FRESH-
WATER
BIOLOGY

NEW YORK · LONDON · JOHN WILEY & SONS, INC.

The late **HENRY BALDWIN WARD**

The late **GEORGE CHANDLER WHIPPLE**

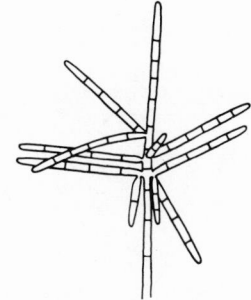

FRESH-
WATER

BIOLOGY

SECOND EDITION

Edited by

W. T. EDMONDSON

Professor of Zoology

University of Washington, Seattle

SECOND EDITION
SECOND PRINTING, JULY, 1963

Library of Congress Catalog Card Number: 59–6781

Printed in the United States of America

To G. EVELYN HUTCHINSON

Authors

M. W. ALLEN, Department of Plant Nematology, University of California, Davis

ALBERT H. BANNER, Department of Zoology and Entomology, University of Hawaii, Honolulu

JOHN LANGDON BROOKS, Osborn Zoological Laboratory, Yale University, New Haven, Connecticut

ROYAL BRUCE BRUNSON, Department of Zoology, Montana State University, Missoula

FENNER A. CHACE, JR., Division of Marine Invertebrates, Smithsonian Institution, U.S. National Museum, Washington, D.C.

B. G. CHITWOOD, Laboratory for Comparative Physiology and Morphology, Kaiser Foundation Research Institute, Richmond, California

WILLIAM J. CLENCH, Museum of Comparative Zoology, Harvard University, Cambridge, Massachusetts

WESLEY R. COE (Emeritus), Department of Biology, Yale University, New Haven, Connecticut

HENRY S. CONARD (Emeritus), Department of Biology, Grinnell College, Grinnell, Iowa

GEORGES DEFLANDRE, Laboratoire de Micropaléontologie de l'Ecole Pratique des Hautes Etudes, Muséum National d'Histoire Naturelle, 13 Place Valhubert, Paris V, France

RALPH W. DEXTER, Department of Biology, Kent State University, Kent, Ohio

FRANCIS DROUET, Department of Biology, New Mexico Highlands University, Las Vegas

W. T. EDMONDSON, Department of Zoology, University of Washington, Seattle

GEORGE F. EDMUNDS, JR., Division of Biology, University of Utah, Salt Lake City

LEONORA K. GLOYD, State Natural History Survey, Urbana, Illinois

CLARENCE J. GOODNIGHT, Department of Biological Sciences, Purdue University, Lafayette, Indiana

ASHLEY B. GURNEY, Entomology Research Division, U.S. Department of Agriculture, Washington, D.C.

OLGA HARTMAN, Allan Hancock Foundation, University of Southern California, Los Angeles

HORTON H. HOBBS, JR., Department of Biology, University of Virginia, Charlottesville

LESLIE HUBRICHT, 1285 Willow Ave., Louisville, Kentucky

H. B. HUNGERFORD, Department of Entomology, University of Kansas, Lawrence

LIBBIE H. HYMAN, American Museum of Natural History, New York, New York

MAURICE T. JAMES, Department of Zoology, Washington State University, Pullman

MINNA E. JEWELL, Thornton Junior College, Harvey, Illinois

E. RUFFIN JONES, Department of Biology, University of Florida, Gainesville

JAMES B. LACKEY, Department of Civil Engineering, University of Florida, Gainesville

HUGH B. LEECH, California Academy of Sciences, Golden Gate Park, San Francisco

FOLKE LINDER, Folkskoleseminariet, Gävle, Sweden

J. G. MACKIN, Department of Oceanography, Agricultural and Mechanical College of Texas, College Station

ERNESTO MARCUS, Universidade de São Paulo, Caixa Postal 6994, São Paulo, Brazil

N. T. MATTOX, Allan Hancock Foundation, University of Southern California, Los Angeles

J. PERCY MOORE (Emeritus), Department of Zoology, University of Pennsylvania, Philadelphia

W. C. MUENSCHER, Department of Botany, Cornell University, Ithaca, New York

IRWIN M. NEWELL, Division of Life Sciences, University of California, Riverside

LOWELL E. NOLAND, Department of Zoology, University of Wisconsin, Madison

SOPHY PARFIN, Division of Insects, Smithsonian Institution, U.S. National Museum, Washington, D.C.

RUTH PATRICK, Academy of Natural Sciences, Philadelphia, Pennsylvania

W. E. RICKER, Fisheries Research Board of Canada, Biological Station, Nanaimo, British Columbia, Canada

MARY DORA ROGICK, College of New Rochelle, New York

HERBERT H. ROSS, State Natural History Survey, Urbana, Illinois

MILTON W. SANDERSON, State Natural History Survey, Urbana, Illinois

WILLIAM W. SCOTT, Department of Biology, Virginia Polytechnic Institute, Blacksburg, Virginia

FREDERICK K. SPARROW, Department of Botany, University of Michigan, Ann Arbor

R. Y. STANIER, Department of Bacteriology, University of California, Berkeley

R. H. THOMPSON, Department of Botany, University of Kansas, Lawrence

WILLIS L. TRESSLER, U.S. Navy, Hydrographic Office, Washington, D.C.

C. B. VAN NIEL, Hopkins Marine Station, Pacific Grove, California

PAUL S. WELCH, Department of Zoology, University of Michigan, Ann Arbor

MILDRED STRATTON WILSON, Division of Marine Invertebrates, Smithsonian Institution, U.S. National Museum, Washington, D.C., and Arctic Health Research Center, Anchorage, Alaska

The late **MIKE WRIGHT**

HARRY C. YEATMAN, Department of Biology, University of the South, Sewanee, Tennessee

Authors of the First Edition

EDWARD ASAHEL BIRGE

NATHAN AUGUSTUS COBB

WESLEY ROSWELL COE

HERBERT WILLIAM CONN

CHARLES BENEDICT DAVENPORT

CHARLES HOWARD EDMONDSON

CARL H. EIGENMANN

HERBERT SPENCER JENNINGS

EDWIN OAKES JORDAN

CHARLES DWIGHT MARSH

JOHN PERCY MOORE

JAMES GEORGE NEEDHAM

EDGAR WILLIAM OLIVE

ARNOLD EDWARD ORTMANN

ARTHUR SPERRY PEARSE

RAYMOND HAINES POND

EDWARD POTTS

JACOB ELLSWORTH REIGHARD

RICHARD WORTHY SHARPE

VICTOR ERNEST SHELFORD

FRANK SMITH

JULIA WARNER SNOW

CAROLINE EFFIE STRINGER

BRYANT WALKER

HENRY BALDWIN WARD

GEORGE CHANDLER WHIPPLE

ROBERT HENRY WOLCOTT

Authors of the First Edition

EDWARD ASAHEL BIRGE

NATHAN AUGUSTUS COBB

WESLEY ROSWELL COE

HERBERT WILLIAM CONN

CHARLES BENEDICT DAVENPORT

CHARLES HOWARD EDMONDSON

CARL H. EIGENMANN

HERBERT SPENCER JENNINGS

EDWIN OAKES JORDAN

CHARLES DWIGHT MARSH

JOHN PERCY MOORE

JAMES GEORGE NEEDHAM

EDGAR WILLIAM OLIVE

ARNOLD EDWARD ORTMANN

ARTHUR SPERRY PEARSE

RAYMOND HAINES POND

EDWARD POTTS

JACOB ELLSWORTH REIGHARD

RICHARD WORTHY SHARPE

VICTOR ERNEST SHELFORD

FRANK SMITH

JULIA WARNER SNOW

CAROLINE EFITH STRINGER

BRYANT WALKER

HENRY BALDWIN WARD

GEORGE CHANDLER WHIPPLE

ROBERT HENRY WOLCOTT

Preface

S ince its publication in 1918, Ward and Whipple's *Fresh-Water Biology* has been the most frequently used single source of information about North American fresh-water fauna and flora, but for many years it has been so out of date as to be essentially useless with most groups of organisms. Because there is still a great need for a concise guide to the North American fresh-water biota, a new edition has been prepared. At the time of its original publication and for some years thereafter *Fresh-Water Biology* served not only as a manual for identification of aquatic plants and animals, but also as one of the few sources available in English for general information about their ecology and habits. Knowledge of all these groups has progressed so much since 1918 that to revise the book today in the same scope as the original would tremendously enlarge it. Other sources for the more general material have meanwhile become readily available. It was obvious that to maintain as convenient a size as possible, a change in the approach would have to be made.

It was therefore decided that the new edition would preserve the major function of the original and be primarily a handbook to aid in the identification of the fauna and flora of inland waters of North America. It can no longer serve as a textbook of general biology, or of the principles of limnology. For the biology of the groups included, many sources of information have become available since 1918, and some of these are referred to in the pertinent chapters. Readers who require an introduction to limnology are referred to the books by Hutchinson, Ruttner, and Welch (see Bibliography of Chapter 1). Moreover, it was decided, in the interests of a more complete treatment of the groups included, to omit a consideration of the strictly internal parasites, since these organisms are not truly aquatic, and are adequately treated elsewhere.

xiii

The free-swimming cercaria larvae of the Trematoda are not included, nor are spiders, some of which live associated with water. Aquatic vertebrates are also omitted, but references are given in Chapter 1 to books that can be used to identify fishes, amphibians, reptiles, and birds found in aquatic habitats. Detailed description of techniques is likewise not within the scope of the present book, but since these matters are so important in connection with identification, some of the major methods of most general application are suggested in the final chapter; special techniques particularly applicable to the various groups are described in the appropriate places.

On the other hand, not all of the changes have resulted in the deletion of material. The bacteria and vascular plants, which were represented by fragmentary discussions in the first edition, have been given considerably more detailed treatment. Fungi, bryophytes, tardigrades, and polychaetes, omitted from the first edition, are now included.

The different groups of organisms are not given equal treatment here. Although the species is ordinarily taken as the taxonomic unit in ecological work, some of the keys in the present book stop at genus. Moreover, the detail of treatment varies somewhat among the chapters that do go to species. The reasons for giving different groups different treatment include the following considerations, in various proportions. Some groups, particularly the insects, but also most of the algae, are so large that keys to the North American species would occupy several volumes. Certain of these are reasonably well treated by monographs, and the present keys to genera serve as a convenient guide to the literature. Thus, sheer bulk requires abbreviated treatment of some groups. But there are other considerations involved also. Some organisms require such special technique and knowledge that identification for serious purposes by nonspecialists is impractical. The determination of characters separating the species is so difficult that secure identification can be made only by persons who have taken the time to acquire a great amount of experience. Such experience must obviously be based on more literature than a general handbook. Although this is true of every group of organisms to some extent, certain ones are much less approachable than others. Further, some groups have been studied much more than others, and a more complete treatment is possible at this time. Finally, it must be recognized that the various authors have reacted differently to the challenge of presenting their groups in a rather artificially delimited space, and have chosen to use their space in somewhat different ways.

In general the geographical coverage is North America north of the Rio Grande, but in some chapters it seemed especially useful to include species reported so far only from Mexico or other adjacent areas, in the expectation that known ranges may be extended by future investigations.

The plan of the volume as it now appears was established in 1951 after much preliminary study and consultation. The length of the interval before completion resulted from a combination of circumstances including the death of two of the prospective authors before they had assembled usable manuscripts, but it largely indicates the time required to prepare such a treatise.

The new edition consists mostly of new material. In certain chapters, some of the illustrations and textual material from the first edition are used, but in many, nothing remains from the old edition. Nevertheless the new book as a whole rests heavily on the efforts of the original authors, and their names are reprinted here to recognize that fact. It is a pleasure to record that two of these authors have revised their sections for the new edition: W. R. Coe and J. Percy Moore.

The figures are of diverse origin. Most of them have been prepared by the authors themselves for this edition; these new figures do not carry specific credit lines. Illustrations which have been copied from the literature are credited to the original author by the phrase "After. . . ." The phrase "By. . ." indicates that the illustration was prepared by the author of the chapter in the first edition. Other situations are indicated in the legends of the figures concerned.

Journal titles are abbreviated according to the system used by Chemical Abstracts.

The present book, then, consists essentially of a series of illustrated keys, each being preceded by an exposition of material that the reader must understand in order to use the key. Since the introductory text is directed toward the problem of identification of material, it should not be taken to be a complete discussion of all aspects of the group concerned. The user of this new edition is earnestly requested to note the restriction that has been placed on its scope. He is expected to bring to the book a general knowledge of the morphology and life history of the group he intends to study. The taxonomy of fresh-water organisms is so complex and vast a subject that a single treatise cannot now include both elementary introductions to the biology of the groups and detailed keys to identification.

This book is intended for a wide audience. It is expected to be used by advanced students of zoology and botany who are interested in aquatic organisms and ecology. Also, there is much here for the professional worker. Systematists will find some fresh approaches, and limnologists, fisheries biologists, and sanitary engineers will be able to use the book in a variety of investigations in which identity of organisms is of central importance. In many cases, nothing will substitute for authoritative identifications by specialists, but at the very least this book will be of value in preliminary identifications and in increasing the limnologist's knowledge of the groups with which he works. The beginner in biology will find the book a helpful guide if he is alert in recognizing the difficulties pointed out in Chapter 1.

Finally, I wish to acknowledge with gratitude the great amount of help and encouragement I have received during the long period between the initiation of the present book and its appearance in print. Discussions of a variety of matters with James E. Lynch, Paul L. Illg, and many others too numerous to mention have been most helpful. John M. Kingsbury supplied useful advice. Mildred S. Wilson has helped far beyond the ordinary bounds of authorship. My wife, Yvette Hardman Edmondson, has been helpful in many ways at all stages in the development of the book. Thanks are due Ernst Mayr, Robert

R. Miller, and R. C. Stebbins for suggesting references to appropriate works on vertebrates, some of which are cited in Chapter 1. The book could not have been brought to completion without the generous cooperation of Arthur W. Martin, Jr., Executive Officer of the Department of Zoology, University of Washington.

The special contribution of Professor G. Evelyn Hutchinson of Yale University to the development of this volume calls for special recognition. His encouragement and counsel given me over a long period of time were most important in providing impetus for the new edition, and he has continued to supply help and sound advice as the work has progressed. Many of the authors of this volume have benefited from Professor Hutchinson's contributions, and they join me in the recognition given on another page.

Most of all, I am grateful to the authors who have faced the problem of selecting a small fraction of their knowledge to present in this volume. Not all of the authors agree with decisions I have made about scope and style; their tolerance is greatly appreciated. We all recognize the very great need for a genuine monographic faunistic and floristic treatment of North America. While this book does not pretend to fill this particular need, we believe that it represents a useful volume in itself and an important preparatory step toward the day when it will be possible to produce such a work.

W. T. EDMONDSON

Seattle, Washington
June, 1959

Contents

1 **INTRODUCTION** 1
    W. T. Edmondson

2 **INTRODUCTION TO THE PROTISTA** 7
    R. Y. Stanier

3 **BACTERIA** 16
    C. B. van Niel
    R. Y. Stanier

4 **FUNGI** 47
    Frederick K. Sparrow
    Key to Fungi Imperfecti   *William W. Scott*

5 **MYXOPHYCEAE** 95
    Francis Drouet

6 **ALGAE** 115
    R. H. Thompson

7 **BACILLARIOPHYCEAE** 171
    Ruth Patrick

8 **ZOOFLAGELLATES** 190
    James B. Lackey

9   **RHIZOPODA AND ACTINOPODA**   232
      Georges Deflandre

10   **CILIOPHORA**   265
      Lowell E. Noland

11   **PORIFERA**   298
      Minna E. Jewell

12   **COELENTERATA**   313
      Libbie H. Hyman

13   **TURBELLARIA**   323
      Introduction   *Libbie H. Hyman*   323
      Tricladida   *Libbie H. Hyman*   326
      Catenulida   *E. Ruffin Jones*   334
      Macrostomida   *E. Ruffin Jones*   338
      Neorhabdocoela   *E. Ruffin Jones*   341
      Alloeocoela   *E. Ruffin Jones*   359

14   **NEMERTEA**   366
      Wesley R. Coe

15   **NEMATA**   368
      B. G. Chitwood
      M. W. Allen

16   **GORDIIDA**   402
      B. G. Chitwood

17   **GASTROTRICHA**   406
      Royal Bruce Brunson

18   **ROTIFERA**   420
      W. T. Edmondson

19   **BRYOZOA**   495
      Mary Dora Rogick

20   **TARDIGRADA**   508
      Ernesto Marcus

21   **OLIGOCHAETA**   522
      Clarence J. Goodnight

22   **POLYCHAETA**   538
      Olga Hartman

23   **HIRUDINEA**   542
      J. Percy Moore

24 **ANOSTRACA** 558
    Ralph W. Dexter

25 **NOTOSTRACA** 572
    Folke Linder

26 **CONCHOSTRACA** 577
    N. T. Mattox

27 **CLADOCERA** 587
    John Langdon Brooks

28 **OSTRACODA** 657
    Willis L. Tressler

29 **FREE-LIVING COPEPODA** 735
    Introduction *Mildred Stratton Wilson and Harry C. Yeatman*   735
    Calanoida *Mildred Stratton Wilson*   738
    Cyclopoida *Harry C. Yeatman*   795
    Harpacticoida *Mildred Stratton Wilson and Harry C. Yeatman*   815

30 **BRANCHIURA AND PARASITIC COPEPODA** 862
    Mildred Stratton Wilson

31 **MALACOSTRACA** 869
    Fenner A. Chace, Jr.
    J. G. Mackin
    Leslie Hubricht
    Albert H. Banner
    Horton H. Hobbs

32 **INTRODUCTION TO AQUATIC INSECTA** 902
    Herbert H. Ross

33 **EPHEMEROPTERA** 908
    George F. Edmunds, Jr.

34 **ODONATA** 917
    Leonora K. Gloyd
    Mike Wright

35 **PLECOPTERA** 941
    W. E. Ricker

36 **HEMIPTERA** 958
    H. B. Hungerford

37 **NEUROPTERA** 973
    Ashley B. Gurney
    Sophy Parfin

38 COLEOPTERA   981
    Hugh B. Leech
    Milton W. Sanderson

39 TRICHOPTERA 1024
    Herbert H. Ross

40 LEPIDOPTERA 1050
    Paul S. Welch

41 DIPTERA 1057
    Maurice T. James

42 ACARI 1080
    Irwin M. Newell
    Parasitengona   1080
    Halacaridae   1108
    Oribatei   1110

43 MOLLUSCA 1117
    William J. Clench

44 BRYOPHYTA 1161
    Henry S. Conard

45 VASCULAR PLANTS 1170
    W. C. Muenscher

46 METHODS AND EQUIPMENT 1194
    W. T. Edmondson

    INDEX 1203

FRESH-
WATER
BIOLOGY

1

Introduction

W. T. EDMONDSON

The scope of this new edition of Ward and Whipple's *Fresh-Water Biology* is specifically described in the Preface. It is hoped that the user of the book will read the Preface before making use of the book. The primary purpose of this Introduction is to facilitate the use of the keys.

Organization

Essentially the book consists of a series of illustrated keys to the fresh-water flora and fauna of North America north of the Rio Grande, each preceded by a statement of information necessary to its correct use. In some, the introductory statement is quite brief because relatively few and obvious features are used or because definitions are given in the body of the key. In others, the identification is based upon study of the details of many organ systems, and a lengthier introductory statement is required. In some, ecological information is especially useful.

Although no key is given to major groups, certain guides have been provided. Chapter 2 introduces the protistan groups. The organization of the

Crustacea is outlined in Chapter 24, and problems of classification are further discussed in Chapter 27. Chapter 32 is devoted to an introduction to insects. Cross references among the chapters have been supplied.

To present a guide to the literature, each chapter is followed by a list of references to the major monographs and compilations, where, in turn, additional references can be found. The final chapter presents a summary of methods of collection and preservation.

The Reader

A statement is perhaps needed of the background required of the user of this book. The book is not intended for the novice in biology. Nevertheless, it is recognized that Ward and Whipple has long been used by elementary students with little background, and will probably continue to be so used. The following introductory remarks are therefore given in more detail than would be necessary for the fully trained professional biologist, but all readers will find them a necessary guide to the book.

The text before each key outlines the features of morphology used in the key but does not pretend to be a complete discussion of all the features of the group. Therefore, the user of the key must obtain some introductory knowledge of zoology or botany, particularly morphology and systematics, and it will help him to be familiar with the organisms in the live condition. Anybody attempting to use the book with insufficient background will be faced with unfamiliar terms and concepts, and will have to find structures the nature of which he may not know. Nevertheless, a good way for a beginner to aid his study of a group is to try to identify a variety of members. Keying the organisms will turn his attention to features he might otherwise not notice, and the necessity of making clear-cut decisions about structural features will teach him morphology and terminology in a highly specific and definite way. References are given to a number of useful introductory and advanced books at the end of this chapter.

Use of the Keys

Almost all the principal keys in the book are of the bracket type. At the beginning of the key are two statements headed **1a** and **1b**. The two statements are intended to be mutually exclusive, and any organism in the group should be correctly described by one or the other of the two statements. The user decides which statement is applicable to the specimen, then proceeds to the couplet indicated by the number at the end of the line. In a few cases it is convenient to have three or even four choices at one point; these are indicated by **c** and **d**. The number in parentheses after the number of the first line of a couplet indicates the number of the previous couplet that leads to that line. This back-track number is often a great convenience since it permits one to work back through the key with a minimum of effort. It will be useful when an erroneous identification appears to have been made, and will permit a check with the key when one has a named specimen or has made a tentative identification merely by examining the figures. In two chapters (3 and 23), additional use is made of tabular keys, and explanations are given in those chapters.

Two different general approaches are taken in these keys. Most are hierarchical keys, meaning that an organism is successively identified as to its class, order, family, genus, and species. Such a key is basically organized as a classification of the group, and the successive couplets which lead to a specific identification can be reassembled to form abbreviated descriptions of the various taxa. Examples of hierarchical keys are found in chapters on tardigrades (20) and malacostracans (31). Such keys work very well in many groups and are satisfactory in that closely related forms occur together in the key. Some groups, however, are such that a strictly hierarchical key is very difficult to use.

Therefore, some of the keys are practical, with little or no attempt to outline the classification. In these keys identification is achieved by the use of characters that are not of fundamental use in establishing the larger taxa, but are much more easily found or specified. Since the key is not organized as a classification of the group, related taxa may be separated by a number of pages. Examples of practical keys are those of the bacteria (3), rotifers (18) and mites (42). An explanation of the kind of reasons that lead to the use of a practical rather than a hierarchical key is given in Chapter 18.

It must be realized that the key characters are highly selected, and that even in hierarchical keys it is not usually possible to reconstitute the full, formal definitions of families and other higher groups. Very often only the most obvious and easily studied characters will be cited in the key, and others, perhaps of greater biological importance, will not be mentioned. The reason for this is that determinative keys are guides to identification, not monographic treatments of groups. For full details, one must, of course, study the appropriate original and compilative literature.

A few hints that will facilitate accurate use of the keys may be useful. In the first place, no identification should be attempted until the introductory text of the section has been studied. Much of the specialized terminology is explained in the introductory text, although some may be given within the body of the key. Users without much experience in morphology may find that they will have to make frequent reference to dictionaries and introductory textbooks.

Sometimes a choice cannot be made on the basis of the material at hand; the animal may be immature; a structure may not be visible, may not be developed, or may not agree with either of the choices offered by the key. Occasionally identification must be abandoned at this point, but often it is worthwhile to track down the organism in two branches of the key. Usually one of the identifications will prove to be obviously wrong and the other can be confirmed by reference to the literature where details are given. Sometimes the choices may seem to be clear enough but the organism at hand may obviously be different from the one finally given by the key. In such cases an error must be suspected and may be located by backtracking through the key. In both these cases, however, the possibility exists that one has an undescribed species, or a species not included in the key because it has not previously been reported from North America, or for some other reason such as extreme rarity.

Some groups, such as gastrotrichs and certain crustaceans, have not been sufficiently studied in this country to permit at this time construction of keys that will identify all species that can easily be collected; many species await description. For that reason, deviation from the description must be taken seriously and questionable identifications in any group confirmed by reference to full descriptions or to a specialist. Ideally these species should not fit the key, but obviously the keys often cannot discriminate between a known form and a similar one that has not yet been described. Moreover, much of our knowledge of some groups is based upon old work done by men who worked before some of the present understanding of the requirements for systematic work was formulated, or who, for other reasons, presented incomplete or inaccurate descriptions. Recent investigators have not yet been able to restudy all the old species from types and new material, and early errors are still preserved and perpetuated. Thus, some chapters of this book have to be regarded as provisional accounts of the published knowledge of the groups, and at the same time as invitations to serious taxonomic work.

Therefore, at this point it is necessary to include a warning about the proper way in which identifications must be made. One should never make identifications important to any scientific investigation on the sole basis of a casual use of this book. If the book is to be used seriously, the user must develop sufficient background and familiarity with the group and the literature that he will recognize deviations of undescribed material from the key and will not force material into conformity with known species simply because he feels he must find a name for the organism. This book will very greatly facilitate the acquisition of the knowledge necessary to make identifications, but it will not substitute for intelligent work and for proper use of the literature.

One sometimes hears complaints that it is too much trouble to find the characters called for in a particular key; one has to count the number of minute spines on the eleventh segment of a small appendage of an organism that has many much larger structures, or has to examine minute and complex internal structures. The point is that these characters are used because the work of generations of systematists shows that they are useful and that certain others, possibly easier to see, are useless or misleading. It should not be assumed that because the key separates two species on an apparently trivial difference, there are no other differences. It may be that in some groups further research will reduce the use of some of the more difficult characters, but in general the trend appears to be in the opposite direction. The plain fact is that in order to identify a flatworm one must cut microtome sections, and that in order to identify a copepod one must be able to remove certain appendages without mangling them. Anybody who is unwilling to do the required work should not try to make identifications, for it is only experienced investigators who can take short cuts. In some cases, a key to a local biota may be made using simpler characters because of the absence of close relatives that can be separated only on the basis of difficult characters.

It must be recognized that a key is only a guide, and in order to make successful identifications, one must do more than merely read a key. A thoroughly sound identification always involves additional work beyond tracking the specimen through a key; the specimen must be compared with a complete description and figure. With specialists, having a thorough knowledge of the material, this last step may sometimes be entirely mental. Nevertheless, any identification that does not make this step in some degree must be regarded as provisional.

When the user of this book decides that one of the keys is unworkable in any part, and if he is sure that he knows enough about the organisms to use the key, he is earnestly requested to inform the editor of his difficulties. It is inconceivable that in a book of this magnitude the combined efforts of authors and editor should not have left errors and points of weakness. If this book is ever to appear in a third edition it will be only because such a book is needed and because an improved version is possible. Improvement obviously can best be achieved by a collaboration of the users and the authors.

References

GENERAL

Borradaile, L. A., et al. 1958. *The Invertebrata*, rev. ed. Macmillan, New York. **Bronn, H. G. 1859–.** *Klassen und Ordnungen des Tier-Reichs*. Winter, Leipzig. (A very extensive treatise, in many volumes, incomplete for some groups.) **Brown, F. A., Jr. (ed.). 1950.** *Selected Invertebrate Types*. Wiley, New York. **Grassé, P. P. (ed.) 1952–.** *Traité de zoologie. Anatomie, systématique, biologie*. Masson, Paris. (Many volumes, still being published.) **Hutchinson, G. Evelyn. 1957.** *A Treatise on Limnology*. Wiley, New York. **Hyman, Libbie H.** *The Invertebrates*. Vol. I, 1940, *Protozoa through Ctenophora*. Vol. II, 1951, *Platyhelminthes and Rynchocoela*. Vol. III, 1951, *Acanthocephala, Aschelminthes and Entoprocta*. Vol. IV, 1955, *Echinodermata*. McGraw-Hill, New York. (More volumes in preparation.) **Kükenthal, W. 1923–.** *Handbook der Zoologie*. Gruyter, Berlin and Leipzig. (Extensive, in many volumes, but less detailed than Bronn.) **Mayr, E., E. G. Linsley, and R. L. Usinger. 1953.** *Methods and Principles of Systematic Zoology*. McGraw-Hill, New York. **Parker, T. J. and W. A. Haswell, revised by O. Lowenstein. 1951.** *A Textbook of Zoology*, 2 vols. Macmillan, London. **Pearse, A. S., (ed.). 1949.** *Zoological Names. A List of Phyla, Classes and Orders*, 4th ed. Amer. Assoc. Adv. Sci., Section F. Duke University, Durham, N. C. **Pennak, R. W. 1953.** *Fresh-Water Invertebrates of the United States*, Ronald, New York. **Ruttner, F. 1953.** *Fundamentals of Limnology*. (Translated by D. G. Frey and F. E. J. Fry.) University of Toronto Press, Toronto. **Welch, P. S. 1952.** *Limnology*, 2nd ed. McGraw-Hill, New York. **1948.** *Limnological Methods*. Blakiston, Philadelphia.

VERTEBRATES

Bishop, Sherman C. 1943. *Handbook of Salamanders*. Comstock, Ithaca, New York. **Blair, W. Frank, Albert P. Blair, Pierce Brodkorb, Fred R. Cagle, and George R. Moore. 1957.** *Vertebrates of the United States*. McGraw-Hill, New York. **Carr, Archie, 1952.** *Handbook* *of Turtles.* Comstock, Ithaca, New York. **Lagler, Karl F. 1957.** *Fresh-Water Fishery Biology.* Brown, Dubuque, Iowa. **Livezey, R. L. and A. H. Wright. 1947.** A synoptic key to the salientian eggs of the United States. *Am. Midland Naturalist,* 37:179–222. **Peterson, R. T. 1941.** *A Field Guide to Western Birds.* Houghton Mifflin, Boston. **1947.** *A Field Guide to the Birds,* 3rd ed. Houghton Mifflin, Boston. **Pettingill, O. S. 1956.** *A Laboratory and Field Manual of Ornithology.* Burgess, Minneapolis. (Appendix lists references to state works on birds.) **Pough, R. H. 1951.** *Audubon Water Bird Guide. Water, Game and Large Land Birds. Eastern and Central North America from Southern Texas to Central Greenland.* Doubleday, Garden City, New York. **Stebbins, Robert C. 1951.** *Amphibians of Western North America.* University of California Press, Berkeley. **Wright, A. H. 1929.** Synopsis and description of North American tadpoles. *Proc. U. S. Natl. Museum,* 74:1–70. **Wright, Anna and A. H. Wright. 1949.** *Handbook of Frogs and Toads.* Comstock, Ithaca, New York.

2

Introduction
to the Protista

R. Y. STANIER

This introduction, a brief survey of the salient properties and possible inter-relationships of the various microbial groups, is included for the benefit of the reader who has little previous acquaintance with microorganisms, and is designed to provide him with a general map of the terrain exposed in detail in the following chapters.

In 1866, Haeckel proposed the recognition of the kingdom Protista as a rational solution of the more and more difficult taxonomic problems being posed by the microorganisms. As knowledge of the microbial forms of life accumulated during the first half of the nineteenth century, it became increasingly evident that these organisms span the gap between the two classical living kingdoms with a multitude of transitional forms, which display a mixture of "plantlike" and "animallike" properties in all possible combinations. Haeckel perceived that this taxonomically awkward situation found its explanation in evolutionary theory: the microorganisms, many of which have preserved a great simplicity of structure throughout evolutionary history, are contemporary organisms that lie on the evolutionary level of the ancestral stems of the higher plants and animals. Therefore, the concepts of "plant" and "animal," which are derived from the consideration of more evolved biotypes, are inapplicable in microbiological context. The most taxonomically workable treatment of these more primitive forms is to group them in a third kingdom, the *Protista*, whose members can be differentiated from the higher plants and animals in terms of their level of biological complexity. An alternative proposal would be to abandon the concept of kingdoms, making phyla the primary division of living organisms.

Despite the compelling logic of Haeckel's proposal, most botanists and zoologists have maintained a somewhat reserved attitude toward the concept of the Protista, and the best dividing lines, certainly arguable, for separating protists from higher plants and animals have never been established by general agreement. There are four central groups which must be assigned to the Protista: algae (Chapters 5, 6, and 7), protozoa (Chapters 8, 9, and 10), fungi (Chapter 4), and bacteria (Chapter 3). Broadly speaking, these organisms are distinguished from the higher forms by a relative lack of differentiation: many of them are either unicellular or coenocytic (bacteria, protozoa, fungi), and even among those that display true multicellularity (e.g., some of the algae) there is little or no tissue differentiation. However, the sponges (Chapter 11) might with almost equal propriety be placed either among the invertebrates or among the protists; and in terms of structural complexity, there is relatively little difference between the most highly developed green algae and the bryophytes (Chapter 44). For the purposes of the present discussion, we shall treat as protists only the four groups of algae, fungi, protozoa, and bacteria.

Within the Protista as just defined, two principal subgroups can be distinguished on the basis of internal cell structure. The cell of the protozoa, fungi, and most algae displays all the major structural features of the cell as we find it in plants and animals. The nucleus in the resting state is surrounded by a distinct membrane, and during division reveals a typical chromosomal organization: both meiosis and mitosis are readily recognizable. The cytoplasm contains plastids: mitochondria and (in the photosynthetic forms) chloroplasts. When examined with an electron microscope, the locomotor organelle of flagellates and ciliates, despite its frequent very considerable specialization, is found to have the same fundamental internal structure as the flagella or cilia of higher plants and animals; it contains a multistranded core, whose two central fibrils are morphologically distinguishable from the nine peripheral ones. There is nothing about the basic construction of the individual cell that serves to distinguish these protists from higher plants and animals; it is, rather, *the relative simplicity of the whole organism* that marks them as protists.

The same is not true of the bacteria and of one group traditionally assigned to the algae, the blue-green algae. The cell in these two groups is constructed on a much less highly differentiated plan. First, bacteria and blue-green algae contain centrally located intracellular bodies with the outstanding chemical property of nuclei (i.e., localization therein of desoxypentose nucleic acid); but despite much careful cytological work, it is still not possible to homologise them with true nuclei. The claims of the occurrence of mitotic division in bacteria made by certain investigators are flatly rejected by others, and in the blue-green algae, the great majority of cytologists who have made a careful study of nuclear structure have failed to observe stages that could be described as typical of mitosis. The existence in bacteria of modes of gene transfer (transduction, type transformation) without counterparts in any other living group can perhaps also be regarded as evidence pointing to a unique mode of organization of the genetic material, although the science of bacterial genetics is still too young to permit a definite conclusion in this respect. Second, typical plastids are lacking in bacteria and blue-green algae. This is most clearly evident in the photosynthetic representatives, which show upon microscopic examination an even distribution of photosynthetic pigments throughout the cytoplasm. Recent analytical studies have shown that there is, in fact, a localization of the photosynthetic pigments of bacteria and blue-green algae in submicroscopic cytoplasmic particles which have been variously termed *grana* or *chromatophores*; but these bodies are far smaller, and apparently far simpler in structure, than typical chloroplasts. Third, the locomotor organelle of bacteria known as the flagellum is not structurally homologous with the flagellum or cilium of plants, animals, and other protists; it consists of a single submicroscopic fibril some 0.03 μ in diameter. Last, it ought to be mentioned that there are two features of the cell whose structural significance is not yet clear, but which are of great practical value in distinguishing bacteria and blue-green algae from other protists. When examined in the living state, the cells of bacteria and blue-green algae have a characteristic and most unusual appearance: the cytoplasm is completely devoid of streaming movements, and there are no vacuoles, at least in young and healthy cells. Thus, on the basis of cell structure the bacteria and blue-green algae constitute an isolated subgroup of the Protista, and may be termed collectively the *lower protists* (sometimes called Monera) in distinction to the other algae, the protozoa, and the fungi (*higher protists*).

General Properties and Interrelationships of the Algae and Protozoa

Among the higher protists, there is much evidence to suggest a common evolutionary stem, and one which must have been characterized by a photosynthetic mode of metabolism. Among present-day forms, the organisms that probably lie closest to this stem are the various groups of unicellular algal flagellates, organisms regarded by the algologist as the most primitive representatives in a series of algal divisions, and placed by the protozoologist in the subclass Phytomastigina of the class Mastigophora (Flagellata)—a taxonomic difference of opinion that speaks eloquently for their ambiguous relationships. Considered from the standpoint of formal taxonomy, the algologist's approach is certainly the better one: the various photosynthetic flagellates fall into a number of sharply defined groups—dinoflagellates, cryptomonads, euglenids, chrysomonads, volvocine flagellates—which differ neatly from one another with respect to flagellar structure, nature of the bounding cellular membrane, photosynthetic pigments, and reserve food materials. Thus, even though these forms no doubt lie closest to the photosynthetic stem group of the higher protists, they constitute a series of variations on the theme of the flagellate unicell, and it is not now possible to perceive their evolutionary interrelationships. They comprise, as it were, the starting points for a number of parallel and in many cases multidirectional evolutionary sequences. The two characteristic dimensions of evolution from the photosynthetic flagellates can be diagrammed as follows:

Multicellularity, reduction of
motility, maintenance of photosynthetic metabolism (higher algae)

↑

Photosynthetic unicellular   ⟶   Unicellularity, maintenance
flagellates   of motility, loss of photo-
                                           synthetic metabolism (protozoa)

Not every group of photosynthetic flagellates shows affinities to multicellular photosynthetic forms. The euglenids, for example, seem not to have progressed beyond a primitive colonialism, represented by one order (the Euglenocapsales) which contains a single genus with but two species. In the dinoflagellate line, multicellularity is represented by a small group of immobile, branching filamentous organisms (the Dinotrichales; brackish water Dinophyceae) whose affinities are revealed by the characteristic dinoflagellate structure of their zoospores. The volvocine flagellates, on the other hand, are related to the whole large and complex class of green algae (Chlorophyta) and, through them, to the higher plants. As G. M. Smith has very aptly pointed out, the relationships of plants, conventionally shown in the form of a much-branched tree, can be more accurately diagrammed as a tree (the Chlorophyta and the higher plants) adjoined by a number of shrubs of varying height (the remaining divisions of algae).

It has already been mentioned that protozoologists classify the photosynthetic flagellates as a subclass of the Mastigophora, a treatment dictated by the close morphological resemblances between many photosynthetic and nonphotosynthetic flagellates. In fact, every group of photosynthetic flagellates has a set of colorless counterparts (sometimes referred to by algologists as *leucophytes*), whose affinities to a particular photosynthetic flagellate group can still be established by characteristic morphological features, such as the number, nature, and arrangement of the flagella. It should be noted that loss of chloroplasts and adaptation to a saprophytic mode of existence has occurred also in nonflagellate algae (e.g., diatoms and nonmotile unicellular green algae). These colorless forms are, however, ignored by the protozoologist. Examples of very closely related photosynthetic-colorless pairs are *Euglena* and *Astasia*, *Chlamydomonas* and *Polytoma*. In certain *Euglena* species, the transformation of a photosynthetic to a colorless flagellate has been achieved in the laboratory; this change, involving as it does the degeneration and loss of the chloroplasts, is an irreversible one. These are thus good grounds for assuming that the entire group of colorless unicellular flagellates placed in the protozoa have been derived, at various times in evolutionary history, from photosynthetic flagellates. In many cases, of course, this primary physiological transformation has been followed by a long history of morphological evolution, so that among present-day colorless flagellates there are specialized groups—for example, the trypanosomes, the poly- and hypermastigotes—which show no distinct morphological affinities to any of the photosynthetic flagellate groups, and whose origin is thus no longer ascertainable (Chapter 8).

The amoeboid protozoa (Sarcodina or Rhizopoda, Chapter 9) show affinities to the flagellates, often possessing flagella at certain stages of development, and can thus also be presumed to have a primary flagellate origin. Indeed, among the photosynthetic flagellates there is one group, the chrysomonads, characterized by naked cells which are strikingly similar to amoeboid protozoa: locomotion is often predominantly pseudopodial, and in the genus *Chrysamoeba* the nutrition is in part holozoic, despite the presence of chloroplasts. Thus the two central protozoan groups, Mastigophora and Sarcodina, probably arose from a variety of forms among the photosynthetic unicellular flagellates. Subsequent morphological evolution in these colorless protozoan groups has in turn given rise to the highly specialized amoeboid and flagellate protozoa, as well as to the sporozoans and the ciliates.

The Origins of the Fungi

In the past, there have been two schools of thought on the origin of fungi: according to one school, they originated from the algae; according to the other, from nonphotosynthetic protists. In terms of the relationships between algae and protozoa described above, these two concepts of fungal phylogeny can be formulated in a slightly different manner. Since the fungi are without exception nonphotosynthetic, the real question at issue is whether they had a direct origin among photosynthetic protists (algae), by loss of photosynthetic pigments in forms that had already developed the coenocytic type of structure characteristic of fungi, or whether they arose at second hand from photosynthetic protists, being derived by further morphological evolution from unicellular, colorless ancestors of the amoeboid-flagellate type. The second alternative is more plausible. Among the lower phycomycetes, the production of flagellate reproductive cells (zoospores, in some cases also gametes) is universal. These organisms can be separated into several groups on the basis of the mode of flagellation. For example, one major group (Chytridiales, Blastocladiales, Monoblepharidales) is characterized by posteriorly uniflagellate reproductive cells. Another (Lagenidiales, Saprolegniales, Leptomitales, Peronosporales) is characterized by biflagellate reproductive cells, the two flagella (whiplash and tinsel, respectively) being distinct from one another in structure and mode of action. Among algae, the number and nature of the flagella is a character of great significance, which accompanies other major group differential characters such as the composition of the photosynthetic pigment system and the nature of the outer bounding cell membrane; hence it seems plausible to assume that the different kinds of flagellation encountered among the lower phycomycetes reflect a polyphyletic origin, from various groups of flagellate ancestors. The mycelial habit of vegetative thallus construction, well-nigh universal in the higher phycomycetes, the Ascomycetes, the Basidiomycetes and the Fungi Imperfecti, is not nearly so conspicuous among the lower phycomycetes. In some of the chytrids, the vegetative structure consists principally of a sporangium, which ultimately undergoes internal cleavage followed by rupture and the liberation of zoospores. To this sporangium is attached a more or less extensive rhizoidal system of branched, tapering threads. In many chytrids the rhizoidal system is nonreproductive, and serves simply to anchor the sporangium to a solid substrate from which nutrients are absorbed through the rhizoids. The chytrids also include forms in which the rhizoidal system has acquired a reproductive function, being capable of indefinite proliferation from the primary sporangial center, and of forming new sporangia at additional points during its growth. This may possibly be the origin of the coenocytic[1] mycelial habit which is the predominant mode of fungal vegetative construction.

The mycelial habit can be regarded as an adaptive modification conditioned by growth on or in solid substrates, characteristic of nearly all the higher groups of fungi (except the unicellular yeasts). The predominantly terrestrial habitat of the higher phycomycetes, the ascomycetes and the basidiomycetes has further resulted in these forms in the permanent loss of flagellate reproductive cells, and the development of reproductive structures (e.g., sporangiospores, conidia, ballistospores) better suited for aerial dispersion. Among the lower protists, the actinomycetes (a largely immotile group of mycelial organisms which are clearly related to the unicellular, motile true bacteria) provide an interesting parallel example of this evolutionary trend. For further discussion of fungi, see Chapter 4.

The Interrelationships of the Lower Protists

As already mentioned, the lower protists (bacteria and blue-green algae) are sharply separated from the other protists by the nature of their cellular construction. As commonly defined by algologists, the blue-green algae are an exclusively photosynthetic group which carry out typical plant photosynthesis. As commonly defined by bacteriologists, the bacteria are largely nonphotosynthetic, the few photosynthetic forms (green and purple bacteria) being characterized physiologically by their photosynthesis, which is not of the plant type; this is shown by the fact that it is never accompanied by the liberation of free oxygen. Thus the two major groups of lower protists are conventionally separated from one another on physiological grounds. Such a separation entails difficulties, however. There are certain colorless, fila-

[1] Even though higher fungi are septate, the mycelium is universally coenocytic. The apparent "transverse walls" of the septate forms are pseudosepta, with a central pore through which free passage of cytoplasm and nuclei can occur. The essentially coenocytic condition of higher fungi is demonstrated by the phenomenon of heterokaryosis (the establishment of two karotypes in a single mycelium, following hyphal fusion between genetically different strains) as well as by the diploidization of the mycelium which is a characteristic feature of the life cycle in basidiomycetes. Hence the notable definition by Langeron of a fungus as "a multinucleate cytoplasmic mass, motile (by cytoplasmic streaming) in a system of tubes."

mentous organisms such as the large "sulfur bacteria," *Beggiatoa* and *Thiothrix*, which have long been known to show striking morphological resemblances to blue-green algae; in fact, *Beggiatoa* is unmistakably a morphological counter-part of the blue-green algae placed in the genus *Oscillatoria*. The whole prob-lem of the relationships between "bacteria" and "blue-green algae" has recently been analyzed by Pringsheim (1949), who has suggested a new and far more satisfactory approach to the matter of their separation. Most, if not all blue-green algae are capable of a characteristic type of locomotion known as *gliding movement*, which occurs only when the cell is in contact with a solid surface, and is accomplished without the aid of detectable locomotor or-ganelles. Among "bacteria" as commonly defined, there are three groups which also show this singular type of gliding movement: the filamentous sulfur bacteria already mentioned above; the colorless, filamentous organisms of the family Vitreoscillaceae; and the unicellular myxobacteria, most of which are further characterized by a complex and unique type of life cycle. Pringsheim regards the mechanism of locomotion as fundamental to the elucidation of relationships, and hence considers these three groups of colorless gliding "bacteria" as having affinities with the blue-green algae: the filamentous forms show relatively close morphological resemblances to filamentous blue-green algae, whereas the myxobacteria are evidently an isolated and highly specialized group, since a life cycle like that of the higher myxobacteria is never found in unicellular blue-green algae. These colorless groups, together with the blue-green algae, comprise, then, a natural assemblage of lower protists which Pringsheim designates as the *gliding organisms*. Quite dis-tinct from these are the so-called true bacteria, which are either immotile or motile by means of flagella, and the spirochaetes, many of which have been shown in recent years to possess flagella. These groups comprise the lower protists which Pringsheim designates collectively as the *swimming organisms*; between them and the gliding forms there are no indications of relationships, and the two large groups may well be of independent origin. The true bac-teria, in addition to the well-known unicellular types, include also filamentous organisms such as *Sphaerotilus*, and mycelial forms, the actinomycetes.

The affinities of both swimming and gliding forms to the higher protists appear to be remote. Among the gliding forms, the photosynthetic represent-atives (i.e., the blue-green algae) resemble one group of higher algae, the Rhodophyta, in the nature of their accessory photosynthetic pigments (phyco-cyanin and phycoerythrin); the red algae are, however, a very highly developed group and the relationship, if it exists, must be remote. The colorless gliding organisms all seem to be terminal evolutionary groups. Insofar as the swim-ming forms and their nonmotile relatives are concerned, two relationships to higher protists have been suggested: a relationship between actinomycetes and true fungi, and a relationship between spirochaetes and protozoa. In each case, the superficial resemblances of vegetative construction which prompted the suggestion are probably better interpreted as a reflection of evolutionary convergence. Despite the existence of flagellar locomotion in the unicellular true bacteria, a relationship to any of the algal flagellates appear unlikely;

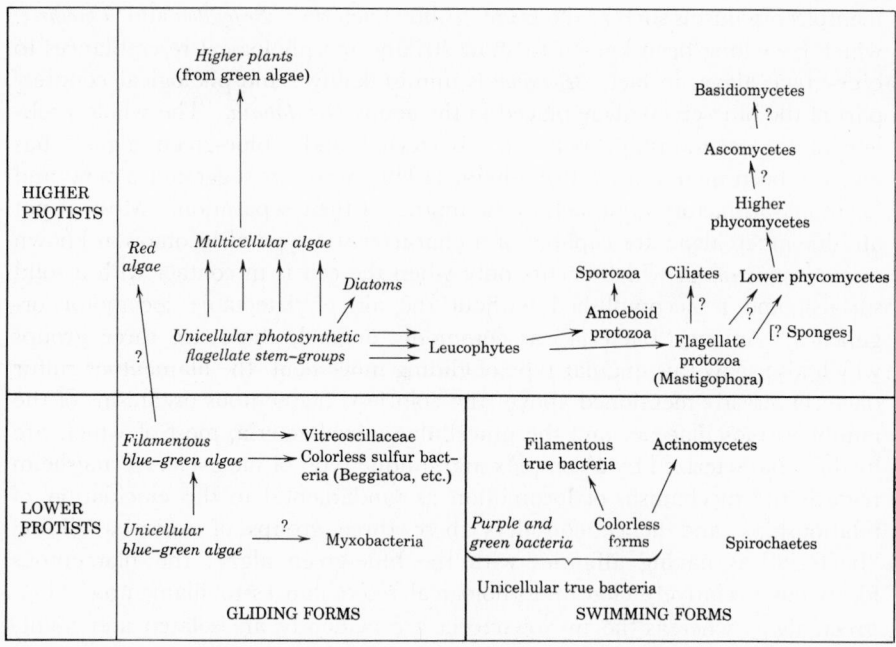

Fig. 2.1. A schematic representation of the constituent groups in the Protista, designed to show possible inter-relationships. Photosynthetic groups are printed in italics, nonphotosynthetic groups in roman letters.

it is improbable that the photosynthetic bacteria could represent a stem group from which the photosynthetic flagellates developed, in view of the unique nature of bacterial photosynthesis, coupled with the existence in these bacteria of special forms of chlorophyll (bacteriochlorophyll, *Chlorobium* chlorophyll) not present in other photosynthetic organisms. Furthermore, an attempt to derive the photosynthetic stem groups among the higher protists from photosynthetic bacteria would necessitate the assumption that plant photosynthesis had arisen independently at two points in evolution, since it also exists in blue-green algae. Hence we look in vain for a plausible origin for the photosynthetic flagellates among existing lower protists.

The conclusions reached from this survey of the Protista are summarized in Fig. 2.1. For those who are interested in a more detailed exploration of the taxonomic and evolutionary questions that have been raised, a brief list of selected references is appended.

References

Benecke, F. 1912. *Bau und Leben der Bakterien.* B. G. Teubner, Leipzig.   **Copeland, H. F. 1938.** The kingdoms of organisms. *Quart. Rev. Biol.,* 13:383–420.   **Doflein, F. and E. Reichenow. 1949.** *Lehrbuch der Protozoenkunde,* 6th ed. G.. Fischer, Jena.   **Fritsch, F. E.** *The Structure and Reproduction of the Algae.* Vol. I, 1935. Vol. II, 1945. Cambridge University

Press, Cambridge. **Langeron, M. 1945.** *Précis de mycologie.* Masson, Paris. **Lwoff, A. 1944.** *L'Évolution physiologique, étude des pertes de fonctions chez les microorganismes.* Hermann, Paris. **Pringsheim, E. G. 1949.** The relationships between bacteria and myxophyceae. *Bacteriol. Rev.,* 13:47–98. **Smith, G. M. 1955.** *Cryptogamic Botany,* 2nd ed. McGraw-Hill, New York. **Sparrow, F. K. 1943.** *Aquatic Phycomycetes.* University of Michigan Press, Ann Arbor. **Stanier, R. Y., M. Doudoroff, and E. A. Adelberg. 1957.** *The Microbial World.* Prentice-Hall. Englewood Cliffs, New Jersey. **Thimann, K. V. 1955.** *The Life of Bacteria.* Macmillan, New York.

3

Bacteria

C. B. VAN NIEL

R. Y. STANIER

Introduction: Some General Remarks on the Definition and Classification of Bacteria

It has never been easy to find satisfactory criteria for defining the group of microorganisms which by common consent is regarded as "bacteria." However, this does not imply that it is, in general, difficult to recognize the majority of such organisms as "bacteria"; rather the reverse is true.

Perhaps the best way to arrive at an answer to the question, "When is an organism considered a bacterium?" is to start by describing the attributes of the various major bacterial groups. Our aim will not be to achieve a *scientific classification*, as this term is understood by taxonomists of the higher plants and animals. Scientific classifications have definite phylogenetic connotations, and present-day knowledge of the bacteria is much too inadequate to permit attempts along such lines. But simply as an aid in the gross identification of the bacteria most frequently encountered in bodies of water, the various groups described below will, we believe, provide a useful general chart of the material.

First, there are the many types of so-called "true bacteria" or "eubacteria," which are small, unicellular organisms of various shapes, equipped with rigid

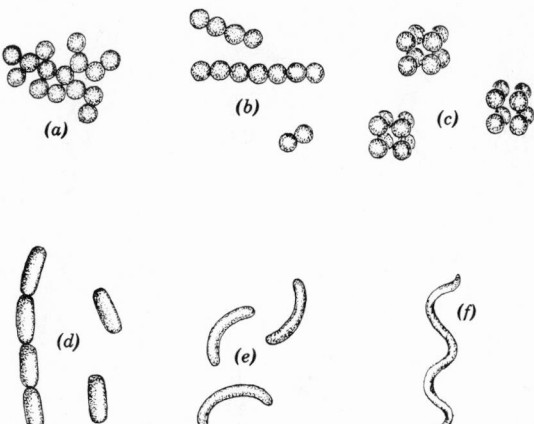

Fig. 3.1. Cellular form and arrangement of unicellular true bacteria. (*a*) *Micrococcus.* (*b*) *Streptococcus.* (*c*) *Sarcina.* (*d*) *Bacterium.* (*e*) *Vibrio.* (*f*) *Spirillum.*

cell walls and often capable of locomotion in a liquid medium. Most uni-cellular true bacteria are from 0.5 to 1.5 μ in diameter and not more than 10 μ in length; a few organisms of considerably larger dimensions are also included, since they can be linked by a series of gradations to the smaller members. Three basic forms occur (Fig. 3.1): spheres ("cocci"); cylinders with square or rounded ends, sometimes assuming an almost ellipsoidal shape ("bacteria" or "bacilli"); and curved rods, consisting of a single half-turn ("vibrios") or one to several complete helices ("spirilla"). Multiplication is always accomplished by binary transverse fission. Sometimes cells remain attached after division, forming small and characteristically arranged aggre-gates. Thus spherical eubacteria may occur as chains (*Streptococcus* type), as tetrads, or as cubical packets (*Sarcina* type). Chain formation also commonly occurs in certain rod-shaped eubacteria, notably some of those that form endospores (see below). Motility is probably universal in the curved eubac-teria, common in the cylindrical forms, and extremely rare in the spherical ones. Its occurrence is associated with the presence of locomotor organelles known as flagella, which are very fine hairlike structures of cytoplasmic origin, composed of fibrous proteins and extending through the cell wall (Fig. 3.2*a*). Each flagellum is of the order of 0.03 μ in diameter and of variable length (up to at least 20 μ). Being well below the limit of resolution of the light microscope, individual flagella cannot be seen on living cells; but during movement they tend to become intertwined in flagellar bundles, which can just be seen under favorable conditions by dark-field or phase microscopy, and with certain large eubacteria even by ordinary light microscopy. Studies by the dark-field technique show that the flagellar bundle is always posteriorly situated on a moving cell, and displacement is achieved by a propellerlike action. There are two principal modes of insertion of the flagella: in some bacteria they are attached at one or both poles of the cell (polar flagellation);

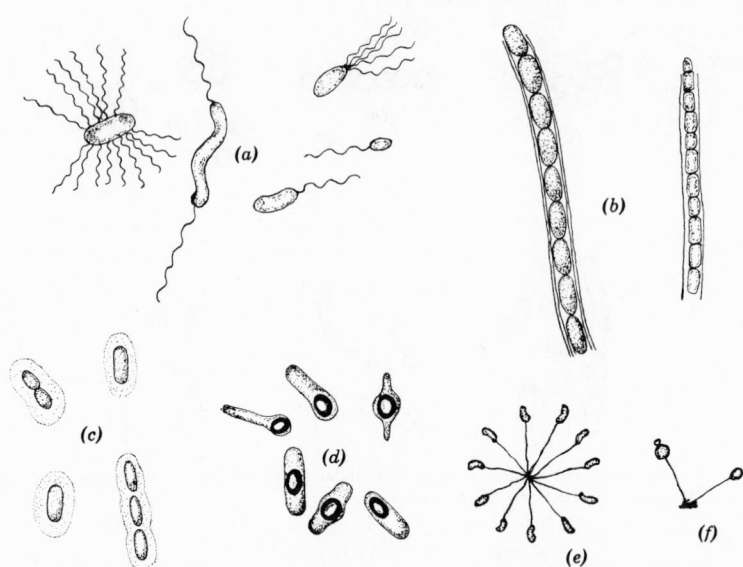

Fig. 3.2. Morphology of true bacteria. (*a*) Various types of flagellar arrangement. (*b*) Sheaths. (*c*) Capsules. (*d*) Endospores. (*e*) Stalks, *Caulobacter* type. (*f*) Stalks, *Blastocaulis* type.

in others, they are attached laterally at a number of points (peritrichous flagellation). Electron-microscopic studies show that the bacterial flagellum is a much simpler structure than the homologous organelle of algae, protozoa, and phycomycetes which bears the same name; it consists of a single fibril, and has been termed by one protein chemist "a macromolecular hair," whereas the flagella of the other microbial groups are always composite and contain many fibrils, often arranged in a fairly complex manner.

A highly characteristic resting structure known as an *endospore* (Fig. 3.2*d*) is produced by many true bacteria. One endospore arises in each vegetative cell by a process of free cell formation, during which part of the cytoplasm is surrounded and cut off from the rest by a highly refractile and impermeable wall. The endospore may then be liberated by the disintegration of the remaining cytoplasm and the surrounding vegetative cell wall. The formation of endospores is fairly common in rod-shaped eubacteria, rare in cocci, and of doubtful occurrence in vibrios and spirilla.

Some true bacteria are capable of secreting extracellular structures known as capsules, stalks, and sheaths (Fig. 3.2*b,c,e,f*). Capsule formation is quite common, and consists in the deposition about the cell of a slime layer which may extend for a distance several times the diameter of the cell itself. The extent, physical sharpness, and permanence of this slime layer are exceedingly variable, being conditioned by such factors as the properties of the capsular material (e.g., its solubility in the medium) and the kind and amount of nutrients available. Stalk formation is also probably not uncommon, particularly in habitats that are low in dissolved nutrients.

The simplest stalked eubacteria (*Caulobacter* type) consist of single cells, attached at one end to the substrate by a holdfast or short stalk, and equipped at the other end with one or more polar flagella; following cell division, the distal cell swims away, and in turn becomes sessile on a suitable surface. There is some evidence to suggest that organisms of the *Caulobacter* type are free-swimming in environments rich in nutrients, and become sessile only when the supply of nutrients falls to a low level.

Some eubacteria form more complex stalks. In these organisms (the *Gallionella* type), the kidney-shaped cell lies at right angles to the stalk, which is a regularly twisted, bandlike structure, and becomes bifurcated as a consequence of cell division; the two daughter cells remain attached to the tips of the resulting branches (Fig. 3.10). By repeated divisions in this fashion, a many-celled colony united to the substrate by a much-ramified stalk is formed. Ensheathed true bacteria (*Leptothrix* type) consist of chains of cells enclosed within a common sheath; reproduction occurs by the liberation of flagellated single cells ("swarmers") from the open end of the sheath. The sheath may lie free, may adhere along its entire length to the substrate, or may be attached terminally to the substrate by a holdfast.

The majority of unicellular true bacteria are nonphotosynthetic, but there are also a few photosynthetic forms; with the single exception of *Rhodomicrobium*, whose morphological peculiarities will be described later, these so-called *green* and *purple eubacteria* are the only organisms commonly regarded as bacteria that are characterized by a photosynthetic type of metabolism. Like the blue-green algae, they show no microscopically observable localization of photosynthetic pigments within the cell (chloroplasts absent).

The eubacterial cell shows little internal structure in the living state: the nuclear bodies are not visible, plastids cannot be detected, and vacuoles are extremely rare—perhaps never present in healthy, young cells. The cytoplasm has a finely granular or hyaline appearance, whose regularity is marred only in some cases by inclusions of sulfur, fat, volutin, or other reserve materials. Protoplasmic streaming has never been observed.

Linked to the unicellular true bacteria by a singularly complete series of transitional forms are the actinomycetes, which have a mycelial vegetative structure. The mycelium is composed of much-branched, nonseptate hyphae seldom more than 0.5 μ in diameter. Some actinomycetes reproduce by massive fragmentation of the mycelium into short elements indistinguishable from rod-shaped true bacteria, each of which can again give rise to a mycelium; others have a permanent mycelium, and reproduction occurs exclusively by the formation of spherical conidia, borne either singly or in chains on the tips of surface hyphae. Permanent immotility is characteristic of most actinomycetes.

A few multicellular filamentous bacteria are known that show unmistakable resemblances in general structure to unicellular eubacteria. The best studied is *Caryophanon*, a very large organism that occurs in cow dung. This organism may reach 5 by 30 μ in size, and consists of a peritrichously flagellated trichome containing 10 to 20 disc-shaped cells, each of which measures about 5 by 1.5

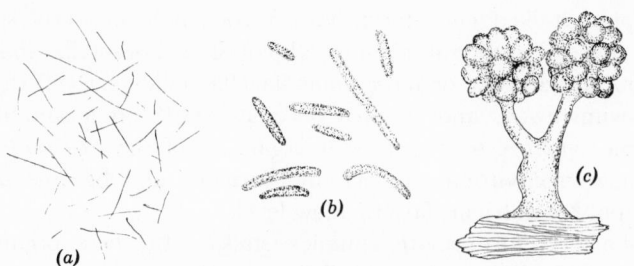

Fig. 3.3. Myxobacteria. (*a*) Habit sketch of cytophagas. × 450. (*b*) Vegetative cells of a fruiting myxobacterium. × 1400. (*c*) Fruiting body of *Chrondrymyces*. × 20.

to 3 μ. It reproduces by binary transverse fission of the entire trichome.

Sharply distinguishable from the true bacteria and their congeners is a second major group, the myxobacteria (Fig. 3.3). Myxobacterial vegetative cells are rods of approximately the same dimensions as rod-shaped true bacteria (0.5 to 1.5 by 5 to 10 μ), and likewise multiply by binary transverse fission (Fig. 3.3). They are very weakly refractile and highly flexible, features attributable to the absence of a rigid cell wall. Motility is universal (although it may be suppressed temporarily under unfavorable environmental conditions), and is of the gliding type also found in blue-green algae; it occurs only when the cell is in contact with a solid surface, and is not associated with the presence of demonstrable locomotor organelles. The simplest myxobacteria (*Cytophaga* type) exist only in the vegetative condition. However, many members of the group form elaborate, spatially localized, many-celled *fruiting bodies* which contain resting cells (Fig. 3.3c). The individual resting cells may be either shortened rods, distinguishable only by their size from vegetative cells, or spherical to oval structures known as *microcysts*, which are surrounded by thick, refractile walls. The mode of formation of microcysts is quite different from that of eubacterial endospores; each microcyst is produced by the rounding up and encystment of an entire vegetative cell. Sometimes many individual resting cells may be enclosed in a much larger common membrane, such structures being termed (macro-) *cysts*. The fruiting body is often elevated from the substrate by a stalk composed of hardened slime and vegetative cells.

Certain gliding, nonphotosynthetic organisms with a filamentous vegetative structure are customarily included among the bacteria. The best-known representatives are the filamentous colorless sulfur bacteria (*Beggiatoa* type), easily recognizable because their trichomes often contain many globules of elementary sulfur (Fig. 3.8). Morphologically, *Beggiatoa* and the other filamentous sulfur bacteria show unmistakable resemblances to blue-green algae belonging to the family Oscillatoriaceae, and they have often been regarded as apochlorotic relatives of these photosynthetic organisms. Other colorless, filamentous gliding organisms (*Vitreoscilla* type) show less certain resemblances

to specific forms of blue-green algae; some of them have very short and un-
stable trichomes and appear to be transitional to the unicellular nonfruiting
myxobacteria of the *Cytophaga* type.

The two major assemblages so far described, for which one may use the general designations of "swimming bacteria" and "gliding bacteria" (Pringsheim,
1949c), include between them the bulk of the bacteria. There remain to be
considered a few small groups which, for one reason or another, cannot be
fitted satisfactorily into either major category, yet have properties that allow
them—perhaps *faute de mieux*—to be regarded as "bacteria." Of these, the
most important are the *spirochaetes:* unicellular organisms with highly flexible
spiral cells, which multiply by binary transverse fission (Fig. 3.4a). The
spirochaetal cell is extremely slender (seldom more than 0.3 μ in diameter).
Two subgroups can be distinguished on the basis of the length of the cell.
Cells of the small forms (*Leptospira* type) rarely exceed 10 μ in length, whereas
cells of the large forms (*Spirochaeta* type) may attain lengths of several hundred
microns. It is by no means evident that these two groups share a common
cell structure, particularly since cytological studies of the large-celled group
are still fragmentary. All spirochaetes are motile in a liquid medium, and
there is good evidence for the occurrence of flagella on some of the small-
celled forms. In spite of this, their slenderness and, above all, their flexibility
can easily serve to distinguish them from spiral eubacteria.

All the organisms so far described multiply by binary transverse fission. A
few bacteriumlike organisms that multiply by budding are known. The
photosynthetic form, *Rhodomicrobium* (Duchow and Douglas, 1949), has small,
oval cells from which thin, filamentous extensions are formed, at whose tips
buds develop which gradually reach the dimensions of the parent cell (Fig.
3.4b). At this stage a transverse wall is laid down in the connecting filament,
thus separating the two cells. Frequently they remain attached, however, and
by continued budding produce loose colonies held together by the connecting
filaments. A roughly similar mode of multiplication occurs in the nonphotosynthetic form, *Hyphomicrobium* (Mevius, 1953).

Last, there is *Crenothrix*, a large, conspicuous, filamentous organism first
described almost one hundred years ago, but still not very well known. The
filament is composed of large cylindrical cells, very variable in size (2 to 5
μ wide, up to 25 μ long), and the entire filament may be almost 1 mm in
length (Fig. 3.10c). It is enclosed in a sheath attached to the substrate,

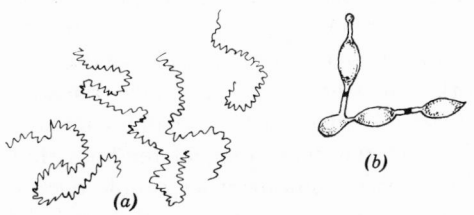

Fig. 3.4. (a) Habit sketch of *Spirochaeta plicatilis.* × 660. (b) *Rhodomicrobium.* × 2000.

whose distal portion is thin and colorless, becoming thicker and yellowish-brown as the basal region is approached (a consequence of the deposition of Fe_2O_3). It reproduces by nonmotile, approximately spherical conidia, which are produced in abundance in the distal portion of the filament.

Apart from the filterable viruses, which are with dubious propriety included among bacteria by certain taxonomists, this terminates the list of major recognizable morphological types. Many attempts have been made to formulate a definition including all these organisms that would permit a simple and clear-cut distinction from other microbial groups; none of these attempts has stood the test of time. The chief stumbling block is the apparent structural simplicity of bacteria; no outstanding *positive* group characteristics are evident. Unpalatable as such a statement may be to the scientific mind, we are forced to admit that at the present time it is mainly because of their *small size* that certain microorganisms are classified as bacteria. For example, the actinomycetes resemble certain imperfect fungi very closely in general morphology, but they are generally considered bacteria because they are distinguishable at a glance by the fineness of their hyphae, which are always at least five times thinner than fungal hyphae. A second feature that helps to distinguish bacteria from most protozoa, fungi, and algae (with the exception of blue-green algae) is the peculiar appearance of the cytoplasm in living cells, notably the absence of vacuolization and of streaming movements. This property will enable a careful observer to differentiate without difficulty the large, nonmotile, rod-shaped true bacteria from the fission yeasts that may be broadly similar in size and mode of vegetative reproduction. The one microbial group whose differentiation from bacteria presents a really difficult problem is the group of blue-green algae, some of whose representatives are morphologically indistinguishable from colorless gliding bacteria—probably a reflection of a close genetic relationship between the two groups. The distinction is arbitrarily made on a physiological basis; but to the microscopist, differentiation between a photosynthetic blue-green algae and a colorless, gliding bacterium may not prove easy, since visual detection of photosynthetic pigments is dependent on the ability to observe a sufficient mass of cell material—a point discussed further in the section on methods.

Methods

The exact identification of most bacteria can be accomplished only by means of more or less elaborate culture techniques and diagnostic tests which must be conducted on pure cultures. The elementary student of fresh-water biology is not likely to possess the special knowledge and equipment necessary for this type of work, unless he is a bacteriologist already—and if such is the case, the present brief and elementary account can be ignored. We shall restrict ourselves here to the description of methods that can be employed by anyone who possesses the equipment for simple microscopy, but it should be emphasized that for a more detailed study of aquatic bacteria cultural and physiological procedures must be adopted.

Direct observation with a good microsope, supplemented by a few simple staining methods and microchemical tests, and coupled with a fair knowledge of environmental factors, permits the approximate identification of numerous bacterial types. Details of shape and certain structures, particularly flagella and stalks, are often more clearly visible by darkfield illumination or by phase-contrast microscopy. Stains are sometimes helpful, but the widely held notion that bacteria are difficult to see in the living state is incorrect, and the observation of living material should always be given first preference. On the basis of the shape of individual cells or cell aggregates, usually the first feature that is noted, a primary distinction can be made among cocci, rod-shaped bacteria, spirilla, spirochaetes, filamentous forms, and mycelial forms. Sometimes cell size may suggest itself as a significant diagnostic character, particularly if it is outside the normal range for bacteria. Furthermore, the presence of conspicuous inclusion bodies, of readily observable cellular pigments, or of colored capsules and sheaths can be used as a means for rapid primary differentiation because of their infrequent occurrence and characteristic association with highly specialized bacterial types.

The inclusion bodies found most frequently in aquatic bacteria are sulfur globules and crystals of calcium carbonate. The former can be recognized by their very high refractility, appearing when small as spherical, black bodies, and when somewhat larger as round structures with an intense and rather wide black border. When cells containing sulfur globules are heated to the boiling point of water, the sulfur melts, and if there are many such bodies in each cell, they fuse; on cooling, the sulfur solidifies, and eventually forms characteristic monoclinic crystals. Crystallization is generally slow, and preparations so treated should be examined for crystals at various times up to a day later. Material suspected of containing sulfur can also be extracted, after careful drying, with carbon disulfide; this treatment causes disappearance of the intracellular globules, and from the carbon disulfide extract it is usually easy to obtain well-developed monoclinic crystals. When a sufficiently large amount of material is available, the presence of sulfur can generally be detected by preliminary drying followed by brief exposure to a flame. The sulfur then burns (blue flame!) and emits SO_2, easily recognizable by its odor.

Inclusion bodies of calcium carbonate are also spherical ("sphaerolites"), and even more dense and opaque than sulfur globules. On treatment of such cells with dilute acid, the crystals dissolve; if the quantity of calcium carbonate is large and the total liquid volume small, dissolution may be accompanied by the visible evolution of gas bubbles (CO_2), but the solubility of CO_2 in water is high, and it is often difficult to observe such a liberation of gas.

Some bacteria are strikingly colored—an obvious diagnostic character of considerable value. However, the amount of pigment in a single cell of bacterial dimensions is generally too small to be observable microscopically, and the unequivocal detection of color is possible only when large masses of bacteria (as clumps) are present, or when the individual cells are exceptionally large.

Although motile bacteria are not apt to be overlooked or confused with inanimate particles, nonmotile bacteria are not so easily recognized by the beginner, particularly if the water sample also contains much fine debris. It is here that stains become useful, since they will often color the cells selectively. Dilute (0.1 per cent) aqueous methylene blue and aqueous carbol erythrosin (phenol 5 per cent, erythrosin 1 per cent) can be recommended; of the two, the latter is perhaps more selective. The material to be stained is dried on a slide at room temperature, "fixed" by passing the slide (preparation upwards) through the flame of a Bunsen or alcohol burner, and flooded for a few minutes with the dye; it is then rinsed, and either dried and examined under oil immersion, or examined wet through a coverslip with a high dry or oil immersion lens. Examination in the wet state is preferable. It should be kept in mind that such stains do not infallibly differentiate cells from debris.

The use of India ink (a suspension of very fine carbon particles) as a so-called "negative stain" is very helpful in revealing capsules, sheaths and stalks. In general, the best method for applying it is to place a small drop on a slide beside a small drop of the material under study, and to lower a coverslip onto the slide in such a way that the two drops merge with the formation of a gradient between the carbon suspension and the sample. Care should be exercised in avoiding an excess of India ink, which renders the material on the slide completely invisible. Aggregation of the carbon particles is likely to occur if the solution being examined is strongly acid or alkaline; in such cases, preliminary neutralization is indicated. In satisfactory preparations capsules, sheaths, and stalks appear as clear areas surrounding or attached to the cells, against the gray background furnished by the carbon particles. An advantage of the India-ink method is that the bacteria need not be killed, so that observations under conditions of increased contrast can be made without destroying vital activities.

Especially useful for the study of *iron bacteria* is a staining procedure based on the formation of Prussian blue. The material to be examined is covered with or dispersed in a solution of potassium ferrocyanide, *after which* a dilute solution of a mineral acid is added. As the acid dissolves the water-insoluble iron compounds ($Fe(OH)_3$ and Fe_2O_3) with which the sheaths and stalks of the iron bacteria are impregnated, the ferric ions are immediately precipitated *in situ* by the ferrocyanide as Prussian blue. If the acid is added at the side of the ferrocyanide-treated material so that it slowly diffuses into the latter, the gradual formation of the Prussian blue can be followed under the microscope, and the staining stopped at the desired moment by washing the slide in water. In this manner more delicately stained preparations can often be obtained. Subsequent staining with phenol-erythrosin provides a means of revealing the presence of bacterial cells; the red-stained cells contrast beautifully with the blue-stained sheaths or stalks.

Observations on living material make possible the detection of motility. Movements caused by flagella are usually rapid (about 30 to 50 μ per second); gliding motility is very much slower (less than 50 μ per minute), and, in contrast to flagellar motility, occurs only when cells are in direct contact with a solid substrate. It is sometimes difficult to distinguish between bacterial motility and Brownian movement; mistaken conclusions are best avoided by a careful comparison of the behavior of the cells in the absence and presence of poisons such as dilute $HgCl_2$. Currents, often the result of evaporation of water around the edges of the coverglass, may also simulate active cellular movement; interference from this source can be avoided by the use of "hanging drops," prepared by inverting a coverglass with a drop of material on it over a depression slide, and sealing the edge with Vaseline, paraffin, or oil.

In true bacteria, the type of motion gives strong indications concerning the arrangement of flagella. Polarly flagellated bacteria exhibit "darting" movements, often accompanied by rapid rotation about the long axis of the cell, whereas peritrichously flagellated bacteria move in a more stately manner, generally with periodic lateral oscillations of the cell. It should be remembered that physico-chemical factors, above all oxygen tension, greatly influence movement. Strictly aerobic bacteria display motility only in the presence of oxygen, and their movement is likely to come to a halt fairly soon if the suspension is dense and the space between slide and coverglass completely filled with liquid. Conversely, strictly anaerobic bacteria are rapidly killed by exposure to oxygen, and become motionless if air has free access to the suspension, as in hanging drops. A simple way to provide both aerobic and anaerobic conditions in one preparation is to include a few small air bubbles in a wet mount; aerobic bacteria remain motile longest in the immediate vicinity of the air bubbles and tend to accumulate around the edge, while anaerobes soon become motionless in those spots and form aggregations where the oxygen tension is lowest. A third type of behavior is shown by the so-called "microaerophilic" motile bacteria, especially the colorless spirilla. These organisms require oxygen but are favored by a low oxygen tension; hence, they accumulate in sharp, narrow bands a short distance away from the edge of the air bubbles, moving ever closer to the air-water interface as the oxygen supply diminishes.

The accumulation of motile bacteria in certain parts of the slide in response to differences in oxygen tension is one manifestation of chemotactic behavior. In general, chemotaxis can be observed when bacteria move through an environment in which there is a sufficiently steep concentration gradient of a response-inducing substance. Passage into an area of unfavorable concentration causes the organisms to reverse the direction of their movement, which leads to a gradual "trapping" of cells in regions of optimal concentration. Many chemicals can induce such taxes; the phenomenon may be simply demonstrated by inserting a capillary filled with a solution of a suitable substance into a drop that contains a suspension of motile bacteria. The concentration gradient established as a result of diffusion causes an accumulation of cells towards (positive taxis) or away from (negative taxis) the source of the solution used.

A few bacteria also display phototactic responses. The photosynthetic purple sulfur and nonsulfur bacteria react to changes in light intensity in the same manner nonphotosynthetic motile bacteria react to changes in the concentration of chemical substances. To observe phototactic movements, it is often sufficient to darken one-half of the microscopic field and watch the behavior of bacteria swimming from the light into the dark region. A sudden reversal of direction as the cell crosses the boundary region that brings it back into the area of higher light intensity is an unmistakable indication of phototaxis, and thereby establishes the organism as a purple bacterium. A more refined, and slightly more elaborate device is the "light trap," generally a piece of opaque material with a small hole which is placed between the light source and the bacterial suspension. Motile purple bacteria eventually all collect in the small illuminated area. As a simple procedure for determining photosynthetic ability in motile bacteria establishment of a phototactic response is unsurpassed and definitive, and of particular value in view of the fact (mentioned above) that the characteristic pigmentation often cannot be observed in single cells.

So far, the procedures discussed have been based on the tacit assumption that the samples to be examined contain sufficient bacteria to be observable by direct microscopy. It must be realized that direct microscopy is feasible only if the bacterial population is at least of the order of 10^7 cells per ml; examined under high power, such a suspension will show, on the average, only about 10 cells per field. Since the population density of clear bodies of water with low content of organic matter (e.g., most lakes and streams) is much below this figure, concentration of the cells by a process such as filtration, is often an essential preliminary to microscopic examination. A filter with a maximum pore diameter of 0.2 to 0.5 μ is required to retain bacteria, and the liquid must be forced through it by pressure or suction. Resuspension of the collected cells is not always easy, but can usually be achieved by passing a small volume of water through the filter in the reverse direction (see Chapter 46).

In environments with low concentrations of dissolved nutrients, only a small proportion of the bacterial population is free-floating; the majority develop characteristically in close physical contact with (and often actually attached to) solid objects—planktonic organisms, sessile algae, and aquatic plants, rocks, etc. This periphytic habit of growth, as it is called, can be partly attributed to the fact that such solid structures furnish the bacteria with nutrient materials (e.g., excreted organic substances from plankton organisms and aquatic plants, the sheath materials of algae). Even when the substrate is not a source of nutrients, attachment still offers substantial advantages to the bacteria. Because of the flow of water past it, a fixed cell is exposed to a far greater amount of dissolved nutrients than is a freely floating one. It is also possible that dissolved organic materials are concentrated on solid substrates by adsorption. A special method, the *submerged-slide technique*, can be used for the study of these periphytic bacteria; they gradually accumulate and grow on the surface of a glass slide immersed in water and they may produce a rich and varied population, often in the form of characteristic microcolonies. Many bacteria that would escape detection by direct examination of water samples grow luxuriantly on such submerged slides, and are revealed in their natural arrangement. Because of its simplicity and far-reaching potentialities, the submerged-slide technique is the method of choice for investigating the microflora of most natural waters, except those (e.g., hot springs, or grossly polluted streams, lakes, and ponds) with very large bacterial populations.

Henrici (1933) recommended attaching the slides at intervals to a line suspended from a float and fastened to an anchor of adequate weight. Rubber-covered copper wire forms a satisfactory line. A piece of adhesive tape is doubled around one end of a slide to form a small projecting tab which can be perforated for attachment to the line. The tab may be labeled as desired, coated with paraffin, and fastened to the line with a short length of rubber-covered wire. A much simpler procedure, useful if one is not interested in studying the microflora at various depths, is to stick a slide or coverslip into a piece of cork, which is then floated on the water, glass down. The use of readily decomposable materials (silk, cotton, iron) or of toxic substances for lines and for slide attachment should be avoided, since they may cause considerable modifications of the developing microflora.

The rapidity with which bacteria develop on such slides is very variable. In Henrici's experiments, principally conducted in Wisconsin and Minnesota lakes, about three to ten days usually elapsed before films of suitable density for microscopic examination were formed. On the other hand, we have often obtained very satisfactory slide cultures in less than twenty-four hours. It is wise to submerge a number of slides at each site and to remove them at intervals. By this means the slides can be examined before they have become too densely covered and something can be learned of the succession of microbial types.

After removal from the water, slides may be examined in wet mounts (the most satisfactory procedure, in our opinion) or fixed and stained by the methods mentioned in earlier paragraphs. Henrici has shown that the slide-culture technique can be extended to the quantitative analysis of bacterial populations in water. For further details, the original papers should be consulted (Henrici, 1933, 1936, 1939; Henrici and Johnson, 1935).

General Considerations on Bacterial Ecology

There is almost no known type of metabolic activity that cannot be found in some bacterial group; and many kinds of metabolic reactions are brought about uniquely by special groups of bacteria. Some bacteria exhibit a high degree of nutritional versatility; it has been shown, for example, that *Pseudomonas* species may be able to satisfy their energy and carbon requirements by oxidizing any one of approximately 100 simple organic compounds. Other bacteria may show an equally high degree of nutritional specialization; for example, the nitrifying bacterium *Nitrosomonas* can obtain its energy only by the oxidation of ammonia to nitrite. With respect to oxygen, every conceivable mode of response may be found in bacteria: some are strict aerobes, some are strict anaerobes, some may grow best in the presence of low concentrations of oxygen (the microaerophiles), and some can develop well in either the presence or the absence of oxygen. The same diversity is shown with respect to hydrogen-ion concentration; although the majority of bacteria grow best under neutral or slightly alkaline conditions, certain sulfur-oxidizing forms flourish in environments with a pH close to 0, while the urea-decomposing bacteria grow well at a pH of 11. The temperature range is equally wide; certain marine bacteria are able to grow at temperatures slightly below the freezing point of pure water, and thermophilic forms can be found in hot springs at a temperature of 80 C. In view of these facts, it is not surprising that the bacteria are widespread in nature; indeed, there is probably no natural environment capable of supporting the development of living organisms in which bacteria cannot be found. Clearly, therefore, the kind of ecological generalization that can be made concerning one of the higher groups of plants or animals would be meaningless here.

Nevertheless, it may be contended that the ecology of the bacteria can be studied with a greater precision and elegance than that of any other living group. It is, however, a microecology, since the significant environment for a given type of bacterium may be contained in a volume of a few cubic microns. A single cellulose fiber undergoing decomposition in mud or soil will support a characteristic and highly specialized microflora, and in closely adjacent regions wholly different microfloras may predominate. Thousands of such microenvironments may lie concealed from the gross ecological eye in a few grams of soil or mud.

For the study of microecology, the microbiologist has devised a special procedure known as the enrichment- or elective-culture technique; this procedure consists in preparing a nutrient medium of defined chemical composition, which is inoculated with a mixed bacterial population (e.g., that contained in a small amount of soil or water) and incubated under defined conditions of temperature, aeration, and illumination. By determining the nature of the predominant resulting microflora, it is then possible to define with great precision the ecological conditions that favor the development of the microbial group in question; by extension, it can be assumed that in a natural environment where such physico-chemical conditions occur, this particular microbial group will come to the fore. As a result of the knowledge derived from enrichment-culture studies conducted over the past half-century, it is now possible to state with a fair degree of certainty the nature of the predominant microflora that will be found in many natural environments where the physico-chemical conditions are relatively well defined. It may be noted in passing that there appear to be no geographical limitations on the distribution of the special physiological groups of bacteria that are amenable to study by enrichment-culture methods; a particular physico-chemical microenvironment, no matter what its geographical location, will lead to the emergence of the same predominant microflora. It is possible that certain physiological groups of bacteria have distinct terrestrial, marine, and freshwater representatives; but this point has still not been established with certainty.

Thanks to their physiological versatility, the bacteria play cardinal roles at a number of different points in the cycle of matter in nature. They are found in almost every situation in which organic matter is formed by the metabolic activities or death of other living organisms; they are the principal agents in the so-called *mineralization process*—the decomposition of this organic matter with eventual liberation of carbon as carbon dioxide, nitrogen as ammonia, phosphorus as inorganic phosphate, and sulfur as hydrogen sulfide. The further oxidations of ammonia to nitrates and of hydrogen sulfide to sulfates are also brought about by special groups of bacteria. Other bacteria may reduce these most highly-oxidized forms of carbon, nitrogen, and sulfur, thereby converting carbonates to methane, nitrates to nitrites, nitrous oxide, or elementary nitrogen, and sulfates to hydrogen sulfide. The fixation of atmospheric nitrogen, an essential reaction in the maintenance of the earth's fixed nitrogen supply, is very largely a bacterial activity, although blue-green algae also participate.

In any large natural environment, such as a body of water, all these bacterial transformations of matter proceed simultaneously, although at any given time and place one particular process may be predominant, thus leading to a temporary mass development of the responsible microbial agents. When several different kinds of bacteria are capable of performing the same chemical transformation, the exact nature of the microflora will be determined by the physical conditions, such as the presence or absence of oxygen and light, or the hydrogen-ion concentration. This point can be illustrated by considering a very simple example—the mineralization of acetate, which is a common product of the breakdown of carbohydrates, fats, and proteins. When acetate is liberated into an anaerobic environment shielded from illumination, it can be oxidized by the representatives of three special groups: the methane bacteria, which couple the oxidation with a reduction of carbon dioxide to methane; the sulfate-reducing bacteria, which couple the oxidation with a reduction of sulfates to hydrogen sulfide; and the nitrate-reducing bacteria, which couple the oxidation with a reduction of nitrates, principally to N_2. Which group actually predominates will be determined by the relative availability of CO_2, sulfates, and nitrates. Under anaerobic conditions where light is available, the nonsulfur purple bacteria, which employ acetate as the oxidant for the photosynthetic reduction of carbon dioxide, will come to the fore. If oxygen is present, other bacteria predominate, notably the *Pseudomonas* types, most of which are strict aerobes and oxidize acetate with molecular oxygen. When the supply of combined nitrogen is limited in an oxygen-rich environment, the *Pseudomonas* types will be supplanted as acetate-oxidizers by nitrogen-fixing bacteria of the *Azotobacter* group.

In the following sections, we shall discuss the application of these general principles to specialized aquatic environments, and indicate, insofar as this can be done, the predominant bacterial types the investigator may expect to encounter.

Clear Lakes and Streams

Clear lakes and streams contain little dissolved organic matter and have a relatively high oxygen content. The total bacterial population is low (of the order of 1 to 1000 organisms per ml), except in close contact with surfaces. From studies by bacteriological techniques such as the plating method and the elective-culture procedure, it appears that the commonest bacteria in such environments are representatives of the *Vibrio* and *Pseudomonas* groups— polarly flagellated true bacteria, strictly aerobic, and capable of oxidizing many simple organic compounds. The detection of these forms is usually not possible by microscopy, however, since their absolute abundance is so low. The slide-culture technique is particularly useful when studying this kind of environment; it reveals in addition the existence of many highly distinctive morphological types, most of which have not yet been studied by the usual bacteriological methods. Some of the common forms likely to be so encountered will be described below.

Stalked Forms

The commonest representatives are unicellular rods or vibrios, attached to the slide by slender stalks and often occurring in rosettelike clusters. Multiplication occurs by binary transverse fission; the outermost daughter cell is set free and swims away until it encounters a new substrate, when it settles down and proceeds to secrete a stalk. Apart from the peculiarity of stalk formation, many of these forms appear to be typical pseudomonads or vibrios; they are commonly referred to as bacteria of the *Caulobacter* type. Other simple, stalked organisms that also multiply by binary transverse fission have distinctly fusiform cells; after division, the outer daughter cell develops a long slender tip, and occasionally pairs of cells with a stalk and holdfast at both extremities can be observed. Hence, it seems probable that the apical cell develops its stalk and becomes anchored to the substrate prior to the completion of cell separation, a conclusion borne out by the observation that bacteria of this type are often found in quite large patches on a slide. These forms may be stalked myxobacteria, but the point has not yet been established with certainty.

Other unicellular, stalked bacteria differ from the above-mentioned organisms in their method of reproduction, which occurs by bud formation at the distal tip of the cell. They are commonly referred to as organisms of the *Blastocaulis* and *Hyphomicrobium* types. The structure of the stalk shows considerable variety; some forms are attached directly to the substrate by an amorphous holdfast, and others have long, slender stalks, which may radiate in whorls from a common center of attachment.

A more complex organization is found in organisms of the *Nevskia* type, which appear to be relatively rare. The organism forms a colony held together by a gummy material arranged in the form of dichotomously branched stalks which arise from a common base; a single cell tips each branch. The cells are rod-shaped and, in contrast to those of the *Caulobacter* type, are set at right angles to the axis of the stalk, so that upon transverse fission a dichotomization of the stalk occurs. The growth habit is similar to that encountered in iron bacteria of the *Gallionella* type.

Unstalked Forms

Many other bacteria, not visibly attached to the substrate by stalks or hold-fasts, will be encountered on slide cultures.   Long, many-celled filaments, frequently enclosed in sheaths, are common.   Many of them are doubtless true bacteria of the *Leptothrix-Sphaerotilus* type, but it is likely that filamentous, gliding organisms of the *Vitreoscilla* group, known to occur in fresh water, will also develop on slides.   A satisfactory diagnosis can best be based on observation of living material for the type of motility.   Microcolonies of rod-shaped or spherical bacteria, sometimes enclosed in mucoid capsules, also occur.   Forms whose capsules contain ferric hydroxide are known as the *Sideromonas* and *Siderocapsa* types (Hardman and Henrici, 1939).   Spiral bacteria (spirilla and spirochaetes) have been found on slide cultures.   The amount of work that has been done by the slide-culture method is relatively restricted, and it is very likely that any observer who undertakes a careful and extensive study by this method will find many additional types not described above.

Sulfur Springs and Sulfide-Containing Environments

Bodies of water containing H_2S have a highly specialized and character-istic microflora.   The restricted character of this microflora is a consequence of the fact that H_2S is poisonous for most living organisms, but can be used as an energy source by a few groups of bacteria.   Sulfide may be formed by volcanic activities (as in sulfur springs), or by the microbiological decomposition of organic matter.   In the former case, the microflora is composed almost exclusively of so-called "sulfur bacteria"; in the latter, these organisms will be accompanied by the various types of bacteria that participate in the decomposition of the organic material and in the formation of H_2S.

The sulfur bacteria, so named by Winogradsky (1887), comprise three groups which can be readily recognized by microscopic examination.   They are:

1. The green and purple (or red) sulfur bacteria.
2. The large, colorless, filamentous sulfur bacteria.
3. The large, colorless, nonfilamentous sulfur bacteria.

Group 1

Morphologically the most conspicuous are the purple *purple sulfur bacteria* (Fig. 3.5).   The majority are relatively large, generally stuffed with sulfur globules, and often so intensely pigmented that even single individuals appear distinctly red.   Frequently they occur in large and dense masses, easily detectable by the naked eye.

Following is a discussion of a variety of differently shaped types that can be distinguished.

Thiopedia.   Nearly spherical, nonmotile bacteria, generally about 1 to 2 μ in diameter, with a strong tendency to remain attached in tetrads or even

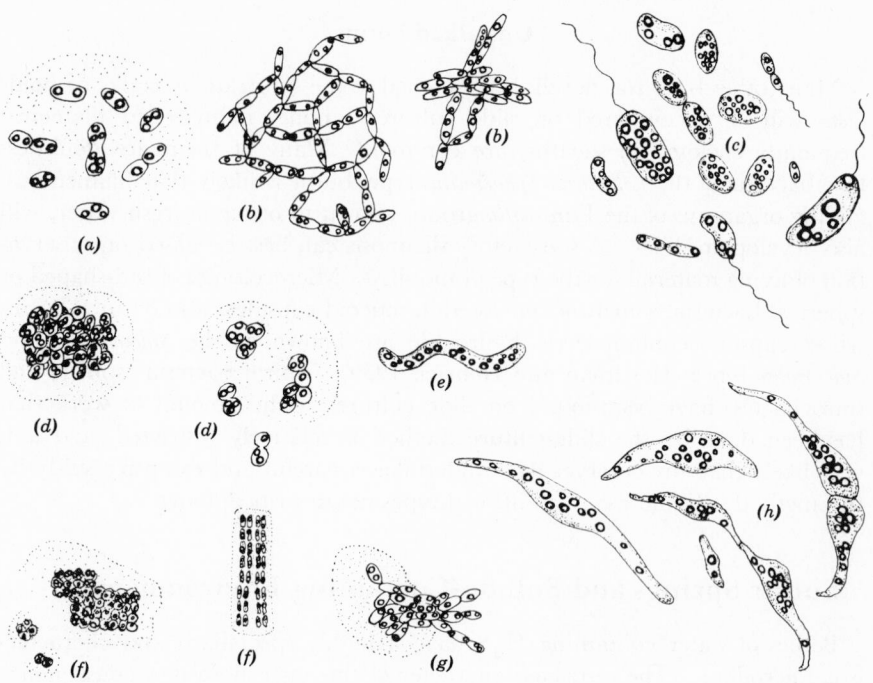

Fig. 3.5. Purple sulfur bacteria. (*a*) *Thiothece*. (*b*) *Thiodictyon*. (*c*) *Chromatium*, various forms. (*d*) *Thiocystis*. (*e*) *Thiospirillum*. (*f*) *Thiopedia*. (*g*) *Amoebobacter*. (*h*) *Rhabdochromatium*.

large sheets one cell thick. The individual cells often have an angular appearance owing to the refractile properties of the sulfur globules. In lakes and pools this type may constitute the vast majority of the purple sulfur bacterial flora, localized in layers where the H₂S concentration, pH, light intensity, etc., are optimal (Fig. 3.5*f*).

Thiosarcina. Similar to *Thiopedia*, but with somewhat larger cells (2 to 3 μ), with a tendency to form regular, three-dimensional packets, containing 8 to 64 cells. Especially when small, the packets may show motility.

Thiocystis, Thiothece, Thiocapsa. Relatively large, spherical to clearly ovoid cells (2.5 to 6 or 7 μ), loosely grouped in small aggregates surrounded by a common slime capsule, in which the individual cells are clearly separated from one another. Single individuals or small groups of cells may break through the common capsule, swim away, and establish a new "colony" elsewhere (Fig. 3.5*a,d*).

Thiopolycoccus, Amoebobacter, Lamprocystis. Somewhat spherical to distinctly rod-shaped bacteria, forming extensive aggregates, but without conspicuous common capsules. *Thiopolycoccus* is described as a nonmotile, spherical bacterium, a little more than 1 μ in diameter, occurring in large, irregular clumps; the cells of *Amoebobacter* and *Lamprocystis* are clearly rod-shaped, from 1 to 3.5 μ wide, and up to 5 or 6 μ long; they are apt to exhibit motility (Fig. 3.5*g*). With this group may be included the purple sulfur bacterium known as *Thiodictyon*, whose rod-shaped cells have a tendency to remain attached together in a reticulate structure somewhat reminiscent of the green alga *Hydrodictyon* (Fig. 3.5*b*).

Chromatium. Motile, short rods or beautifully kidney-shaped cells of various sizes, ranging from about 1 to 10 μ in width, and from 2 to 25 μ in length. There has been a tendency to consider organisms of this general description, and falling within a specified size range, as individual "species," but experiments have shown that the size of individuals in cultures started from single cells may vary enormously with environmental conditions (Fig. 3.5*c*). A form that has been designated as *Rhabdochromatium*, distinguished by a typically elongated cell of very uneven diameter, often with attenuated ends, should probably be regarded as a particular form of *Chromatium* induced by unfavorable environmental conditions, such as too high a sulfide concentration (Fig. 3.5*h*).

Thiospirillum. Motile, spiral cells, often of great size (up to 2.5 μ or more in width, and occasionally over 100 μ long), and generally occurring singly. Mass developments form delicately colored and easily dispersed patches (Fig. 3.5*e*).

Small purple sulfur bacteria. This group of unnamed organisms occurs as short, motile rods or spirals, 0.5 to 1 μ in width, and fails to deposit sulfur globules inside its cells. It is not known whether mass developments of these organisms ever occur in nature. Their recognition is difficult unless special culture experiments are resorted to. It may be confidently asserted, however, that phototactically active, small rods, found in large numbers in an H_2S-containing environment belong to this group.

This entire assemblage of anaerobic photosynthetic bacteria is observed in nature only where H_2S is present. The major photosynthetic pigment is a chlorophyll (*bacteriochlorophyll*), whose green color is masked by intensely red to purplish carotenoids. Since the development of these bacteria depends upon their photosynthetic activity, they do not grow in permanently dark localities. The absorption spectrum of the pigment system is, however, almost exactly complementary to that of the green and blue-green algae, extending into the infrared region of the spectrum up to about 900 μ. Consequently, mass developments of purple bacteria are frequently found directly beneath those of algae.

Green sulfur bacteria. If the sulfide concentration of the habitat is relatively high (more than 50 mg H_2S per liter) these purple bacteria are frequently accompanied by *green sulfur bacteria* (Fig. 3.6). These are small, ovoid to rod-shaped, nonmotile organisms, generally less than 1 μ in diameter, and of a faint yellowish-green color. Usually the pigmentation cannot be clearly discerned unless the organisms occur in aggregates. In spite of the fact that they are dependent on light and H_2S for growth and carry out a vigorous oxidation of sulfide with the production of free sulfur, they have never been observed with sulfur globules in their cells, presumably because of their small size. An exception should be made for *Clathrochloris* (see below). The most effective wave length for photosynthetic activity by these organisms lies

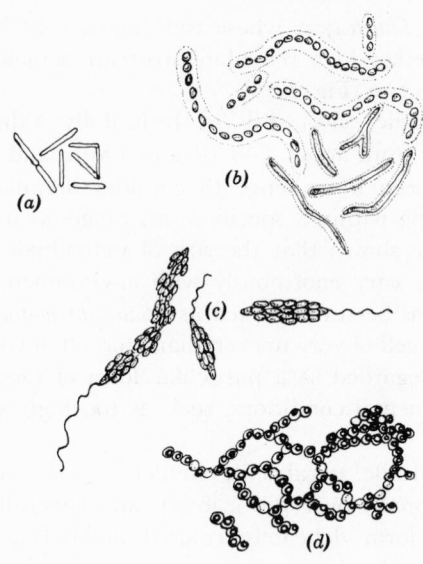

Fig. 3.6. Green sulfur bacteria. (*a*) *Microchloris.* (*b*) *Chlorobium.* (*c*) *Chlorochromatum aggregatum.* (*d*) *Clathrochloris.*

around 730 to 750 millimicrons, where the bacteriochlorophyll of the purple bacteria does not absorb appreciably (Fig. 3.7). Thus, these organisms fill the spectral gap between the algae and purple bacteria, a fact of obvious ecological significance.

There is reason to believe that certain small blue-green algae have been mistaken for green bacteria (Pringsheim, 1953). A sharp distinction between the two groups of organisms is possible if sufficient material is available for a study of the pigments and the photosynthetic metabolism. The algae display absorption maxima at 620 and 680 μ, for which phycocyanin and chlorophyll *a*, respectively, are responsible, but not at 750 μ. Besides, the algae produce oxygen when illuminated, and can thrive in sulfide-free environments, whereas the green sulfur bacteria fail to do so.

The green sulfur bacteria comprise two groups, presumably distinguishable by the shape of the cells when grown under optimal conditions. The first

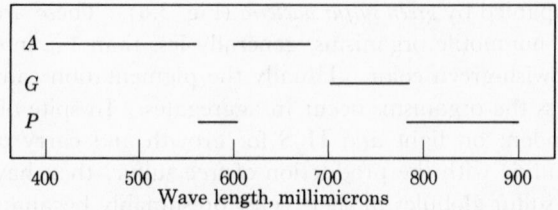

Fig. 3.7. Light-absorption bands for photosynthetic activity of algae, *A*, green sulfur bacteria, *G*, and purple sulfur bacteria, *P*.

group is composed of spherical or ovoid bacteria; the cells of the second are rod-shaped.

In the first group belong the bacteria of the *Chlorobium* type, the best studied among the green sulfur bacteria. The cells are ovoid, about 0.7 by 0.9 to 1.5 μ, occurring singly, in chains, or in clumps. Clumped aggregates of approximately spherical "green bacteria," referred to as *Pelogloea*, have been regarded as growth forms of *Chlorobium* (van Niel, 1948). On the basis of recent studies Pringsheim (1953) believes, however, that *Pelogloea* is actually a blue-green alga. It must be mentioned that in an unfavorable environment *Chlorobium* may produce rod-shaped and long, irregular filamentous growth forms (Fig. 3.6b).

Spherical green sulfur bacteria with sulfur inclusions, and arranged in trellis-shaped aggregates have been described as *Clathrochloris* (Fig. 3.6d).

Normally, rod-shaped green sulfur bacteria range from the small (0.4-μ wide) *Microchloris*, occurring singly, to the aggregate-forming *Pelodictyon* types, which occur as netlike masses, irregular clumps, or bundles of parallel strands of cell chains (Fig. 3.6a).

Some green bacteria form associations with other organisms. The most interesting of these is the complex known as *Chlorochromatium aggregatum* (*Chloronium mirabile*), which is composed of a large, polarly flagellated "inner bacterium," covered with a single layer of green bacteria arranged with their long axes parallel to that of the carrier. Multiplication of the two component organisms appears to be closely synchronized, so that the barrel-shaped structure is maintained during development (Fig. 3.6c). Cultural studies on such complexes have not yet been pursued successfully; it is therefore unknown whether they represent fortuitous aggregates of different types of organisms occurring in the same environment, or typically symbiotic structures, comparable to lichens.

Group 2

The colorless, filamentous sulfur bacteria (Fig. 3.8), known as *Beggiatoa*, *Thiothrix*, *Thioploca*, and *Thiospirillopsis*, are also restricted to places where H_2S is present, but because their metabolism is oxidative instead of photosynthetic, they can flourish only in regions where oxygen is available. Hence, they are generally found in areas where one body of water with an adequate O_2 content borders on another which supplies the H_2S. Here the colorless, filamentous sulfur bacteria may form dense mats of a slightly yellowish-white appearance.

Beggiatoa. Occurs in the form of single filaments of indeterminate length and with a diameter ranging from 1.8 to 16 μ. The filaments exhibit characteristic gliding movements, often accompanied by bending and rotation around the long axis. They are stuffed with sulfur globules if the H_2S supply is adequate (Fig. 3.8a,b).

Thiothrix. Resembles *Beggiatoa* in most respects, but the filaments lack motility, and are attached to a substrate by a small holdfast. The apical cells

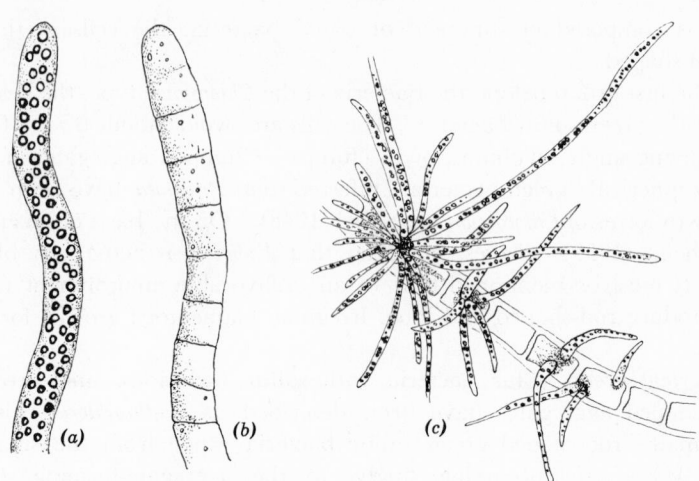

Fig. 3.8. Colorless, filamentous sulfur bacteria. (*a*) *Beggiatoa*, terminal portion of a filament filled with sulfur globules. (*b*) Same, without sulfur globules. (*c*) Tufts of *Thiothrix*, growing on a filamentous alga.

can become detached, glide over the substrate, and reproduce another filament (Fig. 3.8*c*).

Thioploca. Best described as a bundle of *Beggiatoa* filaments enclosed in a common slime sheath.

Thiospirillopsis. Occurs in the form of helical filaments; it appears to be the colorless counterpart of the blue-green algae of the *Spirulina* type. The described *Spirulina* species vary greatly in width, but only thin (*ca.* 1 μ wide) forms of *Thiospirillopsis* have been found so far.

Group 3

Thiovolum. Notably in the immediate neighborhood of decaying algae, there may be found mass developments of nonfilamentous, extremely motile, large, ovoid sulfur bacteria, known as *Thiovulum* (Fig. 3.9*a*). The rapid and whirling motion, accompanied by rotation of the cells, suggests the presence of flagella, but these structures have not yet been detected. Usually the sulfur globules are aggregated in the posterior half of the cell.

Mass developments of *Thiovulum* have a strong tendency to accumulate as a thin veil which is easily dispersed by slight disturbances, but rapidly reforms when the cause of the disturbance is eliminated. This behavior is probably the result of a chemotactic aggregation.

Achromatium. *Achromatium oxaliferum* (poorly named for an organism that does not contain oxalate crystals as was originally suspected) is perhaps the most spectacular of all bacteria (Fig. 3.9*b*). It occurs in the form of large single cells, ovoid to short-cylindrical with round ends, measuring *ca.* 7 to 35 μ in width, and up to 100 μ in length, filled with spherolites of $CaCO_3$, and suggesting bags of marbles. Interspersed between the spherolites are smaller sulfur globules which remain after treatment of the cells with dilute acid. The specific gravity of the bacteria is high as a result of the mineral content; hence, they are found only at the bottom of a body of water, frequently in the upper layer of mud or detritus. Slow gliding movements occur, during which the cell rotates about its long axis. This bacterium can easily be separated from other microorganisms if advantage is taken of its high specific gravity; after a mud sample has been shaken, the $CaCO_3$-containing cells sediment before other bacteria. *A. volutans*, also known as *Thiophysa*, resembles *A. oxaliferum* in all respects, except for the $CaCO_3$ inclusions. Somewhat similar to *A. volutans*, but smaller and nonmotile, are the recently described *Thiogloea* types (Devidé, 1952). They are round, ovoid, or short, rod-shaped bacteria, measuring 0.7 to 6 μ in width by 3 to 10 μ in length; they occur in the form of mucilaginous colonies containing from less than a hundred to many thousands of individuals.

Macromonas. A rod-shaped, polarly-flagellated, colorless sulfur bacterium, about 5 by 10 μ, *Macromonas* (Fig. 3.9*c*) is characterized by its inclusion bodies which, like those of *Achromatium oxaliferum*, are comprised of both $CaCO_3$ spherolites and sulfur globules. It is often found in mud.

Thiobacillus. In addition to the three groups of sulfur bacteria described above, small, colorless, motile, rod-shaped *Thiobacillus* types are generally present, sometimes in vast numbers, in a sulfide-containing environment. The absence of sulfur globules in their cells makes it impossible to recognize them as sulfur bacteria by simple microscopy, but positive chemotaxis toward H_2S, coupled with the absence of a phototactic response, may help to identify them.

In samples collected some time prior to examination, even those sulfur bacteria that normally contain sulfur globules may appear devoid of the characteristic inclusions. In general, it is possible to induce the rapid reappearance of sulfur globules by adding a small amount of a neutral sulfide solution (final concentration not over 20 mg H_2S per l) to the samples. Those sulfur bacteria which are still viable rapidly oxidize the sulfide to sulfur, and will contain sulfur globules within a few hours. With purple sulfur bacteria, simultaneous exposure to light is essential; this is best accomplished by filling a glass-stoppered bottle completely with the sulfide-treated sample, and placing it at a distance of 10 to 20 cm from an ordinary incandescent bulb (25 to

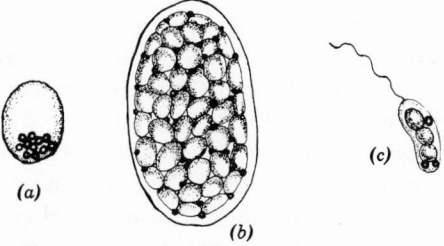

(a)

(b)

(c)

Fig. 3.9. Colorless, nonfilamentous sulfur bacteria. (*a*) *Thiovulum*. (*b*) *Achromatium*. (*c*) *Macromonas*.

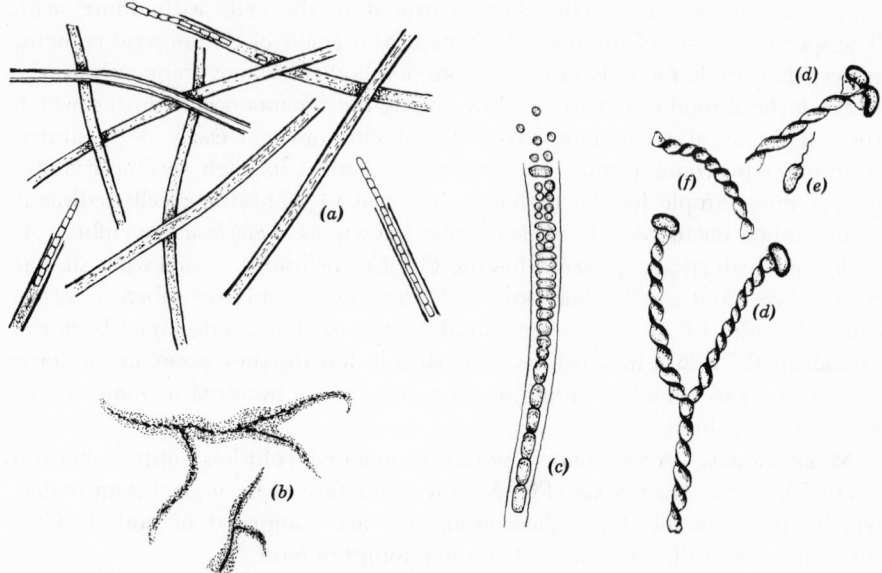

Fig. 3.10. Iron bacteria. (*a*) *Leptothrix ochracea*, habit sketch showing ensheathed filaments and empty sheaths. (*b*) *Leptothrix crassa*. (*c*) Filament of *Crenothrix polyspora*. (*d–f*) *Gallionella*: (*d*) Tips of stalks, showing terminal cells; (*e*) Flagellated swarm cell; (*f*) Fragment of stalk.

60 watt). Light from fluorescent tubes is far less effective, because such light sources have a low emission in the near infrared region.

Iron Springs and Similar Environments

An iron spring is usually recognizable by the fact that an orange-brown precipitate forms in a sample that has stood exposed to air. The explanation for this phenomenon is that the iron, originally present as ferrous ions, oxidizes, and the oxidation product settles out in the form of $Fe(OH)_3$. It is in such environments that the "iron bacteria" (Fig. 3.10) are encountered. The geochemical origin of iron springs is still uncertain. However, water can easily become charged with ferrous ions by seepage through a layer of earth or mud in which reduction processes occur, and it may well be that iron springs obtain their water from such sources. An identical ecological situation is encountered in streams or ditches receiving an inflow from the banks of seepage water charged with ferrous ions. At a short distance from the inflow, the characteristic precipitate of $Fe(OH)_3$ accumulates.

Although ferrous iron is spontaneously oxidized when exposed to air, except in very strongly acid solutions, the oxidation in iron springs and similar habitats is quite generally accompanied by the development of specialized bacteria, and the ecology of these organisms suggests that they may play a part in the oxidation process. Microscopic examination of the flocculent, orange-brown matter reveals that it is partly "organized"; the most frequently encountered and conspicuous structures are tube-shaped (*Leptothrix* sheaths), with regularly twisted flat bands (*Gallionella* stalks) occurring in greater or lesser abundance among them.

Leptothrix. The organism itself is a short, rod-shaped bacterium with a pronounced tendency to chain formation. It excretes a mucilaginous sheath of considerable dimensions which becomes light brown as a result of impregnation with $Fe(OH)_3$. Swarm cells are released at the tip of the filament and, after swimming away, start new tubes. Only in very young and as yet lightly encrusted sheaths can the bacteria themselves be seen by microscopic examination. In a typical deposit very few of the sheaths still contain cells.

By far the most satisfactory preparations of *Leptothrix* are obtained on slides submerged near the outflow of the water containing ferrous iron. On these slides, the actual bacterial filaments can be observed before the sheaths have become too heavily encrusted to obscure the contents. Careful microscopy of the slides may fail to suggest the presence of iron in the sheaths at an early stage; but application of the Prussian-blue staining method will show that even the thin, transparent, and colorless portions of the sheath already contain appreciable quantities of ferric iron.

There are several members of the *Leptothrix* group, distinguishable by the different appearances of their tubes. *L. ochracea* (Fig. 3.10*a*) forms smooth, uniformly thick, and straight to slightly curved or bent tubes, and *L. crassa* (Fig. 3.10*b*) forms much more irregular, often branched tubes, which are, in addition, much darker than the vivid ochre tubes of *L. ochracea*. The dark color of *L. crassa* tubes is caused by the presence of appreciable amounts of Mn_3O_4.

Leptothrix trichogenes forms a sheath which adheres along its entire length to the substrate. The chain of rod-shaped cells moves back and forth inside the sheath, suggesting a railroad train moving within rather wide tracks.

Gallionella. Regularly twisted bands containing $Fe(OH)_3$ had long been observed in iron deposits, and described as an organism (*Gallionella*). In 1926, Cholodny discovered that these structures are stalks, formed by a small kidney-shaped bacterium located at the tip (Fig. 3.10*d–f*). When the bacterium divides, the two daughter cells separate, each one continuing the deposition of its own stalk, with the formation of a branched colony. The connection between the bacterium and the stalk is apparently not a very tight one; the organisms are easily dislodged by slight disturbances. The delicate nature of the *Gallionella* colony makes the slide-culture technique the only suitable method for observing its habit.

Apart from these two major types, three further kinds of iron bacteria have been described. They are:

Crenothrix polyspora. A filamentous bacterium forming a holdfast and secreting a sheath which becomes impregnated with $Fe(OH)_3$, much like *Leptothrix* (Fig. 3.10*c*). It differs from *Leptothrix* by producing in the apical portion of the sheath small, roundish, presumably immotile reproductive cells which are dispersed by water currents.

Siderocapsa. A small, spherical bacterium, found embedded in irregular masses of $Fe(OH)_3$-containing mucus.

Sideromonas. A small, rod-shaped, nonfilamentous bacterium, whose habit is similar to that of *Siderocapsa*.

The last two types are usually encountered on decaying plant residues.

For the sake of completeness it should be mentioned that the characteristic microorganisms in iron springs also include two "iron flagellates" (Pringsheim, 1946), one a colorless, stalked protist (*Anthophysa*, Fig. 8.82), the other a green, euglenoid flagellate which secretes around itself a thick, crusty housing impregnated with $Fe(OH)_3$, within which the organism normally lives and divides (*Trachelomonas*, Fig. 6.11, 6.12).

Studies on the iron bacteria have been fragmentary. Pringsheim (1949a) has recently shown that *Leptothrix* can also live in an environment where ferrous iron is absent. Under these conditions it assumes a somewhat different habit, and is indistinguishable from the characteristic sewage bacterium *Sphaerotilus* (see the following section).

Water Contaminated with Sewage and Industrial Wastes

The chemical composition of the contaminating material determines what types of bacteria predominate in the environments considered here. The microflora is also influenced by the fact that some kinds of waste, especially household sewage, contribute vast numbers of bacteria characteristic of the normal microflora of the human intestinal tract. These bacteria do not grow much in their new environment, but may persist for long periods.

With the recognition that the enteric diseases are water-borne, the development of bacteriological methods for detecting fecal contamination in water became an urgent sanitary problem. During the past fifty years a vast amount of bacteriological research has been devoted to this problem. Since the most abundant and easily recognizable bacterial inhabitant of the intestinal tract of warm-blooded animals is *Escherichia coli*, detection of fecal contamination has been based on the specific detection of this bacterium. Such detection depends on the use of culture techniques, which will not be described here; for further details concerning the theory and practice of sanitary water analysis consult Prescott, Winslow, and McCrady (1946).

In heavily contaminated waters the bacterial population may be so dense as to be observable by direct microscopic examination. Most of the organisms found are eubacterial cocci and rods of various kinds, distinguishable only by isolation in pure culture followed by the application of differential tests. In certain cases the spirilla, conspicuous because of their striking shape and movements, may also abound. The two most distinctive bacteria in such contaminated bodies of water are *Sphaerotilus natans* and *Zoogloea ramigera* (Fig. 3.11).

Sphaerotilus. A filamentous bacterium, usually found attached to the sides of basins, to floating debris, or to dead leaves, forming dense, feltlike

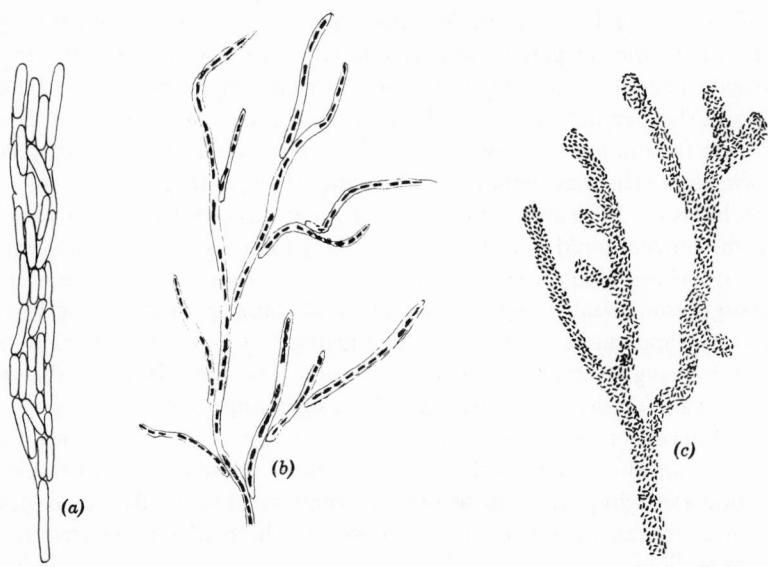

Fig. 3.11. Sewage organisms. (*a*) *Sphaerotilus natans*, typical group of cells. × 1300. (*b*) *Sphaerotilus natans*, Cladothrix growth form. × 570. (*c*) *Zoogloea ramigera*, portion of zoogloeal aggregate. × 340.

masses of a dirty-whitish to brownish shade. *Sphaerotilus* is composed of strings of rod-shaped bacteria, enclosed in a common sheath which is attached to some solid object by means of a holdfast (Fig. 3.11*a,b*).

There is a considerable morphological resemblance between *Sphaerotilus* and the iron bacteria of the *Leptothrix* group. It was long believed that *Leptothrix* could be distinguished from *Sphaerotilus* by the presence of $Fe(OH)_3$ in its sheaths. However, the studies of Cataldi (1939) and of Pringsheim (1949a) have shown convincingly that *Leptothrix* can develop in dilute solutions of organic matter in the absence of ferrous salts. Under such conditions the sheaths do not contain $Fe(OH)_3$ and the organism is then indistinguishable from *Sphaerotilus*. It seems, therefore, that these two organisms are ecological variants of a single type, the presence or absence of ferrous salts determining whether the organism will develop as a *Leptothrix* or as a *Sphaerotilus*.

Other ensheathed, filamentous eubacteria, like *Cladothrix* and *Clonothrix*, have been inadequately characterized. *Cladothrix* shows a pronounced resemblance to *Sphaerotilus*, and some bacteriologists consider the two identical. *Clonothrix* is now generally regarded as synonymous with *Crenothrix* (see p. 39).

Zoogloea ramigera. A small, rod-shaped organism. The cells are held together by slime in irregular masses, which characteristically resemble branched antlers (Butterfield, 1935; Wattie, 1942). From the fact that the individual cells are polarly flagellated and do not form spores, it may be inferred that *Z. ramigera* represents a special growth form of some pseudomonas; the name would thus merely designate a more or less readily recognizable ecotype (Fig. 3.11*c*).

Spirillum. Usually found in decomposing plant material; they thrive particularly where the oxygen tension is relatively low, since they are micro-aerophiles. The representatives of this group range from about 1 to 100 μ in length, and the steepness of the helix, as well as the number of turns, is subject to wide fluctuations. Exceptionally long individuals are probably composed of several cells that remain attached after division (Fig. 3.1f).

Spirochaetes. In heavily contaminated water samples spirochaetes are not infrequently encountered. The only form likely to be seen by ordinary microscopy is *Spirochaeta plicatilis* (Fig. 3.4a); it is a very thin, long, and extremely flexible organism, usually about 0.5 μ wide, and 100 μ or more in length, with characteristic and rapid locomotion. A moving *S. plicatilis* resembles a thin, long metal spring, waving and coiling about. The organism can thrive at very low oxygen tensions, and may even be a strict anaerobe.

The various decomposition processes that give rise to the formation of H_2S and ferrous salts also call forth the development of sulfur and iron bacteria in and around decaying particles of organic matter. For a discussion of the more typical representatives of these two groups the reader is referred to the preceding sections.

Sulfate reduction and methane fermentation. Much of the decomposition of organic debris occurs in the lower strata of a contaminated body of water, particularly in the mud and silt. Here, the most characteristic anaerobic processes are sulfate reduction and methane fermentation.

Sulfate reduction is a process in which simple organic substances (alcohols, fatty acids, etc.) are oxidized by bacterial action with the concomitant reduction of sulfate to sulfide. The sulfate-reducing bacteria are small, motile rods, with a tendency to assume a somewhat spiral shape. They are strict anaerobes and are found in large numbers only in places where sulfate is present in some abundance. It is very probable that by far the largest amount of H_2S produced in nature stems from sulfate reduction rather than from the direct decomposition of sulfur-containing organic materials such as proteins.

The methane fermentation is an oxidative decomposition of simple organic compounds, coupled with the reduction of CO_2 to CH_4. Bacteria of various shapes—cocci, sarcinae, and rods, known respectively as *Methanococcus*, *Methanosarcina*, and *Methanobacterium*—can carry out this type of transformation. These organisms are all immotile and incapable of producing endospores. They are strict anaerobes.

Microscopically, neither sulfate-reducing bacteria nor methane-producing bacteria are unambiguously identifiable; nevertheless, the presence of strictly anaerobic pseudomonas, vibrio, or spirillum types (negative aerotaxis!) found in an environment where H_2S is being generated, and not exhibiting any kind of phototaxis, will be sufficient justification for the tentative conclusion that the organisms are, in fact, sulfate-reducing bacteria (*Desulfovibrio*). Similarly, the presence of typical sarcinae in mud in which gas (CH_4) is being produced may be taken as strong presumptive evidence that the packets represent *Methanosarcina*, especially since the other known anaerobic sarcinae are not likely to be encountered in mud samples in large enough numbers to be detectable by microscopic examination.

In addition to the more or less easily recognizable types of bacteria described above, heavily contaminated water samples are apt to contain various kinds that are striking enough to the microscopist, but about which practically nothing is known. There is here a large field in which much rewarding work remains to be done. The use of elective culture methods should aid immensely in accumulating information of a kind that will eventually lead to a better understanding of the properties of these virtually unknown organisms and their role in the natural cycle of matter.

Keys for the Recognition of Bacterial Types Frequently Encountered in Water and Mud

These are tabular keys in which a series of choices is offered. The best description should be selected, and a further selection made from the set of correspondingly numbered alternatives.

A detailed and comprehensive key for the determination of the generic position of bacteria, by V. B. D. Skerman, is now available in the 7th edition of Bergey's Manual, pp. 987–1032; experience with this key has shown it to be excellent. This same treatise also contains information on the authorities for the generic names and a detailed classification of the recognized genera down to the species level.

For a general consideration of the problems of bacterial classification, reference may be made to van Niel (1955).

KEY I. PRIMARY DIFFERENTIATION BY SHAPE

Spherical to ovoid cells	1
Bean- to rod-shaped cells	2
Spiral cells	3
Filamentous forms	4

1a Purple- to red-colored cells, with sulfur globules; photosynthetic (purple sulfur bacteria):

Cells arranged in tetrads and sheets (Fig. 3.5f)	*Thiopedia*
Cells arranged in cubical packets	*Thiosarcina*
Cells clumped in irregular masses, not conspicuously mucoid	*Thiopolycoccus*
Cells in mucoid colonies	*Thiocapsa*
(Fig. 3.5a)	*Thiothece*
(Fig. 3.5d)	*Thiocystis*

1b Yellow-green, photosynthetic (green sulfur bacteria):

Cells not containing sulfur globules (Fig. 3.6b)	*Chlorobium*
Cells with sulfur globules (Fig. 3.6d)	*Clathrochloris*

1c Colorless, sulfur-containing cells:

Actively motile ("swimming") (Fig. 3.9a)	*Thiovulum*
Gliding motility, with $CaCO_3$ crystals (Fig. 3.9b)	*Achromatium oxaliferum*
Gliding motility, no crystals	*Achromatium volutans* (= *Thiophysa*)
Nonmotile, in gelatinous colonies	*Thiogloea*

1d Cells with secretions of $Fe(OH)_3$ ("iron bacteria") | *Siderocapsa*

1e Members of the *Micrococcus, Streptococcus, Sarcina* types:
　　Spores of *Bacillus* and *Clostridium* types.
　　Cysts of *Azotobacter*, microcysts of *Myxobacteria* types.

2a Purple- to red-colored cells, with sulfur globules; photosynthetic; phototactic when motile (purple sulfur bacteria):
　　Cells in irregular masses (Fig. 3.5*g*). . . . . *Amoebobacter*
　　　　　　　　　　　　　　　　　　　　　　　　　　　　Lamprocystis
　　　　　　　　　　　　　　　　　　　　　　　　　　　Thiopolycoccus
　　Cells in reticulate groups (Fig. 3.5*b*) . . . . *Thiodictyon*
　　Cells single, usually motile (Fig. 3.5*c*) . . . . *Chromatium*
　　(Fig. 3.5*h*) . . . . *Rhabdochromatium*

2b Cells purple- to red-colored, without sulfur globules; photosynthetic; phototactic when motile:
　　Cells unstalked, motile. . . . . *Rhodopseudomonas*
　　Cells stalked (Fig. 3.4*b*) . . . . *Rhodomicrobium*

2c Yellow-green, photosynthetic organisms, in H_2S environments (green sulfur bacteria):
　　Single, small cells (Fig. 3.6*a*) . . . . *Microchloris*
　　Cells in trellis-shaped aggregates . . . . *Pelodictyon*
　　Cells in ovoid group, arranged around an inner bacterium (Fig. 3.6*c*). . . . . .
　　　　　　　　　　　　　　　　　　　　　　　　　　Chlorochromatium

2d Nonphotosynthetic bacteria:
　　Swimming types: Swarmers of stalked or filamentous forms; Members of the *Pseudomonas, Thiobacillus, Vibrio, Bacillus, Clostridium*, etc., groups. When fairly large (5 by 10 μ) and containing $CaCO_3$ crystals (Fig. 3.9*c*) . . . *Macromonas*

2e Gliding types:
　　Members of the Cytophaga and Sporocytophaga groups.
　　Vegetative cells of the higher myxobacteria.

2f Stalked bacteria:
　　Members of the *Caulobacter, Blastocaulis, Hyphomicrobium* types, if stalk encrusted with $Fe(OH)_3$ (Fig. 3.10*d–f*) . . . . *Gallionella*

2g Bacteria with $Fe(OH)_3$-containing capsules. . . . . *Sideromonas*

3a Purple- to red-colored cells, photosynthetic and phototactic:
　　With sulfur globules (Fig. 3.5*e*) . . . . *Thiospirillum*
　　Without sulfur globules . . . . *Rhodospirillum*

3b Colorless cells, nonflexible (Fig. 3.1*f*) . . . . *Spirillum*
　　(Fig. 3.1*e*) . . . . *Vibrio*
　　　　　　　　　　　　　　　　　　　　　　　　　　　　Desulfovibrio

3c Flexible, thin bacteria (Fig. 3.4*a*) . . . . *Spirochaeta*
　　　　　　　　　　　　　　　　　　　　　　　　　　　　Leptospira

4a Colorless, with sulfur globules:
　　Not regularly coiled; gliding movement; single (Fig. 3.8*a,b*) . . . . *Beggiatoa*
　　Not regularly coiled; gliding movement, but filaments together
　　in common sheath . . . . *Thioploca*
　　Not regularly coiled, nonmotile, attached (Fig. 3.8*c*) . . . . *Thiothrix*
　　Regularly coiled . . . . *Thiospirillopsis*

4b Ensheathed, with encrustation of $Fe(OH)_3$ (Fig. 3.10*a,b*) . . . . . . *Leptothrix*
　　(Fig. 3.10*c*). . . . . *Crenothrix*

4c Ensheathed, but without $Fe(OH)_3$ (Fig. 3.11*a,b*). . . . . *Sphaerotilus*

4d Not ensheathed; with gliding movements . . . . *Vitreoscilla*

4e Forming branched mycelium (moldlike) . . . . . **Actinomycetes (*Streptomyces, Nocardia, Micromonospora*, etc.)**

KEY II. PRIMARY DIFFERENTIATION BY CONSPICUOUS FEATURES OTHER THAN CELL SHAPE

Color:

Purple to red . . . . Key I, **1a, 2a, 2b, 3a**

Yellow-green . . . . Key I, **1b, 2c**

Brown, due to $Fe(OH)_3$ . . . . Key I, **1d, 2f, 2g, 4b**; Key III, **2**

Inclusion bodies:

Sulfur globules . . . . Key III, **1**

$CaCO_3$ crystals . . . . Key I, **1c, 2d**

Characteristic cell aggregates:

Cubical packets . . . . Key I, **1a, 1e**

Reticulate masses . . . . Key I, **2a, 2c**

Antler-shaped aggregates (Fig. 3.11*c*) . . . . ***Zoogloea ramigera***

KEY III. PRIMARY DIFFERENTIATION BY HABITAT (ECOLOGY)

Environments containing H_2S: "sulfur bacteria" (p. 31) **1**

Environments containing Fe^{++}: "iron bacteria" (p. 38) **2**

In mud and silt **3**

In heavily contaminated bodies of water (p. 40) **4**

On slide cultures (p. 26) **5**

1a Purple and red sulfur bacteria . . . . Key I, **1a, 2a, 3a**

1b Green sulfur bacteria . . . . Key I, **1b, 2c**

1c Colorless, nonfilamentous sulfur bacteria . . . . Key I, **1c**

1d Colorless, filamentous sulfur bacteria . . . . Key I, **4a**

2a Single-celled, nonstalked . . . . Key I, **1d, 2g** (***Siderocapsa, Sideromonas***)

2b Single-celled, stalked . . . . Key I, **2f** (***Gallionella***)

2c Filamentous, ensheathed . . . . Key I, **4b** (***Leptothrix, Crenothrix***)

3 Bacteria with $CaCO_3$ crystals . . . Key I, **1c, 2d** (***Achromatium, Macromonas***)

4a Single-celled, in antler-shaped, mucilaginous aggregates (Fig. 3.11*c*) . . . . ***Zoogloea ramigera***

4b Filamentous, attached (Fig. 3.11*a, b*) . . . . ***Sphaerotilus***

5a Stalked . . . . Key I, **2f**

5b Filamentous, ensheathed . . . . Key I, **4b, 4c, 4d**

References

Bergey, D. H., R. S. Breed, E. G. D. Murray, and Nathan R. Smith (eds.). 1957. *Bergey's Manual of Determinative Bacteriology,* 7th ed. Williams and Wilkins, Baltimore. **Butterfield, C. T. 1935.** Studies of sewage purification. II. A zooglea-forming bacterium isolated from activated sludge. *Public Health Repts., U. S.,* 50:671–684. **Cataldi, M. S. 1939.** Estudio fisiologico y sistematico de algunas Chlamydobacteriales. Thesis, Buenos Aires. **Cholodny, N. 1926.** *Die Eisenbakterien.* G. Fischer, Jena. **Devidé, Z. 1952.** Zwei neue farblose Schwefelbakterien, *Thiogloea ruttneri, n. gen., n. sp.* und *Thiogloea ragusina n. sp. Schweiz. Z. Hydrol.,* 14:446–455. **Duchow, E. and H. C. Douglas, 1949.** *Rhodomicrobium vannielii,* a new photosynthetic bacterium. *J. Bacteriol.,* 58:409–416. **Hardman, Y. and A. T.**

Henrici. 1939. Studies of freshwater bacteria. V. The distribution of *Siderocapsa treubii* in some lakes and streams. *J. Bacteriol.*, 37:97–104. **Henrici, A. T. 1933.** Studies of freshwater bacteria. I. A direct microscopic technique. *J. Bacteriol.*, 25:277–286. **1936.** Studies of freshwater bacteria. III. Quantitative aspects of the direct microscopic method. *J. Bacteriol.* 32:265–280. **1939.** The distribution of bacteria in lakes. In: *Problems of Lake Biology, Publ. Am. Assoc. Advance. Sci.*, 10:39–64. **Henrici, A. T., revised by C. W. Emmons, C. E. Skinner, and H. M. Tsuchiya. 1947.** *Molds, Yeasts and Actinomycetes*, 2nd ed. Wiley, New York. **Henrici, A. T. and D. C. Johnson. 1935.** Studies of freshwater bacteria. II. Stalked bacteria, a new order of Schizomycetes. *J. Bacteriol.* 30:61–93. **Kucera, S. and R. S. Wolfe. 1957.** A selective enrichment method for *Gallionella ferruginea*. J. Bacteriol. 74:344–349. **Lauterborn, R. 1915.** Die sapropelische Lebewelt. *Verhandl. nat.-med. Vereins Heidelberg*, N. F., 13:394–481. **Mechsner, K. 1957.** Physiologische und morphologische Untersuchungen an Chlorobakterien. *Arch. für Mikrobiol.* 26:32–51. **Mevius, W. 1953.** Beiträge zur Kenntnis von *Hyphomicrobium vulgare* Stutzer et Hartleb. *Arch. Mikrobiol.*, 19:1–29. **Prescott, S. C., C.-E. A. Winslow, and Mac H. McCrady. 1946.** *Water Bacteriology*. Wiley, New York. **Pringsheim, E. G. 1946.** On iron flagellates. *Phil. Trans. Roy. Soc. London, Ser. B*, 232:311–342. **1949a.** The filamentous bacteria *Sphaerotilus, Leptothrix, Cladothrix*, and their relation to iron and manganese. *Phil. Trans. Roy. Soc. London, Ser. B*, 233:453–482. **1949b.** Iron bacteria. *Biol. Revs. Cambridge Phil. Soc.*, 24:200–245. **1949c.** The relationship between bacteria and Myxophyceae. *Bacteriol. Rev.*, 13:47–98. **1952.** Iron organisms. *Endeavour*, 11:208–214. **1953.** Die Stellung der grünen Bakterien im System der Organismen. *Arch. Mikrobiol.*, 19:353–364. **Stanier, R. Y., M. Doudoroff, and E. A. Adelberg. 1957.** *The Microbial World*. Prentice-Hall, Englewood Cliffs, New Jersey. **van Iterson, W. 1958.** *Gallionella ferruginea* Ehrenberg in a different light. *N. V. Noord Hollandische Uitgevers Maatschappij*. **van Niel, C. B. 1955.** Classification and taxonomy of the bacteria and bluegreen algae. In: *A Century of Progress in the Natural Sciences, 1853–1953*. California Academy of Sciences, San Francisco. **1957.** The sub-order *Rhodobacteriineae*. In: R. S. Breed, *et al.* (eds.). *Bergey's Manual of Determinative Bacteriology*, 7th ed. Williams and Wilkins, Baltimore. **Waksman, S. A. 1950.** *The Actinomycetes: Their Nature, Occurrence, Activities and Importance*. Chronica Botanica, Waltham, Massachusetts. **Wattie, E. 1942.** Cultural characteristics of zooglea-forming bacteria isolated from activated sludge and trickling filters. *Public Health Repts. U. S.*, 57:1519–1534. **Williams, M. A. and S. C. Rittenberg. 1957.** A taxonomic study of the genus *Spirillum*. *Intern. Bull. Bacterial Nomenclature and Taxonomy*, 7:49–110. **Williams, R. E. O. and C. C. Spicer (eds.). 1957.** *Microbial Ecology*. Cambridge University Press, Cambridge, England. **Winogradsky, S. 1887.** Ueber Schwefelbacterien. *Botan. Zeitung*, 45:489–507, 513–523, 529–539, 545–559, 569–576, 585–594, 606–610. **1888.** *Beiträge zur Morphologie und Physiologie der Bacterien*. Vol. I, *Zur Morphologie und Physiologie der Schwefelbacterien*. Arthur Felix, Leipzig. **1949.** *Microbiologie du sol. Oeuvres complètes*. Masson, Paris. (The two preceding references are reprinted in this volume in French translation as: Les sulfobactéries, pp. 24–41; and Contribution à la morphologie et à la physiologie des sulfobactéries, pp. 83–126.)

4

Fungi

FREDERICK K. SPARROW

The bulk of the true fungi occurring in fresh water belong to the Phycomycetes although there are a few aquatic Ascomycetes and Fungi Imperfecti.

Phycomycetes (Key on p. 50)

The Phycomycetes as a group are characterized by having an indefinite number of spores borne in a *sporangium*, and by a vegetative system or *mycelium* which is composed of elements or *hyphae* lacking cross walls, except where reproductive structures are delimited. Although this vegetative system in certain higher forms may be visible to the naked eye, where, for example, it may form a cottony halo around a dead fish or insect in the water, in a very large number of species it may be completely within the substratum and, further, may consist of only a few delicate, strongly tapering, very minute *rhizoids*.

Aquatic Phycomycetes exist on a wide variety of substrata. They are found as parasites or saprophytes on algae composing the phytoplankton, on non-planktonic algae, other aquatic fungi, spores of higher plants, vegetable and animal debris, eggs, embryos and adults of microscopic animals, and empty integuments of aquatic insects. Others are wound parasites of larger aquatic

animals such as fish, amphibia, etc., on which they continue to live as saprophytes after the death of the host.   Many, such as the members of the Blastocladiales, Monoblepharidales, and Leptomitales are found on twigs and fruits
that have fallen into the water or have been placed by the investigator in the
water in wire mesh traps.   The Plasmodiophorales (regarded by many as
more related to the Mycetozoa) are obligate parasites of aquatic angiosperms
and other aquatic Phycomycetes.

Parasitic forms must simply be searched for in nature by examining collections of appropriate hosts.   Saprophytes may be trapped by "baiting" appropriate sites, or gross cultures of debris from such sites, with bits of herbaceous
stems, leaves, pollen grains, boiled cellophane (unwaterproofed), pieces of
snake skin, shrimp shell (decalcified), hemp seeds, etc.   Normal soils have
yielded a large number of Phycomycetes.   These are obtained commonly by
placing a teaspoonful of soil in a petri dish, covering it with sterile water,
and baiting with boiled split hemp seed (*Cannabis*), boiled cellophane, etc.

The thallus, composed of the vegetative or nutrient-gathering system and
one or more reproductive rudiments, is extremely varied in character and
extent.   The tubular, much-branched mycelium of the higher Phycomycetes
is probably the most familiar.   This may be both intra- and extramatrical.
There are, however, other types of thalli that may consist solely of the reproductive rudiment (*holocarpic*, Fig. 4.2) or one (*monocentric*, Fig. 4.32) or more
(*polycentric*, Fig. 4.55) reproductive rudiments borne on the vegetative system.   The latter may be *rhizoidal* rather than hyphal in character.   That is,
the nutrient-gathering threads may taper strongly as they radiate from the
reproductive rudiment and be relatively restricted in their growth (Fig. 4.32).
Such a vegetative system is characteristic of the large aquatic order Chytridiales.   In some groups, notably the Saprolegniales, under certain environmental conditions the coenocytic hyphae may form segmented portions, or
gemmae, which may absciss, remain dormant, and eventually germinate to
form new hyphae.

Reproductive rudiments may be transformed at maturity into thin-walled
sporangia, or, either asexually or sexually, into thick-walled resting spores
which germinate after a period of rest.   The sporangia discharge spores
through one or more apertures formed in the sporangium wall.   In the Chytridiales this discharge pore may form upon deliquescence of one or more
papillae (*inoperculate*, Fig. 4.15), or after the dehiscence of a well-defined
specialized circular portion of the sporangium wall (*operculate*, Fig. 4.47).

The *spore*, as might be expected in aquatic groups of the Phycomycetes, is
nearly always motile, being propelled by one or more flagella.   Its body is
somewhat varied in shape, but the important feature is the number and position of the flagella.   This is an absolute essential for detecting relationships.
There are five types of zoospores:

1. The posteriorly uniflagellate type, with more or less spherical body, is
characteristic of the lower forms, such as the Chytridiales, Monoblepharidales,
and Blastocladiales.

2. The anteriorly uniflagellate type is formed by the small order Hyphochytriales which simulate the Chytridiales in their body plan.

3. The heterocont type is one in which there is a short anterior flagellum and a longer backwardly directed one, characteristic of the Plasmodiophorales.

4. The anteriorly biflagellate type is formed in the Saprolegniales. This is the pip-shaped, so-called primary zoospore which usually soon encysts and gives rise to the following type.

5. The laterally biflagellate type. Here, the spore body has been variously termed *bean-shaped, reniform, grape-seedlike*, etc. From a shallow groove in the body at mid-region are attached two equal, oppositely directed flagella. It is a curious circumstance that in the Saprolegniales, which include the common "fish molds," the zoospores emerging from the sporangia are of the anteriorly biflagellate type. These encyst after a varied period of motility and each cyst gives rise to a zoospore of the secondary type. This phenomenon is termed *diplanetism*, more properly, *dimorphism*. Zoospores of the laterally biflagellate type are found in the Saprolegniales, Leptomitales, Lagenidiales, and Peronosporales.

In the Synchytriaceae the sporangia are produced in a group or *sorus* (Fig. 4.6).

In the genus *Ancylistes* we have an undoubted member of a terrestrial group, many of whose members are specialized parasites of insects. Although *Ancylistes* has returned to the aquatic environment, it still maintains the habit of forcefully discharging nonflagellated *conidia*, or externally borne spores, into the air as do its terrestrial relatives in the order Entomophthorales.

Sexual reproduction in the Phycomycetes is extremely varied in type and may be isogamous (like gametes), anisogamous (alike in form but differing in size), or oogamous (egg and sperm). Since determination of the genera does not rest primarily on this phase of the organism, it need not be considered here. It should be mentioned however, that in the Blastocladiales, particularly in *Allomyces*, several types of life cycles exist. For example, if the zygote fails to encyst, and instead germinates at once without meiosis, it results in the interpolation of a new, diploid, sporophyte plant in the life cycle, and alternation of isomorphic generations. Other types of cycles have also been discovered in members of this genus (Sparrow, 1943). In Phycomycetes belonging to other orders, the zygote usually becomes an encysted structure (*resting spore*), which after a quiescent period undergoes meiosis and germinates. Here, the diploid phase is one-celled.

Ascomycetes (Key on p. 87)

In this group the vegetative system is a well-developed mycelium consisting of septate hyphae. These form fruiting structures or *ascocarps* within which the ascospores are borne. Ascospores are formed within narrow saclike structures termed *asci*, eight usually being produced in each ascus. The spores are nonmotile.

The asci in *Vibrissea* and *Apostemidium* are borne within soft and semi-gelatinous sessile or short-stalked fruiting structures (Figs. 4.107, 4.108) formed on twigs in swift-running water. Those of *Mitrula*, which occurs on decaying vegetation, are produced within a fleshy spatulate head terminating a stout stalk (Fig. 4.106), the whole up to 6 cm in height. In *Loramyces*, *Ophiobolus*, and *Ceriospora*, found on decaying parts of higher plants, the asci are enclosed in a flask-shaped structure (*perithecium*) more or less immersed in the substratum (Fig. 4.111).

Fungi Imperfecti (Key on p. 89)

Aquatic Imperfecti have been found thus far primarily in running streams where they inhabit decaying sunken leaves, especially those of alder and oak (Ingold, 1942; Ranzoni, 1953). The variety of types found is truly astonishing and would well repay investigation.

As in the Ascomycetes, the mycelium is septate and the externally formed reproductive structures, termed *conidia*, are usually borne at the apices of the hyphae or on a specialized portion of it called the *conidiophore*. Typically, conidia in aquatic species are produced and disseminated below the water surface. A few species when growing in shallow water produce their conidia just above the water surface and liberate them on the surface film. Members of certain genera have been termed *aeroaquatic*, i.e., their vegetative stage is submerged, whereas their conidia are produced above the water surface.

The key for this group was prepared by William W. Scott.

The figures of Phycomycetes in this chapter are mostly from *Aquatic Phycomycetes* by Frederick K. Sparrow by permission of the publishers, University of Michigan Press. All are highly magnified.

KEY TO GENERA: CLASS PHYCOMYCETES

Spores motile or nonmotile, the motile ones formed in indefinite numbers in a sporangium borne on a coenocytic vegetative system; the nonmotile spores externally and singly formed at the tip of a segmented vegetative system; sexual reproduction iso-, aniso-, or oogamous, or by zygospore formation.

1a		Reproducing asexually by zoospores; thallus varied in character . . .	**2**
1b		Reproducing asexually by forcibly discharged conidia produced singly at the tip of an unbranched conidiophore; thallus segmented. Parasitic in desmids. (Fig. 4.1) . . . . . . Order **Entomophthorales** Family **Ancylistaceae** *Ancylistes* Pfitzer	
2a	(1)	Zoospores posteriorly uniflagellate	**3**
2b		Zoospores anteriorly uniflagellate, usually formed, at least in part, outside sporangium . . . . Order **Hyphochytriales**	**60**
2c		Zoospores biflagellate	**62**
3a	(2)	Thallus either lacking a vegetative system and converted as a whole into a reproductive structure, or forming typically a system of tapering rhizoids (rarely mycelial) and one or more reproductive structures; zoospores usually with a refractive oil globule. . . . . Order **Chytridiales**	**4**
3b		Thallus composed of a more or less well-developed, usually branched mycelium; rhizoids present or absent; cytoplasm often foamy or reticulate in appearance.	**53**

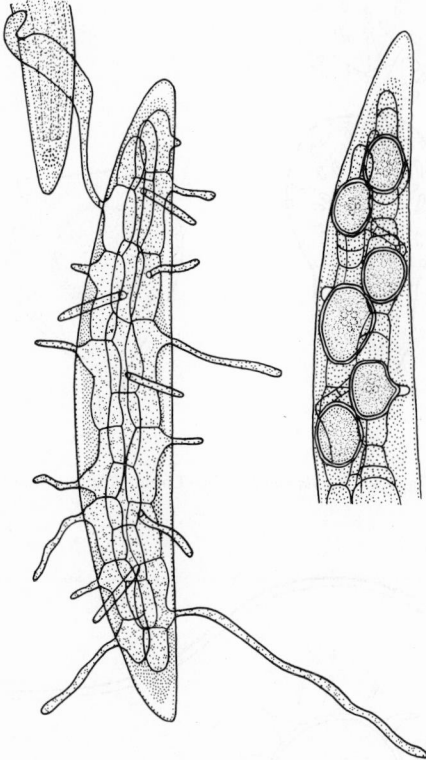

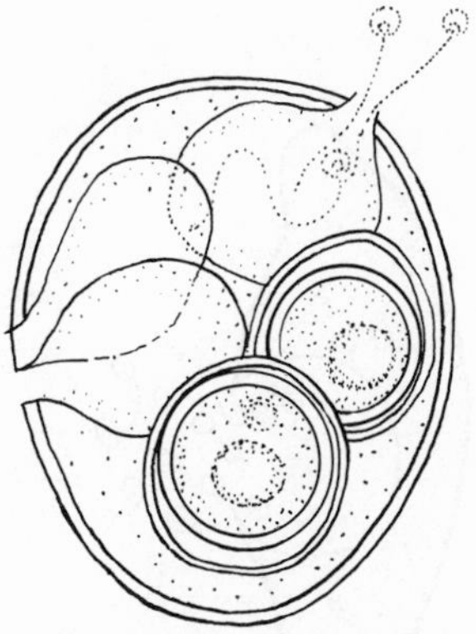

Fig. 4.2. *Olpidium gregarium*
(Nowakowski) Schroeter
in rotifer egg.
(After Sparrow.)

Fig. 4.1. *Ancylistes closterii* Pfitzer in *Closterium*.

4a (3) Sporangium opening by one or more pores upon the deliquescence of papillae . . . . Series **Inoperculatae**   5

4b   Sporangium opening by the dehiscence of an operculum formed at the apex of a discharge papilla . . . . Series **Operculatae**   41

5a (4) Thallus endobiotic, lacking a specialized vegetative system   6

5b   Thallus epi- and endobiotic, or entirely endobiotic; monocentric; reproductive organ epi- or endobiotic . . . . Family **Phlyctidiaceae**   12

5c   Thallus interbiotic, the rhizoids extending along the surface or burrowing somewhat into the substratum; reproductive rudiment lying more or less free, forming a sporangium, prosporangium, or resting spore. . . . . Family **Rhizidiaceae**   26
         See also note under **25a** which applies here.

5d   Thallus polycentric, endo- or interbiotic   37

6a (5) Thallus composed of a single reproductive rudiment which becomes converted into a reproductive organ . . . . . . Family **Olpidiaceae**   7

6b   Thallus forming an endobiotic prosorus which gives rise to a sorus of sporangia. . . . . Family **Synchytriaceae**   9

6c   Thallus composed of a series of unbranched, conjoined segments, each of which becomes a sporangium . . Family **Achlyogetonaceae**   11

7a (6) Thallus lying loosely within the substratum; in algae, plant spores or bodies, or eggs of microscopic animals. (Fig. 4.2) . . . .
                                     Olpidium (Braun) Rabenhorst

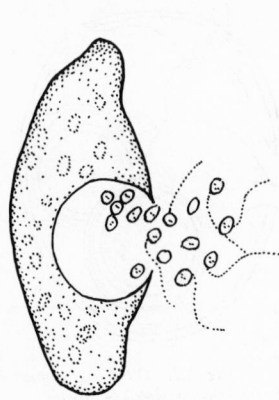

Fig. 4.3. *Sphaerita dangeardii*
Chatton and Brodsky in *Euglena*.
(After Sparrow.)

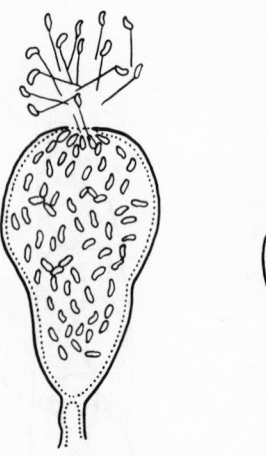

Fig. 4.4. *Rozella rhipidii-spinosi* Cornu. (After Cornu.)

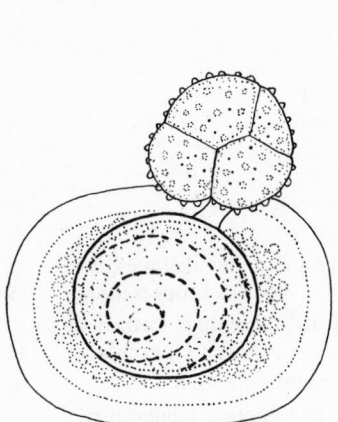

Fig. 4.5. *Micromycopsis cristata* Scherffel.
(After Scherffel.)

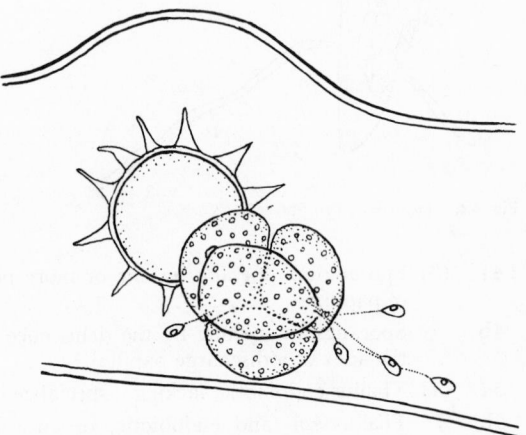

Fig. 4.6. *Micromyces zygogonii* Dangeard. (After Sparrow.)

7b		Thallus completely filling the often hypertrophied host structure . . .	8
8a	(7)	In nuclei of Euglenophyta. (Fig. 4.3). . . . . **Sphaerita** Dangeard	
8b		In other water fungi. (Fig. 4.4) . . . . **Rozella** Cornu	
9a	(6)	Sorus endobiotic	10
9b		Sorus epibiotic, formed at the tip of a discharge tube from the prosorus; sporangia as in *Endodesmidium* (Fig. 4.5) . . . . **Micromycopsis** Scherffel	
10a	(9)	Zoospores freed within the host cell from several large sporangia. (Fig. 4.6) . . . . **Micromyces** Dangeard	
10b		Zoospores freed from numerous small spherical sporangia which emerge more or less amoeboidly through pores from the sorus. (Fig. 4.7) . . . . **Endodesmidium** Canter	

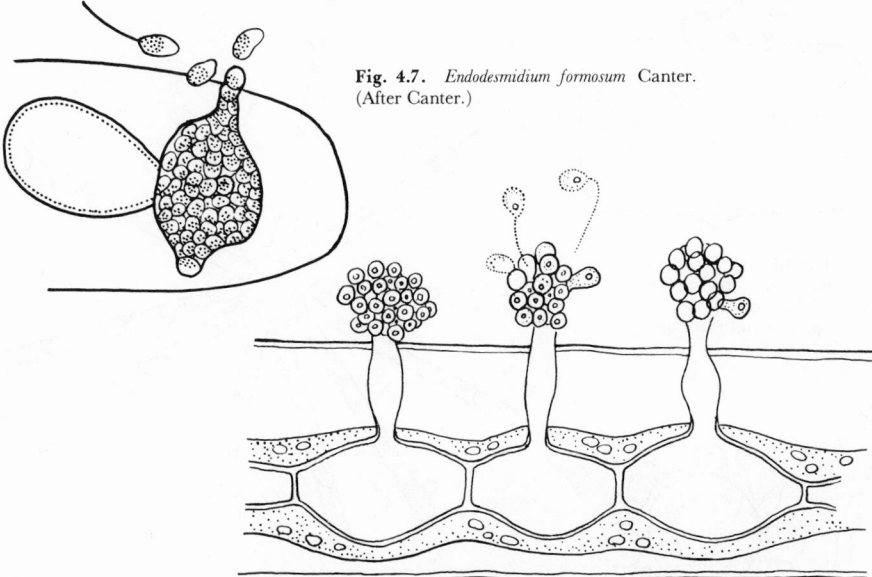

Fig. 4.7. *Endodesmidium formosum* Canter. (After Canter.)

Fig. 4.8. *Achlyogeton entophytum* Schenk in *Cladophora*. (After Schenk.)

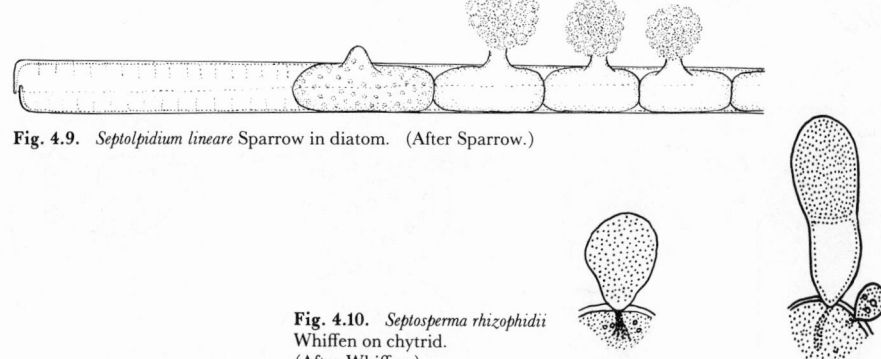

Fig. 4.9. *Septolpidium lineare* Sparrow in diatom. (After Sparrow.)

Fig. 4.10. *Septosperma rhizophidii* Whiffen on chytrid. (After Whiffen.)

11a (6) Zoospores upon discharge encysting at the mouth of the discharge tube. (Fig. 4.8) . . . . **Achlyogeton** Schenk

11b Zoospores temporarily clustering after discharge, eventually swimming away without encystment. (Fig. 4.9) . **Septolpidium** Sparrow

12a (5) Resting spore divided by a septum into an empty basal and a distal fertile part. Parasitic on Phycomycetes. (Fig. 4.10) . . . . **Septosperma** Whiffen

12b Resting spore 1-celled 13

13a (12) Sporangium epibiotic, completely fertile, sterile parts of thallus endobiotic 14

13b Sporangium epibiotic, with a sterile septate base, or knoblike sterile part. 20

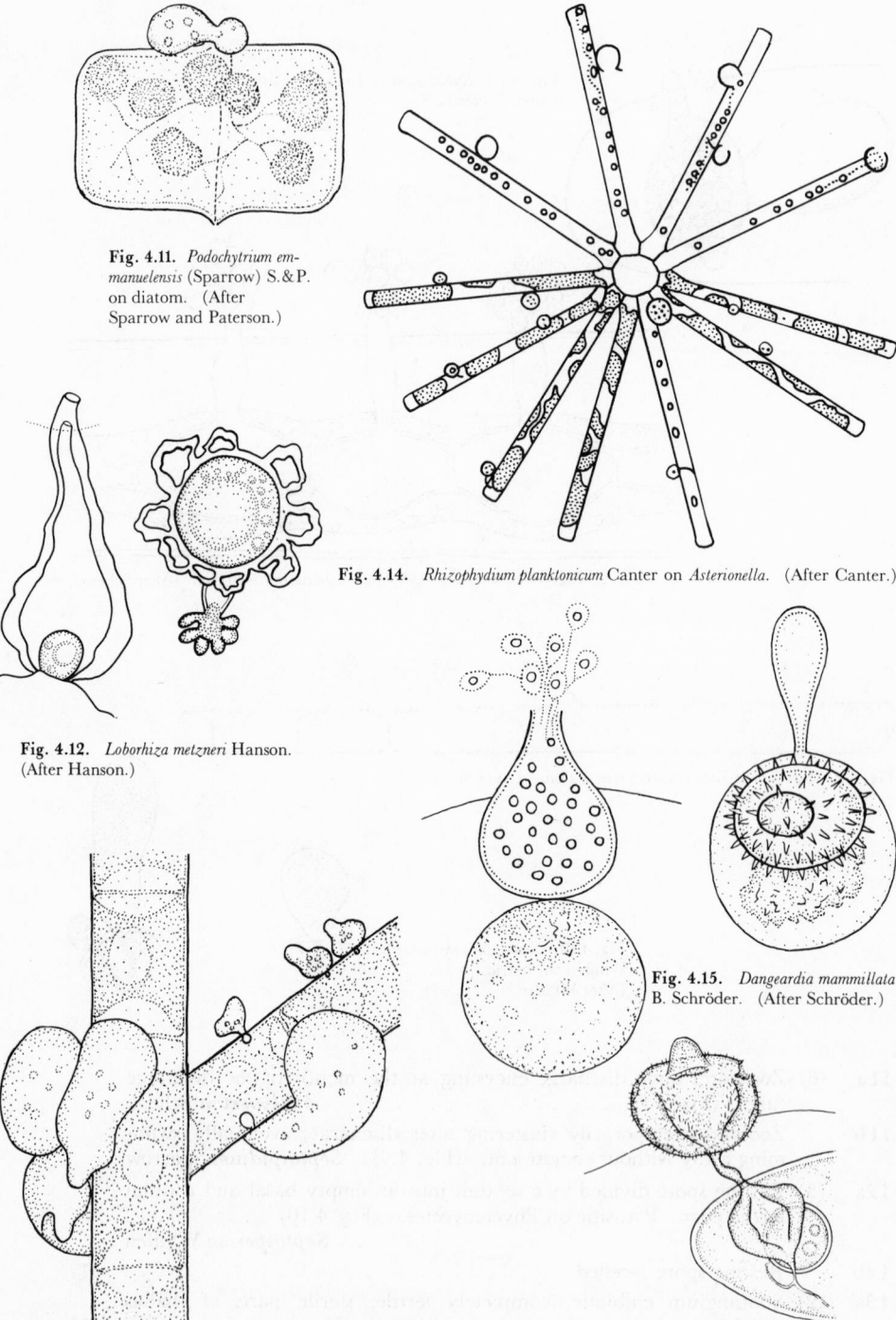

Fig. 4.11. *Podochytrium emmanuelensis* (Sparrow) S.&P. on diatom. (After Sparrow and Paterson.)

Fig. 4.14. *Rhizophydium planktonicum* Canter on *Asterionella*. (After Canter.)

Fig. 4.12. *Loborhiza metzneri* Hanson. (After Hanson.)

Fig. 4.15. *Dangeardia mammillata* B. Schröder. (After Schröder.)

Fig. 4.13. *Phlyctidium anatropum* (Braun) Rabenhorst on green alga. (After Sparrow.)

Fig. 4.16. *Blyttiomyces spinulosis* (Blytt) Bartsch. (After Bartsch.)

13c   Sporangium, resting spore, and sterile parts all endobiotic **23**

14a (13) Whole body of germinating spore enlarging to form a reproductive rudiment **15**

14b   Only a portion of spore body enlarging to form the sporangium, the remainder appearing as a basal cyst on wall of sporangium; not apophysate. On diatoms. (Fig. 4.11) . . . . ***Podochytrium*** Pfitzer
Includes *Rhizidiopsis* Sparrow. See also **20a**.

14c   Body of germinating spore either sessile and enlarging to form a prosporangium; or lying free, not enlarging, and producing a germ tube the tip of which in contact with host expands to form an epibiotic sporangium. **21**

15a (14) Endobiotic part tubular, digitate, or papillate, never a tapering branched or unbranched rhizoid **16**

15b   Endobiotic part a tapering, unbranched or branched rhizoid either arising directly from the tip of the germ tube of encysted zoospore or appearing to arise from a subsporangial apophysis **17**

16a (15) Endobiotic part digitate; sporangia internally proliferating. In *Volvox*. (Fig. 4.12) . . . . ***Loborhiza*** Hanson

16b   Endobiotic part papillate or tubular. (Fig. 4.13) . . . .
                                           Phlyctidium (Braun) Rabenhorst

17a (15) Rhizoids arising from basal region of sporangium **18**

17b   Endobiotic part a subsporangial apophysis from which rhizoids emerge **19**

18a (17) Sporangium and resting spore epibiotic; rhizoids developed to a varying degree. On a wide variety of plant and animal substrata. (Fig. 4.14) . . . . ***Rhizophydium*** Schenk

18b   Sporangium extracellular, with a bushy tuft of rhizoids; resting spore (?) endobiotic. On *Pandorina*. (Fig. 4.15). . . . .
                                             Dangeardia B. Schröder

19a (17) Sporangium bearing a prominent apiculus; resting spore endobiotic. On zygospores of Conjugatae. (Fig. 4.16). . . ***Blyttiomyces*** Bartsch

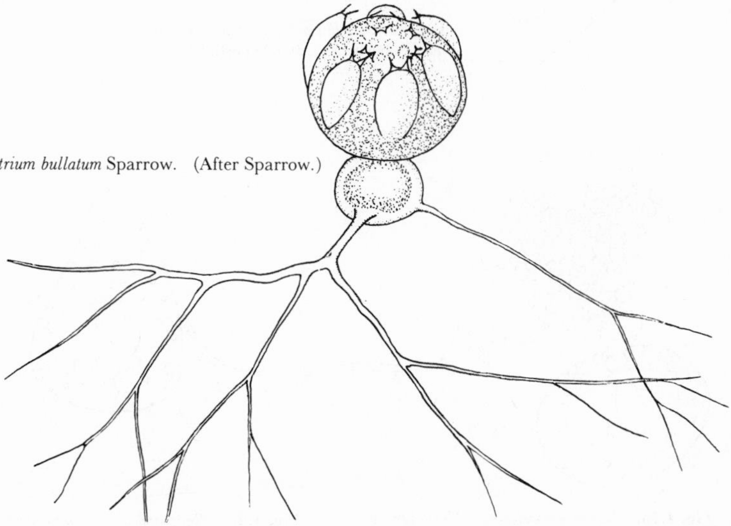

Fig. 4.17. *Phlyctochytrium bullatum* Sparrow. (After Sparrow.)

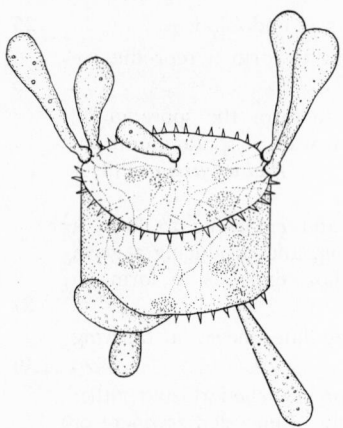

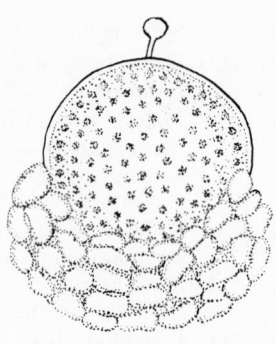

Fig. 4.21. *Scherffeliomyces parasitans* Sparrow on cyst of *Euglena*.

Fig. 4.18. *Podochytrium cornutum* Sparrow on *Stephanodiscus*. (After Sparrow.)

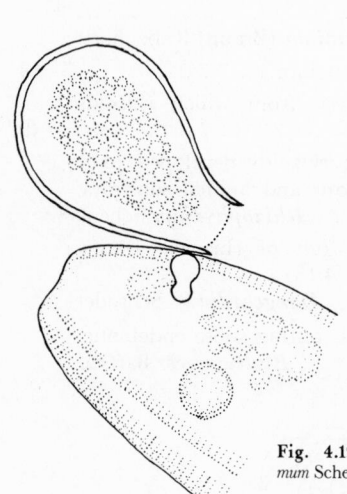

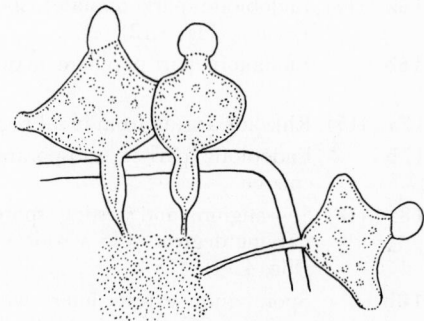

Fig. 4.22. *Coralliochytrium scherffelii* Domján. (After Domján.)

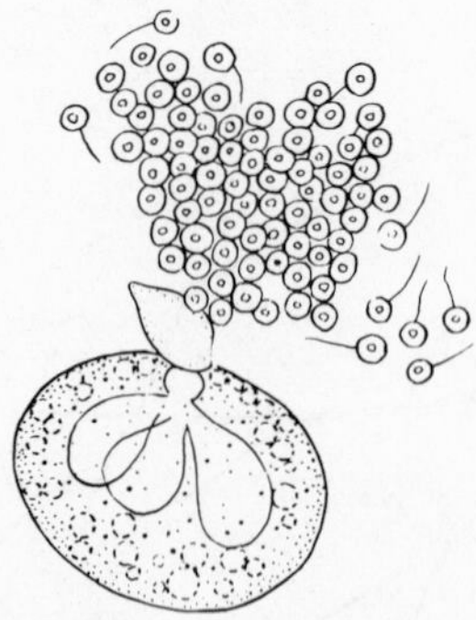

Fig. 4.19. *Physorhizophidium pachydermum* Scherffel. (After Scherffel.)

Fig. 4.20. *Saccomyces endogenus* (Nowakowski) Sparrow. (After Serbinow.)

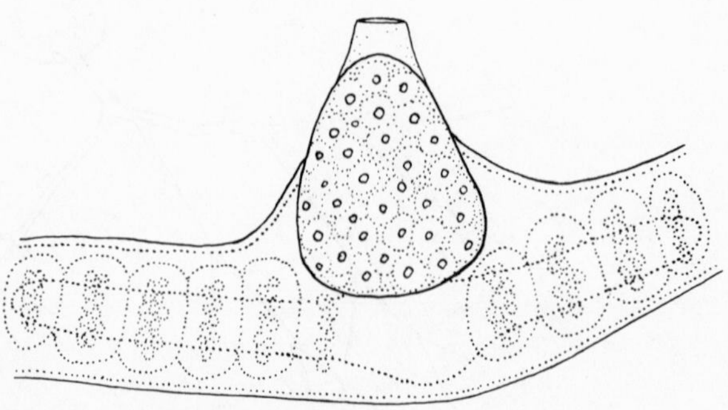

Fig. 4.23. *Rhizosiphon crassum* Scherffel in blue-green alga. (After Scherffel.)

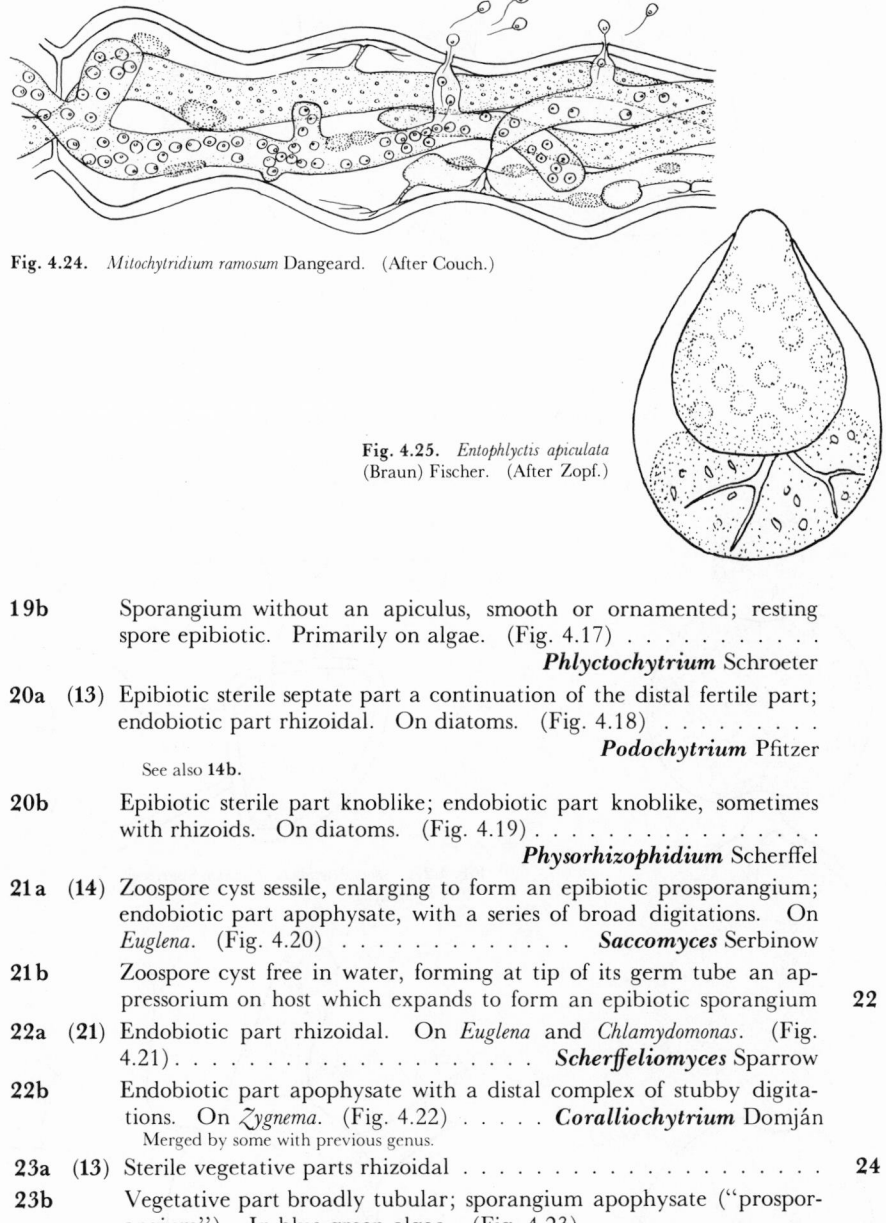

Fig. 4.24. *Mitochytridium ramosum* Dangeard. (After Couch.)

Fig. 4.25. *Entophlyctis apiculata* (Braun) Fischer. (After Zopf.)

19b   Sporangium without an apiculus, smooth or ornamented; resting spore epibiotic. Primarily on algae. (Fig. 4.17) . . . .
                                      Phlyctochytrium Schroeter

20a (13) Epibiotic sterile septate part a continuation of the distal fertile part; endobiotic part rhizoidal. On diatoms. (Fig. 4.18) . . . .
                                      Podochytrium Pfitzer
            See also **14b**.

20b   Epibiotic sterile part knoblike; endobiotic part knoblike, sometimes with rhizoids. On diatoms. (Fig. 4.19) . . . .
                                      Physorhizophidium Scherffel

21a (14) Zoospore cyst sessile, enlarging to form an epibiotic prosporangium; endobiotic part apophysate, with a series of broad digitations. On *Euglena*. (Fig. 4.20) . . . . ***Saccomyces*** Serbinow

21b   Zoospore cyst free in water, forming at tip of its germ tube an appressorium on host which expands to form an epibiotic sporangium   **22**

22a (21) Endobiotic part rhizoidal. On *Euglena* and *Chlamydomonas*. (Fig. 4.21) . . . . ***Scherffeliomyces*** Sparrow

22b   Endobiotic part apophysate with a distal complex of stubby digitations. On *Zygnema*. (Fig. 4.22) . . . . . ***Coralliochytrium*** Domján
            Merged by some with previous genus.

23a (13) Sterile vegetative parts rhizoidal **24**

23b   Vegetative part broadly tubular; sporangium apophysate ("prosporangium"). In blue-green algae. (Fig. 4.23) . . . .
                                      Rhizosiphon Scherffel

24a (23) Sporangium strongly tubular, forming one or more discharge tubes. In the desmid *Docidium*. (Fig. 4.24) . . . ***Mitochytridium*** Dangeard

24b   Sporangium spherical, pyriform, or irregular, not tubular **25**

25a (24) Rhizoids or rhizoidal axes arising directly from body of sporangium. Primarily in algae. (Fig. 4.25) . . . .
                                      Entophlyctis Fischer and ***Phlyctorhiza*** Hanson
            Reproductive rudiment said in some instances to arise from body of encysted spore, or in others, from either body of encysted spore or from tip of a short germ

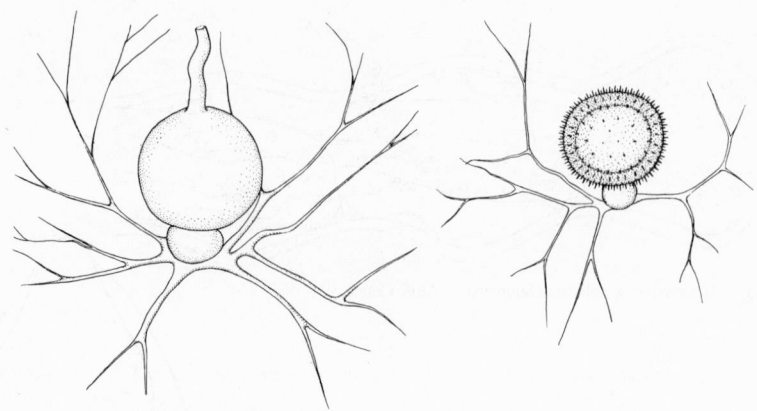

Fig. 4.26. *Diplophlyctis intestina* (Schenk) Schroeter. (After Sparrow.)

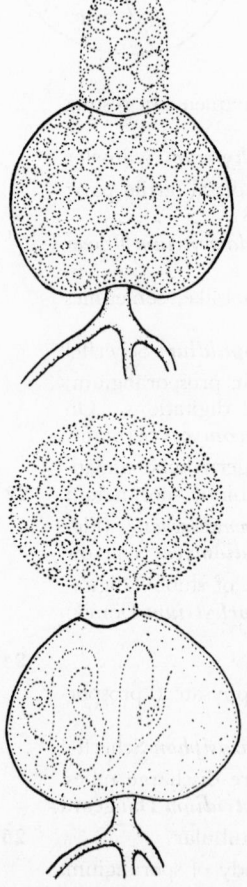

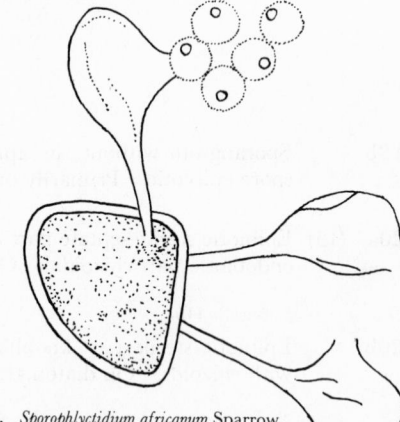

Fig. 4.27. *Sporophlyctidium africanum* Sparrow. (After Sparrow.)

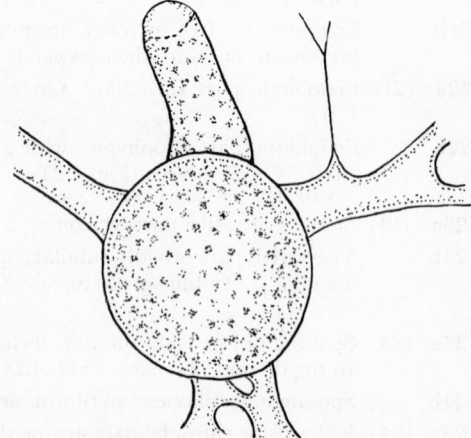

Fig. 4.28. *Rhizidium ramosum* Sparrow. (After Sparrow.)

Fig. 4.29. *Rhizophlyctis petersenii* Sparrow. (After Sparrow.)

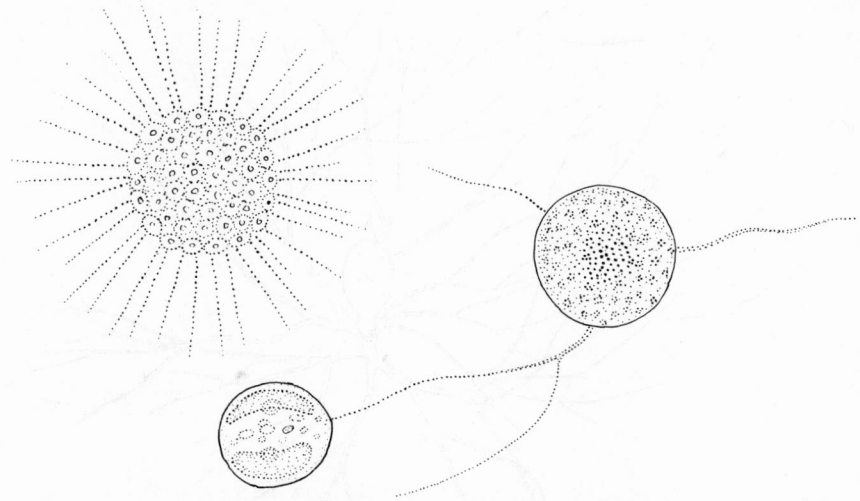

Fig. 4.30. *Nowakowskia hormothecae* Borzi. (After Borzi.)

tube, or as an outgrowth of an apophysis.   In *Phlyctorhiza*, on chitin and keratin, sporangium forms by vesication of rhizoids.

25b Rhizoids arising from an apophysis. In algae. (Fig. 4.26) . . . . . .
Diplophlyctis Schroeter

26a **(5)** Reproductive rudiment converted directly into a sporangium or resting spore **27**

26b Reproductive rudiment forming a prosporangium, either from the body of the encysted zoospore, or from a germ tube produced by the encysted zoospore **35**

27a **(26)** Rhizoids arising from body of sporangium at several points, or from a conspicuous main axis; resting spore mostly asexually formed   **28**

27b Rhizoids arising from a subsporangial apophysis; resting spore sometimes sexually formed **32**

28a **(27)** Vegetative system unbranched or branched, arising from one place on the sporangium. **29**

28b Vegetative system of branched rhizoids arising from more than one place on sporangium. **30**

29a **(28)** Vegetative system an unbranched, double-contoured tube; aplanospores, not zoospores, formed. On *Protoderma*. (Fig. 4.27) . . . . . .
Sporophlyctidium Sparrow

29b Vegetative system of branched rhizoids arising from a more or less prolonged main axis; zoospores formed. On a variety of substrata. (Fig. 4.28) . . . . *Rhizidium* Braun

30a **(28)** Sporangium wall persistent after discharge of zoospores through one or more pores or tubes. On a variety of substrata. (Fig. 4.29) . . .
Rhizophlyctis Fischer

Certain rhizophlyctoid fungi, in which a pellicle covering the cytoplasm at the base of the discharge tube is pushed back at emergence of the zoospores as an "endo-operculum," have been placed in a genus *Karlingia* Hanson.   There is doubt as to the constancy of this endo-operculum under varying environmental conditions.   It does not appear comparable in constancy or origin with the operculum of wall material formed in an operculate chytrid.

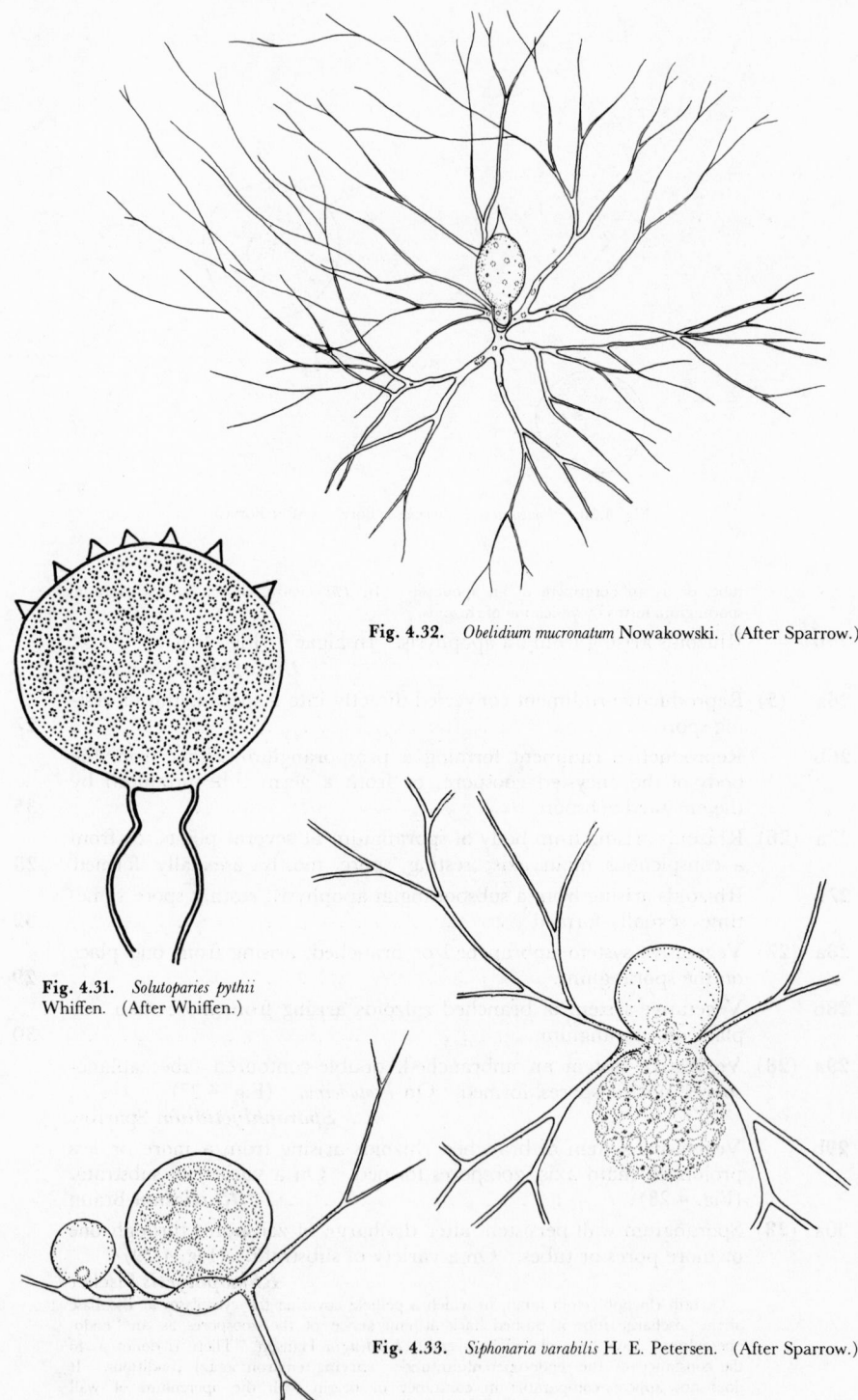

Fig. 4.32.   *Obelidium mucronatum* Nowakowski.   (After Sparrow.)

Fig. 4.31.   *Solutoparies pythii*
Whiffen.   (After Whiffen.)

Fig. 4.33.   *Siphonaria varabilis* H. E. Petersen.   (After Sparrow.)

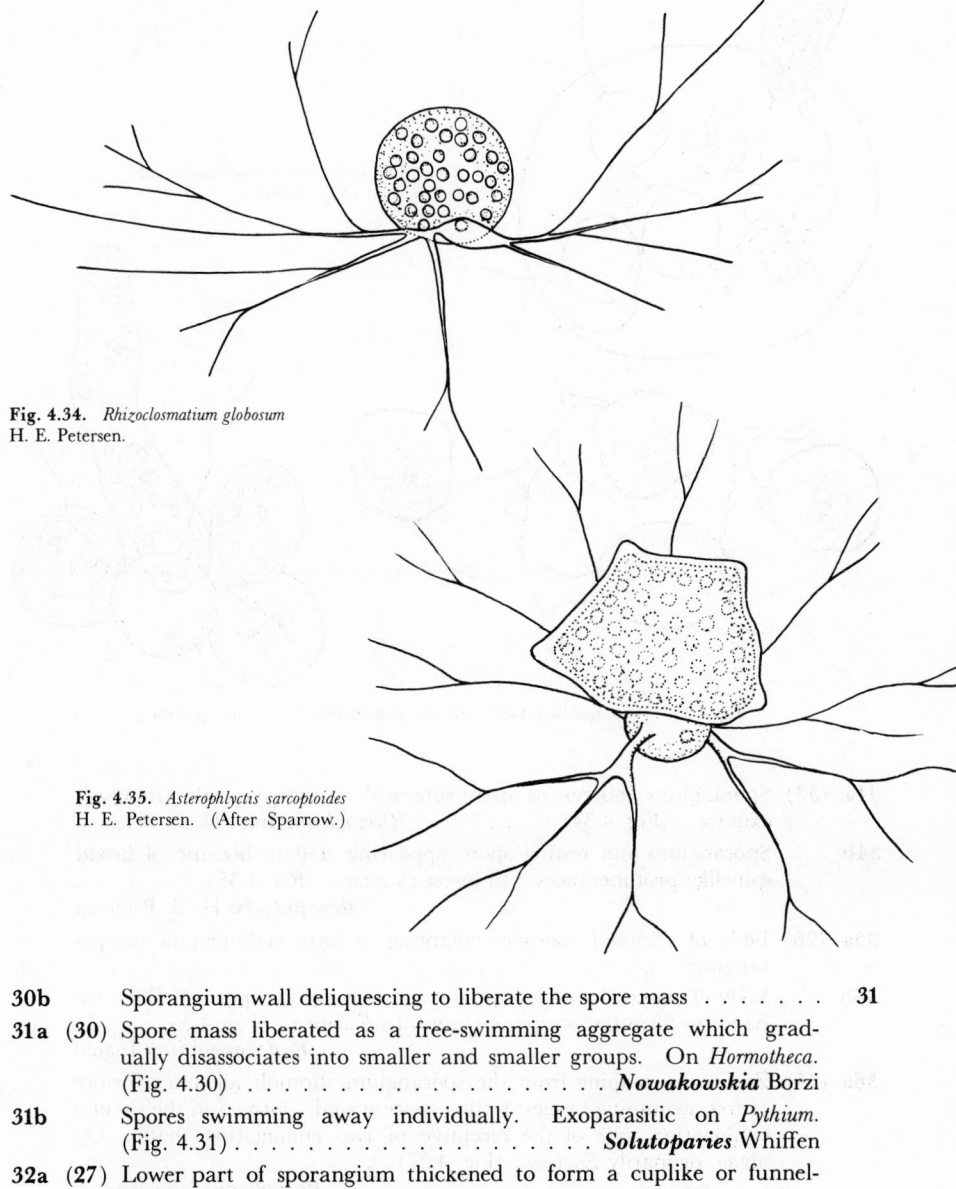

Fig. 4.34. *Rhizoclosmatium globosum* H. E. Petersen.

Fig. 4.35. *Asterophlyctis sarcoptoides* H. E. Petersen. (After Sparrow.)

30b Sporangium wall deliquescing to liberate the spore mass **31**

31a (30) Spore mass liberated as a free-swimming aggregate which gradually disassociates into smaller and smaller groups. On *Hormotheca*. (Fig. 4.30) . . . . *Nowakowskia* Borzi

31b Spores swimming away individually. Exoparasitic on *Pythium*. (Fig. 4.31) . . . . *Solutoparies* Whiffen

32a (27) Lower part of sporangium thickened to form a cuplike or funnel-like base; apophysis inconspicuous; sporangium fusiform with a single apical, solid spine. In insect exuviae. (Fig. 4.32) . . . .
Obelidium Nowakowski

32b Lower part of sporangium not differentiated from remainder; apophysis usually conspicuous; sporangium smooth or bearing spine-like protuberances **33**

33a (32) Rhizoids delicate, wide-lumened only near apophysis, if at all **34**

33b Rhizoids wide-lumened throughout; resting spores sexually formed by conjugation of thalli by means of rhizoidal anastomosis. In insect exuviae. (Fig. 4.33) . . . . *Siphonaria* H. E. Petersen

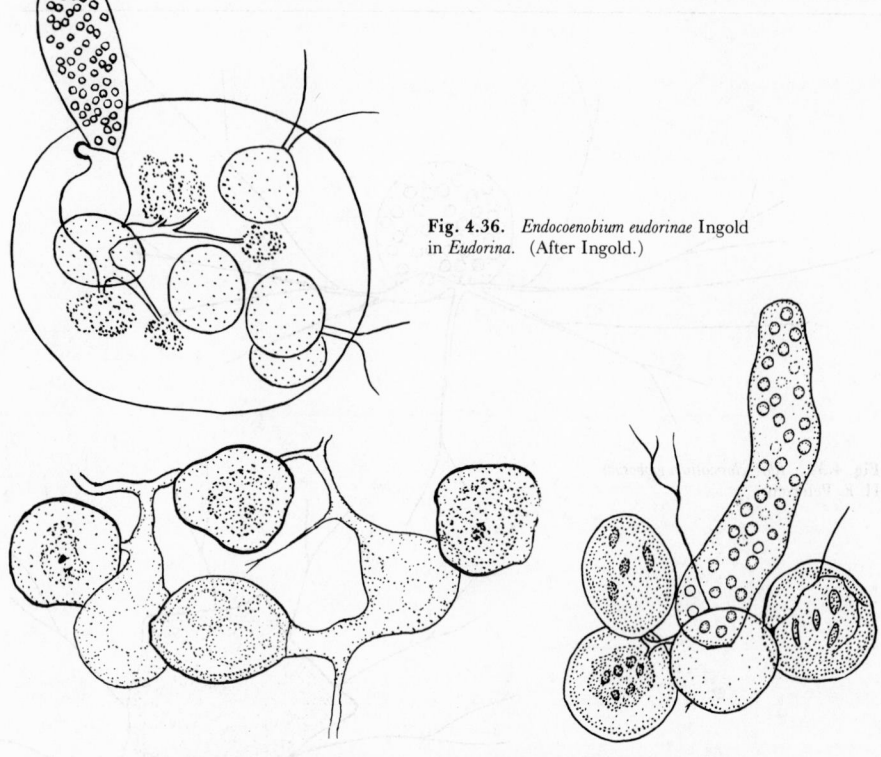

Fig. 4.36. *Endocoenobium eudorinae* Ingold in *Eudorina*. (After Ingold.)

Fig. 4.37. *Polyphagus laevis* Bartsch on encysted euglenoids. (After Sparrow.)

34a (33) Sporangium and resting spore spherical or subspherical. In insect exuviae. (Fig. 4.34) . . . . ***Rhizoclosmatium*** H. E. Petersen

34b Sporangium and resting spore appearing stellate because of broad spinelike protuberances. In insect exuviae. (Fig. 4.35) . . . .
                                          Asterophlyctis H. E. Petersen

35a (26) Body of encysted zoospore enlarging to form rudiment of prosporangium **36**

35b Body of encysted zoospore producing a germ tube, part of which expands to form the prosporangium. In *Eudorina*. (Fig. 4.36) . . . . .
                                            Endocoenobium Ingold

36a (35) Zoospores escaping from the sporangium through an apical orifice as free-swimming bodies; resting spore sexually formed in the tip of a conjugation tube of the receptive of two conjugating thalli. On algae, primarily *Euglena*. (Fig. 4.37) . . . .
                                     Polyphagus Nowakowski

36b Zoospores (aplanospores) escaping through a subapical or lateral pore, devoid of flagellae, or germinating within the sporangium; resting spore sexually formed in the receptive of two conjugating adnate thalli. On *Draparnaldia*. (Fig. 4.38) . . . .
                                     Sporophlyctis Serbinow

37a (5) Zoospores with flagellae **38**

37b Zoospores lacking flagellae, with amoeboid movement only. In jelly of *Chaetophora*. (Fig. 4.39) . . . . ***Amoebochytrium*** Zopf

38a (37) Thallus with septate turbinate cells; vegetative system strongly rhizoidal, somewhat irregular **39**

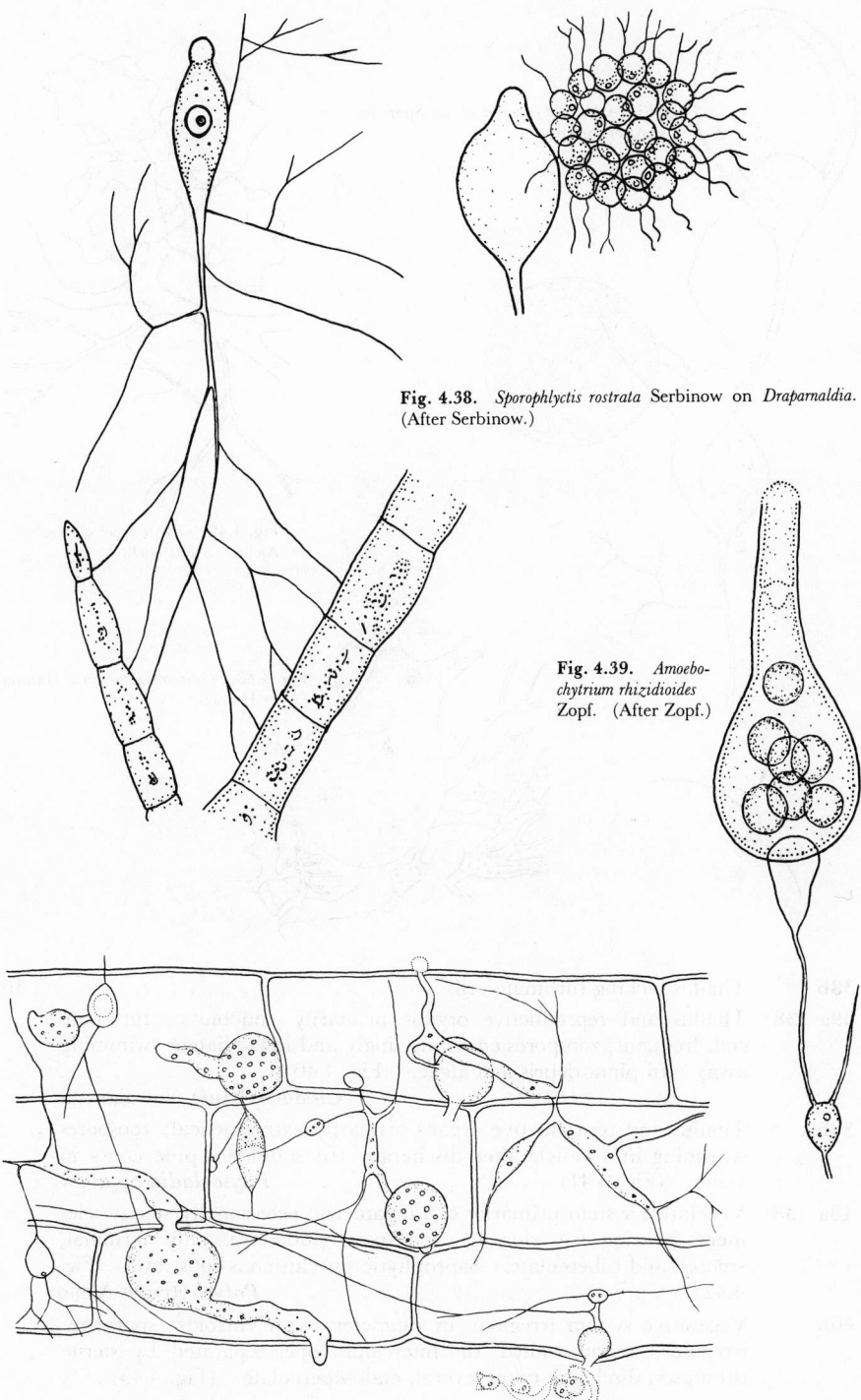

Fig. 4.38. *Sporophlyctis rostrata* Serbinow on *Drapamaldia*. (After Serbinow.)

Fig. 4.39. *Amoebo-chytrium rhizidioides* Zopf. (After Zopf.)

Fig. 4.40. *Cladochytrium tenue* Nowakowski in plant debris. (After Sparrow.)

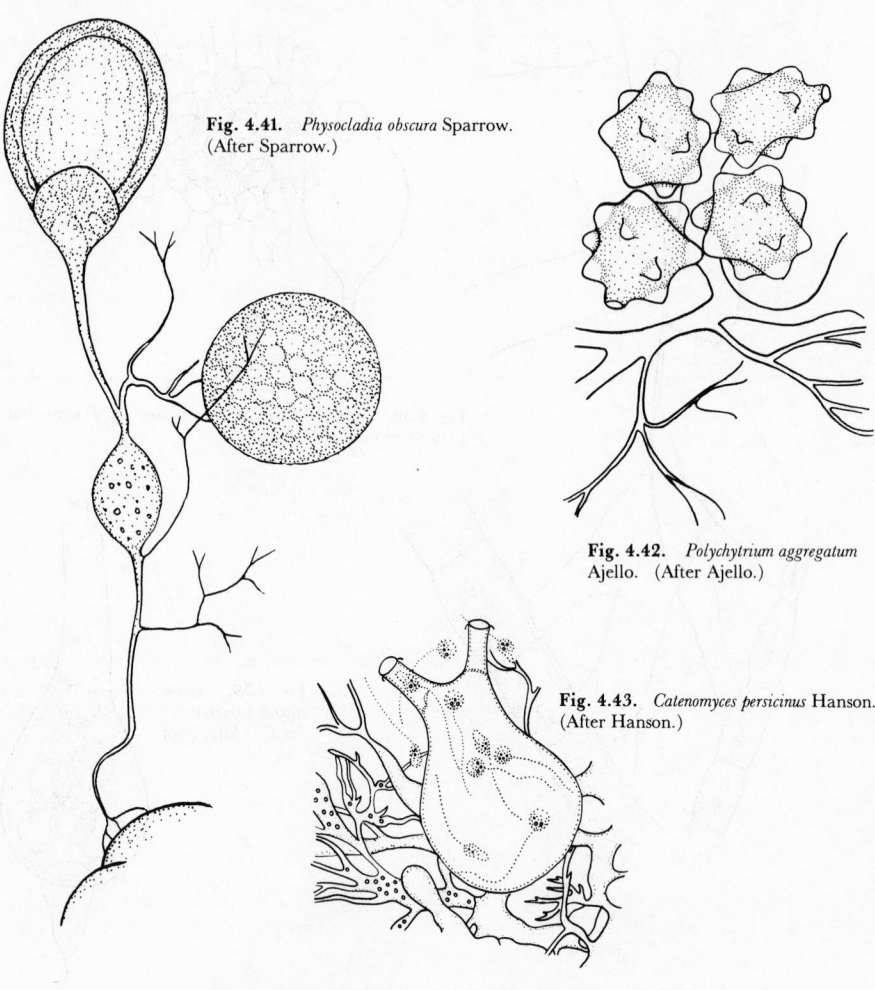

Fig. 4.41. *Physocladia obscura* Sparrow. (After Sparrow.)

Fig. 4.42. *Polychytrium aggregatum* Ajello. (After Ajello.)

Fig. 4.43. *Catenomyces persicinus* Hanson. (After Hanson.)

38b	Thallus lacking turbinate cells	**40**
39a (38)	Thallus and reproductive organs primarily endobiotic; turbinate cells frequent; zoospores emerging singly and immediately swimming away. In plant debris and algae. (Fig. 4.40). . . . . *Cladochytrium* Nowakowski	
39b	Thallus and reproductive organs primarily extramatrical; zoospores swarming in a vesicle after discharge. In staminate pine cones in water. (Fig. 4.41) . . . . *Physocladia* Sparrow	
40a (38)	Vegetative system primarily of isodiametric occasionally septate elements bearing few rhizoids; sporangia globose, usually terminal, smooth and tuberculate. Saprophytic on chitinous substrata. (Fig. 4.42) . . . . *Polychytrium* Ajello	
40b	Vegetative system irregular in diameter, with rhizoids; sporangia irregular, smooth-walled, the intercalary ones separated by sterile isthmuses; discharge tubes several, endo-operculate. (Fig. 4.43) . . . *Catenomyces* Hanson	

Probably better placed in Blastocladiales (**53**).

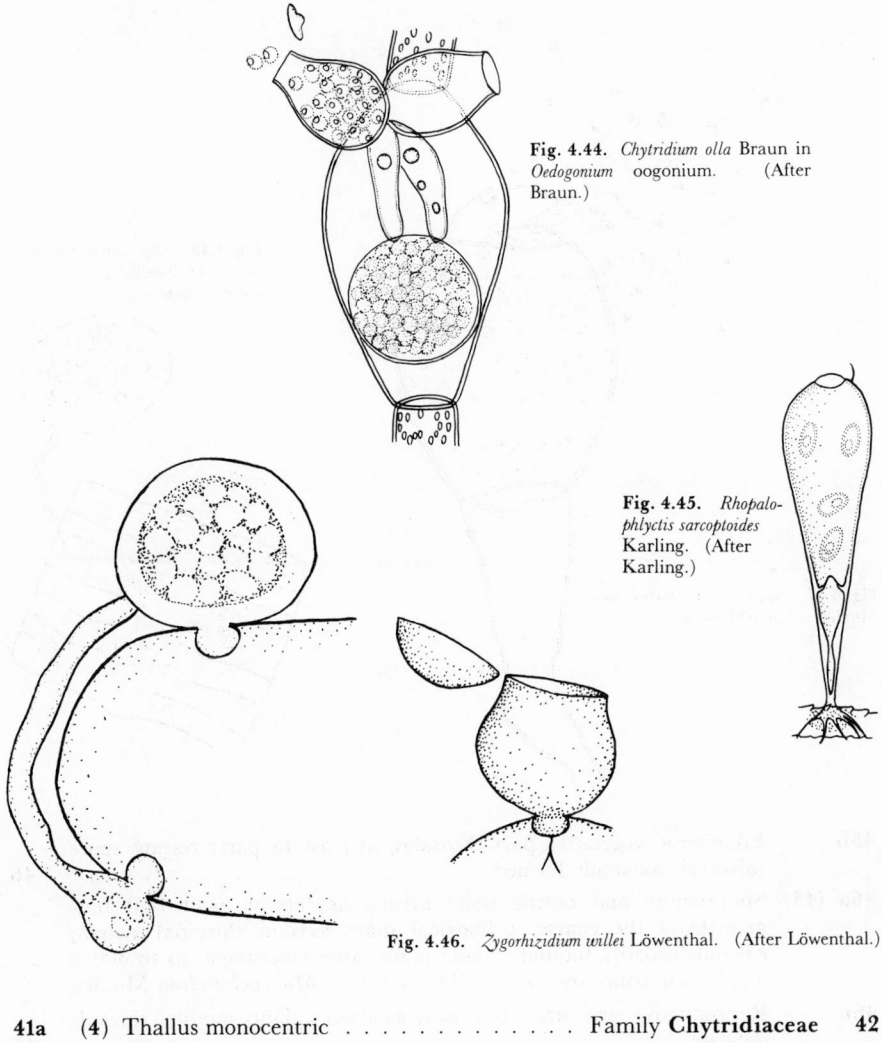

Fig. 4.44. *Chytridium olla* Braun in *Oedogonium* oogonium. (After Braun.)

Fig. 4.45. *Rhopalophlyctis sarcoptoides* Karling. (After Karling.)

Fig. 4.46. *Zygorhizidium willei* Löwenthal. (After Löwenthal.)

41a (4) Thallus monocentric . . . . Family **Chytridiaceae** **42**

41b   Thallus polycentric . . . . Family **Megachytriaceae** **51**

42a (41) Sporangium epibiotic; resting spore endo- or epibiotic **43**

42b   Sporangia and resting spores endobiotic **49**

43a (42) Sporangium epibiotic, resting spore endobiotic; rhizoids not segmented; zoospores swimming directly away from sporangium. Primarily on algae. (Fig. 4.44) . . . . ***Chytridium*** Braun

43b   Sporangium epibiotic, resting spore epibiotic (where known) **44**

44a (43) Sporangium completely fertile **45**

44b   Sporangium continuous or with a sterile basal part. On insect exuviae. (Fig. 4.45) . . . . ***Rhopalophlyctis*** Karling

45a (44) Endobiotic vegetative part a small bulbous structure; resting spore sexually formed after conjugation with a small thallus by an elongate conjugation tube. On Conjugatae. (Fig. 4.46) . . . . ***Zygorhizidium*** Löwenthal

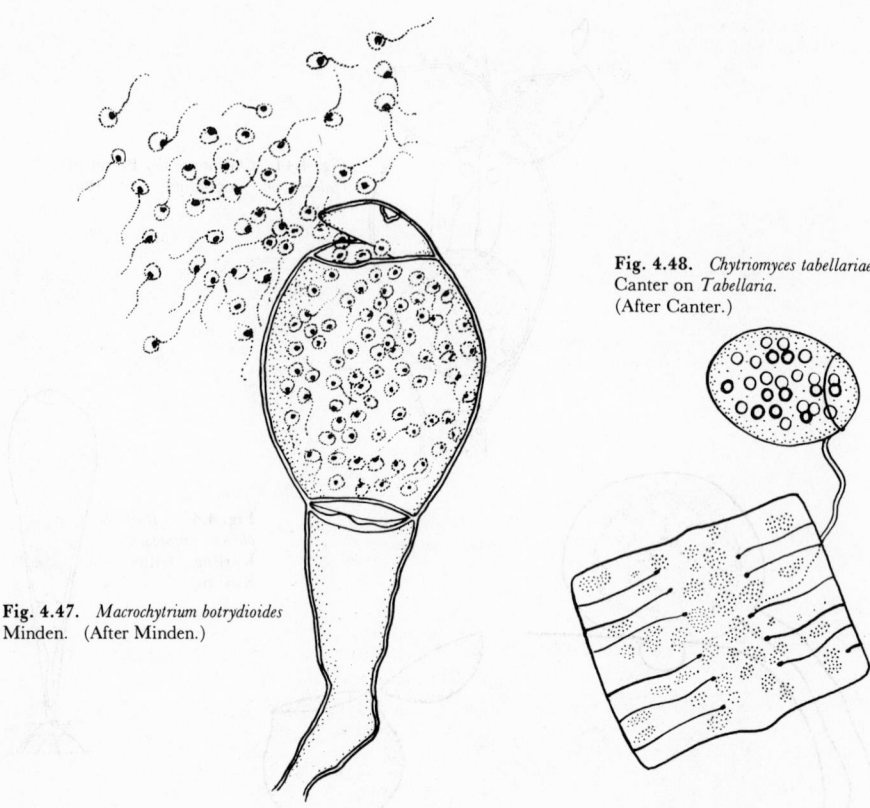

Fig. 4.48. *Chytriomyces tabellariae* Canter on *Tabellaria*. (After Canter.)

Fig. 4.47. *Macrochytrium botrydioides* Minden. (After Minden.)

45b		Endobiotic vegetative part rhizoidal, at least in part; resting spore (always?) asexually formed	**46**
46a	**(45)**	Sporangium and resting spore arising as lateral, walled-off outgrowths of the coarse, cylindrical main axis of rhizoidal system; rhizoids broadly tubular; whole plant large (sporangia up to 800 μ long). On fruits and twigs. (Fig. 4.47). . **Macrochytrium** Minden	
46b		Reproductive structures not arising as above; plants minute, rhizoids delicate.	**47**
47a	**(46)**	Rhizoids arising from base of sporangium or apophysis; sporangium and resting spore extramatrical. On a variety of substrata. (Fig. 4.48) . . . . **Chytriomyces** Karling	
47b		Rhizoids arising from a series of subsporangial catenulate segments.	**48**
48a	**(47)**	Sporangium tubular, entire cyst of zoospore expanding. Saprophytic in vegetable debris. (Fig. 4.49). . . . . **Cylindrochytridium** Karling	
48b		Sporangium more or less globose, a portion of the zoospore cyst persistent on the sporangium. Saprophytic in vegetable debris. (Fig. 4.50). . . . . **Catenochytridium** Berdan	
49a	**(42)**	Rhizoids arising directly from the body of the sporangium	**50**
49b		Rhizoids arising from a subsporangial swelling. In algae and vegetable debris. (Fig. 4.51). . . . . . . **Nephrochytrium** Karling	

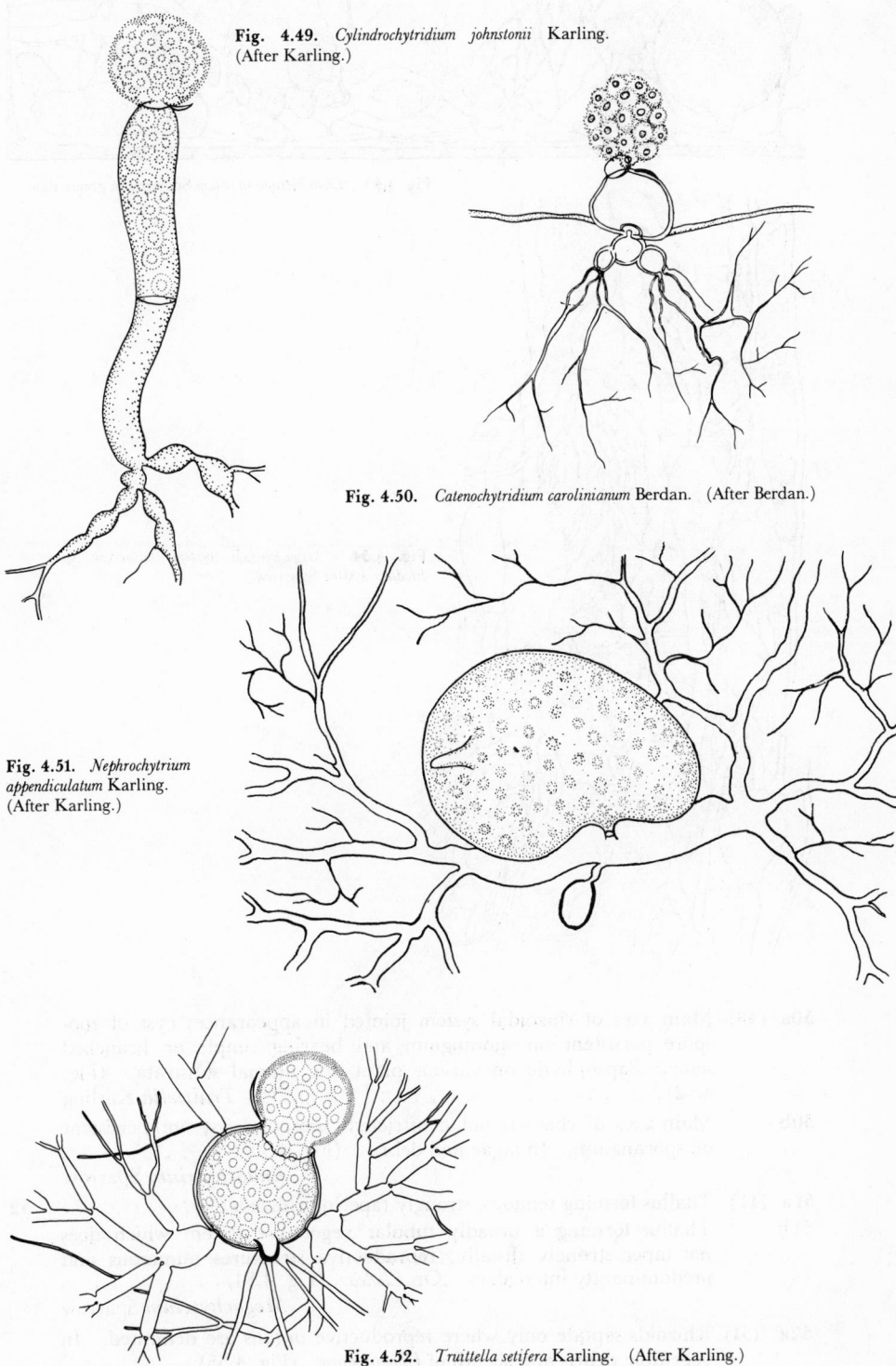

Fig. 4.49. *Cylindrochytridium johnstonii* Karling. (After Karling.)

Fig. 4.50. *Catenochytridium carolinianum* Berdan. (After Berdan.)

Fig. 4.51. *Nephrochytrium appendiculatum* Karling. (After Karling.)

Fig. 4.52. *Truittella setifera* Karling. (After Karling.)

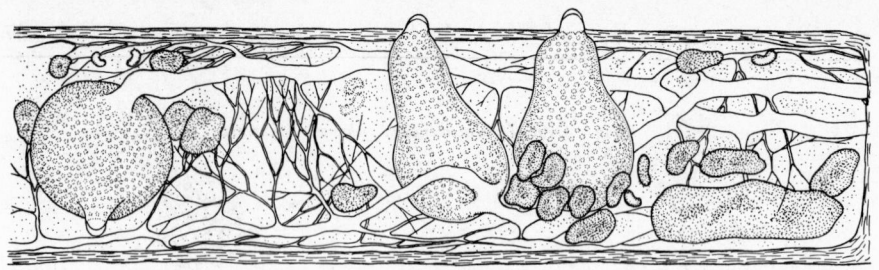

Fig. 4.53. *Endochytrium ramosum* Sparrow in green alga.

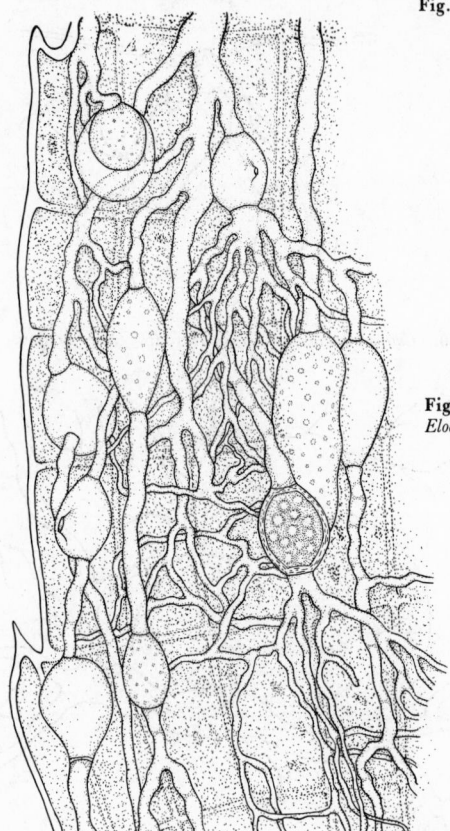

Fig. 4.54. *Megachytrium westonii* Sparrow in *Elodea*. (After Sparrow.)

50a (49) Main axes of rhizoidal system jointed in appearance; cyst of zoospore persistent on sporangium and bearing simple or branched setae. Saprophytic on various plant and animal substrata. (Fig. 4.52)..................................... ***Truittella*** Karling

50b Main axes of rhizoids not constricted; zoospore cyst not persistent on sporangium. In algae and debris. (Fig. 4.53)...........
Endochytrium Sparrow

51a (41) Thallus forming tenuous, strongly tapering rhizoids.......... **52**

51b Thallus forming a broadly tubular vegetative system which does not taper strongly distally; reproductive structures numerous and predominantly intercalary. On *Elodea*. (Fig. 4.54).........
Megachytrium Sparrow

52a (51) Rhizoids septate only where reproductive organs are delimited. In vegetable debris and sheath of *Chaetophora*. (Fig. 4.55) ....
Nowakowskiella Schroeter

68

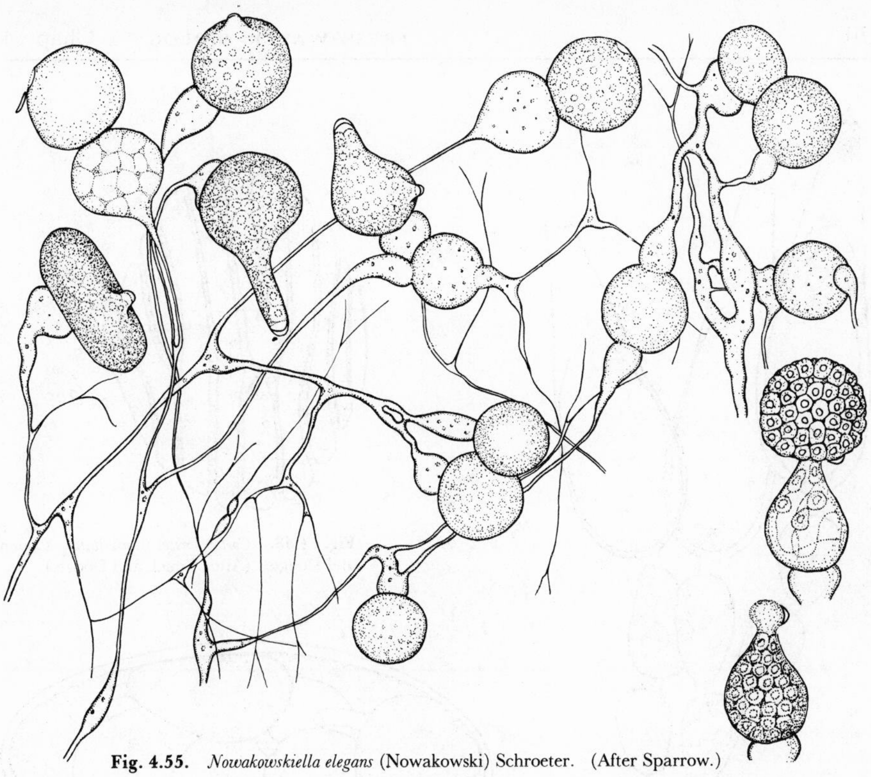

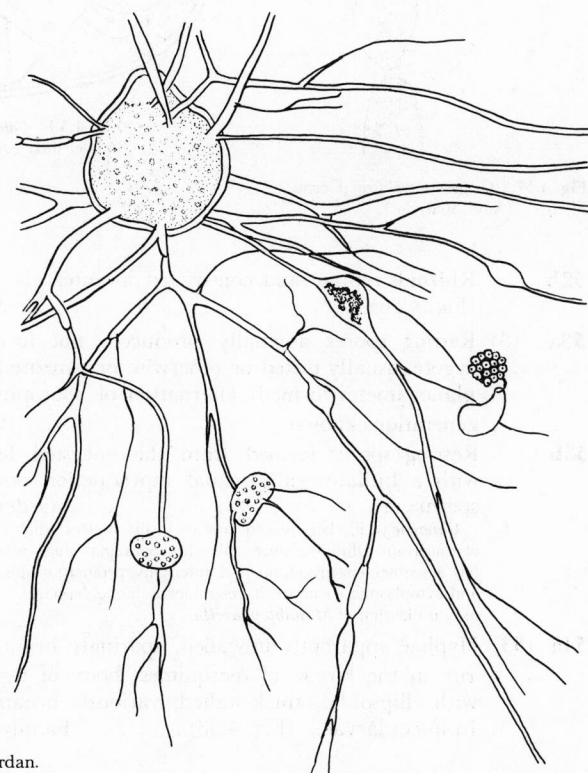

Fig. 4.55. *Nowakowskiella elegans* (Nowakowski) Schroeter. (After Sparrow.)

Fig. 4.56. *Septochytrium variabile* Berdan.
(After Berdan.)

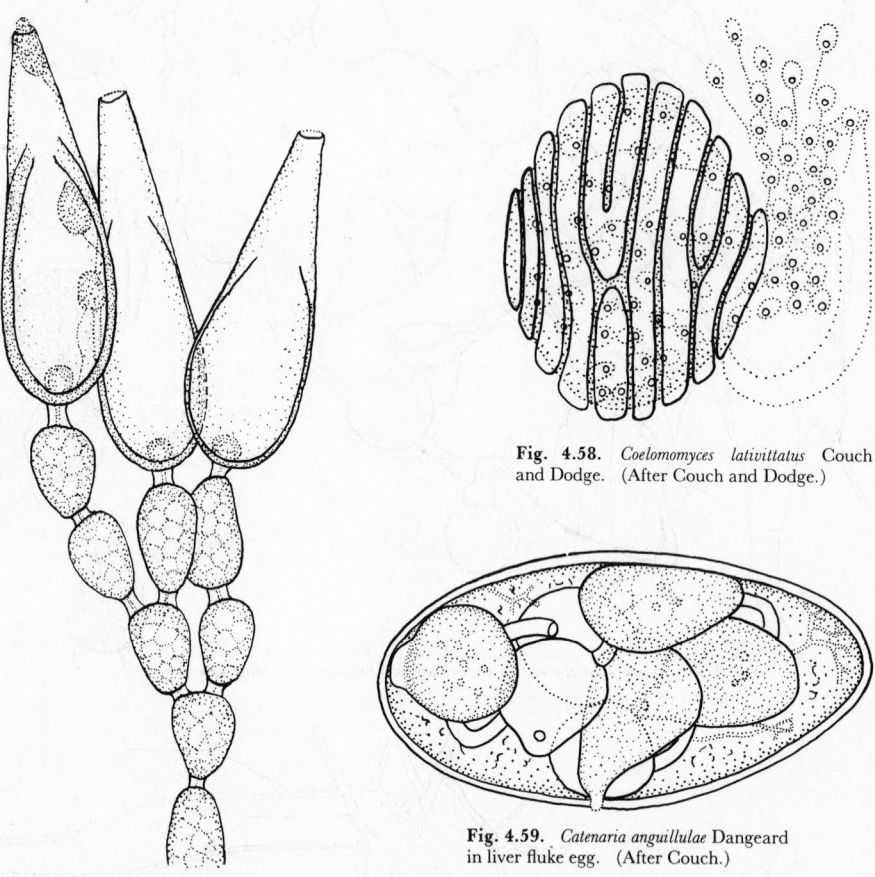

Fig. 4.58. *Coelomyces lativittatus* Couch and Dodge. (After Couch and Dodge.)

Fig. 4.59. *Catenaria anguillulae* Dangeard in liver fluke egg. (After Couch.)

Fig. 4.57. *Gonapodya prolifera* (Cornu) Fischer. (After Sparrow.)

52b   Rhizoids septate and constricted at intervals. In vegetable debris. (Fig. 4.56) . . . . ***Septochytrium*** Berdan

53a   (3) Resting spores asexually produced, not formed directly from a zygote, usually pitted or otherwise ornamented; iso- or anisogamous planogametes formed; alternation of sporophyte and gametophyte generations known . . . . Order **Blastocladiales**   54

53b   Resting spores formed from the encysted fertilized egg; usually with a bullate wall; sexual reproduction oogamous, with motile sperm . . . . Order **Monoblepharidales**   59

> *Gonapodya* Fischer, placed here or in the Blastocladiales, was until recently a genus of uncertain affinities, since only the sporangial stage was known. (Fig. 4.57). It has a segmented mycelium and internally proliferous sporangia. Both in zoosporic and cytoplasmic structure it resembles a monoblepharid. Its newly discovered sexuality is like that of ***Monoblepharella.***

54a   (53) Hyphae apparently unwalled, sparingly branched, parasitic primarily ·in tne larvae of mosquitoes; body of insect filled at maturity with ellipsoidal, thick-walled, variously ornamented resting spores. In insect larvae. (Fig. 4.58) . . . . . . Family **Coelomomycetaceae**
                                                        Coelomomyces Keilin

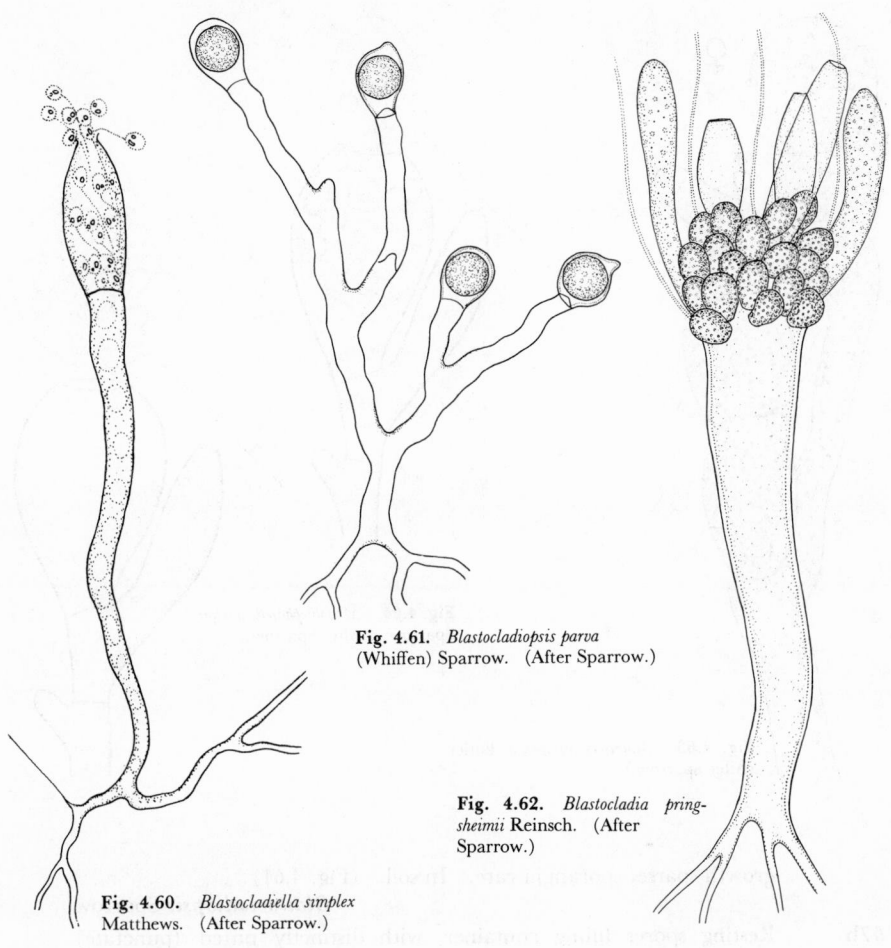

Fig. 4.61. *Blastocladiopsis parva* (Whiffen) Sparrow. (After Sparrow.)

Fig. 4.62. *Blastocladia pring-sheimii* Reinsch. (After Sparrow.)

Fig. 4.60. *Blastocladiella simplex* Matthews. (After Sparrow.)

54b   Hyphae walled, not parasitic in insect larvae **55**

55a (54) Thallus tubular, sparingly if at all branched, at maturity segmenting into alternating reproductive rudiments and sterile portions; rhizoids at intervals along its length. In algae, microscopic animals, and Phycomycetes. (Fig. 4.59) . . . . Family **Catenariaceae**
Catenaria Dangeard
See **40b**.

55b   Thallus composed of a basal cell, anchored in the substratum by rhizoids, which may bear the reproductive structures; or reproductive structures formed on well-developed hyphae which arise distally from the basal cell. . . . . Family **Blastocladiaceae**   **56**

56a (55) Thallus a somewhat spherical or cylindrical basal cell, with rhizoids, bearing a single reproductive rudiment; alternation of sporophyte and gametophyte (isogamous gametes) known in some species. On organic debris in soils. (Fig. 4.60) . . . ***Blastocladiella*** Matthews

56b   Thallus bearing more than one reproductive rudiment **57**

57a (56) Resting spores not filling the container, smooth-walled; vegetative

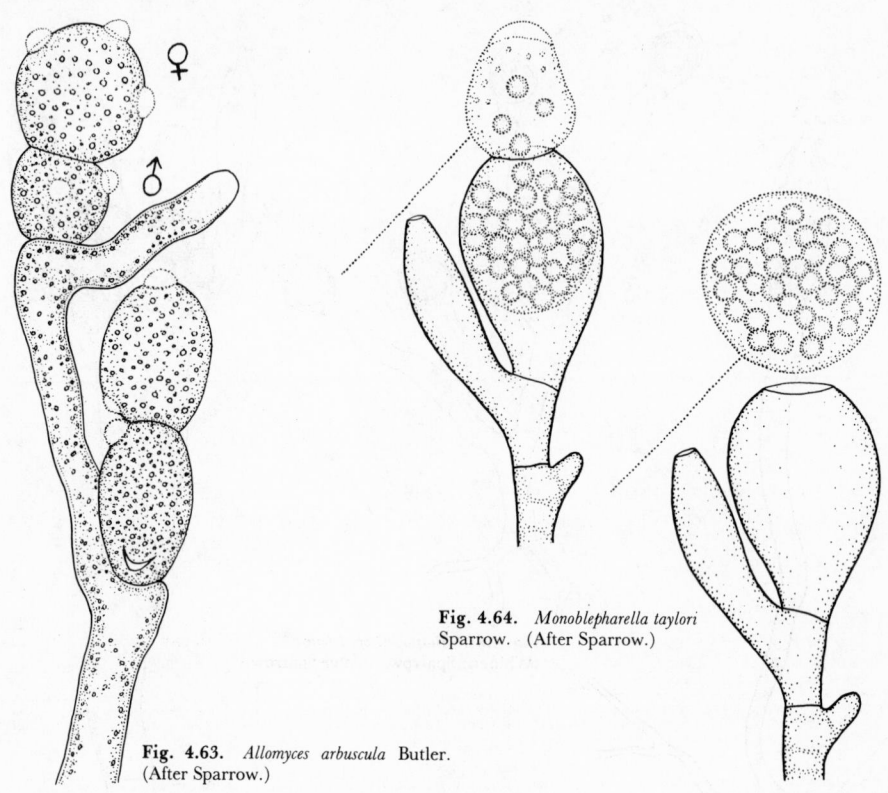

Fig. 4.64. *Monoblepharella taylori* Sparrow. (After Sparrow.)

Fig. 4.63. *Allomyces arbuscula* Butler. (After Sparrow.)

growth sparse; sporangia rare. In soil. (Fig. 4.61) . . . .
                                                      Blastocladiopsis Sparrow

57b   Resting spores filling container, with distinctly pitted (punctate) walls . . . . : **58**

58a **(57)** Reproductive rudiments borne directly on the basal cell or on lobes or short branches; growth limited; without known sexual reproduction. On fruits and twigs. (Fig. 4.62) . . . . **Blastocladia** Reinsch

58b   Reproductive rudiments borne on an extensive hyphal system of unlimited growth, which arises from a basal cell; sexual reproduction iso- or anisogamous, involving alternation of generations. In soil. (Fig. 4.63) . . . . **Allomyces** Butler

59a **(53)** Zygote swimming away from oogonium, propelled by the persistent flagellum of the male gamete; mycelium exceedingly delicate. In tropical soils. (Fig. 4.64) . . . . **Monoblepharella** Sparrow
           See note at **53b**.

59b   Zygote emerging from oogonium, without a flagellum, encysting and forming an oospore at orifice of oogonium to which it remains attached; mycelium less delicate. In cold springs, etc.; on twigs. (Fig. 4.65) . . . . **Monoblepharis** Cornu

60a **(2)** Thallus monocentric **61**

60b   Thallus polycentric, composed of tubular elements and reproductive rudiments. (Fig. 4.66) . . . . Family **Hyphochytriaceae**
                                                       Hyphochytrium Zopf

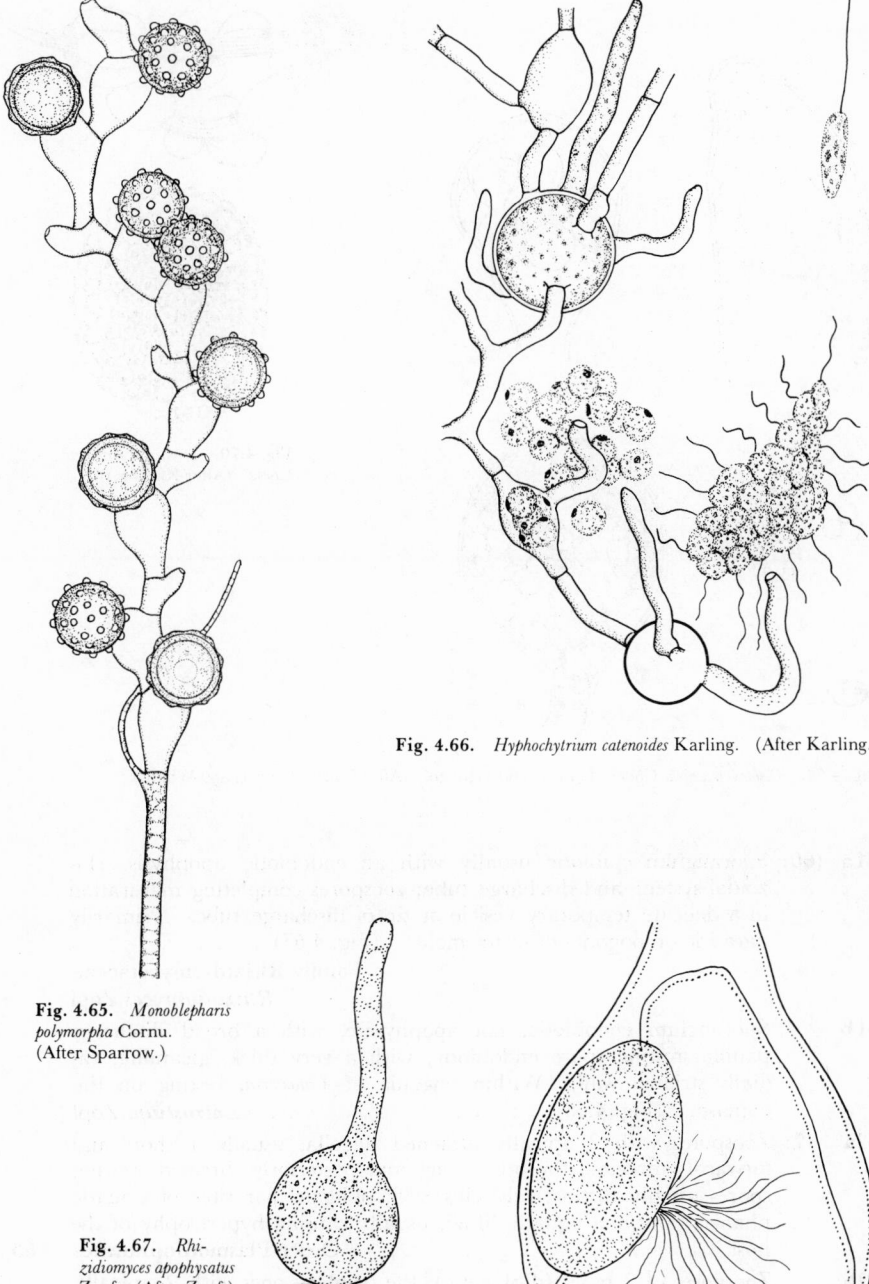

Fig. 4.66.   *Hyphochytrium catenoides* Karling.   (After Karling.)

Fig. 4.65.   *Monoblepharis polymorpha* Cornu. (After Sparrow.)

Fig. 4.67.   *Rhizidiomyces apophysatus* Zopf.   (After Zopf.)

Fig. 4.68.   *Latrostium comprimens* Zopf. (After Zopf.)

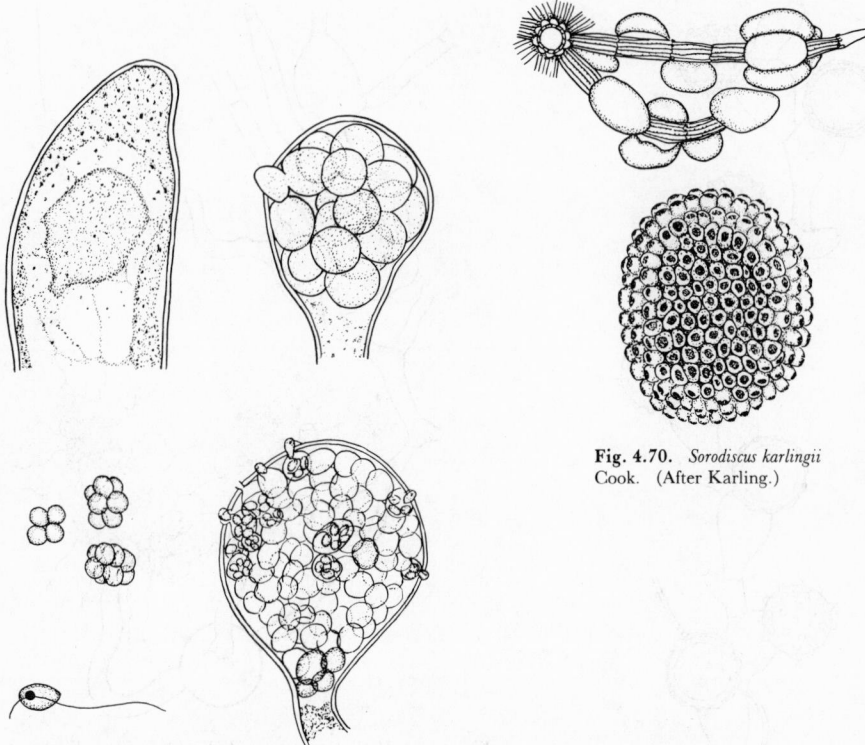

Fig. 4.70. *Sorodiscus karlingii* Cook. (After Karling.)

Fig. 4.69. *Octomyxa achlyae* Couch, Leitner, and Whiffen. (After Couch, Leitner, and Whiffen.)

61a **(60)** Sporangium epibiotic usually with an endobiotic apophysis, rhizoidal system, and discharge tube; zoospores completing maturation in a delicate temporary vesicle at tip of discharge tube. Primarily parasitic on oogonia of water molds. (Fig. 4.67) . . . .
Family **Rhizidiomycetaceae**
Rhizidiomyces Zopf

61b Sporangium endobiotic, not apophysate, with a broad discharge papilla; resting spore endobiotic, with a very thick, gleaming, radially striated wall. Within oogonia of *Vaucheria*, resting on the ooplasm. (Fig. 4.68) . . . . *Latrostium* Zopf

62a **(2)** Zoospores with 2 apically attached flagella, usually 1 short and forwardly directed, the other long and posteriorly directed; resting spores usually distinctly clustered; obligate parasites of aquatic phanerogams and aquatic fungi, usually causing hypertrophy of the host . . . . Order **Plasmodiophorales**   63

62b Zoospores of 2 types (diplanetic), the primary ones with 2 apically attached, equal flagella, the secondary (arising from the primary after a period of encystment) with 2 lateral, oppositely directed flagella   65

63a **(62)** Resting spores in clusters of 8. Parasitic in filamentous aquatic Phycomycetes. (Fig. 4.69) . . . .
Octomyxa Couch, Leitner, and Whiffen

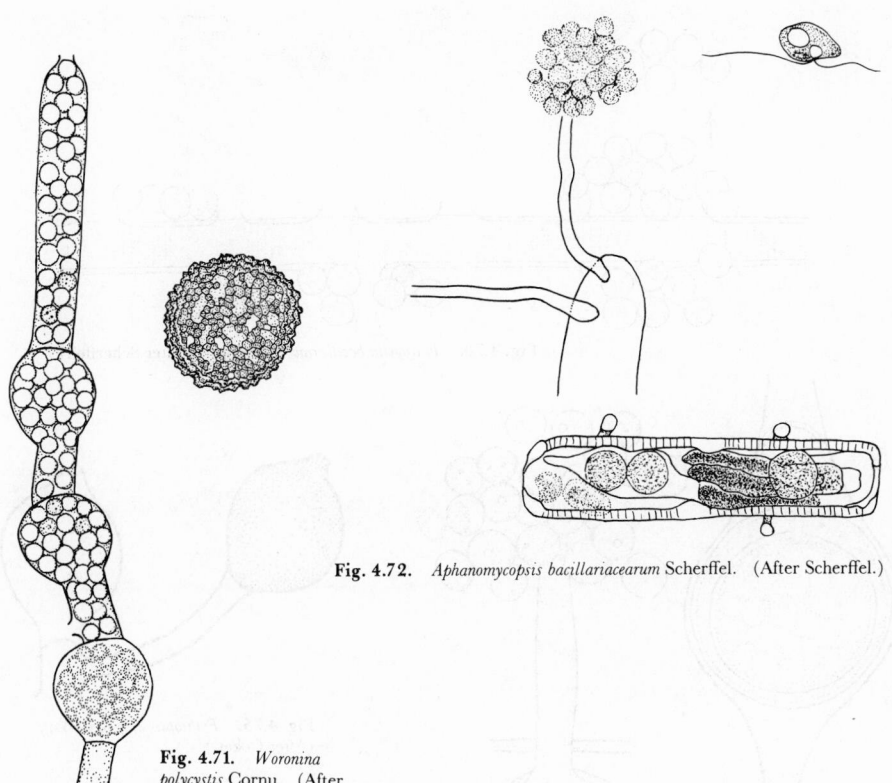

Fig. 4.72. *Aphanomycopsis bacillariacearum* Scherffel. (After Scherffel.)

Fig. 4.71. *Woronina polycystis* Cornu. (After Sparrow.)

63b   Resting spores not in clusters of 8. . . . .   **64**

64a **(63)** Resting spores forming a flat, 2-layered, more or less rounded plate. Parasitic in *Chara, Callitriche,* and *Pythium.* (Fig. 4.70). . . . .
            Sorodiscus Lagerheim and Winge

64b   Resting spores forming a compact, irregular cluster. In the hypertrophied hyphae of filamentous water molds or in algae.   (Fig. 4.71) . . . . *Woronina* Cornu

65 **(62)** Zoospores cleaved out within the sporangium, diplanetic, the primary spore motile or encysting immediately after discharge, or completely suppressed, only secondary zoospores being formed **66**

65b   Zoospores cleaved out either within the sporangium, or partly or wholly formed outside the sporangium, in which case they are usually surrounded by a more or less evanescent vesicle; only secondary type of zoospore formed. **89**

66a **(65)** Thallus holo- or eucarpic; hyphae of eucarpic forms not arising from a basal cell and without constrictions or cellulin plugs; oogonium with one or more eggs . . . . Order **Saprolegniales**   **67**

66b   Thallus always eucarpic, the hyphae bearing constructions at intervals accompanied by cellulin plugs; oogonium usually with a single egg. Saprophytic, primarily on fruits and twigs. . . . .
            Order **Leptomitales**   **83**

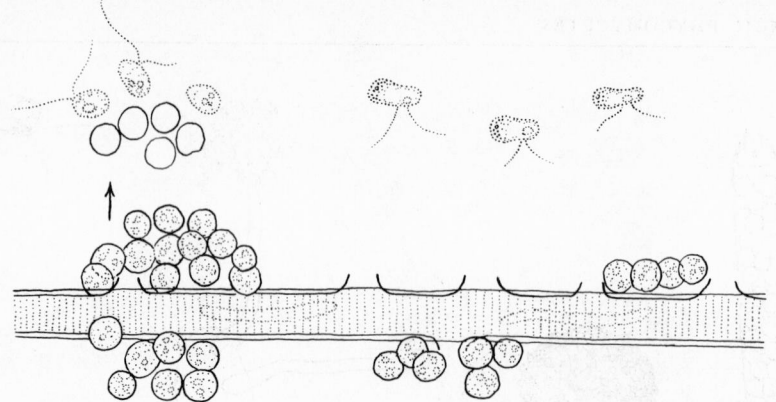

Fig. 4.73. *Ectrogella bacillariacearum* Zopf. (After Scherffel.)

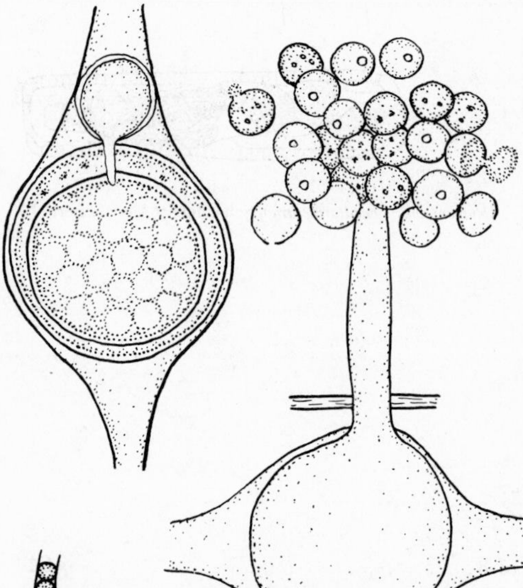

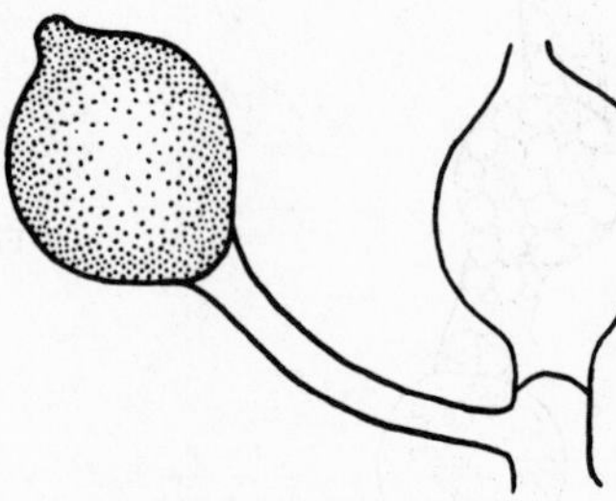

Fig. 4.75. *Pythiopsis cymosa* deBary. (After Coker.)

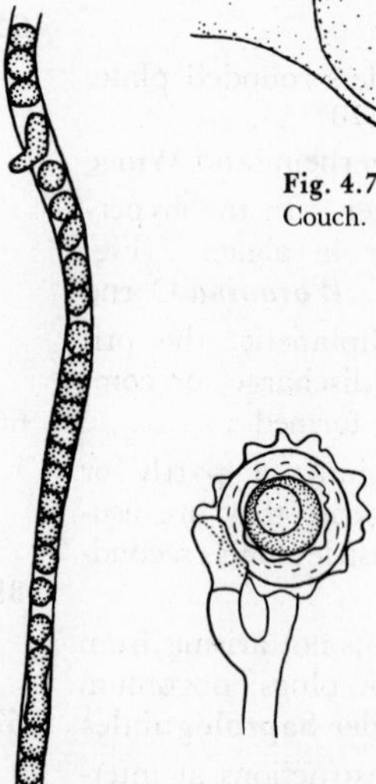

Fig. 4.74. *Pythiella vernalis* Couch. (After Couch.)

Fig. 4.76. *Leptolegnia eccentrica* Coker. (After Coker.)

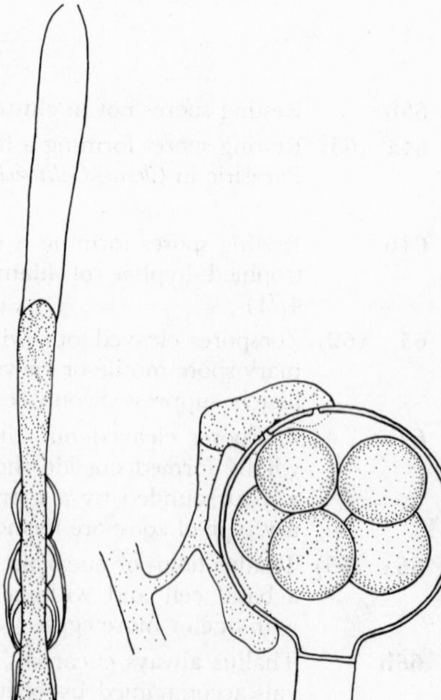

Fig. 4.77. *Saprolegnia ferax* (Gruith.) Thuret. (After Coker.)

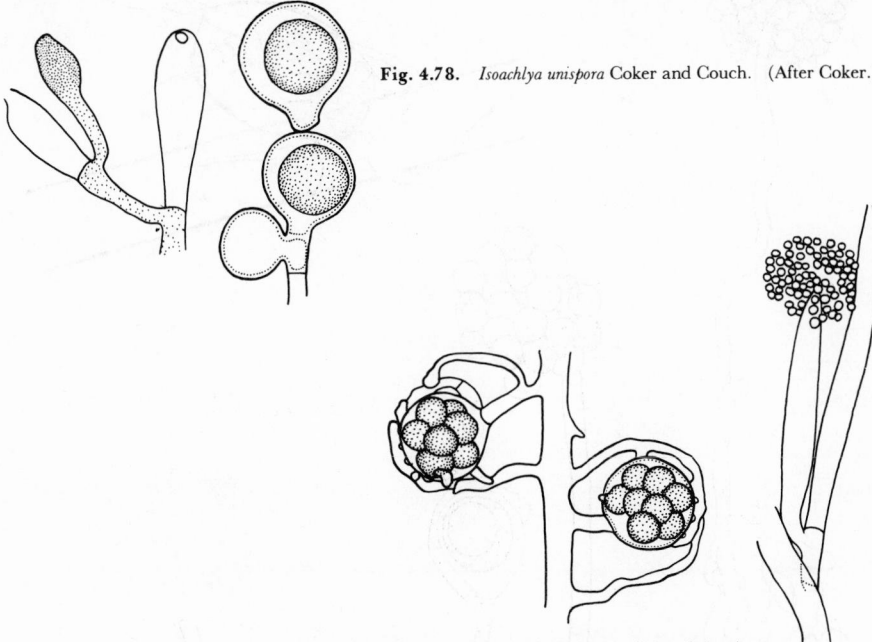

Fig. 4.78. *Isoachlya unispora* Coker and Couch. (After Coker.)

Fig. 4.79. *Achlya americana* Humphrey. (After Coker.)

67a (66) Thallus holocarpic, endobiotic, zoospores usually encysting in a
cluster at the orifice of the discharge tube . . . .
Family **Ectrogellaceae** **68**

67b Thallus eucarpic, composed of hyphae which bear the reproductive
organs . . . . Family **Saprolegniaceae** **70**

68a (67) Sporangium unbranched **69**

68b Sporangium branched, the long discharge tube thick-walled at point
of emergence from host. Parasitic in diatoms. (Fig. 4.72) . . . . . .
Aphanomycopsis Scherffel

69a (68) Sporangium tubular, with one or more short discharge tubes. Parasitic in diatoms. (Fig. 4.73) . . . . *Ectrogella* Zopf

69b Sporangium spherical, with one to several long projecting discharge
tubes. Parasitic in hyphae of *Pythium* in which it causes local
hypertrophy. (Fig. 4.74). . . . . *Pythiella* Couch

70a (67) Zoospores only of the primary (anteriorly biflagellate) type; sporangia globose. Saprophytic. (Fig. 4.75) . . . . *Pythiopsis* deBary

70b Zoospores diplanetic (see **62b**) or showing evidences of diplanetism **71**

71a (70) Primary zoospores emerging as flagellated or nonflagellated structures **72**

71b Primary zoospores encysting within the sporangium. **78**

72a (71) Motile spores of both primary and secondary types produced. **73**

72b Primary zoospores for most part emerging from the sporangia without flagella, encysting at orifice of sporangium, each cyst producing
a secondary zoospore **75**

73a (72) Zoospores in more than one row in the sporangium **74**

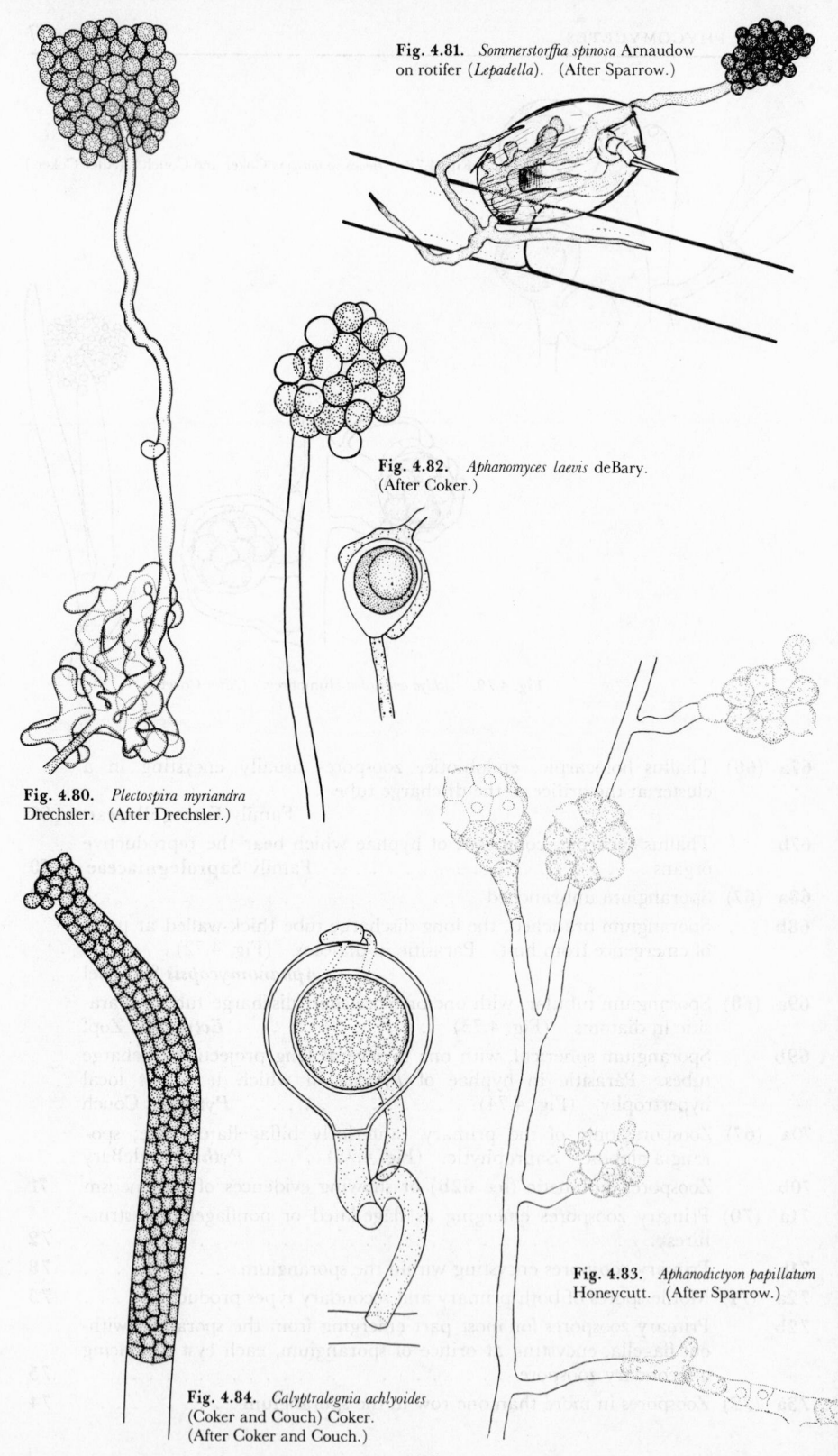

Fig. 4.81. *Sommerstorffia spinosa* Arnaudow on rotifer (*Lepadella*).   (After Sparrow.)

Fig. 4.82. *Aphanomyces laevis* deBary. (After Coker.)

Fig. 4.80. *Plectospira myriandra* Drechsler.   (After Drechsler.)

Fig. 4.83. *Aphanodictyon papillatum* Honeycutt.   (After Sparrow.)

Fig. 4.84. *Calyptralegnia achlyoides* (Coker and Couch) Coker. (After Coker and Couch.)

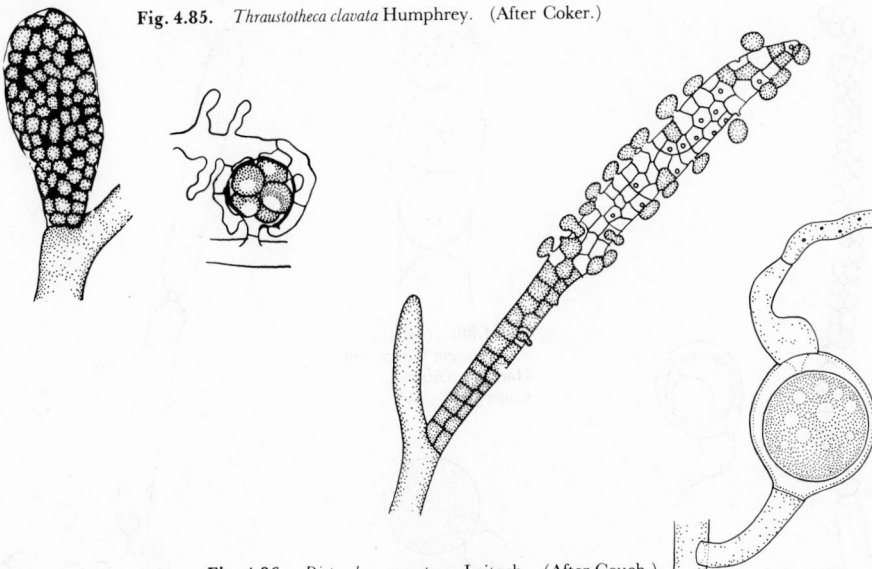

Fig. 4.85. *Thraustotheca clavata* Humphrey. (After Coker.)

Fig. 4.86. *Dictyuchus monosporus* Leitgeb. (After Couch.)

73b		Zoospores in one row in the sporangium. Saprophytic. (Fig. 4.76) . . . . *Leptolegnia* deBary
74a	**(73)**	New sporangia formed within old ones by internal proliferation (rare in *S. parasitica*). Saprophytic and parasitic. (Fig. 4.77) . . . . *Saprolegnia* Nees
74b		New sporangia formed primarily by cymose branching. Saprophytic. (Fig. 4.78) . . . . *Isoachlya* Kauffman
75a	**(72)**	Zoospores formed in more than one row in the nearly cylindrical sporangia. Saprophytic. (Fig. 4.79) . . . . *Achlya* Nees
75b		Spores formed in one row, at least in the cylindrical parts of the sporangia. **76**
76a	**(75)**	Sporangium completely cylindrical **77**
76b		Zoosporangium consisting of a basal portion of inflated lobulations and a narrow cylindrical discharge tube. Saprophytic, and parasitic on roots of higher plants. (Fig. 4.80) . . . . . *Plectospira* Drechsler
77a	**(76)**	Hyphae depauperate, epiphytic on algae, the tips armed with spiny processes for capturing of living rotifers. Predaceous on rotifers. (Fig. 4.81) . . . . *Sommerstorffia* Arnaudow
77b		Hyphae well developed, without spiny capturing processes. Saprophytic and parasitic on algae and fungi. (Fig. 4.82) . . . . *Aphanomyces* deBary
78a	**(71)**	Oogonium bearing several eggs **79**
78b		Oogonium bearing a single egg **81**
79a	**(78)**	Zoosporangia more or less pyriform, scattered laterally along a delicate mycelium. Keratinophilic saprophytes. (Fig. 4.83) . . . . . *Aphanodictyon* Honeycutt

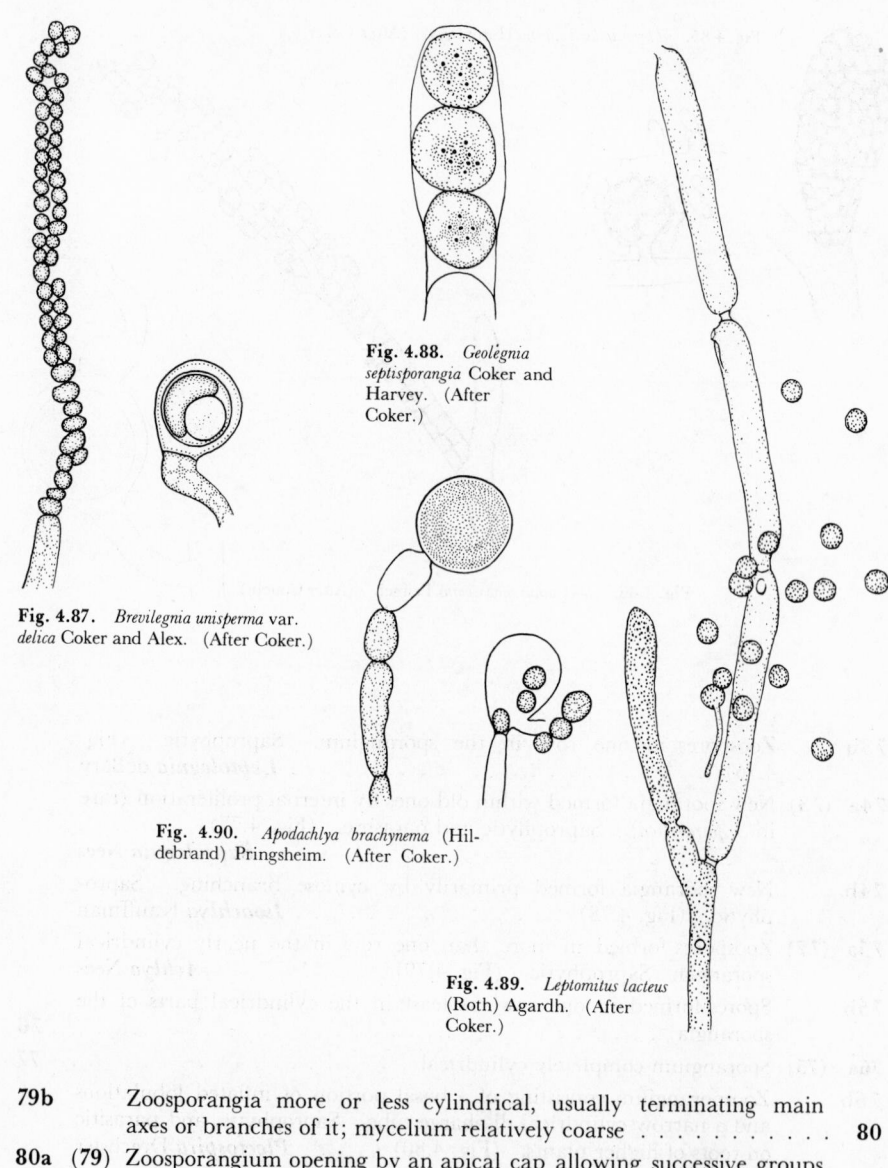

Fig. 4.88. *Geolegnia septisporangia* Coker and Harvey. (After Coker.)

Fig. 4.87. *Brevilegnia unisperma* var. *delica* Coker and Alex. (After Coker.)

Fig. 4.90. *Apodachlya brachynema* (Hildebrand) Pringsheim. (After Coker.)

Fig. 4.89. *Leptomitus lacteus* (Roth) Agardh. (After Coker.)

79b   Zoosporangia more or less cylindrical, usually terminating main
              axes or branches of it; mycelium relatively coarse **80**

80a (79) Zoosporangium opening by an apical cap allowing successive groups
             of encysted primary spores to escape. Saprophytic. (Fig. 4.84) . . .
                                                              Calyptralegnia Coker

80b   Zoosporangium not operculate, the wall bursting or deliquescing to
              liberate the encysted primary spores. Saprophytic. (Fig. 4.85) . . .
                                                             Thraustotheca Humphrey

81a (78) Mycelium well developed, extensive, somewhat sprawling; encysted
             spores usually in several rows, emerging as swimming secondary
             spores through the sporangium wall. Saprophytic. (Fig. 4.86) . . .
                                                             Dictyuchus Leitgeb

81b   Mycelium somewhat depauperate, dense **82**

82a (81) Sporangium wall soon disappearing, the encysted zoospores thin-
             walled, remaining clumped; the zoospores in one or more rows in

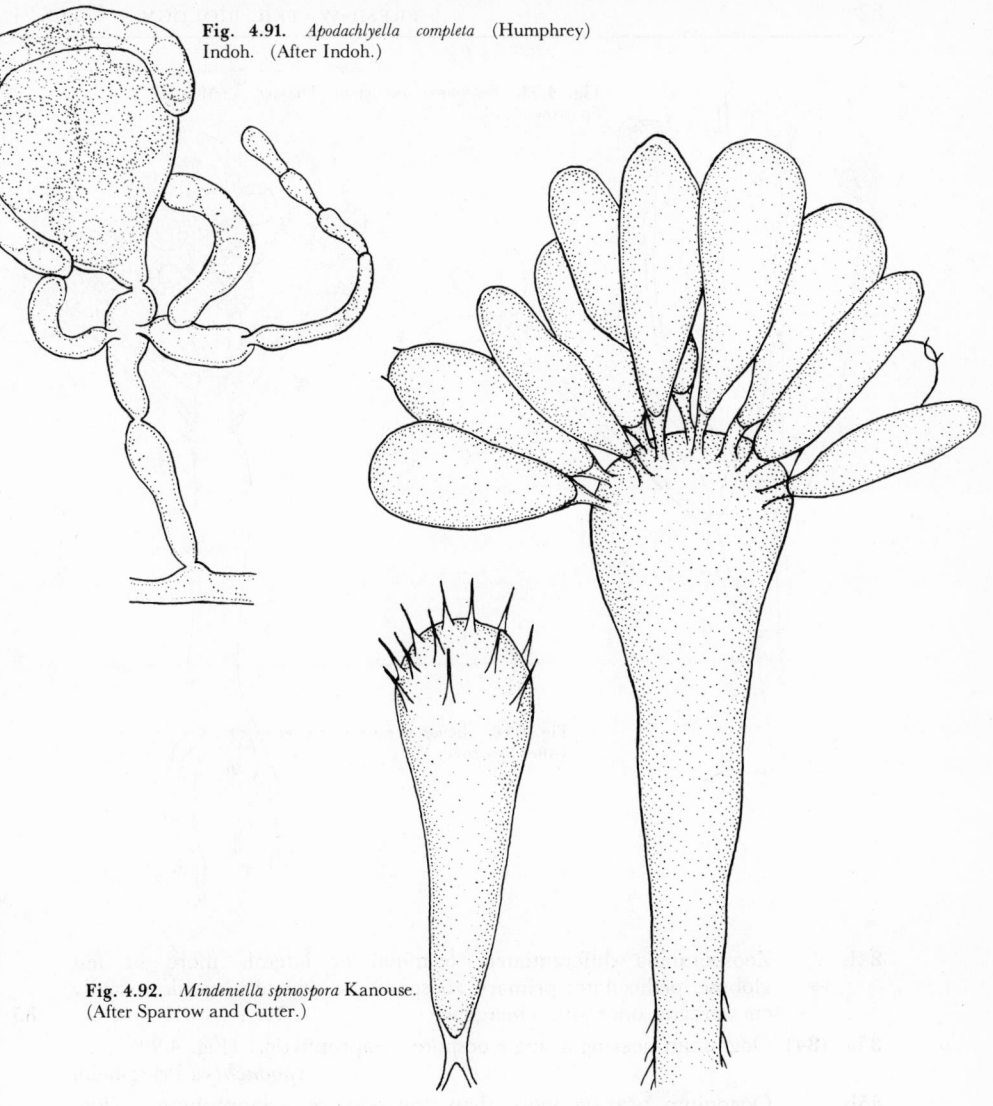

Fig. 4.91. *Apodachlyella completa* (Humphrey) Indoh. (After Indoh.)

Fig. 4.92. *Mindeniella spinospora* Kanouse. (After Sparrow and Cutter.)

sporangium, germinating by a tube or a secondary zoospore. Saprophytic. (Fig. 4.87) . . . . **Brevilegnia** Coker and Couch

82b   Sporangium wall persistent, encysted spores thick-walled, in one row; never swimming. Saprophytic. (Fig. 4.88) . . . .
                                        Geolegnia Coker

83a **(66)** Thallus filamentous throughout, not differentiated into basal cell and hyphal branches; zoospores diplanetic . . . .
                                Family **Leptomitaceae** **84**

83b   Thallus differentiated into a holdfast system, basal cell, and hyphal branches that bear the pedicellate reproductive organs; zoospores only of the secondary type, the primary ones completely suppressed
                                Family **Rhipidiaceae** **86**

84a **(83)** Zoosporangia undifferentiated segments of the mycelium; primary and secondary zoospores motile. Saprophytic. (Fig. 4.89) . . . . .
                                    Leptomitus Agardh

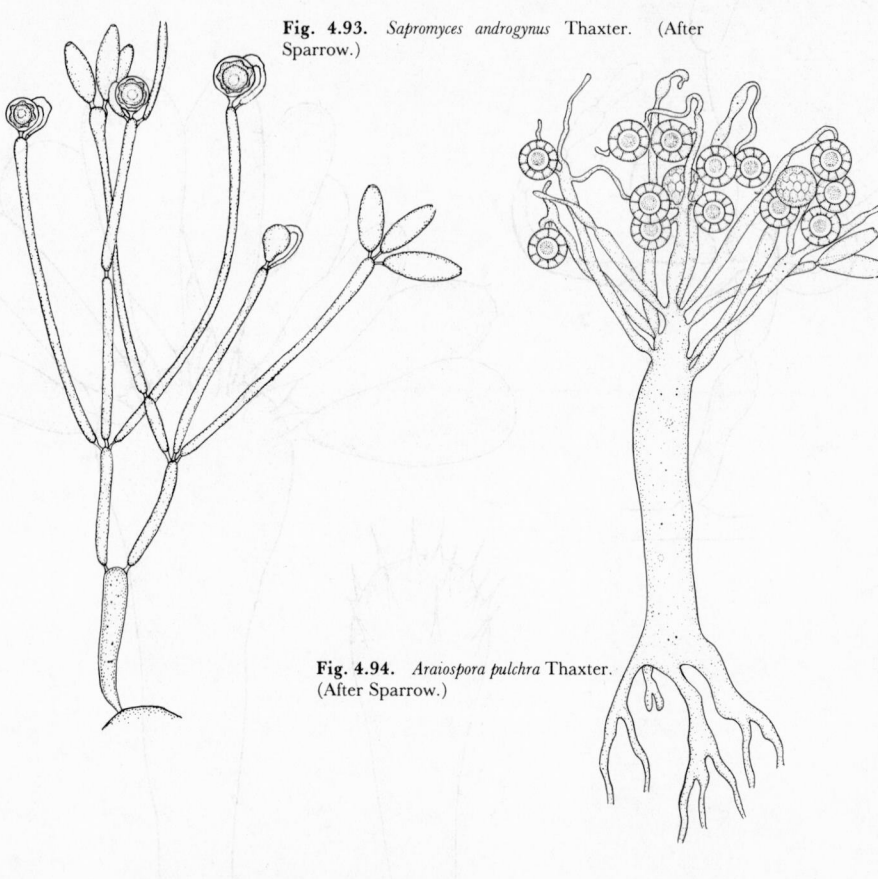

Fig. 4.93. *Sapromyces androgynus* Thaxter. (After Sparrow.)

Fig. 4.94. *Araiospora pulchra* Thaxter. (After Sparrow.)

84b	Zoosporangia differentiated, terminal or lateral, more or less globose, pedicellate; primary zoospores occasionally motile, usually encysting at once after emergence	**85**
85a (84)	Oogonium bearing a single oospore. Saprophytic. (Fig. 4.90) . . . *Apodachlya* Pringsheim	
85b	Oogonium bearing more than one oospore. Saprophytic. (Fig. 4.91). . . . . *Apodachlyella* Indoh	
86a (83)	Reproductive structures arising directly from the basal cell; sporangia spiny and (more commonly) smooth; resting spores spiny, parthenogenetic. Saprophytic on fruit. (Fig. 4.92). . . . . *Mindeniella* Kanouse	
86b	Reproductive structures borne on hyphal branches which arise from apex of basal cell; oospores formed	**87**
87a (86)	Basal cell slender, often poorly defined; sporangia whorled or in umbels, all smooth-walled. Saprophytic, primarily on twigs and fruits. (Fig. 4.93). . . . . *Sapromyces* K. Fritsch	
87b	Basal cell well developed; sporangia smooth-walled, spiny, or both. .	**88**
88a (87)	Zoosporangia smooth and spiny-walled; outer oospore wall cellular. Saprophytic on twigs and fruits. (Fig. 4.94) . . *Araiospora* Thaxter	
88b	Zoosporangia all smooth-walled; outer oospore wall coarsely reticulate. Saprophytic on fruits and twigs. (Fig. 4.95). . . . . *Rhipidium* Cornu	

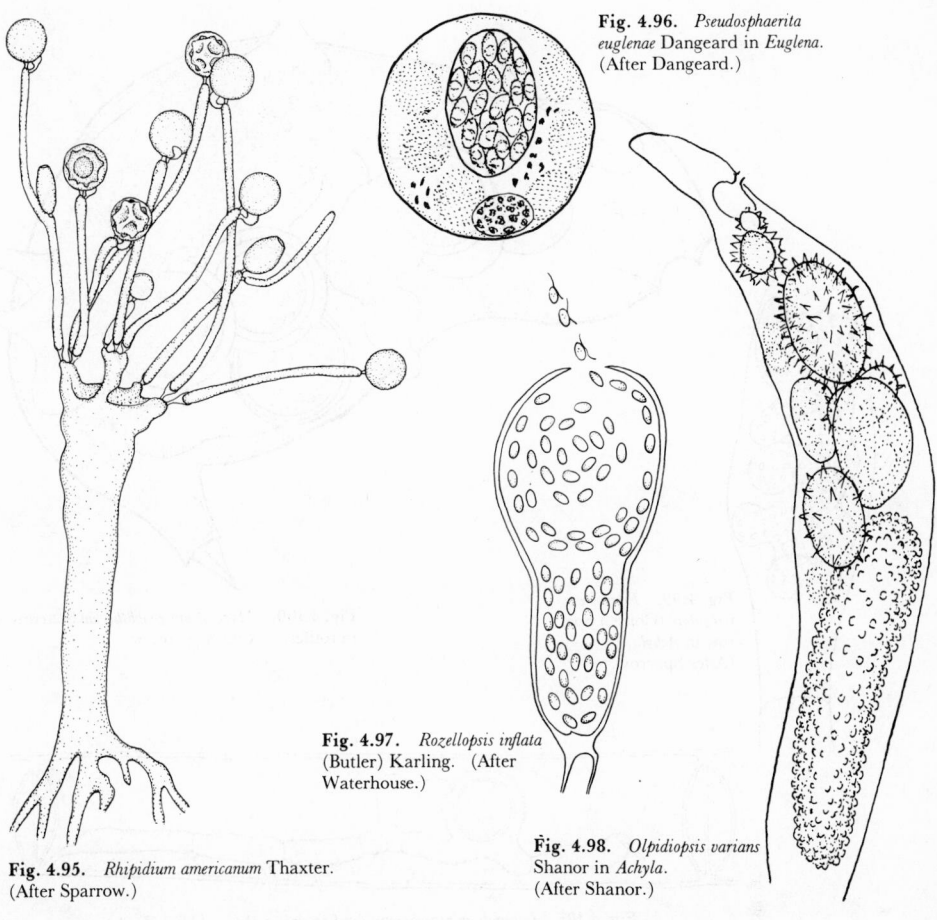

Fig. 4.96. *Pseudosphaerita euglenae* Dangeard in *Euglena*. (After Dangeard.)

Fig. 4.97. *Rozellopsis inflata* (Butler) Karling. (After Waterhouse.)

Fig. 4.95. *Rhipidium americanum* Thaxter. (After Sparrow.)

Fig. 4.98. *Olpidiopsis varians* Shanor in *Achyla*. (After Shanor.)

89a (65) Thallus holocarpic, endobiotic, either unicellular or consisting at maturity of a series of unbranched or occasionally branched segments of limited extent . . . . Order **Lagenidiales** 90

89b Thallus eucarpic, distinctly mycelial and usually both intra- and extramatrical; zoosporangia persistent or deciduous "conidia" . . . . Order **Peronosporales** 95

90a (89) Thallus always 1-celled; zoospores formed within the sporangium; resting spore lying free in the host . . . . Family **Olpidiopsidaceae** 91

90b Thallus predominantly multicellular, occasionally 1-celled, zoospores completing their maturation in a vesicle at the tip of the discharge tube, resting spore formed in a container. . . . . Family **Lagenidiaceae** 94

91a (90) Parasitic in aquatic Phycomycetes and algae. 92

91b Parasitic in Euglenophyceae and Cryptophyceae. (Fig. 4.96) . . . . *Pseudosphaerita* Dangeard

92a (91) Resting spore predominantly sexually formed after conjugation of thalli 93

92b Resting spore apparently asexually formed. Parasitic in aquatic Phycomycetes. (Fig. 4.97). . . . . *Rozellopsis* Karling

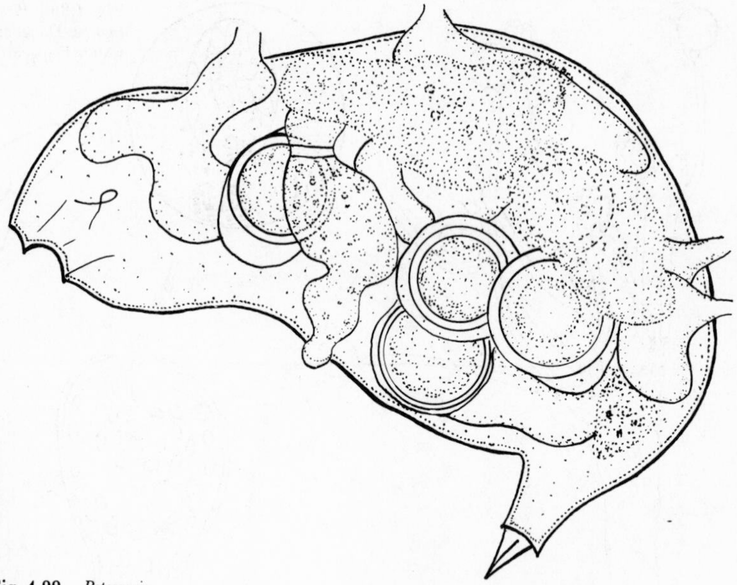

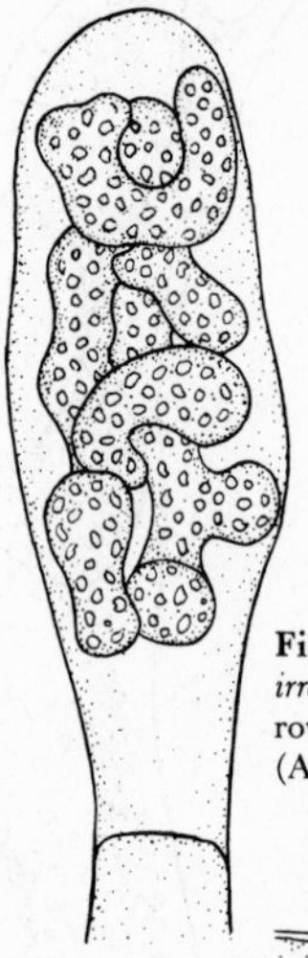

Fig. 4.99. *Petersenia irregulare* (Const.) Sparrow in *Achyla*. (After Sparrow.)

Fig. 4.100. *Myzocytium zoophthorum* Sparrow in rotifer. (After Sparrow.)

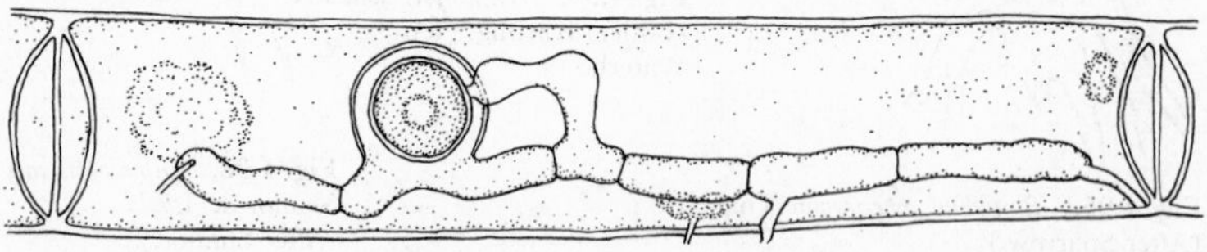

Fig. 4.101. *Lagenidium rabenhorstii* Zopf in green alga. (After Zopf.)

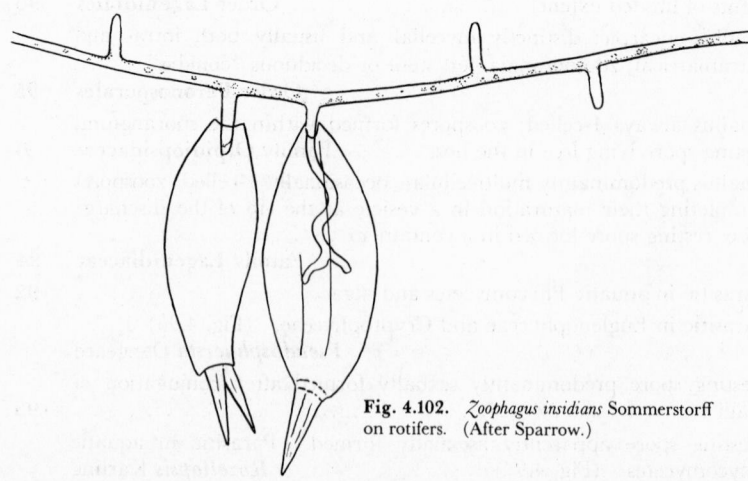

Fig. 4.102. *Zoophagus insidians* Sommerstorff on rotifers. (After Sparrow.)

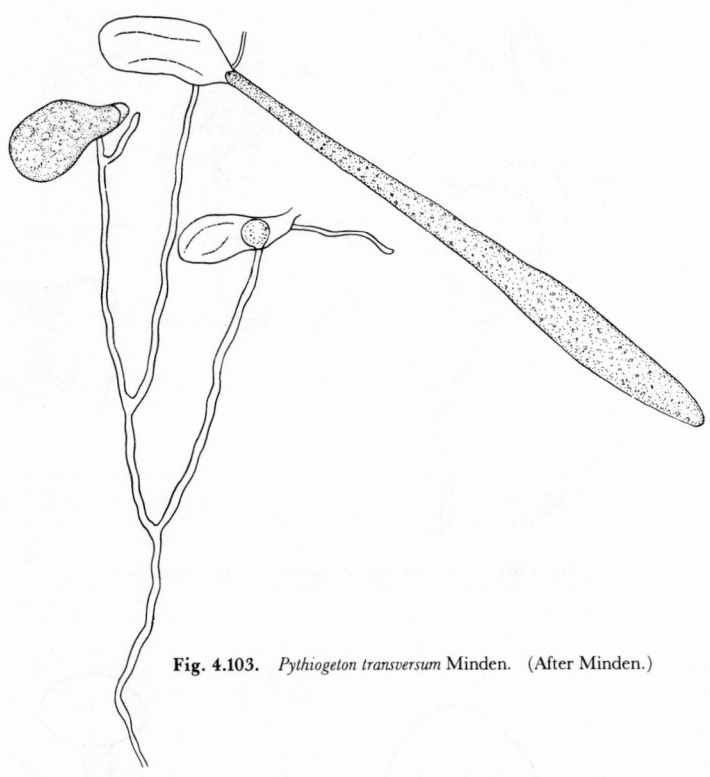

Fig. 4.103. *Pythiogeton transversum* Minden. (After Minden.)

93a (92) Zoosporangia predominantly spherical or ellipsoidal with one or, rarely, more discharge tubes. Parasitic in aquatic Phycomycetes. (Fig. 4.98) . . . . **Olpidiopsis** Cornu
Includes *Pseudolpidium* with resting spores lacking companion cells.

93b Zoosporangia predominantly irregularly lobed or tubular, usually with more than one discharge tube. Parasitic in aquatic Phycomycetes. (Fig. 4.99) . . . . **Petersenia** Sparrow

94a (90) Thallus within one host cell, unbranched, strongly constricted at the cross walls; thallus segments regular, linklike; antheridial cell poorly differentiated. Parasites of fresh-water algae and microscopic animals. (Fig. 4.100) . . . . **Myzocytium** Schenk

94b Thallus unbranched or branched, in one or more cells of host; little or not at all constricted at the cross walls; segments of thallus often irregular; antheridial cell usually well differentiated. Parasites in fresh-water algae, pollen grains; weakly parasitic in mosquito larvae, copepods, *Daphnia*, and rotifer eggs. (Fig. 4.101) . . . .
Lagenidium Schenk

95a (89) Hyphae bearing short, lateral, peglike outgrowths adapted for the capturing of rotifers; zoosporangium filamentous throughout. Predaceous parasites of rotifers. (Fig. 4.102) . . . .
Zoophagus Sommerstorff

95b Hyphae not as above; zoosporangium filamentous, or filamentous with a lobulate base, or somewhat spherical, or bursiform **96**

96a (95) Zoosporangium primarily bursiform, its long axis at an angle to the delicate supporting hypha; oospore with a very thick, gleaming

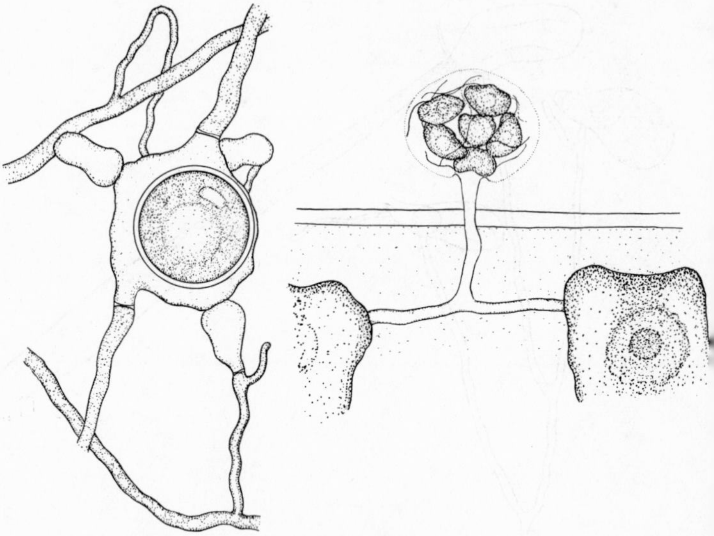

Fig. 4.104. *Pythium angustatum* Sparrow.   (After Sparrow.)

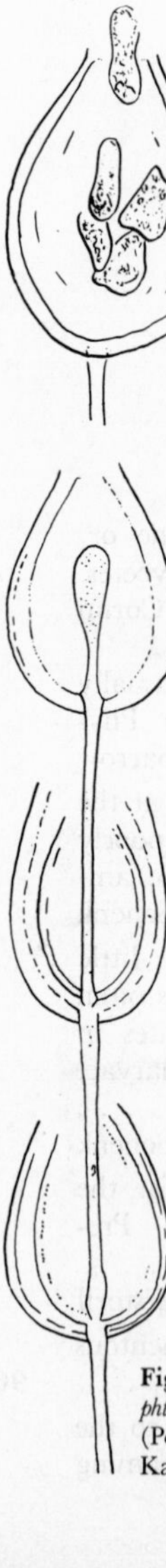

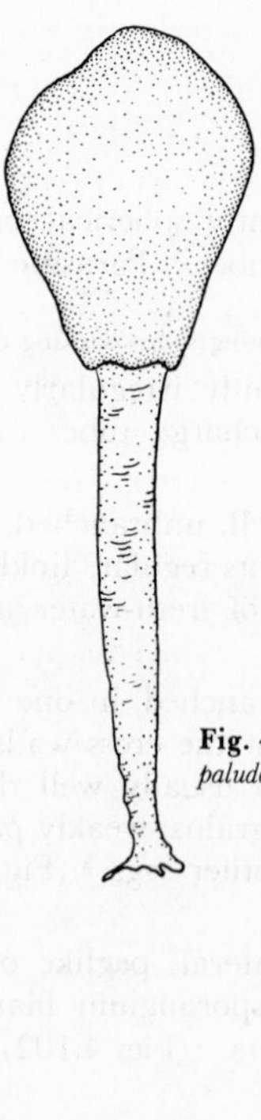

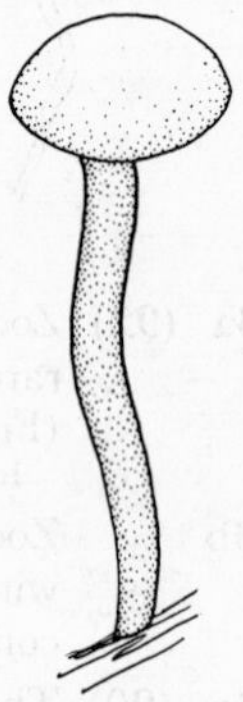

Fig. 4.107. *Vibrissea truncarum* (Albertini and Schweinitz) Fries.

Fig. 4.106. *Mitrula paludosa* Fries.

Fig. 4.105. *Phytophthora gonapodyides* (Pet.) Buisman.   (After Kanouse.)

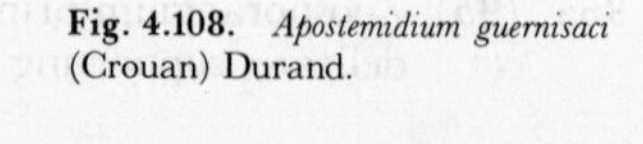

Fig. 4.108. *Apostemidium guernisaci* (Crouan) Durand.

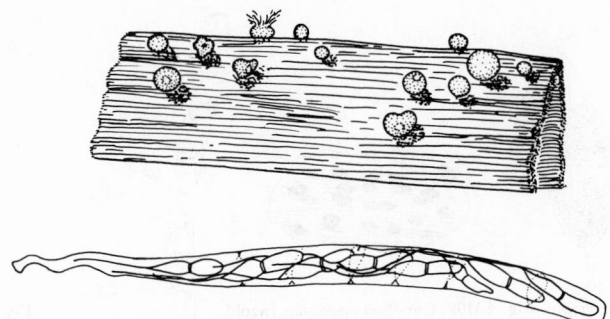

Fig. 4.109. *Loramyces juncicola* Weston on *Juncus*. (After Weston.)

wall; zoospores completing development out in water after quick
disappearance of a long, narrow vesicle. Saprophytic on twigs and
fruits. (Fig. 4.103) . . . . ***Pythiogeton*** Minden

96b Zoosporangium filamentous, or filamentous and lobulate, or somewhat spherical and the long axis parallel with that of the supporting
hypha 97

97a **(96)** Zoospores formed within a vesicle produced at the tip of the evacua-
tion tube. Saprophytic in soil and water. (Fig. 4.104) . . . .
Pythium Pringsheim

97b Zoospores leaving sporangium completely formed; sporangia always
ovoid or citriform. Saprophytic, and parasitic on higher plants.
(Fig. 4.105) . . . . ***Phytophthora*** deBary

KEY TO GENERA: CLASS ASCOMYCETES

1a Fruiting body stalked 2
1b Fruiting body sessile 3
2a **(1)** Fruiting body soft, viscid, stalked, the ascogenous portion being
somewhat ellipsoidal or pyriform in outline and somewhat furrowed,
yellowish, 2–6 cm or more high, with a sharply differentiated white
or pinkish stem. On decaying vegetation. (Fig. 4.106) . . . .
Mitrula Persoon *ex* Fries

2b Fruiting body having the ascogenous portion caplike, yellow or
somewhat orange, 3–5 mm in diameter; stem up to 1.5 cm high,
white, gray, or brownish; whole plant 4–15 mm high; the escaped
spores forming a vibrating cloud around the plant. On twigs in
cold running water. (Fig. 4.107) . . . . ***Vibrissea*** Fries

3a **(1)** Fruiting body a cushionlike structure, sessile on substratum, waxy,
turbinate, or convex; solitary or gregarious; bluish-pallid to orange-
ochraceous, dark brown below, 1–4 mm in diameter, often with a
dimple in center. On twigs in water. (Fig. 4.108) . . . .
Apostemidium Karsten

3b Fruiting body a perithecium, more or less submerged in the
substratum 4

4a **(3)** Perithecium resting on a subiculum; superficial on the substratum;
dark, more or less fleshy, surrounded by a gelatinous layer; ascospores eight, fusiform, once septate, surrounded by a gelatinous

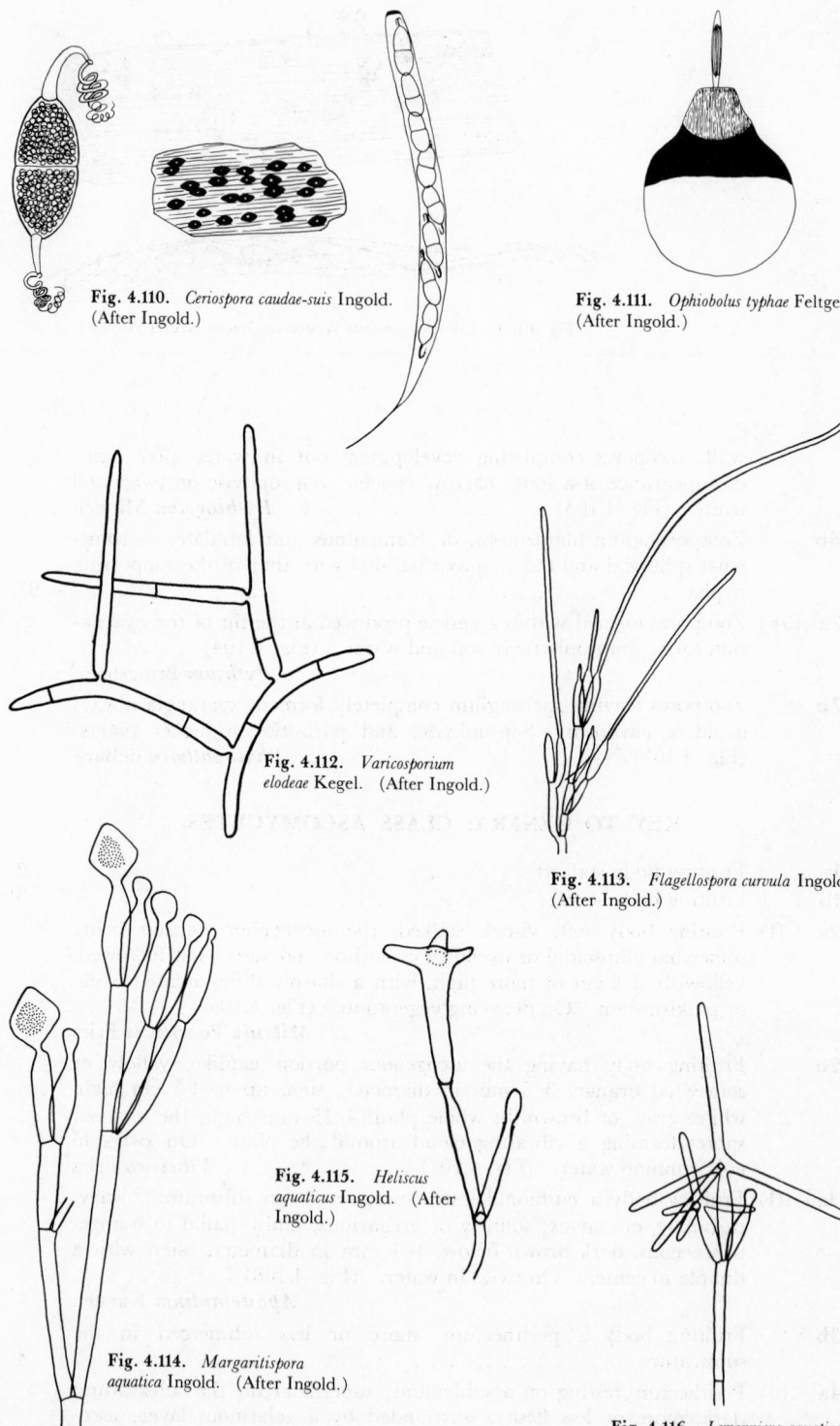

Fig. 4.110. *Ceriospora caudae-suis* Ingold. (After Ingold.)

Fig. 4.111. *Ophiobolus typhae* Feltgen. (After Ingold.)

Fig. 4.112. *Varicosporium elodeae* Kegel. (After Ingold.)

Fig. 4.113. *Flagellospora curvula* Ingold. (After Ingold.)

Fig. 4.115. *Heliscus aquaticus* Ingold. (After Ingold.)

Fig. 4.114. *Margaritispora aquatica* Ingold. (After Ingold.)

Fig. 4.116. *Lemonniera aquatica* De Wildeman. (After Ingold.)

capsule, each bearing a single long, filamentous appendage. On *Juncus*. (Fig. 4.109) . . . . *Loramyces* Weston

4b   Perithecium sunken in the substratum, with a short neck **5**

5a   (4) Ascospores ovoid, once septate, with a long, gelatinous appendage at each end. (Fig. 4.110) . . . . *Ceriospora* Niessl

5b   Ascospores filamentous, 3- or 4-septate, not appendaged. (Fig. 4.111) . . . . *Ophiobolus* Riess

KEY TO GENERA: FUNGI IMPERFECTI
by William W. Scott

1a   Submerged aquatic fungi. Mycelium found in the vascular system of submerged and decaying angiosperm debris. Spores normally formed and dispersed beneath the surface of the water **2**

1b   Aeroaquatic fungi. Mycelium found in submerged and decaying vegetable debris. Conidiophores emergent and conidia produced above the surface of the water.   **16**

2a   (1) Conidia (phialospores) produced in basipetal succession from the apex of a definite, hyaline phialide (a one-celled flasklike structure from the end of which conidia are abstricted). The septum separating the conidium from the phialide not formed until the spore is fully grown. **3**

2b   Conidia (aleuriospore; i.e., spores on a sterigma that has no relation to the growth point of the hypha) formed as a terminal portion of a hypha that is early separated by a septum from the parent hypha . .   **7**

2c   Conidia (radulaspores; i.e., terminal or lateral chlamydospore, like a conidium verum but not deciduous) borne on minute sterigma without any reference to the growing point of the conidiophore. First-formed conidium terminal, remaining conidia lateral; each conidium consisting of a main axis with 3 secondary branches produced laterally from only one side of the main axis. Each lateral branch capable of forming tertiary ramuli in the same one-sided manner. Conidiophores simple. (Fig. 4.112) . . . . . . *Varicosporium* Kegel

3a   (2) Conidia simple, lacking divergent processes, unicellular or 2-septate. Conidiophores branched to form a group of phialides **4**

3b   Conidia possessing 3 or 4 divergent processes. Conidiophores simple or branched **5**

4a   (3) Conidia hyaline, filiform, unicellular, curved or sigmoid, tapering towards the ends. Conidiophore usually branched, forming a group of 2 to 10 phialides. (Fig. 4.113) . . . . *Flagellospora* Ingold

4b   Conidia formed below the surface, unicellular, tetrahedral to subspherical, with a conspicuous glycogen vacuole; those formed on the water surface 2-septate, elongate or broadly fusiform, with a conspicuous vacuole in each cell. (Fig. 4.114) . . . *Margaritispora* Ingold

5a   (3) Conidia consisting of a rodlike, 2-celled main axis with 3 divergent processes developed at the apex. Conidiophores simple or branched to form a group of phialides. (Fig. 4.115) . . . . . *Heliscus* Ingold

5b   Conidia consisting of 4 long, divergent arms, usually becoming septate. Conidiophores simple or branched to form a group of phialides **6**

6a   (5) Four long, divergent arms of the conidium inserted on the phialide at the point of divergence of the 4 arms. (Fig. 4.116). . . . .
Lemonniera De Wildeman

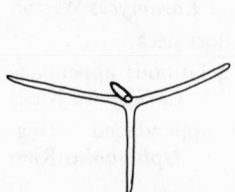

Fig. 4.117. *Alatospora acuminata* Ingold (After Ingold.)

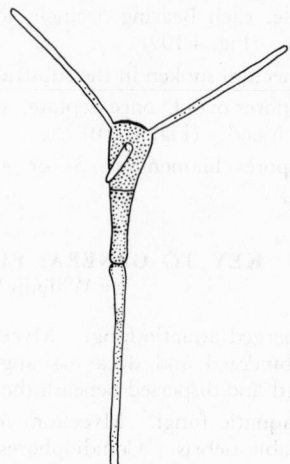

Fig. 4.118. *Clavariopsis aquatica* De Wildeman. (After Ingold.)

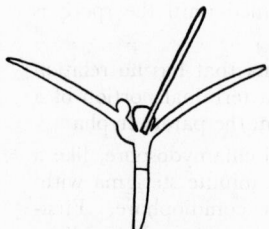

Fig. 4.119. *Tetracladium marchalianum* De Wildeman. (After Ingold.)

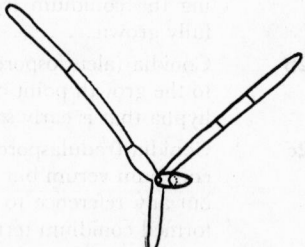

Fig. 4.120. *Articulospora tetracladia* Ingold. (After Ingold.)

6b		Two of the 4 arms of the conidia forming a curved main axis from which arise 2 lateral arms inserted at the middle of the axis. (Fig. 4.117) . . . . **Alatospora** Ingold	
7a	(2)	Conidia tetraradiate.	**8**
7b		Conidia eel-like, crescent-shaped, lemon-shaped, or consisting of a main axis with lateral branches; never tetraradiate	**13**
8a	(7)	Conidia attached to the conidiophore by the tip of one of its arms . .	**9**
8b		Conidia attached to the conidiophore near the point of divergence of the 4 arms.	**12**
9a	(8)	Conidia with a clavate to pyriform main axis from which 3 divergent branches successively arise	**10**
9b		Conidia with a straight or curved, septate main axis, from which arise 3 strongly divergent, septate branches.	**11**
10a	(9)	Main axis of the conidia broadly clavate or narrowly pyriform, 1-septate, with 3 divergent arms arising from the truncate base of the main axis. Conidiophores usually simple. (Fig. 4.118) . . . . **Clavariopsis** De Wildeman	
10b		Main axis of the conidia narrowly clavate, 1- or 2-septate, giving rise to 3 unequal divergent branches with one or more knobs or	

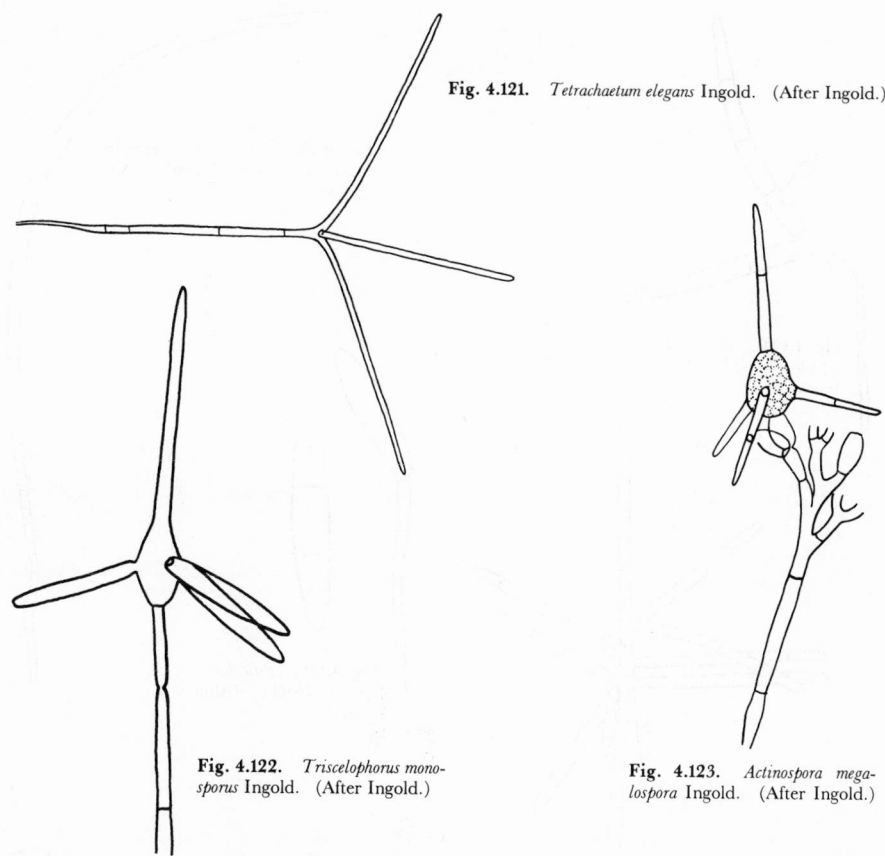

Fig. 4.121.   *Tetrachaetum elegans* Ingold.   (After Ingold.)

Fig. 4.122.   *Triscelophorus mono-sporus* Ingold.   (After Ingold.)

Fig. 4.123.   *Actinospora mega-lospora* Ingold.   (After Ingold.)

fingerlike projections arising on the upper side.   Conidiophores sparingly branched. (Fig. 4.119) . . . **Tetracladium** De Wildeman

11a   **(9)** Conidia with a slender main axis, and 3 somewhat longer divergent branches arising from a common point, a narrow isthmus or constriction found where each arm joins the main axis.   Conidiophores simple or branched. (Fig. 4.120) . . . . . . **Articulospora** Ingold

11b   Conidia with a slender main axis, curved, with 2 lateral branches diverging at the same level from the convex side of the main axis, the 2 lateral arms and the upper and lower portion of the main axis forming a tetraradiate spore.   Conidia liberated by the breakdown of a "separating cell" present at the upper end of the conidiophore. (Fig. 4.121) . . . . **Tetrachaetum** Ingold

12a   **(8)** Conidia consisting of an elongated main axis continuous with the conidiophores, and 3 elongated secondary ramuli forming a whorl of 3 branches near the base of the main axis.   Conidiophores unbranched. (Fig. 4.122). . . . . **Triscelophorus** Ingold

12b   Conidia consisting of a central, spherical part from which radiate 4 (rarely 8) long, straight, hyaline arms.   Conidia large, borne terminally on conidiophores consisting of a straight, simple lower part and an upper, closely branched part with apically swollen branches. (Fig. 4.123) . . . . **Actinospora** Ingold

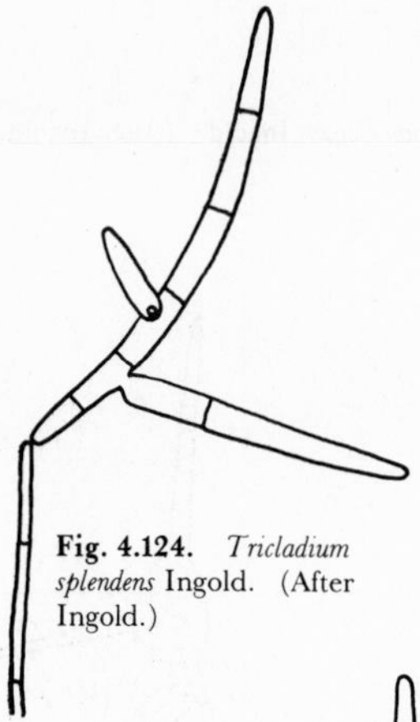

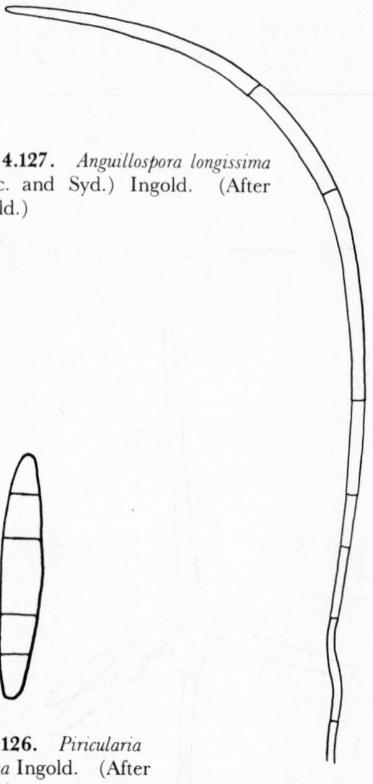

Fig. 4.127. *Anguillospora longissima* (Sacc. and Syd.) Ingold. (After Ingold.)

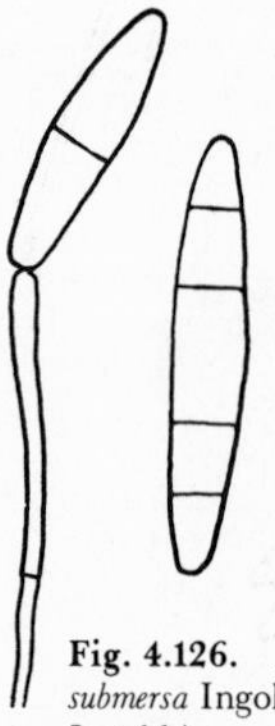

Fig. 4.124. *Tricladium splendens* Ingold. (After Ingold.)

Fig. 4.126. *Piricularia submersa* Ingold. (After Ingold.)

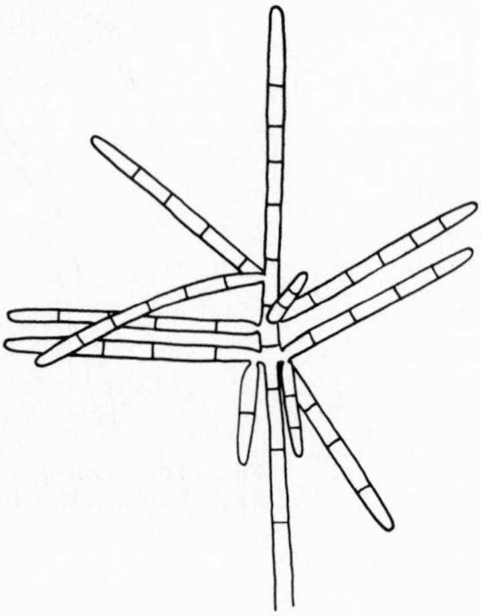

Fig. 4.125. *Dendrospora erecta* Ingold. (After Ingold.)

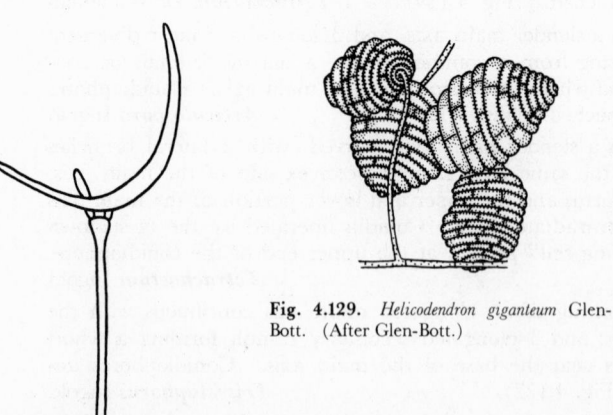

Fig. 4.129. *Helicodendron giganteum* Glen-Bott. (After Glen-Bott.)

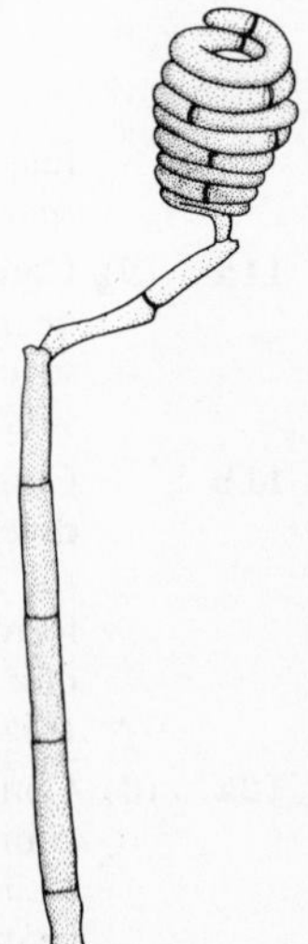

Fig. 4.130. *Helicoon fuscosporum* Linder. (After Linder.)

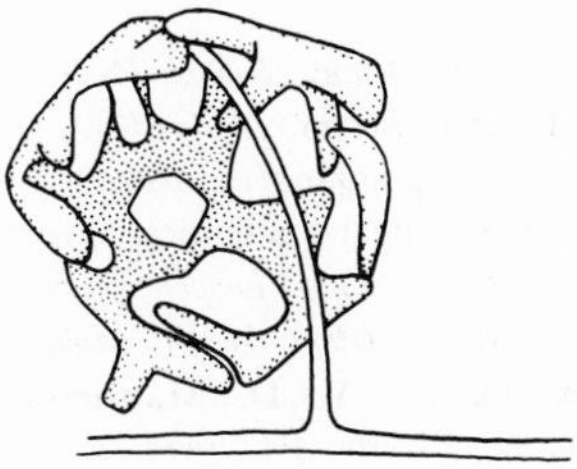

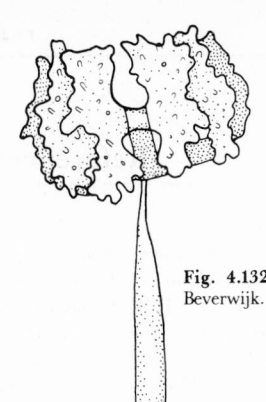

Fig. 4.132. *Candelabrum spinulosum* van Beverwijk. (After van Beverwijk.)

Fig. 4.131. *Clathrosphaerina zalewskii* van Beverwijk. (After van Beverwijk.)

13a	**(7)**	Conidia consisting of a main axis with secondary ramuli, with or without tertiary ramuli arising from it	**14**
13b		Conidia lemon-shaped, elongate and eel-like, or crescent-shaped . . .	**15**
14a	**(13)**	Conidia consisting of a fusiform, 3- to 6-septate main axis from which arise, at different levels, 2 lateral branches, each of which is narrowly constricted at the base. (Fig. 4.124). . **Tricladium** Ingold	
14b		Conidia consisting of a long, slender septate main axis, continuous with the conidiophores, from which arise 6 to 12 long, septate secondary ramuli, narrowly constricted at the base and arising in whorls of 3 or in pairs from the lower part of the main axis; tertiary ramuli when present constricted at the base, arising from the lower most secondary branches. (Fig. 4.125) . . . . . **Dendrospora** Ingold	
15a	**(13)**	Conidia lemon-shaped, nonseptate, or 1-septate when liberated, but becoming 3- to 6-septate later. (Fig. 4.126). . **Piricularia** Saccardo	
15b		Conidia elongate, 6- to 10-septate, curved, or sigmoid, liberated from the conidiophores by the breakdown of a short "separating cell" present at the upper end of the conidiophores. (Fig. 4.127) . . **Anguillospora** Ingold	
15c		Conidia unicellular, crescent-shaped or sigmoid, attached to the conidiophores by a small "stalk cell" at a point along the convex surface. (Fig. 4.128) . . . . **Lunulospora** Ingold	
16a	**(1)**	Spores helicoid consisting of septate filaments coiled to form a three-dimensional helix	**17**
16b		Spores clathroid forming a hollow sphere or network of branching and forking filaments	**18**
17a	**(16)**	Cells of helix proliferating to form secondary helicoid conidia. (Fig. 4.129) . . . . **Helicodendron** Peyronel	
17b		Cells of helix never proliferating to form daughter conidia. (Fig. 4.130) . . . . **Helicoon** Morgan	
18a	**(16)**	Conidia hyaline, multicellular; a spherical network produced by repeated forking and meeting of the forked tips of the filaments. (Fig. 4.131) . . . . **Clathrosphaerina** van Beverwijk	
18b		Conidia hyaline, multicellular, consisting of 4 central cells forming an H-shaped structure each cell of which forms 2 lateral cells at its apex. (Fig. 4.132) . . . . **Candelabrum** van Beverwijk	

References

AQUATIC PHYCOMYCETES

Canter, H. M. 1950. Fungal parasites of the Phytoplankton. I. *Ann. Botany, London* (N.S.), 14:263–289. **1951.** Fungal parasites of the Phytoplankton. II. *Ann. Botany, London* (N.S.), 15:129–156. **Canter, H. M. and J. W. G. Lund. 1948.** Studies on plankton parasites. I. Fluctuations in the numbers of *Asterionella formosa* Hass. in relation to fungal epidemics. *New Phytologist*, 47:238–261. **1951.** Studies on plankton parasites. *Ann. Botany, London* (N.S.) 15:359–371. **Coker, W. C. 1923.** *The Saprolegniaceae, with Notes on Other Water Molds.* University of North Carolina Press, Chapel Hill. **Coker, W. C. and V. D. Matthews. 1937.** Saprolegniales, Saprolegniaceae, Ectrogellaceae, Leptomitaceae. *North Am. Flora*, 2:15–67. **Karling, J. S. 1942.** *The Simple Holocarpic Biflagellate Phycomycetes.* Published by the author, New York. **Middleton, J. T. 1943.** The taxonomy, host range and geographic distribution of the genus *Pythium. Torrey Botan. Club, Mem.*, 20:1–171. **Sparrow, F. K. 1943.** *Aquatic Phycomycetes, Exclusive of the Saprolegniaceae and Pythium.* University of Michigan Press, Ann Arbor. (Second edition in press.) **1951.** *Podochytrium cornutum* n. sp., the cause of an epidemic on the planktonic diatom *Stephanodiscus. Trans. Brit. Mycol. Soc.*, 34:170–173.

AQUATIC ASCOMYCETES

Durand, E. J. 1908. The Geoglossaceae of North America. *Ann. Mycologici*, 6:387–477. **Ingold, C. T. 1951.** Aquatic Ascomycetes: *Ceriospora caudae-suis* n. sp. and *Ophiobolus typhae. Trans. Brit. Mycol. Soc.*, 34:210–215. **Weston, W. H. Jr. 1929.** Observations on Loramyces, an undescribed aquatic ascomycete. *Mycologia*, 21:55–76.

AQUATIC FUNGI IMPERFECTI

Ingold, C. T. 1942. Aquatic Hyphomycetes of decaying alder leaves. *Trans. Brit. Mycol. Soc.*, 25:339–417. (Subsequent volumes of this journal have a number of other papers by this author on aquatic Imperfecti.) **Ranzoni, F. V. 1953.** The Aquatic Hyphomycetes of California. *Farlowia*, 4:353–398.

5

Myxophyceae

FRANCIS DROUET

The Myxophyceae, also called Cyanophyta or blue-green algae, are macroscopic or microscopic, gelatinous, leathery, mealy, or stony; they are attached to various substrata or float in water. Each plant is composed either of individual cells surrounded by gelatinous material (*sheaths*) which they secrete, or of chains (*trichomes*) of cells encased in most species in more or less cylindrical sheaths. The trichome (which becomes branched in the Stigonemataceae) and its sheath comprise the *filament*, which in several families becomes branched where regenerating ends of broken trichomes grow through the parent sheaths and continue development. The cells contain various pigments (phycocyanin, phycoerythrin, carotin, and the chlorophylls), proteid granules, vacuoles, *pseudovacuoles* (irregularly shaped granules, presumably gas bubbles, which appear black in transmitted light and red in reflected light), and oil globules. The cell has a thin outer membrane. Certain cells in several families develop thick walls and become *spores* (reproductive bodies) or *heterocysts;* the spores have walls of equal thickness throughout (Fig. 5.26), and the heterocysts have nodular thickenings inside the walls adjacent to the attached vegetative cells (Fig. 5.16). In the Chamaesiphonaceae, the cells divide internally into endospores (undifferentiated reproductive cells). The trichomes dissociate into several-celled fragments (*hormogonia*, which move by some obscure means) with the death of intervening cells.   Trichomes reproduce by fragmentation or, in a few groups, by spores.

These plants almost invariably develop in masses.   Cell division, enlargement, reproduction, and regeneration take place constantly and often rapidly under widely diverse environmental conditions, and considerable morphological variation can be expected among members of a population.   Numerous individuals should therefore be studied before identification is attempted. Specimens can best be preserved by quick drying on paper, mica, plastic, or glass at room temperatures and in the open air.   Plankton algae should be killed by the addition of commercial formalin before they are dried.   Dried specimens can be studied easily when parts are mounted on a slide in water or in a dilute filtered solution of one of the household detergents.   Calcium carbonate can be dissolved out of these plants in weak solutions of nitric, hydrochloric, or acetic acids.   For differentiating among various species of sheathed Oscillatoriaceae, the reaction of chlor-zinc-iodine is useful: the material should be mounted successively in a strong iodine-potassium iodide solution, in a supersaturated solution of zinc chloride, and again in the iodine solution.

The terminology of the coccoid families (Chroococcaceae, Chamaesiphonaceae, and Clastidiaceae) is given according to the revision of Drouet and Daily (1955).   The major synonyms are given to permit location of the more familiar names.

Most of the illustrations were prepared by Miss Janice F. Bush.

The following key includes the fresh-water species found in North America north of the Rio Grande.

KEY TO SPECIES

1a   Cells becoming completely separated from each other by sheath material during division **2**

1b   Cells attached directly to each other, separated only by their membranes **17**

2a   (1) Cells dividing into daughter cells of equal sizes . . . .
                            Family **Chroococcaceae**   **3**

2b   Cells, at least at the base of the plant, dividing into daughter cells of unequal sizes **14**

3a   (2) Cells before division spherical, ovoid, discoid, cylindrical, or pyriform, never dividing in planes perpendicular to the long axis **7**

3b   Cells before division ovoid to cylindrical, each dividing in a plane perpendicular to the long axis . . . . **Coccochloris** Sprengel
                Formerly *Aphanothece* Nägeli, *Gloeothece* Nägeli, *Synechococcus* Nägeli, and *Rhabdoderma* Schmidle and Lauterborn.

4a   (3) Cells before division ovoid to cylindrico-elliptic, up to 3 times as long as they are broad **5**

4b   Cells before division cylindrical, up to 8 or 10 times as long as they are broad **6**

5a   (4) Cells 7–45 μ in diameter . . . .
                            C. aeruginosa (Nägeli) Drouet and Daily

5b   Cells 5–8 μ in diameter. (Fig. 5.1) . . . . . **C. stagnina** Sprengel

Fig. 5.1. *Coccochloris stagnina.*

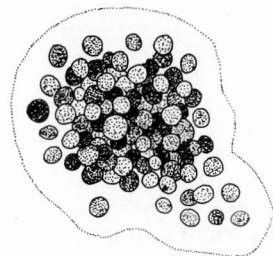

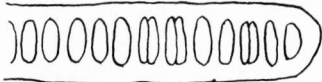

Fig. 5.2. *Johannesbaptistia pellucida.*

Fig. 5.3. *Anacystis cyanea.*

6a (4) Cells 2–6 μ broad, straight, quasi-truncate at the ends. . . . .
  C. elabens (Brébisson) Drouet and Daily

6b   Cells 1–3 μ broad, often curved, rotund or tapering at the
  ends . . . . **C. peniocystis** (Kützing) Drouet and Daily

7a (3) Cells before division spherical, ovoid, cylindrical, or pyriform,
  distributed through a flat or curved surface; cell division proceeding successively in two planes perpendicular to each other **11**

7b   Cells before division discoid, in a single linear series within the
  gelatinous matrix; division proceeding in a single plane through
  the diameter of the cell. . . . . **Johannesbaptistia** J. de Toni
  (Fig. 5.2) . . . . **J. pellucida** (Dickie) Taylor and Drouet

7c   Cells before division spherical, irregularly distributed through the
  gelatinous matrix or in a series of rows in 3 planes perpendicular
  to each other; cell division proceeding successively in 3 planes perpendicular to each other . . . . **Anacystis** Meneghini **8**
    Formerly *Gloeocapsa* Kützing, *Microcystis* Kützing, *Chroococcus* Nägeli, *Eucapsis*
  Clements and Shantz, *Aphanocapsa* Nägeli.

8a (7) Cells without pseudovacuoles, plants not developing as water
  blooms **10**

8b   Cells containing pseudovacuoles, plants developing as water
  blooms; i.e., floating up to surface of a collection **9**

9a (8) Cells 3–7 (rarely 2.5–10) μ in diameter. (Fig. 5.3) . . . .
  A. cyanea (Kützing) Drouet and Daily

9b   Cells 0.5–2 μ in diameter . . . .
  A. incerta (Lemmermann) Drouet and Daily

10a (8) Cells 0.5–2 μ in diameter, sheaths hyaline . . . .
  A. marina (Hansgirg) Drouet and Daily

10b   Cells 2–6 μ in diameter (larger when parasitized); sheaths hyaline
  or becoming brown, yellow, red, or blue . . . .
  A. montana (Lightfoot) Drouet and Daily

10c   Cells 6–12 μ in diameter, sheaths hyaline. . . . .
  A. thermalis (Meneghini) Drouet and Daily

10d   Cells 12–50 μ in diameter, sheaths hyaline . . . .
  A. dimidiata (Kützing) Drouet and Daily

11a (7) Plant spherical to ovoid; cells spherical to ovoid, cylindrical, or
  pyriform, regularly or irregularly arranged . . . .
  Gomphosphaeria Kützing **13**
    Formerly *Coelosphaerium* Nägeli and *Marssoniella* Lemmermann.

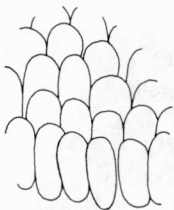

Fig. 5.4. *Microcrocis geminata.*

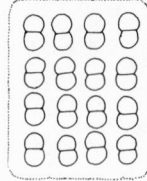

Fig. 5.5. *Agmenellum quadruplicatum.*

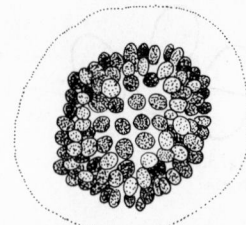

Fig. 5.6. *Gomphosphaeria wichurae.*

Fig. 5.7. *Entophysalis lemaniae.*

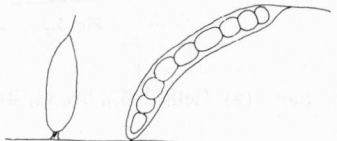

Fig. 5.8. *Clastidium setigerum.*

11b   Plant a flat or curved plate; cells ovoid or cylindrical, irregularly arranged . . . . *Microcrocis* Richter (Fig. 5.4) . . . . *M. geminata* (Lagerheim) Geitler

11c   Plant a flat or curved plate; cells spherical, ovoid, or cylindrical, arranged regularly in series of rows perpendicular to each other . . . . *Agmenellum* Brébisson   12
Formerly *Merismopoedia* Kützing.

12a   (11) Cells 1–3.5 μ in diameter, plants 1- to 64-celled. (Fig. 5.5) . . . . *A. quadruplicatum* (Meneghini) Brébisson

12b   Cells 4–10 μ in diameter, plants larger and often foliose . . . . . . *A. thermale* (Kützing) Drouet and Daily

13a   (11) Cells 3–5 μ in diameter, containing pseudovacuoles; plants developing as water blooms. (Fig. 5.6). . . . . *Gomphosphaeria wichurae* (Hilse) Drouet and Daily

13b   Cells 2–4 μ in diameter, without pseudovacuoles; plants aquatic . . *G. lacustris* Chodat

13c   Cells 4–15 μ in diameter, without pseudovacuoles; plants aquatic or subaerial . . . . *G. aponina* Kützing

14a   (2) Plants basally attached to larger water plants, at first unicellular, each dividing internally into a chain of cells . . . . Family **Clastidiaceae**   16

14b   Plants aquatic, at first unicellular; cells dividing to form cushions of radial structure, the cells below often elongating downward into the substratum; endospores formed in certain cells. . . . . Family **Chamaesiphonaceae** *Entophysalis* Kützing
Formerly *Dermocarpa* Crouan, *Chamaesiphon* A. Braun and Grunow, *Pleurocapsa* Thuret, *Hyella* Bornet and Flahault, and *Radaisia* Sauvageau.

15a   (14) On rocks, wood, shells, etc. . . . . . . *E. rivularis* (Kützing) Drouet

15b   On larger plants. (Fig. 5.7) . . . . *E. lemaniae* (Agardh) Drouet and Daily

16a   (14) Plant terminating above in a hairlike extension of the sheath . . . . *Clastidium* Kirchner (Fig. 5.8) . . . . *C. setigerum* Kirchner

Fig. 5.9. *Stichosiphon sansibaricus.*

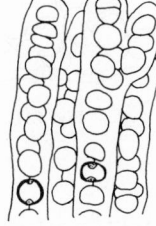

Fig. 5.10. *Capsosira brebissonii.*

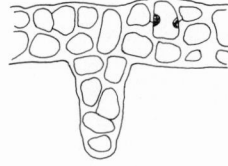

Fig. 5.11. *Stigonema minutum.*

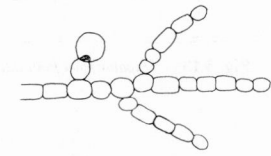

Fig. 5.12. *Nostochopsis lobatus.*

16b		Plant smooth at the apex. . . . . ***Stichosiphon*** Geitler	
		(Fig. 5.9) . . . . . ***S. sansibaricus*** (Hieronymus) Drouet and Daily	
17a	(1)	Trichomes never branching, heterocysts present or absent.	27
17b		Trichomes branching, heterocysts present	
		Family **Stigonemataceae**	18
18a	(17)	Cells of the trichome dividing only in planes perpendicular to the axis except in the basal branches, where parallel divisions also occur. (Fig. 5.12).	23
18b		Cells of the trichome dividing in planes perpendicular to and parallel with the axis throughout the plant. (Fig. 5.11)	19
19a	(18)	Filaments upright and parallel in a cushion; plants aquatic.	
		Capsosira Kützing	
		(Fig. 5.10) . . . . ***C. brebissonii*** Kützing	
19b		Filaments intertwined, or upright and parallel only at the surface of the plant . . . . ***Stigonema*** Agardh	20
20a	(19)	Cells of the ultimate branches chiefly uniseriate	21
20b		Cells of the ultimate branches chiefly multiseriate	22
21a	(20)	Filaments 7–15 μ in diameter, plants subaerial.	
		S. hormoides (Kützing) Bornet and Flahault	
21b		Filaments 20–30 μ in diameter, plants terrestrial.	
		S. panniforme (Agardh) Bornet and Flahault	
21c		Filaments 35–50 μ in diameter, plants aquatic	
		S. ocellatum (Dillwyn) Thuret	
22a	(20)	Filaments 18–29 μ in diameter, plants subaerial and aerial. (Fig. 5.11) . . . . ***S. minutum*** (Agardh) Hassall	
22b		Filaments over 40 μ in diameter, plants rigid, blue-green, aquatic . . . . ***S. mamillosum*** (Lyngbye) Agardh	
22c		Filaments over 40 μ in diameter, plants soft, yellowish, subaerial . . . . ***S. informe*** Kützing	
22d		Filaments 27–30 μ broad; plants velvety, subaerial	
		S. turfaceum (Smith and Sowerby) Cooke	
23a	(18)	Filaments radial in a spherical gelatinous matrix, plants aquatic . . . . ***Nostochopsis*** Wood	
		(Fig. 5.12) . . . . ***N. lobatus*** Wood	

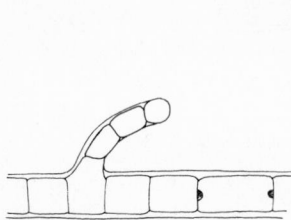

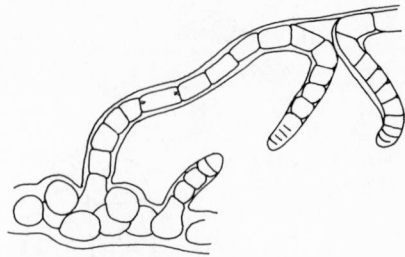

Fig. 5.13. *Hapalosiphon fontinalis.*   **Fig. 5.14.** *Fischerella ambigua.*

23b   Filaments sheathed, but not imbedded in a gelatinous matrix. . . .   **24**

24a   (23)   Plants terrestrial . . . . ***Fischerella*** Gomont   **26**

24b   Plants aquatic . . . . ***Hapalosiphon*** Nägeli   **25**

25a   (24)   In ponds, lakes, and streams. (Fig. 5.13) . . . .
                  H. fontinalis (Agardh) Bornet

25b   In hot springs. . . . . ***H. laminosus*** (Kützing) Hansgirg

26a   (24)   Trichomes of upper filaments cylindrical. (Fig. 5.14). . . . .
                  Fischerella ambigua (Nägeli) Gomont

26b   Trichomes of upper filaments torulose . . . .
                  F. thermalis (Schwabe) Gomont

27a   (17)   Spores absent (except in *Gloeotrichia* and *Aulosira*), cells chiefly
                  cylindrical **60**

27b   Spores formed in most species, cells chiefly spherical to barrel-
                  shaped . . . . Family **Nostocaceae**   **28**

28a   (27)   Heterocysts terminal in the trichome **55**

28b   Heterocysts intercalary in the trichome. **29**

29a   (28)   Individual sheaths distinct, plants aquatic or subaerial . . . .
                  Hydrocoryne Schwabe
                  (Fig. 5.15) . . . . ***H. spongiosa*** Schwabe

29b   Individual sheaths coalesced or inconspicuous **30**

30a   (29)   Trichomes without sheaths or in fragile, coalesced gelatin of in-
                  definite shape **46**

30b   Trichomes included in a common gelatinous matrix of definite
                  shape **31**

31a   (30)   Trichomes parallel in a fragile, hyaline, gelatinous sac; plants
                  aquatic . . . . ***Wollea*** Bornet and Flahault
                  (Fig. 5.16) . . . . ***W. saccata*** (Wolle) Bornet and Flahault

31b   Trichomes curved and intermeshed in a firm matrix which is at
                  first hyaline, becoming yellow or brownish . . . ***Nostoc*** Vaucher   **32**

32a   (31)   Plants terrestrial or on wet substrata **43**

32b   Plants aquatic **33**

33a   (32)   Plants macroscopic, on stones, etc., or unattached **37**

33b   Plants microscopic, epiphytic or endophytic **34**

34a   (33)   Trichomes straight or broadly curving **36**

34b   Trichomes much contorted **35**

35a   (34)   Spores spherical, 8–10 μ in diameter. . . . . ***N. cuticulare*** Brébisson

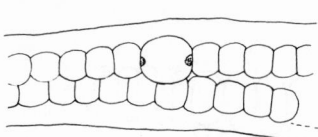

Fig. 5.15. *Hydrocoryne spongiosa.*

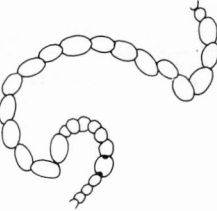

Fig. 5.17. *Nostoc linckia.*

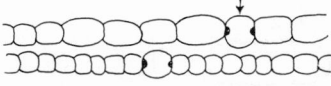

Fig. 5.16. *Wollea saccata.* The arrow
points to heterocyst.

35b		Spores subspherical to oblong, 5–6 × 5–8 μ . . . . *N. hederulae* Meneghini	
35c		Spores spherical to depressed-spherical, 5–6 μ in diameter. . . . . . *N. entophytum* Bornet	
36a	(34)	Spores spherical, 6 μ in diameter . . . . *N. maculiforme* Bornet and Flahault	
36b		Spores oblong, 4 × 6–8 μ. . . . . *N. paludosum* Kützing	
37a	(33)	Plants attached to substrata, in flowing water	42
37b		Plants unattached, floating	38
38a	(37)	Plants firm, spores absent	41
38b		Plants soft, developing spores	39
39a	(38)	Trichomes densely contorted, spores 6–7 × 7–10 μ. (Fig. 5.17). . . *N. linckia* (Roth) Bornet	
39b		Trichomes lax	40
40a	(39)	Spores spherical, 6–7 μ in diameter, with hyaline walls . . . . *N. piscinale* Kützing	
40b		Spores ovoid, 6–8 × 7–10 μ, with brownish walls. . . . . *N. rivulare* Kützing	
40c		Spores ovoid, 6 × 8–10 μ, with hyaline walls. . . . . *N. carneum* (Lyngbye) Agardh	
40d		Spores cylindrical, 6–7 × 10–12 μ, with hyaline walls . . . . *N. spongiiforme* Agardh	
41a	(38)	Plants up to 1 cm in diameter, forming water blooms; trichomes containing pseudovacuoles . . . . *N. caeruleum* Lyngbye	
41b		Plants up to 6 cm in diameter, trichomes without pseudo-vacuoles. . . . . *N. pruniforme* (Linnaeus) Agardh	
42a	(37)	Plants spherical, to 2 cm in diameter; trichomes torulose . . . . . . *N. sphaericum* Vaucher	
42b		Plants spherical or cushion-shaped, to 10 cm in diameter; trichomes cylindrico-torulose . . *N. verrucosum* (Linnaeus) Vaucher	
42c		Plants spherical or cushion-shaped, up to 60 cm in diameter; trichomes torulose. . . . . *N. amplissimum* Setchell	
42d		Plants discoid, trichomes torulose . . . . *N. parmelioides* Kützing	
43a	(32)	Plants firm	45
43b		Plants soft	44

Fig. 5.18. *Aphanizomenon holsaticum.*

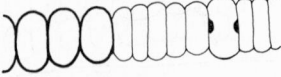

Fig. 5.19.   *Nodularia spumigena.*

Fig. 5.20.   *Anabaena oscillarioides.*

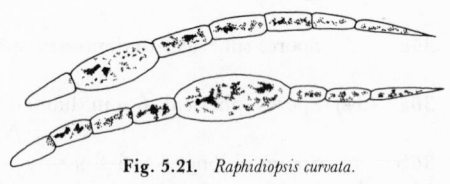

Fig. 5.21.   *Raphidiopsis curvata.*

Fig. 5.22.   *Cylindrospermum licheniforme.*

44a   (43) Cells cylindrical, 4 μ in diameter; spores 6–8 $\times$ 14–20 μ . . . . . .
                        Nostoc ellipsosporum (Desmazières) Rabenhorst

44b   Cells spherical, 3–5 μ in diameter; spores 4–8 $\times$ 8–12 μ . . . .
                        N. muscorum Agardh

44c   Cells spherical, 2–3 μ in diameter; spores 4 $\times$ 6 μ . . . .
                        N. humifusum Carmichael

45a   (43) Plants large, at first globose, becoming laminate or filiform . . . . .
                        N. commune Vaucher

45b   Plants spherical, to 2 cm in diameter; trichomes 5–8 μ in diam-
                eter . . . . ***N. microscopicum*** Carmichael

45c   Plants spherical, microscopic; trichomes 8–9 μ in diameter . . . . .
                        N. macrosporum Meneghini

46a   (30) Terminal cells of the trichome similar to other vegetative cells . . .   48

46b   The several terminal cells at each end of the trichome colorless
                and elongate, the vegetative cells pseudovacuolate; plants develop-
                ing as water blooms . . . . ***Aphanizomenon*** Morren   47

47a   (46) Spores long-cylindrical. (Fig. 5.18). . . .   ***A. holsaticum*** Richter
                Formerly referred to as *A. flos-aquae.*

47b   Spores ovoid . . . . ***A. ovalisporum*** Forti

48a   (46) Vegetative cells spherical, or longer than broad . . ***Anabaena*** Bory   50

48b   Vegetative cells depressed-spherical or discoid . . . .
                        Nodularia Mertens

49a   (48) Cells 8–18 μ in diameter. (Fig. 5.19). . . ***N. spumigena*** Mertens

49b   Cells 6–7 μ in diameter, spores 7–10 μ in diameter. . . . .
                        N. sphaerocarpa Bornet and Flahault

49c   Cells 4–6 μ in diameter, spores 6–8 μ in diameter . . . .
            N. harveyana (Thwaites) Thuret

50a (48, 56) Cells blue-green, without pseudovacuoles; plants aquatic **52**

50b   Cells containing pseudovacuoles, plants developing as water blooms **51**

51a (50) Trichomes 4–8 μ in diameter, spores 7–13 μ in diameter . . . . . .
            Anabaena flos-aquae (Lyngbye) Brébisson

51b   Trichomes 8–10 μ in diameter, spores 16–18 μ in diameter . . . . .
            A. circinalis (Harvey) Rabenhorst

52a (50) Spores spherical or ovoid **53**

52b   Spores cylindrical **54**

53a (52) Spores catenate, 7–9 μ in diameter . . . . . *A. variabilis* Kützing

53b   Spores chiefly solitary and next to the heterocysts, 12–20 μ in diameter . . . . *A. sphaerica* Bornet and Flahault

54a (52) Spores 20–40 μ long, next to and on only one side of the heterocyst . . . . *A. unispora* Gardner

54b   Spores up to 100 μ long, next to and on both sides of the heterocysts. (Fig. 5.20) . . . .   **54′**

54c   Spores 14–17 μ long, remote from the heterocysts . . . .
            A. inaequalis (Kützing) Bornet and Flahault

54d   Spores up to 30 μ long, remote from the heterocysts . . . .
            A. catenula (Kützing) Bornet and Flahault

54′a (54) Vegetative cells 12 μ broad . . . . *A. bornetiana* Collins

54′b   Vegetative cells 3–6 μ broad (Fig. 5.20) . . . .
            A. oscillatorioides Bory

55a (28) Heterocysts developing at both ends of the trichome. **56**

55b   Trichome bearing a heterocyst at one end, tapering to a point at the other end; cells pseudovacuolate; plants developing as water blooms . . . . *Raphidiopsis* Fritsch and Rich (Fig. 5.21) . . . . *R. curvata* Fritsch and Rich

56a (55) Spores remote from the heterocysts . . . . . . *Anabaena* Bory   **50**

56b   Spores adjacent to the heterocysts . . *Cylindrospermum* Kützing   **57**

57a (56) Spores developing in a catenate series . . . . . *C. catenatum* Ralfs

57b   Spores solitary **58**

58a (57) Walls of mature spores rough . . . . *C. majus* Kützing

58b   Walls of mature spores smooth **59**

59a (58) Spores long-cylindrical, up to 40 μ long . . . .
            C. stagnale (Kützing) Bornet and Flahault

59b   Spores oblong, 12–14 × 20–30 μ. (Fig. 5.22) . . . .
            C. licheniforme (Bory) Kützing

59c   Spores ovoid to oblong, 9–12 × 12–20 μ. . . *C. muscicola* Kützing

60a (27) Trichome of the same thickness throughout its length **72**

60b   Trichome thick at the base, tapering above, often into a colorless hair . . . . Family **Rivulariaceae**   **61**

61a (60) Sheaths coalesced; plants spherical or cushion-shaped **68**

61b   Sheaths of the individual trichomes distinct **62**

62a (61) Terminals hairs permanent **63**

62b   Terminal hairs ephemeral, heterocysts absent, trichomes 1–3 μ in diameter, plants aquatic. . . . . *Amphithrix* Kützing (Fig. 5.23) . . . . . *A. janthina* (Montagne) Bornet and Flahault

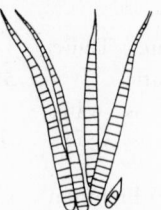

Fig. 5.23. *Amphithrix janthina.*

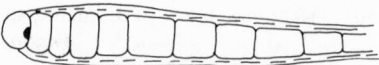

Fig. 5.24. *Calothrix parietina.*

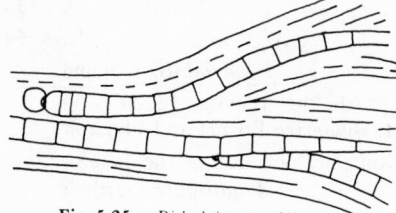

Fig. 5.25. *Dichothrix gypsophila.*

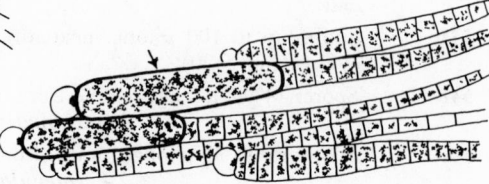

Fig. 5.26. *Gloeotrichia echinulata.* The arrow points to a spore.

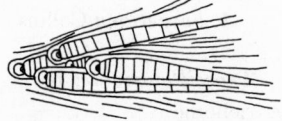

Fig. 5.27. *Rivularia minutula.*

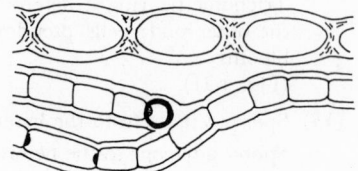

Fig. 5.28. *Aulosira implexa.*

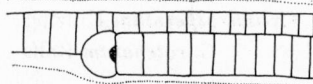

Fig. 5.29. *Desmonema wrangelii.*

63a (62) Filaments branched, the bases of the branches included for a short distance within the parent sheaths . . . . . **Dichothrix** Zanardini   66

63b   Filaments unbranched or branched, the branches not included within the parent sheath . . . . **Calothrix** Agardh   64

64a (63) Heterocysts absent . **C. juliana** (Meneghini) Bornet and Flahault

64b   Heterocysts present   65

65a (64) Filaments 5–10 μ.in diameter in the middle; sheaths at first hyaline, becoming yellow or brown. (Fig. 5.24) . . . .
                                    C. parietina (Nägeli) Thuret

65b   Filaments 18–22 μ in diameter in the middle, sheaths hyaline. . . .
                                    C. adscendens (Nägeli) Bornet and Flahault

66a (63) Upper ends of sheaths enlarged, lamellate, and frayed. (Fig. 5.25). . . . **Dichothrix gypsophila** (Kützing) Bornet and Flahault

66b　　　　Upper ends of sheaths cylindrical or pointed . . . .　**67**

67a　(66)　Basal heterocysts spherical to ovoid, sheaths becoming yellowish, filaments up to 15 μ in diameter in the middle . . . .
　　　　　　　　　　　D. baueriana (Grunow) Bornet and Flahault

67b　　　　Basal heterocysts spherical to ovoid, sheaths becoming dark brown, filaments 10–12 μ in diameter in the middle . . . .
　　　　　　　　　　　D. orsiniana (Kützing) Bornet and Flahault

67c　　　　Basal heterocysts spherical to hemispherical, filaments 40–90 μ in diameter in the middle . . . . *D. inyoensis* Drouet

67d　　　　Basal heterocysts depressed-hemispherical, filaments 25–30 μ in diameter in the middle . . . . *D. hosfordii* (Wolle) Bornet

68a　(61)　Spores absent . . . . ~ . . . . *Rivularia* Agardh　**70**

68b　　　　Thick-walled spores present . . . . *Gloeotrichia* Agardh　**69**

69a　(68)　Spores long-cylindrical; cells pseudovacuolate; gelatin firm; plants microscopic, forming water blooms. (Fig. 5.26). . . . .
　　　　　　　　　　　G. echinulata (Smith and Sowerby) Richter

69b　　　　Spores long-cylindrical; pseudovacuoles absent; gelatin firm; plants up to 1 cm in diameter, attached . . . *G. pisum* (Agardh) Thuret

69c　　　　Spores ovoid to long-ovoid, with stratified walls; pseudovacuoles absent; plants soft-gelatinous, of indefinite size . . . .
　　　　　　　　　　　G. natans (Hedwig) Rabenhorst

69d　　　　Spores short-cylindrical, with unstratified walls; pseudovacuoles absent; plant soft-gelatinous, of indefinite size . . . .
　　　　　　　　　　　G. salina (Kützing) Rabenhorst

70a　(68)　Plants not impregnated with calcium carbonate, spherical; trichomes 4–16 μ in diameter . . . . . *Rivularia bornetiana* Setchell

70b　　　　Plants impregnated with calcium carbonate, at least in part . . . .　**71**

71a　(70)　Plants microscopic, epiphytic; trichomes 4–9 μ in diameter; sheath narrow . . . . *R. dura* Roth

71b　　　　Plants macroscopic, on rocks, etc.; trichomes 9–13 μ in diameter; sheaths thick, lamellate, yellow or brown. (Fig. 5.27) . . . .
　　　　　　　　　　　R. minutula (Kützing) Bornet and Flahault

71c　　　　Plants macroscopic, stony, on rocks, etc.; trichomes 4–8 μ in diameter; sheaths thin . . . *R. haematites* (de Candolle) Agardh

72a　(60)　Trichomes without heterocysts, sheaths present or absent . . . . . .
　　　　　　　　　　　Family **Oscillatoriaceae**　**86**

72b　　　　Trichomes containing heterocysts, sheaths firm, filaments branched
　　　　　　　　　　　Family **Scytonemataceae**　**73**

73a　(72)　Spores absent. . . . .　**74**

73b　　　　All the vegetative cells eventually becoming thick-walled spores, plants aquatic. . . . . *Aulosira* Kirchner　**73'**

73'a　(73)　Trichomes 5–8 μ broad . . . . *A. laxa* Kirchner

73'b　　　　Trichomes 8–10 μ broad. (Fig. 5.28) . . . .
　　　　　　　　　　　A. implexa Bornet and Flahault

74a　(73)　Trichomes solitary within the sheaths. . . . .　**75**

74b　　　　Trichomes several within a common sheath; filaments sub-dichotomously branched above, aquatic . . . .
　　　　　　　　　　　Desmonema Berkeley and Thwaites
　　　　　　　　(Fig. 5.29) . . . . . *D. wrangelii* (Agardh) Bornet and Flahault

75a　(74)　Branches single (issuing from the sheath beside a heterocyst) or absent . . . .　**83**

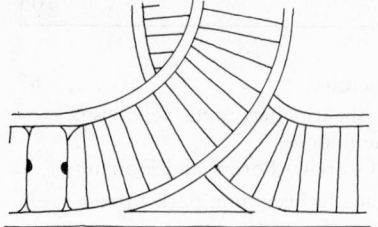

Fig. 5.30. *Scytonema crispum.*

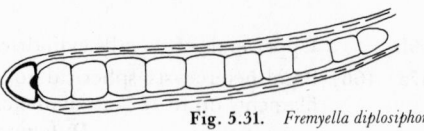

Fig. 5.31. *Fremyella diplosiphon.*

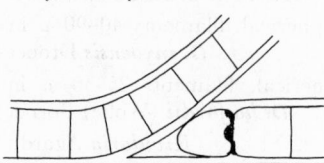

Fig. 5.32. *Tolypothrix lanata.*

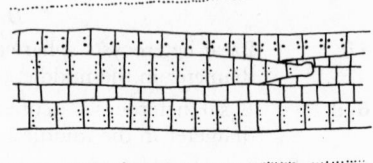

Fig. 5.33. *Microcoleus vaginatus.*

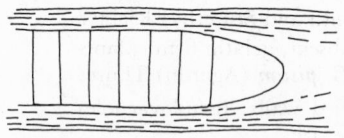

Fig. 5.34. *Porphyrosiphon notarisii.*

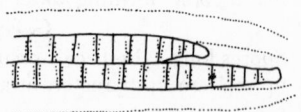

Fig. 5.35. *Hydrocoleum homoeotrichum.*

75b		Branches geminate, at least in the older parts of the plant.	
		Scytonema Agardh	**76**
76a	(75)·	Plants terrestrial, or aerial, on wood, rocks, etc.	**78**
76b		Plants aquatic	**77**
77a	(76)	Cells ½–⅙ as long as broad, filaments 18–30 μ in diameter, sheaths cylindrical. (Fig. 5.30) . . . **S. crispum** (Agardh) Bornet	
77b		Cells ½–2 times as long as broad, filaments 18–24 μ in diameter, sheaths cylindrical . . . . **S. coactile** Montagne	
77c		Cells ½–3 times as long as broad, filaments 10–15 μ in diameter, sheaths thickened and sometimes lamellate at the tips	
		S. tolypotrichoides Kützing	
78a	(76)	Sheaths thick and often lamellate, at least at the tips	**80**
78b		Sheaths cylindrical, of the same thickness throughout	**79**
79a	(78)	Filaments 15–22 μ in diameter, growing parallel with each other above in upright fascicles	
		S. guyanense (Montagne) Bornet and Flahault	
79b		Filaments 7–15 μ in diameter, growing parallel with each other above in upright fascicles. . . . . **S. hofmannii** Agardh	
79c		Filaments 10–15 μ in diameter, not forming upright fascicles	
		S. ocellatum Lyngbye	
80a	(78)	Sheaths much thickened and often gelatinous at the tips	**82**
80b		Sheaths somewhat thickened and lamellate at the tips.	**81**
81a	(80)	Filaments 15–21 μ in diameter. . . . **S. mirabile** (Dillwyn) Bornet	
81b		Filaments 18–36 μ in diameter. . **S. myochrous** (Dillwyn) Agardh	
82a	(80)	Filaments 15–30 μ in diameter, growing parallel with each other above in upright bundles; sheaths lamellose, those of the twin branches coalesced . . . . **S. crustaceum** Agardh	

82b		Filaments 24–40 μ in diameter, not in bundles; sheaths gelatinous throughout . . . . **S. densum** (A. Braun) Bornet
82c		Filaments 24–70 μ in diameter, not in bundles; sheaths gelatinous and ocreate only at the tips . . . . **S. alatum** (Carmichael) Borzi
83a	(75)	Filaments long and branched . . . . . . **Tolypothrix** Kützing 85
83b		Filaments short, usually not branched; heterocysts basal . . . . . . **Fremyella** J. de Toni 84

Formerly *Microchaete* Thuret.

84a	(83)	Filaments 6–7 μ in diameter . . . . **F. tenera** (Thuret) J. de Toni
84b		Filaments 7–10 μ in diameter. (Fig. 5.31) . . . . **F. diplosiphon** (Gomont) Drouet
85a	(83)	Filaments 10–15 μ in diameter, cells shorter than broad . . . . **Tolypothrix distorta** (Hofman-Bang) Kützing
85b		Filaments 8 (rarely to 10) μ in diameter, cells as long as or longer than broad . . . . **T. tenuis** Kützing
85c		Filaments 9–12 μ in diameter, cells as long as or longer than broad. (Fig. 5.32) . . . . **T. lanata** (Desvaux) Wartmann
86a	(72)	Trichomes sheathless or solitary within the sheaths 114
86b		Trichomes more than one within a sheath, at least at the base of the plant 87
87a	(86)	Trichomes few within a sheath 91
87b		Trichomes many within a sheath. . . . . **Microcoleus** Desmazières 88
88a	(87)	Trichomes becoming capitate at the tips. (Fig. 5.33) . . . . 88'
88b		Trichomes becoming sharply long-pointed at the tips . . . . **M. acutissimus** Gardner
88c		Trichomes becoming conical at the tips 89
88'a	(88)	Trichomes 3–4 μ broad, cells longer than broad . . . . **M. monticola** (Kützing) Hansgirg
88'b		Trichomes 4–8 μ broad, cells as long as or shorter than broad . . . **M. vaginatus** (Vaucher) Gomont
89a	(88)	Sheaths turning blue when treated with chlor-zinc-iodine; trichomes 2–4 μ in diameter, constricted at the cross walls. . . . . . . **M. rupicola** (Tilden) Drouet
89b		Sheaths not turning blue when treated with chlor-zinc-iodine. . . . 90
90a	(89)	Trichomes 4–5 μ in diameter, constricted at the cross walls . . . . . **M. lacustris** Farlow
90b		Trichomes 6–10 μ in diameter, not constricted at the cross walls . . **M. paludosus** (Kützing) Gomont
91a	(87)	Sheaths gelatinous, diffluent; plants aquatic . . . . **Hydrocoleum** Kützing 93
91b		Sheaths firm, irregular in outline . . . . . **Schizothrix** Kützing 94
91c		Sheaths firm, cylindrical, lamellated, becoming red; plants sub-aerial or aerial . . . . **Porphyrosiphon** Kützing 92
92a	(91, 123)	Trichomes 3–6 μ in diameter . . . . **P. fuscus** Gomont
92b		Trichomes 8–20 μ in diameter. (Fig. 5.34) . . . . **P. notarisii** (Meneghini) Kützing
93a	(91)	Trichomes 6–8 μ in diameter, becoming attenuate and capitate at the tips; cells ½ as long to as long as broad. (Fig. 5.35) . . . . . . **Hydrocoleum homoeotrichum** Kützing

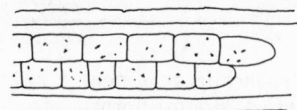

Fig. 5.36. *Schizothrix friesii.*

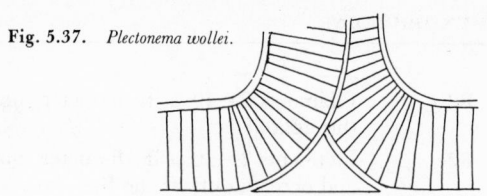

Fig. 5.37. *Plectonema wollei.*

93b		Trichomes 4–8 μ in diameter, becoming short-attenuate and truncate at the tips; cells $\frac{1}{5}$ as long to as long as broad	
		Hydrocoleum groesbeckianum Drouet	
94a	(91)	Plants subaerial or aerial, filaments prostrate and intertwined, at least next to substratum	**99**
94b		Plants aquatic, filaments growing upright and parallel from the base	**95**
95a	(94)	Plants gelatinous, not encrusted with calcium carbonate	**98**
95b		Plants encrusted with calcium carbonate, at least in part	**96**
96a	(95)	Filaments aggregated in fascicles, trichomes 1–3 μ in diameter, plant stony . . . . ***Schizothrix fasciculata*** (Nägeli) Gomont	
96b		Filaments not aggregated in fascicles	**97**
97a	(96)	Plant stony, trichomes 1–2 μ in diameter	
		S. pulvinata (Kützing) Gomont	
97b		Plant mealy, trichomes 2–3 μ in diameter	
		S. vaginata (Nägeli) Gomont	
97c		Plant gelatinous, calcified below; trichomes 1–2 μ in diameter . . .	
		S. lacustris A. Braun	
98a	(95)	Trichomes 5–11 μ in diameter . . . . ***S. rivularis*** (Wolle) Drouet	
98b		Trichomes 3–5 μ in diameter. . ***S. penicillata*** (Kützing) Gomont	
98c		Trichomes 1–2.5 μ in diameter, cells as long as or shorter than broad . . . . ***S. tinctoria*** (Agardh) Gomont	
98d		Trichomes 1–2.5 μ in diameter, cells longer than broad	
		S. arenaria (Berkeley) Gomont	
99a	(94)	Filaments prostrate next to substratum, aggregated into bundles above	**101**
99b		Filaments all prostrate and intertwined, trichomes 1–2 μ in diameter	**100**
100a	(99)	Plants calcified, filaments contorted.	
		S. coriacea (Kützing) Gomont	
100b		Plants calcified, filaments straight or flexuous	
		S. lateritia (Kützing) Gomont	
100c		Plants not calcified, filaments contorted	
		S. calcicola (Agardh) Gomont	
100d		Plants not calcified, filaments straight or flexuous	
		S. lardacea (Cesati) Gomont	
101a	(99)	Sheaths, at least at the surface of the plant, becoming colored . . .	**107**
101b		Sheaths always hyaline	**102**
102a	(101)	Outer membrane of the end cell thin	**104**
102b		Outer membrane of the end cell becoming thickened	**103**
103a	(102)	Trichomes 3–4 μ in diameter, cells longer than broad	
		S. dailyi Drouet	

103b		Trichomes 4–6 μ in diameter, cells chiefly shorter than broad. . . . *S. stricklandii* Drouet	
104a	(102)	Trichomes constricted at the cross walls	**106**
104b		Trichomes not constricted at the cross walls	**105**
105a	(104)	Cells 2–5 μ in diameter, mature end cell truncate-conical . . . . . . *S. californica* Drouet	
105b		Cells 3–5 μ in diameter, mature end cell simply conical . . . . *S. purcellii* W. R. Taylor	
106a	(104)	Cells 1–2 μ in diameter, as long as or shorter than broad; end cell rotund . . . . *S. fragilis* (Kützing) Gomont	
106b		Cells 2–4 μ in diameter, up to 6 times as long as broad; end cell acute-conical . . . . *S. longiarticulata* Gardner	
106c		Cells 3–6 μ in diameter, up to twice as long as broad; end cell blunt-conical. (Fig. 5.36) . . . . . . *S. friesii* (Agardh) Gomont	
107a	(101)	Sheaths becoming red	**112**
107b		Sheaths becoming blue	**111**
107c		Sheaths becoming yellow or brown	**108**
108a	(107)	Ends of trichomes becoming conical	**110**
108b		Ends of trichomes becoming very sharply pointed	**109**
109a	(108)	Trichomes 2–3 μ in diameter. . . . . *S. macbridei* Drouet	
109b		Trichomes 4–6 μ in diameter. . . . . *S. acutissima* Drouet	
110a	(108)	Trichomes 7–13 μ in diameter . . . . *S. muelleri* Nägeli	
110b		Trichomes 4–6 μ in diameter. . . . . *S. wollei* Drouet	
110c		Trichomes 3–4 μ in diameter. . . . . *S. lamyi* Gomont	
110d		Trichomes 2–3 μ in diameter. . . . . *S. fuscescens* Kützing	
111a	(107)	Trichomes 2–3 μ in diameter, end cell becoming blunt-conical . . . *S. heufleri* Grunow	
111b		Trichomes 2–4 μ in diameter, end cell becoming acute-conical . . . *S. giuseppei* Drouet	
111c		Trichomes 4–7 μ in diameter, end cell becoming long and very sharply conical . . . . *S. taylorii* Drouet	
112a	(107)	Outer wall of end cell thin	**113**
112b		Outer wall of end cell conspicuously thickened, trichomes 4–7 μ in diameter . . . . *S. richardsii* Drouet	
113a	(112)	Trichomes 1–2.5 μ in diameter. . . . *S. roseola* (Gardner) Drouet	
113b		Trichomes 4–7 μ in diameter, not constricted at the cross walls; end cell acuminate . . . . *S. acuminata* (Gardner) Drouet	
113c		Trichomes 6–8 μ in diameter, constricted at the cross walls; end cell acute-conical . . . . *S. purpurascens* (Kützing) Gomont	
113d		Trichomes 4–9 μ in diameter, constricted at the cross walls; end cell rotund . . . . *S. thelephoroides* (Montagne) Gomont	
114a	(86)	Sheaths inconspicuous or coalesced	**125**
114b		Trichomes in individual cylindrical sheaths	**115**
115a	(114)	Filaments rarely if ever branched	**117**
115b		Filaments with frequent scytonematoid branching . . . . *Plectonema* Thuret	**116**
116a	(115)	Trichomes 25–50 μ in diameter, plants aquatic. (Fig. 5.37) . . . . *P. wollei* Farlow	

116b   Trichomes 11–22 μ in diameter, plants aquatic . . . .
                         Plectonema tomasinianum (Kützing) Bornet
116c   Trichomes 3 μ in diameter, plants aquatic . . . .
                         P. purpureum Gomont
116d   Trichomes 2 μ in diameter, filaments spiraled, plants subaerial . . .
                         P. cloverianum Drouet
116e   Trichomes 1–2 μ in diameter, in gelatin of subaerial and aquatic
              algae . . . . *P. nostocorum* Bornet
116f   Trichomes 1–2 μ in diameter, plants aquatic in limestone and
              shells . . . . *P. terebrans* Bornet and Flahault
117a (115)   Plants terrestrial or subaerial, the surface filaments growing
              parallel in upright or prostrate fascicles   123
117b   Plants aquatic (only *L. aestuarii* being equally aquatic and ter-
              restrial) . . . . *Lyngbya* Agardh   118
118a (117)   Plants developing on a substratum   120
118b   Plants planktonic   119
119a (118)   Trichomes 15–25 μ in diameter, cells containing pseudovacuoles . .
                         L. birgei Smith
119b   Trichomes 1–2 μ in diameter, cells without pseudovacuoles . . . . .
                         L. contorta Lemmermann
120a (118)   Sheaths turning blue when treated with chlor-zinc-iodine   122
120b   Sheaths not turning blue when treated with chlor-zinc-iodine. . . .   121
121a (120)   Trichomes 8–25 μ in diameter, cells shorter than broad.   (Fig.
              5.38). . . . . *L. aestuarii* (Mertens) Liebmann
121b   Trichomes 1–2 μ in diameter, cells as long as or longer than
              broad . . . . *L. epiphytica* Hieronymus
122a (120)   Trichomes 7–13 μ in diameter, constricted at the cross walls; cells
              as long as or slightly shorter than broad. . . *L. putealis* Montagne
122b   Trichomes 5–10 μ in diameter, constricted at the cross walls; cells
              $\frac{1}{3}$–$\frac{1}{6}$ as long as broad . . . . *L. giuseppei* Drouet
122c   Trichomes 5–10 μ in diameter, not constricted at the cross walls;
              cells $\frac{1}{3}$–$\frac{1}{6}$ as long as broad . . . . *L. patrickiana* Drouet
122d   Trichomes 4–7 μ in diameter, not constricted at the cross walls;
              cells as long as or slightly shorter than broad. . . . .
                         L. taylorii Drouet and Strickland
122e   Trichomes 2–3 μ in diameter, not constricted at the cross walls;
              cells as long as broad . . . . . *L. versicolor* (Wartmann) Gomont
122f   Trichomes 2–3 μ in diameter, constricted at the cross walls; cells
              shorter than broad . . . . *L. diguetii* Gomont
123a (117)   Sheaths red   . . . . *Porphyrosiphon* Kützing   92
123b   Sheaths hyaline . . . . *Symploca* Kützing   124
124a (123)   Trichomes 4–10 μ in diameter, constricted at the cross walls . . . .
                         S. kieneri Drouet
124b   Trichomes 5–8 μ in diameter, not constricted at the cross walls.
              (Fig. 5.39) . . . . *S. muscorum* (Agardh) Gomont
124c   Trichomes 3–4 μ in diameter, not constricted at the cross walls. . .
                         S. muralis Kützing
124d   Trichomes 1–2.5 μ in diameter, plant calcified . . . .
                         S. dubia (Nägeli) Gomont

124e   Trichomes 1–2 μ in diameter, plant not calcified. . . . .
         S. thermalis (Kützing) Rabenhorst

125a **(114)** Sheath material inconspicuous or absent **139**

125b   Trichomes in a gelatinous matrix of coalesced sheaths. . . . .
             Phormidium Kützing **126**

126a **(125)** Ends of trichomes rotund or conical **130**

126b   Ends of trichomes capitate **127**

127a **(126)** Ends of trichomes curved **129**

127b   Ends of trichomes straight **128**

128a **(127)** Cells 4–9 μ in diameter, as long or ½ as long as broad . . . .
         P. favosum (Bory) Gomont

128b   Cells 5–11 μ in diameter, ½–¼ as long as broad . . . .
          P. subfuscum Kützing

129a **(127)** Cells 6–9 μ in diameter, ½–⅓ as long as broad . . . .
         P. uncinatum (Agardh) Gomont

129b   Cells 4–7 μ in diameter, as long or ½ as long as broad. (Fig.
    5.40). . . . . **P. autumnale** (Agardh) Gomont

129c   Cells 4–5 μ in diameter, 1–2 times as long as broad . . . .
         P. setchellianum Gomont

130a **(126)** End cells rotund **136**

130b   End cells blunt-conical **132**

130c   End cells acute-conical **131**

131a **(130)** Cells 2–5 μ in diameter, longer than broad. . . . .
         P. anabaenoides Drouet

131b   Cells 3–7 μ in diameter, up to ⅙ as long as broad . . . .
         P. richardsii Drouet

132a **(130)** Trichomes constricted at the cross walls, 1–2 μ in diameter . . . . .
         P. tenue (Meneghini) Gomont

132b   Trichomes not constricted at the cross walls **133**

133a **(132)** Plant stony, calcified; trichomes 4–5 μ in diameter . . . .
         P. incrustatum (Nägeli) Gomont

133b   Plant gelatinous **134**

134a **(133)** Trichomes 1–2 μ in diameter. . **P. laminosum** (Agardh) Gomont

134b   Trichomes 3–5 μ in diameter **135**

135a **(134)** Cells longer than broad, stratum membranaceous . . . .
         P. inundatum Kützing

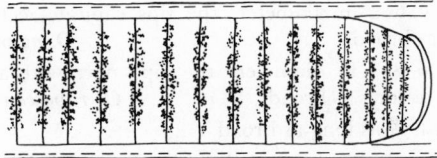

Fig. 5.38. *Lyngbya aestuarii.*

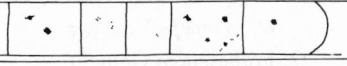
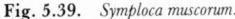

Fig. 5.39. *Symploca muscorum.*

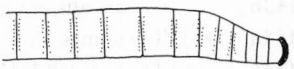

Fig. 5.40. *Phormidium autumnale.*

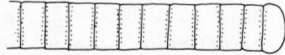 Fig. 5.41. *Oscillatoria tenuis.*

135b   Cells longer than broad, stratum thick and leathery . . . .
                   Phormidium corium (Agardh) Gomont

135c   Cells shorter than broad, stratum thin and tough . . . .
                   P. papyraceum (Agardh) Gomont

136a (130)   Trichomes not constricted at the cross walls **138**

136b   Trichomes constricted at the cross walls **137**

137a (136)   Cells 1–2 μ in diameter, longer than broad; trichomes in gelatin of
                   planktonic algae . . . . **P. mucicola** Naumann and Huber

137b   Cells 1.5–2 μ in diameter, longer than broad . . . .
                   P. luridum (Kützing) Gomont

137c   Cells 2–3 μ in diameter, trichomes moniliform . . . .
                   P. groesbeckianum Drouet

137d   Cells 2–3 μ in diameter, longer than broad . . . .
                   P. molle (Kützing) Gomont

137e   Cells 2–3 μ in diameter, shorter than broad . . . .
                   P. minnesotense (Tilden) Drouet

137f   Cells 4–12 μ in diameter . . . . **P. retzii** (Agardh) Gomont

138a (136)   Cells 0.5–1 μ in diameter, plants of hot springs . . . .
                   P. treleasei Gomont

138b   Cells 2–2.5 μ in diameter, longer than broad . . . .
                   P. valderianum Gomont

138c   Cells 4–6 μ in diameter, up to ¼ as long as broad . . . .
                   P. ambiguum Gomont

139a (125)   Trichomes rigidly spiraled **155**

139b   Trichomes straight or flexuous, not permanently and rigidly
                   spiraled . . . . **Oscillatoria** Vaucher   **140**

140a (139)   Trichomes without pseudovacuoles, not developing as water
                   blooms **142**

140b   Trichomes containing pseudovacuoles, developing as water
                   blooms **141**

141a (140)   Trichomes 2–3 μ in diameter, the tips rotund; cells longer than
                   broad . . . . **O. rileyi** Drouet

141b   Trichomes 2–5 μ in diameter, the tips capitate; cells longer than
                   broad . . . . **O. prolifica** (Greville) Gomont

141c   Trichomes 4–6 μ in diameter, the tips capitate; cells as long or
                   ½ as long as broad . . . . **O. agardhii** Gomont

141d   Trichomes 6–8 μ in diameter, the tips narrowed and truncate;
                   cells ½–⅓ as long as broad . . . . . . **O. rubescens** de Candolle

142a (140)   Cells at least ⅓ as long as broad **148**

142b   Cells at most ⅓ as long as broad; tips of trichomes rounded,
                   truncate, or capitate **143**

143a (142)   Cross walls granulated, at least in part **145**

143b   Cross walls not granulated **144**

144a (143)   Trichomes 16–90 μ in diameter . . . . **O. princeps** Vaucher

144b   Trichomes 12–15 μ in diameter . . . . . **O. proboscidea** Gomont

145a (143)   Trichomes not constricted at the cross walls **147**

145b   Trichomes constricted at the cross walls **146**

146a (145) Trichomes 10–20 μ in diameter, the tips short-attenuate and sub-capitate . . . . ***O. sancta*** (Kützing) Gomont

146b　　　Trichomes 9–11 μ in diameter, the tips slightly attenuate and rounded . . . . ***O. ornata*** (Kützing) Gomont

147a (145) Trichomes 11–20 μ in diameter, the tips straight. . . . .
O. limosa (Roth) Agardh

147b　　　Trichomes 10–17 μ in diameter, the tips curved . . . .
O. curviceps Agardh

147c　　　Trichomes 6–8 μ in diameter, the tips curved and capitate . . . . .
O. anguina (Bory) Gomont

148a (142) Tips of trichomes attenuate, not capitate . . . .　151

148b　　　Tips of trichomes attenuate and capitate . . . .　150

148c　　　Tips of trichomes neither attenuate nor capitate . . . .　149

149a (148) Trichomes 6–11 μ in diameter, granulated but not constricted at the cross walls, the tips straight . . . ***O. irrigua*** (Kützing) Gomont

149b　　　Trichomes 8 μ in diameter, not granulated or constricted at the cross walls, the tips straight . . . . ***O. simplicissima*** Gomont

149c　　　Trichomes 4–10 μ in diameter, granulated and somewhat constricted at the cross walls, the tips curved. (Fig. 5.41) . . . .
O. tenuis Agardh

149d　　　Trichomes 2–4 μ in diameter, constricted at the cross walls, the tips curved . . . . ***O. articulata*** Gardner

149e　　　Trichomes 3–4 μ in diameter, not constricted at the cross walls, the tips straight . . . . ***O. chlorina*** (Kützing) Gomont

149f　　　Trichomes 2–4 μ in diameter, slightly constricted at the cross walls, the tips straight or curved . . . .
O. geminata (Meneghini) Gomont

150a (148) Cells 2–3 μ in diameter, several times longer than broad . . . . . .
O. splendida Greville

150b　　　Cells 2.5–5 μ in diameter, as long as or somewhat longer than broad . . . . ***O. amoena*** (Kützing) Gomont

151a (148) End cells truncate-conical . . . .　154

151b　　　End cells blunt- or rounded-conical. . . . .　153

151c　　　End cells acuminate- or acute-conical . . . .　152

152a (151) Cells 3–5 μ in diameter, longer than broad. . . . .
O. acuminata Gomont

152b　　　Cells 3–4 μ in diameter, shorter than broad . . . .
O. animalis Agardh

152c　　　Cells 4–6 μ in diameter, always shorter than broad . . . .
O. brevis (Kützing) Gomont

152d　　　Cells 6–8 μ in diameter, as long as or shorter than broad . . . . . .
O. boryana (Agardh) Bory

153a (151) Cells 4–6 μ in diameter, as long as or shorter than broad . . . . . .
O. formosa Bory

153b　　　Cells 5.5–8 μ in diameter, as long as or longer than broad . . . . .
O. cortiana (Meneghini) Gomont

153c　　　Cells 8–13 μ in diameter, shorter than broad. . . . .
O. chalybea Mertens

153d　　　Cells 6.5–8 μ in diameter, shorter than broad . . . .
O. okenii Agardh

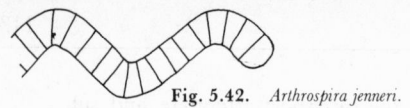
Fig. 5.42. *Arthrospira jenneri.*

Fig. 5.43. *Spirulina major.*

154a (151) Cells 3–5 μ in diameter, as long as or longer than broad . . . . . .
            Oscillatoria granulata Gardner
154b   Cells 4–6.5 μ in diameter, shorter than broad . . . .
          O. terebriformis (Agardh) Gomont
155a (139) Trichomes without evident cross walls. . . . . . **Spirulina** Turpin **158**
155b   Trichomes with conspicuous cross walls. . . . .
          Arthrospira Stizenberger **156**
156a (155) Cells blue-green, 4–8 μ in diameter, without pseudovacuoles; plants not developing as water blooms. (Fig. 5.42) . . . .
         A. jenneri (Kützing) Stizenberger
156b   Cells containing pseudovacuoles, plants developing as water blooms **157**
157a (156) Trichomes 2.5–3 μ in diameter, cells longer than broad . . . .
            A. gomontiana Setchell
157b   Trichomes 3–5 μ in diameter, cells shorter than broad . . . . . .
        A. khannae Drouet and Strickland
157c   Trichomes 6–8 μ in diameter, cells shorter than broad . . . . . .
         A. platensis (Nordstedt) Gomont
158a (155) Turns of the spiral touching each other. **160**
158b   Turns of the spiral far apart. **159**
159a (158) Trichomes 0.5–1 μ in diameter, spiral 1.5–2.5 μ in diameter, 1.2–2 μ between turns. . . . . **Spirulina subtilissima** Kützing
159b   Trichomes 1 μ in diameter, spiral 1.5 μ in diameter, 3–4 μ between turns. . . . . **S. caldaria** Tilden
159c   Trichomes 1–2 μ in diameter, spiral 2.5–5 μ in diameter, 2.5–5 μ between turns. (Fig. 5.43) . . . . **S. major** Kützing
159d   Trichomes 3–5 μ in diameter, spiral 8–12 μ in diameter, 8–12 μ between turns. . . . . **S. princeps** W. and G. S. West
160a (158) Trichomes 1–2 μ in diameter, spiral 3–5 μ in diameter . . . . . .
            S. subsalsa Oersted
160b   Trichomes 1 μ in diameter, spiral 2–3 μ in diameter. . . . .
       S. labyrinthiformis (Meneghini) Gomont

References

Bornet, E., and C. Flahault. **1886–1888.** Révision des Nostocacées hétérocystées. *Ann. Sci. Nat.* VII. *Botan.*, 3:323–381, 4:343–373, 5:51–129, 7:177–262. **Drouet, F. 1951.** Cyanophyta. In: G. M. Smith. *Manual of Phycology*, pp. 159–166. Chronica Botanica, Waltham, Massachusetts. **Drouet, F., and W. A. Daily. 1956.** Revision of the coccoid Myxophyceae. *Butler Univ. Botan. Studies*, 12:1–218. **Forti, A. 1907.** *Sylloge Myxophycearum.* Vol. 5 in: J. B. de Toni. *Sylloge Algarum.* **Geitler, L. 1930–1932.** *Cyanophyceae.* Vol. 14 in: L. Rabenhorst (ed.). *Kryptogamen-Flora von Deutschland, Österreich und der Schweiz.* Akademische Verlagsgesellschaft m. b. H., Leipzig. **1942.** *Schizophyta: Klasse Schizophyceae.* Vol. 1b in: A. Engler and K. Prantl. *Die natürlichen Pflanzenfamilien*, 2nd ed. Duncker and Humblot, Berlin. **Gomont, M. 1892.** Monographie des Oscillariées (Nostocacées homocystées). *Ann. Sci. Nat.* VII. *Botan.*, 15:263–368, 16:91–264. **Tilden, J. E. 1910.** *Minnesota Algae.* Vol. I, *The Myxophyceae of North America.* Minneapolis.

6

Algae[1]

R. H. THOMPSON

The algae are an assemblage of Protista containing unicellular to multi-cellular motile or immobile forms, forms having a certain amount of differentiation and specialization of parts. None of them, however, possesses a multicellular sex organ with a sterile jacket comparable to the archegonium or antheridium of the bryophytes. This one negative feature alone distinguishes lower plants classed as Algae from those classed as Bryophyta. The possession of photosynthetic pigments by all but a very few forms of algae distinguishes them from the Fungi. The problem of general classification is discussed in Chapter 2. Many of the genera included in the Order Mastigophora of protozoological classifications will be found in this key; see also chapter 8.

Most species of algae are widely distributed in North America and do not show strong geographical limitation. The occurrence of particular species is determined more by ecological conditions than by geographical location. Detached plants or fragments travel over and populate the large watersheds of continents. On the feathers and scaly legs of water fowl they move between pond, lake, and river, or traverse a continent during migratory flights. Car-

[1]Exclusive of the Myxophyceae and Bacillariophyceae.

ried by the wind in the dust from intermittant puddles and ponds, their spores leap geographic and climatic barriers that hold in check many of the higher plants. Thus, one may find some kind of alga present at any place provided with suitable moisture and minerals. Only a few algae are apparently restricted to such special habitats as hot springs, salterns, and alpine snowfields.

Collection and Identification

The collecting of algae is simple and requires a minimum of equipment: vials, jars, and newspapers. The only specialized, and invaluable, piece of equipment is a plankton net, which should be at least as fine as 24 mesh.

Portions of attached or floating algal mats and aquatic seed plants furnish the larger forms as well as the small epiphytic forms. Concentrated collections can be made of desmids, diatoms, and many planktonic forms, both motile and free-floating, by squeezing such algal mats or aquatic seed plants. A large handful of the material is held over a wide-mouthed jar and squeezed thoroughly until nothing more drips into the jar. A plankton net is indispensable for obtaining a concentrated sample of the organisms dispersed and floating in open waters.

Examination of a collection should be made as soon as possible, since even with special care in keeping the collection cool, many of the motile forms die and disintegrate. This is particularly true of the Chrysophyceae. Algae also deteriorate rapidly from lack of aeration, so the containers should not be filled more than half full and, if possible, they should be transported unstoppered.

At several points in the key, reference is made to the typical color of the organisms, and a dichotomy is made on that basis. Unfortunately, the color of some organisms may vary somewhat with the conditions in which they have grown. Some species have colors unusual in the group, and many of the colors fade in preserved material. If there is doubt about the color, it is necessary to run the organisms down two branches of the key. If an obvious misidentification is made, one of the first points to check is the color. For the most part, in living material (except in a very few instances relative to the bulk of algal forms) the color differences between algal groups are quite definite. In one season a person can easily learn to appreciate these differences, even when slight, if he periodically makes a collection and attempts to identify everything in it. In this way a feeling is gained not only for color differences but also for chromatophore morphology, which distinguishes the different algal groups.

If possible, material should be identified in the living condition, though this cannot always be done. However, identification of unicellular, motile forms can be done only from living material.

The identification of preserved material should ordinarily be attempted only after one is thoroughly familiar with the different algal groups, with their chromatophore morphology, and with their appearance after preservation. Even then identification of many forms cannot be carried to species. But, with the above background, one is in a position to make an educated guess about characters in the living condition, just as one usually has to do for type of reproduction where that is the only criterion of separation (e.g., *Chlorella* vs. *Chlorococcum*).

Preservation

There are several solutions for the preservation of collections to be found in books on microtechnique (see Chapter 46). The simplest method is to use commercial formalin which varies from 35 to 40 per cent. Enough formalin is added to the jar or vial containing the algae and water to make a one-tenth dilution of the formalin, to 3.5 or 4.0 per cent. The vials or jars should have been marked previously for the amount of commercial formalin to be added. As low as 2.5 per cent formalin gives good preservation with little if any shrinkage. For those who wish to fix and preserve algal material for future staining and permanent mounting, a good universal fixing and preserving solution is FPA: commercial formalin 5 parts, propionic acid 5 parts, 50 per cent ethyl alcohol 90 parts. Glacial acetic acid may be substituted if propionic is unavailable. Preservation in liquid has the disadvantage of slow evaporation and possible irretrievable loss of the material by drying out in the vial. Many algae may also be preserved as dried herbarium specimens. The larger and filamentous forms may be dried on herbarium paper, and smaller forms on thin sheets of mica. Air drying without heat gives the best results.

There is considerable literature on the culture of algae for life history or physiological study. Harold C. Bold's article gives an excellent review and bibliography on this subject (see References).

Classification

The classification used here follows those phycologists who recognize divisional (phylum) rank for the major groups within the Algae. The divisions and classes represented in fresh water are as follows:

Division Euglenophyta, euglenoids
Division Chlorophyta, grass-green algae
  Class Chlorophyceae
  Class Charophyceae, stoneworts
Division Chrysophyta
  Class Xanthophyceae, yellow-green algae
  Class Chrysophyceae, golden or yellow-brown algae
  Class Bacillariophyceae, diatoms (Chapter 7)
Division Phaeophyta, brown algae
Division Pyrrophyta
  Class Desmokontae
  Class Dinophyceae, dinoflagellates
Division Rhodophyta, red algae
Division Cyanophyta, (Myxophyceae) blue-green algae (Chapter 5)

In addition there are two small groups, the *Chloromonadophyceae* and the *Cryptophyceae*, which are poorly known and whose systematic position is problematical. Here the treatment by G. M. Smith is followed, and they are listed as groups of uncertain position. The sequence in which the different divisions are presented is entirely one of convenience and economy, so it should not be construed as showing interrelationship or having any evolutionary significance.

Certain genera form colonies with a gelatinous matrix, and the following list is given as an aid to identification:

Chlorophyceae: *Sphaerocystis, Gloeocystis, Asterococcus, Palmella, Coccomyxa,* (Palmella stages). Xanthophyceae: *Gloeochloris, Gloeobotrys, Botryococcus.* Chrysophyceae: *Chrysocapsa, Phaeosphaera,* (Palmella stages). Euglenophyceae: (Palmella stages). Cryptophyceae: (Palmella stages).

The following key to the genera of fresh-water algae includes lines leading to the *Bacillariophyceae* and to the *Myxophyceae* which are presented in separate chapters. Unless otherwise indicated the illustrations are original, by the author.

KEY TO GENERA

1a		Cells with pigments; motile or immobile.	**6**
1b		Cells colorless; all motile	**2**
		See also Chapter 8.	
2a	(1)	With a median groove (i.e., girdle and sulcus). (Figs. 6.373, 6.380)	**370**
2b		Without a median groove	**3**
3a	(2)	Cells with pharyngeal rods or containing paramylum	**113**
3b		Cells without such rods; starch or leucosin present	**4**

It is not necessary to determine the nature of the stored food to decide this couplet. If the cells contain paramylum, they are of obvious euglenoid morphology (**113**). If they lack pharyngeal rods, they are colorless Chlorophyceae, Chrysophyceae, or Cryptophyceae and, regardless of stored food, are separated in later couplets.

4a	(3)	Equally 2- or 4-flagellate; starch present (iodine is a test for starch)	**14**
4b		Unequally biflagellate. (Fig. 6.412).	**5**
5a	(4)	Cells with a deep gullet. (Figs. 6.414, 6.415).	**403**
5b		Cells without a deep gullet (monads) . . . . (Chapter 8)	
6a	(1)	Pigments contained in chromatophores	**7**
6b		Pigments not contained in chromatophores. Blue-green, olive, red-violet, violet . . . . (Chapter 5) **Myxophyceae**	
7a	(6)	Chromatophores grass green or yellow-green	**8**
7b		Chromatophores blue-green, olive, shades of brown, or red.	**91**

Diatoms may give trouble here; they can be recognized on the basis of morphology. See Chapter 7.

8a	(7)	Chromatophores grass green; starch or paramylum present	**9**
8b		Chromatophores yellow-green; no starch or paramylum present . .	**86**
9a	(8)	Unicellular or colonial (not filamentous).	**10**
9b		Filamentous or parenchymatous-appearing. (Figs. 6.274, 6.275) .	**62**
10a	(9)	Flagellated and motile	**11**

10b		Not flagellated, mostly immobile.	**16**
11a	**(10)**	Unicellular	**12**
11b		Colonial	**15**
12a	**(11)**	Cells containing starch; 1-, 2-, 4-, or 8-flagellate	**13**
12b		Cells containing paramylum; 1-, 2-, or 3-flagellate	**104**
13a	**(12)**	Cells with a lorica. (Fig. 6.54, 6.346)	**146**
13b		Cells without a lorica	**14**
14a	**(4, 13)**	Cells without a cell wall; periplast firm. (Fig. 6.39)	**130**
14b		Cells with a cell wall. (Fig. 6.45)	**135**
15a	**(11)**	Colony with a gelatinous matrix. (Fig. 6.61)	**153**
15b		Colony without a gelatinous matrix	**160**
16a	**(10)**	Unicellular	**17**
16b		Colonial	**40**
17a	**(16)**	Cells with a median constriction, often in pairs	**234**
17b		Cells without a median constriction	**18**
18a	**(17)**	Aquatic	**19**
18b		Terrestrial; vesicular (Fig. 6.128), or siphonaceous (Fig. 6.129) and branched	**200**
19a	**(18)**	Endophytic or endozoic.	**20**
19b		Free-living; attached or free-floating.	**21**
20a	**(19)**	Endophytic in aquatic plants.	**190**
20b		Endozoic	**186**
21a	**(19)**	Free-floating and microscopic	**22**
21b		Attached; sessile or stalked. (Fig. 6.111)	**37**
22a	**(21)**	Cells spheric, subspheric, or subcylindric	**23**
22b		Cells pyriform, cylindric or variously shaped	**28**
23a	**(22)**	With a thick, gelatinous sheath.	**165**
23b		Without such a sheath	**24**
24a	**(23)**	Cells with few to many needlelike setae. (Fig. 6.109)	**191**
24b		Cells without such setae.	**25**
25a	**(24)**	Cells minute, smooth; subspheric or subcylindric	**179**
25b		Cells larger; globose; smooth, warty, sculptured, or with appendages.	**26**
26a	**(25)**	Wall smooth, sculptured, or appearing loculate or beaded	**27**
26b		Wall bearing few to many stout spinelike appendages	**204**
27a	**(26)**	Wall smooth or with a very few buttonlike thickenings.	**188**
27b		Wall sculptured or appearing loculate or beaded	**203**
28a	**(22)**	Cells setose with few to many setae. (Fig. 6.138)	**205**
28b		Without setae; poles may be prolonged as spines	**29**
29a	**(28)**	Cells pyriform, ellipsoid, or cylindric	**30**
29b		Spindle-shaped, acicular, lunate; or, triangular, pyramidal, or irregular	**33**
30a	**(29)**	Pyriform with thickened knoblike places in the wall	**31**
30b		Ellipsoid or cylindric	**32**
31a	**(30)**	Chloroplast single, axial, massive cup-shaped, or with numerous processes.	**189**
31b		Chloroplasts more than 1, parietal; thickened parts stratified	**201**

32a (30) Ellipsoid to subcylindric or panduriform **207**

32b   Fusiform, oblong-cylindric, or long-cylindric **228**

33a (29) Spindle-shaped, acicular, or lunate **34**

33b   Pyramidal, triangular, or irregular. **211**

34a (33) Spindle-shaped; arcuate to sigmoid **35**

34b   Acicular or strongly lunate **36**

35a (34) With a costate, biturbinate envelope (Fig. 6.99); short- or long-pointed **187**

35b   Without such an envelope **180**

36a (34) Cells acicular or linear **209**

36b   Cells lunate **210**

37a (21) Sessile, cells with 1 or more setae. (Fig. 6.269) **287**

37b   Stalked **38**

38a (37) With paramylum; stalk massive **105**

38b   With starch; stalk massive or slender. **39**

39a (38) With a tough gelatinous sheath and stalk **163**

39b   Cell wall of cellulose; stalk ending in a discoid holdfast **193**

40a (16) Colonies branched and epizoic. **41**

40b   Colonies not epizoic; simple filaments or radiate **42**

41a (40) Cells containing paramylum **105**

41b   Cells containing starch **164**

42a (40) Cells with a median constriction. **249**

42b   Cells without a median constriction **43**

43a (42) Microscopic, sessile, few-celled clusters; cells bearing hairlike setae. (Fig. 6.266) **287**

43b   Microscopic or macroscopic colonies; those bearing setae planktonic. **44**

44a (43) Colonies macroscopic and gelatinous, or microscopic with a gelatinous matrix **45**

44b   Colonies microscopic without a wide matrix, or else colony a flat plate. **52**

45a (44) Colonies amorphous, saccate, netlike or branched **169**

45b   Colonies microscopic and more or less spheric. **46**

46a (45) Cells spheric or subspheric, remote or in close groups. **47**

46b   Cells not spheric, or if so, attached by threads **48**

47a (46) Sheaths of individual cells evident **165**

47b   Sheaths not clearly evident or else lacking **167**

48a (46) Cells connected by threads or free; ellipsoid to cylindric **49**

48b   Cells fusiform, spindle-shaped, lunate, or irregularly-curved cylinders. **50**

49a (48) Cells connected by radiating threads; spheric to subcylindric . . . **196**

49b   Cells not connected by threads. **181**

50a (48) Cells fusiform or spindle-shaped to pointed-cylindric **183**

50b   Cells ovate-attenuate, lunate, or irregularly-curved cylinders. . . . **51**

51a (50) Cells ovate-attenuate **216**

51b   Cells lunate, sausage-shaped, or curved cylinders **217**

52a (44) Cells spindle-shaped; joined end to end in branching colonies . . . **215**

52b   Cells variously shaped; not forming branching colonies. **53**

53a	(52)	Colony a flat plate or saclike net	54
53b		Colony not a flat plate or net	57
54a	(53)	Colony a flat plate of 2 to many cells	55
54b		Colony a net, or netlike	198
55a	(54)	With a gelatinous matrix; cells in groups of 4	184
55b		Without a matrix or matrix not evident	56
56a	(55)	Colony cruciate or linear and usually 4-celled	223
56b		Colony of 2 opposed or of 16 to 64 radiately united cells	197
57a	(53)	Cells enclosed by the expanded mother cell wall	218
57b		Cells not so enclosed	58
58a	(57)	Cells with 1 to several setae; spheric or ellipsoid	192
58b		Cells not bearing setae; various shaped	59
59a	(58)	Many-celled compound colonies united by empty cell walls	194
59b		Cells of many-celled colonies not so united	60
60a	(59)	Cells lunate; in groups of 4 to 16 with convex sides opposed	215
60b		Cells not lunate; radiately united or forming a hollow sphere	61
61a	(60)	Cells radiately united; lanceolate to cordate in shape	199
61b		Cells spheric or ovate; united as a hollow sphere	196
62a	(9)	Filamentous; simple or branched	63
62b		Parenchymatous or pseudoparenchymatous	81
63a	(62)	Alga a simple filament of cells	64
63b		Alga branched, sparingly or profusely	73
64a	(63)	Cells with a shallow or a distinct median constriction	249
64b		Cells without a median constriction	65
65a	(64)	Chloroplast a parietal laminate band	66
65b		Chloroplast not a laminate band; parietal or axial	67
66a	(65)	Filaments with a gelatinous sheath	265
66b		Filaments without a gelatinous sheath	266
67a	(65)	Chloroplast parietal	68
67b		Chloroplast axial or massive and filling the cell	71
68a	(67)	Chloroplast a reticulate or perforate sheet	69
68b		Chloroplasts 1 or more; not reticulate	70
69a	(68)	Certain cells with apical caps, appearing cross-striated	294
69b		Without such cells	271
70a	(68)	Chloroplasts ringlike bands or appearing so	289
70b		Chloroplasts not ringlike	259
71a	(67)	Chloroplasts stellate or massive	72
71b		Chloroplast platelike; extending the cell length	263
72a	(71)	With 1 stellate or massive chloroplast per cell	272
72b		With 2 stellate or cushionlike chloroplasts per cell	261
73a	(63)	Without cross walls but constricted at each dichotomy	200
73b		Multicellular	74
74a	(73)	Branching, whorled (Fig. 6.286); plants macroscopic	299
74b		Branching, not whorled; macroscopic or microscopic	75
75a	(74)	Chloroplast laminate; extending completely or incompletely to cell ends	77
75b		Chloroplast a reticulate sheet extending full length of cell	76

76a   (75) Cells bearing a turbinate cap or a bulbous-based seta   295
76b   Cells not bearing caps or setae   296
77a   (75) Cells bearing setae   78
77b   Cells not bearing setae; branches may be attenuated   80
78a   (77) Setae are cytoplasmic hairs with a basal sheath   286
78b   Setae are cellular in nature   79
79a   (78) Seta very long, hollow, continuous with cell bearing it   273
79b   Seta cut off by a cross wall from cell bearing it 282
80a   (77) Endophytic, encrusting or perforating substratum, or ter-
             restrial 274
80b   Free-living; all aquatic 277
81a   (62) Alga consisting of packets of few cells   185
81b   Alga macroscopic or microscopic; consisting of many cells   82
82a   (81) Thallus cylindrical   290
82b   Thallus endophytic or crustose, or an expanded sheet   83
83a   (82) Endophytic in the walls of other algae   276
83b   Free-living, epiphytic, or epizoic   84
84a   (83) Cells bearing hairlike setae   178
84b   Cells not bearing setae   85
85a   (84) Alga a macroscopic expanded sheet of cells   291
85b   Alga crustose; epiphytic or epizoic   284
86a   (8) Cells flagellated and motile; with trichocysts (Fig. 6.403) and
             chromatophores 393
86b   Cells not flagellated and immobile   87
87a   (86) Unicellular or colonial   88
87b   Filamentous or vesicular. (Fig. 6.322)   327
88a   (87) Colonial   89
88b   Unicellular 301
89a   (88) Colonies gelatinous, attached or free   90
89b   Colonies branched and attached   302
90a   (89) Matrix cartilaginous; cells peripheral and in 2's or 4's 331
90b   Matrix gelatinous; cells scattered throughout   303
91a   (7) Chromatophores blue-green, olive, or red to red-violet   97
91b   Chromatophores golden yellow, or olive brown to dark brown   92
92a   (91) Golden yellow to pale olive brown   93
92b   Dark olive brown to dark brown   94
93a   (92) With striated siliceous walls; unicellular or colonial . . . .
                                        (Chapter 7) **Bacillariophyceae**
93b   Without striated siliceous walls   332
94a   (92) Filamentous or parenchymatous (Fig. 6.422) and macro-
             scopic 406
94b   Unicellular   95
95a   (94) Motile   96
95b   Immobile; epiphytic or epizoic on fish   386
96a   (95) Shades of brown; with girdle and sulcus or bivalved theca   369
96b   Greenish to olive brown; without a girdle and sulcus   394
97a   (91) Pigments blue-green or olive   98

97b   Pigments red to red-violet; filamentous or unicellular **405**

98a   (97)   Chromatophores blue-green **99**

98b   Chromatophores olive **101**

99a   (98)   Unicellular and motile **100**

99b   Filamentous; chromatophores asteroid or discoidal **405**

100a   (99)   With a girdle and sulcus **369**

100b   Without a girdle and sulcus **397**

101a   (98)   Unicellular **102**

101b   Filamentous, simple or branched **406**

102a (101)   Motile **103**

102b   Immobile and tetrahedral **394**

103a (102)   With a girdle and sulcus, or with a bivalved theca **369**

103b   Without such structures; with or without a deep gullet. **394**

Division **Euglenophyta** (Mastigophora)

104a   (12)   Cells pigmented **105**

104b   Cells colorless **113**

105a (38, 41, 104)   Flagellated; free-swimming or sedentary **106**

Order **Colaciales**, Family **Colaciaceae**

105b   Encapsulated; solitary or arborescent; epizoic. (Fig. 6.1). . . . . .

                                              Colacium Ehrenberg

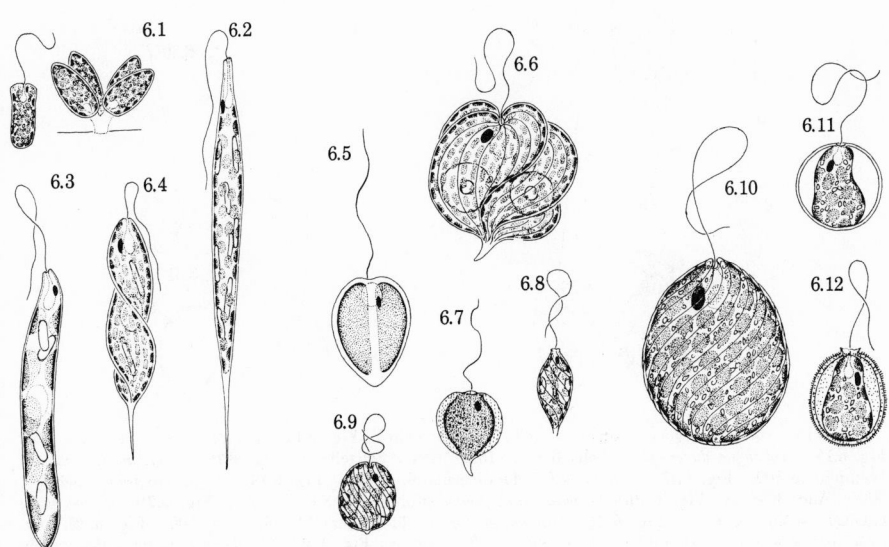

Fig. **6.1.** *Colacium vesiculosum* Ehrenberg. × 390. Fig. **6.2.** *Euglena acus* Ehrenberg. × 500. Fig. **6.3.** *E. deses* Ehrenberg. × 500. Fig. **6.4.** *E. tripteris* (Dujardin) Klebs. × 500. Fig. **6.5.** *Cryptoglena pigra* Ehrenberg. × 1000. Fig. **6.6.** *Phacus quinquemarginatus* Jahn and Shawhan. × 500. Fig. **6.7.** *P. suecica* Lemmermann. × 780. Fig. **6.8.** *Lepocynclis acicularis* France. × 500. Fig. **6.9.** *L. ovum* (Ehrenberg) Lemmermann. × 500. Fig. **6.10.** *L. texta* (Dujardin) Lemmermann. × 500. Fig. **6.11.** *Trachelomonas volvocina* Ehrenberg. × 500. Fig. **6.12.** *T. hispida* (Perty) Stein. × 500.

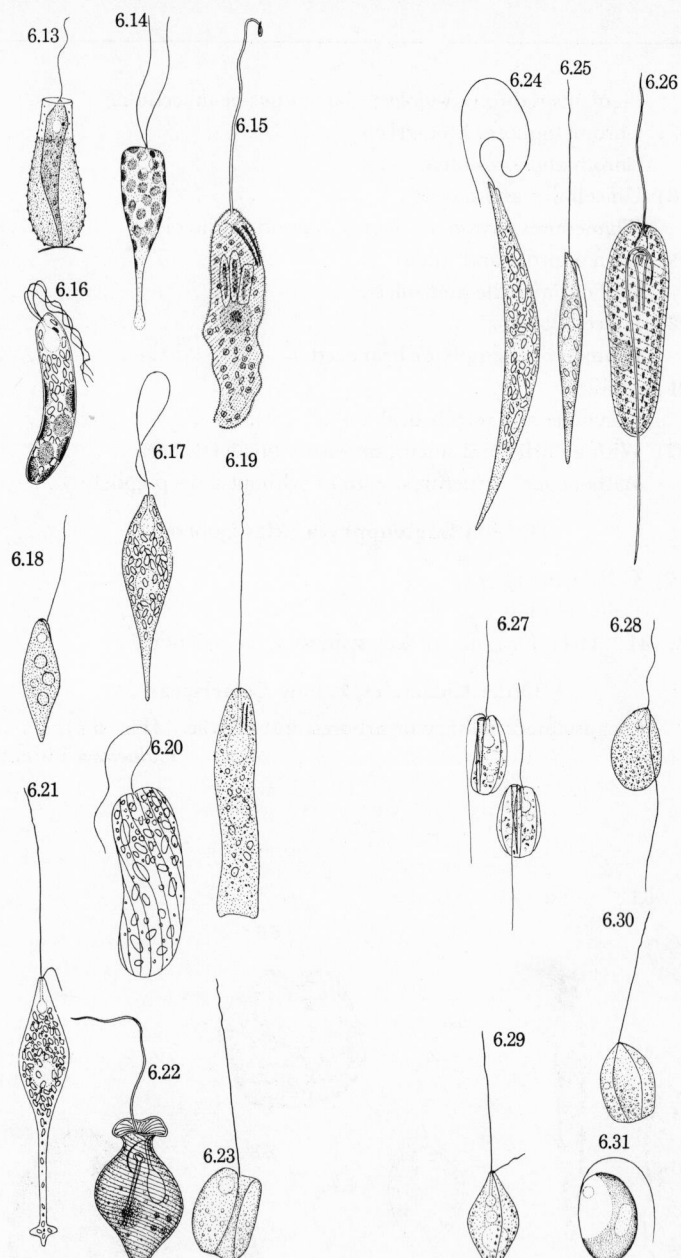

Fig. 6.13. *Ascoglena vaginicola* Stein. × 500. (After Stein.) **Fig. 6.14.** *Eutreptia viridis* Perty. × 390. **Fig. 6.15.** *Jenningsia diatomophaga* Schaeffer. × 162. (After Schaeffer.) **Fig. 6.16.** *Euglenamorpha hegneri* Wenrich. × 500. **Fig. 6.17.** *Astasia klebsii* Lemmermann. × 500. **Fig. 6.18.** *Euglenopsis vorax* Klebs. × 500. (After Klebs.) **Fig. 6.19.** *Peranema trichophorum* (Ehrenberg) Stein. × 585. **Fig. 6.20.** *Rhabdomonas incurva* Fresenius. × 1000. **Fig. 6.21.** *Distigma proteus* (O. F. Müller) Ehrenberg. × 500. **Fig. 6.22.** *Urceolus cyclostomus* (Stein) Mereshkowsky. × 665. (After Senn.) **Fig. 6.23.** *Petalomonas mediocanellata* Stein. × 1170.

Fig. 6.24. *Menoidium falcatum* Zacharias. × 375. **Fig. 6.25.** *Heteronema acus* (Ehrenberg) Stein. × 450. **Fig. 6.26.** *Dinema griseolum* Perty. × 330. (After Lemmermann.) **Fig. 6.27.** *Entosiphon sulcatum* (Dujardin) Stein. × 390. **Fig. 6.28.** *Anisonema ovale* Klebs. × 1000. **Fig. 6.29.** *Sphenomonas quadrangularis* Stein. × 325. (After Stein.) **Fig. 6.30.** *Notosolenus apocamptus* Stokes. × 1000. (After Stokes.) **Fig. 6.31.** *Pedinomonas minor* Korsch. × 4000.

Order **Euglenales**, Family **Euglenaceae**

106a (105) Uniflagellate 107
106b   Bi- or triflagellate. 112
107a (106) With a lorica. (Fig. 6.13) 111
107b   Without a lorica 108
108a (107) Cells slightly to very plastic. (Figs. 6.2, 6.3, 6.4) . . . .
                                                              Euglena Ehrenberg
108b   Cells rigid 109
109a (108) Cells with 2 elongate, platelike chloroplasts. (Fig. 6.5) . . . . . .
                                                           Cryptoglena Ehrenberg
109b   Cells with many small chloroplasts 110
110a (109) Cells strongly compressed. (Figs. 6.6, 6.7) . . . *Phacus* Dujardin
110b   Cells not at all compressed. (Figs. 6.8, 6.9, 6.10) . . . .
                                                            Lepocynclis Perty
111a (107) Free-swimming. (Figs. 6.11, 6.12) . . *Trachelomonas* Ehrenberg
111b   Sedentary and attached. (Fig. 6.13) . . . . *Ascoglena* Stein
112a (106) Cells with 2 equal flagella. (Fig. 6.14) . . . . . . *Eutreptia* Perty
112b   Cells with 3 equal flagella. (Fig. 6.16) *Euglenamorpha* Wenrich
113a (3, 104) Cells triflagellate (see line 112) . . . . *Euglenamorpha* Wenrich
113b   Cells uni- or biflagellate 114
114a (113) Cells uniflagellate 115
114b   Cells biflagellate 123
115a (114) With a pigmented eyespot (see line 108) . . . *Euglena* Ehrenberg
115b   Without an eyespot 116
116a (115) Cells strongly or moderately plastic 117
116b   Cells rigid 121
117a (116) Without pharyngeal rods 118
117b   With pharyngeal rods. (Fig. 6.22) 119
118a (117) Periplast faintly striate; gullet terminal. (Fig. 6.17) . . . .
                                                               Astasia Ehrenberg
118b   Periplast clearly striate; gullet excentric. (Fig. 6.18) . . . .
                                                            Euglenopsis Klebs
119a (117) Cells urceolate or flask-shaped. (Fig. 6.22) . . . .
                                                       Urceolus Mereschkowsky
119b   Cells somewhat cylindric 120
120a (119) Feeding on diatoms only. (Fig. 6.15) . . . *Jenningsia* Schaeffer
120b   Not feeding on diatoms only. (Fig. 6.19) . *Peranema* Dujardin
121a (116) Cells compressed 122
121b   Cells cylindric, spirally ridged. (Fig. 6.20) . . . .
                                                          Rhabdomonas Fresenius
122a (121) Without pharyngeal rods; lunate; triangular in section. (Fig.
            6.24) . . . . *Menoidium* Perty
122b   With pharyngeal rods; ovoid-deltoid-asymmetric. (Fig. 6.23) . . .
                                                          Petalomonas Stein
123a (114) Without pharyngeal rods; trailing flagellum very short 124
123b   With pharyngeal rods (Fig. 4.22); trailing flagellum long or
            short 126
124a (123) Spindle-shaped but continually changing shape. (Fig. 6.21) . . . .
                                                             Distigma Ehrenberg
124b   Spindle-shaped to campanulate with weak costae 125
125a (124) Spindle-shaped; 1 to 4 straight costae; central gelatinous sphere.
            (Fig. 6.29) . . . . *Sphenomonas* Stein
125b   Ovoid to campanulate; refractive bodies posteriorly. (Fig.
            6.30) . . . . *Notosolenus* Stokes

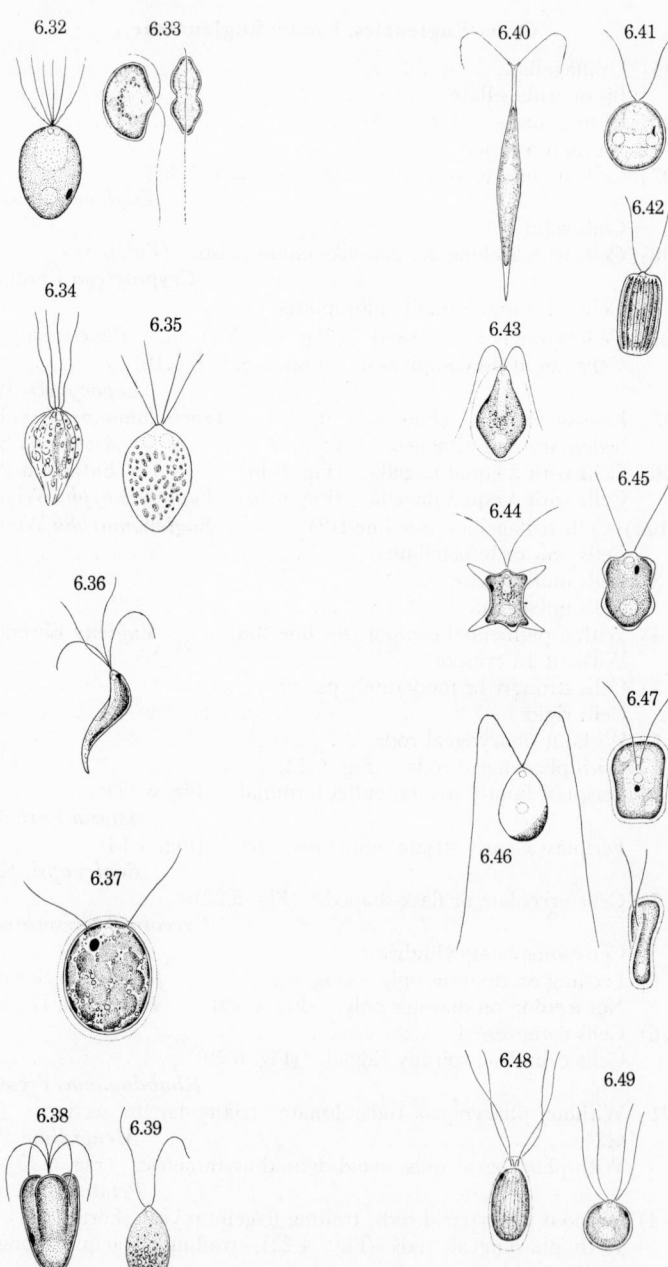

Fig. 6.32. *Polyblepharides singularis* Dangeard. × 1000. **Fig. 6.33.** *Heteromastix angulata* Korsch. × 780. **Fig. 6.34.** *Polytomella citri* Kater. × 390. **Fig. 6.35.** *P. agilis* Aragao. × 1000. (After Aragao.) **Fig. 6.36.** *Spermatozoopsis exultans* Korsh. × 780. **Fig. 6.37.** *Gloeomonas ovalis* Klebs. × 500. **Fig. 6.38.** *Pyramidomonas tetrarhynchus* Schmarda. × 780. **Fig. 6.39.** *Polytoma uvella* Ehrenberg. × 780.

Fig. 6.40. *Chlorogonium elongatum* Dangeard. × 295. **Fig. 6.41.** *Chiamydomonas cingulata* Pascher. × 390. **Fig. 6.42.** *C. penium* Pascher. × 390. **Fig. 6.43.** *Sphaerellopsis fluviatilis* (Stein) Pascher. × 390. **Fig. 6.44.** *Lobomonas pentagona* Hazen. × 520. **Fig. 6.45.** *L. rostrata* Hazen. × 1560. **Fig. 6.46.** *Platychloris minima* Pascher. × 2000. (After Pascher.) **Fig. 6.47.** *Mesostigma viridis* Lauterborn. × 780. **Fig. 6.48.** *Carteria crucifera* Korsh. × 390. **Fig. 6.49.** *C. globulosa* Pascher. × 390.

126a (123) Trailing flagellum the shorter; cells very plastic **127**
126b   Trailing flagellum the longer. **128**
127a (126) Cylindric, striae delicate. (Fig. 6.19) . . . . *Peranema* Dujardin
127b   Cylindric to spindle-shaped, smooth or with elevated spiraled
           ridges. (Fig. 6.25) . . . . *Heteronema* Dujardin
128a (126) Cells markedly compressed; oval, rigid, or plastic. (Fig. 6.28). . .
           Anisonema Dujardin
128b   Cells ellipsoid or ovoid; not compressed; rigid **129**
129a (128) Ellipsoid; periplast tough and spirally striate. (Fig. 6.26) . . . . .
           Dinema Perty
129b   Ovoid; with a cone-shaped pharyngeal apparatus. (Fig. 6.27) . .
           Entosiphon Stein

Division **Chlorophyta**, Class **Chlorophyceae**,
Order **Volvocales**, Family **Polyblepharidaceae**

130a (14) Uni- or biflagellate **131**
130b   With 4 to 8 flagella **132**
131a (130) Uniflagellate; flagellum curved back. (Fig. 6.31). . . . .
           Pedinomonas Korshikov
131b   Biflagellate; with a long swimming and short, curved trailing
           flagellum. (Fig. 6.33) . . . . *Heteromastix* Korshikov
132a (130) With 4 flagella **133**
132b   With 6 to 8 flagella. (Fig. 6.32) . . . *Polyblepharides* Dangeard
133a (132) Cells colorless. (Figs. 6.34, 6.35) . . . . . . *Polytomella* Aragao
133b   Cells pigmented. **134**
134a (133) Cells spindle-shaped and spiraled. (Fig. 6.36) . . . .
           Spermatozoopsis Korshikov
134b   Cells hemispherical or pyriform. (Fig. 6.38) . . . .
           Pyramidomonas Schmarda

Family **Chlamydomonadaceae**

135a (14) Cells biflagellate **136**
135b   Cells with 4 flagella. **144**
136a (135) Cells colorless. (Fig. 6.39) . . . . *Polytoma* Ehrenberg
136b   Cells pigmented. **137**
137a (136) With cytoplasmic processes from cell to envelope **162**
137b   Without such processes **138**
138a (137) Circular in anterior view. **139**
138b   Anterior view lenticular or irregular. **142**
139a (138) Flagella inserted close together **140**
139b   Flagella inserted widely apart. (Fig. 6.37). . . *Gloeomonas* Klebs
140a (139) Cells fusiform or spindle-shaped. (Fig. 6.40) . . . .
           Chlorogonium Ehrenberg
140b   Cells not of the above shapes. **141**
141a (140) Protoplast same shape as and filling its envelope. (Figs. 6.41,
           6.42) . . . . *Chlamydomonas* Ehrenberg
141b   Protoplast variable in shape; not filling its envelope. (Fig.
           6.43) . . . . *Sphaerellopsis* Korshikov
142a (138) Cell envelope with regular or irregular processes. (Figs. 6.44,
           6.45) . . . . *Lobomonas* Dangeard
142b   Cells compressed **143**
143a (142) Ovate in outline; flagella at narrow end. (Fig. 6.46). . . . .
           Platychloris Pascher
143b   Rounded to rectangular; flagella inserted near cell center. (Fig.
           6.47) . . . . *Mesostigma* Lauterborn

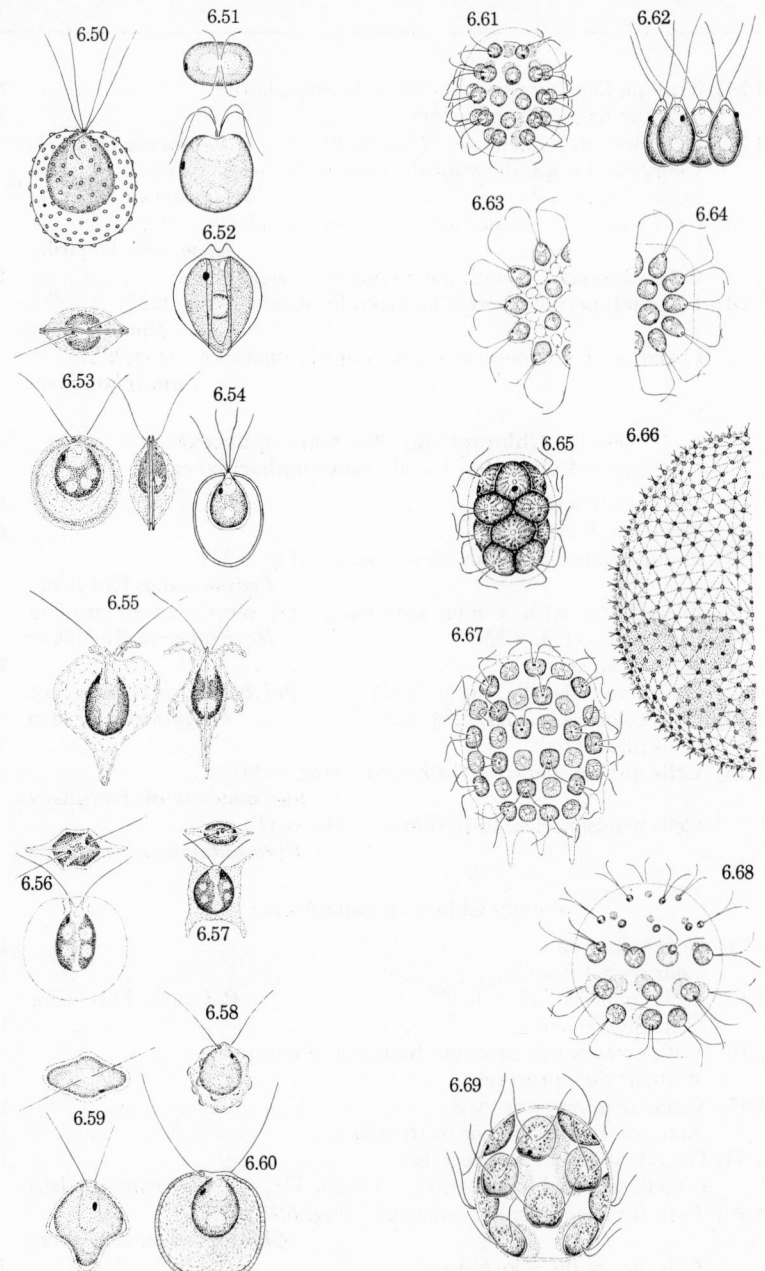

Fig. 6.50. *Pedinopera granulosa* (Playfair) Pascher. × 500. (After Playfair.) **Fig. 6.51.** *Platymonas elliptica* G. M. Smith. × 390. **Fig. 6.52.** *Scherffelia phacus* Pascher. × 1000. (After Pascher.) **Fig. 6.53.** *Phacotus lenticularis* (Ehrenberg) Stein. × 780. **Fig. 6.54.** *Coccomonas orbicularis* Stein. × 500. **Fig. 6.55.** *Wislouchiella planctonica* Skvortzow. × 780. **Fig. 6.56.** *Pteromonas angulosa* (Carter) Lemmermann. × 780. **Fig. 6.57.** *P. aculeata* Lemmermann. × 390. **Fig. 6.58.** *Thoracomonas irregularis* Korsh. × 780. **Fig. 6.59.** *Cephalomonas granulata* Higinbotham. × 1145. (After Higinbotham.) **Fig. 6.60.** *Dysmorphococcus variabilis* Takeda. × 780.

Fig. 6.61. *Eudorina elegans* Ehrenberg. × 150. **Fig. 6.62.** *Gonium sociale* (Dujardin) Warming. × 780. **Fig. 6.63.** *G. formosum* Pascher. × 340. **Fig. 6.64.** *G. pectorale* Müller. × 340. **Fig. 6.65.** *Pandorina charkowiensis* Korsh. × 390. **Fig. 6.66.** *Volvox aureus* Ehrenberg. × 133. (After G. M. Smith.) **Fig. 6.67.** *Platydorina caudata* Kofoid. × 195. **Fig. 6.68.** *Pleodorina californica* Shaw. × 195. **Fig. 6.69.** *Volvulina steinii* Playfair. × 390.

144a (135) Cells compressed **145**
144b   Cells not compressed. (Figs. 6.48, 6.49) . . . . *Carteria* Diesing
145a (144) Weakly compressed; broadly rounded posterior. (Fig. 6.51). . . .
                                                    Platymonas G. S. West
145b   Strongly compressed; posterior pointed. (Fig. 6.52) . . . .
                                                    Scherffelia Pascher

Family **Phacotaceae**

146a (13) Cells biflagellate **147**
146b   Cells with 4 flagella. (Fig. 6.50) . . . . . . *Pedinopera* Pascher
147a (146) Lorica obviously compressed **148**
147b   Lorica not compressed or very slightly so **152**
148a (147) With projections from flattened faces. (Fig. 6.55) . . . .
                                                    Wislouchiella Skvortzow
148b   Without such projections **149**
149a (148) Union of lorica halves clearly evident. (Fig. 6.53) . . . .
                                                    Phacotus Perty
149b   Union of lorica halves not evident **150**
150a (149) Lorica verrucose. (Fig. 6.58) . . . . . *Thoracomonas* Korshikov
150b   Lorica smooth, granulate or pitted **151**
151a (150) Strongly compressed; posterior rounded or emarginate. (Figs.
            6.56, 6.57) . . . . *Pteromonas* Seligo
151b   Weakly compressed; posterior acuminate. (Fig. 6.59) . . . .
                                                    Cephalomonas Higinbotham
152a (147) With a single flagellar pore. (Fig. 6.54). . . . *Coccomonas* Stein
152b   With separate flagellar pores. (Fig. 6.60) . . . .
                                                    Dysmorphococcus Takeda

Family **Volvocaceae**

153a (15) Colony a flat plate of cells **154**
153b   Colony globose to ellipsoid **155**
154a (153) Colony square or circular; flagella from one side. (Figs. 6.62,
            6.63, 6.64) . . . . *Gonium* Müller
154b   Colony oval with posterior processes; flagella from each side.
            (Fig. 6.67) . . . . *Platydorina* Kofoid
155a (153) Over 500 cells per colony. (Fig. 6.66) . . . . . *Volvox* Linnaeus
155b   Not over 256 cells per colony **156**
156a (155) All cells of a colony equal or nearly so in size **157**
156b   Cells of a colony of two distinct sizes. (Fig. 6.68). . . . .
                                                    Pleodorina Shaw
157a (156) Cells with polar processes; haematochrome often present **162**
157b   Cells without such processes; haematochrome not present **158**
158a (157) Cells of vegetative colony closely compacted. (Fig. 6.65) . . . . .
                                                    Pandorina Bory
158b   Cells of vegetative colony remote from each other **159**
159a (158) Cells spheric or nearly so. (Fig. 6.61) . . . . *Eudorina* Ehrenberg
159b   Cells anteriorly-posteriorly flattened. (Fig. 6.69) . . . .
                                                    Volvulina Playfair
160a (15) Cells biflagellate **161**
160b   Cells quadriflagellate. (Fig. 6.72) . . *Spondylomorum* Ehrenberg
161a (160) Colonies with 2 or 4 cells. (Fig. 6.70) . . *Pascheriella* Korshikov
161b   Colonies of more than 4 cells. (Fig. 6.71) . . *Pyrobotrys* Arnoldi

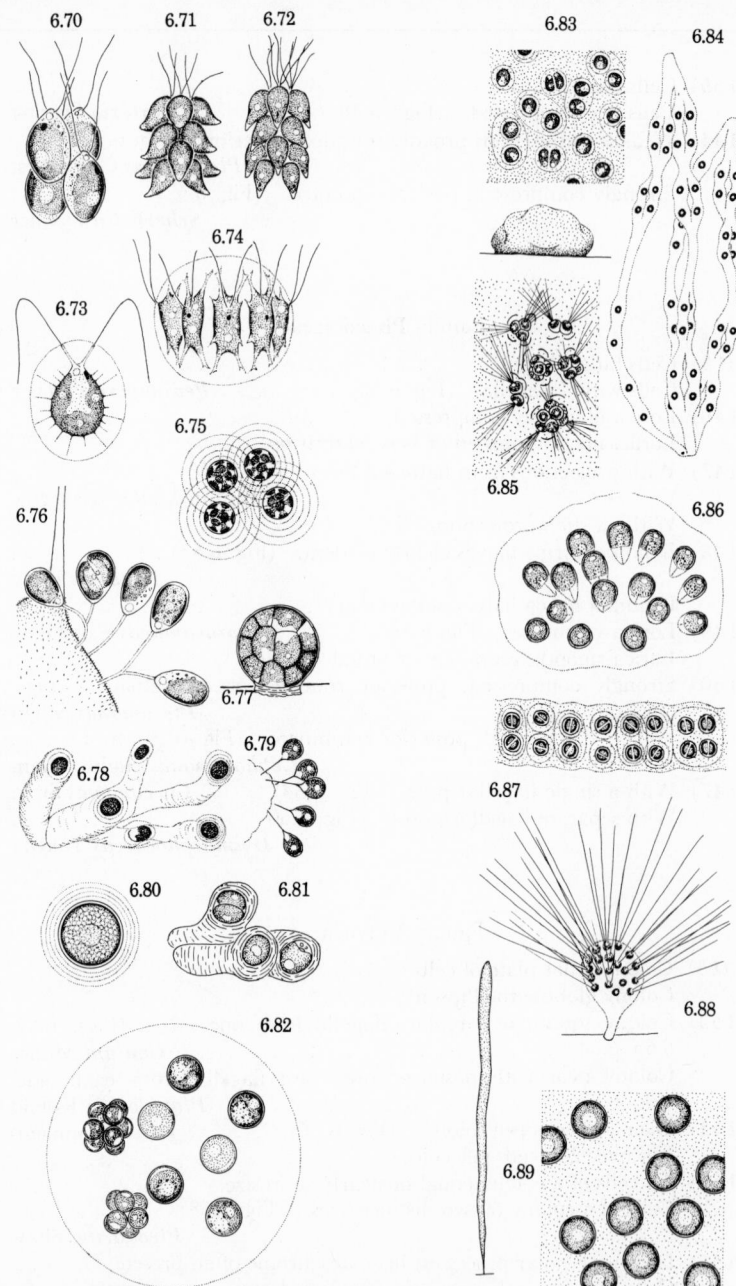

Fig. 6.70. *Pascheriella tetras* Korsh. × 780. **Fig. 6.71.** *Pyrobotrys stellata* Korsh. × 780. **Fig. 6.72.** *Spondylomorum quaternarium* Ehrenberg. × 780. **Fig. 6.73.** *Haematococcus lacustris* (Griod.) Rostaf. × 250. **Fig. 6.74.** *Stephanosphaera pluvialis* Cohn. × 390. **Fig. 6.75.** *Asterococcus superbus* (Cienkowski) Scherffel. × 404. **Fig. 6.76.** *Chlorangium javanicum* Lemmermann. × 780. **Fig. 6.77.** *Malleochloris sessilis* Pascher. × 412. **Fig. 6.78.** *Hormotila mucigena* Borzi. × 250. (After West.) **Fig. 6.79.** *Stylosphaeridium stipitatum* (Bachm.) Geitler. × 390. **Fig. 6.80.** *Gloeocystis gigas* (Kützing) Lagerheim. × 390. **Fig. 6.81.** *Hormotilopsis gelatinosa* Trainor and Bold. × 435. (After Trainor and Bold.) **Fig. 6.82.** *Sphaerocystis schroeteri* Chodat. × 390.

Fig. 6.83. *Palmella miniata* Leibl. × 390. Habit. × 5. **Fig. 6.84.** *Schizodictyon catenatum* Thompson. × 90. **Fig. 6.85.** *Schizochlamys gelatinosa* A. Braun. × 404. **Fig. 6.86.** *Askenasyella chlamydopus* Schmidle. × 404. **Fig. 6.87.** *Palmodictyon viride* Kützing. × 390. **Fig. 6.88.** *Apiocystis brauniana* Nägeli. × 195. **Fig. 6.89.** *Tetraspora cylindrica* (Wahlb.) Agardh. × 390 Habit. × 5.

Family **Haematococcaceae**

162a (137, 157) Unicellular; often bright red. (Fig. 6.73). . . . .
                                                        Haematococcus Agardh
162b   Colonial; of 8 to 16 cells. (Fig. 6.74). . . *Stephanosphaera* Cohn

Order **Tetrasporales,** Family **Chlorangiaceae**

163a   (39) Stalk massive, about as wide as cell. (Fig. 6.77) . . . .
                                                        Malleochloris Pascher
163b   Stalk very slender and pointed. **164**
164a (163, 41) Solitary or gregarious; epiphytic in algal matrices. (Fig.
              6.79) . . . . *Stylosphaeridium* Geitler et Gimesi
164b   Solitary or branched; epizoic on crustacea. (Fig. 6.76) . . . . . .
                                                        Chlorangium Stein
165a (47, 23) With pseudocilia. **177**
165b   Without pseudocilia. **166**

Family **Palmellaceae**

166a (165) With a stellate chloroplast. (Fig. 6.75) . . *Asterococcus* Scherffel
166b   With a massive cup-shaped chloroplast. (Fig. 6.80) . . . .
                                                        Gloeocystis Nägeli
167a (47) Cells with one chloroplast; becoming remote from one another . .   **168**
167b   Cells remaining in pyramidate groups or else with several chloro-
            plasts **213**
168a (167) With a single stellate chloroplast. (Fig. 6.75). . . . .
                                                        Asterococcus Scherffel
168b   With a parietal cup-shaped chloroplast. (Fig. 6.82) . . . .
                                                        Sphaerocystis Chodat
169a (45) Colonies amorphous or saccate.   **174**
169b   Colonies simple, branched, or netlike **170**
170a (169) Cells at distal ends of thick gelatinous branches. **171**
170b   Cells scattered or definitely spaced. **172**
171a (170) Zoospores biflagellate. (Fig. 6.78) . . . . *Hormotila* Borzi
171b   Zoospores quadriflagellate. (Fig. 6.81). . . . *Hormotilopsis* Bold
172a (170) Colony a simple or branched tube of many cells. (Fig. 6.87) . . .
                                                        Palmodictyon Kützing
172b   Colony netlike; cells definitely spaced **173**
173a (172) With a lenticular mesh; an expanded net or in chains. (Fig.
            6.84) . . . . *Schizodictyon* Thompson
173b   With a cubical mesh; cells at the corners **227**
174a (169) Colonies amorphous **175**
174b   Colonies saccate. **177**
175a (174) Cells pyriform, with broad poles outermost. (Fig. 6.86) . . . . . .
                                                        Askenasyella Schmidle
175b   Cells spheric. **176**
176a (175) Sheaths of individual cells distinct; colony crustose. (Fig.
            6.83) . . . . *Palmella* Lyngbye

Family **Tetrasporaceae**

176b   Sheaths indistinct; cell-wall fragments scattered in matrix. (Fig.
            6.85) . . . . *Schizochlamys* A. Braun
177a (165, 174) Microscopic and with distinct pseudocilia. (Fig. 6.88) . . . .
                                                        Apiocystis Nägeli
177b   Macroscopic; pseudocilia evident by staining. (Fig. 6.89) . . . . .
                                                        Tetraspora Link

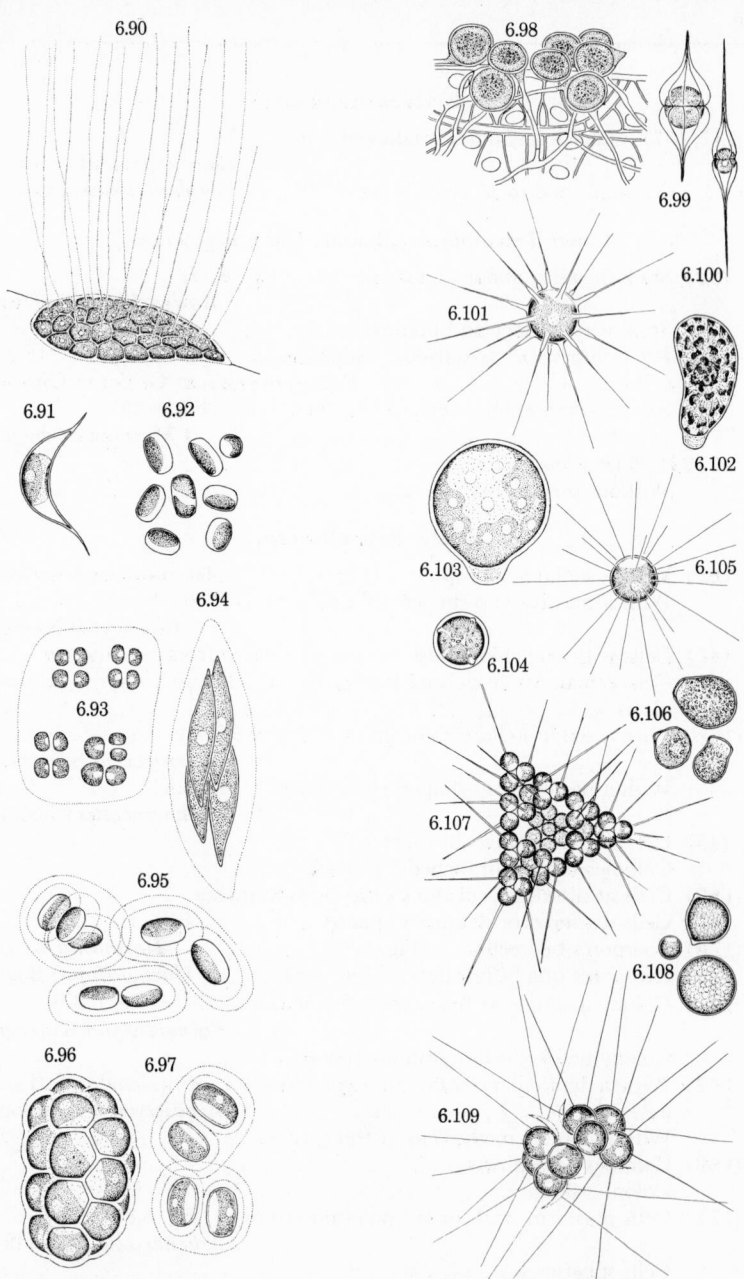

Fig. 6.90. *Chaetopeltis orbicularis* Berthold. × 412. **Fig. 6.91.** *Ourococcus bicaudatus* Grobety. × 808. **Fig. 6.92.** *Nannochloris bacillaris* Naumann. × 1560. **Fig. 6.93.** *Dispora crucigenioides* Printz. × 404. **Fig. 6.94.** *Elakatothrix viridis* (Snow) Printz. × 404. **Fig. 6.95.** *Dactylothece braunii* Lagerheim. × 808. **Fig. 6.96.** *Chlorosarcina minor* Gerneck. × 780. **Fig. 6.97.** *Coccomyxa dispar* Schmidle. × 780.

Fig. 6.98. *Phyllobium sphagnicola* G. S. West. × 187. (After West.) **Fig. 6.99.** *Desmatractum bipyramidatum* (Chodat) Pascher. × 808. **Fig. 6.100.** *D. indutum* (Geitler) Pascher. × 390. **Fig. 6.101.** *Acanthosphaera zackariasi* Lemmermann. × 808. **Fig. 6.102.** *Kentrosphaera bristolae* G. M. Smith. × 412. **Fig. 6.103.** *Chlorochytrium lemnae* Cohn. × 390. **Fig. 6.104.** *Oöphila amblystomatis* Lambert. × 390. **Fig. 6.105.** *Golenkinia radiata* Chodat. × 808. **Fig. 6.106.** *Myrmecia aquatica* G. M. Smith. × 487. (After Smith.) **Fig. 6.107.** *Errerella bornhemiensis* Conrad. × 500. **Fig. 6.108.** *Chlorococcum humicola* (Nägeli) Rabenhorst. × 390. **Fig. 6.109.** *Micratinium pusillum* Fres. × 780.

178a (84) Setae gelatinous threads without a basal sheath. (Fig. 6.90). . . .
Chaetopeltis Berthold

178b Setae cytoplasmic threads with or without a basal sheath **286**

Family **Coccomyxaceae**

179a (25) With 1 chloroplast per cell; reproduction by cell division. (Fig. 6.92) . . . . *Nannochloris* Naumann

179b With 2 to 4 chloroplasts per cell; reproduction by autospores . . . **201**

180a (35) One or both poles sharply attenuated; reproduction by cell division. (Fig. 6.91). . . . . *Ourococcus* Grobety

180b Ending in slender spines; both simple, or one forked **193**

181a (49) Cell sheath evident; solitary or in chains of 2 or 4 cells. (Fig. 6.95) . . . . *Dactylothece* Lagerheim

181b Sheaths indistinct, confluent in indefinite-sized colonies. **182**

182a (181) Wall smooth; with one chloroplast lacking a pyrenoid. (Fig. 6.97) . . . . *Coccomyxa* Schmidle

182b Wall setose; 1 to 4 chloroplasts each with a pyrenoid **214**

183a (50) Cells loosely arranged; reproduction by division. (Fig. 6.94) . . . *Elakatothrix* Wille

183b Cells in parallel groups of 4; reproduction autosporous **214**

184a (55) Cells ellipsoid to rectangular; growth by division; no pyrenoids. (Fig. 9.93) . . . . *Dispora* Printz

184b Cells ellipsoid to triangular; with one pyrenoid; autosporous. . . . **226**

185a (81) Aquatic, free or endophytic; packets ensheathed. (Fig. 6.96) . . . *Chlorosarcina* Gerneck

185b Aerial, cells solitary, paired, or at most in 4-celled packets. **289**

Order **Chlorococcales,** Family **Chlorococcaceae**

186a (20) Within amphibian eggs; inside of cell wall with pits. (Fig. 6.104). . . . . *Oöphila* Lambert ex Printz

186b Within protozoa or hydroids; cell wall even throughout **202**

187a (35) Planktonic or in gelatinous matrix of other algae. (Figs. 6.99, 6.100). . . . . *Desmatractum* W. et G. S. West

187b Green, or red with haematochrome; on moist cliffs, or causing "red snow" **207**

188a (27) Wall with a few buttonlike thickenings; variable in size. (Fig. 6.108). . . . . *Chlorococcum* Fries

188b Wall smooth and evenly thick throughout. **202**

189a (31) Cells globose-pyriform; chloroplast without processes. (Fig. 6.106). . . . . *Myrmecia* Printz

Family **Endosphaeraceae**

189b Cells elongate-pyriform; chloroplast with palisadelike processes. (Fig. 6.102). . . . . *Kentrosphaera* Borzi

190a (20) Cells tubular, simple or branched, with globose tips. (Fig. 6.98) . . . . *Phyllobium* Klebs

190b Cells globose to ellipsoid with localized lamellated thickenings. (Fig. 6.103). . . . . *Chlorochytrium* Cohn

Family **Micractiniaceae**

191a (24) Base of setae thickened for some length. (Fig. 6.101). . . . . *Acanthosphaera* Lemmermann

191b Setae equally slender throughout. (Fig. 6.105) . . . . *Golenkinia* Chodat

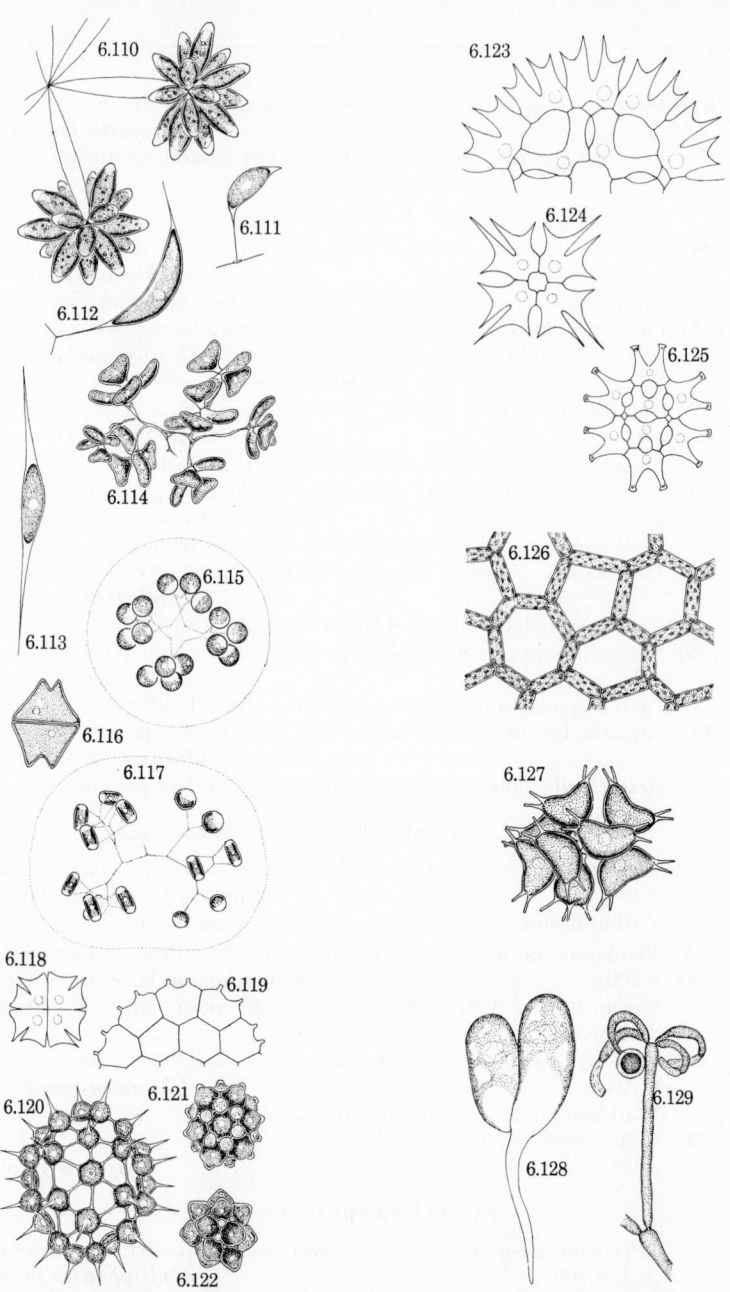

Fig. 6.110. *Actidesmium hookeri* Reinsch. × 390.   **Fig. 6.111.** *Characium ornithocephalum* A. Braun. × 390.
Fig. 6.112. *Schroederia anchora* G. M. Smith. × 404.   **Fig. 6.113.** *S. setigera* (Schröder) Lemmermann. × 390.
Fig. 6.114. *Dimorphococcus lunatus* A. Braun. × 390.   **Fig. 6.115.** *Dictyosphaerium pulchellum* Wood. × 390.
Fig. 6.116. *Euastropsis Richteri* (Schmidle) Lagerheim. × 1000.   (After G. M. Smith.)   **Fig. 6.117.** *Dictyosphaerium planctonicum* Tiffany and Ahls. × 780.   **Fig. 6.118.** *Pediastrum tetras* (Ehrenberg) Ralfs. × 404.
Fig. 6.119. *P. boryanum* (Turpin) Meneghini. × 390.   **Fig. 6.120.** *Coelastrum chodati* Ducell. × 390.   **Fig. 6.121.** *C. cambricum* Arch. × 390.   **Fig. 6.122.** *C. sphaericum* Nägeli. × 390.

Fig. 6.123. *Pediastrum biradiatum* Meyen. × 500.   **Fig. 6.124.** *P. biradiatum* var. *longicornutum* Gutwin. × 500.   **Fig. 6.125.** *P. duplex* Meyen var. *reticulatum* Lagerheim. × 390.   **Fig. 6.126.** *Hydrodictyon reticulatum* (L.) Lagerheim. × 1.5.   **Fig. 6.127.** *Sorastrum spinulosum* Nägeli. × 780.   **Fig. 6.128.** *Protosiphon botryoides* (Kützing) Klebs. × 390.   **Fig. 6.129.** *Dichotomosiphon tuberosus* (A. Braun) Ernst. × 25.   (After Ernst.)

134

192a **(58)** Cells each with one seta; in pyramidal groups of 4. (Fig. 6.107) . . . . ***Errerella*** Conrad

192b With 2 to 7 setae; cells quadrately united. (Fig. 6.109) . . . . . .
Micractinium Fresenius

Family Characiaceae

193a **(39, 180)** Epiphytic on algae and aquatic phanerogams. (Fig. 6.141) . . ***Characium*** A. Braun

193b Planktonic. (Figs. 6.112, 6.113) . . . ***Schroederia*** Lemmermann

194a **(59)** Cells spindle-shaped, in radial clusters. (Fig. 6.110) . . . .
Actidesmium Reinsch

194b Cells not spindle-shaped; in quadrate or linear coenobia **195**

195a **(194)** With 2 reniform and 2 cylindric cells in a colony. (Fig. 6.114) . . . . ***Dimorphococcus*** A. Braun

195b Cells of a colony all the same shape **221**

196a **(49, 61)** With 2 or 4 cells in groups joined by threads. (Figs. 6.115, 6.117) . . . . ***Dictyosphaerium*** Nägeli

Family Coelastraceae

196b With 2 to 64 cells joined laterally to form a hollow sphere. (Figs. 6.120, 6.121, 6.122) . . . . ***Coelastrum*** Nägeli

Family Hydrodictyaceae

197a **(56)** Colony 2-celled; outer side emarginate. (Fig. 6.116) . . . .
Euastropsis Lagerheim

197b Colony 4- to 64-celled. (Figs. 6.118, 6.119, 6.123, 6.124, 6.125). . .
Pediastrum Meyer

198a **(54)** Macroscopic, cells forming a saccate net. (Fig. 6.126) . . . . . .
Hydrodictyon Roth

198b Microscopic; cells at corners of a cubical mesh **227**

199a **(61)** Cells pyriform to cordate-reniform with 1 to 4 stout spinelike processes. (Fig. 6.127). . . . . ***Sorastrum*** Kützing

199b Cells fusiform to cylindric **227**

Family Protosiphonaceae

200a **(18, 73)** Alga a pyriform to cylindric, rarely branched vesicle. (Fig. 6.128) . . . . ***Protosiphon*** Klebs

Family Dichotomosiphonaceae

200b Alga a dichotomously-branched tube constricted at the branches. (Fig. 6.129). . . . . ***Dichotomosiphon*** Ernst

Family Oöcystaceae

201a **(31, 179)** Cells minute, with 2 to 4 parietal chloroplasts. (Fig. 6.132) . .
Palmellococcus Chodat

201b Cells relatively large; wall with localized thickening. (Fig. 6.131). . . . . ***Excentrosphaera*** G. T. Moore

202a **(186, 188)** With one parietal chloroplast; wall moderately thick. (Fig. 6.133). ***Chlorella*** Beijerinck

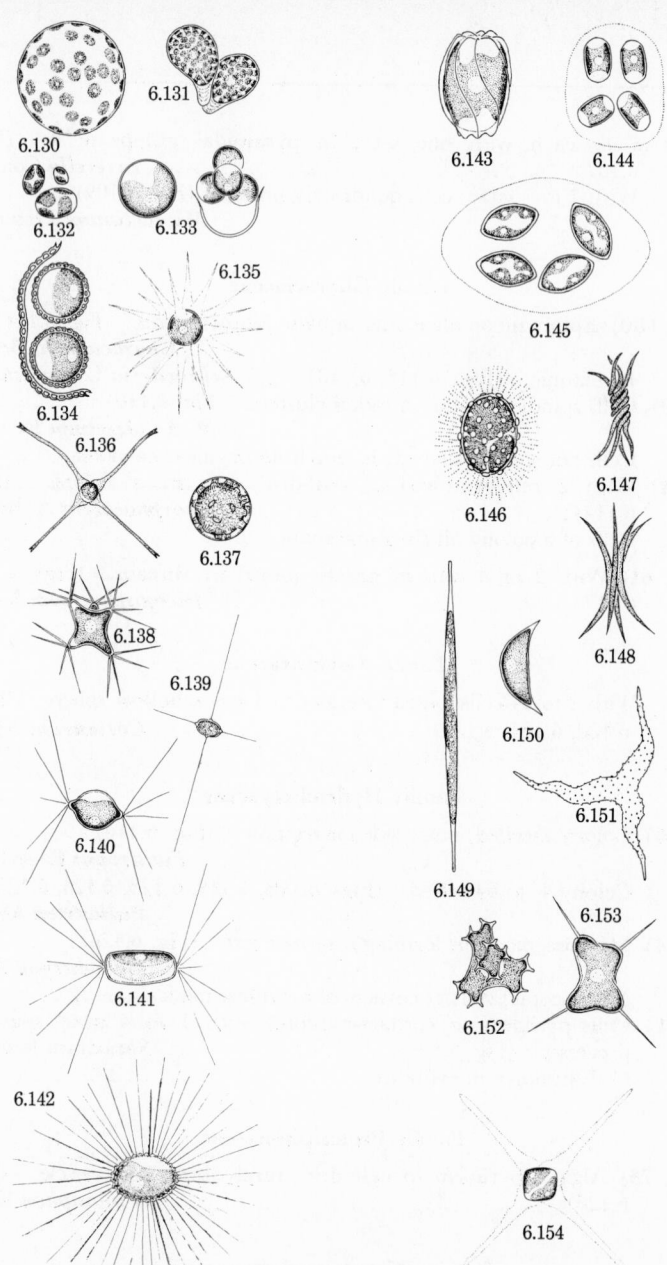

Fig. 6.130. *Eremosphaera viridis* deBary. × 390. **Fig. 6.131.** *Excentrosphaera viridis* G. T. Moore. × 412. **Fig. 6.132.** *Palmellococcus protothecoides.* (Krüg) Chodat. × 390. **Fig. 6.133.** *Chlorella vulgaris* Beijerinck. × 780. **Fig. 6.134.** *Keriochlamys styriaca* Pascher. × 780. **Fig. 6.135.** *Echinosphaerella limnetica* G. M. Smith. × 390. **Fig. 6.136.** *Pachycladon umbrinus* G. M. Smith. × 195. **Fig. 6.137.** *Trochiscia pachyderma* (Reinsch) Hansgirg. × 1000. **Fig. 6.138.** *Polyedriopsis spinulosa* Schmidle. × 390. **Fig. 6.139.** *Chodatella wratislawiensis* (Schröder) Lay. × 390. **Fig. 6.140.** *C. citriformis* Snow. × 390. **Fig. 6.141.** *C. subsalsa* Lemmermann. × 390. **Fig. 6.142.** *Franceia tuberculata* G. M. Smith. × 500.

Fig. 6.143. *Scotiella nivalis* (Chodat) Fritsch. × 780. **Fig. 6.144.** *Oöcystis natans* (Lemmermann) Wille. × 412. **Fig. 6.145.** *O. novae-semliae* Wille. × 412. **Fig. 6.146.** *Siderocelis ornatus* (Fott) Fott. × 780. **Fig. 6.147.** *Ankistrodesmus spiralis* (Turner) Lemmermann. × 390. **Fig. 6.148.** *A. falcatus* (Corda) Ralfs. × 404. **Fig. 6.149.** *Closteriopsis longissima* Lemmermann. × 390. **Fig. 6.150.** *Closteridium lunula* Reinsch. × 390. **Fig. 6.151.** *Cerasterias irregulare* G. M. Smith. × 500. (After Smith.) **Fig. 6.152.** *Tetraëdron lobatum* (Nägeli) Hansgirg var. *subtetraedricum* Reinsch. × 412. **Fig. 6.153.** *T. regulare* Kützing var. *incus* Teiling. × 825. **Fig. 6.154.** *Treubaria crassispina* G. M. Smith. × 390.

202b		Chloroplasts numerous, discoidal; wall very thin. (Fig. 6.130) . . *Eremosphaera* de Bary
203a	**(27)**	Wall sculptured or scrobiculate. (Fig. 6.137) . . . . *Trochiscia* Kützing
203b		Wall thick, appearing beaded or to contain locules. (Fig. 6.134) . . . . *Keriochlamys* Pascher
204a	**(26)**	With 4 stout appendages bifurcate at apex, usually brown. (Fig. 6.136) . . . . *Pachycladon* G. M. Smith
204b		With many long, tapering gelatinous spines. (Fig. 6.135) . . . . . *Echinosphaerella* G. M. Smith
205a	**(28)**	Irregularly tetrahedral; setose at each apex. (Fig. 6.138) . . . . . *Polyedriopsis* Schmidle
205b		Citriform, cylindric, or ellipsoid **206**
206a	**(205)**	Citriform to cylindric; setae at the poles or median. (Figs. 6.139, 6.140, 6.141) . . . . *Chodatella* Lemmermann
206b		Ellipsoid; setae scattered over surface. (Fig. 6.142) . . . . *Franceia* Lemmermann
207a	**(32, 187)**	With spiraled longitudinal ridges. (Fig. 6.143) . . . . *Scotiella* Fritsch
207b		Without such ridges **208**
208a	**(207)**	Wall smooth, often with polar thickenings. (Figs. 6.144, 6.145) . . . . *Oöcystis* Nägeli See also 220.
208b		Wall with many small verrucae. (Fig. 6.146) . . *Siderocelis* Fott
209a	**(36)**	Acicular often sigmoid or arcuate. (Figs. 6.147, 6.148) . . . . . . *Ankistrodesmus* Corda
209b		Straight; with an axial row of many pyrenoids. (Fig. 6.149) . . . *Closteriopsis* Lemmermann
210a	**(36)**	Chloroplast parietal, filling cell, with short polar spines. (Fig. 6.150) . . . . *Closteridium* Reinsch
210b		Chloroplasts axial, separated at the middle of cell **237**
211a	**(33)**	Tetrahedral, each apex with a long stout spine. (Fig. 6.154) . . . *Treubaria* Bernard
211b		Triangular, tetrahedral to irregular; without long spines. **212**
212a	**(211)**	Triangular; cell body tapered into 3 arms. (Fig. 6.151) . . . . . . *Cerasterias* Reinsch
212b		Tetrahedral to variously lobed or irregular. (Figs. 6.152, 6.153) . . . . *Tetraëdron* Kützing
213a	**(167)**	Cells remaining in pyramidal groups of 4. (Fig. 6.155) . . . . . . *Radiococcus* Schmidle
213b		Cells with 2 or more parietal chloroplasts. (Fig. 6.160) . . . . . . *Planktosphaeria* G. M. Smith
214a	**(182, 183)**	Cells in indefinite floccose gelatinous colonies. (Fig. 6.156) . . . . *Bohlinia* Lemmermann
214b		Colonies fusiform with 1 to 4 in coenobia. (Fig. 6.157) . . . . . . *Quadrigula* Printz
215a	**(52, 60)**	Solitary or in chains; poles minutely apiculate. (Fig. 6.159) . . . . *Dactylococcus* Nägeli
215b		Colonial aggregates of few to 100 or more cells. (Fig. 6.158) . . . *Selenastrum* Reinsch
216a	**(51)**	Colony spheric; cells in radiating clusters of 2 or 4. (Fig. 6.161) . . . . *Gloeoactinium* G. M. Smith
216b		Colony irregular; cells paired, adhering to mother cell wall. (Fig. 6.162) . . . . *Dichotomococcus* Korshikov
217a	**(51)**	Cells lunate or sausage-shaped; united. **222**

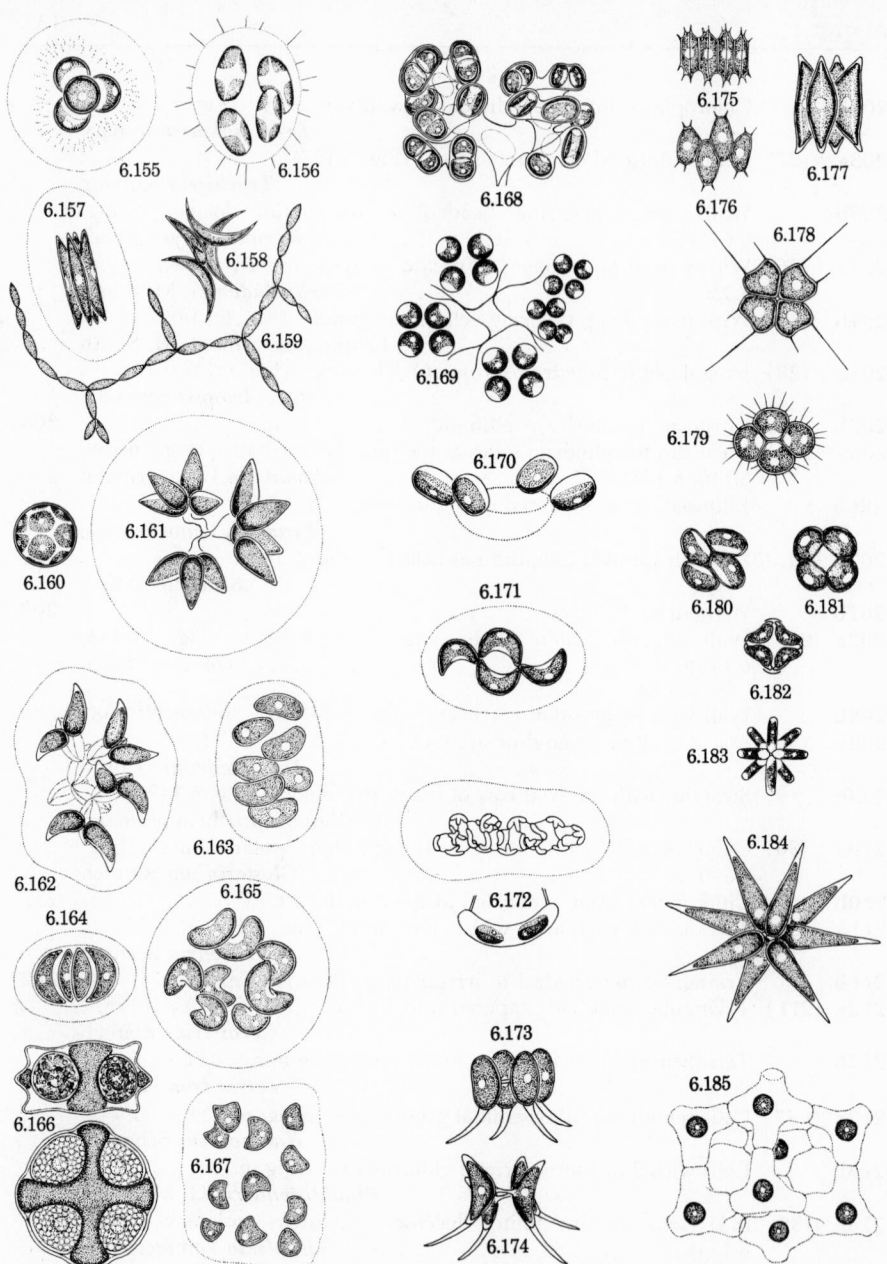

Fig. 6.155. *Radiococcus nimbatus* (de Wildermann) Schmidle. × 780. **Fig. 6.156.** *Bohlinia echidna* (Bohlin) Lemmermann. × 787. (After Prescott and Croasdale.) **Fig. 6.157.** *Quadrigula closterioides* (Bohlin) Printz. × 404. **Fig. 6.158.** *Selenastrum gracile* Reinsch. × 390. **Fig. 6.159.** *Dactylococcus infusionum* Nägeli. × 200. (After G. M. Smith.) **Fig. 6.160.** *Planktosphaeria gelatinosa* G. M. Smith. × 500. **Fig. 6.161.** *Gloeoactinium limneticum* G. M. Smith. × 1560. **Fig. 6.162.** *Dichotomococcus lunatus* Fott. × 780. **Fig. 6.163.** *Nephrocytium agardhianum* Nägeli. × 404. **Fig. 6.164.** *N. lunatum* W. West. × 390. **Fig. 6.165.** *Kirchneriella obesa* (W. West) Schmidle. × 808. **Fig. 6.166.** *Gloeotaenium loitlesbergianum* Hansgirg. × 412. **Fig. 6.167.** *Kirchneriella malmeana* (Bohlin) Wille. × 390.

Fig. 6.168. *Lobocystis dichotoma* Thompson. × 390. **Fig. 6.169.** *Westella botryoides* (W. West) de Wildermann. × 780. **Fig. 6.170.** *Quadricoccus verrucosus* Fott. × 1560. **Fig. 6.171.** *Tetrallantos lagerheimii* Teiling. × 404. **Fig. 6.172.** *Tomaculum catenatum* Whitford. Cell × 325; colony × 70. (After Whitford.) **Fig. 6.173.** *Coronastrum ellipsoideum* Fott. × 780. **Fig. 6.174.** *C. lunatum* Thompson. × 780.

217b   Cells lunate or curved cylinders; in groups of 4 or 8. (Figs. 6.165, 6.167) . . . . ***Kirchneriella*** Schmidle

218a (57) Cells separated by a cruciate gel impregnated with calcite. (Fig. 6.166) . . . . ***Gloeotaenium*** Hansgirg

218b   Cells not so separated. **219**

219a (218) Cells ellipsoid, subcylindric to panduriform **220**

219b   Cells oblong, reniform or lunate; enclosing wall gelatinized. (Figs. 6.163, 6.164). . . . . ***Nephrocytium*** Nägeli

220a (219) Enclosing wall expanded equally throughout. (Figs. 6.144, 6.145) . . . . ***Oöcystis*** Nägeli
       See also **208**.

220b   Enclosing wall bilobed with a daughter cell in each lobe. (Fig. 6.168). . . . . ***Lobocystis*** Thompson

221a (195) Cells touching one another; colony cruciate or linear. (Fig. 6.169). . . . . ***Westella*** de Wildermann

221b   Cells remote, cruciately arranged on the margin of the expanded mother cell wall. (Fig. 6.170) . . . . ***Quadricoccus*** Fott

Family **Scenedesmaceae**

222a (217) Cells blunt-lunate; united into 4-celled colonies. (Fig. 6.171) . . .   ***Tetrallantos*** Teiling

222b   Cells sausage-shaped; many united by strands into a saccate colony. (Fig. 6.172). . . . . ***Tomaculum*** Whitford

223a (56) Cells united at point of mutual contact **224**

223b   Cells united by slender strands; coenobia 4-celled. (Figs. 6.173, 6.174). . . . . ***Coronastrum*** Thompson

224a (223) Cells ellipsoid, fusiform, or cylindric; united laterally **225**

224b   Cells ellipsoid, phaseolate to triangular; cruciately united **226**

225a (224) Colony linear. (Figs. 6.175, 6.176). . . . . ***Scenedesmus*** Meyen

225b   Colony cruciate. (Fig. 6.177) . . . . ***Tetradesmus*** G. M. Smith

226a (184, 224) Outer face of each cell with one or more spines. (Figs. 6.178, 6.179) . . . . ***Tetrastrum*** Chodat

226b   Cells not bearing spines. (Figs. 6.180, 6.181, 6.182) . . . .   ***Crucigenia*** Morren

227a (199, 173, 198) Cells united centrally at points of mutual contact. (Figs. 6.183, 6.184) . . . . ***Actinastrum*** Lagerheim

227b   Colony a gelatinous cubical meshwork; cells at the corners. (Fig. 6.185). . . . . ***Pectodictyon*** Taft

Order **Zygnematales**, Family **Mesotaeniaceae**

228a (32) Chloroplast a spiral band or axial with spiraled ridges. (Fig. 6.186). . . . . ***Spirotaenia*** Brébisson

228b   Chloroplast not spiraled **229**

229a (228) Chloroplasts platelike and axial **230**

229b   Chloroplasts asteroid, or longitudinally ridged **232**

Fig. 6.175. *Scenedesmus brasiliensis* Bohlin. × 404. **Fig. 6.176.** *S. denticulatus* Lagerheim. × 404. **Fig. 6.177.** *Tetradesmus wisconsinensis* G. M. Smith. × 780. **Fig. 6.178.** *Tetrastrum heterocanthum* (Nordstedt) Chodat. × 600. **Fig. 6.179.** *T. staurogeniaeforme* (Schröder) Lemmermann. × 780. **Fig. 6.180.** *Crucigenia alternans* G. M. Smith. × 600. **Fig. 6.181.** *C. lauterbornei* Schmidle. × 600. **Fig. 6.182.** *Crucigenia crucifera* (Wolle) Collins. × 780. **Fig. 6.183.** *Actinastrum hantzschii* Lagerheim. × 780. **Fig. 6.184.** *A. gracillimum* G. M. Smith. × 390. **Fig. 6.185.** *Pectodictyon cubicum* Taft. × 390.

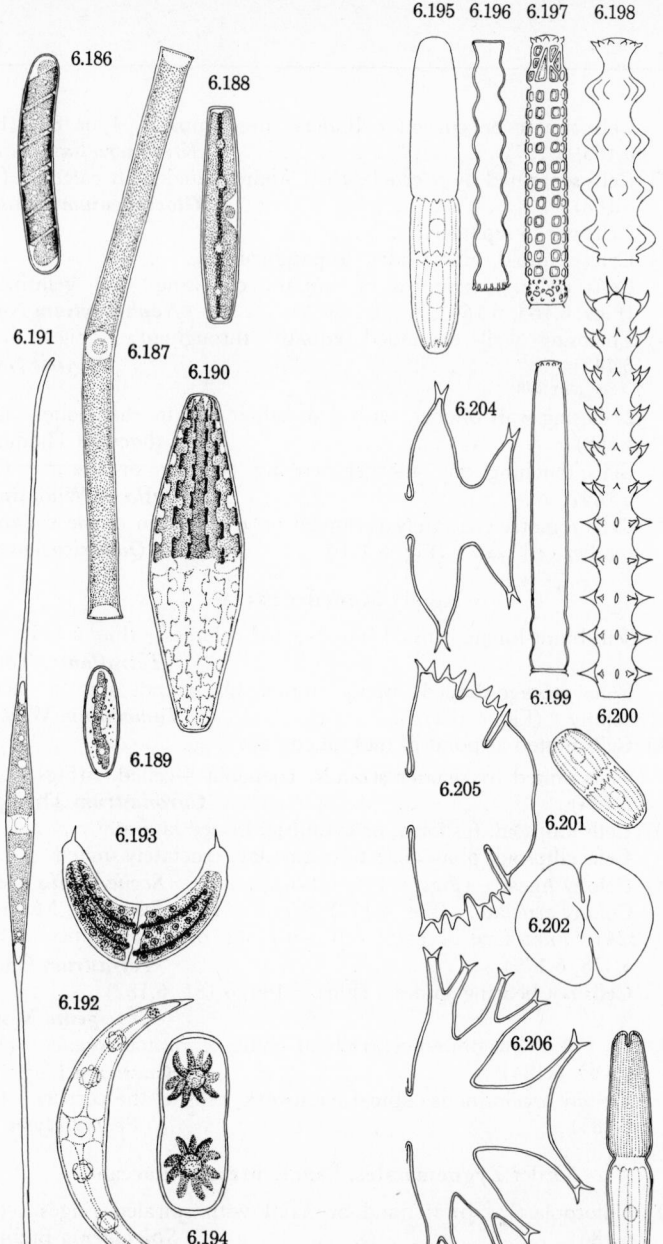

Fig. 6.186. *Spirotaenia condensata* Brébisson. × 195. **Fig. 6.187.** *Gonatozygon kinahani* (Arch.) Rabenhorst. × 404. **Fig. 6.188.** *Roya anglica* G. S. West. × 500. (After Hodgetts.) **Fig. 6.189.** *Mesotaenium aplanosporum* Taft. × 390. **Fig. 6.190.** *Netrium digitus* (Ehrenberg) Itz. and Rothe. × 195. **Fig. 6.191.** *Closterium setaceum* Ehrenberg. × 390. **Fig. 6.192.** *C. leibleinii* Kützing. × 390. **Fig. 6.193.** *Spinoclosterium curvatum* Bernard var. *spinosum* Prescott. × 142. (After Prescott.) **Fig. 6.194.** *Cylindrocystis brébissonii* Meneghini. × 390.

Fig. 6.195. *Penium margaritaceum* (Ehrenberg) Brébisson. × 390. **Fig. 6.196.** *Docidium undulatum* Bail. × 390. **Fig. 6.197.** *Pleurotaenium trochiscum* W. and G. S. West var. *tuberculatum* G. M. Smith. × 195. **Fig. 6.198.** *P. nodosum* (Bail.) Lundell. × 195. **Fig. 6.199.** *P. ehrenbergii* (Brébisson) DBy. × 390. **Fig. 6.200.** *Triploceras gracile* Bailey. × 292. **Fig. 6.201.** *Penium polymorphum* Perty. × 412. **Fig. 6.202.** *Staurastrum orbiculare* Ralfs. × 404. **Fig. 6.203.** *Tetmemorus brébissonii* (Meneghini) Ralfs. × 195. **Fig. 6.204.** *Micrasterias pinnatifida* (Kützing) Ralfs. × 404. **Fig. 6.205.** *M. truncata* (Corda) Brébisson. × 412. **Fig. 6.206.** *M. radiata* Hassall. × 390.

230a (229) Poles of cells truncated; cells long-cylindric. (Fig. 6.187) . . . . .
                                        Gonatozygon de Bary
230b   Poles of cells rounded **231**
231a (230) Chloroplast indented at center; cells cylindric, straight, or arcuate.
              (Fig. 6.188). . . . . *Roya* W. et G. S. West
231b   Chloroplast not indented; cells subcylindric to elliptico-cylindric.
              (Fig. 6.189). . . . . *Mesotaenium* Nägeli
232a (229) With 2 asteroid chloroplasts. (Fig. 6.194). . . . .
                                        Cylindrocystis Meneghini
232b   Chloroplasts longitudinally ridged **233**
233a (232) Without a girdle piece; cells fusiform to cylindric.   (Fig.
              6.190). . . . . *Netrium* Nägeli
233b   With a girdle piece; cells cylindric. **238**

Family **Desmidiaceae**

234a   (17) Unicellular, occasionally in pairs **235**
234b   Colonial; colonies irregular or filamentous **249**
235a (234) Cylindric and straight, or attenuate and lunate. **236**
235b   Compressed or not compressed and with radiating polar
              processes. **243**
236a (235) Cylindric or slightly attenuate; straight **238**
236b   Poles equally attenuate, slightly to markedly lunate **237**
237a (210, 236) Poles of cell each with a stout spine. (Fig. 6.193) . . . .
                                        Spinoclosterium Bernard
237b   Poles of cell not bearing spines. (Figs. 6.191, 6.192) . . . .
                                        Closterium Nitzsch
238a (233, 236) Cells with a median constriction **239**
238b   Cells without a median constriction (Figs. 6.195, 6.201) . . . . . .
                                        Penium Brébisson
              See also **241**.
239a (238) Poles of cell truncate **240**
239b   Poles of cell incised or bearing spinescent processes **242**
240a (239) Basal inflation of each semicell vertically plicate. (Fig. 6.196). . .
                                        Docidium Brébisson
240b   Basal inflation not plicate. **241**
241a (240) Base of semicell inflated; poles usually with tubercules.   (Figs.
              6.197, 6.198, 6.199) . . . . *Pleurotaenium* Nägeli
241b   Base of semicell not inflated; without polar tubercules . . . .
                                        Penium Brébisson
              See also **238**.
242a (239) Apex of cell slightly compressed and incised. (Fig. 6.203) . . . . .
                                        Tetmemorus Ralfs
242b   Apex of cell flattened and bearing 2 spinescent processes. (Fig.
              6.200). . . . . *Triploceras* Bailey
243a (235) Cells compressed **244**
243b   Cells not compressed; poles bearing 3 or more radiating processes.
              (Fig. 6.202). . . . . *Staurastrum* Meyen
244a (243) Semicells much incised or deeply divided. (Figs. 6.204, 6.205,
              6.206). *Micrasterias* Agardh
244b   Semicells entire or at most bilobed. **245**
245a (244) Semicells with 2 long divergent arms. (Figs. 6.207, 6.208). . . . .
                                        Staurastrum Meyen
245b   Semicells without arms **246**
246a (245) Apices of cells clearly once incised. (Figs. 6.209, 6.210) . . . . . .
                                        Euastrum Ehrenberg

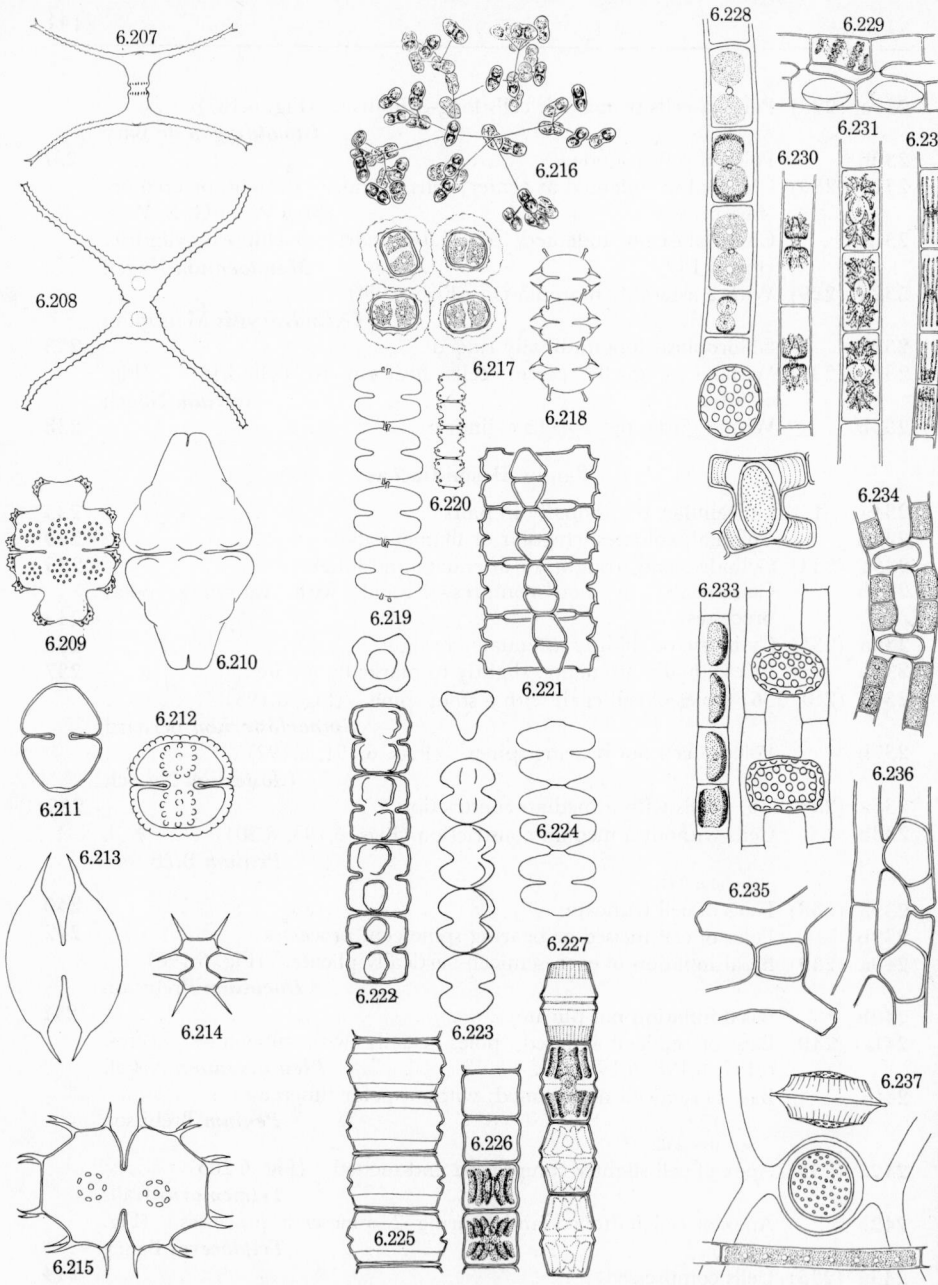

Fig. 6.207. *Staurastrum leptocladium* Nordstedt. var. *denticulatum* G. M. Smith. × 780. **Fig. 6.208.** *S. chaetoceras* (Schröder) G. M. Smith. × 404. **Fig. 6.209.** *Euastrum gemmatum* Brébisson. × 363. **Fig. 6.210.** *E. didelta* (Turpin) Ralfs. × 390. **Fig. 6.211.** *Cosmarium granatum* Brébisson. × 390. **Fig. 6.212.** *C. monomazum* Lundell. × 412. **Fig. 6.213.** *Arthrodesmus convergens* Ehrenberg. × 390. **Fig. 6.214.** *A. octocornis* Ehrenberg. × 404. **Fig. 6.215.** *Xanthidium cristatum* Brébisson var. *uncinatum* Brébisson. × 404.

Fig. 6.216. *Cosmocladium constrictum* Arch. × 195. **Fig. 6.217.** *Oöcardium stratum* Nägeli. × 312. (After Senn.) **Fig. 6.218.** *Onychonema laeve* Nordstedt var. *latum* W. and G. S. West. × 412. **Fig. 6.219.** *Sphaerozosma aubertianum* W. West var. *archerii* (Gutw.) W. and G. S. West. × 390. **Fig. 6.220.** *S. granulatum* Roy et Biss. × 404. **Fig. 6.221.** *Desmidium aptogonum* Brébisson. × 412. **Fig. 6.222.** *Phymatodocis nordstedtiana* Wolle. × 260. (After W. and G. S. West.) **Fig. 6.223.** *Spondylosium moniliforme* Lund. × 200. **Fig. 6.224.** *S. pulchrum* Arch. × 400. **Fig. 6.225.** *Desmidium grevillii* (Kützing) DBy. × 404. **Fig.. 6.226.** *Hyalotheca dissiliens* (J. E. Smith) Brébisson. × 390. **Fig. 6.227.** *Gymnozyga moniliformis* Ehrenberg. × 390.

142

246b   Apices of cells not incised. **247**

247a (246) Cells bearing obvious, rather long spines **248**

247b   Cells smooth to variously embellished, at most with minute barbs. (Figs. 6.211, 6.212). . . . . ***Cosmarium*** Corda

248a (247) Cell wall of equal thickness throughout. (Figs. 6.213, 6.214) . . . ***Arthrodesmus*** Ehrenberg

248b   Wall at mid-face of each semicell thickened, often scrobiclate. (Fig. 6.215). . . . . ***Xanthidium*** Ehrenberg

249a (42, 64, 234) Colony filamentous. **251**

249b   Colony not filamentous. **250**

250a (249) Planktonic; cells united by gelatinous bands. (Fig. 6.216) . . . . . ***Cosmocladium*** Brébisson

250b   Attached to calcareous rocks in flowing water; cells at the extremities of branched gelatinous tubes. (Fig. 6.217) . . . . ***Oöcardium*** Nägeli

251a (249) Cells united by polar processes. **252**

251b   Cells adhering by a contact surface or lobes. **253**

252a (251) Processes long; wide apart and overlapping adjacent cells. (Fig. 6.218). ***Onychonema*** Wallich

252b   Processes short or tuberculate, close together and not overlapping. (Figs. 6.219, 6.220 . . . . ***Sphaerozosma*** Corda

253a (251) Cells triangular or quadrangular in end view. **254**

253b   Cells circular or elliptic in end view. **256**

254a (253) Apices of young semicells replicate. (Fig. 6.221) . . . . ***Desmidium*** Agardh

254b   Apices of young cells not replicate. **255**

255a (254) End view quadrangular. (Fig. 6.222). . . . . ***Phymatodocis*** Nordstedt

255b   End view triangular. (Fig. 6.223) . . . ***Spondylosium*** Brébisson

256a (253) Cells ellipsoid in end view **257**

256b   Cells circular in end view **258**

257a (256) Cells markedly compressed; median constriction relatively deep. (Fig. 6.224). . . . . ***Spondylosium*** Brébisson

257b   Cells not greatly compressed; median constriction shallow and acute to scarcely evident. (Fig. 6.225) . . . ***Desmidium*** Agardh

258a (256) Cells square to rectangular in side view. (Fig. 6.226) . . . . ***Hyalotheca*** Ehrenberg

258b   Cells barrel-shaped, attenuated ends striate. (Fig. 6.227) . . . . . ***Gymnozyga*** Ehrenberg

Family **Zygnemataceae**

259a (70) With 2 opposed, discoidal chloroplasts. (Fig. 6.228) . . . . ***Pleurodiscus*** Lagerheim

259b   With 1 or more spiral, bandlike chloroplasts **260**

260a (259) Usually spiraled more than once; distinct conjugation tube. (Fig. 6.229). . . . . ***Spirogyra*** Link

260b   Rarely spiraled more than half a turn; no conjugation tube. (Fig. 6.232). ***Sirogonium*** Kützing

◄

Fig. 6.228. *Pleurodiscus borinquenae* Tiffany. × 210. (After Tiffany.) **Fig. 6.229.** *Spirogyra varians* (hassall) Kützing. × 85. **Fig. 6.230.** *Zygnemopsis spiralis* (Fritsch) Transeau. × 195. **Fig. 6.231.** *Zygnema insigne* (Hassall) Kützing. × 195. **Fig. 6.232.** *Sirogonium sticticum* (Engl. Bot.) Kützing. × 42. **Fig. 6.233.** *Mougeotiopsis calospora* Palla. × 250. (After Skuja.) **Fig. 6.234.** *Zygogonium ericetorium* Kützing. × 180. **Fig. 6.235.** *Mougeotia laetevirens* (A. Braun) Wittrock. × 202. **Fig. 6.236.** *M. transeaui* Collins. × 202. **Fig. 6.237.** *Debarya ackleyana* Transeau. × 210. (After Transeau.)

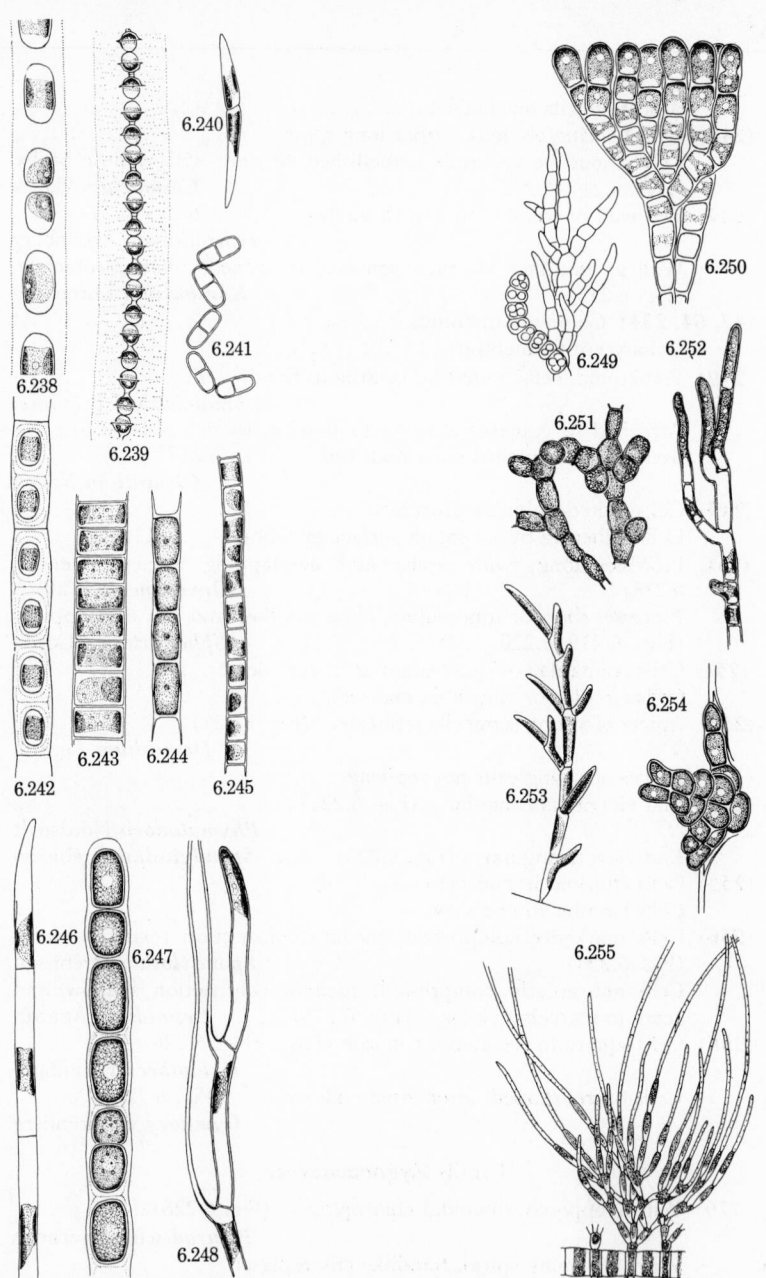

Fig. 6.238. *Geminella interrupta* Turpin. × 500. **Fig. 6.239.** *Radiofilum conjunctivum* Schmidle. × 390. **Fig. 6.240.** *Raphidionema sempervirens* Chodat. × 390. **Fig. 6.241.** *Stichococcus bacillaris* Nägeli. × 85. **Fig. 6.242.** *Binuclearia tatrana* Wittrock. × 500. **Fig. 6.243.** *Ulothrix zonata* (Web. et Mohr) Kützing. × 202. **Fig. 6.244.** *Microspora amoena* (Kützing) Rabenhorst. × 202. **Fig. 6.245.** *Hormidium subtile* (Kützing) Herring. × 390. **Fig. 6.246.** *Uronema elongatum* Hodgetts. × 500. **Fig. 6.247.** *Cylindrocapsa geminella* Wolle. × 390. **Fig. 6.248.** *Fridaea torrenticola* Schmidle. × 250.

Fig. 6.249. *Leptosira mediciana* Borzi. × 195. **Fig. 6.250.** *Gongrosira debaryana* Rabenhorst. × 195. **Fig. 6.251.** *Gomontia lignicola* G. T. Moore. × 195. **Fig. 6.252.** *Chlorotylium cataractum* Kützing. × 195. **Fig. 6.253.** *Microthamnion strictissimum* Rabenhorst. × 412. **Fig. 6.254.** *Entocladia pithophorae* (G. S. West) G. M. Smith. × 195. **Fig. 6.255.** *Draparnaldiopsis alpina* Smith and Klyver. × 325. (After Smith and Klyver.)

144

261a   (72)   With 2 stellate chloroplasts per cell **262**
261b   With 2 massive, cushionlike chloroplasts; terrestrial. (Fig. 6.234). . . . . ***Zygogonium*** Kützing
262a (261)   Gametangium with stratified gel around zygote. (Fig. 6.230) . . . ***Zygnemopsis*** Skuja
262b   Gametangium without gel around zygote. (Fig. 6.231) . . . . . . ***Zygnema*** Agardh
263a   (71)   Chloroplast without pyrenoids. (Fig. 6.233) . . . . ***Mougeotiopsis*** Palla
263b   Chloroplast with pyrenoids. **264**
264a (263)   Gametangium with gel around the zygote. (Fig. 6.237) . . . . . . ***Debarya*** Wittrock
264b   Gametangium without gel. (Figs. 6.235, 6.236) . . . . ***Mougeotia*** Agardh

Order **Ulotrichales,** Family **Ulotrichasceae**

265a   (66)   Cells cylindric; continuous or in pairs. (Fig. 6.238) . . . . ***Geminella*** Turpin
265b   Cells spheric to ellipsoid; continuous. (Fig. 6.239) . . . . ***Radiofilum*** Schmidle
266a   (66)   Filaments of 10 or fewer cells **267**
266b   Filaments of many cells. **268**
267a (266)   Terminal cells pointed. (Fig. 6.240). ***Rhaphidionema*** Lagerheim
267b   Terminal cells broadly rounded. (Fig. 6.241) . . . . ***Stichococcus*** Nägeli
268a (266)   Cells appearing as in pairs. (Fig. 6.242) . . ***Binuclearia*** Wittrock
268b   Cells not in pairs **269**
269a (268)   Cells shorter than wide, or dimensions equal. (Fig. 6.243). . . . . ***Ulothrix*** Kützing
269b   Cells longer than wide **270**
270a (269)   Cells 5 to 10 times longer than wide; end cells pointed. (Fig. 6.246). . . . . ***Uronema*** Lagerheim
270b   Cells half to 2 times longer than wide; end cells not pointed. (Fig. 6.245). . . . . ***Hormidium*** Kützing

Family **Microsporaceae**

271a   (69)   Cells 2 to 3 times as long as wide; walls of H-shaped pieces in optical section. (Fig. 6.244). . . . . ***Microspora*** Thuret
271b   Cells at least 6 times as long as wide; walls not composed of H-shaped pieces **298**

Family **Cylindrocapsaceae**

272a   (72)   With concentric sheaths; cells ellipsoid for most part. (Fig. 6.247). ***Cylindrocapsa*** Reinsch
272b   Without concentric sheaths; cells square to rectangular **293**
273a   (79)   Distal end of many cells prolonged as setae. (Fig. 6.248) . . . . . ***Fridaea*** Schmidle
273b   Cells without such processes **274**
274a (80, 273) Encrusting or perforating shells, rocks, and wood. **275**
274b   Endophytic or terrestrial **276**
275a (274)   Branches perforating shells, rock, or wood. (Fig. 6.251) . . . . . . ***Gomontia*** Bornet et Flahault

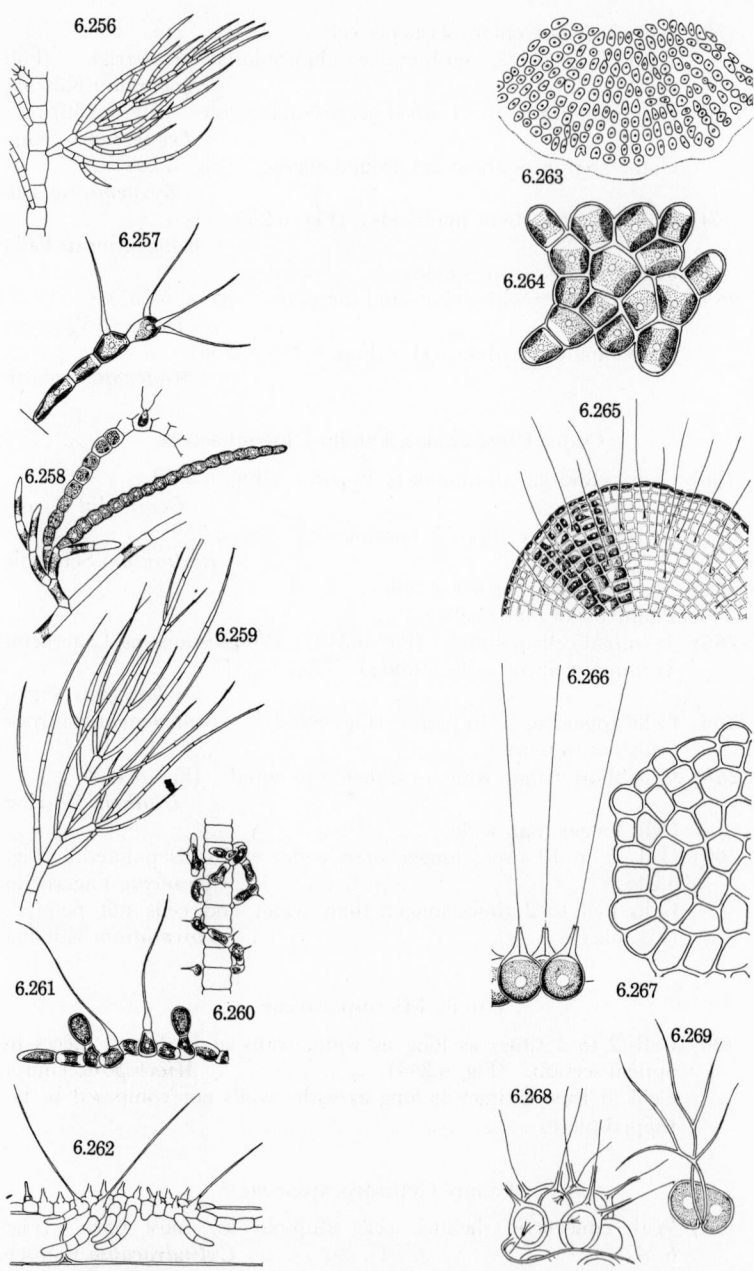

Fig. 6.256. *Draparnaldia plumosa* (Vaucher) Agardh. × 85. **Fig. 6.257.** *Thamniochaete huberi* Gay. × 250.
Fig. 6.258. *Chaetophora incrassata* (Hudson) Hazen. × 195. **Fig. 6.259.** *Stigeoclonium lubricum* (Dillw.)
Fries. × 200. **Fig. 6.260.** *Aphanochaete vermiculoides* Wolle. × 202. **Fig. 6.261.** *Chaetonema irregulare*
Nowakowski. × 412. **Fig. 6.262.** *Aphanochaete repens* A. Braun. × 195.

Fig. 6.263. *Pseudoulvella americana* (Snow) Wille. (After Snow.) **Fig. 6.264.** *Protoderma viride* Kützing.
× 500. **Fig. 2.265.** *Coleochaete scutata* Brébisson. × 85. **Fig. 6.266.** *Chaetosphaeridium globosum* (Nordstedt)
Klebahn. × 404. **Fig. 6.267.** *Dermatophyton radians* Peter. × 166. **Fig. 6.268.** *Conochaete comosa* Klebahn.
× 500. (After Prescott and Croasdale). **Fig. 6.269.** *Dicranochaete reniformis* Hieronymus. × 404.

275b   With erect, sparingly branched filaments; sporangia terminal. (Fig. 6.250). . . . . **Gongrosira** Kützing

276a (274, 83) Terrestrial on mud; freely branched. (Fig. 6.249) . . . .
                                                               Leptosira Borzi

Family **Chaetophoraceae**

276b   Endophytic in walls of other algae and aquatic phanerogams. (Fig. 6.254). . . . . **Entocladia** Reinke

277a (80) Terminal cells of filaments rounded **278**

277b   Terminal cells of filaments attenuated to pointed **279**

278a (277) Filaments with alternating series of long and short cells. (Fig. 6.252). **Chlorotylium** Kützing

278b   Cells nearly same size throughout; terminal cells tapered. (Fig. 6.253). . . . . **Microthamnion** Nägeli

279a (277) Cells of main axis distinctly larger than those of branches **280**

279b   Cells of axes and branches of about same size **281**

280a (279) Main axis of alternate long and short cells. (Fig. 6.255). . . . . .
                             Draparnaldiopsis Smith et Klyver

280b   Cells of main axis same size throughout. (Fig. 6.256) . . . .
                                     Draparnaldia Bory

281a (279) Alga with a firm gel and definite shape. (Fig. 6.258) . . . .
                                   Chaetophora Schrank

281b   Alga with an indistinguishable gel and indefinite in shape. (Fig. 6.259). . . . . **Stigeoclonium** Kützing

282a (79) Filaments procumbent, with or without erect branches. **283**

282b   Filaments erect and sparingly branched. (Fig. 6.257) . . . .
                                   Thamniochaete Gay

283a (282) Procumbent filaments with short, erect branches. (Fig. 6.261) . .
                                   Chaetonema Nowakowski

283b   Filaments and branches wholly procumbent. (Figs. 6.260, 6.262). . . . . **Aphanochaete** A. Braun

284a (85) Epiphytic **285**

284b   Epizoic on turtles, several cells thick at center. (Fig. 6.267) . . . .
                                   Dermatophyton Peter

285a (284) One cell thick throughout. (Fig. 6.264) . . **Protoderma** Kützing

285b   Several cells thick in the middle. (Fig. 6.263) . . . .
                                   Pseudoulvella Wille

Family **Coeleochaetaceae**

286a (78, 178) Growing on or beneath the sheath of other algae. (Fig. 6.265). . . . . **Coleochaete** Brébisson

286b   Epiphytic and solitary or clustered, with or without a gelatinous envelope. **287**

287a (37, 43, 286) Cells reniform; setae branched and without a basal sheath. (Fig. 6.269) . . . . **Dicranochaete** Hieronymus

287b   Cells spheric to ellipsoid; setae simple and with a basal sheath. . . **288**

288a (287) One seta per cell. (Fig. 6.266) . . **Chaetosphaeridium** Klebahn

288b   Two to several setae per cell. (Fig. 6.268) . **Conochaete** Klebahn

Family **Sphaeropleaceae**

289a (70, 185) Cross walls thick and warty; germlings tapered at each end. (Fig. 6.270). . . . . **Sphaeroplea** Agardh

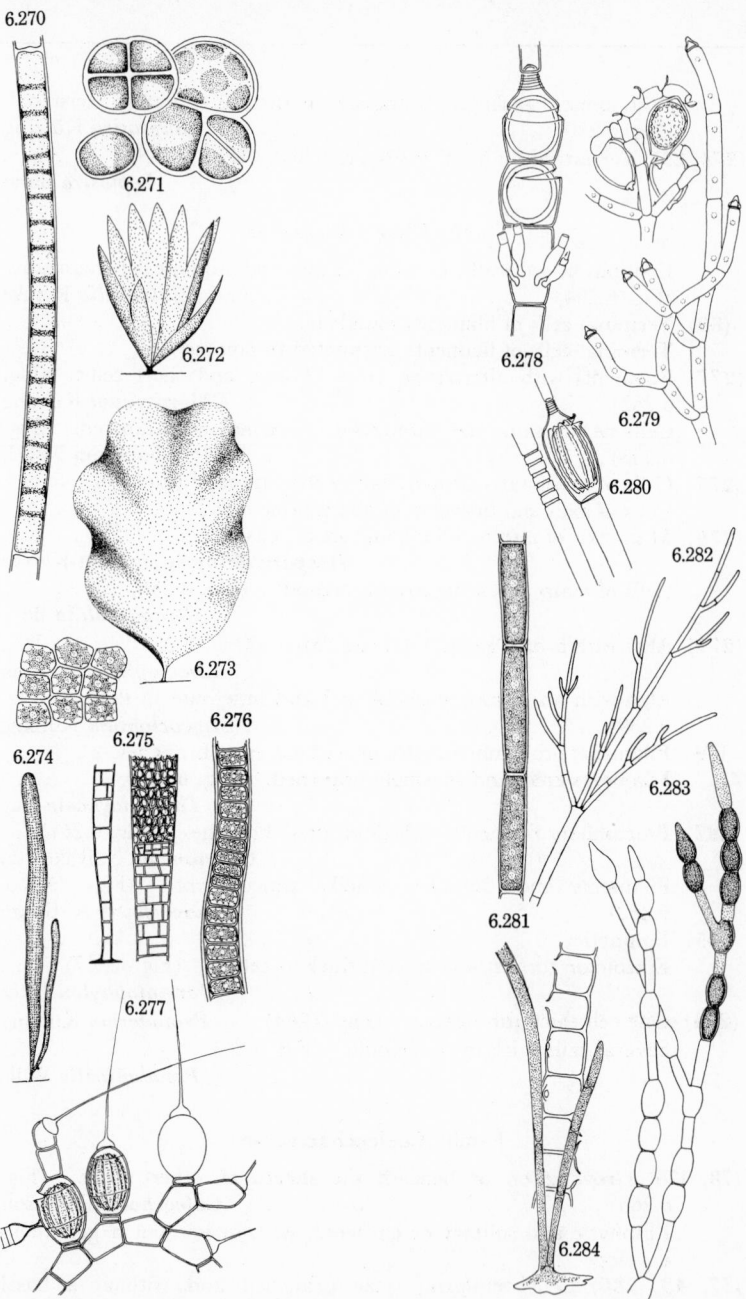

Fig. 6.270. *Sphaeroplea annulina* (Roth) Agardh. × 250. **Fig. 6.271.** *Protococcus viridis* Agardh. × 780. **Fig. 6.272.** *Monostroma quaternarium* (Kützing) Desmaz. × 0.5. **Fig. 6.273.** *Prasiola mexicana* J. A. Agardh. Habit. × 1; detail × 500. **Fig. 6.274.** *Enteromorpha prolifera* (Fl. Dan.) J. G. Agardh. × 97. **Fig. 6.275.** *Schizomeris leibleinii* Kützing. × 150. **Fig. 6.276.** *Schizogonium murale* Kützing. × 250. **Fig. 6.277.** *Bulbochaete varians* Wittrock. × 195.

Fig. 6.278. *Oedogonium hians* Nordstedt and Hirn. × 202. **Fig. 6.279.** *Oedocladium operculatum* Tiffany. × 195. **Fig. 6.280.** *Oedogonium crenulatocostatum* Wittrock forma *cylindricum* Hirn. × 202. **Fig. 6.281.** *Rizoclonium hieroglyphicum* (Agardh) Kützing. × 195. **Fig. 6.282.** *Cladophora glomerata* (L.) Kützing. × 42. **Fig. 6.283.** *Pithophora oedogonia* (Montagne) Wittrock. × 170. **Fig. 6.284.** *Basicladia chelonum* (Collins) Hoffmann and Tilden. × 95; portion with zoosporangia × 101.

Family **Protococcaceae**

289b Characteristically aerial and rarely filamentous. (Fig. 6.271) . . .
Protococcus Agardh

Order **Ulvales,** Family **Ulvaceae**

290a (82) Cylinder hollow; chloroplasts parietal. (Fig. 6.274) . . . .
Enteromorpha Link
290b Cylinder solid **292**
291a (85) Thallus fan-shaped, erect; chloroplasts parietal. (Fig. 6.272) . . .
Monostroma Thuret
291b Thallus irregular, prostrate; chloroplast stellate and axial **293**

Family **Schizomeridaceae**

292a (290) Chloroplast parietal; filamentous below parenchymatous above.
(Fig. 6.275) . . . . *Schizomeris* Kützing
292b Chloroplast axial and stellate **293**

Order **Schizogoniales,** Family **Schizogoniaceae**

293a (272, 291, 292) Usually filamentous, sometimes more than 1 cell wide.
(Fig. 6.276) . . . . *Schizogonium* Kützing
293b Usually an expanded sheet of cells; with marginal rhizoids or
with a stalk. (Fig. 6.273) . . . . *Prasiola* Meneghini

Order **Oedogoniales,** Family **Oedogoniaceae**

294a (69) Filaments simple; attached but frequent in floating masses.
(Figs. 6.278, 6.280) . . . . *Oedogonium* Link
294b Filaments branched; aquatic or terrestrial. **295**
295a (76, 294) Branching unilateral; cells with bulbous-based setae, all
aquatic. (Fig. 6.277) . . . . *Bulbochaete* Agardh
295b Freely branched; cells with displaced caps; mostly terrestrial.
(Fig. 6.279) . . . . *Oedocladium* Stahl

Order **Cladophorales,** Family **Cladophoraceae**

296a (76) Freely branched throughout **297**
296b Branches few, from the base or rhizoidal in form. **298**
297a (296) With numerous dark akinetes. (Fig. 6.283) . . . .
Pithophora Wittrock
297b Without conspicuous akinetes. (Fig. 6.282) . . . .
Cladophora Kützing
298a (296, 271) Filaments simple or with short rhizoidal branches. (Fig.
6.281) . . . . *Rhizoclonium* Kützing
298b Branching from the base; epizoic on turtles. (Fig. 6.284) . . . . .
Basicladia Hoffmann and Tilden

Class **Charophyceae,** Order **Charales,** Family **Characeae**

299a (74) Corona 5-celled; alga with stipulodes, corticated or not. (Figs.
6.285, 6.286) . . . . *Chara* Valliant
299b Corona 10-celled; alga without stipulodes and uncorticated **300**
300a (299) Fertile branches simple or once to repeatedly forked. (Figs. 6.287,
6.288) . . . . *Nitella* Agardh
300b Fertile branches pinnate. (Fig. 6.289) . . . *Tolypella* Leonhardi

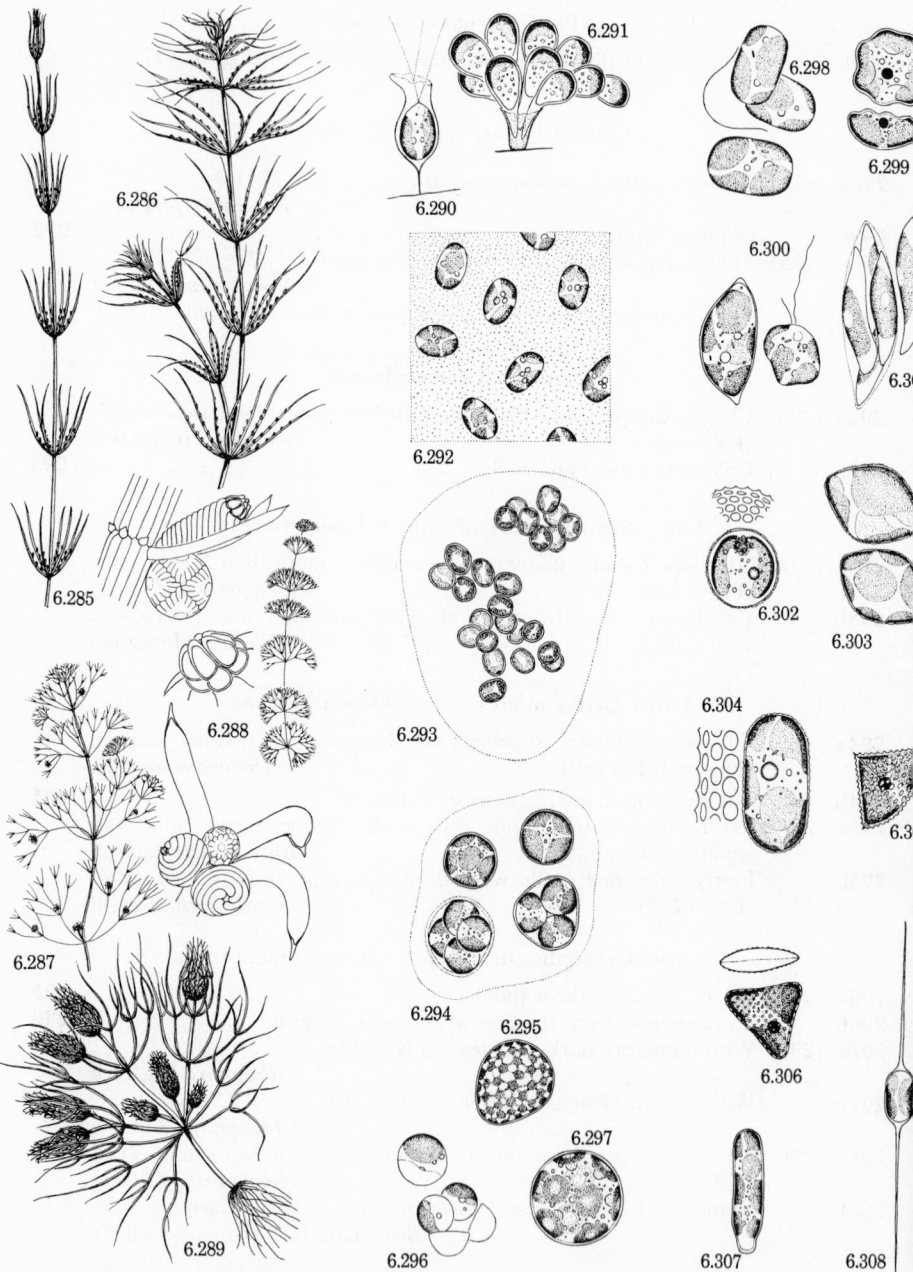

Fig. 6.285. *Chara globularis* Thuill. × 0.5; fertile node × 20. **Fig. 6.286.** *C. sejuncta* A. Braun. × 0.5. **Fig. 6.287.** *Nitella microcarpa* A. Braun var. *megacarpa* (T.F.A.) Nordstedt. × 0.5; fertile branch × 20; corona × 180. **Fig. 6.288.** *N. transilis* Allen. × 0.5. **Fig. 6.289.** *Tolypella glomerata* (Desv.) Leonhardi. × 0.3.

Fig. 6.290. *Stipitococcus urcelolatus* W. and G. S. West. × 1560. **Fig. 6.291.** *Malleodendron caespitosum* Thompson. × 404. **Fig. 6.292.** *Gloeochloris smithiana* Pascher. × 500. (After G. M. Smith.) **Fig. 6.293.** *Gloeobotrys limnetica* (G. M. Smith) Pascher. × 500. (After G. M. Smith.) **Fig. 6.294.** *Chlorobotrys regularis* (West) Bohlin. × 808. **Fig. 6.295.** *Leuvenia natans* Gardner. × 390. **Fig. 6.296.** *Diachros simplex* Pascher. × 1250. (After Pascher.) **Fig. 6.297.** *Botrydiopsis arhiza* Borzi. × 390.

Division **Chrysophyta**, Class **Xanthophyceae**, Order **Rhizochloridales**, Family **Stipitococcaceae**

301a	(88)	Cells naked, within a stalked, urn-shaped lorica. (Fig. 6.290). . . . *Stipitococcus* W. and G. S. West	
301b		Cells with a wall; without a lorica	**306**

Order **Heterocapsales**, Family **Malleodendraceae**

302a	(89)	Cells ovoid to obpyriform; at the extremities of gelatinous branches. (Fig. 6.291). . . . . *Malleodendron* Pascher	
302b		Cells spheric, cylindric or panduriform	**321**
303a	(90)	Cells ellipsoid or subspheric	**304**
303b		Cells regularly spheric or globose	**305**

Family **Chlorosaccaceae**

304a	(303)	Colony spherical to ovoid, attached or free; reproduction by zoospores. (Fig. 6.292) . . . . *Gloeochloris* Pascher	

Order **Heterococcales**, Family **Gloeobotrydiaceae**

304b		Colony amorphous; planktonic; reproduction by autospores. (Fig. 6.293). . . . . *Gloeobotrys* Pascher	
305a	(303)	With few to many cells; mother cell wall not evident. (Fig. 6.294). *Chlorobotrys* Bohlin	

Family **Pleurochloridaceae**

305b		With 4 cells at most; mother cell halves present. (Fig. 6.296) . . . *Diachros* Pascher	
306a	(301)	Cells free-floating	**307**
306b		Cells attached to some substratum	**322**
307a	(306)	Walls of cell smooth	**308**
307b		Walls of cells sculptured	**315**
308a	(307)	Cells spheric to ovoid or pyriform	**309**
308b		Cells cylindric, fusiform, or hemispheric and crenate	**311**
309a	(308)	Cells spheric.	**310**
309b		Young cells spheric; mature cells ovid to pyriform. (Fig. 6.295). *Leuvenia* Gardner	
310a	(309)	Aquatic; temporarily colonial; mother cell halves present. (Fig. 6.296). . . . . *Diachros* Pascher See also **305**.	
310b		Terrestrial or aquatic; solitary and extremely variable in size. (Fig. 6.297). . . . . *Botrydiopsis* Borzi	
311a	(308)	Cells cylindric, straight, or curved	**312**
311b		Cells fusiform or hemispheric and crenate in outline	**313**
312a	(311)	Cell length at most twice width; poles equally rounded. (Fig. 6.298). . . . . *Monallantus* Pascher	
312b		Cell length more than twice the width	**319**
313a	(311)	Cells hemispheric with crenated margin. (Fig. 6.299) . . . . *Chlorogibba* Geitler	
313b		Cells spindle-shaped or fusiform	**314**

◄

Fig. 6.298. *Monallantus brevicylindrus* Pascher. × 1100. (After Pascher.) **Fig. 6.299.** *Chlorogibba trochisciaeformis* Geitler. × 1250. **Fig. 6.300.** *Pleurogaster lunaris* Pascher. × 800. (After Pascher.) **Fig. 6.301.** *Chlorocloster pirenigera* Pascher. × 720. (After Pascher.) **Fig. 6.302.** *Arachnochloris minor* Pascher. × 1250. (After Pascher.) **Fig. 6.303.** *Trachychloron biconicum* Pascher. × 1050. (After Pascher.) **Fig. 6.304.** *Chlorallanthus oblongus* Pascher. × 1440. (After Pascher.) **Fig. 6.305.** *Tetraedriella acuta* Pascher. × 780. **Fig. 6.306.** *Gonochloris sculpta* Geitler. × 780. **Fig. 6.307.** *Bumilleriopsis breve* (Gern.) Printz. × 390. **Fig. 6.308.** *Centritractus belonophorus* (Schmidle) Lemmermann. × 404.

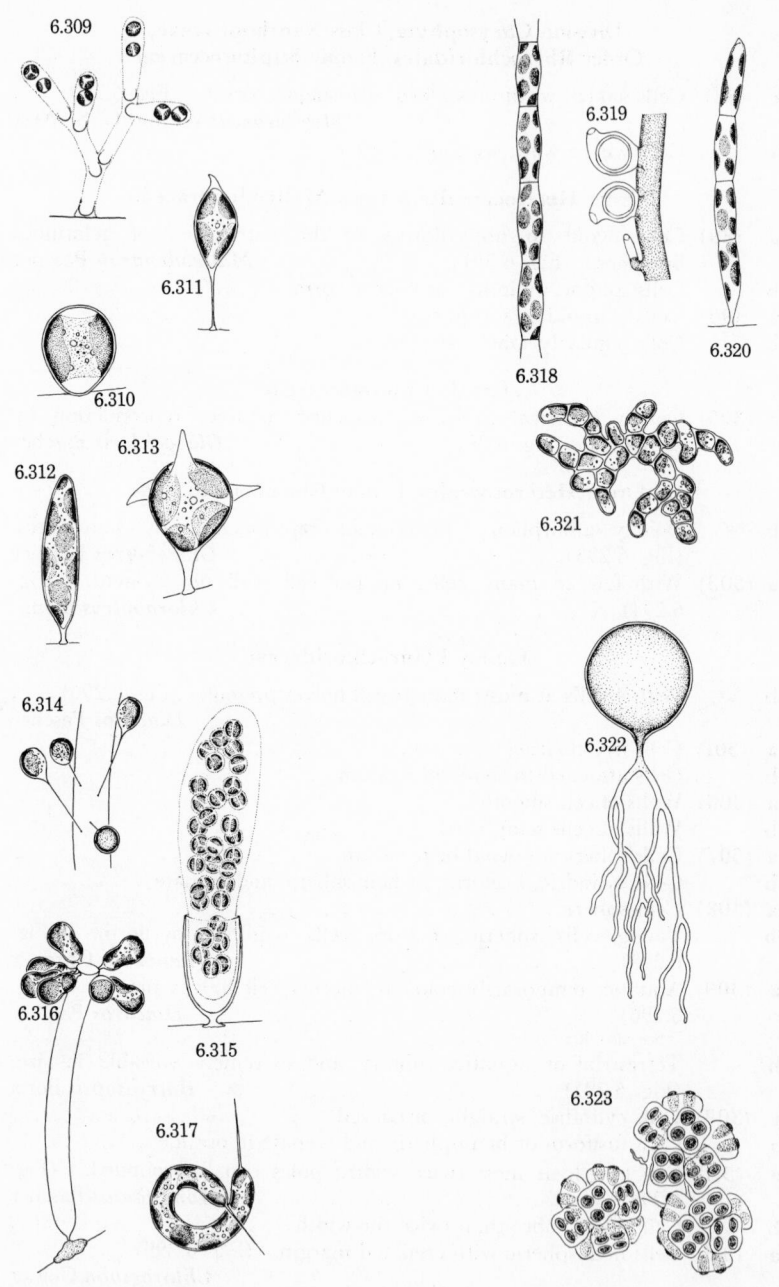

Fig. 6.309. *Mischococcus confervicola* Nägeli. × 390. **Fig. 6.310.** *Lutherella adhaerens* Pascher. × 808. **Fig. 6.311.** *Characiopsis longipes* Borzi. × 404. **Fig. 6.312.** *C. polychloris* Pascher. × 412. **Fig. 6.313.** *Dioxys tricornuta* Thompson. × 780. **Fig. 6.314.** *Peroniella planctonica* G. M. Smith. **Fig. 6.315.** *Chlorothecium pirottae* Borzi. × 404. **Fig. 6.316.** *Ophiocytium arbusculum* (A. Braun) Rabenhorst. × 390. **Fig. 6.317.** *O. capitatum* Wolle. × 390.

Fig. 6.318. *Tribonema bombycinum* (Agardh) Derbes et Solier. × 404. **Fig. 6.319.** *Vaucheria aversa* Hassall. × 45. **Fig. 6.320.** *Bumillaria exilis* Klebs. × 412. **Fig. 6.321.** *Monocilia simplex* Pascher. × 202. **Fig. 6.322.** *Botrydium granulatum* (L.) Grev. × 15. **Fig. 6.323.** *Botryococcus braunii* Kützing. × 180.

314a (313) Poles unevenly attenuated into short, blunt, or long, slender spines. (Fig. 6.300) . . . . ***Pleurogaster*** Pascher
314b Poles evenly attenuated; cells straight, arcuate, or sigmoid. (Fig. 6.301). . . . . ***Chlorocloster*** Pascher
315a (307) Cells spheric or ellipsoid to cylindric. 316
315b Cells triangular to pyramidal. 318
316a (315) Cells ellipsoid to cylindric 317
316b Cells spheric to ellipsoid; with 1 parietal, lobed chromatophore. (Fig. 6.302). . . . . ***Arachnochloris*** Pascher
317a (316) Ellipsoid to biconic; sculpturing finely reticulate. (Fig. 6.303) ***Trachychloron*** Pascher
317b Ellipsoid to cylindric; sculpturing regular rows of pits. (Fig. 6.304). . . . . ***Chlorallanthus*** Pascher
318a (315) Cells pyramidal. (Fig. 6.305). . . . . . . ***Tetraedriella*** Pascher
318b Cells triangular and strongly compressed. (Fig. 6.306). . . . . . . ***Goniochloris*** Geitler

Family **Centritractaceae**

319a (312) Mamillate-capitate at one end; straight or curved. (Fig. 6.307). . ***Bumilleriopsis*** Printz
319b Cells not mamillate-capitate 320
320a (319) Wall with halves overlapping or separated; ending in long spines; straight. (Fig. 6.308) . . . . ***Centritractus*** Lemmermann
320b Arcuate or coiled; with or without terminal spines 326
321a (302) Cells panduriform to cylindric; stipitate. 326

Family **Mischococcaceae**

321b Cells spheric; usually in pairs at the ends of gelatinous branches. (Fig. 6.309). . . . . ***Mischococcus*** Nägeli

Family **Chloropediaceae**

322a (306) Sessile without a stalk; cells ovoid. (Fig. 6.310) . . . . ***Lutherella*** Pascher
322b Cells with a stalk 323

Family **Characiopsidaceae**

323a (322) Cells ovoid to fusiform or lanceolate. (Figs. 6.311, 6.312 . . . . . ***Characiopsis*** Borzi
323b Cells spheric to oval or angular and lobed 324
324a (323) Cells 2- or 3-lobed, with or without processes. (Fig. 6.313) . . . . ***Dioxys*** Pascher
324b Cells rounded at apex; with or without a terminal spine 325
325a (324) Stalk very slender, longer than the cell; without a terminal spine. (Fig. 6.314). . . . . ***Peroniella*** Gobi

Family **Chlorotheciaceae**

325b Stalk shorter than the cell or cell nearly sessile 326
326a (320, 321, 325) Cells spheric, ellipsoid to pyriform; stalk massive. (Fig. 6.315). . . . . ***Chlorothecium*** Borzi
326b Cells panduriform to cylindric; stalk spinelike. (Figs. 6.316, 6.317). . . . . ***Ophiocytium*** Nägeli
327a (87) Filamentous; simple or branched 328
327b Vesicular; terrestrial 331
328a (327) Alga a simple filament 329
328b Alga branched 330

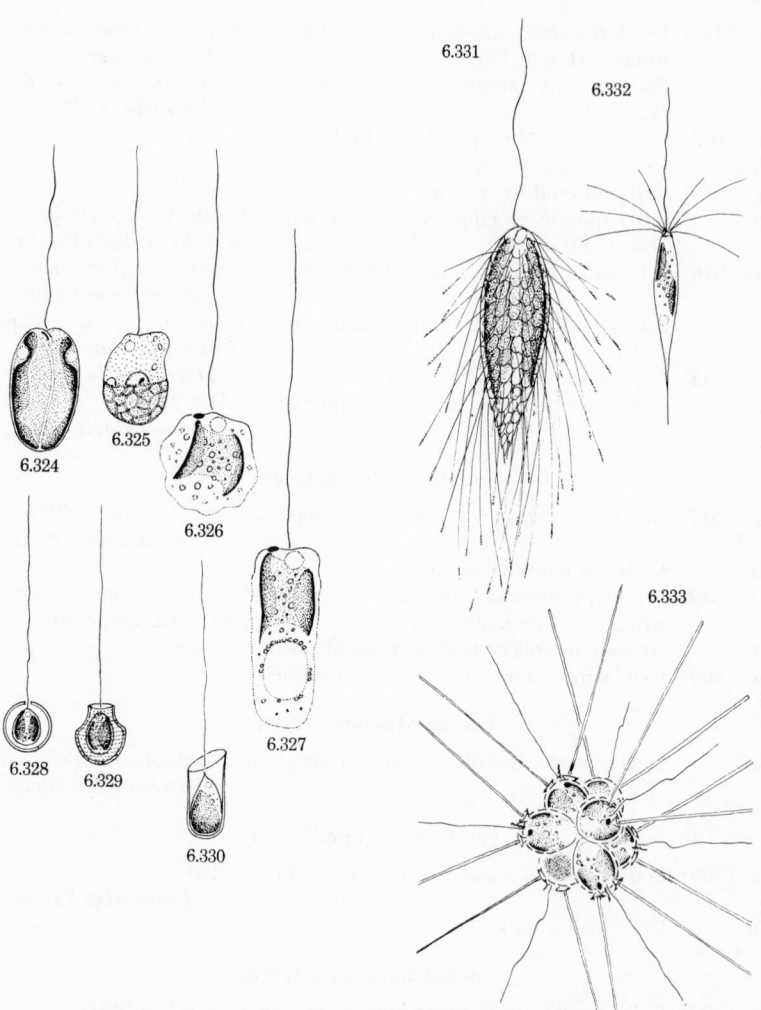

Fig. 6.324. *Amphichrysis compressa* Korshikov. × 539. Fig. 6.325. *Chrysapsis sagene* Pascher. × 667. (After Pascher.) Fig. 6.326. *Chromulina globosa* Pascher. × 1040. Fig. 6.327. *C. flavicans* Butschli. × 1040. Fig. 6.328. *Chrysococcus rufescens* Klebs. × 1040. Fig. 6.329. *C. amphora* Lackey. × 1040. Fig. 6.330. *Kephyrion ovum* Pascher. × 1040. Fig. 6.331. *Mallomonas caudata* Iwanoff. × 539. Fig. 6.332. *M. akrokomos* Ruttner. × 520. Fig. 6.333. *Chrysosphaerella longispina* Lauterborn. × 520.

Order **Heterotrichales,** Family **Tribonemataceae**

329a (328) Cell length 2 or more times width, broadest at the middle. (Fig. 6.318) . . . . **Tribonema** Derbes et Solier

329b Cell length less than twice the width; width equal throughout. (Fig. 6.320) . . . . **Bumillaria** Borzi

Family **Monociliaceae**

330a (328) Alga multicellular; cells with few chromatophores. (Fig. 6.321) . . . . **Monocilia** Gerneck

Order **Heterosiphonales,** Family **Vaucheriaceae**

330b   Alga a nonseptate, branched filament; chromatophores numerous; aquatic or terrestrial. (Fig. 6.319) . . . . *Vaucheria* De Candolle

Family **Botrydiaceae**

331a **(90, 327)** Vesicle globose or branched; with colorless, branched rhizoidal branched. (Fig. 6.322) . . . . *Botrydium* Wallroth

Doubtful **Xanthophyceae**

331b   Cells included or protruding from a cartilaginous gelatinous matrix. (Fig. 6.323) . . . . *Botryococcus* Kützing

Class **Chrysophyceae,** Order **Chrysomonadales** (=Chrysomonadina), Family **Chromulinaceae**

332a **(93)** Cells flagellated; free-swimming or sedentary 333
332b   Cells not flagellated; immobile. 352
333a **(332)** With 1 flagellum 334
333b   With 2 flagella only, or with apparently 3 (Fig. 6.334) 340
334a **(333)** Cells naked 335
334b   Cells within a lorica (Fig. 6.346) or bearing siliceous rods or scales 337
335a **(334)** Cells with 2 anterior-lateral contractile vacuoles. (Fig. 6.324). . .
                                                Amphichrysis Korshikov
335b   Cells with 1 anterior contractile vacuole 336
336a **(335)** Chromatophore reticulate. (Fig. 6.325). . . . *Chrysapsis* Pascher
336b   Chromatophores entire. (Figs. 6.326, 6.327) . . . .
                                                Chromulina Cienkowski
337a **(334)** Cells with a lorica 338
337b   Cells with siliceous rods or scales; unicellular or colonial. 339
338a **(337)** Lorica ovoid or globose; flagellar opening relatively small. (Figs. 6.328, 6.329) . . . . *Chrysococcus* Klebs
338b   Lorica ovoid-cylindric with a wide orifice. (Fig. 6.330) . . . . . .
                                                Kephyrion Pascher

Family **Mallomonadaceae**

339a **(337)** Unicellular; scales of most species with slender needles. (Figs. 6.331, 6.332) . . . . *Mallomonas* Perty
339b   Colonial; cells with 2 to several siliceous rods at the anterior portion of the cell. (Fig. 6.333). . *Chrysosphaerella* Lauterborn
340a **(333)** Cells with 2 flagella only 341

Family **Prymnesiaceae**

340b   Cells with 2 flagella and a flagellumlike hapteron. (Fig. 6.334). .
                                                Chrysochromulina Lackey
341a **(340)** Flagella equal or subequal 342
341b   Flagella markedly unequal 346
342a **(341)** Unicellular 343
342a   Colonial 344

Family **Coccolithophoridaceae**

343a **(342)** With ringlike coccoliths in the cellular envelope. (Fig. 6.335). . .
                                                Hymenomonas Stein

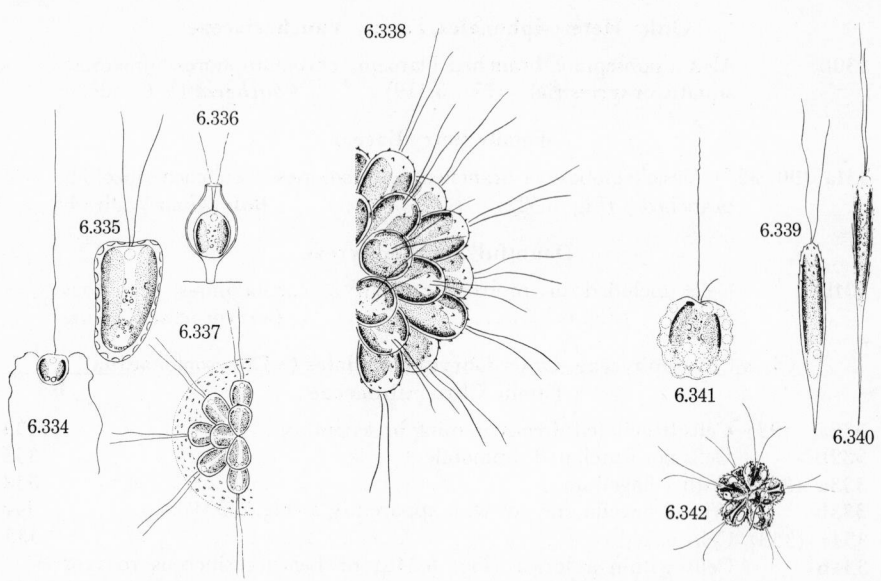

Fig. 6.334. *Chrysochromulina parva* Lackey. × 780. **Fig. 6.335.** *Hymenomonas roseola* Stein. × 500. **Fig. 6.336.** *Derepyxis amphora* Stokes. × 808. **Fig. 6.337.** *Syncrypta volvox* Ehrenberg. (After Stein.) × 454. **Fig. 6.338.** *Synura uvella* Ehrenberg. × 195. **Fig. 6.339.** *S. adamsii* G. M. Smith. Single cell. × 390. **Fig. 6.340.** *S. caroliniana* Whitford. Single cell. × 390. **Fig. 6.341.** *Ochromonas crenata* Klebs. × 600. **Fig. 6.342.** *Skadovskiella sphagnicola* Korsh. × 404.

Family **Syncryptaceae**

343b　　　Cells sedentary and within a lorica. (Fig. 6.336) . . . .
　　　　　　　　　　　　　　　　　　　　　　　　Derepyxis Stokes
344a **(342)** Colony with a gelatinous matrix. (Fig. 6.337) . . . .
　　　　　　　　　　　　　　　　　　　　　　　　Syncrypta Ehrenberg
344b　　　Colony without a gelatinous matrix **345**

Family **Synuraceae**

345a **(344)** Cells with 2 parietal, platelike chromatophores. (Figs. 6.338, 6.339, 6.340) . . . . **Synura** Ehrenberg
345b　　　Cells with 2 axial, platelike chromatophores. (Fig. 6.342) . . . . .
　　　　　　　　　　　　　　　　　　　　　　　　Skadovskiella Korshikov

Family **Ochromonadaceae**

346a **(341)** Cells with a vaselike or cylindrical lorica **350**
346b　　　Cells without a lorica **347**
347a **(346)** Unicellular. (Fig. 6.341) . . . . **Ochromonas** Wysotzki
347b　　　Colonial **348**
348a **(347)** Colony a circle of cells arranged as a shallow cone. (Fig. 6.343) .
　　　　　　　　　　　　　　　　　　　　　　　　Cyclonexis Stokes
348b　　　Colony globular; cells at the periphery of a gelatinous matrix . . . **349**
349a **(348)** Inner pole of cells flat or rounded; smaller flagellum minute. (Fig. 6.344) . . . . **Uroglenopsis** Lemmermann
349b　　　Inner pole of cells attenuate; shorter flagellum $\frac{1}{3}$ to $\frac{1}{2}$ the length of the longer. (Fig. 6.345) . . . . . . **Uroglena** Ehrenberg

350a (346) Solitary or gregarious; sedentary; lorica smooth. (Fig. 6.348) . . .
Epipyxis Ehrenberg

350b   Branched colonial or solitary **351**

351a (350) Colony free-swimming. (Figs. 6.346, 6.347, 6.349) . . . .
Dinobryon Ehrenberg

351b   Epiphytic; lorica with hairlike projections in optical section.
(Fig. 6.350) . . . . *Hyalobryon* Lauterborn

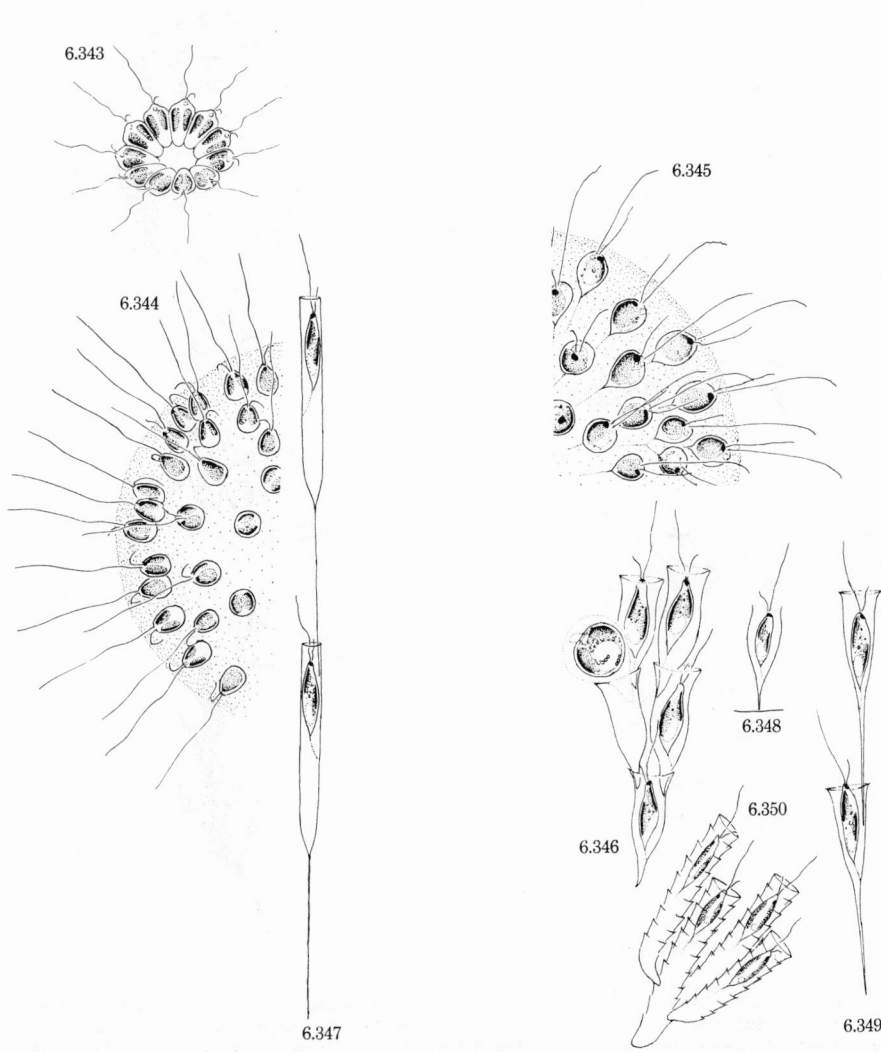

Fig. 6.343. *Cyclonexis annularis* Stokes. × 539.   **Fig. 6.344.** *Uroglenopsis americana* (Calkins) Lemmermann. × 520.   **Fig. 6.345.** *Uroglena volvox* Ehrenberg. × 520.   **Fig. 6.346.** *Dinobryon sertularia* Ehrenberg. × 520.   **Fig. 6.347.** *D. borgei* Lemmermann. var. *elongata* Pascher. × 520.   **Fig. 6.348.** *Epipyxis utriculus* Stein. × 520.   **Fig. 6.349.** *Dinobryon stipitatum* Stein. × 520.   **Fig. 6.350.** *Hyalobryon ramosum* Lauterborn. × 539.

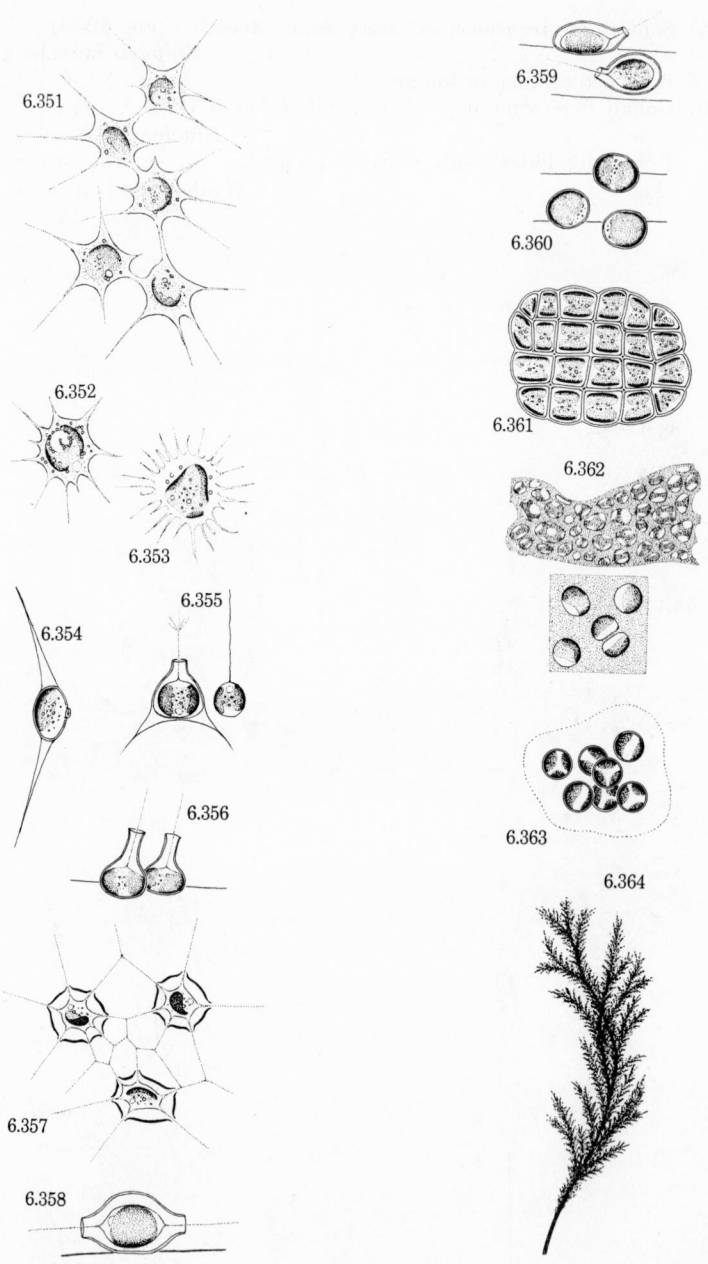

Fig. 6.351. *Chrysidiastrum catenatum* Lauterborn. × 390. **Fig. 6.352.** *Chrysamoeba radians* Klebs. × 404. **Fig. 6.353.** *Rhizochrysis scherffelii* Pascher. × 404. **Fig. 6.354.** *Bitrichia phaseolus* (Fott) Bourelly. × 808. **Fig. 6.355.** *Chrysopyxis bipes* Stein. × 780. **Fig. 6.356.** *Lagynion scherffelii* Pascher. × 780. **Fig. 6.357.** *Heliapsis mutabilis* Pascher. × 780. **Fig. 6.358.** *Chrysoamphitrema ovum* Scherffel. × 1560.

Fig. 6.359. *Kybotion ellipsoideum* Thompson. × 1200. **Fig. 6.360.** *Epichrysis paludosa* (Korsch.) Pascher. × 780. **Fig. 6.361.** *Phaeoplaca thallosa* Chodat. × 808. **Fig. 6.362.** *Phaeosphaera perforata* Whitford. Habit. × 3; detail × 375. (After Whitford.) **Fig. 6.363.** *Chrysocapsa planctonica* (W. and G. S. West) Pascher. × 600. **Fig. 6.364.** *Hydrurus foetidus* (Vill.) Trev. × 1.

Order **Rhizochrysidales,** Family **Rhizochrysidaceae**

352a (332) Cells amoeboid with rhizopodia *353*
352b Cells not amoeboid *361*
353a (352) Cells naked *354*
353b Cells enclosed in a test or lorica *356*
354a (353) Cells in linear colonies. (Fig. 6.351). . . . .
             Chrysidiastrum Lauterborn
354b Cells solitary or in temporary irregular colonies *355*
355a (354) Cells with short, relatively thick pseudopodia. (Fig. 6.352) . . . .
             Chrysamoeba Klebs
355b Cells with long slender rhizopodia. (Fig. 6.353) . . . .
             Rhizochrysis Pascher
356a (353) Free-floating; lorica with tapering polar horns. (Fig. 6.354). . . .
             Bitrichia Woloszynska
356b Attached to some substratum *357*
357a (356) Base of lorica or test flat and broad *358*
357b Base of lorica two-pronged; prongs connected by a fine thread
             around the host filament. (Fig. 6.355) . . . . *Chrysophyxis* Stein
358a (357) Cells with 1 or 2 slender rhizopodia *359*
358b Cells with more rhizopodia, each issuing from a separate pore.
             (Fig. 6.357). *Heliapsis* Pascher
359a (358) Lorica ovoid or ellipsoid, horizontal with terminal pores. *360*
359b Lorica bottle-shaped; erect with a single opening. (Fig. 6.356) . .
             Lagynion Pascher
360a (359) With a short neck and pore at one end. (Fig. 6.359). . . . .
             Kybotion Pascher
360b With a short neck and pore at each end. (Fig. 6.358) . . . .
             Chrysoamphitrema Scherffel
361a (352) Filamentous and microscopic or macroscopic and crustose *368*
361b Not filamentous. *362*
362a (361) Colonial with a gelatinous sheath or matrix. *363*
362b Solitary or colonial, without a sheath *364*
363a (362) Colony with gelatinous or hairlike setae *367*
363b Colony without setae *365*

Order **Chrysosphaerales,** Family **Chrysosphaeraceae**

364a (362) Solitary or in few-celled aggregates; epiphytic. (Fig. 6.360) . . . .
             Epichrysis Pascher

Order **Chrysocapsales,** Family **Chrysocapsaceae**

364b Colony compact, crustose of rarely more than 30 cells. (Fig.
             6.361). . . . . *Phaeoplaca* Chodat
365a (363) Colony of many cells; globose to cylindric; perforated. (Fig.
             6.362). . . . . *Phaeosphaera* W. and G. S. West
365b Colony not perforated; microscopic or macroscopic *366*
366a (365) Colony microscopic and globular; of few cells. (Fig. 6.363) . . . .
             Chrysocapsa Pascher

Family **Hydruraceae**

366b Colony macroscopic; tufted; fetid. (Fig. 6.364). . . . .
             Hydrurus Agardh

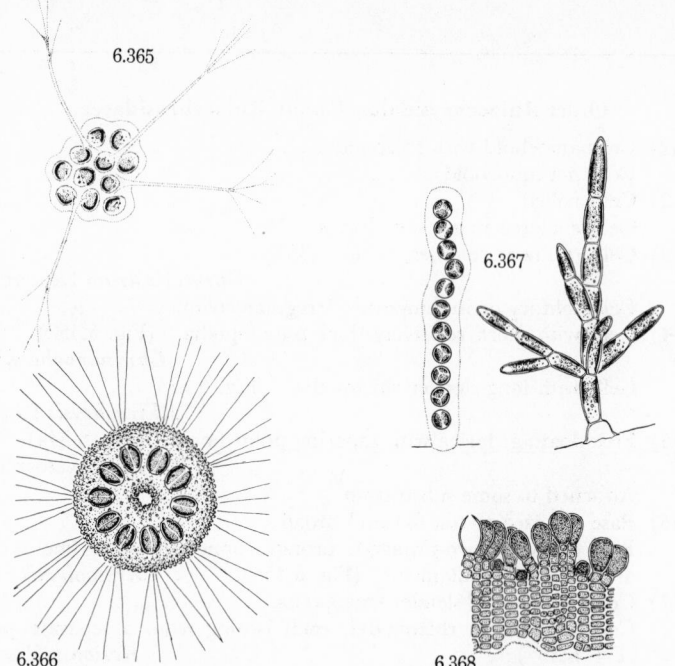

Fig. 6.365. *Naegeliella flagellifera* Correns. × 390. **Fig. 6.366.** *Chrysostephanosphaera globulifera* Scherffel. × 404. **Fig. 6.367.** *Phaeothamnion confervicola* Lagerheim. × 390. (*a*) Palmella. **Fig. 6.368.** *Heribaudiella fluviatilis* (Gom.) Svedelius. (After Flahault.)

Family **Naegeliellaceae**

367a (363) Colony sessile; with multiple gelatinous setae. (Fig. 6.365) . . . .
Naegeliella Correns

367b   Free-floating; cells radial in one plane; setae hairlike. (Fig. 6.366) . . . . ***Chrysostephanosphaera*** Scherffel

Order **Chrysotrichales**, Family **Phaeothamniaceae**

368a (361) Filamentous, branched, and microscopic; palmella stage a branching gelatinous tube containing uniseriate spherical cells. (Fig. 6.367) . . . . ***Phaeothamnion*** Lagerheim

Division **Phaeophyta**

368b   Crustose and macroscopic; growing on rocks in swiftly flowing streams as olive-brown discs. (Fig. 6.368) . . . .
Heribaudiella Gomont

Division **Pyrrophyta**, Class **Desmokontae**, Order **Desmomonadales**, Family **Prorocentraceae**

369a (96, 100, 103) Cells with a bivalved theca without a transverse furrow; flagella issuing from a funnel-shaped pore at the anterior end. (Fig. 6.369) . . . . ***Exuviaella*** Cienkowski

369b   Cells with a transverse furrow or girdle; naked or thecate 370

Class **Dinophyceae**

370a (2, 369) Cells motile 371
370b   Cells immobile 384

371a (370) Girdle completely encircling the cell. **372**
371b   Girdle incompletely encircling the cell. **376**
372a (371) Cells without a theca **373**
372b   Cells with a theca composed of plates **377**

Order **Gymnodiniales,** Family **Gymnodiniaceae**

373a (372) Girdle spiraling steeply to the left; ends displaced 3 to 5 times
their width at the sulcus; sulcus oblique. (Fig. 6.370) . . . .
Gyrodinium Kofoid and Swezy
373b   Girdle straight, or spiraled to left with slight displacement at ends
adjoining the sulcus. **374**
374a (373) Girdle supramedian, median, or inframedian. (Figs. 6.371,
6.372, 6.373, 6.374) . . . . ***Gymnodinium*** Stein

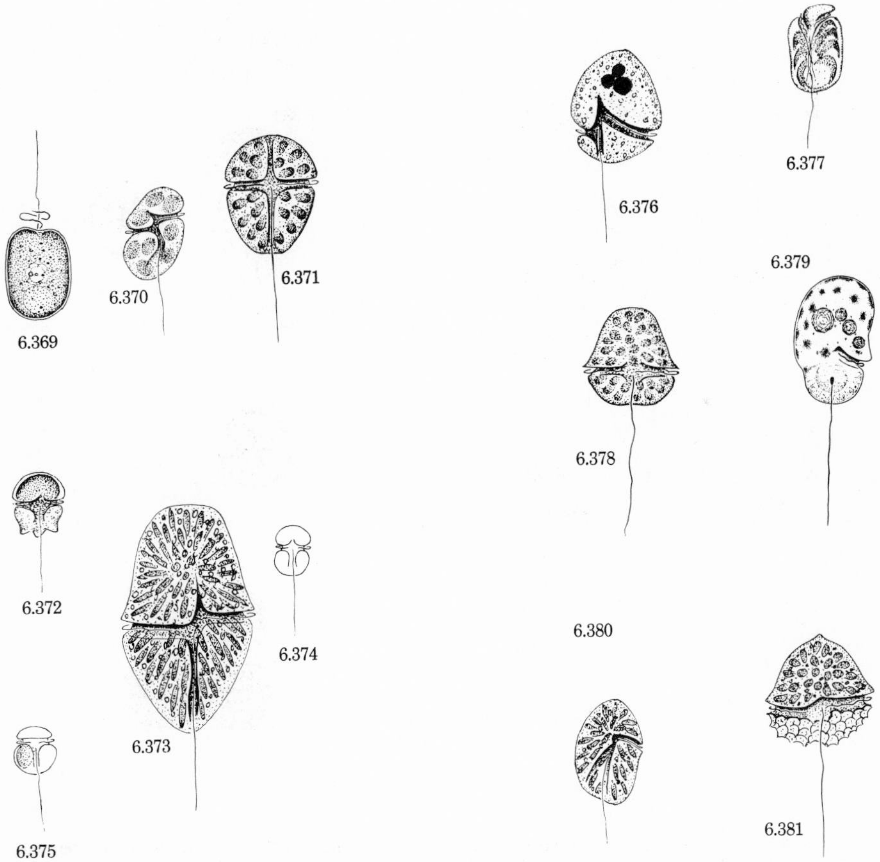

Fig. **6.369.** *Exuviaella compressa* Ostenf. × 539. Fig. **6.370.** *Gyrodinium pusillum* (Schilling) Kof. and Swezy. × 539. Fig. **6.371.** *Gymnodinium aeruginosum* Stein. × 539. Fig. **6.372.** *G. triceratium* Skuja. × 539. Fig. **6.373.** *G. fuscum* (Ehrenberg) Stein. × 539. Fig. **6.374.** *G. albulum* Lindemann. × 539. Fig. **6.375.** *Amphidinium lacustre* Stein. × 520. Fig. **6.376.** *Massartia vorticella* (Stein) Schiller. × 520. Fig. **6.377.** *Amphidinium klebsii* Kof. and Swezy. × 539. Fig. **6.378.** *Massartia musei* (Danysz) Schiller. × 539. **Fig. 6.379.** *Bernardinium bernardinense* Chodat. × 1040. Fig. **6.380.** *Hemidinium nasutum* Stein. × 539. **Fig. 6.381.** *Woloszynskia reticulata* Thompson. × 520.

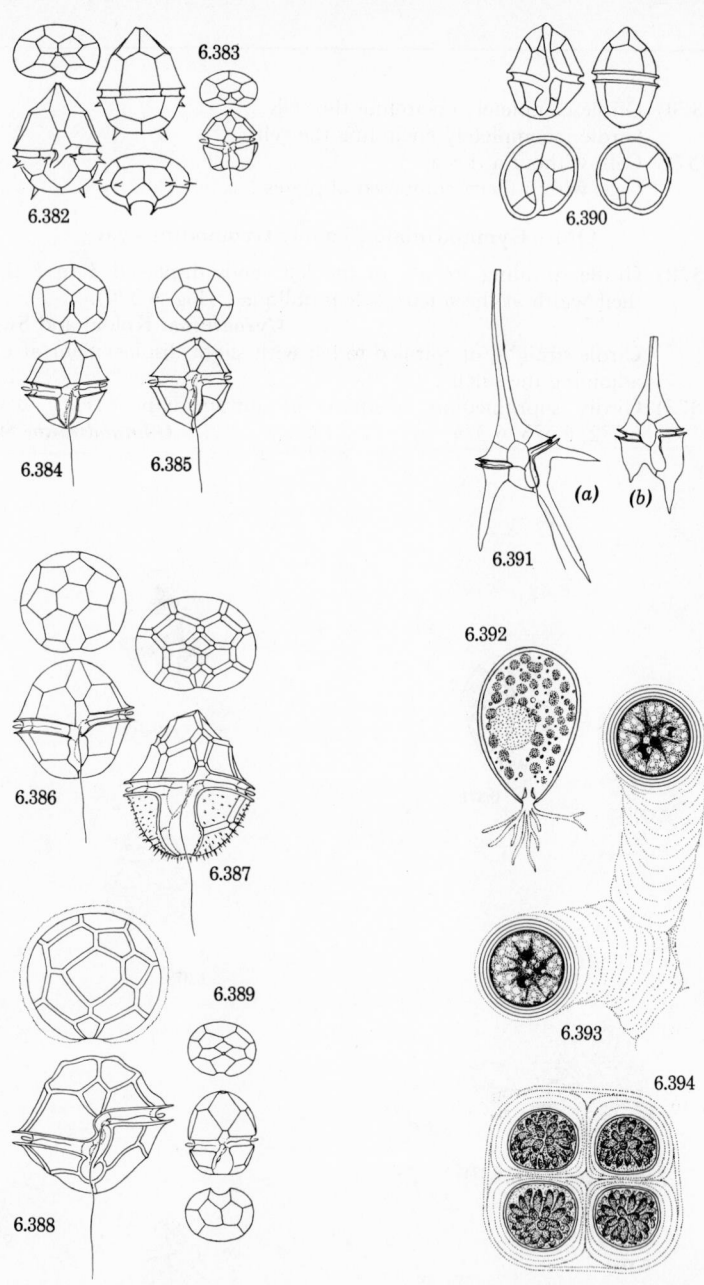

Fig. 6.382. *Glenodinium quadridens* (Stein) Schiller. × 404. **Fig. 6.383.** *G. penardiforme* (Lindemann) Schiller. × 390. **Fig. 6.384.** *G. kulczynskii* (Woloszynska) Schiller. × 390. **Fig. 6.385.** *G. berolinense* (Lemmermann) Lindem. × 390. **Fig. 6.386.** *Sphaerodinium cinctum* (Ehrenberg) Woloszynska × 390. **Fig. 6.387.** *Peridinium palatinum* Lauterborn. × 390. **Fig. 6.388.** *P. gatunense* Nygaard. × 390. **Fig. 6.389.** *P. umbonatum* Stein. × 404.

Fig. 6.390. *Gonyaulax apiculata* (Penard) Entz fil. × 270. (After Entz; not known for the U.S.) **Fig. 6.391.** *Ceratium hirundinella* (O.F.M.) Schrank. × 195. (*a*) Forma *austriacum* Zederb. (*b*) Forma *brachyceras* Daday. **Fig. 6.392.** *Oödinium limneticum* Jacobs. × 312. (After Jacobs.) **Fig. 6.393.** *Urococcus insignis* (Hassall) Kützing. × 195. **Fig. 6.394.** *Gloeodinium montanum* Klebs. × 310.

374b		Girdle near the anterior or the posterior end	**375**

375a (374) Girdle near anterior end; epicone very small. (Figs. 6.375, 6.377) . . . . *Amphidinium* Claparède and Lachmann

375b Girdle near posterior end; hypocone very small. (Figs. 6.376, 6.378) . . . . *Massartia* Conrad

376a (371) Girdle starting in lower third of cell. (Fig. 6.379) . . . . *Bernardinium* Chodat

Order **Peridiniales**, Family **Glenodiniaceae**

376b Girdle starting in upper third of cell. (Fig. 6.380) . . . . *Hemidinium* Stein

377a (372) Theca composed of many (50 or more) small plates. (Fig. 6.381) . . . . / . . . . . . *Woloszynskia* Thompson

377b Theca composed of few (15 to 25) plates **378**

378a (377) Sulcus reaching both apices, usually sigmoid **383**

378b Sulcus entering eitheca ⅓ of the way at most **379**

379a (378) Epitheca produced as a long horn; hypotheca bearing 2 or 3 horns **383**

379b Epitheca not produced as a horn but may be conical. **380**

380a (379) With 5 to 6 apical plates and hypothecal spines. (Fig. 6.382) . . . *Glenodinium* Stein

380b With 2 to 4 apical plates; with or without spines **381**

381a (380) Without anterior intercalary plates except for one species, which has 1, and its variety, which has 2 such plates. (Figs. 6.383, 6.384, 6.385) . . . . *Glenodinium* Stein

381b Characteristically with 2 to 4 anterior intercalary plates **382**

382a (381) With 4 anterior intercalary and 6 postcingular plates. (Fig. 6.386) . . . . *Sphaerodinium* Woloszynska

Family **Peridiniaceae**

382b With 2 or 3 anterior intercalary and 5 postcingular plates. (Figs. 6.387, 6.388, 6.389) . . . . *Peridinium* Ehrenberg

Family **Gonyaulaceae**

383a (378, 379) With one postintercalary plate; girdle ends, at the sulcus, usually displaced 2 or more times width of girdle. (Fig. 6.390) . . *Gonyaulax* Diesing

Family **Ceratiaceae**

383b Without a postintercalary plate; girdle ends, at the sulcus, nearly even and dissipated. (Fig. 6.391) . . . . *Ceratium* Schrank

Order **Dinocapsales**

384a (370) Cells in branched or irregular gelatinous masses **385**

384b Cells solitary; planctonic; epiphytic or epizoic **386**

Family **Gloeodiniaceae**

385a (384) Cells enclosed in concentric gelatinous sheaths. (Fig. 6.394). . . . *Gloeodinium* Klebs

385b Cells at the ends of gelatinous stalks. (Fig. 6.393) . . . . *Urococcus* Kützing

386a (95, 384) Cells free-floating. **388**

386b Cells attached; epiphytic or epizoic **387**

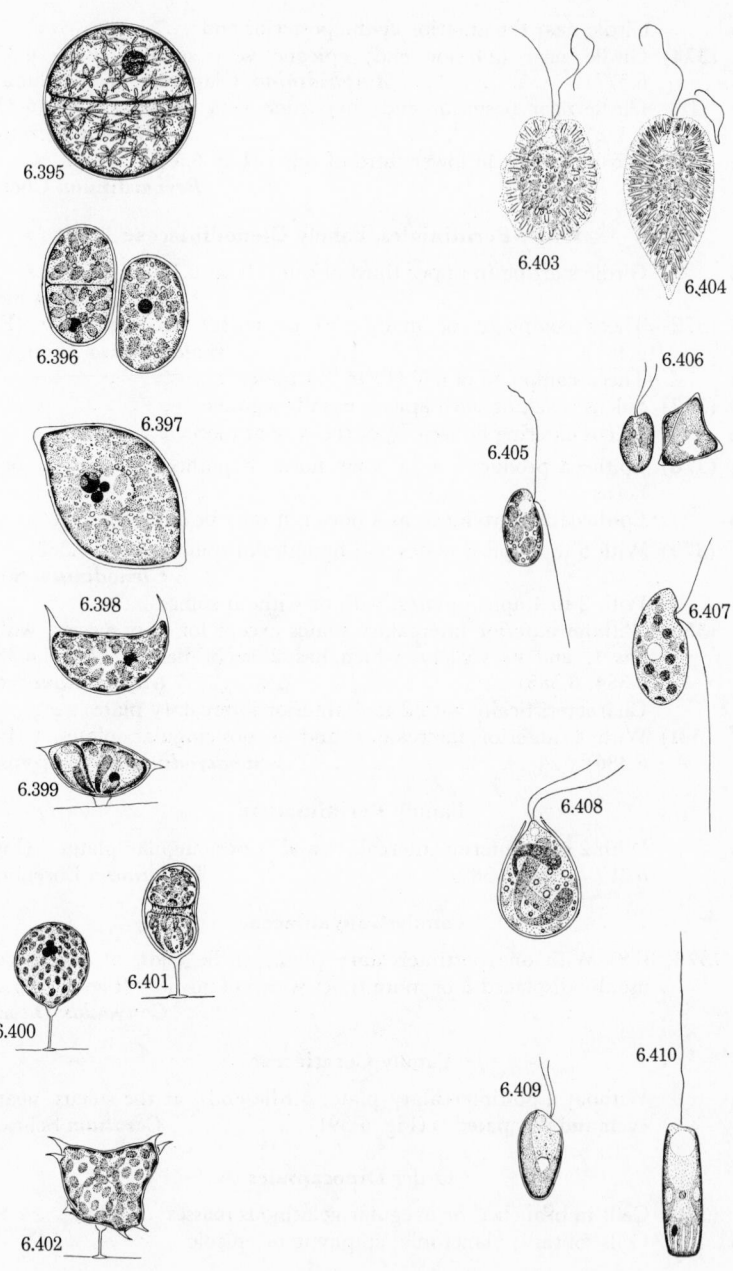

Fig. 6.395. *Hypnodinium sphaericum* Klebs. × 404.   **Fig. 6.396.** *Phytodinium simplex* Klebs.   × 727. **Fig. 6.397.** *Cystodinium bataviense* Klebs. × 363.   **Fig. 6.398.** *C. iners* Geitler. × 363.   **Fig. 6.399.** *Raciborskia oedoginii* (P. Richter) Pascher. × 404.   **Fig. 6.400.** *Stylodinium globosum* Klebs. × 390.   **Fig. 6.401.** *Dinopodiella phaseolus* Pascher. × 808.   **Fig. 6.402.** *Tetradinium javanicum* Klebs. × 404.

Fig. 6.403. *Gonyostomum latum* Iwanoff. × 404.   **Fig. 6.404.** *G. semen* Dies. × 360.   **Fig. 6.405.** *Merotrichia capitata* Skuja. × 390.   **Fig. 6.406.** *Tetragonidium verrucatum* Pascher. × 404.   **Fig. 6.407.** *Cyanomonas coeruleus* Lackey. (After Lackey.)   **Fig. 6.408.** *Cyanomastix morgani* Lackey. × 1000. (After Lackey.)   **Fig. 6.409.** *Chroomonas nordstedtii* Hansgirg. × 1560.   **Fig. 6.410.** *Monomastix opisthostigma* Scherffel. × 808.

Family **Blastodiniaceae**

387a (386) Epizoic on fish; globose to pyriform with rhizopodial attachments. (Fig. 6.392). . . . . *Oödinium* Jacobs
387b   Epiphytic; attached by a discoidal holdfast **390**

Family **Phytodiniaceae**

388a (386) Cells spheric, large; protoplast with girdle, sulcus, and eyespot. (Fig. 6.395). . . . . *Hypnodinium* Klebs
388b   Cells small, lunate to spheric or ellipsoid **389**
389a (388) Cells spheric to ellipsoid; reproduction by autospores. (Fig. 6.396). *Phytodinium* Klebs
389b   Cells lunate; reproduction by zoospores or aplanospores. (Figs. 6.397, 6.398) . . . . *Cystodinium* Klebs
390a (387) Cells lunate-fusiform or bean-shaped. **391**
390b   Cells globose, ovoid or square to tetrahedral **392**
391a (390) Lunate-fusiform with apical "spines"; stalk median to the long axis. (Fig. 6.399) . . . . *Raciborskia* Woloszynska
391b   Bean-shaped; stalk at one end. (Fig. 6.401) . . . . *Dinopodiella* Pascher
392a (390) Cells globose to square; without spines. (Fig. 6.400). . . . . *Stylodinium* Klebs
392b   Cells tetrahedral with spines at free apices. (Fig. 6.402) . . . . . . *Tetradinium* Klebs

Organisms of Uncertain Position

Order **Chloromonadales**

393a (86) Trichocysts scattered; flagella anterior (Figs. 6.403, 6.404) . . . . . *Gonyostomum* Diesing
393b   Trichocysts in needlelike clusters; flagella lateral. (Fig. 6.405) . . *Merotrichia* Mereschkowski

Class **Cryptophyceae,**
Order **Cryptococcales,** Family **Cryptococcaceae**

394a (96, 102, 103) Motile **395**
394b   Immobile and tetrahedral; planctonic. (Fig. 6.406) . . . . *Tetragonidium* Pascher

Order **Cryptomonadales**

395a (394) Cells with flagella inserted anteriorly **396**
395b   Cells with flagella inserted laterally and medially. **404**
396a (395) Cells with a deep gullet; pigmented or colorless. **402**
396b   Cells without a gullet. **397**

Family **Cryptochrysidaceae**

397a (100, 396) Cells with a lorica; chromatophore blue-green. (Fig. 6.408). *Cyanomastix* Lackey
397b   Cells without a lorica. **398**
398a (397) Chromatophores blue-green **399**
398b   Chromatophores not blue-green **400**
399a (398) With several discoidal, parietal chromatophores. (Fig. 6.407). . . . *Cyanomonas* Oltmanns
399b   With 1 parietal, laminate chromatophore. (Fig. 6.409) . . . . . . *Chroomonas* Hansgirg

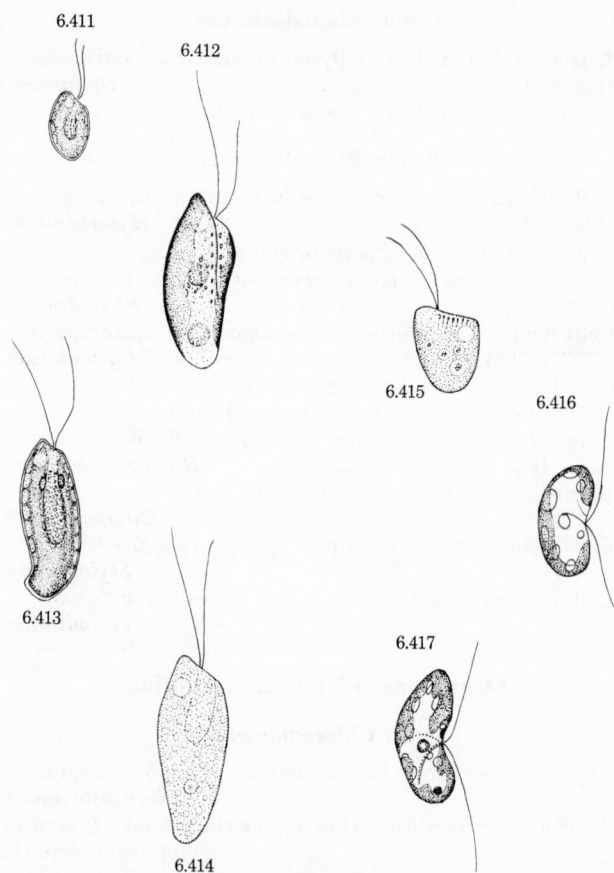

Fig. 6.411. *Rhodomonas lacustris* Pascher. × 520. **Fig. 6.412.** *Cryptochrysis commutata* Pascher. × 1040. **Fig. 6.413.** *Cryptomonas reflexa* (Marsson) Skuja. × 520. **Fig. 6.414.** *Chilomonas paramaecium* Ehrenberg. × 539. **Fig. 6.415.** *Cyathomonas truncata* Fromental. × 1040. **Fig. 6.416.** *Nephroselmis olivacea* Stein. × 667. (After Pascher.) **Fig. 6.417.** *Protochrysis phaeophycearum* Pascher. × 800. (After Pascher.)

400a (398) With 1 flagellum and a posterior group of trichocysts. (Fig. 6.410) . . . . ***Monomastix*** Scherffel
400b With 2 flagella. **401**
401a (400) One chromatophore, olive to brown or red, and with a pyrenoid. (Fig. 6.411) . . . . ***Rhodomonas*** Karsten
401b Two chromatophores; olive and without pyrenoids. (Fig. 6.412) . . . . ***Cryptochrysis*** Pascher

Family **Cryptomonadaceae**

402a (396) Colorless. **403**
402b Olive to brown chromatophores, with or without pyrenoids. (Fig. 6.413) . . . . ***Cryptomonas*** Ehrenberg
403a (5, 402) Cells greatly flattened; anterior end broadly truncate. (Fig. 6.415) . . . . ***Cyathomonas*** Fromentel

403b   Cells slightly flattened; anterior end oblique. (Fig. 6.414) . . . . .
Chilomonas Ehrenberg

Family **Nephroselmidaceae**

404a **(395)** Cells with a gullet; without an eyespot. (Fig. 6.416) . . . .
Nephroselmis Stein

404b   Cells without a gullet; with an eyespot. (Fig. 6.417) . . . .
Protochrysis Pascher

Division **Rhodophyta,** Class **Rhodophyceae**

405a **(97, 99)** Alga unicellular or filamentous **407**
405b   Alga multicellular and simple or branching **406**
406a **(94, 101, 405)** Alga macroscopic, firm or cartilaginous **411**
406b   Alga macroscopic or in macroscopic tufts; lax and collapsing on removal from water **412**

Subclass **Bangioideae**
(of uncertain systematic position)

407a **(405)** Unicellular with a gelatinous sheath; terrestrial. (Fig. 6.418) . . .
Porphyrdium Nägeli

407b   Filamentous **408**

Order **Bangiales**

408a **(407)** Chromatophores stellate and axial **409**
408b   Chromatophores discoidal to strap-shaped, parietal **410**

Family **Goniotrichaceae**

409a **(408)** Miscroscopic; filaments simple or falsely branched; chromatophores bright blue-green. (Fig. 6.420) . . . . . *Asterocytis* Gobi
409b   Micro- or macroscopic; filaments red-violet to red-brown **410**

Family **Bangiaceae**

410a **(408, 409)** Filaments simple below, at least 2 cells thick above. (Fig. 6.419) . . . . *Bangia* Lyngbye
410b   Filaments simple throughout, microscopic; chromatophores olivaceous. (Fig. 6.421) . . . . *Kyliniella* Skuja

Family **Erythrotrichiaceae**

411a **(406)** Freely branched; corticated with a single axial row of large discoidal or short-cylindric cells. (Fig. 6.422) . . . .
Campsopogan Montagne

411b   Branched or cylindric and simple; cells of axial row long-cylindric **415**

Subclass **Florideae,** Order **Nemalionales**

412a **(406)** With a central axis of large cells bearing smaller-celled branches **413**
412b   Cells of axis and branches not markedly different **414**

Family **Batrachospermaceae**

413a **(412)** Carposporophyte of compact filaments. (Fig. 6.423) . . . .
Batrachospermum Roth

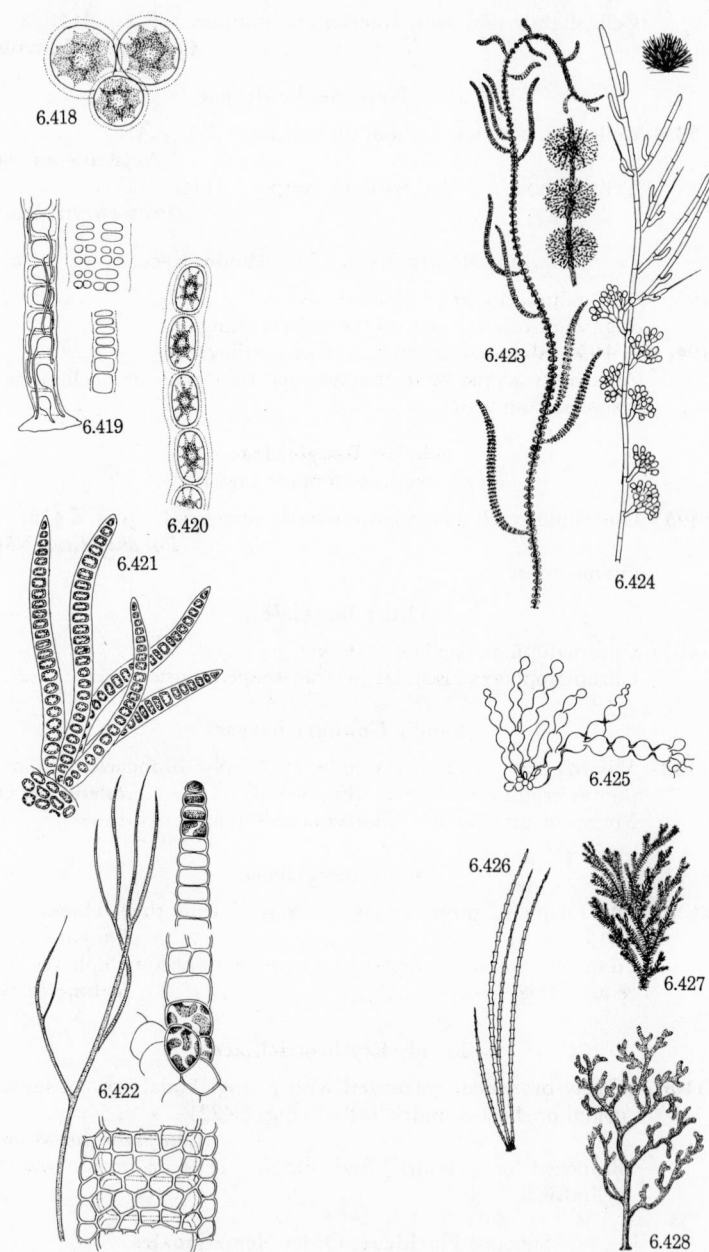

Fig. 6.418. *Porphrydium cruentum* (Smith and Soerly) Nägeli. × 750. **Fig. 6.419.** *Bangia fuscopurpurea* (Dillw.) Lyngbye. Basal portion × 195; detail × 275. **Fig. 6.420.** *Asterocytis smaragdina* (Reinsch) Forti. × 500. **Fig. 6.421.** *Kyliniella latvica* Skuja. × 375. (After Flint.) **Fig. 6.422.** *Campsopogan coeruleus* (Balbis) Mont. × 0.5; detail × 195.

Fig. 6.423. *Batrachospermum boryanum* Sirodot. Habit. × 4; detail × 35. **Fig. 6.424.** *Audouinella violacea* (Kützing) Hamel. Habit × 0.5; detail × 90. **Fig. 6.425.** *Sirodotia californica* Setch. × 195; detail carpogonium and antheridium × 390. **Fig. 6.426.** *Lemanea annulata* Kützing. × 0.5. **Fig. 6.427.** *Thorea ramosissima* Bory. (After Wolle.) **Fig. 6.428.** *Tuomeya fluviatilis* Harvey. × 1.

413b   Carposporophyte of lax, wide-spreading filaments. (Fig. 6.425) . . . . *Sirodotia* Kylin

Family Chantransiaceae

414a   **(412)**   Microscopically branched; dark red-violet; growing as tufts up to ½ in. (Fig. 6.424) . . . . *Audouinella* Bory

Family Thoreaceae

414b   Macroscopic and branched; olive to black; axial core composed of intertwined branches. (Fig. 6.427) . . . . *Thorea* Bory

Family Lemaneaceae

415a   **(411)**   Alga simple-appearing; differentiated into nodes and internodes. (Fig. 6.426) . . . . *Lemanea* Bory

415b   Alga profusely branched; not differentiated into nodes and internotes. (Fig. 6.428) . . . . *Tuomeya* Harvey

References

Ahlstrom, E. H. 1937. Studies on variability in the genus Dinobryon (Mastigophora). *Trans. Am. Microscop. Soc.*, 56:139–159. **Ahlstrom, E. H. and L. H. Tiffany. 1934.** The algal genus Tetrastrum. *Am. J. Botany*, 21:499–507. **Allegre, C. F. and T. L. Jahn. 1934.** A survey of the genus *Phacus* Dujardin (Protozoa; Euglenoidina). *Trans. Am. Microscop. Soc.*, 62:233–244. **Allen, F. M. 1883.** Notes on the American species of Tolypella. *Bull. Torrey Botan. Club*, 10:107–117. **Bold, Harold C. 1942.** The cultivation of algae. *Botan. Rev.*, 8:69–138. **Bourrelly, P. 1957.** Rechèrches sur les Chrysophycées. *Rev. Algologique. Mém. Hors-Série No. 1.* **Carter, N. 1923.** (See West and West.) **Chodat, R. 1926.** Scenedesmus. Étude de génétique, de systématique expérimentale et d'hydrobiologie. *Rev. hydrologie*, 3:71–258. **Collins, F. S. 1909.** The green algae of North America. *Tufts Coll. Studies. Scientific Ser.* 2:79–480 (Stigeoclonium). **Conrad, W. 1933.** Revision du genre *Mallomonas* Perty (1851) incl. *Pseudo-Mallomonas* Chodat (1920). *Mem. musée Roy. hist. nat. Belg.*, 56:1–82. **1934.** Matériaux pour une monographie du genre Lepocynclis Perty. *Arch. Protistenk.*, 82:203–249. **Deflandre, G. 1926.** *Monographie du genre Trachelomonas Ehr.* Nemours, Paris. **Gojdics, M. 1953.** *The genus Euglena.* University of Wisconsin Press, Madison. **Hazen, T. E. 1902.** The Ulotrichaceae and Chaetophoraceae of the United States. *Mem. Torrey Botan. Club.* 11:135–250. **Huber-Pestalozzi, G.** *Das Phytoplankton der Süsswassers, Systematik und Biologie. Die Biennengewässer Einzelderstellungen aus der Limnologie und ihren Nachbargebieten; herausgegeben von Dr. August Thienemann.* Vol. 16: Part 1, 1938, *Cyanophyceen, Blaualgen; Schizomycetes;* Part 2 (1), 1941, *Chrysophyceae; Farblose Flagellaten; Heterokontae;* Part 2 (2), 1942, *Diatomeae;* Part 3, 1950, *Cryptophyceae, Chloromonadinae, Peridineae;* Part 4, 1955, *Euglenophyceae.* E. Schweizerbart'sche verlagsbuchhandlung (Nägle u. Obermiller), Stuttgart. **Kofoid, C. A. and O. Swezy. 1921.** The free-living unarmored Dinoflagellata. *Mem. Univ. Calif.* 5:1–538. **Lackey, J. B. 1938.** Scioto River forms of Chrysococcus. *Am. Midland Naturalist*, 20:619–623. **Lefèvre, M. 1932.** Monographie des especes d'eau douce du genre Peridinium. *Arch. Botan.*, 2:1–210. **Pascher, A.** (ed.). *Die Süsswasserflora Deutschlands, Österreichs und der Schweiz.* Heft 2, 1913, *Flagellatae II*, Chrysomonadinae, Cryptomonadinae, Eugleninae, Chloromonadinae. Heft 3, 1913, *Dinoflagellatae (Peridineae).* Heft 4, 1927, *Volvocales—Phytomonadinae; Flagellatae IV— Chlorophyceae 1.* Heft 5, 1915, *Chlorophyceae II*, Tetrasporales, Protococcales, Einzellige Gottungen Unsicherer Stellung. Heft 6, 1914, *Chlorophyceae III*, Ulotrichales, Microsporales, Oedogoniales. Heft 7, 1921, *Chlorophyceae IV*, Siphonocladiales, Siphonales. Heft 9, 1923, *Zynemales.* Heft 11, 1925, *Heterokontae, Phaeophyta, Rhodophyta, Charophyta.* Heft 12, 1925,

Cyanophyceae, Cyanochloridinae—Chlorobacteriaceae. G. Fischer, Jena. **Pochmann, A. 1942.** Synopsis der Gattung *Phacus. Arch. Protistenk.* 95:81–252. **Prescott, G. W. 1951.** *Algae of the Western Great Lakes Area.* Cranbrook Institute of Science Bulletin No. 31. **Rabenhorst, L.** (ed.). *Kryptogamen-Flora von Deutschland, Österreich und der Schweiz.* Vol. 10 (2), 1930; *Silicoflagellatae, Coccolithineae.* Vol. 10 (3), 1935, *Gymnodiniales;* 1938, *Peridiniales.* Vol. 11, 1939, *Heterokonten (Xanthophyceae).* Vol. 12 (4), 1939, *Oedogoniales.* Vol. 13 (1), 1933, 1935, 1937. *Die Desmidiaceen.* Vol. 13 (2), 1940, *Zygnematales.* Vol. 14, 1932, *Cyanophyceae.* Akademische Verlagsgesllschaft m.b.H., Leipzig. **Robinson, C. B. 1906.** The Chareae of North America. *Bull. N. Y. Botan. Garden.,* 4:244–308. **Skvortzow, B. B. 1925.** Die Euglenaceengattung Trachelomonas Ehrenberg. Eine Systematische Übersicht. *Proc. Sungari River Biol. Sta.,* 1:1–101,. **Smith, G. M. 1916.** A monograph of the algal genus Scenedesmus based on pure culture studies. *Trans. Wisconsin Acad. Sci.,* 18:422–530. **1920.** Phytoplankton of the Inland Lakes of Wisconsin. Part I. Myxophyceae, Phaeophyceae, Heterokontae, and Chlorophyceae exclusive of the Desmidiaceae. *Wisconsin Geol. Nat. Hist. Survey,* Bull. 57 Sci. Ser. 12. **1924.** Phytoplankton of the Inland Lakes of Wisconsin. Part II. Desmidiaceae. *Bull. Univ. Wisconsin,* Serial 1270, General Ser. 1048. **1950.** *The Fresh-water Algae of the United States.* McGraw-Hill, New York. **Tiffany, L. H. 1930.** *The Oedogoniaceae; A monograph.* The Spahr & Glenn Co., Columbus, Ohio. **1937.** Oedogoniales. *North American Flora,* Vol. 11, Part 1, pp. 1–102. **Tilden, J. E. 1910.** *Minnesota Algae.* Vol. I, *The Myxophyceae of North America.* Minneapolis. **Transeau, E. N. 1951.** *The Zygnemataceae.* Ohio State University Press, Columbus. **West, W. and G. S. West.** *A Monograph of the British Desmidiaceae.* Vol. 1, 1904. Vol. 2, 1905. Vol. 3, 1908. Vol. 4, 1912. Vol. 5, 1923, by N. Carter. Ray Society, London. **Wood, R. D. 1948.** A review of the genus *Nitella* (Characeae) of North America. *Farlowia,* 3:331–398.

7

Bacillariophyceae

RUTH PATRICK

Bacillariophyceae, or diatoms, are unicellular algae, usually microscopic, that are characterized by having a cell wall of silica. This wall consists of two valves that are more or less flat surfaces, held together by a band or girdle (Fig. 7.1). The thickness of this siliceous cell wall may vary greatly from very thin in certain typical plankton species to quite thick in certain of the temperate and arctic benthic species. Classification is based on the markings present on the siliceous cell walls.

In some cases the characters useful in taxonomy can be seen in fresh mounts using a high dry objective, but critical identification usually involves examination of cleaned material with an oil immersion objective. Cleaning is accomplished by boiling the material with nitric or sulphuric acid to which an oxidizing agent (usually potassium dichromate) is added. It is then repeatedly washed with distilled water, allowing the frustules to settle before decanting. A small amount of the material is then transferred to a cover glass and dried before being mounted in a highly refractive medium such as Hyrax or euparol.

Diatoms may live singly or form colonies or filaments. All secrete a jelly that more or less covers the siliceous wall; the amount and distribution vary greatly. For example, in some species it causes the cells to adhere to one an-

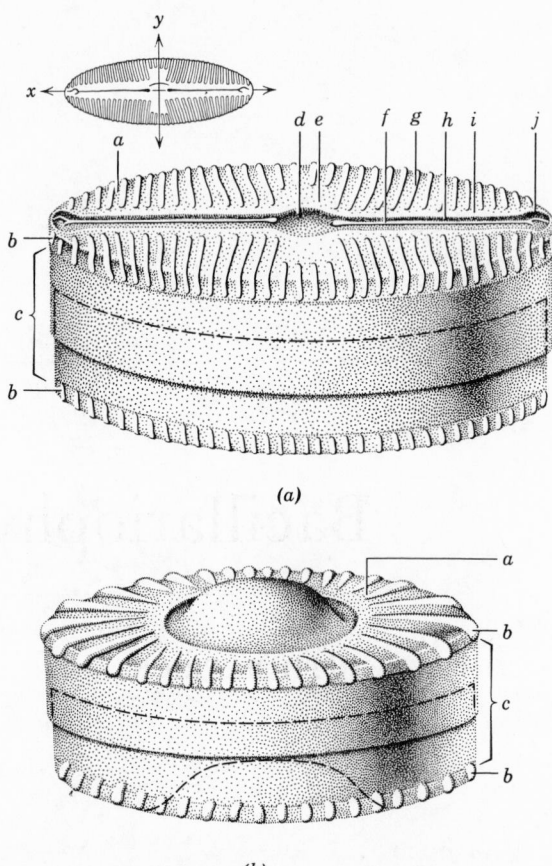

Fig. 7.1. Oblique views of diatoms to show structure. (*a*) Pennate type (*Pinnularia*). (*b*) Centric type (*Cyclotella*). Symbols: *a*, valve; *b*, valve mantle; *c*, girdle; *d*, central nodule; *e*, central area; *f*, raphe; *g*, stria; *h*, *i*, axial area; *j*, terminal nodule; *y*, transverse axis; *x*, apical or longitudinal axis.

other and form filaments or colonies; in other species it forms gelatinous stalks or pads by means of which the diatoms are attached to the substrate. In those forms that are epiphytic or live in fast-moving water this jelly provides many different methods of attachment.

Each diatom cell contains one nucleus. Vegetative cell division is accomplished by mitosis, and reproductive nuclei are formed by meiosis. The chromosomes are easily definable, but due to their great number, they are sometimes hard to count.

The chloroplasts may be numerous or few, and they vary greatly in size and shape. The pigments that have been identified in diatoms are: chlorophyll *a*, chlorophyll *c*, β-carotene, ε-carotene, fucoxanthin, neofucoxanthin A, neofucoxanthin B, diatoxanthin, and diadinoxanthin. It is the abundance of the carotenoid pigments that gives the diatoms their golden-brown color.

The storage products of diatoms are mostly fats and oils, unlike most of the algae, which store various types of carbohydrates.

Reproduction is most commonly vegetative. Auxospore formation, which appears in many cases to be a type of sexual reproduction, has been recorded in numerous diatom genera (Patrick, 1954). Its incidence of occurrence, however, when compared with ordinary vegetative cell division, is relatively rare.

Distribution

Diatoms are widely distributed throughout the fresh and salty waters of the world. They are also found in damp mud and on moist plants such as *Sphagnum* (Patrick, 1948). Although some species are very tolerant of widely varying ecological conditions, many are quite specific in their requirements. For example, certain species will withstand only a very definite concentration of dissolved substances such as chlorides.

The species and their relative abundance have been very useful in indicating the type of water in which the diatom flora lives. Ecologists have used them to indicate the degree of pollution in a body of water, and other variations in ecological conditions. Paleontologists can often ascertain by the diatom flora of sediment cores whether a deep lake or a shallow pond existed in prehistoric times, as well as the temperature and some of the chemical characteristics of the water. In determining environmental conditions a much more reliable estimate can be made if the pattern of species, i.e., kinds of species and relative abundance, is considered, rather than a few so-called indicator species.

It is much more difficult to make precise statements about genera than about species, and only a few generalizations can be made about the type of ecological environment in which some genera find their best development. Most species in the genera *Eunotia* and *Frustulia* are most often found in water low in calcium and magnesium and with a pH below 7.0. Certain species of these genera seem to prefer a range of pH 5 to 6.

Other genera contain species which for the most part seem to avoid acid water and very low concentrations of calcium and magnesium. They are *Mastogloia, Diploneis, Amphipleura, Gyrosigma, Denticula, Epithemia,* and *Rhopalodia.* Genera such as *Cylindrotheca* and most species of *Nitzschia* seem to prefer waters of fairly high ionic content. Such waters may be hard, as in the western central part of the United States, or they may be more or less brackish.

Temperature is another environmental factor that seems to affect the distribution of diatoms. Certain genera such as *Tetracyclus* and *Amphicampa* are usually found in cool mountainous regions. Likewise, the genera *Diatoma* and *Ceratoneis* are more often encountered in cool temperate regions than in warmer southern temperate areas. The species of such genera have their best development in winter, early spring, or late fall if the temperature of the water is warm during the summer months.

Classification

The most generally accepted system of classifying diatoms is that which
Schütt published in 1896. He divided the diatoms into two main groups
which Hustedt (1930) referred to as the Centrales and the Pennales. Recently Hendey (1937) proposed a system of classification which seems much
more natural and logical. It is, in general, the classification I have followed
in this treatment of the diatoms, although a few changes and additions have
been made. As will be seen below, the classification recognizes one order and
several suborders of equal standing. Further studies may indicate that these
suborders should be raised to the rank of orders.

Division Chrysophyta
Class Bacillariophyceae
Order Bacillariales

Suborder Discineae
  Family Coscinodiscaceae
    Subfamily Melosiroideae
      Genus *Melosira*
    Subfamily Coscinodiscoideae
      Genera *Coscinodiscus, Cyclotella,*
      Stephanodiscus

Suborder Biddulphineae
  Family Biddulphiaceae
    Subfamily Biddulphioideae
      Genus *Biddulphia*
    Subfamily Terpsinioideae
      Genera *Terpsinoe, Hydrosera*
  Family Chaetoceraceae
    Subfamily Chaetoceroideae
      Genus *Chaetoceros*

Suborder Soleniineae
  Family Rhizosoleniaceae
    Subfamily Rhizosolenioideae
      Genera *Rhizosolenia, Attheya*

Suborder Araphidineae
  Family Fragilariaceae
    Subfamily Tabellarioideae
      Genera *Tetracyclus, Tabellaria*
    Subfamily Meridionoideae
      Genera *Meridion, Diatoma*
    Subfamily Fragilarioideae
      Genera *Asterionella, Ceratoneis, Centronella,*
      Fragilaria, Opephora, Synedra, Amphicampa

Suborder Raphidioidineae
  Family Eunotiaceae
    Subfamily Peronioideae
      Genus *Peronia*
    Subfamily Eunotioideae
      Genera *Eunotia, Actinella*

Suborder Monoraphidineae
  Family Achnanthaceae
    Subfamily Cocconeioideae
      Genus *Cocconeis*
    Subfamily Achnanthoideae
      Genera *Achnanthes, Rhoicosphenia,*
      Eucocconeis

Suborder Biraphidineae
  Family Naviculaceae
    Subfamily Naviculoideae
      Genera *Amphipleura, Anomoeneis,*
      Brebissonia, Caloneis, Diatomella,
      Diploneis, Frustulia, Gyrosigma,
      Mastogloia, Navicula, Neidium,
      Pinnularia, Stauroneis
    Subfamily Amphiproroideae
      Genus *Amphiprora*
  Family Gomphonemaceae
    Subfamily Gomphonemoideae
      Genera *Gomphonema, Gomphoneis*
  Family Cymbellaceae
    Subfamily Cymbelloideae
      Genera *Cymbella, Amphora*
  Family Epithemiaceae
    Subfamily Epithemioideae
      Genera *Epithemia, Denticula*
    Subfamily Rhopalodioideae
      Genus *Rhopalodia*
  Family Nitzschiaceae
    Subfamily Nitzschioideae
      Genera *Cylindrotheca, Bacillaria,*
      Hantzschia, Nitzschia
  Family Surirellaceae
    Subfamily Surirelloideae
      Genera *Surirella, Cymatopleura,*
      Campylodiscus

Glossary

ALVEOLA. A thin area in the valve wall sometimes with pores; surrounded by a siliceous thickening that usually extends into the interior of the frustule. The alveolae may be relatively simple in structure, or they may be complex with variously formed internal projections of silica and connecting membranes. They are usually more or less circular or hexagonal in shape.

APICAL AXIS. The axis of the valve connecting the two apices. The raphe or the pseudoraphe either lie in this axis or are eccentric to it.

AXIAL AREA. The clear area between the raphe and the ends of the striae.

CANAL RAPHE. A raphe that lies in a groove or channel, usually located in a more or less marked crest or keel. It connects with the internal protoplasm through a series of apertures in the membrane that forms the base of the canal. Terminal nodules of the raphe are usually not evident. The presence of central pores of the raphe is variable.

CENTRAL AREA. The clear area in the center of the valve around the central nodule. Often the central nodule merges into the central area so that they appear as one.

CENTRAL NODULE. A thickened area between the two central pores of the raphe. It may vary considerably in size and shape.

CENTRAL PORES OF RAPHE. The points in the central nodule where the external branch of the raphe connects by a transverse canal with the internal branch of the raphe.

COSTA. A rib or thickening of silica. The thickening may be internal or external to the valve surface.

FRUSTULE. The diatom cell wall, which is composed of two valves joined by a connecting band known as the girdle. A portion of the girdle is joined to the valve mantle of each valve.

GIRDLE. Two bands of silica, one attached to each valve mantle. One of these bands overlaps the other. Thus, the diatom frustule has a boxlike formation. When the diatom divides these two bands of silica separate, and each daughter cell retains one valve and band inside of which a new valve and band is formed.

INTERCALARY BANDS. Bands of silica that often occur between the valve mantle and the girdle of the diatom. They may be few or many in number and vary greatly in width. They sometimes extend into the valve to form a septum.

KEEL. A projection of the valve surface, usually more or less eccentric to the apical axis. In it is enclosed the canal raphe which is characteristic of certain genera of diatoms.

KEEL PUNCTAE. Usually apertures or pores in the plate lying below the canal raphe. In some cases the keel punctae, as commonly recognized, may be the membrane between the apertures or pores rather than the pores themselves.

LOCULE. A division or chamber in the internal septum.

POLE OF VALVE. The end of an axis from which the markings on the valve usually radiate.

PSEUDORAPHE. An axial area without a raphe.

PUNCTAE. Small holes or thin bits of valve wall with pores in them surrounded by a thickening of the wall. The thickening may be only very slight and may extend either inward or to the exterior.

RAPHE. A slit in the valve forming an external and internal canal, often more or less >-shaped, through which the protoplasm of the diatom may flow. By this means the diatom protoplasm is in intimate contact with the environment. The raphe is always found on the valve.

SEPTUM. An internal plate (an extension of an intercalary band) lying parallel to the valves of the diatom. It may extend completely across the valve or it may be only marginal in its development. It usually arises from an intercalary band.

SETA. A long, thin spine.

SPINE. A short, pointed, siliceous outgrowth, tapering from base to point.

STAUROS. A central nodule that extends almost, if not quite, to the margins of the valve.

STRIA. A line of punctae. The close juxtaposition of the punctae may make the stria appear to be a solid line.

TERMINAL OR POLAR NODULE. An enlarged, usually thickened, area of the wall in which the raphe terminates, forming an external and an internal fissure.

TRANSVERSE AXIS. The axis of the valve that connects the two margins of the valve and is perpendicular to the apical axis.

VALVE. The valve is composed of a more or less flattened surface and a mantle. Each diatom frustule is composed of two valves joined by a connecting band known as the girdle.

VALVE SURFACE. The surface of a diatom that possesses most of the markings on which identification is based.

VALVE MANTLE. The portion of the valve that is apparent in girdle view.

WING. A thin projection of the valve surface, much better developed than a keel, often arising near the apical axis of the valve but becoming most apparent at junction of valve with valve mantle. At this point the raised surface of the valve may be distinctly elevated. The canal raphe may be enclosed in the wing.

KEY TO GENERA

1a		Valve with true raphe or pseudoraphe	**2**
1b		Valve without true raphe or pseudoraphe	**48**
2a	(1)	Valve with pseudoraphe on both frustules	**3**
2b		Valve with raphe present on at least one valve; raphe may be present only near ends of valve or enclosed in keel or canal	**14**
3a	(2)	Frustules with septa	**4**
3b		Frustules without septa	**5**
4a	(3)	Septae usually straight; valves without costae. (Fig. 7.2). . . . . *Tabellaria* Ehr.	
4b		Septae curved, valves with costae. (Fig. 7.3) . . . *Tetracyclus* Ralfs	

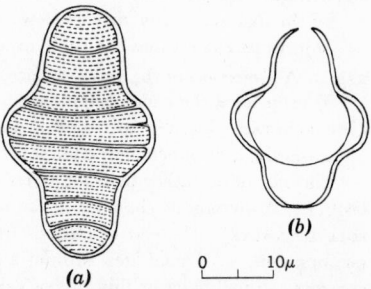

Fig. 7.3. *Tetracyclus lacustris* Ralfs. (a) Valve view. (b) Internal septum. × 1000.

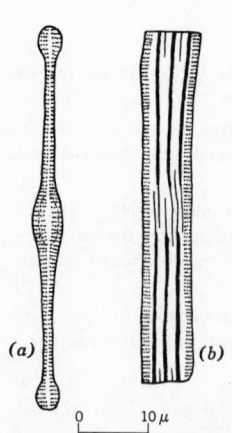

Fig. 7.2. *Tabellaria fenestrata* (Lyngb.) Kütz. (a) Valve view. (b) Girdle view showing septae.

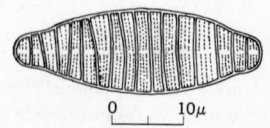

Fig. 7.4. *Diatoma vulgare* Bory. × 1000.

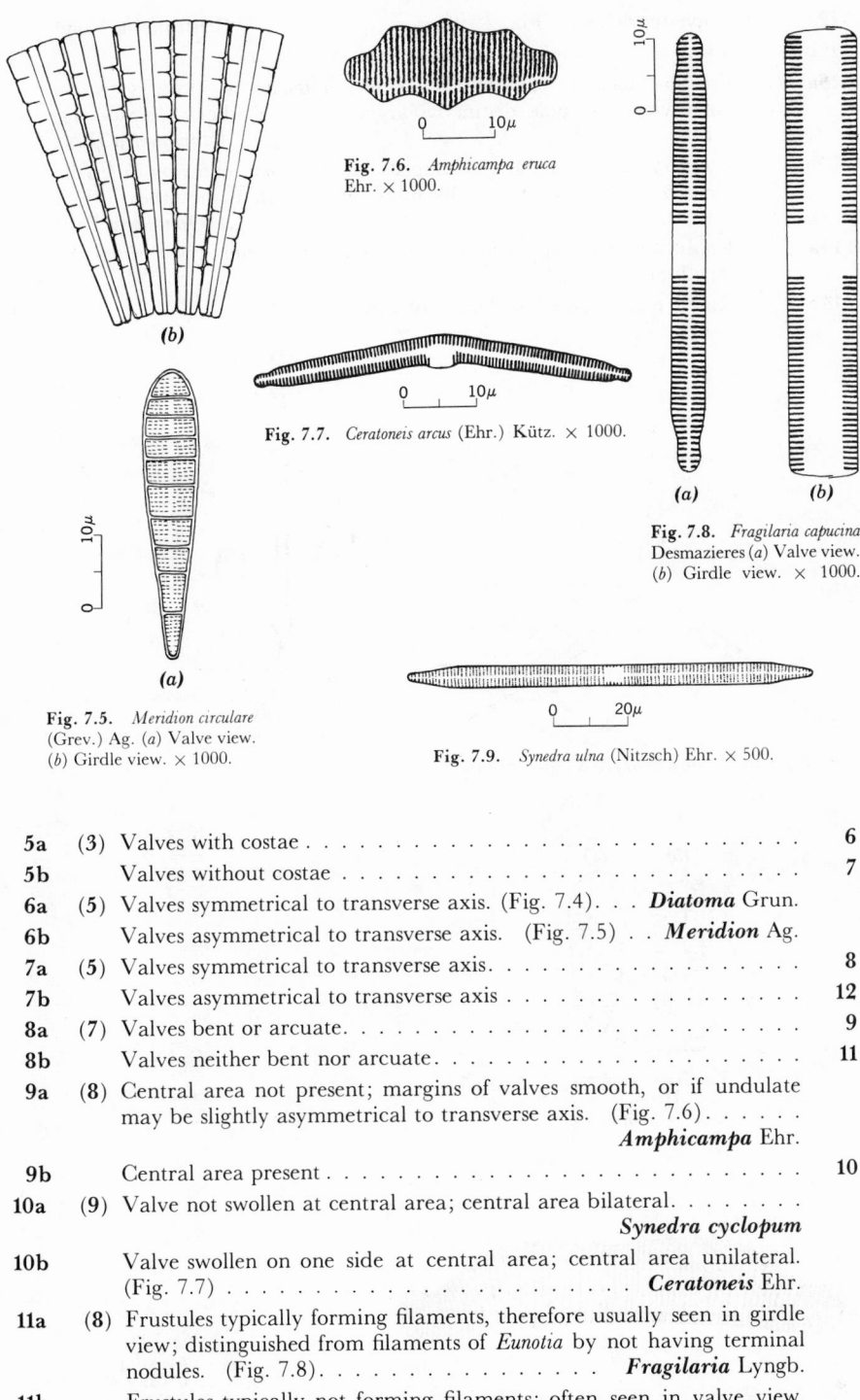

Fig. 7.6. *Amphicampa eruca*
Ehr. × 1000.

Fig. 7.7. *Ceratoneis arcus* (Ehr.) Kütz. × 1000.

Fig. 7.8. *Fragilaria capucina*
Desmazieres (*a*) Valve view.
(*b*) Girdle view. × 1000.

(b)

(a)

(a)   **(b)**

Fig. 7.5. *Meridion circulare*
(Grev.) Ag. (*a*) Valve view.
(*b*) Girdle view. × 1000.

Fig. 7.9. *Synedra ulna* (Nitzsch) Ehr. × 500.

5a   (3) Valves with costae. **6**

5b   Valves without costae **7**

6a   (5) Valves symmetrical to transverse axis. (Fig. 7.4). . . *Diatoma* Grun.

6b   Valves asymmetrical to transverse axis. (Fig. 7.5) . . *Meridion* Ag.

7a   (5) Valves symmetrical to transverse axis. **8**

7b   Valves asymmetrical to transverse axis **12**

8a   (7) Valves bent or arcuate. **9**

8b   Valves neither bent nor arcuate. **11**

9a   (8) Central area not present; margins of valves smooth, or if undulate
            may be slightly asymmetrical to transverse axis. (Fig. 7.6). . . . . .
                                                              Amphicampa Ehr.

9b   Central area present. **10**

10a (9) Valve not swollen at central area; central area bilateral. . . . .
                                                              Synedra cyclopum

10b   Valve swollen on one side at central area; central area unilateral.
            (Fig. 7.7) . . . . *Ceratoneis* Ehr.

11a (8) Frustules typically forming filaments, therefore usually seen in girdle
            view; distinguished from filaments of *Eunotia* by not having terminal
            nodules. (Fig. 7.8). . . . . *Fragilaria* Lyngb.

11b   Frustules typically not forming filaments; often seen in valve view.
            (Fig. 7.9) . . . . *Synedra* Ehr.

12a (7) Valve branched. (Fig. 7.10) . . . . ***Centronella*** Voigt

12b   Valve not branched **13**

13a (12) Frustules forming star-shaped colonies, cuneate in girdle view, in valve view one pole distinctly larger than the other. (Figs. 7.11, 4.14). . . . . ***Asterionella*** Hass.

13b   Frustules not forming star-shaped colonies, not cuneate in girdle view; one end of valve usually much broader than the other. (Fig. 7.12). . . . . ***Opephora*** Petit

14a (2) Raphe apparent only at terminal nodules, not extending throughout length of valve **15**

14b   Raphe extending whole length of valve. **17**

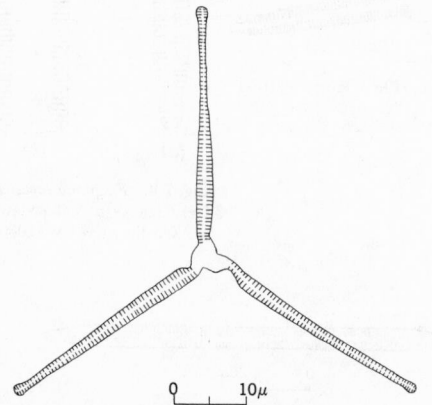

Fig. 7.10. *Centronella reichelti* Voigt. × 1000.

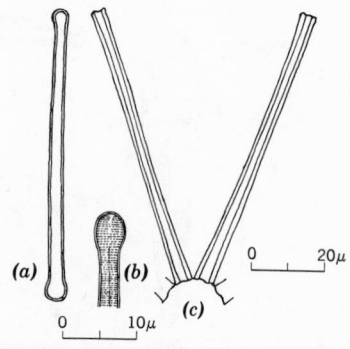

Fig. 7.11. *Asterionella formosa* Hass. (*a*) Valve outline. (*b*) Valve end. (*c*) Frustules in girdle view.

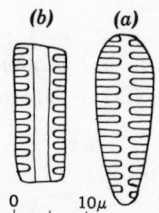

Fig. 7.12. *Opephora martyi* Herib. (*a*) Valve view. (*b*) Girdle view. × 1000.

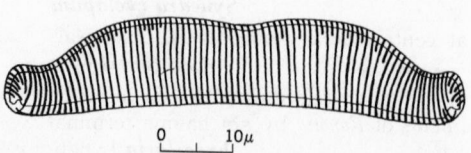

Fig. 7.14. *Eunotia praerupta* var. *bidens* (Ehr.) Grun. × 1000.

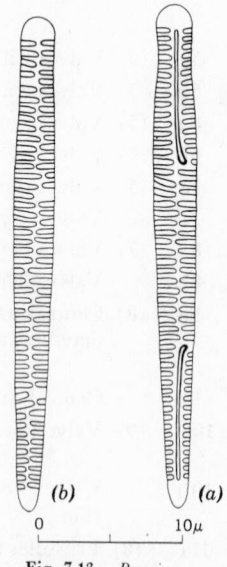

Fig. 7.13. *Peronia erinacea* Bréb. and Arn. (*a*) Valve with raphe. (*b*) Valve without raphe. × 2000.

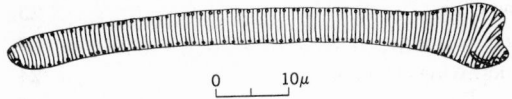

Fig. 7.15. *Actinella punctata* Lewis. × 1000.

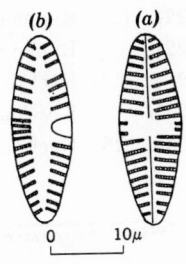

Fig. 7.18. *Achnanthes lanceolata* (Bréb.) Grun.: Cl. and Grun. (*a*) Valve with raphe. (*b*) Valve without raphe. × 1250.

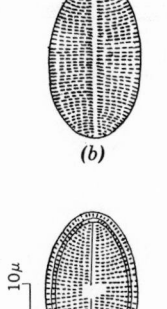

(*b*)

(*a*)

Fig. 7.16. *Cocconeis placentula* var. euglypta (Ehr.) Cl. (*a*) Valve with raphe. (*b*) Valve without raphe. × 1000.

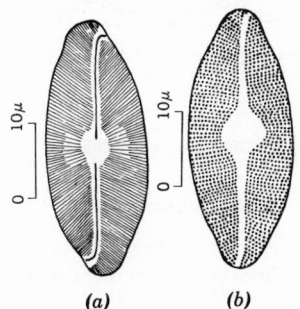

(*a*)   (*b*)

Fig. 7.17. *Eucocconeis flexella* (Kütz.) Cl. (*a*)Valve with raphe. (*b*) Valve without raphe. × 1000.

(*a*)   (*b*)

Fig. 7.19. *Rhoicosphenia curvata* (Kütz.) Grun. (*a*) Valve with complete raphe. (*b*) Valve with rudimentary raphe. × 1000.

15a	(14)	Apical axis straight. (Fig. 7.13) . . . . ***Peronia*** Bréb. and Arn.
15b		Apical axis slightly or distinctly bent **16**
16a	(15)	Valve symmetrical to transverse axis; in girdle view rectangular. (Fig. 7.14) . . . . ***Eunotia*** Ehr.
16b		Valve asymmetrical to transverse axis; in girdle view slightly wedge-shaped. (Fig. 7.15) . . . . ***Actinella*** Lewis
17a	(14)	Raphe on only one valve, the other valve with pseudoraphe or with rudimentary raphe. **18**
17b		Raphe well developed on both valves. **21**
18a	(17)	Valve with raphe having a more or less distinct rim; usually elliptical in outline. (Fig. 7.16) . . . . ***Cocconeis*** Ehr.
18b		Valve with raphe without distinct rim; usually not elliptical in outline **19**
19a	(18)	Raphe sigmoid, sometimes sigmoid only near terminal nodules. (Fig. 7.17). ***Eucocconeis*** Cl.
19b		Raphe straight **20**
20a	(19)	Valve symmetrical to transverse axis. (Fig. 7.18) . . . . ***Achnanthes*** Bory
20b		Valve asymmetrical to transverse axis. (Fig. 7.19) . . . . ***Rhoicosphenia*** Grun.
21a	(17)	Raphe in a canal. **22**
21b		Raphe not in a canal **31**

22a (21) Raphe enclosed in a keel or wing 23
22b   Raphe not enclosed in a keel or wing 29
23a (22) Keel punctae usually distinct, wings lacking 24
23b   Keel punctae lacking, wings distinct 27
24a (23) Keel on margin of valve eccentric or central 25

Fig. 7.20. *Cylindrotheca gracilis* (Bréb.) Grun.: V. H. × 1000.

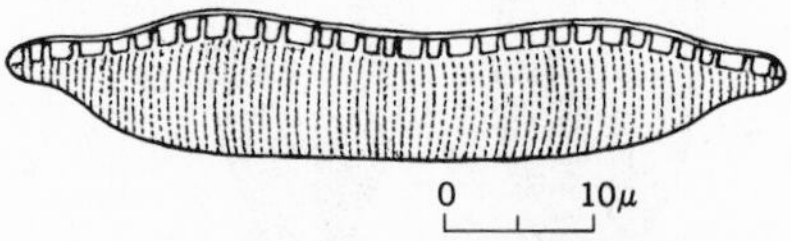

Fig. 7.21. *Hantzschia amphioxys* (Ehr.) Grun.: Cl. and Grun. × 1000.

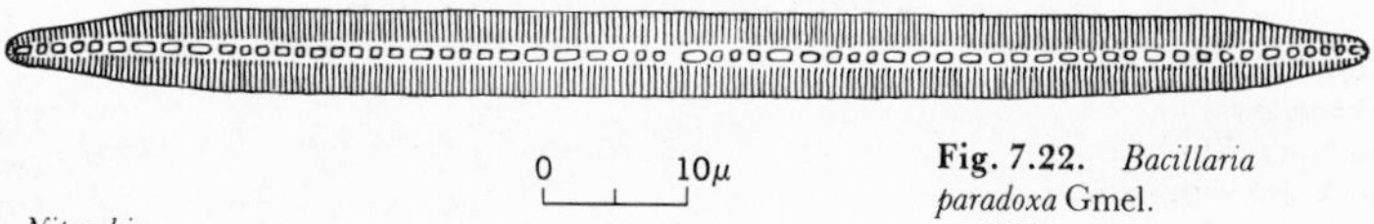

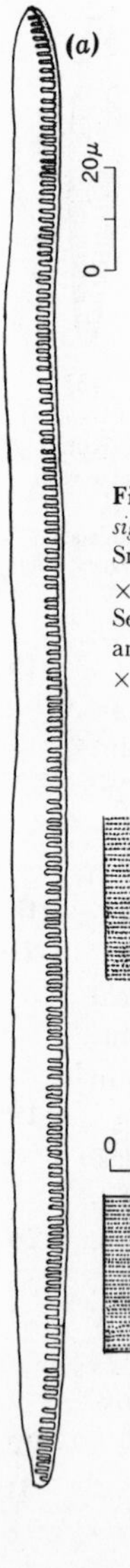

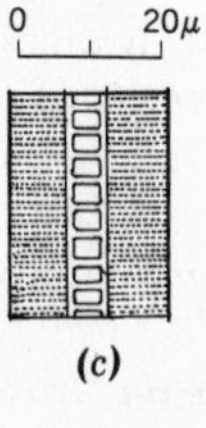

Fig. 7.23. *Nitzschia sigmoidea* (Nitzsch) W. Sm. (*a*) Valve outline, × 500. (*b*) and (*c*) Sections showing striae and keel punctae, × 1000.

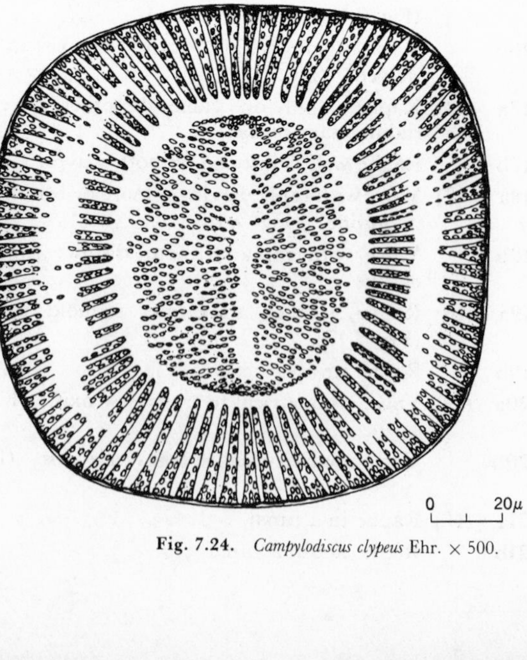

Fig. 7.22. *Bacillaria paradoxa* Gmel. × 1000.

Fig. 7.24. *Campylodiscus clypeus* Ehr. × 500.

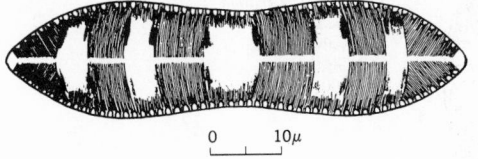

Fig. 7.25. *Cymatopleura solea* (Bréb.) W. Sm. × 1000.

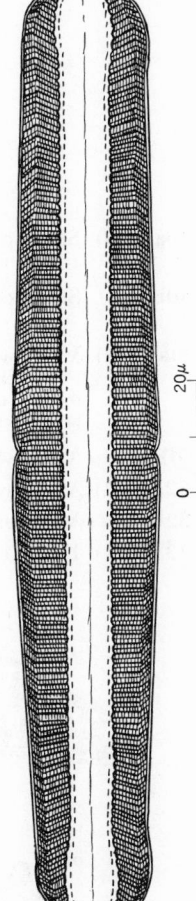

Fig. 7.27. *Rhopalodia gibba* (Ehr.) Müll. × 750.

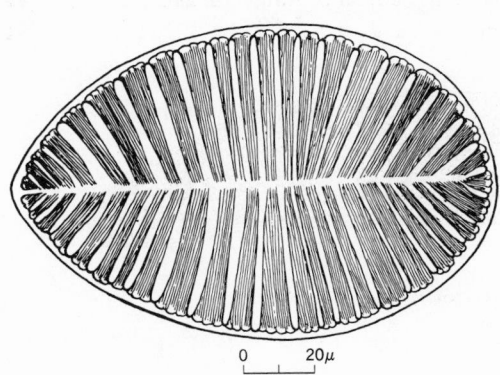

Fig. 7.26. *Surirella striatula* Turp. × 500.

24b	Keel central, valve twisted, thus raphe appears twisted. (Fig. 7.20). . . . . *Cylindrotheca* Raben.	
25a	(24)	Keels on same margins of valves. (Fig. 7.21) . . . *Hantzschia* Grun.
25b		Keels on or near opposite margins of valve, or central 26
26a	(25)	Keels central; cells usually forming a colony. (Fig. 7.22). . . . . *Bacillaria* Gmel.
26b		Keel more or less eccentric, cells not usually forming colonies. (Fig. 7.23). . . . . *Nitzschia* Hass.
27a	(23)	Frustules saddle-shaped. (Fig. 7.24). . . . . . . *Campylodiscus* Ehr.
27b		Frustules not saddle-shaped, sometimes twisted 28
28a	(27)	Surface of valve undulate. (Fig. 7.25). . . . *Cymatopleura* W. Sm.
28b		Surface of valve not undulate; in girdle view sometimes twisted or sigmoid. (Fig. 7.26) . . . . *Surirella* Turp.
29a	(22)	Raphe enclosed in a canal with pores 30
29b		Raphe enclosed in a canal without pores. (Fig. 7.27). . . . . *Rhopalodia* Müll.

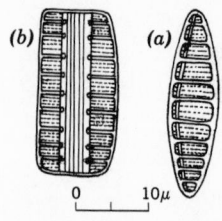

Fig. 7.28. *Denticula thermalis* Kütz.
(*a*) Valve view.
(*b*) Girdle view.
× 1000.

30a (29) Shape of valve symmetrical to apical axis. (Fig. 7.28) . . . .
                                                    Denticula Kütz.

30b Shape of valve asymmetrical to apical axis. (Fig. 7.29) . . . .
                                                   Epithemia Bréb.

31a (21) Valve with wings. (Fig. 7.30) . . . . **Amphiprora** Ehr.
        The genus *Tropodoneis* is very similar to *Amphiprora*, differing mainly in the fact that the frustules are straight, not twisted, in girdle view, and the species are all brackish or marine in habitat.

31b Valve without wings **32**

32a (31) Valve symmetrical to both longitudinal and transverse axes **33**

32b Valve asymmetrical to either longitudinal or transverse axis **45**

33a (32) Frustules with inner septum **34**

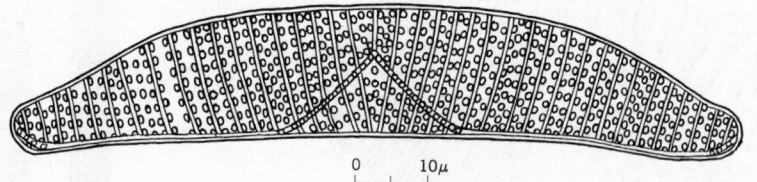

Fig. 7.29. *Epithemia turgida* (Ehr.) Kütz. × 1000.

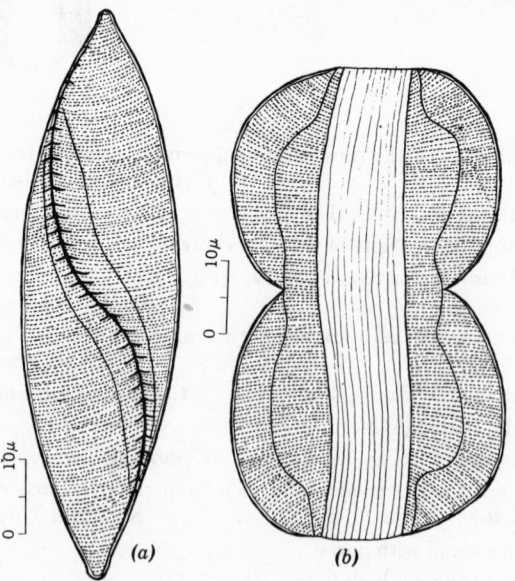

Fig. 7.30. *Amphiprora alata* (Ehr.) Kütz.
(*a*) Valve view.
(*b*) Girdle view.
× 1000.

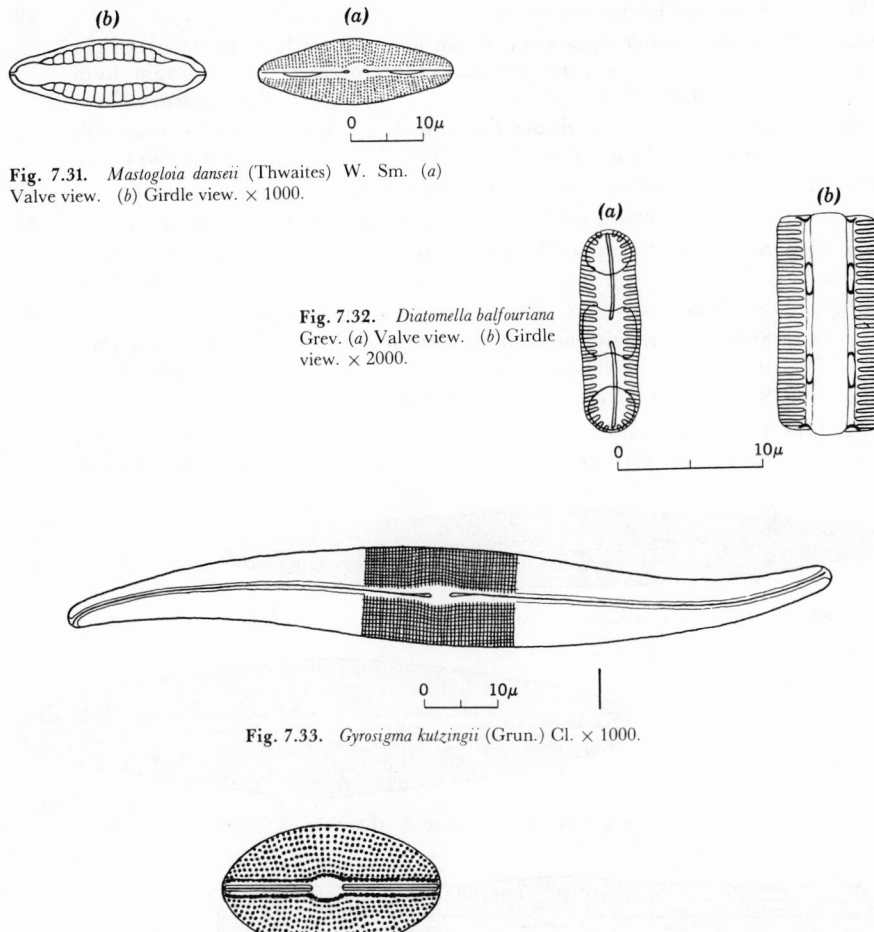

Fig. 7.31. *Mastogloia danseii* (Thwaites) W. Sm. (*a*)
Valve view. (*b*) Girdle view. × 1000.

Fig. 7.32. *Diatomella balfouriana*
Grev. (*a*) Valve view. (*b*) Girdle
view. × 2000.

Fig. 7.33. *Gyrosigma kutzingii* (Grun.) Cl. × 1000.

Fig. 7.34. *Diploneis elliptica* (Kütz.) Cl. × 1000.

33b		Frustules without inner septum	**35**
34a	(33)	Septum with locules on edges of valve. (Fig. 7.31)	

Mastogloia Thw.

34b Septum without locules on edges of valve; with large holes at center and near terminal nodules. (Fig. 7.32). . . . . . *Diatomella* Grun.

35a (33) Frustule sigmoid; longitudinal and transverse striae cross each other at right angles. (Fig. 7.33) . . . . *Gyrosigma* Hass.

> The genus *Pleurosigma*, which is usually found in salt water, is very similar in appearance to *Gyrosigma*. It differs by having the longitudinal and transverse striae cross at an angle other than a right angle.

35b Frustule not sigmoid. **36**

36a (35) Raphe enclosed in a siliceous rib **37**

36b Raphe not enclosed in a siliceous rib **39**

37a (36) Valve costate, one or two rows of alveolae or punctae between costae. (Fig. 7.34). . . . . *Diploneis* Ehr.

37b   Valve not having costae   38

38a   (37)   Central nodule drawn out so that it is at least half the length of the
             valve; no central area present, central pores of raphe distant from
             each other. (Fig. 7.35) . . . . *Amphipleura* Kütz.

38b   Central nodule not drawn out to such a degree; much less than half
             the length of the valve. (Fig. 7.36) . . . . *Frustulia* Grun.

39a   (36)   Striae crossed by one or several longitudinal lines or by a band. . . .   40

39b   Striae not crossed by distinct longitudinal lines or by a band   43

40a   (39)   Striae costate, crossed by a band that is more or less distinct.   (Fig.
             7.37). . . . . *Pinnularia* Ehr.

40b   Striae punctate or costate, crossed by a longitudinal line or lines . . .   41

41a   (40)   Striae indistinctly punctate or appearing costate; longitudinal lines
             near margin of valve. (Fig. 7.38). . . . . *Caloneis* Cl.

41b   Striae punctate, crossed by several longitudinal lines   42

42a   (41)   Longitudinal lines scattered, central pores of raphe turned, if at all,
             in the same direction. (Fig. 7.39). . . . . *Anomoeoneis* Pfitz.

Fig. 7.35. *Amphipleura pellucida* Kütz. × 1000.   0   10μ

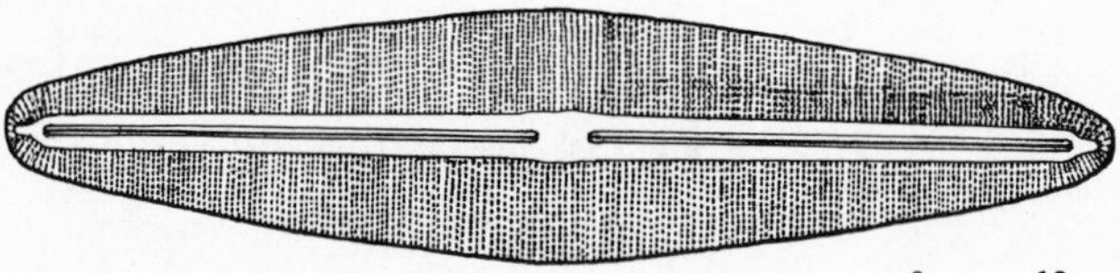

Fig. 7.36. *Frustulia rhomboides* (Ehr.) De T. × 1000.   0   10μ

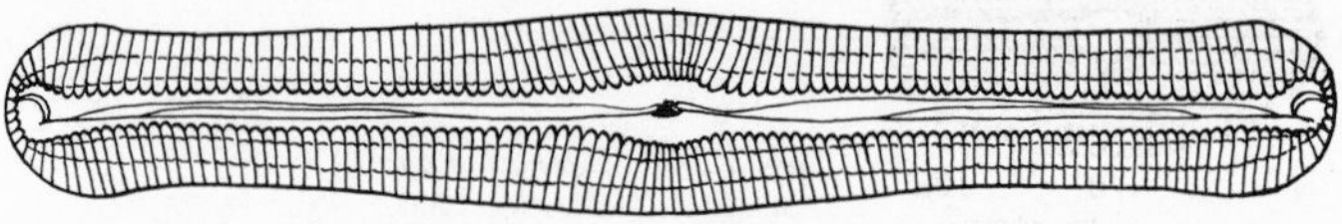

0   10μ   **Fig. 7.37.** *Pinnularia nobilis* Ehr. × 350.

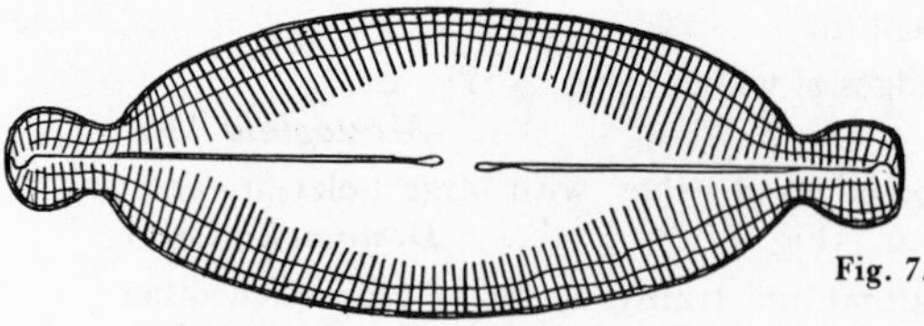

Fig. 7.38. *Caloneis amphisbaena* (Bory) Cl. × 1000.

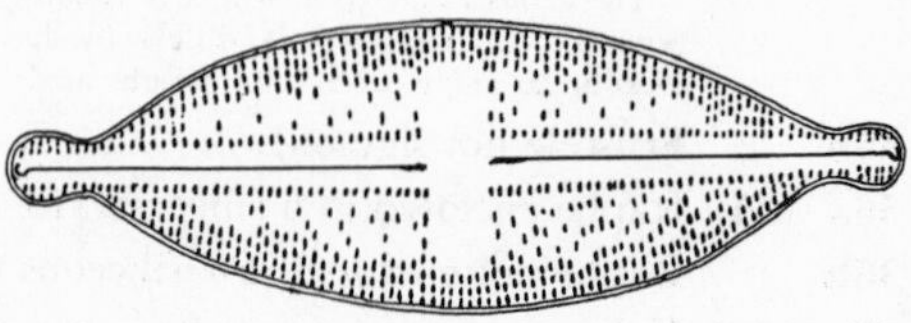

0   10μ

Fig. 7.39. *Anomoeoneis sphaerophora* (Kütz.) Pfitz. × 1000.   0   10μ

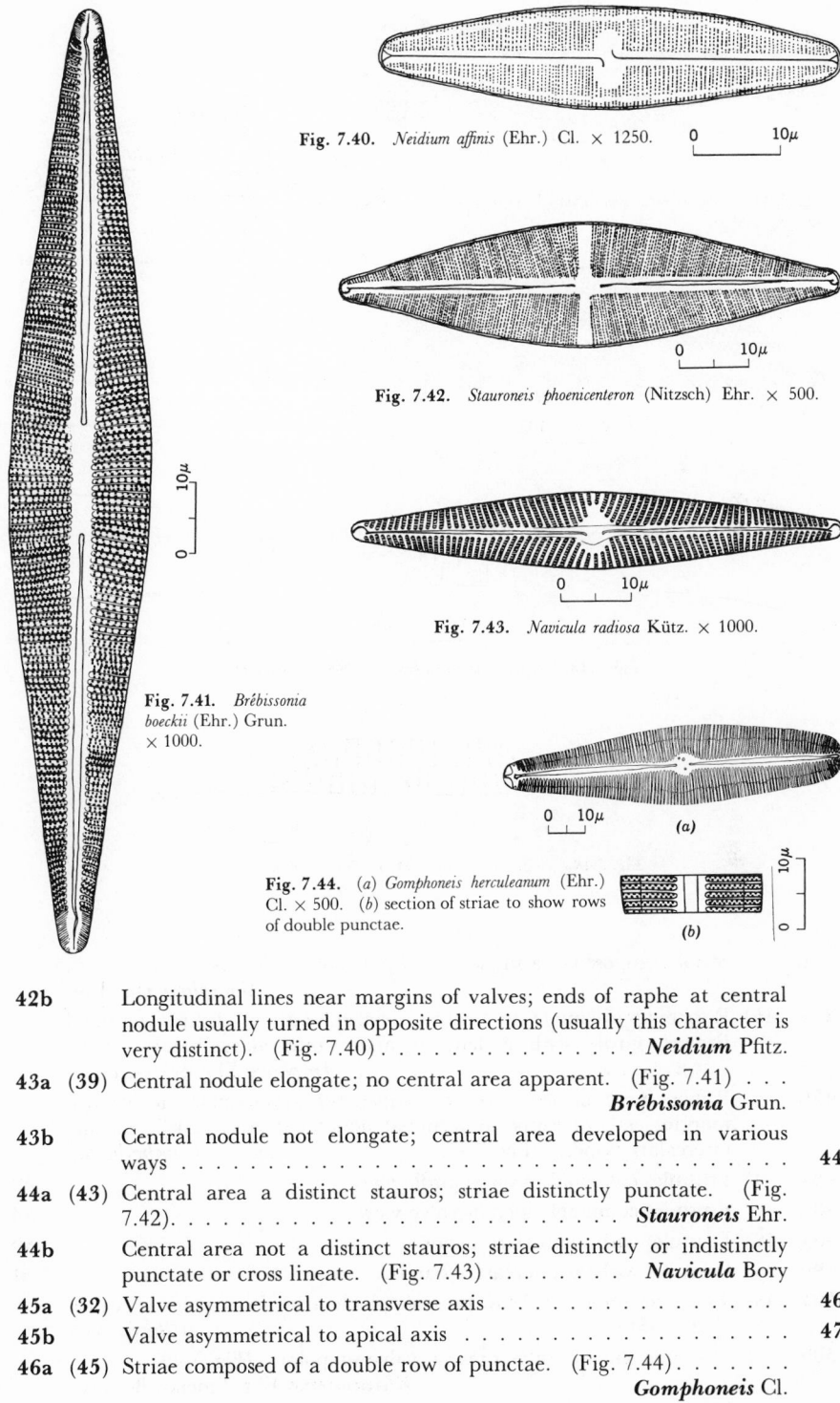

Fig. 7.40. *Neidium affinis* (Ehr.) Cl. × 1250.   0   10μ

Fig. 7.42. *Stauroneis phoenicenteron* (Nitzsch) Ehr. × 500.

Fig. 7.43. *Navicula radiosa* Kütz. × 1000.

Fig. 7.41. *Brébissonia boeckii* (Ehr.) Grun. × 1000.

Fig. 7.44. (*a*) *Gomphoneis herculeanum* (Ehr.) Cl. × 500. (*b*) section of striae to show rows of double punctae.

42b		Longitudinal lines near margins of valves; ends of raphe at central nodule usually turned in opposite directions (usually this character is very distinct). (Fig. 7.40). . . . . **Neidium** Pfitz.
43a	(39)	Central nodule elongate; no central area apparent. (Fig. 7.41) . . . **Brébissonia** Grun.
43b		Central nodule not elongate; central area developed in various ways **44**
44a	(43)	Central area a distinct stauros; striae distinctly punctate. (Fig. 7.42). . . . . **Stauroneis** Ehr.
44b		Central area not a distinct stauros; striae distinctly or indistinctly punctate or cross lineate. (Fig. 7.43) . . . . **Navicula** Bory
45a	(32)	Valve asymmetrical to transverse axis **46**
45b		Valve asymmetrical to apical axis **47**
46a	(45)	Striae composed of a double row of punctae. (Fig. 7.44). . . . . . . **Gomphoneis** Cl.

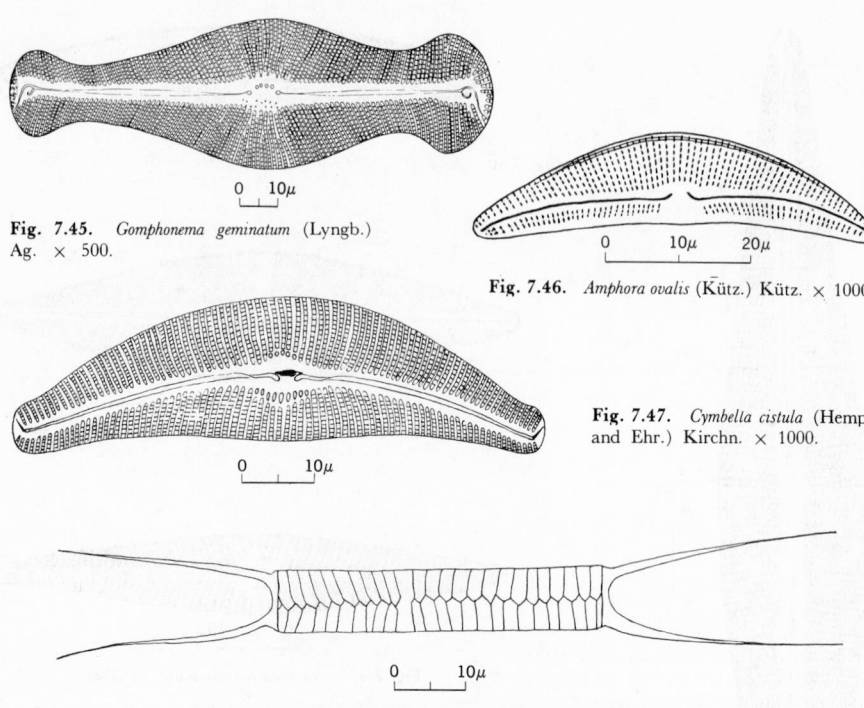

Fig. 7.45. *Gomphonema geminatum* (Lyngb.) Ag. × 500.

Fig. 7.46. *Amphora ovalis* (Kütz.) Kütz. × 1000.

Fig. 7.47. *Cymbella cistula* (Hempr. and Ehr.) Kirchn. × 1000.

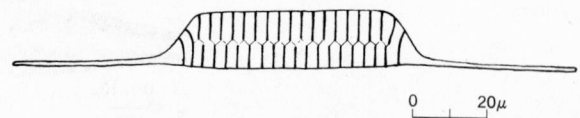

Fig. 7.48. *Attheya zachariasi* Brun. × 500. (After Hustedt.)

Fig. 7.49. *Rhizosolenia eriensis* H. L. Sm. × 500.

46b	Striae composed of a single row of punctae. (Fig. 7.45) . . . . **Gomphonema** Hust.	
47a	(45)	Raphe approximate to ventral margin, without distinct terminal fissures, girdle with a few to many intercalary bands. (Fig. 7.46). . . . . **Amphora** Ehr. emend Kutz.
47b		Raphe more or less eccentric sometimes approximate to ventral margin; ends of raphe at terminal nodules distinct; girdle lacking intercalary bands. (Fig. 7.47). . . . . **Cymbella** Ag.
48a	(1)	Frustules commonly seen in girdle view 49
48b		Frustules commonly seen in valve view. 54
49a	(48)	Frustules with intercalary bands 50
49b		Frustules without intercalary bands. 51
50a	(49)	Two long spines or setae present which arise from poles or valve. (Fig. 7.48) . . . . **Attheya** West
50b		One long spine or setae present, valve unipolar. (Fig. 7.49) . . . . . **Rhizosolenia** Ehr. Emend. Brightw.

51a (49) Frustules forming long filaments **52**

51b   Frustules not forming true filaments **53**

52a (51) Frustules linked by long setae or slender spines extending from valve surface, girdle of frustule not conspicuous. (Fig. 7.50) . . . .
Chaetoceros Ehr.

52b   Frustules not so linked, girdle of frustule conspicuous, joined to valve mantle by groove. (Fig. 7.51). . . . . *Melosira* Kütz.

53a (51) In girdle view transverse costae appearing as musical notes, poles of valve not appearing elevated but capitate. (Fig. 7.52). . . . . . .
Terpsinöe Ehr.

53b   In girdle view valve undulate, poles prolonged into obtuse processes, spines often scattered over central region of valve. (Fig. 7.53) . . . .
Biddulphia Gray

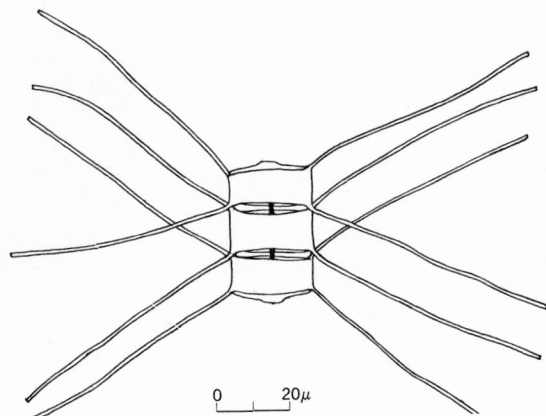

Fig. 7.50.   *Chaetoceros elmorei* Boyer. × 500.

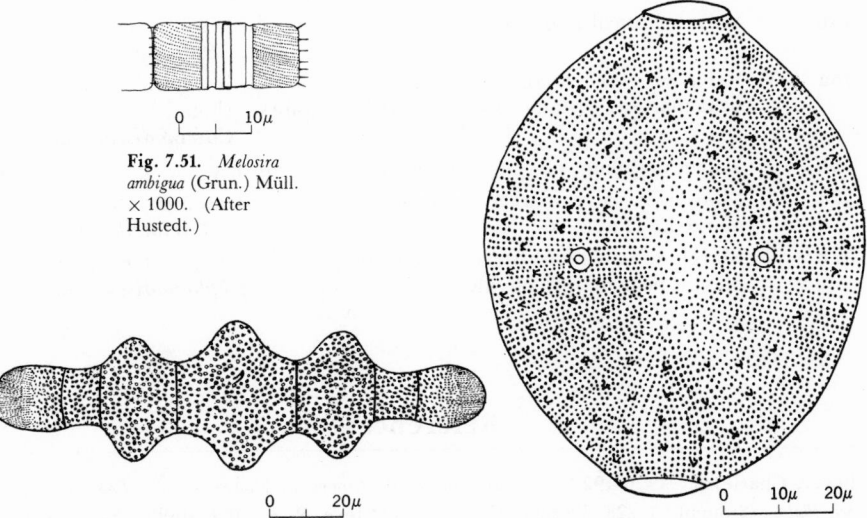

Fig. 7.51. *Melosira ambigua* (Grun.) Müll. × 1000. (After Hustedt.)

Fig. 7.52.   *Terpsinöe musica* Ehr. × 500.

Fig. 7.53.   *Biddulphia laevis* Ehr. × 750.

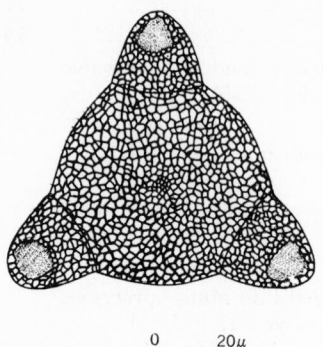

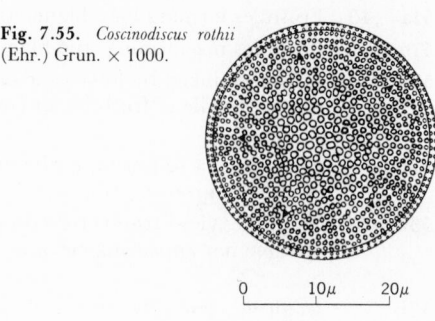

Fig. 7.55. *Coscinodiscus rothii* (Ehr.) Grun. × 1000.

Fig. 7.54. *Hydrosera triquetra* Wall. × 500.

Fig. 7.56. *Cyclotella stelligera* (Cl. and Grun.) V. H. × 1000.

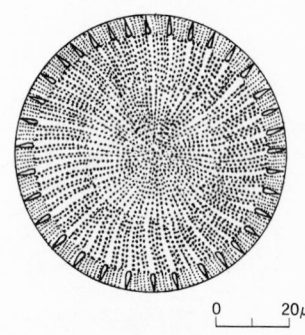

Fig. 7.57. *Stephanodiscus niagarae* Ehr. Note spines. × 500.

54a   **(48)**   Valves having a triangular appearance. (Fig. 7.54) . . . .
                                                 Hydrosera Wallich

54b   Valves circular or elliptical **55**

55a   **(54)**   Valve circular in outline. **56**

55b   Valve elliptical in outline with raised obtuse processes at poles . . . .
                                                 Biddulphia Gray

56a   **(55)**   Valve more or less convex; markings of same type in central areas as near margins. Many species are without spines. (Fig. 7.55). . . . .
                                           Coscinodiscus Ehr.

56b   Valves undulate; marginal and central areas distinct. **57**

57a   **(56)**   Margin of valve marked by coarse costae; central area variously marked. (Fig. 7.56) . . . . ***Cyclotella*** Bréb.

57b   Margin of valve with striae; striae composed of fine punctae; spines present near margin of valve. (Fig. 7.57) . . ***Stephanodiscus*** Ehr.

References

Boyer, Charles S. 1926–1927. Synopsis of North American Diatomaceae. *Proc. Acad. Nat. Sci. Phila.*, 78(Suppl.):1–228, 79(Suppl.)229–583.   **Cleve, P. T. 1894–1895.** Synopsis of the naviculoid diatoms. *Kgl. Svenska Vetenskaps akad. Handl.*, Part I, 26:1–194, Part II, 27:1–219.

Hendey, N. Ingram. 1937. The plankton diatoms of the Southern Seas. *Discovery Repts.*, 16:153–364. **Hustedt, F. 1930.** *Bacillariophyta.* Heft 10 in: A. Pascher (ed.). *Die Süsswasser-Flora Mitteleuropas.* G. Fischer, Jena. **1930–1937.** *Die Kieselalgen.* Vol. 7 (1,2) in: L. Rabenhorst (ed.). *Kryptogamen-Flora von Deutschland, Österreich und der Schwiez,* Akademische Verlagsgesellschaft M.L.H., Leipzig. **Patrick, Ruth. 1948.** Factors effecting the distribution of diatoms. *Botan. Rev.*, 14:473–524. **1954.** Sexual reproduction in diatoms. In: *Sex in Microorganisms*, pp. 82–89. American Association for the Advancement of Science, Washington, D. C. **Schütt, F. 1896.** Bacillariales (Diatomeae) In: A. Engler and K. Prantl *Die natürlichen Pflanzenfamilien,* Vol. 1 (1b), pp. 31–150. Duncker and Humblot, Berlin. **Van Heurck, Henri. 1896.** *A Treatise on the Diatomaceae.* (Translated by Wynne E. Baxter.) Wesley, London. **1880.** *Synopsis des Diatomées de Belgique.* Anvers, Belgium.

8

Zooflagellates

JAMES B. LACKEY

There is no quite satisfactory title for this chapter because some, but not all, organisms of this description are closely related to algae. The making of a key to the colorless flagellate protoza is therefore an ambiguous procedure.

First, such a key should omit those colorless forms which, besides possessing flagella, produce starch and which structurally, except for a lack of chlorophyll, are manifestly Phytomonadida. This includes species such as *Polytoma*. It should also omit many other colorless forms whose affinities are certainly in the orders Cryptomonadida, Dinoflagellida, and Euglenida. These colorless flagellates are treated with the Algae in Chapter 6.

Second, it should include doubtful forms, whose taxonomic status is not yet well defined. It should also include a number of forms which in many reference books are placed in one class or order and in others are placed in a different class or order. This is especially true of certain organisms that have recently been placed in the order Chrysomonadida by such authorities as Fritsch or Grassé. The procedure may lead to some duplication, but it certainly offers the student a better chance for identification, if generic and specific names are not changed.

Third, since the aim is to identify certain organisms, and not to be a taxonomic *tour de force*, the key should follow a recent and well-considered classification. This key follows the classification in Hall's *Protozoology*.

Although the majority are in Hall's Class Zoomastigophorea some of the organisms are in his Class Phytomastigophorea, and the key is not broken down to Class, Order, Suborder and Family. Instead, it aims at identifying 90 genera of colorless flagellates. This includes most of the 29 in the first edition of *Fresh-Water Biology;* most of the free-living genera of the 110 in Pascher and Lemmermann's *Die Süsswasserflora*, Volume I, Flagellatae I; most of those in Grassé's *Traité de Zoologie*, Volume I, which are listed as free-living in his Class "Zooflagelles" and a number of those in his Class "Chrysomonadines"; some doubtful species; and some recently described species. Since no classification finds total acceptance, the user can identify an organism from the key, then place the genus according to the classification he uses. Not all of these genera will be found in Hall, Grassé, Pascher, or any other text, because of divergent views on classification. The key, the drawings, the brief descriptive matter should help to identify any one.

In making a key for the identification of these organisms, dependence has been placed largely on easily recognized structures or characteristics such as pseudopodia; flagella; body shape; shell, test, lorica, or house; collar; and type of colony, if the organism is not solitary.

With pseudopodia, besides the usual protoplasmic extrusion, often blunt or lobelike, there may occur fine, pointed filopodia, or there may be protoplasmic extrusions through which axial processes or fibrils extend, hence the term axopodia. Structures in connection with flagella require more consideration. In most cases flagella and cilia emerge from a minute, stainable granule just beneath the pellicle, termed a blepharoplast. This is certainly a kinetic body, but some textbooks show flagellar fibrils originating from centrioles (in the division of *Oxyrrhis*, for example) which usually form the poles of a mitotic spindle, and which in the resting cell (*Dimorpha mutans*) are embedded in a homogeneous cytoplasmic area near the nucleus, termed a centrosphere. Another kinetic structure is termed a parabasal body, kinetonucleus, or a "nucleus of motion." This may be large or small, but is usually close to the flagellar base, and if there is a blepharoplast an appreciable distance away, the two are connected by a stainable fibril. Where there is one kinetonucleus, there are as many blepharoplasts as flagella. These structures take the usual chromatin stains, but kinetonuclei or parabasals are sometimes visible (because of larger size) in the living cells. The curious band organs or bandelettes, usually visible in the living organisms, belong here, but their function is obscure.

Reserve food materials also may be troublesome. Oil droplets are round, and have a black rim, as does a bubble of air. Paramylum (seemingly confined to euglenoid flagellates) occurs in a variety of shapes, and usually shows a laminated structure. Leucosin and volutin are difficult to separate, are usually irregularly rounded, and under strong light have an optically bluish tinge. The latter frequently may fill half the cell or more, as a single granule. Volutin is a refractive, usually nonstainable, probably nitrogenous material. Deposits of the carbohydrate leucosin are usually round in outline.

KEY TO GENERA

1a		Single set of cell organelles	**2**
1b		Double set of cell organelles.	**82**
2a	(1)	Definitive pseudopodia formed	**3**
2b		Definitive pseudopodia lacking	**11**
3a	(2)	Flagella hardly distinguishable from fine or hairlike pseudopodia. . .	**4**
3b		Flagella distinct; pseudopodia blunt or thickened	**6**
4a	(3)	Cell normally stalked	**5**
4b		Cell normally without stalk; 2 flagella. . . . . . . *Dimorpha* Gruber	

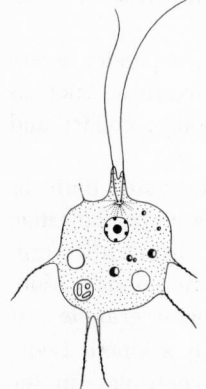

◄ **Fig. 8.1.** *Dimorpha mutans* Gruber. This organism is generally unsatisfactory to classify (Grassé considers it a rhizopod) but is usually flagellated. The body is round to elongate, 10–25 μ in diameter. The pseudopodia have axial filaments and the 2 equal flagella emerge from 1 or 2 bodies, presumably centrioles, since there is an evident centrosphere around them. Vacuoles usually 2, variable in position, nucleus central.

4c   Cells normally without stalk; many flagella . *Multicilia* Cienkowski

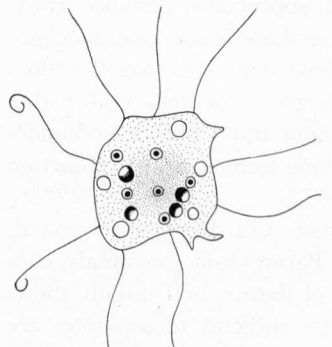

◄ **Fig. 8.2.** *Multicilia lacustris* Lauterborn. Organism generally spherical, but quite amoeboid, with long flagella emerging at any point. These are definitely not axopodia, and are too constant to be regarded as pseudopodia. Cytoplasm clear except for oil drops. Nuclei several (6 in the organism figured) and there are several contractile vacuoles. Large, about 40 μ in diameter. Holozoic.

5a   **(4)** Pseudopodia from any part of cell . . . . ***Actinomonas*** S. Kent

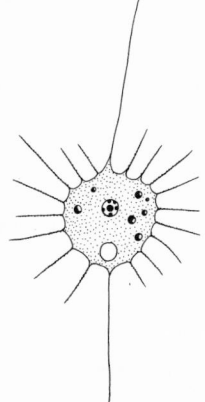

◄ **Fig. 8.3.** *Actinomonas mirabilis* S. Kent. Body spherical, attached or free-swimming, about 15 μ in diameter. Flagellum 20–30 μ. Pseudopodia fine, rough, numerous. Nucleus median, contractile vacuole basal, may be multiple. (Helioflagellida in Hall, 1953.)

5b   Pseudopodia from ring around flagellum base . . . .
Pteridomonas Penard

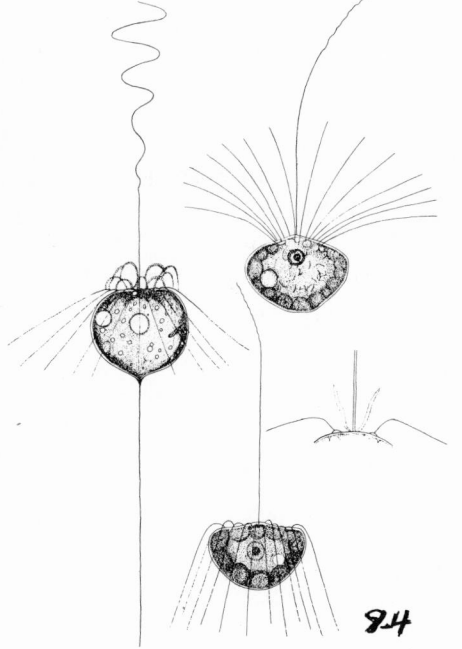

◄ **Fig. 8.4.** *Pteridomonas pulex* Penard. Spherical to hemispherical, 10 μ in diameter. Single, stout, long anterior flagellum vibratile at its tip. Ring of fine axopodia (pointed, fine pseudopodia, with stiffening axial filaments) around flagellum base, which at times is also encircled by short filipodia (narrow, fingerlike pseudopodia.) Nucleus central. Vacuoles variable in number and position. Normally stalked. (After Skuja.)

6a   **(3)** One anterior flagellum **7**
6b   One anterior flagellum, 1 trailing **8**

7a   **(6)** Flagellum traceable to nucleus. . . . . ***Mastigamoeba*** Schulze

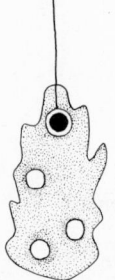

See also Chapter 9, line 11 in key.

◄ **Fig. 8.5.** *Mastigamoeba reptans* Stokes.   Generally teardrop-shaped, no long pseudopodia, 10–15 μ long.   Variable number of vacuoles.   Nucleus anterior or median, large.   Flagellum to 40 μ long, visibly connected to nucleus.

7b   Flagellum not traceable to nucleus . . . . ***Mastigella*** Frenzel

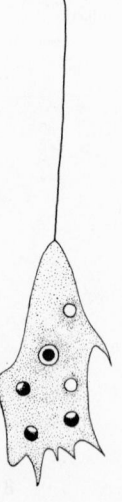

◄ **Fig. 8.6.** *Mastigella simplex* S. Kent.   Elongate, pseudopodia rather short, mostly posterior, body 15 μ long, flagellum to 30 μ.   Nucleus central, not visibly connected to flagella.   Contractile vacuoles variable.   No bacteroids.

8a   (6) Pseudopodia from mouthlike region **9**

8b   Pseudopodia from any point on cell **10**

9a   (8) Pseudopodia from a ring of granules . . . .
Thaumatomonas de Saedeleer

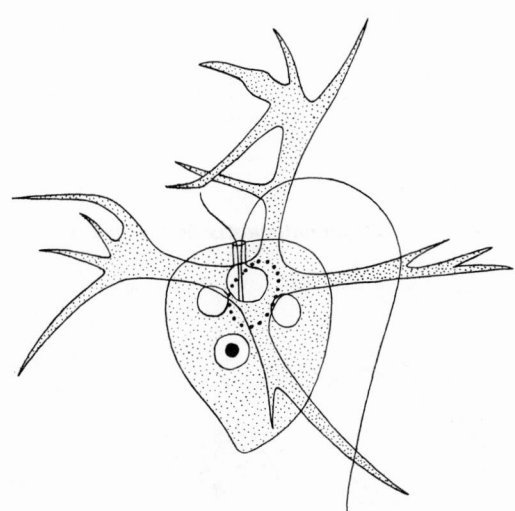

◄ **Fig. 8.7.** *Thaumatomonas lauterborni* de Saedeleer. Monad a flattened oval, or elongate with rounded ends, or the posterior end may be broadly pointed. Cells are 6–10 μ long, and half to two-thirds as wide. A poorly-defined gullet or invagination at the anterior end opens into a reservoir into which 1 or 2 vacuoles feed. This is enclosed by a ring of stainable granules and the flagella emerge from this region. The anterior one is short, the trailing one is about twice the body length. At times branching pseudopodia are produced rapidly from this area, and are rapidly withdrawn. In this respect, and in being found in a decoction of wheat, *Thaumatomonas* is strikingly like *Bodopsis*

9b   Pseudopodia not from a ring of granules . . *Bodopsis* Lemmermann

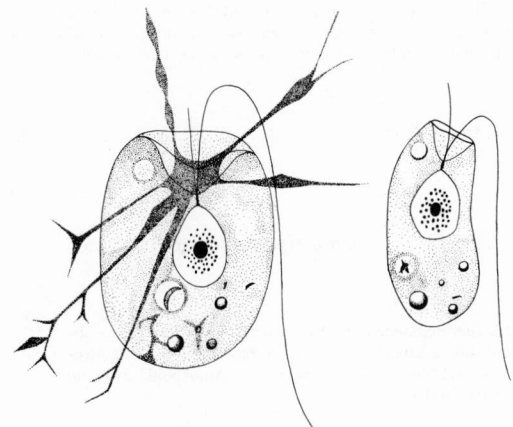

◄ **Fig. 8.8.** *Bodopsis godboldi* Lackey. Cell a flattened or elongate oval, 10–15 μ long, with anterior depression from which at times long, branching pseudopodia are produced. Contractile vacuole anterior; nucleus median. Anterior flagellum short, posterior one long, both originating from a basal body inserted in the bottom of the anterior depression and adjacent to the nuclear membrane. Thus far found only in sewage sludge.

10a (8) Pseudopodia blunt, simple . . . . ***Reckertia*** Conrad

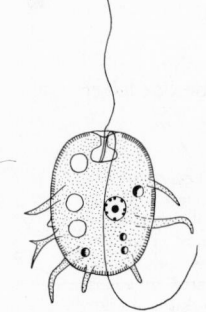

◀ **Fig. 8.9.** *Reckertia sagittifera* Conrad. Body a flattened oval with numerous peripheral trichocysts. Anterior opening into an enlarged reservoir from the floor of which emerge 2 flagella. Anterior one short, posterior one about twice the body length. Vacuoles variable, nucleus median. Pseudopodia as fingerlike lobes from any part of body.

10b   Pseudopodia compound, pointed . . ***Thaumatomastix*** de Saeldeleer

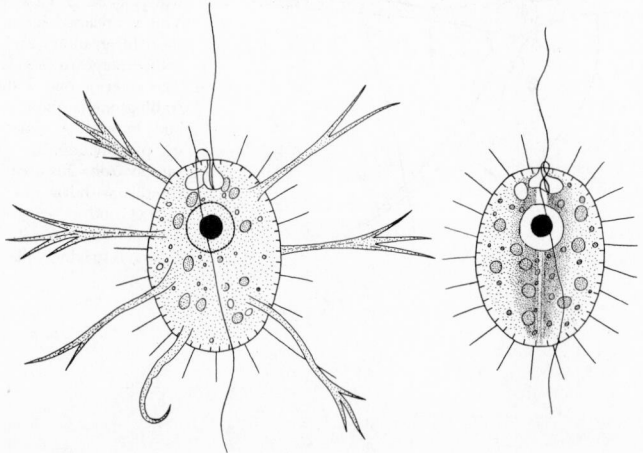

Fig. 8.10. *Thaumatomastix setifera* Lauterborn. Body a flattened oval with trichocysts, 15–30 μ long. An anterior opening into a reservoir, from the floor of which emerge a short (about body length) anteriorly directed flagellum, and a trailing flagellum about twice the body length, both very tenuous. Nucleus in anterior third. One anterior contractile vacuole. Two types of extrusions—hairlike setae which seem to be present all the time, and long, slender pseudopodia which branch repeatedly near their tips and which are extruded only occasionally. Cytoplasm light brown in color. Marine as well as fresh-water. (After Lauterborn.)

11a (2) No flagella; probably secondarily lost. **12**

11b   One or more flagella. **13**

12a (11) Naked . . . . ***Amastigomonas*** de Saedeleer

◀ **Fig. 8.11.** *Amastigomonas debruynei* de Saedeleer. Very small, to 8 μ long. No observed flagellum at any time, but a kinetonucleus at the base of a hooking proboscis. Nucleus central, contractile vacuole posterior. Amoeboid, but no pseudopodia formed. (After de Saedeleer.)

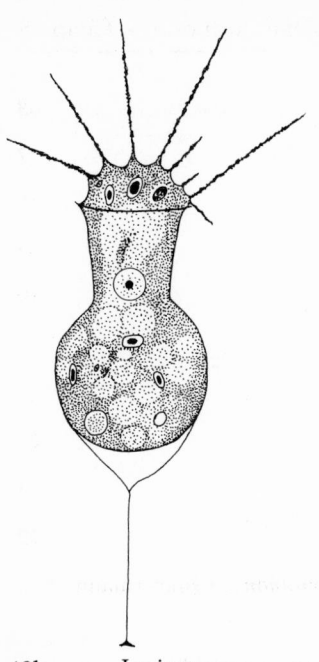

◄ **Fig. 8.12.** *Salpingorhiza pascheriana* Klug. No permanent flagellum. Test vase-shaped, stipitate. Organism not attached to test, but fills it; pseudopodia normally formed at the anterior end. Holozoic. Nucleus in neck region, contractile vacuole posterior. About 15 μ long. (After Klug.)

12b	Loricate . . . . *Salpingorhiza* Klug	
13a	(11) Cell polyaxial . . . . *Actinomonas* S. Kent	
13b	Cell monaxial	**14**
14a	(13) No visible anterior flagella	**15**
14b	One or more visible anterior flagella	**16**
15a	(14) Proboscis present, one trailing flagellum . . . *Rhynchomonas* Klebs	

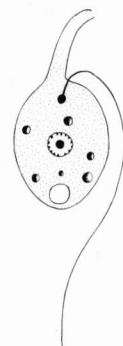

◄ **Fig. 8.13.** *Rhynchomonas nasuta* Klebs. Body a rounded oval. Single trailing flagellum, twice the body length or more, arises from the base of an anterior proboscis, which apparently drags the body forward by hooking movements. Visible kinetonucleus. Nucleus median, protoplasm vacuolate. Cell to 10 μ long.

15b   Proboscis absent, one trailing flagellum . . . . *Clautriavia* Massart

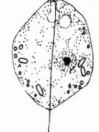

◄ **Fig. 8.14.** *Clautriavia parva* Massart. Cell oval or pointed-oval, flattened, about 6–10 μ long, 5–8μ wide. Shape fixed. No visible anterior flagellum but a trailing flagellum 25–35 μ long extends straight behind as the organism moves ahead with a steady glide. Nucleus lateral, a gullet-reservoir-vacuole system at the anterior end. Inclusions seem to be paramylum, so this is probably a colorless euglenid. Common in sewage at times.

16a   **(14)** More than 1 anterior flagellum **63**

16b   Only 1 anterior flagellum **17**

17a   **(16)** Only anterior flagella present **18**

17b   Anterior and trailing flagella present **51**

18a   **(17)** Collar absent **19**

18b   Collar present **31**

19a   **(18)** Prominent liplike anterior extension present **26**

19b   Prominent liplike anterior extension absent **20**

20a   **(19)** Lorica absent **21**

20b   Lorica present **22**

21a   **(20)** Cell with keel . . . . ***Ancyromonas*** Lemmermann

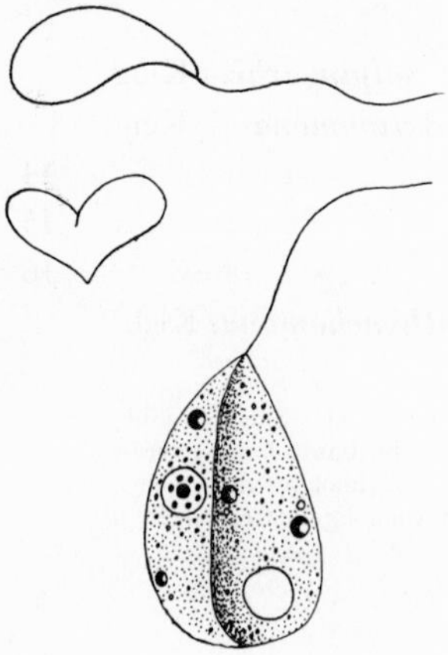

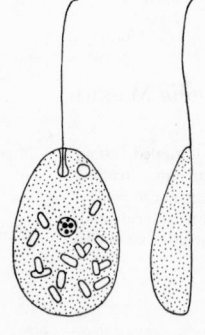

◀ **Fig. 8.15. *Ancyromonas contorta*** Lemmermann. Cell teardrop-shaped in side view, keeled in cross section, oval, but pointed anteriorly. Quite rigid in shape. Nucleus median lateral, contractile vacuole posterior. Single swimming flagellum cell length. Inclusions oil. About 10 μ long. Saprozoic?

Both *Ancyromonas* and *Thylacomonas* (line **21b**) could belong in the Euglenophyceae; *Thylacomonas* almost certainly does, because it contains paramylum granules and apparently has a gullet-reservoir system.

21b   Cell flattened . . . . ***Thylacomonas*** Schewiakoff

◀ **Fig. 8.16.** *Thylacomonas compressa* Schewiakoff. This colorless flagellate has been seen in foul water. It is almost certainly a colorless Euglenid. It is figured here because of its uncertain status. The cell is oval in outline, rounded above, flattened ventrally and rigid in shape due to a thick periplast, which is smooth. A single flagellum emerges from a gulletlike area, and is rigid except for its vibratile tip. Near it is a single contractile vacuole, the arrangment suggesting a gullet-reservoir vacuole system. The cytoplasm is clear except for a number of paramylumlike bodies. Length 20 μ, width 12 μ, flagellum about body length. Rare.

22a   **(20)** Lorica free **23**

22b   Lorica attached **25**

23a (22) Lorica with attached sand grams, urnlike . . ***Domatomonas*** Lackey

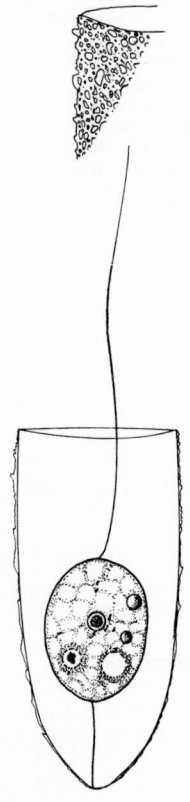

◀ **Fig. 8.17.** *Domatomonas cylindrica* Lackey. Organism very like *Codoneca*, but amoeboid to some extent and smaller, about 10–15 μ in diameter. It has a quite small median nucleus. The lorica is distinctive—a deep urn, transparent but with adherent debris. Lorica 30 by 12 μ, truncate anteriorly, rounded posteriorly. Animal attached by a stipe to lorica base. Nutrition?

23b   Lorica without attached sand grams **24**
24a (23) Lorica tubelike, smooth . . . . ***Aulomonas*** Lackey

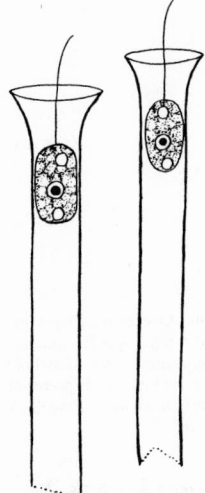

◀ **Fig. 8.18.** *Aulomonas purdyi* Lackey. Organism ovately spherical, without an attaching pedicel, in a thin, transparent tubelike lorica, flaring at the anterior end, of uniform diameter as far back as the broken end. No complete tubes have ever been found. Organism about 15 μ long, 12 μ wide. Nucleus median. Two vacuoles.

24b   Lorica campanulate, with annulae . . . . **Codomonas** Lackey

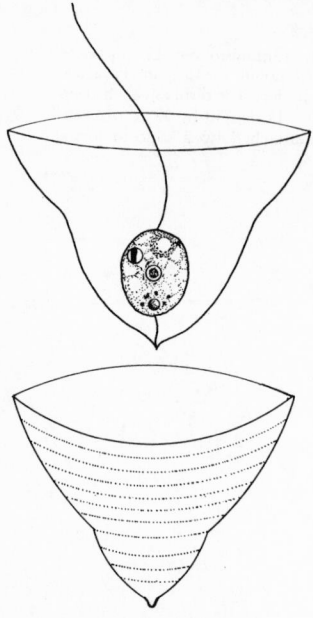

◄ **Fig. 8.19.** *Codomonas annulata* Lackey. Monad almost spherical, colorless, with a median nucleus and a contractile vacuole near the flagellum base. Flagellum 3 times body length, not vibratile, but moves freely as a whole. Holozoic, ingestion near flagellum base. Attached by a short pedicel to the thin transparent lorica which is broadly campanulate and ringed with faint amulae. Absence of a lip precludes terming it a *Bicoeca* as Bourrelly would do. Planktonic.

25a (22) Lorica attached along its side . . . . **Platytheca** Stein

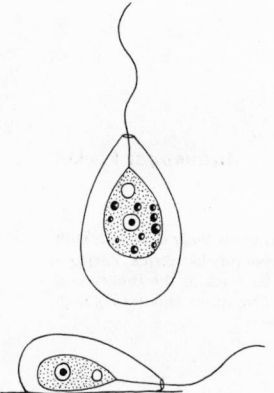

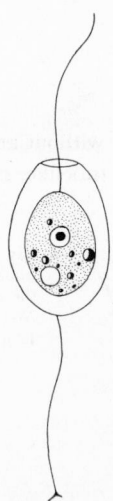

Fig. 8.20. *Platytheca micropora* Stein. Organism flattened, rounded posteriorly, pointed anteriorly, lives within a colorless test of the same shape, sessile upon algae or other plants. One flagellum, cell length or more. Anterior contractile vacuole, median nucleus. Test becomes brown with age, up to 20 μ. Nutrition?

Fig. 8.21. *Condoneca inclinata* S. Kent. Organism, 20 μ long, ovoid, lives in an ovoid test, open at the top. Test is attached by a thin flexible stalk and is transparent. No amoeboid change has been noted in the organism, which has a posterior contractile vacuole and a median nucleus. Flagellum tenuous, about 30 μ long. Saprozoic?

25b   Lorica stalked. . . . . **Codoneca** J. Clark

26a **(19)** No lorica . . . . ***Oicomonas*** S. Kent

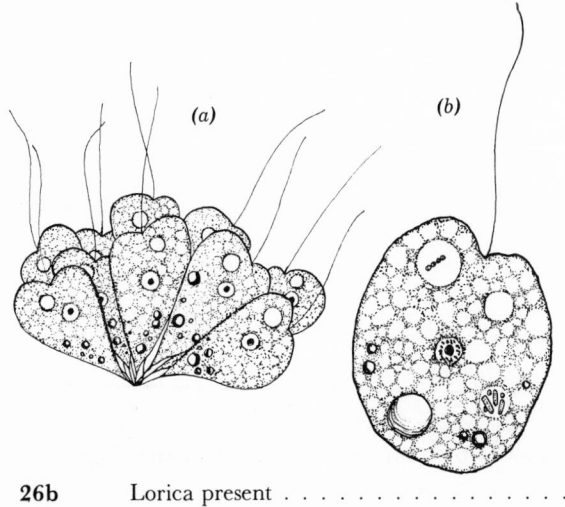

(a)

(b)

◄ **Fig. 8.22.** (*a*) *Oicomonas socialis* Moroff. (*b*) *O. termo* (Ehrenberg) S. Kent. Organisms spherical, or with a pointed posterior end. Short-stalked or free-swimming. Solitary or colonial. Nucleus generally median, contractile vacuole near the base of the flagellum. Food may be ingested near the flagellum base. In *O. ocellata* there is an eyespot near the flagellum base. Size varies, from 5–20 μ in *O. termo*. No lip is formed but there may be a slight depression at the flagellum base. Holozoic.

26b   Lorica present **27**
27a **(26)** Solitary. **28**
27b   Colonial **29**
28a **(27)** Sessile, attached to lorica by contractile stalk . . . .
                                        Bicoeca (J. Clark) Stein

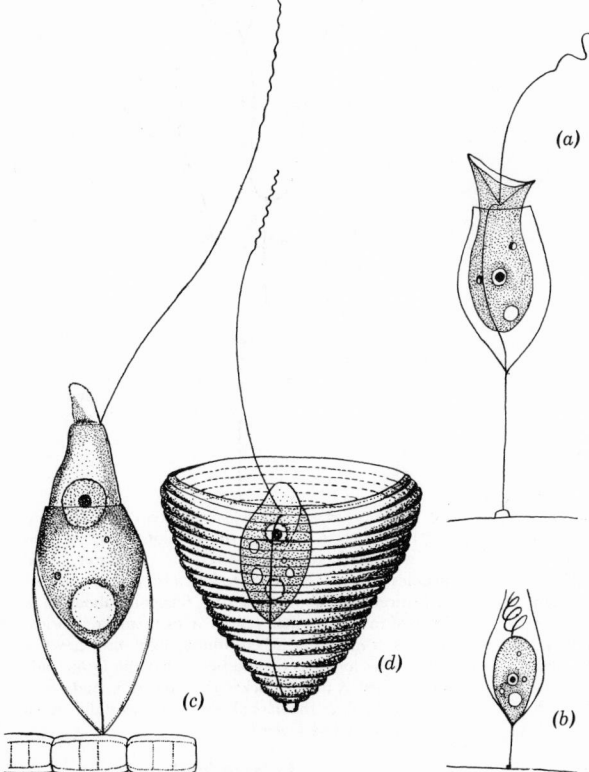

(a)

(c)

(d)

(b)

◄ **Fig. 8.23.** *Bicoeca*. (*a, b*) *B. exilis* Penard. Cell ovoid, but with a flaring protrusible anterior lip. Attached inside lorica by a contractile stalk, which is inserted anteriorly near the single flagellum. Nucleus central, contractile vacuole posterior. Flagellum about twice the body length. Lorica thin, transparent, stipitate, urn-shaped, but with a pointed posterior end and a slight constriction behind the anterior rim. Monad about 12 μ long, 10 μ in diameter. (After Penard.)

(*c*) *B. lacustris* J. Clarke. Cell almost ovoid, similar to *B. exilis*. Lorica ovoid, truncate anteriorly, pointed posteriorly, no necklike constriction. Usually sessile, but may be stalked. Older shells brown. (After Skuja.)

(*d*) *B. multiannulata* Skuja. Monad about like two previous species. Organism planktonic, however, in a brown shell which varies somewhat in shape but always presents a large number of concentric rings. Bourrelly regards it as a variety of *B. planctonica* Kissilew. (After Skuja.)

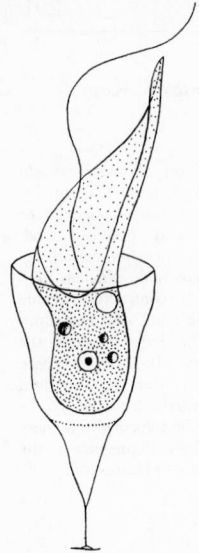

Fig. 8.24. *Histiona aroides* Pascher. Organism with a very pronounced flaring lip. Contractile vacuole anterior, flagellum somewhat longer than body length. Nucleus inconspicuous, median. Shell colorless, thin, with wide anterior opening, narrowing to a foot or a very short stipe behind. Organism has no pedicel attaching it to lorica.

28b   Stalked, not attached to lorica by contractile stalk . . *Histiona* Vogt

29a (27) Colony arboroid 30

29b   Colony a rosette . . . . *Stephanocodon* Pascher

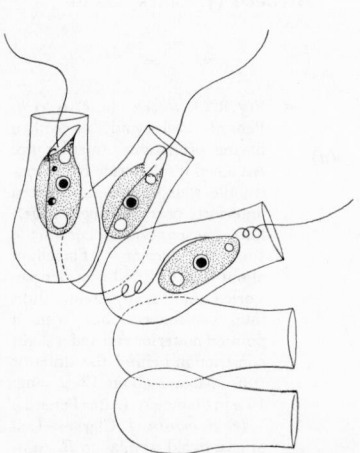

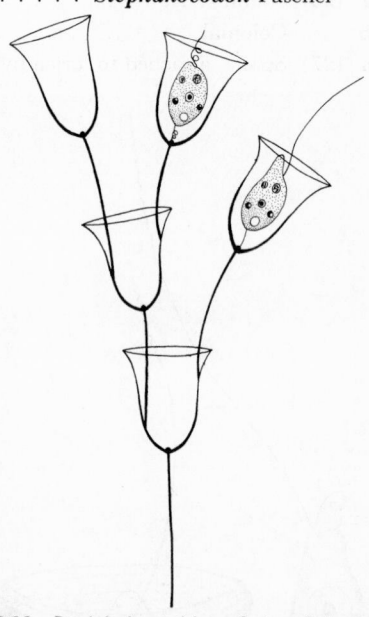

Fig. 8.25. *Stephanocodon socialis* Lauterborn. Organisms ovoid, without a very pronounced lip; a median nucleus and a posterior contractile vacuole. Cell is about 10 μ long, flagellum about 30 μ. Organism attached to base of lorica by a long thin pedicel whose insertion is in the lip area, near the active flagellum, which suggests it may also be a flagellum used as an anchor. Colonial in habit, the colonies resembling flattened rosettes. Loricae are urn-shaped, in *socialis* somewhat distended basally. Shells are exceedingly thin and transparent and their mode of attachment to each other is not apparent.

Fig. 8.26. *Poteriodendron petiolatum* Stein. Cell ovoid, with lip, which is retracted on occasion. Nucleus central, contractile vacuole posterior. Holozoic, food vacuoles present. Attached to lorica by a thin contractile pedicel. Cell about 20 μ long, flagellum about 35 μ. Colonial in habit, each cell in a campanulate lorica, which is transparent when young, light brown when older. The lorica is diagnostic in that the lower part of each shell is much thicker than the rim, and there is a small knob at the base above the heavy stalk. Each stalk arises inside a lower lorica.

30a (29) Lorica campanulate . . . . *Poteriodendron* Stein

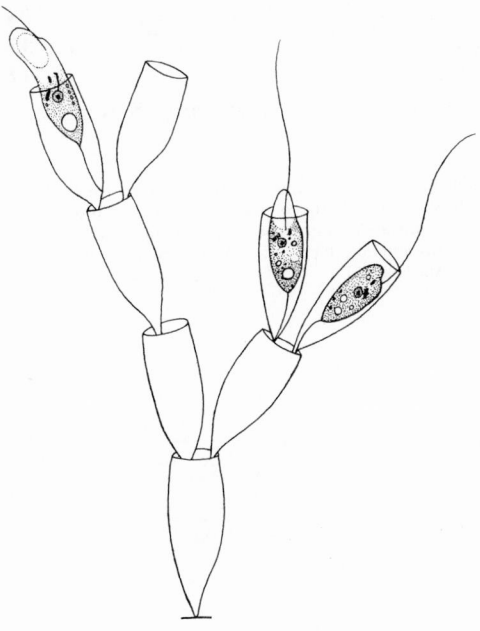

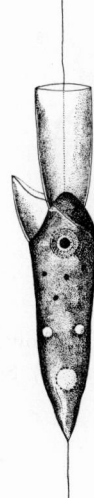

◄ **Fig. 8.27.** *Codonodendron ocellatum* Pascher. Organism ovoid, with extensive lip formation at times, about 25–25 μ long when lip is not extended. Basal contractile vacuole, median nucleus. There is an elongate stigma near the flagellum base, and a row of dots near the nucleus. Colonial, loricate, each lorica being rather urn-shaped and attached very near the rim of a lower lorica. Colony sessile on substrate by direct attachment (no stalk) of oldest lorica. (After Pascher.)

Fig. 8.28. *Codonosigopsis kosmos* Skuja. This species is solitary but the other two ► species are colonial, with several zooids at the end of a long stalk. *C. kosmos* is about 25 μ, collar rim to base. The cell is spindle-shaped, rounded in front, tapering below. Nucleus anterior, 2 or more posterior contractile vacuoles. There is a long pedicel (45 μ, Skuja) and a long flagellum. The upper collar is long and narrow as in the other species. The lower collar is only partly formed and is quite reminiscent of *Histiona*. (After Skuja.)

30b		Lorica urn-shaped	***Codonodendron*** Pascher	
31a	(18)	Collar single		32
31b		Collar double		36
32a	(31)	Solitary		33
32b		Colonial		43
33a	(32)	With lip outside collar	***Codonosigopsis*** Senn	

33b   No lip   34
34a (33) No lorica   35
34b   Loricate   37
35a (34) Collar single; sessile . . . . *Monosiga* S. Kent

◄ **Fig. 8.29.** *Monosiga varians* Skuja. Cell spherical to pointed-ovoidal in shape, usually attached directly to the substrate. At the top is a thin transparent protoplasmic collar within which a single flagellum arises. Cell about 10–15 μ long, flagellum to 30 μ. Nucleus anterior, 1 or 2 contractile vacuoles in the post median part of the cell. (After Skuja.)

35b   Collar double   36
36a (35; 31) Sessile . . . . *Diplosiga* Frenzel

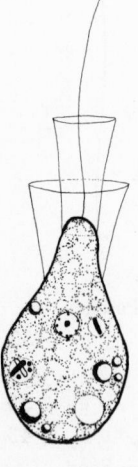

◄ **Fig. 8.30.** *Diplosiga socialis* Frenzel. Cell about 15 μ long, almost pyriform, with an anterior nucleus and a small posterior contractile vacuole. There is a single quite long flagellum and a double collar, the lower wide and short, the upper long and narrow. Holozoic. Sessile. *D. francei* Lemmermann has a short stout stalk.

36b   Stalked . . . .   *Dicraspedella* Ellis

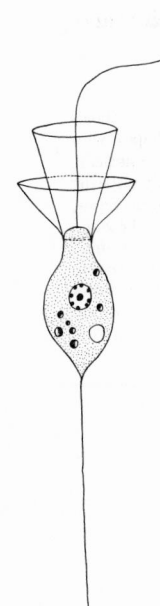

◄ **Fig. 8.31.** *Dicraspedella stokesii* Ellis. A small monad 6–10 μ long, rounded, but with an anterior neck and a pointed posterior end which gives rise to a long thin pedicel. There are 2 collars, the upper long and slightly flaring, the lower short and widely flaring. The flagellum is long, nucleus median, and the contractile vacuole posterior.

37a (34) Lorica double. . . . .   *Diploeca* Ellis

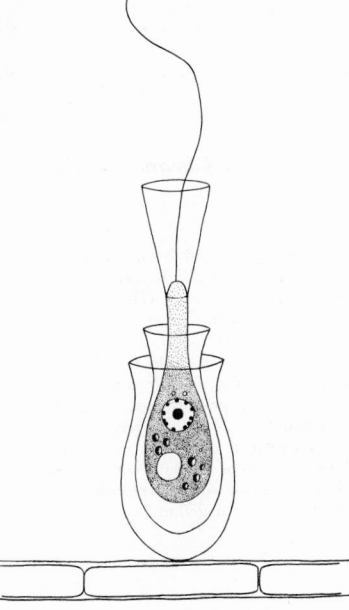

◄ **Fig. 8.32.** *Diploeca placita* Ellis. Zooid with a long anterior neck, rounded posteriorly. Nucleus median, contractile vacuole posterior. Oil or volutin present. Collar narrow but flares. Flagellum about 30 μ, cell about 20 μ. The lorica is double, both parts being urn-shaped. No stalk, the organisms being sessile.

37b   Lorica single **38**

38a **(37)** Collar double . . . . *Diplosigopsis* France

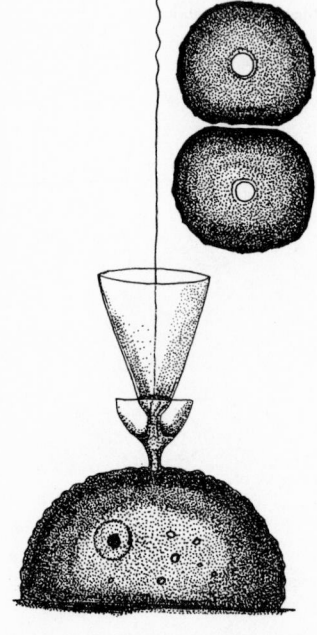

◄ **Fig. 8.33.** *Diplosigopsis siderotheca* Skuja. Zooids rounded or hemispherical (due to shell shape) with median nucleus and one or more posterior contractile vacuoles. Two collars, the lower more widely flaring than the anterior. In *D. siderotheca* the shell is low, rough, brown, and sits flatly on the substrate; the zooid extrudes a narrow neck which gives rise to a small lower collar and a longer upper collar. The flagellum is about 35 μ long, and the zooid from base to rim of upper collar about 20 μ. The 3 remaining species are all sessile, but are either pointed or rounded posteriorly. (After Skuja.)

38b   Collar single **39**

39a **(38)** Collar external to lorica **40**

39b   Collar internal to lorica **41**

40a **(39)** Collar widely flaring; attached . . . . *Choanoeca* Ellis

Fig. 8.34. *Choanoeca perplexa* Ellis. Ellis described this organism in 1929 from brackish water and gave it the species name because no flagellum was present, except temporarily in daughter cells prior to their becoming sedentary. However, the flaring collar is also distinctive. A very similar organism has been found in stagnant water in Florida on a few occasions; a flagellum could not always be seen but was certainly present on occasion in sessile organisms. The cell is ovoid, about 12 μ long and the flagellum about 20 μ to 30 μ. The width of the collar is about 20–24 μ; it is very transparent as is the lorica, which normally has a short stalk; the pyriform cell practically fills the lorica. Nucleus median, contractile vacuole basal. Organisms holozoic. Those seen in Florida are assigned here despite the apparently constant flagellum.

40b   Collar slightly flaring; attached . . . . *Salpingoeca* J. Clark

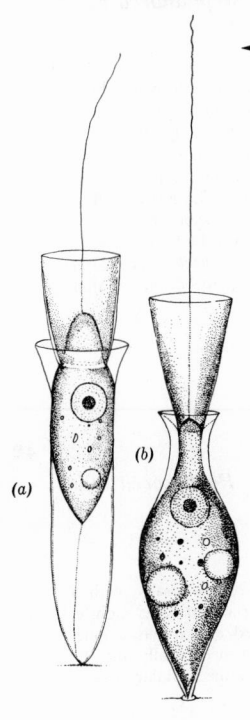

◄ **Fig. 8.35.** (*a*) *Salpingoeca vaginicola* Stein. (*b*) *S. buetschlii* Lemmermann. A large number of species all characterized by a more or less
elongate lorica, constricted either at the mouth or in a neck region,
and all attached—some by stipes, some sessile. In *S. buetschlii* the
organism almost fills the lorica, which is sessile. The nucleus is anterior, the contractile vacuole about median. The lorica is pointed-
vase-shaped. In *S. vaginicola* the lorica is elongate and sessile and
the monad, which fills about half of it, is attached to the base by a
contractile stalk. In some species of *Salpingoeca* the loricae are
brown. Size is rather variable, although few of the monads ever exceed 15 μ in length; most of them are ovoid. (After Skuja.)

40c   Collar slightly flaring; not attached . . . . *Lagenoeca* S. Kent

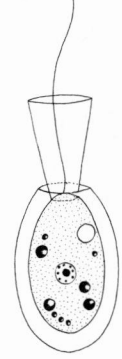

◄ **Fig. 8.36.** *Lagenoeca ovata* Lemmermann. Cell ovoid, 10–15 μ long, in a thin
lorica which is also ovoid. Collar thin, rather short, half cell length. Single
contractile vacuole, posterior, nucleus median. Zooid not attached to shell.
Shell not attached. Grassé considers this genus identical with the genus
Salpingoeca on the basis that *Lagenoeca* has simply lost its attachment. However,
L. cuspidata is stated to have 5 posterior pointed processes on its lorica, and no
such organism is known in *Salpingoeca*. Furthermore, the student should have
some means of identifying unattached as well as attached forms.

41a (39) Lorica expanded to large median or anterior bulb. . . . .

Stephanoeca Ellis

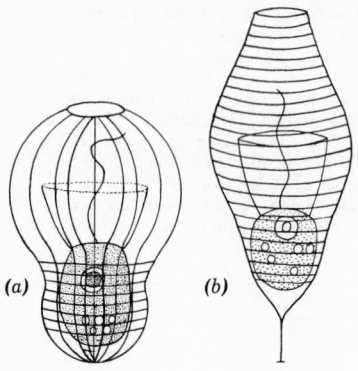

◄ **Fig. 8.37.** (*a*) *Stephanoeca ampulla* (S. Kent) Ellis. (*b*) *S. kenti* Ellis. Organism contained within a thin and bulbous lorica. About 15 μ long, flagellum but little longer. Nucleus median. Posterior end of the cell is vacuolated, but at least one of these is a contractile vacuole. The lorica is diagnostic; it is sessile, or there is a short stalk. It is thin, and annulate top to bottom in *S. kenti* and in the posterior smaller part in *S. ampulla*. The latter also has longitudinal striae top to bottom. (After Ellis.)

(*a*)   (*b*)

41b   Lorica expanded posteriorly. **42**
42a (41) Lorica bulb thick-walled . . . . *Pachysoeca* Ellis

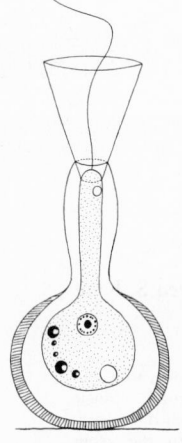

◄ **Fig. 8.38.** *Pachysoeca longicollis* Ellis. Organism round basally, with a long thin neck, retractable, surmounted by a collar which flares slightly. The basal portion of the lorica is likewise rounded and it has a long, slightly distended neck. The lorica is heavily thickened basally and the few specimens seen were so dark brown as to obscure cellular details. The lorica is diagnostic, however.

42b   Lorica bulb thin-walled . . . . *Diphanoeca* Ellis

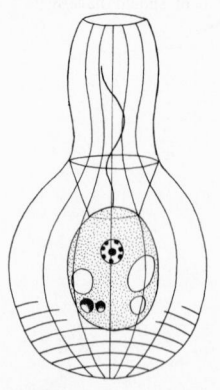

◄ **Fig. 8.39.** *Diphanoeca grandis* Ellis. Zooid almost exactly like those of *Stephanoeca*. These organisms are probably both saprozoic. The lorica of *Diphanoeca* is distinctive—greatly enlarged basally and sessile. The lower part is annulate, and there are longitudinal striae also, which apparently separate ribs; but the very few specimens seen did not show definite ribs. (After Ellis.)

43a **(32)** Colony naked **44**

43b   Colony not naked **47**

44a **(43)** Colony attached to substrate **45**

44b   Colony free. **46**

45a **(44)** Stalk unbranched or branches short . . . .   ***Codonosiga*** Senn

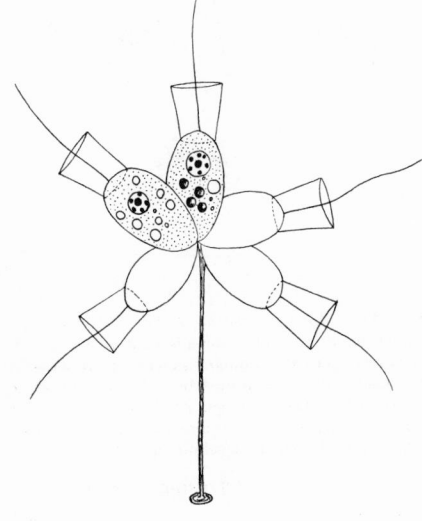

◄ **Fig. 8.40.** *Codonosiga botrytis* (Ehrenberg) S. Kent. Organism ovoid, with collar and flagellum as in Monosiga. Cell 15–20 μ, flagellum 20–30 μ. Cytoplasm vacuolated, but at least one contractile vacuole about median, nucleus anterior. Colonial, about 4 to 10 monads attached directly to a stout stalk. One of the larger collared monads, not uncommon.

45b   Stalk with long branches . . . .   ***Codonocladium*** Stein

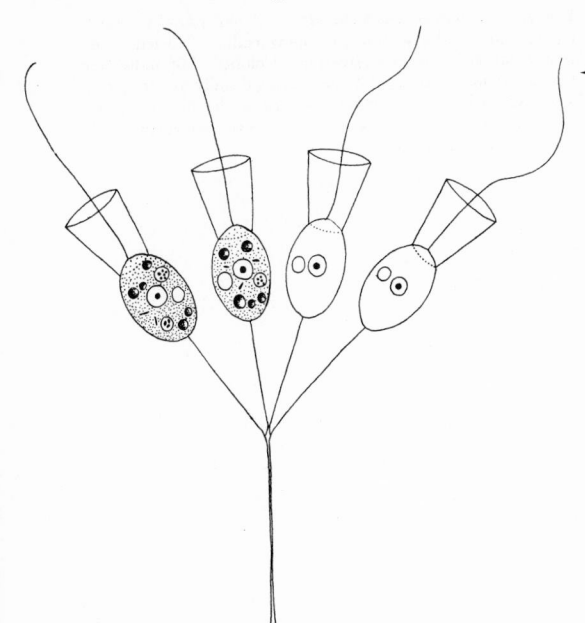

◄ **Fig. 8.41.** *Codonocladium umbellatum* (Tatem) Stein. Monad about like *Codonosiga*, except smaller (10–15 μ). Organisms do have oil drops or volutin, however, and the nucleus is central. Lower part of the stalk is thick, but it branches repeatedly in the upper part; in *corymbosum*, the branching is corymbose; in *umbellatum* all zooids branch from a common point on the stalk.

46a (44) Colony linear . . . . *Desmarella* S. Kent

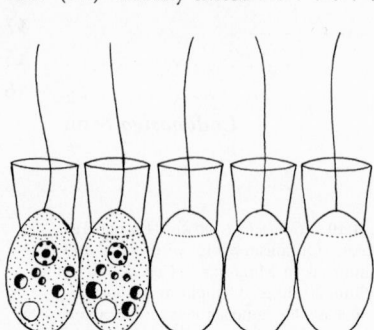

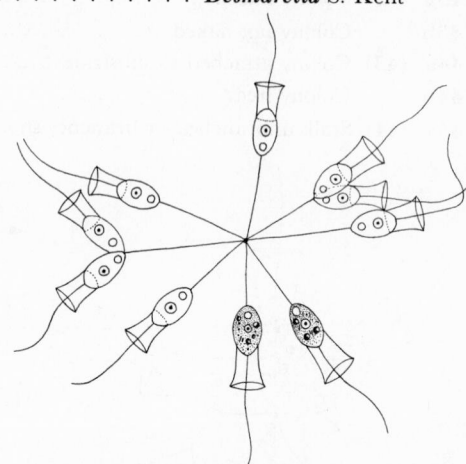

Fig. 8.42. *Desmarella moniliformis* S. Kent. Monads free-living, not stalked, but attached tangentially to each other in linear colonies. A typical ovoid collared monad, sometimes very abundant in stagnant water plankton, each zooid about 10 μ in length. The contractile vacuole is basal, the nucleus slightly anterior, and there is usually a group of oil droplets or volutin granules in the middle of the cell.

Fig. 8.43. *Astrosiga radiata* Zach. Free-floating, radiate colonies of collared monads, each on a stem, all the stems attached at a common center. (In *A. disjuncta* the monads are not stemmed but attached by their pointed bases.) Collar somewhat flaring in its outer third. Nucleus central, contractile vacuole basal. Monads about 15 μ long, flagellum up to 40 μ.

46b   Colony radiate . . . . *Astrosiga* S. Kent
47a (43) Colony loricate. 48
47b   Colony in jelly 49
48a (47) Loricae separate, colony arboroid . . . . *Polyoeca* S. Kent

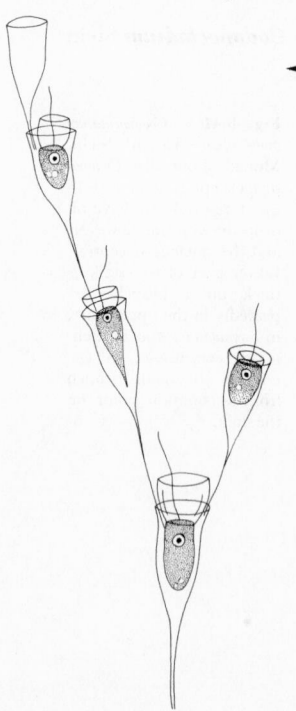

◄ **Fig. 8.44.** *Polyoeca dumosa* Dunkerly. Zooids round to ovoid, truncate anteriorly, with a ballooning collar. Nucleus anterior, contractile vacuole posterior. Colonial, but individual loricae are long campanulate, narrowing distally to a long thin stalk, which is attached inside the rim of the mother lorica. No attaching pedicel for the organism. Size and nutrition not given. (After Dunkerly.)

48b   Common tubular lorica, zooids in ends of branches . . . .
                                                        Stelexomonas Lackey

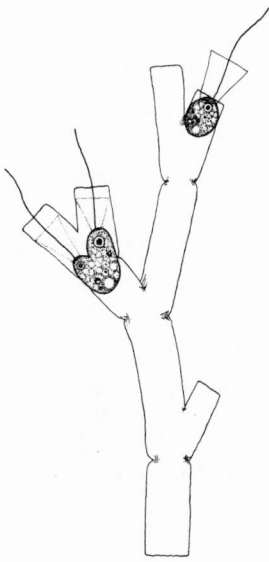

◄ **Fig. 8.45.** *Stelexomonas dichotoma* Lackey. Organisms round to ovoid, about 15 μ long, with an anterior nucleus and a posterior vacuole. Protoplasm vacuolate. Collar slightly flaring, as long as cell. Flagellum about 2½ times cell length. Holozoic. The colonial shell is diagnostic; no attached shells have been seen, but the fragments are transparent, dendroid, and branch dichotomously. They are of uniform diameter, and there is a slight depression at each branching point, which is lightly wrinkled. Organisms show no attaching pedicel, but do retract completely into lorica.

49a (**47**) Colony spherical, large . . . . *Sphaeroeca* Lauterborn

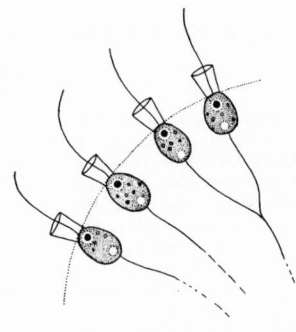

◄ **Fig. 8.46.** *Sphaeroeca volvox* Lauterborn. Colony round, may be as large as 150 μ in diameter, with numerous zooids tangentially arranged, collar exserted. Collar does not flare. Zooids attached to each other by long posterior pedicels, which may branch, between the center and periphery. Jelly sometimes brown. Nucleus anterior, contractile vacuole posterior, little evidence of holozoic nutrition. Cells about 10–15 μ long.

49b   Colony not spherical **50**
50a (**49**) Colony irregular, amorphous . . . . *Proterospongia* S. Kent

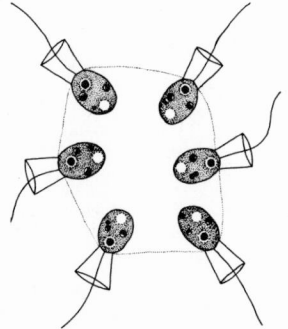

◄ **Fig. 8.47.** *Proterospongia haeckelii* S. Kent. Free-floating collared monads in irregular clumps of jelly, the colony usually flattened, up to 50 μ in diameter, and the monads tangential, but the whole collar exserted. Cells about 10 μ long, ovoid, with an anterior nucleus and a posterior contractile vacuole. Sometimes cited as *Proterospongia*. Some authors maintain that *Protospongia* is nothing more than bits of fresh-water sponge. This organism is discussed in more detail by the author in his paper, Morphology and biology of a species of *Protospongia*, *Trans. Am. Microscop. Soc.*, 1959.

50b Colony arboroid or flattened. . . . . . *Phalansterium* Cienkowski

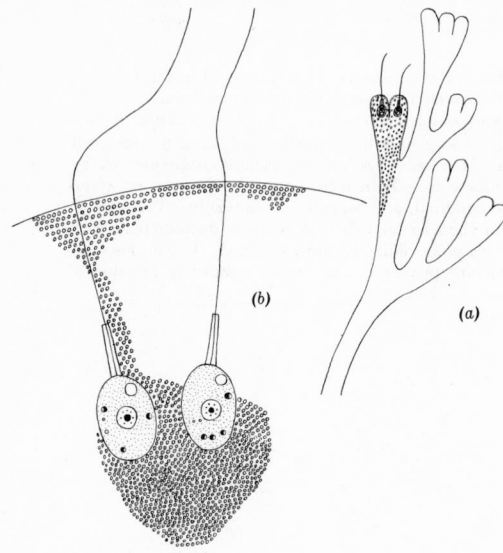

◀ **Fig. 8.48.** (*a*) *Phalansterium digitatum* Stein. (*b*) *P. consociatum* (Fres.) Cienk. Organisms enclosed in a gelatinous mess, usually flattened in *P. consociatum,* but arboroid with the tips rounded in *P. digitatum.* Zooids rounded ovoids with a central nucleus and oil droplets in the cytoplasm. Pascher says there are 2 contractile vacuoles in the posterior end; but the few colonies observed had only 1. The collar is distinctive; narrow at the top, and expanding very slightly towards the base. Both zooid and collar are deeply insunk, the single long, rather stiff flagellum traversing the jelly for as much as one-third of its length. Cells large, up to 20 μ. Jelly brown, due to ferric hydroxide. Colonies may be quite large and are attached.

50c Colony finger-shaped. . . *Cladospongia* Iyengar and Ramanathan

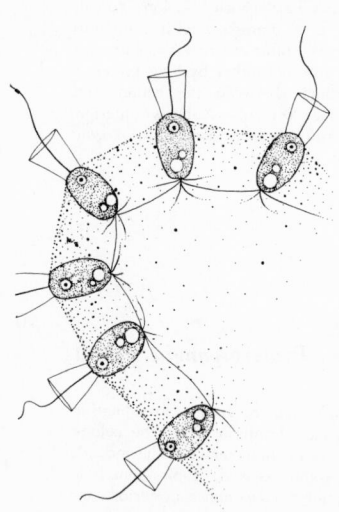

◀ **Fig. 8.49.** *Cladospongia elegans* Iyengar and Ramanathan. Collared cells about 10 μ long, in the periphery of fingerlike mucilaginous matrices, which are themselves attached to some solid substrate. Each cell ovoid but somewhat truncate at the anterior end, where the slightly flaring collar joins it. Collar about as long as the cell, flagellum about 3 to 4 times cell length. One or 2 posterior vacuoles, nucleus anterior. Each cell is connected to neighboring cells by protoplasmic strands; as many as 7 of these pass out from the base of a single cell. (After Iyengar and Ramanathan.)

51a (17) One anterior, 1 trailing flagellum. **52**

51b One anterior, 2 trailing flagella. **61**

52a **(51)** Trailing flagellum attached to fixed point on substrate. Body reniform. . . . . ***Pleuromonas*** Perty

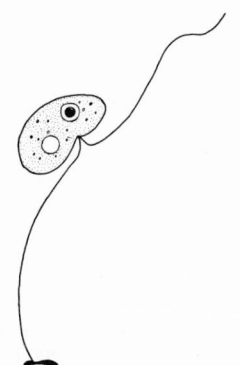

◄ **Fig. 8.50.** *Pleuromonas jaculans* Perty. This small "jumping monad," about 6–10 μ long is reniform in shape and rounded. There is an anterior nucleus and a posterior contractile vacuole. Two flagella emerge from a lateral invagination; the longer, trailing, normally attaches to some object, and the anterior, also quite long, is used to thrash the body about. The organism is very sensitive to oxygen diminution, and will detach and swim toward the edge when under a cover glass. The trailer is then identified. Swimming is a rapid rotation about a straight line.

52b Trailing flagellum attached along body **53**

52c Trailing flagellum free. **55a**

53a **(52)** Body greatly attenuated . . . . . ***Phanerobia*** Hartmann and Chagus

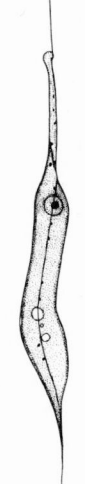

◄ **Fig. 8.51.** *Phanerobia pelophila* Skuja. An extremely long (to 50 μ) organism of rather frequent occurrence in anaerobic sewage. The body may approach 10 μ in thickness; it is an elongate cylinder, pointed behind, but with a neck area about ¹/₄ to ¹/₅ of the entire length. This neck terminates in a rounded knob with a pronounced overhang. From this a swimming flagellum emerges, which is about 35 μ long and vibratile in its distal third; a trailing flagellum also emerges, to which the body clings and which is about 15 μ longer than the body. The organism is quite metabolic, and the cytoplasm is granular with a few small, dark inclusions. The nucleus is in the anterior third and one or more contractile vacuoles are in the posterior third. Only a gliding movement has been observed.

53b Body not attenuated **54**

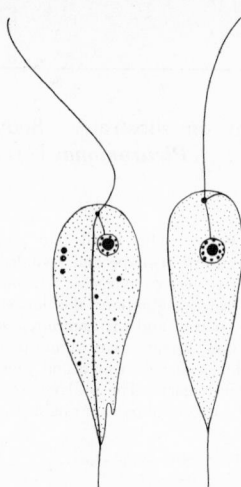

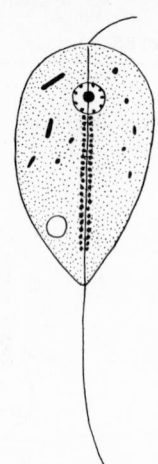

Fig. 8.52. *Cercomastix parva* Hartmann and Chagas. Elongate, broadly rounded anteriorly, tapering to flexible or amoeboid point posteriorly. Flagellum about body length. A stiff process (flagellum?) extending from the anterior end, posteriorly beyond end of body. Nucleus anterior with an often visible fibril (rhizoplast) extending from it to a blepharoplast (the basal granule from which the flagellum originates). Flagellum about body length, or 15–25 μ. No visible vacuoles. Body highly plastic. Cytoplasm homogenous.

Fig. 8.53. *Helkesimastix faecicola* Woodcok and Lapage. Common in sewage sludge, this very small organism (8 by 3 μ) is flattened, rounded anteriorly, somewhat pointed posteriorly. A prominent posterior flagellum, which is in contact with the body so that the organism frequently rolls on it as on an axis, extends somewhat more than body length behind. Bordering this flagellum are 16 to 24 pairs of round bodies, very definite in appearance. No anterior flagellum can ordinarily be seen; it is very short and carried to one side. Progression is steady and rapid, in a straight line. Nucleus anterior, contractile vacuole posterior. No pseudopod formation seen, but nutrition is holozoic.

54a (53) Readily forms pseudopodlike processes . ***Cercomastix*** Lemmermann

54b   Body shape quite constant . . ***Helkesimastix*** Woodcock and Lapage

55a (52) Flagella emerge subterminally or laterally **56**

55b   Flagella emerge anteriorly or from a definite mouth depression. . . . **59**

56a (55) Body rigid, oval. . . . . ***Colponema*** Stein

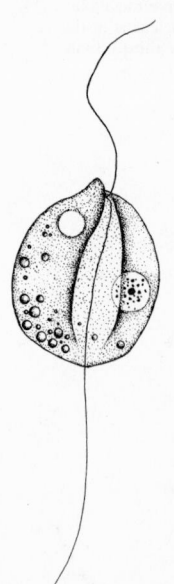

◄ **Fig. 8.54.** *Colponema loxodes* Stein. This somewhat flattened spherical organism is anomalous in position. Hollande would place the species in the Chloromonadida. There is some evidence at times of a gullet to the vacuole which, despite absence of paramylum, might relate it to a colorless euglenid. It may be 25 μ in diameter and 20 μ thick. A deep anteroposterior curved groove bordered by a ridge, which begins at a forward point, slightly eccentric; from the side of this point emerge one forward-directed flagellum about body length and one trailing flagellum about 3 times as long. The contractile vacuole is just behind and lateral to the flagella. The nucleus is central and large. Oil is stored. The body is rigid and the shape is characteristic.

56b   Body plastic **57**

57a **(56)** Body normally attenuate anteriorly . . . . ***Spiromonas*** Perty

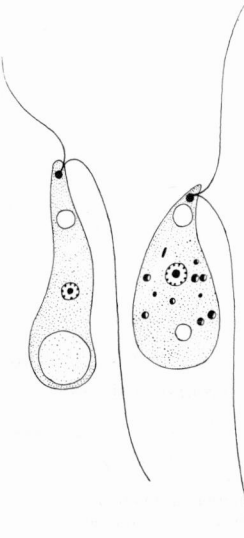

Fig. 8.55. *Spiromonas angusta* (Dujardin) Alexieff. Cell shaped like a curved teardrop, slightly arched, about 10 μ long with 2 subequal flagella inserted just behind the anterior point. Because the organism is primarily a swimmer, the trailing nature of the longer (20 μ) flagellum is rarely evident. There is a small contractile vacuole behind the base of the flagella and a small kinetonucleus anterior to the median nucleus. The organism is extremely active and voracious. It uses its pointed anterior to penetrate and absorb other organisms, which aggregate in a huge posterior food vacuole. One *Spiromonas* will absorb as many as four *Monas sociabilis*, and round up into a trembling mass whose features are no longer distinguishable. The mechanism of swallowing or absorbing other organisms is not yet explained.

57b   Body not attenuate anteriorly **58**

58a **(57)** Body contours usually rounded . . . . ***Bodo*** Ehrenberg

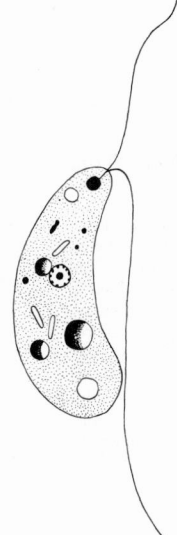

Fig. 8.56. *Bodo caudatus* Dujardin. Free-living, naked cells, flattened or rounded, often curved in an antero-posterior arc. *B. caudatus* is 10–25 μ long, 6–15 μ wide, but *B. minimus* Klebs is only about 5 μ long. The body is plastic and sometimes forms pseudopodia, but each species has a definite shape. All appear to be holozoic, and oil is stored. Nucleus median and large. Anterior to the nucleus is a large refractive body, the kinetonucleus, which is the origin of the flagella. In passing it should be noted that at least 5 genera of Protomastigineae—*Amastigomonas, Rhynchomonas, Bodo, Spiromonas* and *Phyllomitus*—definitely possess this organelle. They are not keyed together, however, despite evident relationship, because the number and kind of flagella have been used to determine their position in the key. Possession of a kinetonucleus by *Amastigomonas* may be regarded as indicating close relationship to *Bodo* and it is believed by some workers that the organism has secondarily lost its flagella.

The flagella of *Bodo* arise from the inside arc of the curve. The anterior flagellum is used in gliding (the commonest movement) in a jerky fashion. An anterior contractile vacuole is common, but sometimes there is a posterior one. Swimming flagellum about body length, trailing flagellum about twice as long.

58b   Body contours usually irregular or amoeboid. . . . .
                                             Cercobodo Krassilotschick

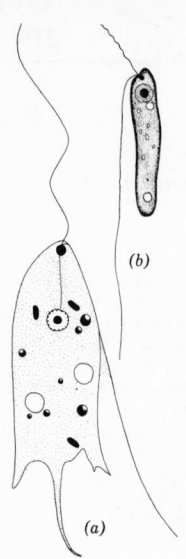

Fig. 8.57. (*a*) *Cercobodo agilis* (Moroff) Lemmermann. (*b*) *C. angustus* Skuja. *C. agilis* is flattened, frequently amoeboid with short, usually posterior pseudopodia. *C. angustus* is usually of fixed shape, rarely forms pseudopodia. Both have a kinetonucleus from which the flagella emerge. *Agilis* shows a peculiarity of the genus; the trailing flagellum frequently adherent to the body for a portion of its length by a fold or stretching of the plasma membrane. Vacuoles usually 2, variable in position, nucleus usually in the anterior third of the cell. (*b* after Skuja.)

59a (55) Kinetonucleus visible **60**

59b No visible kinetonucleus . . . . *Dinomonas* S. Kent

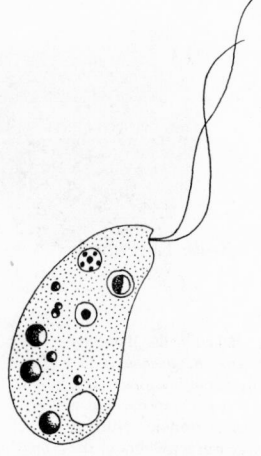

Fig. 8.58. *Dinomonas vorax* S. Kent. A rounded, curved organism about 15 μ long, 8 μ in diameter. Two subequal swimming flagella arising from a small anterior indentation on the concave surface. The contractile vacuole is posterior. Holozoic.

60a (59) Flagella from mouthlike lipped depression . . . . . *Parabodo* Skuja

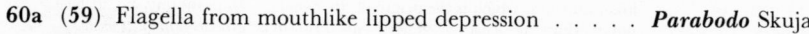

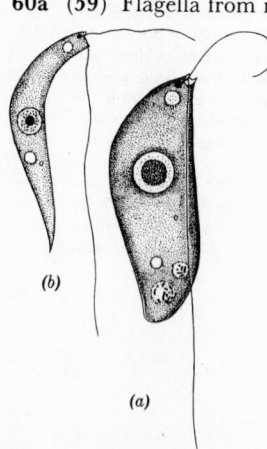

Fig. 8.59. (*a*) *Parabodo nitrophilus* Skuja. (*b*) *P. attenuatus* Skuja. Elongate, somewhat flattened, but oval in cross section, pointed posteriorly and to some extent anteriorly. Two lips partly ring a mouthlike anterior depression from which 2 flagella emerge. Body plastic but does not form pseudopodia. Often slightly S-shaped. Anterior flagellum short, often used in the "hooking" fashion characteristic of the Bodonidae. Trailing flagellum half to twice the length of the body. Both flagella from a kinetonucleus, usually visible. Nucleus central, contractile vacuole anterior, one or more other (food?) vacuoles posterior. Cytoplasm generally uniformly granular. (After Skuja.)

60b   Flagella from mouthlike depression; body saclike . . . .
Phyllomitus Stein

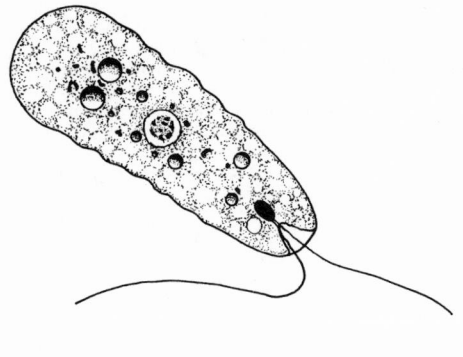

◄ **Fig. 8.60.** *Phyllomitus amylophagus* Klebs. This organism is quite active, very metabolic, and the trailing nature of the longer flagellum is not evident. According to Pascher there is no kinetonucleus; Hollande describes one. The organism figured does not have an elongate kinetonucleus such as Hollande shows (Grassé) but does have a saclike depression with an evident kinetonucleus at its base, from which the flagella arise. Vacuole anterior, nucleus median. Cell about 15 μ long, a flattened plastic cylinder.

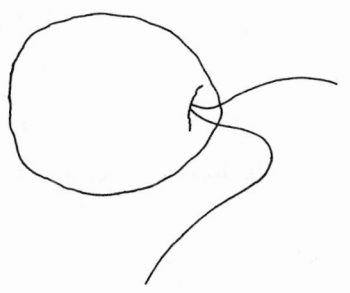

61a (51) Trailing flagella about as long as anterior flagellum. . . . .
Trimastigamoeba Whitmore

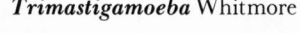

◄ **Fig. 8.61.** *Trimastigamoeba* sp. Whitmore. The figured organism from sewage and stagnant water is not unlike the one described by Whitmore; only a 3-flagellated stage has been seen. Flagella of equal length with a single kinetonucleus. Cell an elongate oval with a median nucleus and a posterior vacuole. Pseudopodia not formed although body is quite amoeboid. Movement deliberate, a slow swimming. 25 by 10 μ.

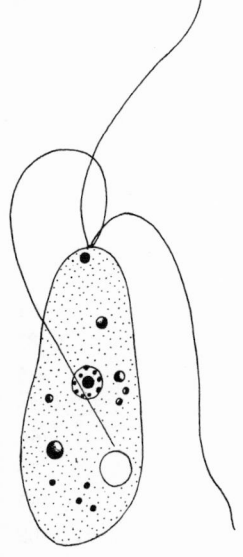

61b   Trailing flagella longer than anterior **62**

62a **(61)** Body flattened, oval . . . . ***Macromastix*** Stokes

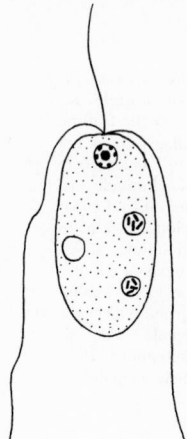

◄ **Fig. 8.62.** *Macromastix lapsa* Stokes. About the same size as *Dallingeria*, but a rather flattened oval. The anterior flagellum is less than body length, and the 2 posterior flagella do not undulate as in *Dallingeria* (or *Prymnesium*). The nucleus is anterior and the vacuole is median. Both these genera have a nucleus in which there is a central endosome with large peripheral chromatin granules.

62b Body attenuate anteriorly . . . . ***Dallingeria*** S. Kent

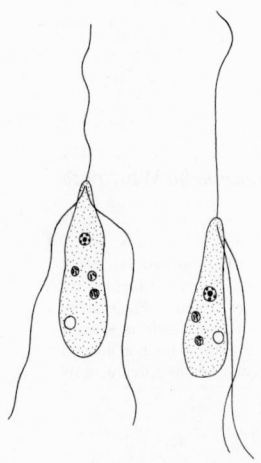

◄ **Fig. 8.63.** *Dallingeria drysdali* S. Kent. Free-living, somewhat elongate, pyriform, but rounded in cross section, bluntly pointed anteriorly, and with decidedly plastic bodies. These organisms are uncommon but are sometimes abundant in polluted water. They are 5–10 μ long, have a posterior contractile vacuole, and 1 forward-directed flagellum about 15 μ long and 2 similar ones posteriorly directed. All 3 emerge just behind the anterior tip, from a practically common insertion. Holozoic.

63a **(16)** Two anterior flagella 64
63b Four or more flagella 81
64a **(63)** Solitary. 65
64b Colonial 73
65a **(64)** Extra pellicular covering absent 66
65b Extra pellicular covering present 70
66a **(65)** Flagella approximately equal 67
66b Flagella subequal 69
67a **(66)** Organisms attached, globose. . . . . ***Amphimonas*** Dujardin

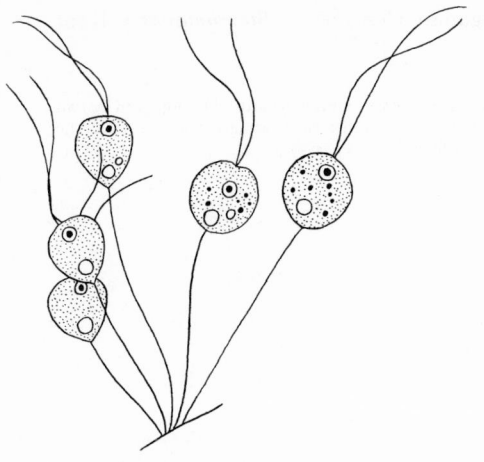

Fig. 8.64. *Amphimonas globosa* S. Kent. Spherical cells, about 15 μ in diameter with 2 long flagella (30 μ) and a tendency to occur in groups. Attached by a long thin stalk. Nucleus median, 1 basal contractile vacuole. Pascher says 2 vacuoles and there are sometimes more, but only 1 is contractile. Holozoic, and with a characteristic swaying motion on their stalks.

67b   Organisms free **68**

68a **(67)** Organisms flattened, oval, with keel. . . . . . . *Streptomonas* Klebs

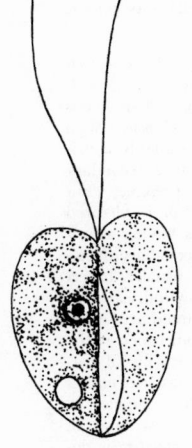

Fig. 8.65. *Streptomonas cordata* (Perty) Klebs.   Cell heart-shaped, flattened but with an antero-posterior keel or lobe. Nucleus median, contractile vacuole posterior and near the periphery. About 20 μ long, flagella about 25 μ. Swims with a rotating movement. Evidently rare, only one or two personal observations. Pascher says the contractile vacuole is anterior but shows it as posterior.

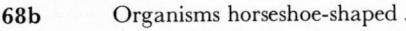

68b   Organisms horseshoe-shaped . . . . *Furcilla* Stokes

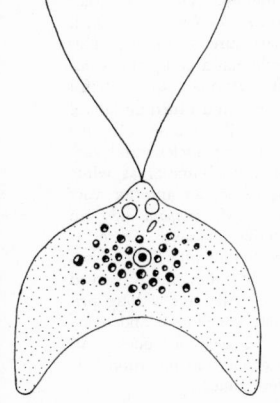

Fig. 8.66. *Furcilla lobosa* Stokes. A horseshoe-shaped cell, with the center enlarged but the arms thinned to a rounded point. Length about 15 μ, breadth across the arms about 20 μ. Shape diagnostic. The flagella are about 15 μ long. Movement a slow rotation. One or more contractile vacuoles at the flagellar base, which is a small papilla. Evidently rare, probably an anaerobe, and saprozoic. Grassé considers this a colorless member of the Volvocales. (After Lemmermann.)

69a (66) Primary flagellum rigid, organism elongate . . **Sterromonas** S. Kent

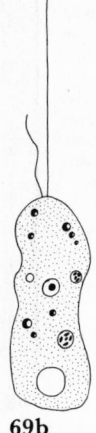

◄ **Fig. 8.67.** *Sterromonas formicina* S. Kent. Grassé regards this organism as an elongate, rather cylindrical *Monas*. The rather rigid primary flagellum and the posterior contractile vacuole hardly justify a separate genus.

69b Primary flagellum rigid, organism spherical, free . . . .
Monas (Ehrenberg) Stein

Monas is quite variable in shape, attachment, and number of flagella—its generic characters need re-examination.

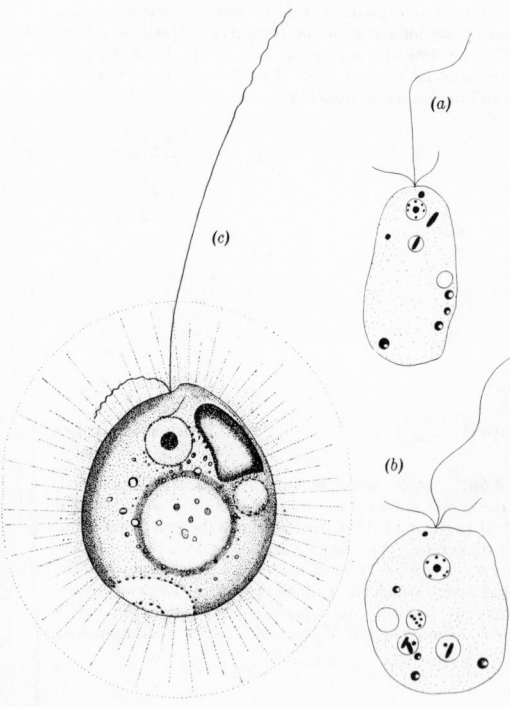

◄ **Fig. 8.68.** (*a, b*) *Monas vivipara* Ehrenberg. (*c*) *M. coronifera* Skuja. This genus separable from *Oicomonas* with difficulty; presumably a primary and 1 or 2 secondary flagella are present. But *Oicomonas ocellata* is pictured by Pascher as having a primary and a secondary flagellum. *Monas* usually has a spherical to cylindrical cell, sometimes pointed at the anterior end, sometimes at the posterior, especially if attached by the posterior end as in *M. socialis*. Single or colonial, with or without an eyespot. *M. coronifera* is described as having a gelatinous mantle, traversed by faint radiations. However, other species of *Monas* show such a structure at times. This species has a conspicuous volutin granule, an anterior nucleus, and a median lateral contractile vacuole. *M. vivipara* has its nucleus and vacuole in the same areas, while *vulgaris* has an anterior vacuole. *M. vivipara* has an anterior eyespot. Size is variable for all of these species, but they rarely exceed 20 μ. All have a long primary flagellum, and may have 2 accessory ones. All have holozoic nutrition. (*c* after Skuja.)

69c   Primary flagellum rigid, organism variable, attached . . . .
                                                              Stomatochone Pascher

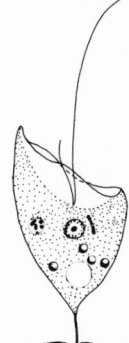

◄ **Fig. 8.69.** *Stomatochone infundibuliformis* Pascher. A genus of small flagellates, 10–15 μ long, normally attached to other plankton organisms. When free, they are normally broadly rounded, but with a rather obliquely truncated anterior end. A long and a short flagellum appear to be constant. Attachment is by a posterior pseudopodlike attenuation; in a median cross section they are round, and the anterior end is oblique, deeply excavate, with a wide flaring thin lobe as an upper border. Contractile vacuole basal, nucleus about median, and a cytoplasmic band organ adjacent to it. Since the characteristics as listed above appear constant, the genus seems well founded, even though the unattached stage strongly suggests *Monas*.

70a (65) Covering a lorica **71**

70b   Covering an investment of spiny plates . . . . . ***Physomonas*** Stokes

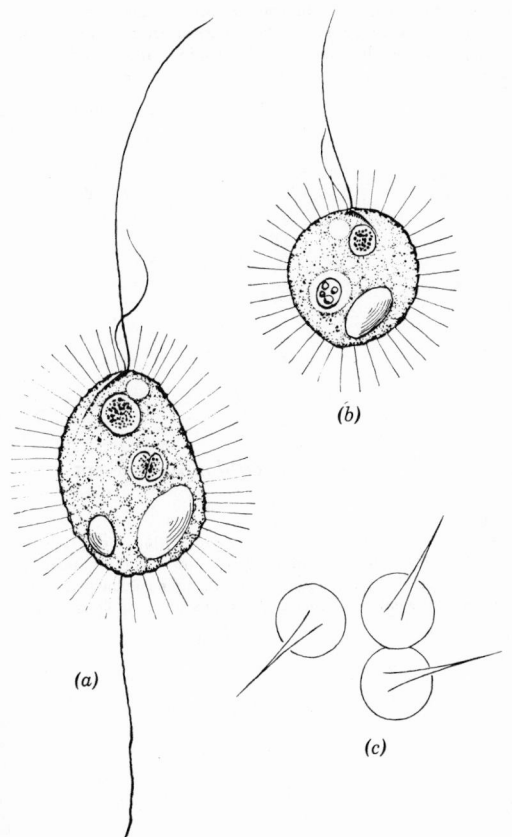

(b)

(a)

(c)

◄ **Fig. 8.70.** An organism very close to *Physomonas vestita* Stokes. (*a*) The attached form, showing 2 subequal flagella, the longer being carried forward in a fairly rigid curve. There is one anterior contractile vacuole; an anterior nucleus, not however, tangent to the blepharoplasts; 1 food vacuole containing blue-green algal cells; and 2 posterior volutin granules. (*b*) The swimming form is essentially similar. Each shows a structure extending from the base of the flagella which is either a shallow groove or a bandelette. The surface of the cell suggests a rough pellicle, but no silicious plates, as shown in (*c*) are demonstrable. (*c* from Grassé, after Korschikov.)

71a **(7υ)** Mouth area deeply excavate . . . . *Stenocodon* Pascher

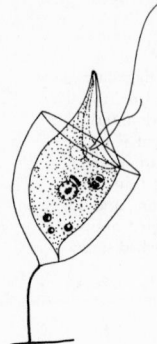

◄ **Fig. 8.71.** *Stenocodon epiplankton* Pascher. This cell, almost precisely like *Stomatochone*, lives attached by its attenuated base inside a lorica so transparent as to be almost invisible. The loricae are open curved cones, 15–20 μ long, with a stalk, and almost always turned sideways. Leucosin, a median nucleus, and an anterior contractile vacuole are present.

71b   Mouth area slightly indented 72
72a **(71)** Lorica cylindrical. . . . . *Stokesiella* Lemmermann

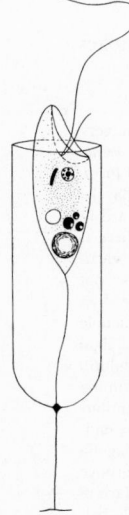

◄ **Fig. 8.72.** *Stokesiella epipyxis* Pascher. The several species are attached by a contractile pedicel to the base of a lorica, which is variously shaped and always stalked. In *epipyxis* the lorica is a straight urn, round at its side and truncate at the top. The cell is typically ovoid, with an anterior lip at whose base emerge 2 flagella. The long one is about 25 μ long, the shorter about 5 μ, and the cell is about 10 μ. There is frequently a large basal leucosin granule, and a basal or median vacuole. A stigma and a band organ are located near the somewhat anterior nucleus.

72b   Lorica globose . . . . *Diplomita* S. Kent

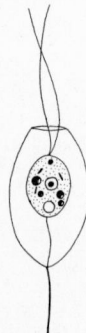

◄ **Fig. 8.73.** *Diplomita socialis* S. Kent. Ovoid cells in an ovoid, thin lorica which has a stout but not long stalk. The lorica is truncate at the top. The organism is about 10 μ long, the 2 equal apical flagella about 20 μ. A stigma is located near the flagellar base, the nucleus is median, and the single contractile vacuole is basal. Holozoic.

73a (64) Lorica present **74**

73b   Lorica absent. **77**

74a (73) Mouth area excavate **75**

74b   Mouth area not excavate **76**

75a (74) Loricae campanulate, attached to other loricae . . . .
                                              Stylobryon Fromentel

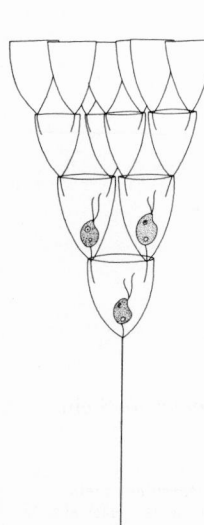

◄ **Fig. 8.74.** *Stylobryon abbotti* Stokes. The colony shape is diagnostic
—the colony stem is very long, but all daughter loricae are attached
near the rims of other loricae by a very short stem. Loricae are
companulate and hyaline to dark brown. Zooids resemble those of
Stokesiella (Grassé) with a median to anterior nucleus, posterior contractile vacuole, and two subequal flagella from the base of a somewhat pronounced lip, the longer about cell length (15 μ). No
personal observations. (After Lemmermann.)

75b   Loricae attached to a common stalk. . . . . . ***Codonobotrys*** Pascher

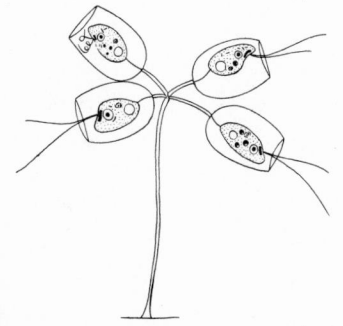

◄ **Fig. 8.75.** *Condonobotrys physalis* Pascher. Zooids living in urn-shaped or campanulate loricae that arise by
short stems from a common point on the thick parent
stem. Organisms small, about 8 μ long, ovoid, obliquely truncate, a small depression in the mouth area
giving rise to 2 subequal flagella, the longer about 10
μ, the shorter about 8 μ. A band organ lies above the
anterior nucleus, the contractile vacuole is basal and
the organisms are definitely holozoic. Loricae usually
colorless. (After Grassé.)

76a   **(74)** Lorica compound, an arboroid tube. . . . . ***Cladomonas*** Stein

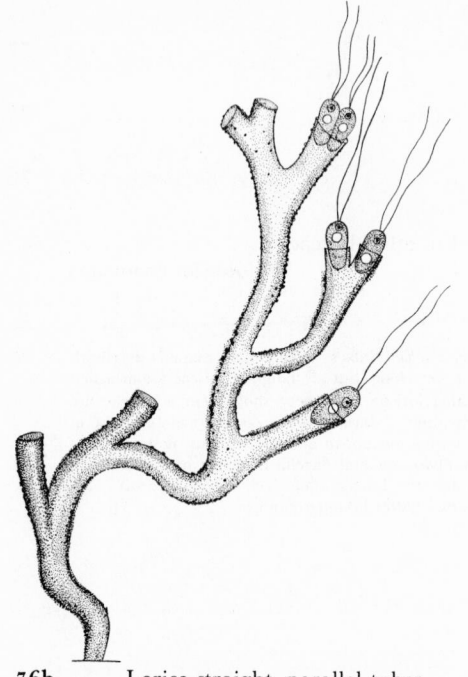

◄ **Fig. 8.76.** *Cladomonas fruticulosa* Stein. This dendroid colony consists of pipe-like, curved, freely-branching tubes, often with attached debris. A single monad usually occupies the end of each tube; about 10 μ long, and ovoid in shape, with subequal flagella about 30–40 μ long, an anterior nucleus and a median contractile vacuole. (After Stein.)

76b   Lorica straight, parallel tubes. . . . . ***Rhipidodendron*** Stein

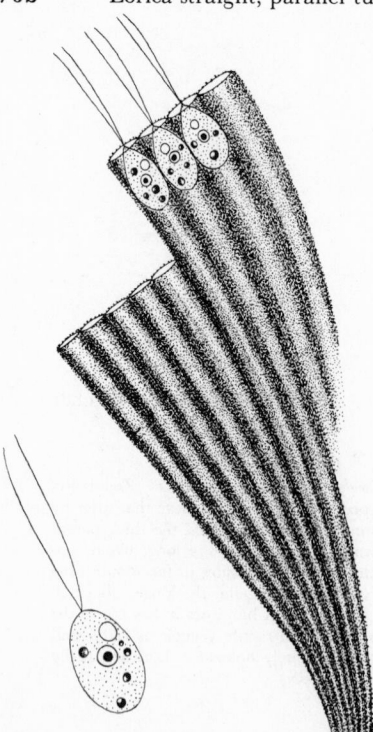

◄ **Fig. 8.77.** *Rhipidodendron splendidum* Stein. Cells 5–8 μ long, ovoid, borne in the ends of long brown tubes attached to each other for much of their length, forming a flat plate resembling organ pipes. Colonies often 2 mm in length and width. The cell contains a median nucleus, just antero-lateral to which is a contractile vacuole. Two subequal flagella are about 15–20 μ long. The organism is most frequently found in rather clear, acid, brown water; its shell is diagnostic.

77a (73) Colony enclosed in gelatinous matrix . . . . **Spongomonas** Stein

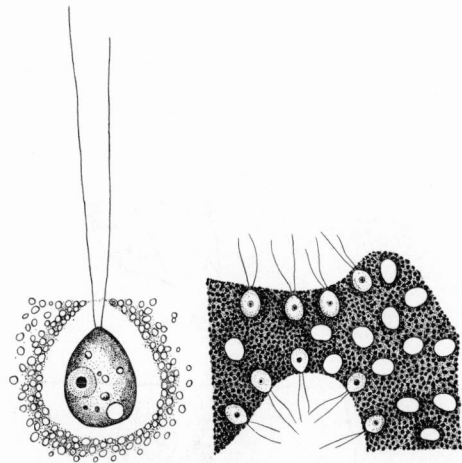

Fig. 8.78. *Spongomonas uvella* Stein.   Rather large colonies, 500 μ or more, attached to debris, plants, etc. The jelly is rarely clear, but brown or gray and with brown or yellow granules.   The cells are isolated in small pits in the jelly, somewhat pyriform or ovoid, but pointed anteriorly.   There is a single contractile vacuole, median to posterior in location, and a large nucleus with a karyosome in a medio-lateral position. Two slightly subequal flagella emerge from the anterior tip of the cell and are about 3 times the length of the cell which is 10–15 μ.   The colonial mass is diagnostic.   According to Lemmermann the 3 species have vacuoles located posteriorly in *S. intestinun*, median in *S. uvella* and anteriorly in *S. sacculus*, which does not quite agree with the condition noted above.

77b   Colony not enclosed **78**

78a (77) Stalk thin, branching repeatedly. . . . . **Cladonema** S. Kent

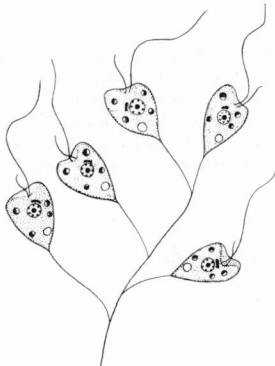

Fig. 8.79.   *Cladonema pauperum* Pascher.   Colonial, but the zooids are borne at the end of short stalks, whereas the secondary stalks in *Dendromonas* are long, except for those near the end, producing a rather umbel-like colony.   The zooids in *Cladonema* are also obliquely truncate.   Nucleus median, contractile vacuole basal.   A band organ lies just above the nucleus, and the accessory flagellum is about 3–4 μ long, the cell about 10 μ.

78b   Stalk thickened. **79**

79a (78) Zooids single at branch ends . . . . **_Dendromonas_** Stein

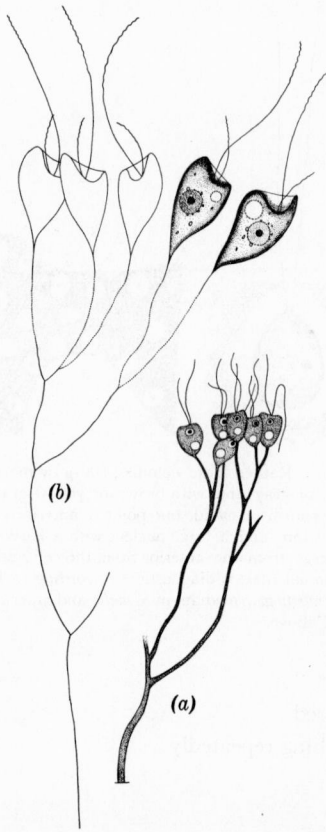

Fig. 8.80. (*a*) *Dendromonas virgaria* (Weisse) Stein. (*b*) *Dendromonas cryptostylis* Skuja. Colonial, cells single at the end of dendroid, rather thick, often brown stalks. Cells round, but transversely truncate at the anterior end, with a slight depression from which emerge 2 long, but subequal flagella. Nucleus anterior, contractile vacuole basal, cytoplasm quite dense. In the few colonies seen, no food particles or vacuoles were ever noted. Cells 10 μ long, equally wide. In *D. cryptostylis* Skuja the stem is thin, the vacuole anterolateral, the nucleus median. *D. distans* (Pascher) has a stigma. (*a* after Doflein; *b* after Skuja.)

79b   Zooids grouped at branch ends **80**

80a **(79)** Epizoic on copepods.   Stalk heavy, rigid . . **Cephalothamnion** Stein

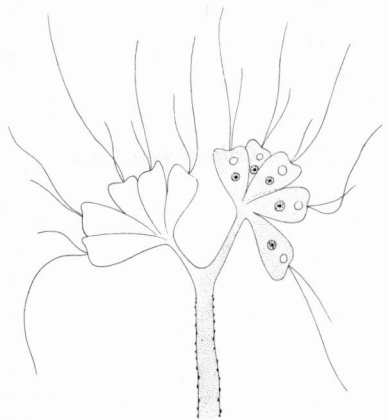

◄ **Fig. 8.81.** *Cephalothamnion cyclopum* Stein. Zooids all clustered at end of a thick, rigid stalk to which they are attached. Cells somewhat conical, but their anterior face is somewhat obliquely truncate or lipped. About 8 μ long, the accessory flagellum about the same length, the principal about 25 μ long. Nucleus median, contractile vacuole posterior. Generally epizoic on copepods.

80b   Not epizoic, stalk flexible, heavy, with attached foreign matter . . . .
                                                        Anthophysa Borg

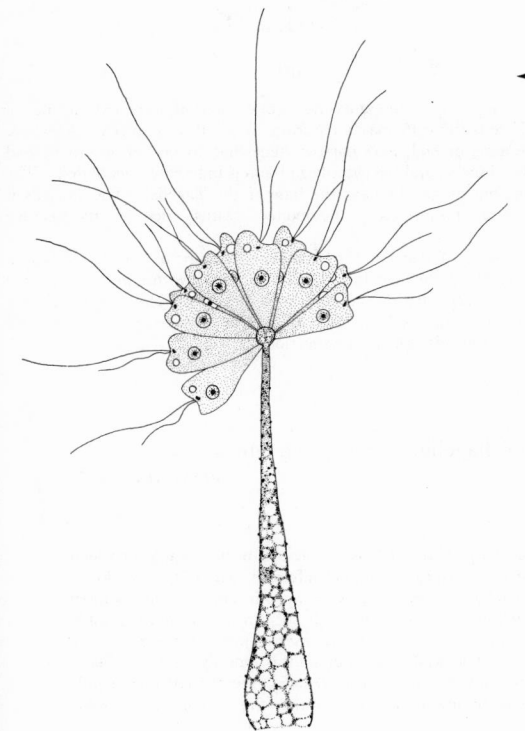

◄ **Fig. 8.82.** *Anthophysa vegetans* (O. F. Müller) Stein.   Cells very similar to those of *Cephalothamnion* except for a stigma borne beneath the flagella. Hollande found a posterior vacuole, but those we have seen have an anterior contractile vacuole, and usually a stigma, which Pascher says is lacking in *vegetans* but present in *steinii*. Since even members of the same colony may vary with regard to the stigma, the existence of 2 species is open to question. The stalk is heavy, but somewhat flexible and it is usually brown. Attached to debris or vegetation.

81a (63) Four flagella **82**

81b   More than 4 flagella **83**

82a (81, 1) All 4 flagella anterior . . . . *Collodictyon* Carter

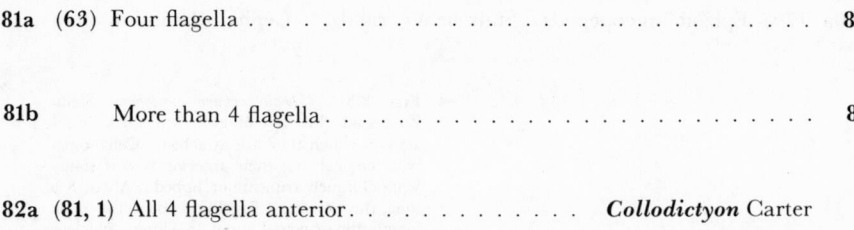
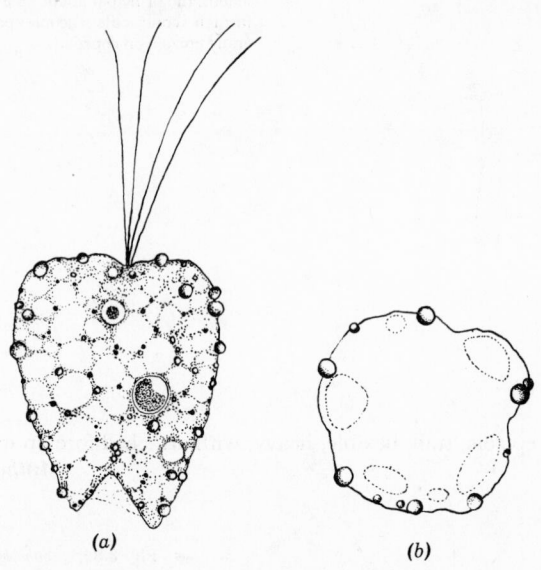

(a)   (b)

Fig. 8.83. *Collodictyon triciliatum* Carter. Cells free-swimming by means of 4 anteriorly directed flagella, of approximately equal length, generally that of the body; cells may vary from 25 to 60 μ in length. They are broadly rounded, but usually truncate at the anterior end, and narrow somewhat to one or several broad points posteriorly. The surface is distinctly lumpy, and the cytoplasm is often markedly vacuolated. The one or more contractile vacuoles are anterior; one is usually near the base of the flagella. The nucleus is generally small and central. This organism is of unsatisfactory taxonomic status, despite its relative abundance.

(*a*) A living cell of *Collodictyon triciliatum* showing 4 flagella, the vacuolated nature of the cytoplasm, a small contractile vacuole just below and to the left of the flagellar base, a nucleus below this, and an ingested *Chlorella* about the median right. Below this is a right peripheral contractile vacuole. The cytoplasm contains numerous refringent spheres. (*b*) shows that these spheres as well as the cytoplasmic vacuoles are peripheral in distribution, when the organism is seen in optical cross section.

82b   Three posterior, 1 anterior flagella. Emergence lateral . . . .
                                                            Tetramitus Perty

◄ **Fig. 8.84.** *Tetramitus pyriformis* Klebs. This organism is quite common in sewage and partially anaerobic organic infusions, where it is important as a consumer of bacteria. Cells concave-convex in cross section, pointed behind and rounded in front. It may reach 20 μ in length and is plastic. Three swimming and 1 trailing flagella are inserted just at the beginning of the concave surface. The trailer is about half as long again as the body, the others less than body length. The nucleus is anterior and the vacuole posterior. There is no undulating membrane; this is not the *Trimastix convexa* of Grassé.

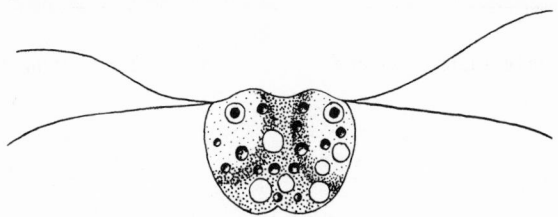

Fig. 8.85. *Gyromonas ambulans* Seligo. Small saddle-shaped organisms of foul water which may range from 5–10 μ in size, but are thin through the middle and on both flanks. The anterior end is rounded but truncate, with a median indentation, and there are several vacuoles, at least one of which is contractile, in the posterior portion. Two nuclei are located, one on each side, near the flagella bases. Two pairs of flagella about 15 μ long arise at the anterolateral juncture. These beat synchronously and rather slowly, so the organism swims with a rather deliberate movement. The flagella are never directed forward; their beat is from the side to rear.

83a   **(81)**   Two sets of 3 flagella . . . . *Trigonomonas* Klebs

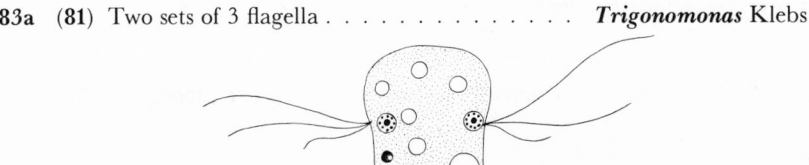

Fig. 8.86. *Trigonomonas compressa* Klebs. Another of the anaerobic or facultative flagellates from foul waters. Lemmermann says the cells may reach a length of 33 μ but personal observation has shown none over 20 μ with most of them about 10 μ. Cell somewhat rectangular in outline with a slight forward thickening and a slight twist. Thickness about ⅓ to ½ length. Cytoplasm vacuolated, at least one contractile vacuole. Two nuclei, located near the flagella bases. Flagella a set of 3, each of different length, on each side. Beat as in *Gyromonas*. This description agrees with that of Lemmermann and also of Grassé, but not with the figure of Klug, shown in Grassé.

83b   More than 6 flagella **84**

84a   **(83)**   Eight flagella **85**

84b   Many flagella . . . . *Paramastix* Skuja

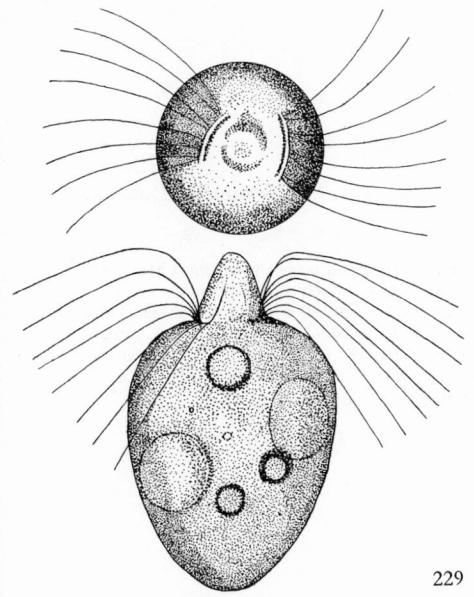

Paramastix may not have a double set of organelles, but its two bands of flagella suggest it has.

◄ **Fig. 8.87.** *Paramastix conifera* Skuja. An organism of questionable affinities, resembling the ciliates *Strombidium* or *Strobilidium*. Ovoid, but truncate anteriorly, with a large papilla at whose base 2 rows of flagella (cilia, Skuja) from 8 to 12 per row, are given off. Flagella not longer than body length (20 μ). Holozoic, but pseudopodia formation or ingestion by amoeboid processes not adequately described. Nucleus large, anterior; contractile vacuole posterior. Food vacuoles present. (After Skuja.)

85a **(84)** Two sets of 4 lateral flagella . . . . *Trepomonas* Dujardin

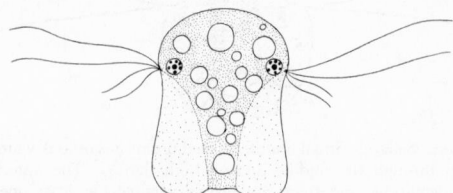

Fig. 8.88. *Trepomonas rotans* Klebs. All broadly oval, flattened in cross section, and with 2 posterior lobe-like inflations. May attain a length of 20 μ, a width of 15 μ and a thickness of 10 μ. Usually quite vacuolate, with at least one posterior contractile vacuole. Two nuclei, lateral, near the flagella bases, which are at the median lateral points where the lobes begin. Flagella on each side, 2 long (20 μ) and 2 short (8 μ), which normally sweep from side to rear. Rotates deliberately in swimming. Foul water species.

85b   Two sets of 3 lateral flagella, 2 posterior flagella **86**

86a **(85)** Posterior body end pointed. . . . . *Urophagus* Klebs

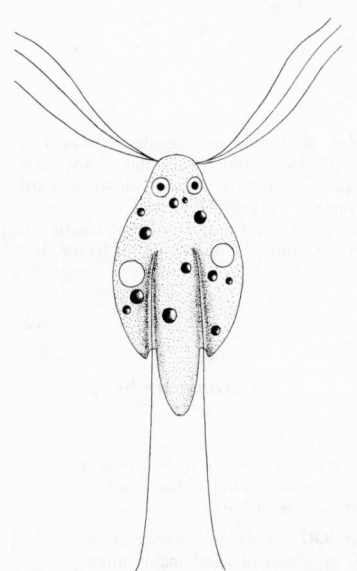

◄ **Fig. 8.89.** *Urophagus rostratus* (Stein) Klebs. Cell spindle-shaped, broadly rounded in front, with a long pointed posterior end. May reach 30 μ in length. Often has vacuoles, 2 of which, on the right and left, are contractile. Holozoic, at times with food vacuoles. Two anterior nuclei near the bases of 2 groups of 3 equal, anteriorly directed flagella. About the mid-point of the cell 2 furrows originate, extending backward alongside the rounded central point. At the same place 2 trailing flagella also emerge and extend back about 20–30 μ. Foul water. Rotates in swimming.

86b   Posterior body end rounded . . . . *Hexamitus* Dujardin

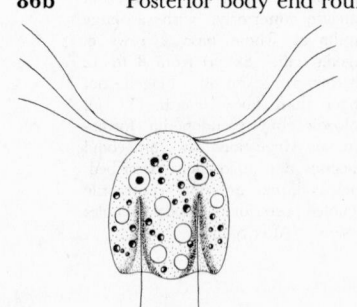

◄ **Fig. 8.90.** *Hexamitus inflatus* Dujardin. Much like *Urophagus*, except the cell is flattened and there is no posterior point. The cell is broadly oval, as long as 30 μ, as wide as 20 μ, about 10 μ thick, and is truncate posteriorly. The 2 nuclei are located between the swimming and trailing flagella. Rotates in swimming. Foul water.

References

Fritsch, F. E. 1948. *The Structure and Reproduction of the Algae.* Cambridge University Press, Cambridge. **Grassé, P. P. (ed.). 1952.** *Traité de Zoologie. Anatomie-systematique biologie.* Tome I, Fascicule I, *Protozoaires (generalites, flagelles).* Masson, Paris. **Hall, R. P. 1953.** *Protozoology.* Prentice-Hall, New York. **Kudo, Richard R. 1947.** *Protozoology,* 3rd ed. Thomas, Springfield, Illinois. **Pascher, A. and E. Lemmermann. 1914.** *Flagellatae.* In: *Die Süsswasserflora Deutschlands, Österreichs und der Schweiz,* Heft 1. G. Fischer, Jena. **Skuja, H. 1948.** Taxonomie des Phytoplanktons einiger Seen in Uppland, Schweden. *Symbolae Botan. Upsaliensis,* 9(3):1–399. **Smith, Gilbert M. 1950.** *The Fresh-Water Algae of the United States,* 2nd ed. McGraw-Hill, New York.

9

Rhizopoda
and Actinopoda

GEORGES DEFLANDRE

Classification

In the general introduction to the Superclass Rhizopoda in the large French *Traité de Zoologie*, the author stated that a *natural* group of Rhizopoda (class or superclass) does not actually exist (Deflandre, 1953, pp. 3–4). The first and main character that was the basis of all the ancient classifications, i.e., the possession of expansions of cytoplasm, the pseudopods, is not restricted to a limited group of Protozoa. It is well known that numerous Protista, and Proto-phyta as well as Protozoa, possess pseudopods that are fundamentally identical in their morphology and often in their physiology. A large number of Flagel-lata, without the flagella, would have the appearance of true *Amoebae* and con-versely many *Amoebae* show a true flagellate stage during their life cycle.

This large and interesting problem cannot be discussed here. Practically and didactically we call Rhizopoda a group of organisms that possesses in common the faculty *de vivre avec des pseudopodes* (to live with pseudopods). Following this definition it is quite natural to take the morphology of the pseudopods as the basis of the classification. This justifies the first couplets of the key.

Concerning the Superclass Actinopoda, similar difficulties arise, and the distinction between certain Flagellata and Heliozoa is often more or less inadequate or questionable (see Deflandre, 1953, pp. 267–268, and Trégouboff, 1953).

Testaceous Rhizopoda

A large number of Rhizopoda possess a shell enclosing the body and provided with an opening for the pseudopodia. Formerly all the Rhizopoda possessing shells were included in the Class or Order *Testacea*. Actually, each of the three Classes *Lobosa*, *Filosa*, and *Granulo-reticulosa* (based upon the morphology of pseudopods) contains a group of testaceous forms, constituting the three Orders *Testacealobosa* (line **3b** in the key), *Testaceafilosa* (line **74b**), and *Thalamia* (line **105b**).

Considered together, there are two classes of shells: those entirely secreted by the amoeba itself, and those built with the aid of extraneous materials. The former may be rigid or more or less flexible. The rigid shells often show scales, generally siliceous, all of one kind or of different kinds (aperture-scales, body-scales, spine-scales). These scales are secreted in the cytoplasm and the sorting out, building up, and cementing together depend on the action of the amoeba itself. In the latter, the substances used are selected by the animal from among its surroundings. One species chooses only fine grains of sand or quartz of nearly equal size, another chooses grains of various dimensions, others prefer diatoms exclusively or mixed with sand or with spherical shells of minute Chrysomonads.

Generally the body of the shell is fairly uniform in size and shape, corresponding to the species. But the spines, horns, and projections may vary more or less considerably within certain limits, either in form, number, or position.

The form of the aperture of the shell and the nature of its border or margin is usually of great importance for specific and eventually for generic discrimination. These characters are usually very constant for each species.

The following key includes all the genera reported from North America. In the larger genera, several representative species likely to be encountered are included.

KEY TO GENERA

1a   Pseudopods without axial filaments   . . . Superclass **Rhizopoda**   2

1b   Pseudopods radiating, with axial filaments . . . .
                                         Superclass **Actinopoda** 109
              Among the Superclass Actinopoda, only one class, the Heliozoa, inhabit fresh water. No central capsule between endoplasm and ectoplasm. Pseudopodia raylike, generally with axial filament.

2a   (1) Pseudopods lobose, fingerlike, rarely anastomosing . . . .
                                           Class **Lobosa**   3

2b   Pseudopods filiform, pointed, branched, and anastomosing . . . . .
                                          Class **Filosa** 74

2c   Pseudopods delicate, reticulate, in which minute granules cir-
        culate . . . . Class **Granulo-reticulosa**   **105**

3a   (2) Without shell . . . . Order **Amoebaea**   **4**

3b   With shell . . . . Order **Testacealobosa**   **12**

4a   (3) Without flagellum . . . . Suborder **Amastigogenina**   **5**

4b   With flagellum, temporary or not . . . Suborder **Mastigogenina**   **11**
        <small>Flagellum appears only during a more or less long stage of the life cycle; when it is absent the organism may be confused with Amastigogenina. Cytological study would be necessary in this case.</small>

5a   (4) With a pellicle . . . . Family **Thecamoebidae**   **6**

5b   Without a pellicle **7**
        <small>The pellicle is a layer, often folded, that generally shows a double outline when observed in optical section.</small>

6a   (5) With a rayed stage when suspended in water . . . .
                                    Rugipes Schaeffer

Two species, of which 1 lives in fresh water.

◄ **Fig. 9.1.** *Rugipes bilzi* Schaeffer. × 300. Broad anterior zone of clear protoplasm. Nucleus single, oval. 1 to 6 contractile vacuoles. Minute bluish granules and crystals. Marshes in Tenn. Length 70 μ. (After Schaeffer.)

6b   Without a rayed stage . . . . ***Thecamoeba*** Fromentel

◄ **Fig. 9.2.** *Thecamoeba verrucosa* (Ehrenberg). × 100. Pseudopods very short, broad lobes. Moves very slowly. Surface of the membrane marked by lines giving a wrinkled appearance. Habitat sphagnous swamp. Size 250–300 μ. A common form, almost identical, living in moss or in the soil is also called *Thecamoeba terricola* (Greeff). (After Leidy.)

7a   (5) Large amoebae with indeterminate pseudopods directing loco-
        motion . . . . Family **Chaosidae**   **8**

7b   Amoebae of medium to large size with determinate conical or
        tapering pseudopods, not directing locomotion. . . . .
                                Family **Mayorellidae**   **10**
        <small>The determinate pseudopods always appear at the same place on the body of the organism. They can vary in shape and size but they retain the main character of the group. The indeterminate pseudopods also retain this character, but they appear, grow, and disappear at any place on the body.</small>

7c   Amoebae of small size (20–60 μ), mostly coprophilous or soil-
        inhabiting. . . . . Family **Hartmannellidae**
        <small>Three genera: **Acanthamoeba** Volkonsky (cyst angular), **Hartmanella** Alexeieff, and **Glaeseria** Volkonsky (cysts spherical). The genera and their species (which comprise a number of amoebae having the shape of *Vahlkampfia limax*) cannot be determined without a detailed knowledge of cytological features. Therefore they are not included in the present key.</small>

7d   Disc-shaped amoebae with granular protoplasm and a margin of
        clear protoplasm; no or indeterminate pseudopods. . . . .
                                Family **Hyalodiscidae**
        <small>*Hyalodiscus* Hertwig and Lesser. Thin ectoplasmic layer with ridges radiating from the central mass. About 3 species, 2 marine and 1 fresh water.</small>

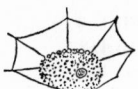

◄ **Fig. 9.3.** *Hyalodiscus rubicundus* Herting and Lesser. × 315. Body discoidal. Endoplasm reddish-yellow in color, enclosing numerous vacuoles and one or more nuclei. Habitat ooze of ponds; scarce. Size 40–60 μ. (After Penard.)

8a (7) One nucleus, or 2 (rarely many) . . . . **Chaos** Linnaeus

(= *Amoeba* Ehrenberg). Terminology according to A. A. Schaeffer, 1926.

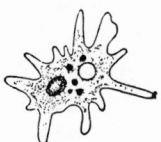

◄ **Fig. 9.4.** *Chaos diffluens* Muller. × 100. (= *Amoeba proteus* Pallas.) A common form, very changeable in the shape of the body; usually with numerous pseudopods, sharply distinguished from the body. Nucleus single, discoid, 40–50 μ. Protoplasm with bipyramid crystals. One contractile vacuole. Habitat both stagnant and clear water. Size, during reptation, 300–600 μ. (After C. H. Edmondson.)

8b Many nuclei; body usually enclosing symbiotic bacteria. . . . . . .
Pelomyxa Greeff

Many species: *P. carolinensis* Wilson and *P. laureata* Penard, apparently without symbiotic bacteria; *P. illinoisensis* Kudo, *P. nobilis* Penard, *P. vivipara* Penard.

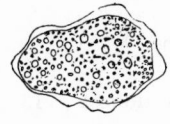

◄ **Fig. 9.5.** *Pelomyxa palustris* Greeff. × 25. Size 1500–3000 μ. Nuclei numerous. Endoplasm enclosing sand, brilliant corpuscles and bacteria simple and double rod. Habitat ooze of ponds, water with putrefying leaves, sphagnous swamps. Modern authors tend to consider as *Pelomyxa* the large or medium-sized amoebae showing a fixed polarity, with a single anterior pseudopod and a posterior urosphaera (= spherical zone ending the body, often provided with a tuft of very delicate pseudopods). (After Penard.)

9a Small, conical, determinate pseudopods very numerous . . . .
Dinamoeba Leidy 10

9b A few large, radiating pseudopods . . . . . **Astramoeba** Vejdovsky

◄ **Fig. 9.6.** *Astramoeba radiosa* (Ehrenberg). × 100. Body spheric. Pseudopods more or less rigid, not withdrawn and reformed rapidly. One nucleus, spheric. Habitat algae; widely distributed. Size 100 μ or less, with pseudopods. (After Leidy.)

10a (7, 9) Protoplasm bluish. . . . . **Dinamoeba horrida** Schaeffer

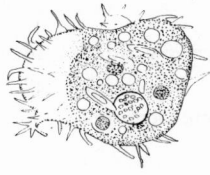

◄ **Fig. 9.7.** *Dinamoeba horrida* Schaeffer. × 330. Shape irregularly ovoidal. Length 80–150 μ. Nucleus large, spheric. Contractile vaculoles numerous. Found in a culture of material from marshes in Tenn. (After Schaeffer.)

10b Protoplasm hyaline . . . . **D. mirabilis** Leidy

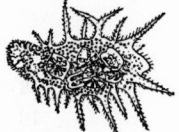

◄ **Fig. 9.8.** *Dinamoeba mirabilis* Leidy. × 100. Very changeable in shape, with many tapering pseudopods frequently branched. Two nuclei, 1 contractile vacuole. Habitat standing water. Size 200 μ. (After Leidy.)

11a   (4) Amoebae of medium or large size (40–130 μ), generally with a permanent flagellum. . . . . Family **Mastigamoebidae**
Mastigamoeba Schulze. One flagellum issuing from the nucleus. This is often difficult to see without fixing and staining the amoebae, but with the help of the phase-contrast microscope, the fine filament connecting the external flagellum to the nucleus is easily seen. Many species not well characterized. See also Chapter 8.

◄ **Fig. 9.9.** *Mastigamoeba longifilum* Stokes. × 1000. Body very changeable in shape, with indeterminate pseudopods. Flagellum long, active. Nucleus anterior, small. Size, 12–30 μ. Habitat standing water among decaying vegetation. (After Conn.)

11b   Amoebae of small size (15–50 μ), producing forms with 2 flagellae. . . . . Family **Vahlkampfiidae**
Vahlkampfia Chatton and Lalung-Bonnaire. Amoeba sluglike, producing flagellate forms with 2 flagella. Numerous forms, commonly called "*Limaxamoebae*" occurring in water, soil, dung, and giving very easily abundant cultures in which the flagellate stage can be observed. The type is *Vahlkampfia limax* (Dujardin). See also Chapter 8.

◄ **Fig. 9.10.** *Vahlkampfia limax* (Dujardin). × 225. (After Penard.)

12a   (3) Pseudopods fingerlike, not anastomosing   . .   Suborder **Eulobosa**   13
12b   Pseudopods often pointed, anastomosing . . . .
                                  Suborder **Reticulo-lobosa**   70
13a (12) Shell membranous   14
        See also line **56.**
13b   Shell with mineral or organic particles, sometimes very minute. . .   31
14a (13) Shell with a definite aperture   15
14b   Shell without a definite aperture. . . . . Family **Cochliopodiidae**
This family constitutes an intermediate link between Amoebaea and Thecamoebae. *Cochliopodium* Hertwig and Lesser. Commonly dome-shaped, but exceedingly flexible and changeable. About 6 to 8 species.

◄ **Fig. 9.11.** *Cochliopodium bilimbosum* Auerbach. × 300. Membranous covering punctulated, capable of great expansion. Pseudopods pointed. One nucleus (*n*). One or 2 contractile vacuoles. Common among algae. Diameter 30–60 μ. (After Leidy.)

15a (14) Shell semirigid, flexible near the aperture . . . .
                               Family **Microcoryciidae**   16
15b   Shell rigid . . . .   Family **Arcellidae**   22
16a (15) Aperture with a membranous frame. . . . *Microcorycia* Cockerell
        About 3 to 4 species.

◄ **Fig. 9.12.** *Microcorycia flava* (Greeff) Cock. × 210. The membranous covering is dome-shaped but very changeable in form. Pseudopods short and thick. Habitat mosses. Diameter 80–100 μ. (After Penard.)

16b   Aperture without a membranous frame   17
17a (16) Shell membrane double   18
17b   Shell membrane simple   19

18a (17) With an external mucilaginous layer . . . . ***Amphizonella*** Greeff

One species.

◄ **Fig. 9.13.** *Amphizonella violacea* Greeff. × 100. Protoplasm violet. One nucleus. Patelliform during locomotion. Habitat mosses and *Sphagnum*. Diameter 125–250 μ. (After Penard.)

18b With an internal hyaline membrane. . . . . ***Diplochlamys*** Greeff

About 3 to 4 species. Hemispherical to cup-shaped, loosely coated with organic and mineral particles.

◄ **Fig. 9.14.** *Diplochlamys fragilis* Penard. × 150. Color gray spotted with black. Thirty to 40 nuclei. Pseudopods short and thick. Diameter 70–125 μ. Habitat mosses; not common. (After Penard.)

19a (18) Shell membrane violet . . . . ***Zonomyxa*** Nüsslin
One species resembling *Amphizonella*, but with many nuclei.

◄ **Fig. 9.15.** *Zonomyxa violacea* Nüsslin. × 100. Body pyriform during loco-motion. Habitat *Sphagnum*. Size 140–160 and to 250 μ in active motion. (After Penard.)

19b Shell membrane hyaline or yellowish **20**

20a (19) Aperture elongate . . . . ***Parmulina*** Penard

One species.

◄ **Fig. 9.16.** *Parmulina cyathus* Penard. × 400. Body navicu-loid with a thick membrane agglutinating foreign particles. Habitat mosses. Diameter 40–60 μ. (After Penard.)

20b Aperture circular **21**

21a (20) One nucleus . . . . ***Microchlamys*** Cockerell
One species.

◄ **Fig. 9.17.** *Microchlamys patella* Claparède and Lachmann. × 310. Shell circular in dorsal view, with large aperture. Pseudopod usually single. One nucleus, 1 contractile vacuole. Habitat mosses in swamps. Diameter 40 μ. (After Penard.)

21b Two nuclei . . . . ***Penardochlamys*** Deflandre

One species, resembling *Arcella*.

◄ **Fig. 9.18.** *Penardochlamys arcelloides* Penard. × 375. Body spheroid, with homogenous membrane. Aperture small. Habitat aquatic vege-tation. Diameter 60 μ. (After Penard.)

22a (15) Aperture equal to or less than ½ of the shell diameter **23**

22b   Aperture nearly as wide as the base of the shell.   Shell with punc-
             tae sometimes indistinct . . . . *Pyxidicula* Ehrenberg   **29**
             Five species, of which 3 are reported from N. A.

23a   (22) One nucleus.   Shell hemispherical, membrane punctate. . . . . . .
                                                    Antarcella Deflandre

◄ **Fig. 9.19.** *Antarcella pseudarcella* Penard. × 430. Nucleus spheric with an eccentric nucleole and' many smaller nucleoles.   Shell dome-shaped; membrane yellowish-brown, very finely punctated, sometimes with siliceous particles at the top of the shell.   Habitat mosses.   Diameter 43–45 μ.   (After Penard.)

23b   Two nuclei or more (8 or 10 to 200)   . . . . ***Arcella*** Ehrenberg   **24**
             About 30 species. Shell generally circular in apical view. Membrane distinctly and
             densely punctated, brown or yellow in color. Aperture central, circular, sometimes
             lobate. Protoplasm united to the inside of the shell by delicate threads, epipodia,
             which are internal pseudopods of a special kind, often retractile, attached to the shell
             by a small and inconspicuous disc.

24a   (23) With 2 nuclei **25**

24b   With 6 to 200 nuclei. Shell flattened, with a wide aperture . . . .
                                                        A. polypora Penard

◄ **Fig. 9.20.** *Arcella polypora* Penard. × 100. Regularly plano-convex, with sharp border. Aperture with a very distinct row of pores. Habitat aquatic vegetation. Diameter 80–150 μ. (After Penard.)

24c   With 36 to 200 nuclei . . . . ***A. megastoma*** Penard

◄ **Fig. 9.21.** *Arcella megastoma* Penard. × 160.   Shell very flattened with a wide aperture, 0.4 to 0.5 of the entire diameter, which varies from 190–365 μ.   Habitat marshes and ponds, among algae. (After Deflandre.)

25a (24) Shell higher than the breadth of the base, mitriform or balloon-shaped. . . . . **A. mitrata** Leidy

◄ **Fig. 9.22.** *Arcella mitrata* Leidy. × 125. Dome mostly inflated, its summit and sides evenly rounded or mamillated (var. *gibbula* Deflandre 1928). Aperture not exactly circular but crenulated. Habitat *Sphagnum* and *Utricularia*. Diameter 100–180 μ, height 100–162 μ. (After Leidy.)

25b Shell not higher than the breadth of the base **26**

26a (25) Shell flattened **27**

26b Shell dome-shaped . . . . **A. vulgaris** Ehrenberg

◄ **Fig. 9.23.** *Arcella vulgaris* Ehrenberg. × 150. Surface smooth or with regular undulations. Pseudopods long and transparent. Many contractile vacuoles. Two nuclei opposite in position. Diameter 100–152 μ. Habitat ponds, among algae and other plants. (After Leidy.)

27a (26) With teeth or dentate border . . . . **A. dentata** Ehrenberg

◄ **Fig. 9.24.** *Arcella dentata* Ehrenberg. × 100. Shell having in lateral view the appearance of a crown when the points are well developed. Apically stellate or dentate, with 8 to 14 teeth. Habitat bogs and swamps. Diameter with the spines 123–184 μ. (After Leidy.)

27b Without teeth **28**

28a (27) With a distinct border. Cytoplasm colored green by chlorellae. . . . . **A. artocrea** Leidy

◄ **Fig. 9.25.** *Arcella artocrea* Leidy. × 170. Dome convex, mamillated or pitted; basal border everted and rising from a quarter to nearly half the height of the test. Protoplasm bright green, pseudopods colorless. Habitat bogs and *Sphagnum* ponds. Diameter 184–216 μ. (After Leidy.)

28b Without a distinct border . . . . **A. discoides** Ehrenberg

◄ **Fig. 9.26.** *Arcella discoides* Ehrenberg. × 175. Shell smooth, with a large circular aperture; in lateral view plano-convex with a rounded border. Two nuclei. Common in pond water. Diameter 90–146 μ. (After Penard.)

29a (22) Shell very small, about 20 μ. **30**

29b   Shell much larger, about 80 μ . . . ***Pyxidicula cymbalum*** Penard

◄ **Fig. 9.27.** *Pyxidicula cymbalum* Penard. × 210. Shell patelliform, brown in color, with distinct punctae. Aperture round, bordered by a narrow rim. Contractile vacuole single. Pseudopodia unknown. Habitat wet moss and *Sphagnum*. Diameter 85–90 μ. Found by E. Penard in material from Summit Lake, Colorado. (After Penard.)

30a   (29) With inverted aperture. Shell discoid, circular . . . .
                                      P. operculata Ehrenberg

◄ **Fig. 9.28.** *Pyxidicula operculata* Ehrenberg. × 925. With a narrow inverted margin, forming a large, circular orifice; membrane smooth, transparent, finely punctate, becoming darker or nearly brown with age. Habitat aquatic vegetation. Diameter 17–21 μ. (After Penard.)

30b   Aperture not invaginate or re-entrant . . . . . ***P. scutella*** Playfair

◄ **Fig. 9.29.** *Pyxidicula scutella* Playfair. × 1000. Shell minute, very depressed, almost saucer-shaped; membrane pale yellow-red, showing a faint punctation, sometimes coarsely scrobiculate. Habitat ponds and lakes, British Columbia. Diameter 16–22 μ. (After Playfair and Wailes.)

31a   (13) Shell with foreign particles, without plates secreted by the cytoplasm. **32**

31b   Shell with plates or scales, rounded or angular, sometimes with foreign mineral particles . . . . Family **Nebelidae**   **53**

32a   (31) Aperture at the extremity of the shell, which possesses an axial symmetry . . . . Family **Difflugiidae**   **42**

32b   Aperture on the side of the shell or ventral; symmetry dorso-ventral. . . . . Family **Centropyxidae**

33a   (32) Aperture triangular; shell hemispherical . . . ***Trigonopyxis*** Penard

One species. This is the *Difflugia arcula* of Leidy.

◄ **Fig. 9.30.** *Trigonopyxis arcula* (Leidy) Penard. × 180. (*Difflugia arcula* Leidy.) Aperture triangular or irregularly trilobed, or roughly quadrangular, never invaginated. Habitat marshy places, among moss or *Sphagnum*. Diameter 90–120 μ. (After Deflandre.)

33b   Aperture not triangular **34**

34a (33) Inferior lip of the aperture extending to the superior lip, sometimes covering it partially; shell hemispherical or elliptical . . . . .
Bullinularia Penard

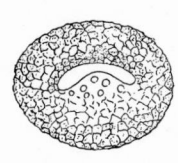

This is *Bullinula* Penard 1911. One species.

◄ **Fig. 9.31.** *Bullinularia indica* Penard. × 120. Shell brownish, of small
siliceous plates and grains, closely cemented upon a chitinous pellicle.
Superior lip with a row of pores. Habitat mosses. Diameter 120–150 μ.
(After Penard.)

34b Inferior lip not extending to the superior lip **35**

35a (34) Aperture linear, lunate; superior lip without pores; shell hemispherical . . . . ***Plagiopyxis*** Penard **36**
Two species.

35b Aperture rounded or angular; shell mostly membranous with en-
crusted foreign particles, or covered with sandy material . . . . . .
Centropyxis Stein **37**
Twenty-eight species.

36a (35) Inferior lip dipping far into the interior of the shell, rounded . . . .
Plagiopyxis callida Penard

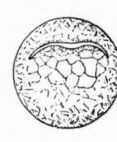

◄ **Fig. 9.32.** *Plagiopyxis callida* Penard. × 150. Shell gray, yellow, or brown in
color, usually smooth and clear. The lips overlap to such an extent that the
aperture is very difficult to observe. One nucleus. Habitat mosses. Diameter
55–135 μ, but usually 90–110 μ. (After Wailes and Penard.)

36b Inferior lip slightly dipping into the interior of the shell, triangular . . . . ***P. labiata*** Penard

◄ **Fig. 9.33.** *Plagiopyxis labiata* Penard. × 155. Brown in color. Smaller than
the preceding species. Pseudopods unknown. Reported by E. Penard from
British Columbia. Diameter 80–88 μ. (After Penard.)

37a (35) With the aperture eccentric **38**

37b With the aperture central **41**

38a (37) With horns or spines **39**

38b Without horns **40**

39a (38) With a row of spines at the border. Shell compressed, cap-
shaped . . . . ***Centropyxis aculeata*** (Ehrenberg) Stein

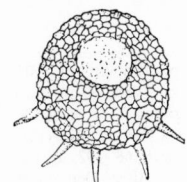

◄ **Fig. 9.34.** *Centropyxis aculeata* (Ehrenberg) Stein. × 150. Membrane
brownish, frequently incrusted with sand grains or diatoms. Spines 2 to
10, rarely more. The animal is very shy, sometimes extending a single
pseudopod. Habitat ponds, lakes, marshes, on algae. Frequent. Diameter 120–150 μ without the spines. (After Leidy.)

39b   With spines irregularly distributed on the dome of the shell. Shell
        almost hemispherical but asymmetric. . . . .
                                    C. hemisphaerica (Barnard) Wailes

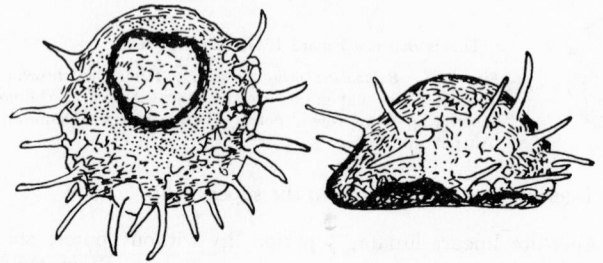

Fig. 9.35. *Centropyxis hemisphaerica* (Barnard) Wailes. × 825. Differs from the preceding species having in
lateral view a much higher shell. Six to 12 large horns, 20–42 μ in length. Habitat aquatic vegetation.
Only known in N. and S. A. Diameter 140–160 μ without the spines. (After Wailes.)

40a (38) Shell of large size, more than 200 μ, discoidal or largely elliptical,
           mostly irregular in outline . . . . . *C. ecornis* (Ehrenberg) Leidy

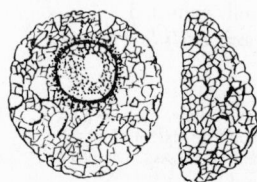

◄ Fig. 9.36. *Centropyxis ecornis* (Ehrenberg) Leidy. × 130.
Large-sized species with a shell covered with quartz sand
grains; color usually grey, sometimes brownish. Aperture
circular or irregularly lobed, not very much eccentric.
Habitat only open water, among plants or mosses. Greatest
diameter, 200–275 μ. (After Leidy.)

40b   Shell of medium size, more than 100 μ, elliptical or ovoid in front
        view, more or less flattened . . *C. constricta* (Ehrenberg) Penard

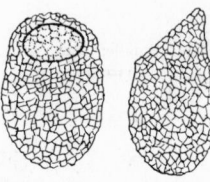

◄ Fig. 9.37. *Centropyxis constricta* (Ehrenberg) Penard. × 80. Aperture at the border of the shell, eccentric, largely elliptic or nearly
circular. Shell covered with closely-set sand grains, giving a grey
color to the test. Habitat mosses and *Sphagnum*. Length 120–150
μ, breadth 75–90 μ. (After Leidy.)

40c   Shell of small size, ovoid, in lateral view pear-shaped . . . .
                                            C. aërophila Deflandre

Fig. 9.38. *Centropyxis aërophila* Deflandre. × 375. Shell chitinous, finely punctate or rough, with or without foreign particles, principally of vegetable origin, hyaline or yellowish, sometimes yellowish-brown.
Aperture nearly semicircular or elliptic. A widely distributed and common species, giving in ecologically
different media a number of adapted varieties which are morphologically stable. Type living especially
among mosses on the trees, length 53–85 μ; var. *sylvatica* Deflandre, on forest moss, length 68–102 μ; var.
sphagnicola Deflandre, nearly circular in outline, among *Sphagnum*, diameter 49–66 μ. (After Deflandre.)

41a (*37*) With a stellate aperture. Shell nearly hemispherical in side view. . . . . *C. stellata* Wailes

◄ **Fig. 9.39.** *Centropyxis stellata* Wailes. × 50. A large species, circular in oral view; test composed of irregularly-shaped siliceous plates, without protuberances. Animal not observed. Reported only from N. A., British Columbia (G. H. Wailes). Habitat spring and stream. Diameter 335–400 μ. (After Wailes.)

41b With a circular aperture. Shell hemispherical with rounded border. *C. arcelloides* Penard

◄ **Fig. 9.40.** *Centropyxis arcelloides* Penard. × 165. Membrane thin, chitinous with small siliceous plates. Aperture circular, faintly invaginated, about half the diameter of the shell in width. Habitat among moss and *Sphagnum*. Diameter 100–110 μ. (After Penard.)

42a (*32*) Shell with internal partition or diaphragm, with deeply constricted neck and transverse perforated partition at the point of constriction . . . . *Pontigulasia* Rhumbler

A widely distributed genus, comprised of about 7 species. The internal partition or diaphragm of *Pontigulasia* may be overlooked in living animals. It appears clearly when the shell is mounted in Canada balsam.

◄ **Fig. 9.41.** *Pontigulasia vas* Leidy. × 100. (= *Pontigulasia spectabilis* Penard.) Resembling *Difflugia oblonga* in appearance except for the deeply constricted neck. Internal partition well observed only in oral view, with 2 to 4 perforations. Sometimes with zoochlorellae (Leidy). Habitat subaquatic mosses at pond sides and in wet *Sphagnum*. Length 125–170 μ. (After Penard.)

42b Shell without internal partition **43**

43a (*42*) Shell transversely polygonal, membranous. Foreign mineral particles scarce or absent. . . . . *Sexangularia* Awerintzew

Three species.

◄ **Fig. 9.42.** *Sexangularia polyedra* Deflandre. × 375. Shell pear-shaped, transparent, spangled with minute quartz grains which are more abundant near the aperture. Pseudopods numerous, straight or forked, active. Habitat *Sphagnum*. Length 60–70 μ. (After Deflandre.)

43b Shell transversely circular or rarely elliptical. **44**

44a (43) Aperture partially closed by a transverse diaphragm. Shell with
a short, annular neck. . . . . **Cucurbitella** Penard
Three species.

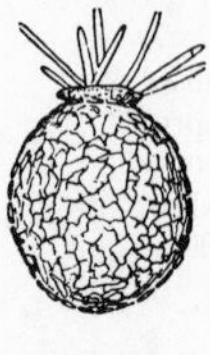

Fig. 9.43. *Cucurbitella mespiliformis* Penard. × 125. Shell ovoid; aperture small, irregularly serrated, surrounded by a 3- or 4-lobed neck with undulating margin. Protoplasm containing green symbiotic chlorellae. Nucleus with a large central nucleole. Pseudopods numerous. Habitat ooze of ponds and lakes, and aquatic vegetation. Length 116–140 μ. (After Penard.)

44b Aperture without diaphragm. Shell varying from globular to
elongated pyriform or acuminate . . . . **Difflugia** Leclerc 45
A large genus comprised of more than 60 species, which should be divided into many genera in the near future.

45a (44) With a more or less marked neck. 46
45b Without a neck 48
46a (45) Neck deeply constricted, with outer margin always recurved . . . .
D. urceolata Carter

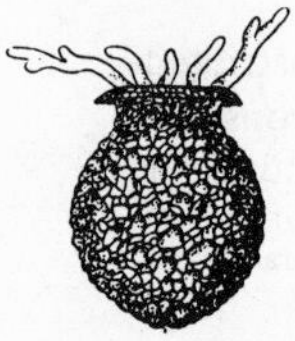

Fig. 9.44. *Difflugia urceolata* Carter. × 75. Shell ovoid or spherical, generally without spines, but a variety, *D. urceolata* var. *olla* Leidy, possesses a few short stubby spines developed from the fundus. Forty to 60 nuclei. Habitat ooze of pond water. Dimensions 200–350 μ, mostly 220–250 μ. (After Leidy.)

46b Neck not deeply constricted. 47
47a (46) Neck sometimes indistinct, aperture large and entire. Shell thin,
transparent, smooth . . . . **D. lebes** Penard

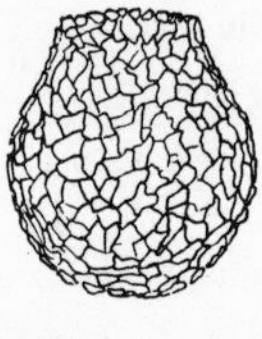

Fig. 9.45. *Difflugia lebes* Penard. × 60. Shell very fragile, covered with silicious flattened particles. Collar straight, rarely recurved (= var. *elongata* Penard). Sometimes more than 100 nuclei. Feeds on large diatoms. Habitat ooze at the bottom of lakes, ponds. One of the greatest *Difflugia*, some reaching 400 μ in length, or more. (After Penard.)

47b Neck distinct, aperture lobated or with undulating margin. Shell
brownish, mamillated in outline. . . . . . . **D. tuberculata** Wallich

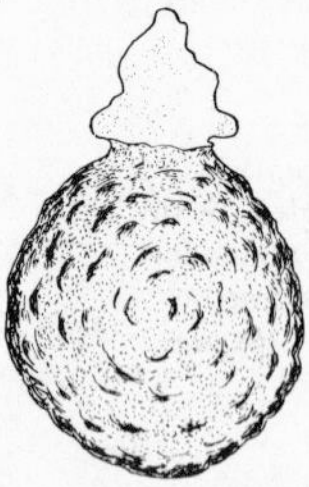

Fig. 9.46. *Difflugia tuberculata* Wallich. × 188. Ovoid, in transverse section circular, with subhemispheric elevations exteriorly and corresponding depressions interiorly, the surface having a mulberry-shaped appearance. Membrane chitinous, covered with irregularly-sized minute sand grains. Aperture hexilobate, with a narrow collar. One nucleus. Habitat marshes and ponds. Length 120–140 μ. (After Deflandre.)

48a (45) Aperture lobed or denticulate. 49
48b Aperture entire, circular. 51
49a (48) With a number of teeth 50

49b   With 3 to 4 blunt lobes. . . . . **D. lobostoma** Leidy

◄ **Fig. 9.47.** *Difflugia lobostoma* Leidy. × 105. Shell ovoid or nearly spheric, usually with a quadrilobate (rarely trilobate) aperture. Pseudopods few. Cytoplasm sometimes with chlorellae. Habitat ponds, ditches, and marshy. places; common. Length 90–150 μ. (After C. H. Edmondson.)

50a **(49)** Margin with large spines. . . . . **D. corona** Wallich

◄ **Fig. 9.48.** *Difflugia corona* Wallich. × 90. Shell ovoid inclining to spheroid, composed of large sand grains or flattened silicious plates, smooth and regular in outline. Teeth usually more than 12 in number, evenly arranged or not. Fundus with 6 to 9 spines, but sometimes with 1 to 4, or none at all. Nucleus single. One of the best known species of *Difflugia*, used by Jennings in his genetical research. Common in ooze of ponds. Length without spines 180–230 μ. (After Leidy.)

50b   Margin with minute spines, sometimes not well marked, or without spines . . . . **D. rubescens** Penard

◄ **Fig. 9.49.** *Difflugia rubescens* Penard. × 330. Piriform, yellowish in color; membrane chitinous, encrusted with foreign particles and diatoms. Aperture circular, bordered by an incurved crenulate margin; protoplasm containing numerous granules of a brick-red color; cyst equally red or brownish-red in color. Habitat aquatic vegetation and among *Sphagnum* or moss. Length 65–105 μ. (After Leidy.)

51a **(48)** Usually covered with diatoms. Shell elongate more or less vial-shaped. . . . . **D. bacillifera** Penard

◄ **Fig. 9.50.** *Difflugia bacillifera* Penard. × 180. Shell variable in form, usually with a rounded fundus tapering into a narrow, often cylindrical neck; membrane hyaline, covered wholly or in part with diatoms, which are sometimes scattered with sand particles or foreign bodies such as minute tests of siliceous flagellates or rhizopoda, e.g. *Euglypha laevis*. Habitat *Sphagnum* pools and mosses. Length 145–160 μ. (After Penard.)

51b   Covered with quartz sand grains **52**
52a **(51)** Pear-shaped with posterior border usually rounded . . . .
                                                    D. oblonga Ehrenberg
This is *Difflugia pyriformis* Perty.

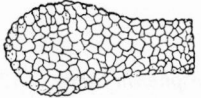

◄ **Fig. 9.51.** *Difflugia oblonga* Ehrenberg. × 60. (= *Difflugia pyriformis* Perty.) A very common species that seems to be exceedingly variable in form and size, but that really comprises a number of elementary species and varieties not yet discriminated. The type here drawn is 100–300 μ in length. Habitat ooze of ponds, lakes, and among aquatic vegetation. (After Leidy.)

52b   Elongate, more or less cylindrical, somewhat enlarged posteriorly
              and acuminate . . . . *D. acuminata* Ehrenberg

◄ **Fig. 9.52.** *Difflugia acuminata* Ehrenberg. × 125. Shell surface rarely
smooth, frequently carrying an incrustation of sand grains. End of fundus
acute or with a knoblike process. Widely distributed, usually associated
with the preceding species. Habitat ooze of sands and marshy ground.
Length from 100–300 μ. (After Leidy.)

53a (31) Shell spiral, more or less compressed, largely composed of minute
              curved, rodlike plates. . . . . *Lesquereusia* Schlumberger   **54**
              Eight to 10 species.

53b   Shell not spiral. **56**

54a (53) Shell without sand grains **55**

54b   Shell primarily of sand grains, few plates . . *L. modesta* Rhumbler

◄ **Fig. 9.53.** *Lesquereusia modesta* Rhumbler. × 125. Often with a *Difflugia*-like
appearance; neck short and broad, slightly turned to one side. Nucleus single.
Pseudopods few, large, and long. Habitat mostly bogs, among *Sphagnum*, also
ponds in the ooze with the large *Difflugia*. Length 100–150 μ. (After Penard.)

55a (54) Plates thick, short, shell slightly compressed . . . .
                                          L. epistomium Penard

◄ **Fig. 9.54.** *Lesquereusia epistomium* Penard. × 150. Neck very sharply distinguished from the rounded shell and very abruptly turned to one side. Rods
disposed as in *L. spiralis*, but reniform and stouter. Habitat *Sphagnum* swamps;
not frequent. Length 110–125 μ. (After Penard.)

55b   Plates slender, elongate . . . . *L. spiralis* Ehrenberg

◄ **Fig. 9.55.** *Lesquereusia spiralis* Ehrenberg. × 125. Shell homogenous
with a meshwork of short curved rods (vermiform pellets of Leidy), transparent. Neck prominent, sharply turned to one side. Nucleus single,
situated posteriorly. Pseudopods as in *Difflugia*. A handsome rhizopod,
widely distributed throughout the world. Habitat marshes, among
Sphagnum and other subaquatic vegetation. Length 110–140 μ. (After
Penard.)

56a (53) Scales or plates very apparent.   **59**

56b   Scales indistinct, membrane seeming structureless . . . .
                                          Hyalosphenia Stein   **57**
              Eight to 10 species. See also **13.**

57a **(56)** Surface of shell without undulations **58**

57b Surface of shell with undulations . . . . **H. elegans** Leidy

◄ **Fig. 9.56.** *Hyalosphenia elegans* Leidy. × 250. Shell flask-shaped, compressed, brownish or yellowish in color, very transparent. Two minute pores, opposite each other, are in the base of the neck. Nucleus single. Pseudopods few. Habitat bogs among *Sphagnum* and moss. Length 75–100 μ. (After Penard.)

58a **(57)** Shell exceedingly transparent, without pores through the fundus. . . . . **H. cuneata** Stein

◄ **Fig. 9.57.** *Hyalosphenia cuneata* Stein. × 300. Shell shorter and broader than that of the preceding species, greatly compressed, colorless. Without zoochlorellae. A rare species. Habitat clear water and among *Sphagnum*. Length 60–75 μ. *n*, nucleus. (After Leidy.)

58b With 2 or more distinct opposite pores through the fundus . . . . . **H. papilio** Leidy

◄ **Fig. 9.58.** *Hyalosphenia papilio* Leidy. × 200. Shell ovoid or pyriform, compressed, yellowish in color. Protoplasm not filling the shell but attached to the inner surface by threads (epipodia) always containing chlorellae. Pseudopods often numerous, active. Generally 2, but sometimes 4 to 6 pores about the border of the fundus. Habitat *Sphagnum;* common in swamps. Length 110–140 μ. (After Leidy.)

59a **(56)** Plates quadrangular **60**

59b Plates not quadrangular **61**

60a **(59)** Shell normally pyriform, with siliceous plates . . . . **Quadrulella** Cockerell

About 4 to 5 species. This genus can be considered a subgenus of *Nebela*.
Note: The distinction between siliceous and calcareous plates is very easily seen with the polarizing microscope. The endogenous siliceous plates always remain black (or extinct) in crossed nicols.

◄ **Fig. 9.59.** *Quadrulella symmetrica* Wallich. × 175. Plates very transparent, usually regularly arranged in transverse and longitudinal series. Many varieties, some being short, others long, others curved. Habitat *Sphagnum* and moss, generally submerged. Length 68–150 μ. (After Leidy.)

60b   Shell normally globular, more or less compressed, with calcareous
plates . . . . *Paraquadrula* Deflandre
About 3 to 4 species or more. Very easy to identify in polarized light; in crossed
nicols the plates are beautifully illuminated.

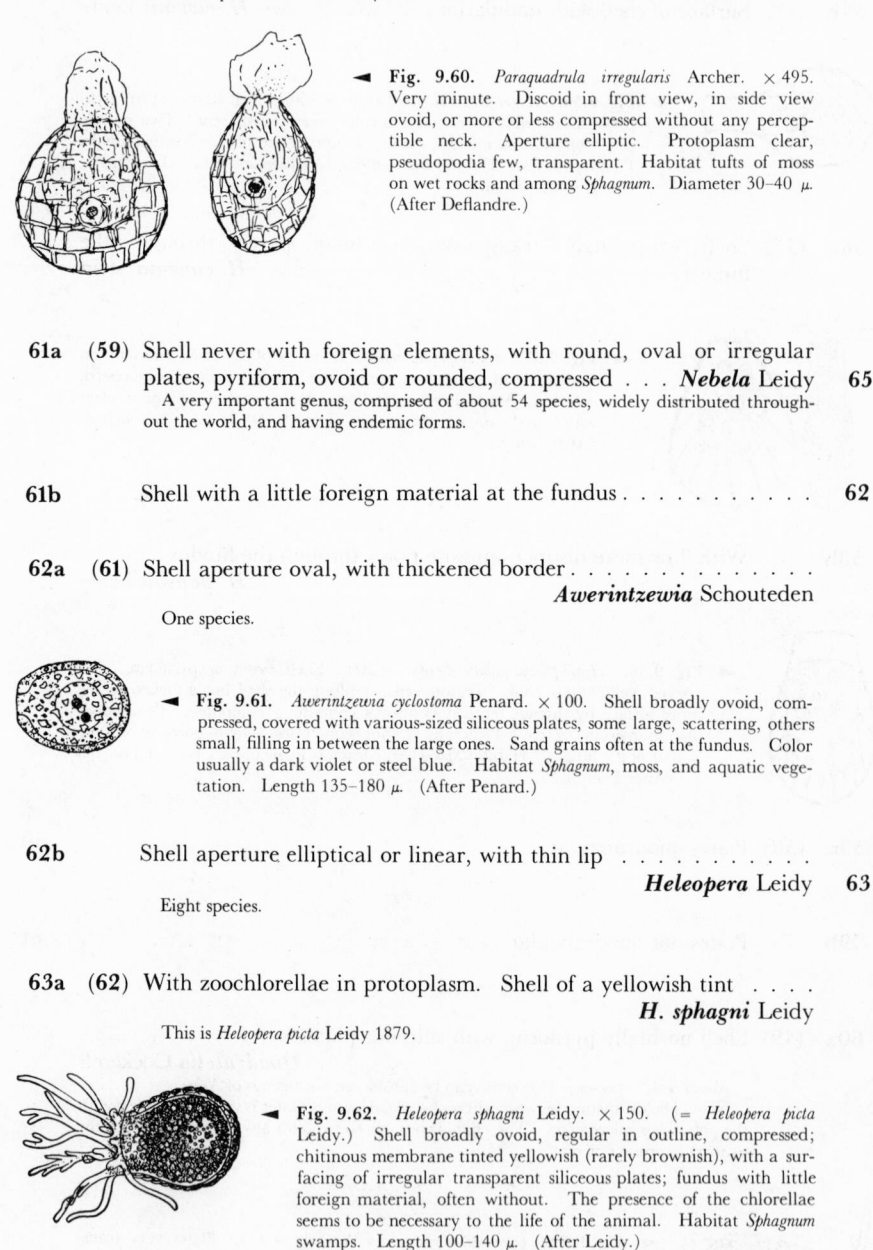

◄ **Fig. 9.60.** *Paraquadrula irregularis* Archer. × 495.
Very minute. Discoid in front view, in side view
ovoid, or more or less compressed without any perceptible neck. Aperture elliptic. Protoplasm
clear, pseudopodia few, transparent. Habitat tufts of moss
on wet rocks and among *Sphagnum*. Diameter 30–40 μ.
(After Deflandre.)

61a   (59) Shell never with foreign elements, with round, oval or irregular
plates, pyriform, ovoid or rounded, compressed . . . *Nebela* Leidy   **65**
A very important genus, comprised of about 54 species, widely distributed throughout the world, and having endemic forms.

61b   Shell with a little foreign material at the fundus **62**

62a   (61) Shell aperture oval, with thickened border . . . .
Awerintzewia Schouteden
One species.

◄ **Fig. 9.61.** *Awerintzewia cyclostoma* Penard. × 100. Shell broadly ovoid, compressed, covered with various-sized siliceous plates, some large, scattering, others
small, filling in between the large ones. Sand grains often at the fundus. Color
usually a dark violet or steel blue. Habitat *Sphagnum*, moss, and aquatic vegetation. Length 135–180 μ. (After Penard.)

62b   Shell aperture elliptical or linear, with thin lip . . . .
Heleopera Leidy   **63**
Eight species.

63a   (62) With zoochlorellae in protoplasm. Shell of a yellowish tint . . . .
H. sphagni Leidy
This is *Heleopera picta* Leidy 1879.

◄ **Fig. 9.62.** *Heleopera sphagni* Leidy. × 150. (= *Heleopera picta*
Leidy.) Shell broadly ovoid, regular in outline, compressed;
chitinous membrane tinted yellowish (rarely brownish), with a sur-
facing of irregular transparent siliceous plates; fundus with little
foreign material, often without. The presence of the chlorellae
seems to be necessary to the life of the animal. Habitat *Sphagnum*
swamps. Length 100–140 μ. (After Leidy.)

63b   Without zoochlorellae **64**

64a **(63)** Shell greyish or amethyst, strongly compressed, especially near the aperture. . . . . **H. petricola** Leidy

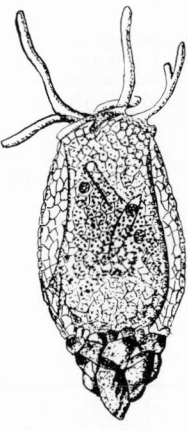

> ◄ **Fig. 9.63.** *Heleopera petricola* Leidy. × 220. Covered with amorphous scales or silicious plates which form a loose reticulation; with the fundus provided with sand grains. Outline subject to much variation, but generally the aperture is convex or, when cut straight across, more or less rounded at the corners. A common species. Habitat *Sphagnum* or moss in boggy places. Length 80–100 μ. Var. *amethystea* Penard is larger than the type (length 115–120 μ) and of a pure amethystine tint. (After Leidy.)

64b Shell vinous red, compressed, but not so strongly near the aperture as in *H. sphagni* . . . . **H. rosea** Penard

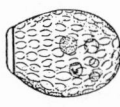

> ◄ **Fig. 9.64.** *Heleopera rosea* Penard. × 150. More robust and broader than *H. petricola* var. *amethystea;* corners of the aperture obtusely angular; shell vinous or rose-colored, lips yellow or sometimes lightish brown. Habitat mosses and *Sphagnum* in swamps. Length 90–135 μ. (After Penard.)

65a **(61)** Aperture smooth with well-defined lips. **67**
65b Aperture more or less irregular, bordered by scales, without lips **66**
66a **(65)** Fundus without spines. Shell ovoid; border of the aperture crenulate . . . . **Nebela dentistoma** Penard

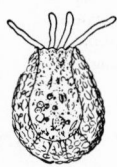

> ◄ **Fig. 9.65.** *Nebela dentistoma* Penard. × 160. Shell transparent, with polygonal rounded plates not jointing but united by minute chitinous bridges. Habitat *Sphagnum* and mosses. Length 70–115 μ. (After Penard.)

66b Fundus provided with spines. . . . . **N. caudata** Leidy

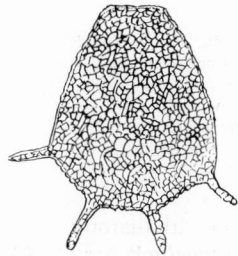

> ◄ **Fig. 9.66.** *Nebela caudata* Leidy. × 330. Shell ovoid with a rounded or angulous fundus ornated with 3 to 5 hollow horns (generally 4); covered with polygonal or circular scales, jointing or rarely overlapping. Habitat *Sphagnum* swamps. Length 78–90 μ. (After Leidy.)

67a   (65)   Shell flask-shaped with a distinct tubular, long, narrow neck. Plates rounded . . . . *N. lageniformis* Penard

◄ **Fig. 9.67.** *Nebela lageniformis* Penard. × 175. Body of shell broad, ellipsoid, prolonged into a tubular neck. Large elliptic or polygonal rounded plates, sometimes intermixed with small polygonal scales. Habitat mosses and *Sphagnum*, in forests or peaty bogs. Length 125–130 μ. (After Penard.)

67b   Shell pear-shaped or rounded **68**

68a   (67)   Shell broader than long, rounded, transversally elliptical . . . . . . *N. flabellulum* Leidy

◄ **Fig. 9.68.** *Nebela flabellulum* Leidy. × 150. The transverse diameter usually equals or exceeds the length. A very short but distinct cylindric neck is present. Habitat *Sphagnum* swamps. Length 72–96 μ, breadth 89–100 μ. (After Leidy.)

68b   Shell more or less elongate **69**

69a   (68)   Shell elongate pyriform, with 2 lateral pores . . . . *N. militaris* Penard

◄ **Fig. 9.69.** *Nebela militaris* Penard. × 330. Two minute pores opposite each other are situated at the first third of the shell; shell transparent chitinous, and yellowish in color, with minute plates sometimes scarcely distributed or wanting. Habitat *Sphagnum* swamps and moss on forest soil. Length 50–72 μ. (After Penard.)

69b   Without pores . . . . *N. collaris* Ehrenberg

◄ **Fig. 9.70.** *Nebela collaris* Ehrenberg. × 150. Shell pyriform with a convex aperture, in side view oblong, aperture always notched. A very common species comprised of many varieties. Often difficult to separate from *N. bohemica* Taranek in which the aperture is not notched. Habitat sphagnous swamps, mosses, aquatic vegetation in peaty bogs. Length 115–130 μ. (After Leidy.)

70a   (12)   Aperture at the extremity of the shell, axially situated. **71**

70b   Aperture on the side of the shell, ventrally situated . . . . *Wailesella* Deflandre

One species.

◄ **Fig. 9.71.** *Wailesella eboracensis* Wailes. × 800. Shell entirely chitinous, brownish or dark brown; ovoid in front view, dissymmetric in side view; aperture circular without any invagination. Nucleus single. Pseudopodia few, pointed. A world-wide, but surely overlooked, species. Found in Alaska and British Columbia. Habitat *Sphagnum* and mosses, subaquatic. Length 20–28 μ. (After Wailes.)

71a   (70)   Shell membranous, densely covered with sand grains or with diatoms or other foreign elements . . . . *Phryganella* Penard   **72**
          Four species.

71b   Shell membranous, without foreign elements, or with few **73**

72a   (71)   Large size (165–220 μ), foreign elements large.   Shell hemispherical, usually of rough contour . . . . **P. nidulus** Penard

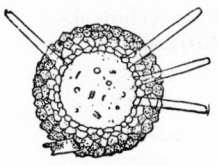

◄ **Fig. 9.72.** *Phryganella nidulus* Penard. × 90.   Aperture large. Pseudopods slender, but often accompanied by broad lobes of protoplasm extending radially in all directions.   Animal very shy; without an examination of the living organism, the identification of the shell alone is not possible.   Habitat the ooze of ponds and lakes.   Diameter 180–200 μ.   (After Penard.)

72b   Small size (30–50 μ), foreign elements small (diatoms and sand grains).   Shell hemispherical. . . . . . . **P. hemisphaerica** Penard

◄ **Fig. 9.73.** *Phryganella hemisphaerica* Penard. × 250.   Shell yellowish or brownish.   Aperture large without any invagination, sometimes bordered with greater scales or grains. Pseudopodia as in the preceding species.   Habitat the ooze of ponds and lakes.   Diameter 40–55 μ.   (After Penard.)

73a   (71)   Shell compressed, ovoid, or pear-shaped . . . .
Cryptodifflugia Penard

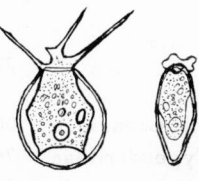

Two to 3 species.

◄ **Fig. 9.74.** *Cryptodifflugia compressa* Penard. × 660.   In side view strongly compressed.   Aperture circular.   Pseudopodia straight, linear, often pointed at the extremity and anastomosing near the base.   Habitat aquatic vegetation.   Length 16–18 μ.   (After Penard.)

73b   Shell not compressed . . . . *Difflugiella* Cash
Shell membranous, smooth; hyaline or yellowish; rarely with any foreign elements; 6 to 8 species.

◄ **Fig. 9.75.** *Difflugiella oviformis* Penard. × 450.   Shell ovoid, chitinous and transparent, yellowish or brownish in color, without foreign elements.   Pseudopodia rarely active.   Habitat marshes, aquatic vegetation.   Length 60–20 μ.   (After Penard.)

74a   (2)   Without shell . . . .   Order **Aconchulina**
Naked amoebae, with pseudopods filiform and pointed.   Our knowledge of this group is deficient.   Only 1 genus with 4 species . . . . *Penardia* Cash

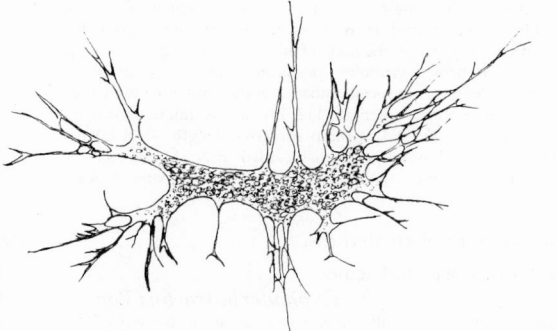

◄ **Fig. 9.76.** *Penardia mutabilis* Cash. × 200. Body when at rest roughly ovoid; pseudopodia projecting from the surface at various points, slender and pointed, branching and anastomosing, ultimately forming a widely-spreading network; colorless. Habitat swampy ground. Found at Cortes Island, British Columbia, by G. H. Wailes. Diameter when at rest, 90–100 μ; fully extended, 300–400 μ. (After Cash.)

74b   With shell . . . . Order **Testaceafilosa**   75
75a (74) With a single aperture   77
75b   With 2 opposite apertures . . . . . . Family **Amphitremidae**   76
76a (75) Shell without foreign particles, strongly compressed, yellowish-
       brown in color . . . . **Ditrema** Archer
       One species.

◄ **Fig. 9.77.** *Ditrema flavum* Archer. × 255. Shell homogeneous; in broad view with nearly parallel sides and rounded ends. In side view, long elliptic; end views oval, each with a central small oval aperture; protoplasm always enclosing green zoochorellae. Nucleus single. Pseudopodia straight, unbranched. Habitat *Sphagnum*. Length 45–77 μ. (= *Amphitrema*.) (After Penard.)

76b   Shell with foreign particles, more or less compressed, hyaline or
       brownish . . . . **Amphitrema** Archer
       Four species.

◄ **Fig. 9.78.** *Amphitrema wrightianum* Archer. × 215. In outline similar to the preceding form, but always possessing scantily distributed foreign particles; siliceous sand grains, diatoms. The apertures provided with short tubelike external collars. Zoochlorellae present. Habitat *Sphagnum*. Length 61–95 μ. (After Penard.)

77a (75) With distinct rounded or elongated plates   78
77b   Without distinct plates, often with foreign elements . . . .
                                         Family **Gromiidae**   100
78a (77) Shell retort-shaped . . . . Family **Cyphoderiidae**   79
78b   Shell not retort-shaped, straight   82
79a (78) Rounded scales very distinct. Neck gently recurved, never furnished with a disc-shaped collar. Shell generally brownish in
       color . . . . **Cyphoderia** Schlumberger   80
       Five to 6 species.
79b   Scales very minute, giving a punctate appearance and eventually
       with foreign elements. Neck with a delicate, transparent disc-
       shaped collar, perpendicular to the aperture. . . **Campascus** Leidy
       Four species.

◄ **Fig. 9.79.** *Campascus cornutus* Leidy. × 150. Shell transverse section trigonal with rounded angles; lateral processes developed from the fundus. Plates small, round, more or less covered by foreign particles. In common with *Cyphoderia*, the body of all species of *Campascus* enclose minute yellow or brown granules (pheosomes) very resistant to reagents. *C. cornutus* is a rare species inhabiting the ooze of lakes (China Lake, Uinta Mountains). Length 112–140 μ. A minute form, *C. minutus* Penard, retort-shaped, without processes (length 50–60 μ) not infrequent in Europe has not yet been recorded in N. A. and may be overlooked at this time. *cv*, contractile vacuole. (After Leidy.)

80a (79) Membrane composed of nonimbricated discs   81
80b   Membrane composed of imbricated scales . . . .
                                  Cyphoderia trochus Penard
       The character imbricated (or not) is generally given by the study of the optical section of the empty shell.

(a)

(b)

◀ **Fig. 9.80.** *Cyphoderia trochus* Penard. × 1300. (*a*) The circular bi-convex imbricated plates are arranged in diagonal rows. In optical section the test is quite different from *C. ampulla*, (*b*). The true type (shell conical) seems to be restricted to deep-water lakes. The varieties are found in *Sphagnum* and other aquatic vegetation. Length of the commonest form, var. *amphoralis* Wailes, 87–150 μ. (After Wailes and Penard.)

81a (80) Fundus rounded or mamillate . . . . *C. ampulla* Ehrenberg

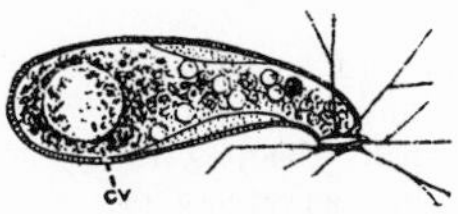

◀ **Fig. 9.81.** *Cyphoderia ampulla* Ehrenberg. × 160. Plates rounded or oval, cemented together in diagonal rows, presenting a hexagonal appearance. Pseudopodia few but very long. Protoplasm with minute granules and crystals. Habitat mosses, ooze of ponds and lakes. Length 61–195 μ. Several varieties of this species are known, sometimes confused with *C. trochus*. *cv*, contractile vacuole. (After Leidy.)

81b With a more or less sharply rounded fundus . . . .
C. ampulla var. *papillata* Wailes

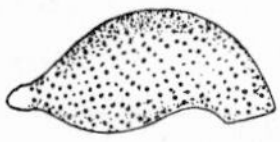

◀ **Fig. 9.82.** *Cyphoderia ampulla* var. *papillata* Wailes. × 150. This variety resembles the type except in the outline of the fundus. Plates never overlapping. Habitat *Sphagnum* and aquatic vegetation. Length 113–135 μ. (After C. H. Edmondson.)

82a (78) With circular or elliptical scales . . . . . . Family **Euglyphidae** 83
82b With very elongated, curved scales . . . . Family **Paulinellidae**

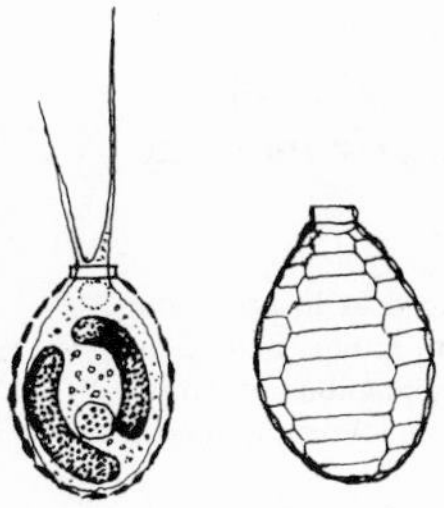

Paulinella Lauterborn, 1 species. Shell small, oviform.

◀ **Fig. 9.83.** *Paulinella chromatophora* Lauterborn. × 600. Shell colorless or of a pale lemon color, not compressed, with a small oval aperture provided with a short neck; scales arranged in alternating transverse rows. Protoplasm devoid of food particles, with 2 sausage-shaped curved symbiotic blue-green algae. These Cyanelles are minute Cyanophyceae, having the same function as the zoochorellae (Chlorophyceae). Habitat submerged vegetation. Length 20–32 μ. A small form that is not rare, but often overlooked. (After Deflandre.)

83a (82) Aperture terminal 84
83b Aperture subterminal 97
84a (83) Aperture in oral view circular or oval 85
84b Aperture elliptical, elongate, or linear 93
85a (84) With a short or long tail composed of scales. Shell ovoid, elliptical, or pyriform. . . . . *Pareuglypha* Penard

One species.

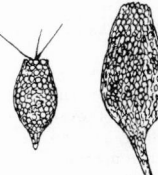

◀ **Fig. 9.84.** *Pareuglypha reticulata* Penard. × 300 (left) and × 500 (right). Very variable in outline. Transverse section circular, aperture circular, bordered by a chitinoid laciniated short collar; scales generally circular, sometimes elliptic, variable in size, becoming smaller near the tail. Nucleus single, contractile vacuoles 1, 2, or more. Habitat aquatic vegetation and submerged mosses. Found in British Columbia by G. H. Wailes. Length 55–70 μ. (After Wailes and Penard.)

85b   Without a tail **86**

86a (85) Aperture bordered by scales. Plates round or oval; margin of aperture with prominent denticles which are specialized scales; spines often developed . . . . **Euglypha** Dujardin **87**
Seventeen species.

86b   Aperture bordered by a dentate neck without scales. With a distinct hyaline collar, denticulate or laciniate . . . .
Tracheleuglypha Deflandre

One species.

◄ **Fig. 9.85.** *Tracheleuglypha dentata* Vejdovsky. × 310. (= *Sphenoderia dentata* Penard.) Shell ovoid or pyriform; aperture furnished with a neck, straight or everted, with more or less numerous teeth. Body scales circular or elliptic, imbricating and frequently giving the appearance of a hexagonal design. Habitat *Sphagnum*, moss, and clear water. Length 42–52 μ. (After Penard.)

87a (86) Aperture circular **88**

87b   Aperture oval **91**

88a (87) Spines at fundus only **89**

88b   Spines not at fundus only **90**

89a (88) Shell not compressed, of large size (100–140 μ). One or 2 spines . . . . **Euglypha mucronata** Leidy

◄ **Fig. 9.86.** *Euglypha mucronata* Leidy. × 165. The elliptic imbricating plates are arranged in longitudinal alternating rows; the spines are modified scales with an elliptic embase. Habitat sphagnous swamps. Length 108–140 μ. (After Leidy.)

89b   Shell elongate, very little or not at all compressed, of small size (30–50 μ). Spines in a tuft . . . . **E. cristata** Leidy

◄ **Fig. 9.87.** *Euglypha cristata* Leidy. × 425. Plates arranged as in preceding species; the fundus is furnished with a tuft of divergent spines, 3 to 8 in number. Pseudopodia rarely extended. Habitat sphagnous swamps. Length 33–84 μ, but usually ranging between 40 and 55 μ. Varieties occur with 1, 2, or no spines. (After Leidy.)

90a (88) Shell elongate-oviform, not compressed. Spines usually absent, scattered when present . . . . **E. tuberculata** Dujardin

◄ **Fig. 9.88.** *Euglypha tuberculata* Dujardin. × 375. Plates round or oval, imbricating, presenting a regular hexagonal design. Aperture scales finely serrated, 8 to 12 in number. Nucleus large, spheric. Pseudopodia numerous, long, fine, radiating, generally straight, seldom branched. Habitat mosses, *Sphagnum*, and submerged vegetation; generally distributed. Length 45–100 μ. (After Edmondson.)

90b   Spines lateral . . . . **E. brachiata** Leidy

◄ **Fig. 9.89.** *Euglypha brachiata* Leidy. × 180. Shell elongate and cylindric, slightly constricted near the center; plates circular or oval, imbricating in a regular manner. From 2 to 6 large, long, and recurved spines situated among the 3 rows adjoining the aperture. Habitat submerged *Sphagnum*. Length 92–128 μ. (After Leidy.)

91a   (87)   Shell with spines **92**

91b   Always destitute of spines. . . . .   *E. laevis* Ehrenberg

◄   **Fig. 9.90.**   *Euglypha laevis* Ehrenberg. × 460.   Shell small, oviform, glabrous; transverse section and aperture elliptic to subcircular; aperture bordered by a single row of pointed scales; body scales oval, slightly imbricated. Habitat mosses and *Sphagnum;* world-wide distribution. Length 22–55 μ. (After Wailes.)

92a   (91)   Shell of medium size (40–90 μ), oviform.   Transverse section of shell elliptical . . . .   *E. ciliata* Ehrenberg

◄   **Fig. 9.91.**   *Euglypha ciliata* Ehrenberg. × 250.   Moderately compressed. Plates oval, imbricated.   Needlelike spines are produced from the entire surface or in a line around the lateral border of the shell, occasionally wanting. Habitat mosses *Sphagnum,* and aquatic vegetation.   Length 40–90 μ.   (After Penard.)

92b   Transverse section of shell denticulate with acute margin . . . . . .

                                      E. compressa Carter

◄   **Fig. 9.92.**   *Euglypha compressa* Carter. × 225.   Shell large, broadly oviform, greatly compressed.   Body scales elliptic, imbricating, and presenting a hexagonal design.   Numerous spines on the margin only, singly or in tufts of 2 or 3.   Habitat *Sphagnum* and aquatic vegetation. Length 70–132 μ.   (After Leidy.)

93a   (84)   Shell globular or ovoid, more or less compressed but always at least moderately so, with a distinct, hyaline neck . . . .

                               Sphenoderia Schlumberger   **94**

         Eight species.

93b   Without a neck, only with narrow lip **95**

94a   (93)   With numerous small scales as in other Euglyphidae.   Shell globular or ovoid . . . .   *S. lenta* Schlumberger

◄   **Fig. 9.93.**   *Sphenoderia lenta* Schlumberger. × 300.   Body scales subcircular or broadly oval, regularly imbricated; aperture linear, formed by a thin chitinous collar which is constituted of minute scales very difficult to define. General outline and neck very variable. Possibly several species confused. Habitat, *Sphagnum* and submerged vegetation. Length 30–64 μ. *cv,* contractile vacuole. (After Leidy.)

94b   With a pair of large plates on each side and a number of other smaller plates . . . .   *S. macrolepis* Leidy

◄   **Fig. 9.94.**   *Sphenoderia macrolepis* Leidy. × 330.   Shell small, oviform; the 2 large plates are elliptic and transversally disposed; aperture linear, neck conical in side view.   Habitat *Sphagnum.*   Not common, but a very distinct and pretty species. Length 27–45 μ.   (After Wailes.)

95a **(93)** Border of the aperture very thin and finely dentate. Shell covered
with elongate-elliptical plates, usually brown in color . . . .
Assulina Ehrenberg   **96**

Three species

95b   Border of the aperture not dentate. . . . . **Placocista** Leidy
Five species.

◄ **Fig. 9.95.** *Placocista spinosa* Leidy. × 170. Shell broadly oval, compressed, with a convex and smooth or undulate aperture; plates oval,
imbricating in a regular manner; margin of the shell furnished at
regular intervals with long lanceolate or awl-shaped spines, which are
movably articulated. Habitat *Sphagnum.* Length 100-170 μ. (After
Leidy.)

96a **(95)** Of small size (28–58 μ), oval. Brown in color, but clearer than
the following species . . . . **Assulina muscorum** Greeff

◄ **Fig. 9.96.** *Assulina muscorum* Greeff. × 300. Shell oviform, compressed.
Plates usually arranged in alternating diagonal rows, or occasionally irregularly.
Habitat mosses and *Sphagnum;* common. Length 28–50 μ. Empty tests are
numerous nearly everywhere, but living animals are not common. (= *A. minor*
Penard.) (After Penard.)

96b   Of large size (60–90 μ), and with rounded shell . . . .
A. seminulum Ehrenberg

◄ **Fig. 9.97.** *Assulina seminulum* Ehrenberg. × 290. Adult forms are chocolate
brown in color. Transverse section lenticular. Nucleus large. Pseudopodia
few, straight, slender. Habitat *Sphagnum* and moss in marshy places; common.
Length 60–90 μ, and up to 150 μ. (After Leidy.)

97a **(83)** Shell elongate-oval, usually compressed; aperture subterminal.
Scales circular or oval . . . . **Trinema** Dujardin   **98**
Four species.

97b   Shell very transparent, shaped as *Trinema,* but with elongate
plates. Scales elongate, irregularly disposed . . **Corythion** Taranek
Four to 5 species.

◄ **Fig. 9.98.** *Corythion dubium* Taranek. × 375. Shell subcircular or oviform;
aperture oval, rarely circular; plates small, oval, elongate, nonimbricated,
often irregularly disposed. Habitat mosses and *Sphagnum;* common. Length
30–45 μ. A particular species, *C. acutum* Wailes, with a short spine at the
fundus, is known only from British Columbia. (After Penard.)

98a **(97)** Oral extremity narrow **99**

98b   Oral extremity broad. . . . . . . **Trinema complanatum** Penard

◄ **Fig. 9.99.** *Trinema complanatum* Penard. × 500. Shell in broad view of
nearly uniform width, with semicircular extremities; body scales circular,
imbricated. Aperture circular, oblique, invaginated, seeming oval in front
view. Habitat mosses and *Sphagnum.* Length 25–60 μ. (After Penard.)

99a **(98)** Shell of large size with distinct plates, oviform, compressed anteriorly . . . . **T. enchelys** Ehrenberg

◄ **Fig. 9.100.** *Trinema enchelys* Ehrenberg. × 310. Body scales circular. Aperture circular, oblique, invaginated, and surrounded by a number of rows of very minute scales. Pseudopodia few, fine, and long. Habitat mosses, *Sphagnum* and aquatic vegetation; common. Length 40–100 μ. (After Penard.)

99b   Shell of small size, plates indistinct . . . . **T. lineare** Penard

◄ **Fig. 9.101.** *Trinema lineare* Penard. × 500. Usually the plates are distinguishable only about the edges where they may appear as minute undulations. Aperture round. Habitat mosses, *Sphagnum*, aquatic vegetation, moistened soils. Generally distributed and apparently the commonest form of all Rhizopoda. Length 16–30 μ. (After Penard.)

100a **(77)** Shell smooth, destitute of foreign particles **101**
100b   Shell with sparsely distributed foreign elements **103**
100c   Shell covered with foreign elements. **104**
101a **(100)** Shell thick, spherical, rigid or slightly flexible, body filling the envelope. . . . . **Gromia** Dujardin
         Two species.

◄ **Fig. 9.102.** *Gromia fluviatilis* Dujardin. × 25. Envelope spheric or subspheric, seldom changing shape; protoplasm habitually covering the surface of the envelope. Pseudopodia numerous, anastomosing. Habitat aquatic plants. Diameter 90–250 μ. A very doubtful form, which may be identical with *G. terricola* Leidy, but the latter possesses granulo-reticulate pseudopodia and is covered with foreign particles. (After Leidy.)

101b   Shell thin, hyaline, and flexible. Aperture circular or oval, elastic . . . . **Lecythium** Hertwig and Lesser **102**
         Five species.

101c   Shell thin, hyaline, but rigid. Aperture circular, not deformable . . . . **Chlamydophrys** Cienkowsky

Nine species.

◄ **Fig. 9.103.** *Chlamydophrys minor* Belar. × 1000. Shell rigid, ovoid, with a small aperture. Protoplasm filling the shell, with a very distinct spheric nucleus placed posteriorly, and a transversal layer of brilliant granules. Pseudopodia straight, numerous, radiating. Habitat aquatic vegetation. Length 16–20 μ. (After Belar.)

102a **(101)** Envelope hyaline, spherical, or pyriform. Aperture large and capable of great dilatation . . . . **Lecythium hyalinum** Ehrenberg

◄ **Fig. 9.104.** *Lecythium hyalinum* Ehrenberg. × 260. The protoplasm is clear and colorless, with a large spherical nucleus; contractile vacuole single. Pseudopodia numerous, straight and pointed, sometimes branched. Habitat, clear water and submerged *Sphagnum*. Diameter 30–48 μ. *cv*, contractile vacuole. (After Leidy.)

102b   Envelope yellow, pyriform; transverse section lenticular or arcuate.
Aperture small, as elastic as in *L. hyalinum* . . . ***L. mutabile*** Bailey

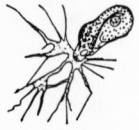

◄ **Fig. 9.105.** *Lecythium mutabile* Bailey. × 165. Shell very changeable in form, often distorted by the ingestion of large diatoms. Protoplasm enclosing brilliant granules. Nucleus large, spheric. Contractile vacuoles 1 or 2. Habitat clear water, among aquatic vegetation. Diameter 40–70 μ and up to 140 μ. (= *Pamphagus*.) (After Penard.)

103a   (100)   Shell ovoid, membranous, hyaline or brown.   Aperture linear . . .
Capsellina Penard

Four species.

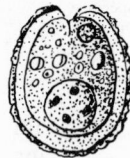 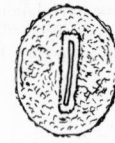

◄ **Fig. 9.106.** *Capsellina bryorum* Penard. × 530. Shell elliptic in transverse section; aperture linear, deeply invaginated. Membrane grayish or slightly brownish, rough and with scarcely distributed siliceous particles. Nucleus single. Habitat mosses. Length 35–40 μ. Found in British Columbia by G. H. Wailes. (After Penard.)

103b   Shell ovoid, flexible, compressed.   Aperture rounded . . . .
Plagiophrys (Claparède and Lachmann) Penard

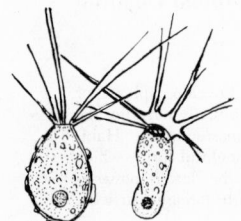

One species.

◄ **Fig. 9.107.** *Plagiophrys parvipunctata* Penard. × 260. Membrane finely and regularly punctate with a few minute scales. Nucleus single, small. Pseudopodia numerous, straight. Habitat among aquatic plants. Length 50 μ. Found in British Columbia by G. H. Wailes. (After Penard.)

104a   (100)   Shell covered with minute particles and short bristles . . . .
Diaphoropodon Archer

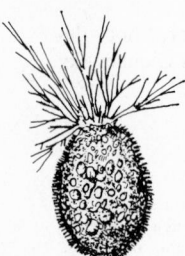

One species.

◄ **Fig. 9.108.** *Diaphoropodon mobile* Archer. × 260. Shell membranous, flexible, brown in color, more or less ovoid, covered with hairlike cils, rigid and of a chitinous nature. Pseudopodia long, numerous, branching but not anastomosing. Habitat aquatic vegetation. Length 60–113 μ. (After Hoogenraad.)

104b   Shell covered with sand grains and dirt particles. . . . .
Pseudodifflugia Schlumberger

Nine species.

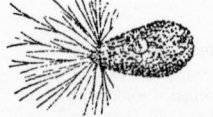

◄ **Fig. 9.109.** *Pseudodifflugia gracilis* Schlumberger. × 250. Shell ovoid, not compressed, light brown or yellowish in color, covered with fine quartz sand grains. Aperture circular, devoid of neck. Pseudopodia numerous, long, delicate, straight or forked. Habitat ooze of ponds, lakes, aquatic vegetation. Length 20–65 μ. *n.* nucleus. Other species of *Pseudodifflugia* are known in N. A. (After Leidy.)

105a **(2)** Without shell . . . . Order **Athalamia**

Biomyxa Leidy, 2 to 3 species. Body without a covering; pseudopodia formed from any part of the surface.

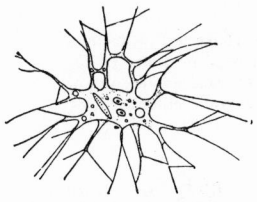

◄ **Fig. 9.110.** *Biomyxa vagans* Leidy. × 65. The body moves slowly but continuously; no distinction between ectoplasm and endoplasm observed. Pseudopodia long, branching and anastomosing, with a perceptible circulation of minute granules along the filaments. Nucleus granular. Habitat *Sphagnum* swamp. Dimensions extremely variable; large individuals may measure 480 μ between the tips of the pseudopodia. (After Penard.)

105b With shell . . . . Order **Thalamia** 106

106a **(105)** Shell with a single aperture 107

106b Shell with 2 or more apertures . . . . Family **Microcometesidae** 108

107a **(106)** Shell of medium or large size, not forming colonies, never attached at the substratum . . . . Family **Allogromiidae**

The family Allogromiidae is comprised of 8 genera, one of which is well known, the others being rarely met with and insufficiently studied. The well-known genus is *Lieberkühnia* Claparède and Lachmann, 3 species. Envelope very flexible, changeable in shape.

◄ **Fig. 9.111.** *Lieberkühnia wageneri* Claparède and Lachmann. × 130. Shell pyriform or elongate. Pseudopodia long, anastomosing, extending from a protoplasmic peduncle at the aperture, and laterally situated. Nuclei as many as 200. Contractile vacuoles numerous. Habitat mosses. Length 90–100 μ. (After Penard.)

107b Shell of small size, often attached laterally and eventually forming colonies . . . . Family **Microgromiidae**

The 4 genera of the Family Microgromiidae actually known contain minute forms that present real affinities with amoeboid loricated species belonging to the great group of the *Chrysomonadina*, Family Chrysamoebae (see Chapter 6, key line **355a** in Family Rhizochrysidaceae.) *Microgromia* Hertwig, 5 species. Shell globular, aperture with an internal lamella laterally situated.

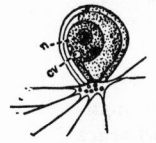

◄ **Fig. 9.112.** *Microgromia socialis* Hertwig. × 545. Shell rigid, with a short neck. Pseudopodia long, anastomosing, arising from a peduncle at the aperture. This peduncle is precisely limited by the internal lamella. Sometimes colonies are formed. Habitat standing water. Length 20 μ. *n.* nucleus; *cv*, contractile vacuole. (After Hertwig.)

108a **(106)** Shell rounded or more or less polyhedral, with 3 to 5 apertures . . . . *Microcometes* Cienkowsky

Two species.

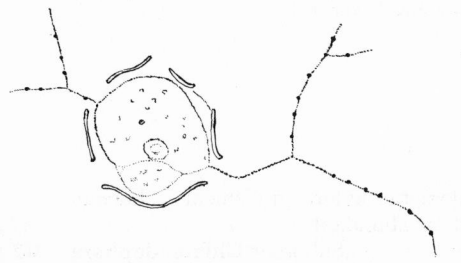

◄ **Fig. 9.113.** *Microcometes paludosa.* Cienkowsky. × 1330. Shell irregularly hemispherical, yellowish or dark brown; apertures with a narrow neck. Nucleus single. Contractile vacuoles 3 to 6. Habitat aquatic vegetation. Diameter 12–22 μ. (After DeSaedeleer.)

108b   Shell spherical, with 2 apertures . . . . *Diplophrys* Barker
         Two species.

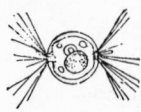

◄ **Fig. 9.114.** *Diplophrys archeri* Barker. × 1200. The characteristic pseudopodia, which are long and branched, extend from opposite poles of the
envelope. The protoplasm always encloses a large spherical gobule, yellow
or brown in color. Nucleus single. One or more contractile vacuoles.
Diameter 8–20 μ. (After Penard.)

109a   (1) No central corpuscle.   Central nucleus from which the axial
         filaments arise. . . . . Order **Actinophrydia**   110

109b   Nucleus eccentric, generally a central corpuscle from which the
         axial filaments arise   . . . . Order **Centrohelidia**   111
         See also **116**.

110a   (109) Nucleus single . . . . *Actinophrys* Ehrenberg
         Four species.

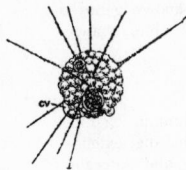

◄ **Fig. 9.115.** *Actinophrys sol* Ehrenberg. × 245. Body spherical with
protoplasm highly vacuolated. Usually 1 contractile vacuole which
rises and pushes out the surface as a rounded globule before bursting.
Pseudopodia extending from all parts of the body, with axial filaments
arising from the membrane of the single nucleus. Habitat pond water
among aquatic plants; very common. Diameter 40–50 μ. *cv*, contractile vacuole. (After Leidy.)

110b   Nuclei many . . . . *Actinosphaerium* Stein
         Two species.

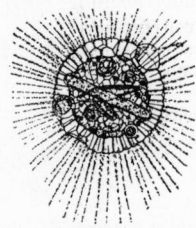

◄ **Fig. 9.116.** *Actinosphaerium eichhornii* Ehrenberg. × 40. Protoplasm
vacuolated with very large vacuoles pressed together about the
periphery. Nuclei numerous, disseminated in the endoplasm near its
periphery; 2 or more (up to 14) contractile vacuoles. Pseudopodia
extending from all parts of the body, but axial filaments ending free in
the protoplasm. Habitat aquatic vegetation, in ponds and lakes; not
infrequent. Average diameter 200–300 μ; a variety, var. *majus* Penard,
measures 600–780 μ and up to 1000 μ. *cv*, contractile vacuole.
(After Leidy.)

111a   (109) With no skeleton or foreign elements.   A single nucleus or many
         nuclei . . . . Suborder **Aphrothoraca**
         One genus with 2 species reported in N. A. . . . . *Actinolophus* Schultze

◄ **Fig. 9.117.** *Actinolophus minutus* Walton. × 350. Pseudopodia very short, extending
from all parts of the body. Nucleus single, posteriorly situated. Habitat river water.
Diameter 12 μ, length of pedicel 70 μ. (After Walton.)

111b   With a mucilaginous shell aglomerating foreign elements.   Foreign
         elements scarce and indistinct, or abundant . . . .
                                          Suborder **Chlamydophora**   112

111c With a mucilaginous shell provided with spicules. Spicules chitinous or siliceous . . . . Suborder **Chalarothoraca** 113

112a (111) Surface of the mucilagenous investment with delicate extraneous particles and bacteria. Body investment spherical, concentric . . .
Astrodisculus Greeff

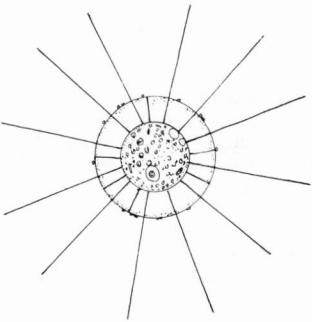

Three to 6 species.

◄ **Fig. 9.118.** *Astrodisculus radians* Greeff. × 500. Very minute form with a thick, colorless mucilagenous investment. Pseudopodia not numerous, of moderate length; nucleus not large, placed eccentrically. One contractile vacuole. Habitat pools and ditches; not common. Diameter 13–17 μ. Found in British Columbia by G. H. Wailes. (After Penard.)

112b Outer envelope composed of siliceous grains and diatoms . . . . . .
Elaeorhanis Greeff

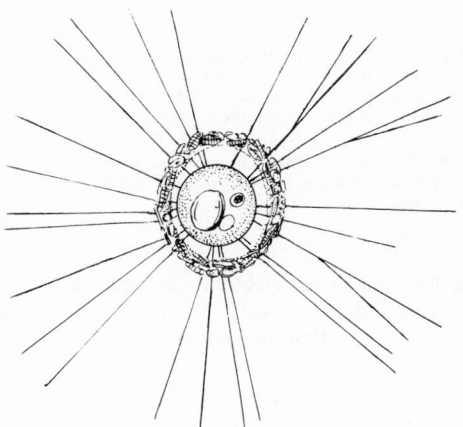

One species.

◄ **Fig. 9.119.** *Elaeorhanis oculea* Archer. × 400. Body enclosed in a spheric membrane; outer investment spheric or ellipsoid. Protoplasm bluish in color containing a large yellow oil globule. Nucleus single, eccentrically placed. Habitat lakes and moorland pools; not common. Diameter 50–60 μ. Found in British Columbia by G. H. Wailes. (After Penard.)

113a (111) Numerous chitinous spicules radiating between the pseudo-podia . . . . *Heterophrys* Archer
Six species.

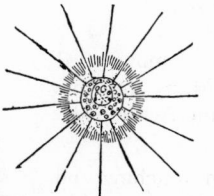

◄ **Fig. 9.120.** *Heterophrys myriopoda* Archer. × 190. Outer border of the envelope presenting a villous appearance due to the arrangement of the spicules. Raylike pseudopodia penetrate the envelope, their axes arising from the central granule. Nucleus single, eccentric. Protoplasm usually green, crowded with zoochlorellae. Contractile vacuole usually absent. Habitat marshes and standing water. Diameter 65–80 μ. (After Penard.)

113b Spicules siliceous **114**

114a   (113) Spicules all of one kind **115**

114b   Spicules of 2 kinds: siliceous in the form of plates and delicate
                    radiating spines. . . . . **Acanthocystis** Carter

Seventeen to 18 species. Species in addition to the one figured of this important
genus are known in N. A. *A. aculeata* Hertwig and Lesser possesses radial spines
nail-headed in form, with the base enlarged. *A. brevicirrhis* Perty has short, straight
or slightly curved spines. *A. myriospina* Penard is provided with long, tapering, very
numerous spines.

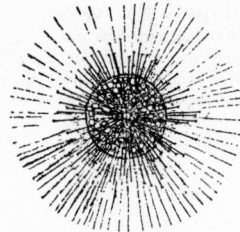

◄ **Fig. 9.121.** *Acanthocystis chaetophora* Schrank. × 250.ᐟ The
skeletal plates are oval, arranged tangentially and slightly imbricated. The spinous rays are of 2 kinds, the numerous long
ones acutely forked, the less numerous short ones widely forked
at the distal extremities. Nucleus large, eccentric; no contractile vacuole. Endoplasm green in color, enclosing
zoochlorellae. Habitat lakes, ponds, and moorland pools.
Diameter of the body 35–60 μ, rarely up to 100 μ. (After
Leidy.)

115a   (114) Spicules scattered through the envelope, surrounding the base of
                    the pseudopodia . . . . **Raphidiophrys** Archer

Twelve species.

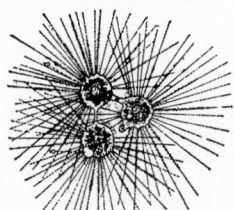

◄ **Fig. 9.122.** *Raphidiophrys elegans* Hertwig and Lesser. × 150.
Subcircular; disc-shaped spicules having thickened edges and
forming elongate cone-shaped accumulations around the pseudopodia. Nucleus single, placed eccentrically. One contractile
vacuole. Green zoochlorellae sometimes present. Often numbers of these individuals are grouped into colonies, joined by
protoplasmic processes. Habitat aquatic plants. Diameter
30–40 μ. (After Leidy.)

115b   Spicules closely united. Skeletal elements globular, forming a
                    compact envelope completely surrounding the body . . . .
                                                                Pompholyxophrys Archer

Three species.

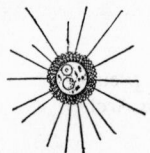

◄ **Fig. 9.123.** *Pompholyxophrys punicea* Archer. × 200. The siliceous spherical
globules usually in 3 rows about the body. Endoplasm usually reddish, containing numerous colored granules and vegetable food particles. Nucleus
spherical, large, eccentric. Pseudopodia very tenuous and indistinct.
Habitat ponds and swamps. Diameter 25–30 μ. (After Penard.)

116   (109) Pseudo-Heliozoa. Modern cytological works have demonstrated
                 that the following forms are not truly Heliozoa. Many are now
                 considered as bridging the gap between Rhizopoda and Actinopoda.

116a   Pseudopodia raylike, soft, and anastomosing when touching; no
                 envelope . . . . Order **Proteomyxida**   **117**

116b   Pseudopodia radiating, forked or branched, with a solid envelope, perforated, sometimes stalked . . . . Order **Desmothoraca**

Envelope or capsule spherical, homogenous, pedunculate. *Clathrulina* Cienkowsky, 4 species.

◄ **Fig. 9.124.** *Clathrulina elegans* Cienkowsky. × 130. Envelope more or less chitinous, perforated by numerous large openings usually regularly placed. Protoplasm not filling the test. Nucleus single, placed centrally. One or more contractile vacuoles. Pseudopodia very delicate, without axial filaments. Habitat *Sphagnum* swamps and among aquatic plants; very common in some localities. Diameter of envelope 60–90 μ. (After Leidy.)

117a   **(116)** Endoplasm colorless. Body amoeboid, normally spherical . . . . .
                                       Nuclearia Cienkowsky

One species.

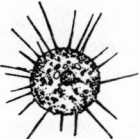

◄ **Fig. 9.125.** *Nuclearia simplex* Cienkowsky. × 250. The body is capable of changing shape. Pseudopodia arising from all parts of the body. Nucleus central; contractile vacuoles, more than one. Diameter 20–50 μ. Reported only by Conn from Connecticut, this form seems to be doubtful. (After Conn.)

117b   Endoplasm red or brown. Body spherical or elongated . . . .
                                       Vampyrella Cienkowsky

Twenty species, a number of which are doubtful.

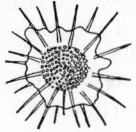

◄ **Fig. 9.126.** *Vampyrella lateritia* Cienkowsky. × 250. Pseudopodia arising from all parts of the body; numerous straight, elongated, and filamentous pseudopodia are intermixed with a variable number of shorter and stouter capitate processes which elongate and contract incessantly. Nucleus and contractile vacuole usually concealed by the content of the protoplasm. Habitat shallow bog-pools, among algae upon which it feeds. Diameter 30–40 μ and up to 80 μ. (After Conn.)

References

Bovee, E. C. 1953. Oscillosignum nov. gen. proboscidium n. sp. (and dakotaensis n. sp.), family Mayorellidae. *Trans. Am. Microscop. Soc.*, 72:328–336. **1954.** Morphological identification of free-living Amoebida. *Proc. Iowa Acad. Sci.*, 60:599–615. **Cash, J., G. H. Wailes, and J. Hopkinson. 1905–1921.** *The British Fresh-Water Rhizopoda and Heliozoa*, 5 Parts. Ray Society, London. **Chatton, E. 1953.** Ordre des Amoebiens nus. In: P. P. Grassé (ed.). *Traité de Zoologie*, Tome I, Fascicule 2, pp. 5–91. Masson, Paris. **Cockerell, T. D. A. 1911.** The fauna of Boulder County, Colorado. *Univ. Colo. Studies*, 8:227–256. **Conn, H. W. 1905.** A preliminary report on the Protozoa of the fresh-water fauna of Connecticut. *Conn. State Geol. and Nat. Hist. Survey Bull.*, 2:1–67. **Decloître, L. 1953.**

Recherches sur les Rhizopodes Thécamoebiens de l'A.O.F. *Mém. I. F. A. N.*, No. 31:1–249. **1956.** *Les Thécamoebiens de l'Eqe (Groenland).* Hermann et Cie, Paris. (A.S.I. No. 1242). **Deflandre, G. 1928.** Le genre *Arcella* Ehr. Morphologie. Biologie. Essai phylogénétique et systématique. *Arch. Protistenk.*, 64:152–287. **1929.** Le genre *Centropyxis* Stein. *Arch. Protistenk.*, 67:323–374. **1936.** Étude monographique sur le genre *Nebela* Leidy (Rhizopoda-Testacea). *Ann. Protistol.*, 5:201–286. **1947.** *Microscopie pratique. Le microscope et ses applications. La faune et la flore microscopiques des eaux. Les microfossiles.* Lechevalier, Paris. (31 plates with 536 figures concern fresh-water biology.) **1953.** Ordres des Aconchulina, des Athalamia et des Thécamoebiens. In: P. P. Grassé (ed.). *Traité de Zoologie*, Tome I, Fascicule 2, pp. 92–148. Masson, Paris. **Edmondson, C. H. 1906.** The protozoa of Iowa. *Proc. Davenport Acad. Sci.*, 11:1–124. **1912.** Protozoa of High Mountain Lakes in Colorado. *Univ. Colo. Studies*, 9:65–74. **Gauthier-Lièvre, L. 1953.** Les genres *Nebela, Paraquadrula et Pseudonebela* (Rhizopodes testacés) en Afrique. *Bull. Soc. Hist. Nat. Afrique du Nord*, 44:324–346. **Grandori, R. and E. 1934.** Studî sui Protozoi del terreno. *Boll. Lab. Zool. agr. Milano*, 5:1–340. **Grospietsch, T. 1953.** Rhizopodenanalytische Untersuchungen an Mooren Ostholsteins. *Arch. Hydrobiol.*, 47:321–452. **1958.** Wechseltierchen (Rhizopoden). *Ed. Mikrokosmos, Stuttgart:* 1:82. (Up-to-date, good introductory work.) **Hempel, A. 1898.** A list of the Protozoa and Rotifera found in Illinois River and adjacent lakes at Havana, Ill. *Bull. Illinois State Lab. Nat. Hist.*, 5:301–388. **Hoogenraad, H. R. and De Groot, A. A. 1940.** Zoetwaterrhizopoden en Heliozoen. *Fauna van Nederland*, Afl. 9:1–303. **Johnson, Percy L. 1934.** Concerning the genus Ouramoeba (Leidy), *Ann. Protistol.*, 4:25–29. **Kirby, Harold. 1941.** Organisms living on and in Protozoa. Chapter XX in: G. N. Calkins and F. M. Summers (eds.). *Protozoa in biological research.* Columbia University Press, New York. **Leidy, J. 1879.** Fresh-water Rhizopods of North America. *U. S. Geol. Survey Territ.*, 12:1–324. **Mote, R. F. 1954.** A study of soil protozoa on an Iowa virgin prairie. *Proc. Iowa Acad. Sci.*, 61:570–592. **Oye, P. Van. 1956a.** Rhizopoda Venezuelas, mit besonderer Berücksichtigung ihrer Biogenographie. *Ergeb. deutsch. Limnol. Venezuela-Exped. 1952*, 1:329–360. **1956b.** On the Thécamoeban fauna of New Zealand, with description of four new species and biogeographical discussion. *Hydrobiologia*, 8:16–37. **Penard, E. 1902.** *Faune rhizopodique du bassin du Léman.* H. Kündig, Geneva. **1904.** *Les Héliozoaires d'eau douce.* Geneva. **1905.** *Les Sarcodinés des grands lacs.* Georg et Cie, Geneva. **1911.** Rhizopodes d'eau douce. *British Ant. Exp.*, 1:203–262. **1935.** Rhizopodes d'eau douce. Récoltes, préparation et souvenirs. *Bull. Soc. franç. microscop.*, 4:57–73. **1938.** *Les infiniment petits dans leurs manifestations vitales.* Geneva. **Ray, D. L. 1951.** Agglutination of bacteria: a feeding method in the soil ameba Hartmannella sp. *J. Expl. Zool.*, 118:443–466. **Schaeffer, A. A. 1926.** Taxonomy of the amebas with description of thirty-nine new marine and fresh-water species. *Papers Dept. Marine Biol. Carnegie Inst.*, 26:1–116. **Stepanek, M. 1952.** Testacea of the pond of Hradek at Kunratice (Prague). *Acta Musei nation.* Pragae, 8:1–55. **1956.** Amoebina and amoebic stages of flagellata freely living in garden soil. *Universitas Carolina Biologica*, 2:125–159. **Trégouboff, G. 1953.** Classe des Héliozoaires. In: P. P. Grassé. *Traité de Zoologie*, Tom. I, Fascicule 2, pp. 465–489. Figs. **Wailes, G. H. 1912.** Fresh-water Rhizopods and Heliozoa from the States of New York, New Jersey and Georgia, U. S. A., with supplemental notes on Seychelles species. *J. Linnean Soc. London Zool.*, 32:121–161. **1928.** Fresh-water and marine Protozoa from British Columbia with description of new species. *Museum Notes Vancouver*, 3:25–37. **1930.** Protozoa and Algae, Mount Ferguson B. C. *Museum Notes Vancouver*, 5:160–165. **1931.** Munday Lake and its Ecology. *Museum Notes Vancouver*, 6:34–39. **1932.** Protozoa and Algae from Lake Tenquille, B. C. *Museum Notes Vancouver*, 7:19–23. **Wailes, G. H. and E. Penard. 1911.** Rhizopoda. *Proc. Roy. Irish Acad.*, Clare Island Survey, Part 65, 64 pp.

10

Ciliophora

LOWELL E. NOLAND

Everyone who will use this chapter should know the meaning of such terms as *cilia* (Fig. 10.1*a*), *macronucleus* (the larger nucleus), *micronucleus* (the smaller one), *pellicle* (cell membrane), *ectoplasm* (firmer peripheral cytoplasm), and *endoplasm* (more fluid inner cytoplasm). Certain other technical terms, however, may need definition, such as: *syncilium* (Fig. 10.1*c*), two or a very few cilia fused into one; *pectinelles* (Fig. 10.1*e*), short rows of strong cilia, not fused, occurring in series; *membranelles* (Fig. 10.1*b*), short double or triple rows of cilia fused into triangular, pennantlike blades, commonly occurring in a series called the *adoral zone* (Fig. 10.1*g*) leading to the mouth; *cirrus* (Fig. 10.1*f*), conical group of cilia fused into a single organelle like the hairs of a wet brush; *membrane* or *undulating membrane* (Fig. 10.1*d*), row of cilia fused to form a sheet with a free edge that moves in a wavy motion; *caudal cilia*, long posterior tactile cilia, few in number; *tactile setae*, stiff bristlelike cilia, sensory in nature. In hypotrichs cirri are named according to location, as shown in Fig. 10.1*g*.

Since the mouth region provides important criteria for classification, it is desirable to know the meaning of certain terms applied to it, such as: *peristome*, a differentiated external area adjacent to the mouth associated with food-getting; *oral groove*, a linear depression or furrow, usually ciliated, leading to the mouth; *buccal cavity*, a food-conducting space or tube, commonly ciliated, open to the

265

outside by way of the functional *mouth* or buccal aperture, and communicating internally through the true mouth or *cytostome* with the *gullet* (cytopharynx), the latter being an unciliated passage leading from the true mouth into the cytoplasm, nearly always closed except when feeding. The mouth is often armed with special supporting or food-getting structures, such as: *toxicysts*, minute poison sacs for killing or paralyzing prey; *trichocysts*, tiny rodlike bodies that can shoot out slender threads into the water when appropriately stimulated; *trichites*, slender supporting rods in the gullet wall or elsewhere. Since trichites, trichocysts, and toxicysts often look very much alike, it is not always evident which is which when they are seen inactive in the cell. Trichocysts are not limited to the mouth region, whereas toxicysts and trichites are more commonly found there than elsewhere.

The terms *right* and *left* in the key always refer to the organism's right and left, not the observer's. In a semitransparent ciliate like a hypotrich, ventral ciliary structures can often be seen through the body when viewed dorsally, and then seem to have right and left sides reversed.

There is a large group of marine ciliates, and another considerable group of parasitic species. These are not treated in this key, except as they are likely to be found free in fresh water or in plain view on the exterior of fresh-water organisms. Limitations of space have made it necessary to omit some of the rarer genera, but most of those found in fresh water are included. Anyone wishing to be sure of species must consult more complete monographs, such as that of Kahl. The approximate number of species in each genus is noted in the key.

Since the group covered by this key is large and complicated, each group above genus is given a separate line, which serves as a heading and will facilitate finding the group, e.g., line 11.

KEY TO GENERA

1a		Cilia present throughout active (unencysted) life; no suctorial tentacles . . . . Class **Ciliata**	**2**
1b		Cilia present during free-swimming juvenile stages only; suctorial tentacles present . . . . Class **Suctoria**	**282**
2	(1)	Class **Ciliata**	**3**
3a	(2)	All nuclei of same type; no mouth; parasitic in digestive tract of amphibians . . . . Subclass **Protociliata** Not treated in this key; see such monographs as that of Metcalf for this group.	
3b		Two types of nuclei (macro- and micro-); mouth usually present; mostly free-living . . . . Subclass **Euciliata**	**4**
4	(3)	Subclass **Euciliata**	**5**
5a	(4)	Peristome either lacking or, if present, provided with simple cilia or membranes; no cirri or membranelles	**6**
5b		Peristome present, bordered by an adoral zone of membranelles leading clockwise to the mouth . . . . . . Order **Spirotrichida**	**159**
6a	(5)	Peristome an apical, spirally-rolled funnel, leading clockwise to	

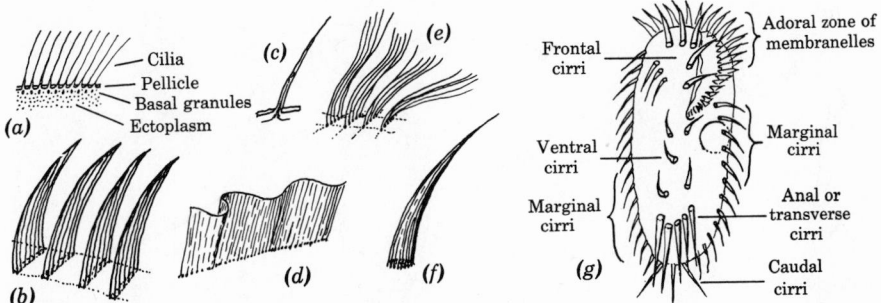

Fig. 10.1. Diagrams illustrating the different kinds of ciliary structures and their distribution. (*a*) Cilia. (*b*) Membranelles. (*c*) Syncilium. (*d*) Undulating membrane. (*e*) Pectinelles. (*f*) Cirrus. (*g*) Ventral view of a hypotrich showing the different groups of cirri.

		mouth, ciliated inside; body otherwise unciliated, rigid; ectozoic on amphipods . . . . Order **Chonotrichida**	281
6b		Peristome not a spirally-rolled apical funnel	7
7a	(6)	Peristome circular, bordered by membranes running counterclockwise around it and into a vestibule containing mouth, anus, and contractile-vacuole pore; body otherwise unciliated except that an aboral circlet is present in free-swimming stages or species. . . . Order **Peritrichida**	242
7b		With the general body surface wholly or partly ciliated; mouth, anus, and contractile-vacuole pore opening independently . . . . . Order **Holotrichida**	8
8	(7)	Order **Holotrichida**	9
9a	(8)	Cytostome at or near surface; no extensive peristome; buccal cavity, if present, does not contain cilia or membranes; gullet closed except when feeding . . . . . . Suborder **Gymnostomina**	11
9b		Cytostome at the inner end of a buccal cavity open to the outside and having cilia or membranes in it	10
10a	(9)	Simple cilia in buccal cavity; no membranes. . . . . Suborder **Trichostomina**	85
		Often difficult to see; when in doubt follow through both alternatives of the key.	
10b		Undulating membranes within or leading into the buccal cavity . . Suborder **Hymenostomina**	113
11	(9)	Suborder **Gymnostomina**	12
12a	(11)	Mouth at or very near the anterior end, though in a few genera it continues down the side as a slit with slight swollen margins usually provided with trichites . . . . Tribe **Prostomata**	14
12b		Mouth lateral or ventral (*Teuthophrys*, at line **69** in the key, belongs here, though its mouth is apparently anterior).	13
13a	(12)	Mouth a nearly invisible lateral slit on convex side of tapering front end, or a lateral opening at base of an anterior proboscis . . . Tribe **Pleurostomata**	58
		In doubtful cases consult Family Spathidiidae at line **49** in the key, as this family is intermediate between Prostomata and Pleurostomata.	
13b		Mouth ventral; no proboscis; often with trichites in the gullet wall . . . . Tribe **Hypostomata**	72
14	(12)	Tribe **Prostomata**	15

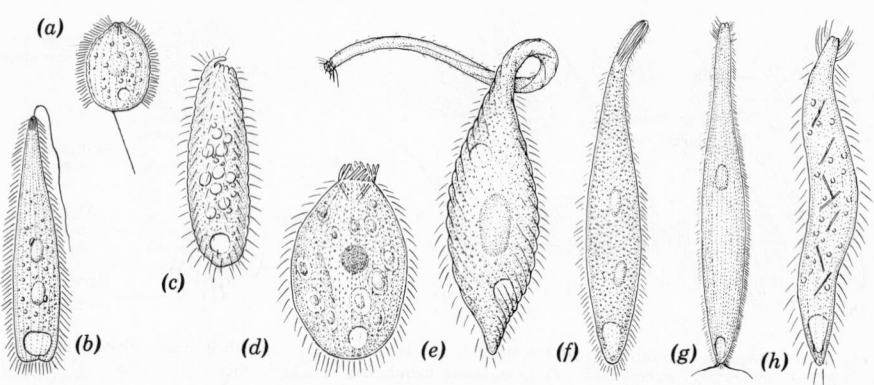

Fig. 10.2. (*a*) *Urotricha farcta* Caparède and Lachmann, 22 μ. × 500. (*b*) *Ileonema ciliata* (Roux), 75 μ. × 400. (*c*) *Chilophrya utahensis* (Pack), 50 μ. × 500. (*d*) *Spasmostoma viride* Kahl, 60 μ. × 360. (*e*) *Lacrymaria olor* O. F. Müller, 120 μ (body only). × 250. (*f*) *Trachelophyllum apiculatum* (Perty), 516 μ. × 80. (*g*) *Urochaenia ichthyoides* Savi, size not recorded. (*h*) *Chaenea teres* (Dujardin), 200 μ. × 220. (*b* after Roux; *c* after Pack; *d* and *h* after Kahl; *g* after Savi.)

15a	**(14)**	With tentacles extending radially from the cell in all directions when at rest, but retracted and hardly visible when swimming . . . Family **Actinobolinidae**	**40**
15b		Tentacles absent or localized or not completely retractile	**16**
16a	**(15)**	An ectoplasmic armor present, consisting of many small plates of translucent organic material . . . . Family **Colepidae**	**41**
16b		No armor of this type present	**17**
17a	**(16)**	Anterior end hollowed out in cuplike fashion; animals sedentary in a secreted shell or jelly . . . . Family **Metacystidae**	**42**
17b		Not as just described.	**18**
18a	**(17)**	Mouth anterior, round, or slightly elongated; if slitlike, very short, not over ⅓ body width	**19**
18b		Mouth a long slit, beginning at anterior end, sometimes extending down one side of the cell . . . . Family **Spathidiidae**	**49**
19a	**(18)**	Mouth at anterior pole, surrounded by an unciliated area which is often raised as a blunt cone; cell body encircled by 1 to 3 rings of strong cilia or pectinelles; elsewhere naked or with short cilia . . . . Family **Didiniidae**	**45**
19b		Cell uniformly ciliated around the mouth, and usually over the entire body . . . . Family **Holophryidae**	**20**
20	**(19)**	Family **Holophryidae**.	**21**
21a	**(20)**	Rear third unciliated, except for one or a few long caudal cilia; 12 species. (Fig. 10.2*a*) . . . . *Urotricha* Claparède and Lachmann	
21b		Body uniformly ciliated at the rear	**22**
22a	**(21)**	Mouth with processes (flaps, whips, or tentacles)	**23**
22b		Mouth without such processes.	**26**
23a	**(22)**	With short oral flaps that can be bent down toward the mouth. . .	**24**
23b		Oral processes long, whiplike, or tentaclelike.	**25**
24a	**(23)**	With 1 oral flap; 1 species. (Fig. 10.2*c*) . . . . *Chilophrya* Kahl	
24b		Several oral flaps; 1 species. (Fig. 10.2*d*). . . *Spasmostoma* Kahl	

25a (23) Oral process long, whiplike; 3 species. (Fig. 10.2b). . . . .
Ileonema Stokes

25b   Oral processes shorter, tentaclelike **26**

26a (22) Tapering anteriorly into a proboscis, which has a ciliated tip set off by a groove and a circle of longer cilia; 10 species. (Fig. 10.2e) . . . . *Lacrymaria* Ehrenberg

26b   Without the type of proboscis tip described above **27**

27a (26) Body over 4 times as long as wide, tapering anteriorly **28**

27b   Less than 4 times as long as wide **30**

28a (27) Body nearly circular in cross section **29**

28b   Body somewhat flattened; glides with one side against substratum; 7 species. (Fig. 10.2f) . . . .
Trachelophyllum Claparède and Lachmann

29a (28) Two caudal processes; 1 species. (Fig. 10.2g) . . . .
Urochaenia Savi

29b   No caudal processes; 8 species. (Fig. 10.2h). . . . .
Chaenea Quennerstedt

30a (27) Mouth a subapical slit; body slightly bent in front; 50–100 μ. . . .   **31**

30b   Mouth terminal, sometimes oval, but not slitlike. **33**

31a (30) Front end much bent over; 6 species. (Fig. 10.3a) . . . .
Platyophrya Kahl

31b   Body bent over only very slightly in front **32**

32a (31) Mouth slit short, its right and left margins much the same; 4 species. (Fig. 10.3b). . . . . *Microregma* Kahl

32b   Mouth slit longer, right margin higher with stronger cilia (syncilia?); 10 species. (Fig. 10.3c). . . . .
Plagiocampa Schewiakoff

33a (30) Cell glass clear except for food; rigid body; 50–100 μ **34**

33b   Cell not so transparent; more pliable; size various **36**

34a (33) Cilia rows spiral; postoral suture runs back to an indentation near the cell middle; 6 species. (Fig. 10.3d) . . . . *Placus* Cohn

34b   Cilia rows straight, longitudinal; no postoral suture **35**

35a (34) Cell tapers toward the front; no caudal cilium; 4 species. (Fig. 10.3e) . . . . *Rhopalophrya* Kahl

35b   Bluntly rounded in front; with caudal cilium; 4 species. (Fig. 10.3f) . . . . *Pithothorax* Kahl

36a (33) Fig-shaped, tapering somewhat at front end **37**

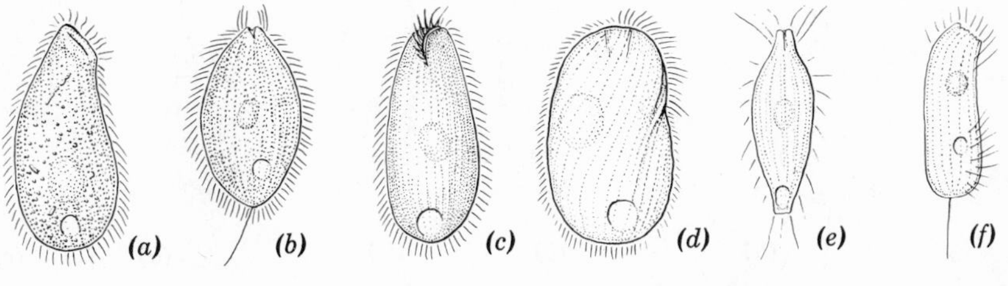

Fig. 10.3. (a) *Platyophrya vorax* Kahl, 60 μ. × 375. (b) *Microregma auduboni* (Smith), 50 μ. × 360. (c) *Plagiocampa mutabilis* Schewiakoff, 40 μ. × 530. (d) *Placus luciae* (Kahl), 50 μ. × 430. (e) *Rhopalophrya sulcata* Kahl, 50 μ. × 365. (f) *Pithothorax rotundus* Kahl, 30 μ. × 580. (b after Smith; e and f after Kahl.)

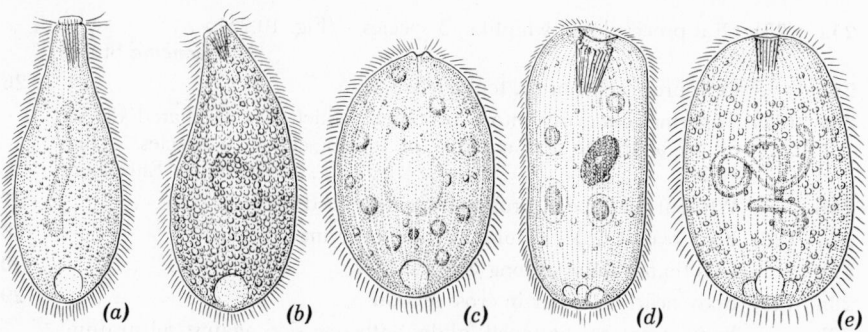

Fig. 10.4. (a) *Enchelyodon elegans* Kahl, 180 μ. × 200. (b) *Enchelys simplex* (Kahl), 85 μ. × 410. (c) *Holophrya simplex* Schewiakoff, 60 μ. × 1000. (d) *Prorodon teres* Ehrenberg, 200 μ. × 175. (e) *Pseudoprorodon ellipticus* Kahl, 120 μ. × 280. (*a*, *d*, and *e* after Kahl; *c* after Schewiakoff.)

36b		Tapering little or none at either end	**38**
37a	(36)	With oral papilla projecting from center of mouth; 15 species. (Fig. 10.4*a*) . . . . **Enchelyodon** Claparède and Lachmann	
37b		No oral papilla evident; 17 species. (Fig. 10.4*b*) . . **Enchelys** Hill	
38a	(36)	Body a regular ellipsoid; mouth round, at anterior pole; gullet wall thin, with delicate trichites or none; no adoral row of bristles; 22 species. (Fig. 10.4*c*) . . . . . . **Holophrya** Ehrenberg	
38b		Body usually a bent ellipsoid; mouth usually elliptical, often slightly displaced from anterior pole; a row of inconspicuous short bristles extends backward from mouth on one side.	**39**
39a	(39)	Gullet trichites double, conspicuous, their external ends slightly below surface level; nucleus compact; 30 species. (Fig. 10.4*d*) . . . **Prorodon** Ehrenberg	
39b		Gullet trichites simple, visible, reaching surface; nucleus usually long; 12 species. (Fig. 10.4*e*) . . . . **Pseudoprorodon** Blochmann	
40	(15)	Family **Actinobolinidae**; 1 genus, 2 species. (Fig. 10.5*a*) . . . . . **Actinobolina** Strand	
41	(16)	Family **Colepidae**; 1 genus in fresh water, 16 species. (Fig. 10.5*b*) . . . . **Coleps** Nitzsch	

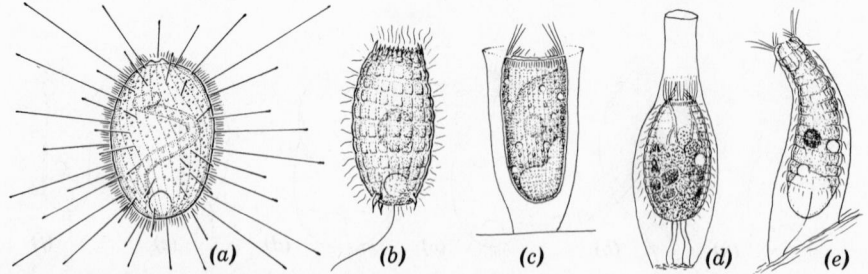

Fig. 10.5. (a) *Actinobolina radians* (Stein), 125 μ. × 150. (b) *Coleps hirtus* (O. F. Müller), 60 μ. × 320. (c) *Pelatractus grandis* (Penard), 160 μ. × 120. (d) *Vasicola ciliata* Tatem, 100 μ. × 150. (e) *Metacystis recurva* Penard, 50 μ. × 420. (*c* and *e* after Penard; *d* after Kahl.)

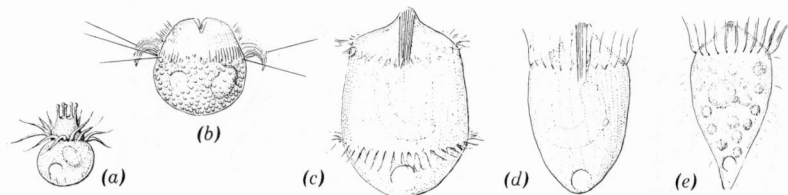

Fig. 10.6. (a) *Mesodinium acarus* Stein, 22 μ. × 500. (b) *Askenasia volvox* (Claparède and Lachmann), 50 μ. × 250. (c) *Didinium nasutum* (O. F. Müller), 90 μ. × 280. (d) *D. balbianii* (Fabre-Domergue), 80 μ. × 280. (e) *Acropisthium mutabile* Perty, 80 μ. × 280. (b after Blochmann; d and e after Schewiakoff.)

42	**(17)**	Family **Metacystidae**	**43**
43a	**(42)**	With 1 or more long caudal cilia	**44**
43b		No long caudal cilia; 2 species. (Fig. 10.5c). . . ***Pelatractus*** Kahl	
44a	**(43)**	No bulging vesicle at rear; 5 species. (Fig. 10.5d). . . . . ***Vasicola*** Tatem	
44b		Bulging vesicle at rear; 11 species. (Fig. 10.5e) . . . . ***Metacystis*** Cohn	
45	**(19)**	Family **Didiniidae**	**46**
46a	**(45)**	With equatorial groove and oral tentacles; body length less than 40 μ; 3 species. (Fig. 10.6a). . . . . ***Mesodinium*** Stein	
46b		No equatorial groove or oral tentacles; over 50 μ	**47**
47a	**(46)**	Two differing ciliary wreaths in front of equator; none behind; 3 species. (Fig. 10.6b). . . . . ***Askenasia*** Blochmann	
47b		One or more wreaths, all of same type (pectinelles)	**48**
48a	**(47)**	Cell rounded at rear; no body cilia besides those of the wreaths; 7 species. (Fig. 10.6c,d) . . . . ***Didinium*** Stein	
48b		Cell pointed at rear; body cilia present; 1 species. (Fig. 10.6e) . . . . ***Acropisthium*** Perty	
49	**(18)**	Family **Spathidiidae**	**50**
50a	**(49)**	Wormlike, more than 5 times as long as wide; 3 species. (Fig. 10.7a). . . . . ***Homalozoon*** Stokes	
50b		Not over 4 times as long as wide	**51**

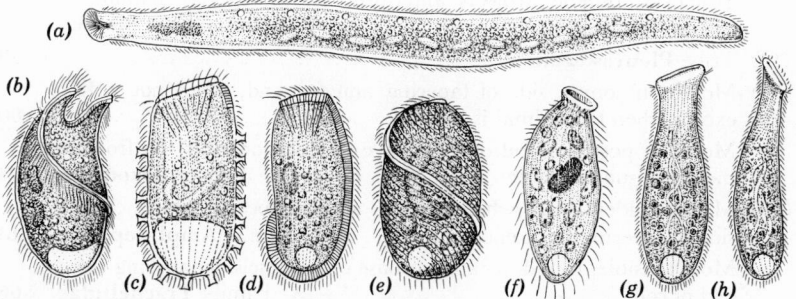

Fig. 10.7. (a) *Homalozoon vermiculare* (Stokes), 450 μ. × 200. (b) *Diceras bicornis* Kahl, 200 μ. × 100. (c) *Legendrea bellerophon* Penard with retracted tentacles, 140 μ. × 200. (d) *Penardiella interrupta* (Penard), 100 μ. × 250. (e) *Perispira ovum* Stein, 95 μ. × 250. (f) *Enchelydium virens* Kahl, 87 μ. × 300. (g) and (h) *Spathidium spathula* (O. F. Müller), 120 μ. × 240. (b and f after Kahl; c and d after Penard.)

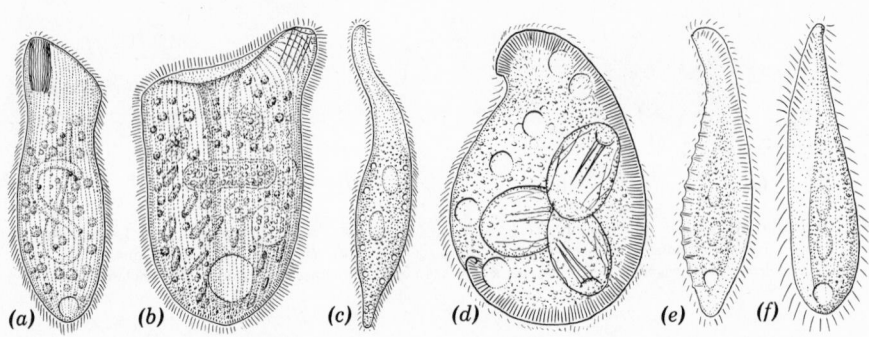

Fig. 10.8. (a) *Cranotheridium taeniatum* Schewiakoff, 170 μ. × 210. (b) *Spathidioides sulcatum* Brodsky, 75 μ. × 500. (c) *Amphileptus claparedei* Stein, 250 μ. × 150. (d) *Bryophyllum lieberkühni* Kahl, 415 μ. × 90. (e) *Loxophyllum helus* (Stokes), 220 μ. × 160. (f) *Acineria incurvata* Dujardin, 120 μ. × 300. (a after Schewiakoff; b and f after Kahl.)

51a	**(50)**	Front end not projecting in pointed lobes **52**
51b		Front end of cell projecting as 2 pointed lobes; 1 species. (Fig. 10.7b) . . . . *Diceras* Eberhard
52a	**(51)**	With extensible and retractile toxicyst-bearing tentacles; 3 species. (Fig. 10.7c) . . . . *Legendrea* Fauré-Fremiet
52b		No such tentacles present **53**
53a	**(52)**	Mouth ridge very long, extending to rear end or beyond **54**
53b		Mouth ridge no longer than width of body. **55**
54a	**(53)**	Cilia rows straight; 3 species. (Fig. 10.7d) . . . *Penardiella* Kahl
54b		Cilia rows spiral; 3 species. (Fig. 10.7e) . . . . . *Perispira* Stein
55a	**(53)**	Mouth slit closed except when feeding **56**
55b		Mouth remains open; 6 species. (Fig. 10.7f) . . . . *Enchelydium* Kahl
56a		One large trichocyst bundle at anterior end of mouth **57**
56b		No single trichocyst bundle, oral trichocysts evenly spaced or in small groups; 66 species. (Fig. 10.7g,h) . . . *Spathidium* Dujardin
57a	**(56)**	Trichocyst bundle without any special papilla; 2 species. (Fig. 10.8a) . . . . *Cranotheridium* Schewiakoff
57b		Trichocyst bundle projects anteriorly into a special papilla; 4 species. (Fig. 10.8b) . . . . *Spathidioides* Brodsky
58	**(13)**	Tribe **Pleurostomata** **59**
59a	**(58)**	Mouth on convex side of tapering anterior end; often not visible except when the animal is feeding **60**
59b		Mouth at posterior end of a concave indentation near the front of the cell; usually visible . . . . Family **Loxodidae** **71**
60a	**(59)**	Mouth evident only when feeding, slitlike, located along convex side of tapering front end . . . . Family **Amphileptidae** **61**
60b		Mouth visible, round, located at base of proboscis or tapering front end of cell . . . . Family **Tracheliidae** **66**
61	**(60)**	Family **Amphileptidae** **62**
62a	**(61)**	Entire body surface covered with cilia **63**
62b		Ciliated ventrally; few or no dorsal cilia; flattened **64**

63a (62) Mouth not extending back of cell middle; trichocysts not evident along mouth; 4 species. (Fig. 10.8c) . . *Amphileptus* Ehrenberg

63b   Mouth extending back beyond cell middle; trichocysts along the entire mouth; 10 species. (Fig. 10.8d) . . . . *Bryophyllum* Kahl

64a (62) Mouth not extending back of cell middle; lateral margins not extended and thin; no trichocysts on aboral margin **65**

64b   Mouth extending back of cell middle; body with thin wide lateral margins; often with trichocysts on aboral as well as oral margin; 22 species. (Fig. 10.8e) . . . . *Loxophyllum* Dujardin

65a (64) Aboral margin rolled upward, carrying 4 rows of cilia to dorsal surface; 1 species. (Fig. 10.8f) . . . . *Acineria* Dujardin

65b   Not as above; 37 species. (Fig. 10.9a) . . *Lionotus* Wrzesniowski

66   (60) Family **Tracheliidae** **67**

67a (66) Body flat, bladelike in front; ectocommensal on isopods and amphipods; 2 species. (Fig. 10.9b) . . . . . *Branchioecetes* Kahl

67b   Body not flattened; free-living. **68**

68a (67) Body ovoid or short ellipsoid; proboscis short, fingerlike; 3 species. (Fig. 10.9c) . . . . *Trachelius* Schrank

68b   Shape not as above; 1 or more long proboscides **69**

69a (68) Three proboscides; mouth central between bases of proboscides; 1 species. (Fig. 10.9d) . . . *Teuthophrys* Chatton and Beauchamp

69b   Only 1 proboscis; mouth lateral at its base **70**

70a (69) Body, not including proboscis, about twice as long as wide; 4 species. (Fig. 10.9e) . . . . *Paradileptus* Wenrich

70b   Body much longer, at least 3 times as long as wide; 20 species. (Fig. 10.9f) . . . . *Dileptus* Dujardin

71   (59) Family **Loxodidae**; 4 species. (Fig. 10.9g) . . *Loxodes* Ehrenberg

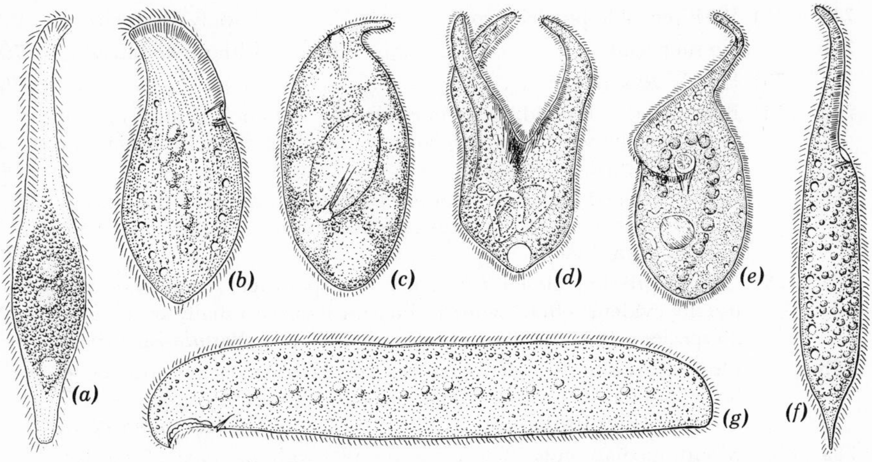

Fig. 10.9. (a) *Lionotus fasciola* (Ehrenberg), 140 μ. × 370. (b) *Branchioecetes gammari* (Penard), 125 μ. × 280. (c) *Trachelius ovum* Ehrenberg, 330 μ. × 100. (d) *Teuthophrys trisulca* Chatton and Beauchamp, 270 μ. × 120. (e) *Paradileptus robustus* Wenrich, 350 μ. × 100. (f) *Dileptus anser* (O. F. Müller), 155 μ. × 350. (g) *Loxodes magnus* Stokes, 500 μ. × 140. (b after Penard; d and e after Wenrich.)

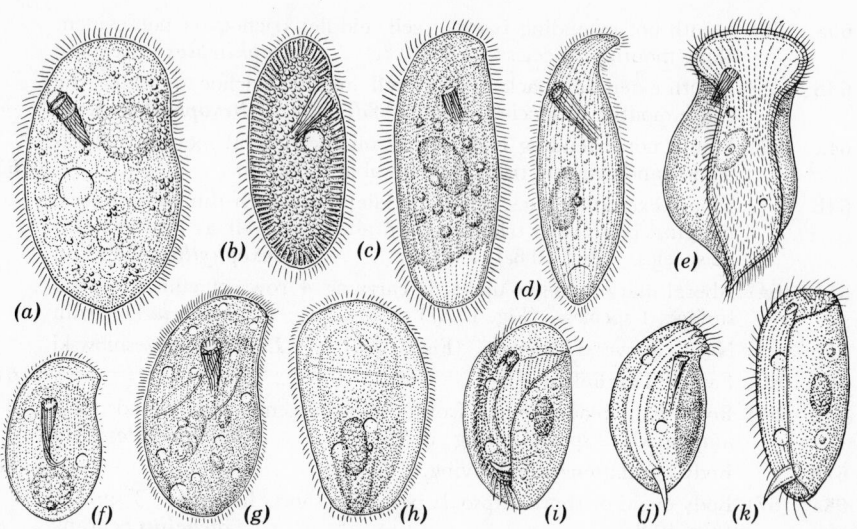

Fig. 10.10. (a) *Nassula ornata* Ehrenberg, 125 μ. × 265. (b) *Cyclogramma trichocystis* (Stokes), 60 μ. × 475. (c) *Chilodontopsis depressa* (Perty), 74 μ. × 450. (d) *Orthodon hamatus* Gruber, 180 μ. (e) *Phascolodon vorticella* Stein, 100 μ. × 320. (f) *Chilodonella uncinata* (Ehrenberg), 42 μ. × 500. (g) *C. cucullula* (O. F. Müller), 125 μ. × 215. (h) *Gastronauta membranacea* Engelmann, 52 μ. × 500. (i) *Trochilioides recta* (Kahl), 40 μ. × 560. (j) *Trochilia minuta* (Roux), 30 μ. × 700. (k) *Dysteria navicula* Kahl, 40 μ. × 650. (c after Blochmann; d after Gruber; e after Stein; i and k after Kahl.)

72	**(13)**	Tribe **Hypostomata**	**73**
73a	**(72)**	Body usually flattened; ventrally ciliated; dorsal side bare, or at most with a few tactile cilia	**74**
73b		Body roundish in cross section, sometimes slightly flattened ventrally; ciliated all over . . . . Family **Nassulidae**	**75**
74a	**(73)**	With movable posterior spur . . . . Family **Dysteriidae**	**82**
74b		No such spur . . . . Family **Chlamydodontidae**	**79**
75	**(73)**	Family **Nassulidae**	**76**
76a	**(75)**	Body nearly ellipsoid, sometimes slightly flattened ventrally; oral depression present; external ends of gullet trichites lie somewhat below cell surface	**77**
76b		Body flattened ventrally, rounded dorsally; no oral depression; gullet trichites reach cell surface at mouth; anterior end bent slightly toward left	**78**
77a	**(76)**	Oral depression closed externally by sphincter; trichocysts not usually evident; often swims rolling on its axis; usually over 70 μ; 20 species. (Fig. 10.10a) . . . . **Nassula** Ehrenberg	
77b		Oral depression widely open to the exterior; heavy trichocyst layer; animal commonly glides on substratum; less than 70 μ; 7 species. (Fig. 10.10b) . . . . **Cyclogramma** Perty	
78a	**(76)**	Mouth median; gullet lies nearly parallel with main axis; 8 species. (Fig. 10.10c) . . . . **Chilodontopsis** Blochmann	
78b		Mouth located at right side; gullet at about 45° angle to main axis; 4 species. (Fig. 10.10d) . . . . **Orthodon** Gruber	
79	**(74)**	Family **Chlamydodontidae**	**80**

80a (79) Body not flattened; planktonic; anterior ciliated area continues down ventral surface as a strip with raised edges; 2 species. (Fig. 10.10*e*) . . . . **Phascolodon** Stein

80b Body distinctly flattened; ciliated ventrally; glides on substratum; rarely swims free. 81

81a (80) Mouth opening round; gullet trichites form a tube; no oral membrane; 30 species. (Fig. 10.10*f,g*) . . . . . . **Chilodonella** Strand

81b Mouth linear, transverse; anterior oral cilia fused into a membrane; 1 species. (Fig. 10.10*h*) . . . . . **Gastronauta** Engelmann

82 (74) Family **Dysteriidae** 83

83a (82) Besides frontal and right lateral cilia there is a ciliated area left of mouth; 5 species. (Fig. 10.10*i*) . . . . **Trochilioides** Kahl

83b No ciliated area left of mouth. 84

84a (83) Lateral margins of the ventral ciliated strip not raised as keels; 8 species. (Fig. 10.10*j*) . . . . **Trochilia** Dujardin

84b Margins of ventral ciliated groove raised as keels; mostly marine; 3 fresh-water species. (Fig. 10.10*k*). . . . . . . **Dysteria** Huxley

85 (10) Suborder **Trichostomina** 86

86a Peristome a circular furrow around a median anterior protuberance bearing longer cilia; live in secreted gelatinous tubes or cases . . . . Family **Marynidae** 101

86b Peristome not as described above; no gelatinous case 87

87a (86) Body small, flattened, rigid; ciliation sparse, borne on pellicular ridges; one side more convex than the other; mouth usually lateral or far back on straighter side; 2 vacuoles near mouth . . . . Family **Trichopelmidae** 109

87b Not having all the above characteristics 88

88a (87) Rear part of cell unciliated except for some long caudal cilia on pointed rear end . . . . Family **Trimyemidae** 100

88b Body uniformly ciliated 89

89a (88) With a row of longer cilia spiralling forward from rear end of body to mouth. 90

89b With no such spiral row of longer cilia 91

90a (89) Rear end rather pointed, bearing a few long caudal cilia . . . . . . Family **Spirozonidae** 98

90b Rear end rounded and without long caudal cilia . . . . Family **Trichospiridae** 99

91a (89) Mouth area shallow; no definite oral groove; no ciliated pharyngeal tube . . . . Family **Clathrostomidae** 95

91b Mouth area sunken in, with oral groove leading to it, or with ciliated pharyngeal tube, or both 92

92a (91) With single posterior contractile vacuole 93

92b Contractile vacuole not posterior, or more than 1 94

93a (92) Cell indented near middle of one side, where oral groove passes around to mouth . . . . Family **Colpodidae** 103

93b Oral groove transverse, in anterior fourth of cell; with mouth at its left end . . . . Family **Plagiopylidae** 108

94a (92) Ectozoic, living usually on outside of clams and snails. . . . . . . Family **Conchophthiridae** 107

94b Free-living; not on host animal . . . . . . Family **Paramecidae** 96

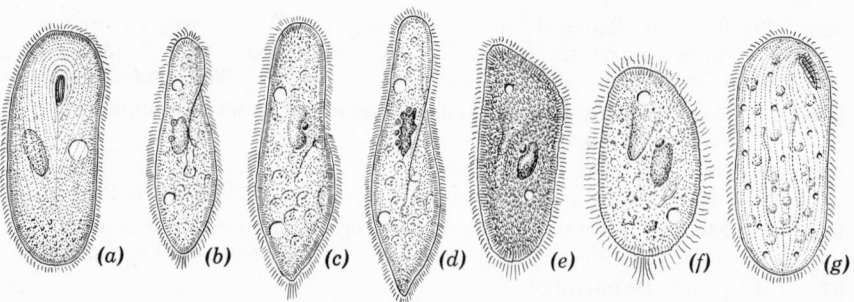

Fig. 10.11. (a) *Clathrostoma viminale* Penard, 127 μ. × 220. (b) *Paramecium aurelia* Ehrenberg, 165 μ. × 160. (c) *P. caudatum* Ehrenberg, 210 μ. × 150. (d) *P. multimicronucleatum* Powers and Mitchell, 275 μ. × 125. (e) *P. bursaria* (Ehrenberg), 130 μ. × 200. (f) *P. trichium* Stokes, 90 μ. × 250. (g) *Physalophrya spumosa* (Penard), 240 μ. × 120. (a and g after Penard.)

95   (91) Family **Clathrostomidae;** 1 genus, 3 species. (Fig. 10.11*a*) . . . .
                                         Clathrostoma Penard

96   (94) Family **Paramecidae** **97**

97a (96) Two contractile vacuoles, 1 in front, 1 in rear half of cell; 15
                species. (Fig. 10.11*b–f*) . . . . *Paramecium* Hill
                Work by Fauré-Fremiet on the oral silver-line system of *Paramecium* suggests that this
                genus may perhaps belong more properly in the suborder Hymenostomina.

97b   One or many contractile vacuoles; macronucleus vermiform or
                moniliform; 2 species. (Fig. 10.11*g*) . . . . . *Physalophrya* Kahl

98   (90) Family **Spirozonidae,** with only 1 genus and 1 species. (Fig.
                10.12*a,b*) . . . . *Spirozona* Kahl

99   (90) Family **Trichospiridae,** with only 1 genus and 1 species. (Fig.
                10.12*c*) . . . . *Trichospira* Roux

100 (88) Family **Trimyemidae,** with only 1 genus and 1 fresh-water species.
                (Fig. 10.12*d*) . . . . *Trimyema* Lackey

101 (86) Family **Marynidae** **102**

102a (101) Peristome completely surrounding anterior prominence; no long
                caudal cilia; 1 species. (Fig. 10.12*e,f*) . . . . . . *Maryna* Gruber

102b   Anterior prominence not completely surrounded by peristomial
                groove; tuft of long caudal cilia present; 2 species. (Fig.
                10.12*g*) . . . . *Mycterothrix* Lauterborn

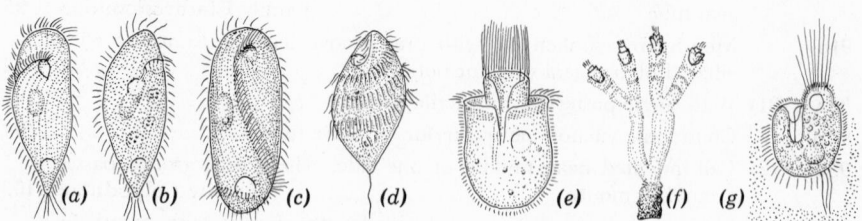

Fig. 10.12. (a) *Spirozona caudata* Kahl, 80 μ. × 250. (b) The same, lateral view. (c) *Trichospira inversa* (Claparède and Lachmann), 90 μ. × 250. (d) *Trimyema compressum* Lackey, 35 μ. × 500. (e) *Maryna socialis* Gruber, 150 μ. × 110. (f) Colony of the same in gelatinous tubes. × 25. (g) *Mycterothrix erlangeri* Lauterborn in gelatinous cases, 55 μ. × 210. (a–d after Kahl; e–f after Gruber; g after Penard.)

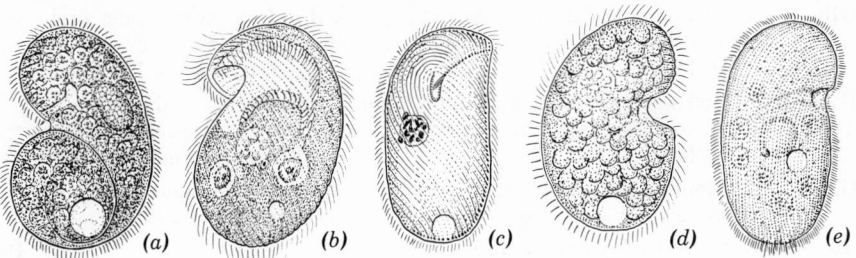

Fig. 10.13. (a) *Tillina magna* Gruber, 135 μ. × 200. (b) *Bresslaua vorax* Kahl, 90 μ. × 300. (c) *Bryophrya bavariensis* Kahl, 100 μ. × 260. (d) *Colpoda cucullus* O. F. Müller, 87 μ. × 300. (e) *Conchophthirus anodontae* (Ehrenberg), 135 μ. × 200. (b and c after Kahl.)

103 (93) Family **Colpodidae** **104**

104a (103) Large (around 200 μ); pharynx is long and curved; contractile vacuole lies in a posterodorsal protuberance; 5 species. (Fig. 10.13a) . . . . *Tillina* Gruber

104b Pharynx short; contractile vacuole not as above **105**

105a (104) Oral groove deeply excavated; gives cell hollow look in oral region; 2 species. (Fig. 10.13b) . . . . *Bresslaua* Kahl

105b Oral groove not greatly hollowed out. **106**

106a (105) Mouth mid-ventral; oral indentation is shallow and indistinct; 4 species. (Fig. 10.13c) . . . . *Bryophrya* Kahl

106b Mouth near lateral margin; oral notch noticeable to quite distinct; 17 species. (Fig. 10.13d). . . . . *Colpoda* Müller

107 (94) Family **Conchophthiridae,** with a single genus; 8 species. (Fig. 10.13e) . . . . *Conchophthirus* Stein

108 (93) Family **Plagiopylidae,** with only 1 certain fresh-water genus; 3 species. (Fig. 10.14a) . . . . *Plagiopyla* Stein

109 (87) Family **Trichopelmidae** **110**

110a (109) Mouth in front half of cell, with delicate pharyngeal trichites running inward from it **111**

110b Mouth at or posterior to middle of cell; no trichites evident in the pharynx **112**

111a (110) Cell nearly oval, with little indication of oral area at margin; 2 species. (Fig. 10.14b) . . . . *Pseudomicrothorax* Mermod

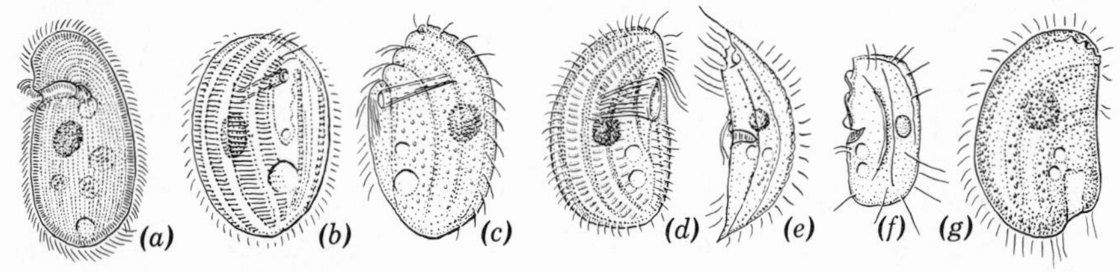

Fig. 10.14. (a) *Plagiopyla nasuta* Stein, 125 μ. × 160. (b) *Pseudomicrothorax agilis* Mermod, 54 μ. × 360. (c) *Trichopelma sphagnetorum* (Levander), 33 μ. × 600. (d) *T. euglenivora* (Kahl), 47 μ. × 400. (e) *Drepanomonas dentata* Fresenius, 50 μ. × 400. (f) *D. revoluta* Penard, 35 μ. × 400. (g) *Microthorax pusillus* Engelmann, 32 μ. × 600. (a, c, d, and f after Kahl; b and e after Penard.)

111b   Cell obliquely truncate or slightly beaked in front at one side; oral area more definitely visible at the cell margin; 5 species. (Fig. 10.14*c,d*) . . . . *Trichopelma* Levander

112a   (110) Cell semilunar or semicircular; mouth near middle of the concave or straighter side; aboral side strongly convex; 5 species. (Fig. 10.14*e,f*). . . . . *Drepanomonas* Fresenius

112b   Mouth near the posterior end of the cell on the straighter side; 14 species. (Fig. 10.14g) . . . . *Microthorax* Engelmann

113   (10) Suborder **Hymenostomina**   114

114a   (113) Mouth small, at cell surface, C-shaped; pharyngeal tube runs perpendicularly inward; small refractile body often visible near mouth . . . . Family **Ophryoglenidae**   150

114b   Not as just described.   115

115a   (114) Preoral peristomal furrow linear, running from front end to mouth, bordered on right by membrane   116

115b   No preoral furrow bordered by membrane, but often with membranes in mouth . . . . Family **Frontoniidae**   119

116a   (115) With double undulating membrane on right side of peristomal furrow . . . . Family **Cohnilembidae**   152

116b   Membrane at right of oral furrow not double   117

117a   (116) Bottom of peristomal furrow not ciliated; membrane curves around rear of mouth before entering; no internal ciliated pocket near mouth . . . . Family **Pleuronematidae**   153

117b   Bottom of preoral furrow ciliated; membrane runs directly into mouth, curving later; an internal ciliated pocket beside mouth. . . . . Family **Philasteridae**   151

119   (115) Family **Frontoniidae**   120

120a   (119) Mouth huge, occupying at least ⅔ of the ventral surface; 3 species. (Fig. 10.15*a*) . . . . *Lembadion* Perty

120b   Mouth not over ½ the cell length   121

121a   (121) Mouth ⅓ to ½ the cell length   122

121b   Mouth not exceeding ⅓ the cell length.   123

122a   (121) Front end tapering to a blunt point; rear end rounded; 1 species. (Fig. 10.15*b*) . . . . *Leucophrys* Ehrenberg

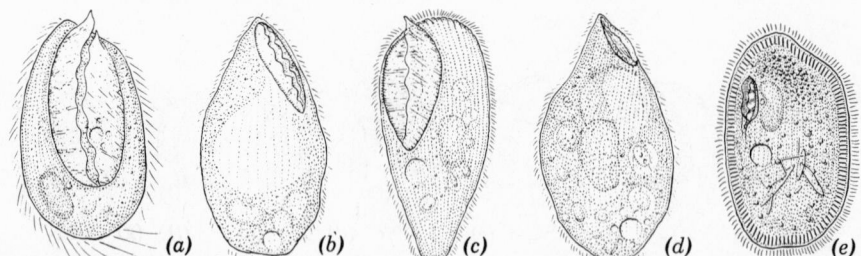

Fig. 10.15. (*a*) *Lembadion magnum* (Stokes), 62 μ. × 430. (*b*) *Leucophrys patula* Ehrenberg, 180 μ. × 160. (*c*) *Turania vitrea* Brodsky, 150 μ. × 200. (*d*) *Leucophrydium putrinum* Roux, 130 μ. × 230. (*e*) *Frontoniella complanata* Wetzel, 110 μ. × 235. (*c* after Brodsky; *d* after Roux; *e* after Wetzel.)

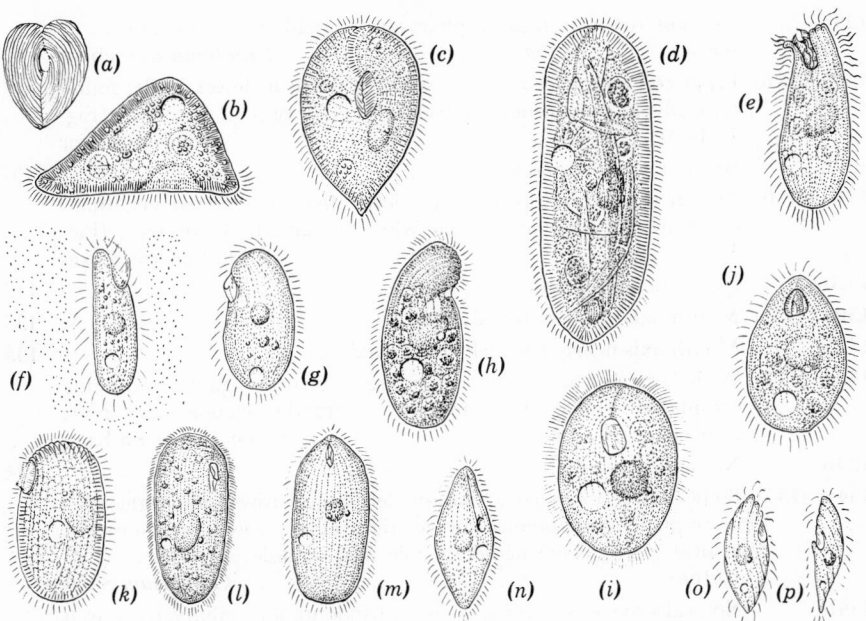

Fig. 10.16. (*a*) *Stokesia vernalis* Wenrich, ventral view, 150 μ. × 90. (*b*) Same, side view, 150 μ. × 170. (*c*) *Disematostoma bütschlii* Lauterborn, 150 μ. × 150. (*d*) *Frontonia leucas* Ehrenberg, 600 μ. × 60. (*e*) *Espejoia culex* (Smith), 50 μ. × 450. (*f*) *Cyrtolophosis mucicola* Stokes, 25 μ. × 350. (*g*) *Pseudoglaucoma muscorum* Kahl, 27 μ. × 600. (*h*) *Colpidium colpoda* (Ehrenberg), 100 μ. × 225. (*i*) *Glaucoma scintillans* Ehrenberg, 50 μ. × 420. (*j*) *Tetrahymena pyriformis* (Ehrenberg) 50 μ. × 400. (*k*) *Dichilum platessoides* Fauré-Fremiet, 135 μ. × 145. (*l*) *Malacophrys rotans* Kahl, 45 μ. × 450. (*m*) *Monochilum frontatum* Schew., 75 μ. × 280. (*n*) *Bizone parva* Lepsi, 54 μ. × 320. (*o*) *Aristerostoma minutum* Kahl, ventral view, 26 μ. × 550. (*p*) Same, side view. (*a, c, g, m, o*, and *p* after Kahl; *b* after Wenrich; *k* after Fauré-Fremiet; *l* after Schewiakoff; *n* after Lepsi.)

122b		Rear end of the cell tapering to a point; front end blunt; 1 species. (Fig. 10.15*c*) . . . . **Turania** Brodsky	
123a	**(121)**	One or more long caudal cilia present	**140**
123b		No such long caudal cilia	**124**
124a	**(123)**	Mouth outline pointed in front	**125**
124b		Mouth rounded or blunt in front	**130**
125a	**(124)**	Mouth about ¼ the body length, starting at front end; 1 species. (Fig. 10.15*d*) . . . . **Leucophrydium** Roux	
125b		Mouth back a little way from front end	**126**
126a	**(125)**	With heavy trichocyst layer; cell usually more than 150 μ.	**127**
126b		Trichocysts not evident; cell usually less than 120 μ	**130**
127a	**(126)**	With funnel-like pharynx leading from back of mouth into the cytoplasm; 1 species. (Fig. 10.15*e*) . . . . . **Frontoniella** Wetzel	
127b		With no such pharynx evident back of mouth	**128**
128a	**(127)**	Dorsal side low-conical; ventrally flat; preoral suture not well developed; 2 species. (Fig. 10.16*a,b*) . . . . . . **Stokesia** Wenrich	
128b		Body ellipsoid or somewhat flattened; preoral suture present, though hard to see	**129**
129a	**(128)**	With posterodorsal median suture; many trichites in pharynx; 2 species. (Fig. 10.16*c*) . . . . **Disematostoma** Lauterborn	

129b   No posterodorsal suture; pharyngeal trichites few or absent; 7 species. (Fig. 10.16*d*) . . . . **Frontonia** Ehrenberg

130a **(126)** Front end obliquely truncate, with mouth at its lower angle; found typically in jelly around insect and snail eggs; 2 species. (Fig. 10.16*e*) . . . . **Espejoia** Bürger

130b   Not as described above **131**

131a **(130)** Mouth near front, opening anteriorly; lives in secreted gelatinous case into which it jerks back when disturbed; 4 species. (Fig. 10.16*f*) . . . . **Cyrtolophosis** Stokes

131b   Not as described above **132**

132a **(131)** Mouth axis oblique to body axis **133**

132b   Mouth axis nearly parallel to body axis **135**

133a **(132)** With 1 membrane on left oral margin, beating against projecting ectoplasmic rim on right margin; contractile vacuole near rear; 2 species. (Fig. 10.16*g*) . . . . **Pseudoglaucoma** Kahl

133b   Not as described above **134**

134a **(133)** Body outline indented at mouth; left ciliary rows meet right ones at an angle along preoral suture; right ciliary rows curve outward at oral groove; contractile vacuole near middle; 2 species. (Fig. 10.16*h*) . . . . **Colpidium** Stein

134b   No oral groove and usually no oral indentation; ciliary rows arch smoothly over mouth, show no deflection on right, contractile vacuole about ⅔ of the way down the cell; 13 species. (Fig. 10.16*i*) . . . . **Glaucoma** Ehrenberg

135a **(132)** Mouth roughly triangular, with 1 membrane on right margin, 3 inside; 5 species. (Fig. 10.16*j*) . . . . **Tetrahymena** Furgason

135b   Mouth not as described above **136**

136a **(135)** Two membranes evident in mouth, 1 overarching the other in front like a hood; 1 species. (Fig. 10.16*k*) . . . .
                                  Dichilum Schewiakoff

136b   Only 1 membrane evident in the mouth **137**

137a **(136)** Mouth narrow, slitlike, only a little way down from anterior pole; 2 species. (Fig. 10.16*l*) . . . . **Malacophrys** Kahl

137b   Mouth some distance down from anterior pole **138**

138a **(137)** With pharyngeal funnel extending back from mouth; body oval or ellipsoid; 5 species. (Fig. 10.16*m*) . . . **Monochilum** Schewiakoff

138b   No pharyngeal funnel evident; both ends pointed **139**

139a **(138)** With inconspicuous preoral groove; contractile vacuole near middle; 1 species. (Fig. 10.16*n*). . . . . **Bizone** Lepsi

139b   No preoral groove; contractile vacuole near end of cell; 2 species. (Fig. 10.16*o,p*) . . . . **Aristerostoma** Kahl

140a **(123)** Cilia in 1 or more zones encircling the cell. **141**

140b   Cilia not in girdles; usually in longitudinal rows. **142**

141a **(140)** With 1 zone of short rows of cilia encircling the cell at middle; front and rear ends bare except for the long caudal cilium; 1 species. (Fig. 10.17*a*) . . . . **Urozona** Schewiakoff

141b   Two encircling bands of cilia, with oral groove between; swim revolving on a mucus thread spun from tuft of caudal cilia; 1 species. (Fig. 10.17*b*) . . . . **Urocentrum** Nitzsch

142a **(140)** Mouth at or in front of the middle **143**

142b		Mouth in rear half of the cell	**149**
143a	(142)	Body slightly compressed laterally; mouth about halfway down the cell and indented slightly	**144**
143b		Body round in cross section or dorsoventrally flattened	**146**

144a (143) Anterior pole with small, unciliated frontal plate; narrow preoral groove leads from the frontal plate to mouth at middle; 4 species. (Fig. 10.17c) . . . . *Uronema* Dujardin

144b No frontal plate, no preoral groove. **145**

145a (144) Membranes of mouth not closed behind to form a pocket; the contractile vacuole lies at the posterior end; 1 species. (Fig. 10.17d) . . . . *Dexiotrichides* Kahl

145b Oral membranes closed behind to form a pocket, contractile vacuole located a little distance forward from the rear end; 1 species. (Fig. 10.17e) . . . . *Uronemopsis* Kahl

146a (143) Body considerably flattened dorsoventrally. **147**

146b Body only slightly flattened dorsoventrally **148**

147a (146) Oral membranes meet at rear margin of mouth to form a pocket; 7 species. (Fig. 10.17f) . . . . *Saprophilus* Stokes

147b Mouth small; 1 oral membrane which does not form a pocket at mouth; 2 species. (Fig. 10.17g) . . . . . . *Platynematum* Kahl

148a (146) Both ends of cell with unciliated terminal projections; 2 species. (Fig. 10.17h) . . . . *Balanonema* Kahl

148b With bare frontal plate only at anterior end; no posterior projection; 9 species. (Fig. 10.17i) . . . . . . *Loxocephalus* Eberhard

149a (142) Body flattened; mouth lateroposterior, receiving oral groove from rear; 2 species. (Fig. 10.17j) . . . . *Cinetochilum* Perty

149b Body spindle-shaped; with anterior unciliated frontal plate; 1 species. (Fig. 10.17k) . . . . *Homalogastra* Kahl

150 (114) Family **Ophryoglenidae**, including only 1 genus; 12 species. (Fig. 10.18a) . . . . *Ophryoglena* Ehrenberg

151 (117) Family **Philasteridae**, including only 1 fresh-water genus; 1 species. (Fig. 10.18b) . . . . *Philasterides* Kahl

152 (116) Family **Cohnilembidae**, including 1 fresh-water genus; 7 species. (Fig. 10.18c) . . . . *Cohnilembus* Kahl

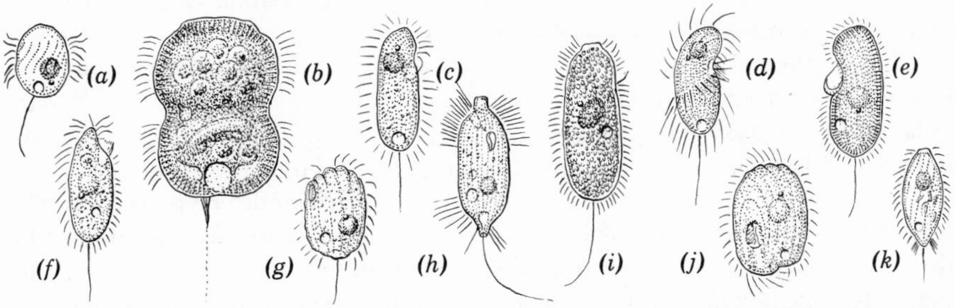

Fig. 10.17. (a) *Urozona bütschlii* Schewiakoff, 23 μ. × 350. (b) *Urocentrum turbo* (O. F. Müller), 78 μ. × 250. (c) *Uronema griseolum* (Maupas), 42 μ. × 320. (d) *Dexiotrichides centralis* (Stokes), 35 μ. × 340. (e) *Uronemopsis kenti* (Kahl), 80 μ. × 180. (f) *Saprophilus agitatus* Stokes, 40 μ. × 310. (g) *Platynematum sociale* (Penard), 40 μ. × 250. (h) *Balanonema biceps* (Penard), 44 μ. × 350. (i) *Loxocephalus plagius* (Stokes), 54 μ. × 320. (j) *Cinetochilum margaritaceum* Perty, 38 μ. × 320. (k) *Homalogastra setosa* Kahl, 30 μ. × 350. (d, e, g, and k after Kahl; h after Penard.)

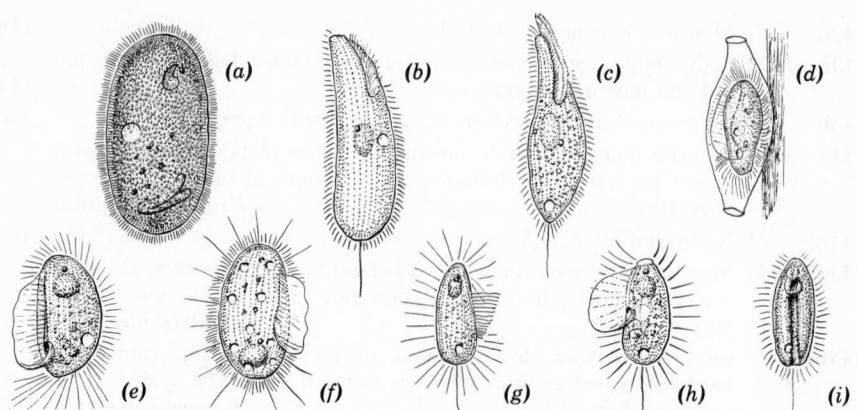

Fig. 10.18. (a) Ophryoglena atra Lieberkühn, 285 μ. × 85. (b) Philasterides armata (Kahl), 75 μ. × 335. (c) Cohnilembus pusillus (Quennerstedt), 47 μ. × 500. (d) Calyptotricha pleuronemoides Phillips, 25 μ. × 470. (e) Pleuronema coronatum Kent, 55 μ. × 275. (f) Histiobalantium natans (Claparède and Lachmann), 68 μ. × 240. (g) Ctedoctema acanthocrypta Stokes, 25 μ. × 500. (h) Cyclidium glaucoma O. F. Müller, 28 μ. × 460. (i) Cristigera phoenix Penard, 45 μ. × 290. (b after Kahl; d after Phillips; f and g after Stokes.)

153   (117) Family **Pleuronematidae**. **154**

154a (153) Living in a case which it secretes; about 50 μ long; 1 species. (Fig. 10.18d) . . . .   ***Calyptotricha*** Phillips

154b   Not case-dwelling; size various **155**

155a (154) Large forms, from 70–180 μ in length **156**

155b   Smaller species, less than 70 μ in length **157**

156a (155) One contractile vacuole; long tactile cilia at rear only; 4 species. (Fig. 10.18e) . . . .   ***Pleuronema*** Dujardin

156b   Many contractile vacuoles; long cilia interspersed all over body; 2 species. (Fig. 10.18f) . . . . ***Histiobalantium*** Stokes

157a (155) Membrane on right margin of preoral groove swings around rear of mouth to form a pocket. **158**

157b   Preoral groove slightly oblique; membrane low at mouth end; no pocket; 2 species. (Fig. 10.18g) . . . . ***Ctedoctema*** Stokes

158a (157) The preoral peristomal furrow ends posteriorly at mouth; 22 species. (Fig. 10.18h) . . . . ***Cyclidium*** O. F. Müller

158b   Peristomal furrow continues back of mouth to end of cell; 5 species. (Fig. 10.18i) . . . . ***Cristigera*** Roux

159   (5) Order **Spirotrichida** **160**

160a (159) No locomotor cirri present **161**

160b   Locomotor cirri on ventral surface; also cilia in some genera . . . . Suborder **Hypotrichina** **205**

161a (160) Body surface ciliated . . . . Suborder **Heterotrichina** **163**

161b   Cilia usually sparse or absent on the body **162**

162a (161) Body flattened, rigid, often with bizarre excisions or spurs; 8 oral membranelles . . . . Suborder **Ctenostomina** **196**

162b   Body circular in cross section; membranelles numerous in circle at anterior end . . . . Suborder **Oligotrichina** **186**

163 (161) Suborder **Heterotrichina** **164**

164a (163) Peristome deeply sunken into cell . . . . . Family **Bursariidae** 183
164b Peristome on surface level, not deeply sunken in 165
165a (165) Peristome linear, narrow; straight or spiralling 167
165b Peristome broad; often roughly circular in outline 166
166a (165) Peristome unciliated and with a well-developed membrane on its right margin . . . . Family **Condylostomidae** 180
166b Peristome ciliated; no membrane on right margin . . . . Family **Stentoridae** 181
167a (165) Peristome oblique or spiral . . . . Family **Metopidae** 169
167b Peristome longitudinal, parallel to body axis 168
168a (167) No visible pharyngeal cavity at mouth end of adoral zone except when feeding . . . . Family **Reichenowellidae** 173
168b Persistent pharyngeal cavity at mouth end of the adoral zone of membranelles . . . . Family **Spirostomidae** 175
169 (167) Family **Metopidae** 170
170a (169) General body ciliation fairly uniform 171
170b Body ciliation limited to the spiral peristome region; long tail spine; 7 species. (Fig. 10.19*a*). . . . . ***Caenomorpha*** Perty
171a (170) Body rigid; pellicle thrown up into several spiral keels; 1 species. (Fig. 10.19*b*) . . . . ***Tropidoatractus*** Levander
171b Body fairly flexible, without rigid keels on pellicle 172
172a (171) Body somewhat compressed; peristome ventral, oblique, ⅓ cell width; 2 species. (Fig. 10.19*c*) . . . . ***Bryometopus*** Kahl
172b Body twisted; peristome narrow, running backward spirally; 49 species. (Fig. 10.19*d*) . . . . . ***Metopus*** Claparède and Lachmann
173 (168) Family **Reichenowellidae** 174
174a (173) Body elongate; with a single terminal contractile vacuole; 1 species. (Fig. 10.19*e*) . . . . ***Reichenowella*** Kahl
174b Body oval; with 2 or more contractile vacuoles; 2 species. (Fig. 10.19*f*) . . . . ***Balantidioides*** Penard
175 (168) Family **Spirostomidae**. 176
176a (175) Small (less than 60 μ); with short peristome and few membranelles; 1 species. (Fig. 10.20*b*). . . . . ***Protocrucia*** Da Cunha

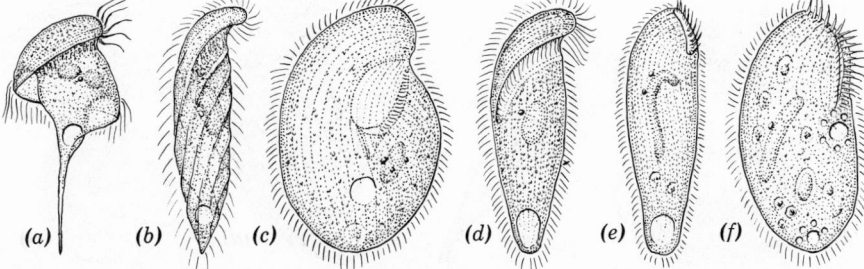

Fig. 10.19. (*a*) *Caenomorpha medusula* Perty, 87 μ. × 340. (*b*) *Tropidoatractus acuminatus* Levander, 100 μ. × 300. (*c*) *Bryometopus sphagni* (Penard), 80 μ. × 375. (*d*) *Metopus es* (O. F. Müller), 135 μ. × 220. (*e*) *Reichenowella nigricans* Kahl, 250 μ. × 120. (*f*) *Balantidioides bivacuolata* Kahl, 100 μ. × 300. (*c* after Penard; *e* and *f* after Kahl.)

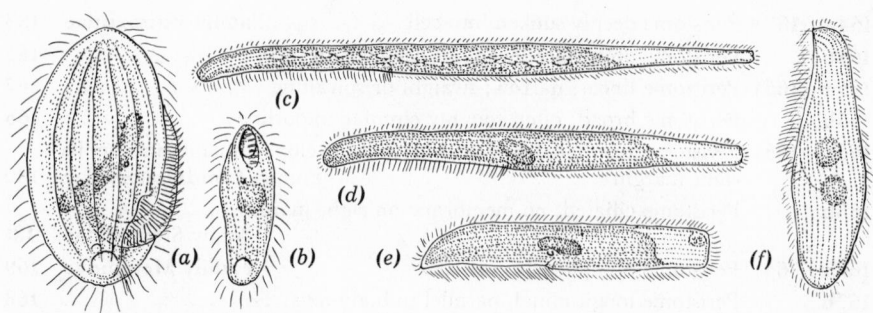

Fig. 10.20. (*a*) *Phacodinium metchnikoffi* (Certes), 100 μ. × 315. (*b*) *Protocrucia sp.* (from soil and dung cultures), 35 μ. × 550. (*c*) *Spirostomum minus* Roux, 900 μ. × 75. (*d*) *S. teres* Claparède and Lachmann, 500 μ. × 100. (*e*) *Pseudoblepharisma crassum* Kahl, 200 μ. × 180. (*f*) *Blepharisma lateritum* (Ehrenberg), 180 μ. × 200. (*a* and *e* after Kahl.)

176b		Larger with long peristome and many membranelles	**177**
177a	**(176)**	Body flattened, rigid; longitudinal ridges on dorsal surface; 1 species. (Fig. 10.20*a*) . . . . **Phacodinium** Prowazek	
177b		Body more elongate, flexible, no marked ridges	**178**
178a	**(177)**	With well-developed undulating membrane on right side of peristome; 12 species. (Fig. 10.20*f*). . . . . . **Blepharisma** Perty	
178b		No undulating membrane on right side of peristome	**179**
179a	**(178)**	Very elongate, wormlike, highly contractile ciliates; 7 species. (Fig. 10.20*c,d*) . . . . **Spirostomum** Ehrenberg	
179b		Moderately elongate, but not very contractile ciliates; 2 species. (Fig. 10.20*e*) . . . . **Pseudoblepharisma** Kahl	
180	**(166)**	Family **Condylostomidae**. With 1 genus, mostly marine; 3 fresh-water species. (Fig. 10.21*a*) . . . . **Condylostoma** Bory	
181	**(166)**	Family **Stentoridae**	**182**
182a	**(181)**	Body flattened; peristome ventral; contractile vacuole posterior; 3 species. (Fig. 10.21*b*) . . . . **Climacostomum** Stein	
182b		Body trumpet-shaped; peristome at free end; contractile vacuole near mouth; 12 species. (Fig. 10.21*c*). . . . . **Stentor** Oken	
183	**(164)**	Family **Bursariidae**	**184**

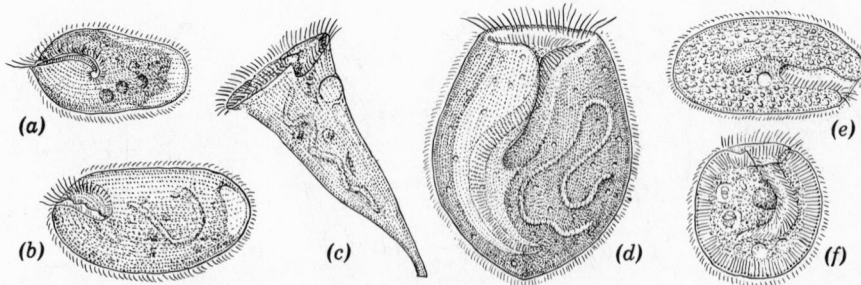

Fig. 10.21. (*a*) *Condylostoma tardum* Penard, 200 μ. × 100. (*b*) *Climacostomum virens* (Ehrenberg), 280 μ. × 90. (*c*) *Stentor roeseli* Ehrenberg, 540 μ half extended. × 60. (*d*) *Bursaria truncatella* O. F. Müller, 775 μ. × 40. (*e*) *Thylacidium truncatum* Schewiakoff, 80 μ. × 300. (*f*) *Bursaridium schewakoffi* Lauterborn, 250 μ. × 65. (*a* after Penard; *e* after Roux; *f* after Lauterborn.)

184a (183) Peristomal funnel turns left posteriorly 185

184b Peristome continues toward animal's right at its posterior end; 3 species. (Fig. 10.21*f*) . . . . **Bursaridium** Lauterborn

185a (185) Large (500 μ and over); peristome partly divided by longitudinal fold; 1 species. (Fig. 10.21*d*) . . . . **Bursaria** O. F. Müller

185b Smaller (100 μ and less); with undivided peristomal cavity; 1 species. (Fig. 10.21*e*) . . . . **Thylacidium** Schewiakoff

186 (162) Suborder **Oligotrichina**. 187

187a (186) Case-bearing species . . . . Family **Tintinnidae** 192

187b Naked species 188

188a (187) Adoral zone of membranelles closed, forming a circle with the mouth inside it. . . . . Family **Strobilidiidae** 191

188b Adoral zone not closed, the mouth breaking the circle at the margin . . . . Family **Halteriidae** 189

189 (188) Family **Halteriidae**. 190

190a (189) No tactile cilia; 4 species. (Fig. 10.22*a*) . . . .
Strombidium Claparède and Lachmann

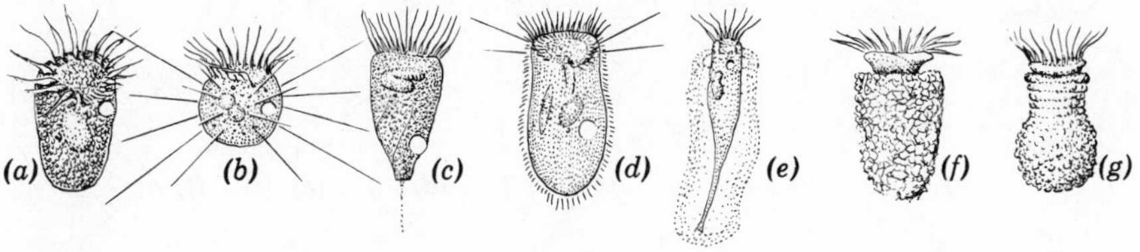

Fig. 10.22. (*a*) *Strombidium viride* Stein, 105 μ. × 125. (*b*) *Halteria grandinella* (O. F. Müller), 27 μ. × 300. (*c*) *Strobilidium gyrans* (Stokes), 60 μ. × 200. (*d*) *Strombidinopsis setigera* Stokes, 80 μ. × 200. (*e*) *Tintinnidium fluviatile* Stein, 180 μ. × 100. (*f*) *Tintinnopsis cylindrata* Kofoid and Campbell, 45 μ. × 125. (*g*) *Codonella cratera* (Leidy), 63 μ. × 150. (*e* after Entz.)

190b Lateral body surface with long tactile cilia; can dart off very swiftly; 5 species. (Fig. 10.22*b*) . . . . **Halteria** Dujardin

191 (188) Family **Strobilidiidae.** With only 1 genus; 4 species. (Fig. 10.22*c*) . . . . **Strobilidium** Schewiakoff

192 (187) Family **Tintinnidae**. 193

193a (192) Case delicate, gelatinous or mucoid, with some attached particles; sometimes found without a case. 194

193b Case firmer, pseudochitinous, with adherent or incorporated foreign particles 195

194a (193) Case very delicate or even absent; body surface ciliated; 2 species. (Fig. 10.22*d*) . . . . **Strombidinopsis** Kent

194b Case more definite; only narrow ciliary band below adoral zone; 2 species. (Fig. 10.22*e*) . . . . **Tintinnidium** Stein

195a (193) Case cylindrical with no neck portion at the open end; 4 fresh-water species. (Fig. 10.22*f*) . . . . **Tintinnopsis** Stein

195b Case vaselike, with bulk and neck; 3 fresh-water species. (Fig. 10.22*g*) . . . . **Codonella** Haeckel

196 (162) Suborder **Ctenostomina**. 197

197a (196) Anterior row of cilia on left side; 4 short rows posteriorly on left side; at least 2 short rows posteriorly on right side . . . .
Family **Epalcidae** 199

197b No anterior row on left side; no posterior rows on right; cilia on left long and in cirruslike groups 198

198a (197) Middle ciliary band long, extending far over on both right and left sides . . . . Family **Discomorphidae** 204

198b Middle ciliary band short   . . . . Family **Myelostomidae** 202

199 Family **Epalcidae** 200

200a (199) Two short posterior rows of cilia on right side; no intermediate rows; 1 species. (Fig. 10.23*a*) . . . . . .   *Pelodinium* Lauterborn

200b Four rows of cilia on right side 201

201a (200) Some of posterior points bear thorns set off by constrictions; 6 species. (Fig. 10.23*b*) . . . . *Saprodinium* Lauterborn

201b No such thorns set off; 7 species. (Fig. 10.23*c*) . . . .
Epalxis Roux

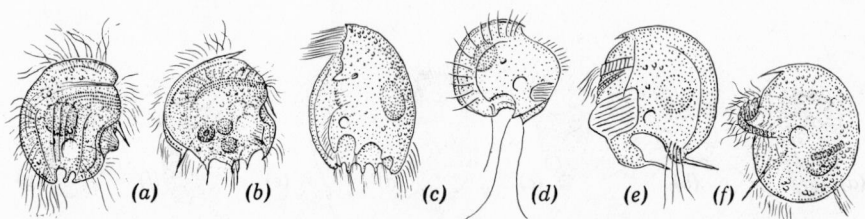

Fig. 10.23. (*a*) *Pelodinium reniforme* Lauterborn, 45 μ. × 330. (*b*) *Saprodinium dentatum* Lauterborn, 70 μ. × 200. (*c*) *Epalxis mirabilis* Roux, 40 μ. × 440. (*d*) *Myelostoma flagellatum* (Penard), 25 μ. × 500. (*e*) *Atopodinium fibulatum* Kahl, 45 μ. × 390. (*f*) *Discomorpha pectinata* Levander, 75 μ. × 225. (*a, b,* and *f* after Wetzel; *c* after Roux; *d* after Penard; *e* after Kahl.)

202 (198) Family **Myelostomidae** 203

203a (202) Rear border shows no excisions or only one; 6 species. (Fig. 10.23*d*) . . . . *Myelostoma* Kahl

203b Two such excisions; 1 species. (Fig. 10.23*e*) . . *Atopodinium* Kahl

204 (198) Family **Discomorphidae**. With only 1 genus; 1 species. (Fig. 10.23*f*) . . . . *Discomorpha* Poche

205 (160) Suborder **Hypotrichina** 206

206a (205) Adoral zone of membranelles well developed and easily seen; dorsal tactile bristles often present 207

206b Adoral membranelles few and concealed in lateral groove; no dorsal bristles . . . . Family **Aspidiscidae** 241

207a (206) Longitudinal rows of cirri or cilia commonly present, especially marginal rows . . . . Family **Oxytrichidae** 208

207b Usually no longitudinal rows; frontal, anal, and caudal groups of cirri well developed . . . . Family **Euplotidae** 240

208 (207) Family **Oxytrichidae** 209

209a (208) Anal (transverse) cirri absent 210

209b Anal (transverse) cirri present (sometimes hard to see) 219

CILIOPHORA 287

210a (209) Ventral and marginal rows of cirri run spirally **216**
210b These rows straight or but slightly oblique **211**
211a (210) Three strong frontal cirri somewhat set off from the others **215**
211b Not as above; frontal cirri but little larger than the ventral cirri, and often in continuous rows with them **212**
212a (211) Small (50–100 μ); not elongated; cirri long, sparse, no special frontals; 2 species. (Fig. 10.24a) . . . . *Psilotricha* Stein
212b Long elliptical, frontal cirri differentiated **213**
213a (212) All frontal cirri in short oblique rows (front right to rear left); 1 species. (Fig. 10.24b) . . . . *Eschaneustyla* Stokes
213b Frontal cirri in rows that curve to the left toward the peristome as they run forward **214**
214a (213) Two rows of ventral cirri; 2 species. (Fig. 10.24c) . . . . *Paraholosticha* Kahl
214b Numerous rows; 2 species. (Fig. 10.24d) . *Hemicycliostyla* Stokes

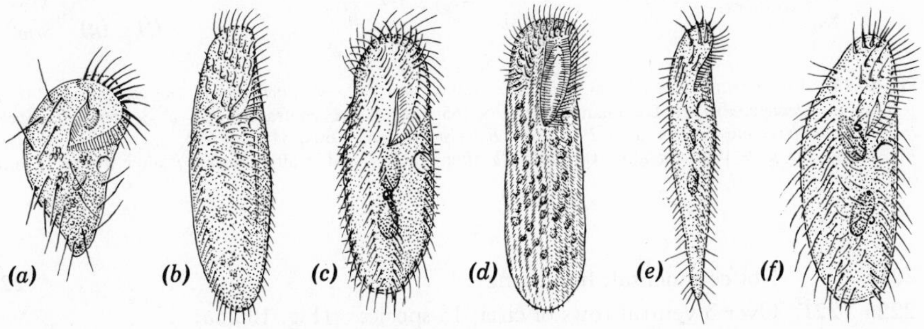

Fig. 10.24. (a) *Psilotricha acuminata* Stein, 90 μ. × 230. (b) *Eschaneustyla brachytona* Stokes, 190 μ. × 170. (c) *Paraholosticha herbicola* Kahl, 170 μ. × 175. (d) *Hemicycliostyla trichota* Stokes, 420 μ. × 80. (e) *Uroleptus piscis* (O. F. Müller), 200 μ. × 155. (f) *Kahlia acrobates* Horvath, 150 μ. × 200. (a after Stein; b and d after Stokes; c after Kahl; f after Horvath.)

215a (211) Elongate, tapering posteriorly; usually 2 rows of ventral cirri; 15 species. (Fig. 10.24e) . . . . *Uroleptus* Ehrenberg
215b Long oval, with 5 to 8 rows of ventral cirri, similar to the marginal cirri; 2 species. (Fig. 10.24f) . . . . *Kahlia* Horvath
216a (210) Plump, planktonic forms, with broad frontal area, pointed rear end; 1 species. (Fig. 10.25a) . . . . . *Hypotrichidium* Ilowaisky
216b Slender forms with narrowed front end **217**
217a (216) Front end narrowed, but not drawn out into a long and slender form; 8 species. (Fig. 10.25b) . . . . *Strongylidium* Sterki
217b Front end long and slender, over ¼ body length **218**
218a (217) Attenuated front end markedly extensile and contractile; 4 species. (Fig. 10.25c, d) . . . . *Chaetospira* Lachmann
218b Not contractile; 5 species. (Fig. 10.25e) . . . *Stichotricha* Perty
219a (209) Anterior cirri in ventral rows enlarged a little to form frontals but no separate specialized frontals **220**

219b   With separate specialized frontal cirri, the anterior 3 as a rule
             being especially large **224**

220a (219) Small (35–80 μ); long, widely spaced cirri; 2 marginal, 1 ventral
             row; 5 species. (Fig. 10.25f) . . . . *Balladyna* Kowalewsky

220b   Not as described above **221**

221a (220) Commensal on hydra; 1 species. (Fig. 10.25g) . . . .
                                                            Kerona Ehrenberg

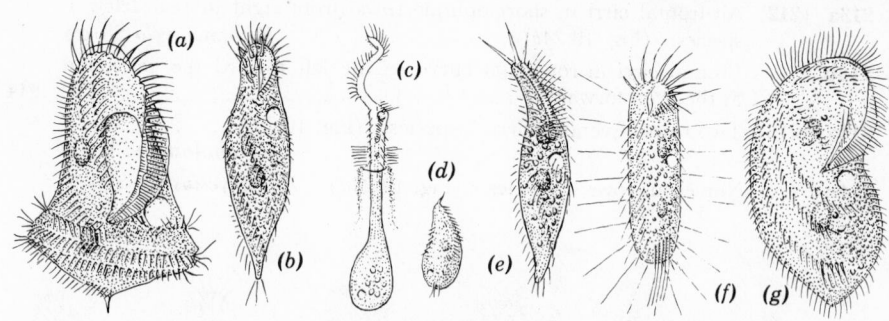

Fig. 10.25. (a) *Hypotrichidium conicum* Ilowaisky, 120 μ. × 250. (b) *Strongylidium crassum* Sterki, 145 μ.
× 200. (c) *Chaetospira mülleri* Lachmann, 200 μ. × 165. (d) Same, contracted, 67 μ. × 165. (e) *Stichotricha aculeata* Wrzesniowski, 94 μ. × 290. (f) *Balladyna elongata* Roux, 34 μ. × 650. (g) *Kerona polyporum*
Ehrenberg, 165 μ. × 180. (a after Ilowaisky; b after Kahl; c and d after Sterki; f after Roux; g after
Stein.)

221b   Not commensal; free-living **222**

222a (221) Over 3 ventral rows of cirri; 15 species. (Fig. 10.26a) . . . .
                                                            Urostyla Ehrenberg

222b   One to 3 ventral rows of cirri **223**

223a (222) Three rows of ventral cirri; 2 species. (Fig. 10.26b) . . . .
                                                            Trichotaxis Stokes

223b   Two rows of ventral cirri; 8 species. (Fig. 10.26c) . . . .
                                                            Keronopsis Penard

224a (219) Ventral cirri all in rows, not large and specialized **225**

224b   Some ventrals in groups, often large and specialized **228**

225a (224) Over 3 rows of ventral cirri; 15 species. (Fig. 10.26a) . . . .
                                                            Urostyla Ehrenberg

225b   One to 3 rows of ventral cirri **226**

226a (225) With tail at rear; 5 species. (Fig. 10.26d) . . . . *Paruroleptus* Kahl

226b   Body not drawn out posteriorly into a tail **227**

227a (226) With only 1 row of ventral cirri; 1 species in fresh water. (Fig.
             10.26e) . . . . *Amphisiella* Gourret and Roeser

227b   Two rows of ventral cirri; 10 species. (Fig. 10.26f) . . . .
                                                            Holosticha Wrzesniowski

228a (224) Some or all ventral cirri in continuous long rows; postoral and
             posterior cirrus groups present **229**

228b   No continuous long rows of cirri; only groups **232**

229a (228) Cirrus rows parallel to longitudinal axis of cell **230**

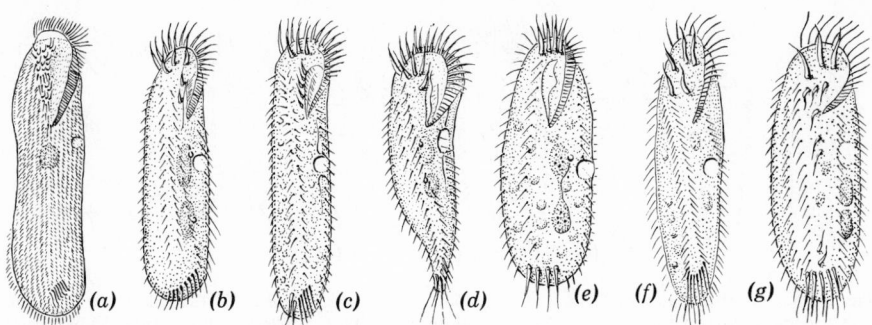

Fig. 10.26. (a) *Urostyla grandis* Ehrenberg, 350 μ. × 100. (b) *Trichotaxis fossicola* Kahl, 160 μ. × 190. (c) *Keronopsis muscorum* Kahl, 250 μ. × 135. (d) *Paruroleptus magnificus* Kahl, 430 μ. × 70. (e) *Amphisiella oblonga* Schewiakoff, 160 μ. × 190. (f) *Holosticha vernalis* Stokes, 180 μ. × 150. (g) *Onychodromopsis flexilis* Stokes, 100 μ. × 310. (a after Stein; b, c, and d after Kahl; e after Schewiakoff; f and g after Stokes.)

229b	Rows of cirri definitely oblique to longitudinal axis	**231**
230a (229)	The 5 anal cirri are uniform and in 1 row; 1 species. (Fig. 10.26*g*) . . . . ***Onychodromopsis*** Stokes	
230b	The 2 right anal cirri are larger and more posterior in position; 2 species. (Fig. 10.27*a*) . . . . ***Pleurotricha*** Stein	
231a (229)	A long oblique row of cirri cuts across ventral surface; 3 species. (Fig. 10.27*b*) . . . . ***Gastrostyla*** Engelmann	
231b	Rows of ventral cirri short, rarely extending beyond mouth; 2 species. (Fig. 10.27*c*) . . . . ***Gonostomum*** Sterki	
232a (228)	Rear end of cell bears a long, very contractile stalk; 1 species. (Fig. 10.27*d*) . . . . ***Ancystropodium*** Fauré-Fremiet	
232b	Rear end without such a stalk.	**233**
233a (232)	With 12 to 15 strong cirri on the frontal field; 4 macronuclei; 1 species. (Fig. 10.27*e*) . . . . ***Onychodromus*** Stein	
233b	Fewer frontal cirri, usually 8 in 3 groups; nearly always 2 macronuclei, rarely 1 or 4	**234**
234a (233)	With the posterior end of the cell tapering to form a tail-like extremity; 5 species. (Fig. 10.27*f*) . . . . . ***Urosoma*** Kowalewsky	
234b	Posterior end not attenuated	**235**

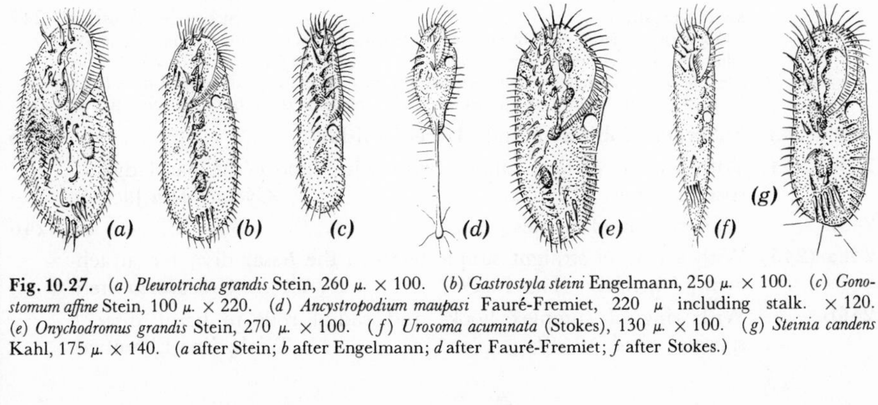

Fig. 10.27. (a) *Pleurotricha grandis* Stein, 260 μ. × 100. (b) *Gastrostyla steini* Engelmann, 250 μ. × 100. (c) *Gonostomum affine* Stein, 100 μ. × 220. (d) *Ancystropodium maupasi* Fauré-Fremiet, 220 μ including stalk. × 120. (e) *Onychodromus grandis* Stein, 270 μ. × 100. (f) *Urosoma acuminata* (Stokes), 130 μ. × 100. (g) *Steinia candens* Kahl, 175 μ. × 140. (a after Stein; b after Engelmann; d after Fauré-Fremiet; f after Stokes.)

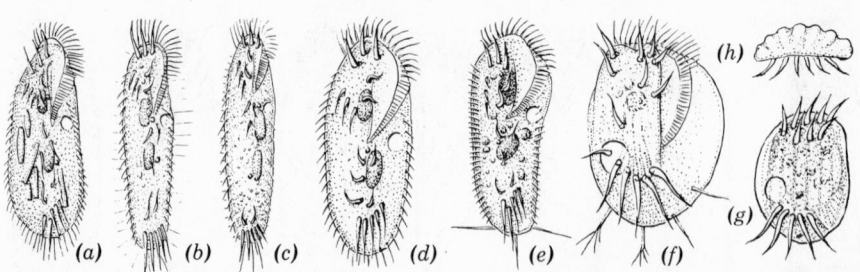

Fig. 10.28. (a) *Oxytricha fallax* Stein, 150 μ. × 160. (b) *Tachysoma pellionella* (O. F. Müller), 100 μ. × 250. (c) *Opisthotricha procera* Kahl, 110 μ. × 225. (d) *Histrio histrio* (O. F. Müller), 160 μ. × 165. (e) *Stylonychia mytilus* Ehrenberg, 150 μ. × 160. (f) *Euplotes patella* (O. F. Müller), 118 μ. × 190. (g) *Aspidisca costata* (Dujardin), 32 μ. × 500. (h) Same, end view, showing dorsal ridges of this species. (a after Stein; b after Roux; c after Kahl.)

235a (234) Right border of peristome curves left, or spirals into a pit in peristome; 9 species. (Fig. 10.27g) . . . . **Steinia** Diesing

235b Right border of peristome curved but little or none 236

236a (235) Cell quite flexible, bends easily to right or left 237

236b Cell stiff, no right-left, slight dorsoventral bending 239

237a (239) Right and left marginal cirri meet as continuous row at rear end; 11 species. (Fig. 10.28a). . . . . **Oxytricha** Ehrenberg

237b Marginal cirri interrupted at rear, near caudal cirri. 238

238a (237) No caudals at rear; 5 species. (Fig. 10.28b) . . . **Tachysoma** Stokes

238b Caudals present; 9 species. (Fig. 10.28c) . . . **Opisthotricha** Kent

239a (236) Right and left marginal cirri meet as continuous row at rear end; 7 species. (Fig. 10.28d) . . . . **Histrio** Sterki

239b Marginal cirri interrupted at rear near the caudal cirri; 12 species. (Fig. 10.28e) . . . . **Stylonychia** Ehrenberg

240 (207) Family **Euplotidae;** only 1 genus in fresh water; 8 species. (Fig. 10.28f) . . . . **Euplotes** Ehrenberg

241 (206) Family **Aspidiscidae;** 7 species. (Fig. 10.28g, h) . . . . **Aspidisca** Ehrenberg

242 (7) Order **Peritricha** 243

243a (242) Ectozoic, with aboral ring of cilia used for gliding about on surface of host . . . . Suborder **Mobilia** 244

243b Sessile or (rarely) free-swimming though many possess a motile swarmer stage . . . . Suborder **Sessilia** 247

   Apparently a few species of vorticellids remain permanently free swimming in the telotroch stage, as, for instance, *Opisthonecta henneguyi* Fauré-Fremiet (Fig. 10.29a), but since the possibility remains that a sessile stage may sometime be discovered these forms have not been included in the present key. For more on this subject see Kahl, p. 663.

244 (243) Suborder **Mobilia;** Family **Urceolariidae** 245

245a (244) With a circle of long ciliary bristles just above the basal disc; 7 species. (Fig. 10.29b, c) . . . . **Cyclochaeta** Jackson

245b No such ring of bristles 246

246a (245) With a ring of straight simple teeth in the basal disc, for attachment; 4 species. (Fig. 10.29d, e) . . . . **Urceolaria** Stein

246b With ring of complex hooks and radial rods in basal disc; 8 species. (Fig. 10.29f, g) . . . . **Trichodina** Ehrenberg

247 (243) Suborder **Sessilia** **248**
248a (247) Cell not protected by secreted case   . . . .   Tribe **Aloricata**   249
248b   Rigid secreted case present . . . .   Tribe **Loricata**   271
249 (248) Tribe **Aloricata** **250**
250a (249) Rear end drawn out into a point bearing 1 or 2 bristlelike
            spines . . . .   Family **Astylozoonidae** 254
250b   Rear end attached directly or by stalk to substratum **251**
251a (250) Basal end of body bulblike, oral end lengthened; contractile
            vacuole near middle of cell and connecting with vestibule by
            long canal . . . .   Family **Ophrydiidae** 270
251b   Not as described above **252**
252a (251) Stalk absent or very short; body tapers basally serving as a
            stalk . . . .   Family **Scyphidiidae** 257
252b   With definite stalk secreted by aboral tip (scopula) **253**
253a (252) Stalk contractile, containing a specialized contractile band
            (spasmoneme) . . . .   Family **Vorticellidae** 266
253b   Stalk not contractile . . . .   Family **Epistylidae** 260
254 (250) Family **Astylozoonidae** **255**
255a (254) Surface of cell below peristome smooth or annulate, but without
            spinelike processes   256
255b   Pellicle below peristome bears 2 to 4 rows of large spinelike
            processes; 2 species. (Fig. 10.29*h*). . . . . . .   ***Hastatella*** Erlanger
256a (255) Without a secreted gelatinous layer on the cell surface; 4 species.
            (Fig. 10.29*i*) . . . .   ***Astylozoon*** Engelmann
256b   With this layer; 1 species. (Fig. 10.29*j*) . . . . .   ***Geleiella*** Stiller
257 (252) Family **Scyphidiidae** **258**
258a (257) Ectozoic on fishes; nucleus an inverted rounded cone; 5 species.
            (Fig. 10.30*a*). . . . .   ***Glossatella*** Bütschli
258b   Not as described above **259**

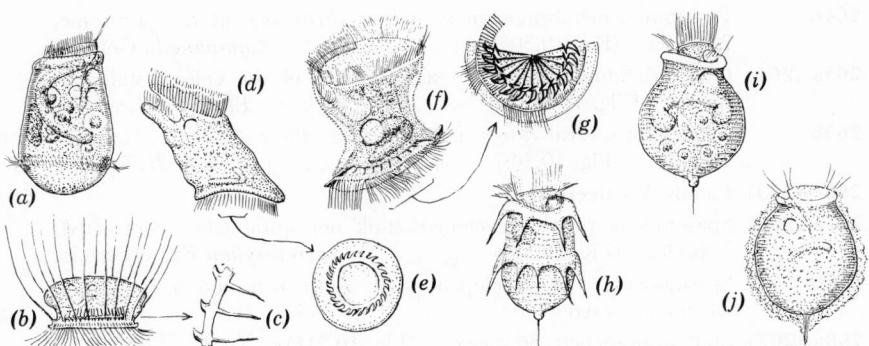

Fig. 10.29. (*a*) *Opisthonecta henneguyi* Fauré-Fremiet, 160 μ. × 115. (*b*) *Cyclochaeta spongillae* Jackson, 60 μ. × 215. (*c*) Same, hooks of basal disc. (*d*) *Urceolaria mitra* (Siebold), 120 μ. × 180. (*e*) Same, hooks of basal disc. (*f*) *Trichodina pediculus* Ehrenberg, 60 μ. × 300. (*g*) Same, hooks of basal disc. (*h*) *Hastatella radians* Erlanger, 40 μ. × 400. (*i*) *Astylozoon faurei* Kahl, 40 μ. × 450. (*j*) *Geleiella vagans* Stiller, presumably about 40 μ. × 400. (*a*, *h*, and *i* after Fauré-Fremiet; *b* and *c* after Jackson; *d* and *e* after Claparède and Lachmann; *f* and *g* after James-Clark; *i* from Kahl after Stiller.)

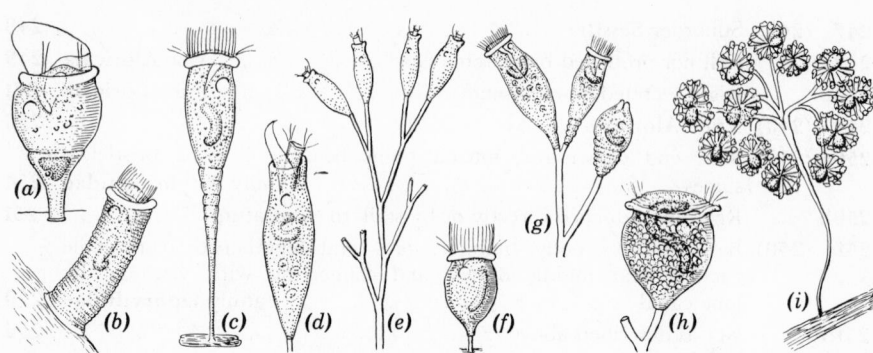

Fig. 10.30. (a) *Glossatella tintinnabulum* (Kent), 40 μ. × 450. (b) *Scyphidia physarum* Lachmann, 90 μ. × 210. (c) *Paravorticella crassicaulis* (Kent), 90 μ. × 400. (d) *Pyxidium cothurnioides* Kent, 56 μ. × 400. (e) *Opercularia coarctata* Claparède and Lachmann, 50 μ (one cell). × 160. (f) *Rhabdostyla pyriformis* (Perty), 30 μ. × 400. (g) *Epistylis plicatilis* Ehrenberg, 150 μ (one cell). × 100. (h) *Campanella umbellaria* Linnaeus, 180 μ. × 80. (i) *Systylis hoffi* Bresslau, 200 μ (one cell). × 25. (a after Penard; b after Lachmann; c and f after Kent; g after Roux; i after Bresslau.)

259a (258) Aboral end of cell tapers very little, bears attachment disc; 13 species. (Fig. 10.30b) . . . . *Scyphidia* Dujardin

259b Aboral (basal) region gradually narrowing to stalklike end; 3 species. (Fig. 10.30c) . . . . *Paravorticella* Kahl

260 (253) Family **Epistylidae** **261**

261a Peristomial furrow shallow; disc not set off from border by deep incision **263**

261b Peristomial furrow deep, separating disc and border. **262**

262a (261) Stalk unbranched; 15 species. (Fig. 10.30d) . . . *Pyxidium* Kent

262b Stalk branched; 35 species. (Fig. 10.30e). . . . *Opercularia* Stein

263a (261) Stalk unbranched; 30 species. (Fig. 10.30f) *Rhabdostyla* Kent

263b Stalk branched. **264**

264a (263) Peristomial membranes make only a little more than 1 turn around the peristome **265**

264b Peristomial membranes make 4 to 6 turns around the peristome; 2 species. (Fig. 10.30h) . . . . *Campanella* Goldfuss

265a (264) One individual at end of each branch of the colony stalk; 44 species. (Fig. 10.30g) . . . . *Epistylis* Ehrenberg

265b Clusters of several dozen individuals at the end of each branch; 1 species. (Fig. 10.30i) . . . . *Systylis* Bresslau

266 (253) Family **Vorticellidae** **267**

267a (266) Spasmoneme poorly developed; stalk not spiral when contracted; 7 species. (Fig. 10.31a) . . . . *Intranstylum* Fauré-Fremiet

267b Spasmoneme well developed; stalk is thrown into a spiral coil when contracted **268**

268a (267) Stalk unbranched; 86 species. (Fig. 10.31b). . . . .
                                                      Vorticella Linnaeus

268b Stalk branched; colonial species. **269**

269a (268) At each branching point of the stalk the spasmoneme also branches, so that the whole colony contracts as a unit; 16 species. (Fig. 10.31c,d,e) . . . . *Zoothamnium* Bory

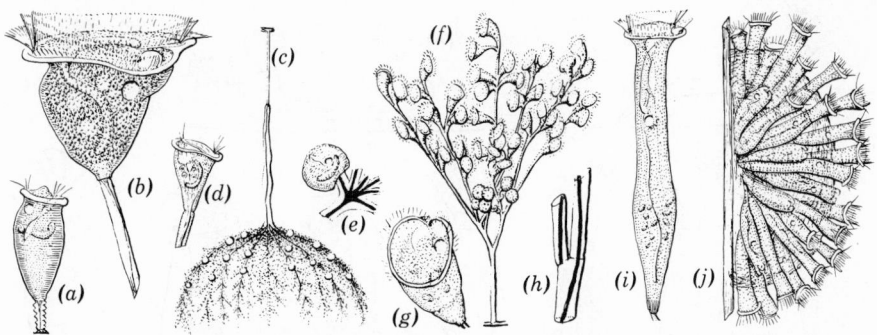

Fig. 10.31. (a) *Intranstylum invaginatum* Stokes, 40 μ. × 350. (b) *Vorticella campanula* Ehrenberg, 100 μ. × 130. (c) *Zoothamnium arbuscula* Ehrenberg, colony, 4 mm. × 9. (d) Same, one individual, 50 μ. × 125. (e) Same, branching point in a developing colony showing spasmoneme continuous in the branches. (f) *Carchesium polypinum* Linnaeus, colony, 1.2 mm. × 30. (g) Same, one individual, 100 μ. × 150. (h) Same, branching point showing spasmoneme starting anew in one branch. (i) *Ophrydium eichhorni* Ehrenberg, 260 μ. × 135. (j) Same, colony. × 45. (a after Kahl; c, d, and e after Wesenberg-Lung; f after Ehrenberg; j after Kent.)

269b At branching points spasmoneme continues into 1 branch only, a new spasmoneme starting in the other branch; 8 species. (Fig. 10.31*f, g, h*) . . . . **Carchesium** Ehrenberg

270 **(251)** Family **Ophrydiidae;** 12 species. (Fig. 10.31*i, j*). . . . . **Ophrydium** Bory

271 **(248)** Tribe **Loricata**. **272**

272a **(271)** Peristome border free from margin of case; body attached by basal end to case inside . . . . Family **Vaginicolidae** **273**

272b Peristome border attached to inwardly reflected margin of case aperture; only the peristomal disc protrudes when feeding . . . . . Family **Lagenophryidae** **280**

273 **(272)** Family **Vaginicolidae**. **274**

274a **(273)** Case broadly adherent to substratum on flat side, with ascending neck; 19 species. (Fig. 10.32*a*) . . . . **Platycola** Kent

274b Case largely free of substratum, attached basally **275**

275a **(274)** Case aperture remains open during contraction **276**

275b Aperture closed by lid or stopper during contraction **277**

276a **(275)** No stalk on case; 30 species. (Fig. 10.32*b*). . **Vaginicola** Lamarck

276b Case has a stalk; 27 species. (Fig. 10.32*c*) . . . . **Cothurnia** Ehrenberg

277a **(275)** Case closed by a lid that is hinged to the case **278**

277b Case closed by lid or stopper borne by animal **279**

278a **(277)** The lid is attached to the case inside below the level of the aperture; 4 species. (Fig. 10.32*d*) . . . . **Thuricola** Kent

278b Lid is attached to the free margin of the aperture of the case; 2 species. (Fig. 10.32*e*) . . . . **Caulicola** Stokes

279a **(277)** Lid is a pseudochitinous secreted structure, fastened to the body; 7 species. (Fig. 10.32*f*) . . . . **Pyxicola** Kent

279b The "stopper" is protoplasmic, an extension of the cell body; 1 species. (Fig. 10.32*g*) . . . . **Pachytrocha** Kent

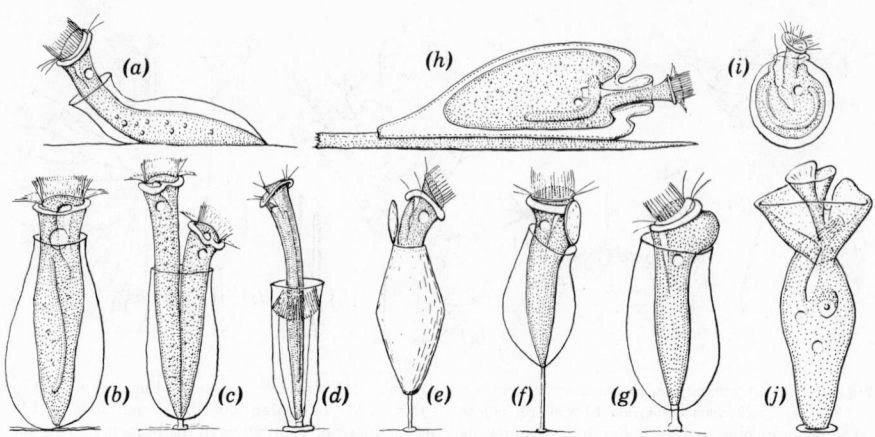

Fig. 10.32. (*a*) *Platycola longicollis* Kent, 125 μ. × 225. (*b*) *Vaginicola ingenita* (O. F. Müller), 55 μ. × 500. (*c*) *Cothurnia imberbis* Ehrenberg, 77 μ. × 380. (*d*) *Thuricola folliculata* (O. F. Müller), 300 μ. × 100. (*e*) *Caulicola pyxidiformis* (D'Udekem), 160 μ. × 150. (*f*) *Pyxicola affinis* Kent, 100 μ. × 220. (*g*) *Pachytrocha cothurnoides* Kent, 60 μ. × 440. (*h*) *Lagenophrys vaginicola* Stein, 48 μ. × 700. (*i*) *L. nassa* Stein, 65 μ. × 180. (*j*) *Spirochona gemmipara* Stein, 100 μ. × 330. (*a, f,* and *g* after Kent; *d* and *h* after Kahl; *e* after D'Udekem; *i* and *j* after Penard.)

280a (272) Family **Lagenophryidae;** 1 fresh-water genus; about 25 species. (Fig. 10.32*h, i*) . . . . *Lagenophrys* Stein

281 (6) Order **Chonotrichida;** Family **Spirochonidae;** 1 fresh-water genus; 9 species. (Fig. 10.32*j*). . . . . *Spirochona* Stein

282 (1) Class **Suctoria; Order Tentaculiferida** 283

283a (282) Body extended into lobes or arms on the tips of which the tentacles are borne. 286

283b Body not extended into lobes or arms. 284

284a (283) Stalk thick; pellicle thick; no case; body somewhat compressed; tentacles fairly thick with cuplike or flattened tips . . . . Family **Discophryidae** 302

284b Stalk, when present, usually slender; pellicle thin; case present or not; tentacles capitate like pinheads 285

285a (284) Tentacles usually not in fascicles; embryos budded off externally, size like parent . . . . Family **Podophryidae** 287

285b Tentacles commonly in fascicles; embryos budded off internally, smaller than parent . . . . Family **Acinetidae** 291

286a (283) Tentacles fine, threadlike . . . . Family **Dendrosomidae** 306

286b Tentacles are merely the retractile tips of arm branches; ectozoic . . . . Family **Dendrocometidae** 310

287 (285) Family **Podophryidae** 288

288a (287) With case or lorica protecting the body 290

288b No case or lorica around the body 289

289a (288) With stalk; 5 species. (Fig. 10.33*a*). . . . *Podophrya* Ehrenberg

289b No stalk; 9-species. (Fig. 10.33*b*). . . . . *Sphaerophrya* Claparède and Lachmann

290a (288) Case cuplike, supporting body on its rim; tentacles scattered; 3 species. (Fig. 10.33*c*) . . . . *Paracineta* Collin

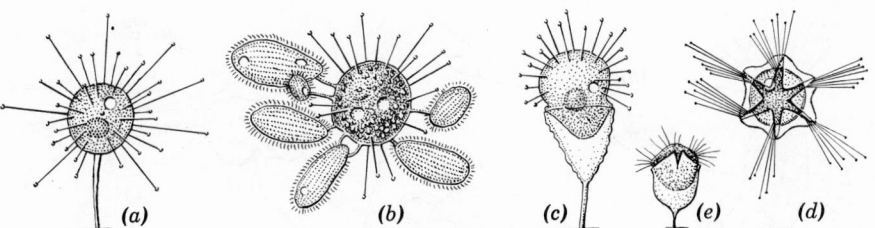

Fig. 10.33. (a) *Podophrya fixa* (O. F. Müller), 42 μ. × 200. (b) *Sphaerophrya magna* Maupas, 38 μ. × 250. The animal is figured with six captured ciliates, immobilized and held at the tips of the tentacles. (c) *Paracineta crenata* (Fraipont), 42 μ. × 260. (d) *Metacineta mystacina* (Ehrenberg), 80 μ. × 100. (e) Same, side view of case and body. (a after Cienkowsky; b after Maupas; c after Collin; d and e after Kent.)

290b		Case with valvelike openings; tentacles in groups; 2 species. (Fig. 10.33d, e) . . . . ***Metacineta*** Bütschli	
291	**(285)**	Family **Acinetidae**	292
292a	**(291)**	No tentacles; endozoic in ciliates; 1 species. (Fig. 10.34a, b) . . . . ***Endosphaera*** Engelmann	
292b		Tentacles present; free-living or ectozoic	293
293a	**(292)**	With neither a stalk nor a case	294
293b		With stalk or case or both.	295
294a	**(293)**	Pellicle ridged; 1 species. (Fig. 10.34c) . . . . ***Anarma*** Goodrich and Jahn	
294b		Pellicle smooth; 2 species. (Fig. 10.34d) . . . . . . ***Hallezia*** Sand	
295a	**(293)**	With stalk, but without case.	296
295b		Case present, with or without stalk	299
296a	**(295)**	Stalk heavy; body often accumulates debris about it; 1 species. (Fig. 10.34e) . . . . ***Squalorophrya*** Goodrich and Jahn	
296b		Stalk slender	297

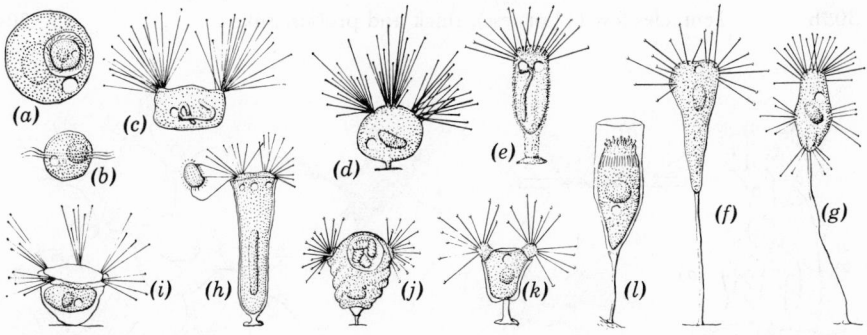

Fig. 10.34. (a) *Endosphaera engelmanni* Entz, 36 μ. × 310. (b) Same, free swimming swarmer stage. (c) *Anarma brevis* Goodrich and Jahn, 125 μ. × 75. (d) *Hallezia brachypoda* (Stokes), 40 μ. × 150. (e) *Squalorophrya macrostyla* Goodrich and Jahn, 90 μ. × 120. (f) *Tokophrya lemnarum* (Stein), 115 μ. × 140. (g) *Multifascicula-tum elegans* Goodrich and Jahn, 70 μ. × 150. (h) *Periacineta buckei* Kent, 110 μ. × 170. (i) *Solenophrya inclusa* Stokes, 43 μ wide. × 200. (j) *Acineta tuberosa* Ehrenberg, 60 μ. × 150. (k) *A. limnetis* Goodrich and Jahn, 45 μ. × 160. (l) *Thecacineta cothurnioides* Collin, 45 μ. × 300. (a and b after Lynch and Noble; c, e, g, and k after Goodrich and Jahn; d and i after Stokes; f, h, and l after Collin; j after Maupas.)

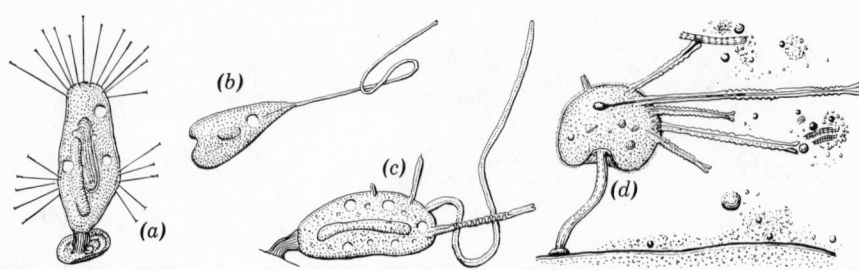

Fig. 10.35. (a) *Discophrya elongata* (Claparède and Lachmann), 85 μ. × 210. (b) *Rhyncheta cyclopum* Zenker, 140 μ. × 100. (c) *Rhynchophora palpans* Collin, 68 μ. × 260. (d) *Choanophrya infundibulifera* Hartog, 50 μ. × 250. (a, c, and d after Collin; b after Zenker.)

297a (296)   Stalk short; 2 species. (Fig. 10.34*d*) . . . . ***Hallezia*** Sand

297b   Stalk moderate to long 298

298a (297)   Tentacles apical; 10 species. (Fig. 10.34*f*) . . ***Tokophrya*** Bütschli

298b   Tentacles both at end of cell and along the sides; 1 species. (Fig. 10.34*g*) . . . . ***Multifasciculatum*** Goodrich and Jahn

299a (295)   Case directly attached to the substratum 300

299b   Case attached by a stalk to the substratum 301

300a (299)   Case tapering, narrowed where it is attached to substratum; 3 species. (Fig. 10.34*h*) . . . . ***Periacineta*** Collin

300b   Case broadly attached to the substratum at its base; 4 species. (Fig. 10.34*i*) . . . . ***Solenophrya*** Claparède and Lachmann

301a (299)   Body adherent to lorica but partly exposed; often compressed; 15 species. (Fig. 10.34*j, k*) . . . . ***Acineta*** Ehrenberg

301b   Body lies completely within the case, only tentacles emerging; 10 species. (Fig. 10.34*l*) . . . ., . ***Thecacineta*** Collin

302 (284)   Family **Discophryidae** 303

303a (302)   Tentacles numerous, in fascicles, scattered all over the cell; 10 species. (Fig. 10.35*a*) . . . . ***Discophrya*** Lachmann

303b   Tentacles few (12 or less), thick and proboscislike 304

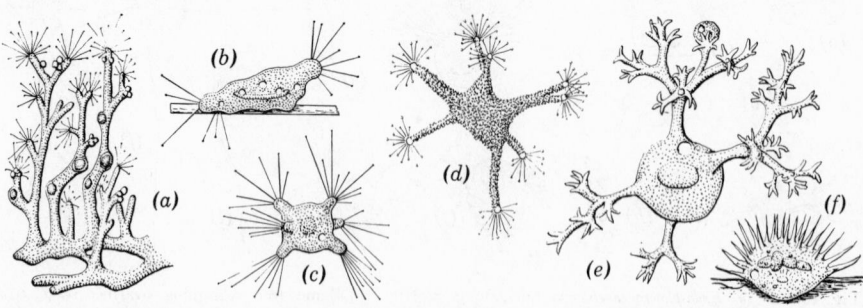

Fig. 10.36. (a) *Dendrosoma radians* Ehrenberg, part of colony. × 20. (b) *Trichophrya epistylidis* (Claparède and Lachmann), about 200 μ wide. × 80. (c) *Staurophrya elegans* Zacharias, 50 μ. × 160. (d) *Astrophrya arenaria* Awerinzew, body 170 μ. × 400. (e) *Dendrocometes paradoxus* Stein, body 50 μ. × 200. (f) *Stylocometes digitatus* (Stein), 100 μ wide. × 120. (a and b after Kent; c after Zacharias; d after Awerinzew; e after Wrzesniowski; f after Plate.)

304a (303)   No stalk; 3 species. (Fig. 10.35*b*) . . . .   *Rhyncheta* Zenker
304b   Provided with a stalk 305
305a (304)   Stalk less than body length; body elongate; 1 species. (Fig.
             10.35*c*) . . . . *Rhynchophora* Collin
305b   Stalk 1 to 2 times body length; body roughly spherical; 1 species.
             (Fig. 10.35*d*) . . . . *Choanophrya* Hartog
306 (286)   Family **Dendrosomidae** 307
307a (306)   Lobes or arms of body do not have secondary branches 308
307b   Lobes or arms of the body branch again into secondary branches;
             1 species. (Fig. 10.36*a*) . . . .   *Dendrosoma* Ehrenberg
308a (306)   Body lobes short and blunt; 7 species. (Fig. 10.36*b*) . . . .
                                     Trichophrya Claparède and Lachmann
308b   Body lobes elongate 309
309a (308)   Body with 6 lobes of medium length; 1 species. (Fig. 10.36*c*) . . .
                                                  Staurophrya Zacharias
309b   Body with 8 longer, armlike lobes; body covered with sand grains;
             1 species. (Fig. 10.36*d*) . . . .   *Astrophrya* Awerinzew
310 (286)   Family **Dendrocometidae** 311
311a (310)   Arms branched; 1 species. (Fig. 10.36*e*) . . *Dendrocometes* Stein
311b   Arms not branched; 1 species. (Fig. 10.36*f*) . . . .
                                                  Stylocometes Stein

References

Collin, B. 1912. Étude monographique sur les Acinétiens. *Arch. zool. exp. et gén.*, 51: 1–457.   **Fauré-Fremiet, E.** A monograph on ciliates and suctorians is in preparation, in Vol. II P. P. Grassé *Traité de Zoologie*, Masson, Paris.   **Guilcher, Yvette.** 1951. Contribution à l'étude des ciliés gemnipares, chonotriches et tentaculifères. *Ann. sci. nat. zool. et biol. animale*, Ser. 11, 13:33–132.   **Hall, R. P.** 1953. *Protozoology.* Prentice-Hall, New York. **Jahn, T. L. and F. F. Jahn.** 1949. *How to Know the Protozoa.* Brown, Dubuque, Iowa. **Kahl, Alfred.** 1930–1935. *Wimpertiere oder Ciliata.* Vol. I of F. Dahl *Die Tierwelt Deutschlands.* (The most complete existing monograph of the ciliates.)   **Kent, W. Sayville.** 1881–1882. *A manual of the Infusoria*, 3 vols. D. Bogue, London.   (Old but still extremely useful.) **Kudo, R. R.** 1954. *Protozoology*, 4th ed. Thomas, Springfield, Illinois.   **Penard, E.** 1922. *Étude sur les Infusoires d'eau douce.* Georg et Compagnie, Geneva.

11

Porifera

MINNA JEWELL

All the North American fresh-water sponges at present known can be assigned to the Family Spongillidae Gray 1867, belonging to the Order Haplosclerina Topsent. The Spongillidae are usually characterized by their fresh-water habitat and the formation of peculiar reproductive structures known as gemmules. Because they appear to be much less influenced by environmental conditions than the vegetative parts of the sponge, these gemmules are important in taxonomy. In fact most of our species cannot be definitely identified if gemmules are absent.

The sponge gemmule consists of an inner protoplasmic part surrounded by one or more membranous coats. Outside of this is usually a pneumatic coat or granular crust which often acts as a float, in which the gemmule spicules are wholly or partially embedded. In most gemmules there is an opening through the wall to allow the escape of cellular material. At this point the membrane usually extends through the crust forming a foraminal tube, which may or may not project beyond the outer surface of the crust.

The spicules of silica, which make up the recticular skeletal structure of the sponge as well as the outer coat of most gemmules, may be classified according to shape, and also according to position in the sponge. In general, two shapes of spicules are recognized, acerate and birotulate. The acerate spicule is typically long and slender, cylindric except near the ends where it may narrow more or less abruptly to points (Figs. 11.1, 11.2). The terminations of acerate spicules may be oxi (sharp points, Fig. 11.4), strongyli (rounded or abruptly terminated, Fig. 11.4), or clavate (inflated or club-shaped, Fig. 11.22); hence spiculi alike at both ends may be designated as amphioxi, amphistrongyli, or biclavi. A common modification of the acerate spicule is the fusiform-acerate, which is largest near the center and slopes gradually to the ends. Acerate spicules may be either straight or curved.

A birotulate spicule consists of a cylindrical shaft enlarged at each end to form a disc or wheel-like structure, the rotula, like an axle with a wheel at each end (Figs. 11.10, 11.11). The rotulae may have entire (smooth) margins or may be more or less incised, in some cases appearing as a circle of spines arranged like the spokes of a wheel.

Spicules are classified as skeletal spicules, gemmule spicules, and flesh or dermal spicules, depending upon their position. Skeletal spicules, usually more or less bound together in fasicles, make up the recticular network that supports and gives shape to the sponge. These spicules are always of the acerate type. Since they are the largest of the spicules they are often called "macroscleres," in contrast to the smaller spicules, "microscleres" found associated with the gemmules or free in the tissues of the sponge. Gemmule spicules are found in the walls of the gemmules. They show the greatest variety in shape and arrangement, hence are most useful in taxonomy. Flesh spicules are spicules that are neither bound together in forming the skeleton nor associated with the gemmules, but are scattered through the living tissue of the sponge. When such spicules are confined to the delicate dermal membrane they are known as dermal spicules. Since, however, the same type of spicule is often found in both the dermal membrane and the underlying tissues, the terms "flesh spicule" and "dermal spicule" are commonly used interchangeably. Flesh or dermal spicules are frequently entirely lacking.

Spicules are prepared for microscopic examination by disintegration of the surrounding tissues with concentrated nitric acid. This can be done quickly by boiling a small fragment of sponge in a large drop of nitric acid on a slide, but a much better and cleaner preparation is made by macerating a small piece of sponge, including gemmules, for 12 or more hours in a small test tube containing about 5 cc of the nitric acid. Frequent shaking accelerates the process. The spicules settle to the bottom, whereas the organic matter not in solution tends to float and can be decanted off. The spicules can then be freed from acid by washing in several changes of water, allowing them to settle or using the centrifuge each time. The water is then replaced by alcohol, and a drop of the alcohol, containing spicules, is ignited on a cover glass. The spicules will be found uniformly distributed over the cover glass, and can be mounted on balsam or damar. This preparation will give an excellent view of the individual spicules, but not of their position or arrangement, so a fragment of the sponge, including part of the dermal layer and several gemmules, one or more of which have been cut in half or crushed, should also be cleared and mounted for examination.

In the following key, a number of the figures and descriptive paragraphs have been retained from that of the first edition by Edward Potts.

KEY TO SPECIES

1a   Gemmules with acerate spicules only . . . .
                  Subfamily **Spongillinae** Vedjovsky 1883   **2**

1b   Gemmules with birotulate spicules arranged at right angles to their surfaces. . . . . Subfamily **Meyeninae** Vedjovsky 1883   **9**

2a   (1) All spicules of the acerate type . . . . . . **Spongilla** Lamarck 1815   **3**

2b   Dermal spicules minute birotulates **8**

3a   (2) Dermal spicules present, minute acerates **4**

3b   Dermal or flesh spicules lacking. **5**

4a   (3) Dermal spicules spined . . . . **S. lacustris** (Linnaeus) 1759

        Sponge usually green, frequently giving off long cylindrical fingerlike branches. Skeleton spicules smooth amphioxi; straight or slightly curved. Dermal spicules slender amphioxi, completely microspined. Gemmule spicules subcylindrical; ends pointed or rounded, variably curved or straight; spined; arranged variably from tangential to nearly erect depending on the thickness of the gemmule crust; sometimes absent. Gemmules yellow to orange in color, distributed throughout the sponge; foraminal aperture infundibular. Strong odor resembling garlic or fish present even in fresh specimens. Distribution world wide.

        Coincident with its ability to live in all types of waters, from soft to brackish, and in many climates, this sponge shows variations in growth form, firmness of texture, robustness of spicules, and thickness of granular crust which, in the past, have been described as distinct species, but which are now regarded as responses to environmental conditions.

◄ **Fig. 11.1.** Skeleton and gemmule spicules of *Spongilla lacustris*, var. *montana*. × 100. (After Potts.)

4b   Dermal spicules smooth . . . . **S. aspinosa** Potts 1880

        Sponge evergreen, encrusting, thin, sending out numerous long, slender waving branches from a relatively thick basal membrane. Gemmules few, in scattered branches. Skeleton spicules smooth, straight or slighly curved, rather abruptly pointed. Dermal spicules minute, smooth, straight or curved, slender, gradually pointed. From clear standing water in N. J. and Va.

◄ **Fig. 11.2.** Spicules of *Spongilla aspinosa*. Four types of spicules figured: ordinary skeleton spicules abruptly pointed at both ends; skeleton spicules abruptly pointed at both ends; skeleton spicule, acute or rounded at one end; malformations of skeleton spicules, with processes at or near one end; small smooth dermal spicules. Globular or discoidal masses of silica frequently observed in this species. × 100. (After Potts.)

5a   **(3)**   Gemmules forming a pavement layer with foraminal tubes turned upward, or in groups with foraminal tubes turned outward **6**

5b   Gemmules associated in groups surrounded by a parenchyma layer. Foraminal tubes turned inward or toward the substratum. **7**

6a   **(5)**   Skeleton spicules smooth . . . . **S. fragilis** Leidy 1851

Sponge encrusting in subcircular patches, thin at edges, occasionally 1 or more inches thick near the middle. In the most varied situations, apparently preferring standing water, though also in running water. Abundant. Gemmules abundant, primarily in 1 or more pavement layers. Also in compact groups surrounded by a cellular parenchyma charged with subcylindrical spined acerates. Skeleton spicules smooth, slightly curved, rather abruptly pointed. True dermals wanting. Found in most of the U. S.

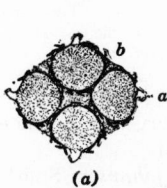 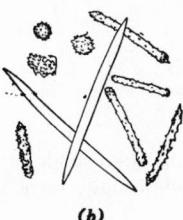

(a)   **(b)**

Fig. 11.3. *Spongilla fragilis.* (*a*) Section of group of gemmules; *a*, foraminal tubules, always curved outward; *b*, envelope with acerate spicules. × 12. (*b*) Three types of spicules figured: skeleton spicules, smooth, abruptly pointed; variable parenchymal spicules, subcylindrical, subspined; spined, spherical forms frequently seen throughout the species. × 100. (After Potts.)

6b   Skeleton spicules profusely spined except at tips . . . .

S. heterosclerifera Smith 1918

Sponge thin, encrusting; skeleton spicules fairly stout amphioxi; closely microspined except on smooth terminal parts. No true flesh spicules. Gemmule spicules on the foraminal side stout, cylindrical, strongly-spined amphistrongyli varying to long, slender amphistrongyli, and, on the side opposite the foramen, to slender, smooth, or sparsely-spined amphioxi. Gemmules abundant, forming a pavement layer on the substratum, surrounded and bound together by a cellular pneumatic layer which is crowded with spicules, mostly of the short, stout, amphistrongylous type. Oneida Lake, N. Y.

Whether gemmules may form groups as in *S. fragilis* is not known.

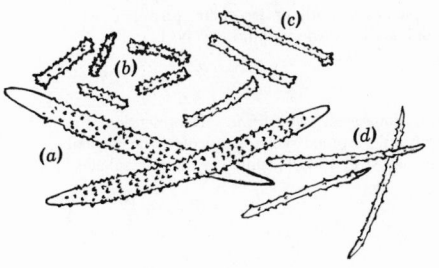

◄ **Fig. 11.4.** *Spongilla heterosclerifera.* (*a*) Skeletal spicules. (*b*) Gemmule amphistrongyli from foraminal side of gemmule. (*c*) Gemmule amphistrongyli from substrate side of gemmule. (*d*) Gemmule amphioxi from substrate side of gemmule. (After Smith.)

7a   **(5)**   Gemmules in groups with foraminal tubes turned inward. Surrounding parenchyma heavily charged with spicules. . . . .

S. mackayi Carter 1885

Sponge brown, thin, encrusting. Gemmules in compact hemispherical groups of 8 to 12 or more, resting on the flat side, surrounded by a parenchyma of unequal cells, charged with numerous coarsely-spined spicules nearly as long as the rather few, less strongly-spined skeleton spicules. Widely distributed in northern states and Canada . . . .

Synonym, *S. iglooformis* Potts 1887

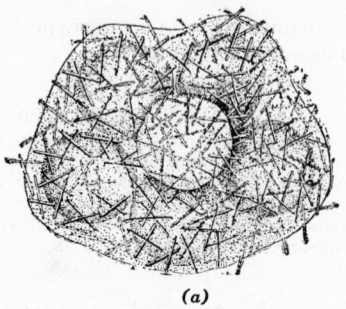

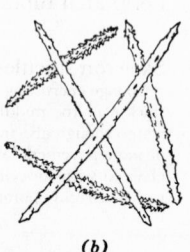

(a)   (b)

Fig. 11.5. *Spongilla mackayi.* (a) Lateral view of dome-shaped group of gemmules. (Foraminal tubules open inward and are invisible.) × 25. (b) Two types of spicules figured: skeleton spicules, weakly spined; "parenchymal spicules" nearly equally long, but more spinous. × 100. (After Potts.)

7b   Gemmules in a single layer surrounded by a very thin parenchyma.
Foramina usually turned toward the substratum . . . .
                                              S. johanseni Smith 1930

Sponge forming a thin unbranched layer. Skeletal structures almost entirely of vertical fibers or columns of spicules, few of which branch or connect with other columns by transverse fibers. These vertical columns widen at the base, where their spicules change from parallel arrangement to diverge from each other and merge with the basal membrane, the whole resembling the buttress of a tree. Skeletal spicules small, very slender, straight, with numerous small spines. Gemmule spicules similar to skeletals but ¾ as long, slightly fusiform, and with longer spines. No flesh spicules observed. Gemmules abundant in groups of 10 to 20, surrounding parenchyma a very thin layer of flattened cells. Foraminal apertures bordered by a flaring bowl-shaped membrane. Tundra lake near Shippigan, New Brunswick. (No figures published, and type specimen not available.)

8a   (2) Only one type of birotulate flesh spicule present . . . .
                                       Corvospongilla Annandale 1911

Type species *C. loricata.* Spongillidae in which the gemmule spicules are without trace of rotules and the flesh spicules have slender cylindrical shafts that bear a circle of strong recurved spines at or near either end.

Only American species yet described, **C. novae-terrae** (Potts) 1886: Sponge encrusting, gemmules rather numerous, very large, crust absent or inconspicuous. Skeleton spicules relatively few, slender, gradually pointed, smooth or microspined. Dermal spicules very abundant, minute, birotulate. Gemmule spicules smooth or irregular, furnished with long spines. Found only in shallow water of lakes in Newfoundland (48° N. L.).

 ◄ **Fig. 11.6.** Spicules of *Corvospongilla novae-terrae.* Representing the slender, smooth or sparsely microspined skeleton spicules; the dermal spicules, birotulates of unequal size; and the spinous gemmule spicules. × 100. (After Potts.)

8b   More than one type of birotulate flesh spicule present . . . .
                                        Parameyenia Jewell 1952

Type species **P. discoides,** formerly **Spongilla discoides** Penney 1933. Sponges with 2 or more types of birotulate flesh spicules. Gemmules usually without spicules. Only 1 species as yet known:

P. discoides (Penney) 1933. Sponge with 2 types of skeletal spicules and 2 types of birotulate flesh spicules in addition to birotulate dermal spicules. Gemmules naked, biconvex discs, without apertures, and with 2 chitinous coats which are separated around the periphery of the disc by air cells. Granular crust restricted to the margins of the discs, and without spicules. Skeletal spicules of the more abundant type slender, entirely-spined amphioxi; those of the less abundant type somewhat shorter amphioxi bearing large, almost cylindrical spines which subdivide at their extremities to form 2 or 3 smaller spines. Dermal spicules slender birotulates resembling those of *Corvospongilla* and *Corvomeyenia*. Flesh spicules abundant, birotulates of 2 distinct classes, resembling the gemmule spicules of *Heteromeyenia*, except that the longer type with hooked rotule rays is the more abundant. Horseshoe Pond, Lexington Co., S. C.

The position of this species is at present uncertain. Penney (1957) stated that the bodies he described as gemmules in 1933 are now known to be bryozoan statoblasts, so the gemmules of the species are unknown. If, upon further study, this sponge is found to produce gemmules of the *Myenia* type (in which the rotules are arranged against the gemmule crust with the shaft perpendicular to the gemmule surface), the species should then be assigned to the genus *Corvomeyenia*, as suggested by Penney, and become *C. discoides* (Penney). If, however, gemmules are produced which are not of the *Myenia* type, of if further study should establish that this species does not produce gemmules, the species will remain as *Paramyenia discoides* (Penney), type of the genus.

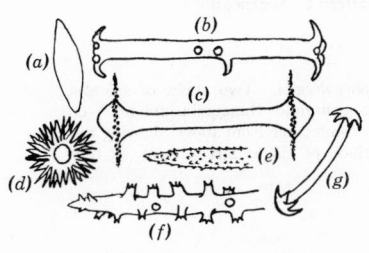

◄ **Fig. 11.7.** *Paramyenia discoides.* (a) Gemmule, side view, × 23. (b) and (c) Flesh spicules. (d) Rotule of flesh spicule like c. (e) and (f) Ends of 2 types of skeletal spicules. (g) Dermal spicule. All spicules × 500. (After Penney.)

9a (1) Dermal spicules acerate or lacking **10**

9b Dermal spicules stellate **28**

9c Dermal spicules birotulate . . . . ***Corvomeyenia*** Weltner 1913

Type species and only American species yet described, **C. everetti**: Sponge green consisting entirely of slender filaments, little more than $\frac{1}{16}$ in. in diameter. Gemmules few, but usually large with a thick crust. Skeleton spicules slender, cylindrical, smooth. Dermal spicules, minute birotulates with slender cylindrical shafts and caplike rotules notched into 5 or 6 hooks. Gemmule birotulates long and clublike; shafts smooth and slender; rotules formed of 5 or 6 stout, recurved, acuminate hooks. Berkshire Co., Mass., Nova Scotia, and Wis.

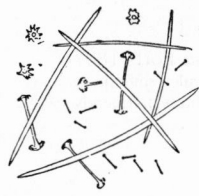

◄ **Fig. 11.8.** Spicules of *Corvomeyenia everetti.* Four types of spicules figured: smooth, skeleton spicules; gemmule birotulates; end view of rotule formed of hooked rays; minute dermal birotules. × 100. (After Potts.)

10a (9) Foraminal tubes of gemmules not greatly elongated, no filamentous appendages. **11**

10b Foraminal tubes of gemmules greatly elongated or with conspicuous filamentous appendages . . . . ***Carterius*** Petr 1886 **27**

The gemmules possess a long foraminal tube, the outer end of which carries an irregularly-lobed disc or is provided with long filaments. Type species C. tentaspermus. The genus *Carterius* is sometimes confused with the genus *Heteromeyenia*. The latter, however, has gemmule spicules of *two distinct classes*, differing in both length and shape, whereas the gemmule spicules of *Carterius* are all of one class as to shape of rotules, and usually show complete intergradation as to length.

11a (10) Gemmule birotulates of a single class. **12**

11b   Gemmule birotulates of 2 distinct classes differing in both length and shape . . . . *Heteromeyenia* Potts 1881   **21**

Type species, *H. argyrosperma* (Potts) 1880. Gemmules surrounded by intermingled birotulates of 2 classes, generally differing in form, and whose shafts are of unequal lengths. The proximal discs of all rest upon the chitinous coat; the outer extremities of the less numerous class project beyond the others.

12a   **(11)**   Rotules of gemmule birotulates equal or nearly equal **13**

12b   Rotules of gemmule birotulates unequal, the proximal rotule being the larger . . . . *Tubella* Carter 1881

Type species *T. paulula* (Bowerbank). Only American species known, *T. pennsylvanica* Potts 1882: Sponge minute, encrusting, on stones or timbers in shallow water. Gemmules very numerous, small. Skeleton spicules very variable in length and curvature, entirely spined; spines large, conical. Dermal spicules wanting. Birotulates of gemmules numerous with a large rotule next to the coat and a small distal rotule, varying from the diameter of the shaft to that of the proximal rotule. Margin of large rotule usually entire but margin of small often angular and notched. Shaft smooth. Averse to light and found as a rule under stones and roots. Eastern U. S. generally.

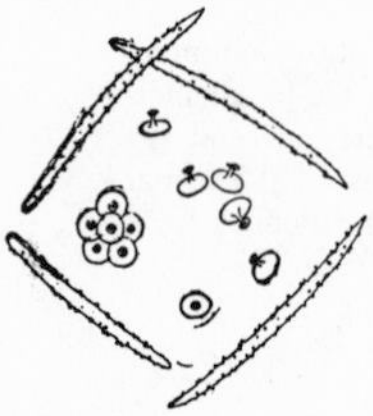

◀ **Fig. 11.9.** Spicules of *Tubella pennsylvanica.* Two types of spicules figured: spined skeleton spicules; gemmule "inaequibirotulates," or trumpet-shaped spicules; group of rotules seen from above, showing the relative sizes of the rotules; surface of single large rotule. × 100. (After Potts.)

13a   **(12)**   Margins of rotules serrated or incised   . . . . *Meyenia* Carter 1881   **14**

Type species *M. fluviatilis:* Sponges with crust of gemmules charged with birotulate spicules arranged perpendicularly around the chitinous coat, so that one disc is applied to the latter, while the other forms a part of the surface of the gemmule.

13b   Margins of rotules entire   . . . . . *Trochospongilla* Vejdovsky 1883   **20**

Type species *T. erenaceus* (Ehrenberg apud Lieberkuhn) 1856: Sponges having birotulate gemmule spicules with shafts short, usually less than the diameter of a rotule, and perpendicular to the surface of the gemmule. Rotulae approximately equal in size and with entire margins.

14a   **(13)**   Rays and spines of birotulates entire **15**

14b   Rays and spines of birotulates subdivided and microspined . . . . . .   *Meyenia subdivisa* Potts 1887

Sponge massive, encrusting, compact. Gemmules few. Skeleton spicules smooth or microspined, abruptly pointed. Birotulates very numerous, robust, shafts frequently spined; rays short but subdivided. From St. Johns River near Palatka, Fla.

◀ **Fig. 11.10.** Spicules of *Meyenia subdivisa.* Three types of spicules figured: smooth and spined skeleton spicules; long, massive gemmule birotulates, spined and subspined; rotules of same. × 100. (After Potts.)

15a   **(14)**   Margins of rotules coarsely dentate **16**

15b   Margins of rotules very finely serrate. . . . . . *M. millsii* Potts 1887

Sponge encrusting. Gemmules small. Skeleton spicules nearly straight, slender, rather abruptly pointed, entirely microspined. Gemmule birotules very numerous, very symmetrical, their shafts usually smooth. Rotules sometimes microspined. From Sherwood Lake, near Deland, Fla.

◄ **Fig. 11.11.** Spicules of *Meyenia millsii*. Three types of spicules figured: microspined skeleton spicule; mature gemmule birotulates with smooth shafts; probably immature forms with less notching on the rotules; face of rotulates lacinulate or delicately notched, and without rays. × 100. (After Potts.)

16a **(15)** Length of birotulate spicules not more than twice the diameter of a rotule **17**

16b Length of birotulate spicules exceeding twice the diameter of a rotule **19**

17a **(16)** Length of birotulate spicules greater than the diameter of a rotule. Skeletal spicules smooth **18**

17b Length of birotulate spicules less than the diameter of a rotule. Skeletal spicules usually microspined except at extremities . . . . . .

M. mülleri (Lieberkuhn) 1856

Sponge cushionlike or with hepatiform lobes; gemmules abundant in tissues of older parts. Skeleton spicules varying from abundantly microspined except at extremities to sparsely spined or smooth, more than one type often occurring in the same specimen. Dermal spicules wanting. Gemmule birotulates shorter than the diameter of the rotule, rotules deeply cleft, the rays sharply pointed. Granular crust varying greatly in thickness so that in some specimens the gemmules may be almost transparent with birotulates sparsely scattered, whereas other specimens have 2 or even 3 layers of closely crowded birotulates surrounding the gemmule. Widely distributed in N. A. and Europe.

◄ **Fig. 11.12.** Spicules of *Meyenia mulleri*. Three types of spicules figured: skeleton spicules, × 120; birotulate gemmule spicules; same malformed; group of rotulae; single rotules showing an ordinary distribution of the rays. × 250. (After Potts.)

18a **(17)** Shafts of birotulates generally smooth or with a single spine . . . . .

M. fluviatilis Carter 1881

Sponge massive, often lobate, gemmules numerous throughout; skeleton spicules smooth, curved, fusiform amphioxi; dermal spicules wanting. Shafts of birotulates smooth or with a single large spine. Rotules frequently umbonate, the margins irregularly denticulate, the depth of denticulation showing considerable variation. Widely distributed in both Europe and N. A.

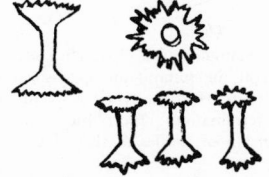

◄ **Fig. 11.13.** *Meyenia fluviatilis*. Gemmule birotulates. × 125. (Upper figures after Carter, lower figures after Weltner.)

18b   Shafts of birotulates with enormous spines . . **M. robusta** Potts 1887

Sponge massive, encrusting, thin. Gemmules scarce. Skeleton spicules pointed, smooth. Birotulates large, generally malformed. Shafts abounding in spines as long as rays of the rotules. Collected near Susanville, Calif., Mexico, and Central Kan.

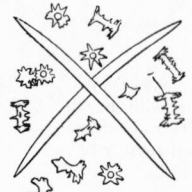

◄ **Fig. 11.14.** Spicules of *Meyenia robusta*. Three types of spicules figured: smooth skeleton spicules; coarsely-spined gemmule birotulates; single rotules; exceedingly misshapen forms. × 100. (After Potts.)

19a **(16)** Length of birotulate spicules 2 to 3 times the diameter of a rotule. Shafts smooth . . . . **M. subtilis** (Weltner) 1895

Sponge thin, encrusting. Skeleton needles extremely slender, scantily covered with short spines. Dermal spicules wanting. Gemmules small, spherical; foramen a simple pore, or a very short tube. Birotulates delicate, slender, of variable length; shaft thin, smooth, long. Rotules small, split nearly to the center, with 10 to 20 blunt rays. Kissimee Lake, Fla.

No figure yet published.

19b   Length of birotulate spicules many times the diameter of a rotule. Shafts spined . . . . **M. crateriformis** Potts 1882

Sponge encrusting, thin. Gemmules small, white, very numerous. Granular crust of gemmules extremely thick, the foraminal tube standing at the center of a craterlike depression. Skeleton spicules slender, sparsely-spined amphioxi. Gemmule birotulates extremely long and slender, shafts abundantly spined. Rotules of 3 to 6 short recurved hooks. Due to their extreme length and the poor support given by their very small rotules, these spicules frequently lose their parallel arrangement when handled, and fall over crossing each other irregularly. Usually in shallow rapidly flowing streams.

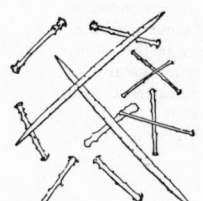

◄ **Fig. 11.15.** Spicules of *Meyenia crateriformis*. Three types of spicules figured: slender microspined skeleton spicules; mature gemmule birotulates with short hooked rays; supposed immature forms. × 100. (After Potts.)

20a **(13)** Skeleton spicules smooth. . . . .
Trochospongilla leidyi (Bowerbank) 1863

Sponge of a peculiar light-gray or drab color, encrusting thin, persistent. Gemmules numerous, each surrounded by a capsule of skeleton spicules. Skeleton spicules short, smooth, robust. Dermal spicules wanting. Gemmule spicules short, birotulate, margins entire and exflected. From La. and Tex. as well as original field of discovery near Philadelphia. Generally distributed in the lower Illinois River.

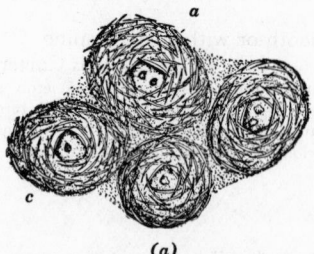
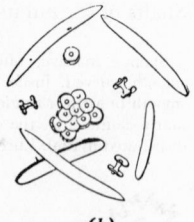

(a)   **(b)**

Fig. 11.16. *Trochospongilla leidyi.* (a) Upper surface of portion of a layer of gemmules, each of which is surrounded by a lattice capsule, *c*, of spicules resembling those of the skeleton; at the summit an open space around the foraminal aperture, *a*, more than one sometimes being present. × 50. (b) Four types of spicules figured: smooth skeleton spicules, abruptly pointed; same, with rounded terminations; short birotulates with entire margins; same with rotule twisted or exflected; face of rotule; group of rotules as they appear upon the surface of the gemmules. × 100. (After Potts.)

20b   Skeleton spicules strongly spined. . . . . . ***T. horrida*** (Weltner) 1893
Sponge encrusting, white, gray, yellow, or brown. No gemmule spicules except birotulates which are smooth-margined low, small. Lives in standing or flowing water. Rare. Reported from Ill. to Fla., and Tex.

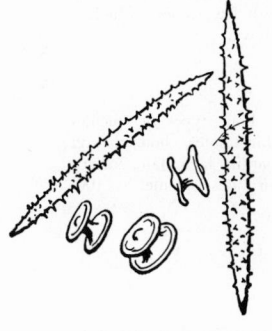

◄ **Fig. 11.17.** *Trochospongilla horrida.* Spinous skeleton spicules. × 180. Birotulate gemmule spicules. × 400. (After W. Kükenthal.)

21a (11) Dermal spicules lacking **22**

21b   Dermal spicules present, small microspined acerates **26**

22a (21) Rotules of gemmule spicules of smaller class with finely serrated margins. Shafts smooth or with few spines **23**

22b   Rotules of gemmule spicules of smaller class coarsely dentate, shafts abundantly spined **25**

23a (22) Skeletal spicules smooth or sparsely spined with smooth ends **24**

23b   Skeletal spicules heavily and entirely microspined, sometimes terminated by one large spine . . . .
Heteromeyenia pictouensis Potts 1885
Sponge massive, nonfasiculated, encrusting; texture very compact. Gemmules few; skeletal spicules short, robust amphistrongyli, entirely spined or with one or both ends terminating in a single large spine. Gemmule birotulates of longer class numerous, one-third their length longer than short birotulates; shafts mostly smooth, conspicuously fusiform, frequently with one or more spines near the middle; rotules of 3 to 6 irregularly placed rays recurved at their extremities. Birotulates of shorter class compactly arranged, shafts smooth or occasionally with a single spine; rotulae large, almost flat, margins finely lacinulate, occasionally microspined.

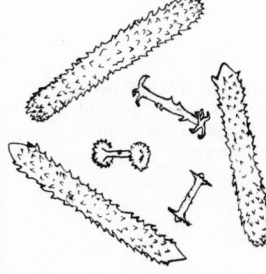

◄ **Fig. 11.18.** *Heteromeyenia pictouensis.* Skeletal and gemmule spicules. × 125 (After Potts.)

24a (23) Gemmule birotulates of smaller class spool-shaped with outer surfaces flat. . . . . **H. ryderi** Potts 1882

 Sponge massive, often hemispherical. Gemmules numerous, crust thick, foramina short and inconspicuous. Skeleton spicules gradually pointed, entirely spined except at the tips. Dermal spicules wanting. Shafts of long birotulates spined, rotules of 3 to 6 short recurved hooks, sometimes umbonate. Rotules of small birotulates nearly as great in diameter as the length of their shafts. Shafts smooth or with few spines. Shallow flowing water, Fla. to Nova Scotia, and inland at least as far as Iowa.

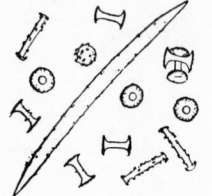

◀ **Fig. 11.19.** Spicules of *Heteromeyenia ryderi.* Four types of spicules figured: skeleton spicule; long gemmule birotulates, hooked and spined; short birotulates; surface of rotules, margins lacinulate, surface microspined or granulated; spherical amorphous spicule. × 100. (After Potts.)

24b Gemmule birotulates of smaller class with both outer and inner surfaces broadly conical . . . . **H. conigera** Old 1931

 Colonies thin, flat, irregular; gemmules more or less abundant, free or resting on substratum; foraminal tubes short, inconspicuous. Skeleton spicules amphioxi, spined except at extremities, not fasciculated. Dermal spicules wanting. Gemmule birotulates of larger class with spined cylindrical shafts; rotules of 3 to 6 short hooklike recurved rays. Birotulates of shorter class approximately 0.6 length of longer birotulates, shafts smooth, rarely spinous, length of shaft between rotules less than the diameter of rotule. Rotules broadly conical on both inner and outer surfaces, their margins thin and finely serrated. Occasionally birotulates of smaller class malformed with shafts thick, irregular, and with broadly conical spines. Early Co., Ga.

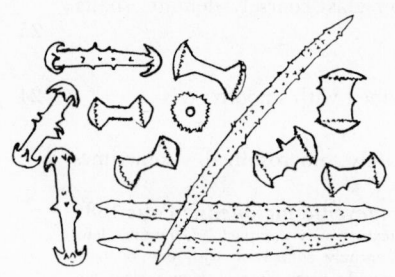

◀ **Fig. 11.20.** *Heteromeyenia conigera.* Gemmule birotulates of longer and of shorter classes × 250. Skeletal spicules × 125. Irregular gemmule spicules of shorter class × 250. (After Old.)

25a (22) Rays of gemmule birotulates of the larger class, 1 to 4 strong clawlike recurved hooks . . . . **H. argyrosperma** (Potts) 1880

 Sponge usually minute, encrusting, gray. Gemmules abundant, large. Foraminal tubes somewhat prolonged. Skeleton spicules rather slender, sparsely spined amphioxi. Dermal spicules wanting. Shafts of long birotulates sparsely spined; rays of rotules few, long, stout, and clawlike. Short birotulates much smaller, shafts abundantly spined, rotules very irregular. Widely distributed in eastern U. S.

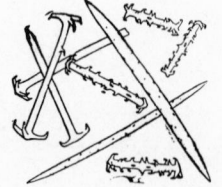

◀ **Fig. 11.21.** Spicules of *Heteromeyenia argyrosperma.* Three types of spicules figured: sparsely microspined skeleton spicules; gemmule birotulates of the longer class with 1 to 3 hooked rays; spined birotulates of the shorter class. × 100. (After Potts.)

25b   Rays of gemmule birotulates of larger class wanting, the rotules represented by clavate enlargements at the ends of shafts . . . .
H. biceps Lendenschmidt 1950

Sponge green to yellow, encrusting, texture compact. Gemmules abundant in basal regions; spherical; foramen inconspicuous; foraminal tube short. Skeleton spicules straight or slightly curved amphioxi; smooth to microspined except at extremities; fasciculated. Dermal spicules lacking. Gemmule birotulates of shorter class fragile; shafts slightly longer than diameter of the rotule, slender, smooth, or spined, occasionally extending beyond the rotule; rotules flat, irregular, deeply serrate, the rays often subdivided. Birotulates of larger class with stout cylindrical shafts, smooth or spined; the rotules without disc or rays, represented by coarsely spined rounded enlargements at the extremities of the shaft. From inlet and outlet of Douglas Lake, Cheboygan Co., Mich.

◄ **Fig. 11.22.** *Heteromeyenia biceps*. Gemmule spicules of two types. × 600. (After Lindenschmidt.)

26a (21) Rotules of gemmule spicules of the larger class domeshaped with ends of the recurved rays still father incurved like the letter J. Short birotulates about ⅔ the length of longer. (See Fig. 11.24) . . . . . .
H. repens (Potts) 1880

Sponge encrusting, thin. Gemmules not abundant. Skeleton spicules rather slender, sparsely microspined, gradually pointed. Dermal spicules nearly straight, entirely spined. Gemmule birotulates of longer class comparatively few; shafts, smooth or with one or a few conspicuous spines often irregularly bent. Rotules domeshaped, rays incurved like fish hooks. Small birotulates very numerous, about ⅔ the length of the large ones. Quiet, almost stagnant water, N. J., Pa., and Mich.

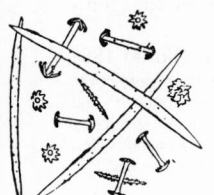

◄ **Fig. 11.23.** Spicules of *Heteromeyenia repens*. Five types of spicules figured: microspined skeleton spicules; gemmule birotulates of the longer class, with recurved hooked rays; birotulates of the shorter class with less pronounced rays; rotules of same; small dermal spicules, coarsely spined; amorphous spicule. × 100. (After Potts.)

26b   Rotules of gemmule spicules of the larger class with rays strongly reflexed, the ends slightly but not markedly incurved. Birotulates of the shorter class about ¾ the length of the longer . . . .
H. baileyi (Bowerbank) 1863

Sponge encrusting, gemmules abundant. Skeleton spicules slender, subfusiform amphioxi. Dermal spicules fusiform acerates, entirely spined; spines of the middle cylindrical, truncated, very long, and large; abundant in both dermal and interstitial membranes. Rotules of larger birotulates conical. Rotules of smaller birotulates mushroom-shaped. Shafts entirely spined.

Because the original description of *H. baileyi* failed to mention 2 classes of birotulates, its taxonomic position was, for a long time, uncertain. It is probable that specimens of this species collected in the past have been identified as *H. repens*, which it closely resembles. In fact opinions still differ as to whether 2 distinct species are actually involved; some authors consider *H. repens* a variety or even an ecological modification of *H. baileyi*. The accompanying figure, drawn from type materials, brings out the main differences between their gemmule birotulates.

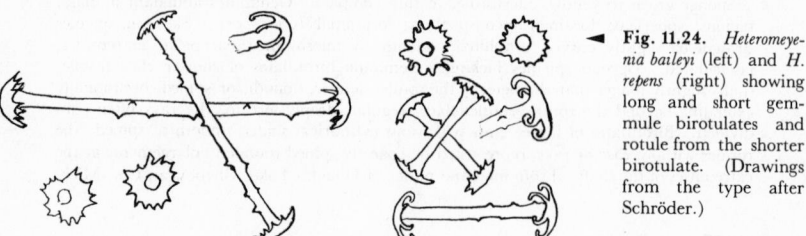

◄ **Fig. 11.24.** *Heteromeyenia baileyi* (left) and *H. repens* (right) showing long and short gemmule birotulates and rotule from the shorter birotulate. (Drawings from the type after Schröder.)

27a **(10)** Length of foraminal tube one-half to once the diameter of the gemmule. Tendrils short, irregular, waving . . . .
Carterius tubispermus (Mills) 1881

Sponge massive. Gemmules numerous. Skeleton spicules rather slender, gradually pointed, sparsely spined. Dermal spicules long, slender, entirely spined. Gemmule birotulates abundant, irregular in length, shaft smooth or with 1 or more spines, rotules arched, rays numerous, long, incurved. Widely distributed.

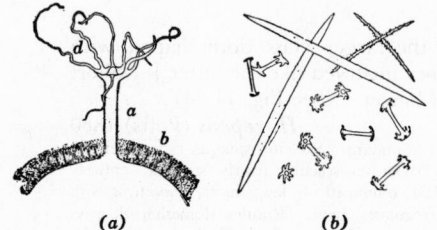

(a)   (b)

◄ **Fig. 11.25.** *Carterius tubispermus.* (a) Partial section of gemmule: *a*, Foraminal aperture prolonged into a long tubule flaring and funnel-shaped at its extremity, and divided into several short tendrils, *d*, or cirrous appendages; *b*, birotulate spicules. × 50. (After Potts.) (b) Three types of spicules figured: skeleton spicules; gemmule birotulates; face of rotule; long-spined slender dermal acerates. × 100. (After Potts.)

27b Length of foraminal tube about ¼ to ½ the diameter of the gemmule. Tendrils 1 or 2, long, enveloping the tube . . . .
C. latitentus (Potts) 1881

Sponge often encrusting stones in rapidly running water. Gemmules numerous. Cirrous appendages at first flat and ribbonlike, becoming slender and rounded, and occasionally subdividing. Skeleton spicules smooth or sparsely microspined, gradually pointed. Dermal spicules long, entirely spined. Birotulates stout, shafts with numerous long pointed spines. Rays of rotules deeply cut and sometimes recurved. In Pa., western N. Y., and Illinois River.

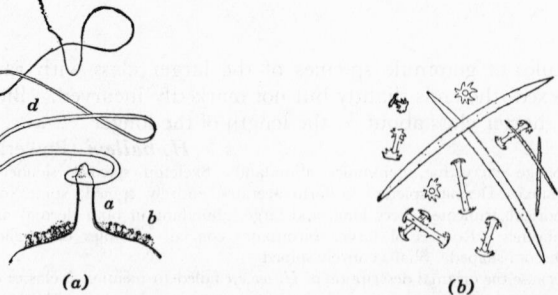

(a)   (b)

Fig. 11.26. *Carterius latitentus.* (a) Partial section of gemmule: *a*, foraminal tube short; *b*, birotulate spicules; *d*, 1 or 2 long and broad, ribbonlike cirrous appendages. × 30. (After Potts.) (b) Three types of spicules figured: skeleton spicules; gemmule birotulates variable in length; face of rotule; spined dermals. × 100. (After Potts.)

27c   Length of foraminal tube about ¼ the diameter of the gemmule.
Tendrils 3 to 5, very long, twisted, frequently branched. . . . .
C. tentaspermus (Potts) 1880
Sponge forming irregular masses creeping upon and around water plants and roots, less
frequently encrusting stones. Gemmules rather numerous. Tendrils as much as half an
inch long. Skeleton spicules slender, very sparsely microspined, gradually pointed.
Dermal spicules slender, nearly straight, entirely spined. Birotulates with cylindrical
shafts, abundantly spined, rotules often irregular. N. J., eastern Pa., and Wis.

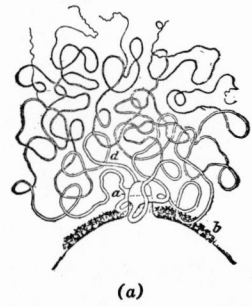 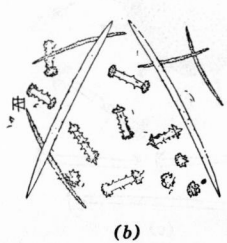

(a)   (b)

Fig. 11.27. *Carterius tentaspermus.* (a) Section of gemmule: *a*, short tubule; *b*, birotulate spicules; *d*, long,
slender cirrous appendages. × 35. (b) Three types of spicules: skeleton spicules; spined gemmule birotulates
with burrlike rotules; ends of same; long, spinous, acerate dermal spicules. × 100. (After Potts.)

28a   (9)   Gemmule birotulates of 2 distinct classes . . . .
Asteromeyenia Annandale 1909   **29**
Type species *A. radiospiculata.* Spongillidae having birotulate gemmule spicules of 2
types and flesh spicules in the form of anthasters.

28b   Gemmule birotulates of a single class . . . . . . *Dosilia* Gray 1867
Type species *Dosilia plumosa* (Carter) 1849.
Only species yet reported in the United States ***Dosilia palmeri*** (Potts) 1885: Sponge
massive, subspherical, lobate. Skeleton spicules sparsely microspined, curved, gradually
pointed. Dermal spicules starshaped, consisting of a variable number of arms of various
lengths, radiating from a large smooth globular body; arms spined throughout. Gemmule birotulates with long spined shafts, rotules notched. From Colorado River, 60
miles below Fort Yuma, Fla., and Mexico.

◄   **Fig. 11.28.** Spicules of *Dosilia palmeri.* Five types of spicules
figured: robust, microspined skeleton spicule; spined gemmule
birotulates; rotules of same, irregularly notched; substellate dermal
spicules; imperfect form of same with only two rays; amorphous
"Scotch terrier" forms. × 100. (After Potts.)

29a   (28)   Rotule spines of longer gemmule birotulates with a simple curve . . .
Asteromeyenia plumosa (Weltner) 1895
Sponge massive, though brittle and friable. Skeleton spicules slender, smooth,
sharply pointed at both ends, nearly straight. Shaft of long birotulates almost smooth,
slender, straight; rotules a circle of curved hooks, joined at the base. Short birotulates
with stouter shafts, profusely, irregularly, and strongly spined; rotules not markedly
convex in profile, irregularly, narrowly, and deeply serrated. Free spicules very minute,

abundant, resembling those of *Dosilia*. Gemmules large, spherical, with single, very small aperture having short, straight foraminal tubule. From Pinto Creek, Kinney County, Tex., and Shreveport, La.; one specimen measured 29 × 25 cm.

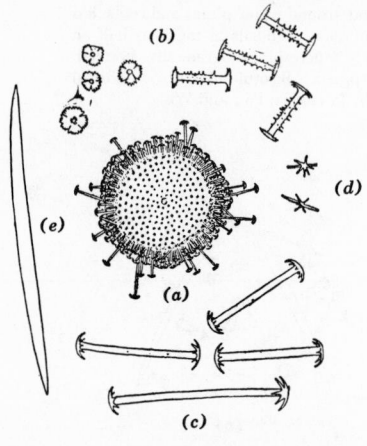

◄ **Fig. 11.29.** *Asteromeyenia plumosa.* (*a*) Gemmule showing aperture in center. × 35. (*b*) Short birotulates. × 120. (*c*) Long birotulates. × 120. (*d*) Free microscleres. × 120. (*e*) Skeleton spicule. × 120. (After Annandale.)

29b   Rotule spines of longer gemmule spicules distinctly recurved, their tips incurved . . . . ***A. radiospiculata*** (Mills) 1888

Resembles *A. plumosa*. In profile the rays of the longer gemmule spicule have almost the form of a J. Ohio and Ill. At Granite City, Ill., specimens were taken from settling tanks of the city water works measuring 42 × 12 × 8 cm.

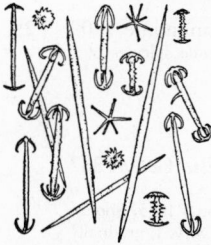

◄ **Fig. 11.30.** Spicules of *Asteromeyenia radiospiculata.* × 100. (From mount.)

References

Annandale, N. 1911. *Fresh-Water Sponges, Hydroids and Polyzoa. Fauna British India.* Taylor & Francis, London.   **Arndt, W. 1926.** Die Spongillidenfauna Europas. *Arch. Hydrobiol.,* 17:337–365.   **Bowerbank, J. S. 1863.** A Monograph of the Spongillidae. *Proc. Zool. Soc. London 1863,* 440–472.   **Carter, H. J. 1881.** History and classification of the known species of Spongilla. *Ann. Mag. Nat. Hist.,* 7:77–107.   **DeLaubenfels, M. W. 1936.** A discussion of the sponge fauna of the Dry Tortugas with material for a revision of the families and orders of the Porifera. *Carnegie Inst. Wash. Publ.,* 467:1–225.   **Gee, N. Gist. 1931.** A contribution toward an alphabetic list of the known fresh-water sponges. *Peking Nat. Hist. Bull.,* 5:31–52.   **Jewell, Minna E. 1935.** An ecological study of the fresh-water sponges of northern Wisconsin. *Ecol. Monographs,* 5:461–504.   **Jewell, M. E. 1952.** The genera of North American fresh-water sponges. Parameyenia new genus. *Trans. Kansas Acad. Sci.,* 85:445–457.   **Old, Marcus C. 1932.** Taxonomy and distribution of the fresh-water sponges (Spongillidae) of Michigan. *Papers Mich. Acad. Sci.,* 28:205–259.   **Potts, Edward. 1887.** Fresh-water sponges; a monograph. *Proc. Acad. Nat. Sci. Phila.,* 39:158–279.   **Smith, Frank. 1921.** Distribution of the fresh-water sponges of North America. *Bull. Illinois Nat. Hist. Survey,* 14:9–22.   **Weltner, W. 1895.** Spongillidenstudien. III. Katalog und Verbreitung der bekannten Susswasserschwamme. *Arch. Naturgeschichte* (pt. I), 61:114–144.

12

Coelenterata

LIBBIE H. HYMAN

The phylum Cnidaria or Coelenterata is almost exclusively marine, being represented in fresh water by only a few forms: the colonial hydroid *Cordylophora lacustris*, some medusae and their polyp stages, and the hydras. All belong to the class Hydrozoa and the order Hydroida, and are undoubtedly derived from marine forebears.

Cordylophora lacustris has the appearance of a marine hydroid, forming a true colony consisting of a basal hydrorhizal network cemented to stones, sticks, tree roots, and similar objects, and one or more stems arising from the hydrorhiza and bearing the hydranths, in an alternating arrangement. The hydranths are of the naked athecate type and are strewn irregularly with filiform tentacles. The gonophores are of the reduced type known as sporosacs which occur as prominent bulbous bodies on the stems just below the hydranths. Medusae are therefore wanting; development takes place inside the sporosacs to the planula stage and the planulae escape and attach to found new colonies. *Cordylophora lacustris* has been found sporadically in a number of separated localities in the United States, in both fresh and brackish water. It seems to thrive best in brackish water where colonies may reach a height of 60 mm, whereas in fresh water, colonies are often but 10 to 15 mm in height. A recent good reference on this hydroid is that of Hand and Gwilliam (1951); photographs of living colonies appear in Roch (1924).

There is but one fresh-water medusa in North America, *Craspedacusta sowerbyi*[1]. This when newly released from the polyp stage is a tiny creature with a tall bell bearing eight tentacles; but it grows to a diameter of 15 to 20 mm with many tentacles in three sets. It has been found sporadically but often in great numbers in almost every state east of the Great Plains and also in several western localities, in ponds, lakes, rivers, reservoirs, and artificial bodies of water of various kinds. Usually all the medusae in any one locality are of the same sex. The hydroid stage is a minute colony of a few naked polyps devoid of tentacles; these reproduce asexually by giving off planula-like bodies that develop into new colonies, and they also bud off the minute medusae. The hydroid stage was named *Microhydra ryderi* and was so called in the literature until its relationship to *Craspeducusta sowerbyi* was discovered; but it must now be called by the latter name, as this has priority. Some of the many references about the medusa and its hydroid in the United States are: Payne (1924), Bennitt (1932), Breder (1937), Causey (1938), Schmitt (1939), Woodhead 1943), Dexter *et al.* (1949), and Arnold (1951). Figures of young and mature medusae appear in Hyman (1940), p. 460.

The hydras are familiar objects of zoological study as the only easily obtainable fresh-water coelenterates. They are small creatures common the world over in lakes, ponds, and streams; they are attached to vegetation, fallen leaves, stones, tree roots, and other objects. Although the hydras have a general resemblance to marine hydroids they lack all trace of a medusoid stage and bear the gonads directly. They are also solitary under normal conditions. The hydra body consists of a short distal region and a long column. The distal region is formed of a conical eminence, the *hypostome*, bearing the mouth at its tip and encircled at its base by a set of hollow, very extensile tentacles, varying in number from three or four to eight or nine, but usually numbering five or six. The column is divisible into a longer stouter distal *stomach* region in which food is digested and absorbed, a shorter proximal *stalk*, and a basal or *pedal disc* for attachment.

Hydras reproduce asexually by budding at the base of the stomach region; successive buds appear in a spiral sequence in the distal direction. Normal hydras do not undergo fission, but longitudinal or transverse fission may result from certain conditions as a means of regulation.

Sexual reproduction is generally related to season, i.e., temperature, occurring in autumn or early winter in most species but in spring and early summer in the green hydra. Hydras may be either dioecious or hermaphroditic; the majority of North American species are dioecious. The gonads develop in the epidermis from cell aggregations. They cover the entire stomach region in dioecious species but in hermaphroditic species the testes are distal, the ovaries proximal on the stomach region. The testes are conical to pumpkinlike elevations provided with a nipple except in *Hydra oligactis*. The ovaries are rounded protuberances eventually containing one egg that goes through an amoebalike shape before assuming its final rounded form.

---

[1]The original spelling of the trivial name is *sowerbii* but as the medusa was named after a Mr. Sowerby, it appears more proper to correct the spelling (see Russell, 1953).

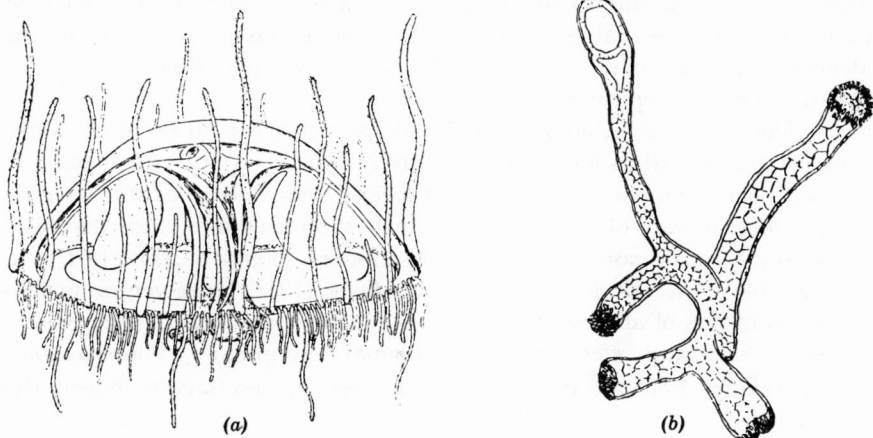

Fig. 12.1. *Craspedacusta sowerbyi.* (*a*) Medusa. (*b*) Hydroid. (By permission from *The Invertebrates*, Vol. 1. by Libbie H. Hyman. Copyright 1940. McGraw-Hill Book Company, Inc.)

When ripe it ruptures through the epidermis and must then be fertilized shortly. When fertilized it begins to develop and the later embryo secretes around itself a shell or theca, probably chitinous. The form of this theca is an important taxonomic character. These thecated embryos either drop to the substratum or are fastened to objects while the chitin is in a sticky phase.

All hydras are armed with four different kinds of nematocysts: stenoteles, holotrichous isorhizas, atrichous isorhizas, and desmonemes. They are illustrated in the accompanying figures, in the undischarged state. The shape, relative and absolute sizes of the nematocysts, and the manner of coiling of the filament or thread inside the isorhizas constitute important taxonomic characters that must be determined on fresh, undischarged nematocysts.

Hydras may be obtained by bringing in submersed vegetation from unpolluted ponds and lakes and covering it with suitable water, whereupon the hydras come to the surface and may be picked off. Search should also be made on the under sides of fallen leaves and of stones in spillways. For the laboratory cultivation of hydras and of *Daphnia* or brine shrimps as their food consult Gatenby and Beams (1950), Chapter 44, Hyman (1930) and Loomis (1953). Hydras may be satisfactorily fixed for whole mounts by letting them expand in a small amount of water in a finger bowl and then suddenly flooding them with hot Bouin's fluid. Sexual reproduction may be induced in many species by placing cultures in a refrigerator for a couple of weeks; feeding is continued.

The identification of hydras to species is a very difficult matter and should be attempted only by those willing to spend time and effort on it. Preserved hydras cannot be identified except by the embryonic theca in some cases. To identify species the hydra must first be cultivated to determine normal form and proportions; fresh undischarged nematocysts of all four kinds must be measured under oil immersion and the details of the filament in undischarged isorhizas noted.   Sexual specimens are usually necessary to complete the identification, especially embryonic thecae.   New species should not be described unless complete data are at hand.

The following key includes all the valid species known to occur in North America, but as hydras have been intensively studied in only small parts of the continent, a number of undescribed species undoubtedly exist.   Users of the key must beware of forcing a fit of their specimens into those described in the key.   In all cases the original descriptions should be consulted as it is impossible to give all details in the key.   Ewer (1948) has given summaries of the characters of all described species throughout the world, together with the synonymies and references to the original descriptions.   Therefore only references after 1948 are given here, as it seems unnecessary to repeat this material.

Colors are mentioned in the following key but are of no great value for identification as they often depend upon the color of recently ingested food. The colors given are those usually seen in the species in question.

KEY TO SPECIES

1a   Medusae. (Fig. 12.1*a*) . . . ***Craspedacusta sowerbyi*** Lankester 1880
        Very small with 8 tentacles when newly released; to 20 mm or more across with many
        tentacles when mature; sporadic in lakes, ponds, rivers, reservoirs, artificial bodies of
        water, and aquaria; recorded from most states east of the Great Plains; also in Okla.,
        Tex., Calif., Ore., Wash., British Columbia.

1b   Hydroids   2

2a   (1) Hydroid without tentacles. (Fig. 12.1*b*) . . . .
                                                        Hydroid stage of *C. sowerbyi*
        Forms minute colony of a few tentacleless polyps without hydrorhizal attachment;
        buds off medusae, also asexual buds; formerly called *Microhydra ryderi* before its relation
        to *C. sowerbyi* was discovered.

2b   Hydroid with tentacles   3

3a   (2) Hydroid colonial, immovably attached to objects by a hydrorhizal
             network . . . .   ***Cordylophora lacustris*** Allman 1871

Forms fair-sized colonies of alternating hy-dranths with sporosacs on their stems, in brackish or fresh water.   Recorded from Newport, R. I.; Woods Hole region; Cambridge, Mass.; Philadelphia; Baltimore; Illinois River near Havana; Mississippi River in Ill. and Ark.; La.; San Joaquin River, Calif., Panama Canal.

◄ **Fig. 12.2.** *Cordylophora lacustris.* (*a*) Branch from colony. (*b*) Tip of branch.

(*a*)   (*b*)

3b   Hydroid solitary, capable of locomotion, attached by pedal disc . . .
                                                                Hydras   **4**

4a (3) Gastrodermis green from symbiotic zoochlorellae . . . .
           Chlorohydra viridissima (Pallas) 1766
> Green hydra, small, hermaphroditic, theca spherical, formed of polygonal plates; common throughout the U. S.; but it has not been settled that all green hydras constitute one species.

4b   Gastrodermis not green, without zoochlorellae . . . . . . ***Hydra***   5

5a (4) Tentacles shorter than the column; stenoteles large, to 20 to 25 μ; holotrichous isorhizas broadly oval; small, pale species   6

5b   Tentacles longer than the column, drooping; stenoteles less than 20 μ long; holotrichous isorhizas narrowly oval except in *utahensis*   7

6a (5) Tentacles held erect . . . .   ***H. americana*** Hyman 1929
> White hydra, dioecious, theca spherical with long spines; common in the eastern U. S., also recorded from Okla. This species was formerly misidentified as *H. vulgaris;* the latter does not exist in the western hemisphere and it is impossible that *americana* could be identical with it, as still maintained by some.

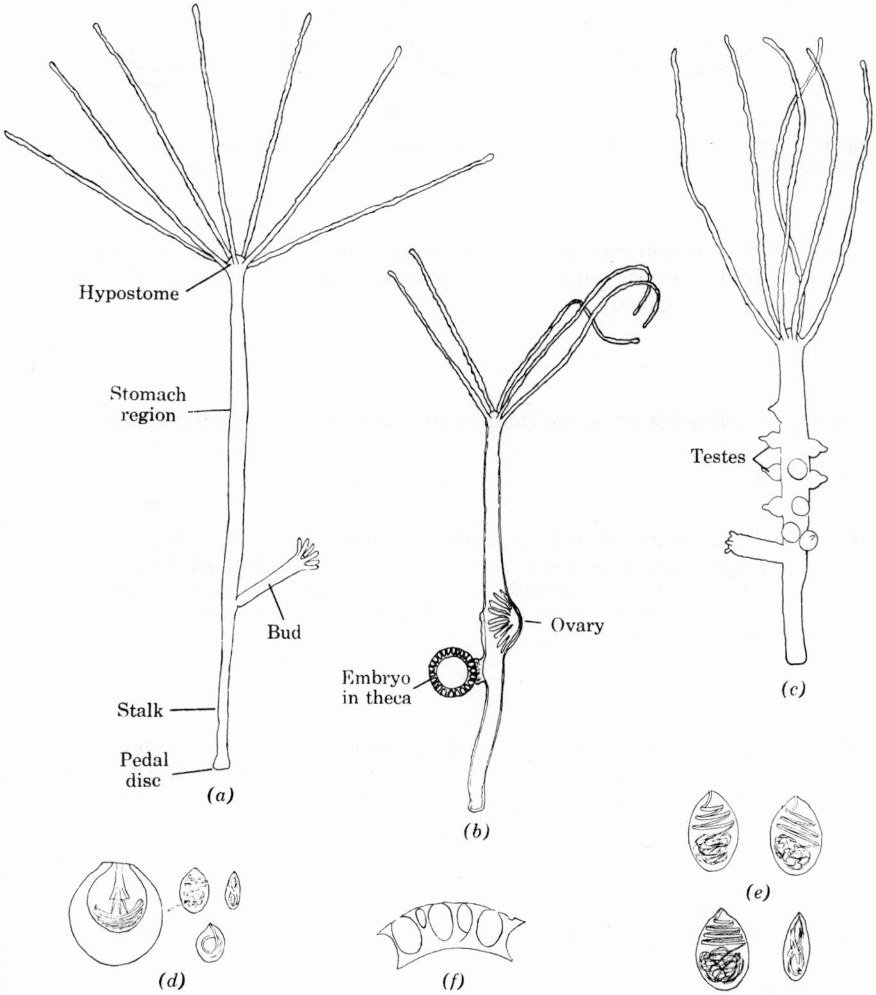

Fig. 12.3. *Hydra americana.* (a) Asexual form with bud. (b) Female with ovary to right, embryo inclosed in theca to left. (c) Male with several testes, also bud. (d) Four kinds of nematocysts, to scale; stenoteles shown in size limits. (e) Enlarged view of isorhizas. (f) Enlarged view of piece of theca. (After Hyman.)

6b   Tentacles held irregularly . . ***H. hymanae*** Hadley and Forrest 1949
          Jersey hydra, pale to brown, hermaphroditic, theca planoconvex, with short, scattered
          spines: N. J. only.

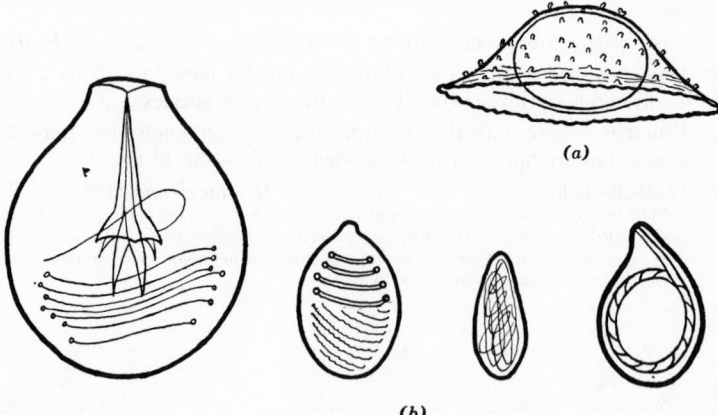

(a)

(b)

Fig. 12.4. *Hydra hymanae.* (*a*) Embryo in theca. (*b*) Four kinds of nematocysts to scale. (After Hadley
and Forrest.)

7a **(5)** Stalk region very distinct from stomach region; dioecious; very large,
          brown species with long drooping tentacles; theca spherical, thin,
          scarcely spined. **8**

7b   Stalk not at all or not conspicuously distinct from stomach region . .   **9**

8a **(7)** Filament in holotrichous isorhizas coiled lengthwise; testes without
          nipples; stenoteles small . . . . ***H. oligactis*** Pallas 1766
          Brown hydra, widely distributed in the U. S., except probably southeastern states; dis-
          tinguished from all other species by the lengthwise coils of the filament in the holo-
          trichous isorhizas.

8b   Filament in holotrichous isorhizas coiled transversely; testes with
          nipples; stenoteles larger . . . . ***H. pseudoligactis*** Hyman 1931
          False brown hydra, common in the north central states; cannot be distinguished
          from *H. oligactis* except by the nematocysts or the testes.

9a **(7)** Hermaphroditic species, small. **10**

9b   Dioecious species, larger, theca always spherical **11**

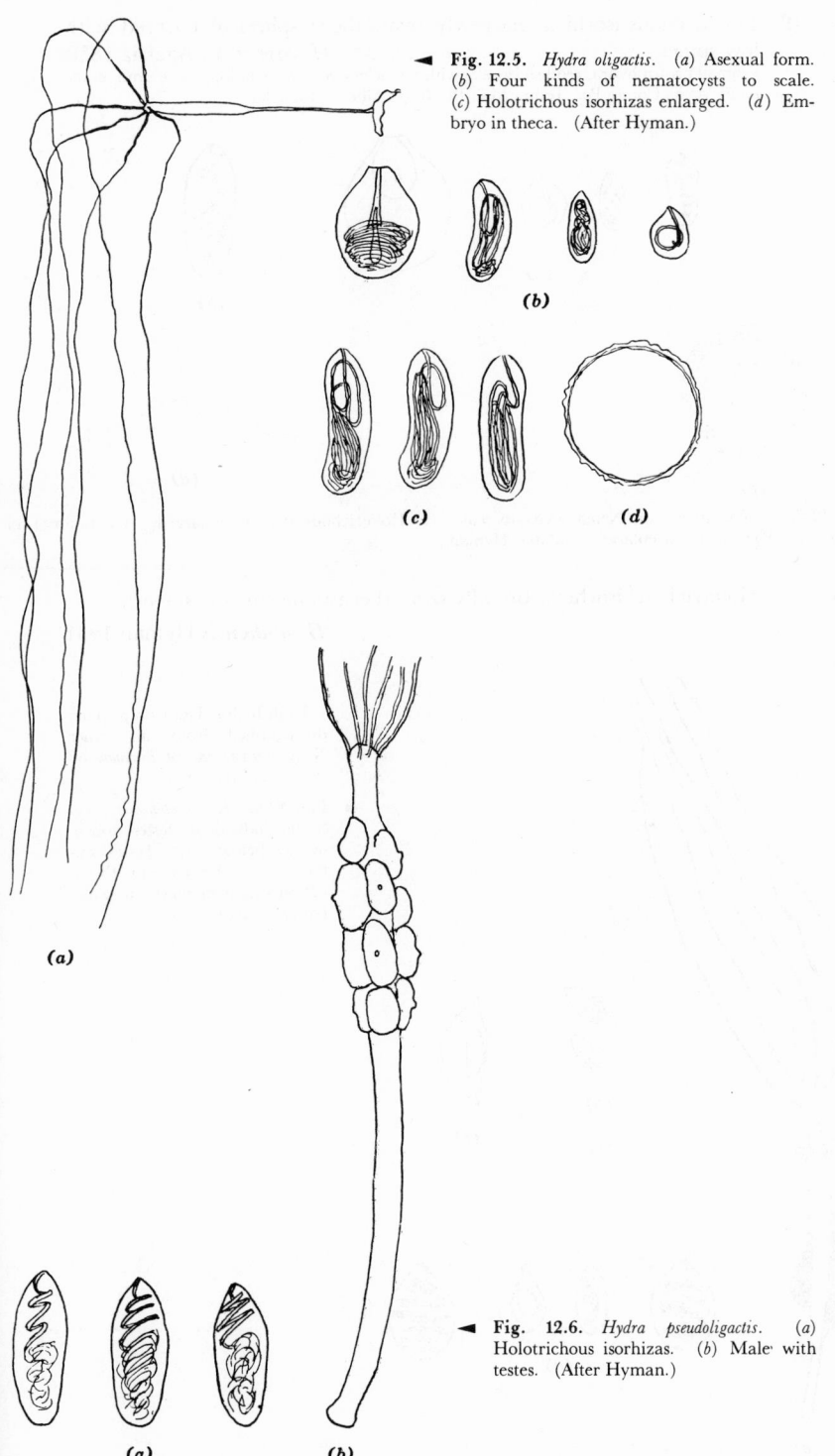

Fig. 12.5. *Hydra oligactis.* (*a*) Asexual form. (*b*) Four kinds of nematocysts to scale. (*c*) Holotrichous isorhizas enlarged. (*d*) Embryo in theca. (After Hyman.)

(b)

(c)   (d)

(a)

Fig. 12.6. *Hydra pseudoligactis.* (*a*) Holotrichous isorhizas. (*b*) Male with testes. (After Hyman.)

(a)   (b)

10a (9) Holotrichous isorhizas narrowly ovai; theca spherical covered with
low spines . . . . **H. carnea** L. Agassiz 1850

<small>Small brown hydra, reddish brown, with tentacles about twice as long as column, com-
mon; Mass., Conn., Pa., Tenn., Mo., Ill., Neb., Alberta, Canada.</small>

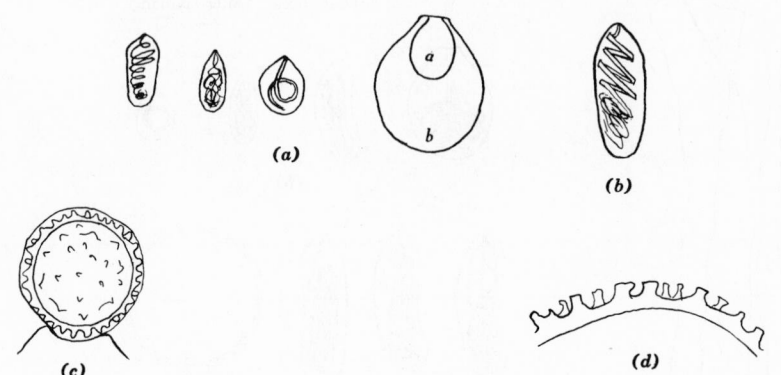

Fig. 12.7. *Hydra carnea.* (*a*) Nematocysts to scale. (*b*) Holotrichous isorhiza enlarged. (*c*) Embryo in
theca. (*d*) Piece of theca enlarged. (After Hyman.)

10b   Holotrichous isorhizas broadly oval, theca planoconvex, smooth . . .

H. utahensis Hyman 1931

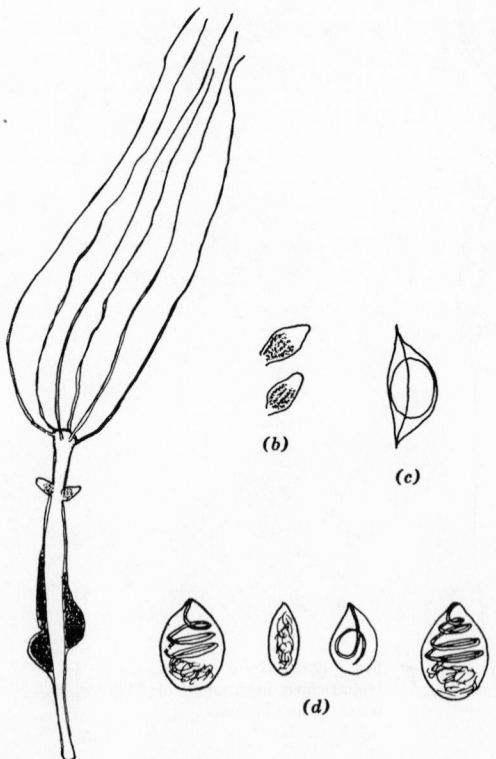

Utah hydra, Utah only, pale,
distinguished from all other
N. A. hydras except *hymanae* by
the form of the theca.

◄ **Fig. 12.8.** *Hydra utahensis.* (*a*)
Sexual individual, testes above,
ovaries below. (*b*) Testes enlarged. (*c*) Embryo in theca.
(*d*) Smaller nematocysts to scale.
(After Hyman.)

11a   (9) Stalk not indicated externally, or not much so **12**

11b   With evident stalk . . . .   ***H. cauliculata*** Hyman 1938

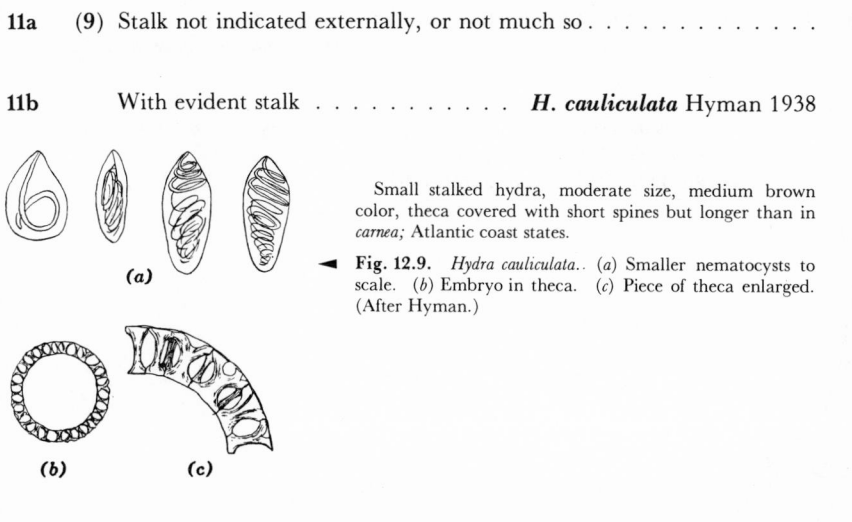

Small stalked hydra, moderate size, medium brown color, theca covered with short spines but longer than in *carnea;* Atlantic coast states.

◄ **Fig. 12.9.** *Hydra cauliculata.·* (*a*) Smaller nematocysts to scale. (*b*) Embryo in theca. (*c*) Piece of theca enlarged. (After Hyman.)

12a   (11) Tentacles very long, 3 to 5 times the column length **13**

12b   Tentacles not exceptionally long, not more than twice the column
                 length . . . .   ***H. littoralis*** Hyman 1931
                 Swift-water hydra, pinkish-orange or greenish in nature, brown in cultivation, theca
covered with long spines, typically found under stones in spillways or along shores subject to wave action; N. Y., N. J., Ill., Pa., Okla., probably widespread.

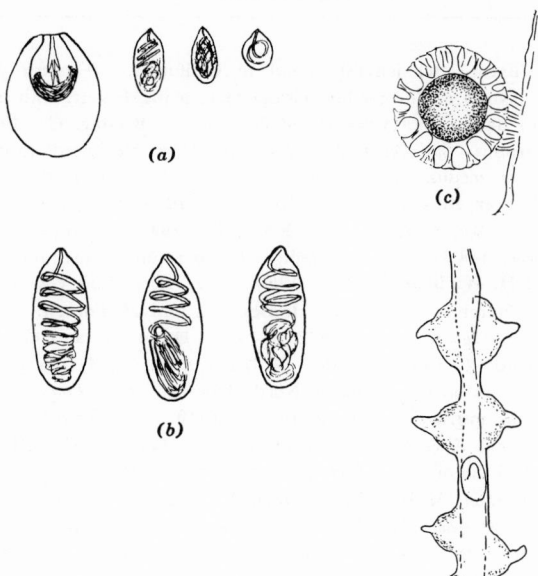

Fig. 12.10. *Hydra littoralis.* (*a*) Nematocysts to scale.· (*b*) Holotrichous isorhizas enlarged. (*c*) Embryo with theca. (*d*) Testes. (After Hyman.)

13a   **(12)** Theca very thin and delicate with a few low spines . . . .
                                        H. canadensis Rowan 1930

Alberta hydra, Alberta, Canada, only; colorless, very large, original description omits nematocysts.

◄ **Fig. 12.11.** *Hydra canadensis.* (After Rowan.)

13b   Theca very thick with long forked spines . . . .
                                        H. oregona Griffin and Peters 1939
              Oregon hydra, near Portland, Ore., only; small, pink or orange, differs from all other
              species in transverse coils of the filament in the atrichous isorhizas; published without any
              figures.

References

Arnold, J. **1951.** Fresh-water jellyfish found in California. *Wasmann J. Biol.*, 9:81–82.
Bennitt, R. **1932.** Notes on the medusa Craspedacusta in Missouri with a summary of the
American records to date. *Am. Naturalist*, 66:287–288.   **Breder, C.** **1937.** Fresh-water
jellyfish at the Aquarium. *Bull. N. Y. Zool. Soc.*, 40. (Lists all records to date.)   **Causey,**
D. **1938.** Fresh-water medusa in Arkansas. *Science*, 88:13. (Also other records in south-
central U. S.)   **Dexter, R., et al.** **1949.** Recent records of Craspedacusta from Ohio and
Pennsylvania.· *Ohio J. Sci.*, 49:235–241.   **Ewer, R.** **1948.** A review of the Hydridae.
Proc. Zool. Soc. London, 118:226–244. (Complete bibliography of hydra taxonomy to date.)
Gatenby, J. B. and H. W. Beams (eds.). **1950.** *The Microtomist's Vade-Mecum* (*Bolles Lee*),
11th ed. J. and A. Churchill, London.   **Hadley, C., and H. Forrest.** **1949.** Description
of *Hydra hymanae.* *Am. Museum Novitates*, 1423:1–14.   **Hand, C., and G. Gwilliam.** **1951.**
New distribution records for Cordylophora. *J. Wash. Acad. Sci.* 41:206–209.   **Hyman, L.**
1930. Taxonomic studies on the hydras of North America. II. *Trans. Am. Microscop. Soc.*,
49:322–333. (Contains instructions for culture.)   **1940.** *The Invertebrates*, Vol. I, p. 460.
McGraw-Hill, New York. (Figures of Craspedacusta.)   **Loomis, W.** **1953.** Cultivation of
hydra under controlled conditions. *Science*, 117:565–566.   **Payne, F.** **1924.** A study of
Craspedacusta. *J. Morphol.*, 38:387–430.   **Roch, F.** **1924.** Experimentelle Untersuchungen
an Cordylophora. *Z. Morphol. Ökol. Tiere*, 2:350–426, 667–670. (Photographs of *Cordylophora*
colonies.)   **Russell, F. S.** **1953.** *The medusae of the British Isles.* Cambridge University
Press, Cambridge. (Bibliography of Craspedacusta.)   **Schmitt, W.** **1939.** Freshwater
jellyfish records since 1932. *Am. Naturalist*, 73:83–89.   **Woodhead, A.** **1943.** Around the
calendar with Craspedacusta. *Trans. Am. Microscop. Soc.*, 62:379–381.

13

Turbellaria

LIBBIE H. HYMAN

E. RUFFIN JONES

INTRODUCTION

Libbie H. Hyman

The class Turbellaria of the phylum Platyhelminthes includes all of the free-living members of the phylum as well as a few epizoic or possibly ecto-parasitic forms (none in the fresh-water fauna of North America). The fresh-water Turbellaria range in size from microscopic forms just visible to the naked eye to the paludicolous planarians, which are up to 30 mm in length. The body is vermiform, that is, longer than it is wide, often much so, and is usually dorsoventrally flattened but not necessarily so; cross sections may be circular or oval but are generally convex dorsally, flattened ventrally. Among conspicuous external characters are eyes, ciliated pits or grooves, and sensory hairs or bristles—but these are not necessarily present. Statocysts occur in a few fresh-water turbellarians. The epidermis is always ciliated in whole or in part, generally only partially in the larger forms. It contains rod-shaped bodies known as *rhabdoids* which are secretions of epidermal or subepidermal gland cells. There are two main types of rhabdoids: *rhabdites*, which are rods shorter in length than the height of the epidermis, and *rhammites*, much longer and often sinuous. Rhabdites occur throughout the class but rhammites are of limited distribution.

The digestive tract lacks an anus (although anal pores occur in a few marine turbellarians). The mouth is ventrally located anywhere along the mid-ventral line from about the middle of the body to near the anterior end. It leads either directly into a muscular pharynx or into a pharyngeal cavity that houses the pharynx. The pharynx occurs in three grades of structure: simple, bulbous, and plicate. The simple pharynx or *pharynx simplex*, characteristic of the fresh-water orders Catenulida and Macrostomida, is a short tube of inturned ciliated epidermis without special musculature, but underlain by the ordinary layers of subepidermal muscles. In the *bulbous* type of pharynx the distal part of the pharyngeal tube remains as a pharyngeal cavity, but the proximal part is greatly thickened into a glandulo-muscular bulb whose tip projects into the pharyngeal cavity. The bulbous pharynx is slightly protrusible; there are two main types, the dolioform (often spelled doliiform) and the rosulate. The former, characteristic of the dalyellioid rhabdocoels, is cask-shaped and oriented parallel to the body axis. The rosulate type, characteristic of the typhloplanoid rhabdocoels, is spheric and oriented at right angles to the body axis. A variant of the dolioform pharynx, termed *variable*, is found among the lower alloeocoels and differs in a less rigid shape and various alterations of the arrangement of the muscle layers. The plicate pharynx, typical of triclads and higher alloeocoels, is a glandulo-muscular cylinder projecting into the large pharyngeal cavity and in the groups mentioned is attached to the anterior end of this cavity, hence projects freely downward or backward. It is protrusible through the mouth by muscular elongation.

From the pharynx the intestine proceeds without the intervention of any definite esophagus or by way of a very short esophagus. The intestine consists of a glandular and absorptive epithelium of tall cells, mostly without any muscular investment, or with an inconspicuous one. The intestine ranges in form from a simple sac to a greatly diverticulated tube, in correlation with the size and shape of the animal. Polypharyngy is not uncommon in fresh-water triclads. The space between the digestive tract and the surface epidermis is filled with a mesenchyme (parenchyma) and contains muscle layers and fibers.

The excretory system takes the form of protonephridial tubules that give off branches ending in flame bulbs. A single median tubule is present in the order Catenulida; in all other Turbellaria the tubules are at least paired and in triclads there are 1 to 4 tubules on each side.

In fresh-water turbellarians the nervous system is sunk into the mesenchyme and lies just below the subepidermal musculature. It is composed of longitudinal cords, of which a ventral pair is generally the most prominent, and of an anterior brain or cerebral ganglia to which the cords are connected. The brain consists of a pair of more or less fused nervous masses and gives off nerves to the head margins and sensory organs of the head.

Asexual reproduction is not uncommon in the fresh-water Turbellaria. The families Catenulidae and Microstomidae regularly reproduce by fissioning into chains of zoids, and fission, also fragmentation, is not uncommon in the Planariidae. However, reproduction is generally sexual, involving copulation and mutal exchange of sperm, as all fresh-water turbellarians are hermaphroditic. The reproductive system is highly complicated and as identification is generally based on the details of this system, these details must be ascertained by the use of serial sections before an identification can be attempted. It therefore must be understood at the start that identification of Turbellaria is excessively difficult and requires expert knowledge of the details of the reproductive system.

In the male system there are one to many testes and one pair of sperm ducts (or a single duct in case of one testis) connecting to the copulatory apparatus. The paired ducts may or may not unite to a common duct before entering the copulatory complex. There are one or two ovaries, each with a duct proceeding to the copulatory complex; this duct is called ovovitelline duct when it receives the yolk glands, otherwise oviduct. In some Turbellaria the yolk glands are combined with the ovary, which is then termed *germovitellarium;* but usually they form separate compact or follicular structures. There may be separate male and female copulatory complexes each with a gonopore, but usually the two are combined into one complex mostly with one common gonopore. The parts of the male copulatory complex are the *seminal vesicle*, which is a muscular chamber for the accumulation of sperm, the *prostatic vesicle*, which supplies a glandular secretion ejaculated with the sperm, and the *penis*. When seminal and prostatic vesicles are combined in one structure this is termed *penis bulb*. The penis can be a fleshy protrusible papilla or a hard simple to complex stylet. Main parts of the female complex besides the terminations of the oviducts are one or two sacs, also called bursae, for the reception of sperm received at copulation. Further details will be found in the introductory remarks to the orders.

In line with the disease prevalent in systematics today, that of raising the ranks of groups, the former three orders of fresh-water Turbellaria, Rhabdocoela, Alloeocoela, and Tricladida, have been increased to at least five and by some authors to eight by the simple process of raising suborders to orders. We here reluctantly admit the breaking up of the old order Rhabdocoela into three orders, but are not prepared to accept the similar dissolution of the Alloeocoela. We positively decline to accept now or hereafter the reduction of Tricladida to a subgroup under Alloeocoela Seriata.

KEY TO ORDERS OF TURBELLARIA

1a	Pharynx of the simplex type; intestine sacciform	2
1b	Pharynx bulbous or plicate or some variant thereof; intestine sacciform or branched	3
2a	(1) With single median protonephridium . . (p. 334) Order **Catenulida**	
2b	With a pair of protonephridia . . . . (p. 338) Order **Macrostomida**	
3a	(1) Pharynx bulbous; intestine sacciform or with but slight lateral outpouchings . . . . (p. 341) Order **Neorhabdocoela**	
3b	Pharynx not bulbous, generally plicate	4

4a   **(3)** Intestine 3-branched, 1 branch extending forward, the other 2
              backwards . . . . (p. 326) Order **Tricladida.**
4b   Intestine not 3-branched, sacciform or laterally diverticulated, or in
              one genus (*Bothrioplana*) forked around the pharynx but single again
              behind this . . . . (p. 359) Order **Alloeocoela**

ORDER TRICLADIDA

Libbie H. Hyman

The triclads, or planarians, because of their size, are the most familiar
representatives of the free-living flatworms or Turbellaria. They are fairly
common in ponds, lakes, springs, and vernal waters among vegetation, beneath stones, or crawling over the bottom. They are the largest of the fresh-
water Turbellaria, ranging from about 5 to 30 mm in length. All are of
flattened elongated form, with a more or less obvious head bearing two or
more eyes, except in cavernicolous species, which are eyeless. The head
varies from a markedly triangular shape with a pair of obvious earlike projections, the *auricles*, to a truncate form with inconspicuous auricles. All
members of the family Kenkiidae and most members of the Dendrocoelidae
have an *adhesive organ* in the center of the ventral face of the anterior margin,
in the form of a cushion supplied with gland cells, nerves, and muscle fibers.
Colors are limited to white, gray, shades of brown, and black.

The digestive tract is uniform throughout the group and hence offers no
characters of taxonomic value. A central pharyngeal cavity opening ventrally
by the mouth is occupied by the cylindrical pharynx, a glandulo-muscular
tube attached to the anterior end of the pharyngeal cavity and capable of
being protruded through the mouth by muscular elongation. At its attached
end it leads into the intestine, which is three-branched; one branch extends
anteriorly to or into the head; the other two proceed posteriorly one to either
side of the pharyngeal cavity. Because of the form of the intestine, the
planarians are also known as *triclads*. Each branch gives off numerous side
branches or diverticula.

Generic and specific identifications among the triclads depend almost
wholly on the details of the reproductive system, especially of the copulatory
apparatus. Most species become sexually mature and breed with reference to
season but some are seldom found in the sexual state or may form genetically
distinct asexual races. All fresh-water triclads are hermaphroditic but cross
fertilization is necessary for development of the eggs. There are a few to
numerous testes, situated laterally either anterior to the pharynx or extending
all or most of the body length. They connect with a pair of sperm ducts that
course posteriorly alongside the pharyngeal cavity. As they approach the
copulatory complex the sperm ducts widen into tubular or sacciform *spermi-*

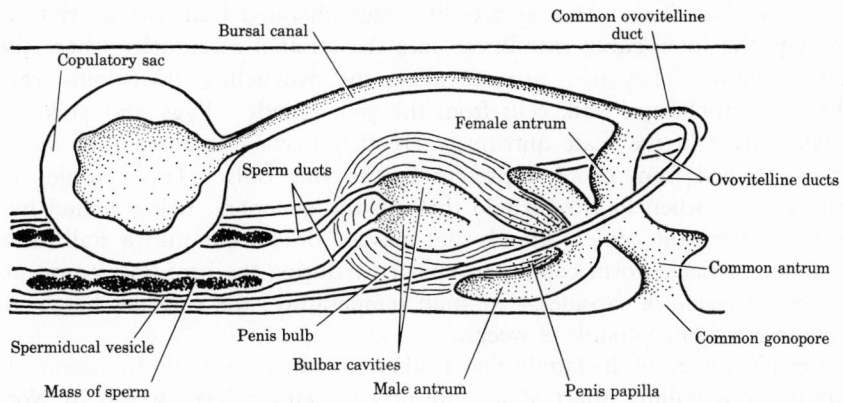

Fig. 13.1. Sagittal view of the copulatory apparatus of *Dugesia tigrina*, also illustrating terminology. (From F. A. Brown, *Selected Invertebrate Types*, Wiley, New York.)

ducal vesicles in which the sperm are stored. There is a single pair of ovaries situated shortly behind the head from which a pair of *ovovitelline ducts* proceed posteriorly close to the sperm ducts. The ovovitelline ducts collect from the *yolk glands*, which are cell clusters between the intestinal diverticula.

The copulatory complex lies immediately behind the pharyngeal cavity with the male part anterior to the female part (Fig. 13.1). The sperm ducts continue from the spermiducal vesicles as narrow tubes that may or may not unite to a common sperm duct before entering the male copulatory apparatus. This consists typically of a rounded or cylindrical glandulo-muscular body, the *penis bulb*, from which a conical muscular *penis papilla* projects posteriorly into a cavity, the *male antrum*. Inside the penis bulb is a single or paired *bulbar cavity* that continues as the *ejaculatory duct* along the penis papilla to its tip. In some planarians (only one North American species), one or more penislike bodies known as *adenodactyls* occur in conjunction with the male apparatus. The female apparatus consists chiefly of a sac, the *copulatory sac* or *copulatory bursa*, that lies between the pharyngeal cavity and the penis bulb. It receives the sperm in copulation. From it a canal, the *bursal canal*, proceeds posteriorly above the male apparatus, then curves ventrally to open into the female antrum, a small chamber continuous anteriorly with the male antrum. Both antra open ventrally by a common *gonopore* situated not far behind the mouth. The ovovitelline ducts open separately or after union into the curve of the bursal canal or after union into the roof of the male antrum, sometimes into the female antrum. The female antrum and terminal parts of the ovovitelline ducts are liberally supplied with eosinophilous *cement glands*.

Copulation occurs by two worms adhering by the posterior part of their ventral surfaces and inserting the penis papilla into the bursal canal of the partner. Copulation is mutual and each worm discharges sperm into the copulatory sac of the partner. The sperm remain but a short time in this sac, soon passing out by the bursal canal and up the ovovitelline ducts to their beginnings at the ovaries where they remain in a small expansion that acts as a *seminal receptacle*. During breeding eggs liberated from the ovaries pass through the intervening membrane into the seminal receptacles where they are fertilized. They then proceed along the ovovitelline ducts into which there are discharged yolk cells from the yolk glands. Eggs and yolk cells finally arrive in the male antrum where they become inclosed in a *shell* or *capsule* formed from droplets present in the yolk cells. This capsule, also called *cocoon*, when finished is laid through the gonopore, being coated by a secretion from the cement glands that may be drawn out into a stalk of attachment. Each worm copulates repeatedly during the breeding season and lays a succession of capsules. At mild temperatures the capsules hatch into minute worms in a couple of weeks.

Certain genera of the family Planariidae reproduce asexually by fission; this consists of a pulling apart at a more or less definite level which in North American species lies behind the pharynx. Members of the other families of fresh-water planarians do not undergo fission. *Phagocata velata* and *vernalis* multiply by fragmentation into a number of small pieces that wall themselves in mucous cysts and therein undergo a process of reorganization into small worms.

Some planarians are best collected by gathering up submersed vegetation from ponds and lakes and packing it into vessels with just enough water to cover, whereupon the planarians come to the surface and can be picked off. Other species have to be sought on the undersides of stones in streams, and along the shores of ponds and lakes. Some can be successfully baited by placing narrow strips of raw beef along the edges of springs and other sites of running water. Planarians often collect in hordes on such bait and can be washed off into jars. Planarians may be kept in earthenware, glass, or enameled dishes or pans in shallow water in dim light. Covering the containers is desirable. They may be fed three times a week by placing in the pans small bits of vertebrate liver, or cut up mealworms, earthworms, clams, and so on. After an hour all traces of food should be removed and the water changed. Planarians may be killed in a flat extended condition by dropping on them 2 per cent nitric acid as they are crawling in a minimum amount of water. This is not equally successful with all species (best with *Dugesia*) and the percentage of nitric acid may need to be varied. Immediately after death the worms should be flooded with a fixative consisting of 0.7 per cent sodium chloride saturated with corrosive sublimate. After half an hour this should be thoroughly washed out in several changes of water and the worms then run up to 70 per cent alcohol to which iodine should be added until it is certain that all corrosive sublimate has been removed. Any whole mount stain may be applied to worms fixed in this manner; they are also suitable for sectioning.

A taxonomic determination of an unknown planarian cannot be made without recourse to sagittal serial sections of a sexual specimen. The genera cannot be determined without such sections. After species have been worked out by a specialist some of them can then be recognized by external characters such as head shape (determinable only on live specimens), eye arrangement, and, to a less extent, color. However, anyone seriously interested in identifying planarians must realize that sagittal serial sections of the copulatory region of sexually mature specimens are indispensable. Juvenile specimens usually cannot be identified, nor can members of species or races that propagate indefinitely by the asexual method. In all attempts at identification the descriptions in original articles should be consulted. Valuable photographs of living worms of the more common eastern species appear in Kenk (1935).

The following keys give all species known to occur in North America but as large parts of the continent have not been well studied, additional species will probably be discovered in the future. Extensive descriptions of North American species will be found in the references cited, and there is a more detailed synopsis of the families, genera, and species in Hyman's study XII (1951). The Kenkiidae are omitted here because of lack of space; they are treated in Hyman's studies VIII (1937b), X (1939), XI (1945), and XIII (1954).

KEY TO SPECIES

1a   Inner muscular zone of the pharynx of distinct circular and longitudinal layers. **2**

◄ **Fig. 13.2.** Scheme of arrangement of the layers of the inner muscular zone of the pharynx. Left, circular and longitudinal layers separate as in Planariidae and Kenkiidae; right, circular and longitudinal fibers intermingled, as in Dendrocoelidae.

1b   Circular and longitudinal fibers of the inner muscular zone of the pharynx intermingled; usually with an adhesive organ; usually with eyes. . . . . Family **Dendrocoelidae**   **19**
In N. A. most colored planarians and all planarians with triangular heads belong to the Planariidae. The family to which white planarians belong (all white planarians in N. A. have truncate heads) cannot be determined without examining cross sections through the pharynx.

2a   (1)   Without an adhesive organ in the center of the anterior margin; nearly always with eyes; colored or white. . . . Family **Planariidae**   **3**
2b   With an adhesive organ; white, eyeless cave dwellers . . . .
                                           Family **Kenkiidae**
       Not treated here. See Hyman (1937b, 1939, 1945, and 1954).

3a   (2)   Eyes 2 (sometimes wanting in *Phagocata subterranea*), accessory eyes sometimes present as abnormalities **4**
3b   Eyes numerous. **17**
4a   (3)   Head triangular; colored **5**
4b   Head truncate or nearly so; colored or white **9**
5a   (4)   Head very triangular, with prominent auricles; testes numerous, extending the body length. . . . . *Dugesia* Girard   **6**
5b   Head of low triangular form, with low auricles; testes few, prepharyngeal. **8**

6a   **(5)** Auricles narrow, pointed; ovovitelline ducts entering the bursal
canal separately . . . . **Dugesia dorotocephala** (Woodworth)

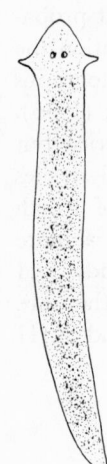

Largest N. A. planarian, to 25 mm or more, uniformly dark brown to black,
sometimes with a light mid-dorsal stripe. Pa. and Va., west to the Pacific coast;
usually in springs or spring-fed waters. Includes *agilis*. See Hyman (1925).

◄ **Fig. 13.3.** *Dugesia dorotocephala.*

6b   Auricles broad, blunt; ovovitelline ducts unite at entrance into
bursal canal **7**

7a   **(6)** Copulatory sac large. (Figs. 13.1, 13.4). . . . . **D. tigrina** (Girard)

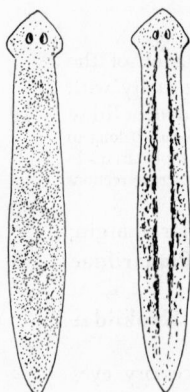

Moderate size, 15–18 mm long, often smaller, color very variable,
most commonly spotted brown and white or brown with a wide white
mid-dorsal streak; most common N. A. triclad. Everywhere in ponds,
lakes, rivers, on vegetation, under stones. Breeds spring and early
summer; capsules stalked, under stones. Old name, **Planaria
maculata.** See Kenk (1944).

◄ **Fig. 13.4.** Two main color patterns of *Dugesia tigrina.*

7b   Copulatory sac small . . . . **D. microbursalis** Hyman 1931
Small, dark, almost black, 10–12 mm long, shape same as *D. tigrina.* Conn., Mass.;
under stones, ponds and streams. Breeds late summer and early autumn. See
Hyman (1931b).

8a   **(5)** Penis normal with bulb and papilla . . . . *Cura foremanii* (Girard)

> Short, broad, plump, to 15 mm long; colored uniformly seal brown or dark gray to black; auricles with a slanting white dash (auricular sense organ); capsules spherical, stalked; will lay capsules continuously in well-fed laboratory cultures. New England and Canada, west into Mich., south into Tenn. and N. C.; in cool creeks and rivers. Old name, *Planaria simplicissima.* See Kenk (1935).
>
> ◄ **Fig. 13.5.** *Cura foremanii.*

8b   Penis degenerate without bulb and with very small papilla . . . . . .
                                Hymanella retenuova Castle 1941

> Small, to 14 mm long but usually less; grayish; capsule oval, not stalked, retained for a long time in the male antrum. Mass., Del., N. C.; vernal ponds and spring-fed swamps.
>
> ◄ **Fig. 13.6.** *Hymanella retenuova.* (After Castle.)

9a   **(4)** Copulatory complex with an adenodactyl; colored . . . .
                                Planaria dactyligera Kenk 1935

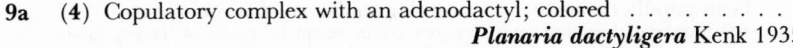

> Small, slender, to 13 mm long; uniformly dark brown or gray to almost black; cannot be distinguished from species of *Phagocata* except by the adenodactyl, observable only in sections; capsules spherical or slightly oval, not stalked. Va.; springs, spring-fed swamps and ponds.
>
> ◄ **Fig. 13.7.** *Planaria dactyligera.* (After Kenk.)

9b   Copulatory complex without an adenodactyl; colored or white;
      capsules always spherical, not stalked . . . . *Phagocata* Leidy   **10**

10a **(9)** Polypharyngeal **11**

10b   Monopharyngeal. (Fig. 13.8) **13**

11a **(10)** Colored. **12**

◄ **Fig. 13.8.**   General appearance of monopharyngeal species of *Phagocata*.

11b   White . . . . *Phagocata subterranea* Hyman 1937
               Small, about 5 mm in length; sometimes eyeless. Caves; Ind.

12a   **(11)** Penis papilla long and pointed . . . *P. gracilis gracilis* (Haldeman)

Relatively large, broad, 15–20 mm long; margins of head expanded; uniformly
colored dark gray to brownish or grayish black. Pa. and Va., westward to Mo.;
mostly in springs; common. See Kenk (1935).

◄ **Fig. 13.9.**   *Phagocata gracilis.*

12b   Penis papilla truncate . . . . . *P. gracilis woodworthi* Hyman 1937
               Cannot be distinguished from *P. gracilis gracilis* except by details of the copulatory
               apparatus. New England, west to the Delaware River. See Hyman (1937a).

13a   **(10)** Colored. **15**

13b   White. **14**

14a   **(13)** Testes extending to posterior end . . . . *P. nivea* Kenk 1953
               Small, delicate, to 8 mm. Alaska only.

14b   Testes extending to level of mouth . . . .
                                          P. morgani (Stevens and Boring)
               Larger, 10–17 mm. Springs and creeks throughout the Appalachian region, also
               Canada, Wis., Mich. Old name, *Planaria truncata.* See Kenk (1935). See also **17b.**

15a   **(13)** Ejaculatory duct with ventral blind sac . . . . . *P. velata* (Stringer)
               Slender, to 15 mm, usually smaller; uniformly gray, whitish when young or senile;
               fragments into small pieces that encyst and regenerate into tiny worms. Mississippi
               Valley, Mich. and Ontario, west of Neb., possibly Colo., south into Mo., also N. Y.
               state; in springs, creeks, and spring-fed ponds and marshes. Seldom found in the sexual
               state. See Castle and Hyman (1941).

15b   Ejaculatory duct without ventral blind sac. **16**

16a   **(15)** Copulatory sac sacciform . . . .
                                          P. gracilis monopharyngea Hyman 1945
               Externally indistinguishable from the other subspecies of *P. gracilis* but mono-
               pharyngeal; 15 mm or more. Iowa, in a drain.

16b   Copulatory sac U-shaped . . . . *P. vernalis* Kenk 1944
               Indistinguishable from *P. velata* except by details of the copulatory apparatus; also

has habit of fragmentation and encystment. North central states; temporary ponds, winter and spring. Seldom sexual.

◄ **Fig. 13.10.** (*a*) Penis papilla of *Phagocata vernalis* without ventral diverticulum of ejaculatory duct. (*b*) Penis papilla of *Phagocata velata* with such diverticulum. (After Kenk.)

(*a*)   (*b*)

17a   (3)   Eyes in a band around anterior end . . . . . . *Polycelis* Ehrenberg   **18**

17b   Eyes in 2 groups in usual site; white . . . .
Phagocata morgani polycelis Kenk 1935

Identical with *P. morgani* except for the eyes. See **14b.** Va.; springs and creeks.

18a   (17)   Penis papilla short, truncate; sperm ducts entering penis bulb asymmetrically. . . . . *Polycelis coronata* (Girard)

Large, 15–20 mm long, uniformly dark brown or black, head rounded with projecting auricles. Hill and mountain streams; Black Hills of S. Dak., westward to the Pacific Coast. See Hyman (1931a).

◄ **Fig. 13.11.** Head of *Polycelis coronata.*

18b   Penis papilla elongate, pointed; sperm ducts symmetrical . . . .
Polycelis borealis Kenk 1953

Light to dark brown, practically indistinguishable externally from *Polycelis coronata.* Streams; Alaska.

19a   (1)   Eyes in 2 longitudinal groups in usual position. . . . .
Sorocelis americana Hyman 1937

Each eye group with 6 to 20 eyes; white, with small adhesive organ; slender, 12–15 mm long. Ozark region; caves, also springs outside of caves. Very abundant when present. See Hyman (1937a, 1939).

◄ **Fig. 13.12.** Head of *Sorocelis americana.*

19b   Eyes otherwise or absent. **20**

20a   (19)   White, with or without adhesive organ **22**

20b   Colored, with adhesive organ, eyes 2 . . . .   **21**

21a   (20)   Penis papilla wanting . . . . . . *Rectocephala exotica* Hyman 1954
Black, 14 mm long; bursal canal very long. Known only from lily pond, Washington, D. C., possibly imported.

21b   With long pointed penis papilla. . . . .
Dendrocoelopsis piriformis Kenk 1953
Brown or brownish gray, may be striped; bursal canal of usual length. Alaska; streams.

22a   (20)   With normal male copulatory apparatus **24**

22b   With massive penis bulb and reduced penis papilla; ejaculatory duct runs ventrally in penis bulb . . . . *Procotyla* Leidy   **23**

23a (22) Adhesive organ present; eyes irregular in usual sites . . . .
Procotyla fluviatilis Leidy

Large, broad, thin, to 20 mm long. New England, west to Wis. and Ill., also Canada, south to N. C.; most common white planarian of the eastern U. S.; springs, ponds, streams, lakes. See Hyman (1928).

◄ **Fig. 13.13.** *Procotyla fluviatilis.* (After Buchanan.)

23b   Without adhesive organ; eyes 2 or wanting . . . .
P. typhlops Kenk 1935
Small, slender, to 12 mm long; rare. Springs, subterranean habitats; Va., Fla.

24a (22) Adhesive organ evident; white   25
24b   Adhesive organ only microscopically determinable; penis papilla very short . . . . *Dendrocoelopsis alaskensis* Kenk 1953
Streams; Alaska.

25a (24) Without eyes; penis bulb surrounded by large eosinophilous mass
Macrocotyla glandulosa Hyman 1956
Stream below cave; Mo.

25b   With 2 eyes; penis bulb not with eosinophilous mass. . . . .
Dendrocoelopsis vaginatus Hyman 1935
Large, 15–20 mm long. Flathead Lake, Mont.

ORDER CATENULIDA

E. Ruffin Jones

The Catenulida are small, slender, threadlike worms, usually white in color, which are quite common in stagnant water. Ciliated pits or grooves are generally present and light-refracting bodies occur frequently in the genus *Stenostomum*, but statocysts are the rule in *Catenula*. The intestine is non-diverticulated and the parenchyma is in the form of a loose network. The protonephridium opens to the exterior in the caudal region. The male reproductive system has an antero-dorsal gonopore and includes a dorsal testis connected by a sperm duct to a penis. The female system, which has no ducts and no permanent gonopore, is represented only by one or more ovaries in a median ventral position near the middle of the body. Eggs are of the entolecithal type with the yolk stored within the egg rather than in special yolk cells.   These systems are rarely seen since the Catenulida usually reproduce asexually, forming chains of two to many zooids.

KEY TO SPECIES

1a   With a preoral circular groove separating the preoral lobe from the remainder of the body . . . . Family **Catenulidae**   **2**
Brain ovoid at base or near middle of preoral zone, and not distinctly subdivided into anterior and posterior lobes.

1b   Without a preoral circular groove . . . . . . Family **Stenostomidae**   **7**
Brain near mouth and clearly divided into 2 anterior and 2 posterior lobes; intestine fills most of body laterally but varies in extent caudally; parenchyma scanty.

2a   (1)   Preoral lobe not subdivided, without longitudinal furrows; body usually composed of 2 to many zooids . . . . *Catenula* Dugès   **3**
No rhabdites; intestine generally short leaving posterior half of body filled with parenchymatous cells; statocyst usually present.

2b   Preoral lobe more or less subdivided by 2 semicircular constrictions extending laterally from the ventral surface and with its basal portion swollen and channelled with longitudinal furrows; body seldom composed of zooids and if zooids are present they rarely exceed 2 in number . . . . *Suomina* Marcus

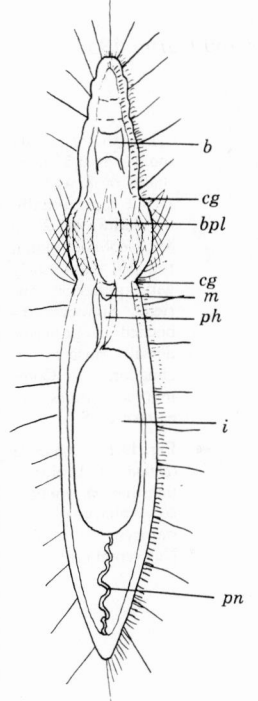

Rhabdites present; intestine almost fills body laterally, but a conspicuous portion of caudal region is left free; usually without statocyst.   Although this genus has been reported several times from the U. S. no species identification has been given. *S. turgida* (Zach) 1902 probably occurs in N. A. as may other species.

◀ **Fig. 13.14.**   Ventral view of *Suomina turgida*. × 160.   *b*, brain; *bpl*, basal section of preoral lobe; *cg*, ciliated groove; *i*, intestine; *m*, mouth; *ph*, pharynx; *pl*, preoral lobe; *pn*, protonephridium.   (After Marcus.)

3a   (2)   Length of cephalic region not more than twice its width **4**
3b   Length of the cephalic region more than twice its width **6**
4a   (3)   Cephalic region approximately one-half the length of the first zooid; chains of zooids suggestive of totem pole in appearance . . . .
Catenula lemnae (Anton Dugès) 1832

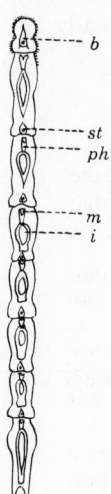

Length of single zooids up to 1 mm, but chains of 32 or more individuals reach a length of more than 6 mm. Protonephridium not readily obvious. Eastern U. S., probably cosmopolitan.

◄ **Fig. 13.15.** *Catenula lemnae.* *b*, brain; *i*, intestine; *m*, mouth; *ph*, pharynx; *st*, statocyst. (After Nuttycombe.)

4b   Cephalic region less than half the length of the first zooid; chains of zooids do not suggest totem pole **5**

5a   (4) With a median dorsal ciliated pit in the preoral zone . . . .
                                                        C. virginia Kepner and Carter 1930

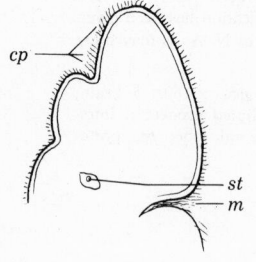

Length of single zooids 0.5 mm, chains of 1 to 4 zooids usual. Eastern U. S., D. C. to Ga.

◄ **Fig. 13.16.** Sagittal section through mouth and dorsal pit of *Catenula virginia.* × 250. *cp*, dorsal ciliated pit; *m*, mouth; *st*, statocyst. (After Kepner and Carter.)

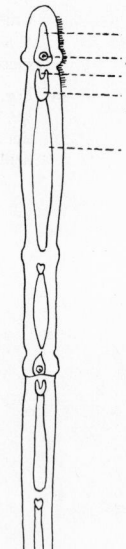

Length of single zooid to 0.5 mm, chains of 8 as long as 3.2 mm; usually in chains of 2 to 4. Cephalic region rounded anteriorly with posterior auricles. Statocyst embedded in posterior dorsal surface of ganglion. Commonest species in eastern U. S.

◄ **Fig. 13.17.** *Catenula confusa.* *b*, brain; *i*, intestine; *m*, mouth; *ph*, pharynx; *st*, statocyst. (After Nuttycombe.)

5b   Without dorsal ciliated pit in preoral zone . . . .
                                                        C. confusa Nuttycombe 1956

6a   (3) Cephalic region elongated anteriorly into a point; with lobes immediately anterior to the preoral groove . . . .
                                                        C. leptocephala Nuttycombe 1956

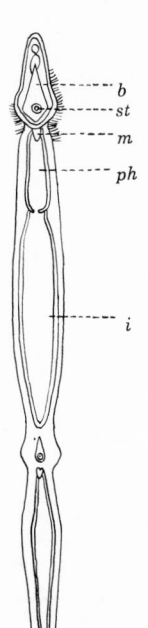

Length rarely exceeding 1 mm and not more than 4 zooids in a chain; intestine almost fills body laterally; Ga.

◄ **Fig. 13.18.** *Catenula leptocephala.* *b*, brain; *i*, intestine; *m*, mouth; *ph*, pharynx; *st*, statocyst. (After Nuttycombe.)

6b Cephalic region rounded anteriorly; lobes midway in cephalic region. . . . . **C. sekerai** Beauchamp 1919

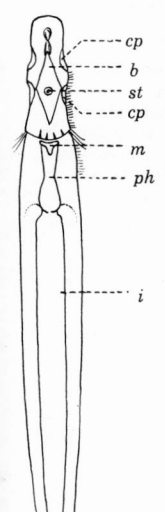

Length of single zooids 0.55 mm, maximum of 2 zooids in a chain with a length of 0.95 mm. Lateral walls of enteron almost in contact with epidermis, lateral ciliated pits present. Ga.

◄ **Fig. 13.19.** *Catenula sekerai.* *b*, brain; *cp* ciliated pit; *i*, intestine; *m*, mouth; *ph*, pharynx; *st*, statocyst. (After Nuttycombe.)

7a (1) Body rarely composed of zooids except in larval and immature stages; anterior end more or less elongated into snoutlike projection . . . . **Rhynchoscolex** Leidy

Single species known from N. A. . . . . . . **Rhynchoscolex simplex** Leidy 1851

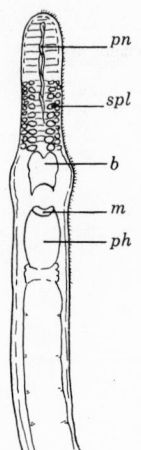

Length to 9 mm; body very slender, coils and twists in snakelike fashion; preoral region elongated into proboscislike prostomium containing pseudo metamerically arranged sensory plates; true ciliated pits and rhabdites lacking. *Stenostomum coluber* Leydig 1854 and the European species **R. vejdovski** Sekera 1930 are both probably identical with *R. simplex*. Eastern U. S., Mass. to Fla.

◄ **Fig. 13.20.** Ventral view of anterior end of *Rhynchoscolex simplex*. × 95. *b*, brain; *m*, mouth; *ph*, pharynx; *pn*, protonephridium; *spl*, sensory plates. (After Marcus.)

7b   Body usually consisting of chain of 2 or more zooids; anterior end showing little or no tendency to snout formation. . . . .
                                            Stenostomum O. Schmidt
Length of single zooids 0.3–2.5 mm; ciliated pits open and usually conspicuous; light-refracting bodies often present; cosmopolitan; at least 22 valid species known from U. S. but identification often difficult.   For key to species see Nuttycombe and Waters (1938).

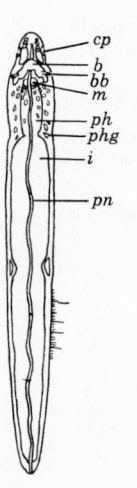

◄ **Fig. 13.21.** Dorsal view of *Stenostomum virginianum*. × 55. *b*, brain; *bb*, bowl-shaped body; *cp*, ciliated pit; *i*, intestine; *m*, mouth; *ph*, pharynx, *phg*, pharyngeal gland; *pn*, protonephridium. (After Nuttycombe and Waters.)

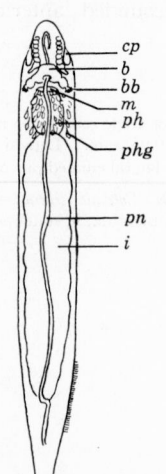

◄ **Fig. 13.22.** Dorsal view of *Stenostomum tenuicaudatum*. × 130. *b*, brain, *bb*, bowl-shaped body; *cp*, ciliated pit; *i*, intestine; *m*, mouth; *ph*, pharynx, *phg*, pharyngeal gland; *pn*, protonephridium. (After Nuttycombe and Waters.)

ORDER MACROSTOMIDA

E. Ruffin Jones

In the Macrostomida the body may be flattened or cylindric in shape, and is white or only slightly pigmented.   Ciliated pits and pigmented eyes are usually present, but statocysts do not occur in fresh-water species.   The intestine may have small lateral diverticula or a preoral blind sac. In the Microstomidae asexual reproduction is common but in the Macrostomidae only sexual reproduction is known. The male reproductive system consists of paired testes with sperm ducts which lead to the copulatory apparatus. This latter consists of a seminal vesicle, a prostate vesicle, and a cuticular penis stylet which is tube- or funnel-like. There is a male antrum and a male gonopore. The Macrostomidae have a pair of ovaries, each with an oviduct leading to the female antrum which opens to the exterior by way of the female gonopore. In the Microstomidae, ovary and oviduct are single. Both male and female gonopores are located in the posterior part of the body. Eggs are of the entolecithal type as in the Catenulida.

KEY TO SPECIES

1a   Without preoral enteric blind sac; body dorsoventrally compressed; posterior end usually a broadened adhesive disc; no asexual reproduction . . . . Family **Macrostomidae**
Single genus reported from fresh water in N. A. . . . . . ***Macrostomum*** O. Schmidt

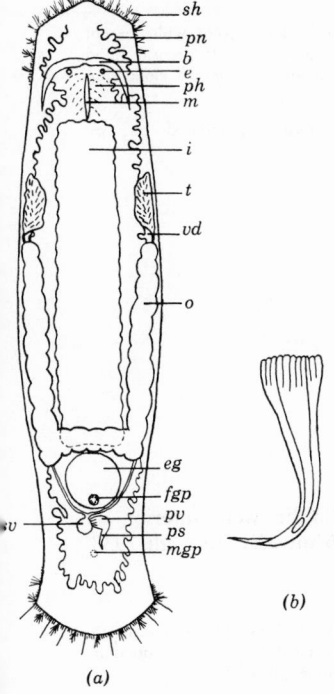

Length 0.8–2.5 mm; body flattened, both ends more or less round; mouth behind brain and eyes; rhabdites present; nephridiopores variable in number and location. More than 25 species known from the U. S., many cosmopolitan in distribution. For key to species see Ferguson (1940).

◄ **Fig. 13.23.** *Macrostomum appendiculatum.* (*a*) Dorsal view. *b*, brain; *e*, eye; *eg*, egg; *fgp*, female genital pore; *i*, intestine; *m*, mouth; *mgp*, male genital pore; *o*, ovary; *ph*, pharynx; *pn;* protonephridium; *ps*, penis stylet; *pv*, prostate vesicle; *sh*, sensory hair; *sv*, seminal vesicle; *t*, testis; *vd*, vas deferens. × 65. (After Ferguson.) (*b*) Penis stylet. × 440.

1b   With preoral blind sac; body cylindric, tapering at posterior end and often composed of zooids or segments . . . . Family **Microstomidae**   2
Single genus reported from fresh water in N. A. . . . . . ***Microstomum*** O. Schmidt
With the characters of the family; the integument often contains nematocysts obtained from feeding on the coelenterate *Hydra*.

2a   (1) With 2 reddish-yellow pigmented eye spots. . . . .
M. lineare (Müller) 1774

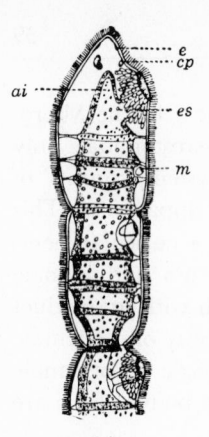

Single individuals up to 1.8 mm; chains of 16 to 18 zooids 9–11 mm in length; very active; body slender, spindle-shaped; color yellowish to grayish-brown, rarely rose-colored; preoral blind sac short; cuticular stylet of penis with curved point. Eastern U. S., N. Y., Pa., Va.

◄ **Fig. 13.24.** *Microstomum lineare.* (*a*) Anterior portion of a chain; *ai,* preoral portion of intestine; *cp,* ciliated pit; *e,* eye; *es,* esophagus; *m,* mouth. × 10. (After Graff.) (*b*) Chitinous stylet of penis. Much enlarged. (After Schultz.)

(a)   *(b)*

2b   Without eyes **3**

3a   (2) With caudal appendage; brain lies under enteric blind sac . . . . . .
                                    M. caudatum Leidy 1850

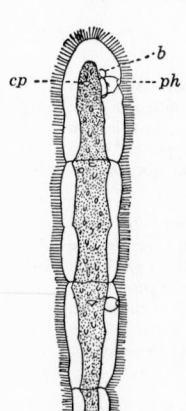

Length 1.5–3.0 mm in chains of 2 to 8 zooids; color white except for intestinal contents; anterior end bluntly rounded; preoral blind sac wide and blunt; caudal appendage a distinctly set-off elongation of dorsal body surface. N. Y., Pa., Mich., Va.

◄ **Fig. 13.25.** *Microstomum caudatum* *b,* brain; *cp,* ciliated pit; *ph,* pharynx. (After Silliman.)

3b   Without caudal appendage; cerebral ganglia lie well anterior to preoral blind sac. . . . . **M. bispiralis** Stirewalt 1937

Length 3–4 mm in chains usually of 4 zooids; body spindle-shaped and white to yellowish in color; preoral blind sac short and conical in shape. Cuticular stylet of copulatory organ bent into a rather deep double spiral. Va.

◄ **Fig. 13.26.** *Microstomum bispiralis,* penis stylet. × 500. (After Stirewalt.)

ORDER NEORHABDOCOELA

E. Ruffin Jones

The order Neorhabdocoela includes more species of fresh-water Turbellaria than any of the other orders. Body color ranges from white to highly pigmented. Eyes are common but by no means universal, and ciliated pits and grooves are frequently lacking. The nephridia may open to the body surface or into the pharyngeal cavity. In the suborder Dalyellioida the mouth is near the anterior end and the pharynx is dolioform, and in the suborders Typhloplanoida and Kalyptorhynchia the mouth is more posterior and the pharynx is usually of the rosulate type. The Kalyptorhynchia have in addition to the pharynx a protrusible muscular proboscis at the anterior end which secretes a sticky material used in capturing prey. Asexual reproduction does not occur in the Neorhabdocoela, and the identification of species is based largely on the details of the reproductive systems, although other characteristics are also important. In the male system of most species, the testes are paired and compact, and a sperm duct connects each testis with the copulatory apparatus. This apparatus is very variable but often includes seminal and prostatic vesicles and a penis, penis bulb, or cirrus which may or may not be armed with cuticular structures. The basic parts of the female system are the ovary and yolk gland, either or both of which may be paired, and their associated ducts. Ovary and yolk gland are usually distinct but may be combined to form a germovitellarium. Other structures such as a seminal receptacle, a copulatory bursa, and a uterus may be added. Uterus and seminal receptacle are often paired, but the copulatory bursa is rarely so.

KEY TO SPECIES

1a   Without a proboscis **2**

1b   With a protrusible or eversible proboscis in a sheath at the anterior end and with a rosulate pharynx . . . Suborder 3 **Kalyptorhynchia**
Section 1—all marine, the Schizokalyptorhynchia.
Section 2—includes all fresh-water Kalyptorhynchia . . . .
                                        Eukalyptorhynchia   **36**
Proboscis apical or subapical, undivided, opens to exterior, and lacks lateral gland sacs. Ovary or germovitellarium paired or unpaired; separate male and female gonopores or a common sex opening; usually with uterus.

2a   (1) Pharynx dolioform; mouth terminal or ventral and near anterior tip; gonopore single; no rhammite tracts . . . .
                                  Suborder 1 **Dalyellioida**   **3**

2b   Pharynx usually rosulate; mouth ventral and usually some distance back of anterior tip; 1 or 2 gonopores; rhammite tracts often present. . . . . Suborder 2 **Typhloplanoida**
       Single family reported from fresh water in N. A. . . . . **Typhloplanidae**   **7**

3a   (2) Dalyellioida with paired (seldom unpaired) germovitellarium or

paired ovary and yolk gland; penis usually with a tube-shaped, needlelike stylet. . . . . Family **Provorticidae**
Gonopore near posterior end of enteron; testes paired.　Single genus from fresh water in N. A. . . . . *Provortex* Graff
Single species known from N. A.　. . . . . *P. virginiensis* Ruebush and Hayes 1940

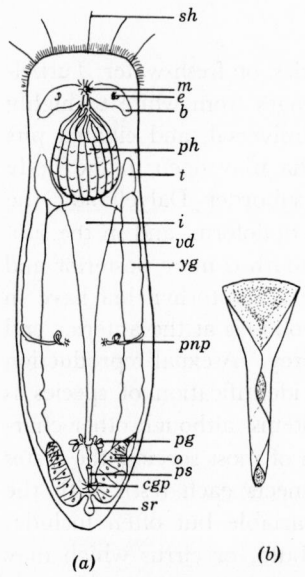

Length 0.5–0.7 mm; unpigmented; 6 to 8 sensory hairs at anterior end; eyes black, ovaries pear-shaped postero-lateral in body, ventral to genital antrum; yolk glands dorsolateral; testes anterior, either side of pharynx; vasa deferentia long; copulatory organ small, ovoid, with small funnel-shaped cuticular tube. Va., Tenn.

◄ **Fig. 13.27.** *Provortex virginiensis.* (*a*) Ventral view. × 90. *b*, brain; *cgp*, common genital pore; *e*, eye; *i*, intestine; *m*, mouth; *o*, ovary; *pg*, prostate gland; *ph*, pharynx; *pnp*, excretory pore; *ps*, penis stylet; *sh*, sensory hair; *sr*, seminal receptacle; *t*, testis; *vd*, vas deferens; *yg*, yolk gland. (*b*) Stylet. × 670. (After Ruebush and Hayes.)

(*a*)　　　　　　　(*b*)

3b　　Dalyellioida with small unlobed, unpaired ovary separated from the paired, branched yolk glands; male copulatory organ usually with complex cuticular apparatus . . . . Family **Dalyelliidae**　　**4**
Mouth in first third of body; gonopore in last third of body; testes paired, copulatory bursa present.

4a　**(3)**　Dalyelliidae without a separate pocket for the chitinous part of the copulatory organ. . . . .　　**6**

4b　　Dalyelliidae with a separate pocket for the chitinous part of the copulatory organ; with 2 black eyes, each consisting of 2 pigment spots connected by a narrower pigment band; body usually heavily pigmented . . . . *Castrella* Fuhrmann　　**5**

5a　**(4)**　Chitinous stylet with 2 stalks or handles, usually connected by 8 to 10 cross commissures and bearing at their distal ends a rodlike cross piece provided with a row of 12 end spines. . . *C. graffi* Hayes 1945
Length about 1.5 mm; color brown to reddish; bursa a simple sac; no locules in female genital canal. This species was identified by Graff in 1911 as *C. pinguis*, but Hayes demonstrated that it is a distinct species. N. Y.

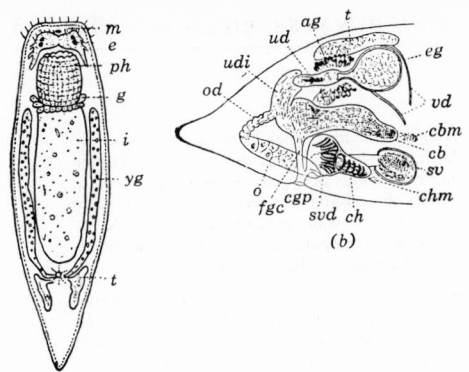

Fig. 13.28. *Castrella graffi.* (*a*) Entire: *e*, eye; *g*, gland; *i*, intestine; *m*, mouth; *ph*, pharynx; *t*, testis; *yg*, yolk gland. × 30. (After Silliman.) (*b*) Sexual organs from animal compressed from side: *ag*, accessory glands; *cb*, copulatory bursa; *cbm*, retractor muscles of same; *cgo*, common genital pore; *ch*, pocket that contains stylet; *chm*, one of 4 muscles of same; *eg*, egg; *fgc*, female genital canal; *o*, ovary; *od*, oviduct; *sv*, seminal vesicle; *svd*, duct from same; *t*, testis; *ud*, duct of uterus; *udi*, uterus diverticulum of antrum; *vd*, vas deferens × 60. (After Graff.)

5b Chitinous stylet with a single, long cylindrical handle, bearing at its distal end a plate provided with a variety of spines . . . .
C. marginata (Leidy) 1847

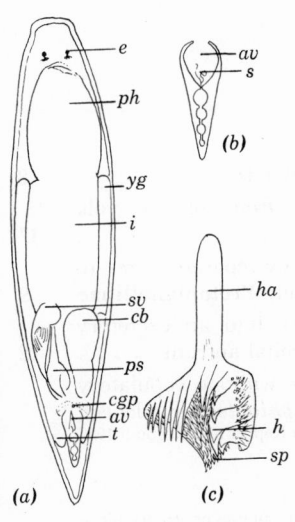

Length up to 2 mm; color brownish to dark gray or black; cuticular stylet more or less broom-shaped with 5 or more different types of spines; spermatophores present, bursa flask-shaped. Female genital canals contain 4 locules of constant structure. Formerly called **C. pinguis** Silliman 1884. N. Y., Pa., Wis., Va.

Fig. 13.29. *Castrella marginata.* (*a*) Ventral view. × 52. *av*, antral vestibule; *cb*, copulatory bursa; *cgp*, common genital pore; *e*, eye; *i*, intestine; *ph*, pharynx; *ps*, penis stylet; *sv*, seminal vesicle; *t*, testis; *yg*, yolk gland. (After Kepner and Gilbert.) (*b*) Antral vestibule enlarged. *av*, antral vestibule; *s*, tail of sperm. (After Kepner and Gilbert.) (*c*) Penis stylet × 190. *h*, hook; *ha*, handle; *sp*, spine. (After Hayes.)

6a (**4**) Usually large forms without a uterus and with zoochlorellae in the parenchyma. **Dalyellia** Fleming
Single species known from N. A.. . . . . **D. viridis** (G. Shaw) 1791

Similar in appearance to *Microdallyellia*. Length 5.0 mm; unpigmented or with brownish pigment, but body green in color due to presence of zoochlorellae. Eyes saucer-shaped; testes in anterior half of body; eggs when present more than 3 in number and scattered through parenchyma. N. Y., N. C.

Fig. 13.30. *Dalyellia viridis.* Chitinous stylet: *ea*, terminal branch; *st*, two-parted stalk. Much enlarged. (After Graff.)

6b Usually small forms with a uterus and usually without zoochlorellae.
Microdalyellia Gieysztor

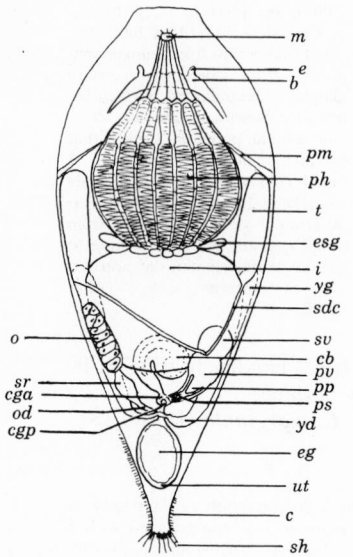

Length to 3.5 mm; color brownish, yellowish, or reddish, often heavily pigmented; testes in anterior or posterior half of body; eggs when present usually not more than 3 and in uterus in posterior body half. At least 17 valid species reported from N. A. Distribution cosmopolitan. For species identification see Ruebush and Hayes (1937).

◄ **Fig. 13.31.** *Microdalyellia gilesi.* Ventral view. × 20, slightly compressed. *b*, brain; *c*, cilia; *cb*, copulatory bursa; *cga*, common genital antrum; *cgp*, common genital pore; *e*, eye; *eg*, egg; *esg*, esophageal gland; *i*, intestine; *m*, mouth; *o*, ovary; *od*, oviduct; *ph*, pharynx; *pm*, retractor muscle of pharynx; *pp*, penis papilla; *ps*, penis stylet; *pv*, prostate vesicle; *sdc*, common sperm duct (vas deferens); *sh*, sensory hairs; *sr*, seminal receptacle; *sv*, seminal vesicle; *t*, testis; *yd*, yolk duct; *yg*, yolk gland; *ut*, uterus. (After Jones and Hayes.)

7a (2) Typhloplanidae with testes ventral to the yolk glands 8
7b   Typhloplanidae with testes dorsal to or in front of the yolk glands 17
8a (7) The 2 end stems of the excretory system open by separate pores to the surface of the body . . . . Subfamily **Protoplanellinae** 9
8b   The end stems of the excretory system empty into an excretory vesicle combined with the mouth, or into the genital antrum 12
9a (7) Protoplanellinae with simple sexual apparatus without copulatory bursa . . . . *Amphibolella* Findenegg
      Single species known from N. A.. . . . . *A. virginiana* Kepner and Ruebush 1937

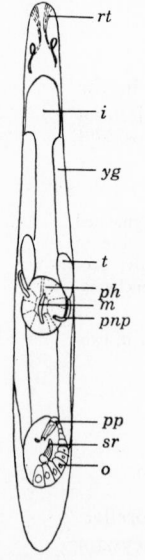

Length 0.7–1.0 mm; slender; unpigmented and without eyes or sensory pit at anterior end; penis simple, short, conical; ovary single, short, on left side beneath yolk glands; oviduct opens into left yolk duct; yolk glands slender, cylindric, paired. Pa., Va.

◄ **Fig. 13.32.** *Amphibolella virginiana.* Ventral view. × 80. *i*, intestine; *m*, mouth; *o*, ovary; *ph*, pharynx; *pnp*, excretory pore; *pp*, penis; *rt*, rhabdite tract; *sr*, seminal receptacle; *t*, testis; *yg*, yolk gland. (After Kepner and Ruebush.)

9b   Protoplanellinae with copulatory bursa **10**

10a   (9) Without sensory pit at anterior end; mouth posterior to anterior
          body third . . . . *Krumbachia* Reisinger   **11**
          <small>Copulatory organ without stylet; ejaculatory duct cuticularized; without copulatory
          antrum and without toothlike structures.</small>

10b   With anterior sensory pit and paired ciliated pits; mouth in anterior
          body third . . . . *Prorhynchella* Ruebush
          <small>Single species known from N. A. . . . . *P. minuta* Ruebush 1939</small>

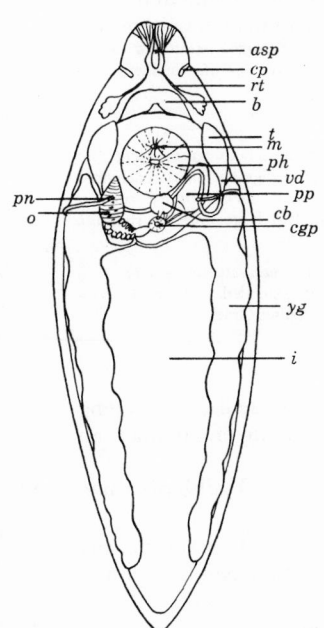

Length 0.6–1.0 mm; unpigmented; gonopore near posterior end of anterior body third; testes ventral, on either side of pharynx; copulatory organ an elongated muscular sac on left side just posterior to pharynx; ovary pear-shaped, ventral, on right side; yolk glands paired, lateral; copulatory bursa a muscular sac. Conn.

◄ **Fig.13.33,** *Prorhynchella minuta.* Ventral view. × 100. *asp*, anterior sensory pit; *b*, brain; *cb*, copulatory bursa; *cgp*, common genital pore; *cp*, ciliated pit; *i*, intestine; *m*, mouth; *o*, ovary; *ph*, pharynx; *pn*, protonephridium; *pp*, penis; *rt*, rhabdite tract; *t*, testis; *vd*, vas deferens; *yg*, yolk glands. (After Ruebush.)

11a   (10) Very small forms with large eyes and colorless bodies . . . .
          Krumbachia minuta Ruebush 1938

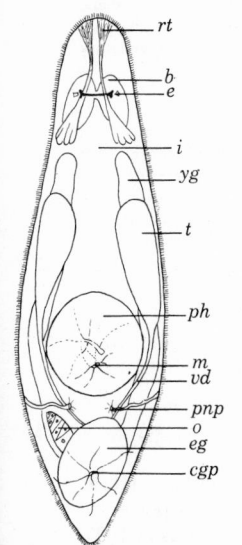

Length 0.3–0.6 mm; gonopore ventral near posterior end; testes paired, ventrolateral, anterior to pharynx; penis a muscular bulb anterior to gonopore; ovary club-shaped; ventral on right side; yolk glands paired, somewhat ribbon-shaped, lateral; copulatory bursa a muscular sac near genital pore; smallest sexually mature rhabdocoele described from N. A. so far. Va.

◄ **Fig. 13.34.** *Krumbachia minuta.* Ventral view. × 100. *b*, brain; *cgp*, common genital pore; *e*, eye; *eg*, egg; *i*, intestine; *m*, mouth; *o*, ovary; *ph*, pharynx; *pnp*, excretory pore; *rt*, rhabdite tract; *t*, testis; *vd*, vas deferens; *yg*, yolk gland. (After Ruebush.)

11b   Larger forms; body opaque; eyes lacking or if present quite small.
               Krumbachia virginiana Kepner and Carter 1931

    Length 2–3 mm; slender with little or no pigment, but with opaque material in extensive enteron; gonopore near anterior end of posterior body third; reproductive organs much as in *K. minuta* (above) except that copulatory bursa has small anterior pear-shaped region and much larger posterior region. Va.

12a **(8)** Anterior end of body developed into a retractile telescoping sensory
               projection; end stems of excretory system open into genital antrum. .
               Subfamily **Rhynchomesostominae**

Single genus . . . . ***Rhynchomesostoma*** Luther
Single species. . . . . ***R. rostrata*** (Müller) 1774

Length to 5 mm; body transparent but with yellowish-red pigment in parenchyma; ventral surface flat, dorsal surface convex; pharynx small, two large anterior rhabdite glands. N. Y., Va.

◄ **Fig. 13.35.** *Rhynchomesostoma rostrata.* (*a*) Proboscis partly extended. (*b*) Fully contracted. × 40. (After Graff.)

(b)

(a)

12b   Anterior end of body not developed into a retractile, telescoping
               sensory projection; end stems of excretory system empty into an
               excretory vesicle combined with the mouth . . . .
               Subfamily **Typhloplaninae**   **13**

13a **(12)** With copulatory bursa, copulatory antrum, seminal receptacle and
               paired uteri; without dermal rhabdites. . . . . ***Castrada*** O. Schmidt   **14**

13b   Without copulatory bursa or copulatory antrum. **16**

14a **(13)** With zoochlorellae in mesenchyme. . ***C. hofmanni*** (M. Braun) 1885

    Length 1.5 mm; unpigmented but may appear green due to zoochlorellae; shape cylindrical; without eyes; large rhabdoids in tracts; gonopore just posterior to pharynx; testes paired, elongate pear-shaped; yolk glands deeply lobed; penis and copulatory bursa enclosed by muscular wall of copulatory antrum. N. Y.

◄ **Fig. 13.36.** *Castrada hofmanni.* Penis, copulatory bursa, and antrum. Diagram from preparation subjected to pressure. *ac*, Copulatory antrum; *ed*, ejaculatory duct; *pgs* prostate secretions; *rm*, circular muscle; *sp*, sperm in copulatory bursa; *sv*, seminal vesicle; *ts*, toothlike spine. Much enlarged. (After Luther.)

14b   Without zoochlorellae in mesenchyme **15**

15a **(14)** With well-developed copulatory bursa outside of muscular mantle
               of penis and in addition to blind sac of penis; gonopore in posterior
               body third . . . ***C. virginiana*** Kepner, Ruebush and Ferguson 1937

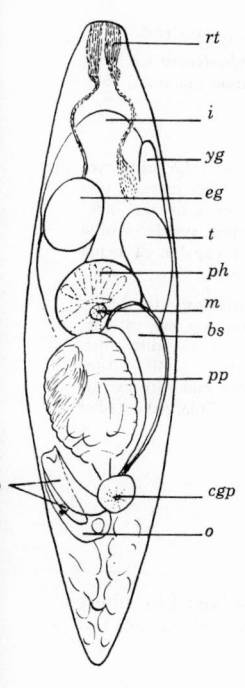

Length 1.5–2.0 mm; color apple green due to pigment in mesenchyme; no eyes; testes large, paired, anterolateral to pharynx; penis and blind sac enclosed in weak muscular mantle; ejaculatory duct cuticularized; blind sac lined by spine bearing membrane; ovary single, just posterior to gonopore; paired uteri extend forward from genital antrum as does single, muscular, armed, copulatory bursa. Va.

◄ **Fig. 13.37.** *Castrada virginiana.* Ventral view. × 50. *bs*, blind sac of penis; *cb*, copulatory bursa; *cgp*, common genital pore; *eg*, egg; *i*, intestine; *m*, mouth; *o*, ovary; *ph*, pharynx; *pp*, penis; *rt*, rhabdite tracts; *t*, testis; *yg*, yolk gland. (After Kepner, Ruebush, and Ferguson.)

15b With well-developed penis blind sac, but without a second blind sac arising from the copulatory antrum; gonopore in second fifth of body just behind mouth . . . .
C. lutheri Kepner, Stirewalt and Ferguson 1939

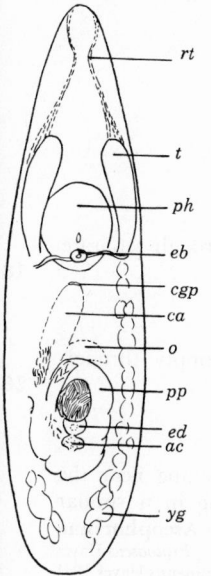

Length 1.5 mm; color pale green; no eyes; penis blind sac enclosed in muscular mantle of penis bulb and divided into 3 regions by the structure of its spines; genito-intestinal duct present; otherwise reproductive systems similar to those of *C. virginiana*. Va.

◄ **Fig. 13.38.** *Castrada lutheri.* Ventral view. × 65. *ac*, copulatory antrum; *ca*, common genital antrum; *cgp*, common genital pore; *eb*, pharyngeal excretory beaker; *ed*, ejaculatory duct; *o*, ovary; *ph*, pharynx; *pp*, penis; *rt*, rhabdite tract; *t*, testis; *yg*, yolk gland. (After Kepner, Stirewalt, and Ferguson.)

16a   **(13)** Usually with eyes; without zoochorellae; uterus, if present, paired . . .
Strongylostoma Örsted
Single species reported from N. A. . . . . *S. gonocephalum* (Silliman) 1884

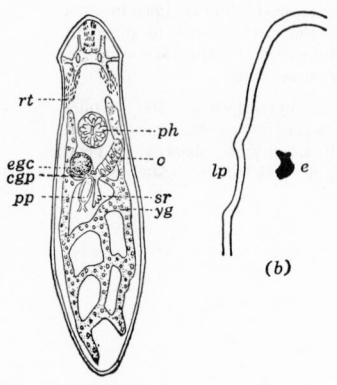

Length 1.2 mm; parenchyma yellowish; eyes carmine red; small dermal rhabdoids present, two shallow oval pits just behind eyes at the side; seminal receptacle an independent vesicle arising from the ovovitelline duct and capable of being closed by special sphincter muscles.

◄ **Fig. 13.39.** *Strongylostoma gonocephalum.* (a) Entire animal: *cgp*, common genital pore; *egc*, egg capsule; *o*, ovary; *ph*, pharynx; *pp*, penis; *rt*, rhabdite tract; *sr*, seminal receptacle; *yg*, yolk gland. × 40. (After Silliman.) (b) Outline of anterior end with eye (e) and shallow pit (*lp*) of one side. Enlarged. (After Graff.)

16b   Without eyes; with zoochlorellae; uterus paired . . . .
Typhloplana Ehrenberg
Single species reported from N. A. . . . . *T. viridata* (Abildgaard) 1790

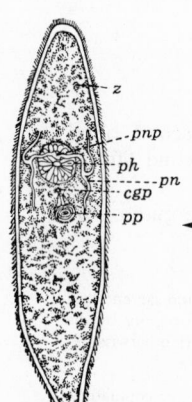

Length 0.5–1.0 mm; transparent, but zoochlorellae in parenchyma produce grass-green color and obscure the internal organs; body tapers toward both ends; gonopore posterior to pharynx; testes small, pear-shaped to elongate, near pharynx; male genital canal with small spines; ejaculatory duct cuticularized, penis pear-shaped; summer eggs develop in body of parent; winter eggs yellowish-brown in color and up to 10 in number. Ill., Mich., N. Y., and Va.

◄ **Fig. 13.40.** *Typhloplana viridata.* *cgp*, common genital pore; *ph*, pharynx; *pn*, protonephridium; *pnp*, excretory pore; *pp*, penis; *z*, zoochlorellae. × 70. (After Graff.)

17a   **(7)** The 2 main stems of the excretory system empty through separate pores **18**

17b   The 2 main stems of the excretory system do not empty through separate pores **26**

18a   **(17)** With an ascus (a tubelike muscular gland-organ opening near the genital pore) or at least with special glands opening in a similar position . . . . Subfamily **Ascophorinae**
Single genus known from fresh water in N. A. . . . . *Protoascus* Hayes
Single species known. . . . . *P. wisconsinensis* Hayes 1941

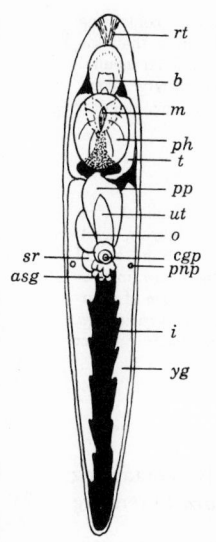

Length 0.4–1.5 mm; body slender, circular or almost so in cross section, colorless to light yellow; no eyes; gonopore near middle of body; testes laterodorsal to pharynx; penis, seminal vesicle, and prostatic vesicle enclosed in heavy muscular sheath; ejaculatory duct cuticularized; copulatory antrum present; ovary lying to right of ejaculatory duct and posterior to penis bulb, yolk glands follicular; uterus present, sometimes tending to be paired; seminal receptacle spherical in shape and with 2 pores. Wis.

◄ **Fig. 13.41.** *Protoascus wisconsinensis. asg,* ascus gland; *b,* brain; *cgp,* common genital pore; *i,* intestine; *m,* mouth; *o,* ovary; *ph,* pharynx; *pnp,* excretory pore; *pp,* penis; *rt,* rhabdite tract; *sr,* seminal receptacle; *t,* testis; *ut,* uterus; *yg,* yolk gland. (After Hayes.)

18b Without an ascus and without special glands opening near the genital pore **19**

19a (18) Yolk glands much branched, anastomosing; mouth near anterior end; body usually triangular in cross section . . . .
Subfamily **Phaenocorinae** **20**
Gonopore near middle of body; pharynx cask-shaped or somewhat cone-shaped, directed anteriorly; pear-shaped appendages in genital antrum.

19b Yolk glands smooth or slightly lobed, rarely branched; mouth not near anterior end; body rounded or oval in cross section . . . .
Subfamily **Olisthanellinae**
Single genus known from fresh water in N. A. . . . . *Olisthanella* Voigt

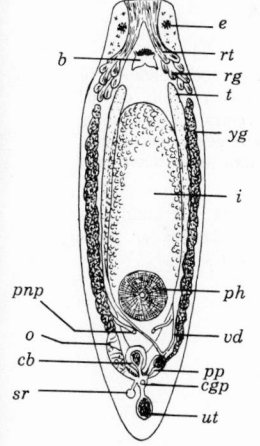

Length 1.0–6.0 mm; unpigmented, but perivisceral fluid or gut contents may give some color; testes simple; yolk glands elongated, slightly indented; no cuticular stylet but ejaculatory duct with cuticular covering. Although this genus has been reported several times from N. A. no definite species has been listed. Wis., Va., Mich.

◄ **Fig. 13.42.** *Olisthanella truncula nassonoffii.* Ventral view. × 14. *b,* brain; *cb,* copulatory bursa; *cgp,* common genital pore; *e,* eye; *i,* intestine; *o,* ovary; *ph,* pharynx; *pnp,* excretory pore; *pp,* penis; *rg,* rhabdite gland; *rt,* rhabdite tract; *sr,* seminal receptacle; *t,* testis; *ut,* uterus; *vd,* vas deferens; *yg,* yolk gland. (After Graff.)

20a (19) Relatively large animals, up to 4.8 mm in length; without zoochlorellae and to date reported only from sulphur springs. . . . . . .
Pseudophaenocora Gilbert
Single species known. . . . . *P. sulfophila* Gilbert 1938
Similar in appearance to *Phaenocora.* Body unpigmented; gonopore in anterior body

half; testes arborescent, extensive, dorsal to yolk glands; penis small, containing "prostatic bladder" as well as prostatic vesicle, and with single wall rather than sac within sac construction; blind sac arises from superior genital antrum; inferior genital antrum extensive with large papilla projecting into lumen; intestinal bursa large, communicating with intestine and female genital canal by long ducts; ovary long, slender; yolk glands paired, branched, and anastomosing posteriorly. Va.

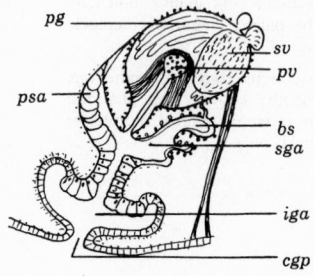

◄ **Fig. 13.43.** *Pseudophaenocora sulfophila.* Sagittal view of male copulatory organ and genital antrum semidiagrammatic and much enlarged. *bs*, blind sac of genital antrum; *cgp*, common genital pore; *iga*, inferior genital antrum; *pg*, prostate gland; *psa*, edge of pear-shaped appendages of bursa; *pv*, prostate vesicle; *sga*, superior genital antrum; *sv*, seminal vesicle. (After Gilbert.)

20b　　Smaller animals, 1.0–3.0 mm in length; frequently containing zoochlorellae. . . . . ***Phaenocora*** Ehrenberg　　**21**

21a　(20)　Male copulatory organ built on a simple sac within a sac plan . . . .　　**22**

21b　　Male copulatory organ not built on a simple sac within a sac plan. . . . .　　**24**

22a　(21)　Ejaculatory duct armed with spines. . . . .　　**23**

22b　　Ejaculatory duct not armed with spines . . . .
　　　　　　　　　　　　　　　　　　　　P. virginiana Gilbert 1935

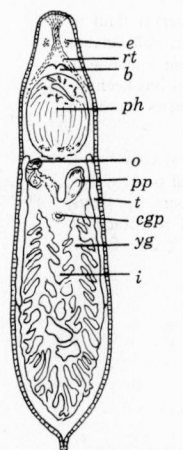

Length 0.7–2.0 mm; color usually dark green due to zoochlorellae, but anterior end of body often heavily pigmented; eyes reddish; dermal rhabdites numerous; pharynx large; yolk glands ventral; anastomosing from base of pharynx to posterior end of body; female genital canal long, warty in appearance, and with simple unicellular glands whose ducts enter the canal singly; intestinal bursa a simple sac. Va.

◄ **Fig. 13.44.** *Phaenocora virginiana.* Ventral view. × 28. *b*, brain; *cgp*, common genital pore; *e*, eye; *i*, intestine; *o*, ovary; *ph*, pharynx; *pp*, penis; *rt*, rhabdite tract; *t*, testis; *yg*, yolk gland. (After Gilbert.)

23a　(22)　With irregular strands of brownish pigment superficial in position on dorsal side of body, and with zoochlorellae. . . . .
　　　　　　　　　　　　　　　　　　　　P. falciodenticulata Gilbert 1938

Length 1.0 mm; inferior genital antrum small; only one pear-shaped lobe in superior genital antrum; female genital canal relatively long; intestinal bursa small; oviduct enters proximal end of female genital canal; male copulatory organ simple; denticles of ejaculatory duct long, slender, considerably curved. Va.

◄ **Fig. 13.45.** *Phaenocora falciodenticulata.* Somewhat diagrammatic and much enlarged view of evaginated male copulatory organ showing spines. (After Gilbert.)

23b Without pigment and without zoochlorellae . . . .
                                                              P. agassizi Graff 1911
Length 1.0–2.0 mm; unpigmented, but intestinal contents may give greenish-yellow color to body; eyes reddish yellow; rhabdites present; intestine more or less deeply indented; testes small, slender, lateral to intestine near middle of body. N. Y.

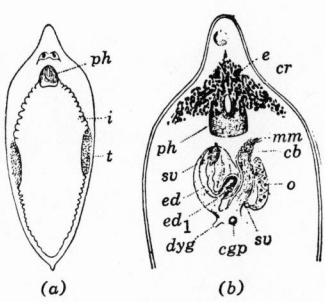

◄ **Fig. 13.46.** *Phaenocora agassizi.* (*a*) Slightly compressed: *i*, intestine; *ph*, pharynx; *t*, testis; × 45. (*b*) Anterior part enlarged: *cb*, copulatory bursa; *cgo*, common genital pore; *cr*, crystalloids (clear or slightly colored); *dyg*, duct of yolk gland; *e*, eye; *ed* and *ed₁*, ejaculatory duct; *mm*, muscles of bursa; *o*, ovary; *sr*, seminal receptacle; *sv*, seminal vesicle; × 70. (After Graff.)

(*a*)   (*b*)

24a (21) Proximal portion of copulatory organ containing sperm and prostatic secretions, separated from distal portion containing long ejaculatory duct by a muscular diaphragm. **25**

24b Portion of copulatory organ containing ejaculatory duct may be retracted into space partially occupied by prostatic secretion, ejaculatory duct short and straight . . ***P. highlandense*** Gilbert 1935

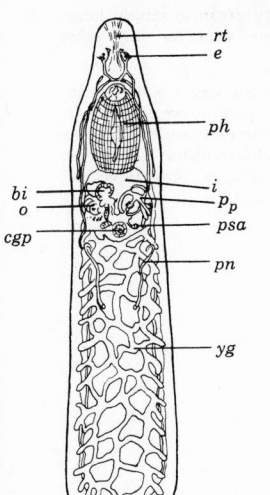

Length 3.0–3.7 mm; robust; pharynx large; red pigment extends from anterior end to base of pharynx; granular pigment confined to eyes; rhabdites few, scattered; yolk glands with dorsal extensions and anastomosing freely; testes tubular, branched; copulatory organ small, denticles confined to distal end of evaginated penis; no male genital canal; female genital canal short, thick, without giant unicellular glands; intestinal bursa divided into dorsal and ventral chambers; zoochlorellae often present. Va.

◄ **Fig. 13.47.** *Phaenocora highlandense.* Ventral view. × 20. *bi*, intestinal bursa; *cgp*, common genital pore; *e*, eye; *i*, intestine; *o*, ovary; *pp*, penis; *ph*, pharynx; *pn*, protonephridium; *psa*, pear-shaped appendage; *rt*, rhabdite tract; *yg*, yolk gland. (After Gilbert.)

25a (24) Without a pseudo ejaculatory duct . . . . . ***P. kepneri*** Gilbert 1935

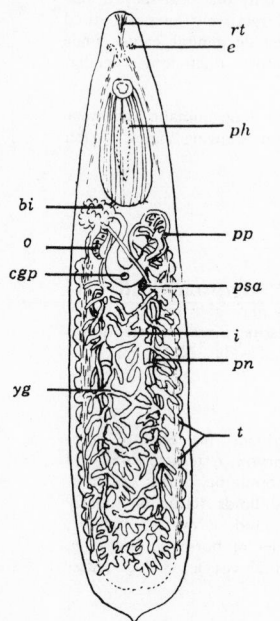

Length 2.0–2.5 mm; unpigmented except for eyes; yolk glands ventral; testes follicular, ejaculatory duct long, lined with cuticular layer, distal portion with denticles of 2 sizes; male copulatory organ very large, $\frac{1}{7}$ body length; well-developed male genital canal present; female canal long and tapering; intestinal bursa large, folded; unicellular glands that empty into female genital canal highly developed with ducts grouped together in sheaflike fashion; zoochlorellae usually present. Middle Atlantic states.

◄ **Fig. 13.48.** *Phaenocora kepneri.* Ventral view. × 50. *bi*, intestinal bursa; *cgp*, common genital pore; *e*, eye; *i*, intestine; *o*, ovary; *pp*, penis; *ph*, pharynx; *pn*, protonephridium; *psa*, pear-shaped appendages; *rt*, rhabdite tract; *t*, testis; *yg*, yolk gland. (After Gilbert.)

25b With a pseudo ejaculatory duct in addition to the true ejaculatory duct . . . . ***P. lutheri*** Gilbert 1937

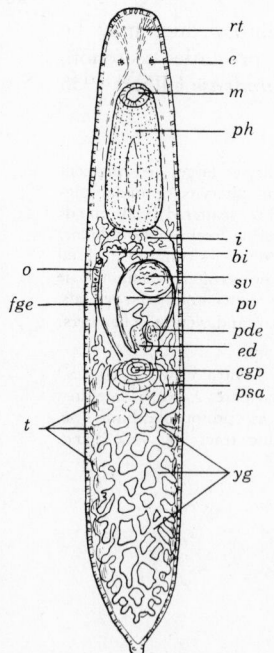

Length 1.75–3.0 mm; color grayish, anterior end rose-colored; zoochlorellae never observed; unpigmented except for eyes; dermal rhabdites present; female genital organs typical for genus; male copulatory organ extremely large; spines of ejaculatory duct larger than for any other American species. Va.

◄ **Fig. 13.49.** *Phaenocora lutheri.* Ventral view. × 40. *bi*, intestinal bursa; *cgp*, common genital pore; *e*, eye; *ed*, ejaculatory duct with denticles; *fgc*, female genital canal; *i*, intestine; *m*, mouth; *o*, ovary; *pde*, pseudo-ejaculatory duct; *ph*, pharynx; *psa*, pear-shaped appendages; *pv*, prostate vesicle; *rt*, rhabdite tract; *sv*, seminal vesicle; *t*, testis; *yg*, yolk gland. (After Gilbert.)

26a **(17)** The 2 main excretory stems empty into an excretory basin combined with the mouth; with typical rosulate pharynx. . . . .
Subfamily **Mesostominae** 27

26b The 2 main stems of the excretory system empty through a common excretory pore located between the mouth and the gonopore; pharynx elongated and its free end directed posteriorly . . . .
Subfamily **Opistominae**

A single genus . . . . **Opistomum** O. Schmidt
A single species . . . . **O. pallidum** O. Schmidt, 1848
Length to 4.5 mm; transparent but often with greenish mesenchyme fluid and yellow to red gut contents; body cylindrical, tapering towards both ends, without rhabdites; eyes lacking; gonopore slightly posterior to mouth; testes elongate, ribbonlike, lateral to intestine; seminal vesicle enclosed in sheath with penis; penis with a crown of 16 spines; ovary club-shaped, lying between mouth and gonopore; uterus simple, lateral to pharynx; yolk glands elongate narrow ribbons, irregularly lobed; animal avoids sunlight and will not stand much heat. Va.

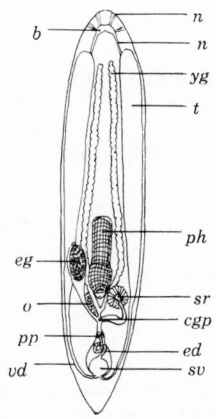

◄ **Fig. 13.50.** *Opistomum pallidum.* Ventral view. × 15. (Intestine and excretory systems not shown.) *b*, brain; *cgp*, common genital pore; *ed*, ejaculatory duct; *eg*, egg; *n*, nerves; *o*, ovary; *ph*, pharynx; *pp*, penis; *sr*, seminal receptacle; *sv*, seminal vesicle; *t*, testis; *vd*, vas deferens; *yg*, yolk gland. (After Graff.)

27a **(26)** Usually without a ventral pit and without insemination canal . . . .
Mesostoma Ehrenberg 28

27b With a ventral pit (sensory pouch) and with an insemination canal joining the copulatory bursa and the ovovitelline duct . . . .
Bothromesostoma Braun

Single species known from N. A. . . . . ***B. personatum*** (O. Schmidt) 1848
Length to 7.0 mm; body colored by clear brown pigment which, combined with the dark color of the intestinal contents, often produces a dark-brown to bluish-black color in the region of the intestine; eyes oval, about as far from the sides of the body as from each other, visible only in lightly pigmented specimens; ventral pit posterior to eyes; common opening for mouth, excretory, and genital systems approximately in middle of ventral surface; summer and winter eggs produced. Mich., Va.

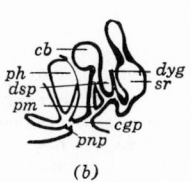

(b)

◄ **Fig. 13.51.** *Bothromesostoma personatum.* (*a*) Entire animal. × 5. (After Schmidt.) (*b*) Diagram of sexual organs: *cb*, copulatory bursa; *cgp*, common genital pore; *dsp*, spermatic duct; *dyg*, duct of yolk gland; *ph*, pharynx; *pm*, opening of penis; *pnp*, excretory pore; *sr*, seminal receptacle. Much enlarged. (After Luther.)

(a)

28a (27) With a single copulatory bursa **29**

28b   With 2 copulatory bursae . . **Mesostoma californicum** Hyman 1957

Length 5 mm; color dark gray; body elongate slender; a conspicuous lateral depression on each side at anterior end. Eyes present but not visible in living animal. Gonopore just behind mouth. Both copulatory bursae quite large. Calif.

◄ **Fig. 13.52.** *Mesostoma californicum*. From life. *sp*, sensory pit. (After Hyman.)

29a (28) Seminal vesicle and prostatic vesicle distinct but not separated from one another, and both enclosed in penis bulb **32**

29b   Seminal vesicle and prostatic vesicle separate and not enclosed together in penis bulb **30**

30a (29) Anterior end usually not truncate and without a median pit **31**

30b   Anterior end broad, truncate, and with a median pit which receives rhammite tracts . . . . **M. arcticum** Hyman 1938

Length 4.0–5.0 mm; color grayish-brown to brown; copulatory bursa spherical with slender duct; large prostatic vesicle between seminal vesicle and penis bulb; mouth and gonopore combined. Manitoba, Canada; Wyo.

◄ **Fig. 13.53.** *Mesostoma arcticum*. Dorsal view. × 20. *ap*, anterior pit; *cga*, common genital antrum; *e*, eye; *eg*, egg; *i*, intestine; *l*, lip; *m*, mouth; *ph*, pharynx; *rt*, rhabdite tract; *yg*, yolk gland. (After Hyman.)

31a (30) Body very transparent; uteri with anterior extensions producing H-shape; without a ventral pit . . . . . . **M. ehrenbergii** (Focke) 1836
Length to 10 mm or sometimes more; color pale yellow to brown; intestinal contents usually yellowish-brown; shape thin and leaflike, tapering at both ends; conspicuous rhabdite tracts in anterior end; eyes black; shallow pit on each side of dorsal surface at anterior end; gonopore posterior to mouth; ejaculatory duct lacks cuticular lining; both summer and winter eggs produced; summer eggs develop and young embryos may be seen moving around in uterus of parent. According to Hyman, **M. macropenis** Hyman 1939 is at best a variant of this species. Cosmopolitan.

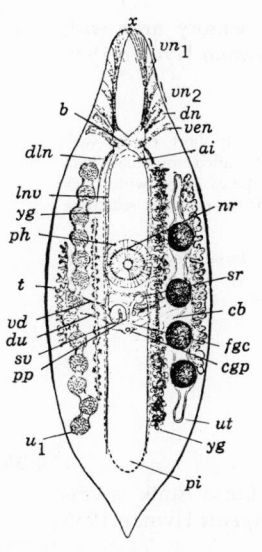

◄ **Fig. 13.54.** *Mesostoma ehrenbergii.* Diagram from ventral side showing nervous, digestive, and reproductive systems. Left side shows summer eggs, the right, winter eggs: *ai*, anterior branch of intestine; *b*, brain; *cb*, copulatory bursa; *cgp*, common genital pore; *dln*, dorsal longitudinal nerve; *dn*, dorsal nerve of brain; *du*, duct of uterus; *fgc*, female genital canal; *lnv*, ventral longitudinal nerve; *nr*, pharyngeal nerve ring; *o*, ovary; *ph*, pharynx; *pi*, posterior branch of intestine; *pp*, penis; *sr*, seminal receptacle; *sv*, seminal vesicle; *t*, testis; *ut*, uterus; *vd*, vas deferens; *ven*, ventral nerve of brain; vn_1 and vn_2, the 2 pairs of anterior nerves of brain; *x*, chiasma of anterior nerves; *yg*, yolk gland. × 6. (After Graff, Vogt, Fuhrmann, and Luther.)

31b   Body opaque; dark brown in color; uteri without anterior extensions; with a ventral pit . . . . . . **M. macroprostatum** Hyman 1939

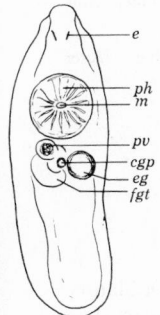

Length 2.0–2.5 mm; anterior end narrowed, posterior end rounded, rectangular in cross section; prostatic vesicle large; male genital canal long; penis papilla large; lumen of penis papilla separated from prostatic vesicle by valve; gonopore some distance posterior to mouth. Position uncertain— probably does not belong in this genus. Wyo.

◄ **Fig. 13.55.** *Mesostoma macroprostatum.* Dorsal view. × 20. *cgp*, common genital pore; *e*, eye; *eg*, egg; *fgt*, female genital tract; *m*, mouth; *ph*, pharynx; *pv*, prostate vesicle. (After Hyman.)

32a **(29)** Gonopore posterior to and distinctly separate from mouth **33**

32b   Gonopore combined with mouth or so close as to be almost combined **34**

33a **(32)** With a small blind sac in the penis papilla; no sensory pits . . . . . .
                M. virginianum Kepner, Ferguson, and Stirewalt 1938

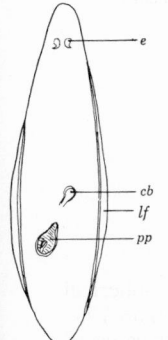

Length 3.0 mm; body chocolate brown; oblanceolate, bulky, smallest at anterior end, frequently forms pliable lateral flaps; one pair black eyes; bursa flask-shaped with long stalk; uteri sac-shaped; short dermal rhabdites as well as rhammites present; testes large anastomosing along mid-dorsal line. Va.

◄ **Fig. 13.56.** *Mesostoma virginianum.* Dorsal view. × 15. *cb*, copulatory bursa; *e*, eye; *lf*, lateral flaps; *pp*, penis. (After Kepner, Ferguson, and Stirewalt.)

33b   Without a blind sac in the penis; well-developed sensory area with
             pit on each side at anterior end . . . *M. columbianum* Hyman 1939

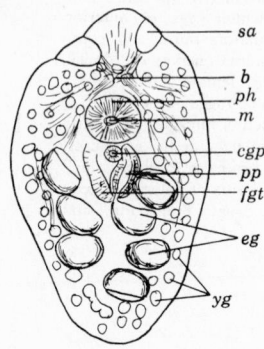

Length 1.3 mm; color dark gray; body short, plump with
round ends; no eyes; bursa stalk thick-walled, muscular, with
sphincter at entrance into antrum; penis large with thick
muscular wall and central passage for the prostatic secretion.
D. C.

◀ **Fig. 13.57.** *Mesostoma columbianum.* Dorsal view. × 35. *b*,
brain; *cgp*, common genital pore; *eg*, egg; *fgt*, female genital
tract; *m*, mouth; *ph*, pharnyx; *pp*, penis; *sa*, sensory area; *yg*,
yolk glands, (After Hyman.)

34a (32) Penis papilla not curved or beaklike **35**
34b   Penis papilla distinctly curved and beaklike; bursa stalk asym-
             metrical with reference to bursa sac. . . *M. curvipenis* Hyman 1955

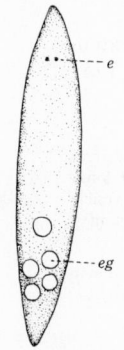

Length 3-4 mm; color grayish-brown. Shape fusiform; copulatory bursa
somewhat squarish, sac-shaped; bursa stalk curves around it and enters from
the rear. Dela.

◀ **Fig. 13.58.** *Mesostoma curvipenis.* From life. *e*, eye; *egc*, egg capsule. (After
Hyman.)

35a (34) Penis papilla mammiform with inner cuticularized tube; copulatory
             bursa cylindric. . . . . *M. vernale* Hyman 1955

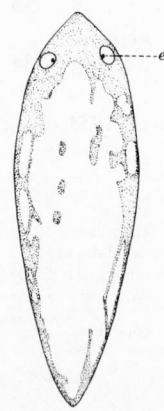

Length 3.5 mm; color opaque brown to black; body plump, convex
dorsally, flattened ventrally; eyes conspicuous—widely separated with
large white areas around them; penis bulb divided diagonally into seminal
and prostatic portions.

◀ **Fig. 13.59.** *Mesostoma vernale.* From life. *e*, eye. (After Hyman.)

35b   Penis papilla very small, cuticularized; copulatory bursa spherical
             with long broad canal . . . . *M. andrewsi* Hyman 1957
             Length 2 mm or more; color dark brown, pigmentation more pronounced on ventral

side; eyes poorly developed; no sensory depressions; penis oblong with sinuous cuticular-ized ejaculatory duct; prostatic portion of penis bulb only slightly developed. Alaska.

36a (1) Eukalyptorhynchia with special protractor muscles of the proboscis **38**

36b   Eukalyptorhynchia without special protractor muscles for the
proboscis . . . . Family **Koinocystidae** **37**

37a (36) Proboscis well developed with terminal opening; gonads paired . . .
Koinocystis Meixner

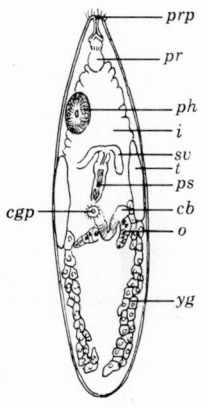

Pronounced sphincter at base of end cone of proboscis; seminal vesicles paired. This genus has been reported from Va., Mich., and Wis., but no species identification appears in the literature.

◄ **Fig. 13.60.** *Koinocystis tvaerminnensis.* Ventral view. × 14. *cb*, copulatory bursa; *cgp*, common genital pore; *i*, intestine; *o*, ovary; *ph*, pharynx; *pr*, proboscis; *prp*, proboscis pore; *ps*, penis stylet; *sv*, seminal vesicle; *t*, testis; *yg*, yolk gland. (After Karling.)

37b   Proboscis small with subterminal opening; single ovary; left testis
much reduced. . . . . *Microkalyptorhynchus* Kepner and Ruebush
Single species . . . . *M. virginianus* Kepner and Ruebush 1935

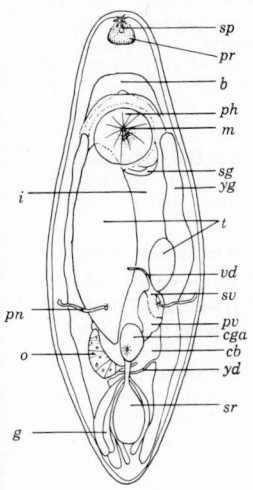

Average length 1.0 mm; body spindle-shaped, gonopore in anterior part of last third of body; excretory pores paired, slightly anterior to gonopore; right testis large, penis large, club-shaped; enclosing seminal and prostatic vesicles and a funnel-shaped ejaculatory duct lined by a short cuticular tube; yolk glands large, club-shaped, located laterally; large seminal receptacle, copulatory bursa rudimentary. Although placed in the Koinocystidae by Kepner and Ruebush, this form is probably not a Kalyptorhychid. Va.

◄ **Fig. 13.61.** *Microkalyptorhynchus virginianus.* Ventral view. × 65. *b*, brain; *cb*, copulatory bursa; *cga*, common genital antrum; *g*, accessory glands; *i*, intestine; *m*, mouth; *o*, ovary; *ph*, pharynx; *pn*, protonephridium; *pr*, proboscis (?); *pv*, prostate vesicle; *sg*, salivary gland; *sp*, sensory pit; *sr*, seminal receptacle; *sv*, seminal vesicle; *t*, testis; *vd*, vas deferens; *yd*, yolk duct, *yg*, yolk gland. (After Kepner and Ruebush.)

38a (36) With unpaired gonads and the male gonopore posterior to the
female . . . . Family **Gyratricidae**
Single genus . . . . *Gyratrix* Ehrenberg
Single species reported from fresh water in N. A.. . . . .
G. hermaphroditus Ehrenberg 1831
Length to 2.0 mm; body colorless, rounded in cross section; very contractile and may round up into a ball or elongate into a thin cylinder; rhabdites only in proboscis; mouth just anterior to middle of body; the male pore, the dorsal opening of the copulatory

bursa, and a ventral pore for the passage of eggs, all posterior; testis usually on right; yolk glands on left; copulatory bursa large, median in position, posterior to large median uterus; prostatic vesicle with chitinous tube protruding into chitinous tip of ejaculatory duct and both enclosed in a common sheath. One subspecies, **G. herma-phroditus hermaphroditus** Ehrenberg reported from fresh water in N. A. with 2 red or black eyes and with a hook on the end of the chitinous sheath of the ejaculatory duct. Cosmopolitan.

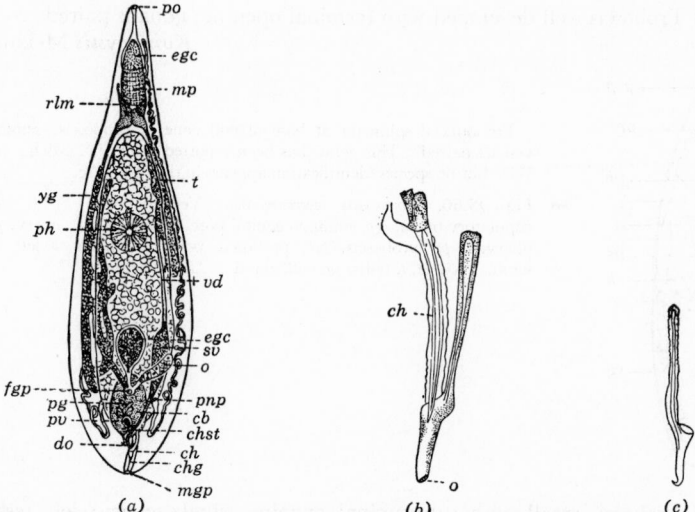

Fig. 13.62. *Gyratrix hermaphroditus.* (*a*) Ventral view of compressed specimen. *cb*, copulatory bursa; *ch*, chitinous tube; *chg*, chitinous stylet leading from prostate vesicle; *chst*, stalk of chitinous tube; *do*, dorsal opening of copulatory bursa; *ecp*, end cone of proboscis; *egc*, egg capsule in uterus; *fpg*, female genital pores; *mgp*, male genital pore; *mp*, muscular portion of proboscis; *o*, ovary; *pg*, prostate glands; *ph*, pharynx; *pnp*, external opening of protonephridium; *po*, external opening of proboscis sheath; *pv*, prostate vesicle; *rlm*, attachment of the long proboscis retractor muscles; *sv*, seminal vesicle; *t*, testis; *vd*, vas deferens; *yg*, yolk gland; × 30. (After Graff.) (*b*) stylet sheath with straight tube. *ch*, chitinous stylet; *o*, opening of stylet sheath. Much enlarged. (After Hallez.) (*c*) *Gyratrix hermaphroditus hermaphroditus.* Stylet sheath with curved point. Much enlarged. (After Graff.)

38b   With paired gonads and a single gonopore. . . . Family **Polycystidae**   **39**
            Mouth and pharynx anterior to middle of body; seminal vesicle paired or unpaired, completely separated from prostatic vesicle or fused only posteriorly and not surrounded by a common muscular mantle; with or without bursa.

39a (38) Seminal vesicle entirely separate from prostatic vesicle; stylet lying in duct connecting prostatic vesicle with male genital canal and receiving material only from prostatic vesicle . . . *Polycystis* Kölliker
            Single species known from fresh water in N. A. . . . . *P. goettei* Breslau 1906
            Length to 2.0 mm; body colorless anteriorly, reddish posteriorly, often appearing striated; no seminal receptacle; yolk glands branched and anastomosing; proboscis opening terminal; otherwise similar in shape, location, and appearance of organs, etc. to *Klattia virginiensis* (see below). *P. roosevelti* Graff (1911) appears to be identical with the European *P. goetti.* N. Y.

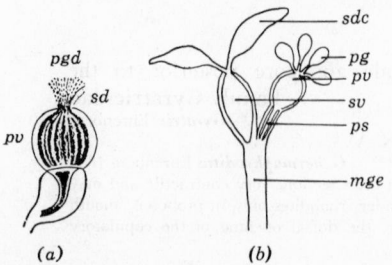

◄ **Fig. 13.63.** *Polycystis goettei.* (*a*) Chitinous stylet. *pgd*, duct leading from the prostate glands; *pv*, prostate vesicle; *sd*, sperm ducts. × 400. (After Graff.) (*b*) Diagram of appendages of male genital canal of *Polycystis.* *mgc*, male genital canal; *pg*, prostate glands; *ps*, chitinous stylet; *pv*, prostate vesicle; *sdc*, spermiducal vesicle; *sv*, seminal vesicle. (After Graff.)

39b   Seminal vesicle fused posteriorly with prostatic vesicle; stylet lying
in male genital canal and receiving materials from both the seminal
vesicle and the prostatic vesicle . . . .
Klattia Kepner, Stirewalt and Ferguson
Single species. . . . . *K. virginiensis* Kepner, Stirewalt, and Ferguson 1939

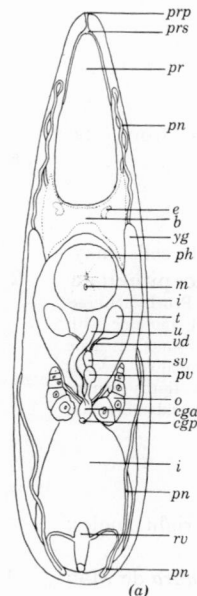

(a)

Length 1.0–1.5 mm; color white to muddy; proboscis opening subterminal; excretory pore near posterior end of body; eyes black, just
behind proboscis; testes small, pear-shaped, one on either side behind
pharynx; seminal vesicle and prostatic vesicle separated anteriorly and
not surrounded by a common muscular mantle; one ovary on either
side of gonopore; yolk glands long, club-shaped, lateral in position;
uterus large, extending from gonopore to posterior border of pharynx;
seminal receptacle present. Va., Pa.

◀ **Fig. 13.64.** *Klattia virginiensis.* (*a*) Ventral view. × 50. *b*, brain;
cga, common genital antrum; *cgp*, common genital pore; *e*, eye; *i*,
intestine; *m*, mouth; *o*, ovary; *ph*, pharynx; *pn*, protonephridium; *pr*,
proboscis; *prp*, opening of proboscis sheath; *prs*, proboscis sheath;
pv, prostate vesicle; *rv*, renal vesicle; *sv*, seminal vesicle; *t*, testis,
u, uterus; *vd*, vas deferens; *yg*, yolk gland. (*b*) Diagram of appendages
of male genital canal of *Klattia. mgc*, male genital canal; *pg*, prostate
gland; *ps*, cuticular stylet; *pv*, prostate ves-
icle; *sv*, seminal vesicle; *vd*, vas deferens.
(Both drawings after Kepner, Stirewalt, and
Ferguson.)

(b)

ORDER ALLOEOCOELA

E. Ruffin Jones

The Alloeocoela are for the most part marine, but there are a few which
occur quite commonly in fresh water. Although smaller than the triclads,
they are usually larger than other fresh-water Turbellaria. The shape and
color of the body are quite variable. The intestine tends to be more or less diverticulated, and the pharynx is of the variable or plicate types in all the fresh-
water forms from North America. The protonephridia are often branched
and may have a number of external openings. Eyes are frequently lacking,
but ciliated pits or grooves are quite common. A statocyst occurs in marine
forms but is rather rare in fresh-water species. The testes are usually
follicular rather than compact and the penis is commonly unarmed or has
only a simple cuticular stylet. The ovary is generally paired and distinct

from the yolk glands, which are often follicular, but ovary and yolk gland may be combined to form a germovitellarium.   The male gonopore may be combined with the mouth or with the female gonopore, or there may be as many as three distinct gonopores, the third being the opening of the seminal bursa. The various accessory reproductive structures that occur in the Neorhabdocoela may also occur in the Alloeocoela.

KEY TO SPECIES

<table>
<tr><td>1a</td><td>Alloeocoela with yolk glands more or less distinct from the ovaries . . . .</td><td>6</td></tr>
</table>

<table>
<tr><td>1b</td><td>Alloeocoela without distinct yolk glands although yolk cells may surround ova; stylet present . . . . . . Suborder Lecithoepitheliata
Only 1 family known from fresh water in N. A. . . . . . . Family Prorhynchidae
Pharynx large, usually cylindrical, opening at or near anterior end; intestine with lateral diverticula more or less well developed; well developed brain; no statocyst; male copulatory organ with cuticular stylet opening into mouth tube; unpaired median ventral germo-vitellarium; female gonopore ventral in second body third; genito-intestinal duct present but no other female accessory apparatus; may be found in brackish or fresh water, or may be terrestrial.</td><td>2</td></tr>
</table>

<table>
<tr><td>2a</td><td>(1)</td><td>Prorhynchidae with penis stylet bent approximately at right angles; testis follicles more or less paired and scattered. . . . .
 Geocentrophora de Man </td><td>3</td></tr>
</table>

<table>
<tr><td>2b</td><td>Prorhynchidae with penis stylet straight or only slightly curved and with unpaired testis masses consisting of thickly accumulated follicles . . . . Prorhynchus Schultze
Single species known from N. A. . . . . P. stagnalis M. Schultze 1851</td></tr>
</table>

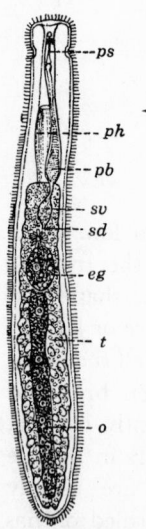

Length to 6 mm; body slender with little or no pigment, threadlike; numerous pear-shaped glands in integument. Cosmopolitan.

◄ **Fig. 13.65.** *Prorhynchus stagnalis.* *eg*, mature egg; *o*, ovary; *pb*, bulb of ejaculatory duct; *ph*, pharynx; *pa*, penis stylet; *sd*, sperm duct; *sv*, seminal vesicle; *t*, testis follicle. × 15. (After Graff.)

3a **(2)** With eyes. **4**

3b Without eyes **5**

4a **(3)** Head region distinctly broadened or with auriclelike appendages, remainder of body somewhat ovoid in shape . . . .
Geocentrophora sphyrocephala (de Man) 1876

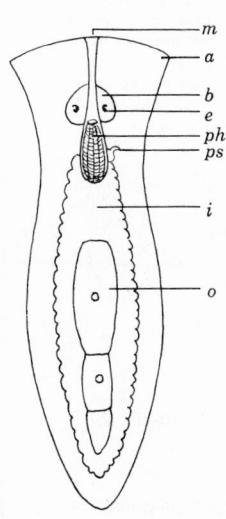

Length 0.6–3.0 mm; broadened head region followed by distinct narrow neck, unpigmented; pharynx ⅓ body width; diverticula of gut inconspicuous; eyes greenish-yellow to dark brown; 6 to 12 testis follicles; seminal vesicle at base of pharynx; bulb of ejaculatory duct well developed; male sex organs usually present in all large individuals. Va.

◄ **Fig. 13.66.** *Geocentrophora sphyrocephala.* × 20. *a*, auricle; *b*, brain; *e*, eye; *i*, intestine; *m*, mouth; *o*, ovary; *ph*, pharynx; *ps*, penis stylet. (After Steinbock.)

4b Head region very little broader than rest of body and without auricles or earlike projections; body more or less ribbonlike, tapering slightly toward posterior end . . . . . *G. applanata* (Kennel) 1888
Length to 4.0 mm; unpigmented; no neck; pharynx ½ to ⅔ body width; diverticula of intestine conspicuous; eyes greenish-yellow to dark brown or red; testis follicles variable but averaging 7 to 8; seminal vesicle near base of pharynx; bulb of ejaculatory duct small; male sex organs seldom developed. Cosmopolitan.

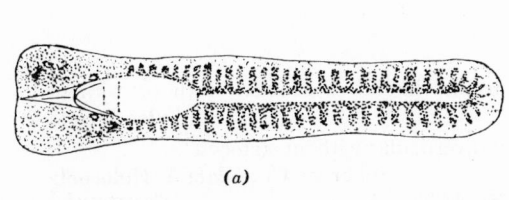

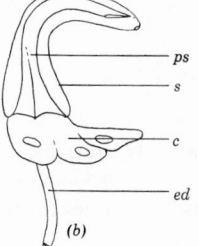

(*a*)

(*b*)

Fig. 13.67. *Geocentrophora applanata.* (*a*) From life. × 20. (After Kennel.) (*b*) Penis much enlarged. *c*, basal cells; *ed*, ejaculatory duct; *ps*, penis stylet; *s*, penis sheath. (After Jones.)

5a   (3) Body black in color. . . . . **G. tropica** Hyman 1941

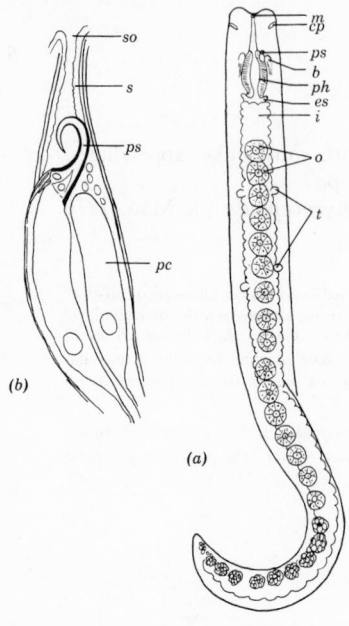

Length to 5.0 mm; heavily pigmented; long, slender, bandlike with sides of body parallel in anterior body half but tapering gradually in posterior half; testis follicles few in number; copulatory organ lined with 4 very elongate cells; stylet more strongly curved than in other species. Canal Zone.

◀ **Fig. 13.68.** *Geocentrophora tropica.* (*a*).Dorsal view. × 20. *b*, brain; *cp*, ciliated pit; *es*, esophagus; *i*, intestine; *m*, mouth; *o*, ovary; *ph*, pharynx; *ps*, penis stylet; *t*, testis. (*b*) Sagittal section of male copulatory apparatus. Much enlarged. *pc*, large cells lining lumen of copulatory organ;; *ps*, penis stylet; *s*, penis sheath; *so*, opening of penis sheath into pharyngeal pocket. (After Hyman.)

5b   Body unpigmented . . . . **G. baltica** (Kennel) 1883

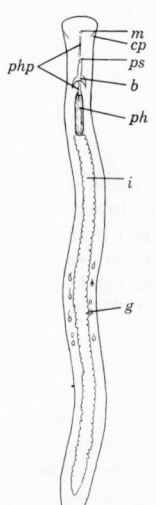

Length to 10.0 mm; width to 0.5 mm; ribbon-shaped; intestinal diverticula fairly pronounced; glands unusually abundant throughout body with several (usually 5 to 6) conspicuous clusters of glands on each side of the intestine near the middle of the body; testis follicles numerous, usually 20 or more; .seminal vesicle elongate, sausage-.shaped; penis bulb well developed; female gonad extending from middle of body to posterior end of intestine; female genital pore in anterior half of second body third. Va.

◀ **Fig. 13.69.** *Geocentrophora baltica* (Habit. Sketch. × 8). *b*, brain; *cp*, ciliated pit; *g*, gland packets; *i*, intestine; *m*, mouth; *ph*, pharynx; *php*, pharynx pocket; *ps*, penis stylet. (After Steinbock.)

6a   (1) Intestine usually with distinct lateral diverticula; with or without a statocyst . . . . Suborder **Seriata**   7

6b   Intestine without distinct diverticula; without statocyst . . . .
                               Suborder **Cumulata** or **Holocoela**

Single family known from fresh water in N. A. . . . . **Plagiostomidae**
Mouth in anterior part of body; gonopore posterior; penis unarmed; yolk glands separate from ovaries and both paired; no accessory female apparatus.

Single genus known from fresh water in N. A. . . . . **Hydrolimax** Haldeman
Single species . . . . **H. grisea** Haldeman 1842

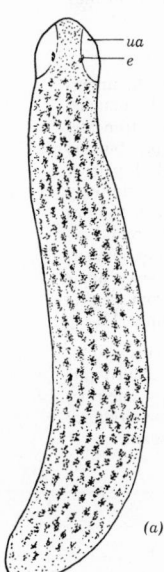

Length 13–15 mm; color white below, gray above except for white area near eyes; body plump, sluglike in appearance; mouth conspicuous on ventral side of head, gonopore ventral near posterior end; pharynx bulbous, excretory pore in mid-dorsal line; testes numerous dorso-lateral to anterior part of intestine; seminal vesicle large, oval near middle of body; prostate vesicle enormous, cylindric, opening into penis bulb; penis papilla small; ovaries follicular, lateral to intestine; yolk glands numerous, follicular, occupying most of peripheral zone of lateral body region; no copulatory sac or seminal receptacle present. Thickness and opaqueness of body prohibit much study of internal structure of living animal. N. J., Pa., Va.

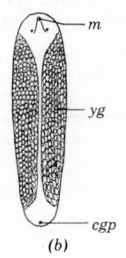

◄ **Fig. 13.70.** *Hydrolimax grisea.* (*a*) Dorsal view. × 6. *e*, eye; *ua;* unpigmented area. (*b*) Ventral view to show distribution of yolk glands. *cgp*, common genital pore; *m*, mouth; *yg*, yolk gland. (*a* from original drawing by Hyman. After Hyman.)

7a   (6) With a statocyst. . . . . Family **Otomesostomidae**
Single genus . . . . *Otomesostoma* Graff
Single species . . . . *O. auditivum* (Du Plessis) 1874

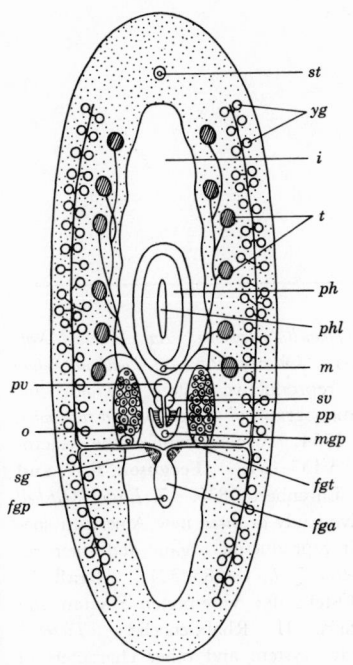

Length to 5.0 mm, nearly half as wide as it is long; color white below, yellow-brown above; body rounded at each end; with paired ciliated pits in addition to the statocyst; no adhesive papillae; 3 pairs of excretory stems with several dorsal and ventral pores; intestine not clearly diverticulated; pharynx a short tube directed ventrally; testes follicular; ovaries behind the pharynx; male gonopore anterior to the female; no bursa present. Va., Calif.

◄ **Fig. 13.71.** *Otomesostoma auditivum.* Dorsal view. × 20. *fga*, female genital antrum; *fgp*, female genital pore; *fgt*, female genital tract; *i*, intestine; *m*, mouth; *mgp*, male genital pore; *o*, ovary; *ph*, pharynx; *phl*, pharynx lumen; *pp*, penis; *pv*, prostate vesicle; *sg*, shell gland; *st*, statocyst; *sv*, seminal vesicle; *t*, testis; *yg*, yolk gland. (After Bresslau.)

7b   Without a statocyst. . . . . Family **Bothrioplanidae**
Single genus . . . . **Bothrioplana** Braun
Single species. . . . . **B. semperi** M. Braun 1881

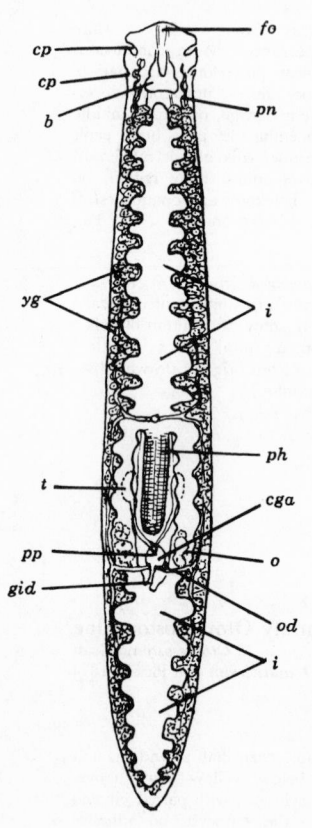

Length 2.5–8.0 mm; width about $\frac{1}{7}$ length; unpigmented and colorless except for intestinal contents; ribbon-shaped with anterior end almost truncate, posterior end rounded; 2 pairs of ciliated pits but no statocyst; adhesive papillae at posterior end; 1 pair of excretory stems opening by ventral pore in mid-body anterior to mouth; anterior and posterior gut sections joined by 2 lobes lateral to and forming a ring around cylindric, posteriorly directed pharynx; testes 2 in number, small, compact; ovaries posterior to pharynx; single gonopore; genito-intestinal duct extends from genital antrum to posterior unpaired section of intestine. Va.

◄ **Fig. 13.72.** *Bothrioplana semperi.* Ventral view. × 18. *b*, brain; *cga*, common genital antrum; *cp*, ciliated pit; *fo*, frontal organ; *gid*, genito-intestinal duct; *i*, intestine; *o*, ovary; *od*, oviduct; *ph*, pharynx; *pn*, protonephridium; *pp*, penis; *t*, testis; *yg*, yolk gland. (After Bresslau.)

References

Castle, W. 1941. The morphology and life history of *Hymanella retenuova*. *Am. Midland Naturalist*, 26:85–97.   **Castle, W. and L. H. Hyman. 1934.** Observations on *Fonticola velata* (Stringer), including a description of the anatomy of the reproductive system. *Trans. Am. Microscop. Soc.*, 53:154–171.   **Ferguson, F. F. 1940.** A monograph of the genus *Macrostomum*, O. Schmidt 1848. Part VII. *Zool. Anz.*, 192:120–145.   **1954.** A monograph of the Macrostomine Worms of Turbellaria. *Trans. Am. Microscop. Soc.*, 73:137–164.   **Ferguson, F. F. and W. J. Hayes. 1941.** A Synopsis of the Genus *Mesostoma*, Ehrenberg 1835. *J. Elisha Mitchell Sci. Soc.*, 57:1–52.   **Gilbert, C. M. 1935.** A comparative study of three new American species of the genus *Phaenocora* with especial reference to their reproductive organs and their relationships with the other described forms of the genus. *Acta Zool.*, 16:283–384.   **Graff, L. Von. 1911.** Acoela, Rhabdocoela and Alloeocoela des Ostens der Vereinigten Staaten von Amerika. *Z. wiss. Zool.*, 99:321–428.   **1913.** Turbellaria. II. Rhabdocoelida. *Tierreich Lief*, 35:1–484.   **Hyman, L. H. 1925.** The reproductive system and other characters of

Planaria dorotocephala Woodworth. *Trans. Am. Microscop. Soc.*, 44:51–89. **1928**. Studies on the morphology, taxonomy, and distribution of North American triclad Turbellaria. I. *Procotyla fluviatilis*, commonly but erroneously known as *Dendrocoelum lacteum*. *Trans. Am. Microscop. Soc.*, 47:222–255. **1931a**. Studies, etc. III. On *Polycelis coronata* (Girard). *Trans. Am. Micro. Soc.*, 50:124–135. **1931b**. Studies, etc. V. Description of two new species. *Trans. Am. Micro. Soc.*, 50:336–343. **1935**. Studies, etc. VI. A new dendrocoelid from Montana, *Dendrocoelopsis vaginatus*, n. sp. *Trans. Am. Microscop. Soc.*, 54:338–345. **1937a**. Studies, etc. VII. The two species confused under the name *Phagocata gracilis*, the validity of the generic name *Phagocata* Leidy 1847, and its priority over *Fonticola* Komarek 1926. *Trans. Am. Microscop. Soc.*, 56:298–310. **1937b**. Studies, etc. VIII. Some cave planarians of the United States. *Trans. Am. Microscop. Soc.*, 56:457–477. **1939**. North American triclad Turbellaria. X. Additional species of cave planarians. *Trans. Am. Microscop. Soc.*, 58:276–284. **1945**. North American triclad Turbellaria. XI. New, chiefly cavernicolous, planarians. *Am. Midland Naturalist*, 34:475–484. **1951a**. *The Invertebrates*. Vol. II, *Platyhelminthes and Rhynchocoela*. McGraw-Hill, New York. **1951b**. North American triclad Turbellaria. XII. Synopsis of the known species of fresh-water planarians of North America. *Trans. Am. Microscop. Soc.*, 70:154–167. **1953**. North American triclad Turbellaria. XIV. A new, probably exotic, dendrocoelid. *Am. Museum Novitates*, No. 1629:1–6. **1954**. North American triclad Turbellaria. XIII: Three new cave planarians. *Proc. U. S. Natl. Museum*, 103:563–573. **1955**. Descriptions and records of fresh-water Turbellaria from the United States. *Am. Museum Novitates*, No. 1714:1–36. **1956**. North American triclad Turbellaria. XV. Three new species. *Am. Museum Novitates*, 1808:1–14. **1957**. North American Rhabdocoela and Alloeocoela. VI. A Further Study of Mesostoma. *Am. Museum Novitates*, No. 1829:1–15. **Kenk, R. 1935**. Studies on Virginian triclads. *J. Elisha Mitchell Sci. Soc.*, 51:79–125. **1944**. The fresh-water triclads of Michigan. *Misc. Publ. Mus. Zool. Univ. Mich.* 51:9–44. **1953**. The fresh-water triclads (Turbellaria) of Alaska. *Proc. U. S. Natl. Museum*, 103:164–186. **Marcus, E. 1945a**. Sobre Microturbellarios do Brasil. *Com. Mus. Hist. Nat. Montevideo*, 1:1–60. **1945b.** Sobre Catenulida Brasileiros. *Univ. São Paulo Fac. filosol. ciênc. e letras Bd. Zool*, 10:3–133. **1946**. Sobre Turbellaria limnicos brasileiros. *Univ. São Paulo Fac. filosol. ciênc. e letras Bol. Zool.*, 11:5–254. **Nuttycombe, J. W. 1956**. The Catenula of the eastern United States. *Am. Midland Naturalist*, 55:419–433. **Nuttycombe, J. W. and A. J. Waters. 1938**. The American species of the genus *Stenostomum*. *Proc. Am. Phil. Soc.*, 79: 213–301. **Ruebush, T. K. 1941**. A key to the American fresh-water Turbellarian genera, exclusive of the Tricladida. *Trans. Am. Microscop. Soc.*, 60:29–40. **Ruebush, T. K. and W. J. Hayes. 1939**. The genus Dalyellia in America. II. *Zool. Anz.*, 128:136–152.

14

Nemertea

Creeping slowly upon the vegetation in pools, streams, and lakes, or on the debris at the bottom are often found small, slender, usually brightly colored worms with ciliated, unsegmented bodies, belonging to the phylum Nemertea (also called Rhynchocoela). They are easily recognized by the long, protrusible proboscis armed with a calcareous central stylet and two pouches of accessory stylets (Fig. 14.1). They are seldom more than 18 mm long and 1.5 to 2 mm in diameter. The head bears two or three pairs of small ocelli. The colors are red, orange, or grayish-green. The young are whitish or pale yellow.

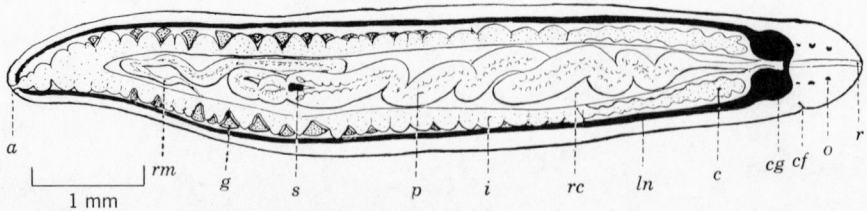

Fig. 14.1. *Prostoma rubrum* (Leidy). Diagram of living individual flattened beneath a cover glass, showing internal anatomy: *a*, anus; *c*, pyloric cecum; *cf*, cephalic furrow; *cg*, cerebral ganglia; *g*, gonad; *i*, intestine; *ln*, lateral nerve; *o*, ocellus; *p*, proboscis; *r*, rhynchodaeum; *rc*, rhynchocoel; *rm*, retractor muscle of proboscis; *s*, central stylet and basis.

In the United States fresh-water nemerteans are widely distributed, being found from New England to Florida and westward to Ohio, Nebraska, Washington, and California. They have been given a variety of names but all those that have been examined from these widely separated localities appear to belong to a single species, *Prostoma rubrum* (Leidy). They are readily carried to new localities on water plants and presumably on the feet of birds.

They feed on minute worms, crustaceans, insects, and unicellular organisms, and are in turn devoured by larger individuals of the same invertebrate phyla and by fishes.

Many individuals are protandric, functioning first as males and later changing to the female phase. They then deposit mucous capsules, each containing 30 to 50 or more ova which undergo direct development, usually after being fertilized by another worm. Self-fertilization occurs occasionally. The nemerteans can be kept for several generations in a balanced aquarium if they are occasionally fed a little liver.

References

Coe, W. R. 1943. Biology of the nemerteans of the Atlantic coast of North America. *Trans. Conn. Acad. Arts and Sci.*, 35:129–328. Leidy, J. 1850. Description of new genera of Vermes. *Proc. Acad. Nat. Sci. Phila.*, 5:124–126.

15

Nemata

B. G. CHITWOOD

M. W. ALLEN

The phyletic name Nemates was proposed in 1919 for the organisms we commonly call nematodes or Nematoda. In this chapter the emended name Nemata is used. Rudolphi (1809) originally proposed the name Nematoidea as an order. Thereafter the term was contracted to Nematoda, from which the word *nematode* was derived. With increased knowledge the group was promoted to phyletic rank, ignoring Cobb's (1931) diagnosis of the group and his article (1932) on the English word "nema." We ourselves are partially to blame for this situation and if future generations will adopt the terms Nemata, nema, etc., the names used for these organisms will become stabilized in the literature.

Nemas are bilaterally symmetric, cylindroid, unsegmented, bisexual, triploblastic organisms which were formerly placed in the phylum Nemathelminthes in most textbooks. This placement has been disputed by comparative anatomists from the time of Bastian (1866) and Huxley (1856, 1864) to this day. The presence of a body cavity not lined by a definite mesodermal layer placed this group in the Aschelminthes (Grobben), the placement followed by Hyman (1951). We formerly accepted the Aschelminthes as a series including the phyla Nematoda, Rotifera, Gastrotricha, Echinodera, Nematomorpha, and possibly the Acanthocephala. However, the last group presents so many similarities to the Platyhelminthes (Cestoda) that the direct placement of all unsegmented worms in the Subkingdom Scolecida seems to be the most logical procedure.

Identification of nemas requires very detailed study of many minute anatomical features. It is suggested that after a knowledge of basic anatomy is acquired, the figures (15.1–15.23) be studied in conjunction with the following discussion before identifications are attempted.

External Covering

Externally nemas are covered by a noncellular layered cuticle composed of various scleroproteins; this cuticle is molted four times in the course of the life history of the individual. The cuticle commonly bears minute to coarse transverse striations and may in addition bear longitudinal lateral ridges or cuticular thickenings marked by grooves or incisures. In a few forms the longitudinal ridges may extend all the way around the body, but this is exceptional rather than the rule. The lateral ridges apparently serve as stiffening structures, since movement is for the most part confined to a dorsoventral plane. In addition to the foregoing, minute internal markings termed punctations commonly form characteristic patterns in some species, and other cuticular layers may give the cuticle a cross-hatched appearance. The cuticle commonly bears various setae and glands treated elsewhere.

Beneath the cuticle the epidermis usually takes the form of four *chords*, one dorsal, two lateral, and one ventral connected by a thin sheet of protoplasm which is anucleate. Among the fresh-water nemas the chords are cellular rather than syncytial. The lateral chords contain the hypodermal glands which are commonly present as two sublateral rows of unicellular glands with minute individual pores through the cuticle.

The lateral chords also contain cell bodies of those structures going to make up the somatic papillae or setae if such are present as well as the cells forming paired genital papillae or setae in the male. Aside from the function of hypodermis, i.e., formative tissue of the cuticle, the chords also serve as carriers for the somatic nerves and lateral canals of the excretory system whenever these are present. Other glands are also apparently of hypodermal origin. These include the *amphidial glands*, which are paired unicellular structures the nuclei of which are usually to be found posterior and dorsolateral to the nerve ring; the orifices of these glands are situated dorsolaterally in the cephalic or postcephalic regions and are termed lateral organs or *amphids* (Figs. 15.8, 15.13, 15.16). Among the fresh-water nemas the external amphids are commonly highly modified in appearance: sometimes circular, question mark, spiral, or pocketlike in shape; and among the soil nemas and parasitic forms, the external amphids are more commonly porelike.

Posteriad, usually in the mid-region of the tail, a pair of unicellular glands open laterally in a few fresh-water nemas; these are called *phasmids*. Male fresh-water nemas may bear a median series of few to many supplementary organs; such structures may or may not be sclerotized and have often been found to be connected with unicellular adhesive glands (see *Anaplectus granulosus*). However, the most characteristic hypodermal glands of the aquatic nemas are the *caudal glands* (Figs. 15.20, 15.21). These consist of three unicellular glands, the cell bodies of which are commonly located in the upper part of the tail, immediately behind the anus. Each of the clavate gland cells has an elongate process leading toward the tip of the tail; usually the three glands unite in an *ampulla* some little distance from the caudal extremity. The ampulla is connected with a terminal *pore* through a minute needle valve termed the *spinneret* (Figs. 15.9, 15.13). Through this structure an adhesive thread is spun in much the same manner as in spiders; with the aid of this thread, fresh-water nemas attach themselves to the substratum—rocks, algae, twigs, gills of crayfish, and an unlimited number of other locations. They are thus able to maintain a position even in the most rapid currents.

The somatic musculature of nemas consists of four submedian longitudinal bands of spindle-shaped muscle cells attached to the hypodermis throughout the length of the body. These cells together with the cuticle, hypodermis, and chords make up the body wall. Between two chords there may be from four to sixteen or more muscle cells in a cross section; if the smaller number is present the nema is said to be *meromyarian*, and if the larger number is present it is said to be *polymyarian*.

Digestive System

The oral opening is at the anterior extremity of nearly all nemas. It is usually surrounded by six or three lips bearing sensory organs termed *papillae* or *setae*. Sometimes there are specialized structures having to do with ingestion such as *jaws*, *probolae*, or *circumoral rugae* (Figs. 15.3, 15.5, 15.8, 15.10). Posterior to this structure, between the oral opening and the definite esophagus we find the secondary stomodeum, commonly called the *stoma* (pharynx of Cobb, buccal capsule of various other authors). Basically this region is a cylindric sclerotized tube formed from the modified hypodermal head cells (*arcade*) which lie posteriad in the body cavity. However, it may bear one or more teeth and may contain or be transformed into a protrusible hollow *stylet* (spear) which functions like a hypodermic needle (Figs. 15.2, 15.23).

Of all the structures in nemic anatomy the stoma shows the greatest diversification, which is understandable, since it is the structure in primary contact with food. Basically there are three parts: *cheilostom*, *protostom*, and *telostom;* of these the first and last are short, and the second is an elongate more or less triradiate tube. The protostom is subdivisible into prostom and mesostom (rarely a metastom is distinctly separate). The walls of the corresponding parts are called *rhabdions*, i.e., cheilorhabdions, protorhabdions (prorhabdions, mesorhabdions, metarhabdions), and telorhabdions. In each section the dorsal and two subventral rhabdions may or may not be distinctly separate. *Teeth* or *denticles* may be formed at any level in the stoma and may be symmetric or asymmetric. The stoma is commonly surrounded by esophageal tissue to a greater or lesser extent. There are two primary means of *stylet* (spear) formation in the Nemata: first, through the transformation of the stomato-rhabdions themselves, forming a stomatostyl (Order Tylenchida); and second, through the movement of a subventral tooth to an axial position (Order Dorylaimida). In the latter, the new stylet is formed in the subventral wall of the esophagus, and in the former, the new stylet is formed *in situ* prior to molts.

Posterior to the stoma is the esophagus and esophago-intestinal valve. This structure represents the primary stomodeum and corresponds to the pharynx and esophagus of platyhelminthes, gastrotrichs, and tardigrades, or to the mastax and esophagus of rotifers. It is lined with cuticle and is triradiate with two rays of the lumen directed subdorsad and one ray directed ventrad. Basically the esophagus is divisible into *corpus* (*procorpus* and *metacorpus*), *isthmus*, and *bulbar* regions, this condition being preserved in saprozooic groups. However, the isthmus tends to become obscure in fresh-water nemas, and in many forms the esophagus becomes cylindric or two-part cylindroid. Such form is usually correlated with feeding habits involving larger substrata. Internally the esophagus is a syncytium characterized by constant numbers of nuclei. Opposite the distal ends of the esophageal radii we find the marginal nuclei embedded in noncontractile fibrils, which apparently act as stabilizers; the marginal tissue probably represents ectoderm forming the lining of the esophagus. In the dorsal and subventral walls of the esophagus there are transverse and oblique radial muscles with their accompanying nuclei; these also are constant in position and number. The salivary glands of nemas, called esophageal glands, ramify the posterior part of the esophagus and extend anteriad, each having a separate pore into the lumen. As a general rule there are three unicellular glands, one dorsal and two subventral, but in some groups the subventral glands are duplicated, and there are two pairs in tandem. Esophageal gland development and position of gland orifices is correlated with feeding habits, and is used as a taxonomic character.

The esophago-intestinal valve is actually a part of the esophagus, but since it is also a connecting piece with the intestine it is commonly treated separately. Like the esophagus it is basically triradiate but it may acquire dorsoventral or lateral symmetry secondarily. The size, symmetry, and extent of involvement with intestinal tissue varies with the taxonomic group.

The intestine or mesenteron is a straight tube composed of low, cuboidal to high, columnar epithelial cells of endodermal origin. Internally these cells are usually lined with a bacillary layer. The cells may be uniform in size and character throughout the length of the intestine, or specialized individual cells or differentiated areas may be present. Various types of cell inclusions are the most conspicuous features of the intestinal cells, and the coloring or refraction due to these cell inclusions is a useful means of recognizing groups as well as species. Unfortunately, much more intense work must be done on the cell inclusions before keys can be constructed making use of them. For the present, we can state that nemic intestinal cell inclusions are sometimes birefringent, sometimes not. Oil globules, protein globules, and glycogen have been identified as stored food, and plant pigments and various inorganic and organic crystals have been identified as waste products.

Like the fore-gut, the hind-gut or rectum is formed as an invagination and is lined with cuticle. Developmentally it is considered the proctodeum. The anus or cloacal opening is ventral, usually a considerable distance from the end of the tail (Fig. 15.7). The hind-gut is termed rectum or cloaca according to the sex, since in the male the reproductive system opens into the rectum from the ventral side. Between intestine and rectum there is a sphincter termed the intestino-rectal valve, posterior to which the dorsal and two subventral unicellular rectal glands also open into the rectum. These structures are confined to certain groups of nemas, primarily the saprozoic and animal parasitic forms; supposedly they are homologues of the malpighian tubules of tardigrades and insects, and in a few representatives these glands are present as multiples of three, showing some evidence of transition from unicellular to multicellular glands. The function is unknown, but their absence in tissue parasites (filarioids) and tylenchoids, both of which are apparently restricted to a liquid diet, may be a clue.

Nervous System

The nervous system of nemas is much too complicated to discuss with any degree of completeness in a work of this nature. Briefly stated, the nervous system consists of a circum-esophageal commissure with associated ganglia forming a brain, eight anterior cephalic nerves, and four posterior somatic nerves, together with various sensory organs, minor commissures, specialized vaginal nerves, and an esophago-sympathetic system and a recto-sympathetic system. The nerve cells of the cephalic papillary nerves are connected with the anterior side of the nerve ring; two subdorsal, two lateral, and two subventral nerves extend anteriad to the labial region, where each nerve divides into three branches, one to a papilla or seta of the internal circle of cephalic sensory organs, and the others to papillae or setae of the external circle. The lateral papillary nerves innervate only the lateral papilla of the internal circle and one papilla or seta of the external circle. Presumably the lateral papillary nerves also innervate any subcephalic setae or papillae, but this matter has not been worked out.

The number, arrangement, and degree of development of the *cephalic sensory organs* is always given considerable weight in taxonomic work. In addition to the cephalic papillary nerves there are two dorsolateral amphidial nerves extending anteriad from the nerve ring. These innervate the amphids which are apparently chemoreceptors, flushed by the amphidial glands previously mentioned. Animal parasitic and saprozoic nemas also commonly have a pair of cervical papillae, the deirids, which are innervated in much the same way as the amphids. Although ocelli are commonly present in aquatic nemas, we know much too little about them to make any very definite statements. When ocelli are present they are subdorsal, in the anterior esophageal region; they may be contained inside or outside the wall of the esophagus and may consist of diffuse pigments spots, well-defined pigment spots, or clear pigment cups with accompanying lenses. Nothing is known about their innervation.

Excretory System

The excretory system of nemas is highly varied. When present, it nearly always opens through a single ventral pore in the anterior part of the body (Fig. 15.3). Among the aquatic nemas this pore usually leads into a rather delicate protoplasmic canal connected with a single ventral gland cell situated free in the body cavity posterior to the base of the esophagus. This cell is sometimes called the renette cell (Figs. 15.10, 15.13). Plectids are exceptional in that there is a distinctly sclerotized terminal duct. Saprozoic soil nemas and parasitic nemas nearly always have a well-developed sclerotized terminal excretory duct and two or more protoplasmic lateral canals situated in the lateral chords. These lateral chords have been considered homologues of the protonephridial system of other invertebrates, even though there are apparently no flame cells.

Reproductive System

The reproductive system usually consists of one or two tubular gonads originating from a genital primordium in the ventral mid-region of the body. If paired gonads are present, they are usually opposed with the blind ends or germinal regions *outstretched* or *reflexed* back toward the mid-region of the body (Fig. 15.9). In the female the genital opening or *vulva* is ventral, commonly situated near the middle of the body, but it may be located anywhere from the esophageal region to the preanal region; the vagina is lined with cuticle of ectodermal origin (Fig. 15.3). In free-living nemas the female reproductive system tends to be less complicated than in parasitic forms, though distinct germinal and growth zones of the ovary, sometimes a distinct oviduct, and single or paired simple uteri each with a distinct seminal receptacle may be recognized. The eggs are capsuliform, covered with a chitinous shell and various other membranes; externally the shell may bear spines or other ornamentation. The male reproductive system is very similar to the female, except that there may be paired seminal vesicles emptying into a common vas deferens. The vas deferens is commonly subdivisible into two sections; the part near the gonads having little or no musculature is commonly quite glandular, and the part extending to the cloaca may be covered with a rather well-developed muscle sheath, in which case there is little glandular development and it is termed an ejaculatory duct. Sperm cells vary from amoeboid to flagellate.

The male copulatory organs, paired sclerotized spicules, are formed as an evagination of the dorsal wall of the cloaca (Figs. 15.7, 15.14, spiculum). They are inserted and withdrawn from the female as a result of specialized spicular muscles which extend from the spicules to the body wall. Posteriorly the spicules are guided by another sclerotized development of the cloacal wall. This is termed the *gubernaculum*.

Repiratory and circulatory systems are absent. These organisms are so constructed that the colorless fluid is aerated without the need of special vessels. The movements of the body serve to propel the body fluid irregularly about the body cavity and among the organs.

Faunistic Separation

All nemas, even those of soil are more or less aquatic, since in the active stage they require a moisture film in which to move and through which to breath. Fresh water commonly contains the washings from land, and plant parasites and free-living stages of land-vertebrate parasites are found in run-off waters. As fresh-water streams approach salt water there are some forms adapted to brackish water to be found and a very few species are capable of adapting themselves to a salt or fresh environment during the life of a single individual.

A few economically important plant pathogenic nemas were first described from fresh water. Thus, *Criconemoides simile* is a terrestrial plant pathogen, though first described from specimens found in filter beds (also on roots of grapevine) by Cobb (1918), and *Dolichodorus heterocephalus* is an aquatic plant pathogen originally described from Douglas Lake, Michigan (Cobb 1914), and Silver Springs, Florida, now known to attack many aquatic plants in various parts of the country and commonly found to cause damage to economic crops in lowlands in various parts of the country.

Mosses, algae, diatoms, and aquatic plants are primarily inhabited by adenophore nemas of the orders Chromadorida, Monhysterida, Enoplida, and Dorylaimida. Some are herbivorous, some carnivorous, a few saprozoic (plectids). Considerable ecologic work is being done at present.

Snails, crayfish, and other aquatic animals have their nemic commensals and parasites as do all other types of living organisms.

The peculiar lack of a brackish-water fauna has stimulated some investigators to unsuccessfully attempt acclimatization of fresh-water nemas to increasing saline concentrations. Thus, Kreis (1927) reports *Dorylaimus stagnalis* died in 0.1 per cent NaCl after 4 to 6 hours, in 0.25 per cent after 60 to 80 minutes, and in 1.5 per cent after 10 minutes. Considering adaptation studies in connection with excretion or osmotic balance in protozoans and turbellarians, parallel studies might well be made on a series of aquatic secernents and adenophores.

Life History

The life cycle of rhabditids usually requires from 2 to 14 days, that of cephalobids 14 to 21 days. Of the remaining nemas encountered in fresh water relatively little is known. Steiner and Heinly (1922) made the most thorough study available on the life cycle of *Mononchus papillatus*. These authors found that the females were hermaphroditic and produced only 1 or 2 eggs at a time, which were deposited daily. Reproduction occurred between the ages of 6 to 10 weeks, after which the female might continue life for 8 more weeks. The eggs are spinulate and apparently adapted to entanglement in debris, thus preventing their reaching an anaerobic environment. Embryonic development required 6 to 7 days, postembryonic development 6 to 7 weeks. During the latter period three molts were observed; just prior to each molt the animals became sluggish, apparently as a result of some specialized physiological process.

Nielsen (1949) determined the time required for development from egg to egg in a number of fresh-water nemas as follows: *Alaimus primitivus*, *Prismatolaimus dolichurus*, and *Plectus cirratus*, 20 to 30 days; *Achromadora dubia*, *Wilsonema auriculatum*, and *Plectus parvus*, 20 days; *Anaplectus granulosus*, 25 days; *Tripyla setifera*, 30 to 40 days. However, the length of life cycle is not a sound index of the reproductive rate, since some organisms produce many eggs and others produce very few. A female of *Plectus parvus* produced 96 eggs in 15 days, and a female of *Panagrolaimus elongatus* produced 209 eggs in 22 days. Nielsen computes the weight of the latter female as 1 γ and the eggs produced as 500 per cent of the weight of the mother. It seems probable that certain saprozoic rhabditids would present a still higher figure of egg production. The eggs of the more typically aquatic nemas tend to be somewhat larger in proportion to the body and a smaller number are generally produced at one time. Among some of the groups there also appear to be rather distinct seasonal fluctuations in the population, so that at certain periods only larvae of a single stage are encountered.

Collection and Technique

Although the fresh-water Nemata are so widespread, and so abundant, it is not always easy to isolate them without the use of special methods. Few of these nemas exceed 2 to 3 mm in length and they are so slender and transparent as to make it practically impossible to locate them without the use of a dissecting microscope. With special methods they can be easily collected in considerable quantity.

A few centigrams of mud or sand may be placed in a beaker, thoroughly roiled, allowed to settle momentarily and the supernatant fluid poured into a second beaker and allowed to settle. The upper fluid may then be carefully decanted and samples from the sediment drawn off with a pipette and examined in syracuse dishes. Samples of algae may be squeezed directly into a beaker or syracuse dish for examination. Larger samples usually require either the Baermann or screening techniques.

The Baermann apparatus, first invented for the collection of larval parasites of vertebrates such as hookworms, is highly adapted to the collection of soil and plant parasitic nemas. It may also be used, though perhaps not quite as successfully, for the collection of aquatic species. The equipment necessary for this procedure is a glass funnel, commonly 500 cc; a 2- to 3-in. length of rubber tubing; a hose clamp, ring stand, or other suitable holder for the funnel; and a piece of ordinary screen wire, with the addition of cheesecloth, linen, or facial tissue if the material to be sampled is fine. The apparatus is assembled and water is placed in the lower part of the funnel; the wire gauze is used as a basket in the upper part of the funnel and the sample material is added to the basket. The water level is then raised in the funnel to the height of the sample material. The nemas become active, swim out from the sample, and gradually sink to the bottom of the funnel. Portions of the liquid may be drawn off after 1 to 24 hours and examined in syracuse dishes. Nielsen has recently modified the Baermann technique for the quantitative extraction of nemas and rotifers from soil and moss. He places the entire apparatus in a plywood box supplied with an electric bulb and states that after an hour the temperature rises to 30 C. This temperature causes paralysis of the organisms so that after 12 hours most of the nemas and rotifers are to be found in the rubber tubing. When they are removed from the tube and the temperature is reduced they again become active and are easily located.

Screening procedures for the collection of nemas are based primarily on the methods described by Cobb (1918). The use of screens and settling is highly adaptable to the materials to be sampled. In fact good screening and settling may be developed to the level of an art. A full set of equipment consists of U. S. standard sieves Nos. 20, 60, 200, and 325, and two or three buckets. With a sandy bottom with a minimum of debris, only the buckets and 200 or 325 screens need be used. One bucket should be scraped along the bottom until it contains a sediment accumulation of 2 to 3 in., then the bucket, half full of water, is roiled vigorously with the hands, the sand permitted to settle (perhaps 30 seconds), and the supernatant poured into the second bucket. The first bucket is roiled again, then permitted to settle, and another sample is added to the second bucket. This process may be repeated several times using a series of sedimentation buckets. Thus the sand in the first or collection bucket may be nearly completely freed of microorganisms. The supernatant from the sedimentation buckets which have been settling during this period (15 to 30 minutes), may then be decanted with great care not to disturb the bottom. Usually the residue in these buckets contains nemas, rotifers, oligochaetes, etc., in some fine sand and perhaps a bit of clay or silt. Such materials may be examined directly, or the sediment may be concentrated by pouring the material through the No. 200 or 325 screen; while pouring, the screen should be shaken by holding with one hand and hitting the side with the other hand. Additional clean water may be used to rinse very fine materials through the screen. When the liquid passing through the screen takes the appearance of clean water, the screen residue is rinsed into a beaker and samples are taken for study. Other screens are provided in order to remove coarser debris and larger organisms if this is desired and the material to be sampled so indicates. The presently marketed U. S. sieves (primarily used for soil analysis) are so fitted that they may be used at the same time in series. When this is done one obtains a graded series of organisms according to the screen on which they are caught. However, when they are used in this way for the collection of aquatic nemas, considerable care must be taken not to overload one of the finer sieves and cause the water to run out the side.

Once one has so processed the material as to obtain an abundance of organisms for study, one has a choice of either fixing the material in bulk or picking out the individual nemas. In order to identify the species, they must be picked out and mounted on slides, and it is easier to do this while they are alive and motile. However, circumstances may prevent such ideal procedure, in which case, we have found that addition of commercial formalin to make a 4 per cent solution in the concentrated collection is satisfactory.

Picking may be accomplished either by the use of a fine capillary pipette or with a very fine needle. A watchmaker's fine pivot broach (obtainable from any watch repairman) may be inserted in a wooden handle. They break rather easily and must often be replaced or sharpened on a stone. Any hard wood can also be sharpened to a very fine point and used for this purpose. It is then used to pick up the individual nema in much the same way one might spear spaghetti with a sharp knitting needle. Specimens so picked may be placed directly on a slide in a small drop of water or 4 per cent formalin; two or three pieces of fine glass wool about the same size as the nemas should be placed in the formalin drop as supports. An 18 mm round cover slip may then be added, the slide placed on a turntable and ringed with suitable sealing material. Paraffin from a small birthday candle (pink or white) may be used; another procedure is to apply with a brush a mixture of half paraffin half vaseline, or hot paraffin. In order to make such slides more durable a second ringing with Zut, lactophenol gum, or fingernail polish may be applied. Specimens so mounted are excellent for immediate morphologic study, as they show details with greater clarity than any of the more permanent preparations. With care in selection of glass wool supports it is possible to use the weight of the cover slip to hold living nemas in place, and plain water or intravitam stains may be substituted for the formalin drop on the slide.

The importance of studying living specimens cannot be overemphasized in gaining a sound understanding of morphology and physiology. Cold 5 per cent formalin-fixed material is the best substitute for the living organisms. Hot fixation changes cell inclusions, causing fat droplets to coalesce, and it also causes other types of changes in the organisms that destroy the natural appearance. Intravitam stains such as methylene blue, neutral red, and ammoniacal carmine together with intramortem stains such as crystal violet bring out some types of morphology in a striking manner.

In order to prepare more permanent mounts of fresh-water nemas one usually picks them out and places them in a small drop of water in a BPI watch crystal or embryological watch glass. After a fair number have been placed in such a drop, hot (50 C) FAAGO[1] may be added. The dish with its contents is then placed in an evaporation chamber and left there until the

[1]FAAGO: Commercial formalin 5 cc, 50 per cent alcohol 90 cc, glycerine 1.5 cc, acetic acid 2 cc, trace of osmic acid.

nemas are covered by a thin film of glycerin on the bottom of the dish. The evaporation process must be very slow to avoid shrinkage. A small amount of pure glycerin is then added and the nemas are picked out and placed in a drop of glycerin on a slide. Thereafter supports and cover slip are added and sealed into place with suitable ringing material.

Lactophenol[2] is considered a satisfactory mounting medium. Specimens may be fixed in FAAGO, then transferred to a mixture of formalin 5 per cent, lactophenol 5 per cent, and evaporated to lactophenol. In this case evaporation can be done quite rapidly in an oven at 40 C. Specimens are mounted in pure lactophenol and ringed with fingernail polish or Zut.

Another acceptable technique is to add a trace of acid fuchsin or cotton blue to the formalin-lactophenol mixture. After evaporation, specimens may appear somewhat overstained but when mounted in plain lactophenol they will destain slightly over a period of years.

Measurement and Identification

The first step in the identification of a nema is to make a low-power outline camera-lucida drawing. In this sketch one marks the base of the stoma or stylet, the nerve ring, and excretory pore (if visible), the base of the esophagus, position of vulva, extent of gonads, size of eggs, position of anus, and length of spicules. Diameters of the body are given at each level noted. In addition, one then makes oil immersion camera-lucida sketches of the head region, esophageal region, and caudal region in both male and female. Full-length camera-lucida sketches are very nice to have but are extremely time-consuming. It is suggested that students prepare at least one such drawing at a magnification that will produce a final drawing of 30 to 60 in. This may be done by making a series of short-length sketches, noting positions of over-lap, then piecing them together. By so doing one is forced to take note of such points as shape and size of intestinal cells, form and number of oöcytes, and other details that would otherwise be overlooked.

Almost all descriptions of aquatic nemas make use of some type of formula to express the various measurements. In general there have been two basic formulae proposed, those of Cobb and de Man. The former is not in use at the present time, but since so many descriptions of aquatic nemas were published using this formula it is essential to understand it. The illustration (Fig. 15.1) is self-explanatory for the most part. All figures are given as percentages of the total body length. Percentages of the distance from the anterior end are given above the line and below the line diameters of the body are given at the corresponding levels. Cobb commonly indicated whether the ovaries were reflexed or outstretched by superscriptions at the position of the vulva, thus '50- would mean anterior ovary reflexed and posterior ovary outstretched with vulva at 50 per cent of the body length. The extent of the gonads as a percentage of the body length was also some-times indicated in the same manner, thus $'^{25}50^{25-}$ would mean each of the

[2]Lactophenol: lactic acid 46 cc, phenol 60 cc, glycerin 96 cc.

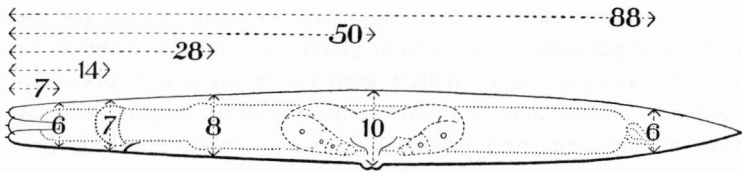

Fig. 15.1. Diagram in explanation of the Cobb system of descriptive formula used for nemas; 6, 7, 8, 10, 6 are the transverse measurements, and 7, 14, 28, 50, 88 are the corresponding longitudinal measurements. The formula in this case is:

$$\frac{7. \quad 14. \quad 28. \quad 50. \quad 88.}{6. \quad \ \ 7. \quad \ \ 8. \quad 10. \quad \ \ 6.}$$

The measurements are simply percentages of the length, and the formula, as printed in the key, may be regarded as somewhat in the nature of a conventionalized sketch of the nema with dimensions attached.

The measurements are taken with the animal viewed in profile; the first is taken at the base of the stoma, the second at the nerve ring, the third at the cardiac constriction (base of the "neck"), the fourth at the vulva in females and at the middle (M) in males, the fifth at the anus. (After Cobb.)

gonads measured 25 per cent of the body length. He also included schematic indications of lateral ridges, striations, form of stoma, position of excretory pore, shape of head, and spicules in some of his formulae. Any questionable uses can usually be understood by a study of Cobb's various papers including his chapter in the original edition of this book (1918).

The formula system proposed by de Man is currently more in vogue. According to this system one gives the length in microns or millimeters followed by a series of ratios:

$a =$ length/greatest breadth
$b =$ length/length of esophagus from anterior end
   (In forms with esophageal glands overlapping intestine, b usually indicates length of body divided by distance to base of median bulb, but individual authors differ in usage.)
$c =$ length/length of tail from anus
$V =$ distance of vulva from anterior end as a per cent of total length
$G\,1 =$ extent of anterior gonad as per cent of body length
$G\,2 =$ extent of posterior gonad as per cent of body length

In this scheme a, b, and c correspond to α, β, γ as found in the literature. Length of stoma, spicules, gubernaculum, setae, and size of eggs are usually expressed in microns. However, relative length of these structures or the position of amphids may be expressed in terms of number of corresponding body diameters. Cephalic setae and amphids may be described in terms of cephalic body diameter, and length of tail and spicules may be described in terms of anal body diameter.

In order to identify aquatic nemas to group, one looks for certain rather minute structures. If they are observed, one is treading on firm ground, but since some of these structures may be overlooked or misinterpreted, negative findings are hardly significant. For that reason, one looks for a whole series of structures. If one or more are found the identification can proceed with some hope of success. In order to make major separations, we look for phasmids, caudal glands (spinneret), hypodermal glands, sclerotized terminal excretory duct, lateral excretory canals (rarely seen except in living specimens under oil immersion), caudal alae or preanal supplementary organs, setae or papillae, form of esophagus and esophago-intestinal valve, amphids (externally modified and large or porelike and probably not observed), ovary number and whether or not the ovaries are reflexed or outstretched, and, finally, form of cuticular markings.

The accompanying key covers only the common genera. Great advances have been made in the taxonomy of these forms in recent years. The student desiring to identify a form to proper genus and species is advised to consult the literature very thoroughly, going over the Zoological Record and Helminthological Abstracts to work up a full bibliography.

KEY TO GENERA

Only the most commonly encountered fresh-water genera and families are included. This key should place to genus, subfamily, or family over 90 per cent of the specimens encountered. Recent involved generic splitting has, for the most part, been omitted.

1a   Excretory canals (protoplasmic) present (often hard to see, easiest in living specimens or sections); terminal excretory duct cuticularized; caudal glands absent; phasmids usually present (rudimentary or apparently absent in many members of the Tylenchida); amphids usually minute, porelike and cephalic in position (Fig. 15.5*b*,*c*), rarely postcephalic and transversely oval; sensory organs papilloid, seldom setose (*Butlerius, Atylenchus, Eutylenchus*); hypodermal glands absent; male with or without caudal alae; preanal supplementary organs not as a ventral series (at most one papilloid) . . . . Class **Secernentea**   2

1b   Excretory canals absent; terminal excretory duct not cuticularized (except in Plectidae and Diphtherophoridae as well as a few marine nemas); caudal glands present or absent; phasmids absent (at most *phasma*, areas without apparent internal connection in some marine nemas); amphids usually externally modified, spiral (Fig. 15.10*c*), circular, pocketlike or modifications thereof, usually postcephalic in position; cephalic sensory organs commonly setose and some postcephalic in position; hypodermal glands commonly present; caudal alae and spinneret present or absent; male usually without caudal alae; preanal supplementary organs commonly as a ventral series, papilloid or tuboid   . . . . . . Class **Adenophorea**   39

2a   (1) Stylet present (Fig. 15.2); only one excretory canal; rectal glands absent; esophagus basically 3 part, namely, corpus (often with bulb), isthmus, and nonvalved glandular region . . . .
                                              Order **Tylenchida**   3

2b   Stylet absent; paired excretory canals, rectal glands usually present; esophagus of varied form   . . . . . .   Order **Rhabditida**   29

3a   (2) Dorsal gland orifice in median bulb . . . .
                                   Superfamily **Aphelenchoidea**   4

3b   Dorsal gland orifice in procorpus, usually near base of stylet . . . .
                                    Superfamily **Tylenchoidea**   7

4a (*3*) Tail conically attenuated to filiform; stylet without knobs or knobs very weak. (Usually carnivorous, about 9 species) . . . . . .
Seinura Fuchs

4b Tail conoid to bluntly rounded 5

5a (*4*) Esophageal glands not overlapping intestine, but enclosed in esophageal wall. (A small genus). . . *Paraphelenchus* Micoletzky

5b Esophageal glands overlapping intestine 6

6a (*5*) Female tail bluntly rounded. (A small genus of fungivorous forms) . . . . *Aphelenchus* Bastian

6b Female tail conoid. (This is a very large genus including plant pathogens, algivorous, and fungivorous species.) (Fig. 15.2). . . .
Aphelenchoides Fischer

7a (*3*) Anterior part of stylet (prorhabdions) usually greatly elongated; cuticular striations moderate to coarse, even spinate. . . . .
Family **Criconematidae**   8

7b Anterior part of stylet (prorhabdions) not usually greatly elongated; striations fine to coarse but never spinate. 13

8a (*7*) Bodies of both sexes definitely elongate, cylindroid; females with 2 ovaries; posterior part of corpus (metacorpus) not greatly enlarged or elongated . . . . Subfamily **Dolichorinae**   9

8b Bodies of both sexes not definitely elongate, cylindroid; females with 1 ovary; posterior part of corpus (metacorpus) greatly enlarged, elongated 10

9a (*8*) Glandular region of esophagus distinctly set off from intestine. (Aquatic to terrestrial, 2 species.) (Fig. 15.3) . . . .
Dolichodorus Cobb

9b Glandular region of esophagus not distinctly set off from intestine. (Usually terrestrial, 1 named species) . . . . *Belonolaimus* Steiner

10a (*8*) Annulation coarse, often with spines, at least in females. . . . . . .
Subfamily **Criconematinae**   11

10b Annulation moderate, never with spines . . . .
Subfamily **Paratylenchinae**
(A large genus, primarily terrestrial) . . . .
Paratylenchus Micoletzky

11a (*10*) Cuticle of females with scales or spines. (A large genus, primarily terrestrial.) (Fig. 15.4) . . . . *Criconema* Hofmänner and Menzel

11b Cuticle of females never with scales or spines. 12

12a (*11*) Cuticle of female very coarsely striated, annules somewhat overlapping; females without sheath. (A large genus, primarily terrestrial). . . . . *Criconemoides* Taylor

12b Cuticle of female moderately to coarsely striated, annules not overlapping; females with sheath. (A large genus, primarily terrestrial) . . . . *Hemicycliophora* de Man

13a (*7*) Esophageal glands overlapping intestine, not enclosed in esophageal wall . . . . Family **Hoplolaimidae**

13b Esophageal glands not overlapping intestine, enclosed in esophageal wall. 21

14a (*13*) Females pear-shaped to lemon-shaped; males without highly specialized caudal alae (merely continuation of lateral ridges) . . .
Subfamily **Heteroderinae**   15

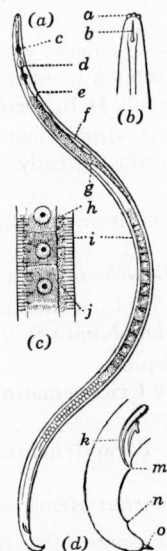

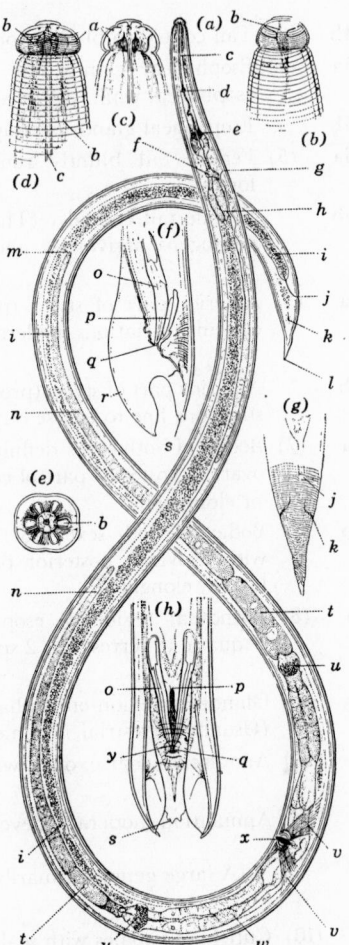

Fig. 15.2. *Aphelenchoides microlaimus.* (*a*) Lateral view. × 130. (*b*) Anterior end. × 350. (*c*) Middle section of body. × 450. (*d*) Posterior end. × 450.

a, lips; *b*, stylet; *c*, sucking-bulb; *d*, nerve ring; *e*, excretory pore; *f*, excretory gland; *g*, blind end of testicle; *h*, intestine; *i*, cuticula or skin; *j*, spermatozoon; *k*, right spiculum or penis; *m*, anus; *n*, papilla; *o*, terminus. (After Cobb.)

Fig. 15.3. *Dolichodorus heterocephalus.* (*a*) Nearly side view of a female. × 88. (*b*) Lateral view of surface of head, more highly enlarged. × 460. (*c*) Sagittal section of head. × 460. (*d*) Dorsoventral view of head. × 460. (*e*) Front view of head. × 460. (*f*) Side view, posterior extremity of male. × 230. (*g*) Ventral view, posterior extremity of female. × 230. (*h*) Ventral view, posterior extremity of male. × 230.

a, papilla; *b*, cephalic organ of unknown significance; *c*, spear; *d*, base of stylet; *e*, median bulb; *f*, nerve ring; *g*, excretory pore; *h*, cardiac swelling; *i*, intestine; *j*, anus; *k*, lateral caudal pores; *l*, terminus; *m*, blind end of posterior ovary; *n*, ovary; *o*, left spiculum; *p*, gubernaculum; *q*, distal end of accessory piece; *r*, left flap of bursa; *s*, terminus of male; *t*, ovum; *u*, spermatozoa; *v*, vaginal muscles; *w*, uterus; *x*, vulva; *y*, anus. (After Cobb.)

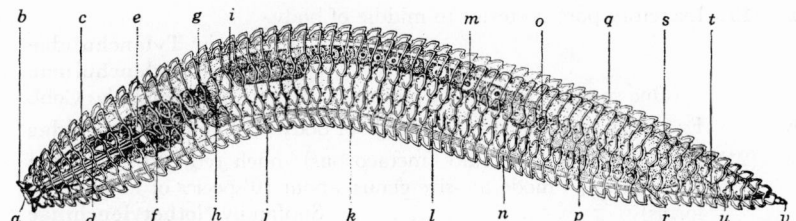

Fig. 15.4. *Criconema octangulare.* × 292. *a*, mouth opening; *b*, lip region; *c*, spear muscles; *d*, shaft of stylet; *e*, base of stylet; *f*, cuticular tube of esophagus; *g*, nerve ring; *h*, posterior portion of esophagus; *i*, flexure in ovary; *j*, body muscles; *k*, cuticula; *l*, one of the 8 longitudinal rows of modified cuticula; *m*, ovum; *n*, muscles of body wall; *o*, sublateral modification of the cuticula; *p*, uterus; *q*, subdorsal modification of the cuticula; *r*, vulva; *s*, muscles of the body wall; *t*, rectum; *u*, anus; *v*, terminus. (After Cobb.)

14b	Females not pear-shaped to lemon-shaped; males with specialized caudal alae.	**17**
15a	(14) Females pear-shaped, body wall not modified, but cuticle with involved striation patterns; vulva terminal near anus; males with lateral cheeks and without numerous cephalic striae. . . . . *Meloidogyne* Goeldi	
15b	Females pear-shaped to lemon-shaped; males without lateral cheeks but with 4 or more cephalic striae	**16**
16a	(15) Vulva and anus rather widely separated; body wall transformed somewhat but not to a persistent cyst. (One species described, terrestrial). . . . . . . *Meloidodera* Chitwood, Hannon, and Esser	
16b	Vulva and anus very closely approximated; body wall transformed to a persistent cyst. (A large genus of over 20 species) . . . . *Heterodera* Schmidt	
17a	(14) Female tails elongate, 2 or more times anal body diameter . . . . . Subfamily **Pratylenchinae**	**18**
17b	Female tails quite short, 1½ or less times as long as anal body diameter . . . . Subfamily **Hoplolaiminae**	**19**
18a	(17) Female with 1 ovary; males with caudal alae surrounding tail tip. (A large genus, about 20 named species, primarily terrestrial) . . . . *Pratylenchus* Filipjev	
18b	Female with 2 ovaries; males with caudal alae not surrounding tail tip. (A moderate-size genus with about 8 species, primarily aquatic, secondarily terrestrial) . . . . *Radopholus* Thorne	
19a	(17) One or more annules of cephalic region longitudinally divided into numerous plates. (A small genus of about 4 species, primarily terrestrial). . . . . *Hoplolaimus* Daday	
19b	Cephalic annules not longitudinally divided into numerous plates.	**20**
20a	(19) Dorsal esophageal gland orifice less than ⅓ stylet length posterior to base of stylet. (A moderate-size genus, about 10 species, primarily terrestrial) . . . . *Rotylenchus* Filipjev	
20b	Dorsal esophageal gland orifice ⅓ or more stylet length posterior to base of stylet. (A small genus, about 3 species, primarily terrestrial). . . . . *Helicotylenchus* Steiner	

21a (13) Excretory pore posterior to middle of body. . . . .
Family **Tylenchulidae**
Subfamily **Tylenchulinae**
(One species, primarily terrestrial) . . . . . . *Tylenchulus* Cobb

21b Excretory pore anterior to middle of body . . Family **Tylenchidae**   22

22a (21) Median esophageal bulb (metacorpus) much reduced, valve not apparent. (A moderate-size genus, about 10 species of fungivorous forms). . . . . Subfamily **Nothotylenchinae**
Nothotylenchus Thorne

22b Median esophageal bulb (metacorpus) not much reduced, valve apparent . . . . Subfamily **Tylenchinae**   23

23a (22) With 4 cephalic setae, primarily aquatic species   24

23b Cephalic sensory organs papilloid.   25

24a (23) Cuticle of body with numerous longitudinal broken ridges. (One named species) . . . . *Atylenchus* Cobb

24b Cuticle of body without numerous longitudinal ridges. (One named species) . . . . *Eutylenchus* Cobb

25a (23) Tails of both sexes attenuated to filiform   26

25b Tails of both sexes elongate-conoid to nearly attenuated   27

26a (25) Distance from anterior end to middle of median bulb (metacorpus) less than distance from same to base of esophagus. (A large genus of about 30 species, commonly aquatic, secondarily terrestrial) . . .
Tylenchus Bastian

26b Distance from anterior end to middle of median bulb (metacorpus) greater than distance from same to base of esophagus. (A moderate-size genus, about 8 species, commonly aquatic, secondarily terrestrial). . . . . *Psilenchus* Cobb

27a (25) Female with 1 ovary. (A large genus, about 30 species, sometimes aquatic, usually terrestrial). . . . . *Ditylenchus* Filipjev

27b Female with 2 ovaries   28

28a (27) Without sclerotized cephalic framework; tails of both sexes acute or subacute. (A small genus of about 4 species, commonly aquatic, sometimes terrestrial) . . . . *Tetylenchus* Filipjev

28b With sclerotized cephalic framework; tails usually not acute. (A large genus, about 40 species, primarily terrestrial, occasionally aquatic). . . . . *Tylenchorhynchus* Cobb

29a (2) Posterior part of esophagus glandular, with no rudiments of valved terminal bulb   30

29b Posterior part of esophagus muscular, with well-developed to rudimentary pigeon-wing valved bulb   32

30a (29) Metacorpus distinct with a median bulb of varied degrees of development . . . . Family **Diplogasteridae**
(Primarily soil, secondary fresh water, about 300 species.)
(Fig. 15.5) . . . . *Diplogaster* Schulze

30b Metacorpus indistinct, no evidence of a median bulb   31

31a (30) Stoma rudimentary, female with one ovary . Family **Cephalobidae**
Subfamily **Daubayliinae**
(Parasites of aquatic snails, 3 species). . . . .
Daubaylia Chitwood and Chitwood

31b Stoma greatly elongated, cylindric (usually in soil). . . . .
Family **Cylindrocorporidae**
Cylindrocorpus Goodey

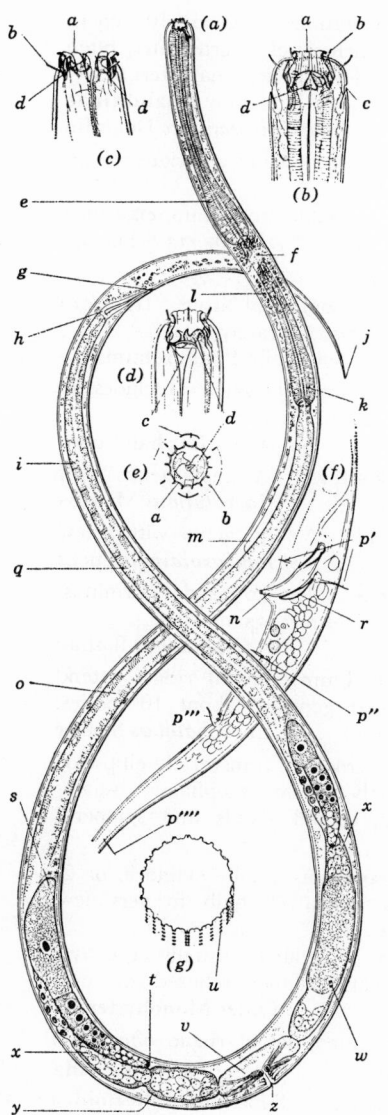

Fig. 15.5. *Diplogaster fictor.* (*a*) Side view of female. × 205. (*b*) Head of the same, seen in dorsoventral view, lips nearly closed. (*c*) Head of the same, lateral view, lips nearly wide open. × 585. (*d*) Head of the same, lateral view, lips partially closed. × 585. (*e*) Front view of mouth, partially closed. × 585. (*f*) Lateral view, posterior portion of a male specimen. × 400. (*g*) Somewhat diagrammatic perspective view showing markings of the cuticula. × 585.

a, one of the lips; *b*, one of the 6 cephalic setae; *c*, amphid; *d*, one of the 2 more or less evertible pharyngeal hook-shaped teeth; *e*, median esophageal bulb; *f*, nerve ring; *g*, anus; *h*, rectum; *i*, intestine; *j*, terminus; *k*, posterior esophageal bulb; *l*, nerve cells; *m*, renette cell (?); *n*, left spiculum; *o*, lumen of the intestine; *p'*, preanal male seta; *p''*, *p'''*, *p''''*, postanal male setae and papillae; *q*, one of the cells of the intestine; *r*, accessory piece; *s*, flexure in anterior ovary; *t*, blind end of anterior ovary; *v*, vagina; *w*, synapsis in egg in the anterior uterus; *x*, one of the spermatozoa in the vagina; *y*, uterus, *z*, vulva. (After Cobb.)

32a (29) Stoma cylindric, terminated by a glottoid apparatus . . . .
                              Family **Rhabditidae**   **33**

32b   Stoma not truly cylindric, not terminated by a glottoid apparatus . . . . Family **Cephalobidae**   **35**

33a (32) Head with paired outwardly acting mandibles. (Usually in soil) . . . . Subfamily **Diploscapterinae**
                              Diploscapter Cobb

33b   Head without paired outwardly acting mandibles **34**

34a (33) Cuticle with asymmetric ornamentations. (Usually inhabit moss and decaying logs, about 7 genera and 35 species) . . . .
                              Subfamily **Bunonematinae**
                              Bunonema Jägerskiold

34b   Cuticle without asymmetric ornamentations. (About 10 genera and 300 species, saprozoic and commensals of invertebrates, often associated with decaying plants, a few species characteristically aquatic). . . . . Subfamily **Rhabditinae**
Rhabditis sensu lato Dujardin

35a   **(32)**   Female with 2 ovaries. (Three genera; usually associates of insects; a few species associates of snails) . . . .
Subfamily **Alloionematinae**
Alloionema Schneider

35b   Female with 1 ovary. **36**

36a   **(35)**   Stoma not completely surrounded by esophageal tissue. (Includes over 11 genera; usually saprozoic, terrestrial, rarely aquatic) . . . .
Subfamily **Panagrolaiminae**   **37**

36b   Stoma completely surrounded by esophageal tissue. (Includes at least 9 genera, usually in decaying organic matter) . . . .
Subfamily **Cephalobinae**   **38**

37a   **(36)**   Some cephalic sensory organs (6) setose. (A small genus with several aquatic species). . . . . *Macrolaimus* Maupas

37b   All cephalic sensory organs papilloid. (A large genus with a few aquatic species.) (Fig. 15.6). . . . . *Panagrolaimus* Fuchs

38a   **(36)**   Female tail bluntly rounded, lateral ridges to caudal terminus. (A large genus, usually terrestrial, sometimes aquatic) . . . .
Cephalobus Bastian

38b   Female tail attenuated (tip rarely blunt), lateral ridges extend only to phasmids. (A moderate-size genus of about 10 species, usually terrestrial, sometimes aquatic) . . . *Eucephalobus* Steiner

39a   **(1)**   Amphids circular, spiral, shepherd's crook, or transversely ellipsoid; cephalic sensory organs commonly setose; esophagus usually terminated by a bulbar swelling; caudal glands and spinneret present unless otherwise noted.   **40**

39b   Amphids pocketlike to porelike; esophagus usually cylindric or 2-part cylindric, terminated by a nonvalved bulb in very few genera **70**

40a   **(39)**   Esophago-intestinal valve usually elongated (usually greatly), labial region without rugae; ovaries either reflexed or outstretched . . . . Order **Monohysterida**   **41**

40b   Esophago-intestinal valve not elongated, labial rugae often well developed; ovaries usually reflexed   . . . Order **Chromadorida**   **62**

41a   **(40)**   Ovaries reflexed   . . . . Superfamily **Plectoidea**   **42**

41b   Ovaries outstretched **56**

42a   **(41)**   Posterior part of esophagus elongated, glandular; amphids posterior to 4 cephalic setae . . . . Family **Camacolaimidae**
Subfamily **Aphanolaiminae**   **43**

42b   Posterior part of esophagus muscular **44**

43a   **(42)**   Stoma rudimentary. (Aquatic, about 15 species.) (Fig. 15.7) . .
Aphanolaimus de Man

43b   Stoma small but distinct. (Aquatic, 2 named species) . . . . . .
Paraphanolaimus Micoletzky

44a   **(42)**   Esophagus greatly elongated, esophago-intestinal valve muscular, bulbar; 10 cephalic setae. . . . . Family **Bastianiidae**
(Fresh water, about 5 species.) (Fig. 15.8). . *Bastiania* de Man

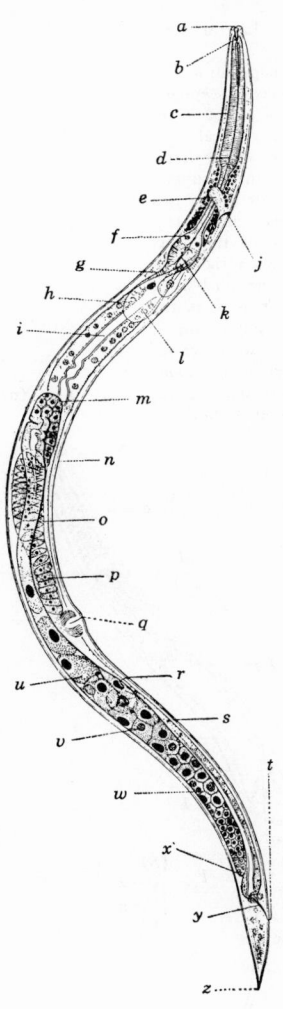

Fig. 15.6. *Pangrolaimus sub-elongatus*. Lateral view of a female. × 250. *a*, lips; *b*, stoma; *c*, anterior portion of esophagus; *d*, posterior extremity of anterior portion of esophagus; *e*, nerve ring; *f*, cardiac bulb; *g*, beginning of intestine; *h*, one of the cells of the intestine; *i*, lumen of the intestine; *j*, excretory pore; *k*, esophago-intestinal valve; *l*, sinus cell; *m*, flexure in single ovary; *n*, cuticula; *o*, ovary; *p*, spermatozoon in uterus; *q*, vulva; *r*, nucleus in ovum; *s*, body cavity; *t*, anus; *u*, ripe ovum; *v*, unripe ovum; *w*, oocyte; *x*, blind end of ovary; *y*, rectum; *z*, terminus. (After Cobb.)

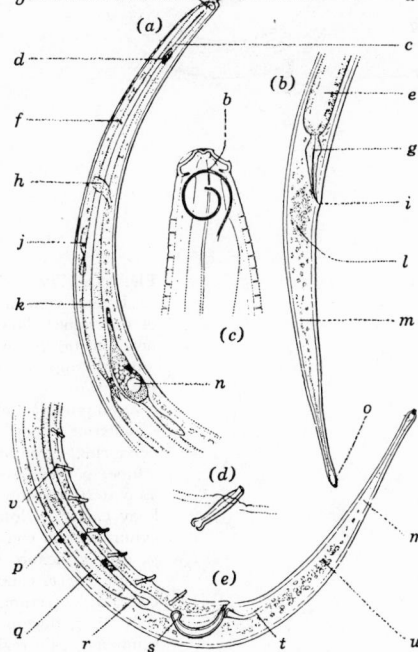

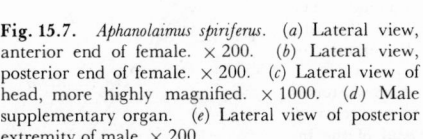

Fig. 15.7. *Aphanolaimus spiriferus*. (*a*) Lateral view, anterior end of female. × 200. (*b*) Lateral view, posterior end of female. × 200. (*c*) Lateral view of head, more highly magnified. × 1000. (*d*) Male supplementary organ. (*e*) Lateral view of posterior extremity of male. × 200.

a, mouth opening; *b*, amphid; *c*, lumen of esophagus; *d*, pigmented eye-spots (?); *e*, intestine; *f*, nerve cell; *g*, rectum; *h*, nerve ring; *i*, anus; *j*, glands; *k*, esophagus; *l*, caudal gland; *m*, duct of caudal gland; *n*, renette cell; *o*, spinneret; *p*, ejaculatory duct; *q*, intestine; *r*, anterior end of cloaca; *s*, right spiculum; *t*, backward pointing accessory piece; *u*, nerve cells (?); *v*, male supplementary organs. (After Cobb.)

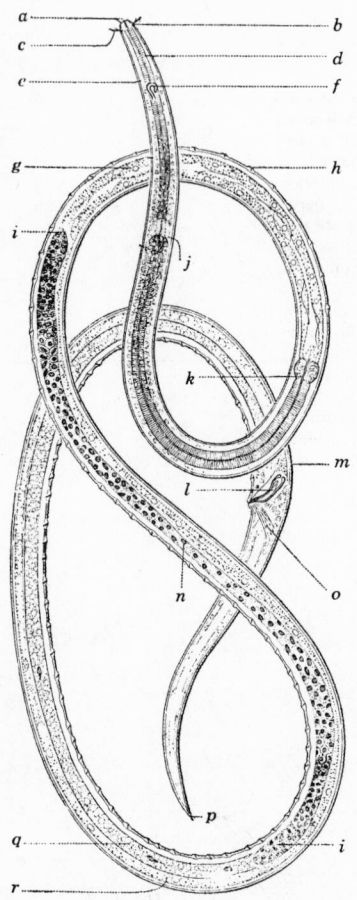

Fig. 15.8. *Bastiana exilis.* Lateral view of a male specimen. × 350.

a, one of the 6 cephalic papillae; *b,* one of the posterior set of 4 submedian cephalic setae; *c,* one of the anterior set of 6 cephalic setae; *d,* esophagus; *e,* cervical seta; *f,* amphid; *g,* one of the cells of the intestine; *h,* one of the numerous male supplementary organs; *i,* blind ends of the two testes; the two testes join each other at *n,* the complete development of the spermatozoa taking place between the locations indicated by *i* and *n;* the junction of the testes with the vas deferens is on the far side of the body and is not shown; *j,* nerve ring; *k,* esophago-intestinal valve; *l,* left spiculum; *m,* cuticula; *n,* spermatozoon; *o,* anal muscle; *p,* terminus; *q,* vas deferens; *r,* intestine. (After Cobb.)

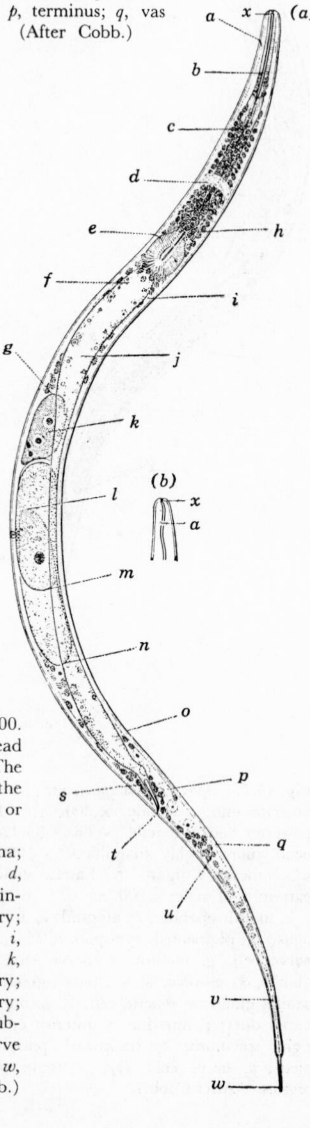

Fig. 15.9. *Rhabdolaimus minor.* × 500. (*a*) Lateral view of female. (*b*) Head of the same, showing amphid. The head in (*a*) is twisted, so that the amphid appears as if ventral, or nearly so.

a, amphid; *b,* long, narrow stoma; *c,* anterior group of nerve cells; *d,* nerve ring; *e,* bulb; *f,* wall of the intestine; *g,* flexure in anterior ovary; *h,* posterior group of nerve cells; *i,* body cavity; *j,* lumen of intestine; *k,* ovum; *l,* blind end of posterior ovary; *m,* egg; *n,* flexure in posterior ovary; *o,* cuticula; *p,* caudal glands; *q,* subcuticula; *s,* rectum; *t,* anus; *u,* nerve cells (?); *v,* duct of caudal glands; *w,* spinneret; *x,* lip region. (After Cobb.)

44b   Esophagus usually (but not always) greatly elongated; esophago-intestinal valve usually elongated but not bulbar; cephalic setae not in form of 10 in one circle. **45**

45a (44) Stoma much elongated, in form of a narrow tube, often armed at anterior end and difficult to distinguish posteriorly from esophageal lumen; cephalic sensory organs variable from setae to papillae, and from spiral or circular to transversely oval or pore-like cervical amphids . . . . Family **Leptolaimidae**   **46**

45b   Stoma prismoid to elongate-conoid; cephalic sensory organs all papillae or with 4 cephalic setae . . . . Family **Plectidae**   **50**

46a (45) Stoma long and narrow; anteriorly armed or unarmed; amphids circular to minute, transversely oval or porelike, subcephalic; esophagus terminated by muscular swelling, chromadoroid; esophago-intestinal valve elongated with modification to intestine; supplementary organs apparently absent . . . .
                       Subfamily **Rhabdolaiminae**
        (Aquatic, about 4 species.) (Fig. 15.9). . . . .
                       Rhabdolaimus de Man

46b   Stoma usually long and narrow, sometimes collapsed; sometimes short and wide to capsuliform; amphids various modifications of unispiral, position usually subcephalic but not necessarily so; esophagus terminated by a more or less distinctly set off muscular bulb; supplementary organs present or absent **47**

47a (46) Esophagus with inconspicuous anterior and conspicuous posterior bulb; esophago-intestinal valve more or less elongated, flattened; amphids postcephalic, unispire to circular; stoma short or long . . .
                       Subfamily **Haliplectinae**
        (A small genus, about 6 species; brackish, fresh-water, or marine) . . . . *Haliplectus* Cobb

47b   Esophagus without trace of anterior bulb; posterior bulb development varied, esophago-intestinal valve varied; amphids circular to unispiral; postcephalic to cervical; stoma varied **48**

48a (47) Stoma short and wide, jointed; esophagus terminated by clavate swelling; esophago-intestinal valve elongated; supplementary organs tuboid, usually numerous. (Aquatic, about 3 species.) (Fig. 15.10). *Anonchus* Cobb

48b   Stoma otherwise; esophago-intestinal valve varied; supplementary organs varied but not numerous tuboid. **49**

49a (48) Stoma short and more or less conoid; amphids circular; one mid-ventral papilloid supplement. (One species, primarily soil) . . . .
                       Domorganus Goodey

49b   Stoma elongated, collapsed, possibly 2-part in tandem; amphids unispiral; male unknown. (One species, aquatic). . . . .
                       Paraplectonema Strand

50a (45) Anterior end with ornamental or specialized dental equipment . . .
                       Subfamily **Wilsonematinae**   **51**

50b   Anterior end without special modifications. . Subfamily **Plectinae**   **54**

51a (50) Head with 6 large sclerotized inwardly acting structures; stoma short and more or less conoid; amphids postcephalic; unispiral to circular; esophago-intestinal valve short; caudal glands and spinneret absent. (Aquatic to terrestrial, 5 to 10 species). . . . . .
                       Teratocephalus de Man

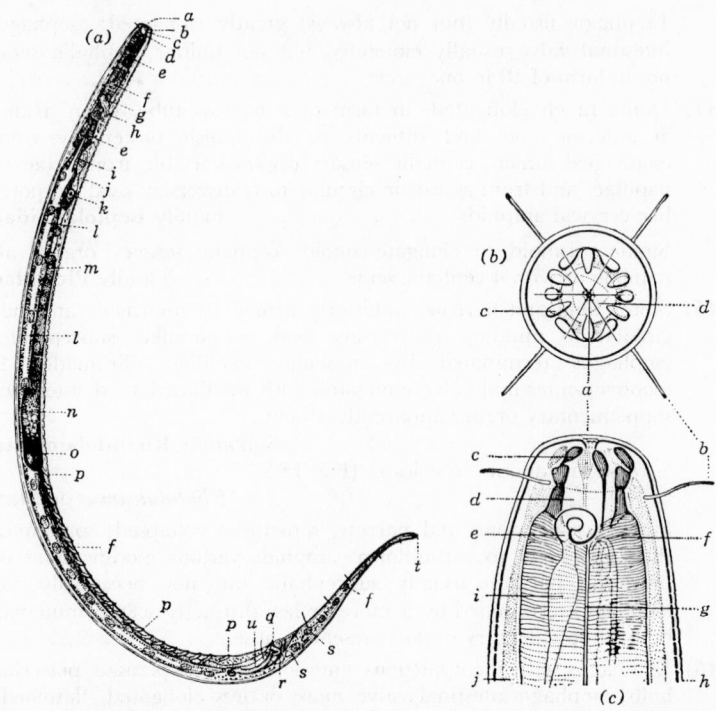

Fig. 15.10. *Anonchus monhystera.* (*a*) Lateral view. × 139.

a, mouth opening; *b*, stoma; *c*, cephalic seta; *d*, amphid; *e*, esophagus; *f*, sublateral hypodermal gland; *g*, nerve ring; *h*, excretory pore; *i*, cardia (esophago-intestinal valve); *j*, anterior end of intestine; *k*, renette cell; *l*, lumen of intestine; *m*, blind end of testicle; *n*, testicle; *o*, spermatozoa; *p*, one of the numerous supplementary organs; *q*, anus; *r*, gubernaculum; *s*, one of the caudal glands; *t*, terminus; *u*, right spiculum. (After Cobb.)

(*b*) Anterior view. × 1000. (*c*) Lateral view of anterior end. × 1000.

a, mouth opening; *b*, cephalic seta; *c*, sclerotized element, anterior portion of stoma; *d*, stoma; *e*, spiral amphid; *f*, radial musculature of esophagus; *g*, lumen of esophagus; *h*, cuticula; *i*, ampulla of amphidial gland (?); *j*, body wall. (After Cobb.)

51b		Head otherwise; amphids varied; esophago-intestinal valve usually elongated	**52**
52a	(51)	Head with 6 externally curved appendages. (One species, fresh water) . . . . **Anthonema** Cobb	
52b		Head otherwise armed	**53**
53a	(52)	Head with 4 stout, incurved, nonbranched setae and 4 submedian cushionlike cuticular swellings. (One species, fresh water) . . . . . **Tylocephalus** Crossman	
53b		Head with branched lamellar expansions and cushions; 4 enclosed setae. (Semi-aquatic, saprozoic, 6 species) . . . **Wilsonema** Cobb	
54a	(50)	Stoma short and conoid; esophagus greatly elongated, esophago-intestinal valve elongated; 4 cephalic setae; preanal supplements tuboid. (Aquatic, 2 species.) (Fig. 15.11) . . **Chronogaster** Cobb	
54b		Stoma cylindroid; esophagus not greatly elongated; probably 4 cephalic setae; supplements papilloid to tuboid	**55**

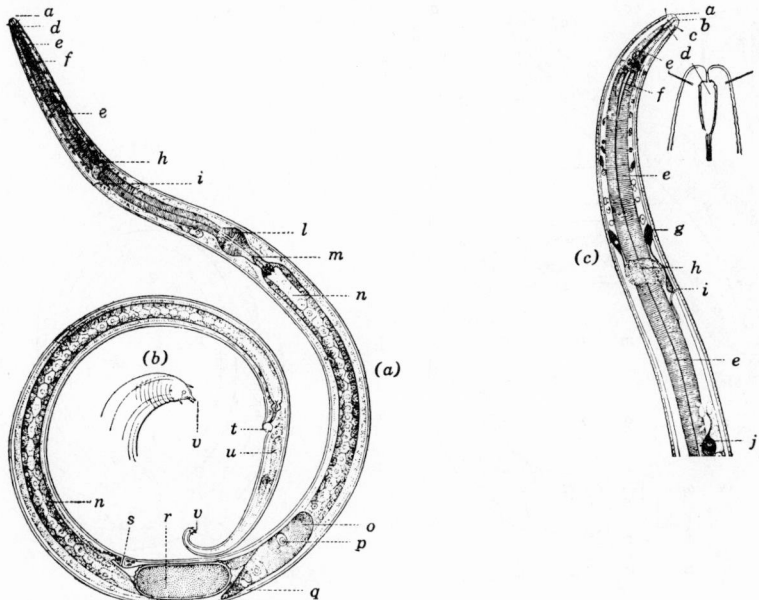

Fig. 15.11. *Chronogaster gracilis.* (*a*) Lateral view. × 150. (*b*) Posterior end. × 985. (*c*) Anterior end. × 300 and (inset) tip, greatly enlarged.

a, lips; *b,* papilla; *c,* cephalic seta; *d,* stoma; *e,* esophagus; *f,* esophageal lumen; *g,* problematical organs; *h,* nerve ring; *i,* excretory pore; *j,* renette cell; *l,* valvular apparatus in bulb; *m,* cardia (esophago-intestinal valve); *n,* intestine; *o,* flexure in ovary; *p,* nucleus of ovum; *q,* blind end of ovary; *r,* egg; *s,* vulva; *t,* anus; *u,* caudal gland; *v,* spinneret. (After Cobb.)

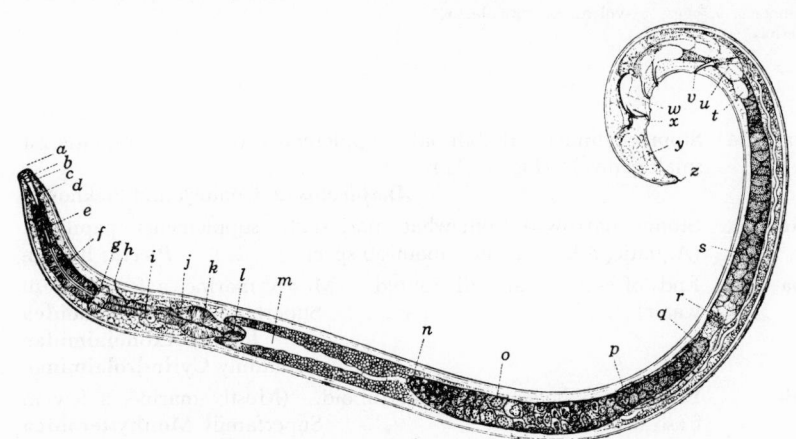

Fig. 15.12. *Anaplectus granulosus.* Male. × 220.

a, mouth; *b,* papillalike cephalic setae; *c,* lateral organ; *d,* stoma; *e,* posterior chamber of stoma; *f,* esophagus; *g,* nerve ring; *h,* excretory pore; *i,* renette cell; *j,* glandular (?) cell; *k,* bulb; *l,* esophago-intestinal valve; *m,* intestine; *n,* blind end of anterior testicle; *o,* spermatocyte; *p,* flexure in posterior testicle; *q,* blind end of posterior testicle; *r,* junction of testicles; *s,* vas deferens; *t,* glandular (?) organ; *u,* muscle to one of the three supplementary organs; *v,* anterior supplementary organ; *w,* spiculum; *x,* anus; *y,* one of the caudal papillae; *z,* spinneret. (After Cobb.)

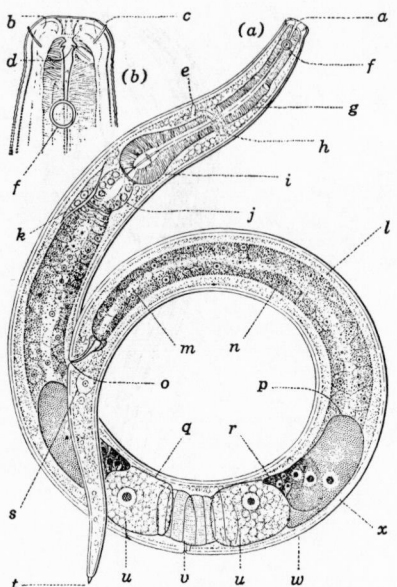

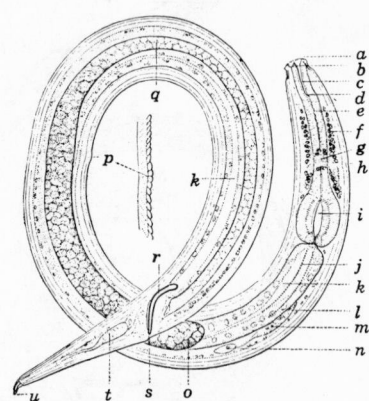

Fig. 15.13. *Microlaimus fluviatilis.* (*a*) Lateral view of a female. × 250. (*b*) Head of the same. × 700.

a, mouth opening; *b*, one of the 6 cephalic papillae; *c*, one of the 4 cephalic setae; *d*, one of the small stomatal teeth; *e*, excretory pore; *f*, circular amphid; *g*, esophagus; *h*, nerve ring; *i*, cardiac bulb; *j*, esophago-intestinal valve; *k*, renette cell; *l*, body cavity; *m*, lumen of intestine; *n*, one of the cells of the intestine; *o*, anus; *p*, flexure in posterior ovary; *q*, uterus; *r*, blind end of posterior ovary; *s*, one of the 3 caudal glands; *t*, spinneret; *u*, eggs; *v*, vulva; *w*, cuticula; *x*, epidermis. (After Cobb.)

Fig. 15.14. *Chromadora canadensis.* × 300.

a, mouth opening; *b*, dorsal tooth; *c*, stoma; *d*, base of the stoma; *e*, esophagus; *f*, nerve cells; *g*, nerve ring; *h*, excretory pore; *i*, valvular apparatus of the bulb; *j*, longitudinal row of cuticular markings characteristic of the genus; *k*, intestine, *l*, renette cell; *m*, nucleus of renette cell; *n*, cell accessory to the renette cell; *o*, blind end of testicle; *p*, reversal of the striations of the cuticula; *q*, vas deferens; *r*, spiculum; *s*, anus; *t*, caudal gland; *u*, spinneret. (After Cobb.)

55a	(54)	Stoma cylindric, rhabditoid; supplements tuboid. (Aquatic, a small genus.) (Fig. 15.12) . . . . *Anaplectus* de Coninck and Stekhoven
55b		Stoma narrowed somewhat posteriad; supplements papilloid. (Aquatic, a large genus, about 50 species) . . . . *Plectus* Bastian
56a	(41)	Ends of esophageal radii tuboid. (Mostly marine, a few in fresh water) . . . . Superfamily **Axonolaimoidea** Family **Axonolaimidae** Subfamily **Cylindrolaiminae** 57
56b		Ends of esophageal radii not tuboid. (Mostly marine, a few in fresh water) . . . . Superfamily **Monhysteroidea** 58
57a	(56)	Male with 3 papilloid preanal supplements. (One species, fresh water) . . . . *Aulolaimus* de Man
57b		Male with 1 preanal papilloid supplement. (Fresh water, 10 species) . . . . *Cylindrolaimus* de Man
58a	(56)	Stoma elongated, usually with 1 to 3 teeth at anterior joint . . . . . Family **Microlaimidae** 59

58b   Stoma short funnel-shaped, teeth apparently absent. (A large
genus, primarily marine, with many fresh-water species) . . . . . .
Family **Monhysteridae**
Monhystera Bastian

59a   **(58)**   One anterior outstretched ovary; 6 small cephalic setae; amphids
postcephalic, circular; esophageal bulb small; esophago-intestinal
valve inconspicuous. (Aquatic, few species) . . . .
Monhystrella Cobb

59b   Two outstretched opposed ovaries; cephalic setae 0, 4, 6, or 10;
amphids variable in position; esophago-intestinal valve varied . . .   **60**

60a   **(59)**   Esophago-intestinal valve elongate, flattened; 4 or 6 short plus 4
long cephalic setae; amphids perfect or broken circle, variable in
position; supplements apparently absent. (A large genus, primarily
marine, secondarily fresh water and moist soil.) (Fig. 15.13). . . .
Microlaimus de Man

60b   Esophago-intestinal valve not particularly elongated **61**

61a   **(60)**   Cephalic setae 4; amphids postcephalic, circular to spiral;
esophago-intestinal valve large, wide. (A small genus, fresh
water). . . . . *Microlaimoides* Hoeppli

61b   Cephalic setae 10; amphids small, circular, postcephalic; esophago-
intestinal valve short and wide. (A small genus, fresh water) . . .
Rogerus Hoeppli

62a   **(40)**   Amphids nearly labial in position, usually minute or ellipsoid.
(Mostly marine, some in fresh water) . Family **Chromadoridae**   **63**

62b   Amphids postlabial, circular to spiral . Family **Cyatholaimidae**   **66**

63a   **(62)**   Cuticular punctation interrupted laterally. (Extremely large
genus, mostly marine.) (Fig. 15.14) . . . . *Chromadora* Bastian

63b   Cuticular punctations not interrupted laterally **64**

64a   **(63)**   Teeth 3, subequal. (Fresh water and marine). . . . .
Prochromadorella Micoletzky

64b   Dorsal tooth large, subventrals weak or absent. **65**

65a   **(64)**   Dorsal tooth heavily sclerotized. . . . . *Punctodora* Filipjev

65b   Dorsal tooth weakly sclerotized . . . . . . *Chromadorita* Filipjev

66a   **(62)**   Only 4 cephalic setae . . . . Subfamily **Ethmolaiminae**   **67**

66b   Head with 10 cephalic setae . . . . Subfamily **Cyatholaiminae**   **69**

67a   **(66)**   Stoma cylindroid, 3 teeth at anterior end of stoma (bulb massive,
globoid, divided, all fresh water). (Fig. 15.15) . . . .
Ethmolaimus de Man

67b   Stoma not cylindroid or without 3 equal teeth at anterior end;
bulb not massive **68**

68a   **(67)**   One ovary, anterior, reflexed; 1 small anterior dorsal tooth. . . . .
Monochromadora Schneider

68b   Two ovaries, opposed, reflexed. (Fresh water, about 10 species) . . . . *Prodesmodora* Micoletzky

69a   **(66)**   Esophagus with distinct bulb. (Fresh water, 9 species.) (Fig.
15.16). . . . . *Achromadora* Cobb

69b   Esophagus clavate to cylindroid. (Mostly marine, 2 species fresh
water). . . . . *Paracyatholaimus* Micoletsky

70a   **(39)**   Stylet absent in all stages of development . . . . Order **Enoplida**
Cuticle at head not reduplicated. (Most in fresh water) . . . . .
Superfamily **Tripyloidea**   **71**

70b   Stylet present, at least in ovic larva; cephalic sensory organs always papilloid; caudal glands absent **86**

71a (70) Stoma usually in form of heavily sclerotized buccal capsule, armed with 1 or more teeth, often with denticles; setae absent . . . .
                Family **Mononchidae**   **72**

71b   Stoma not in form of heavily sclerotized buccal capsule; setae present or absent. **78**

72a (71) Stoma narrow, cylindroid, with small ventral tooth at base . . . . .
                Bathyodontus Fielding

72b   Stoma otherwise **73**

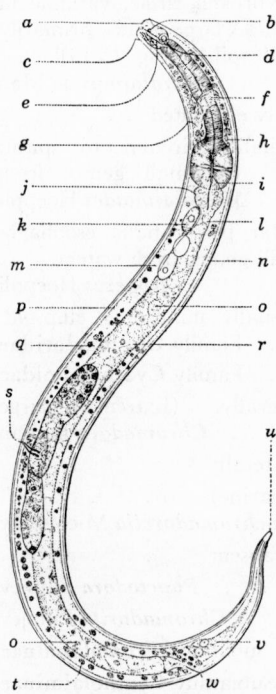

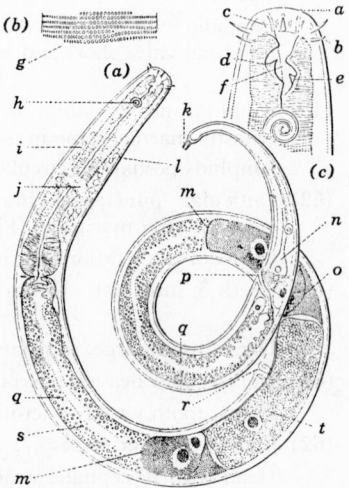

Fig. 15.15. *Ethmolaimus americanus.* Lateral view of a female. × 238.

a, lips; b, minute dorsal and ventral stomatal teeth; c, one of the 4 cephalic setae; d, amphid; e, stoma; f, nerve ring; g, excretory pore; h, nerve cells; i, cardiac bulb; j, esophago-intestinal valve; k, renette cell (?); l, beginning of main portion of the intestine; m, one of 2 pairs of unicellular organs of unknown significance; n, cuticula; o, one of the cells of the intestine; p, subcuticula; q and r, body cavity; s, vulva; t, nucleus of one of the muscle cells; u, spinneret; v, one of the caudal glands; w, anus. (By Cobb.)

Fig. 15.16. *Achromadora minima.* (a) Lateral view of a female. × 400. (b) Lateral view, cuticular markings. × 737. (c) Lateral view of head. × 737.

a, cephalic papilla; b, cephalic seta; c, one of the ribs of the stoma; d, dorsal stomatal tooth; e, subventral (?) stomatal tooth; f, stoma, g, cuticular markings; h, amphid; i, nerve cell; j, nerve ring; k, spinneret; l, excretory pore; m, flexure of ovary; n, one of the caudal glands; o, blind end of posterior ovary; p, anus; q, intestine; r, vulva; s, one of the granules of the intestine; t, egg. (After Cobb.)

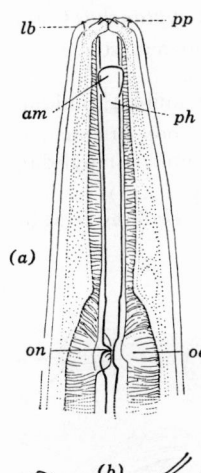

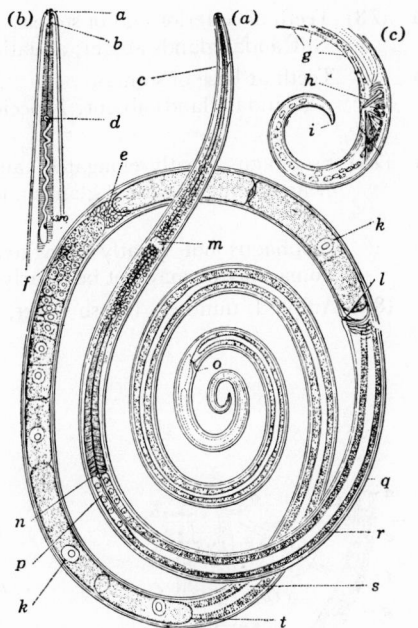

Fig. 15.17. *Cryptonchus nudus.* (*a*) Anterior end, lateral view. × 286. (*b*) Posterior end, lateral view. × 181.

lb, lip region; *pp,* labial papillae; *am,* amphid; *ph,* stoma; *on,* onchus or tooth; *oe,* esophagus; *sp,* spinneret. (After Cobb.)

Fig. 15.18. *Alaimus simplex.* (*a*) Lateral view of a female. × 200. (*b*) Anterior extremity, lateral view. × 370. (*c*) Posterior extremity of a male, lateral view. × 260.

a, lip region; *b,* stoma; *c,* amphid; *d,* amphid, enlarged; *e,* group of spermatozoa at posterior portion of ovary; *f,* blind end of ovary; *g,* male supplementary papillae; *h,* left spiculum; *i,* terminus; *k,* egg; *l,* vulva; *m,* nerve ring; *n,* posterior extremity of esophagus; *o,* rectum; *p,* modified cells of anterior intestine; *q,* cuticula; *r,* wall; *s,* lumen of intestine; *t,* flexure in single ovary. (After Cobb.)

73a	(**72**)	Stoma conoid, with small dorsal tooth and 1 large and 1 small ventral tooth . . . . ***Mononchulus*** Cobb	
73b		Stoma not conoid	74
74a	(**73**)	Ventral stomatal ridge bearing longitudinal row of denticles . . . . ***Prionchulus*** Cobb	
74b		Ventral stomatal ridge absent	75
75a	(**74**)	Numerous subventral denticles present . . . . ***Mylonchulus*** Cobb	
75b		Without denticles	76
76a	(**75**)	Dorsal tooth usually massive, anterior to mid-stomatal . . . . ***Monochus*** Bastian	
76b		Dorsal tooth small or absent, all 3 teeth poststomatal	77
77a	(**76**)	Teeth retrorse . . . . ***Anatonchus*** Cobb	
77b		Teeth not retrorse . . . . ***Iotonchus*** Cobb	
78a	(**71**)	Stoma cylindric . . . . Family **Ironidae**	79
78b		Stoma not cylindric	80

79a (78) Teeth at anterior end of stoma. . . . . Subfamily **Ironinae**
(Caudal glands absent; aquatic, 16 species). . . . *Ironus* Bastian

79b Teeth at base of stoma . . . . Subfamily **Cryptonchinae**
(Caudal glands absent; 1 species, aquatic.) (Fig. 15.17) . . . . .
Cryptonchus Cobb

80a (78) Esophagus greatly elongated, anterior end attenuated, stoma not
clearly defined; caudal glands and spinneret absent . . . .
Family **Alaimidae**   81

80b Esophagus not greatly elongated, anterior end not attenuated,
stoma may or may not be clearly defined . . Family **Tripylidae**   83

81a (80) Amphids minute. (Fresh water, 16 species.) (Fig. 15.18) . . . . .
Alaimus de Man

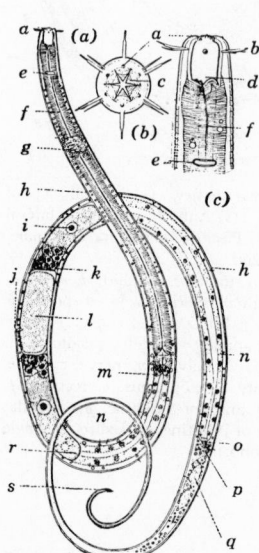

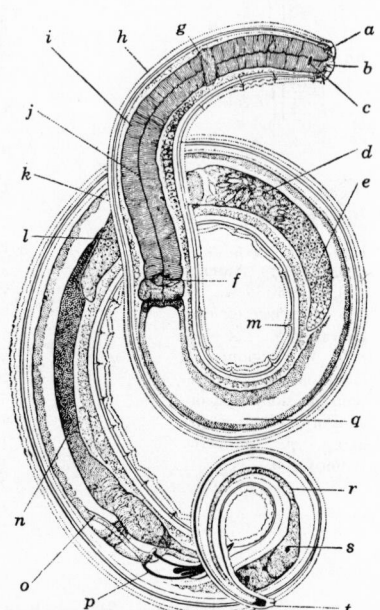

Fig. 15.19. *Prismatolaimus stenurus.* (a) Lateral view of a female. × 200. (b) Front view of head. × 565. (c) Side view of head. × 565.
a, one of the 6 cephalic papillae; b, one of the 10 cephalic setae; c, one of the 6 thin lips; d, stoma; e, amphid; f, lumen of the esophagus; g, nerve ring; h, cuticula; i, nucleus of ovum; j, vulva; k, blind end of posterior ovary; l, egg; m, beginning of the intestine; n, one of the cells of the wall of the intestine; o, rectum; p, anus; q, one of the caudal glands; r, flexure in anterior ovary; s, spinneret. (After Cobb.)

Fig. 15.20. *Tripyla lata.* × 100.
a, labial papilla; b, lip; c, amphid; d, spermatozoon; e, spermatocyte of anterior testis; f, base of esophagus, pseudo-bulb; g, nerve ring; h, cuticula; i, esophagus; j, lining of esophagus; k, intestine; l, posterior testis; m, male supplementary organ; n, vas deferens; o, retractor muscle of spiculum; p, right spiculum; q, intestine; r, duct of caudal gland; s, caudal gland; t, spinneret. (After Cobb.)

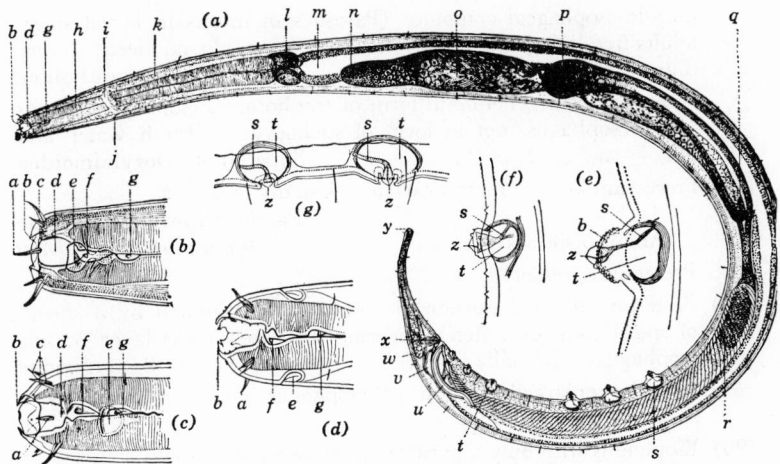

Fig. 15.21. *Trilobus longus.* (a) Male. (b) Head, lateral view. (c) Head, lateral view. (d) Head, ventral view. (e) Anterior supplementary organ. (f) Posterior supplementary organ. (g) Two supplementary organs from an exceptional female.

a, lateral seta; b, papilla; c, submedian seta; d, stoma; e, lateral organ; f, tooth; g, tooth; h, esophagus; i, nerve ring; j, excretory pore; k, body muscles; l, esophago-intestinal valve; m, intestine; n, blind-end anterior testicle; o, testicle; p, junction of testicles; q, blind-end posterior testicle; r, vas deferens; s, nerve of supplementary organ; t, cavity of supplementary organ; u, left spiculum; v, gubernaculum; w, the 3 caudal glands; x, anus; y, spinneret; z, apex of supplementary organ. (After Cobb.)

81b		Amphids distinct, not minute	**82**
82a	(81)	Amphids transverse slits. (Fresh water, 11 species) . . . . *Amphidelus* Thorne	
82b		Amphids ellipsoid, transverse. (Fresh water, 1 species). . . . . . . *Adorus* Cobb	
83a	(80)	Stoma cylindroid. (Fresh water, about 15 species.) (Fig. 15.19) . . . . *Prismatolaimus* de Man	
83b		Stoma conoid to obscure	**84**
84a	(83)	Lips massive, deeply cut, dividing head into 3 jaws . . . . *Trischistoma* Cobb	
84b		Lips not massive, not dividing head into jaws	**85**
85a	(84)	Stoma collapsed, male with papilloid preanal supplements. (Fresh water, 14 species.) (Fig. 15.20) . . . . . . *Tripyla* Bastian	
85b		Stoma rather conoid, male with bubblelike preanal supplements. (Fresh water, 22 species.) (Fig. 15.21) . . . . . *Trilobus* Bastian	
86a	(70)	Adult male with muscular caudal sucker. (Parasites of vertebrates). . . . . Order **Dioctophymatida**	
86b		Adult male without muscular caudal sucker . . . . Order **Dorylaimida**	**87**
87a	(86)	Esophageal glands in form of a single row of stichocytes, outside contour of esophagus. (Parasites of vertebrates). . . . . Superfamily **Trichuroidea**	
87b		Esophageal glands not as above.	**88**
88a	(87)	Intestine degenerate, in form of trophosome and growing anterior to base of esophagus, esophageal glands in 2 rows of stichocytes,	

outside esophageal contour.   (Parasites of insects in larval stage, adults free-living, commonly in fresh water, do not feed; a very difficult group) . . . . Superfamily **Mermithoidea**

88b   Intestine not degenerate in form of trophosome, esophageal glands inside esophagus, not in form of stichocytes. (Fresh water and soil) . . . . Superfamily **Dorylaimoidea**   89

89a **(88)** Prerectum absent; meromyarian. (Usually in soil) . . . .
            Family **Diphtherophoridae**
            (About 6 species). . . . . *Diphtherophora* de Man

89b   Prerectum present   90

90a **(89)** Posterior enlarged portion of esophagus surrounded by a sheath of spiral muscles (often appearing as a refractive layer outside esophagus). (Usually in soil) . . . . Family **Belondiridae**

90b   Posterior enlarged portion of esophagus not surrounded by a sheath of spiral muscles   91

91a **(90)** Esophagus with only a pyriform or elongate basal bulb . . . .
            Family **Leptonchidae**   92

91b   Esophagus with enlarged posterior third or more . . . .
            Family **Dorylaimidae**   98

92a **(91)** Stoma armed with a mural tooth . . . Subfamily **Campydorinae**
            Campydora Cobb

92b   Stoma armed with an axial stylet   . . . Subfamily **Leptonchinae**   93

93a **(92)** Tail extremely long and filiform; stylet with distinct, rather knob-like extensions. (One species) . . . . . *Aulolaimoides* Micoletzky

93b   Tail short to conically attenuated.   94

94a **(93)** Esophageal bulb not set off by a constriction.   95

94b   Esophageal bulb set off by a constriction.   97

95a **(94)** Bulb length 2 or more times neck diameter. (Seven species) . . . .
            Dorylaimoides Thorne and Swanger

95b   Bulb length less than twice neck diameter   96

96a **(95)** Bulb lining divided into 2 sections. (One species). . . . .
            Tyleptus Thorne

96b   Bulb lining not divided into 2 sections. (Three species). . . . . . .
            Leptonchus Cobb

97a **(94)** Stylet with additional stiffening piece. (Six species) . . . .
            Tylencholaimellus Cobb

97b   Stylet without additional stiffening piece. (Three species) . . . . .
            Doryllium Cobb

98a **(91)** Stoma heavily sclerotized, enlarged, often dentate. . . . .
            Subfamily **Actinolaiminae**
            (Fresh water, about 31 species.) (Fig. 15.22) . . . .
            Actinolaimus Cobb

98b   Stoma not heavily sclerotized, enlarged, or dentate   99

99a **(98)** Stoma armed with a mural tooth   . . . Subfamily **Nygolaiminae**   100

99b   Stoma armed with an axial stylet.   102

100a **(99)** Esophagus with definitely set off basal portion. (Usually in soil, about 21 species) . . . . *Nygolaimus* Cobb

100b   Esophagus without definitely set off basal portion   101

101a **(100)** Lips 6. (One species, moist soil near irrigation ditch) . . . .
            Oionchus Cobb

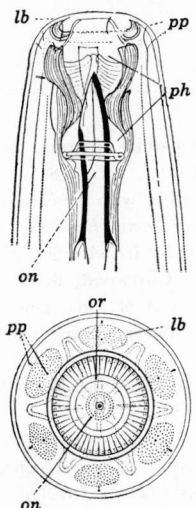

Fig. 15.22. *Actinolaimus radiatus.* (a) Anterior end in lateral view. × 285. (b) Anterior view. × 514.

lb, lip region; *pp*, innervated papillae; *ph*, stoma; *on*, onchus or spear; *or*, mouth opening. (After Cobb.)

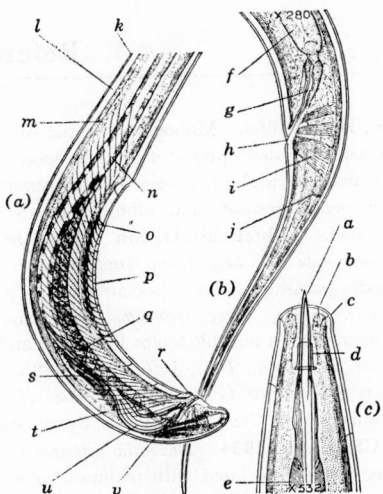

Fig. 15.23. *Dorylaimus fecundus.* (a) Tail end of a male. × 188. (b) Tail end of a female. × 280. (c) Head end of a female. × 532.

a, apex of spear, showing oblique opening; *b*, papilla of the anterior circlet; *c*, papilla of the posterior circlet; *d*, guiding ring for the spear; *e*, commencement of the esophagus; *f*, prerectum; *g*, rectum; *h*, anus; *i*, anal muscles; *j*, caudal papilla; *k*, outer cuticula; *l*, inner cuticula; *m*, muscular layer; *n*, prerectum; *o*, one of the numerous oblique copulatory muscles; *p*, one of the ventral series of male supplementary organs; *q*, ejaculatory duct; *r*, pair of preanal papillae; *s*, retractor muscles of the spicula; *t*, muscular layer; *u*, right spiculum; *v*, accessory piece. (After Cobb.)

101b		Lips 3. (One species in water pools) . . . . . ***Enoplochilus*** Kreis
102a	**(99)**	Stylet not greatly attenuated, at most with simple basal extensions . . . . Subfamily **Dorylaiminae** (Over 300 species, usually in soil, some species characteristically in fresh water.) (Fig. 15.23) . . . . ***Dorylaimus*** Dujardin
102b		Stylet greatly attenuated or distinctly compound with knobbed or flanged basal extensions. (Usually in soil about plant roots, rarely found in fresh water) . . . Subfamily **Tylencholaiminae** 103
103a	**(102)**	Stylet with extensions less than twice as long as labial diameter. (Usually in soil, about 15 species) . . . . ***Tylencholaimus*** de Man
103b		Stylet with extensions more than twice as long as labial diameter 104
104a	**(103)**	Stylet very long with extensions long and flanged. (Usually parasitic on roots of terrestrial plants, about 15 species) . . . . ***Xiphinema*** Cobb
104b		Stylet shorter, extensions rodlike or broadly flanged. (About 15 species). . . . . ***Enchodelus*** Thorne

References

Bastian, H. C. 1865. Monograph on the Anguillulidae, or free nematoids, marine, land and fresh water; with descriptions of 100 new species. *Trans. Linn. Soc. London*, 25:73–184. **1866.** On the anatomy and physiology of the nematoids, parasitic and free; with observations on their zoological position and affinities to the echinoderms. *Phil. Trans. Roy. Soc. London*, 156:545–638. **Buetschli, O. von. 1873.** Beiträge zur Kenntniss der freilebenden Nematoden. *Nova Acta Ksl. Leop.-Carol. Deutsch. Akad. Naturf.*, 36:1–144. **Chitwood, B. G. 1935.** Nematodes parasitic in and associated with, Crustacea, and descriptions of some new species and a new variety. *Proc. Helminthol. Soc. Wash., D.C.*, 2:93–96. **1949a.** Ring nematodes (Criconematidae) a possible factor in decline and replanting problems of peach orchards. *Proc. Helminthol. Soc. Wash., D.C.*, 16:6–7. **1949b.** Root-knot nematodes—Part I. A revision of the genus *Meloidogyne* Goeldi, 1887. *Proc. Helminthol. Soc. Wash., D.C.*, 16:90–104. **1958.** The English word "Nema" revised. Systematic zool., 6:184–186. **Chitwood, B. G. and M. B. Chitwood. 1934.** *Daubaylia potomaca* n. sp., a nematode parasite of snails with a note on other nemas associated with molluscs. *Proc. Helminthol. Soc. Wash., D.C.*, 1:8–9. **1938.** Notes on the "culture" of aquatic nematodes. *J. Wash. Acad. Sci.*, 28:455–460. **1937–1950.** *An Introduction to Nematology.* Monumental Printing Co., Baltimore. **Chitwood, B. G. and A. McIntosh. 1934.** A new variety of *Alloionema* (Nematoda: Diplogasteridae), with a note on the genus. *Proc. Helminthol. Soc. Wash., D.C.*, 1:37–38. **Cobb, M. V. 1915.** Some freshwater nematodes of the Douglas Lake region in Michigan. *Trans. Am. Microscop. Soc.*, 34:21–47. **Cobb, N. A. 1893.** Nematodes, mostly Australian and Fijian. *Macleay Mem. Vol. Linn. Soc.*, N. S. Wales. Dept. Agric. Misc. Publ. No. 13. **1913.** New nematode genera found inhabiting fresh water and non-brackish soils. *J. Wash. Acad. Sci.*, 3:432–444. **1914.** North American free-living fresh water nematodes. *Trans. Am. Microscop. Soc.*, 33:35–99. **1918a.** Free-living nematodes. In: Ward and Whipple. *Fresh-Water Biology*, 1st ed., pp. 499–505. Wiley, New York. **1918b.** Filter-bed nemas. Nematodes of the slow sand- and filter-beds of American cities. *Contrib. Sci. Nemat.*, No. 7:189–213. **1920.** One hundred new nemas. *Contrib. Sci. Nemat.*, No. 9:217–343. **1931.** Some recent aspects of Nematology. *Science*, 73:22–29. **1932.** The English word "Nema." *J. Am. Med. Assoc.*, 98:75. **Filipjev, I. N. and J. H. S. Stekhoven. 1941.** *A Manual of Agricultural Helminthology.* E. J. Brill, Leiden. **Goodey, T. 1947.** *Domorganus macronephriticus* n.g., n.sp., a new cylindrolaimid free-living soil nematode. *J. Helminthol.*, 21:175–180. **1951.** *Soil and Freshwater Nematodes.* Wiley, New York. **Hoeppli, R. 1926.** Studies of free-living nematodes from the thermal waters of Yellowstone Park. *Trans. Am. Microscop. Soc.*, 45:234–255. **Huxley, T. 1856.** Lectures on general natural history. *Med. Times, London*, N.S., 13:27–30. **Kreis, H. A. 1927.** Über die Bedeutung der geographischen Verbreitung der freilebenden marinen und Süsswassernematoden. *Verhandle. schweiz. Naturforsch. Ges.*, 2:196–197. **Hyman, L. H. 1951.** *The Invertebrates.* Vol. III, *Acanthocephala, Aschelminthes and Entoprocta.* McGraw-Hill, New York. **Man, J. G. de. 1876.** Onderzoekingen over vrij in de aarde levende Nematoden. *Tijdschr. Ned. Dierk. Ver.*, 2:78–196. **1880.** Die einheimischen, frei in der reinen Erde und im süssen Wasser lebende Nematoden. *Tijdschr. Ned. Dierk. Ver.*, 5:1–104. **1884.** *Die frei in der reinen Erde und im süssen Wasser lebenden Nematoden.* E. J. Brill, Leiden. **1904.** Nematodes libres. Résultats du voyage du S. Y. Belgica. *Exped. Antarctique Belge.* **1921.** Nouvelles recherches sur les Nématodes libres terricoles de la Hollande. *Capita Zool.*, 1:1–62. **Micoletzky, H. 1913.** Die freilebenden Süsswasser-nematoden der Ostalpen I, II. *Sitzber. Akad. Wiss. Wien Math. naurtw. Kl.*, 122:111–122, 543–548. **1914a.** Oekologie ostalpiner Süsswasser-Nematoden. *Intern. Rev. gen. Hydrobiol. Hydrog.* **1914b.** Freilebende Süsswasser-Nematoden der Ost-Alpen mit besonderer Berücksichtigung des Lunzer Seengebietes. *Zool. Jahrb. Abt. Syst.*, 36:331–546. Nachtrag, 1915. 38:246–274. **1922.** Freie Nematoden aus dem Grunschlamm norddeutscher Seen (Madüund Plönersee). *Arch. Hydrobiol.*, 13:532–560. **1925.** Die freilebenden Süsswasser und Moornematoden Dänemarks. *Kgl. Danske*

Videnskab. Selskabs. Scrifter. Sect. Sci., 8 s., 10:57–310. **1925b.** Zur Kenntnis tropischer freilebender Nematoden aus Surinam, Trinidad und Ostafrika. *Zool. Anz.*, 64:1–28. **Overgaard, C. 1948a.** An apparatus for the quantitative extraction of nematodes and rotifers from soil and moss. *Natura Jutlandica*, 1:271–278. **1948b.** Studies on the soil microfauna. I. The moss inhabiting nematodes and rotifers. Detl Laerde Selskabs Skrifter. Ser. Sc. Nat. I:1–98. **Nielsen, C. Overgaard. 1949.** Studies on the soil microfauna. II. The soil inhabiting nematodes. Naturhist. Mus. Aarhus, 131 pp. **Steiner, G. and Heinly, H. 1922.** The possibility of control of *Heterodera radicicola* and other plant-injurious nemas by means of predatory nemas, especially by *Mononchus papillatus* Bastian. J. Wash. Acad. Sc. V. 12 (16):367–386. **Thorne, G. 1937.** A revision of the nematode family Cephalobidae Chitwood and Chitwood. *Proc. Helminthol. Soc. Wash., D.C.*, 4:1–16. **1939.** A monograph of the nematodes of the superfamily Dorylaimoidea. *Capita Zool.*, 8:1–261. **Thorne, G. and H. H. Swanger. 1936.** A monograph of the nematode genera *Dorylaimus* Dujardin, *Aporcelaimus* n.g., *Dorylaimoides* n.g., and *Pungentus* n.g. *Capita Zool.*, 6:1–156.

16

Gordiida

B. G. CHITWOOD

The organisms commonly known as horsehair worms are at present placed in the Order Gordiida of the Phylum or Class Nematomorpha Vejdosky. This group is supposed to contain another order, the marine Nectonematida. The soundness of the association of the two orders and the status of the phylum itself may be questioned. In the adult stage gordiids are free-living, unsegmented, cylindroid worms from 4 to 40 or more in. long, inhabiting fresh water. The body is hard, even wiry, and is colored from tannish to dark brown and sometimes has blackish pigment, particularly near the head. They are particularly common in warm pools or sluggish streams between May and September. In the larval stages they are parasitic in Arthropods (crustaceans and insects) or molluscs.

The body is covered externally by a noncellular, multilayered cuticle, which commonly contains many types of specialized structures such as bristles, tubercles, araeolae, and pore-canals (Figs. 16.1, 16.2). These structures make up the greater part of the taxonomic characters in use at the present time. In addition, the shape of the posterior end is used in identification. The only copulatory apparatus is the forking of the tail, relatively common in the male, less common in the female. The tail may be extremely abrupt and blunt or it may be forked in two or three large blunt lobes (Fig. 16.1).

Adults are commonly fixed for microscopic study in formalin, or in alcohol then cleared in lactophenol, beech-wood creosote, or similar materials. Color markings in the living specimen may fade to a minor extent in clearing fluid. Such matters as taper of body, squareness of head, shape of tail, color markings at head and sometimes on other parts of the body are noted. One usually cuts out a section from the mid-region of body, boils it in lactophenol, slits it open and scrapes out the internal tissues; then the skin or cuticle is mounted on a slide, external side up, and a cover slip is added. A careful study of the cuticular pattern is then made. The form of the male tail, cloacal groove, and tubercles or hairs in this area are also noted.

KEY TO GENERA

1a   Cuticle smooth, at most with faint squares but no distinct separation into sections known as araeolae. Bristles, if present, arising from fibrous cuticle. Mouth not connected with intestine. Ovaries not enclosed by mesenchyme. Male tail bifurcate, with crescentlike transverse postcloacal ridge; no spines or warts. Female with bluntly rounded tail . . . . Family **Gordiidae**
Single genus . . . . *Gordius* Linneaus

1b   Cuticle not smooth, with araeolae, cuticular papillae and other ornamentations. Mouth connected with esophagus. Ovaries enclosed by mesenchyme, double mesenteries in female. Male tail bifurcate or simple, without transverse ridge. Female tail bluntly rounded or trifurcate . . . . Family **Chordodidae**   2

2a   (1) Female tail trilobed, male tail in 2 long parts, with hair groups around cloaca . . . . Subfamily **Paragordiinae**
Single genus. (Fig. 16.1) . . . . *Paragordius* Camerano

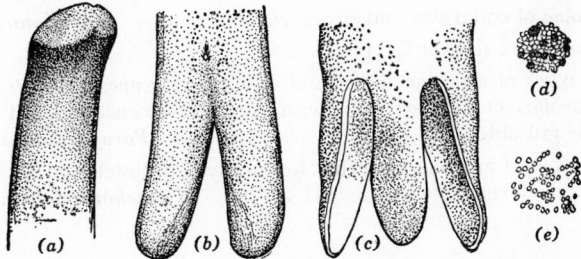

Fig. 16.1. *Paragordius varius.* (a) Lateral aspect of head. × 25. (b) Ventral view of tail. × 25. (c) Dorsal view of tail, female. × 25. (d) Surface view of cuticle of male. Highly magnified. (e) Surface view of cuticle of female. Highly magnified. (a, b, by Ward in first edition of this book; c, after Stiles; d, e, after Montgomery.)

2b   Female tail simple, male tail bifurcate or simple   3

3a   (2) Cuticle without araeolae but with paired ridges and inter-ridge structures, male tail with weak furrow, cloacal opening terminal . . .
Subfamily **Chordodiolinae**
Single genus . . . . *Chordodiolus* Heinze

3b   Cuticle with araeolae, male tail various **4**

4a (3) Male tail simple, with weak furrow, no copulatory ornamentations; araeolae usually of 2 heights; female tail simple. . . . .

                                     Subfamily **Chordodinae**   **5**

4b   Male tail bifurcate, no postcloacal slit, groups of hairs at sides of cloaca; cuticle with 1 or 2 types of araeolae; female tail simple . . . .

                               Subfamily **Parachordodinae**   **8**

5a (4) Two types of araeolae; pore canals, no interaraeolar bristles . . . . . .

                                Pseudochordodes Carvalho

5b   Only 1 type of araeolae. **6**

6a (5) Male with deep groove from tail tip anterior to cloaca; no pore canals between large araeolae. Araeolae not grouped. (Fig. 16.2) . . . . *Chordodes* Creplin

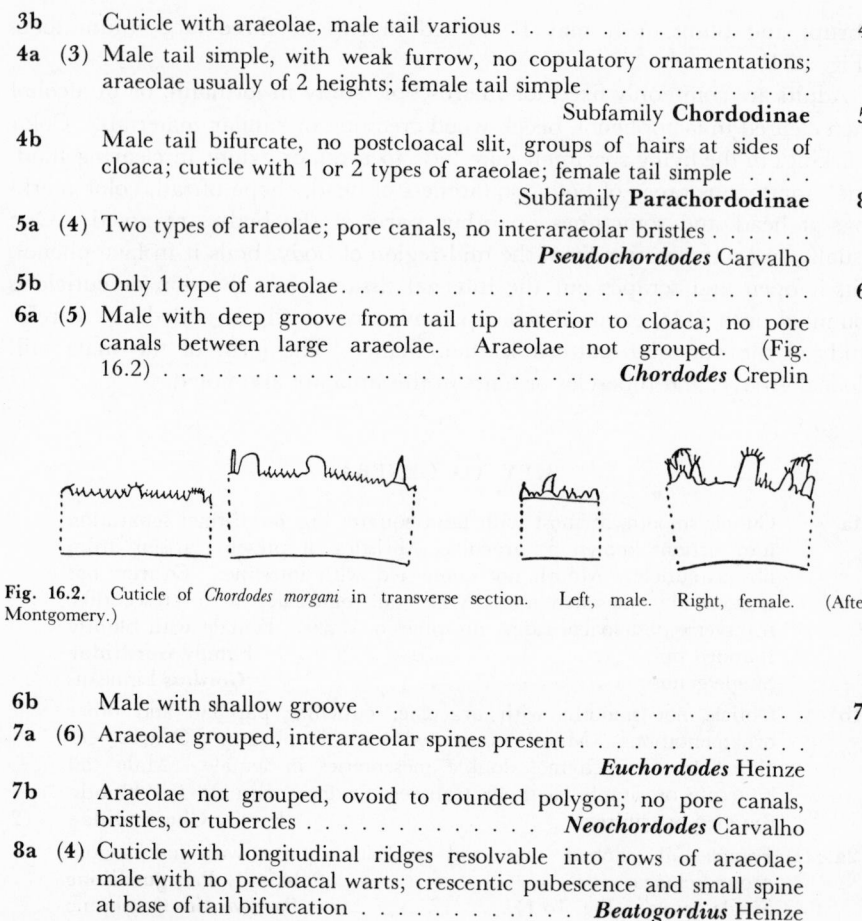

Fig. 16.2.   Cuticle of *Chordodes morgani* in transverse section.   Left, male.   Right, female.   (After Montgomery.)

6b   Male with shallow groove **7**

7a (6) Araeolae grouped, interaraeolar spines present . . . .

                                   Euchordodes Heinze

7b   Araeolae not grouped, ovoid to rounded polygon; no pore canals, bristles, or tubercles . . . .   *Neochordodes* Carvalho

8a (4) Cuticle with longitudinal ridges resolvable into rows of araeolae; male with no precloacal warts; crescentic pubescence and small spine at base of tail bifurcation . . . .   *Beatogordius* Heinze

8b   Otherwise **9**

9a (8) Araeolae of equal size and structure . . . .   *Gordionus* Müller

9b   Araeolae of 2 sizes or 2 types. **10**

10a (9) Two types of araeolae, large and small, larger ones in pairs; no interaraeolar setae; male with warts and pubescence around cloaca; female tail obtuse . . . .   *Paragordionus* Heinze

10b   Two types of araeolae differing in density; no postcloacal slit; female cloaca dorsoventrally slit . . . .   *Parachordodes* Camerano

References

Carvalho, J. C. M. 1942. Studies on some Gordiacea of North and South America. *J. Parasitol.*, 28:213–222.   **Filipjev, I. N. and J. H. S. Stekhoven. 1941.** *A Manual of Agricultural Helminthology.* E. J. Brill, Leiden.   **Goodey, T. 1951.** *Soil and Freshwater Nematodes.* Wiley, New York.   **Hyman, L. 1951.** *The Invertebrates.* Vol. III, *Acanthocephala,*

Aschelminthes and Endoprocta. McGraw-Hill, New York.   **Leidy, J. 1851.** On the Gordiaceae. *Proc. Acad. Nat. Sci. Phila.*, 5:262–263, 266, 275.   **1856.** A synopsis of Entozoa and some of their ectocongeners observed by the author. *Proc. Acad. Nat. Sci., Phila.*, 8:42–58.   **1870.** The *Gordius*, or hair-worm. *Am. Ent.*, 2:193–197.   **Thorne, G. 1940.** The hair-worm, *Gordius robustus* Leidy, as a parasite of the Mormon cricket, *Anabrus simplex* Haldeman. *J. Wash. Acad. Sci.*, 30:219–231.

17

Gastrotricha

ROYAL BRUCE BRUNSON

Identification

The Gastrotricha, which have a certain similarity to the Rotifera, have been variously classified as a phylum or as a class of the Aschelminthes.

One of the primary characteristics that separates genera of gastrotrichs is the presence or absence of caudal furca, or caudal prongs. These appear as posterolateral extensions of the body and may be simple, jointed, or in a few cases scaled. Not to be confused with true furca are various types of spines that project from the posterior end, and those types of rounded or odd-shaped protuberances found in the genera *Neogossea* and *Stylochaeta* (Figs. 17.1, 17.2). These protuberances usually have several small spines or long trailing spines projecting from them, and they lack adhesive glands. The normal position of the furca in the living animal is as a posterior extension of the body; however, in disturbed individuals they may be spread outward or upward or folded underneath. The folded position is often characteristic of preserved animals that contracted during the killing and fixing process.

Generic differentiation is also determined by the type of body covering. The genus *Ichthydium* is characterized by a cuticle devoid of scales or spines. However, animals of this genus (as well as of other genera) may have either or both anterior and posterior pairs of tactile bristles (see Fig. 17.9). These should not be confused with spines. In *Ichthydium* as well as other genera that possess some areas of smooth surface, the cuticle is so transparent that there is no doubt about the absence of scales. In those genera characterized by scales, the scales can easily be seen under high dry magnification (440 ×). Exceptions occur in such species as *Lepidodermella trilobum*. Because of their small size, oil immersion is necessary to discern individual scales, but on high dry magnifications the surface of the animal has a definite "scaly" appearance. Those species of *Chaetonotus* that possess scales in addition to spines are indicated as such in the key only if the scales are easily distinguished under high dry magnification.

The type and shape of the head is diagnostic in specific determination. A cephalic shield may or may not be present. This shield may be only a thickened portion of the cuticle on the front of the animal (easily seen with high dry magnification) or it may project slightly outward from the head contours. The lobes of the head, which are also diagnostic, can usually be seen only on living specimens. However, the tactile ciliary tufts may aid in determining the number of lobes present; i.e., in the case of five-lobed heads, there are two pairs of tufts which are somewhat evenly spaced in the hemisphere forming the anterior portion of the head. Also helpful in specific determination are the comparative sizes and types of lobes. These are illustrated in the key on the drawings of the species that possess them.

Identification of species of *Chaetonotus* is largely determined by the number and kinds of spines present. In some animals only long spines are present, and these originate either from a limited area on the back or from the whole trunk area. Evidence obtained to date indicates that the number of spines for any given species is constant, but care should be taken in diagnosis, because the spines may be broken off quite easily. Apparently they are not regenerated. Some spines are smooth, and many are either singly or doubly bifurcate; i.e., there is a small accessory barb or forking that can be seen under high power.

In some animals the spines may all be very much the same size; in others, the spines may be short on the head and neck and long on the trunk; and in still others, the spines on the head are short, and successive posterior spines are longer, so that there is a gradual increase in the length from those on the head to those above the caudal furca. In the latter case the spines above the furca are at least twice the length of those on the head.

Although it is not definitely known just how much variation occurs in the total length of the members within a species, evidence indicates that size is diagnostic. Proportions are particularly important; i.e., ratio of length of pharynx or length of caudal furca to the total length. These measurements should be taken on the living animal or on a relaxed specimen.

Distribution

Although gastrotrichs are world wide in their distribution, cosmopolitanism of individual species is not yet an established fact. Gastrotrichs can be found in most natural waters. Shallow ponds, beach pools, *Sphagnum* bogs, and the psammolittoral areas of lakes produce greater numbers of individuals and species than other types of habitats. Apparently, most species exhibit habitat specificity, with pH possibly a determining factor. Most gastrotrichs are benthic, although some occur as adventitious plankters.

Lepidodermella occurs in a variety of habitats, whereas *Chaetonotus* generally is found in association with some plant, such as *Spirogyra*, *Myriophyllum*, *Sphagnum*, *Utricularia*. Most species of *Ichthydium* are found in the thin surface layer of bottom mud, mud on the basal leaves of *Nymphea* (*Nuphar*), and in areas of low oxygen concentration. *Polymerurus* and *Stylochaeta* have been found in warm-water beach pools, and the only place *Dasydytes* has been found was in the grassy, overflow area of a permanent pond. Reportedly *Dichaetura* is a swamp-dweller. Some species occur in great numbers on the under side of *Lemna* and *Spirodela*.

Collection

Collection of plant-dwelling species is best made by squeezing the plants so that the water collects in a small container. Many such squeezings will concentrate the individuals. Similarly, scooping or siphoning the surface mud will limit larger areas to the surface area of the container, thus concentrating the specimens. Examination of material by means of a dissecting microscope equipped with 15 × wide field oculars greatly facilitates the search. The transfer of the specimen to a slide is made by means of a pipette. Methods of relaxing and mounting are discussed in Chapter 46,[1] and by Brunson (1950) and Pennak (1953).

The following key covers all species thus far reported from North America, except for the few species for which information was either inadequate or was known to be inaccurate. Inasmuch as so little work has been done on the Gastrotricha of North America, the average worker will undoubtedly find forms not covered by the key, and he should always be cognizant of that fact when using it. The figures show dorsal views (unless otherwise noted), and in some cases enlarged views of scales.

KEY TO SPECIES

1a	True caudal furca or prongs absent; on the posterior end there may be rounded protuberances which bear projecting or trailing spines	2
1b	Caudal furca present as posterolateral extensions of the body, usually with adhesive glands	5
2a	(1) Club-shaped tentacles present; long spines limited to posterior end	

---

[1]Henry W. Decker working at the University of Washington has found that gastrotrichs are better relaxed with crystalline cocaine than with solutions. In this method, the gastrotrich is confined in a small drop of water and a minute bit of cocaine is applied with a fine needle dipped into powdered cocaine. As soon as the animal stops moving, formalin or other fixative is added. This method has resulted in fixed specimens almost identical in appearance with live ones, and presumably it will be found to be useful with other organisms as well.

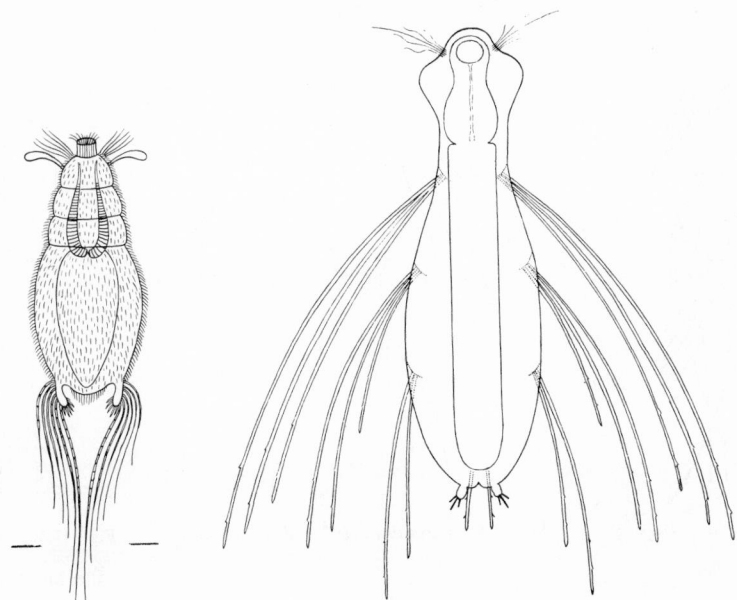

Fig. 17.1. *Neogossea fasciculata.* (After Daday.)

Fig. 17.2. *Stylochaeta scirteticus.* (After Brunson.)

of body; short spines may be present on body; posterior protuberances bear trailing spines. . . . . Family **Neogosseidae** (Fig. 17.1). . . . . *Neogossea* Remane 1927
tacles present; long trailing spines originating on posterior protuberances; body with short spines; pharynx with posterior half enlarged. Two species reported from U. S., *N. fasciculata* and *N. sexiseta.* [Robin C. Krivanek and Jerome O. Krivanek. 1958. A new and a redescribed species of *Neogossea* (Gastrotricha) from Louisiana. *Trans. Am. Microscop. Soc.*, 77:423–428.]

2b   Tentacles absent; spines occur in tufts on body surface; most of rest of body smooth . . . . Family **Dasydytidae**   **3**

3a   (2) Small posterior protuberances present. (Fig. 17.2) . . . .
          Stylochaeta Hlava 1904
          Distinct head, neck, and trunk regions; body covering smooth; 3 sets of spines originating on sides–anterior 3 spines longest, middle 3 shorter, posterior 2 shortest; posterior protuberances short, bearing spines; pharynx pear-shaped; total body length, 160 μ; pharynx length, 40 μ; animal "jumps." One species in U. S. *S. scirteticus* Brunson 1950.

3b   Posterior protuberances absent; long trailing spines cross in back of body . . . . *Dasydytes* Gosse 1851   **4**

4a   (3) Four long spines projecting from each side of body. (Fig. 17.3) . . .
          D. saltitans Stokes 1887
          Distinct head, neck, and trunk regions; body oval with smooth covering; 4 long spines originate on each side and may cross on back; 2 trailing curved spines and 2 long straight spines project from posterior; head with 2 rows of cilia; total body length, 85 μ

4b   Six long spines projecting from each side of body. (Fig. 17.4) . . . .
          D. oöeides Brunson 1950
          Distinct head, neck and trunk regions; trunk egg-shaped with smooth covering; 6 long spines originate on each side and are usually carried folded next to the trunk; 2 long trailing spines, which cross in back of the body, and 2 shorter spines project from posterior end; pharynx pear-shaped; total body length, 88 μ.

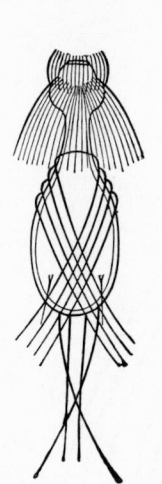

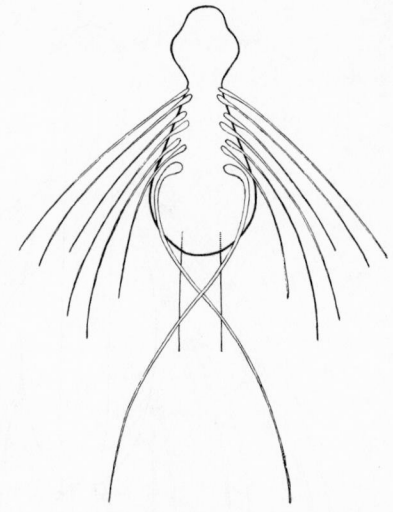

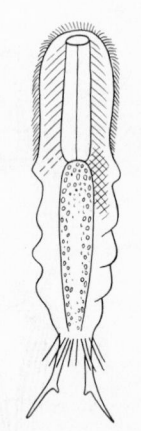

Fig. 17.3. *Dasydytes saltitans.* (After Stokes.)

Fig. 17.4. *Dasydytes ooeides.* (After Brunson.)

Fig. 17.5. *Dichaetura capricornia.* (After Metchnikoff.)

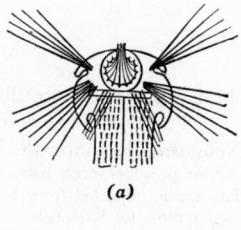

(a)

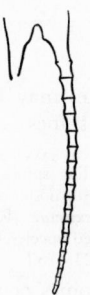

(b)

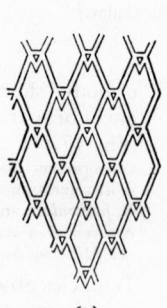

(c)

Fig. 17.6. *Polymerurus rhomboides.* (a) Ventral view of head. (b) Caudal branch. (c) Scales enlarged. (After Stokes.)

5a	(1) Caudal furca branched . . . . Family **Dichaeturidae**	
	(Fig. 17.5) . . . . *Dichaetura* Lauterborn 1913	

Caudal furca forked; head length more than ⅓ total length; body covering in folds; a single transverse row of spines or tactile bristles above caudal furca; total length, 150 μ. Not yet reported from N. A.

5b	Caudal furca not branched . . . . Family **Chaetonotidae**	6
6a	(5) Caudal furca segmented, nearly as long as body	
	Polymerurus Remane 1927	7
6b	Caudal furca not segmented, much shorter than body	8
7a	(6) Body covered with rhomboid scales. (Fig. 17.6)	
	P. rhomboides (Stokes) 1887	

Caudal furca jointed; lobes of head projecting in a hooklike posterior extension; 2 pairs of tactile ciliary tufts; body covered with overlapping rhomboidal scales; total length, 295 μ.

7b   Body surface covered with small, pointed excresences. (Fig. 17.7) . .
                                                    P. callosus Brunson 1950

Caudal furca jointed; lobes of head not hooklike as in *P. rhomboides;* body covering with numerous small, pointed excresences arranged in 12 to 18 alternating longitudinal rows; light-refracting granules in gut; total length, 267 μ; furca length, 107 μ; pharynx length, 47 μ.

8a (6) Body covering smooth; no scales or spines present except 1 or 2 pairs of tactile bristles . . . . ***Ichthydium*** Ehrenberg 1830   **9**

8b   Body covering of scales or spines or both **17**

9a (8) Head single-lobed. (Fig. 17.8) **10**

9b   Head 3- or 5-lobed. (Figs. 17.12, 17.14). **13**

10a (9) Furca $\frac{1}{5}$ or more of total length **11**

10b   Furca less than $\frac{1}{8}$ of total length **12**

11a (10) Head rounded; pharynx oval; only posterior tactile bristles present. (Fig. 17.8) . . . . ***I. monolobum*** Brunson 1950

Head and pharynx oval; body covering smooth; caudal furca curved like a pair of ice tongs; posterior tactile bristles above caudal furca; total length, 137 μ; pharynx length, 33 μ; furca length, 30 μ.

11b   Head and pharynx rectangular; anterior and posterior tactile bristles present. (Fig. 17.9) . . . . ***I. cephalobares*** Brunson 1947

Head roundly rectangular from both dorsal and side views; body covering clear, smooth; caudal furca curved like a pair of ice tongs; both anterior and posterior tactile bristles present; total length, 143 μ; pharynx length, 37 μ; furca length, 30 μ.

12a (10) Length of head and pharynx equal to $\frac{1}{3}$ of total length; head wider than body; posterior tactile bristles only; furca less than $\frac{1}{20}$ of total length. (Fig. 17.10) . . . . ***I. brachykolon*** Brunson 1947

Head single-lobed, oval; posterior end of body abbreviated; posterior tactile bristles present; ventral cilia abundant and apparently longer than width of body; total length, 127 μ; pharynx length, 43 μ; furca length, 5 μ.

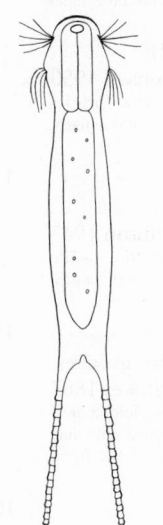
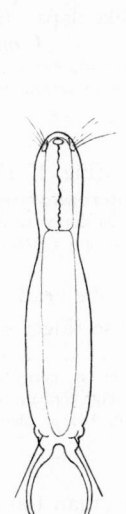
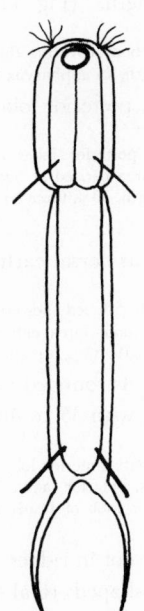
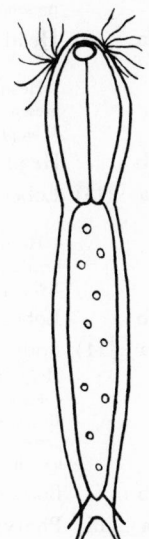

Fig. 17.7. *Polymerurus callosus.* (After Brunson.)

Fig. 17.8. *Ichthydium monolobum.* (After Brunson.)

Fig. 17.9. *Ichthydium cephalobares.* (After Brunson.)

Fig. 17.10. *Ichthydium brachykolon.* (After Brunson.)

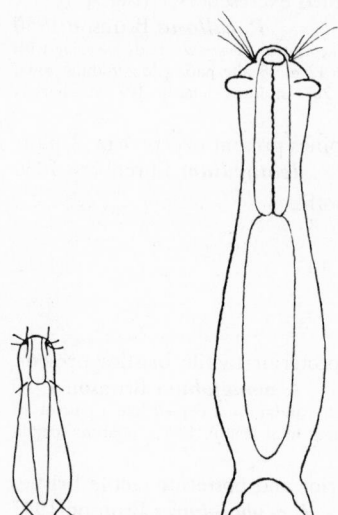
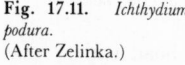
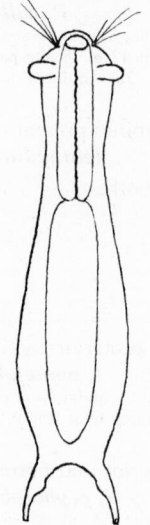
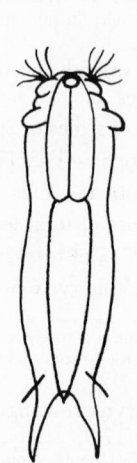
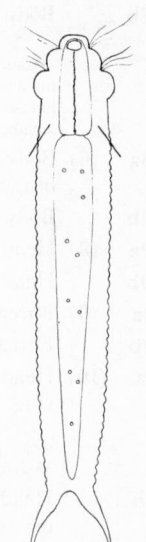

Fig. 17.11. *Ichthydium podura.* (After Zelinka.)

Fig. 17.12. *Ichthydium auritum.* (After Brunson.)

Fig. 17.13. *Ichthydium macropharyngistun.* (After Brunson.)

Fig. 17.14. *Ichthydium sulcatum.* (After Stokes.)

12b   Head and pharynx less than $\frac{1}{3}$ of total length; head not wider than body; anterior and posterior tactile bristles present; furca more than $\frac{1}{20}$ of total length. (Fig. 17.11) . . . .

Icthydium podura (O. F. Müller) 1773

Head rounded; body covering thin and pliable; anterior and posterior tactile bristles present; total length, 75 μ; pharynx length, 19 μ; furca length, 9 μ.

13a   **(9)** Head 3-lobed, posterior lobes as earlike flaps. (Fig. 17.12). . . . . .

I. auritum Brunson 1950

Head 3-lobed; posterior lobes as small, dorsal, earlike flaps; no tactile bristles present; caudal furca carried at higher plane than ventral body surface; total length, 87 μ; pharynx length, 30 μ; furca length, 11 μ.

13b   Head 5-lobed **14**

14a   **(13)** Lobes present as dorsal earlike flaps. (Fig. 17.13) . . . .

I. macropharyngistum Brunson 1947

Head indistinctly 5-lobed; posterior 4 lobes as dorsal earlike flaps; posterior tactile bristles present; caudal furca enlarged at base; gut tapering and dark; total length, 100 μ; pharynx length, 33 μ; furca length, 17 μ.

14b   Lobes normal, as rounded portions of the head **15**

15a   **(14)** Body covering with 35 to 40 transverse ridges separated by grooves. (Fig. 17.14) . . . . *I. sulcatum* Stokes 1887

Head 5-lobed, with anterior lobe distinctly set off; cuticula transparent, folded into 40 transverse folds; anterior tactile bristles present; cephalic shield (thickened cuticula) present on anterior lobe of head; total length, 183 μ; pharynx length, 37 μ; furca length, 22 μ.

15b   Body covering not in ridges **16**

16a   **(15)** Pharynx pear-shaped; total length less than 100 μ. (Fig. 17.15) . . .

I. minimum Brunson 1950

Head indistinctly 5-lobed, pointed; posterior tactile bristles present; pharynx pear-shaped; short cilia, thickly covering ventral surface; total length, 70 μ; pharynx length, 30 μ; furca length, 11 μ.

16b   Pharynx oval; total length more than 100 μ.   (Fig. 17.16) . . . . . .
                                                I. leptum Brunson 1947
               Head distinctly 5-lobed; cuticula smooth and pliable; body slender; base of caudal
               furca enlarged; total length, 170 μ; pharynx length, 33 μ; furca length, 20 μ.

17a   **(8)** Only scales present; no spines, although 1 or 2 pairs of tactile bristles
               may be present. **18**

17b   Surface of body with spines or scales and spines . . . .
                                          Chaetonotus Ehrenberg 1830   **21**

18a   **(17)** Body covered with complex stalked scales, each with a basal plate, a
               stalk, and an end plate.   (Fig. 17.17). . . *Aspidiophorus* Voigt 1904
               Head irregular, appears to be rounded; no neck constriction; caudal furca normal; 2
               pairs of tactile ciliary tufts; body covered with elaborate scales which are composed of a
               rounded basal plate and a stalk supporting a rhomboid end plate; length, 100–350 μ.
               Not yet reported from N. A.

18b   Scales not stalked **19**

19a   **(18)** Each scale with minute ridge or keel.   (Fig. 17.18) . . . .
                                        Heterolepidoderma Remane 1927
               Head irregular, appears to be rounded; usually 2 pairs of tactile ciliary tufts;
               posterior tactile bristles present; body covered with small scales, each of which is
               keeled; total length, 80–200 μ.   Not yet reported from N. A.

19b   Scales not keeled . . . . *Lepidodermella* Blake 1933   **20**
               The former name, *Lepidoderma* Zelinka (1889) was preoccupied.

20a   **(19)** Head 5-lobed; scales can be seen to project from surface of body
               when viewed along any margin.   (Fig. 17.19) . . . .
                                             L. squamatum (Dujardin) 1841
               Head distinctly 5-lobed; body distinctly scaled, scales occurring in alternating rows
               and seemingly pointed anteriorly; 2 pairs of tactile ciliary tufts; total length, 150
               (110–220) μ; pharynx length, 48 μ; furca length, 24 μ.   Apparently, there are
               ecological races of this species, with some races much smaller than others.

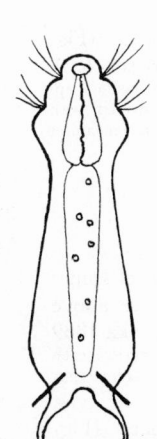

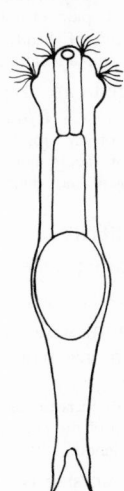

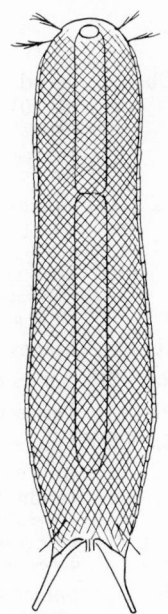

Fig. 17.15.   *Ichthydium minimum.* (After Brunson.)

Fig. 17.16.   *Ichthydium leptum.* (After Brunson.)

Fig. 17.17.   *Aspidiophorus marinus.* (After Remane.)

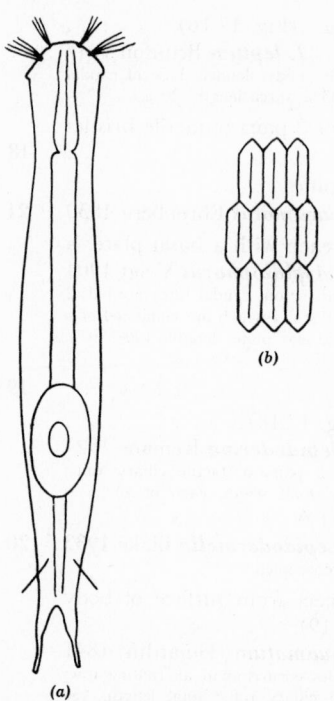

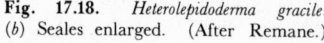

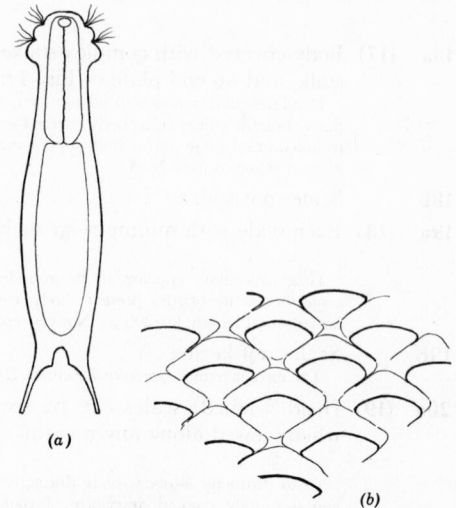

Fig. 17.18. *Heterolepidoderma gracile.* (b) Seales enlarged. (After Remane.)

Fig. 17.19. *Lepidodermella squamatum.* (b) Scales enlarged. (After Brunson.)

20b   Head 3-lobed; scales minute (diameter less than 3 μ). (Fig. 17.20) . . . . **Lepidodermella trilobum** Brunson 1950
Head distinctly 3-lobed, with 1 pair of tactile ciliary tufts; ventral cilia thick: cuticula thickened on anterior end of head; scales tiny (3 μ), hexagonal; total length, 177 μ; pharynx length 50 μ; furca length, 20 μ.

21a   **(17)** Spines elaborately enlarged at the base; animal large. (Fig. 17.21) . . . . **Chaetonotus robustus** Davison 1938
Largest known fresh-water gastrotrich—total length, 585-615 μ; head irregular; appears to be rounded; 2 pairs of tactile ciliary tufts; body thickly covered with elaborate spines which are winged and pouchlike; scale with its spine measures 50-60 μ.

21b   Spines not elaborately modified   22
22a   **(21)** Spines originate from a distinct scale   23
22b   Distinct scales not present   24
23a   **(22)** Spines with small bifurcation; scales enlarged around base of spine; spines increase gradually in size from those on head to those above caudal furca. (Fig. 17.22). . . . . **C. similis** Zelinka 1889
Head indistinctly 5-lobed; spines increase gradually in size toward posterior; each spine embedded in a raised portion of the cuticula, and bifurcate distally; animal has dishevelled appearance; total body length, 120-220 μ.

23b   Spines simple, nearly of same size over the body; scales flat. (Fig. 17.23) . . . . **C. brevispinosus** Zelinka 1889
Head indistinctly 5-lobed, appears rounded; about 11 rows of alternating spines, slightly longer posteriorly; each spine originates from a basal scale; total length, 95-149 μ; pharynx length, 23 μ.

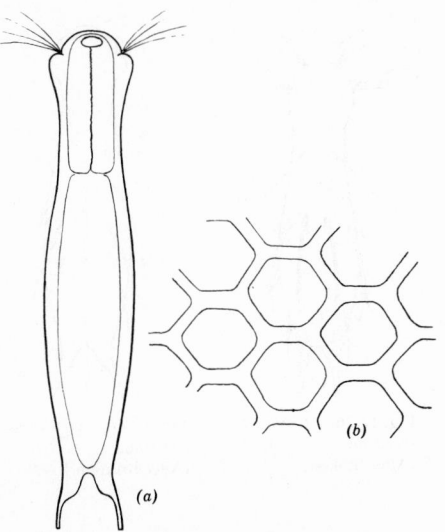

Fig. 17.20. *Lepidodermella trilobum.* (*b*) Scales enlarged. (After Brunson.)

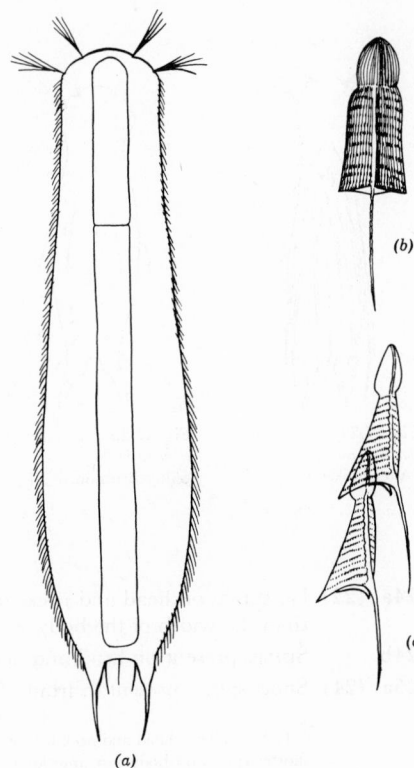

Fig. 17.21. *Chaetonotus robustus.* (*b*) Dorsal view of body scales. (*c*) Side view of body scales. (After Davison.)

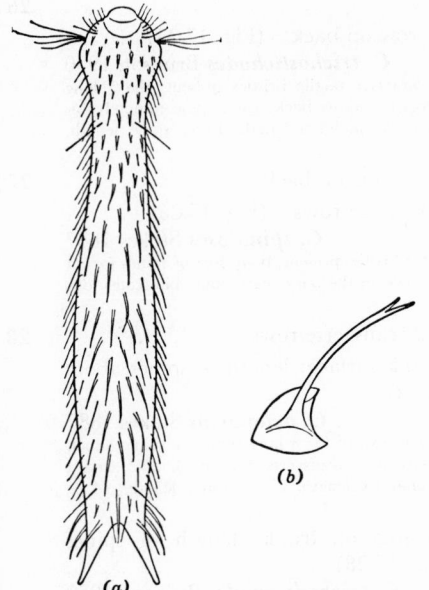

Fig. 17.22. *Chaetonotus similis.* (*b*) Side view of body spine. (After Zelinka.)

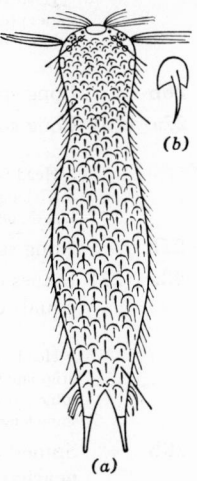

Fig. 17.23. *Chaetonotus brevispinosus.* (*b*) Spinose scale. (After Zelinka.)

 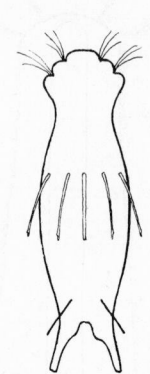 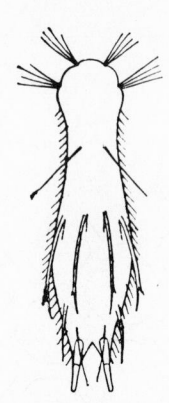 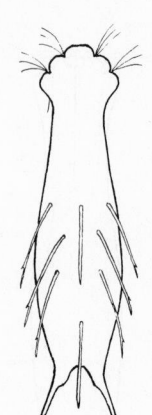

Fig. 17.24. *Chaetonotus longispinosus.* (After Stokes.)

Fig. 17.25. *Chaetonotus trichostichodes.* (After Brunson.)

Fig. 17.26. *Chaetonotus spinulosus.* (After Stokes.)

Fig. 17.27. *Chaetonotus octonarius.* (After Brunson.)

24a **(22)** No spines on head and neck; some spines about as long as or longer than the width of the body **25**

24b Spines present on head and neck **29**

25a **(24)** Short spines present in front of the long spines. (Fig. 17.24) . . . . .
Chaetonotus longispinosus Stokes 1887
Head 5-lobed; head and neck free of spines; 8 long dorsal spines in 2 transverse rows; shorter spines on body just anterior to long spines; long spines bifurcate; total length, 74 μ.

25b No short spines present **26**

26a **(25)** Long spines 5, forming transverse row on back. (Fig. 17.25) . . . . .
C. trichostichodes Brunson 1950
Head 5-lobed; body covering smooth; posterior tactile bristles present; the 5 long spines present are arranged in a transverse row across back; each spine minutely bifurcate; total length, 87 μ; pharynx length, 27 μ; furca length, 13 μ; spine length, 17 μ.

26b Long spines 7 or 8, not in transverse row on back **27**

27a **(26)** Long spines 7, originating in 2 transverse rows. (Fig. 17.26) . . . . .
C. spinulosus Stokes 1887
Head 5-lobed; anterior and posterior tactile bristles present; body free of spines except for 7 long spines which form 2 transverse rows on the back; each spine bifurcate; total length, 68–89 μ.

27b Long spines 8, not originating in 2 transverse rows **28**

28a **(27)** Spines originating over most of trunk surface; length of spines about equal to the body width. (Fig. 17.27) . . . .
C. octonarius Stokes 1887
Head indistinctly 5-lobed; body free of spines except for 8 long bifurcate spines which originate from the posterior half of the body in 4 transverse rows of 3, 2, 2, and 1 spines, respectively; total length, 100 μ; pharynx length, 27 μ; furca length, 14 μ; spine length, 22 μ.

28b Spines originating from limited area on trunk; length of spines nearly twice the body width. (Fig. 17.28) . . . .
C. trichodrymodes Brunson 1950
Head 5-lobed; body covering smooth except for 8 bifurcate spines which originate from a triangular area in the middle of the trunk; total length, 107 μ; pharynx length, 33 μ; furca length, 16 μ; spine length, 48 μ.

29a **(24)** Some spines present which are as long as or longer than width of
body **30**

29b No spines as long as the width of body **31**

30a **(29)** Long spines many, originating over most of trunk; short spines
present only on head and neck. (Fig. 17.29) . . . .
C. acanthophorus Stokes 1887
Head 5-lobed; head and neck covered with short spines; trunk covered with
elongated spines which form 4 transverse and 5 longitudinal rows; total length, 100 μ;
pharynx length, 30 μ; furca length, 16 μ; short spine length, 4 μ; long spine length,
24 μ.

30b Long spines 7, originating from limited area on trunk; other spines
increase gradually in length from those on the head to those above
caudal furca. (Fig. 17.30). . . . . *C. anomalus* Brunson 1950
Head 5-lobed; 6 to 8 longitudinal rows of spines (of 8 to 10 spines each) which increase
in size posteriorly; in addition, 7 long spines, each twice bifurcate, originate in a hexagonal area on the trunk and project beyond other spines; total length, 147 μ; pharynx
length, 43 μ; furca length, 27 μ; length of long spines, 45–60 μ.

31a **(29)** Head 3-lobed; all spines short. (Fig. 17.31) . . . .
C. formosus Stokes 1887
Head 3-lobed, with thickened cuticula anteriorly—an indistinct cephalic shield;
posterior tactile bristles present; retractable oral bristles; 8 to 12 longitudinal, alternating rows of about 30 short spines each; total length, 167 μ; pharynx length, 47 μ;
furca length, 24 μ; spine length, 7 μ.

31b Head 5- or apparently single-lobed; spines short or show a gradual
increase in size from those on head to those above caudal furca. . . . **32**

32a **(31)** Spines with 3-pronged base; body length greater than 400 μ. (Fig.
17.32) . . . . *C. gastrocyaneus* Brunson 1950
Head irregular, appears to be rounded or flattened anteriorly; cephalic shield present;
2 pairs of tactile ciliary tufts; 10 to 16 longitudinal rows of spines; each spine bifurcate, bent, and with 3-pronged base; gut wall usually a pale sky-blue color; total

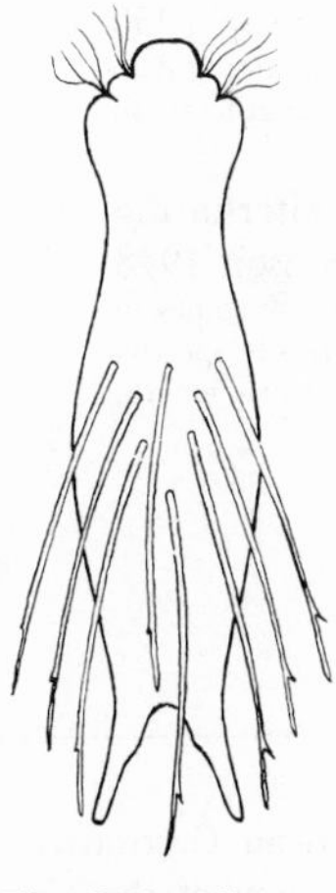

Fig. 17.28. *Chaetonotus trichodrymodes.* (After Brunson.)

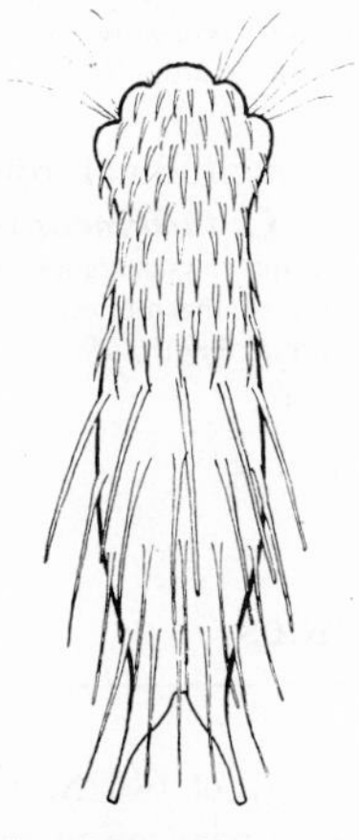

Fig. 17.29. *Chaetonotus acanthophorus.* (After Brunson.)

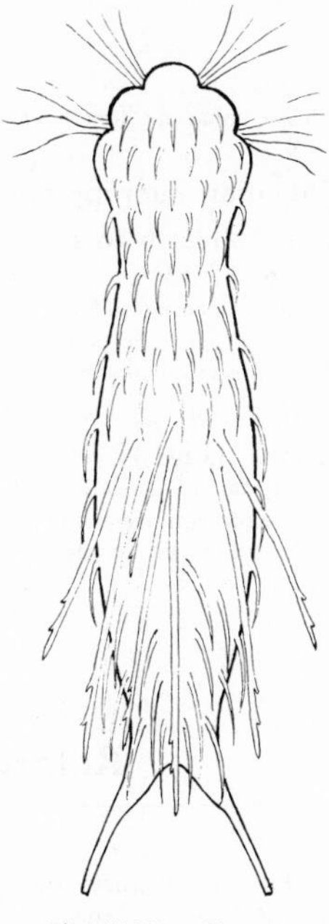

Fig. 17.30. *Chaetonotus anomalus.* (After Brunson.)

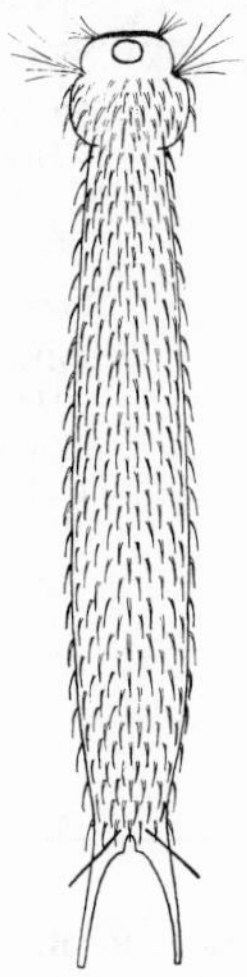

Fig. 17.31. *Chaetonotus formosus.* (After Brunson.)

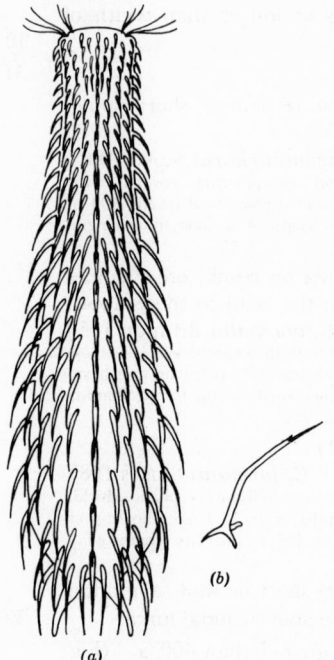

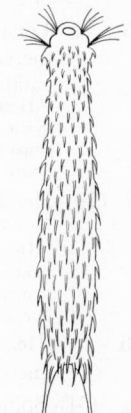

Fig. 17.32.   *Chaetonotus gastrocyaneus.*
(*b*) One spine enlarged.
(After Brunson.)

Fig. 17.33.   *Chaetonotus vulgaris.*
(After Brunson.)

Fig. 17.34.   *Chaetonotus tachyneusticus.*
(After Brunson.)

length, 347–485 μ; pharynx length, 83 μ; furca length, 42 μ; length of posterior spines, 50 μ.

32b   Spines as cuticular outcroppings; body length less than 400 μ. **33**

33a **(32)** Size less than 150 μ; 6 to 8 longitudinal rows of alternating spines on body. (Fig. 17.33) . . . . . **Chaetonotus vulgaris** Brunson 1950
Head indistinctly 5-lobed; 6 to 8 longitudinal rows of spines, each row containing 8 to 9 spines; spines increase in size posteriorly; total length, 103 μ; pharynx length, 30 μ; furca length, 13 μ; posterior spine length, 7 μ.

33b   Size more than 150 μ; 10 to 12 longitudinal rows of alternating spines on body. (Fig. 17.34) . . . . **C. tachyneusticus** Brunson 1948
Head 5-lobed; 12 to 16 longitudinal, alternating rows of spines with 20 to 30 spines in each row; spines increase in size posteriorly; swimming motion characteristically speedy; total length, 203 μ; pharynx length, 60 μ; furca length, 29 μ; posterior spine length, 12 μ.

References

Brunson, R. B. 1949. The life history and ecology of two North American Gastrotrichs. *Trans. Am. Microscop. Soc.*, 68:1–20.   **1950.** An introduction to the taxonomy of the Gastrotricha with a study of eighteen species from Michigan. *Trans. Am. Microscop. Soc.*, 69:325–352.   **De Beauchamp, P. M. 1934.** Sur la morphologie et l'ethologie des Neogossea.

Bull. Soc. Zool. France, 58:321–342.   **Grünspan, T.   1910.**   Die Süsswasser-Gastrotrichen Europas. Eine zusammenfassende Darstellung ihrer Anatomie, Biologie und Systematik. *Ann. Biol. Lacustre*, 4:211–365.   **Hyman, L. H.   1951.**   *The Invertebrates*. Vol. III, *Acanthocephala, Aschelminthes, and Entroprocta*, pp. 151–170.   McGraw-Hill, New York.   **Murray, J.   1913.** Gastrotricha. *J. Quekett Microscop. Club*, 12:211–238.   **Packard, C. E.   1936.**   Observations on the Gastrotricha indigenous to New Hampshire. *Trans. Am. Microscop. Soc.*, 55:422– 427.   **Pennak, R. W.   1953.**   *Fresh-Water Invertebrates of the United States*, pp. 148–158. Ronald, New York.   **Remane, A.   1935–1936.**   Gastrotricha (Gastrotricha und Kinorhyncha). In: Bronns *Klassen und Ordnungen des Tierreichs*. Band IV, Abt. II, Buch 1, Teil 2, Lfrg 1–2, pp. 1–242. Akademische Verlagsgesellschaft, Leipzig.   **Stokes, A. C.   1887.**   Observations on Chaetonotus. *Microscope*, 7:1–9, 33–43. .   **Zelinka, C.   1889.**   Die Gastrotrichen. Eine Monographische Darstellung ihrer Anatomie, Biologie und Systematik. *Z. wiss. Zool.*, 49:209–384.

18

Rotifera

W. T. EDMONDSON

The Rotifera, also called Rotatoria or wheel animalcules, are a group of small, usually microscopic, pseudocoelomate animals which have been variously regarded either as a class of the Phylum Aschelminthes, or as a separate phylum (Hyman, 1951, p. 54). They are characterized by the possession of both a corona, which is either a ciliated area or a funnel-shaped structure at the anterior end, and a specialized pharynx called the mastax, with its cuticular lining differentiated into trophi, a series of pieces that act as jaws.

The rotifers have attracted much attention from microscopists because of their wide-spread distribution in waters of all kinds, the great abundance in which they frequently occur, and the striking beauty of some of the species (Hudson and Gosse, 1886, pp. 3–4). For a general account of the group, the excellent work by Hyman (1951) is recommended. More detail is given in the unfinished monograph by Remane (1929–1933). The emphasis in the following paragraphs will be upon those aspects of morphology of greatest use in determinative work.

General Anatomy

Eosphora najas (Fig. 18.1), although not especially common, is a convenient example since it shows well many of the structures it is necessary to know to use the key. *Epiphanes* (= *Hydatina*) *senta*, often cited as a typical rotifer, is less useful.

The body is covered with a thin, flexible cuticle overlying the thin syncitial hypodermis. The animal is weakly differentiated into several regions; *head, neck, trunk,* and *foot,* separated from each other by folds. In some rotifers the regions are seen only as gradual changes in diameter of the body, and often a separate neck is not present (Figs. 18.76, 18.86). In many rotifers there are permanent transverse folds or grooves in the cuticle dividing the body into sections (Fig. 18.116). Although the sections are often called segments or joints, there is no true metamerism.

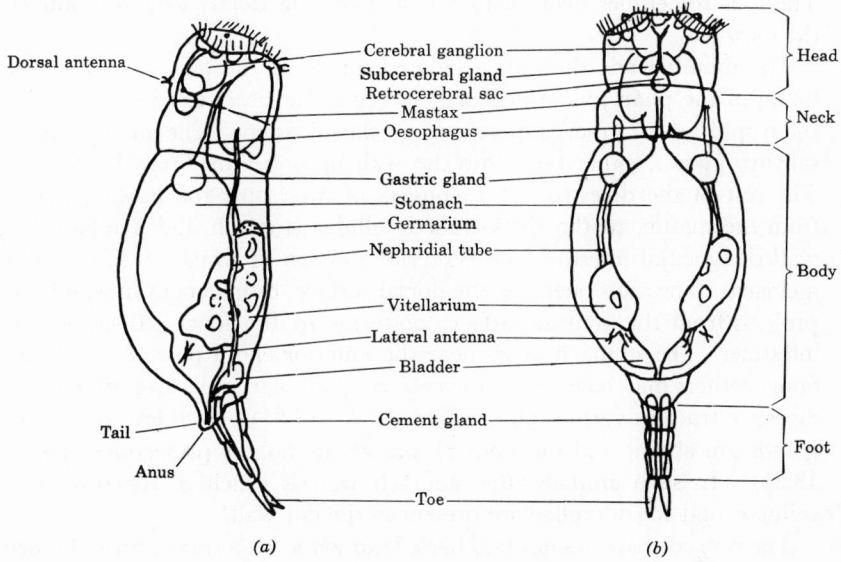

Fig. 18.1. Anatomy of a representative illoricate rotifer (Monogononta). (*a*) Lateral. (*b*) Dorsal. (Based on *Eosphora najas,* modified from Harring and Myers.)

The *foot* is a prolongation of the body posterior or ventral to the anus, and it is not properly called a tail, that term being restricted to a fold or prolongation dorsal or anterior to the anus (Figs. 18.1, 18.87). The foot bears at its end two conical *toes,* and contains within two cement glands which secrete a sticky material through ducts opening at the tips or near the base of the toes.

At the anterior end is an oblique ciliated area, the *corona,* with a mouth near the ventral or posterior border.

There are glands of unknown function in the head that are of use in identifying some rotifers. These are an unpaired glandular *retrocerebral sac* with a long forked duct opening on the apical area of the corona, and a pair of *subcerebral glands* lying beside the duct of the retrocerebral sac, also with ducts leading to the apical area. Together, the sac and glands are called the retrocerebral organ. Rotifers may have both sac and glands, either one, or neither. In some species the ducts are vestigial. A few rotifers (*Lindia,*

Synchaeta) have a small retrocerebral sac filled with a dark red pigment which is sometimes mistaken for the eye. An interesting relation is that in a general way, and with certain exceptions, these structures are most fully developed in families with the least elaborate protonephridia.

There are three prominent tactile sense organs on the body: two *lateral antennae* located on the sides toward the posterior part, and a *dorsal antenna* on the mid-line of the head. The dorsal antenna represents fusion of two antennae; it receives two nerves, and a few rotifers actually possess two dorsal antennae lying close together (Figs. 18.62, 18.63). The corona may bear stiff cirri, evidently derived from cilia, which are said to have a tactile function. The foot sometimes bears a tactile seta on the dorsal side just anterior to the toes.

The mouth leads through a ciliated gullet to the *mastax*. The cuticular lining of the pharynx is thickened, forming the *trophi* or jaws; the details of the trophi are of great importance in classification. The mastax may have salivary glands, either built into the wall or projecting from it (Fig. 18.16). The rest of the digestive tract consists of an esophagus leading posteriorly from the mastax to the thick-walled cellular stomach and a relatively thin-walled, syncitial intestine, the posterior portion of which is differentiated as a cloaca. The anus opens on the dorsal surface, or in genera in which the foot projects from the ventral surface, posterior to the foot. *Asplanchna* has no intestine. The stomach bears near the anterior end a pair of gastric glands. Some rotifers may have several accessory glandular structures attached to the digestive tract in various places (Fig. 18.73, 18.93). In a few groups, gastric glands are absent and the stomach has cecae, hollow projections (Figs. 18.8, 18.81). In such animals, the stomach wall is syncitial, digestion is intracellular, and zoochlorellae are present in the gut wall.

The reproductive system is of basic taxonomic importance since the primary systematic division of the group into classes is made on the basis of the number of ovaries and the development of males. The female reproductive system lies ventral to the digestive tract and is made up of a vitellogermarium (often called simply the ovary) which consists of a small germarium that supplies nuclei to eggs, and a massive yolk gland or vitellarium. An oviduct leads to the cloaca. In many of the Monogononta, the yolk gland has eight large nuclei, although there are fewer in some species, and the Flosculariceae and Collothecidae regularly have many more.

The rotifers ordinarily encountered are amictic females; that is, they reproduce parthenogenetically by diploid eggs. The eggs may be laid free in the water, or they may be attached to plants or other surfaces, or to the cuticle of the mother where they are carried until they hatch (Sudzuki, 1955d, 1957). Most planktonic species carry their eggs. Some rotifers are viviparous, carrying one or more developing embryos in the greatly enlarged oviduct (Fig. 18.20). In response to adverse environmental conditions, some females, called mictic, may lay haploid eggs which then develop into males. When such haploid eggs are fertilized, the resulting eggs have a thicker shell than the parthenogenetic ones, are resistant to drying, and generally take a relatively long time to develop; such eggs are called resting eggs and always give rise to females.   There are reports of resting eggs being produced parthenogenet-ically, but little is known about them.   Ordinarily mictic and amictic females look alike and can be distinguished only by the eggs they lay, but in a few species there are morphological differences between the two.

Males have not been described for most species, although they have been found in many Monogononta.   Males are apparently completely absent in the Bdelloids, but do occur in the Seisonidae.   In a few genera the males look like small caricatures of the females, except for the reproductive system (Figs. 18.38, 18.75), but in most there is great sexual dimorphism, the males lacking a digestive system and bearing little resemblance to the females (Figs. 18.5, 18.6, 18.12, 18.102).   The occurrence of males in most species where they are known is restricted to short periods of the year, although they may become very abundant while they are present.   No attempt is made to key males in this chapter.   Descriptions and illustrations of many males are given by Wesenberg-Lund (1923), Wisznieski (1934b) and Sudzuki (1955–1956).   The male reproductive system of the Monogononta ordinarily consists of a testis and a ciliated sperm duct with glands.   The tip of the duct is evertible and may be lined with thickened cuticle, forming a penis.   Copulation may be cloacal, or, more frequently, the sperm are injected through the cuticle of the female into the body cavity.

The rest of the organ systems are little used in identification, although they may be of importance in classification.   The excretory system consists of a pair of protonephridial tubes with flame bulbs.   Most commonly the tubes lead to a bladder which has a duct leading to the cloaca, but in some, the nephridia lead directly to the distensible cloaca.   The muscular system con-sists of a series of longitudinal and circular bands which move the body, net-works of muscle in the viscera, and a few muscles between the viscera and body wall.   In some rotifers, certain of the more powerful muscles are prom-inently striated, as in the lateral muscles of *Polyarthra* (Fig. 18.4), the muscles of the appendages of *Hexarthra* (Fig. 18.6), and the hypopharyngeal muscle of *Synchaeta* (Fig. 18.59).   The nervous system consists of a dorsal cerebral gang-lion, accessory ganglia in mastax and foot regions, and various nerves.   Most rotifers have prominent eyes, usually red.   There may be a single eye lying on the cerebral ganglion, or a pair of frontal eyes connected with the ganglion by optic nerves, or both.   In some, the eyes are provided with clear, spheric lenses embedded in the red pigment mass.   The body cavity contains a loose syncytial network of amoeboid cells.   They may be very difficult to see, but in species with large body cavities they may be prominent (*Asplanchna*).

Variations of the Body Wall

In many rotifers, the cuticle is thin and flexible throughout, and the body very mobile.   These rotifers, when disturbed or preserved without special treatment, contract with the corona and foot retracted within the body (Figs. 18.20, 18.102, 18.105).   In many other rotifers the cuticle may be stiffened in places forming relatively inflexible plates. Such a structure is called a *lorica*, and the details are of great use in identification; it is therefore necessary to determine its presence and to recognize the structure. In some rotifers the lorica is simple, and consists merely of a general thickening of the cuticle of the body, forming a boxlike structure without any obvious differentiation within the different regions (Fig. 18.76). The presence of such a lorica is made evident by the fact that the outline of the animal does not change as it moves about. In preserved material, the presence of the lorica is recognized by the firm, definite appearance of the body wall. More commonly, the lorica consists of a dorsal arched plate and a ventral flat plate joined together at the edges. In such cases, the anterior part of the ventral plate may be connected with the dorsal plate by a flexible cuticular membrane which permits expansion when the head is retracted (Fig. 18.11). In some, the dorsal and ventral plates are joined together by lateral strips of thin cuticle folded inward, forming deep *lateral sulci* (Figs. 18.41, 18.46). In *Mytilina* there is a dorsal median sulcus of a similar nature, the ventral plate being firmly joined to the dorsolateral plates (Fig. 18.40), and in *Diplois*, both lateral and dorsal sulci exist (Fig. 18.39). The lateral sulci may be quite shallow (Fig. 18.28).

With animals like *Euchlanis*, *Lepadella*, and especially the spiny Brachionids, (Figs. 18.10, 18.11, 18.29) there is little doubt about the presence of a lorica. However, there are some animals in which the plates are fairly flexible and little differentiated from the connecting areas of cuticle; such a structure is usually called a *semilorica*. There may be some difficulty in identifying semi-loricate animals. Missing the point in the key will ordinarily lead to obviously erroneous identifications, and if such misidentification is reached, one of the first points to check is the lorica. Some of the difficult genera are keyed out in two places.

Most loricate animals have the head covered with thin, flexible cuticle. There are some in which the head, too, is loricate, with plates forming a head sheath (Fig. 18.26).

Many of the animals with spines on the anterior edge of the lorica show great variability in the length of spines, and there has been a tendency to use trinomial or even quadrinomial nomenclature with varieties and forms being designated. While this kind of nomenclature has been a convenience in faunal lists, the variety no longer has sanction as a preservable taxon. Some of the loricate species show distinct cyclomorphosis, with a succession of morphological variations occurring during the year.

Although the lorica is much used in identification of rotifers, it is of little use in making the major taxonomic divisions. Evidently it has evolved independently in the various families, and even within some genera it shows great variation; *Cephalodella* contains species with no lorica and some with a well-developed spined lorica (Fig. 18.38).

Other important modifications of the body wall are movable setae and paddles, quite different from the spines that are projections of the lorica. These may take the form of long, slender filaments (*Filinia*, Fig. 18.5), flat paddles (*Polyarthra*, Fig. 18.4), or hollow outgrowths of the body furnished with setate spines and powered by large striated muscles (*Hexarthra*, Fig. 18.6). The paddles serve for jumping locomotion, but may also serve as protection against being swallowed by somewhat larger animals.

Variations of the Foot

The foot may be highly modified from the example described. There may be only one toe, representing a fusion of the two (Figs. 18.42, 18.43). In some there is an additional toelike structure, a dorsal spur (Fig. 18.65). In most of the Trichocercidae, the toes are long, unequal, slender filaments, with the cement glands opening at the base. There may be several basal spurs (Fig. 18.22). The free-swimming Flosculariaceae with a foot have a ciliated cup at the end (Fig. 18.90). In the sessile Flosculariaceae and Collothecaceae the foot ends in a *peduncle* (Fig. 18.102), and the cement glands are represented by cushions of hypodermis distributed throughout the foot. Most of the sessile rotifers secrete a gelatinous tube from the foot (Fig. 18.102) which may be supplemented by fecal pellets (*Ptygura pilula*, Fig. 18.95; *Floscularia janus*) or pellets constructed in a special apparatus just posterior to the chin (Fig. 18.97b). In *Atrochus*, the foot is reduced to a hemispheric cushion without toes (Fig. 18.104). The foot of the actively crawling bdelloids and many Monogononta is jointed, and in other forms it may be wrinkled or quite smooth. The foot may be capable of being withdrawn into the body or lorica, but in some genera is not (*Platyias*, Fig. 18.30).

The foot is a very small, insignificant structure in many rotifers, and in some it is absent altogether. Reduction of the foot seems to have evolved independently several times in the rotifers, and there are footless genera in six families. In general, the foot is small or absent in planktonic rotifers.

The foot is used for permanent attachment by most Flosculariaceae and Collothecaceae, and for temporary attachment by many other groups. Most of the bdelloids and some *Testudinella* will attach temporarily to a plant and may build up a tube of debris around themselves.

The bdelloids have cement glands in the rostrum, and a very characteristic mode of locomotion in this group is to attach by the foot, extend the body, attach by the rostrum, and by contracting the body bring the foot up close to the rostrum. This inchworm or leechlike locomotion serves as a certain recognition of a bdelloid, but most can also swim with extended corona.

Variations of the Corona

The ciliated corona is used for locomotion and obtaining food. The structure of the corona is of basic importance in the classification of rotifers, but since the head is usually retracted when animals are disturbed or preserved without special methods, coronal characters are used as little as possible in the present key. Nevertheless, it is necessary to understand the basic structure and its variations. There is a relation between the morphology of the corona and the swimming and feeding habits of the animal.

The basic corona type, often described as the primitive condition, can be regarded as a ciliated band around the head, enclosing a bare apical area and prolonged at the posterior on the ventral side (Fig. 18.2a). This corona, therefore, consists of a buccal field and a circumapical band, and other types of corona can be described as modifications of it. The primitive pattern is found in a good many Notommatidae, although the circumapical band is very small or even missing in some. Some of these animals have tufts of elongated cilia at the sides of the buccal area, and a further development of this type has lateral ciliated projections, *auricles*, from the head, the ciliation being

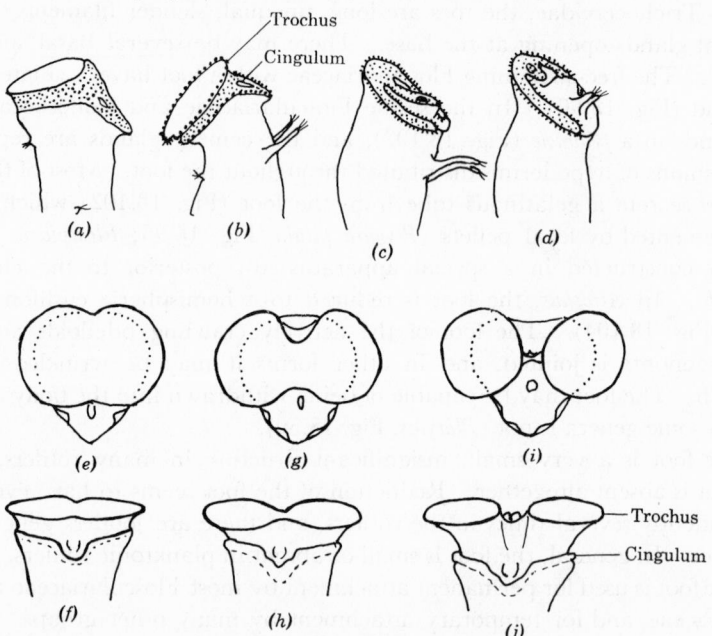

Fig. 18.2. Structure of corona. (*a*) Basic, primitive type. (*b*) *Ptygura*. (*c*) *Conochiloides*. (*d*) *Conochilus*. (*e, f*) Basic type in frontal and ventral view. (*g, h*) Flosculariacean type. (*i, j*) Bdelloid (*Philodina*) type. (After de Beauchamp.)

either continuous with the buccal area or limited to the tips (Fig. 18.87). Notommatid auricles are introvertable, and are usually extended only when the animal is swimming. The buccal field may project as a prominent chin directed toward the posterior (Fig. 18.87).

Other variations involve reduction of the circumapical band to a very narrow extent and reduction of the buccal field so that the mouth lies at its posterior border. In the free-swimming Brachionidae (*Brachionus, Euchlanis, Epiphanes*, others), there is little ciliation posterior to the mouth, and the buccal field is highly modified (Fig. 18.76). It is a triangular area anterior to the mouth, bordered on each side by a line of enlarged, fused cilia, the pseudotroch, and anteriorly or dorsally by a series of tufts of stiff cirri. A narrow line of cilia, evidently the remains of the circumapical band, leads from the mouth completely around the head.

In many actively swimming rotifers the corona is reduced to a thin course of cilia around the head, enclosing a large apical area, and leading to a small buccal area. The ciliation may be broken by gaps (Asplanchnidae, Synchaetidae, Gastropidae, Trichocercidae, and some Notommatidae; Figs. 18.20, 18.59).

A projecting rostrum may be developed from the apical area. In the bdelloids the rostrum is a very prominent structure.

In an extensive series of variations found in the Flosculariidae and Bdelloidea, the mouth is quite close to the ventral border of the reduced buccal field, and the cilia on the anterior and posterior margins of the ciliated areas are elongated. The anterior line of larger cilia is called the trochus, the posterior the cingulum (Figs. 18.2b, 18.95b, 18.96). Earlier workers erroneously homologized these bands with the preoral and postoral bands of the trochophore-type larva. In most, but not all, of the Flosculariidae there is a dorsal gap in the ciliation of the circumapical band, where the trochus leads into the cingulum (Fig. 18.2b). In this group, the edges of the corona may be thrown into lobes, the number of which is of taxonomic significance (Figs. 18.96, 18.97). The bdelloids can be regarded as the extreme development of this type, for the dorsal gap has become so deeply indented toward the ventral side that it has cut through, breaking the trochus into two circles; it is these circles that give the illusion of rotating wheels in many bdelloids (Fig. 18.2e–j). The trochal circles are often incomplete medially (Fig. 18.116).

The corona of the Conochilidae looks rather different from the closely related Flosculariidae, since it is terminal and the mouth lies toward the dorsal side (Figs. 18.2d, 18.91). There is some doubt about the homology, but the outer part can most probably be regarded as equivalent to a circumapical band with a wide dorsal gap, but reflexed to the ventral side where it surrounds the buccal area. Thus, what appears superficially to be the apical area is actually part of the ventral surface of the body and accounts for the presence of the lateral antennae on the corona of *Conochilus* (Fig. 18.2d).

The corona of the Collothecidae represents a great modification. In *Collotheca*, the anterior part of the body is drawn out into a wide bowlshaped structure, the margin of which is usually lobed and beset with very long, fine filaments, generally called setae, although they are not stiff (Fig. 18.102). Within the bowl there is a ventral shelflike diaphragm which is usually ciliated. There may be vibratile cilia within the bowl, and in some species on the margin as well, but most do not have coronal cilia. Adult members of some of the Collothecidae lack coronal cilia completely except for a small area at the bottom of the coronal bowl. The *Collotheca* corona takes advantage of the thigmotactic responses of many protozoa and motile algae. When such organisms swim by chance into contact with the long filaments, they tend to follow them until they reach the corona, when the lobes fold together and snap the prey through the mouth. Usually capture is aided by a lashing movement of the filaments.

In some rotifers the ciliation is greatly reduced to a small, evenly ciliated area around the mouth, and in the bdelloid *Philodinavus* all that remains is a tuft of cilia projecting from the mouth (Fig. 18.121). Two genera, *Drilophaga* and *Wulfertia*, are remarkable in having the mouth some distance posterior to the small corona (Figs. 18.71, 18.72). Adult *Cupelopagis*, *Atrochus*, and *Acyclus* have no coronal cilia or setae (Figs. 18.21, 18.103, 18.104).

Variations of the Trophi

The trophi of rotifers are important systematic features used as criteria at all taxonomic levels. Not only are the major variations useful in separating classes, orders, and families, but some types of trophi vary in such a way that species can be recognized on the basis of details of trophi alone (Fig. 18.55).

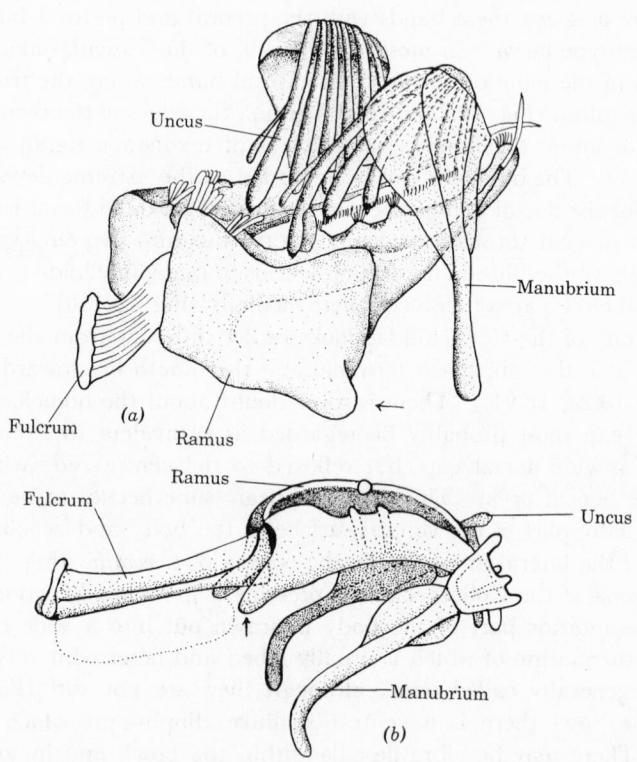

Fig. 18.3. Trophi in oblique view to show structure. (*a*) Malleate (submalleate). (*b*) Generalized virgate. The dotted line shows the outline of the fully contracted hypopharyngeal muscle. Each arrow points to an alula. [*a* based on *Euchlanis pellucida*, modified from Stoszberg; *b* original (J. B.).]

Naturally, there is a relation between the variations of the trophi and the investing musculature which together make up the mastax, but it is ordinarily necessary to examine only the hard parts for purposes of identification.

Almost all rotifer trophi consist of seven separate basic pieces arranged in three sections (Fig. 18.3). Thus, to the unpaired, median, ventral *fulcrum* is attached a pair of *rami* which move in opposition to each other like a pair of forceps. Together, these three pieces are called the *incus*. The rami are pulled apart by muscles attached to the fulcrum and to the outer part of the bases of the rami. In many trophi this part of the ramus is expanded into an *alula*. Lying adjacent to each ramus is a *malleus* consisting of two pieces, a toothed *uncus* and a *manubrium* hinged together, and moving in a plane perpendicular to the rami. The manubrium generally has an expanded *head*, which is the part attached to the uncus, and a handlelike *cauda*. In addition to these seven standard parts, there may be *subunci*, rods connecting each ramus to its corresponding uncus (Figs. 18.3*a*, 18.102*d*), and an epipharynx, one or more rods in the wall of the gullet serving to support the mouth (Fig. 18.79). Other extra supporting members may be developed in various places.

Several types of trophi can be recognized. They are named according to the relative development of the parts. Naturally there are transitions, and some aberrant ones that do not fit well into any of the main types. The development of the various parts and the associated muscles are related to the feeding habits.

Malleate trophi. This type can be taken as the basic one to which the others may be compared (Figs. 18.3*a*, 18.46). All parts are well developed and functional. The rami are massive, generally triangular, at most slightly curved. The inner margin may be toothed. The unci have several large teeth. There are subunci and there may be an epipharynx. Malleate trophi are ordinarily symmetrical although some have a slight asymmetry (Fig. 18.46). The malleate mastax functions to grind, grasp, and pump. Primarily it seems to grind material brought down by the ciliated gullet by pounding the rami and tips of the unci together. The tips of the rami may be thrust forward to grasp relatively large pieces in the gullet and pull them in. To some extent the mastax may act merely as a pump or valve and relatively large organisms can be drawn in and passed through unbroken into the stomach.

The malleate type is found in some very commonly encountered rotifers, the Brachionidae and Proalidae. In the latter family the trophi show modifications suggesting the virgate type, described next. The details of structure are used as specific characters in the Proalidae and to some extent in *Euchlanis*. In *Euchlanis* the manubria and fulcrum are slightly longer and more slender than in the others, and a separate subtype, *submalleate*, has been proposed for them. No use of this division will be made in the key. The malleate type is often recognizable on the basis of the general shape of the mastax (Figs. 18.75, 18.76). Shape and orientation of the fulcrum alone often permit ready recognition.

In some malleate mastaces there is a hypopharyngeal muscle in the ventral part of the mastax wall which, by contraction, enlarges the lumen and serves a pumping function. This muscle becomes very important in the next type.

Virgate trophi. This type (Fig. 18.3*b*) is specialized for pumping, and the trophi are considerably modified from those of the malleate type, although transitions exist. In general, the rami are thin, curved plates which together form a hemispheric dome. The mallei tend to lie more nearly parallel to the fulcrum than at right angles, because the rami are curved or because the fulcrum is attached at an angle to the axis of the rami (Fig. 18.85). Alulae are often well developed. The fulcrum is greatly elongated and tends to point toward the posterior rather than ventral. Subunci are well developed in many. Some virgate trophi are asymmetric (Figs. 18.22, 18.87). The powerful hypopharyngeal muscle inserts posteriorly on the fulcrum, and anteriorly on thin cuticle of the floor of the mastax cavity. By contracting, it enlarges the mastax lumen and functions in pumping. This muscle is a single piece in most species, but in *Synchaeta* and some others it is a prominent V-shaped striated band (Fig. 18.59).

Virgate trophi are of various kinds and might well be divided into subclasses, although a system has not been proposed. One of the kinds of variation is fusion of parts to form a firm support against the action of the hypopharynx. The extreme shows in *Tylotrocha* where all the parts are fused into a dome (Fig. 18.62). The virgate type is widely distributed, being found in most Notommatidae, Trichocercidae, Synchaetidae, and Gastropidae, and has transitions to the malleate type in the Gastropidae. Details of the trophi are used as generic and specific characters. Virgate trophi can usually be recognized in entire animals on the basis of the long fulcrum and the orientation.

In the usual manner of operation, the tips of the unci are used to pierce some organism and the contents are pumped out as described. Some species of *Notommata* and *Trichocerca* will work down a filament of alga such as *Mougeotia* or *Spirogyra*, sucking out the contents of the cells one by one (Fig. 18.22). Many possessors of virgate trophi are predaceous. *Synchaeta* and *Ploesoma* eat other rotifers, and large *Cephalodella* eat small *Cephalodella*. Organisms small enough are swallowed whole, others are pumped empty.

Cardate trophi. This type is modified for pumping, but the hypopharyngeal muscle is not developed. Rather, during pumping the lumen is widened by a rolling motion of the trophi. This action gives a characteristic appearance, and one can learn to recognize the type under low magnification by the motion alone. In preserved material the type can be recognized in entire specimens by the presence of a dorsal outgrowth from the head of the manubrium (Fig. 18.74). Cardate trophi are found only in the Lindiidae, and variations are of specific value.

Forcipate trophi. This type is modified for grasping (Fig. 18.56). The rami are flattened and lie in the same plane as the fulcrum. They have sharp tips and are often toothed along the medial border. The mallei lie very nearly in the same plane as the incus. The unci consist of only one or few teeth. The trophi are thrust through the mouth with a simple, direct motion, and small organisms are dragged back by the tips of the unci and rami. This type is limited to the Dicranophoridae, and variations are of use in separating genera and species. It can be recognized easily when the animals are seen feeding, for these animals look very different from those possessing incudate trophi, the only other type that can also be thrust well out of the mouth. The characteristic shape can easily be seen in lightly compressed animals.

Incudate trophi. This type is also specialized for grasping, but the modification are quite different from those of the forcipate type (Fig. 18.20). It is characterized by great development of the incus with large, toothed rami and small, practically vestigial mallei. In operation, the trophi are rotated from their resting position with fulcrum pointing ventrally, more than 90° around a transverse axis, and this brings them outside the mouth. In addition, the mastax performs a sucking function by means of a dorsal, expansible sac. Thus, prey are caught by the everted rami, pulled to the mouth, and sucked in by the operation of the dorsal sac.

Fully developed incudate trophi are possessed only by *Asplanchna* and *Asplanchnopus*. The type is easily recognized in sufficiently compressed specimens. *Harringia*, the third genus of the Asplanchnidae, has incudate trophi that somewhat resemble the malleate type, in that the mallei are relatively much larger, particularly the manubrium (Fig. 18.61).

Malleoramate trophi. This type is related to the malleate from which it differs chiefly in the shape and orientation of the manubria and unci. The manubrium is reduced to an elongate crescentic object bordering the lateral edge of the uncus; the cauda is missing. The rami are long and narrow, and completely overlain by the unci. The unci have a great many fine teeth, often with the ventral ones somewhat enlarged (Fig. 18.93, 18.95). This type is restricted to the Flosculariaceae. It is easily recognized, in live animals particularly, by the incessant pounding in a characteristic motion. Even in dead specimens, the shape of the mastax is characteristic. Just how much grinding is actually done by the trophi is questionable, for only very small organisms are able to pass down the long, thin gullet.

Ramate trophi. This type is very similar to the malleoramate, representing a somewhat more extreme departure from the malleate in the same direction as the malleoramate. The rami are semicircular, multitoothed plates, the teeth being thickenings of the plates. Usually several teeth near the middle are enlarged (Fig. 18.106). The name ramate is based on a misconception since it is the unci, not the rami, that are most developed. The fulcrum is very small, and even absent in *Abrochtha* (Fig. 18.119). The ramate mastax is usually associated with a long gullet. In *Abrochtha* the gullet is short and the tips of the unci may be projected through the mouth and used in gnawing, but the usual function appears to be crushing the food by rolling the unci together. The ramate type is limited to the Bdelloidea. Ordinarily there is no difficulty in recognizing it, even in preserved specimens, although compression may be required to render the trophi visible. Its motion in live animals somewhat resembles that of the malleoramate type.

Uncinate trophi. This type has a very different development of the unci from that of the ramate type (Figs. 18.21c, 18.102c). The unci have but few teeth, often one large tooth with several small accessory teeth, but even these may be absent. The manubria are vestigial. The unci are attached to the rather large bowed rami by elongate subunci.

The greatest difference in the mastax from the other types is in the structure of the muscular part. Most of the pharynx is expanded to form a large, thin-walled chamber (proventriculus, Fig. 18.102d). At the anterior end, a hollow *pharyngeal tube* projects into it from the anterior part of the system, the *infundibulum*. Prey are snapped through the highly distensible tube into the proventriculus from which there is no escape. The presence of the easily seen tube is diagnostic of this type. The trophi lie at the posterior end of the proventriculus.

The trophi may function to tear and rip food and push the pieces into the stomach, but apparently in many species they are too weak to perform the former function. Apparently some digestion may take place in the proventriculus of large *Collotheca* and of *Cupelopagis*, for one often observes several empty loricas of minute *Trichocerca* and other rotifers, or intact cell walls of Dinoflagellates.

Uncinate trophi are limited to the Collothecaceae. In animals stuffed with food it may be very difficult to see the trophi, but the unique structure of the anterior part of the digestive system and the corona usually makes examination of the trophi unnecessary for identification.

One other type of trophi is known in the rotifers, the *fulcrate*. This type is aberrant, not fully understood, and is limited to the Order Seisonidae, marine epizootic animals that are not treated in this book.

Identification of type of trophi. As has been indicated in connection with the descriptions, the type of trophi can ordinarily be recognized in whole animals, untreated, or at most subjected to light pressure. Occasionally the animal must be crushed to force the trophi out. For critical work, and in doubtful cases where details must be seen, the trophi must be removed intact from the animal. Critical taxonomic work at the species level should never be done with crushed preparations; it is necessary to have the parts free and in natural orientation. This is most easily accomplished with sodium hypochlorite (commercial Chlorox). To do many such examinations routinely, one needs a Watson-type compressor (Fig. 46.1), but in the absence of that valuable tool, the rotifer may be put into a depression on a slide, the solution added, and covered in such a way that the animal is kept in the angle between the slide and cover. The specimen must be found quickly with high power before the soft parts dissolve, for otherwise the trophi will be very difficult to locate. The technique is described by Myers (1938).

Nomenclature

Because many of the names first applied to the rotifers were published in obscure books or journals, many synonyms came into wide use during the nineteenth century. In 1913 Harring published a revision of nomenclature in which the principle of priority was applied consistently. The names in Harring's Synopsis have been adopted by most modern workers, except in certain cases shown to be wrong, but many of the old names persist. In the present key these synonyms are given, but only for the purpose of convenience in reading the literature; they should not be used.

Distribution

Rotifers are so easily transported that their distribution is usually regarded as potentially cosmopolitan. Many species are, in fact, nearly world wide in their distribution, and the occurrence of most species in a given locality seems to be controlled mostly by the environmental features of that locality, not by its location. Nevertheless, some species do show a distinctly limited geographical range and as careful studies are made, more cases are being revealed. For instance, *Trichocerca platessa*, *Pseudoploesoma formosa*, and *Kellicottia bostoniensis* seem to be limited to North America. One specimen of the last species has been reported from Sweden (Carlin, 1943), but this was in a location that might have received material from North American ships. Other species, common in Europe, have not yet been reported from other continents. *Keratella cochlearis* is generally absent from the tropics, although it is probably the commonest planktonic species in temperate regions. A number of brachionids are restricted to warmer climates. In general, the fauna of high altitudes and high latitudes is composed of relatively few species of wide distribution. Although some genera have not yet been reported from North America, there is as yet no reason for thinking that a limitation of distribution extends to the generic level, and for that reason all genera are mentioned.

The genera that are consistently caught most commonly and in greatest abundance with a plankton net in the open water of lakes are *Keratella*, *Kellicottia*, *Polyarthra*, *Synchaeta*, *Filinia*, and *Asplanchna*. Less common genera that nevertheless are frequently collected in great numbers are *Hexarthra*, *Conochilus*, *Conochiloides*, *Ploesoma*, *Pompholyx*, and *Gastropus*. Pelagic species of *Collotheca* are frequently found in some locations but are not as characteristic of the plankton as those listed. Such a list for the littoral fauna would be too long for practical use.

Preservation

Cocaine, neosynephrin, and other relaxing agents have been used successfully to obtain expanded specimens of rotifers (Chapter 46) when followed by a rapid fixative such as osmic acid. A much easier method involving hot water and no relaxing agent works remarkably well with many species, and has produced some of the best museum specimens the author has seen. A large number of organisms are placed in a Syracuse dish somewhat less than half full. An equal amount of boiling water is suddenly poured into the middle of the dish after the organisms have had time to become extended. Many of the organisms in the dish will either be distorted by high temperature or not killed because they were at the edges which remain cool, but an annular region within the dish is likely to contain well-fixed specimens. This method works with free-swimming and sessile organisms both. The latter are placed in the dish on small pieces of their substrate. Preservative is added to the dish and the material stored on vials or mounted on slides. Mild centrifugation may be effective (Edmondson, 1952).

Use of Key

The key is strictly practical, with no attempt made to key out families before genera. This has the disadvantage of separating closely related genera, but makes identification much easier.

Rotifer orders and families are established very largely on the character of the corona and trophi, two of the more difficult structures to study. The easiest characters to ascertain have to do with the presence or absence of the foot, movable cuticular spines and paddles, and, when the lorica is well developed, its structure. Therefore, the foot and cuticle are used early in the key, avoiding the necessity of making decisions about trophi and corona. The foot apparently has been lost independently in six different familes, and movable, cuticular structures appear in two families. Use of trophi cannot be avoided, but by using the easier lorica and foot characters first, it is minimized. The use of trophi might be reduced even further, but this would necessitate the use of characters that are difficult or impossible to use in preserved material.

Most fully loricate animals can be better identified from contracted than expanded material; although genera can be identified either way, specific characters often cannot be seen with the head expanded. Soft-bodied rotifers with virgate or forcipate trophi are relatively easy. There are few rotifers likely to be common in the plankton of lakes that cannot be identified to species in the contracted state when simply preserved with formalin. The Bdelloids can very easily be recognized as such on the basis of the ramate trophi, but well-behaved live specimens and prolonged study are ordinarily necessary for generic and specific identifications. It is difficult to fix bdelloids extended well enough to permit this, although it can be done; the hot water method is often very effective. Most of the sessile rotifers belonging to the Collothecaceae and Floscularicaceae can be identified only from expanded living specimens. For many of the Floscularicaceae well-fixed material is adequate, but even the best specimens of Collothecaceae usually are not good enough. Some of the species of Flosculariidae are so characteristic that they can be recognized when contracted, and some of the specific characters of *Limnias* are best seen in contracted specimens. Nevertheless, generic recognition depends upon the fully expanded corona. Most species of the Collothecaceae can be placed in the genus even though contracted, but for specific identification, live animals are ordinarily necessary. The soft corona of *Collotheca* is so easily distorted that only the more characteristic species can be named on the basis of fixed material.

With sufficient experience, one can learn to recognize many rotifers on the basis of informal characters that are not suitable for use in a key, or are difficult to describe accurately. The manner of locomotion, the way in which the trophi are moved, even when the parts cannot be seen, or other features permit one to recognize certain genera or species. Some species have unique structures that permit ready recognition, but they cannot be used in a key to genera; the tail of *Notommata copeus* (Fig. 18.87*h*), and the tube of *Limnias melicerta* (Fig. 18.96) are examples.

It is impractical within the confines of this book to provide a key to all the species, even limited to those that have been reported from North America. The key includes all the fresh-water genera, whether or not yet reported from North America. In monotypic genera, the book obviously serves to identify species as well, as long as one does not have under examination an undescribed second species of the genus. An attempt has been made to figure or provide information about particularly common species, especially planktonic ones. For small, common genera, especially if planktonic, critical information is generally provided on all known species so that species identification is possible. But some of these genera are due for a revision that will change the status of the described species; e.g., *Conochilus*, line 95a. Nevertheless, there are vast numbers of species that cannot even be mentioned. Illustrations of several species in large genera are provided as aids to generic identification, but it should not be assumed that a species at hand belongs to one of the species illustrated even if there is a close resemblance. In short, this key can be relied upon for specific identifications in only some genera, and in any case identifications should be confirmed by reference to full descriptions if they are to be used in scientific work.

For secure identification of species, extensive use must be made of the literature. The original literature on rotifers is very scattered, but fortunately there are some very useful compilations. The well-known Synopsis of the Rotatoria by Harring (1913) provides a catalog of all species described to that year with synonymy and bibliography. Wiszniewski (1954) issued what amounts to a new edition of the Synopsis, covering all species described through 1939. Important nomenclatorial changes were explained in an earlier paper (Wiszniewski, 1953). These catalogs, supplemented by the *Zoological Record*, permit one to locate all the known species; however, some of the synonymy may be changed by further research. The comprehensive compilation by Voigt (1957) will be invaluable, containing as it does extensive keys to species and many illustrations. Although only European species are included in the key, all known species are listed. Even those who do not read German will find the plates and bibliography useful.

Papers by the following authors list most of the species known to occur in North America and provide information about a great many of them: Ahlstrom, Burger, Carlin-Nilson, Edmondson, Harring, Jennings, Myers. The European literature cannot be ignored with a group as ubiquitous as the rotifers. The book by Hudson and Gosse (1889) is still useful in providing descriptions and figures of many species commonly found in North America, although the nomenclature is outdated and many genera were inadequately treated. In addition, papers by the following authors are especially useful

not only in describing new species, but in providing good illustrations and
amplified descriptions of common species: Bartoš, de Beauchamp, Bērzinš,
Carlin, Donner, Hauer, Wiszniewski, Wulfert.

In the following key, an attempt has been made to cite, in connection with
each genus, important papers published after the period covered by Wiszniewski's catalog, not only papers describing new species, but also those with an
abundance of figures and redescriptions. A large literature is developing on
the bdelloid fauna of damp soil and moss, but no attempt has been made to
cover it.

The fauna of hard waters contains many fewer species than that of soft, acid
waters, and most American investigators will not see many of the species
described originally by Harring and Myers from Northern Wisconsin, New
Jersey, and Maine. Information on the distribution of species with relation
to pH was given by Myers (1931, 1942).

Classification

Since the key is practical, a classification is provided indicating the location of each genus in the key by the line number. The classification is slightly
modified from that of Remane, and is organized on the basis of Phylum status
for the Rotifera. To convert this to a classification on the basis of Class
status for the rotifers, strike out the first two lines headed "Class," change
Monogononta to order, and list the following orders as suborders. The
Seisonidae and the Bdelloidea are sometimes grouped as subdivisions of the
Digononta, but they are sufficiently different that they more properly are
given coordinate designation with the Monogononta, even though the resulting groups are not differentiated into a number of orders, as is the Class
Monogononta. This classification differs from that of Remane largely in the
distribution of genera he included in the Lecanidae, and it departs from
Pennak's modification (1953) largely in that some of his subfamilies of the
Notommatidae are here kept as full families, and the Flosculariaceae are
differently divided. Some of the divisions are quite arbitrary, and some
of the families might well be organized into subfamilies. The subfamily
Brachioninae as given here is very heterogeneous and needs to be divided.
However, a useful revision of rotifer classification must await the accumulation of more information; for example, Sudzuki (1956b) suggested on the basis
of male morphology that *Ascomorpha* may be related to the Trichocercidae.

Class Seisonidea

Order Seisonida
  Family Seisonidae
      Genus *Seison* is an epibiont on marine crustaceans and is, therefore, not in-
      cluded in the key.

Class Bdelloidea

Order Bdelloida

Family Habrotrochidae

Genera *Ceratotrocha* 114b, *Habrotrocha* 115b, *Otostephanus* 115a, *Scepanotrocha* 114a

Family Philodinidae

Genera *Anomopus* (omitted, see 110), *Didymodactylus* 117a, *Dissotrocha* 122a, *Embata* 121a, *Macrotrachela* 119b, *Mniobia* 116b, *Philodina* 121b, *Pleuretra* 122b, *Rotaria* 119a, *Zelinkiella* (omitted, see 110)

Family Adinetidae

Genera *Adineta* 125a, *Bradyscela* 125b

Family Philodinavidae

Genera *Abrochtha* 123a, *Henoceros* 124a, *Philodinavus* 124b

Bdelloid of uncertain position: *Synkentronia* (see 110, Fig. 18.124)

Class Monogononta

Order Ploima

Family Brachionidae

Subfamily Brachioninae

Genera *Anuraeopsis* 8a, *Brachionus* (includes *Schizocerca*) 31a, *Cyrtonia* 81b, *Dipleuchlanis* 46a, *Diplois* 42a, *Epiphanes* 81a, *Euchlanis* (includes *Dapidia*) 46b, *Kellicottia* 11a, *Keratella* 12a, *Lophocharis* 38b, *Macrochaetus* 25a, *Manfredium* 28a, *Mikrocodides* 68a, *Mytilina* 42b, *Notholca* (includes *Pseudonotholca* and *Argonotholca*) 12b, *Platyias* 31b, *Proalides* 15b, *Rhinoglena* 79a, *Trichotria* 26a, *Tripleuchlanis* 45a, *Wolga* 38a

Subfamily Colurinae

Genera *Colurella* 24a, *Lepadella* 32a, *Paracolurella* 24b, *Squatinella* 36a

Family Lecanidae

Genera *Lecane* 44a, *Monostyla* 44b

Family Proalidae

Genera *Bryceela* 72a, *Proales* 82b, *Proalinopsis* 82a, *Wulfertia* 75b

Family Notommatidae

Genera *Cephalodella* (includes *Metadiaschiza*) 40b, *Dorria* 84a, *Dorystoma* 85a, *Drilophaga* 75a, *Enteroplea* (includes *Pseudoharringia* 92b) 76a, *Eosphora* 89b, *Eothinia* 87a, *Itura* 84b, *Monommata* 69a, *Notommata* 91b, *Pleurotrocha* 92a, *Pseudoploesoma* 35b, *Resticula* 89a, *Rousseletia* 86a, *Scaridium* 28b, *Sphyrias* 70a, *Taphrocampta* 73a

Family Lindiidae

Genus *Lindia* 77a

Family Birgeidae

Genus *Birgea* 71a

Family Trichocercidae

Genera *Ascomorphella* 15a, *Elosa* 14b, *Trichocerca* (includes *Diurella*) 22a

Family Gastropidae

Genera *Ascomorpha* 9a, 18b, *Chromogaster* 9b, *Gastropus* 34a

Family Dicranophoridae

Genera *Albertia* 55b, *Aspelta* 59a, *Balatro* 54a, *Dicranophorus* 56b, *Encentrum* (including *Parencentrum*) 52b, *Erignatha* 58a, *Myersinella* 60a, *Paradicranophorus* 56a, *Pedipartia* 49a, *Streptognatha* 57a, *Wigrella* 51a, *Wierzejskiella* 52a

Family Tylotrochidae (new designation)

Genus *Tylotrocha* 64a

Family Tetrasiphonidae

Genus *Tetrasiphon* 65a

Family Asplanchnidae

Genera *Asplanchna* 18a, *Asplanchnopus* 63a, *Harringia* 63b

Family Synchaetidae
Genera *Ploesoma* 35a, *Polyarthra* 4a, *Synchaeta* (includes *Parasynchaeta*) 61a
Family Microcodonidae
Genus *Microcodon* 67a
Order Flosculariaceae
Family Testudinellidae
Genera *Filinia* (includes *Tetramastix*, *Fadeewella*) 5a, *Horaella* 17b, *Pompholyx*
14a, *Testudinella* 94a, *Trochosphaera* 17a, *Voronkowia* 99a
Family Hexarthridae
Genus *Hexarthra* 5b
Family Flosculariidae
Genera *Beauchampia* 101a, *Floscularia* 105a, *Lacinularia* 100a, 106a, *Limnias*
104a, *Octotrocha* 105b, *Ptygura* (includes *Pseudoecistes*) 99b, 103a, *Sinantherina*
100b, 106b
Family Conochilidae
Genera *Conochiloides* 96a, *Conochilus* 95a
Order Collothecaceae
Family Collothecidae
Genera *Acyclus* 109a, *Atrochus* 109b, *Collotheca* (including *Hyalocephalus*) 108b,
Cupelopagis 19b, *Stephanoceros* 108a

Rotifers of uncertain position (Fig. 18.125)
(Not reported from N. A.)

Cordylosoma perlucidum Voigt cannot be placed, since the trophi were not seen and
details of the corona are uncertain.
Cypridicola parasitica Daday has ramate trophi, one ovary, and some unique characteristics, but is incompletely described. It lives attached to appendages of ostracods in salt water.
Vanoyella globosa Evens. Described from contracted material, this species is incompletely characterized and cannot be assigned to a family.

In the key, a fairly conservative nomenclature has been used, although
various proposed changes have been indicated. Authorities and dates for
specific names can be found in the catalogs by Wiszniewski (1954) and
Harring (1913).

Size is not critical in identifying genera of rotifers, and is helpful only
occasionally. The sizes stated in the key represent the range of predominant
approximate total length of the commonest species.

Most of the illustrations have been prepared especially for this new edition.
Many of the new drawings, identified by the initials J. B., were made by Miss
Janice F. Bush; those not so attributed were made by the author. Figures
based on drawings of specimens by the author are indicated as *Original*.
Otherwise, the figures have been redrawn from the source indicated. Most
of the stippled drawings show the rotifers as if by reflected light to make the
structure clear; many of the tone values are different with transmitted light.

KEY TO GENERA

1a	With 1 ovary. Trophi of any type but ramate . Class **Monogononta**	**2**
1b	With 2 ovaries (Fig. 18.116). (**Digononta**) Trophi ramate (Figs. 18.106, 18.116) . . . . Class **Bdelloidea**	**110**

It is usually unnecessary to see the ovaries or to make a close examination of the trophi to decide this couplet with live specimens. Most of the common aquatic bdelloid rotifers can usually be recognized at a glance under low power because of the characteristic "2-wheeled" appearance when swimming or feeding, or by the method of crawling on a substrate. *Adineta* (Fig. 18.122) is the chief exception. Relatively few Monogononta have two prominent "wheels" and those that do can usually be recognized by other features. Most of the bdelloids may crawl in inchworm or leech fashion on surfaces with the corona withdrawn. The ramate trophi can usually be recognized in entire animals (Figs. 18.116, 18.106). In preserved material, the bdelloids are easily distinguished from the loricate or sessile Monogononta by general appearance.

The marine epibiont *Seison*, belonging to the Seisonidae of the Digononta, is not included in this key.

a (1) With a foot, or with an attachment disc on the ventral surface (Fig. 18.21) **19**
A posterior median spine on the lorica should not be mistaken for a foot (Figs. 18.10, 18.12a).

b Without foot or attachment disc **3**
Genera with a completely retractible foot may give trouble, but the foot opening can be found, or the foot itself seen inside the body; e.g., *Brachionus*, *Testudinella*, *Gastropus*, and *Ploesoma*.

a (2) Body with movable, cuticular appendages, either filiform or flattened paddles, or hollow outgrowths bearing setae and containing muscles **4**
Fully loricate animals, with stiff, immovable spines are not included here (Figs. 18.10, 18.29).

b Body without such appendages **6**
There may be a lorica with spines.

a (3) Body with flattened cuticular appendages ("paddles," attached in 4 groups to dorsolateral and ventrolateral surfaces near anterior end). Trophi virgate. (Fig. 18.4) . . . . *Polyarthra* Ehrenberg
The older literature lists 2 species, *P. trigla* and *P. platyptera*, regarded by Carlin

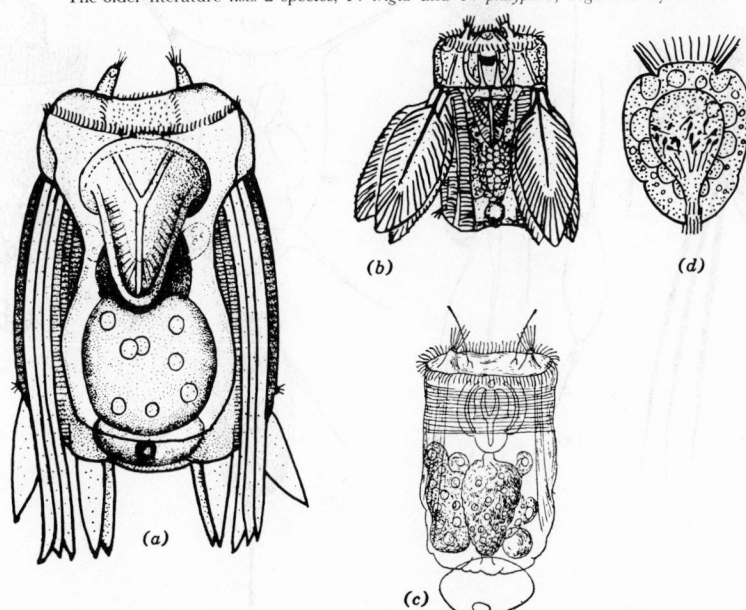

(b)

(d)

(a)

(c)

Fig. 18.4. *Polyarthra.* (*a*) *P. vulgaris*, ventral. × 400. (*b*) *P. euryptera*, dorsal. × 130. (*c*) Apterous form from resting egg ("*Anarthra*"). × 220. (*d*) Male [a original (J. B.); *b* modified from Bartoš; *c* after Hood; *d* after Wesenberg-Lund.]

as unrecognizable. At present 10 species are recognized: *P. euryptera* (Wierzejski) 1891, *P. remata* (Skorikow) 1896, *P. major* (Burckhardt) 1900, *P. minor* (Voigt) 1904, *P. dolichoptera* (Idelson) 1925, *P. longiremis* Carlin 1943, *P. vulgaris* Carlin 1943, *P. proloba* Wulfert 1941, *P. dissimulans* Nipkow 1952, *P. bicera* Wulfert 1956. All but the last one are included in a key by Nipkow (1952), and all but the last three are described by Carlin (1943). Both authors give photographic illustrations. Gillard (1952) gave outline sketches of all but the last two, and a key in Dutch. Bartoš (1951d) gave very useful illustrations and a key to 8 species in Czech.

The specific characters include: shape, length, and structure of paddles, number of nuclei in vitellarium, location of lateral antennae, and presence of an extra pair of small, cuticular appendages on the ventral side. All but *P. bicera* have 3 paddles in each of the 4 groups. *P. euryptera* is recognizable by having 12 nuclei in the vitellarium; *P. minor* and *P. remata* have 4, all the others 8 (including *bicera*?). *P. minor* differs from *P. remata* and all others in having the appendages of the right side distinctly shorter than those of the left. *P. bicera* is unique in having 2 setiform projections from the posterior surface. *P. proloba* is recognizable by a large ventral lobe containing the posterior part of the mastax, but Pejler (1957a) questions the validity of this character. The remaining 5 species can be separated only by careful study of appendages and other features. Of them, only *P. vulgaris* has the lateral antennae lying well anterior to the posterior corners of the body (Fig. 18.4a). *P. euryptera* and *P. major* have paddles much broader than those of the other species (Fig. 18.4b). Additional useful information by Donner (1954) and Sudzuki (1955). The possibility of hybridization was discussed by Pejler (1956).

The individuals that hatch from resting eggs do not have paddles; such forms were originally referred to as **Anarthra,** now apparently an invalid genus (Fig. 18.4c). *Polyarthra* is very common in the plankton of lakes and ponds, and may become very abundant.

4b   Appendages arranged otherwise. Trophi malleoramate **5**

5a   **(4)** Appendages are setiform extensions of cuticle. (Fig. 18.5) . . . . .

                                                 Filinia Bory de St. Vincent

Wiszniewski's catalog lists: *F. aseta* Fadeew 1925, *F. brachiata* (Rousselet) 1901, *F. camascela* Myers 1938, *F. cornuta* (Weisse) 1847, *F. longiseta* (Ehrenberg) 1834,

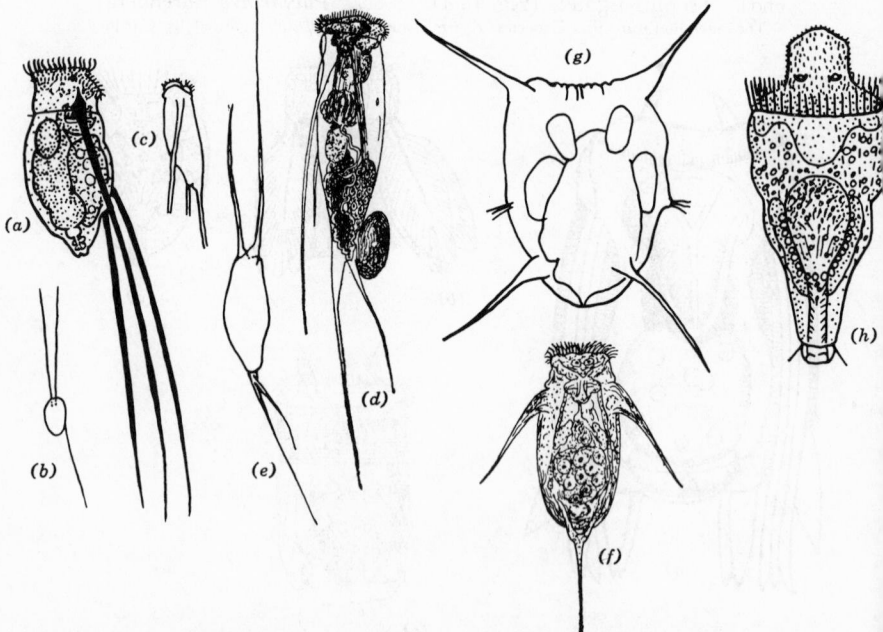

Fig. 18.5. *Filinia.* (*a*) *F. longiseta*, lateral. × 160. (*b*) *F. longiseta*, contracted. (*c*) *F. terminalis.* (*d*) *F* (= *Tetramastix*) *opoliensis*, lateral. × 150. (*e*) *F. opoliensis*, contracted. (*f*) *F. brachiata.* × 200. (*g*) *F.* (*Fadeewella*) *minuta*, contracted. (*h*) *F. longiseta*, male. (*a* modified from Donner; *c* after Edmondson and Hutchinson; *d, f* after Rousselet; *g* after Smirnov; *h* after Wesenberg-Lund; *b, e*, original.)

F. passa (Müller) 1786, *F. terminalis* (Plate) 1886, *F.* (as *Fadeewella*) *minuta* (Smirnow) 1928, *F.* (as *Tetramastix*) *opoliensis* (Zacharias) 1898.

Some authors synonymize *F. limnetica* Fadeew and *F. maior* (Colditz) with *F. longiseta* (Ehrenberg), as above, and others regard them as good species (Carlin, 1943, Donner 1954.) Pejler (1957a) regards *F. maior* as a synonym of *F. terminalis*. These species need further critical study. Additional information by Donner (1954). Hauer (1953) described a form of *F. longiseta* without a posterior spine. Formerly *Triarthra*. Remane includes *Tetramastix* in *Filinia*, presumably on the basis that a second, small, posterior spine is insufficient for generic separation. Possibly additional differences will be found. On the same basis, *Fadeewella*, not yet found in N. A., should probably be included (Fig. 18.5*g*). Length of body, up to 175 μ. Very common in plankton, especially *F. longiseta*.

5b   Appendages are hollow outgrowths of body wall, containing striated muscles and bearing setae. (Fig. 18.6) . . . .
Hexarthra Schmarda

Key to 8 species by Bartoš (1948). The specific characters include the details of setation of the appendages, presence of cylindric processes on posterior surface, presence of lower lip and number of teeth in uncus. Additional information by Hauer (1941, 1953, 1957) and Sládeček (1955). Length of body about 400 μ. Common in plankton, including that of saline and alkaline lakes. This genus was formerly known as *Pedalia* Barrois (originally *Pedalion* Hudson), but the priority of *Hexarthra* is recognized (Neal, 1951, Hemming, 1955).

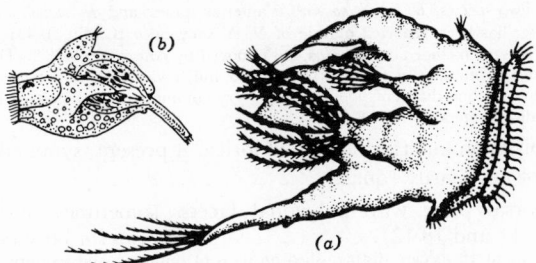

Fig. 18.6. *Hexarthra mira.* (*a*) Lateral. (*b*) Male. [*a* modified from Hudson and Gosse (J.B.); *b* after Wesenberg-Lund.]

6a   (**3**) With lorica or semilorica **7**
6b   Without lorica **16**
7a   (**6**) Lorica composed of 2 plates, dorsal and ventral, joined by flexible cuticle, forming sulci. (Figs. 18.7, 18.9) **8**
7b   Lorica otherwise **10**

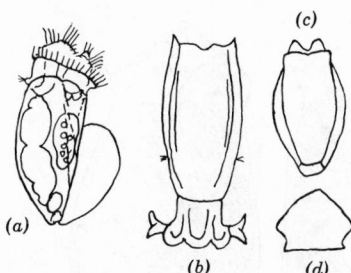

Fig. 18.7. *Anuraeopsis fissa.* (*a*) Lateral, with egg. × 55. (*b*) Dorsal, with posterior structure (homolog of foot?) extended, *A. fissa* var. *navicula.* (*c*) Ventral. (*d*) Optical cross section. (*a, b* after Donner; *c, d* after Ahlstrom.)

8a   (**7**) Dorsal plate arched, ventral plate almost flat. (Fig. 18.7) . . . . .
Anuraeopsis Lauterborn

Four species. 115 μ. Only *A. fissa* (Gosse) has been reported from U. S. where it is common in plankton and littoral. Additional information by Wulfert (1956).

8b   Lorica otherwise **9**

9a   (8) Dorsal plate less than ¾ the width of ventral plate. (Fig. 18.8) . .
                                                        Ascomorpha Perty
          Four species, up to 200 μ. Common in ponds and in lake plankton. *A. minima*,
          80 μ, is the smallest known metazoan. *A. ecaudis* is illoricate (Fig. 18-8*d-g*).
          Additional information by Hauer (1937) and de Beauchamp (1932).

9b   Ventral plate slightly narrower than dorsal plate. (Fig. 18.9) . . .
                                                        Chromogaster Lauterborn
          Two species have been recognized, but Carlin (1943) proposed to unite and place
          them in *Ascomorpha*. Formerly *Anapus*.

10a   (7) Lorica a boxlike structure composed primarily of 2 plates closely
          joined at the sides. The anterior and posterior ends of the ventral
          plate may be connected with the dorsal plate by flexible mem-
          branes, permitting extrusion of the head and eggs. (Fig. 18.11) . .   11
          Almost all species in this series have 4 or 6 spines on the anterior dorsal margin
          of the lorica.

10b   Lorica otherwise   13
          None of the animals in this series has a series of spines on the anterior dorsal
          margin of the lorica.

11a   (10) Spines on the anterior margin of lorica asymmetrically unequal in
          length. (Fig. 18.10) . . . . ***Kellicottia*** Ahlstrom
          Two species, *K. longispina* with 6 anterior spines, and *K. bostoniensis* with 4. The
          latter has been reported outside of N. A. only once (Carlin, 1943). Up to 1 mm.
          Originally included in *Notholca*, but removed by Ahlstrom (1938). There is evidently
          a certain amount of cyclomorphosis, and much variation in size and proportion of
          spines among different populations. Very common in plankton of lakes, less so in
          shallow waters.

11b   Spines on anterior margin of lorica, if present, symmetric, although
          not necessarily equal   12

12a   (11) Dorsal plate with polygonal facets, sometimes obscure. (Figs.
          18.11 and 18.12) . . . . ***Keratella*** Bory de St. Vincent
          About 15 species, distinguished on basis of spination and sculpture of dorsal plate
          (compare Fig. 18.12*a* with *j, j* with *k*). In some the sculpture may be difficult to see,
          but it can be brought out by partial drying. *K. cochlearis* is probably the commonest

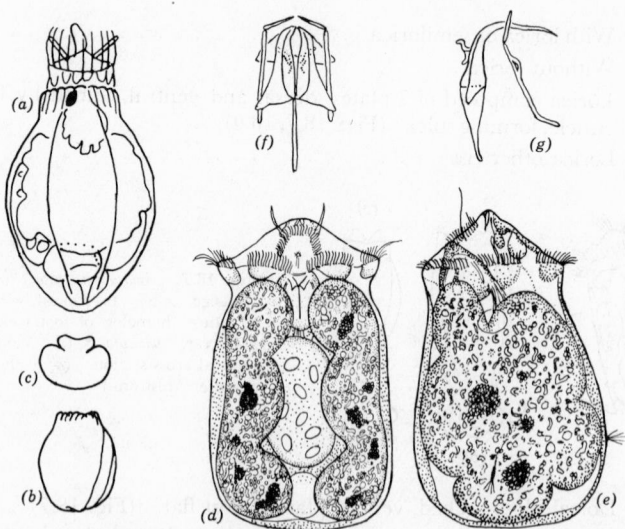

Fig. 18.8. *Ascomorpha. A. saltans:* (*a*) Dorsal, corona extended. × 250. (*b*) Contracted. (*c*) Optical cross section. *A. ecaudis:* (*d*) Ventral. (*e*) Lateral. × 200. (*f*) Trophi, ventral. (*g*) Trophi, lateral. [*a-c, f, g* after de Beauchamp; *d, e* modified from Remane (J.B.).]

planktonic rotifer. Carlin (1943) proposed to throw the familiar name *K. cochlearis* into synonymy with *K. stipitata* Ehrenberg on the basis of a presumed error by Ehrenberg in drawing a *K. quadrata* dorsal pattern in a *K. cochlearis* outline. Some authors have accepted Carlin's suggestion, but there has also been vigorous objection (Bērzinš, 1954b), and it may be hoped that *K. cochlearis* will not be replaced. There is uncertainty about the connection of Ehrenberg's original *K. stipitata* with the animal recorded as *stipitata* from South America by Zelinka (Fig. 18.12*h*) or the lower Elbe of Germany by Remane. These animals may be assignable to *K. americana* Carlin (= *gracilenta* Ahlstrom), but there are differences in the dorsal sculpture from Ehrenberg's figure. Formerly **Anuraea**.

Ahlstrom's monograph (1943) will serve for identification of species likely to be collected in America except *K. canadensis*. The following species not listed in Ahlstrom's key have been described or recognition proposed: *ahlstromi* Russell 1951, *canadensis* Bērzinš 1954c, *hiemalis* Carlin 1943, *lenzi* Hauer 1953, *sancta* Russell 1944, *testudo* (Ehrenberg) Carlin 1943, *ticinensis* (Callerio) Carlin 1943. *K. serrulata* is a synonym of *K. faculata* (Gillard) 1948, and *carinata* Russell 1950 is a synonym of *K. javana* Hauer (Russell) 1952. Additional papers containing useful information, bibliographies, descriptions of varieties, and figures are: Bērzinš (1954b, 1955), Klement (1955), Hauer (1954a,b), Pejler (1957c), Ruttner-Kolisko (1949), Tafall (1942), and Wulfert (1956). Despite much work, the systematics of this genus is not entirely satisfactory, and further nomenclatorial changes are to be expected. Cyclomorphosis is pronounced.

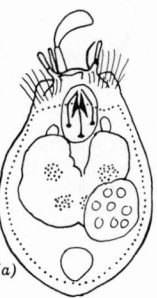

Fig. 18.9. *Chromogaster. C. ovalis:* (*a*) Dorsal. × 175. (*b*) Ventral, contracted. *C. testudo:* (*c*) Ventral, contracted. (Combined from several sources.)

(*a*)

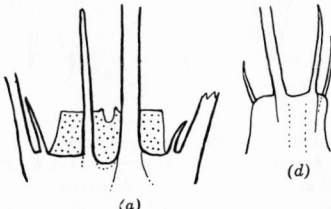

(*a*)

(*d*)

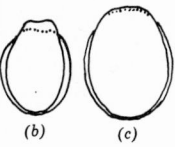

(*b*) (*c*)

Fig. 18.10. The 2 species of *Kellicottia* with corona retracted. *K. longispina:* (*a*) Dorsal view of anterior part of lorica. Anterior end of ventral plate is stippled. (*b*) Dorsal view of specimen from Imikpuk at Point Barrow, Alaska. × 55. (*c*) Specimen from Hall Lake, Washington. × 55. *K. bostoniensis:* (*d*) Dorsal view of anterior part of lorica. (*e*) Specimen from Hall Lake, Washington. × 55. (*f*) Dorsal view of specimen from Boston. *a–e* original; *f* after Rousselet.

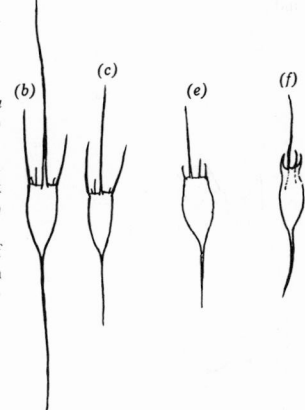

(*b*) (*c*) (*e*) (*f*)

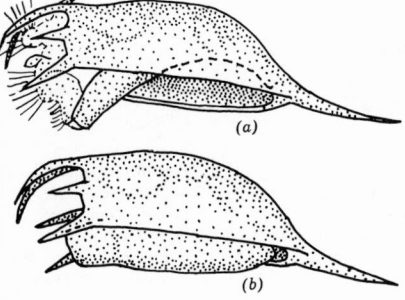

(*a*)

(*b*)

Fig. 18.11. *Keratella cochlearis* in lateral view to show structure of lorica. (*a*) Head extended, pushing anterior end of ventral plate down and drawing posterior end of ventral plate up. The anterior end of the membrane connecting ventral and dorsal plates is left unstippled. (*b*) Head contracted, ventral plate straightened out. The specimen was slightly compressed, making the ventral plate bulge out. Ordinarily, the venter is concave. (Original.)

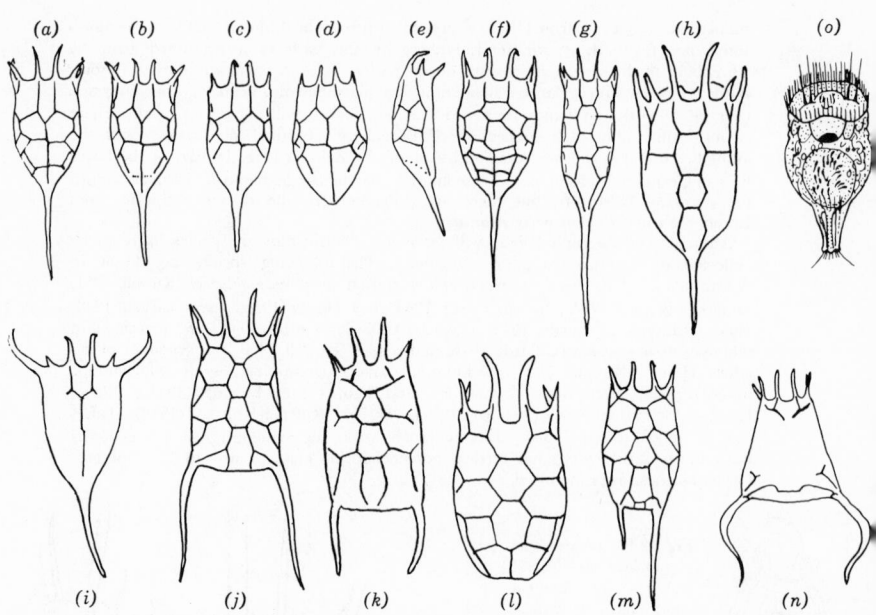

Fig. 18.12. *Keratella.* (*a–d, f–m*) Lorica in dorsal view. (*e*) Lorica in lateral view. (*o*) Male of *K. cochlearis.* (*a–e*) *K. cochlearis.* (*f*) *K. earlinae.* (*g*) *K. americana.* (*h*) *K. stipitata* (see text). (*i*) *K. taurocephala.* (*j*) *K. quadrata.* (*k*) *K. hiemalis.* (*l*) *K. serrulata.* (*m*) *K. valga.* (*n*) *K. canadensis.* Various magnifications about × 140. (*a–g, i, j, l–m* after Ahlstrom; *h*, after Zelinka; *k*, after Carlin; *n*, after Bērziņš; *o*, after Wesenberg-Lund.)

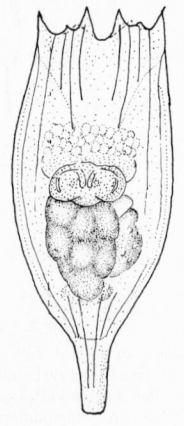

Fig. 18.13. *Notholca acuminata,* dorsal, head retracted. × 250. [Original (J.B.).]

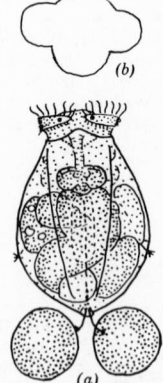

Fig. 18.14. *Pompholyx sulcata.* (*a*) Dorsal view of specimen carrying 2 eggs. × 230. (*b*) Optical cross section. (Combined from various sources.)

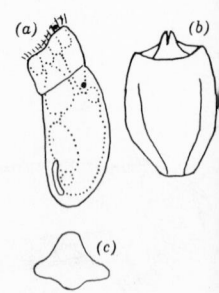

Fig. 18.15. *Elosa woralli.* (*a*) Lateral. × 225. (*b*) Dorsal, contracted. (*c*) Optical cross section. (*a, b* modified from Wiszniewski; *c* after Lord.)

12b   Dorsal plate smooth, with pustules, or with longitudinal striations,
             but not with facets. (Fig. 18.13) . . . . **Notholca** Gosse
             About 15 species, up to 180 μ. Common in ponds and other small waters, not
             common in lake plankton. Some authors recognize a subgenus **Argonotholca** for the
             species with ventral keel. See also Pejler 1957a. The genus needs revision. See
             Carlin (1943), Gillard (1948).

13a (10) Body lobed or flat in cross section. **14**

13b   Body more or less circular in cross section **15**

14a (13) Body in cross section with 4 approximately equal lobes or greatly
             flattened; 2 frontal eyes; trophi malleoramate. (Fig. 18.14) . . . . .
             . . . . **Pompholyx** Gosse
             Three species. *P. sulcata*, 120 μ, very common in ponds and lakes. *P. complanata*,
             distinguishable by very depressed body, and *P. trilobata* Pejler 1957c, with 3-lobed
             cross section, neither known in N. A.

14b   Body with 3 large lobes and an inconspicuous ventral lobe. Lorica
             with posterior, ventral, crescentic or oval opening. A large cer-
             ebral eye and a small frontal eye. Trophi virgate, asymmetric.
             (Fig. 18.15) . . . . **Elosa** Lord
             Two species, up to 90 μ. Common in psammon and *Sphagnum*.

15a (13) Lorica very poorly developed, rather flexible; with a few longi-
             tudinal folds but no transverse plications. Trophi virgate. (Fig.
             18.16) . . . . **Ascomorphella** Wiszniewski
             One species, 120 μ. *A. volvocicola* is one of the 3 species of rotifers found in *Volvox*
             colonies. It has previously been referred to as *Ascomorpha* or *Hertwigiella*. There is a
             rudimentary foot so minute it is often overlooked.

15b   Lorica fairly stiff, plicate, with several transverse folds. With
             cylindric posterior structure to which egg is attached. Trophi
             modified malleate. (Fig. 18.17) . . . . . **Proalides** de Beauchamp
             Three species. Not yet reported in N. A. Additional species by Rodewald 1940.

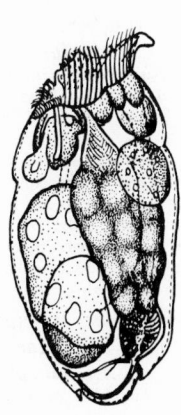

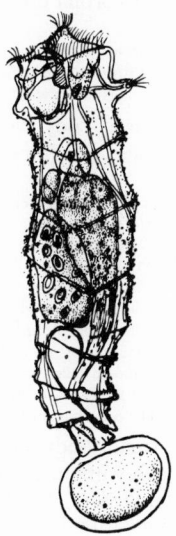

Fig. 18.16. *Ascomorphella
volvocicola*, lateral. × 240.
[Modified from de Beau-
champ (J.B.).]

Fig. 18.17. *Proalides verrucosus*, lateral. [Modified
from de Beauchamp (J.B.).]

16a   (6) Trophi malleoramate. (Fig. 18.95). **17**
16b   Trophi otherwise, incudate or virgate **18**
         Atrochus will appear at this point and fail to key further if the foot was not observed (see **109b**).

17a   (16) Corona a band of cilia encircling body.   (Fig. 18.18) . . . .
                                                    Trochosphaera Semper
         Two species; *T. aequatorialis* Semper has the band of cilia around the middle, *T.
         solstitialis* has the band somewhat toward one end.   Of rare occurrence, but may be
         very abundant when found.   In small, usually polluted waters, Midwest.

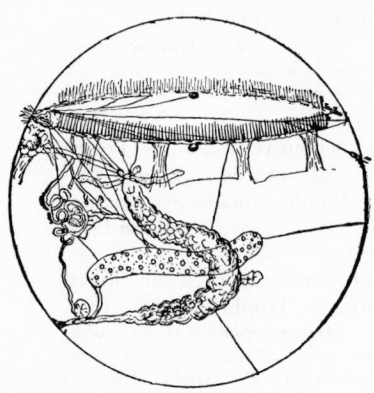

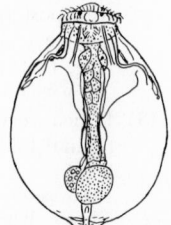

Fig. 18.18. *Trochosphaera solstitialis*, lateral.
× 80. (After Rousselet.)

Fig. 18.19.   *Horaella
brehmi*, dorsal.   × 110.
(Modified from Donner.)

17b   Circular corona borne on short neck.   (Fig. 18.19) . . . .
                                                    Horaella Donner
         One species, not yet reported from the U. S. (Donner, 1949a).
         Note: *Filinia* hatched from resting eggs, without setae, will appear at this point
         (see **5a**), as probably would *Cypridicola* (p. 438).

18a   (16) With large body cavity, the stomach lying well away from the
         epidermis.   No intestine.   Trophi incudate.   (Fig. 18.20). . . . . .
                                                    Asplanchna Gosse
         A difficult, variable genus (de Beauchamp, 1951). Wiszniewski's catalog includes:
         A. priodonta Gosse 1850, *A. herricki* de Guerne 1888, *A. brightwelli* Gosse 1850, *A.
         girodi* de Guerne 1888, *A. intermedia* Hudson 1886, *A. sieboldi* (Leydig) 1854, *A. sil-
         vestri* Daday 1902.
         The ovary in the first two is nearly spheric, band- or horseshoe-shaped in the rest.
         A. herricki is distinguished by the presence of a 2-celled glandular structure near the
         cloaca.   Other specific characters include details of trophi.   Some species have large,
         lateral extensions of the body wall which develop cyclomorphotically.   *A. priodonta* is
         very common in lake plankton and is the species most often collected there in N. A.
         Up to 1500 μ.   Predatory, swallowing whole rotifers and crustaceans as well as colonial
         algae.   The anus has shifted, and lies on the same side as the mouth.

18b   With small body cavity, the stomach lying close to epidermis.
         Stomach without gastric glands, with diverticula and algae in wall.
         Trophi virgate, very slender . . . . *Ascomorpha* Perty
         Most *Ascomorpha* have a semilorica (**8b**) but *A. ecaudis* is illoricate and will key here
         (Fig. 18.8).   See also *Ascomorphella* (**15a**).

         Individuals of *Polyarthra* that hatch from resting eggs have no paddles and may come
         to this point in the key, but may be recognized easily by differences in trophi, gut, and
         corona.   (See **4a**.)

19a   (2) With a foot, rather than ventral attachment disc **20**

19b　　　With attachment disc on ventral surface of ovate body.　Corona
　　　　　a large bowl-shaped structure without marginal cilia or setae.
　　　　　(Fig. 18.21) . . . . *Cupelopagis* Forbes
　　　　　　　Probably only 1 species, up to 800 μ.　Generally attached to flat leaves of aquatic
　　　　　　　plants.
　　　　　　　Formerly *Apsilus*.　Other members of the Collothecaceae key out under **107**.

20a　**(19)**　Foot ends in 1 or 2 toes . . . .　**21**
　　　　　　　If there is 1 toe it is tapering and ends in a point or rounded tip; it is not a
　　　　　　　peduncle which is cylindrical and ends bluntly or in an expanded disc.　(Figs. 18.95,
　　　　　　　18.102).　No truly sessile species belongs here.

20b　　　Foot ends otherwise, either in a peduncle, attachment disc, ciliated
　　　　　cup, or without a specialized structure . . . .　**93**
　　　　　　　Foot may be a hemispheric cushion at end of body, without toes.

21a　**(20)**　With lorica or semilorica . . . .　**22**

21b　　　Without lorica . . . .　**47**

22a　**(21)**　Body twisted in segment of helix, asymmetric.　Toes unequal.
　　　　　Trophi virgate, asymmetric.　(Fig. 18.22). . *Trichocerca* Lamarck
　　　　　　　About 90 species, of which most are littoral, but some are common in lake plankton.
　　　　　　　Length of the lorica 100–500 μ.　The genus includes species formerly in *Diurella*
　　　　　　　(Edmondson, 1935), which included species with equal toes and those in which the
　　　　　　　short toe was more than ⅓ the length of the other.　*Trichocerca*, as presently defined,
　　　　　　　is a heterogenous genus, and possibly should be split, but not on the basis of toe length.
　　　　　　　The monograph by Jennings (1903) is out of date, but includes most of the common
　　　　　　　North American species.　Formerly *Rattulus*.　Many species have been described since
　　　　　　　Wiszniewski's catalog (1954): Donner (1950a, 1953), Edmondson (1948b), Hauer
　　　　　　　(1952a), Myers (1942), Wulfert (1940, 1956).　Further useful information by Carlin
　　　　　　　(1939), Wulfert (1939a, 1956).

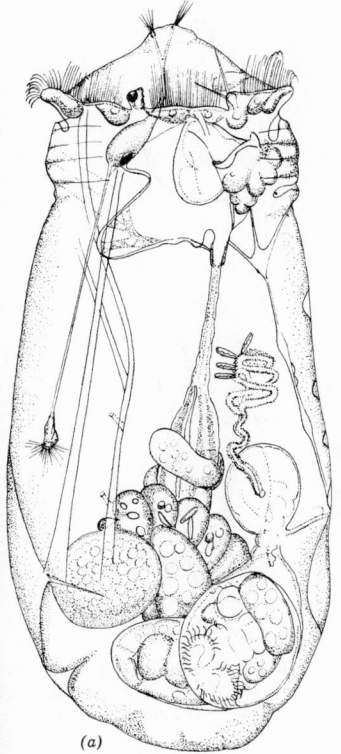
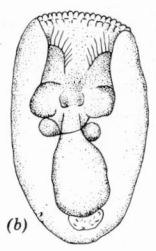
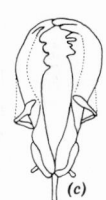

(b)

(c)

(a)

Fig. 18.20.　*Asplanchna priodonta.*　(*a*) Lateral view of specimen
with several embryos. × 100.　(*b*) Dorsal view of contracted
specimen.　(*c*) Trophi.　(*a* original drawing supplied by E. R.
Hollowday; *b*, *c* original.)

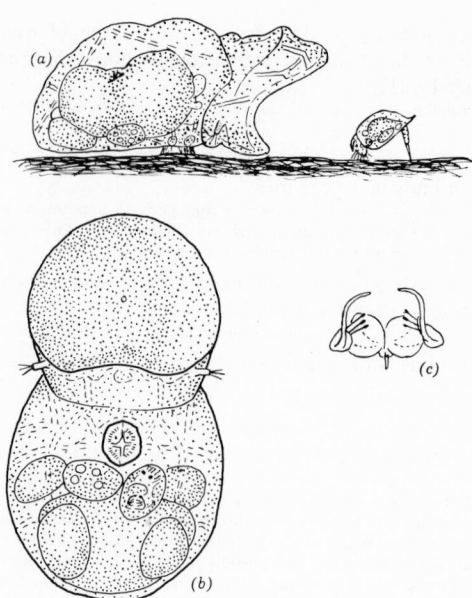

Fig. 18.21. *Cupelopagis vorax.* (*a*) Lateral, about to catch a *Colurella.* × 60. (*b*) Ventral, specimen removed from substrate. (*c*) Trophi of advanced embryo. (Original.)

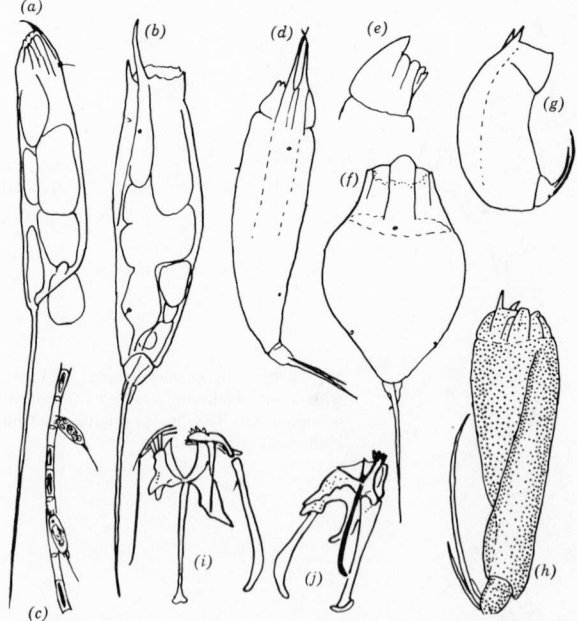

Fig. 18.22. *Trichocerca.* (*a*) *T. cylindrica.* × 130. (*b*) *T. longiseta.* (*c*) Two *T. longiseta* pumping cell contents from *Mougeotia.* (*d*) *T. similis.* × 380. (*e*) Head of *T. capucina.* (*f*) *T. multicrinis.* (*g*) *T. porcellus.* (*h*) *T. insignis,* showing twisted shape of lorica. (*i*) Dorsal view of trophi of *T. bicornis.* (*j*) Lateral view of same. (*a, b, d–g* after Jennings; *c, h* original; *i, j* after de Beauchamp.)

22b   Body not twisted **23**
         Note: Some of the smaller *Trichocerca* are not conspicuously twisted and may give
         trouble at this point, but they have asymmetric trophi and cannot easily be confused
         with any other genera in couplets **23–48**; a superficial resemblance to small *Cephalodella*
         (Fig. 18.38) may be briefly confusing. See Fig. 18.22*d, g*.

23a **(22)** Body compressed (laterally narrowed), and jointed foot originates
         near posterior end of body in longitudinal ventral cleft in lorica.
         (Fig. 18.23) **24**
23b   Body not compressed, or if somewhat compressed, foot other-
         wise. **25**

24a **(23)** Lorica continuous across dorsum. (Fig. 18.23) . . . .
                                           Colurella Bory de St. Vincent
         About 15 species, 50–150 μ. No joint of foot longer than toes. The head carried
         a semicircular shield dorsal to corona which is retractable within lorica. Littoral,
         browsing over plants, scraping up small organisms with head shield (Fig. 18.21*a*).
         Very common. Formerly *Colurus*. A variable, difficult genus. Additional species
         and figures by Hauer (1924), Carlin (1939), Bērziņš (1949).

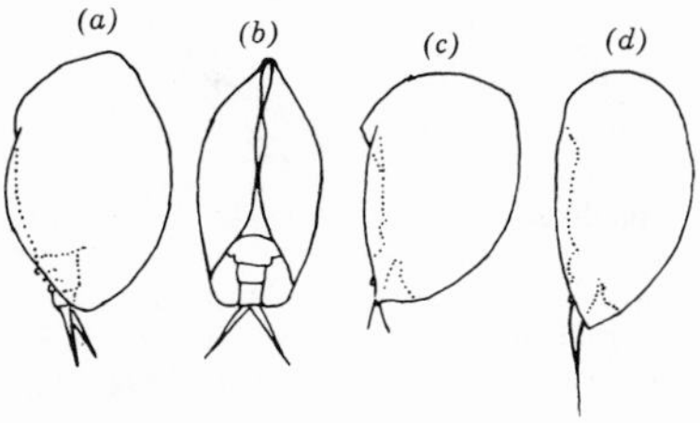

Fig. **18.23.** *Colurella.* (*a, b*) *C. obtusa*, in
lateral and ventral views. × 225. (*c*) *C. bi-
cuspidata*. × 225. (*d*) *C. adriatica*. ×225. (*a,
b* after Hauer; *c, d* after Carlin.) See also Fig.
18.21*a*.

Fig. **18.24.**   *Paracolurella
aemula,* lateral. × 200.
(After Myers.)

24b   Lorica with longitudinal dorsal sulcus. (Fig. 18.24) . . . .
                                           Paracolurella Myers
         Three species, 300 μ. Terminal joint of foot longer than toes. Littoral. Rare.
         Includes *Mytilina pertyi*.

25a **(23)** Dorsal surface of lorica with pairs of long spines placed symmetri-
         cally with respect to mid-line. (Fig. 18.25). . *Macrochaetus* Perty
         Seven species, up to 200 μ. Littoral, swimming among plants. Remane (1933)
         proposed to include these species in *Trichotria* (see line **25b**). Common. Formerly
         Polychaetus.

25b   Dorsal surface without spines, or at most with but 1 or 2 spines
         on mid-line. **26**

26a **(25)** First joint of foot with heavy dorsal spines. (Fig. 18.26) . . . . . .
                                           Trichotria Bory de St. Vincent
         About 10 species. Up to 400 μ. Littoral, common. Formerly *Dinocharis*. Addi-
         tional species by Myers (1942). Figures of commonest species by Wulfert (1956). See
         also *Wolga* (line **38a**).

26b   Foot, if jointed, without such spines on first joint **27**
27a **(26)** Foot and toes together longer than lorica. **28**
27b   Foot and toes together shorter than lorica **29**
28a **(27)** Lorica pear-shaped, dorsum bulging. Trophi malleate. (Fig.
         18.27) . . . . *Beauchampiella* Remane
         One species, 540 μ. Ponds. Very rare. Formerly placed in *Scaridium* to which it
         has only a superficial resemblance. Named *Eudactylota* by Manfredi, but this name
         was preoccupied (Gallagher, 1957). Description and figure by Wulfert (1940).
         Remane (1929) used *Beauchampiella* for this in his text, but retained *Eudactylota* in the
         formal statement of classification. It would appear that *Beauchampiella* should be used
         rather than *Manfredium*. Gallagher.

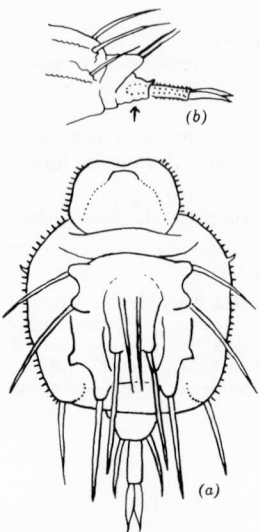

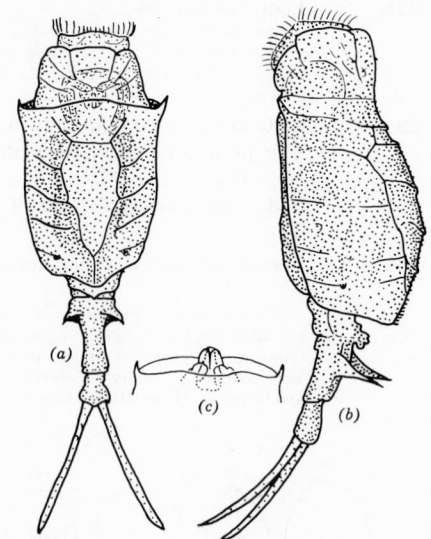

Fig. 18.25. (*a*) *Macrochaetus near subquadratus*, dorsal. (*b*) *M. collinsi*, lateral view of posterior end. × 350. The arrow points to first joint of foot. (*a* original; *b* after Myers.)

Fig. 18.26. *Trichotria tetractis*. (*a*, *b*) Dorsal and lateral views of expanded specimens. × 190. (*c*) Dorsal view of anterior end of contracted animal showing head sheath folded. The specimen shown in *b* was somewhat compressed. (Original.)

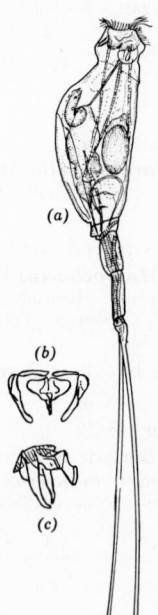

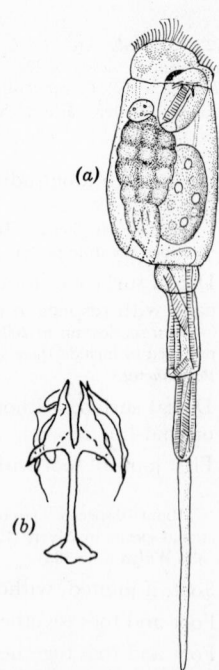

Fig. 18.27. *Manfredium eudactylotum*. (*a*) Lateral. × 150. (*b*) Trophi dorsal. (*c*) Trophi lateral. [After de Beauchamp (J.B.).]

Fig. 18.28. *Scaridium longicaudum*. (*a*) Lateral. × 110. (*b*) Trophi. [*a* original (J.B.); *b* after Beauchamp.]

28b   Lorica cylindric, dorsum not bulging.   Trophi modified virgate.
(Fig. 18.28) . . . . ***Scaridium*** Ehrenberg
<blockquote>One species known, but probably more exist.   Up to 425 μ.   Littoral.   The very thin lorica may be overlooked.   Common.   Description and figures by Donner (1943a).   Some species of **Monommata** would come to this point if the fairly stiff cuticle were regarded as a semilorica.   See **69a**.</blockquote>

29a   **(27)**   Lorica of 1 piece, either continuous around body or an arched dorsal plate with venter of flexible cuticle, or of dorsal and ventral plates firmly fused at the edges with or without dorsal sulcus **30**

29b   Lorica otherwise, in most cases consisting of dorsal and ventral plates separated by flexible membranes which in many form sulci. (Figs. 18.41, 18.45*b*) **39**
<blockquote>Some of these animals have a poorly developed lorica of rather flexible plates.</blockquote>

30a   **(29)**   With an even number of spines projecting forward from the dorsal anterior margin of the lorica. **31**
<blockquote>*Brachionus* belongs here, although 2 species deviate.   See **38b**.</blockquote>

30b   Without such spines, or with uneven number of short spines. (Fig. 18.33*c*) **32**

31a   **(30)**   Foot annulated, retractible within body. (Fig. 18.29) . . . .
Brachionus Pallas
<blockquote>About 25 species, mostly littoral, some planktonic in lakes.   Characteristically found in hard water.   Some species abundant in salt waters.   Very common.   Monograph by Ahlstrom (1940).   Includes **Schizocerca**, not known from N. A.   See also Bērziņš (1943) and Hauer (1952a, 1953).</blockquote>

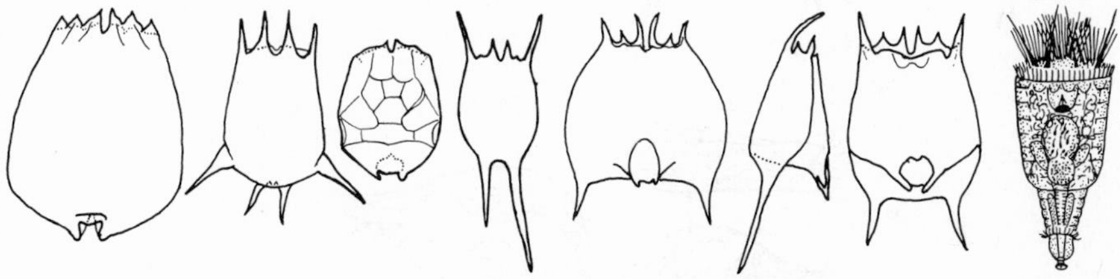

Fig. 18.29.   *Brachionus.*   (*a–g*) Loricae.   (*h*) Male.   (*a*) *B. plicatilis.*   (*b*) *B. calyciflorus.*   (*c*) *B. angularis.*   (*d*) *B. havanaensis* (described from Havana, Illinois).   (*e*) *B. quadridentata,* ventral.   (*f*) Same, with longer spines, lateral.   (*g*) *B. bidentata.*   Various magnifications × 45–100; *h* × 200.   (*a–g* after Ahlstrom; *h* after Wesenberg-Lund.)

31b   Foot jointed, not retractible within body. (Fig. 18.30) . . . .
Platyias Harring
<blockquote>Four species; 3 in Ahlstrom's monograph (1940), additional 1 by Wulfert (1956) not reported from N. A.   Wiszniewski (1954) admits only *P. quadricornis,* placing the others in **Brachionus.**   350 μ.   Formerly **Noteus.**</blockquote>

32a   **(30)**   Ventral plate of lorica with opening for foot about halfway between middle and posterior end.   Foot jointed; at rest lies in groove extending back from foot opening. (Fig. 18.31) . . . .
Lepadella Bory de St. Vincent
<blockquote>Over 50 species, littoral.   Most 100–200 μ.   Very common. Harring's monograph is incomplete, but contains the species that are likely to be found commonly in hard waters.   Formerly **Metopidia.**   Additional information by Donner (1943c), Wulfert (1956).</blockquote>

32b   Foot otherwise **33**

33a   **(32)**   Foot attached near middle of ventral surface, and typically extends at approximately right angles to longitudinal axis of body . .   **34**

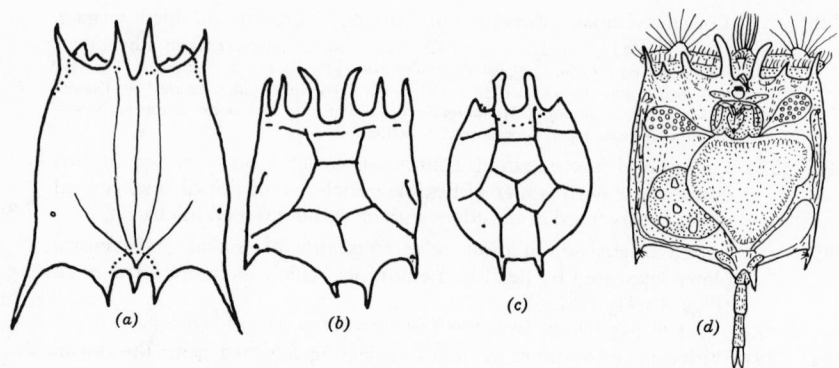

Fig. 18.30. The 3 species of *Platyias* reported from N. A. (*a–c*) Dorsal views of loricae. (*d*) Expanded specimen showing foot and corona. × 120. (*a*) *P. polyacanthus.* (*b, d*) *P. patulus.* (*c*) *P. quadricornis.* Wiszniewski (1954) regards the first two as belonging in *Brachionus.* (*a–c* after Ahlstrom; *d* original.)

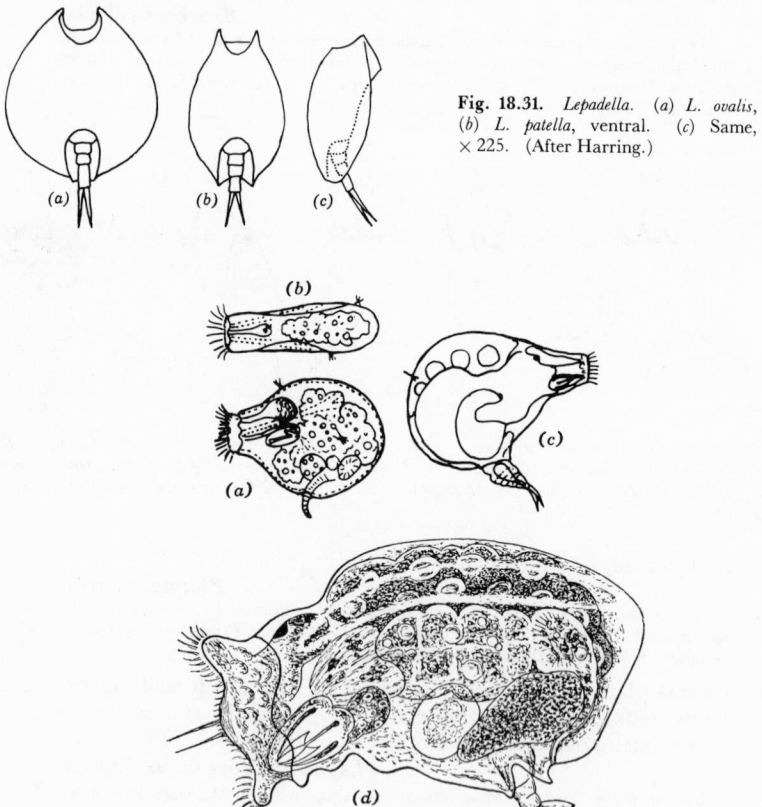

Fig. 18.31. *Lepadella.* (*a*) *L. ovalis*, ventral. (*b*) *L. patella*, ventral. (*c*) Same, lateral. × 225. (After Harring.)

Fig. 18.32. The 3 species of *Gastropus.* (*a*) *G. stylifer*, lateral. × 130. (*b*) Same, dorsal. (*c*) *G. minor*, lateral. (*d*) *G. hyptopus.* × 170. (*a, b* after Rousselet; *c* after Wesenberg-Lund; *d* after Hudson and Gosse.)

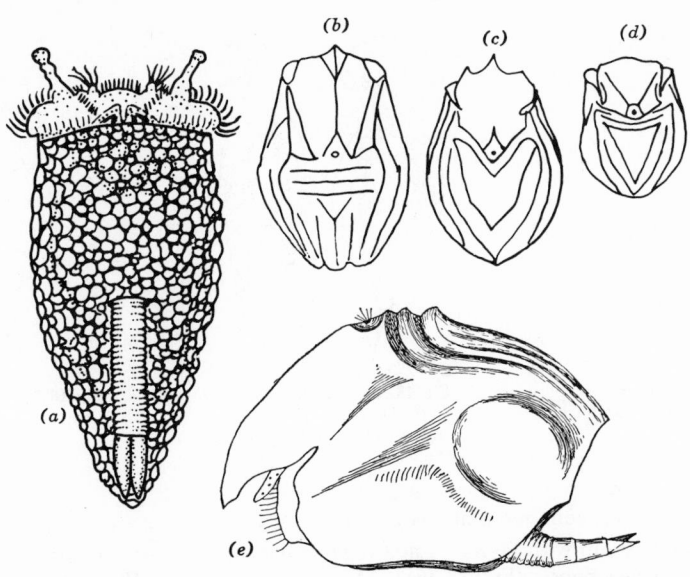

Fig. 18.33. The 4 species of *Ploesoma*. (*a*) *P. hudsoni*, ventral. × 110. (*b–d*) Dorsal views of loricae. (*b*) *P. lenticulare*. × 150. (*c*) *P. triacanthum*. × 175. (*d*) *P. truncatum*. × 80. (*e*) *P. lenticulare*, lateral. × 300. (*a* modified from Wesenberg-Lund; *b–d* after Brauer; *e* after Wierzejski and Zacharias.)

33b		Foot terminal or attached nearer to posterior end than to middle of ventral surface.	**36**
34a	(**33**)	Lorica thin, transparent, without obvious sculpture, body compressed. (Fig. 18.32) . . . . ***Gastropus*** Imhof	

Three species, 90–370 μ. littoral and planktonic. *G. stylifer*, common in lake plankton is the most colorful rotifer; violet epidermis, blue stomach studded with orange fat globules, and red eye. Wiszniewski (1953) places the other 2 species in ***Postclausa*** Hilgendorff 1899. Figures of *G. hyptopus* by Wulfert (1939a), of *G. stylifer* by Wulfert 1956.

34b		Lorica with sculpture or areolations	**35**
35a	(**34**)	Corona without prominent processes lateral to mouth. Dorsal antenna on lorica. Lorica heavily marked with ridges and grooves or with areolations. (Fig. 18.33) . . . . ***Ploesoma*** Herrick	

Four species, up to 500 μ. Ponds and lake plankton. *P. hudsoni* has a frothy-appearing epidermis which gives the thin, flexible lorica an areolated appearance. Wiszniewski (1953) removes this species to the monotypic genus ***Bipalpus*** Wierzejski and Zacharias 1893. Figure by Wulfert (1939a, 1956).

35b		Corona with large prominences lateral to mouth. Anterior dorsal margin of lorica notched permitting emergence of antenna. Lorica marked with small areolations and several light ridges but not grooves. (Fig. 18.34) . . . . ***Pseudoploesoma*** Myers	

One species, 215 μ. Littoral in acid water.

36a	(**33**)	Corona covered by dorsal semicircular, nonretractible shield. Head not retractible. Body somewhat depressed. (Fig. 18.35) . . . ***Squatinella*** Bory de St. Vincent	

Fourteen species, up to 220 μ. Littoral. Formerly ***Stephanops***. Additional information by Carlin (1939), Wulfert (1936, 1939a, 1950, 1956).

36b		Without such head shield	**37**
37a	(**36**)	Lorica without true ventral plate, or if a plate is present, is merely a somewhat stiffer area in the flexible cuticle across venter (Fig. 18.45*d*). See *Euchlanis* (**46b**).	

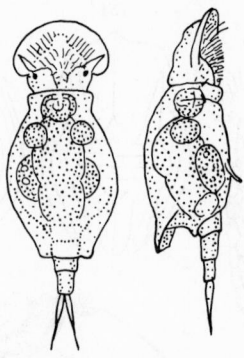

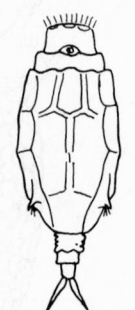

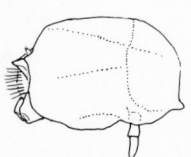

Fig. 18.34. *Pseudoploesoma*
formosum, lateral. × 100.
(Modified from Myers.)

Fig. 18.35. *Squatinella mutica.*
(*Left*) Dorsal. (*Right*) Lateral. × 300. (Modified from
Wulfert.)

Fig. 18.36. *Wolga spinifera.* (After Western.)

37b   Lorica continuous across venter **38**

38a   (37) Dorsum with facets. Lateral antennae arise under small spinous
            projections. (Fig. 18.36) . . . . **Wolga** Skorikov
            One species. Littoral. Rare. Remane (1929) includes this in **Trichotria** (see
            26a). Figures and discussion by Bērziņš (1954).

38b   Dorsum without facets. Lateral antennae unprotected by spines.
            (Fig. 18.37) . . . . **Lophocharis** Ehrenberg
            Four species, up to 175 μ. Littoral.
            Epiphanes mollis will come to this point if its somewhat stiff cuticle is regarded as a
            semilorica. (Fig. 18.76). *Brachionus dimidiatus* var. *inermis*, the only *Brachionus*
            without spines, and the anomalous *B. tridens*, will also key out here (see **31a**).
            They have not been reported from N. A. Species of *Brachionus* with very short spines
            will key here if the spines were not recognized. *Ascomorphella* will key out here if
            the minute foot was observed (see **15a**). *Mikrocodides* will come to this point if the
            rather stiff cuticle was regarded as a lorica. (See Fig. 18.65.)

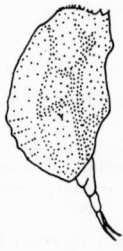

Fig. 18.37. *Lophocharis*
salpina, lateral, head retracted. × 180. (After
Harring.)

39a   (29) Lorica fairly flexible, weakly developed **40**

39b   Lorica stiff, well developed. **41**

40a   (39) Trophi forcipate (Figs. 18.48–18.58) . . . Family **Dicranophoridae**   **49**
            Some Dicranophoridae have a semilorica.

40b   Trophi virgate. (Fig. 18.38) . . **Cephalodella** Bory de St. Vincent
            Over 100 species described; probably many synonyms created. Very common in
            littoral. A varied genus, some with no perceptible lorica (see **91a**), most with weak
            lorica, some with well-developed lorica. Key by Wulfert (1938a). Additional in-
            formation by Donner (1950b), Margaleff (1948), Myers (1942), Wulfert (1940, 1943,
            1956). Many species common in N. A. described by Harring and Myers (1924) and
            Myers (1934a). Formerly **Diaschiza, Metadiaschiza,** and **Furcularia** in part.

41a   (39) Lorica with dorsal longitudinal sulcus (Fig. 18.40*b*) **42**

41b   Lorica without dorsal sulcus **43**

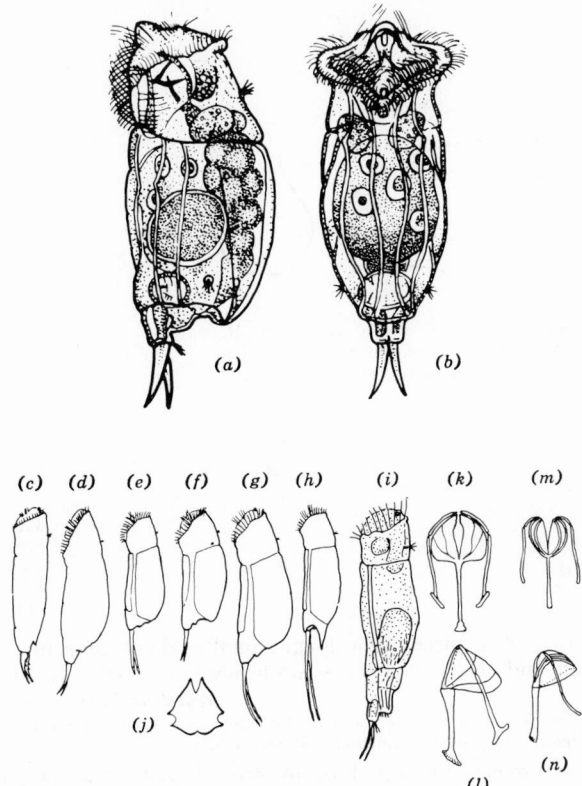

Fig. 18.38. *Cephalodella.* (*a*) *C. auriculata*, lateral. (*b*) Same, ventral. × 330. (*c*) *C. forficula.* (*d*) *C. megalocephala.* (*e*) *C. intuta.* (*f*) *C. exigua.* (*g*) *C. gibba.* × 80. (*h*) *C. mucronata.* (*i*) Male of *C. gibba.* (*j*) Optical cross section of *C. plicata* showing the 4 plates of the lorica connected by thin membranes. (*k*) Trophi of *C. gibba*, dorsal. (*l*) Same, lateral. (*m*) Trophi of *C. megalocephala*, dorsal. (*n*) Same, lateral. [*a, b, i* after Dixon-Nuttall and Freeman (J.B.); *j* after Remane, *c–h, k–n* after Harring and Myers.]

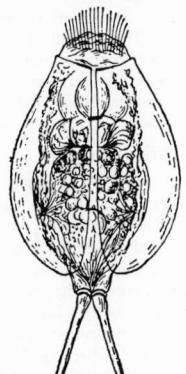

Fig. 18.39. *Diplois da-viesiae*, dorsal. × 87. (After Weber.)

42a **(41)** Lorica composed of 3 separate plates separated by sulci; 1 ventral plate and 2 dorsolateral plates. (Fig. 18.39) . . . .
Diplois Gosse

One species, *Sphagnum* bogs, 500 μ. Rare.

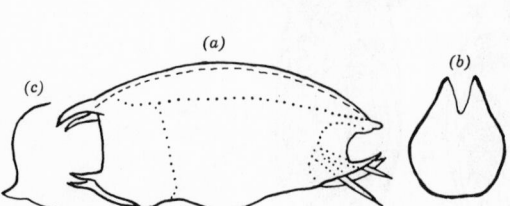

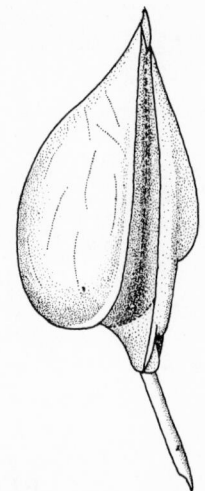

Fig. 18.40. *Mytilina.* (*a*) Lateral view of contracted *M. mucronata.* × 140. (*b*) Optical cross section of same showing dorsal sulcus. (*c*) Lateral view of anterior end of *M. ventralis.* (Original.)

Fig. 18.41. Lateral view of *Lecane aspasia* showing structure of lorica. [After Myers (J.B.).]

42b   Lorica of 1 piece, split longitudinally along dorsum. Ventral plate and dorsolateral plates are firmly fused. (Fig. 18.40). . . . .
                                Mytilina Bory de St. Vincent
Nine species, up to 250 μ. Littoral. Common, especially in hard water. Formerly **Salpina**. Additional information by Wulfert (1939a).

43a   **(41)**   Foot projects through hole in ventral plate near posterior end (Figs. 18.41, 18.42) **44**

43b   Foot projects between plates of lorica (Fig. 18.45*a*) **45**

44a   **(43)**   With 2 toes, separate or barely fused at base. (Figs. 18.41, 18.42). . . . . ***Lecane*** Nitzsch
Over 100 species, littoral, up to 300 μ, most smaller. Very common. No recent monograph exists, but the paper by Harring and Myers (1926) includes many common species. Additional species and information by Ahlstrom (1938), Bartoš (1951b), Bērziņš (1943, 1949b), Carlin (1939), Hauer (1929, 1940, 1952a), Russell (1953), Varga (1945), Yamamoto (1951, 1955). Probably many synonyms have been created because the details of shape of the anterior margin depend somewhat upon the state of contraction (see Donner, 1954). Some synonymy was proposed by Wiszniewski (1954). (See *Monostyla*, **44b**.)

44b   With 1 toe which may be split toward distal end. (Fig. 18.43). . . .
                                  Monostyla Ehrenberg
Nearly 100 species, up to 300 μ, most smaller. Littoral. Very common. Edmondson (1935) proposed to fuse *Monostyla* with *Lecane* because of the great similarity in structure of lorica, and the existence of species of *Lecane* with partly fused toes and *Monostyla* with partly divided toes (Fig. 18.42). *L. elasma* may have the toes completely free or partly fused at the base. The proposal has not met with complete adoption, largely because of the unwieldly nature of the large genus resulting, and the fact that intergrading species are few, so that the names ordinarily have a clearcut meaning. Additional species and information by Bērziņš (1943), Hauer (1952a), Myers (1942), Varga (1945), Wulfert (1939b). See also the comments about *Lecane*, **44a**.

45a   **(43)**   Lateral sulci wide, with stiff flange projecting laterally. (Fig. 18.44)   ***Tripleuchlanis*** Myers
See **Euchlanis** (**46b**).

45b   Lateral sulci without flange **46**

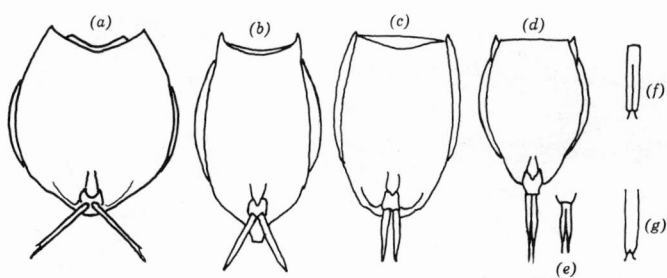

Fig. 18.42. *Lecane*, lorica in ventral view. (*a*) *L. luna*. (*b*) *L. ohioensis*. (*c*) *L. depressa*. (*d*) *L. elasma*, with separate toes. (*e*) *L. elasma*, toes partly fused at base. (*f*) *L. sympoda*, toes. (*g*) *Monostyla furcata*, toe. Various magnifications, about × 200. (*a–d* after Harring and Myers; *e–g* after Hauer.)

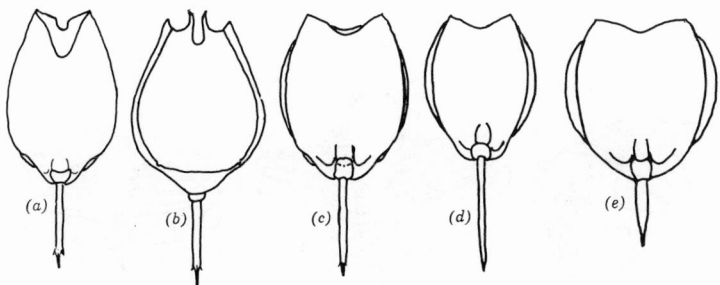

Fig. 18.43. *Monostyla*, lorica in ventral view. (*a*) *M. bulla*. × 180. (*b*) *M. quadridentata*. (*c*) *M. lunaris*. (*d*) *M. lunaris* (= *crenata*). (*e*) *M. closterocerca*. × 225. *b*, dorsal, others ventral. See also Fig. 19.42*f, g*. (After Harring and Myers.)

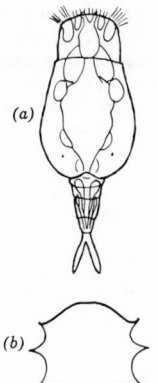

Fig. 18.44. *Tripleuchlanis plicata.* (*a*) Dorsal. × 115. (*b*) Optical cross section. (After Myers.)

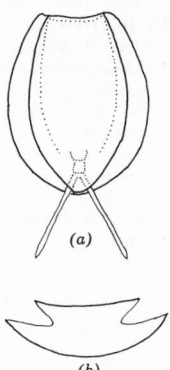

Fig. 18.45. *Dipleuchlanis propatula.* (*a*) Dorsal view. × 140. (*b*) Optical cross section. (After Myers.)

46a (**37, 45**) Dorsal plate flat, narrower than the arched ventral plate. (Fig. 18.45) . . . . *Dipleuchlanis* de Beauchamp
    Two species. Littoral. See *Euchlanis* (**46b**).

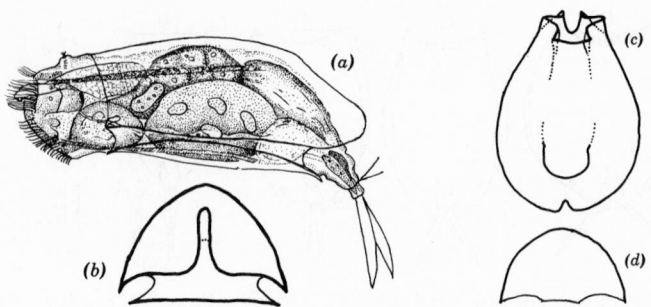

Fig. 18.46. *Euchlanis dilatata.* (*a*) Lateral. × 200. (*b*) Optical cross section, *E.* (*Dapidia*) *calpidia.* × 100. (*c*) Ventral view of lorica. (*d*) Optical cross section. [After Myers (J.B.).]

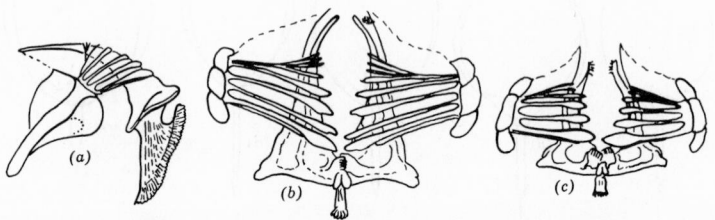

Fig. 18.47. Trophi of *Euchlanis.* (*a*) *E.* (*Dapidia*) *calpidia,* trophi, lateral view. (*b*) Same, frontal view. (*c*) *E.* (*Euchlanis*) *dilatata,* trophi, frontal view. (After Myers.)

46b	Dorsal plate arched, wider than the flat ventral plate. (Figs. 18.46, 18.47). . . . . **Euchlanis** Ehrenberg	

Myers (1930) proposed to reduce the genera **Euchlanis, Dapidia, Dipleuchlanis,** and **Tripleuchlanis** to subgenera of the genus *Euchlanis,* but in writing continued to use the names in the manner of generic names. *Dipleuchlanis* (**46a**) and *Tripleuchlanis* (**45a**) are separated on the basis of the structure of the lorica. There are 2 species of *Dipleuchlanis,* and 1 of *Tripleuchlanis,* in brackish water. *Dapidia* was distinguished by the absence of the minute comb of teeth near the tips of the rami (Fig. 18.47*a*). *Dapidia* and *Euchlanis* each contain species with and without ventral plate and lateral sulci. It is probably preferable to leave *Dipleuchlanis* and *Tripleuchlanis* as genera and combine *Dapidia* with *Euchlanis* as done by Wiszniewski (1954).

There are 18 species. Myers (1930) described all the species likely to be found commonly in the U. S. Additional species and information by Carlin (1939), Wulfert (1939b, 1950, 1956), Hauer (1930). *E. triquetra* is a synonym of *E. incisa.* Common in littoral, but some species may occur in plankton. Length of dorsal plate up to 500 μ.

47a	**(21)**	Foot and toes together longer than rest of body.		**69**
47b		Foot and toes together shorter than rest of body		**48**
48a	**(47)**	Trophi forcipate (Figs. 18.48–18.58) . . Family **Dicranophoridae**		**49**

This point can usually be decided without dissolving out the trophi, by examining the lightly compressed animal in dorsal or ventral view. Most of the members of this group have a cylindric body, simple terminal, oblique, or prone corona (Fig. 18.55), and relatively short toes. Most are active swimmers. Useful compilations by Harring and Myers (1928) and Wulfert (1936).

48b	Trophi otherwise.	**61**

Note: Many of the genera included in couplets **49–92** are extensively treated in the papers of Harring and Myers (1922, 1924, 1928). It is recommended that the plates of these papers be studied before identification of notommatid and dicranophorid

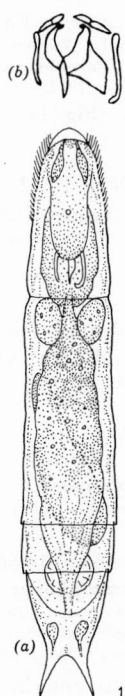

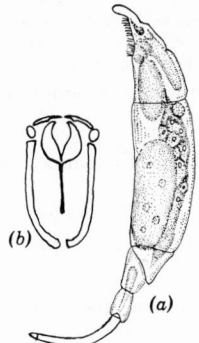

Fig. 18.49. *Wigrella depressa.* (*a*) Lateral. × 290. (*b*) Trophi. [After Wiszniewski (J.B.).]

Fig. 18.48. *Pedipartia gracilis.* (*a*) Dorsal. × 375. (*b*) Trophi, ventral. (After Myers.)

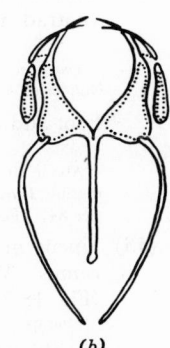

Fig. 18.50. *Wierzejskiella ricciae.* (*a*) Lateral. × 220. (*b*) Trophi. [After Harring and Myers (J.B.).]

rotifers is attempted. In addition, those working in acid-water districts should examine the appropriate papers by Myers.

In this section of the key it is necessary to use the characters of the trophi to separate some of the genera. Some species in these genera may be reliably recognized on the basis of external characters (Fig. 18.87*h*), but external characters cannot be consistently used for sure separation of genera.

49a (40, 48) Left ramus with large, subsquare alula, toes rudimentary. (Fig. 18.48) . . . . **Pedipartia** Myers
  1 species, 200 μ. Psammon and *Sphagnum.*

49b   Without such alula. 50

50a (49) With small pieces intercalated between manubrium and uncus (Fig. 18.51), or with manubrium attached to ramus (Fig. 18.50) . . . 51

50b   Without such pieces, manubrium attached directly to uncus 53

51a (50) Body greatly depressed. (Fig. 18.49) . . . . **Wigrella** Wiszniewski
  Two species. Psammon.

51b   Body nearly cylindric 52

52a (51) Foot more than ¼ of total length. (Fig. 18.50) . . . .
  . . . . **Wierzejskiella** Wiszniewski
  Four species. Mostly in psammon and *Sphagnum.* Up to 350 μ. Includes species formerly placed in **Encentrum.**

52b   Foot much shorter. (Fig. 18.51) . . . . . . ***Encentrum*** Ehrenberg
Over 50 species, many in brackish water; common in littoral. Wiszniewski (1953) removed 5 species with stiff, transversely plicated cuticle to **Parencentrum**. Additional species and information by Donner (1943b), Wulfert (1936, 1939a).

53a **(50)** Joint between uncus and ramus near tips of both (Fig. 18.55) . . .   **54**

53b   Otherwise **57**

54a **(53)** Posterior part of body expanded transversely. (Fig. 18.52). . . . .
Balatro Claparède
Two species, attached to Oligochaetes. Placed in ***Albertia*** by Wiszniewski (1954).

54b   Posterior part of body not expanded **55**

55a **(54)** With 2 toes. **56**

55b   With 1 short conical toe arising from rudimentary foot. (Fig. 18.53) . . . . ***Albertia*** Dujardin
Nine species, 150 μ. Mostly parasitic in or on Oligochaetes. *A. typhylina* is found free. (Harring and Myers, 1922.)

56a **(55)** Two very slender toes arising from rudimentary foot, directed ventrad; ridges around body. Body very stout. (Fig. 18.54). . . .
Paradicranophorus Wiszniewski
Two species, 470 μ, living on bottom, not swimming. Additional information by Donner (1943b), Wulfert (1939a).

56b   Foot and toes otherwise, without furrows. (Fig. 18.55, 18.56) . . .
Dicranophorus Nitzsch
About 60 species, up to 600 μ, mostly smaller. Littoral. Common. **Dorria** and possibly **Itura** will key out here if the modified virgate trophi were considered forcipate (see **84**). Formerly **Diglena**. Additional information by Wulfert (1936, 1939a).

57a **(53)** Uncus hinged at its posterior or lateral end to anterior tip of ramus. With large extra sclerotized pieces attached to tips of unci. (Fig. 18.57a) . . . . ***Streptognatha*** Harring and Myers
One species, 240 μ. Littoral. Rare.

57b   Trophi otherwise. **58**

58a **(57)** Uncus hinged at its posterior or lateral tip to ramus as well as to manubrium, forming a triple joint. (Fig. 18.57b) . . . .
Erignatha Harring and Myers
Seven species, up to 175 μ. Littoral. Fairly common.

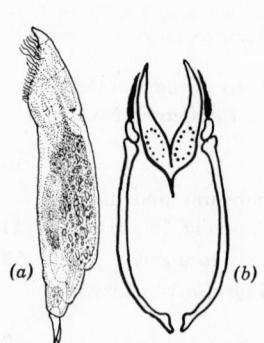

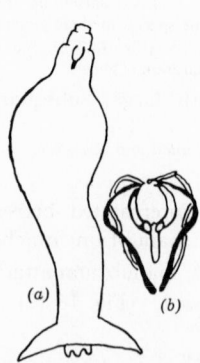

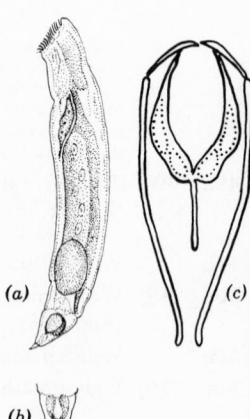

Fig. 18.51. *Encentrum felis.* (*a*) Lateral. × 320. (*b*) Trophi. [After Harring and Myers (J.B.).]

Fig. 18.52. *Balatro.* (*a*) *B. calvus,* dorsal. (*b*) Trophi of *B. anguiformis.* (*a* after Claparède, *b* after Issel.)

Fig. 18.53. *Albertia typhylina.* (*a*) Lateral. × 290. (*b*) Foot, dorsal. (*c*) Trophi. [After Harring and Myers (J.B.).]

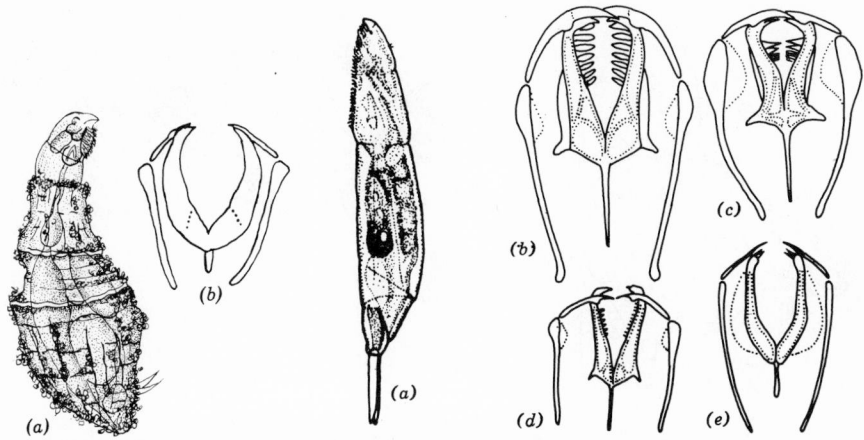

Fig. 18.54. *Paradicranophorus hudsoni.*
(a) Lateral. × 100. (b) Trophi.
[Combined from Wiszniewski and
Wulfert (J.B.).]

Fig. 18.55. *Dicranophorus.* (a) *D. forcipatus,* lateral. × 170.
(b–e) Trophi. (b) *D. forcipatus.* (c) *D. lütkeni.* (d) *D. hercules.*
(e) *D. caudatus.* [After Harring and Myers (J.B.).]

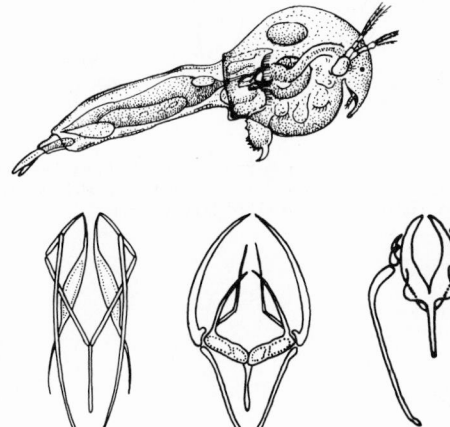

Fig. 18.56. *Dicranophorus isothes* attacking a
Chydorus. Most of the dicranophorids eat only
organisms that are small enough to be pulled in
through the mouth by the trophi. *D. isothes* is an
exception. [After Myers (J.B.).]

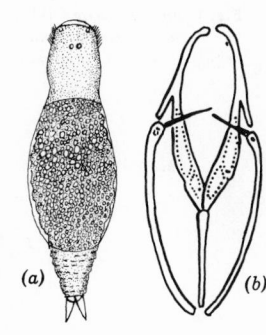

Fig. 18.57. Dicranophorid trophi. (a) *Streptognatha
lepta.* (b) *Erignatha clastopis.* (c) *Aspelta aper.* (After
Harring and Myers.)

Fig. 18.58. *Myersinella tetra-
glena.* (a) Dorsal. (b) Trophi.
[After Wiszniewski (J.B.).]

58b   Uncus otherwise   59

59a   (58) Uncus irregular in shape, attached to ramus near mid-length.
          (Fig. 18.57c) . . . . **Aspelta** Harring and Myers
          Over 15 species, up to 255 μ. Littoral. Additional information and species, Bērziņš
          (1949b), Myers (1942).

59b   Uncus otherwise   60

60a   (59) Uncus a small needle lying across ramus at its mid-length.   (Fig.
          18.58) . . . . **Myersinella** Wiszniewski
          One species, 130 μ. Psammon.

60b   Uncus hinged near its mid-length to tip of ramus. . . . . **Itura**
          Itura, a member of the Notommatidae, will key out here if the trophi were regarded
          as forcipate. Harring and Myers (1928) described the trophi as modified virgate. See
          84b.

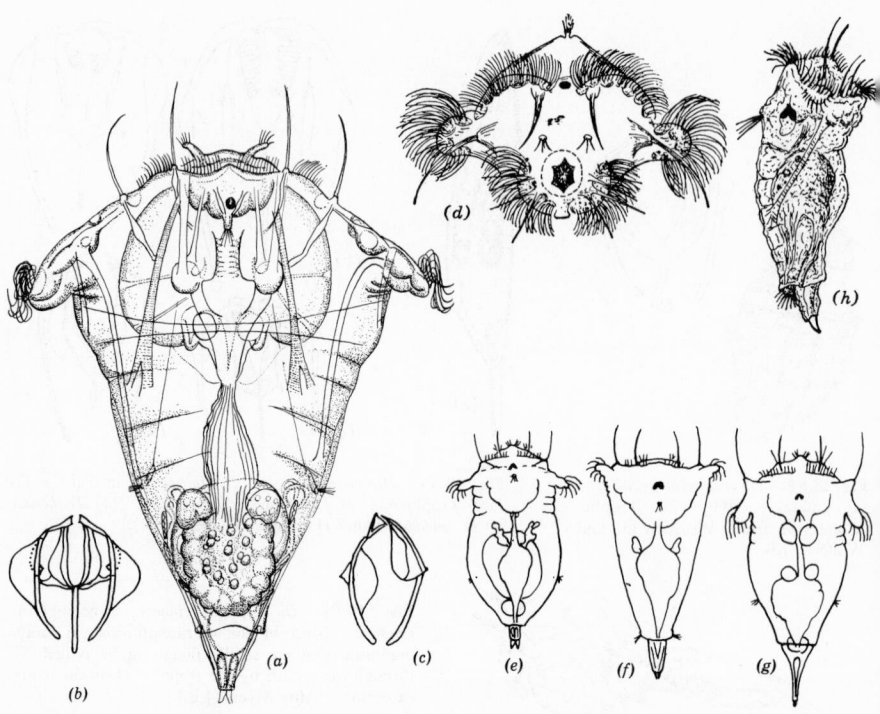

Fig. 18.59. *Synchaeta.* (*a*) *S. pectinata*, dorsal. × 300. (*b*) Same, trophi, ventral. (*c*) Same, trophi, lateral. (*d*) *S. baltica*, frontal, showing corona. (*e*) *S. oblonga.* (*f*) *S. tremula.* (*g*) *S. stylata.* (*h*) Male of *S. tremula.* × 300. (*a* original drawing supplied by E. S. Hollowday; *b–h* modified from Rousselet.)

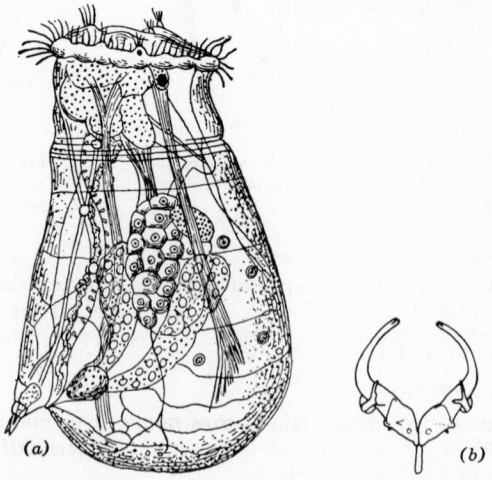

Fig. 18.60. *Asplanchnopus myrmeleo.* (*a*) Lateral. × 80. (*b*) Trophi. (*a* after Weber; *b* original.)

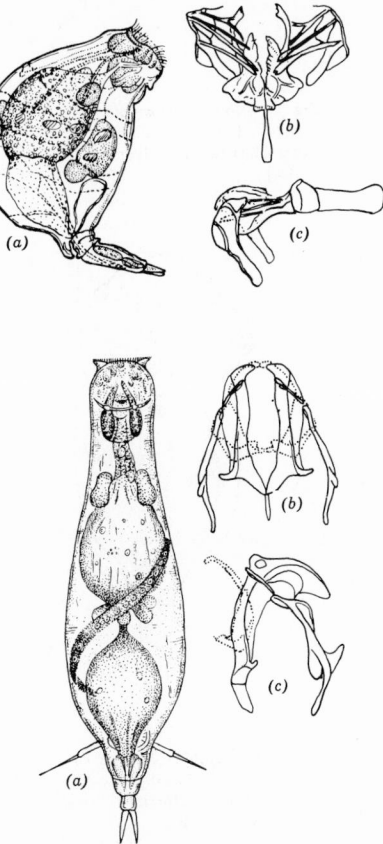

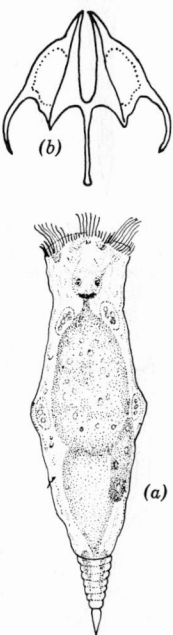

Fig. 18.61. *Harringia rousseleti.* (*a*) Lateral view of somewhat distorted specimen. (*b*) Trophi, frontal. (*c*) Trophi, lateral. [After de Beauchamp (J.B.).]

Fig. 18.63. *Tetrasiphon hydrocora.* (*a*) Dorsal. × 70. (*b*) Trophi, ventral. (*c*) Trophi, lateral. [After Harring and Myers (J.B.).]

Fig. 18.62. *Tylotrocha monopus.* (*a*) Dorsal. × 270. (*b*) Trophi. [After Harring and Myers (J.B.).]

61a (48) Trophi virgate, with prominent V-shaped striated hypopharyngeal muscle, visible in ventral view of intact animal. Corona with auricles and setae. (Fig. 18.59) . . . . **Synchaeta** Ehrenberg
About 30 species. 200–600 μ. Largely planktonic, many in salt water. *S. pectinata* is particularly frequent in lake plankton, but others are often collected. No recent monograph, but Rousselet (1902) described most of the species likely to be collected in fresh water. See Pejler (1957c) for an additional species.

61b Trophi otherwise; if virgate, not with V-shaped muscle 62

62a (61) Trophi incudate (Fig. 18.60), eversible through mouth 63

62b Trophi otherwise. 64

63a (62) Intestine absent. (Fig. 18.60) . . . . . . **Asplanchnopus** duGuerne
Three species, up to 1 mm. Littoral. Not common. Summary of species by Myers (1934b).

63b Intestine present. (Fig. 18.61) . . . . . **Harringia** de Beauchamp
Two species. Littoral. Not common.

64a (62) Trophi very highly modified virgate; parts fused to form triangular dome. (Fig. 18.62) . . . . **Tylotrocha** Harring and Myers
One species, 250 μ. In ponds. Not common.

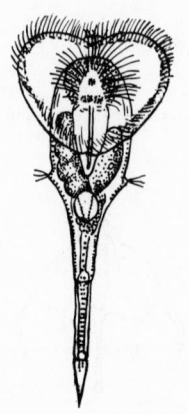

Fig. 18.66. *Monommata grandis*, lateral. × 120. [After Harring and Myers. (J.B.).]

Fig. 18.64. *Microcodon clavus*, ventral. × 250. [Modified from Hyman (J.B.).]

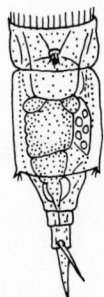

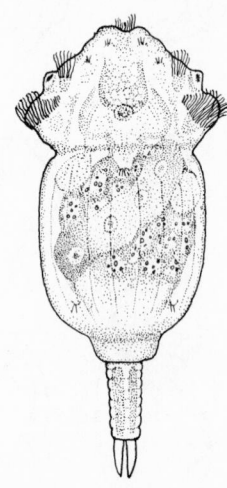

Fig. 18.65. *Mikrocodides chlaena*, dorsal. × 170. [After Weber.]

Fig. 18.67. *Sphyrias lofuana*, dorsal. × 200. [After Harring and Myers (J.B.).]

64b	Trophi otherwise.	**65**
65a	**(64)** With long knobbed lateral antenna near foot, 2 dorsal antennae, and highly modified virgate trophi. (Fig. 18.63)	

Tetrasiphon Ehrenberg

One species, up to 1000 μ. Fairly common in acid water ponds.

65b	Otherwise	**66**
66a	**(65)** With 1 toe (may have additional dorsal spur on foot)	**67**
66b	With 2 toes.	**69**
67a	**(66)** Foot about half of total length, slender, jointed. With purple plates just anterior to mastax. (Fig. 18.64)	

Microcodon Ehrenberg

One species, 200 μ. Mostly littoral, but may be found in plankton.

67b	Foot shorter than half of total length; without such plates.	**68**
68a	**(67)** With prominent spur on dorsal side of foot just anterior to toe, and plications in cuticle of dorsum. (Fig. 18.65)	

Mikrocodides Bergendal

One or 2 species, 250 μ. Littoral. Misspelled by Wiszniewski (1954). Additional species by Rodewald 1940.

68b	Otherwise	**70**

Some species of *Proales* and *Pleurotrocha* have 1 toe, although the majority have 2 (see 82 and 92).

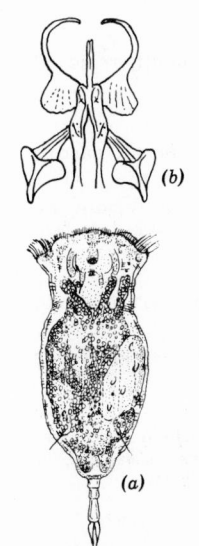
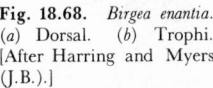
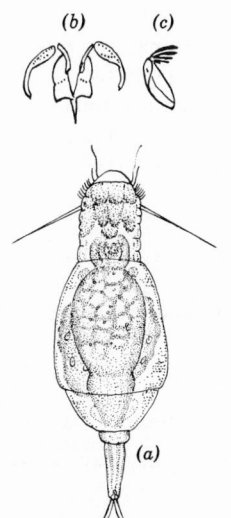
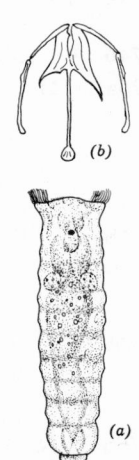

Fig. 18.68. *Birgea enantia.*
(a) Dorsal. (b) Trophi.
[After Harring and Myers
(J.B.).]

Fig. 18.69. *Bryceela
tenella.* (a) Dorsal.
× 160. (b) Trophi.
(c) Malleus, oblique.
[After Wiszniewski
(J.B.).]

Fig. 18.70. *Taphro-
campta annulosa.* (a)
Dorsal. × 250. (b)
Trophi. [After Har-
ring and Myers (J.B.).]

69a (47, 66) Toes longer than rest of body. (Fig. 18.66) . . . .
Monommata Bartsch
Over 20 species, 250–700 μ. Littoral. Most easily recognized by the long, unequal toes, but 1 species has equal toes. Formerly **Furcularia** in part. Additional species and information by Bērziņš (1949a), Myers (1937b), Varga (1954). The weak lorica of **Scaridium** may be overlooked. See **28b**.

69b Toes shorter than rest of body. 70

70a (68, 69) Head broader than body, separated from it by groove. Corona consists of several rows of cilia on bulging surface. Two eyes in lateral protuberances on corona. Trophi modified. (Fig. 18.67) . .
Sphyrias Harring
One species, 300 μ. Littoral. Eats rotifers. Easily distinguishable by corona and shape of body.

70b Otherwise 71

71a (70) Body barrel-shaped, with very narrow cylindric foot set off from body. Trophi highly modified. (Fig. 18.68) . . . .
Birgea Harring and Myers
One species, 275 μ. Littoral. Stomach with zoochlorellae, no gastric glands.

71b Otherwise 72

72a (71) Corona a ventral oval disc with long cirri resembling those on hypotrich protozoa. With 2 long styli directed toward posterior. Head much depressed, set off by constriction from body. (Fig. 18.69) . . . . **Bryceela** Remane
Three species, 130 μ. Littoral. Review by Rodewald (1935). Additional information by Wulfert (1940).

72b Otherwise 73

73a (72) Body annulated. Foot rudimentary. Small toes directed ventrad. (Fig. 18.70) . . . . **Taphrocampta** Gosse
Three species, all described by Harring and Myers (1924). 200 μ. Littoral.

73b   Otherwise **74**

74a (73) Corona a simple circumapical ring of cilia with mouth removed some distance toward posterior **75**

74b   Mouth not removed from corona **76**

75a (74) Corona on a more or less cylindric, retractible prominence, narrow relative to body. Manubria expanded at posterior tips. (Fig. 18.71) . . . . *Drilophaga* Vejdowsky

Three species, only 1 from N. A., up to 275 μ. The European species have been reported as ectoparasites on Oligochaetes. The North American species is apparently free-living. Review by Pawlowski (1934).

75b   Corona wider, nearly as wide as body. Posterior part of body widened. Manubria not expanded. (Fig. 18.72) . . . .

Wulfertia Donner

One species, not yet reported from N. A.

76a (74) Stomach with large band-shaped, forked gastric glands and 2 pairs of other attached structures. (Fig. 18.73) . . *Enteroplea* Ehrenberg

One species, 600 μ. Littoral. Remane proposed to include *Pseudoharringia*, but was not followed by Wiszniewski (1954). See **92b**. Formerly *Triphylus*.

76b   Otherwise **77**

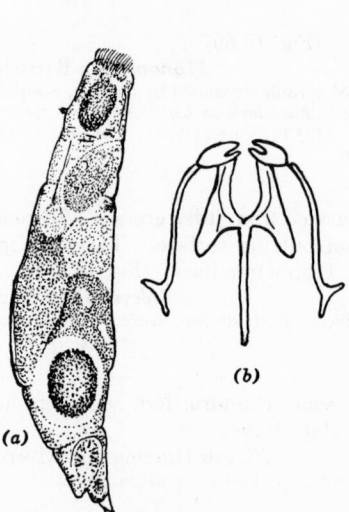

Fig. 18.71. *Drilophaga judayi.* (a) Lateral. × 170. (b) Trophi. [After Harring and Myers (J.B.).]

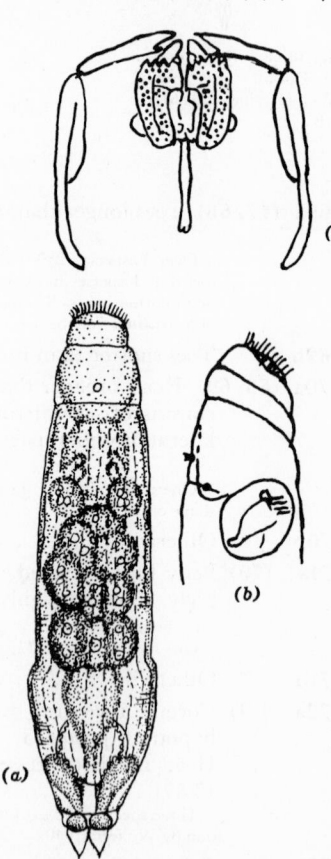

Fig. 18.72. *Wulfertia ornata.* (a) Dorsal. × 320. (b) Anterior end, lateral. (c) Trophi. [After Donner (J.B.).]

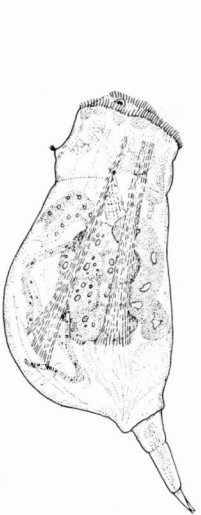

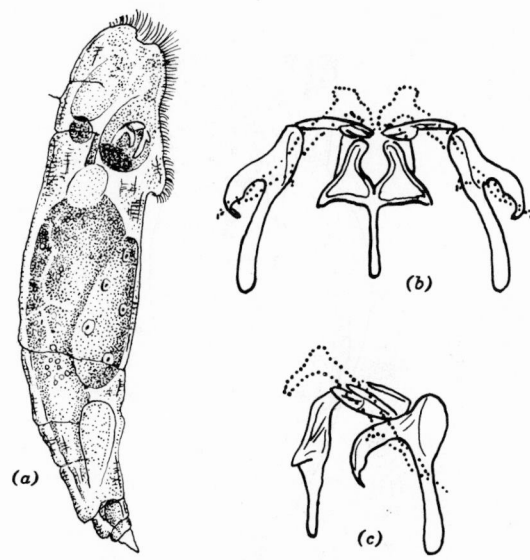

(a)

(b)

(c)

Fig. 18.73. *Enteroplea lacustris*, lateral. × 150. [After Harring and Myers (J.B.).]

Fig. 18.74. *Lindia truncata*. (a) Lateral. × 150. (b) Trophi, ventral. (c) Trophi, lateral. [After Harring and Myers (J.B.).]

77a (76) Trophi cardate (determinable in lateral view of intact animal). Body spindle-shaped, toes directed ventrad. (Fig. 18.74) . . . . . .
Lindia Dujardin

About 18 species, 300–500 μ. Mostly littoral. Two species removed to *Halolindia*. See also Běrziņš (1949a), Donner (1954), Russell (1947).

77b Otherwise 78

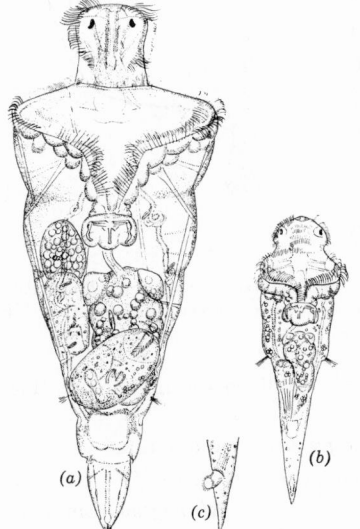

(a)

(b)

(c)

Fig. 18.75. *Rhinoglena frontalis*. (a) Ventral. × 200. (b) Male. (c) Penis. (Original supplied by E. S. Hollowday.)

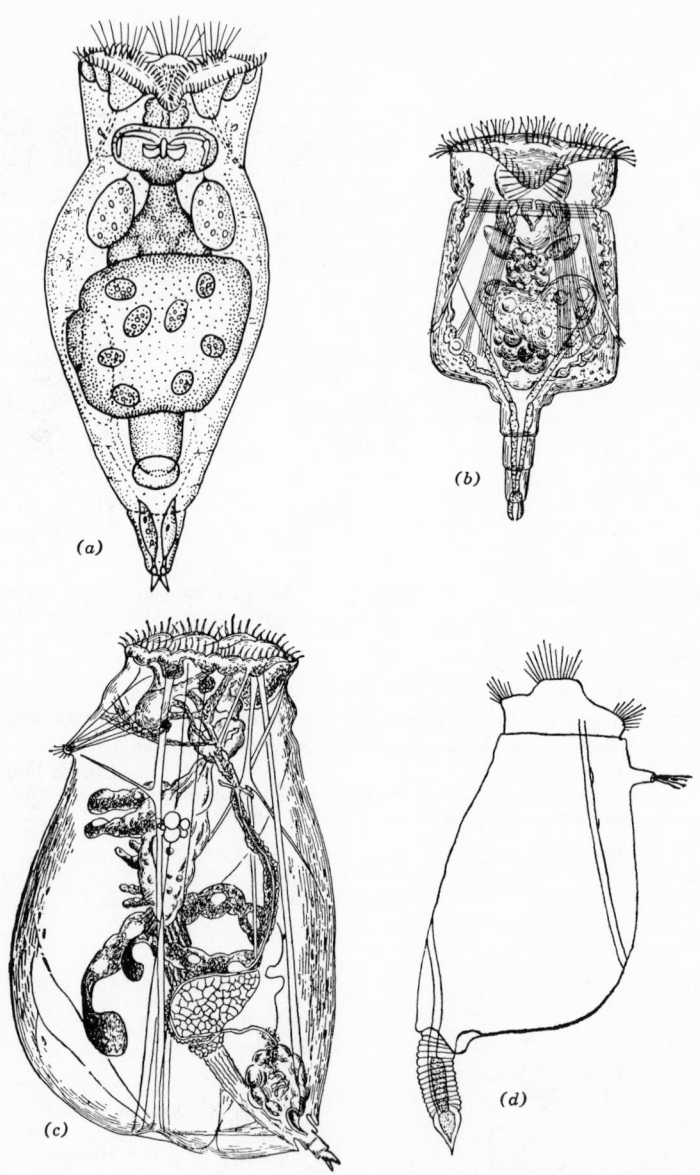

Fig. 18.76. *Epiphanes.* (*a*) *E. senta,* ventral. × 150. (*b*) *E. brachionus,* ventral. (*c*) *E. clavulata,* lateral. (*d*) *E. macroura* (=*mollis*), lateral. (*a* original; *b* after Weber; *c* after Hudson and Gosse; *d* after Hempel.)

78a (77) Trophi clearly malleate or weakly modified toward virgate (Figs. 18.78, 18.79). 79

78b   Trophi virgate, or a nonmalleate modification of virgate 83

79a (78) Corona with proboscis containing 2 eyes. (Fig. 18.75) . . . .
                                                    Rhinoglena Ehrenberg
            Two species, 300 μ. Littoral. Formerly ***Rhinops.***

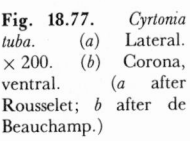

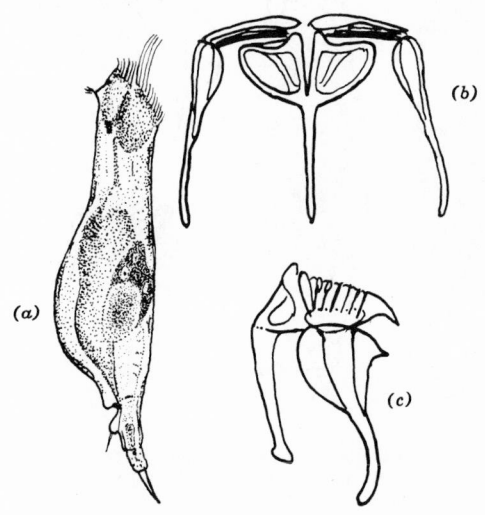

Fig. 18.77. *Cyrtonia tuba.* (a) Lateral. × 200. (b) Corona, ventral. (a after Rousselet; b after de Beauchamp.)

Fig. 18.78. *Proalinopsis caudatus.* (a) Lateral. × 180. (b) Trophi, ventral. (c) Trophi, lateral. [After Harring and Myers (J.B.).]

79b   Corona without such a proboscis 80

80a (79) Corona of *Epiphanes* type (Figs. 18.76, 18.77) 81

80b   Corona otherwise 82

81a (80) Body conic, cylindric, or sacform; toes short. (Fig. 18.76) . . . . .
                                                      Epiphanes Ehrenberg

        Five species, 500 μ. Littoral. Found commonly in pools receiving drainage from cow barns. *E. senta (Hydatina)* is one of the most widely known rotifers because of its use in elementary texts, but is by no means the commonest. The species of the genus formerly were distributed among **Hydatina, Notops,** and **Brachionus.**

81b   Body arched, its longitudinal axis sigmoid, tapering to foot with long, slender toes. (Fig. 18.77). . . . . **Cyrtonia** Rousselet
        One species, 360 μ. Littoral. Often attaches to plants by thread spun out from toes. Cilia of corona unusually long.

82a (80) With setate papilla just posterior to anus. (Fig. 18.78) . . . .
                                                  Proalinopsis Weber

        Seven species, 200 μ. Littoral. Fairly common. Trophi have some virgate characteristics.

82b   Without such papilla. (Fig. 18.79) . . . . **Proales** Gosse
        About 40 species, 200–400 μ. Littoral. Some act as scavengers, eating out the contents of dead cladocera and insect larvae. Additional species and information by Běrzinš (1949c), Donner (1954), Edmondson (1948), Myers (1942), Wulfert (1939a, 1940, 1950). Compilation of common species by Harring and Myers (1922, 1924). See also the related **Wulfertia (75b).**

        Cordylosoma perlucidum, which has not been found since its discovery by Voigt, might come to this part of the key. The corona apparently is different from any described type, but the trophi were not observed. (See "Rotifers of uncertain position," p. 438, Fig. 18.125a,b.)

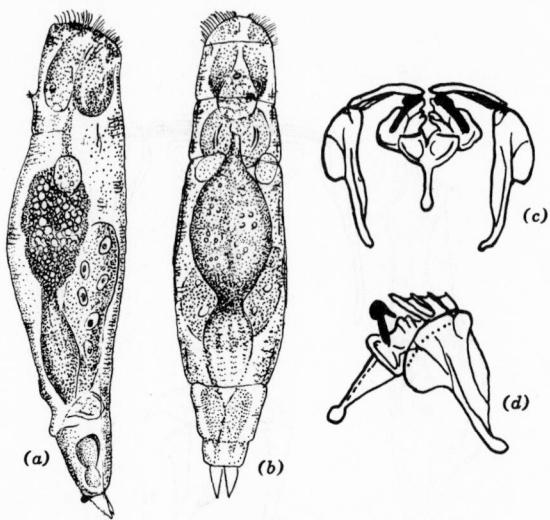

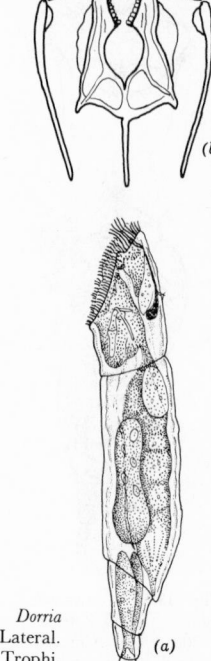

Fig. 18.79. *Proales.* (*a*) *P. fallaciosa*, lateral. × 145. (*b*) *P. decipiens*, dorsal. (*c*) *P. decipiens*, trophi, ventral. (*d*) Same, lateral. The epipharyngeal pieces are shown in solid black. [After Harring and Myers (J.B.).]

Fig. 18.80. *Dorria dalecarlica.* (*a*) Lateral. × 300. (*b*) Trophi. [After Myers (J.B.).]

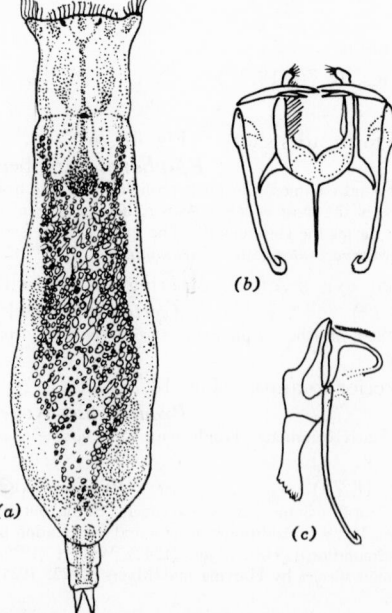

Fig. 18.81. *Itura aurita.* (*a*) Dorsal. × 260. (*b*) Trophi, ventral. (*c*) Trophi, lateral. [After Harring and Myers (J.B.).]

Fig. 18.82. *Dorystoma caudata.* (*a*) Lateral. × 225. (*b*) Trophi, lateral. [After Voigt (J.B.).]

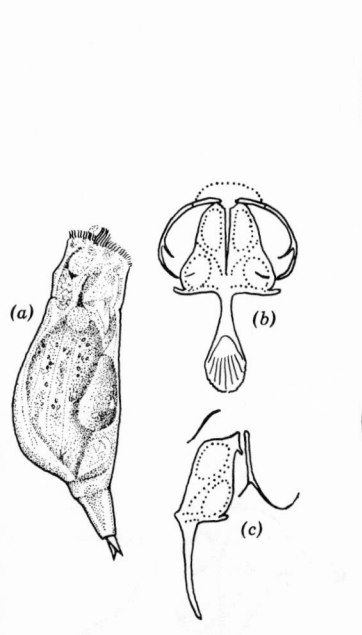
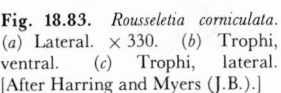
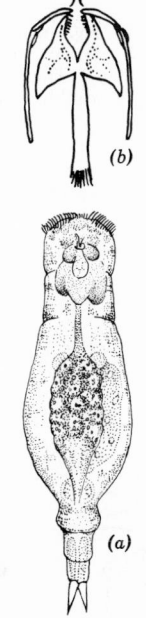
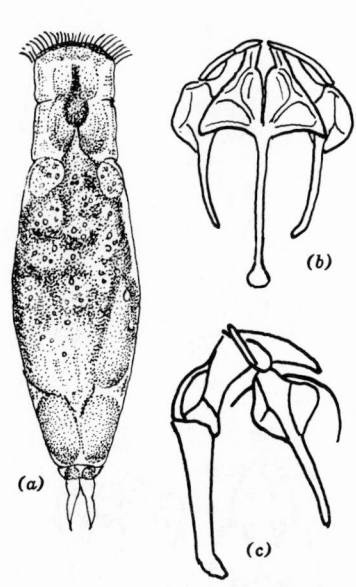

Fig. 18.83. *Rousseletia corniculata.* (*a*) Lateral. × 330. (*b*) Trophi, ventral. (*c*) Trophi, lateral. [After Harring and Myers (J.B.).]

Fig. 18.84. *Eothinia elongata.* (*a*) Dorsal. × 140. (*b*) Trophi. [After Harring and Myers (J.B.).]

Fig. 18.85. *Resticula melandocus.* (*a*) Dorsal. × 170. (*b*) Trophi, ventral. (*c*) Trophi, lateral. [After Harring and Myers (J.B.).]

83a (78) Trophi virgate, strongly modified toward forcipate type. **84**

83b   Trophi otherwise. **85**

84a (83) With gastric glands. Toes short, blunt. Foot glands unusually long, reaching forward into body cavity. (Fig. 18.80) . . . .
                                                                                         Dorria Myers
           One species, 200 μ. In moss in rapidly flowing brooks.

84b   Without gastric glands, but with anterior cecae instead. With 2 frontal and 1 cerebral eyes. (Fig. 18.81). . . . .
                                                                              Itura Harring and Myers
           Six species, up to 400 μ. Littoral. Stomach with zoochlorellae. Additional information by Donner (1954), Wulfert (1935).

85a (83) Manubrium forming a nearly circular plate with small handle projecting at posterior margin. With prominent supra-anal spine. (Fig. 18.82) . . . . ***Dorystoma*** Harring and Myers
           Two species. Littoral. Not known from N. A. See Wulfert (1939b).

85b   Otherwise **86**

86a (85) Uncus absent. Manubrium a slender forked rod. (Fig. 18.83) . .
                                                                              Rousseletia Harring
           One species, 130 μ. Littoral.

86b   Uncus present. Manubrium otherwise. **87**

87a (86) Inner margin of both rami finely toothed, symmetric. (Fig. 18.84) . . . . ***Eothinia*** Harring and Myers
           Nine species, 400 μ. Littoral. Additional species by Bērziņš (1949c), Myers (1933c).

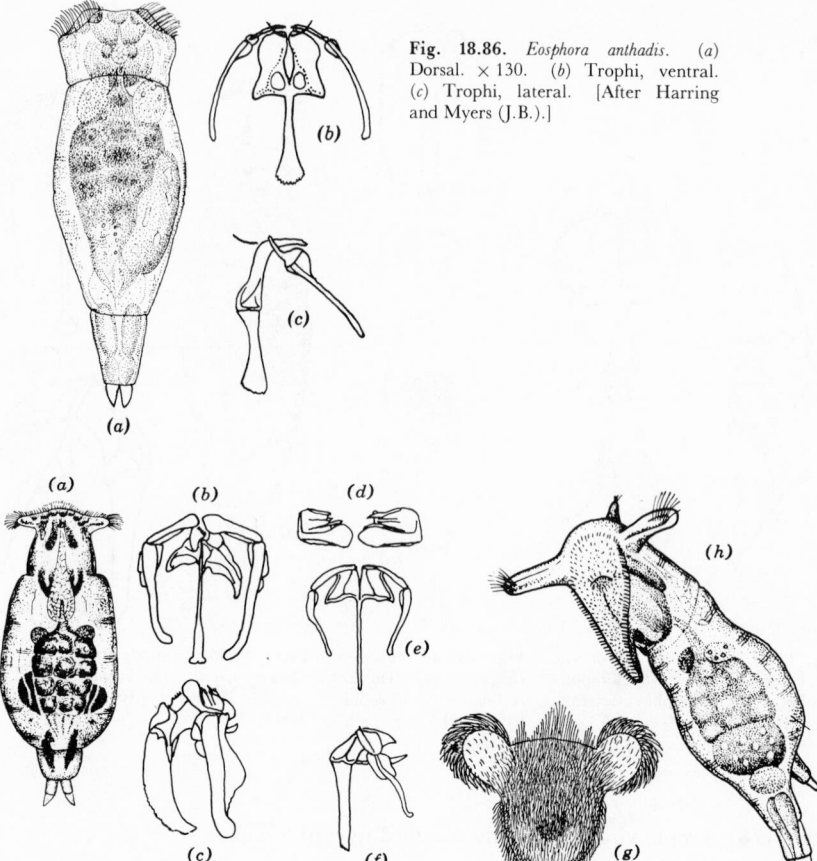

Fig. 18.86. *Eosphora anthadis.* (*a*) Dorsal. × 130. (*b*) Trophi, ventral. (*c*) Trophi, lateral. [After Harring and Myers (J.B.).]

Fig. 18.87. *Notommata pachyura:* (*a*) Dorsal. × 75. (*b*) Trophi, ventral. (*c*) Trophi, lateral. (*d*) Unci, frontal. *N. pseudocerbus:* (*e*) Trophi, ventral. (*f*) Trophi, lateral. (*g*) *N. aurita,* corona. (*h*) *N. copeus,* oblique view. × 60. [*a* combined from various sources (J.B.); *b–f* after Harring and Myers; *g* after Wesenberg-Lund; *h* based on photograph of glass model in the American Museum of Natural History (J.B.).]

87b   Otherwise **88**

88a (87) Rami with a nearly right angle bend at mid-length (Figs. 18.85*c*, 18.86*c*) **89**

88b   Otherwise **90**

89a (88) Uncus has 1 principal tooth with 1 to 5 preuncial teeth. Retrocerebral organ consisting of sac alone, no glands. (Fig. 18.85) . . .
                                                          Resticula Harring and Myers
                Seven species, up to 400 μ. Littoral. See Wulfert (1935, 1939b).

89b   Uncus has 1 principal tooth without preuncial teeth. Fulcrum short, broad. Both sac and glands present. (Fig. 18.1, 18.86) . . .
                                                          Eosphora Ehrenberg
                Seven species, 500 μ. Littoral. See Wulfert (1935).

90a (88) Rami strongly arched, hemispheric plates **91**

90b   Rami otherwise **92**

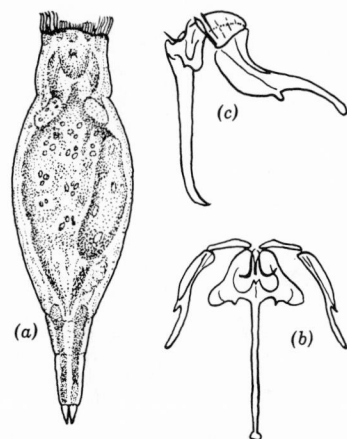

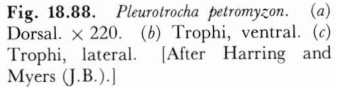

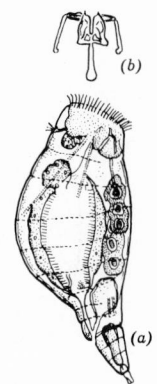

Fig. 18.88. *Pleurotrocha petromyzon.* (*a*) Dorsal. × 220. (*b*) Trophi, ventral. (*c*) Trophi, lateral. [After Harring and Myers (J.B.).]

Fig. 18.89. *Pseudoharringia similis.* (*a*) Lateral. (*b*) Trophi. [After Fadeew (J.B.).]

91a (90) Uncus consists simply of rodlike tooth. (Fig. 18.38) . . . .
Cephalodella Bory de St. Vincent
Most species of the genus have a soft lorica. See **41b.**

91b Uncus a platelike structure with 1 or 2 principal teeth and in some cases small accessory teeth. (Fig. 18.87) . **Notommata** Ehrenberg
About 50 species, 200–1000 μ. Littoral. A large and varied genus. Remane (1933) defined a number of species groups. Additional species and information by Bērziņš (1949c), Donner (1944, 1954), Myers (1942), Wulfert (1935, 1939a,b, 1940, 1950). Most of the species of **Copeus,** no longer recognized, are now included in *Notommata.*

92a (90) Manubrium with small ventral projection. (Fig. 18.88) . . . . . .
Pleurotrocha Ehrenberg
Twelve species, 250 μ. Littoral.

92b Manubrium without such projection. Uncus with 5 teeth. (Fig. 18.89) . . . . **Pseudoharringia** Fadeew
Two species. Littoral. Not known from N. A. Incompletely described. See **Enteroplea, 76a.**

93a (20) Digestive system of Collotheca type (Fig. 18.102). Trophi incudate . . . . Order **Collothecaceae** 107
Corona of most with long setae, marginally ciliated or bare in but very few of the species in this group. Most of these animals are sessile.

93b Digestive system otherwise. Trophi malleoramate (Fig. 18.95). Corona always marginally ciliated . . . . . Order **Flosculariaceae** 94

94a (93) With distinct lorica. Retractible annulated foot ends in ciliated cup. 2 frontal eyes. (Fig. 18.90) . . . .
Testudinella Bory de St. Vincent
About 25 species, up to 250 μ. Littoral. The species are distinguished on the basis of shape, size, and position of foot opening, position of lateral antennae, and outline of lorica in dorsal view and cross section. Foot opening ventral in most species, terminal in some. The commonest hard-water species is *T. patina,* circular in dorsal view. *T. caeca* may attach itself temporarily and form a rough tube of debris. Formerly **Pterodina.** Additional species and information by Bartoš (1951b), Carlin (1939), Gillard (1947, 1952), Myers (1942), Yamomoto (1951).

94b Without lorica 95

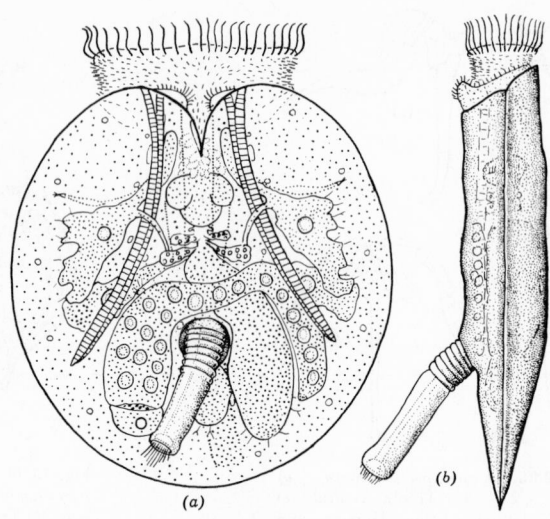

Fig. 18.90. *Testudinella patina.* (*a*) Ventral. × 300. (*b*) Lateral. [Combined from various sources, chiefly Seehaus (J.B.).]

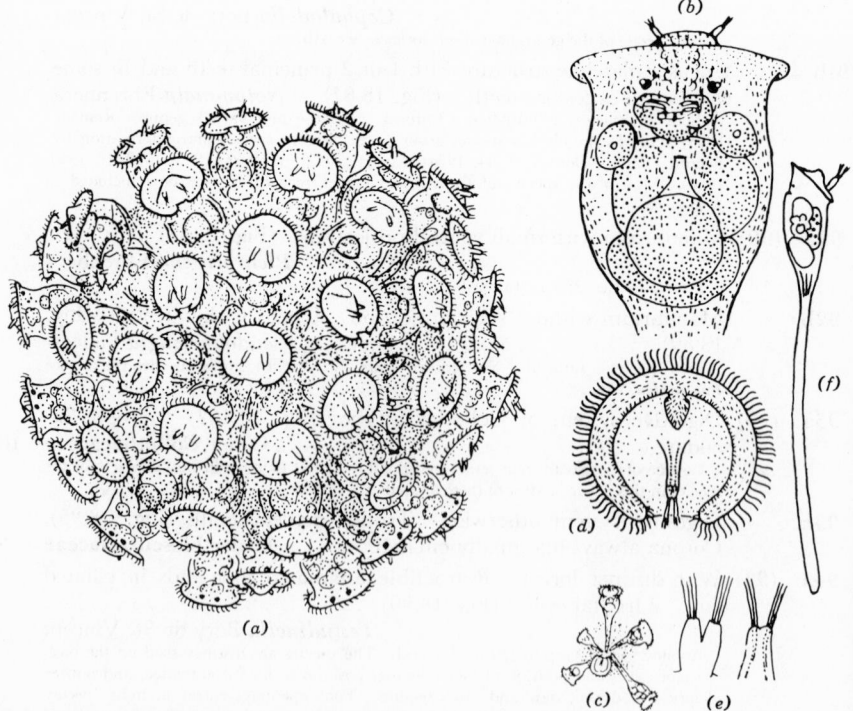

Fig. 18.91. *Conochilus.* (*a*) Colony of *C. hippocrepis.* × 25. (*b*) *C. hippocrepis*, dorsal view of anterior part, cilia omitted. × 100. (*c*) Colony of *C. unicornis.* (*d*) *C. unicornis*, frontal view of corona. (*e*) Variations in structure of antennae of *C. unicornis.* (*f*) *C. unicornis* (= *norvegicus*). × 44. (*a–d* original; *e*, *f* after Burckhardt.)

95a (94) One or 2 cylindric (lateral) antennae within coronal disc (Fig. 18.2*d*). (Fig. 18.91) . . . . **Conochilus** Hlava

Two species are generally recognized: *C. hippocrepis* (Schrank) with 2 antennae, forming colonies of about 100 animals (Fig. 18.91*a*), common in ponds; and *C. unicornis* with 1 antenna, representing 2 fused, with a relatively shorter foot, forming colonies of only a few individuals (Fig. 18.91*c*), and common in lake plankton. Burckhardt (1943) designated a very large unicorn form (1300 μ) with greatly elongated foot as *C. norvegicus*, but this might represent a case of gigantism and allometric distortion in a population adapted to low temperature. The total length of *hippocrepis* varies up to about 800 μ and of *unicornis* up to 450 μ. Possibly several species are combined in *unicornis*. Sometimes unicorn animals with *hippocrepis* body shape are found in large colonies. The structure of the antenna varies in different populations (Fig. 18.91*e*). *C. hippocrepis* is often found with unequal antennae, the left being as small as half the length of the right. The stomach is partially divided into 2 unequal chambers by an asymmetrically placed paramedian partition (Edmondson, 1940). This species may accumulate in great numbers in shadows of lily pads and sticks in quiet ponds. Detailed discussion of possibility of hybridization by Pejler (1956).

95b   Antennae on body, posterior to corona **96**

96a (95) Lateral antennae elongate, close together on ventral surface of body, or fused for part of length. Corona of *Conochilus* type (Fig. 18.2*c*). (Fig. 18.92) . . . . **Conochiloides** Hlava

Four species, planktonic. Ordinarily solitary, but young may attach to tube of mother, forming small groups. Ahlstrom (1938) figured the trophi of all.

96b   Antennae otherwise; may be long or short, but located on sides of body, not close together on venter. Corona of *Ptygura* type (Fig. 18.2*b*). **97**

Most of these animals are sessile, and of the free-swimming ones, most are colonial. Only 4 species swim solitarily.

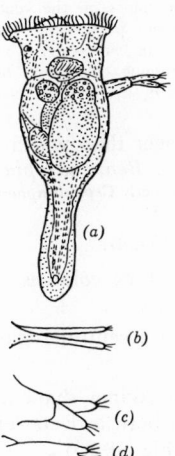

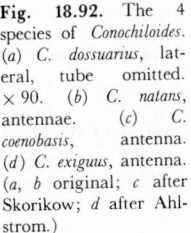

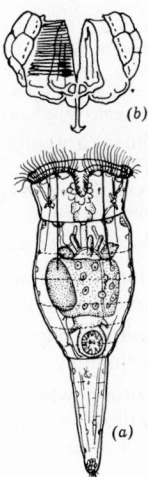

Fig. 18.92. The 4 species of *Conochiloides*. (*a*) *C. dossuarius*, lateral, tube omitted. × 90. (*b*) *C. natans*, antennae. (*c*) *C. coenobasis*, antenna. (*d*) *C. exiguus*, antenna. (*a, b* original; *c* after Skorikow; *d* after Ahlstrom.)

Fig. 18.93. *Voronkowia mirabilis.* (*a*) *Ventral.* (*b*) Trophi, left uncus omitted. [After Fadeew (J.B.).]

97a (96) Animals free-swimming, alone or in colonies.   98
97b   Animals attached to substrate.   101
98a (97) Animals solitary, not forming definite colonies.   99
98b   Animals forming spheric or ellipsoidal colonies, ordinarily of many
           individuals.   100
99a (98) Foot ends in tuft of cilia. (Fig. 18.93) . . . . *Voronkowia* Fadeew

           A speculation by F. J. Myers that *Voronkowia* is really the free-swimming larva of
           Octotrocha speciosa has been quoted in the literature (Wulfert 1939a, Donner, 1949).
           This identity is made unlikely by the difference in trophi and the fact that Fadeew's
           figure shows an egg maturing (see **105b** and Fig. 18.98). Likewise, it is probably not
           a larval **Sinantherina** (Voigt, 1957). Solitary, free-swimming larvae of Flos-
           culariidae will key out here, but in general they may be distinguished by the relatively
           much larger size of the mastax, as compared to the body width. *Voronkowia* seems
           closest to *Ptygura*, but there are anatomical differences. Bogoslowski (1958) has pre-
           sented evidence that **Voronkowia** is the larva that hatches from the resting egg,
           Sinantherina socialis.

99b   Foot ends otherwise. (Fig. 18.95) . . . . *Ptygura* Ehrenberg

           Over 20 species, up to 500 μ, of which 3 may swim as adults. See **103a** for sessile
           species. Key by Edmondson (1949). *Pseudoecistes rotifer* should be included in
           the genus. Formerly **Oecistes.**

100a (98) Colony in a gelatinous matrix. (Fig. 18.99). . . . .

                                              Lacinularia Schweigger

           Nine species, of which 2 form free-swimming colonies. Of these 2 only *L. ismailo-
           viensis* has been reported from N. A. Over 2 mm in diameter. Easily distinguishable
           from *Conochilus* by the position of antennae, and from *Sinantherina* by presence of
           gelatinous matrix. Frequent in hard water. See **106a** for sessile species.

100b   Colony not in gelatinous matrix. (Fig. 18.100) . . . .

                                      Sinantherina Bory de St. Vincent

           Five species, of which 2 form free-swimming colonies as adults. Of these, 2, *S.
           spinosa* is distinguished by the presence of many fine spines on the venter, and *S.
           semibullata* by 2 dark wartlike structures on the anterior part of the venter. (See
           Canella, 1952). Over 2 mm in diameter. Frequent in hard water. See **106b** for
           sessile species. Formerly **Megalotrocha.** Free-swimming colonies of larvae of *S.
           socialis* will key out here. They may be distinguished by the presence of 4 warts on
           the venter; the larvae do not carry eggs.

101a (97) Dorsal antenna a prominent cylinder, longer than width of body.
           (Fig. 18.94) . . . . *Beauchampia* Harring

           One species, up to 500 μ, usually much smaller. Formerly **Cephalosiphon.** See also
           Donner (1954).

101b   Dorsal antenna at most a short cylinder or knob.   102
102a (101) Animals solitary, or at most forming branching colonies by young
           attaching to tubes of old animals 103
102b   Animals regularly colonial, radiating from a central region of at-
           tachment. 106
103a (102) Corona when fully expanded circular or kidney-shaped, with a
           shallow ventral indentation, but not with both a deep ventral in-
           dentation and a wide, deep dorsal gap. (Fig. 18.95) . . . .

                                              Ptygura Ehrenberg

           Over 20 species, of which most are sessile. Nearly all make definite tubes which are
           usually gelatinous, sometimes with a fibrous structure. One species (*P. pilula*) supple-
           ments its tube with fecal pellets. Most species about 250 μ, some up to 1 mm. Key
           by Edmondson (1949). See also **Lacinularia** (**106a**). Formerly **Oecistes.** See also
           Běrzinš (1949b, 1950b), Donner (1954), Wright (1957).

103b   Corona definitely lobed   104
104a (103) Corona with 2 lobes formed by wide dorsal gap and deep ventral
           depression. (Fig. 18.96) . . . . *Limnias* Schrank

           Six species, all sessile. May be solitary or in branching groups. May be over 1 mm
           long. Tube-building described by Wright (1954).

104b   Corona with more than 2 lobes.   105

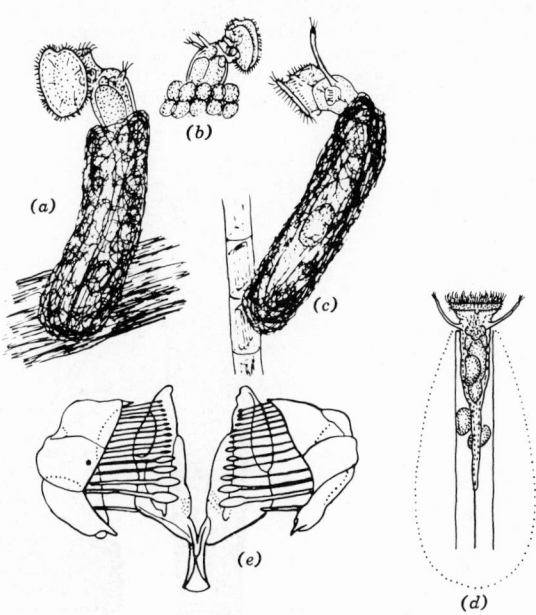

(b)

(a)

(c)

(e)

(d)

Fig. 18.94. *Beauchampia crucigera,* lateral. × 110. (Original.)

Fig. 18.95. *Ptygura.* (*a*) *P. crystallina,* oblique. × 65. (*b*) *P. pilula,* oblique. (*c*) *P. longicornis bispicata,* lateral. × 125. (*d*) The planktonic *P. libera,* ventral. × 100. (*e*) *P.* (= *Pseudoecistes*) *rotifer,* trophi. (*a–c, e* original; *d* after Myers.)

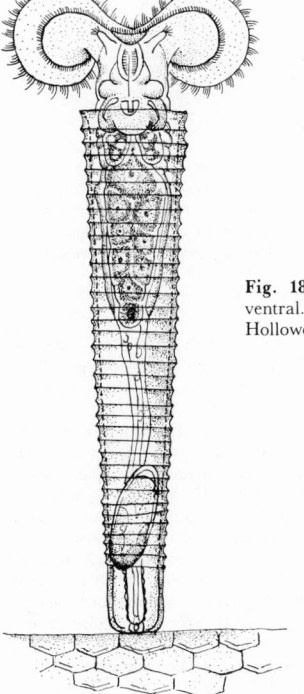

Fig. 18.96. *Limnias melicerta,* ventral. × 60. [Modified from Hollowday (J.B.).]

105a (104) Corona with 4 lobes. (Fig. 18.97) . . . . ***Floscularia*** Cuvier

Six species, all sessile; 5 make brown tubes of pellets. Of these, 4 species use pellets made in a special structure on the head, the other uses fecal pellets. Tube-building of *F. ringens* described by Wright (1950). May be solitary or in branching colonies. May be over 1 mm long. Formerly ***Melicerta.***

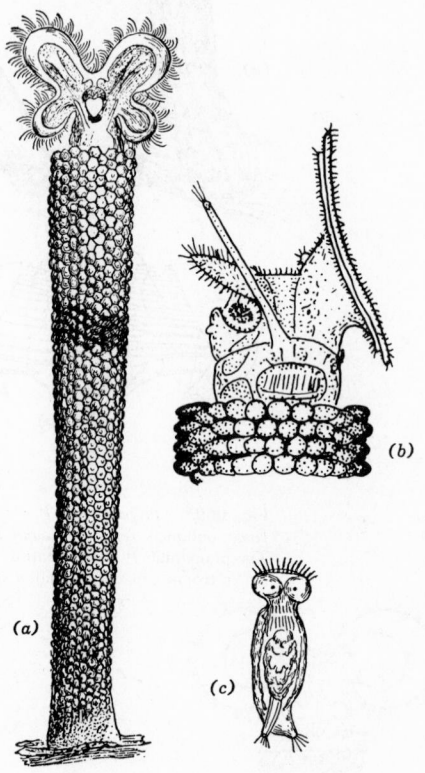

(b)

(a)

(c)

Fig. 18.97. *Floscularia.* (a) *F. ringens,* ventral view of animal in its tube. × 60. (b) *F. conifera,* lateral view showing chin and pellet cup. (c) Male. (a after Hudson and Gosse; b original; c after Weber.)

105b   With more than 4 lobes. (Fig. 18.98) . . . . . ***Octotrocha*** Thorpe

One species, *O. speciosa,* was described from China, and has been reported from a number of localities in the United States. However, there is some doubt about the specific and even the generic identity of the American form with Thorpe's. His description states that there is a wide dorsal gap (Fig. 18.98a), but the American form has a small, distinct, dorsal lobe in place of the gap (Fig. 18.98b). Thorpe's figure gives the distinct impression of being made from a partly contracted specimen, and it is possible that the small dorsal lobe was missed. In anterior view, the lobe is not seen. The American form would be more properly described as having 3 pairs of lobes than 4, but specimens have been observed to stand with the large ventral lobes partly folded in such a way as to give the appearance of 2, as in Thorpe's drawing. The trophi of the American form (Fig. 18.98e), while clearly malleoramate, are unique in the reduction of the small teeth and the enlargement of the large ones. Thorpe's figure of trophi also shows 3 large teeth. It differs in detail, but he evidently drew a crushed preparation, and small details cannot be trusted. The animal is sessile; either solitary, or crowded together in large groups when common. May be confused with *Lacinularia* if the corona is not observed.

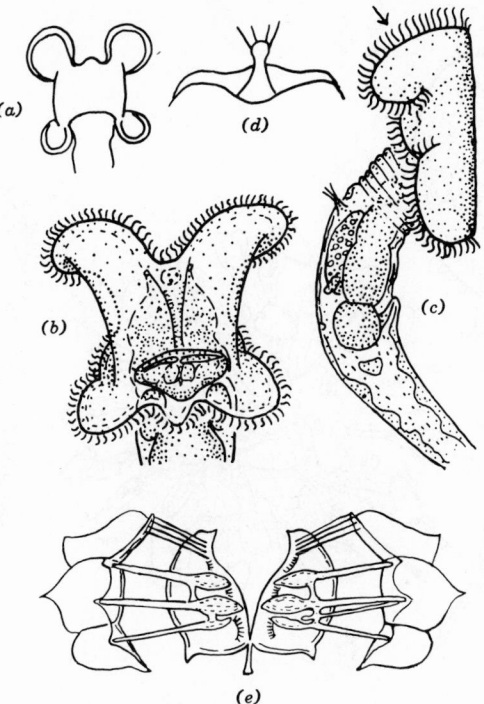

Fig. 18.98. *Octotrocha.* (*a*) *O. speciosa,* dorsal view of corona of specimen from China. The rest are figures of American specimens. (*b*) Dorsal view of corona. × 90. (*c*) Lateral. (*d*) Ventral view of contracted animal showing dorsal antenna. (*e*) Trophi. (*a* from Thorpe; *b–d* original; *e* unpublished original by F. J. Myers.)

106a (102) Colony in a gelatinous matrix. (Fig. 18.99) . . . .
Lacinularia Schweigger

Nine species, of which 6 form sessile colonies, attaching either to the substrate directly or to a penduncle secreted by the rotifers. *L. flosculosa* is the only sessile species yet reported from N. A. It occasionally is found as a solitary individual and may be confused with *Ptygura*. It may also be briefly confused with *Octotrocha* (**105b**). See **100a** for free-swimming species.

106b Colony not in a gelatinous matrix. (Fig. 18.100) . . . .
Sinantherina Bory de St. Vincent

Five species, of which 3 form sessile colonies. Eggs are attached to specialized structure on foot (oviferon) found only in this genus. The most common species, *S. socialis,* has 4 dark, ovate, wartlike structures on the ventral anterior surface. (Figures by Wulfert, 1939a). *S. procera,* not yet reported from N. A., also has 4 warts, but has a much longer and more slender foot, and the corona is of a different shape from that in Fig. 18.100*a. S. ariprepes* has an oviferon consisting of 2 humps, and no warts. See **100b** for free-swimming species. Formerly *Megalotrocha.* See also Edmondson (1939, 1940).

The larvae of *S. socialis* aggregate into a spheric, free-swimming colony. After a period of exploring substrates, the colony disbands, and the individuals swim around in contact with the substrate. They then reaggregate and attach close together (Surface, 1906). Thus the members of a colony are of nearly the same age. Colonies of *S. ariprepes* apparently grow by accretion; aggregates of free-swimming larvae have not been seen. (Edmondson, 1939).

107a (93) Margin of corona furnished either with long, very fine setae, or,

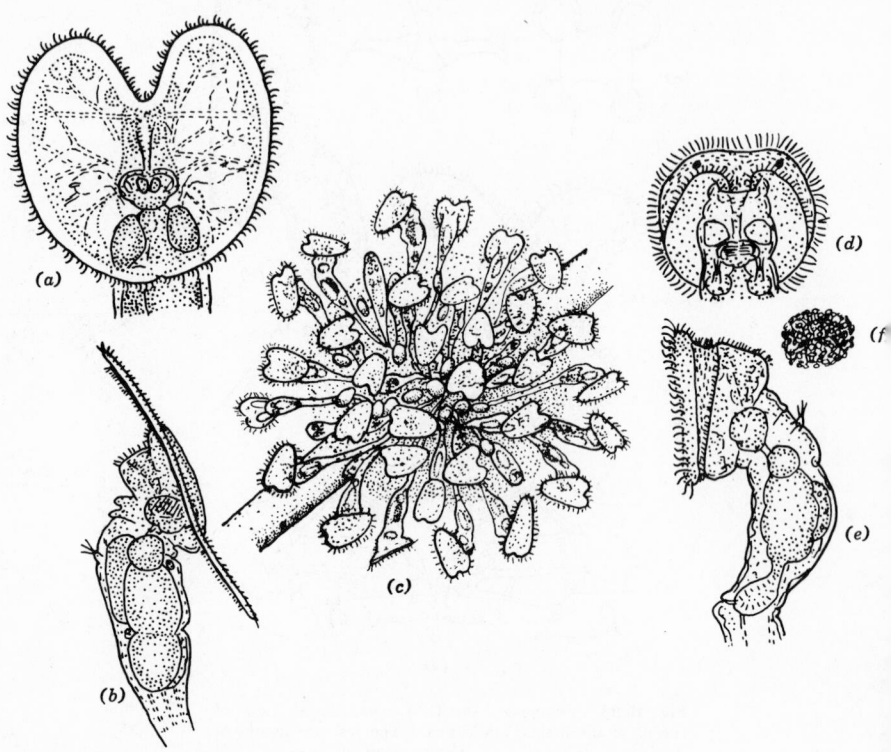

Fig. 18.99. The 2 species of *Lacinularia* known from North America. *L. flosculosa:* (*a*) Corona, dorsal. × 60. (*b*) Lateral, foot omitted. (*c*) Colony. *L. ismailoviensis:* (*d*) Corona, ventral. × 230. (*e*) Lateral, foot omitted. (*f*) Colony. × 12. [*a, b, d–f* original; *c* modified from Hudson and Gosse (J.B.); *d* and *e* based on fixed specimens.]

	in a few cases, ciliated (Figs. 18.101, 18.102). (If bare, margin not lobed.) Free-swimming larvae of Collothecids will appear here . . .	**108**
107b	Margin of corona bare and with at least 1 lobe	**109**
108a	**(107)** Lobes of corona about as long as body, and setae arranged in whorls. (Fig. 18.101) . . . . **Stephanoceros** Ehrenberg	

One species, over 1 mm. See also *Collotheca*, **108b.**

108b   Lobes of corona short, or corona unlobed. If lobes elongated, setae not arranged in whorls. (Fig. 18.102). . **Collotheca** Harring

Over 50 species, mostly sessile, in clear, gelatinous tubes. Five species are free-swimming and may be found in lake plankton. There is question about the proper generic position of *C. millsii* and *evansoni*, which have very elongate lobes but whose setae are arranged differently from those of *Stephanoceros fimbriatus*. Key to species by Bērziņš (1951). See also Donner, 1954. Formerly **Floscularia**, but that name must be used for the genus once known as **Melicerta**. **Hyalocephalus trilobus** Lucks probably belongs in this genus.

109a **(107)** Corona with long, flexible dorsal lobe. Foot cylindric. (Figs. 18.103, 18.100) . . . . **Acyclus** Leidy

One species, living attached in colonies of *Sinantherina*, the larvae of which it eats. Over 1 mm.

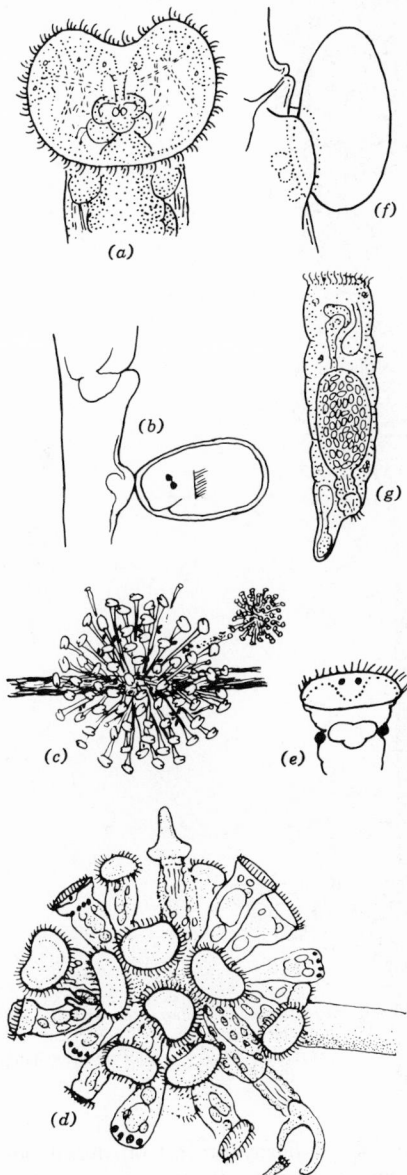

Fig. 18.100. *Sinantherina.* *S. socialis:* (*a*) Corona, dorsal. × 50. (*b*) Part of foot with oviferon and attached egg, lateral. (*c*) Colony of adults with colony of larvae leaving. (*d*) Small colony, with two *Acyclus inquietus* (see **109a**). × 15. (*e*) Anterior end of larva showing lateral warts and relative size of mastax. *S. ariprepes.* (*f*) Oviferon and attached egg, lateral. (*g*) Male, lateral. × 115. [Original (*c*, after surface J.B.).]

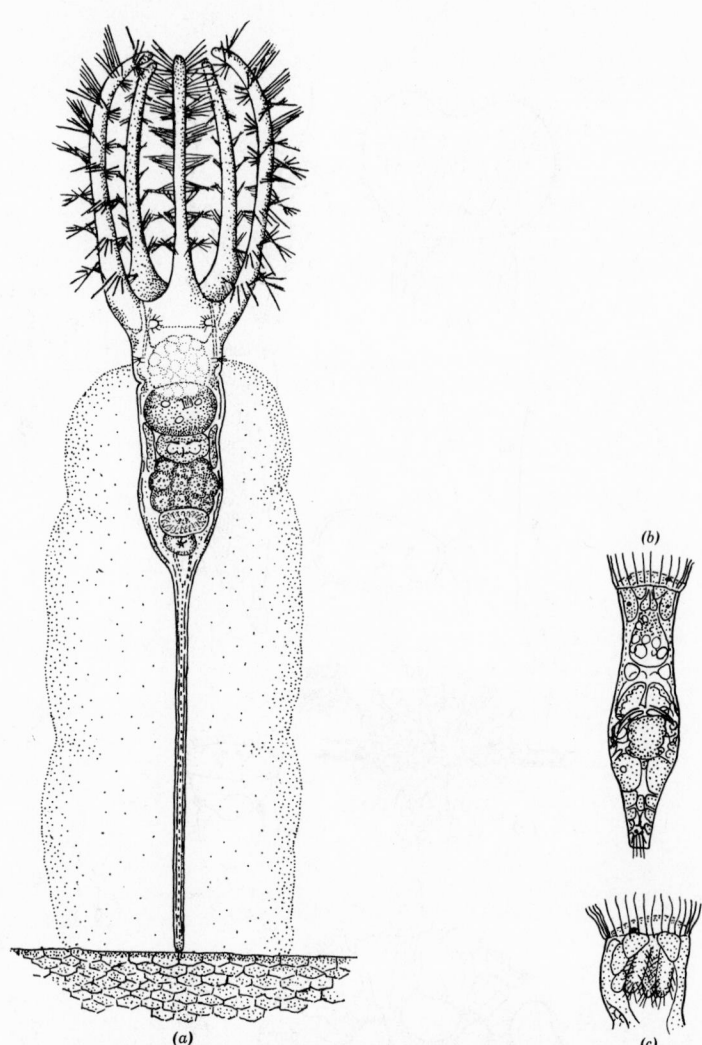

Fig. 18.101. *Stephanoceros fimbriatus.* (*a*) Dorsal. × 80.
(*b*) Free-swimming larva, dorsal. (*c*) Somewhat older
larva, lateral. [*a* original (J.B.); *b*, *c* after de Beauchamp.]

109b   Corona with short, horny process on dorsal side. Foot reduced
to a hemispheric projection at the posterior end. (Fig. 18.104) ⋮ .
                                                    Atrochus Wierzejski

One species, 1500 μ. Lying on bottom or on plants.

110   **Bdelloidea**

The bdelloids are among the most difficult rotifers to investigate. Much of the
identification of genera and species is based on coronal characters which can be seen
only when the animal is attached and feeding, with corona expanded. They may
spend much time crawling with the corona contracted or swimming. They are
difficult to fix extended well enough for identification, although sometimes the hot

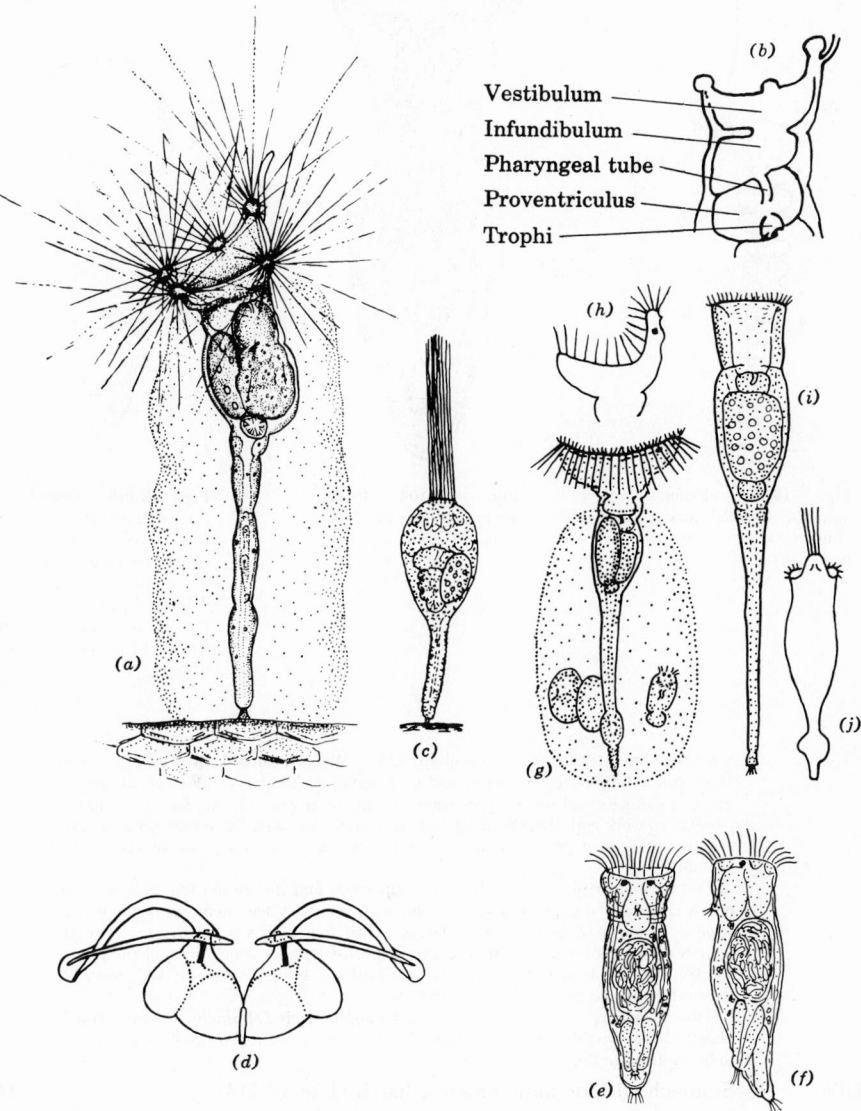

Vestibulum
Infundibulum
Pharyngeal tube
Proventriculus
Trophi

Fig. 18.102. *Collotheca.* (*a*) *C. ornata* var. *cornuta,* oblique. × 130. Note the short peduncle at tip of foot. (*b*) Diagram of anterior part of digestive system of *Collotheca.* (*c*) Contracted *Collotheca* showing bundle of setae and the invisibility of coronal characteristics in a contracted specimen. (*d*) Trophi of *C. judayi.* Subunci are shown in solid black. (*e*) Male of *C. gracilipes,* dorsal. (*f*) Same, lateral. × 370. The rest of the figures show the planktonic species reported from North America. (*g*) *C. mutabilis* in swimming position, lateral. Note hatching egg. × 135. (*h*) Corona of *C. mutabilis* in feeding position, the animal lying motionless in the water. (*i*) *C. pelagica,* ventral. × 180. (*j*) *C. libera,* dorsal. × 200. [*a–c, e–i* original (*a* J.B.); *d* after an unpublished drawing by F. J. Myers; *j* after Zacharias.]

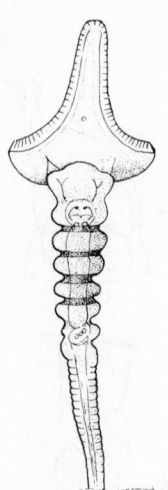 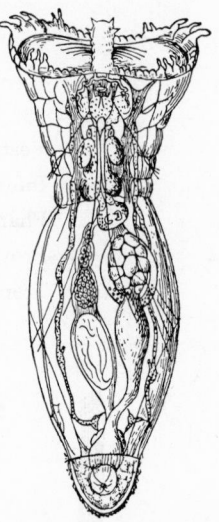 

Fig. 18.103. *Acyclus inquietus*, dorsal, tube omitted. × 75. [Original (J.B.).]

Fig. 18.104. *Atrochus tentaculatus*, dorsal. × 35. (After Wierzejski.)

Fig. 18.105. Two contracted bdelloids, to show that many taxonomic characters cannot be seen in contracted material. The two projections from the lower animal are spurs, not toes. (Original.)

water method is strikingly successful (p. 433). When disturbed, they generally contract, concealing specific, generic, and even familial characters. (Fig. 18.105). The group is distinguished by the possession of ramate trophi. It requires considerable patience to work with these animals, but since there are some very interesting systematic and biological problems, the effort is rewarding (see for example Hsu, 1955, 1956a,b).

The vast majority of species live in damp moss, and the species that will be commonly encountered in ponds and lakes will be a relatively few members of the genera *Rotaria*, *Philodina*, *Dissotrocha*, and *Adineta*. The following key to the group relies heavily on the comprehensive monograph of Bartoš (1951c) which keys all the known species. See also Hauer (1939) and Dobers (1915). It is necessary to distinguish between the toes and pedal spurs (Fig. 18.113).

Two marine epibionts are omitted, *Anomopus* and **Zelinkiella**. **Synkentronia** Hauer (1938) is epibiontic on fresh-water arthropods, but is too insufficiently known to be included in the key (Fig. 18.124).

| 110a | (1) | Stomach with definite lumen, ciliated (Fig. 18.114) . . . . | 111 |

| 110b | | Stomach without lumen, the syncitial mass containing food vacuoles, may have a frothy appearance (Fig. 18.110) . . . . |

Family **Habrotrochidae** 113

This fact was discovered by Burger (1948) in one species of *Habrotrocha*. It is assumed that the entire family lacks a lumen. The previous definition, based on misinterpretation, stated that the lumen of the stomach was wide, and that food was compressed into pellets.

| 111a | (110) | Corona with 2 separate trochal circles on pedicels (Figs. 18.2*i*, 18.2*j*, 18.116) . . . . Family **Philodinidae** | 116 |

| 111b | | Corona otherwise . . . . | 112 |

In 1 genus, the corona is bilobed, but without upper and lower lips (Fig. 18.119). In others the "wheel" appearance is completely lacking.

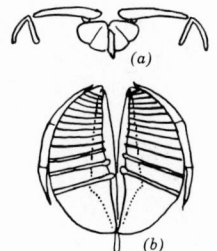

Fig. 18.106. Ramate
trophi. (a) Cross sec-
tion. (b) Frontal.
(After de Beauchamp.)

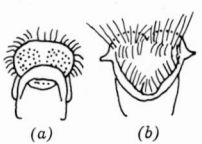

Fig. 18.107. *Scepano-
trocha* corona. (a) *S.
rubra.* (b) *S. corniculata.*
(After Bryce.)

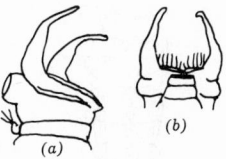

Fig. 18.108. *Ceratotrocha
cornigera,* anterior end.
(a) Lateral. (b) Dorsal.
(After Murray.)

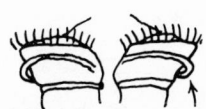

Fig. 18.109. *Otoste-
phanos auriculatus,* co-
rona, ventral. The
arrow points to the
shelf-like membrane.
(After Milne.)

112a	**(111)**	Corona reduced to small ciliated areas near mouth and on ros-trum. Pharynx short. Trophi partly exsertible for chewing . . . . Family **Philodinavidae** **123**
112b		Corona 2 flat ciliated areas on ventral side of head, separated by narrow longitudinal groove (Fig. 18.122) . . . Family **Adinetidae** **125**
113a	**(110)**	Corona with prominent outgrowths of dorsal or ventral lip **114**
113b		Corona without such outgrowths **115**
114a	**(113)**	Corona with upper lip forming a wide, transparent lobe covering corona. (Fig. 18.107). . . . . ***Scepanotrocha*** Bryce Six species, in moss.
114b		Corona with corners of lower lip drawn out into 2 long horns. (Fig. 18.108) . . . . ***Ceratotrocha*** Bryce Three species, up to 350 μ. In moss, especially *Sphagnum*.
115a	**(113)**	Sides of coronol pedicels with shelflike membrane. (Fig. 18.109) ***Otostephanos*** Milne Seven species, about 400 μ. In drying mosses. Not yet reported from N. A.
115b		Without such thickenings. (Fig. 18.110) . . . ***Habrotrocha*** Bryce Over 100 species, 300 μ. Most common in submerged moss, but occurring in dry-ing moss. Many secrete flask-shaped capsules in which they live. Includes species formerly known as ***Callidina.***
116a	**(111)**	Foot with toes **117**
116b		Foot without toes, with attachment disc. (Fig. 18.111) . . . . ***Mniobia*** Bryce Over 40 species, up to 1000 μ, most smaller. In drying mosses, some epizotic.
117a	**(116)**	Foot with 2 toes. (Fig. 18.112) . . . . . . ***Didymodactylus*** Milne One species. Not yet reported from N. A.
117b		Foot with 3 or 4 toes. **118**
118a	**(117)**	Foot with 3 toes **119**
118b		Foot with 4 toes . . . . . . : **120**

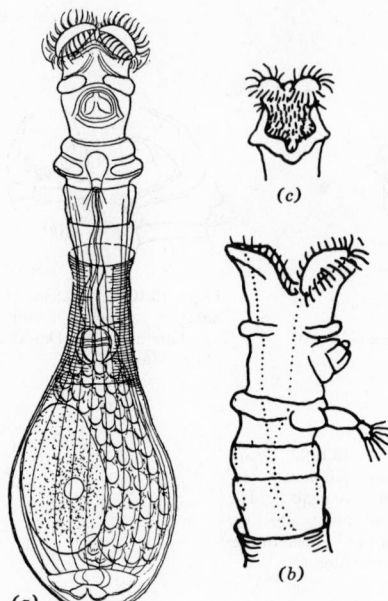

Fig. 18.111. *Mniobia circinnata*, spurs and attachment disc. (After Murray.)

Fig. 18.110. *Habrotrocha*: (*a*) *H. angusticollis*, ventral view, in tube. × 325. Note food vacuoles in stomach. (*b*) Same, corona in lateral view. (*c*) Corona of *H. spicula*, ventral. × 400. (*a*, *b* after Murray; *c* after Bryce.)

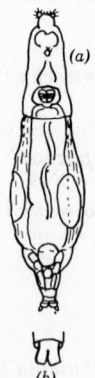

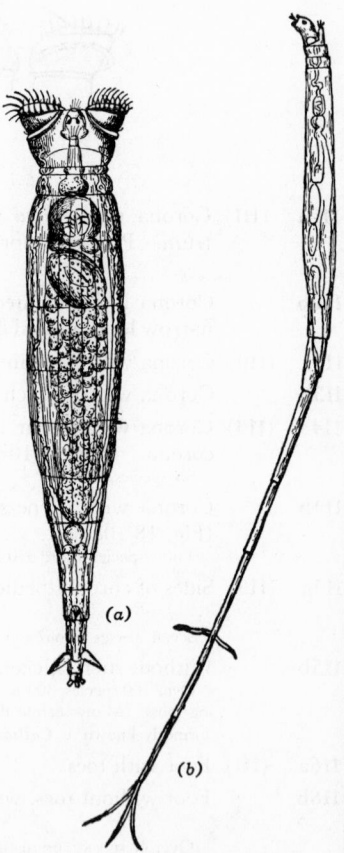

Fig. 18.112. *Didymodactylus carnosus*. (*a*) Dorsal, showing spurs; Toes retracted. (*b*) Toes. (After Milne.)

Fig. 18.113. (*a*) *Rotaria citrinus*, dorsal. × 170. (*b*) *R. neptunia*, the most elongate rotifer known, lateral. × 100. In both figures note the 3 terminal toes and the 2 spurs on the foot. (After Weber.)

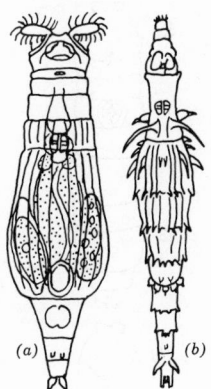

Fig. 18.114. (a) Macro-trachela quadricornifera, dorsal. × 140. (b) M. multispinosa, dorsal. (After Bartoš.)

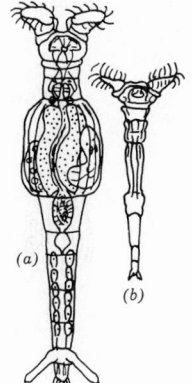

Fig. 18.115. (a) Emabata commensalis, dorsal. × 90. (b) E. laticeps. (a after Bartoš, b after Murray.)

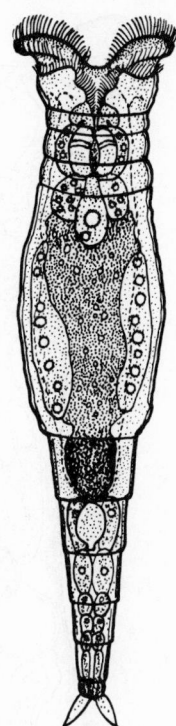

Fig. 18.116. Philodina roseola, ventral. × 230. Note the 2 spurs; toes are not shown. [After Hickernell (J.B.).]

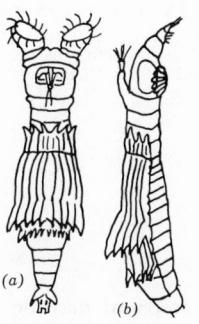

Fig. 18.118. Pleuretra brycei. (a) Dorsal, feeding. (b) Lateral, crawling. (After Bar-toš.)

Fig. 18.120. Henoceros falcatus, head, ventral. (After Milne.)

Fig. 18.117. Dissotrocha. D. aculeata: (a) Lateral view of crawling animal. (b) Spurs. D. macrostyla: (c) Lateral view of end of foot. (After Weber.)

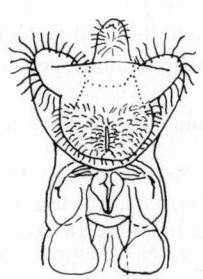

Fig. 18.119. Abrochtha intermedia, head, ven-tral. × 250.

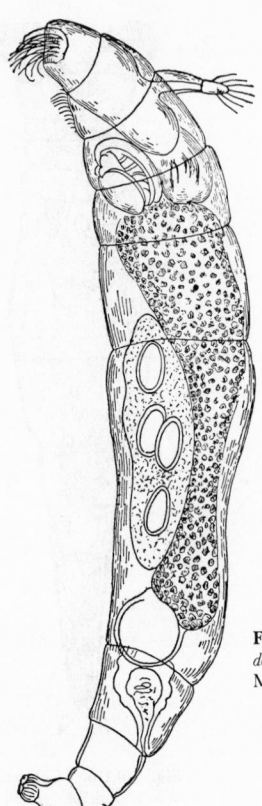

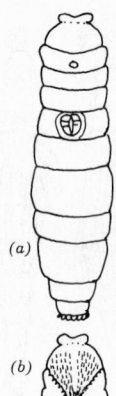

Fig. 18.122. *Adineta vaga*, ventral. × 80. (After Hudson and Gosse.)

Fig. 18.123. *Bradyscela clauda*. (*a*) Dorsal. (*b*) Corona. (After Bryce.)

Fig. 18.121. *Philodinavus paradoxus*, lateral × 400. (After Murray.)

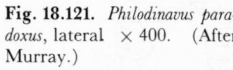

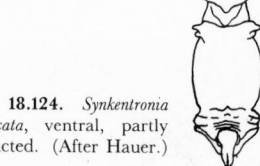

Fig. 18.124. *Synkentronia complicata*, ventral, partly contracted. (After Hauer.)

119a (118) Eyes, if present, in rostrum; may be absent. Viviparous. (Fig. 18.113). . . . . ***Rotaria*** Scopoli
Over 20 species, up to 1500 μ, most much smaller. Littoral, some epizotic. *Rotaria rotatoria* commonly becomes abundant in laboratory cultures. Formerly **Rotifer.**

119b Eyes always absent. Oviparous. (Fig. 18.114 . . . .
Macrotrachela Milne
Nearly 70 species, up to 600 μ, most smaller. Mostly in drying mosses. Some species with elaborate cuticular spination.
The species of *Rotaria*, as compared with those of *Macrotrachela*, tend to have elongate spurs and toes. Only in *Macrotrachela* is there a development of elaborate cuticular spines on the body.

120a (118) Cuticle of trunk quite thin and flexible. 121

120b Cuticle of trunk stiff, in most species having ridges, plications or spines. 122

121a (120) Spurs long, flat, and wide. Foot half or more of total length. (Fig. 18.115) . . . . ***Embata*** Bryce
Five species, 600 μ. Mostly epizootic on aquatic arthropods.

121b Spurs short, or if elongate, not flat and wide. Foot somewhat less than half total length. (Fig. 18.116) . . . . ***Philodina*** Ehrenberg
Over 40 species, 500 μ. Mostly littoral. Commonly becomes abundant in laboratory cultures.

122a (120) Spurs several times as long as width at base. Viviparous. (Fig. 18.117) . . . . ***Dissotrocha*** Bryce
Six species, 500 μ. Mostly littoral, and in submerged *Sphagnum*. Some species were formerly included in **Philodina.**

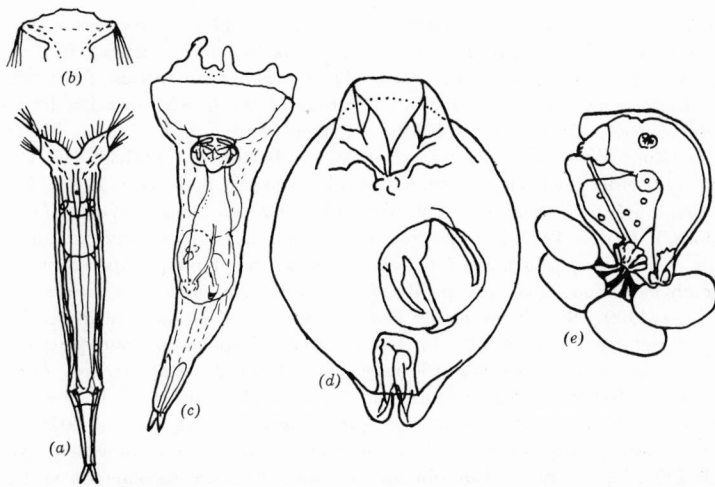

Fig. 18.125. Rotifers of uncertain position. (*a*) *Cordylosoma perlucidum*, dorsal, fully expanded. (*b*) Same, corona in swimming position. (*c*) Rotifer that may belong to this genus. Drawing based on specimen from Germany in American Museum of Natural History, and sketch provided by J. Hauer, who sent the specimen to F. J. Myers. (*d*) *Vanoyella globosa*, contracted specimen. (*e*) *Cypridicola parasitica*, lateral view of contracted specimen. (*a, b* from Voigt; *c* original; *d* redrawn from Evens; *e* redrawn from Daday.)

122b		Spurs shorter. Oviparous. (Fig. 18.118) . . . . . ***Pleuretra*** Bryce
		Ten species, 400 μ. On mosses, some in submerged *Sphagnum*.
123a	**(112)**	Corona with 2 ciliated areas on short pedicels, but without upper and lower lips. (Fig. 18.119) . . . . ***Abrochtha*** Bryce
123b		Corona more reduced 124
124a	**(123)**	Corona consists of 2 small, circular areas, elevated above ventral surface of head only at anterior margin. (Fig. 18.120) . . . . ***Henoceros*** Milne
		Two species littoral. Not yet reported from N. A.
124b		Corona even more greatly reduced; cilia limited to small area around mouth and tip of rostrum. (Fig. 18.121) . . . . ***Philodinavus*** Harring
		One species, littoral, or running water. Not yet reported from N. A. Formerly **Microdina**.
125a	**(112)**	Spindle-shaped. Head wider than neck. (Fig. 18.122) . . . . ***Adineta*** Hudson
		Eleven species, 500 μ. Littoral and in mosses, including *Sphagnum*.
125b		Cylindric. Head not wider than neck. (Fig. 18.123) . . . . ***Bradyscela*** Bryce
		Two species. Not yet reported from N. A.

References

Ahlstrom, E. H. 1932. Plankton Rotatoria from Mexico. *Trans. Am. Microscop. Soc.*, 51:242–251. **1934.** Rotatoria of Florida. *Trans. Am. Microscop. Soc.*, 53:251–256. **1938.** Plankton Rotatoria from North Carolina. *J. Elisha Mitchell Sci. Soc.*, 54:88–110. **1940.** A revision of the rotatorian genera *Brachionus* and *Platyias* with descriptions of one new species and two new varieties. *Bull. Am. Museum Nat. Hist.*, 77:143–184.   **1943.** A revision of the rotatorian genus *Keratella* with descriptions of three new species of five new varieties. *Bull. Am. Museum Nat. Hist.*, 80:411–457.   **Bartoš, E.   1948.** On the Bohemian species of the genus *Pedalia* Barrois. *Hydrobiologia*, 1:63–77.   **1951a.** Rotatoria of the Czechoslowakian Iceland-expedition. *Hydrobiologia*, 3:244–250.   **1951b.** Ceskoslovenske druhy rodu Testudinella a Pompholyx (Rot.) *Sbornik Kl. Prir. Brně.*, 29:10–20.   **Bartoš, E.   1951c.** The Czechoslovak Rotatoria of the order Bdelloidea. Vestnik Ceskosl. Zoologicke Spolecnost. 15:241–244, 345–500. **1951d.** Klíč k určváníi vířníků rodu Polyarthra Ehrb. *Čas. nár. Musea, přírod. oddíl.*, 118:82–91. **Beauchamp, P. D. de.   1907.** Morphologie et variations de l'appareil rotateur dans la series des Rotifères. *Arch. zool. exp. et gén. Ser. 4*, 6:1–29.   **1908.** Sur l'interprétation morphologique et la valeur phylogénique du mastax des Rotifères. *Compt. rend. assoc. Franc. avance. sci.*, 36 sess., 2:649–652.   **1909.** Rechèrches sur les Rotifères: les formation tégumentaires et l'appareil digestif. *Arch. zool. exp. et gén., Ser. 4*, 10:1–410.   **1912a.** Rotifères communiqués par M. M.-K. Harring: *Scaridium eudactylotum* Gosse et le mastax des *Dinocharis. Bull. soc. zool. France*, 37:182–187.   **1912b.** Rotifères communiqués par M. H. K. Harring et C. F. Rousselet: contribution à l'étude des Atrochides. *Bull. Soc. Zool. France*, 37:242–254.   **1912c.** Sur deux formes inférieures D'Asplanchnidés (avec description d'une espèce nóuvelle). *Soc. Zool. de France*, 36:223–233.   **1932.** Contribution à l'étude du genre Ascomorpha et des processus digestifs chez les rotifères. *Bull. Soc. Zool. France*, 57:428–449.   **1940.** Croisière du Bougainville aux Iles Australes Françaises. XII. Tubellariés et Rotifères. *Mém. muséum nat. hist. nat. Paris, N.S.*, 14:313–326.   **Bĕrzinš, B.   1943.** Systematisch-faunistisches Material über Rotatorien Lettlands. *Folia Zool. Hydrobiol. Riga*, 12:210–244.   **1949a.** Einige neue *Monommata*-Arten (*Rotatoria*) aus Schweden. *Arkiv Zool.*, 42A:1–6.   **1949b.** Taxonomic notes on some Swedish Rotatoria. *J. Quekett Microscop. Club, Ser. 4*, 3:25–36.   **1949c.** Einige neue Notommatidae-Arten (Rotatoria) aus Schweden. *Hydrobiologica*, 1:312–321.   **1950a.** Einige neue Bdell. Rotatorien aus dem Aneboda-Gebiet (Schweden). *Kgl. Fysiogrof. Sällskap. Lund. Förh.*, 20:1–6.   **1950b.** Observations on rotifers on sponges. *Trans. Am. Microscop. Soc.*, 69:189–193.   **1951.** On the Collothecacean Rotatoria. *Arkiv Zool., Ser. 2*, 1:565–592. **1953.** Zur Kenntnis der Rotatorien aus West-Australien. *Lunds Univ. Arsskr., NF 2*, 49:3–12. **1954a.** Zur Rotatorienfauna Siziliens. *Hydrobiologica*, 6:309–320.   **1954b.** Nomenklatorische Bemerkungen an einigen planktischen Rotatorien-Arten aus der Gattung Keratella. *Hydrobiologia*, 6:321–327.   **1954c.** A new rotifer, *Keratella canadensis. J. Quekett Microscop. Club*, 4:113–115.   **1955.** Taxonomie and Verbreitung von *Keratella valga* und verwandten Formen. *Arkiv Zool., Ser. 2*, 8:549–559.   **Budde, E.   1925.** Die parasitischen Rädertiere mit besonderer Berücksichtigung der in der Umgegend von Minden I. W. beobachteten Arten. *Z. Morphol. Ökol. Tiere*, 3:706–784.   **Burckhardt, G.   1943.** Hydrobiologische Studien an Schweizer Alpenseen. *Z. Hydrol.*, 9:354–385.   **Burger, A.   1948.** Studies on the moss dwelling bdelloids (Rotifera) of Eastern Massachusetts. *Trans. Am. Microscop. Soc.*, 67:111–142. **Canella, M. F.   1952.** Osservazioni su Sinantherina semibullata (Thorpe) e su altri Flosculariacea. *Ann. Univ. Ferrara. N. S.*, 1:171–257.   **Carlin-Nilson, B.   1935.** Rotatorien aus Mexico. *Kgl. Fysiograf. Sällskap. Lund. Förh*, 5:1–11.   **Carlin, B.   1939.** Über die Rotatorien einiger Seen bei Aneboda. *Medd. Lunds Univ. Limn. Inst.*, 2:1–68.   **1943.** Die Planktonrotatorien des Motalaström. *Medd. Lunds Univ. Limn. Inst.*, 5:1–255.   **Dobers, E.   1915.** Über die Biologie der Bdelloidea. *Intern. Rev. 7* (Suppl. 1):1–128.   **Donner, Josef Von. 1943a.** Zur Rotatorienfauna Südmährens. Mit Beschreibung der neuen Gattung *Wulfertia. Zool. Anz.*, 143:21–33.   **1943b.** Zur Rotatorienfauna Südmährens. II. *Zool. Anz.*, 143:63–75.   **1943c.** Zur Rotatorienfauna Südmährens. III. *Zool. Anz.*, 143:173–179. **1950a.** Zur Rotatorienfauna Südmährens. IV. *Zool. Anz.*, 145:139–155.   **1944b.** Rotatorien einiger Teiche um Admont. *Mitt. naturw. Ver. Steiremark*, 78:1–9.   **1949.** *Horaëlla brehmi* nov. gen. nov. spec. *Hydrobiologia*, 2:134–140.   **1950.** Rädertiere der Gattung Cephalodella aus Südmähren. *Arch. Hydrobiol.*, 42:304–328.   **1953.** *Trichocerca (Diurella) ruttneri* nov. spec., ein Rädertier aus Insulinde, Indien und dem Neusiedlersee. *Österr. zool. Z.*, 4:19–22.   **1954.** Zur Rotatorienfauna Südmährens. (Abschluss). *Österr. zool. Z.*, 5:30–117. **Edmondson, W. T.   1935.** Some Rotatoria from Arizona. *Trans. Am. Microscop. Soc.*, 54:301–306.   **1936.** New Rotatoria from New England and New Brunswick. *Trans. Am. Microscop. Soc.*, 55:214–222.   **1938.** Three new species of Rotatoria. *Trans. Am. Microscop. Soc.*,

57:153–157. **1939.** New species of Rotatoria, with notes on heterogonic growth. *Trans. Am. Microscop. Soc.*, 58:459–472. **1940.** The sessile Rotatoria of Wisconsin. *Trans. Am. Microscop. Soc.*, 59:433–459. **1944.** Ecological studies of sessile Rotatoria. Part I. Factors affecting distribution. *Ecol. Monographs*, 14:31–66. **1945.** Ecological studies of sessile Rotatoria. Part II. Dynamics of populations and social structures. *Ecol. Monographs*, 15:141–172. **1948a.** Rotatoria from Penikese Island, Mass., with description of *Ptygura agassizi*, n. sp. *Biol. Bull.* 94:169–173. **1948b.** Two new species Rotatoria from sand beaches. *Trans. Am. Microscop. Soc.*, 67:149–152. **1949.** A formula key to the rotatorian genus *Ptygura. Trans. Am. Microscop. Soc.*, 68:127–135. **1950.** Centrifugation as an aid in examining and fixing rotifers. *Science*, 112:49. **Gallagher, J. J. 1957.** Generic classification of the Rotifera. *Proc. Penn. Acad. Sci.*, 31:182–187. **Gillard, A. 1947.** Testudinella vanoyei n. sp., un nouveau Rotateur. *Ann. soc. roy. zool. Belg.*, 73:24–25. **1948.** De Branchionidae (Rotatoria) van België met Beschouwingen over de Taxonomie van de Familie. *Natuurw. Tijdschr.*, 30:159–218. **1952.** Raderdieren van Katanga. *Med. Landbouw. Opzoekingsstat. Gent.*, 17:333–352. **Graaf, F. de. 1956.** Studies on Rotatoria and Rhizopoda from the Netherlands. *Biol. Jaarboek*, 145–216. **Harring, H. K. 1913.** Synopsis of the Rotatoria. *Bull. U. S. Natl. Museum*, 81:1–226. **1914.** A list of the Rotatoria of Washington and vicinity, with descriptions of a new genus and ten new species. *Proc. U. S. Natl. Museum*, 46:387–405. **1916.** A revision of the Rotatorian genera *Lepadella* and *Lophocharis* with descriptions of five new species. *Proc. U. S. Natl. Museum*, 51:527–568. **Harring, H. K. and F. J. Myers. 1922.** The rotifer fauna of Wisconsin. *Trans. Wisconsin Acad. Sci.*, 20:553–662. **1924.** The rotifer fauna of Wisconsin, II. A revision of the notommatid rotifers, exclusive of the Dicranophorinae. *Trans. Wisconsin Acad. Sci.*, 21:415–549. **1926.** The rotifer fauna of Wisconsin, III. A revision of the genera *Lecane* and *Monostyla. Trans. Wisconsin Acad. Sci.*, 22:315–423. **1928.** The rotifer fauna of Wisconsin, IV. The Dicranophorinae. *Trans. Wisconsin Acad. Sci.*, 23:667–808. **Hauer, J. 1924.** Zur Kenntnis des Rotatorien genus *Colurella* Bory de St. Vincent. *Zool. Anz.*, 59:177–189. **1929.** Zur Kenntnis der Rotatorien-genera Lecane und Monostyla. *Zool. Anz.*, 83:143–164. **1930.** Zur Rotatorienfauna Deutschlands, I. *Zool. Anz.*, 92:219–222. **1931a.** Zur Rotatorienfauna Deutschlands. II. *Zool. Anz.*, 93:7–13. **1931b.** Zur Rotatorienfauna Deutschlands. III. *Zool. Anz.*, 94:173–184. **1935.** Zur Rotatorienfauna Deutschlands. IV. *Zool. Anz.*, 110:260–264. **1936a.** Zur Rotatorienfauna Deutschlands. V. *Zool. Anz.*, 113:154–157. **1936b.** Rädertiere aus dem Naturschutzgebiet Weingartener Moor. *Beitr. naturkund. Forsch. in Südwestdeutschland*, 1:129–152. **1936c.** Zur Rotatorienfauna Deutschlands. VI. *Zool. Anz.*, 115:334–336. **1937.** Zur Kenntnis der Rotatorienfauna des Eichener Sees. *Beitr. naturkund. Forsch. Südwestdeutschland*, 2:165–173. **1937–38.** Die Rotatorien von Sumatra, Java, und Bali nach den Ergebnissen der Deutschen Limnologischen Sunda-Expedition. *Arch. Hydrobiol., Suppl.*, 7:296–384, 507–602. **1938.** Zur Rotatorienfauna Deutschlands. VII. *Zool. Anz.*, 123:213–219. **1939.** Zur Kenntnis der Russelrädertiere (*Bdelloidea*) des Schwarzwaldes. *Beitr. naturkund. Forsch. Südwestdeutschland*, 4:163–173. **1940.** Beitrag zur Kenntnis der Rotatorien warmer Quellen Deutschlands. *Zool. Anz.*, 130:156–158. **1941.** Rotatorien aus dem "Zwischengebiet Wallacea". *Intern. Rev. ges. Hydrobiol. Hydrog.*, 41:177–203. **1952a.** Rotatorien aus Venezuela und Kolumbien. *Ergeb. deutsch. limnol. Venezuela-Exped.*, 1:277–314. **1952b.** Pelagische Rotatorien aus dem Windgfällweiher, Schluchsee und Titisee im südlichen Schwarzwald. *Arch. Hydrobiol. Suppl.-Bd.*, 20:212–237. **1953.** Zur Rotatorienfauna von Nordostbrasilien. *Arch. Hydrobiol.*, 48:154–172. **1957.** Rotatorien aus dem Plankton des Van-Sees. *Arch. Hydrobiol.*, 53:23–29. **Hemming, F. (ed.). 1955.** Validation, under the plenary powers, of the generic name "Hexarthra" Schmarda 1854 (Class Rotifera) and matters incidental thereto. *Intern. Comm. Zool. Nomenclature. Opinions and Declarations.* Opinion 326. 9:269–281. **Hickernell, L. M. 1917.** A study of desiccation on the rotifer *Philodina roseola*, with special reference to cytological changes accompanying desiccation. *Biol. Bull.*, 32:343–407. **Hsu, W. S. 1956a.** Oogenesis in the Bdelloidea rotifer Philodina roseola Ehrenberg. *Cellule rec. cytol. histo.*, 57:283–296. **1956b.** Oogenesis in Habrotrocha tridens (Milne). *Biol. Bull.*, 111:364–374. **Hudson, C. T. and P. H. Gosse. 1886.** *The Rotifera; or wheel-animalcules, both British and foreign*, 2 vols. Longmans, Green,

London. **1889** Supplement. **Hyman, L. H. 1951.** *The Invertebrates.* Vol. III, *Acanthocephala, Aschelminthes and Entoprocha.* McGraw-Hill, New York. **Jennings, H. S. 1894.** A list of the Rotatoria of the Great Lakes and some of the inland lakes of Michigan. *Bull. Mich. Fish Comm.*, 3:3–34. **1900.** Rotatoria of the United States, with especial reference to those of the Great Lakes. *Bull. U. S. Fish Comm.*, 1900:67–104. **1903.** Rotatoria of the United States. II. A monograph of the Rattulidae. *Bull. U. S. Fish Comm.*, 1902:272–352. **Klement, V. 1955.** Über eine Missbildung bei dem Rädertier *Keratella cochlearis* und über eine neue Form von *Keratella quadrata.* *Zool. Anz.*, 155:321–324. **Leissling, R. 1924.** Zur Kenntnis von *Pompholyx sulcata* Hudson. *Zool. Anz.*, 59:88–100. **Lucks, R. 1929.** Rotatoria. Rädertiere. *Biol. Tiere Deutschlands*, 10:1–176. **Margalef, R. 1947.** Notas sobre algunos Rotiferos. *Publ. inst. biol. apl. Barcelona*, 4:135–148. **Myers, F. J. 1930.** The rotifer fauna of Wisconsin. V. The genera *Euchlanis* and *Monommata.* *Trans. Wisconsin Acad. Sci.*, 25:353–411. **1931.** The distribution of Rotifera on Mount Desert Island. *Am. Museum Novitiates*, 494:1–12. **1933a.** A new genus of rotifers (*Dorria*). *J. Roy. Microscop. Soc.*, 53:118–121. **1933b.** The distribution of Rotifera on Mount Desert Island. II. New Notommatidae of the genera *Notommata* and *Proales.* *Am. Museum Novitiales*, 659:1–26. **1933c.** The distribution of Rotifera on Mount Desert Island. III. New Notommatidae of the genera *Pleurotrocha, Lindia, Eothinia, Proalinopsis*, and *Encentrum.* *Am. Museum Novitiales*, 660:1–18. **1934a.** The distribution of Rotifera on Mount Desert Island. IV. New Notommatidae of the genus *Cephalodella.* *Am. Museum Novitiates*, 699:1–14. **1934b.** The distribution of Rotifera on Mount Desert Island. V. A new species of the Synchaetidae and new species of Asplanchnidae, Trichocercidae and Brachionidae. *Am. Museum Novitiates*, 700:1–16. **1934c.** The distribution of Rotifera on Mount Desert Island. VI. New Brachionidae of the genus *Lepadella.* *Am. Museum Novitiates*, 760:1–10. **1934d.** The distribution of Rotifera on Mount Desert Island. VII. New Testudinellidae of the genus *Testudinella*, and a new species of Brachionidae of the genus *Trichotria.* *Am. Museum Novitiates*, 761:1–8. **1936a.** Psammolittoral rotifers of Lenape and Union lakes, New Jersey. *Am. Museum Novitiates*, 830:1–22. **1936b.** Rotifers from the Laurentides National Park with descriptions of two new species. *Can. Field-Natur.* 50:82–85. **1937a.** A method of mounting rotifer jaws for study. *Trans. Am. Microscop. Soc.*, 56:256–257. **1937b.** Rotifera from the Adirondack region of New York. *Am. Museum Novitiates*, 903:1–17. **1938.** New species of Rotifera from the collection of the American Museum of Natural History. *Am. Museum Novitiates*, 1011:1–17. **1940.** New species of Rotatoria from the Pocono Plateau, with note on distribution. *Notulae Naturae Acad. Nat. Sci. Phila.*, 51:1–12. **1941.** Lecane curvicornis var. miamiensis, new variety of Rotatoria, with observations on the feeding habits of Rotifers. *Notulae Naturae Acad. Nat. Sci. Phila.*, 75:1–8. **1942.** The rotatorian fauna of the Pocono plateau and environs. *Proc. Acad. Nat. Sci. Phila.*, 94:251–285. **Neal, M. 1951.** Application for the stabilisation of the name for the genus of the Class Rotifera formerly known as "Pedalion." *Bull. Zool. Nomencl.*, 6:73–78. **Nipkow, F. 1952.** Die Gattung *Polyarthra* Ehrenberg im Plankton des Zürichsees und einiger anderer Schweizer Seen. *Schweiz. Z. Hydrol.*, 14:135–181. **Pawlowski, L. K. 1934.** Drilophaga bucephalus Vejdovský, ein parasitisches Rädertier. *Mém. acad. polon. Sci. Classe sci. math et nar.*, Ser. B, 1934:95–104. **Pejler, B. 1956.** Introgression in planktonic Rotatoria with some points of view on its causes and conceivable results. *Evol.*, 10:246–261. **1957a.** On variation and evolution in planktonic Rotatoria. *Zool. Bidr. Uppsala*, 32:1–66. **1957b.** Taxonomical and ecological studies on planktonic Rotatoria from northern Swedish Lapland. *Kgl. Svenska Vetenskapskad. Handl.*, Ser. 4, 6:1–68. **1957c.** Taxonomical and ecological studies on planktonic Rotatoria from Central Sweden. *Kgl. Svenska Vetenskapskad. Handl.*, Ser. 4, 6:1–52. **Pennak, R. W. 1953.** Fresh-water invertebrates of the United States. Ronald Press, New York. **Peters, F. 1931.** Untersuchungen über Anatomie and Zellkonstanz von *Synchaeta.* *Z. wiss. Zool.*, 139:1–119. **Remane, A. 1929–33.** Rotatorien. In: Bronns *Klassen und Ordnungen des Tierreichs*, B. and IV, Abt. II, Buch 1, pp. 1–4. Akademische Verlagsgesellschaft, Leipzig. **1929.** Rotatoria. In: *Die Tierwelt der Nord- und Ostsee.* Viie:1–156. **1933.** Zur Organisation der Gattung *Pompholyx* (Rotatoria). *Zool. Anz.*, 103:188–193. **Rodewald, L. 1935.** Rädertierfauna Rumäniens. I. Neue Rädertiere aus den Hochmooren der Bukowina, nebst Bemerkungen zur Gattung Bryceela Remane. *Zool. Anz.*, 111:225–233. **1940.** Rädertierfauna Rumäniens. IV. *Zool. Anz.*, 130:272–289. **Rousselet, C. F. 1902.** The genus *Synchaeta*: a monographic study with descriptions of five new species. *J. Roy. Microscop. Soc.*, 1902:269–290, 393–411. **Russell, C. R. 1944.** A new rotifer from New Zealand. *J. Roy. Microscop. Soc.*, 64:121–123. **1947.** Additions to the Rotatoria of New Zealand. I. *Trans. Roy. Soc. New Zealand*, 76:403–408. **1950.** Additions to the Rotatoria of New Zealand. III. *Trans. Roy. Soc. New Zealand*, 78:161–166. **1951.** The Rotatoria of the Upper Stillwater Swamp. *Records Canterbury Museum New Zealand*, 5:245–251. **1952.** Additions to the Rotatoria of New Zealand. 4. *Trans. Roy. Soc. New Zealand*, 80:59–62. **1953.** Some Rotatoria of the Chatham Islands. *Records Canterbury Museum New Zealand*, 6:237–244. **Ruttner-Kolisko, A. 1949.** Zum Formwechsel- und Artproblem von Anuraea aculeata (*Keratella quadrata*). Hydrobiologica, 1:425–468. **Seehaus, W. 1930.** Zur morphologie der Rädertiergattung *Testudinella* Bory de St. Vincent (= *Pterodina* Ehrenberg). *Z. wiss. Zool.*, 137:175–273. **Sládeček, V. 1955.** A note on the occurrence of *Hexarthra fennica* Levander in Czechoslovakian Oligohaline waters. *Hydrobiologica*, 7:64–67. **Sudzuki, M. 1955a.** On the general structure and the seasonal occurrence of the males in some Japanese rotifers. I. *Zool. Mag. (Dobutsugaku Zasshi)*, 64:126–129. **1955b.** On the general structure and the seasonal occurrence of the males in some Japanese rotifers. II. *Zool. Mag. (Dobutsugaku Zasshi)*, 64:130–136. **1955c.** On the general structure and the seasonal occurrence of the males in some Japanese rotifers. III. *Zool. Mag. (Dobutsugaku Zasshi)*, 64:189–193. **1955d.** Studies on the egg-carrying types in Rotifera. I. Genus *Pompholyx*. *Zool. Mag. (Dobutsugaku Zasshi)*, 64:219–224. **1955e.** Life history of some Japanese rotifers. I. *Polyarthra trigla* Ehrenberg. *Sci. Repts. Tokyo Kyoiku Daigaku*, Sect. B., 8:41–64. **1956a.** On the general structure and the seasonal occurrence of the males in some Japanese rotifers. IV. *Zool. Mag. (Dobutsugaku Zasshi)*, 65:1–6. **1956b.** On the general structure and the seasonal occurrence of the males in some Japanese rotifers. V. *Zool. Mag. (Dobutsugaku Zasshi)*, 65:329–334. **1956c.** On the general structure and the seasonal occurrence of the males in some Japanese rotifers. VI. *Zool. Mag. (Dobutsugaku Zasshi)*, 65:415–421. **1957a.** Studies on the egg-carrying types in Rotifera. II. Genera *Brachionus* and *Keratella*. *Zool. Mag. (Dobutsugaku Zasshi)*, 66:11–20. **1957b.** Studies on the egg-carrying types in Rotifera. III. Genus *Anuraeopsis*. *Zool. Mag. (Dobutsugaku Zasshi)*, 66:407–415. **Surface, F. M. 1906.** The formation of new colonies of the rotifer *Megalotrocha alboflavicans* Ehr. *Biol. Bull.*, 11:183–192. **Tafall, B. F. O. 1942.** Rotiferos planctonicos de Mexico. *Soc. Mex. His. Nat.*, 3:23–79. **Thomasson, K. 1953.** Studien über das südamerikanische Süsswasserplankton. *Arkiv Zool.*, Ser. 2, 6:189–194. **Thorpe, V. G. 1893a.** The Rotifera of China. *J. Roy. Microscop. Soc.*, 1893:145–152. **1893b.** Pond life in China. *J. Quekett Microscop. Club*, Ser. 2, 5:226–227. **Varga, L. 1945.** Die Sommer-Rotatorien des Kis-Balatons. *Külön. Mag. Biol. Kutat. Munkaibol.*, 16:36–102. **Voigt, M. 1957.** Rotatoria. Die Rädertiere Mitteleuropas, 2 vols. Borntraeger, Berlin. **Weber, E. F. 1898.** Fauna rotatorienne du basin de Léman. *Rev. suisse zool.*, 5:263–785. **Wesenberg-Lund, C. 1923.** Contributions to the biology of the Rotifera. I. The males of the Rotifera. *Mém. acad. roy. sci. Danemark*, Ser. 8, 4:189–345. **Wiszniewski, J. 1929.** Zwei neue Rädertierarten: Pedalia intermedia n. sp. und Paradicranophorus limosus n. g. n. sp. *Bull. intern. acad. polon. sci.* Ser. B., 2:137–153. **1934a.** Les rotifères psammiques. *Ann. mus. zool. polon.* 10:339–399. **1934b.** Les mâles des Rotifères psammiques. *Mém. acad. polor. sci. Classe sci. math. et nat. Sér. B.*, 1934:143–164. **1953.** Les rotifères de la faune polonaise et des regions avoisinantes. *Polskie Arch. Hydrobiol.*, 1:317–490. (Polish, French summary.) **1954.** Matériaux relatifs à la nomenclature et à la bibliographie des Rotifères. *Polskie Arch. Hydrobiol.*, 2:7–260. **Wright, H. G. S. 1950.** A contribution to the study of *Floscularia ringens*. *J. Quekett Microscop. Club.* Ser. 4, 3:103–116. **1954.** The ringed tube of *Limnias melicerta* Weisse. *Microscope*, 10:13–19. **1957.** The rotifer fauna of East Norfolk. *Trans. Norfolk and Norwich Nat. Soc.*, 18:1–23. **Wulfert, K. 1935.** Beiträge zur Kenntnis der Rädertierfauna Deutschlands. I. *Arch. Hydrobiol.*, 28:583–602. **1936.** Beiträge zur Kenntnis der Rädertierfauna Deutschlands. II. *Arch. Hydrobiol.*, 30:401–437. **1937a.** Beiträge zur Kenntnis der Rädertierfauna Deutschlands. III. *Arch. Hydrobiol.*, 31:592–635. **1937b.** Zur Kentniss der Lebensgemeinschaften der Restlochgewässer des Braunkohlenbergbaues. *Z. Naturwiss.*, 91:56–69. **1938a.** Die Rädertiergattung *Cephalodella* Bory de St. Vincent. *Arch. Naturgeschichte*, 7:137–152. **1938b.** Die Tierwelt der Quellen. 3. Die Rädertiere des Goldlochs bei Eifersdorf. *Beitr. Biol. Glatzer Schneeberges*, 4:384–394. **1939a.** Beiträge zur Kenntnis der Rädertierfauna Deutschlands. IV. *Arch. Hydrobiol.*, 35:563–624. **1939b.** Einige neue Rotatorien aus Brandenburg und Pommern. *Zool. Anz.*, 127:65–75. **1940.** Rotatorien einiger ostdeutscher Torfmoore. *Arch. Hydrobiol.*, 36:552–587. **1942.** Neue Rotatorienarten aus deutschen mineralquellen. *Zool. Anz.*, 137:187–200. **1943.** Rädertiere aus dem Salzwasser von Hermannsbad. *Zool. Anz.*, 143:164–172. **1944.** Bericht über Rotatorien aus einiger Düngerproben. *Z. Morphol. Ökol. Tiere*, 40:377–388. **1950.** Das Naturschutzgebiet auf dem Glatzer Schneeberg. *Arch. Hydrobiol.*, 44:441–471. **1956.** Die Rädertiere des Teufelssees bei Friedrichshagen. *Arch. Hydrobiol.*, 51:457–495. **Yamamoto, K. 1951.** On six new Rotatoria from Japan. *Annotationes Zool. Japon.*, 24:157–162. **1953.** Preliminary studies on the Rotatorian fauna of Korea. *Pacific Sci.*, 7:151–164. **1955.** A new Rotifer (Order Ploima) from Japan. *Annotationes Zool. Japon.*, 28:33–34.

19

Bryozoa

MARY DORA ROGICK

Whether Ectoprocta and Entoprocta should be regarded as two independent phyla, or as classes or subphyla under the Bryozoa or Polyzoa, is still an unsettled controversy. It will remain so until much more profound studies are made of their morphology and development, on which data are very incomplete at present. Hyman (1951) favors the two-phyla view, but some bryozoologists, including the present author, favor the single-phylum view.

Bryozoa are sessile, although young colonies of *Cristatella*, *Lophopodella*, *Pectinatella*, and possibly *Lophopus* have a capacity for a slight movement over the substratum. The growth habit distinguishes most of the fresh-water genera. The genera are readily distinguished with the naked eye but the compound microscope is required for correct identification of most of the species and for a study of anatomical details of the zoids. *Pectinatella magnifica*, the jelly ball, forms gelatinous, slimy, yet quite firm masses larger than a human head. The mossy, brownish, firm tubular branches of *Plumatella* and *Fredericella* form mats, sometimes of very loose texture, other times densely packed, in some species completely encrusting, in others attached at the base

with some branches soon rising free from the substratum. *Paludicella* forms
delicate traceries of much finer texture than the Plumatellas.

Bryozoan colonies consist of more or less elongate individuals called zoids,
which are small and numerous. The zoids have a crown or lophophore bearing numerous long, ciliated tentacles. In the Ectoprocta (Bryozoa) the
lophophore is completely retractile into a tentacular sheath which can be
pulled into the body cavity of the zoid. In the Entoprocta the tentacles
merely curve inward over the top of the vestibule of the calyx and the lopho
phore flap or rim puckers up around them, and the tentacles are not retractile
into the zoid's body. When the water containing the colonies is disturbed,
the tentacles are withdrawn or folded inward with great speed, and remain so
until the disturbance is over. The shape of the Entoproct lophophore is
elliptical or nearly circular, and in the fresh-water Ectoprocts it is generally
horseshoe-shaped, except in a few species such as those of *Fredericella*, *Paludicella*,
and *Pottsiella*, in which it ranges from very broadly elliptical to circular.

The fresh-water ectoproct zoids consist of body wall and polypide.

The body wall has two major layers, the outer ectocyst or cuticula and the
inner laminated endocyst. In some genera (*Hyalinella*, *Lophopodella*, *Lophopus*)
the ectocyst is colorless, delicate, transparent. In other genera (*Fredericella*,
Plumatella, *Stolella*) the ectocyst may vary from almost transparent to opaque
and also may be encrusted with minute particles, debris, or coarse sand grains,
depending on the species and the habitat, but it is generally firmer or more
rigid than the ectocyst of the previous group of genera.

The polypide includes the tentacular crown, the digestive tract, and associated musculature. The polypide is suspended in a large body cavity and
attached to the body wall by a funiculus from the digestive tract and by the
musculature that retracts the lophophore or tentacular crown into the body
cavity. The digestive tract is U-shaped, consisting of a mouth surrounded by
the tentacles, a pharynx, esophagus, stomach, rectum, and anus. A fold of
tissue, the epistome, overhangs the mouth of the Phylactolaemata, an order of
the Class Ectoprocta. A nerve net supplies the body wall and other parts of
the body. A fairly conspicuous ganglion is located dorsally, under the epistome, just back of the mouth and pharynx. From this ganglion large nerve
trunks go into the lophophore. The excretory system in the Lophopodidae
consists of two ciliated canals which occur in the epistomeal region and which
empty their contents into one or more of the median tentacles. Polypides
may degenerate into so-called "brown bodies," shrunken, rounded, yellow to
orange to brown balls which may either be extruded or which may regenerate
into new zoids.

Reproduction

Both sexual and asexual reproduction occur. Spermatozoa are developed
on the funiculus, ova on the body wall. Ciliated larvae are formed by sexual
reproduction inside colonies, e.g., in *Hyalinella* the larvae rotate freely for a

time inside the parental body cavity and then are released from the zooecial tubes to swim about actively for a short time before metamorphosis into a new colony occurs. Asexual reproduction may take the form of production of buds, hibernacula, or statoblasts (sessoblasts and floatoblasts). Budding is common to all the fresh-water species. Zoids bud off similar zoids from their body wall. Another form of asexual reproduction is the formation of chitin-covered germinating bodies called hibernacula and statoblasts, which upon germination give rise to new colonies. Hibernacula are irregular in shape, sometimes even of a somewhat jagged outline, brown in color, and are attached to the substratum by a cementing substance. They are produced by *Paludicella* and *Pottsiella*. Statoblasts are of much more regular shape and are produced by all the Phylactolaemata. Their shape may vary from a circular to an elliptical disc. Conspicuous spines are present on statoblasts of *Pectinatella*, *Cristatella*, *Lophopodella*, and *Lophopus*, but are lacking in statoblasts of *Fredericella*, *Hyalinella*, *Plumatella*, and *Stolella*. Two types of statoblasts are produced, the sessoblasts and the floatoblasts. Sessoblasts are often called resting or attached statoblasts, because they remain cemented to the zooecial tube or to the substratum after the zooecial tubes have disintegrated. They are thick and elliptical, and some are provided on the free face with a thin, chitinous rim, the vestigial annulus, which may contain a suggestion of "cell" markings but which does not have genuine air cells. Floatoblasts are generally called floating or free, because they are produced on the funiculus in great numbers and are first released to float freely in the zoid's body cavity, and then released from the zoid tubes into the surrounding water. A disc-shaped, dark reddish-brown, chitin-covered central capsule containing germinal tissue from which new zoids will develop is characteristic of both sessoblasts and floatoblasts. A light amber-colored gas-filled float or annulus of "air cells" is characteristic of floatoblasts. It usually makes them light enough to float to the surface. Some genera produce both sessoblasts and floatoblasts, others only one of the two types. Floatoblasts bearing conspicuous spines are sometimes called spinoblasts. The floatoblasts are of extreme importance in identification, the sessoblasts less so. Sometimes it is possible to identify a species from only a single floatoblast. The entire significance of the various types of statoblasts is not fully understood. It is known that stato-blasts are produced in summer and fall, and that they germinate in the spring, summer, and fall, and indoors at such times as temperature and other ecological conditions are right for germination. It is also known that statoblasts may tide the species over periods of drought or winter long after the parent colonies have disintegrated or disappeared. Bryozoa can be raised at any time during the year in the laboratory from statoblasts that have been stored either dry or in water in the refrigerator for a number of months. *Lophopodella* statoblasts have germinated 50 months after dry storage. Germination takes place a few days after the statoblasts are put into the water at room temperature. Why the same species of *Plumatella* should produce both sessoblasts and floatoblasts is not very clear.

Colonies of *Cristatella*, *Lophopodella*, and *Pectinatella* occasionally undergo fission and move apart. If a *Hyalinella* or *Plumatella* colony is severed or damaged, some of the zoids remain alive to carry on colony expansion.

Distribution

Bryozoa may spread from one place to another by various means. Water animals (birds, mammals) and wind may transport statoblasts from one body of water to another. Brown (1933, p. 308). found some statoblasts still viable and capable of germination after passing through the digestive tracts of frogs, salamanders, turtles, and mallard ducks.

The fresh-water Bryozoa, although very common in lakes, ponds, and rivers are not of much economic importance. However, some, like *Plumatella*, *Fredericella*, and *Paludicella* have been known to clog or greatly reduce the diameter of water pipes by forming a thick lining mat inside the pipes. Others, like *Lophopodella*, when crushed in their vicinity are toxic to some fish. This lethal effect was observed in 1948 by the Pennsylvania Fish Commission and confirmed by experiments with damaged colonies of *Pectinatella gelatinosa* by Dr. Shujitsu Oda of Tokyo.

Bryozoa live in various types of fresh waters of pH range 5.3 to 8.0, and up to elevations of 3950 meters. They occur at various depths from shore-line level to 214 meters. They grow on submerged materials of many kinds, such as plant leaves and stems, fallen trees, mussel and snail shells, rocks, and similar objects, and in close association with sponges. Some tendipedid larvae are associated with Bryozoa.

Fresh-water Bryozoa are cosmopolitan in distribution, some of the species occurring on several continents. Their diet consists of microscopic organisms, both plant and animal. The Bryozoa in turn may be eaten by fishes, snails, and possibly other animals.

Identification and Preservation

There are two chief difficulties in preserving fresh-water Bryozoa. One is to kill them in an expanded condition, with tentacles spread out. The second is to keep the soft gelatinous forms like *Pectinatella* from disintegrating. The first problem is met by narcotization of zoids by gradual addition of crystals of chloral hydrate until the polypides do not react to touch, and then plunging them into either 70 per cent alcohol or 10 per cent formalin. The second problem has not yet been solved completely. The usual practice is to put the soft forms into 10 per cent formalin.

Since identification of some species depends upon the number of tentacles and shape of the lophophore, both of which are difficult to determine after retraction or ordinary preservation, anyone interested in identification and preservation of Bryozoa should make a count of the tentacles on several zoids and note the shape of the lophophore, whether horseshoe-shaped, circular, or elliptical, while the colonies are still alive, before narcotization or preservation are begun.

The following classification is based on the works of leading authorities in the group:

Class (or Phylum) Entoprocta
    Family Urnatellidae
        Genus *Urnatella* Leidy 1851
Class (or Phylum) Ectoprocta
  Order Gymnolaemata
    Family Paludicellidae
        Genus *Paludicella* Gervais 1836
    Family Victorellidae
        Genus *Pottsiella* Kraepelin 1887
  Order Phylactolaemata
    Family Cristatellidae
        Genus *Cristatella* Cuvier 1798
    Family Fredericellidae
        Genus *Fredericella* Gervais 1838
    Family Lophopodidae
        Genera *Lophopodella* Rousselet 1904, *Lophopus* Dumortier 1835, *Pectinatella* Leidy 1851
    Family Plumatellidae
        Genera *Hyalinella* Jullien 1885, *Plumatella* Lamarck 1816, *Stolella* Annandale 1909

KEY TO SPECIES

1a   Individual or colony stalked. Stalk topped by a distinct head or calyx which has a circle of tentacles that curl or roll inward toward the lophophore base but are not retractable into the interior of the calyx. Anus and mouth inside the tentacular circle. Calyx contains all the body systems . . . . Class (or Phylum) **Entoprocta**
        Only 1 species known from N. A. . . . . **Urnatella gracilis** Leidy 1851

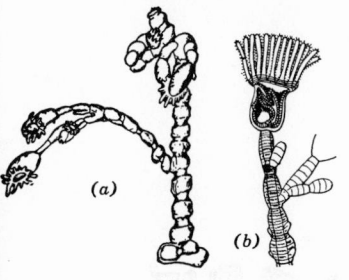

Colony consists of a basal plate from which arise from 1 to 6 segmented or beaded chitinized stalks which are tipped by one or more branches and calyces. Number of segments in stalk varies from 1 to 18. Colonies grow on rocks, mussel shells, and other objects. From Schuylkill River, Pa.; Scioto River and Lake Erie, Ohio; Grand and Clinton Rivers, Mich.; Tippecanoe River, Ind.; Licking River, Ky.; Lake Dallas, Tex.; Fairport, Ia.; James River, Va.

◄ **Fig. 19.1.** *Urnatella gracilis.* (*a*) Colony from Havana, Illinois River. × 13. (*b*) Single polyp. (*a* after Davenport; *b* after Leidy.)

1b   Colonies branching, usually "plantlike" in fresh-water forms. Stalks not topped by a distinct calyx, but a crown of tentacles may protrude either from the tip or from a special tube near the zoid tip. Anus outside tentacular crown. Lophophore or tentacular crown shape ranges from circular to ellipsoid to horseshoe-shaped, depending upon the species. Tentacular crown completely retractile into the zooecial tube or body cavity . . . . . Class (or Phylum) **Ectoprocta**   **2**

2a   **(1)** Tentacular crown circular, mouth in center and not overhung by a lip or epistome. Species mostly marine. No statoblasts produced, but hibernacula may be formed by some . . . Class **Gymnolaemata**   **3**

2b   Tentacular crown usually horseshoe-shaped, occasionally ellipsoid to circular. Mouth protected by an overhanging lip or epistome. Statoblasts produced . . . . Class **Phylactolaemata**   **4**

3a   **(2)** Colony consists of stolons from which erect, single, more or less cylindrical, hyaline zoids, with a pentagonal terminal orifice arise at intervals . . . . *Pottsiella erecta* (Potts) 1884

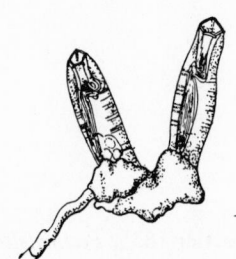

Zooecia arise from cylindrical and sometimes from long stolons. Septa present in stolons near place where the zoids originate. Zoids taper slightly at both ends. Zoid height 1.44–5.6 mm, zoid width 0.187–0.216 mm. Lophophore circular; 19 to 21 tentacles. Photophil, on stones and sponges. From Tacony Creek, Schuylkill and Delaware Rivers, Pa.; Lake Dallas, Tex.; James River, Va.; and Loosahatchie River, Tenn., Dr. Harold Harry, coll.

◄ **Fig. 19.2.** *Pottsiella erecta.* × 25. (After Kraeplin.)

3b   Colony not stolonate but formed of straight lines of zooecia, from which arise secondary and tertiary, etc. straight lines of zoids. Secondary branches arise oppositely from the swollen part of the primary line of zooecia. Orifice not terminal but at the end of a tube which projects frontally or laterally a short distance below the distal end of the zoid. Orifice square or quadrilateral. Zoids slender, club-shaped . . . . *Paludicella articulata* (Ehrenberg) 1831
Colony a pale yellow threadlike tracery on rocks, shells, etc. Zoids have 16 to 18 tentacles. Zoid length 1.09–2.0 mm, width 0.174–0.243 mm. Zoid wall thin, chitinous, transparent to translucent, varying in color with age from pale to deep yellow. Hibernacula irregular in shape, dark brown when mature. Widely distributed species in N. A. and on other continents.

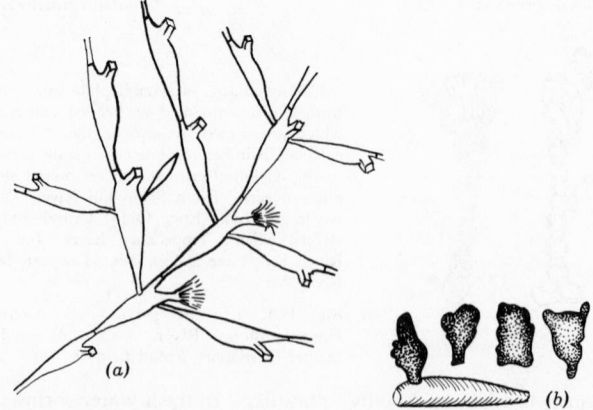

Fig. 19.3. *Paludicella articulata.* (*a*) Growth habit. Note squared orifices frontally placed. (*b*) Four variously-shaped hibernacula, one attached to a zoid.

4a **(2)** Statoblasts of only one kind, sessoblasts, attached directly to the zoid wall or substratum and not provided with float (annulus) of air cells, or with hooked processes or spines of any kind . . . .
Fredericella H. Milne-Edwards   **5**

4b Statoblasts usually of two types; one the sessoblast (attached either to zoid wall or to substratum), the second a floatoblast or free-floating type which does not adhere to the substratum but is provided with a float of gas-filled cells. Tentacles borne on horseshoe-shaped lophophore **6**

5a **(4)** Sessoblasts longer than wide, generally kidney-shaped, but may be oval to slightly angular. Shape rather variable. Tentacle crown circular. From 17 to 27 tentacles, with 20 to 22 being the most common number . . . . **F. sultana** (Blumenbach) 1779

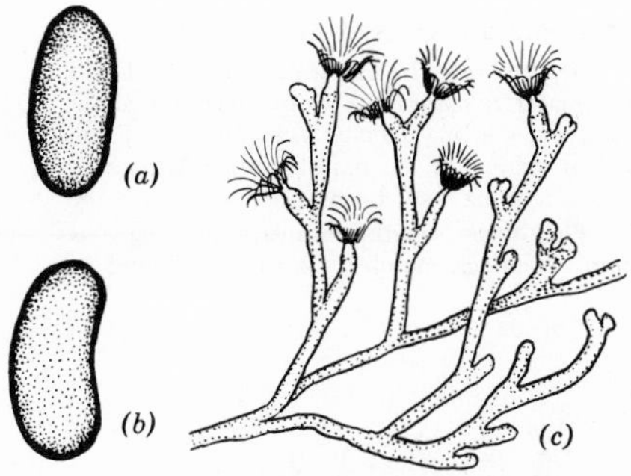

Colony tan to brown, branching antler-like and open, zooecial tubes cylindrical, orifices terminal, ectocyst encrusted lightly with debris, algae, or sand. Sessoblast length 0.27–0.57 mm, width 0.139–0.37 mm. Zooecial tube length 1.73–4.8 mm, width 0.16–0.35 mm. Very widely distributed in N. A. and abroad.

◄ **Fig. 19.4.** *Fredericella sultana.* (*a, b*) Sessoblast shapes. (*c*) Colony growth habit. Note circular lophophore bearing tentacles.

5b Sessoblasts broad, circular or nearly so. From 24 to 30 tentacles. . .
F. australiensis subsp. **browni** Rogick 1945

Branching of colony antlerlike, open as in *F. sultana*. Ectocyst a light tan color, rather opaque, encrusted with sand grains and debris and of considerable rigidity and firmness. Zooecial tubes wider than in *sultana* (0.391–0.576 mm). Sessoblasts usually nearly circular, 0.331–0.461 mm long and 0.266–0.367 mm wide. Polypides short and stubby. From an alkali pond in Uinta County, Wyo.

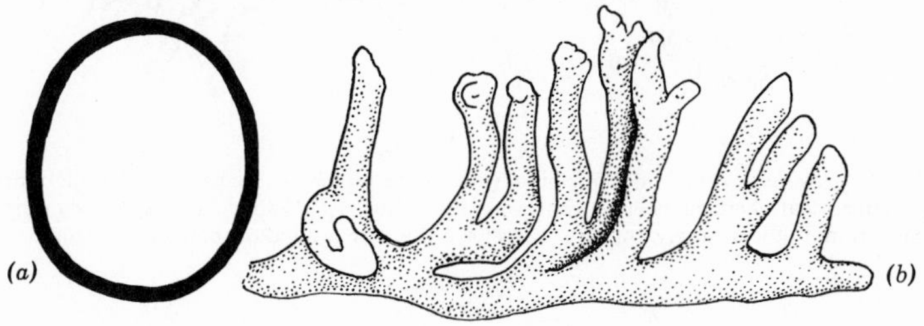

Fig. 19.5. *Fredericella australiensis* subsp. *browni.* (*a*) Sessoblast. (*b*) Colony growth habit.

6a **(4)** Floatoblasts provided with processes or hooks or spines **7**

6b Floatoblasts not provided with processes, hooks, or spines; circular or ellipsoid **10**

7a **(6)** Floatoblasts with processes, hooks, or spines at the two opposite poles only. Sessoblasts absent. **8**

7b Floatoblasts with processes, hooks, spines peripherally arranged . . .   **9**

8a **(7)** Floatoblast spindle-shaped, with a single process at each pole; i.e., with the two polar ends prolonged into a short, acute point. . . . . .
Lophopus crystallinus (Pallas) 1766

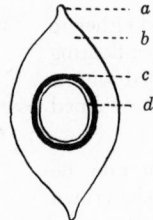

Colony soft, light-colored, shaped like a sac, erect; sometimes lobed by indentations of the surface, and looking like a glove. Ectocyst transparent, colorless, delicate, gelatinous. Endocyst soft, yellowish. Clusters of zoids divided into lobes. About 60 tentacles. From Schuylkill and Illinois Rivers.

◄ **Fig. 19.6.** *Lophopus crystallinus* statoblast. *a*, polar process or spine; *b*, float; *c*, heavy circle or line indicates the extent and rim of statoblast's capsule; *d*, thin line represents the extent of the encroachment or overlapping of the float upon the statoblast capsule. (After Kraepelin.)

8b   Floatoblasts ellipsoid, with several hooked processes at each pole . . .
Lophopodella carteri (Hyatt) 1866

Colony very closely resembles that of *Lophopus crystallinus*. Colony measures about 6.5 by 13.0 mm and may have about 45 polypides in it. Floatoblasts ellipsoid. From each of the two poles of the statoblast's float arise 6 to 20 multibarbed processes or spines. The number of tiny curved barbs on a spine may be as high as 22 but the average number per spine is about 11. Floatoblast length, exclusive of spines, is 0.84–1.009 mm and width is 0.64–0.78 mm. Tentacle number 52 to 82. Found in Lake Erie, Ohio; Ill.; N. J.; Pa.

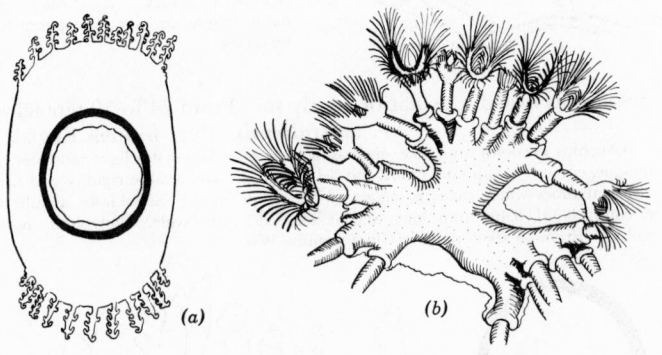

Fig. 19.7. *Lophopodella carteri.* (*a*) Spinoblast with barbed spines at each pole. Capsule represented by a heavy black line, float encroachment by thin, wavy inner line. (*b*) Colony habit sketch, showing a few of the zoids expanded and others simply cut off. Ectocyst is thin film around the bulk of the colony.

9a **(7)** Floatoblasts with a single row of about 11 to 26 large hooked spines rising from the periphery of the float . . . .
Pectinatella magnifica Leidy 1851

Colonies hyaline, in the form of a small rosette, lobed, with horizontal tubes only, secreting a gelatinous base or substratum of immense size, sometimes as large as or larger than a human head. This jellylike base may be from very watery to quite solid; it is translucent and often inhabited by insect larvae. The colonies form starlike single-layered patches over the surface of this jelly. Statoblasts are circular to subrectangular, with broad float. In side view they are shaped like a shallow hat. Statoblast diameter from 0.79 to more than 1 mm. Peripheral spines around statoblast end in hooks. Lophophore horseshoe-shaped, with 50 to 84 tentacles. Very common in stagnant water. Found in many states.

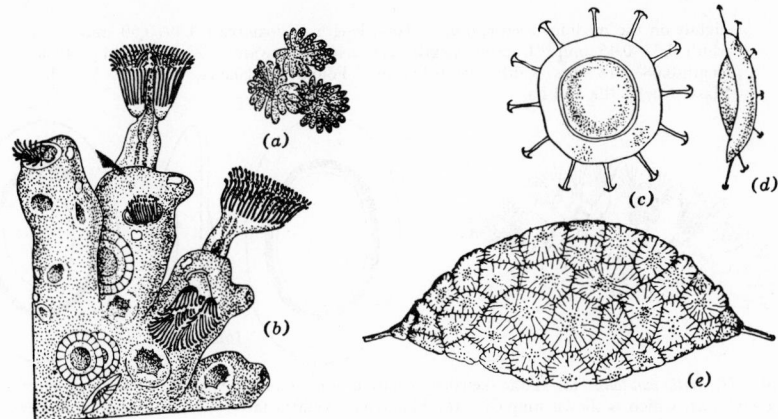

Fig. 19.8. *Pectinatella magnifica.* (a) Young colony. × 1. (b) Section highly magnified. (c) Statoblast, ventral view. (d) Statoblast, profile. ˙(e) Colony on plant stem × 0.5. (After Kraeplin.)

9b   Floatoblasts with a double row of hooked processes, one row arising from each statoblast face in the region of the capsule, outlining the capsule and extending over and beyond the float . . . .
Cristatella mucedo Cuvier 1798

Colony ribbonlike, unbranched, soft, gelatinous, whitish, with a flat "sole.". All polypides contract into a common cavity. From 55 to 99 tentacles. Floatoblasts about 1 mm in diameter (0.75–1.25 mm) and provided on the dorsal side with 9 to 34 hooked spines and on the ventral side with 20 to 65 hooked spines. A cluster of several sharply curved hooks occurs at the tip of each statoblast spine. Found in Ohio, Pa., and R. I.

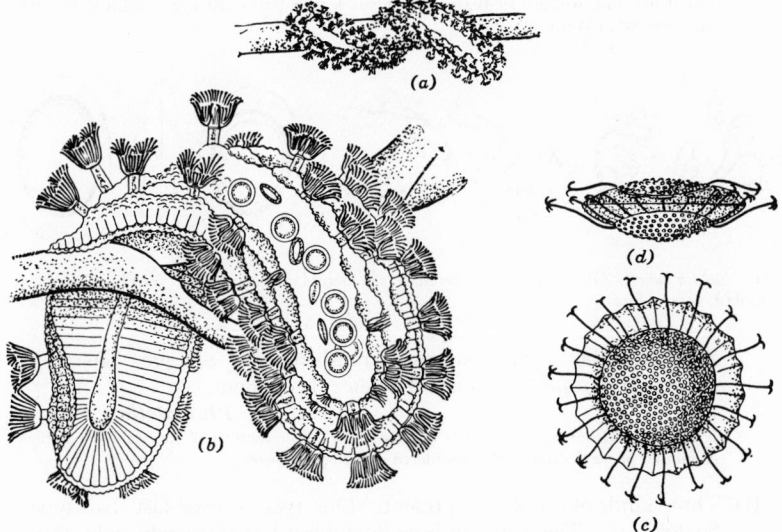

Fig. 19.9. *Cristatella mucedo.* (a) Colony. × 1. (b) Colony much enlarged. (c) Statoblast, ventral view. (d) Statoblast, profile. × 25. (After Allman.)

10a   (6) Zooecia provided with a baggy, clear, transparent ectocyst. Endocyst rather soft and fairly transparent. . . . .
Hyalinella punctata (Hancock) 1850

The colony resembles rows of colorless or faintly yellow vesicles. Zooecia not keeled. From 35 to 55 tentacles. Floatoblasts ellipsoid, with large float which encroaches slightly on the medium-sized capsule. Total length of floatoblast 0.49–0.60 mm, total width 0.33–0.45 mm. Capsule length 0.31–0.40 mm, width 0.25–0.31 mm. Float length 0.08–0.18 mm, width 0.04–0.13 mm. Found in Canada, Ohio, N. Y., Me., Mass., Mich., Ill., and Pa.

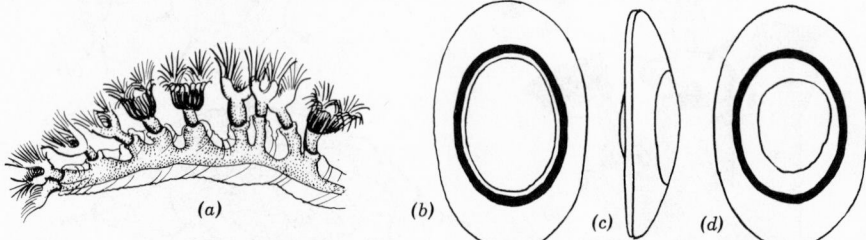

Fig. 19.10. *Hyalinella punctata.* (*a*) Habit sketch of a part of the colony. Baggy transparent ectocyst encloses the endocyst, which is shown stippled. (*b*) Floatoblast, ventral face. (*c*) Floatoblast, side view. (*d*) Floatoblast, dorsal face. Capsule outlined by very heavy line, float encroachment by thin inner one.

10b   Zooecia provided with a closely fitting colored cylindrical or tubular ectocyst which varies in appearance from translucent to opaque and is often encrusted. It is somewhat firmer than in the preceding species **11**

11a   (10)   Zooecial tubes narrowest at point of origin, widening noticeably toward the distal end, therefore looking somewhat club-shaped. Ectocyst grayish, opaque, lightly encrusted, wrinkled, and flexible . .

                                                  Stolella indica Annandale 1909

      Colonies form coarse, openly branched, sometimes loosely tangled mats of grayish, very long, curved, club-shaped zoids. Interzooecial septa not complete. Tentacles number about 30 to 35. Floatoblasts similar to those of *Hyalinella punctata* in shape. Floatoblast total length 0.35–0.41 mm, total width 0.24–0.28 mm. Capsule length 0.26–0.305 mm, width 0.19–0.24 mm. Float length 0.05–0.10 mm, width 0.03–0.06 mm. Found at Westtown, Pa.

Fig. 19.11. *Stolella indica.* (*a*) Colony habit sketch. (*b*) Floatoblast, dorsal face. (*c*) Floatoblast, side view. (*d*) Floatoblast, ventral face.

11b   Zooecial tubes firm, brownish, translucent to opaque, often encrusted and of fairly uniform diameter throughout. . . . .

                                          Plumatella Lamarck   **12**

      The classification of this genus is unsatisfactory. The number of species is debatable. A number of species have been considered varieties of *P. repens.*

12a   (11)   Three kinds of statoblasts present. One type a sessoblast, two types floatoblasts. The distinguishing floatoblast has extremely pale, thin-walled valves, yellow in color. It is almost twice as long as wide, with capsule extending the whole length of the statoblast. The other type of floatoblast is a typical sturdy, dark-colored body . . . . . .

                                                P. casmiana Oka 1907

      Plumatella casmiana colonies and the second type of floatoblast can hardly be distinguished from *P. emarginata.* The above described thin-walled pale statoblast is the only truly distinctive feature between the two species. Found in Lake Erie (Ohio and Canada), Ind., Mich.

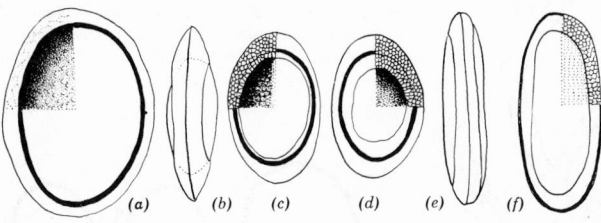

Fig. 19.12. *Plumatella casmiana.* (*a*) Sessoblast. One sector filled in to show faintly marked, light-colored vestigial annulus and the very large dark capsule. (*b*) Floatoblast of the ordinary (*Plumatella*) type, side view. (*c*) Floatoblast of ordinary type, ventral face; one sector filled in to show float (annulus) of fully-developed air cells and the extent to which the float encroaches on the darker central capsule. (*d*) Floatoblast of the ordinary type, dorsal face. (*e*) Statoblast of third type, special to *casmiana* species, side view. (*f*) Statoblast of the third type, special to *casmiana* species; one face, showing the rudimentary flat float and the capsule, which occupies the whole length and width of the floatoblast.

12b   Statoblasts of two kinds, one a dark sessoblast with vestigial annulus and the second a deeply colored brown or reddish-brown floato-blast **13**

13a (12) Floatoblasts long and narrow, more than twice as long as wide, with a medium-sized capsule. . . . . *P. fruticosa* Allman 1844

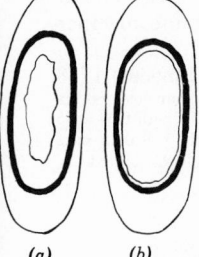

(a)   (b)

Colony brown in color, irregularly and openly branched, attached only at its origin. Zoid tubes cylindrical, furrowless, obscurely keeled. The distinctive features are the very elongate floatoblasts whose total length is 0.41–0.57 mm and total width 0.17–0.27 mm. Capsule length 0.22–0.27 mm and width 0.15–0.17 mm. Found in Ill. and Pa.

◀ **Fig. 19.13.** *Plumatella fruticosa* floatoblasts. (*a*) Dorsal face, where a larger part of the capsule is covered by the float. (*b*) Ventral, less covered face. (After Toriumi.)

13b   Floatoblasts range from nearly circular to ellipsoid, but the proportion of total length to total width is less than 2 to 1 . . . .   **14**

14a (13) Floatoblasts with float covering considerably more of one side of capsule than of the other. Floatoblasts noticeably flatter on more covered side . . . . *P. emarginata* Allman 1844

Has been regarded as a variety of *repens*. Zoids adherent along the greater part of their length; long, cylindrical, keeled, and furrowed, the furrow continuous with the notched emargination of the orifice. Encrusted ectocyst varies from a sandy to a very dark brown color, with the tips always considerably lighter. From 30 to 54 tentacles. Floatoblast total length 0.37–0.5 mm, total width 0.21–0.31 mm. Capsule length 0.23–0.29 mm, width 0.18–0.23 mm. Sessoblast total length 0.40–0.58 mm, width 0.27–0.36 mm. *P. emarginata* is widely distributed and has been reported from Ill., Ind., Mich., N. Y., Ohio, Pa., U., Wyo.

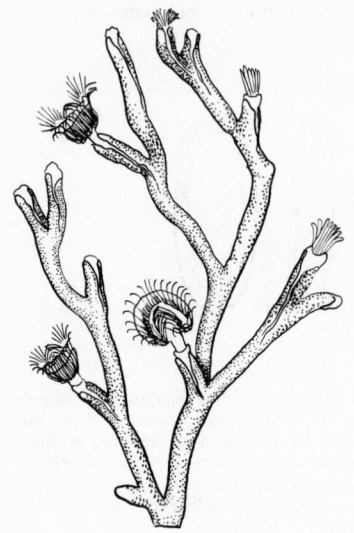

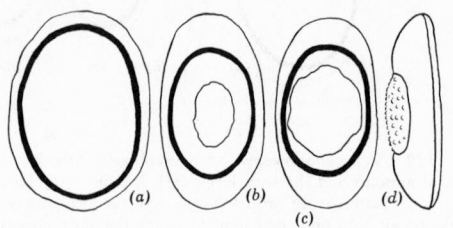

◄ **Fig. 19.14.** *Plumatella emarginata* colony showing the growth habit. Three tentacular crowns are fully extruded, three only partly so. (After Kraepelin.)

Fig. 19.15. *Plumatella emarginata* statoblasts. (*a*) Sessoblast, with large capsule and vestigial annulus. (*b*) Floatoblast, dorsal, more covered face. (*c*) Floatoblast, ventral, less covered face. (*d*) Floatoblast, side view.

14b   Floatoblasts with float covering approximately the same amount of capsule on each side. Float not noticeably flatter on one side than on the other **15**

15a   **(14)**   Colony with translucent, pale yellowish to deep amber-colored ectocyst. Zooecia adherent throughout only part of their length. Colony generally branches loosely, the zooecia do not fuse together into a solid mass of parallel tubes . . . .
                                   Plumatella repens (Linnaeus) 1758
      Zooecial tubes slender, cylindrical, pellucid. Keel generally absent from free zooecia but a faint one may be present on older basal ones. Floatoblast total length 0.34–0.45 mm; total width 0.22–0.29 mm; capsule length 0.24–0.31 mm, width 0.18–0.25 mm. Sessoblast total length 0.42–0.51 mm, total width 0.28–0.38 mm. Widely distributed.

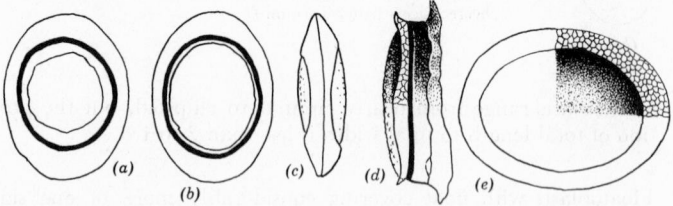

Fig. 19.16. *Plumatella repens* statoblasts. (*a*) Floatoblast, dorsal face. (*b*) Floatoblast, ventral face. (*c*) Floatoblast, side view. (*d*) Sessoblast, side view, showing the thick capsule, the vestigial annulus, and the uneven basal cementing substance at right. (*e*) Sessoblast, view of unattached face, showing the large dark capsule and its surrounding vestigial annulus.

15b   Colony with firm reddish to brown to dark gray ectocyst. Zooecia soon rise free from the substratum into closely cemented parallel vertical tubes. . . . . ***P. fungosa*** (Pallas) 1768

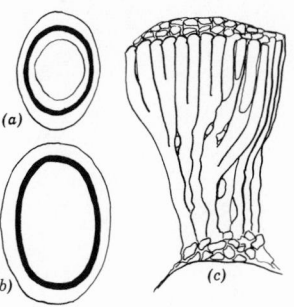

Has been regarded as a variety of *repens*. Colony forms compact firm masses around objects. Floatoblasts broadly ellipsoid, varying in total length 0.34–0.53 mm and in total width 0.21–0.4 mm. Sessoblasts 0.48 mm long, 0.39 mm wide. Found in Ill., Me., Mass.

◄ **Fig. 19.17.** *Plumatella fungosa.* (*a*) Floatoblast. (*b*) Sessoblast. (*c*) Habit sketch of part of colony, showing many zooecial tubes cemented together. (*a*, and *b* after Kraepelin; *c* after Allman.)

References

Allman, G. J. 1856. *A Monograph of the Fresh-Water Polyzoa.* Ray Society, London. **Brown, C. J. D. 1933.** A limnological study of certain fresh-water Polyzoa with special reference to their statoblasts. *Trans. Am. Microscop. Soc.*, 52:271–316. **Davenport, C. B. 1904.** Report on the fresh-water Bryozoa of the U. S. *Proc. U. S. Natl. Museum*, 27:211–221. **Hozawa, S. and M. Toriumi. 1940.** Some fresh-water Bryozoa found in Manchoukuo. *Rept. Limnobiol. Survey Kwantung and Manchoukuo*, 3:425–434. (In Japanese.) **Hyatt, A. 1866–1868.** Observations on Polyzoa, Suborder Phylactolaemata. *Commun. Essex Inst.*, 4:167–228; 5:97–112, 145–160, 193–232. **Hyman, L. H. 1951.** *The Invertebrates.* Vol. III, *Acanthocephala, Aschelminthes, and Entoprocta.* The pseudocoelomate Bilateria. McGraw-Hill, New York. **Kraepelin, K. 1887.** Die Deutschen Süsswasser-Bryozoen. Eine Monographie. I. Anat.-Syst. Teil. *Abhandl. Naturw. Verein, Hamburg*, 10:1–168. **Leidy, J. 1851.** On some American fresh-water Polyzoa. *Proc. Acad. Nat. Sci. Phila.*, 5:320–321. **Marcus, E. 1934.** Über *Lophopus crystallinus* (Pall.). *Zool. Jahrb. Abt. Anat. u. Ontog. Tiere*, 58:501–606. **1940.** Mosdyr (Bryozóa eller Polyzóa). Danmarks Fauna, Handbøger over den Danske Dyreverden. *Dansk Naturhist. Forening*, 46:1–401. **Rogick, M. D. 1940.** Studies on freshwater Bryozoa. IX. *Trans. Am. Microscop. Soc.*, 59:187–204. **1945a.** Studies on fresh-water Bryozoa. XV. *Ohio J. Sci.*, 45:55–79. **1945b.** Studies on fresh-water Bryozoa. XVI. *Biol. Bull.*, 89:215–228. **Rogick, M. D. and H. van der Schalie. 1950.** Studies on fresh-water Bryozoa. XVII. *Ohio J. Sci.*, 50:136–146. **Toriumi, M. 1942.** Studies on fresh-water Bryozoa of Japan. III. Freshwater Bryozoa of Hokkaidô. *Sci. Repts. Tôhoku Imp. Univ. Fourth Ser.*, 17:197–205. **1943.** Studies on freshwater Bryozoa of Japan, V. The variations occurring in the statoblasts and in the number of the tentacles of *Cristatella mucedo* Cuv. *Sci. Repts. Tôhoku Imp. Univ., Fourth Ser.*, 17:247–253. **1951.** Taxonomical study of freshwater Bryozoa, I. *Fredericella sultana* (Blum.). *Sci. Repts. Tôhoku Univ., Fourth Ser.*, 19:167–178.

20

Tardigrada

ERNESTO MARCUS

There is some question about whether the Tardigrades are to be regarded as more closely related to the Arthropoda or to the Annelida; some authors place them in a separate phylum. In this chapter they will be regarded as a class of the Arthropoda.

They are generally no more than 1 mm long, with a head and four trunk segments (Figs. 20.10, 20.11). The skin is cuticularized, as are the fore- and hind-gut. There are four pairs of legs, set off from the trunk, with claws, fingers, or disclike endings. Smooth muscles occur in metamerical groups for head, trunk, and legs. The nervous system has a brain (Figs. 20.1, 20.26, *br*) and two longitudinal ventral nerve cords united in four ganglia (*v*), with peripherical nerves. Many species have eyes. The mouth (Fig. 20.1, *mt*) is terminal or ventroterminal. The fore-gut is provided with two secretory and excretory glands (*ng*), two calcareous protrusible stylets (*st*), and a sucking pharynx (*ph*), and the mid-gut (*md*) has muscles. The hind-gut opens with ventral anus (Fig. 20.5, *an*). The gonad (*ov*) is an unpaired sac; the gonoducts open with ventral pore (*h*) or into the rectum (Fig. 20.1), which bears three glands (*rt*) in Meso- and Eutardigrada. Storage cells exist (Fig. 20.31, *cs*) in the body cavity. There are no respiratory or circulatory systems. Sexes are separate and the females are oviparous (Figs. 20.20, 20.21, 20.27). Development is direct (Fig. 20.19), the cuticle being molted.

The first description of the "little water bear" was given by J. Goeze (1773). The resistance to dryness of "il tardigrado" was studied by L. Spallanzani (1776), and the name "Tardigrades" was applied in L. Doyère's fundamental memoirs on the class (1840–1842).

There are nearly 350 species in mosses, lichens, and among algae. The animals are chiefly herbivorous. There are about a dozen marine species, the others are terrestrial or aquatic, generally without well-defined ecological limits. During desiccation of their habitat most species contract, become barrel-shaped (tuns Fig. 20.2), and are able to survive in a state of extremely diminished animation for many years, even during prolonged exposure to temperatures far below the freezing point of water. Some populations of generally lacustrine species do not survive desiccation, either as individuals or as eggs. Dry tuns and eggs may be dispersed by wind, and this causes the very wide distribution of many species. The dry stage differs from the cyst (Fig. 20.3) that is formed within the old cuticle (*ce*) and occurs in aquatic and terrestrial Eutardigrada. The latter are more frequent in water than Heterotardigrada. All Tardigrades crawl and do not swim.

Distribution

From North America north of the Mexican border, 36 nonmarine species are listed (Mathews 1938; Ramazzotti 1956), of which 18 are recorded from the United States. That is a very small number compared with Scotland, Sweden, France, Germany, Switzerland, and Italy, each of which has at least 40 to 50 species. Because many species have a wide range of distribution, the present key includes all nonmarine genera and 76 species and forms frequently or occasionally found in fresh water, although most of them are not yet reported from North America. Many more may be found in water, where probably all Tardigrades washed into this habitat by rains can live. Before considering a fresh-water Tardigrade not covered by the key as new, one should consult the literature.

Identification

Systematically important characters are the toes of the Arthrotardigrada (Fig. 20.4, *t*); the cephalic appendages (Figs. 20.10, 20.12) that are absent in the Eutardigrada (Figs. 20.1, 20.35); the sensory papillae of *Milnesium* (Fig. 20.14); the plates and their appendages in the Scutechiniscidae (Figs. 20.7–20.11); the length of the gullet (Fig. 20.36, *g*), the cuticular apophyses (*ap*) and placoids (*m*, *mc*) in the pharynx (Fig. 20.17); the claws (Figs. 20.28–20.30) and the eggs (Figs. 20.19–20.25) of the Macrobiotidae.

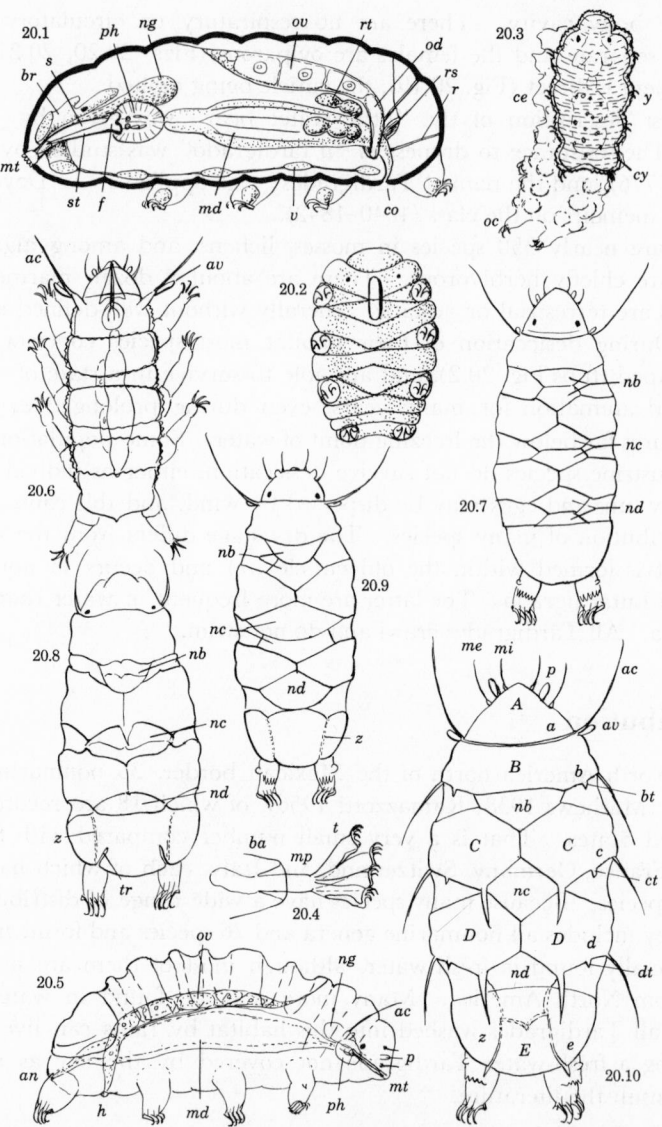

Fig. 20.1. General anatomy of a female Tardigrade (Genus *Macrobiotus*). **Fig. 20.2.** Dry stage of a *Macrobiotus* with bubble of air in the mouth tube. **Fig. 20.3.** Cyst of *Hypsibius nodosus*. **Fig. 20.4.** Leg of an Arthrotardigrade (*Styraconyx sargassi*). **Fig. 20.5.** *Parechiniscus chitonides*. (After Thulin.) **Fig. 20.6.** *Oreella mollis*. (After Murray.) **Fig. 20.7.** Plates of *Echiniscus* (*Bryochoerus*). **Fig. 20.8.** Plates of *Echiniscus* (*Hypechiniscus*). (After Murray and Thulin.) **Fig. 20.9.** Plates of *Echiniscus* (*Bryodelphax*). **Fig. 20.10.** Plates and appendages of *Echiniscus* (*Echiniscus*).

Fig. 20.11. *Pseudechiniscus suillus*. **Fig. 20.12.** Head of *Pseudechiniscus tridentifer*. (After Bartoš.) **Fig. 20.13.** *Mopsechiniscus imberbis*. (After du Bois-Reymond Marcus.) **Fig. 20.14.** Head of *Milnesium tardigradum* with its sensory papillae. **Fig. 20.15.** Claws of *Macrobiotus islandicus*. **Fig. 20.16.** Claws of *Macrobiotus pullari*. **Fig. 20.17.** Buccal apparatus of a Eutardigrade (Family Macrobiotidae.) **Fig. 20.18.** Claws of *Macrobiotus ambiguus*. **Fig. 20.19.** *Macrobiotus hufelandii* hatching from the egg. **Fig. 20.20.** Egg of *Macrobiotus richtersii*. **Fig. 20.21.** Egg of *Macrobiotus hastatus*. **Fig. 20.22.** *Macrobiotus occidentalis*, process of egg shell. **Fig. 20.23.** *Macrobiotus furciger*, process of egg shell. (After Murray.) **Fig. 20.24.** *Macrobiotus ampullaceus*, processes of egg shell. (After Thulin.) **Fig. 20.25.** *Macrobiotus dispar*, processes of egg shell.

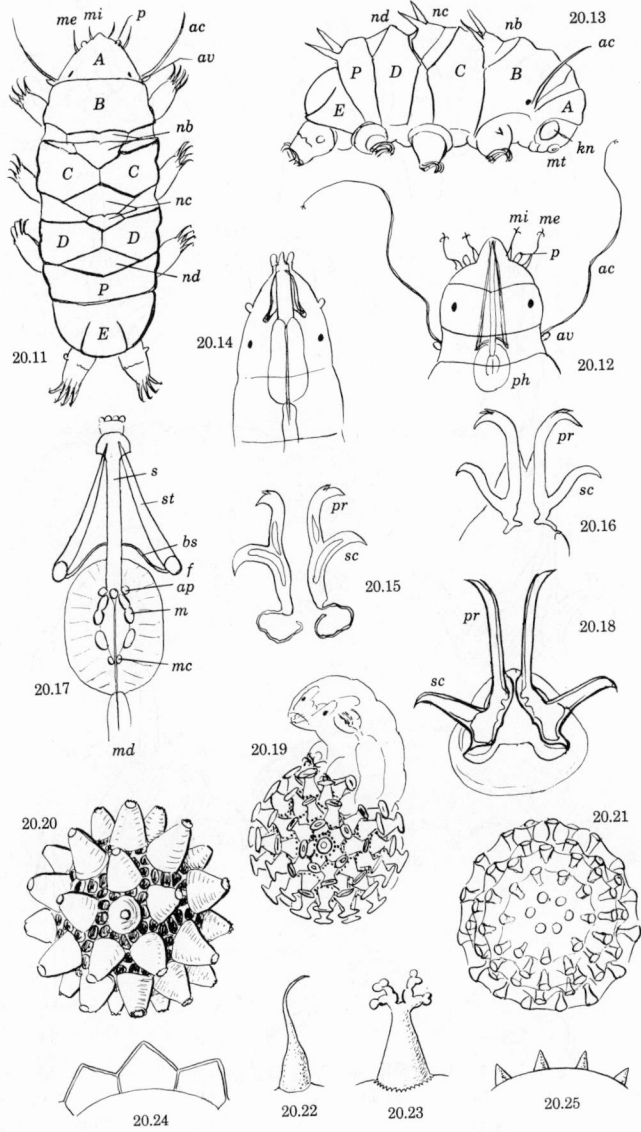

A, head plate; *a*, hind corner of head plate; *ac*, lateral cirrus; *an*, anus; *av*, clava; *B*, first segmental or shoulder plate; *b*, hind corner of shoulder plate; *ba*, basal part of leg; *br*, brain; *bs*, bearer of stylet; *bt*, dorsolateral spicule over *b;* *C*, second segmental plates; *c*, hind corner of 2nd plates; *cd*, dorsal spine over *c; ce*, old cuticle; *co*, orifice of cloaca; *cs*, storing cells; *ct*, dorsolateral spicule over *c; cy*, cuticle of cyst; *D*, third segmental plates; *d*, hind corner of 3rd plates; *dd*, dorsal spine over *d; dt*, dorsolateral spicule over *d; E*, end plate; *e*, hind corner of end plate; *f*, furca; *h*, gonopore; *kn*, cephalic knob; *m*, macroplacoid; *mc*, microplacoid; *md*, mid-gut; *me*, external medial cirrus; *mi*, internal medial cirrus; *mp*, middle part of leg; *mt*, mouth; *n*, claws; *nb*, first intersegmental plate; *nc*, second intersegmental plate; *nd*, third intersegmental plate; *ng*, buccal gland; *oc*, old claw; *od*, oviduct; *ov*, ovary; *P*, pseudosegmental plate; *p*, cephalic papilla; *ph*, pharynx; *pr*, principal branch of claw; *r*, rectal muscles; *rs*, receptaculum seminis; *rt*, rectal glands; *s*, mouth tube; *sc*, secondary branch of claw; *st*, stylet; *t*, toe with claw (distal part of leg); *tr*, furrow of trefoil; *v*, ganglion of ventral nerve cord; *y*, young claw; *z*, facet.

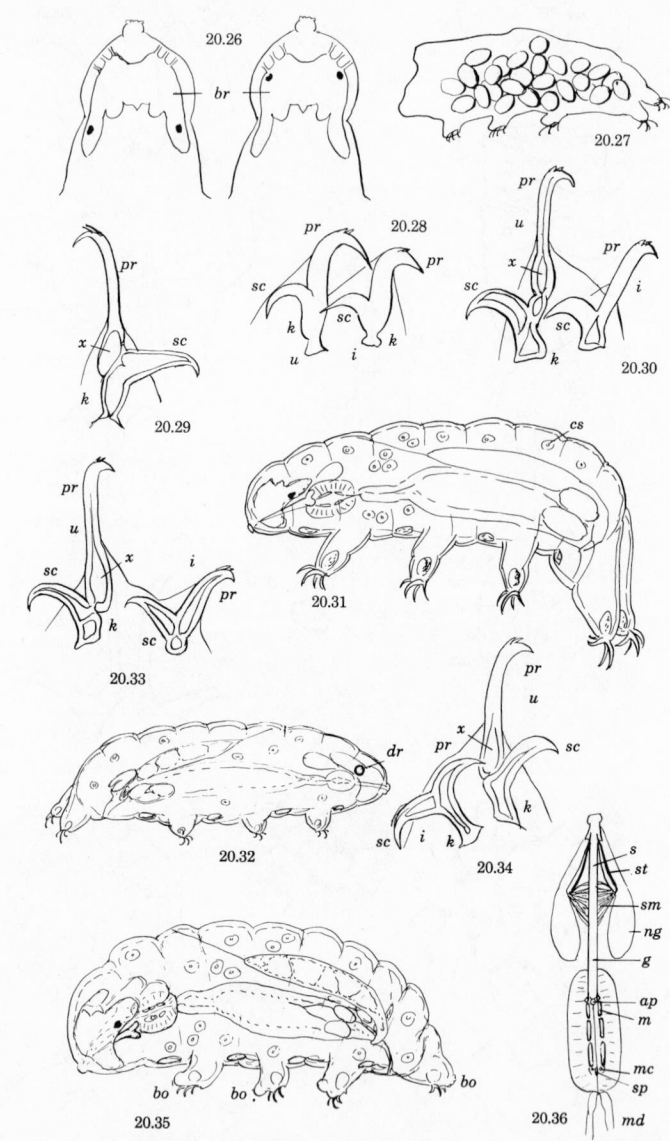

Fig. 20.26. Heads of *Macrobiotus hufelandii* with posterior and *Macrobiotus furciger* with anterior eyes. (After Thulin.)   **Fig. 20.27.** Eggs of *Hypsibius augusti* laid in the moulted cuticle.   **Fig. 20.28.** *Hypsibius (Calohypsibius) verrucosus,* claws of the fourth right leg.   **Fig. 20.29.** *Hypsibius (Isohypsibius) myrops,* outer (posterior) claw.   **Fig. 20.30.** *Hypsibius (Diphascon) recamieri,* claws.   **Fig. 20.31.** *Hypsibius (Isohypsibius) augusti.*   **Fig. 20.32.** *Hypsibius (Isohypsibius) myrops.*   **Fig. 20.33.** *Hypsibius (Hypsibius) dujardini,* claws of the fourth right leg.   **Fig. 20.34.** *Hypsibius (Hypsibius) convergens,* claws of the fourth left leg. **Fig. 20.35.** *Hypsibius (Hypsibius) evelinae.*   **Fig. 20.36.** Buccal apparatus of *Hypsibius (Diphascon) scoticus.*

ap, apophyses; *bo,* boss above the claws; *br,* brain; *cs,* storing cells; *dr,* fat droplet; *g,* gullet; *i,* inner (anterior) claw; *k,* common base of claw; *m,* macroplacoid; *mc,* microplacoid; *md,* mid-gut; *ng,* buccal gland; *pr,* principal branch of claw; *s,* mouth tube; *sc,* secondary branch of claw; *sm,* muscles of stylet; *sp,* septulum (cuticular disc in pharynx); *st,* stylet; *u,* outer (posterior) claw; *x,* flexible piece of claw.

For classification adult animals with symmetrical appendages should be used. Young Scutechiniscidae, recognizable by having only two instead of four claws (Fig. 20.10, *n*), may have the appendages of the plates still incomplete. Molting young and adult Macrobiotidae cast off the cuticular parts (placoids, stylets, etc.) of their buccal apparatus, and such "simplex-stages" cannot be identified. Eggs squeezed out of the body of Macrobiotidae may have incomplete processes; young animals pressed out of the eggs may show placoids and claws different from those of the adults.

The texture of the plates of Scutechiniscidae as well as the placoids and claws of Eutardigrada are examined with high power in water, lightly compressed below a cover slip. For distinguishing between *Isohypsibius* (Fig. 20.29) and *Hypsibius* or *Diphascon* (Fig. 20.30) an outer (posterior) claw must be laid so that it shows the secondary branch (*sc*) exactly in profile. This is easiest with one of the six anterior legs. In *Isohypsibius* a right angle is formed between the dorsal margin of the secondary branch and the base (*k*), and these margins form an even arch in *Hypsibius* and *Diphascon*.

KEY TO SPECIES

1a		With a lateral cirrus (Figs. 20.10, 20.11, *ac*) on each side of the head			**2**
1b		Without a lateral cirrus . . . . Order **Eutardigrada**			**25**
2a	(1)	Pharynx with placoids (as in Figs. 20.17, 20.36) . . . . Order **Mesotardigrada**			
		One species from hot springs in Japan: *Thermozodium esakii* Rahm 1937.			
2b		Pharynx without placoids . . . . Order **Heterotardigrada**			**3**
3a	(2)	Claws, fingers, or adhesive discs joined to the legs by toes (Fig. 20.4, *t*) . . . . Suborder **Arthrotardigrada**			
		Marine; about 9 species.			
3b		Claws (Fig. 20.10, *n*) immediately inserted in legs, without intermediate toes . . . . Suborder **Echiniscoidea**			**4**
4a	(3)	Cuticle of the back segmented by metamerical thickenings, pads, or plates . . . . Family **Scutechiniscidae**			**5**
4b		Cuticle of the back smooth or papillose, without metamerical thickenings, pads, or plates . . . . Family **Nudechiniscidae**			
		One genus marine (1 species) and another, ***Oreella*** J. Murray 1910 (Fig. 20.6), with 2 species, terrestrial, but possibly sometimes in fresh water.			
5a	(4)	Dorsal cuticle with platelike metamerical thickenings behind the first pair of legs . . . . Genus ***Parechiniscus*** Cuénot 1926			
		One species; Fig. 20.5.			
5b		The whole back including the head with distinct plates			**6**
6a	(5)	End plate (Fig. 20.10, *E*) behind the third segmental plates (*D*) or separated from them by 1 or 2 intersegmental plates (*nd*). . . . . . . Genus ***Echiniscus*** S. Schultze 1840			**7**
		About 122 species.			
6b		End plate (Fig. 20.11, *E*) adjacent to the fourth paired or unpaired pseudosegmental plates (*P*)			**22**
7a	(6)	More than 3 intersegmental plates result from bipartition of first and second, or first, second, and third intersegmental plates			**8**

7b Not more than 3 intersegmental plates (Fig. 20.10, *nb, nc, nd*) . . . .
Subgenus *Echiniscus* Thulin 1928 **10**

8a (7) First, second, and third intersegmental plates (Fig. 20.7, *nb, nc, nd*) each divided in two. . . . . . . Subgenus *Bryochoerus* Marcus 1936
One species, in terrestrial mosses and lichens.

8b First and second intersegmental plates divided in two **9**

9a (8) End plate faceted (Fig. 20.8, *z*) and trefoliate (*tr*) . . . .
Subgenus *Hypechiniscus* Thulin 1928
One species: *E.* (*H.*) *gladiator* J. Murray 1905, in inundated mosses (Scottish Lochs); also in land moss. British Columbia.

9b End plate faceted (Fig. 20.9, *z*) but not trefoliate; i.e., without lateral furrows . . . . Subgenus *Bryodelphax* Thulin 1928
Two species, in terrestrial mosses and lichens.

10a (7) The only lateral appendages are the hairs at *a* (lateral cirri) **11**

10b Besides the hair at *a* there are more lateral appendages **14**

11a (10) Head plate almost or entirely smooth . . . . **11'**

11b Head plate sculptured . . . . **11''**

11'a (11) With third intersegmental plate . . . .
Echiniscus (*Echiniscus*) *mauccii* Ramazzotti 1956
Wis.

11'b Without third intersegmental plate . . . .
E. (*E.*) *phocae* du Bois-Reymond Marcus 1944
Wis.

11''a (11) Claws (Fig. 20.10, *n*) 12–18 μ long **12**

11''b Claws 25 μ long. **13**

12a (11') End plate faceted (as *E* in Fig. 20.10). . . . .
E. (*E.*) *reticulatus* J. Murray 1905

12b End plate not faceted (as *E* in Fig. 20.11). . . . . **'** . .
E. (*E.*) *wendti* Richters 1903

13a (11') Plates unsculptured, olivaceous with darker green spots; cephalic papilla (Fig. 20.10, *p*) much shorter than cephalic medial cirri (*me, mi*) . . . . *E.* (*E.*) *viridis* J. Murray 1910

13b Plates sculptured, grayish pink; cephalic papilla as long as cephalic medial cirri . . . . *E.* (*E.*) *calvus* Marcus 1931

14a (10) Except the lateral cirri, all lateral appendages are spines . . . .
E. (*E.*) *spiniger* Richters 1904

14b Besides the lateral cirri (hairs at *a*), with other setalike lateral appendages. **15**

15a (14) With third intersegmental plate (Fig. 20.10, *nd*) **16**

15b Without third intersegmental plate **20**

16a (15) A tiny spiculum (Fig. 20.10 at *e*. See also Fig. 20.8, *tr*) in the furrow of the trefoil . . . . *E.* (*E.*) *granulatus* (Doyère) 1840

16b A seta or longer spine at *e*. **17**

17a (16) End plate faceted (as *E* in Fig. 20.10) **18**

17b End plate not faceted (as *E* in Fig. 20.10) **19**

18a (17) With seta at *b* . . . .
E. (*E.*) *merokensis* Richters forma *suecica* Thulin 1911

18b Without seta at *b* . . . . *E.* (*E.*) *merokensis* Richters 1904
Wash.

19a (17) With furcate seta at *a* or *d* or both, on one or both sides of animal . . . .
E. quadrispinosus Richters forma *fissispinosa* J. Murray 1907

19b		No furcate appendages	
		E. (E.) quadrispinosus Richters forma *cribrosa* J. Murray 1907	
20a	**(15)**	Lateral appendages ending with a knob	
		E. (E.) tympanista J. Murray 1911	
20b		Lateral appendages without terminal thickenings	**21**
21a	**(20)**	A seta at *e*. . . . . *E. (E.) oinonnae* Richters 1903	
		British Columbia, Wis.	
21b		No seta at *e*, only a spiculum (Fig. 20.10, *e*) in the furrow of the trefoil . . . . *E. (E.) spitsbergensis* Scourfield 1897	
22a	**(6)**	Head with 2 lateral cirri (Fig. 20.13, *ac*) and 2 broad, flat knobs (*kn*) . . . . . Genus *Mopsechiniscus* du Bois-Reymond Marcus 1944	
		One species.	
22b		Head with 2 lateral cirri (Fig. 20.11, *ac*), 2 clavae (*av*), 2 external (*me*), 2 internal (*mi*) cirri and 2 cephalic papillae (*p*)	
		Genus *Pseudechiniscus* Thulin 1911	**23**
		About 16 species.	
23a	**(22)**	External and internal medial cirri tridentate at their points (Fig. 20.12, *me*, *mi*); the filaments at *a* (lateral cirri) and *e* with 3 small denticles. Aquatic . . . . *P. tridentifer* Bartoš 1936	
23b		Without tridentate cirri or appendages	**24**
24a	**(23)**	Lateral cirrus a hair; pseudosegmental plate unpaired (Fig. 20.11, *P*).	**24′**
24b		Lateral cirrus bladelike; pseudosegmental plates paired; hygrophilous . . . . *P. cornutus* (Richters) 1906	
24′a	**(24)**	Posterior border of pseudosegmental plate straight	
		P. suillus (Ehrenberg) 1853	
		Occasionally in submerged mosses. Alaska, Calif., Vt.	
24′b		Posterior border of pseudosegmental plate with two paramedian lobes . . . . *P. ramazzottii* Maucci 1952	
		Wis.	
25a	**(1)**	With 6 sensory papillae around the mouth and 2 a little farther behind (Fig. 20.14). . . . . Family **Milnesiidae**	
		One species: *Milnesium tardigradum* Doyère 1840, occasionally limnetic. Ill., Wash., Wis., British Columbia, Canada.	
25b		Without sensory papillae around the mouth	
		Family *Macrobiotidae*	**26**
26a	**(25)**	Two unbranched claws on each leg; each claw with a bifid point. . .	
		Genus *Haplomacrobiotus* May 1948	
		One species.	
26b		The two claws of every leg (Figs. 20.15, 20.30) composed of a principal (*pr*) and a secondary (*sc*) branch each	**27**
27a	**(26)**	The principal branches face each other, the secondary branches opposed to each other (Fig. 20.15)	
		Genus *Macrobiotus* S. Schultze 1834	**28**
		About 94 species.	
27b		The principal branches of the claws parallel and directed forwards, the secondary branches parallel and directed backwards (Figs. 20.28, 20.30)	**54**
28a	**(27)**	Cuticle at least partially bossed, granulated, or distinctly dotted . . .	**29**
28b		Cuticle smooth, at most slightly dotted	**33**
29a	**(28)**	Two dorsal cones between third and fourth legs; usually aquatic . . .	
		M. dispar J. Murray 1907	
29b		Without isolated cones.	**30**

30a (29) Texture most striking on the legs . . *M. echinogenitus* Richters 1904
British Columbia, Canada.

30b   Texture distinct on the whole back   31

31a (30) Three macroplacoids (Fig. 20.36, *m*) . . *M. furcatus* Ehrenberg 1859

31b   Two macroplacoids (Fig. 20.17, *m*)   32

32a (31) Processes of egg shell straight, rigid cones or spines; diameter of egg
(without processes) 0.09–01 mm . . . . *M. islandicus* Richters 1904

32b   Processes of egg shell ending with soft, undulate bristles (Fig. 20.22);
diameter of egg (without processes) 0.058 mm . . . .
M. occidentalis J. Murray 1910

33a (28) Egg shell smooth (as in Fig. 20.27), aquatic . . . .
M. macronyx Dujardin 1851

33b   Egg shell with processes   34

34a (33) Processes embedded in a hyaline outer zone of the shell (Fig. 20.21);
hygrophilous, in turf mosses . . . . . . *M. hastatus* M. Murray 1907

34b   No common cuticular mantle around the processes   35

35a (34) Three macroplacoids (Fig. 20.36, *m*)   36

35b   Two macroplacoids (Fig. 20.17)   43

36a (35) Placoids as long as broad; mouth tube (Fig. 20.17, *s*) narrow. . . . .
M. intermedius Plate 1888
Canada.

36b   Placoids longer than broad; mouth tube wide   37

37a (36) Egg shell with polygonal depressions around the processes (Fig.
20.20) . . . . *M. richtersii* J. Murray 1911

37b   Egg shell not areolated   38

38a (37) Processes of egg shell ramified (Fig. 20.23) . . . .
M. furciger J. Murray 1907

38b   Processes of egg shell not ramified, sometimes slightly fringed.   39

39a (38) Branches of claw united about half way (Fig. 20.15)   40

39b   Branches of claw diverging from the base, V-shaped (Fig. 20.16)   42

40a (39) Processes of egg shell like upside down egg cups (Fig. 20.19) . . . . .
M. hufelandii S. Schultze 1833
Mich., Wash., D. C., Wis.

40b   Processes of egg shell without distal discs   41

41a (40) Processes bulbous or conical . . . .   41'

41b   Processes hemispherical . . . . *M. montanus* J. Murray 1910

41'a (41) Bases of processes surrounded by dots . . . .
M. harmsworthi var. *coronata* Barros 1942
Calif.

41'b   No dots around bases of processes . . . .   41''

41''a (41') Second placoid as long as first and third . . . .
M. harmsworthi J. Murray 1907
Canada.

41''b   Second placoid half as long as first and third, nearly contiguous
with first. . . . . *M. tonollii* Ramazzotti 1956
Wis.

42a (39) Pharynx long-oval; stylets much dilated in basal half . . . .
M. dubius J. Murray 1907

42b   Pharynx short-oval; stylets only with the usual broadened furca
(Fig. 20.1, *f*); aquatic . . . . *M. pullari* J. Murray 1907

43a (35) Secondary branch of claw much shorter than the principal one and forming a nearly right angle with it (Fig. 20.18) **44**

43b Principal and secondary branch of claw only little different in size, either united about half way (Fig. 20.15) or diverging from common base (Fig. 20.16) **47**

44a (43) Smooth eggs laid in the molted skin (Fig. 20.27); aquatic . . . .
M. macronyx Dujardin 1851

44b Single eggs with processes on the shell **45**

45a (44) Processes separate from each other (Fig. 20.25); usually aquatic. . . .
M. dispar J. Murray 1907

45b Processes touching each other at their bases **46**

46a (45) Claws long and fine, principal branch slightly curved; frequently in water . . . . *M. ambiguus* J. Murray 1907

46b Claws short and thick, principal branch strongly curved; in permanently wet and in drying mosses (Fig. 20.24) . . . .
M. ampullaceus Thulin 1911

47a (43) Stylets much dilated in basal half . . . . *M. dubius* J. Murray 1907

47b Stylets with only the usual broadened furca (Fig. 20.1, *f*) **48**

48a (47) Egg shell with polygonal fields around the processes . . . .
M. grandis Richters 1911

48b Egg shell not areolated **49**

49a (48) Eyes in front of the constriction of the head (Fig. 20.26) . . . .
M. furciger J. Murray 1907

49b Eyes behind the constriction of the head or on its level **50**

50a (49) Processes of egg shell end with undulate bristles (Fig. 20.22) . . . . .
M. occidentalis J. Murray 1910

50b Processes of egg shell rigid, not with bristlelike terminations **51**

51a (50) Pharynx without microplacoid **52**

51b Pharynx with microplacoid (Fig. 20.17, *mc*) **53**

52a (51) Principal (Fig. 20.15, *pr*) and secondary (*sc*) branch of claw united about half way . . . . *M. islandicus* Richters 1904

52b Principal (Fig. 20.16, *pr*) and secondary (*sc*) branch of claw diverging from common base, claws V-shaped; aquatic . . . .
M. pullari J. Murray 1907

53a (51) Processes of egg shell end with discs (Fig. 20.19) . . . .
M. hufelandii S. Schultze 1833
See 76.

53b Processes of egg shell pointed or rounded cones . . . .
M. echinogenitus Richters 1904
See 57.

54a (27) Pharynx always without placoids . . . Genus *Itaquascon* Barros 1938
One species.

54b Pharynx with placoids that are absent only during the molt ("simplex stage") . . . . Genus *Hypsibius* Ehrenberg 1848 **55**
About 98 species.

55a (54) A flexible piece (Figs. 20.29, 20.30, *x*) within the outer (posterior) claw of each leg unites the principal (*pr*) and secondary (*sc*) branches **59**

55b No flexible piece in any claw (Fig. 20.28) . . . .
Subgenus *Calohypsibius* Thulin 1928 **56**

56a **(55)** Cuticle smooth; hygrophilous . . . .
Hypsibius (Calohypsibius) ornatus (Richters) forma *caelata*
Marcus 1928

56b   Cuticle with spines, warts, or nodules. **57**

57a **(56)** Spines present . . . . *H. ornatus* (Richters) 1900

57b   Without spines **58**

58a **(57)** Round nodules of equal size forming transverse rows on the back;
hygrophilous . . . .
H. (C.) ornatus (Richters) forma *caelata* Marcus 1928

58b   Frequently angular warts of irregular size and partly coalescent,
not in rows . . . . *H. (C.) verrucosus* (Richters) 1900

59a **(55)** The secondary branch (Fig. 20.29, *sc*) of the outer (posterior) claw
forms a right angle with the common base (*k*) of the claw . . . . . .
Subgenus *Isohypsibius* Thulin 1928   **60**

59b   The secondary branch (Fig. 20.30, *sc*) of the outer claw (*u*) continues the common base (*k*) of the claw evenly arched **73**

60a **(59)** Back with granules, papillae, warts, or spines **61**

60b   Cuticle of back, not always that of the legs, smooth **69**

61a **(60)** Body slender, legs long (Fig. 20.31). **62**

61b   Body stumpy, legs short (Fig. 20.32) **64**

62a **(61)** Two macroplacoids; frequently in water . . . .
Hypsibius (Isohypsibius) annulatus (J. Murray) 1905

62b   Three macroplacoids **63**

63a **(62)** Back with hemispheric bosses or regular granules, ventral side and
legs smooth; claws of fourth leg nearly equal in size . . . .
H. (I.) asper (J. Murray) 1906

63b   Small irregular granules all over the body and the legs; claws of
fourth leg very different in size; aquatic, but also in drying mosses . .
H. (I.) granulifer (Thulin) 1928

64a **(61)** With spines or pointed processes **65**

64b   No acuminate processes **67**

65a **(64)** Warts with spines between them . . . *H. (I.) sattleri* (Richters) 1902

65b   Mamillary pointed processes **66**

66a **(65)** Processes small, separated at their bases, not covering the whole
back . . . . *H. (I.) papillifer* (J. Murray) 1905

66b   Processes large, touching each other at their bases, covering the
whole back; aquatic . . . .
H. (I.) papillifer (J. Murray) forma *bulbosa* Marcus 1928

67a **(64)** Cuticular granulation consists of refractive areolae; aquatic. . . . . . .
H. (I.) baldii Ramazzotti 1945

67b   Cuticle bossed **68**

68a **(67)** Ten transverse rows of bosses, the first and tenth with an odd number (5) of them . . . . *H. (I.) tuberculatus* (Plate) 1888
British Columbia.

68b   Seven transverse rows of bosses, all with even numbers, so that the
dorsal mid-line is free of bosses . . . *H. (I.) nodosus* (J. Murray) 1907

69a **(60)** Legs long; aquatic. (Fig. 20.31) . *H. (I.) augusti* (J. Murray) 1907
Me.

69b   Legs short (Fig. 20.32) **70**

70a	**(69)**	Mouth terminal	**71**
70b		Mouth ventral, subterminal	**72**
71a	**(70)**	Posterior eyes; pharynx oval, with apophyses (Fig. 20.36, *ap*), 3 thick macroplacoids and microplacoid . . . . *H. (I.) prosostomus* (Thulin) 1928	
71b		Fat droplets (Fig. 20.32, *dr*) in the place of anterior eyes (see Fig. 20.26); pharnyx longish oval, without apophyses and microplacoid, with 3 thin macroplacoids ; aquatic. . . . . *H. (I.) myrops* du Bois-Reymond Marcus 1944	
72a	**(70)**	Outer (posterior) and inner (anterior) claw differ slightly in length . *H. (I.) tetradactyloides* (Richters) 1907	
72b		Outer claw much longer than inner claw. . . . . *H. (I.) schaudinni* (Richters) 1909 Texas.	
73a	**(59)**	Gullet not longer than half the length of the pharynx (Fig. 20.17) . . Subgenus **Hypsibius** (Thulin) 1928	**74**
73b		Gullet (Fig. 20.36, *g*) never less than half the length of the pharynx (*ph*) . . . . Subgenus **Diphascon** (Plate) 1888	**80**
74a	**(73)**	Two macroplacoids deeply 2-lobed on outer side . . . . *Hypsibius (Hypsibius) zetlandicus* (J. Murray) 1907	
74b		No unilateral outer notch of the 2 macroplacoids, at most a constriction in the first.	**75**
75a	**(74)**	Egg shell with processes	**76**
75b		Egg shell smooth (Fig. 20.27)	**77**
76a	**(75)**	Processes of egg shell short pegs embedded in a hyaline outer zone of the shell . . . . *H. (H.) arcticus* (J. Murray) 1907 British Columbia.	
76b		Processes of egg shell conical or hemispherical, projecting over the shell . . . . *H. (H.) oberhaeuseri* (Doyère) 1840 Calif., Wis., British Columbia.	
77a	**(75)**	Legs smooth, without bosses.	**78**
77b		Legs with bosses (Fig. 20.35, *bo*) dorsally over the claws. . . . . *H. (H.) evelinae* (Marcus) 1928	
78a	**(77)**	Pharynx nearly as broad as long . . *H. (H.) pallidus* (Thulin) 1911	
78b		Pharynx distinctly longer than broad	**79**
79a	**(78)**	Macroplacoids thin rods; the branches of both claws united at the bases only (Fig. 20.33), and much longer than the common bases; aquatic. . . . . *H. (H.) dujardini* (Doyère) 1840	
79b		Macroplacoids broad rods, or the second nutlike; secondary branch (Fig. 20.34, *sc*) of inner claw (*i*) as long as the common base (*k*). . . . . . *H. (H.) convergens* (Urbanowicz) 1925 Niagara Falls.	
80a	**(73)**	Pharynx with 2 macroplacoids	**81**
80b		Pharynx with 3 macroplacoids (Fig. 20.36, *m*)	**85**
81a	**(80)**	With eyes.	**82**
81b		Without eyes	**84**
82a	**(81)**	Eight pairs of dorsal bosses; cuticle on back and sides closely beset with granules; aquatic. . . . . *Hypsibius (Diphascon)* (Bartoš) 1937	
82b		No bosses, cuticle smooth	**83**

83a (82) Gullet about as long as pharynx . . . .
Hypsibius (Diphascon) conjungens (Thulin) 1911
83b   Gullet almost twice as long as pharynx . . . .
H. (D.) recamieri (Richters) 1911
84a (81) Posterior part of gullet annulate; no microplacoid . . . .
H. (D.) angustatus (J. Murray) 1905
84b   Whole gullet smooth; with microplacoid . . . . 84′
84′a (84) Pharynx twice as long as broad . . . .
H. (D.) spitzbergensis (Richters) 1903
84′b   Pharynx spherical or short oval . . . .
H. (D.) oculata forma *vancouverensis* Thulin 1911
British Columbia, Calif.
85a (80) Gullet longer than pharynx 86
85b   Gullet as long as pharynx 87
86a (85) Body small, short, broad; macroplacoids equal in size; outer and inner claw of one leg differ little . . *H. (D.) chilenensis* (Plate) 1888
Canada.
86b   Body big, longish, narrow; macroplacoids increase in length and diameter from first to third; outer and inner claw very different in size. . . . . *H. (D.) alpinus* (J. Murray) 1906
Canada.
87a (85) Mouth terminal (Fig. 20.32); series of placoids hardly longer than half the length of the pharynx . *H. (D.) prorsirostris* (Thulin) 1928
87b   Mouth ventral, subterminal (Fig. 20.35); series of placoids longer than half the length of the pharynx. . . . .
H. (D.) scoticus (J. Murray) 1905
British Columbia, Canada.

References

Barros, R. de. 1942–43. Tardigrados do Estado de S. Paulo. *Rev. Bras. Biol.*, 2:257–269; 2:373–386; 3:1–10. **Bartoš, E. 1941.** Studien über die Tardigraden des Karpathengebietes. *Zool. Jahrb. Syst.*, 74:435–472. **Cuénot, L. 1932.** Tardigrades. *Faune de France*, 24:1–96. **du Bois-Reymond Marcus, E. 1944.** Sobre Tardigrados brasileiros. *Commun. Zool. Mus.*, 1:1–19. **Englisch, H. 1936.** Uber die lateralen Darmanhangsdrüsen und die Wohnpflanzen der Tardigraden. *Zool. Jahrb. Syst.*, 68:325–352. **Heinis, F. 1910.** Systematik und Biologie der moosbewohnenden Rhizopoden, Rotatorien und Tardigraden der Umgebung von Basel, etc. *Arch. Hydrobiol.*, 5:1–115. **Marcus, E. 1929.** Tardigrada. In: Bronns *Klassen und Ordnung*, Band IV, Abt. IV, Buch 3. Akademische Verlagsgesellschaft, Leipzig. **1936.** *Tardigrada.* In: *Das Tierreich*, 66. Gruyter, Berlin and Leipzig. **Mathews, G. B. 1938.** Tardigrada from North America. *Am. Midland Naturalist*, 19:619–627. **May, R. M. 1948.** *La vie des Tardigrades.* J. Rostand (ed.). *Histoires naturelles*, Librairie Gallimard, Paris. **Müller, J. 1935.** Zur vergleichenden Myologie der Tardigraden. *Z. wiss. Zool.*, 147:171–204. **Murray, J. 1910.** *Tardigrada.* British Antarctic Expedition 1907–09, Vol. 1, Part 5, pp. 81–185. London. **Pennak, R. W. 1940.** Tardigrada. Ecology of the microscopic Metazoa, etc. *Ecol. Monographs*, 10:572–579. **Petersen, B. 1951.** The Tardigrade fauna of Greenland. *Medd. Grønland*, 150:1–94. **Pigoń, A. and B. Weglarska. 1953.** The respiration of Tardigrada: a study in animal anabiosis. *Bull. acad. polon. sci. Classe II*, 1:69–72. **1955a.** Anabiosis in Tardigrada. *Bull. acad. polon. sci. Classe II*, 3:31–34. **1955b.** Rate of metabolism in tardigrades during active life and anabiosis. *Nature*, 176:121–122. **Rahm, G. 1937.** Eine neue Tardigraden-Ordnung, etc. *Zool. Anz.*, 120:65–

71.   **Ramazzotti, G. 1945.** I tardigradi d'Italia. *Mem. ist. ital. idrobiol. Dott. Marco De Marchi*, 2:29–166.   **1954.** Nuove tabelle di determinazione dei generi *Pseudechiniscus* ed *Echiniscus* (Tardigrada). *Mem. ist. ital. idrobiol. Dott. Marco De Marchi*, 8:177–204.   **1956.** Tre nuove specie di Tardigradi ed altre specie poco comuni. *Atti soc. ital. sci. nat.*, 95:284–291.   **1958.** Nuove tabelle di determinazione dei generi *Macrobiotus* e *Hypsibius* (Tardigradi). *Mem. ist. ital. idrobiol. Dott. Marco De Marchi*, 10:69–120.   **Rodriguez-Roda, J. 1952.** Tardigrados de la fauna española. *Trabajos Mus. Ci. Nat. new ser. Zool.*, 1:1–86.   **Thulin, G. 1911.** Beiträge zur Kenntnis der Tardigradenfauna Schwedens. *Arkiv Zool.*, 7:1–60.   **1928.** Ueber die Phylogenie und das System der Tardigraden. *Hereditas*, 11:207–266.

21

Oligochaeta

CLARENCE J. GOODNIGHT

Except for a relatively small number of marine species, the oligochaetes are fresh-water and terrestrial animals. The body is clearly divided into segments which are separated externally by distinct grooves. Anterior and dorsal to the mouth is the prostomium, a small extension which is variously modified in different species and has some diagnostic value. The mouth is contained in segment 1, the peristomium; posteriorad from this, the segments are numbered consecutively. The boundaries between the segments are indicated as 5/6, 6/7, etc.

The bristlelike setae which aid the worms in locomotion are easily seen. Their shape and arrangement are often of taxonomic importance. They occur in all fresh-water oligochaetes but the branchiobdellids. In general, two types are recognized: the hair-setae, which are slender and flexible, and the needle setae, which are heavier and less flexible. The needle setae are variously shaped, varying from simple, straight needles to the highly modified "crotchet" setae (Fig. 21.1). The "crotchet" setae are greatly curved and commonly end in a double prong. Many setae possess a swelling, the nodulus. Another modification of diagnostic value is the variation of the distal end of

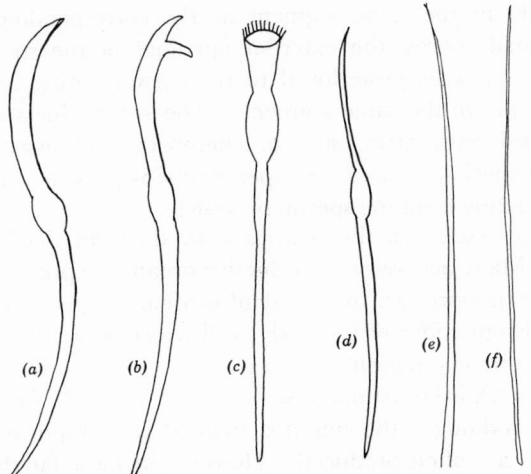

Fig. 21.1. Types of setae. (*a*) Single pointed crotchet. (*b*) Double pointed crotchet. (*c*) Pectinate. (*d*) Needle with nodulus. (*e*) Needle without nodulus. (*f*) Hair.

the setae. They may terminate in a simple point, be bifid, or serrate; and some show intermediate teeth between the distal bifurcations.

Some aquatic oligochaetes possess gills either at the posterior end or along the sides of the segments. The number and arrangements of the gills are of importance.

Identification

For the correct identification of many species or even genera, it is necessary to study the arrangement of the internal organs, particularly the reproductive system. This is especially true of the large opaque worms of the families Tubificidae and Lumbriculidae. These observations may be made with the aid of careful dissections or, preferably, with serial sections.

Internally the segmentation is evidenced by septa which divide the segments. More or less in the center of the coelom is the digestive tract which is modified into different organs in the various families. Nephridia or excretory organs are typically paired and present in all segments except a few anterior and posterior ones. In a few species, the nephridia are, however, highly modified and may develop asymmetrically. There is usually a dorsal pulsating blood vessel, several ventral blood vessels, and many connecting ones. The ventral nerve cord is conspicuous in the ventral portion of the coelem.

The reproductive organs of these hermaphroditic animals are of great use in the identification of species. One or more pairs of testes are attached to the anterior septa of certain segments and extend into the coelomic cavity of the segment. The sperm ducts possess internal openings or spermiducal funnels which are usually in the same segment as the corresponding testes. The spermiducal or male pores, the external openings of the sperm ducts, are usually on some segments posteriorad to the corresponding testes; but in a few species, they are in the same segment. The sperm ducts are commonly variously modified with atria, storage chambers, and associated prostate glands. In some species, an accessory sperm tube is present. This consists of a blind tube extending from the spermatic vesicle.

Posterior to the testes are the ovaries with their modified oviducts and oviducal pores. Many accessory reproductive organs are present. One which is of considerable importance in the identification of species is the spermathecae, saclike invaginations of the body wall which serve for the temporary storage of sperm after copulation.

In the families Opisthocystidae, Aeolosomatidae, and Naididae, asexual reproduction by budding is the common method of multiplication. In these forms long chains are often produced. However in most families, sexual reproduction is the usual method; and in all families it may occur at more or less definite seasons of the year. In the species of many families, conspicuous external glandular swellings occur during the period of sexual reproduction. These swellings are the clitellum which secretes the egg capsule. Among some forms, the segments on which it occurs are of importance in identification.

Distribution

Many of the smaller worms, such as members of the Aeolosomatidae and Naididae, are widely distributed and may be found in nearly any part of the country; others are more restricted in their distribution. Locality data are given for the restricted ones; however, it must be remembered that our present knowledge of aquatic oligochaetes is very limited and many new species and even genera remain to be identified.

KEY TO SPECIES

1a	Without setae; with posterior sucker; pharynx with dorsal and ventral chitinous jaws; commensal on crayfish	
	Family **Branchiobdellidae**	2
1b	With well-developed setae on most segments; without suckers or chitinous jaws.	26
2a	(1) With one pair of testes in segment 5. . . ***Branchiobdella*** Odier 1823	3
2b	With two pairs of testes in segments 5 and 6	4
3a	(2) Dorsal and ventral jaws dissimilar with a 5–4 dental formula; peristomium entire. . . . . ***B. americana*** Pierantoni 1912 Eastern U. S. and Tex.	
3b	Dorsal and ventral jaws similar, with a 4–4 dental formula; peristomium bilobed . . . . ***B. tetradonta*** Pierantoni 1906 Calif.	
4a	(2) Body with appendages.	5
4b	Body without appendages	8

5a   (4) Appendages in the form of blunt cylindrical projections along the median dorsal line of the body . . . . ***Pterodrilus*** Moore 1895   **6**

5b   Appendages in the form of pointed bands encircling the dorsal surface of the body . . . . ***Cirrodrilus*** Pierantoni 1905
One species, **C. *thysanosomus*** (Hall) 1914 from the Great Basin. (Fig. 21.2).

6a   (5) Segments 7 and 8 with funnel-shaped enlargements of the dorsal portions; funnel of 8 excavated dorsally so its dorsal margin bears two small "horns". . . . . ***Pterodrilus durbini*** Ellis 1919
Ind. and Mich.

6b   Without these funnel-shaped enlargements. **7**

7a   (6) Dorsal appendages on segments 2 to 8 inclusive . . . .
P. distichus Moore 1895
N. Y., Ohio, Ind., and Ill.

7b   Dorsal appendages on segments 3, 4, 5, and 8. (Fig. 21.3). . . . . .
P. alcicornus Moore 1895
N. C. and Va.

7c   Dorsal appendages on segment 8 only; a simple, 4-horned appendage . . . . ***P. mexicanus*** Ellis 1919
Ark.

8a   (4) Accessory sperm tube present. (Fig. 21.4). **9**

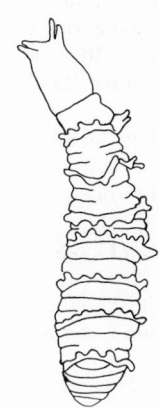

Fig. **21.2.** *Cirrodrilus thysanosomus.* (After Hall.)

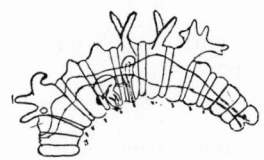

Fig. **21.3.** *Pterodrilus alcicornus.* × 50. (After Moore.)

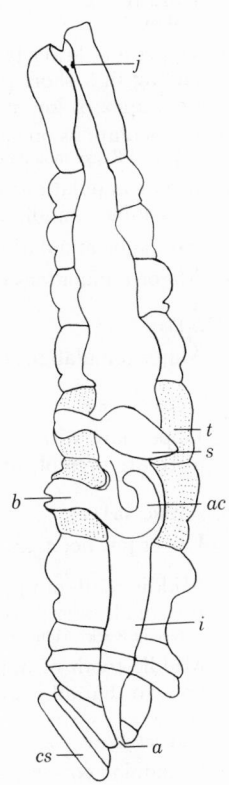

Fig. **21.4.** *Cambarincola elevata.* *j,* jaws; *t,* testes; *s,* spermatheca; *b,* male bursa; *ac,* accessory sperm tube; *i,* intestine; *a,* anus; *cs,* caudal sucker.

8b   Accessory sperm tube absent **19**

9a   (8) Body cylindrical, not flattened, posterior end not conspicuously en-
               larged, anterior nephridia opening to the outside through a single
               pore . . . . **Cambarincola** Ellis 1912   **10**
               Several other uncommon species have been described, including **C. meyeri** Goodnight
               1942 (Ky.), **C. macbaini** Holt 1955 (Ky.), **C. branchiophila** Holt 1954 (Va.), and
               C. gracilis Robinson 1954 (West Coast).

9b   Body flattened, posterior end enlarged so that the body is racket-
               or spatula-shaped, anterior nephridia opening to the outside through
               separate pores . . . . **Xironogiton** Ellis 1919   **17**

10a   (9) Upper lip composed of 4 subequal lobes **11**

10b   Upper lip entire except for a small median emargination **13**

11a (10) Major annulations of body segments distinctly and visibly elevated
               over minor annulations . . . **Cambarincola chirocephala** Ellis 1919
               East of Rocky Mountains.

11b   Major annulations of body segments not elevated over minor an-
               nulations **12**

12a (11) Upper and lower jaws similar, appearing as large triangular blocks
               terminating in a sharp tooth, without lateral teeth, but with uneven
               margins . . . . **C. macrocephala** Goodnight 1943
               Wyo.

12b   Upper and lower jaws dissimilar, upper jaw in the form of a triangle,
               ending in a short point with several minute denticulations on each
               side, apex of lower jaw bifurcated into two points with two minute
               denticulations on each side . . . . . . **C. philadelphica** (Leidy) 1851
               East of Rocky mountains.

13a (10) Major annulations of some segments visibly and distinctly elevated
               over minor annulations **14**

13b   No major annulations distinctly elevated over minor ones. **15**

14a (13) Major annulations of segments 2 to 8 elevated . . . .
                                               C. floridana Goodnight 1941
               Fla.

14b   Major annulations of segment 8 only elevated. (Fig. 21.4) . . . . . .
                                               C. elevata Goodnight 1940
               Central states.

15a (13) Upper jaw with 3 prominent teeth; if as, in a few specimens, 5
               teeth are present, the 2 lateral ones are very small. . . . .
                                               C. inversa Ellis 1919
               Wash. and Ore.

15b   Upper jaw not as above, but with 5 noticeable teeth. **16**

16a (15) Middle tooth of upper jaw long and prominent when compared with
               the small, lateral teeth . . . . **C. macrodonta** Ellis 1912
               East of Rocky Mountains.

16b   Middle tooth of upper jaw longer than the other 4 teeth but small
               enough that all 5 teeth may be considered subequal . . . .
                                               C. vitrea Ellis 1919
               Middle West.

17a   (9) Glandular concave discs near the lateral margin of the ventral sur-
               face of segments 8 and 9; body usually spatula-shaped . . . .
                                               Xironogiton occidentalis Ellis 1919
               Wash. and Ore.

17b   No conspicuous glandular discs near the lateral margin of the ven-
               tral surface of segments 8 and 9; body usually flask-shaped **18**

18a   (17) Two teeth of the longest pair in the upper jaw separated by only
            1 tooth; if 2 long teeth are contiguous, the inner one is the longer.
            (Fig. 21.5) . . . . *X. instabilius instabilius* (Moore) 1894
            Eastern states. Holt (1949) believes that this form does not possess a true accessory
            sperm tube.

18b   Two teeth of the longest pair in the upper jaw separated by 2 teeth;
            if 2 long teeth are contiguous, the outer one is usually the longer.
            (Fig. 21.5). . . . . *X. instabilius oregonensis* Ellis 1919
            Ore., Wash., Calif.

Fig. 21.5.   *Xironogiton instabilius.*
(After Ellis.)

19a   (8)   Without a pair of large clear glands in each of the 9 postcephalic
            segments; spermatheca not bifid   20

19b   With a pair of large clear glands in each of the 9 postcephalic seg-
            ments; spermatheca bifid . . . . *Bdellodrilus* Moore 1895
            One species, *B. illuminatus* (Moore) 1894; from the eastern states.

20a   (19) Major annulations of body segments secondarily divided, especially
            noticeable in the median segments. . . *Triannulata* Goodnight 1940   21

20b   Major annulations of body segments not secondarily divided   22

21a   (20) Lips entire except for a slight median emargination . . . .
                                              T. magna Goodnight 1940
            Wash. and Ore.

21b   Lips divided into lobes . . . . *T. montana* Goodnight 1940
            Wash. and Ida.

22a   (20) Body flattened; sucker ventral . . . . *Xironodrilus* Ellis 1919   23

22b   Body not flattened; sucker terminal . . . .
                                              Stephanodrilus Pierantoni 1906
            One species, *S. obscurus* Goodnight 1940; from Calif.

23a   (22) Upper and lower jaws each with 3 teeth   24

23b   Upper and lower jaws each with more than 3 teeth   25

24a   (23) Median teeth of upper and lower jaws shorter than lateral teeth
                                    Xironodrilus pulcherrimus (Moore) 1894
            N.C.

24b		Median tooth of upper and lower jaws longer than the lateral ones
		Zironodrilus appalachius Goodnight 1943
		N. C.
25a	**(23)**	Median tooth of the upper jaw the longest tooth if teeth are odd in number; if teeth are even in number, the median pair is the longest . . . . ***X. formosus*** Ellis 1919
		Central states.
25b		Median tooth of the upper jaw shorter than either of the 2 teeth adjoining it, if odd in number; if even in number, one of median pair is shorter . . . . ***X. dentatus*** Goodnight 1940
		Mo., Okla., and W. Va.
26a	**(1)**	Reproduction chiefly by budding; clitellum when present on one or more of segments 5 to 8 (or on segments 21 to 23 in *Opistocysta*); size small, usually less than 25 mm in length. **27**
26b		Reproduction mostly sexual, never by budding; clitellum ordinarily posterior to segment 8. Usually larger in size **64**
27a	**(26)**	Septa imperfectly developed; ventral and dorsal setae bundles containing hair setae; prostomium usually broad and ventrally ciliated; mostly with oil globules in the integument . . . .
		Family **Aeolosomatidae** **28**
		With 1 genus, *Aeolosoma* Ehrenberg 1831.
27b		Septa well developed; ventral setae "crotchet," with a swelling, the nodulus; without ventral hair setae **34**
28a	**(27)**	Oil globules colorless, light yellow, blue, or green **29**
28b		Oil globules orange or red **33**
29a	**(28)**	Prostomium not wider than the following segments; oil droplets colorless . . . . ***Aeolosoma niveum*** Leydig 1865
29b		Prostomium wider than the following segments **30**
30a	**(29)**	Crotchet or needle setae occur between the bundles of hair setae . . . **31**
30b		Without crotchets or needle setae. **32**
31a	**(30)**	Without crotchet setae in the first 3 setigerous segments; crotchet setae of the following segments bifid; oil globules yellow . . . .
		A. tenebrarum Vejdovsky 1860
31b		With crotchet setae on the first 3 setigerous segments; crotchet setae not bifid; oil globules pale green. . . . . ***A. leidyi*** Cragin 1887
32a	**(30)**	Without nephridia in the esophageal region; oil globules colorless, yellowish, or yellow-green . . . . ***A. variegatum*** Vejdovsky 1884
32b		With nephridia in the esophageal region; oil globules greenish or bluish. . . . . · . . . . ***A. headleyi*** Beddard 1888
33a	**(28)**	Prostomium not wider than the following segments. . . . .
		A. quaternarium Ehrenberg 1831
33b		Prostomium wider than the following segments. (Fig. 21.6) . . . . .
		A. hemprichi Ehrenberg 1831
34a	**(27)**	Penis well developed; gonads in segments 21 and 22; spermathecae posterior to the gonads; lateral commissural blood vessels in all body segments . . . . Family **Opisthocystidae**
		Only one genus and species known, *Opisthocysta flagellum* (Leidy) (Fig. 21.7). This species resembles a *Pristina* with its anterior proboscis and the dorsal seta beginning in the second segment. It is easily distinguished by its peculiar posterior gills.
34b		Penis lacking, gonads in some segments between 4 and 8; spermathecae in the same segments as the testes; lateral commissural blood vessels only in the anterior body segments . . . . . Family **Naididae** **35**

35a (34) Without dorsal setae. . . . . *Chaetogaster* K. Von Baer 1827   **36**
35b   With both dorsal and ventral setae **40**
36a (35) Prostomium well developed, pointed, with long sensory hairs, length
         0.5–5 mm . . . . *C. diastrophus* (Gruithuisen) 1828
36b   Prostomium indistinct **37**
37a (36) Length of worm 2.5–25 mm (ordinarily 10–15 mm); longest setae of
         segment 2 usually more than 200 μ . . . .
                                      C. diaphanus (Gruithuisen) 1828
37b   Length of worm 7 mm or less **38**
38a (37) Esophagus very short; ordinarily found living in the mantle cavity
         of pulmonate snails. (Fig. 21.8). . . . *C. limnaei* K. von Baer 1827
38b   Esophagus well developed, almost as long as the pharynx; free-
         living **39**
39a (38) Commissural blood vessels ordinarily rudimentary on the pharynx;
         with an incision on the anterior margin of the prostomium (at times
         difficult to see) . . . . *C. cristallinus* Vejdovsky 1883
39b   Commissural blood vessels ordinarily well developed on the pharynx;
         without an incision on the anterior margin of the prostomium . . . .
                                      C. langi Bretscher 1896

40a (35) Without hair setae in dorsal bundles **41**
40b   With hair setae in dorsal bundles **43**

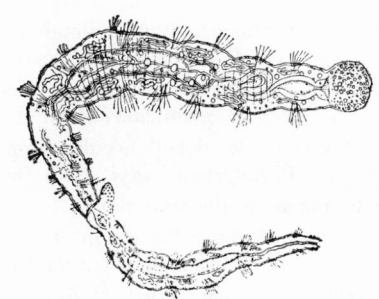

Fig. 21.6. *Aeolosoma hemprichi.* × 20. (After Lankester.)

Fig. 21.7. Posterior end of *Opisthocysta flagellum.* × 16. (After Leidy.)

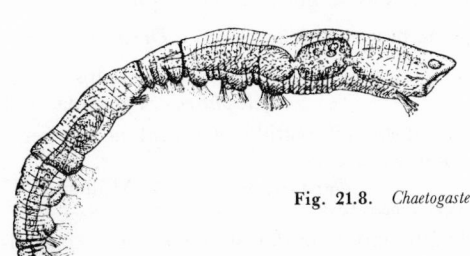

Fig. 21.8. *Chaetogaster limnaei.* × 40. (After Lankester.)

41a (40) Segment 3 much longer than the other segments; dorsal setae begin
in segment 3 . . . . *Amphichaeta* Tauber 1879
One species *A. americana* Chen 1944.

41b Segment 3 not longer than the other segments; dorsal setae begin in
segment 5 or 6 . . . . **42**

42a (41) Setae of all dorsal bundles crotchet, begin in segment 5 . . . .
Paranais Czerniavsky 1880
One species, *P. litoralis* (Müller) 1784 from the New England coast, probably also
occurs in adjacent fresh waters.

42b Dorsal setae nearly straight, slightly toothed or simple pointed, begin
in segment 6 with 1 per bundle . . . . . . *Ophidonais* Gervais 1838
One species *O. serpentina* (Müller) 1773 easily recognized by the small irregularly
distributed dorsal setae, the 4 large transverse pigmented areas on the anterior region,
and the relatively large size. (Fig. 21.9.)

43a (40) First anterior dorsal setae on segment 2 or 3 . . . .
Pristina Ehrenberg 1828   **44**

43b First anterior dorsal setae on segment 4, 5, or 6 . . . . **50**

44a (43) Prostomium rounded, not elongated to form a proboscis **45**

44b Prostomium elongated to form at least a short proboscis **46**

45a (44) Dorsal hair setae of segments 3 and 4 very elongate . . . .
P. bilongata (Chen 1944)

45b Dorsal hair setae of all segments approximately equal in length
P. osborni (Walton 1906)

46a (44) Dorsal hair setae not serrate. **47**

46b Dorsal hair setae serrate. **48**

47a (46) Dorsal hair setae 1 per bundle; short proboscis, living animal flesh
red with white dots . . . . *P. breviseta* Bourne 1891

47b Dorsal hair setae 2 or 3 per bundle; long proboscis; living animal
not colored as above . . . . *P. schmiederi* Chen 1944

48a (46) Dorsal hair setae of segment 3 very elongate; dorsal needles simple
pointed. . . . . *P. longiseta leidyi* Smith 1896

48b No dorsal hair setae very elongate; dorsal needle setae bifid **49**

49a (48) Dorsal needle setae with nodulus and with distal bifurcation formed
of unequal teeth; short proboscis . . . . . *P. plumiseta* Turner 1935

49b Dorsal needle setae without nodulus and with distal bifurcation
formed of equal, fine teeth; well-developed proboscis . . . .
P. aequiseta Bourne 1891

50a (43) Posterior end modified into a gill bearing repiratory organ, the
branchial area; usually living in tubes **51**

50b Posterior end not modified into a gill-bearing respiratory organ . . . **54**

51a (50) Branchial area without long processes or palps . . . *Dero* Oken 1815   **52**

51b Ventral margin of branchial area with long processes or palps . . . .
Aulophorus Schmarda 1861   **53**

52a (51) Normally 4 pairs of gills; distal bifurcation of dorsal needle setae
longer than proximal. (Fig. 21.10) . . . .
Dero digitata (O. F. Müller) 1773
According to Sperber (1948) *D. limosa* Leidy is a synonym of this species.

52b Normally 3 pairs of gills; bifurcations of dorsal needle setae approximately equal . . . . *D. obtusa* d'Udekem 1855

53a (51) First dorsal setae on segment 5; 3 or 4 pairs of well-developed gills.
(Fig. 21.11) . . . . *Aulophorus furcatus* (O. F. Müller) 1773

53b First dorsal setae on segment 6; only slightly developed gills . . . . .
A. vagus Leidy 1880

54a **(50)** Prostomium elongated to form a proboscis . . . .
Stylaria Lamarck 1816 **55**

54b Prostomium not elongated to form a proboscis. **56**

55a **(54)** Proboscis projects from the apex of the prostomium. (Fig. 21.12*b*) .
S. fossularis Leidy 1852

55b Proboscis inserted in a notch in the prostomium. (Fig. 21.12*a*) . . .
S. lacustris (Linnaeus) 1767

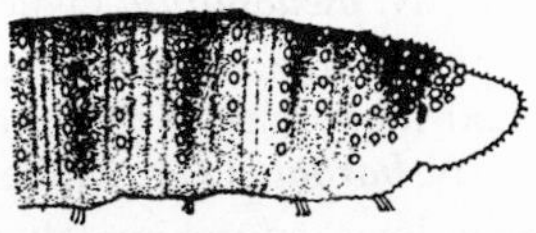

Fig. 21.9. Anterior end of *Ophidonais serpentina.* × 40. (After Piguet.)

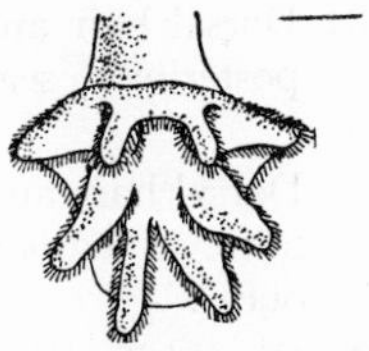

Fig. 21.10. Posterior end of *Dero digitata.* × 25. (After Bousefield.)

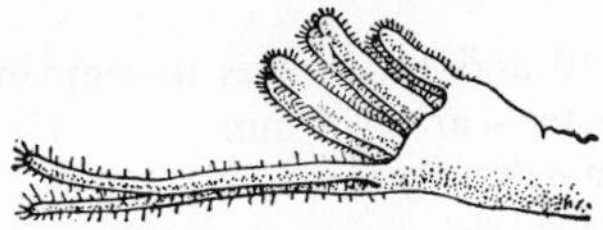

Fig. 21.11. Posterior end of *Aulophorus furcatus.* × 40. (After Bousefield.)

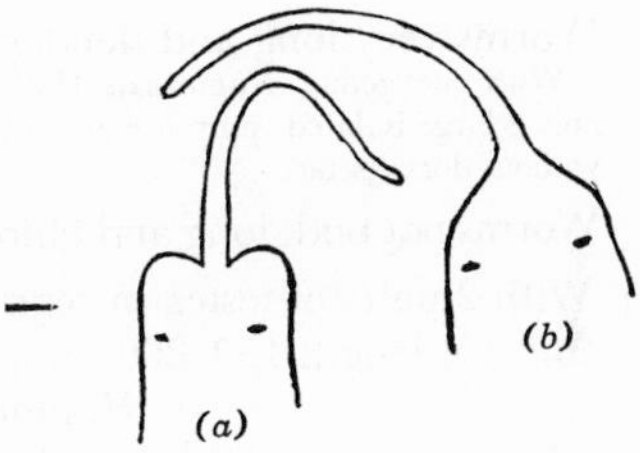

Fig. 21.12. Prostomium and proboscis of *Stylaria.* (*a*) *S. lacustris.* (*b*) *S. fossularis.* × 40. (By Smith.)

56a **(54)** One or more of the dorsal hair setae of segment 6 much longer than those of the other segments and equal to 3 or 4 times the diameter of the body . . . . *Slavina* Vejdovsky 1883
One species, *S. appendiculata* (d'Udekem) 1855 has body surface studded with sensory papillae and with foreign bodies.

56b Dorsal hair setae of segment 6 similar in length to those of other segments **57**

57a **(56)** Dorsal hair setae serrate along one or both borders; 2 to 3 secondary annulations in each segment . . *Vejdovskyella* Michaelsen 1903
One species, *V. comata* (Vejdovsky) 1883.

57b Dorsal hair setae not serrate along borders; without secondary annulations on segments **58**

58a (57) Eyes normally present, ventral setae of segments 2 to 5 mostly well
          differentiated from those of the more posterior segments . . . .
          Nais Müller 1773   **59**

58b   Without eyes, ventral setae of segments 2 to 5 only slightly different
          from those of the more posterior segments . . *Allonais* Sperber 1948
          One species *A. paraguayensis* (Michaelsen) 1905.

59a (58) Dorsal needle setae single pointed. **60**

59b   Dorsal needle setae bifid **62**

60a (59) Dorsal needle setae with short, blunt tips . *Nais simplex* Piguet 1906

60b   Dorsal needle setae with long, sharp tips. **61**

61a (60) Dorsal hair and needle setae with 1 to 3 per bundle; ventral setae
          posterior to segment 5 thin, not strongly curved . . . .
          N. pseudobtusa Piguet 1906

61b   Dorsal hair and needle setae up to 5 per bundle; ventral setae pos-
          terior to segment 5 heavier, shorter, and more curved than anterior
          ones . . . . *N. barbata* O. F. Müller 1773

62a (59) Distal bifurcations of dorsal needle setae long, approximately paral-
          lel; ventral setae with distal bifurcations twice as long as proximal
          N. elinguis, O. F. Müller 1773

62b   Distal bifurcations of dorsal needle setae short, diverging; posterior
          ventral setae with bifurcations of approximately the same length **63**

63a (62) Stomach dilates abruptly; ventral setae of segments 2 to 5 with distal
          tooth longer than proximal . . . . *N. variabilis* Piguet 1906

63b   Stomach dilates gradually; ventral setae of segments 2 to 5 very similar
          to more posterior ones . . . . *N. communis* Piguet 1906

64a (26) Worms very long and slender (filiform) . . . . Family **Haplotaxidae**   **65**
          With one genus, **Haplotaxis** Hoffmeister 1843. The members of this family have
          only 2 large isolated ventral setae and 2 small dorsal setae per segment; many segments
          without dorsal setae.

64b   Worms not both long and filiform **66**

65a (64) With 2 pairs of testes in segments 10 and 11, ovaries in segments 12
          and 13, length 150–200 mm, diameter scarcely 1 mm . . . .
          Haplotaxis gordioides (G. L. Hartmann) 1821
          Found throughout the U. S., subterranean in habit.

65b   With 1 pair of testes in segment 10, ovaries in segments 15 and 16,
          length 100–150 mm, diameter 0.6–0.7 mm . . *H. forbesi* Smith 1918
          From Ill., subterranean in habit.

66a (64) Ordinarily with not more than 2 well-developed setae per bundle;
          male pores ordinarily on some other segment than 11 or 12 . . . . . .   **67**

66b   Ordinarily with more than 2 well-developed setae per bundle in some
          segments; male pores on segment 11 or 12 . . . .   **80**

67a (66) Male pores on one or more segments anterior to segment 12; ovaries,
          1 to 3 pairs in the region of segments 9 to 13; male pores on the seg-
          ment that contains the most posterior pair of testes . . . .
          Family **Lumbriculidae**   **68**

67b   Male pores exceptionally on segments 12 or 13, commonly more pos-
          terior. Ovaries in segment 13. Male pores opening posterior to the
          segments that contain the testes . . . . **Earthworms**   **79**
          Members of 2 families include but a few aquatic forms. Some few terrestrial species
          inhabit flood plains and may be found in water during flooding.

68a (67) Some setae bifid **69**

68b   Setae all single pointed. **71**

69a **(68)** Prostomium not elongated into a proboscis; all setae bifid, with distal tooth smaller than proximal . . . . ***Lumbriculus*** Grube 1844   **70**

69b   Prostomium elongated into a proboscis; setae anterior to the clitellum clearly but not deeply bifid; setae posterior to clitellum single pointed . . . . ***Kincaidiana*** Altman 1936
        One species, **K. hexatheca** Altman 1936; from Wash.

70a **(69)** Male pores on segment 8 (occasionally on 7). . . . .
        Lumbriculus variegatus (O. F. Müller) 1774
        Entire U. S.

70b   Male pores on segment 10, occasionally on 11 . . . .
        L. inconstans (F. Smith) 1895
        Entire U. S.

71a **(68)** Large, single unpaired spermatheca with many tubular diverticula in segment 8; prostomium elongated into a proboscis . . . .
        Sutroa Eisen 1888   **72**

71b   Spermatheca without tubular diverticula, prostomium may or may not be elongated into proboscis **73**

72a **(71)** Spermatheca with 2 or 3 simple diverticula, rarely bifid at ends
        S. rostrata Eisen 1888
        Calif.

72b   Spermatheca with many branched diverticula . . . .
        S. alpestris Eisen 1893
        Calif.

73a **(71)** Unpaired median male pore in segment 10; spermatheca 1 or 2 in number, asymmetrical in position, openings in segments 9, or 8 and 9; prostomium elongated into a proboscis . . . .
        Mesoporodrilus F. Smith 1896   **74**

73b   Male pores paired; spermathecae paired, symmetrical in position **75**

74a **(73)** With 1 pair of testes and funnels in segment 10 . . . .
        M. asymmetricus F. Smith 1896
        Ill.

74b   With 2 pairs of testes and funnels in segments 9 and 10 . . . .
        M. lacustris (Verrill) 1871
        Lake Superior, Mich.

75a **(73)** With 1 pair of male pores in segment 9; prostomium elongated into a proboscis . . . . ***Premnodrilus*** F. Smith 1900
        One species, **P. palustris** F. Smith 1900, from Fla.

75b   With male pores in segment 10. . . . .   **76**

76a **(75)** Spermathecal pores in segment 8 or 8 and 9; prostomium elongated into a proboscis . . . . ***Rhynchelmis*** Hoffmeister 1843   **77**

76b   Spermathecal pores posterior to segment 8 . . . .   **78**

77a **(76)** Nephridia alternate or irregular, without prominent ventral glands
        R. elrodi F. Smith and Dickey 1918
        Mont.

77b   Nephridia 1 pair per segment, posterior to segment 13. Five pairs of prominent median glands composed of prostates in the region of the ventral nerve cord in segments 2 to 6 . . . .
        R. glandula Altman 1936
        Wash.

78a **(76)** Spermathecal pores in segment 9; prostomium rounded . . . .
        Eclipidrilus Eisen 1881
        One species, **E. frigidus** Eisen 1881; from Calif.

78b   Spermathecal pores in segment 11, or 11 and 12. Prostomium elongate . . . . ***Trichodrilus*** Claparède 1862
One species, *T. allobrogum* Claparède 1862; from Ill.

79a **(67)** Clitellum beginning on segments 14 to 16 and extending over 10 to 12 segments; male pores on segments 18/19 or on 19, recognizable only in section; few or no dorsal pores; without well-developed gizzard . . . . ***Sparganophilus*** Benham 1892
Family **Glossoscolecidae;** several species of genus known from various regions of the U. S.

79b   Clitellum beginning on segments 18 to 23 and extending over 4 to 6 segments; male pores on segment 12, 13, or 15, conspicuous; gizzard limited to segment 17; first dorsal pores on segments 4/5 . . . .
Eiseniella Michaelsen 1900
Family **Lumbricidae;** one highly variable species, *E. tetraedra* (Savigny) 1826 is widely distributed.

80a **(66)** Setae simple pointed and usually straight, spermathecae open between segments 4/5 or 3/4 and 4/5. Usually whitish in appearance and seldom more than 25 mm in length . . . Family **Enchytraeidae** **81**
Many genera and species described in this family, but are incompletely known. The genera included have been recorded from fresh water. Ice worms live in melting ice of glaciers. Most species belong to the Enchytraeidae.

80b   Ventral setae ordinarily cleft; spermathecae if present usually open on segment 10; usually reddish in appearance, commonly more than 25 mm in length, many live in tubes . . . . . . Family **Tubificidae** **86**

81a **(80)** Setae arranged in 6 bundles per segment, 2 subdorsal, 2 lateral, and 2 ventral . . . . ***Chirodrilus*** Verrill 1871

81b   Setae arranged in 4 bundles per segment **82**

82a **(81)** Esophagus expanding abruptly into the intestine . . . .
Henlea Michaelsen 1889

82b   Esophagus gradually merging into intestine **83**

83a **(82)** Setae straight. . . . . ***Enchytraeus*** Henle 1837

83b   Setae sigmoid **84**

84a **(83)** Head pore generally at the apex of the prostomium, nephridia plurilobed; peneal bulb with muscular strands . . . .
Mesenchytraeus Eisen 1878

84b   Head pore between prostomium and segment 1; nephridia not plurilobed; peneal bulb as a rule without muscular strands, but covered by an investment of muscle **85**

85a **(84)** Testes deeply divided, forming a number of distinct lobes. . . . . . .
Lumbricillus Örsted 1884

85b   Testes undivided and massive. . . . . . ***Marionina*** Michaelsen 1889

86a **(80)** Segments of the posterior portion of the body each with a dorsal and a ventral gill . . . . ***Branchiura*** Beddard 1892
One introduced species, *B. sowerbyi* Beddard 1892; widely distributed. (Fig. 21.13.)

86b   Segments of body without gills. **87**

87a **(86)** Body surface covered with many cuticular papillae . . . .
Peloscolex Leidy 1852 **88**

87b   Body surface not covered with many cuticular papillae **89**

88a **(87)** Each segment with a prominent papilla and many smaller ones
P. variegatus (Leidy) 1852
Pa.

88b   Each segment with 2 rows of papillae, one in the setae zone, the

other between the segments, occasionally a third irregular row. (Fig. 21.14) . . . . ***P. multisetosus*** (F. Smith) 1900
Ill.

89a **(87)** Dorsal bundles without hair setae. **90**

89b   Dorsal bundles with hair setae **95**

90a **(89)** With a single spermathecal pore in the median portion of segment 10; sperm ducts without definite prostate glands . . . .
Rhizodrilus F. Smith 1900
One species, **R. lacteus** F. Smith 1900; from Ill. Chen (1940) considers **Rhizodrilus** a synonym of **Monopylephorus.**

90b   With paired spermathecal pores in segment 10; sperm ducts with definite prostate glands **91**

91a **(90)** Setae indistinctly cleft and sometimes simple pointed; 10 or more small definite prostates on each sperm duct . . . .
Telmatodrilus Eisen 1879   **92**

91b   Setae distinctly cleft; sperm ducts each with one definite prostate gland. (Fig. 21.15) . . . . ***Limnodrilus*** Claparède 1862   **93**
A number of species not included here also described from Calif. by Eisen (1885); **L. gracilis** Moore 1906, from the Great Lakes, was described from immature specimens.

92a **(91)** Spermathecae opening in front of and between the ventral fascicles of setae in segment 10 . . . . . ***Telmatodrilus vejdovskyi*** Eisen 1879
Calif.

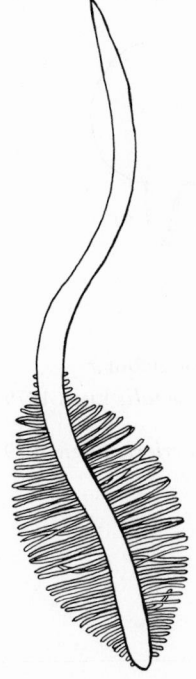

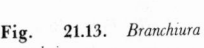

Fig. 21.13. *Branchiura sowerbyi.*

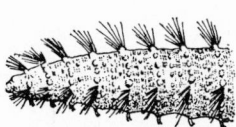

Fig. 21.14. *Peloscolex multisetosus.* (By Smith.)

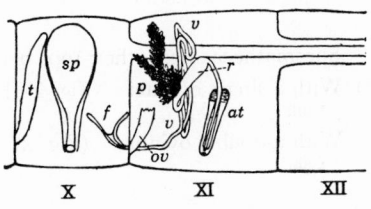

Fig. 21.15. Reproductive organs of *Limnodrilus sp.* *t*, spermary; *sp*, spermathecá; *f*, spermiducal funnel; *v*, *v*, sperm duct; *p*, prostate gland; *r*, atrium; *at*, penis and penis sheath; *ov*, ovary. × 20. (After Moore.)

92b Spermathecae opening in front and between the ventral and lateral fascicles of setae in segment 10 . . . .
Telmatodrilus mcgregori Eisen 1900
Calif.

93a (91) Body· with more than 140 segments . . . .
Limnodrilus hoffmeisteri Claparède 1862
Entire U. S.

93b Body with less than 140 segments 94

94a (93) Penis sheath about 4 to 6 times as long as broad; proximal prongs of setae much reduced . . . . *L. udekemianus* Claparède 1861
Entire U. S.

94b Penis sheath about 20 to 30 times as long as it is broad; proximal prongs with setae only slightly reduced . . . .
L. claparedianus Ratzel 1869
Entire U. S.

95a (89) Two lateral teeth of dorsal pectinate setae not widely divergent; length of atrium and penis combined at least ⅔ that of the remainder of the sperm duct . . . . *Ilyodrilus* Eisen 1879 96

95b Two lateral teeth of dorsal pectinate setae widely divergent; length of atrium and penis combined much shorter than the remainder of the sperm duct . . . . *Tubifex* Lamarck 1816
One common species, **T. tubifex** (O. F. Müller) 1774; from the entire U. S.; very common in polluted waters. The anterior end is imbedded in mud tubes while the posterior end waves about.

96a (95) Spermathecae not bent, but globular and inflated. (Fig. 21.16c). . . .
Ilyodrilus fragilis Eisen 1879
Calif.

21.16. Spermathecae of *Ilyodrilus* (a) *I. perrierii.* (b) *I. sodalis.* (c) *I. fragilis.* (After Eisen.)

96b Spermathecae bent, their tops being saclike and not globular 97

97a (96) With a single oviduct. (Fig. 21.16b) . . . . . . *I. sodalis* Eisen 1879
Calif.

97b With a double oviduct. (Fig. 21.16a) . . . . *I. perrierii* Eisen 1879
Calif.

References

Altman, Luther Clare. 1936. Oligochaeta of Washington. *Univ. Wash. Publ. Biol.*, 4:1–137.
Cernosvitov, L. 1936. Oligochaeten aus Südamerika. Systematische stellung der Pristina flagellum Leidy. *Zool. Anz.*, 113:75–84. **Chen, Y. 1940.** Taxonomy and faunal relations

of the limnitic Oligochaeta of China. *Contr. Biol. Lab. Sci. Soc. China Zool. Ser.*, 14:1–131. **1944.** Notes on Naidomorph Oligochaeta of Philadelphia and vicinity. *Notulae Naturae Acad. Nat. Sci. Phila.*, No. 136:1–8. **Eisen, G. 1885.** Oligochaetological researches. *Rept. U. S. Fish Comm. for 1883*, 11:879–964. **1896.** Pacific Coast Oligochaeta. II. *Mem. Calif. Acad. Sci.*, 2:123–198. **1905.** Enchytraeidae of the West Coast of North America. *Harriman Alaska Expedition*, 12:1–166. **Ellis, Max M. 1919.** The Branchiobdellid worms in the collection of the United States National Museum with descriptions of new genera and new species. *Proceedings U. S. Natl. Museum*, 55:241–265. **Galloway, T. W. 1911.** The common fresh water Oligochaeta of the United States. *Trans. Am. Microscop. Soc.*, 30:285–317. **Goodnight, C. J. 1940.** The Branchiobdellidae (Oligochaeta) of North American crayfishes. *Illinois Biol. Monographs*, 17:5–75. **Michaelsen, W. 1900.** *Oligochaeta. Das Tierreich*, 10. Gruyter, Berlin and Leipzig. **Moore, J. P. 1893.** On some leech-like parasites of American crayfishes. *Proc. Acad. Nat. Sci., Phila.*, 419:428. **1906.** Hirudinea and Oligochaeta collected in the Great Lakes Region. *Bull. U. S. Bur. Fisheries*, 21:153–171. **Pierantoni, U. 1912.** Monografia dei Discodrilidae. *Ann. Mus. Zool. Univ. Napoli N. S.*, 3:1–28. **Smith, Frank. 1900a.** Notes on species of North American Oligochaeta. III. *Bull. Illinois State Lab. Nat. Hist.*, 5:441–458. **1900b.** Notes on species of North American Oligochaeta. IV. *Bull. Illinois State Lab. Nat. Hist.*, 5:459–478. **1905.** Notes on species of North American Oligochaeta. V. *Bull. Illinois State Lab. Nat. Hist.*, 7:45–51. **Smith, Frank and Welch, P. S. 1913.** Some new Illinois Enchytraeidae. *Bull. Illinois State Lab. Nat. Hist.*, 9:615–636. **Sperber, C. 1948.** A taxonomical study of the Naididae. *Zool. Bidrag Fran Uppsala*, 28:1–296. **Stephenson, J. 1930.** *The Oligochaeta*. Oxford University Press, London. **Walton, L. B. 1906.** Naididae of Cedar Point, Ohio. *Am. Naturalist*, 40:683–706. **Welch, P. S. 1914.** Studies on the Enchytraeidae of North America. *Bull. Illinois State Lab. Nat. Hist.*, 10:123–212.

22

Polychaeta

OLGA HARTMAN

Polychaetes are only seldom inhabitants of fresh water; more often they are euryhaline (tolerant to variable salinities to brackish or fresh water) and in some instances they can be experimentally cultured in salt-free water. Nine species are known from North America. With few exceptions they inhabit streams or lakes which have, or have had, recent connection with the sea. Most of the species are small (a few mm long) and thus escape detection. They are members of widely dispersed families and are presumed to have originated separately in their unusual (nonmarine) habitats. Some differ from their nearest relatives in having modified methods of reproduction or physiological peculiarities. Some fresh-water polychaetes are members of monotypic genera, but many others are representatives of marine genera. Other geographic areas, notably South America (Correa, 1948) and southern Asia (Feuerborn, 1932), are known to have a more diversified fresh-water polychaete fauna.

In North America most species belong to the family Nereidae. *Neanthes limnicola* (including, presumably, *N. lighti*) was first recorded from Lake Merced, a reservoir in the water supply system of San Francisco; it is known from San Francisco northward, along the Russian River and some of its tributaries to Coos Bay, Oregon. The cosmopolitan, euryhaline *N. succinea* (including *N. saltoni*) is common in the highly fluctuating brackish Salton Sea in California, which sometimes receives much fresh water from the Colorado River. The same species is found along temperate shores of eastern and western North America, especially in brackish to nearly fresh water.

Laeonereis culveri has been experimentally maintained in fresh water (Johnson, 1903), but is typically a brackish species along much of the eastern and southeastern United States, where it occupies extensive sand flats and provides an important source of food for shore birds. *Namanereis hawaiiensis* is recorded from a spring near Honolulu, Hawaii, and *Lycastoides alticola* is known only from Lower California, Mexico, where it inhabits a mountain stream at an elevation of 7000 ft. A species of *Lycastopsis*, with three, instead of four, pairs of peristomial tentacles, inhabits the highest fringes of intertidal zones along much of the western shores of the Americas (North American records unpublished), where considerable freshening from creeks and streams occurs, especially in areas overgrown with sparse vegetation; its habitat approaches a terrestrial condition.

The family Sabellidae is represented by one species of the fresh-water genus *Manayunkia*. *M. speciosa* inhabits the brackish to fresh-water tidal streams of Pennsylvania, New Jersey, and adjacent shores attached to stones. As *M. eriensis* it is recorded from Lake Erie in 55 feet of water; it extends west possibly to Duluth Harbor, Lake Superior. The Serpulidae are known from North America through a single species, *Mercierella enigmatica*, common in Lake Merritt, Oakland, California and tributaries along the western end of the Gulf of Mexico. This survives in both fresh and sea water, but reproduces in brackish conditions. It is a fouling organism on the bottoms of ships, piers, and other floating objects.

Most fresh-water polychaetes are colonial, associated with sand and mud. Nereids construct burrows or transient tubes; sabellids occupy more or less permanent tubes of mucus and particles of detritus; and serpulids form tube masses that resemble coral heads.

Some fresh-water polychaetes have direct development in which the typical trochophore is modified or somewhat suppressed. *Neanthes limnicola* has an intracoelomic development, with the 20- or more-segmented young emerging from the hermaphroditic parent (as *N. lighti*, Smith, 1950) at intersegmental furrows. Hermaphroditism with ova and sperm produced in the same or successive segments is known, though it is not limited to fresh-water polychaetes. Giant ova, few in number, are described for *Lycastopsis* (Feuerborn, 1932). *Neanthes succinea* in Salton Sea develops normally, the dioecious adults developing into swarming, epitokal adults, giving rise to trochophoral, planktonic larvae which develop gradually into the settling young. *Mercierella enigmatica* in Lake Merritt sheds enormous numbers of minute ova into the surrounding water, where fertilization and succeeding development occur. The trochophore is typical and planktonic for a normal period of time.

KEY TO SPECIES

1a   With head end exposed (Fig. 22.1) **2**

1b   With head end covered by tentacular processes (Fig. 22.2) **3**

2a   **(1)** Parapodia distinctly biramous (Fig. 22.3) **4**

2b   Parapodia uniramous (Fig. 22.4) **6**

3a   **(1)** In calcereous tubes; tentacular crown accompanied with operculum
                                        Merciella enigmatica (Faurel) 1923

3b   In sandy, mucoid tubes; tentacular crown without operculum . . . . .
                             Manyunkia speciosa (Leidy) 1858
       Includes *M. eriensis* (Krecker) 1939.

4a   **(2)** Dorsal cirri of middle and posterior segments are increasingly small
       and inserted at upper base of notopodium **5**

4b   Dorsal cirri of middle and posterior segments large and inserted near
       distal end of notopodium (Fig. 22.3); proboscidial organs dark, horny;
       on both sides of N. A. . . . .
                Neanthes succinea (Frey and Leuckhart) 1849
       Includes *N. saltoni* Hartman 1936.

5a   **(4)** Organs on the proboscis in the form of tufts of pale, soft papillae; in-
       habits eastern and southern shores of U. S. . . . .
                  Laeonereis culveri (Webster) 1879

5b   Organs on the proboscis dark, chitinized; inhabits tidal streams of
       western U. S. and Lake Merced . ***Neanthes limnicola*** (Johnson) 1903
       Includes *N. lighti* Hartman 1936.

6a   **(2)** Prostomium clearly bilobed in front; tentacular cirri articled (jointed);
       inhabits mountain stream of Lower California . . . .
                Lycastoides alticola (Johnson) 1903

6b   Prostomium not bilobed in front; tentacular cirri smooth **7**

7a   **(6)** Anterior end with four pairs of peristomial tentacles; inhabits stream
       near Honolulu . . . . ***Namanereis hawaiiensis*** (Johnson) 1903

7b   Anterior end with 3 pairs of peristomial tentacles; inhabits intertidal
       fringe zones of western America . . . . ***Lycastopsis*** sp.

References

Correa, D. D. **1948.** A Polychaete from the Amazon Region. *Bol. Fac. Univ. Sao Paulo Zool.*, No. 13:245–257. **Feuerborn, H. J.** **1932.** Eine Rhizocephale und zwei Polychaeten aus dem Süsswasser von Java und Sumatra. *Intern. Ver. Limnol. Stuttgart Verhandl.*, 5:618–660. **Johnson, H. P.** **1903.** *Fresh-Water Nereids from the Pacific Coast and Hawaii, with Remarks on Fresh-Water Polychaeta in General*, pp. 205–222. Edward Laurens Mark Anniv. Vol. Holt, New York. **1908.** *Lycastis quadraticeps*, an hermaphrodite nereid with gigantic ova. *Biol. Bull. Marine Biol. Lab.*, 14:371–386. **Leidy, J.** **1883.** *Manayunkia speciosa*. *Proc. Acad. Nat. Sci. Phila.*, 35:204–212. **Pettibone, M. H.** **1953.** Fresh-water polychaetous annelid, *Manayunkia speciosa* Leidy, from Lake Erie. *Biol. Bull.*, 105:149–153. **Smith, R. I.** **1950.** Embryonic development in the viviparous nereid polychaete, *Neanthes lighti* Hartman. *J. Morphol.*, 87:417–466. **1953.** The distribution of the polychaete *Neanthes lighti* in the Salinas River estuary, California, in relation to salinity, 1948–1952. *Biol. Bull.*, 105:335–347.

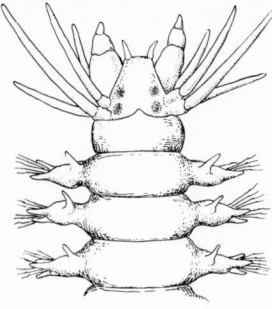

Fig. 22.1. *Neanthes limnicola*, anterior end in dorsal view. × 10. (After Johnson.)

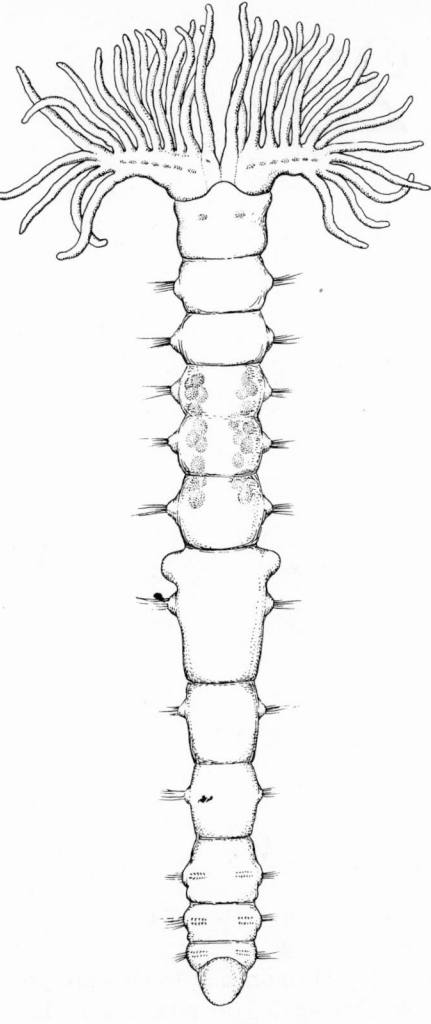

Fig. 22.2. *Manayunkia speciosa*, entire animal in dorsal view. × 25. (After Leidy.)

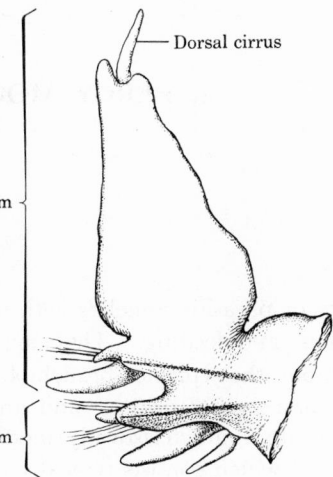

Dorsal cirrus

Notopodium

Neuropodium

Fig. 22.3. *Neanthes succinea*, biramous parapodium. × 20. (After Ehlers.)

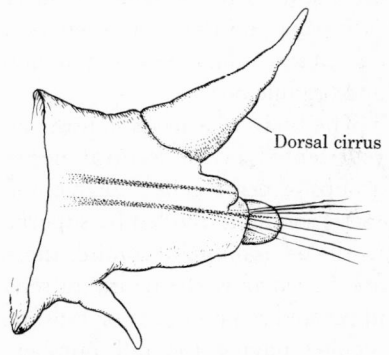

Dorsal cirrus

Fig. 22.4. *Lycastis hawaiiensis*, uniramous parapodium. × 40. (After Johnson.)

23

Hirudinea

J. PERCY MOORE

The Hirudinea or leeches are predatory or parasitic annelids with terminal suckers serving in attachment, locomotion, and feeding. They are closely related to the Oligochaeta, and resemble the epizoic Branchiobdellidae (p. 524) in the possession of suckers, median genital orifices, and analogous jaws, and in the absence of setae. They are characteristically modified for procuring and digesting their peculiar food, which consists typically of blood and other animal juices, but also of smaller annelids, snails, insect larvae, and organic ooze.

The body of a leech consists of 34 somites designated I to XXXIV, each represented in the central nerve cord by a ganglion usually formed of 6 groups of nerve cells. Segment I is much reduced. Externally, each neuromeric somite is divided by superficial furrows into 2 to 16 rings or annuli. In the 3- or 5-annulate somite, the middle annulus is aligned with the ganglion and is known as the neural or sensory annulus. It bears usually 3 or 4 pairs of integrated sense organs or sensillae on the dorsum and 3 pairs on the venter. Somites having the full number of annuli characteristic of the genus are termed complete or perfect. Incomplete or abbreviated somites occur at both ends of the body and may have any number less than the complete somites (Fig. 23.6) into which they grade. Recognizing the triannulate somite as basic, and that more complex ones may be derived by repeated binary division of the annuli, the following symbols are used for precise designation of the annuli. Counting from the head end, those of the triannulate somite are *a1*, *a2*, and *a3*, with *a2* the neural or sensory annulus. These bisected become *b1* to *b6;* repeated subdivisions give tertiary annuli *c1* to *c12*, and quaternary annuli *d1* to *d24*. But only very few of the fourth order are ever developed, and the neural annulus is usually less divided than the others. The annular composition of complete somites is usually characteristic of genera, and the composition of incomplete somites is frequently characteristic of species. Sensillae are named from the axis of the body as dorsal or ventral paramedian, intermediate, paramarginal (supra- and submarginal), and marginal, and the fields in between them respectively as median, paramedian, intermediate and paramarginal.

Differentiated longitudinal body regions are: head, somites I to VI; precitellum, VII to X*a2*; clitellum XI, XII and contiguous annuli of X and XIII; middle body (region of complete somites, gastric caeca,and testesacs), XIII to XXIV; anal region, XXV to XXVII; and caudal (sucker), XXVIII to XXXIV. Among the Piscicolidae the caudal end of the clitellum more or less distinctly separates the trachelosome (neck) and the urosome (abdomen), and somite XIII may bear a collar enclosing the end of the clitellum. The oral or head sucker surrounds the mouth, forming little more than the lips in Hirudidae and Erpobdellidae, but being more or less widely expanded about the proboscis pore (mouth) in the Rhynchobdellida, especially the Piscicolidae. The oral sucker also bears the principal eyes. The caudal or subanal sucker is the most compacted part of the body; somites XXVIII to XXXIV have lost most of their distinctness and their ganglia are crowded into one mass. It is larger than the head; complexly muscular; discoid, shallowly concave, or more or less deeply cupped; and supported on a central, often narrow peduncle which gives the body much freedom of motion when attached. When detached the sucker is capable of great changes in shape.

Surface Organs

In addition to the sensillae, which are confined to the sensory annuli or their morphological equivalents, there are other surface organs—eyes, papillae, and tubercles. The simplest eyes are ocelli consisting of a single visual cell largely embedded in pigment. They may be found singly in any part, including the caudal sucker, and aggregations of them in a common network of chromatophores generally form the eyes of Piscicolidae. True integrated eyes are confined to the head, and are segmental, and usually replace sensillae of the paramedian and intermediate series. They reach their highest development in the Hirudidae, where they consist of a central optic nerve encased in a cylindrical, transparent core of "glass" or visual cells, and these are again surrounded, except at the summit, by a dense black cover of chromatophores. Such eyes are simple when single, compound when one or two small ones are attached to the main one. The term papilla is limited here to the minute protrusive sense organs (Bayer's organs, etc.) which are scattered in large numbers over the surface. Tubercles are the larger projections which include the deeper dermal tissues and muscle; they are often covered with papillae and as both tubercles and their papillae are retractile, their appearance varies greatly with the state of the leech. When the leech is hungry, they are likely to be prominent and rough, when the stomach is full they are lower and smoother. Very small tubercles having a diameter of less than one-third the width of the annulus can be termed tuberculae; very large ones with a diameter exceeding two-thirds of the width of the annulus are termed tuberosities. The latter are likely to bear a rosette of papillae at the summit. No North American fresh-water leech possessing gills is known, but several have pulsatile marginal vesicles.

Digestive Tract

The digestive tract (Figs. 23.2, 23.5) is a tube from mouth to anus, and is divided into buccal chamber, pharynx, esophagus, stomach or crop, intestine, and rectum. In the Hirudidae the buccal chamber commonly includes 3 compressed muscular jaws bearing serial teeth on the ridge. The teeth may be in one or two rows (mono- or distichodont). The pharynx is a muscular suction bulb (Hirudidae), a long straight tube (distichous Hirudidae and Erpobdellidae), or a slender exsertile proboscis moving within a sheath. Unicellular salivary or proboscis glands open on the jaws or at the tip of the proboscis. They are diffuse (Fig. 23.2) or in compact lobes (Fig. 23.4). The esophagus is well developed only in some Rhynchobdellida, in which it is a narrow tube with a vertical S-fold when retracted, and in some bears a pair of pouches or glands. In the Hirudidae the esophagus is scarcely distinguishable from the stomach and often bears small, irregular caeca which may contain food. In the Erpobdellidae, the extension of the pharynx far into somite XII practically eliminates the esophagus. The large stomach or crop, which in many leeches is practically a storage tank in which blood may last for many weeks, extends from somite XIII to XIX or XX. It may be a simple straight tube more or less camerated, or moniliform, with intersegmental constrictions; or it may be complicated by from 1 to 14 pairs of lateral caeca, 1 or 2 pairs in each somite. These may be simple or variously lobed and branched. The last or postcaeca is the most constant and largest; it is reflexed, extending even to the anus, and has lateral lobes of the same pattern as the other gastric caeca (sanguivorous Hiruididae and Glossiphoniidae). In the Piscicolidae the 2 postcaeca may be united in varied degree and may even be completely coalesced; in the predacious Hirudidae they may be the only caeca, and in the Erpobdellidae, caeca may be entirely absent. The intestine is a narrow tube, acaecate or with 4 pairs of usually simple caeca (most Glossiphoniidae, Fig. 23.2), which connects with the dorsal anus by a short rectum.

Reproduction

Leeches are hermaphroditic. The gonopores are median on the ventral face of the clitellum, the male always preceding, although the two outlets may be united. Testes (properly testesacs) are confined to the middle body segments and usually occur in pairs alternating with the gastric enlargements; usually there are 6 or 5 (Rhynchobdellida), 9 or 10 (Hirudidae), and numerous very small saccules in grape-bunch clusters on each side (Erpobdellidae). A capillary vas deferens or sperm duct enlarges anterior to the testes into a seminal vesical which continues in an ejaculatory duct opening into the atrium or terminal male organ. The seminal vesicle is a long posterior loop, a simple coil, a closely folded epididimal knot, or other shape. The atrium consists essentially of three parts (Fig. 23.12), a thin-walled eversible bursa, a thick-walled glandular and muscular median chamber, and a pair of horns or cornua opening into the latter and of similar structure. They receive the ejaculatory ducts at their tips and with the median chamber constitute in the Rhynchobdellida and Pharyngobdellida the spermatophore organ, effective in hypodermic insemination. In the Hirudidae, which inseminate by copulation, the organ becomes much more complex and presents many generic modifications. The most complex are found in the distichodont Hirudidae (Fig. 23.10), in which the atrium is greatly elongated, houses a filiform evaginable penis, and has an enlarged head covered by prostate glands in which the greatly reduced horns are embedded and concealed. Ovaries like the testes, are enclosed in pairs in coelomic sacs. Although they take various forms they are mostly simple, the ovisacs terminating in ducts which join to form a common duct or vagina. The most complex modifications occur in the distichodont Hirudidae, in which they follow the plan of elongation that reciprocally adapts them to the male organs (Fig. 23.10b). In the Piscicolidae there are other complications, most of which are glandular, or facilitate insemination by hypodermic injection.

A most striking distinction from other annelids is the great reduction of the coelom into a system of sinuses and lacunae, the extent and arrangement of which vary with genera and families, but these are too difficult to work out for the use in ordinary identification. Nephridia are not used for the same reason. In the Rhynchobdellida there is also a separate vascular system with colorless blood; in the other orders it is united with the lymph lacunae, and hemolymph is red.

Color is derived chiefly from three sources: blood or hemolymph of the Rhynchobdellida; reserve cells which form the white bands and spots, the latter often metameric; and amoeboid excretophores, mostly wandering, which carry the true pigments of various colors and are arranged between muscle strands and other organs to give the characteristic patterns.

Preservation and Identification

Taxonomic and anatomical study of leeches is greatly facilitated by properly prepared material and equally hampered by faulty preparations. The preserved leeches should be straight, moderately extended, and undistorted. They should be well fixed and preserved in fluids strong enough to prevent maceration or softening and not so strong as to render them overhard and brittle. They are completely ruined when they dry out, which often happens to museum specimens or when collectors are busy with other things.

As leeches contract excessively and irregularly on contact with irritating preserving fluids, the first step in good preparation is to stupefy or anesthetize them. Many narcotics or other drugs will accomplish this, but the best, if available, is carbon dioxide, such as in soda water in siphons or bottles, or other carbonated waters. Other good anesthetics are chloroform or ether fumes, chloral, chlorotone, cocaine hypchlorate (of about 0.1 per cent strength), a very weak nicotine or tobacco decoction, magnesium sulphate, or alcohol added very gradually, or very weak acids like lemon juice, in which the leeches usually die extended. The leeches are placed in a small covered vessel or stoppered bottle nearly filled with clean water to prevent escape from the stupefying agent. They are very sensitive to nicotine, and in the field one of the best and simplest methods is to drop a few shreds of tobacco into the water with them, just enough to give the water a very faint tint. The leeches will usually be completely narcotized and relaxed in 30 to 60 minutes. If the decoction is made too strong they will die quickly, and contracted.

When the leeches no longer respond to pinching with a forceps or similar stimulation they are rapidly drawn between the thumb and fingers to remove excess mucus and are laid in a flat dish, extended, side by side, and in contact with each other. To keep them in place and prevent distortion when the fixing fluid is poured on, a piece of muslin or other thin cloth, or tissue paper, or filter paper moistened with the fluid, may be placed on them. The fixing fluid is then gently poured on. Usually not quite enough to cover them is poured on at first in order to prevent floating and disarrangement. After allowing a few minutes (longer for large leeches) for them to harden partially, sufficient fluid is added to completely immerse them, care being taken to prevent floating. For ordinary museum or taxonomic purposes 50 per cent alcohol or 2 per cent formaldehyde will answer perfectly, the latter being preferable because it is less likely to cause the cuticle to separate. After the fluid has thoroughly penetrated and the leeches have fully stiffened, they are transferred to stronger solutions and are finally preserved in generous quantities of 85 per cent alcohol or 4 per cent formaldehyde. They should be placed in wide-mouthed bottles or vials of sufficient length and diameter to keep them straight and to avoid crowding and distortion. Wads of absorbent cotton or cheesecloth may be used to hold them in place if necessary. When time is short or stupefiers are lacking, a bath of hot water (about 170° F) will usually cause the leeches to die extended.

For special purposes of study, suitable methods of fixation should be used and noted on the label. Most of the methods recommended for earthworms are applicable to leeches. For a general fixative, a saturated solution of corrosive sublimate is excellent. Flemming's fluid used with the customary precautions is one of the best for faithful preservation of histological detail. Bouin's fluid is even better as it ensures good staining. To bring out more clearly the sensillae and surface markings weak chromic acid (about 0.5 per cent aqueous solution, or better, equal parts of 0.25 per cent chromic acid and 0.25 per cent platinum chloride), or Perenyi's fluid are useful. But too strong solutions, too long exposure, or failure to wash out very thoroughly before preservation in alcohol, are likely to be disastrous, since chromic acid renders the tissues very brittle and Perenyi's fluid or any strong acid causes the connective tissues to swell and histolize.

As some of the coloring matters of leeches are freely dissolved or altered by the fixing and preserving fluids it is very desirable that the living colors be noted on the label. Every additional ecological or other fact added to the record increases the value of the specimens. To ensure thorough study many specimens of each lot are needed. Single specimens, especially when poorly preserved, are frequently an annoyance.

The following key is intended as a guide to the determination of the species of fresh-water leeches of North America north of Mexico, and may not serve for species outside of that area. Because of the plasticity of leeches which change in appearance with physiological and environmental conditions and with methods of preservation, recourse to full description is advised in case of doubt. Leeches should be studied both living and preserved, and in preservation they should be carefully stupefied and arranged in natural positions of moderate extension before fixation. Names in brackets [ ] are synonyms used in the first edition of this book.

Leeches should not be identified on the basis of host or habitat. Only the morphology will lead to secure identifications.

For each species, a reference number is given in parenthesis which will lead to a more complete description.

KEY TO SPECIES

1a   Mouth a small pore on the head sucker through which the pharyngeal proboscis may be protruded; no jaws; no denticles; blood colorless . . . . Order **Rhynchobdellida**   **2**

1b   Mouth large, opening from behind into entire sucker cavity; pharynx fixed; blood red **12**

2a   (1) Body not divided externally into trachelosome and urosome (Fig. 23.3), usually flattened and much wider than head (except some Helobdella and Actinobdella); head sucker not freely expanded; eyes mostly integrated, confined to head (except *Placobdella hollensis*); complete somites mostly 3-annulate . . . .
                                              Family **Glossiphoniidae**   **3**

2b   Body (especially in contraction) more or less distinctly divided at XIII into trachelosome and urosome (Fig. 23.7), usually long and narrow and little flattened; head sucker usually freely expanded;

eyes are ocelli or aggregations of ocelli on head, neck, or caudal
sucker; somites usually more than 3-annulate . . . .
Family **Piscicolidae**   9

3a   (2) Mouth within sucker cavity on II or III (Fig. 23.2B); body not
excessively flattened; cutaneous tubercles smooth and mostly few or
absent; salivary glands diffuse; sperm duct without closely folded
epididymis (seminal vesicle).   4

3b   Mouth apical or subapical on sucker rim (Fig. 23.2P); body often
excessively flattened; eyes 1 pair on III (except *P. hollensis*); gastric
caeca 7 pairs; salivary glands usually compact; esophagus pouches
often present   6

4a   (3) Eyes 4 pairs on inner paramedian lines of II to V; conspicuously
spotted on dorsum; after egg-laying translucent and gelatinous.
(Fig. 23.1) . . . . *Theromyzon* Philippi 1884
   A. Gonopores separated by 2 annuli, ♂ XI/XII, ♀ XIIa2/a3. (Ref. 1) . . . .
                 T. meyeri (Livahow) 1902
   B. Gonopores separated by 3 annuli, ♂ XIa2/a3, ♀ Xa2/a3. (Refs. 1, 14) . . . .
                 T. rude (Baird) 1863
   C. Gonopores separated by 4 annuli, ♂ XIa2/a3, ♀ XII/XIII. (Refs. 1. 10) . . . . . .
             T. tessulatum (O. F. Müller) 1774

4b   Eyes fewer than 4 pairs; body mostly opaque, not gelatinoid   5

5a   (4) Eyes 3 pairs; gastric caeca 6 pairs . . . . *Glossiphonia* Johnson 1816
   A. Eyes in 2 paramedian rows on II to IV; a pair of dark brown stripes more or less
broken by pale spots and low, rounded tubercles on sensory annuli; gonopores separated by 2 annuli, ♂ XI/XII, ♀ XIIa2/a3; testisacs 10 pairs. (Refs. 1, 5, 10) . .
             G. complanata (Linnaeus) 1758
   B. Eyes in a roughly triangular pattern of 3 groups of 2 each on III to V; a median but
no paired brown stripes or none; gonopores united at XIIa1/a2; testisacs 6 pairs.
(Refs. 1, 5) . . . . *G. heteroclita* (Linnaeus) 1758

5b   Eyes 1 or 2 pairs; gastric caeca 7 pairs; salivary glands diffuse; mouth
in III at level of eyes, which are simple. (Fig. 23.2) . . . .
        Batrachobdella Viguier 1879
   A. Eyes 2 pairs, on III smaller and nearly in contact, on IV widely separated.
(Refs 1, 2) . . . . *B. paludosa* (Carena) 1827
   B. Eyes 1 pair on III, pigment united or slightly separated; no white bar on VIa3.
(Refs. 1, 2) . . . . *B. picta* (Verrill) 1872
   C. Eyes as in B, in white area, pigment usually completely merged; a dense white
bar across VIa3. (Ref. 4) . . . . *B. phalera* Graf 1899
            [*Placobdella phalera* (Graf)]

5c   Eyes 1 pair, simple, well separated; mouth in III or IV; gastric caeca
1 to 6 pairs, may change with amount of contained food; salivary
glands diffuse; size small. (Fig. 23.3) . . . *Helobdella* E. Blanchard
   A. Nuchal gland and scute on dorsum of VIII; ♂ and ♀ pores separated by 1 annulus
(XIIa2). (Refs. 1, 10) . . . . *H. stagnalis* (Linnaeus) 1758
   A.A. No nuchal gland or scute b.
   B. Form rounded and in extension almost filamentous; unpigmented and translucent;
♂ and ♀ pores separated by XIIa2; postcaeca only. (Refs. 3, 4) . . . .
           H. elongata (Castle) 1899
          [*Glossiphonia nepheloidea* (Graf)]
    BB. Form moderately flattened and wider postclitellum; gonopores separated by
XIIa2; gastric caeca normally 6 pairs but anterior pairs may be completely
retracted when starving.
   C. Tubercles absent or limited to median line of posterior somites; color coffee brown
with 6 or 7 large white spots on sensory annuli. (Refs. 3, 10) . . . .
           H. fusca (Castle) 1900
   D. Tubercles absent or nearly so; color a conspicuous pattern of longitudinal brown
stripes and transverse rows of white spots. (Ref. 11) . . . .
         H. punctata-lineata Moore 1939
   E. Tubercles small, smooth, and conical, deeply pigmented and often double, mostly

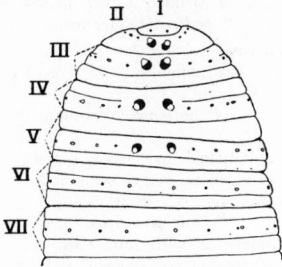

Fig. 23.1. *Theromyzon.* Dorsal view of anterior 7 somites showing annuli, eyes, and sensillac. × 20. (Original.)

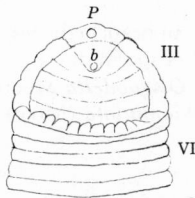

Fig. 23.2. Position of mouth in Glossiphiidae: *P, Placobdella parasitica* (Say); *b, Batrachobdella picta* (Verrill). (Original.)

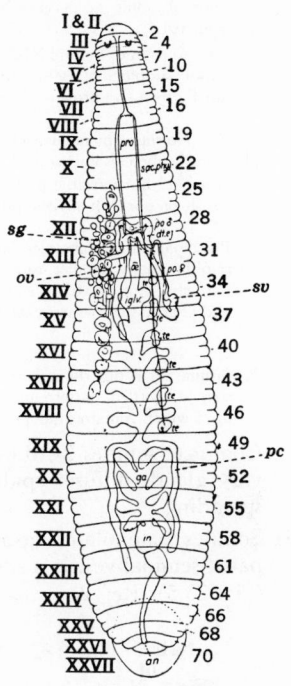

Fig. 23.3. *Helobdella fusca* (Castle). External form with segmentation and annulation and general internal anatomy. I–XXVII, somites; 2–70, annuli; *an*, anus; *dt.ej*, ductus ejaculatorius; *ga*, caecate intestine; *iglv*, stomach crop; *in*, bulbous intestine; *ov*, ovisac; *pc*, postcaecum; *po* ♂ and *po* ♀, gonopores; *pro, oe*, esophagus; *sac phy*, pharyngeal sac; *sg*, diffuse salivary gland; *sv*, seminal vesicle; *te*, testisacs. × 20. (Modified from Castle.)

in posterior middle field; many fine longitudinal light and dark lines, no large white spots. (Refs. 12, 16) . . . . *H. lineata* (Verrill) 1874
[*G. fusca* (Castle) part]

F. Tubercles prominent and numerous in 5 to 7 or 9 long series on neural annuli, smooth, conical, and black or dark brown; general color dark yellowish-brown. (Refs. 3, 10) . . . . *H. papillata* Moore 1906
[*G. fusca* (Castle) part]

6a **(3)** Complete somites 3-annulate, *a3* and *a1* may be faintly subdivided; resting form usually very broad and flat; eyes on III appear as one pair united in a common pigment mass and may be compound; salivary glands compact; epididymis a tight knot. (Fig. 23.4)
Placobdella Blanchard 1896

A. Somites I–V distinctly widened to form a discoid head; dorsum of body with 3 prominent tuberculated ridges; eyes separated by their diameter. (Ref. 10) . . . . .
P. montifera Moore 1912

AA. Without widened head and tuberculated ridges *B.*

B. Anus of adult at XXIII/XXIV; postanal somites a slender sucker penducle. (Ref. 10) . . . . *P. pediculata* Heminway 1912

BB. Anus in or behind XXVII; postanal somites normal.

C. Simple supplementary eyes in series following pair on head; annulus *a2* darkened with much green and brown pigment; size medium. (Refs. 1, 10) . . . .
P. hollensis (Whitman) 1872

CC. No supplementary eyes; large and very flat.

D. Dorsal tubercles low, smooth domes, often suppressed; opaque and heavily pigmented in a bold but variable pattern of brown, green, and yellow, venter with about 12 bluish or purplish stripes; dorsal and ventral furrows aligned. (Refs. 1, 10) . . .
P. parasitica (Say) 1824

E. Principal dorsal tubercles large, elevated, and roughened with numerous sensory papilla, many smaller tubercles and papillae; integument translucent and color pattern a fine mixture, venter without stripes; margins of body very thin, dorsal and ventral furrows not exactly aligned. (Refs. 10, 16) . . . .
P. ornata (Verrill) 1872
[*P. rugosa* (Verrill)]

F. Similar to E but the tubercles more uniform and smaller and the color in life a striped mottled pattern of brown and green underlain by about 30 dark brown lines which remain after preservation. (Ref. 13) . . *P. multilineata* Moore 1953

6b Complete somites 2- to 6-annulate; resting form not foliacious; salivary glands diffuse; epididymis loosely folded and the duct often spreading. . . . .

7a (6) Somites biannulate; gastric caeca 7 pairs, unbranched; testisacs 5 pairs, seminal vesicle a simple enlarged fold of sperm duct; 1 species. (Fig. 23.5) (Ref. 2) . . . . *Oligobdella* Moore 1918
O. biannulata (Moore) 1900

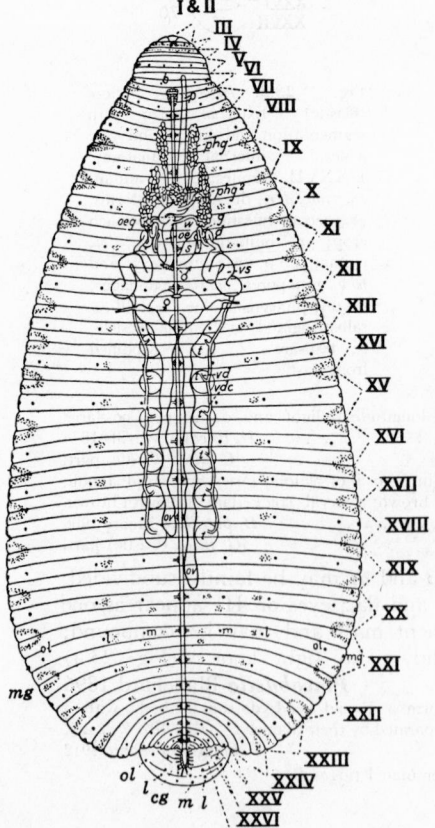

Fig. 23.4. *Placobdella parasitica* (Say). External form, segmentation and markings, and internal anatomy except stomach, intestine, etc. I–XXVII, somites; *b*, brain; *cg*, caudal ganglion; *d*, ductus ejaculatorius; *g*, epididymis; *m, l, ol, mg*, paramedian, intermediate, supramarginal and marginal sensillae; *oe*, esophagus, *oeg* and *w*, esophageal pouch ducts; *ov*, ovisacs; *p*, proboscis; *phg¹²*, compact salivary glands; *s*, atrial cornu; *t*, testisacs; *vd* vas efferens; *vdc*, sperm duct or vas deferens; *vs*, vesicules seminalis; ♂ and ♀, male and female pores. × 2. (Modified from Whitman.)

7b   Complete somites 3- to 6-annulate **8**

8a (7) Complete somites 3-annulate; caudal sucker without a circle of retractile papillae; eyes well separated; gonopores united, XII *a2/a3*;
gastric caeca 5 or 6 pairs; testisacs 5 pairs. (Refs. 2, 9). . . . .
Oculobdella Autrum 1936
O. lucida Meyer and Moore 1954

8b   Caudal sucker with a circle of glands and retractile digitate processes;
eyes 1 pair on III; gastric caeca 7 or 6 pairs; salivary glands diffuse.
(Fig. 23.6) . . . . *Actinobdella* Moore 1901

A. Somites 3-annulate; sucker processes conical, about 30; dorsal tubercles prominent,
   in 5 series. (Ref. 2) . . . . *A. triannulata* Moore 1924
B. Somites 6-annulate; sucker processes about 60; dorsal tubercles in 5 series.
   (Ref. 2) . . . . *A. annectens* Moore 1906
C. Somites of 6 unequal annuli; sucker glands and processes about 30; median
   dorsal tubercles only. (Refs. 2, 10) . . . . *A. inequiannulata* Moore 1901

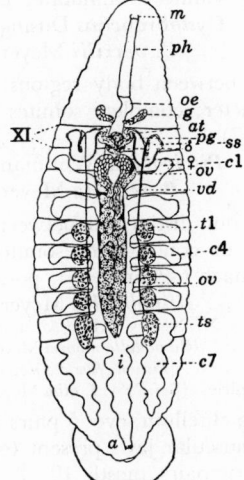

Fig. 23.5. *Oligobdella biannulata* Moore.
General form and anatomy, boundaries of
middle somite indicated; eyes and diffuse
salivary glands indicated. *a*, anus; *at*,
atrium with cornua; *c* 1–4, gastric caeca;
c 7, postcaecum; *g*, esophageal pouch; *i*,
intestine; *m*, mouth (proboscis pore); *oe*
esophagus; *ov*, ovisac; *pg*, prostate glands;
ph, pharynx; *ss*, seminal vesicle; *t* 1–5,
testisacs; *vd*, vas deferens; ♂ and ♀ pores.
× 20. (After Moore.)

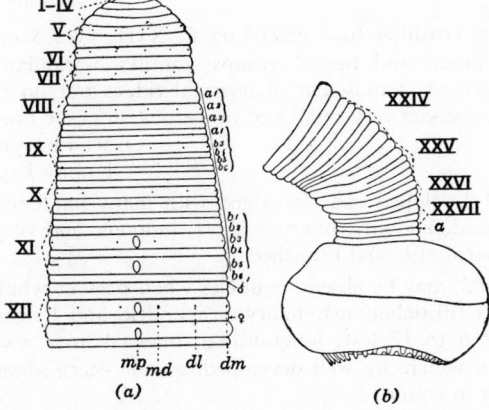

Fig. 23.6. *Actinobdella inequiannulata* Moore. (*a*) Segmentation, annulation, sensillae, and median tubercles of
somites I–XII, *md*, *dl*, *dm*, paramedian, intermediate and
supramarginal sensillae; *mp*, median tubercles. ♂ and ♀,
male and female pores. × 20. (*b*) Lateral view of posterior
end with sucker and its papillae. (After Moore.)

9a   (2) Pulsatile vesicles on margins of sensory annuli of urosome. **10**

9b   No pulsatile vesicles **11**

10a (9) Pulsatile vesicles 11 pairs (XIII to XXIII), small, on 2 tertiary annuli; on preserved leeches usually contracted and obscure; somites 14-annulate; postcaeca completely united into one; testisacs 6 pairs .
Pisicola Blainville 1818

A. No ocelli on caudal sucker; cephalic eyes 2 (or 1) pairs; gonopores separated by 4 tertiary annuli. (Refs. 1, 7). . . . . *P. punctata* (Verrill) 1871
AA. Ocelli on caudal sucker.
B. Caudal ocelli 8 to 10; crescentic; gonopores separated by 2 annuli; sperm duct much convoluted. (Ref. 8) . . . . *P. salmonsitica* Meyer 1946
C. Caudal ocelli punctiform 10 or 12; cephalic eyes 2 pairs but variable; sperm duct simply looped; gonopores by 2 annuli. (Refs. 1, 7). . . . *P. milneri* (Verrill) 1871
D. Caudal ocelli punctiform, 12 or 14; cephalic eyes same as C; gonopores by 3 annuli; sperm duct simply looped. (Refs. 1, 5) . . . . *P. geometra* (Linnaeus) 1758

10b Pulsatile vesicles 11 pairs, larger, covering 2 secondary annuli and persistent after preservation; complete somites 7-annulate; caudal sucker very large. (Refs. 1, 7) . . . . . *Cystobranchus* Dusing 1859
C. verrilli Meyer 1940

11a (9) Form clavate with no sharp limitation between body regions; both suckers much smaller than body diameter; complete somites 3-annulate; mouth central in sucker. (Ref. 7) . . . .
Piscicolaria Whitman 1889
P. reducta Meyer 1940

11b Form of body and suckers as in last, but distinction between trachelosome and urosome more evident in contraction; somites 14-annulate; mouth central in sucker; stomach 6-chambered; postcaeca completely fused. (Fig. 23.7). . . . . *Illinobdella* Meyer 1940
A. Gonopores separated by 4 secondary annuli, ♂ XI/XII, ♀ XII*b4/b5*; a distinct seminal vesicle. (Ref. 7) . . . . *Illinobdella moorei* Meyer 1940
[*Piscicola punctata* Verrill] part
B. Gonopores separated by 2 annuli; no seminal vesicle. (Ref. 7) . *I. alba* Meyer 1940

12a (1) Pharynx a suction bulb not extending to clitellum; eyes 5 pairs in an arch on II to VI; somites 5-annulate; muscular jaws present (except *Haemopis* sp.); testisacs large, in metameric pairs, mostly 10. . . . . . .
Order *Gnathobdellida*
Family **Hirudidae**   **13**

12b Pharynx a crushing tube extending to XIII; eyes 3 or 4 pairs in separate labial and buccal groups; somites 5-annulate but often further divided; 3 muscular pharyngeal ridges but no true jaws or denticles; testisacs very small and numerous, in grape-bunch arrangement . . . . Order **Pharyngobdellida**
Family **Erpobdellidae**   **16**

13a (12) Jaws well developed and saw-edged with many fine denticles in one row (monostichodont); pharynx short, bulbous, and very muscular; gastric caeca large and branched **14**

13b Jaws varied, may be absent, denticles when present wholly or partly in 2 rows (distichodont); pharynx cylindric and longer, with thin walls and 6 to 12 low, longitudinal, internal folds; except for the postcaeca, which are well developed, gastric caeca absent or vestigial except in young. **15**

14a (13) Gastric caeca 1 pair per somite of middle region. No copulatory glands or pores. A species introduced through medical practice . . .
Hirudo medicinalis (Linnaeus) 1758

14b Copulatory glands and pores in linear patterns in one group behind

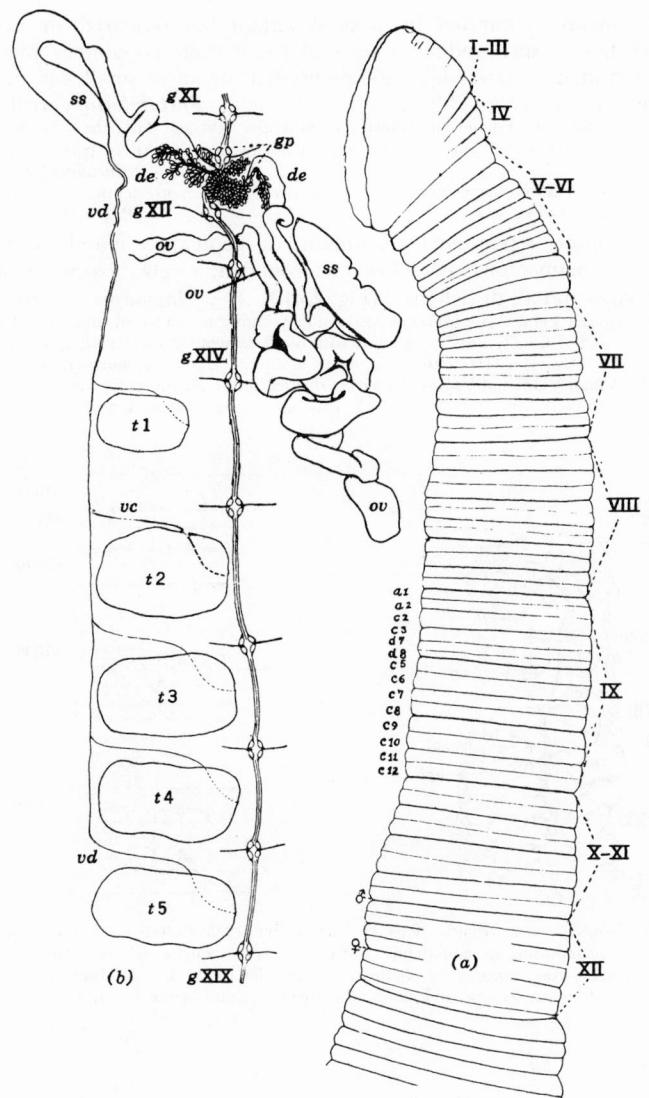

Fig. 23.7. *Illinobdella moorei* Meyer. (*a*) External morphology of somites I to XIII, showing complete annulation with symbols for IX, position of ♂ and ♀ pore and eye. × 35. (*b*) Dissected reproductive organs in relation to nerve cord, *g* XI–XIX. *de*, ejaculatory duct; *gp*, prostate glands; *ov*, ovisac; *ss*, seminal vesicle; *vd*, vas deferens; *ve*, vas efferens; *t* 1–5, testisacs. (After Moore.)

gonopore. (Fig. 23.8) . . . . ***Macrobdella*** Verrill 1872

A. A median dorsal series of about 21 bright red spots; gonopores separated by 5 annuli, ♂ XI/XII, ♀ XII/XIII; copulatory gland pores in a square figure in furrows XIII/XIV and XIV*b1/b2;* denticles about 65. (Refs. 1, 10) . . . .
M. decora (Say) 1824

B. Red spots as in A; gonopores separated by 2½ annuli, ♂ XII*b1,* ♀ XII*a2/b5;* copulatory gland pores in 4 transverse rows of 6 each on annuli XIII*b6,* XIV*b1* and XIV*b2;* denticles *ca.* 40. (Ref. 1). . . . . *M. sestertia* Whitman 1884

C. No red spots; gonopores separated by 2 annuli, ♂ XI/XII; copulatory gland pores in 2 rows of 4 in furrows XIII/XIV and XIV*b1/b2;* denticles about 50. (Ref. 13) . . .
M. ditetra Moore 1953

15a **(13)** Gonopores separated by 3 or 4 annuli but obscured in mature leeches by surrounding systems of counterpart copulatory pits and prominences; jaws high and compressed; denticles small and in part distichous. (Fig. 23.9) . . . . ***Philobdella*** Verrill 1874

    A. Median dorsal stripe dark brown, paired stripes light and dark, the supra marginal sometimes broken but no discrete spots; denticles about 20. (Refs. 1, 16). . . . . .

                                          P. floridana Verrill 1874

    B. Median dorsal stripe light yellow; dorsolateral brown spots; denticles about 40 (35–48). (Ref. 1) . . . . ***P. gracilis*** Moore 1901

15b Gonopores separated by 5 annuli; no copulatory glands or pores; penis filamentous; jaws low and rounded, rarely absent; denticles coarse and all distichous. (Fig. 23.10) . . . ***Haemopis*** Savigny 1820

    A. Annuli VII*a3* and VIII*a1* enlarged but only faintly or not at all divided; color variable but usually heavily blotched with black, brown, and yellowish-gray; denticles 12–16 pairs. (Refs. 1, 10) . . . . ***H. marmorata*** (Say) 1824

    B. Annuli VII*a3* and VIII*a1* completely divided into *b* annuli; color uniformly gray or plumbous, a median dorsal black stripe and orange marginal stripes, rarely with a

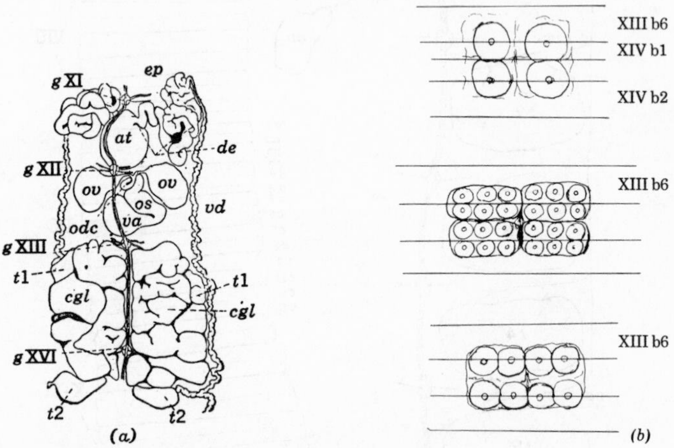

Fig. 23.8. (*a*) *Macrobdella decora* (Say). Part of reproductive organs dissected. *at*, atrium; *cgl*, copulatory glands; *de*, ductus ejaculatorius; *ep*, epididymis; *g* XI–XIV, nerve ganglia; *od*, *odc*, oviduct; *os*, ovisac; *ov*, ovary; *t* 1, 2, first and second testisacs; *va*, vagina; *vd*, vas deferens. × 3. (*b*) *Macrobdella*. Diagrams of copulatory pores of *M. decora*, *sestertia*, and *ditetra*. Enlarged. Annuli symbols indicated. (*a* after Moore; *b* original.)

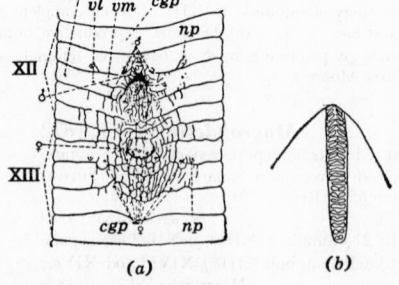

Fig. 23.9. *Philobdella gracilis* Moore. (*a*) External genital region. ♂, ♀ orifices with their respective systems of gland pores *cgp* ♂ and *cgp* ♀; *np*, nephropores; *sbm*, *vl*, *vm*, submarginal, intermediate, and paramedian ventral sensillae. × 3 ½. (*b*) Outline of cross profile of jaw with incomplete series of denticles. × 35. (After Moore.)

few black spots; denticles 20 to 25 pairs; very large. An aquatic and a terrestrial
form. (Refs. 1, 10). . . . . *H. lateralis* (Say) 1824

  C. Annuli VII*a3* and VIII*a1* like *marmorata;* size and general form like *plumbea*, with a
median dorsal stripe like *lateralis;* denticles 9 to 12 pairs. Young with metameric
dark bands. (Ref. 6) . . . .
                                              H. kingi Mather 1954

  D. Annuli VII*a3* and VIII*a1* enlarged and divided; ♂ pore XI*b5/b6;* ♀ XII*b5/b6;* color
variable shades of dull green, always more or less blotched with black, but no median
dorsal stripe; jaws and denticles absent or vestigial. Our largest leech. (Refs. 10,
16). . . . . *H. grandis* (Verrill) 1874

  E. VII*a3* and VIII*a1* less enlarged and divided than *grandis;* gonopores near middle of
XI*b6* and XII*b6*. Resembles *grandis* most closely in general structure and *marmorata*
in reproductive organs. (Ref. 10) . . . . *H. plumbea* Moore 1912

16a (12) Ejaculatory duct with long preatrial loop reaching to ganglion XI. .   **17**

16b   Ejaculatory duct without preatrial loop; median atrium relatively
large and cornua directed chiefly laterad; annulus *b6* enlarged and
subdivided as in *Dina*. (*Dina* and *Mooreobdella* may be regarded as
subgenera of *Erpobdella*. (Fig. 23.13) . **Mooreobdella** Pawlowski 1955

  A. Atrium ellipsoidal, wider than long with horns shorter than diameter of median
atrium; gonopores normally by 3 annuli, ♂ XII*b2/a2,* ♀ XII/XIII; eyes 3 pairs on
III and IV, length about 1¼ inches. (Refs. 1, 15). . *M. microstoma* (Moore) 1901

  B. Atrium globoid with prominent horns longer than its diameter; gonopores by 2 annuli, ♂ XII*a1/a2,* ♀ X*b5/b6;* eyes 3 or 4 pairs, 1 on III, 2 on IV; color in life pale red;
length to 2 inches. (Refs. 15, 16) . . . . *M. fervida* (Verrill) 1874

  C. Atrium as in B; gonopores by 2–2½ annuli, ♂ XII*a2* to XII*a2/b5,* ♀ XII/XIII; eyes 3
pairs on III and IV; length to 1¼ inches. (Refs. 13, 15) . . . .
                                                 M. bucera (Moore) 1949

17a (16) Atrial cornua with sheep-horn spiral coil; annulus *b6* of complete
somites enlarged and partially divided. (Fig. 23.11) . . . .
                                        Nephelopsis Verrill 1872

Eyes 4 pairs, 2 labial, 2 buccal; gonopores by 2 annuli; postclitellar region wider and
flatter; color yellowish with black spots, or plain. (Refs. 1, 10) . . . .
                                          N. obscura Verrill 1872

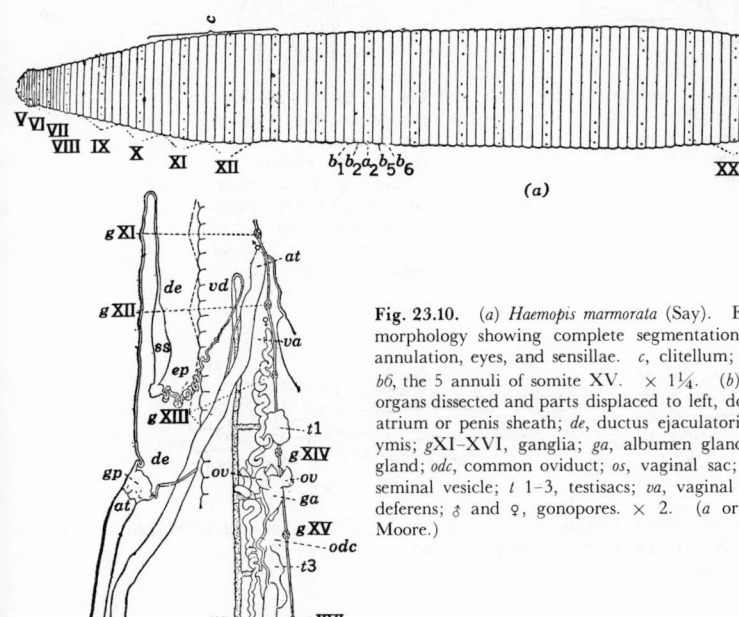

(a)

Fig. 23.10. (a) *Haemopis marmorata* (Say). External dorsal
morphology showing complete segmentation (V–XXVII),
annulation, eyes, and sensillae. *c*, clitellum; *b1, b2, a2, b5,
b6*, the 5 annuli of somite XV. × 1¼. (b) Reproductive
organs dissected and parts displaced to left, dorsal view. *at*,
atrium or penis sheath; *de*, ductus ejaculatorius; *ep*, epididymis; *g*XI–XVI, ganglia; *ga*, albumen gland; *gp*, prostate
gland; *odc*, common oviduct; *os*, vaginal sac; *ov*, ovisac; *ss*,
seminal vesicle; *t* 1–3, testisacs; *va*, vaginal stalk; *vd*, vas
deferens; ♂ and ♀, gonopores. × 2. (a original; b after
Moore.)

(b)

17b   Atrial cornua simply curved conic **18**
18a   **(17)**   Annulus *b6* not appreciably enlarged or subdivided.   (Fig. 23.12) . .
                                                        Erpobdella Blainville 1818
>   Eyes 3 pairs, 1 labial, 2 buccal; several color phases but mostly with 2 or 4 broad
>   longitudinal stripes of brown spotted with ashy grey. A heavily black-barred form
>   (*annulata* Moore 1922) occurs in Vancouver and the Northwest border states. (Refs.
>   1, 10). . . . . ***E. punctata*** (Leidy) 1870

18b   Annulus *b6* (and somtimes *a2*) of complete somites distinctly en-
                larged and subdivided; size medium or small and color in life more
                or less reddish from blood . . . . ***Dina*** E. Blanchard 1892
>   A. Eyes absent; gonopores separated by 2 annuli (♂ X*b5/b6*, ♀ XII*b1/b2*); color longi-
>   tudinal stripes. (Ref. 1) . . . . ***D. anoculata*** Moore 1898
>   B. Eyes 4 pairs; gonopores by 3 to 3½ annuli, ♂ XII*a2* to *a2/b5*, ♀ XIII*b1/b2;* nearly
>   pigmentless or with a few dark spots; length extended 1 inch. (Ref. 10) . . . . . .
>   ***D. parva*** Moore 1912
>   C. Eyes and gonopores (♂ XII*a2*) like *parva*, but size 2 inches or less and color heavily
>   blotched with a median dorsal dark stripe. (Ref. 14) . . . .
>   ***D. dubia*** Moore and Meyer 1951
>   D. Eyes 3 pairs, 1 on II, 2 on IV–V; gonopores by 2 annuli, ♂ XII*b1/b2*, ♀ XII*b5/b6;*
>   liver color in life, size 2 inches or less. (Refs. 1, 12) . . . ***D. lateralis*** (Verrill) 1871

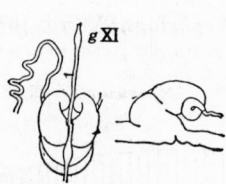

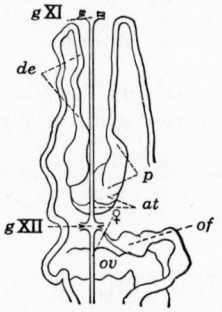

Fig.   23.11. *Nephelopsis*
obscura (Verrill). Dorsal
and lateral views of
atrium *in situ.* × 3.
(Original.)

Fig.   23.12.   *Erpobdella*
punctata (Leidy). Atrium
and neighboring parts of
reproductive organs.   *at*,
atrium, *de*, ductus ejaculatorius; *g*XI–XII, ganglia; *of*, closed inner end of
ovisac; *ov*, ovary; *p*, atrial
horn. × 7½. (After Moore.)

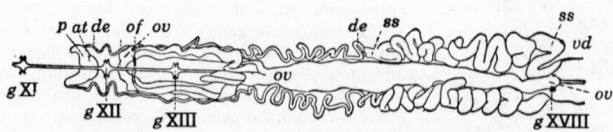

Fig. 23.13. *Mooreobdella fervida* (Verrill). Reproductive organs except testisacs. *at*, atrial cornua; *de*, ductus
ejaculatorius; *g*XI–XVIII, ganglia; *of*, closed inner end of ovisac; *ss*, seminal vesicle; *vd*, vas deferens.
× 3 ¼. (After Moore.)

References

The code numbers before each species name in the key will serve as guides to full descriptions and additional references. Only the most important early (before 1938) papers are cited; all others are listed in Number 1.

(1) **Autrum, H. 1939.** Hirudineen. In: Bronns *Klassen und Ordnungen des Tierreichs.* Band IV, Abt. III, Buch 4, Teil 2. Akademische Verlagsgesellshaft, Leipzig. (Literature to 1938.) (2) **1936.** Hirudineen. In: Bronns *Klassen und Ordnungen des Tierreichs.* Band IV, Abt. III, Buch 4, Teil 1, pp. 1–96. Akademische Verlagsgesellschaft, Leipzig. (Glossiphoniidae only.) (3) **Castle, E. 1900.** Some North American fresh-water Rhynchobdellidae. *Bull. Mus. Comp. Zool.,* 36:18–64. (4) **Graf, Arnold. 1899.** Hirudinienstudien. *Abhandl. Kaiserl. Leop. Carol. Deutschen Akad. Naturf.,* 72:217–404. (5) **Harding, W. A. 1910.** A revision of British leeches. *Parasitology,* 3:130–201. (6) **Mather, C. K. 1954.** Haemopis kingi, new species. *Am. Midland Naturalist,* 52:460–468. (7) **Meyer, M. C. 1940.** A revision of the leeches (Piscicolidae) living on fresh-water fishes of North America. *Trans. Am. Microscop. Soc.,* 59:354–376. (8) **1946.** A new leech *Piscicola salmositica. J. Parasitol.,* 32:467–476. (9) **Meyer, M. C. and J. P. Moore. 1954.** Notes on Canadian leeches. *Wasmann J. Biol.,* 12:63–96. (10) **Moore, J. P. 1912.** The leeches of Minnesota. *Geol. Nat. Hist. Survey Zool. Ser.,* No. 5 Pt. 3. (11) **1939.** Helobdella punctato-lineata. *Puerto Rico J. Public Health Trop. Med.,* 1939:422–428. (12) **1952.** Professor Verrill's freshwater leeches. *Notulae Naturae Acad. Nat. Sci. Phila.,* No. 245:1–15. (13) **1953.** Three undescribed North American leeches. *Notulae Naturae, Acad. Nat. Sci. Phila.,* No. 250:1–13. (14) **Moore, J. P. and M. C. Meyer. 1951.** Leeches (Hirudinea) from Alaskan and adjacent waters. *Wasmann J. Biol.,* 9:11–77. (15) **Pawlowski. 1955.** Revision des genres Erpobdella et Dina. *Bull. Soc. Sci. et Let. Lodz Cl. III,* 6:1–15. (16) **Verrill, A. E. 1874.** Synopsis of American Fresh Water Leeches. *Report of Commissioner of Fisheries for 1872–73,* Pt. II:666–689.

24

Anostraca

RALPH W. DEXTER

INTRODUCTION TO THE FRESH-WATER CRUSTACEA

Crustaceans are a class of the phylum Arthropoda specifically characterized by the presence of two pairs of antennae. Respiration is accomplished by means of gills or directly through the body surface. The vast majority of Crustacea are aquatic, and a large number of species are found in fresh water. Three orders, the Anostraca, Notostraca, and Conchostraca, are confined to inland waters.

Following is a synoptic outline of fresh-water crustacea:

Phylum Arthropoda. With paired, jointed appendages on a body nearly always segmented (segmentation is obscured in some Crustacea, especially smaller ones).

Subphylum Mandibulata. Mouth parts always include a pair of mandibles.

Class Crustacea. Two pairs of antennae; respiration by gills or body surface.

Subclass Branchiopoda (phyllopods). Many pairs of flattened appendages on thorax serving for both locomotion and respiration. (The name Euphyllopoda was formerly used for an order to include the suborders Anostraca, Notostraca and

Conchostraca. See Chapter 27 for a discussion of the problems of classifying the Branchiopoda.)

Order Anostraca (fairy shrimps). Eleven to 17 pairs of thoracic appendages; elongate, cylindrical body without a carapace; eyes stalked (this chapter).

Order Notostraca (tadpole shrimps). Forty to 60 pairs of thoracic appendages; body depressed and partly covered by a dorsal shieldlike carapace; eyes sessile (Chapter 25).

Order Conchostraca (clam shrimps or claw shrimps). Ten to 28 thoracic appendages; body compressed and completely enclosed within a bivalve carapace; eyes sessile (Chapter 26).

Order Cladocera (water fleas). Four to 6 pairs of thoracic appendages; body compressed, all except head usually enclosed within a bivalve carapace; second antennae used for locomotion; single compound eye (Chapter 27).

Subclass Ostracoda (seed shrimps). Two (or 3) pairs of thoracic appendages (p. 660); body compressed and entirely enclosed within a bivalve carapace (Chapter 28).

Subclass Copepoda (copepods). Five or 6 pairs of thoracic appendages, the first 4 pairs being biramous; body small, cylindrical, and divided into a metasome and a urosome. Parasitic forms greatly modified (Chapters 29 and 30).

Subclass Branchiura (fish lice). With suction cups on maxillae, body depressed. Ectoparasitic on fish, sometimes swim freely. Formerly considered an order of the copepoda (Chapter 30).

Subclass Malacostraca. Body consisting of 20 segments, 5 in head, 8 in thorax, and 7 in abdomen (Chapter 31).

ANOSTRACA

The crustacean order Anostraca, or fairy shrimps, are among the most primitive crustaceans of our recent fauna. They are elongated, somewhat cylindrical, delicate animals without the carapace characteristic of the other groups of Branchiopoda. Hence they bear the name Anostraca (without a shield). The head is large and prominent, and bears the sensory organs, clasping organs of the male, and the mouthparts. The compound *eyes* are large and stalked. In the middle of the forehead there is a single, sessile eyespot known as the *ocellus*. The *first antennae* are slender and inconspicuous. Often they are unsegmented. The *second antennae* are large and swollen, and in the male they are enormously developed (Fig. 24.8a,c). Often they have *antennal appendages* (Fig. 24.3), processes, tubercules, or spines. Each species has developed unique characteristics of the second antennae, used in classification. Sometimes there is a *frontal appendage* (Fig. 24.23b) attached to the head between the second antennae, which presumably aids them as claspers. This, too, is important in the taxonomy of fairy shrimps. The mouthparts consist of an overhanging labrum, a pair of mandibles, and two pairs of maxillae. The buccal cavity, vertical esophagus, globular stomach, and digestive glands are located within the head. A long, straight intestine runs to the extreme posterior end of the animal.

In North American forms the thorax contains either 11 or 17 segments, each bearing a pair of foliaceous appendages which serve for locomotion, food getting, and respiration. The appendages, which are biramous, lobed, and setose, are much alike, although the last pair is not usually as fully developed as the others. Laterally there are one or two pre-epipodites at the base of each appendage; often these pre-epipodites are fused to varying degrees. Basally on the appendage there is the epipodite, which never has spines, and distally there are the exopodite and endopodite. Along the medial margin are five endites bearing setae as well as filaments.

Following the limb-bearing thoracic segments, sometimes called the pregenital region or trunk, are two partly fused thoracic segments which contain the reproductive organs. In the male there are two penes, the position and structure of which are of taxonomic value. In the female there is a conspicuous ovisac in which the eggs develop. Posterior to the genital segments are seven abdominal or postgenital segments without jointed appendages, but the last one, the telson, carries two *cercopods* (Fig. 25.8a). These are usually elongated and armed with spines or long filaments.

North American fairy shrimps range in size from 5 to 6 mm (*Eubranchipus floridanus* Dexter) to 60 to 100 mm (*Branchinecta gigas* Lynch). Most of the species average about 20 to 25 mm in total length. Sexes are separate and internal fertilization is ordinarily necessary for development. Often males are less abundant than females, but usually not as uncommon as is generally believed. Sometimes males are more abundant than the females. The gonads of both sexes are paired, tubular organs. The fertilized eggs develop in the egg sac and may be released into the water or remain in the sac at the death of the female.

Fairy shrimps are local and sporadic in occurrence. One pool may have an abundant population and another nearby may not have any. Some regions have a great many pools inhabited by them and others have few, if any. While many parts of North America remain to be explored for specimens, there are some regions, especially in the southeastern United States, where field collectors have never found them. Not only is the pattern of geographical distribution irregular, but from year to year their occurrence may fluctuate tremendously in distribution and abundance. During certain favorable years the fairy shrimps may be widespread and abundant, and at other times they may be rare, if present at all, in the same bodies of water. Also, the duration of active existence varies greatly from year to year depending on climatic factors. In one pond studied by the writer a generation was completed in three weeks (April 22 to May 13, 1944). In that same pond another generation existed for 24 weeks (November 8, 1951 to April 26, 1952). Usually there is but one generation a year in each pond. However, on occasions a pond may dry out and refill before temperatures get too high for hatching and another hatch may then take place. This is not necessarily a second generation in the usual sense of the word, since many eggs require a resting period and do not hatch for some time even though they may have been soaked in water several different times. For the most part fairy shrimps live in temporary pools and ponds of fresh water, except the brine shrimp, *Artemia salina* L., which inhabits the Great Salt Lake, saline bodies of water, and salt evaporating basins. This species can tolerate even saturated saline solutions. Variations in this species have been correlated with the salinity of the medium in which they developed. In spite of such salinity tolerance, there is no marine species of Anostraca. One species, *Branchinecta shantzi* Mackin, is known to inhabit alpine lakes. Ordinarily fairy shrimps are not found in bodies of water where fish are likely to live. Fairy shrimps can usually withstand predation from amphibians and carnivorous insects, but are soon eradicated in the presence of fish. Rain pools and temporary ponds that form from melting snow and ice are the usual habitats of fairy shrimps. The eggs are distributed by winds and by the transport of mud carried by animals visiting the ponds. There is a nauplius type of larva (Fig. 24.8*d*). Many molts take place in development. The hatching of certain species, the number of molts, and rate of development depend upon the temperature. Some species hatch only in cold water. Hatching also seems to require a previous period of drying. At least drying is a stimulus to the hatching of many species. Freezing also seems to be a stimulus, but apparently is not always required. Experimentally a small number of eggs of a few species have been hatched without either drying or freezing, but hatching is much more successful after drying of the eggs, and in nature this almost always happens. Eggs of some species are very resistant to desiccation.

Adults are found only in the spring in the northeastern United States. However, those species that occur in northern latitudes and high altitudes are found as adults in the summer season. Southern and western species appear whenever sufficient water is present in the temporary pools to produce a population.

The fairy shrimps swim gracefully on their backs. The appendages are always faced toward the source of light. Mature specimens frequently rest on their backs on the bottom sediments. Color is extremely variable, differing from place to place and from one life stage to another. Color is sometimes determined by the kind of food ingested. They feed on microorganisms and detritus. The food is concentrated in a ventral groove between the bases of the appendages, on a mucilaginous string which is continuously moved forward to the mouth by the action of the appendages.

With the name of each species given in the key will be found a list of the states, provinces, and localities in North America from which the species has been reported in the literature. Present knowledge of distribution is very incomplete and much work remains to be done. As a group, the Anostraca is found from sea level to high alpine tundra, and from Mexico to the Arctic plains.

There are six families of Anostraca known from North America thus far. They, with North American genera, are as follows: Polyartemiidae (*Polyartemiella*), Artemiidae (*Artemia*), Branchinectidae (*Branchinecta*), Streptocephalidae (*Streptocephalus*), Thamnocephalidae (*Thamnocephalus, Branchinella*), Chirocephalidae (*Pristicephalus, Chirocephalopsis, Eubranchipus, Artemiopsis*). Altogether, 27 species have thus far been recorded, all of which are included and illustrated in the following key.

Linder (1941) believes the classification of families should rest upon the structure and arrangement of the male reproductive system and the pre-epipo-dites of the thoracic appendages; and the configurations of the second an-tennae with their appendages, and of the frontal appendages when present, should be used to separate the lower categories.

For the following key, an attempt has been made to combine characteristics of the reproductive system, the thoracic appendages, and head appendages. The key is based on male specimens, and all figures are of males unless other-wise indicated. In making identifications, most of the necessary observations can be made on whole specimens with a dissecting microscope, using trans-mitted or reflected light as necessary. In some cases it may be necessary to sever the head from the body in order to make critical observations of the shape of frontal or antennal appendages.

Acknowledgment is made to Drs. Folke Linder, Walter G. Moore, J. G. Mackin, and N. T. Mattox for their constructive criticism of the key.

KEY TO SPECIES

1a　　Pregenital swimming appendages 17 to 19 pairs . . . .
　　　　　　　　　　　　　　　　　　　Family ***Polyartemiidae***　　**2**
　　One genus Polyartemiella Daday 1909.

1b　　Pregenital swimming appendages 11 pairs. . . . .　**3**

2a　(1) Tuberculiform frontal appendage. Male clasping antenna (includ-ing antennal appendage) quadriramose. (Fig. 24.1) . . . .
　　　　　　　　　　　　　Polyartemiella hazeni (Murdoch) 1874
　　Coastal plains of Alaska, Yukon Territory, Northwest Territories.

2b　　No frontal appendage. Male clasping antenna (including antennal appendage) triramose. (Fig. 24.2) . . . . . . ***P. judayi*** Daday 1909
　　Pribiloff Islands and Alaska.

3a　(1) Terminal segment of male second antenna with complex cheliform terminal segment (Fig. 24.5) . . . . Family **Streptocephalidae**　**4**
　　*One genus, **Streptocephalus** Baird 1852.*

3b　　Terminal segment simple. . . . .　**8**

4a　(3) Cercopods of mature male curve inward with long bristles on proxi-mal portion and short curved spines on distal portion . . . .　**5**

4b　　Cercopods of mature male straight with long bristles along entire length. . . . .　**6**

5a　(4) Medial portion of terminal branch on male clasping antenna bears 2 teeth on anterior margin near proximal end. (Fig. 24.3). . . . . .
　　　　　　　　　　　　　Streptocephalus seali Ryder 1879
　　N. Y., N. J., Md., Va., Ill., Mo., Minn.,.N. D., N. C., S. C., Fla., Ala., Miss., La., Tex., Okla., Kan., Colo., Neb., Ariz., Calif., Ore., Mont., Alberta, Vera Cruz.

5b　　Medial branch of terminal segment on male clasping antenna bears 3 teeth. (Fig. 24.4) . . . . ***S. similis*** Baird 1852
　　Santo Domingo, Puerto Rico, Jamaica.

6a　(4) Lateral branch of terminal segment on male clasping antenna spi-nous. Cercopods blunt. (Fig. 24.5) . . . ***S. antillensis*** Mattox 1950
　　Puerto Rico.

Fig. 24.1. *Polyartemiella hazeni.* Posterior view of anterior part of head. (After Daday.)

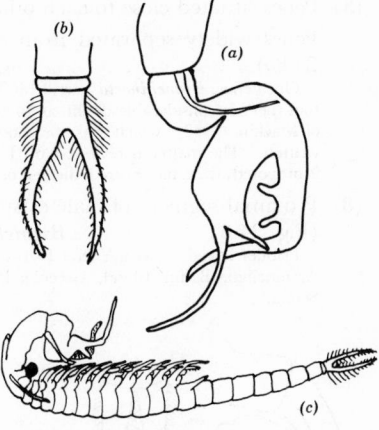

Fig. 24.3. *Streptocephalus seali.* (*a*) Second antenna. (*b*) Cercopods. (*c*) Lateral view in swimming position. (After Creaser.)

Fig. 24.2. *Polyartemiella judayi.* Dorsal view of head. (After Daday.)

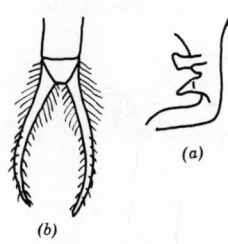

Fig. 24.4. *Streptocephalus similis.* (*a*) Diagnostic portion of second antenna showing the 3 teeth. (*b*) Cercopods. (Modified from Creaser.)

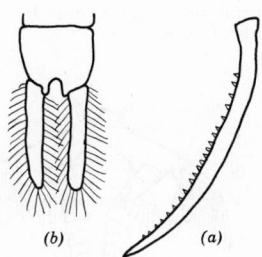

Fig. 24.5. *Streptocephalus antillensis.* (*a*) External branch of terminal segment of second antenna. (*b*) Cercopods. (After Mattox.)

6b　　　Lateral branch not spinous. Cercopods elongated and pointed. . . . 　7

7a　(6) Medial branch of terminal segment of the male clasping antenna bears a process near the end. Posterior spur of lateral branch sabre-shaped. (Fig. 24.6). . . . . **S. texanus** Packard 1871
Neb., Kan., Mo., Colo., Okla., Tex., N. M., Ariz., Utah, Calif., Mont., San Luis Potosi (Mexico), Fla.

7b　　　Medial branch does not have a process near the end. Posterior spur of lateral branch shaped something like a miniature human foot in outline at the end. (Fig. 24.7). . . **S. dorothae** Mackin 1942
N. M., Tex., Okla.

8a   (3)   Penes situated close to each other on the ventral surface **14**

8b   Penes widely separated from each other on lateral surface. (Fig. 24.8*a*) . . . . Family **Branchinectidae**   **9**

> One genus, **Branchinecta** Verrill 1869. Some of the species were so poorly described that there is considerable doubt as to their identity. The nomenclature used is that of Mackin (1952), which was the most recent revision at the time this chapter was written. The matter is reviewed by Lynch (1958), and restoration of certain names is proposed; these names are indicated below.

9a   (8)   Proximal segment of male clasping antenna serrate on inner margin. (Fig. 24.8) . . . . **Branchinecta paludosa** (O. F. Müller) 1788

> Pribilof Island, Alaska, Yukon Territory, coastal plains of Arctic Ocean and Canadian Archipeligo, Baffin Island, Greenland, Labrador, Quebec, Manitoba, Wyo., Nova Scotia.

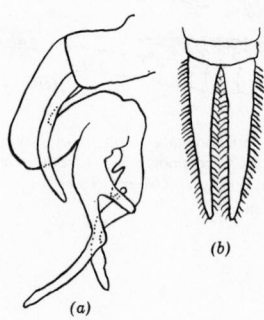

Fig. 24.6. *Streptocephalus texanus.* (*a*) Second antenna, lateral view. (*b*) Cercopods. (After Creaser.)

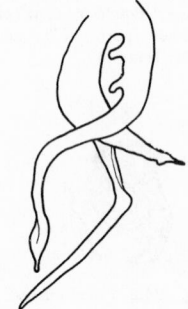

Fig. 24.7. *Streptocephalus dorothae.* Medial view of terminal segment of antenna. (After Mackin.)

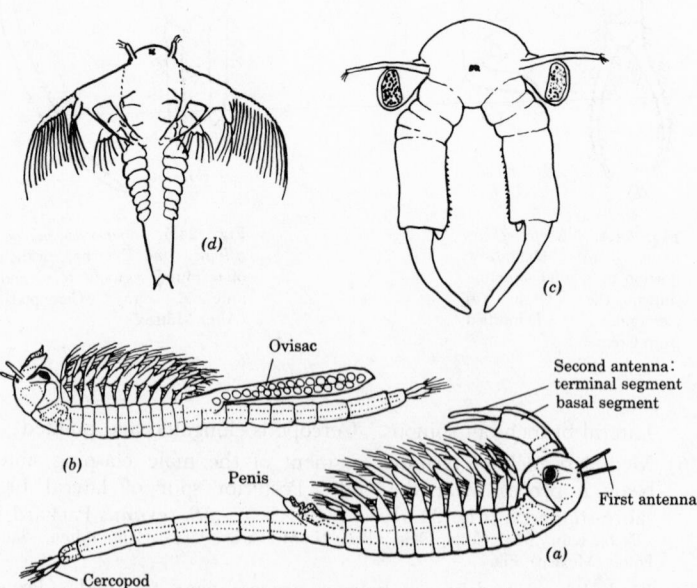

Fig. 24.8. *Branchinecta paludosa.* (*a*) Lateral view of male in swimming position, about twice natural size. (*b*) Female, same view. (*c*) Head of male, anterior. (*d*) Larva. (After Sars.)

9b   Proximal segment not serrate on inner margin. **10**

10a   (9) A knob or spur process at base of male clasping antenna **11**

10b   No knob or spur process at base of male clasping antenna **13**

11a (10) Proximal segment of male clasping antenna bears a rounded knob at
              base on inner margin which also has a swollen spinous area near
              middle of segment. Tips of distal segment are recurved. (Fig.
              24.9) . . . . ***B. shantzi*** Mackin 1952
              <small>Wyo., Colo., Nev., Calif., Ore., Utah. See Mackin (1952) for synonymy. Formerly
              B. coloradensis.</small>

11b   Proximal segment bears a spurlike process at base of inner margin.
              (Fig. 24.6). No elevated prominence on segment **12**

12a (11) Proximal segment of male clasping antenna bears a fingerlike pro-
              cess with a tuberculated tip near the inner angle of the segment
              distally from the spur. (Fig. 24.10) . . . . ***B. lindahli*** Packard 1883
              <small>Colo., Okla., Tex., Wyo., Kan., Neb., N. M., Ariz. (Formerly *B. packardi* Pearse
              1913.)</small>

12b   No such fingerlike process distally from the spur, but inconspicuous
              spines near distal end. Tips of distal segment are not recurved.
              (Fig. 24.11). . . . . ***B. mackini*** Dexter 1956
              <small>Nev., Wash., Calif.</small>

13a (10) Body size very large (50–90 mm). Male cercopods with short,
              widely spaced spines. (Fig. 24.12) . . . . ***B. gigas*** Lynch 1937
              <small>Wash., Mont., Nev., Utah.</small>

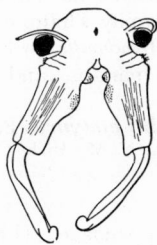

Fig. 24.9. *Branchinecta schantzi.* Head, front view. (After Shantz.)

Fig. 24.10. *Branchinecta lindahli.* Basal segment of second antenna. (By Pearse.)

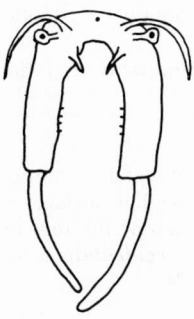

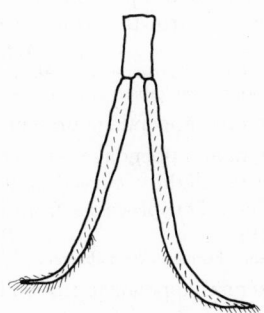

Fig. 24.11. *Branchinecta mackini.* Head, front view. (After Dexter.)

Fig. 24.12. *Branchinecta gigas.* Cercopods. (After Lynch.)

13b   Body size medium (25–36 mm). Male cercopods with long fil-
               aments. Uncornified bulge on anterior margin of proximal segment
               of male antenna. . . . . *Branchinecta cornigera* Lynch 1958
               Wash.

13c   Body size small (10–12 mm). Male cercopods with long filaments.
               No bulge on anterior margin of proximal segment of male antenna.
               (Fig. 24.13). . . . . *B. coloradensis* Packard 1874
               Calif., Ia., Kan., Neb., N. D., Wyo., Colo., N. M., Utah, Ariz., Nev., Wash.
               Formerly *B. lindahli.*

14a **(8)** Distal segment of male second antenna compressed, triangular, and
               blade-shaped. (Fig. 24.14). . . . . Family **Artemiidae**
               One genus, *Artemia* Leach 1819. One species, *A. salina* (Linnaeus) 1758.
               Conn., Utah, Wash., Ore., Calif., Lower Calif., Nev., N. D., Saskatchewan, Santo
               Domingo, Puerto Rico.
               The form of European and Asian animals varies greatly according to salinity (Gajew-
               ski, 1922), but American material does not (Bond, 1932).

14b   Distal segment not so **15**

15a **(14)** Genital segments very much swollen, containing large vesiculae
               seminales, mostly visible from the outside, no loop upwards of the
               vas deferens visible from the outside . . . . Family **Chirocephalidae**   **18**
               N. A. forms have antennal appendages except *Artemiopsis stephanssoni* which has
               an outgrowth on the labrum instead. N. A. forms lack a frontal appendage. Three
               genera: *Chirocephalopsis* Daday 1910, *Eubranchipus* Verrill 1870 and *Pristicephalus*
               Daday 1910.

15b   Genital segments not much swollen, no vesiculae seminales, but vas
               deferens makes a loop upwards, visible through the cuticle. N. A.
               forms have a frontal appendage. . . . . Family **Thamnocephalidae**   **16**
               Two genera: *Thamnocephalus* Packard 1887 and *Branchinella* Sayche 1903.

16a **(15)** Abdomen flattened, with membranous margins, and continuous
               with triangular cercopods. (Fig. 24.15). . . . .
                              Thamnocephalus platyurus Packard 1879
               Neb., Mo., Kan., Colo., Okla., Tex., Ariz., Nev., N. M., Utah, Calif., San Luis
               Potasi (Mexico).

16b   Abdomen and cercopods not so modified **17**

17a **(16)** Two terminal branches at end of frontal appendage. (Fig. 24.16) . .
                              Branchinella lithaca (Creaser) 1940
               Ga. Formerly *Chirocephalus lithacus* Creaser 1940.

17b   Three terminal branches at end of frontal appendage. (Fig.
               24.17). . . . . *B. alachua* Dexter 1953
               Fla.

18a **(15)** No antennal appendage, but labrum contains a rounded, verru-
               ciform outgrowth. (Fig. 24.18). . . . .
                              Artemiopsis stephanssoni Fr. Johansen 1922
               Northwest Territories and Alaska. The distal segment of the second antenna has
               2 thorn-shaped projections. Linder (1933) described a variety with but one thorn.

18b   Antennal appendage present **19**

19a **(18)** Antennal appendage of male clasping antenna a blunt, hornlike
               process, slightly curved and armed on medial surface with short
               spines. The processes from both sides touch at the mid-line. (Fig.
               24.19) . . . . *Pristicephalus occidentalis* (Dodds) 1923
               Calif. Formerly *Branchinecta occidentalis* Dodds 1923.

19b   Antennal appendage ribbonlike. **20**

20a **(19)** Antennal appendage very long, narrow, and uniformly tapering
               with very small serrations on margins. (Fig. 24.20). . . . .
                              Chirocephalopsis bundyi (Forbes) 1876
               Alaska, Yukon Territory, Alberta, Manitoba, Ontario, Quebec, Mass., N. Y., Mich.,

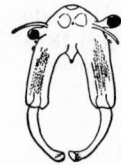

Fig. 24.13. *Branchinecta coloradensis.* Head, front view. (After Shantz.)

Fig. 24.14. *Artemia salina.* Head, dorsal view. (After Daday.)

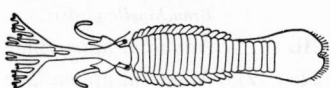

Fig. 24.15. *Thamnocephalus platyurus.* Dorsal view. (After Packard.)

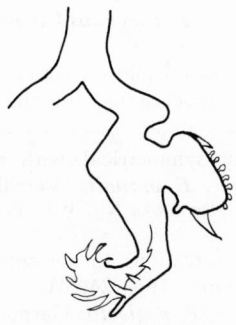

Fig. 24.16. *Branchinella lithaca.* Frontal append-age. (After Creaser.)

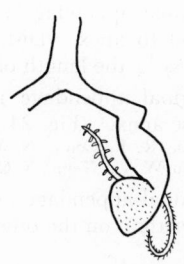

Fig. 24.17. *Branchinella alachua.* Frontal append-age. (After Dexter.)

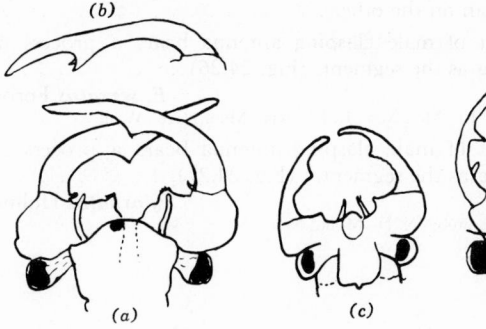

Fig. 24.18. *Artemiopsis stefanssoni:* (*a*) Head, dorsal. (*b*) Distal segment of second antenna of same show-ing the two thorn-shaped processes. (*c*) *A. stefanssoni* var. *groenlandicus:* head, ventral. (*d*) *A. bungei:* head, dorsal (not recorded from N. A.). (After Linder.)

Wis., Ohio, Ind., Wyo., N. H., Vt., Ill., Minn., S. D., Neb., Utah, Wash. Formerly ***Branchipus gelidus*** Hay 1889.

20b   Antennal appendage short or moderate in length with conspicuous serrations on margins **21**

21a **(20)** Antennal appendage in form of corkscrew which can be unrolled. Male clasping antennae sickle-shaped and thin. (Specimens in

southern U. S. have body armature but those in northern U. S. do not.). (Fig. 24.21) . . . . . . ***Eubranchipus holmani*** (Ryder) 1879
Conn., N. Y., N. J., Md., Va., Pa., Ohio, Minn., Tenn., La., Ga., N. C. Synonym, ***Branchinella gissleri.***

21b   Antennal appendage flat or coiled cylindrically **22**

22a (21) Antennal appendage short, not extending beyond proximal segment of clasping antenna **23**

22b   Antennal appendage may be extended beyond proximal segment of clasping antenna. **25**

23a (22) Antennal appendage with large toothlike serrations which extend to a sharp pointed apex. Distal segment of male clasping antenna has a process less than ¼ the length of the segment. (Fig. 24.22) . . . .
E. oregonus Creaser 1930
Okla., Ore., British Columbia, Wash.

23b   Antennal appendage with small serrations on margin which do not extend to apex. Distal segment of male clasping antenna has a process ¼ the length of the segment **24**

24a (23) Antennal appendage nearly bilaterally symmetrical with slightly obtuse apex. (Fig. 24.23) . . . . ***E. vernalis*** (Verrill) 1869
Mass., R. I., Conn., N. Y., N. J., Del., Md., Pa., Ohio, Ky., Ind., Ill., Mich., Ontario, W. Va., Tenn., N. C.

24b   Antennal appendage asymmetrical, the serrations on one side much longer than on the other. Apex very blunt. (Fig. 24.24) . . . . . .
E. neglectus Garman 1926
Ky., Ohio.

25a (22) Antennal appendage narrowly elliptical with margins studded with spinous papillae. Distal segment of male clasping antenna bent at nearly right angle and has no process. (Fig. 24.25). . . . .
E. floridanus Dexter 1953
Fla.

25b   Antennal appendage broad and asymmetric with serrations longer on one side than on the other **26**

26a (25) Distal segment of male clasping antenna bears a process near its base ½ as long as the segment. (Fig. 24.26) . . . .
E. serratus Forbes 1876
Md., Wis., Ind., Ill., Mo., Neb., Kan., Okla., Mont., Ore., Wash.

26b   Distal segment of male clasping antenna bears a process near its base ⅛ as long as the segment. (Fig. 24.27). . . . .
E. ornatus Holmes 1910
Wis., Minn., Manitoba, N. D., Mont., Neb.

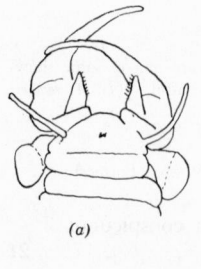

(a)

(b)

Fig. 24.19. *Pristicephalus occidentalis.* (a) Head, dorsal. (b) Head, lateral. (After Linder.)

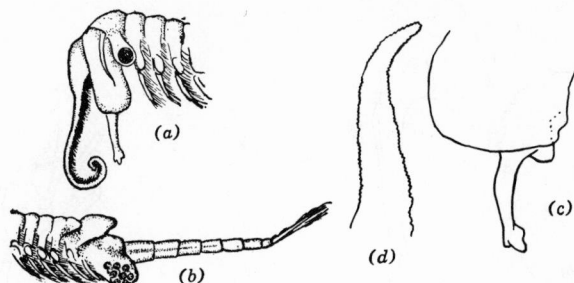

Fig. 24.20. *Chirocephalus bundyi.* (*a*) Lateral view of head of male. (*b*) Lateral view of posterior part of female. (*c*) Male second antenna. (*d*) Antennal appendage. (*a, b* by Pearse; *c, d* after Creaser.)

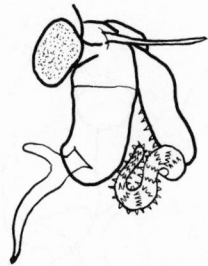

Fig. 24.21. *Eubranchipus holmani.* Head, lateral view. (After Mattox.)

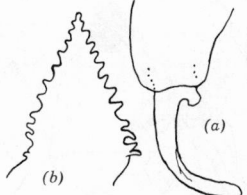

Fig. 24.22. *Eubranchipus oregonus.* (*a*) Second antenna. (*b*) Antennal appendage. (After Creaser.)

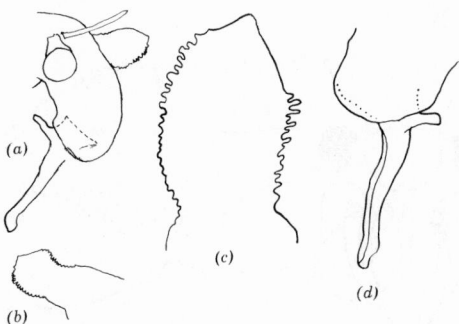

Fig. 24.23. *Eubranchipus vernalis.* (*a*) Head, lateral. (*b, c*) Antennal appendage. (*d*) Second antenna. (*a, b* after Mattox; *c, d* after Creaser.)

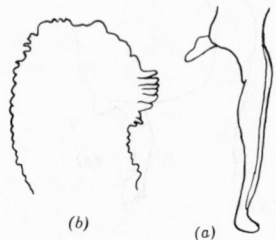

Fig. 24.24. *Eubranchipus neglectus.* (*a*) Second antenna. (*b*) Antennal appendage. (After Creaser.)

Fig. 24.25. *Eubranchipus floridanus.* Ventral view of head, antennae thrown forward. (After Dexter.)

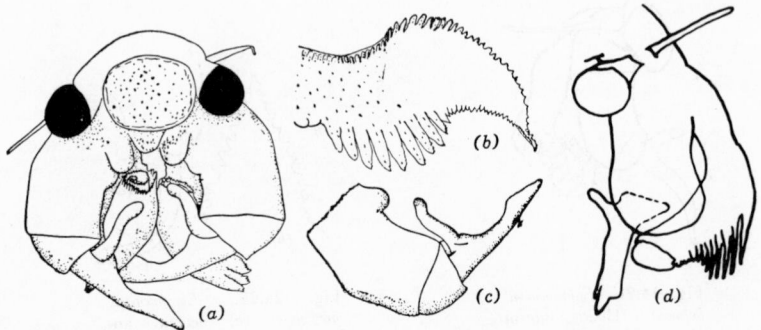

Fig. 24.26. *Eubranchipus serratus.* (*a*) Posterior view of head severed from body. (*b*) Antennal appendage. (*c*) Second antenna. (*d*) Lateral view of head. (*a, b, c* by Pearse, *d* original.)

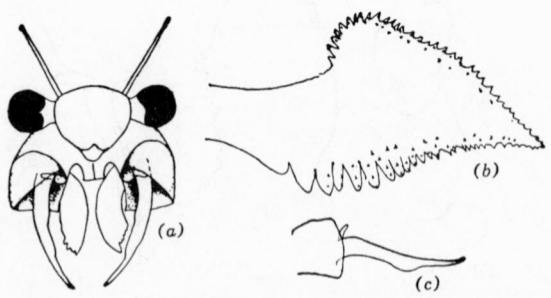

Fig. 24.27. *Eubranchipus ornatus.* (*a*) Posterior view of head severed from body. (*b*) Antennal appendage. (*c*) Second antenna. (After Holmes.)

References

Bond, R. M. 1932. Observations on Artemia *"franciscana"* Kellogg, especially on the relation of environment to morphology. *Intern. Rev. ges. Hydrobiol. Hydrog.*, 28:117–125. **Coopey, R. W. 1946.** Phyllopods of southeastern Oregon. *Trans. Amer. Microscop. Soc.*, 65:338–345. **Creaser, E. P. 1929.** The Phyllopoda of Michigan. *Papers Mich. Acad. Sci.*, 11:381–388. **1930a.** Revision of the phyllopod genus Eubranchipus, with the description of a new species. *Occasional Papers Mus. Zool. Univ. of Mich.*, No. 208:1–13. **1930b.** The North American phyllopods of the genus *Streptocephalus*. *Occasional Papers Mus. Zool. Univ. of Mich.*, No. 217:1–15. **Daday, Eugène. 1910.** Monographie systématique des Phyllopodes Anostracés. *Ann. sci. nat. (N. S.) Zool.*, 11:91–492. **Dexter, R. W. 1943.** Collecting fairy shrimps for teaching and research. *Turtox News*, 21:1–4. **1946.** Further studies on the life history and distribution of Eubranchipus vernalis (Verrill). *Ohio J. Sci.*, 46:31–44. **1953.** Studies on North American fairy shrimps with the description of two new species. *Am. Midland Naturalist*, 49:751–771. **1956.** A new fairy shrimp from western United States, with notes on other North American species. *J. Wash. Acad. Sci.*, 46:159–165. **Dexter, R. W. and M. S. Ferguson. 1943.** Life history and distributional studies on Eubranchipus serratus Forbes (1876). *Am. Midland Naturalist*, 29:210–222. **Dexter, R. W. and C. H. Kuehnle. 1951.** Further studies on the fairy shrimp populations of northeastern Ohio. *Ohio J. Sci.*, 51:73–86. **Gajewski, N. 1922.** Über die Variabilität bei Artemia salina. *Intern. Rev. ges. Hydrobiol. Hydrog.*, 10:139–159, 299–309. **Heath, Harold. 1924.** The external development of certain phyllopods. *J. Morphol.*, 38:453–483. **Johansen, Fritz. 1922.** Euphyllopod Crustacea of the American Arctic. *Rept. Canadian Arctic Exped. 1913–18.* 7: part G, 34 pp. **Linder, Folke. 1933.** Die Branchipoden des arktische Gebietes. *Fauna Arctica*, 6:183–204. **1941.** Contributions to the morphology and the taxonomy of the Branchiopoda Anostraca. *Zool. Bidrag Från Uppsala*, 20:101–302. **Lochhead, J. H. 1941.** Artemia, the "brine shrimp." *Turtox News*, 19:41–45. **Lynch, J. E. 1937.** A giant new species of fairy shrimp of the genus Branchinecta from the state of Washington. *Proc. U. S. Natl. Mus.*, 84:555–562. **1958.** Branchinecta cornigera, a new species of anostracan phyllopod from the state of Washington. *Proc. U. S. Natl. Mus.*, 108:25–37. **Mackin, J. G. 1939.** The identification of the species of Phyllopoda of Oklahoma and neighboring states. *Proc. Oklahoma Acad. Sci.*, 19:45–47. **1952.** On the correct specific names of several North American species of the phyllopod genus Branchinecta Verrill. *Am. Midland Naturalist*, 47:61–65. **Mathias, Paul. 1937.** Biologie des Crustacés Phyllopodes. *Actualités sci. et ind.*, No. 447:1–107. **Moore, W. G. 1951.** Observations on the biology of *Streptocephalus seali*. *Proc. Louisiana Acad. Sci.*, 14:57–65. **1955.** The life history of the spiny-tailed fairy shrimp in Louisiana. *Ecology*, 36:176–184. **Packard, A. S. 1883.** A monograph of the phyllopod Crustacea of North America with remarks on the order Phyllocarida. *Twelfth Ann. Rept. U. S. Geol. and Geog. Surv. Terr. for 1878*, Sect. 2:295–592. **Shantz, H. L. 1905.** Notes on the North American species of Branchinecta and their habitats. *Biol. Bull.*, 9:249–264. **Van Cleave, H. J. and Sister S. M. Hogan. 1931.** A comparative study of certain species of fairy shrimps belonging to the genus Eubranchipus. *Trans. Illinois State Acad. Sci.*, 23:284–290. **Verrill, A. E. 1870.** Observations on phyllopod Crustacea of the family Branchipidae with descriptions of some new genera and species from America. *Proc. Am. Assoc. Adv. Sci. 18th meeting, 1869*, pp. 230–247.

25

Notostraca

FOLKE LINDER

The Notostraca are an order of the subclass Branchiopoda. The segmentation of these animals is rather different from the other suborders in that some rings bear more than one pair of appendages, and there is no direct relation between the number of chitinized rings and segments. Therefore, the term *body ring* will be used instead of segment.

The first eleven body rings form the thorax; the thorax is followed by the abdomen, which consists of two series of parts of segments—the series of body rings and the series of legs united to each other—plus the telson. The number of legs is much greater than the number of body rings, but the legs diminish in size rapidly toward the caudal end, leaving some caudal body rings free of legs. There is no fixed number of legs on any of the leg-bearing abdominal rings; the placement of the legs is independent of the boundaries between the rings, and the number of legs varies a little within the limits of a species, as do the number of leg-bearing rings, the number of legless rings, and the total number of rings. The series of legs may stop at any place under a ring, and it is sometimes necessary to count half leg-bearing rings, but only approximately. A short formula is useful, e.g., $11 + (9.5-11.5) + (4-5.5) = 25-27$ describes *Lepidurus couesii* as having 9.5 to 11.5 abdominal leg-bearing rings 4 to 5.5 legless rings and a total number of at least 25, at most 27 rings, not counting the telson. Incomplete rings should be counted; they may be small pieces or almost complete rings, and may be marked with an *i* in the formula. The size and number of spines at the rings are, generally, not good characters for taxonomy, nor is the number of rings exposed behind the carapace because of varying contraction of the specimens. The length of the telson is worth attention, but the length of the supra-anal plate projecting from the telson in the genus *Lepidurus*, and the arrangment of spines at its dorsal side are especially important. Bilobation of this plate is not a valid character. In the genus *Triops*, the pattern of spines at the dorsal side of the telson shows an unbroken series of variations and will not distinguish species, at least not in American forms. Total length of body cannot be given with accuracy because of the highly varying state of contraction, especially in preserved specimens. Length of carapace, measured in the mid-line, gives a good idea of the size of the specimen. The carapace, more flattened in males than in females, may or may not be furnished with small spines on its dorsal side, and this is no character of species. Size and pattern of spines at the posterior emargination of carapace are worth attention, though the number of these spines is subject to great variation within the limits of a species. The paired eyes are rather uniform within the group, with the exception that they sometimes are unusually small (Fig. 25.4). The *nuchal organ* is situated just behind the eyes, in the mid-line (Fig. 25.4). Its shape is consistent in the American forms of *Lepidurus*, where it is rounded, as seen from above, as well as in American forms of *Triops*, where it is triangular. As for the legs, details useful for the taxonomy are shape and length of endites of the first pair of legs, which are usually longer in the genus *Triops* than in *Lepidurus*. Because of the smallness of the last pairs, it may be difficult to make out the exact number, but great accuracy is not generally necessary for purposes of identification. There is some variation in this respect within the limits of a species. The coxal lobe is counted as the first endite. Females may be recognized by ovisacs attached to the eleventh pair of appendages; otherwise, the sexes are difficult to distinguish.

More detailed treatments with full references to the literature are given by Linder (1952) and Longhurst (1955). The illustrations are from the author's 1952 paper.

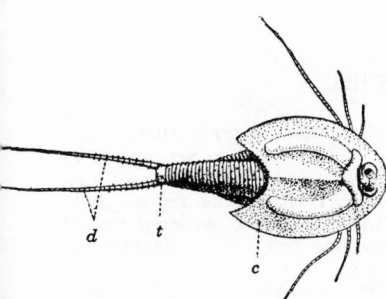

Fig. 25.1.   *Triops longicaudatus*. *c*, carapace; *t*, telson; *d*, cercopods. (After Packard.)

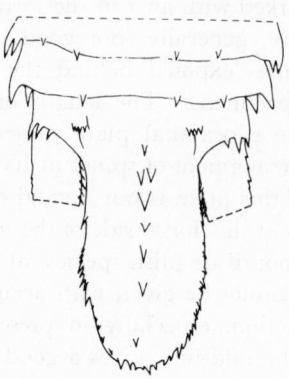

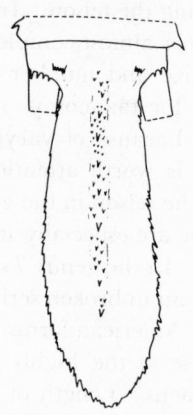

Fig. 25.2. Dorsal view of telson
of a female *Lepidurus packardi*.

Fig. 25.3. Dorsal view
of telson of a male
Lepidurus couesii.

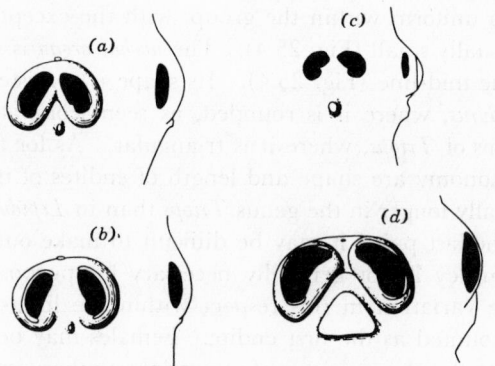

Fig. 25.4. Eyes and nuchal organ. (*a*) Female holo-
type of *Lepidurus packardi*. (*b*) Male paratype of *L.
couesii*. (*c*) Female paratype of *L. lynchi*. (*d*) Male of
Triops longicaudatus.

KEY TO SPECIES

1a   No supra-anal plate on the telson; total number of body rings 34 +
i to 44 (American forms) . . . . **Triops** Schrank 1803

      Apus Schaeffer 1756 is a frequently used synonym, but this name belongs to a bird (Hol-
thuis and Hemming, 1956). Single species in America, *T. longicaudatus* LeConte (1846)
(Figs. 25.1, 25.4*d*). Synonyms: *A. aequalis* Packard (1871); *A. newberryi* Packard
(1871); *A. lucasanus* Packard (1871); *A. oryzaphagus* Rosenberg (1947); *A. biggsi* Rosenberg 1947.

      Distribution: Mont., Ore., Wyo., Calif., Nev., Utah, Colo., Neb., Ariz., N. Mex., Kan.,
Okla., Tex.; Galapagos Islands; Hawaiian Islands; Mexico; Haiti; St. Vincent Islands;
Argentina.

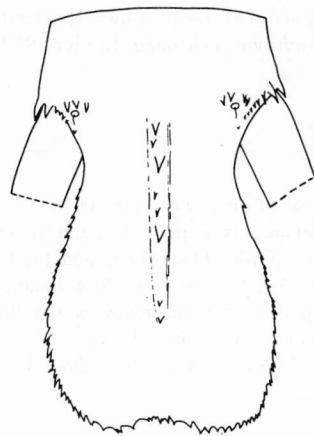

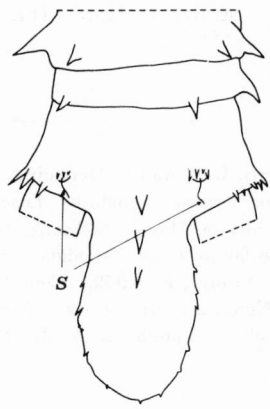

Fig. 25.5. Dorsal view of telson of female *Lepidurus bilobatus*.

Fig. 25.6. Dorsal view of telson and last two body rings of female *Lepidurus lynchi* var. *echinatus*. (*S*, dorsal sensory setae.)

1b   A supra-anal plate projecting from telson between caudal filaments, total number of body rings 25 to 34 . . . . ***Lepidurus*** Léach   **2**

2a   (1)   Total number of rings 25 to 29; abdominal leg-bearing rings 9.5 to 13; pairs of legs 34 to 46. . . . .   **3**

2b   Total number of rings 30 to 34; abdominal leg-bearing rings 14.5 to 18; pairs of legs 60 to 71. . . . .   **5**

3a   (2)   Endites 3 to 5 of first leg rather similar in size, projecting very little or not at all beyond margin of carapace, supra-anal plate only 7 to 13 per cent of length of carapace. . . . . ***L. arcticus*** Pallas 1793
          Synonym: ***Apus glacialis*** Kryer.   Pribilof Islands, Alaska, King William Land, Labrador, Greenland.

3b   Endites 3 to 5 of first leg very dissimilar in size, fifth endite of first leg clearly projecting beyond margin of carapace, supra-anal plate 17 to 44 per cent of length of carapace **4**

4a   (3)   Mediodorsal spines on supra-anal plate not on a keel, few in number, rather similar in size. (Figs. 25.2, 25.4*a*) . . ***L. packardi*** Simon 1886
          Synonym: ***L. patagonicus*** Berg?   Calif., Patagonia (?)

4b   Mediodorsal spines on supra-anal plate on a distinct keel, numerous (20 to 100), and highly variable in size. (Figs. 25.3, 25.4*b*) . . . . . .
                                         L. couesii Packard 1875
          Synonym: ***L. macrurus*** Lilljeborg 1877.   Alberta, Saskatchewan, Manitoba in Canada; Mont., N. D., Ore., Ida., Ut., in U. S. A.; Russia; Northern Siberia, and Turkestan.

5a   (2)   Anterior part of nuchal organ between posterior part of eye tubercles, not far from posterior margin of eyes (Fig. 25.4*a*), mediodorsal spines on supra-anal plate placed on a keel. (Fig. 25.5) . . . .
                                        L. bilobatus Packard 1883
          Colo., Utah?

5b   Nuchal organ considerably behind eye tubercles (Fig. 25.4*c*), mediodorsal spines on supra-anal plate not placed on a keel (Fig. 25.6) . . .   **6**

6a   (5)   Posterior part of lateral margins of carapace with minute spines. (Fig. 25.4*c*) . . . . ***L. lynchi*** Linder 1952
          Wash., Nev.

6b   Posterior part of lateral margins of carapace with large spines, directed
          straight outward. (Fig. 25.6) . . . *L. lynchi* var. *echinatus* Linder 1952
          Ore.

References

Holthuis, L. B. and F. Hemming. 1956. Proposed use of the plenary powers (a) to validate
the generic name "Lepidurus" Leach 1819 and to designate a type species for, and to determine
the gender of "Triops" Schrank, 1803 (Class Crustacea, Order Phyllopoda) and (b) to validate the family name "Apodidae" Hartert, 1897 (Class Aves). *Bull. Zool. Nomenclature*, 12:67–
85.   **Linder, F. 1952.** Contributions to the morphology and taxonomy of the Branchiopoda Notostraca, with special reference to the North American Species. *Proc. U. S. Natl. Mus.*,
102:1–69.   **Longhurst, A. R. 1955.** A review of the Notostraca. *Bull. Brit. Mus. Zool.*,
3:1–57.

26

Conchostraca

N. T. MATTOX

The phyllopods belonging to the order Conchostraca are all characterized by being enveloped by a bivalve shell which completely covers the more or less compressed body. This shell is not attached directly to the trunk somites, but is held by a strong adductor muscle which passes through the dorso-anterior portion of the body. In all but a few forms the shell is marked by a varying number of concentric lines of growth. In many species these lines of growth are numerous and increase in number as long as the animal lives. Each line of growth apparently represents an ecdysis.

The body of the conchostracans is composed of two major divisions, a head and the postcephalic body or trunk. The head of all conchostracans is very conspicuous. The anterior portion is attenuated into a rostrum that in many species is pointed, spatulate, and notched in profile view. The compound eyes are sessile, dorsal, and close together. The ventral ocellus is triangular in lateral view. The first antennae are always shorter than the second antennae and are usually provided with a series of dorsal, sensory papillae. The second antennae are biramous, with two long flagella arising from a basal scape. The flagella are variously segmented in the different groups and are used as swimming organs by all species.   In the family Limnadiidae there is a frontal organ on the mid-dorsal surface of the head.   This organ is of a pyriform shape attached by the narrower end of the appendage.

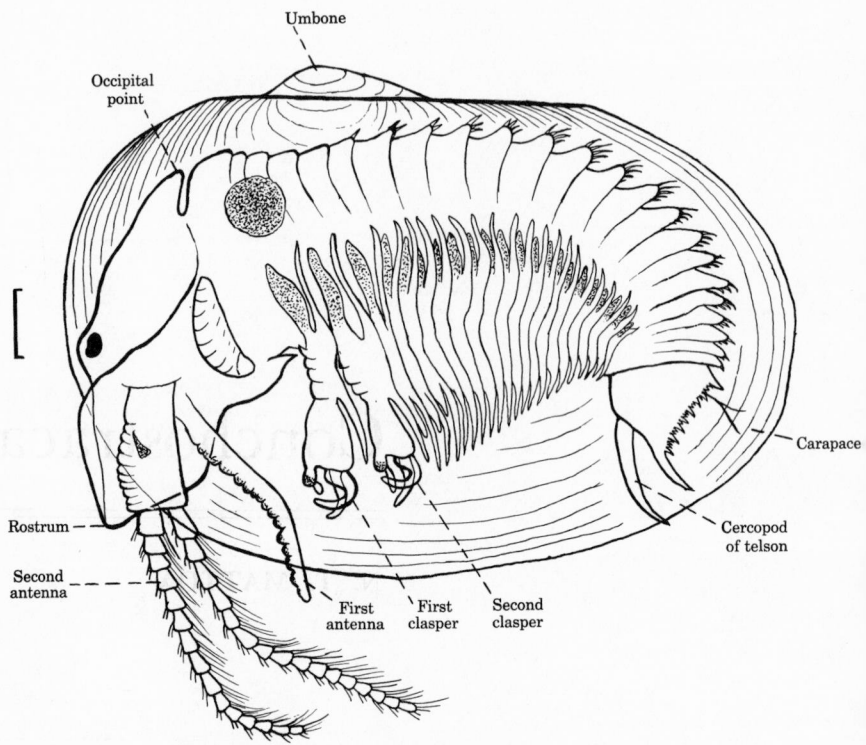

Fig. 26.1.   *Cyzicus mexicanus*, male, with left valve of carapace removed.   Scale in this and subsequent figures, 1 mm.

The trunk appendages vary in number from 10 to 28 pairs and are very uniform in appearance, except for the sexual dimorphism of the first one or two pairs in the males.   The typical appendage is a flattened, foliaceous swimming leg.   There are six endite divisions variously lobed with one exopodite, a branchial flabellum and epipodite on each appendage.   In the females the epipodites of one or more appendages on the ninth to eleventh somites are greatly extended for the purpose of carrying eggs in the carapace chamber.   The eggs are cemented to these epipodites and carried in raftlike clusters until the death of the female or until ecdysis.   The first and second postcephalic appendages of the males are modified into pincerlike prehensile organs used for grasping the female shell during copulation.   The fourth endite is broad with a thumblike extension, the fifth endite forms the claw, and the sixth endite of these appendages is variously extended as a digitiform process.   The form of these male appendages is used in the classification of the different groups, also they are responsible for one of the common names given to the group, "claw shrimp." Another common name, given because of the shell, is "clam shrimp."

The trunk of the conchostracans terminates in a broad, truncate telson. The typical telson terminates in a pair of elongated ventral spines or cercopods. These spines, or telson claws, are variously spined or are smooth in the different groups. The dorsal surface of the telson possesses two lateral ridges surmounted by a series of spines on each ridge (Fig. 26.4). Arising from between these ridges there is usually a biramous filament that varies in position in the different species. The form of the telson, number of pairs of dorsal spines, and the form of the terminal spines are used as taxonomic characters.

The Conchostraca are commonly found as freely swimming, littoral animals in lakes, ponds, and temporary fresh-water pools. As a group they have a wide geographic distribution, but many species are very local, like certain species of the genus *Eulimnadia* which are known only from their type locality. It appears as though many species are restricted to one locality, and others such as *Lynceus brachyurus* and *Cyzicus mexicanus* have a very extensive distribution. The conchostracans typically are found in warmer waters than most of the Anostraca. Conchostracans are usually found during the late spring and summer months of the year in the temperate zones.

The history of the classification of the conchostracan phyllopods is a devious and confusing one. The family Limnadiidae was created by W. Baird (1849) to include all of the group. A. S. Packard (1874) erected the second family Estheriidae to include the "limniids" and the "estheriids." In 1900 G. O. Sars added the family Cyclestheriidae to include a new genus, *Cyclestheria*. It was Stebbing (1910) who correctly established the family name Cyzicidae applying to most of the "estherid" group. Later Daday (1915), apparently unaware of Stebbing's work, proposed a revision of the family Estheriidae substituting the families Caenestheriidae and Leptestheriidae, and at the same time upholding the family Lynceidae, instead of Limnetidae, as created by Stebbing and Sayce. The latter family is based on the priority of the generic name *Lynceus* Müller (1776) over *Limnetis* Lovén (1846).

The "estheriids" have been much misunderstood, with the generic name *Estheria* being used until recent times. *Estheria* was first used to designate these animals by Rüppell (1837). This name must be disregarded as applying to the conchostraca on the basis of two rules, priority and synonymy. First, the name *Estheria* was originally used by Robineau-Desvoidy in 1830 for a genus of Diptera, hence on the basis of priority the name as applied to the conchostracans cannot be used. Second, the name *Estheria* as applied to the conchostracans must be considered in synonymy with *Cyzicus* as proposed by M. Audouin (1837). In present-day usage the name *Cyzicus* applies to only a small number of those animals formerly known as *Estheria*, and to only one genus in the family Cyzicidae. The four genera in the latter family are: *Caenestheria* Daday, *Caenestheriella* Daday, *Eocyzicus* Daday, and *Cyzicus*

Audouin, all of which are represented in the North American fauna except *Caenestheria*. In the family Leptestheriidae Daday there is found one North American genus of "estheriids," the genus *Leptestheria* G. O. Sars.

KEY TO SPECIES

1a Valves without lines of growth; first antennae 2-segmented; only the first pair of postcephalic limbs prehensile in the male . . . .
    Family **Lynceidae** Stebbing 1902
    One genus . . . . *Lynceus* Müller 1776 (= *Limnetis*)   2

1b Valves typically with one or more lines of growth; first antennae not segmented, with a series of dorsal sensory papillae; first and second postcephalic limbs of male prehensile.   5

2a (1) Shell subspherical; frontal ridge of head does not reach tip of rostrum; rostrum not truncate at tip.   3

2b Shell suboval; frontal ridge extends to end of rostrum; in male anterior termination of rostrum truncate and broadly expanded; second antennae 29-segmented; shell length 4–6 mm; claw digit short (brachydactyl). (Fig. 26.2). . . . . . . *L. brevifrons* (Packard) 1877
    Kan., Colo., N. M., and Mexico.

3a (2) Right and left claspers of male equal in form   4

3b Right and left claspers of male not equal in form, claw (endite 6) of right clasper thick and heavy, claw of left slender; length of shell up to 5.5 mm; rostrum of both male and female very broad; flagella of second antennae 20-segmented . . . .
    L. gracilicornis (Packard) 1871
    Tex.

4a (3) Length of shell up to 4.5 mm; head in profile beak-shaped; rostrum broad in male, pointed in female; flagella of second antennae 16-segmented; claw of claspers regularly and smoothly curved, sickle-shaped. (Fig. 26.3) . . . .
    L. brachyurus O. F. Müller 1785
    Wide distribution; reported from Montreal, Quebec, Ottawa, Mass., N. H., R. I., N. Y., Ohio, Ill., Mich., Colo., Ore., Wash., and Alaska. Also found in Europe and Asia. (=*L. gouldi* Baird)

Fig. 26.2. *Lynceus brevifrons*. Profile of head of male.

Fig. 26.3. *Lynceus brachyurus*. Profile and frontal view of head of male.

4b   Length of shell up to 5 mm; head very broad in profile; broad rostrum in male, in female strongly mucronate; flagella of second antennae 14- and 17-segmented; claw of claspers large, as long as width of the "hand" of the clasper, with terminal knob on edge of claw; twelfth pair of trunk appendages of male terminating in diagnostic large, strong, recurved hook . . . . . . *L. mucronatus* (Packard) 1875
Mont., Kan., and Alberta

5a   (1)   With a pedunculate, pyriform frontal organ on the mid-dorsal surface of the head . . . . Family **Limnadiidae** Sars 1896   6

5b   No frontal organ on the head   17

6a   (5)   Shell broadly oval, compressed; ventral surface of telson at point of articulation of terminal claws (cercopods) without a spine . . . . . .
Limnadia Brongniart 1820
Only one species, **L. lenticularis** (Linnaeus) 1761 (=**L. americana** Morse). Shell 12.5 by 9 mm, broad, ovate, and with 18 lines of growth; first antennae shorter than scape of second antennae; flagella of second antennae with 12 to 14 segments; 22 pairs of legs; only females known.
Mass. and in coastal lakes of the Arctic Ocean.

6b   Shell narrow, ovate; usually (except *E. alineata*) 1 to 5 lines of growth, up to 12; conspicuous ventral spine on telson at base of terminal spines (cercopods); 18 pairs of legs; first antennae variable in length; second antennae flagella with 9 segments. . . . .
Eulimnadia Packard 1874   7

7a   (b)   Shell with lines of growth   8

7b   Shell with no lines of growth; 4.2 by 2.6 mm; 9 to 12 telson spines; forked filament of telson between spines 3 and 4. . . . .
E. alineata Mattox 1953
In ricefields at Stuttgart, Ark.

8a   (7)   Shell with 1 to 4 lines of growth, elongate, not strongly convex dorsally.   9

8b   Shell with 5 to 12 growth lines, usually ovate and slightly convex dorsally.   13

9a   (8)   Telson with 9 or 10 dorsal spines   10

9b   Telson with 12 to 16 dorsal spines   11

10a   (9)   Shell 5–6 mm long by 3–4 mm wide; 1 to 4 growth lines; rostrum rounded; forked filament arises between telson spines 3 and 4. . . . .
E. antillarum (Baird) 1852
La. and Mexico.

10b   Shell averages 4.3 by 2.5 mm 1 to 4 growth lines; rostrum pointed; first antennae extend to fifth segment of second antennae in male; forked filament arises between telson spines 2 and 3. (Fig. 26.4)
E. francesae Mattox 1953
Md.

11a   (9)   Shell normally with 2 growth lines; shell of male averages 4.2 by 2.5 mm; front of head slightly convex; first antennae of male extend to fourth segment of second antennae; forked filament of telson between spines 3 and 4. . . . . *E. diversa* Mattox 1937
Ill.

11b   Shell with 3 or 4 lines of growth; 6 to 7.5 mm in length.   12

12a   (11)   Telson with 12 dorsal spines; forked filament arising between telson spines 1 and 2; rostrum not pointed and inflected; shell size average 6.2 by 3.8 mm; first antennae of male does not extend beyond scape of the second antennae . . . . *E. agassizii* Packard 1874
Mass.

12b   Telson with 16 dorsal spines, forked filament arising between spines
             6 and 7; rostrum strongly pointed and inflected; shell size average
             7.3 by 4.3 mm; first antennae of male extend to fourth segment of
             second antennae. (Fig. 26.5). . . *Eulimnadia inflecta* Mattox 1939
             Ill. and Ohio.

13a   (8)   Average of 5 lines of growth; mature size less than 8 mm in length   14
13b   Lines of growth 7 to 12; mature size more than 8 mm   15
14a   (13)   Shell size averages 5 by 3 mm; 7 to 9 telson spines; male first antennae
             slightly longer than female's; rostrum of male extended and sharply
             pointed . . . . *E. antlei* Mackin 1940
             Okla.

14b   Shell size averages 7 by 4 mm; telson with 16 to 20 spines; male
             first antennae extend to third segment of second antennae, in female
             shorter; rostrum rounded. (Fig. 26.6) . . . *E. texana* Packard 1871
             Tex., Kan., Neb., and Okla.

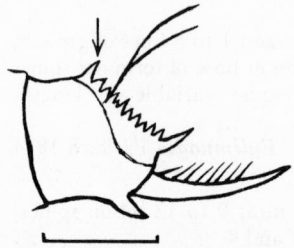

Fig. 26.4. *Eulimnadia francesae.*
Lateral view of telson, characteristic of genus. The arrow
points to the most anterior dorsal
spine.

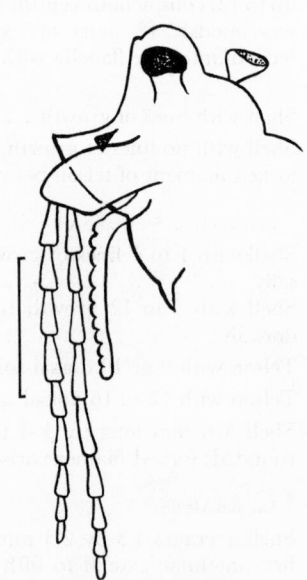

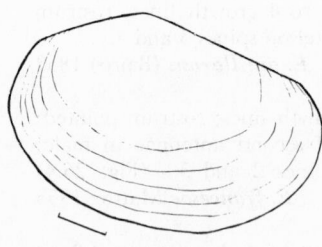

Fig. 26.5. *Eulimnadia inflecta.*
Lateral view of head of male
showing characters of the genus.

Fig. 26.6. *Eulimnadia texana.* Lateral
view of shell of male.

15a   (13)   Male first antennae extend beyond scape of second antennae; forked
             filament of telson arises between spines 3 and 4   16
15b   Male first antennae extend only to end of scape of second antennae;
             forked filament of telson arising between spines 5 and 6; 14 pairs of

telson spines; average of 10 growth lines; shell size averages 8.5 by 6 mm. . . . . ***E. stoningtonensis*** Berry 1926
Conn.

16a **(15)** Male rostrum not attenuated anteriorly. Male first antennae extend to second segment of second antennae; 14 to 20 pairs of telson spines; average size of shell 8.1 by 5.5 mm; average of 7 growth lines . . . . ***E. thompsoni*** Mattox 1939
Ill.

16b Male rostrum attenuated to sharp point. Male first antennae extend to third segment of second antennae; 14 to 16 telson spines; average of 10 growth lines, up to 12; shell size averages 8.4 by 5.2 mm. (Fig. 26.7). . . . . ***E. ventricosa*** Mattox 1953
Md., Va., and Ga.

16c The male rostrum pointed but not greatly inflected; occipital notch on head conspicuous; 14 to 20 dorsal telson spines; telson cercopods longer than dorsal telson margin; 9 to 12 growth lines, crowded; shell size average 6.8 by 4.2 mm; second antennal scape extends ½ length beyond rostrum. . . . . ***E. oryzae*** Mattox 1954
From rice fields at Stuttgart, Ark.

17a **(5)** Rostrum at anteroventral extremity armed with a conspicuous spine
Family **Leptestheriidae** Daday 1915
One genus, **Leptestheria** Sars 1898 (=*Estheria* pro parte). Shell long and narrow, 11 by 6 mm; 6 to 15 lines of growth; 18 to 25 small, variable telson spines; male rostrum spatulate in profile, female acuminate. (Fig. 26.8) . . . .
Leptestheria compleximanus (Packard) 1877
Kans., Calif., Colo., Tex., Ut., and Mexico.

17b Rostrum apex without a spine. . . . Family **Cyzicidae** Stebbing 1910   **18**

18a **(17)** Occipital part of head greatly produced, acute, with a deep occipital notch **19**

18b Occipital part of head rounded, occipital notch shallow and not conspicuous. . . . . ***Eocyzicus*** Daday 1915 (=*Estheria* pro parte)   **24**

19a **(18)** Rostrum of male and female terminating acutely; flagellum of second antennae 14- or 15-segmented . . . .
Caenestheriella Daday 1915 (=*Estheria* pro parte)   **20**

19b Rostrum of male broadly spatulate in profile; rostrum of female terminating acutely; flagella of second antennae 16 to 22 segments
Cyzicus Audouin 1837 (=*Estheria* pro parte)   **22**

20a **(19)** Shell compressed; umbones not prominent, anteriorly located **21**

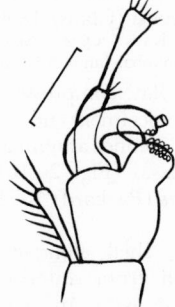

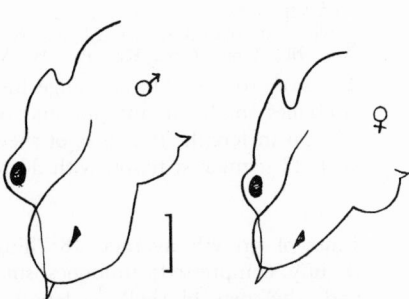

Fig. 26.7. *Eulimnadia ventricosa.* First clasper of male.

Fig. 26.8. *Leptestheria compleximanus.* Lateral view of heads of male and female showing rostral spines.

20b   Shell very thick, globose; umbones large, median; lines of growth
             21 to 35; shell high and thick; average shell size 7.5 mm long,
             6 mm wide, 3.8 mm thick; maximum length 9 mm; second anten-
             nae 14 and 15 segments; male first antennae extend to fifth seg-
             ment of second; telson with 17 to 25 spines. . . . .
             Caenestheriella belfragei (Packard) 1871
             Tex., Okla., and Kans.

21a **(20)** Lines of growth 13 to 21, average 15; maximum length about
             8 mm; umbones ⅓ length from anterior end; male first antennae
             extend to tenth segment of second; telson with 11 to 15 spines.
             (Fig. 26.9). . . . . *C. setosa* (Pearse) 1912
             Neb., Okla., Ore., S. D., Tex., and Mo.

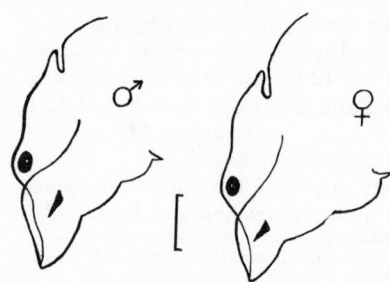

Fig. 26.9. *Caenestheriella setosa.* Lateral views of heads of male and female.

21b   Lines of growth 15 to 26, average 20; maximum length about
             11 mm; umbones ⅕ length from anterior end; telson with 20 to
             25 spines; only females known. . . . . *C. gynecia* Mattox 1950
             Ohio.

22a **(19)** Lines of growth distinct; shell not greatly swollen; umbo near an-
             terior end of shell **23**

22b   Lines of growth crowded, indistinct, 35 or more; shell very globose;
             average shell size 12.2 mm long, 7 mm wide, 6 mm thick; first male
             antennae extending to segment 7 of second antennae; telson spines
             numerous and small . . . . *Cyzicus morsei* (Packard) 1871
             S. D., Neb., Okla., Ia., and N. D.

23a **(22)** Lines of growth 25 to 35; shell moderately swollen; short, straight
             hinge line; umbones prominent, anterior to ¼ length of shell from
             anterior end; mature length 9–12 mm; thickness of shell ⅓ the
             length; second antennae with 16 and 17 segments; telson with 40
             to 50 spines . . . . *C. mexicanus* (Claus) 1860
             Widely distributed species; Mexico, N. Mex., Tex., Ariz., Kans., Okla., Neb.,
             Ark., Ill., Tenn., Ohio, Ky., Pa., W. Va., Va., Md., Manitoba, and Alberta.

23b   Lines of growth 15 to 25; hinge line rounded; shell flat, compressed;
             umbones small, at anterior end of shell; mature specimens up to
             16 mm in length; thickness of shell ¼ the length; second antennae
             with 22 segments; telson with 30 to 50 dorsal spines. (Fig. 26.10)
             C. californicus (Packard) 1874
             Calif.

23c   Lines of growth average 18; hinge line straight; shell elongate,
             slightly compressed; umbones small, at ⅕ length from anterior
             end; thickness of shell ⅕ length; second antennae with 16 seg-
             ments each; telson average of 31 dorsal spines . . . .
             C. elongatus Mattox 1957
             Several localities in southern part of Calif.

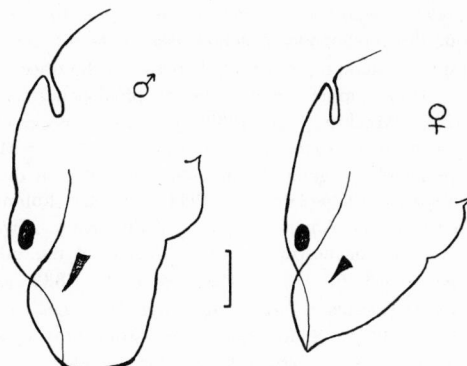

Fig. 26.10. *Cyzicus californicus*. Profile views of heads of male and female.

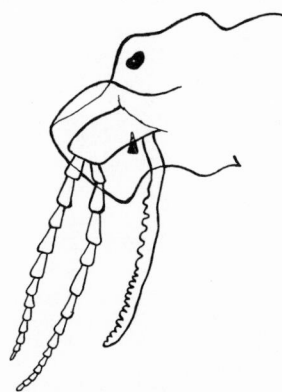

Fig. 26.11. *Eocyzicus concavus*. Profile view of head of male.

24a **(18)** Shell with 18 to 22 lines of growth; umbones $\frac{1}{5}$ length from anterior end; shell $\frac{1}{2}$ as thick as wide; telson with 16 to 17 spines; first antennae with 15 to 20 dorsal sensory papillae; second antennae with 12 and 14 segments; male rostrum very broad in profile. (Fig. 26.11). . . . . ***Eocyzicus concavus*** (Mackin) 1939
Tex.

24b Shell with 14 to 16 lines of growth; umbones $\frac{1}{4}$ length of shell from anterior end; shell $\frac{2}{3}$ as thick as wide; telson with 12 to 15 spines; first antennae with 13 to 18 dorsal sensory papillae; second antennae with 12 and 13 segments . . . . ***E. digueti*** (Richard) 1895
Calif. and Lower California

References

Audouin, M. V. **1837.** Séance du Ier Février. *Ann. Soc. Ent. France,* 6, Bull. Ent. 9–11.
Baird, W. **1849.** Monograph of the family Limnadiidae, a family of entomostracous Crustacea. *Proc. Zool. Soc. London,* Part 17:84–90. **Daday de Dees, E.** **1915.** Monographie systématique des phyllopodes conchostraces. *Ann. Sci. Nat. Zool. Ser. 9,* 20:39–330. **1923.** Monographie

systématique des phyllopodes conchostraces. *Ann. Sci. Nat. Zool. Ser. 10*, 6:255–390. **1925.** Monographie systématique des phyllopodes conchostraces. *Ann. Sci. Nat. Zool. Ser. 10*, 8:143–184. **1926.** Monographie systématique des phyllopods conchostraces. *Ann. Sci. Nat. Zool. Ser. 10*, 9:1–81. **1927.** Monographie systématique des phyllopodes conchostraces. *Ann. Sci. Nat. Zool. Ser. 10*, 10:1–112. **Mackin, J. G. 1939.** Key to the species of phyllopoda of Oklahoma and neighboring states. *Proc. Oklahoma Acad. Sci.*, 19:45–47. **Mattox, N. T. 1939.** Description of two new species of the genus Eulimnadia and notes on the other phyllopoda of Illinois. *Am. Midland Naturalist*, 22:642–653. **1954.** A new Eulimnadia from the rice fields of Arkansas with a key to the American species of the genus. *Tulane Studies Zool.*, 2:3–10. **1957.** A new estheriid conchostracan with a review of the other North American forms. *Am. Midland Naturalist*, 58:367–377. **Packard, A. S. 1883.** A monograph of the phyllopod crustacea of North America. *Twelfth Ann. Rept. U. S. Geol. and Geog. Surv. for 1878 (Hayden)*, section 1:295–592. **1874.** Synopsis of the fresh-water Phyllopod Crustacea of North America. *U. S. Geol. and Geog. Surv. Terr. (Hayden) for 1873*:618. **Rüppell, E. 1837.** Über *Estheria dahalcensis* Rüppell, neue Gattung aus der Familia der Daphniden. (in Strauss-Durcheim). *Abhandl. senckenberg. Mus.*, 2:117–128. **Sars, G. O. 1900.** On some Indian Phyllopoda. *Arch. Math. Naturvidenskab.*, 22 (6). **Stebbing, T. R. R. 1910.** General catalogue of South African Crustacea. *Ann. South African Mus.*, 6:281–599.

27

Cladocera[1]

JOHN LANGDON BROOKS

Classification

The most acceptable modern concept of the position of the Cladocera within the subclass Branchiopoda differs from that held when the original edition was prepared and deserves comment. The earlier classification divided the Branchiopoda into the Phyllopoda and the Cladocera. The unnaturalness of this scheme was apparent to Calman (1909). He indicated that the differences between the three major groups of the Phyllopoda (Anostraca, Notostraca, and Conchostraca) were as great as the differences between the Conchostraca and the Cladocera and proposed that the subclass Branchiopoda be divided into four orders, Anostraca, Notostraca, Conchostraca, and Cladocera. To these Scourfield would add a fifth order, the Lipostraca, for the Devonian *Lepidocaris rhyniensis* Scourfield. But not all of these five orders show a similar degree of kinship, and some grouping of the orders to indicate

[1]The key is essentially that which Birge devised for the original edition and parts of the text and most of the illustrations have been retained. That his key still suffices is both a tribute to him, and an indication that the group has received relatively little subsequent attention by students in this country. The chief changes in the key concern, on the one hand, the arrangement of the major groups to accord with modern views of the systematics of the order, and on the other hand, revisions of the genera *Camptocercus* and *Daphnia*. The North American representatives of these two genera are treated in accordance with the studies of Mackin (1930) on *Camptocercus* and Brooks (1957) on *Daphnia*.

their closeness seems desirable.   For example, the fossil Lipostraca are very much more like the Anostraca than they are like the other orders.   Of more immediate concern is the close similarity of the Conchostraca and Cladocera which is not evident in Calman's scheme.

A reasonable way to indicate these relationships in the classification is to group the similar orders into superorders.   Gerstaecker in 1866 proposed the name Diplostraca to include the Cladocera and Conchostraca, and Eriksson in 1934 suggested the name Onychura for the same group.   Although Onychura has been adopted by Brown (1950), there appears to be no reason for not using the earlier name, Diplostraca, for the superorder embracing these two of Calman's orders.   (The orders Anostraca and Lipostraca would then constitute an equivalent superorder.   The Notostraca alone would comprise the third.)   The Diplostraca are characterized by the possession of a bivalve carapace enclosing body and appendages, and an abdomen with the end (together with the post-abdomen) bent ventrally and forward.   The abreptor thus formed is provided with spines, and serves to cleanse unwanted large particles from the median space between the legs.   The orders are as defined by Calman (1909).   The close similarity between the structure of adult Cladocera and the larvae of certain conchostracans strongly suggests that the Cladocera are neotenic (paedomorphic) derivatives of some early conchostracan.

In the classification Birge used in the original edition, the major dichotomy of the Cladocera was based upon the relative size of the carapace and the nature of the thoracic appendages.   In the group known as the Calyptomera the carapace enclosed the body and feet, and in the Gymnomera, the carapace did not so enclose the body, being only a brood sac on the dorsal side of the body.   The thoracic appendages of the Calyptomera are flattened for filtering and respiratory exchange of a sort characteristic of the subclass, and, indeed, providing the basis of its name, Branchiopoda.   The legs of the Gymnomera are not flattened but are composed of subcylindrical joints.   The Gymnomera comprise two rather different kinds of animals: the Polyphemidae with six genera (a total of about eighteen species) two of which are primarily marine, and the sole species of *Leptodora* which is given its own family, the Leptodoridae.   There is cogent evidence, however, against any close relationship between these two families. The pelagic, predaceous lives of the representatives of these families are similar, and the similarity of both carapace and feet is probably the convergent result of adaptation to this existence.

Sars' groups Gymnomera and Calyptomera must therefore be abandoned as unnatural, but his grouping of the families into four well-marked "tribes," Ctenopoda (Sididae, Holopedidae), Anomopoda (Daphnidae, Bosminidae, Macrothricidae, Chydoridae), Onychopoda (Polyphemidae), and Haplopoda (Leptodoridae) still seems reasonable.   The distinctness of the representatives of these four groups is evident; their relationship is not.   The most isolated is certainly *Leptodora* (Haplopoda).   Its large size (up to 18 mm long), lack of branchial appendages on the legs, and aberrant body organization set it off from the others, but the most significant feature is that the winter eggs hatch into a nauplius (or metanauplius) larva.   All other cladocerans develop directly from both parthenogenetic and fertilized eggs. This retention of a nauplius larva makes the possibility of the derivation of *Leptodora* from any of the other groups, all of which have lost all semblance of a nauplius, seem rather remote. In order to express adequately the distinctness of *Leptodora* from the other Cladocera, the scheme of Eriksson (1934) is followed, in which the Haplopoda are set off from all other Cladocera. The group comprising all of the Cladocera except *Leptodora kindtii* (Focke) is named Eucladocera by Eriksson, and in some ways *Leptodora* is more like an aberrant conchostracan than a derivative of the Eucladocera. The Eucladocera, then, includes Sars' "tribes" Ctenopoda, Anomopoda, and Onychopoda. However, the term "tribe" when used today as a taxonomic category denotes a group of genera. Sars' "tribes" are superfamilies and must be named in accordance with the rules for formulating family-group names. Ctenopoda is replaced by Sidoidea, Anomopoda by Chydoroidea, and Onychopoda by Polyphemoidea. A synopsis of the classification of the Cladocera appears below. The numbers in parentheses are those of the key lines where the groups appear.

Subclass: Branchiopoda (as defined by Calman 1909)
Superorder: Diplostraca Gerstaecker (=Onychura Eriksson)
Order: Cladocera Calman
Suborder: Haplopoda Sars (1)
  1. Family: Leptodoridae Lilljeborg (1)
Suborder: Eucladocera Eriksson (1)
  A. Superfamily Sidoidea, superfam. n. (=Ctenopoda Sars) (3)
    2. Family: Sididae (Baird) (4)
    3. Family: Holopedidae Sars (4)
  B. Superfamily Chydoroidea, superfam. n. (=Anomopoda Sars) (3)
    4. Family: Daphnidae (Straus) (14)
    5. Family: Bosminidae Sars (14)
    6. Family: Macrothricidae Norman and Brady (14)
    7. Family: Chydoridae Stebbing (13)
  C. Superfamily Polyphemoidea, superfam. n. (=Onychopoda Sars) (2)
    8. Family: Polyphemidae Baird (2)

Structure and Behavior

All of the Cladocera of North America, with the exception of two species, although exhibiting considerable variation, present a general pattern of structure and behavior which is described below. The exceptional species are *Polyphemus pediculus*, the only North American fresh-water representative of the superfamily Polyphemoidea and *Leptodora kindtii*, the sole member of the Haplopoda. Their characteristics are noted above and in the key. The vast majority of cladocera range in size from about 0.2 to 3.0 mm, or even more. All have a distinct head and a body covered by a fold of the cuticle, which extends backward and downward from the dorsal side of the head and constitutes a bivalve *carapace*. The junction of head and body is sometimes marked by a depression, the *cervical sinus* or *notch* (Figs. 27.4, 27.31, 27.49).[2]

[2]The figures referred to are designed to give the specific characters rather than the anatomy, which is shown only incidentally.

Cladocera have two light-sensitive organs in the head, the large compound *eye* and the smaller *ocellus* (Fig. 27.1). The eye has numerous or few lenses (Figs. 27.9, 27.33, 27.3) and is capable of being rotated by three muscles on each side. It is a most conspicuous organ, by its size, its dark pigment, and its constant motion during life.

The compound eye is usually present; the ocellus is more variable. The ocellus is sometimes absent (*Diaphanosoma, Daphnia retrocurva, Daphnia longiremis*); sometimes rudimentary (many forms of *Daphnia*); sometimes larger than the eye (*Leydigia, Dadaya*); and may be the sole organ sensitive to light as in *Monospilus*.

In the head are also the brain, the optic ganglion with its numerous nerves to the eye, the antennal muscles, and the anterior part of the digestive tract. The head bears two pairs of sensory appendages: (1) *First antennae*, or the *antennules* as they are usually called in this group (Figs. 27.4, 27.36, 27.73, 27.5), which carry sense-hairs, the olfactory setae (usually placed at the end),

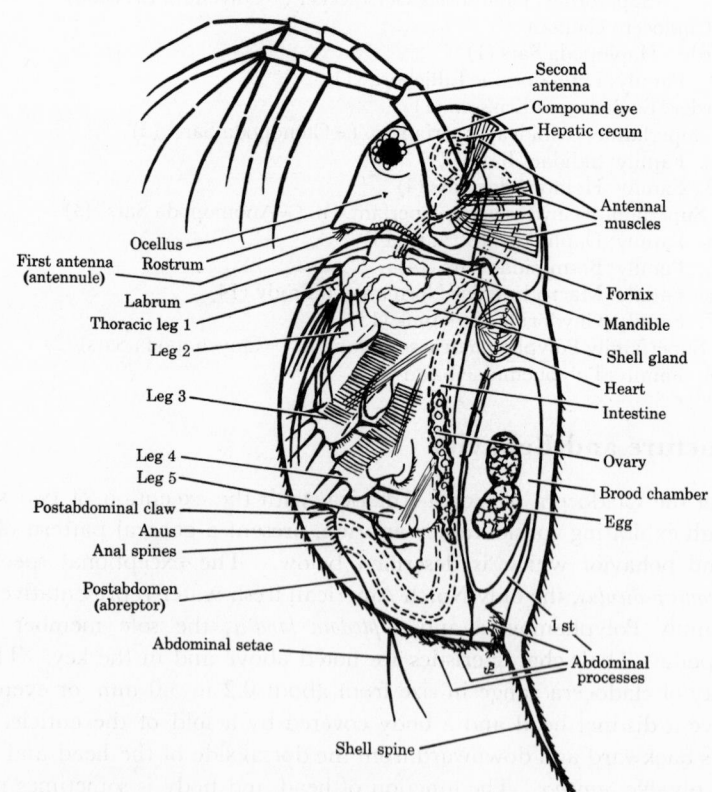

Fig. 27.1. Lateral view of female *Daphnia similis* showing features important in cladoceran taxonomy. (Modified from Claus.)

and also ordinarily have one or more lateral sense-hairs. (2) *Second antennae*, usually called merely *antennae* in the Cladocera, which are the main organs of locomotion; they are large, swimming appendages, with a stout basal joint bearing two branches or rami, which, in turn, carry long plumose setae. The antennae are moved by powerful muscles, which may occupy a great part of the interior of the head (Fig. 27.1). The number of the antennary setae can be expressed by a formula which shows the number of the setae on each joint of each branch of the antenna starting with the proximal joint; the numbers for the dorsal branch occupying the place of the numerator of a fraction. The formula thus constructed reads for *Daphnia* (Fig. 27.1), $\frac{0-0-1-3}{1-1-3}$; that for *Sida* (Fig. 27.4), $\frac{0-3-7}{1-4}$.

The type of locomotion depends on the size of the antennae, the length and number of the *setae*, and on the size of the antennal muscles. Sidids (Fig. 27.5) progress by powerful strokes of the broad antennae, and the smaller daphnids (Fig. 27.36) move by more numerous and less vigorous strokes. The heavier forms of this family (Fig. 27.32) with smaller antennae have a rotating, unsteady motion, produced by rapid strokes. In the Macrothricidae and Chydoridae (Figs. 27.68, 27.78, 27.93) the postabdomen is often an efficient aid to locomotion. *Drepanothrix* (Fig. 27.60), whose antennae bear saberlike setae, scrambles and pushes itself about, and the mud-haunting *Ilyocryptus* (Fig. 27.68) crawls about among the weeds pulling with its antennae and pushing with its postabdomen. The members of the large family of the Chydoridae have small antennae and move them very rapidly; their progress varies from a rapid whirling motion, as in *Chydorus* (Fig. 27.108), to a slower wavering and tottering progress, as in *Acroperus* (Fig. 27.81).

The head also bears the mouthparts: the mandibles, maxillules, maxillae, and the upper (labrum) and lower (paragnath) lips (Fig. 27.1). The mandibles are stout, strongly chitinized organs, made in one piece and without a palpus. Their opposing faces are toothed and ridged and they grind the food very efficiently. The maxillules, a pair of minute appendages, lie concealed on the ventral surface of the body between the mandibles and the median, conical paragnath. Each is a small, pointed structure, bearing several curved setae. These appendages work like a pair of hands to push the food mass between the mandibles. These were considered the maxillae before their true nature was elucidated by Cannon and Leak (in Cannon, 1933). The maxillae are minute or absent. They, or their rudiments, carry the opening of the excretory gland but play no part in feeding. The labrum is large and extends posteriorly covering the other mouthparts ventrally. In many of the Macrothricidae and Chydoridae the labrum bears a keel or projection which is of taxonomic value (Figs. 27.4, 27.12, 27.62, 27.94).

The axis of the head may continue that of the body (*extended*, Fig. 27.56), or it may be bent ventrad (*depressed*, Fig. 27.42). That part in front of the eye is known as the *vertex*. There is usually a beak in front of, or between, the antennules, which is known as the *rostrum*, whose size and shape has taxonomic value. There is commonly a ridge above the insertion of the antenna, which helps to stiffen the side of the head and to support the pull of the antennary muscles. This is the *fornix*, whose shape and extent may form an important taxonomic character (Figs. 27.15, 27.39).

The *carapace*, though called bivalve, is really in one piece, bent along the back, but never shows a division or joint at this place. It has very different shapes: as seen from the side, it may appear nearly square, oval, or round. It may be marked in the most various fashions. It may bear hairs or spines, along the ventral edge. There may be a single *spine* at the superoposteal angle, prolonging the junction of the valves, as in *Daphnia*, or each valve may have one or more spines at the inferoposteal angle (Fig. 27.33). This angle in the Chydoridae may be acute or rounded, smooth or toothed, and these configurations are of taxonomic value. The inner wall of the carapace is far thinner than the outer, and through it some respiratory exchange occurs. The blood circulates freely in the space between the walls.

The heart, an oval or elongated sac (Figs. 27.1, 27.4, 27.44) whose rapid pulsations are easily seen in the living animal, lies just back of the head, on the dorsal side. It receives the blood from the hemocoel through a pair of lateral ostia and expels it from its anterior end. There are no blood vessels, but the circulation passes along definite courses through a complex series of passages all over the body. The movements of the blood corpuscles may be readily seen in transparent Cladocera.

Respiration is not served by any single organ. The legs and the inner wall of the valves are the main surfaces for the exchange of gases.

In the anterior part of the valves lies an organ whose structure is not readily apparent. This is the maxillary or shell gland (Figs. 27.1, 27.4, 27.10), a flattened glandular tube with several loops, which subserves the functions of excretion and osmoregulation.

The body lies free within the valves and is divided into the appendage-bearing thorax, which is not plainly segmented, and the abdomen (plus a pygidium or *postabdomen*). In the Eucladocera the segmentation of neither thorax nor abdomen is clearly marked. The gut leads through the body. Along the sides of the middle portion of the gut lie the simple reproductive organs. Attached to the ventral side of the body are the ordinarily five, sometimes six, pairs of legs. These are mainly flattened structures, each with several parts, bearing numerous hairs and long setae (Figs. 27.1, 27.101). Their structure is too complex to describe here (cf. Eriksson, 1934, for a detailed account of both structure and function). In the Sidoidea all the legs are similar and flattened (foliaceous). Their use is to create a current of water through the valves, bringing in oxygen for respiration and particles of food. The latter consists chiefly of algae, though nothing edible that the current brings in is rejected. The food particles collect in the ventral food groove which runs between the bases of the legs and passes forward toward the mouth. At the anterior end of the food groove the food particles are entangled in the viscid labral secretion. These masses of food and secretion are pushed between the mandibles by the maxillules, the mandibles grind them up, and the comminuted material passes into the esophagus. Cladocera are normally eating all of the time. In the Chydoroidea the feet differ in structure, the first pair being more or less prehensile and having other functions besides the main one of drawing in water. Many species, especially among the macrothricids and chydorids live chiefly among the weeds and coarse detritus, and the hooks and spines of the first leg aid them in locomotion and also in picking up large food particles.

Most members of the Sidoidea feed by filtering particles from the water, and most Chydoroidea are equipped for picking up fine particulate material from the bottom. Although the trophic ecology of the majority of each superfamily can thus be characterized, there are within each group forms that deviate considerably from the general behavior. Within the Sidoidea, for example, *Latona* has taken to feeding on the bottom, and within the Chydoroidea the Daphnidae and Bosminidae filter particles from the water. All of these are secondary modifications away from the structural and functional organization that can be said to characterize each superfamily (Eriksson, 1934). The Polyphemoidea feed primarily by seizing relatively large particles with their prehensile legs. *Polyphemus* lives chiefly in marshes and in the weedy margins of ponds and lakes, but may also be found in the open waters of large lakes. It feeds largely on protozoa, rotifers, and minute crustacea. *Leptodora* utilizes similar prey.

In the more transparent species the full extent of the digestive tract can be seen. The narrow esophagus (Figs. 27.1, 27.4, 27.54) widens suddenly into the stomach, which lies in the head and whose posterior end passes imperceptibly into the intestine. Attached to the stomach in many species are two sacs, often long and curved (Figs. 27.1, 27.6, 27.12, 27.56). These are the *hepatic caeca*, which store and possibly digest food. The stomach and intestine have a muscular wall, a lining of dark-colored glandular cells, and an inner peritrophic membrane. The cavity is ordinarily filled with food. The intestine has a direct course in the first four families of Eucladocera. In the Macrothricidae it is sometimes direct (Fig. 27.62), and sometimes convoluted, and there is often a *caecum* attached to the ventral side near the posterior end (Figs. 27.81, 27.100). The terminal part of the intestine, the *rectum*, is always transparent and the muscles that open and close it can easily be seen. The anus usually lies either at or near the end of the postabdomen in the first five families of Eucladocera, or on the dorsal side in the Chydoridae and in some forms of the other families (Figs. 27.44, 27.49, 27.56, 27.67).

On the dorsal surface of the abdomen there are often one to several finger-like projections, the *abdominal processes* (Fig. 27.1). These often function to retain the eggs in the brood chamber. They are numbered in order from the anterior end.

The *abreptor* (often called *postabdomen*) is ordinarily jointed to the rest of the body and is bent forward; hence its dorsal side may come to be ventral in position. Proximally on the dorsal side it bears the two *abdominal seta* which are often very long (Fig. 27.45). At the end of the postabdomen are two *terminal claws*. The concave side of the claw is provided with spines and teeth of various sizes and arrangements. Their patterns are often of taxonomic importance and several terms are used to indicate the several general patterns into which claw spination falls. Where there are only a few large spines, they occur near the base of the claw and are referred to as *basal spines* (Figs. 27.11, 27.66, 27.85). Where the spines are minute and of the same length along the greater part of the claw, that part of the claw is said to be *denticulate* (Figs. 27.13, 27.74). When the spines are intermediate in size between these two extremes they are usually grouped into a row called a *pecten* or comb. A claw may bear a pecten of intermediate-size teeth between large basal spines and distal denticulation (Fig. 27.105), or as in *Daphnia* there may be *three pectens* with teeth of the same or different sizes (Figs. 27.19, 27.25 27.26). The morphologically dorsal (although it may seem posterior or ventral) surface of the postabdomen usually bears spines arranged in rows. In the Sididae, Holopedidae, and Daphnidae the single row of spines on either side of the postabdomen is called the *anal spines*. The Bosminidae lack these spines and in the Macrothricidae (Fig. 27.66) and Chydoridae (Figs. 27.86, 27.89) there may be one or more rows of *lateral spines* in addition to the marginal anal spines. These spines and teeth may have the most diverse shape and structure (squamae, fascicles, etc.), and furnish important taxonomic characters. Their main function seems to be to comb the legs and keep them clean and free from foreign matter and from parasites which might otherwise readily attach themselves.

Reproduction

The reproduction of the Cladocera is noteworthy. During most of the year the females produce eggs which develop without being fertilized into more parthenogenetic females. These eggs may number only two, the usual number in the Chydoridae, or there may be on occasion more than twenty as in the larger Daphnidae. The eggs are deposited in the cavity bounded by the dorsal part of the valves and the upper side of the body—the *brood chamber* Here they develop and hatch in a form quite like that of the parent except smaller. Hence there are no free-living larval forms of Cladocera, such as are so abundant in the Copepoda (except in *Leptodora*, the ephippial eggs of which hatch as metanauplii). The eggs of most Cladocera are well provided with yolk and will develop normally outside of the chamber. In *Polyphemus*, however, the eggs are small and have little yolk. The hypodermal cells on the dorsum which form the floor of the completely closed brood chamber secrete a fluid which apparently nourishes the developing eggs. In other genera the brood chamber may be partially closed behind by the abdominal processes.

This parthenogenesis will continue until conditions become unfavorable for the cladoceran. When the food supply fails or the ponds begin to dry (the

resultant overcrowding may exhaust the food), the production of parthenogenetic eggs dwindles. Some of these eggs develop into males instead of
females. At the same time the adult females of the population produce a
different kind of egg. The cytoplasm of these eggs not only has a very different appearance, being opaque and dark in color, but typically the nucleus is
haploid, requiring fertilization. In "sexual" females producing these eggs,
the carapace enclosing the brood chamber begins to thicken and darken. In
the Chydorinae the entire carapace is altered. In the Daphnidae (Figs. 27.31,
27.36, 27.51), a semielliptical portion of the dorsal region of each valve becomes greatly altered to form the *ephippium*, so called for its resemblance to a
saddle. In either case after the eggs have been extruded into the brood
chamber and fertilized, the altered carapace closes around the eggs and at the
next molt eggs and carapace (or ephippium) are shed as a unit. The early
embryo into which the eggs have developed will lie dormant, often withstand-
ing freezing and drying, until the return of conditions suitable for the continuation of development.

This process of sexual reproduction, which occurs at different times of the
year in different species, involves a shift from the formation of parthenogenetic
eggs to the formation of the so-called "resting" eggs, as well as the development of a larger percentage of the parthenogenetic eggs into males. The
results of the several attempts to determine experimentally the environmental
conditions responsible for these processes have not been very clear-cut, although they do indicate that the conditions under which sexual eggs are produced are not necessarily those that initiate male production, and vice versa.
Much evidence points to a sudden decrease in the amount of food a female
gets as being important in initiating sexual egg production. Males usually
occur when a population is dense or just after it has been: over-crowding is
believed to increase the percentage of males. The ability to produce ephippial
eggs parthenogenetically is known to have evolved within one species of
Daphnia, and may occur sporadically elsewhere in the Cladocera.

The males are smaller than the females and usually of similar form. They
are distinguished by larger antennules; the postabdomen is usually somewhat
modified (Fig. 27.102); the first foot is frequently armed with a stout hook
which serves to clasp the females. In some genera, *Moina* for example, this
function is performed by the very large antennules (Fig. 27.50).

Distribution

The Cladocera are found in all sorts of fresh waters. Lakes and ponds contain a much larger number of forms than rivers do. The shallow, weedy
backwaters of a lake whose level is fairly permanent harbor a greater variety
of species than does any other kind of locality. Here are found almost all of
the Chydoridae and Macrothricidae, as well as most of the representatives of
the other families. While by far the greater number of species belong to the
littoral region, living among the weeds and feeding on algae and similar

organisms, a few species live near the mud, although not specially adapted to a life in the mud; such are *Alona quadrangularis* and *Drepanothrix*. The genera *Ilyocryptus* and *Monospilus* live regularly on the bottom; their structure is adjusted to a life in the mud and their shells are often overgrown by algae. These forms may and do swim, but more often scramble about on the bottom, pulling with their antennae and pushing with the postabdomen. In both forms the old shell is not cast off in molting, the new and larger shell appearing beneath it (Figs. 27.68, 27.75).

The species of *Moina* are found most commonly in muddy pools, such as those in brick-yards, though not confined to such waters (some species live in saline lakes). *Daphnia* are likely to be found in temporary pools of clear and weedy water, in small ponds, and in lakes.

The limnetic region of the inland lakes has a cladoceran population large in number of individuals but not rich in species. *Chydorus sphaericus* is almost the only chydorid that is ever abundant here, though any species may be present as an accidental visitor. The regularly limnetic species belong chiefly to the genera *Bosmina*, *Diaphanosoma*, *Daphnia*, and *Holopedium*. Apart from transparency and a general lightness of build, the limnetic forms generally have no peculiar characters. *Holopedium* forms a conspicuous exception to this statement, as its globular gelatinous case is unique in the group and indeed in the Crustacea.

Certain species are intermediate in character between the limnetic and the littoral forms. Such is *Ophryoxus gracilis* (Fig. 27.56), which paddles about in the open waters between weeds, and such also is *Sida crystallina* (Fig. 27.4). Both of these forms are transparent, but they are never present in large numbers in the open water, nor are they likely to be found far out from the weedy margin.

The Cladocera are sometimes cited as a group in which a study of geographic distribution promises little of interest because the species are so widely distributed. A few species, of which *Chydorus sphaericus* is noteworthy, appear to be truly cosmopolitan animals. The range of many species, probably the majority, includes several continents. A significant number are, however, restricted to parts of a single continent. Therefore, careful studies of taxonomy and distribution should yield data of considerable zoogeographical interest.

Although many of the bottom-dwelling chydorids and macrothricids can be collected almost anywhere on the North American continent, some of these and many of the sidids and daphnids are restricted to particular regions. Many species are found only in the southern part of the continent, in the southern United States and southward. Most of these southern species also occur in South America, in fact 23 species are known to be common to the two continents (cf. Birge, 1910). (For example, *Latonopsis fasciculata*, *Pseudosida bidentata*, *Holopedium amazonicum*, *Daphnia laevis*, *Ceriodaphnia rigaudi*, *Moinodaphnia macleayii*, *Grimaldina brazzai*, *Alona karua*). Certain species are limited to the northern part of the continent, Alaska, Canada, northern

United States (*Daphnia middendorffiana, Daphnia longiremis, Eurycercus glacialis*). The three last-named species are clearly circumpolar in distribution. Some of the species restricted to this continent have a limited distribution, and others may occupy large areas. *Daphnia retrocurva*, for example, is a common plankter in the lakes of the entire northern half of the continent and is restricted to this region. Careful distributional studies within genera, the taxonomy of which are reasonably sound, would provide valuable information about the history of the fresh-water fauna of the New World.

Preservation and Identification

Formalin or alcohol are the preservatives usually used for Cladocera. Formalin is convenient for field use because a relatively small volume must be added to bring the final strength of the fluid up to 15 to 20 per cent of formaldehyde. It does, however, distort some of the soft-bodied species. For *Pseudosida, Latona, Latonopsis, Moina,* and *Diaphanosoma* strong alcohol (95 per cent) often gives better preservation. For careful histological or cytological studies, regular procedures for fixing and staining should be followed.

The transparent bodies of the Cladocera require only clearing and a mounting medium of low refractive index to reveal nearly all structural detail. Most Cladocera have to be examined microscopically, often with oil immersion objectives, so that liquid mounting media are inadequate. Mounting in glycerine jelly, using the double cover slip method, gives excellent results and is highly recommended. With this method the specimen, which has been run up to half or full strength glycerine (as desired), is mounted in glycerine jelly on a small cover slip (½-inch circles are convenient), oriented as it is to be seen in the final preparation. This cover slip with jelly and specimen is inverted and placed in the center of a larger cover slip (¾- or 1-inch square or circle) on a warming plate. When the jelly has melted so that it reaches the rim of the small cover slip, the preparation is cooled. This preparation of the specimen between the two cover slips is finally mounted in some resin (damar, Canada balsam) with the larger cover slip uppermost. Care is taken that the space under the small cover slip and under the edges of the large cover slip are completely filled with resin, thus sealing in the glycerine jelly. Such double cover slip preparations are relatively permanent; some have been kept without any signs of deterioration for more than ten years.

It is necessary to count legs and recognize specialization of the first two pairs to identify Cladocera. Ordinarily, observations can be made through the carapace, but it may be necessary to make dissections in some cases. The material should be brought into glycerine, and the carapace and legs removed with very fine needles. Number 00 insect pins mounted in dowel handles can be used for this purpose.

The following key includes all and illustrates most species known to occur in North America.

KEY TO SPECIES

1a   Large (adult female 7–18 mm long), with body and legs not covered by bivalve carapace. Carapace reduced to small brood sac. Legs not flattened, but with cylindrical joints; without branchial appendages. . . . . Suborder **Haplopoda** Sars

Sole Family . . . . **Leptodoridae** Lilljeborg
Head elongated, slender; eye filling anterior end. Body 4-jointed, the first part (head and thorax) bearing the 6 legs and dorsal brood sac; abdomen clearly divided into 3 segments. Postabdomen not reflexed, with 2 short stylets. Antennules small, freely movable. Antennae with very large basal joint; rami 4-jointed; with numerous setae. Mandibles long, slender, pointed, with 3 spines near apex. Six pairs of legs, first pair very long; all prehensile, without branchial appendages. Esophagus very long, stomach in last abdominal segment. ♀ with very long antennules. The young from winter eggs hatch as a metanauplius.

Sole genus . . . . *Leptodora* Lilljeborg 1860
Sole species . . . . *L. kindtii* (Focke) 1844
This remarkable, transparent form is the largest of the Cladocera, the ♀ reaching a length of 18 mm. Predaceous, though its weak mandibles prevent it from devouring any tough plankters.

Not uncommon in lakes of northern U. S. and northward.

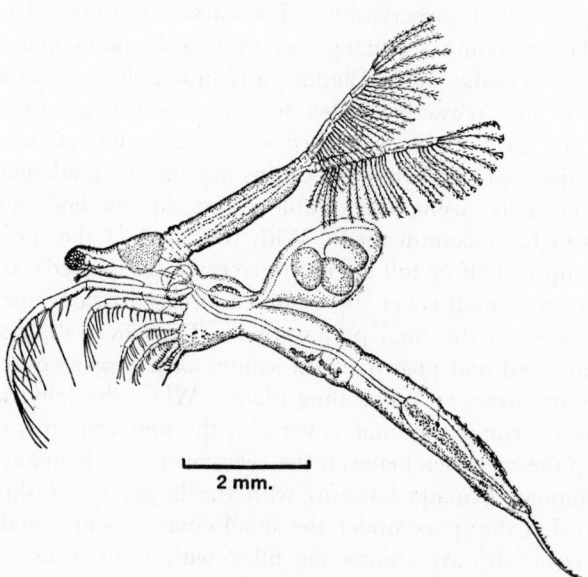

2 mm.

Fig. 27.2. *Leptodora kindtii.* (By Birge.)

1b   Small, less than 6 mm in length (usually much less). Body and legs enclosed in bivalve carapace (*Polyphemus*, Fig. 27.3, is sole exception). Legs usually with flattened branchial appendages . . . Suborder **Eucladocera** Eriksson   **2**

2a   (1) Five or 6 pairs of flattened appendages **3**

2b   Four pairs of jointed appendages with subcylindrical joints . . . . .
Superfamily **Polyphemoidea** (= Tribe **Onychopoda** Sars)

Sole North American family . . . . **Polyphemidae** Baird
Body very short. Carapace converted into large globular brood sac. Caudal process long, slender, with 2 long caudal stylets or setae. Rami of antennae with 3 and 4 joints. Eye very large; no ocellus. Labrum large. Two small hepatic caeca.

Sole genus in inland waters of North America . **Polyphemus** O. F. Müller 1785
Sole species . . . . **P. pediculus** (Linné) 1761

Brood sac globular, with 20 to 25 young in full grown specimens. Antennules very small, on ventral surface of head. Head large, filled in front by huge movable eye. Antennae with 7 setae on each ramus. Legs stout, with strong claws, and branchial appendages; fourth pair very small. Length, ♀ measured to back of brood sac, up to 1.5 mm; ♂ 0.8 mm.

Common throughout northern part of continent, in pools, marshes, and margins of lakes.

◄ **Fig. 27.3.** *Polyphemus pediculus.* (By Birge.)

0.5 mm.

3a   (2) Six pairs of legs, all similar except most posterior, all flattened . . .
                Superfamily **Sidoidea** (= Tribe **Ctenopoda** Sars)   **4**

3b   Five or 6 pairs of legs. First and second pair more or less prehensile with cylindrical joints, others leaflike . . . .
              Superfamily **Chydoroidea** (= Tribe **Anomopoda** Sars)   **13**

4a   (3) Carapace of usual type, without gelatinous mantle. Antenna biramous in female, rami flattened, the dorsal bearing numerous setae, both lateral and terminal . . . . Family **Sididae** Baird   **5**
    Head large; cervical sinus present. Antennules large, movable, with 9 olfactory setae. Ventral ramus of antennae with terminal setae only. Eye large, with numerous lenses; ocellus small or absent. Intestine simple, usually with more or less distinct median caecum or enlargement at anterior end; rarely with 2 hepatic caeca. Heart elongated. Male usually with characteristic antennule; the flagellum united with the base into one structure, long, tapering, with a row of fine spinules toward apex; usually with grasping organ on first foot and copulatory organs on postabdomen. (Figs. 27.4–27.11.)

4b   Carapace enclosed in gelatinous mantle, formed by carapace. Antenna of female uniramous, cylindrical, with 3 terminal setae. (Fig. 27.9) . . . . Family **Holopedidae** Sars
    Animal enclosed in a large, globular, transparent, gelatinous case, open ventrally and forming 2 valves. Body much compressed, shell of head and body very thin and high, as seen from side, leaving uncovered the mouth parts, the ends of the legs, and the hind part of the body. Antennule small, fixed; with 6 olfactory setae and lateral sense-hair, but no flagellum. Antennae in ♀ long; basal joint curved, annulated; the single ramus 2-jointed; antennae of ♂ biramous. Postabdomen large, fleshy, not bent forward; with rather long, curved anal spines and clusters of very fine spinules; abdominal setae long, set on single, long conical projection. Claws large, curved, denticulate, not set off from body by distinct joint. Eye small, with numerous lenses; ocellus small. Intestine simple with 2 hepatic caeca. Branchial sac on second to fifth legs. Color transparent. Swims on its back.   ·
    Sole genus . . . . *Holopedium* Zaddach 1855   **12**

5a   (4) Dorsal ramus of antenna 3-jointed; rostrum present . . . .
                             Sida Straus 1820
    Head with large gland on dorsal side; pointed rostrum; no fornices. Antennules of ♀ attached to side of rostrum, short, truncate, with short flagellum. Ventral ramus of antennae 2-jointed. Antennules of ♂ very long; no copulatory organ; first leg with hook.
    Only one known species . . . . **S. crystallina** (O. F. Müller) 1875

Color yellow-hyaline, sometimes with brilliant blue spots. Length, ♀ 3.0–4.0 mm; ♂ 1.5–2.0 mm.

Common in lakes and ponds among weeds.

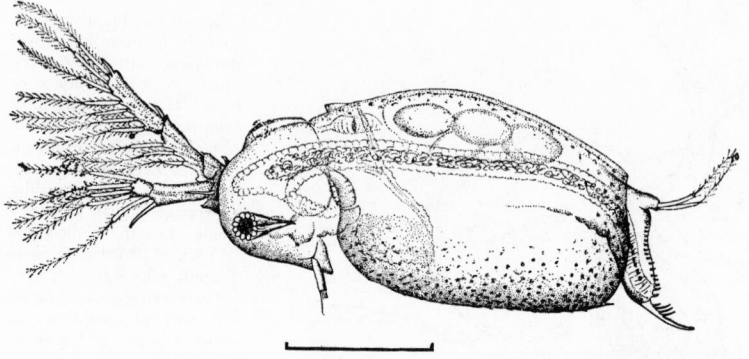

1 mm.

Fig. 27.4.   *Sida crystallina.*   (By Birge.)

5b   Dorsal ramus of antenna 2-jointed **6**

6a   (5) With lateral expansion on basal joint of dorsal ramus of antenna. . . . . *Latona* Straus 1820   **7**
    Large, tongue-shaped projection on ventral side of head, its ventral surface concave
    (Figs. 27.5, 27.6). Ventral ramus of antennae 3-jointed. Long setae on posterior
    margin of valves. Eye dorsal, far from optic ganglion.   ♂ with copulatory organ; no
    hook on first leg.

6b   Without lateral expansion of antenna. **8**

7a   (6) Antennary expansion very large; no hepatic caeca. . . . .
                                    L. setifera (O. F. Müller) 1785
    Antennules of both sexes alike, bent, with large, hairy flagellum set on at angle,
    looking like continuation of base. Color yellow; not transparent; old ♀ often with
    brilliant colors in late autumn. Length, ♀ 2.0–3.0 mm, ♂ *ca.* 1.5 mm.
    Widely distributed, but rarely abundant, among weeds in ponds and lakes.

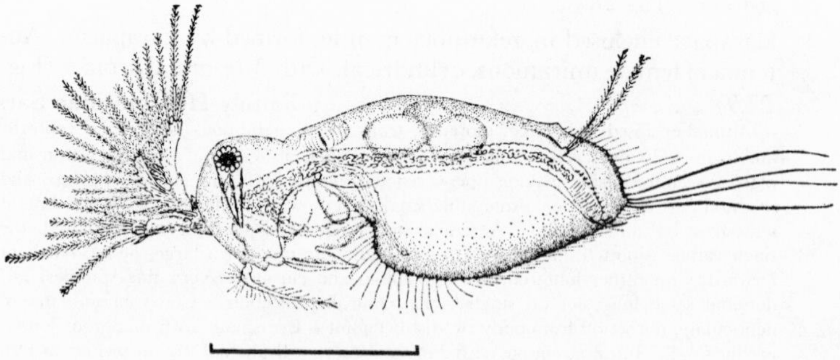

1 mm.

Fig. 27.5.   *Latona setifera.*   (By Birge.)

7b   Antennary expansion small; hepatic caeca present. . . . .
                                    L. parviremis Birge 1910
    Antennule of ♀ with basal part and long slender flagellum, like *Latonopsis;* of ♂
    very long, like other Sididae. Color yellow. Length, ♀ to 2.5 mm; ♂ 0.8 mm.
    Me. to Wis. in weedy waters of lakes.

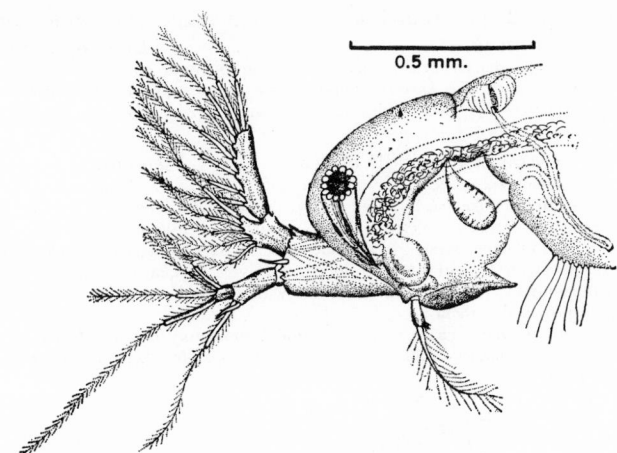

Fig. 27.6. *Latona parviremis.* (By Birge.)

8a **(6)** No anal spines on postabdomen. . . . ***Diaphanosoma*** Fischer 1850   **9**

No rostrum, fornix, or ocellus. Antennule small, truncated; olfactory setae terminal, with slender flagellum. Dorsal ramus of antennae 2-jointed; ventral 3-jointed. Claws with 3 basal spines. ♂ with long antennule; copulatory organ; hook on first foot.

8b   Anal spines present on postabdomen **10**

9a **(8)** Reflexed antenna (i.e., antenna held against carapace) not reaching posterior margin of valves . . . ***D. brachyurum*** (Liéven) 1848

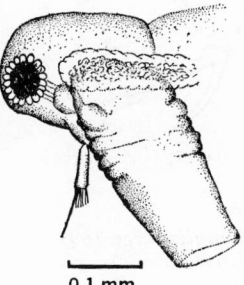

Eye pigment large; eye filling end of head. Color yellowish-transparent. Length ♀ 0.8–0.9 mm; ♂ *ca.* 0.4 mm. Common in marshes and weedy margins of lakes. Very probably the next species is merely a limnetic variety of this.

◄ **Fig. 27.7.** *Diaphanosoma brachyurum.* (By Birge.)

0.1 mm.

9b   Reflexed antenna reaching or exceeding posterior margin of valves . . . . ***D. leuchtenbergianum*** Fischer 1850

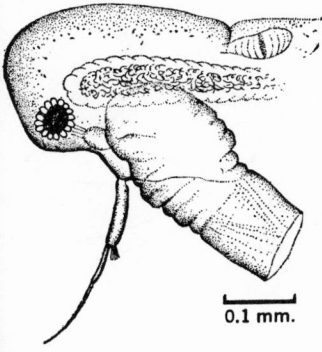

Eye not filling end of head, pigment small. Color hyaline. Length, ♀ 0.9–1.2 mm; ♂ to 0.8 mm. Common in open waters of lakes.

◄ **Fig. 27.8.** *Diaphanosoma leuchtenbergianum.* (By Birge.)

0.1 mm.

10a   (8) Eye dorsal, far from insertion of antennule and optic ganglion.
No rostrum. . . . . *Latonopsis* Sars 1888   **11**

> No tongue-shaped process on ventral side of head, or antennary expansion.
> Otherwise much like *Latona parviremis*. Posterior margin of valves with very long
> setae (often lost). Male with long antennule, copulatory organ, and hook on
> first foot.

10b   Eye ventral or in middle of head . . . . . *Pseudosida* Herrick 1884

> Sole known species . . . . *P. bidentata* Herrick 1884
> General form like *Sida* but head more depressed and dorsum more arched. Rostrum
> present; no fornix or cervical glands. Antennules attached as in *Sida*, long basal
> part with olfactory setae on each side, and long flexible flagellum. Dorsal ramus of
> antennae with 2, ventral with 3, joints; setae very unequal in length. Postabdomen
> with about 14 clusters of spinules; claws with 2 large basal spines and a very small
> spine proximal to them.
>
> ♂ with antennule characteristic of family; copulatory organs. Complex grasping
> apparatus on first leg. Color yellowish, semi-transparent. Length, ♀ to 1.8 or 2.0 mm;
> ♂ 0.9 mm.
>
> Southern U. S. in pools and lakes.

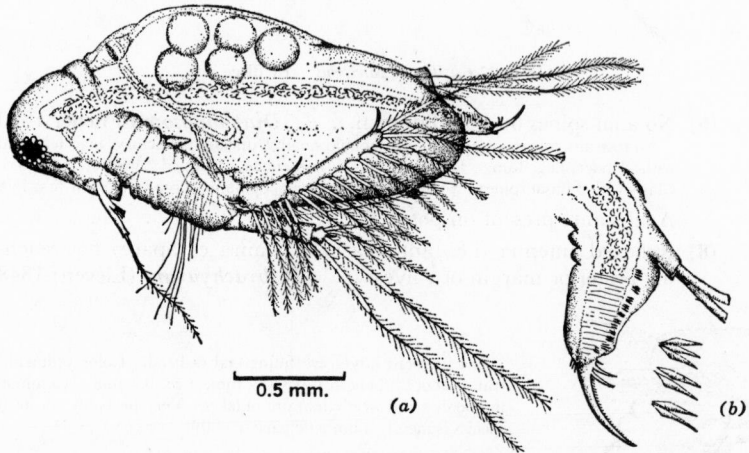

Fig. 27.9. *Pseudosida bidentata.* (*a*) Lateral. (*b*) Postabdomen. (*a* after Foster; *b* after Birge.)

11a   (10) Shell gland drawn out into very long posterior loop . . . .
Latonopsis occidentalis Birge 1891

> Postabdomen with about 9 small anal spines. Color yellowish-transparent. Length,
> ♀ to 1.8 mm; ♂ *ca.* 0.6 mm.
> New England to Colo. and Tex.; in weedy pools and lakes.

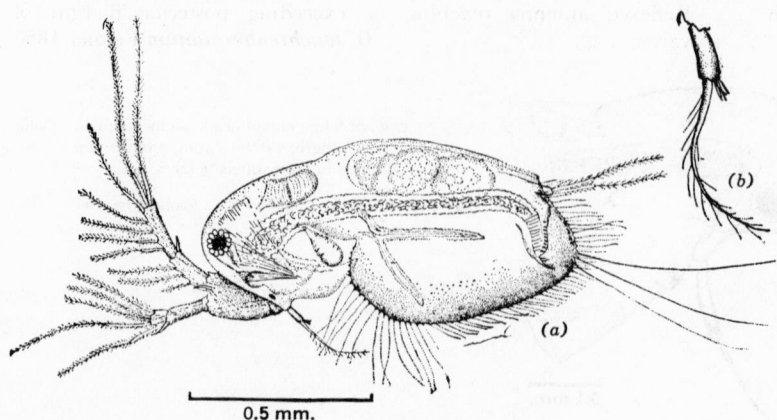

Fig. 27.10. *Latonopsis occidentalis.* (By Birge.)

11b   Shell gland without long posterior loop. . . . .
                                    L. fasciculata Daday 1905

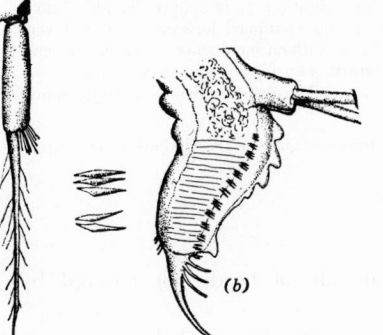

Postabdomen with projections on dorsal (posterior) margin and 12 to 14 clusters of 2 to 3 lancet-shaped anal spines. Color yellowish. Length, ♀ to 2.0 mm; ♂ to 1.0 mm.
La., Tex., in weedy pools and lakes.

◄ **Fig. 27.11.** *Latonopsis fasiculata.* (*a*) Antennule. (*b*) Postabdomen. (By Birge.)

12a   **(4)** Ventral margin of valves with fine spines. . . . .
                             Holopedium gibberum Zaddach 1855

Postabdomen elongated (*ca.* ⅓ length of body) and tapering; anal spines numerous, up to 20. Claws with 1 basal spine.   Length, ♀, 1.5–2.2 mm; ♂ 0.5–0.6 mm.
This species is not uncommon in open water in lakes of the northern part of U. S. and northward.   South in mountains of west to Calif., Colo.

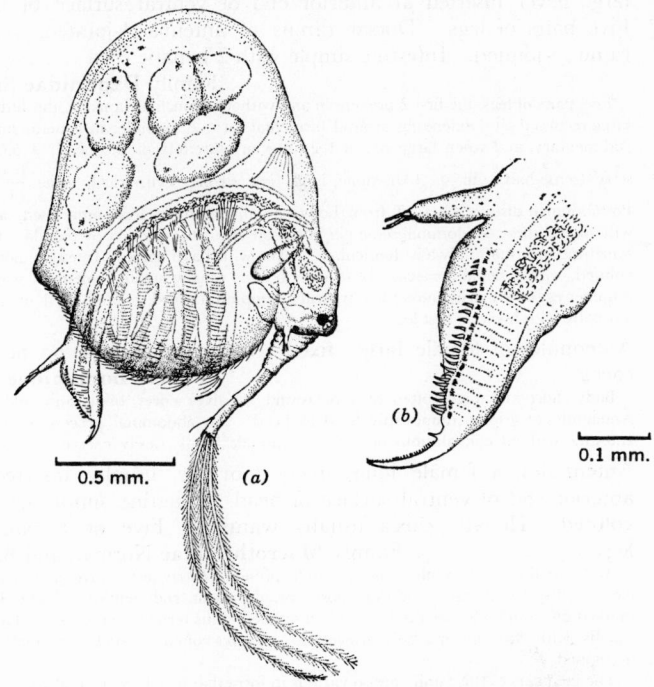

0.5 mm.   (*a*)

(*b*)

0.1 mm.

Fig. 27.12.   *Holopedium gibberum*   (*a*) Lateral (gelatinous case not shown).   (*b*) Postabdomen.   (By Birge.)

12b   Ventral margin of valves without spinules . . . .
                                    H. amazonicum Stingelin 1904

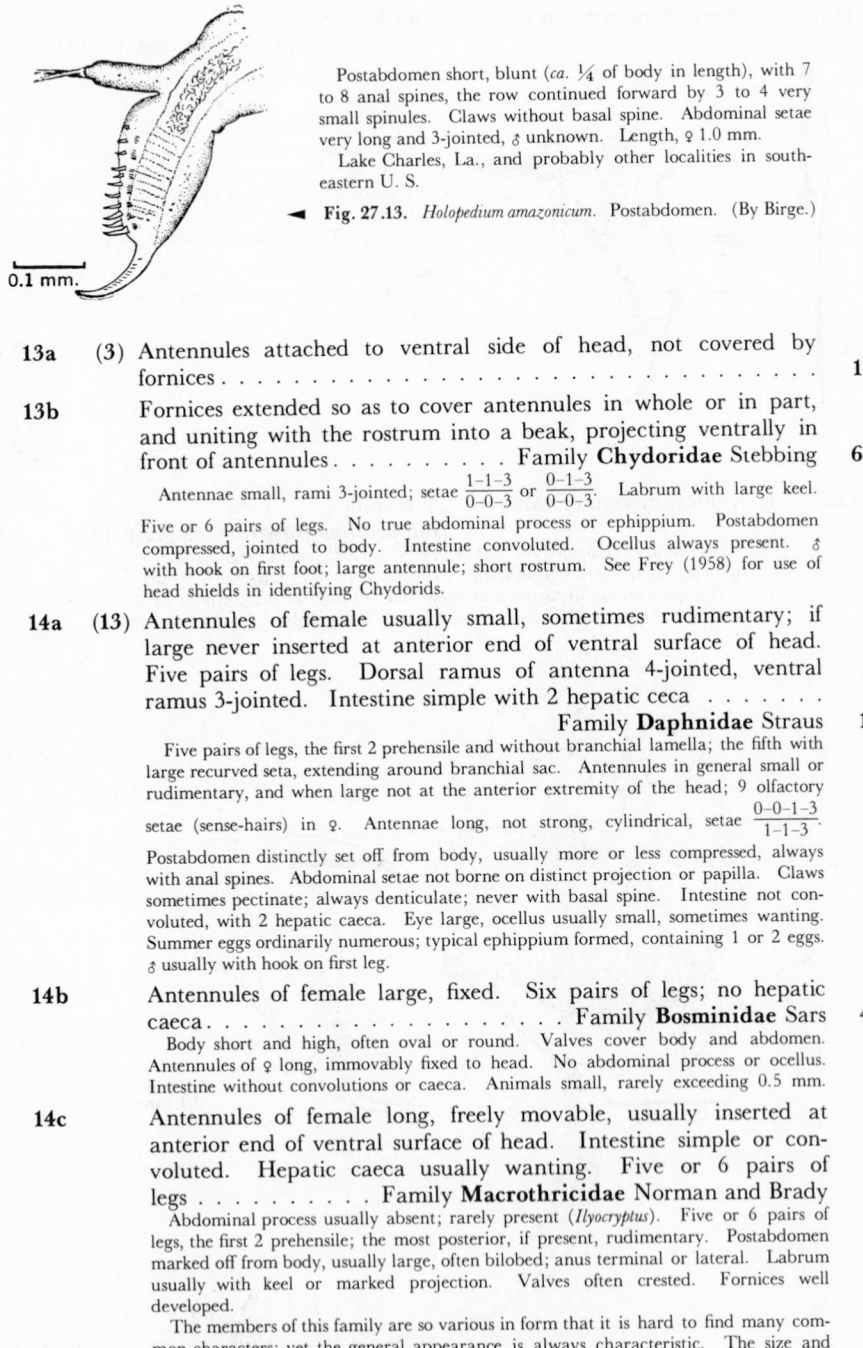

Postabdomen short, blunt (*ca.* ¼ of body in length), with 7 to 8 anal spines, the row continued forward by 3 to 4 very small spinules. Claws without basal spine. Abdominal setae very long and 3-jointed, ♂ unknown. Length, ♀ 1.0 mm.
Lake Charles, La., and probably other localities in southeastern U. S.

◄ **Fig. 27.13.** *Holopedium amazonicum.* Postabdomen. (By Birge.)

0.1 mm.

13a (3) Antennules attached to ventral side of head, not covered by fornices 14

13b Fornices extended so as to cover antennules in whole or in part, and uniting with the rostrum into a beak, projecting ventrally in front of antennules . . . . Family **Chydoridae** Stebbing 67

Antennae small, rami 3-jointed; setae $\frac{1-1-3}{0-0-3}$ or $\frac{0-1-3}{0-0-3}$. Labrum with large keel. Five or 6 pairs of legs. No true abdominal process or ephippium. Postabdomen compressed, jointed to body. Intestine convoluted. Ocellus always present. ♂ with hook on first foot; large antennule; short rostrum. See Frey (1958) for use of head shields in identifying Chydorids.

14a (13) Antennules of female usually small, sometimes rudimentary; if large never inserted at anterior end of ventral surface of head. Five pairs of legs. Dorsal ramus of antenna 4-jointed, ventral ramus 3-jointed. Intestine simple with 2 hepatic ceca . . . . Family **Daphnidae** Straus 15

Five pairs of legs, the first 2 prehensile and without branchial lamella; the fifth with large recurved seta, extending around branchial sac. Antennules in general small or rudimentary, and when large not at the anterior extremity of the head; 9 olfactory setae (sense-hairs) in ♀. Antennae long, not strong, cylindrical, setae $\frac{0-0-1-3}{1-1-3}$.

Postabdomen distinctly set off from body, usually more or less compressed, always with anal spines. Abdominal setae not borne on distinct projection or papilla. Claws sometimes pectinate; always denticulate; never with basal spine. Intestine not convoluted, with 2 hepatic caeca. Eye large, ocellus usually small, sometimes wanting. Summer eggs ordinarily numerous; typical ephippium formed, containing 1 or 2 eggs. ♂ usually with hook on first leg.

14b Antennules of female large, fixed. Six pairs of legs; no hepatic caeca . . . . Family **Bosminidae** Sars 49

Body short and high, often oval or round. Valves cover body and abdomen. Antennules of ♀ long, immovably fixed to head. No abdominal process or ocellus. Intestine without convolutions or caeca. Animals small, rarely exceeding 0.5 mm.

14c Antennules of female long, freely movable, usually inserted at anterior end of ventral surface of head. Intestine simple or convoluted. Hepatic caeca usually wanting. Five or 6 pairs of legs . . . . Family **Macrothricidae** Norman and Brady 51

Abdominal process usually absent; rarely present (*Ilyocryptus*). Five or 6 pairs of legs, the first 2 prehensile; the most posterior, if present, rudimentary. Postabdomen marked off from body, usually large, often bilobed; anus terminal or lateral. Labrum usually with keel or marked projection. Valves often crested. Fornices well developed.

The members of this family are so various in form that it is hard to find many common characters; yet the general appearance is always characteristic. The size and position of the antennules will show the membership of every genus except *Ilyocryptus;* and there is no trouble in recognizing that genus as belonging to the family.

15a (14) Rostrum present 16

15b Rostrum absent, cervical sinus present 34

16a (15) Cervical sinus absent. Valves with posterior spine. Usually with crest on anterior surface of head . . . ***Daphnia*** O. F. Müller 1785   **17**

Form oval or elliptical, except as modified by crest on head (helmet) in some species. Body always compressed, often greatly so. Valves reticulated; dorsal and ventral margins rounding over toward each other and provided with spinules along posterior part. Rostrum well marked in ♀ and pointed. Antennules small or rudimentary, not movable, placed behind rostrum. Abdominal processes 3 to 4, all ordinarily developed; the anterior especially long, tongue-shaped and bent forward. Ephippium with 2 eggs. Summer eggs often very numerous.

Head of ♂ without rostrum; antennules large, movable, ordinarily with long, stout, anterior seta or flagellum; first leg with hook and long flagellum.

See Brooks (1957) for a monographic treatment of the highly variable species of *Daphnia* inhabiting N. A.

16b   Cervical sinus present. No crest **30**

17a   **(16)**   Carapace continues anteriorly onto head along mid-dorsal line as median strip between halves of head shield (view specimen dorsally). Heavy-bodied forms with little lateral compression. (Figs. 27.14*a*, 27.15*b*) . . . .

Subgenus ***Ctenodaphnia*** Dybowski and Grochowski 1895   **18**

(a)

(b)

Large forms with bodies tending to be spherical; along mid-dorsal line carapace projects forward onto head shield. Fornices high and often with a lateral keel on the valve continuous with the fornix. The ephippial eggs lie parallel or oblique to the dorsal edge of ephippium; dorsal edge of ephippium continued forward as toothed spine; teeth of proximal and middle pectens of postabdominal claw always larger than teeth of distal pecten.

◄   **Fig. 27.14.** Dorsal views of representatives of (*a*) Subgenus *Ctenodaphnia* (*Daphnia similis*) and (*b*) Subgenus *Daphnia* (*Daphnia pulex*). See also Fig. 27.15*b* for dorsal view of another *Ctenodaphnia*.

17b   Apex of head shield projects posteriorly along mid-dorsal line onto carapace. (Fig. 27.14*b*). . . . . . . Subgenus ***Daphnia*** sensu stricto   **19**

Large or small forms with laterally compressed bodies; along mid-dorsal line the head shield projects backward onto carapace; fornices low, and never with lateral keel on valve. The ephippial eggs lie at right angles to the dorsal edge of ephippium; dorsal edge of ephippium never continued forward as toothed projection. Teeth of all 3 pectens of postabdominal claw may be of same size, or those of proximal and middle pectens may be larger than those of distal.

18a   **(17)**   Postabdomen with deeply sinuate posterior margin . . . .

D. magna Straus 1820

This species can be readily distinguished from all other *Daphnia* on this continent by the deeply sinuate posterior (dorsal) margin of the postabdomen. The head shield bears a pair of longitudinal ridges on either side of the median keel. Posterior end of fornix either rounded or pointed. Lateral keel on valve appears to be a continuation of fornix. Male with large spinulate genital papilla at base of postabdominal claw. Distal portion of antennular flagellum setulate. Length, ♀ to 5.0 mm; ♂ 2.0 mm or more.

Neb., N. D., Saskatchewan; Calif. to British Columbia. Ponds, small lakes.

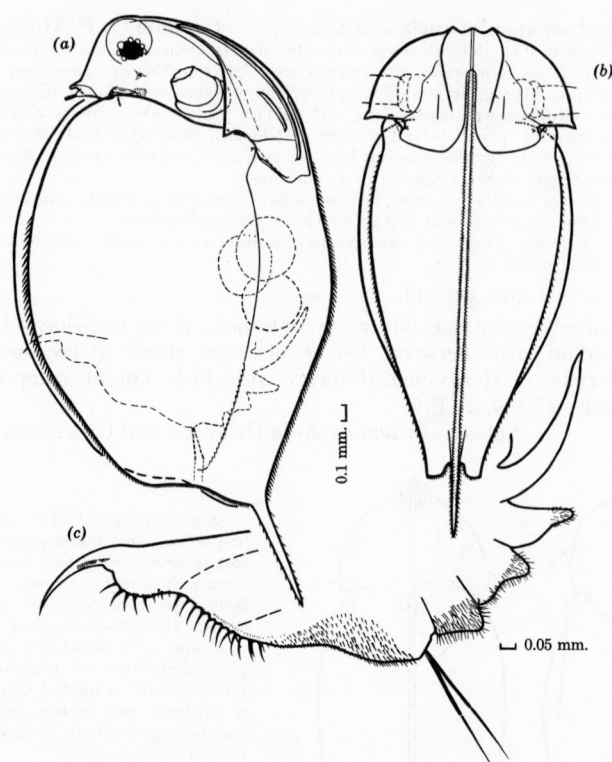

Fig. 27.15. Female of *Daphnia magna*. (a) Lateral view. (b) Dorsal view. (c) Postabdomen.

18b   Posterior postabdominal margin not sinuate . . . .
D. similis Claus 1876

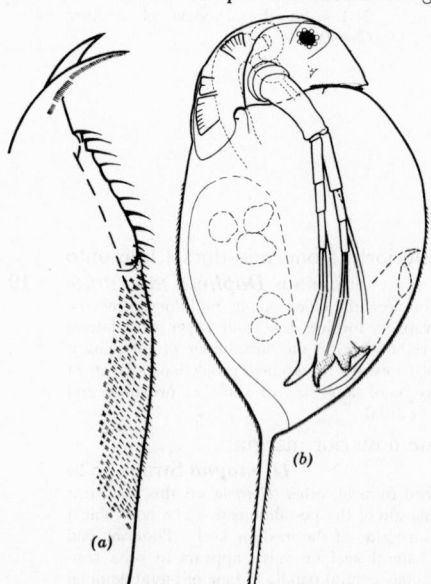

Posterior ends of fornix produced into long points. Lateral shell keel well developed (see Fig. 27.1). Usually with long shell spine. Head shape variable, depending on width of mid-dorsal extension of carapace. Dorsal margin of postabdomen not sinuate; 10 to 14 anal spines.

Male without genital papilla at base of postabdominal claws. Length, ♀ to 2.8 mm; ♂ to 1.8 mm.

Pools; Calif., N. D., Neb., and Saskatchewan.

◄ **Fig. 27.16.** Female of *Daphnia similis*. (a) Postabdomen. (b) Lateral view. See Fig. 27.1 for another form of female; also Fig. 27.14a for dorsal view.

19a (17) Swimming hairs of reflexed antenna never reaching posterior margin of valves in adult female **20**

19b Swimming hairs of reflexed (i.e., folded against body) antenna reaching posterior margin of valves . . . . ***D. longiremis*** Sars 1861

Valves broadly oval; shell spine long and slender. Head with well-developed rostrum and rounded crest, often with elongate, even retrocurved helmet. Tip of swimming hairs of antennae usually extend beyond posterior margin of valves when antennae are reflexed. Seta which arises from the apex of first joint of the 3-jointed (ventral) ramus does not reach end of that ramus. Second abdominal process about size of third. Nine to 11 anal spines; first 2 usually much longer than remainder. Teeth of all 3 combs of postabdominal claw small, of nearly the same size. Small, length of ♀ (head and carapace) 0.8–1.2 mm. ♂ extremely rare.

Northern N. A. south to northern U. S. Limnetic, confined to hypolimnion of lakes during stratification.

◄ **Fig. 27.17.** *Daphnia longiremis.*

0.1 mm.

20a (19) Teeth of all 3 pectens of postabdominal claw small and inconspicuous, of about the same length (cf. Fig. 27.19). Ocellus present. . . **21**

20b Teeth of middle and proximal pectens of about same size, somewhat larger than teeth of distal comb (cf. Fig. 27.24). Ocellus inconspicuous or absent **26**

20c Teeth of middle pecten distinctly larger than teeth of either proximal or distal pectens (cf. Fig. 27.25). Ocellus present. **27**

21a (20) Small, head and valves 1 mm or less in length. . . . .
D. ambigua Scourfield 1947
In lateral view the valve of this species appears almost circular and the head relatively small. The head often drawn out into a small point anteriorly. Occasion-

ally in adolescent female the point is sufficiently long that the head in lateral view resembles an equilateral triangle. Rostrum moderately developed. Shell spine always less than ½ length of carapace; sometimes considerably less. Postabdomen with 7 to 10 anal spines, gradually decreasing in length away from claw. Claw with 3 pectens of fine teeth.

Male with large antennules, longer than head. Flagellum slightly shorter than basal joint and 3 to 4 times as long as olfactory setae. Length, ♀ 0.75–1.0 mm; ♂ 0.9 mm.

In ponds and the deep water of stratified lakes. Southern part of continent; Central America north to New England, Ohio, Wash.

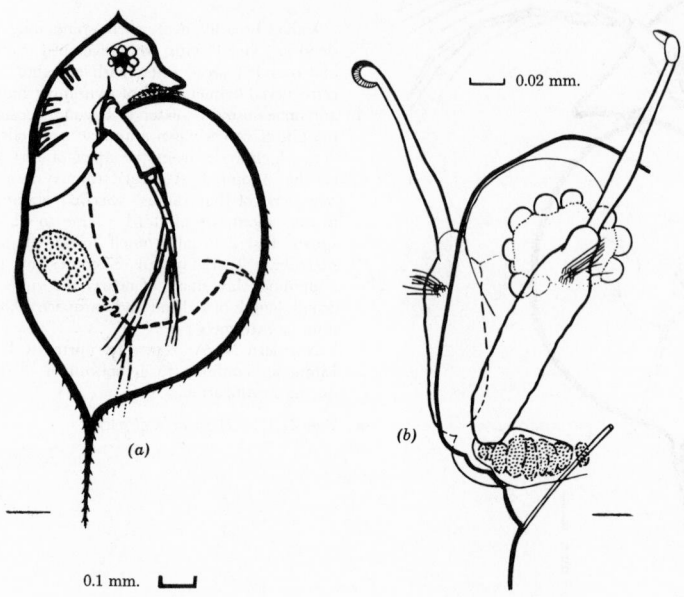

Fig. 27.18. *Daphnia ambigua.* (*a*) Female. (*b*) Head of male.

21b   Large, head and valves more than 1.25 mm long **22**

22a (**21**) Second abdominal process in mature female much smaller than first, about ¼ length of first. **23**

22b   Second abdominal process intermediate in length between first and third, at least ½ length of first **24**

23a (**22**) Spinulation extends over slightly more than ½ ventral margin of valve. Anterior margin of head with rounded crest. Head as deep as long . . . . *D. laevis* Birge 1879

This species and the next are easily distinguished from all other North American species by their elongate form. In lateral view the valve is at least 1½ times, often twice, the width (dorsoventral). The shell spine is especially long and slender, at least ¾ as long as valve, often its equal in length. In both species, the second abdominal process is much smaller than either first or third, being in this respect unlike any other species on this continent. Postabdomen with 9 to 14 (usually about 10) anal spines. Teeth of all 3 pectens on claw of about the same length.

♂: Basal joint of antennule relatively short, about ½ diameter of eye. Flagellum about length of olfactory setae.

Length (head and carapace), ♀ 1.2–1.7 mm; ♂ 1.0 mm.

This species can be distinguished morphologically from *D. dubia* on the basis of head shape: in *laevis* the head is about as long as it is deep at level of rostrum. Anterior margin rounded, or if slightly pointed, apex is always in mid-line.

Lives primarily in ponds, often temporary ones. Southern N. A., north to southern Conn., Okla., Calif.

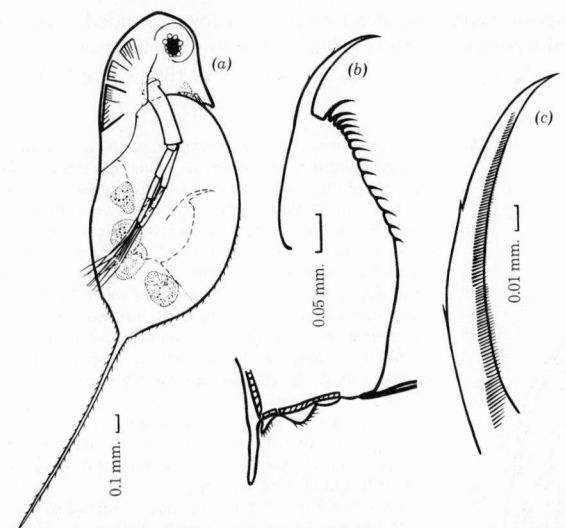

Fig. 27.19. *Daphnia laevis.* (*a*) Lateral. (*b*) Postabdomen. (*c*) Postabdominal claw.

23b Spinulation extends over posterior ¾ of ventral margin of valve, at least. Anterior margin of head produced into pointed helmet with apex well dorsal of mid-line; often helmet is retrocurved . . . . . .

D. dubia Herrick 1895

General description as indicated above. This species can be distinguished from *dubia* on the basis of its head shape. Except in very early spring, head with pointed helmet, with apex well dorsal to mid-line. Helmet sometimes marked by retrocurved apex. Presence of an ocellus, as well as its general body form, readily distinguish it from *retrocurva*.

Length, ♀ 1.2–1.8 mm; ♂ 1.0 mm.

Common in lakes in narrow belt from New England, west to Wis.

◄ **Fig. 27.20.** *Daphnia dubia.*

24a (22) Anterior margin of head often with low, rounded crest, but never
produced into helmet. Head twice as deep as long . . . .
                                    D. rosea Sars 1862 emend. Richard 1896

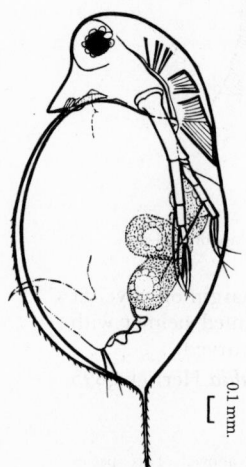

This species frequently bears a toothed crest on the anterodorsal margin of the head in young specimens. The anterior
margin of the head, although usually crested, is never produced
into a helmet such as characterizes *galeata mendotae*, to which
rosea is similar in many respects. The optic vesicle is very
close to anteroventral margin of head, never farther removed
than the diameter of a lens of the eye. Shell spine, slender
and weak, ⅓ to ½ length of valves (often broken). The relative size of the first abdominal process increases with age
after maturity: in the first mature instar, the first process
is about as long as the second; in large females it may be
twice as long. Second and third about same size. Postabdomen
with 12 to 15 anal spines. Teeth in all 3 pectens on claw
of about the same size. Synonym: *dentifera*.

Male usually with toothed crest similar to that of young
females. Basal joint of antennule relatively short and flagellum
about length of olfactory setae.

Length, ♀ usually 1.2–1.6 mm, may be up to 2 mm; male 0.8 mm.

Ponds, small lakes of western N. A. from Alaska to Calif., east
to Alberta, Colo.

◄ **Fig. 27.21.** *Daphnia rosea.*

[0.1 mm.

24b   Anterior margin produced into helmet so that head is always
longer than ½ greatest depth **25**

25a (24) Dorsal margin of head never with concavity at level of most anterior antennal muscle. Helmet usually pointed. . . . .
                            D. galeata Sars 1864 **mendotae** Birge 1918

A large species in which the head is produced anteriorly into a broad helmet except
in the very early spring. Helmet may be sharply or bluntly pointed, of various
shapes. As a consequence of this development, the optic vesicle is well removed
from the ventral margin of the head. Valve a broad oval, less than 1½ times as long
as wide. Shell spine at least ½ as long as valves. First abdominal process longer
than second, and second longer than third. Postabdomen with 9 to 11 anal spines.
Claw as in preceding species.

Male with pointed helmet; otherwise as in *rosea*. Length (head and carapace), ♀ 1.3–
3.0 mm; ♂ 1 mm.

In lakes of northern part of continent, especially common in lakes of glaciated
regions. Infrequent in mountainous regions of U. S., Canada, but present in some
mountain lakes of Central America.

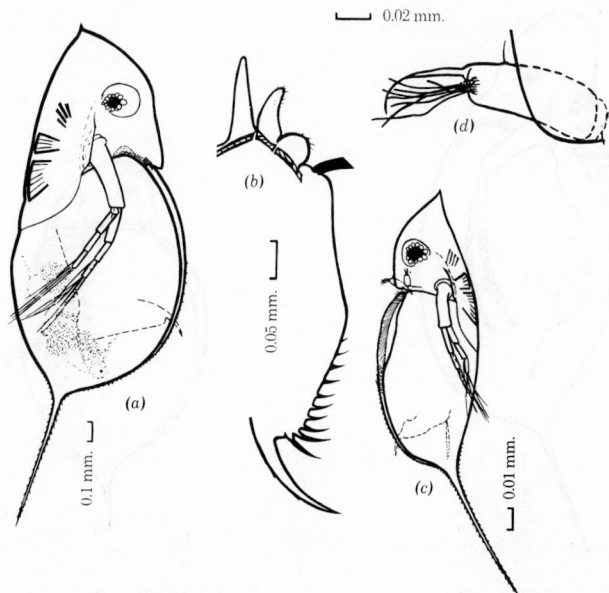

Fig. 27.22. *Daphnia galeata.* (*a*) Female. (*b*) Postabdomen, female. (*c*) Male. (*d*) Antennule of male.

25b   Dorsal margin of head with concavity at level of most anterior antennal muscle. Helmet rounded. (Fig. 27.23) . . . .
D. thorata Forbes 1893

Much like preceding species except that valves, viewed laterally, form elongate oval, at least 1½ times as long as wide. Head with elongate helmet always broadly rounded. ♂ usually without helmet, otherwise like ♂ of galeata mendotae. Length, ♀ 1.3–1.8 mm; ♂ 1.0 mm.

Restricted to large lakes of Wash., Ida., Mont., southern British Columbia, and Alberta.

26a (20)   Anterior margin of head with broadly rounded crest longest in mid-line. (Fig. 27.24) . . . . **D. parvula** Fordyce 1901

This species closely resembles *D. retrocurva* except for the degree of development of the helmet and rostrum; *parvula* always has a small rostrum and at most a rounded helmet which lengthens the head by no more than the diameter of the eye. ♂ with flagellum of antennule slightly longer than olfactory setae. Second abdominal process of very small, smaller than third. Length, ♀ 0.75–1.0 mm, occasionally up to 1.2 mm; ♂ 0.6 mm.

In ponds and small lakes in southern part of continent; from Central America north to southern New England, southern Saskatchewan, Wash.

26b   Anterior margin of head produced into helmet, apex of which is always dorsal to mid-line. (Fig. 27.25) . . . .
D. retrocurva Forbes 1882

The large, usually retrocurved helmet, together with the lack, or minute size of the ocellus, serves to differentiate this species. In winter or very early spring the helmet is smaller, but always present, with its greatest extension dorsal to the mid-line. This distinguishes *retrocurva* from *parvula*, in which the helmet, when present, is rounded. Shell spine long, *ca.* ⅔ length of carapace. Postabdomen with 6 to 10 (usually 8) anal spines. Teeth of middle pecten, 6 to 20 in number, all about the same size. Teeth of proximal pecten about as long as those of middle, but finer and usually slightly less numerous. ♂ with helmet less well developed than in ♀ of same body size. Antennule and abdominal processes much as in *parvula*.

Length of carapace of mature ♀, 0.8–0.9 mm; length of head and carapace may be up to 1.6 mm; length, ♂, 0.8 mm.

In lakes of northern N. A. except Alaska and Arctic Canada, south to New England, Wis., Wash.

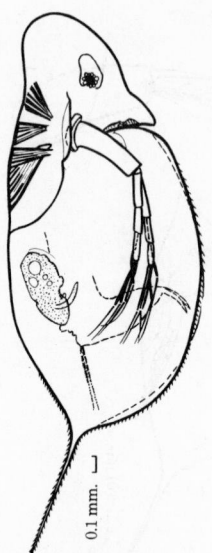

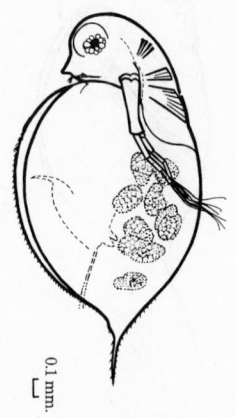

Fig. 27.23.   *Daphnia thorata.*

Fig. 27.24.   *Daphnia parvula.*

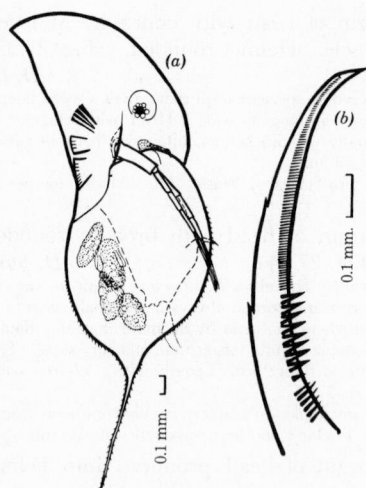

Fig. 27.25.   *Daphnia retrocurva.*   (a) Lateral.
(b) Postabdominal claw.

27a (20) Ventral margin of head concave; optic vesicle contiguous with
           margin in lateral view.   28
27b   Ventral margin of head sinuate or more or less straight, never
           strongly concave; optic vesicle usually separated from margin in
           lateral view.   29
28a (27) Head longest over optic vesicle; posterior margin of ventral part of

head separated from anterior margin of valve by wide gap; exoskeleton of dorsal part of head often distinctly brown . . . .

D. middendorffiana Fischer 1851

One of the largest of North American *Daphnia*. Females often 2.5–3.0 mm in length.
Cuticle of head dorsal to fornix usually light brown. Pigmentation may extend to
basal joint of antenna. Optic vesicle fills anteriormost part of head. Rostrum of
variable size and shape, but ventral margin of head always distinctly concave.
Dorsal margin of head without crest; bulging over attachment of anterior antennal
muscles. Shell spine slender, usually ⅓ to ½ length of carapace. First abdominal
process long, nearly twice as long as second. Postabdomen long and narrow;
posterior margin straight or with concavity under middle of spinate portion. Anal
spines 12 to 14, decreasing gradually in length away from claw. Teeth of middle
pecten of postabdominal claw 5 to 7 in number, separated at their bases, and about
twice as long as teeth of proximal pecten. Males are rare in high latitude populations
and ephippial eggs are usually made parthenogenetically (Edmondson 1955). ♂ antennule is distinctive as flagellum is expanded into cup-shaped tip. Synonym *D. pulex*
var. *tenebrosa* Sars.

Alaska, Northern Canada, south mostly in mountains to Calif. Ponds, lakes.

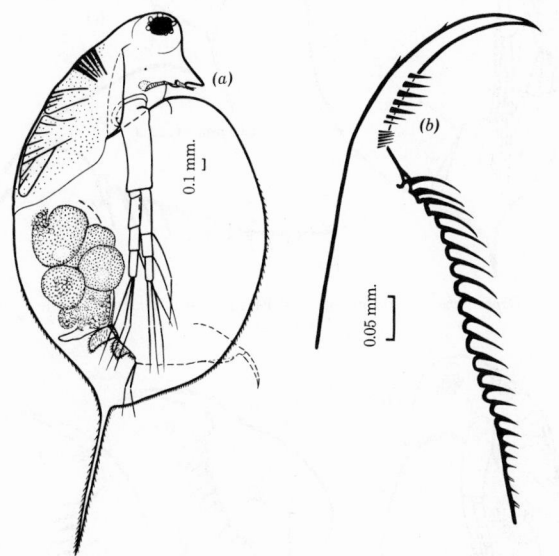

Fig. 27.26. *Daphnia middendorffiana.* (*a*) Lateral. (*b*) End of postabdomen.

28b   Head longest in mid-line or between mid-line and optic vesicle
with anterior margin of head more or less straight perpendicular
to body axis; posterior margin of ventral part of head close to
anterior margin of valves; exoskeleton of dorsal part of head never
distinctly brown . . . ***D. pulex*** Leydig 1860 emend. Richard 1896

Head and carapace form broad oval. Head length about ½ the depth at level of
rostrum. Anterior margin of head broadly rounded, sometimes almost a straight line
normal to body axis, in lateral view. Ventral margin of head concave. Tip of
rostrum directed posteroventrally. Eye large; ocellus of moderate size. Sometimes
in males and young females the crest between anterior and posterior adductors is
denticulate. Median carina on posterior surface of head continued into a mound between tips of antennules. Valves, oval in outline, become almost circular in large
females. Spinulation on ventral margin never more than covers posterior half. Shell
spine ⅓ to ⅕ of carapace length in mature specimens. Abdominal processes
gradually decreasing in length; the second about ¾ as long as first. Anal spines 10 to
16 in number (usually 12 to 14) gradually decreasing in length away from claw.
Teeth of middle pecten of postabdominal claw with 5 to 9 teeth, contiguous at their
bases; tooth length decreasing proximally. Four to 8 teeth in proximal comb. Teeth
usually about ½ as long as those of middle comb (may be nearly as long).

Length, ♀ 1.3–2.2 mm, occasionally even larger, ♂ 1.1 mm.

♀: Length of basal portion of antennule about equal to diameter of eye. Flagellum slightly shorter than basal joint; about twice as long as olfactory setae. Second abdominal process extending beyond base of anal setae in fully mature males.

A widespread and variable species, living in both ponds and lakes, entire continent.

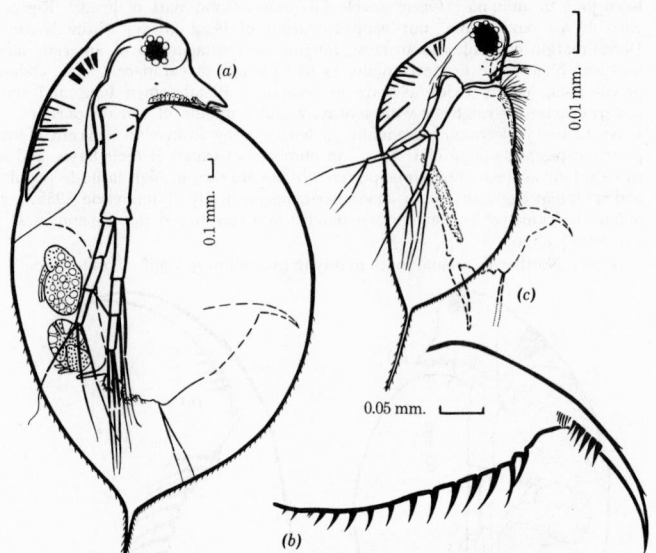

Fig. 27.27. *Daphnia pulex.* (*a*) Female. (*b*) End of postabdomen of female. (*c*) Male.

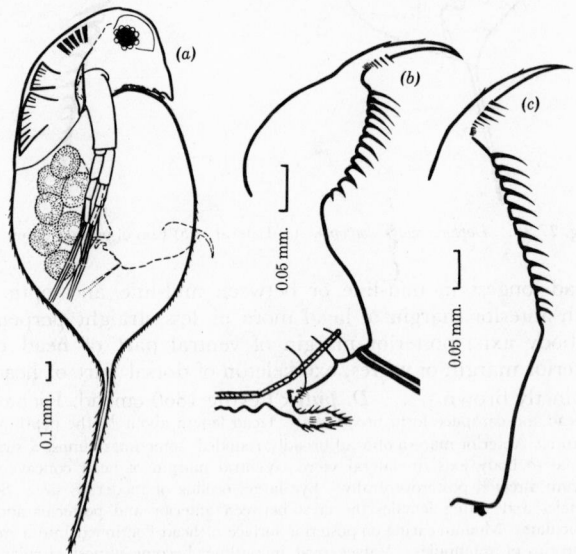

Fig. 27.28. *Daphnia schødleri.* (*a*) Female. (*b*) Postabdomen, male. (*c*) Postabdomen, female.

29a **(27)** Spinules on dorsal margin of valves large, interspinule distance less than 1½ times spinule length; shell spine more than ½ as long as valves, arising in mid-line . . . . **D. schødleri** Sars 1862

Dorsal and ventral margins of both head and valves tend to be subparallel to the body axis. Ventral margin of head never strongly concave; often nearly straight. Tip of rostrum directed posteriorly. Eye of moderate size, ocellus small. Anterior margin with broadly rounded crest, occasionally a similarly rounded helmet. Median carina on posterior surface of head never continued as mound between tips of antennules. Spinulation on ventral margin of valves extends forward onto anterior half. Dorsal margin of body straight over junction of head and carapace, subparallel with body axis. Shell spine long, ½ the length of valves or more. Postabdomen with 13 to 16 anal spines, all of nearly equal length. Five to 7 (usually 5 or 6) teeth in middle pecten of claw. Teeth show marked decrease in size proximally so that most proximal tooth is often very small. Proximal comb with 5 to 9 teeth, about ¼ to ⅓ as long as longest in middle pecten.

♂: Length of basal joint of antennule slightly less than diameter of eye; slightly curved. Flagellum of antennule slightly shorter than basal joint and about twice as long as olfactory setae (much as in *pulex*). Deep bay on dorsal surface of postabdomen. Second abdominal process extends about halfway to base of anal setae.

Length (head and carapace) ♀ 1.5–2.0 mm; ♂ 1 mm.

Western N. A., Alaska to Calif., east to Great Lakes region, Tex. Lakes, large ponds.

29b   Spinules on dorsal margin of valves small, interspinule distance at least twice, often 3 times spinule length . . ***D. catawba*** Coker 1926

A small species, 1.0–1.5 mm long, in which the ventral edge of the large optic vesicle never reaches the margin of the head. Head always with at least a slight rounded crest. Eye and ocellus of moderate size. Median carina on posterior surface of head continues between tips of antennules, and is nearly as high as, or higher than, tips of antennules as it passses between them. Valves broadly oval, sometimes nearly circular. Spinulation on ventral edge continues well forward onto anterior half. Posteriorly spinulation stops before base of shell spine. Spinules on both dorsal and ventral margins widely spaced, with spaces between at least twice spinule length. Shell spine long, ⅓ to ½ of carapace length. First to fourth abdominal processes of gradually decreasing length. Second and third with very sparse pubescence. Postabdomen with 8 to 11 anal spines, with teeth in distal half of series being much the longer. Middle pecten of claw with 2 to 4 large, widely separated teeth. Five to 10 teeth in proximal comb, ¼ to ⅕ as long as longest teeth of middle comb, and only slightly longer than those of distal comb.

♂: Flagellum of antennule relatively short, only 1½ times the length of the olfactory setae. Second abdominal process of mature male, rudimentary, smaller than third. Bay on dorsal surface not well marked.

Pond and lakes. Southeastern U. S. north to New England, southern Saskatchewan.

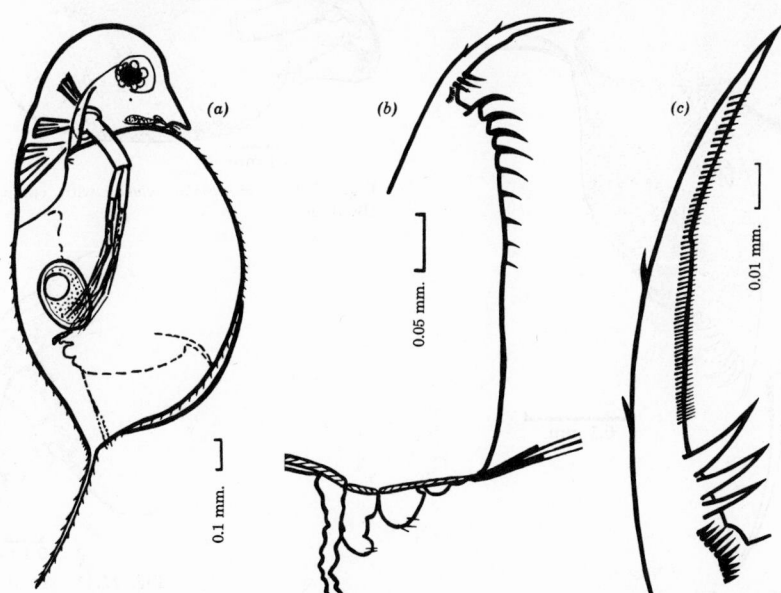

Fig. 27.29. *Daphnia catawba.* (*a*) Female. (*b*) Postabdomen, female. (*c*) Postabdominal claw.

30a   (16)   Ventral margin of valve merges into posterior margin in broad curve . . . . *Simocephalus* Schødler 1858   **31**

> Body large and heavy; shell thick. Head and rostrum small. Valves large, somewhat quadrate, with rounded angles and sometimes a posterior spine; marked with oblique striae, anastomosing irregularly and with cross-connections. Two abdominal processes developed, placed far apart. Postabdomen large, broad, truncate, posterior and emarginate and bearing the anal spines. Claws rather straight with fine teeth along entire length. Teeth near base may be enlarged to form the proximal pecten. Summer eggs numerous; ephippium large, triangular, with one egg. Antennules of ♂ like those of ♀ but with 2 lateral sense-hairs. First leg without flagellum and with small hook. Poor swimmer; often swims on its back. Color yellow to yellow-brown.

30b   Posterior end of ventral margin of valve extended into a point or spine (inferoposteal angle) . . . . . . *Scapholeberis* Schødler 1858   **33**

> Body not compressed; shape more or less quadrate. Cervical sinus deep. Fornices and rostrum well developed. Head small, depressed. Valves almost rectangular, the inferoposteal angle produced into a longer or shorter spine; ventral margin with short, fine setae. Claws denticulate, not pectinate. One abdominal process developed. Antennules small, about alike in both sexes, borne behind rostrum. Summer eggs numerous; one ephippial egg. ♂ much like ♀; hook on first leg.

31a   (30)   Postabdominal claw with proximal pecten. Ocellus rhomboidal or round. (Fig. 27.30). . . *Simocephalus exspinosus* (Koch) 1841

> Vertex with obtuse or rounded angle. No posterior spine on valves. Postabdomen slightly narrower toward apex; anal spines up to 12 in number, evenly curved, not bent; claw with pecten of 8 to 12 teeth at its base and with row of fine teeth distal to the pecten. Length, ♀ to 3.0 mm; ♂ to 1.3 mm.
> Not common, but occurring over most of continent.

31b   Postabdominal claw without proximal pecten, all teeth on claw small, of same length throughout. Ocellus elongated, triangular or rhomboidal **32**

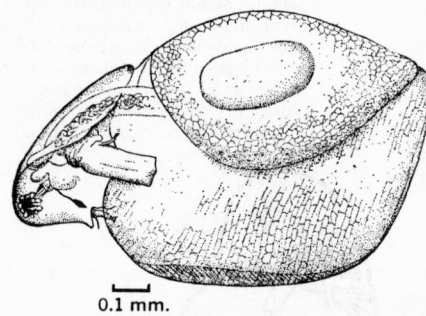

0.1 mm.

Fig. 27.31. *Simocephalus vetulus*, with ephippium. (By Birge.)

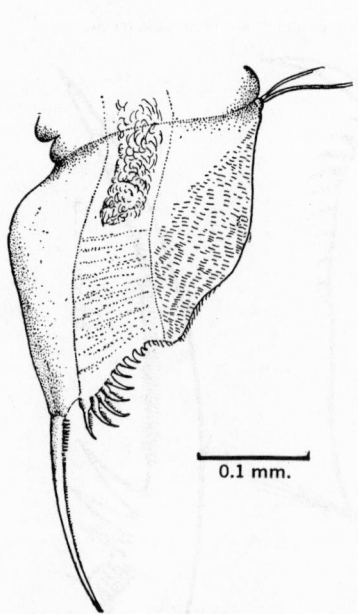

0.1 mm.

Fig. 27.30. *Simocephalus exspinosus.* Postabdomen. (By Birge.)

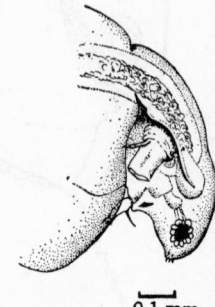

0.1 mm.

Fig. 27.32. *Simocephalus serrulatus.* Anterior end. (By Birge.)

32a (31) Vertex rounded over. No posterior spine on valves. Ocellus usually elongated. (Fig. 27.31) . . . . . *S. vetulus* Schødler 1858

Ocellus large, elongated, rarely rhomboidal. No spine on valves, though there may be a blunt posterior angle. Postabdomen very broad, deeply emarginate; anal spines about 10, decreasing from the claws; the larger bent and ciliate at the base. Claws long, slender, nearly straight, with fine teeth along entire length. Length, ♀ to 3.0 mm; ♂ *ca*. 1.0 mm.

Not very abundant, but found everywhere in weedy water.

32b Vertex angulate, spinous. Blunt, rounded posterior spine on valves of older individuals. Ocellus rhomboidal or triangular, rarely elongated. (Fig. 27.32) . . . . . *S. serrulatus* (Koch) 1841

Shape, and degree of spination of vertex extremely variable. Otherwise much like *vetulus* except for presence of blunt posterior spine in *serrulatus*. Length, ♀ 2.8–3.0 mm; ♂ to 0.8 mm.

Common everywhere among weeds.

33a (30) Color usually dark, often nearly black. (Fig. 27.33) . . . .
Scapholeberis kingi Sars 1903

Valves arched dorsally in old specimens; posterior and ventral margins straight; at their junction a spine often short, but often very long. Antennules very small, almost immovable, set behind beak. Postabdomen short and broad, rounded at posterior end; 5 to 6 anal spines. Length, ♀ 0.8–1.0 mm; ♂ *ca*. 0.5 mm.

Forms with a frontal spine or horn have been found in arctic regions of N. A. Common everywhere in pools and lakes in weedy water, or swimming on its back near or at the surface.

This species has often, and erroneously, been designated *S. mucronata* (O. F. Müller) 1785.

33b Color whitish or greenish; transparent or opaque, not black. (Fig. 27.34) . . . . *S. aurita* (Fischer) 1849

Head larger than in *kingi*, rostrum long, lying against margin of valves. Antennules behind rostrum, conical, large, and movable; sense-hair about middle. Valves with blunt projection at inferoposteal angle, obscurely striate and reticulate in front, and with small elevations elsewhere. Length, ♀ *ca*. 1.0 mm; ♂ 0.5 mm.

Not common; in weedy pools and margins of lakes. Northern part of continent.

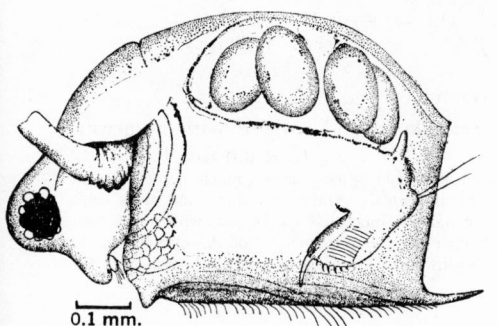

0.1 mm.

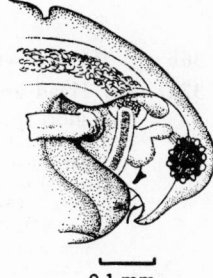

0.1 mm.

Fig. 27.33. *Scapholeberis kingi.* (By Birge.)

Fig. 27.34. *Scapholeberis aurita.* Anterior end. (By Birge.)

34a (15) Head small and depressed (Fig. 27.36). Antennules small. Valves oval or round. No postanal extension of postabdomen . . .
Ceriodaphnia Dana 1853   **35**

General form rounded or oval; size small, rarely exceeding 1 mm. Vertex a rounded or angular projection, usually nearly filled by eye. Valves oval or round to subquadrate, usually ending in a sharp dorsal angle or short spine. Antennules not very freely movable. One abdominal process ordinarily developed. Postabdomen large, of various shapes. Ephippium triangular, with one egg placed longitudinally. Antennules of ♂ with long, stout seta, a modification of flagellum; first leg with hook and long flagellum. Free swimming; motion saltatory.

34b   Head large and usually extended (Fig. 27.49).   Antennules large
            and freely movable. Postabdomen with postanal extension. **42**

35a (34) Head with a short spine or horn over eye on anterior margin. . . .
                                                    C. rigaudi Richard 1894

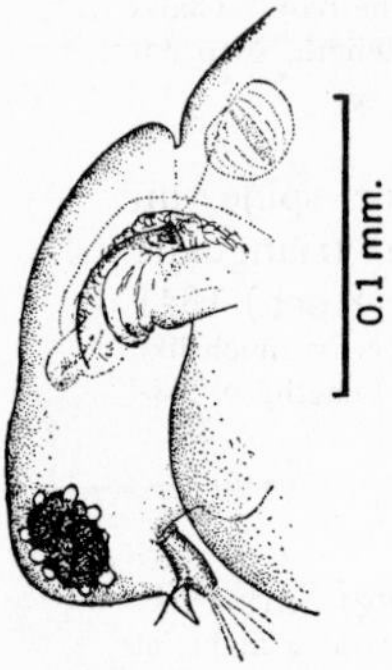

Valves reticulated. Head produced in front of antennules into a
short, conical, sharp-pointed, hornlike process.   Two abdominal
processes. Postabdomen with 5 to 6 anal spines. Claws smooth or
denticulate. Antennules rather slender; lateral sense-hair somewhat distal to middle. Length, ♀ 0.4–0.5 mm; ♂ (South American)
0.38 mm.

Pools. La., Tex. The form with horn on vertex also is found in
S. A., mingled with typical *C. rigaudi*. Probably both forms should
be included in *C. cornuta* Sars.

◄ **Fig. 27.35.** *Ceriodaphnia rigaudi.* Anterior end. (By Birge.)

35b   Head without horn **36**
36a (35) Claws with proximal pecten . . . . . . **C. reticulata** (Jurine) 1820

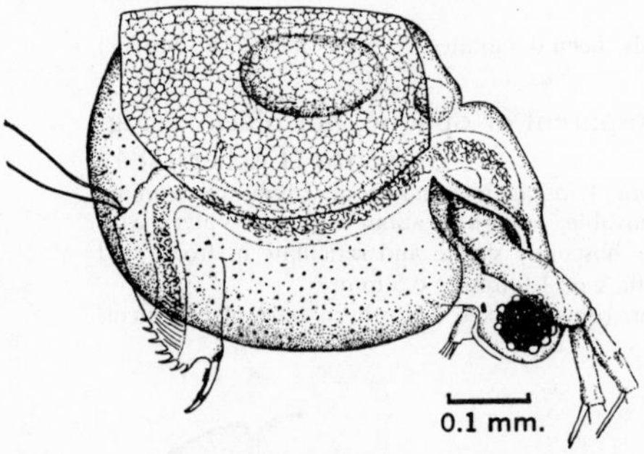

Head rounded or obtusely angulated
in front of antennules. Valves reticulated, ending in spine or angle. Antennules small with sense-hair near
apex. Anal spines 7 to 10. Claws
with pecten of 6 to 10 teeth and denticulate. Color variable, shades of red
and yellow. Length, ♀ 0.6–1.4 mm;
♂ 0.4–0.8 mm.

Common, widely distributed.

◄ **Fig. 27.36.**   *Ceriodaphnia reticulata*, with
ephippium. (By Birge.)

36b   Claws without proximal pecten **37**
37a (36) Head and valves strongly reticulated and covered with numerous
            short spinules . . . . **C. acanthina** Ross 1897
            General shape rotund with well-developed spine. Head much depressed, not
            angulated in front of antennules or at vortex. Antennules short and thick with
            sense-hair near apex. Postabdomen narrow, much like *quadrangula*, with 7 to 9 anal
            spines. Claws denticulate, the denticles in the proximal $\frac{2}{5}$ of the claw obviously
            longer than the remainder. Color whitish-transparent to very dark. Length, ♀ to 1.0
            mm; ♂ unknown.
            Manitoba, in weedy slough.

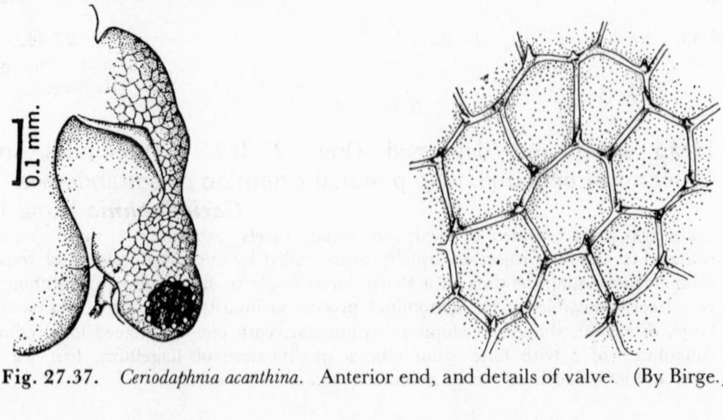

Fig. 27.37.   *Ceriodaphnia acanthina.* Anterior end, and details of valve. (By Birge.)

37b Valves not spinulate **38**

38a (37) Postabdomen abruptly incised near apex. Margin serrate above, with spines below . . . . ***C. megalops*** Sars 1861

Head angulated in front of antennules; valves striated. Antennules with sense-hair near apex. Postabdomen broad, with an angle near apex, cut into below angle, finely serrate above and with 7 to 9 slender anal spines below. Claws not pectinate. Length, ♀ 1.0–1.5 mm; ♂ 0.6–0.8 mm.

Widely distributed but not common.

◄ **Fig. 27.38.** *Ceriodaphnia megalops.* Postabdomen. (By Birge.)

0.1 mm.

38b Postabdomen of ordinary form; not incised. **39**

38c Postabdomen very broad, obliquely truncate. **41**

39a (38) Fornices projecting into spinous processes. Eye small . . . .
C. lacustris Birge 1893

0.1 mm.

Head angulated in front of antennule; vertex with fine spinules. Fornices very broad, triangular; with spines at tip. Valves with stout, short posterior spine, sometimes divided, but usually with 3 to 4 spinules. Postabdomen like *C. quadrangula.* ♂ unknown. Color yellow, transparent. Length, ♀ 0.8–0.9 mm.

Most of U. S. Limnetic in lakes.

◄ **Fig. 27.39.** *Ceriodaphnia lacustris.* Lateral and dorsal views. (By Birge.)

39b Fornices of ordinary form; eye large **40**

40a (39) Head inflated in front of antennules. Small species not exceeding 0.7 mm. . . . . ***C. pulchella*** Sars 1862

Form of type characteristic of genus. Head rounded in front; inflated in region behind eye, angulated in front of antennules. Valves reticulated but not plainly so. Postabdomen not sinuate above anal spines, which number 7 to 10. Length, ♀ 0.4–0.7 mm; ♂ 0.5 mm.

Found among weeds and limnetic in lakes and in pools; reported from most regions of continent. (Occasional specimens will be difficult to assign either to this species or to *quadrangula,* yet will agree closely with these descriptions. These variations should be more carefully studied.)

0.1 mm.

◄ **Fig. 27.40.** *Ceriodaphnia pulchella.* Anterior end. (By Birge.)

40b   Head angulate but not inflated in front of antennules.   Length to
1.0 mm . . . . *C. quadrangula* (O. F. Müller) 1785
General form like *reticulata*. Valves reticulated, often not plainly marked. Postabdomen narrowing toward apex, often, but not always, sinuate above anal spines,
which number 7 to 9. Claws large, denticulate.   ♂ antennules with long flagellum,
hooklike at tip. Color transparent to pinkish opaque. Length, ♀ to 1.0 mm; ♂ to
0.6 mm.
Common in all regions, found among weeds, also limnetic.

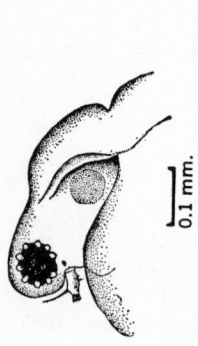
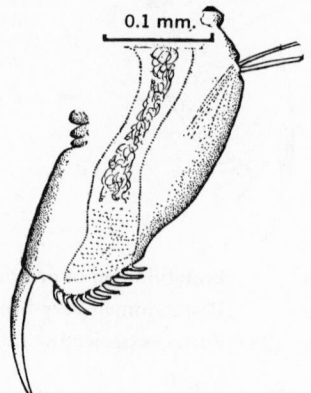

Fig. 27.41.   *Ceriodaphnia quadrangula.* Anterior end and postabdomen.   (By Birge.)

41a   (38) Vertex evenly rounded without spines.   Antennules of moderate
length. . . . . *C. laticaudata* P. E. Müller 1867

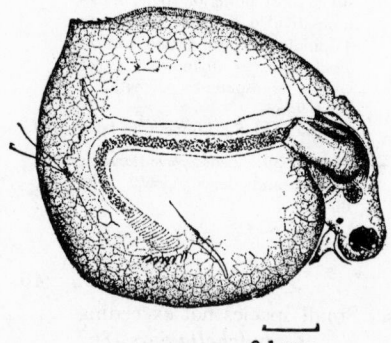

General form round. Valves ventricose below. Postabdomen large, dilated near middle,
obliquely truncated and bearing 8 to 11 spines
on lower margin. Claws long, denticulate.
Color transparent or opaque, through red and
red-brown to nearly black. Length, ♀ to 1.0
mm, but not seen larger than 0.7 mm in U. S.
♂ to 0.7 mm.
Reported from most of U. S. west of Rockies.

◄ Fig. 27.42.   *Ceriodaphnia laticaudata.*   (By
Birge.)

0.1 mm.

41b   Vertex angulate, with spines. Antennules long . . . .
C. rotunda Sars 1862

General form much like that of preceding species. Head angulate at
vertex, with spines. Antennules long and slender. Postabdomen somewhat enlarged, but not so much as in *laticaudata*, tapering toward apex,
obliquely truncate, with 7 to 9 anal spines. Color yellowish or brown,
not transparent. Length, ♀ to 1.0 mm; ♂ to 0.6 mm.
Rare, Wisc. Both this species and the preceding one live among weeds.

◄ Fig. 27.43.   *Ceriodaphnia rotunda.*   Anterior end.   (After Lilljeborg.)

42a   (34) Body compressed.   Valves elliptical, crested dorsally, completely

covering body. Ocellus present. Fornix and abdominal process
well developed. . . . . **Moinodaphnia** Herrick 1887
    Cervical sinus present; no cervical gland. Valves tumid in posterodorsal region;
crested; minute spines on ventral margin; sharp angle, but no spine, at junction of
dorsal and ventral margins; marked with oblique striae, usually invisible in preserved specimens. Antennules attached on ventral surface of head, sense-hair about
middle; olfactory setae small. One large abdominal process, broad, concave in
front, somewhat saddle-shaped, forming a transition to the condition in *Moina*.
Postabdomen as in *Moina*, with slender postanal projection bearing about 10 finely
ciliated spines and a much longer distal spine with 2 unequal prongs, the *bident* (Fig.
27.52). Claws denticulate. Summer eggs numerous. Male (South American) much
like *Moina*, with large curved antennules.
    Only one certain species . . . . **M. macleayii** (King) 1853

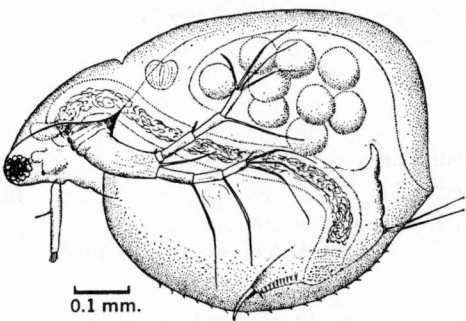

Color yellowish, transparent.
Length, ♀ *ca.* 1.0 mm.
  La. and southward. In weedy
pools and lakes.

◄ **Fig. 27.44.** *Moinodaphnia macleayii.*
(By Birge.)

0.1 mm.

42b   Body thick and heavy. Valves somewhat rhomboidal, not wholly
covering body. Fornix small. Ocellus absent. Abdominal process
represented by horseshoe-shaped fold. (Fig. 27.49) . . . .
                             Moina Baird 1850   **43**
    Cervical sinus present. Valves thin, obscurely reticulated or striated; no posterior
spine. Head large, thick, rounded in front; sometimes with deep depression above eye;
no rostrum. Antennules long, spindle-shaped, freely movable; lateral sense-hair about
middle. No regular abdominal projection, but in old ♀ a horseshoe-shaped ridge
which closes the brood cavity. Postabdomen extended into conical postanal part,
bearing ciliated spines and bident. Claws small; abdominal setae very long. Summer
eggs numerous; ephippium oval, with 1 or 2 eggs. Antennule of ♂ very long and stout,
modified into clasping organ; denticulate, with small recurved hooks at apex. First
leg with hook.
    The species of *Moina* ordinarily inhabit muddy pools and similar places. They are
soft-bodied, weak animals; likely to be much distorted by preserving fluids. The
species are much alike and often hard to distinguish unless ♂ and ephippial ♀ are
present.

43a **(42)** Fewer than 8 postanal spines. Animal small, about 0.5 mm
long . . . . **M. micrura** Kurz 1874

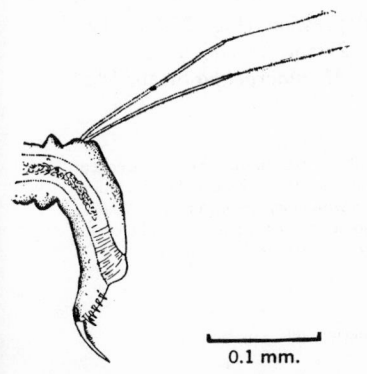

Small, transparent; head relatively very large;
deep cervical sinus; supraocular depression small or
absent. Terminal portion of postabdomen small
with 4 to 6 spines and a much longer bident.
Claws pectinate. Male unknown. Length, ♀ 0.5–
0.6 mm.
  Ill., Ark.

◄ **Fig. 27.45.** *Moina micrura.* Postabdomen. (By
Birge.)

0.1 mm.

43b More than 8 postanal spines. Animal larger, about 1.0 mm or more **44**

44a (43) Dorsal surface of head hairy. . . . . *M. irrasa* Brehm 1937

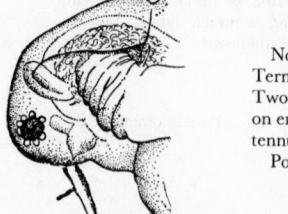

Group of prominent hairs on back of head. Antennule relatively small and short. Postabdomen with about 11 ciliated spines. Claw pectinate with about 10 teeth in pecten. Length, ♀ 0.9 mm. (There is possibly another North American species of *Moina* slightly larger than *irrasa* also with hair on its head. In the larger species, also from Carson Sink, Nev., the lateral sense-hair arises from the middle of the relatively large antennule, rather than from the proximal third as in *irrasa*, cf. Brehm, 1937.)

◄ **Fig. 27.46.** *Moina irrasa*, showing hair on head. (After Brehm.)

44b Dorsal surface of head without hairs **45**

45a (44) Distal anal spine large, forked. (Fig. 27.52). **46**

45b Distal anal spine not forked, and of variable size. . . . .
M. hutchinsoni Brehm 1937

Head depressed. Eye relatively small. Seven to 10 ciliated anal spines. Postabdominal claw without pecten (but with fine hairs at base of claw proximal to usual site of pecten) see Fig. 27.47. Distal anal spine not forked as is characteristic of rest of genus. Instead it is a single, nonciliated spine, variable in length, usually shorter than first ciliated spine. Length, ♀ 1.6 mm.

Alkaline lakes, western U. S. and Canada.

◄ **Fig. 27.47.** *Moina hutchinsoni,* end of postabdomen. (After Brehm.)

46a (45) Claws pectinate; supraocular depression present (Fig. 27.49); no flagellum on first leg of male. **47**

46b Claws not pectinate; no supraocular depression (Fig. 27.53); first leg of male with long flagellum. (Figs. 27.50c, 27.51c) . . . .
M. macrocopa Straus 1820

Not very transparent; yellowish or greenish. Head extended. Terminal part of postabdomen long, with 10 to 12 spines besides bident. Two ephippial eggs (Fig. 27.51). ♂ with elongated head; 5 to 6 hooks on end of antennule, sense-hairs somewhat proximal to middle of antennules (Fig. 27.50). Length, ♀ to 1.8 mm; ♂ 0.5–0.6 mm.

Pools; widely distributed.

◄ **Fig. 27.48.** *Moina macrocopa.* Anterior end. (By Birge.)

0.2 mm.

47a (46) Two ephippial eggs; antennule of male with sense-hair in middle . . . . **M. brachiata** (Jurine) 1820

Body stout, heavy, greenish, not transparent. Head ordinarily much depressed, so that vertex often lies almost on level of ventral margin of valves. Deep supraocular depression. Valves faintly reticulated. Postanal spines, 7 to 11 besides bident; claws pectinate. Antennules of ♀ with 4 hooks at tip; first leg without flagellum. Length, ♀ to 1.5 mm; ♂ unknown in U. S.

Widely distributed; in pools.

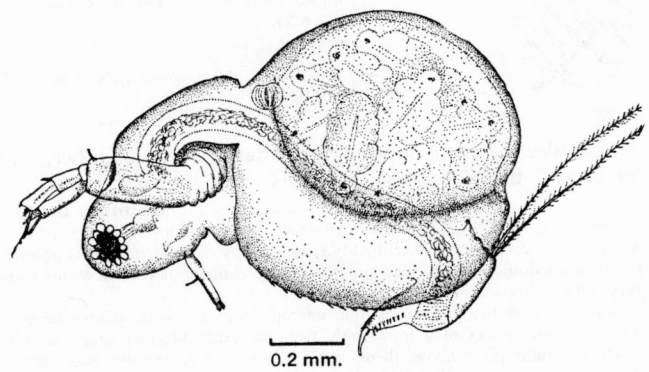

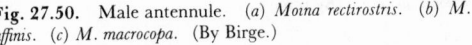

Fig. 27.49. *Moina brachiata.* (By Birge.)

47b One ephippial egg (Fig. 27.51*a,b*) **48**

48a (47) Valves smooth; ephippium reticulated around edges, smooth in middle (Fig. 27.51*a*); antennules of male with sense-hair near middle (Fig. 27.50*a*) . . . . **M. rectirostris** (Leydig) 1860

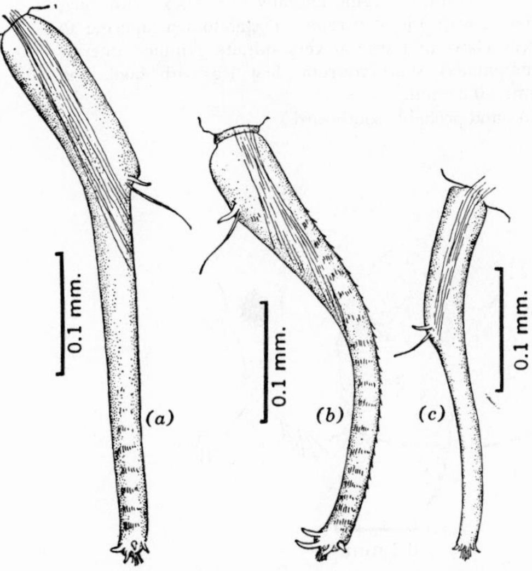

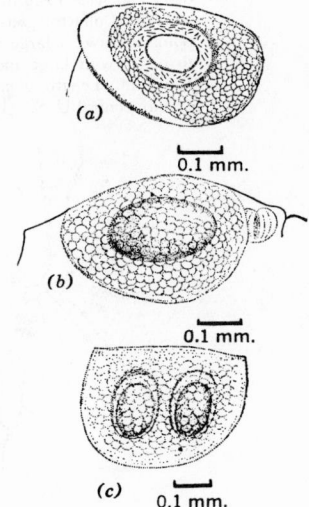

Fig. 27.50. Male antennule. (*a*) *Moina rectirostris.* (*b*) *M. affinis.* (*c*) *M. macrocopa.* (By Birge.)

Fig. 27.51. Ephippium. (*a*) *Moina rectirostris.* (*b*) *M. affinis.* (*c*) *M. macrocopa.* (*a, c* after Lilljeborg; *b* by Birge.)

Colorless, or with bluish cast. Head extended or little depressed; deep cervical and supraocular depressions. Postabdomen with long projection and 10 to 15 postanal spines and bident. Claws pectinate. Antennules of ♂ with 5 to 6 hooks at apex. Length, ♀ 1.0–2.0 mm; ♂ 0.4–0.6 to 1.0 mm.

Widely distributed in muddy pools.

48b   Valves striate; ephippium reticulated all over (Fig. 27.51*b*); antennules of male with sense-hair near base (Fig. 27.50*b*) . . . . .
                                                                 M. affinis Birge 1893

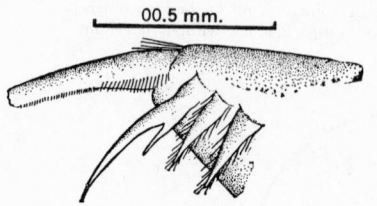

00.5 mm.

Much like *M. rectirostris*, from which the young ♀♀ are hardly distinguishable. Antennules of ♂ broad, fringed with fine hairs on inner margin; 4 to 6 hooks at end. Length, ♀ 0.8–1.0 mm; ♂ 0.3–0.6 mm.
Wis. to La.

◄ **Fig. 27.52.** *Moina affinis,* apex of postabdomen. (By Birge.)

49a **(14)** Antennules of female approximately parallel to each other, curving backward, fixed to head; olfactory setae on side, usually near base . . . . ***Bosmina*** Baird 1845   **50**

    The taxonomy of the North American representatives of this genus is very confused; many "races" are commonly named but the validity of these entities as geographical subspecies is dubious. The least unsatisfactory treatment at present seems to be the very conservative one followed here.

    Animal usually hyaline; valves thin; inferoposteal angle with spine—the *mucro* (pl. *mucrones*). Antennules of ♀ immovably fixed to head; olfactory setae on side with small triangular plate above them; distal portion of antennules looks segmented. Antenna with 3- and 4-jointed rami. Postabdomen somewhat quadrate; anus terminal; spines small and inconspicuous; claws set on a cylindrical process.

    ♂ smaller than ♀, with short, blunt rostrum; large free antennules; hook and long flagellum on first leg.

49b   Antennules united at base, and diverging at apex; numerous long olfactory setae on their ventral side. . . ***Bosminopsis*** Richard 1895

    Sole American species. . . . . ***B. deitersi*** Richard 1895

    In general much like *Bosmina* (see **50**). Basal part of antennules united with each other and head to form very long rostrum; diverging laterally near apex, with long, straggling, olfactory setae. Antenna with 3-jointed rami. Postabdomen tapering to point at claws, 1 large spine near claws and several very minute spinules anterior to it. ♂ with large movable antennules; short rostrum; first leg with hook and flagellum. Length, ♀ *ca.* 0.35 mm; ♂ 0.25 mm.

    South-central U. S. (La., Okla., and probably southward.)

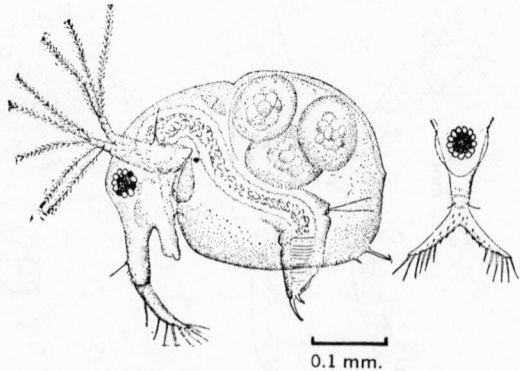

0.1 mm.

Fig. 27.53. *Bosminopsis deitersi.* (By Birge.)

50a **(49)** Proximal pecten of postabdominal claw with 3 (or 4) very large spines, and distal pecten with 2 to 6 small spines continuing distally into minute spinules . . . .
                                Bosmina longirostris (O. F. Müller) 1785

    The small sense-hair is usually nearer to the center of the space between the eye and the base of the antennule than to the base of the latter. The mucro is short. Eye usually large. Postabdomen with 2 pectens. The 3 or 4 teeth in the proximal set are very large, increasing in length distally. The most proximal 2 to 6 teeth of the distal pecten are enlarged and easily visible. Their size decreases distally, being minute near tip of claw.

Common in ponds and lakes throughout continent.

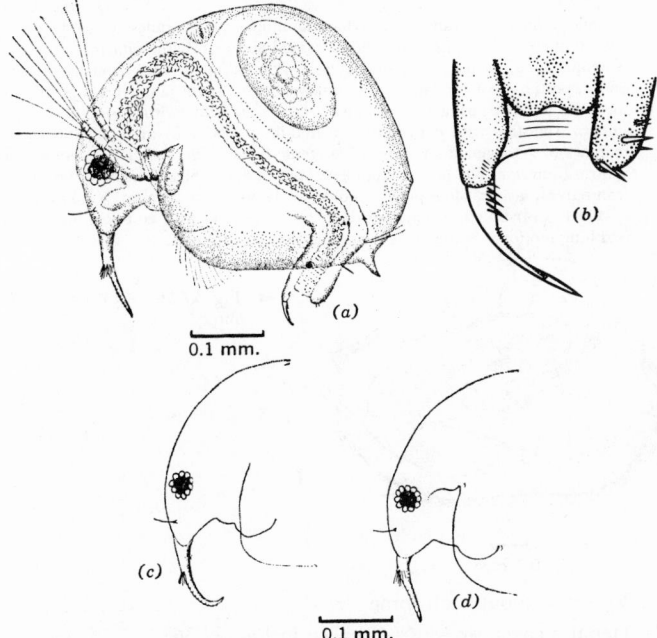

Fig. 27.54. *Bosmina longirostris.* (*a*) Typical specimen. (*b*) Postabdomen more highly magnified. Rostrum. (*c*) var. *cornuta.* (*d*) var. *brevicornis.* (*a, c, d* by Birge; *b* after Austin.)

50b   Proximal pecten of postabdominal claw with 5 to 6 large spines, and with numerous very fine spinules in distal pecten . . . .
                                          B. coregoni Baird 1857

The sense-hair is usually near the base of the antennule. Body form very variable; dorsal margin of carapace usually marked by a hump. Mucrones usually longer than in *longirostris*, but very variable, as are antennules. Postabdominal claw has 5 to 6 large spines which increase in length distally. The distal pecten composed of long, very fine spinules, often difficult to see. Generally 2 rows of 4 to 8 spinules on the body of the postabdomen morphologically ventral to anus. Common in ponds and lakes throughout continent.

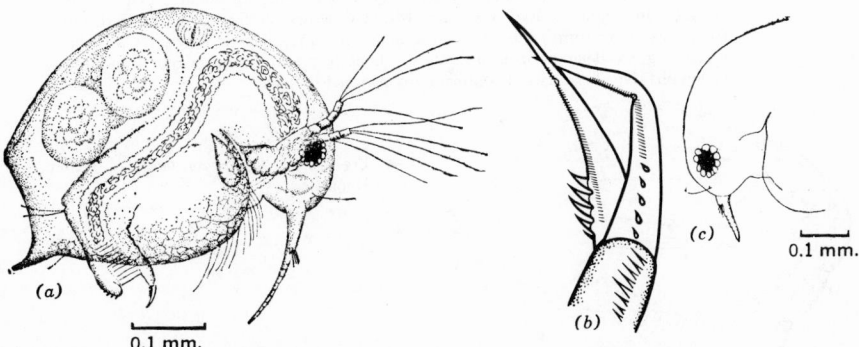

Fig. 27.55. *Bosmina coregoni.* (*a*) Lateral. (*b*) Tip of postabdomen. (*c*) Head of young. (*a, c* by Birge. *b* after Austin.)

51a   **(14)** Intestine convoluted. (Fig. 27.56) **52**

51b   Intestine simple. (Fig. 27.62) **57**

52a   **(51)** Valves with spine at superoposteal angle. Small hepatic caeca. . .

Ophryoxus Sars 1861

      Sole species. . . . . *O. gracilis* Sars 1861

      General form elongated, somewhat daphnid. Antennules long, slender, fringed with numerous hairs behind, lateral sense-hair near base; olfactory setae unequal. Antennae long, weak. Six pairs of legs. Postabdomen long, tapering at apex, anus dorsal, postanal portion large with numerous short, blunt ciliated spines, the proximal mere elevations bearing fine spinules. Claws straight, with (usually) 2 stout basal spines. Intestine with convolution in middle of body; 2 small hepatic caeca. Antennules of ♂ longer than those of ♀; sense-hairs longer. Vasa deferentia open on ventral (anterior) side of postabdomen about middle. Strong hook on first leg. Color transparent, last leg often purple in old ♀♀. Length, ♀ to 2.0 mm; ♂ 1.0 mm.

      Widely distributed, among weeds in lakes. Swims with constant but rather feeble paddling motion. Spine longer in young than in adult.

◄ **Fig. 27.56.** *Ophryoxus gracilis.* (By Birge.)

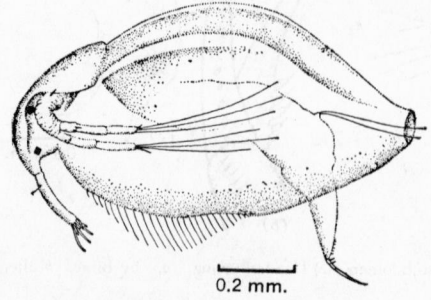

0.1 mm.

52b   Valves without such spine **53**

53a   **(52)** Hepatic caeca present (cf. caeca in Fig. 27.56). **54**

53b   Hepatic caeca absent, setae $\dfrac{0-0-0-3}{1-1-3}$ **56**

54a   **(53)** Antennary setae ♀, $\dfrac{0-0-0-3}{0-0-3}$; ♂, $\dfrac{0-0-0-3}{1-1-3}$; valves narrowed behind and prolonged into short tube . . . . ***Parophryoxus*** Doolittle 1909

      Sole species . . . . *P. tubulatus* Doolittle 1909

      Form elongated oval; narrow crest on head and valves. Head rounded, rostrum well marked; cervical sinus present. Valves thin, transparent; unmarked or faintly reticulated; prolonged behind into a sort of tube, best seen from above; ventral margin with moderate setae. Postabdomen elongated, triangular; postanal part long and slender, narrowed toward apex somewhat as in *Ophryoxus;* bearing a few very small spines. Claws long, rather straight; with 2 basal spines. Antennules cylindrical, slender; with basal sense-hair and 3 conspicuously long olfactory setae. Antennae long, slender; basal joint annulated; setae not conspicuously dissimilar. Legs, 6 pairs; the last rudimentary. Eye moderate, with few lenses; ocellus large, some distance from apex of rostrum. Intestine convoluted, with small hepatic caeca. ♂ with hook on first leg; vas deferens opens near claws. Length, ♀ to 1.2 mm. Color transparent-yellowish. Northern part of continent among weeds in lakes.

◄ **Fig. 27.57.** *Parophryoxus tubulatus.* (After Doolittle.)

0.2 mm.

54b Setae $\dfrac{0\text{--}0\text{--}1\text{--}3}{1\text{--}1\text{--}3}$; animal small, spherical. (Fig. 27.58) . . . .

Streblocerus Sars 1862   **55**

Body round-oval, not compressed or crested. Labrum with large, serrate, acute process. Antennules large, flat, bent, or rather twisted, broadened in distal part; with lateral sense-hair near base, several hairs on posterior face, rows of fine hairs, and subequal olfactory setae. Postabdomen bilobed (Fig. 27.59); the preanal part compressed, semicircular; the anal part rounded, with fine spines or hairs. Claws small, curved, with several equal minute denticles on concave edge. Five pairs of legs. Intestine convoluted, with small hepatic caeca. ♂ (European, of *S. serricaudatus*) small, triangular, much like ♀; first leg without hook.

55a **(54)** Dorsal margin of valves smooth . . *S. serricaudatus* (Fischer) 1849

Preanal part of postabdomen with serrate margin and bearing rows of fine hairs. Anterior margin of antennule somewhat toothed. Color whitish-opaque to yellowish. Length, ♀ *ca.* 0.5 mm; ♂ *ca.* 0.25 mm.

Rare but widely distributed over continent in weedy pools and margins of lakes.

0.1 mm.

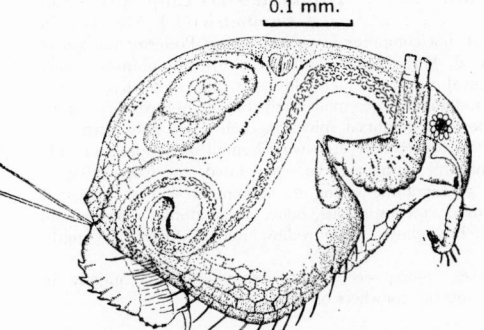

0.05 mm.

Fig. 27.58. *Streblocerus serricaudatus.* (By Birge.)   **Fig. 27.59.** *Streblocerus pygmaeus.* (By Birge.)

55b   Valves reticulate, the edges of the reticulations making scalelike ridges which give the dorsal margin a serrate appearance. . . . . . .

S. pygmaeus Sars 1901

Preanal part of postabdomen not serrate, with 4 to 5 rows of fine hairs. Color grayish-white, opaque, to nearly black in ephippial ♀. ♂ unknown.

Length, ♀ 0.2–0.25 mm. The smallest member of the family and one of the smallest of the group.

La., in weedy pools, with *S. serricaudatus*.

56a **(53)** Convolution of intestine in middle of body. Valves crested, with a strong tooth on crest . . . . *Drepanothrix* Sars 1861

Sole species . . . . *D. dentata* (Eurén) 1861

Valves reticulated; dorsal margin arched, crested, with conspicuous, short, backward-pointing tooth about middle. Antennules broad, flat, twisted, though not so much as in *Streblocerus;* postabdomen compressed but not extended into a thin edge; almost quadrate as seen from side. Margin with 2 rows of small spines, about 20, and with several rows of hairs besides scattered groups; apex truncate, emarginate, with anus in depression. Claws short, broad, crescentic, smooth, or denticulate; 5 pairs of legs. ♂ much like young ♀; hook on first leg; postabdomen without spines; vasa deferentia open in front of claws. Color whitish to yellowish; opaque or transparent. Length, ♀ *ca.* 0.7 mm; ♂ *ca.* 0.4 mm.

Not commonly collected though widely distributed and probably not very rare in shallow waters of lakes; on bottom or among weeds. Most of northern part of continent.

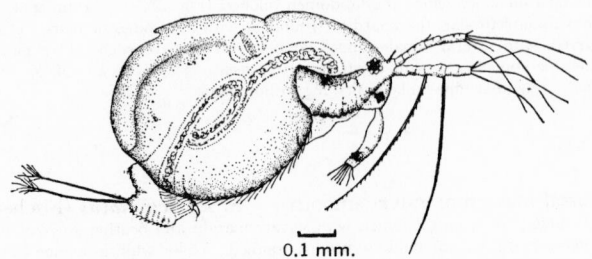

Fig. 27.60.   *Drepanothrix dentata.*   (By Birge.)

56b   Convolutions of intestine in hind part of body and in post-abdomen. No dorsal tooth . . . . *Acantholeberis* Lilljeborg 1853

Sole species . . . . *A. curvirostris* (O. F. Müller) 1776

Form in general angular-oval, not compressed, without crest. Posterior margin of valves rounded over into ventral, both fringed with long, close-set, plumose setae. Labrum with long, slender, conical process. Antennules large, flat, somewhat curved, expanded toward apex. Postabdomen large, moderately broad, not compressed or divided, hairy, with 20 or more small dorsal spines in each row; anus terminal. Claws short, stout, broad, curved, denticulate, and with 2 small basal spines set side by side. Six pairs of legs. Intestine without caeca, convoluted, the loops lying in great part in postabdomen. ♂ resembling young ♀; antennules with 2 proximal sense-hairs; first leg with small, inconspicuous hook, postabdomen emarginate dorsally; vasa deferentia open behind claws. Color yellow, not transparent. Length, ♀ to 1.8 mm; ♂ 0.5–0.7 mm.

In pools and margins of lakes among weeds; reported especially frequently in *Sphagnum* bogs. Most of U. S., possibly elsewhere on continent.

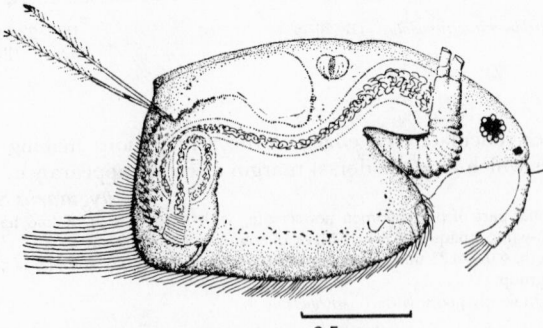

Fig. 27.61.   *Acantholeberis curvirostris.*   (By Birge.)

57a   **(51)** Hepatic caeca present; postabdomen bilobed (Fig. 27.62); anten-nary setae $\dfrac{0-0-1-3}{1-1-3}$ **58**

57b   No hepatic caeca; postabdomen various **59**

58a   **(57)** Postabdomen very large, with few spines . . . .
                                                          Grimaldina Richard 1892

Sole species . . . . *G. brazzai* 1892

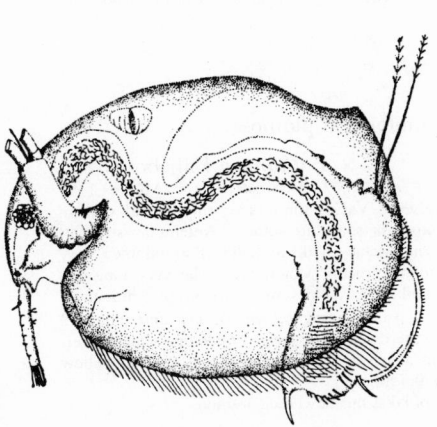

Body compressed, somewhat quadrangular with all margins of valves slightly convex. Postabdomen enormous, much compressed, roughly semielliptical in form; the preanal portion divided by a notch into 2 parts, of which the anterior is the smaller; a long spine in the notch which marks junction of anal and preanal parts; on anal part 2 lateral rows of small, slender spines, about 7 in anterior, and 5 in posterior row. Claws small, denticulate, with 1 small basal spine. Ephippium rounded-quadrangular; egg-chambers reniform with concave sides toward each other. ♂ (South American) small, like immature ♀; antennules with 2 basal sense-hairs; small hook on first leg. Color reddish-brown. Length, ♀ to 0.9 mm; ♂ 0.5 mm.

La. and southward. Weedy pools of clear water.

◄ **Fig. 27.62.** *Grimaldina brazzai.* (By Birge.)

58b Postabdomen of moderate size; numerous spines on preanal part, clusters of hairs on anal part . . . . **Wlassicsia** Daday 1903
Sole American species . . . . **W. kinistinensis** Birge 1910
Form oval, not compressed. Valves crested; with spines on ventral margin; marked by very delicate transverse striae which anastomose, forming fine vertical meshes. Olfactory setae subequal. Two rounded projections at base of labrum on ventral surface of head. Labrum with strong conical projection pointing backward and a second projection just in front of small terminal lobe. Postabdomen with fine spines and hairs. Abdominal setae very long, not set on projection. Claws with very small basal spines. Five pairs of legs; branchial sacs on all legs. ♂ with large antennule; small keel on labrum; hook on first leg. Color yellow. Length, ♀ 0.8 mm; ♂ 0.4 mm. Northern U. S. and Southern Canada.

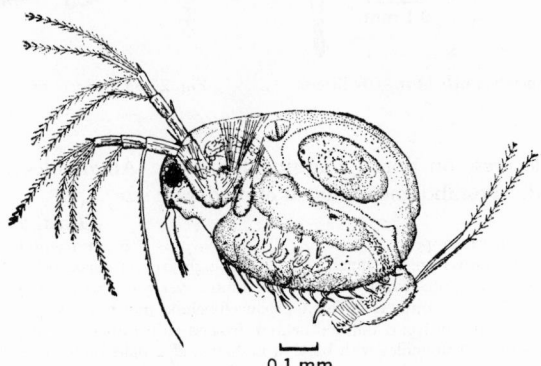

0.1 mm.

Fig. 27.63. *Wlassicsia kinistinensis.* (By Birge.)

59a **(57)** Antennary setae $\dfrac{0-0-0-3}{1-1-3}$ 60

59b Antennary setae $\dfrac{0-0-1-3}{1-1-3}$; basal seta of 3-jointed ramus stout and stiff (Fig. 27.69). . . . . **Macrothrix** Baird 1843 63
Shape oval or rotund, somewhat compressed, with dorsal crest. Head large, ordinarily not depressed; vertex evenly or abruptly rounded; rostrum short. Ventral margin of valves ordinarily with long, stout, movable spines, which project in several

directions. Antennules large; lateral sense-hair near base. Antennae large; the proximal seta of 3-jointed ramus long, stiff, spinous; the others sparsely plumose or partly spinous. Five pairs of legs. No abdominal process. Postabdomen large; often bilobed. Claws small. Intestine simple, no caeca. ♂ with large antennules; hook on first leg.

59c   Antennary setae $\dfrac{0\text{-}1\text{-}1\text{-}3}{1\text{-}1\text{-}3}$; all similar and plumose. . . . .

Lathonura Lilljeborg 1853

Sole species (Fig. 27.64) . . . . **L. rectirostris** (O. F. Müller) 1785

General form long-oval, not compressed. Valves unmarked; the ventral margin with short, close-set, smooth, lancet-shaped or spatulate spines. Antennules straight, with sense-hair near base; 2 pairs of sense setae in distal half. Postabdomen very small, extended behind into a long conical process, which bears the very long abdominal setae; covered with fine spines and setae. Claws small; smooth or denticulate. Summer eggs, 2 to 10; 1 ephippial egg. ♂ like young ♀, with larger antennules; 2 lateral sense-hairs, the additional one—the distal—the larger; olfactory setae longer. First leg with hook. Vas deferens opens at claws. Color transparent to clear yellow or greenish. Length, ♀ to 1.0 mm; ♂ ca. 0.5 mm.

Widely distributed in weedy margins of lakes but nowhere common.

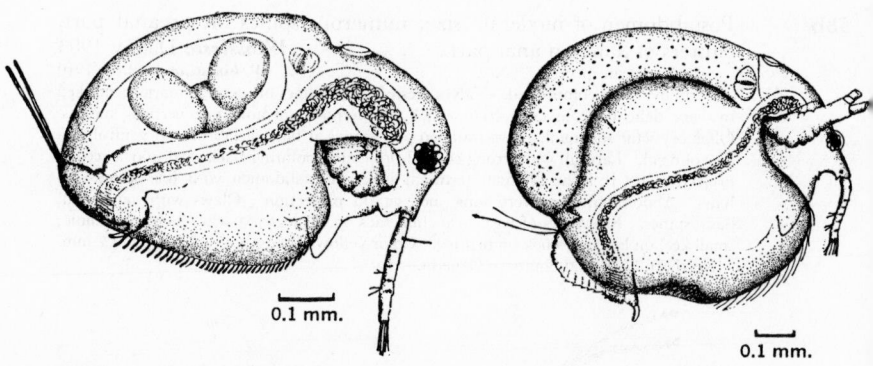

0.1 mm.

0.1 mm.

Fig. 27.64. *Lathonura rectirostris.* (By Birge.)   **Fig. 27.65.** *Bunops serricaudata.* (By Birge.)

60a **(59)** Wide crest on dorsal margin of valves. Antennules at apex of head. Postabdomen bilobed of moderate size . . . .

Bunops Birge 1893

Sole American species . . . . **B. serricaudata** (Daday) 1888

General form rounded, much compressed; high keel on dorsal side. Front of head flat, somewhat kite-shaped, with boss or umbo over eye. Strong triangular keel on labrum. Valves faintly reticulated, produced behind into rounded projection; ventral margin gaping in front, inflexed behind, fringed with rather long straggling hairs or weak setae. Antennules with basal sense-hair and 2 pairs of sense setae near apex; olfactory setae somewhat unequal. Postabdomen much like *Streblocerus:* bilobed, preanal portion flattened, semicircular, with 7 to 8 notches or teeth on the dorsal margin and 3 to 4 rows of fine hairs; anal portion with fine hairs and 3 to 4 spines. Color transparent, tinged with yellow. ♂ unknown. Length of ♀ to 1.0 mm.

Very local in distribution, but not rare when present; northern U. S.

60b   Dorsal crest on valves absent or small. Vertex of head forming a sharp angle in front of insertion of antennules. Postabdomen very large, with numerous long spines. (Fig. 27.68) . . . .

Ilyocryptus Sars 1861   **61**

General form oval-triangular, the head forming the apex of the triangle, while the enormously dilated ventral and posterior edges of the valves round into each other; these have long, close-set, fixed setae, usually branched and fringed. Antennules long, freely movable, 2-jointed, basal joint very small, attached to ventral side of head behind vertex; olfactory setae unequal. Antennae short, powerful; basal joint annulated nearly to apex; with long sense setae; motor setae not plumose, either smooth or with sparse hairs. Abdominal process long, tongue-shaped, hairy. Postabdomen large, broad, compressed; anus on side or near apex; many spines on dorsal margin; numerous, long, curved, lateral spines and setae; fine spinules near base of claws. Claws long, straight, denticulate, and with 2 slender basal spines. Intestine simple, no caeca, but enlarged near rectum. Six pairs of legs. ♂ with larger antennules than ♀, bearing 2 sense-hairs; no hook on first leg.

In most species the old shells are not cast off in molting but overlie the youngest in several layers. The species live in mud, creep about among weeds, though they can and do swim; are often greatly loaded with mud and vegetable growths, nearly concealing structure.

61a **(60)** Anus opening on dorsal margin of postabdomen (Fig. 27.67); molting imperfect **62**

61b   Anus at end of postabdomen; molting complete . . . .
                                                        I. acutifrons Sars 1862

Postabdomen not emarginate; about 8 small spines near claws, shortest next to claw; about 6 long, curved, lateral spines, about 8 marginal spines corresponding to preanals of other species; the proximal 2 directed forward; from distal spine of this set a series of very small marginals to anus. Antennule club-shaped, hairy. Ocellus nearer eye than insertion of antennules. Claws as in *I. sordidus*. Three to 4 summer eggs. ♂ unknown. Color reddish or yellowish. Length, ♀ *ca.* 0.7 mm.
Widely distributed.

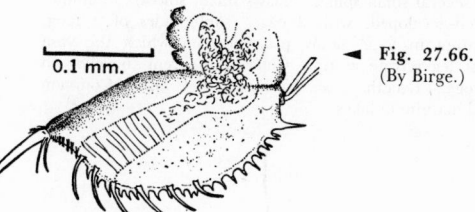

0.1 mm.

◄ **Fig. 27.66.** *Ilyocryptus acutifrons.* Postabdomen. (By Birge.)

62a **(61)** Eight or more preanal spines; antennary setae short . . . .
                                                        I. sordidus (Liéven) 1848

Postabdomen emarginate where anus opens; 8 to 14 preanal marginal spines; lateral postanal spines about 8 to 10; marginal row of numerous smaller spines. Ocellus nearer base of antennule than eye. Six to 8 summer eggs.
Color red, but often so loaded with debris as to be opaque. Length, ♀ *ca.* 1.0 mm; ♂ 0.42 mm.
Not very common but widely distributed in weeds on muddy bottoms.

0.1 mm.

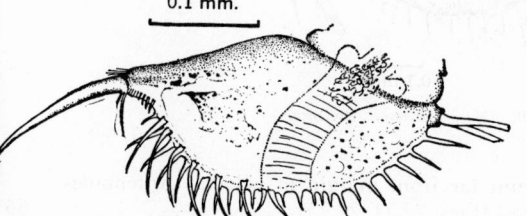

◄ **Fig. 27.67.** *Ilyocryptus sordidus.* Postabdomen. (By Birge.)

62b   Five to 7 preanal spines; antennary setae ordinarily very long . . .
I. spinifer Herrick 1884

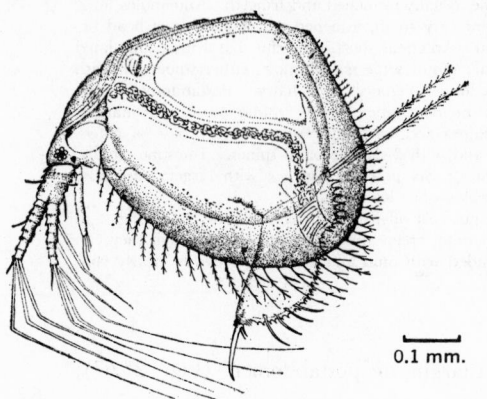

Anus opens in depression on dorsal margin of postabdomen; 5 to 7 preanal spines; 4 to 8 postanal lateral spines in outer row. Antennary setae usually long, sometimes equaling length of valves; although in some specimens they are short, apparently because of wear. Eight to 10 summer eggs; true ephippium formed and cast off (Sars). ♂ unknown. Color yellow or reddish. Length, ♀ to 0.8 mm.

Not uncommon, throughout U. S.

0.1 mm.

◄ **Fig. 27.68.** *Ilyocryptus spinifer.* (By Birge.)

63a **(59)** Dorsal margin of head evenly rounded (Fig. 27.71) **64**

63b   Dorsal margin of head curved abruptly in front of eye. Antennules slender . . . . **Macrothrix rosea** (Jurine) 1820
Form broadly ovate. Valves reticulated, crested, not serrate. Head large; its dorsal margin rounding over abruptly into anterior margin. Antennules long, slender, not enlarged near apex; lateral sense-hair near base on small elevation; olfactory setae unequal. Postabdomen extended into blunt process, on which abdominal setae are borne, preanal part semielliptical, with numerous spinules along convex edge and many fine hairs; anal part with several small spines. Claws small, smooth. Summer eggs numerous; ephippium well-developed, with 2 eggs. Antennules of ♂ long, curved. Postabdomen terminating in long, fleshy projection on which the vasa deferentia open. Hook of first leg serrate at tip. Color transparent to yellowish or sometimes with a ruddy tinge. Length, ♀ *ca.* 0.7 mm; ♂ 0.4 mm. Common everywhere in marshy pools and margins of lakes.

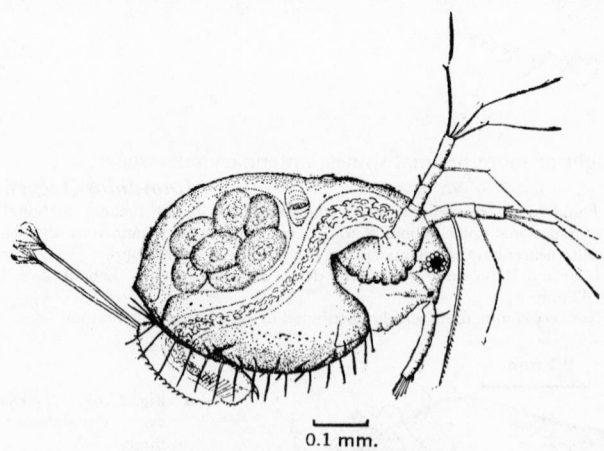

0.1 mm.

Fig. 27.69. *Macrothrix rosea.* (By Birge.)

64a **(63)** Head extended; rostrum far from margin of valves. Antennules enlarged near distal end (Figs. 27.71–27.73) **65**

64b   Head much depressed; rostrum close to margin of valves.   Antennules slender; not enlarged near distal end . . . .
M. borysthenica Matile 1890

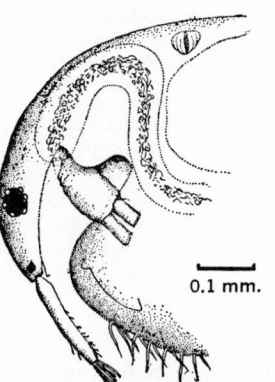

Dorsal margin of head evenly rounded over into that of valves, without sinus. Front of head recurved so that rostrum is very close to valves. Antennules with a few scattered fine hairs; olfactory setae small, equal. Postabdomen elongated, bilobed; with numerous fine spinnules and hairs on both lobes. Claws small. Eye moderate; ocellus at rostrum. Color transparent. Length, ♀ to 1.1 mm.

Southwestern U. S.

0.1 mm.   ◄ **Fig. 27.70.** *Macrothrix borysthenica.* Anterior end. (After Matile.)

65a (**64**) Postabdomen bilobed (cf. Fig. 27.69). **66**

65b   Postabdomen not bilobed . . . . *M. laticornis* (Jurine) 1820
Form round-ovate. Valves crested, the dorsal edge serrate with fine teeth. Head evenly rounded. Labrum with large triangular process. Antennule broader distally; a setiferous projection on posterior margin near apex; anterior margin with several fine incisions and clusters or rows of hairs; olfactory setae conspicuously unequal. Postabdomen with numerous fine spines and hairs; anus terminal. Claws small. Color grayish-white or yellowish. Length, ♀ 0.5–0.7 mm; ♂ 0.3–0.4 mm.
Widely distributed; found in all parts of the country but nowhere very abundant.

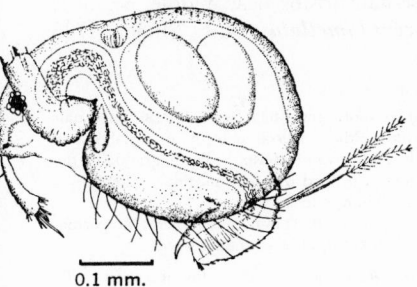

◄ **Fig. 27.71.** *Macrothrix laticornis.* (By Birge.)

0.1 mm.

66a (**65**) Conspicuous fold or folds of shell of head at cervical sinus. . . . . .
M. montana Birge 1904

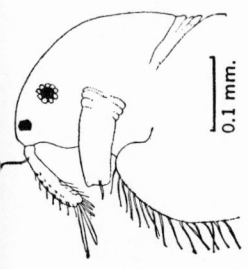

Form ovoid. Head large; dorsal margin evenly rounded; the shell extended into collarlike folds in front of cervical sinus. Antennules stout, large, enlarged near apex, about 6 anterior cross-rows of hairs and 3 to 4 stouter posterior setae; olfactory setae unequal. Postabdomen bilobed. Claws hardly larger than spines. Color transparent in preserved specimens. ♂ unknown. Length, ♀ *ca.* 0.55 mm.

Rocky Mountains.

◄ **Fig. 27.72.** *Macrothrix montana.* Anterior end. (By Birge.)

66b   No such folds . . . . . . *M. hirsuticornis* Norman and Brady 1867

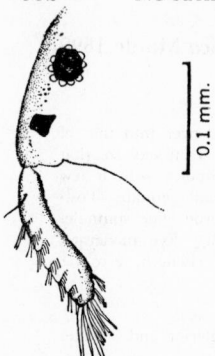

Form broadly ovate, not very different from *M. laticornis*. Antennules broad, flat, bent, varying in form but always enlarged distally; with 6 to 8 rows of stiff hairs on anterior side; sometimes stout setae on posterior side; olfactory setae unequal. Postabdomen large, broad, bilobed; preanal part not flattened and without projection for abdominal setae; numerous small spines and hairs on both anal and preanal parts. ♂ unknown. Length, ♀ 0.55 mm.

Widely distributed.

◄ **Fig. 27.73.** *Macrothrix hirsuticornis.* Anterior end. (By Birge.)

67a **(13)** Anus terminal (Fig. 27.74). Two hepatic caeca. Summer and ephippial eggs numerous . . . . . Subfamily **Eurycercinae** Kurz   **68**

Sole genus . . . . *Eurycercus* Baird 1843

Body stout and heavy, a broad oval in lateral view. Ventral margin of valves more or less concave. Antennae short and powerful. Postabdomen very large, flattened, general form quadrangular; anus terminal, in depression; dorsal margin with numerous (80 to 120), sawlike teeth. Claws on spiniferous projection with 2 basal spines; denticulate. Six pairs of legs. Intestine with hepatic caeca and convolutions. ♂ like young ♀; hook on first leg; vas deferens opens at base of claw on ventral (anterior) side.

67b Anus on dorsal side of postabdomen, postanal portion of which bears denticles (Fig. 27.79). No hepatic caeca. Two summer eggs; 1 ephippial egg. ♂ with strong hook on first leg . . . .

Subfamily **Chydorinae**   **69**

68a **(67)** Antennule short and thick; sense-hair arising near middle. . . . . . *Eurycercus lamellatus* (O. F. Müller) 1785

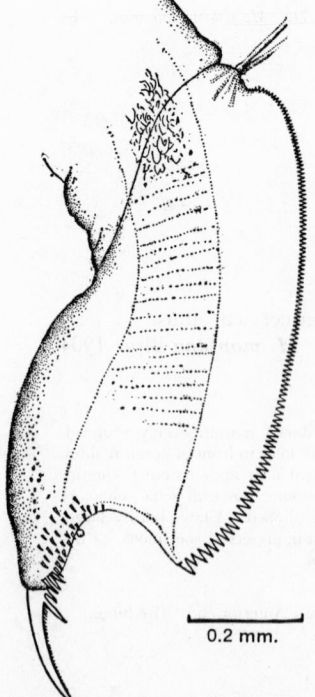

Antennules short and thick with lateral sense-hair arising near middle. Dorsal margin of postabdomen with nearly 100 or more teeth. Teeth gradually decrease in length toward base of anal setae. Length, ♀ to 3.0 mm or more; ♂ to 1.4 mm.

Absent in far north, common elsewhere; in permanent pools or margins of lakes among weeds.

◄ **Fig. 27.74.** *Eurycercus lamellatus.* Postabdomen. (By Birge.)

0.2 mm.

68b Antennule elongate; lateral sense-hair arising near apex. (Not figured) . . . . *E. glacialis* Lilljeborg 1887

Dorsal margin of postabdomen usually with 80 to 90 teeth of nearly same length throughout the series. Lateral sense-hair of antennule less stout than in *lamellatus*, and arises near apex. One of the largest Cladocera, ♀♀ may attain 6 mm, often 4–5 mm; ♂♂ up to 2 mm.

Northern Alaska and northern Canada.

69a (67) Compound eye present 70

69b Compound eye absent; ocellus only . . . . . *Monospilus* Sars 1861

Sole species . . . . *M. dispar* Sars 1861

Form oval or round. Shell not cast in molting, as in *Ilyocryptus*. Valves nearly round with fine setae along ventral edge. Head very small, depressed, movable. Keel of labrum with about 4 scallops on ventral edge. Postabdomen broad, short, with about 5 to 7 marginal denticles and numerous clusters of fine hairs. Eye lacking; ocellus large. Antennules short, not reaching apex of rostrum. ♂ with hook on first leg; postabdomen tapering, triangular, somewhat resembling that of *Graptoleberis*. Color brown-yellow. Length, ♀ *ca.* 0.5 mm; ♂ *ca.* 0.4 mm.

Northern U. S. and Canada; rare.

◄ **Fig. 27.75.** *Monospilus dispar.* (By Birge.)

70a (69) Compound eye and ocellus of ordinary size; antennules do not project beyond rostrum, though olfactory setae may (see, for example, Figs. 27.78, 27.82) 71

70b Compound eye and ocellus very large; antennules project far beyond rostrum (Fig. 27.76) . . . . *Dadaya* Sars 1901

Sole species . . . . *D. macrops* (Daday) 1898

Form rounded-oval; not compressed. Head small, much depressed; tumid above eye; rostrum short and broad. Antennules long, moderately stout, projecting far beyond rostrum. Postabdomen of moderate size, compressed, somewhat broadened behind anus, slightly narrowing toward apex; angle rounded; about 14 to 18 marginal denticles. Claws small, one small basal spine. Eye very large, with few lenses; ocellus nearly as large, crowded down into rostrum. ♂ unknown. Color dark brown. Length, ♀ *ca.* 0.3 mm.

Southern U. S.

◄ **Fig. 27.76.** *Dadaya macrops.* (By Birge.)

71a (70) Posterior margin of valves not greatly less than maximum height (see, for example, Figs. 27.78, 27.82) 72

No species of *Pleuroxus* belong in this section, although some individuals of *P. striatus*, **108b**, and *P. hamulatus*, **110b**, may seem to.

71b Posterior margin of valves considerably less than maximum height (see, for example, Figs. 27.100, 27.111). 96

All species of *Pleuroxus* belong here; also *Alonella excisa* and *exigua*.

72a (71) Body compressed; claws with secondary tooth in middle (Fig. 27.79). 73

72b   Body not greatly compressed; claws with one basal spine (Fig.
27.89), or rarely none (Fig. 27.86).
(For all species with 2 spines on terminal claws, see **96** ff.) **79**

73a   **(72)**   Antennal formula $\dfrac{0-1-3}{0-0-3}$ . . . . *Camptocercus* Baird 1843   **74**

Form oval; greatly compressed; often with crest on head and back.   Valves with
angles rounded, ventral margin concave in middle; small teeth at inferoposteal
angle; longitudinally striated.   Postabdomen very long, slender, with numerous
marginal denticles and lateral squamae.   Claws long, straight, with 1 basal spine;
a series of small denticles terminating in a larger one about the middle of claw;
extremely fine teeth thence to apex.   Five pairs of legs.

73b   Antennal formula $\dfrac{1-1-3}{0-0-3}$ **75**

74a   **(73)**   Postabdomen with 45 to 65 minute marginal denticles . . . .
C. oklahomensis Mackin 1930
Head without keel, eye relatively large, situated less than its own diameter from
the margin of the head.   Length, ♀ to 1.0 mm; ♂ smaller, with strong hook on first leg.
Okla., Kan. in shallow, temporary pools.

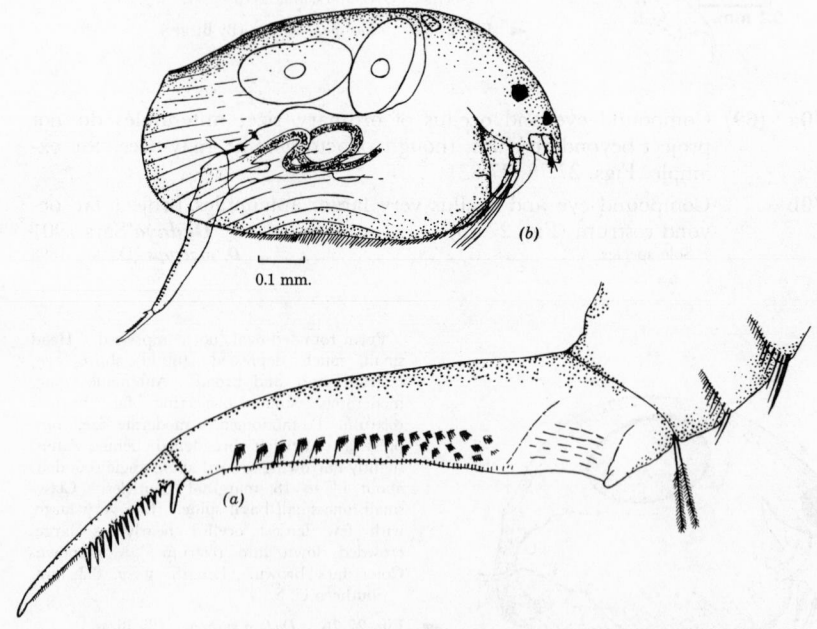

Fig. 27.77.   *Camptocercus oklahomensis.*   (*a*) Female.   (*b*) Postabdomen.   (After Mackin.)

74b   Postabdomen with 20 to 30 marginal denticles.   (Not figured) . . .
C. macrurus (O. F. Müller) 1785
Head and valves with crest.   Much like *C. rectirostris.*
Rare, but reported from most regions in the U. S.

74c   Postabdomen with 15 to 17 marginal denticles. . . . .
C. rectirostris Schødler 1862
Head extended or depressed.   Crest (keel) on head and valves.   ♂ without denticles
on postabdomen.   Color yellow-transparent.   Length, ♀ to 1.0 mm.
Common everywhere among weeds in margins of lakes, etc.

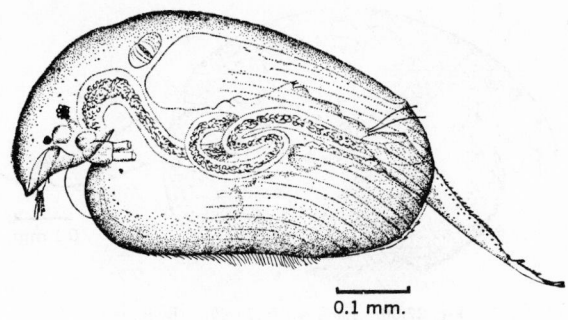

Fig. 27.78. *Camptocercus rectirostris.* (By Birge.)

75a *(73)* Six pairs of legs, the last small. . . . . *Alonopsis* Sars 1862   **76**
General form resembling *Acroperus* but less compressed and without crest. Keel of labrum moderate or small, almost triangular. Valves obliquely striated but striae often inconspicuous. Postabdomen long, broad; with well-developed marginal denticles. Six pairs of legs, the last very small. ♂ with usual characteristics. Color yellow.

75b Five pairs of legs. **77**

76a *(75)* Postabdomen with 15 to 17 marginal denticles. . . . .
A. elongata Sars 1861

Minute tooth at inferoposteal angle of valves. Postabdomen with lateral fascicles. Length, ♀ *ca.* 0.8 mm. Rare.

◄ **Fig. 27.79.** Postabdomen. (*a*) *Alonopsis elongata.* (*b*) *Alonpsis aureola.* (After Doolittle.)

a)   *(b)*

0.1 mm.   0.1 mm.

76b Postabdomen with 11 marginal denticles. . . . .
A. aureola Doolittle 1912
No lateral fascicles or inferoposteal tooth. Length, ♀ *ca.* 1.9 mm; ♂ unknown. Both species in margins of lakes and ponds among weeds. Rare; reported only from Me.

77a *(75)* Valves flattened laterally; vas deferens of male debouching at the tip of the postabdomen near the base of the claw **78**

77b Valves gaping anteriorly; vas deferens of male debouching near middle of the anterior (ventral) margin of the postabdomen . . . .
Euryalona Sars 1901
Sole American species . . . . *E. occidentalis* Sars 1901
General form resembling *Kurzia* (**78b**), but less compressed; no crest. Valves gaping in front, tumid in inferoanterior region; marked obscurely with concentric lines; dorsal margin arched. Keel of labrum angled behind but not prolonged. Postabdomen very long, slender, lobed at apex; with about 20 marginal and very fine lateral denticles. Claws straight, armed about as in *Camptocercus.* Five pairs of legs; hook on first leg of ♀. ♂ with strong hook; vas deferens opens on upper (ventral) side of postabdomen about middle. Color dark brown-yellow. Length, ♀ to 1.0 mm; ♂ 0.7 mm.
Southern U. S. and southward; not uncommon in weedy pools and lakes.

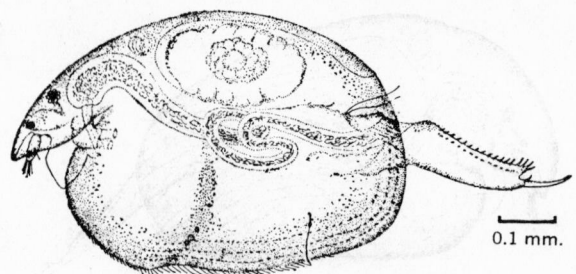

0.1 mm.

Fig. 27.80. *Euryalona occidentalis.* (By Birge.)

78a **(77)** Marginal denticles of postabdomen absent . **Acroperus** Baird 1843
  Body thin, compressed; crest on head and back. Valves subquadrate, obliquely striated; inferoposteal angle rounded or acute, usually with teeth. Postabdomen large, compressed; without marginal denticles but with lateral row of squamae. Claws long, straight, with 1 basal spine and secondary denticles, much as in *Camptocercus*. Intestine with large intestinal cecum. Eye larger than ocellus. Color yellow-transparent. Considerable variation in relative size of crest.
  Sole American species . . . . *A. harpae* Baird 1843

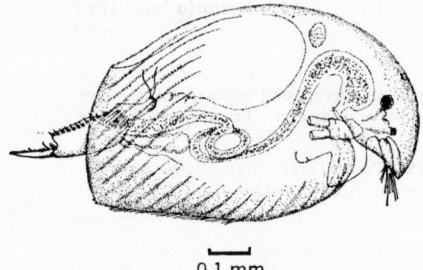

Rostrum acute. Eleven to 12 groups of fine spinules on postabdomen. Length, ♀ to 0.9 mm; ♂ to 0.6 mm.
Common everywhere, among weeds, in relatively open water; not in muddy pools. (Forms with a well-developed crest on head are here considered forms of *A. harpae*, rather than a separate species, *A. angustatus* Sars 1863.)

0.1 mm.

◄ **Fig. 27.81.** *Acroperus harpae.* (By Birge.)

78b   Marginal denticles present . . . .
                  Kurzia Dybowski and Grochowski 1894
  This genus is *Alonopsis* (part) of older authors; *Pseudalona* Sars.
  Sole American species . . . . *K. latissima* (Kurz) 1874

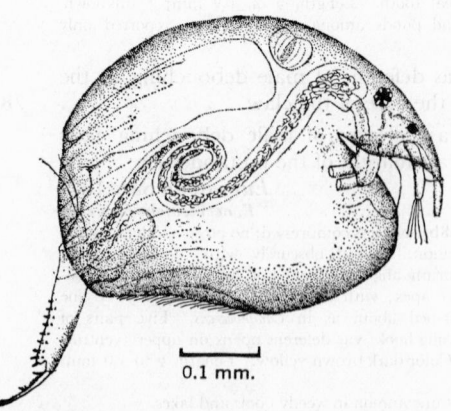

General form subquadrate; greatly compressed; but with only slight crest on back, none on head. Head small, the rostrum reaching not much below middle of valves, though longer than antennules. Postabdomen long, slender; lower angle usually produced into a lobe; 10 to 12 marginal denticles. Claws of *Camptocercus* type. ♂ like ♀; rostrum shorter; postabdomen with small denticles; vas deferens opens on ventral (upper) side; strong hook on first leg. Color yellowish, transparent. Length, ♀ 0.6 mm; ♂ 0.4 mm.
Found in all regions of continent among weeds in pools or lakes.

0.1 mm.

◄ **Fig. 27.82.** *Kurzia latissima.* (By Birge.)

79a	(72)	Rostrum not greatly exceeding antennules (Fig. 27.90)	80
79b		Rostrum considerably exceeding antennules (Fig. 27.121).	95
80a	(79)	Rostrum pointed	81
80b		Rostrum broad, semicircular (Fig. 27.83)	

Graptoleberis Sars 1863

Sole species . . . . *G. testudinaria* (Fischer) 1848
Posterior margin with 2 strong teeth at inferoposteal angle; valves and head with conspicuous reticulation. Head large; fornix very broad, forming a semicircular rostrum, covering antennules and extending down as far as ventral margin of valves. Postabdomen bent at the sharp preanal angle, tapered toward claws, so that form is nearly triangular; marginal spines small; lateral fascicles minute, sometimes wanting. Claws small, with 1 minute basal spine, sometimes wanting. ♂ with long, slender postabdomen, without spines; vas deferens open on ventral side; claws very minute; hook of leg slender.

Color gray to yellow-white; sometimes opaque. Length, ♀ 0.5–0.7 mm; ♂ 0.5 mm. or less.

Common among weeds or on bottom of pools and margin of lakes.

◄ **Fig. 27.83.** *Graptoleberis testudinaria.* (By Birge.)

0.1 mm.

| 81a | (80) | Postabdomen with numerous clusters of large spines (Fig. 27.86) . . | |

Leydigia Kurz 1874   86

General shape oval, much compressed but not crested. Head small, extended; keel of labrum rhomboidal with angles blunt or rounded. Postabdomen very large, compressed, semielliptical in form; postanal part much expanded, with numerous clusters of spines; spines in distal clusters very long. Claws long and slender. Eye smaller than ocellus. ♂ with blunt rostrum; process on upper (ventral) side of postabdomen, on which vas deferens opens; postabdomen with spines. Color yellow.

81b		Postabdomen without numerous clusters of large spines	82
82a	(81)	Postabdomen with marginal and lateral denticles (Fig. 27.89 *a–c*,	
		e–g)	83
82b		Postabdomen with marginal denticles only (Figs. 27.85*d*, 27.118) . . .	85
83a	(82)	Postabdomen relatively long and narrow; marginal denticles numerous, longer distally. Basal spine stout and long.	

Oxyurella Dybowski and Grochowski 1894   84

In general like *Alona*. Postabdomen long, slender; with marginal and lateral denticles, the former numerous and ending in a group of large denticles at angle of postabdomen. Terminal claw straight, with one large basal spine, attached some way distal to base of claw. Color yellow or yellow-brown.

This genus is the same as *Odontalona* Birge.

| 83b | | Postabdomen not noticeably narrow; distal denticles not conspicuously larger. Basal spine small. . . . . *Alona* Baird 1850 | 87 |

Most species will key here. General form subquadrate; compressed, not crested. Valve with superoposteal angle rounded or wellmarked; inferoposteal angle rounded, with or without small teeth. Fornices broad; rostrum short and blunt, little exceeding the apex of the antennules. Antennules short, thick; olfactory setae equal. Keel of labrum large, ordinarily rounded; the posterior angle not acuminate. Legs, 5 pairs, rarely 6; the sixth, if present, rudimentary. Postabdomen broad, compressed, with various armature. Claws with 1 basal spine; denticulate. Color some shade of yellow, varying from light to dark, with shade of brown in large species. All species littoral.

84a   **(83)** With 12 to 15 marginal denticles . . . .
                                   Oxyurella tenuicaudis (Sars) 1862

Marginal denticles very small near anus; the distal 4 to 5 much longer; the penultimate the largest. Length, ♀ *ca.* 0.5 mm; ♂ 0.4 mm.
  Widely distributed throughout U. S. but not abundant anywhere.

◄ **Fig. 27.84.** *Oxyurella tenuicaudis.* Apex of postabdomen. (See also Fig. 27.88*b*.) (By Birge.)

84b   With about 16 marginal denticles . . . *O. longicauda* (Birge) 1910

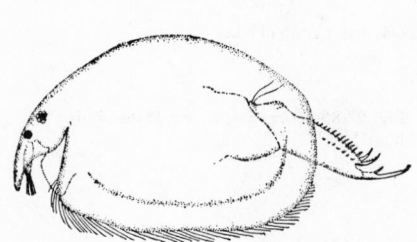

Between *Alona* and *Euryalona* in form. Valves with concentric marking. About 16 marginal denticles, larger distally; the penultimate much larger, and the ultimate larger still and serrate on concave side. Basal spine stout, attached about ⅓ of way from base of claw. ♂ unknown. Length, ♀ 0.5–0.6 mm.
  Rather rare, among weeds; southern U. S. and southward.

◄ **Fig. 27.85.** *Oxyurella longicaudis.* (By Birge.)

85a   **(82)** Postabdomen large, denticles very small (Fig. 27.118) . . . .
                                   Alonella diaphana   **12€**
            Turn to key at number indicated, where the species named is discussed.

85b   Postabdomen of moderate size; denticles of usual size (Fig. 27.85*a*) . . . . *Alona guttata*   **9€**
            Turn to key at number indicated, where the species named is discussed.

86a   **(81)** Valves without markings. (Not figured) . . . .
                                   Leydigia quadrangularis (Leydig) 1860
            Keel of labrum with minute setae. Claws with basal spine. Length, ♀ to 0.9 mm; ♂ *ca.* 0.7 mm.
            In all regions of the continent; not common; found singly among weeds.

86b   Valves striated longitudinally . *L. acanthocercoides* (Fischer) 1854
            Keel of labrum with long cilia. Claws without basal spine. Length, ♀ to 1.0 mm or more; ♂ (European) 0.7 mm.
            Rare; southern U. S. and southward.

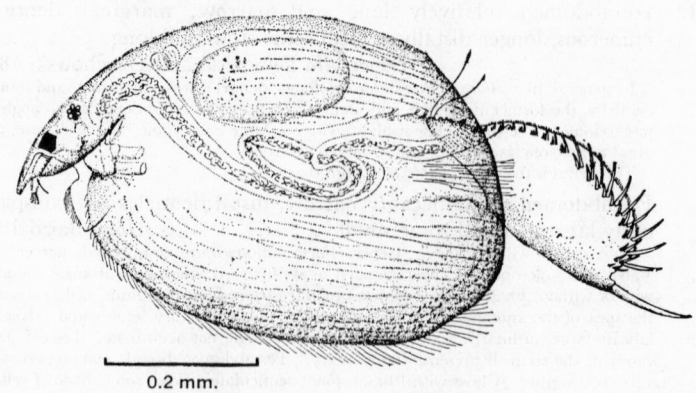

├─── 0.2 mm. ───┤

Fig. 27.86. *Leydigia acanthocercoides.* (By Birge.)

87a (83) Inferoposteal angle with 1 to 4 small teeth **88**

87b Inferoposteal angle without teeth **89**

88a (87) Basal spine of terminal claw very long, about ⅓ length of claw itself . . . . *Alona monacantha* Sars 1901

In general form not unlike *A. intermedia.* Valves with distinct longitudinal striae; inferoposteal angle with 1 to 3 small teeth. Postabdomen with 9 to 10 denticles; claws with very long basal spine. Keel of labrum produced into posterior angle. Length, ♀ 0.35–0.4 mm.

Southern U. S. and southward; in weedy pools.

◄ **Fig. 27.87.** *Alona monacantha.* (By Birge.)

0.1 mm.

88b Basal spine of terminal claw short, less than ¼ length of claw itself . . . . *A. karau* King 1853
Similar to above species. However, striae on valves are oblique; inferoposteal angle with 1 to 4 minute teeth. Postabdomen broad, expanded behind anus; apex rounded; with usually 8 margin denticles and as many larger lateral fascicles. Keel of labrum rounded, not produced into posterior angle. Color yellow, transparent. Length, ♀ 0.45 mm, in South America 0.23 mm.
Southern U. S. and southward; not rare; in pools and lakes.

General shape like that of *Alona* and easily taken for a member of that genus (see **87**). Valves with oblique striae; inferoposteal angle with 1 to 4 minute teeth. Postabdomen broad, expanded behind anus; apex rounded; with about 8 minute marginal denticles and as many larger lateral fascicles. Claws with 1 small basal spine. Color yellow, transparent. Length, female 0.45 mm; male (South American), 0.23 mm.

Southern U. S. and southward; not rare, in pools and lakes.

0.1 mm.

◄ **Fig. 27.88.** *Alona karua.* (By Birge.)

89a (87) Postabdomen long, narrow; distal marginal denticles very long. . . *Oxyurella* **84**
Turn to key at number indicated where genus is discussed.

89b Postabdomen not noticeably long **90**

90a (89) Postabdomen with marginal denticles only. (Fig. 27.89d) . . . . . *A. guttata* Sars 1862
Form much like *A. costata,* but usually smaller and dorsal margin less arched. Valves smooth, striate, or tuberculate. Postabdomen short, broad, slightly tapering toward apex; truncate, angled, with longest marginal denticles at angle; denticles 8 to 10, pointed, small; no squamae. Claws with small basal spine. Postabdomen of ♂

without spines; vas deferens opens behind claws, without any projection. Length, ♀ *ca.* 0.4 mm; ♂ 0.3–0.35 mm.
Not uncommon, everywhere on continent.

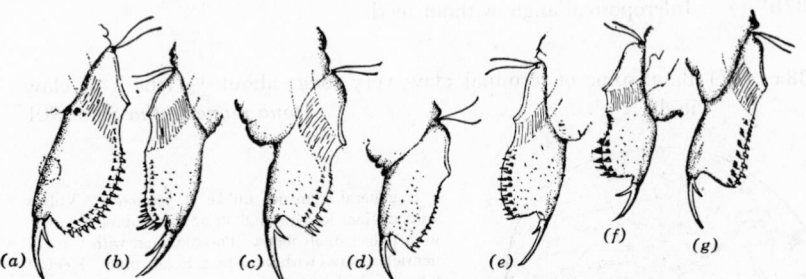

Fig. 27.89. Postabdomens. (*a*) *Alona quadrangularis.* (*b*) *Oxyurella tenuicaudis.* (*c*) *Alona costata.* (*d*) *A. guttata.* (*e*) *A. rectangula.* (*f*) *A. rectangula* var. *pulchra.* (*g*) *A. intermedia.* These figures are not drawn to the same scale. (By Birge.)

90b   Postabdomen with both marginal and lateral denticles · **91**

91a   (90)   Size large; postabdomen with 14 or more marginal denticles **92**

91b   Size moderate or small; postabdomen with fewer than 14 denticles **93**
       Many unstudied species unlisted.

92a   (91)   Cluster of fine spinules at base of claw . . *A. affinis* (Leydig) 1860

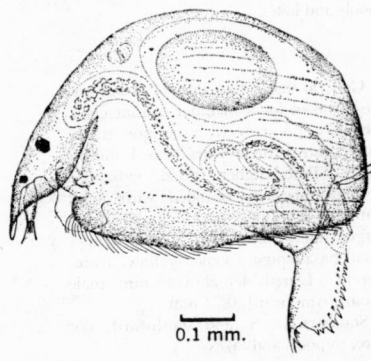

Greatest height usually near middle of valves. Valves longitudinally striated or reticulated, often not plainly marked. Labrum with rhomboidal keel; its corners often angulated, sometimes rounded. Postabdomen large, not widened behind anus; with 14 to 16 serrate marginal denticles and a lateral row of small squamae. Claws long, denticulate; with long basal spine and 4 to 5 spinules inside of basal spine. Six pairs of legs, the last rudimentary. Length, ♀ to 1.0 mm; ♂ to 0.7 mm.
    The largest species of the genus; very abundant in all parts of continent, in margin of ponds and lakes, among weeds.

0.1 mm.   ◀ **Fig. 27.90.** *Alona affinis.* (By Birge.)

92b   No spinules at base of claw. (Fig. 27.89*a*) . . . .
                              A. quadrangularis (O. F. Müller) 1785
    Greatest height usually posterior to middle of valves. Valves usually plainly striated, sometimes conspicuously so, with a reticulated area in inferoanterior region. Labrum with large keel of variable form; often quadrate or with rounded angles. Postabdomen large, flattened, dorsal margin dilated; with 15 to 18 serrate marginal denticles and row of lateral squamae. Claws large, with long basal spine; no spinules on inside of basal spine. Length, ♀ to 0.9 mm; ♂ to 0.6 mm.
    In localities similar to those where preceding species is found; also on bottom of open water.

93a   (91)   Lateral fascicles or squamae do not extend beyond dorsal margin of postabdomen (Fig. 27.89*c*) . . . . *A. costata* Sars 1862
    Evenly arched or greatest height behind middle; posterior margin convex. Valves striated or smooth. Postabdomen short, broad; with straight dorsal (lower) margin, tapering toward apex, with about 12 subequal denticles and a row of fine squamae. Postabdomen of ♂ tapering; no marginal denticles; very fine squamae; vas deferens opens at apex of process extending out ventral to (above) claws; claws without basal spine. Length, ♀ 0.5 mm or more; ♂ 0.4 mm.
    Found everywhere, very abundant.

93b   Lateral fascicles long, extending beyond dorsal margin (Fig.
       27.89e,f)   94
94a (93) Postabdomen not broadened toward apex (Fig. 27.89e,f) . . . . . .
                                        A. rectangula Sars 1861
       Body evenly arched; general form like that of *A. guttata*. Valves striated, reticulated,
       or smooth, rarely tuberculate; ventral margin usually somewhat convex. Postab-
       domen short, slightly enlarged toward apex, angle rounded; with 8 to 9 marginal
       denticles or bundles of setae and about as many fascicles, the distal long enough to
       project beyond margin of postabdomen. Intestine without caecum, enlarged at junc-
       tion of intestine and rectum, somewhat as in *Ilyocryptus*. Length, ♀ 0.35–0.42 mm.
       Common everywhere.

94b   Postabdomen broader toward apex (Fig. 27.89g) . . . .
                                        A. intermedia Sars 1862
       Body evenly arched but not very high. Postabdomen long, broad, enlarged toward
       apex, with rounded angle; the 8 to 9 marginal denticles rather small and thick; the
       lateral denticles or fascicles much more conspicuous, consisting of bundles of fine
       setae. The distal seta in each bundle is the largest and the size of setae increases
       toward apex of postabdomen. The distal bundles project beyond margin of post-
       abdomen. Length, ♀ ca. 0.4 mm.
       Rare. (Specimens closely agreeing with Lilljeborg's description and figures found in
       Wis. Possibly not Sars' *intermedia*, as his figure of that species resembles some
       varieties of Lilljeborg's *rectangula*.)

◄ **Fig. 27.91.** *Alona intermedia.* (By Birge.)

0.1 mm.

95a (79) Postabdomen with marginal denticles only . . . . *Alonella* 129
       Turn to key at number indicated, where 2 species are discussed.
95b   Postabdomen with 2 to 4 marginal denticles, and long series of
       lateral denticles. Rostrum very long, recurved . . . .
                                        Rhynchotalona Norman 1903
       Sole species . . . . *R. falcata* (Sars) 1861

       General form like that of *Alona*. Rostrum
       very long, slender, and recurved under the
       head. Postabdomen stout, thick, bent at
       anus, truncate at apex; with about 4 rather
       stout marginal denticles near apex, and a
       lateral series of very fine spinules continued
       in an unbroken row nearly to anus. In-
       testine with caecum. ♂ (European) with
       long rostrum, bilobed at apex; postabdomen
       tapering and armed with hairs only; ordinary
       hook on first leg. Color yellow or greenish.
       Length, ♀ 0.5 mm; ♂ 0.4 mm.
       Northern U. S.

◄ **Fig. 27.92.** *Rhynchotalona falcata.* (By Birge.)

0.1 mm.

96a (71) Body elongated, form not spherical (Figs. 27.99, 27.100)   97
96b   Body spherical or broadly ellipsoidal (Figs. 27.104, 27.115)   112
97a (96) Lower part of posterior margin of valve excised or crenulated
       (Figs. 27.122, 27.123) . . . . *Alonella excisa, A. exigua* 130

97b   Posterior margin of valve with numerous teeth along entire
           length . . . . *Pleuroxus procurvus, P. truncatus* 105, 106
           Turn to key at numbers indicated, where 2 species are discussed.

97c   Posterior margin of valve with teeth (if any) only at inferoposteal
           angle   98

98a (97) Inferoposteal angle well marked, ordinarily with teeth (Fig.
           27.117)   99

98b   Inferoposteal angle rounded, with or without teeth (Figs. 27.93,
           27.99)   100

99a (98) Rostrum long (most species) . . . . *Pleuroxus* 103
           Take up the key at the number indicated, where the genus is discussed as a unit.

99b   Rostrum short (Fig. 27.117) . . . . *Alonella dentifera* 125
           Take up key at number indicated.
           N.B. If rostrum is broad, semicircular at end, see 80.

100a (98) Inferoposteal angle with well-marked tooth or teeth   101

100b   Inferoposteal angle without teeth, or tooth very small; rostrum
           long or short   103

101a (100) Rostrum long, recurved . . . . *Pleuroxus striatus* 108
           Take up key at number indicated.

101b   Rostrum short . . . . *Dunhevedia* King 1853   102
           General shape rounded. Valves tumid, gaping below; obscurely reticulated; infero-
           posteal angle rounded, with 1 or 2 teeth on ventral margin in front of angle. Post-
           abdomen bent abruptly behind anus; postanal part thick, somewhat foot-shaped as
           seen from side, its dorsal (lower) margin lying parallel to ventral margin of valves;
           with many fine denticles and setae. Claws short, curved, with 1 basal spine. ♂ with
           usual characters; postabdomen same shape as in ♀, with fine hairs only.

102a (101) Body short and high, as dorsal margin is much arched. (Fig.
           27.93) . . . . *D. crassa* King 1853
           Keel of labrum produced into a somewhat tonguelike form, its ventral margin
           smooth. Color yellow. Length, ♀ to 0.5 mm; ♂ *ca.* 0.36 mm.
           Throughout U. S. Not common; among weeds.

102b   Body more elongate, as dorsal margin is little arched. (Fig.
           27.94) . . . . *D. serrata* Daday 1898
           Usually 2 teeth at inferoposteal angle, a very small posterior one and a larger
           anterior one. Keel of labrum serrate in anterior part, smooth behind; about 10 to 12
           serrations, pointing backward. ♂ unknown. Color yellow. Length, ♀ *ca.* 0.7 mm.
           Southern U. S.; in pools and lakes, among weeds; not abundant.

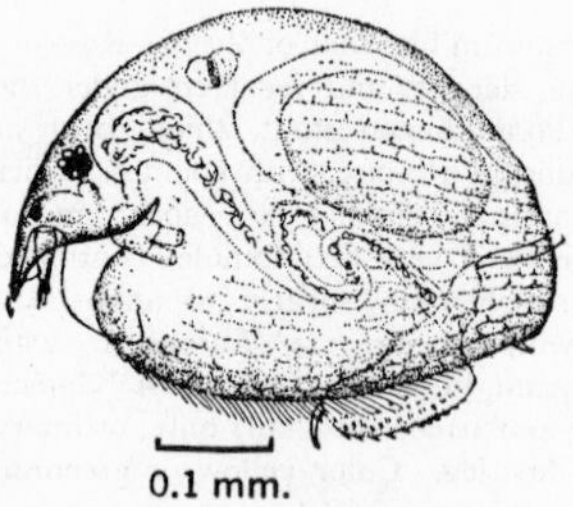

Fig. 27.93. *Dunhevedia crassa.* (By   Fig. 27.94. *Dunhevedia serrata.* (a) Labrum. (b)
Birge.)   Postabdomen. (By Birge.)

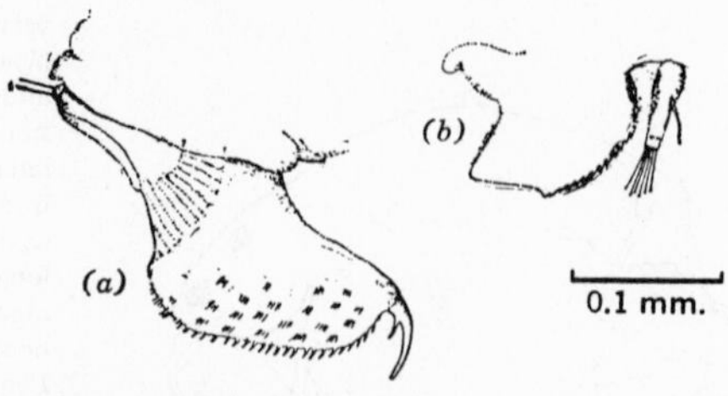

103a (100) Postabdominal claws with 2 basal spines . . . .
                                                  Pleuroxus Baird 1843   104
           Rostrum long and pointed, rarely bent forward. Dorsal margin much arched;
           posterior margin short, usually less than ½ height, rarely toothed along entire
           length; inferoposteal angle rarely rounded, usually sharp and toothed. Keel of labrum
           large, usually tongue-shaped; posterior angle prolonged. Postabdomen with

marginal denticles only. ♂ smaller than ♀, with usual characters; postabdomen varied in different species.

Three types of form are distinguishable in the genus: (1) Relatively long and low species: *striatus* type (*P. striatus, hastatus, hamulatus*). (2) Short, high-arched forms: *denticulatus* type (*P. denticulatus, aduncus, trigonellus, truncatus*). (3) Like (2) with rostrum bent forward: (*P. procurvus, uncinatus*). All species littoral.

03b   Postabdominal claws with 1 basal spine. . . *Alonella,* most species   **123**
       Take up key at number indicated, where one subgenus is discussed.

04a (103)   Rostrum bent up in front (Figs. 27.95, 27.96) **105**

04b   Rostrum not bent forward (Fig. 27.99). **106**

05a (104)   Rostrum bent sharply into a hook; teeth along whole posterior margin of valves (Fig. 27.95) . . . ***Pleuroxus procurvus*** Birge 1878
       General form and markings like those of *P. denticulatus*. Posterior margin of valves with 7 to 8 teeth along the whole length. Postabdomen like *P. denticulatus* but slightly more broadened behind anus. ♂ unknown. Color yellowish, transparent, or opaque. Length, ♀ *ca.* 0.5 mm.
       Northern U. S., common in weedy waters. (Although Birge refers to this species both in 1891 and 1910 as *procurvatus* 1878, the original spelling was *procurvus*.)

05b   Rostrum merely curved forward; teeth at inferoposteal angle only. (Fig. 27.96) . . . . ***P. uncinatus*** Baird 1850
       Inferoposteal angle with 2 to 4 rather long, curved teeth, sometimes branched Rostrum long, acute, bent forward. Postabdomen like that of *P. trigonellus*, broad, somewhat tapered toward apex; about 13 sizable marginal denticles. Color dirty gray, or with green or yellow tinge. Length, ♀ 0.7–0.9 mm; ♂ (European) 0.56 mm.
       This species is very close to *P. trigonellus*, separated by procurved rostrum and large teeth at inferoposteal angle.
       Northern U. S.

06a (104)   Numerous teeth along entire posterior margin of valve. (Fig. 27.97). . . . . ***P. truncatus*** (O. F. Müller) 1785
       Posterior margin with numerous (more than 20) close-set teeth; valves striated, the striae on middle of valves nearly longitudinal, the others oblique. Postabdomen much like that of *P. trigonellus*, slightly tapering toward apex, angle rounded; 12 to 14 marginal denticles, increasing in size distally. Color yellow-brown. Length, ♀ *ca.* 0.6 mm; ♂ (European) 0.45 mm. (Note: this species is assigned by some authors to its own genus **Peracantha** Baird.)
       Northern U. S.

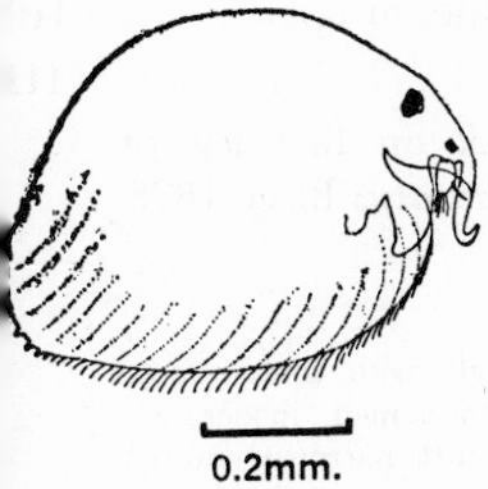

0.2mm.   0.1 mm.   0.1 mm.

Fig. 27.95. *Pleuroxus procurvus.* (By Birge.) 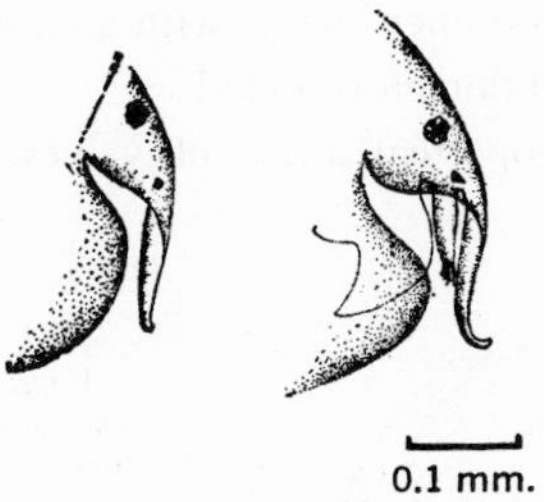   **Fig. 27.96.** *Pleuroxus unicinatus.* European specimens. (By Birge.)   **Fig. 27.97.** *Pleuroxus truncatus.* European specimen. (By Birge.)

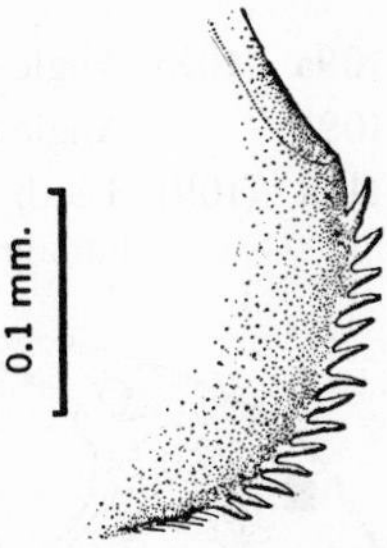

06b   Posterior margin of valve with teeth at inferoposteal angle only . .   **107**

07a (106)   Postabdomen long, slender, convex on ventral (upper) side (Figs. 27.99, 27.100) **108**

07b   Postabdomen of moderate length; ventral (upper) margin straight, or nearly so; greatest width behind anus (Figs. 27.100–27.103) . . .   **109**

108a (107) Superoposteal angle sharp, but not projecting; inferoposteal angle
             a sharp point . . . . *P. hastatus* Sars 1862

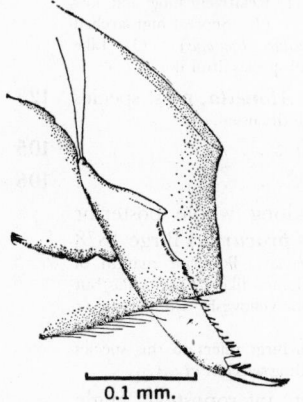

Inferoposteal angle a sharp point, with a very small
tooth; valves reticulated, longitudinal marks often more
distinct, giving appearance of striation. 16 to 18 marginal
denticles. Color yellow, transparent or opaque; not black
unless ephippial. Length, ♀ *ca.* 0.6 mm; ♂ *ca.* 0.45 mm.
Rather rare; throughout U. S.

◄ **Fig. 27.98.** *Pleuroxus hastatus.* Posterior end. (By Birge.)

0.1 mm.

108b   Superopo teal angle overhanging; inferoposteal angle rounded with
              small tooth in front of it . . . . *P. striatus* Schødler 1863

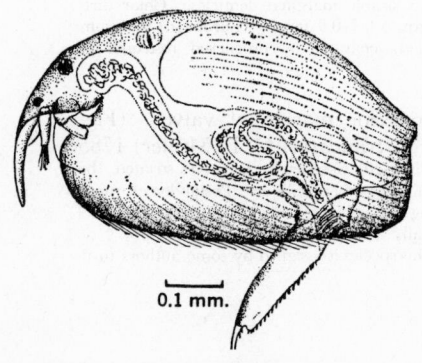

General shape much like that of *P.
hastatus* but never so high-arched as this
may be. Valves obviously striated. Postabdomen long, slender, with 20 or more
marginal denticles. Color dark, especially
opaque on dorsal side, often nearly black.
Length, ♀ *ca.* 0.8 mm; ♂ *ca.* 0.6 mm.
In all parts of U. S.; common among
weeds.

◄ **Fig. 27.99.** *Pleuroxus striatus.* (By Birge.)

0.1 mm.

109a (107) Angle of postabdomen sharp, with a cluster of spines at apex   110
109b   Angle of postabdomen rounded   111
110a (109) Teeth at inferoposteal angle of valves; no hook on first leg of
             female . . . . *P. denticulatus* Birge 1878

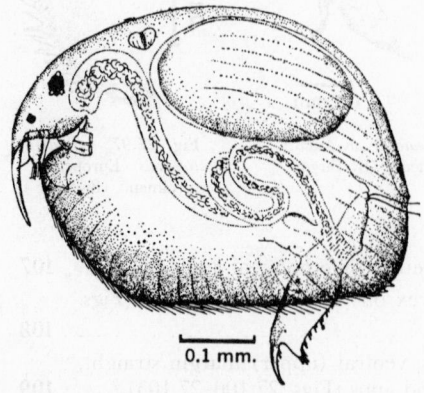

Inferoposteal angle with small toothlike spines. Postabdomen moderately
long, straight, very little narrowed toward
apex; length of postanal part 1.5 times, or
more, that of anal emargination; apex
truncate; with cluster of fine, straight
denticles at apex and 8 to 12 anterior to
these. Color greenish or yellowish, usually transparent. Length, ♀ 0.5-0.6 mm;
♂ 0.36 mm.

Common everywhere, in weedy water.

◄ **Fig. 27.100.** *Pleuroxus denticulatus.* (By
Birge.)

0.1 mm.

10b Inferoposteal angle rounded, without teeth; first leg of female with
stout hook . . . . **P. hamulatus** Birge 1910

Inferoposteal angle rounded, without teeth, valves reticulated; also marked by very
fine striae, with run nearly longitudinally. Rostrum long, recurved. Keel of labrum
small, rounded, prolonged. Postabdomen much like that of *P. denticulatus,* but with
apex more rounded and denticles not so crowded there. Denticles about 12 to 14. ♂
unknown. Color horn-yellow, often dark on dorsal side like *P. striatus.* Length, ♀
ca. 0.6 mm.

New England and southern U. S.; probably a southern form; not reported from
north-central region. Common in pools and weedy waters.

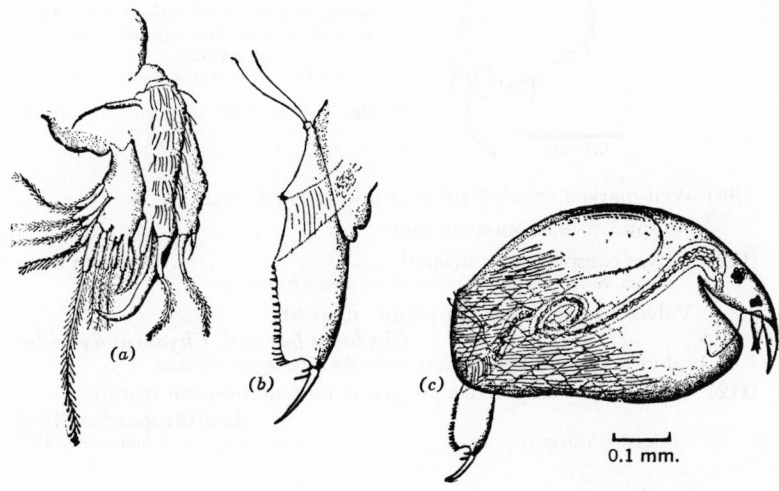

Fig. 27.101. *Pleuroxus hamulatus.* (*a*) First leg. (*b*) Postabdomen. (*c*) Lateral. (By Birge.)

111a **(109)** Series of marginal denticles longer than anal emargination; postabdomen of male broadened in middle of postanal part with
crescentic dorsal margin . . . . **P. trigonellus** (O. F. Müller) 1785

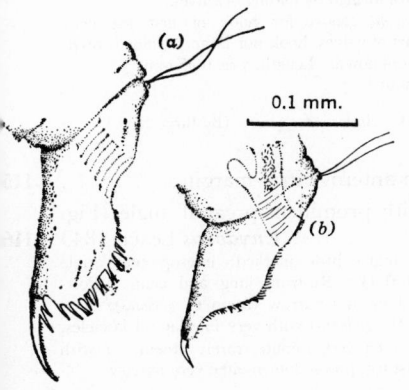

Form of *P. denticulatus* type. Inferoposteal
angle with 2 or 3 small teeth, often minute,
sometimes wanting. Postabdomen much as
in *P. denticulatus;* but dorsal margin slightly
convex, broader behind anus; apex rounded;
14 to 16 marginal denticles, longer toward
apex, but not distinctly clustered there. ♂
postabdomen is characteristic; broadened behind anus into a semielliptical plate, bearing
thick-set hairs, no spines; greatly narrowed
toward apex, forming a slender prolongation.
Color yellowish, transparent; postabdomen
often dark. Length, ♀ 0.6 mm; ♂ 0.4 mm.

Not common; widely distributed in U. S.,
Canada.

◄ **Fig. 27.102.** *Pleuroxus trigonellus.* Postabdomen: (*a*) Female. (*b*) Male. (By
Birge.)

111b   Row of marginal denticles about as long as anal emargination;
                male postabdomen not crescentic . . . . *P. aduncus* (Jurine) 1820

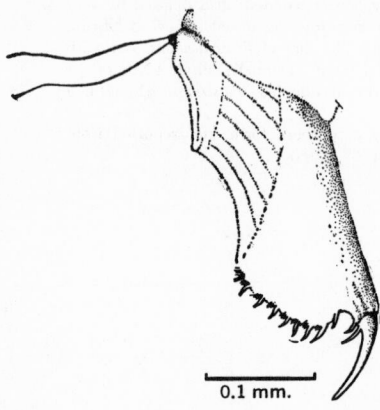

♀ very closely resembling *P. trigonellus*, but
differing as follows: Valves striated; inferoposteal angle usually without teeth. Postabdomen shorter, the length of postanal part
hardly exceeding anal emargination; dorsal
margin slightly arched, with 9 to 12 marginal
denticles; apex rounded. ♂ postabdomen very
different from that of *P. trigonellus;* narrower
than that of ♀, tapered toward claws; no dorsal
enlargement or apical prolongation. Color
horn-yellow, sometimes opaque. Length, ♀
ca. 0.6 mm; ♂ *ca.* 0.45 mm.
          Western U. S.; among weeds or in pools.

◄ **Fig. 27.103.** *Pleuroxus aduncus.* (By Birge.)

0.1 mm.

112a   (96) Well-marked or small spine at inferoposteal angle **113**
112b   No spine at inferoposteal angle **114**
113a (112) Valves conspicuously striated. . . . . *Alonella nana* **128**
                  Turn to key at number indicated, where species is discussed.
113b   Valves reticulated or not plainly marked. . . . .
                            Chydorus barroisi, Chydorus hybridus **122**
                  Turn to key at number indicated, where the 2 species are discussed.
114a (112) Valves with conspicuous projection on anteroventral margin . . . .
                              Anchistropus Sars 1862
                  Sole American species . . . . *A. minor* Birge 1893

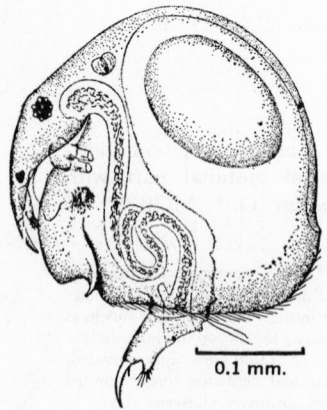

Form globular. Ventral region tumid anteriorly;
ventral margins of valves bent sharply away from each
other about ⅓ of way from front, and valve folded out
into a hollow groove and tooth, which contains the
strong hook of the first leg. Head large, bulging over
eye, the fornices broad and forming a sort of flaplike
rostrum, which can be closely pressed to the valves.
Postabdomen broad at base, preanal angle overhanging; narrowing sharply toward apex, which is prolonged
into a lobe; a few marginal spines. Claws with long,
slender basal spine; denticulate or smooth. First leg
of ♀ with strong hook, toothed on concave side, which
lies in groove formed by folding of valves.
     In *A. minor*, groove for hook of first leg near
anterior part of valves; hook not large. Color brown-
yellow. ♂ unknown. Length, ♀ *ca.* 0.35 mm.
     Throughout U. S.

◄ **Fig. 27.104.** *Anchistropus minor.* (By Birge.)

0.1 mm.

114b   Valve without such projection on anteroventral margin **115**
115a (114) Postabdomen ordinarily short with prominent preanal angle (Fig.
                27.111) . . . . *Chydorus* Leach 1843 **116**
                  Shape spherical or ovate. Posterior angles little marked; inferoposteal angle
                  usually unarmed. Antennules short and thick. Rostrum long and acute. Post-
                  abdomen usually short and broad, rarely long and narrow (as in *C. globosus*); apex
                  rounded; with marginal denticles only or (*C. globosus*) with very fine lateral fascicles.
                  Claws with 2 basal spines, the proximal often very minute, rarely absent. ♂ with
                  short rostrum, thick antennule, hook on first leg, postabdomen often very narrow.

115b   Postabdomen large, preanal angle ordinarily not prominent (Figs.
            28.116, 27.118). . . . . *Alonella* Sars 1862   **123**
    Although the species of this genus are diverse in many respects, they are similar
    in the shape and surface sculpturing of the head shield.   There are 3 groups of
    species within this genus that might well constitute separate genera:
        (1) *Alonella* proper.   Rostrum long, slender, recurved; usually conspicuously so;
    postabdomen with marginal denticles only; claws with 1 basal spine.   *A. rostrata,
    dadayi, nana.* (2) *Paralonella.* Rostrum short, hardly exceeding antennules; post-
    abdomen with very small marginal denticles, with or without lateral fascicles; claws
    with 1 basal spine.   *A. dentifera, diaphana, globulosa.*   (3) *Pleuroxalonella.* Ros-
    trum moderate; postabdomen with marginal denticles only; claws with 2 basal
    spines.   *Pleuroxus*-like.   *A. excisa, exigua.*

116a **(115)** Postabdomen long, narrow, *Pleuroxus*-like. . . . .
                                        Chydorus globosus Baird 1850

Almost spherical; valves smooth or reticulated, sometimes
striated in front.   Postabdomen with small preanal angle;
numerous marginal denticles and very fine lateral fascicles.
Claws with 2 basal spines, the distal very long and slender.
Color bright yellow to dark brown, usually with dark spot in
center of valve.   Length, ♀ to 0.8 mm; ♂ 0.6 mm.
    Everywhere; in lakes and ponds, among weeds, but never
present in large numbers.
    C. globosus might well be a type of a separate genus.   The
other species fall into 3 groups: (1) The *sphaericus* group or
Chydorus proper (*C. sphaericus, gibbus, piger, latus, ovalis*).   (2) The
faviformis group, similar to (1) but with greatly developed
cuticular structures (*C. faviformis, bicornutus*).   (3) The *barroisi*
group, with toothed labrum; denticles of postabdomen shortest in
middle of row (*C. barroisi, hybridus, poppei*).

◄ **Fig. 27.105.** *Chydorus globosus.* Postabdomen. (By Birge.)

116b   Postabdomen short, broad; preanal angle marked **117**

117a **(116)** Valve covered with deep polygonal cells . . . .
                                            C. faviformis Birge 1893

Much like *sphaericus* in form and size.   ♂ unknown.   Color yellow to light brownish.
Length, ♀ 0.5–0.6 mm.
    U. S. and Canada; not common.

◄ **Fig. 27.106.** *Chydorus faviformis.* Cast shell.
(By Birge.)

0.1 mm.

117b   Valve with deep polygonal cells and cuticular ridges . . . .
                                     C. bicornutus Doolittle 1909

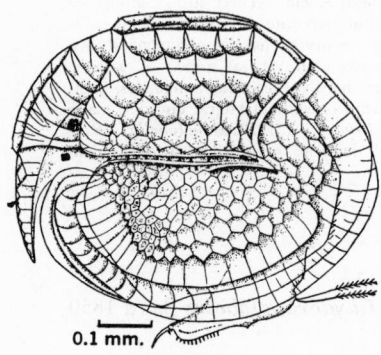

Like *faviformis* in having deep polygonal cuticular cells; but distinguished by the development of an extraordinary and complex system of thin cuticular ridges, which extend far beyond the ordinary cells. A long horn extends laterally from the middle dorsal region of each valve, from which radiate some of the ridges. ♂ unknown. Color yellow. Length, ♀ to 0.7 mm.
Northern U. S. and Canada.

◄ **Fig. 27.107.**   *Chydorus bicornutus.*   (After Doolittle.)

0.1 mm.

117c   Valve of ordinary type. **118**
118a (117) Ventral edge of keel of labrum smooth (Fig. 27.111) **119**
118b   Ventral edge of keel of labrum with 1 or more teeth (Figs. 27.113,
              27.114). **122**
119a (118) Anterodorsal surface of valves and head flattened . . . .
                                     C. gibbus Lilljeborg 1880

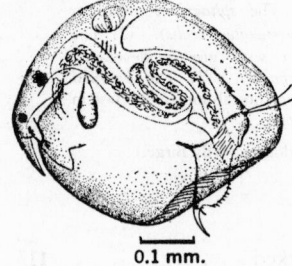

The curve of the dorsal surface somewhat flattened, both in front and behind, making a sort of hump in center of dorsal margin. Valves reticulated. Head small; rostrum projects from valves in characteristic way. Postabdomen with 8 to 10 marginal denticles. Color yellowish to brown. Length, ♀ 0.5 mm.
Northern U. S. and Canada; rare.

◄ **Fig. 27.108.**   *Chydorus gibbus.*   (By Birge.)

0.1 mm.

119b   Dorsal surface not flattened; form usually spherical or broadly
              ovate **120**
120a (119) Small forms, not exceeding 0.5 mm, usually less **121**
120b   Larger forms, to 0.8 mm. Antennules short and thick with all
              olfactory setae terminal. . . . . *C. latus* Sars 1862

Much like *sphaericus*, but larger. Mandible attached some distance back of junction of head and valve. Denticles of postabdomen 10 to 12. Claws sometimes with only 1 basal spine. Color dark yellow-brown. Length, ♀ to 0.7–0.8 mm.
Rare; Canada, near Lake Erie.

◄ **Fig. 27.109.**   *Chydorus latus.*   (By Birge.)

0.1 mm.

651

120c   Forms about 0.5 mm long.   Antennule with 1 olfactory seta
proximal to cluster at end . . . . *C. ovalis* Kurz 1874

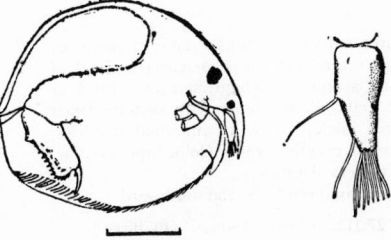

Form round or broad oval. Postabdomen with rounded apex; 12 to 15 marginal denticles. Claws with 2 basal spines, the proximal minute. Color yellow, transparent. Length, ♀ to 0.6 mm; ♂ (European), 0.5 mm.
Rare; northern U. S.

◄ **Fig. 27.110.** *Chydorus ovalis.* Entire specimen and antennule. (By Birge.)

0.2 mm.

121a (120) Fornices gradually narrowing into rostrum.   All olfactory setae on
end of antennule . . . . *C. sphaericus* (O. F. Müller) 1785

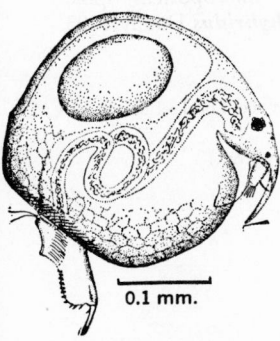

Spherical or broadly elliptical. Shell usually reticulated, sometimes smooth (var. *nitidus* Schødler), sometimes punctate (var. *punctatus* Hellich), or with elevations (var. *coelatus* Schødler). Postabdomen with 8 to 9 marginal denticles. Claws small; proximal basal spine very minute. ♂ with postabdomen much emarginate. Color light yellow to dark brown. Length, ♀ 0.3–0.5 mm; ♂ 0.2 mm. Small limnetic forms constitute var. *minor* Lilljeborg.
The commonest of all Cladocera; found all over the world.

◄ **Fig. 27.111.** *Chydorus sphaericus.* (By Birge.)

0.1 mm.

121b   Fornices abruptly narrowed into rostrum.   Two olfactory setae
on side of antennule . . . . *C. piger* Sars 1862
General form much like that of *C. sphaericus.* Ventral margin of valves densely ciliated; valves ordinarily marked by oblique striae, sometimes smooth. Fornices abruptly narrowed at rostrum. Antennule with usual lateral sense seta and 2 olfactory setae on side. Postabdomen with 8 to 9 rather long marginal denticles. Claws with 2 basal spines, the proximal one minute. ♂ postabdomen narrow, but not excavated. Color light to dark yellow. Length, ♀ *ca.* 0.4 mm.
Rare; reported only from Me.

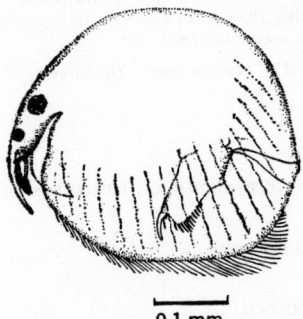

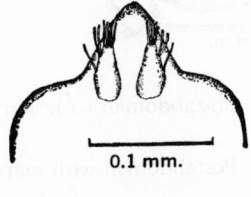

0.1 mm.

0.1 mm.

Fig. 27.112. *Chydorus piger.* Entire specimen and lower side of rostrum with antennules. (By Birge.)

122a (118) Ventral edge of labrum with several teeth; short spine at inferoposteal angle of valves . . . . **C. barroisi** (Richard) 1894

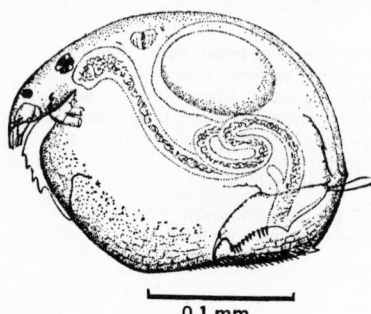

Form and size much like those of *sphaericus*,
though ventral margin is less curved. Keel of
labrum acuminate behind; serrate; with 4 or
more teeth. Postabdomen with well-developed
preanal angle; 10 to 12 marginal denticles,
shortest in middle of row. Color brown-yellow.
Length, ♀ *ca.* 0.4 mm.
    Rare; southern U. S. and southward.
◄ **Fig. 27.113.** *Chydorus barroisi.* (By Birge.)

0.1 mm.

122b   Ventral edge of labrum with one tooth; inferoposteal spine
present . . . . **C. hybridus** Daday 1905

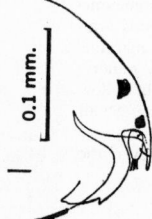

0.1 mm.

Similar to *barroisi* but with only 1 tooth on keel of labrum.
Rare; U. S.
◄ **Fig. 27.114.** *Chydorus hybridus.* (By Birge.)

122c   Ventral edge of labrum with 1 tooth; no spines on valves . . . . . .
C. poppei Richard 1897

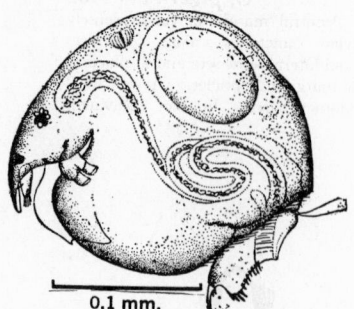

Like *hybridus* but without spine at inferoposteal
angle. Tooth on labrum sometimes small or
obsolescent.
    La., Calif.; rare.
    Very probably the last 2 species should be listed
as varieties of *barroisi*. These species were first
placed in *Pleuroxus*, but have no very close affinity
with either *Pleuroxus* or *Chydorus;* they might well
be made a separate genus.
◄ **Fig. 27.115.** *Chydorus poppei.* (By Birge.)

0.1 mm.

123a (115) Postabdomen with marginal and lateral denticles; rostrum short . . **124**

123b   Postabdomen with marginal denticles only **126**

124a (123) Valves with inferoposteal angle toothed **125**

124b   No inferoposteal teeth; form rotund . . . .
Alonella globulosa Daday 1898

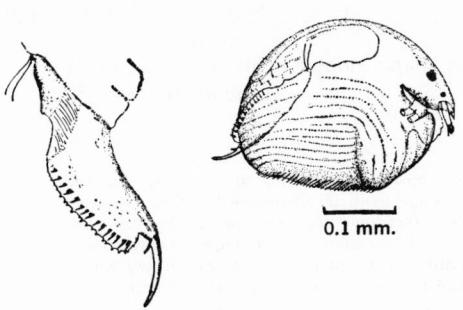

Small, shape oval-rotund; head reaching about to middle of valves. Valves striated; all margins rounded and without teeth. Postabdomen long, narrow; broadest near anus; about 12 minute marginal denticles and as many slender lateral fascicles. Keel of labrum with 2 notches. Color yellow-brown. Length, ♀ 0.3 0.4 mm. (= *A. sculpta* Sars.)

Southern U. S. and southward, among weeds.

◄ **Fig. 27.116.** *Alonella globulosa.* (By Birge.)

125a (124) Inferoposteral angle with 1 to 4 fine teeth; valves striated . . . . . .
See *Alona karau* **88**

125b Inferoposteal angle with 1 to 3 strong teeth, valves reticulated . . .
Alonella dentifera Sars 1901

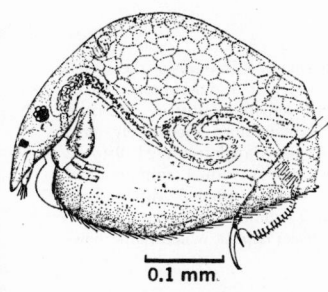

Back high-arched; inferoposteal angle acute, with 1 to 3 fairly strong teeth. Rostrum reaches nearly to ventral margin of valves. Postabdomen large, broad, somewhat expanded behind anus; apex rounded; with about 12 minute marginal denticles, and as many very minute lateral fascicles. Claws with 1 very long basal spine. Color yellow-brown. Length, ♀ *ca.* 0.4 mm; ♂ 0.35 mm.

Southern U. S. and southward; not rare, in pools and lakes.

◄ **Fig. 27.117.** *Alonella dentifera*, with developing ephippium. (By Birge.)

126a (123) Marginal denticles on postabdomen minute; postabdomen large, bent behind anus; no inferoposteal tooth on valves . . . .
A. diaphana (King) 1853

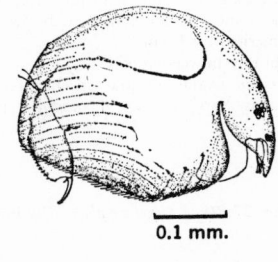

Head short, rostrum not reaching more than ⅔ distance toward ventral margin. Valves striated, sometimes passing into reticulation, often inconspicuous; inferoposteal angle rounded, without teeth. Postabdomen long, slightly enlarged behind anus; with numerous, very minute marginal denticles and no other spines. Claws long; 1 basal spine. Length, ♀ 0.5 mm; ♂ 0.4 mm. Color yellow, transparent.

Southern U. S. and southward; in pools and lakes; rare.

◄ **Fig. 27.118.** *Alonella diaphana.* (By Birge.)

126b Marginal denticles of ordinary size; inferoposteal tooth present . . . **127**
127a (126) Claws with 1 basal spine . . . . ◄ **128**
127b Claws with 2 basal spines; posterior margin of valves excised near inferoposteal angle **130**

128a (127) Rostrum long, recurved **129**

128b   Rostrum short or moderate; shape globose; valves conspicuously
                  striated. . . . . **A. nana** (Baird) 1850

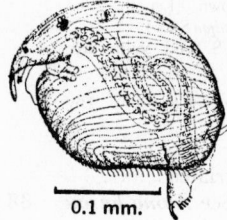

Very minute; *Chydorus*-like. Valves coarsely and conspicuously striated; minute tooth in inferoposteal region. Rostrum variable, usually rather long, recurved, considerably exceeding antennules. Postabdomen short; preanal angle strongly projecting; apex rounded; about 6 marginal denticles. Claws with 1 small spine. Color brownish, usually opaque. Length, ♀ 0.2–0.28 mm; ♂ 0.25 mm.

Northern U. S. and Canada; rare. The smallest member of the family.

0.1 mm.

◄ **Fig. 27.119.** *Alonella nana.* (By Birge.)

129a (128) Shape an elongated oval; valves striated . . . .
                           A. acutirostris (Birge) 1878

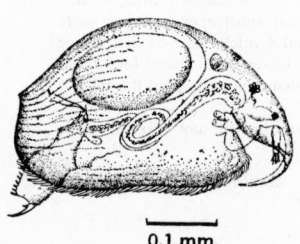

General form not unlike that of a *Pleuroxus* of the *striatus* type. Valves striated or reticulated; inferoposteal angle rounded with minute tooth, sometimes absent. Rostrum long, slender, recurved. Postabdomen moderately long, somewhat tapering toward apex; angle rounded; 9 to 12 small, marginal denticles. Claws with 1 minute basal spine. Color yellow or brown, usually rather dark. Length, ♀ *ca.* 0.5 mm; ♂ *ca.* 0.4 mm.

Note: Birge called this species *rostrata*, which name however is pre-empted (Frey).

Rather rare; most of U. S.

0.1 mm.

◄ **Fig. 27.120.** *Alonella acutirostris.* (By Birge.)

129b   Shape, a short oval; valves strongly reticulated . . . .
                           A. dadayi Birge 1910

Shape oval-rotund. Valves strongly reticulated all over; inferoposteal angle rounded, with several minute teeth. Rostrum long, pointed, recurved. Keel of labrum acuminate behind, and its margin with 1 projection. Postabdomen short, wide; preanal angle strongly marked, as in *Chydorus;* with numerous small denticles; apex rounded. Claws with 1 basal spine. Color yellow to brown, often opaque. Length, ♀ 0.25–0.3 mm; ♂ (South American) 0.2 mm.

Southern U. S. and southward; not rare in weedy pools.

(This species is *Leptorhynchus dentifer* Daday, whose specific name has to be changed on removing to *Alonella,* as Sars' species *A. dentifera* preoccupies the name.)

◄ **Fig. 27.121.** *Alonella dadayi.* (By Birge.)

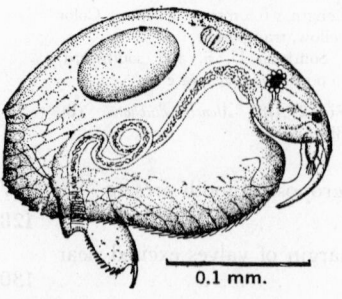

0.1 mm.

130a (127) Postabdomen fairly long; angled at apex; valves reticulated and
with fine striae . . . . ***A. excisa*** (Fischer) 1854
General appearance *Pleuroxus*-like. Rostrum moderate to long, neither so prolonged
nor so recurved as in *A. rostrata;* longer in southern forms. Inferoposteal angle
marked, sometimes produced into a point; posterior margin above it excised, sometimes crenulated. Postabdomen long, narrow, not narrowing much toward apex; apex
angled; with about 9 to 10 small marginal denticles. Color yellow to brown. Length,
♀ to 0.5 mm; ♂ 0.28 mm.
Not uncommon throughout continent; in weedy pools and lakes.

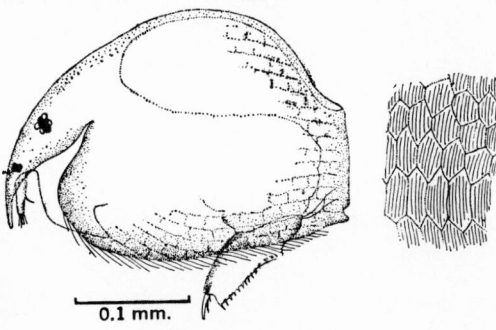

◄ **Fig. 27.122.** *Alonella exisa.*
Entire specimen and details of
markings of valve. (By
Birge.)

`0.1 mm.`

130b Postabdomen short; rounded at apex; valves without fine striae . .
A. exigua (Lilljeborg) 1853

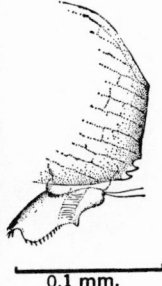

Much like the preceding species but smaller. About 6 to 8 small
marginal denticles. Color yellow, not very transparent. Length, ♀ 0.35
mm; ♂ 0.28 mm.
Northern U. S. and Canada; rare.

◄ **Fig. 27.123.** *Alonella exigua.* Posterior end. (By Birge.)

`0.1 mm.`

References

Austin, T. S. 1942. The fossil species of *Bosmina. Am. J. Sci.,* 240:325–331. **Birge, E. A.
1891.** Notes on Cladocera. II. List of Cladocera from Madison, Wisc. *Trans. Wisconsin Acad.
Sci.,* 8:379–398. **1893.** Notes on Cladocera. III. Descriptions of new and rare species.
Trans. Wisconsin Acad. Sci., 9:275–317. **1910.** Notes on Cladocera. IV. Descriptions of new
and rare species chiefly southern. *Trans. Wisconsin Acad. Sci.,* 16:1018–1066. **Brehm, V.
1937.** Zwei neu Moina-Formen aus Nevada, U. S. A. *Zool. Anz.,* 117:91–96. **Brooks, J. L.
1957.** The Systematics of North American *Daphnia. Mem. Conn. Acad. Arts Sci.,* 13:1–180.
Brown, F. A., Jr. (ed). 1950. *Selected Invertebrate Types.* Wiley, New York. **Calman, W. T.
1909.** Crustacea. Third Fascicle, Part VII in: Lankester, *Treatise of Zoology.* A. and C. Black,
London. **Cannon, H. G. 1933.** On the feeding mechanism of the Branchiopoda. *Phil. Trans.
Roy. Soc. London, Ser. B.,* 222:267–352. **Edmondson, W. T. 1955.** Seasonal life history of *Daphnia* in an arctic lake. *Ecology,* 36:439–455. **Frey, D. G. 1958.** The taxonomic and phyloge-
netic significance of the head pores of the Chydoridae. (Cladocera). *Intern. Rev.,* 44:27–50.
Eriksson, S. 1934. Studien über die Fangapparate der Branchiopoden. *Zool. Bidrag Uppsala,* 15:

24–287.   **Lilljeborg, W.   1900.**   Cladocera Sueciae. *Nova Acta. Regiae Soc. Sci. uppsaliensis*, Ser. 3, 19:1–101.   **Lockhead, J. H. 1950.**   *Daphnia magna.* In: E. A. Brown (ed.). *Selected Invertebrate Types.* Wiley, New York.   **Mackin, J. G.   1930.**   Studies on the Crustacea of Oklahoma. I. *Camptocercus oklahomensis* n. sp. *Trans. Am. Microscop. Soc.,* 49:46–53.   **Richard, J.   1894.**   Revision des Cladocères. Part I. Sididae. *Ann. Sci. Nat. Zool.,* 7:279–389.   **1896.**   Revision des Cladocères. Part II. Daphnidae. *Ann. Sci. Nat. Zool.,* 8:187–363.   **Sars, G. O.   1901.**   Contributions to the knowledge of the freshwater Entomostraca of South America. Part I. Cladocera. *Arch. Math. Naturridenskab,* 23:1–102.   **Wagler, E.   1927.**   Branchiopoda, Phyllopoda. In: Kükenthal and Krumbach, *Handbuch der Zoologie,* III Bande, 1ste Hälfte, pp. 303–398. Gruyter, Berlin and Leipzig.

28

Ostracoda

WILLIS L. TRESSLER

The Ostracoda are small, bivalved crustaceans which are found in both fresh-water and marine environments. In size they average about 1 mm, but in fresh water they range in length from 0.35 mm to about 7 mm. They can be easily recognized and distinguished from the Conchostraca by the absence of lines of growth on the valves and the smaller number of appendages, and are distinguishable from the small clams, such as the Sphaeridae, by the absence of lines of growth and by the distinct arthropodan structure. There are something over 1700 species of known Ostracoda, of which about one-third are found in fresh water. They inhabit a wide variety of environments, being found almost everywhere in all types of fresh water; in lakes, pools, swamps, streams, cave waters, heavily polluted areas, etc. They are all free-living with the exception of some commensal forms; one entire genus lives a more or less parasitic or commensal existence on the gills of various species of fresh-water crayfish (*Entocythere*). Other ostracods have been reported from the intestinal tracts of fish and amphibia; undoubted parasitism, however is unknown.

In 1748 Linneaus, in his *Systema Naturae*, published a note on a species of crustacean he called "*Monoculus concha pedata*," and for many years the term "Monoculus" was used to refer to all entomostracans. In 1853, Baker de-

scribed a form which is now recognizable as a *Cypris* and in 1776, O. F
Müller established this genus as the type form of the Ostracoda.   To the
general biologist, the genus *Cypris* is all inclusive as far as the ostracods
are concerned, but actually, this genus has been so restricted within recent
years that it now includes only a few species characterized by extreme obesity.

Taxonomic Anatomy

It will be sufficient to take up the general structure of those parts useful in
taxonomic work without going into a detailed study of the internal anatomy of
the ostracod body and shell.   The following characters are considered of
primary taxonomic importance and will be most stressed in the key to the
species that forms the bulk of this chapter: size and shape of the shell, character of shell surface, presence and length of the natatory setae of the second
antennae, segmentation of the second antennae, form and number of spines of
the maxillary process, armature of the third thoracic leg, shape and armature
of the caudal furca (Fig. 28.1).

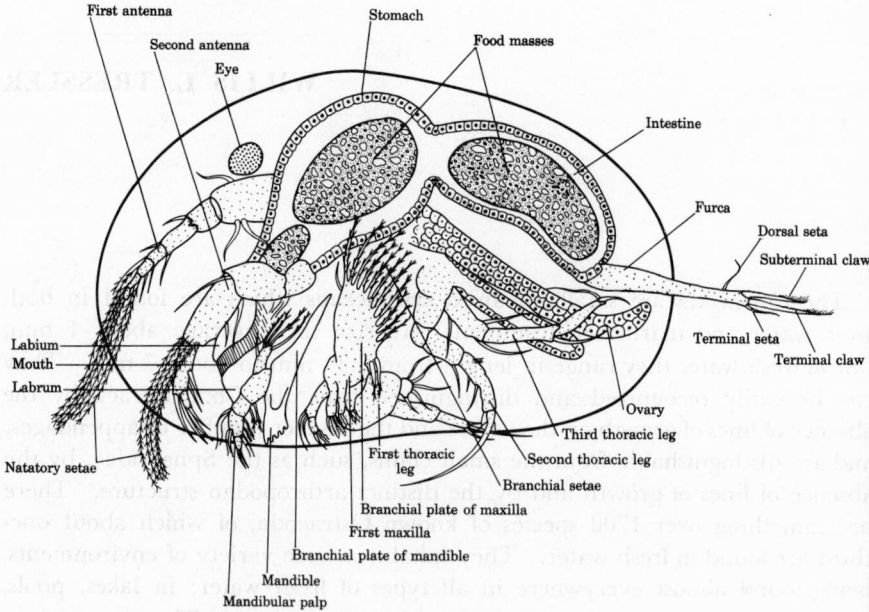

Fig. 28.1.   Diagram of anatomy of female ostracod, lateral view.

The general shape of fresh-water ostracods is much more uniform than that
of the marine species and there is far less tendency to ornamentation of the
shell surface with spines, projections, pits, or tubercles.   Shell shape is commonly regular, and the surface of valves is comparatively smooth.   Colors
tend toward greenish, yellowish, or whitish hues; some forms are brown in
color, others tend toward a bluish shade of green.   The amount of hairiness is

variable and often of specific importance. The valve margins in many forms, particularly at the extremities, exhibit short radiating canals, the so-called "pore canals." The valve edges are frequently tuberculated, or less frequently serrated. Valve surfaces may have striations or complex patterns of anastomosing lines, pits, or tubercles. A hinge, similar to that found in the Mollusca, is present in all fresh-water Ostracoda. An eye, either single or double, may be present, and will be visible through the valves. Scars indicating the attachment of the adductor valve muscles are present in the center of the valves and are of taxonomic importance.

There are seven pairs of appendages in the family Cypridae, to which the great majority of fresh-water ostracods belong. These are, from anterior to posterior, the first antenna (antennule), second antenna, mandible, first maxilla, first thoracic leg, second thoracic leg, and third thoracic leg. The body terminates in a pair of caudal furca (Fig. 28.1). The first and second antennae are provided with natatory setae, which are ordinarily well developed on the first antenna, but whose presence and degree of development are variable on the second antenna. Both pairs of antennae commonly extend anteriorly through the valve aperture where they are vibrated rapidly back and forth in swimming. A labrum and labium are present followed by the mandibles (Fig. 28.2), which are chitinous, elongated bodies, each bearing a

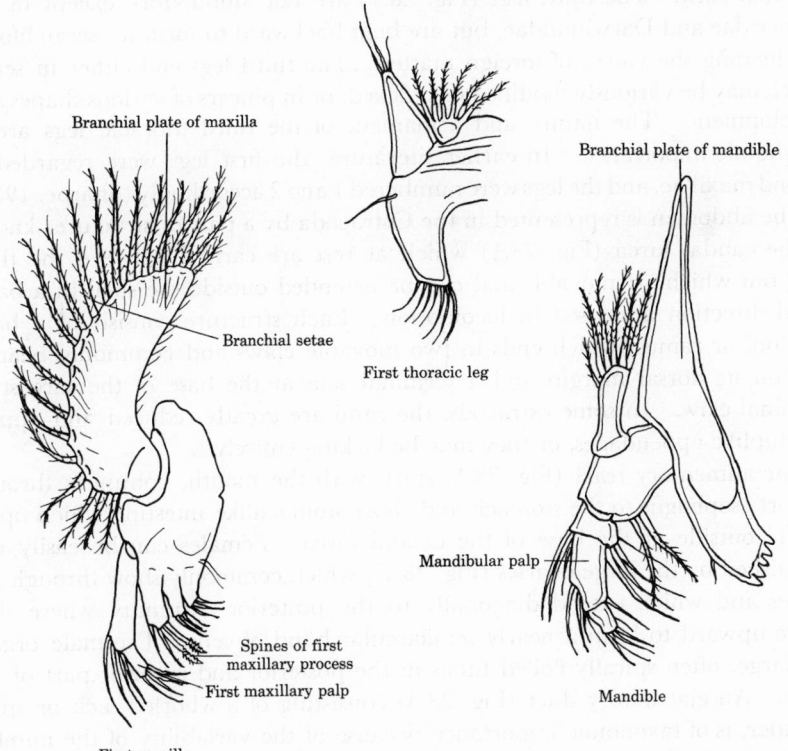

Fig. 28.2. Appendages of ostracod.

well-developed pediform palp, located immediately behind the base of the second antenna. The body of the mandible is equipped with a cutting edge, which is divided into several strong, bifurcate teeth. The palp has on its outer side a narrow branchial appendage which bears a number of plumose setae.

The first maxilla (Fig. 28.2) is composed of a thick, muscular basal part from which four processes originate, the larger of which is jointed and movable and is known as the maxillary palp; the other three processes are the true masticatory lobes of the maxilla. The first masticatory lobe is commonly armed with two strong spines, which may be plain or toothed and which are of specific importance. A large flattened plate, the branchial plate, bears a number of plumose setae. The branchial plate functions as in other crustacea in creating a current of water within the shell cavity.

The first thoracic appendages or legs, sometimes referred to as the second maxillae, are not well developed, although they contain the same parts as the first maxillae (Fig. 28.2). A single masticatory lobe is present. In the male, the palps of the first legs are often modified to form prehensile structures (Fig. 28.3) for grasping the female during copulation. A branchial plate is present but is usually greatly reduced. The second thoracic appendages, called leg 2 in the key (Fig. 28.1), are pediform and usually armed with a long, tapering terminal claw. The third legs (Fig. 28.1) are not ambulatory except in the Cytheridae and Darwinulidae, but are bent backward to form a "scratchfoot" for cleaning the valves of foreign matter. The third legs end either in setae, which may be variously modified or reflexed, or in pincers of various shapes and development. The nature and armament of the third thoracic legs are of taxonomic importance. In earlier literature, the first legs were regarded as second maxillae, and the legs were numbered 1 and 2 accordingly (Sharpe, 1918).

The abdomen is represented in the Ostracoda by a pair of structures known as the caudal furca (Fig. 28.1) which at rest are carried between the third legs, but which are movable and can be extended outside the shell in a backward direction and used in locomotion. Each structure consists of a basal portion, or ramus, which ends in two movable claws and commonly bears a seta on its dorsal margin and a terminal seta at the base of the largest or terminal claw. In some ostracods, the rami are greatly reduced and appear as whiplike appendages, or they may be lacking entirely.

The alimentary tract (Fig. 28.1) starts with the mouth, continues through a short esophagus to the stomach and short stomachlike intestine which opens to the outside at the base of the caudal furca. Females can be easily distinguished by the large ovaries (Fig. 28.1), which commonly show through the valves and which extend diagonally to the posterior extremity where they curve upward to form a nearly semicircular band of cells. The male organs are large, often spirally coiled tubes in the posterior and anterior part of the body. An ejaculatory duct (Fig. 28.3) consisting of a whorled sack or spiny cylinder, is of taxonomic importance because of the variability of the number of whorls of spines in different species. In many ostracods the spermatozoa are longer than the body of the adult. Many Ostracoda are parthenogenetic

and reproduce entirely by unfertilized eggs. The eggs hatch into larvae
which resemble the adult and then molt a number of times before reaching
maturity, in the course of which the size and shape of the shell and the body
appendages change greatly. Unless the stages of development are well known
for the particular species, it is usually impossible to determine the species in

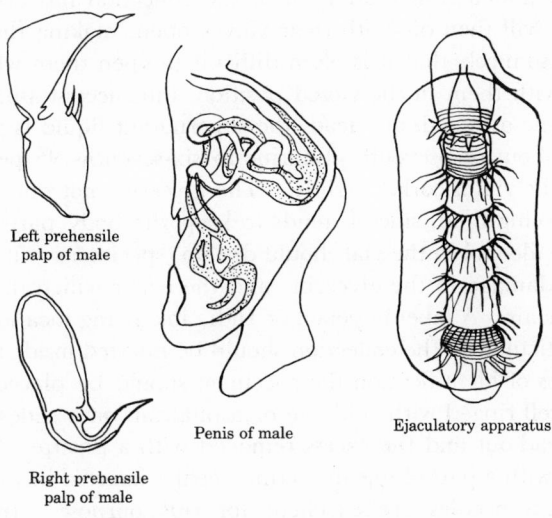

Left prehensile
palp of male

Right prehensile
palp of male

Penis of male

Ejaculatory apparatus

Fig. 28.3. Reproductive organs of male ostracod.

immature forms. Immature forms can be recognized by their size, few adult
ostracods being less than 0.50 mm in length, and by a general roundness and
undeveloped appearance of the appendages. The shell shape also tends to be
similar to the cyprid larvae type found in barnacles, with a high anterior
portion and sloping posterior dorsal margin.

Methods of Collection

Fresh-water ostracods may be found in almost any type of aquatic habitat
with the possible exception of the pure waters of springs. They are rarely
caught in plankton nets because most of them live on or near the bottom
and only occasionally swim upward far enough to be netted. Where high
vegetation exists, these little creatures may climb up along the stems of
plants and thus be secured by nets drawn through the water plants. A Birge
or cone net, which is a small net protected at the opening by a coarse wire screen,
is commonly used either unmodified or with runners which make it easier to
drag it over the bottom. A small cone net attached to a pole is useful for
dragging through vegetation, and an ordinary naturalist's dredge of small
size will collect many bottom forms, particularly those that burrow in the
topmost layer of the substratum.

Preservation and Identification

The collected sample is usually brought into the laboratory alive and placed in flat dishes or watch glasses. The organisms can be picked out under the low powers of a binocular dissecting microscope with a pipette. The living ostracods should be placed in a vial containing a little water to which 95 per cent alcohol is added slowly after all of the collection has been transferred. The ostracods will then die with their valves open, making dissection easier. The valves fit so nicely that it is often difficult to open them when the organism has died with them in the closed position, thus necessitating crushing the shell. When the ostracods are dead, the supernatant liquid is poured off and a mixture of about one-seventh glycerine to six-sevenths 95 per cent alcohol is added and the vial is corked tightly. The glycerine not only makes a more suitable preserving fluid, since it tends to keep the body parts soft and flexible, but if the alcohol in the vial should dry up, specimens will remain a long time without damage in the glycerine and the water which the glycerine extracts from the air. A label in pencil or India ink giving location, depth, date, and other particulars of the collection should be inserted inside the vial.

For purposes of identification the specimen should be placed in a drop of glycerin in a cell ringed with gold size or asphaltum on a slide. The glycerin should be spread out and the excess removed with a pipette. The valves are first removed with a pair of fine dissecting needles inserted in wooden handles; the finest insect needles are excellent for this purpose. In dissection, a binocular dissecting microscope should be used with as low a power as possible. The valves when removed are placed to one side in the ring, or if too large or thick to go under the cover glass, they should be placed in a drop of balsam at the other end of the slide from the label. The appendages are then systematically removed from the body and spread out in the glycerine for future examination. The cell is covered with a coverglass and later sealed. Examination should be made under a compound microscope. A useful and inexpensive addition to the ordinary microscope is a plankton counting square, which can be inserted within the ocular, calibrated with a stage micrometer, and then used against a set of values for measurements of length, height, and thickness of the specimens.

A record card should be made of each dissection or species from a collection, which should include the number on the slide label, the species name, date and location of collection, name of collector, sex of specimen, number of specimens in collection, length, height, and if possible width of specimen, plus a camera lucida outline drawing of the valves and any other pertinent features. A second card should be made for each species and filed according to taxonomic position, giving collection locations for that species and the number of the record card and slide. A slide cabinet having a number of horizontal trays is suitable for storage of slides of dissected specimens. Drawings should be made with a camera lucida to obtain the outline and should include a side and, where possible, a dorsal view of one or both of the valves and the appendages which show taxonomic features of importance in the particular species.

A few workers advocate clearing and staining of ostracods before dissection and identification. This is not advised unless special studies are contemplated, because it greatly extends the time required to identify specimens, which is already considerable when compared with that required for other entomostraca, and because staining and clearing add little if anything to the clarity or definition of chitinous appendages.

Literature

A list of the more important references helpful in identifying North American fresh-water Ostracoda is given at the end of this chapter. Among the references which should prove of most value are Sars' Ostracoda of Norway (1928), Furtos' Ostracoda of Ohio (1933), Dobbin's Ostracoda from Washington (1941) and Hoff's Illinois Ostracoda (1942b). Müller's comprehensive synopsis of the group (1912) is an excellent reference and starting point, because it summarizes all literature through 1908. The drawings to be found in the present key have been largely taken from those authors' works, and where possible from the original drawings illustrating the new species. The worker in the field will also find the above mentioned reports, especially those of Furtos and Hoff, excellent for anatomical, ecological, bibliographical, and other information on ostracods not covered in the present chapter because of lack of space. Sharpe's chapter on Ostracoda in the 1918 edition of Ward and Whipple also gives an excellent summary of this phase of the study of ostracods.

The following key includes and illustrates all species known to occur in North America through 1953.

KEY TO SPECIES

1a	Second antenna 2-branched; 1 branch (exopodite) rudimentary, immobile; endopodite with at least 7 segments and long natatory setae. Marine . . . . Suborder **Myodocopa**	
1b	Second antenna with endopodite and exopodite both well developed and used in swimming. Marine . . . . . . Suborder **Cladocopa**	
1c	Second antenna with both branches flattened, foliacious. Marine . . . . Suborder **Platycopa**	
1d	Second antenna with well-developed endopodite armed with claws at tip; exopodite either absent or present as a rudimentary scale, or simple, long seta. Mostly fresh water . . . Suborder **Podocopa**	**2**
2a	(1) Exopodite of second antenna simple, long and setiform . . . . Family **Cytheridae**	**166**
2b	Exopodite of second antenna absent or reduced to a platelike scale	**3**
3a	(2) Abdomen without furca; third legs ambulatory and similar to first legs, ventrally directed. . . . . Family **Darwinulidae**	

One genus, **Darwinula** Brady and Norman 1889.
Leg 1 with well-developed anteriorly directed masticatory process; endopodite of 3 segments; large branchial plate.
One species, **D. stevensoni** (Brady and Robertson) 1870. (Fig. 28.4).
Length 0.68–0.75 mm, height 0.27–0.31 mm, width 0.24–0.25 mm. Valves weakly

arched; surface smooth, sparsely hairy and of a pearly luster. Antenna very short, stout, without natatory setae. ♂ not definitely known in America. Muddy bottoms of large lakes. Apr. to Sept. Ohio (Lake Erie), Ga., Ill., Mass., Mich., Yucatan.

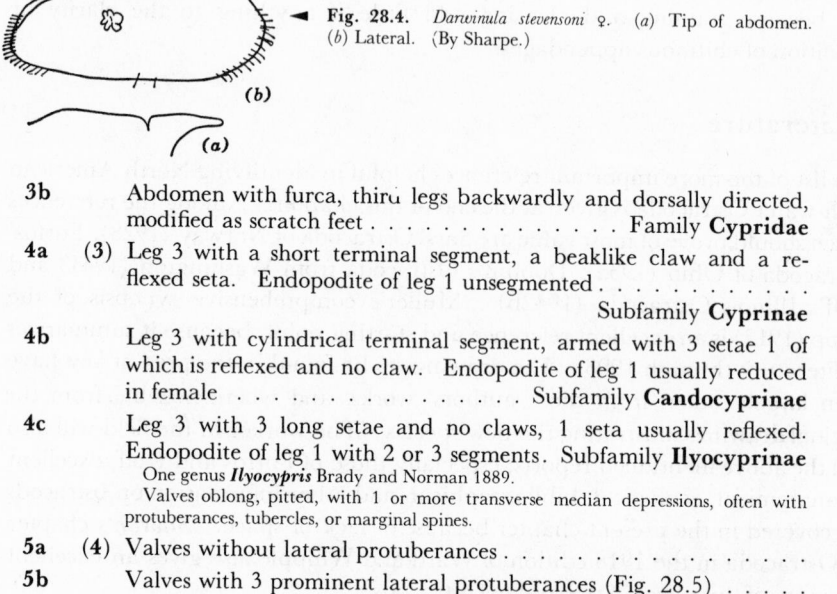

◄ **Fig. 28.4.** *Darwinula stevensoni* ♀. (a) Tip of abdomen. (b) Lateral. (By Sharpe.)

3b   Abdomen with furca, third legs backwardly and dorsally directed, modified as scratch feet . . . . Family **Cypridae**   **4**

4a   (3)   Leg 3 with a short terminal segment, a beaklike claw and a reflexed seta. Endopodite of leg 1 unsegmented . . . . Subfamily **Cyprinae**   **77**

4b   Leg 3 with cylindrical terminal segment, armed with 3 setae 1 of which is reflexed and no claw. Endopodite of leg 1 usually reduced in female . . . . Subfamily **Candocyprinae**   **7**

4c   Leg 3 with 3 long setae and no claws, 1 seta usually reflexed. Endopodite of leg 1 with 2 or 3 segments. Subfamily **Ilyocyprinae**   **5**
One genus *Ilyocypris* Brady and Norman 1889.
Valves oblong, pitted, with 1 or more transverse median depressions, often with protuberances, tubercles, or marginal spines.

5a   (4)   Valves without lateral protuberances **6**

5b   Valves with 3 prominent lateral protuberances (Fig. 28.5) . . . . .
                           Ilyocypris gibba (Ramdohr) 1808
Length 0.85–0.95 mm. Shell much tuberculated anteriorly and posteriorly, decidedly furrowed anteriorly. Two prominent tubercles just back of the eye spot. Furca nearly straight; terminal claws nearly equal in length and plain. Terminal setae of furca about $^2/_5$ length of terminal claw. Swampy regions in mud. Spring. Colo., Ill., Ohio.

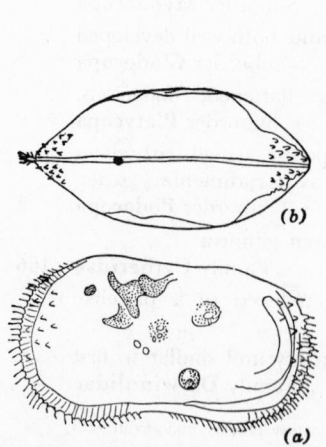

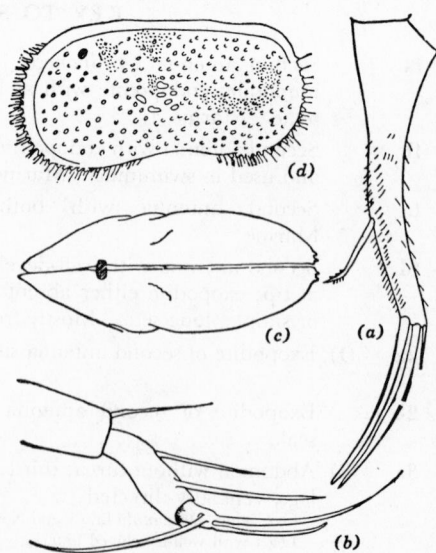

Fig. 28.5. *Ilyocypris gibba* ♀. (a) Lateral. (b) Dorsal. (By Sharpe.)

Fig. 28.6. *Ilyocypris bradyii* ♀. (a) Furca. (b) End of leg 3. (c) Dorsal. (d) Lateral. (By Sharpe.)

6a　**(5)** Natatory setae rudimentary; not extending beyond penultimate
segment (Fig. 28.6) . . . . *I. bradyii* Sars 1890
　　Length 0.95 mm, height 0.50 mm, width 0.32 mm.　Shell weakly tuberculate and
without furrows.　Furca strongly curved with a much broadened base; 10 times as long
as narrowest width; dorsal setae plumose and bent near tip.　Distal half or dorsal
margin of furca with fine hairs.　Running waters and pools left by streams. Colo., Ill.,
Wisc., Ohio (Lake Erie).

6b　　Natatory setae of second antenna well developed, extending to tips
of claws or beyond (Fig. 28.7) . . . . . *I. biplicata* (Koch) 1838
　　Length 1.10 mm, height 0.58 mm.　Valves with 2 transverse folds which are
sharply marked off.　Color opaque, whitish gray with brownish tinge.　Surface of
valves coarsely granular, with densely set small pits.　Furca slightly curved; terminal
claws thin.　♂ slightly smaller than ♀ but otherwise similar.　Ejaculatory tubes with
18 whorls of radiating spikes. Ponds, ditches. Tex.

7a　**(4)** Natatory setae of second antenna always present, usually extending
beyond terminal claws; occasionally rudimentary . . . .
Tribe **Cyclocyprini**　**8**

7b　　Natatory setae of second antenna always absent . . . .
Tribe **Candonini**　**40**

8a　**(7)** Natatory setae of second antenna well developed, extending
beyond tips of terminal claws . . . .　**9**

8b　　Natatory setae of second antenna rudimentary . . . .
Candocypria Furtos 1933
　　One species, *C. osborni* Furtos 1933 (Fig. 28.8).
　　Length 0.97 mm, height 0.55 mm, width 0.38 mm.　Pore canals conspicuous. Hyaline borders at anterior and posterior end.　Left valve longer than right.　Surface
smooth with a few hairs and puncta.　Color yellow-brown with dark brown on mid-
lateral surface and along dorsal margin.　With 5 natatory setae on antenna 2, extending only to proximal third of penultimate segment.　Leg 3 with terminal segment with
unequal terminal setae.　Furca almost straight, 10 times narrowest width; dorsal
margin smooth; terminal seta short.　♂ similar to ♀ but shorter; ejaculatory duct
cylindrical with 5 whorls of spines distinctly separated.　Small streams.　May and
early June. Ohio.

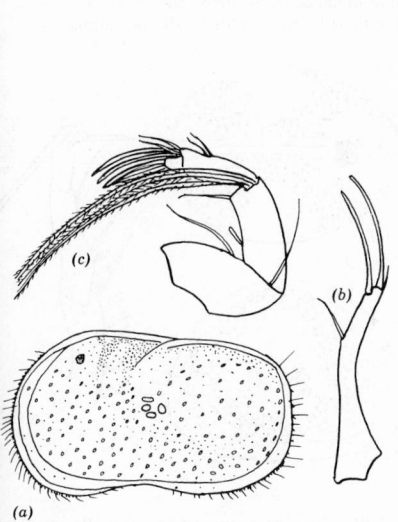
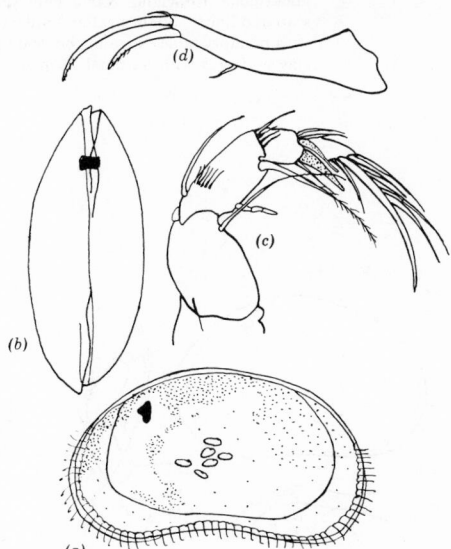

Fig. 28.7.　*Ilyocypris biplicata* ♀.　*(a)* Lateral.
(b) Furca.　*(c)* Antenna 2.

Fig. 28.8.　*Candocypria osborni.*　*(a)* Lateral ♀.
(b) Dorsal ♀.　*(c)* Antenna 2 ♂.　*(d)* Furca ♂.

9a   (8) Valves compressed. Terminal segment of leg 3 small, with 2 short
           claws and a long, reflexed seta   20
9b   Valves very tumid. Terminal segment of leg 3 otherwise.   10
10a   (9) Leg 3 with 1 long, reflexed seta and 2 clawlike setae . . . .
                                                    Cyclocypria Dobbin 1941
           One species, *C. kinkaidia* Dobbin 1941 (Fig. 28.9).
           Length 0.45 mm. Valves equal, oval, rounded evenly dorsally. Rather long hairs
           on anterior and posterior margins of valves. Natatory setae of antenna 2 reach beyond
           tips of terminal claws by 1½ times the length of terminal claws. Furca with terminal
           claw more than half the length of ramus. ♂ unknown. Lakes, ponds. May through
           July. Wash.

10b   Leg 3 with 1 long, reflexed seta, 1 shorter reflexed seta, and 1
           short seta. . . . . *Cyclocypris* Brady and Norman 1889   11
11a (10) Length of valves less than 0.5 mm   18
11b   Length of valves greater than 0.5 mm   12
12a (11) Dorsal seta of furca present, well developed   13
12b   Dorsal seta of furca rudimentary or absent (Fig. 28.10) . . . . . .
                                                        C. cruciata Furtos 1935
           Length of ♂ 0.53 mm, height 0.37 mm, width 0.37 mm. Color light with dark blue
           bands forming an "X" when viewed from above. Left valve longer and higher than
           right. Anterior margin of right valve with conspicuous flange. Surface of valves
           smooth with a few scattered hairs. Natatory setae of antenna 2 extend beyond tips of
           terminal claws by not quite length of claws. Terminal segment of leg 3 three times
           longer than wide, the short terminal seta gently curved, ⅓ length of segment. Furca
           straight, 12½ times narrowest width, dorsal margin smooth, dorsal seta absent, re-
           placed by a papilla. ♀ unknown. Ponds, lakes. June, Aug. Mass., N. Y.
           (Chautauqua Lake.)

13a (12) Length of valves less than 0.70 mm.   16
13b   Length of valves greater than 0.70 mm.   14
14a (13) Length of valves less than 0.80 mm.   15
14b   Length of valves greater than 0.80 mm (Fig. 28.11). . . . .
                                                          C. globosa (Sars) 1862
           Length of ♀ 0.88 mm; ♂ slightly larger. Height equal to ⅔ the length. Width
           about equal to height. Color light yellowish-brown. Surface of valves smooth with
           scattered hairs. Slight hyaline borders in anterior and posterior regions. Antenna 2
           with natatory setae extending beyond tips of terminal claws by almost the length of the
           claws. Leg 3 with terminal segment greater than ½ the length of the penultimate

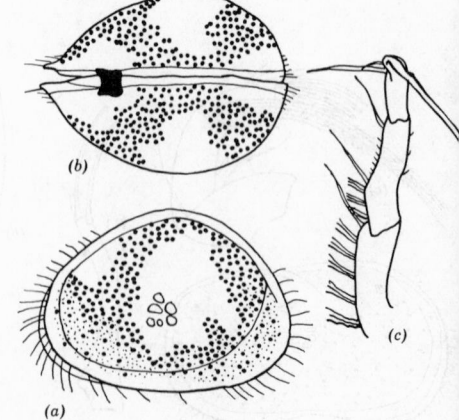

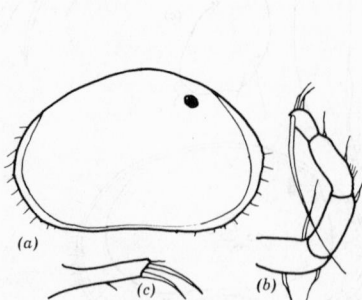

Fig. 28.9. *Cyclocypria kinkaidia* ♀. (a) Right
valve. (b) Leg 3. (c) Furca.

Fig. 28.10. *Cyclocypris cruciata* ♂. (a) Right valve.
(b) Dorsal. (c) Leg 3.

segment. Furca long and fairly straight with minute spinules on most of the dorsal edge. Dorsal and terminal setae of about the same length. Claws short and stout. Lakes. June. Northwest Territories (Dolphin and Union Strait).

15a   **(14)** Shell evenly rounded on dorsal margin (Fig. 28.12) . . . .
                                     C. nahcotta Dobbin 1941
      Length 0.71 mm. Height $2/3$ length. Surface of valves covered with puncta and scattered hairs. Long, coarse hairs on extremities. Antenna 2 with natatory setae reaching beyond tips of claws by more than the length of the antenna. Terminal claw of leg 2 long and slender. Terminal segment of leg 3 equal to $2/3$ length of penultimate segment. Furca with long, slender claws. ♂ present. Lakes, pools. May, Aug., Nov. Wash. (Lake Washington).

15b   Shell boldly arched, seen laterally; highest about the middle (Fig. 28.13) . . . . ***C. ampla*** Furtos 1933
      Length 0.74 mm, height 0.52 mm, width 0.52 mm. Color glossy yellow-brown, densely speckled with small dark-brown spots. Surface of valves with numerous long hairs and a few scattered puncta. Natatory setae of antenna 2 extend beyond tips of terminal claws by twice the length of the claws. Terminal segment of leg 3 $2/3$ as long as penultimate segment. Furca curved, 15 times longer than narrowest width; distal half of dorsal margin with sparse fine hairs. ♂ slightly smaller than ♀. Ponds, marshes, small lakes, cold streams. Feb. to Apr., and Nov. Ohio.

16a   **(13)** Terminal claws of furca strongly bent at tips. **17**

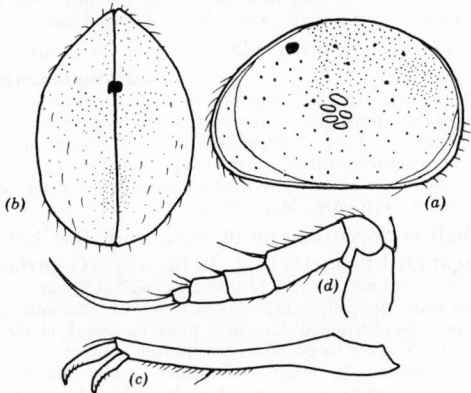

Fig. 28.11. *Cyclocypris globosa* ♀. (*a*) Lateral. (*b*) Dorsal. (*c*) Furca. (*d*) Leg 2.

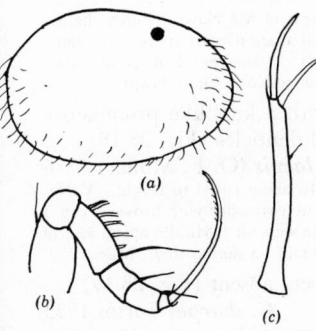

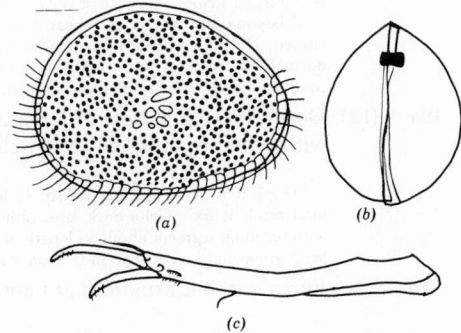

Fig. 28.12. *Cyclocypris nahcotta* ♀. (*a*) Right valve. (*b*) Leg 2. (*c*) Furca.

Fig. 28.13. *Cyclocypris ampla.* (*a*) Right valve ♀. (*b*) Dorsal ♀. (*c*) Furca ♂.

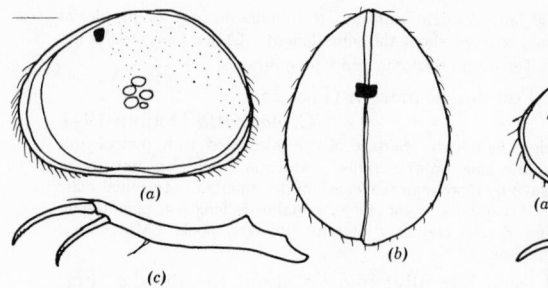

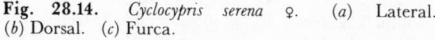

Fig. 28.14. *Cyclocypris serena* ♀. (*a*) Lateral. (*b*) Dorsal. (*c*) Furca.

Fig. 28.15. *Cyclocypris washingtonensis* ♀. (*a*) Left valve. (*b*) Leg 3. (*c*) Furca.

16b   Terminal claws of furca slightly curved at tips (Fig. 28.14) . . . . .
                                                C. serena (Koch) 1841

Length 0.60 mm. Height nearly ¾ length. Color dark, olivaceous brown. Surface of valves smooth and polished with delicate hairs at extremities. Leg 3 with terminal segment scarcely ½ as long as penultimate segment. Furca slightly curved, dorsal margin nearly smooth; claws slender, the terminal claw equal to about ½ length of ramus. ♂ similar to ♀. Mont., Mich., Pribilof Islands.

17a **(16)** Dorsal margin of furca plain; valves without tubercles (Fig. 28.15) . . . . **C. washingtonensis** Dobbin 1941

Length 0.59 mm. Height ⅗ length; very tumid. Surface of valves with scattered hairs; longer hairs at margins. Natatory setae reaching the length of the antenna beyond tips of terminal claws. Terminal claw of leg 2 1½ times length of 3 terminal segments and prominently toothed. Terminal segment of leg 3 ⅔ length of penultimate segment. Furca rather stout; dorsal and terminal setae about equal in length. ♂ present. Ponds. Feb., Apr., May. Wash.

17b   Distal half of dorsal margin of furca with fine hairs; valves with small, scattered tubercles (Fig. 28.16) . . . **C. forbesi** Sharpe 1897

Length 0.55–0.62 mm, height 0.39–0.47 mm, width 0.36 mm. Color, sepia brown. Occasional hairs, especially along margins of valves. Natatory setae of antenna 2 reach beyond tips of terminal claws by 3 times the length of the claws. Terminal segment of leg 3 ⅗ the length of the penultimate segment. Furca somewhat bent, terminal claw about ½ the length of the ramus. ♂ very similar in shape and size to ♀. Lakes, ponds (among vegetation). Apr., Aug. Ill., Mass., S. C.

18a **(11)** Dorsal seta of furca absent or rudimentary 1

18b   Dorsal seta of furca present (Fig. 28.17) . . **C. ovum** (Jurine) 1820

Length 0.49 mm, height 0.33 mm, width 0.33 mm. Color bright chestnut brown with lighter areas which give it a striped appearance when viewed from above. This species can be distinguished from *C. sharpei* by having a deeper color and not so pronounced stripes. Surface of valves slightly hairy. Natatory setae of antenna 2 extend beyond tips of claws by nearly twice the length of the claws. Furca slightly curved, 11 times as long as narrowest width; dorsal margin with sparse fine hairs; dorsal and terminal seta of about the same length. ♂ smaller than ♀, otherwise similar. Ponds, edges of shallow lakes, marshes. Apr. to Nov. Ohio, Wash.

19a **(18)** Dorsal margin of base of furca armed with a knoblike prominence with 3 points, followed by 4 well-marked denticles (Fig. 28.18). . .
                                  C. laevis (O. F. Müller) 1776

Length 0.50 mm. Height nearly ¾ length; width about equal to height. Valves moderately hairy. Color dark, brownish gray, delicately speckled with brown. Leg 3 with terminal segment about ½ length of penultimate segment. Middle apical seta of leg 3 somewhat longer than in *C. ovum;* distinctly sigmoid. ♂ similar to ♀. Mass.

19b   Furca without armature at base; dorsal seta absent (Fig. 28.19) . .
                                  C. sharpei Furtos 1933

Length 0.47 mm, height 0.33 mm, width 0.33 mm. Color chestnut brown with lighter areas in ocular region and posterior to the middle. Surface of valves with long hairs and a few scattered pucta. Natatory setae of antenna 2 extend beyond tips of terminal claws by the length of the claws. Furca slightly curved, 10 times narrowest width in length, and with a smooth dorsal margin. ♂ slightly smaller than ♀, otherwise similar. Ponds, lakes, marshes. Feb. to Nov. Ohio, Ill., N. Y., Fla., Iowa, S. C., La., New Brunswick.

20a	**(9)**	Margins of one valve tuberculated . . . . ***Physocypria*** Vavra 1897	**31**
20b		Margins of both valves smooth . . . . ***Cypria*** Zenker 1854	**21**
21a	**(20)**	Surface of shell without striations	**22**
21b		Surface of shell with closely set, parallel, anastomosing striations (Fig. 28.20) . . . . ***C. turneri*** Hoff 1942	

Length 0.55 mm, height 0.35 mm. Width slightly less than ½ the length. Anterior and posterior margins with thin hyaline border. Color yellow with occasional lightbrown markings. A few scattered hairs along the margins. Natatory setae of antenna 2 reach beyond tips of terminal claws by about 3 times the length of the longest terminal claw but vary considerably in length. Furca stout and curved; ventral margin 7 times narrowest width of ramus. ♂ similar to ♀ but slightly smaller. Furca more curved; prehensile palps dissimilar and unequal. This species has long been confused with the European *C. exculpta* to which it is closely related. Ponds

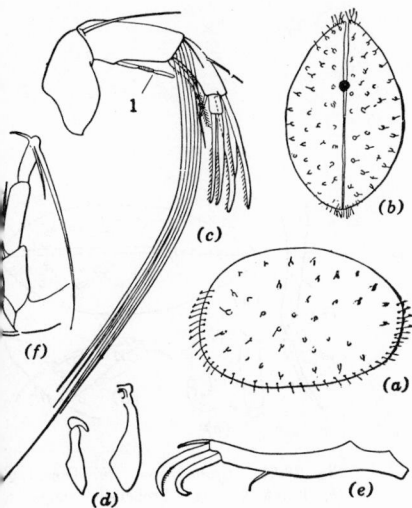

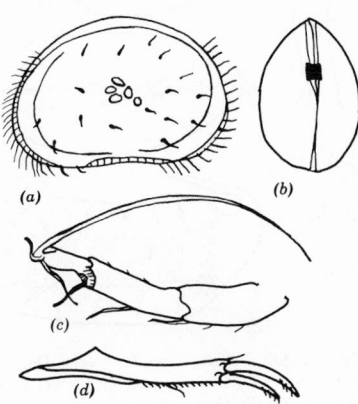

ig. 28.16. *Cyclocypris forbesi* ♂. (*a*) Right valve.) Dorsal. (*c*) Antenna 2. (*d*) Maxillary palps.) Furca. (*f*) Leg 3. (By Sharpe.)

Fig. 28.17. *Cyclocypris ovum.* (*a*) Left valve ♀. (*b*) Dorsal ♀. (*c*) Leg 3 ♂. (*d*) Furca ♂.

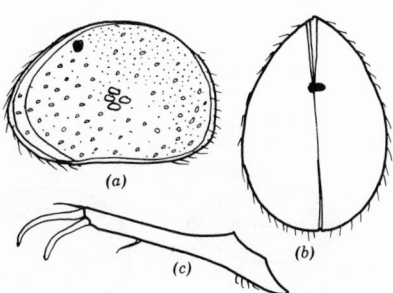

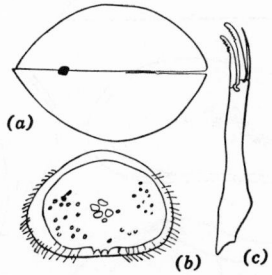

Fig. 28.18. *Cyclocypris laevis* ♀. (*a*) Lateral. (*b*) Dorsal. (*c*) Furca.

Fig. 28.19. *Cyclocypris sharpei* ♀. (*a*) Left valve. (*b*) Dorsal. (*c*) Furca. (By Sharpe.)

and lakes, often associated with algae, water plants, and grass. Very abundant from Mar. to June. Newfoundland, Dela., Ill., Ala., Ohio, Tenn., Miss., Va., S. C., Wis., Mich., Utah, Wash., Alaska.

22a (21) Shell more than 0.70 mm in length 23
22b   Shell less than 0.70 mm in length 24
23a (22) Terminal claw of furca ⅗ length of ventral margin of furca; ventral margin of shell almost straight (Fig. 28.21) . . . .
$$\textbf{\textit{C. obesa}}\text{ Sharpe 1897}$$
Length 0.74–0.86 mm, height 0.44–0.52 mm. Color brownish. Width about ⅖ length. Natatory setae of antenna 2 extend beyond tips of terminal claws by twice the length of the claws. Furca with terminal seta about 1½ times narrowest width of ramus in length; dorsal seta equal to about ⅓ length of subterminal claw. ♂ slightly smaller than ♀ but otherwise similar. Penis triangular with 2 subequal terminal lobes. Lakes, pools, springs in shallow water associated with grasses and water plants. Summer. Ill., Ohio, D. C., Mich., S. C.

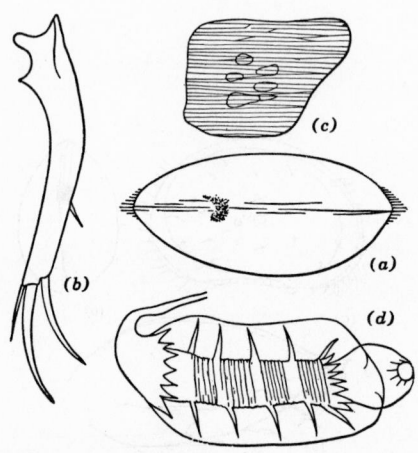

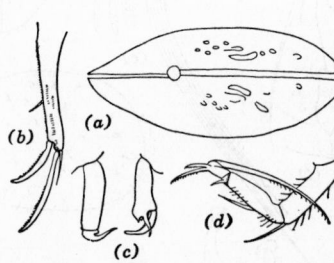

Fig. 28.20. *Cypria turneri.* (a) Dorsal ♀. (b) Furca ♀. (c) Striations on shell. (d) Ejaculatory duct ♂. (By Sharpe.)

Fig. 28.21. *Cypria obesa.* (a) Dorsal ♀. (b) Furca ♀. (c) Prehensile palps ♂. (d) Leg 3 ♀. (By Sharpe.)

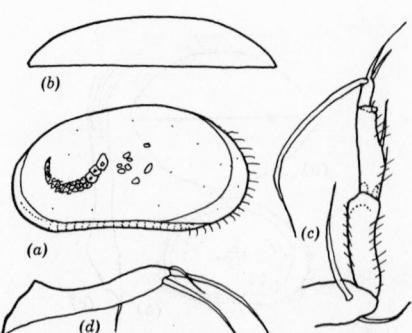

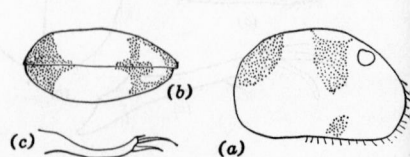

Fig. 28.22. *Cypria mediana* ♀. (a) Right valve. (b) Left valve from below. (c) Leg 3. (d) Furca.

Fig. 28.23. *Cypria inequivalva* ♀. (a) Lateral. (b) Dorsal. (c) Furca. (By Sharpe.)

23b   Terminal claw of furca almost as long as ventral margin of furca; shell elongate; ventral margin of shell concave (Fig. 28.22) . . . . .
         C. mediana Hoff 1942
     Length 0.94 mm, height 0.50 mm. Width about $^2/_5$ length. Color, light with a few flakes of pigment on the anterior and posterior slopes of the valves. Natatory setae of antenna 2 reach beyond tips of terminal claws by about the length of the claws. Distal setae on terminal segment of leg 3 unequal, the longer being about twice the length of the shorter. Furca slightly curved. About 7 times longer than narrowest width of ramus. ♂ similar to ♀ but smaller. Testes large. Right valve with slight sinuation at both ends of dorsal margin. Claw of leg 2 relatively longer than in ♀   Pools among plants. June. Ill.

24a   **(22)**   Dorsal seta of furca well developed **25**
24b   Dorsal seta of furca absent or rudimentary (Fig. 28.23) . . . .
         C. inequivalva Turner 1893
     Length 0.42–0.55 mm, height 0.35–0.38 mm, width 0.26–0.28 mm. Surface of valves glossy with fine hairs. Shell light color with anterior and posterior band of darker color. Natatory setae of antenna 2 extend beyond tips of terminal claws by 3 times the length of the claws. Furca strongly curved; dorsal seta absent or rudimentary. ♂ similar. Ejaculatory duct with 5 whorls closely crowded together. Shallow ponds among algae. Ohio, Ga., S. C.

25a   **(24)**   Dorsal seta of furca less than ½ length of subterminal claw **26**
25b   Dorsal seta of furca at least ½ length of subterminal claw **27**
26a   **(25)**   Three color blotches on shell, one anterior, one posterior, and one behind the eye (Fig. 28.24) . . . . ***C. maculata*** Hoff 1942
     Length 0.48–0.56 mm, height 0.33–0.39 mm. Width greater than ½ the length. Surface of shell with a few scattered hairs. Natatory setae of antenna 2 extend beyond tips of terminal claws by 3 times the length of the claws. Furca somewhat curved; length of ventral margin about 8 times narrowest width of ramus. ♂ with shell of same size and shape as ♀. Propodus of right prehensile claw of ♂ much enlarged distally. Clear water in ponds, lake shores among algae. Ill.

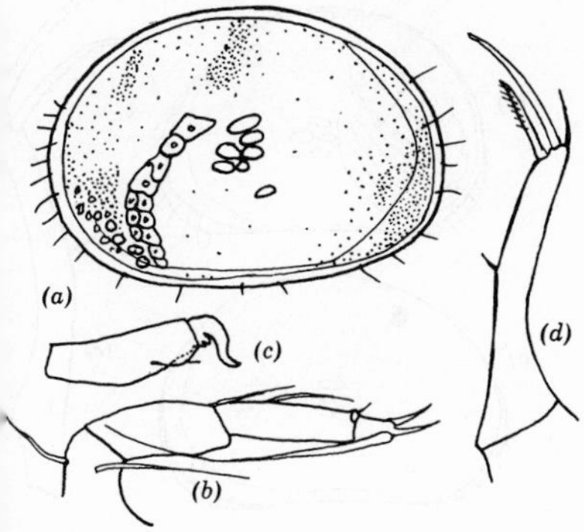

Fig. 28.24. *Cypria maculata.* (a) Left valve ♀. (b) Leg 3 ♀. (c) Right prehensile palp ♂. (d) Furca ♀.

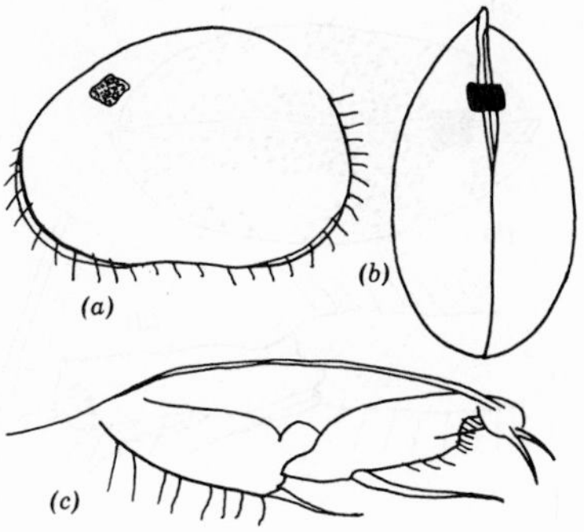

Fig. 28.25. *Cypria pellucida* ♀. (a) Lateral. (b) Dorsal. (c) Leg 3.

26b   Color blotches absent, shell hyaline yellow (in alcohol) (Fig. 28.25) . . . . ***C. pellucida*** Sars 1901
     Length 0.58 mm, height 0.39 mm, width 0.31 mm. Valves transparent without striations. Left valve projects beyond right valve at anterior end. Leg 3 with 2 short, equal setae on terminal segment, equal in length to 1⅔ length of the segment. Furca straight; dorsal seta about ⅓ length of subterminal claw. Terminal seta slightly less than ½ length of terminal claw. ♂ present. Marshes. Ohio.

27a   **(25)**   Margins of valves without prominent pore canals **28**

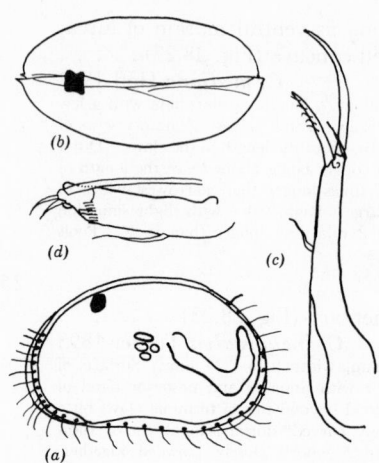

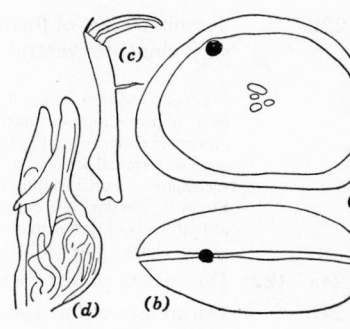

Fig. 28.27. *Cypria ophthalmica.* (a) Latei ♀. (b) Dorsal ♀. (c) Furca ♀. (d) Pen (By Sharpe.)

Fig. 28.26. *Cypria pseudocrenulata* ♀. (a) Left valve. (b) Dorsal. (c) Furca. (d) Leg 3 distal portion.

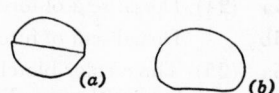

Fig. 28.29. *Cypria mons.* (a) Lateral. (b) Dorsal. (After Chambers.)

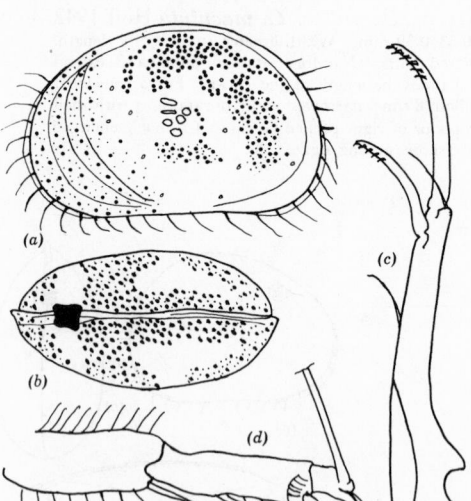

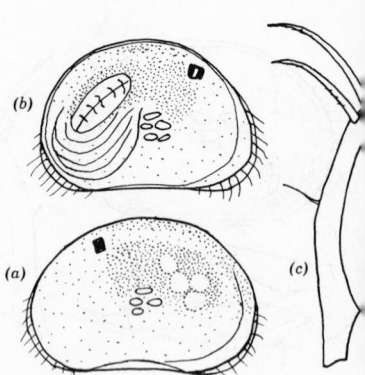

Fig. 28.28. *Cypria palustera.* (a) Left valve ♀. (b) Dorsal ♂. (c) Furca ♂. (d) Leg 3 ♀.

Fig. 28.30. *Cypria lacustris.* (a) Lateral (b) Lateral ♂. (c) Furca ♀.

27b   Margin of each valve with a row of short, thick pore canals which resemble tubercles (Fig. 28.26) . . **C. pseudocrenulata** Furtos 1936
    Length 0.55 mm, height 0.35 mm, width 0.29 mm. Surface of valves smooth with a few marginal hairs. Well-developed hyaline border along anterior margin of both valves. Left valve projects beyond right at anterior end. Natatory setae of antenna 2 project beyond tips of terminal claws by 3 times length of claws. Furca slightly curved; 13 times longer than narrowest width. ♂ smaller than ♀, otherwise similar. Creeks. Aug. Fla.

28a **(27)** Dorsal margin of furca with fine hairs   2

28b   Dorsal margin of furca plain   3

29a **(28)** Valves nearly equal in length and highly arched when seen laterally; surface of valves evenly covered with small specks of brown pigmentation (Fig. 28.27). . . . . . . *C. ophthalmica* (Jurine) 1820

Length 0.56–0.61 mm, height 0.36–0.44 mm. Width about ⅖ the length. Surface of valves smooth, with small dark spots and with hyaline border at anterior and posterior ends. Natatory setae of antenna 2 extend beyond tips of terminal claws by more than 3 times the length of the terminal claws. Terminal segment of leg 3 almost as wide as long; the 2 shorter terminal setae slightly unequal. Furca stout and slightly curved; dorsal margin finely toothed. ♂ slightly smaller than ♀ but otherwise similar in shape and coloration. In waters containing an abundance of decaying material; tolerant of acid conditions. Nova Scotia, Ga., S. C., Ill., Tenn., Mont., Fla., Utah, Wash., Mich., Alaska.

29b Left valve larger than right; valves elongate; color light brown with dark brown areas (Fig. 28.28) . . . *C. palustera* Furtos 1935

Length 0.63 mm, height 0.39 mm, width 0.29 mm. Surface of valves smooth. Hyaline border well developed. Natatory setae of antenna 2 extend beyond tips of terminal claws by 3 times the length of the claws. Terminal setae of leg 3 equal in length and 1½ times length of terminal segment. Furca slightly curved, 10 times longer than narrowest width. Dorsal margin smooth. ♂ smaller than ♀ otherwise similar. Ponds and marshes. June. Mass.

30a **(28)** Valves with numerous, almost confluent puncta. (Fig. 28.29) . . . *C. mons* Chambers 1877

Length 0.70 mm. This ostraced is not well described and is of doubtful validity. Colo. (Mt. Elbert) at 11,000 feet elevation.

30b Valves without puncta (Fig. 28.30) . . . . . *C. lacustris* Sars 1890

Length 0.60 mm, height 0.40 mm, width 0.21 mm. Color transparent white with faint yellowish tinge. Valves very pellucid with smooth and polished surfaces. Broad hyaline borders at both ends. Furca curved, terminal claw about ½ the length of ramus. ♂ smaller than ♀. Posterior portion of the shell broader and more deflexed. Lakes. June. Mich., Tex., Northwest Territories.

31a **(20)** Margin of right valve tuberculated **32**

31b Margin of left valve tuberculated (Fig. 28.31) . . . .
Physocypria dentifera (Sharpe) 1898

Length 0.69 mm, height 0.38 mm, width 0.26 mm. Color brownish-yellow with 4 dark-brown spots. Surface of valves smooth. Natatory setae of antenna 2 reach beyond tips of terminal claws by the length of the antenna. Terminal segment of the leg 3 ⅓ longer than broad; distal setae equal. Furca straight, 10 times longer than narrowest width of ramus; dorsal seta ⅓ the length of subterminal claw. ♂ unknown. Weedy ponds. Aug. Ohio, N. J., Ill.

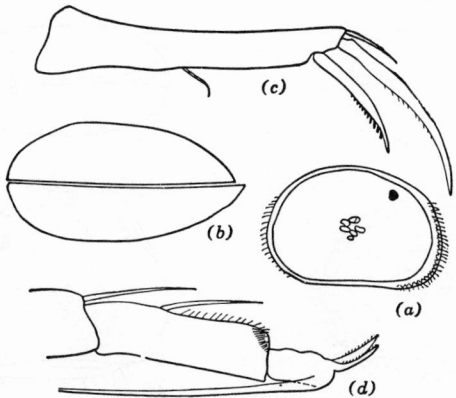

Fig. 28.31. *Physocypria dentifera* ♀. (*a*) Lateral. (*b*) Dorsal. (*c*) Furca. (*d*) Leg 3. (By Sharpe.)

32a (31) Length less than 0.70 mm **33**

32b Length greater than 0.70 mm (Fig. 28.32) . . . .

P. posterotuberculata Furtos 1935

    Length 0.72 mm, height 0.45 mm, width 0.40 mm. Color light with narrow chestnut-brown band encircling the margin, and surrounding the eye. Posterior extremity of right valve with a distinct row of tubercles, which extend beyond the margin. Natatory setae of antenna 2 extend beyond tips of terminal claws by 4 times the length of the claws. Terminal setae of leg 3 nearly equal. Furca curved, $9\frac{1}{2}$ times longer than narrowest width; dorsal margin smooth. ♂ unknown. Ponds. August. Mass.

33a (32) Terminal claw of furca at least $\frac{1}{2}$ length of ramus **35**

33b Terminal claw of furca less than $\frac{1}{2}$ length of ramus. **34**

34a (33) Right valve larger than left (Fig. 28.33) . . . .

P. denticulata (Daday) 1905

    Length 0.65 mm, height 0.44 mm, width 0.33 mm. Valves sparsely hairy, with scattered puncta. Natatory setae of antenna 2 extend beyond tips of terminal claws by $3\frac{1}{2}$ times length of claws. Terminal setae of leg 3 almost equal. Furca slightly curved and 10 times narrowest width of ramus, in length. Dorsal margin of furca smooth. ♂ smaller than ♀, otherwise similar; ejaculatory duct with 5 crowns of spines. Pools. July, Aug. Yucatan.

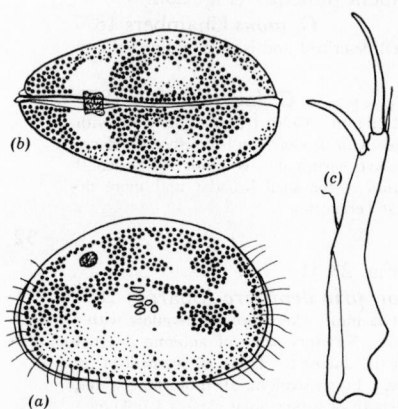 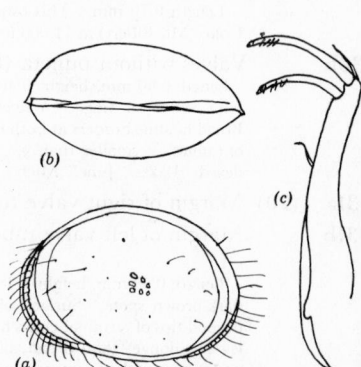

Fig. 28.32. *Physocypria posterotuberculata* ♀. (*a*) Lateral. (*b*) Dorsal. (*c*) Furca.

Fig. 28.33. *Physocypria denticulata.* (*a*) Left valve ♀. (*b*) Dorsal ♀. (*c*) Furca ♂.

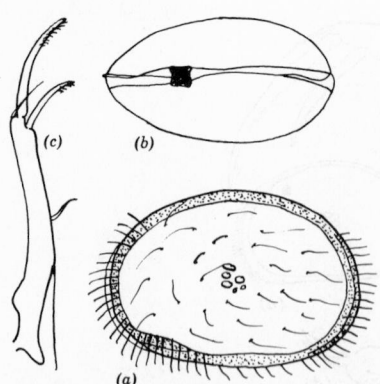 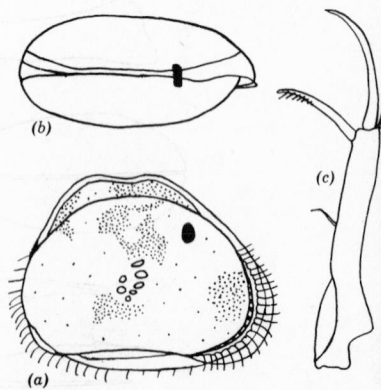

Fig. 28.34. *Physocypria fadeewi* ♀. (*a*) Left valve. (*b*) Dorsal. (*c*) Furca.

Fig. 28.35. *Physocypria inflata.* (*a*) Lateral ♀. (*b*) Dorsal ♀. (*c*) Furca ♂.

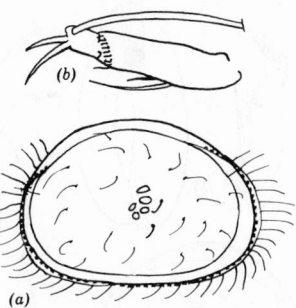

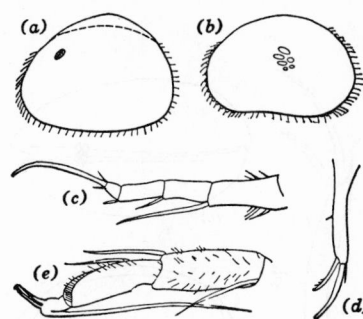

Fig. 28.36. *Physocypria exquisita* ♀. (*a*) Right valve. (*b*) Distal portion of leg 3.

Fig. 28.37. *Physocypria pustulosa* ♀. (*a*) Left valve. (*b*) Right valve. (*c*) Leg 2. (*d*) Furca. (*e*) Leg 3. (By Sharpe.)

34b   Left valve larger than right (Fig. 28.34) . . . .
                                P. fadeewi Dubowsky 1927
          Length 0.65 mm, height 0.45 mm, width 0.32 mm. Marginal denticles are absent from mid-ventral region of right valve. Surface of valves smooth with scattered puncta bearing long, curved hairs. Natatory setae of antenna 2 reach beyond tips of terminal claws by 4 times length of claws. Terminal setae of leg 3 unequal, the longer seta being 1½ times the length of the terminal segment. Furca gently curved and 14 times length of narrowest part of ramus. ♂ not yet reported from America. Rivers. Aug. Fla.

35a   (33)   Left valve with dorsal margin convexly rounded. **36**

35b   Left valve with distinctly sinuated dorsal (Fig. 28.35) . . . .
                                P. inflata Furtos 1933
          Length 0.55–0.62 mm, height 0.41–0.46 mm, width 0.21–0.22 mm. ♂ smaller than ♀, otherwise similar. Dorsal margin of left valve with a distinct sinuation in the middle which gives the appearance of two humps. Pore canals prominent. Color white with brown spots. Surface of valves smooth, sparsely hairy. Natatory setae of antenna 2 extend beyond tips of terminal claws by 3 times length of claws. On mud bottoms in lakes, shallow and deep water. May to Nov. Ohio (Lake Erie).

36a   (35)   Right valve with dorsal flange and higher than left valve **39**

36b   Left valve with dorsal flange, or at least larger than right valve   . .   **37**

37a   (36)   Surface of valves smooth, without pits **38**

37b   Surface of valves with ovoid pits between which are scattered puncta bearing hairs (Fig. 28.36) . . . . . **P. exquisita** Furtos 1936
          Length 0.63 mm, height 0.42 mm. Marginal hairs long. Natatory setae of antenna 2 reach beyond tips of terminal claws by 4 times the length of claws. Terminal setae of leg 3 unequal, the longer seta being almost twice the length of the distal segment. Furca with subterminal claw equal in length to about ½ the length of ramus; terminal seta ⅓ length of terminal claw. ♂ unknown. Aug. Fla.

38a   (37)   Posteroventral margin of right valve with 2 to 4 tubercles (Fig. 28.37) . . . . **P. pustulosa** Sharpe 1898
          Length 0.62 mm, height 0.45 mm. ♂ small, otherwise similar. Color light brown with large dark-brown spots. Surface of valves smooth, moderately hairy with a few puncta. Dorsal flange on left valve. Natatory setae of antenna 2 reach beyond tips of terminal claws by 3 times the length of claws. Terminal setae of leg 3 almost equal. Furca curved, about 9½ times as long as narrowest width of furca; dorsal seta equal to ⅓ the length of subterminal claw; terminal seta ⅓ the length of terminal claw. Lakes along shallow stony bars, rock pools, weedy inlets. May to Nov. S. C., Md., Ohio, Ill., Mich.

38b   Posteroventral margin of right valve with 3 large tubercles followed by many small ones (Fig. 28.38) . . . . . . **P. globula** Furtos 1933
          Length 0.63 mm, height 0.41 mm, width 0.43 mm. Very broad when viewed from above; left valve projects considerably beyond right at anterior end. Color bluish

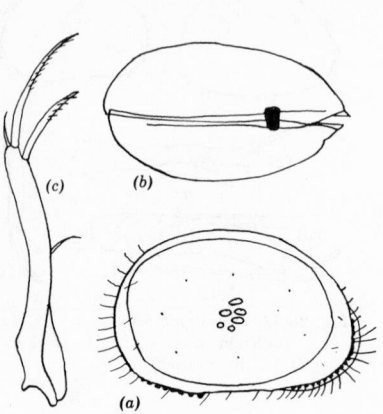

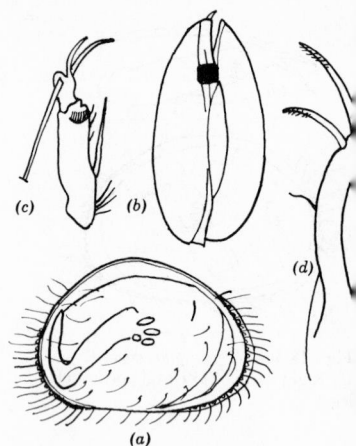

Fig. 28.38. *Physocypria globula.* (*a*) Right valve ♀. (*b*) Dorsal ♀. (*c*) Furca ♂.

Fig. 28.39. *Physocypria gibbera.* (*a*) Rig valve ♀. (*b*) Dorsal ♀. (*c*) Distal portion leg 3. (*d*) Furca ♂.

gray to yellow with large, conspicuous reddish-brown spots. Natatory setae of antenna 2 reach beyond tips of terminal claws by 3 times the length of claws. Terminal setae of leg 3 slightly unequal. Furca curved, 11 times narrowest width of ramus. Dorsal seta of furca ⅓ the length of subterminal claw; terminal claw smooth and longer than ½ length of ramus; terminal seta ⅓ length of terminal claw. ♂ smaller than ♀, otherwise similar. Ponds, small lakes. Mar. to Oct. Ohio.

39a (36) Leg 3 with shorter claw ¾ the length of longer claw (Fig. 28.39) .

                                                P. gibbera Furtos 1936

Length 0.60 mm, height 0.40 mm, width 0.37 mm. Surface of valves minutely punctate and covered with slender, long hairs. Right valve with distinct humplike dorsal flange. Natatory setae extend beyond tips of terminal claws by 3 times the length of the claws. Terminal setae of leg 3 unequal. Furca straight, 10 times narrowest width. ♂ shorter and higher than ♀, ejaculatory duct with 5 crowns of spines. Pools, lakes. Aug. Fla.

39b Leg 3 with two very unequal terminal claws, the longer 2½ times the length of the shorter (Fig. 28.40) . . . .

                                       P. xanabanica Furtos 1936

Length 0.53 mm, height 0.34 mm, width 0.25 mm. Similar to *P. denticulata* but smaller. Marginal tubercles extend further dorsally along anterior and posterior margins. Surface of valves smooth and hairless with scattered puncta. Natatory setae of antenna 2 extend beyond tips of terminal claws by 2½ times the length of the claws. Terminal setae of leg 3 very unequal. Furca curved, 8 times narrowest width; dorsal margin smooth; dorsal seta ⅔ length of subterminal claw, and situated in about the middle of the ramus. ♂ smaller, otherwise similar; ejaculatory duct with 5 crowns of spines. Cenotes. June. Yucatan.

40a (7) Surface of valves smooth.   4

40b Surface of valves tuberculated . . . . . ***Paracandona*** Hartwig 1899

One species, ***P. euplectella*** (Brady and Norman) 1889 (Fig. 28.41).

Small tumid forms; penultimate segment of leg 3 divided, each division with a strong distal seta. Surface of valves reticulated and covered with small, scattered tubercles.

Length 0.68 mm, height 0.32 mm, width 0.32 mm. Valves equal; surface conspicuously reticulated with small scattered tubercles, each with a long stiff hair. Furca slightly curved, 10 times as long as narrowest width; dorsal seta larger than subterminal claw; terminal seta short. ♂ 0.77 mm in length, otherwise similar to ♀. Lakes. Apr. Ohio, N. J.

41a (40) Dorsal seta of furca absent; shell laterally compressed; valves thin . . . . ***Candonopsis*** Vavra 1891

One species, ***C. kingsleii*** (Brady and Robertson) 1870 (Fig. 28.42).

Leg 3 with two unequal terminal setae, the shorter one less than ½ the length of the longer.

Length 0.95 mm, height 0.50 mm, width 0.20 mm. Valves smooth and shiny with a few fine hairs. Antennae 1 and 2 both slender. Mandibular palp with terminal segment narrowly produced. Furca very narrow and slightly curved; dorsal seta absent, claws without strong teeth. ♂ slightly larger than ♀; ejaculatory tubes large and conspicuous through the valve. In leaf cups of bromeliads (muddy creeks in Europe). Dec. Puerto Rico.

41b	Dorsal seta of furca well developed . . . . . . ***Candona*** Baird 1842	**42**
	For species described since this key was finished, see Tressler (1954, 1957).	
42a (41)	Length greater than 1.50 mm.	**43**
42b	Length less than 1.50 mm.	**47**
43a (42)	Height either greater or less than ½ the length	**44**
43b	Height equal to ½ the length (Fig. 28.43)	

C. hyalina Brady and Robertson 1870

Length 1.50 mm, height 0.75 mm. Surface of valves sparsely hairy. Dorsal margin

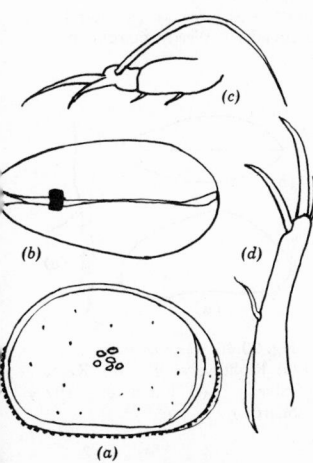

Fig. **28.40.** *Physocypria xanabanica.*
(a) Right valve ♀. (b) Dorsal ♀.
(c) Leg 3 ♂. (d) Furca ♂.

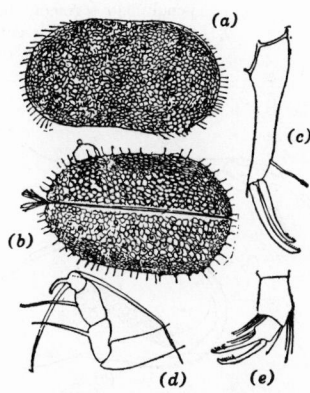

Fig. **28.41.** *Paracandona euplectella* ♀.
(a) Lateral. (b) Dorsal. (c) Furca.
(d) Leg 3. (e) Mandibular palp. (By Sharpe.)

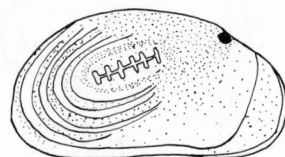

Fig. **28.42.** *Candonopsis kingsleii.*
Lateral ♀.

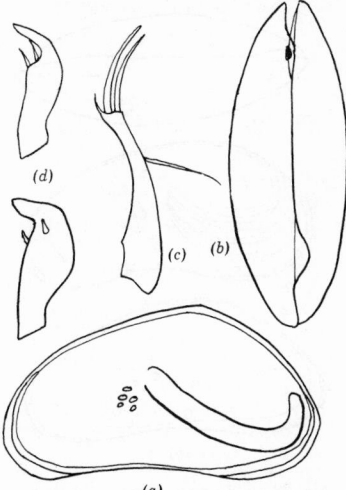

Fig. **28.43.** *Candona hyalina.* (a) Lateral ♀.
(b) Dorsal ♀. (c) Furca ♀. (d) Prehensile palps ♂.

with a rounded point about midway of valve. Ventral margin weakly indented. Furca strongly curved; dorsal margin smooth; dorsal seta about the same length as subterminal claw; terminal seta very short. ♂ about the same size as ♀, but slightly higher; posterior portion of ventral border of valves weakly sinuated. Lake Superior.

44a   **(43)**   Height less than ½ the length **45**

44b   Height greater than ½ the length (Fig. 28.44). . . . .

                                                 C. uliginosa Furtos 1933

           Length 1.50 mm, height 0.85 mm, width 0.70 mm. Left valve extends beyond right anteriorly. Surface sparsely hairy with scattered puncta which resemble minute tubercles near the extremities. Leg 3 with penultimate segment divided; shortest distal seta of terminal segment 4 times the length of the segment. Furca gently curved, 16 times narrowest width; distal third of dorsal margin with sparse fine hairs. ♂ unknown. Temporary marsh. Nov. Ohio.

45a   **(44)**   Length less than 2 mm **46**

45b   Length about 2 mm (Fig. 28.45) . . . . . ***C. ohioensis*** Furtos 1933

           Length 1.78–2.00 mm, height 0.82–0.88 mm, width 0.66 mm. Left valve projects beyond right at both ends. Surface of valves sparsely hairy. Leg 3 with divided penultimate segment; shortest distal seta on terminal segment 3 times the length of the segment. Furca curved, 15 times narrowest width. ♂ as large as ♀; ventral margin of valves more deeply sinuated; posterior end broader. Weedy margins of lakes. Nov. Ohio.

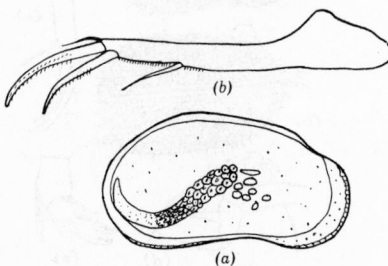

Fig. 28.44.   *Candona uliginosa* ♀.   (*a*) Right valve. (*b*) Furca.

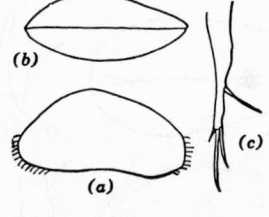

Fig. 28.46.   *Candona crogmaniana.* (*a*) Right valve ♀.   (*b*) Right valve ♂.   (*c*) Furca ♀.   (By Sharpe.)

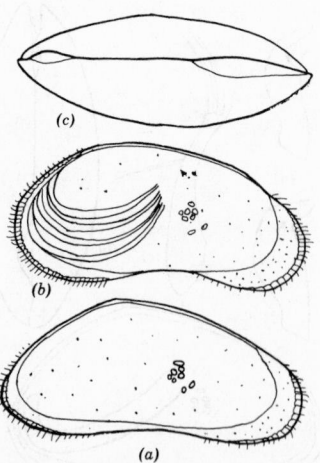

Fig. 28.45.   *Candona ohioensis.* (*a*) Right valve ♀. (*b*) Right valve ♂. (*c*) Dorsal ♀.

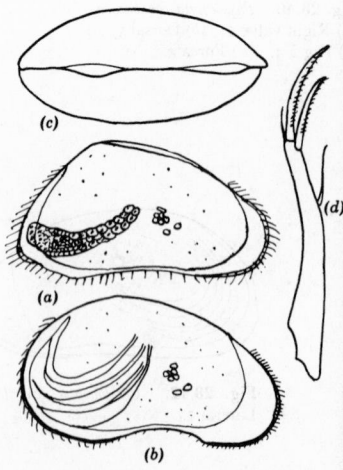

Fig. 28.47.   *Candona intermedia.* (*a*) Right valve ♀. (*b*) Right valve ♂. (*c*) Dorsal ♀. (*d*) Furca ♀.

46a **(45)** Terminal seta of furca only about ⅓ the length of subterminal claw (Fig. 28.46) . . . . **C. crogmaniana** Turner 1894

Length 1.45 mm, height 0.68 mm, width 0.60 mm, elongated; height less than ½ the length. Surface of valves sparsely hairy. Eye black and unusually prominent for the genus *Candona*. Leg 3 with the penultimate segment divided; terminal segment with the shortest distal seta 4 times the length of the segment. Furca slightly curved, 19 times narrowest width. ♂ slightly longer and higher than ♀; dorsal margin of valves more boldly arched. Permanent and temporary ponds, lakes, rivers. Mar. to May, Nov. Ga., Ohio, Ill., Wis., Mich., Kan., Great Slave Lake. Also known as fossil.

46b Terminal seta of furca equal to ½ the length of subterminal claw (Fig. 28.47) . . . . **C. intermedia** Furtos 1933

Length 1.7 mm, height 0.92 mm, width 0.73 mm. Left valve with exterior sinuation on dorsal margin. Left valve projects slightly beyond right at both extremities. Surface sparsely hairy. Leg 3 with penultimate segment divided; terminal segment with shortest distal seta 4 times the length of the terminal segment. Furca slender, 16 times narrowest width. ♂ slightly longer than ♀; ventral margin of left valve with a prominent convex hump near the anterior end. Muddy bottoms, cold, clear streams. May, June. Ohio, Tex.

47a **(42)** Length less than 1.00 mm **48**

47b Length between 1.00 and 1.50 mm **61**

48a **(47)** Height equal to ½ the length **49**

48b Height either greater or less than ½ the length **50**

49a **(48)** Shell sparsely hairy (Fig. 28.48) . . . . **C. parvula** Sars 1926

Length 0.56 mm, height 0.28 mm. Surface smooth and shining with a pearly luster. Leg 3 with penultimate segment undivided; shorter terminal seta of terminal segment short and curved. Furca slightly curved, 8 times narrowest width of ramus. ♂ unknown. Pools in meadows. May. Quebec.

49b Shell with many fine hairs (Fig. 28.49) . . **C. fluviatilis** Hoff 1942

Length 0.68–0.76 mm, height 0.33–0.38 mm, width less than height. Valves covered with pits separated by raised areas. Leg 3 with penultimate segment undivided; shorter terminal seta of terminal segment 2½ to 3 times length of terminal segment. Furca rather stout and little curved, 6 times narrowest width. ♂ unknown. Small streams. Spring. Ill.

50a **(48)** Height less than ½ the length **51**

50b Height greater than ½ the length **56**

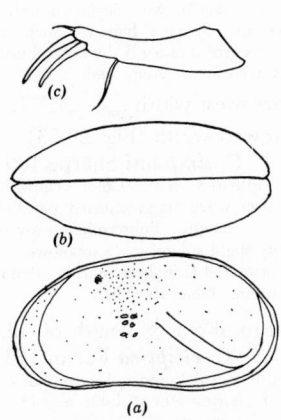

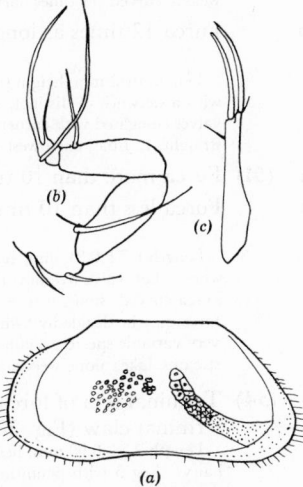

Fig. 28.48. *Candona parvula* ♀. (*a*) Lateral. (*b*) Dorsal. (*c*) Furca.

Fig. 28.49. *Candona fluviatilis* ♀. (*a*) Right valve. (*b*) Leg 3. (*c*) Furca.

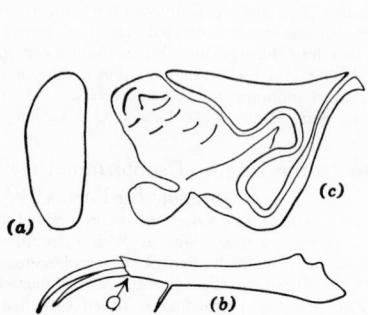

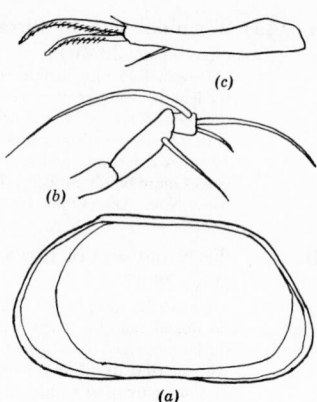

Fig. 28.50. *Candona peirci.* (a) Lateral ♀.
(b) Furca ♂. (c) Penis ♂. (By Sharpe.)

Fig. 28.51. *Candona marengoensis* ♀.
(a) Left valve. (b) Leg 3. (c) Furca.

51a (50) Penultimate segment of leg 3 distinctly or indistinctly divided . . . 5!

51b Penultimate segment of leg 3 undivided . . . . 5·

52a (51) Terminal seta of furca present. . . . . 5!

52b Terminal seta of furca absent (Fig. 28.50) . . . .
C. peirci (Turner) 1895
Length 0.70–0.79 mm, height 0.33–0.37 mm, width 0.22–0.31 mm. Color white tinged with yellow. Shell smooth, much compressed. Furca nearly straight, 12 times as long as width of ramus at narrowest point; subterminal claw more than ⅔ the length of terminal claw. ♂ with ejaculatory duct of 7 whorls of spines. Shallow, weedy ponds. June. Ga.

53a (52) Furca 11 times as long as wide (Fig. 28.51) . . . .
C. marengoensis Klie 1931
Length 0.62 mm, height 0.32 mm, width 0.24 mm. Color light brown. Sparsely haired. Valves with distinct corners, unrounded. Surface of valves with network of distinct, small, polygonal markings. Penultimate segment of leg 3 divided. Furca weakly curved, 11 times narrowest width. ♂ unknown. Cave waters. Aug. Ind.

53b Furca 12 times as long as narrowest width (Fig. 28.52) . . . .
C. jeanneli Klie 1931
Length 0.62 mm, height 0.29 mm, width 0.21 mm. Color brown. Surface of valves with a network of distinct, small, polygonal markings. Sparsely haired. Shape of valves elongated with corners rounded. Penultimate segment of leg 3 divided. Furca straight, 12 times narrowest width. ♂ unknown. Cave waters. Aug. Ind.

54a (51) Furca more than 10 times as long as narrowest width . . . . 5!

54b Furca less than 10 times as long as narrowest width (Fig. 28.53) . .
C. simpsoni Sharpe 1897
Length 0.72–0.81 mm, height 0.33–0.38 mm, width 0.37 mm. Color yellowish-white. Left valve overlaps the right; upper and lower valve margins nearly parallel. Furca curved, stout, 6 to 8 times narrowest width in length. Subterminal claw of furca may be decidedly S-shaped or may show only slight sinuosity. ♂ unknown. A very variable species; synonomous with *C. exilis* Furtos and *C. reflexa* Sharpe. Ponds, streams, lakes along weedy shores. Spring and autumn. Ohio, Ill., Mich.

55a (54) Terminal seta of furca very short; equal to about ⅑ length of sub-terminal claw (Fig. 28.54) . . . . *C. elliptica* Furtos 1933
Length 0.78–0.90 mm, height 0.35–0.37 mm, width 0.20–0.25 mm. Valves sparsely hairy. Leg 3 with penultimate segment undivided; shortest seta of distal segment 2 times narrowest width; dorsal seta less than ½ the length of subterminal claw; terminal claw about ½ the length of ramus; terminal seta short. ♂ larger, otherwise similar to ♀. Muddy, weedy bottoms of lakes and ponds. Mar. to Aug. Ohio, S. C.

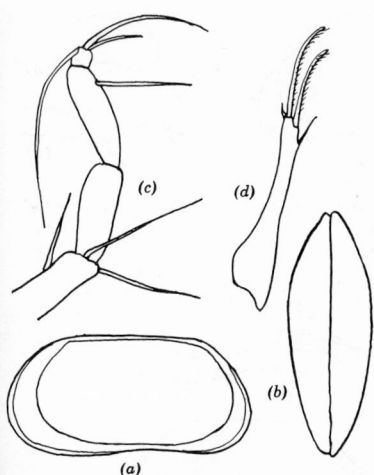

Fig. 28.52. *Candona jeanneli* ♀. (*a*) Left valve. (*b*) Dorsal. (*c*) Leg 3. (*d*) Furca.

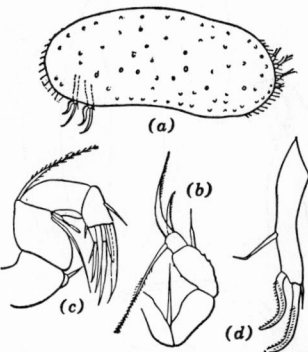

Fig. 28.53. *Candona simpsoni* ♀. (*a*) Lateral. (*b*) Leg 3. (*c*) Antenna 2. (*d*) Furca. (By Sharpe.)

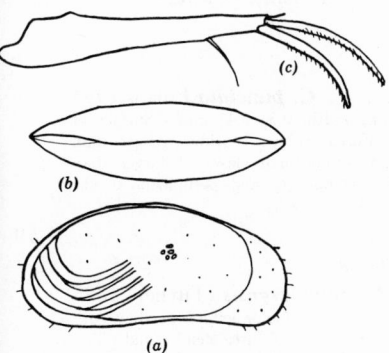

Fig. 28.54. *Candona elliptica* ♀. (*a*) Right valve. (*b*) Dorsal. (*c*) Furca.

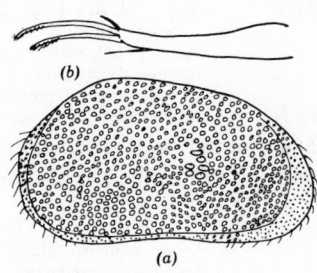

Fig. 28.55. *Candona foveolata.* (*a*) Right valve ♂. (*b*) Furca ♀.

55b   Terminal seta of furca equal to about $\frac{1}{5}$ length of subterminal claw (Fig. 28.55). . . . . **C. foveolata** Dobbin 1941
    Length 0.65 mm, height 0.26 mm. Valves covered with large pits; sparsely hairy. Leg 3 with penultimate segment undivided; the 2 longer terminal seta of terminal segment equal in length; shortest seta $\frac{1}{3}$ the length of longer. Furca 11 times narrowest width; dorsal seta $\frac{1}{2}$ the length of terminal claw. ♂ slightly larger than ♀, otherwise similar. Pools. Feb. Washington.

56a **(50)** Penultimate segment of leg 3 distinctly or indistinctly divided . . . **58**
56b   Penultimate segment of leg 3 undivided **57**
57a **(56)** Valves with few hairs (Fig. 28.56) . . . . . . **C. annae** Mehes 1913
    ♂: length 0.90 mm, height 0.46 mm, width 0.31 mm. Surface of valves smooth and glistening, without puncta and hairless except for marginal hairs, color bluish-irridescent. Leg 3 with penultimate segment undivided; distal setae very unequal, one very short and curved. Furca straight, 14 times narrowest width; dorsal margin smooth. Ejaculatory duct with 5 crowns of spines, widely separated. ♀ unknown from N. A. Ditches. June, Aug. Fla., Mass.

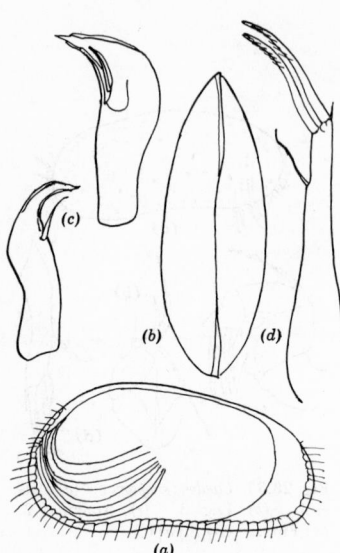

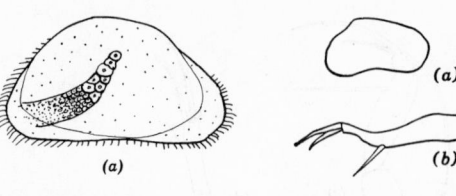

Fig. 28.57. *Candona punctata*
♀. Right valve.

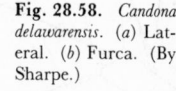

Fig. 28.58. *Candona delawarensis*. (*a*) Lateral. (*b*) Furca. (By Sharpe.)

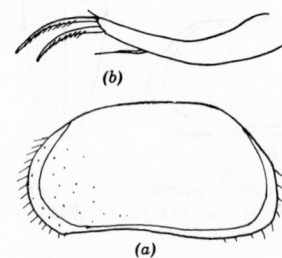

Fig. 28.56. *Candona annae* ♂. (*a*) Left valve. (*b*) Dorsal. (*c*) Prehensile palps. (*d*) Furca.

Fig. 28.59. *Candona balatonica* ♀. (*a*) Left valve. (*b*) Furca.

57b   Valves hairy (Fig. 28.57) . . . . **C. punctata** Furtos 1933
Length 0.85–0.90 mm, height 0.54–0.51 mm, width, 0.37–0.42 mm. Surface of valves pitted, with many, long, stiff hairs. Furcal ramus 13 times longer than narrowest width; dorsal seta $\frac{2}{5}$ the length of subterminal claw. ♂ larger than ♀, with posterior portion of valves broader. Temporary and permanent ponds, marshes, and lakes. Mar. to May, Nov. Ohio, Ill.

58a **(56)** Terminal seta of furca present. **59**

58b   Terminal seta of furca absent (Fig. 28.58) . . . .
                                              C. delawarensis (Turner) 1895
Length 0.95 mm, width 0.43 mm, height 0.54 mm. Color greenish-yellow with brown splotches. Maxillary spines plain. Terminal claws of furca slender and plain. Furca slender and much curved. Creeks. Mar. Dela. (This has not been well described and is a very doubtful species.)

59a **(58)** Furcal ramus less than 10 times narrowest width **60**

59b   Furcal ramus longer than 10 times narrowest width (Fig. 28.59) . .
                                              C. balatonica Daday 1894
Length 0.96 mm, height 0.48 mm. Valves sparsely hairy. Leg 3 with penultimate segment indistinctly divided; shorter terminal seta of distal segment $\frac{5}{8}$ the length of longer, and 4 times the length of segment. Furca curved, 11 times longer than narrowest width. ♂ unknown in N. A. Ponds. Sept. Fla., Alaska.

60a **(59)** Valves hairy (Fig. 28.60). . . . . **C. albicans** Brady 1864
Length 0.78–0.85 mm, height 0.42–0.48 mm. Valves with small pits especially in anterior and posterior portions. Third thoracic leg with penultimate segment divided; shortest distal seta of terminal segment 2 times length of segment. Furca nearly straight, 8 times narrowest width. ♂ similar but larger. Pools, ditches with muddy bottoms, marshes. Spring and early summer. Ill., Ohio, Colo., Calif., Nova Scotia.

60b   Valves nearly hairless with a few bristle-bearing puncta scattered
                over the surface (Fig. 28.61) . . . . **C. biangulata** Hoff 1942
Length 0.70–0.73 mm, height 0.35–0.38 mm. Leg 2 with a large, heavy second segment; claw $1\frac{1}{2}$ times the length of the last 3 segments. Leg 3 with penultimate segment clearly divided; shortest terminal seta of distal segment more than 3 times the length of the segment. Furca relatively straight, 7 times narrowest width. ♂ unknown. Temporary streams and ponds with muddy bottoms. Spring. Ill.

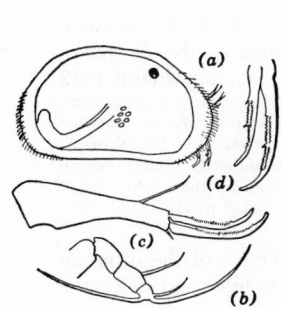

Fig. 28.60. *Candona albicans* ♀.
(*a*) Right valve. (*b*) Leg 3.
(*c*) Furca. (*d*) Terminal claws
of furca. (By Sharpe.)

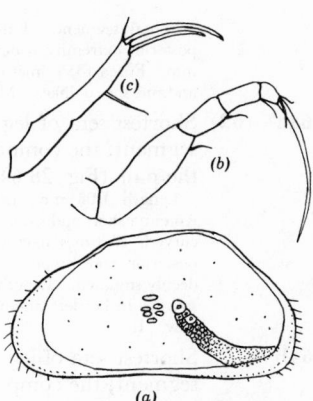

Fig. 28.61.   *Candona biangulata* ♀.
(*a*) Right valve.   (*b*) Leg 2.
(*c*) Furca.

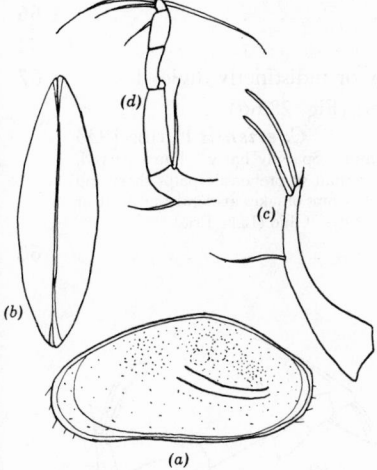

Fig. 28.62. *Candona subgibba* ♀. (*a*) Lateral.
(*b*) Dorsal. (*c*) Furca. (*d*) Leg 3.

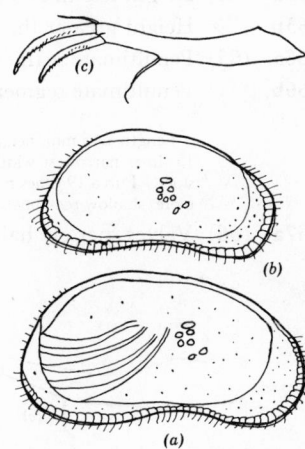

Fig. 28.63. *Candona scopulosa.*
(*a*) Right valve ♂. (*b*) Right valve ♀.
(*c*) Furca ♀.

61a	**(47)**	Height equal to ½ the length	**62**
61b		Height either greater or less than ½ the length	**65**
62a	**(61)**	Length of furcal ramus greater than 10 times narrowest width . . .	**63**
62b		Length of furcal ramus not more than 10 times narrowest width . .	**64**
63a	**(62)**	Terminal seta of furca less than ⅓ length of subterminal claw (Fig. 28.62) . . . . *C. subgibba* Sars 1926	

Length 1.35 mm, height 0.65 mm, width 0.43 mm. Valves thin, semipellucid, almost bare of hairs. Leg 3 with penultimate segment distinctly divided. Furca curved, 13 times narrowest width; long and slender. ♂ unknown. Brackish ponds. Aug. Alaska.

63b		Terminal seta of furca greater than ⅓ length of subterminal claw (Fig. 28.63) . . . . *C. scopulosa* Furtos 1933

Length 0.93–1.00 mm, height 0.47–0.50 mm, width 0.40–0.47 mm. Leg 3 with penultimate segment faintly divided; shortest terminal seta 3½ times length of

terminal segment.   Furca curved, 11 times narrowest width.   ♂ larger than ♀;
posterior extremity wider; ventral margin more deeply sinuated.   Length of ♂ 1.15
mm.   Furca 13½ times narrowest width.   Rocky shores and rock pools, weedy inlets
and marshes of lakes.   May to Aug.   Ohio (Lake Erie).

64a   **(62)**   Shortest seta of leg 3 not over 3 times the length of the ultimate
segment; the companion seta is over twice as long as the shorter of
the pair (Fig. 28.64) . . . .   ***C. acuta*** Hoff 1942
Length 1.08 mm, height 0.55 mm, width 0.57 mm.   A few scattered hairs.
Antenna short and stout.   Leg 3 with penultimate segment divided.   Furca somewhat
curved, 10 times narrowest width.   ♂ larger and with valves of different shape;
posterior end broadly rounded; anterior end narrowly rounded; ventral margin
deeply sinuated.   Length of ♂ up to 1.31 mm.   Left valve longer than right.   Small
streams in borders among weeds and decaying vegetation; small ponds.   May, July,
Nov.   Ill.

64b   Shortest seta of leg 3 more than 3 times the length of the ultimate
segment; the companion seta about equal to or less than twice the
length of the shortest distal seta (Fig. 28.65) . . . .
                                                     C. indigena Hoff 1942
Length 0.96–1.10 mm, height 0.55–0.65 mm, width 0.52 mm.   Valves with a few
scattered puncta bearing hairs.   Leg 3 with a divided penultimate segment.   Furca
straight, 9 times narrowest width.   Males with a more concave ventral margin and
more rounded posterior margin.   Temporary ponds among decaying vegetation.
Spring.   Ill., Tenn., Mich.

65a   **(61)**   Height less than ½ the length.   **66**

65b   Height greater than ½ the length. **71**

66a   **(65)**   Penultimate segment of leg 3 distinctly or indistinctly divided . . .   **67**

66b   Penultimate segment of leg 3 undivided (Fig. 28.66) . . . .
                                                     C. eriensis Furtos 1933
Length 1.15 mm, height 0.46 mm, width 0.40 mm.   Sparsely hairy.   Furca curved,
15 times narrowest width.   ♂ higher and broader than ♀; prehensile palps short and
stout.   Furca 19 times narrowest width.   Muddy bottoms of lakes at depths of 25 ft or
more, shallow rock pools and weedy inlets.   June, July.   Ohio (Lake Erie).

67a   **(66)**   Valves sparsely haired **68**

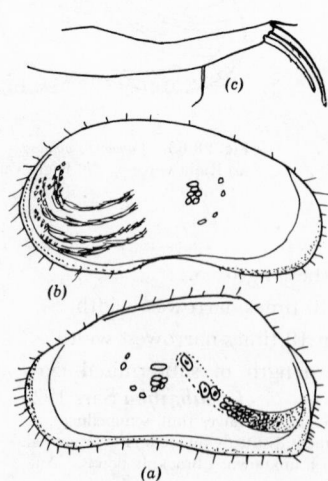

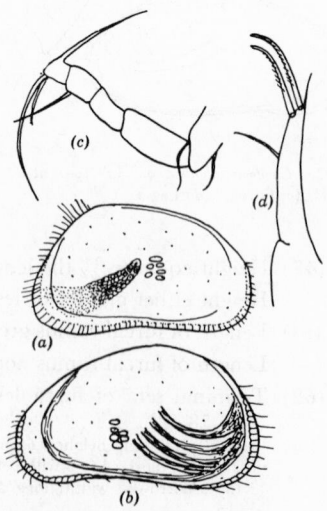

Fig. 28.64.   *Candona acuta.*   (*a*) Right
valve ♀.   (*b*) Left valve ♂.   (*c*) Furca ♀.

Fig. 28.65.   *Candona indigena.*   (*a*) Right
valve ♀.   (*b*) Left valve ♂.   (*c*) Leg 3 ♀.
(*d*) Furca ♀.

67b   Valves hairy (Fig. 28.67). . . . . **C. suburbana** Hoff 1942

Length 1.08 mm, height 0.51 mm, width 0.40 mm. Surface of valves with a considerable number of short, weak hairs set on papillae. Conspicuous pore canals at both ends and along ventral margin. Leg 3 with penultimate segment divided. Furca nearly straight, 11 times longer than narrowest width. ♂ slightly larger than ♀; ventral margin more deeply sinuated; furca more slender. Temporary ponds among plants along the shore. May, June. Ill.

68a (67) Length of furca more than 10 times narrowest width **69**

68b Length of furca less than 10 times narrowest width (Fig. 28.68) . . **C. sharpei** Hoff 1942

Length 1.00 mm, height 0.47 mm, width 0.49 mm. Shell yellowish, transparent, strongly compressed, the left valve overlapping the right at both extremities. Dorsal flanges present. Furca 10 times length of narrowest width. Small ponds. Mar., Apr., Sept. Ga., Ill., S. C.

69a (68) Terminal seta of furca less than ⅓ the length of subterminal claw **70**

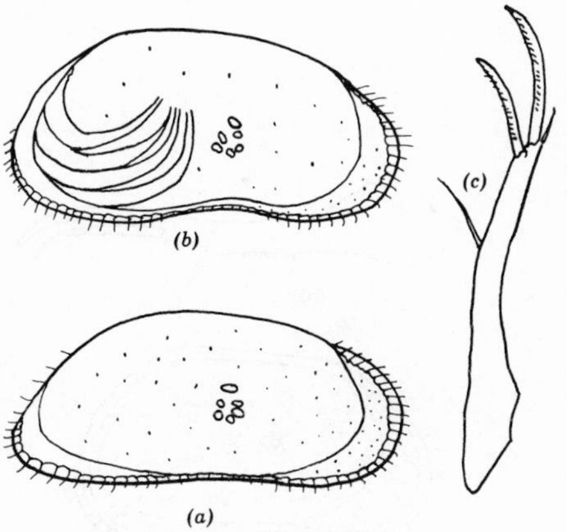

Fig. 28.66. *Candona eriensis.* (a) Right valve ♀. (b) Right valve ♂. (c) Furca ♀.

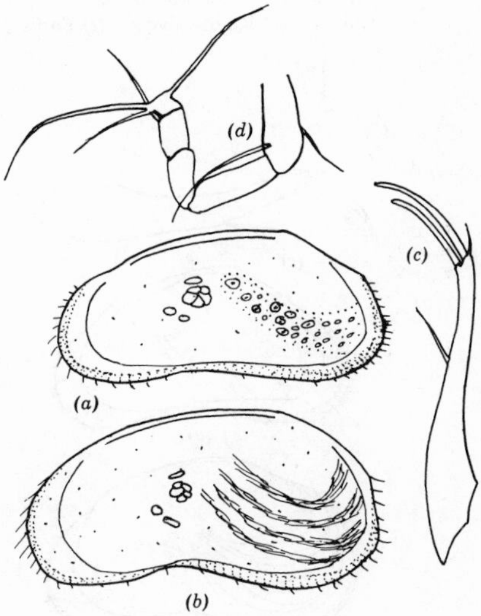

Fig. 28.67. *Candona suburbana.* (a) Right valve ♀. (b) Right valve ♂. (c) Furca ♂. (d) Leg 3 ♂.

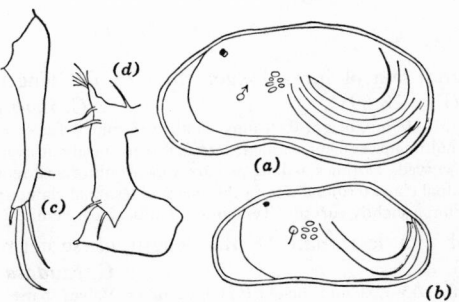

Fig. 28.68. *Candona sharpei.* (a) Lateral ♂. (b) Lateral ♀. (c) Furca ♀. (d) Right maxillary palp ♂. (By Sharpe.)

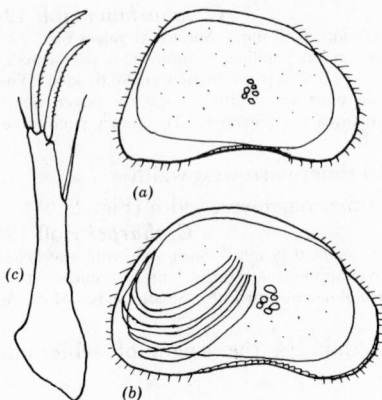

Fig. 28.69. *Candona inopinata.* (*a*) Right valve ♀. (*b*) Right valve ♂. (*c*) Furca ♀.

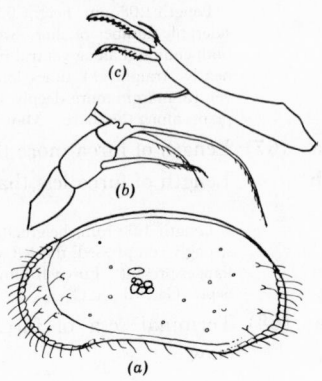

Fig. 28.70. *Candona caudata* ♀. (*a*) Left valve. (*b*) Leg 3. (*c*) Furca.

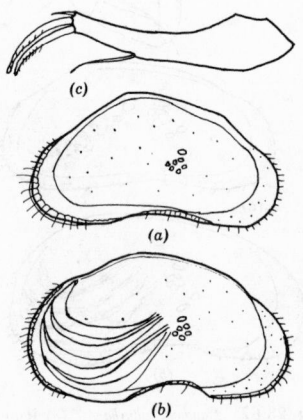

Fig. 28.71. *Candona distincta.* (*a*) Right valve ♀. (*b*) Right valve ♂. (*c*) Furca ♀.

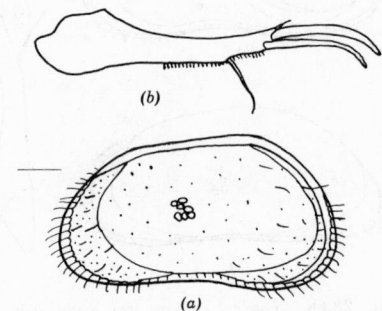

Fig. 28.72. *Candona stagnalis.* (*a*) Right valve ♀. (*b*) Furca.

69b   Terminal seta of furca longer than ⅓ the length of subterminal claw (Fig. 28.69) . . . . *C. inopinata* Furtos 1933

Length 1.40 mm, height 0.60 mm, width 0.53 mm. Valves sparsely hairy. Leg 3 with penultimate segment divided; terminal segment about as long as broad. Furca slightly curved, 11 times as long as narrowest width; dorsal seta slightly longer than subterminal claw; terminal seta ¾ the length of terminal claw. ♂ higher and broader than ♀; furca slightly curved. Temporary ponds. Nov. Ohio.

70a (69) Dorsal seta less than ½ the length of subterminal claw (Fig. 28.70) . . . . *C. caudata* Kaufmann 1900

Length 1.13–1.34 mm, height 0.51–0.63 mm. Valves sparsely haired and marked by minute elevated areas separated by fine grooves. Pore canals conspicuous. Antennae 1 and 2 stout and heavy. Leg 3 with divided penultimate segment. Furca 11 times as long as narrowest width. ♂ unknown. Canals, lakes, ponds, among grass and weeds. June. Ill., Mass., Wash., Mont.

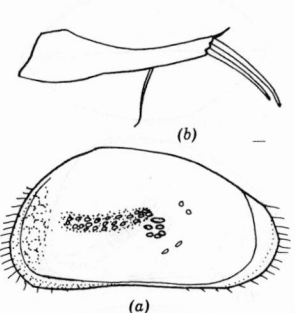

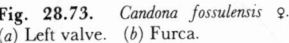

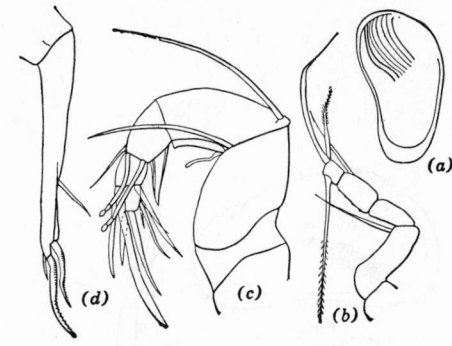

Fig. 28.73. *Candona fossulensis* ♀.
(*a*) Left valve. (*b*) Furca.

Fig. 28.74. *Candona sigmoides* ♂. (*a*) Lateral. (*b*) Leg 3. (*c*) Antenna 2. (*d*) Furca. (By Sharpe.)

70b   Dorsal seta equal in length to length of subterminal claw (Fig. 28.71) . . . . *C. distincta* Furtos 1933
   Length 0.95–1.15 mm, height 0.35–0.55 mm, width 0.38–0.43 mm. Valves very transparent, shining, sparsely hairy. Leg 3 with penultimate segment divided. Furca curved, 13 times as long as narrowest width. ♂ larger than ♀. Left valve larger than right; prehensile palps stout and dissimilar. Marshes, canal basins, small lakes. Mar. to June, Nov. Ohio, Ill.

71a (65) Penultimate segment of leg 3 distinctly divided 72
71b   Penultimate segment of leg 3 undivided (Fig. 28.72) . . . .
                C. stagnalis Sars 1890
   Length 1.06 mm, height 0.55 mm. Surface of valves sparsely hairy with a few refractive tubercles. Furca slightly curved, 11 times as long as narrowest width. ♂ of the same size as ♀; and anterior sinuation in the dorsal margin is present. Ponds. May. Ohio.

72a (71) Valves sparsely haired 73
72b   Valves hairy (Fig. 28.73) . . . . *C. fossulenis* Hoff 1942
   Length 1.00–1.06 mm, height 0.54–0.57 mm, width considerably less than height. Valves with many fine hairs; pore canals obliterated. Irregular, square and diamond-shaped markings on posterior portion of valves. Antennae 1 and 2 slender. Furca gently curved, 9 times as long as narrowest width. ♂ similar to ♀ but slightly longer (up to 1.20 mm in length). Ditches, vernal ponds. Apr. Ill.

73a (72) Furca greater than 10 times longer than narrowest width 74
73b   Furca less than 10 times longer than narrowest width (Fig. 28.74) . . . . *C. sigmoides* Sharpe 1897
   Length 1.06 mm, height 0.54 mm. Shell smooth with few hairs. Antenna 1 stout, antenna 2 short and stout. Penultimate segment of leg 3 divided. Furca considerably curved, 9 times as long as narrowest width. ♂ larger (up to 1.25 mm in length and 0.63 mm in height). Furca longer (12 ×) and more slender. Lakes, rivers, along shore among plants. June, Aug. Ill.

74a (73) Length of furca less than 15 times narrowest width 75
74b   Length of furca greater than 15 times narrowest width (Fig. 28.75) . . . . *C. decora* Furtos 1933
   Length 1.18–1.30 mm, height 0.63–0.70 mm. Surface of valves faintly reticulated, sparsely hairy. Leg 3 with penultimate segment divided. Furca curved, 16 times as long as narrowest width. ♂ larger and wider than ♀; ventral margin sharply sinuated. Furca straight and 18 times as long as narrowest width. Ponds, lakes. Apr. Ohio, Mass., Mich., Great Slave Lake.

75a (74) Terminal seta of furca less than ⅓ the length of subterminal claw 76

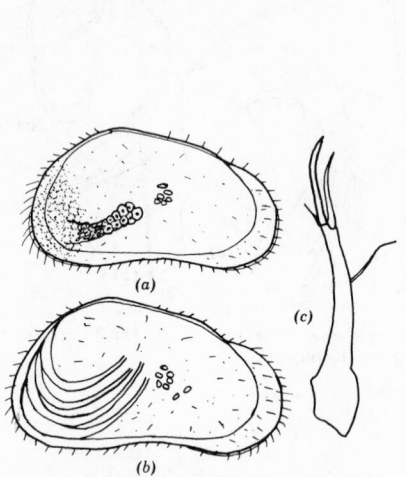

Fig. 28.75. *Candona decora.* (*a*) Right valve ♀.
(*b*) Right valve ♂. (*c*) Furca ♀.

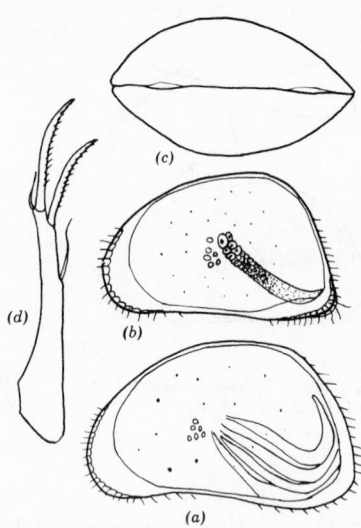

Fig. 28.76. *Candona truncata.* (*a*) Left valve ♂.
(*b*) Left valve ♀. (*c*) Dorsal ♀. (*d*) Furca ♀.

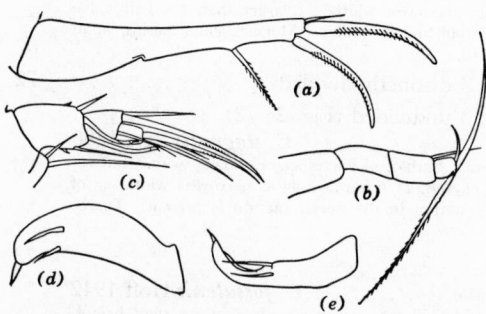

Fig. 28.77. *Candona recticauda* ♂. (*a*) Furca. (*b*) Leg 3.
(*c*) End of antenna 2. (*d*) Right maxillary palp. (*e*) Left
maxillary palp. (By Sharpe.)

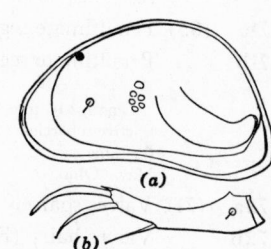

Fig. 28.78. *Candona candida* ♀.
(*a*) Lateral. (*b*) Furca. (By
Sharpe.)

75b   Terminal seta of furca greater than ⅓ length of subterminal claw
            (Fig. 28.76). . . . . **C. truncata** Furtos 1933
      Length 1.00–1.25 mm, height 0.56–0.74 mm, width 0.53–0.70 mm. Sparsely hairy.
      Leg 3 with divided penultimate segment. Furca slightly curved, 12½ times as long as
      narrowest width. ♂ longer and higher; posterior end broadly rounded; furca straight.
      Ponds and marshes. Feb. to May and Nov. Ohio, Mich.

76a **(75)** Penultimate segment of leg 3 very distinctly divided (Fig. 28.77). . .
                                            C. recticauda Sharpe 1897
      Length of ♂ 1.18 mm, height 0.70 mm. Shell covered with scattered papillar eleva-
      tions. Spermatozoa show through valves as 4 bands. Furca straight, 14 times as long
      as narrowest width; dorsal seta ³⁄₅ the length of subterminal claw; terminal seta
      about ¼ the length of terminal seta. ♀ unknown. Bottoms of ponds. Feb. Ill.

76b   Penultimate segment of leg 3 indistinctly divided (Fig. 28.78) . . .
                                            C. candida (O. F. Müller) 1776
      Length 1.05–1.20 mm, height 0.60 mm. Shell prominently arched dorsally, highest
      point in middle. Furca 5 times as long as narrowest width and decidedly curved.
      Shallow, temporary ponds and ditches. Apr. and Sept. Mass., Ore., Mont.

77a **(4)** Masticatory process of maxilla with 6 equally developed setae . . .   **78**

77b   Masticatory process of maxilla with 2 or 3 spinelike setae **80**

78a (77) Second antenna with penultimate segment divided; distal 3 setae of leg 3 unmodified; terminal seta of furca absent . . . .
Notodromus Lilljeborg 1853

One species, *N. monacha* (O. F. Müller) 1776 (Fig. 28.79).
Length 1.18 mm, height 0.85 mm, width 0.75 mm. Color brownish-yellow. Antenna 2 6-segmented. Leg 3 terminating in 3 setae, 1 of which is reflexed. Active swimmers, resembling Cladocera in their movements. ♂ larger than ♀; prehensile palps similar; furca more curved than in ♀. Permanent ponds and lakes among algae. Spring and summer. Ind., Ill., Minn., Alaska.

78b Second antenna with penultimate segment undivided; seta on distal end of leg 3 formed into one claw and a reflexed seta; terminal seta of furca present . . . . *Cyprois* Zenker 1854 **79**

79a (78) Length of shell less than 1.5 mm (Fig. 28.80) . . . .
C. occidentalis Sars 1926

Length 1.00 mm, height 0.65 mm. Valves very thin and transparent, without sculpturing, with hyaline borders at both ends. Appendages similar to those of *C. marginata*. ♂ smaller than ♀; shells more compressed and higher in proportion to the length. Prehensile palps distinctly unequal in size and shape; ejaculatory tubes comparatively short. Pools. May. Ontario.

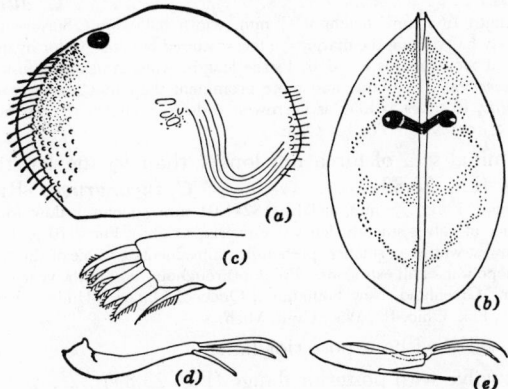

Fig. 28.79. *Notodromus monacha.* (*a*) Lateral ♂. (*b*) Dorsal ♂. (*c*) Maxillary spines. (*d*) Furca ♀. (*e*) End of leg 3.

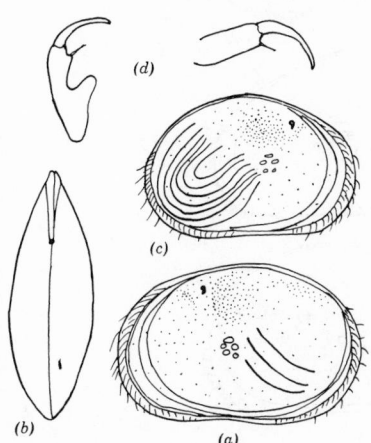

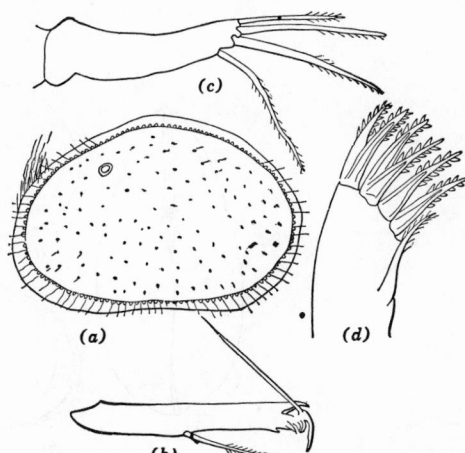

Fig. 28.80. *Cyprois occidentalis.* (*a*) Lateral ♀. (*b*) Dorsal ♀. (*c*) Lateral ♂. (*d*) Prehensile palps of maxillipeds ♂.

Fig. 28.81. *Cyprois marginata.* (*a*) Lateral ♀. (*b*) End of leg 3. (*c*) Furca ♀. (*d*) Maxillary spines. (By Sharpe.)

79b   Length of shell more than 1.5 mm (Fig. 28.81) . . . .
                                                 C. marginata (Strauss) 1821
              Length 1.53 mm, height 0.96 mm, width 0.75 mm. Color uniform yellow. Nata-
              tory setae of antenna 2 extend to tips of claws. Furcal ramus little curved; dorsal seta
              modified, clawlike. ♂ smaller than ♀; posterior margin of shell more rounded in ♂.
              Prehensile palps nearly equal in shape and size. Furca much more curved. Vernal
              ponds. Apr. to July. Ill., Ohio, Mich.

80a   (77) Furca rudimentary; reduced to a long setae or flagellum   150
80b   Furca well developed   81
81a   (80) Margin of one valve tuberculated . . . . . *Cyprinotus* Brady 1885   82
81b   Margins of both valves smooth   96
82a   (81) Length greater than 1.5 mm   83
82b   Length less than 1.5 mm   88
83a   (82) Height greater than ½ the length of valves   84
83b   Height not more than ½ the length of valves   85
84a   (83) Terminal seta of furca nearly as long as terminal claw (Fig.
              28.82) . . . . *C. aureus* Sars 1895
              Length 1.65 mm, height 0.97 mm, width 0.85 mm. Surface of valves smooth,
              sparsely hairy along the margins; a few scattered puncta. Color hyaline yellow. Seen
              from above, width is less than ½ the length, which will distinguish this species from
              C. incongruens. Eye large and more prominent than in *C. incongruens.* Furca almost
              straight; 10 times as long as narrowest width. ♂ smaller than ♀ but not as yet re-
              ported from N. A. Ponds. Ohio.

84b   Terminal seta of furca not longer than ½ the length of terminal
              claw (Fig. 28.83) . . . . *C. incongruens* (Ramdohr) 1808
              Length 1.40–1.75 mm, height 0.82–1.04 mm. Color yellow to brownish-yellow.
              Surface of valves smooth; left valve overlaps right. Furca 10 or 11 times as long as
              narrowest width. ♂ smaller; prehensile palps unequal. One of the most common and
              cosmopolitan of all ostracods. Present throughout most of the year in all kinds of fresh
              water. Greenland, New Foundland, Quebec, Ontario, Hudson Bay, Penn., D. C.,
              N. C., Fla., Ohio, Ill., Wis., Utah, Mich.

85a   (83) Left valve without posterior flange   86
85b   Left valve with posterior flange (Fig. 28.84) . . . .
                                                 C. inconstans Furtos 1933
              Length 1.50 mm, height 0.76 mm, width 0.60 mm. Surface of valves smooth, with
              scattered puncta bearing short hairs; margins moderately hairy. Natatory setae of
              antenna 2 extend beyond tips of terminal claws by ⅓ the length of claws. Both

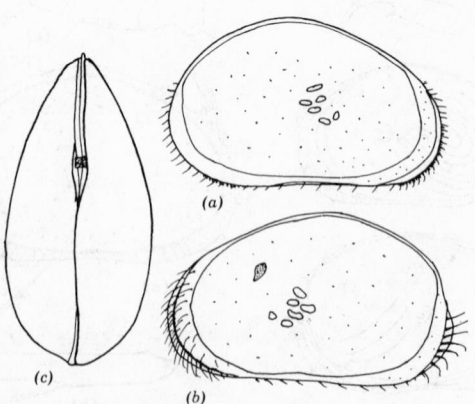

Fig. 28.82. *Cyprinotus aureus* ♀. (*a*) Right valve. (*b*) Left
valve. (*c*) Dorsal.

maxillary spines bluntly toothed. Furca weakly S-shaped; 20 times narrowest width; the only species of this genus with a posterior flange on left valve. ♂ unknown. Ponds, pools. June. Yucatan.

86a (85) Left valve without posterior spine. **87**

86b Left valve with prominent spine on posterior end (Fig. 28.85) . . .
C. unispinifera Furtos 1936

Length 1.70 mm, height 0.80 mm, width 0.68 mm. Surface very faintly pitted with large scattered puncta. Sparsely haired except as margins. Maxillary spines coarsely toothed. Furca moderately S-shaped, 25 times narrowest width. ♂ similar to ♀ but somewhat smaller; ejaculatory duct with 32 crowns of spines. July. Yucatan.

87a (86) Natatory setae of second antenna reach beyond tips of terminal claws; length of furcal ramus 16 times longer than narrowest width (Fig. 28.86) . . . . *C. dentatus* Sharpe 1918

Length 1.35–1.60 mm. Right valve with tuberculate margins as a rule. Color brownish-yellow. Furca gently curved, 16 times narrowest width. Both maxillary spines toothed. ♂ common. Temporary ponds, lakes. Tenn., Neb., Tex.

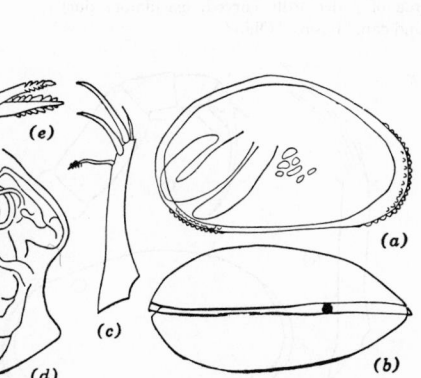

28.83.   *Cyprinotus incongruens.* (*a*) Right valve ♀. Dorsal ♀. (*c*) Furca ♀. (*d*) Penis ♂. (*e*) Maxillary s. (By Sharpe.)

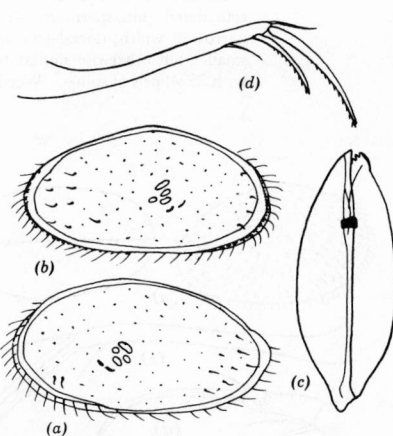

Fig. 28.84.   *Cyprinotus inconstans* ♀. (*a*) Left valve. (*b*) Right valve. (*c*) Dorsal. (*d*) Furca.

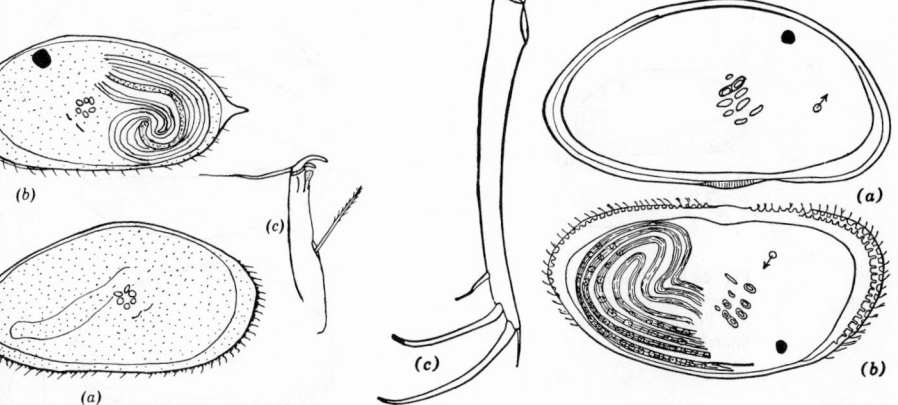

28.85.   *Cyprinotus unispinifera* ♀. (*a*) Right ♀. (*b*) Left valve. (*c*) Leg 3.

Fig. 28.86.   *Cyprinotus dentatus.* (*a*) Left valve from within ♀. (*b*) Right valve from within ♀. (*c*) Furca ♀. (By Sharpe.)

87b   Natatory setae of second antenna extend only to tips of terminal claws; furcal ramus 20 times longer than narrowest width (Fig. 28.87) . . . . *C. americanus* Cushman 1905
           Length 1.50 mm, height 0.80 mm, width 0.70 mm. Valves colorless. Translucent, surface smooth; right valve larger than left. Terminal segment of leg 3 constricted in the middle. Furca nearly straight; 20 times narrowest width. Ponds and ditches. Apr. Mass.

88a (82)   Right valve tuberculated **89**

88b   Left valve tuberculated (Fig. 28.88). . . . *C. scytoda* Dobbin 1941
           Length 1.35 mm. Height about ½ the length. Surface of valves with rather rough crepe or fine-grained leather appearance. Sparsely hairy. Both spines of third maxillary process toothed. Furca 10 times narrowest width. ♂ unknown. Ponds. Mar. Cal.

89a (88)   Spines on maxillary process with teeth **90**

89b   Spines on maxillary process without teeth (Fig. 28.89) . . . .
                                    C. crenatus Turner 1893
           Length 1.23 mm, height 0.63 mm, width 0.60 mm. Color yellowish-green. Valves reticulated, thin, spermaries show through. Furca gently curved, 18 times longer than narrowest width; dorsal seta about ½ the length of subterminal claw. ♂ common, smaller but otherwise similar to ♀. Furca of ♂ decidedly curved; ejaculatory duct with 25 whorls of spines. Weedy ponds and canal basins. Ohio.

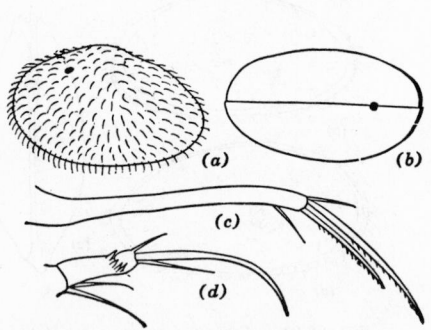

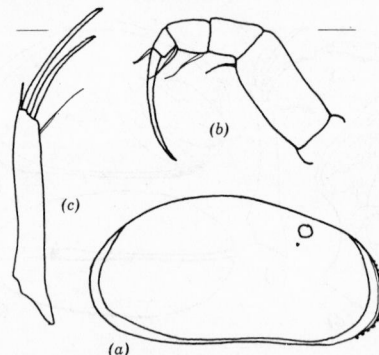

Fig. 28.87. *Cyprinotus americanus* ♂. (*a*) Lateral. (*b*) Dorsal. (*c*) Furca. (*d*) Leg 2. (By Sharpe.)

Fig. 28.88. *Cyprinotus scytoda* ♀. (*a*) Lateral. (*b*) Leg 2. (*c*) Furca.

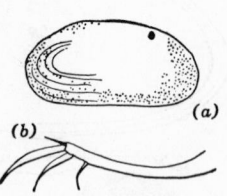

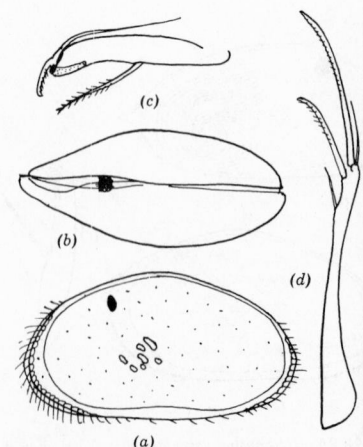

Fig. 28.89. *Cyprinotus crenatus.* (*a*) Lateral ♀. (*b*) Furca ♂. (By Sharpe.)

Fig. 28.90. *Cyprinotus fluviatilis* ♀. (*a*) Lateral. (*b*) Dorsal. (*c*) Leg 3. (*d*) Furca.

90a (89) Spines on maxillary process with well-developed teeth on one spine,
the other spine only faintly tooched. **91**

90b   Spines on maxillary process both toothed. **92**

91a (90) Natatory setae of second antenna extend only slightly beyond tips
of terminal claws; length of furcal ramus 19 times narrowest
width (Fig. 28.90). . . . . *C. fluviatilis* Furtos 1933
Length 1.35 mm, height 0.72 mm, width 0.56 mm. Surface of valves smooth, with
scattered puncta and marginal hairs. Furca slightly curved; 19 times longer than
narrowest width; dorsal seta $\frac{1}{2}$ the length of subterminal claw; terminal seta $\frac{2}{5}$
the length of terminal claw. ♂ smaller but otherwise similar to ♀. Ejaculatory
duct with 31 whorls of spines. Creeks, rivers, lakes. July and Aug. Del., Ohio,
Ariz., Wash.

91b   Natatory setae of second antenna extend beyond tips of terminal
claws by $\frac{2}{3}$ their length; length of furcal ramus 11 times narrowest
width (Fig. 28.91). . . . . *C. glaucus* Furtos 1933
Length 1.15 mm, height 0.58 mm, width 0.50 mm. Color gray. Surface of valves
smooth, scattered puncta, moderately hairy along the extremities. Furcal ramus
straight, 11 times narrowest width; dorsal margin with sparse hairs. ♂ smaller, more
triangular (length 1.00 mm). Ejaculatory ducts with 26 whorls of spines. Stony bars,
weedy inlets, rock pools. May to Nov. Ohio (Lake Erie).

92a (90) Dorsal seta of furca about the length of subterminal claw **93**

92b   Dorsal seta of furca shorter than subterminal claw. **94**

93a (92) Shell height less than $\frac{1}{2}$ the length (Fig. 28.92) . . . .
C. pellucidus Sharpe 1897
Length 1.20 mm, height 0.75 mm. Valves usually transparent and covered with a

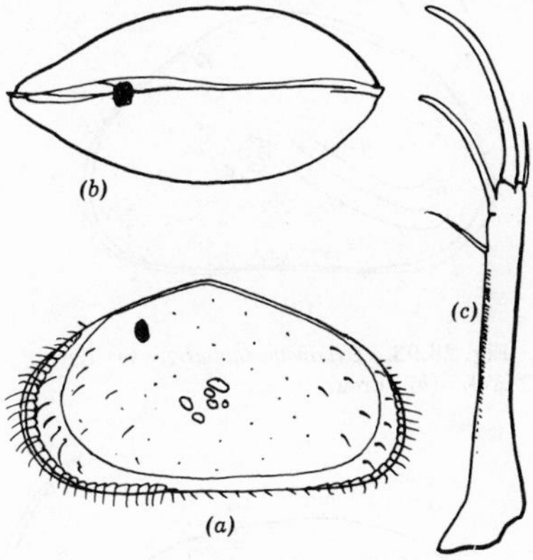

Fig. 28.91. *Cyprinotus glaucus.* (*a*) Left valve ♀.
(*b*) Dorsal ♀. (*c*) Furca ♂.

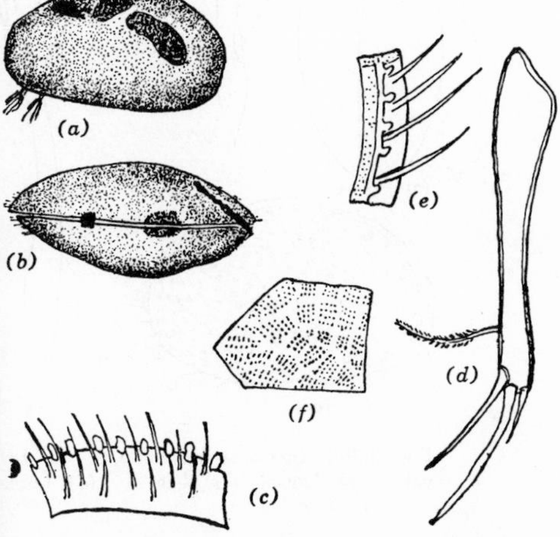

Fig. 28.92. *Cyprinotus pellucidus.* (*a*) Lateral ♀.
(*b*) Dorsal ♀. (*c*) Ventroanterior margin of righ†
valve. (*d*) Furca ♀. (*e*) Inner margin of left shell.
(*f*) Markings on valves. (By Sharpe.)

regular arrangement of dotted lines. Right shell tuberculated and larger than left. Color clear uniform yellowish. Maxillary spines both toothed. Dorsal seta of furca about equal to length of subterminal claw. Shallow ponds and pools. Apr. to Sept. Ill., Idaho, Wash., Mexico.

93b   Shell height greater than ½ the length (Fig. 28.93) . . . .
                                                              C. salinus (Brady) 1862
    Length 1.2 mm, height about ⅔ the length, width about ½ the length. Right valve tuberculated. Furca about straight, terminal seta about ¼ the length of terminal claw; dorsal seta about as long as the subterminal claw. ♂ unknown. Tex.

94a (92) Furca long and slender; 16 times as long as narrowest width **95**
94b   Furca short and stubby; 11 times as long as narrowest width (Fig. 28.94) . . . . ***C. fretensis*** (Brady) 1870
    Length 1.27 mm, height 0.75 mm. Right valve tuberculated. Left valve overlaps right. Surface of valves covered with delicate hairs. Spines of third maxillary process delicately toothed. Furca with terminal claw exceeding ½ the length of ramus; dorsal seta ¾ the length of subterminal claw. ♂ unknown. Wash.

95a (94) Dorsal margin of valves with a distinct angle slightly behind the middle (Fig. 28.95) . . . . ***C. putei*** Furtos 1936
    Length 1.33 mm, height 0.76 mm, width 0.57 mm. Surface of valves conspicuously pitted and with scattered dark spinous processes, each with a short, blunt hair. Color light, with irregular reddish-brown patches. Natatory setae of antenna 2 reach slightly beyond tips of terminal claws. Both maxillary spines with broad, flat teeth. Furca gently curved, 16 times longer than narrowest width. ♂ smaller than ♀ but otherwise similar; ejaculatory duct with 30 crowns of spines. Pools. Aug. Yucatan.

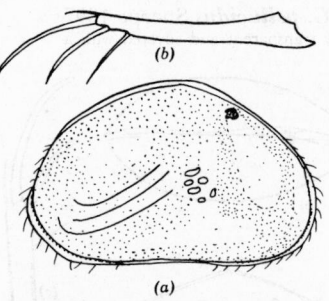

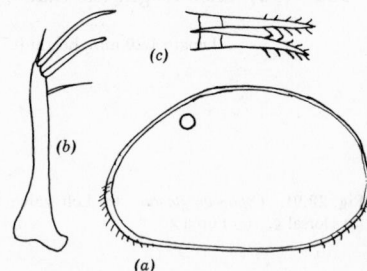

Fig. 28.93. *Cyprinotus salinus* ♀. (*a*) Lateral. (*b*) Furca.

Fig. 28.94. *Cyprinotus fretensis* ♀. (*a*) Lateral. (*b*) Furca. (*c*) Third maxillary process spines.

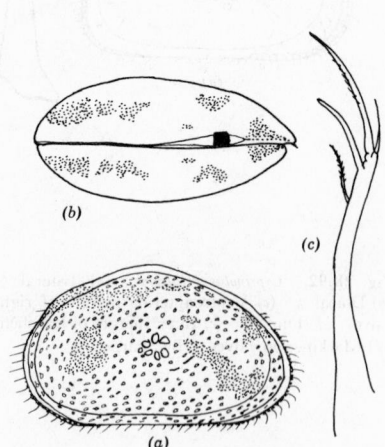

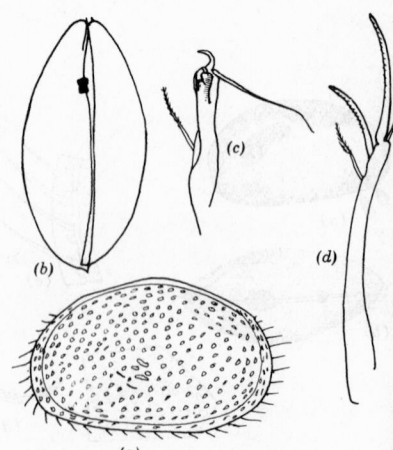

Fig. 28.95. *Cyprinotus putei.* (*a*) Right valve ♀. (*b*) Dorsal ♀. (*c*) Furca ♂.

Fig. 28.96. *Cyprinotus symmetricus* ♀. (*a*) Left valve. (*b*) Dorsal. (*c*) Leg 3. (*d*) Furca.

95b   Dorsal margin of valves broadly arched and rounded (Fig. 28.96) . . . . **C. symmetricus** (G. W. Müller) 1898
Length 1.15 mm, height 0.71 mm, width 0.58 mm. Surface of valves faintly pitted and with scattered puncta bearing hairs. Natatory setae of antenna 2 extend slightly beyond tips of terminal claws. Both maxillary spines with large, broad teeth. Furcal ramus gently curved, 16 times as long as narrowest width; dorsal seta ¾ the length of subterminal claw; terminal seta less than ½ the length of terminal claw. ♂ not reported from N. A. July. Yucatan.

96a **(81)** Natatory setae of second antenna usually not reaching to tips of terminal claws **112**

96b   Natatory setae of second antenna reaching at least to tips of terminal claws **97**

97a **(96)** Valves oblong, compressed; anterior margins with a conspicuous irregular canal system . . . . . . **Stenocypria** G. W. Müller 1901
One species, *S. longicomosa* Furtos 1933 (Fig. 28.97).
Length 1.40 mm, height 0.65 mm, width 0.40 mm. Surface of valves clothed with stiff hairs which become fewer and longer toward the posterior extremity where two hairs are extremely long and prominent. Ventral margin slightly sinuated. Natatory setae extend beyond tips of terminal claws by less than their length. Maxillary palp with terminal segment slightly longer than broad; maxillary spines smooth. Furca strongly curved, 13 times longer than narrowest width. ♂ unknown. Lakes on mudbottoms. July, Nov. Ohio (Lake Erie).

97b   Valves of normal shape; margins without irregular canal system . .   **98**

98a **(97)** Furcal rami at least ½ as long as valves **126**

98b   Furcal rami usually less than ½ the length of valves **99**

99a **(98)** Second segment of leg 2 with two well-developed setae on ventral margin . . . . **Chlamydotheca** Saussure 1858 **120**

99b   Second segment of leg 2 with the usual single seta **100**

00a **(99)** Anterior margin of each valve with a row of radiating septa . . . . **Cypretta** Vavra 1895 **146**
Small forms less than 1 mm in length; dorsal margin boldly arched; tumid when viewed from above.

00b   Valves without radiating septa at anterior margins **101**

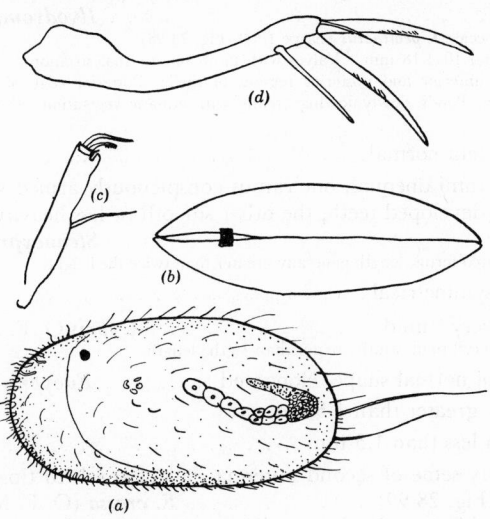

Fig. 28.97. *Stenocypria longicomosa* ♀. (*a*) Lateral.
(*b*) Dorsal. (*c*) Leg 3. (*d*) Furca.

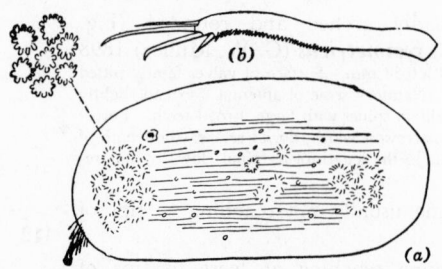

Fig. 28.98. *Ilyodromus pectinatus* ♀. (*a*) Lateral, with detail of shell. (*b*) Furca. (By Sharpe.)

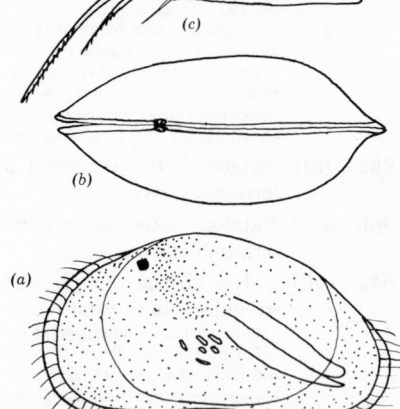

Fig. 28.99. *Eucypris crassa* ♀. (*a*) Lateral. (*b*) Dorsal. (*c*) Furca.

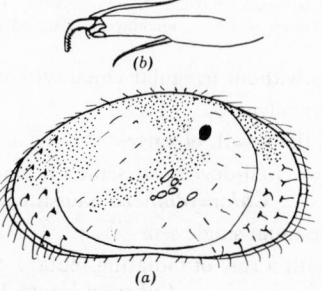

Fig. 28.100. *Eucypris fuscatus* ♀. (*a*) Lateral. (*b*) Leg 3.

101a (100) Dorsal seta of furca heavily developed and similar to subterminal claw. Terminal segment of maxillary palp about as long as broad . . . . *Ilyodromus* Sars 1895
One species, *I. pectinatus* Sharpe 1908 (Fig. 28.98).
Length 1.10–1.18 mm. Valves with faint longitudinal striations. Reticulate patterns on anterior and posterior regions of shell. Posterior edge of furca decidedly pectinate. Ponds, slowly flowing streams with aquatic vegetation (*Typha, Iris, Chara*). S. C.

101b   Dorsal seta normal. **102**

102a (101) Furcal rami unequal, one ramus conspicuously armed with a series of well-developed teeth, the other smooth or less heavily armed. . . *Stenocypris* Sars 1890 **145**
Elongated forms, length generally greater than twice the height.

102b   Furca symmetrical **103**

103a (102) Shells very tumid. . . . . *Cypris* O. F. Müller 1776 **114**
Shell very broad, width greater than ½ the length.

103b   Shells of normal shape, elongated . . . . . . *Eucypris* Vavra 1891 **104**

104a (103) Length greater than 1.5 mm **105**

104b   Length less than 1.5 mm **108**

105a (104) Natatory setae of second antenna do not reach to tips of terminal claws (Fig. 28.99) . . . . *E. crassa* (O. F. Müller) 1785
Length 1.90 mm. Height about ½ the length. Width slightly less than ½ the length. Color greenish. Valves thin with well-marked pellucid zone; surface smooth and shining. In dorsal view, both ends greatly attenuated; the posterior end produced

into a beaklike point. Maxillae with maxillary lobes less attenuated than in other species of the genus; palp with distal joint comparatively shorter and broader. Furca strongly built with long, slender terminal claw. ♂ unknown. Pools in grassy swamps. Spring. Ontario, Va.

105b   Natatory setae of second antenna reach to tips of terminal claws or slightly beyond. **106**

106a (105) Spines of maxillary process smooth **107**

106b   Spines of maxillary process toothed (Fig. 28.100) . . . .

E. fuscatus (Jurine 1820)

Length 1.72 mm, height 0.95 mm, width 0.94 mm. Color light green with 3 dorso-lateral dark-blue bands on the sides. Surface of valves with a few wartlike tubercles near the extremities; a few short, delicate hairs. Furca slightly curved; 25 times as long as narrowest width. ♂ unknown. Temporary ponds. Spring. Mass., N. Y., Ohio, Ill., Mexico.

107a (106) Length less than 2.00 mm (Fig. 28.101) . . *E. hystrix* Furtos 1933

Length 1.55 mm, height 0.88 mm, width 0.88 mm. Surface of valves smooth, thickly covered with long hair. Color light green, brilliantly banded with dark bluish-green. Natatory setae of antenna 2 extend to tips of terminal claws. Furca almost straight, 25 times as long as narrowest width. ♂ unknown. Small weedy temporary ponds, in woods. May. Ohio.

107b   Length greater than 2.00 mm (Fig. 28.102) . . . .

E. virens (Jurine) 1820

Length 1.70–2.30 mm, height 0.90–1.00 mm, valves dull green with two yellow spots in region of eye spot. Left valve slightly overlaps right. Surface of valves covered

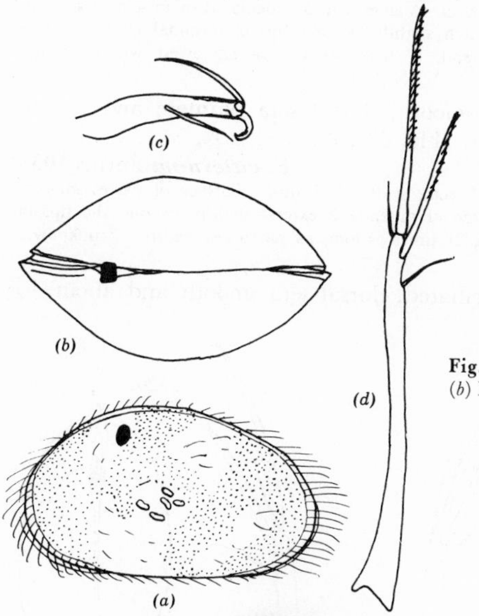

Fig. 28.101. *Eucypris hystrix* ♀. (*a*) Lateral. (*b*) Dorsal. (*c*) Leg 3. (*d*) Furca.

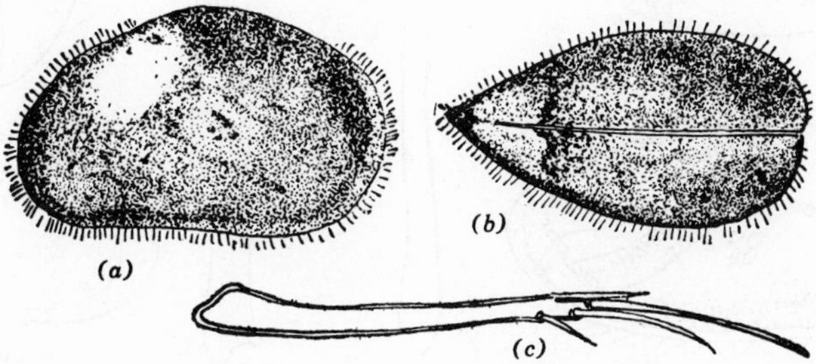

Fig. 28.102. *Eucypris virens* ♀. (*a*) Lateral. (*b*) Dorsal. (*c*) Furca. (By Sharpe.)

with short hairs. Natatory setae of antenna 2 reach to tips of terminal claws. Both maxillary spines smooth. Furca weakly S-shaped, 20 times as long as narrowest width. ♂ unknown. Apr. to July. Weedy ponds. Wis., Ohio, Mass., Mexico.

108a (104) Length greater than 1.00 mm 109

108b Length less than 1.00 mm (Fig. 28.103) . **E. arcadiae** Furtos 1936
Length 0.82 mm, height 0.45 mm, width 0.38 mm. Color light green with a series of delicate, longitudinal dark-blue stripes on lateral surface. Surface of valves, smooth, hairless. Natatory setae of antenna 2 extend beyond tips of terminal claws. Furca straight, 21 times as long as narrowest width; terminal seta ½ the length of terminal claw. ♂ unknown. Aug. Fla.

109a (108) Terminal seta of furca less than ½ the length of terminal claw . . . 110

109b Terminal seta of furca equal to ½ the length of terminal claw (Fig. 28.104) . . . . **E. rava** Furtos 1933
Length 1.26 mm, height 0.66 mm, width 0.67 mm. Color yellowish-brown, with an oval light area in the ocular region of each valve. Surface of valves smooth, moderately hairy, a few scattered puncta. Maxillary spines bluntly toothed. Furca gently curved, slender, 18 times as long as narrowest width. ♂ unknown. Small streams. May. Ohio.

110a (109) Surface of valves without reticulations; both maxillary spines toothed 111

110b Surface of valves definitely reticulated; one spine of maxillary process toothed, the other with only faint indications of teeth (Fig. 28.105) . . . . **E. reticulata** (Zaddach) 1844
Length 1.10–1.30 mm, height 0.72 mm, width 0.65 mm. Color dark green with two light patches in the region of the eyes. Valves with a reticulated or tasselated surface. Natatory setae of antenna 2 reach slightly beyond tips of terminal claws. Furca straight, weakly bent near the end, 12 times as long as narrowest width. Small temporary grassy pools. Mass., N. Y., N. J., Ill.

111a (110) Dorsal margin of furca smooth; dorsal seta plumose and ⅔ the length of subterminal claw (Fig. 28.106) . . . .
E. cisternina Furtos 1936
Length 1.41 mm, height 0.75 mm, width 0.72 mm. Surface of valves smooth, moderately hairy. Natatory setae of antenna 2 extend slightly beyond the tips of terminal claws. Furca straight, 21 times as long as narrowest width. ♂ unknown. July. Yucatan.

111b Dorsal margin of furca ciliated; dorsal seta smooth and about ½

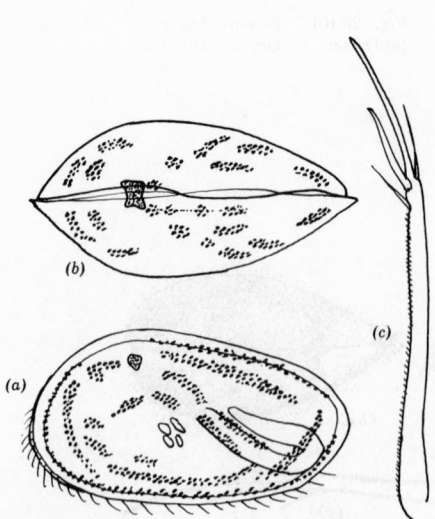

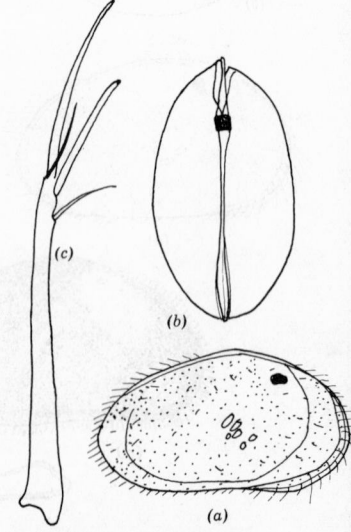

Fig. 28.103. *Eucypris arcadiae* ♀. (*a*) Left valve. (*b*) Dorsal. (*c*) Furca.

Fig. 28.104. *Eucypris rava* ♀. (*a*) Right valve. (*b*) Dorsal. (*c*) Furca.

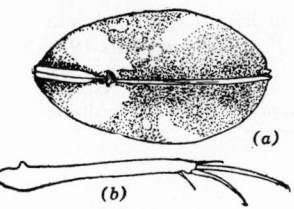

Fig. 28.105. *Eucypris reticulata* ♀.
(a) Dorsal. (b) Furca. (By Sharpe.)

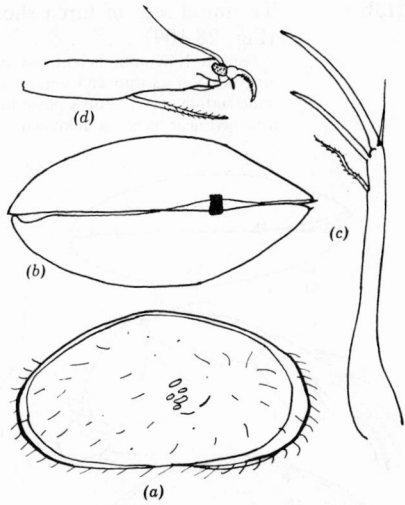

Fig. 28.106. *Eucypris cisternina* ♀. (a) Right valve. (b) Dorsal. (c) Furca. (d) Leg 3.

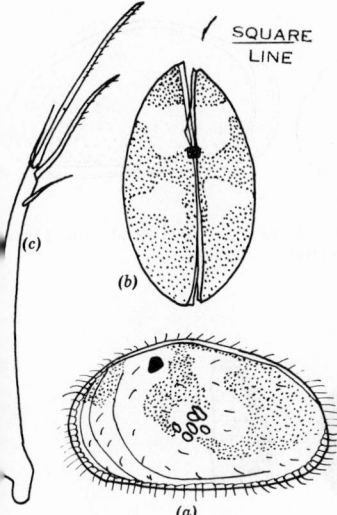

SQUARE LINE

Fig. 28.107. *Eucypris affinis hirsuta* ♀.
(a) Lateral. (b) Dorsal. (c) Furca.

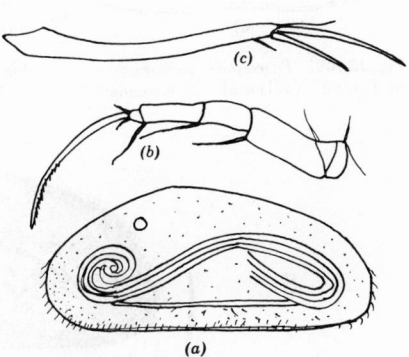

Fig. 28.108. *Prionocypris longiforma*. (a) Left valve ♂. (b) Leg 2 ♀. (c) Furca ♀.

the length of the subterminal claw (Fig. 28.107). . . . .
             E. affinis hirsuta (Fischer) 1851

Length 1.35 mm, height 0.77 mm, width 0.67 mm. Color light green with three dorsolateral dark-brown or dark-blue bands. Surface of valves smooth with a few short hairs. Natatory setae of antenna 2 extend to tips of terminal claws. Furca gently curved, 24 times as long as narrowest width. ♂ have been reported but not described from N. A. Temporary ponds and marshes. Spring. Ohio, Ill., Mexico.

112a (96) Natatory setae of second antenna reaching at least halfway to tips of terminal claws. 115

112b Natatory setae of second antenna greatly reduced and barely evident . . . . **Prionocypris** Brady and Norman 1896 113

113a (112) Terminal seta of furca 2½ times length of dorsal seta (Fig. 28.108) . . . . **P. longiforma** Dobbin 1941

Length 1.25 mm, height ⅖ the length. Left valve overlaps right slightly at both extremities. Prominent pore canals at valve margins. Surface valves with numerous papillar projections and fine hairs. Antenna 1 with 7 segments. Natatory setae rudimentary. Terminal segment of maxillary palp longer than broad. Furca long and slender, 15 times as long as narrowest width. Males with testes showing through shell. Streams. Apr. Wash.

113b   Terminal seta of furca short, about equal in length to dorsal seta
             (Fig. 28.109) . . . . **P. canadensis** Sars 1926
             Length 1.40 mm, height not fully attaining ½ the length, width about ²⁄₅ the
             length. Valves thin and semipellucid; surface smooth and sparsely hairy. Natatory
             setae rudimentary. Furca powerfully developed and straight. Dorsal seta small and
             attached near apex. ♂ unknown. Brooks. Aug. Alberta.

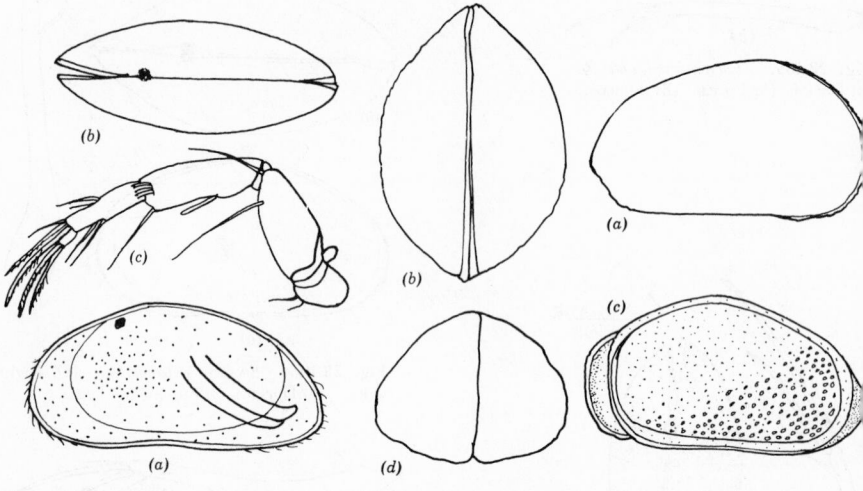

Fig. 28.109. *Prionocypris canadensis* ♀.   Fig. 28.110. *Cypris subglobosa* ♀. (*a*) Lateral. (*b*) Dorsal
(*a*) Lateral. (*b*) Dorsal. (*c*) Antenna 2.   (*c*) Right valve from inside. (*d*) End view.

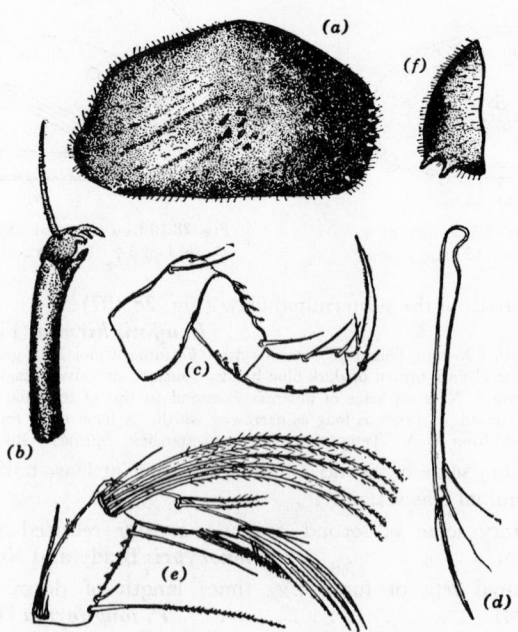

Fig. 28.111. *Cypris pubera* ♀. (*a*) Lateral. (*b*) End of leg 3.
(*c*) Leg 1. (*d*) Furca. (*e*) Antenna 2. (*f*) Posteroventral
portion of shell. (By Sharpe.)

114a (103) Breadth at least ¾ the length (Fig. 28.110) . . . .
Cypris subglobosa Sowerby 1840

Length 1.2 mm, height about ⅗ the length, width ¾ the length. Surface of valves strongly punctured with small pits, the pattern resembling a thimble. Color green. Lateral portion of carapace is prominantly swollen. Dorsal margin convex; ventral margin concave and sinuated. La., Mich.

114b Breadth less than ¾ the length (Fig. 28.111). . . . .
C. pubera (O. F. Müller 1776)

Length 2.10 mm, height 1.25 mm, width 1.20 mm. A dark patch on the highest central, lateral portion of valve. Anterior and posterior margins with prominent external tubercles. Valves sparsely hairy. Two prominent tubercles at posteroventral margin of valve. Leg 1 4-segmented, third and fourth segments united. Furca nearly straight; 24 times as long as narrowest width. ♂ unknown. Ponds. Apr. to June. Ore., Wash., Wyo., Ontario.

115a (112) Teeth of maxillary spines smooth **116**

115b Teeth of maxillary spines denticulated; shell elongate, dorsal margin scarcely arched at all; surface of valves with hair-bearing puncta between which are hairless puncta . . . .
Herpetocypris Brady and Norman 1889 **142**

For a species described since this key was completed, see Tressler (1954).

116a (115) Distal segment of maxillary palp cylindrical . . . .
Cypriconcha Sars 1926 **118**

116b Distal segment of maxillary palp broadened distally . . . .
Candonocypris Sars 1895 **117**

For a species described since this key was completed, see Tressler (1954).

117a (116) Posterior margins of valves serrate (Fig. 28.112) . . . .
C. pugionis Furtos 1936

Length 3.90 mm, height 1.90 mm, width 1.70 mm. Surface of valves smooth, with numerous puncta bearing short, straight hairs. Color light with dark extremities and a dark patch on either side of ocular region. Natatory setae extend to tips of terminal claws. Maxillary spines very coarsely toothed. Terminal claw of leg 2 straight, daggerlike; distal half with short denticles. Furca straight, 24 times as long as narrowest width. ♂ unknown. Pools. Aug. Fla.

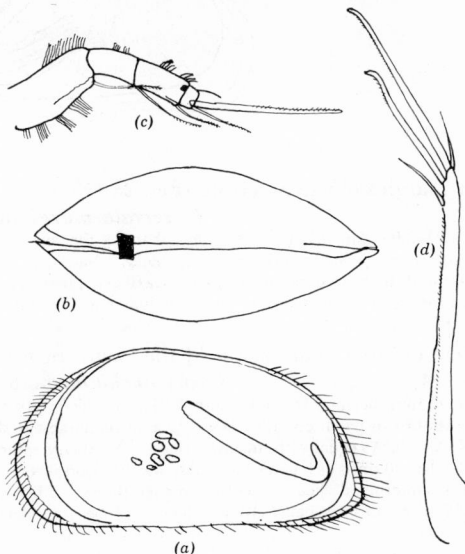

Fig. 28.112. *Candonocypris pugionis* ♀. (*a*) Lateral. (*b*) Dorsal. (*c*) Leg 2. (*d*) Furca.

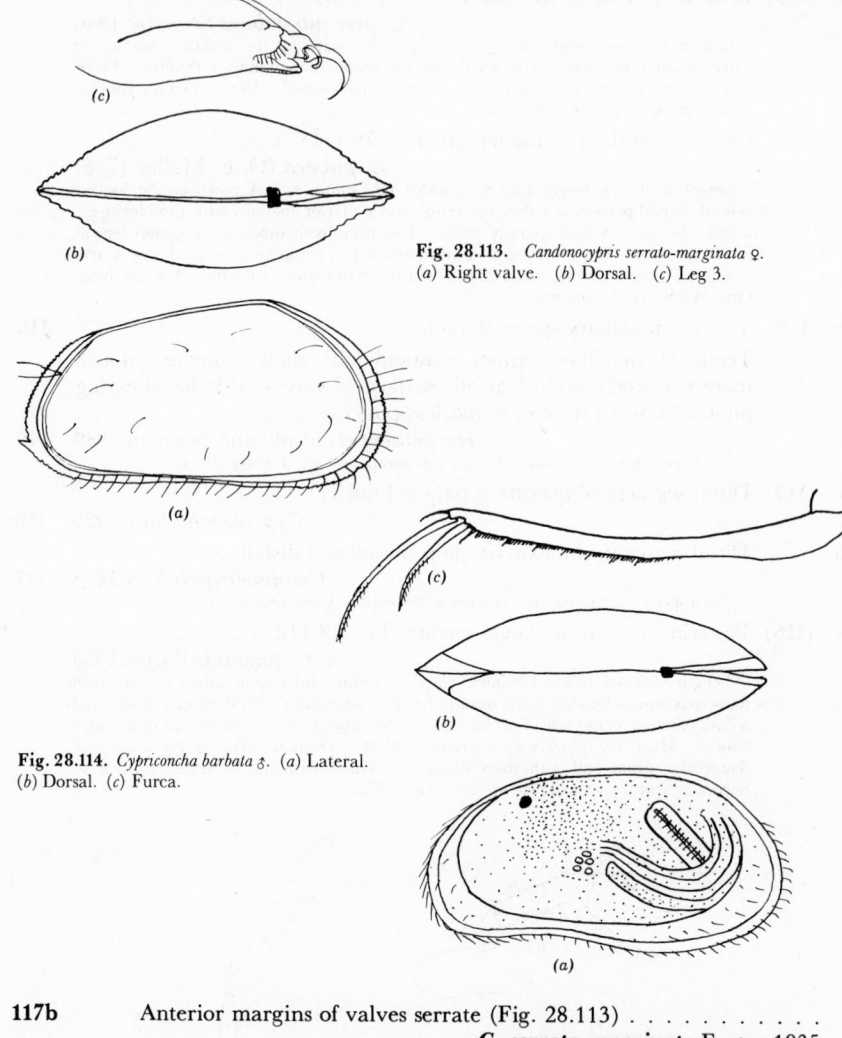

Fig. 28.113. *Candonocypris serrato-marginata* ♀.
(*a*) Right valve. (*b*) Dorsal. (*c*) Leg 3.

Fig. 28.114. *Cypriconcha barbata* ♂. (*a*) Lateral.
(*b*) Dorsal. (*c*) Furca.

117b   Anterior margins of valves serrate (Fig. 28.113) . . . .
                                    C. serrato-marginata Furtos 1935
    Length 1.71–2.05 mm, height slightly less than ½ the length; width about ⅖
length. Surface of valves smooth, sparingly hairy. Natatory setae of antenna 2
extend as far as distal third of terminal claws. Maxillary spines smooth. Furca gently
curved, 18 times as long as narrowest width. ♂ similar to ♀. Only immature ♂ have
been reported from N. A. Cenotes. July, Aug. Fla., Yucatan.

118a (116) Dorsal seta of furca small, about ¼ the length of subterminal claw
                 (Fig. 28.114) . . . . *Cypriconcha barbata* (Forbes) 1893
    Length 4.00 mm, height 2.00 mm, width 1.60 mm. Color dirty yellowish-brown
with a reddish-brown patch on either side. One of the largest fresh-water ostracods
found in N. A. Valves equal with smooth surfaces. Maxillary spines smooth. Furca
long and slender, 20 times as long as narrowest width. Dorsal seta of furca small, ¼
the length of subterminal claw. ♂ slightly smaller than ♀, otherwise similar except
for slight differences in shell shape. Rivers, sloughs. June, July, Aug. Wyo., Alberta,
Great Slave Lake.

118b   Dorsal seta at least ½ the length of subterminal claw **119**

119a (118) Length about 4.00 mm (Fig. 28.115) . . **C. gigantea** Dobbin 1941
    Length 4.00 mm. Right valve slightly longer than left at posterior. Surface of
valves smooth with short hairs. Natatory setae reach a little more than halfway to

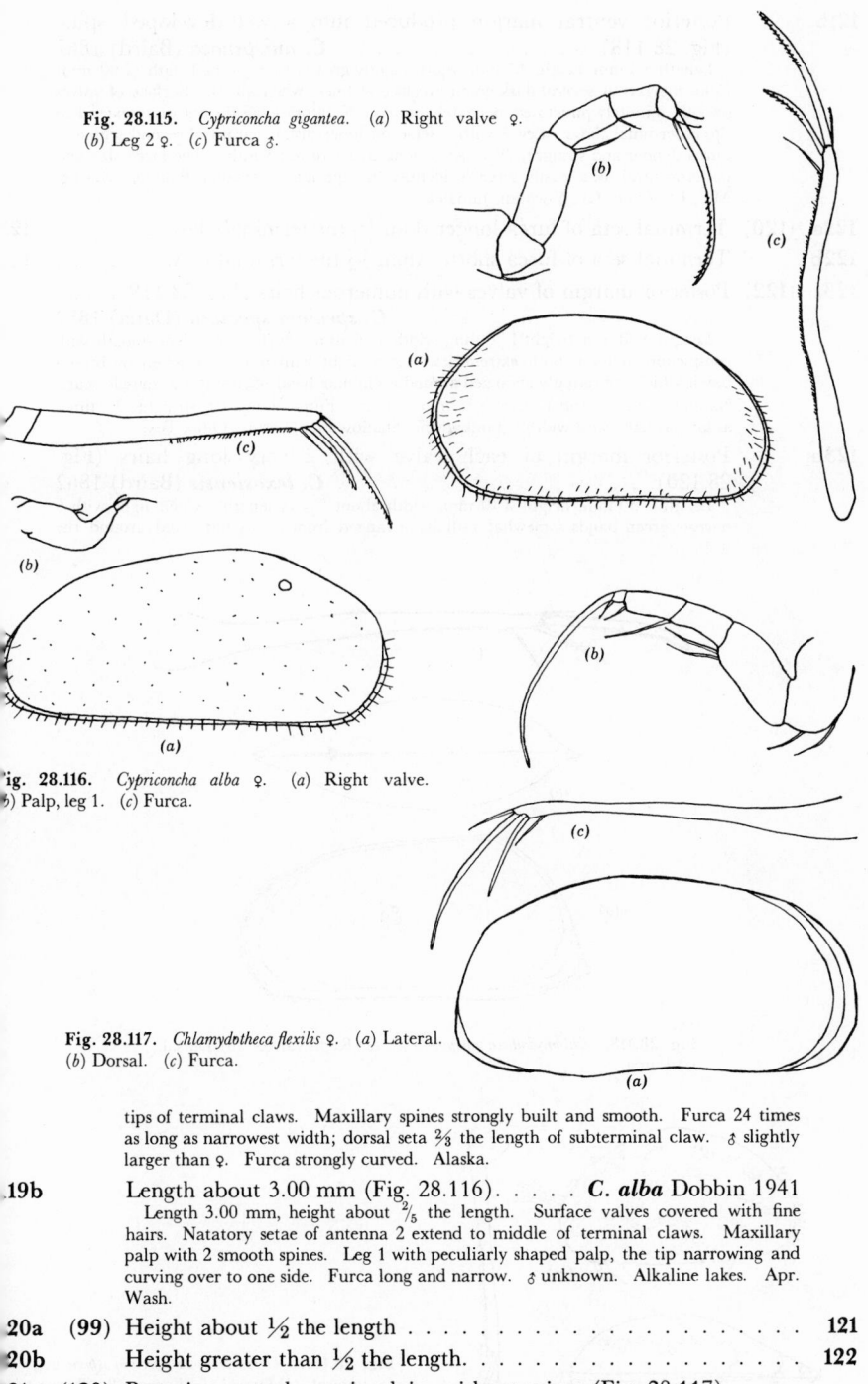

Fig. 28.115. *Cypriconcha gigantea.* (*a*) Right valve ♀.
(*b*) Leg 2 ♀. (*c*) Furca ♂.

(*b*)

(*c*)

(*a*)

(*c*)

(*b*)

(*a*)

ig. 28.116. *Cypriconcha alba* ♀. (*a*) Right valve.
b) Palp, leg 1. (*c*) Furca.

(*b*)

(*c*)

Fig. 28.117. *Chlamydotheca flexilis* ♀. (*a*) Lateral.
(*b*) Dorsal. (*c*) Furca.

(*a*)

tips of terminal claws. Maxillary spines strongly built and smooth. Furca 24 times
as long as narrowest width; dorsal seta ⅔ the length of subterminal claw. ♂ slightly
larger than ♀. Furca strongly curved. Alaska.

19b   Length about 3.00 mm (Fig. 28.116). . . . . . *C. alba* Dobbin 1941
       Length 3.00 mm, height about ⅖ the length. Surface valves covered with fine
       hairs. Natatory setae of antenna 2 extend to middle of terminal claws. Maxillary
       palp with 2 smooth spines. Leg 1 with peculiarly shaped palp, the tip narrowing and
       curving over to one side. Furca long and narrow. ♂ unknown. Alkaline lakes. Apr.
       Wash.

20a   (99)   Height about ½ the length **121**

20b   Height greater than ½ the length. **122**

21a   (120)   Posterior ventral margin plain, without spines (Fig. 28.117) . . . .
                       Chlamydotheca flexilis (Brady) 1862
       Length 4.00 mm, width about ⅓ the length, height almost ½ the length. Furca
       slender, almost straight; dorsal margin finely ciliated; dorsal seta about ½ the length
       of subterminal claw. ♂ unknown. Pools in sand dunes. July. La. (Grand Isle).

121b   Posterior ventral margin produced into a well-developed spine
(Fig. 28.118) . . . . **C. unispinosa** (Baird) 1862

Length 5.6 mm, height 2.5 mm, width slightly greater than ⅓ the length (2.00 mm).
Color marked by several dark-green streaks and lines (when alive). Surface of valves
smooth, minutely punctated, delicately haired. Natatory setae do not quite extend to
tips of terminal claws. Leg 2 with 2 setae on inner distal margin of second segment.
Furca slender and straight, 29 times as long as narrowest width. The large size and
posteroventral spine easily serve to identify this species. ♂ smaller than ♀. Spring.
Md., Ill., Ohio, La., Yucatan, Jamaica.

122a **(120)** Terminal seta of furca longer than ½ the terminal claw **123**

122b   Terminal seta of furca shorter than ½ the terminal claw **124**

123a **(122)** Posterior margin of valves with numerous hairs (Fig. 28.119). . . .
                                             C. speciosa speciosa (Dana) 1852

Length 3.00 mm, height 1.70 mm, width 1.60 mm. Surface of valves smooth with
conspicuous hairs at both extremities. Color light with 6 narrow green or brown
bands which are radially arranged around a circular band enclosing the muscle scars.
Natatory setae extend to tips of terminal claws. Furca elongated, straight, 20 times
as long as narrowest width. ♂ unknown. Shallow canal basin. Ohio, Tex.

123b   Posterior margin of each valve with 2 very long hairs (Fig.
28.120). . . . . **C. texasiensis** (Baird) 1862

Length 3.30 mm, height 2.10 mm, width about $^{8}/_{15}$ of length. Color light with 6
narrow green bands somewhat radially arranged from a circular band around the

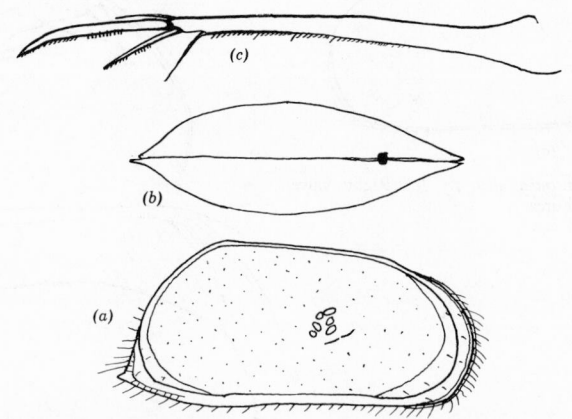

Fig. 28.118. *Chlamydotheca unispinosa* ♀. (*a*) Right valve. (*b*) Dorsal.
(*c*) Furca.

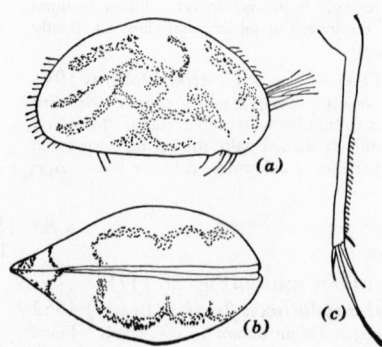

Fig. 28.119. *Chlamydotheca speciosa speciosa* ♀.
(*a*) Lateral. (*b*) Dorsal. (*c*) Furca. (By Sharpe.)

muscle marks. Surface of valves smooth with scattered puncta bearing delicate hairs. Natatory setae do not reach to tips of terminal claws. The 2 setae of second segment of leg 2 elongated and extending to base of terminal claw. Furca slightly curved, 19 times as long as narrowest width. ♂ unknown. Pools. Tex., La. (Grand Isle), Yucatan.

24a (122) Shorter than 3.00 mm **125**

24b Longer than 3.00 mm (Fig. 28.121). . . **C. azteca** (Saussure) 1858

Length 3.30 mm, height 2.00 mm, width 1.80 mm. Color yellowish-gray. Shell with no special markings. Natatory setae reach to tips of terminal claws. Furca almost straight, about 18 times as long as narrowest width; dorsal margin faintly pectinate. ♂ unknown. Ditches and pools. October. Tex., Mexico.

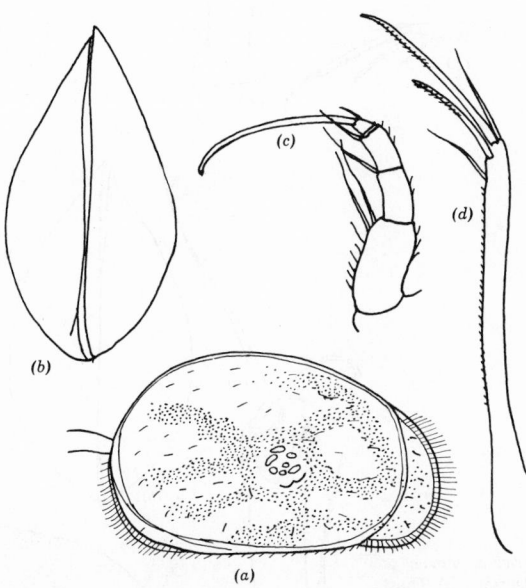

Fig. 28.120. *Chlamydotheca texasiensis* ♀. (*a*) Right valve. (*b*) Dorsal. (*c*) Leg 2. (*d*) Furca.

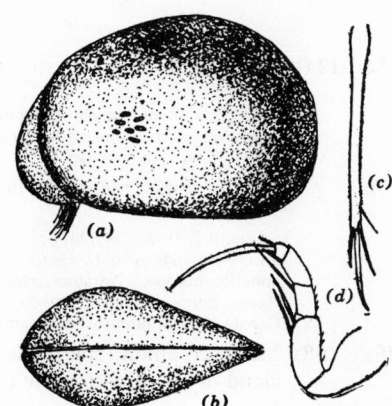

Fig. 28.121. *Chlamydotheca azteca* ♀. (*a*) Lateral. (*b*) Dorsal. (*c*) Furca. (*d*) Leg 2. (By Sharpe.)

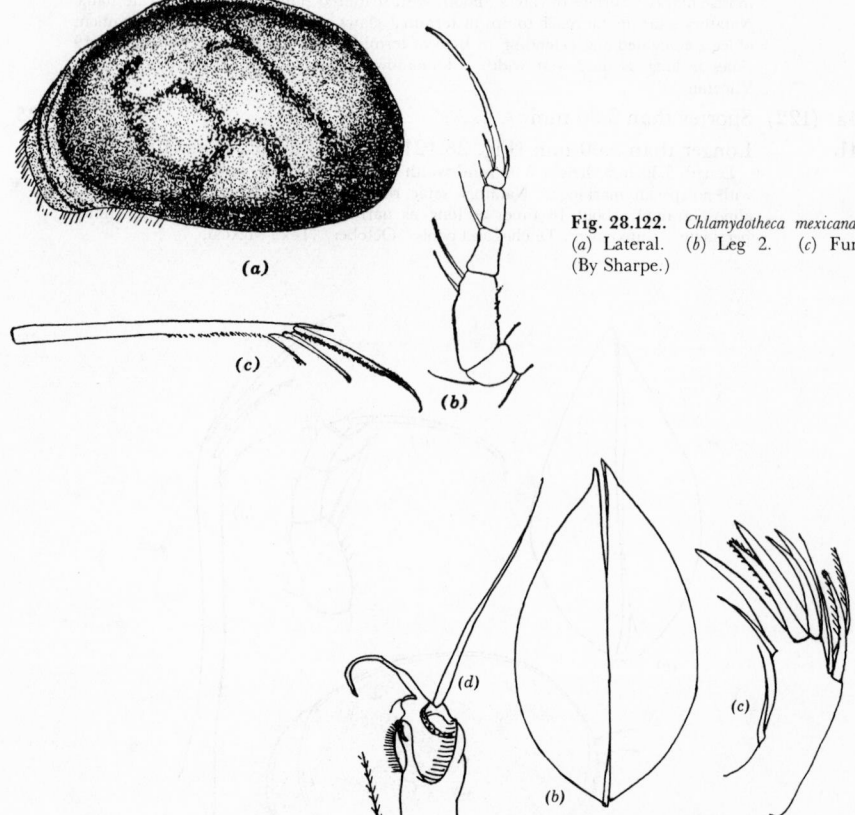

Fig. 28.122. *Chlamydotheca mexicana* ♀.
(*a*) Lateral.   (*b*) Leg 2.   (*c*) Furca
(By Sharpe.)

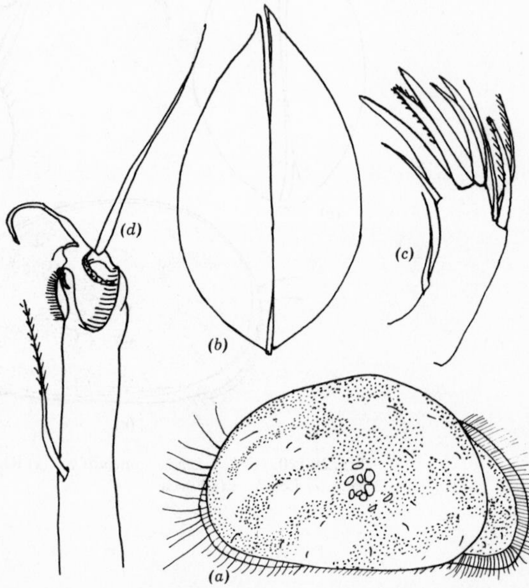

(*a*)

(*c*)

(*b*)

Fig. 28.123. *Chlamydotheca arcuata* ♀.
(*a*) Right valve. (*b*) Dorsal. (*c*) Third
maxillary process. (*d*) Leg 3.

(*d*)

(*c*)

(*b*)

(*a*)

125a (124) Distal half only of dorsal margin of furca pectinated (Fig.
28.122) . . . . **C. mexicana** Sharpe 1903
   Length 2.75 mm, height 1.55 mm, width 1.60 mm. Color light with 2 or 3 greenish
bands irregularly arranged on shell. Furca straight, 20 times as long as narrowest
width. Ponds. September. Mexico (Durango).

125b   Most of dorsal margin of furca pectinated (Fig. 28.123) . . . .
                                                        C. arcuata (Sars) 1902
   Length 2.70 mm, height 1.56 mm, width 1.15 mm. Color light with 6 radially
arranged bands as in *C. speciosa*. Valves smooth with conspicuous hairs along the
posterior margin. Natatory setae of antenna 2 do not extend to tips of terminal
claws. Furca elongated, straight, 20 times as long as narrowest width. ♂ unknown.
Canal basins. Summer and autumn. Ohio, Fla., La., Mexico.

126a (98) Males common; spermatic vessels in the male forming a dense
            cloud or spiral within the anterior part of shell . . . .
                                                Cypricercus Sars 1895 **125**

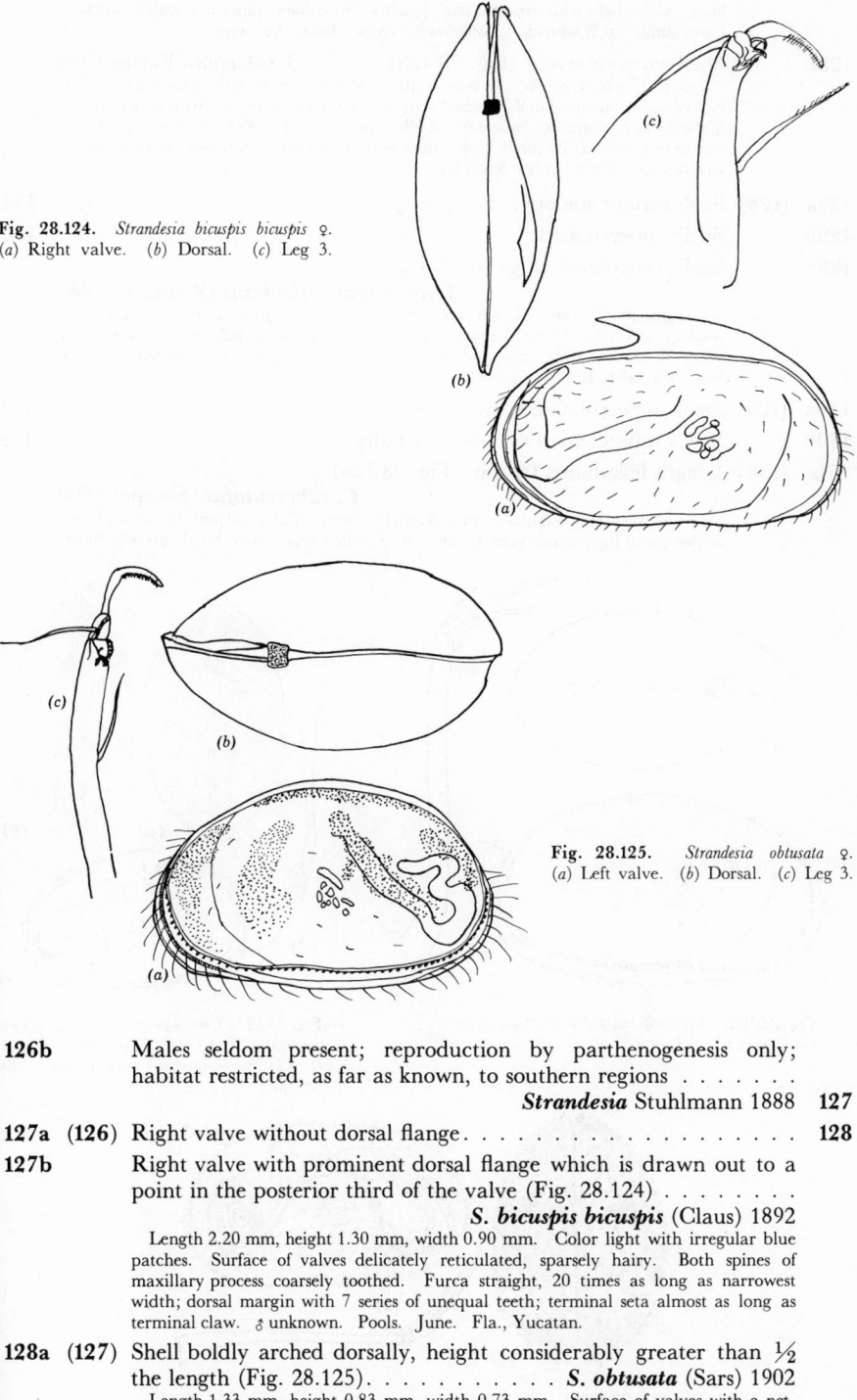

Fig. 28.124. *Strandesia bicuspis bicuspis* ♀.
(*a*) Right valve. (*b*) Dorsal. (*c*) Leg 3.

Fig. 28.125. *Strandesia obtusata* ♀.
(*a*) Left valve. (*b*) Dorsal. (*c*) Leg 3.

126b   Males seldom present; reproduction by parthenogenesis only; habitat restricted, as far as known, to southern regions . . . .
                                     Strandesia Stuhlmann 1888   **127**

127a   **(126)**   Right valve without dorsal flange. **128**

127b   Right valve with prominent dorsal flange which is drawn out to a point in the posterior third of the valve (Fig. 28.124) . . . .
                                         S. bicuspis bicuspis (Claus) 1892

Length 2.20 mm, height 1.30 mm, width 0.90 mm. Color light with irregular blue patches. Surface of valves delicately reticulated, sparsely hairy. Both spines of maxillary process coarsely toothed. Furca straight, 20 times as long as narrowest width; dorsal margin with 7 series of unequal teeth; terminal seta almost as long as terminal claw. ♂ unknown. Pools. June. Fla., Yucatan.

128a   **(127)**   Shell boldly arched dorsally, height considerably greater than ½ the length (Fig. 28.125). . . . . ***S. obtusata*** (Sars) 1902

Length 1.33 mm, height 0.83 mm, width 0.73 mm. Surface of valves with a network of heavy-walled, polygonal reticulations with scattered puncta bearing short hairs. Color light with irregular dark patches. Maxillary spines moderately toothed. Furca similar to *S. intrepida*. ♂ unknown. Pools. June. Yucatan.

128b   Shell elongate-ovoid (Fig. 28.126) . . . . **S. intrepida** Furtos 1936

Length 1.76 mm, height 0.98 mm, width somewhat greater than ½ the length. Surface of valves longitudinally striated with a reticulated network; hairless except for sparsely haired margins. Spines of maxillary process both moderately toothed. Furca very gently curved, 20 times longer than narrowest width. ♂ smaller than ♀, otherwise similar. Pools. July. Yucatan.

129a (126)   Shell surface smooth **134**

129b   Shells tuberculated. **130**

129c   Shells reticulated (Fig. 28.127) . . . .

                    Cypricercus reticulatus (Zaddach) 1844

Length 1.0–1.3 mm. Height somewhat greater than ½ the length. Width somewhat greater than ½ the length. Surface of valves with polygonal reticulations. Furca slender, straight in ♀ and S-shaped in ♂; dorsal margins toothed in distal half. Va., Md., Ill.

130a (129)   Shells tuberculated, with few hairs **131**

130b   Shells tuberculated and densely hairy. **132**

131a (130)   Length less than 1.00 mm (Fig. 28.128) . . . .

                          C. tuberculatus (Sharpe) 1908

Length 0.93 mm, height 0.53 mm, width 0.70 mm. Color purplish-brown with one or two dorsal lighter transverse bands. Shell surface very tuberculated, sparsely hairy.

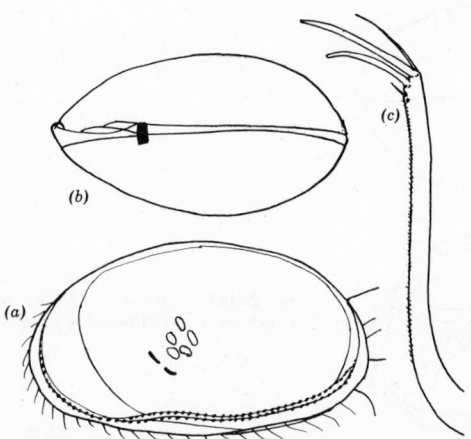

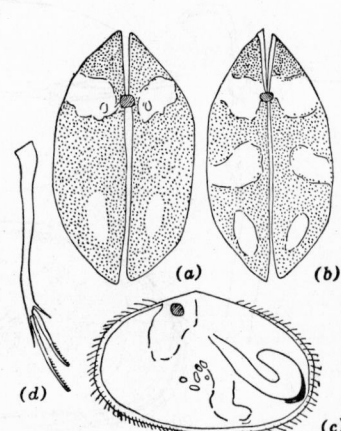

Fig. 28.126. *Strandesia intrepida.* (*a*) Left valve ♂. (*b*) Dorsal ♂. (*c*) Furca ♀.

Fig. 28.127. *Cypricercus reticulatus.* (*a*) Var. *major* dorsal ♀. (*b*) Var. *minor* dorsal. (*c*) Var. *major* lateral. (*d*) Furca. (By Sharpe.)

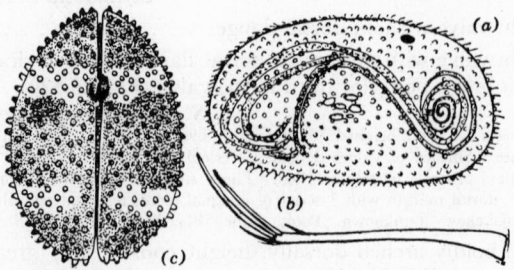

Fig. 28.128. *Cypricercus tuberculatus* ♂. (*a*) Lateral. (*b*) Furca. (*c*) Dorsal. (By Sharpe.)

Plump forms when seen from above. Natatory setae extend slightly beyond tips of terminal claws. Furca straight, 32 times longer than narrowest width. Shallow, weedy, and swampy ponds. Spring. Ill., Ind., Mich.

131b Length greater than 1.00 mm (Fig. 28.129) . . . .
C. tincta Furtos 1933

Length 1.50 mm, height 0.85 mm, width 0.74 mm. Color light green, brilliantly banded with three dorsolateral dark-blue bands. Natatory setae extend to tips of terminal claws. Both maxillary spines sharply toothed. Furca gently curved, 26 times as long as narrowest width. ♂ smaller than ♀, otherwise similar. Temporary marshes, pools. May. Ohio, Mich.

132a (130) Terminal claw of furca equal to at least ½ the length of ramus. . . **133**

132b Terminal claw of furca less than ½ the length of ramus (Fig. 28.130) . . . . *C. horridus* Sars 1926

Length 1.05 mm, height somewhat more than ½ the length, width about ⅔ the length. Left valve overlaps right. Surface of valves very uneven being everywhere covered with short, stout spikes, and densely distributed fine hairs. Furca straight. ♂ unknown. Canals, pools. June. Ontario.

133a (132) Natatory setae of second antenna do not reach to tips of terminal claws (Fig. 28.131) . . . . *C. fuscatus* (Jurine) 1820

Length 1.50 mm, height somewhat greater than ½ the length, width about equal to the height. Surface of valves with numerous small knobs and rather densely hairy. Color light yellowish-brown with a conspicuous chocolate band extending obliquely down either valve. Furca about 25 times longer than narrowest width. ♂ unknown. Shallow, grassy ponds and swamps. Spring and early summer. Ohio, Ill., Ga., Dela., Mass.

133b Natatory setae of second antenna reach beyond tips of terminal claws (Fig. 28.132) . . . . *C. splendida* Furtos 1933

Length 1.75 mm, height 0.98 mm, width 0.88 mm. Color olive green brilliantly banded with bluish-green. Surface of valves covered with conspicuous wartlike tubercles. Natatory setae reach to tips of claws. Maxillary spines both toothed. ♂ smaller than ♀, otherwise similar; ejaculatory duct with 30 whorls of spines. Temporary ponds, marshes, ditches rich in vegetation. Feb. to May and Oct. to Nov. Ohio, Mass., N. Y.

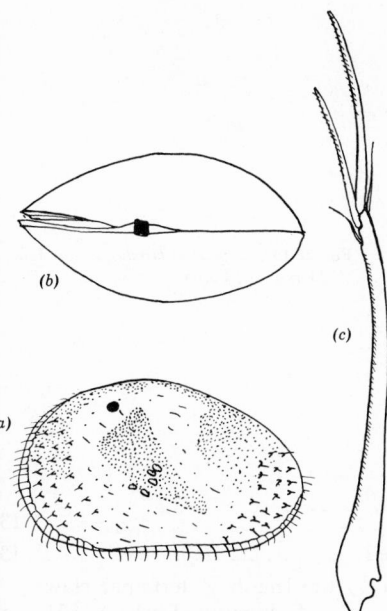

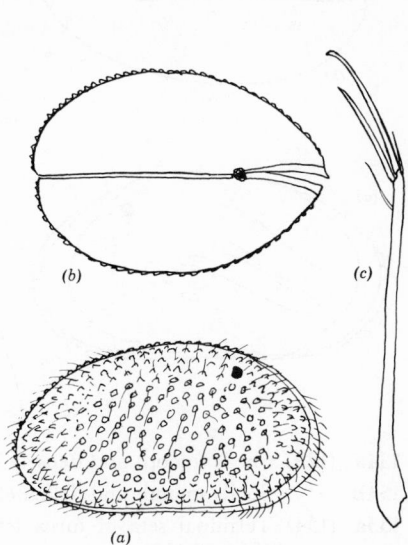

Fig. 28.129. *Cypricercus tincta* ♀. (*a*) Lateral. (*b*) Dorsal. (*c*) Furca.

Fig. 28.130. *Cypricercus horridus* ♀. (*a*) Lateral. (*b*) Dorsal. (*c*) Furca.

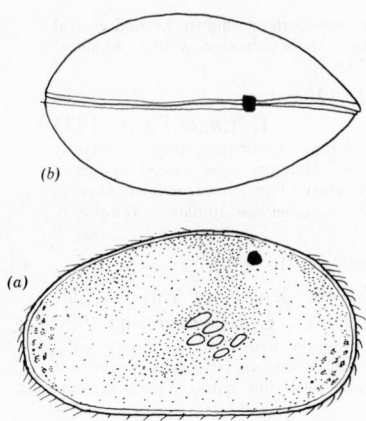

Fig. 28.131. *Cypricercus fuscatus* ♀. (*a*) Lateral. (*b*) Dorsal.

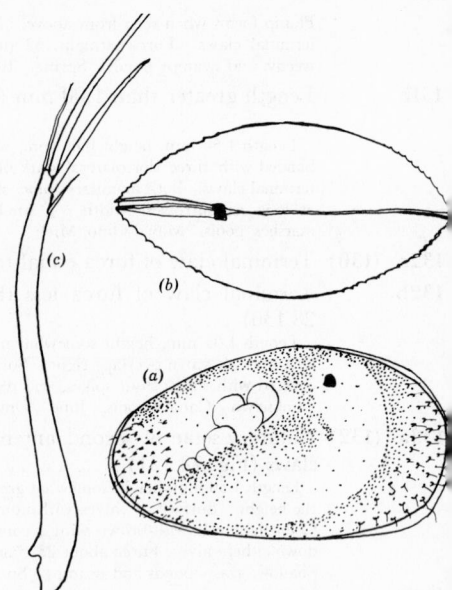

Fig. 28.132. *Cypricercus splendida* ♀. (*a*) Latera (*b*) Dorsal. (*c*) Furca.

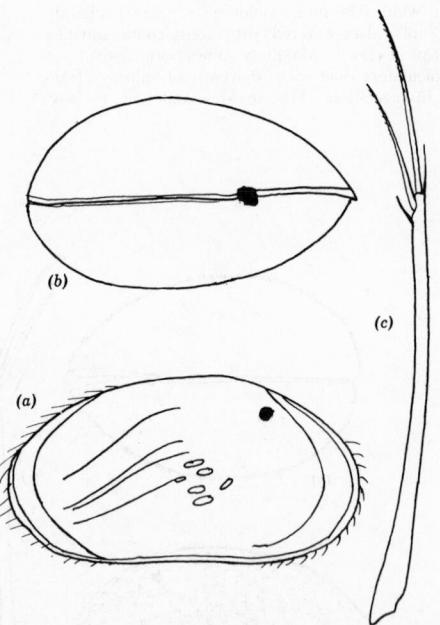

Fig. 28.133. *Cypricercus hirsutus* ♀. (*a*) Lateral (*b*) Dorsal. (*c*) Furca.

134a (129) Shells smooth and densely haired 135
134b Shells smooth and delicately haired. 136
135a (134) Terminal seta of furca less than ½ the length of terminal claw (Fig. 28.133) . . . . **C. hirsutus** (Fischer) 1851
Length 1.10 mm, height about ½ the length, width somewhat greater than height. Left valve conspicuously larger than right and overlapping it anteriorly. Color dark

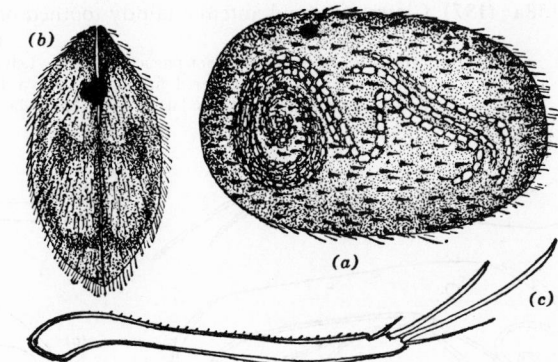

Fig. 28.134. *Cypricercus passaica* ♂.
(a) Lateral. (b) Dorsal. (c) Furca.
By Sharpe.)

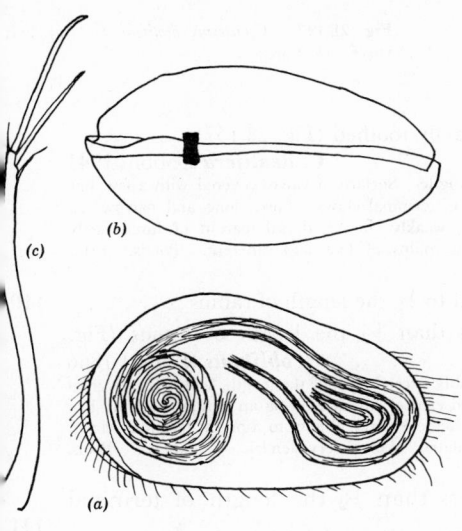

Fig. 28.135. *Cypricercus mollis* ♂. (a) Left valve.
(b) Lateral. (c) Furca.

bluish-green. Surface of valves rather densely hairy. Maxillary spines smooth.
Furca long and slender. ♂ unknown. Pribilof Islands.

135b   Terminal seta of furca longer than ½ the length of terminal claw
(Fig. 28.134) . . . . **C. passaica** (Sharpe) 1903
     Length 1.60 mm, height 0.80 mm, width 0.82 mm. Color brownish with dark-blue
patches laterally and dorsally. Natatory setae reach slightly beyond tips of terminal
claws. Furca about ½ the length of shell, 23 times as long as wide; dorsal margin
weakly pectinate. Weedy ponds. Spring. Mass., N. J.

136a   **(134)**   Length greater than 1.00 mm. 137

136b   Length less than 1.00 mm (Fig. 28.135) . . . .
                               C. mollis Furtos 1936
     Length (of ♂) 0.80 mm, height 0.44 mm, width 0.40 mm. Surface of valves smooth,
sparsely hairy. Maxillary spines toothed. Prehensile palps unequal, 2-segmented.
Furca feebly S-shaped, 23 times longer than narrowest width. Ejaculatory duct with
13 rings of spines. ♀ unknown. Pools. Aug. Fla.

137a   **(136)**   Length greater than 1.50 mm. 138

137b   Length less than 1.50 mm. 139

138a (137) Claws of second antenna faintly toothed or smooth (Fig. 28.136). .
　　　　　　　　　　　　　　　　　　　　　　　C. elongata Dobbin 1941
Length 1.82 mm. Shell elongate and narrow. Left shell slightly higher than right. Surface of valves with scattered fine hairs. Furca 19 times longer than narrowest width. ♂ similar to ♀. Ponds, lakes. Feb., Apr., May. Wash.

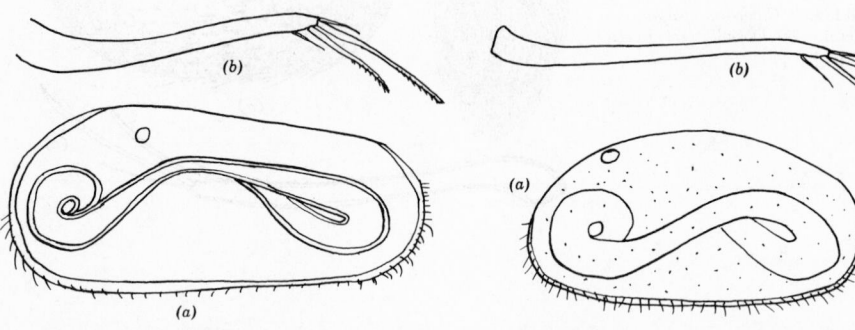

Fig. 28.136. *Cypricercus elongata.* (*a*) Left valve ♂. (*b*) Furca ♀.

Fig. 28.137. *Cypricercus dentifera* ♂. (*a*) Left valve. (*b*) Furca.

138b Claws of second antenna heavily toothed (Fig. 28.137) . . . .
　　　　　　　　　　　　　　　　　　　　　　　C. dentifera Dobbin 1941
Length 2.00 mm. Shell rather fragile. Surface of valves covered with a few fine hairs. Natatory setae reach to tips of terminal claws. Furca long and narrow, 22 times longer than narrowest width, weakly curved; dorsal margin of furca finely toothed. ♂ slightly smaller than ♀; palps of first legs different. Ponds. Feb., Mar. Wash.

139a (137) Terminal claw of furca equal to ½ the length of ramus **140**

139b Terminal claw of furca less than ½ the length of ramus (Fig. 28.138) . . . . *C. obliqvus* (Brady) 1866
Length 1.26 mm, height considerably exceeding half the length, width not as great as height. Color light greenish. End view of valves showing an oblique junction line between the two shells. Natatory setae extend almost to tips of terminal claws. Maxillary spines indistinctly denticulated. Furca very slender. ♂ unknown. Pools, small lakes. D. C., Md., Ariz.

140a (139) Terminal seta of furca less than ½ the length of terminal claw **141**

140b Terminal seta of furca equal to ½ the length of terminal claw (Fig. 28.139). . . . . *C. columbiensis* Dobbin 1941
Length 1.27 mm. Surface of valves with a few fine hairs. Right valve slightly larger than left. Furca long and slender, 21 times as long as narrowest width, slightly curved; dorsal margin faintly toothed. ♂ slightly smaller than ♀. Ponds. Apr., May. Wash.

141a (140) Length less than 1.25 mm (Fig. 28.140) . . . .
　　　　　　　　　　　　　　　　　　　　　　　C. serratus Tressler 1950
Length 1.08 mm, height 0.69 mm. Hyaline margins narrow, with a few scattered hairs. Valves smooth, with a few scattered polygonal markings, sparsely hairy. Testes with marked spiral coil in anterior valve chamber. Ejaculatory duct long and slender with 24 whorls of spines. ♀ unknown. July. Ponds. Wyo. (Medicine Bow Mts., elev. 10,200 ft).

141b Length greater than 1.25 mm (Fig. 28.141) . . . .
　　　　　　　　　　　　　　　　　　　　　　　C. affinis (Fischer) 1851
Length 1.35 mm, height 0.77 mm, width 0.67 mm. Color light green with three dorsolateral dark-brown or dark-blue bands. Maxillary spines toothed. Furca gently curved, 24 times as long as narrowest width. ♂ present. Temporary ponds and marshes. Apr. and May. Ohio, Ill., Ontario, Mexico, Alaska.

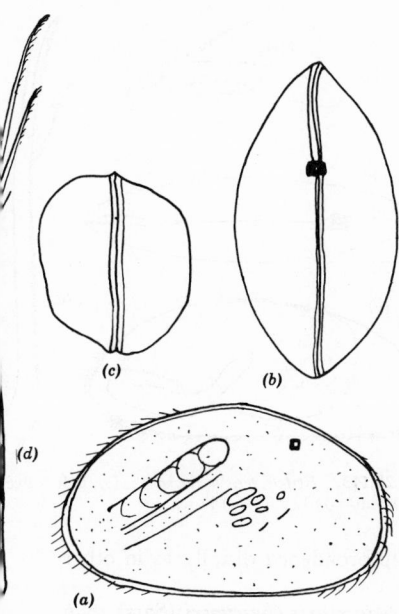

Fig. 28.138. *Cypricercus obliqvus* ♀. (a) Lateral.
(b) Dorsal. (c) End view. (d) Furca.

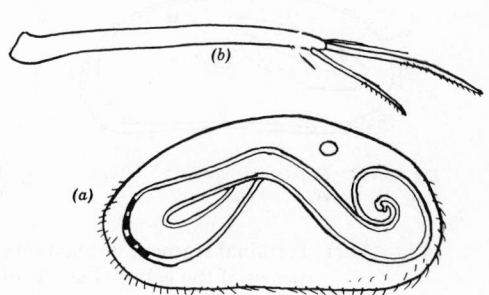

Fig. 28.139. *Cypricercus columbiensis.* (a) Right
valve ♂. (b) Furca ♀.

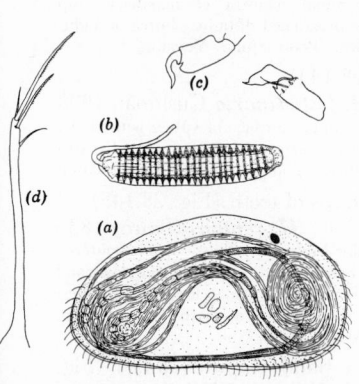

Fig. 28.140. *Cypricercus serratus* ♂. (a) Lateral. (b) Ejaculatory duct. (c) Prehensile
palps. (d) Furca.

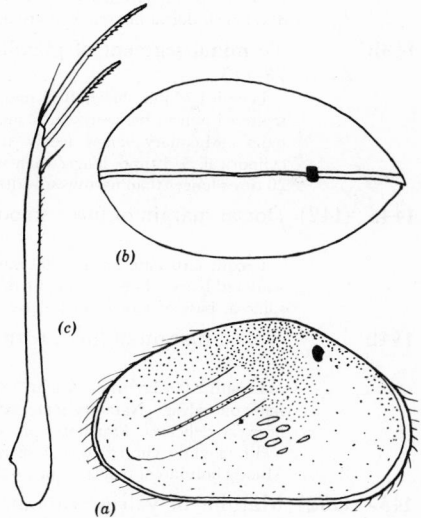

Fig. 28.141. *Cypricercus affinis* ♀. (a) Lateral.
(b) Dorsal. (c) Furca.

142a **(115)** Natatory setae of second antenna extend to tips of terminal
claws **143**

142b Natatory setae of second antenna do not reach to tips of terminal
claws **144**

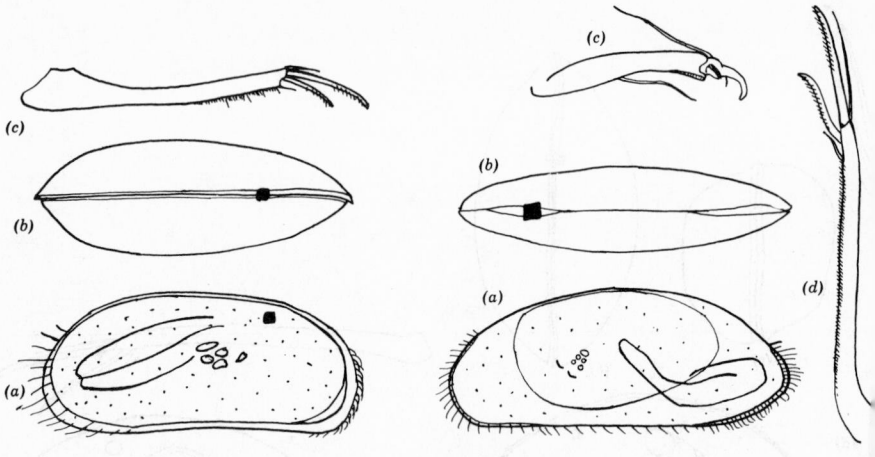

Fig. 28.142. *Herpetocypris chevreuxi* ♀. (*a*) Lateral. (*b*) Dorsal. (*c*) Furca.

Fig. 28.143. *Herpetocypris meridana* ♀. (*a*) Left valve. (*b*) Dorsal. (*c*) Leg 3. (*d*) Furca.

143a (142) Terminal segment of maxillary palp broadened distally as in other species of the genus (Fig. 28.142) . . . .
Herpetocypris chevreuxi (Sars) 1896
Length 2.30 mm, height 0.95 mm, width ⅓ the length, olivaceous. Clouded with dark green. Surface of valves smooth and polished; few hairs except at posterior extremity. Valves unequal, left overlapping right at both ends. Natatory setae extend to tips of terminal claws but are thin and not plumose. Furca powerfully developed; dorsal margin with groups of small teeth. ♂ unknown. Utah.

143b Terminal segment of maxillary palp not broadened distally (Fig. 28.143) . . . . *H. meridana* Furtos 1936
Length 1.37 mm, height 0.55 mm, width 0.35 mm. Surface of valves smooth with scattered puncta between which smaller puncta are not evident. Delicate marginal hairs. Maxillary spines feebly toothed. Terminal segment of maxillary palp cylindrical, 3½ times longer than wide and not broadened distally. Furca straight, 20 times longer than narrowest width. ♂ unknown. Pools. July. Yucatan.

144a (142) Dorsal margin of furca smooth (Fig. 28.144). . . . .
H. testudinaria Cushman 1908
Length 2.10 mm, height 1.00 mm, width 0.80 mm. Surface of valves with short, scattered hairs. Furca about 14 times longer than narrowest width; a small, extra spine at base of subterminal claw; claws untoothed. Ponds. May. Newfoundland.

144b Dorsal margin of furca with 4 or 5 groups of teeth (Fig. 28.145) . .
H. reptans (Baird) 1835
Length 2.00–2.50 mm, height 0.80 mm. Color brownish-yellow. A few scattered hairs on valves. Natatory setae extend barely halfway to tips of terminal claws and are very delicate. Maxillary spines toothed. Furca 16 times as long as narrowest width, slightly curved; dorsal margin with 5 rows of coarse teeth. ♂ unknown. Muddy bottoms of ponds. Apr. to Sept. Cal., Wash.

145a (102) Margins of valves without prominent band of pore canals (Fig. 28.146) . . . . *Stenocypris fontinalis* Vavra 1895
Length 1.51 mm, height 0.55 mm, width 0.44 mm. Surface of valves smooth, with scattered puncta. Margins less hairy than in *S. malcolmsoni*. Natatory setae extend slightly beyond tips of terminal claws. Maxillary spines faintly toothed. Broader furcal ramus curved, 16 times narrowest width, heavily armed with teeth along dorsal margin. Straight ramus with few teeth. ♂ unknown. Ponds. June. Yucatan, Trinidad.

145b Margin of valves with a broad, striated band of pore canals (Fig. 28.147). . . . . *S. malcolmsoni* (Brady) 1859
Length 2.00 mm, height 0.78 mm, width 0.60 mm. Surface of valves with scattered puncta, otherwise smooth. Sparsely hairy except along posterior margin where hairs

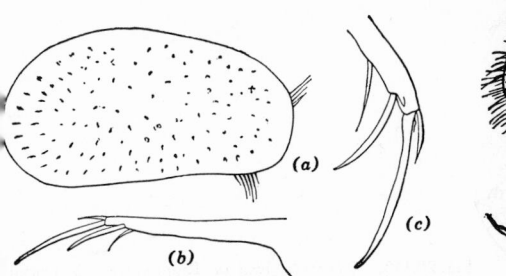

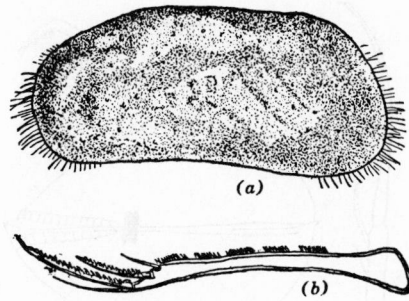

Fig. 28.144. *Herpetocypris testudinaria* ♀. (*a*) Lateral. (*b*) Furca. (*c*) End of furca showing extra spine. (By Sharpe.)

Fig. 28.145. *Herpetocypris reptans* ♀. (*a*) Lateral. (*b*) Furca.

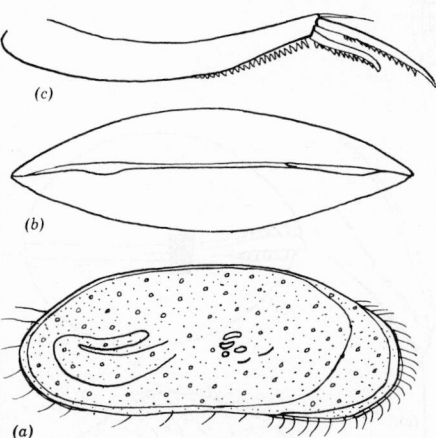

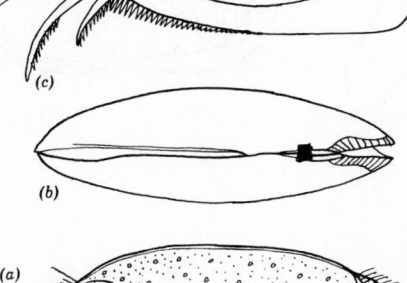

Fig. 28.146. *Stenocypris fontinalis* ♀. (*a*) Right valve. (*b*) Dorsal. (*c*) Furca.

Fig. 28.147. *Stenocypris malcolmsoni* ♀. (*a*) Right valve. (*b*) Dorsal. (*c*) Furca.

increase in length. Maxillary spines toothed. Natatory setae extend to tips of terminal claws. Furca 15 times longer than narrowest width. One ramus curved, the other straight; curved ramus with dorsal margin armed with a row of very stout subequal teeth; claws also stoutly armed with teeth. ♂ unknown. Pools. July. Yucatan, Trinidad.

146a	**(100)**	Terminal seta of furca present. **147**
146b		Terminal seta of furca absent (Fig. 28.148) . . . .

Cypretta turgida (Sars) 1895

Length 0.80 mm, height 0.53 mm, width 0.56 mm. Surface of valves sparsely hairy, very faintly pitted. Natatory setae extend slightly beyond tips of terminal claws. Maxillary spines with slender, delicate teeth. Furca almost straight, 20 times narrowest width; dorsal margin smooth. ♂ unknown. Fish ponds with flowing water. July. N. C.

147a	**(146)**	Length of valves greater than 0.80 mm. **148**
147b		Length of valves less than 0.80 mm (Fig. 28.149) . . . .

C. intonsa Furtos 1936

Length 0.55 mm, height 0.40 mm, width 0.44 mm. Color light with four dark spots, one situated on antero- and another on the posterolateral surface of each valve. Sur-

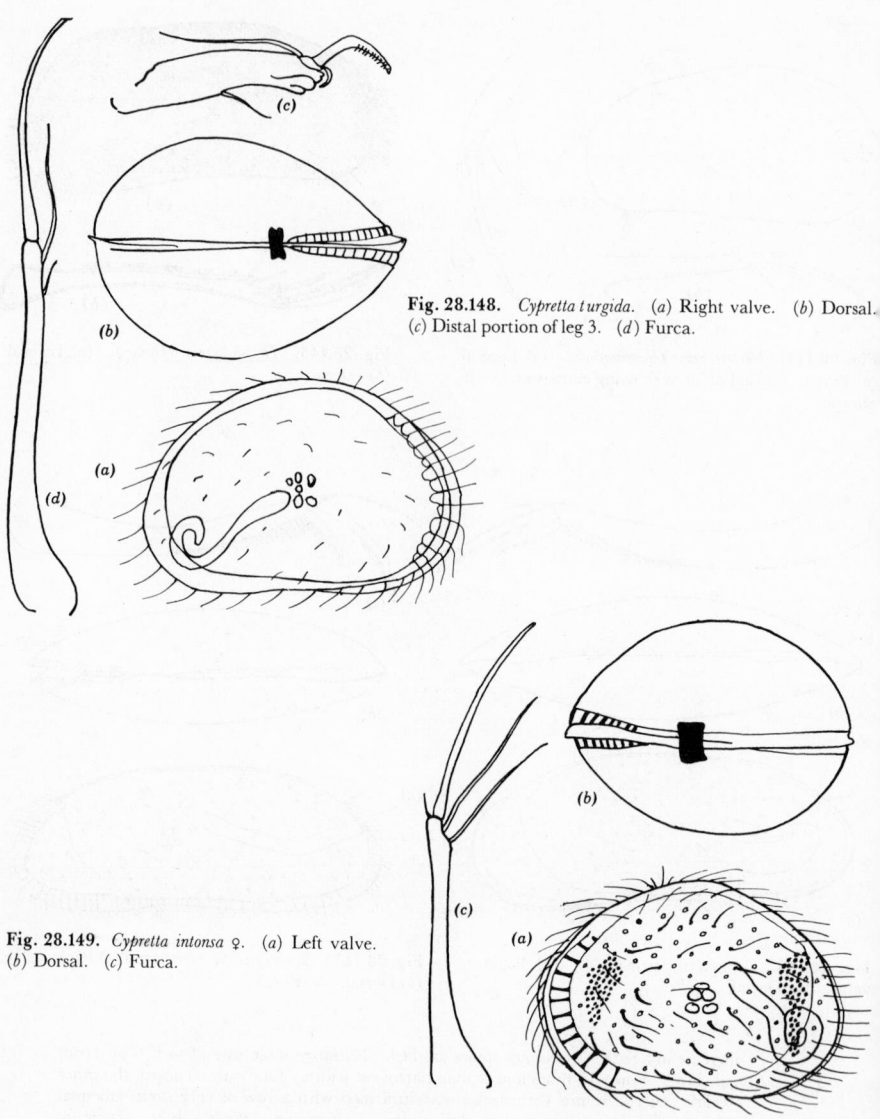

Fig. 28.148. *Cypretta turgida.* (*a*) Right valve. (*b*) Dorsal. (*c*) Distal portion of leg 3. (*d*) Furca.

Fig. 28.149. *Cypretta intonsa* ♀. (*a*) Left valve. (*b*) Dorsal. (*c*) Furca.

face of valves delicately pitted; numerous long, slender hairs and short, stiff spines. Natatory setae extend slightly beyond tips of terminal claws. Maxillary spines smooth. Furca 16 times longer than narrowest width; dorsal margin smooth. ♂ unknown. Ditches, pools. Aug. Fla.

148a (147) Valves not uniformly dark-pigmented **149**

148b   Valves uniformly dark-pigmented (Fig. 28.150) . . . .

                                                       C. nigra Furtos 1936

       Length 0.92 mm, height 0.70 mm, width 0.76 mm. Surface of valves smooth, covered with slender, curved hairs; margins hairy without a hyaline border. Color dark blue except on marginal areas and ocular region. Natatory setae extend to tips of terminal claws or slightly beyond. Maxillary spines smooth. Terminal claw of leg 2 strong, of striking brown color. Furca slender, 16 times as long as narrowest width. ♂ unknown. Pools. Aug. Fla.

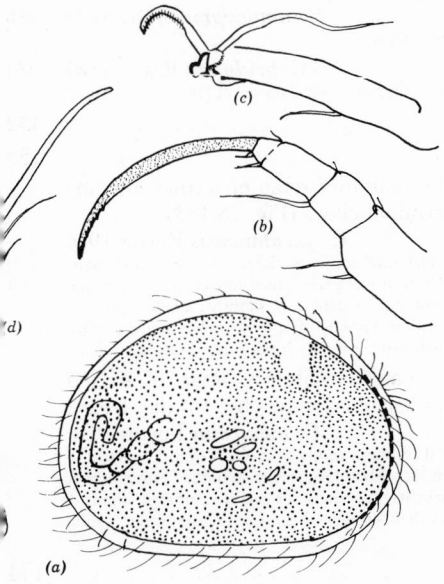

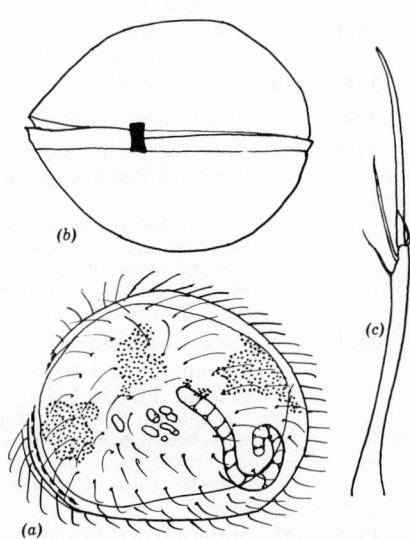

Fig. 28.150. *Cypretta nigra* ♀. (*a*) Right valve. (*b*) Leg 2. (*c*) Distal portion of leg 3. (*d*) Furca.

Fig. 28.151. *Cypretta brevisaepta* ♀. (*a*) Left valve. (*b*) Dorsal. (*c*) Furca.

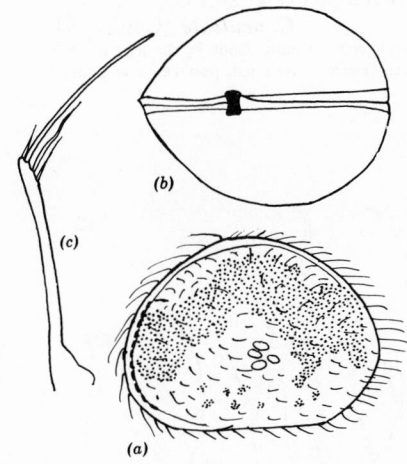

Fig. 28.152. *Cypretta bilicis.* (*a*) Left valve ♀. (*b*) Dorsal ♀. (*c*) Furca ♂.

149a (148) Valves spotted (Fig. 28.151) . . . . . . **C. brevisaepta** Furtos 1936

Length 0.85 mm, height 0.67 mm, width 0.70 mm. Surface of valves smooth with numerous short, blunt spinelike processes, each bearing a strong, curved hair. Color light yellow, with scattered dark-blue patches. Natatory setae extend to tips of terminal claws. Maxillary spines smooth. Furca curved slightly, 15 times longer than narrowest width. ♂ smaller than ♀, otherwise similar. Ejaculatory duct barrel-shaped with what appear to be 18 crowns of spines. Pools. Aug. Fla.

149b Each valve with a longitudinal dark band (Fig. 28.152). . . . . . .
. . . . **C. bilicis** Furtos 1936

Length 0.94 mm, height 0.70 mm, width 0.72 mm. Color light yellow with a very conspicuous longitudinal dark band extending in an undulating fashion on each valve. Surface of valves smooth with slender hairs of moderate length. Natatory setae extend slightly beyond tips of terminal claws. Maxillary spines smooth. Furca slightly curved, 18 times as long as narrowest width. ♂ smaller than ♀, otherwise similar. Ejaculatory duct with 16 crowns of spines. Pools. Aug. Fla.

150a **(80)** Valves compressed . . . . *Potamocypris* Brady 1870 **160**
Branchial plate of leg 1 with only 1 or 2 setae.

150b Valves tumid . . . . *Cypridopsis* Brady 1867 **151**
For species described since this key was completed, see Tressler (1954).

151a **(150)** Length less than 0.50 mm **152**

151b Length greater than 0.50 mm **153**

152a **(151)** Valves smoothly arched dorsally, natatory setae of second antenna
barely extend beyond tips of terminal claws (Fig. 28.153) . . . . . .
C. yucatanensis Furtos 1936
Length 0.35 mm, height 0.25 mm, width 0.28 mm. Surface of valves pitted, hairless. Maxillary spines smooth. Flagellum of furca 2⅔ times longer than base and
somewhat separated from it; dorsal seta not evident. ♂ slightly smaller than ♀,
otherwise similar. Ejaculatory duct with 6 crowns of spines. The smallest adult
fresh-water ostraced ever described. Pools, cave waters. June, July. Yucatan.

152b Valves with angulated dorsal margin; natatory setae of second
antenna extend beyond tips of terminal claws by ½ the length
of setae (Fig. 28.154) . . . . *C. mexicana* Furtos 1936
Length 0.35–0.38 mm, height 0.25–0.26 mm, width 0.25–0.26 mm. Surface of
valves smooth, hairless, with a few puncta. Maxillary spines smooth. Furca consist
of a slender base terminating in a delicate flagellum. ♂ slightly smaller than ♀,
otherwise similar. Ejaculatory duct with 6 crowns of spines. Cave water. July.
Yucatan.

153a **(151)** Surface of valves pitted **154**

153b Surface of valves without pits **157**

154a **(153)** Surface of valves without spines but fairly hairy **155**

154b Surface of valves with spines between pits (Fig. 28.155) . . . .
C. aculeata (Costa) 1847
Length 0.65–0.75 mm. Height $^7/_{10}$ the length. Width about ½ the length. Surface thickly covered with rounded pits. Anterior end rounded, posterior end pointed.
♂ unknown. Pools. Ontario.

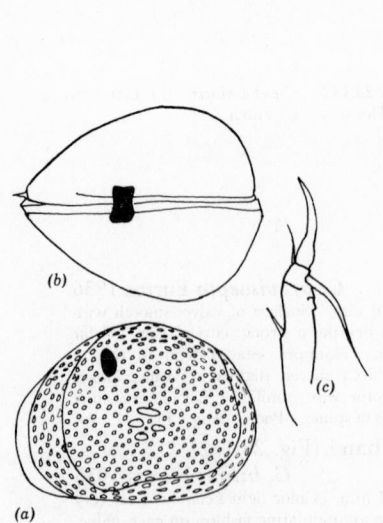

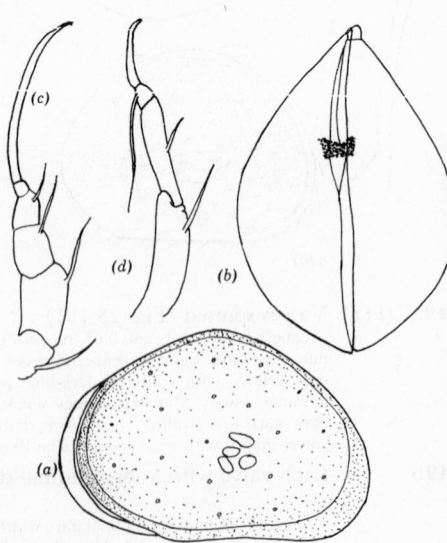

Fig. 28.153. *Cypridopsis yucatanensis* ♀. (*a*) Left
valve. (*b*) Dorsal. (*c*) Leg 3.

Fig. 28.154. *Cypridopsis mexicana* ♀. (*a*) Left valve.
(*b*) Dorsal. (*c*) Leg 2. (*d*) Leg 3.

155a (154) Dorsal bristle present on furca **156**
155b   Furca without dorsal bristle (Fig. 28.156) . . . .
                                                 C. rhomboidea Furtos 1936
    Length 0.68 mm, height 0.40 mm, width 0.43 mm. Surface of valves pitted and hairy. Natatory setae of antenna 2 extend well beyond tips of terminal claws. Maxillary spines bluntly toothed near tips. Flagellum of furca slightly more than twice length of base and not clearly separated from it. ♂ unknown. Ponds, pools. June. Yucatan.

156a (155) Natatory setae of second antenna extend beyond tips of terminal claws by about ½ the length of the setae (Fig. 28.157) . . . .
                                                 C. okeechobei Furtos 1936
    Length 0.64 mm, height 0.40 mm, width 0.42 mm. Color light yellow with dark-green stripes similarly arranged to those of *C. vidua.* Surface of valves pitted with short, curled hairs. Natatory setae extend beyond tips of terminal claws by ½ the length of claws. A smaller leg 3 and the presence of ♂ distinguish this species from *C. vidua.* Furca with base clearly separated from flagellum. ♂ common, smaller than ♀ but otherwise similar. Ejaculatory duct with 14 whorls of spines. Lakes, rivers. Aug. Fla.

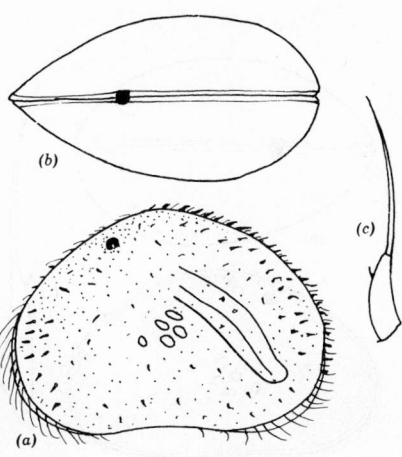

Fig. 28.155. *Cypridopsis aculeata* ♀. (*a*) Lateral. (*b*) Dorsal. (*c*) Furca.

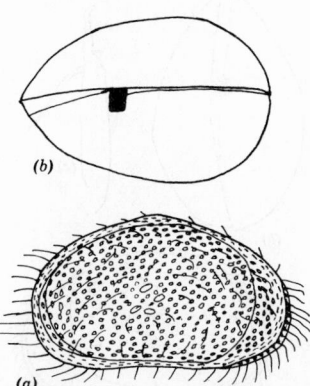

Fig. 28.156. *Cypridopsis rhomboidea* ♀. (*a*) Right valve. (*b*) Dorsal.

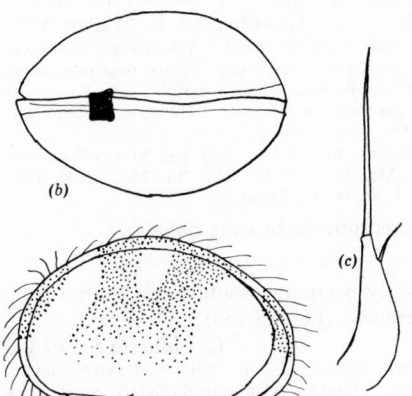

Fig. 28.157. *Cypridopsis okeechobei* ♀. (*a*) Left valve. (*b*) Dorsal. (*c*) Furca.

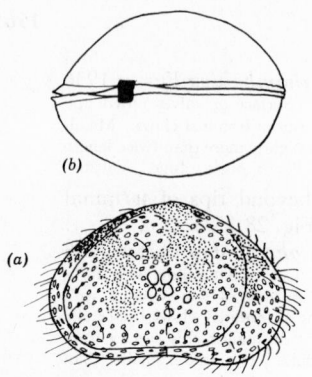

Fig. 28.158. *Cypridopsis vidua* ♀.
(*a*) Lateral. (*b*) Dorsal.

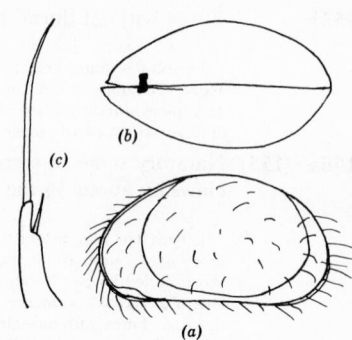

Fig. 28.159. *Cypridopsis viduella* ♀. (*a*) Left
valve. (*b*) Dorsal. (*c*) Furca.

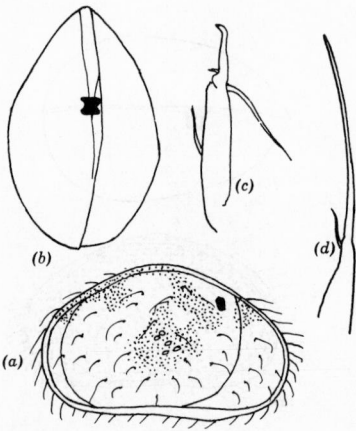

Fig. 28.160. *Cypridopsis niagrensis* ♀. (*a*) Right
valve. (*b*) Dorsal. (*c*) Leg 3. (*d*) Furca.

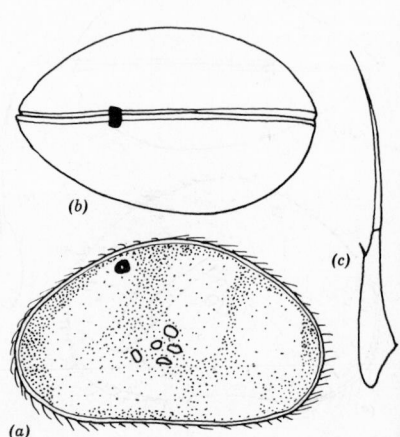

Fig. 28.161. *Cypridopsis helvetica* ♀. (*a*) Lateral.
(*b*) Dorsal. (*c*) Furca.

156b   Natatory setae of second antenna barely extend beyond tips of
terminal claws (Fig. 28.158). . . . . . **C. vidua** (O. F. Müller) 1776
Length 0.60–0.75 mm. Height equal to ⅔ the length. Width equal to height.
Plump, tumid forms. Surface of valves pitted and hairy. Three prominent dark-
green or dark-brown bands extend laterally from a similar dorsal band. Furca with
flagellum 2½ times the length of the base. ♂ unknown. The most common and
widely distributed of fresh-water Ostracoda. Often develops in aquaria. A variety, *C. vidua obesa*, lacks the dark bands. Apr. to Dec., Feb. and Mar. Newfoundland, Ontario, Mass., N. Y., D. C., Md., S. C., Tenn., Ohio, Ill., Mo., Mich., Wis.,
Fla., La., Miss., Ore., Tex., Mont., Utah, Wash., Trinidad.

157a (153) Surface of valves with scattered puncta bearing hairs **158**
157b   Surface of valves smooth. **159**
158a (157) Terminal claw of leg 3 strongly pectinate; slightly less than ⅓ the
length of the penultimate segment (Fig. 28.159) . . . .
                                                      C. viduella Sars 1896
Length 0.60 mm, height 0.33 mm, width 0.33 mm. Surface of valves smooth,
with scattered puncta bearing hairs. Natatory setae extend slightly beyond tips
of terminal claws. Maxillary spines smooth. Flagellum of furca 1½ times as long
as base and separated from it. ♂ unknown. Pools. July. Yucatan.

58b   Terminal claw of leg 3 approximately straight with tip bent, smooth, ⅓ as long as penultimate segment (Fig. 28.160) : . . . . .
C. niagrensis Furtos 1936

Length 0.58 mm, height 0.35 mm, width 0.36 mm. Surface of valves smooth, with scattered puncta bearing short hairs. Natatory setae extend beyond tips of claws by ½ the length of the claws. Maxillary spines with suggestion of denticulation at tips. Furca with clearly separated flagellum; dorsal seta short, slender. ♂ unknown. Pools. July. Yucatan.

59a (157)   Natatory setae of second antenna barely extend beyond tips of terminal claws (Fig. 28.161) . . . . . *C. helvetica* Kaufmann 1893

Length 0.65 mm, height 0.38 mm, width 0.39 mm. Color, similar to that of *C. vidua* but with ground color of pale whitish and bands almost black, also a fourth indistinct stripe at the posterior part. Surface of valves smooth and finely haired. Furca with long, slender flagellum. ♂ unknown. Pools. Ontario, Utah, Okla.

59b   Natatory setae extend beyond tips of terminal claws by about ½ the length of the setae (Fig. 28.162) . . . . . *C. inaudita* Furtos 1936

Length 0.72 mm, height 0.49 mm, width 0.32 mm. One maxillary spine slightly toothed at distal end. Terminal claw of leg 3 exceptionally long for the genus. Flagellum of furca 1⅔ the length of base and distinctly separated from it; dorsal seta short. ♂ slightly smaller than ♀; ejaculatory duct with 14 whorls of spines. Pools, ponds. June, July. Yucatan.

60a (150)   Valves pitted **162**

60b   Valves without pits or with weak, shallow depressions only **161**

61a (160)   Valves with few hairs (Fig. 28.163) . . . .
Potamocypris pallida Alm 1914

Length 0.71 mm, height 0.37 mm, width 0.27 mm. Color green except for a clear area around the eye. Surface with only a few scattered puncta and hairs. Natatory setae extending only to distal third of penultimate segment. Furca with dorsal seta; flagellum separated from the base and twice as long. ♂ unknown. Cold Springs. Spring and summer. Ohio.

61b   Valves hairy (Fig. 28.164) . . . . . . *P. islagrandensis* Hoff 1943

Length 0.58–0.64 mm, height 0.35–0.38 mm. Width not quite ½ the length. Natatory setae extend beyond tips of terminal claws by about ⅓ the length of setae. Furca with dorsal seta. Similar to *P. smaragdina* but differs in having nearly equal valves, a weak hyaline flange on anterior border and poorly developed pits and hairs on the valves. ♂ smaller but otherwise similar to ♀. Temporary ponds. La. (Grand Isle).

62a (160)   Height of shell less than ⅔ the length **163**

62b   Height of shell equal to ⅔ the length. **165**

63a (162)   Natatory setae of second antenna extend beyond tips of terminal claws by about ⅓ the length of the setae. **164**

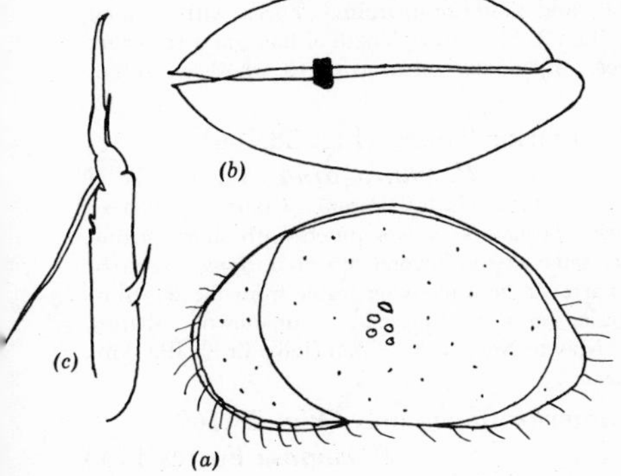

Fig. 28.162. *Cypridopsis inaudita* ♀. (*a*) Left valve. (*b*) Dorsal. (*c*) Leg 3.

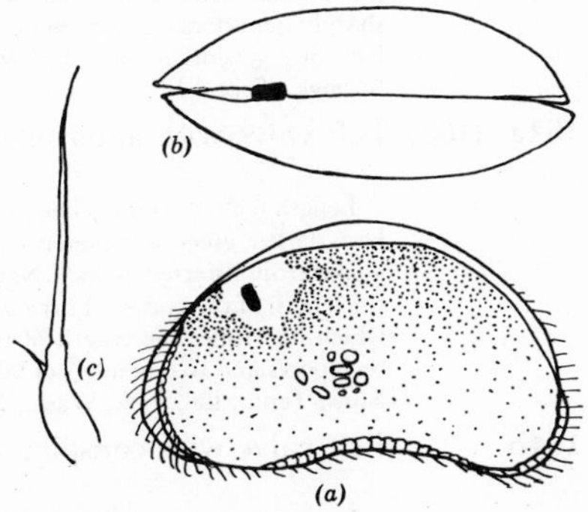

Fig. 28.163. *Potamocypris pallida* ♀. (*a*) Lateral. (*b*) Dorsal. (*c*) Leg 3.

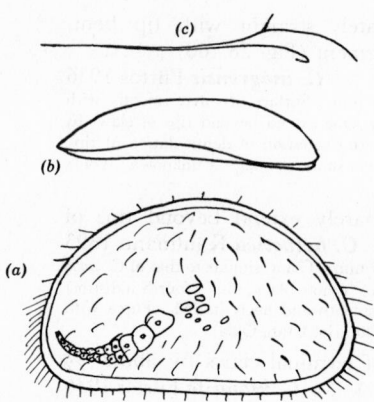

Fig. 28.164. *Potamocypris islagrandensis* ♀.
(a) Left valve. (b) Dorsal. (c) Furca.

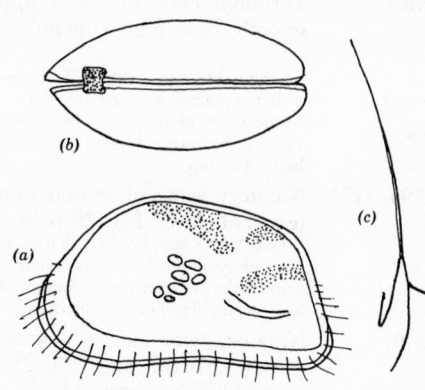

Fig. 28.165. *Potamocypris illinoisensis.* (a) Lateral ♀.
(b) Dorsal ♂. (c) Furca ♀.

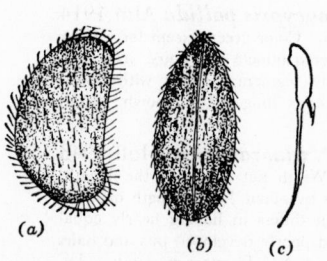

Fig. 28.166. *Potamocypris smaragdina* ♀.
(a) Lateral. (b) Dorsal. (c) Furca.

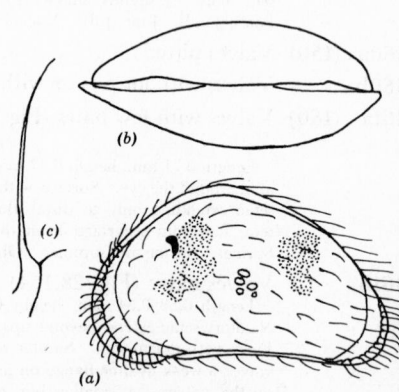

Fig. 28.167. *Potamocypris comosa* ♀. (a) Lateral.
(b) Dorsal. (c) Furca.

163b   Natatory setae of second antenna greatly reduced, the longest
hardly longer than the width of the antepenultimate segment (Fig.
28.165) . . . . **P. illinoisensis** Hoff 1943
   Length 0.56 mm, height 0.30 mm.   Width from $^3/_7$ to almost $^1/_2$ the length.   Hyaline border around anterior, ventral, and posterior margins.   Furca with a long,
sharply bent dorsal seta present; flagellum 2½ to 3 times length of base, and separated
from it.   ♂ similar to ♀ but smaller.   Ejaculatory duct with 10 whorls of spines.
Springs.   Sept.   Ill.

164a (163) Left valve with anterior and posterior flanges (Fig. 28.166). . . . . .
                    P. smaragdina (Vavra) 1891
   Length 0.60–0.70 mm, height 0.33–0.40 mm, width 0.28 mm.   Color yellow-green
with darker green dorsolateral stripes.   Surface of valves pitted with short, strong,
backwardly directed hairs.   Natatory setae extend beyond tips of terminal claws by
the length of the claws.   Furca with narrow base scarcely separable from the flagellum
which is 3 times the length of the base; dorsal seta present.   ♂ unknown.   Shallow
rock pools and weedy inlets of lakes.   May to Nov.   S. C., Ohio (Lake Erie), Ill., Mo.,
Miss., Tenn., La., Tex., Wash., Mexico.

164b   Left valve with conspicuous anterior flange only (Fig. 28.167) . . .
                    P. comosa Furtos 1933
   Length 0.68 mm, height 0.37 mm, width 0.27 mm.   Color light green with two
conspicuous bright-green dorsolateral stripes.   Surface of valves pitted with numerous
long, strong hairs.   Furca with dorsal seta, flagellum nearly 4 times as long as base
and clearly separated from it.   ♂ smaller than ♀ but otherwise similar; ejaculatory

duct with 11 whorls of spines. Weedy inlets and rock pools of lakes. May to Sept. Ohio (Lake Erie).

165a (162) Surface without spines between the hairs of the posterior portion (Fig. 28.168) . . . . . . *P. variegata* (Brady and Norman) 1889
Length 0.55 mm, height 0.32 mm, width 0.20 mm. Color and markings as in *P. elegantula.* Natatory setae extend beyond tips of terminal claws but not quite the length of the claws. Terminal claw of leg 3 longer than in *P. elegantula.* Furca with dorsal seta; flagellum 3 times length of base. ♂ unknown. Small ponds. Aug. Ohio.

165b Surface with many exceedingly short spines between the hairs of the posterior portion (Fig. 28.169) . . . *P. elegantula* Furtos 1933
Length 0.55 mm, height 0.38 mm, width 0.25 mm. Color light green with two dark-green dorsolateral stripes. Surface of valves moderately hairy, pitted. Natatory setae extend beyond tips of terminal claws by the length of the claws. Furcal ramus with dorsal seta; flagellum 3 times longer than base. ♂ unknown. Small, weedy ponds. Sept. and Mar. Ohio.

166a (2) Shells of normal width, hinge line not toothed **168**

166b Shells very broad, right valve with anterior and posterior portions of hinge line toothed . . . . *Metacypris* Brady and Robertson 1870
For a species described since this key was completed, see Tressler (1956).

167a (166) Valves without pits (Fig. 28.170) . *M. maracaoensis* Tressler 1941
Length 0.78 mm, height 0.39 mm, width 0.64 mm. Color gray with a much darker area in the anterior half of the valve. Surface with a pattern of polygonal areas in anterior half, with few hairs. Valves smooth. Spine of first antenna reaches to middle of fourth segment. Exopodite of antenna 2 reaches to tips of terminal claws. Eight mandibular teeth, not split. ♂ unknown. Bromeliads. Jan., July, Dec. Fla., Puerto Rico.

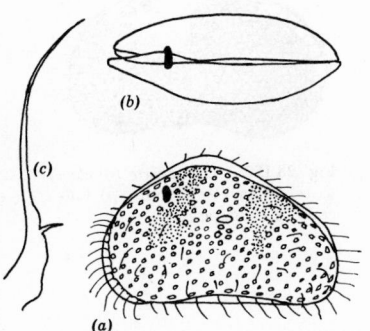

Fig. 28.168. *Potamocypris variegata* ♀. (*a*) Lateral. (*b*) Dorsal. (*c*) Furca.

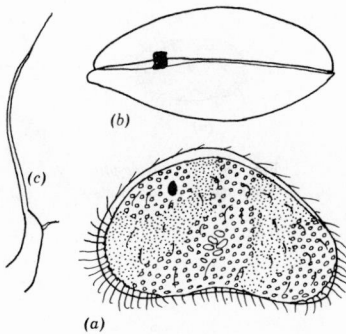

Fig. 28.169. *Potamocypris elegantula* ♀. (*a*) Lateral. (*b*) Dorsal. (*c*) Furca.

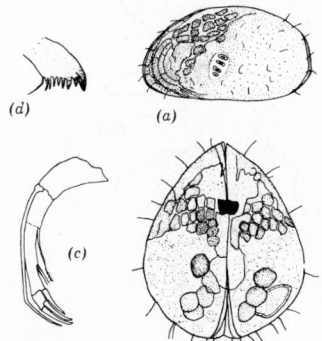

Fig. 28.170. *Metacypris maracaoensis* ♀. (*a*) Lateral. (*b*) Dorsal. (*c*) Antenna 2. (*d*) Mandibular teeth.

167b   Valves with pits and a few long, stiff hairs (Fig. 28.171). . . . .
                                                      M. americana Furtos 1936
              Length 0.55 mm, height 0.31 mm. Very tumid when seen from above. Color
              gray. Surface of valves pitted, with a few long stiff hairs. Spine of antenna 1
              reaches to middle of terminal claw. Exopodite of antenna 2 reaches to middle of
              terminal claws. Seven mandibular teeth, each split. ♂ unknown. Pools. July.
              Yucatan.

168a (166) Parasitic on gills of Crustacea; terminal claws of legs with 4 large
              teeth. ***Entocythere*** Marshall 1903   175

168b   Free-living forms; terminal claws of legs with not more than 2
              teeth, or plain . . . . ***Limnocythere*** Brady 1868   169
              For a species described since this key was completed, see Tressler (1957).

169a (168) Shell with conspicuous, well-developed protuberances.   172

169b   Shell without well-developed protuberances   170

170a (169) Shell with 1 dorsolateral furrow; shell sculpturing inconspicuous . .   171

170b   Shell with 2 dorsolateral furrows; surface sculpturing conspicuous
              (Fig. 28.172) . . . . ***L. reticulata*** Sharpe 1897
              Length 0.66–0.77 mm, height 0.36 mm, width 0.25 mm. Color grayish-white.
              Shell conspicuously marked with a network of honeycombed, polygonal reticulations

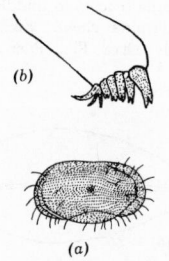

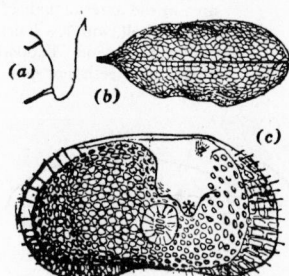

Fig. 28.171. *Metacypris*
americana ♀. (*a*) Lateral. (*b*) Mandibular
teeth.

Fig. 28.172. *Limnocythere reticulata*
♀. (*a*) Furca. (*b*) Dorsal. (*c*) Lateral. (By Sharpe.)

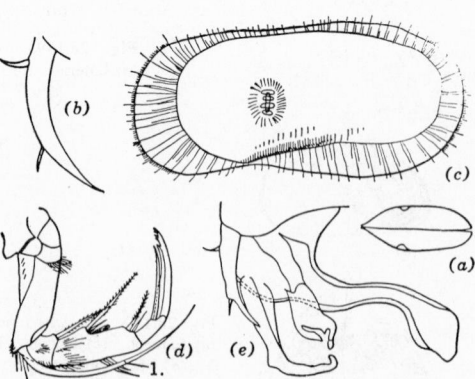

Fig. 28.173.   *Limnocythere illinoisensis.*   (*a*) Dorsal ♀.
(*b*) Furca ♀.   (*c*) Lateral ♀.   (*d*) Antenna 2 ♀.   (*e*) Sexual
organs ♂.   (By Sharpe.)

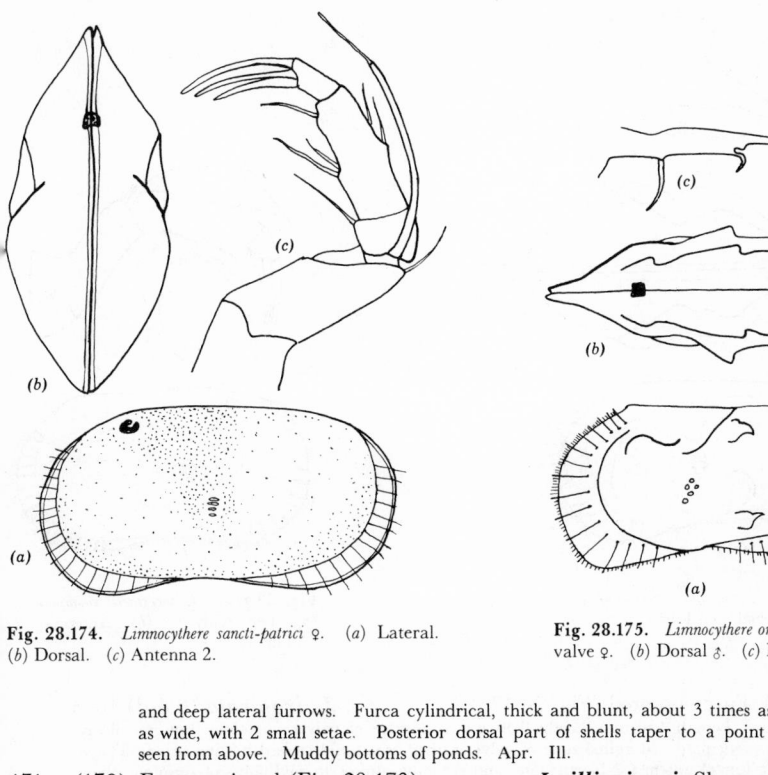

Fig. 28.174. *Limnocythere sancti-patrici* ♀. (*a*) Lateral. (*b*) Dorsal. (*c*) Antenna 2.

Fig. 28.175. *Limnocythere ornata*. (*a*) Left valve ♀. (*b*) Dorsal ♂. (*c*) Furca ♂.

and deep lateral furrows. Furca cylindrical, thick and blunt, about 3 times as long as wide, with 2 small setae. Posterior dorsal part of shells taper to a point when seen from above. Muddy bottoms of ponds. Apr. Ill.

171a (170) Furca pointed (Fig. 28.173) . . . . . . ***L. illinoisensis*** Sharpe 1897
  Length 0.88 mm, height 0.40 mm, width 0.29 mm. Color dark grayish-white. Shell faintly reticulate. Flagellum of antenna 2 2-segmented. Furca cylindrical; 7 times as long as wide. ♂ with well-developed grasping organs; terminal claw of antenna 2 of ♂ armed with 3 or 4 strong teeth at tip. Sandy bottoms, bayous, and lake shores. May. Ill.

171b Furca blunt (Fig. 28.174) . . . .
  L. sancti-patrici Brady and Robertson 1869
  Length 0.79 mm, height 0.42 mm, width 0.36 mm. Color light yellowish-brown. Valves thin and pellucid; surface faintly reticulated with scattered hairs at each extremity. Furca directed downwards and club-shaped; apical bristle slightly longer than lateral bristle. ♂ slightly larger than ♀; shell narrower and more elongated. Lakes. Mich., Tex.

172a (169) Length less than 0.90 mm 173
172b Length greater than 0.90 mm (Fig. 28.175) . . . .
  L. ornata Furtos 1933
  Length 0.88 mm, height 0.43 mm. Color gray. Surface of valves coarsely reticulated; margins with short brushlike hairs with a few longer hairs, which arise from slender, straight processes in the marginal zone. Furca elongated and terminating in a strong spine; terminal abdominal seta missing. ♂ larger than ♀; posterior extremity narrower. Lakes. July, Nov. Ohio (Lake Erie).

173a (172) Dorsal seta of furca arises directly from the base without the intervention of a papilla 174
173b Dorsal seta of furca arises from a papilla (Fig. 28.176) . . . .
  L. verrucosa Hoff 1942
  Length 0.54–0.60 mm, height 0.30 mm. ♂ slightly longer and not as high as ♀. Width about equal to height. Surface of valves covered with small raised areas giving a reticulated appearance. Protuberances very marked and giving the shell an inflated, bulging look. Furca with trilobed base; posterior lobe small and papillalike with a long seta. Lakes. Aug. Ill.

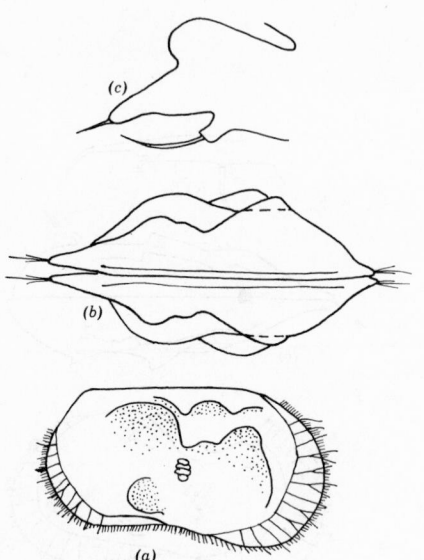

Fig. 28.176. *Limnocythere verrucosa* ♀. (*a*) Right valve. (*b*) Dorsal. (*c*) Furca.

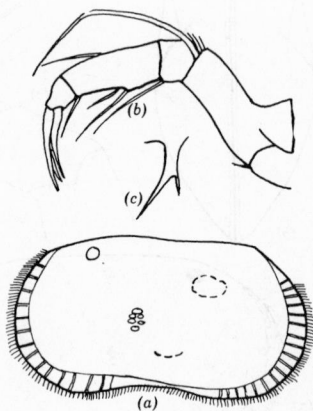

Fig. 28.177. *Limnocythere inopinata* ♂. (*a*) Left valve. (*b*) Antenna 2. (*c*) Furca.

174a **(173)** Furca pointed (Fig. 28.177) . . . . **L. inopinata** (Baird) 1843

    Length 0.56 mm, height 0.30 mm. Surface of valves sometimes but not always reticulate. Marginal zone of valves rather broad and crossed by fine striae. Flagellum of antenna 2 biarticulate, and reaching almost to mid-point of terminal claws. Furca of conical shape, directed slightly forward. ♂ slightly larger than ♀, otherwise similar. Alkaline lakes. Oct. Wash.

174b Furca blunt (Fig. 28.178) . . . . **L. glypta** Dobbin 1941

    Length 0.60 mm, height 0.30 mm. Surface reticulated with very small pits. Very slight marginal zone present, which feature serves to distinguish this species from *L. ornata* and *L. reticulata.* Posteroventral margin of left valve with numerous small teeth. Furca 4 times as long as broad; both setae of nearly equal length. ♂ with prominent copulatory apparatus. Lake bottoms. July. Wash.

175a **(168)** Length 0.50 mm or more **177**

175b Length less than 0.50 mm **176**

176a **(175)** Length less than 0.40 mm **183**

176b Length between 0.40 and 0.50 mm **188**

177a **(175)** Terminal end of distal portion of base of male copulatory complex forming a long, narrow process, extending nearly parallel to the clasping apparatus (Fig. 28.179) . . **Entocythere serrata** Hoff 1944

    Length of ♂ 0.48 mm, height 0.28 mm. Surface of valves with very few short, fine hairs. Dorsal part of valves with dark brown flecks. Eye well developed. Maxillary palp with spine on the convex surface. Respiratory plate of mandible represented by 3 setae. Clasping apparatus very simple; distal end of base narrowed and extended to form a long, gently curved process. ♀ similar to ♂ in shell structure. Parasitic on gills of crayfish *Cambarus diogenes diogenes* living in burrows in swamps, spring-fed meadows. Aug., June. Ill.

177b Terminal end of distal portion of the base of male copulatory complex not forming such a process **178**

178a **(177)** The border at the juncture of the 2 rami forming nearly a simple right angle or merely rounded **179**

178b Juncture of the vertical and horizontal rami forms a pronounced angle, the external border being extended to form a pointed projection (Fig. 28.180) . . . . **E. illinoisensis** Hoff 1942

Length of ♂ 0.54–0.61 mm, height 0.28–0.29 mm. ♀ slightly larger than ♂. Surface of shell leatherlike and transparent. Color yellowish-brown in preserved specimens. Eyes fused and located far anteriorly. Antenna 1 with 5 segments, mandibular palp with 2 segments; these two characteristics distinguish the species from *E. cambaria*. Parasitic on gills of *Orconectes* (= *Cambarus*) *propinques*, *O. virilis*, and *O. immunis*. Ill., Ohio.

179a (178) Shell greatly enlarged anteriorly and ventrally (Fig. 28.181) . . . .
E. claytonhoffi Rioja 1942

Length of ♂ 0.50–0.57 mm, height 0.27–0.28 mm. ♀ with accessory structures on last segment of antenna 1. Clasping appendage stout and L-shaped. Parasitic on *Procambarus* (= *Cambarus*) *blandingii cuevachicae*. Cueva Chica, Mexico.

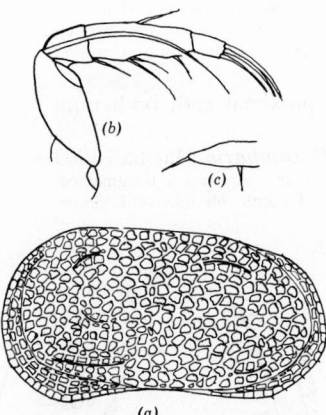

Fig. 28.178. *Limnocythere glypta* ♀. (*a*) Left valve. (*b*) Antenna 2. (*c*) Furca.

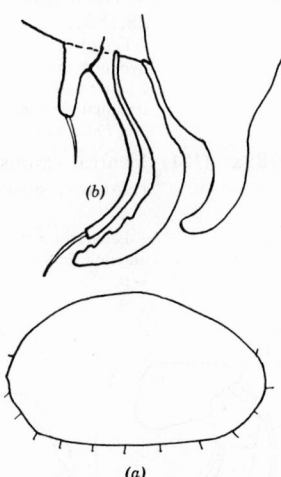

Fig. 28.179. *Entocythere serrata* ♂. (*a*) Right valve ♂. (*b*) Copulatory complex showing clasping apparatus.

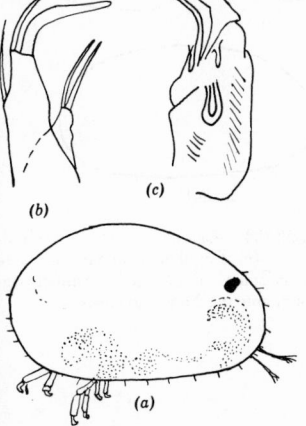

Fig. 28.180. *Entocythere illinoisensis.* (*a*) Lateral ♀. (*b*) Maxillary process and end of maxillary palp. (*c*) Copulatory complex ♂ showing clasping appendage.

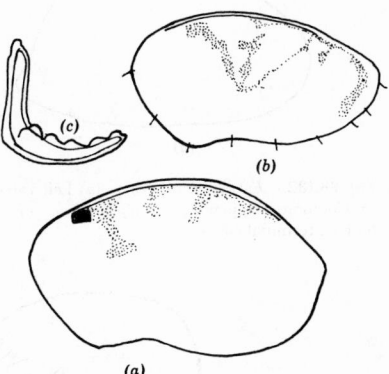

Fig. 28.181. *Entocythere claytonhoffi.* (*a*) Lateral ♀. (*b*) Lateral ♂. (*c*) Clasping appendage.

179b   Shell not so enlarged. **180**

180a (179) Clasping apparatus stout, swollen at juncture of the 2 rami; the 2
                rami never joined at an angle much greater than a right angle . . . **181**

180b   Clasping apparatus slender, little if any swollen at the juncture of
                the vertical and horizontal rami (Fig. 28.182) . . . .
                                              E. columbia Dobbin 1941
   Length of ♀ 0.60 mm. ♂ considerably smaller than ♀. Valves, thin, delicate, transparent. Antenna 1 6-jointed. Mandibular palp 4-segmented. Antenna 2 ends in 3
   stout claws. Copulatory appendage large and conspicuous with a long curved protuberance. Parasitic on crayfish gills. Jan., June, Sept. Wash.

181a (180) The 2 rami approximately equal in length **182**

181b   Vertical ramus much longer than the horizontal ramus (Fig.
                28.183) . . . . **E. elliptica** Hoff 1944
   Length of ♂ 0.51–0.56 mm, height 0.25–0.28 mm. Shell usually poorly pigmented.
   Eye large and conspicuous. Distal portion of base of copulatory apparatus elongated.
   ♀ larger and not so regularly elliptical as ♂. Antenna 2 with accessory structures on
   terminal segment. Parasitic on the gills of several species of *Orconectes* (= *Cambarus*)
   and *Procambarus*. Oct., Nov. Fla., Ga., S. C.

182a (181) Ventral ramus wider in center than at proximal end; both rami
                relatively stout throughout (Fig. 28.184) . . . .
                                              E. cambaria Marshall 1903
   Length of ♂ 0.60 mm. Shell thin, fragile, transparent. Antenna 1 6-segmented;
   antenna 2 4-segmented. Flagellum unsegmented. Parasitic on gills of *Cambarus*.
   Wis.

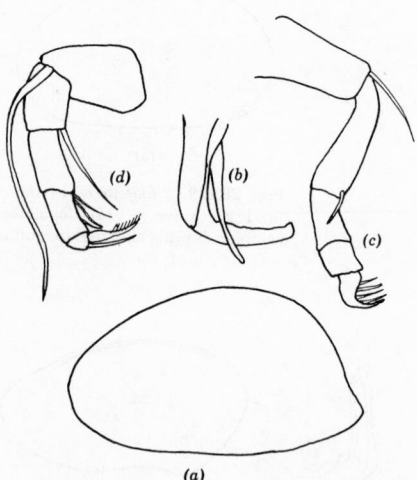

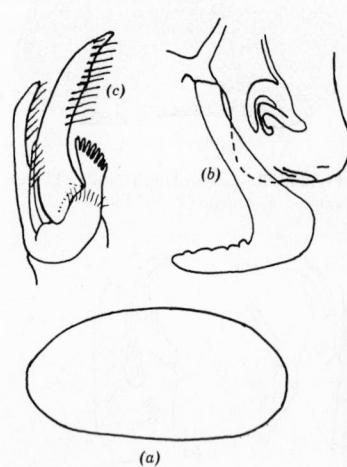

Fig. 28.182. *Entocythere columbia.* (*a*) Left valve ♀.
(*b*) Copulatory appendage ♂. (*c*) Leg 3 ♀. (*d*) Antenna 2 terminal claws ♂.

Fig. 28.183. *Entocythere elliptica.* (*a*) Lateral ♂. (*b*) Copulatory apparatus ♂, to
show clasping organ. (*c*) Terminal segment of antenna 2 and end claws, ♀.

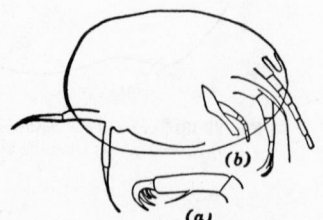

Fig. 28.184.   *Entocythere cambaria.*
(*a*) End of leg 3. (*b*) Lateral. (After
Marshall.)

182b   Vertical ramus proximally wider than in center of ramus; both rami relatively slender near the ends of clasping apparatus (Fig. 28.185) . . . . *E. mexicana* Rioja 1944
Length of ♂ 0.51–0.56 mm, height 0.27–0.32 mm. Length of ♀ 0.41–0.51 mm, height 0.21–0.28 mm. Shell transparent, hyaline. Violet pigment in central dorsal region. Parasitic on gills of *Paracambarus* and *Procambarus*. Villa Juarez (Puebla), Mexico.

183a **(176)** Teeth of internal border of horizontal ramus of clasping apparatus grouped together some distance from the distal margin, dorsal margin of shell evenly rounded (Fig. 28.186). . . . .
E. equicurva Hoff 1944
Length of ♂ 0.30–0.34 mm, height 0.17–0.19 mm. Valves well pigmented, especially in central portion of the dorsal half of each valve. Clasping apparatus sickle-shaped and more or less equally curved throughout. ♀ slightly larger than ♂. Parasitic on gills of crayfishes of the genera *Procambarus* and *Orconectes*. Ga., Fla. Ala.

183b   Teeth of internal border well spaced; dorsal margin of shell not evenly rounded. **184**

184a **(183)** Clasping apparatus C-shaped; distal margin with 2 teeth (Fig. 28.187) . . . . *E. dobbinae* Rioja 1944
Length of ♂ 0.27–0.30 mm, height 0.18–0.19 mm. Length of ♀ 0.30–0.35 mm, height 0.18–0.19 mm. Valves hyaline, violet pigment in dorsal median part of body. Shell ovoid. Parasitic on gills of *Paracambarus* and *Procambarus*. Villa Juarez (Puebla), Mexico.

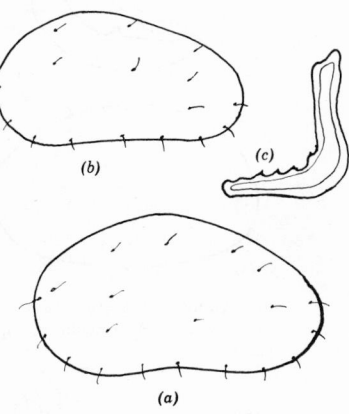

Fig. 28.185. *Entocythere mexicana.* (*a*) Lateral ♀. (*b*) Lateral ♂. (*c*) Copulatory appendage ♂.

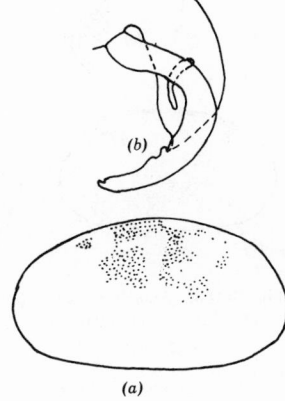

Fig. 28.186. *Entocythere equicurva* ♂. (*a*) Right valve. (*b*) Copulatory apparatus to show clasping appendage.

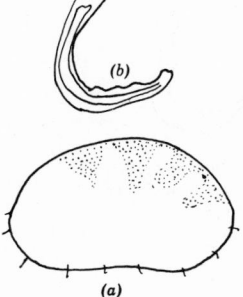

Fig. 28.187. *Entocythere dobbinae.* (*a*) Lateral ♀. (*b*) Copulatory appendage ♂.

184b   Clasping apparatus not distinctly C-shaped; distal margin with
more than 2 teeth **185**

185a **(184)** A talon or teeth present on external border of the clasping apparatus of male **186**

185b   External border of clasping apparatus of male entire (Fig.
28.188). **E. riojai** Hoff 1943
Length of ♂ 0.32–0.37 mm, height 0.18–0.19 mm. Anterior third of valves unmarked and transparent; posterior two-thirds marked with very fine pigment flecks.
Eye large and prominent. Ventral margin of valves slightly convex. Penultimate
segment of antenna 1 divided. Copulatory organ consists of a base and 3 accessory
pieces. End of clasping apparatus blunt and marked distally by 3 or 4 teeth. ♀
similar to ♂. Parasitic on gills of *Orconectes virilis*, *O. propinquus*, *O. palmeri longimanus*,
and *O. meeki*. Streams and creeks. Ill., Ohio, Ark.

186a **(185)** External border of horizontal ramus with 2 teeth of which the
proximal tooth is enlarged to form a talon, or with only a talon . . **187**

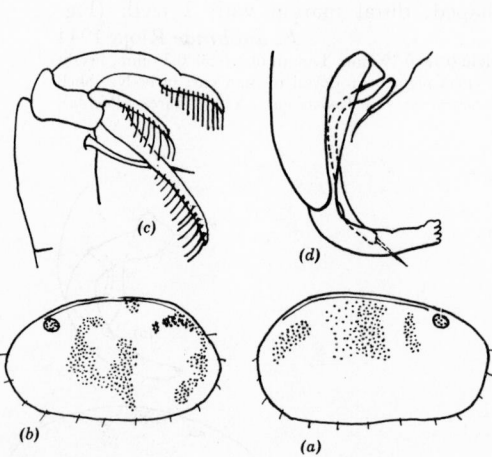

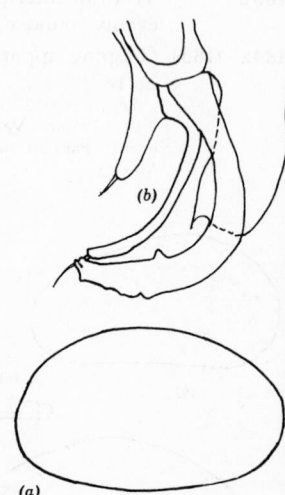

Fig. 28.188. *Entocythere riojai.* (*a*) Lateral ♀. (*b*) Lateral ♂.
(*c*) Distal portion of antenna 1 ♂. (*d*) Copulatory organ ♂.

Fig. 28.189. *Entocythere talulus* ♂.
(*a*) Lateral. (*b*) Copulatory apparatus.

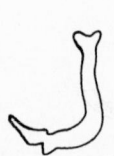

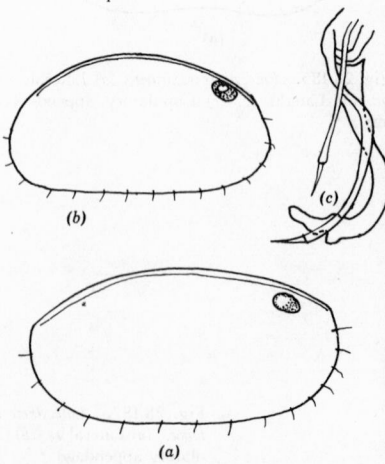

Fig. 28.190. *Entocythere
sinuosa* ♂. Copulatory
appendage.

Fig. 28.191. *Entocythere heterodonta.* (*a*) Lateral ♀.
(*b*) Lateral ♂. (*c*) Copulatory apparatus ♂.

186b   Two teeth on external border of horizontal ramus nearly equal in
size (Fig. 28.189) . . . . *E. talulus* Hoff 1944
     Length of ♂ 0.34 mm, height 0.21 mm. Dorsal and ventral margins of shell convex.
     Valves not heavily pigmented. Antenna 2 with 5 distal setae. Proximal 3 segments of
     mandibular palp fused. Respiratory plate of maxilla with numerous long distal setae
     on protopodite. ♀ similar to ♂; ventral margin of shell concave. Parasitic on
     gills of *Procambarus alleni*, burrows under boards in marsh. Fla.

187a (186) Vertical ramus proximally straightened (Fig. 28.190) . . . .
                                                                E. sinuosa Rioja 1940
     Length of ♂ 0.31 mm, height 0.16 mm. Shell transparent. Ventral margin of shell
     nearly straight in ♂ but convex in ♀. Large prominent eye. ♀ slightly larger than
     ♂. Similar to *E. heterodonta* except for shape of copulatory appendage of ♂. Para-
     sitic on gills of *Procambarus* (=*Cambarus*) *blandingii cuevachicae*. Cueva Chica, San
     Luis Potosi, Mexico.

187b   Clasping apparatus evenly curved throughout (Fig. 28.191) . . . .
                                                                E. heterodonta Rioja 1940
     Length of ♂ 0.31 mm, height 0.16 mm. ♀ slightly larger than ♂. Valves very trans-
     parent. Ventral margin of shell nearly straight in ♂ but convex in ♀. Eye large and
     prominent. Parasitic on gills of *Cambarellus* (=*Cambarus*) *montezumae*. Mexico.

188a (176) First antenna composed of 6 segments and with 4 or more distal
setae **189**

188b   First antenna composed of 7 segments and with 3 distal setae
(Fig. 28.192). . . . . *E. insignipes* (Sars) 1926
     Length of ♀ 0.45 mm. Height nearly ⅔ the length. Valves very thin with smooth
     surface devoid of hairs. Width of shell not quite ⅓ the length. ♂ unknown. Canada.

189a (188) External border of clasping apparatus of male entire **191**

189b   A talon or teeth present on the external border of the male clasping apparatus **190**

190a (189) Talon short, little recurved, and not extending parallel to the horizontal ramus (Fig. 28.193) . . . . *E. copiosa* Hoff 1942
     Length of ♀ 0.38–0.44 mm, height 0.21–0.24 mm. Eye usually conspicuous. Sur-
     face of valves smooth and almost hairless. Ventral margin distinctly sinuate. All
     segments of maxillary palp fused. ♂ similar but ventral margin of shell is often
     straight. Distal end of antenna 2 with 3 claws. Parasitic on several species of
     Orconectes (=*Cambarus*). Ill.

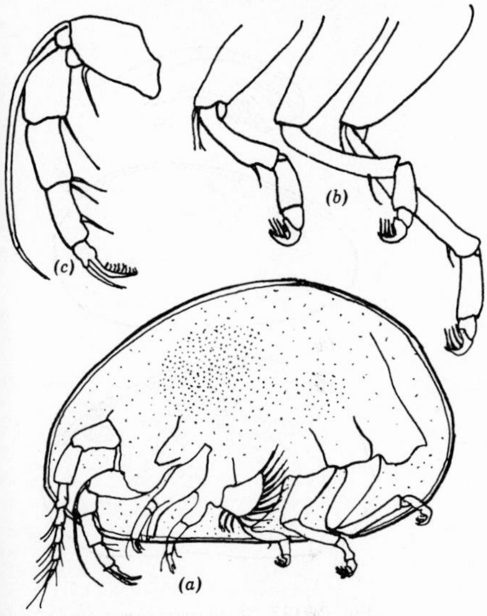

Fig. 28.192. *Entocythere insignipes* ♀. (*a*) Right valve with enclosed body. (*b*) Right series of legs. (*c*) Antenna 1.

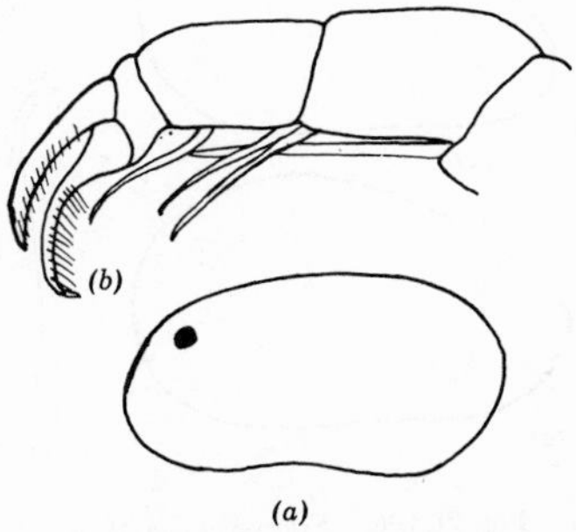

Fig. 28.193. *Entocythere copiosa* ♀. (*a*) Lateral. (*b*) Distal end of antenna 2.

190b   Talon long, recurved and extending in its distal part parallel to the post-talon portion of the clasping apparatus (Fig. 28.194). . . .
E. hobbsi Hoff 1944

Length of ♂ 0.34–0.40 mm, height 0.21–0.25 mm. Ventral margin of valves straight; left valve considerably higher than right. Valves with dorsal pigmented areas. Vertical ramus of clasping apparatus very long. ♀ similar to ♂ in shell shape and size. Parasitic on _Procambarus advena_ and other species of the genus. S. C., Ga., Fla.

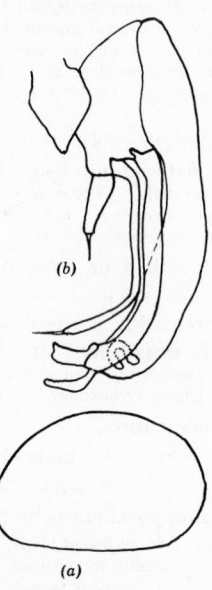

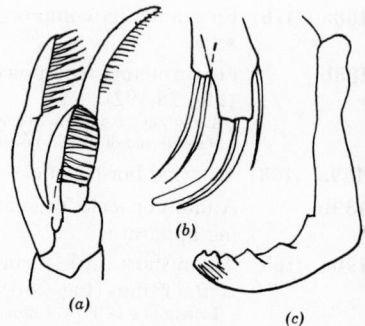

Fig. 28.194. _Entocythere hobbsi_ ♂. (_a_) Lateral. (_b_) Copulatory complex.

Fig. 28.195. _Entocythere dorsorotunda_ ♂. (_a_) Claws of antenna 2. (_b_) Palp and base of maxilla. (_c_) Clasping apparatus.

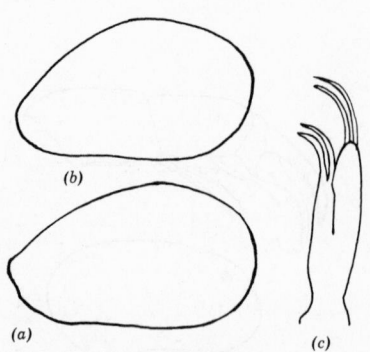

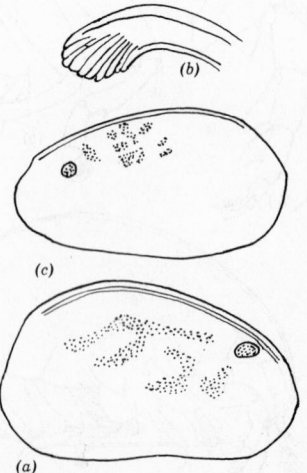

Fig. 28.196. _Entocythere donaldsonensis._ (_a_) Lateral ♀. (_b_) Lateral ♂. (_c_) Maxilla ♀.

Fig. 28.197. _Entocythere humesi._ (_a_) Lateral ♀. (_b_) End of ♂ clasping apparatus. (_c_) Lateral ♂.

191a (189) Horizontal ramus very short, reduced to a mere lobe at the end of the vertical ramus **192**

191b Horizontal ramus over ⅓ the length of the vertical ramus (Fig. 28.195) . . . . *E. dorsorotunda* Hoff 1944

Length of ♂ 0.42–0.46 mm, height 0.23–0.26 mm. All margins of shell convex. Antenna 1 composed of 6 segments. Clasping apparatus with a nearly straight vertical ramus joined at an angle greater than a right angle by the horizontal ramus forming an L-shaped clasping apparatus. ♀ similar in valve shape to ♂ except for a nearly straight ventral margin. Parasitic on several species of *Procambarus*. Ga., Fla.

192a (191) End of clasping apparatus with acute teeth, male antenna reported as having 2 distal claws only. Right shell longer than left (Fig. 28.196). . . . . *E. donaldsonensis* Klie 1931

Length 0.46 mm, height 0.26 mm. Valves smooth without hairs. Right valve longer than left. Antenna 1 sparsely haired; end segment 5 times as long as it is wide. Antenna 2 with 2 well-developed end claws; natatory setae extend beyond tips of terminal claw. ♂ similar to ♀. Parasitic on crayfish found in caves. Ind.

192b End of clasping apparatus fan-shaped, corrugated with rounded or blunt teeth; male antenna with 3 distal claws (Fig. 28.197) . . . *E. humesi* Hoff 1943

Length of ♂ 0.45 mm, height 0.19 mm. Ventral margin nearly straight sometimes with a slight concavity in anterior portion. Clasping appendage sickle-shaped, curved more distally than proximally and formed of a highly chitinized bar distally widened and fan-shaped. ♀ similar to ♂ except concavity of anterior portion of ventral margin of valves more marked. Antenna with 4 terminal setae; respiratory plate represented by 2 respiratory setae. Parasitic on gills of *Cambarus robustus* (= *C. bartonii robustus*). Streams. N. Y., Ohio.

References

Blake, Charles H. 1931. Two freshwater ostracods from North America. *Bull. Museum Comp. Zool. Harvard*, 72:279–292. **Dobbin, Catherine N.** 1941. Freshwater Ostracoda from Washington and other western localities. *Univ. Wash. Publs. Biol.*, 4:174–246. **Ferguson, Edward.** 1952. A preliminary report on the freshwater ostracods of Orangeburg County, South Carolina, *Trans. Am. Microscop. Soc.*, 71:272–276. 1957. Ostracoda (Crustacea) from the northern lower peninsula of Michigan. *Trans. Am. Microscop. Soc.*, 76:212–218. **Furtos, Norma C.** 1933. The Ostracoda of Ohio. *Ohio Biol. Survey*, 5:411–524. 1935. Freshwater ostracods from Massachusetts. *J. Wash. Acad. Sci.*, 25:530–544. 1936a. Freshwater Ostracoda from Florida and North Carolina. *Am. Midland Naturalist*, 17:491–522. 1936b. On the ostracods from the Cenotes of Yucatan and vicinity. *Carnegie Inst. Wash. Publ.*, No. 457:89–115. **Hoff, C. Clayton.** 1942a. The subfamily Entocytherinae, a new subfamily of freshwater Cytherid Ostracoda, with descriptions of two new species of the genus Entocythere. *Am. Midland Naturalist*, 27:63–73. 1942b. The ostracods of Illinois, their biology and taxonomy. *Illinois Biol. Monograph*, 19:1–196. 1943a. The Cladocera and Ostracoda of Reelfoot Lake. *J. Tenn. Acad. Sci.*, 18:49–107. 1943b. The description of a new ostracod of the genus *Potamocypris* from Grand Isle, Louisiana, and records of ostracods from Mississippi and Louisiana. *Occasional Papers Marine Lab. La. State Univ.*, 3:2–11. 1943c. Two new ostracods of the genus *Entocythere* and records of previously described species. *J. Wash. Acad. Sci.*, 33:276–286. 1944. New American species of the ostracod genus *Entocythere*. *Am. Midland Naturalist*, 32:327–357. **Klie, Walter.** 1931. Campagne spéologique de C. Bolivar et R. Jeannel dans l'Amerique du Nord (1928). 3 Crustaces Ostracodes. *Arch. zool. exp. et gén.*, 71:333–344. **Müller, G. W.** 1900. Deutschlands Süsswasser Ostracoden. *Zoologica (Stuttgart)*, 30:1–112. 1912. Ostracoda. *Das Tierreich*, 31. Gruyter, Berlin and Leipzig. **Rioja, Enrique.** 1940. Morfologia de un ostracode epizoario observado sobre *Cambarus (Cambarellus) montezumae* Sauss. de Mexico, *Entocythere heterodonta* n. sp. y descripcion de algunos de sus estados larvarios.

Anales inst. biol. Univ. Méx., 11:593–609. **1942a.** Descripcion de una especia y una sub-specie nuevas del genero *Entocythere* Marshall, procedentes de la Gueva Chica (San Luis Potosi Mexico). *Ciencia, Mex.*, 3:201–204. **1942b.** Consideraciones y datos acerca del genero *Entocythere* (Crust. Ostracodos) y algunas de sus especies, con descripcion de una nueva. *Anales. inst. biol. Univ. Méx.*, 13:685–697. **1944.** Nuevos datos de los *Entocythere* (Crus. ostracodos) de Mexico. *Anales. inst. biol. Univ. Méx.*, 15:1–22. **Sars, Georg O. 1926.** Freshwater Ostracoda from Canada and Alaska. *Rept. Can. Arctic Exped., 1913–1918*, 7:1–22. **1928.** *Ostracoda. An Account of the Crustacea of Norway*, 9. Cammermeyer, Oslo. **Sharpe, Richard W. 1918.** The Ostracoda. In: Ward and Whipple. *Fresh-Water Biology*, pp. 790–827. Wiley, New York. **Tressler, Willis L. 1941.** Ostracoda from Puerto Rican Bromeliads. *J. Wash. Acad. Sci.*, 31:263–269. **1950.** A synopsis of the ostracod genus *Cypricercus*, with a description of one new species from Wyoming. *J. Wash. Acad. Sci.*, 40:291–295. **1954.** Freshwater Ostracoda from Texas and Mexico. *J. Wash. Acad. Sci.*, 44:138–149. **1956.** Ostracoda from bromeliads in Jamaica and Florida. *J. Wash. Acad. Sci.*, 46:333–336. **1957.** The Ostracoda of Great Slave Lake. *J. Wash. Acad. Sci.*, 47:415–423.

29

Free-Living Copepoda

MILDRED STRATTON WILSON

HARRY C. YEATMAN

The free-living copepods, together with the parasitic copepods, constitute the Order Copepoda of the Class Crustacea in the Phylum Arthropoda. The three suborders of free-living copepods found in fresh and other inland water bodies are the same as those found in marine waters—the Calanoida, the Cyclopoida, and the Harpacticoida.

The segmented body of the copepods of these suborders has two noticeable divisions separated by a major articulation. The point of this articulation is easily determined in the Calanoida and Cyclopoida and in some Harpacticoida because of a noticeable difference in the widths of the two parts (Fig. 29.1). In the Calanoida, the articulation occurs between the somite of the fifth leg and the genital segment; in the Cyclopoida and the Harpacticoida, it occurs between the somites of the fourth and fifth legs. Two terms of convenience are frequently used to refer to these anterior and posterior divisions of the copepod body—*metasome* and *urosome*. In the literature, there are various interpretations of the particular somites that should be included in these terms. To avoid confusion herein, they are considered as containing the same body somites in each suborder.

The metasome (sometimes called cephalothorax) can be divided into two regions—the head with five pairs of appendages (first antennae or antennules,

second antennae, mandibles, maxillules, and maxillae); and the thorax with
six pairs of appendages (maxillipeds, four pairs of well-developed swimming
legs referred to as legs 1 to 4, and one pair, leg 5, which may be modified or
vestigial). The first body segment, referred to as the *cephalic segment*, is composed of the head and a thoracic somite (bearing the maxillipeds); sometimes
a second thoracic somite bearing leg 1 is also fused with the head. The last
two segments of the metasome may be partially or entirely fused. The metasome may thus have six, five, or four segments, depending on the genus or
species.

The term urosome as used herein includes the genital segment and the succeeding abdominal segments. In the male, the urosome usually has five distinct segments. In the female, the first abdominal somite is fused with the
genital somite and the whole is referred to as the *genital segment;* other fusions
of somites may reduce the number of segments in the urosome to three or
two (Table 29.1). In the Cyclopoida and Harpacticoida, the genital segment usually bears a pair of vestigial legs (leg 6) more developed in the male
than in the female. The urosome ends in paired processes called *caudal rami*
bearing terminal and lateral *caudal setae*.

The appendages are basically constructed on the biramous plan, with a
1- or 2-segmented basal portion (*basipodite* or *basipod*) which bears an outer
ramus or branch (*exopodite* or *exopod*) and an inner ramus (*endopodite* or *endopod*). One ramus is always lacking in the first antenna, sometimes in the
second antenna, mandible, and fifth leg.

The differentiation of the structure and habit of the three suborders of free-
living, fresh-water copepods is shown in Fig. 29.1 and Table 29.1. The sexes
may be distinguished from one another by the characters of the urosome, first
antennae, and fifth legs, as given in the table. Females are easily recognized

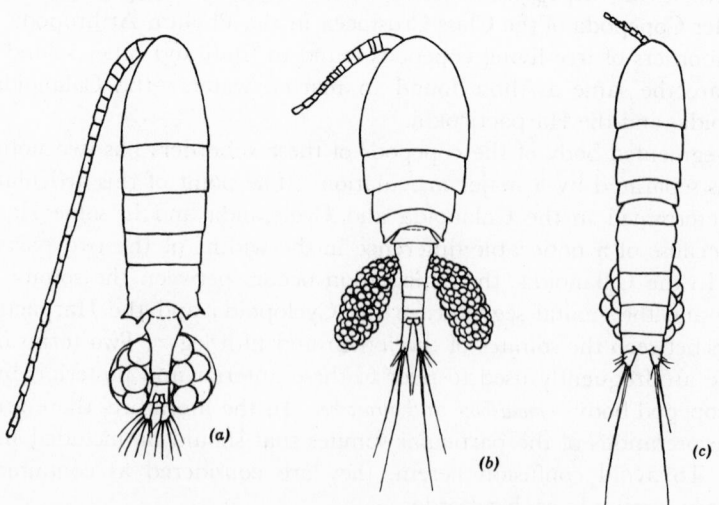

Fig. 29.1. Diagrams of the three types of free-living, fresh-water copepods, habitus of female in dorsal
view. (*a*) Calanoid. (*b*) Cyclopoid. (*c*) Harpacticoid. (*a, b* original, Yeatman; *c* after Sars.)

Table 29.1. Characteristics of the Suborders of Copepoda

Calanoida	Cyclopoida	Harpacticoida
	Habitus	
Anterior part of body much broader than posterior.	Anterior part of body much broader than posterior.	Anterior part of body usually little broader than posterior.
Marked constriction between somite of 5th leg and genital segment.	Marked constriction between somites of 4th and 5th legs.	Slight, or no constriction between somites of 4th and 5th legs.
Urosome ♀ 2-, 3-, or 4-segmented.	Urosome ♀ 4-segmented.	Urosome ♀ 4-segmented.
Urosome ♂ 5-segmented.	Urosome ♂ 5-segmented.	Urosome ♂ 5-segmented.
Caudal setae equal or not in length.	Caudal setae unequal in length.	Caudal setae unequal in length.
1 egg sac, carried medially.	2 egg sacs, carried laterally.	Usually 1 egg sac, carried medially.
Spermatophore elongate; 1 or more may be attached to ♀ genital protuberance.	Spermatophore kidney-shaped; 2 may be attached to ♀ genital protuberance.	Spermatophore elongate; 1 or more may be attached to ♀ genital protuberance.
	First Antennae	
Reach from near end of metasome to near end of caudal setae.	Reach from proximal third of cephalic segment to end of metasome.	Reach from proximal fifth to end of cephalic segment.
♀ 23 to 25 segments.	♀ 6 to 17 segments.	♀ 5 to 9 segments.
♂ left, similar to ♀.	♂ both left and right geniculate.	♂ both left and right geniculate.
♂ right, geniculate or not.		
	Leg 5	
Similar to other legs, or modified.	Not like other legs, vestigial. 1-, 2-, 3-segmented.	Not like other legs, vestigial. 2 segments or segments fused.
Basal portion 2 segments.	Basal segment not enlarged on inner margin.	Basal segment enlarged on inner margin into broad expansion usually bearing spines and setae.
Exopod 2 or 3 segments.		
Endopod present or not, 3-segmented or modified.	Endopod lacking.	
♀ symmetrical.	Symmetrical, alike in ♀ and ♂.	Symmetrical, more developed in ♀ than in ♂.
♂ asymmetrical.		
	Habit	
Planktonic rarely littoral.	Littoral, few species planktonic.	Extremely littoral.
Lakes, ponds, ditches.	Mostly around shores of lakes, or in ponds and ditches.	Shores of lakes, ponds, ditches; on vegetation, in mud, debris, wet moss above water, between sand grains on damp beaches.

when carrying egg sacs (Fig. 29.1) or spermatophores, the sperm capsules attached to the genital area by the male during copulation (Figs. 29.11a 29.64a, 29.150b).

In development, the copepods pass by molting through five or six nauplius stages, and six copepodid stages, of which the last is the adult. The body form of the copepodid is similar to that of the adult. Immature copepods constantly give the novice trouble in making identifications (Fig. 29.7). The early copepodid stages (I–III) can be recognized because all pairs of legs are not developed, or are very rudimentary. Stages IV and V in which four or five pairs of legs are present may be confused with the adult stage. Instructions for recognizing these are included under each suborder.

In the keys, distribution data refer to the presently known range of the species. Future collecting will undoubtedly extend the ranges of many species or disclose species not yet recorded for North America. For those who are interested in checking their specimens further, a bibliography is listed at the end of the chapter. These in turn include many other references. Developmental forms and structure are well summarized by Gurney (1931).

CALANOIDA

Mildred Stratton Wilson

Classification

The Calanoida of North America belong to four families, of which all except the Diaptomidae include both marine and fresh-water genera, or genera that have both marine and inland species, as indicated:

Family Pseudocalanidae
   Senecella. A single species found in fresh water (N. A., Asia) and in coastal marine and brackish waters (Asia).
Family Centropagidae
   Limnocalanus. Marine coastal and fresh-water species.
   Osphranticum. Fresh-water species.
Family Temoridae
   Eurytemora. Marine, coastal brackish, fresh-water, or euryhaline species.
   Epischura. Fresh-water species.
   Heterocope. Fresh-water species.
Family Diaptomidae
   Acanthodiaptomus. Fresh-water species.
   Diaptomus. Fresh- and inland saline water species.

The diaptomids are the most common calanoid copepods in all types of inland water bodies (lakes, permanent or seasonal ponds, roadside ditches). Only one species of *Acanthodiaptomus* occurs on the continent, but a large number of species of *Diaptomus* referable to several subgenera are known (Table 29.2). Those groups listed as occurring in Eurasia were split off as

Table 29.2. North American Diaptomidae

Genera and Subgenera	Distribution by Continent	Species Known in N. A. (through 1958) (Number indicates position in key)
Acanthodiaptomus Kiefer	Eurasia; N. A.	denticornis (65a, 89a)
Diaptomus Westwood Subgenera:	All continents	
(Diaptomus)	Eurasia; N. A. (incl. Greenland)	castor (49b), glacialis (49a)
(Arctodiaptomus) Kiefer	Eurasia; Africa; N. A. (incl. West Indies)	bacillifer (46a), arapahoensis (32a, 46b), kurilensis (58a, 105a), saltillinus (59a), floridanus (59b), dorsalis (57a), dampfi (57b), asymmetricus (55a)
(Mixodiaptomus) Kiefer	Eurasia; N. A.	theeli (47a)
(Eudiaptomus) Kiefer	Eurasia; N. A.	gracilis (54b)
(Sinodiaptomus) Kiefer	Eurasia; N. A.	sarsi (54a) (introduced?)
(Nordodiaptomus) M. S. Wilson	Asia; N. A.	alaskaensis (48a)
(Hesperodiaptomus) Light	N. A.; Asia	shoshone (44a), novemdecimus (45a), kenai (34a), hirsutus (43a), caducus (41a), nevadensis (37a), eiseni (40b), arcticus (40c), breweri (40a), scheffeli (39a), wardi (35a), franciscanus (50a), augustaensis (33a), unisonae (37b), victoriaensis (39b)
(Aglaodiaptomus) Light	N. A.	stagnalis (45b), leptopus (24a), clavipes (24b), clavipoides (26a), spatulocrenatus (19a), conipedatus (23b), pseudosanguineus (21a), marshianus (20a), dilobatus (22a), saskatchewanensis (23a), lintoni (17b), forbesi (17a)
(Mastigodiaptomus) Light	N. A. (incl. West Indies)	purpureus (28a), albuquerquensis (31a), amatitlanensis (30a), texensis (29a), montezumae (31b)
(Leptodiaptomus) Light	N. A.; Asia	tyrrelli (67a, 92a), pribilofensis (66a, 93b), coloradensis (67b, 93a), sicilis (72a, 97a), siciloides (78a, 101a), connexus (78b, 101b), ashlandi (72b, 98a), judayi (83a, 99b), insularis (77a, 99a), spinicornis (70a, 96a), signicauda (84a, 104a), novamexicanus (64a, 82a, 102a), nudus (82b, 103a), moorei (84b, 104b), mexicanus (94a), trybomi (62a), minutus (61a)
(Onychodiaptomus) Light	N. A.	sanguineus (86a, 109a), louisianensis (88b, 110a), virginiensis (88a, 110b), hesperus (87a, 107a), birgei (85a, 106a)
(Skistodiaptomus) Light	N. A.	oregonensis (74b, 113a), pygmaeus (73a, 113b), pallidus (74a, 112a), reighardi (73a, 115a), mississippiensis (79a, 114a), bogalusensis (80b, 116b), sinuatus (80a, 116a)
(Microdiaptomus) Osorio Tafall	N. A.	cokeri (61b)
(Prionodiaptomus) Light	N. A.; S. A.	colombiensis (51a)

genera by Kiefer (1932) from the well-established and nearly cosmopolitan genus *Diaptomus*. Light (1938, 1939), using similar morphological criteria, divided North American species into named groups but considered that the structural range of difference was not sufficient to give the rank of genus to most of the Eurasian or North American groups. There is considerable justification for Light's viewpoint, particularly since these and groups subsequently named by Kiefer and others are all referable to a basic and essentially narrow definition of the genus *Diaptomus*, and do not exhibit the structural gaps that are so characteristic of genera in the Calanoida. Moreover, the taxonomic soundness and usefulness of many of the groups, even as subgenera, are dependent upon more precise diagnosis, and upon greater knowledge, evaluation and interpretation of structure, variation, and distribution than have been presented in the literature. As is often true when a large, widely distributed genus is split up into groups of related species, the exact definition, validity, and status of these diaptomid groups can be realized only from a critical, comparative study of all the groups over their entire range of distribution. Until such study has been completed, confusion in the literature is avoided if these groups are considered as subgenera. In using the names given to diaptomid groups, however, it must be realized that with the exception of some African-Eurasian groups referred by Kiefer to a subfamily Paradiaptominae, the Eurasian and North American groups are taxonomic equivalents (as shown in Table 29.2).

Distribution

The present key includes the species described through 1958 that are known to occur in fresh- or inland saline-water bodies of North America, including Central America, Greenland, and the West Indies. Except for the euryhaline species of *Eurytemora*, species of marine genera that may be found in coastal waters of varying salinity (*Pseudodiaptomus, Acartia, Centropages, Tortanus*) are not included. Two species of *Diaptomus* that occur in the Canal Zone have been omitted because they are essentially a part of the South American fauna (see Marsh, 1913, 1929). Mexico, Central America, and the West Indies are relatively unknown regions—only nine valid fresh-water calanoid species have been recorded from Mexico and from Central America. Most of these species are an extension of the northern fauna; only one (*D. colombiensis*) is also known from South America. Three species are known from the West Indies; of these, one (*D. dorsalis*) occurs in the southeastern United States; the others belong to North American subgenera of *Diaptomus*.

The most recent summaries of distribution of North American Calanoida are those of Marsh (1929, 1933). These papers included fifty species (exclusive of currently recognized synonyms) from the same area covered by the present key with ninety-two species. Of the forty-two additional species, Marsh had considered four as doubtful or synonyms (*Epischura massachusettsensis, Diaptomus arapahoensis, D. pribilofensis, D. pygmaeus*) and one had been overlooked (*D. pseudosanguineus*). Seven are species known from Eurasia and since found in North America (*Eurytemora composita, Acanthodiaptomus denticornis, Diaptomus glacialis, D. theeli, D. gracilis, D. kurilensis, D. sarsi*). The last species may have been introduced into California with exotic water lilies, but the other species have come to light because of studies in the hitherto neglected areas of Alaska and western Canada.

Only one fresh-water genus, *Osphranticum*, is endemic to North America. Seventeen species are currently known to be common to the continental masses of North America (including Greenland) and Eurasia; of these, twelve are diaptomids, four of which are very widely distributed in Europe and Asia (*Acanthodiaptomus denticornis, Diaptomus bacillifer, D. gracilis, D. theeli*), the others being mostly northern species found also in the neighboring areas of Asia. Three other species are closely related to Asian species and may with more knowledge be found to be synonyms or subspecies (*Diaptomus alaskaensis, D. arapahoensis, D. pribilofensis*); these relationships are pointed out in the key. A subgenus of *Diaptomus* (subgenus *Arctodiaptomus*), widely distributed in Eurasia, is represented on the continent and the West Indies by eight species; of these, five are endemic.

The distribution summary given for each species in the key frequently includes records unpublished at the time of compilation. The previously known distribution will therefore appear greatly extended for many species. The distribution picture of North American calanoids is more complete than for the other two groups of fresh-water copepods, but there are still many areas that have been only sparsely collected. Some species are little known and may actually be rare or localized, but such terms are inappropriate in the current state of knowledge.

Literature

Descriptions and distribution of the North American calanoids exclusive of the Diaptomidae can be found in a publication by Marsh (1933). References to other reliable descriptions in the literature are included in the key. The monographs of Schacht (1897) and of Marsh (1907) contain much basic information on the species of *Diaptomus* known through 1907. References to other descriptions or to species described since 1907 are included in the key and bibliography.

Characters Used in the Key

Figures 29.2–29.6 should be studied for illustration of the parts of the body and appendages referred to in the key before attempting to use it. Though these figures specifically illustrate the diaptomids, the terms are applicable to the other genera. Each genus can be determined from examination of the whole specimen of either sex, according to the simple habitus characters given in the key; where necessary, verification may be made by examination of the dissected fifth leg.

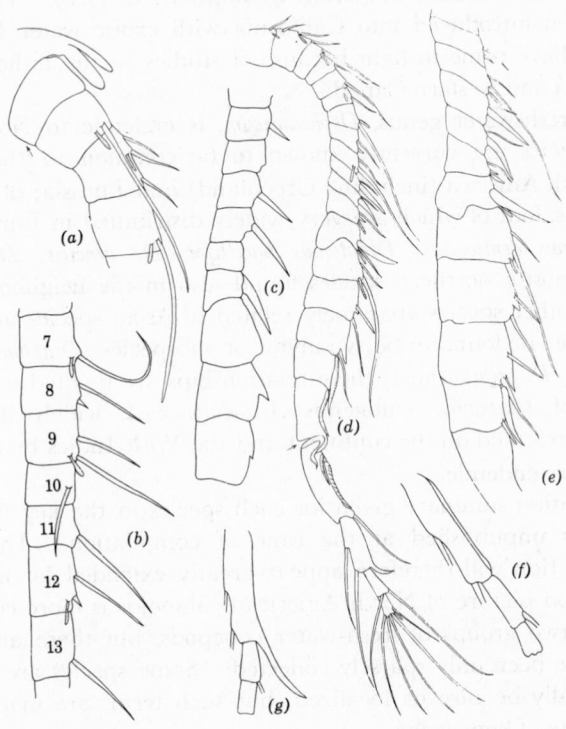

Fig. 29.2. *Diaptomus*, first antennae. See also Table 29.3. (*a*) *D. schefferi*, basal portion of first antenna ♀ (segments 1 to 3) showing rostral filaments and protrusion (or papilla) of ventral face to which antenna is attached. This papilla, indicated by arrow, may sometimes be separated from the body in dissection; it does not bear setae and should not be considered in counting segments. Segment 1 always has an aesthete and 1 seta, the length of which may have specific or group value. Segment 2 usually has 3 setae and an aesthete; segment 3 has 1 seta and an aesthete.

(*b*) *Diaptomus* (generalized), detail of segments 7 to 13 (numbered from base) showing arrangement of setae, aesthetes, and spines on ♀ and left ♂ antennae. Aesthetes (short, foliate structures) are shown on segments 7, 9, and 12, and short spines on segments 8 and 12. Seta of segment 10 is usually eccentrically placed as shown; if 2 setae are present, which is rare, the extra seta is aligned with those of the other segments. See Table 29.3 for complete armature of the total 25 segments.

(*c*) *D. bogalusensis*, ♂ right antenna, segments 8 to 16, showing spines on segments 8, 10, 11, and 13, and spinous processes on segments 15 and 16. In this figure and in those given throughout the key, the setae and aesthetes are omitted. It has become customary in literature to give such an outline form of the spines and processes of these segments because they are often of group or specific importance.

(*d*) *D. dilobatus*, ♂ right antenna. Arrow points to the strong specialized joint or *geniculation* between segments 18 and 19. Segments 1 to 9 are like those of the left antenna. The rest of the appendage is modified through an increase in the number of spines (on segments 10, 11, 13); structural modification of some setae; enlargement of segments 14 to 18 (amount of enlargement specifically variable); and the fusion of segments 19 to 21, and of segments 22 and 23. The segment resulting from fusion of segments 22 and 23 is referred to in the key as segment 23; it has the apex produced into a variously developed process, or has a membrane along the margin of the segment, or is unarmed. Segments 17, 18, and fused segment 19–21 bear peculiar depressed processes.

(*e*) *D. dilobatus*, ♂ right antenna, showing detail of armature of segments 8 to 17. (Compare with *c*.)

(*f*) *D. bogalusensis*, apical segments of ♂ right antenna, showing narrow hyaline membrane along margin of fused segments 22 and 23 (referred to in key as segment 23).

(*g*) *D. forbesi*, apical segments of ♂ right antenna, showing strongly developed process of segment 23. Such processes are often extremely variable within a species, and so have limited taxonomic value.

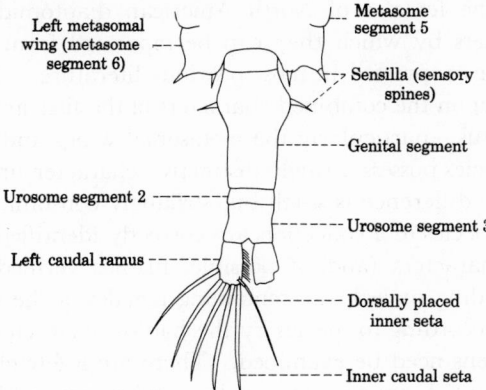

Fig. 29.3. *Diaptomus gracilis* ♀, distal part of metasome and urosome, dorsal, showing parts referred to in key. Females of Calanoida may carry egg sacs (Fig. 29.1a) and spermatophores (Figs. 29.11a and 29.64a) attached to ventral side of the genital segment.

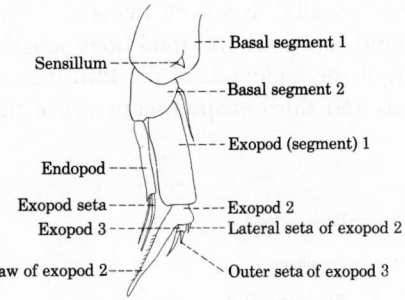

Fig. 29.4. *Diaptomus nevadensis*, ♀ right leg 5, showing parts referred to in key. This appendage is symmetrical in all the species in the key, and only one leg of a pair is illustrated as in this figure. Fig. 29.9b shows a complete pair of legs and also illustrates a type in which endopods are lacking and basal segment 1 of the right and left sides are completely fused.

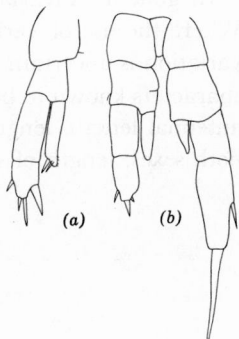

Fig. 29.5. *Diaptomus* sp., showing form of leg 5 in copepodid stage V, posterior view. (a) Female, left leg. (b) Male, both legs.

The species are identifiable from either sex, and the characters of both are included in regular sequence in the key. In order that the most easily observed or least variable characters might be used, it has been necessary to separate the sexes of some species; there are cross references to all of these. Key characters and figures have been largely determined and illustrated from examination of specimens, which have been available for all but four species (*Diaptomus pseudosanguineus, D. dampfi, D. cokeri, D. montezumae*).

The fact that the females of North American diaptomid species possess individual characters by which they can be separated from one another is contrary to opinions expressed in most previous literature. Their identification is dependent upon the combined characters of the first antennae, the fifth legs, and the habitus—particularly the metasomal wings and the genital segment. Several species possess a single distinctive character unlike that of any other; in most, the difference is small but strikingly constant. Once individual females of all species in a collection are correctly identified on the basis of their combined characters (and, if possible, further verified by association with the corresponding males), the remaining females of the sample can usually be separated according to species by the habitus characters alone, so that only whole specimens need be examined. There are a few closely allied species that are not, in the present state of knowledge, separable with absolute certainty. These are: the four species of the *Diaptomus oregonensis* group (key numbers 73–74); *D. forbesi* and *D. lintoni* (17); *D. breweri*, *D. eiseni*, and *D. arcticus* (40); *D. kiseri* and *D. victoriaensis* (39); *D. dorsalis* and *D. dampfi* (57). The lack of distinct differences between the closely allied members of these groups of species seldom presents difficulty, however, since they seem to be infrequently associated in the same body of water.

In general, invariable or little variable characters are used in the key. Where the use of variable characters is unavoidable, or where the range of variation is uncertain, the qualifying word "usually" has been inserted. The characters known to be variable in diaptomid copepods are: total body length, antennal length, length and segmentation of the endopods of the fifth legs of both sexes, length of setae of the endopods and third exopod segments of the

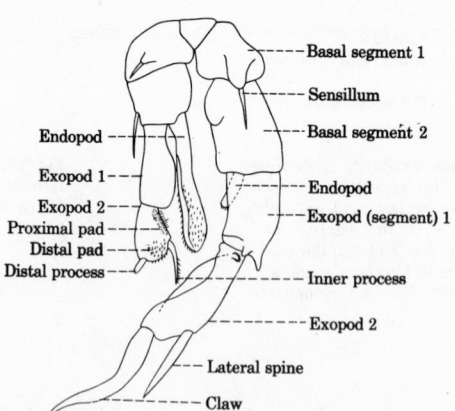

Fig. 29.6. *Diaptomus marshianus*, ♂ leg 5, posterior view, showing parts referred to in key. In the Calanoida, this appendage is asymmetrical in the male, and both legs of the pair are figured in the key. Most of the figures show the posterior view as illustrated above, with the right leg at the right of the figure and the left leg at the left of the figure. The legs are in reverse position in figures labeled anterior view.

female fifth legs, length and shape of the process of the twenty-third segment of the male right first antenna, the segmentation of the female urosome (separation of second and third segments), and the dorsal process of the distal metasome segments found in females of several subgenera. Where these characters are used in the key, supplemental characters in the text and figures should always be checked. The extent of variability of antennal length relative to body length is not known, but it is possible that it may be reliably determined for a species as it is found in a single body of water. In the female diaptomid, this might serve as a means of separating closely or distantly related species of similar habitus form that occur together.

Probably almost any structure in copepods is subject to abnormal development. Anomalies are comparatively rare in calanoid copepods, but the possibility of their occurrence should be kept in mind in identification. Probably the most common form of anomaly is the multiplication of setae, claws, or other structures, placed either on the same or another segment. A diaptomid may, for instance, have the claw of the fifth leg doubled in either sex. In diaptomids having two setae on segment 11 of the first antennae, there may occasionally be present an extra seta on one of segments 13 to 19 on which there is normally only one seta. In the female, this has been observed on only one of the two antennae, and the resulting asymmetry of the two appendages makes it recognizable as an anomalous condition.

Developmental Stages

The key refers only to the adult stage of the species. Isolated developmental forms of few species can be identified with certainty even in the subadult stage (copepodid stage V). The few diaptomid copepods that have a distinctive setation of the first antenna, such as *Diaptomus caducus* and *D. shoshone*, can be identified in stage V from the antennae of the female and the left antenna of the male, since the setation is the same as in the adult. In general, developmental forms should be associated with the species that are known to occur in any body of water that is being studied. Genera can be recognized in late copepodid stages by the characters of the caudal rami and the caudal setae which are similar to those given in the key for the adult. The form of the fifth leg is distinctive for each genus in copepodid stage V (Figs. 29.5, 29.8, 29.9, 29.16, 29.19), and where known in stage IV.

The copepodid stages appear to follow a similar pattern of development in the Calanoida, so that the stage can be recognized by a basic formula. In North American fresh-water calanoids, the males all have five segments in the urosome of the adult, and apparently four segments in stage V. Segmentation of the urosome of the adult female ranges from two to four, and sometimes is specifically variable. To avoid confusion, it is therefore best to consider body segmentation related to each species, rather than to genera. The copepodid stages are best recognized by the number of segmented legs. Legs 1 and 2 first appear in the nauplius and legs 3 to 5 in the copepodids as rudimentary "buds" (Fig. 29.7). The bud is flattened and unsegmented, al-

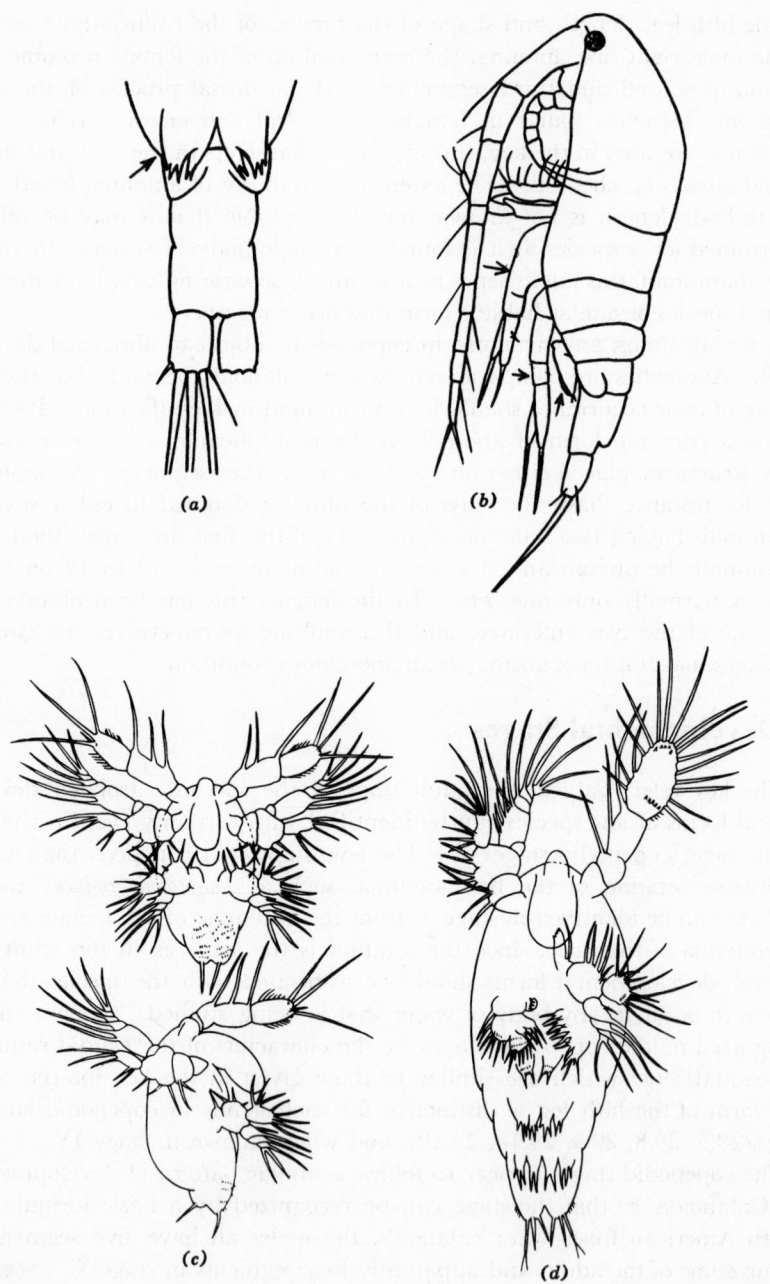

Fig. 29.7. Developmental stages of calanoid copepods. (*a*) *Epischura massachusettsensis*, ventral view, distal part of metasome and urosome, copepodid stage I. Arrow indicates "bud" of leg 3. (*b*) *Diaptomus* (generalized), lateral view, copepodid stage I. Arrows indicate segmented legs 1 and 2, and "bud" of leg 3. (*c*) *Diaptomus*, nauplius stage II, ventral and lateral. (*d*) *Diaptomus*, nauplius stage VI, ventral. Only one of each pair of cephalic appendages is shown: on the right (animal's left), the first antenna, mandible, and maxilla; on the left, the second antenna and maxillule. The paired buds represent the maxillipeds and legs 1 and 2. (*a* after Humes; *b* modified from Gurney; *c*, *d* modified from Wüthrich.)

though it is biramous and obviously represents parts that will develop into the basal portion and two rami (exopod and endopod). In the first copepodid stage (Fig. 29.7b), the two pairs of buds of the last nauplius stage have developed these parts and can be recognized as legs 1 and 2, although usually the exopods and endopods are unsegmented. In addition, the bud of the third leg has appeared. In the two successive molts, the bud changes to a segmented leg and the bud of the next leg appears. By stage IV, all five legs are segmented, and the sexes can first be distinguished by the slight differences in the fifth leg. By stage V, the differences between the sexes in this leg is usually very noticeable (Fig. 29.5).

The following summary is a basic formula for recognition of the copepodid stages of Calanoida; it probably applies to all genera but this has not been entirely determined for *Osphranticum* and *Senecella*, nor for every species of the other genera. It must be remembered that the fifth leg is lacking in the adult female of *Senecella*.

Copepodid Stage	Segmented Legs	Bud
I	Legs 1, 2	Leg 3
II	1, 2, 3	4
III	1, 2, 3, 4	5
IV	1, 2, 3, 4, 5	
V	1, 2, 3, 4, 5	
VI (adult)	1, 2, 3, 4, 5	

References to studies of developmental stages for different genera are included in the References (Gurney, 1931; Davis, 1943; Humes, 1955; Humes and Wilson, 1951; Juday, 1925; Wuthrich, 1948).

Eggs are carried in egg sacs by most of the fresh-water calanoid species of North America (as shown for *Diaptomus* in Fig. 29.1a), but in a few genera the eggs are apparently laid free in the water (*Senecella*, *Limnocalanus*, and most *Epischura*).

Technique of Identification

Two or more genera or several species of one genus of Calanoida may occur together. In a collection, both sexes of a species may not be present. In sorting and in identification, therefore, attention should be given to both sexes and to individuals differing in size and body shape.

In comparing whole specimens or appendages with the figures given in the key, it must be constantly remembered that if the material being examined is not in the same position as that illustrated, it will not appear identical. Cover-glass pressure will cause distortion of tumid protrusions, which have been illustrated as nearly as possible without distortion. In addition, specimens may have parts of the body or appendages contracted, expanded, twisted, or folded "out of line." The last is particularly common with the claw of the right fifth leg of male diaptomids. The first antennae of diaptomids are likewise sometimes vexing to study, because in the final mount the appendages may have become twisted or the setae may not lie in a favorable position. The unmodified female and male antennae are, however, relatively large so the segments can be distinguished from one another at comparatively low magnification. Critical segments can be located by using certain segments as guides (see Table 29.3). The simplicity of structure and small number of setae and aesthetes make it possible, with a little practice, to ascertain the setation of even poorly mounted appendages.

Table 29.3. Summary of Setation Pattern of Diaptomid First Antenna of Female and Left Side of Male

(s = seta   sp = spine   a = aesthete)

Segment	Armature		Segment	Armature		Segment	Armature	
1	s	a	10	s*		19	1(2)s	a
2	3s*	a	11	1(2)s		20	s	
3	s	a	12	s,sp	a	21	2s	
4	s		13	1(2)s		22	2s	
5	s	a	14	1(2)s	a	23	2s	
6	s*		15	1(2)s		24	2s	
7	s	a	16	1(2)s	a	25	5s	a
8	s,sp		17	1(2)s				
9	2s	a	18	1(2)s				

*Only one species (*D. caducus*) has been recorded in literature as having more than 3 setae on segment 2 or more than 1 seta on segment 6; and only two species (*D. caducus* and *D. hirsutus*) have been recorded as having more than 1 seta on segment 10.

As shown by the arrangement of species in the key, those with 2 setae on segment 11 may have 1 or 2 setae on some or all of segments 13 to 19; those with 1 seta on segment 11 have 1 seta on segments 13 to 19. In most species, the left antenna of the male has the same number of setae as in the female, but in a few, as indicated in the key, the armature of the male differs from that of the female.

In locating segments on the mounted appendages, segments 8 and 12 with their invariably present short spines, or segment 9 with its invariable number of 2 setae, may serve as guides. Often it is easiest to locate segments 13 to 19 by counting backwards from the apex (segment 25). Also see Fig. 29.2.

The dorsal aspect of whole specimens should be examined in preservative (formalin or 70 per cent alcohol) or in glycerine, without cover glass, or in built-up mounts for detail of metasomal wings, urosome, and caudal rami. For determination of fine detail, such as hairs on the margins of the caudal rami, the compound microscope must usually be used. If specimens are not straight, it may be necessary to separate the urosome from the rest of the body.

Appendages can be accurately studied only when they have been dissected from the body of the copepod. Dissection (using a pair of mounted no. 12 needles) is not difficult in the Calanoida. For each of the genera, dissect the fifth legs of both sexes. For the diaptomids, remove also the right and left antennae of each sex. Check the setation of both antennae of the female to be certain that they are identical. As mentioned above, an additional, anomalous seta may very rarely be present on one antenna of a pair. Do not confuse broken-off setae with anomaly. The position of a lost seta is indicated by a characteristic indentation of the cuticle. See Fig. 29.2 and Table 29.3 for setal arrangement.

The appendages can usually be most satisfactorily studied in flat glycerine mounts (cover glass supported only by mounting medium), permitting examination with high-power objectives. A combination of techniques is usually advisable for the male fifth legs. Cover-glass pressure may distort tumid protrusions or membranous structures of these appendages, particularly of complex fifth legs of diaptomids. Before mounting, the fifth legs of all calanoids should be examined from several positions under both stereoscopic and compound microscopes, so that the nature and position of protrusions and accessory armature can be understood. For critical study under high magnification of detailed structures such as the left exopod of male diaptomids, it is usually necessary to prepare flattened mounts. Otherwise, the cover of mounts should be supported in some manner such as by bits of cover glass or paper. Most of the outlines of the figures illustrating the male fifth legs in the key have been made from unmounted appendages, using high-power oculars (15X, 20X); where necessary, detail has been added from mounts with unsupported covers studied with high-power objectives.

KEY TO SPECIES

Note: On many of the figures, arrows indicate the characters referred to in the key. For brevity, references to habitat are limited to the two terms lakes and ponds. These are inclusive of all types; particularly the term pond, which refers to permanent and seasonal ponds, pools, or ditches. The figures are original except those for which a source is indicated.

1a   Caudal ramus ♀ ♂ with 4 well-developed setae (plus slender outer seta frequently directed across the others as in Fig. 29.8a, and a slender inner seta). Leg 5 ♀ lacking. Neither right nor left first antenna ♂ geniculate . . . . *Senecella calanoides* Juday 1923

      Family **Pseudocalanidae.** Only species known in the genus. Urosome ♀ 4-segmented. First antennae ♀ ♂ 25-segmented. Oral appendages adult ♂ reduced; those of ♀ normal. Endopods legs 1 to 4 with 1,2,3,3 segments. Length: ♀ 2.65–2.9 mm; ♂ 2.45–2.55 mm. Lakes. Northeastern U. S. west to Great Lakes area; eastern Canada west to Great Slave Lake area; northern Asia (marine, brackish, fresh waters). (Descr. Juday, 1925.)

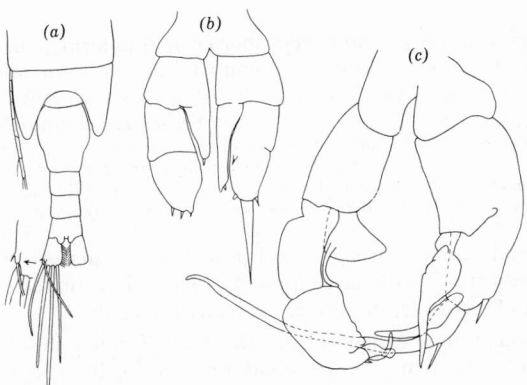

Fig. 29.8. *Senecella calanoides.* (*a*) Distal part of metasome and urosome ♀ with detail of outer margin of caudal ramus, dorsal. (*b*) Leg 5 ♂, late copepodid stage. (*c*) Leg 5 ♂, adult, posterior view.

1b Caudal ramus ♀ ♂ with 3 or 5 well-developed setae. Leg 5 ♀
    present. Right first antenna ♂ geniculate (Fig. 29.2*d*) **2**

2a (1) Caudal ramus ♀ ♂ with 3 well-developed terminal setae and a
    reduced or spiniform outer seta (plus slender, dorsally placed
    inner seta) (Fig. 29.10*a*) **3**

2b Caudal ramus ♀ ♂ with 5 well-developed setae (plus slender, dor-
    sally placed inner seta) (Figs. 29.3, 29.17*a,c*) **7**

3a (2) Caudal ramus ♀ ♂, outer seta slender and setiform, its length
    about equal to that of ramus; urosome ♂ symmetrical. Leg 5 ♀
    and left leg 5 ♂, apex with an elongate spine (as long as or longer
    than last segment in ♀, about as long as in ♂) . . . .
         Heterocope septentrionalis Juday and Muttkowski 1915

      Family **Temoridae**. Only species of genus known in N. A. Urosome ♀ 3-
      segmented. First antennae ♀, ♂ left, 25-segmented. Endopods legs 1 to 4 of
      genus with 1 segment. Leg 2 ♂ of this species asymmetrical, spines of exopod
      segments of right leg modified (enlarged, twisted, or with eccentrically placed
      spinules). Length: ♀ 3.0–4.0 mm; ♂ 3.0–3.8 mm. Ponds, lakes. Alaska;
      northern and western Canada.

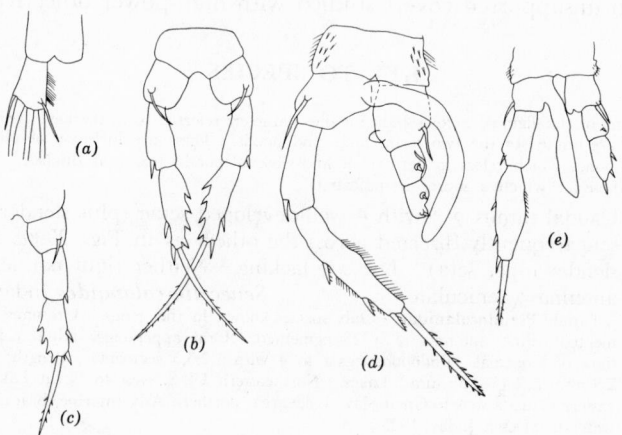

Fig. 29.9. *Heterocope septentrionalis.* (*a*) Caudal ramus and bases of caudal setae ♀. (*b*) Leg 5 ♀, adult.
(*c*) Leg 5 ♀, copepodid stage V. (*d*) Leg 5 ♂, adult, posterior view. (*e*) Leg 5 ♂, copepodid stage V.

3b Caudal ramus ♀ ♂, outer seta shorter or spiniform (length less than
    that of ramus); urosome ♂ asymmetrical, the right side with vari-
    ous processes. Leg 5 ♀ and left leg ♂, apex without an elongate
    apical spine . . . . ***Epischura*** (Family **Temoridae**) **4**

      Urosome ♀ 3-segmented. First antennae ♀, ♂ left, 25-segmented. Endopods
      legs 1 to 4 with 1 segment. Four species are listed in the key. A fifth species,
      E. fluviatilis Herrick 1883, described from Ala., is probably a distinct form of
      this or another genus, but is not sufficiently known to be placed in the key (see
      Herrick, 1895; Marsh, 1933).

4a (3) Caudal ramus ♀ broad, length less than 2 times width. Urosome
    ♂, segment 2 with large process (width equaling or more than
    that of segment), distally or ventrally directed **5**

4b Caudal ramus ♀ slender, length about 2 times the width. Uro-
    some ♂, segment 2 with small process (width less than that of
    segment) **6**

5a (4) Urosome ♀ usually twisted, and the 3 long apical setae of left
    caudal ramus more or less enlarged and decreasing in size from

outer to inner seta. Urosome ♂, segment 5 with a ventral process ending in serrate knob (process usually visible in dorsal aspect). . . .
Epischura lacustris S. A. Forbes 1882

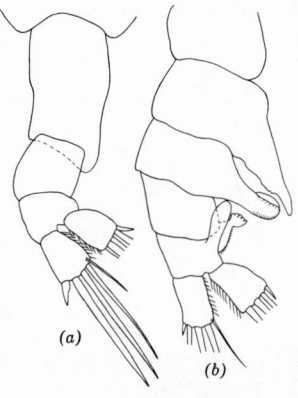

Degree of enlargement of, caudal setae very variable; those of right caudal ramus enlarged or not; leg 5 ♀ similar to that of *E. nevadensis*. Leg 5 ♂, inner process of left basal segment 2 recurved distally so that it overlies the apical segment of left leg in posterior view. Length: ♀ 1.78–2.0 mm; ♂ 1.38–1.6 mm. Lakes. Northeastern coastal states and provinces, west to Minn. and Northwest Territories (see **5b**).

◄ **Fig. 29.10.** *Epischura lacustris.* (*a*) Urosome ♀, dorsal. (*b*) Urosome ♂, dorsal.

5b   Urosome ♀ usually straight and the apical caudal setae more or less uniform in size. Urosome ♂, segment 5, ventral process somewhat triangular, its edge minutely denticulate (process best observed when segments 4 and 5 are separated). (Fig. 29.11*d*) . . . .
E. nevadensis Lilljeborg 1889

Leg 5 ♂, inner process of left basal segment 2 more outwardly curved than in *E. lacustris*, so that it usually overlies the right leg in posterior view. Length: ♀ 1.3–2.5 mm; ♂ 1.27–2.1 mm. Lakes. Pacific coast Alaska to Calif., east to Rocky Mountain region in U. S., to Manitoba in Canada. Associated with *E. lacustris* in northern Canadian lakes.

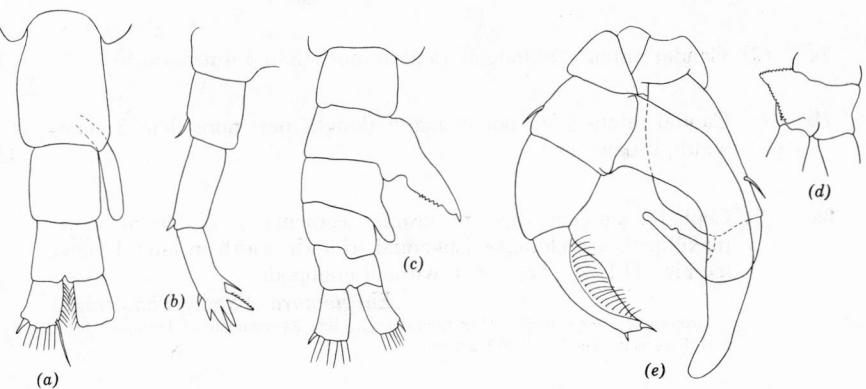

Fig. 29.11. *Epischura nevadensis.* (*a*) Urosome ♀ with spermatophore, dorsal. (*b*) Leg 5 ♀. (*c*) Urosome ♂, dorsal. (*d*) Urosome segment 5 ♂, ventral. (*e*) Leg 5 ♂, posterior view.

6a   (4) Leg 5 ♀, apical segment with 3 equidistant spinous processes placed along inner margin, beginning at about proximal fourth of segment. Urosome ♂ with process on first segment; leg 5 ♂, left basal segment 2, inner process short (about as long as or little longer than width of segment). (Fig. 29.12). . . . .
E. massachusettsensis Pearse 1906

Length: ♀ 3.27–3.55 mm; ♂ 2.9–3.34 mm. Ponds. Mass. (Descr. Humes, 1955.)

6b   Leg 5 ♀, none of spinous processes of inner margin placed above middle of segment. Urosome ♂ without process on first segment; leg 5 ♂, inner process longer than width of segment. (Fig. 29.13) .
                        E. nordenskiöldi Lilljeborg 1889
Length: ♀ 1.64–1.99 mm; ♂ 1.1–1.6 mm. Lakes, ponds. Quebec east to coast, south to N. C.

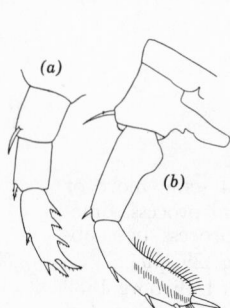

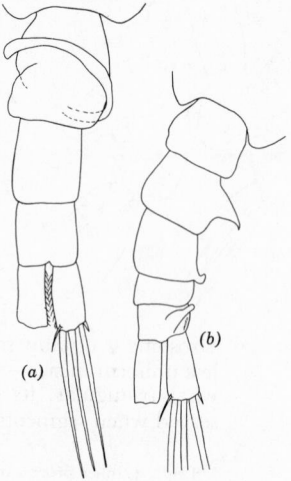

Fig. 29.12. *Epischura massa-chusettsensis.* (*a*) Leg 5 ♀. (*b*) Leg 5 ♂, left. (After Humes.)

Fig. 29.13. *Epischura nordenskiöldi.* (*a*) Urosome ♀, dorsal, showing spermatophore coiled around genital segment. (*b*) Urosome ♂, dorsal.

7a   (2) Caudal ramus ♀ ♂ elongate (length more than 3 times width). . . .   8

7b   Caudal ramus ♀ ♂ not elongate (length not more than 3 times width).   13

8a   (7) Cephalic segment (first metasome segment) ♀ ♂, lateral view, maxillipeds not elongate (subequal to body width in lateral view; see Fig. 29.15*a*). Leg 5 ♀ ♂, without endopods . . . .
                        Eurytemora (Family **Temoridae**)
Urosome ♀ 3-segmented. First antennae ♀, ♂ left, 24-segmented. Endopods leg 1 with 1 segment, legs 2 to 4 with 2 segments.

8b   Cephalic segment ♀ ♂, lateral view, maxillipeds elongate (about 2 times body width in lateral view; see Fig. 29.18*a*). Leg 5 ♀ ♂, with endopods . . . . . . ***Limnocalanus*** (Family **Centropagidae**)   12
Urosome ♀ 3-segmented. First antennae ♀, ♂ left, 24-segmented. Endopods legs 1 to 4 with 3 segments.

9a   (8) Leg 5 ♀, width of inner process of exopod 1 (third segment of leg) about ½ the total length of its segment. Leg 5 ♂, right leg 5-segmented (the apical claw divided into 2 segments) . . . .
                        Eurytemora canadensis Marsh 1920

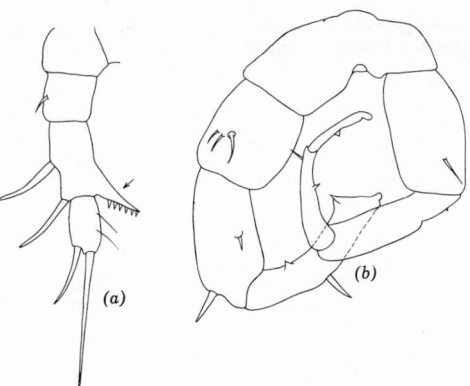

Genital segment ♀ without prominent lateral protrusions. Length: ♀ 1.9–2.25 mm; ♂ 1.9–2.1 mm. Fresh and brackish tundra ponds and lakes., Bering and Arctic coasts, Alaska and Canada. Probably includes *E. tolli* Rylov 1922 from Siberia.

◄ **Fig. 29.14.** *Eurytemora canadensis.* (*a*) Leg 5 ♀ (arrow indicates inner process of exopod 1). (*b*) Leg 5 ♂, posterior view.

9b   Leg 5 ♀, width of inner process more than ½ the length of its segment. Leg 5 ♂, right leg 4-segmented (apical claw not divided) . .   **10**

10a   (9) Leg 5 ♀, inner process of exopod 1 strongly directed backwards. Leg 5 ♂, apex of left leg with 2 stout digitiform processes . . . . . .
                                    E. affinis (Poppe) 1880

       Length: ♀ 1.1–1.5 mm; ♂ 1.0–1.5 mm. A variable marine form found in lakes and ponds of the Atlantic, Pacific, and Gulf of Mexico coastal areas. Most American records of *E. hirundoides* are probably this species.

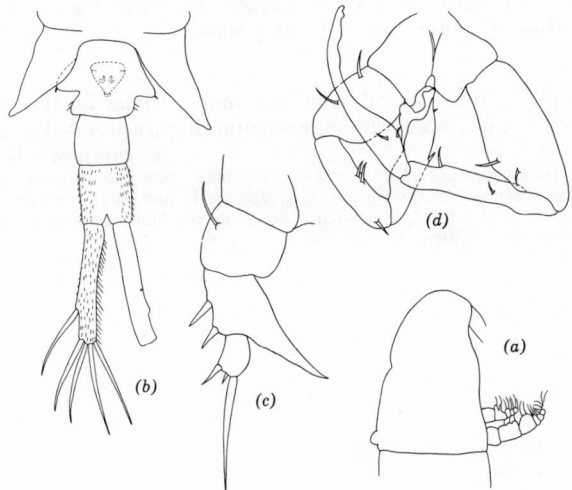

Fig. 29.15. *Eurytemora affinis.* (*a*) Lateral view cephalic segment ♀. (*b*) Metasomal wings and urosome ♀, dorsal, with outline of ventral genital operculum on genital segment. (*c*) Leg 5 ♀. (*d*) Leg 5 ♂, posterior view. (From specimens from Lake Providence, La.)

10b   Leg 5 ♀, inner process of exopod 1 directed inwards. Leg 5 ♂, apex of left leg without digitiform protrusion or with only one **11**

11a   (10) Leg 5 ♀, inner apical spine nearly 4 times the length of the outer. Leg 5 ♂, left, apex without digitiform protrusion. . . . .
                                  E. yukonensis M. S. Wilson 1953

Protrusions of genital segment ♀ somewhat pointed.   Dorsal surfaces of urosome and
caudal rami of both sexes lacking hairs or spinules.   Length: ♀ 1.6 mm; ♂ 1.3 mm.
Fresh-water lakes, ponds.   Alaska.

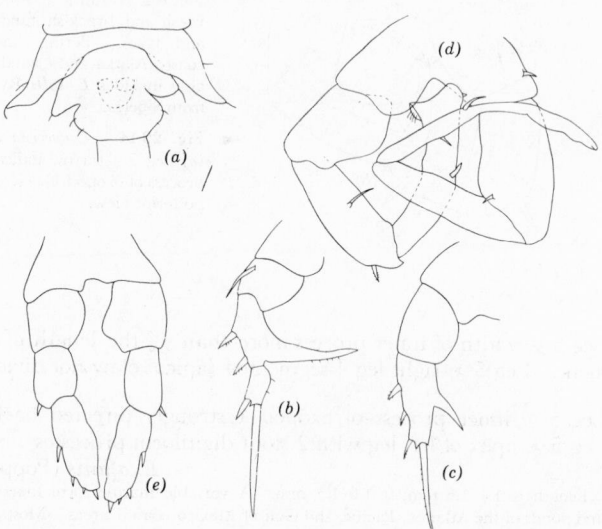

Fig. 29.16. *Eurytemora yukonensis.*   (*a*) Metasomal wings and genital segment ♀ with outline of ventral genital operculum.   (*b*) Leg 5 ♀, adult.   (*c*) Leg 5 ♀, copepodid stage V.   (*d*) Leg 5 ♂, adult, posterior view.
(*e*) Leg 5 ♂, copepodid stage V, posterior view.   (After M. S. Wilson.)

11b   Leg 5 ♀, inner apical spine less than 2 times longer than outer.
                 Leg 5 ♂, left, apex with outer digitiform protrusion (Fig. 29.17) . .
                                                    E. composita Keiser 1929

Protrusions of genital segment ♀ rounded lobes.   Spinulose hairs on dorsal surfaces
of ♀ urosome segments 2 and 3 caudal rami, and of ♂ urosome segment 5.
Length: ♀ 1.2–1.4 mm; ♂ 1.07 mm.   Fresh and brackish ponds, lakes.   Alaska; Asia.
(Descr. M. S. Wilson, 1953*b*.)

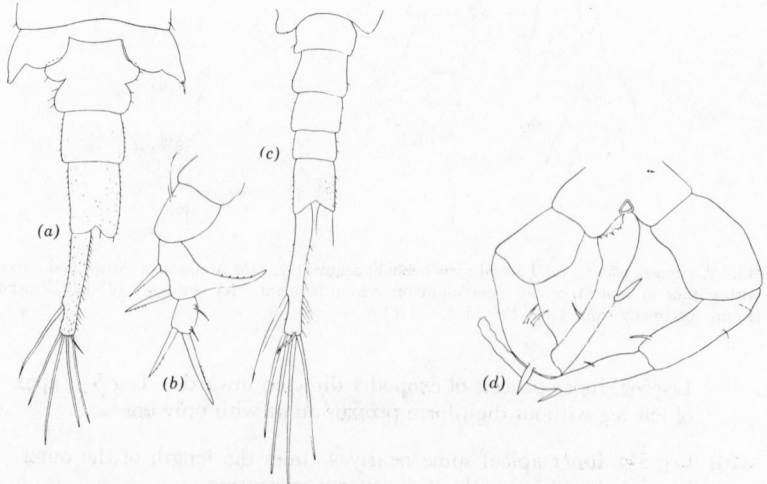

Fig. 29.17. *Eurytemora composita.*   (*a*) Metasomal wings and urosome ♀, dorsal.   (*b*) Leg 5 ♀.   (*c*) Urosome
♂, dorsal.   (*d*) Leg 5 ♂, posterior view.   (After M. S. Wilson.)

12a (8) Cephalic segment ♀ ♂, lateral view, anterior part with dorsal depression. Urosome ♀ ♂, caudal ramus 6 to 7 times longer than wide. (Fig. 29.18) . . . . *Limnocalanus macrurus* Sars 1863
Length: ♀ 2.2–3.15 mm; ♂ 2.2–2.78 mm. Lakes. Northeastern U. S. west to Wis.; Canada; Alaska; Greenland; northern Europe. (Descr. Sars, 1901–1903; Gurney, 1931.)

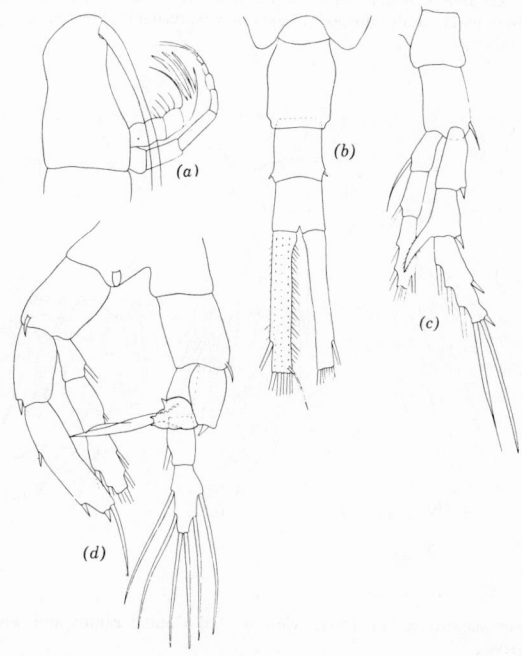

Fig. 29.18. *Limnocalanus macrurus.* (*a*) Lateral view cephalic segment ♀. (*b*) Urosome ♀, dorsal. (*c*) Leg 5 ♀. (*d*) Leg 5 ♂, posterior view.

12b Cephalic segment ♀ ♂, lateral view, without depression. Caudal ramus ♀ ♂ about 4 times longer than wide . . . .
                          L. johanseni Marsh 1920

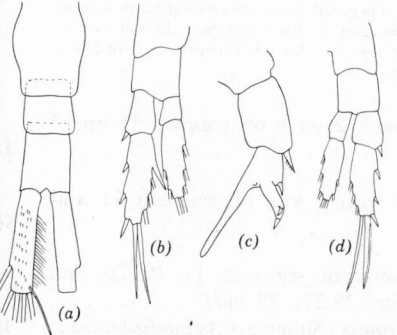

Leg 5 ♀ as in *L. marcrurus*. Leg 5 ♂ differs from that of *L. macrurus* in having the apical outer portion of the right exopod extended considerably beyond the base of the inner process. Length: 2.2–2.9 mm; ♂ 2.0–2.6 mm. Tundra lakes, ponds. Arctic Canada; Alaska.

◄ **Fig. 29.19.** *Limnocalanus johanseni.* (*a*) Urosome ♀, dorsal. (*b*) Leg 5 ♀, copepodid stage V. (*c*) Leg 5 ♂, apex of right exopod. (*d*) Leg 5 ♂, copepodid stage V, posterior view (leg symmetrical in this stage).

13a (7) Caudal ramus ♀ ♂, setae unequal in length, the fourth from outer margin longer and stouter than the others. Leg 5 ♀ ♂, endopods

3-segmented, apical segment with 6 setae; leg 5 ♂, right leg not
ending in single claw . . . .
Osphranticum labronectum S. A. Forbes 1882
Family **Centropagidae.** Only species known in the genus. Urosome ♀ 4-
segmented. First antennae ♀, ♂ left, 24- or 23-segmented. Endopods legs 1 to 4
with 3 segments. Length: ♀ 1.7–2.5 mm; ♂ 1.4–2.3 mm. Ponds, shallow lakes.
Great Lakes area, Canada and U. S. north central and southeastern states, west into
Tex.; Guatemala. Late copepodid stages may be recognized by the unequal length of
the caudal setae.

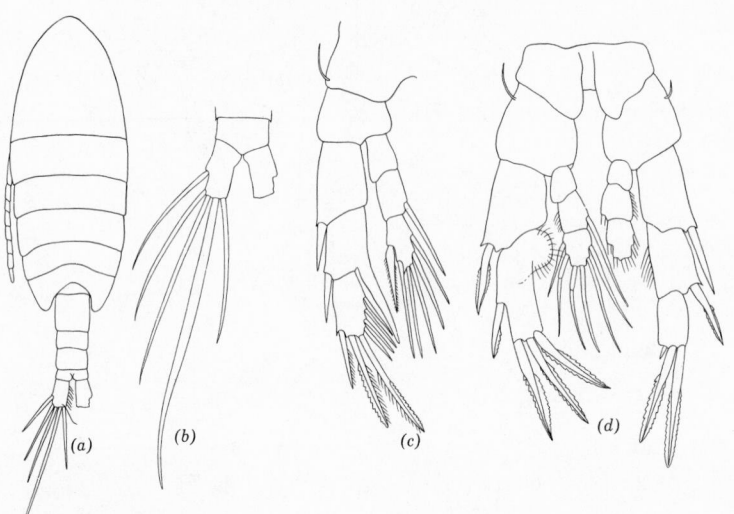

Fig. 29.20. *Osphranticum labronectum.* (*a*) Dorsal view ♀.   (*b*) Caudal ramus and setae ♀.   (*c*) Leg 5 ♀.
(*d*) Leg 5 ♂, posterior view.

13b   Caudal ramus ♀ ♂, setae nearly equal in length, the fourth not
different from the others. Leg 5 ♀ ♂, endopods modified, 1- or
2-segmented, with 0 to 2 apical setae; leg 5 ♂, right leg ending
in single claw . . . . Family **Diaptomidae**   14
The key goes directly to species rather than to genera or subgenera, though where
possible, the species are arranged in related groups. Two genera of this family are
included in the key: *Acanthodiaptomus*, of which one species is known in N. A.; and
Diaptomus with numerous species referable to the several subgenera listed in the key
and in Table 29.2. See Figs. 29.2–29.7 for copepodid stages and explanation of terms
used in key. Table 29.3 gives details of setation of first antennae. In this family,
endopods of leg 1 have 2 segments, those of legs 2 to 4 have 3 segments. Urosome ♀
2- or 3-segmented in North American groups.

14a   (13) First antennae ♀ and ♂ left side, with 2 setae on segment 11 and 1
or 2 on segments 13 to 19   15

14b   First antennae ♀ and ♂ left side, with 1 seta on segment 11 and
1 on segments 13 to 19.   50

15a   (14) First antennae ♀ and ♂ left side, setae on segments 17, 19, 20, and
22 with the end stiffly hooked (Figs. 29.21*c*, 29.24*d*). . . . .
Diaptomus (Subgenus *Aglaodiaptomus*)   16

15b   These setae not hooked. (Do not confuse with curved or twisted
ends that straighten out under cover-glass pressure.).   25

16a **(15)** First antennae ♀, ♂ left side, 2 setae on some of segments 13 to 19
(2 on 16) **17**
Note: check both antennae of ♀; a second anomalous seta may occasionally occur on
a segment of one side.

16b One seta on all these segments **18**

17a **(16)** Leg 5 ♀, endopod usually as long as or very little longer than inner
margin of exopod 1. Leg 5 ♂, right, apical claw longer than
exopod 2 (about 1.6:1). (Fig. 29.21). . . . .
Diaptomus forbesi Light 1938
Metasomal wings ♀ rounded, symmetrical; urosome 3-segmented, symmetrical.
See Fig. 29.2 for apex ♂ right antenna. Length: ♀ 1.3–1.9 mm; ♂ 1.1–1.4 mm.
Ponds, lakes. Pacific coast states; western Canada, east to Saskatchewan.

17b Leg 5 ♀, endopod usually longer than exopod 1 by ⅕ to ¼ of its
own length. Leg 5 ♂, right, claw a little shorter than exopod 2
(about 0.85:1). (Fig. 29.22) . . . . . **D. lintoni** S. A. Forbes 1893
Metasomal wings and urosome ♀ very similar to those of *D. forbesi*; females of
these two species may not be separable with certainty. Leg 5 ♂, left endopod with
very stout dentitions. Length: ♀ 1.72–2.5 mm; ♂ 1.5–2.0 mm. Lakes. Rocky
Mountains, Mont. to Colo.

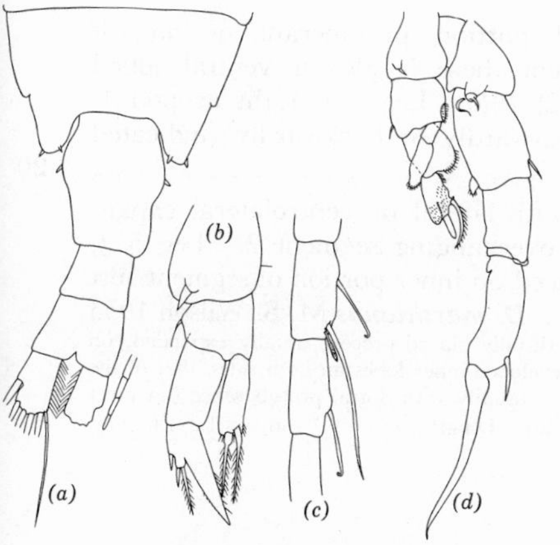

Fig. 29.21. *Diaptomus forbesi.* (*a*) Metasome segments 5 and 6 and urosome ♀, dorsal, with apex of first antenna. (*b*) Leg 5 ♀. (*c*) Segments 16 and 17 of first antenna ♀. (*d*) Leg 5 ♂, posterior view.

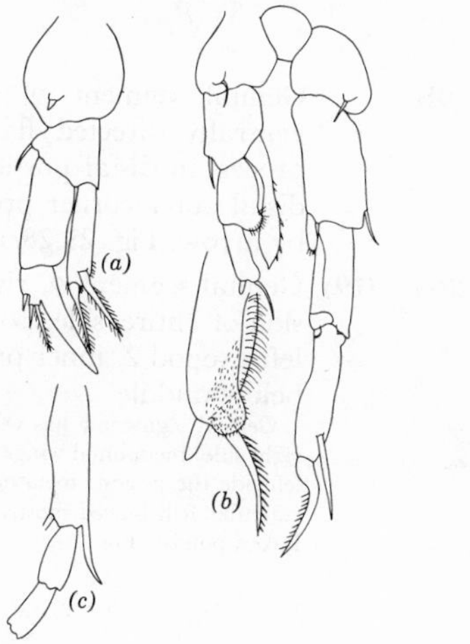

Fig. 29.22. *Diaptomus lintoni.* (*a*) Leg 5 ♀. (*b*) Leg 5 ♂ with detail left exopod 2. (*c*) Apex right antenna ♂.

18a **(16)** Metasomal wings ♀ (dorsal view) with well-developed lobe on
inner side of one or both wings (Fig. 29.24*b*). Leg 5 ♂ (posterior
view), left exopod 2, processes not closely set to one another **19**

18b Metasomal wings ♀ without well-developed inner lobes (very small
lobe may be present). (Fig. 29.29*a*). Leg 5 ♂, left exopod 2,
processes closely set **24**

19a **(18)** Genital segment ♀, proximal portion only slightly protuberant
laterally but right side showing a ventrally directed flange in
dorsolateral view, and lacking a ventral lobed process in distal

portion. Leg 5 ♂, right exopod 1, sclerotized area of distal outer
corner of segment not produced outwardly or backwardly . . . . .
                                      D. spatulocrenatus Pearse 1906

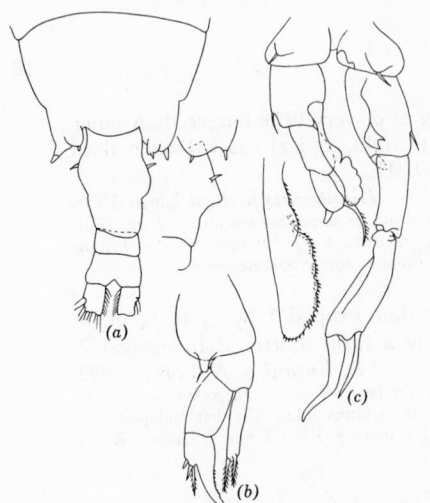

Length: ♀ 1.2–1.6 mm; ♂ 1.1–1.3 mm.
Ponds, lakes. Quebec east to coast; south
in Atlantic coastal states to Md. (Descr.
Marsh, 1929; Kiefer, 1931.)

◄ **Fig. 29.23.** *Diaptomus spatulocrenatus.*
(*a*) Metasome segments 5 and 6 and urosome ♀, dorsal, and dorsolateral view of
flange of right side of genital segment.
(*b*) Leg 5 ♀. (*c*) Leg 5 ♂, posterior view,
with detail of left endopod.

19b   Genital segment ♀, proximal portion protuberant or not, if
              ventrally directed flange present there is also a ventral lobed
              process in distal portion (Fig. 29.26*b*). Leg 5 ♂, right exopod 1,
              distal outer corner produced outwardly or backwardly (indicated
              by arrow, Fig. 29.28*c*) **20**

20a **(19)** Genital segment ♀, right side with lateral or ventrolateral expan-
              sion of entire side, sometimes overhanging segment 2. Leg 5 ♂,
              left exopod 2, inner process placed on inner portion of segment just
              below middle . . . . **D. marshianus** M. S. Wilson 1953
                  Genital segment ♀ has ventrolateral, distally placed process, usually expanded, on
              right side; metasomal wings with well-developed inner lobes on both sides, that of the
              left side the largest; metasome segment 5 usually with dorsal protuberance (an erect
              cuticular frill placed mostly on right side). Length: ♀ 1.5–1.9 mm; ♂ 1.3–1.6 mm.
              Lakes, ponds. Fla.

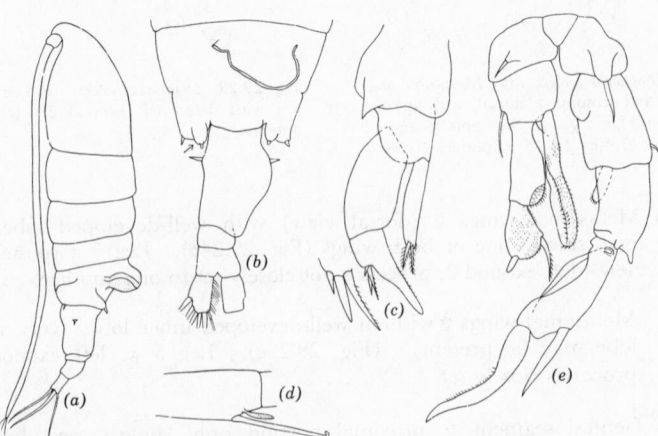

Fig. 29.24. *Diaptomus marshianus.* (*a*) Lateral view ♀. (*b*) Metasome segments 5 and 6 and urosome ♀, dorsal (arrow indicates inner lobe of metasomal wing). (*c*) Leg 5 ♀ with detail exopod setae. (*d*) Segment 19
of first antenna ♀ showing hooked seta and aesthete. (*e*) Leg 5 ♂, posterior view.

20b   Genital segment ♀, right side with or without lateral expansion, if present, only in proximal portion and not overhanging segment 2. Leg 5 ♂, left exopod 2, inner process placed terminally or subterminally **21**

21a   (20)   Genital segment ♀ with a ventrally placed process distad to the genital protuberance at approximately the mid-point of the segment. Leg 5 ♂, left leg reaching beyond the right exopod 1 . . . .
                                      D. pseudosanguineus Turner 1921

         An inadequately known species; no specimens available for illustration. Length: ♀ 2.0 mm; ♂ 1.6 mm. Type locality, St. Louis, Mo.

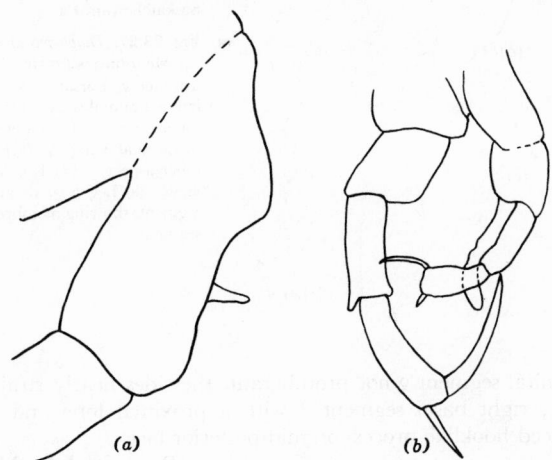

Fig. 29.25. *Diaptomus pseudosanguineus.* (a) Genital segment ♀, lateral view, showing process. (b) Leg 5 ♂, anterior view. (After Turner.)

21b   Genital segment ♀ with or without ventral process, if present, it is placed near distal end of segment. Leg 5 ♂, left leg not reaching beyond the right exopod 1. . . . . **22**

22a   (21)   Genital segment ♀, proximal portion of right side laterally protuberant with ventrally directed flange, and with a small lobed process distally. Leg 5 ♂, right basal segment 2, inner margin expanded into 2-lobed flange . . . ***D. dilobatus*** M. S. Wilson 1958

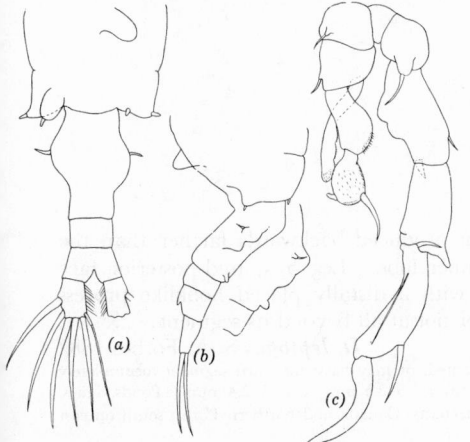

Metasomal wings ♀ asymmetrical, left wing with large inner lobe; dorsal cuticular protuberance of metasome may be present or absent. Length: ♀ 1.8–1.88 mm; ♂ 1.65–1.71 mm. Ponds. La.

◄ **Fig. 29.26.** *Diaptomus dilobatus.* (a) Metasome segments 5 and 6 (with dorsal process) and urosome ♀, dorsal. (b) Lateral view of same, and dorsolateral view of right side of genital segment showing profile view of ventrally directed flange and distal lobe of genital segment. (c) Leg 5 ♂, posterior view. See Fig. 29.2*d,e* for right antenna.

22b   Genital segment ♀ without ventral flange or process. Leg 5 ♂,
               right basal segment 2, inner margin not expanded as above. 2:

23a (22)   Genital segment ♀ slightly protuberant on each side. Leg 5 ♂,
               right basal segment 2 with narrow hyaline membrane on inner
               mid-portion. . . . . **D. saskatchewanensis** M. S. Wilson 1958

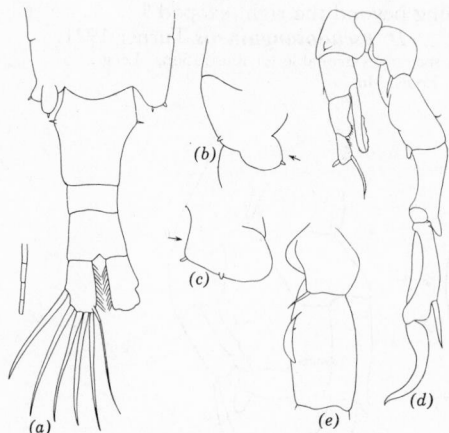

Left metasomal wing ♀ with large
inner lobe. Length: ♀ 1.2–1.79 mm;
♂ 1.18–1.56 mm. Lakes, ponds.
Saskatchewan; La.

◄ **Fig. 29.27.** *Diaptomus saskatchewanensis.*
(*a*) Metasome segments 5 and 6 and
urosome ♀, dorsal. (*b*) Lateral view
left metasomal wing ♀ (arrow indicates
inner aspect). (*c*) Lateral view right
metasomal wing ♀ (arrow indicates
inner aspect). (*d*) Leg 5 ♂, posterior
view. (*e*) Leg 5 ♂, detail right basal
segments showing membrane of second
segment.

23b   Genital segment ♀ not protuberant, the sides nearly straight. Leg
               5 ♂, right basal segment 2 with a proximal lobe and a distally
               placed hooklike process on mid-posterior face . . . .
                                          D. conipedatus Marsh 1907

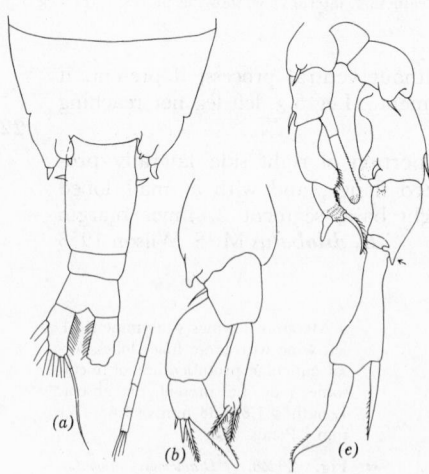

Metasomal wings ♀, each side with
large inner lobe. Length: ♀ 1.5 mm;
♂ 1.3 mm. Ponds. La.

◄ **Fig. 29.28.** *Diaptomus conipedatus.*
(*a*) Metasome segments 5 and 6 and
urosome ♀, dorsal, with apex of first
antenna. (*b*) Leg 5 ♀. (*c*) Leg 5 ♂,
posterior view (arrow indicates backwardly produced, outer distal corner
of exopod 1).

24a (18)   Left metasomal wing ♀ not produced backwards farther than the
               right; with a very small inner lobe. Leg 5 ♂, mid-posterior face
               of right basal segment 2 with a distally placed hooklike process
               that reaches only slightly or not at all beyond its segment. . . . . .
                                          D. leptopus S. A. Forbes 1882
Urosome ♀ may be 2- or 3-segmented, or may have the third segment incompletely
separated from the second. Length: ♀ 1.5–2.5 mm; ♂ 1.25–2.4 mm. Ponds, lakes.
Occurs at all altitudes, east to west coasts, Canada and northern U. S., south on east

coast to Va., on west coast to Ore.; throughout Rocky Mountains; eastern Alaska. Includes **D. piscinae** Forbes 1893 and **D. manitobensis** Arnason 1950.

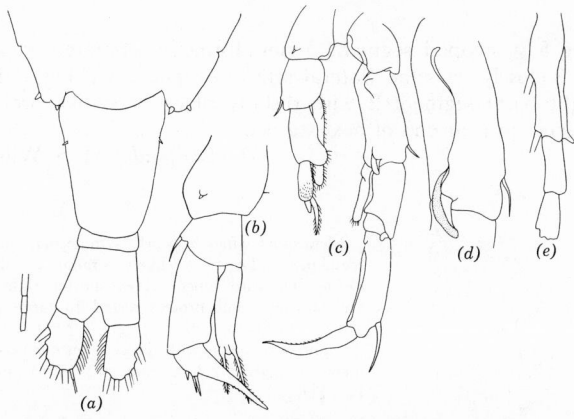

ig. 29.29. *Diaptomus leptopus.* (*a*) Metasomal wings and urosome ♀, dorsal, with apex of first antenna.) Leg 5 ♀. (*c*) Leg 5 ♂, posterior view. (*d*) Leg 5 ♂, detail process of right basal segment 2. (*e*) Apex right ntenna ♂ showing hyaline membrane of fused segment 22–23. (From typical specimens from Mass.)

24b   Left metasomal wing ♀ produced backwards farther than the right; without inner lobe. Leg 5 ♂, right basal segment 2 with distally placed hooklike process that reaches to near end of first exopod segment . . . . **D. clavipes** Schacht 1897

           Urosome ♀ 2-segmented. Length: ♀ 1.37–2.5 mm; ♂ 1.28–2.2 mm. Ponds, lakes. Rocky Mountains U. S., east to Mississippi River; Mexico; Alberta, Manitoba. Includes **D. nebraskensis** Brewer 1898.

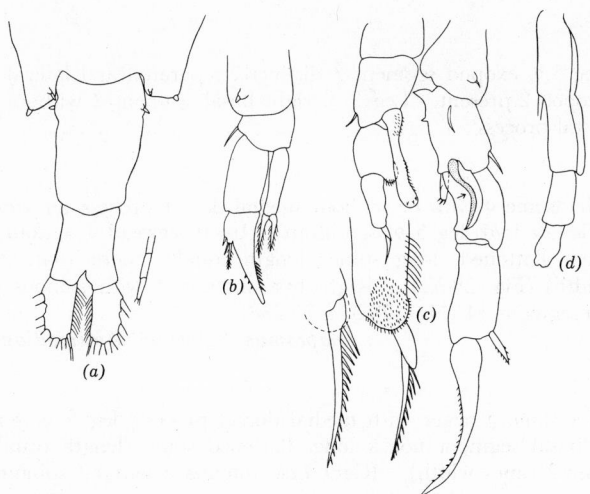

ig. 29.30. *Diaptomus clavipes.* (*a*) Metasomal wings and urosome ♀, dorsal, with apex of first antenna.) Leg 5 ♀. (*c*) Leg 5 ♂, posterior, with details of processes of left exopod 2 (arrow on right exopod 1 indi- ates hooklike process of basal segment 2). (*d*) Apex right antenna ♂.

25a   **(15)** First antennae ♀, ♂ left, 1 seta on all of segments 13 to 19 (see **16b** for note on anomalous setae) **26**

25b   First antennae ♀, ♂ left, 2 setae on some or all of segments 13 to 19 (see also **34a**) **41**

26a (25) Leg 5 ♀, exopod segment 3 not distinctly separated, represented by 2 closely set setae (lateral seta of exopod 2 lacking). Leg 5 ♂, right basal segment 2 with distally placed hooklike process that reaches to near end of next segment . . . .
D. clavipoides M. S. Wilson 1955

Metasomal wings ♀ nearly symmetrical; urosome 2-segmented. Leg 5 ♂ closely similar to that of *D. clavipes* but with longer claw, shorter right endopod, and lacking small process distad to large protruding process of proximal inner margin of right basal segment 2; right antenna ♂ with process on twenty-third segment. Length: ♀ 2.3–2.5 mm; ♂ 2.0–2.13 mm. Ponds. La., Fla.

◄ **Fig. 29.31.** *Diaptomus clavipoides.* (*a*) Leg 5 ♀ with detail exopod setae. (*b*) Leg 5 ♂, posterior. (*c*) Apex right antenna ♂. (After M. S. Wilson.)

26b   Leg 5 ♀, exopod segment 3 distinctly separated and lateral seta of exopod 2 present. Leg 5 ♂, right basal segment 2 without such a distal process **27**

27a (26) Metasome ♀ with or without medial dorsal process on segment 5 (Fig. 29.35*b*); leg 5 ♀, sensillum of basal segment 1 a stout, somewhat flattened, long spine (length usually more than 2 times width) (Fig. 29.32*b*). Right first antenna ♂ with spinous process on segment 14 (Figs. 29.32*d*, 29.33*d*) . . . .
Diaptomus (Subgenus *Mastigodiaptomus*) **28**

27b   Metasome ♀ never with medial dorsal process; leg 5 ♀, sensillum of basal segment not a long, flattened spine (length usually less than 2 times width). Right first antenna ♂ without spinous process on segment 14 . . . . ♀ **33**, ♂ **32**

28a (27) Caudal ramus ♀ with hairs on inner margin only; metasome never with dorsal process. Leg 5 ♂, left exopod 2, distal process a stout broadened continuation of the segment . . . .
D. purpureus Marsh 1907

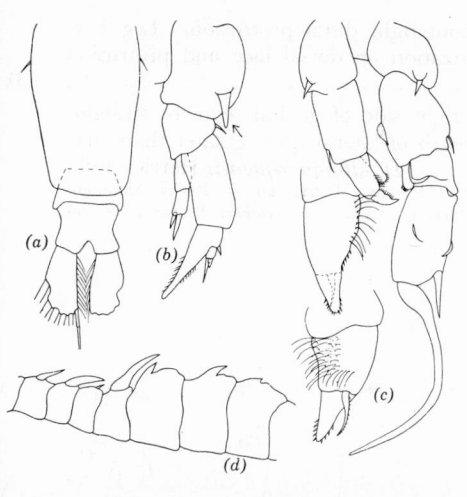

Length: ♀ 2.5 mm; ♂ 2.2 mm. Cuba.

◄ **Fig. 29.32.** *Diaptomus purpureus.* (*a*) Urosome ♀, dorsal. (*b*) Leg 5 ♀ (arrow indicates elongate, flattened sensillum of basal segment 1). (*c*) Leg 5 ♂, posterior, with detail of posterior and anterior aspects of left exopod 2. (*d*) Right antenna ♂ showing spines and processes of segments 10 to 16.

28b   Caudal ramus ♀ with hairs on both margins; metasome usually with dorsal process (a character of all 4 following species, but may be absent in some individuals of a sample). Leg 5 ♂, left exopod 2, distal process narrower than the segment **29**

29a **(28)** Genital segment ♀ without lateral protrusions. Leg 5 ♂, left exopod, segment 1 from 2½ to 3 times longer than segment 2 . . . . .
                                 D. texensis M. S. Wilson 1953
Length: ♀ 1.5–1.6 mm; ♂ 1.4–1.6 mm. Ponds. Tex.; Mexico.

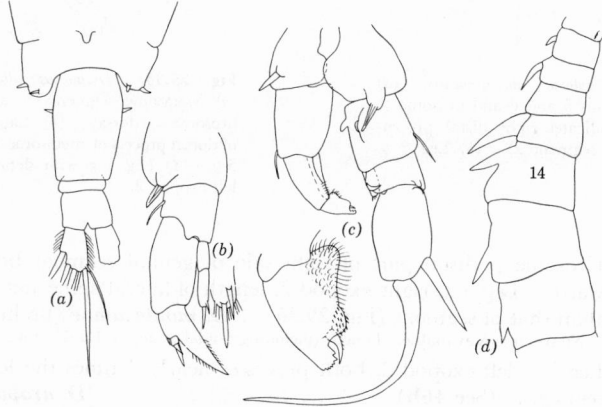

Fig. 29.33. *Diaptomus texensis.* (*a*) Metasome segments 5 and 6 and urosome ♀, dorsal. (*b*) Leg 5 ♀ with detail apex of endopod. (*c*) Leg 5 ♂, posterior, with detail left exopod 2. (*d*) Right antenna ♂ showing spines and processes of segments 10 to 16. (After M. S. Wilson.)

29b   Genital segment ♀ with lateral protrusions. Leg 5 ♂, left exopod, segment 1 subequal to or only a little longer than segment 2 . . . .   **30**

30a **(29)** Urosome ♀, segment 2 with right distal protrusion. Leg 5 ♂, right exopod 2 without sclerotization on dorsal face or protrusion on proximal outer margin. (Fig. 29.34) . . . .
                             D. amatitlanensis M. S. Wilson 1941
Length: ♀ 1.4–1.5 mm; ♂ 1.25–1.4 mm. Lakes. Guatemala.

30b   Urosome ♀, segment 2 without right distal protrusion. Leg 5 ♂, right exopod 2 with sclerotization on dorsal face and protrusion on proximal outer margin. **3**

31a (30) Urosome ♀, distal part of right side of genital segment straight. Leg 5 ♂, right exopod 2, length of lateral spine greater than that of its segment. (Fig. 29.35). . . . ***D. albuquerquensis*** Herrick 1895
Length: ♀ 1.08–1.7 mm; ♂ 0.96–1.5 mm. Ponds, lakes. Rocky Mountain states, Utah south into Central America. Includes ***D. lehmeri*** Pearse 1904 and ***D. a. patzcuarensis*** Kiefer 1938.

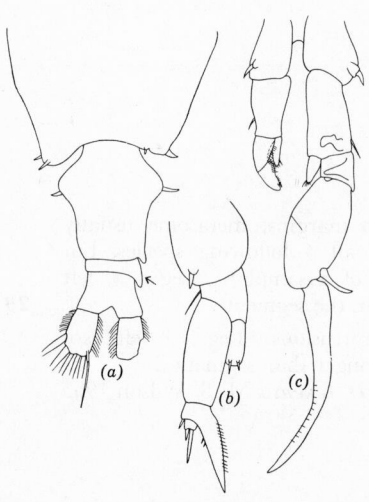

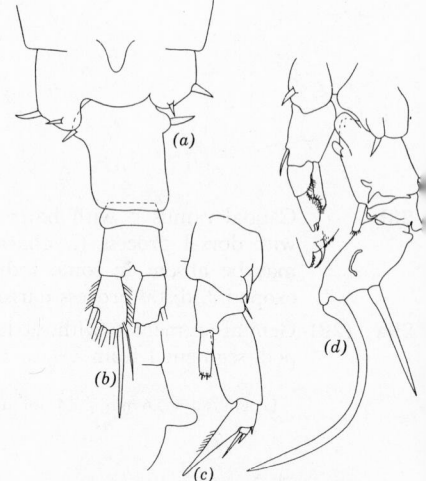

Fig. 29.34. *Diaptomus amatitlanensis.* (*a*) Metasome segments 5 and 6 and urosome ♀, dorsal (arrow indicates right distal protrusion of urosome segment 2). (*b*) Leg 5 ♀. (*c*) Leg 5 ♂.

Fig. 29.35. *Diaptomus albuquerquensis.* (*a*) Metasome segments 5 and 6 and urosome ♀, dorsal. (*b*) Lateral profile of dorsal process of metasome ♀. (*c*) Leg 5 ♀. (*d*) Leg 5 ♂ with detail processes left exopod 2.

31b   Urosome ♀, distal part of right side of genital segment bent outwards. Leg 5 ♂, right exopod 2, length of lateral spine not greater than that of segment (Fig. 29.36). . ***D. montezumae*** (Brehm) 1955
No specimens available. Length (including caudal setae): ♀ 1.6–1.7 mm. Mexico.

32a (27) Leg 5 ♂, left exopod 2, both processes nearly 2 times the length of segment. (See **46b**) . . . . ***D. arapahoensis***

32b   Leg 5 ♂, left exopod 2, processes not longer than segment. . . . . .
Diaptomus (Subgenus ***Hesperodiaptomus***)   **33**

33a (27,32) First antennae ♀, seta of segment 1 and first seta of segment 2 exceedingly elongate (Fig. 29.37*a*). Leg 5 ♂, left exopod 2, inner process setiform and closely set to the base of digitiform distal process (Fig. 29.37*c*) . . . . ***D. augustaensis*** Turner 1910
Leg 5 ♀ has an accessory denticle on the medial surface of the claw. Length: ♀ 1.8 2.3 mm; ♂ 1.5–2.3 mm. Ponds. Ga., N. C., La.

33b   First antennae ♀, seta of segment 1 elongate or not, but first seta of segment 2 shorter or not much longer than the other 2 setae of segment. Leg 5 ♂, left exopod 2, base of inner process not closely set to the base of digitiform distal process but arising anterior to the base and toward the medial portion of distal pad, varying in shape from broad-based, flattened structure with spiniform tip to shortened, spiniform seta **34**

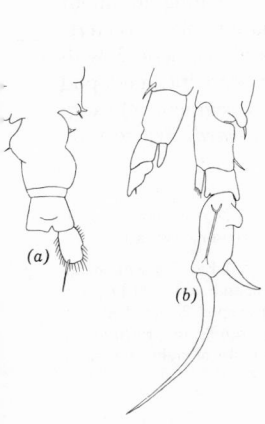

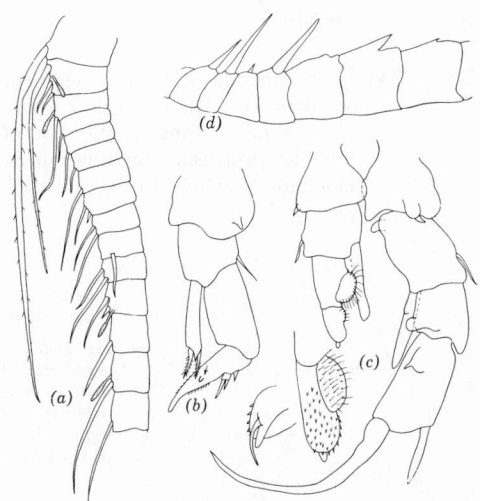

Fig. **29.36.** *Diaptomus monte-umae.* (*a*) Metasomal wings and urosome ♀, dorsal. (*b*) Leg 5 ♂. (After Brehm.)

Fig. **29.37.** *Diaptomus augustaensis.* (*a*) Segments 1 to 15 of first antenna ♀. (*b*) Leg 5 ♀ (arrow indicates accessory denticle). (*c*) Leg 5 ♂ with details posterior aspect left exopod 2 and anterior aspect of processes. (*d*) Segments 10 to 16 of right antenna ♂. (From La. specimens.)

34a (*33*) Leg 5 ♀, exopod 2 broad from the base to near the end of claw where it is abruptly tapered (greatest width about ½ the length). Leg 5 ♂, right basal segments 1 and 2 without inner protrusion or accessory cuticular outgrowth. . . . . **D. kenai** M. S. Wilson 1953

Leg 5 ♂, left exopod 2, the inner process has a very broad base that tapers to a long spinous point, variable in length. Rarely the antennae may have 2 setae on some of segments 13 to 19. Length: ♀ 2.0–3.0 mm; ♂ 1.8–2.5 mm. Common in mountain lakes and ponds, rarer at low altitudes. Alaska to Calif.

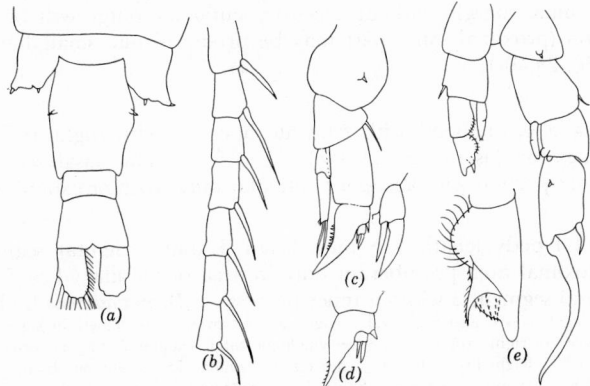

Fig. **29.38.** *Diaptomus kenai.* (*a*) Metasomal wings and urosome ♀, dorsal. (*b*) First antenna ♀ showing setae of segments 13 to 19. (*c*) Leg 5 ♀ with detail setae of exopod 3. (*d*) Leg 5 ♀, variation of exopod. (*e*) Leg 5 ♂ with detail anterior aspect left exopod 2. (After M. S. Wilson.)

34b   Leg 5 ♀, exopod 2 elongate and narrow, tapering gradually
            throughout its length. Leg 5 ♂, right basal segment 1 with inner
            protrusion (Fig. 29.40e) and basal segment 2 with or without a
            lobelike protrusion or accessory cuticular outgrowth on inner
            margin **35**

35a (34) First antennae ♀, seta of segment 1 elongate, reaching to about
            segment 11. Right leg 5 ♂ with this combination of characters:
            claw about as long as the rest of the leg; basal segment 2 with
            lobelike protrusion on inner proximal face of segment; endopod
            elongate (reaching to beyond middle of exopod segment 2) (Fig.
            29.39b) . . . . **D. wardi** Pearse 1905

Length: ♀ 1.3–2.0 mm; ♂ 1.24–1.6 mm. Ponds.
Wash., Mont. (Descr. M. S. Wilson, 1953a.)

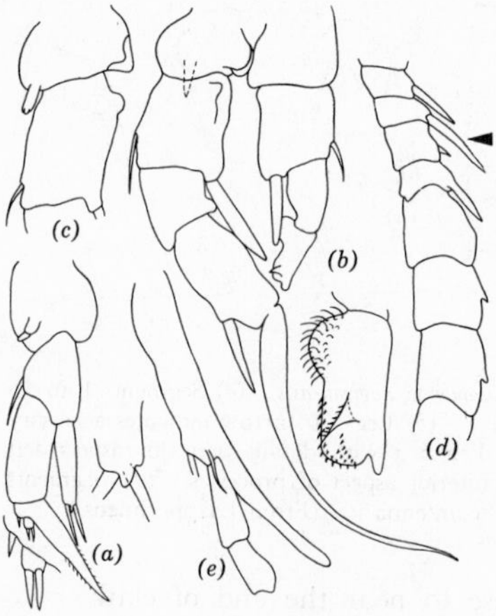

Fig. 29.39. *Diaptomus wardi.* (a) Leg 5 ♀ with detail
lateral exopod setae and apex of endopod. (b) Leg 5 ♂,
anterior view, with detail left exopod 2. (c) Leg 5 ♂,
right basal segments, showing profile of protrusion of
segment 2. (d) Segments 10 to 16 of right antenna ♂.
(e) Apex right antenna ♂.

35b   First antennae ♀, seta of segment 1 not reaching beyond segment
            4. Right leg 5 ♂ with one or two of these characters, but not with
            all. **36**

36a (35) Leg 5 ♀, endopod with prominent inner prolongation of apex,
            produced to point (Fig. 29.40b, g). Leg 5 ♂, right basal segment
            2, inner margin without accessory cuticular outgrowth or expan-
            sion (proximal inner part may be produced into small lobe as in
            Fig. 29.40e). **37**

36b   Leg 5 ♀, endopod with truncate apex or with slight inner pro-
            longation (as in Fig. 29.44a). Leg 5 ♂, right basal segment 2,
            inner portion with accessory cuticular outgrowth or expansion . . .   **38**

37a (36) Total body length ♀ ♂ more than 3 mm. Genital segment ♀,
            proximal area protuberant only in area of sensilla. Leg 5 ♂, left
            basal segment 2 without inner process . . **D. nevadensis** Light 1938
            Leg 5- ♀, see also Fig. 29.4. Right leg ♂ variable as shown in figures; inner
            process of right exopod 1 may be whole or bifid, exopod 2 may be broadened or
            narrow. Right first antenna ♂, process of segment 15 present or absent. Length:
            ♀ 3.85–4.05 mm; ♂ 3.5 mm. Saline lakes and ponds. Nev. (type locality), eastern
            Wash., Mont., N. D.; Saskatchewan. Includes **Hesperodiaptomus dentipes** Kincaid
            1956, corresponding to variation shown in Fig. 29.40f. (Descr. M. S. Wilson, 1953a.)

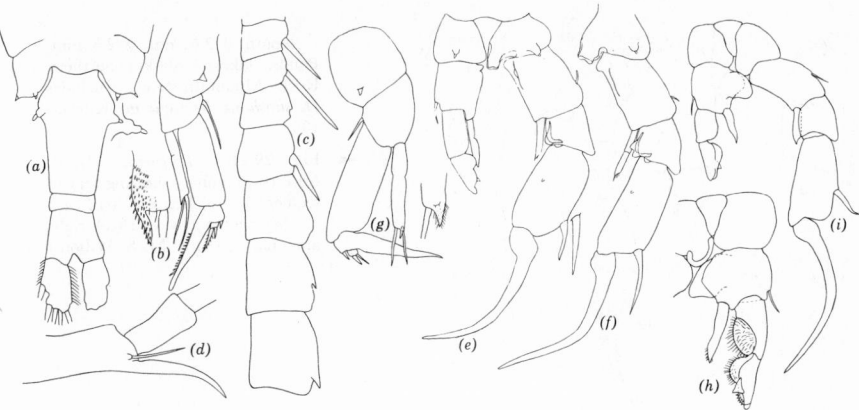

Fig. 29.40. *Diaptomus nevadensis:* (*a*) Metasomal wings and urosome ♀, dorsal. (*b*) Leg 5 ♀ with detail apex of endopod. (*c*) Right antenna ♂ showing spines and processes of segments 10 to 16. (*d*) Right antenna ♂ showing process of segment 23. (*e*) Leg 5 ♂, typical form from Nevada with variation in lateral spine of right exopod 2. (*f*) Right leg 5 ♂, variation (from Saskatchewan specimen). *D. wilsonae:* (*g*) Leg 5 ♀ with detail apex of endopod. (*h*) Left leg 5 ♂, anterior aspect. (*i*) Leg 5 ♂ posterior.

37b   Total body length ♀ ♂ less than 3 mm. Genital segment ♀, entire proximal area of both sides protuberant. Leg 5 ♂, left basal segment 2 with conspicuous, inwardly expanded membranous process (Fig. 29.40*h,i*) . . . . *D. wilsonae* Reed 1958

First antennae ♀, seta of segment 1 reaches to about end of segment 3. Right first antenna ♂, segment 15 with prominent process. Length: ♀ 1.6–1.7 mm; ♂ 1.5 mm. Ponds. Western Hudson Bay region of Canada.

38a (36) First antennae ♀, seta of segment 1 reaching to end of segment 3 or 4 as shown in Fig. 29.41*a* (see also *D. wilsonae*, **37b**). Leg 5 ♂, inner margin of right basal segment 2 with 1 or 2 small, unornamented cuticular outgrowths (rounded or somewhat rectangular in shape) **39**

38b   First antennae ♀, seta of segment 1 not reaching beyond middle of segment 2. Leg 5 ♂, inner margin of right basal segment 2 with a spinulose, denticulate, or serrate protrusion **40**

39a (38) Genital segment ♀ without lateral protrusions. Leg 5 ♂, right basal segment 2 with 1 cuticular outgrowth placed at mid-point of inner margin, and right exopod 2 with regularly curved inner margin . . . . **D. schefferi** M. S. Wilson 1953

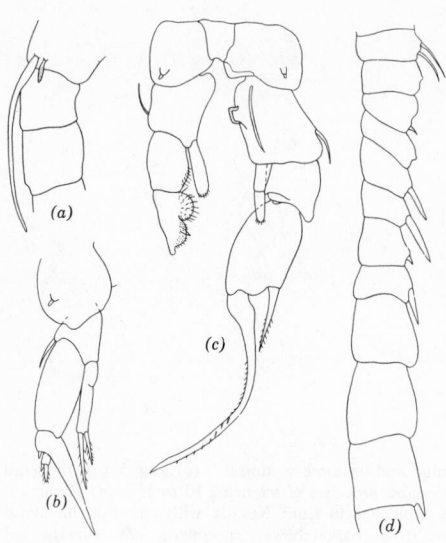

Length: ♀ 2.6 mm; ♂ 2.5 mm.
Ponds, lakes.   Alaska; northern
Rocky Mountain states.   Includes
D. shoshone beringianus Kincaid
1953.

◄ **Fig. 29.41.**   *Diaptomus schefferi.*
(*a*) First antenna ♀ showing seta of
segment 1.   (*b*) Leg 5 ♀.   (*c*) Leg 5
♂.   (*d*) Segments 6 to 16 of right
antenna ♂.   (After M. S. Wilson.)

39b   Genital segment ♀ with lateral protrusions, and metasomal wings
laterally compressed.   Leg 5 ♂, right basal segment 2 with 1
cuticular outgrowth on inner margin, and right exopod 2 with
irregularly expanded inner margin (Fig. 29.42c) . . . .
                                          D. victoriaensis Reed 1958
    Protrusions of genital segment ♀ similar to those of *D. kiseri;* not known whether
metasomal wings ♀ may sometimes be expanded as in *D. kiseri.* Leg 5 ♂, cuticular
outgrowth of right basal segment 2 very narrow, placed distad to a more prominent proximal·lobelike expansion of the segment.   Length: ♀ 2.8 mm; ♂ 2.6 mm.
Known only from pond, Victoria Island, Northwest Territories.

39c   Genital segment ♀ with lateral protrusions, and metasomal wings
expanded as in Fig. 29.42a.   Leg 5 ♂, right basal segment 2 with
2 cuticular outgrowths on inner margin . . **D. kiseri** Kincaid 1953
    Length: ♀ 2.8 mm; ♂ 2.4 mm.   Ponds.   Eastern Wash.; Saskatchewan.

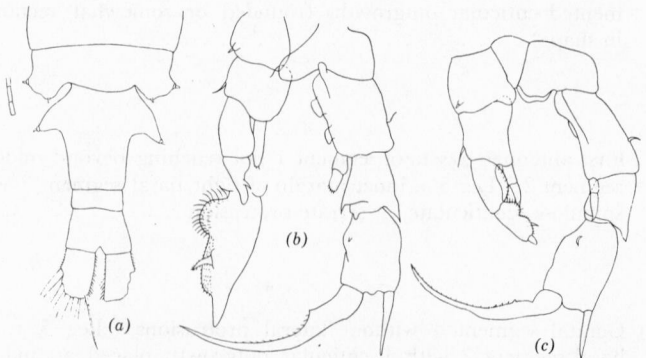

Fig. 29.42.   *Diaptomus kiseri:* (*a*) Metasome segments 5 and 6 and urosome ♀, dorsal, with apex of first antenna.   (*b*) Same, leg 5 ♂, posterior, with detail of left exopod 2, anterior view.   *D. victoriaensis:* (*c*) Leg
♂, posterior.

40a (38) (For characters of ♀, **40a–c**, see legend under each species.) Leg 5 ♂, right basal segment 2 little expanded inwardly and armed only with a medially placed spinulose protrusion . . . .
D. breweri M. S. Wilson 1958

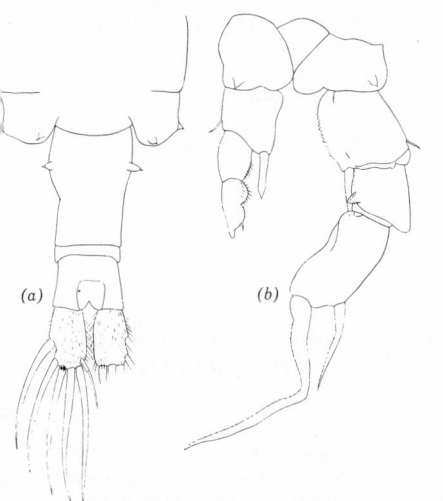

Leg 5 ♀ similar to that of *D. eiseni*. Length: ♀ 4.2–4.5 mm; ♂ 3.2–4.0 mm. Ponds. Neb.; Saskatchewan (type locality).

◄ **Fig. 29.43.** *Diaptomus breweri.* (a) Metasome segments 5 and 6 and urosome ♀, dorsal. (b) Leg 5 ♂, posterior view.

(a)   (b)

40b Leg 5 ♂, right basal segment 2 with anterior portion expanded inwardly and bearing the spinulose or denticulate protrusion, on the mid-posterior face a lengthwise ridge or shelf and a prominent sclerotized structure, usually consisting of single large spine or denticle; right exopod 2 without spinule on posterior face . . . .
D. eiseni Lilljeborg 1889

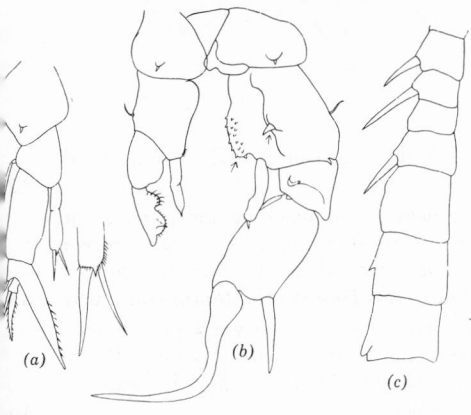

Leg 5 ♀, inner seta of exopod 3 usually reaching beyond middle of claw of exopod 2. Leg 5 ♂, the structure on right basal segment 2 may be large sclerotization of 2 or more points; claw of right leg usually irregular in shape with one or more distinct angles. A highly variable species. Length: ♀ 2.77–4.0 mm; ♂ 2.6–3.5 mm. Ponds, lakes. Calif. (type locality), north to Alaska, east to Labrador.

◄ **Fig. 29.44.** *Diaptomus eiseni.* (a) Leg 5 ♀ with detail apex of endopod. (b) Leg 5 ♂ (arrows indicate denticulate inner protrusion and the spiniform, sclerotized structure of mid-posterior face). (c) Segments 10 to 16 of right antenna ♂. (From specimens of typical form, central Calif.)

(a)   (b)   (c)

40c Leg 5 ♂, right basal segment 2 as in **40b** except that sclerotized structure of posterior face is a very small spinous point placed at about mid-point of lengthwise ridge; right exopod 2 with spinule on posterior face . . . . **D. arcticus** Marsh 1920

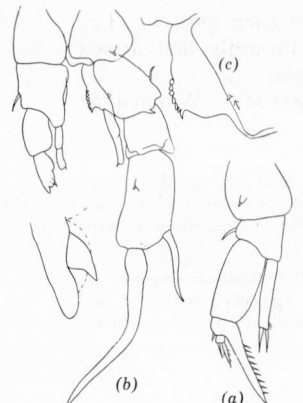

Leg 5 ♀, inner seta of exopod 3 usually not reaching beyond middle of claw. Leg 5 ♂, right, claw curved and without distinct angles though sometimes twisted. Length: ♀ 2.5–3.5 mm; ♂ 2.2 3.0 mm. Lakes, ponds. Arctic coast of Canada (type locality), western Canada; Alaska; Pacific coast and northern Rocky Mountain states. Described from Asia as *D. eiseni occidentalis* Rylov 1922.

◄ **Fig. 29.45.** *Diaptomus arcticus.* (*a*) Leg 5 ♀. (*b*) Leg 5 ♂ with detail processes of left exopod 2. (*c*) Leg 5 ♂, detail inner part of right basal segment 2 (arrow indicates spinous point). (From specimens of type lot, Arctic Canada.)

41a (25) First antennae ♀, ♂ left and right, segment 2 with 4 setae (and aesthete) . . . . *D. caducus* Light 1938

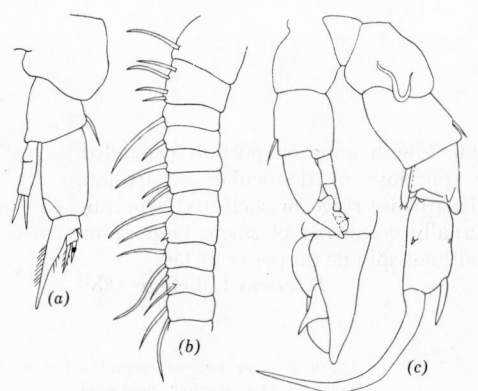

Subgenus *Hesperodiaptomus.* First antennae ♀, ♂ left, with 2 setae on all of segments 13 to 19, and on segments 6 and 10. Length: ♀ 2.8–3.3 mm; ♂ 2.4–2.6 mm. Ponds. British Columbia and Pacific coast states.

◄ **Fig. 29.46.** *Diaptomus caducus.* (*a*) Leg 5 ♀. (*b*) First antenna ♀ showing setae of segments 1 to 11. (*c*) Leg 5 ♂ with outline of anterior aspect of left exopod 2.

41b   First antennae ♀, ♂ left and right, segment 2 with 3 setae (and aesthete) **42**

42a (41) First antennae ♀, seta of segment 1 not reaching beyond end of segment 3 (usually shorter). Right first antenna ♂ with process on segment 23 and leg 5 ♂, left exopod 2, processes not longer than segment . . . . . . Subgenus *Hesperodiaptomus* continued   **43**

42b   First antennae ♀, seta of segment 1 reaching beyond end of segment 3 and only some of segments 13 to 19 with 2 setae. Right first antenna ♂ with process on segment 23 and leg 5 ♂, left exopod 2, processes longer than the segment **46**

42c   First antennae ♀, seta of segment 1 reaching beyond end of segment 3 and all of segments 13 to 19 with 2 setae. Right first antenna ♂ without process on segment 23 (membrane may be present along margin or on apex); leg 5 ♂, left exopod 2, processes either longer or shorter than segment. **47**

43a (42) First antennae ♀, ♂ left, seta of segment 3 elongate, that of ♀ reaching to segment 10, that of ♂ to segment 8 (Fig. 29.47c) . . . . *D. hirsutus* M. S. Wilson 1953

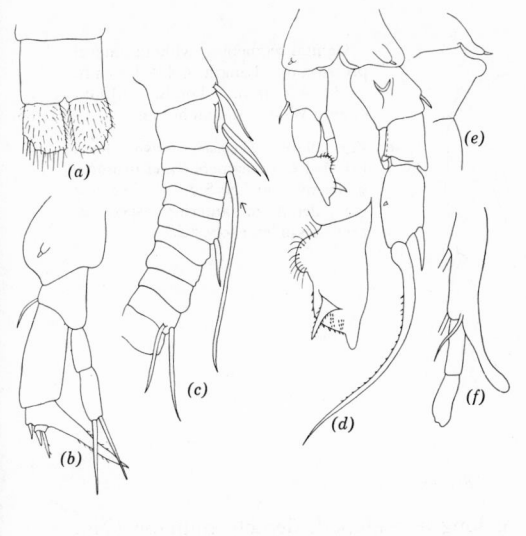

Setation first antennae ♀: 2 setae on 10, 11, 13 to 19. ♂ left differing from ♀: 1 on 10 and 13, 2 on 14 to 19. Caudal ramus ♀ with hairs on dorsal surface. Leg 5 ♂, right basal segment 2 with outwardly produced lobe on posterior face; claw about as long as the rest of the leg. Length: ♀ 1.88 mm; ♂ 1.79 mm. Mountain lakes and ponds, Calif.

◄ **Fig. 29.47.** *Diaptomus hirsutus.* (a) Urosome segment 3 and caudal rami ♀, dorsal. (b) Leg 5 ♀. (c) Segments 1 to 11 of first antenna ♀, showing setae of segments 1, 2, 3, 6 and 10 (arrow indicates elongate seta of segment 3). (d) Leg 5 ♂ with detail of anterior aspect of left exopod 2. (e) Leg 5 ♂, profile view of sensillum of right basal segment 1 and protrusion of posterior face of segment 2. (f) Apex right antenna ♂. (After M. S. Wilson.)

43b   First antennae ♀, ♂ left, seta of segment 3 not so elongate **44**

44a **(43)** First antennae ♀, ♂ left, 2 setae on segments 14, 16, 18 . . . .
                                        D. shoshone S. A. Forbes 1893
      Genital segment ♀ with lateral protrusions. Leg 5 ♂, right, claw swollen at base, shorter than the rest of the leg; left exopod 2, inner process a long, slender spine (Fig. 29.48*e*). Length: ♀ 3.1–4.0 mm; ♂ 2.59–3.3 mm. Lakes and ponds, high altitudes. Rocky Mountains, Alberta to Colo.; Sierra Nevada Mountains, Calif. (Descr. M. S. Wilson, 1953a.)

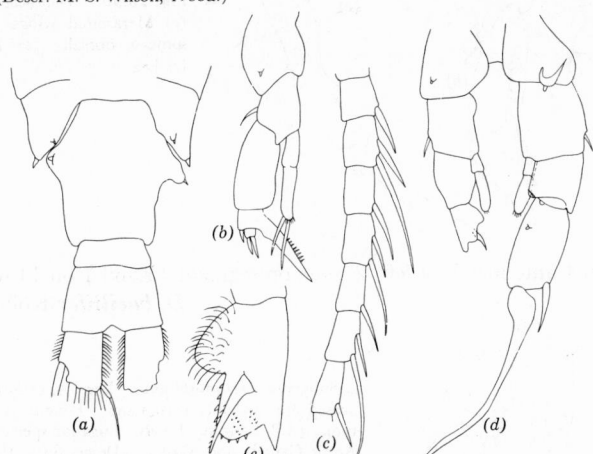

Fig. 29.48. *Diaptomus shoshone.* (a) Metasomal wings and urosome ♀, dorsal. (b) Leg 5 ♀. (c) First antenna ♀ showing setae of segments 13 to 19. (d) Leg 5 ♂. (e) Leg 5 ♂, anterior aspect of left exopod 2. (After M. S. Wilson.)

44b   First antennae ♀, ♂ left, 2 setae on segments 14, 16, 18, and 19 . . . **45**

45a **(44)** Leg 5 ♀, endopod setae shorter than endopod, not plumose. Leg
          5 ♂, right, claw about as long as the rest of the leg. . . . .
                      D. novemdecimus M. S. Wilson 1953

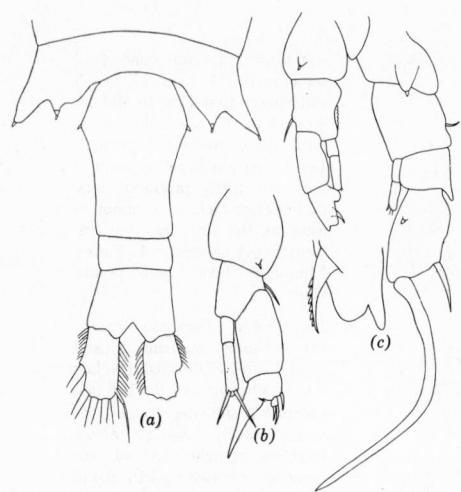

Genital segment ♀ without lateral protrusions. Length: ♀ 4.0–4.5 mm; ♂ 3.8–4.0 mm. Ponds. Mont., eastern Wash.; Saskatchewan.

◄ **Fig. 29.49.** *Diaptomus novemdecimus.* (*a*) Metasome segment 6 and urosome ♀, dorsal. (*b*) Leg 5 ♀. (*c*) Leg 5 ♂ with detail of anterior aspect of processes of left exopod 2.

45b   Leg 5 ♀, endopod setae as long as endopod, densely plumose (Fig. 29.50*b*).   Leg 5 ♂, right, claw shorter than exopod . . . .
D. stagnalis S. A. Forbes 1882

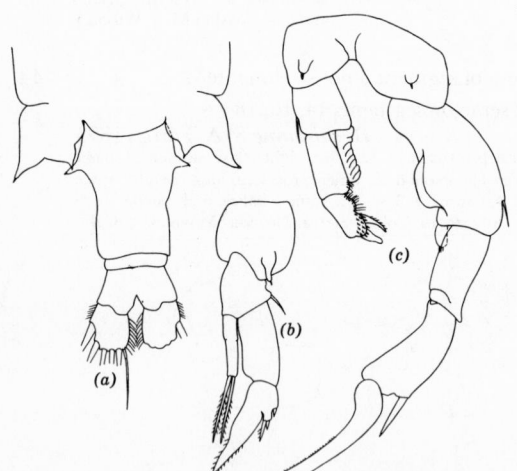

Subgenus **Aglaodiaptomus** (see **15–26**). Length: ♀ 3.2–4.5 mm; ♂ 3.0–4.0 mm. Ponds. Manitobia, Saskatchewan, north central states south to Gulf of Mexico states, east to Va., Ga. (Descr. Marsh, 1929.)

◄ **Fig. 29.50.** *Diaptomus stagnalis.* (*a*) Metasomal wings and urosome ♀, dorsal. (*b*) Leg 5 ♀. (*c*) Leg 5 ♂.

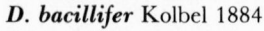

46a   (42)   First antennae ♀, ♂ left, 2 setae on segment 13 and 1 on 14 to 19 . .
D. bacillifer Kolbel 1884

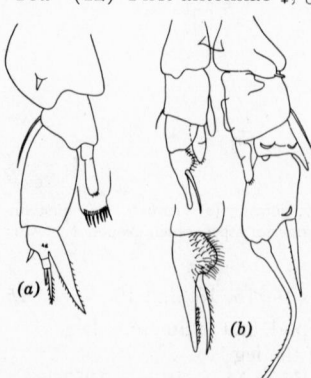

Subgenus **Arctodiaptomus.** Leg 5 ♀, endopod short, armed apically only with hairs. Length: ♀ 1.2–1.45 mm. ♂ 1.2–1.4 mm. Ponds. Eurasian species found in Arctic Canada and Alaska. (Descr. Sars, 1901–1903; Marsh, 1920.)

◄ **Fig. 29.51.** *Diaptomus bacillifer.* (*a*) Leg 5 ♀. (*b*) Leg 5 ♂ with detail left exopod 2. (From specimens from Pribilof Islands, Alaska.)

46b   First antennae ♀, 2 setae on segments 13, 15, and 17. (♂ keys out
              to **32a**). . . . . **D. arapahoensis** Dodds 1915

Subgenus **Arctodiaptomus**. Left antenna ♂ differs from ♀, having 1 seta
on all of segments 13 to 19. Leg 5 ♂, right exopod 2 with large spinous
process on midposterior face (indicated by arrow, Fig. 29.52). Leg 5 ♀ as
in *D. bacillifer*. Length: ♀ 1.6–2.0 mm; ♂ 1.3–1.7 mm. Lakes. Rocky
Mountains, Canada, U. S. Closely related to *D. acutilobatus* Sars 1903
from Siberia. (Descr. M. S. Wilson, 1953a.)

◄ **Fig. 29.52.** *Diaptomus arapahoensis.* Leg 5 ♂ (arrow indicates spinous
process on midposterior face of right exopod 2). (From type lot, Colo.)

46c   First antennae ♀, setation otherwise. **47**

47a (42, 46) First antennae ♀, 2 setae on segments 15, 16, and 17. Leg 5 ♂,
              right exopod 2, length of lateral spine less than width of seg-
              ment . . . . **D. theeli** Lilljeborg 1889

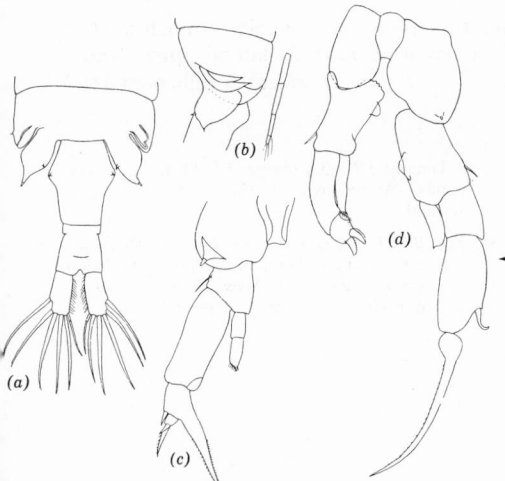

Subgenus **Mixodiaptomus**. Leg
5 ♀, endopod shorter than inner
margin exopod 1, may be 2-
segmented, armed apically with
spinules or hairs. First antennae ♀,
seta of segment 1 reaches to about
segment 11. Length: ♀ 1.5–2.0
mm; ♂ 1.4–1.6 mm. Alaska;
Europe; Asia.

◄ **Fig. 29.53.** *Diaptomus theeli.*
(*a*) Metasomal wings and urosome
♀, dorsal. (*b*) Lateral view right
metasomal wing with apex of first
antenna. (*c*) Leg 5 ♀. (*d*) Leg 5
♂, posterior view. (From specimens from Arctic slope of Alaska.)

47b   First antennae ♀, 2 setae on all of segments 13 to 19. Leg 5 ♂,
              right exopod 2, length of lateral spine as great as or more than
              width of segment. **48**

48a (47) Leg 5 ♀, exopod 3, inner seta about same length as outer. Leg 5
              ♂, left leg reaching to near base of right claw . . . .
                                              D. alaskaensis M. S. Wilson 1951

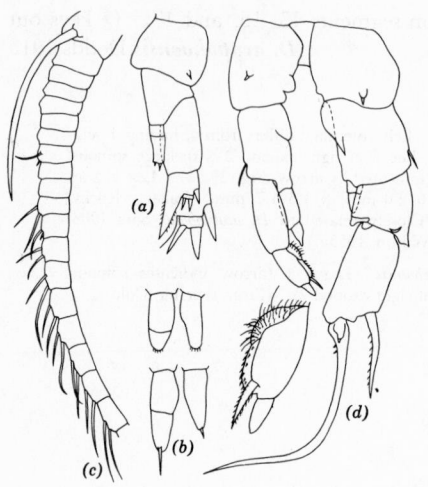

Subgenus **Nordodiaptomus**. Leg 5 ♀, endopod asymmetrical in length and armature, variable, with or without setae. First antennae ♀, seta of segment 1 reaching to segments 11 or 12; antenna ♂ left, setation differing from ♀: 2 on 11, 16, and 19, 1 on 13 to 15, 17, 18. Length: ♀ 1.65 mm; ♂ 1.44 mm. Ponds. Alaska. Closely related to *D. siberiensis* M. S. Wilson 1951 (= *D. rylovi* Smirnov 1930) from Siberia.

◄ **Fig. 29.54.** *Diaptomus alaskaensis.* (a) Leg 5 ♀ with detail exopod setae. (b) Leg 5 ♀, variation in endopod. (c) Segments 1 to 19 of first antenna ♀ showing setae of segments 1, 8, and 11 to 19. (d) Leg 5 ♂ with detail anterior aspect of left exopod 2. (After M. S. Wilson.)

48b		Leg 5 ♀, exopod 3, inner seta at least twice the length of outer. Leg 5 ♂, left leg not reaching to near base of right claw . . . . *Diaptomus* (Subgenus *Diaptomus*)	**49**

49a	(48)	Metasomal wings ♀, outer part produced posteriorly much farther than the inner. Leg 5 ♂, right exopod 2, lateral spine below middle of segment . . . . *D. glacialis* Lilljeborg 1889

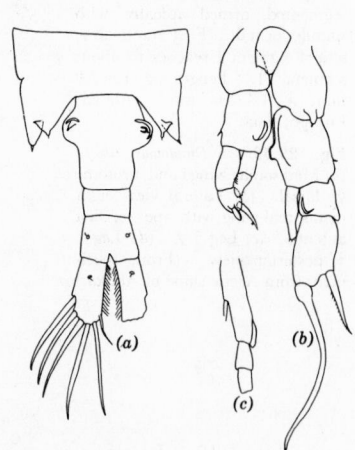

Length: ♀ 2.6–3.0 mm; ♂ 2.5–2.8 mm. Lakes ponds. Arctic slope of Alaska; Arctic Eurasia, Iceland.

◄ **Fig. 29.55.** *Diaptomus glacialis.* (a) Metasome segments 5 and 6 and urosome ♀, dorsal. (b) Leg 5 ♂, posterior view. (c) Apex right antenna ♂. (From specimens from Arctic slope of Alaska.)

49b		Metasomal wings ♀, outer part not produced beyond the inner. Leg 5 ♂, this spine above middle of segment . . . . *D. castor* (Jurine) 1820

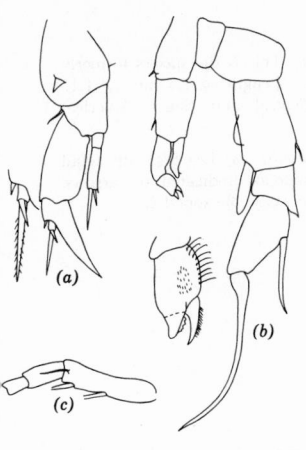

Length: ♀ 1.8–2.5 mm; ♂ 1.75–2.3 mm. Ponds. European species occurring in Greenland. (Descr. Sars, 1901–1903; Gurney, 1931.)

◀ **Fig. 29.56.** *Diaptomus castor.* (*a*) Leg 5 ♀ with detail of exopod setae. (*b*) Leg 5 ♂ with detail left exopod 2. (*c*) Apex right antenna ♂. (From European specimens.)

50a (14) Leg 5 ♀, exopod 3 distinctly separated from exopod 2 and endopod with 2 long setae (at least ½ or more of endopod length). Leg 5 ♂, left exopod 2, distal pad divided medially into lobed portions (Fig. 29.57*b*) . . . . **D. franciscanus** Lilljeborg 1889

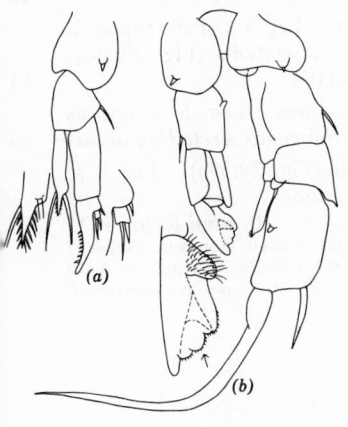

Subgenus **Hesperodiaptomus.** Leg 5 ♀, apex of endopod with distinctive triangular prolongation, spinulose on both margins (Fig. 29.57*a*). Leg 5 ♂, distal pad of left exopod 2 transparent; the lobes may be folded rather than expanded as in Fig. 29.57*b*, but careful focusing makes their outlines visible. Length: ♀ 1.2–2.0 mm; ♂ 1.1–2.0 mm. Lakes, ponds. Southeastern Alaska to Calif. Includes **D. bakeri** Marsh 1907.

◀ **Fig. 29.57.** *Diaptomus franciscanus.* (*a*) Leg 5 ♀ with detail exopod setae and apex of endopod. (*b*) Leg 5 ♂ with detail left exopod 2 (arrow indicates lobes of distal pad).

50b Leg 5 ♀, exopod 3 separated (Fig. 29.58*a*) or not (Fig. 29.66*c*); if separated, the endopod with shorter setae or with none. Leg 5 ♂ not as above **51**

51a (50) Leg 5 ♀, endopod with 2 short setae, the innermost placed laterally above the other and near the beginning of an inner lengthwise hairy groove (Fig. 29.58*a*). Leg 5 ♂, right exopod 2, lateral spine with stout marginal dentitions . . . .
D. colombiensis Thiebaud 1912

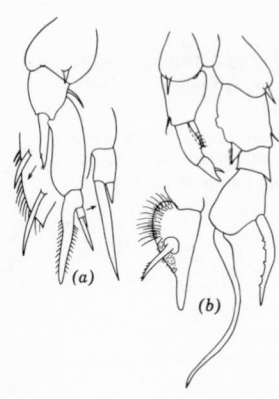

Subgenus *Prionodiaptomus.* This is the species formerly called *D. marshi* Juday 1914. Length: ♀ 1.3 mm; ♂ 1.15 mm. Lakes, reservoirs. Central and South America. (Descr. Marsh, 1913.)

◄ **Fig. 29.58.** *Diaptomus colombiensis.* (a) Leg 5 ♀ with detail of exopod 3 and apex of endopod indicated by arrows. (b) Leg 5 ♂ with detail anterior aspect of exopod 2.

51b		Leg 5 ♀, endopod setae present or not; if present, apical or sub-apical in position (Fig. 29.60a). Leg 5 ♂, lateral spine without stout dentitions. **52**
52a	**(51)**	Leg 5 ♀, exopod 3 separated (see *D. sarsi,* **54a,** for variation). Right first antenna ♂, segment 14 with spinous process (Fig. 29.59d) **53**
52b		Leg 5 ♀, exopod 3 not separated. Right first antenna ♂, segment 14 without spinous process **60**
53a	**(52)**	Leg 5 ♀, endopod with 1 or 2 short setae. Leg 5 ♂, left exopod 2, distal process stout and with crosswise corrugations (Fig. 29.59c); or the right endopod very stout (Fig. 29.60b) **54**
53b		Leg 5 ♀, endopod usually with hairs only. Leg 5 ♂ not as above . . . . *Diaptomus* (Subgenus *Arctodiaptomus*) **55**
54a	**(53)**	Metasome ♀ with dorsal spinous process (Fig. 29.59a). Leg 5 ♂, left exopod 2, distal process with corrugations . . . .

D. sarsi Rylov 1923

Subgenus **Sinodiaptomus.** Leg 5 ♀, endopod has a single short seta; exopod 3 may not always be distinctly developed. Length: ♀ 1.9 mm; ♂ 1.7 mm. An Asian species known only from lily pond in Monrovia, Calif. May have been introduced.

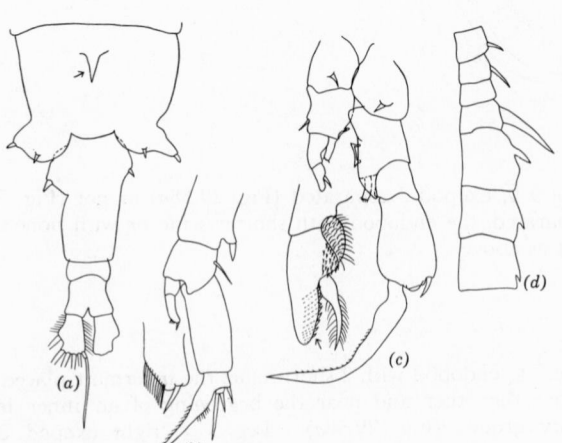

54b   Metasome ♀ without dorsal process. Leg 5 ♂, right endopod long
            and stout . . . . ***D. gracilis*** Sars 1863

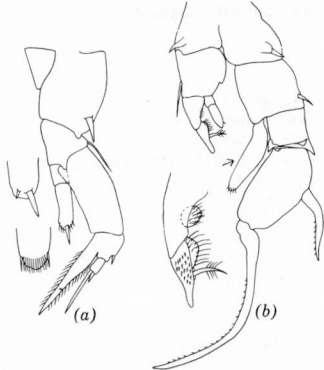

Subgenus ***Eudiaptomus.*** Leg 5 ♀, the inner
endopod seta may be reduced and covered by the
subapical fringe of hairs (Fig. 29.60*a*). Length:
♀ 1.2–1.4 mm; ♂ 1.2–1.3 mm. Lakes, ponds.
Eurasian species common in Alaska. (Fig. 29.3
shows metasomal wings and urosome ♀.) (Descr.
Sars, 1901–1903; Gurney, 1931.)

◄ **Fig. 29.60.** *Diaptomus gracilis.* (*a*) Leg 5 ♀ with
detail apex of endopod (setae and fringe of hairs).
(*b*) Leg 5 ♂ with detail left exopod (arrow indicates
stout endopod). (From Alaskan specimens.)

55a **(53)** Genital segment ♀, distal portion of right side swollen into large
            lateral lobe. Leg 5 ♂, right exopod 2, length of lateral spine less
            than that of segment . . . . ***D. asymmetricus*** Marsh 1907
            Length: ♀ 1.39 mm; ♂ 1.16 mm. Cuba.

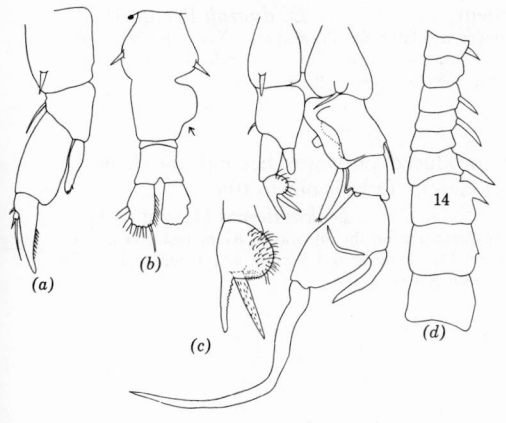

◄ **Fig. 29.61.** *Diaptomus asymmetricus.* (*a*) Leg 5 ♀. (*b*) Urosome ♀, dorsal (arrow indicates
distal lobe of genital segment).
(*c*) Leg 5 ♂ with detail left exopod
2. (*d*) Right antenna ♂ showing
spines and processes of segments
8 to 16.

55b   Genital segment ♀, right side without distal lobe. Leg 5 ♂, length
            of lateral spine about equal to or greater than that of segment . . .   **56**
56a **(55)** Leg 5 ♀, endopod as long as or nearly as long as the inner margin
            of exopod 1. Leg 5 ♂, right exopod 2, lateral spine at middle of
            segment. **57**
56b   Leg 5 ♀, endopod ½ to ¾ of the length of the inner margin of
            exopod 1. Leg 5 ♂, right exopod 2, lateral spine near end of
            segment . . . . ♀ **58**, ♂ **59**

◄───

Fig. 29.59. *Diaptomus sarsi.* (*a*) Metasome segments 5 and 6 and urosome, dorsal (arrow indicates dorsal
spinous process of metasome). (*b*) Leg 5 ♀ with detail of apex of endopod. (*c*) Leg 5 ♂ with detail left
exopod 2 (arrow indicates crosswise corrugations of distal process). (*d*) Segments 10 to 16 of right antenna
♂. (From specimens from Calif.)

57a **(56)** Metasome ♀, segment 5 usually with medial dorsal protuberance (may be single or double as in Fig. 29.62a). Leg 5 ♂, right exopod 2, lateral spine longer than segment (about 1.4:1) . . . .
D. dorsalis Marsh 1907

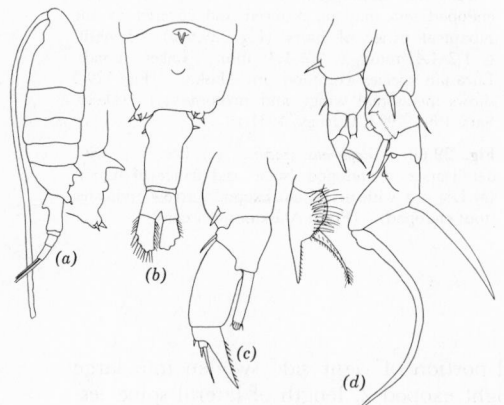

Length: ♀ 1.13 mm; ♂ 1.06 mm. Lakes, ponds. States in Gulf of Mexico region; West Indies. Includes **D. proximus** Kiefer 1936.

◄ **Fig. 29.62.** *Diaptomus dorsalis.* (a) Lateral view ♀ with detail of dorsal process of metasome. (b) Metasome segments 5 and 6 and urosome ♀, dorsal. (c) Leg 5 ♀. (d) Leg 5 ♂ with detail left exopod 2.

57b Metasome ♀ without dorsal protuberance. Leg 5 ♂, lateral spine about same length as segment. . . . . **D. dampfi** Brehm 1932
Inadequately known; no specimens available for illustration. Very closely allied to *dorsalis* of which it may be a variation. Leg 5 ♀ as in *dorsalis*. Length: ♀ 1.2 mm; ♂ 1.0 mm. Lake Peten, Guatemala. (Descr. Brehm, 1939.)

58a **(56)** Right metasomal wing ♀ produced outwards beyond the lateral margin of the body (dorsal view). (♂ keys out to **105**.) . . . .
D. kurilensis (Kiefer) 1937
Leg 5 ♀ may sometimes have a single seta on the endopod. Right first antenna ♂ lacks a spinous process on segment 14. Length: ♀ 1.5 mm; ♂ 1.4 mm. Ponds. Aleutian Islands, Alaska; Asia. (Descr. Kiefer, 1938b.)

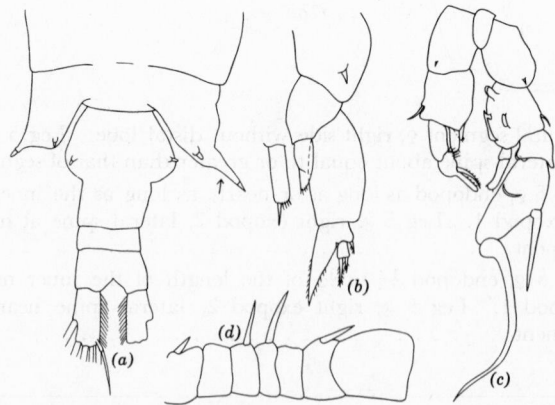

Fig. 29.63. *Diaptomus kurilensis.* (a) Metasome segments 5 and 6 and urosome ♀, dorsal (arrow indicates production of right metasomal wing). (b) Leg 5 ♀ (and endopod of other leg with seta). (c) Leg 5 ♂. (d) Right antenna ♂, segments 8 to 14. (From specimens from Aleutian Islands, Alaska.)

58b   Right metasomal wing ♀ not produced outwards beyond margin of
body **59**

59a (56, 58)   Caudal ramus ♀ usually with hairs on only inner margin. Leg
5 ♂, right basal segment 2 with small hyaline process on inner
margin . . . . **D. saltillinus** Brewer 1898

> ♀: metasome may or may not have medial dorsal process; urosome segment 2 is
> distinct; first antennae usually reach to near end of caudal rami; leg 5, inner seta of
> exopod 3 usually does not reach beyond middle of claw of exopod 2. Length: ♀ 1.3–
> 1.5 mm; ♂ 1.25 mm. Ponds. Neb. south to Tex.

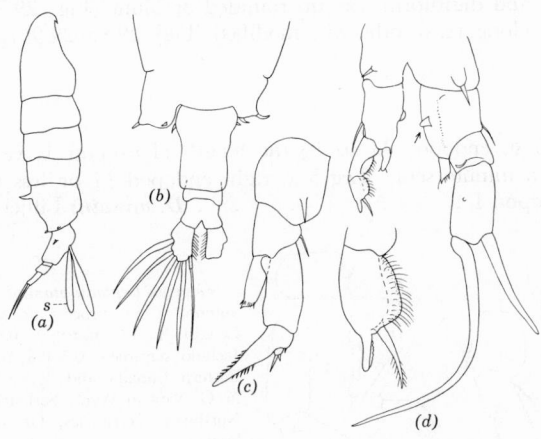

Fig. 29.64. *Diaptomus saltillinus.* (*a*) Lateral view of female with attached spermatophores (*S*). (*b*) Metasome segments 5 and 6 and urosome ♀, dorsal (specimen lacking dorsal process). (*c*) Leg 5 ♀. (*d*) Leg 5 ♂
with detail left exopod 2 (arrow indicates process on inner margin of right basal segment 2).

59b   Caudal ramus ♀ usually with hairs on both margins. Leg 5 ♂,
right basal segment 2 without inner hyaline process . . . .
D. floridanus Marsh 1926

> ♀: metasome may or may not have medial dorsal process; urosome segment 2
> distinct or entirely telescoped within margins of genital segment; first antennae
> usually reach beyond caudal rami; leg 5, inner seta of exopod 3 usually reaches
> beyond middle of claw of exopod 2. Length: ♀ 1.0–1.1 mm; ♂ 0.9 mm. Lakes,
> ponds. Fla., Ga.

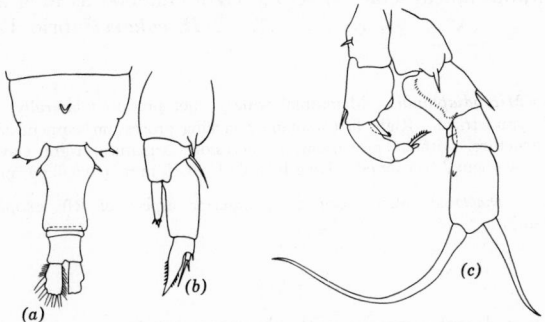

Fig. 29.65. *Diaptomus floridanus.* (*a*) Metasome segments 5 and 6 and urosome ♀, dorsal. (*b*) Leg 5 ♀.
(*c*) Leg 5 ♂.

60a **(52)** Leg 5 ♀, endopod from ¼ to about ⅔ the length of exopod 1, usually without apical setae. Leg 5 ♂, left exopod 2, inner process about as long as distal, flattened, sharply pointed (Figs. 29.66d, 29.67)... **61**

60b Leg 5 ♀, endopod reaching to near end of exopod 1 or beyond, always with 2 apical setae. Leg 5 ♂, left exopod 2, inner process short and digitiform, the tip rounded or blunt (Fig. 29.70c), setiform, elongate, or otherwise modified (Figs. 29.85–29.96)...... **62**

61a **(60)** Leg 5 ♀, endopod ¼ to ½ the length of exopod 1, very rarely with a minute seta. Leg 5 ♂, right endopod ¼ or less of length of exopod 1 .... ***D. minutus*** Lilljeborg 1889

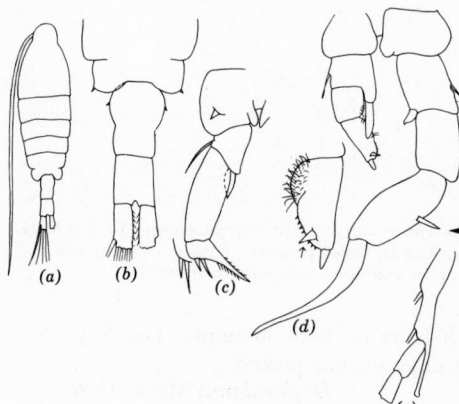

Subgenus ***Leptodiaptomus?*** Right first antenna ♂ has process on segment 23. Length: ♀ 1.0 mm; ♂ 0.9–1.0 mm. Iceland specimens 0.5–0.8 mm. Lakes. Eastern Canada and U. S., south to N. C., west to Wyo.; Saskatchewan and Northwest Territories; Greenland; Iceland.

Fig. 29.66. *Diaptomus minutus.* (a) Dorsal view ♀. (b) Metasomal wings and urosome ♀, dorsal. (c) Leg 5 ♀ with detail setae of exopod 3. (d) Leg 5 ♂ with detail anterior aspect left exopod 2 (arrow indicates flattened, sharply pointed inner process of left exopod 2). (e) Apex right antenna ♂.

61b Leg 5 ♀, endopod about ⅔ the length of exopod 1, with 1 or 2 or without apical setae. Leg 5 ♂, right endopod as long as exopod 1................... ***D. cokeri*** Osorio Tafall 1942

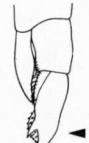

Subgenus ***Microdiaptomus.*** Metasomal wings ♀ not produced laterally; urosome 3-segmented, symmetrical. Right first antenna ♂ lacking process on segment 23. Leg 5 ♂ simple in structure, without protrusions or accessory armature; right claw nearly as long as basal segment 2 + exopod. Length: ♀ 0.615–0.75 mm; ♂ 0.6–0.65 mm. Mexico.

Fig. 29.67. *Diaptomus cokeri.* Leg 5 ♂, anterior aspect of left exopod. (After Osorio Tafall.)

62a **(60)** Leg 5 ♀, basal segment with elongate spiniform sensillum at least as long as the segment itself (Fig. 29.68c). Leg 5 ♂, right basal segment 2 swollen distally into inner lobe and the endopod arising medially............... ***D. trybomi*** Lilljeborg 1889

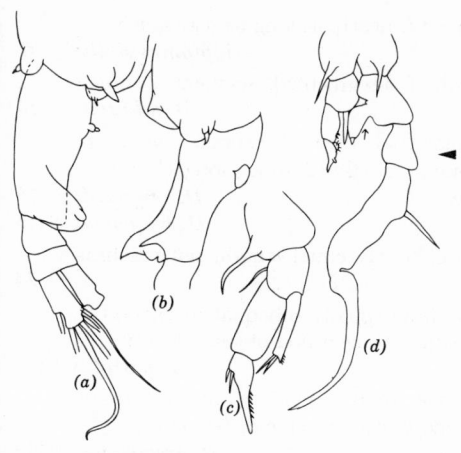

Subgenus *Leptodiaptomus?* Length: ♀
1.5 mm; ♂ 1.4 mm. Known only from
the original collection, Multnomah Falls,
Ore.

◀ **Fig. 29.68.** *Diaptomus trybomi.* (*a*) Metasomal wings and urosome ♀, dorsal. (*b*)
Metasome 6 and genital segment, lateral.
(*c*) Leg 5 ♀. (*d*) Leg 5 ♂ (arrow indicates
endopod). (*a-c* original from cotype specimens in Illinois Natural History Survey;
d after Lilljeborg.)

62b   Leg 5 ♀, sensillum not so elongate.   Leg 5 ♂, right basal segment
2 not as above . . . . ♀ **63,** ♂ **89**
Note: 63–88 refer only to females.   The figures illustrating the females are included with those of the males to which cross reference is made for each species.

63a (62) Leg 5 ♀, exopod with 3 lateral (outer marginal) setae **64**

63b   Leg 5 ♀, exopod with 2 lateral setae **68**

64a (63) Genital segment ♀ with distal process on right side. . . . .
D. novamexicanus ♂ **102**

64b   Genital segment ♀ without distal process on right side **65**

65a (64) Leg 5 ♀, inner apex of endopod produced into sharp point . . . . .
Acanthodiaptomus denticornis ♂ **89**

65b   Leg 5 ♀, inner apex of endopod rounded **66**

66a (65) Genital segment ♀, sensilla borne at ends of lateral protrusions
that are shorter than the sensilla or scarcely developed . . . .
D. pribilofensis ♂ **93**

66b   Genital segment ♀, sensilla borne on protrusions as long as or
longer than the sensilla **67**

67a (66) Right metasomal wing ♀, outer margin usually rounded and expanded laterally, its length less than $1\frac{1}{2}$ times its width. . . . . .
D. tyrrelli ♂ **92**

67b   Right metasomal wing ♀, outer margin hardly or not at all
rounded, not expanded laterally, projecting backwards, its length
more than $1\frac{1}{2}$ times its width. . . . . *D. coloradensis* ♂ **93**

68a (63) Caudal ramus ♀ with hairs on inner margin only **69**

68b   Caudal ramus ♀ with hairs on both margins **85**

69a (68) Genital segment ♀, proximal lateral areas only slightly expanded
and rounded, without lateral protrusions or well defined lobes . . .   **70**

69b   Genital segment ♀, proximal lateral areas with protrusions or
lobes **75**

70a (69) Genital segment ♀ with distal process on right side. . . . .
D. spinicornis ♂ **96**

70b   Genital segment ♀ without such a process **71**

71a (70) Leg 5 ♀, inner apex of endopod rounded **72**

71b   Leg 5 ♀, inner apex of endopod produced into sharp point **73**

72a   (71) Urosome ♀ 3-segmented, segment 2 nearly as long as segment 3 . .
                                                                    Diaptomus sicilis ♂ 97
72b   Urosome ♀ 2- or 3-segmented; if 3-segmented, segment 2 much
             shorter than segment 3 . . . . *D. ashlandi* ♂ 98
73a   (71) Leg 5 ♀, of the 2 lateral marginal setae of exopod, the inner
             noticeably longer than the outer (usually 2 times longer), reach-
             ing beyond the middle of claw. . . . . *D. reighardi* ♂ 115
                                                                    D. pygmaeus ♂ 113
73b   Leg 5 ♀, these 2 lateral setae of nearly equal length, not reaching
             beyond middle of claw.   74
74a   (73) Leg 5 ♀, exopod 2 (to tip of claw) usually subequal to exopod 1,
             and first antennae usually reaching to near end of caudal rami . . .
                                                                    D. pallidus ♂ 112
74b   Leg 5 ♀, exopod 2 usually longer than exopod 1, and first anten-
             nae usually reaching to near end of caudal setae or beyond . . . . .
                                                                    D. oregonensis ♂ 113
75a   (69) Genital segment ♀ without distal process on right side.   76
75b   Genital segment ♀ with distal process on right side   81
76a   (75) Urosome ♀, segment 2 completely separated or not, if so, not
             more than ¼ of length of segment 3   77
76b   Urosome ♀, segment 2 completely separated, more than ¼ of
             length of segment 3   79
77a   (76) Genital segment ♀, protrusion of left side a rounded lobe . . . . . .
                                                                    D. insularis ♂ 99
77b   Genital segment ♀, protrusion of left side, if present, pointed   78
78a   (77) Metasome ♀, greatest width at about the middle; protrusions of
             genital segment at proximal third of segment . . . . . *D. siciloides* ♂ 101
78b   Metasome ♀, greatest width in cephalic segment; protrusions of
             genital segment near middle of segment . . . . *D. connexus* ♂ 101
79a   (76) Genital segment ♀, left side not expanded, right side with short
             protrusion. . . . . *D. mississippiensis* ♂ 114
79b   Genital segment ♀, both sides expanded or with protrusions.   80
80a   (79) Genital segment ♀ with nearly symmetrical lobes on each side . . .
                                                                    D. sinuatus ♂ 116
80b   Genital segment ♀ asymmetrical, the left side with medial swell-
             ing, the right with outwardly directed protrusion that reaches be-
             yond edge of metasomal wing. . . . . *D. bogalusensis* ♂ 116
81a   (75) Left metasomal wing ♀ with well-developed inner lobe extending
             posteriorly beyond the outer portion of the wing.   82
81b   Left metasomal wing ♀ without such a well-developed inner lobe. .   83
82a   (81) Left metasomal wing ♀ conspicuously larger than the right . . . . .
                                                                    D. novamexicanus ♂ 102
82b   Left metasomal wing ♀ not conspicuously larger than the right . . .
                                                                    D. nudus ♂ 103
83a   (81) Right metasomal wing ♀, outer portion not posteriorly produced . .
                                                                    D. judayi ♂ 99
83b   Right metasomal wing ♀, outer portion produced posteriorly   84
84a   (83) Genital segment ♀, distal process usually large (reaching to near
             middle of last segment or beyond) . . . . *D. signicauda* ♂ 104
84b   Genital segment ♀, distal process small, not reaching middle of
             last segment . . . . *D. moorei* ♂ 104

85a (68) Genital segment ♀ (lateral view) with ventral lobed process distad to genital protuberance . . . . *D. birgei* ♂ **106**

85b Genital segment ♀ without ventral lobed process. **86**

86a (85) Metasomal wings ♀ with greatly elongate outer sensilla (length of that on left side ⅓ or more of margin of segment) . *D. sanguineus* ♂ **109**

86b Metasomal wings ♀, sensilla not so elongate **87**

87a (86) Leg 5 ♀, length of endopod setae more than ½ that of endopod . .
D. hesperus ♂ **107**

87b Leg 5 ♀, length of endopod setae not more than ½ that of endopod **88**

88a (87) Metasome segment 5 ♀, right side with posteriorly directed lobed protrusion. . . . . *D. virginiensis* ♂ **110**

88b Metasome segment 5 ♀ protruberant or not on right side; if so, the protrusion directed laterally . . . . *D. louisianensis* ♂ **110**

89a (62) Right first antenna ♂, apex of last segment with outwardly produced process (Fig. 29.69*d*) . . . .
Acanthodiaptomus denticornis (Wierzejski) 1888

♀ keys out to **65a**. Leg 5 ♂, left exopod 2, proximal process directed outwardly instead of mesially as in the genus *Diaptomus*. Length: ♀ 1.7–1.9 mm; ♂ 1.5 mm. Lakes, ponds. Eurasian species found in Alaska, western Canada, south in Rocky Mountains to Wyo. (Descr. Sars, 1901–1903.)

◄ **Fig. 29.69.** *Acanthodiaptomus denticornis.* (*a*) Metasomal wings and urosome ♀, dorsal, with apex of first antenna. (*b*) Leg 5 ♀. (*c*) Leg 5 ♂ with detail left exopod 2. (*d*) Apex right antenna ♂ (arrow indicates process of last segment). (From Alaskan specimens.)

89b Right first antenna ♂, apex of last segment without process **90**

90a (89) Leg 5 ♂, left exopod 2, processes similar to one another, digitiform, with rounded or blunt tips, the outer (distal) distinctly separated from segment, both pads medial in position with slight constriction between them; lateral spine of right exopod 2 not inserted on same plane as that of segment, directed backwards (in mounted specimens it may appear in different positions as shown in figures) (Fig. 29.75*c*). . *Diaptomus* (Subgenus *Leptodiaptomus*) **91**

90b Leg 5 ♂, processes and pads of left exopod 2 and the lateral spine of right exopod 2 not as above **105**

91a (90) Right first antenna ♂, segment 23 without process (membrane may be present as in Fig. 29.70*d*). **92**

91b Right first antenna ♂, segment 23 with process (Figs. 29.73*b*, 29.82*c*) **94**

92a (91) Leg 5 ♂, left exopod 2, distal process longer than the inner, about ½ the length of the outer margin of the segment . . . .
D. tyrrelli Poppe 1888

♀ keys out to **67a**. Lateral protrusions ♀ genital segment may be directed laterally or ventrally. Right antenna ♂, segment 23 has small variable membrane; spine of segment 8 nearly as long as that on 10; segments 15 and 16 with processes. Length: ♀ 1.2–1.9 mm; ♂ 1.1–1.8 mm. Ponds, lakes. Common in mountains, rarer at low elevations. Alaska, east to Labrador; Rocky Mountains, west to Pacific coast; Asia. **D. lighti** M. S. Wilson 1941 may be synonym.

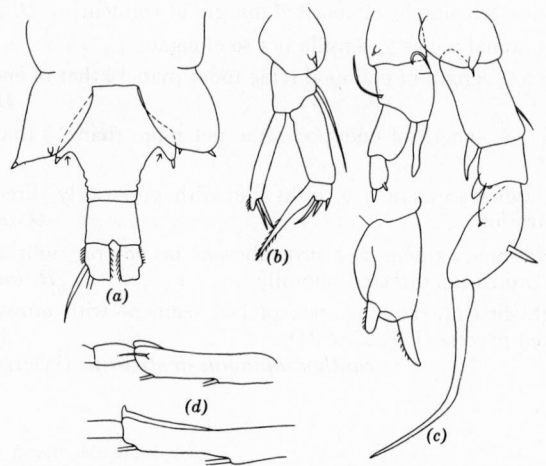

Fig. 29.70. *Diaptomus tyrrelli.* (*a*) Metasomal wings and urosome ♀, dorsal (arrows indicate lateral protrusions of genital segment). (*b*) Leg·5 ♀. (*c*) Leg 5 ♂ with outline anterior aspect left exopod 2. (*d*) Right antenna ♂, showing variation in membrane of segment 23.

92b   Leg 5 ♂, left exopod 2, processes subequal, the distal not more
than ¼ the length of outer margin of segment **93**

93a **(92)** Leg 5 ♂, right exopod 1 with 2 prominent hyaline processes, on
inner margin and on posterior face . . **D. coloradensis** Marsh 1911

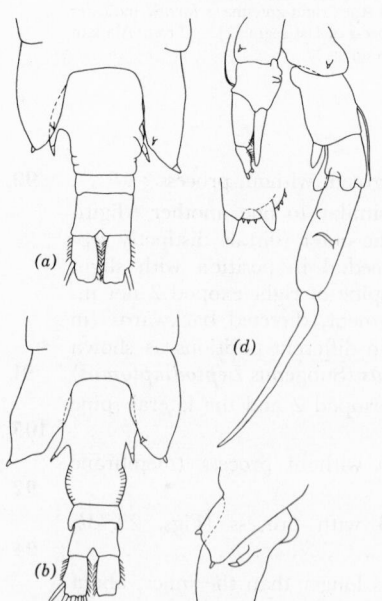

♀ keys out to **67b.** Right antenna ♂, segment 23 without membrane; spines and processes of segments 8 to 16 similar to those of *D. tyrrelli.* Length: ♀ 1.2–1.4 mm; ♂ 1.1–1.3 mm. Ponds, lakes. High altitudes in Rocky Mountains, Colo., Utah.

◄ **Fig. 29.71.** *Diaptomus coloradensis.* (*a*) Metasomal wings and urosome ♀, dorsal (from specimen with distal part of genital segment contracted). (*b*) Metasomal wings and urosome ♀, dorsal (from specimen with fully expanded genital segment). (*c*) Right metasomal wing ♀ and proximal part of genital segment, lateral. (*d*) Leg 5 ♂ with detail processes left exopod 2, posterior.

93b   Leg 5 ♂, right exopod 1 with inconspicuous hyaline processes, that
              of distal inner margin minute or not developed, that of posterior
              face very small, crescent-shaped (Fig. 29.72*b*) . . . .
                          D. pribilofensis Juday and Muttkowski 1915

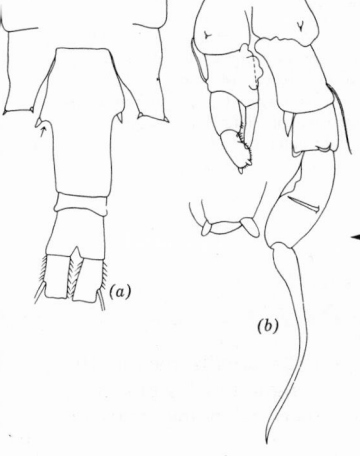

♀ keys out to **66a.** Right antenna ♂ as in *D. coloradensis.* Leg 5 ♂, inner margin of left basal segment
2 has a distinctive 3-lobed hyaline membranous expansion; this tends to collapse or become distorted
in specimens that are not well preserved or from
cover-glass pressure. Length: ♀ 1.1–1.8 mm; ♂ 1.0–
1.6 mm. Ponds, lakes. Common in Alaska and
northwestern Canada; Wis. Closely allied to the
Asian species *D. angustilobus* Sars 1898.

◄ **Fig. 29.72.** *Diaptomus pribilofensis.* (*a*) Metasomal
wings and urosome ♀, dorsal (arrow indicates lateral
protrusion of genital segment). (*b*) Leg 5 ♂ with
detail processes left exopod 2, anterior.

94a **(91)** Right first antenna ♂, segment 15 with prominent spinous process
              (Fig. 29.73*a*) . . . . ***D. mexicanus*** Marsh 1929

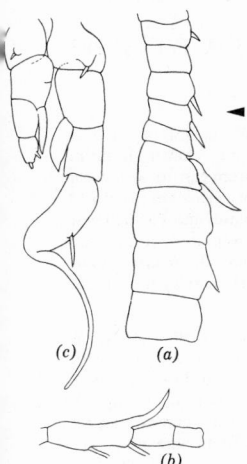

♀ unknown. ♂ known from single specimen taken near Mexico
City, Mexico. Length: 1.2 mm.

◄ **Fig. 29.73.** *Diaptomus mexicanus.* (*a*) Right antenna ♂ showing
spines and processes of segments 8 to 16. (*b*) Apex right antenna ♂.
(*c*) Leg 5 ♂. (From type specimen.)

94b   Right first antenna ♂, segment 15 usually without process; if
              present, it is not prominent **95**

95a **(94)** Right first antenna ♂, process of segment 23 reaching to middle of
              segment 24 or beyond, the tip swollen (Fig. 29.75*d*) or pointed
              (Fig. 29.74*d*), not outcurved or hooklike **96**

95b   Right first antenna ♂, process of segment 23 usually not reaching
              beyond middle of segment 24, usually shorter, the tip outcurved
              or hooklike (Figs. 29.79*f*, 29.82*c*) **100**

96a **(95)** Right first antenna ♂, process of segment 23 usually tapering to
              needlelike point (Fig. 29.74*d*); leg 5 ♂, right basal segment 1 with

greatly enlarged inner protrusion extending posteriorly between the right and left legs (Fig. 29.74c) . . . **D. spinicornis** Light 1938

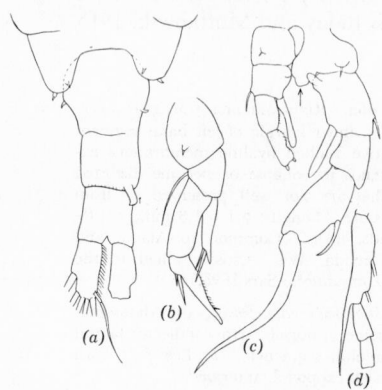

♀ keys out to **70a.** Length: ♀ 1.0–1.3 mm; ♂ 1.0–1.1 mm. Lakes, ponds. Western U. S., Nev., Calif., Wash.

◄ **Fig. 29.74.** *Diaptomus spinicornis.* (a) Metasomal wings and urosome ♀, dorsal (arrow indicates distal process of genital segment). (b) Leg 5 ♀. (c) Leg 5 ♂ (arrow indicates inner protrusion of right basal segment 1). (d) Apex right antenna ♂.

96b   Right first antenna ♂, process of segment 23 usually having the apex swollen, blunt or rounded; leg 5 ♂, right basal segment 1, inner protrusion present or not, but not extending between right and left legs **97**

97a (96) Leg 5 ♂, right basal segment 2, proximal inner portion without protrusion or membranous process; right exopod 1 with large rounded hyaline process on inner posterior face . . . .
                                        D. sicilis S. A. Forbes 1882

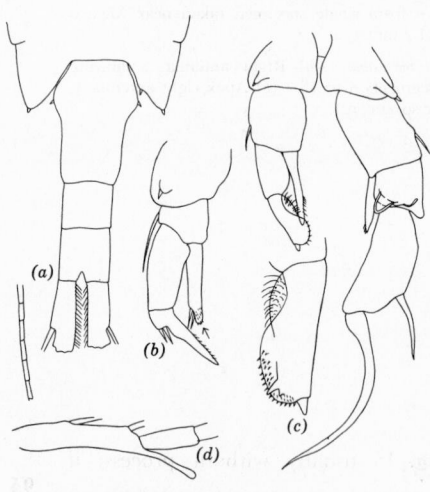

♀ keys out to **72a.** Length: ♀ 1.19–1.9 mm; ♂ 1.1–1.5 mm. Lakes, fresh to saline. Common from east to west coast in Canada and northern states; south to Mo.; Rocky Mountains and Pacific coast states; Alaska. Reported from San Salvador. Includes **D. tenuicaudatus** Marsh 1907 and **D. natriophilus** Light 1938.

◄ **Fig. 29.75.** *Diaptomus sicilis.* (a) Metasomal wings and urosome ♀, dorsal, with apex of first antenna. (b) Leg 5 ♀ (arrow indicates rounded apex of endopod). (c) Leg 5 ♂ with detail anterior aspect left exopod 2. (d) Right antenna ♂, segments 23 and 24. (From Lake Erie specimens.)

97b   Leg 5 ♂, right basal segment 2, proximal inner portion with protrusion or process; right exopod 1, if process present, not as above. **98**

98a (97) Leg 5 ♂, right exopod 2, proximal part of segment widened and forming distinct angle at point of insertion of lateral spine; spine stout, its length 1½ to 2½ times the greatest width of the segment. . . . . **D. ashlandi** Marsh 1893

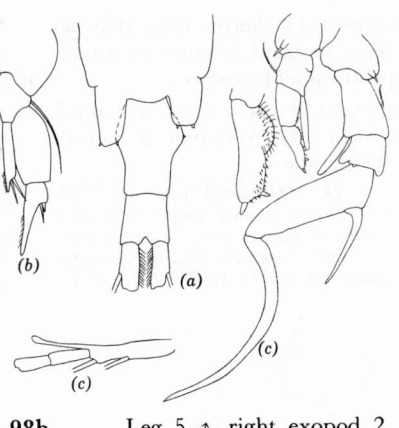

♀ keys out to **72b**. Length: ♀ 0.93–1.4 mm; ♂ 0.9–1.2 mm. Lakes. Common from east to west coast in Canada and northern states.

◄ **Fig. 29.76.** *Diaptomus ashlandi.* (*a*) Metasome segments 5 and 6 and urosome ♀, dorsal. (*b*) Leg 5 ♀. (*c*) Leg 5 ♂ with detail left exopod 2. (*d*) Apex right antenna ♂.

98b   Leg 5 ♂, right exopod 2, proximal portion of segment more or less widened but without distinct angle; spine subequal to or less than greatest width of segment **99**

99a   (98)   Leg 5 ♂, right, distal inner margin exopod 1 with small rounded hyaline protrusion or process and proximal part of exopod 2 conspicuously broadened. (Fig. 29.77). . . . .
                                    D. insularis (Kincaid) 1956
            ♀ keys out to **77a**. Length: ♀ 1.2–1.75 mm; ♂ 1.0–1.5 mm. Alaska.

99b   Leg 5 ♂, right, distal inner margin exopod 1 with elongate, pointed hyaline process and proximal part of exopod 2 not broadened. (Fig. 29.78). . . . . **D. judayi** Marsh 1907
            ♀ keys out to **83a**. Urosome ♀ may be 2- or 3-segmented, or segments 2 and 3 may be indistinctly divided; the distal process of genital segment is very small. Length: ♀ 0.93 mm; ♂ 0.9 mm. Lakes. Rocky Mountains.

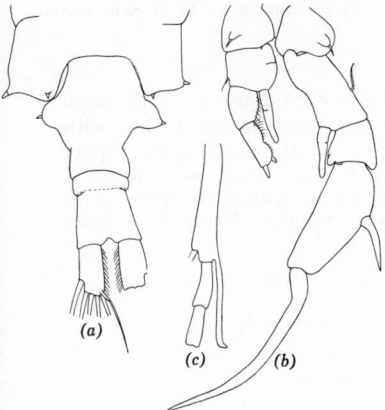

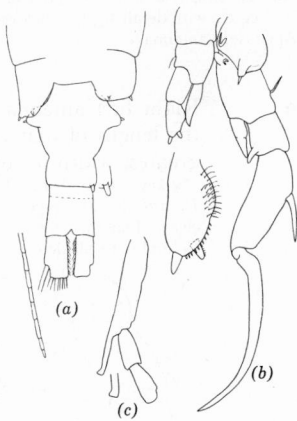

Fig. 29.77. *Diaptomus insularis.* (*a*) Metasomal wings and urosome ♀, dorsal. (*b*) Leg 5 ♂. (*c*) Apex right antenna ♂.

Fig. 29.78. *Diaptomus judayi.* (*a*) Metasomal wings and urosome ♀, dorsal, with apex of first antenna and detail of process of genital segment. (*b*) Leg 5 ♂ with detail left exopod 2. (*c*) Apex right antenna ♂ with detail apex of process of segment 23.

100a   (95)   Right first antenna ♂, spine of segment 11 subequal to or longer than that of 13 (Fig. 29.79*e*). Leg 5 ♂, right basal segment 2, inner proximal portion without upwardly projecting protrusion or process **101**

100b Right first antenna ♂, spine of segment 11 shorter than that of
13 (Fig. 29.84c). Leg 5 ♂, right basal segment 2, inner proximal
portion with upwardly projecting protrusion or process **102**

101a (100) Right first antenna ♂, spine of segment 8 usually not enlarged
(about same length as that on segment 12); metasome in dorsal
view with greatest width at about the middle . . . .
                                       D. siciloides Lilljeborg 1889
♀ keys out to **78a.** Metasome width as in ♂. Leg 5 ♂, right exopod 1, rectangular or squarish hyaline process on distal portion of inner margin. Length: ♀ 1.0–
1.3 mm; ♂ 1.0–1.1 mm. Lakes, ponds. Known from most of the continent except the
extreme north and the east coast. ***D. cuauhtemoci*** Osorio Tafall 1941 may be
synonym.

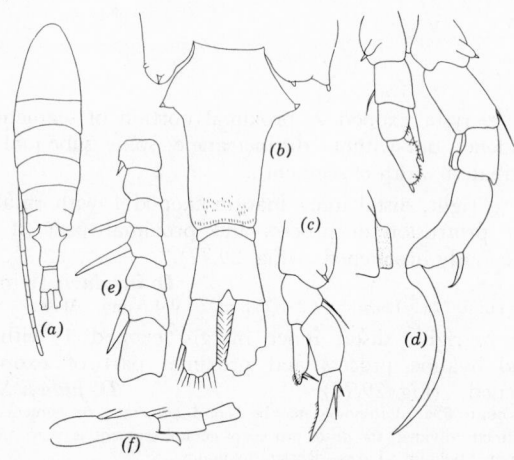

Fig. 29.79. *Diaptomus siciloides.* (a) Dorsal view ♀. (b) Metasomal wings and urosome ♀, dorsal. (c) Leg 5
♀. (d) Leg 5 ♂ with detail hyaline process of right exopod 1. (e) Segments 8 to 14 of right antenna ♂.
(f) Apex right antenna ♂.

101b Right first antenna ♂, spine of segment 8 enlarged (at least twice
the length of that on segment 12); metasome in dorsal view with
greatest width in cephalic segment . . . . ***D. connexus*** Light 1938
♀ keys out to **78b.** Metasome width as in ♂. Leg 5 ♂ very similar to that of
D. siciloides; the type of taxonomic relationship between these two species is not
clear. Length: ♀ 0.9–1.5 mm; ♂ 0.9–1.5 mm. Ponds, lakes. Western U. S., British
Columbia to Mexico; southwestern U. S., east to N. M.

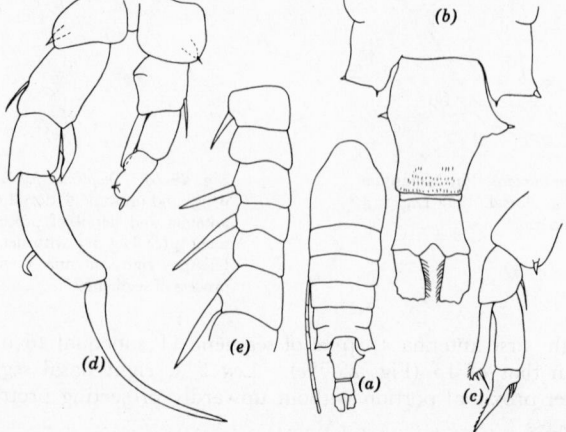

102a (100) Right first antenna ♂, segment 23, base of process starting at middle of segment (Fig. 29.81*c*) . **D. novamexicanus** Herrick 1895
  ♀ keys out to **64a** and **82a**. Leg 5 ♀ may occasionally have 3 instead of the usual 2 setae on the outer lateral margin of the exopod, on one or both legs of a pair. Length: ♀ 1.0–2.0 mm; ♂ 0.98–1.7 mm. Lakes. Rocky Mountains, Utah south into Mexico; Pacific coast region, British Columbia to Calif. Includes **D. washingtonensis** Marsh 1907 and **D. garciai** Osorio Tafall 1942.

102b Right first antenna ♂, segment 23, process entirely apical **103**

103a (102) Leg 5 ♂, right exopod 2, lateral spine placed above middle of segment. (Fig. 29.82) . . . . **D. nudus** Marsh 1904
  ♀ keys out to **82b**. Length: ♀ 1.1–1.3 mm; ♂ 1.1 mm. Lakes, ponds. Rocky Mountain states; Alaska east to Manitoba and Hudson Bay region.

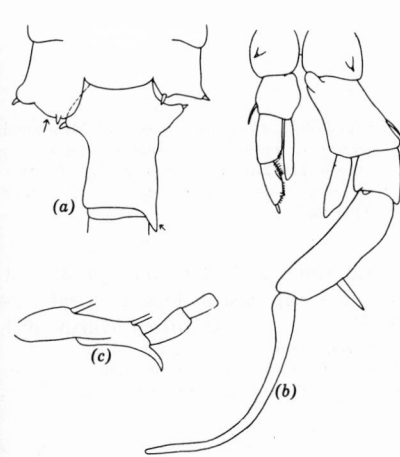

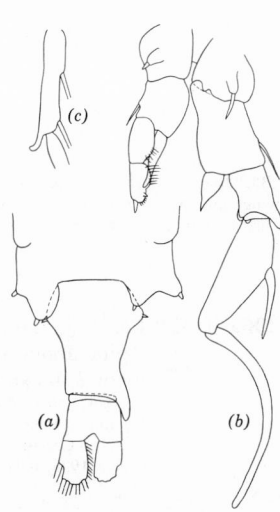

Fig. 29.81. *Diaptomus novamexicanus.* (*a*) Metasomal wings and genital segment ♀, dorsal (arrows indicate enlarged lobe of left metasomal wing and distal process of genital segment). (*b*) Leg 5 ♂. (*c*) Apex right antenna ♂.

Fig. 29.82. *Diaptomus nudus.* (*a*) Metasomal wings and urosome ♀, dorsal. (*b*) Leg 5 ♂. (*c*) Apex right antenna ♂.

103b Leg 5 ♂, right exopod 2, lateral spine placed below middle of segment. **104**

104a (103) Leg 5 ♂, right exopod 1 with rounded, distally directed hyaline process on inner margin. (Fig. 29.83) . . . .
                          D. signicauda Lilljeborg 1889
  ♀ keys out to **84a**. Length: ♀ 1.1–1.5 mm; ♂ 1.0–1.3 mm. Lakes. Rocky Mountains; very common in mountains of Pacific coast region; reported from Ia.

104b Leg 5 ♂, right exopod 1 with subrectangular, mesially directed process. (Fig. 29.84) . . . . **D. moorei** M. S. Wilson 1954
  ♀ keys out to both **84b**. Leg 5 ♂, inner portions of both right and left basal segment 2 protuberant. Length: ♀ 1.27–1.32 mm; ♂ 1.15 mm. Ponds. Fla., La., Tex.

105a (90) Leg 5 ♂, right basal segment 2 with 3 mesially directed processes on posterior face. (See **58a** for ♀ and figures.). . . . .
                          D. kurilensis (Kiefer) 1937

105b Leg 5 ♂, right basal segment 2, never more than 2 processes, if present **106**

Fig. 29.80. *Diaptomus connexus.* (*a*) Dorsal view ♀. (*b*) Metasomal wings and urosome ♀, dorsal. (*c*) Leg 5 ♀. (*d*) Leg 5 ♂, anterior view. (*e*) Segments 8 to 13 of right antenna ♂.

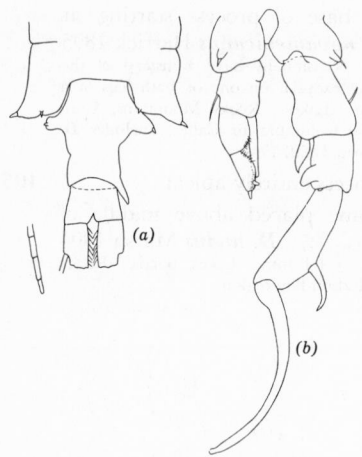

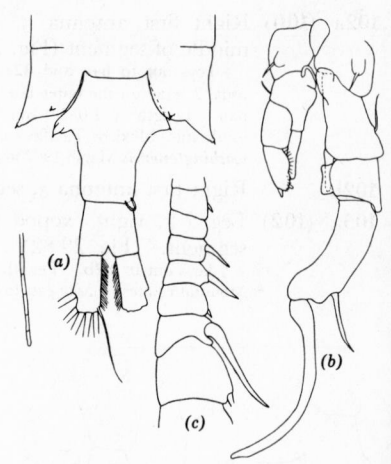

Fig. 29.83. *Diaptomus signicauda.* (*a*) Meta-somal wings and urosome ♀, dorsal, with apex of first antenna. (*b*) Leg 5 ♂.

Fig. 29.84. *Diaptomus moorei.* (*a*) Metasomal segments 5 and 6 and urosome ♀, dorsal, with apex of first antenna. (*b*) Leg 5 ♂. (*c*) Segments 8 to 14 of right antenna ♂. (After M. S. Wilson.)

106a (105) Leg 5 ♂, right exopod 2, lateral spine and claw very prominent (spine about as long as segment, claw at least as long as basal segment 2 + exopod). (Fig. 29.85) . . . . . . **D. birgei** Marsh 1894
♀ keys out to **85a.** Subgenus *Onychodiaptomus.* Length: ♀ 1.3–1.6 mm; ♂ 1.1–1.3 mm. Lakes, ponds. Known from scattered localities in northern states and southern Canada; on East Coast, north to New Brunswick, south to Ga., La. (Descr. Coker, 1926; Kiefer, 1931.)

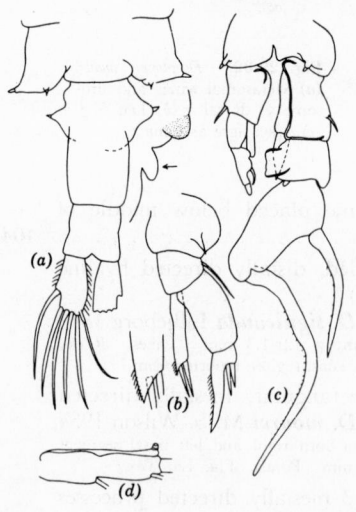

◄ Fig. 29.85. *Diaptomus birgei.* (*a*) Metasomal wings and urosome ♀, dorsal, with lateral profile of genital protuberance and lobed process (indicated by arrow). (*b*) Leg 5 ♀ with detail of endopod apex. (*c*) Leg 5 ♂. (*d*) Segment 23 of right antenna ♂.

106b   Leg 5 ♂, spine and claw not so prominent 107

107a (106) Leg 5 ♂, left exopod 2, both pads prominent and protruding mesially (Fig. 29.86*c*) . . **D. hesperus** M. S. Wilson and Light 1951
♀ keys out to **87a.** Subgenus *Onychodiaptomus.* Urosome ♀ 2-segmented. Length: ♀ 1.2–1.5 mm; ♂ 1.06–1.1 mm. Lakes, ponds. British Columbia to Ore. Includes **D. pugetensis** Carl, 1940; Kincaid, 1953 (*nomen nudum*).

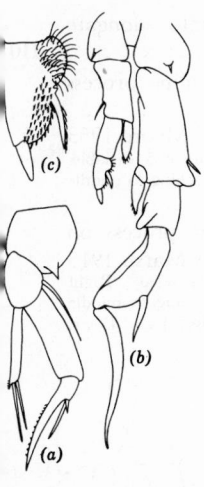

◄ **Fig. 29.86.** *Diaptomus hesperus.* (*a*) Leg 5 ♀. (*b*) Leg 5 ♂. (*c*) Leg 5 ♂, left exopod 2.

107b   Leg 5 ♂, left exopod 2, proximal pad protruding mesially; the distal pad, if present, confined largely to posterior face (Fig. 28.87*c*) **108**

108a **(107)**   Leg 5 ♂, right, claw shorter than exopod . . . .
                           Diaptomus (Subgenus ***Onychodiaptomus*** continued)   **109**

108b   Leg 5 ♂, right, claw as long as or longer than exopod . . . .
                             Diaptomus (Subgenus ***Skistodiaptomus***)   **111**

109a **(108)**   Leg 5 ♂, outer distal part of right basal segment 2 produced into elongate spiniform process, usually backwardly directed. . . . . . .
                                D. sanguineus S. A. Forbes 1876

      ♀ keys out to **86a**. ♀ may show in lateral view a dorsal hump in distal part of metasome; this character variable within a sample; spines of metasomal wings variable. Leg 5 ♂, process of right basal segment 2 variable in length. Right antenna ♂, spine of segment 8 as large as that on 10. Length: ♀ 1.42–2.1 mm; ♂ 1.0–2.1 mm. Ponds. East to west coasts in Canada and northern states, south to Va., Mississippi Valley south to Gulf of Mexico states. (Descr. Schacht, 1897; Humes and M. S. Wilson, 1951.)

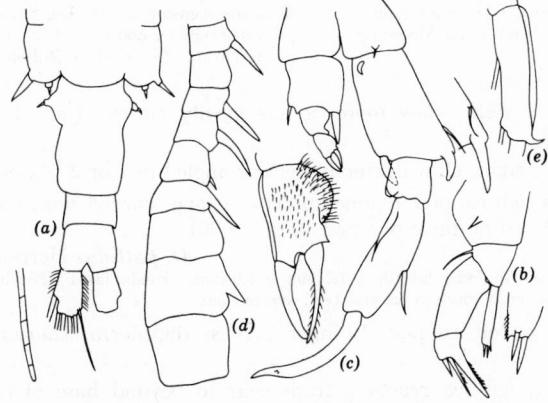

Fig. 29.87. *Diaptomus sanguineus.* (*a*) Metasomal wings and urosome ♀, dorsal, with apex of first antenna. (*b*) Leg 5 ♀ with detail endopod apex. (*c*) Leg 5 ♂ with detail left exopod 2, and variation in process of right basal segment 2. (*d*) Right antenna ♂, segments 8 to 16. (*e*) Right antenna ♂, segment 23.

109b   Leg 5 ♂, right basal segment 2 not produced into elongate
process **110**

110a **(109)** Leg 5 ♂, right exopod 1 with narrow rectangular hyaline process
on inner margin. (Fig. 29.88) . . . .
D. louisianensis M. S. Wilson and Moore 1953
♀ keys out to **88b**. Extent of lateral protrusion of metasome segment 5 variable.
Right first antenna ♂, spine of segment 8 not enlarged, that of 13 reaching to middle
of 14. Length: ♀ 1.85 mm; ♂ 1.33 mm. Ponds. La., Fla.

110b   Leg 5 ♂, right exopod 1 with prominent triangular process on
inner margin. (Fig. 29.89) . . . . . . **D. virginiensis** Marsh 1915
♀ keys out to **88a**. Leg 5 ♂, right basal segment 2 greatly expanded. Right
antenna ♂, spines of segments 8, 10, and 11 reduced, that of 13 reaching to middle
of 15. Length: ♀ 1.36–1.6 mm; ♂ 1.24–1.3 mm. Ponds. Va., Miss., La. (Descr.
M. S. Wilson and Moore, 1953a).

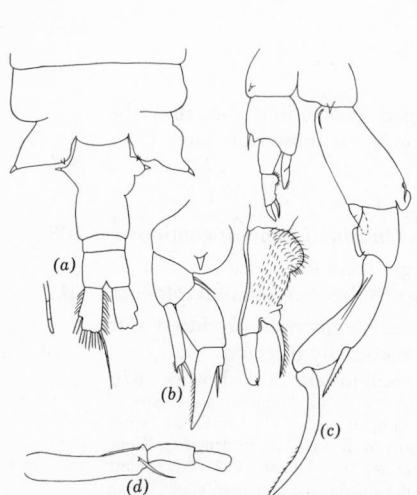

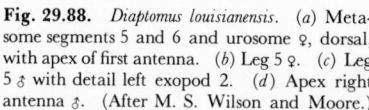

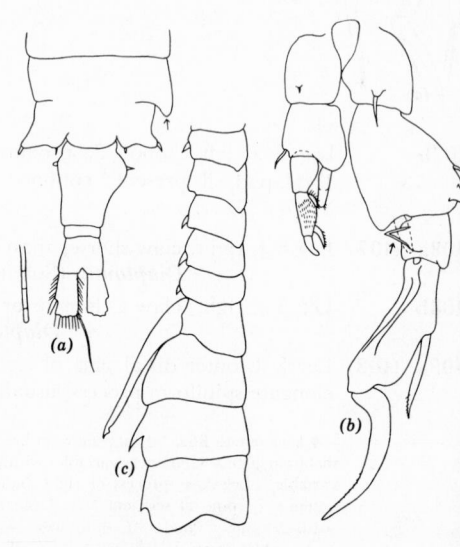

Fig. 29.88. *Diaptomus louisianensis.* (*a*) Meta-some segments 5 and 6 and urosome ♀, dorsal, with apex of first antenna. (*b*) Leg 5 ♀. (*c*) Leg 5 ♂ with detail left exopod 2. (*d*) Apex right antenna ♂. (After M. S. Wilson and Moore.)

Fig. 29.89. *Diaptomus virginiensis.* (*a*) Metasome segments 5 and 6 and urosome ♀, dorsal, with apex of first antenna (arrow indicates protrusion of right metasome segment 5). (*b*) Leg 5 ♂ (arrow indicates process of right exopod 1). (*c*) Right antenna ♂, segments 8 to 16. (After M. S. Wilson and Moore.)

111a **(108)** Leg 5 ♂, right, claw more or less evenly curved (not distinctly
angled) **112**

111b   Leg 5 ♂, right, claw distinctly bent or angled in 1 or 2 places. . . .   **114**

112a **(111)** Leg 5 ♂, left exopod 2, inner process a long, curved seta, reaching
beyond end of distal process. (Fig. 29.90) . . . .
D. pallidus Herrick 1879
♀ keys out to **74a**. Length: ♀ 1.2 mm; ♂ 1.0 mm. Ponds, lakes. North central
and plains states, south to La. and Tex., west to Colo.

112b   Leg 5 ♂, left exopod 2, inner process digitiform, shorter than
distal . . ·. **113**

113a **(112)** Leg 5 ♂, left leg reaching from near to beyond base of claw of
right leg; left exopod, length of segment 2 (measured to base of
distal process) from about ½ to ¾ that of segment 1. (Fig.
29.91) . . . . **D. oregonensis** Lilljeborg 1889
♀ keys out to **74b**. Length: ♀ 1.25–1.5 mm; ♂ 1.25–1.4 mm. Lakes, occasionally in ponds. Common Great Lakes area U. S. and Canada, east and west to

coasts; north to Northwest Territories, south in Rocky Mountains to Colo.; some
published records from northeastern coastal states are *D. pygmaeus*, long considered
a synonym.

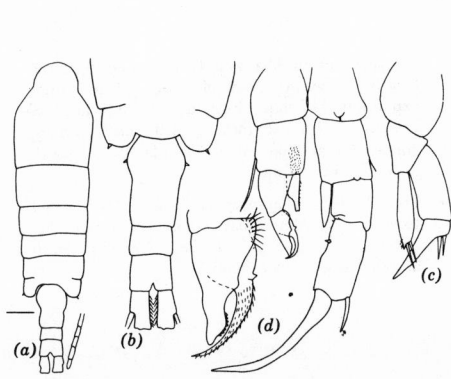

Fig. 29.90. *Diaptomus pallidus.* (*a*) Dorsal view ♀
with apex of first antenna. (*b*) Metasomal wings and
urosome ♀, dorsal. (*c*) Leg 5 ♀. (*d*) Leg 5 ♂ with
detail left exopod 2.

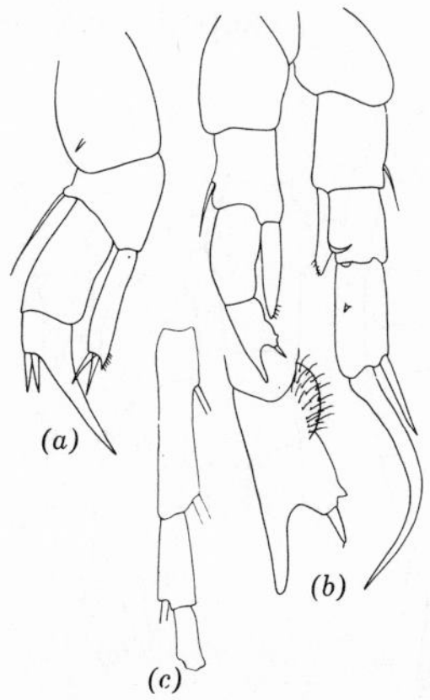

Fig. 29.91. *Diaptomus oregonensis.*
(*a*) Leg 5 ♀. (*b*) Leg 5 ♂ with detail
left exopod 2. (*c*) Apex right antenna ♂. (From Wis. specimens.)

113b   Leg 5 ♂, left leg not reaching beyond middle of right exopod 2;
left exopod, length of segment 2 about ⅓ that of segment 1.
(Fig. 29.92) . . . . **D. pygmaeus** Pearse 1906
         ♀ keys out to **73a;** variation studies are necessary to establish reliable characters
to separate the female from *D. reighardi*. The claw of the right leg 5 ♂ is somewhat
irregular in the proximal part, but without a distinct bend as in *D. reighardi*.
Length: ♀ 1.0–1.1 mm; ♂ 0.97–1.0 mm. Lakes, ponds. Northeastern U. S.
Formerly considered a synonym of *D. oregonensis*.

114a   **(111)**   Leg 5 ♂, right endopod attached to the medial margin of basal
segment 2, greatly enlarged. (Fig. 29.93) . . . .
                           D. mississippiensis Marsh 1894
         ♀ keys out to **79a.** Leg 5 ♂, right, the width of the endopod and the extent of
the expansion of inner part of right exopod 2 are variable. Length: ♀ 1.2–1.3 mm;
♂ 1.1 mm. Lakes, ponds. Southeastern U. S.

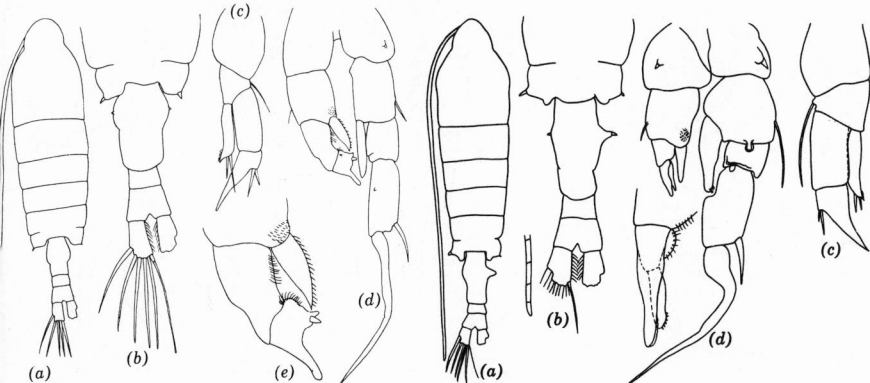

Fig. 29.92. *Diaptomus pygmaeus.* (*a*) Dorsal view ♀.
(*b*) Metasomal wings and urosome ♀, dorsal. (*c*) Leg
5 ♀. (*d*) Leg 5 ♂, posterior view. (*e*) Leg 5 ♂, detail
left exopod and endopod. (From specimens from
Lake Kezar, Me.)

Fig. 29.93. *Diaptomus mississippiensis.* (*a*) Dorsal
view ♀. (*b*) Metasomal wings and urosome ♀, dorsal,
with apex of first antenna. (*c*) Leg 5 ♀. (*d*) Leg 5 ♂
with detail left exopod 2.

114b   Leg 5 ♂, right endopod attached to the posterior margin of basal
            segment 2, not greatly enlarged 11

115a (114) Leg 5 ♂, left exopod 2, distal (outer) process shorter than the total
            exopod (measured to base of process) . . **D. reighardi** Marsh 1895

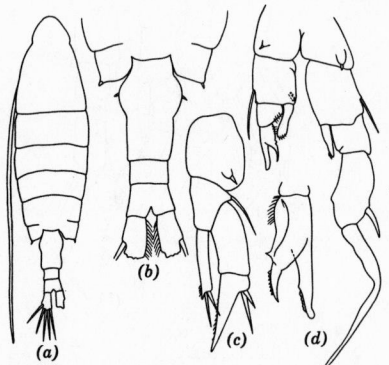

♀ keys out to **73a** (see **113b**). Leg 5 ♂, the
length of inner process of left exopod 2 and inner
expansion of right exopod 2 are variable.
Length: ♀ 1.1–1.2 mm; ♂ 1.0 mm. Lakes,
ponds. Great Lakes area, Canada, U. S., south
to S. C., La., Tenn.

◄ **Fig. 29.94.** *Diaptomus reighardi.* (a) Dorsal
view ♀. (b) Metasomal wings and urosome ♀,
dorsal. (c) Leg 5 ♀. (d) Leg 5 ♂ with detail
anterior aspect left exopod 2.

115b   Leg 5 ♂, left exopod 2, distal process longer than the total
            exopod 11

116a (115) Leg 5 ♂, left endopod reaching to about end of exopod (not in-
            cluding processes). (Fig. 29.95) . . . . . **D. sinuatus** Kincaid 1953
            ♀ keys out to **80a.** Length: ♀ 1.25 mm; ♂ 1.1 mm. Ponds. Fla.

116b   Leg 5 ♂, left endopod reaching beyond exopod by about ½ its
            own length. (Fig. 29.96) . . . .

D. bogalusensis M. S. Wilson and Moore 1953
♀ keys out to **80b.** See Fig. 29.2 for ♂ right antenna. Length: ♀ 1.32 mm;
♂ 1.3 mm. Ponds. La.

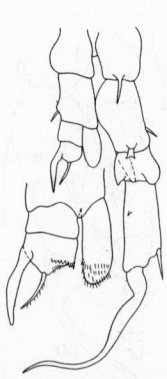

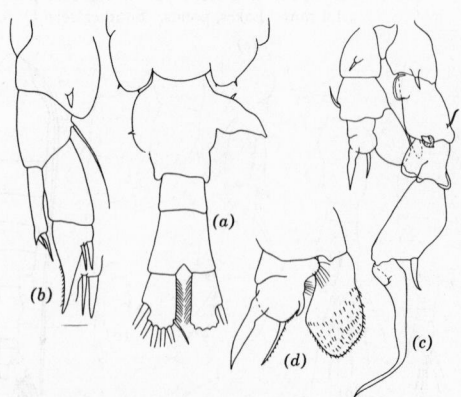

Fig. 29.95. *Diaptomus sinuatus.*
Leg 5 ♂ with detail left exopod
and endopod. (From speci-
men in type lot.)

Fig. 29.96. *Diaptomus bogalusensis.* (a) Metasomal wings
and urosome ♀, dorsal. (b) Leg 5 ♀ with detail of exopod setae. (c) Leg 5 ♂. (d) Leg 5 ♂, left exopod and
endopod. (After M. S. Wilson and Moore.)

CYCLOPOIDA

Harry C. Yeatman

The fresh-water Cyclopoida belong to the family Cyclopidae. Females are more useful in distinguishing the species because the male sometimes fails to show some of the structural features on which separation of one species from another depends. For instance, the male of *Eucyclops agilis* lacks the comblike row of spinules on the outer side of each caudal ramus which is so prominent in the female (Fig. 29.106). Sexes can be distinguished by using characters indicated in Table 29.1. When female specimens are lacking, males can often be successfully identified by the key, since the fifth legs are similar in both sexes.

Immature cyclopoids (copepodid stages) with short first antennae and 2-segmented exopods and endopods of legs 1 to 4 are sometimes incorrectly identified as *Cyclops bicolor*, *Cyclops varicans*, etc. Females with egg sacs are always mature, but often these sacs are absent. In the adult the last urosomal segment (bearing the caudal rami) is shorter than the preceding segment. In the immature copepod it is much longer than (usually twice as long as) the preceding segment, since it has not divided transversely.

Some cyclopoid copepods are parasitic but their free-swimming developmental copepodid stages and those of other parasitic copepods may be collected in plankton tows. One species of the parasitic cyclopoid genus *Ergasilus* has never been collected from a host, and the adults of both sexes are found free-swimming in North American lakes. (See Chapter 30.)

Specimens of Cyclopoida can be fixed in dilute formalin and then placed with a drop of water in a drop of glycerine on a slide. The water evaporates and leaves only the glycerine and specimen which can be examined and dissected under a binocular microscope. First, using needles, the body is easily separated between the fourth and fifth thoracic somites (metasome segments 4 and 5). The important fifth legs on the underside of the fifth thoracic somite are now visible, as are usually the first antennae, caudal rami, and often the swimming legs. If the last are too closely applied to each other, they must be dissected off.

Permanent mounts can be made by placing specimens and dissected parts from glycerine into a drop or two of melted glycerine jelly on a slide. Support round cover slip with broken cover slip pieces and seal with Murrayite cement. See Chapter 46 for more detail.

KEY TO SPECIES

1a   Leg 5 not distinct from fifth metasomal segment and armed with 2 strong inner spines and an outer seta; first antenna usually of 11 segments, but sometimes 9 or 10 (Fig. 29.97c) . . . .
      Ectocyclops Brady

Only 1 species known in N. A., *E. phaleratus* (Koch) 1838 (Fig. 29.97). Length: ♀ 0.90–1.26 mm; ♂ about 0.90 mm. Not common but widespread in U. S. and southern Canada.

1b   Leg 5 consisting of 1 or more distinct segments; first antenna of 6 to 17 segments (Figs. 29.98*b*, and 29.99*d*) . . . .

2a   (1) Leg 5 consisting of 3 distinct segments; first antenna of 16 segments (Fig. 29.98*b*). . . . . ***Orthocyclops*** E. B. Forbes
Only 1 species known in this genus, *O. modestus* (Herrick) 1883 (Fig. 29.98). Length: ♀ 0.80–1.25 mm. ♂ 0.75–0.90 mm. Widespread in U. S. and southern Canada.

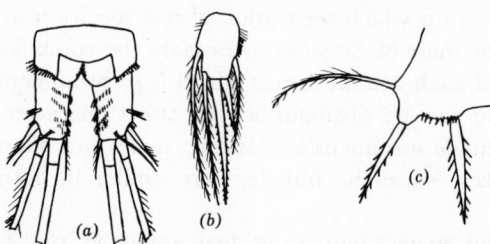

Fig. 29.97. *Ectocyclops phaleratus.* (*a*) Dorsal view of last body segment and caudal rami. (*b*) Terminal segment of endopod of leg 4. (*c*) Leg 5.

Fig. 29.98. *Orthocyclops modestu* (*a*) Dorsal view of last body segmen and caudal rami. (*b*) Leg 5.

2b   Leg 5 consisting of 1 or 2 distinct segments; first antenna of 6 to 17 segments (Figs. 29.99*d* and 29.111*e*) . . . .

3a   (2) Leg 5 consisting of 1 distinct, broad segment and armed with an inner spine and 2 outer setae (Figs. 29.99*d* and 29.106*d*) . . . . . .

3b   Leg 5 consisting of 2 distinct segments (except in some few species which have the basal segment protruding but not separated by a distinct joint from fifth thoracic segment and armed with an outer seta) (Figs. 29.111*e* and 29.136*c*) . . . .

4a   (3) First antenna of 17 segments; large, robust species 1.77–2.88 mm in length. . . . . ***Macrocyclops ater*** (Herrick) 1882
This species differs from other members of the genus *Macrocyclops* in having the fifth leg of only one distinct segment instead of two, so is best included separately in this key. Length: ♀ 1.77–2.88 mm. Not common but probably widespread. Wis., Ill., Minn., Mississippi Valley, N. C.; southern Canada.

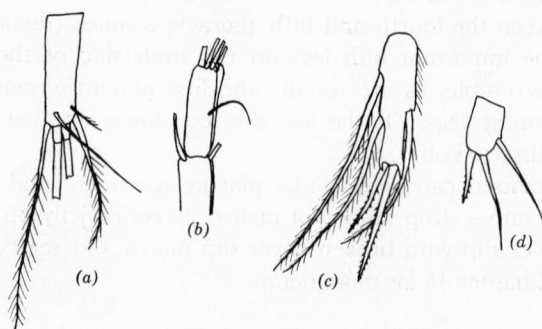

Fig. 29.99. *Macrocyclops ater.* (*a*) Dorsal view of caudal ramus. (*b*) Terminal segment of first antenna (*c*) Terminal segment of endopod of leg 4. (*d*) Leg 5.

4b   First antenna of less than 17 segments; small species, usually under 1.40 mm in length. . . . .

5a   (4) First antenna of 12 segments (in North American species). . . . . .

5b First antenna of 11 or less segments (usually 8 segments in North American species) . . . . *Paracyclops* Claus 10

6a (5) Caudal ramus of female with spinules on outer margin; caudal ramus at least 4 times as broad in males and females (Fig. 29.106*b*). . . . *Eucyclops* Claus 12

6b Caudal ramus of female without spinules on outer margin; caudal ramus about 3 times as long as broad in males and females (Fig. 29.107*a*). . . . . *Tropocyclops* Kiefer 15

7a (3) Distal segment of leg 5 *small* and armed with 2 spines or setae (or 2 spines and 1 seta in the rare *Cyclops venustoides bispinosus*) (Fig. 29.123*d,e*). 8

7b Distal segment of leg 5 *broad* and armed with 3 or more spines or setae (Fig. 29.149*d*) 9

8a (7) Distal segment of leg 5 armed with an apical seta and usually a short or moderately long inner lateral or subapical spine (or with an apical seta and an inner and outer subapical spine in the rare *Cyclops venustoides bispinosus*) (Fig. 29.123*e*); first antenna rarely with hyaline membrane. . . . . *Cyclops* O. F. Müller 16

8b Distal segment of leg 5 armed with an apical seta and a long terminal or subterminal inner spine or seta; last 2 segments of first antenna usually with a hyaline membrane (Fig. 29.140*b,c*). . . . *Mesocyclops* Sars 47

9a (7) Distal segment of leg 5 armed with 2 long spines and a median seta; first antenna of 17 segments (Fig. 29.149*d*). . . . . *Macrocyclops* Claus 53

9b Distal segment of leg 5 armed with 4 or 5 setae or spines; first antenna of 6 segments . . . . *Halicyclops* Norman

Species are usually found in brackish water. Known from Atlantic, Pacific, and Gulf of Mexico coasts. Most North American records listed under the names of the European species *H. aequoreus* and *H. magniceps* are questionable. The literature should be consulted for identification of species of this genus. The most recent key is that of Lindberg (1957). A paper by Kiefer (1936) includes American records to that date and describes West Indian species. A review of the records of the genus in North America and a bibliography of all species described to 1958 are included in M. S. Wilson (1958).

◄ **Fig. 29.100.** *Halicyclops* sp. (*a*) Dorsal view (antennal setae not shown). (*b*) Leg 5 ♀.

0a (5) First antenna of 11 segments (Fig. 29.101) . . . . *Paracyclops affinis* (Sars) 1863

Length: ♀ 0.60–0.85 mm; ♂ 0.56 mm. This rare, creeping species is found in weeds in shallow water and also in water of pitcher plant leaves in Quebec.

0b First antenna of 8 segments 11

1a (10) Caudal ramus 4 to 6 times as long as wide with short transverse rows of spinules next to lateral seta (Fig. 29.102) . . . . *P. fimbriatus* (Fischer) 1853

Length: ♀ 0.70–0.90 mm; ♂ 0.74–0.85 mm. May not occur in N. A., most of the records definitely refer to *P. fimbriatus poppei*.

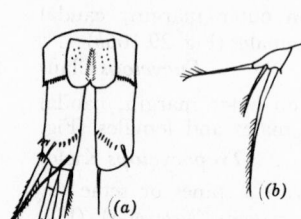

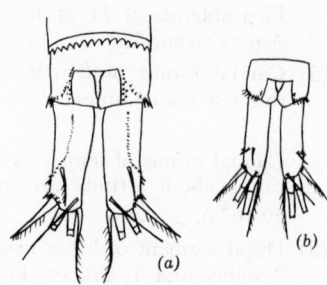

Fig. 29.101. *Paracyclops affinis.* (*a*) Dorsal view of last body segment and caudal rami. (*b*) Leg 5. (After Gurney.)

Fig. 29.102. (*a*) *Paracyclops fimbriatus poppei,* dorsal view of last body segment and caudal rami. (*b*) *P. fimbriatus,* dorsal view of last body segment and caudal rami.

11b　　Caudal ramus 3 to 4 times as long as wide with longitudinal dorsal row of spinules (Fig. 29.102*a*) . . . .
　　　　　　　　　　　　　　　　　P. fimbriatus poppei (Rehberg) 1880
　　　　Length: ♀ 0.70–0.90 mm; ♂ 0.70–0.85 mm. A creeping species found in debris of shallow water. Common and widespread in N. A.

12a　(6)　First antenna not reaching to hind margin of first body segment (Fig. 29.104) 1³

12b　　First antenna reaching beyond hind margin of first body segment and usually to hind margin of second body segment (Fig. 29.106*a*) . . 14

13a　(12)　Caudal ramus 8 to 9 times as long as broad, without marginal saw, but with group of 4 or 5 obliquely arranged spinules at insertion of lateral seta (Fig. 29.103) . . . .
　　　　　　　　　　　　　　　　　Eucyclops macrurus (Sars) 1863
　　　　Length: ♀ 1.10–1.40 mm; ♂ 0.80 mm. Found in S. A., may occur in states bordering on Mexico.

13b　　Caudal ramus 4 times as long as broad, with prominent saw on outer margin (Fig. 29.104) . . . . *E. prionophorus* Kiefer 1931

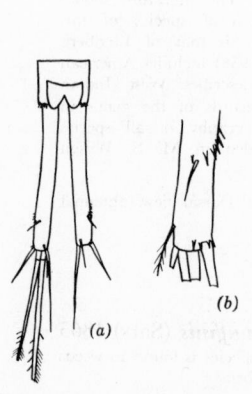

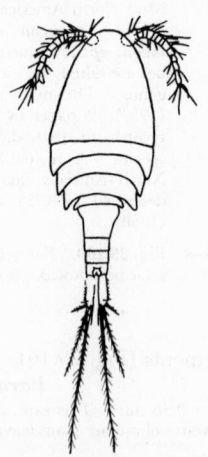

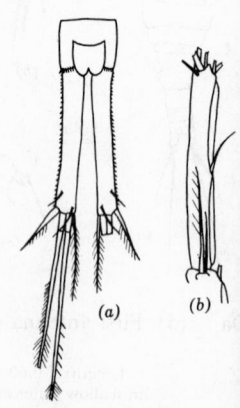

Fig. 29.103. *Eucyclops macrurus.* (*a*) Dorsal view of last body segment and caudal rami. (*b*) Dorsal view of terminal end of caudal ramus. (After Gurney.)

Fig. 29.104. *Eucyclops prionophorus.* Dorsal view ♀.

Fig. 29.105. *Eucyclops speratus.* (*a*) Dorsal view of last body segment and caudal rami ♀. (*b*) Last segment of first antenna ♀. (After Gurney.)

Length: ♀ 0.70–0.94 mm; ♂ 0.80 mm. In small ponds and slow-moving creeks. Conn., N. C., Tenn.

14a (12) Caudal ramus usually more than 5 times as long as broad, lateral spinules very small; inner corner seta usually shorter than ramus in male and female (Fig. 29.105). . . . ***E. speratus*** (Lilljeborg) 1901

Length: ♀ 1.0–1.6 mm; ♂ 0.75–0.80 mm. Found in shallow water. Not common but widespread in N. A. The incompletely described ***Cyclops serrulatus elegans*** Herrick is probably this species.

4b Caudal ramus usually not more than 5 times as long as broad, lateral spinules conspicuous; inner corner seta slightly longer to considerably longer than caudal ramus in male and often in female . . . . ***E. agilis*** (Koch) 1838

Length: ♀ 0.80–1.5 mm; ♂ 0.68–0.8 mm. Probably the commonest littoral cyclopoid copepod in N. A. Often called ***Cyclops serrulatus*** Fischer 1851, but this name has been frequently used for any of the various species bearing spinules on the outer margin of the caudal ramus. A subspecies having much shorter caudal rami and being smaller in size than the typical *E. agilis* is known as ***E. agilis montanus*** (Brady).

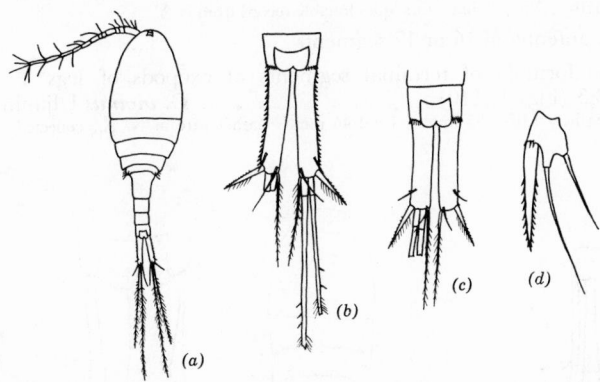

Fig. 29.106. *Eucyclops agilis.* (*a*) Dorsal view ♀. (*b*) Dorsal view of last body segment and caudal rami ♀. (*c*) Dorsal view of last body segment and caudal rami ♂. (*d*) Leg 5. (After Gurney.)

15a (6) Dorsal caudal seta less than twice as long as outermost terminal caudal seta; inner terminal spine of inner end segment of leg 4 less than twice as long as segment (Fig. 29.107) . . . .

Tropocyclops prasinus (Fischer) 1860

Length: ♀ 0.50–0.90 mm; ♂ 0.55–0.60 mm. A very common and widespread limnetic species in N. A.

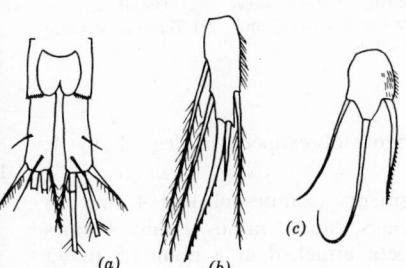

Fig. 29.107. *Tropocyclops prasinus.* (*a*) Dorsal view of last body segment and caudal rami. (*b*) Terminal segment of endopod, leg 4. (*c*) Leg 5. (*a*, *c* after Gurney; *b* after Kiefer.)

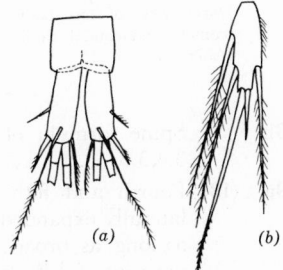

Fig. 29.108. *Tropocyclops prasinus mexicanus.* (*a*) Dorsal view of last body segment and caudal rami. (*b*) Terminal segment of endopod, leg 4. (After Kiefer.)

15b   Dorsal caudal seta more than twice as long as outermost terminal caudal seta; inner terminal spine of inner end segment of leg 4 more than twice as long as segment (Fig. 29.108) . . . .
                        Tropocyclops prasinus mexicanus Kiefer 1938
         Length: ♀ 0.50–0.90 mm; ♂ 0.50–0.60 mm. Limnetic in Mexican lakes, may occur in states bordering on Mexico.

16a   **(8)** Three distal segments of first antenna with row of fine hyaline spines (not conspicuous); caudal ramus usually with longitudinal, dorsal ridge and inner margin hairy; second segment of leg 5 with apical seta and a large spine attached at middle of inner side of segment (Figs. 29.113*d* and 29.111*b,e*) . . ***Cyclops*** (Subgenus ***Cyclops***)   1

16b   Distal segments of first antenna without hyaline spines; caudal ramus without dorsal ridge and inner margin with or without hairs; second segment of leg 5 with apical seta and small or slender spine on the inner side, usually near apex (Fig. 29.114*d*) 2

17a **(16)** First antenna of 14 segments (Fig. 29.109) . . ***C. insignis*** Claus 1857
         Length: ♀ 2.5–5.0 mm. One questionable record from N. Y.

17b   First antenna of 16 or 17 segments 1

18a **(17)** Spine formula of terminal segments of exopods of legs 1 to 4: 2,3,3,3 (Fig. 29.110) . . . . ***C. vicinus*** Uljanin 1875
         Length: ♀ 1.07–1.85 mm; ♂ 1.0–1.46 mm. Rather rare in N. A., collected only in Alaska.

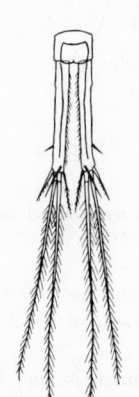 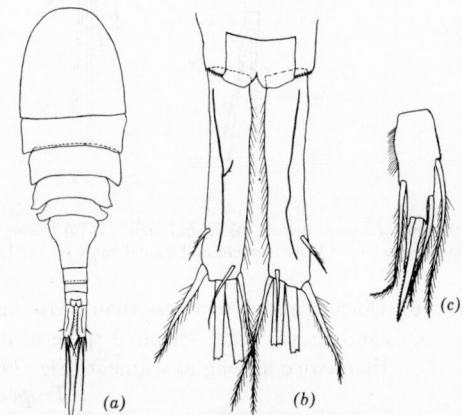

Fig. 29.109. *Cyclops insignis.* Dorsal view of last body segment and caudal rami. (After Sars.)

Fig. 29.110. *Cyclops vicinus.* (*a*) Dorsal view ♀. (*b*) Dorsal view, caudal rami. (*c*) Terminal segment of endopod, leg 4.

18b   Spine formula of terminal segments of exopods of legs 1 to 4: 3,4,3,3 1

19a **(18)** Fourth and fifth metasomal segments (somites of legs 4 and 5) laterally expanded into pointed wings; caudal ramus usually 4 times as long as broad; outer lateral seta attached at a point 65 to 73 per cent of distance from base to apex of caudal ramus . . . .
                                 C. scutifer Sars 1863
         Length: ♀ 1.29–1.9 mm; ♂ 1.0–1.4 mm. A locally common limnetic species. Canada; Alaska; N. Y.

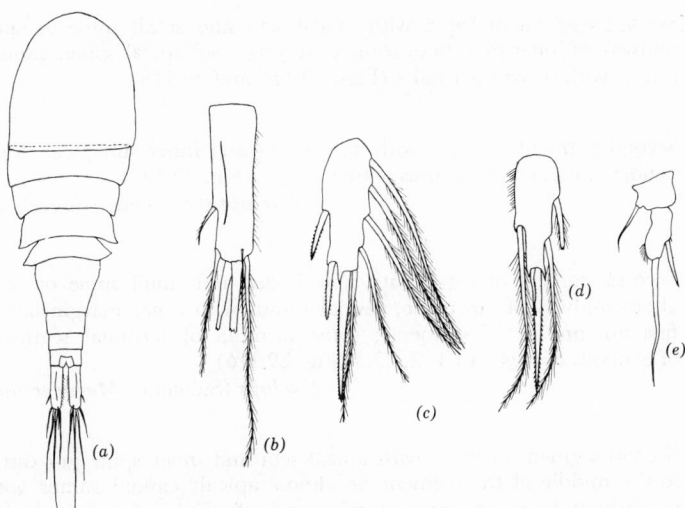

ig. 29.111. *Cyclops scutifer.* (*a*) Dorsal view. (*b*) Dorsal view of caudal ramus. (*c*) Terminal segment of xopod, leg 4. (*d*) Terminal segment of endopod, leg 4. (*e*) Leg 5.

9b   Fourth and fifth metasomal segments not laterally expanded, but usually with small projections at posterolateral angles of these segments; caudal ramus usually at least 5 to 7 times as long as broad; outer lateral seta attached at a point 73 to 87 per cent of distance from base to apex of caudal ramus . . . . . **C. strenuus** Fischer 1851
         Length: ♀ 1.42–2.35 mm; ♂ 1.28–1.56 mm. Rather rare, but collected in Alaska. Most records of this species in N. A. are of **C. scutifer.**

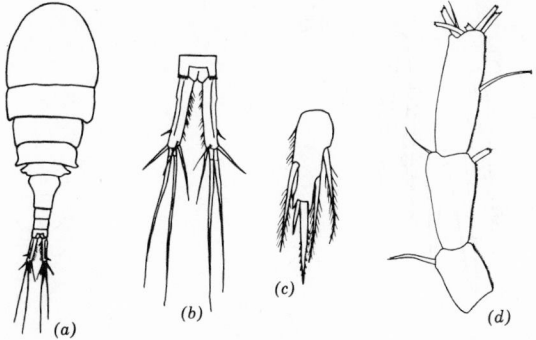

ig. 29.112. *Cyclops strenuus.* (*a*) Dorsal view ♀. (*b*) Dorsal view of last body segment and caudal rami. *c*) Terminal segment of endopod, leg 4. (*d*) Last three segments of first antenna.

20a   (16)   Leg 5 with 2 segments **21**

20b   Leg 5 with 1 segment, the usual basal segment not being separated from fifth metasomal segment; the distinct or partly distinct segment of leg 5 with an apical seta and with or without an inner spine; legs 1 to 4 with rami of 2 segments (Fig. 29.138) . . . .
                          Cyclops (Subgenus *Microcyclops*)   **43**

21a **(20)** Second segment of leg 5 with apical seta and small spine or spur midway of inner margin or somewhat longer subapical spine; caudal ramus with or without hairs (Figs. 29.116 and 29.118) **22**

21b Second segment of leg 5 with apical seta and inner subapical, long slender spine; caudal ramus without hairs (Fig. 29.125). . . . .
Cyclops (Subgenus ***Diacyclops***)   **32**

22a **(21)** Second segment of leg 5 with apical seta and small spine or spur about midway of inner side; caudal ramus with inner margin hairy; first antenna of 17 segments; spine formula of terminal segments of exopods of legs 1 to 4: 2,3,3,3 (Fig. 29.116) . . . .
Cyclops (Subgenus ***Megacyclops***)   **23**

22b Second segment of leg 5 with apical seta and inner spine just distal to the middle of the segment or almost apical; caudal ramus with or without hairs on inner margin; spine formula of terminal segments of exopods of legs 1 to 4; 2,3,3,3 or 3,4,4,4 or very variable (Fig. 29.118) . . . . *Cyclops* (Subgenus ***Acanthocyclops***)   **27**

23a **(22)** Inner terminal spine of endopod of leg 4 shorter than outer terminal spine . . . . *Cyclops magnus* Marsh 1920

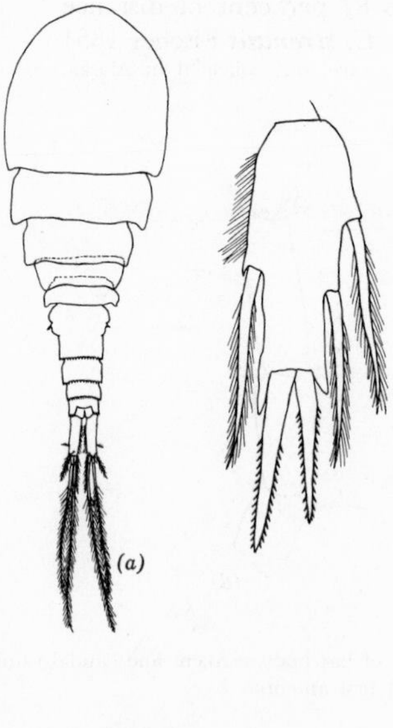

Length: ♀ 1.85–2.55 mm; ♂ somewhat smaller. Not common. Canada; Alaska.

◄ **Fig. 29.113.** *Cyclops magnus.* (*a*) Dorsal view ♀. (*b*) Terminal segment of endopod leg 4.

23b Inner terminal spine of endopod of leg 4 longer than outer terminal spine **24**

24a **(23)** Setae of terminal segment of endopod of leg 4 not extending to distal end of inner terminal spine **25**

24b   Setae of terminal segment of endopod of leg 4 extending beyond
distal end of inner terminal. spine **26**

25a (24) Innermost terminal caudal seta much longer than ramus . . . .
C. viridis (Jurine) 1820

Length: ♀ 1.5–3.0 mm; ♂ 1.4–1.6 mm. Most records refer to other species and
particularly to *C. vernalis,* but possibly occurs in N. A.

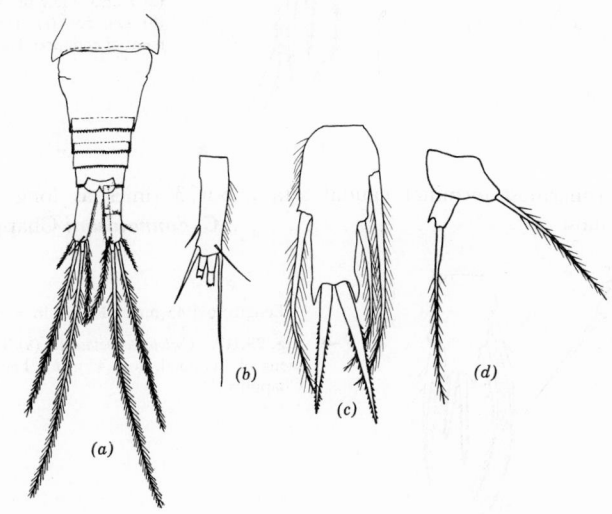

Fig. 29.114. *Cyclops viridis.* (*a*) Dorsal view of urosome and caudal rami ♀. (*b*) Dorsal view of caudal
ramus. (*c*) Terminal segment of endopod, leg 4. (*d*) Leg 5.

25b   Innermost terminal caudal seta shorter than ramus or about the
same length as ramus . . . . *C. gigas* Claus 1857

Length: ♀ 2.0–4.20 mm; ♂ smaller. Rather rare. Alaska.

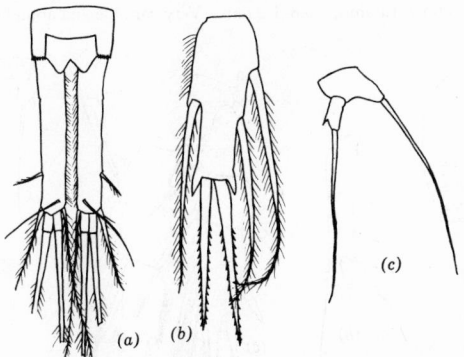

Fig. 29.115. *Cyclops gigas.* (*a*) Dorsal view of last body segment and caudal rami. (*b*) Terminal segment of
endopod, leg 4. (*c*) Leg 5.

26a (24) Innermost terminal caudal seta not quite twice as long as outermost . . . . *C. latipes* Lowndes 1927

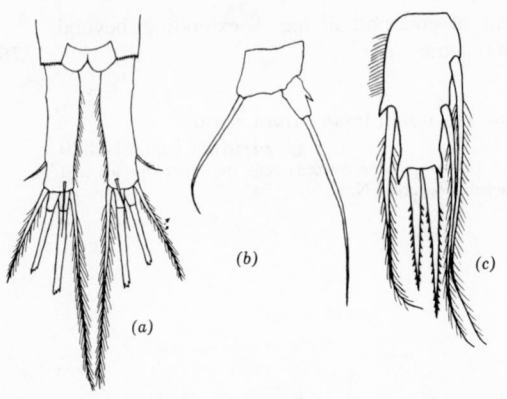

Length: ♀ 1.85–2.5 mm; ♂ 1.5 mm. N. C., Mich., Tenn. Probably more common than present distribution indicates. The incompletely described **C. ingens** Herrick may be identical with this species.

◄ **Fig. 29.116.** *Cyclops latipes.* (a) Dorsal view of caudal rami. (b) Leg 5. (c) Terminal segment of endopod, leg 4.

26b   Innermost terminal caudal seta about 3 times as long as outermost . . . . **C. donnaldsoni** Chappuis 1929

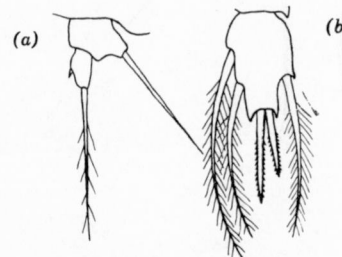

Length: ♀ 1.45 mm. Found in a cave in Ind.

◄ **Fig. 29.117.** *Cyclops donnaldsoni.* (a) Terminal segment of endopod, leg 4. (b) Leg 5. (After Chappuis.)

27a (22) First antenna of 17 segments (occasionally 18) **28**

27b   First antenna of less than 17 segments **29**

28a (27) Inner margin of caudal ramus without hairs . . . .
C. vernalis Fischer 1853

Length: ♀ 0.99–1.8 mm; ♂ 0.8–1.5 mm. Very variable and abundant in N. A.

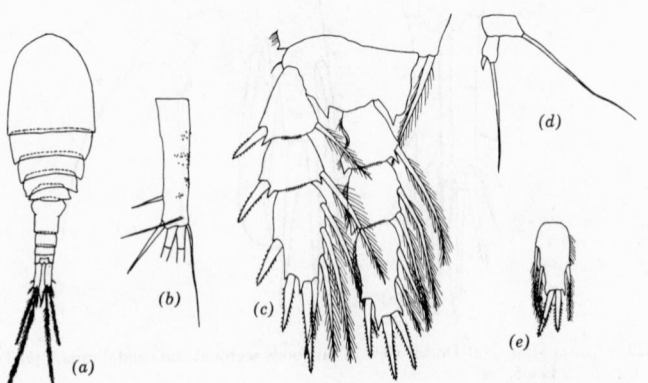

Fig. 29.118. *Cyclops vernalis.* (a) Dorsal view ♀. (b) Dorsal view of caudal ramus (tiny spinules on inner margins, as shown here are not always present). (c) Leg 4. (d) Leg 5. (e) Variation of terminal segment of endopod, leg 4.

8b Inner margin of caudal ramus with tufts of fine hair (Fig. 29.119) . .
              C. carolinianus Yeatman 1944
  Length: ♀ 0.8–1.5 mm; ♂ about 0.76 mm. Not common. N. C.

9a (27) First antenna of 12 segments **30**

9b First antenna of 11 segments (Fig. 29.120) . . . **C. exilis** Coker 1934
  Length: ♀ 0.78–0.88 mm; ♂ 0.7 mm. Not common. In small streams in N. C. and N. Y.

0a (29) Inner margin of caudal ramus without hairs (Fig. 29.121) . . . . . .
              C. capillatus Sars 1863
  Length: ♀ 1.8–2.2 mm. Rare. Alaska; Quebec.

0b Inner margin of caudal ramus with small hairs **31**

1a (30) Setal formula of terminal segments of exopods of legs 1 to 4: 5,5,5,5; posterolateral angles of next to last thoracic segment produced (as in *C. vernalis*) (Fig. 29.122) . . . **C. venustus** Norman and Scott 1906
  Length: ♀ 1.0–1.3 mm; ♂ 0.9 mm. May not occur in N. A. Quebec record is of *C. venustoides*.

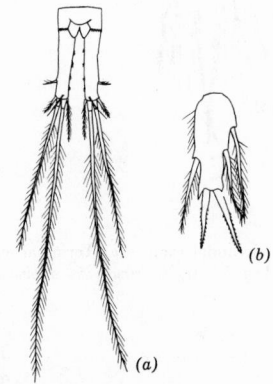

Fig. 29.119. *Cyclops carolinianus.* (a) Dorsal view of last body segment and caudal rami. (b) Terminal segment of endopod, leg 4.

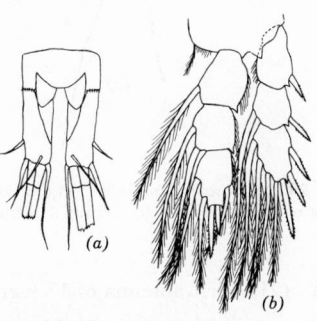

Fig. 29.120. *Cyclops exilis.* (a) Dorsal view of last body segment and caudal rami. (b) Leg 4.

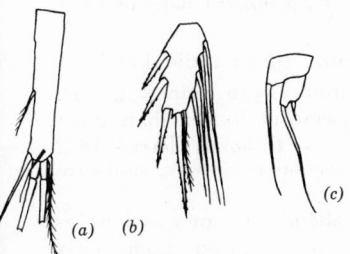

Fig. 29.121. *Cyclops capillatus.* (a) Dorsal view of caudal ramus. (b) Terminal segment of exopod, leg 4. (c) Leg 5.

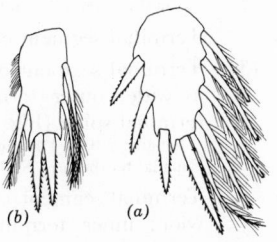

Fig. 29.122. *Cyclops venustus.* (a) Terminal segment of exopod, leg 4. (b) Terminal segment of endopod, leg 4.

31b   Setal formula of terminal segments of exopods of legs 1 to 4: 4,4,4,4; posterolateral angles of next to last thoracic segment not produced . . . . **C. venustoides** Coker 1934

           Length: ♀ 0.82–1.56 mm; ♂ 0.9 mm. Not common. N. C., Ohio, Alaska. According to Kiefer, his **C. pilosus,** bearing hairs on outer as well as inner margins of caudal rami, is this species. A subspecies **C. venustoides bispinosus** Yeatman 1951 (often with an outer as well as an inner subapical spine on the second segment of leg 5) has been found in Ohio and Quebec; length: ♀ 1.6–1.9 mm; ♂ 1.56 mm.

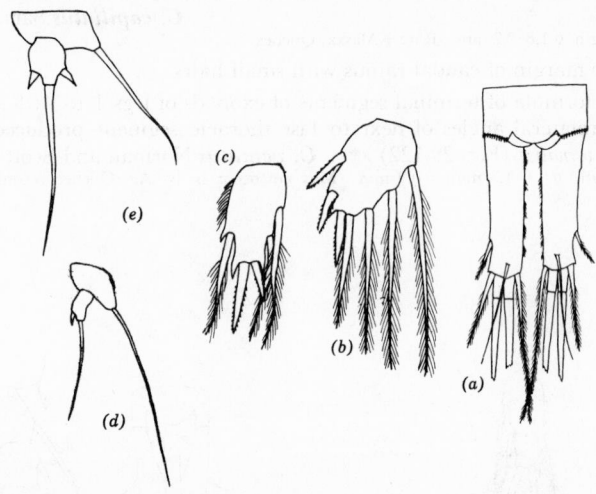

Fig. 29.123. *Cyclops venustoides.* (*a*) Dorsal view of last body segment and caudal rami. (*b*) Terminal segment of exopod, leg 4. (*c*) Terminal segment of endopod, leg 4. (*d*) Leg 5. (*e*) *C. venustoides bispinosus,* leg 5.

32a **(21)** First antenna of 17 segments **33**

32b   First antenna of less than 17 segments **39**

33a **(32)** Outer lateral caudal seta attached at a point $\frac{3}{4}$ to $\frac{4}{5}$ of the distance from base to apex of ramus (Fig. 29.125*a*) **34**

33b   Outer lateral caudal seta attached at a point $\frac{1}{2}$ to $\frac{2}{3}$ of the distance from base to apex of ramus (near middle of ramus) (Fig. 29.130*b*) . . . **37**

34a **(33)** Terminal segment of endopod of leg 4 with long inner seta (not spine) and short outer spine at distal end (Fig. 29.124) . . . .
                                   C. jeanneli Chappuis 1929

           Length: ♀ about 0.90 mm. From a cave in Ind.

34b   Terminal segment of endopod of leg 4 with 2 spines at distal end. . . **35**

35a **(34)** Terminal segment of endopod of leg 4 from $2\frac{1}{2}$ to 3 times as long as wide; outer terminal spine of this segment longer than inner terminal spine (Fig. 29.125). . . . . **C. navus** Herrick 1882
           Length: ♀ 0.90–1.16 mm; ♂ about 0.86 mm. Temporary ponds, wells, small lakes. Canada; northern U. S., N. C.

35b   Terminal segment of endopod of leg 4 about $1\frac{1}{2}$ times as long as wide; inner terminal spine of this segment longer than outer terminal spine **36**

36a **(35)** A seta on outer side of terminal segment of endopod of leg 4; caudal ramus 5 to 7 times as long as wide (Fig. 29.126). . . . .
                                    C. bisetosus Rehberg 1863

           Length: ♀ 0.84–1.51 mm; ♂ 0.80–1.0 mm. Rare in N. A. Quebec.

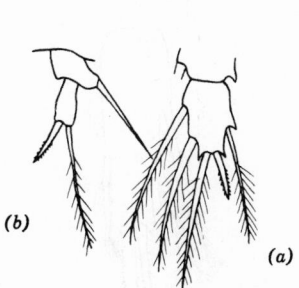

Fig. 29.124.   *Cyclops jeanneli.*
(a) Terminal segment of endopod,
leg 4.   (b) Leg 5.   (After
Chappuis.)

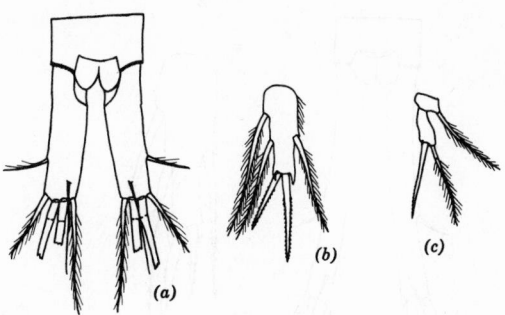

Fig. 29.125. *Cyclops navus.* (a) Dorsal view of last body segment and caudal rami. (b) Terminal segment of endopod, leg
4. (c) Leg 5.

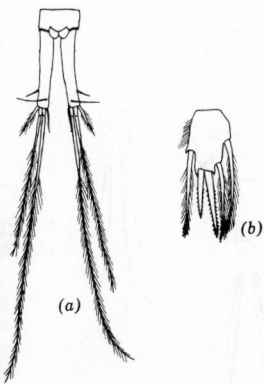

Fig. 29.126.   *Cyclops bisetosus.*
(a) Dorsal view of last body
segment and caudal rami ♀.
(b) Terminal segment of endopod, leg 4.

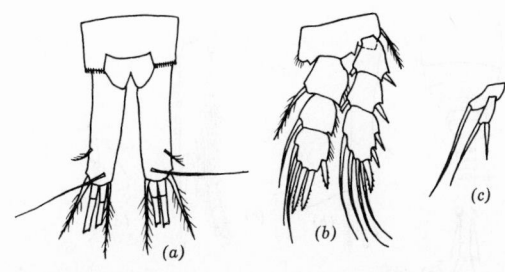

Fig. 29.127. *Cyclops nearcticus.* (a) Dorsal view of last body
segment and caudal rami ♀. (b) Leg 4. (c) Leg 5.

36b A spine on outer side of terminal segment of endopod of leg 4;
    caudal ramus about 4 times as long as wide (Fig. 29.127). . . . . . .
                     C. nearcticus Kiefer 1934
      Length: ♀ 0.50–0.80 mm; rare. In small streams and wells. Mass., N C., Texas.

37a (33) Inner terminal spine of endopod of leg 4 longer than outer terminal
    spine (Fig. 29.128) . . . . **C. haueri** Kiefer 1931
      Length: ♀ 1.15–1.40 mm. Rare. Temporary pools. Conn., Ohio.

37b Inner terminal spine of endopod of leg 4 shorter than outer. **38**

38a (37) Outer terminal spine of endopod of leg 4 about 1½ times as long as
    inner terminal spine; next to last thoracic segment with postero-
    lateral angles rounded (Fig. 29.129) . . . **C. bicuspidatus** Claus 1857
      Length: ♀ 0.95–1.57 mm; ♂ about 1.0 mm. One record from Mass. The subspecies
    C. b. thomasi is the common American form.

38b Outer terminal spine of endopod of leg 4 about twice as long as

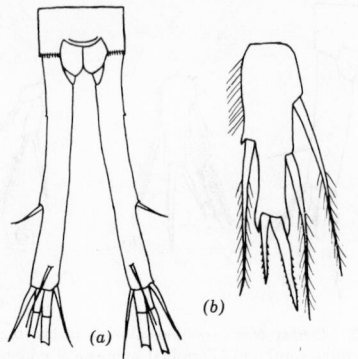

Fig. 29.128. *Cyclops haueri.* (*a*) Dorsal view of last body segment and caudal rami ♀. (*b*) Terminal segment of endopod, leg 4.

Fig. 29.129. *Cyclops bicuspidatus.* Terminal segment of endopod, leg 4.

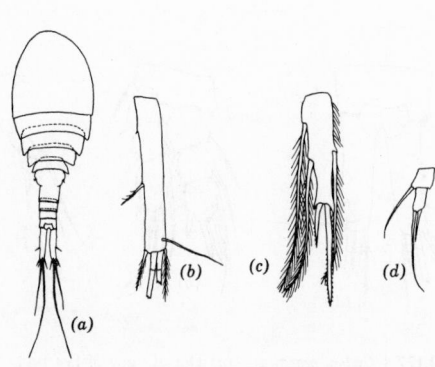

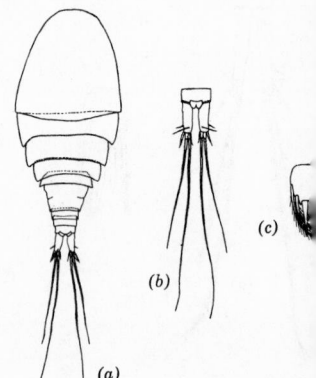

Fig. 29.130. *Cyclops bicuspidatus thomasi.* (*a*) Dorsal view ♀. (*b*) Dorsal view of caudal ramus. (*c*) Terminal segment of endopod, leg 4. (*d*) Leg 5.

Fig. 29.131. *Cyclops crassicaudis brachycercus.* (*a*) Dorsal view ♀ (*b*) Dorsal view of caudal rami (*c*) Terminal segment of endopod leg 4.

inner terminal spine; next to last thoracic segment with papilliform posterolateral process (Fig. 29.130) . . . .
C. bicuspidatus thomasi S. A. Forbes 1882
<small>Length: ♀ 0.9–1.17 mm; ♂ about 0.8 mm. A widely distributed and common limnetic species in N. A.</small>

39a **(32)** First antenna of 14 segments . ***C. bicuspidatus lubbocki*** Brady 1868
<small>Identical with ***C. bicuspidatus*** except for number of segments of first antenna. Length: ♀ about 1.0 mm; ♂ about 0.8 mm. One questionable record from N. Y.</small>

39b First antenna of less than 14 segments 40

40a **(39)** First antenna of 12 segments (Fig. 29.131) . . . .
C. crassicaudis brachycercus Kiefer 1929
<small>Length: ♀ 0.72–1.10 mm; ♂ about 0.75 mm. In stagnant, temporary puddles and in wells. N. C., N. Y., Quebec. Uncommon. Wiley's ***C. bissextilis*** is this species.</small>

40b First antenna of 11 segments 41

1a (40) Terminal segment of endopod of leg 4 with long inner seta and short outer spine at distal end; inner most terminal caudal seta longer than outer . . . . **C. jeanneli putei** Yeatman 1943
Length: ♀ 0.76–0.86 mm; ♂ about 0.70 mm. Rare. In wells in N. C.

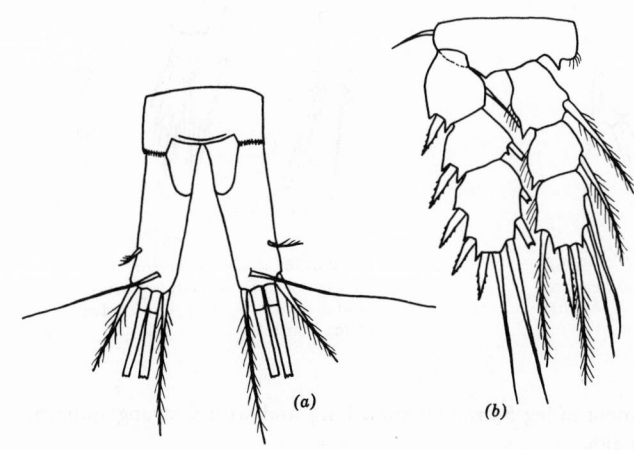

Fig. 29.132. *Cyclops jeanneli putei.* (*a*) Dorsal view of last body segment and caudal rami ♀. (*b*) Leg 4.

1b Terminal segment of endopod of leg 4 with long inner spine and shorter outer spine at distal end; innermost terminal caudal seta shorter than outer **42**

2a (41) Lateral caudal seta at about middle of ramus (Fig. 29.133) . . . . . . **C. nanus** Sars 1863
Length: ♀ 0.45–0.9 mm; ♂ 0.40 mm. Very rare in N. A. From bottom material from lake in N. C.

2b Lateral caudal seta at distal third of ramus (Fig. 29.134) . . . . **C. languidoides** Lilljeborg 1901
Length: ♀ 0.51 mm; ♂ 0.46 mm. Found in debris from spring in Quebec.

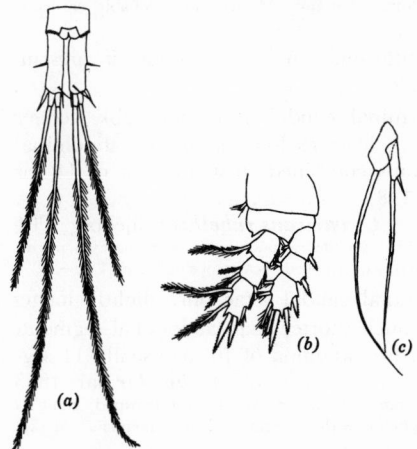

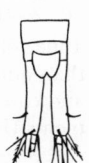

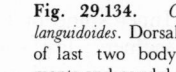

Fig. 29.133. *Cyclops nanus.* (*a*) Dorsal view of last body segment and caudal rami ♀. (*b*) Leg 4. (*c*) Leg 5.

Fig. 29.134. *Cyclops languidoides.* Dorsal view of last two body segments and caudal rami.

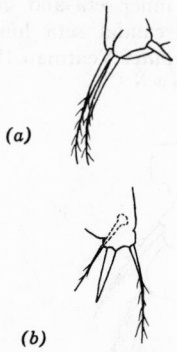

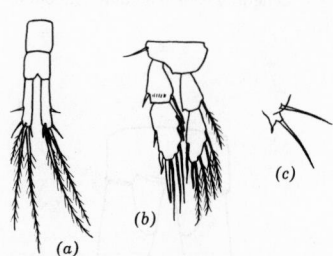

Fig. 29.135.   Cyclops
dimorphus. (a) Leg 5 ♀.
(b) Leg 5 ♂. (After
Kiefer.)

Fig. 29.136.   Cyclops panamensis. (a) Dorsal view of last two body segments and
caudal rami. (b) Leg 4. (c) Leg 5.
(After Marsh.)

43a (20) Free segment of leg 5 broader than long and with a strong spine on
           the inner side. **44**

43b   Free segment of leg 5 longer than broad and with or without a
           tiny spine on the inner side **45**

44a (43) No inner seta on first exopod segment of leg 4; body robust. (Fig.
           29.135). . . . . *C. dimorphus* Kiefer 1934
           Length: ♀ 1.10–1.20 mm; ♂ 0.95 mm.   Known only from Salton Sea, Calif.

44b   An inconspicuous seta on first exopod segment of leg 4; body very
           slender. (Fig. 29.136) . . . . *C. panamensis* Marsh 1913
           Length: ♀ 0.61–0.69 mm.   Panama; Mexico; Fla.; probably occurs in states bordering
           on Mexico.

45a (43) Free segment of leg 5 bearing small spine near distal end of inner
           side that protrudes beyond end of segment and curves towards base
           of terminal seta (Fig. 29.137) . . . . . *G. dentatimanus* Marsh 1913
           Length: ♀ 0.82 mm; ♂ 0.54–0.59 mm.   Panama; Mexico, and probably occurs in
           states bordering on Mexico.

45b   Free segment of leg 5 without spine on inner side or, if present,
           attached to middle of inner side.   **46**

46a (45) Inner of the two middle terminal caudal setae noticeably longer
           than outer median seta and at least as long as all the abdominal
           segments and the caudal ramus combined; first antenna of 11, or
           usually, 12 segments (Fig. 29.138) . . . .
                                          C. varicans rubellus Lilljeborg 1901
           Length ♀: 0.51–0.96 mm; ♂ 0.50 mm   In debris and weeds near shore in ponds,
           etc. Widely distributed in N. A. but easily overlooked because of small size.

46b   Inner of the two middle terminal caudal setae only slightly longer
           than outer median seta and much shorter than abdominal segments
           and caudal ramus combined; first antenna of 10, or usually 11 seg-
           ments (Fig. 29.139) . . . . *C. bicolor* Sars 1863
           Length: ♀ 0.60–0.80 mm; ♂ 0.50 mm.   A rather rare littoral copepod; most of
           the records of this species are *C. v. rubellus* with 11-segmented first antennae. Alaska;
           Mass., Wyo., Ill., Wisc.

47a (8)   Inner spine of leg 5 at middle or just beyond middle of second
           segment; last segment of first antenna bearing hyaline plate with
           one or more distinct notches (Fig. 29.140) **48**

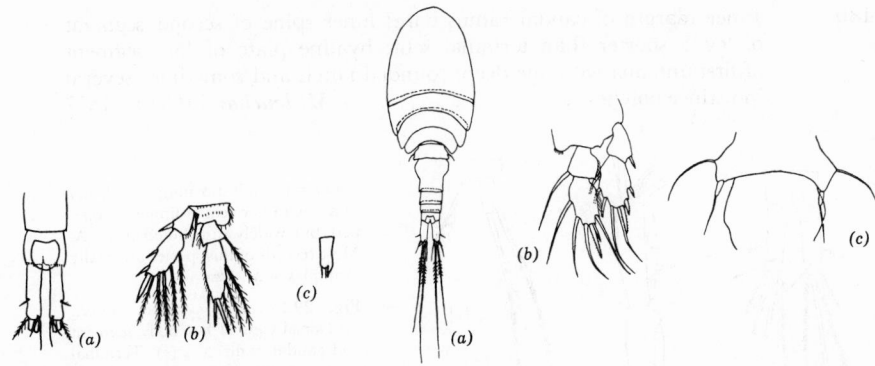

Fig. 29.137. *Cyclops dentatimanus.* (*a*) Dorsal view of last body segment and caudal rami. (*b*) Leg 4. (*c*) Tip of distinct segment of leg 5. (*a*, *b* after Marsh; *c* after Comita.)

Fig. 29.138. *Cyclops varicans rubellus.* (*a*) Dorsal view ♀. (*b*) Leg 4. (*c*) Ventral view of fifth thoracic somite and leg 5.

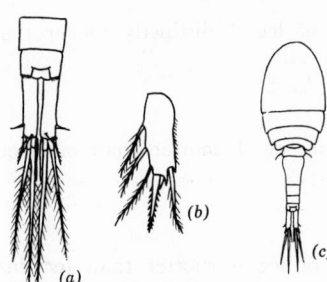

Fig. 29.139. *Cyclops bicolor.* (*a*) Dorsal view of last two body segments and caudal rami. (*b*) Terminal segment of endopod, leg 4. (*c*) Dorsal view of female. (*c* after Sars.)

47b   Inner spine of leg 5 apical or subapical in position on second segment; last segment of first antenna without hyaline plate or with unnotched hyaline plate (Fig. 29.142*c*) **49**

48a (47) Inner margin of caudal ramus with hairs; inner spine of second segment of leg 5 longer than terminal seta; hyaline plate of last segment of first antenna with a number of sharp notches . . . .
                     Mesocyclops edax (S. A. Forbes) 1891
     Length: ♀ 1.0–1.5 mm; ♂ 0.75–0.9 mm. A very common widespread limnetic copepod.

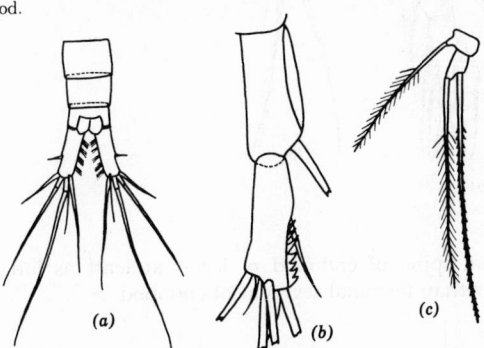

Fig. 29.140. *Mesocyclops edax.* (*a*) Dorsal view of last three body segments and caudal rami ♀. (*b*) Terminal segments of first antenna. (*c*) Leg 5.

48b   Inner margin of caudal ramus bare; inner spine of second segment of leg 5 shorter than terminal seta; hyaline plate of last segment of first antenna with one deep, rounded notch and sometimes several indistinct notches . . . . **M. leuckarti** (Claus) 1857

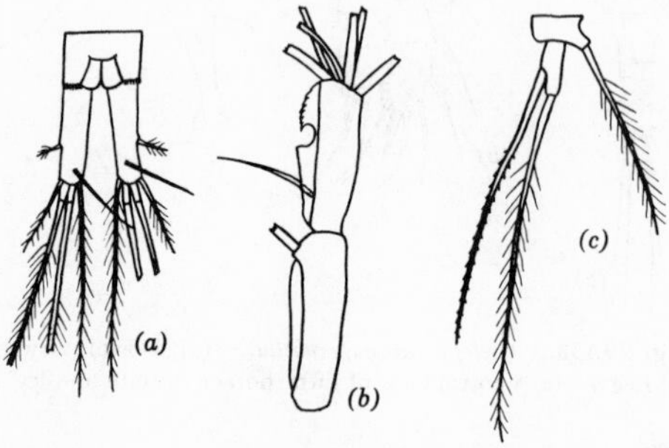

Length: ♀ 0.9–1.3 mm; ♂ 0.75–0.9 mm. A rather scarce limnetic copepod, but widely distributed in N. A. Most records of this species are really examples of *M. edax.*

◄ **Fig. 29.141.** *Mesocyclops leuckarti.* (*a*) Dorsal view of last body segment and caudal rami ♀. (*b*) Terminal segment of first antenna ♀. (*c*) Leg 5.

49a (47) Inner terminal spine of endopod of leg 4 distinctly longer than outer terminal spine **50**

49b   Inner terminal spine of endopod of leg 4 shorter than or about same length as outer terminal spine. **52**

50a (49) Inner terminal spine of endopod of leg 4 shorter than terminal segment of endopod . . . . **M. hyalinus** (Rehberg) 1880

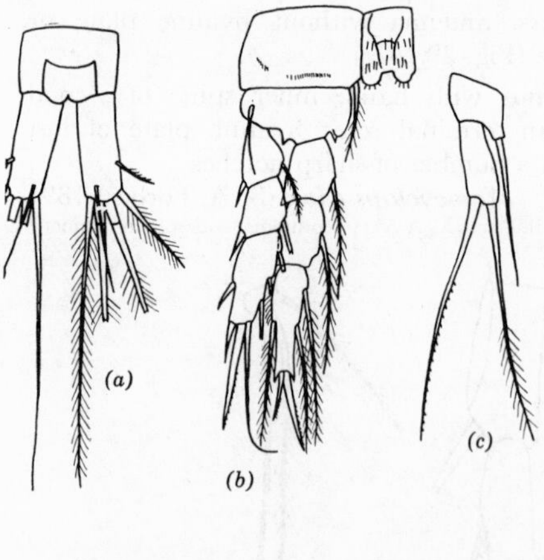

Length: ♀ 0.8–0.9 mm. ♂ 0.6 mm. Central America; may occur in states bordering on Mexico.

◄ **Fig. 29.142.** *Mesocyclops hyalinus.* (*a*) Dorsal view of last body segment and caudal rami ♀. (*b*) Leg 4. (*c*) Leg 5. (After Gurney.)

50b   Inner terminal spine of endopod of leg 4 at least as long as and usually longer than terminal segment of endopod **51**

51a (50) Inner terminal spine of endopod of leg 4 about 5 times as long as outer terminal spine . . . . **M. oithonoides** (Sars) 1863

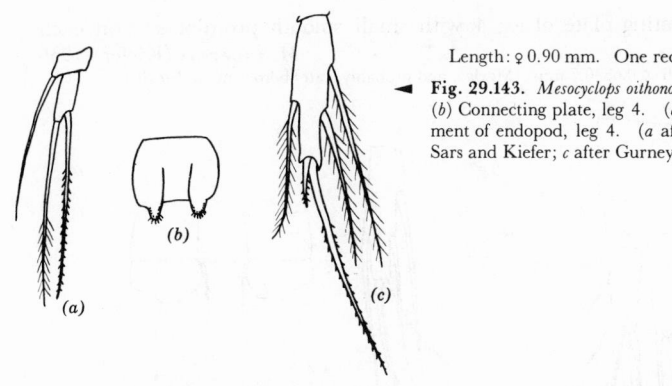

Length: ♀ 0.90 mm. One record from Minn.

◄ **Fig. 29.143.** *Mesocyclops oithonoides.* (*a*) Leg 5. (*b*) Connecting plate, leg 4. (*c*) Terminal segment of endopod, leg 4. (*a* after Sars; *b* after Sars and Kiefer; *c* after Gurney.)

51b   Inner terminal spine of endopod of leg 4 about twice as long as outer terminal spine . . . . **M. tenuis** (Marsh) 1909
Length: ♀ 1.10 mm. Rare. Ariz. and probably other states bordering on Mexico.

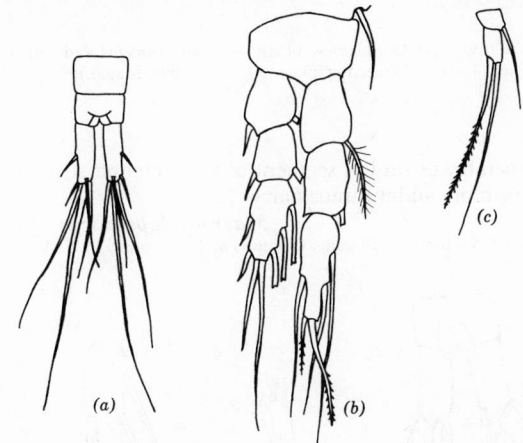

Fig. 29.144. *Mesocyclops tenuis.* (*a*) Dorsal view of last two body segments and caudal rami. (*b*) Leg 4. (*c*) Leg 5.

52a (49) Connecting plate between right and left leg 4 without prominences. . .
M. dybowskii (Lande) 1890

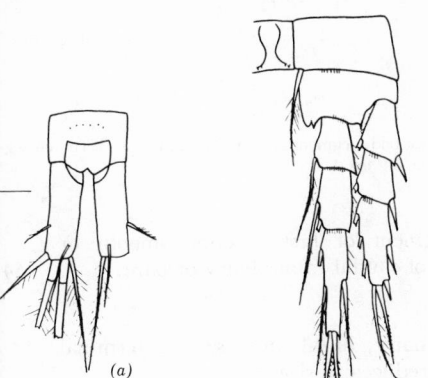

Length: ♀ 0.9–1.1 mm; ♂ 0.65 mm. A rare limnetic species found in Wyo. and Ill.

◄ **Fig. 29.145.** *Mesocyclops dybowskii.* (*a*) Dorsal view of last body segment and caudal rami. (*b*) Leg 4. (After Gurney.)

52b   Connecting plate of leg 4 with small smooth prominences on each
            side . . . . **M. inversus** (Kiefer) 1936
            Length: ♀ 0.68–0.7 mm. Mexico, and probably states bordering on Mexico.

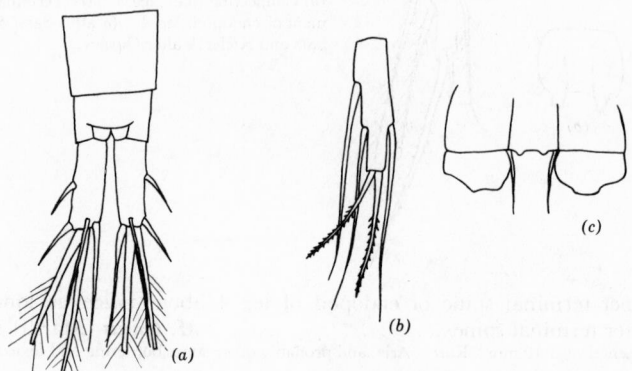

Fig. 24.146. *Mesocyclops inversus.* (*a*) Dorsal view of last two body segments and caudal rami. (*b*) Terminal segment of endopod, leg 4. (*c*) Connecting plate, leg 4. (*c* after Marsh.)

53a   (9) Hyaline membrane on last segment of first antenna strongly toothed;
               inner margin of caudal ramus hairy . . . .
                                    Macrocyclops fuscus (Jurine) 1820
               Length: ♀ 1.8–4.0 mm; ♂ 1.19 mm. Common and widespread in N. A.

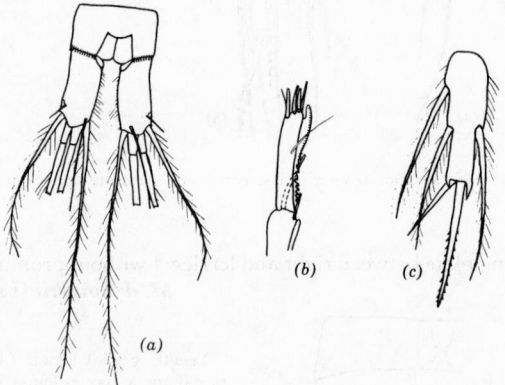

Fig. 29.147. *Macrocyclops fuscus.* (*a*) Dorsal view of last body segment and caudal rami. (*b*) Terminal segment of first antenna ♀. (*c*) Terminal segment of endopod, leg 4.

53b   Hyaline membrane on last segment of first antenna smooth or
            minutely serrated; inner margin of caudal ramus hairy or bare. . . .   **54**

54a (53) Inner margin of caudal ramus hairy; distal inner seta of terminal
               segment of endopod of leg 4 not reduced in size . . . .
                                    M. distinctus (Richard) 1887

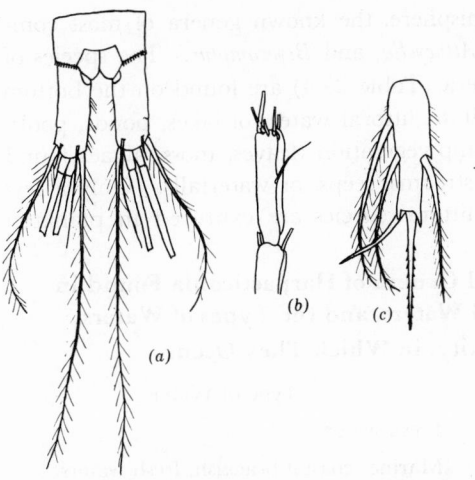

Length: ♀ 2.0–2.2 mm; ♂ 1.2–1.4 mm. No authentic record in N. A., but should be looked for because of its rather wide distribution in the world.

◄ **Fig. 29.148.** *Macrocyclops distinctus.* (*a*) Dorsal view of last body segment and caudal rami. (*b*) Terminal segment of first antenna ♀. (*c*) Terminal segment of endopod, leg 4.

54b   Inner margin of caudal ramus without hairs; distal inner seta of terminal segment of endopod of leg 4 much reduced in size . . . . . .
                                                        M. albidus (Jurine) 1820

The hyaline membrane on the terminal segment of the first antenna is not easily seen when the antenna is in the normal position. Length: ♀ 1.5–2.5 mm; ♂ 0.96 mm. One of the commonest of North American copepods. *Cyclops bistriatus* Koch is a synonym.

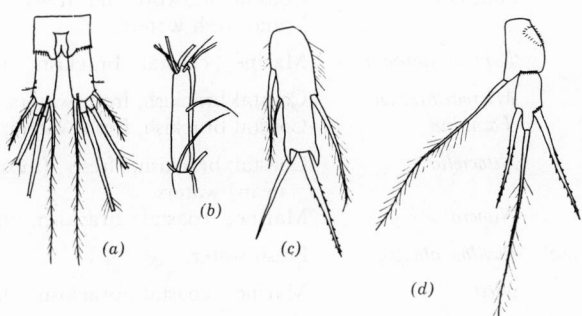

Fig. 29.149. *Macrocyclops albidus.* (*a*) Dorsal view of last body segment and caudal rami. (*b*) Last segment of first antenna ♀. (*c*) Terminal segment of endopod, leg 4. (*d*) Fifth foot.

HARPACTICOIDA

Mildred Stratton Wilson and Harry C. Yeatman

The most frequently encountered species of Harpacticoida of inland waters belong to the family Canthocamptidae, a cosmopolitan group in which about three hundred fresh-water forms have been described. In North America, as in other parts of the Northern Hemisphere, the known genera of most common occurrence are *Canthocamptus*, *Attheyella*, and *Bryocamptus*. The species of these and other canthocamptid genera (Table 29.4) are found on the bottom or among the vegetation of the shallow, littoral waters of lakes, ponds, pools, and ditches. Many forms live in damp vegetation (leaves, moss, algae) found in or along the margins of springs, streams, seeps, or waterfalls. Only a few species or occasional specimens of littoral species are captured in plankton

Table 29.4. The Families and Genera of Harpacticoida Found in North American Continental Waters, and the Types of Waters, with Respect to Salinity, in Which They Occur.

Family	Genus	Type of Water
Parastenocaridae	*Parastenocaris*	Fresh water.
Harpacticidae	*Harpacticus*	Marine; coastal brackish, fresh waters.
	Tigriopus	Marine; coastal brackish, fresh; inland fresh waters.
Cylindropsyllidae	*Stenocaris*	Marine; coastal fresh water.
Laophontidae	*Onychocamptus*	Marine; coastal and inland brackish, saline and fresh waters.
	Heterolaophonte	Marine; coastal brackish, fresh waters.
	Pseudonychocamptus	Marine; coastal fresh water.
Diosaccidae	*Schizopera*	Coastal brackish and fresh; inland saline and fresh waters.
Thalestridae	*Paradactylopodia*	Marine; coastal brackish, fresh waters.
Tachidiidae	*Microarthridion*	Coastal brackish, fresh waters.
	Tachidius	Coastal brackish, fresh waters.
Ameiridae	*Nitocrella*	Coastal brackish, fresh (mostly subterranean) waters.
	Nitocra	Marine; coastal brackish, fresh waters.
Phyllognathopidae	*Phyllognathopus*	Fresh water.
Metidae	*Metis*	Marine; coastal brackish, fresh waters.
Cletodidae	*Cletocamptus*	Marine; coastal and inland brackish, saline and fresh waters.
	Nannopus	Coastal and inland brackish and fresh waters.
	Huntemannia	Marine; coastal brackish; inland fresh waters.
Canthocamptidae	*Mesochra*	Marine; coastal brackish, fresh; inland fresh waters.
	Epactophanes	Fresh water (rarely in brackish or saline).
	Moraria	
	Maraenobiotus	
	Paracamptus	
	Canthocamptus	
	Attheyella	
	Elaphoidella	
	Bryocamptus	

tows in deep or open water. The sand of lake or pond shores may also harbor canthocamptids, but more characteristic of this little-studied habitat is the genus *Parastenocaris*, known from only a few records in North America. Another little-explored habitat is that of subterranean waters, which have yielded large numbers of specimens and forms in Europe.

The remainder of the harpacticoid copepods of continental waters include a diverse group of genera. This is because many families have one or a few species with a wide range of salinity tolerance (Table 29.4). Most of these are found in brackish or fresh-water coastal bodies, but some genera such as *Tigriopus*, *Mesochra*, and *Huntemannia* occur in inland fresh-water lakes. In addition, some genera and species of marine families have become adapted to inland saline waters. Such species as *Onychocamptus mohammed* and *Cletocamptus albuquerquensis* are characteristic faunal elements of some inland saline lakes of North America.

Distribution

The continental harpacticoid fauna of North America has been little investigated so that no region can be said to be well known. Most reports are from localized areas. The studies of Willey in Quebec, of Chappuis and Coker in New York, of Coker in North Carolina, of Carter in Virginia, of C. B. Wilson in Massachusetts, and of M. S. Wilson in Alaska have made these areas, or parts of them, the best known. From one to five species have been reported from some other states or provinces, but for most of the continent, there are no records. What constitutes a rare or common species, the variations within species, or the relationship of distribution to taxonomy are therefore unknown.

Few of the known species groups seem to be restricted to North America. Some cosmopolitan and several species known also from Europe and Asia occur in North America, particularly in the northern portions. Many of these

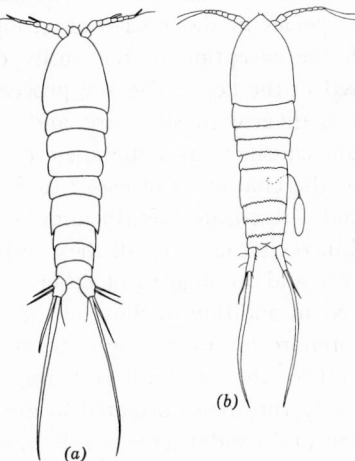

Fig. 29.150. Dorsal view of females of two genera of Canthocamptidae. Types of body form of other genera are shown throughout key. (*a*) *Moraria duthiei*. (*b*) *Canthocamptus robertcokeri* with attached spermatophore.

(*b*)

(*a*)

seem to be identical to the Eurasian forms. Others show close affinities tc
Eurasian species or groups, and may replace them on this continent. Ir
addition, some groups such as the *Bryocamptus minutus* complex, appear to be
represented in North America by a greater number of species and forms thar
are found in Eurasia.

Classification and Literature

The classification followed in the key is that outlined in Lang (1948). This
is the basic work for study of the Harpacticoida, but it should be used with
the knowledge that it is based on literature only through the year 1938. In
some fresh-water genera of the Canthocamptidae and in the genus *Parastenocaris,*
numerous new species have been described since Lang's work. Many of the
new species of Canthocamptidae will be found in Borutsky (1952). Chappuis
(1956 and 1958) has issued revised keys to the species of *Elaphoidella* and oi
Parastenocaris. In addition to these references, recent journals and the *Zoologi-*
cal Record should also be consulted for species that cannot be placed in the key
given here.

All important references to descriptions of species from American waters are
given in the bibliography. For many species, further references to reliable
descriptions are included in the key. A large number of species that are found
in North America have been well described by Gurney (1932) and by Sars
(1903–1911). These, along with Lang's monograph and papers by Coker
(1934) and M. S. Wilson (1956–1958), are valuable basic works for the investigator of North American continental harpacticoid copepods.

Characters and Use of the Key

The cephalic appendages, particularly those of the oral area, are highly
significant in classification at the familial and generic levels in the Harpacticoida. Since these are technically difficult for the nonspecialist to study, and
are not necessary for identification of species in the restricted group found
in fresh water, these appendages with the exception of the easily dissected
and observed first antenna, are not used in the key. The key proceeds from
the characters of the first leg, which is diverse in structure and armature
in the Harpacticoida, and is a significant classificatory appendage throughout
the free-living Copepoda. Beyond this, the characters of legs 2 to 5, the first
antenna, and the habitus are employed. Accurate identification of species
often requires consideration of the combined characters of all these appendages
and the habitus. Therefore, the figures and summaries of other characters
given for each species should be checked in addition to those of the couplets.

To use the key, it is necessary to prepare for microscopic examination a
slide or slides of dissected appendages (first antenna with rostrum, and legs
1 to 5), and the posterior part of the body (urosome) oriented in dorsal view
so that the anal operculum, caudal rami, and caudal setae can be adequately
observed. Dissection (using a pair of mounted no. 12 needles or *minuten*

Nadeln) requires practice but is far from an insurmountable task. Many harpacticoids as found in preserved collections, have the body bent as shown in Fig. 29.151. For beginners, fluid glycerine is recommended as the most satisfactory medium in which to dissect and examine specimens.

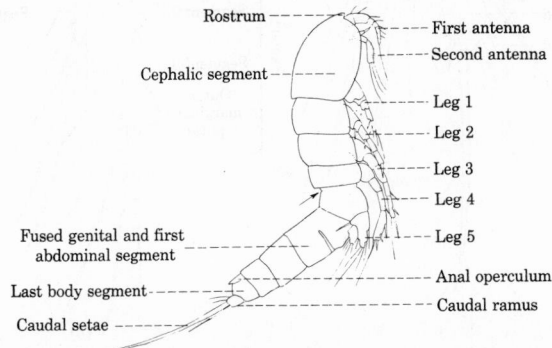

Rostrum
First antenna
Second antenna
Cephalic segment
Leg 1
Leg 2
Leg 3
Leg 4
Fused genital and first abdominal segment
Leg 5
Last body segment
Anal operculum
Caudal ramus
Caudal setae

Fig. 29.151. *Bryocamptus hiemalis* ♀, lateral view (modified from Coker). Arrow indicates place of initial separation of body in dissection. All body parts and appendages are indicated by terminology used in key.

To start dissection, first separate the body into two parts at the point indicated by the arrow in Fig. 29.151 (the flexible attachment of the anterior and posterior portions located between the somites of legs 4 and 5). It is then usually possible to remove leg 5 from its segment, or the segment may be separated from the urosome segments and mounted in ventral aspect. This last method is necessary in dealing with species in which the fifth leg is exceedingly reduced. Legs 1 to 4 may likewise be removed by separation from the segment. An easier method, especially in contracted specimens, is first to separate the segments from one another. When this is done, it is usually advisable to remove the legs from the segment, or to split the segment, so that the legs will lie in a favorable position in the final mount.

The structures of the habitus and appendages referred to in the key are relatively simple and the terminology necessary to distinguish them is not extensive. Figs. 29.151–29.154 illustrate these and should be studied before

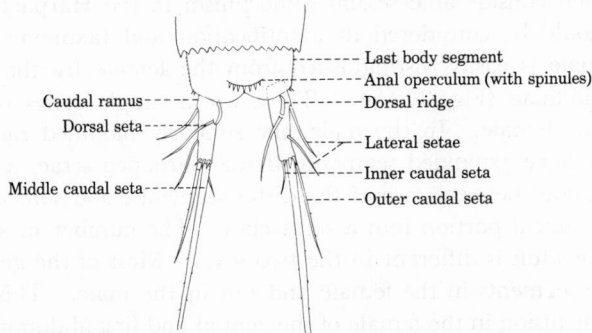

Last body segment
Anal operculum (with spinules)
Caudal ramus
Dorsal ridge
Dorsal seta
Lateral setae
Inner caudal seta
Middle caudal seta
Outer caudal seta

Fig. 29.152. Dorsal view last body segment and caudal rami, *Canthocamptus robertcokeri*, with terminology used in key. (Caudal rami in slightly different aspects.)

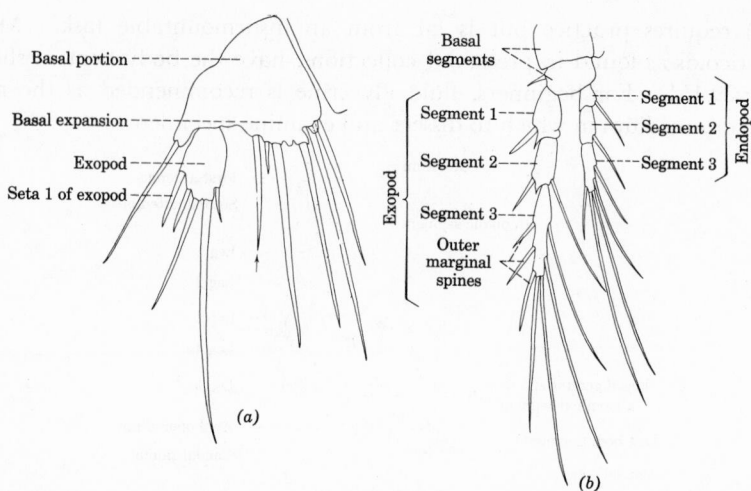

Fig. 29.153. *Canthocamptus oregonensis* ♀, indicating terminology used in key for legs 1 to 5. (a) Left leg 5 and inner portion of base of right leg 5. (In the key, setae are numbered from the outer margin as shown for exopod seta 1. References to the number of setae on the basal portion include only those of the inner part labeled "Basal expansion"; the seta of the outer margin (marked by arrow) is not considered. This figure is an example of a segmented leg; unsegmented legs in which exopod and basal portion are fused are shown in Figs. 29.160, 29.165, and 29.185.) (b) Left leg 3. (This example has an outer spine and inner seta on exopod segment 2, and a total of 7 spines and setae on exopod segment 3. Endopod segment 3 (or apical segment) has total of 5 setae. References in key reading "exopod segment 3 with 6,7,7 spines and setae" refer respectively to the total number on exopod segment 3 of legs 2, 3, and 4.)

attempting to use the key. The types of leg 1 found in the included species are shown in Fig. 29.155 and throughout the key. Length measurements refer to the dorsal mid-line from the rostrum base to the end of the caudal ramus. These have been compiled from literature as well as personal observation. They have little taxonomic value. Harpacticoids are difficult to measure accurately because of the tendency of the body segments to retract within one another, particularly in preserved specimens. In the genus *Attheyella*, extreme examples of such retractions are often found in collections, some specimens appearing to be only half of the length of the others.

There is often considerable sexual dimorphism in the Harpacticoida, and both sexes should be considered in identification and taxonomy. In most genera, the male is easily distinguished from the female by the differences in the first antennae (Fig. 29.154). These are a simple series of setiferous segments in the female. In the male, the similarly modified right and left antennae may have expanded segments, densely grouped setae, a specialized joint (geniculation) between two of the distal segments, and sometimes modification of the apical portion into a stout claw. The number of segments in the body of the adult is different in the two sexes. Most of the genera in the key have nine segments in the female and ten in the male. This difference results from the fusion in the female of the genital and first abdominal segment (Fig. 29.151), so that there are four segments posterior to the somite of the fifth pair of legs in the female, and five in the male. In the Canthocamptidae, there are frequently ventral, lateral, or even complete heavily sclerotized bands on the surface that give this fused segment the appearance of being incompletely or completely separated (as shown for *Attheyella alaskaensis*, Fig. 29.197*b*).

Other sexual differences are more diversely expressed throughout the genera. The caudal rami may be strikingly different in the two sexes of some species of a genus, but in other species they may be identical. Only slight differences, if any, appear in the first leg. The second to the fourth legs are very similar in the two sexes of some genera, but in most fresh-water genera there are structural differences in one or all of these legs. Some occur in the exopod, particularly affecting the third segment, but more commonly there are differences in the endopods, all or some of which may be modified in the male. *Canthocamptus* and *Moraria* are examples of genera in which the endopods of legs 2 to 4 of the male are all different from those of the female; these are illustrated for *Canthocamptus assimilis* in Fig. 29.194. The fifth legs are never alike, those of the male being reduced in size and having fewer setae on the inner basal expansion. The male may also have a well-developed sixth leg (Figs. 29.173, 29.177); in the female this leg is represented by one or two setae associated with the genital field.

Variation and Anomaly

Both variation and anomaly are common in harpacticoids. An attempt has been made in the key to point out the known variations, but it is impossible

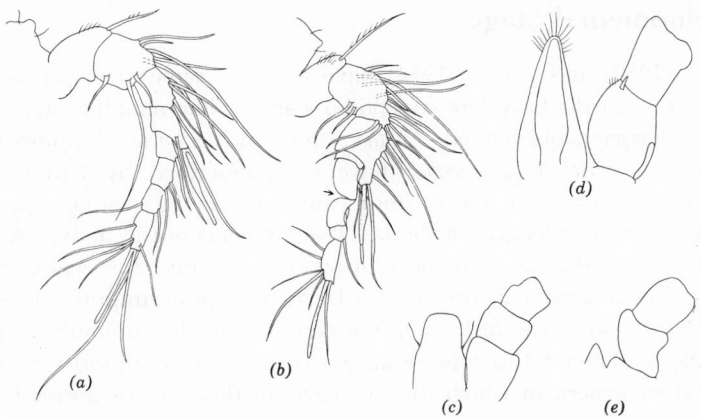

Fig. 29.154. *First antenna and rostrum.* (*a*) *Canthocamptus robertcokeri* ♀, small rostrum with apical papilla and 8-segmented first antenna. (*b*) *C. robertcokeri* ♂, rostrum and modified, geniculate first antenna (arrow indicates point of geniculation). (*c*) *Moraria duthiei*, large rostrum characteristic of genus and antennal base. (*d*) *Huntemannia lacustris*, large protruding rostrum and antennal base. (*e*) *C. oregonensis*, very small pointed rostrum and antennal base.

to cover this at all well when dealing with such an inadequately studied fauna. It is to be expected that individual specimens will be encountered that do not entirely agree with the setation given for the legs, or with the structural characteristics of the caudal rami. For the most part, these are fairly constant characters, but in some species, as indicated, variation of armature is the rule. Usually the segmentation of the first antennae and that of the rami of the legs is constant, but in various genera there is a tendency in some species for variation in the separation of certain segments of these appendages. Sometimes fusion of segments appears as an anomaly rather than a variation, inasmuch as it may affect only one ramus or appendage of a pair and be accompanied by eccentric development of some parts. Other anomalies that will be found are asymmetry of the caudal rami in which one ramus is much shorter than the other; and the "twinning" of setae, in which two setae, particularly on the inner margin of the segment of a leg, arise from the same cuticular indentation.

In the key, two incompletely known groups that seem to be widely distributed and variable, have been referred to respectively as the *Bryocamptus hiemalis* and *B. minutus* complexes. These may include several species, subspecies, local races, or be merely variants of a single species. Before a sound taxonomic interpretation of their status can be made, far more knowledge of their distribution and variation over the entire continent must be available. This is also probably true of some other groups of closely related species in the key, such as *Moraria laurentica* and *M. mrazeki; Attheyella illinoisensis* and *A. nordenskioldii;* and the three species of the *Attheyella dentata* group (*A. dentata, A. americana, A. dogieli*). In the present state of knowledge, it is possible to separate only the nominate forms of these groups.

Developmental Stages

Gurney (1932) and Coker (1934) have included some descriptions of immature forms, particularly late copepodid stages. The nauplius stages of an American harpacticoid called *Canthocamptus northumbricus* (presumably = *Attheyella dentata* or *A. americana*) have been described by Ewers (1930). Late copepodid stages of most canthocamptids can be distinguished from adults by the relative lengths of the last two segments of the body. As in the Cyclopoida, the last body segment is longer than the preceding segment in the immature, but is shorter in the adult. In many canthocamptids, the separation of the exopod of the fifth leg is not complete in the subadult stage, and care should be taken not to confuse an immature canthocamptid with mature forms of other genera in which the two parts of this leg are normally fused.

Figures not otherwise credited are originals by M. S. Wilson.

KEY TO SPECIES

1a   Leg 1 ♀ ♂, distal exopod segment armed apically with 3 to 5 short, clawlike spines, all much shorter than segment (Fig. 29.155*a*)   8

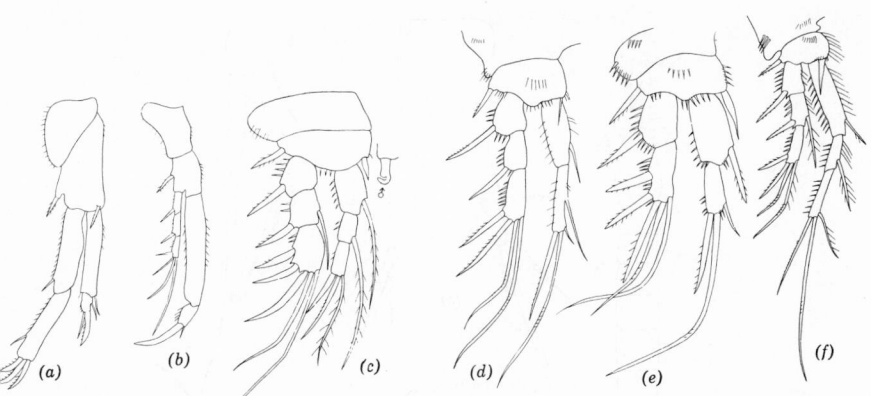

Fig. 29.155. Types of leg 1 illustrating characters referred to in couplets 1–7 of key. Each figure shows the two basal segments, the exopod (at left) and the endopod (at right) of the left leg of the first pair. (*a*) *Harpacticus chelifer*. (*b*) *Heterolaophonte stromi*. (*c*) *Nitocra lacustris* ♀, with detail of modified inner basipodal spine of ♂. (*d*) *Moraria duthiei*. (*e*) *Maraenobiotus insignipes*. (*f*) *Canthocamptus oregonensis*. (*a, b* after Sars.)

1b Leg 1 ♀ ♂, distal exopod segment armed with normal spines and setae, some of which are longer than segment (Fig. 29.155*b–f*) **2**

2a (1) Leg 1 ♀ ♂, exopod 2-segmented (Figs. 29.155*e*, 29.159*e*, 29.173*a*) . . . **9**

2b Leg 1 ♀ ♂, exopod 3-segmented (Fig. 29.155*b–d,f*) **3**

3a (2) Leg 1 ♀ ♂, exopod segment 3 with total of 5 to 6 spines and setae (Fig. 29.155*c*). **15**

3b Leg 1 ♀ ♂, exopod segment 3 with total of 4 spines and setae **4**

4a (3) Leg 1 ♀ ♂, exopod segment 2 with inner seta (Fig. 29.155*f*) **7**

4b Leg 1 ♀ ♂, exopod segment 2 without inner seta (Fig. 29.155*b,d*) . . . **5**

5a (4) Leg 1 ♀ ♂, exopod segment 2 with outer spine (Fig. 29.155*b,d*) **6**

5b Leg 1 ♀ ♂, exopod segment 2 without outer spine (Fig. 29.156) . . . .
Parastenocaris Kessler

Cosmopolitan genus with many species, found in sand and vegetation of all types of fresh-water habitats, including subterranean waters. Genus may be recognized by characters of leg 1, slender vermiform habitus, 2-segmented exopod of leg 3 ♀, reduced endopods of legs 2 to 4 ♀ (1-segmented or a single spine), modified leg 3 ♂ (Fig. 29. 156*f*), and reduced platelike form of leg 5 of both sexes. See Chappuis (1958) for a key to the males of the species known to 1957. Four species have been recorded from N. A.:

P. brevipes Kessler, from sandy beaches of Wisconsin lakes; a new record is from bog mats in Michigan. An uncertain record is that of C. B. Wilson (1932) from Massachusetts (=*P. wilsoni* Borutsky 1952). (Descr. Kessler, 1913.)

P. starretti Pennak (Fig. 29.156), from sandy beach of Wisconsin lake. (Descr. Pennak, 1939.)

P. lacustris Chappuis and *P. delamarei* Chappuis, from the Canadian banks of Lake Erie. (Descr. in Chappuis and Delamare Debouttteville, 1958.)

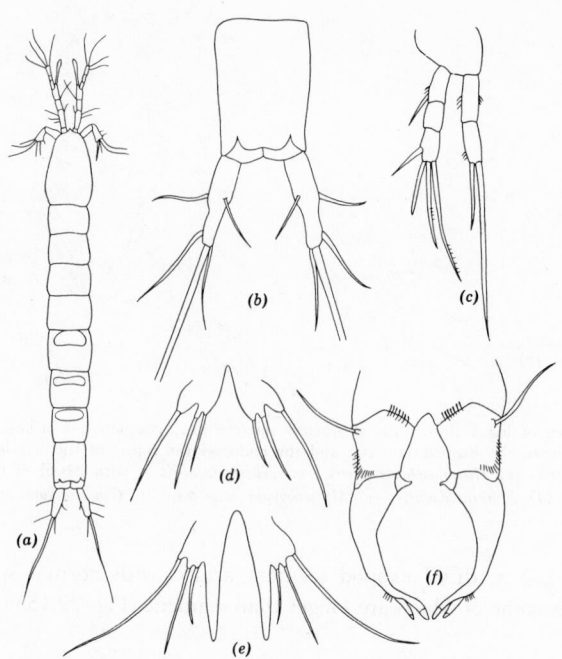

Fig. 29.156. *Parastenocaris starretti.* (*a*) Dorsal view ♀. (*b*) Dorsal view last body segment and caudal rami. (*c*) Leg 1 ♀. (*d*) Leg 5 ♂ (pair). (*e*) Leg 5 ♀ (pair). (*f*) Leg 3 ♂. (After Pennak.)

6a	(5)	Leg 1 ♀ ♂, endopod 1-segmented (Fig. 29.173)	**22**
6b		Leg 1 ♀ ♂, endopod 2-segmented (Fig. 29.155*b,d*)	**10**
6c		Leg 1 ♀ ♂, endopod 3-segmented (Fig. 29.155*c,f*)	**20**
7a	(4)	Leg 1 ♀ ♂, endopod 2-segmented (Fig. 29.155*b,d*)	**31**
7b		Leg 1 ♀ ♂, endopod 3-segmented (Fig. 29.155*c,f*)	**34**

8a (1) Leg 5 ♀, basal expansion with 3 or 4 setae; leg 5 ♂, basal portion not expanded inside, without seta. Leg 2 ♂, endopod segment 2 produced into outer marginal process. . . . .
Harpacticus Milne-Edwards

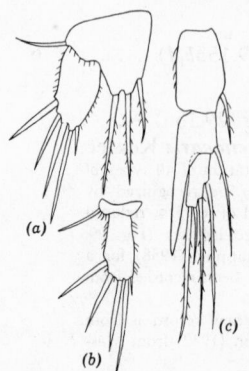

Two brackish-water species recorded from fresh water, Mass. (C. B. Wilson, 1932). ***H. chelifer*** (O. F. Müller) 1776 (Figs. 29.155*a*, 29.157) has 3 claws and no setae on apex of exopod of leg 1; leg 5 ♀ has 5 setae on exopod and 3 on basal expansion. Length ♀ ♂ about 0.9–1.0 mm. Record of ***H. gracilis*** Claus is questionable. (Descr. both species, Sars, 1903–1911.)

◄ **Fig. 29.157.** *Harpacticus chelifer.* (*a*) Leg 5 ♀. (*b*) Leg 5 ♂. (*c*) Endopod leg 2 ♂. (After Sars.)

8b Leg 5 ♀, basal expansion with 5 or 6 setae; leg 5 ♂, basal portion slightly expanded inside, with 1 seta. Leg 2 ♂, endopod segment 2

produced into both inner and outer marginal processes . . . .
Tigriopus Norman

T. californicus (Baker) 1912 (Fig. 29.158), known from tidal pools of varying salinity, Pacific coast (Vancouver Island to Calif.); from inland fresh waters, S. A. and Africa. Leg 5 ♀, basal expansion with 5 setae; leg 5 ♂, exopod with 5 setae. Length: ♀ 0.96–1.4 mm; ♂ 0.94–1.1 mm. Includes **T. triangulus** Campbell 1930. (Descr. Monk, 1941.)

◄ **Fig. 29.158.** *Tigriopus californicus.* (*a*) Leg 5 ♀. (*b*) Leg 5 ♂. (*c*) Endopod leg 2 ♂. (After Campbell.)

9a (2) Leg 1 ♀ ♂, endopod shorter than or about as long as exopod. **22**

9b Leg 1 ♀ ♂, endopod much longer than exopod (about 2 times) . . . . **Onychocamptus** Daday

Genus has legs 2 to 4 ♀ ♂ with 3-segmented exopods; endopods ♀ 2-segmented. Two species known from fresh and inland saline waters in N. A.

O. mohammed (Blanchard and Richard) 1891. First antenna ♀ 5-segmented. Legs 2 to 4: exopod segment 3 ♀ with 6,6,5 spines and setae, ♂ with 6,6,6; endopod segment 2 ♀ with 4,6,3 setae; endopod leg 3 ♂ differs from ♀, 3-segmented, apical segment with 4 setae. Leg 5 ♀ ♂ as in Fig. 29.159. ♀ with 1 or 2 egg sacs. Length range ♀ ♂ 0.6–0.8 mm. Quebec, Saskatchewan; N. D., Mass.; Yucatan; Europe; Asia; Africa. Recorded in literature as **Laophonte mohammed;** includes **L. calamorum** Willey 1923. (Descr. Gurney, 1932.)

O. talipes (C. B. Wilson) 1932, known from coastal fresh water, Mass.; distinguished by modified leg 4 ♀ (apical exopod and endopod segments lamellar, without normal spines or setae).

◄ **Fig. 29.159.** *Onychocamptus mohammed.* (*a*) Dorsal view ovigerous ♀. (*b*) Dorsal view last body segment and caudal rami ♀. (*c*) Leg 5 ♀. (*d*) Leg 5 ♂. (*e*) Leg 1. (After Gurney.)

10a (6) Leg 1 ♀ ♂, endopod not reaching beyond exopod segment 2 . . . . . . **21**

10b Leg 1 ♀ ♂, endopod reaching from about middle of exopod segment 3 to a little beyond end of segment 3 . . . . **11**

10c Leg 1 ♀ ♂, endopod reaching beyond exopod segment 3 by half of apical segment or more. **12**

11a (10) Leg 5 ♀ ♂ segmented (exopod segment distinctly separated from basal expansion) **24**

11b Leg 5 ♀ ♂ unsegmented (Fig. 29.160) . . . . **Stenocaris** Sars

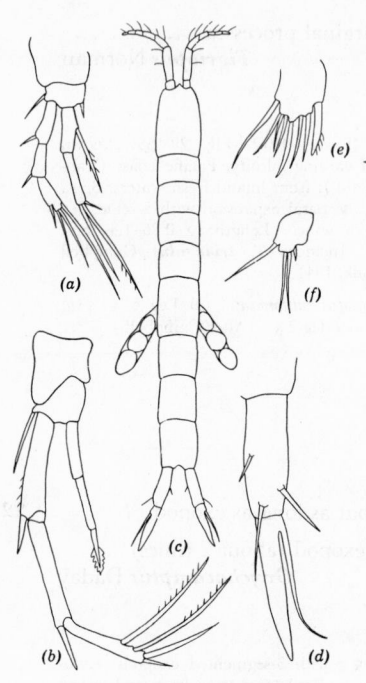

S. *minor* (T. Scott) 1892, marine-brackish species reported from fresh water, Mass. (C. B. Wilson, 1932). Length: ♀ 0.72–0.9 mm; ♂ 0.77 mm. (Descr. ♀ Sars, 1903–1911; ♂ Kunz, 1938).

Fig. 29.160. *Stenocaris minor.* (*a*) Leg 1. (*b*) Leg 4 ♀. (*c*) Dorsal view ovigerous ♀. (*d*) Caudal ramus. (*e*) Leg 5 ♀. (*f*) Leg 5 ♂. (♀ after Sars; ♂ after Kunz.)

12a	**(10)**	Leg 1 ♀ ♂, apex of endopod with single stout spine; if seta present, it is minute or hairlike. **13**
12b		Leg 1 ♀ ♂, apex of endopod with stout spine and longer seta, or with 2 or more well-developed setae **14**
13a	**(12)**	Legs 2 to 4 ♀, endopods 2-segmented and segment 2 with 4,6,4 setae. Legs 2 to 3 ♂, exopod 3 modified (directed inward and spines very stout as in Fig. 29.161a); leg 4 ♂ like ♀; leg 5 ♂ unsegmented . . . . .

Heterolaophonte strömi (Baird) 1837

Euryhaline species reported from coastal ponds, Mass. See Fig. 29.155*b* for leg 1. Length: ♀ about 0.88 mm; ♂ about 0.77 mm. (Descr. Sars, 1903–1911, as *Laophonte strömi.*)

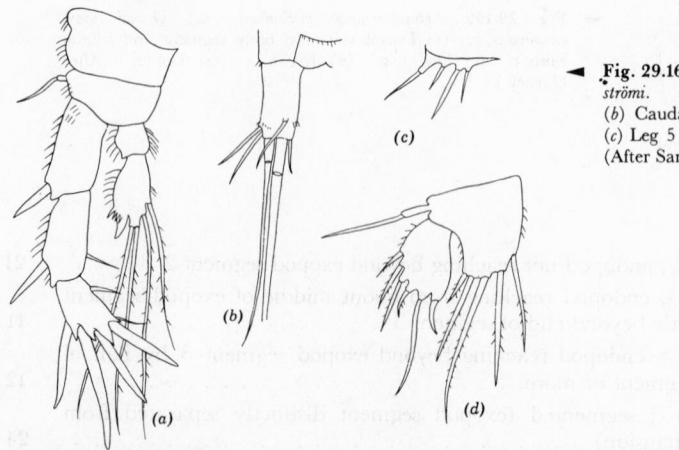

Fig. 29.161. *Heterolaophonte strömi.* (*a*) Leg 3 ♂. (*b*) Caudal ramus, dorsal. (*c*) Leg 5 ♂. (*d*) Leg 5 ♀. (After Sars.)

13b	Legs 2 to 4 ♀, endopods 2-segmented and segment 2 with 4,5,4 setae. Legs 2 to 4 ♂, exopod 3 modified; leg 4 ♂, endopod 1-segmented;

leg 5 ♂, exopod separated . . . .

Pseudonychocamptus proximus (Sars) 1908

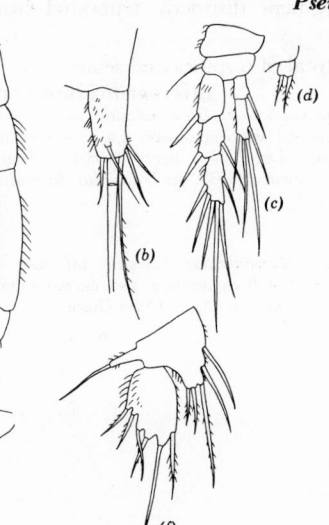

Euryhaline species reported from fresh
water, Mass. Length: ♀ 0.65–0.9 mm; ♂ 0.6
mm.   (Descr. ♀ Sars, 1903–1911 as
Laophonte proxima. ♂ Klie, 1929.)

◄ **Fig. 29.162.** *Pseudonychocamptus proximus.*
(*a*) Leg 1. (*b*) Caudal ramus, dorsal.
(*c*) Leg 4 ♀. (*d*) Endopod leg 4 ♂. (*e*) Leg 5
♂. (*f*) Leg 5 ♀. (♀ after Sars; ♂ after Klie.)

14a (12, 20) Legs 2 to 4 ♀, endopods 3-segmented.   Leg 2 ♂, endopod 2-
segmented, segment 2 with outer or apical modified spine (expanded
and enlarged as articulated spine or as enlarged process) . . . .

Schizopera Sars

Not yet reported from North American continental waters, but species are known from
coastal and inland fresh and saline waters over much of the world.   Two new brackish
species described by Kiefer (1936) from Haiti, *S. triacantha* and *S. haitiana.*

14b Legs 2 to 4 ♀, endopods 2-segmented or lacking (may be 1-segmented
in leg 4).   Leg 2 ♂, endopod if present, 2-segmented and with
normal setae. . . . . ***Paracamptus*** Chappuis   **25**

15a (3) Leg 1 ♀ ♂, exopod segment 3 about same length as segment 2. . . . .   **16**

15b Leg 1 ♀ ♂, exopod segment 3 reduced, about ⅓ the length of exopod
segment 2 . . . . ***Paradactylopodia*** Lang

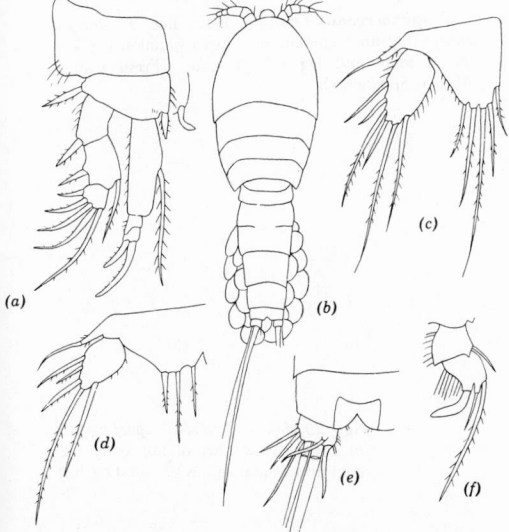

P. brevicornis (Claus) 1866,
euryhaline species known from
Atlantic coast.   Length: ♀ ♂
0.5–0.65 mm.   (Descr. Sars,
1903–1911, as ***Dactylopusia
brevicornis.***)

◄ **Fig. 29.163.** *Paradactylopodia
brevicornis.* (*a*) Leg 1. (*b*) Dorsal view ovigerous ♀. (*c*) Leg 5
♀. (*d*) Leg 5 ♂. (*e*) Dorsal
view last body segment and
caudal ramus. (*f*) Endopod
leg 2 ♂. (After Sars.)

16a   (15) Leg 5 ♀ ♂ unsegmented (Figs. 29.164–29.166) . . . .   1?

16b   Leg 5 ♀ ♂ segmented (exopod segment distinctly separated from
                basal expansion) . . . .   1?

17a   (16) Leg 1 ♀ ♂, exopod segment 3 with total of 6 spines and setae . . . . .
                                                             Microarthridion Lang

> *M. littorale* (Poppe) 1881, euryhaline species known from Atlantic coast. First
> antenna ♀ 6-segmented. Legs 1 to 4 ♀ ♂: exopod segment 3 with 6,6,6,5 spines and
> setae; endopod segment 3 with 5,5,6,5 setae. Leg 5 ♀ ♂ very reduced. Length:
> ♀ 0.45–0.57 mm; ♂ 0.4–0.46 mm. (Descr. Gurney, 1932, as *Tachidius littoralis*.)

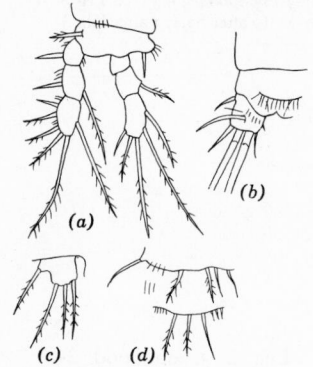

◄ **Fig. 29.164.** *Microarthridion littorale.* (*a*) Leg 1.
(*b*) Dorsal view last body segment and caudal ramus.
(*c*) Leg 5 ♀. (*d*) Legs 5 and 6 ♂. (After Gurney.)

17b   Leg 1 ♀ ♂, exopod segment 3 with total of 5 spines and setae . . . . .
                                                             Tachidius Lilljeborg

Two species known from North American coastal bodies of water.

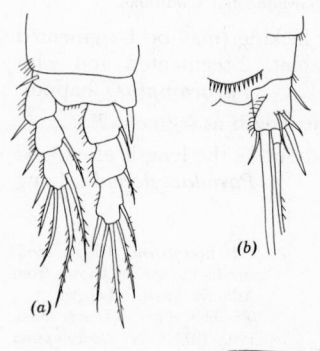

T. discipes Giesbrecht 1881 (= *T. brevicornis*),
euryhaline species reported from Atlantic and
Canadian Arctic coasts. First antenna ♀ 7-segmented.
Legs 1 to 4: exopod segment 3 ♀ ♂ with 5,6,6,5 spines
and setae; endopod 3 ♀ with 5,5,5,5 setae; leg 2 ♂,
endopod modified, segment 3 with 3 apical setae.
Leg 5 ♀ with 9 setae; leg 5 ♂ with 7 setae. Length:
♀ about 0.6 mm; ♂ 0.5 mm. (Descr. Gurney, 1932.)

T. spitzbergensis Oloffson 1917, like *T. discipes*
except that anal operculum ♀ lacks spinules, leg 5 ♀
has 8 setae and leg 5 ♂ 6 setae. Fresh water.
Alaska; Spitsbergen.

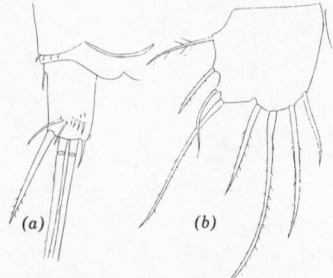

Fig. 29.165. *Tachidius discipes.*
(*a*) Leg 1. (*b*) Dorsal view last
body segment and caudal ramus.
(*c*) Leg 5 ♀. (*d*) Leg 5 ♂. (After
Sars.)

Fig. 29.166. *Tachidius spitzbergensis.*
(*a*) Dorsal view part of last body segment and caudal ramus ♀. (*b*) Leg 5 ♀.

8a   **(16)**   Legs 2 to 3 ♀ ♂, exopod segment 3 with 3 outer marginal spines . . .
                                         Nitocra Boeck   **19**

8b   Legs 2 to 3 ♀ ♂, exopod segment 3 with 2 outer marginal spines . . .
                                 Nitocrella Chappuis

Genus with variable species for which many important characters are incompletely known. Leg 1 may have 4 or 5 spines and setae on exopod segment 3; and exopod segment 2 may have inner seta or not. Legs 2 to 4, endopods 1-, 2-, or 3-segmented; setal formula of legs 1 to 5 may vary within species. Most described species are from subterranean brackish and fresh waters of Europe. Two species reported from American waters have characters of leg 1 that fit into this portion of key.

N. subterranea (Chappuis) 1928, subterranean European species, reported from brackish and fresh-water caves in Yucatan. Legs 2 to 4 ♀ ♂: exopod segment 3 with 5,5,5 spines and setae; endopods 3-segmented, segments 1 and 2 with inner seta, segment 3 with 2,3,2 setae. Leg 5 ♀ ♂ as in Fig. 29.167. Length unknown.

N. incerta (Chappuis) 1933, from fresh water of Curacao, Caribbean Sea; may occur in southern N. A. Original description as *Nitocra lacustris incerta*. Leg 1 as in *Nitocra lacustris* (Fig. 29.155c). Legs 2 to 4 ♀: exopod segment 3 with 6,5,6 spines and setae; endopods 3-segmented, segment 1 without seta, segment 3 with 2,4,4 setae. Leg 5 ♀, exopod with 6, basal expansion with 4 setae. Length: ♀ 0.3 mm. ♂ unknown.

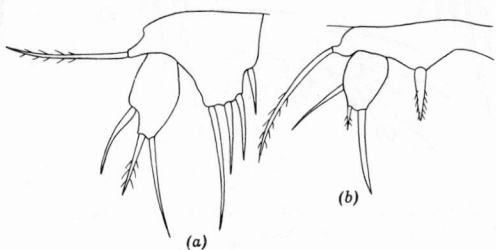

Fig. 29.167. *Nitocrella subterranea.* (*a*) Leg 5 ♀. (*b*) Leg 5 ♂. (After Chappuis.)

9a   **(18)**   Leg 1 ♀, ♂, endopod segment 1 shorter than total exopod and leg 2 ♀, ♂, endopod segment 3 with 3 setae (Figs. 29.155c and 29.168) . .
                       Nitocra lacustris (Schmankewitsch) 1875

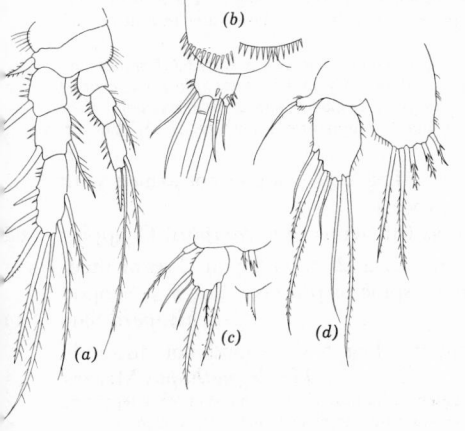

See Fig. 29.155c for leg. 1. Legs 2 to 4 ♀ ♂: exopod segment 3 with 7,7,7 spines and setae; endopod 1, no inner seta; endopod 2, inner seta; endopod 3 with 3,5,5 setae. Leg 5 ♀, exopod with 6 or 5 setae, basal expansion 5 setae; leg 5 ♂, exopod 6 or 7 setae, basal expansion 2 setae. Length: ♀ 0.45–0.62 mm; ♂ 0.4–0.45 mm. Mexico; Tex. (Descr. Gurney, 1932.)

◄ **Fig. 29.168.** *Nitocra lacustris.* (*a*) Leg 2. (*b*) Dorsal view part of last body segment and caudal ramus. (*c*) Leg 5 ♂ with variation of setae of basal expansion. (*d*) Leg 5 ♀. (From Tex. coast specimens.)

19b   Leg 1 ♀, ♂ as in **19a** and leg 2 ♀, ♂, endopod segment 3 with 4 setae . . . . ***Nitocra spinipes*** Boeck 1864
      N. spinipes Boeck 1864. Legs 2 to 4 ♀ ♂: exopod segment 3 with 7,7,7 spines and setae; endopod 1 and 2 with inner seta; endopod 3 with 4,5,5 setae. Leg 5 ♀, exopod

5 or 6 setae, basal expansion 5 setae; leg 5 ♂, exopod 6 setae, basal expansion 3 to 5 setae. Length: ♀ and ♂ 0.66–0.8 mm. Hudson Bay; Atlantic coast; Mexico; Alaska. (Descr. Gurney, 1932.)

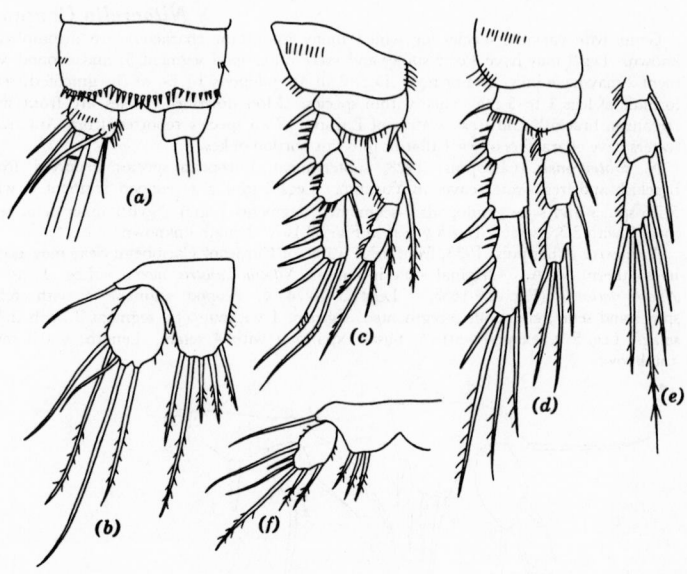

Fig. 29.169. *Nitocra spinipes.* (*a*) Dorsal view last body segment and caudal rami. (*b*) Leg 5 ♀. (*c*) Leg 1 (*d*) Leg 2. (*e*) Endopod leg 3. (*f*) Leg 5 ♂. (After Sars.)

19c   Leg 1 ♀ ♂, endopod segment 1 reaching about to end of exopod . . .
                                                                    Nitocra sp.
         Species of *Nitocra* with this character of leg 1 not reported from fresh water in N. A., but that may occur:
         N. hibernica (Brady) 1880. Legs 2 to 4 ♀ ♂: exopod segment 3 with 5,5,7 spines and setae; endopod 1, no inner seta; endopod 2 with inner seta; endopod 3 with 2,3,3 setae. Leg 5 ♀ ♂, exopod with 6 setae, basal expansion 5 setae. Length: ♀ 0.56–0.75 mm; ♂ 0.51–0.6 mm. Widely distributed species of fresh and brackish water in Europe and Asia. (Descr. Gurney, 1932.)
         N. typica Boeck 1864. Legs 2 to 4 ♀ ♂: exopod segment 3 with 6,6,7 spines and setae; endopod 1 and 2 with inner seta; endopod 3 with 4,5,5 setae. Leg 5 ♀, exopod with 6 setae, basal expansion 5 setae; leg 5 ♂, exopod 6 setae, basal expansion 4 or 3 setae. Length: ♀ 0.6–0.7 mm; ♂ 0.5 mm. Coastal brackish waters N. A., Europe, Africa. (Descr. Gurney, 1932.)

20a (6) Leg 2 ♀ ♂, endopod 2-segmented. Leg 2 ♂, endopod not armed with outer apical modified spine or process . . . .
                                Bryocamptus (subgenus *Limocamptus*) Chappuis   62

20b   Leg 2 ♀, endopod 3-segmented. Leg 2 ♂, endopod 2-segmented, armed with outer apical modified spine or process. Leg 5 ♀, exopod separated . . . . *Schizopera* Sars   14

20c   Leg 2 ♀ ♂, endopod 3-segmented. Leg 5 ♀, exopod not distinctly separated . . . . *Phyllognathopus* Mrazek
         One variable species in genus: *P. viguieri* (Maupas) 1892. Somite of leg 1 separated from cephalic segment. Caudal setae variously developed, may be swollen in ♀ or normal and elongate. Legs 1 to 4 ♀ ♂, exopod segment 2 without inner seta; leg 4, exopod segment 2 without outer spine, endopod 2 or 3 segmented; endopods ♂ like ♀ or differing slightly. Leg 5 ♀ ♂ variable, exopod ♂ separated or not. Length: ♀ 0.35–0.57 mm; ♂ 0.31–0.57 mm. Found in wide variety of habitats: margins of lakes (both in water and sandy banks), pools, springs, subterranean waters, in vegetation; recorded from aquaria and from Bromeliaceous and other plants. Reported from N. J., Mich., Wis. Europe, Africa, Pacific tropical islands. (Descr. Gurney, 1932.)

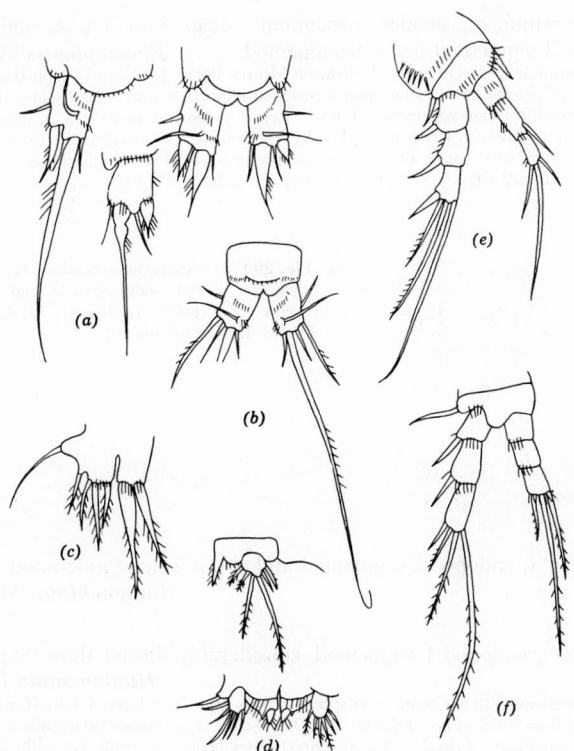

Fig. 29.170. *Phyllognathopus viguieri.* (*a*) Variations of caudal rami and setae ♀. (*b*) Caudal ramus ♂. (*c*) Leg 5 ♀. (*d*) Leg 5 ♂, 2 variations. (*e*) Leg 1. (*f*) Leg 3. (After Gurney.)

21a (10) Fore part of body ♀ ♂ expanded, gradually tapered posteriorly. Legs 2 to 4 ♀ ♂, endopods 3-segmented . . . . ***Metis*** Philippi

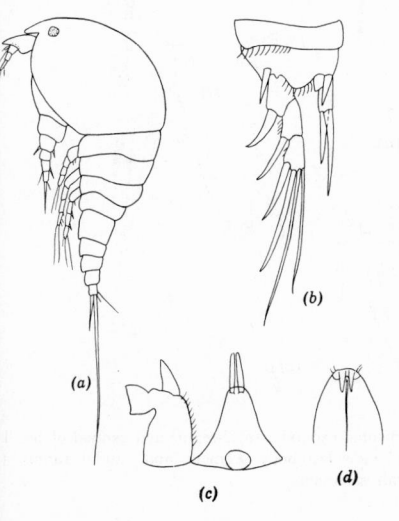

M. jousseaumei (Richard) 1892, euryhaline species known from brackish and fresh water of Atlantic coast. Rostrum ♀ ♂ with 2 apical spines. Legs 2 to 4: exopod segment 3 with 5,6,6 spines and setae; endopod segment 3 with 4,4,4 setae. Leg 5 a reduced plate in both sexes. Length: ♀ ♂ 0.35–0.59 mm. (Descr. Sharpe, 1910 as *M. sarsi;* Gurney, 1927; Nicholls, 1941.)

◄ **Fig. 29.171.** (*a*) *Metis* sp. lateral view ♀. *M. jousseaumei:* (*b*) Leg 1. (*c*) Rostrum, dorsal. (*d*) Rostrum, ventral, with spines turned under. (*a* modified from Sars; *b* after Sharpe; *c, d* after Gurney.)

21b   Body vermiform, slender throughout. Legs 2 to 3 ♀ ♂, endopods
usually 2-segmented, leg 4 1-segmented . . . . ***Epactophanes*** Mrazek
One variable species in genus: *E. richardi* Mrazek 1893. Legs 2 to 4: exopod segment
3 ♀ ♂ with 5,5,5 spines and setae; apical endopod segment ♀ with 1 or 2 setae; endopod
leg 3 ♂ modified and extremely variable. Leg 5 ♀, exopod with 3 to 5 setae, basal
expansion usually with 4 or 5 setae. Leg 5 ♂ unsegmented, represented by 3 short setae.
Length: ♀ ♂ 0.3–0.45 mm. Fresh water, usually in moss. Cosmopolitan species known
from Canada; Alaska, N. Y. (Descr. Gurney, 1932; Lang, 1934.)

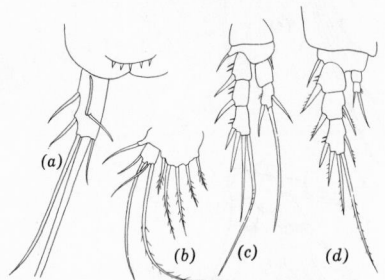

◀ **Fig. 29.172.** *Epactophanes richardi.* (*a*) Dorsal
view part of last body segment and caudal
ramus. (*b*) Leg 5 ♀. (*c*) Leg 1. (*d*) Leg 3 ♀.
(From Alaskan specimens.)

22a (6, 9)   Leg 1 ♀ ♂, endopod 2-segmented and about as long as exopod. . . .
Maraenobiotus Mrazek   23

22b   Leg 1 ♀ ♂, endopod 1-segmented, considerably shorter than exopod
Huntemannia Poppe
Marine genus with one fresh-water species known from Bear Lake, Utah: *H. lacustris*
M. S. Wilson 1958 (Figs. 29.154*d*, 29.173). Leg 1 ♀ ♂, exopod segments 2 and 3
fused or separated. Legs 2 to 4 ♀ ♂: exopods 2-segmented, segment 2 ♀ with 5,6,6 or
5,5,6 spines and setae; segment 2 ♂ with 5,8,7 spines and setae; endopods ♀ ♂ 1-
segmented or papilliform, with 1 or 2 setae. Leg 5 ♀, exopod, length 2 times width,
with 5 setae; basal expansion with 4 setae; leg 5 ♂ unsegmented, with total of 9 setae.
Two egg sacs. Length: ♀ 0.8–0.86 mm; ♂ 0.8–0.95 mm.   Closely related to the marine
species, *H. jadensis* Poppe.

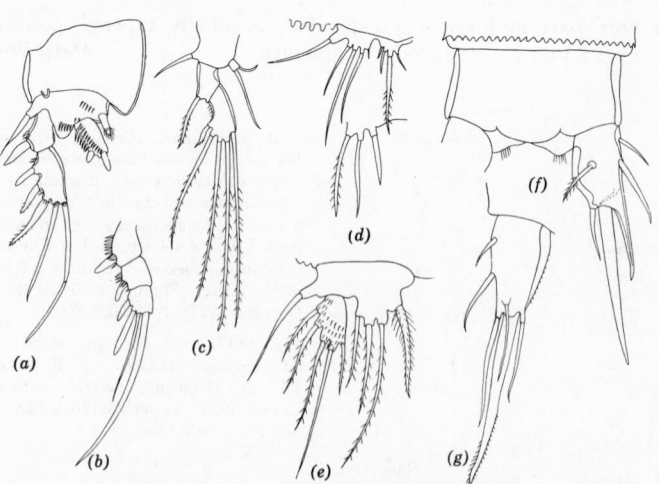

Fig. 29.173. *Huntemannia lacustris.* (*a*) Leg 1 with 2-segmented exopod. (*b*) 3-segmented exopod of leg 1
(*c*) Leg 2 ♂. (*d*) Legs 5 and 6 ♂. (*e*) Leg 5 ♀. (*f*) Dorsal view last body segment and caudal ramus ♂.
(*g*) Ventral view caudal ramus ♀. (From Bear Lake, Utah specimens.)

23a   **(22)**   Legs 1 to 4 ♀ ♂, apical exopod segment with 5,5,6,5 spines and setae . . . . ***Maraenobiotus insignipes*** (Lilljeborg) 1902

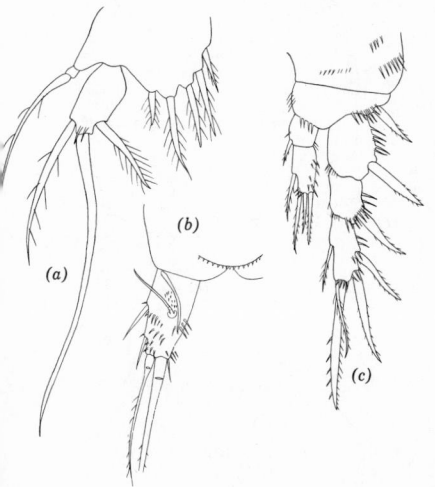

(b)

(a)

(c)

Legs 1 to 4 ♀, endopod segment 2 with 3,4,5,5 setae (see Fig. 29.155e for leg 1). Leg 5 ♂, exopod with 4 setae, basal expansion 2 setae. Widely distributed and variable species. Fresh water, margins of lakes, ponds, waterfalls, usually in vegetation. Length: ♀ 0.6–0.75 mm; ♂ 0.45–0.57 mm. Alaska; western Canada; Greenland; Europe; Asia. (Descr. Oloffson, 1917.)

◄ **Fig. 29.174.** *Maraenobiotus insignipes.* (a) Leg 5 ♀. (b) Dorsal view part of last body segment and caudal ramus ♀. (c) Leg 2. (From Alaskan specimens.)

23b   Legs 1 to 4 ♀ ♂, apical exopod segment with 5,4,5,5 spines and setae . .
                               M. brucei (Richard) 1898

Caudal ramus, setation endopods legs 1 to 4 ♀, leg 5 ♀ ♂, length and habitat similar to those of *M. insignipes.* Alaska; Europe; Asia; Africa. (Descr. Oloffson, 1917.)

◄ **Fig. 29.175.** *Maraenobiotus brucei.* Leg 2. (From Alaskan specimen.)

24a   **(11)**   Legs 2 to 4 ♀ ♂, exopod segment 2 without inner seta (Fig. 29.183c,e) . . . . ***Moraria*** T. and A. Scott   **26**

24b   Legs 2 to 4 ♀ ♂, exopod segment 2 with inner seta . . . .
                           (in part) ***Bryocamptus*** Chappuis   **57**

25a   **(14)**   Legs 2 to 4 ♀ and legs 2 and 4 ♂ without endopods . . . .
                           Paracamptus reductus M. S. Wilson 1956
Caudal ramus of this genus has only middle caudal seta well developed. Legs 2 to 4 ♀ ♂: exopod segment 3 with 5,5,4 spines and setae; leg 3 ♂, endopod well developed with 2 apical setae. Length: ♀ 0.55–0.7 mm; ♂ 0.46–0.47 mm. Margins of fresh-water lakes, Alaska.

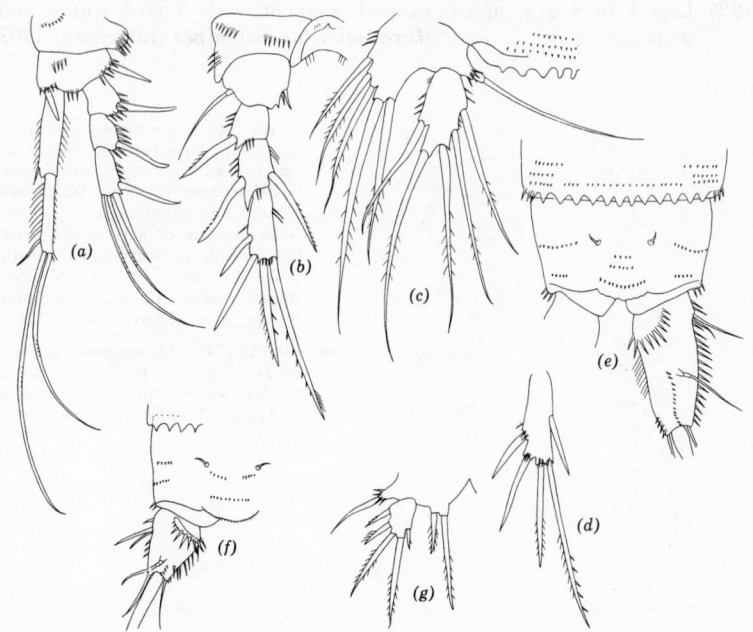

Fig. 29.176. *Paracamptus reductus.* (a) Leg 1. (b) Leg 4 ♀. (c) Leg 5 ♀. (d) Leg 2, exopod segment 3. (e) Dorsal view last body segment and caudal ramus ♀. (f) Same ♂. (g) Leg 5 ♂. (After M. S. Wilson.)

25b   Legs 2 to 4 ♀ and legs 2 and 4 ♂ with endopods . . . .
                              P. reggiae M. S. Wilson 1958

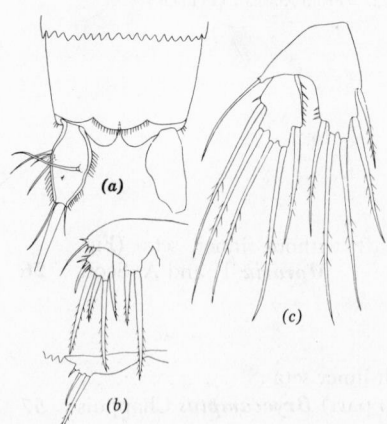

Legs 1 to 4 ♀ ♂, exopods as in *P. reductus;* endopods 2-segmented, apical segment ♀ with 2 setae. Length: ♀ 0.68 mm; ♂ 0.6 mm. Vegetation of ponds and pools, Alaska. Closely related to *P. schmeili* (Mrazck) of Europe.

◄ **Fig. 29.177.** *Paracamptus reggiae.* (a) Dorsal view last body segment and caudal ramus ♀. (b) Legs 5 and 6 ♂. (c) Leg 5 ♀. (From Alaskan specimens.)

26a (24) Legs 2 to 4 ♀, endopod segment 2 with 3,3,3 or 3,4,4 setae, or variants of these numbers. Leg 2 ♂, where known, endopod segment 2 with 2 setae (apical) **27**

26b   Legs 2 to 4 ♀, endopod segment 2 with 4,5,4 setae.   Leg 2 ♂, endopod segment 2 with 3 setae (2 apical, 1 inner) . . . .
Moraria duthiei (T. and A. Scott) 1896
See Figs. 29.151*a*, 29.154*c*, 29.155*d*. First antenna ♀ 8-segmented. Length: ♀ 0.79–
0.9 mm; ♂ 0.7–0.8 mm. Margins of lakes, Alaska, Canada. (Descr. Gurney, 1932.)

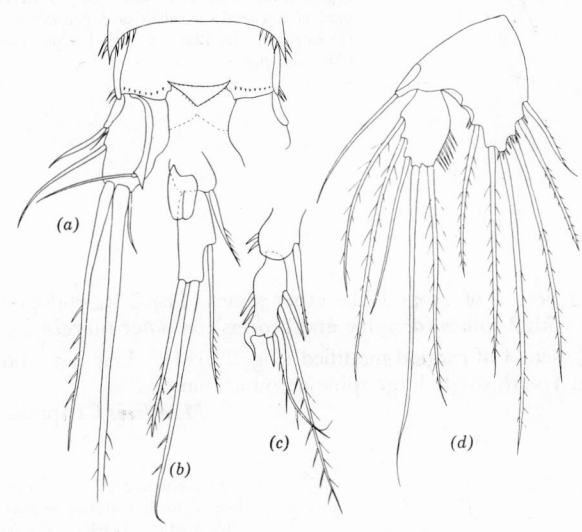

Fig. 29.178. *Moraria duthiei.* (*a*) Dorsal view last body segment and caudal ramus ♀.   (*b*) Endopod leg 2
♂.   (*c*) Endopod leg 4 ♂.   (*d*) Leg 5 ♀. (From specimens from Lake Kluane, Y. T., Canada.)

27a   **(26)** Anal operculum ♀ ♂ rounded distally **28**

27b   Anal operculum ♀ (and presumably that of ♂) pointed distally . . . .
M. virginiana Carter 1944

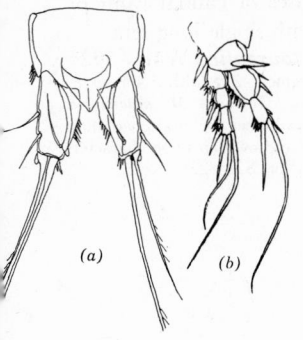

♂ unknown. First antenna ♀ 8-segmented. Legs 2 to
4, endopod segment 2 with 3 setae. Length: 0.45 mm.
Springs. Va.

◀ **Fig. 29.179.** *Moraria virginiana.* (*a*) Dorsal view last
body segment with operculum, and caudal rami ♀.
(*b*) Leg 1. (After Carter.)

28a   **(27)** Posterior margin of body segments ♀ ♂ not serrate **29**

28b   Posterior margin of body segments ♀ ♂ serrate . . . .
M. cristata Chappuis 1929

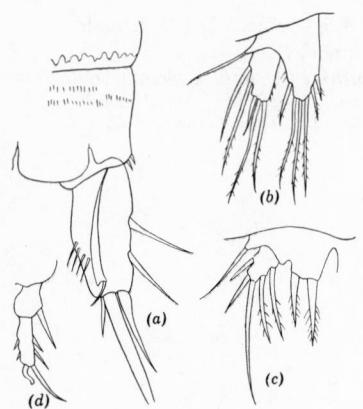

First antenna ♀ 7-segmented. Legs 2 to 4, endopod segment 2 ♀ with 3 setae. Length: ♀ ♂ about 0.6 mm. Ind., Ohio. (Descr. Chappuis, 1931.)

◄ **Fig. 29.180.** *Moraria cristata.* (*a*) Dorsal view of part of last body segment and caudal ramus ♀. (*b*) Leg 5 ♀. (*c*) Leg 5 ♂. (*d*) Endopod leg 4 ♂. (After Chappuis.)

29a (28) Leg 5 ♀, seta 4 of exopod like other setae. Leg 2 ♂, endopod segment 1 with 2 spines (or spine and process) on outer margin **30**

29b Leg 5 ♀, seta 4 of exopod modified (Fig. 29.181). Leg 2 ♂, endopod segment 1 with single large spine on outer margin. . . . .
  M. affinis Chappuis 1927

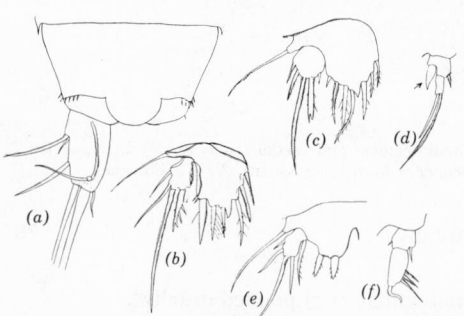

First antenna ♀ 7- or 8-segmented. Legs 2 to 4, endopod segment 2 ♀ with 3 setae. Length: ♀ ♂ 0.4–0.45 mm. N. Y., Alaska.

◄ **Fig. 29.181.** *Moraria affinis.* (*a*) Dorsal view last body segment and caudal ramus ♀. (*b*) Leg 5 ♀ (Alaska). (*c*) Leg 5 ♀ (N. Y.). (*d*) Endopod leg 2 ♂ (arrow indicates spine of outer margin). (*e*) Leg 5 ♂. (*f*) Endopod leg 4 ♂. (*c–f* after Chappuis.)

30a (29) Anal operculum ♀ ♂ enlarged, reaching to bases of caudal rami or beyond. Leg 4 ♂, inner margin of endopod with single long seta. . .
  M. laurentica Willey 1927
First antenna ♀ 7-segmented. Legs 2 to 4, endopod segment 2 ♀ with 3,4,4 setae. Length: ♀ 0.44 mm; ♂ 0.42 mm. Typical form, Quebec. Includes **M. americana** Chappuis 1927 from N. J. (Fig. 29.182*c–e*). It is not known whether the anal operculum and setation of the endopods of legs 2 to 4 ♀ are variable and overlap these characters in *M. mrazeki.* See comment under Variation and Anomaly, pp. 821–822.

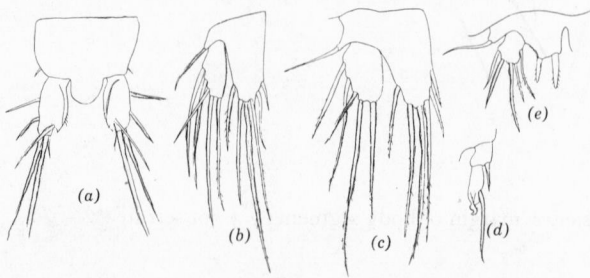

Fig. 29.182. *Moraria laurentica.* (*a*) Dorsal view last body segment with operculum, and caudal rami ♀. (*b*) Leg 5 ♀ (Canada). (*c*) Leg 5 ♀ (N. J.). (*d*) Endopod leg 4 ♂. (*e*) Leg 5 ♂. (*a, b* after Willey; *c–e* after Chappuis.)

30b Anal operculum ♀ ♂ not enlarged (Fig. 29.183). Leg 4 ♂, inner margin of endopod with more than one seta, or with seta and short spines . . . . **M. mrazeki** T. Scott 1902
  First antenna ♀ 7-segmented. Legs 2 to 4, endopod segment 2 with 3,4,4 setae or variable. Length: ♀ 0.33–0.62 mm. Canada; Alaska; Greenland; Europe. (Descr. Gurney, 1932.) (See Comment under **30a.**)

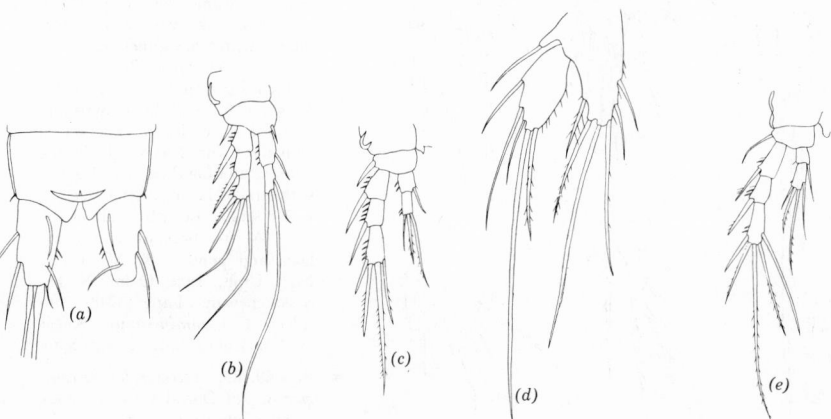

Fig. 29.183. *Moraria mrazeki.* (*a*) Dorsal view last body segment with operculum, and caudal rami ♀. (*b*) Leg 1. (*c*) Leg 2 ♀. (*d*) Leg 5 ♀. (*e*) Leg 4 ♀. (From Alaskan specimens.)

31a (7) Leg 1 ♀ ♂, endopod reaching to near end of exopod or beyond 32

31b Leg 1 ♀ ♂, endopod not reaching beyond exopod segment 2. . . . . . .
  Nannopus Brady

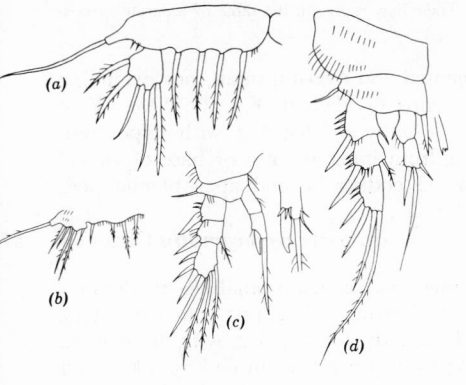

N. palustris Brady 1880, brackish-water species that may occur in coastal fresh waters. First antenna ♀ 5-segmented. Legs 2 to 4 ♀ ♂: exopod segment 3 with 6,7,7 spines and setae; legs 2 to 3, endopods 2-segmented, leg 4, 1-segmented, apical segments with 3,3,2 setae; leg 3 ♂, endopod slightly modified. Length: ♀ ♂ 0.55–0.7 mm. ♀ with 1 or 2 egg sacs. Canada. Includes *N. littoralis* Willey 1923. (Descr. Gurney, 1932; Sars, 1903–1911.)

◄ **Fig. 29.184.** *Nannopus palustris.* (*a*) Leg 5 ♀. (*b*) Leg 5 ♂. (*c*) Leg 3 ♂ with detail apex of endopod. (*d*) Leg 1 with variation on inner seta of basal segment 2. (*a* after Sars; *b, c* after Gurney; *d* modified from Sars and Gurney.)

32a (31) Leg 5 ♀ ♂, exopod segment distinctly separated from basal expansion 34

32b Leg 5 ♀, and usually that of ♂, exopod segment separated from basal expansion only by notch or gap . . . .
  Cletocamptus Schmankewitsch (= **Marshia** Herrick) 33

33a (32) Caudal ramus ♀ ♂, outer and middle caudal setae fused at base. . . .
                              C. albuquerquensis (Herrick) 1895

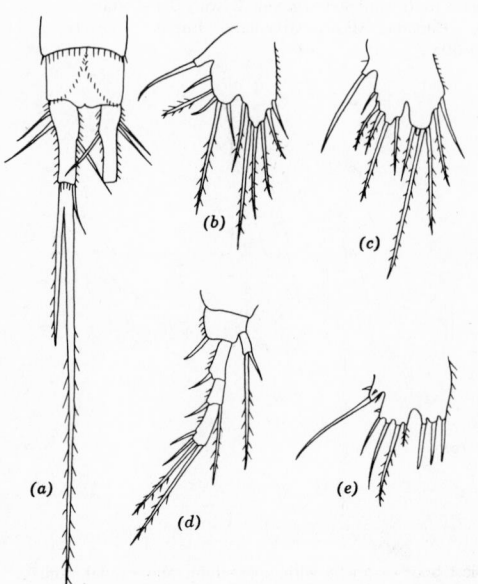

Commonly recorded in literature under genus *Marshia*. First antenna ♀ 6-segmented. Legs 2 to 4 ♀ ♂ (so far as known): exopod segment 3 with 5,5,4 spines and setae; legs 2 to 3 endopods 2-segmented, segment 2 ♀ with 3 and 5 setae, leg 3 ♂ segment 2 with hypophysis and 2 apical setae, leg 4 endopod 1-segmented with 2 setae. Including outer seta of basal portion, leg 5 ♀ with total of 11 or 12 setae; leg 5 ♂ with 8 setae. Length: ♀ about 0.6 mm. Mostly brackish and saline lakes and ponds. N. M., Colo., Nev., Utah, Tex., Okla., N. D.; Saskatchewan. Lang (1948) considers *C. dominicanus* Kiefer (1936), from Haiti, a synonym.

◄ **Fig. 29.185.** *Cletocamptus albuquerquensis.* (*a*) Dorsal view last body segment and caudal rami ♀. (*b, c*) Leg 5 ♀ variations. (*d*) Leg 4 ♀. (*e*) Leg 5 ♂. (After Herrick.)

33b   Caudal ramus ♀ ♂, outer and middle caudal setae separate from one
           another . . . . *Cletocamptus* spp.
           In this group, 3 inadequately known forms have been recorded from N. A. and the
           West Indies: *C. brevicaudata* (Herrick) 1895, N. M., Mass.; *C. bicolor* (C. B. Wilson)
           1932 (=*Attheyella bicolor*), Mass.; and *C. deitersi* (Richard) 1897, Haiti, Guatemala
           (references and descr. Kiefer, 1936). These may represent the same or separate species.

34a (7, 32) Leg 1 ♀ ♂, endopod 2-segmented and rostrum small, not extending
           to end of segment 1 of first antenna or beyond (Fig. 29.154*e*). First
           antenna ♀ 7- or 8-segmented (usually 8); leg 3 ♂ with hypophysis
           (stout spiniform process arising from inner margin or base of second
           segment or its equivalent and extending beyond apex of endopod,
           Figs. 29.208*d*, 29.212*b*, 29.213*e*) . . . .
                              (in part) **Bryocamptus** Chappuis   57

34b   Leg 1 ♀ ♂, endopod 3-segmented and rostrum small to moderately
           developed, not extending beyond segment 1 of first antenna. First
           antenna ♀ 7- to 9-segmented (usually 8); leg 3 ♂ with hypophysis
           (stout spiniform process arising from inner margin or base of second
           segment or its equivalent and extending beyond apex of endopod,
           Figs. 29.190*c*, 29.194*c*, 29.196*c*, 29.202*d*, 29.204*c*)   37

34c   Leg 1 ♀ ♂, endopod 2- or 3-segmented and rostrum greatly enlarged,
           extending to end of segment 1 of first antenna or beyond (Figs.
           29.186, 29.187). First antenna ♀ 6- or 7-segmented; leg 3 ♂, endo-
           pod segment 2 or its equivalent with simple, articulated spine on
           inner margin . . . . *Mesochra* Boeck   35

35a (34) Leg 1 ♀ ♂, endopod 3-segmented   36

35b   Leg 1 ♀ ♂, endopod 2-segmented . . . . . *M. lilljeborgi* Boeck 1864

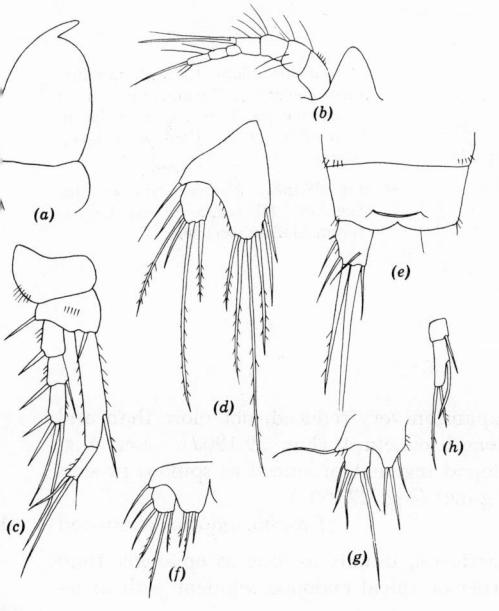

The only euryhaline species yet known from N. A. with this character of leg 1. First antenna ♀ 7-segmented. Leg 1 ♀ ♂, segment 1 reaches beyond exopod. Legs 2 to 4: exopod segment 3 ♀ ♂ with 6,7,7 spines and setae; endopods 2-segmented except leg 3 ♂ which may have segments 2 and 3 fused or separated; endopod 2 ♀ with 5,5,5 setae; apical endopod segment ♂ with 5,2,5 or 4 setae. Length: ♀ 0.42–0.73 mm; ♂ 0.5–0.6 mm. Reported from brackish coastal ponds, Mass., Fla.; Europe. (Descr. Gurney, 1932; Sars, 1903–1911.)

◄ **Fig. 29.186.** *Mesochra lilljeborgi.* (a) Lateral view cephalic segment with rostrum. (b) Rostrum and first antenna ♀. (c) Leg 1. (d) Leg 5 ♀. (e) Dorsal view last body segment and caudal ramus ♀. (f, g) Leg 5 ♂. (h) Endopod leg 3 ♂. (a, g after Gurney; b–f, h after Sars.)

36a   (35)   Leg 1 ♀ ♂, length of endopod segment 1 less than that of total exopod (about 0.8:1) . . . . *M. rapiens* (Schmeil) 1894

Anal operculum ♀ of typical form without spinules; present in ♂ and in Alaskan females. First antenna ♀ usually 7-segmented (sometimes 6). Legs 2 to 4: exopod segment 3 ♀ ♂ with 6,7,7 spines and setae; leg 3 ♂, exopod 3 modified (similar to Fig. 29.188a); endopods 2-segmented except leg 3 ♂; endopod 2 ♀ with 5,5,5 setae, apical endopod segment ♂ with 5,2,4 setae. Leg 5 ♀ exopod and basal expansion with 5 or 6 setae. Length: ♀ 0.5–0.65 mm; ♂ 0.38–0.47 mm. Pacific and Bering Sea coastal brackish and fresh-water ponds, sloughs, lakes; also known from an inland lake (Bear Lake, Utah). Europe; Asia. (Descr. Gurney, 1932.)

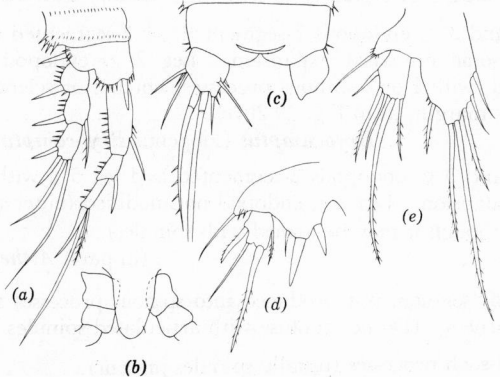

Fig. 29.187. *Mesochra rapiens.* (a) Leg 1 (Alaska). (b) Rostrum and base of first antenna (Bear Lake, Utah). (c) Dorsal view last body segment and caudal ramus (Bear Lake, Utah). (d) Leg 5 ♂ (Alaska). (e) Leg 5 ♀ (Alaska).

36b   Leg 1 ♀ ♂, length of endopod segment 1 more than that of total
exopod (about 1.25:1). . . . . **M. alaskana** M. S. Wilson 1958

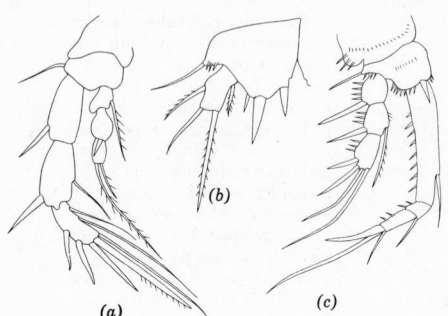

Anal operculum ♀ ♂ with spinules.
First antenna ♀ 7-segmented. Body
length and legs 2 to 5 ♀ ♂ similar to
those of *M. rapiens.* Fresh-water lakes,
Alaska.

◄ **Fig. 29.188.** *Mesochra alaskana.* (*a*)
Leg 3 ♂. (*b*) Leg 5 ♂. (*c*) Leg 1.
(From Alaskan specimens.)

37a **(34)** Leg 5 ♀, seta 2 of basal expansion very reduced, not more than and
usually less than ¼ the length of seta 1 (Fig. 29.190*d*). Leg 4 ♂,
outer corner of apical endopod segment produced as spinous process
(not set off as articulated spine) (Fig. 29.191*c*) . . . .
                                                Canthocamptus Westwood   **41**

37b   Leg 5 ♀, this seta not so reduced, usually as long as or longer than
seta 1. Leg 4 ♂, outer corner of apical endopod segment with articulated spine or seta **38**

38a **(37)** Leg 1 ♀ ♂, endopod segment 1 reaching to middle of exopod segment 3 or beyond . . . . (in part) **Attheyella** Brady   **47**

38b   Leg 1 ♀ ♂, endopod segment 1 not reaching beyond exopod segment 2 . . . .   **39**

39a **(38)** Legs 2 to 4 ♀ ♂, exopod segment 3 with 2 outer marginal spines
(total spines and setae: 5,6,6). Leg 5, basal expansion with 3 or 4
setae in ♀, with none in ♂ . . . . **Elaphoidella** Chappuis   **55**

39b   Legs 2 to 4 ♀ ♂, exopod segment 3 with 3 outer marginal spines
(total spines and setae: 6,7,7 or rarely 6 on leg 4). Leg 5, basal expansion with 5 or 6 (rarely 4) setae in ♀, usually with 2 setae in ♂ . .   **40**

40a **(39)** Legs 2 and 3 ♀, endopods 3-segmented; or 2-segmented and leg 5 ♀
with 5 setae on basal expansion. Leg 2 ♂, endopod segment 2
modified (with 1 or 2 spinous processes, notches or sclerotized knobs
on outer margin, as in Fig. 29.209*c*). . . . .
                                  Bryocamptus (subgenus **Bryocamptus**) Chappuis   **65**

40b   Legs 2 and 3 ♀, endopods 2-segmented and leg 5 ♀ with 6 setae on
basal expansion. Leg 2 ♂, endopod not modified (outer margin continuous though it may be armed with spinules). . . . .
                                                (in part) **Attheyella** Brady   **55**

41a **(37)** Last body segment ♀ ♂ produced into spinous processes at the distal
outer corners. (Do not confuse with articulated spinules.). **42**

41b   Without such processes (usually spinules present). **43**

42a **(41)** Caudal setae ♀ ♂ not jointed at bases; leg 3 ♂, endopod with very
reduced apical setae (shorter than last segment) . . . .
                                    Canthocamptus staphylinus (Jurine) 1820

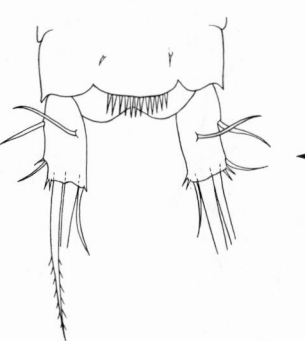

Eurasian species questionably reported from N. A.
Caudal ramus ♀ ♂ without inner marginal spinules. Sper-
matophore not flask-shaped. Leg 5 ♂, exopod with 6 setae.
Length: ♀ ♂ 0.7–1.0 mm. (Descr. Gurney, 1932.)

◀ **Fig. 29.189.** *Canthocamptus staphylinus*, dorsal view last
body segment and caudal rami. (After Gurney.)

42b   Both outer and middle caudal setae jointed at bases; leg 3 ♂, endo-
         pod with well-developed apical setae (at least one longer than total
         endopod). . . . . *C. oregonensis* M. S. Wilson 1956

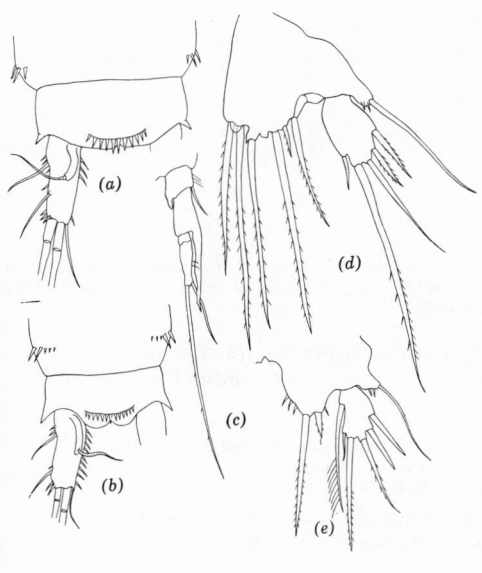

See Figs. 29.153, 29.154*e*,
29.155*f*. Arrangement and number of spinules on inner margin of
caudal ramus variable. Leg 2 ♀ ♂
with 2 setae on distal segment or
portion of endopod. Leg 4 ♀,
outer apical seta of endopod 2
longer than outer spine. Length:
♀ 0.87–0.88 mm; ♂ 0.82 mm.
Roadside ditches, ponds. Ore.,
Calif.

◀ **Fig. 29.190.** *Canthocamptus oregonensis*. (*a*) Dorsal view last body
segment and caudal ramus ♀. (*b*)
Same of ♂. (*c*) Endopod leg 3 ♂.
(*d*) Leg 5 ♀. (*e*) Leg 5 ♂. (After
M. S. Wilson.)

43a (41) Leg 2 ♀ ♂, endopod with 2 setae on distal inner margin of apical
         segment (Fig. 29.191*d,e*). **44**

43b   Leg 2 ♀ ♂, endopod with 1 distal inner seta (Fig. 29.194*b,h*) **45**

44a (43) Caudal ramus ♀ ♂, inner margin with spinules . . . .
                                          C. staphylinoides Pearse 1905

Caudal ramus ♀ ♂, length usually about 2½ times width (may range to 4 times),
dorsomedial portion with proximal lobelike swelling; spinules of inner margin variable
in number and distribution. Outer caudal seta typically slender, setiform, about ¼ the
length of stout middle seta. Anal operculum with 5 to 8 spinules, sometimes spinules
lacking. Leg 4 ♀, outer apical seta of endopod 2 longer than outer spine. Leg 5 ♀,
mid-portion of basal expansion bearing setae 2 to 4 conspicuously produced beyond
bases of setae 1 and 5. Length: ♀ ♂ 0.7–1.0 mm. Lakes, ponds, seasonal pools, and
ditches. Type locality in Neb. Probably widely distributed over most of continent.

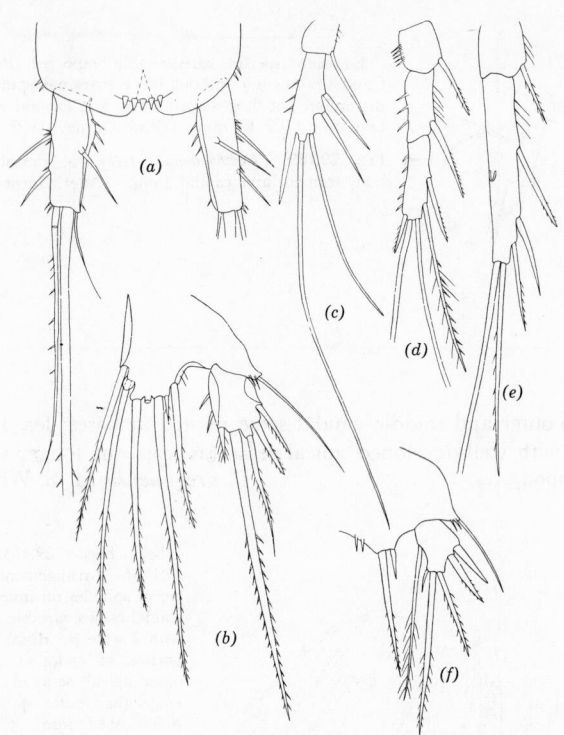

Fig. 29.191. *Canthocamptus staphylinoides.* (*a*) Dorsal view last body segment and caudal rami (rami in slightly different aspects). (*b*) Leg 5 ♀. (*c*) Endopod leg 4 ♂. (*d*) Endopod leg 2 ♀. (*e*) Endopod leg 2 ♂. (*f*) Leg 5 ♂. (From western Canadian specimens.)

44b   Caudal ramus ♀, inner margin without spinules; present or not in
male . . . . **C. sinuus** Coker 1934

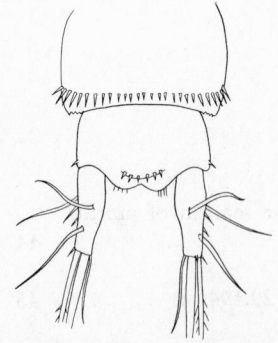

Very similar to *C. staphylinoides* of which it is probably sub-species or local variety. Length: ♀ 0.95 mm; ♂ 0.71 mm. Type locality in N. C., Conn., N. J.

◄ **Fig. 29.192.** *Canthocamptus sinuus*, dorsal view last body segments and caudal rami. (After Coker.)

45a (43)   Caudal ramus ♀ ♂, outer apical seta stout, spiniform. Leg 5 ♂,
where known, exopod with 6 setae **46**

45b   Caudal ramus ♀ ♂, outer apical seta extremely slender. Leg 5 ♂,
exopod with 5 setae . . . . **C. robertcôkeri** M. S. Wilson 1958
See Figs. 29.150*b*, 29.152, 29.154*a,b*. Distal membrane of body segments ♀ ♂ coarsely
serrate (Fig. 29.193*a*) to smooth. Caudal ramus ♀ ♂ length 4 to 5 times width, with or
without inner marginal spinules. Outer caudal seta setiform, shorter to a little longer
than ramus, about ⅕ the length of very stout middle seta. Length: ♀ 0.65-0.8 mm;

♂ about 0.62 mm. Type locality: Lake Erie. Known from lakes in Utah, N. C.; ponds and ditches, La., Kan.

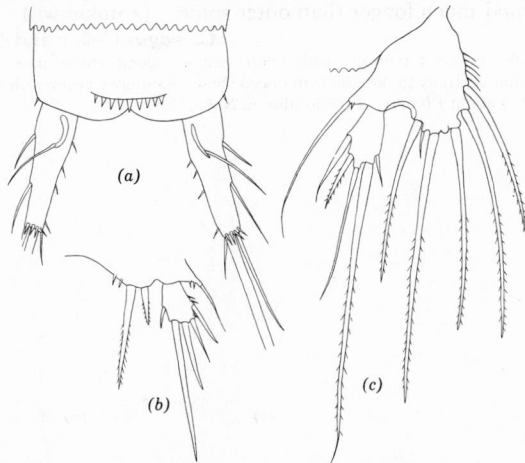

Fig. 29.193. *Canthocamptus robertcokeri.* (*a*) Dorsal view last body segment and caudal rami ♀. (*b*) Leg 5 ♂. (*c*) Leg 5 ♀. (From specimens from Lake Erie.)

46a (45) Leg 5 ♀, mid-portion of basal expansion hardly produced beyond the rest of the segment. Leg 4 ♀, outer apical seta of endopod shorter than or only a little longer than outer spine . . . .
                                                C. assimilis Kiefer 1931

Caudal ramus ♀ ♂, length about 2 to 2½ times width, with or without inner marginal spinules. Anal operculum usually broad, with 5 to 8 spinules. Outer caudal seta about ⅙ the length of middle seta. Length: ♀ ♂ 0.65–0.8 mm. Lakes, ponds, ditches. Type locality in Conn.; Alaska; western Canada; Neb., La., Calif.

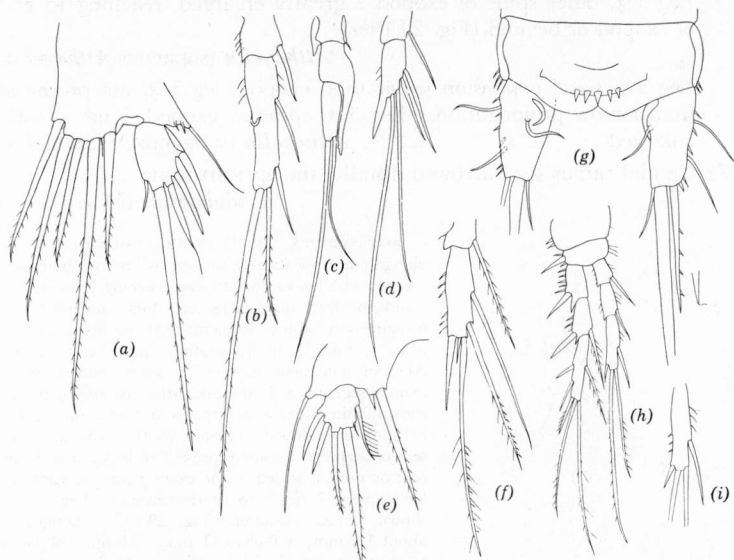

Fig. 29.194. *Canthocamptus assimilis.* (*a*) Leg 5 ♀. (*b*) Endopod leg 2 ♂. (*c*) Endopod leg 3 ♂. (*d*) Endopod leg 4 ♂. (*e*) Leg 5 ♂. (*f*) Endopod leg 4 ♀. (*g*) Dorsal view last body segment and caudal rami (rami in slightly different aspects). (*h*) Leg 2 ♀. (*i*) Detail apex of endopod leg 2 ♀. (From Alaskan specimens.)

46b Leg 5 ♀, mid-portion of basal expansion bearing setae 2 to 4 produced beyond the rest of the segment. Leg 4 ♀, outer apical seta of endopod much longer than outer spine. (♂ unknown) . . . .

        C. vagus Coker and Morgan 1940

Caudal ramus ♀ typically with small spinules along entire inner margin, placed somewhat ventrally so not visible in dorsal view. Anal operculum with 8 to 12 spinules. Length ♀ about 1.0 mm. Type locality in N. C., Wash

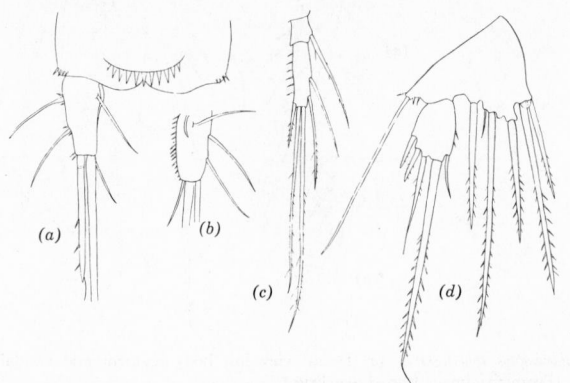

Fig. 29.195. *Canthocamptus vagus.* (*a*) Dorsal view last body segment and caudal ramus. (*b*) Part of caudal ramus, turned so as to show ventromedial spinules. (*c*) Endopod leg 4 ♀. (*d*) Leg 5 ♀. (From typical N. C. specimens.)

47a (38) Leg 5 ♀, both exopod and basal expansion elongate, of nearly same width, the basal part reaching to near end of exopod (Fig. 29.197*d*). Leg 5 ♂, basal expansion produced into narrow, elongate portion that reaches to proximal fourth or third of exopod (Fig. 29.196*b*). Leg 3 ♂, outer spine of exopod 2 greatly enlarged, reaching to end of exopod or beyond (Fig. 29.196*c*) . . . .

        Attheyella (subgenus ***Attheyella***)   **48**

47b Leg 5 ♀, basal expansion wider than exopod; leg 5 ♂, not produced into narrow prolongation. Leg 3 ♂, spine of exopod 2 not greatly enlarged . . . . ***Attheyella*** (subgenus ***Mrazekiella***)   **49**

48a (47) Caudal ramus ♀ ♂ narrowed distally, the apex truncate . . . .

        A. idahoensis (Marsh) 1903

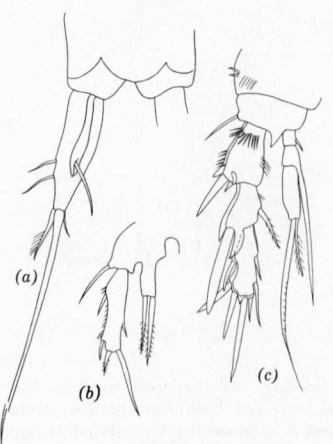

Body segments coarsely serrate. Caudal ramus ♀ ♂ elongate-narrow (nearly as long as last two body segments), with heavy dorsal ridge reaching from base to middle or distal third of ramus. First antenna ♀ 7 or 8-segmented (apical segments may be fused or separate). Legs 2 to 4: setation imperfectly known. Alaskan specimens have 6,7,6 spines and setae on exopod segment 3; Idaho and Montana specimens are shown in literature with 6 spines and setae on leg 3 ♂. Setation of endopods appears to be variable, inner seta of segment 1 usually present in legs 2 and 3, absent or present in leg 4; the exact range of variation for segment 2 needs to be determined. Leg 5 ♀ as shown for *A. alaskaensis* (Fig. 29.197). Length: ♀ about 1.0 mm; ♂ 0.68–0.72 mm. Margins of lakes, moss of streams. Ida., Mont.; Alaska. (Descr. Coker, 1934.)

◀ **Fig. 29.196.** *Attheyella idahoensis.* (*a*) Dorsal view part of last body segment and caudal ramus ♀. (*b*) Leg 5 ♂. (*c*) Leg 3 ♂. (From Alaskan specimens.)

48b   Caudal ramus ♀ (and presumably that of ♂) hardly at all narrowed distally, the apex rounded. (♂ unknown.) . . . .
A. alaskaensis M. S. Wilson 1958

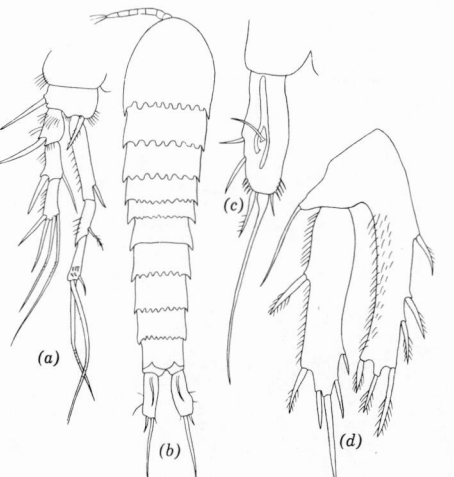

First antenna 7-segmented (may be variable). Legs 2 to 4 ♀: exopod segment 3 with 6,7,6 spines and setae; legs 2 to 3 with inner seta on endopod segment 1, without seta in leg 4; endopod segment 2 with 3,4,3 setae. Length: ♀ 1.04 mm. Known only from margin of Lake Tikchik, Alaska.

◄ **Fig. 29.197.**   *Attheyella alaskaensis.* (*a*) Leg 1. (*b*) Dorsal view ♀. (*c*) Caudal ramus. (*d*) Leg 5 ♀. (From Alaskan specimens.)

49a (47)   Leg 5 ♀, basal expansion produced to middle of exopod segment or beyond.   Leg 5 ♂, seta 3 of exopod nonplumose and more slender than others. (Legs 2 to 3 ♀, endopods typically 3-segmented). **50**

49b   Leg 5 ♀, basal expansion hardly at all produced.   Leg 5 ♂, seta 3 usually similar to others. (Legs 2 to 3 ♀, endopods typically 2-segmented.) **51**

50a (49)   Caudal ramus ♀, outer margin constricted in distal half, typical form with outer apical seta strongly outbent at base, and that of ♂ inbent.   Leg 4 ♂, exopod 3 with outer distal and apical spines modified (Fig. 29.198*c*) . . . . . *A. nordenskioldii* (Lilljeborg) 1902
First antenna ♀ 8- or 9-segmented. Legs 2 to 4: exopod segment 3 ♀ ♂ with 6,7,7 spines and setae; endopods legs 2 to 3 ♀ usually 3-segmented, but segments 2 and 3 sometimes fused or incompletely divided, segment 3 with 4 and 5 setae; endopod leg 4 ♀ 2-segmented, segment 2 with 5 setae; endopods legs 2 and 4 ♂ 2-segmented, segment 2 with 4 or 5 setae; leg 3 ♂, endopod with stout hypophysis. This form appears to have local races with less modification of the caudal ramus than in the typical form; whether it grades into *A. illinoisensis* is unknown. Length: ♀ ♂ 0.85–1.0 mm. Typical form widely distributed Alaska, Canada, western and plains states. Includes *C. hyperboreus* Willey 1925. (Descr. Pearse, 1905, as *C. illinoisensis.*)

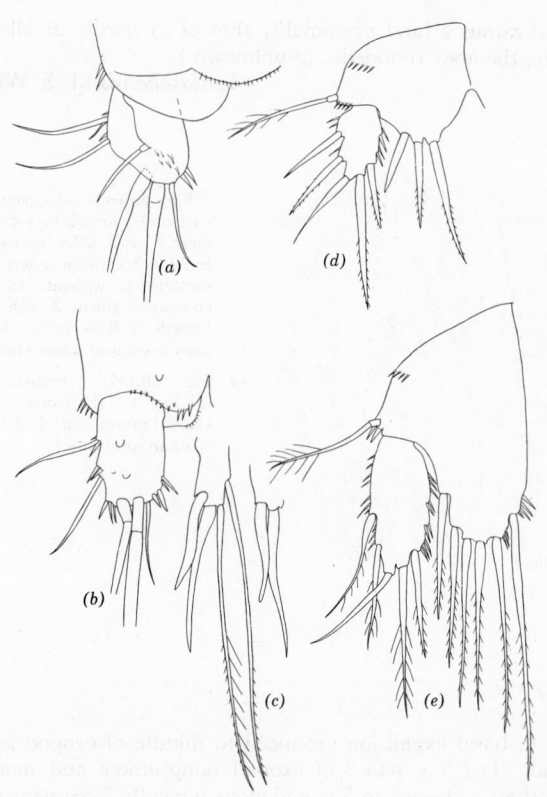

Fig. 29.198. *Attheyella nordenskioldii.* (a) Dorsal view part of last body segment and caudal ramus ♀. (b) Caudal ramus ♂, ventral. (c) Leg 4 ♂, spines of exopod segment 3, different aspects. (d) Leg 5 ♂. (e) Leg 5 ♀. (From Alaskan specimens.)

50b   Caudal ramus ♀, outer margin usually not constricted in distal half, outer caudal seta ♀ ♂ not modified. Leg 4 ♂, exopod 3, spines not modified . . . . *A. illinoisensis* (S. A. Forbes) 1882

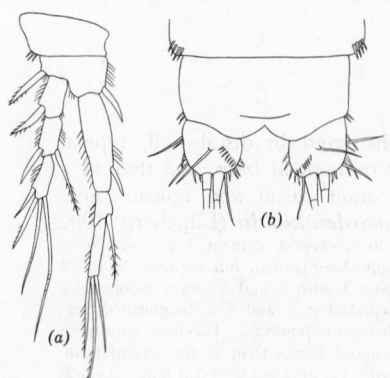

First antenna, setation legs 2 to 4 as in *A. nordenskioldii.* Length: ♀ 0.88–1.0 mm; ♂ 0.64–0.79 mm. Central states, north- and southeastern states (type locality in Ill.). (Descr. Coker, 1934.)

◄ **Fig. 29.199.** *Attheyella illinoisensis.* (a) Leg 1. (b) Dorsal view last body segments and caudal rami ♀. (After Coker.)

51a (49) Leg 5 ♀, basal expansion with 3, 4, or 5 setae. Caudal ramus ♀ ♂ elongate, length at least 2 times width, usually more, and greater than that of last body segment; with hairs or spines on inner margin or dorsal face **52**

51b Leg 5 ♀, basal expansion with 6 setae. Caudal ramus ♀ ♂ short, length not greater than 2 times width, usually less, and subequal to or shorter than last body segment; without inner marginal or dorsal surface hairs or spinules **53**

52a (51) Leg 5 ♀, basal expansion with 3 or 4 setae; leg 5 ♀ ♂, length of exopod segment about 2 times greatest width. Caudal ramus ♀ ♂ somewhat flattened dorsally and ornamented with 2 or more longitudinal rows of spinules . . . . *A. carolinensis* Chappuis 1932

Body segments coarsely serrate. First antenna ♀ 7- or 8-segmented. Rostrum rather prominent, broad. Legs 2 to 4: exopod segment 3 ♀ ♂ with 6,7,7 spines and setae; endopod segment 1 ♀ with inner seta; endopod segment 2 ♀ with 5,6,5 setae, endopod 2 leg 4 ♂ with 5 setae. Length: ♀ 0.7–0.8 mm; ♂ 0.58 mm. N. C., Va. (Descr. Coker, 1934.)

◄ **Fig. 29.200.** *Attheyella carolinensis.* (*a*) Caudal ramus ♂, lateral. (*b*) Dorsal view last body segments and caudal ramus ♀. (After Coker.)

52b Leg 5, basal expansion ♀ with 5 setae; exopod ♀ ♂, length less than 2 times greatest width. Caudal ramus ♀ ♂ ornamented with 2 or 3 oblique inner rows of hairs . . . . *A. pilosa* Chappuis 1929

Body segments serrate. First antenna ♀ 8-segmented. Legs 2 to 4: setation exopod 3 unknown; endopod segment 1 ♀ with inner seta; endopod segment 2 ♀ with 5,6,5 setae; legs 2 and 4 ♂, endopod 2 with 4 and 5 setae. Length: ♀ ♂ about 0.75 mm. Cave waters, Ind., Ken. (Descr. Chappuis, 1931.)

◄ **Fig. 29.201.** *Attheyella pilosa.* (*a*) Caudal ramus, ventral. (*b*) Leg 5 ♀. (*c*) Leg 5 ♂. (After Chappuis.)

53a (51) Caudal ramus ♀ ♂, lateral setae of outer margin usually not arising from same place, and the outer distal edge of ramus not produced as sclerotization. (See comment under Variation and Anomaly, pp. 821–822.) **54**

53b Caudal ramus ♀ ♂, lateral setae placed close together and the outer

distal edge of ramus armed with rounded sclerotization overlying
bases of apical setae . . . . **A. dogieli** (Rylov) 1923

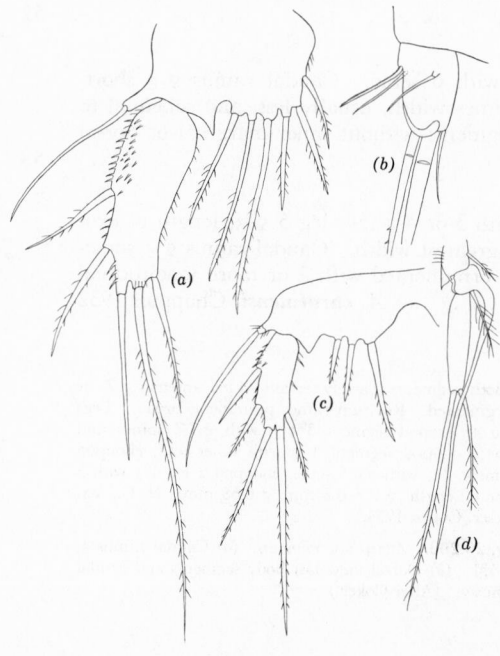

Body segments weakly serrate. First antenna ♀ 8-segmented. Legs 2 to 4: exopod segment 3 ♀ ♂ with 6,7,7 spines and setae; endopod segment 1 ♀ with inner seta; endopod segment 2 of ♀ with 6,6,5 setae, of legs 2 and 4 ♂, 5 and 5 setae; leg 3 ♂, endopod with strongly outcurved and elongate hypophysis. Leg 5 ♀ ♂, exopod elongate, length up to 3 times width. Length ♀ 0.8–1.0 mm. ♂ 0.7 mm. Lakes, ponds. Alaska; western Canada; Wash.; Asia.

◄ **Fig. 29.202.** *Attheyella dogieli.* (a) Leg 5 ♀. (b) Caudal ramus, dorsal. (c) Leg 5 ♂. (d) Endopod leg 3 ♂. (From specimens from Lake Kluane, Y. T., Canada.)

54a (53) Caudal ramus ♀ ♂ about as long as last body segment; on dorsal face of ♀ ramus a prominent triangular or semirectangular sclerotization distad to insertion of dorsal seta. Leg 5 ♀, exopod greatly broadened in basal portion, its length less than 2 times width. Leg 3 ♂, hypophysis of endopod strongly developed (enlarged and outcurved at base and reaching beyond apex of endopod by about half of its own length, as shown in Fig. 29.202*d*) . . . .
A. dentata (Poggenpol) 1874

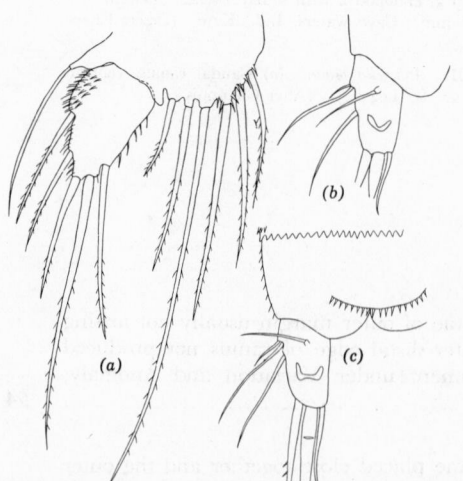

Antenna ♀ and setation legs 2 to 4 as in *A. dogieli.* Leg 5 ♂, exopod with 5, basal expansion with 3 or 4 setae. Length: ♀ 0.56–0.73 mm; ♂ 0.54–0.69 mm. Usually in pools, ponds. Alaska; Canada; Europe. Recorded in literature as *A. northumbrica* (Brady). (Descr. Gurney, 1932.)

◄ **Fig. 29.203.** *Attheyella dentata.* (a) Leg 5 ♀. (b,c) Caudal rami of different specimens showing slight differences in placement of lateral setae and shape of dorsal sclerotization. (From Alaskan specimens.)

4b   Caudal ramus ♀ ♂ shorter than last body segment, about as broad as long, without such sclerotization on dorsal face (do not confuse with narrow ridge that may be associated with dorsal seta). Leg 5 ♀, exopod not greatly broadened in basal portion, its length about 2 times width. Leg 3 ♂, hypophysis of endopod weakly developed (hardly enlarged and not outcurved at base, reaching beyond endopod by only about ⅓ of its length) . . . .
A. americana (Herrick) 1884
Antenna ♀; legs 2 to 4 as above, except leg 2 ♂ like ♀. Leg 5 ♂, basal expansion with 2 or 3 spines. Length: ♀ 0.56–0.73 mm; ♂ 0.54–0.69 mm. Small lakes, ponds, pools. Type locality in Minn. Known from middle and eastern U. S. and Canada. Recorded in literature as subspecies of *A. northumbrica* or *dentata*. (Descr. Coker, 1934; taxonomic status M. S. Wilson, 1958b).

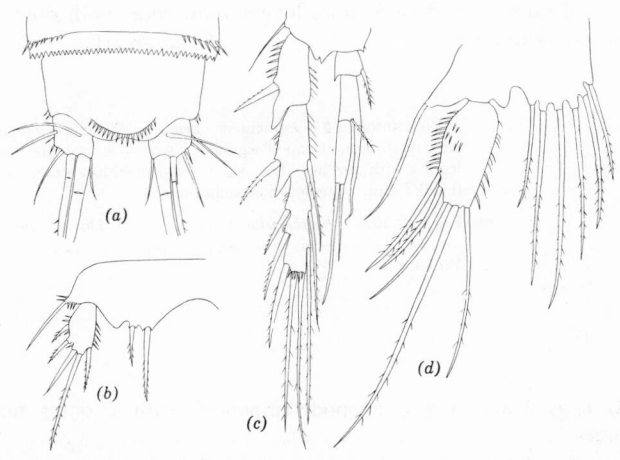

Fig. 29.204. *Attheyella americana.* (a) Dorsal view of last body segments and caudal ramus. (b) Leg 5 ♂. (c) Leg 3 ♂. (d) Leg 5 ♀ (Fla.) (a after Coker.)

55a (39, 40) Leg 5 ♀, basal expansion with 4 setae. Leg 5 ♂, basal expansion lacking setae **56**

55b   Leg 5 ♀, basal expansion with 6 setae. Leg 5 ♂, basal expansion with 2 setae . . . . *A. obatogamensis* (Willey) 1925

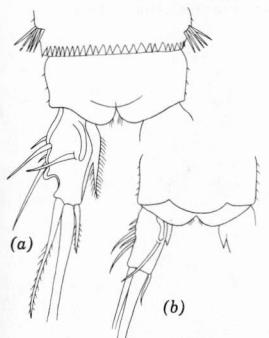

Subgenus *Attheyella*. Caudal ramus ♀ with prominent, ornamented inner process; process of ♂ less prominent and unornamented. First antenna ♀ 8-segmented. Legs 2 to 4: exopod segment 3 ♀ ♂ with 6,7,7 spines and setae; endopod segment 1 ♀ ♂ with inner seta; endopod segment 2 ♀ with 5,6,5 setae; legs 2 and 4 ♂, endopod segment 2 with 4 and 3 setae; leg 3 ♂, outer spine exopod 2 greatly enlarged. Leg 5 ♀, exopod and basal expansion with extremely long setae, most about 3 times as long as the segment. Length: ♀ 0.52–0.57 mm; ♂ 0.39–0.49 mm. Quebec; N. Y. (Descr. Willey, 1934; Coker, 1934, as *A. wierzejskii.*)

◄ Fig. 29.205. *Attheyella obatogamensis.* (a) Dorsal view last body segments and caudal ramus ♀. (b) Same ♂. (After Coker.)

56a (55) Caudal ramus ♀ only a little longer than wide, with prominent dorsal hook. (♂ of American form unknown.). . . . .
Elaphoidella bidens coronata (Sars) 1904

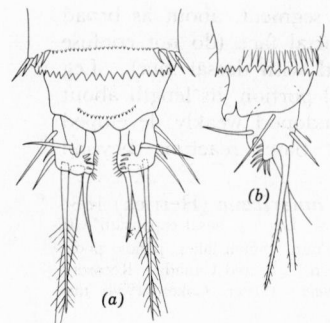

(b)

(a)

Cosmopolitan, parthenogenetic species for which males have been rarely found. First antenna ♀ 8-segmented. Legs 2 to 4 exopod segment 3 with 5,6,6 spines and setae; endopods 2-segmented, segment 1 with inner seta on legs 2 and 3, without on leg 4; segment 2 with 5,6,4 setae. Leg 5 ♀, exopod with 5 setae. Length: ♀ 0.43 mm. Lakes, pools. N. C., Pa., Va., La. (Descr. Coker, 1934.)

◄ **Fig. 29.206.** *Elaphoidella bidens coronata.* (a) Dorsal view last body segments and caudal rami ♀. (b) Lateral view of same. (After Coker.)

56b   Caudal ramus ♀ ♂, 2 to 3 times longer than wide, with dorsal ridge but not with hook . . . . **E. subgracilis** (Willey) 1934

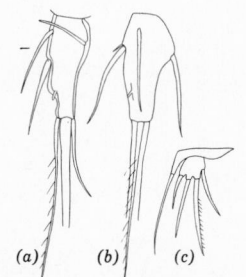

(a)   (b)   (c)

First antenna ♀ 8-segmented. Legs 2 to 4: endopod segment 1 ♀ with inner seta, endopod segment 2 ♀ with 4,5,4 setae; endopod 2 leg 2 ♂ with 3 setae. Leg 5 ♀ ♂, exopod with 4 setae. Length: ♀ 0.5–0.7 mm. Quebec, Saskatchewan.

◄ **Fig. 29.207.** *Elaphoidella subgracilis.* (a) Dorsal view caudal ramus. (b) Lateral view caudal ramus. (c) Leg 5 ♂. (After Willey.)

57a (24, 34)   Legs 2 and 3 ♀ ♂, exopod segment 3 with 2 outer marginal spines. 5⁸

57b   Legs 2 and 3 ♀ ♂, exopod segment 3 with 3 outer marginal spines . . **Bryocamptus zschokkei** (Schmeil) 1893

Subgenus *Bryocamptus*. Highly variable, widely distributed species for which many varietal names have been proposed. First antenna ♀ 8-segmented. Legs 1 to 4: exopod segment 3 ♀ ♂ with 4,6,7,7 spines and setae; endopods 2-segmented (except modified leg 3 ♂), endopod segment 2 ♀ with 4,5,6,5 setae and spines; endopod 2 legs 2 and 4 ♂ with 4 setae. Leg 3 ♂, apical endopod seta modified (broad and flattened) in at least some North American forms (**B. z. alleganiensis** Coker 1934 and populations from Alaska). Length: ♀ 0.49–0.6 mm; ♂ 0.36–0.45 mm. Includes the incompletely known **B. frigidus** (Willey) 1925. N. Y., N. C., Va.; Quebec; Alaska; Europe; Asia. (Descr. Gurney, 1932; Coker 1934.)

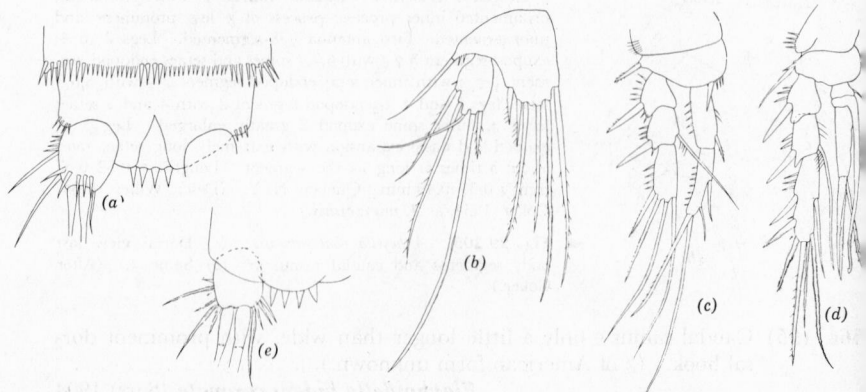

(a)

(b)

(c)   (d)

(e)

58a (57) Legs 2 and 3 ♀ ♂, exopod segment 2 with well-developed inner seta (reaching at least to near end of exopod). . . . .
                                                            B. pygmaeus (Sars) 1862

Subgenus *Bryocamptus*. Caudal ramus ♀ ♂ about as broad as long, overhung dorsally by last body segment. First antenna ♀ 8-segmented. Legs 1 to 4: exopod segment 3 ♀ ♂ with 4,5,6,6 spines and setae, with enlarged distal spine in leg 2 ♂; endopods 2-segmented (except leg 3 ♂), segment 2 ♀ with 3,4,5,5 setae; segment 2 leg 2 ♂ with 3 setae, leg 4 ♂ with 2 to 4 setae. Length ♀ 0.4–0.7 mm; ♂ 0.38–0.5 mm. N. Y.; Europe; Africa. (Descr. Gurney, 1932; Sars, 1903–1911.)

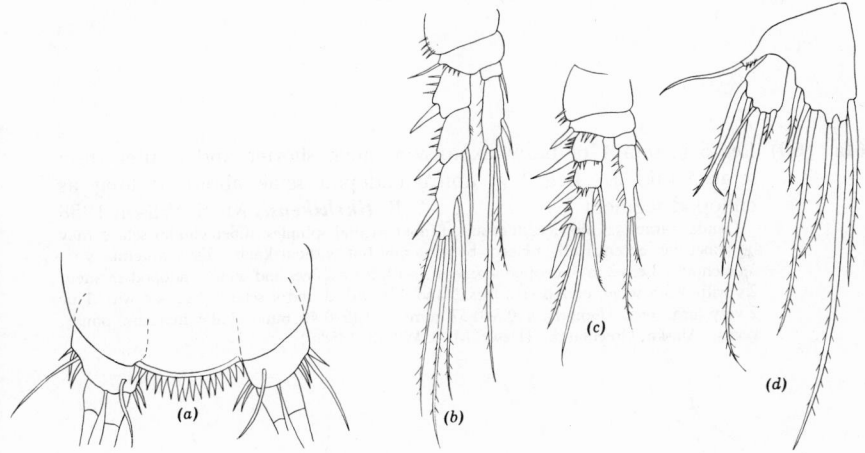

Fig. 29.209. *Bryocamptus pygmaeus.* (*a*) Dorsal view last body segment, operculum, and caudal rami. (*b*) Leg 3 ♀. (*c*) Leg 2 ♂. (*d*) Leg 5 ♀. (*b* after Sars; *a, c, d* after Gurney.)

58b   Legs 2 and 3 ♀ ♂, exopod segment 2 with short inner seta (rarely reaching beyond middle of exopod segment 3). . . . .
                  Bryocamptus (subgenus ***Arcticocamptus***) Chappuis   **59**

59a (58) Leg 3 ♀ ♂, exopod segment 3 with total of 5 spines and setae; caudal setae ♀ not overlying one another **60**

59b   Leg 3 ♀ ♂, exopod segment 3 with total of 6 spines and setae; outer caudal seta ♀ overlying middle seta . . ***B. cuspidatus*** (Schmeil) 1893

Caudal ramus ♀ without medial or dorsal group of spinules, inner seta with basal knob; caudal setae ♂ placed terminally, not modified. First antenna ♀ 8-segmented. Legs 2 to 4, exopod segment 3 ♀ ♂ with 5,6,6 spines and setae; endopod segment 2 ♀ with 4,5,5 setae; endopod 2 legs 2 and 4 ♂ with 3 and 4 setae. Leg 5 ♀ similar to that of *B. tikchikensis* (Fig. 29.212*a*). Length: ♀ 0.53–0.64 mm; ♂ about 0.4 mm. Questionable record from Quebec; Greenland; Europe. (Descr. Gurney, 1932.)

◄ **Fig. 29.210.** *Bryocamptus cuspidatus*, ventral view caudal ramus and setae ♀. (After Gurney.)

60a (59) Caudal ramus ♀ ♂ with crest of medial or dorsal spinules **61**

60b   Caudal ramus ♀ ♂ without crest of spinules. . . . .
                                               B. subarcticus (Willey) 1925

◄────

Fig. 29.208. *Bryocamptus zschokkei.* (*a*) Ventral view last body segments and caudal ramus ♀. (*b*) Leg 5 ♀. (*c*) Leg 1. (*d*) Leg 3 ♂. (*e*) Dorsal view operculum and caudal ramus ♂. (From Alaskan specimens.)

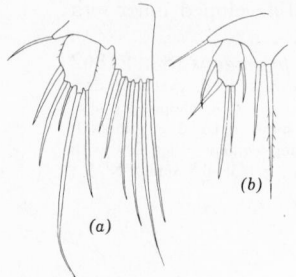

Leg 1 ♀ ♂, exopod segment 2 without inner seta in type specimens. Legs 2 to 4: exopod segment 3 with 5,5,6 spines and setae; endopod segment 2 ♀ with 4,5,5 (or 4) setae; endopod 2 legs 2 and 4 ♂ with 3 and 4 setae. Length: ♀ 0.47–0.48 mm; ♂ 0.38–0.43 mm. Margin of lake, Quebec. (Descr. as *Attheyella subarctica* Willey 1925.)

◄ **Fig. 29.211.** *Bryocamptus subarcticus.* (*a*) Leg 5 ♀. (*b*) Leg 5 ♂. (After Willey.)

61a (60) Leg 5 ♀, seta 4 of basal expansion much shorter and stouter than setae 3 and 5. Leg 3 ♂, apical endopod setae about as long as endopod segment 3 . . . . **B. tikchikensis** M. S. Wilson 1958
Caudal ramus ♀ ♂ alike, with group of inner medial spinules, inner caudal seta ♀ may be sometimes eccentrically placed, bent basally but without knob. First antenna ♀ 8-segmented. Legs 2 to 4: exopod segment 3 with 5,5,6 spines and setae; endopod segment 2 ♀ with 4,5,5 setae; endopod 2 legs 2 and 4 ♂ with 3 and 4 setae. Egg sac with 1 or 2 very large ova. Length: ♀ 0.5–0.57 mm; ♂ 0.45–0.48 mm. Lake margins, ponds, pools. Alaska; Greenland. (Descr. M. S. Wilson, 1958c.)

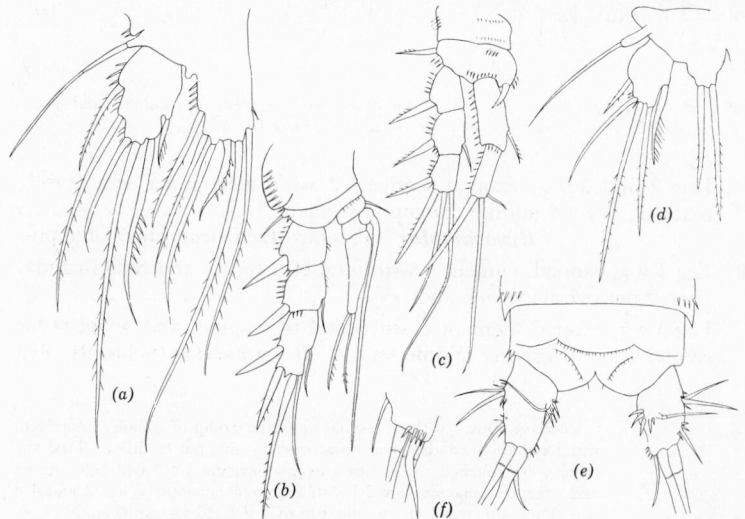

Fig. 29.212. *Bryocamptus tikchikensis.* (*a*) Leg 5 ♀. (*b*) Leg 3 ♂. (*c*) Leg 1. (*d*) Leg 5 ♂. (*e*) Dorsal view last body segment and caudal rami ♀ (rami in slightly different aspects). (*f*) Ventral view caudal setae ♀. (After M. S. Wilson.)

61b Leg 5 ♀, this seta similar in length and slenderness to seta 5. Leg 3 ♂, apical endopod setae much shorter than endopod segment 3 . . . . **B. arcticus** (Lilljeborg) 1902

First antenna ♀, ova, and setation legs 2 to 4 ♀ as in *B. tikchikensis.* Leg 4 ♂, inner apical spine of exopod segment 3 with stout spinules. Length: ♀ 0.4–0.77 mm; ♂ 0.5–0.6 mm. In vegetation of pools, ponds. Essentially arctic-alpine in distribution. Alaska; Europe; Asia. (Descr. Sars, 1903–1911; M. S. Wilson, 1958c.)

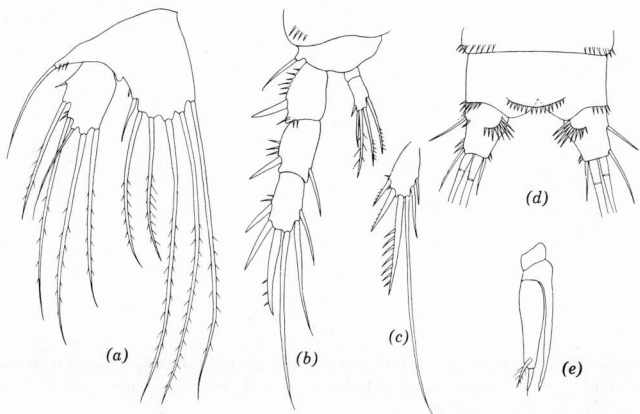

Fig. 29.213. *Bryocamptus arcticus.* (*a*) Leg 5 ♀. (*b*) Leg 4 ♀. (*c*) Leg 4 ♂, exopod segment 3. (*d*) Dorsal view last body segments and caudal rami. (*e*) Endopod leg 3 ♂. (*d* after Sars; *a–c, e* after M. S. Wilson from Alaskan specimens.)

62a   **(20)**   Caudal ramus ♀ (and presumably that of ♂) lacking inner group of spinules. Leg 3 ♀, apical endopod segment with 3 or 4 setae; leg 4 ♀, apical endopod segment with 4 setae. (♂ unknown) . . . .
                                              B. morrisoni (Chappuis) 1929

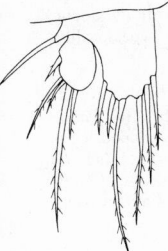

Inadequately known species. Caudal ramus as wide as long. First antenna ♀ 8-segmented. Leg 1, endopod reaching to end of exopod. Legs 2 to 4, setation exopod segment 3 unknown; endopod segment 2 ♀ with 4,4,4 setae. Length: ♀ 0.55 mm. Type locality in Ind. **B. m. elegans** Chappuis 1929, based on single female specimen from Kentucky has 3 setae on endopod 2 of leg 3. (Descr. Chappuis, 1931.)

◀ **Fig. 29.214.** *Bryocamptus morrisoni,* leg 5 ♀. (After Chappuis.)

62b   Caudal ramus ♀ ♂ with distal, inner group of spinules. Legs 3 and 4 ♀, endopod segment 2 with 5 setae . . . . . . **B. hiemalis** complex   **63**
    See Figs. 29.215, 29.216. A group of closely related forms of uncertain taxonomic status, widely distributed in N. A. and known from Asia. Leg 5 ♀ ♂ as in Fig. 29.215. Legs 2 to 4 ♀ ♂: exopod segment 3 with 6,7,7 spines and setae; endopods 2-segmented, except modified leg 3 ♂ (3-segmented and with hypophysis); endopod 2, leg 2 ♀ with 4 or 5 setae; endopod 2 legs 2 and 4 ♂ with 3 setae. The range of variation within and between populations is unknown and the named forms may represent species, subspecies, or merely variations of a single species to which the name *B. hiemalis* is applicable.

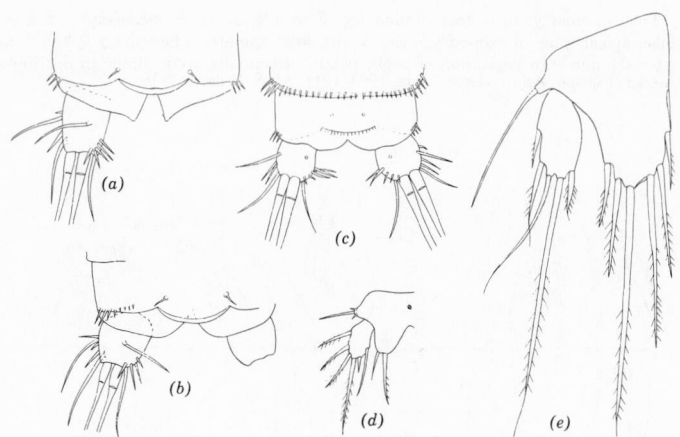

Fig. 29.215. *Bryocamptus hiemalis* group.   (a) *B. hiemalis*, dorsal view last body segment with operculum and caudal ramus ♀ (Utah).   (b) *B. nivalis*, same (Alaska).   (c) *B. h. brevifurca*, same.   (d) *B. h. brevifurca*, leg 5 ♂.   (e) Leg 5 ♀ (Utah).   (c, d after Coker.)

63a (62)   Leg 4 ♀, middle apical seta of endopod segment 2 shorter than outer
              spine (Fig. 29.216*e,f*) **64**

63b   Leg 4 ♀, middle apical seta of endopod segment 2 longer than outer
              spine (Fig. 29.216*c,d*) . . . . **B. hiemalis** (Pearse) 1905

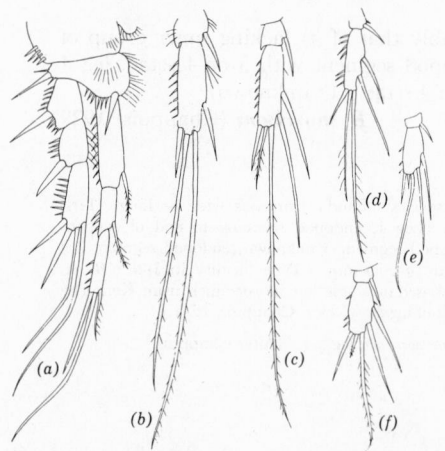

Caudal ramus ♀ ♂ longer than wide.
Leg 2 ♀, endopod segment 2 with 2
setae on inner margin (total 5), or with
1 seta and group of proximal hairs.
Leg 4 ♀, endopod segment 1 without
seta in typical form, but this probably
variable character.   Length: ♀ about
0.77 mm; ♂ 0.55 mm.   Type locality
in Neb.; Mont., Utah.

◄ **Fig. 29.216.** *Bryocamptus hiemalis* group.
(a) *B. hiemalis*, leg 1 (Utah).   (b) Same,
endopod leg 2 ♀.   (c–f) endopod leg 4
♀: (c) *B. hiemalis* (Utah).   (d) *B.
hiemalis* (Mont.).   (e) *B. nivalis*.   (f) *B.
h. brevifurca* (N. C.)   (d, f after Coker;
e after Willey.)

64a (63)   First antenna ♀ 8-segmented . . . . **B. nivalis** (Willey) 1925
              Presumably includes *B. hiemalis brevifurca* Coker 1934.   Caudal ramus ♀ ♂ about
              as long as wide.   Length: ♀ 0.54–0.77 mm; ♂ 0.44–0.58 mm.   Type locality *B. nivalis*
              in Quebec; of *B. h. brevifurca* in N. C.; N. Y., Mich., Alaska.

64b   First antenna ♀ 7-segmented . . . . **B. douwei** (Willey) 1925
              Described as **Canthocamptus douweanus** n. n. Willey 1934.   Inadequately known,
              based on single female specimen.   Caudal ramus longer than wide.   Leg 1, endopod
              described as "not prolonged."   Quebec.

65a (40)   Leg 1 ♀ ♂, endopod segment 1 without inner seta **66**

65b   Leg 1 ♀ ♂, endopod segment 1 with inner seta **67**

66a (65) Anal operculum ♀ ♂ with spinules. Legs 2 to 4 ♀, endopods 2-segmented . . . . **B. hiatus** (Willey) 1925

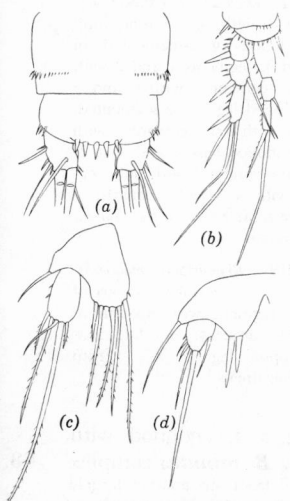

First antenna ♀ 8-segmented. Legs 2 to 4: exopod segment 3 ♀ ♂ with 6,7,7 spines and setae; endopod segment 2 ♀ with 4,5,4 setae; endopod segment 2 leg 2 ♂ with 3 or 4 setae, of leg 4 ♂ with 4 setae. Leg 5 ♀ with 5 setae on exopod and basal expansion. Leg 5 ♂, exopod with 6 setae in typical form. Length: ♀ 0.49–0.55 mm; ♂ 0.41–0.51 mm. Type locality in Quebec. Includes **B. australis** Coker from N. C. (Descr. Willey, 1934; Coker, 1934.)

◄ **Fig. 29.217.** *Bryocamptus hiatus.* (a) Dorsal view last body segments and caudal rami ♀. (b) Leg 1. (c) Leg 5 ♀. (d) Leg 5 ♂. (After Coker.)

66b   Anal operculum ♀ ♂ without spinules. Legs 2 to 3 ♀, endopods 3-segmented; leg 4 ♀, endopod 2-segmented . . . .
B. newyorkensis (Chappuis) 1927
First antenna ♀ 8-segmented. Legs 2 to 4: exopod segment 3 ♀ ♂ with 6,7,7 spines and setae; leg 2: endopod ♀ without inner seta on segment 1, with seta on segment 2, with 4 setae on segment 3, endopod ♂ 2-segmented, without inner seta on segment 1, with 4 setae on segment 2; leg 3 ♀, endopod with inner seta on segments 1 and 2, with 5 setae on segment 3; leg 4 ♀ ♂, endopod without seta on segment 1, segment 2 with 4 setae in ♀, with 3 in ♂. Leg 5 ♂, exopod with 6 setae. Length: ♀ 0.5 mm; ♂ 0.45 mm. N. Y., La.; Siberia.

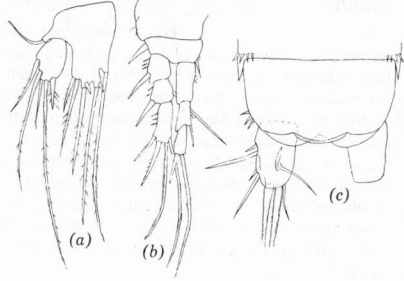

Fig. 29.218. *Bryocamptus newyorkensis.* (a) Leg 5 ♀. (b) Leg 1. (c) Dorsal view last body segment with operculum and caudal ramus. (From Louisiana specimens.)

67a (65) Leg 5 ♀, basal expansion with 5 setae, of which the first is very reduced (and presumably may be sometimes absent). Leg 3 ♂, endopod with one of its apical setae modified (broad, flattened, bifid) . . . . **B. umiatensis** M. S. Wilson 1958

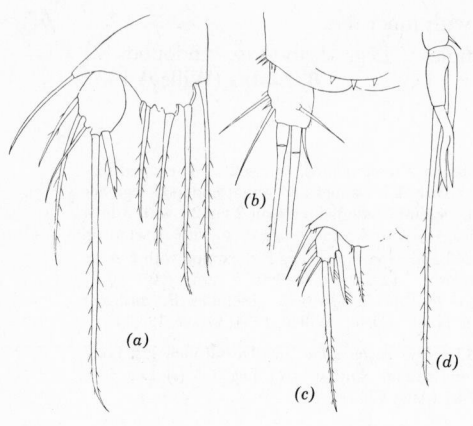

(a)

(b)

(c)

(d)

Anal operculum ♀ with 3 to 5 widely spaced spinules; ♂ with about 8 spinules. First antenna ♀ 8-segmented. Legs 2 to 4: exopod segment 3 ♀ ♂ with 6,7,7 spines and setae; endopods ♀ 3-segmented in legs 2 and 3, segments 1 and 2 with inner seta, segment 3 with 4 and 5 setae; endopod leg 4 ♀ 2-segmented, segment 1 with seta, segment 2 with 5 setae; endopod legs 2 and 4 ♂ 2-segmented, segment 1 with seta, segment 2 with 4 setae. Length: ♀ about 0.68 mm; ♂ 0.63 mm. Lakes, pools. Alaska.

◄ **Fig. 29.219.** *Bryocamptus umiatensis.* (a) Leg 5 ♀. (b) Dorsal view part of last body segment with operculum and caudal ramus ♀. (c) Leg 5 ♂. (d) Endopod leg 3 ♂. (From Alaskan specimens.)

67b Leg 5 ♀, basal expansion with 6 setae. Leg 3 ♂, endopod with normal apical setae . . . . **B. minutus** complex   **68**

A group of species of world-wide distribution, distinguished from one another largely by differences of the caudal rami and setae of the female; the modifications present in the female not found in the male of most species. First antenna ♀ 8-segmented. The species all have similar legs 1 to 5. Legs 2 to 4: exopod segment 3 ♀ ♂ with 6,7,7 spines and setae (except leg 4 of typical *B. minutus*); endopods legs 2 to 3 ♀ 3-segmented, leg 4 ♀ 2-segmented, legs 2 and 4 ♂ 2-segmented, 3-segmented and with hypophysis in leg 3 ♂; apical endopod segment with 4,5,5 setae in ♀; with 4 setae in leg 2 ♂, with 3 or 4 setae in leg 4 ♂. Leg 5 of similar structure in all species, exopod with 5 setae in ♀, 6 in ♂; basal expansion with 6 setae in ♀, 2 in ♂. (Since it is usually not possible to identify males except when associated with the female, the couplets of the key distinguish only the females.)

68a (67) Caudal ramus ♀ with the usual 3 caudal setae present, inserted apically side by side, or the outer more or less underlying the longer middle seta **69**

68b Caudal ramus ♀ with 2 apically placed caudal setae (the outer seta lacking, the long middle seta usually greatly enlarged, the inner seta reduced as usual) . . . . **B. vejdovskyi** (Mrazek) 1893

It appears that an incompletely known group of forms with this character is widely distributed in N. A. These exhibit variations in the form of the female caudal ramus, and spinules of the anal operculum. The typical form has normal operculum spinules, the outer margin of the caudal ramus produced as a spinous point; ramus of ♂ shorter, without spinous point and with 3 normally placed apical setae; leg 4 ♂, endopod segment 2 with 4 setae; known from Alaska, Wash.; Europe; Asia. One inadequately known form in which the outer margin of caudal ramus ♀ lacks the spinous process has been named **B. minusculus** (Willey) 1925 (Fig. 29.220c); Quebec; N. Y. The name **B. vejdovskyi** forma **minutiformis** Kiefer 1934 has been given to otherwise typical specimens from Conn. with bifid anal operculum spinules; also known from Mich. Length: ♀ 0.64–0.8 mm; ♂ 0.6–0.7 mm.

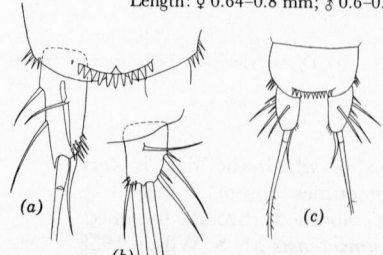

(a)

(b)

(c)

◄ **Fig. 29.220.** *Bryocamptus vejdovskyi.* (a) Dorsal view last body segment with operculum and caudal ramus ♀. (b) Caudal ramus ♂. (c) *B. minusculus*, dorsal view last body segment and caudal rami ♀. (a, b from Alaskan specimens. c after Willey.)

69a (68) Caudal ramus ♀ without apical process at base of caudal setae **70**

69b Caudal ramus ♀ with an apical, outwardly or somewhat dorsally placed process (more or less spiniform) at base of caudal setae . . . .
B. hutchinsoni Kiefer 1929

A variable species. Caudal ramus ♀ ♂ not alike; that of ♀ has outer caudal seta inserted ventrally below the larger middle seta; insertion of setae apical in ♂. The typical form from Conn. (as described by Kiefer) has caudal ramus ♀ little longer than wide, whorl of spinules on inner margin at middle of ramus; anal operculum with nonbifid spinules; leg 2 ♀, endopod segment 1 without inner seta; leg 4 ♂, endopod segment 2 with 3 setae; this includes **B. minutus** forma **simplicidentata** Willey 1934 from Quebec. A variation from Va. has both outer and inner caudal setae ♀ greatly enlarged at base (Carter, 1944); leg 2 ♀, endopod segment 1 with inner seta. A widely distributed form found in Alaska, western Canada, and western U. S., has longer caudal ramus with the whorl of spinules placed distally; bifid opercular spinules; leg 2 ♀ (so far as known), endopod segment 1 with inner seta; leg 4 ♂ (so far as known), endopod segment 2 with 4 setae. Length: ♀ about 0.8 mm; ♂ about 0.65–0.7 mm.

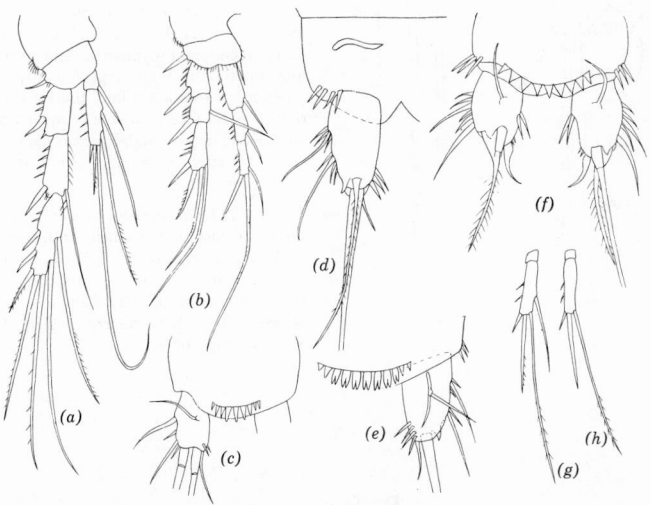

Fig. 29.221. *Bryocamptus hutchinsoni.* (*a*) Leg 4 ♀ (Alaska). (*b*) Leg 1 (Alaska). (*c*) Dorsal view last body segment with operculum and caudal ramus ♂ (Alaska). (*d*) Ventral view part last body segment and caudal ramus ♀ showing typical placement of outer caudal seta below middle seta (Alaska). (*e*) Dorsal view part last body segment with operculum and caudal ramus ♀, variety with bifid opercular spinules (Alaska). (*f*) Dorsal view last body segment and caudal rami (Va.). (*g*) Endopod leg 4 ♂ (Alaska). (*h*) Endopod leg 4 ♂ (Conn.). (*f* after Carter; *h* after Kiefer.)

70a (69) Caudal ramus ♀ with caudal setae placed apically . . . .
B. washingtonensis M. S. Wilson 1958

Anal operculum ♀ ♂ with large, nonbifid spinules. Caudal rami ♀ ♂ alike, elongate (length from 2.3 to 2.8 times width in ♀; about 2 times in ♂). Leg 5 ♀, exopod seta 2 more than 2 times length of outer seta. Leg 4 ♂, endopod segment 2 with 4 setae. Length: ♀ ♂ 0.7–0.77 mm. Ponds, ditches. Wash., Ore.

◄ **Fig. 29.222.** *Bryocamptus washingtonensis.* (*a*) Leg 5 ♀. (*b*) Dorsal view last body segment and caudal rami ♀ (rami in slightly different aspects). (*c*) Leg 5 ♂.

70b Caudal ramus ♀ with outer caudal seta inserted somewhat eccentrically on ventral face (may or may not completely underlie the larger middle seta) . . . . *B. minutus* Claus 1863

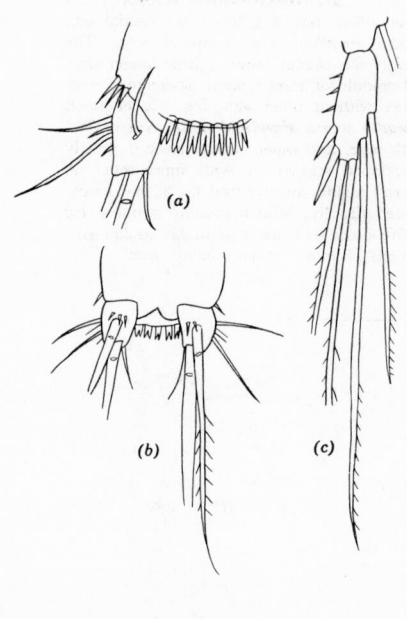

Typical European form has total of 6,7,6 spines and setae on exopod segment 3 of legs 2 to 4 (leg 4 with 2 outer spines); leg 4 ♂, endopod segment 2 with 4 setae; spinules of anal operculum bifid. Length: ♀ 0.5–0.7 mm; ♂ 0.4–0.55 mm. (Descr. Gurney, 1932). Reported from several localities in N. A., but records uncertain. *C. minnesotensis* Herrick has been called subspecies of *B. minutus*, but its identity is uncertain (see M. S. Wilson, 1956). A form referable to *B. minutus*, and which may replace the European form in N. A., has 3 spines on exopod 3 of leg 4 (total 7); leg 4 ♂ with 3 setae on endopod segment 2; and spinules of anal operculum bifid; reported from N. Y. As found in Ind. and Ohio, this form has the outer caudal seta completely underlying the middle seta; the caudal ramus lacks prominent spinules or spinule group as in *B. hutchinsoni*.

◀ **Fig. 29.223.** *Bryocamptus minutus.* (*a*) Dorsal view part last body segment with operculum and caudal ramus ♀. (*b*) Ventral view caudal rami showing eccentric placement of outer caudal seta. (*c*) Leg 4 ♀, exopod segment 3. (After Gurney, from typical European specimens.)

References

CALANOIDA

Brehm, Vincenz. 1939. La Fauna microscopica del Lago Peten, Guatemala. *Escuela Nac. Cienc. biol. Anal.*, 1:173–202. **1955.** Mexicanische Entomostraken. *Österr. zool. Z.*, 6:412 420. **Coker, Robert E. 1926.** Plankton collections in Lake James, North Carolina—copepods and Cladocera. *J. Elisha Mitchell Sci. Soc.*, 41:228–258. **Davis, Charles C. 1943** The larval stages of the calanoid copepod, *Eurytemora hirundoides* (Nordquist). *Chesapeake Biol. Lab. Publ.*, No. 58:1–52. **Dodds, Gideon S. 1915.** Descriptions of two new species of Entomostraca from Colorado, with notes on other species. *Proc. U. S. Natl. Museum*, 49:97–102. **Gurney, Robert. 1931.** *British Fresh-Water Copepoda.* Vol. 1. Ray Society, London. **Herrick, C. L. 1895.** Copepoda of Minnesota. In: C. L. Herrick and C. H. Turner, *Synopsis of Entomostraca of Minnesota.* Zool. Ser. II, Geol. Nat. Hist. Survey Minnesota. **Humes, Arthur G. 1955.** The postembryonic developmental stages of a fresh-water calanoid copepod, *Epischura massachusettsensis* Pearse. *J. Morphol.*, 96:441–472. **Humes, Arthur G. and Mildred Stratton Wilson. 1951.** The last copepodid instar of *Diaptomus sanguineus* Forbes (Copepoda). *J. Wash. Acad. Sci.*, 41:395–399. **Juday, Chancey. 1925.** *Senecella calanoides*, a recently described fresh-water copepod. *Proc. U. S. Natl. Museum*, 66, art. 4:1–6. **Juday, Chancey and R. A. Muttkowski. 1915.** Entomostraca from St. Paul Island, Alaska. *Bull. Wisconsin Nat. Hist. Soc.*, N. S., 13:23–31. **Kiefer, Friedrich. 1931.** Zur Kenntnis der freilebenden Süsswassercopepoden, insbesondere der Cyclopiden Nordamerikas. *Zool. Jahrb. Abt. Syst. Ök.*

Geog. Tiere, 61:579–620. **1932.** Versuch eines systems der Diaptomiden (Copepoda Calanoida). *Zool. Jahrb. Abt. Syst. Ok. Geog. Tiere*, 63:451–520. **1938a.** Ruderfusskrebse (Crust. Cop.) aus Mexiko. *Zool. Anz.*, 123:274–280. **1938b.** Freilebende Süsswassercopepoden von den Nordkurilen. *Bull. Biogeog. Soc. Japan*, 8:75–94. **Kincaid, Trevor. 1953.** *A Contribution to the Taxonomy and Distribution of the American Fresh-Water Calanoid Crustacea.* Privately printed by the author. Calliostoma Co., Seattle. **1956.** *Notes and Descriptions of American Fresh-Water Calanoid Crustacea.* Privately printed by the author. Calliostoma Co., Seattle. **Light, S. F. 1938.** New subgenera and species of diaptomid copepods from the inland waters of California and Nevada. *Univ. California Berkeley Publs. Zool.*, 43:67–78. **1939.** New American subgenera of *Diaptomus* Westwood (Copepoda, Calanoida). *Trans. Am. Microscop. Soc.*, 58:473–484. **Lilljeborg, W. 1889.** In: deGuerne and Richard, Révision de Calanides d'eau douce. *Mém. soc. zool. France*, 2:53–181. **Marsh, C. Dwight. 1907.** A revision of the North American species of *Diaptomus*. *Trans. Wisconsin Acad. Sci.*, 15:381–516. **1911.** On a new species of *Diaptomus* from Colorado. *Trans. Wisconsin Acad. Sci.*, 17:197–199. **1913.** Report on freshwater Copepoda from Panama, with descriptions of new species. *Smithsonian Inst. Publs. Misc. Collections*, 61(3):1–31. **1915.** A new crustacean, *Diaptomus virginiensis*, and a description of *Diaptomus tyrelli* Poppe. *Proc. U. S. Natl. Museum*, 49:457–462. **1920.** The fresh-water Copepoda of the Canadian Arctic Expedition 1913–18. *Rept. Canadian Arctic Exped. 1913–1918*, 7(P. J):1–24. **1926.** On a collection of Copepoda from Florida with a description of *Diaptomus floridanus*, new species. *Proc. U. S. Natl. Museum*, 70, art. 10:1–4. **1929.** Distribution and key of the North American copepods of the genus *Diaptomus*, with the description of a new species. *Proc. U. S. Natl. Museum*, 75, art. 14:1–27. **1933.** Synopsis of the calanoid crustaceans, exclusive of the Diaptomidae, found in fresh and brackish waters, chiefly of North America. *Proc. U. S. Natl. Museum*, 82, art. 18:1–58. **Osorio Tafall, B. F. 1942.** *Diaptomus (Microdiaptomus) cokeri*, nuevos subgenero y especie de Diaptomido de las cuevas de la region de Valles (San Luis Potosi, Mexico) (Copep., Calan.). *Ciencia (Mex.)*, 3:206–210. **Pearse, A. S. 1906.** Freshwater Copepoda of Massachusetts. *Am. Naturalist*, 40:241–251. **Reed, Edward B. 1958.** Two new species of *Diaptomus* from arctic and subarctic Canada (Calanoida, Copepoda). *Canadian J. Zool.*, 36:663–670. **Rylov, V. M. 1923.** On the Eucopepodean fauna of Manchuria. *Ann. Museum Zool. Acad. Sci. Russia*, 24:52–95. **Sars, G. O. 1901–1903.** *An Account of the Crustacea of Norway.* Vol. 4, *Copepoda Calanoida.* Bergen Museum, Bergen. **Schacht, F. W. 1897.** The North American species of *Diaptomus*. *Bull. Illinois State Lab. Nat. Hist.*, 5:97–208. **Turner, C. H. 1910.** Ecological notes on the Cladocera and Copepoda of Augusta, Georgia, with descriptions of new or little known species. *Trans. Acad. Sci. St. Louis*, 19:151–176. **1921.** Ecological studies of the Entomostraca of the St. Louis District. Part I. *Diaptomus pseudosanguineus* sp. nov. and a preliminary list of the Copepoda and Cladocera of the St. Louis district. *Trans. Acad. Sci. St. Louis*, 24(2):1–25. **Wilson, Mildred Stratton. 1941.** New species and distribution records of diaptomid copepods from the Marsh collection in the United States National Museum. *J. Wash. Acad. Sci.*, 31:509–515. **1951.** A new subgenus of *Diaptomus* (Copepoda: Calanoida), including an Asiatic species and a new species from Alaska. *J. Wash. Acad. Sci.*, 41:168–179. **1953a.** New and inadequately known North American species of the copepod genus *Diaptomus*. *Smithsonian Inst. Publs. Misc. Collections*, 122(2):1–30. **1953b.** New Alaskan records of *Eurytemora* (Crustacea, Copepoda). *Pacific Sci.*, 7:504–512. **1954.** A new species of *Diaptomus* from Louisiana and Texas with notes on the subgenus *Leptodiaptomus*. *Tulane Studies Zool.*, 2:49–60. **1955.** A new Louisiana copepod related to *Diaptomus (Aglaodiaptomus) clavipes* Schacht. *Tulane Studies Zool.*, 3:35–47. **1958.** New records and species of calanoid copepods from Saskatchewan and Louisiana. *Canadian J. Zool.*, 36:489–497. **Wilson, Mildred Stratton and S. F. Light. 1951.** Description of a new species of diaptomid copepod from Oregon. *Trans. Am. Microscop. Soc.*, 70:25–30. **Wilson, Mildred Stratton and Walter G. Moore. 1953a.** New records of *Diaptomus sanguineus* and allied species from Louisiana, with the description of a new species (Crustacea: Copepoda). *J. Wash. Acad. Sci.*, 43:121–127. **1953b.** Diagnosis of a new species of diaptomid copepod from Louisiana. *Trans. Am. Microscop. Soc.*, 72:292–295. **Wuthrich, Marguerite. 1948.** Etude du developpement des nauplii de *Diaptomus gracilis*, O. Sars, et *Diaptomus laciniatus*, Lilljeborg. *Rev. suisse zool.*, 55:427–445.

CYCLOPOIDA

Coker, Robert E. 1943. *Mesocylops edax* (S. A. Forbes) *M. leuckarti* (Claus) and related species in America. *J. Elisha Mitchell Sci. Soc.*, 59:181–200. **Forbes, E. B. 1897.** A contribution to the knowledge of North American freshwater Cyclopidae. *Bull. Illinois State Lab. Nat Hist.*, 5:27–83. **Gurney, Robert. 1933.** *British Fresh-Water Copepoda*, vol. 3. Ray Society London. **Johnson, Martin W. 1953.** The copepod *Cyclops dimorphus* Kiefer from the Saltor Sea. *Am. Midland Naturalist*, 49:188–192. **Kiefer, Friedrich. 1929.** *Crustacea Copepoda*, II Cyclopoida Gnathostoma. *Das Tierreich*, 53:51–102. Gruyter, Berlin and Leipzig. **1936** Frielebende Suss- und Salzwassercopepoden von der Insel Haiti. Mit einer Revision der Gattung Halicyclops Norman. *Arch. Hydrobiol.*, 30:263–317. **Lindberg, K. 1957.** Cyclopide (Crustaces copepodes) de la Cote d'Ivoire. *Bull. Inst. fran. Afr. noire, ser. A*, 19:134–179 **Marsh, Charles Dwight. 1910.** A revision of the North American species of *Cyclops*. *Trans Wisconsin Acad. Sci.*, 16:1067–1135. **Price, J. L. 1958.** Cryptic speciation in the *vernali.* group of Cyclopidae. *Canadian J. Zool.*, 36:285–303. **Sars, G. O. 1918.** An account of the Crustacea of Norway. Copepoda Cyclopoida. *Bergens Museums Skrifter*, 6:1–225. **Wilson Mildred Stratton. 1958.** The copepod genus *Halicyclops* in North America, with description o a new species from Lake Pontchartrain, Louisiana, and the Texas coast. *Tulane Studies Zool.* 6:000–000. **Yeatman, Harry C. 1944.** American cyclopoid copepods of the *viridis-vernali.* group, (including a description of *Cyclops carolinianus* n. sp.). *Am. Midland Naturalist*, 32:1–90

HARPACTICOIDA

Borutsky, E. V. 1952. Harpacticoida presnykh vod. Fauna SSSR, Rakoobraznye, III(4) *Zool. Inst. Akad. Nauk SSSR, n. s.*, No. 50, 425 pp. (In Russian.) **Campbell, Mildred H 1930.** Some free-swimming copepods of the Vancouver Island region. II. *Trans. Roy. Soc Can.*, 3rd Ser., 24:177–182. **Carter, Marjorie Estelle. 1944.** Harpacticoid copepods o the region of Mountain Lake, Virginia. *J. Elisha Mitchell Sci. Soc.*, 60:158–166. **Chappuis. P. A. 1927.** Freilebende Süsswasser-Copepoden aus Nordamerika. 2. Harpacticiden. *Zool Anz.*, 74:302–313. **1928.** Nouveaux Copépodes cavernicoles. *Bull. Soc. Sci. Cluj Roumanie* 4:20–34. **1931.** Campagne spéologique de C. Bolivar et R. Jeannel dans l'Amérique du Nord (1928). 4. Crustacés Copépodes. *Arch. zool. exp. et gén.*, 71:345–360. **1933.** Zoologische Ergebnisse einer Reise nach Bonaire, Curacao und Aruba im Jahre 1930. No. 6 Süss- und Brackwasser-Copepoden von Bonaire, Curacao und Aruba. I. Harpacticoida *Zool. Jahrb. Abt. Syst. Ok. Geog. Tiere*, 64:391–404. **1956.** Notes sur les Copépodes. 23 Le genre *Elaphoidella* Chappuis. *Notes Biospéologiques*, 11:61–71. **1958.** Le genre *Parastenocaris* Kessler. *Vie et Milieu* (for 1957), 8:423–432. **Chappuis, P. A. and Cl. Delamare Deboutteville. 1958.** Recherches sur la faune interstitielle littorale du Lac Érié. Le problème des glaciations quaternaires. *Vie et Milieu* (for 1957), 8:366–376. **Coker, Robert E 1934.** Contribution to knowledge of North American freshwater harpacticoid copepod Crustacea. *J. Elisha Mitchell Sci. Soc.*, 50:75–141. **Coker, Robert E. and Juanita Morgan 1940.** A new harpacticoid copepod from North Carolina. *J. Wash. Acad. Sci.*, 30:395–398 **Ewers, Lela A. 1930.** The larval development of freshwater Copepoda. *Ohio State Univ Franz Theo. Stone Lab.*, Contrib. No. 3, 43 pp. **Gurney, Robert. 1927.** Zoological result of the Cambridge Expedition to the Suez Canal, 1924. XXXIII. Report on the Crustacea Copepoda (littoral and semi-parasitic). *Trans. Zool. Soc. London*, 22:451–577. **1932. Th** *British Fresh-Water Copepoda*. Vol. 2. Ray Society, London. **Herrick, C. L. 1895.** Copepoda of Minnesota. In: C. L. Herrick and C. H. Turner. Synopsis of Entomostraca of Minnesota. *Zool. Ser. II, Geol. Nat. Hist. Survey Minnesota.* **Kessler, E. 1913.** *Parastenocari brevipes* nov. gen. et nov. spec., ein neuer Susswasserharpacticide. *Zool. Anz.*, 42:514–520 **Kiefer, F. 1929.** Zur Kenntnis der freilebenden Copepoden Nordamerikas. *Zool. Anz.*, 86:97– 100. **1931.** Zur Kenntnis der freilebenden Süsswassercopepoden, insbesondere der Cyclopiden Nordamerikas. *Zool. Jahrb. Abt. Syst. Ok. Geog. Tiere*, 61:579–620. **1936.** Freilebende Süss- und Salzwassercopepoden von der Insel Haiti. Mit einer Revision der Gattung *Halicyclops* Norman. *Arch. Hydrobiol.*, 30:263–317. **Klie, W. 1929.** Die Copepoda Harpacticoida der südlichen und westlichen Ostee mit besonderer Berucksichtigung der Sandfauna der Kieler Hafen. *Zool. Jahrb. Abt. Syst. Ok. Geog. Tiere*, 57:329–386. **Kunz, H. 1938**

Die Sandbewohnenden Copepoden von Helgoland. I Teil. *Kiel. Meeresforsch.*, 2:223–254. **Lang, Karl. 1934.** Studien in der Gattung *Epactophanes* (Copepoda Harpacticoida). *Arkiv Zool.*, 28A(11):1–27. **1948.** *Monographie der Harpacticiden*, 2 vols. H. Ohlsson Lund. **Marsh, C. Dwight. 1903.** On a new species of *Canthocamptus* from Idaho. *Trans. Wisconsin Acad. Sci.*, 14:112–114. **Monk, Cecil R. 1941.** Marine harpacticoids from California. *Trans. Am. Microscop. Soc.*, 60:75–99. **Nicholls, A. G. 1941.** The developmental stages of *Metis jousseaumei* (Richard) (Copepoda, Harpacticoida). *Ann. Mag. Nat. Hist.*, Ser. 11, 7:317–328. **Oloffson, Ossian. 1917.** Beitrag zur Kenntnis der Harpacticiden-Familien Ectinosomidae, Canthocamptidae (Gen. *Maraenobiotus*) und Tachidiidae nebst Beschreibungen einiger neuen und wenig bekannten, arktischen Brackwasser -und Süsswasser-Arten. *Zool. Bidrag. Uppsala*, 6:1–39. **Pearse, A. S. 1905.** Contributions to the copepod fauna of Nebraska and other states. *Proc. Am. Microscop. Soc.*, 26:145–160. Reprinted in: *Stud. Zool. Lab. Univ. Nebraska*, No. 65. **Pennak, Robert W. 1939.** A new copepod from the sandy beaches of a Wisconsin lake. *Trans. Am. Microscop. Soc.*, 58:224–227. **Rylov, V. M. 1923.** On the eucopepodean fauna of Mantshuria. *Ann. Museum Zool. Acad. Sci. Russia*, 24:52–95. **Sars, G. O. 1903–1911.** *An Account of the Crustacea of Norway.* Vol. 5, *Copepoda Harpacticoida.* Bergen Museum, Bergen. **Sharpe, Richard W. 1910.** Notes on the marine Copepoda and Cladocera of Woods Hole and adjacent regions, including a synopsis of the genera of the Harpacticoida. *Proc. U. S. Natl. Museum*, 38:405–436. **Willey, Arthur. 1925.** Northern Cyclopidae and Canthocamptidae. *Trans. Royal Soc. Can.*, 3rd Ser., 19:137–158. **1927.** Description of a new species of fresh-water copepod of the genus *Moraria* from Canada. *Proc. U. S. Natl. Museum*, 71:1–12. **1934.** Some Laurentian copepods and their variations. *Trans. Roy. Can. Inst.*, 20:77–98. **Wilson, Charles B. 1932.** The copepods of the Woods Hole Region, Massachusetts. *U. S. Natl. Museum Bull.*, No. 158, 1–635. **Wilson, Mildred Stratton. 1956a.** North American Harpacticoid Copepods. 1. Comments on the known fresh-water species of the Canthocamptidae. 2. *Canthocamptus oregonensis*, n. sp. from Oregon and California. *Trans. Am. Microscop. Soc.*, 75:290–307. **1956b.** North American Harpacticoid Copepods. 3. *Paracamptus reductus*, n. sp., from Alaska. *J. Wash. Acad. Sci.*, 46:348–351. **1958a.** North American Harpacticoid Copepods. 4. Diagnoses of new species of fresh-water Canthocamptidae and Cletodidae (Genus *Huntemannia*). *Proc. Biol. Soc. Wash.*, 71:43–48. **1958b.** North American Harpacticoid Copepods. 5. The status of *Attheyella americana* (Herrick) and the correct name for the subgenus *Brehmiella*. *Proc. Biol. Soc. Wash.*, 71:49–52. **1958c.** North American Harpacticoid Copepods. 6. New records and species of *Bryocamptus* (subgenus *Arcticocamptus*) from Alaska. *Trans. Am. Microscop. Soc.*, 77:320–328.

30

Branchiura
and Parasitic Copepoda

MILDRED STRATTON WILSON

Knowledge of North American fresh-water branchiurans and parasitic copepods stems largely from the works of C. B. Wilson (1903–1917). Taxonomic studies, with revised descriptions of known species and integration with the recent work of other continents, as well as further investigation of occurrence (hosts and geographic distribution), habits, and development are much needed. The necessary taxonomic research would be both encouraged and facilitated if investigators of fish parasites would deposit collections of these crustaceans in museums, instead of allowing them to remain geographically scattered or become lost to science.

The summaries presented here are based largely on literature with no attempt to make taxonomic decisions. Characters and figures are given for the recognition of genera. Descriptions of the species known to 1957 can be found by reference to cited literature and Table 30.1.

Branchiura

In modern classification, the Branchiura, represented in North America only by the genus *Argulus* (Fig. 30.1) are recognized as a subclass of the

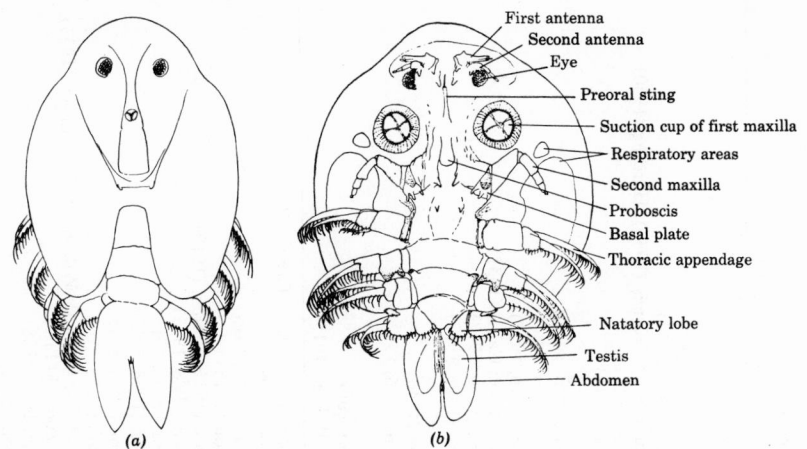

Fig. 30.1. *Argulus.* (*a*) Dorsal view, female, *A. stizostethii* Kellicott. (*b*) Ventral view, male, *A. japonicus* Thiele. (*a* after Wilson, *b* after Meehean.)

Crustacea, not as a subdivision of the Copepoda. *Argulus* is parasitic on fresh- and salt-water fishes, and has been found on amphibians.

Recognition Characters of *Argulus*

Body depressed dorsoventrally; carapace laterally and distally expanded, incised posteromedially so that all or a portion of the segmented thorax and abdomen is free. Ventrally (Fig. 30.1*b*), a pair of prominent suction cups and four setose thoracic appendages (legs). Sexes differing in armature of legs; male usually smaller than female. Total length range of females of known American species, about 5–25 mm.

Literature on *Argulus*

The most recent reviews of North American *Argulus* are those of Meehean (1940) and C. B. Wilson (1944). Meehean's paper is a valuable basic reference because it includes a summary of external anatomy, development, taxonomy, and literature, with an excellent key to the known species. It should be used together with Wilson's paper because of unresolved differences of opinion on synonymy. There is need for studies like that of Hsiao (1950) which demonstrated that the common, widespread *A. japonicus* is a single variable species, so that *A. trilineatus* falls into its synonymy as indicated by Meehean.

Parasitic Copepoda

The four genera known to occur on North American fresh-water fishes and amphibians can be distinguished by the habitus outlines shown in Figs. 30.2–

Table 30.1. Fresh-Water Parasitic Copepoda

List of Species and References to Descriptions

Ergasilus

Synonymy, distribution, hosts: Smith (1949)

caeruleus C. B. Wilson	Wilson (1911, 1916);
(includes confusus Bere)	Bere (1931); Mueller (1936)
centrarchidarum Wright	Wilson (1911)
chautauquaensis Fellows	Wilson (1911); Fig. 30.2 herein
coti Kellicott	Kellicott (1892)
elegans C. B. Wilson	Wilson (1916); Mueller (1936)
elongatus C. B. Wilson	Wilson (1916)
lanceolatus C. B. Wilson	Wilson (1916)
luciopercarum Henderson	Henderson (1926)
megaceros C. B. Wilson	Wilson (1916)
nigritus C. B. Wilson	Wilson (1916)
osburni Tidd and Bangham	Tidd and Bangham (1945a)
versicolor C. B. Wilson	Wilson (1911); Mueller (1936)

Achtheres

Key: C. B. Wilson (1915)

ambloplitis Kellicott	Wilson (1915)
coregoni (Smith)	Wilson (1915)
corpulentus Kellicott	Wilson (1915)
lacae Krøyer	Wilson (1915)
micropteri Wright	Wilson (1915)
pimelodi Krøyer	Wilson (1915, 1916)

Lernaea

Keys: C. B. Wilson (1917); Harding (1950)*

carassii Tidd	Tidd (1933, 1934)
(synonym of cyprinacea*)	
catostomi (Krøyer)	Wilson (1916, 1917)
cruciata (LeSueur)	Wilson (1916, 1917)
cyprinacea Linnaeus	Wilson (1917); Gurney (1933); Harding (1950)
dolabrodes C. B. Wilson	Wilson (1917)
insolens (C. B. Wilson	Wilson (1917)
(= anomala C. B. Wilson)	
pectoralis Kellicott	Wilson (1916, 1917)
(synonym of catostomi*)	
pomotidis (Krøyer)	Wilson (1916, 1917)
ranae Stunkard and Cable	Stunkard and Cable (1931)
tenuis (C. B. Wilson	Wilson (1916, 1917)
tortua Kellicott	Wilson (1916, 1917)
(synonym of catostomi*)	
tortua coquae Dolley	Dolley (1940)
variabilis (C. B. Wilson)	Wilson (1916, 1917)

Salmincola

Key: C. B. Wilson (1915)

beani (C. B. Wilson)	Wilson (1908, 1915); Fasten (1921)
bicauliculata (C. B. Wilson)	Wilson (1908, 1915)
californiensis (Dana)	Wilson (1915, 1916)
carpenteri (Packard)	Wilson (1915)
edwardsii (Olsson)	Wilson (1915); Fasten (1919)
extumescens (Gadd)	Wilson (1908, 1915)
falculata (C. B. Wilson)	Wilson (1908, 1915)
gibber (C. B. Wilson)	Wilson (1908, 1915)
inermis (C. B. Wilson)	Wilson (1908, 1915)
oquassa C. B. Wilson	Wilson (1915)
salmonea (Linnaeus)	Bere (1930) (record); Wilson (1915); Gurney (1933); Friend (1941)
salvelini Richardson	Richardson (1938)
siscowet (Smith)	Wilson (1915)
thymalli (Kessler)	(New record, District of Keewatin, Canada); Gurney (1933); Wilson (1915)
wisconsinensis Tidd and Bangham	Tidd and Bangham (1945b)

*See text, p. 866.

30.5, and by the recognition characters given in the text.   The known species with references to keys and descriptions are listed in Table 30.1.

Recognition Characters of Genera

Family Ergasilidae.   *Ergasilus* (Figs. 30.2, 30.3).   Copepodids (developmental stages and adult) of cyclopoid form; cephalothorax more or less inflated in attached adult female.   First antenna six-segmented; second antenna variously developed from species to species, terminating in strong claw. Maxillipeds absent in female; present and conspicuous in male (Fig. 30.2c). Legs 1 to 4 biramous; leg 5 uniramous, reduced.   Two ovisacs.   Total length,

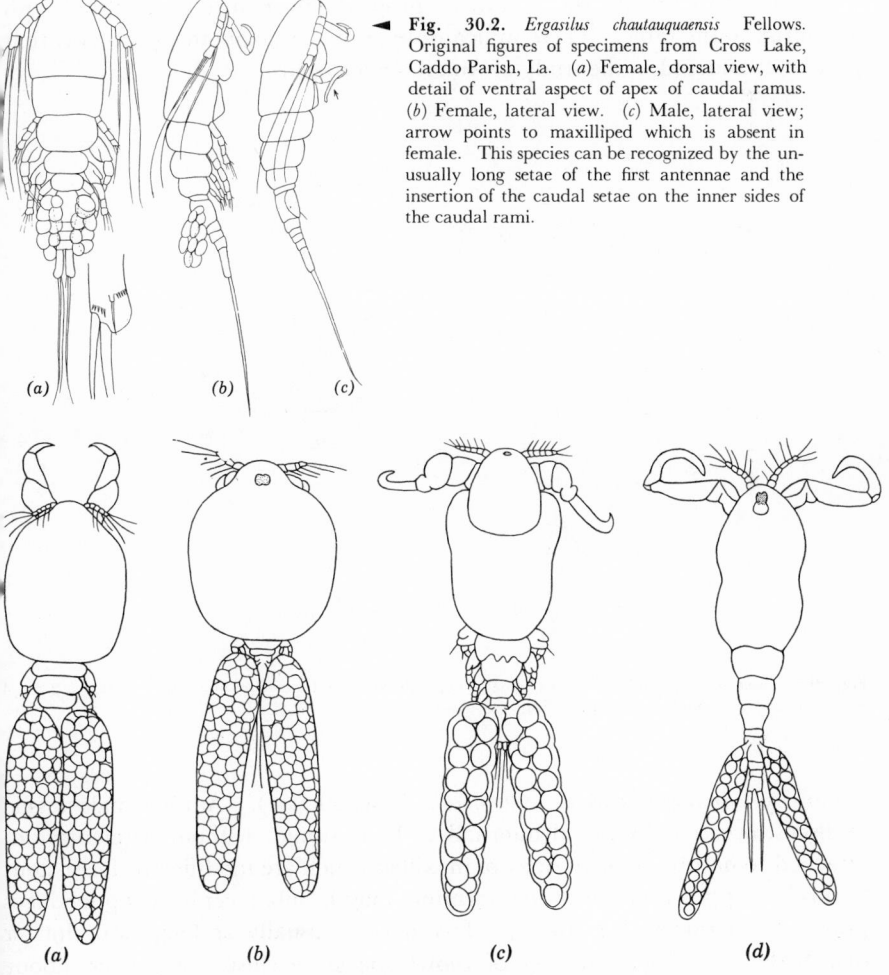

◄ **Fig. 30.2.** *Ergasilus chautauquaensis* Fellows. Original figures of specimens from Cross Lake, Caddo Parish, La.   (*a*) Female, dorsal view, with detail of ventral aspect of apex of caudal ramus. (*b*) Female, lateral view.   (*c*) Male, lateral view; arrow points to maxilliped which is absent in female.   This species can be recognized by the unusually long setae of the first antennae and the insertion of the caudal setae on the inner sides of the caudal rami.

Fig. 30.3.   Dorsal views of ovigerous females of some species of *Ergasilus* to show differences in habitus and development of second antennae.   (*a*) *E. labracis* Krøyer.   (*b*) *E. centrarchidarum* Wright.   (*c*) *E. caeruleus* C. B. Wilson.   (*d*) *E. versicolor* C. B. Wilson.   (Modified from C. B. Wilson.)

1 mm or less. Developmental stages free-swimming; adult female usually attached to gill filaments of host; adult male free-swimming. Adults of *E. chautauquaensis* (Fig. 30.2), frequently collected in plankton tows, have not yet been found associated with a host.

Family Lernaeidae. *Lernaea* (Fig. 30.4). Mature female vermiform; cephalothorax with hornlike processes anchored in tissue of fish or amphibian host, or attached to fish scale; posterior part of body protruding from host. Two ovisacs. Length range of known species from about 5–23 mm. Developmental copepodid stages of male and female (Fig. 30.4b) and of adult male not modified, body form cyclopoid; found on host or free-swimming.

In the most recent taxonomic revision of the genus (Harding, 1950), specimens of North American species were not compared with those of other continents. For this reason, the synonyms proposed by Harding are listed with an asterisk in the table. These indicate some of the taxonomic problems that must be considered in the study of American lernaeids.

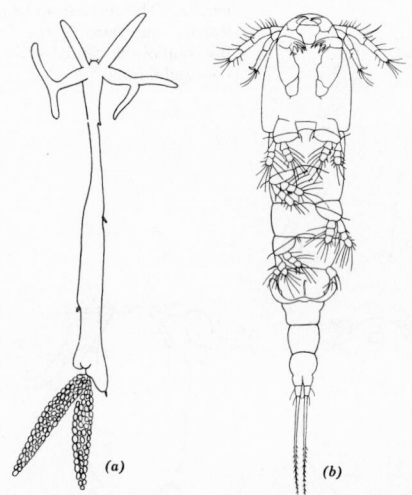

(a)   *(b)*

Fig. 30.4. *Lernaea.* (a) Adult ovigerous female, generalized. (b) Free-swimming female. (Legs 2 to 4 shown on one side only.) (Modified from Gurney.)

Family Lernaeopodidae. *Salmincola* (Fig. 30.5a–c). Adult female sessile on fishes; occasionally found internally. Body tumid, without segmentation; attached to host by bulla at apex of maxillae which are modified to form stout "arms." Upper and lower lips forming mouth tube; cephalic appendages present but reduced; legs absent. Two ovisacs, usually as long as or longer than body. Total length range of known species, exclusive of ovisacs, about 3–8 mm. Both sexes with one free-swimming copepodid stage (Fig. 30.5b) which undergoes modification after attachment to host. Adult male minute, length 1.4 mm or less; attaches to female at time of fertilization and may remain attached throughout life; maxillae not modified as "arms"; maxillipeds usually with stout claws (Fig. 30.5c).

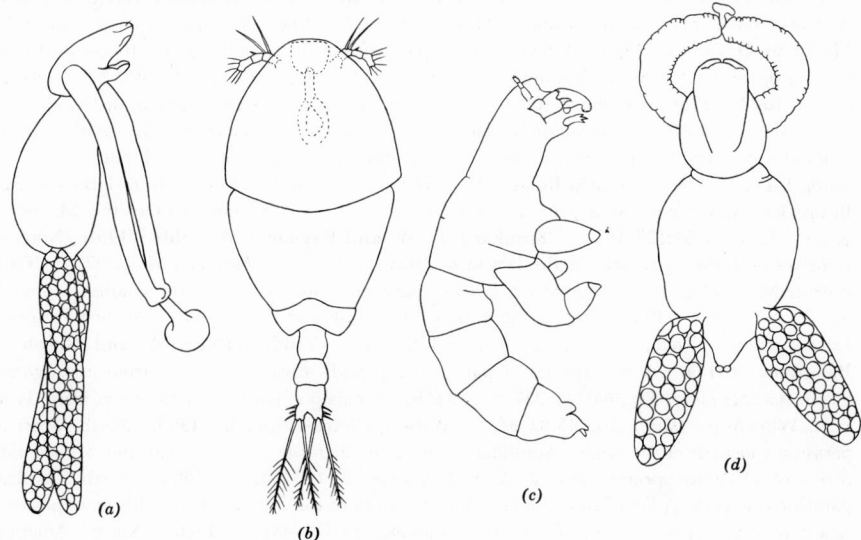

Fig. 30.5. Genera of Lernaeopodidae. (*a*) *Salmincola edwardsii* (Olsson), adult ovigerous female, lateral view. (*b*) *S. edwardsii*, free-swimming copepodid larva, dorsal view. (*c*) *S. thymalli* (Kessler), adult male, lateral view. (*d*) *Achtheres ambloplitis* Kellicott, adult ovigerous female, dorsal view. (*a, d* after Wilson; *b* after Fasten; *c.* after Gurney.)

It is considered questionable whether the genus *Salmincola* should be taxonomically separated from *Achtheres*.

Achtheres (Fig. 30.5*d*). Like *Salmincola* except that body of female is more or less segmented, head has dorsally sclerotized ridges ("carapace" of C. B. Wilson), and abdomen is short. Male hardly different from *Salmincola*. Total length range of known species: female, about 2.8–8 mm, male, 1–2.5 mm.

References

Bere, Ruby. 1930. The parasitic copepods of the fish of the Passamaquoddy region. *Contrib. Can. Biol. and Fisheries*, (N.S.), 5:423–430. **1931.** Copepods parasitic on fish of the Trout Lake region, with descriptions of two new species. *Trans. Wisconsin Acad. Sci.*, 26:427–436. **Dolley, John S. 1940.** A new lernaean (parasitic copepod) from minnows in Lafayette County, Mississippi. *Trans. Am. Microscop. Soc.*, 59:70–77. **Fasten, Nathan. 1919.** Morphology and attached stages of first copepodid larva of *Salmincola edwardsii* (Olsson) Wilson. *Publ. Puget Sound Biol. Sta. Univ. Wash.*, 2:153–181. **1921.** Another male copepod of the genus *Salmincola* from the gills of the chinook salmon. *Biol. Bull.*, 41:121–124. **Friend, G. F. 1941.** The life-history and ecology of the salmon gill-maggot *Salmincola salmonea* (L.) (Copepod Crustacean). *Trans. Roy. Soc. Edinburgh*, 60:503–541. **Gurney, Robert. 1933.** *British Fresh-Water Copepoda.* Vol. 3. Ray Society, London. **Harding, J. P. 1950.** On some species of *Lernaea* (Crustacea, Copepoda: parasites of freshwater fish). *Bull. Brit. Museum (Nat. Hist.) Zool.*, 1:1–27. **Henderson, Jean T. 1926.** Description of a copepod gill-parasite of Pike-Perches in lakes of northern Quebec, including an account of the free-swimming male and some developmental stages. *Contrib. Can. Biol. and Fisheries* (N.S.), 3:235–246.   **Hsiao, Sidney C.   1950.**   Copepods from Lake Erh Hai, China.   *Proc. U. S. Natl. Museum*, 100:161–200.   **Kellicott, David S.   1892.** A crustaceous parasite of the "Miller's Thumb" (*Cottus*).   Proc. *Am. Microscop. Soc.*, 14:76–79. **Meehean, O. Lloyd.   1940.**   A review of the parasitic Crustacea of the genus *Argulus* in the collections of the United States National Museum. *Proc. U. S. Natl. Museum*, 88:459–522.   **Mueller, Justus F.   1936.**   Notes on some parasitic copepods and a mite, chiefly from Florida fresh water fishes.   *Am Midland Naturalist*, 17:807–815.   **Richardson, Laurence R.   1938.**   An account of a parasitic copepod, *Salmincola salvelini* sp. nov., infecting the Speckled Trout.   *Can. J. Research*, 16D:225–229.   **Smith, Roland. F.   1949.**   Notes on *Ergasilus* parasites from the New Brunswick, New Jersey, area, with a check list of all species and hosts east of the Mississippi River. *Zoologica*, 34:127–182.   **Stunkard, H. W. and Raymond M. Cable.   1931.**   Notes on a species of *Lernaea* parasitic in the larvae of *Rana clamitans*.   *J. Parasitol.*, 18:92–97.   **Tidd, Wilbur M.   1933.**   A new species of *Lernaea* (parasitic Copepoda) from the goldfish.   *Ohio J. Sci.*, 33:465–468.   **1934.**   Recent infestations of goldfish and carp by the "Anchor Parasite," *Lernaea carassii*. *Trans. Am. Fisheries Soc.*, 64:176–180.   **.Tidd, Wilbur M. and Ralph V. Bangham.   1945a.**   A new species of parasitic copepod, *Ergasilus osburni*, from the burbot. *Trans. Am. Microscop. Soc.*, 64:225–227.   **1945b.**   A copepod parasite of the cisco from Trout Lake, Wisconsin.   *Ohio J. Sci.*, 45:82–84.   **Wilson, Charles Branch.   1903.**   North American parasitic copepods of the family Argulidae, with a bibliography of the group and a systematic review of all known species.   *Proc. U. S. Natl. Museum*, 25:635–742.   **1908.**   North American parasitic copepods: A list of those found upon the fishes of the Pacific Coast, with descriptions of new genera and species.   *Proc. U. S. Natl. Museum*, 35:431–481.   **1911.**   North American parasitic copepods belonging to the family Ergasilidae.   *Proc. U. S. Natl. Museum*, 39:263–400. **1915.**   North American parasitic copepods belonging to the Lernaeopodidae, with a revision of the entire family.   *Proc. U. S. Natl. Museum*, 47:565–729.   **1916.**   Copepod parasites of freshwater fishes and their economic relations to mussel glochidia.   *Bull. Bur. Fisheries*, 34:331–374. **1917.**   The economic relations, anatomy, and life history of the genus *Lernaea*. *Bull. Bur. Fisheries*, 35:165–198.   **1944.**   Parasitic copepods in the United States National Museum.   *Proc. U. S. Natl. Museum*, 94:529–582.

31

Malacostraca

FENNER A. CHACE, JR.

J. G. MACKIN

LESLIE HUBRICHT

ALBERT H. BANNER

HORTON H. HOBBS, JR.

The malacostracan crustaceans differ from those belonging to the other four crustacean subclasses in having a definite and fixed number of body segments or somites. Of the twelve orders of the Malacostraca, only four have fresh-water representatives in North America. These orders may differ rather noticeably from one another (Fig. 31.1), but all the species involved are alike in having 19 pairs of appendages, exclusive of the eyes.

General Characteristics

The *mysids* or opposum shrimps (Fig. 31.1c) superficially resemble the true decapod shrimps but they are generally regarded as the lowest of the four orders considered here. They have stalked eyes and a carapace covering most of the thoracic somites, but the carapace is not fused with the thorax posteriorly, the thoracic limbs are little differentiated and have natatory exopods, and there are no gills in most of the species. Adult females are characterized by the development of a brood pouch or marsupium composed of lamellate endites attached to the bases of the pereiopods; in this brood pouch the eggs are hatched and the young sheltered until they are able to shift for themselves, hence the name sometimes given to these forms. Males are distinguished by the absence of a brood pouch and often by the pronounced development of the exopod of the fourth pleopod.

The *isopods* or sow-bugs (Fig. 31.1*a*) lack a carapace, are somewhat flattened dorsoventrally, and have the pleopods modified as respiratory organs.

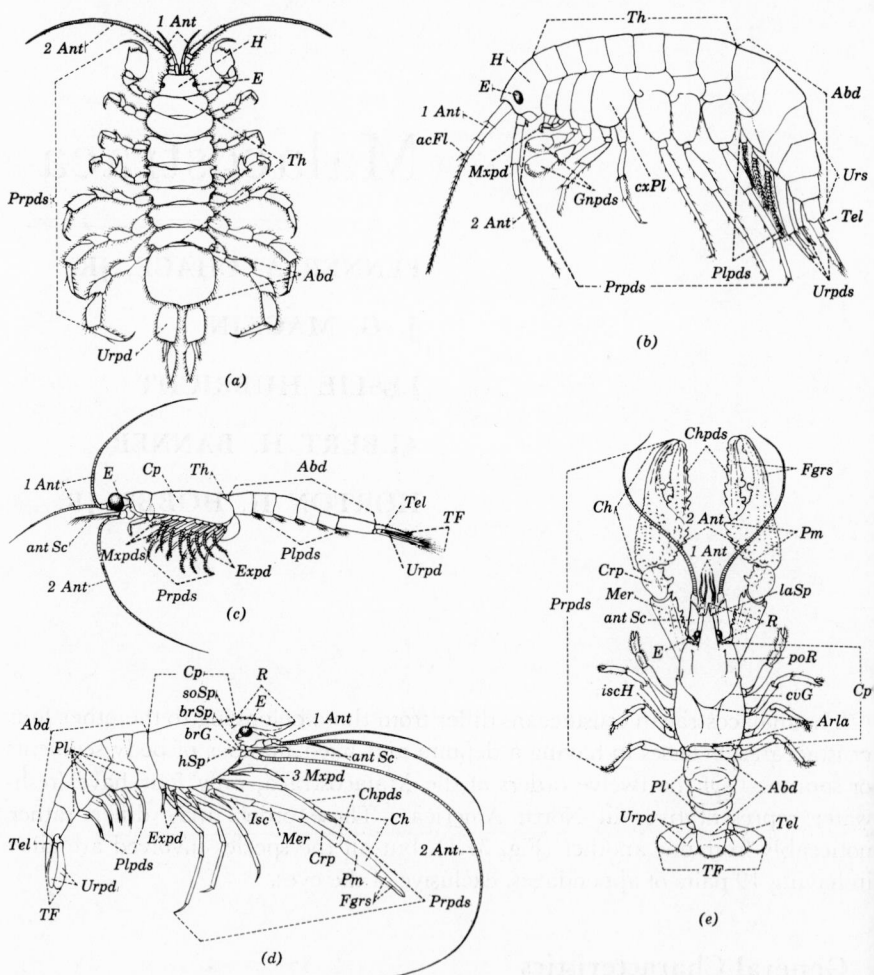

Fig. 31.1. Examples of malacostracan crustaceans and structures. (*a*) An isopod, *Asellus*, sp. (*b*) An amphipod, *Gammarus lacustris*. (*c*) A mysid, *Mysis relicta*. (*d*) A generalized shrimp. (*e*) A crayfish, *Procambarus spiculifer*.

Abd, abdominal somites; *acFl*, accessory flagellum of first antenna; *1Ant*, first antenna (antennule); *2Ant*, second antenna (antenna); *antSc*, antennal scale; *Arla*, areola; *brG*, branchiostegal groove; *brSp*, branchiostegal spine; *Ch*, chela; *Chpds*, chelipeds; *Cp*, carapace; *Crp*, carpus; *cvG*, cervical groove; *cxPl*, coxal plate; *E*, eye; *Expd*, exopod; *Fgrs*, fingers of chela; *Gnpds*, gnathopods; *H*, head; *hSp*, hepatic spine; *Isc*, ischium; *iscH*, hook on ischium; *laSp*, lateral spine of rostrum; *Mer*, merus; *Mxpd*, maxilliped; *3Mxpd*, third maxilliped; *Pl*, pleura of abdominal somites; *Plpds*, pleopods; *Pm*, palm of chela; *poR*, postorbital ridge; *Prpds*, pereiopods; *R*, rostrum; *soSp*, supraorbital spine; *Tel*, telson; *TF*, tail fan; *Th*, thoracic somites; *Urpd*, uropod; *Urs*, urosome somites. (*a–c* After Smith.)

A brood pouch is developed in adult females between the bases of the pereiopods. In males, the inner ramus of the second pleopod bears a slender copulatory stylet.

The *amphipods* or scuds (Fig. 31.1*b*) are compressed laterally, and the respiratory organs are attached to the pereiopods. As in the last two groups, the
adult female has a brood pouch formed by oostegites attached to the bases of
the pereiopods. In separating the sexes of amphipods, care must be taken to
distinguish between the oostegites and the gills which may originate close together. Males can be recognized by the presence of sexual organs near the
base of the last pereiopods, and often also by more obvious secondary characters involving the antennae and gnathopods.

Among the *decapods*, the caridean shrimps or prawns (Fig. 31.1*d*) can be
distinguished by the relatively thin, rather translucent shell, the laterally compressed rostrum, and the presence of pincers on the first two pairs of pereiopods only. Males have two slender stylets on the inner margin of the inner
branch of the second pair of pleopods, whereas there is but one such stylet in
this position in the females.

The crayfishes or crawfishes (Fig. 31.1*e*) have a somewhat heavier and more
densely pigmented shell, a dorsally flattened rostrum, and pincers on the first
three pairs of pereiopods. Males can be readily distinguished from females by
the modification of the first two pairs of pleopods; these function as gonopods
and are normally flexed against the lower surface of the animal between the
last pair of pereiopods. The male openings can also often be discerned on the
coxae of the fifth pereiopods and the female openings on the third pereiopods.
In the crayfishes found east of the Rocky Mountains (Cambarinae) there is an
annulus ventralis functioning as a seminal receptacle in the female between the
bases of the fifth pereiopods, and the gonopods of the male are especially complex. Of particular interest in this subfamily are the two distinctly different
and usually alternating morphological forms exhibited by adult males. These
are designated as "first" and "second" form males. The first form is in the
breeding stage and can be recognized by the corneous condition of the well-
defined terminal elements of the first pleopod, whereas in the second form male
the terminal elements of the first pleopod are not so well defined, are usually
blunt, and are never corneous. Moreover, the hooks on the ischia of the pereiopods are strongly developed and sometimes corneous in the first form, whereas
in the second form they are usually reduced and seldom corneous. The shape
of the rostrum is used in identification. The *acumen* is the cephalic subtriangular apex of the rostrum, frequently delimited at its base by small spines
or an interrupted margin of the rostrum proper. Measurements of the carapace length used in the key to crayfishes are made from the tip of the rostrum
to the posterior margin in the dorsal midline.

Distribution

Most of the malacostracan crustaceans live in water less than three or four
feet deep in springs, ditches, streams, rivers, ponds, and lakes, where there are

weeds, stones, or debris to afford concealment in the daytime. Isopods, amphipods, shrimps, and crayfishes are often encountered in subterranean streams
and pools, and some crayfishes live in burrows leading to ground water in
areas where there may be no surface water. Few species are found north of
southern Canada, but mysids and amphipods occur all the way to the Arctic
Ocean. In the United States, malacostracans are most abundant in the eastern, central, and southern states, and west of the coastal ranges along the
Pacific coast; except for a few species of amphipods and crayfishes, they are
rare or absent in the vast area of the western Great Plains and the Rocky
Mountains.

None of the true fresh-water crabs of the family Potamonidae occurs north
of Mexico and the West Indies, but crabs of at least three normally marine
families may be found in fresh waters of the United States. The brackish-
water mud crab, *Rhithropanopeus harrisii* (Gould) 1841, may invade fresh-water
streams along the Atlantic and Gulf coasts, and it has recently been introduced along the Pacific shores of California and Oregon. The common edible
blue crab of the Atlantic coast, *Callinectes sapidus* Rathbun 1896, is sometimes
found in fresh water where it may attain a very large size. Finally, the river
crab of eastern Mexico, *Platychirograpsus typicus* Rathbun 1914, has been introduced into the Hillsboro River near Tampa, Florida, where it lives in holes in
the banks above water level. As none of these crabs is of general occurrence
in the fresh waters of North America, they are not included in the following
key. Several brackish-water isopods and amphipods which may be found
occasionally in fresh-water portions of tidal rivers and streams have also been
omitted.

The division of the chapter among the authors is as follows: Banner, Mysidacea; Chace, Caridea and Introductory text; Hobbs, Nephropsidea; Hubricht, Amphipoda; Mackin, Isopoda.

The key mentions all recognized species described through 1957. Of course
there may be difference of opinion about synonymy, and it is known that
many undescribed species exist. References are given either to original descriptions or to useful redescriptions in compilations.

KEY TO GENERA AND SPECIES

1a		Without carapace. Head united with first thoracic somite. Seven free thoracic somites. Eyes sessile or absent . . . . Orders **Isopoda** and **Amphipoda**	**2**
1b		With carapace which covers most or all of thoracic somites. Eyes stalked . . . . Orders **Mysidacea** and **Decapoda**	**19**
2a	(1)	Body more or less flattened dorsoventrally. Telson fused with last abdominal somite. Branchii or branchial plates, if present, at- tached to abdominal appendages. First 3 pairs of pleopods variously modified but never with multiarticulate rami. (Fig. 31.1*a*) . . . . Order **Isopoda**	**3**

2b Body laterally compressed. Telson small but distinctly separated from last abdominal somite by a suture. Branchial appendages attached to basal segments of pereiopods or ventral surface of thoracic somites. Three anterior pairs of abdominal appendages with multiarticulate rami; rami of last 3 pairs with 1 or 2 joints. (Fig. 31.1*b*) . . . . Order **Amphipoda**   8
See also Bousfield (1958). (There has not yet been an opportunity to evaluate the new species proposed in this paper.)

3a **(2)** Uropods absent. Parasitic. Body of female markedly asymmetrical (Fig. 31.2); male symmetrical and situated beneath abdomen of female. . . . . Family **Bopyridae**
Only one genus and species in fresh waters of North America . . . .
Probopyrus bithynis Richardson 1904

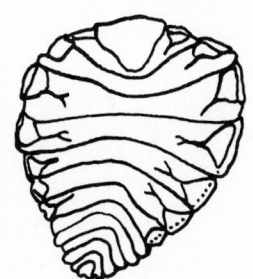

This species is parasitic in the gill chambers of shrimps of the genera *Macrobrachium* and *Palaemonetes* from streams flowing into the Gulf of Mexico.

◀ **Fig. 31.2.** *Probopyrus bithynis.* A female specimen showing asymmetry. (After Richardson.)

3b Uropods present. Free-living. Body symmetrical.   4

4a **(3)** Uropods attached to last abdominal somite in lateral position near base. (Fig. 31.3*a*)   5

4b Uropods attached to last abdominal somite in terminal or subterminal position . . . . Family **Asellidae**   7
This family contains 2 genera and numerous species, and is by far the most important and most characteristic of the North American fresh-water isopods.

5a **(4)** Abdomen composed of 2 segments . . . . . Family **Sphaeromidae**   6

5b Abdomen composed of 6 segments. . . . . . . Family **Cirolanidae**
Only one genus and species in fresh water north of Mexico and the West Indies. . .
Cirolanides texensis Benedict 1896
This species inhabits springs and caverns of the Balcones escarpment of Tex. and cavernous areas to the west as far as Rock Springs.

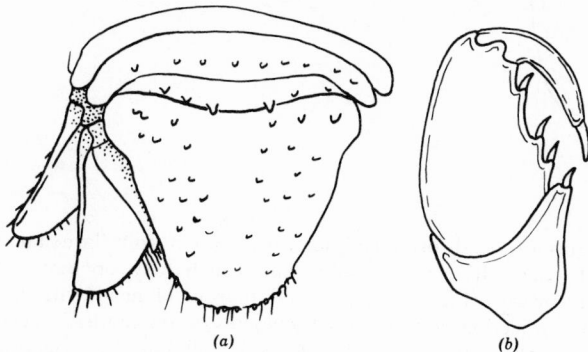

(a) *(b)*

Fig. 31.3. (*a*) *Cirolana* sp. Van Name. Last abdominal segment with laterally attached uropoda. (*b*) First pereiopod of *Cirolanides texensis.*

6a   (5) Outer margins of uropods serrate . . . . ***Sphaeroma*** Latreille

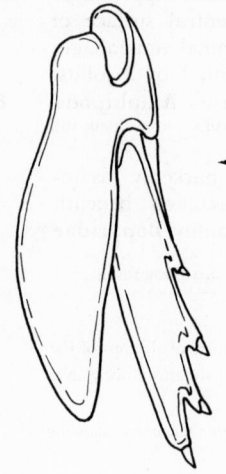

Species of this genus are able to roll into a ball. One species, *S. terebrans* Bate 1866, bores into wood pilings, boat hulls, and other wood structures in brackish and fresh water. Reported from Fla. and probably widely distributed around Gulf Coast. Common in marsh areas of La.

◄ **Fig. 31.4.** *Sphaeroma terebraus.* Uropod. Dense setae covering the uropoda, which are characteristic of the species of *Sphaeroma*, are not drawn.

6b   Outer margins of uropods smooth . . . . . ***Exosphaeroma*** Stebbing

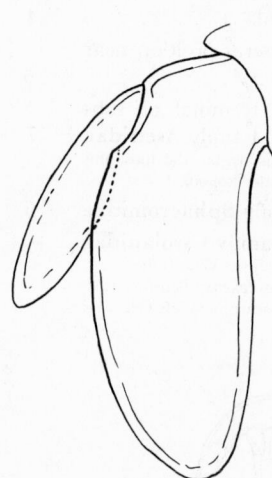

Several species have been reported from fresh-water habitats. *E. thermophilum* (Richardson) 1897 is found in warm springs in N. M. *E. insulare* Van Name 1940, is reported from fresh water of San Nicholas Island, Calif. Other primarily marine species occasionally invade fresh water in Alaska and N. Y.

◄ **Fig. 31.5.** *Exosphaeroma thermophilum.* Uropod.

7a   (4) Margins of head produced laterally into a thin lamellar plate which covers base of mandible and which may or may not be deeply incised (Fig. 31.6*a*); frontal margin of head with median carina projecting between basal segments of antennules. Terminal segment of exopod of third pleopod (gill cover or operculum) triangular or half-moon-shaped (Fig. 31.6*b*); suture between

ultimate and penultimate segments beginning at posterior median angle and traversing gill cover obliquely forward and laterally . . .
Lirceus Rafinesque

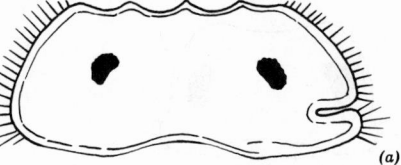
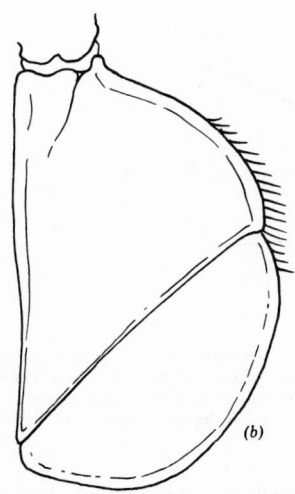

This genus (= *Mancasellus* Harger) contains no blind species although several may enter caves. The following 13 species and several subspecies, all confined to the area east of the great plains from Canada to the Gulf of Mexico, are keyed by Hubricht and Mackin (1949):

L. lineatus (Say) 1818; *L. fontinalis* Rafinesque 1820; *L. brachyurus* (Harger) 1876; *L. hoppinae* (Faxon) 1889; *L. louisianae* (Mackin and Hubricht) 1938; *L. alabamae* Hubricht and Mackin 1949; *L. bicuspidatus* Hubricht and Mackin 1949; *L. bidentatus* Hubricht and Mackin 1949; *L. garmani* Hubricht and Mackin 1949; *L. hargeri* Hubricht and Mackin 1949; *L. megapodus* Hubricht and Mackin 1949; *L. richardsonae* Hubricht and Mackin 1949; *L. trilobus* Hubricht and Mackin 1949.

◄ **Fig. 31.6.** (*a*) *Lirceus* sp., outline of head. The right margin is shown as incised, the left margin is entire. The median frontal carina is characteristic of the genus. (*b*) *Lirceus* sp., exopodite of the third pleopod (gill cover, or operculum).

7b   Lateral margins of head without lamellar expansion; no median frontal carina (Fig. 31.7*b*). Terminal segment of third pleopods quadrangular, suture between last two segments beginning somewhat anterior to posterior median tip and running nearly straight across operculum (Fig. 31.7*c*). . . . . . **Asellus** Geoffrey St. Hillaire

The eyed, surface-water species of this genus include the following:
A. communis Say 1818; *A. brevicaudus* Forbes 1876; *A. intermedius* Forbes 1876; *A. tomalensis* Harford 1877 (Richardson 1904); *A. militaris* Hay 1878; *A. attenuatus* Richardson 1900; *A. dentadactylus* Mackin and Hubricht 1938; *A. montanus* Mackin and Hubricht 1938.

Those species in which the eyes are absent or reduced and the body pigment pattern is absent or reduced are as follows. Mostly subterranean species:
A. stygius (Packard) 1871; *A. nickajackensis* (Packard) 1881; *A. smithi* (Ulrich) 1902; *A. alabamensis* (Stafford) 1911; *A. tridentatus* (Hungerford) 1922; *A. antricolus* (Creaser) 1931; *A. californicus* Miller 1933; *A. macropropodus* (Chase and Blair) 1937; *A. hobbsi* Maloney 1939; *A. acuticarpus* (Mackin and Hubricht) 1940; *A. adentus* Mackin and Hubricht 1940; *A. dimorphus* (Mackin and Hubricht) 1940; *A. oculatus* (Mackin and Hubricht) 1940; *A. packardi* (Mackin and Hubricht), 1940;

A. spatulatus (Mackin and Hubricht) 1940; *A. stiladactylus* (Mackin and Hubricht) 1940; *A. pricei* Levi 1949.

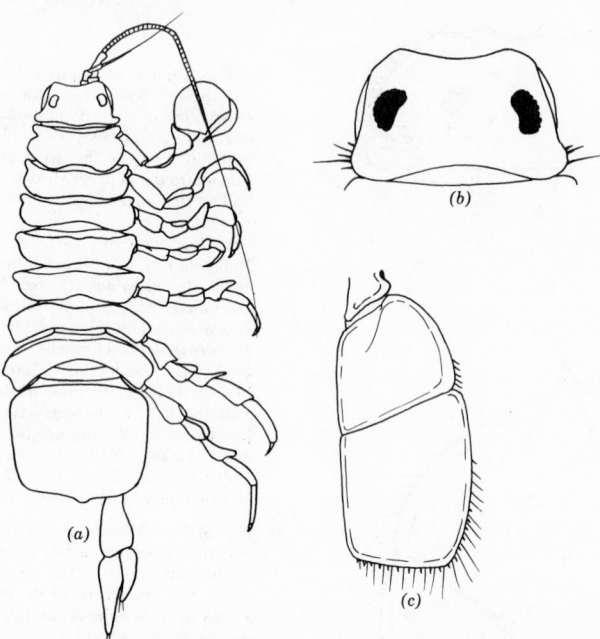

Fig. 31.7. (*a*) *Asellus militaris* Hay, a fresh-water isopod. This is the most common species in the eastern U. S., ranging from the eastern edge of the prairies to the Atlantic coast, and from the Great Lakes to the Gulf. In temporary waters, ponds and swamps. (*b*) *Asellus* sp., outline of head. Lack of lateral lamellae and median frontal carina distinguishes this genus from *Lirceus*. (*c*) *Asellus* sp. exopodite of third pleopod (gill cover, or operculum).

8a (2) First antenna without accessory flagellum. Mandible without palp. First maxilla with small, 1-jointed palp . . . .
Family **Talitridae**
Only one genus and species in fresh waters of North America . . . .
Hyalella azteca (Saussure) 1858
This species [= *H. **knickerbockeri*** (Bate) 1862] is widely distributed in permanent bodies of water with submerged vegetation.

8b First antenna with accessory flagellum. Mandible with palp. First maxilla with 2-jointed palp 9

9a (8) First antenna shorter than second antenna; flagellum in female very short, in mature male, very long. Seventh pereiopod much shorter than sixth, with second segment much expanded . . . . . .
Family *Haustoriidae*
Only one genus and species in North American fresh waters . . . .
Pontoporeia affinis Lindstrom 1855
This species (=*P. hoyi* Smith 1871 = *P. filicornis* Smith 1874) is found in the Great Lakes and other deep-water lakes in the northern U. S. and Canada, as well as in northern Europe.

9b First antenna usually longer than second; no sexual difference in length of flagellum. Pereiopods more or less slender with second segment of fifth to seventh pairs only moderately expanded; seventh pereiopod about as long as or longer than sixth . . . .
Family **Gammaridae** 10

10a (9) Accessory flagellum of first antenna with 3 to 7 segments. Urosome somites bearing dorsal spines. Without sternal gills . . . **11**

10b Accessory flagellum of first antenna with 1 long and 1 short segment. Urosome somites without dorsal spines. Sternal gills present or absent. **12**

11a (10) First gnathopod of male smaller than second. Coxal gills without cylindrical appendages. Inner ramus of third uropod more than ½ as long as outer. . . . . *Gammarus* Fabricius

Six species are known from North American fresh water. Two species are rather local in distribution: *G. troglophilus* Hubricht and Mackin 1940, caves and springs in southern Ill. and eastern Mo.; and *G. acherondytes* Hubricht and Mackin 1940, from caves in southern Ill. The 4 more widely distributed species may be distinguished as follows:

G. lacustris Sars 1865. (=*G. limnaeus* S. I. Smith). (Fig. 31.1*b*). With sensory organs on the second antenna of the male only. Inner ramus of the third uropod about ¾ as long as the outer. Terminal segment of outer ramus of third uropod with plumose setae. Length of adult males more than 12 mm. Found throughout Canada and Alaska and the northern U. S., extending southward in the Rocky Mountain region to Nev. and N. M. and eastward into Okla.

G. pseudolimnaeus Bousfield 1958. Similar to *G. lacustris* but more slender and without the plumose setae on the terminal segment of the outer ramus of the third uropod. Length of adult males more than 12 mm. Springs and small streams in the Great Lakes region and along the Mississippi River as far south as northeastern Ark.

G. minus Say 1818. (=*G. propinquus* W. P. Hay, =*G. purpurascens* W. P. Hay). Similar to *G. pseudolimnaeus*, but smaller, with the inner ramus of the third uropod only ⅔ as long as the outer. Found south of the southern limit of Pleistocene glaciation from eastern Pa. to Okla, and south into northern Ala.

G. fasciatus Say 1818. Without sensory organs on the second antenna of either sex. Found in the Great Lakes (probably introduced) and in streams along the Atlantic coast from Me. to Fla.

11b First gnathopod of male larger than second. Coxal gills with cylindrical appendages. Inner ramus of third uropod very small. .
Anisogammarus Derjavin

Two known species: *A. ramellus* (Weckel) 1907, found in streams and ponds from British Columbia to northern Calif.; and *A. oregonensis* Shoemaker 1944, from Ore.

12a (10) Third uropod with 2 rami, inner one very small. **13**

12b Third uropod with 1 small ramus, usually free, but sometimes fused to peduncle **15**

13a (12) Outer ramus of third uropod with 2 to 12 segments in mature male . . . . *Allocrangonyx* Schellenburg

Only one species in North American fresh waters . . . . *A. pellucidus* (Mackin) 1935

Found in caves, springs, and seeps in the Ozark and Arbuckle Mountains.

13b Outer ramus of third uropod with only one segment **14**

14a (13) Outer ramus of third uropod much longer than peduncle. First 4 coxal plates deeper than their segments. Second antenna of mature male armed with paddle-shaped sensory organs. . *Crangonyx* Bate

Eleven described species, 7 of which have somewhat restricted ranges as follows: *C. anomalus* Hubricht 1943, springs in southern Ohio and central Ky.; *C. serratus* (Embody) 1910, streams and ponds, Va. to Fla.; *C. forbesi* (Hubricht and Mackin) 1940, springs, Ozark region; *C. occidentalis* (Hubricht and Harrison) 1941, streams and ponds, British Columbia and Wash.; *C. hobbsi* Shoemaker 1941, caves in northern Fla.; *C. antennatus* (Packard) 1881, caves in southwestern Va. to northern Ala.; *C. dearolfi* Shoemaker 1942, caves in eastern Pa.

The 4 species which are found over most of eastern U. S. may be distinguished as follows: *C. obliquus* (Hubricht and Mackin) 1940, palmar margins of the propodi of the gnathopods 1½ to 2 times as long as the posterior margins and armed on each side with more than 25 notched spines of uniform size, size medium to large; *C. shoemakeri* (Hubricht and Mackin) 1940, palmar margins of the propodi of the gnathopods concave and armed with heavy notched spines, lower margins of the dactyli of the female armed with 10 to 12 teeth, a medium-size species. *C. gracilis gracilis* Smith

1871, palmar margins of the propodi of the gnathopods of the female convex and armed with weak, minutely notched spines, except at heel, lower margin of the dactyli smooth or with only a few setae, size small; *C. gracilis packardii* Smith 1888, differs from typical *gracilis* in having degenerate eyes, found in wells and caves.

14b   Outer ramus of third uropod not longer than peduncle. First coxal plates shallower than their segments. Second antenna of male without sensory organs . . . . ***Bactrurus*** W. P. Hay

Three blind species found in subterranean waters: *B. brachycaudus* Hubricht and Mackin 1940, Ozark region; *B. hubrichti* Shoemaker 1945, eastern Kan. and Okla.; *B. mucronatus* (Forbes) 1876, eastern Iowa and Mo. to Ohio.

15a **(12)** Second antenna of mature male with paddle-shaped sensory organs. Eyes well developed . . . . ***Synurella*** Wrzesniewski

Four known species: *S. dentata* Hubricht 1943, southern Ohio to northern Tenn.; *S. bifurca* (O. P. Hay), 1882, Gulf states, Ark. to Ala.; *S. chamberlaini* (Ellis) 1941, Md. to S. C.; *S. johanseni* Shoemaker 1920, Alaska.

15b   Second antenna of mature male without sensory organs. Without eyes 16

16a **(15)** Sixth and seventh thoracic somites without bifurcate sternal gills 17

16b   Sixth and seventh thoracic somites with bifurcate sternal gills. . . . 18

17a **(16)** Size small, not exceeding 10 mm in length. Urosome somites usually free . . . . ***Stygobromus*** Cope

Ten described species, found in caves, wells, and seeps: *S. mackini* Hubricht 1943, southwestern Va.; *S. vitreus* (Cope) 1872, Ky. to Ala.; *S. spinosus* (Hubricht and Mackin) 1943, western Va.; *S. putealis* (Holmes) 1908, Wis.; *S. heteropodus* Hubricht 1943, eastern Mo.; *S. exilis* Hubricht 1943, Ky. to Ala.; *S. onondagaensis* (Hubricht and Mackin) 1940, Ozark region; *S. iowae* Hubricht 1943, Ia.; *S. smithi* Hubricht 1943, eastern Tenn. to Ala.; *S. hubbsi* Shoemaker 1942, Ore.

17b   Size large, exceeding 10 mm in length. Urosome somites fused . . ***Stygonectes*** W. P. Hay

Two subterranean species: *S. flagellatus* (Benedict) 1896 and *S. balconis* Hubricht 1943, from the Edwards Plateau of Tex.

18a **(16)** Ramus of third uropod not fused to peduncle . ***Synpleonia*** Creaser

Six described species known from caves, seeps, and wells: *S. emarginata* Hubricht 1943, W. Va.; *S. pizzinii* Shoemaker 1938, eastern Pa. to Va.; *S. clantoni* Creaser 1934, western Mo. and Ark., eastern Kan. and Okla.; *S. americana* (Mackin) 1935, Ozark region southwestward into northeastern Tex.; *S. alabamensis* (Stout) 1911, Ala.; *S. tenuis* (S. I. Smith) 1874 (=*S. hayi* Hubricht and Mackin), N. Y. and Conn. to Va.

18b   Ramus of third uropod very small, fused to peduncle . . . . ***Apocrangonyx*** Stebbing

Two described species known from caves, seeps, and wells: *A. subtilis* Hubricht 1943, southern Ill.; *A. lucifugus* (O. P. Hay) 1882, reported from wells in Abingdon, Ill., but possibly in error for Abingdon, Va.

19a **(1)** Carapace coalesced dorsally with no more than first 3 thoracic somites. Only first or first 2 pairs of thoracic appendages modified as maxillipeds, and none of following pairs chelate . . . .
Order **Mysidacea**   20

19b   Carapace coalesced dorsally with all thoracic somites. First 3 pairs of thoracic appendages modified as maxillipeds and at least 2 following pairs chelate . . . . Order **Decapoda**   22

20a **(19)** Telson with terminal indentation. Antennal scale rounded at tip. Third pleopod of male various but composed of at least 2 articles. Fourth pleopod of male with 7 articles in exopod 21

20b   Telson with tip truncate. Antennal scale acute at tip. Third

pleopod of male a simple, undivided plate. Fourth pleopod of
male with 2 articles in exopod . . . .
Neomysis awatchensis (Brandt) 1851

This species (previously known as *Neomysis mercedis* Holmes 1897) is primarily a
brackish-water form occurring in shallow bays near the outlets of streams along the
Pacific coasts of N. A. and Asia. It is sometimes found in isolated bodies of fresh
water, as in lakes behind the sand dunes along the Oregon coast, and in those fresh-
water basins which are still narrowly connected with the sea, such as Lakes Washington
and Union at Seattle. Specimens reach a maximum length of about 15 mm.

21a **(20)** Antennal scale only slightly longer than antennular peduncle.
Body of mandible with strong tooth on lateral surface. Third
pleopod of male with a single, unsegmented branch borne on basal
article. . . . . *Taphromysis louisianae* Banner 1953

This species may be widespread through the Gulf Coast region and may even be
found in brackish water. Mature individuals reach 8 mm in length.

21b Antennal scale almost $1\frac{1}{2}$ times as long as antennular peduncle.
Mandible without tooth on lateral surface. Third pleopod of male
biramous, with outer branch composed of 5 articles. (Fig.
31.1*c*) . . . . *Mysis relicta* Lovén 1861

This species is considered a relict of the great glaciers of the Pleistocene. As
these continental glaciers advanced they carried with them the brackish-water, cir-
cumarctic mysid, *M. oculata* (Fabricius) 1780. When the southern edge of the
glaciers withdrew, the animals were left stranded in a series of fresh-water lakes, and
slight differences in the form of the animals were fixed so that they are now known as
a separate species. *M. relicta* is reported from the lakes of the Great Lakes system and
the Finger Lakes of New York, and thence northward to bodies of fresh and brackish
water lying near the Arctic Ocean. It is also known from northern Europe. Mature
specimens reach a length of 15 to 25 mm. For more detailed discussion see Holmquist
(1949) and Tattersall and Tattersall (1951).

22a **(19)** Body and rostrum laterally compressed. Third pair of pereiopods
not chelate . . . . Section **Caridea** 23

22b Body subcylindrical anteriorly; rostrum and abdomen dorsally de-
pressed. First 3 pairs of pereiopods chelate . . . .
Section **Nephropsidea**
Family **Astacidae** 33

23a **(22)** A supraorbital spine on either side of base of rostrum. Fingers of
chelae with terminal tufts of hair. . . . . Family **Atyidae** 24

23b No supraorbital spines. Fingers of chelae not tufted . . . .
Family **Palaemonidae** 26

24a **(23)** Eyes reduced, unpigmented. Carpus of second cheliped excavated
for reception of proximal end of chela. Exopods on all pereiopods. . . . . *Palaemonias ganteri* Hay 1901

(a)

(c)   (b)

This species is thus far recorded only from
Mammoth Cave in Ky.

◄ **Fig. 31.8.** *Palaemonias ganteri.* (*a*) Anterior part of
body. (*b*) Carpus and chela of first pereiopod. (*c*)
Carpus and chela of second pereiopod. × 8.5.
(Modified after Fage.)

24b Eyes well developed, pigmented. Carpus of second cheliped not
excavated for reception of proximal end of chela. Fifth pereiopod
without exopod. . . . . *Syncaris* Holmes 25

25a **(24)** Rostrum normally with one or two dorsal teeth. Exopods on first 4 pereiopods in adults . . . . **S. pacifica** (Holmes) 1895

Recorded from streams in Sonoma, Marin and Napa counties, Calif.

◄ **Fig. 31.9.** *Syncaris pacifica.* (*a*) Anterior part of body. (*b*) Carpus and chela of first pereiopod. (*c*) Carpus and chela of second pereiopod. × 5.

25b Rostrum without dorsal teeth. Fourth pereiopod without exopod . . . . **S. pasadenae** (Kingsley) 1896

Recorded from Los Angeles, San Bernardino, and San Diego counties, Calif. The species is probably now extinct in many localities where it was formerly abundant.

◄ **Fig. 31.10.** *Syncaris pasadenae.* Anterior part of body. × 6.

26a **(23)** Lower lateral spine on carapace branchiostegal (placed near anterior margin of carapace and below branchiostegal groove). Mandible without palp. Small species . . . . **Palaemonetes** Heller 27

26b Lower lateral spine on carapace hepatic (placed far from anterior margin of carapace and at posterior end of branchiostegal groove). Mandible with palp. Large species . . . . . **Macrobrachium** Bate 30

27a **(26)** Eyes reduced and unpigmented 28

27b Eyes well developed and pigmented 29

28a **(27)** Rostrum unarmed ventrally. Both pairs of chelipeds similar in size and shape. **Palaemonetes (Alaocaris) antrorum** Benedict 1896

This species is known only from subterranean waters near San Marcos, Tex.

◄ **Fig. 31.11.** *Palaemonetes (Alaocaris) antrorum.* Anterior part of body. × 11. (After Holthuis.)

28b Rostrum armed with teeth on both margins. Two pairs of chelipeds different in size and shape . . . .
Palaemonetes (Palaemonetes) cummingi Chace 1954
This species is known only from a cave in Alachua County, Fla.

29a **(27)** Branchiostegal spine on anterior margin of carapace just below branchiostegal groove. Posterior pair of dorsal spines on telson placed midway between anterior pair and end of telson . . . .
P. (P.) paludosus (Gibbes) 1850

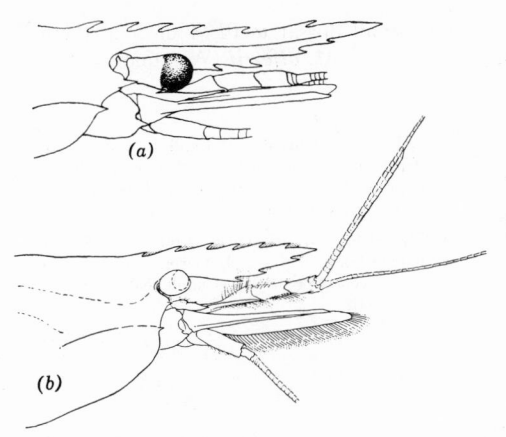

This species (=*P. exilipes* Stimpson 1871) is relatively common from N. J. to Fla., occasional and probably introduced west of the Alleghenies.

◀ **Fig. 31.12.** (*a*)   *Palaemonetes* (*Palaemonetes*) *paludosus.* Anterior part of body. × 5.   (*b*) *P.* (*P.*) *cummingi.* × 4.   (*a* after Holthuis, *b* after Chace.)

29b   Branchiostegal spine placed distinctly behind anterior margin of carapace and some distance below branchiostegal groove. Posterior pair of dorsal spines of telson placed much closer to end of telson than to anterior spines, often in same row as posterior marginal spines . . . . **P. (P.) kadiakensis** Rathbun 1902

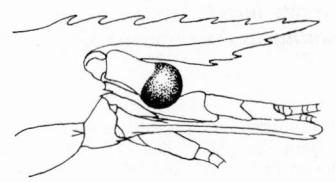

This is the common fresh-water shrimp west of the Alleghenies from southern Ontario and the Great Lakes to the Gulf coast and northeastern Mexico.

◀ **Fig. 31.13.** *Palaemonetes* (*Palaemonetes*) *kadiakensis.* Anterior part of body. × 3. (After Holthuis.)

30a   (26) Carpus of second cheliped distinctly shorter than merus. Rostrum arched over eyes; tip directed upward; 4 to 6 of dorsal teeth placed behind level of posterior margin of orbit . . . . **Macrobrachium carcinus** (Linnaeus) 1758

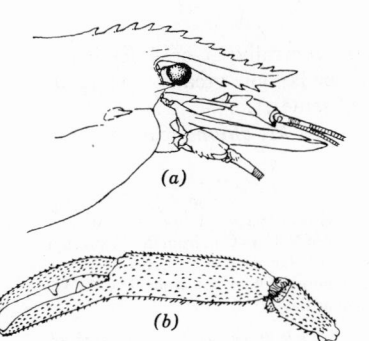

This large West Indian shrimp (=*M. jamaicensis* (Herbst) 1792) is found in the U. S. only in Fla. from St. Augustine to Miami and at Big Pine Key, and in Tex. from Matagorda Bay to Mexico.

◀ **Fig. 31.14.** *Macrobrachium carcinus.* (*a*) Anterior part of body. × 1.   (*b*) Carpus and chela of second pereiopod. (*b* after Holthuis.)

30b   Carpus of second cheliped as long as, or longer than merus **31**

31a   (30) Second chelipeds dissimilar, larger one much more robust, with inflated palm; fingers of smaller one arched, gaping.   Rostrum

nearly straight, with 12 to 15 small dorsal teeth, 4 or 5 of them
behind orbital margin . . . . *M. olfersii* (Wiegmann) 1836

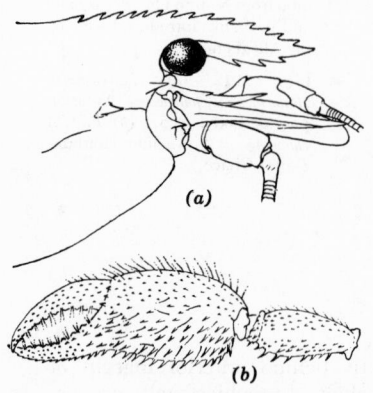

(a)

(b)

This is primarily a Central and South
American species which has been found
occasionally in the St. Johns River system
and at St. Augustine, Fla.

◄ **Fig. 31.15.** *Macrobrachium olfersii.* (a) Anterior
part of body. × 3.5. (b) Carpus and chela of
larger second pereiopod of adult male. (b after
Hedgpeth.)

31b   Second chelipeds similar, subequal, with straight fingers. Rostrum
somewhat arched, with 9 to 13 dorsal teeth, 2 to 4 of them behind
orbital margin **32**

32a **(31)** Dorsal teeth of rostrum continued to tip, 2 placed behind orbital
margin. Second chelipeds long and strong; fingers covered with
velvet or feltlike pubescence. . . *M. acanthurus* (Wiegmann) 1836

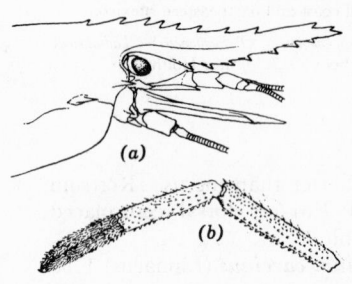

(a)

(b)

This subtropical shrimp invades the coastal
regions of Ga., eastern Fla., Miss., La., and Tex.
It is sometimes found in brackish and salt water.

◄ **Fig. 31.16.** *Macrobrachium acanthurus.* (a) Anterior
part of body. × 1.1. (b) Carpus and chela of
second pereiopod. (b after Holthuis.)

32b   Tip of rostrum unarmed dorsally and ventrally; 3 or 4 teeth behind orbital margin. Second chelipeds not greatly enlarged;
fingers naked or with scattered tufts of setae . . . .
                              M. ohione (Smith) 1874

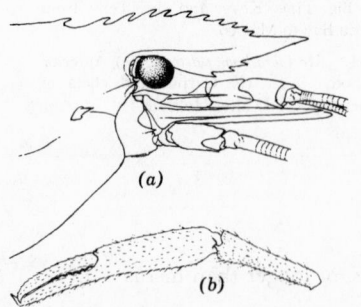

(a)

(b)

M. ohione has been recorded from the Atlantic
coastal plain from N. C. to Ga.; from the Mississippi
drainage system as far north as St. Louis, Mo., and
Washington County, Ohio; and from Tex. as far
south as Aransas Bay.

◄ **Fig. 31.17.** *Macrobrachium ohione.* (a) Anterior
part of body. × 2.7. (b) Carpus and chela of
second pereiopod of adult male. (b after Holthuis.)

33a **(22)** Distal portion of first pleopod of male tubular and without tubercles
or spines. Second, third, and fourth pereiopods of male without

hooks on ischia. Gills present on last thoracic somite. Females without an annulus ventralis. (Fig. 31.23) . . . .

Subfamily **Astacinae**

Pacifastacus Bott

The following 7 species and subspecies have been recognized along the Pacific slope from British Columbia to Calif., and one of them, *P. gambelii*, has crossed the divide into the upper Missouri drainage: *P. gambelii* (Girard) 1852 (2 subspecies), *P. nigrescens* (Stimpson) 1857 (2 subspecies), *P. leniusculus* (Dana) 1852, *P. trowbridgii* (Stimpson) 1857, and *P. klamathensis* (Stimpson) 1857.

33b Distal portion of first pleopod of male terminating in 2 or more distinct parts. Second, third, or fourth pereiopods of male with hooks on ischia. Gills absent on last thoracic somite. Females with an annulus ventralis . . . . Subfamily **Cambarinae** 34

34a (33) Teeth lacking on opposable margins of ischia of third maxillipeds. Albinistic . . . . *Troglocambarus* Hobbs

Only one species known . . . . *T. maclanei* Hobbs 1942
Inhabits subterranean waters in peninsular Fla.

34b Teeth present on opposable margins of ischia of third maxillipeds. Albinistic or pigmented 35

35a (34) First pleopod terminating in 2 parts. (Figs. 31.31–31.41) 36

35b First pleopod terminating in 3 or more parts. (Figs. 31.18–31.22 and 31.24–31.30) 37

NOTE: The remaining portion of the key is based on "first form" males (see introductory remarks).

36a (35) Terminal elements of first pleopod short and bent caudad at right angles to main shaft of appendage (Figs. 31.40, 31.41), except in *Cambarus obeyensis*. (Fig. 31.39) . . . . *Cambarus* Erichson 138

Forty-three species and subspecies ranging from New Brunswick to Tex.; most species between the Blue Ridge and the Mississippi River.

36b Terminal elements of first pleopod short or long, but if short, never bent at right angles to main shaft of appendage (Figs. 31.31–31.38). . . . . *Orconectes* Cope 95

Fifty-nine species and subspecies ranging from Me. to Tex.; most species in the Central Basin.

37a (35) Males with hooks on ischia of third or third and fourth pereiopods (Figs. 31.24–31.30) . . . . *Procambarus* Ortmann 42

Ninety-seven species and subspecies ranging from New England and the Great Lakes to Mexico, Guatemala, Honduras, and Cuba; 68 species in the U. S., most of which occur in the southeast.

37b Males with hooks on ischia of second and third pereiopods. All species small, less than 2 in. in length . . . *Cambarellus* Ortmann 38

Fifteen species and subspecies ranging from Fla. and Ill. to Veracruz and Michoacan. Five species in the U. S. (Figs. 31.18–31.22).

38a (37) Terminal elements of first pleopod straight. (Fig. 31.18) . . . . . .

C. shufeldtii (Faxon) 1884

Mississippi basin.

38b Terminal elements of first pleopod curved 39

39a (38) Central projection of first pleopod extending farther caudad than other terminal elements. (Fig. 31.22) . . . . *C. ninae* Hobbs 1950

Southern Tex.

39b Central projection of first pleopod never extending farther caudad than other terminal elements 40

40a (39) Areola (Fig. 31.1) 5 or 6 times longer than broad. (Fig. 31.19) . .

C. puer Hobbs 1945

Ark., La., Tex.

40b Areola 2 to 4 times longer than broad 41

41a **(40)** Hook on ischiopodite of second pereiopod bituberculate. (Fig.
31.20) . . . . ***C. schmitti*** Hobbs 1942
Ala. and Fla.

41b Hook on ischiopodite of second pereiopod simple. (Fig. 31.21) . .
C. diminutus Hobbs 1945
Southern Ala.

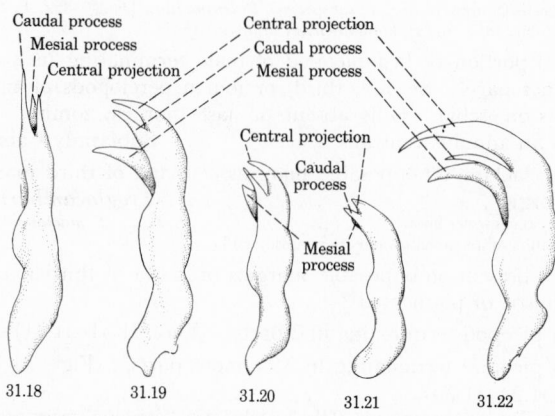

Fig. 31.18. *Cambarellus shufeldtii.* Lateral view of first left pleopod of first form male. **Fig. 31.19.** *Cambarellus puer.* Lateral view of first left pleopod of first form male. **Fig. 31.20.** *Cambarellus schmitti.* Lateral view of first left pleopod of first form male. **Fig. 31.21.** *Cambarellus diminutus.* Lateral view of first left pleopod of first form male. **Fig. 31.22.** *Cambarellus ninae.* Lateral view of first left pleopod of first form male.

42a **(37)** Albinistic. Eyes reduced **43**

42b Pigmented. Eyes normal **45**

43a **(42)** Hook on ischiopodite of fourth pereiopod bituberculate . . . . . .
Procambarus acherontis (Lönnberg) 1894
Peninsular Fla. (Hobbs, 1942).

43b Hook on ischiopodite of fourth pereiopod simple. **44**

44a **(43)** Margins of rostrum tapering toward acumen. . . . .
P. pallidus (Hobbs) 1940
Peninsular Fla. (Hobbs, 1942).

44b Margins of rostrum subparallel or convex (2 subspecies) . . . . . .
P. lucifugus (Hobbs) 1940
Peninsular Fla. (Hobbs, 1942a).

45a **(42)** Two spines on each side of carapace just caudad of cervical
groove **46**

45b Only 1 spine or none on each side of carapace just caudad of
cervical groove. **53**

46a **(45)** Small spine on mesial margin of coxa of cheliped . . . .
P. versutus (Hagen) 1870
Ala., Ga., Fla. (Hobbs, 1942a).

46b No spine on mesial margin of coxa of cheliped. **47**

47a **(46)** Cephalic process of first pleopod absent **48**

47b Cephalic process of first pleopod present **49**

48a **(47)** Central projection of first pleopod more than twice as long as
broad. (Fig. 31.24, 31.1*d,e*) . . . . . ***P. spiculifer*** (LeConte) 1856
Ala., Ga., Fla. (Hobbs, 1942a).

48b Central projection of first pleopod less than twice as long as broad . . . . *P. raneyi* Hobbs 1953
Ga., S. C.

49a **(47)** Cephalic process of first pleopod mesial to central projection . . . .
P. suttkusi Hobbs 1953
Ala., Fla.

49b Cephalic process of first pleopod cephalic or lateral to central projection **50**

50a **(49)** Cephalic process of first pleopod with apex directed distally or cephalodistally . . . . *P* **51**

50b Cephalic process of first pleopod with apex directed caudo-distally **52**

51a **(50)** Apex of mesial process of first pleopod distad of that of other terminal elements . . . . *P. natchitochae* Penn 1953
Ark., La.

51b Apex of mesial process of first pleopod never distad of that of central projection . . . . *P. penni* Hobbs 1951
La., Miss.

52a **(50)** Terminal end of cephalic process of first pleopod pointed . . . . . .
P. vioscai Penn 1946
La., Miss., Tenn.
P. echinatus Hobbs 1956
S. C.

52b Terminal end of cephalic process of first pleopod rounded . . . . .
P. dupratzi Penn 1953
Ark., La.

53a **(45)** First left pleopod with a prominent angular shoulder (occasionally rounded in *P. clarkii*) on cephalic margin at base of distal third of appendage. (Fig. 31.25) **54**

53b First left pleopod without angular shoulder on cephalic margin at base of distal third of appendage **55**

54a **(53)** Cephalic process of first pleopod consisting of a broad rounded lobe, the caudodistal margin of which may be angular or rounded . . . . *P. troglodytes* (LeConte) 1856
North of Altamaha River in Ga. and S.C.
(Fig. 31.25) . . . . *P. clarkii* (Girard) 1852
Tex. to Escambia County, Fla., and North to Ark. and Ky. Introduced, Nev., Calif.
P. okaloosae Hobbs 1942
Between Yellow and Perdido Rivers in Ala. and Fla.

54b Cephalic process acute . . . . *P. paeninsulanus* (Faxon) 1914
Fla. and southern Ga.
P. howellae Hobbs 1952
Tributaries of Altamaha River in Ga.

55a **(53)** Length of chela less than twice its greatest width except in *P. pygmaeus* where the length is approximately twice the width. **56**

55b Length of chela more than twice its greatest width **60**

56a **(55)** Hooks on ischia of third and fourth pereiopods . . . .
P. geodytes Hobbs 1942
St. Johns drainage system in Fla.

56b Hooks on ischia of third pereiopods only **57**

57a **(56)** Corneous central projection of first pleopod platelike and directed laterad (3 subspecies) . . . . *P. rogersi* (Hobbs 1938)
Panhandle of Fla. (Hobbs, 1942a).

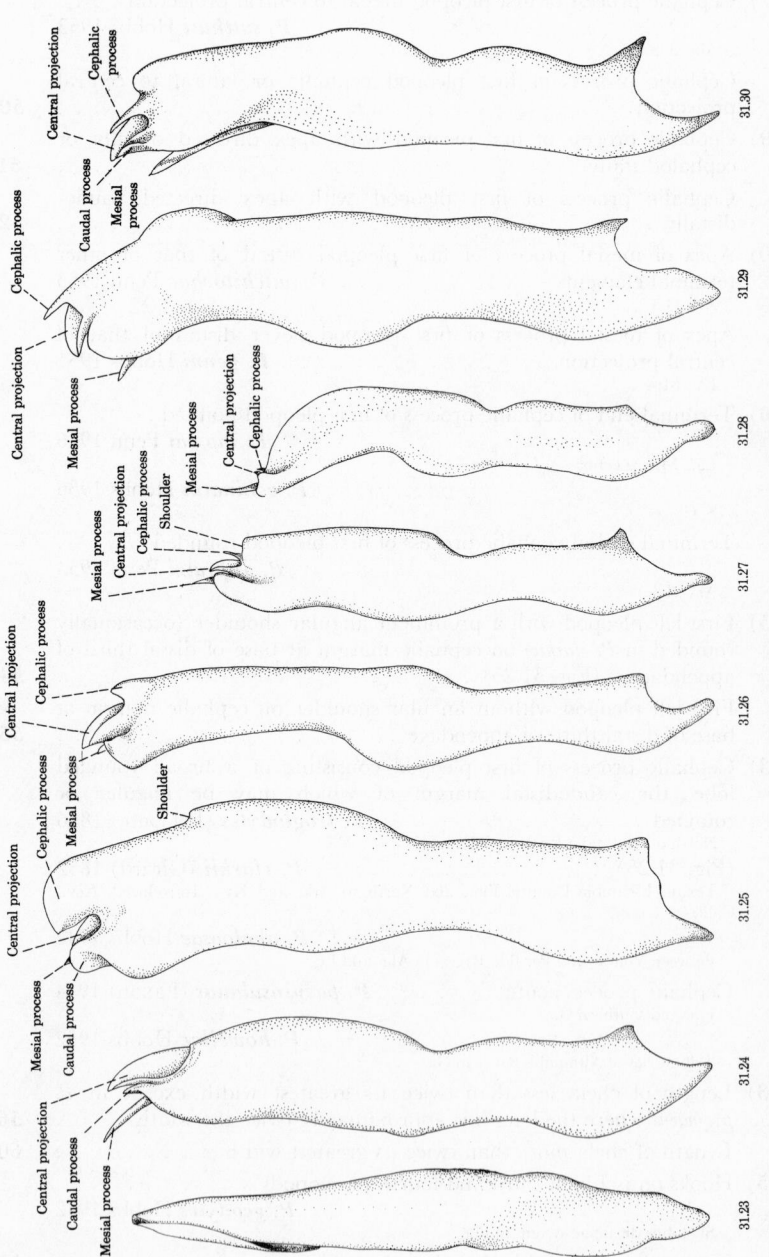

Fig. 31.23. *Pacifastacus gambelii gambelii.* Lateral view of first left pleopod of first form male. **Fig. 31.24.** *Procambarus clarkii.* Lateral view of first left pleopod of first form male. **Fig. 31.25.** *Procambarus spiculifer.* Lateral view of first left pleopod of first form male. **Fig. 31.26.** *Procambarus advena.* Lateral view of first left pleopod of first form male. **Fig. 31.27.** *Procambarus gracilis.* Lateral view of first left pleopod of first form male. **Fig. 31.28.** *Procambarus barbatus.* Lateral view of first left pleopod of first form male. **Fig. 31.29.** *Procambarus enoplosternum.* Lateral view of first left pleopod of first form male. **Fig. 31.30.** *Procambarus blandingii blandingii.* Lateral view of first left pleopod of first form male.

57b	Corneous central projection of first pleopod beaklike and directed caudad or caudodistad.	**58**
58a (**57**)	Mesial process of first pleopod directed caudad at a right angle to main shaft of appendage . . ***Procambarus truculentus*** Hobbs 1954 North of Altamaha River in Ga.	
58b	Mesial process of first pleopod directed caudodistad	**59**
59a (**58**)	Cephalic process of first pleopod present although at times rudimentary; integument never green with scarlet markings. (Fig. 31.26) . . . . ***P. advena*** (Le Conte 1856) Southern Ga. and Northern Fla. (Hobbs, 1942*a*).	
59b	Cephalic process of first pleopod absent; integument green with scarlet markings . . . . ***P. pygmaeus*** Hobbs 1942a Southern Ga. and Northern Fla.	
60a (**55**)	Areola obliterated at least at mid-length. (Fig. 31.27) . . . . ***P. gracilis*** (Bundy) 1876 Wis. to Okla. (Williams, 1954). ***P. hagenianus*** (Faxon) 1884 Ala., Miss. (Faxon, 1914).	
60b	Areola not obliterated	**61**
61a (**60**)	Rostrum without lateral spines and not emarginate	**62**
61b	Rostrum with lateral spines or emarginate	**75**
62a (**61**)	Chela with a conspicuous tuft of setae on inner margin of palm . .	**63**
62b	Chela without conspicuous setae on inner margin of palm	**66**
63a (**62**)	Caudal process of first pleopod forming a sharp sinuous ridge on caudolateral portion of appendage . . . . ***P. shermani*** Hobbs 1942 Extreme western Fla. to Miss.	
63b	Caudal process of first pleopod never forming a sinuous ridge. . . .	**64**
64a (**63**)	Cephalic margin of first pleopod with an angular prominence just proximad of terminal elements. . . . . ***P. pubischelae*** Hobbs 1942 Southeastern Ga. and northern Fla. ***P. tulanei*** Penn 1953 Southern Ark. and northern La.	
64b	Cephalic margin of first pleopod with no such prominence	**65**
65a (**64**)	Caudal process of first pleopod forming a corneous bladelike structure along caudolateral tip of appendage. . . . . ***P. hubbelli*** (Hobbs) 1940 Choctawhatchee drainage system in Ala. and Fla. (Hobbs, 1942a).	
65b	Caudal process of first pleopod rounded and noncorneous. (Fig. 31.28) . . . . ***P. barbatus*** (Faxon) 1890 Southeastern S. C. and Ga. north of Altamaha River. (Hobbs 1942a). ***P. escambiensis*** Hobbs 1942 Extreme western Fla.	
66a (**62**)	Cephalic process of first pleopod lacking . . . . ***P. mancus*** Hobbs and Walton 1957 Eastern Miss.	
66b	Cephalic process of first pleopod present	**67**
67a (**66**)	Cephalic process of first pleopod on mesial side of appendage and in no way hooding the central projection.	**69**
67b	Cephalic process hooding the central projection at least in part. . .	**68**
68a (**67**)	Terminal portion of central projection directed at no more than a 45° angle to main shaft of appendage. . . . . ***P. jaculus*** Hobbs and Walton 1957 Central Miss.	

68b   Terminal portion of central projection directed at approximately
               a 90° angle to main shaft of appendage. . . . .
                                    Procambarus pearsei (Creaser 1934)
               Eastern N. C.
                                         P. planirostris Penn 1953
               Western La. and southern Miss.
                                       P. hybus Hobbs and Walton 1957
               Eastern Miss.
69a   (67)   Hooks on ischia of third pereiopods only (2 subspecies) . . . .
                                         P. simulans (Faxon 1884)
               Kan. and Ark. to Mex.   (Williams, 1954).
                                        P. rathbunae (Hobbs 1940)
               Panhandle of Fla.   (Hobbs, 1942a).
69b   Hooks on ischia of third and fourth pereiopods   70
70a   (69)   Tip of first pleopod terminating in 4 acute elements. . . . .
                                         P. viae-viridis (Faxon) 1914
               Northeastern Ark. to Miss. and Ala.
70b   Tip of first pleopod never terminating in more than 3 acute
               elements   71
71a   (70)   Central projection of first pleopod long and scythelike, and the
               most conspicuous of the terminal elements. . ***P. tenuis*** Hobbs 1950
               Ark., Okla.   (Williams, 1954).
71b   Central projection short and never the most conspicuous of the
               terminal elements   72
72a   (71)   Mesial process of first pleopod conspicuously large, subspatulate
               and bent caudad at an angle of 45–90° to the main shaft of the
               appendage. . . . . ***P. kilbyi*** (Hobbs) 1940
               Northern Fla.   (Hobbs, 1942a).
72b   Mesial process never subspatulate; always slender   73
73a   (72)   Mesial process spiculiform and arched . . . .
                                       P. latipleurum Hobbs 1942
               Gulf County, Fla.
73b   Mesial process slender but not needlelike and never strongly
               arched   74
74a   (73)   Caudal process of first pleopod thumblike . . . .
                                      P. apalachicolae Hobbs 1942
               Gulf to Walton County, Fla.
74b   Caudal process rounded but not thumblike. . . . .
                                       P. econfinae Hobbs 1942
               Bay County, Fla.
75a   (61)   Areola less than 5 times as long as its least width, except occa-
               sionally in *P. seminolae*   76
75b   Areola more than 5 times as long as its least width   84
76a   (75)   Length of inner margin of palm of chela greater than length of
               dactyl.   77
76b   Length of inner margin of palm of chela less than length of
               dactyl.   78
77a   (76)   Length of acumen as long as or longer than rest of rostrum. . . . .
                                         P. youngi Hobbs 1942
               Panhandle of Fla.
77b   Length of acumen much shorter than rest of rostrum . . . .
                                       P. hinei (Ortmann) 1905
               Southern La. and Tex.   (Penn, 1956).

78a (76) Cephalic process of first pleopod entirely lateral to central
projection . . . . ***P. lepidodactylus*** Hobbs 1947
Northeastern S. C.

78b Cephalic process of first pleopod cephalic to central projection . . .   79

79a (78) Cephalic surface of first pleopod with an angular hump at base of
cephalic process . . . . ***P. litosternum*** Hobbs 1946
Canoochee, Ogeechee, and Newport Rivers in Ga.

79b Cephalic surface of first pleopod without an angular hump at base
of cephalic process   80

80a (79) Cephalic process of first pleopod spiculiform and directed
distad   81

80b Cephalic process acute but not needlelike and usually inclined
caudodistad   82

81a (80) Areola less than 4 times longer than least width . . . .
P. pubescens (Faxon) 1884
Between Savannah and Altamaha Rivers in Ga.

81b Areola more than 4 times longer than least width . . . .
P. seminolae Hobbs 1942
Southern Ga. and northern Fla.

82a (80) Areola less than 4 times longer than least width . . . .
P. pictus (Hobbs) 1940
Northeastern Fla. (Hobbs, 1942a).

82b Areola more than 4 times longer than least width   83

83a (82) Cephalic process of first pleopod distinctly hooding the central
projection. (Fig. 31.29) . . . . . . ***P. enoplosternum*** Hobbs 1946
Tributary of Ohoopee River in Ga.

83b Cephalic process small and not hooding the central projection . . .
P. angustatus (LeConte) 1856
Southern Ga.

84a (75) Hooks on ischia of fourth pereiopods bituberculate   85
84b Hooks on ischia of fourth pereiopods simple   86

85a (84) Mesial process of first pleopod extends far distad beyond other
terminal elements . . . . ***P. alleni*** (Faxon) 1884
Peninsular Fla.

85b Mesial process not extending distad beyond other terminal
elements but directed caudodistad . . . . ***P. bivittatus*** Hobbs 1942
Western Fla. to La.

86a (84) Caudal process of first pleopod vestigial or absent   87
86b Caudal process of first pleopod well developed   90

87a (86) Mesial process of first pleopod flattened, bladelike . . . .
P. fallax (Hagen) 1870
Southeastern Ga. and peninsular Fla.

87b Mesial process rounded in section, not bladelike   88

88a (87) Mesial process of first pleopod extending much farther distad than
minute central projection . . . . . ***P. pycnogonopodus*** Hobbs 1942
Western panhandle of Fla.

88b Mesial process not extending farther distad than the well-
developed central projection   89

89a (88) Caudal process of first pleopod represented by a minute tooth at
caudal base of central projection . . . . . ***P. leonensis*** Hobbs 1942
Eastern panhandle of Fla.

89b Caudal process of first pleopod absent . . . . ***P. lunzi*** (Hobbs 1940)
Southeastern S. C.

90a (86) Cephalic process of first pleopod directed at a right angle to the
           main shaft of the appendage   91
90b   Cephalic process directed at an angle less than a right angle to the
           main shaft of the appendage   93

91a (90) Distal fourth of lateral surface of first pleopod with a distinct
           longitudinal ridge. . . . . *Procambarus acutissimus* (Girard) 1852
           Kemper County, Miss.
91b   Distal fourth of lateral surface of first pleopod without a longi-
           tudinal ridge but with a deep excavation or a knob or both near
           base of terminal elements   92

92a (91) Caudal process of first pleopod contiguous with central pro-
           jection . . . . *P. hayi* (Faxon) 1884
           Western Ala. and eastern Miss.
92b   Caudal process of first pleopod not touching central projec-
           tion . . . . *P. lecontei* (Hagen) 1870
           Southern Ala., southern Miss.

93a (90) Knob or rounded prominence present on cephalic or cephalo-
           lateral side of first pleopod at base of terminal elements.   94
93b   No knob or rounded prominence present. . . . .
                                                  P. evermanni (Faxon) 1890
           Southern Miss. to western panhandle of Fla. (Hobbs, 1942a).

94a (93) Shoulder on cephalodistal portion of first pleopod broadly arched
           over distal third of appendage. . . . . . *P. verrucosus* Hobbs 1952
           Eastern Ala.
94b   Shoulder on cephalodistal or cephalolateral portion of first
           pleopod forming a knob. (Fig. 31.30) . . . .
                                                  P. blandingii (Harlan) 1830
           Three subspecies. New England to Mexico.

95a (36) First pleopod with central projection more than twice as long as
           mesial process (Fig. 31.32) . . . . . . Subgenus *Faxonella* Creaser   96
95b   First pleopod with central projection less than 2 times as long as
           mesial process (Figs. 31.33–31.38) . . Subgenus *Orconectes* Cope   97

96a (95) Central projection of first pleopod more than 3 times as long as
           mesial process. (Fig. 31.32). . . . . *O. (F.) clypeatus* (Hay) 1899
           Ark. and La. to Ga. and Fla.
96b   Central projection of first pleopod less than 3 times as long as
           mesial process. . . . . *O. (F.) beyeri* Penn 1950
           N. W. La.

97a (95) Albinistic. Eyes reduced   98
97b   Pigmented. Eyes normal   99

98a (97) Mesial process of first pleopod directed caudodistad . . . .
                                                  O. (O.) inermis Cope 1872
           Southern Ind.
98b   Mesial process of first pleopod directed distad . . . .
                                            O. (O.) pellucidus (Tellkampf) 1844
           Four subspecies. Ind. to N. Ala.

99a (97) Areola obliterated at least at mid-length   100
99b   Areola broad or narrow but never obliterated   104

100a (99) Length of acumen ¾ that of entire rostrum . . . .
                                                  O. (O.) lancifer (Hagen) 1870
           Ill. to La.
100b   Length of acumen less than ¾ that of entire rostrum   101

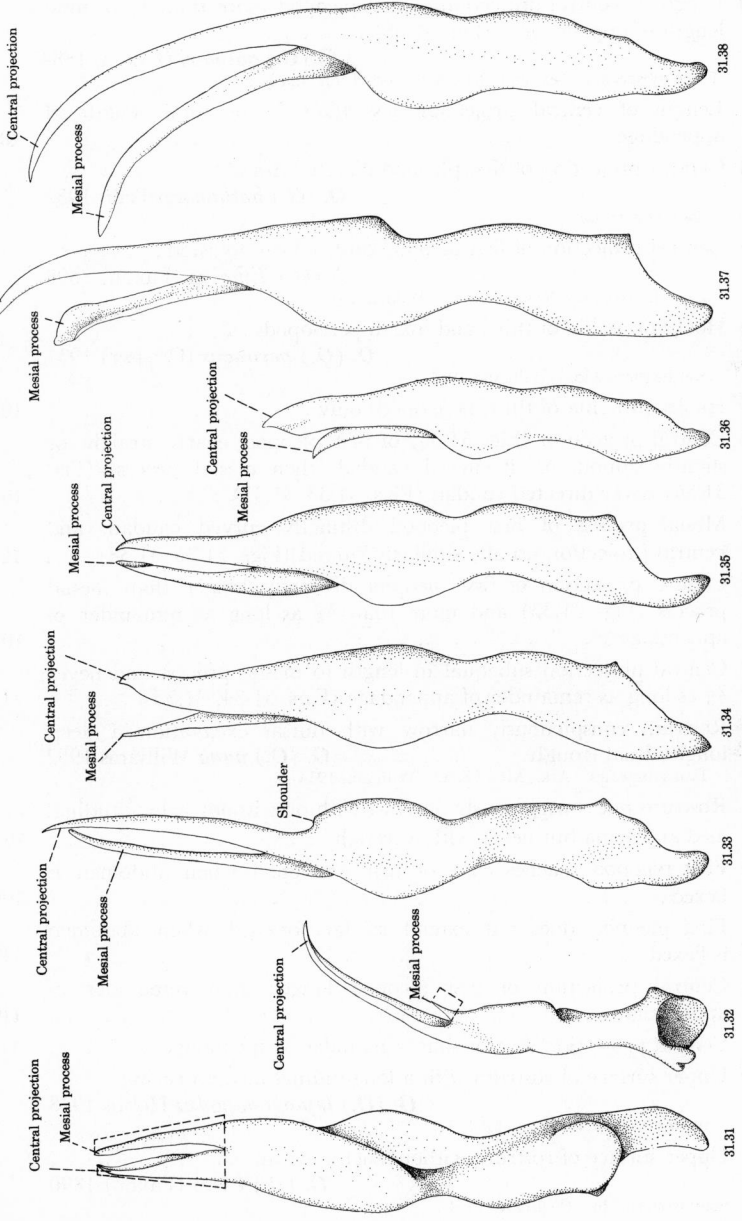

Fig. 31.31. *Orconectes (Orconectes) propinquus.* Mesial view of first left pleopod of first form male. **Fig. 31.32.** *Orconectes (Faxonella) clypeatus.* Cephalic view of first left pleopod of first form male. **Fig. 31.33.** *Orconectes (Orconectes) juvenilis.* Lateral view of first left pleopod of first form male. **Fig. 31.34.** *Orconectes (Orconectes) limosus.* Lateral view of first left pleopod of first form male. **Fig. 31.35.** *Orconectes (Orconectes) propinquus.* Lateral view of first left pleopod of first form male. **Fig. 31.36.** *Orconectes (Orconectes) sloani.* Lateral view of first left pleopod of first form male. **Fig. 31.37.** *Orconectes (Orconectes) virilis.* Lateral view of first left pleopod of first form male. **Fig. 31.38.** *Orconectes (Orconectes) palmeri.* Lateral view of first left pleopod of first form male.

101a **(100)** Rostrum without lateral spines . . . .
Orconectes (Orconectes) mississippiensis Faxon 1884
Miss. (Faxon, 1885).

101b Rostrum with lateral spines **102**

102a **(101)** Length of central projection of first pleopod more than ⅓ of total length of appendage. (Fig. 31.38) . . . .
O. (O.) palmeri (Faxon) 1884
Three subspecies. Tex. and Okla. to western Tenn. and La.

102b Length of central projection less than ⅓ of total length of appendage **103**

103a **(102)** Central projection of first pleopod directed distad . . . .
O. (O.) hathawayi Penn 1952
Southeastern La.

103b Central projection of first pleopod directed caudodistad . . . .
O. (O.) difficilis (Faxon) 1898
Eastern Okla., northwestern Ark. (Williams, 1954).

104a **(99)** Hooks on ischia of third and fourth pereiopods . . . .
O. (O.) peruncus (Creaser) 1931
Southeastern Mo. (Williams, 1954).

104b Hooks on ischia of third pereiopods only **105**

105a **(104)** Central projection (Fig. 31.36) of first pleopod nearly straight or slightly sinuous or, if curved caudad, then mesial process (Fig. 31.36) never directed caudad (Figs. 31.33–31.35) **106**

105b Mesial process of first pleopod distinctly curved caudad, and central projection usually similarly curved (Figs. 31.36–31.38) . . . **124**

106a **(105)** Central projection of first pleopod distinctly longer than mesial process (Fig. 31.33) and more than ⅔ as long as remainder of appendage **107**

106b Central projection subequal in length to mesial process and never ⅔ as long as remainder of appendage (Figs. 31.34, 31.35) **115**

107a **(106)** Rostrum conspicuously narrow with dorsal excavation a deep longitudinal trough . . . . *O. (O.) nana* Williams 1952
Two subspecies. Ark., Mo., Okla. (Williams, 1954).

107b Rostrum not conspicuously narrow, with or without a longitudinal median carina but never with a trough. **108**

108a **(107)** First pleopod reaches coxa of first pereiopod when abdomen is flexed **109**

108b First pleopod does not extend so far forward when abdomen is flexed. **112**

109a **(108)** Central projection of first pleopod longer than remainder of appendage **110**

109b Central projection shorter than remainder of appendage **111**

110a **(109)** Upper surface of rostrum with a longitudinal median carina . . . .
O. (O.) leptogonopodus Hobbs 1948
Western Ark. (Williams, 1954).

110b Upper surface of rostrum without such a carina . . . .
O. (O.) hylas (Faxon) 1890
Southwestern Mo. (Williams, 1954).

111a **(109)** Mesial margin of dactyl of chela with raised tubercles forming a serrate margin along proximal portion . . . .
O. (O.) ozarkae Williams 1952
Southern Mo., northern Ark. (Williams, 1954).
(Fig. 31.33). . . . . *O. (O.) juvenilis* (Hagen) 1870
Eastern tributary of Mississippi River in Ala., Va., Ky., Tenn.

111b　　　　Mesial margin of dactyl of chela never serrate . . . .
　　　　　　　　　　　　　　　　　　O. (O.) medius (Faxon) 1884
　　　　　　　　Southeastern Mo.　(Williams, 1954).

112a　(108)　Margins of rostrum concave laterally . . . .
　　　　　　　　　　　　　　　　　　O. (O.) rusticus (Girard) 1852
　　　　　　　　Five subspecies.　Eastern tributary of Mississippi River and southern Great Lakes
　　　　　　　　drainage as far south as Ala. and Tenn.

112b　　　　Margins of rostrum never concave laterally . . . .　**113**

113a　(112)　Margins of rostrum subparallel and upper surface with a median
　　　　　　　carina. . . . . **O. (O.) neglectus** (Faxon) 1885
　　　　　　　　Two subspecies.　Tex. and Colo. to Mo.　(Williams, 1954).

113b　　　　Margins of rostrum convergent; carina may or may not be
　　　　　　　present . . . .　**114**

114a　(113)　First pleopod reaching base of second pereiopod; branchiostegal
　　　　　　　spine absent . . . . **O. (O.) menae** (Creaser) 1933
　　　　　　　　Western Ark.　(Williams, 1954).

114b　　　　First pleopod reaching base of third pereiopod; branchiostegal
　　　　　　　spine present . . . . **O. (O.) luteus** (Creaser) 1933
　　　　　　　　Eastern Kan., southern Mo., northern Ark.　(Williams, 1954).

115a　(106)　Terminal elements of first pleopod widely separated distally
　　　　　　　(Fig. 31.34) . . . .　**116**

115b　　　　Terminal elements subparallel and not widely separated distally
　　　　　　　(Fig. 31.35) . . . .　**119**

116a　(115)　Terminal elements of first pleopod distinctly divergent (central
　　　　　　　projection directed cephalodistad and mesial process caudo-
　　　　　　　distad) . . . .　**117**

116b　　　　Central projection directed distad, never cephalodistad . . . .　**118**

117a　(116)　Lateral surface of carapace with only one spine . . . .
　　　　　　　　　　　　　　　　　　O. (O.) indianensis (Hay) 1896
　　　　　　　　Southern Ind.

117b　　　　Lateral surface of carapace with more than one spine.　(Fig.
　　　　　　　31.34) . . . . **O. (O.) limosus** (Rafinesque) 1817
　　　　　　　　Me. to Va.　(Ortmann, 1906).

118a　(116)　Margins of rostrum concave laterally; dactyl of chela more than
　　　　　　　twice as long as inner margin of palm. **O. (O.) shoupi** Hobbs 1947
　　　　　　　　Cumberland drainage in Tenn.

118b　　　　Margins of rostrum convex or converging; dactyl of chela less than
　　　　　　　twice as long as inner margin of palm . . . .
　　　　　　　　　　　　　　　　　　O. (O.) wrighti Hobbs 1948
　　　　　　　　Southern Tenn.

119a　(115)　First pleopod with an angular shoulder at base of central pro-
　　　　　　　jection . . . . **O. (O.) obscurus** (Hagen) 1870
　　　　　　　　N. Y. to W. Va. and Va.　(Ortmann, 1906).

119b　　　　First pleopod without such a shoulder . . . .　**120**

120a　(119)　Mesial process of first pleopod never extending quite so far distad
　　　　　　　as central projection . . . .　**121**

120b　　　　Mesial process extending at least as far distad as central pro-
　　　　　　　jection . . . .　**122**

121a　(120)　Rostrum with a median longitudinal carina . . . .
　　　　　　　　　　　　　　　　　　O. (O.) rafinesquei Rhoades 1944
　　　　　　　　Central Ky.

121b　　　　Rostrum without a median longitudinal carina . . . .
　　　　　　　　　　　　　　　　　　O. (O.) virginiensis Hobbs 1951
　　　　　　　　Southeastern Va.

122a (120) Tip of central projection of first pleopod lies in a groove on the
           mesial process 123
122b   Tip of central projection mesial to mesial process . . . .
                          Orconectes (Orconectes) tricuspis Rhoades 1944
           Western Ky.
123a (121) Carapace covered with conspicuous punctations . . . .
                                O. (O.) eupunctus Williams 1952
           Southern Mo., northern Ark. (Williams, 1954).
123b   Carapace more sparsely studded with shallow punctations. (Figs.
           31.31, 31.35) . . . . *O. (O.) propinquus* (Girard) 1852
           Three subspecies. Ontario to Wis. and south to Ky., W. Va., Pa., N. Y. (Rhoades,
           1944).
124a (105) Length of central projection of first pleopod less than ¼ of total
           length of pleopod 125
124b   Length of central projection more than ¼ of total length of
           pleopod. 127
125a (124) Rostrum with a median longitudinal carina. (Fig. 31.36) . . . . .
                                      O. (O.) sloani (Bundy) 1876
           Ind. and Ohio.
125b   Rostrum without a median longitudinal carina 126
126a (125) Distal end of central projection of first pleopod blunt . . . .
                                O. (O.) kentuckiensis Rhoades 1944
           Western Ky.
126b   Distal end of central projection acute. . . . .
                                    O. (O.) harrisoni (Faxon) 1884
           Southwestern Mo. (Williams, 1954).
127a (124) Opposable margin of dactyl of cheliped with a distinct excision
           near base. 128
127b   Opposable margin of dactyl without such an incision 130
128a (127) Terminal elements of first pleopod recurved so that their apices are
           directed perpendicular to the main shaft of the appendage . . . . .
                                    O. (O.) immunis (Hagen) 1870
           Mass. to Colo. (Williams, 1954).
128b   Terminal elements recurved but slightly 129
129a (128) Areola very narrow with room for only 1 row of punctations . . . .
                                      O. (O.) nais (Faxon) 1885
           Great Plains and Ozark region. (Williams, 1954).
129b   Areola broader with room for 2 or more rows of punctations.
           (Fig. 31.37) . . . . *O. (O.) virilis* (Hagen) 1870
           Colo., Saskatchewan, and Manitoba to Ontario (introduced into Md.).
130a (127) Dactyl of chela at least 3 times as long as inner margin of
           palm. . . . . *O. (O.) longidigitus* (Faxon) 1884
           Southern Mo., northern Ark. (Williams, 1954).
130b   Dactyl of chela less than 3 times as long as inner margin of
           palm 131
131a (130) Distal portion of mesial process recurved so that its apex is
           directed perpendicular to the main shaft of the appendage 132
131b   Distal portion of mesial process not so strongly recurved 136
132a (131) Rostrum with a median longitudinal carina 133
132b   Rostrum without such a carina 134
133a (132) Carapace with both lateral and branchiostegal spines . . . .
                                  O. (O.) alabamensis (Faxon) 1884
           Northern Ala., southern Tenn. (Faxon, 1885).

133b   Carapace without lateral and branchiostegal spines . . . .
                    O. (O.) compressus (Faxon) 1884
             Ky., Tenn., Miss., Ala.

134a   (132)   Areola more than 20 times longer than broad . . . . . : . . . . . .
                       O. (O.) hobbsi Penn 1950
             Pontchartrain watershed in La.

134b   Areola less than 20 times longer than broad **135**

135a   (134)   Central projection of first pleopod gently recurved throughout its
             entire length . . . . ***O. (O.) validus*** (Faxon) 1914
             Northern Ala. and southern Tenn.

135b   Central projection almost straight to base of distal third where
             it is suddenly recurved . . . . ***O. (O.) rhoadesi*** Hobbs 1949
             Nashville Basin, Tenn.

136a   (131)   Terminal elements of first pleopod less than $\frac{1}{3}$ of total length
             of appendage. . . . . ***O. (O.) marchandi*** Hobbs 1948
             Northern Ark., southern Mo. (Williams, 1954).

136b   Terminal elements more than $\frac{1}{3}$ of total length of appendage . . .   **137**

137a   (136)   Cephalic margin of first pleopod with an angular shoulder at base
             of central projection . . . . . ***O. (O.) punctimanus*** (Creaser) 1933
             Southern Mo., northern Ark. (Williams, 1954).

137b   Cephalic margin of first pleopod without an angular shoulder . . .
                       O. (O.) meeki (Faxon) 1898
             Two subspecies. Northern Ark., eastern Okla. (Williams, 1954).

138a   (36)   Albinistic **139**

138b   Pigmented **143**

139a   (138)   Chela conspicuously setose . . . . . ***Cambarus setosus*** Faxon 1899
             Southwestern Mo. and northeastern Okla. (Williams, 1954).

139b   Chela not conspicuously setose **140**

140a   (139)   Rostrum with a longitudinal median carina . . . .
                       C. cahni Rhoades 1941
             Northern Ala.

140b   Rostrum without a longitudinal median carina **141**

141a   (140)   Lateral spines on rostrum . ***C. hamulatus*** Cope and Packard 1881
             Southeastern Tenn. and northern Ala.

141b   No lateral spines on rostrum **142**

142a   (141)   Antennal scale more than twice as long as broad . . . .
                       C. hubrichti Hobbs 1952
             Southern Mo.

142b   Antennal scale less than twice as long as broad . . . .
                       C. cryptodytes Hobbs 1941
             Panhandle of Fla.

143a   (138)   Rostrum with lateral spines or minute corneous teeth **144**

143b   Rostrum without lateral spines or teeth. **146**

144a   (143)   Antennal flagellum strongly compressed and bearded on inner
             margin . . . . ***C. cornutus*** Faxon 1884
             Green River, Edmundson County, Ky.

144b   Antennal flagellum normal **145**

145a   (144)   Rostrum less than $\frac{1}{4}$ total length of carapace . . . .
                    C. rusticiformis Rhoades 1944
             Cumberland drainage in Ky. and Tenn.
                    C. hubbsi Creaser 1931
             Southern Mo. (Williams, 1954).

145b   Rostrum more than ¼ total length of carapace . . . .
                      Cambarus extraneus Hagen 1870
Eastern Tenn., northern Ala., northwestern Ga. (Ortmann, 1931).
                          C. spicatus Hobbs 1956
Fairfield and Richland Counties, S. C.

146a **(143)** Areola obliterated at least at mid-length **147**

146b   Areola broad or narrow but never obliterated at mid-length **150**

147a **(146)** Central projections of first pleopods crossed . . . .
                           C. dissitus Penn 1955
Northern La.

147b   Central projections of pleopods subparalled, never crossed **148**

148a **(147)** Dactyl of cheliped with a distinct excision on opposable margin. (Fig. 31.40). . . . . **C. fodiens** (Cottle) 1863
Ontario, Mich., Ill., Ind., Ohio.
                        C. uhleri Faxon 1884
Coastal plain, Md. to S. C.
                        C. byersi Hobbs 1941
Coastal plain, Miss. to Fla.
                        C. hedgpethi Hobbs 1948
Tex. and La.

148b   Dactyl of cheliped without a distinct excision on opposable margin **149**

149a **(148)** Inner margin of palm of chela with a row of prominent tubercles . . . . **C. ortmanni** Williamson 1907
Southern Ind.

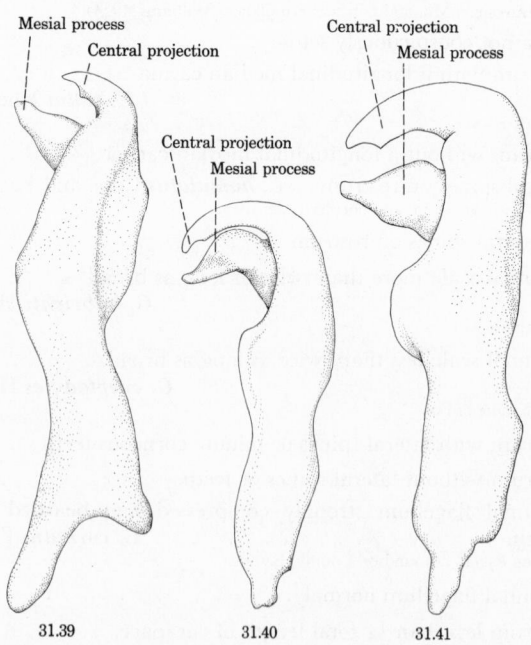

31.39   31.40   31.41

Fig. 31.39. *Cambarus obeyensis.* Lateral view of first left pleopod of first form male. **Fig. 31.40.** *Cambarus fodiens.* Lateral view of first left pleopod of first form male. **Fig. 31.41.** *Cambarus bartonii bartonii.* Lateral view of first left pleopod of first form male.

149b		Inner margin of palm of chela with tubercles appressed	

C. diogenes Girard 1852

Two subspecies. Minn. to Tex. and N. J. to panhandle of Fla. except in mountains.

150a (146) Chela flattened, broadly subtriangular, with 2 or 3 rows of well-defined tubercles along inner margin of palm **151**

150b Chela not strongly flattened, not broadly subtriangular, and never with more than 2 rows of tubercles along inner margin of palm . . **152**

151a (150) Areola more than 8 times longer than wide and usually constituting more than 38 per cent of entire length of carapace (including rostrum) . . . . *C. striatus* Hay 1902

Southern Ala. to Ky.

C. reduncus Hobbs 1956

N. C. and S. C.

C. floridanus Hobbs 1941

Middle panhandle of Fla.

151b Areola less than 8 times longer than wide and constituting less than 38 per cent of entire length of carapace. . . . .

C. latimanus LeConte 1856

N. C. to Fla.

152a (150) Inner margin of palm of chela with a single row of tubercles (only occasionally with a second row above) which proximally form a cristiform row or are obviously fused to form a ridge **153**

152b Inner margin of palm of chela with or without 1 or 2 rows of tubercles which never form a cristiform row or ridge proximally. **159**

153a (152) Chela with many conspicuous long setae . . . .

C. asperimanus Faxon 1914

Mountains of N. C. and S. C.

C. friaufi Hobbs 1953

Eastern Highland Rim in Tenn.

C. brachydactylus Hobbs 1953

Western Highland Rim in Tenn.

153b Chela not conspicuously setose **154**

154a (153) Areola more than 8 times longer than broad. **155**

154b Areola less than 8 times longer than broad. **156**

155a (154) Outer margin of chela subserrate and costate . . . .

C. carolinus Erichson 1846

Mountains from Pa. to S. C. (Ortmann, 1931).

155b Outer margin of chela smooth and rounded . . . .

C. monongalensis Ortmann 1905

Mountains in Pa. and W. Va. (Ortmann, 1931).

156a (154) Antennal scale broader than ½ its length . . . .

C. cristatus Hobbs 1955

Miss.

156b Antennal scale narrower than ½ its length. **157**

157a (156) Central projection of first pleopod directed at less than a right angle to main shaft of appendage. (Fig. 31.39) . . . .

C. obeyensis Hobbs and Shoup 1947

Cumberland Plateau in Tenn.

157b Central projection of first pleopod directed at approximately a right angle to main shaft of appendage. **158**

158a (157) Central projection of first pleopod strongly arched. . . . .

C. parvoculus Hobbs and Shoup 1947

Cumberland Plateau in Tenn. and southwestern Va.

158b   Central projection almost straight, not arched . . . .
                    Cambarus distans Rhoades 1944
                    Upper Cumberland River in Ky. and Tenn.

159a **(152)** Length of areola greater than 38 per cent of carapace length (including rostrum). . . . . **C. tenebrosus** Hay 1902
                    Ky., Tenn., Ala.

159b   Length of areola less than 38 per cent of carapace length **160**

160a **(159)** Margins of rostrum angular at base of acumen. . . . .
                    C. sciotensis Rhoades 1944
                    Kanawha and Ohio drainage system in Ohio, W. Va., Va.

160b   Margins of rostrum rounded or tapering at base of acumen **161**

161a **(160)** Palm of chela without tubercles and fingers without a longitudinal ridge above. . . . . **C. longulus** Girard 1852
                    Two subspecies. Va., W. Va., N. C., Tenn. (Ortmann, 1931).

161b   Palm of chela with tubercles and fingers with a longitudinal ridge above. **162**

162a **(161)** Margins of rostrum tapering to tip of acumen . . . .
                    C. montanus Girard 1852
                    Three subspecies. W. Va. and Tenn., and Md. to Ga. (Ortmann, 1931).
                    C. robustus Girard 1852
                    Ohio, and Mich. to Pa. (Crocker, 1957).

162b   Acumen of rostrum somewhat distinctly delimited at base. (Fig. 31.41) . . . . **C. bartonii** (Fabricius) 1798
                    Five subspecies. New Brunswick to Ga., and Ind. to Ala. (Ortmann, 1931).

References

ISOPODA

Benedict, J. E. 1896. Preliminary description of a new genus and three new species of Crustacea from an artesian well at San Marcos, Texas. *Proc. U. S. Natl. Museum*, 18:615–617.   **Chase, H. D. and A. P. Blair. 1937.** Two new blind isopods from northeastern Oklahoma. *Am. Midland Naturalist*, 18:220–224.   **Cope, E. D. 1872.** On the Wyandotte Cave and its fauna. *Am. Naturalist*, 6:406–422.   **Cope, E. D. and A. S. Packard, Jr. 1881.** The fauna of Nickajack Cave. *Am. Naturalist*, 15:877–882.   **Creaser, E. P. 1931.** A new blind isopod of the genus *Caecidotea*, from a Missouri cave. *Occasional Papers Museum Zool. Univ. Mich.*, 222:1–7.   **Forbes, S. A. 1876.** List of Illinois Crustacea with description of new species. *Bull.' Illinois Museum Nat. Hist.*, 1:3–25.   **Hatchett, S. P. 1947.** Biology of the Isopoda of Michigan. *Ecol. Monographs*, 17:47–79.   **Hay, O. P. 1878.** Description of a new species of Asellus. *Illinois State Lab. Nat. Hist. Bull.*, 2:90–92.   **1882.** Notes on some fresh-water Crustacea, together with descriptions of two new species. *Am. Naturalist*, 16:143–146, 241–243.   **Hay, W. P. 1902.** Observations on the crustacean fauna of Nickajack Cave, Tennessee, and vicinity. *Proc. U. S. Natl. Museum*, 25:417–439.   **Hubricht, Leslie and J. G. Mackin. 1949.** The freshwater isopods of the genus *Lirceus* (Asellota, Asellidae). *Am. Midland Naturalist*, 42:334–349.   **Hungerford, H. B. 1922.** A new subterranean isopod from Kansas. *Kansas Univ. Sci. Bull.*, 14:175–181.   **Levi, H. W. 1949.** Two new species of cave isopods from Pennsylvania. *Notulae Naturae Acad. Nat. Sci., Phila.*, 220:1–6.   **Mackin, J. G. and Hubricht, Leslie. 1938.** Records of distribution of species of isopods in central and southern United States, with descriptions of four new species of Mancasellus and Asellus (Asellota, Asellidae). *Am. Midland Naturalist*, 19:628–637.   **1940.** Descriptions of seven new species of Caecidotea (Isopoda, Asellidae) from the central United States. *Trans. Am. Microscop. Soc.*, 59:383–397.   **Maloney, J. O. 1939.** A new cave isopod from Florida. *Proc. U. S. Natl. Mus.*, 86:457–459.   **Miller, M. A. 1933.** A new blind isopod, *Asellus californicus*, and a revision of the subterranean asellids. *Univ. Calif. Berkeley Publs. Zoöl.*, 39:97–110.   **Packard, A. S., Jr. 1871.** The Mammoth Cave and its inhabitants. *Am.*

Naturalist, 5:739–761.   **1888.**   The cave fauna of North America, with remarks on the anatomy of the brain and origin of the blind species. *Mem. Natl. Acad. Sci.*, 4:1–156.   **Richardson, Harriet. 1900.**   Synopses of North American invertebrates. VIII. The Isopoda. *Am. Naturalist*, 34:295–309.   **1904.**   Isopod crustaceans of the northwest coast of North America. In: *Harriman Alaska Expedition*, 10, pp. 211–230. Doubleday, Page, New York.   **1905.**   A monograph on the isopods of North America. *Bull. U. S. Natl. Museum*, 54:i–liv, 1–727.   **Say, Thomas. 1818.**   An account of the crustacea of the United States. *F. Acad. Nat. Sci. Phila.*, 1: 374–401.   **Smith, S. I. 1874a.**   The crustacea of the fresh waters of the United States. *Rept. U. S. Comm. Fisheries*, 2:637–665.   **Stafford, B. E. 1911.**   A new subterranean isopod. *Pomona Coll. J. Entomol.*, 3:572–575.   **Ulrich, C. J. 1902.**   A contribution to the subterranean fauna of Texas. *Trans. Am. Microscop. Soc.*, 23:83–100.   **Van Name, W. G. 1936.**   The American land and fresh-water isopod Crustacea. *Bull. Am. Museum Nat. Hist.*, 71:i–viii, 1–535.   **1940.**   A supplement to the American land and fresh-water isopod Crustacea. *Bull. Am. Museum Nat. Hist.*, 77:109–142.   **1942.**   A second supplement to the American land and fresh-water isopod Crustacea. *Bull. Am. Museum Nat. Hist.*, 81:299–329.

AMPHIPODA

Adamstone, F. B. 1928.   Relict amphipods of the genus *Pontoporeia*. *Trans. Am. Microscop. Soc.*, 47:366–371.   **Bate, C. Spence. 1862.**   *Catalogue of the amphipodous Crustacea in the British Museum.* London.   **Benedict, J. E. 1896.**   Preliminary description of a new genus and three new species of crustacea from an artesian well at San Marcos. *Proc. U. S. Natl. Museum*, 18:615–617.   **Bousfield, E. L. 1958.**   Fresh-water amphipod crustaceans of glaciated North America. *Can. Field-Naturalist*, 72:55–113.   **Cope, E. D. and A. S. Packard, Jr. 1881.**   The fauna of Nickajack Cave. *Am. Naturalist*, 15:877–882.   **Creaser, E. P. 1934.**   A new genus and species of blind amphipod with notes on parallel evolution in certain amphipod genera. *Occasional Papers Museum Zool. Univ. Mich.*, 282:1–7.   **Derjavin, A. N. 1927.**   The Gammaridae of the Kamchatka Expedition, 1908–1909. *Russ. Hydrobiol. Z.*, 6:8.   **Ellis, T. Kenneth. 1940.**   A new amphipod of the genus *Crangonyx* from South Carolina. *Charleston Museum Leaf.*, 13:1–8.   **1941.**   A new fresh-water amphipod of the genus *Stygobromus* from South Carolina. *Charleston Museum Leaf.*, 16:1–8.   **Embody, George C. 1910.**   A new fresh-water amphipod from Virginia, with some notes on its biology. *Proc. U. S. Natl. Museum*, 38:299–305.   **Fage, Louis. 1931.**   Crustacés amphipodes et décapodes. Campagne spéologique de C. Bolivar et R. Jeannel dans l'Amerique du Nord (1928). *Arch. zool. exp. et gén.*, 71:361–374.   **Forbes, S. A. 1876.**   List of Illinois Crustacea with description of new species. *Bull. Illinois Museum Nat. Hist.*, 1:3–25.   **Hay, O. P. 1882.**   Notes on some freshwater Crustacea, together with descriptions of two new species. *Am. Naturalist*, 16:143–146, 241–243.   **Hay, W. P. 1902a.**   Observations on the crustacean fauna of the region about Mammoth Cave, Kentucky. *Proc. U. S. Natl. Museum*, 25:223–236.   **1902b.**   Observations on the crustacean fauna of Nickajack Cave, Tennessee, and vicinity. *Proc. U. S. Natl. Museum*, 25:416–439.   **Holmes, S. J. 1908.**   Description of a new subterranean amphipod from Wisconsin. *Trans. Wisconsin Acad. Sci.*, 16:77–80.   **Hubricht, Leslie. 1943.**   Studies on the nearctic freshwater Amphipoda. III. Notes on the freshwater Amphipoda of eastern United States, with descriptions of ten new species. *Am. Midland Naturalist*, 29:683–712.   **Hubricht, Leslie and J. G. Mackin. 1940.**   Descriptions of nine new species of fresh-water amphipod crustaceans with notes and new localities for other species. *Am. Midland Naturalist*, 23:187–218.   **Hubricht, Leslie and C. H. Harrison. 1941.**   The fresh-water Amphipoda of Island County, Washington. *Am. Midland Naturalist*, 26:330–333.   **Mackin, J. G. 1935.**   Studies on the Crustacea of Oklahoma. III. Subterranean amphipods of the genera *Niphargus* and *Boruta*. *Trans. Am. Microscop. Soc.*, 54:41–51.   **Packard, A. S. 1888.**   The cave fauna of North America, with remarks on the anatomy of the brain and origin of the blind species. *Mem. Natl. Acad. Sci.*, 4:1–156.   **Sars, G. O. 1863.**   Beretning om i Sommeren 1862 foretagen zoologisk Reise i Christianias og Trondhjems Stifter. *Nyt. Mag. Natur.*, 12:193–252.   **Saussure, H. de. 1858.**   Mémoir sur divers crustacés nouveaus du Mexique et des Antelles. *Mém. soc. phys. hist. nat. Geneva*, 14:417–496.   **Say, Thomas. 1818.**   An account of the Crustacea of the United States. *J. Acad. Nat. Sci. Phila.*, 1: 374–401.   **Schellenberg, A. 1937.**   Die Amphipodengattungen um *Crangonyx*, ihre Verbreitung und ihre Arten. *Mitt. Zool. Mus. Berlin*, 22:31–44.   **Shoemaker, C. R. 1920.**   Crustacea. Part E. Amphipoda. *Rept. Can. Arctic Exped. 1913–1918*, 7:1–30. **1938.** A new species of fresh-water amphipod of the genus *Synpleonia*, with remarks on related genera. *Proc. Biol. Soc. Wash.*, 51:137–142. **1940.** Notes on the amphipod *Gammarus minus* Say and description of a new variety, *Gammarus minus* var. *tenuipes*. *Jour. Wash. Acad. Sci.*, 30:388–394. **1941.** A new subterranean amphipod of the genus *Crangonyx* from Florida. *Charleston Museum Leaf.*, 16:9–14. **1942a.** Notes on some American fresh-water amphipod crustaceans and descriptions of a new genus and two new species. *Smithsonian Inst. Publs. Misc. Collections*, 101:1–31. **1942b.** A new cavernicolous amphipod from Oregon. *Occasional Papers Univ. Mich.*, 466:1–6. **1944.** Description of a new species of Amphipoda of the genus *Anisogammarus* from Oregon. *J. Wash. Acad. Sci.*, 34:89–93. **1945.** Notes on the amphipod genus *Bactrurus* Hay, with description of a new species. *J. Wash. Acad. Sci.*, 35:24–27. **Smith, S. I. 1874a.** The Crustacea of the fresh waters of the United States. *Rept. U. S. Comm. Fisheries*, 2:637–665. **1874b.** Report on the amphipod crustaceans. (In Hayden's) *Rept. U. S. Geol. Survey Territ. 1873*, 7:608–611. **1875.** The crustaceans of the caves of Kentucky and Indiana. *Am. J. Sci. Arts*, III. 9:476–477. **1888.** *Crangonyx vitreus* and *packardii*. *Mem. Natl. Acad. Sci.*, 4:34–36. **Stebbing, T. R. R. 1899.** Amphipoda from the Copenhagen Museum and other sources. Part II. *Trans. Linn. Soc. London, II (Zool.)*, 7:395–432. **1906.** Amphipoda. I. Gammaridea. *Das Tierreich*, 21. Friedländer, Berlin. **Stout, V. R. 1911.** A new subterranean freshwater amphipod. *Pomona Coll. J. Entomol.*, 3:570–571. **Ulrich, C. J. 1902.** A contribution to the Subterranean fauna of Texas. *Trans. Am. Microscop. Soc.*, 23:83–100. **Weckel, A. L. 1907.** The fresh-water Amphipoda of North America, *Proc. U. S. Natl. Museum*, 32:25–58. **Wrzesnioski, A. 1877.** Über die anatomie der Amphipoden. *Z. wiss. Zool.*, 28:403–406.

MYSIDACEA

Banner, A. H. 1948. A taxonomic study of the Mysidacea and Euphausiacea (Crustacea) of the Northeastern Pacific. Part II. Mysidacea, from tribe Mysini through subfamily Mysidellinae. *Trans. Roy. Can. Inst.*, 27:65–125. **1953.** On a new genus and species of mysid from southern Louisiana (Crustacea, Malacostraca). *Tulane Studies Zool.*, 1:1–8. **1954a.** A supplement to W. M. Tattersall's review of the Mysidacea of the United States National Museum. *Proc. U. S. Natl. Museum*, 103:575–583. **1954b.** New records of Mysidacea and Euphausiacea from the northeastern Pacific and adjacent areas. *Pacific Sci.*, 8:125–139. **Holmquist, C. 1949.** Über eventuelle intermediäre Formen zwischen *Mysis oculata* Fabr. und *Mysis relicta* Lovén. *Lunds Univ. Årskr. N. F.*, 2. 45:1–26. **Smith, S. I. 1874a.** The crustacea of the fresh waters of the United States. *Rept. U. S. Comm. Fisheries*, 2:637–665. **Tattersall, W. M. 1951.** A review of the Mysidacea of the United States National Museum. *Bull. U. S. Natl. Museum*, 201:i–x, 1–292. **Tattersall, W. M. and Olive S. Tattersall. 1951.** *The British Mysidacea.* Ray Society. London.

DECAPODA

Benedict, J. E. 1896. Preliminary description of a new genus and three new species of Crustacea from an artesian well at San Marcos, Texas. *Proc. U. S. Natl. Museum*, 18:615–617. **Bouvier, E.-L. 1925.** Recherches sur la morphologie, les variations, la distribution géographique des crevettes de la famille des Atyidés. *Encyclopédie entomologique.* Ser. A, Vol. 4. **Chace, F. A., Jr. 1954.** Two new subterranean shrimps (Decapoda: Caridea) from Florida and the West Indies, with a revised key to the American species. *J. Wash. Acad. Sci.*, 44:318–324. **Cope, E. D. and A. S. Packard, Jr. 1881.** The fauna of Nickajack Cave. *Am. Naturalist*, 15:877–882. **Cope, E. D. 1872.** On the Wyandotte cave and its fauna. *Am. Naturalist*, 6:406–422. **Creaser, E. P. 1931a.** The Michigan decapod crustaceans. *Papers Mich. Acad. Sci.*, 13:257–276. **1932.** The decapod crustaceans of Wisconsin. *Trans. Wisconsin Acad. Sci., Arts and Letters*, 27:321–338. **Creaser, E. P. and A. I. Ortenburger. 1933.** The decapod crustaceans of Oklahoma. *Univ. Oklahoma Publ. Biol. Survey*, 5:14–47. **Crocker, Denton W. 1957.** The crayfishes of New York State. *N. Y. State Museum and Sci. Ser. Bull.* No. 355:1–97. **Fage, Louis. 1931.** Crustacés amphipodes et décapodes. Campagne speologique de C. Bolivar et R. Jeannel dans l'Amerique du Nord (1928). *Arch. zool. exp. et gén.*, 71:361–374. **Faxon, Walter. 1884.** Descriptions of new species of Cambarus; to which is added a synonymical list of the known species of *Cambarus* and *Astacus*. *Proc. Am. Acad. Arts Sci.*, 20:107–158. **1885.** A revision of the Astacidae. *Mem. Museum Comp. Zool.*

Harvard, 10:1–186.   **1890.** Notes on North American crayfishes, Family Astacidae. *Proc. U. S. Natl. Museum*, 12:619–634.   **1898.** Observations on the Astacidae in the United States National Museum and in the Museum of Comparative Zoology, with descriptions of new species and subspecies to which is appended a catalogue of the known species and subspecies. *Proc. U. S. Natl. Museum*, 20:643–694.   **1914.** Notes on the crayfishes in the United States National Museum and the Museum of Comparative Zoölogy with descriptions of new species and subspecies to which is appended a catalogue of the known species and subspecies. *Mem. Museum Comp. Zool. Harvard*, 40:347–427.   **Forbes, S. A. 1876.** List of Illinois Crustacea with description of new species. *Bull. Illinois Museum Nat. Hist.*, 1:3–25.   **Hagen, H. A. 1870.** Monograph of the North American Astacidae. *Illustrated Cat. Museum Comp. Zool. Harvard*, 3:1–109.   **Harris, J. A. 1903.** An ecological catalogue of the crayfishes belonging to the genus *Cambarus*. *Univ. Kansas Sci. Bull.*, 2:51–187.   **Hay, O. P. 1882.** Notes on some freshwater crustacea, together with descriptions of two new species. *Am. Naturalist*, 16:143–146, 241–243.   **Hay, W. P. 1896.** The crayfishes of the state of Indiana. *Ann. Rept. Indiana Geol. Survey*, 20:475–507.   **1899.** Synopsis of North American invertebrates. 6. The Astacidae of North America. *Am. Naturalist*, 33:957–966.   **1902a.** Two new subterranean crustaceans from the U. S. *Proc. Biol. Soc. Wash.*, 14:179–180.   **1902b.** Observations on the crustacean fauna of Nickajack Cave, Tennessee, and vicinity. *Proc. U. S. Natl. Museum*, 25:417–439.   **1902c.** Observations on the crustacean fauna of the region about Mammoth Cave, Kentucky. *Proc. U. S. Natl. Museum*, 25:223–236.   **Hedgpeth, J. W. 1949.** The North American species of Macrobrachium (River Shrimp). *Texas J. Sci.*, 1:28–38.   **Holthuis, L. B. 1952.** The subfamily Palaemoninae. A general revision of the Palaemonidae (Crustacea Decapoda Natantia) of the Americas. *Allan Hancock Found. Publ., Occasional Papers*, 12:1–396.   **Hobbs, H. H., Jr. 1942a.** The crayfishes of Florida. *Univ. Florida Publs. Biol. Sci. Ser.*, 3:i–vi, 1–179.   **1942b** A generic revision of the crayfishes of the subfamily Cambarinae (Decapoda, Astacidae) with the description of a new genus and species. *Am. Midland Naturalist*, 28:334–357.   **1942c.** On the first pleopod of the male Cambari (Decapoda, *Astacidae*). *Proc. Florida Acad. Sci.*, 5:55–61.   **1945a.** Notes on the first pleopod of the male Cambarinae (Decapoda, *Astacidae*). *Quart. J. Florida Acad. Sci.*, 8:67–70.   **1945b.** Two new species of crayfishes of the genus *Cambarellus* from the Gulf coastal states, with a key to the species of the genus (Decapoda, *Astacidae*). *Am. Midland Naturalist*, 34:466–474.   **1948.** On the crayfishes of the *Limosus* section of the genus *Orconectes* (Decapoda: *Astacidae*). *J. Wash. Acad. Sci.*, 38:14–21.   **1952.** A new crayfish of the genus *Procambarus* from Georgia with a key to the species of the Clarkii subgroup. *Quart. J. Florida Acad. Sci.*, 15:165–174.   **Ortmann, A. E. 1905.** The mutual affinities of the species of the genus *Cambarus*, and their dispersal over the United States. *Proc. Am. Phil. Soc.*, 44:91–136.   **1906.** The Crawfishes of the state of Pennsylvania. *Mem. Carnegie Museum*, 343–523.   **1931.** Crawfishes of the southern Appalachians and the Cumberland Plateau. *Ann. Carnegie Museum*, 20:61–160.   **Packard, A. S., Jr. 1871.** The Mammoth Cave and its inhabitants. *Am. Naturalist*, 5:739–761.   **1888.** The cave fauna of North America, with remarks on the anatomy of the brain and origin of the blind species. *Mem. Natl. Acad. Sci.*, 4:1–156.   **Pearse, A. S. 1910.** The crawfishes of Michigan. *Mich. State Biol. Survey Publ.*, 1:9–22.   **Penn, G. H. 1950.** The genus *Cambarellus* in Louisiana. *Am. Midland Naturalist*, 44:421–426.   **1952.** The genus *Orconectes* in Louisiana. *Am. Midland Naturalist*, 47:743–748.   **1956.** The genus *Procambarus* in Louisiana. *Am. Midland Naturalist*, 56:406–422.   **Rhoades, R. 1944.** The crayfishes of Kentucky, with notes on variation, distribution and descriptions of new species and subspecies. *Am. Midland Naturalist*, 31:111–149.   **Say, Thomas. 1818.** An account of the crustacea of the United States. *F. Acad. Nat. Sci. Phila.*, 1:374–401.   **Smith, S. I. 1874a.** The crustacea of the fresh waters of the United States. *Rept. U. S. Comm. Fisheries*, 2:637–665.   **Ulrich, C. J. 1902.** A contribution to the Subterranean fauna of Texas. *Trans. Am. Microscop. Soc.*, 23:83–100.   **Villalobos, A. 1955.** Cambarinos de la Fauna Mexicana (Crustacea Decapoda). Tesis presentada para aspirar al Grado de Doctor en Ciencias Biológicas, Departamento de Biologia, Facultad de Ciencias, Universidad National Autónoma de México.   **Williams, A. B. 1952.** Six new crayfishes of the genus *Orconectes* (Decapoda: Astacidae) from Arkansas, Missouri and Oklahoma. *Trans. Kansas Acad Sci.*, 55:330–351.   **Williams, A. B. 1954.** Speciation and distribution of the crayfishes of the Ozark Plateaus and Quichita Provinces. *Univ. Kansas Sci. Bull.*, 36 (Pt. 2):803–918.   **Williams, A. B. and A. B. Leonard. 1952.** The crayfishes of Kansas. *Univ. Kansas Sci. Bull.*, 34 (Pt. 2):960–1012.

32

Introduction to
Aquatic Insecta

HERBERT H. ROSS

Although the great majority of insects are terrestrial, a number of them are aquatic, occurring almost entirely in fresh water. Five insect orders are entirely aquatic, and seven others contain families or genera that are primarily aquatic. In all, about 5000 species of insects are found in fresh-water habitats in North America. These insects are distinguished from other aquatic arthropods by possessing a 3-segmented thorax typically bearing three pairs of legs, and, in the adult, two pairs of wings. The head is also distinctive, forming a definite capsule bearing one pair of antennae and three pairs of mouthparts, the latter consisting of paired mandibles and maxillae and the labium, which is composed of the fused second maxillae. These parts are frequently highly modified and difficult to identify. In the Hemiptera, for instance, certain structures of the mouthparts form long stylets which fit together to form a piercing-sucking beak.

In a few insects the young change little in general structure from first instar to adult. The wingless orders Thysanura and Collembola are examples of such orders. All the orders of winged insects and their allies pass through developmental stages which are different in appearance. In many groups the wings develop externally as pads which increase in size at each molt. In this type of development, called gradual metamorphosis, the young are designated as *nymphs*. Hemiptera, Odonata, Plecoptera, and Ephemeroptera are aquatic insects of this type. In other orders the wings develop internally for several instars, then are everted as external pads in the last preadult instar. In this type of development, known as complete metamorphosis, the early instars without wing pads are designated as *larvae* and the preadult instar with wing pads as the *pupa*. Common examples among aquatic insects include the orders Coleoptera, Neuroptera, Megaloptera, Trichoptera, and Diptera.

In the main, each group of insects is most abundant in certain types of fresh-water habitats. Damselflies and dragonflies are predominantly pond or shallow lake species, although some forms occur in running water. Stoneflies and mayflies are predominantly running water forms, although certain groups of the latter order are chiefly pond dwellers. Caddisflies are abundant in both streams and lakes, but require well-aerated water. Beetles, spongeflies, and the other groups also occur in both streams and lakes.

In only the aquatic beetles and bugs are both adults and nymphs or larvae adapted for living in the water. In the other groups the immature stages live in the water and the adults (sometimes the pupae also) are terrestrial.

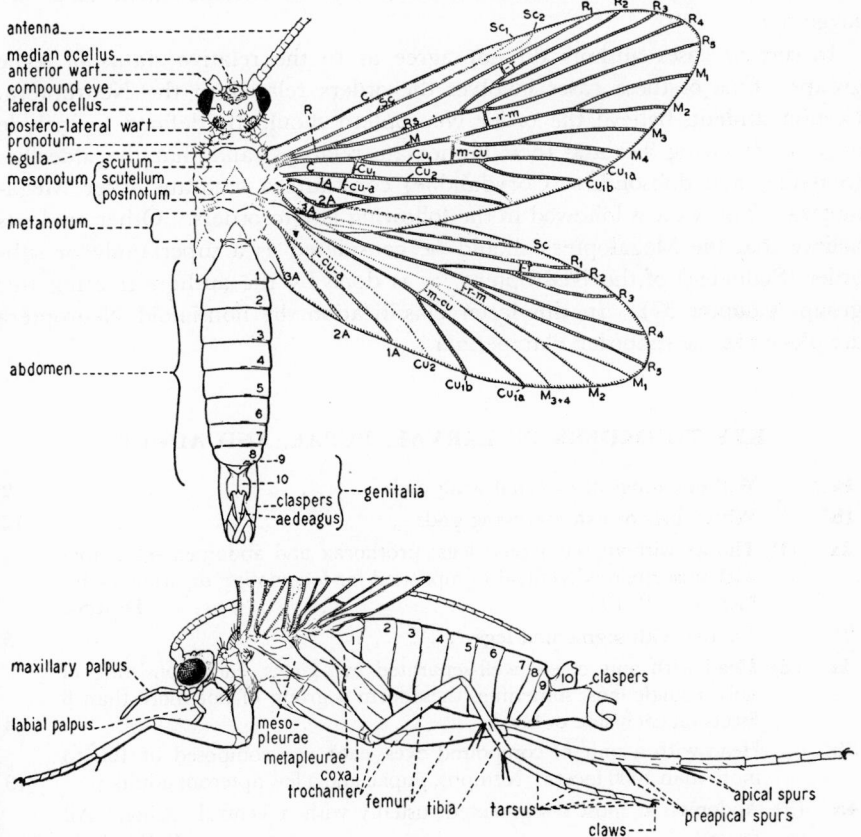

Fig. 32.1. Dorsal and lateral views of an adult caddisfly, *Rhyacophila lobifera*, illustrating terminology of parts. (From Illinois Natural History Survey.)

Certain water bugs, such as the water-striders, live on the surface of the water and are termed semiaquatic. A few species of springtails (Collembola) are also semiaquatic. Some parasitic wasps (Hymenoptera) are truly aquatic. The adults swim under water and lay their eggs on the host eggs of Hemiptera, Odonata, and possibly of other aquatic insects.

The principal groups of aquatic insects constitute an important part of the biota of fresh-water communities. They are important economically both as natural fish food and as indicator organisms for identifying the ecological characteristics of streams.

The insect structures and body regions most frequently used in taxonomic keys are illustrated in Fig. 32.1. The illustration is of a caddisfly, and in some other insects it may be difficult to be certain of homologies of parts. Assistance in ascertaining such homologies can be obtained from the introductory textbooks listed at the end of this chapter. The key to the orders of the adults treats only species that habitually occur in or on the water. In collections made at lights or by sweeping vegetation along water, species of many terrestrial orders will be caught. To sort and orient material from general catches such as these the more extensive ordinal keys in entomological texts are suggested.

In certain cases authors do not agree as to the relative status of some groups. One of these cases involves the orders related to the Neuroptera. Certain students believe the order Neuroptera should be defined to include only the lacewing flies and their immediate relatives, and that the alderflies (Sialidae) and dobsonflies (Corydalidae) constitute a separate order Megaloptera. This view is followed in the following key to orders. Other students believe that the Megaloptera should be considered as a superfamily or suborder (Sialoidea) of the Neuroptera, as is done by the authors treating this group (Chapter 37). In this latter classification the nonsialoid Neuroptera are placed in the suborder Planipennia.

KEY TO ORDERS OF LARVAE, PUPAE, AND ADULTS

1a Without wings or external wing pads. 2
1b With wings or external wing pads 12
2a (1) Thorax without segmented legs; prothorax and abdomen sometimes with unsegmented ventral humps used for locomotion or attachment. Larvae. (P. 1057). . . . . **Diptera**
2b Thorax with segmented legs. 3
3a (2) Head with only a few well-separated eye facets, each consisting of only a single lens; sometimes these form a group of not more than 8 facets on each side of head 4
3b Head with a pair of compound eyes, each eye composed of 100 to more than 1000 facets. Nymphs, pupae and a few apterous adults. . . 13
4a (3) Abdomen at most 6-segmented, usually with a ventral spring. All stages. . . . . **Collembola**
    This group is not treated further here. See Usinger (1956).
4b Abdomen with more than 6 segments, never with a ventral spring. Larvae. (Couplets **5–11**) 5

5a (**4**) Mouthparts forming a very long, slender, projecting beak; feeding in sponges (*Sisyridae*). (P. 973) . . . . **Neuroptera**

5b Mouthparts not forming a beak **6**

6a (**5**) Venter of abdomen having several pairs of appendages, each one having a ring of fine hooks or crochets around it. (P. 1050) . . . . . . **Lepidoptera**

6b Venter of abdomen without appendages bearing crochets **7**

7a (**6**) Apex of abdomen ending in 1 or 2 appendages or lobes bearing sclerotized hooks **8**

7b Apex of abdomen without terminal, hooked appendages **10**

8a (**7**) Apex of abdomen with only one mesal, fingerlike lobe bearing 4 hooks (*Gyrinidae*). (P. 981) . . . . **Coleoptera**

8b Apex of abdomen with 2 lateral lobes or appendages, each bearing 1 or 2 hooks. **9**

9a (**8**) Each terminal leg bearing only 1 hook, which is divided into several sclerites. (P. 1024) . . . . **Trichoptera**

9b Each terminal leg bearing 2 or more hooks, each hook simple, undivided (*Corydalidae*). (P. 973, 904) . . . . **Megaloptera**

10a (**7**) Each tarsus ending in a single claw. (P. 981) . . . . **Coleoptera**

10b Each tarsus ending in 2 claws of about equal length **11**

11a (**10**) Apex of abdomen ending in a long, slender process or "tail" (*Sialidae*). (P. 973, 904) . . . . **Megaloptera**

11b Apex of abdomen not ending in a long tail. (P. 981) . . . **Coleoptera**

12a (**1**) Wings large and functional, or if small pivoting on a special area of small sclerites which provide an articulation with the body. Adults **24**

12b Wings forming flaps whose epidermis flows into that of the body with no break or articulation (Fig. 32.2); never functional; often the adult wing can be seen pleated inside the immature wing **13**

13a (**3, 12**) Abdomen with a paired series of sclerous plates, each plate bearing a series of sharp hooks or teeth (Fig. 32.2). Pupae. (P. 1024) . . . . . **Trichoptera**

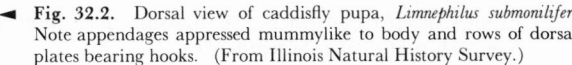

◄ **Fig. 32.2.** Dorsal view of caddisfly pupa, *Limnephilus submonilifer*. Note appendages appressed mummylike to body and rows of dorsal plates bearing hooks. (From Illinois Natural History Survey.)

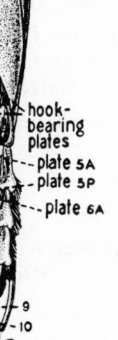

13b		Abdomen occasionally with a series of hooks or teeth on margins of segments, but never on pairs of distinctive plates.	**14**
14a	(13)	Active forms with freely movable legs and antennae, not found in cases, cocoons, or cells in wood or earth. Nymphs and a few apterous adults. (Couplets **15–18**)	**15**
14b		Body mummylike (Fig. 32.2), often in a case, cocoon, or earthen cell, or within a hard covering formed by a larval skin (puparium); the appendages held close to the body and sometimes seemingly fused with the wall of the body. Pupae. (Couplets **19–23**)	**19**
15a	(14)	Labium or lower lip forming an extensile, elbowed, grasping organ. (P. 917) . . . . **Odonata**	
15b		Labium not extensile and not forming such an organ	**16**
16a	(15)	Sides of abdomen having a series of platelike or leaflike gills. (P. 908) . . . . **Ephemeroptera**	
16b		Sides of abdomen without gills; in some forms a few ventral clusters of filamentous gills occur at the extreme base of the abdomen	**17**
17a	(16)	Antennae at most with 10 segments; mouthparts forming a long tubular or short triangular beak. (P. 958) . . . . **Hemiptera**	
17b		Antennae with 25 to 100 segments; mouthparts not forming a beak	**18**
18a	(17)	Tarsi each with a maximum of 3 segments. Nymphs and a few apterous adults. (P. 941) . . . . **Plecoptera**	
18b		Tarsi each 5-segmented. Apterous adults. (P. 1024) . **Trichoptera**	
19a	(14)	Body having only one pair of wing pads. (P. 1057) . . . . **Diptera**	
19b		Body having 2 pairs of wing pads.	**20**
20a	(19)	Appendages and body heavily sclerotized and dark, the appendages appearing to be fused with the body. (P. 1050) . . . . **Lepidoptera**	
20b		Appendages and body chiefly submembranous, the appendages free from the body	**21**
21a	(20)	Base of abdomen constricted to a narrow waist. (Not included) . . . **Hymenoptera**	
21b		Base of abdomen not constricted to a waist.	**22**
22a	(21)	Front wing pads thickened; antennae with less than 15 segments. (P. 981). . . . . **Coleoptera**	
22b		Front wing pads no thicker or massive than hind wing pads; antennae with more than 20 segments	**23**
23a	(22)	Size small, 10 mm or less. (P. 973, 904) . . . . **Neuroptera**	
23b		Size larger, 12 mm or more. (P. 973, 904) . . . . . . **Megaloptera**	
24a	(12)	Front wings hard, opaque, and without venation, the two forming a strong cover over the abdomen; hind wings sometimes lacking. (P. 981). . . . . **Coleoptera**	
24b		Front wings chiefly membranous, usually with a well-developed and branching venation, or absent.	**25**
25a	(24)	Having only 1 pair of wings.	**26**
25b		Having 2 pairs of wings	**27**
26a	(25)	Metanotum having a pair of knobbed balancing organs or halteres. (P. 1057). . . . . **Diptera**	
26b		Metanotum without halteres	**27**
27a	(25, 26)	Mouthparts forming a piercing-sucking beak which is either	

		tubular and jointed, directed posteriorly, or broad, and triangular, pointed downward. (P. 958) . . . . **Hemiptera**	
27b		Mouthparts not forming a beak.	**28**
28a	(27)	Antennae short, usually little longer than width of head, and all but the base hairlike	**29**
28b		Antennae much longer, threadlike or even more massive	**30**
29a	(28)	Front and hind wings about equal in length and area. (P. 917) . . . **Odonata**	
29b		Hind wings much smaller than front wings, in some species entirely lacking. (P. 908). . . . . **Ephemeroptera**	
30a	(28)	Middle and hind legs with tarsi 3-segmented. (P. 941) . . . . **Plecoptera**	
30b		Middle and hind legs with tarsi 5-segmented.	**31**
31a	(30)	Wings and body densely clothed with scales. (P. 1050). . . . . **Lepidoptera**	
31b		Wings and body clothed chiefly with hair, sometimes the wings having patches of scales	**32**
32a	(31)	Each wing at most with 7 or 8 crossveins posterior to those along front margin	**33**
32b		Each wing with a large number of crossveins scattered over all regions of the wing.	**34**
33a	(32)	Wings either covered with dense hair which obscures the venation, or venation composed chiefly of long, relatively parallel veins (Fig. 32.1). (P. 1024) . . . . **Trichoptera**	
33b		Wings never with hair obscuring the venation, the venation either restricted to anterior part of wing or composed of a meshlike arrangement of veins and crossveins. (Not included). . . . . **Hymenoptera**	
34a	(32)	Pronotum large and shieldlike. (P. 973, 904) . . . . . **Megaloptera**	
34b		Pronotum small and comparatively inconspicuous. (P. 973, 904) . . **Neuroptera**	

References

Borror, D. J. and D. M. DeLong. 1954. *An Introduction to the Study of Insects.* Rinehart, New York. **Brues, C. T., A. L. Melander, and F. M. Carpenter. 1954.** Classification of insects. *Bull. Museum Comp. Zool. Harvard*, 108:1–917. **Ross, H. H. 1956.** *A Textbook of Entomology*, 2nd ed. Wiley, New York. **Usinger, R. L. (ed.). 1956.** *Aquatic Insects of California with Keys to North American Genera and California species.* University of California Press, Berkeley and Los Angeles.

33

Ephemeroptera

GEORGE F. EDMUNDS, JR.

The nymphs of Ephemeroptera can be immediately distinguished from all other aquatic insects by the presence of tracheal gills on the abdomen, unpaired tarsal claws, and an enlarged mesothorax. The bodies of many forms are either flattened or fish-shaped. The head is variable in shape and in the burrowing forms usually bears a frontal process. The mandibles of some bear a tusklike projection. The maxillary and labial palpi are primitively of three segments, but either or both may be reduced to two segments. The tarsal claws bear denticles in many genera, and the claws are particularly long among some of the sand-inhabiting genera. Tracheal gills occur on abdominal segments 1 to 7, or are reduced or missing from either end of the series; they occur in a wide variety of forms and are often diagnostic of genera or groups of genera. In counting the abdominal segments to determine the segment from which any gill arises, it is best to count forward from the terminal tenth segment. The tracheal gills arise from the pleura and are usually carried dorsally or laterally, but may be ventral in position. Frequently one of the gills serves the purpose of a protective cover or operculum for the gills caudad of it. There are always two or three tails arising from the tenth abdominal segment.

The key to the family Baetidae should be used with caution. Young nymphs frequently cannot be determined with certainty. The metathoracic

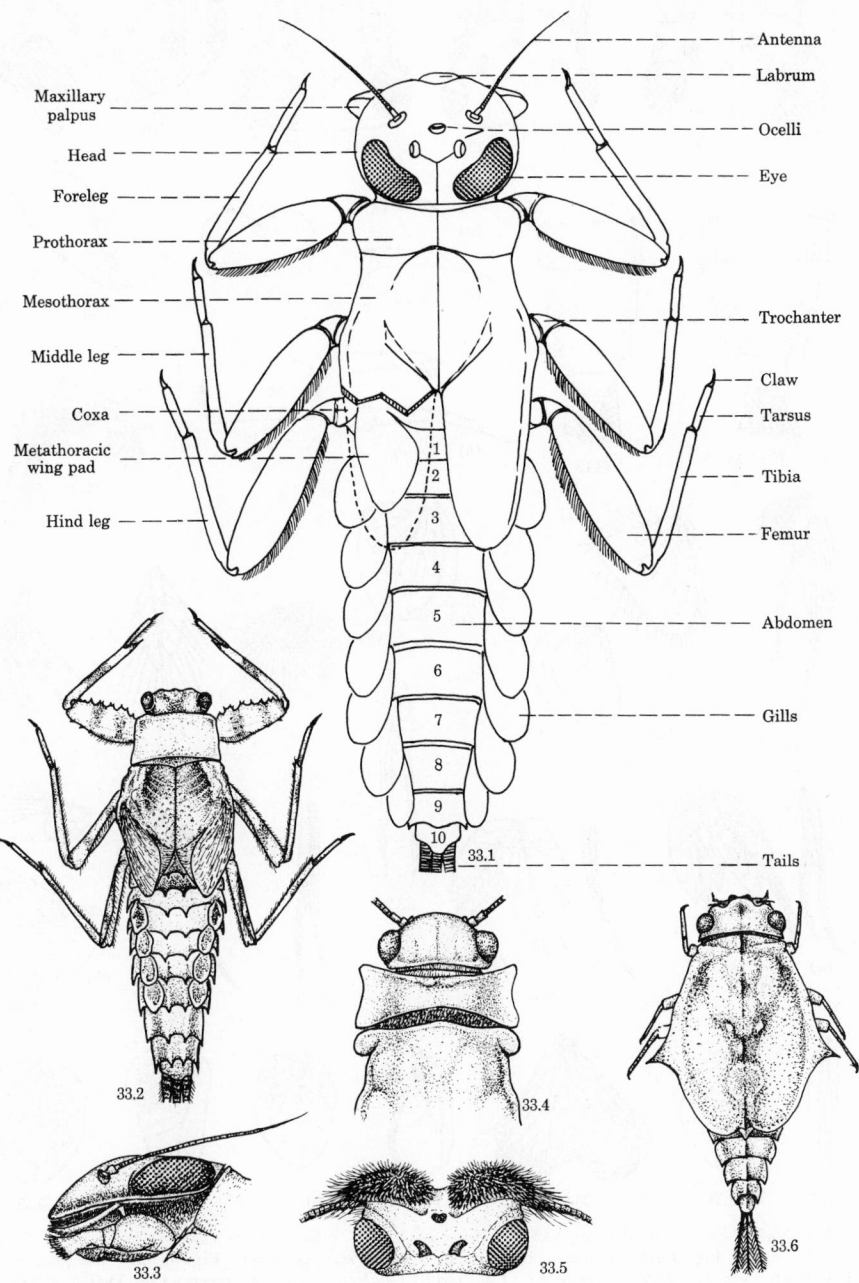

Fig. 33.1. *Cinygmula*, with principal parts labeled. **Fig. 33.2.** *Ephemerella*. **Fig. 33.3.** *Heptagenia*, head, lateral view. **Fig. 33.4.** *Neoephemera*, head and part of thorax. **Fig. 33.5.** *Dolania*, head. **Fig. 33.6.** *Baetisca*.

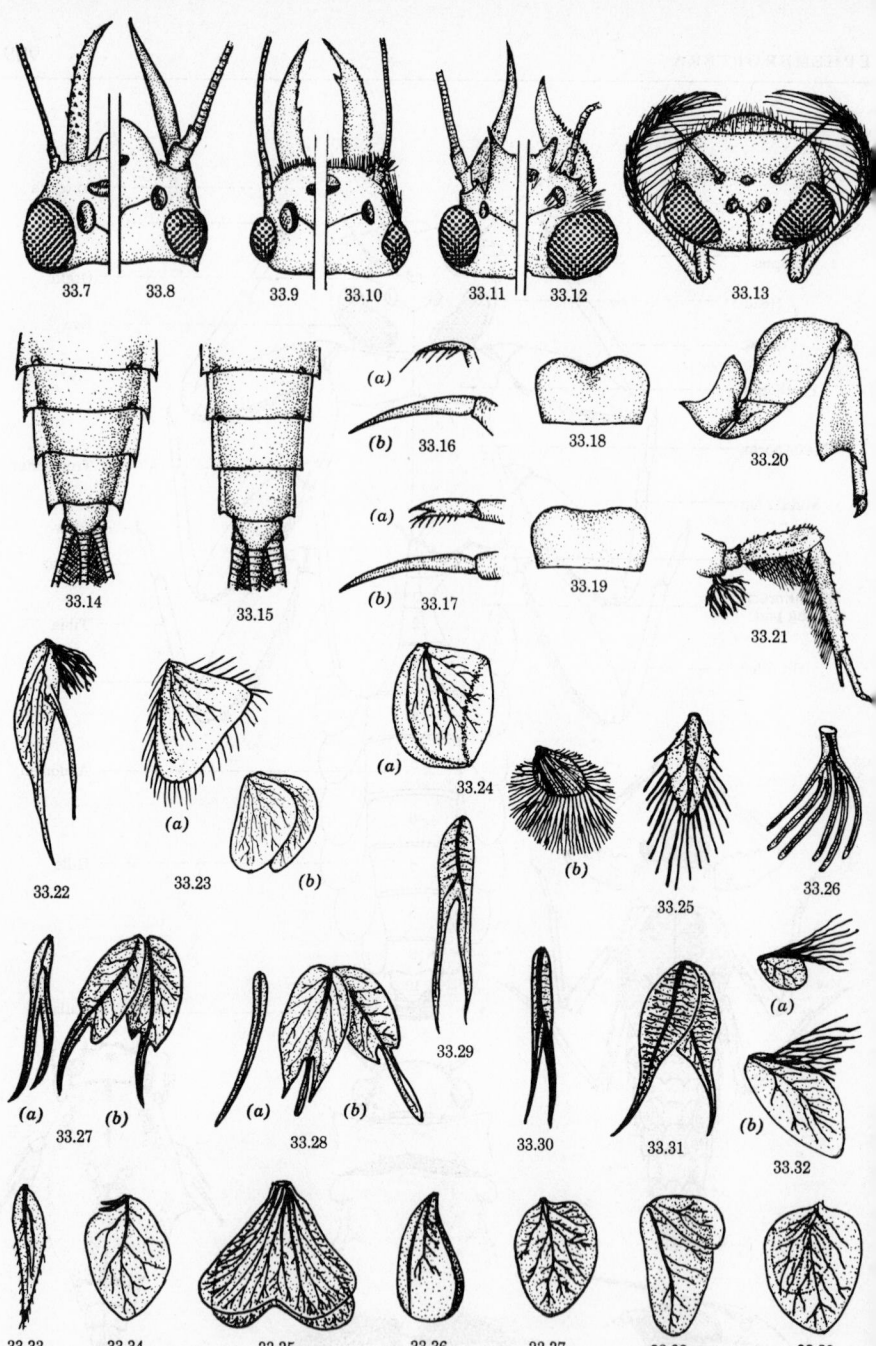

Fig. 33.7. *Ephoron*, head. **Fig. 33.8.** *Hexagenia*, head. **Fig. 33.9.** *Campsurus*, head. **Fig. 33.10.** *Tortopus*, head. **Fig. 33.11.** *Ephemera*, head. **Fig. 33.12.** *Pentagenia*, head. **Fig. 33.13.** *Arthroplea*, head. **Fig. 33.14.** *Siphlonurus*, apex of abdomen. **Fig. 33.15.** *Callibaetis*, apex of abdomen. **Fig. 33.16.** *Ametropus*. (a) Foreclaw. (b) Hind claw. **Fig. 33.17.** *Siphloplecton*. (a) Foreclaw. (b) Hind claw. **Fig. 33.18.** *Habrophlebiodes*, labrum. **Fig. 33.19.** *Paraleptophlebia*, labrum. **Fig. 33.20.** *Hexagenia*, foreleg. **Fig. 33.21.** *Isonychia*, foreleg. **Fig. 33.22.** *Pseudiron*, gill 3. **Fig. 33.23.** *Tricorythodes*. (a) Gill 2. (b) Gill 3. **Fig. 33.24.** *Caenis*. (a) Gill 2. (b) Gill 3. **Fig. 33.25.** *Traverella*, anterior lamella of gill 3. **Fig. 33.26.** *Habrophlebia*, gill 3. **Fig. 33.27.** *Leptophlebia*. (a) Gill 1. (b) Gill 2. **Fig. 33.28.** *Choroterpes*. (a) Gill 1. (b) Gill 2. **Fig. 33.29.** *Habrophlebiodes*, gill 3. **Fig. 33.30.** *Paraleptophlebia*, gill 3. **Fig. 33.31.** *Thraulodes*, gill 3. **Fig. 33.32.** *Cinygma*. (a) Gill 1. (b) Gill 2. **Fig. 33.33.** *Stenonema*, gill 7. **Fig. 33.34.** *Cinygmula*, gill 3. **Fig. 33.35.** *Siphlonurus*, gill 1. **Fig. 33.36.** *Ameletus*, gill 3. **Fig. 33.37.** *Parameletus*, gill 3. **Fig. 33.38.** *Centroptilum*, gill 1. **Fig. 33.39.** *Callibaetis*, gill 3.

wing pad can be seen only by lifting the mesothoracic wing. Mayfly nymphs are best preserved in the field in vials of 95 per cent alcohol immersed in a wide-mouth jar. The vials are plugged with cotton to exclude air bubbles. The diagnostic gills, legs, and tails are soon broken when transporting partly filled vials of liquid. Such larval preservatives as KAAD should be avoided as they cause the flat gills to become distorted by excessive swelling.

KEY TO GENERA

1a		Mandibles with large tusks projecting forward and usually visible from above the head (Figs. 33.7–33.12)	2
1b		Mandible without such tusks (Figs. 33.1–33.6, 33.13)	6
2a	(1)	Tusks depressed; gills tuning-fork-shaped (Fig. 33.30), without fringed margins. . . . . Family **Leptophlebiidae** (Fig. 33.19). . . . . ***Paraleptophlebia*** Lestage	
2b		Tusks variable (Figs. 33.7–33.12), not depressed; gills with fringed margins.	3
3a	(2)	Gills dorsal, curving up over the abdomen; foretibiae fossorial (Fig. 33.20), burrowing nymphs	4
3b		Gills lateral, outspread at sides of abdomen; foretibiae slender, sub-cylindrical, not fossorial; sprawling nymphs . . . . Family **Potamanthidae** ***Potamanthus*** Pictet	
4a	(3)	Head with a conspicuous frontal process between bases of antennae (Figs. 33.7, 33.8, 33.11, 33.12)	5
4b		Front of head rounded, no conspicuous frontal process, clypeal area may be expanded (Figs. 33.9, 33.10) . . . . Family **Polymitarcidae**, Subfamily **Campsurinae**	21
5a	(4)	Mandibular tusks curve downward apically as viewed laterally, upper surface with 20 or more tubercles . . . . Family **Polymitarcidae**, Subfamily **Polymitarcinae** (Fig. 33.7) . . . . ***Ephoron*** Williamson	
5b		Mandibular tusks curve upward apically as viewed laterally, upper surface with hairs or spines, but no tubercles (Figs. 33.8, 33.11, 33.12) . . . . Family **Ephemeridae**	22
6a	(1)	Mesonotum enlarged into a carapacelike structure which encloses the gills; abdominal tergites 6 to 10, only, exposed (Fig. 33.6) . . . . Family **Baetiscidae** (Fig. 33.6) . . . . ***Baetisca*** Walsh	
6b		Mesonotum not as above, gills exposed; all abdominal tergites exposed (Figs. 33.1, 33.2)	7
7a	(6)	Anterolateral angles of head with conspicuous crowns of setae (Fig. 33.5); gills ventral, with fringed margins . . . Family **Behningiidae** ***Dolania*** Edmunds and Traver	
7b		Anterolateral angles of head without such setae; gills either dorsal or without fringed margins.	8
8a	(7)	Forelegs with a dense row of long setae on the inner surface (Fig. 33.21); a tuft of gills at the base of each maxilla, gills may be present at base of each forecoxa.	9
8b		Forelegs with setae other than above; gill tufts wanting on maxillae and forecoxae	10
9a	(8)	Gills dorsal on first abdominal segment; gill tufts present at base of	

each forecoxa (Fig. 33.21) . . . . Family **Siphlonuridae**
Subfamily **Isonychiinae**
(Fig. 33.21) . . . . *Isonychia* Eaton

9b Gills ventral on first abdominal segment; no gill tufts at base of each forecoxa. . . . . Family **Oligoneuriidae** 24

10a (8) Gills on segment 2 operculate or semioperculate covering the succeeding pairs (Figs. 33.23a, 33.24a), those on segment 1 rudimentary or absent 11

10b Gills on segment 2 neither operculate nor semioperculate, either similar to those on following segments (Fig. 33.1) or wanting (Fig. 33.2) 13

11a (10) Gills on segment 2 triangular (Fig. 33.23a) or oval; gills on succeeding segments never with fringed margins (Fig. 33.23b) . . . . . .
Family **Tricorythidae**
(Fig. 33.23) . . . . *Tricorythodes* Ulmer

The genus *Leptohyphes,* now known to occur in Texas (Burks, 1953), will key here; the narrowly oval gill on segment 2, the scalelike structures along the dorsal edge of the femora and the lateral expansion of the abdominal tergites should distinguish *Leptohyphes.* The presence of several unidentified and aberrant relatives of this complex in N. A. makes it undesirable to place *Leptohyphes* in the key at present.

11b Gills on segment 2 quadrate (Fig. 33.24a); gills on succeeding segments with fringed margins (Fig. 33.24b). 12

12a (11) Mesonotum with a distinct rounded lobe on the anterolateral corner (Fig. 33.4); operculate gills fused at median line; metathoracic wing pads present. . . . . Family **Neoephemeridae**
(Fig. 33.4) . . . . *Neoephemera* McDunnough

12b Mesonotum without an anterolateral lobe; operculate gills not fused at median line; metathoracic wing pads wanting . . . . **Caenidae** 25

13a (10) Gills on segment 2 wanting, gill on segment 1 wanting or minute, gills on segment 3 present or absent, gills on segments 3 or 4 may be operculate, semioperculate or similar to those on 4 or 5 to 7; paired spines often present dorsally on head, thorax, and/or abdomen
Family **Ephemerellidae**
(Fig. 33.2) . . . . *Ephemerella* Walsh

13b Gills on segments 1 to 7 (Fig. 33.1); rarely on 1 to 5 only. 14

14a (13) Nymph distinctly depressed; head with eyes *and* antennae dorsal (Figs. 33.1, 33.3, 33.13) 15

14b Nymph not depressed; eyes and/or antennae lateral, anterolateral, or on front of head (Fig. 33.2) 17

15a (14) Gills forked (Figs. 33.29, 33.30), in clusters of filaments (Fig. 33.26), or bilamellate with margins fringed (Fig. 33.25), or terminating in a filament or point (Figs. 33.27b, 33.28b, 33.31); mandibles often visible from above; labial palpi 3-segmented . . . .
Family **Leptophlebiidae** 26

15b Gills of a single lamella usually with a fibrilliform tuft at or near the base (Figs. 33.22, 33.32a,b, 33.34); mandibles not visible dorsally (Figs. 33.1, 33.3); labial palpi 2-segmented . . . .
Family **Heptageniidae** 16

16a (15) Gills with a fingerlike projection extending from near the middle of the lamella (Fig. 33.22); claws greatly elongated; maxillary palpi 3-segmented. . . . . Subfamily **Pseudironinae**
(Fig. 33.22) . . . . *Pseudiron* McDunnough

16b Gill lamellae without such a fingerlike projection (Figs. 33.32a,b, 33.34); claws not noticeably elongated; maxillary palpi 2-segmented . . . . Subfamily **Heptageniinae** 33

17a (14) Claws of middle and hind legs long and slender, about as long as the short tibia; claws of foreleg differ from others in structure (Figs. 33.16a,b, 33.17a,b) 18

17b Claws on all legs similar, usually sharp pointed, rarely spatulate; if claws of middle and hind legs are long and slender, the claws of the foreleg are merely shorter, not different in structure 19

18a (17) Claws on forelegs slender, curved, bearing several long spines (Fig. 33.16a), a fleshy appendage attached to the inner margin of each forecoxa . . . . Subfamily **Ametropodinae** (Fig. 33.16) . . . . *Ametropus* Albarda

18b Claws on forelegs bifid (Fig. 17a); no such fleshy appendage attached to the inner margin of each forecoxa . . **Metretopodinae** 42

19a (17) Gills forked (Figs. 33.29, 33.30), in clusters of filaments (Fig. 33.26), or bilamellate with margins fringed (Fig. 33.25) or terminating in a filament or point (Figs. 33.27b, 33.28b, 33.31); head prognathous or hypognathous . . . . Family **Leptophlebiidae** 26

19b Gills of single or double lamellae, margins entire except in *Acanthametropus* where they are fringed on the inner side only, not as above; head hypognathous 20

20a (19) Posterolateral angles of segments 8 and 9, and usually also of the preceding pairs, produced into distinct flattened spines (Fig. 33.14); if spines are weakly produced the antennae are short, the length being clearly less than twice the width of the head. . . . . Family **Siphlonuridae** 43

20b Posterolateral angles of segments 8 and 9 usually without such spines (Fig. 33.15); if spines are weakly produced, the antennae are long, the length being clearly more than twice the width of the head. . . . . Family **Baetidae** 48

21a (4) Mandibular tusks with a single prominent subapical tooth on the inner margin (Fig. 33.10) . . . . . *Tortopus* Needham and Murphy

21b Mandibular tusks with several teeth on the inner margin (Fig. 33.9). . . . . *Campsurus* Eaton

22a (5) Frontal process of head bifid (Figs. 33.11, 33.12) 23

22b Frontal process of head entire, either truncate, rounded, or conical (Figs. 33.8, 33.20) . . . . *Hexagenia* Walsh

23a (22) Mandibular tusks crenate on outer or upper margin (Fig. 33.12); labial palpi 2-segmented . . . . *Pentagenia* Walsh

23b Mandibular tusks smooth on margins (Fig. 33.11); labial palpi 3-segmented. *Ephemera* Linnaeus

24a (9) Lamellate portion of gills on segments 2 to 7 oval, fibrilliform portion well developed, ventral first gill similar to others, coxae of hind legs much smaller than femora. . . . . *Lachlania* Hagen

24b Lamellate portion of gills on segments 2 to 7 lanceolate, fibrilliform portion absent, ventral first gill greatly enlarged; coxae of hind legs larger than femora . . . . *Homoeoneuria* Eaton

25a (12) Three prominent ocellar tubercles on head; maxillary and labial palpi 2-segmented . . . . *Brachycercus* Curtis

25b No ocellar tubercles present on head; maxillary and labial palpi 3-segmented. (Fig. 33.24). . . . . *Caenis* Stephens

26a (15, 19) Labrum fully as broad as head; gills with fringed margins (Fig. 33.25) . . . . *Traverella* Edmunds

26b   Labrum much narrower than head; gills not as above. **27**

27a   **(26)**   Each of the gills on segments 2 to 6 consist of two clusters of slender filaments (Fig. 33.26). . . . . *Habrophlebia* Eaton

27b   Each gill forked or bilamellate, not as above. **28**

28a   **(27)**   Gills on segment 1 differ in structure from those of the succeeding pairs (Figs. 33.27*a,b*, 33.28*a,b*) **29**

28b   Gills on segment 1 similar in structure to those of the succeeding pairs **30**

29a   **(28)**   Gills on segment 1 forked (Fig. 33.27*a*). . . *Leptophlebia* Westwood

29b   Gills on segment 1 a single linear lamella (Fig. 33.28*a*) . . . .
Choroterpes Eaton

30a   **(28)**   Gills on middle segments as in Fig. 33.29; well-developed postero-lateral spines on abdominal segments 8 and 9 . . . .   **31**

30b   Without the above combination of characters, gills either narrower, without lateral branches on trachea, or more deeply cleft; or posterolateral spines either poorly developed or absent on segment 8 . . . .   **32**

31a   **(30)**   Labrum rather deeply emarginate on fore margin (Fig. 33.18); spinules on posterior margins of segments 6 or 7 to 10. (Fig. 33.29) . . . . *Habrophlebiodes* Ulmer

31b   Labrum only shallowly emarginate on fore margin (Fig. 33.19); spinules on posterior margins of segments 1 to 10. (Fig. 33.30) . . .
Paraleptophlebia Lestage

32a   **(30)**   Gills tuning-fork-shaped (Fig. 33.30), or more deeply divided into two linear lamellae . . . . *Paraleptophlebia* Lestage

32b   Gills bilamellate, each lamella broadly lanceolate (Fig. 33.31) . . . .
Thraulodes Ulmer

33a   **(16)**   Nymph with 2 tails only . . . . *Epeorus* Eaton   **34**

33b   Nymph with 3 tails **36**

34a   **(33)**   A double row of submedian spines present on the abdominal tergites . . . . *Epeorus* Subgenus *Ironodes* Traver

34b   No spines present on the abdominal tergites **35**

35a   **(34)**   Anterior margin of head slightly emarginate; a dense row of long setae on the median line of the abdominal tergites; anterior pair of gills meet beneath body, gills thick and darkened, the anterior and outer margins darkest . . . *Epeorus* Subgenus *Ironopsis* Traver

35b   Anterior margin of head entire; median row of hairs poorly developed or absent; anterior pair of gills variable, gills usually thin and light, anterior and outer margin not noticeably darker . . . . . .
Epeorus Subgenus *Iron* Eaton

36a   **(33)**   Gills inserted ventrally; fibrilliform portion of gills large, longer than gill lamellae . . . . (Probably) *Anepeorus* McDunnough

36b   Gills dorsal or lateral; fibrilliform portion of gills smaller or absent, usually shorter than gill lamellae **37**

37a   **(36)**   Distal segment of maxillary palpi at least 4 times as long as the galea-lacinia (Fig. 33.13). . . . . *Arthroplea* Bengtsson

37b   Distal segment of maxillary palpi much shorter than above. **38**

38a   **(37)**   Gills of first and seventh pairs enlarged, meeting beneath body to form a ventral disc . . . . *Rhithrogena* Eaton

38b   Gills of first and seventh pairs not as above, usually smaller than intermediate pairs **39**

39a (38) Gills of seventh pair reduced to a single slender filament; trachea, if present in this pair of gills, with few or no lateral branches (Fig. 33.33). . . . . ***Stenonema*** Traver

39b Gills of seventh pair similar to preceding pairs, but smaller; trachea in this pair with lateral branches **40**

40a (39) Fibrilliform portion of gills wanting or reduced to a few tiny threads (Fig. 33.34); front of head distinctly emarginate, maxillary palpi at least partly visible at sides of head in dorsal view (Fig. 33.1) . . . . . ***Cinygmula*** McDunnough

40b Fibrilliform portion of gills present on segments 1 to 6, may be wanting on 7; front of head entire or feebly emarginate; maxillary palpi normally not visible in dorsal view **41**

41a (40) First pair of gill lamellae decidedly smaller than the second pair (Figs. 33.32*a,b*); fibrilliform portion of gill 1 longer than lamella; labrum narrow extending not more than ¼ the distance along the anterior margin of the head . . . . ***Cinygma*** Eaton

41b First pair of gill lamellae only slightly shorter than the second pair, often narrower; labrum broad, extending ⅔ to ¾ the distance along the anterior margin of the head. (Fig. 33.3). . . ***Heptagenia*** Walsh

42a (18) Apical segment of labial palpi expanded and truncate apically; tarsi subequal to tibiae. (Fig. 33.17) . . . . ***Siphloplecton*** Clemens

42b Apical segment of labial palpi rounded apically; tarsi longer than tibiae . . . . ***Metretopus*** Eaton

43a (20) Lateral margins of head, pronotum, and mesonotum with conspicuous spines; a row of median spines on the abdominal tergites . . . . . ***Acanthametropus*** Tshernova

43b Head, pronotum, and abdomen without such spines. **44**

44a (43) Gill lamellae double on segments 1 and 2 (Fig. 33.35), sometimes on 3 to 7 also **45**

44b Gill lamellae single on all segments (Figs. 33.36, 33.37). **46**

45a (44) Gills oval, the posterior lamellae on segments 1 and 2 about ⅔ as large as the anterior lamellae, meso- and metathoracic claws almost twice as long as the prothoracic claws . . . . . . ***Edmundsius*** Day

45b Gills with a retuse apical margin, the posterior lamellae on segments 1 and 2 as large as the anterior lamellae (Fig. 33.35); meso- and metathoracic claws not noticeably longer than the prothoracic claws. (Fig. 33.14) . . . . ***Siphlonurus*** Eaton

46a (44) Gills obovate with a sclerotized band along the ventral margin and usually a similar band near or on the dorsal margin (Fig. 33.36); maxillae with a crown of pectinate spines, posterolateral abdominal spines may be feebly developed. . . . . ***Ameletus*** Eaton

46b Gills cordate or subcordate (Fig. 33.37); maxillae not as above; posterolateral abdominal spines well developed **47**

47a (46) Abdominal segments 5 to 9 greatly expanded laterally; meso- and metathorax with midventral spines; labial palpi not as below. . . . . . ***Siphlonisca*** Needham

47b Abdominal segments not greatly expanded laterally; no spines on venter of thorax; last two segments of labial palpi formed into a pincerlike process. (Fig. 33.37) . . . . . . ***Parameletus*** Bengtsson

48a (20) Each gill consists of a simple, flat lamella without additional ventral or dorsal flap, lamellae never double **49**

48b   Each gill consists of double lamellae or of a single lamella with recurved ventral or dorsal flap (Figs. 33.38, 33.39) **56**

49a **(48)** With 2 well-developed tails only, the median tail wanting or represented by a rudiment no longer than the tenth tergite **50**

49b   With 3 well-developed tails, although the median may be shorter and thinner than the laterals, it is much longer than the tenth tergite **52**

50a **(49)** Gills on segments 1 to 5 only; the tails bare or with only a few inconspicuous setae . . . . *Baetodes* Needham and Murphy

50b   Gills on segments 1 to 7; the tails with a noticeable fringe of setae on the inner margin **51**

51a **(50)** Metathoracic wing pads present, though they may be minute. . . . . *Baetis* Leach

51b   Metathoracic wing pads wanting. . . . . . . *Pseudocloeon* Klapalek

52a **(49)** Median tail shorter and thinner than lateral ones; claws with denticles; distal segment of labial palpi rounded apically . . . . . . *Baetis* Leach

52b   Median tail subequal to lateral ones in length and thickness; claws with or without denticles; distal segment of labial palpi variable . . .   **53**

53a **(52)** Tracheae in gills with well-developed lateral branches on the inner side. **54**

53b   Tracheae in gill with only a few poorly developed lateral branches. .   **55**

54a **(53)** Metathoracic wing pads present . . . . *Centroptilum* Eaton

54b   Metathoracic wing pads wanting . . . . *Neocloeon* Traver

55a **(53)** Claws a little more than ½ the length of tarsi, with denticles; maxillary palpi 2-segmented; known only from Calif. and Puerto Rico. . . *Paracloeodes* Day

55b   Claws subequal to tarsi, without denticles; maxillary palpi 3-segmented; known only from Calif. . . . . *Apobaetis* Day

56a **(48)** Gills as in Fig. 33.38; tracheal branches usually on the inner side only; a small dorsal flap at the base. . . . . . . *Centroptilum* Eaton

56b   Gills variable; tracheal branches pinnate; palmate; or primarily on the outer side. **57**

57a **(56)** Metathoracic wing pad present; the smaller lamella or flap on the ventral surface of the gill (Fig. 33.39) . . . . *Callibaetis* Eaton

57b   Metathoracic wing pad wanting; the smaller lamella on the dorsal surface of the gill . . . . *Cloeon* Leach

References

Berner, L. 1950. The mayflies of Florida. *Univ. Florida Studies, Biol. Sci. Series*, 4(4):1–267. **Burks, B. D. 1953.** The mayflies, or Ephemeroptera, of Illinois. *Bull. Illinois Nat. Hist. Surv.*, 26 (Art. 1):1–216. **Day, W. C. 1956.** "Ephemeroptera" in Aquatic Insects of California, Univ. Calif. Press, Berkeley and Los Angeles, pp. 79–105. **Edmunds, G. F., Jr. 1954.** Eatonia (Mimeo., University of Utah, current bibliography of Ephemeroptera), Nos. 1–3, continuing. **Edmunds, G. F., Jr. and R. K. Allen. 1957.** A checklist of the Ephemeroptera of North America North of Mexico. *Ann. Entomol. Soc. Am.* 50:317–324. **Edmunds, G. F., Jr. and J. R. Traver. 1954.** An outline of a reclassification of the Ephemeroptera. *Proc. Entomol. Soc. Wash.*, 56:236–240. **Needham, J. G., J. R. Traver, and Yin-chi Hsu. 1935.** The Biology of Mayflies. Comstock, Ithaca, New York.

34

Odonata

LEONORA K. GLOYD
MIKE WRIGHT

The order Odonata, exclusive of fossil forms, is represented in North America by two suborders, the Anisoptera and the Zygoptera. Their larvae, or nymphs, are distinguished from those of all other insects by the form of the labium, or underlip, which has developed into a protractile organ for grasping living prey. In fact, this organ is unique in the whole animal kingdom. It is composed of two major parts, the prementum and the postmentum, commonly called mentum and submentum respectively, which are hinged together by an elbowlike joint and are also hinged at the base of the postmentum to the underside of the head (Figs. 34.4 and 34.52). At the distal end of the prementum there are two movable palpal lobes each of which has a movable hook and other devices for holding and cleaning prey. The labium is held in a closed position but is ever ready to be thrust out at lightening speed by a forward swing of the postmentum.

The nymph in both suborders is further characterized by having a chewing type of mouthparts; compound eyes; two pairs of wing pads on the dorsal surface of the thorax (fourth instar and later ones) which are directed either caudad or obliquely laterad; three pairs of thoracic legs with two- or three-segmented tarsi; a ten-segmented abdomen; internal caudal gills (supplemented by external gills in the Zygoptera); and five anal appendages consisting of a

pair of dorsolateral cerci, an epiproct (above the anus), and a pair of paraprocts (one at each side and extending below the anus).

The anal appendages differ greatly in appearance in the two suborders. In
many zygopterous nymphs the cerci, not evident in the earlier instars, are
small and easily overlooked when hidden by the projecting edge of the tenth
abdominal segment or by the lateral caudal gills. In North American species
the epiproct and the paraprocts have large, conspicuous outgrowths known as
median and lateral caudal gills or lamellae (Figs. 34.3 and 34.5), each of which
is longer than the combined lengths of the last five abdominal segments. In
the anisopterous nymphs the internal gills in the respiratory chamber of the
intestine (colon) are well developed and there are no external gills. The
respiratory chamber is also used as a means of locomotion by jet propulsion
when the nymph causes the water to be suddenly and forcefully ejected from
it. The anal appendages are usually shorter than the combined lengths of the
last three abdominal segments, and when held closely together form the anal
pyramid. The cerci, epiproct, and paraprocts have commonly been termed,
respectively, the lateral, superior, and inferior appendages. Since the relationship of these parts is different in the adults of the Zygoptera and the
Anisoptera, and the terms superior and inferior appendages do not apply to
the corresponding structures in both nymph and adult (Figs. 34.1 and 34.2),
preference is here given, for the nymphs, to terms commonly used for comparable parts in insects of other orders. Our use of these terms is also in accordance with the recent work of Snodgrass (1954, pp. 33–35).

RELATIONSHIP OF ANAL APPENDAGES IN ZYGOPTERA AND ANISOPTERA

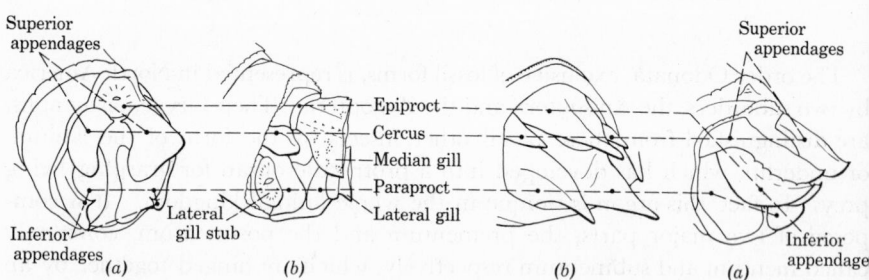

Fig. 34.1. Tenth abdominal segment and anal
appendages of *Ischnura verticalis* (Say).
(a) Adult. (b) Nymph.

Fig. 34.2. Same, *Libellula luctuosa* (Burmeister)
(a) Adult. (b) Nymph.

The nymphal cerci give rise to the superior appendages of the adult, which
in the male are usually enlarged and specialized, and during the last instar
often outgrow the confines of these structures and become recessed in the tenth
segment. The epiproct gives rise to the inferior appendage of the adult male
in the Anisoptera, but is usually represented only by a rather small plate or
process in the female and in both sexes of the Zygoptera. The paraprocts
retain their relative positions in the adult but are considerably changed in
shape. In most Anisoptera and in the female of the Zygoptera they apparently

function as anal plates only, but in the male of the Zygoptera they are modified to form the inferior appendages.

The sex of a nymph in the last instar, and sometimes in earlier instars, can be determined by the external modifications found on the ventral side of certain abdominal segments and on the dorsal side of the epiproct. In the male, the developing accessory genitalia are usually visible through the cuticle and their position may be marked externally by a mid-ventral patchlike area on one or both of segments 2 and 3; the small genital pore on the mid-ventral line of segment 9 is discernible in most species, and the genital valves in the Zygoptera appear as two long, well-separated, sharp-pointed processes (one on either side of the pore) that are free at the posterior end (Fig. 34.17), but in most Anisoptera these valves are indicated only by two leaf-shaped outlines or slightly raised areas; the developing inferior anal appendage of the adult may be seen in outline through the cuticle on the dorsal side of the epiproct, or it may be quite obvious as a hump or thickening, especially in the Aeshnidae and Petaluridae (Figs. 34.70, 34.71, 34.73, 34.74). In the females of the Zygoptera, Aeshnidae, and Petaluridae, the developing ovipositor and its adjacent valves on the ventral side of segment 9 appear as four posteriorly directed processes (Figs. 34.18, 34.19, 34.72), the free ends of which may extend beyond the apex of segment 10. In the Cordulegastridae only two long processes on segment 9 are visible in ventral view, as they cover the two mesal processes (ovipositor). In the Gomphidae the vulvar laminae are represented by two flat, triangular structures which seem to arise from the suture between segments 8 and 9 and lie against the body. In the Macromiidae and Libellulidae the developing female structures can sometimes be discerned under the cuticle on segments 8 and 9 if they are large in the adult, otherwise the only apparent modifications consist of a pair of tiny papillae or triangular knobs at the base of segment 9, framed by the suture and an arc-like demarcation (Corduliinae), or there may be only a small mid-ventral notch in the integument beneath the cuticle at the apex of segment 8. In these two families there is also a pair of spots, on on either side of the mid-ventral line, not found in other anisopterous females. They are similar in size and appearance to the spot for the genital pore in the males and correspond in position to the two papillae found in the same region of the adult female. There is usually no obvious modification of the epiproct in the female nymph, but in a few genera such as *Cordulegaster*, in which the adult has a well-developed epiproct, there is a swelling very similar to that in the male nymph but it is not as broad at the tip. In the figures, some abdominal segments are identified by Roman numerals.

The common name for all insects of the order Odonata is dragonflies. The introduction and use of the name "damselflies" for the Zygoptera and "true dragonflies" for the Anisoptera has proven misleading and confusing even among odonatologists because the "true" was dropped from "dragonflies" and the same name used for both the Anisoptera and the entire order. Furthermore, the French word *demoiselle*, from which "damselflies" may have originated, also applies to all Odonata. Such names as "devil's darningneedle," "mosquito-hawk," "snake doctor," "snake feeder," and "horse stinger" have been used for various species and groups of species but none of these i used consistently or exclusively for either suborder.

Because the proportions and relative lengths of various parts of the nymph as well as the number of setae, vary in different instars of a given species o genus, and because these differences are known only for a few species, the fol lowing key is, of necessity, based on nymphs of the last few instars. It also employs characteristics which will most readily distinguish the suborders and families as represented in North America, for example the Zygoptera and th Macromiidae, and not necessarily characteristics that would be useful on a world-wide basis.

Acknowledgments

In writing the section on Odonata begun by the late Dr. Mike Wright, I have had the generous assistance of everyone from whom it was sought. Dr. E. J Kormondy forwarded all of Dr. Wright's notes, partially written keys, and penciled drawings. Dr. Alvah Peterson kindly sent me all the original draw-ings he had made for "A key to the genera of anisopterous dragonfly nymphs of the United States and Canada (Odonata, suborder Anisoptera)" in joint authorship with Dr. Wright, and gave me permission to use them in any way desired; these drawings are acknowledged in the legends by the phrase "From Wright and Peterson." Dr. H. L. Dietrich, Dr. Alice Ferguson Beatty, the late Dr. J. G. Needham, and Dr. M. J. Westfall, Jr., either gave or loaned specimens of little known nymphs not represented in the Illinois Natura History Survey collection. Dr. E. M. Walker, whom I have consulted fre-quently, and members of the Illinois Natural History Survey staff have given me much good advice and encouragement, both of which are greatly appreciated.

KEY TO GENERA

1a		External gills present in the form of 3 flat and vertical or triquetral caudal lamellae, the longest of which is more than $\frac{1}{3}$ the length of the abdomen (Figs. 34.3, 34.5); abdomen cylindrical, not wider posteriorly than at base. . . . . Suborder **Zygoptera**
1b		External gills absent; the longest caudal appendage less than $\frac{1}{3}$ the length of the abdomen; abdomen more or less flattened dorsoventrally and widened posteriorly from base to mid-length or beyond (Figs. 34.23–34.27) . . . . Suborder **Anisoptera** **4**
2a	(1)	First antennal segment as long as, or longer than, the remaining 6 segments combined (Fig. 34.7); prementum with the distal edge of the ligula deeply and openly cleft medianly (Fig. 34.6) . . . . (p. 923) Family **Calopterygidae**
2b		First antennal segment much shorter than the others combined (Figs. 34.3–34.5, 34.14, 34.15); prementum with the distal edge of the ligula entire or with a closed short median cleft (Figs. 34.8, 34.10–34.13). **3**

3a   **(2)**   Prementum spoon-shaped, greatly narrowed in proximal half, in repose its base reaching to the level of the mesothoracic coxae or beyond (Fig. 34.4); ligula with a closed median cleft (Figs. 34.10, 34.11); 2 or 3 raptorial setae on the movable hook of each palpal lobe; each caudal gill with numerous small tracheae, branched near the margins but otherwise subparallel and almost perpendicular to the main axis (Fig. 34.16) . . . . (p. 924) Family **Lestidae**

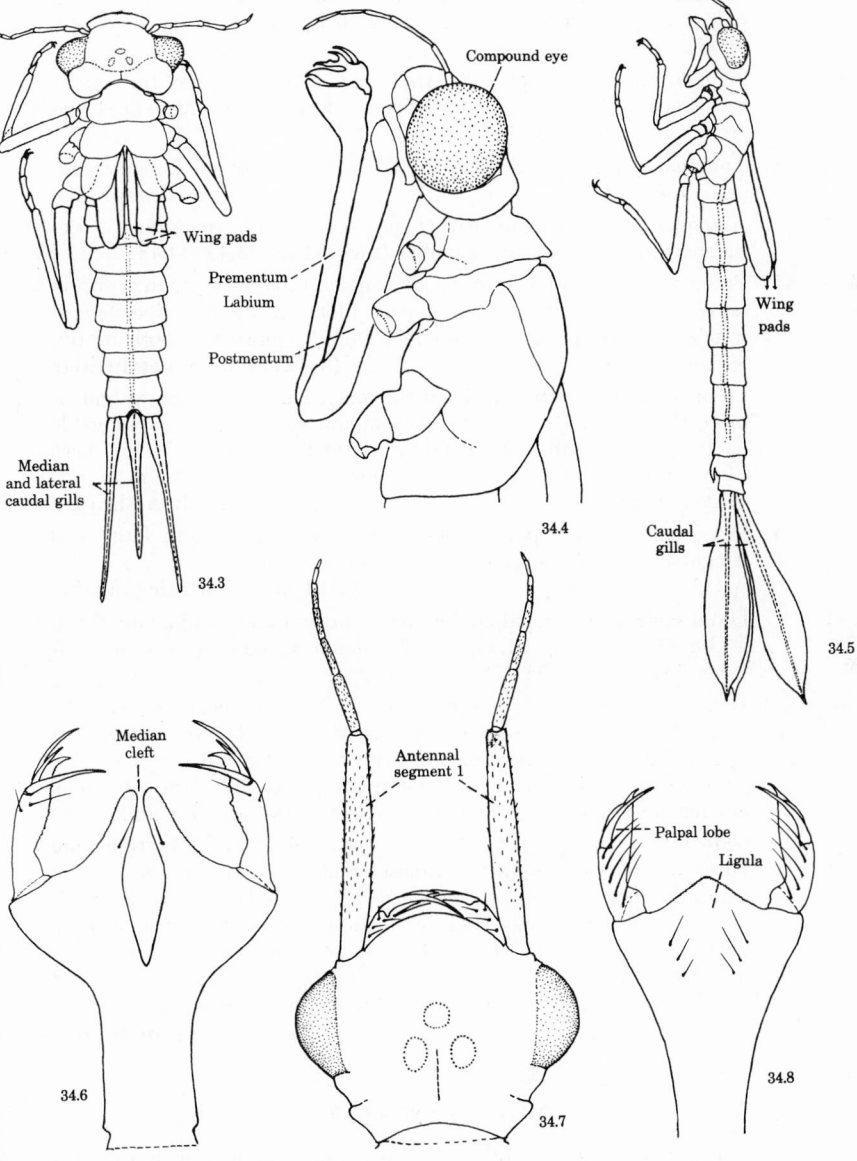

ZYGOPTERA: **Fig. 34.3.** *Argia*, dorsal view. **Fig. 34.4.** *Lestes*, lateral view of head and thorax. **Fig. 34.5.** *Enallagma*, lateral view. **Fig. 34.6.** *Calopteryx*, prementum of labium, dorsal view. **Fig. 34.7.** *Calopteryx*, head, dorsal view. **Fig. 34.8.** *Enallagma*, prementum, dorsal view. (All figures from Wright and Peterson.)

3b   Prementum flat, not greatly narrowed in proximal half, its base reaching to or slightly beyond the level of the prothoracic coxae (Fig. 34.5); ligula without a median cleft (Figs. 34.8, 34.12, 34.13); no raptorial setae on the movable hook of palpal lobe; each caudal gill with well-branched small tracheae, diverging from the main axis at much less than right angles (Figs. 34.18, 34.19) . . . .
(p. 924) Family Agrionidae (Coenagrionidae)

4a (1) Prementum flat, or nearly so, dorsal surface without long stout palpal or premental setae (raptorial bristles) (Figs. 34.24, 34.41, 34.45). **5**

4b   Prementum spoon-shaped, covering the face to the base of the antennae (Figs. 34.27, 34.52, 34.55), dorsal surface with long palpal and premental setae (Figs. 34.44, 34.46, 34.47). **7**

5a (4) Antennae 4-segmented (Figs. 34.28–34.32); mesotarsi 2-segmented; ligula not cleft (Fig. 34.41) . . . . (p. 926) Family **Gomphidae**

5b   Antennae 6- or 7-segmented (Figs. 34.33–34.35); mesotarsi 3-segmented; ligula with a median cleft (Figs. 34.42, 34.43, 34.45) **6**

6a (5) Prementum with sides subparallel in apical three-fifths, then abruptly narrowed near basal hinge; each palpal lobe with a stout dorsolateral spur at base of movable hook (Fig. 34.45); epiproct smooth at tip, without 2 sharp points (Fig. 34.70) . . (p. 929) Family **Petaluridae**

6b   Prementum widest in apical fourth, much narrower in basal half or more (Figs. 34.42, 34.43); no dorsolateral spur at base of movable hook; epiproct with 2 sharp points at tip (Figs. 34.72, 34.73), very small or worn off in *Nasiaeschna* (Fig. 34.74) and absent in one species of *Boyeria* . . . . (p. 931) Family **Aeshnidae**

7a (4) Distal edge of each palpal lobe deeply incised, its teeth large and irregular (Fig. 34.38); ligula with a median cleft . . . .
(p. 934) Family Cordulegastridae

7b   Distal edge of each palpal lobe smooth or evenly crenulate (Figs. 34.39, 34.40, 34.44, 34.46, 34.47); ligula without a median cleft (Figs. 34.44, 34.46, 34.47). **8**

8a (7) Head with a prominent, almost erect, thick frontal horn between the bases of the antennae, its width at base distinctly less than its length (Figs. 34.51, 34.52); metasternum with a broad mesal tubercle (best seen in lateral view) near posterior margin; legs very long, the apex of each hind femur reaching to or beyond the apex of abdominal segment 8 . . . . (p. 934) Family **Macromiidae**

8b   Head without a prominent, almost erect, thick frontal horn—the triangular frontal shelf of *Neurocordulia molesta* (Walsh) (Figs. 34.54, 34.55) is ·flattened, only slightly upcurved, with its width at base greater than its length (Figs. 34.53–34.62); metasternum without a tubercle near posterior margin; legs shorter, the apex of each hind femur usually not reaching to the apex of abdominal segment 8 . . .
(p. 934) Family Libellulidae

Suborder Zygoptera

    The problem regarding the names to be used for two of the families of the Zygoptera has been presented to the International Commission on Nomenclature. Because the evidence is strongly in favor of the names Calopterygidae for one and Agrionidae (replaced by Coenagrionidae or Coenagriidae by some taxonomists) for the other, these names are used here.

Family Calopterygidae

Nymphs of the family Calopterygidae are found only in clear streams or spring-fed woodland pools. These stiff, long-legged, slender, sticklike creatures cling to submerged roots and vegetation.

1a   Ligula cleft medianly almost halfway to base of prementum (Fig. 34.6); body color usually dark . . . . **Calopteryx** Leach

1b   Ligula cleft medianly only to the level of the articulation of the palpal lobes (Fig. 34.9); body color usually light . . . . . **Hetaerina** Hagen

Distribution: Species of the genus *Calopteryx* occur mostly in the eastern half of the U. S. and southern Canada, but their range has a westward spur that includes Minn., Colo., Saskatchewan, and Wash. to Calif. *Hetaerina* is represented in southeastern Canada, eastern and southern U. S., and Mexico.

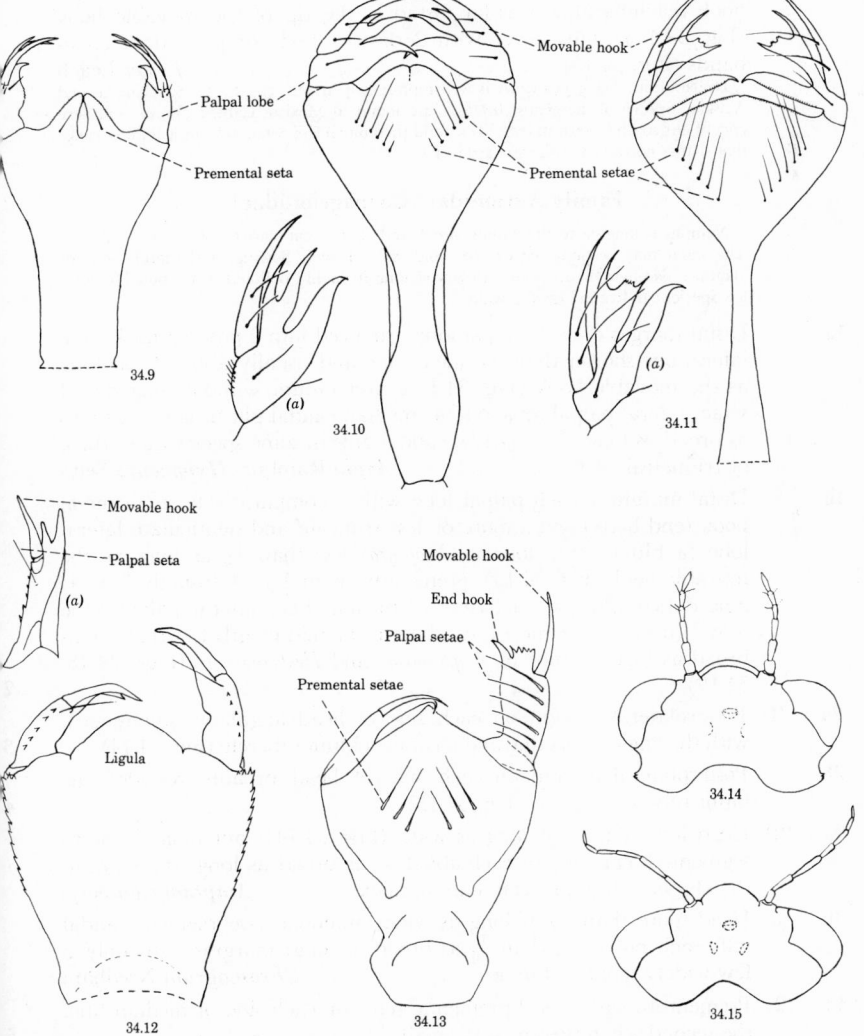

ZYGOPTERA. DORSAL VIEW OF PREMENTUM AND (a) OF PALPAL LOBE: Fig. 34.9. *Hetaerina.* **Fig. 34.10.** *Archilestes.* **Fig. 34.11.** *Lestes.* **Fig. 34.12.** *Argia.* **Fig. 34.13.** *Agrion.* (After Kennedy.) **DORSAL VIEW OF HEAD: Fig. 34.14.** *Amphiagrion.* **Fig. 34.15.** *Ischnura.*

Family Lestidae

The nymphs of the genus *Lestes* are usually plentiful in ponds, small lakes, or streams that have marshy or boggy margins. *Archilestes* nymphs occur primarily in streams having pockets of quiet waters and supporting growths of tall aquatic plants near their margins or on islands in mid-stream.

1a   Distal margin of each palpal lobe divided into 3 sharply pointed processes, the middle one about as long as the inner one, and the outer one distinctly shorter than the movable hook (Fig. 34.10*a*), each caudal gill with 2 well-defined dark crossbands . . . .
Archilestes Selys

1b   Distal margin of each palpal lobe divided into 4 processes — 3 sharply pointed hooks and a short, serrated, truncate protuberance between the smaller outer 2 hooks, with the tip of the outermost hook reaching almost as far distad as the tip of the movable hook (Fig. 34.11*a*); gills never with 2 distinct and complete dark crossbands (Fig. 34.16) . . . . *Lestes* Leach

Distribution: The genus *Lestes* is well represented in Mexico, the U. S., Canada, and Alaska. Species of the genus *Archilestes* are found in Mexico, Calif. to Wash., southern part of our southwestern states to Tex., and then much less commonly in a limited range that swings northward and eastward to Pa.

Family Agrionidae (Coenagrionidae)

Nymphs belonging to this family are found in a great variety of aquatic habitats. The water may be quite hot or very cold, still or swift flowing, and slightly acid or alkaline. No nymphs are known to survive long in highly polluted water but those of a few species can live in brackish water.

1a   Distal margin of each palpal lobe produced into 2 pointed hooks, the lateral one shorter than the mesal one and usually about $\frac{1}{2}$ as long as the movable hook (Fig. 34.12); prementum without long dorsal setae; lateral palpal setae 0 to 4; median caudal gill usually $\frac{1}{3}$ to $\frac{1}{2}$ as broad as long (Fig. 34.17); caudal gills in some species quite thick or triquetral . . . . *Argia* Rambur, *Hyponeura* Selys

1b   Distal margin of each palpal lobe with a comparatively small mesal hook (end hook) and a more or less truncate and denticulate lateral lobe (a blunt acute lobe in *Neoneura*) less than $\frac{1}{3}$ as long as the movable hook (Fig. 34.13); prementum with 1 to 4 (usually 3 or 4) long dorsal setae on each side of median line; lateral palpal setae 3 to 7 (usually 5 or more); caudal gills at mid-length less than $\frac{1}{3}$ as broad as long (except in *Amphiagrion* and *Hesperagrion*) (Figs. 34.18, 34.19). **2**

2a   (1) Posterolateral margin on each side of head angulate (square-cut), with the angle projecting and forming a blunt tubercle (Fig. 34.14) . .   **3**

2b   Posterolateral margin on each side of head broadly rounded, no blunt tubercle present (Fig. 34.15) **4**

3a   (2) Head less than $\frac{1}{2}$ as long as wide (Fig. 34.14); antennae 5- or 6-segmented; caudal gills each about $\frac{1}{3}$ as broad as long, the margins thickly set with setae from base to apex . . . . . *Amphiagrion* Selys

3b   Head more than $\frac{1}{2}$ as long as wide; antennae 7-segmented; caudal gills each not more than $\frac{1}{6}$ as broad as long; margins with only a few widely separated setae . . . . *Chromagrion* Needham

4a   (2) Prementum with 1 or 2 premental setae on each side of median line, the second when present very small. **5**

4b   Prementum with 3 to 7 dorsal setae on each side (Fig. 34.13). **6**

5a   (4) Caudal gills each obliquely divided at $\frac{2}{3}$ the length into a thick laterally carinate basal part and a thin leaflike apical portion; each palpal lobe with 3 setae; short lobe on distal margin of palpal lobe, between the end hook and the movable hook, bluntly pointed (Needham, 1939, p. 243.) (Fig. 34.4) . . . . **Neoneura** Selys

5b   Caudal gills not obliquely divided into two parts differing markedly in appearance; each palpal lobe with 6 lateral setae; short lobe on distal margin of palpal lobe, between end hook and the movable hook, truncate and bearing 3 or less small denticles . . . .
Nehalennia Selys

6a (4) Caudal gills without dark pigment except on the axes and tracheae .   **7**
6b   Caudal gills each with a dark pigmented pattern of spots or crossbands. **11**

7a (6) Tip of each caudal gill broadly rounded, without a terminal point; the terminal angle of the apical sixth of gill more than 90° (Fig. 34.20); median gill slightly more than $\frac{1}{3}$ as broad as long . . . .
Hesperagrion Calvert

7b   Tip of each caudal gill tapered to a point and either bluntly or acutely angled; terminal angle of apical sixth of gill less than 90° (Figs. 34.18, 34.19, 34.21, 34.22) **8**

ZYGOPTERA. EXTERNAL CAUDAL GILLS: **Fig. 34.16.** *Lestes.* **Fig. 34.17.** *Argia.* **Fig. 34.18.** *Agrion (Coenagrion).* (After Kennedy.) **Fig. 34.19.** *Ischnura,* apical sixth of lateral gill marked. METHOD OF MEASURING THE TERMINAL ANGLE OF APICAL SIXTH OF A CAUDAL GILL: **Fig. 34.20.** *Hesperagrion.* (After Needham.) **Fig. 34.21.** (*a*), (*b*), *Enallagma* spp. **Fig. 34.22.** *Ischnura.*

8a (7) Apical sixth of lateral caudal gill with a terminal angle of 30° or less (Figs. 34.19, 34.22) . *Ischnura* Charpentier, *Anomalagrion* Selys

8b Apical sixth of lateral caudal gill with a terminal angle of 60° or more (Fig. 34.21*a*)........................ 9

9a (8) Second segment of each antenna twice as long as the first and ½ as long as the third; prementum long and slender; abdominal segment 10 ½ as long as segment 8.......... *Teleallagma* Kennedy

9b Relative lengths of first 3 antennal segments otherwise, or if same, the prementum is almost as wide as long............... 10

10a (9) Antennae 6-segmented (Fig. 34.5)...... *Enallagma* Charpentier

10b Antennae 7-segmented....... *(Coenagrion) Agrion* Fabricius

11a (6) Each caudal gill with 6 transverse dark pigmented bands, the basal 4 joined along the tracheal axis.................... 12

11b Each caudal gill with fewer bands or marked otherwise........ 13

12a (11) Antennae 7-segmented .... *Zoniagrion* Kennedy

12b Antennae 6-segmented .... *Enallagma* Charpentier

13a (11) Caudal gills with long tapering points, the apical sixth of each gill with a terminal angle of 45° or less (Fig. 34.21*b*) 14

13b Caudal gills with short tapering points, the apical sixth of each gill with a terminal angle of nearly 60° or more (Fig. 34.21*a*)....... 15

14a (13) Antennae 6-segmented .... *Enallagma* Charpentier

14b Antennae 7-segmented (Fig. 34.15) ...... *Ischnura* Charpentier

15a (13) Palpal setae 6 to 7; 2 indistinct arcuate transverse bands at about mid-length of median caudal gill; length of body about 11 mm, caudal gills each about 3 mm .... *Telebasis* Selys

15b Palpal setae 4 to 5; more than 2 transverse bands, or pigment arranged in a different pattern; length of body more than 13 mm, caudal gills each 8 mm or longer .... *Enallagma* Charpentier

> Distribution: Species belonging to the genera *Argia, Enallagma,* and *Ischnura* are widely distributed throughout the U. S., Mexico, and southern Canada, with *Enallagma* represented as far north as Alaska. Species of *Agrion (Coenagrion)* are found in Alaska, Canada, and in the northern part and higher elevations of the U. S. *Amphiagrion* has species that are distributed over the northern three-fourths of the U. S. and across southern Canada. *Chromagrion* is known from one species limited to the eastern half of the U. S. and southeastern Canada; species of *Nehalennia* occur in the same region but in southern Canada one species is found from Nova Scotia to British Columbia. *Anomalagrion* is represented by a single species that occurs in the eastern half of the U. S., Tex., southern Ariz., and Mexico. The one species of *Teleallagma* is found in scattered localities in the eastern U. S. only. *Telebasis* is represented by species occurring from Fla. to Calif., as far north as Kan., and south throughout Mexico. Species of *Hesperagrion* and *Hyponeura* are found in the southwestern U. S. from Tex. to Calif. and in Mexico. Species of *Neoneura* are found in Mexico and one has been reported from southern Tex. *Zoniagrion* is known from a single species found in Calif. only.

Suborder Anisoptera

Family Gomphidae

> The nymphs of this family are a part of the bottom fauna of streams, large and small lakes, spring-fed pools, and more or less permanent ponds. They burrow just below the bottom sediment, sand, or mud, and when the water is clear can sometimes be located by the trails left behind when they move from place to place.

1a Abdominal segment 10 more than twice as long as segments 8 + 9 (Fig. 34.69) .... *Aphylla* Selys

1b Abdominal segment 10 shorter than segments 8 + 9 (Figs. 34.23, 34.25, 34.65, 34.67) 2

2a **(1)** Mesothoracic coxae closer together at bases than prothoracic coxae
(Fig. 34.63); fourth antennal segment elongate, as long as the first
(Fig. 34.32) . . . . ***Progomphus*** Selys

2b Mesothoracic coxae and prothoracic coxae approximately the same
distance apart at their bases (Fig. 34.64); fourth antennal segment
never as above, usually a small rounded knob (Figs. 34.28–34.31) . . **3**

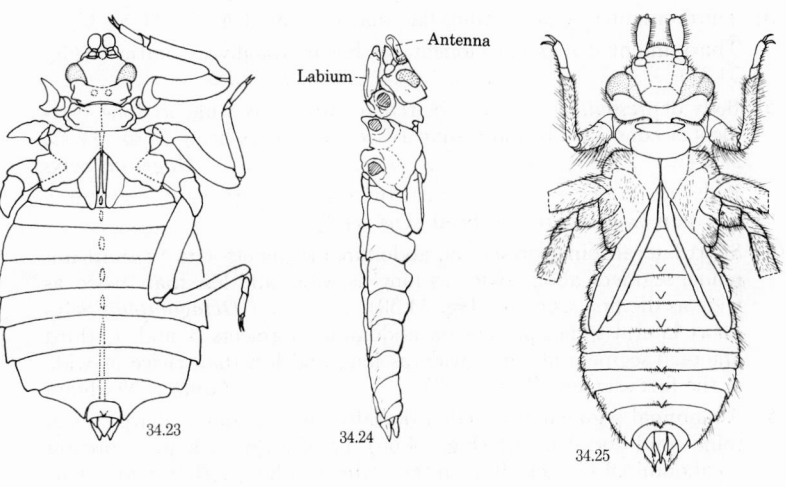

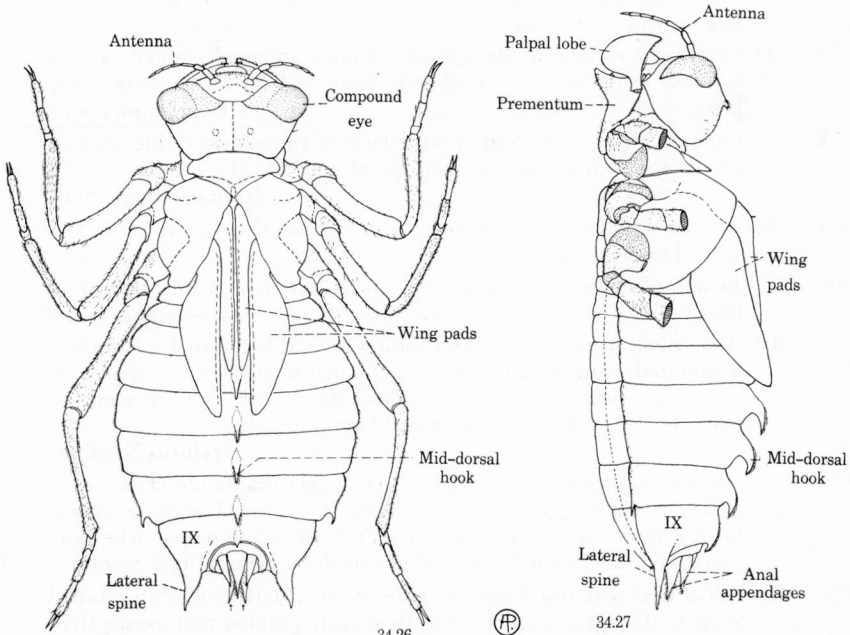

ANISOPTERA: **Fig. 34.23.** *Hagenius*, dorsal view. **Fig. 34.24.** Same, lateral view. **Fig. 34.25.** *Ophio-gomphus*, dorsal view. **Fig. 34.26.** *Epicordulia*, dorsal view. **Fig. 34.27.** Same, lateral view. (Figs. 34.23, 34.25–34.27 from Wright and Peterson.)

3a (2) Wing pads divergent, not parallel with meson (Fig. 34.25) **4**
3b   Wing pads parallel with meson (Fig. 34.23) **5**
4a (3) Dorsal hooks present on abdominal segments 2 to 9 or 3 to 9, those on the posterior segments hooklike (Fig. 34.25) *Ophiogomphus* Selys
4b   Dorsal hooks present only on abdominal segments 2 to 4; segments 8 and 9 may have slight thickenings on the mid-dorsal line, but these never as described above . . . . *Erpetogomphus* Selys
5a (3) Third antennal segment thin, flat, and suboval (Figs. 34.28–34.30). . **6**
5b   Third antennal segment elongate or linear, usually cylindrical (Fig. 34.31). **8**
6a (5) Body depressed, abdomen subcircular, almost as wide as long, segment 5 considerably more than twice as wide as head (Figs. 34.23, 34.24). . . . . *Hagenius* Selys
6b   Body not so depressed; abdomen twice as long as wide, segment 5 less than twice as wide as head (Fig. 34.25) **7**
7a (6) Short lateral spines present on abdominal segments 7 to 9; third antennal segment about twice as long as wide, and less than twice as wide as the first segment (Fig. 34.30). . . . . . . *Octogomphus* Selys
7b   Short lateral spines present on abdominal segments 8 and 9; third antennal segment about as wide as long and less than twice as wide as the first segment (Fig. 34.29) . . . . *Lanthus* Needham
8a (5) Abdominal segment 9 rounded dorsally and without a sharp apical spine (mid-dorsal hook) (Fig. 34.66), or, if a dorsal hook is present on abdominal segment 9, then the segment is longer than wide at its base . . . . *Gomphus* Leach s. lat. **10**
8b   Abdominal segment 9 with an acute mid-dorsal ridge ending in a spine (mid-dorsal hook) at the apex (Fig. 34.67); segment 9 never as long as wide at its base **9**
9a (8) Cerci each as long as the epiproct; fourth antennal segment a conspicuous upturned conic rudiment about as long as the second segment. . . . . *Gomphoides* Selys
9b   Cerci each not more than ¾ the length of the epiproct (Fig. 34.67); fourth antennal segment a small round structure (Fig. 34.31). . . . . *Dromogomphus* Selys
10a (8) Length of abdominal segments 9 + 10 greater than width of 9 at its base (Fig. 34.65). **11**
10b   Length of abdominal segments 9 + 10 less than width of 9 at its base . . . . *Gomphus* Leach
11a (10) Abdominal segment 10 shorter than wide and less than ½ as long as abdominal segment 8; width of abdominal segment 6 less than 1 ½ times the width of the head; palpal lobes with 2 to 4 sagittate, truncate, or stairlike teeth on mesal lateral margin . . . . *Stylurus* Needham
11b   Abdominal segment 10 (dorsal aspect) longer than wide (Fig. 34.65), and more than ½ as long as segment 8; width of abdominal segment 6 more than 1½ times the width of head; palpal lobe with 5 to 8 sagittate, truncate, or stairlike teeth on mesal lateral margin . **12**
12a (11) Palpal lobe with distal half or more of mesal lateral margin toothed to apex, the distalmost 3 or 4 teeth sharply pointed and longer than width at base, and the terminal tooth which appears to be the pointed tip of the palpal lobe about equal in size to the penultimate tooth (Westfall in letter of Feb. 17, 1956, modified and reworded) . . . . . *Arigomphus* Needham

2b   Palpal lobe with the distal half or more of the mesal lateral margin
not toothed all the way to the apex, the pointed tip of the lobe thus
appearing as a strong, smooth, spurlike end hook; the distalmost 3 or
4 teeth truncate and as wide as or wider than long (*australis* and
cavillaris). . . . . **Gomphus** Leach

>   Distribution: The genera *Erpetogomphus*, *Gomphus*, *Ophiogomphus*, and *Stylurus* are repre-
> sented in most of the U. S. Of these, all but *Erpetogomphus* have species that have been
> recorded from Canada, and those of *Stylurus* and *Erpetogomphus* are also known from
> Mexico. *Progomphus* is represented in Mexico and in most of the U. S. except the
> northern states from Wis. to Wash. The species of *Aphylla* and *Gomphoides* range from
> Mexico to Tex., and the former is also found east to Fla. and north to N. C. Species of
> *Arigomphus*, *Dromogomphus*, *Hagenius*, and *Lanthus* occur in southeastern Canada, eastern
> half of the U. S., and west to Kan., Okla. and eastern Tex. *Octogomphus* has a single spe-
> cies whose range is limited to the West Coast region as far north as British Columbia
> and east to Nev.

Family Petaluridae

>   The family Petaluridae includes 5 archaic genera, 2 of which, *Tanypteryx* and *Tachop-
> teryx* are represented in the U. S. The nymphs live in the muck of spring-fed seepage or

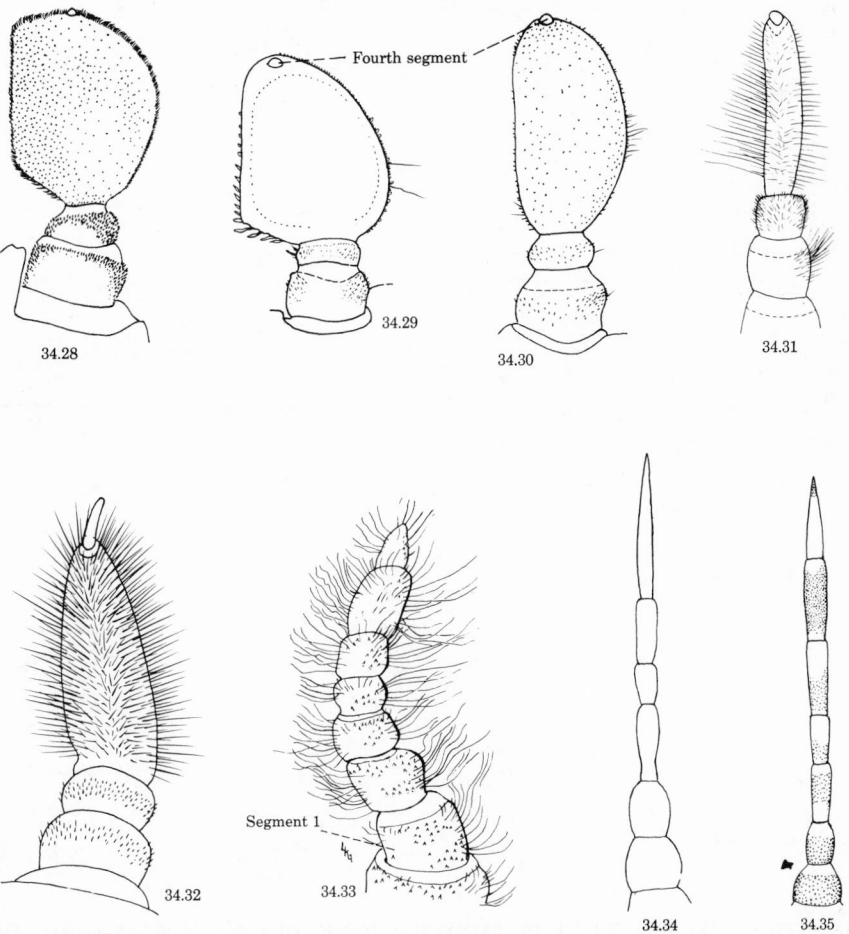

ANISOPTERA. ANTENNAE: **Fig. 34.28.** *Hagenius.* **Fig. 34.29.** *Lanthus.* **Fig. 34.30.** *Octogomphus.* **Fig.
34.31.** *Dromogomphus.* **Fig. 34.32.** *Progomphus.* **Fig. 34.33.** *Tachopteryx.* **Fig. 34.34.** *Aeshna.* **Fig.
34.35.** *Pachydiplax.* (Figs. 34.28–34.32, 34.34, 34.35 from Wright and Peterson.)

boggy areas on wooded hillsides and isolated mountain valleys. The nymph of *Tanypteryx* was first known by that of the Japanese species *pryeri* (Selys) described by Asahina and Okumura in 1949.

1a   Antennae 6-segmented, the third to fifth segments each longer than wide; cerci each more than ½ as long as the epiproct . . . .
Tanypteryx Kennedy

1b   Antennae 7-segmented, the third to fifth segments each wider than long (Fig. 34.33); cerci each less than ½ as long as the epiproct (Fig. 34.71) . . . . *Tachopteryx* Selys
Distribution: *Tachopteryx thoreyi* (Hagen) is known from the eastern half of the U. S. and Tex. *Tanypteryx hageni* (Selys) occurs in the mountainous regions of northern Calif. to British Columbia and in Nev.

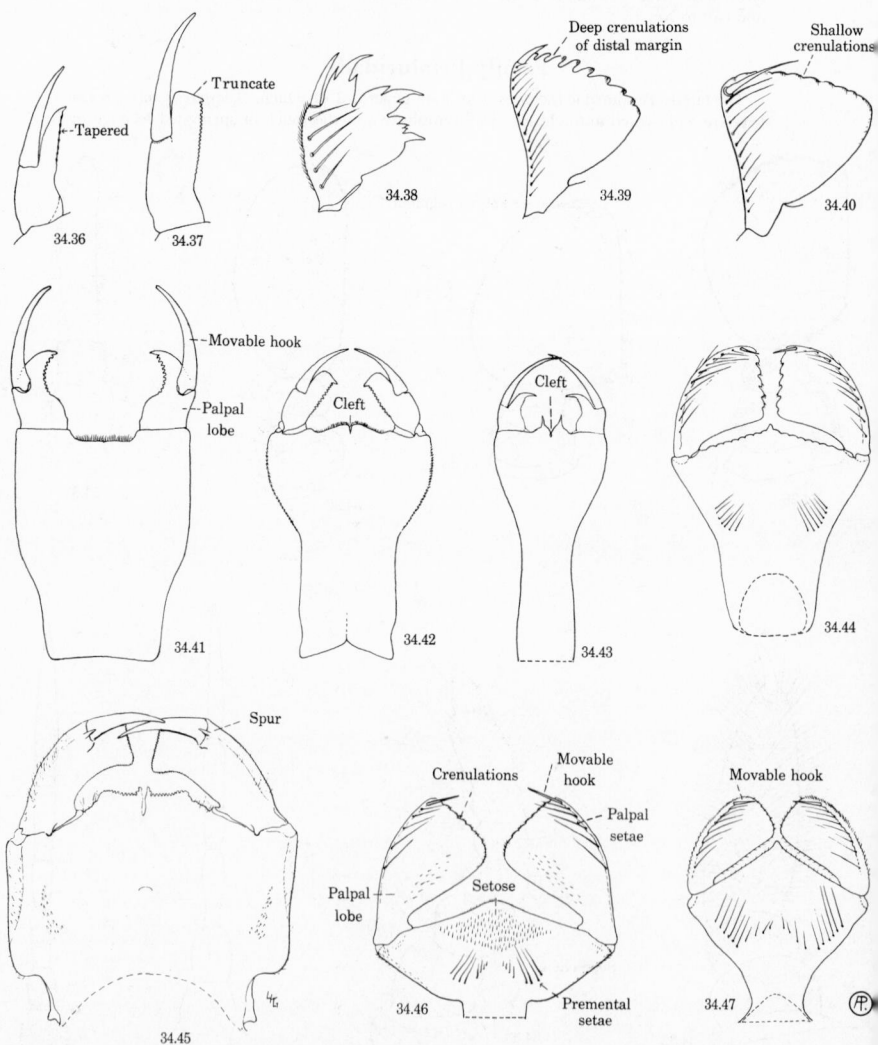

ANISOPTERA. LEFT PALPAL LOBE OF PREMENTUM, DORSAL VIEW: Fig. 34.36. *Basiaeschna*. Fig. 34.37. *Boyeria*. Fig. 34.38. *Cordulegaster*. Fig. 34.39. *Pantala*. Fig. 34.40. *Tramea*. PREMENTUM OF LABIUM, DORSAL VIEW: Fig. 34.41. *Dromogomphus*. Fig. 34.42. *Aeshna*. Fig. 34.43. *Coryphaeschna*. Fig. 34.44. *Plathemis*. Fig. 34.45. *Tachopteryx*. Fig. 34.46. *Epicordulia*. Fig. 34.47. *Libellula*. (Figs. 34.36–34.44, 34.46, 34.47 from Wright and Peterson.)

Family Aeshnidae

The nymphs of the Aeshnidae are climbers and live among reeds and other aquatic vegetation in still waters from a few inches to 1½ ft. deep (Walker 1912, p. 51). They are all of slender build and have long tapering abdomens which are widest a little beyond mid-length.

1a		Palpal lobes with stout lateral setae.	**2**
1b		Palpal lobes without any lateral setae (Figs. 34.42, 34.43)	**3**
2a	(1)	Lateral palpal setae nearly uniform in length . ***Triacanthagyna*** Selys	
2b		Lateral palpal setae very unequal in length, diminishing to very small ones at proximal end of row . . . . ***Gynacantha*** Rambur	
3a	(1)	Protarsi 2-segmented, meso- and metatarsi 3-segmented. . . . . ***Gomphaeschna*** Selys	
3b		All tarsi 3-segmented	**4**
4a	(3)	Ligula with a long marginal spine on each side of, and adjacent to the median cleft (Fig. 34.43). . . . . . . ***Coryphaeschna*** Williamson	
4b		Ligula not so armed (Fig. 34.42) or spines short and distant from cleft.	**5**
5a	(4)	Lateral spines present on abdominal segments 7 to 9 (sometimes a very small one on segment 6)	**6**
5b		Lateral spines present on abdominal segments 4, 5, or 6 to 9	**7**
6a	(5)	Prementum 2 or more times as long as width at base . . ***Anax*** Leach	
6b		Prementum less than 1½ times as long as width at base (*sitchensis*) ***Aeshna*** Fabricius	
7a	(5)	Caudolateral margins of head from dorsal view each with 2 large, well-developed tubercles (Fig. 34.49); eyes small, occupying only ⅓ or less of the lateral margins of the head	**8**
7b		Caudolateral margins of head from dorsal view never with 2 tubercles (Fig. 34.50); eyes large, each occupying about ½ of the lateral margin of the head.	**9**
8a	(7)	Low but distinct mid-dorsal hooks (best seen from lateral view) on abdominal segments 7 to 9; cerci each less than ½ as long as the epiproct (Fig. 34.74); apex of palpal lobe broadly rounded . . . . . . ***Nasiaeschna*** Selys	
8b		No mid-dorsal hooks on abdominal segments; cerci each more than ½ the length of the epiproct; apex of palpal lobe truncate . . . . . . ***Epiaeschna*** Hagen	
9a	(7)	Lateral spines on abdominal segments 6 to 9. . . . ***Aeshna*** Fabricius	
9b		Lateral spines on abdominal segments 4 to 9 or 5 to 9.	**10**
10a	(9)	Rear of head rounded or obtusely angulate	**11**
10b		Rear of head almost rectangular (Fig. 34.50)	**12**
11a	(10)	Epiproct and paraprocts subequal in length and about the same length as abdominal segments 9 + 10; tips of paraprocts strongly incurved . . . . ***Oplonaeschna*** Selys	
11b		Epiproct about ⅘ the length of a paraproct; paraprocts distinctly longer than abdominal segments 9 + 10; rear of head sometimes obtusely angulate (*eremita*). . . . . ***Aeshna*** Fabricius	
12a	(10)	Cerci each about ⅔ the length of the epiproct (Fig. 34.73); tips of paraprocts straight; no moundlike protuberance on each side of mesothorax at about mid-height; palpal lobe tapered to a point at tip (Fig. 34.36) . . . . ***Basiaeschna*** Selys	
12b		Cerci each ⅓ the length of the epiproct or less; tips of paraprocts	

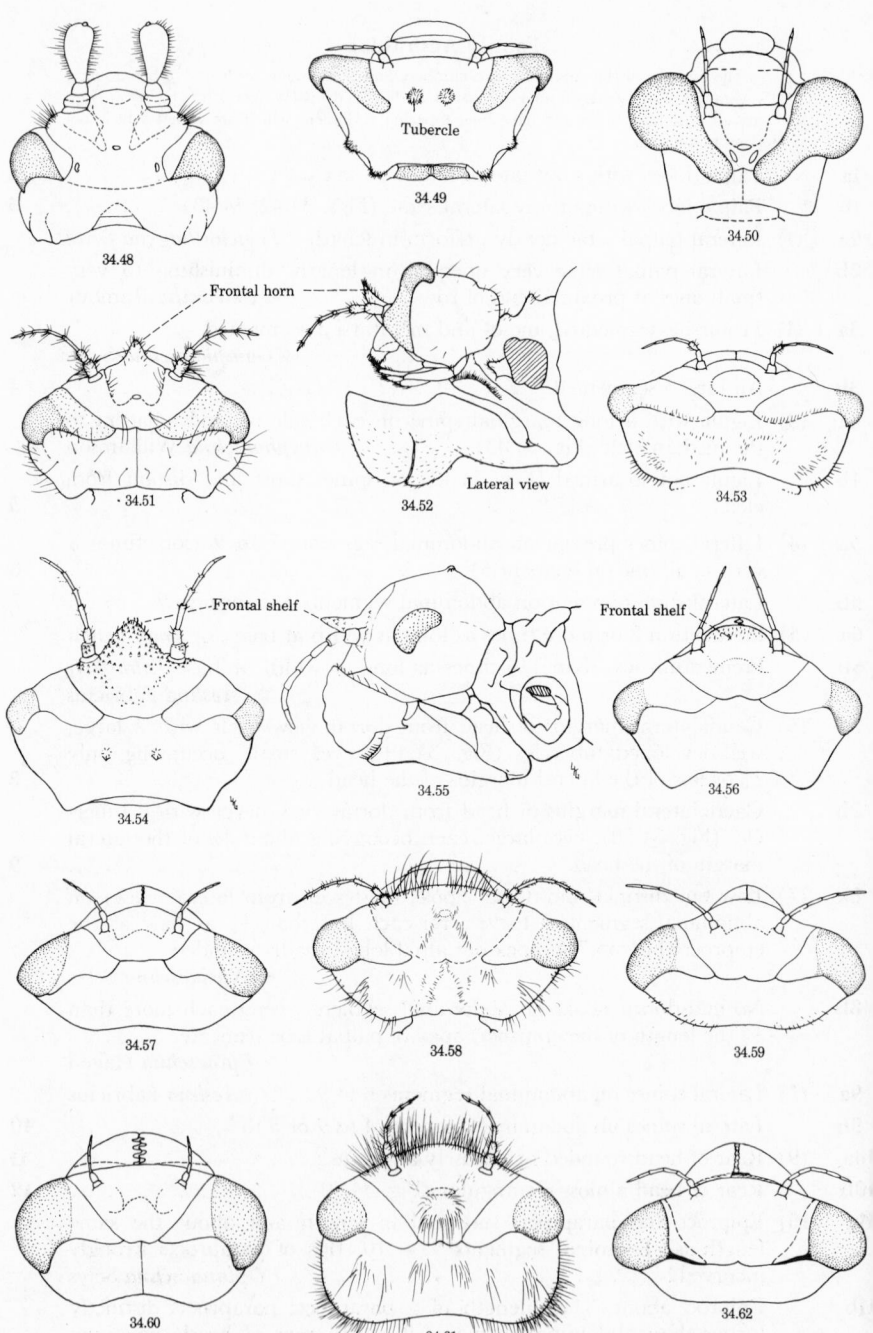

ANISOPTERA. HEAD: **Fig. 34.48.** *Octogomphus.* **Fig. 34.49.** *Nasiaeschna.* **Fig. 34.50.** *Basiaeschna.* **Fig. 34.51.** *Macromia,* dorsal view. **Fig. 34.52.** Same, lateral view. **Fig. 34.53.** *Somatochlora.* **Fig. 34.54.** *Neurocordulia molesta* (Walsh), dorsal view. **Fig. 34.55.** Same, lateral view. **Fig. 34.56.** *Neurocordulia* (other species). **Fig. 34.57.** *Celithemis.* **Fig. 34.58.** *Libellula.* **Fig. 34.59.** *Leucorrhinia.* **Fig. 34.60.** *Paltothemis.* **Fig. 34.61.** *Plathemis.* **Fig. 34.62.** *Pachydiplax.* (Fig. 34.48 after Kennedy; Figs. 34.49–34.52, 34.56–34.62 from Wright and Peterson; Fig. 34.53 redrawn from Walker by Wright and Peterson.)

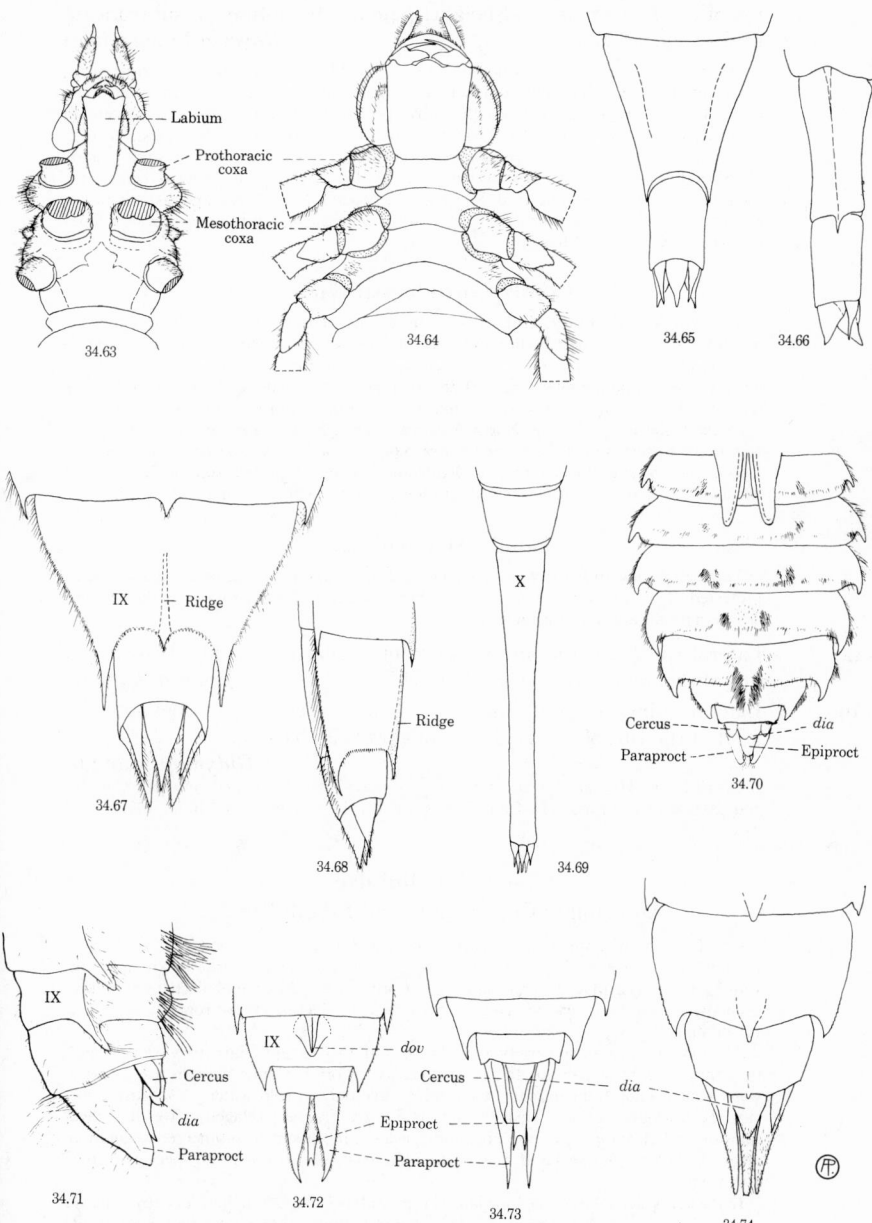

ANISOPTERA. HEAD AND THORAX, VENTRAL VIEW: **Fig. 34.63.** *Progomphus.* **Fig. 34.64.** *Gomphus.* TERMINAL ABDOMINAL SEGMENTS AND ANAL APPENDAGES: **Fig. 34.65.** *Gomphus (Arigomphus),* dorsal view. **Fig. 34.66.** Same, lateral view. **Fig. 34.67.** *Dromogomphus,* dorsal view. **Fig. 34.68.** Same, lateral view. **Fig. 34.69.** *Aphylla,* dorsal view. **Fig. 34.70.** *Tachopteryx* ♂, dorsal view segments 4–10. **Fig. 34.71.** Same, lateral view segments 8–10. **Fig. 34.72.** *Boyeria* ♀, ventral view. **Fig. 34.73.** *Basiaeschna* ♂, dorsal view. **Fig. 34.74.** *Nasiaeschna* ♂, dorsal view. (Figs. 34.63–34.70, 34.72–34.74 from Wright and Peterson.) *dia,* developing inferior appendage of adult male within the nymphal epiproct; *dov,* developing ovripositor and valves of adult.

incurved (Fig. 34.72); a moundlike protuberance on each side of the
mesothorax at about mid-height; palpal lobe obtuse or subtruncate
at tip (Fig. 34.37) . . . . *Boyeria* MacLachlan

Distribution: *Aeshna* is represented in most of N. A. Adults of *Anax* are great wan-
derers and have been reported from Alaska, southern Canada, the U. S., and southward
in N. A. but there is no evidence that any species of this genus passes the nymphal
stages north of southern Canada (Walker, Feb. 1956). Species of *Basiaeschna, Boyeria,
Epiaeschna, Gomphaeschna,* and *Nasiaeschna* occur in southeastern Canada and eastern
U. S. and, except *Gomphaeschna,* range west to Ia., Okla., and eastern Tex., with
those of *Epiaeschna* extending into Mexico. *Coryphaeschna, Gynacantha,* and *Triacanthagyna*
are represented in southeastern U. S. and Mexico. *Oplonaeschna* has one species known
from Ariz., N. M., and Mexico.

Family Cordulegastridae

The nymphs of Cordulegastridae live in the shallow mud or silt in small woodland or
mountain streams. Their bodies are hairy and usually so coated with silt and mud as
to be easily overlooked. Their rather slender 7-segmented antennae, deeply and irregularly incised palpal lobes (Fig. 34.38), divergent wing pads, and lack of mid-dorsal
hooks on their abdomens, serve to distinguish them from all other anisopterous nymphs.

Of the 9 species in the one North American genus *Cordulegaster* Leach, 6 are found
only in the eastern half of the United States, 3 of these ranging into southeastern Canada;
1 occurs only west of the Continental Divide and has been reported from as far north as
Alaska; 2 are known from Mexico, the range of 1 extending into Ariz. and Utah.

Family Macromiidae

The nymphs live in shallow parts of lakes and in the larger streams having wide beds.
Protected by their mottled coloration or a thin covering of silt they sprawl on the bottoms
of such habitats and wait for their prey.

1a   Lateral spines of abdominal segment 9 reach less than halfway to
tips of the anal appendages (Fig. 34.76) . . . . . *Macromia* Rambur

1b   Lateral spines of abdominal segment 9 reach to or beyond the
level of the tips of the anal appendages (Fig. 34.75) . . . .
                                 Didymops Rambur

Distribution: *Macromia* is represented in most parts of the U. S. and southern Canada,
and *Didymops* in southeastern Canada, eastern U. S., and west to Minn., Okla., and
Tex.

Family Libellulidae
Subfamilies Corduliinae and Libellulinae

On the basis of adult characteristics, the genera *Cordulia, Dorocordulia, Epicordulia,
Helocordulia, Neurocordulia, Somatochlora, Tetragoneuria,* and *Williamsonia* (nymph unknown) have been placed in the subfamily Corduliinae. No nymphal characters have
been discovered that can be used to distinguish all its members from those of the
Libellulinae.

The nymphs live in a variety of habitats and their adaptability to various conditions may account in part for the great diversity of forms. Some nymphs are found in
water which is hot to the human touch, others live only in cold waters. At least 2 species, *Erythrodiplax berenice* (Drury) and *Macrodiplax balteata* (Hagen), live in brackish water. The nymphs of some libellulid species climb about in submerged vegetation,
others are bottom dwellers and may accumulate coats of silt, but in no species are they
burrowers.

In the key which follows, the long lateral raptorial setae on each palpal lobe are referred
to as the palpal setae (Fig. 34.46). The number of premental setae (also raptorial) on the
dorsal surface of the prementum refers to the number on each side of the median line.

1a   Abdomen with mid-dorsal hooks, spines, or knobs on 1 or more
segments (Figs. 34.26, 34.27)   2

1b   Abdomen with mid-dorsal hooks, spines, or knobs absent on all segments   26

2a   (1) A mid-dorsal hook, spine, or knob on abdominal segment 9 (Figs.
34.77, 34.78, 34.84)   3

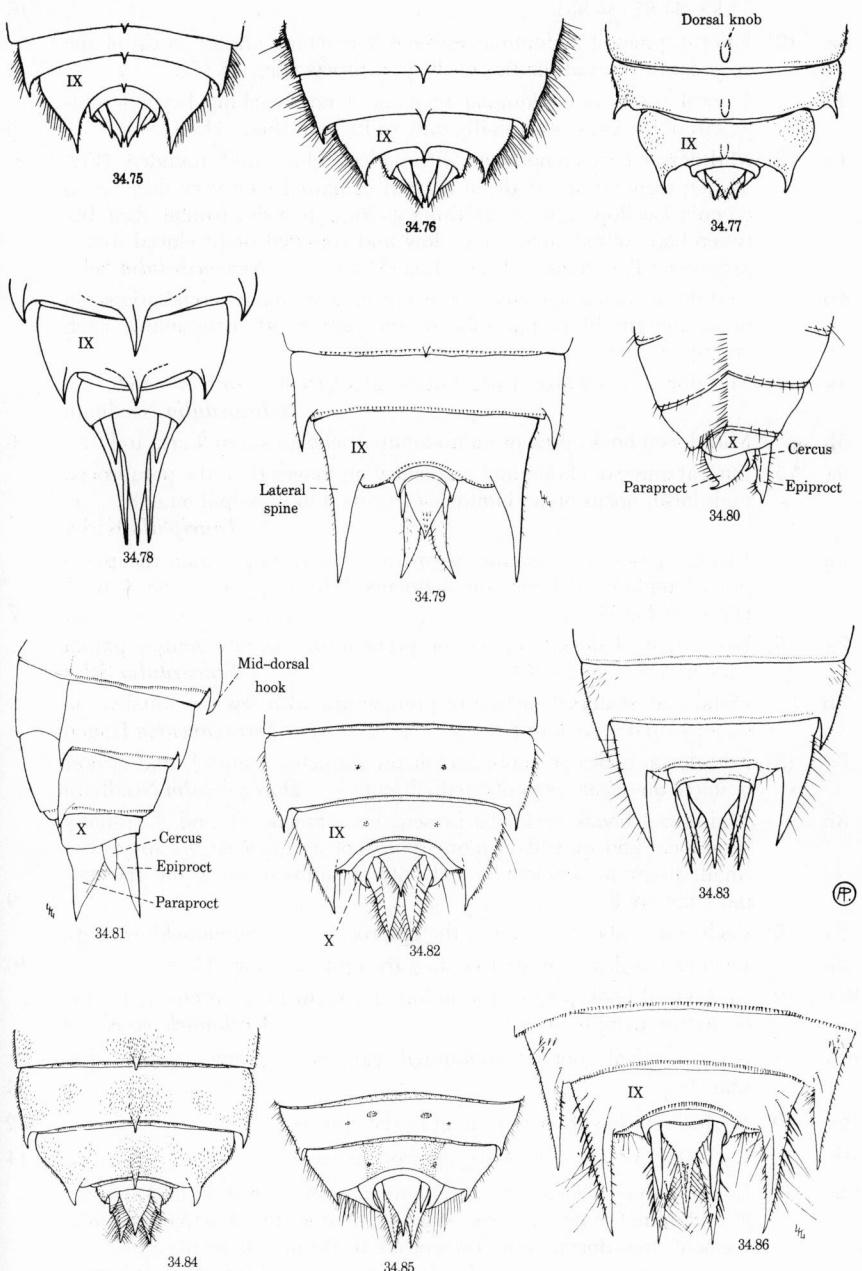

ANISOPTERA. TERMINAL ABDOMINAL SEGMENTS AND ANAL APPENDAGES: Fig. 34.75. *Didymops.* **Fig. 34.76.** *Macromia.* **Fig. 34.77.** *Neurocordulia.* **Fig. 34.78.** *Cannacria.* **Fig. 34.79.** *Celithemis.* **Fig. 34.80.** *Erythemis,* lateral view. **Fig. 34.81.** *Libellula,* lateral view. **Fig. 34.82.** *Pachydiplax.* **Fig. 34.83.** *Pantala.* **Fig. 34.84.** *Perithemis.* **Fig. 34.85.** *Sympetrum.* **Fig. 34.86.** *Tramea.* (Figs. 34.75–34.77, 34.79, 34.80, 34.82–34.85 from Wright and Peterson; Fig. 34.78 redrawn from figure by Byers.)

2b   No mid-dorsal hook, spine, or abdominal segment 9 (Figs. 34.79–
               34.83, 34.85, 34.86) **16**

3a   (2)   Lateral spines of abdominal segment 9 reaching almost to tip of the
               epiproct or beyond the tips of the paraprocts (Fig. 34.77). **4**

3b   Lateral spines of abdominal segment 9 not reaching beyond mid-
               length of the epiproct, usually only to its base (Figs. 34.78, 34.84) . . **8**

4a   (3)   Mid-dorsal hooks knoblike, with apices blunt and rounded (Fig.
               34.77); crenulations on distal margin of palpal lobe very deep, each
               crenula (scallop) 2 or more times as long as wide; frontal shelf be-
               tween bases of antennae either low and rounded or produced into a
               prominent flat triangle (Figs. 34.54–34.56) . . . ***Neurocordulia*** Selys

4b   Mid-dorsal hooks spinelike, with apices acuminate; crenulations on
               distal margin of palpal lobe of medium depth or shallow, each
               crenula as long as, or shorter than, wide **5**

5a   (4)   Mid-dorsal hooks absent on abdominal segments 1 to 5 or 1 to 6. . . .
                                                      Helocordulia Needham

5b   Mid-dorsal hooks present on abdominal segments 2 to 9 or 3 to 9 . .   **6**

6a   (5)   Lateral spines of abdominal segment 9 no longer than the paraprocts;
               mid-dorsal hooks on abdominal segments 3 to 9; palpal setae 8. . . . .
                                                      Tauriphila Kirby

6b   Lateral spines of abdominal segment 9 much longer than the para-
               procts; mid-dorsal hooks on segments 2 to 9; palpal setae 4 to 7
               (Figs. 34.26, 34.27) **7**

7a   (6)   Distal half of dorsal surface of prementum heavily setose; palpal
               setae 4 or 5 (Fig. 34.46) . . . . ***Epicordulia*** Selys

7b   Distal half of dorsal surface of prementum with few or, usually, no
               setae; palpal setae 6 to 8 . . . . ***Tetragoneuria*** Hagen

8a   (3)   Mid-dorsal hooks or knobs absent on segments 3 and 4, and almost
               obsolete on distal segments; palpal setae 7 . ***Dorocordulia*** Needham

8b   Mid-dorsal hooks or knobs present on segments 3 and 4, usually
               prominent and spinelike (in one species of *Somatochlora* the spines are
               small, absent on segment 3, and only a rudiment on 5, but the pal-
               pal setae are 8). **9**

9a   (8)   Each cercus about as long as the epiproct . . . . ***Somatochlora*** Selys

9b   Each cercus distinctly shorter than the epiproct (Fig. 34.78) **10**

10a (9)   Mid-dorsal hook present on abdominal segment 2; premental setae
               of mature nymph 14 to 15 . . . . ***Brechmorhoga*** Kirby

10b   No mid-dorsal hook on abdominal segment 2; premental setae less
               than 14 . . . .   **11**

11a (10)   Each cercus less than ½ as long as the epiproct **12**

11b   Each cercus ½ as long as the epiproct or more **14**

12a (11)   Epiproct much shorter than abdominal segments 8 + 9; lateral spine
               of abdominal segment 9 less than ⅓ of mid-dorsal length of same
               segment; mid-dorsal hooks present on abdominal segments 3 to 9 . .
                                                      Macrothemis Hagen

12b   Epiproct longer than mid-dorsal length of abdominal segments 8 +
               9; lateral spine of abdominal segment 9 at least ½ as long as mid-
               dorsal length of same segment; mid-dorsal hooks present on abdom-
               inal segments 3 to 10 . . . .   **13**

13a (12)   Each lateral spine of abdominal segment 9 longer than mid-dorsal length of that segment; hind legs longer than over-all length of rest of nymph . . . . *Idiataphe* Cowley

13b   Each lateral spine of abdominal segment 9 shorter than mid-dorsal length of that segment; hind legs considerably shorter than over-all length of rest of nymph . . . . *Cannacria* Kirby

14a   (11) Mid-dorsal hooks present on abdominal segments 3 to 10; lateral spines of abdominal segment 9 each less than $\frac{1}{2}$ the mid-dorsal length of that segment . . . . *Brachymesia* Kirby

14b   Mid-dorsal hooks present on abdominal segments 3 to 9; lateral spines of abdominal segment 9 each $\frac{1}{2}$ or more the mid-dorsal length of that segment (Fig. 34.84). **15**

15a   (14) Crenulations of distal margin of each palpal lobe deep; palpal setae 5. . . . . *Perithemis* Hagen

15b   Crenulations of distal margin of each palpal lobe obsolete; palpal setae 7 to 10. . . . . *Dythemis* Hagen

16a   (2) Each cercus $\frac{2}{3}$ to equal to the length of the epiproct **17**

16b   Each cercus about $\frac{1}{2}$ the length of the epiproct **20**

17a   (16) Lateral spines of abdominal segment 9 each longer than that segment mid-dorsally (Fig. 34.83) **18**

17b   Lateral spines of abdominal segment 9 each $\frac{1}{3}$ or less of that segment mid-dorsally **19**

18a   (17) Mid-dorsal hooks or spines on abdominal segments 3 to 4 (sometimes also present on 2 or 5, or both) small and papillate, erect in older nymphs; usually a small spine or angular process at apex of segment 10; palpal setae 14 to 15; crenulations on distal margin of each palpal lobe deep (Fig. 34.39). . . . . *Pantala* Hagen

18b   Mid-dorsal hooks or spines on abdominal segments 2 to 8 prominent, increasing in size posteriorly; lateral palpal setae 7; crenulations on distal margin of palpal lobe very shallow, less pronounced than those of *Tramea* (Fig. 34.40) . . . . *Miathyria* Kirby

19a   (17) Mid-dorsal hooks on abdominal segments 2 to 6 low but spinelike, decreasing in size posteriorly; palpal setae 9. . . *Paltothemis* Karsch

19b   Mid-dorsal hooks on abdominal segments 5 to 8 represented by small knoblike prominences; palpal setae 7 . **Dorocordulia** Needham

20a   (16) Paraprocts noticeably longer than the epiproct (Figs. 34.79, 34.85) .   **21**

20b   Paraprocts and epiproct subequal in length (Fig. 34.81) **22**

21a   (20) Lateral spines of abdominal segment 9 long and straight, reaching to the tips of the paraprocts (Fig. 34.79); mid-dorsal hook always absent on segment 8 . . . . *Celithemis* Hagen

21b   Lateral spines of abdominal segment 9 short, not reaching beyond tips of the cerci (Fig. 34.85); if the lateral spines are long (usually not extending beyond tip of the epiproct) there is a dorsal hook on segment 8 . . . . *Sympetrum* Newman

22a   (20) Premental setae 0 to 3 short, all inconspicuous . . *Ladona* Needham

22b   Premental setae 7 to 21, all prominent (Figs. 34.44, 34.47) **23**

23a   (22) Body smooth, not covered with long hairlike setae; eyes large and prominent, occupying $\frac{1}{2}$ or more than $\frac{1}{2}$ the length of the head (Fig. 34.59) **24**

23b   Body extremely hairy; eyes usually small, occupying less than $\frac{1}{2}$ the length of the head (Figs. 34.58, 34.61) **25**

24a **(23)** Premental setae 18 to 21 (young nymphs may have 16 or 17, Bick, 1955) . . . . **Macrodiplax** Brauer

24b Premental setae 10 to 15 . . . . **Leucorrhinia** Brittinger

25a **(23)** Width of head across eyes less than 1 ¼ width of prothorax across dorsolateral ridges; labium with distal margin of ligula crenulate (Fig. 34.44); abdominal segments 7 to 9 with brown or black, shining mid-dorsal ridges . . . . **Plathemis** Hagen

25b Width of head across eyes more than 1 ¼ width of prothorax across dorsolateral ridges; labium with distal margin of ligula evenly contoured, not obviously crenulate (Fig. 34.47); abdominal segments 7 to 9 without dark, mid-dorsal ridges . . . . **Libellula** Linné

26a **(1)** Apical third of cerci and paraprocts strongly decurved (Fig. 34.80) . 27

26b Apical third of cerci and paraprocts straight or only slightly decurved (Fig. 34.81) 28

27a **(26)** A minute lateral spine on each side of abdominal segment 9; palpal setae 11 or 12 . . . . **Lepthemis** Hagen

27b No lateral spines on abdomen; palpal setae 7 or 8 **Erythemis** Hagen

28a **(26)** Lateral spines on abdominal segment 8 (Figs. 34.82, 34.86). 29

28b No lateral spines on abdominal segment 8 39

29a **(28)** Abdominal segment 9 with each lateral spine about twice the mid-dorsal length of the segment; segment 8 with each lateral spine as long as, or longer than, the mid-dorsal length of the segment (Fig. 34.86). . . . . **Tramea** Hagen

29b Abdominal segment 9 with each lateral spine equal to, or less than, the mid-dorsal length of the segment; segment 8 with each lateral spine less than the mid-dorsal length of the segment (Fig. 34.82) . . . 30

30a **(29)** Tips of lateral spines of abdominal segment 9 extending farther caudad than tip of the epiproct (Fig. 34.82) . . . . **Pachydiplax** Brauer

30b Tips of lateral spines of abdominal segment 9 not extending beyond tip of the epiproct 31

31a **(30)** Each cercus ½ or less than ½ the length of a paraproct 32

31b Each cercus more than ½ to as long as a paraproct 35

32a **(31)** Eyes small, protuberant, directed more upward than outward (front view), and extend well above level of top of head; premental setae 8 to 10, the outer 3 or 4 large and prominent, inner group small and indistinct . . . . **Orthemis** Hagen

32b Eyes large, little or not elevated above level of top of head (Fig. 34.59); premental setae 11 to 15 (or if 10, abdominal segments 6 to 9 each have a heavy mid-dorsal tuft of hair) 33

33a **(32)** Lateral spines of abdominal segments 8 and 9 subequal in length; body hairy . . . . **Erythrodiplax** Brauer

33b Lateral spines of abdominal segment 9 about twice the length of those of segment 8; body smooth 34

34a **(33)** Epiproct as long as, or slightly shorter than a paraproct . . . .
                                                        Leucorrhinia Brittinger

34b Epiproct extends caudad to ⅔ or less than ⅔ the length of a paraproct . . . . **Micrathyria** Kirby

35a **(31)** Crenulation of distal margin of palpal lobe obsolete or shallow with each crenula (scallop) less than ¼ as deep as broad 36

35b Crenulation of distal margin of palpal lobe of medium depth, each crenula ⅓ to ½ as deep as broad. 38

36a (**35**) Palpal setae 6; lateral spines of abdominal segments 8 and 9 incurved; length of full-grown nymph, 10 mm . . **Nannothemis** Brauer

36b Palpal setae 7 to 11; lateral spines of abdominal segments 8 and 9 straight; length of full-grown nymph 20 mm or more **37**

37a (**36**) Sides of head convergent posteriorly (similar to Fig. 34.59); length of epiproct less than ½ apical width of abdominal segment 9 . . . . .
Pseudoleon Kirby

37b Sides of head subparallel (similar to Fig. 34.61); length of epiproct more than ½ apical width of abdominal segment 9 . **Belonia** Kirby

38a (**35**) A dark longitudinal stripe present along the dorsolateral margin of the thorax. . . . . **Cordulia** Leach

38b Thorax unicolored, no such stripe . . . . **Somatochlora** Selys

39a (**28**) Crenulations on distal margin of palpal lobe deep; cerci each about equal in length to the epiproct . . . . **Somatochlora** Selys

39b Crenulations on distal margin of palpal lobe obsolete; cerci each ⅔ the length of the epiproct or slightly less . . . .
Sympetrum Newman (subgenus *Tarnetrum* Needham and Fisher)

Distribution: *Libellula* and *Sympetrum* are rather generally represented throughout all of N. A. Species of *Pachydiplax* and *Plathemis* are found in southern Canada, most of the U. S., and in Mexico. Species of *Celithemis*, *Erythrodiplax*, *Pantala*, *Perithemis*, and *Tramea* are recorded from southeastern Canada, most of the U. S. except the northwestern part, and from Mexico; *Erythemis* has the same distribution but one of its species also occurs in northwestern U. S. *Leucorrhinia* is a circumpolar genus and in N. A. its species are known from Alaska, Canada, and the northern part of the U. S. with the southern limits in Calif., Colo., Tenn., Ohio, and N. Y. *Ladona* is represented in southern Canada, and in the northern border and eastern half of the U. S. The one species of *Nannothemis* is very local in distribution and is found in southeastern Canada and the eastern half of the U. S. *Dythemis* and *Orthemis* are represented in the southern U. S. and Mexico, one species of the latter genus being recorded from as far north as Okla., and Utah. *Belonia* is represented in western U. S. and Mexico. Genera having species that occur in the southern part of N. A. only are: *Brachymesia*, *Cannacria*, *Lepthemis*, *Macrodiplax*, and *Miathyria*—Fla. to Tex. and in Mexico; *Macrothemis* and *Micrathyria*—Tex. and Mexico; *Idiataphe* and *Tauriphila*—Fla.

References

For the determination of nymphs to species the following references are recommended either as indices to the literature, or for keys and descriptions, as indicated by their titles.

Asahina, Syoziro and Teiichi Okumura. 1949. The nymph of *Tanypteryx pryeri* Selys (Odonata, Petaluridae). *Mushi*, 19:(7), pp. 37–38. **Bick, G. H. 1955.** The nymph of Macrodiplax balteata (Hagen). *Proc. Entomol. Soc. Wash.*, 57:191–196. **Hayes, W. P. 1941.** A bibliography of keys for the identification of immature insects. Part II. Odonata. *Entomol. News*, 52:52–55, 66–69, 93–98. **Muttkowski, Richard A. 1910.** Catalogue of the Odonata of North America. *Bull. Public Museum Milwaukee*, 1:1–207. **Needham, James G. 1939.** Nymph of the Protoneurine genus *Neoneura* (Odonata). *Entomol. News*, 50:241–245. **Needham, James G. and Hortense Butler Heywood. 1929.** *A Handbook of the Dragonflies of North America.* Charles C. Thomas, Springfield, Illinois, and Baltimore. **Needham, James G. and Minter J. Westfall, Jr. 1955.** *A Manual of the Dragonflies of North America (Anisoptera) Including the Greater Antilles and the Provinces of the Mexican Border.* University of California Press, Berkeley and Los Angeles. **Smith, R. F. and A. E. Pritchard. 1956.** Chapter 4, Odonata. In: R. L.

Usinger (ed.). *Aquatic Insects of California with Keys to North American Genera and California Species*, pp. 106–153. University of California Press, Berkeley and Los Angeles.   **Snodgrass, R. E. 1954.** The dragonfly larva. *Smithsonian Inst. Publ. Misc. Collections.* 123:1–38.   **Walker, E. M. 1912.** The North American Dragonflies of the Genus Aeshna. *Univ. Toronto Studies, Biol. Ser.* No. 11:1–213.   **Walker, E. M. 1953.** *The Odonata of Canada and Alaska.* Vol. 1, Part I, General; Part II, Zygoptera. University of Toronto Press, Toronto.   **Whitehouse, Francis C. 1948.** Catalogue of the Odonata of Canada, Newfoundland and Alaska. *Trans. Roy. Can. Inst.*, 27:3–56.   **Wright, M. and A. Peterson. 1944.** A key to the genera of anisopterous dragonfly nymphs of the United States and Canada. (Odonata, suborder Anisoptera.) *Ohio J. Sci.*, 44:151–166.

35

Plecoptera

W. E. RICKER

Nymphs of Plecoptera do not deviate greatly from primitive orthopteroid structure. The most obvious features of gross anatomy are the usual division of the body into head, 3 thoracic segments and 10 abdominal segments, the long antennae, and the many-segmented cerci at the end of the body. The "two-tailed" appearance would be a distinctive mark of the order in aquatic habitats, except that nymphs of a few genera of mayflies lack the central caudal filament. Stonefly nymphs are most abundant in rapid stony parts of streams, less common in quiet reaches. They seem to avoid bare mud bottoms completely, but are often numerous on logs or stumps which protrude above the mud. A few species occur along the stony shores of large temperate lakes, and in both large and small arctic or alpine lakes.

Collecting nymphs in streams is done most easily by turning and washing rocks, sticks, or vegetation and catching the nymphs in a dip-net or screen held downstream from the operation. When disturbed or removed from the water, most stoneflies actively run away, but pteronarcids usually roll themselves into a circle like the "woolly-bear" caterpillar of *Isia isabella*. Cast nymphal skins, or *exuviae*, of many species can be found on logs or rocks or under bridges; these are almost as good as actual nymphs for identification, particularly if they are soaked in warm water so that any gills can be distinguished. Collectors usually wish to obtain imagoes also, and, especially early in the morning, these too are most easily obtained on or under bridges whose piers stand in the water.

External Structure

Good accounts of the general external structure of stonefly nymphs are available in the works of Wu (1923, *Nemoura* only) and Hynes (1941). The review below mentions only the characters most useful for identification.

Head. The *head* is often somewhat bent downward, and is extremely and habitually so in the Peltoperlidae. The *eyes* are usually large and posterior (Fig. 35.7a, etc.), but in two genera in each of two different families they are reduced in size and shifted to a more forward position, relatively, by extension of the rear of the head (Fig. 35.7c, j, k). *Ocelli* are usually 3, but the anterior one is lost in Peltoperlidae (Fig. 35.7b) and in several genera of Perlidae (Fig. 35.7c, d).

Mouthparts. *Mouthparts* most clearly distinguish the two suborders of stoneflies. In the Holognatha (as redefined by Frison, 1935) the *glossa* of the labium is about equal in length to the paraglossa (Fig. 35.1a, b), the *lacinia* of the maxilla terminates in one to several blunt, often chisel-shaped teeth, and the *mandibles* are of the thick "herbivorous" type. In Systellognatha the size of the paraglossa greatly exceeds that of the glossa (Fig. 35.1c, d), the lacinia is nearly always tipped by one or more slender spines (Fig. 35.10), and the mandibles are much thinner. Within the Systellognatha, the shape of the *lacinia* is variable, and the differences are useful for finer classification in the family Perlodidae and to some extent in Chloroperlidae; there is a *major spine* or cusp, and almost always a *minor spine* and a variable number of *marginal hairs* or spinules.

Thorax. The shapes and positions of the *thoracic plates* vary a great deal within the order, but no detailed comparative study is available for nymphs. The most conspicuous aberrations are the long, thin posterior extensions of the pronotum, wing pads, and sterna in Peltoperlidae (Fig. 35.2c). More use could be made of thoracic structure in a key, but it does not lend itself particularly well to brief description. The sutures or internal ridges of the sterna, particularly of the mesosternum (Fig. 35.9), have proved useful in subgeneric differentiation of Perlodidae. Dorsally, the *wing pads* are conspicuous (Fig. 35.8); the axes of the hind pair may be either parallel to that of the body, or at a considerable angle to it.

Abdomen. The *abdomen* most nearly of primitive type is perhaps that of Pteronarcidae, in which both the first and the tenth sternites are fairly well developed and in direct connection with the tergite (Fig. 35.2a). In other holognathous families the metacoxae separate the first tergite from the *first sternite.* The latter is still fairly large and quite recognizable in Nemouridae, being separated by a suture from the metasternum, but in Peltoperlidae I cannot distinguish it. In Systellognatha the first sternite apparently became quite narrow before its fusion with the metasternum; usually it is not recognizably distinct from the latter, but a groove which apparently marks the line of division can be seen in *Oroperla* (Fig. 35.2*d*). The *tenth sternite*, already considerably narrowed in Pteronarcidae (Fig. 35.2*a*), becomes further reduced in Peltoperlidae, where it is practically always completely hidden by the ninth sternite (Fig. 35.2*c*). In other Holognatha it narrows to a thin, unsclerotized band at the mid-line but is broad and sclerotized at the side of the sternite (Fig. 35.6*a–b*). In Systellognatha the tenth sternite remains of normal length (Fig. 35.2*g*).

No separate pleurite is recognizable in Plecoptera, but between the tergite and sternite there may be an *unsclerotized fold* (Fig. 35.5). This is presumably primitive; it occurs on segments 1 to 9 of Capniinae and on a varying smaller number of segments in other Nemouridae. In Peltoperlidae and in all Systellognatha segment 2 always has this fold (also segment 1 when it is recognizable), but the more posterior segments are fully sclerotized, except for a partial or complete separation of the third in several of our perlodid subgenera, of the fourth in the extralimital *Perlodes*, and of the 7 gill-bearing segments in *Oroperla*. Pteronarcids also maintain the separation of tergite and sternite as far as there are abdominal gills, though it is absence of spinules rather than really weak sclerotization that here marks the "fold" line; on segments behind the gill-bearing ones the sides are uniformly sclerotized and armed.

At the end of the abdomen, the *subanal lobes* offer some diagnostic characteristics. Commonly they are separate, but in some Capniinae and Leuctrinae they are fused along part or the whole of their mesal edges, especially in last-instar males (Fig. 35.6). The *cerci* vary in length from about half as long as to much longer than the abdomen.

Body surface. The body surface of a nymph and its appendages may be almost naked, or they may bear a variety of *spines, spinules, hairs*, or an appressed pile called *clothing hair*. Relatively few nymphs have long *swimming hairs* on the cerci, of the type common among mayflies, but these are regularly developed in *Isocapnia* (Fig. 35.6*f*). The arrangement of spinules on the back of the head is of value in separating the genera and subgenera of Perlidae, and those of the thorax, legs, etc., will undoubtedly prove to have peculiarities useful for species recognition, in most genera of the order.

Gills. Although a majority of stonefly nymphs lack external *gills*, these are present in several branches of the order, and on various parts of the body. The primitive condition appears to have been one of rather profuse development of simple or once-branched gills. *Oroperla* (Fig. 35.2*d*) is the best North American representative of this condition, though it is unfortunately too scarce to become widely known, and in any event it probably does not possess the complete original complement. Subsequent gill evolution has consisted of a reduction in the number of gills, or a branching of the filaments into finely dissected tufts (Fig. 35.2*a, g*). In a few instances what are probably "new" gills have appeared. Homologies among the different familes are sometimes not certain, but in this chapter gills in similar positions are given the same names, as follows:

M: *Mental gills* occur between the mentum and submentum, at either side. They are known only in *Visoka*, a subgenus of *Nemoura* (Fig. 35.4c).

Sm: *Submental gills.* A simple filament or knob on either side of the submentum, found in many Perlodidae (Fig. 35.4d-g).

Ce (= AT_1 + **AP**): The *cervical gills* (prosternal gills of Hynes) are on the anterior part of the prosternum. There are two sets, of which the outer appears serially homologous with the *anterior thoracic* (AT) gills of the sterna behind, but the inner or *anterior prosternal* (AP) gill has no homologue on other segments. The AP filaments are mesad of and slightly posterior to the AT_1, rising from a different fold of the nymphal skin. Both AT_1 and AP occur in Pteronarcidae, AT_1 being divided to the base in older nymphs and then appearing as two quite separate gills (Fig. 35.2a). Both AT_1 and AP occur also in three of our subgenera of *Nemoura*, where they may be either simple or considerably branched (Fig. 35.4a, b); in the Old-World subgenus *Protonemura*, AT_1 is divided into two separated gills, as in Pteronarcidae. AT_1 occurs undivided and without AP in one subgenus of *Peltoperla* and in two subgenera of *Arcynopteryx* (Figs. 35.2d, 35.4d).

AT_2, AT_3: *Anterior thoracic gills,* found at the lateral anterior corners of the thoracic segments. The occurrence of AT_1 was described above. AT_2 and AT_3 occur in a few subgenera of the genus *Arcynopteryx*, being double in *Oroperla* (Fig. 35.2d) and single in *Megarcys*, *Setvena* and *Perlinodes* (Fig. 35.2e). Both pairs are invariably present in Pteronarcidae (Fig. 35.2a) and Perlidae (Fig. 35.2g), where they are finely branched and tend to divide into two or three separate stems in the larger nymphs. (Hynes calls these gills and the MTA of Perlidae *intersegmental* gills; however the general scheme of insect morphology suggests that each pair should be associated with one segment or the other, and external appearance, particularly the lateral view, points to the association used here.)

PT_1, PT_2, PT_3: *Posterior thoracic gills.* In Pteronarcidae, gills of this series are found near the hind margin of all 3 sterna, inside of the coxae, hence more toward the mid-line than the AT series (Fig. 35.2a). (The AP gill has a corresponding position on the anterior part of the prosternum.) They are finely branched but not double. In *Peltoperla* s.s. there is a ventral unbranched gill, quite probably homologous with PT_3 of pteronarcids, which has its origin underneath the produced metasternum, one on either side (Fig. 35.2c). (For the most posterior metasternal gills of Perlidae, see MTA below.)

ASC_1: *Anterior supracoxal gills* are found on the prothorax only; they occur in *Pteronarcella* (but not *Pteronarcys*), in a few Perlodidae (Fig. 35.2d, e), and in all Perlidae (Fig. 35.2g). They are situated on the body wall above and in front of the coxa. They are single in Perlodidae, in *Pteronarcella* they have one to a few filaments, and in Perlidae they are finely branched.

PSC_1, PSC_2, PSC_3: *Posterior supracoxal gills.* These are found on all 3 segments of *Pteronarcella*, though sometimes reduced to a few filaments; PSC_1 is not as close to the coxa as are the other 2 and may not be strictly homologous. In *Pteronarcys* I have seen gills of this series only on the prothorax (PSC_1), the tufts being small compared to those of the AT or PT series.

In Peltoperlidae gills of this group may be found either on all 3 segments or only on the meso- and metathorax (Fig. 35.2c); they may be single or, if double, they may appear as 2 separate gills.

Posterior supracoxal gills are present on all 3 segments of most Perlidae (Fig. 35.2g). In *Acroneuria*, *Claassenia*, *Neoperla*, *Paragnetina*, etc., PSC_2 has 2 stems, PSC_1 and PSC_3 have single stems, but all are finely branched. *Anacroneuria* lacks PSC_2 and PSC_3, at least in the species studied.

Co_1, Co_2, Co_3: *Coxal gills*, found on the inner surface of the coxae. Known only in *Taeniopteryx* (Fig. 35.2*b*).

MTA: *Thoracico-abdominal gill.* A large, double, finely branched gill occurs at the lateral posterior angle of the metasternum in Perlidae (Fig. 35.2*g*). Since the first abdominal sternite is apparently completely fused to the metasternum to form a combined sternum-sternite, this gill is almost certainly homologous with A_1 of *Oroperla*: it lies right next to the pleural fold of abdominal segment 2. However, to avoid the paradox of referring to an *abdominal* gill on a *thoracic* segment, I have given it a distinctive name.

A_1–A_7: *Abdominal gills* arise from the fold between the tergite and sternite (*Oroperla*) or just mesad of this fold (Pteronarcidae). A_1 and A_2 occur in *Pteronarcys* (Fig. 35.2*a*), and A_1–A_3 in *Pteronarcella*—much dissected in both cases. The only North American perlodid stonefly with abdominal gills is *Oroperla*, which has 7 pairs, unbranched (Fig. 35.2*d*). (For the abdominal gill of Perlidae, see MTA, above.)

SL: *Subanal gills.* The subanal lobe of the cercus bears an unbranched gill in *Peltoperla* s.s. (Fig. 35.2*c*), and a finely divided gill in many Perlidae (Fig. 35.2*b*).

R: Retractile *rectal gills* have been described in the subgenus *Zealeuctra* and may possibly occur much more extensively. They are difficult to demonstrate in preserved material.

Identification

With stoneflies as with other insects, the principal difficulty in preparing a key to nymphs is that they often show no trace of important structural differences that distinguish the adults. The converse proposition is also true, though to a less extent; and a really satisfactory "natural" key can be constructed only on the assumption that its user has both nymphs and male imagoes available simultaneously. At present only the *families* can be infallibly identified by fundamental structural differences in both the mature and the immature stages. The artificiality of the present key is evident in the several places where the same genus appears in different couplets, and in the rather trivial characteristics upon which some separations are based. It is not likely that a natural key to nymphs alone will ever be practical, but additional study will undoubtedly permit improvements. It is not too much to expect that well-grown nymphs of all genera, subgenera, and perhaps even species will eventually be recognizable. At present, species identification is restricted mainly to genera that have good color patterns in Pteronarcidae, Perlodidae, and Perlidae.

In general, the closer a nymph is to its final instar, the easier it is to identify. In really young nymphs body proportions and color patterns are different, and gills tend to have fewer branches or to be lacking entirely. Keys for the young stages of European nymphs have been made (e.g., Hynes, 1941), but most American species have not yet been given such close scrutiny.

A student who wishes to verify an identification, or carry it beyond the limits of this key, will consult principally the works of Claassen (1931), Frison (1935, 1942), Ricker (1952), Harden and Mickel (1952) and Jewett (1956). The last three papers use substantially the generic arrangement of the present key. A guide to the generic usage of Claassen and Frison follows:

This Key	Claassen	Frison
Acroneuria	*Acroneuria* (except *depressa*)	*Acroneuria*
Arcynopteryx	*Perlodes*	*Perlodes*
Brachyptera	*Taeniopteryx fasciata, californica*	*Strophopteryx* (*fasciata*, 1935) *Brachyptera* (1942)
Chloroperla	...	*Chloroperla* (1942)
Claassenia	*Acroneuria "depressa"*	*Claassenia*
Diura	...	*Dictyopterygella*
Hastaperla	*Chloroperla*	*Chloroperla* (1935) *Hastaperla* (1942)
Isogenus	*Isogenus* *Perla aestivalis, bilobata, expansa, hastata, verticalis*	*Isogenus* *Hydroperla* *Diploperla* *Isoperla duplicata* (1935)
Isoperla	*Isoperla* *Clioperla*	*Isoperla*
Nemocapnia	...	*Capnia "vernalis"* (1935) *Nemocapnia* (1942)
Paracapnia	*Capnia "vernalis"*	*Capnia opis* (1942)
Paragnetina	*Perla media, immarginata*	*Togoperla*
Phasganophora	*Perla capitata*	*Neophasganophora*
Taeniopteryx	*Taeniopteryx nivalis, maura, parvula*	*Taeniopteryx*

Distribution

There are three principal kinds of ranges of North American stoneflies.

Northern transcontinental. This group is rather few in numbers, its representatives ranging south into the region of the Great Lakes, the Atlantic provinces of Canada and the New England states, and a few south in the mountains to Georgia. In the West these transcontinental forms mostly occur no farther south than central British Columbia, *Pteronarcys dorsata* and *Arcynopteryx compacta* being the only exceptions.

Eastern. These species occupy the region from the eastern plains and northern Ontario to the Atlantic coast. Within this area there can be distinguished species of general distribution, species limited to Canada or practically so, species confined to the Appalachian mountain chain (especially its southern portion) and species characteristic of the lowlands on either side of these mountains.

Cordilleran. These species are found in the mountain and intermountain regions of the West, mixing with the transcontinental series in central and northern British Columbia and Alaska. The Sierra Nevada and Cascade ranges, north to the Columbia River or a little beyond, contain a number of genera and subgenera not known elsewhere. In the eastern or Rocky Mountain part of the cordillera, central Montana is a convenient boundary for dividing a northern from a southern province. The southern ranges, centering on Colorado, are to a considerable degree isolated from the rest of the cordillera, and this has led to some endemism, but it rarely reaches the sub-generic level. The northern Rocky Mountain region is less well separated from the ranges to the west, and its fauna is not greatly different from that of the coastal region of British Columbia. The high plains tend to have few species of stoneflies, and these are mostly related to the closest major province —cordilleran, northern, or eastern. However, little plains collecting has been done, especially southward. Southern Texas has a few species not known elsewhere, and the tropical genus *Anacroneuria* has been taken there.

KEY TO GENERA

1a   Tips of the glossae produced nearly as far forward as the tips of the paraglossae, or farther (Fig. 35.1*a*, *b*); tip of the lacinia with blunt teeth; middle portion of the tenth abdominal sternite ⅓ as long as the ninth (Fig. 35.2*a*) or less, usually unsclerotized and covered by the ninth (Figs. 35.2*c*, 35.6*a–f*) . . . .

                Suborder **Holognatha** Enderlein (Filipalpia Klapálek)   **2**

1b   Tips of the glossae situated much behind the tips of the paraglossae (Fig. 35.1*c*, *d*); lacinia almost always tipped with 1 or more sharp spines (Fig. 35.10); tenth sternite more than ⅓ as long as the ninth and not covered by it (Fig. 35.2*d*, *g*) . . . .

              Suborder **Systellognatha** Enderlein (Setipalpia Klapálek)   **15**

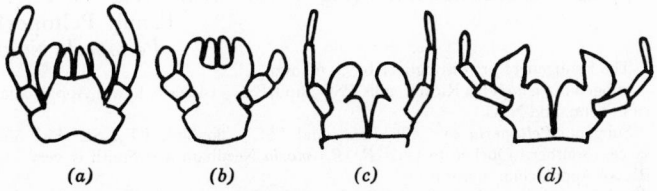

      (a)   *(b)*   *(c)*   *(d)*

Fig. 35.1. Labia. (*a*) *Pteronarcys*. (*b*) *Taeniopteryx*. (*c*) *Perlesta*. (*d*) *Arcynopteryx*. (The *glossa* is the inner of the two terminal divisions of the labium in *a* and *b*, the outer one being the *paraglossa*. In *c* and *d* the glossa appears to be merely a process on the inner side of the paraglossa.)

2a (1) Finely dissected gills present on the ventral side of the thorax and of abdominal segments 1 and 2, sometimes also 3 (Fig. 35.2*a*) . . . .

                            Family **Pteronarcidae**   **3**

2b   Gills absent from abdominal segments 1 to 3; if present on the thorax, they are not dissected (except sometimes the cervical gills) . .   **4**

3a (2) Abdominal segment 3 with gills similar to those on 1 and 2; gills ASC_1, PSC_2, and PSC_3 present in well-grown nymphs; corners of the pronotum broadly rounded . . . . **Pteronarcella** Banks

        Two species; occurs throughout the cordillera, abundant southward.

3b   Abdominal segment 3 without gills (Fig. 35.2*a*), gills ASC_1, PSC_2, and PSC_3 absent; anterior corners of the pronotum usually produced into horns or processes (Fig. 35.2*a*), sometimes *narrowly rounded* . . . . **Pteronarcys** Newman

        Subgenus *Allonarcys* Needham and Claassen: abdomen with lateral processes (Fig. 35.2*a*); 4 species, southern Quebec to Ga.

        Subgenus *Pteronarcys* s.s.: abdomen without processes; 4 species, transcontinental except for the southern plains region (Kan. south). *P. (P.) dorsata* Say is abundant across the northern half of the continent, south to Mont., Minn., and in the Appalachians to Ga.

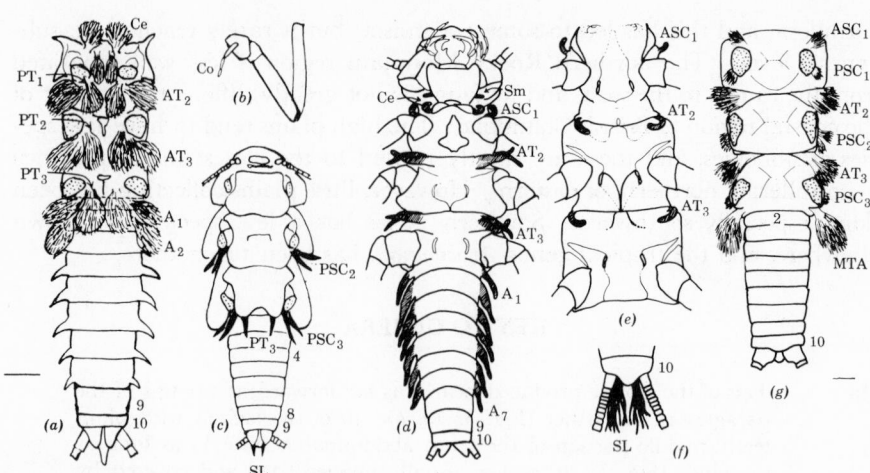

Fig. 35.2. Ventral surface of nymphs, showing thoracic and abdominal gills. (*a*) *Pteronarcys* (*Allonarcys*). (*b*) *Taeniopteryx* (gill on coxa). (*c*) *Peltoperla* (s.s.). (*d*) *Arcynopteryx* (*Oroperla*). (*e*) *Arcynopteryx* (*Megarcys*). (*f*) *Neoperla* (dorsal view). (*g*) *Acroneuria internata*. (Symbols for names of gills are explained in the text.)

4a (2) Thoracic sterna not produced; ocelli 3 (Fig. 35.7*a*) . . . .
            Family **Nemouridae**   5

4b Thoracic sterna produced posteriorly into thin plates broadly overlapping the segment behind (Fig. 35.2*c*); ocelli 2 (Fig. 35.7*b*). . . . .
            Family **Peltoperlidae**
            Peltoperla Needham

The 5 subgenera are distinguishable as follows:

Subgenus *Viehoperla* Ricker: gills PSC$_2$ and PSC$_3$ (single); 1 rare Appalachian species (Tenn. and N. C.).

Subgenus *Peltoperla* s.s.: gills PSC$_2$ and PSC$_3$ (double), PT$_3$, SL (Fig. 35.2*c*); 5 species, southern Quebec to Ga. **P. (P.) maria** Needham and Smith is very common in cool Appalachian streams.

Subgenus *Yoraperla* Ricker: gills Ce, PSC$_1$–PSC$_3$ (double); 2 cordilleran species, **P. (Y.) brevis** Banks is abundant, except in the southern Rockies.

Subgenus *Sierraperla* Jewett: gills Ce, PSC$_1$ and PSC$_2$ (double, each branch forked), PSC$_3$ (single, forked); **P. (S.) cora** Needham and Smith is the only species; Calif., Nev.

Subgenus *Soliperla* Ricker: gills PSC$_2$ and PSC$_3$ (single); 3 species; southern cordillera, uncommon.

5a (4) Second tarsal segment, as seen in side view, about as long as the first (Fig. 35.3*a*) . . . . Subfamily **Taeniopteryginae**   6

5b Second tarsal segment much shorter than the first (Fig. 35.3*b*)   7

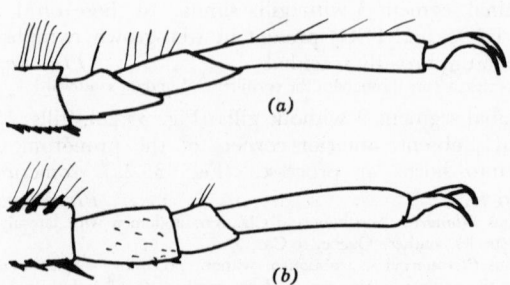

Fig. 35.3. Feet. (*a*) *Taeniopteryx*. (*b*) *Nemoura*.

6a (5) Single gills present on the inner side of each coxa (Fig. 35.2*b*); ninth sternite only slightly produced . . . . **Taeniopteryx** Pictet

Three species, almost wholly eastern. *T. maura* Pictet is an abundant winter stone-fly from southern Labrador, the Great Lakes, and Mississippi drainages to the Gulf of Mexico; it occurs also in the Willamette valley of Ore.

6b Gills absent; ninth sternite much produced, about twice as long as the tergite. . . . . **Brachyptera** Newport

Four subgenera and 15 species; transcontinental, except for the tundra and the southern plains; abundant in the cordillera. *B. fasciata* is abundant in the East.

7a (5) Hind wing pads nearly parallel to the long axis of the body (Fig. 35.8*a–c*); hind legs, when extended, barely reach the tip of the body **8**

7b Hind wing pads strongly diverging from the axis of the body (Fig. 35.8*d*); extended hind legs much surpass the tip of the body . . . . .

Subfamily **Nemourinae**

Nemoura Pictet

Few of our 12 subgenera can be easily distinguished as nymphs. Those with gills are as follows:

Subgenus *Amphinemura* Ris: gills 4, cervical, with 5 or more branches each (Fig. 35.4*a*); 7 species, some abundant; mostly eastern, but *N.* (*A.*) *venusta* Banks is common from Wyo. into Mexico.

Subgenus *Malenka* Ricker: gills as in *Amphinemura;* 9 species, cordilleran: *N.* (*M.*) *californica* Claassen is one of the most widespread and abundant.

Subgenus *Visoka* Ricker: gills 2, mental, branched (Fig. 35.4*c*); 1 species, *N.* (*V.*) *cataractae* Neave; cordillera south to Mont. and Wash., rarely Ore.

Subgenus *Zapada* Ricker: gills 4, cervical, simple (Fig. 35.4*b*) or in *cinctipes* Banks with usually 3 or 4 branches each; 5 western species, of which *cinctipes* is extremely common and widespread; 2 eastern species, uncommon.

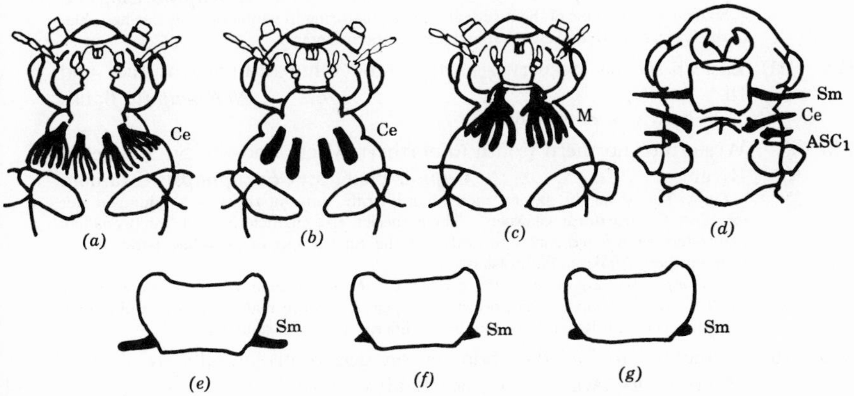

Fig. 35.4. Gills on the head and neck. (*a*) *Nemoura* (*Amphinemura*). (*b*) *Nemoura* (*Zapada*). (*c*) *Nemoura* (*Visoka*). (*d*) *Arcynopteryx* (*Perlinodes*). (*e*) *Isogenus* (*Hydroperla*). (*f*) *Isogenus* (*Malirekus*). (*g*) *Isogenus* (*Pictetia*). (*e–g* show the submentum only; symbols for names of gills are explained in the text.)

8a (7) Dorsal and ventral sclerotized portions of abdominal segments 1 to 9 divided by a membranous fold ventrolaterally (Fig. 35.5*a*) . . . . . .

Subfamily **Capniinae**   **9**

8b Only the first 7 abdominal segments, or fewer, divided by a membranous lateral fold (Fig. 35.5*b, c*) . . . . . . Subfamily **Leuctrinae**   **13**

9a (8) Cerci with mesal and lateral fringes of long silky swimming hairs, several hairs to a segment (Fig. 35.6*f*); subanal lobes not fused mesally in either sex . . . . **Isocapnia** Banks

About 10 species, southern Alaska and central British Columbia to northern Calif. and southern Colo. Most species are rather large for this subfamily. *I. grandis* Banks is largest, most widespread, and most abundant.

9b   Cerci lacking long swimming hairs; subanal lobes fused mesally in the mature male nymph (Fig. 35.6*a*, *b*). **10**

10a   (9)   Bristles on the body slender and inconspicuous (Fig. 35.6*b*, *c*), sometimes scarce or almost lacking.   **11**

10b   Most of the body and appendages densely covered with stout, conspicuous bristles (Fig. 35.6*a*) . . . . **Paracapnia** Hanson
         Two eastern species; *P. opis* Newman is abundant, Minnesota to Newfoundland, south to N. C. and the Ozarks.

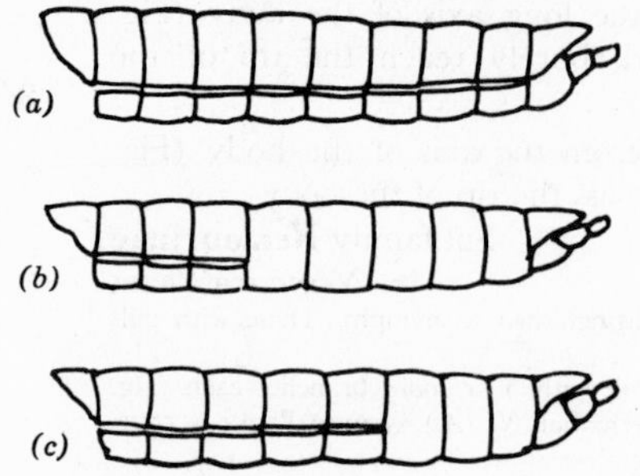

(*a*)

(*b*)

(*c*)

◀ **Fig. 35.5.** Sclerotization of pleurites (the narrow lateral band indicates reduced sclerotization). (*a*) *Isocapnia*. (*b*) *Leuctra* (s.s.). (*c*) *Leuctra* (*Zealeuctra*).

11a   (10)   Inner margin of the hind wing pad with a notch about halfway from base to tip (Fig. 35.8*b*) **12**

11b   Inner margin of the hind wing pad notched very close to the tip, if at all (Fig. 35.8*a*—pad sometimes absent); eastern . . . .
                                    Allocapnia Claassen
         Sixteen species, some of them very abundant; maturing in winter or early spring. Fla. to the Ozarks, north to Newfoundland and northern Ontario.

12a   (11)   Eastern species, occurring north to the Ohio drainage of Ind. and Ill. . . . . **Nemocapnia** Banks
         One species, *N. carolina* Banks.

12b   Western or northern (south to northern Mich. and the St. Lawrence River) . . . . **Capnia** Pictet and **Eucapnopsis** Okamoto
         Capnia has about 40 species, mostly cordilleran, some of them very abundant, for example, *C. gracilaria* Claassen. Three species are characteristic of the Canadian plains and the forested zone eastward, and the tundra species *nearctica* Banks occurs from northern Alaska to Baffin Island.
         Eucapnopsis has 2 species, both cordilleran. *E. brevicauda* Claassen is one of the smallest stoneflies, smaller than nearly all *Capnia* (a mature male nymph is 5 mm long, excluding cerci); it is widespread in the cordillera, and often abundant.

13a   (8)   Segments 1 to 7 of the abdomen divided ventrolaterally by a membranous fold; large species, nearly all western **14**

13b   Fewer abdominal segments divided (Fig. 35.5*b*, *c*). **Leuctra** Stephens
         The 5 subgenera can be distinguished as follows:

	Leuctra (s.s.)	*Despaxia* Ricker	*Zealeuctra* Ricker	*Moselia* Ricker	*Paraleuctra* Hanson
Segments divided laterally	1–4	1–5	1–6	1–6	1–6
Abdominal bristles	very sparse	absent	absent	numerous	sparse
Subanal lobes	separate	separate	half-fused	separate	separate
Number of species	15	1	1	1	4
Distribution	east	west	east	west	general

         Leuctra s.s. is typically eastern Canadian and Appalachian, but is found in spring streams south through Ill. and Ind. *L.* (*Zealeuctra*) *claasseni* Frison occurs through the Mississippi and Ohio Valleys, south to Ky. and Okla. *L.* (*Despaxia*) *glabra* and *L.* (*Moselia*) *infuscata* are known from central British Columbia to central Calif.

Paraleuctra has 4 cordilleran species, the commonest of which, **L.** (**P.**) *sara* Claassen, also occurs from northern Ind. and southern Ontario to Mass. and south to Ga.

14a (13) Body covered by rather coarse appressed pile, whose individual hairs are about $\frac{1}{5}$ as long as a middle abdominal segment; subanal lobes of the male nymph completely fused, the fused plate much produced and rounded in mature specimens, not notched (female unknown) (Fig. 35.6d) . . . . *Megaleuctra* Neave
    One Appalachian species (Tenn.) and 2 or 3 cordilleran species (southern British Columbia to Ore.), all rare.

14b Body with only extremely fine pile, appearing naked; subanal lobes of both sexes fused mesally for $\frac{1}{2}$ or $\frac{2}{3}$ of their length, leaving a notch at the tip (Fig. 35.6e) . . . . *Perlomyia* Banks
    Two cordilleran species, central British Columbia south to Ore. and Colo.; often abundant.

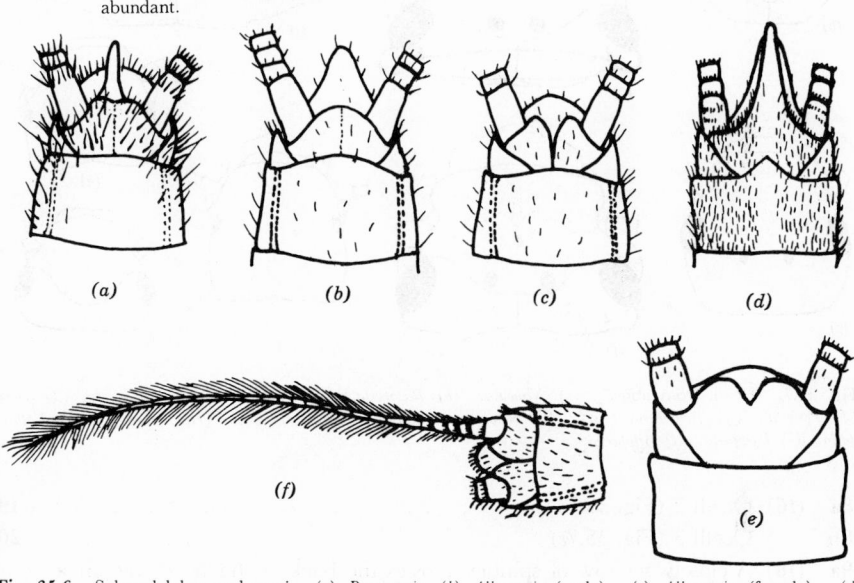

Fig. 35.6. Subanal lobes and cerci. (a) *Paracapnia*. (b) *Allocapnia* (male). (c) *Allocapnia* (female). (d) *Megaleuctra* (male). (e) *Perlomyia*. (f) *Isocapnia*.

15a (1) Profusely branched gills present at the corners of the thoracic sterna and above the front coxae, usually also above the other coxae (Fig. 35.2g); paraglossa broadly rounded (Fig. 35.1c); galea with a transverse suture near the middle. . . . . Family **Perlidae** 16

15b Thoracic gills, when present, single or double (Fig. 35.2d, e), usually entirely lacking; paraglossa pointed (Fig. 35.1d); galea not divided near the middle 25

16a (15) Eyes situated much anterior to the hind margin of the head (Fig. 35.7c). 17

16b Eyes situated normally, close to the hind margin of the head (Fig. 35.7d–g) 18

17a (16) Anterior ocellus absent (Fig. 35.7c); body uniformly colored . . . . . *Atoperla* Banks
    A. *ephyre* Newman is the only species; Minn., Mass., and southward.

17b Anterior ocellus present though small, indistinct in small nymphs; body boldly patterned . . . . *Perlinella* Banks
    P. *drymo* Newman is the only species; Minn. to Nova Scotia, and southward.

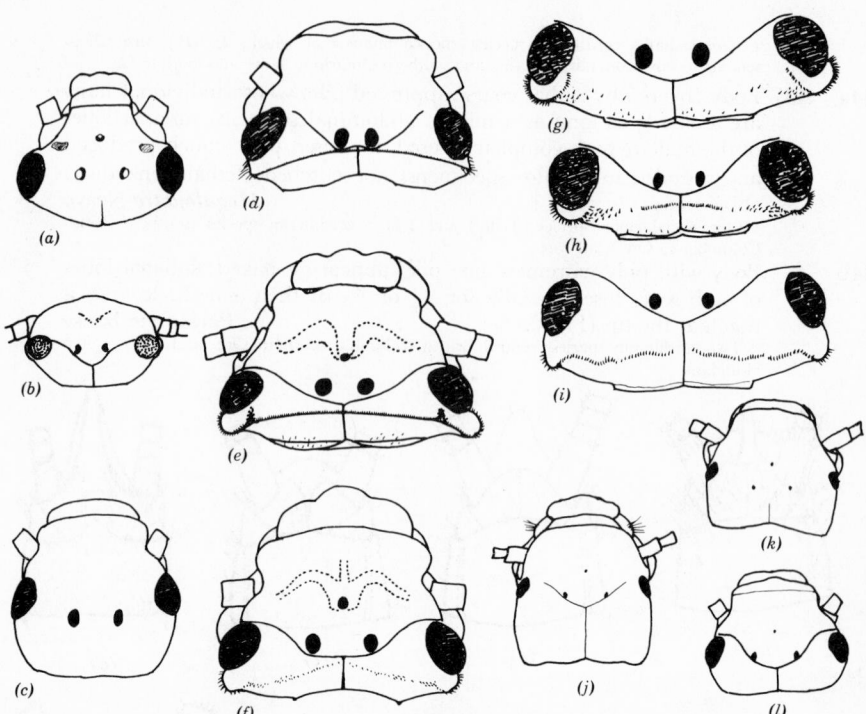

Fig. 35.7. Head, from above. (*a*) *Nemoura*. (*b*) *Peltoperla*. (*c*) *Atoperla*. (*d*) *Neoperla*. (*e*) *Claassenia*. (*f*) *Perlesta*. (*g*) *Acroneuria* (s.s.). (*h*) *Acroneuria* (*Calineura*). (*i*) *Acroneuria* (*Hesperoperla*). (*j*) *Kathroperla*. (*k*) *Paraperla*. (*l*) *Alloperla*.

18a	**(16)**	Ocelli 2 (Fig. 35.7*d*)	**19**
18b		Ocelli 3 (Fig. 35.7*e*)	**20**

19a (18) A closely set row of spinules crosses the back of the head, set on a low occipital ridge (Fig. 35.7*d*); subanal gills present (Fig. 35.2*f*); all supracoxal gills present . . . . **Neoperla** Needham

Two species; *N. clymene* Newman is very abundant from the plains eastward, north to Minn. and New Brunswick.

19b No row of spinules across the back of the head; subanal gills absent (Fig. 35.2*g*); gills PSC_2 and PSC_3 absent . . . .

Anacroneuria Klapálek

Numerous species in Mexico; once recorded from Tex.

20a (18) A closely set regular row of spinules completely crosses the back of the head, inserted on a low occipital ridge (Fig. 35.7*e*) **23**

20b Occipital ridge absent; spinules on the back of the head present mainly at the sides (Fig. 35.7 *g*), or else arranged in a transverse row of varying completeness but always at least a little bit wavy or irregular (Fig. 35.7*f*) **21**

21a (20) Abdomen without frecklelike spots **22**

21b Abdomen covered with freckles; occipital spinules in an irregular line which is nearly complete across the head (Fig. 35.7*f*); subanal gills present (Fig. 35.2*f*) . . . . **Perlesta** Banks

Two species, eastern. *P. placida* Hagen is very abundant from southern Manitoba and the Great Lakes region southward.

22a **(21)** Spinules present somewhere in the central half of the occiput (Fig. 35.7*h, i*) . . . . (in part) *Acroneuria* Pictet

> Subgenus *Hesperoperla* Banks: line of occipital spinules broadly discontinuous at the midline (Fig. 35.7*i*); subanal gills present. *A.* (*H.*) *pacifica* Banks is the only species; abundant throughout the cordillera southward from central British Columbia.
>
> Subgenus *Calineuria* Ricker: line of occipital spinules continuous across the midline but broken at the sides (Fig. 35.7*h*); subanal gills absent; 2 cordilleran species; *A.* (*C.*) *californica* Banks is abundant.
>
> Subgenus *Attaneuria* Ricker: line of occipital spinules practically continuous from one side of the head to the other; subanal gills absent. *A.* (*A.*) *ruralis* Hagen is the only species; eastern, north to the St. Lawrence and Ottawa Rivers.

22b Spinules absent from the central region of the occiput (except a few very close to the hind margin), present mainly in patches near the eyes (Fig. 35.7*g*) . . . . (in part) *Acroneuria* Pictet

> Ten eastern species, including 8 of *Acroneuria* s.s. and 1 each of the subgenera *Eccoptura* Klapálek and *Beloneuria* Needham and Claassen. Abundant in most streams and rivers and along the stony shores of larger lakes. *A. lycorias* Newman occurs from Fla. to Hudson Bay (Churchill, Manitoba).

23a **(20)** Subanal gills present (Fig. 35.2*f*) **24**

23b Subanal gills absent (Fig. 35.2*g*) . . . . *Paragnetina* Klapálek

> Five species, eastern prairie border to northern Saskatchewan and east to the Atlantic. *P. immarginata* Say is abundant in the Appalachians, *P. media* Walker northward.

24a **(23)** Abdominal segments yellow, broadly bordered with black . . . . . . *Phasganophora* Klapálek

> *P. capitata* Pictet, the only species, is abundant and variable; eastern, north to the St. Lawrence valley and Minn. (one Mont. record).

24b Abdominal segments almost wholly brown . . . . . . *Claassenia* Wu

> *C. sabulosa* Banks is the only species; cordillera from N. M. to northern British Columbia, and eastward north of the prairies to the western and southern tributaries of Hudson Bay.

25a **(15)** Hind wing pads set at an angle to the axis of the body (Fig. 35.8*e*); cerci usually at least as long as the abdomen; body almost always pigmented in a distinct pattern, on some part or other . . . .

Family **Perlodidae** **26**

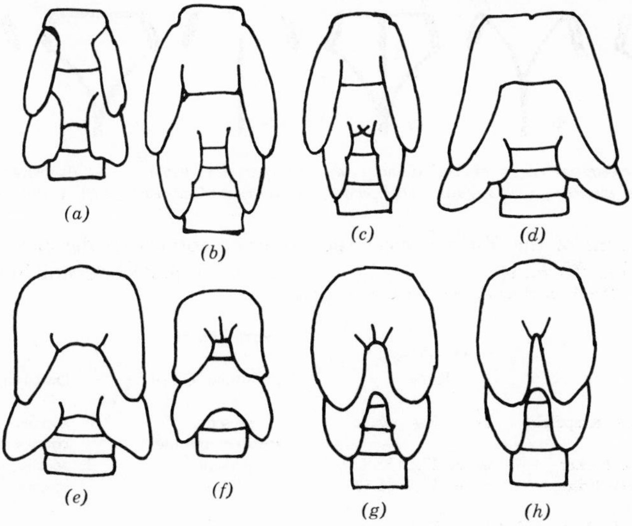

Fig. 35.8. Wing pads. (*a*) *Allocapnia*. (*b*) *Capnia*. (*c*) *Leuctra*. (*d*) *Nemoura*. (*e*) *Isoperla*. (*f*) *Kathroperla*. (*g*) *Alloperla*. (*h*) *Hastaperla*.

25b   Axis of the pads of the hind wings nearly parallel to that of the body (Fig. 35.8 g, h) except in mature *Kathroperla* (Fig. 35.8 f); cerci not more than $\frac{3}{4}$ as long as the abdomen; body almost uniformly brown, without a distinct pattern . . . . . . Family **Chloroperlidae**   38

26a **(25)** Gills absent from the thorax.   27

26b   Gills present on the thorax (Fig. 35.2d, e) . . . .

                     (in part) *Arcynopteryx* Klapálek

Subgenus *Oroperla* Needham: gills Sm, Ce, ASC_1, AT_2 and AT_3 (these double), A_1–A_7 (Fig. 35.2d); mesosternal ridges as in Fig. 35.9a; 1 species, *A. (O.) barbara* Needham; Sierra Nevada, rare.

Subgenus *Perlinodes* Needham and Claassen: gills Sm, Ce, ASC_1, AT_2, AT_3 (Fig. 35.4d); mesosternal ridges as in Fig. 35.9a; 1 species, *A. (P.) aurea* Smith; western cordillera north to Wash.

Subgenus *Megarcys* Klapálek: gills Sm, ASC_1, AT_2, AT_3, (Fig. 35.2e); mesosternal ridges as in Fig. 35.9b; 5 species, cordillera generally, sometimes abundant.

Subgenus *Setvena* Ricker: gills Sm, AT_2, AT_3; mesosternal ridges as in Fig. 35.9a; 2 species; Wash. and southern British Columbia; abundant near timber line.

27a **(26)** Submental gills present, at least twice as long as their greatest width (Fig. 35.4e).   28

27b   Submental gills about as long as wide (Fig. 35.4f, g), or absent. . . .   29

28a **(27)** Arms of the Y-ridge of the mesosternum meet or approach the anterior corners of the furcal pits (Fig. 35.9b, f) . . . .

                     (in part) *Arcynopteryx* Klapálek

Subgenus *Arcynopteryx* s.s.: arms of the Y-ridge approach but do not meet the furcal pits (Fig. 35.9f); denticles of the major cusp of the right mandible inconspicuous or absent. *A. (A.) compacta* MacLachlan is the only species; arctic, boreal, and alpine (south to northwestern Montana, the Michigan shore of Lake Superior, and the White Mountains of N. H.).

Subgenus *Frisonia* Ricker: arms of the Y-ridge meet the furcal pits (Fig. 35.9b); both mandibular cusps with conspicuous denticles; spine of the lacinia $\frac{1}{2}$ as long as the whole lacinia (measured along the outer curvature); 2 cordilleran species; *A. (F.) parallela* Frison is very abundant.

Subgenus *Skwala* Ricker; like *Frisonia*, except that the spine of the lacinia is only $\frac{1}{3}$ of the total lacinial length. *A. (S.) picticeps* Hanson is the only species; southern British Columbia to Ore., rare.

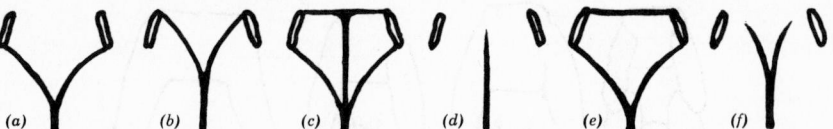

   (a)   (b)   (c)   (d)   (e)   (f)

Fig. 35.9. Mesosternal ridges of Perlodidae. (*a*) *Arcynopteryx* (*Setvena*). (*b*) *Arcynopteryx* (*Mergarcys*). (*c*) *Isogenus* (*Isogenoides*). (*d*) *Isogenus* (*Diploperla*). (*e*) *Isogenus* (*Malirekus*). (*f*) *Arcynopteryx* (s.s.).

28b   Arms of the Y-ridge meet the posterior corners of the furcal pits (Fig. 35.9a, c). . . . . (in part) *Isogenus* Newman

The 3 subgenera are distinguished as follows:

	Mesosternal Ridges	Serrations of Outer Mandibular Cusps	Distribution
Isogenoides Klapálek	as in Fig. 35.9c	present, sometimes minute	7 species, trans-continental
Helopicus Ricker	as in Fig. 35.9a	present	2 species, eastern
Hydroperla Frison	as in Fig. 35.9a	absent	2 species, eastern

I. (Isogenoides) frontalis Newman is abundant in large rivers across the continent northward, south to central Ore. and southern Colo. in the West, and to central N. Y. in the East. Other species are locally abundant. The commonest *Helopicus, I. (H.) subvarians* Banks, occurs from central Ontario to Ga. The 2 species of *Hydroperla* occur from Ind. to Okla. or Tex.

29a (27) Lacinia with only a single cusp and lacking hairs or spinules (Fig. 35.10*d*)............................... 30

29b Lacinia with a minor spine at least ½ as long as the major cusp, and usually also with spinules or hairs along the mesal margin (Fig. 35.10*e, f*)............................. 31

30a (29) Mesosternal ridges as in Fig. 35.9*e* .... ***Rickera*** Jewett
One species, ***R. venusta*** Jewett; southern Ore. and northern Calif., rare.

30b ·Mesosternal ridges as in Fig. 35.9*a* . . . (in part) ***Isogenus*** Newman
Subgenus ***Kogotus*** Ricker: 3 species, cordilleran.
Subgenus ***Remenus*** Ricker: *I.* (*R.*) *bilobatus* Needham and Claassen is the only species; Appalachian.

31a (29) Lacinia with an angle or knob, bearing a tuft of hairs or spinules, situated just below the insertion of the smaller spine (Fig. 35.10*c*) . . 32

31b Mesal margin of the lacinia tapering gradually from the smaller spine to the base, lacking an angle or knob (Fig. 35.10*a, b*)....... 34

32a (31) Submental gills present, short-conical (Fig. 35.4*f*); abdominal tergites dark, each with a transverse row of 8 small but distinct white spots; eastern .... (in part) ***Isogenus*** Newman
Subgenus ***Malirekus*** Ricker: a single species, *I.* (*M.*) *hastatus* Banks; Gaspé peninsula to Ga.; abundant in cool Appalachian streams.

32b Submental gills, if present, short and rounded (Fig. 35.4*g*); abdomen with or without small white spots, but with larger white areas as well.................................. 33

33a (32) Mid-line of head in front of the anterior ocellus wholly dark in color; submental gills absent; arctic or alpine ....
(in part) ***Diura*** Billberg
Subgenus ***Diura*** s.s.: 2 species; *D.* (*D.*) *bicaudata* Linné is abundant in large arctic rivers; *D.* (*D.*) *nanseni* Kempny is known from Mt. Washington, N. H., and Mt. Albert, Gaspé Peninsula.

33b Mid-line of head in front of the anterior ocellus wholly light in color, or almost so; submental gills present or absent; southern Appalachian species .... (in part) ***Isogenus*** Newman
Subgenus ***Yugus*** Ricker: 2 species, southern Appalachians.

34a (31) Submental gills present, about as long as wide (Fig. 35.4*g*) 35
34b Submental gills absent..................... 37

35a (34) Major lacinial spine short, about 0.3 of the total lacinial length as measured along the outer curvature (Fig. 35.10*a*); mesosternal ridges as in Figure 35.9*a* .... (in part) ***Diura*** Billberg
Subgenus ***Dolkrila*** Ricker: *D.* (*D.*) *knowltoni* Frison is the only species; cordillera, Yukon to Ore. and Colo.

35b Major lacinial spine 0.4 to 0.5 of the total lacinial length (Fig. 35.10*b*) .... (in part) ***Isogenus*** Newman
Subgenus ***Diploperla*** Needham and Claassen: mesal margin of the lacinia with only 3 to 4 hairs; mesosternal ridges as in Fig. 35.9*d*. *I.* (*D.*) *duplicatus* Banks is the only species; eastern, north to the Ohio Valley.
Subgenus ***Pictetia*** Banks: mesal margin of the lacinia with 10 or more hairs (Fig.

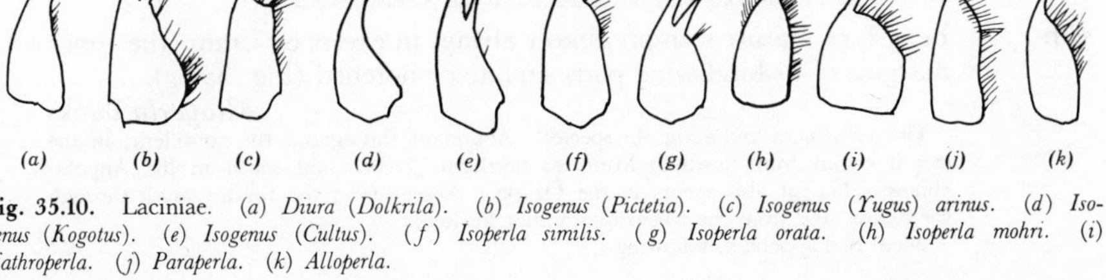

(a)   (b)   (c)   (d)   (e)   (f)   (g)   (h)   (i)   (j)   (k)

Fig. 35.10. Laciniae. (*a*) *Diura* (*Dolkrila*). (*b*) *Isogenus* (*Pictetia*). (*c*) *Isogenus* (*Yugus*) *arinus*. (*d*) *Isogenus* (*Kogotus*). (*e*) *Isogenus* (*Cultus*). (*f*) *Isoperla similis*. (*g*) *Isoperla orata*. (*h*) *Isoperla mohri*. (*i*) *Kathroperla*. (*j*) *Paraperla*. (*k*) *Alloperla*.

35.10*b*); mesosternal ridges as in Fig. 35.9*a*. *I. (P.) expansus* Banks is the only species; Mont. to Colo.
Note: See appendix, p. 957.

36a (34) Tips of the Y-ridge of the mesosternum reaching the anterior corners of the furcal pits (as in Fig. 35.9*b*), *and* a transverse ridge joins the anterior corners of the furcal pits; lacinia with only 1 or 2 marginal hairs proximad of the minor spine; abdominal segments dark above, with a narrow median yellow stripe. . . . . . (in part) *Isogenus* Newman
Subgenus *Osobenus* Ricker: the single species, *I. (O.) yakimae* Hoppe, inhabits the western part of the cordillera from central British Columbia to Calif.

36b Tips of the Y-ridge reach or approach the posterior corners of the furcal pits (Fig. 35.9*a, e*); transverse ridge usually lacking; lacinia with at least 2 marginal hairs, usually many **37**

37a (36) Each abdominal tergite dark on its anterior half, and light posteriorly except sometimes along the very margin, giving a transversely banded appearance; lacinia with only 2 or 3 marginal hairs (Fig. 35.10*e*) . . . . (in part) *Isogenus* Newman
Subgenus *Cultus* Ricker: *I. (C.) decisus* Walker occurs from Hudson Bay to the Gulf of St. Lawrence and in the mountains to Tenn.; there are also 3 rather common cordilleran species.

37b Abdominal pattern usually consisting of longitudinal stripes, less often plain or spotted; when rarely banded, the posterior half of each tergite is dark and the anterior half light; lacinia variable, rarely like Fig. 35.10*e* . . . . *Isoperla* Banks
A genus of truly transcontinental distribution, from Labrador to Fla.; from Alaska to Baja California, Mexico; and from Great Bear Lake to Tex. A number of subgenera have been proposed, but no adequate revision is yet available. Figure 35.10*f–h* shows several types of lacinia; others approach 35.10*i*. About 45 described species in N. A., and many undescribed.
Note: See appendix, p. 957.

38a (25) Sides of the head, behind the eyes, parallel to the long axis of the body; eyes small (Fig. 35.7*j, k*) . . . . . . Subfamily **Paraperlinae**   **39**
38b Sides of the head, behind the eyes, immediately deflected inward; eyes normal (Fig. 35.7*l*) . . . . Subfamily **Chloroperlinae**   **40**

39a (38) Cerci about 0.7 of the length of the abdomen; lacinia slightly broader than the length of its terminal spine (Fig. 35.10*i*) . . . .
Kathroperla Banks
K. perdita Banks is the only species; occurs in the cordillera except the Rocky Mountain region south of Mont.

39b Cerci about 0.5 of the length of the abdomen; breadth of the lacinia about 0.7 of the length of its spine (Fig. 35.10*j*) . . . .
Paraperla Banks
P. frontalis Banks occurs from southern British Columbia and Alberta to Calif. and N. M., and is often abundant; another species occurs north to the Yukon.

40a (38) Length of mature nymph 5–7 mm (excluding the cerci); the inner margins of the hind wing pads almost straight (Fig. 35.8*h*) . . . . . .
Chloroperla Newman and *Hastaperla* Ricker
Chloroperla: *C. terna* Frison is our only species; southern Appalachians, rather rare. *Hastaperla*: 3 species; *H. brevis* Banks is abundant across the continent northward, south to Ga. and to central British Columbia. *H. chilnualna* Ricker occurs in the western cordillera from Vancouver Island to the Sierra Nevada.

40b Length of mature nymph almost always in excess of 7 mm; the inner margins of its hind wing pads sinuate or notched (Fig. 35.8*g*). . . . .
Alloperla Banks
Five subgenera and about 45 species. Abundant throughout the cordillera; in the east it occurs from northern Minn. to northern Quebec and south in the Appalachians to Ga.; it also occurs in the Ozarks. Absent from the tundra south through the plains, and from the Mississippi Valley lowlands.
Note: See appendix, following.

Appendix. Genera and Subgenera Whose Nymphs Are Unknown

Isogenus, subgenus *Chernokrilus* Ricker. Two rare species, western Ore. and northern Calif. Will probably key to couplet **35.**

Calliperla Banks. A single species in western Ore. and Calif. Assigned to the Isoperlinae, it is most likely to key to couplet **37.**

Utaperla Ricker. A single rare species, northern Utah to the Yukon. Although adult characters place it in Paraperlinae, the nymphal structure will probably be closer to that of *Alloperla* than to that of *Paraperla.*

References

The papers and monographs listed include the more important of those that describe nymphs of American species, and a few major works concerning European species.

Aubert, Jacques. 1946. Les plécoptères de la Suisse romande. *Mitt. schweiz. entomol. Ges.,* 20:1–128. **Brink, Per. 1949.** Studies on Swedish stoneflies (Plecoptera). *Opuscula Entomologica,* Suppl. 11:1–250. **Claassen, P. W. 1931.** Plecoptera nymphs of America (north of Mexico). *Thos. Say Foundation of the Entomol. Soc. Am. Publ., No.* 3:1–199. **Clark, Robert L. 1934.** The external morphology of *Acroneuria evoluta* Klapálek. *Ohio J. Sci.,* 34:121–128. **Frison, T. H. 1929.** Fall and winter stoneflies, or Plecoptera, of Illinois. *Bull. Illinois Nat. Hist. Survey,* 18:345–409. **1935.** The stoneflies, or Plecoptera, of Illinois. *Bull. Illinois Nat. Hist. Survey,* 20:281–471. **1942.** Studies of North American Plecoptera, with special reference to the fauna of Illinois. *Bull. Illinois Nat. Hist. Survey,* 22:231–355. **Harden, Philip H. 1942.** The immature stages of some Minnesota Plecoptera. *Ann. Entomol. Soc. Am.,* 35:318–331. **Harden, Philip H., and Clarence E. Mickel. 1952.** The stoneflies of Minnesota (Plecoptera). *Univ. Minnesota Agr. Exp. Sta., Tech. Bull.,* No. 201:1–84. **Holdsworth, R. P. 1941a.** The life history and growth of *Pteronarcys proteus* Newman (Pteronarcidae: Plecoptera). *Ann. Entomol. Soc. Am.,* 34:494–502. **1941b.** Additional information and a correction concerning the growth of *Pteronarcys proteus* Newman (Pteronarcidae: Plecoptera). *Ann. Entomol. Soc. Am.,* 34:714–715. **Hynes, H. B. N. 1941.** The taxonomy and ecology of the nymphs of British Plecoptera, with notes on the adults and eggs. *Trans. Roy. Entomol. Soc. London,* 91:459–557. **1948.** The nymph of *Anacroneuria aroucana* Kimmins (Plecoptera, Perlidae). *Proc. Roy. Entomol. Soc. London* (A), 23:105–110. **Jewett, S. J., Jr. 1954a.** New stoneflies from California and Oregon. *Pan-Pacific Entomologist,* 30:167–179. (Nymphs of *Soliperla* and *Sierraperla.*) **1954b.** New stoneflies (Plecoptera) from western North America. *J. Fisheries Research Board Can.,* 11:543–549. (Nymph of *Megaleuctra.*) **1955.** Notes and descriptions concerning western North American stoneflies (Plecoptera). *Wasmann J. Biol.,* 13:145–155. (Nymphs of *Osobenus* and *Rickera.*) **1956.** Plecoptera. In: R. L. Usinger (ed.). *Aquatic Insects of California, with Keys to North American Genera and California Species,* pp. 155–181, University of California Press, Berkeley and Los Angeles. **Kühreiber, J. 1934.** Die Plekopterenfauna Nordtirols. *Berichte naturwiss.-med. Vereines Innsbruck,* 43/44:1–219. **Lestage, J. A. 1921.** Plecoptera. In: Rousseau, *Les larves et nymphes aquatiques des insectes d'Europe,* Vol. 1, pp. 274–320. **Neave, Ferris. 1934.** Stoneflies from the Purcell Range, B. C. *Can. Entomologist,* 66:1–6. (Nymph of *Kathroperla.*) **Needham, J. G. 1933.** A stonefly nymph with paired lateral abdominal appendages. *J. Entomol. and Zool.,* 25:17–19. (Nymph of *Oroperla.*) **Ricker, W. E. 1943.** Stoneflies of southwestern British Columbia. *Indiana Univ. Publ. Sci. Ser.,* No. 12:1–145. **1950.** Some evolutionary trends in Plecoptera. *Proc. Indiana Acad. Sci.,* 59:197–209. **1952.** Systematic studies in Plecoptera. *Indiana Univ. Publ., Sci. Ser.,* No. 18: 1–200. **Wu, C. F. 1923.** Morphology, anatomy and ethology of *Nemoura.* *Bull. Lloyd Library* Cincinnati, No. 23:(Ent. Ser., No. 3) 1–81.

36

Hemiptera[1]

H. B. HUNGERFORD

The water bugs are characterized by a segmented beak. Many species are dimorphic and may be entirely wingless, possess short wings, or have fully developed hemelytra and flight wings which often make specific determination difficult. Of the fourteen families treated here, three live on the shore (Saldidae, Ochteridae, and Gelastocoridae); five stride over the water surface or floating vegetation (Gerridae, Veliidae, Hebridae, Mesoveliidae, and Hydrometridae); six live beneath the surface (Naucoridae, Belostomatidae, Nepidae, Notonectidae, Pleidae, and Corixidae). The first eight families belong to the semiaquatic and the last six to the aquatic Hemiptera. All of the true aquatics have their antennae hidden beneath the head, and some of the others that now and then submerge themselves, have the antennae either hidden, as do the Gelastocoridae (toad bugs), or shortened, as do the Ochteridae and the hebrid genus *Merragata* which spends considerable time beneath the surface film. General features are shown in Fig. 36.5. The antennae are 3-, 4-, or 5-segmented; subsegments, if they occur, are not counted. The tarsi may be one-, two-, or three-segmented and if claws are present, they may be terminal or inserted well before the tip as they are in Gerridae and Veliidae. The following key covers the fauna of North, Central, and Insular America, not including Trinidad.

[1]Contribution No. 818, Department of Entomology, University of Kansas.

Distribution

Although the key gives an indication of the number of species of a given genus in the territory covered, it gives no indication of the area covered by that genus. Therefore, the following information may be helpful:

Hydrometridae. *Hydrometra martini* Kirkaldy is the common species found over most of the U. S. The number of species increases as one goes south. The genus *Limnobatodes* Hussey is known from a single specimen from Honduras.

Veliidae. The genus *Macrovelia* Uhler is represented by 1 species, *M. hornii* Uhler now known from Calif., Ariz., N. M., Colo., S. D., and Ore. The genus *Rhagovelia* Mayr is widespread, the species usually occur in numbers skating over flowing waters. They are rapid in their movements. The genus *Microvelia* Westwood includes a considerable number of very small striders and is widely distributed from Canada south. Some species have a vast range. *M. hinei* Drake, for example, is found from Ore. to Conn. and south to Fla. and west to Ariz. *M. pulchella* Westwood and its subspecies occurs from Mich. to Peru. The genus *Velia* contains few species but some of them are the largest in the family. *V. brachialis* Stål is common in Fla. and westward through Tex. and Ariz. and south through Mexico to Peru and Brazil.

Gerridae. The best known genus is the widespread *Gerris* Fabricius. The large *G. remigis* Say is common from eastern Canada and Me. to Calif. and Mexico, and *G. marginatus* Say, a smaller species, is reported from many states. *Limnogonus* Stål with 12 species in the western hemisphere has only 2 in the U. S., *L. hesione* Kirkaldy in Ohio and southeast to Fla. and Cuba, and *L. franciscanus* Stål in southern Tex. to Calif. *Potamobates* Champion extends from Mexico to Peru. *Limnometra* Champion (not Mayr) is not found north of Mexico; although this genus is recorded as a synonym of *Tenagogonus* Stål, an older name, this New World group has recently received a generic name of its own, *Tachygerris*. There are several genera belonging to the subfamily *Halobatinae* that are small striders. *Metrobates hesperius* Uhler is found from N. Y. to Fla. and west to Kan. and Minn. *Telmatometra* Begroth is found in British Honduras, Costa Rica, and Mexico. *Brachymetra* Mayr is found in Central and Insular America. *Trepobates pictus* (H. S.) is a little striped species with a range from Me. to Fla. and west to Ariz. *Rheumatobates rileyi* Bergroth in which the males have grasping antennae and curiously modified hind legs is widespread over the eastern states and west to Kan. and Minn.

Mesoveliidae. *Mesovelia mulsanti* B. White is the common species everywhere in the U. S. and ranges from Canada to Brazil.

Hebridae. These very small striders are widespread, yet seldom collected. Their curiously reduced wing venation is often figured in texts.

Saldidae. The common and widespread *Saldula pallipes* (Fabricius) is an Old World species that is found in both N. A. and S. A. Most of our species belong to the genus *Saldula* Van Duzee. The genus *Pentacora* Reuter contains our largest saldids and *P. signoretti* Guérin is most colorful, being black spotted with yellow. It is widespread from N. Y. to Fla., west to Calif., and from

Canada to Mexico; it is very common in Tex. and N. M. The curious genus *Saldoida* Osborn is represented by 1 species in the Philippine Islands, another in Formosa, and 2 in the U. S. These were both described from Fla., and *S. slossoni* Osborn is recorded also from Va., Ala., and Tex.

Ochteridae. Represented in N. A. by the genus *Ochterus* Latreille. Uhler described *O. americanus* and said it occurs on margins of brooks and ponds from Mass. to Tex. It occurs in Kan. and Neb. but is not commonly collected. *O. banksi* Barber winters as fourth instar nymph in Va. (M. L. Bobb, 1951).

Gelastocoridae. The common toad bug in the U. S. is *Gelastocoris oculatus* (Fabricius). It has been taken in many states from coast to coast and from Canada to Mexico. The genus *Nerthra* has 32 species in the New World but only 3 of these appear in the U. S. *N. martini* Todd occurs in Calif., Ariz., and Ga. *N. usingeri* Todd in Calif., and *N. stygica* Say in Fla.

Belostomatidae. This family contains our largest water bugs. Those of the genera *Benacus* Stål and *Lethocerus* Mayr are called "giant water bugs" and "electric light bugs"; the smaller species belong to the genera *Belostoma* Latreille and *Abedus* Stål and are often called "toe biters." *Benacus griseus* Say is a common species from Kan. eastward to N. J., and from Mich. southward to Fla., and westward through Tex. and Ariz. into Mexico and Cuba. *Lethocerus americanus* Leidy is our commonest giant water bug and is known from Me. westward to Ore. and Calif. It is abundant in northern Mich. and Kan. *Belostoma flumieneum* Say, often taken with eggs on the back of the male, is our best-known species, and ranges from Me. westward to Calif., and from Quebec, south and west to Tex., N. M., and Ariz. The genus *Abedus* is largely confined to the southwestern U. S., from N. M. westward to California and southward through Mexico to Panama. The males are often taken with their backs covered with eggs.

Nepidae. *Nepa apiculata* Uhler is our only species in this genus. It is found from Quebec and Ontario, south to Ga. and west to Kan. *Curicta* Stål is confined to Tex. and southwest. In the genus *Ranatra* Fabricius our well-known and most widely distributed species is *Ranatra fusca* Palisot de Beauvois (syn. *Ranatra americana* Montd.); it ranges from Quebec and Ontario south to Fla. and west to Tex., and is very common in Kan. and Mich.

Pleidae. The tiny *Plea striola* Fieber is common from Mich. east to N. J., south to Fla., and west to Calif.

Notonectidae. The genus *Notonecta* Linné is found all over N. A. and S. A. The common black and white backswimmer found across Canada and the U. S. from Ky. north is *Notonecta undulata* Say. Below Ky. it may be replaced by the darker *Notonecta indica* Linné, which is found from Md. to Calif. and south through Mexico to Colombia; it is common in the West Indies. The genus *Buenoa* consists of slender species that often swim in schools in open water at some distance beneath the surface. They are our only water bugs with hydrostatic balance and they possess oxyhemoglobin. Perhaps *B. margaritacea* Torre Bueno is our best-known species ranging from Mich. south to Ga., and from N. Y. and Va. west to Calif.

Naucoridae. *Pelocoris* Stål is our only genus in the East; *P. femoratus* Palisot de Beauvois is the name usually given to a species that is collected from Wis. east to Mass., south to Fla., and west to Kan. The genus *Ambrysus* Stål is represented by species from Tex., Colo., and Wyo., west to Calif., and south to Mexico.

Corixidae. This large family is represented by 20 genera in the area covered by this chapter. The genus *Sigara* Fabricius with its 48 species contains by far the largest number of species and is represented everywhere. *S. alternata* (Say) is our most common and widespread species. None of the species is large. The genus *Hesperocorixa* Kirkaldy contains the large corixids that are found across the U. S. from Me. to Wash. In the West we have *H. laevigata* (Uhler), and elsewhere the common species is *H. vulgaris* (Hungerford) which also occurs on the West Coast. The genus *Trichocorixa* Kirkaldy contains only small shining corixids having the male strigil on the left side of the abdomen, and it is widespread and frequently found in saline waters. *Corisella* Lundblad is a genus of moderate to small species found mostly in the western U. S., although *C. tarsalis* (Fieber) has been taken in Manitoba, Pa., and N. Y. *Cenocorixa* Hungerford is a northwestern genus with *C. utahensis* (Hungerford) ranging east to Iowa; also, *C. bifida* (Hungerford) has been taken in R. I. if Uhler's labels are correct. Our commonest species of *Palmacorixa* Abbott is *P. buenoi* Abbott, which is found from N. Y. to Fla. and from Minn. to Tex. It is common in streams and often brachypterous. The genus *Ramphocorixa* Abbott is represented by *R. acuminata* (Uhler), which is found from Washington, D. C. west to Mexico and from Minn. to Ga. It is found in pasture ponds and lays its eggs on crayfish. In the north we find the genera *Callicorixa* White, *Cymatia* Flor, *Glaenocorisa* Thomson, *Dasycorixa* Hungerford, and *Arctocorisa* Wallengren. *Callicorixa alaskensis* Hungerford ranges from Noorvik, Alaska south to Utah and in the east to Pa. and N. H. *C. audeni* Hungerford, the only species lacking the black spot on the hind tarsus, is common across Canada from British Columbia to Newfoundland and in the states adjoining. *Cymatia americana* Hussey, our only species, extends from Alaska to St. Paul, Minn. and has been taken in S. D. and Cheboygan County, Mich. *Glaenocorisa quadrata* Walley is known only from Quebec and Newfoundland. *Dasycorixa* Hungerford contains 2 species from Canada and *D. hybrida* (Hungerford) described from Minn. *Arctocorisa sutilis* (Uhler) is found from Kodiak, Alaska to Pingree Park, Colo., where *A. lawsoni* Hungerford also occurs. The other species are Canadian. In the southwestern U. S. and Mexico we find the genera *Graptocorixa* Hungerford, *Neocorixa* Hungerford, *Pseudocorixa* Jaczewski, *Morphocorixa* Jaczewski, and *Centrocorisa* Lundblad. *Graptocorixa abdominalis* (Say) is a large species with a red band on the abdomen, and is common from Tex. west to Nev. and south into Mexico. *Pseudocorixa beameri* (Hungerford) from Ariz. is our only representative of the genus in the U. S. *Morphocorixa compacta* (Hungerford) is common in Tex. and extends into Mexico. *M. lundbladi* (Jaczewski), a Mexican species has been taken in Ariz. *Centrocorisa nigripennis* (Fabricius), a common plump-bodied species in Insular America and Mexico, has been taken in Tex. The other species, *C. kollari* (Fabricius) is a South American species that extends north to Lower Calif. In Mexico we find the genera *Krizousacorixa* Hungerford, *Trichocorixella* Jaczewski, and very small *Tenegobia* Bergroth.

KEY TO GENERA

1a		Antennae exposed	**2**
1b		Antennae hidden.	**37**
2a	(1)	Head short and broad; eyes close to or at base of head	**4**
2b		Head very long and slender, at least nearly 3 times as long as broad across the eyes; eyes distant from base of head. (Marsh treaders) . . Family **Hydrometridae**	**3**
3a	(2)	Antennae 4-segmented. (17 species) . . . . . *Hydrometra* Lamarck	
3b		Antennae 5-segmented. (1 species) . . . . . . *Limnobatodes* Hussey	
4a	(2)	Claws inserted before apex of tarsus (except in *Macrovelia*) and winged forms without exposed scutellum	**5**
4b		Claws apical, or winged forms with exposed scutellum	**21**
5a	(4)	Hind femora not long, scarcely if at all exceeding tip of abdomen (except in *Microvelia longipes*). Vertex with a median longitudinal groove. (Broad-shouldered water striders). . . . . Family **Veliidae**	**17**
5b		Hind femora very long, exceeding apex of abdomen. Vertex without a median longitudinal groove, except in Rheumatobates. (Water striders) . . . . Family **Gerridae**	**6**
6a	(5)	Inner margins of eyes sinuate or concave behind the middle; body and abdomen comparatively long and narrow . Subfamily **Gerrinae**	**7**
6b		Inner margins of eyes convexly rounded; body and abdomen comparatively stout and broad . . . . Subfamily **Halobatinae**	**10**
7a	(6)	Metasternum divided by a transverse suture so that abdominal venter appears 7-segmented. (4 species) . . . . *Potamobates* Champion	
7b		Metasternum entire, not divided by a transverse suture	**8**
8a	(7)	Basal tarsal segment of front leg about $\frac{1}{2}$ as long as second. (3 species) . . . . *Limnogonus* Stål	
8b		Basal tarsal segment subequal with second	**9**
9a	(8)	Beak long, tip reaching middle of mesosternum. Antennae long, reaching apex of hind coxae, last segment longest. (2 species) . . . . (Syn. *Tenagogonus* auths. nec Stål) *Tachygerris* Drake	
9b		Beak short. Antennae shorter than above, last segment usually shorter than the first. (25 species). . . . . *Gerris* Fabricius	
10a	(6)	Tibia and first tarsal segment of middle leg with a fringe of long hairs. (Marine) . . . . *Halobates* Eschscholtz	
10b		Tibia and first tarsal segment of middle leg not as above	**11**
11a	(10)	Middle femur longer than hind femur. Very small striders. The males often with antennae and hind legs strangely modified. (15 species). . . . . *Rheumatobates* Bergroth	
11b		Middle femur shorter than hind femur. Plump-bodied striders . . .	**12**
12a	(11)	First tarsal segment of front leg at least $\frac{1}{2}$ the length of second. . . .	**13**
12b		First tarsal segment of front leg less than $\frac{1}{2}$ the length of second . . .	**14**
13a	(12)	Thorax and abdomen with heavy black longitudinal stripes. (1 species) . . . . *Eobates* Drake and Harris	

13b	Thorax and abdomen with no heavy markings, may have faint bars or stripes. (2 species). . . . . ***Brachymetra*** Mayr	
14a	**(12)** Body dorsoventrally flattened	**15**
14b	Body not dorsoventrally flattened.	**16**
15a	**(14)** Middle femur comparatively short, not more than $\frac{2}{3}$ the length of hind femur. Interocular space broader than long. (6 species). . . . ***Metrobates*** B-White	
15b	Middle femur long, nearly as long as hind femur. Interocular space narrower than long. (2 species) . . . . . ***Platygerris*** B-White	
16a	**(14)** First antennal segment longest. (7 species) . . . . ***Trepobates*** Uhler	
16b	First antennal segment not longest. (3 species) . . . . ***Telmatometra*** Bergroth	
17a	**(5)** Ocelli distinct. Hemelytra with 6 closed cells. (1 species) . . . . . . ***Macrovelia*** Uhler	
17b	Ocelli absent or represented by depressed pits only. Hemelytra with less than 6 closed cells, usually 4	**18**
18a	**(17)** Middle tarsi deeply cleft, with leaflike claws and plumose hairs arising from base of cleft. (50 species) . . . . . . ***Rhagovelia*** Mayr	
18b	Middle tarsi not deeply cleft and without plumose hairs arising from base of cleft.	**19**
19a	**(18)** Middle and hind tarsi with claws modified, 4 broad leaflike structures arising from a brief cleft at middle. (1 species) . . . . ***Veloidea*** Gould	
19b	Middle and hind tarsi with claws similar to those of front tarsi	**20**
20a	**(19)** Front tarsus 3-segmented. Antenna not inserted close to the eye. (14 species) . . . . ***Velia*** Latreille	
20b	Front tarsus appearing 1-segmented. Antenna inserted close to the eye. (35 species) . . . . ***Microvelia*** Weston	
21a	**(4)** Wingless, or if winged, without veins in the membrane of hemelytra (except in Mesoveloidea)	**22**
21b	Winged, with veins in the membrane of hemelytra	**25**

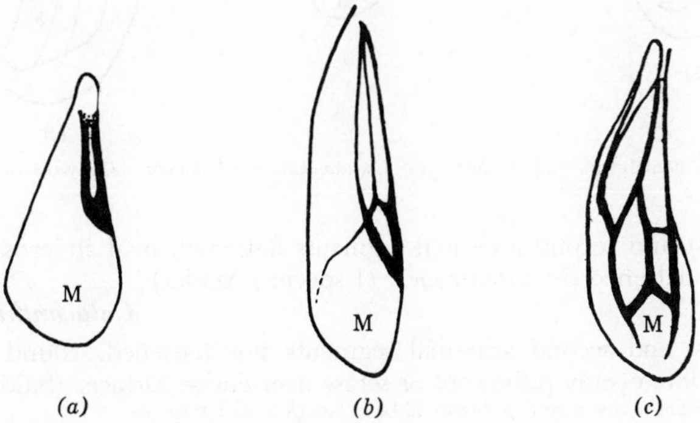

Fig. 36.1. Right hemelytra. (*a*) *Hebrus.* (*b*) *Mesovelia.* (*c*) *Microvelia.* M, membrane.

22a	**(21)** Bucculae small, not forming a longitudinal groove beneath the head . . . . Family **Mesoveliidae**	**23**
22b	Bucculae very large, forming a distinct longitudinal groove, which extends to base of head . . . . Family **Hebridae**	**24**

23a (22) Tarsal claws anteapical. (1 species) . . . *Mesoveloidea* Hungerford

23b   Tarsal claws apical. (6 species) . . . . *Mesovelia* B-White

24a (22) Antennae 5-segmented. (16 species). . . . . *Hebrus* Laporte

24b   Antennae 4-segmented. (4 species) . . . . *Merragata* B-White

25a (21) Antennae longer than head and pronotum; membrane of hemelytra with 4 or 5 closed cells (areolae). (Shore bugs) . . Family **Saldidae** **26**

25b   Antennae shorter than head and pronotum; membrane of hemelytra not as above **35**

26a (25) Anterior lobe of pronotum produced into a pair of prominent dorsally directed conical or thornlike processes. (2 species) . . . . *Saldoida* Osborn

26b   Anterior lobe of pronotum slightly convex to greatly arched, not produced into paired processes **27**

27a (26) Membrane of hemelytra with 4 cells **28**

27b   Membrane of hemelytra with 5 cells of which the fourth may be reduced. **34**

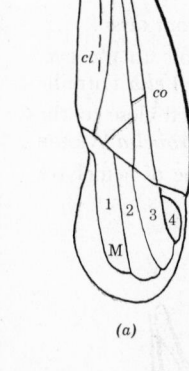 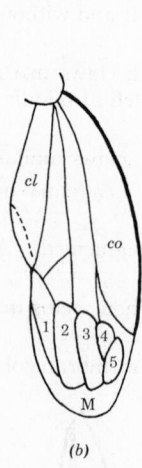 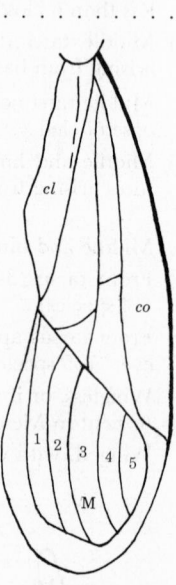

(a)   (b)   (c)

Fig. 36.2.   Right hemelytra.   (a) *Saldula*.   (b) *Chiloxanthus*.   (c) *Pentacora*.   *cl*, clavus; *co*, corium; M, membrane.

28a (27) First and second antennal segments flattened, oval in cross section, the flattened sides glabrous. (1 species; Alaska). . . . . *Calacanthia* Reuter

28b   First and second antennal segments not flattened, round in cross section, evenly pubescent or setose over entire surface. (Salda group) **29**
           Perhaps these 6 genera should all be in Genus *Salda* Fabricius.

29a (28) Anterior lobe of pronotum slightly convex **30**

29b   Anterior lobe of pronotum strongly convex. **33**

30a (29) The membrane with base of cell one prolonged $\frac{2}{5}$ or $\frac{1}{2}$ beyond base of cell 2. (4 species). . . . . *Salda* Fabricius

30b   The membrane with base of cell one prolonged slightly or not more than $\frac{1}{3}$ of its length beyond cell 2 . . . . **31**

31a (30) Corium with 2 entirely distinct veins, the inner one furcate toward apex, its branches touching the suture of the membrane. Apex of cell 1 usually touching or nearly touching apex of cell 2. (41 species) . . .
Saldula Var Duzee

31b Corium with at least the inner vein destroyed toward apex **32**

32a (31) Corium with its 2 veins wholly obsolete. (7 species) . . . .
Micracanthia Reuter

32b Corium with the outer vein and often even with the inner vein distinct at base. (2 species). . . . . *Teloleuca* Reuter

33a (29) Sides of pronotum straight. Callus by no means touching lateral margins, leaving a rather broad margin extended a little behind the middle. Ocelli slightly distant. (5 species) . . . *Ioscytus* Reuter

33b Sides of pronotum more or less sinuate. Callus occupying all or nearly all of the width of pronotum leaving only a narrow margin. Ocelli a little distant. (1 species) . . . . . . *Lampracanthia* Reuter

34a (27) Membrane of hemelytron with 5 oblong cells; the fourth cell not triangular, completely separating the third and fifth cells. (5 species) . . . . *Pentacora* Reuter

34b Membrane of hemelytron with 4 oblong cells; the fourth cell triangular, the third and fifth cells touching beyond apex of fourth cell. (1 species). . . . . *Chiloxanthus* Reuter

35a (25) Ocelli present. Bugs living along shore **36**

35b Ocelli absent. Aquatic bugs **39**

36a (35) Antennae exposed but shorter than head and thorax. (10 species). .
Family **Ochteridae** Latreille

36b Antennae hidden beneath head **37**

37a (1, 36) Ocelli present. (Toad bugs) . . . . Family **Gelastocoridae** **38**

37b Ocelli absent **39**

38a (37) Front tarsus not fused to tibia, articulate; 2 well-developed tarsal claws on front leg in adult; rostrum arising from apex of head, stout, recurved posteriorly. (8 species) . . . . *Gelastocoris* Kirkaldy

38b Front tarsus fused to tibia, not articulate; 1 well-developed tarsal claw on front leg in adult; rostrum appearing to rise on ventral surface of head, slender, projecting anteriorly or ventrally. (16 species) . .
(syn. *Mononyx* Laporte) *Nerthra* Say

39a (37) Hind tarsi with 2 distinct claws **40**

39b Hind tarsi without distinct claws **50**

40a (39) Abdomen with apical appendages **41**

40b Abdomen without apical appendages **47**

41a (40) Apical appendages of abdomen, 'short and flat, retractile. (Giant water bugs and toe biters) . . . . Family **Belostomatidae** **44**

41b Apical appendages of abdomen long and slender, not retractile. (Water scorpions) . . . . Family **Nepidae** **42**

42a (41) Body slender and elongate. (15 species) . . . . . *Ranatra* Fabricius

43a (42) Body elongate oval, width of hemelytra at middle subequal to width at base. (5 species). . . . . *Curicta* Stål

43b Body oval and flat, width of hemelytra at middle distinctly greater than at base. (1 species) . . . . *Nepa* Linné

44a (41) Anterior femur not sulcate. (1 species). . . . . *Benacus* Stål

44b Anterior femur sulcate. **45**

45a (44) Head not conically produced, rostrum short stout. First segment short. (8 species) . . . . *Lethocerus* Mayr

45b Head conically produced, rostrum long, thin. First segment long . . **46**

46a (45) Membrane of hemelytra large. (13 species) . . *Belostoma* Latreille

46b Membrane of hemelytra much reduced. (12 species). . *Abedus* Stål

47a (40) Very small, almost hemispherical bugs that swim on their backs . . .
Family **Pleidae**
(Tiny back swimmers; 4 species) . . . . *Plea* Leach

47b More or less flat bugs, mostly of moderate size with greatly enlarged front femora. (Creeping water bugs) . . . . . . Family **Naucoridae** **48**

48a (47) Rostrum slender and at least as long as front femur. (1 species) . . .
Potamocoris Hungerford

48b Rostrum very broad at base, tapering toward apex, much shorter than front femur **49**

49a (48) Anterior margin of pronotum deeply emarginate behind interocular space. (Fig. 36.3b). **50**

49b Anterior margin of pronotum straight or scarcely concave behind interocular space. (Fig. 36.3a) **52**

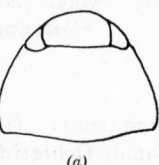 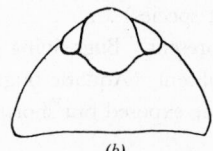

(a)   (b)

Fig. 36.3. Head and pronotum. (a) *Pelocoris*. (b) *Ambrysus*.

50a (49) Propleurae medially produced and platelike near posterior portion of prosternum, subcontiguous at middle and completely covering this portion of the prosternum. Abdominal venter densely pubescent, interrupted by small holes at spiracular openings, and by a transverse row of small holes behind each spiracle. Not strongly dimorphic **51**

50b Prosternum completely exposed, separated from the flattened pleura by simple sutures and not at all produced mesad as above. Abdominal venter naked and with a disclike area near each spiracle. Strongly dimorphic, the brachypterous forms most common and elongate oval in form. (4 species) . . . . . . *Cryphocricos* Signoret

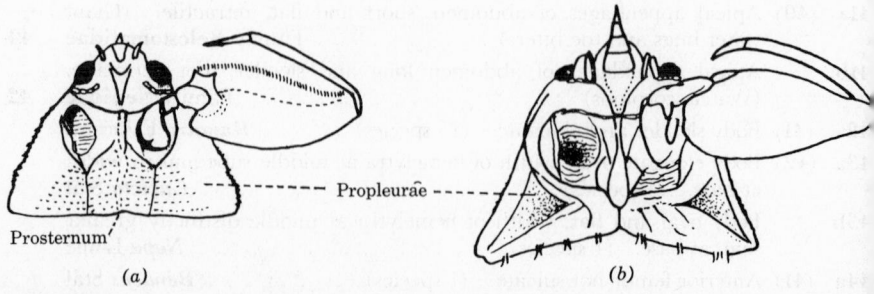

Prosternum

Propleurae

(a)   (b)

Fig. 36.4. Ventral view of head and prothorax. (a) *Cryphocricos*. (b) *Cataractocoris*. (after Usinger.)

51a **(50)** Surface covered with scattered papillalike granules. Eyes distinctly elevated and subglobose. Ventral surface margined by a glabrous area, especially on abdomen. (2 species) . . *Cataractocoris* Usinger

51b Surface in great part shagreened, sometimes highly polished. Head and pronotum often distinctly punctate but not granulate as above. Eyes scarcely elevated above the flattened surface of the head; subtriangular, their inner margins very long and outer margins rounded. Ventral surface of abdomen pubescent almost or quite to lateral margins. (37 species). . . . . *Ambrysus* Stål

52a **(49)** Front tarsi 2-segmented with 2 claws which are often very inconspicuous. (1 species) . . . . *Heleocoris* Stål

52b Front tarsi 1-segmented and with or without a minute, scarcely distinguishable claw **53**

53a **(52)** Inner margins of eyes anteriorly divergent. Meso- and metasterna bearing prominent longitudinal carinae which are broad and foveate or otherwise excavated along middle. Body broadly oval and subflattened **54**

53b Inner margins of eyes anteriorly convergent. Meso- and metasterna without longitudinal carinae at middle. Body strongly robust. (5 species). . . . . *Pelocoris* Stål

54a **(53)** Embolium (Fig. 36.5*d*) quite remarkable, laterally expanded and terminating posterolaterally in a long, sharp, posteriorly curved spine. (1 species) . . . . *Usingerina* LaRivers

54b Embolium not terminating posteriolaterally in a long, sharp, posteriorly curved spine. (6 species) . . . . *Limnocoris* Stål

55a **(39)** Base of head overlapping apex of pronotum; bugs without a distinctly segmented beak; swim normally. (Water boatmen). . . . . . Family **Corixidae**   **58**

55b Base of head not overlapping apex of pronotum; bugs with a distinctly segmented beak; back swimmers . . . . Family **Notonectidae**   **56**

56a **(55)** Hemelytral commissure without a definite hair-lined pit at anterior end **57**

56b Hemelytral commissure with a definite hair-lined pit at anterior end. (26 species). . . . . *Buenoa* Kirkaldy

57a **(56)** Intermediate femur without an anteapical pointed protuberance. (3 species) . . . . *Martarega* B -White

57b Intermediate femur with an anteapical pointed protuberance. (29 species) . . . . *Notonecta* Linné

58a **(55)** Scutellum exposed, covered by pronotum only at anterior margin. Very small corixids. (3 species) . . . . *Tenegobia* Bergroth

58b Scutellum covered by pronotum (rarely with apex visible). See Fig. 36.5 . . . .   **59**

59a **(58)** Rostrum with transverse sulcations absent. (1 species) . . . . *Cymatia* Flor

59b Rostrum with transverse sulcations **60**

60a **(59)** Eyes protuberant with inner anterior angles broadly rounded. **61**

60b Eyes not protuberant, inner anterior angles normal **62**

61a **(60)** Pronotum and clavus strongly rastrate. Mesosternum not medially produced. (1 species). . . . . *Glaenocorisa* Thomson

61b Pronotum and clavus not strongly rastrate. Mesosternum medially produced. (3 species) . . . . *Dasycorixa* Hungerford

62a (60) With rather thick, well-developed apical claw on front tarsus in both sexes; pala of both narrowly digitiform **63**

62b With apical claw on front tarsus spinelike, usually resembling the spines along lower margin of the palm; pala not narrowly digitiform **64**

63a (62) Male abdomen sinistral, strigil absent. Female abdomen slightly asymmetrical. Face of female slightly concave. (2 species) . . . . .
Neocorixa Hungerford

63b Male abdomen dextral, strigil present. Female abdomen normal and face not concave. (14 species) . . . . *Graptocorixa* Hungerford

64a (62) Small shining corixids less than 5.6 mm long; males with sinistral assymetry and with pala short, triangular, the tibia produced apically over it; females with the apices of clavi not exceeding a line drawn through the costal margins of the hemelytra at the nodal furrows. (13 species) . . . . *Trichocorixa* Kirkaldy

64b Not as above **65**

65a (64) Inner posterior angle of eye sharply right angulate to acutely produced; lower posterior angle of front femur of male produced and bearing several rows of stridulatory pegs **66**

65b Not as above **67**

66a (65) Inner posterior angle of eye acutely produced; with a pruinose area on corial side of claval suture. (2 species) . . . .
Krizousacorixa Hungerford

66b Inner posterior angle of eye sharply right angulate, occasionally slightly produced; with a pruinose area on corial side of claval suture. (1 species) . . . . *Trichocorixella* Jaczewski

67a (65) Rugulose species with rear margin of head sharply curved, embracing a very short pronotum; interocular space much narrower than the width of an eye; dorsal median lobe of the seventh abdominal segment of the male bearing a hooklike projection. (3 species) . . . . *Palmocorixa* Abbott

67b Not as above **68**

68a (67) Smooth, shining insects never more than faintly rugulose, ranging in size from 4–8.4 mm long; lateral lobe of prothorax typically with sides tapering to a narrowly rounded apex; all but 2 small species with hind femur pubescent ventrally only at the base; male pala triangular, about equal in length to tibia, with a row of pegs near dorsal margin and another in or near the upper palmer row of bristles. (7 species) . . . . *Corisella* Lundblad

68b Combination of characters not as above **69**

69a (68) Length of pruinose area along claval suture less than twice the length of the distance between the shining basal apices of the corium and clavus; with the postnodol pruinose area (as measured from the cubital angle) shorter than or barely equal to the meron; males without a strigil or a strigilar stalk. (2 species) . . . .
Centrocorisa Lundblad

69b Not as above **70**

70a (69) Short, broad corixids, more than $\frac{1}{3}$ as broad as long; distal portion of corium semihyaline with color pattern often effaced; length of pruinose area along claval suture less than twice the length of the distance between the shining basal apices of the clavus and corium. . **71**

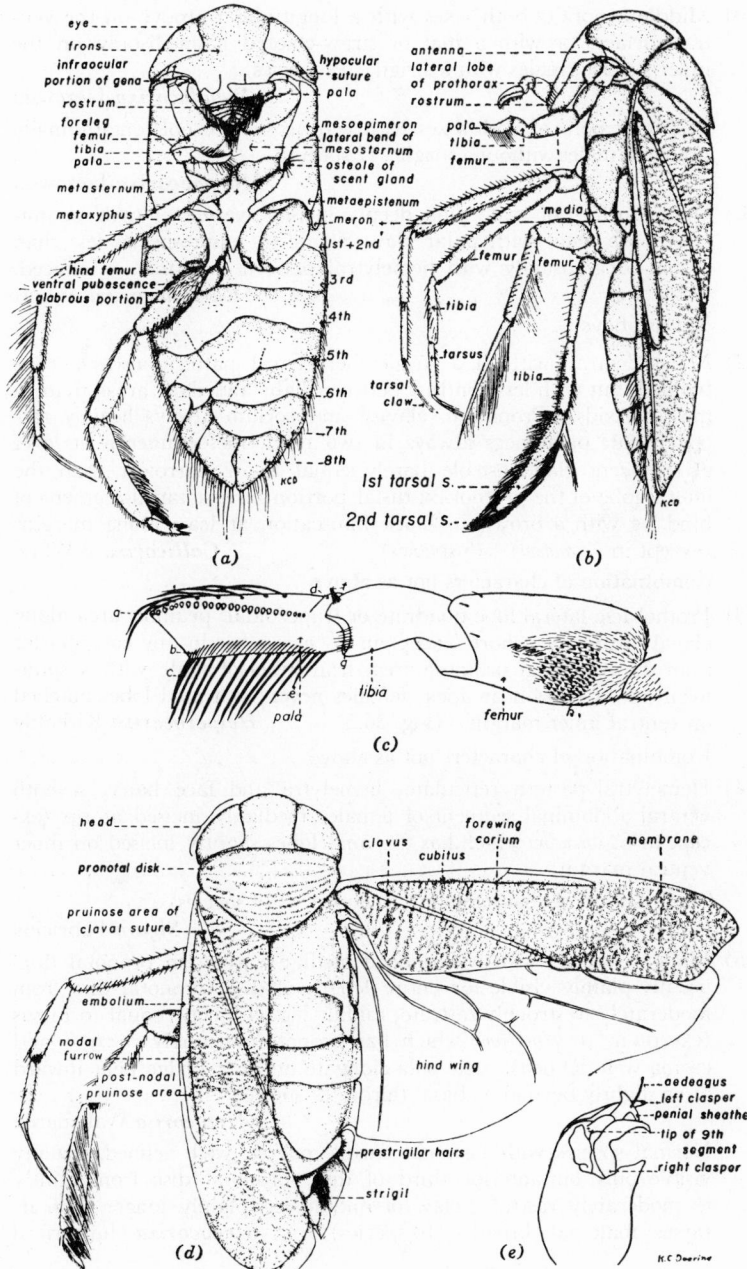

Fig. 36.5. *Hesperocorixa obliqua* (Hungerford), showing principal structures. (*a*) Ventral view of ♂. (*b*) Lateral view of ♀. (*c*) Foreleg of ♂. *a*, peg row; *b*, upper palmar bristles; *c*, claw; *d*, dorsal carina; *e*, lower palmar bristles; *f*, spiniform hair bundle; *g*, pad; *h*, stridular area. (*d*) Dorsal view of ♂. (*e*) Genital capsule.

70b   Not as above . . . . 　**72**

71a   (70)   Middle femora of both sexes with a longitudinal groove on the ventral surface, or with a mat or straw-colored hairs distally on the inner surface; males with a strigil. (5 species) . . . .
                             Pseudocorixa Jaczewski

71b   Middle femora of both sexes with ventral surface not longitudinally grooved; males without a strigil. (2 species) . . . .
                             Morphocorixa Jaczewski

72a   (70)   Upper surface of male pala deeply incised; vertex of male acuminate; both sexes with palar claw serrate at base; species less than 7 mm long, usually with hemelytral pattern indistinct or effaced. (2 species) . . . . ***Ramphocorixa*** Abbott

72b   Not as above . . . . 　**73**

73a   (72)   Males dextral, without a strigil; hemelytral pattern always cross-banded, but with less contrast between light and dark areas than in most corixids. Pronotum, clavus, and corium always heavily rastrate; male palar pegs always in two rows; mesoepimeron at level of the scent gland osteole barely equal to or narrower than the lateral lobe of the prothorax; distal portion of first tarsal segment of hind leg with a brown to black infuscation at least along margins (except in *C. audeni*). (5 species) . . . . ***Callicorixa*** B-White

73b   Combination of characters not as above . . . . 　**74**

74a   (73)   Prothoracic lateral lobe quadrate or trapezoidal; pruinose area along claval suture very short (except in *H. minorella*), in any case shorter than the postnodal pruinose area; front tibia of male with a spiniform tuft of hairs near apex; females never with anal lobes notched on ventral inner margin. (Fig. 36.5) . . . . ***Hesperocorixa*** Kirkaldy

74b   Combination of characters not as above . . . . 　**75**

75a   (74)   Hemelytral pattern reticulate; hemelytra and face hairy; seventh ventral abdominal segment of females medianly incised at tip (except in *C. sorensoni* which has the anal lobes slightly incised on inner ventral margin) . . . . 　**76**

75b   Combination of characters not as above. (49 species) . . . .
                             Sigara Fabricius

76a   (75)   Elongate species with well-defined median carina on pronotal disc, usually plainly visible for entire length of disc; pronotal disc from moderately to strongly rastrate; middle leg with claw equal to tarsus (except in *A. planifrons* which has pronotum strongly rastrate and carina well defined); male pala elongate, its dorsal edge bent inward at or slightly beyond its basal third. (5 species) . . . .
                             Arctocorisa Wallengren

76b   Normal species with pronotal carina not so well defined, usually visible only on anterior third of disc; pronotal disc from faintly to moderately rastrate; claw of middle leg plainly longer than its tarsus; male pala broad. (10 species) . . . . ***Cenocorixa*** Hungerford

References

GENERAL

Blatchley, W. S. 1926. *Heteroptera or True Bugs of Eastern North America.* The Nature Publishing Co., Indianapolis. (Taxonomy out of date.) **Hungerford, H. B. 1920.** The biology and ecology of aquatic and semiaquatic Hemiptera. *Univ. Kansas Sci. Bull.*, 11:1–328. (Taxonomy out of date). **1957.** Some interesting aspects of the world distribution and classification of aquatic and semiaquatic hemiptera. *Proc. Tenth Int. Congress of Entomol.*, Montreal 1956. **Hungerford, H. B., Paul J. Spangler, and Neil A. Walker. 1955.** Subaquatic light traps for insects and other animal organisms. *Trans. Kansas Acad. Sci.*, 58:387–407. **Usinger, R. L. 1956.** Aquatic Hemiptera. In: R. L. Usinger (ed.). *Aquatic Insects of California, with Keys to North American Genera and California Species*, pp. 182–228. University of California Press, Berkeley and Los Angeles.

Papers Containing Keys or Species Lists.

Belostomatidae: **Cummings, Carl. 1933.** The giant water bugs. *Univ. Kansas Sci. Bull.*, 21:197–219. **DeCarlo, José A. 1938.** Los Belostomidos Americanos. *Anales museo arg. cienc nat. "Bernardino Rivadavia" Buenos Aires*, 39:189–260. **1948.** Revision del Genero "Abedus." *Comuns. inst. nacl. invest. circ. nat. museo argentino cienc. nat. "Bernardino Rivadavia" ser. cienc. zool.*, 5:1–24. **Hidalgo, Jose. 1935.** The genus *Abedus*. *Univ. Kansas Sci. Bull.*, 22:493–519.

Corixidae: **Griffith, Melvin E. 1945.** The environment, life history and structure of the Water Boatman *Ramphocorixa acuminata*. *Univ. Kansas Sci. Bull.*, 30:241–365. **Hungerford, H. B. 1948.** The Corixidae of the Western Hemisphere. *Univ. Kansas Sci. Bull.*, 33:1–827. **Sailer, Reece I. 1948.** The genus *Trichocorixa*. *Univ. Kansas Sci. Bull.*, 32:289–407.

Gelastocoridae: **Martin, C. H. 1929.** An exploratory survey of characters of specific value in the genus *Gelastocoris* Kirkaldy and some new species. *Univ. Kansas Sci. Bull.*, 38:351–369. **Melin, D. 1929.** Hemiptera from South and Central America, I. *Zool. Bidrag Uppsala*, 12:151–194. **Todd, Edward L. 1955.** A taxonomic revision of the family Gelastocoridae. *Univ. of Kansas Sci. Bull.*, 37:277–475.

Gerridae (Gerrinae): **Drake, D. J. and H. M. Harris. 1934.** Gerrinae of the Western Hemisphere, *Ann. Carnegie Museum*, 23:179–240. **Hungerford, H. B. 1954.** The genus Rheumatobates Bergroth. *Univ. Kansas Sci. Bull.*, 36:529–288. **Kuitert, Louis C. 1942.** Gerrinae in the University of Kansas Collections. *Univ. Kansas Sci. Bull.*, 28:113–143.

Gerridae (Halobatinae): **Anderson, L. D. 1932.** A monograph of the genus *Metrobates*. *Univ. Kansas Sci. Bull.*, 20:297–311. **Drake, C. J. and H. M. Harris. 1932a.** A synopsis of the genus *Metrobates*. *Ann. Carnegie Museum*, 21:83–89. **1932b.** A survey of Trepobates. *Bull. Brooklyn Entomol. Soc.*, 27:113–123. **1945.** Concerning the genus *"Metrobates."* *Rev. brasil. biol.*, 5:179–180. **Drake, C. J. and F. C. Hottes. 1951.** Notes on the genus *Rheumatobates*. *Proc. Biol. Soc. Wash.*, 64:147–158. **Kenaga, Eugene E. 1941.** The genus *Telmatometra*. *Univ. Kansas Sci. Bull.*, 27:169–183. **Shaw, J. Gilbert. 1933.** A study of the genus *Brachymetra*. *Univ. Kansas Sci. Bull.*, 21:221–233.

Hebridae: **Drake, C. J. and H. M. Harris. 1943.** Notas Sobre Hebridae. *Notas del Museo de la Plata*, 8(zool):41–58.

Hydrometridae: **Hungerford, H. B. and N. E. Evans. 1934.** The Hydrometridae. *Ann. Musei Natl. Hungarici*, 28:31–112. **Sprague, Isabelle Baird. 1956.** The biology and morphology of *Hydrometra martini* Kirkaldy. *Univ. of Kan. Sci. Bull.*, 38:579–693.

Mesoveliidae: **Drake, Carl J. 1948.** Two new Mesoveliidae with check list of American species. *Bol. Entomol. Venezolana*, 7:145–147. **Jaczewski, T. 1930.** Notes on the American species of the genus Mesovelia. *Ann. Musei Zool. Polonici*, 9:3–12. **Neering, Thomasine. 1954** Morphological variations in *Mesovelia mulsanti*. *Univ. of Kan. Sci. Bull.*, 36:125–148.

Naucoridae: **LaRivers, Ira. 1951.** A revision of the genus *Ambrysus* in the United States. *Univ. Calif. Berkeley Publs. Entomol.*, 8:277–338. **1958.** The Ambrysus of Mexico. *Univ. Kansas Sci. Bull.*, 35: Pt. II, No. 10, 1279–1349. **Usinger, Robert L. 1941.** Key to the subfamilies of Naucoridae with a generic synopsis of the new subfamily Ambrysinae. *Ann. Entomol. Soc. Am.*, 34:5–16.

Nepidae: **DeCarlo, Jose A. 1951.** Nepidos de America. *Rev. inst. nacl. invest. cienc. nat. y museo arg. cienc. nat. "Bernardino Ravadavia," cienc. zool.*, 1:385–421. **Hungerford, H. B. 1922.** The Nepidae of North America. *Univ. Kansas Sci. Bull.*, 14:425–469.

Notonectidae: **Hungerford, H. B. 1933.** The genus *Notonecta* of the world. *Univ. Kansas Sci. Bull.*, 21:5–195. **Hutchinson, G. Evelyn. 1942.** Note on the occurrence of *Buenoa elegans* (Fieb.) (Notonectidae, Hemiptera-Heteroptera) in the early postglacial sediment of Lyd

Hyt Pond. *Am. J. Sci.*, 240:335–338.   **1945.** On the species of *Notonecta* (Hemiptera-Heteroptera) inhabiting New England. *Trans. Conn. Acad. Arts and Sci.*, 36:599–605.   **Truxal, Fred S. 1949.** A study of the genus *Martarega. J. Kansas Entomol. Soc.*, 22:1–24.   **1953.** A revision of the genus *Buenoa. Univ. Kansas Sci. Bull.*, 35:1351–1523.

Ochteridae: **Bobb, M. L. 1951.** Life history of *Ochterus banksi. Bull. Brooklyn Entomol. Soc.* 46:92–100.   **Drake, Carl J. 1952.** Concerning American Ochteridae. *Florida Entomologist*, 35:72–75.   **Schell, Dorothydean Viets. 1943.** The Ochteridae of the Western Hemisphere. *J. Kans. Ent. Soc.*, 16:29–46.

Saldidae: **Drake, Carl J. and Ludvik Hoberlandt. 1950.** Catalogue of genera and species of Saldidae. *Acta Entomol. Musei Natl. Pragae*, 26:1–12.

Veliidae: **Bacon, John A. 1956.** A taxonomic study of the genus *Rhagovelia* of the Western Hemisphere. *Univ. Kansas Sci. Bull.*, 38:695–913.   **Drake, C. J. and R. F. Hussey. 1955.** Concerning the Genus *Microvelia* Westwood with descriptions of two new species and a checklist of the American forms. *Florida Entomologist*, 38:95–115.   **Gould, George E. 1931.** The Rhagovelia of the Western Hemisphere. *Univ. Kansas Sci. Bull.*, 20:5–61.

37

Neuroptera

ASHLEY B. GURNEY

SOPHY PARFIN

The most recent trend by some specialists in the systematics of the holometabolous Order Neuroptera since its division into the Orders Megaloptera, Neuroptera, and Raphidiodea in 1903, is to again group these insects into the one Order Neuroptera, this time as suborders (Sialodea, Planipennia, and Raphidiodea). Although the division into orders is based primarily on the striking differences in the biologies and larvae of the three suborders, the great similarity in the adult morphology of both fossil and recent forms tends to bring them together into one group. The adults of the entire order are terrestrial, with two of the suborders, the Sialodea (Megaloptera) and Planipennia, having some larvae which are aquatic. The aquatic larvae of both suborders are campodeiform, have six well-developed thoracic legs, and can be separated by the following key.

KEY TO SUBORDERS

1a   Larvae with chewing mouthparts; no cocoons . . . . (p. 975) **Sialodea**
1b   Larvae with piercing and sucking mouthparts; cocoons of silk . . . . .
                                                 (p. 979) **Planipennia**

Suborder Sialodea

There are two families in this more primitive suborder, which may be recognized by the characters given in the key. The Nearctic Sialidae (alderflies) are included in one genus, *Sialis*, a homogeneous group of about twenty species.

The limits of genera and species in the Nearctic Corydalidae are in a somewhat unsettled state, with a taxonomic revision necessary. At present, about six genera are recognized: *Corydalus* (dobsonflies, hellgrammites), with the well-known species *cornutus* (Linnaeus), and some others (as *cognata* Hagen); *Chauliodes* (fishflies), with two well-known species, *pectinicornis* (Linnaeus) and *rastricornis* Rambur; *Nigronia*, with two well-known species, *fasciata* (Walker) and *serricornis* (Say); *Neohermes*, with at least two species, *filicornis* (Banks) and *californicus* (Walker); *Dysmicohermes*, with three species, *crepusculus* Chandler, *disjunctus* (Walker), and *ingens* Chandler; *Protochauliodes*, with six species, *aridus* Maddux, *infuscatus* (Caudell), *minimus* (Davis), *montivagus* Chandler, *simplus* Chandler and *spenceri* Munroe. The last four genera are closely related to *Chauliodes*, and it is possible that other genera and species may be represented among Nearctic species of the *Chauliodes* complex, and that the taxonomy may be changed in other ways.

The larvae of all Sialodea are aquatic or semiaquatic. The large, blackish hellgrammite is well known as fish bait, the smaller fishfly larva less so. Full-grown larvae vary in length from about 20 mm (alderflies) to 40 mm (fishflies), to over 80 mm (dobsonflies). They are vicious predators, which feed primarily on immature insects by means of heavy, well-toothed mandibles. The labium has segmented palpi, and the maxilla is separate from the mandible. The larvae are likely to be confused only with certain coleopterous larvae, but none of the latter has the same combination of toothed mandible, lateral abdominal appendages, and terminal abdominal structures that occurs in Sialodea. Sialodean larvae have five-segmented legs with paired claws, and seven to nine pairs of lateral abdominal appendages. The distinguishing characters of many genera and species are unknown in the larval stages, and further correlation of larvae and adults is highly desirable. One method of determining larval characters in this group is to preserve the cast larval skins found in the pupal cells and associate them with the resulting adults.

The habitat range of the larvae is from small streams and ponds to large rivers and lakes, and may be equally variable for one species (*Sialis*). The larvae of *Corydalus* and other members of the *Chauliodes* complex may be found under stones in well-aerated streams and rivers. Some species of the *Chauliodes* complex may occur under stones among debris near the banks of streams or ponds. *Sialis* has also been reported burrowing into the mud and detritus of a lake or stream bottom, particularly where the floor is covered with vegetation such as *Phragmites*. Full-grown larvae leave the water and usually pupate in a cell in the soil, frequently under a stone on the banks of the body of water (*Corydalus*, *Chauliodes*), or in a dry stream bed (*Neohermes*, *Nigronia*, *Protochauliodes*). Pupae of *Chauliodes* and *Nigronia* have also been found in rotten logs. There is no cocoon. *Corydalus* larvae may sometimes wander 78 ft or more from the water before pupating. A complete life cycle may require about three years, with almost one month spent in the pupal period; this varies with the species. The number of instars is unknown, although as many as ten have been estimated for *Sialis*, and four to six suggested for *Corydalus*. Eggs are laid in masses near the water; those of *Corydalus* are best known; they are chalky-white patches an inch or less in diameter and are frequently deposited on stones along streams.

The following key has been modified from Chandler (1956), p. 232.

KEY TO GENERA OF SIALODEA (LARVAE)

1a   Abdomen with a long terminal median process tapering to a fine point; paired segmented lateral appendages on abdominal segments 1 to 7; no anal prolegs or abdominal tufts of gills; body length up to about 20 mm. (Figs. 37.1a, 37.3). (Cosmopolitan) . . . .
                                                                               Family **Sialidae**
                                                                                 Sialis Latreille

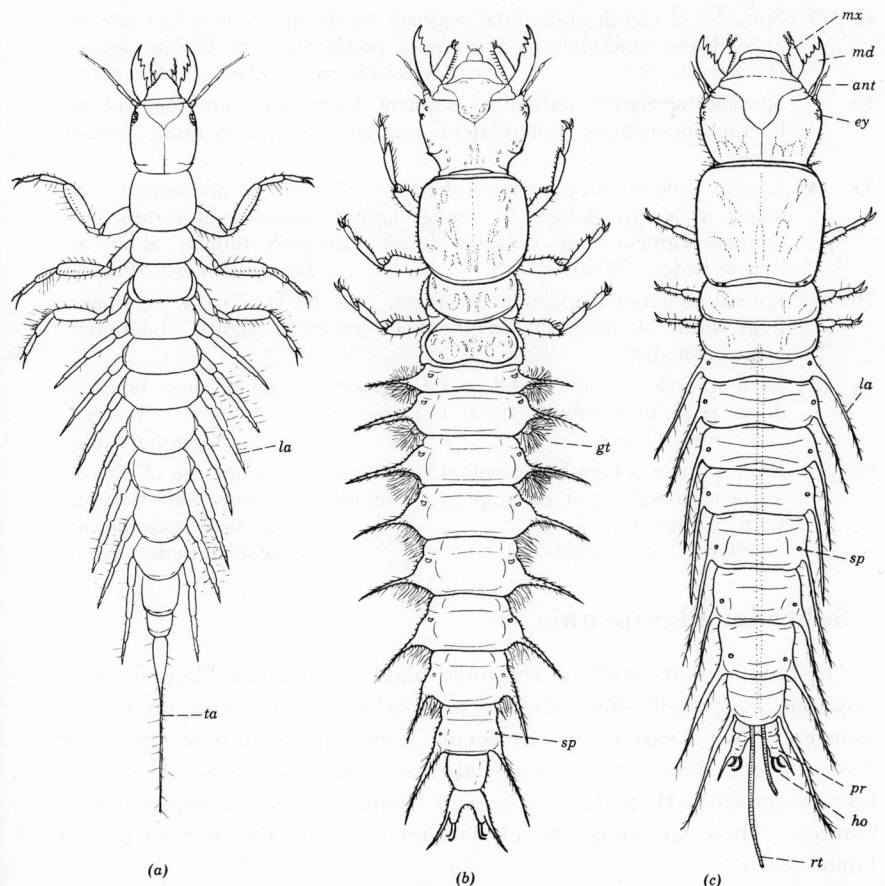

Fig. 37.1. Larvae of Sialodea. (*a*) *Sialis*. (*b*) *Corydalus*. (*c*) *Chauliodes*.

ant, antenna; *ey*, eye; *gt*, gill tuft; *ho*, hooks of proleg; *la*, lateral abdominal appendage; *md*, mandible; *mx*, maxilla; *pr*, proleg; *rt*, respiratory tubule; *sp*, spiracle; *ta*, terminal appendage.

1b   Abdomen with a pair of anal prolegs, each bearing 2 strong hooks; paired, lateral, usually unsegmented appendages on abdominal segments 1 to 8 and 10; tufts of gills sometimes present; body length up to more than 80 mm. (Fig. 37.1*b,c*) . . . . Family **Corydalidae**   2

2a **(1)** Abdomen with tufts of tracheal gills at bases of lateral appendages on segments 1 to 7 (Fig. 37.1*b,gt*), gills lacking in very young larvae; eighth abdominal segment without dorsal respiratory tubules, but with spiracles close to lateral appendages. (Western Hemisphere, possibly elsewhere) . . . . **Corydalus** Latreille

2b   Abdomen without tracheal gill tufts; eighth abdominal segment with pair of spiracles sessile or at ends of dorsal respiratory tubules   3

3a **(2)** Each spiracle of eighth abdominal segment at end of long contractile dorsal respiratory tubule extending to or beyond hooks of anal prolegs. (Eastern and central U. S., Canada) . . . . **Chauliodes** Latreille
<small>See R. D. Cuyler. 1958. The larvae of *Chauliodes* Latreille (Megaloptera: Corydalidae). *Ann. Entomol. Soc. Am.*, 51:582–586.</small>

3b   Each spiracle of eighth abdominal segment sessile or at end of short dorsal respiratory tubule, not extending beyond center of ninth segment.   4

4a **(3)** Spiracles of eighth abdominal segment small, anteromesad of lateral appendages; anterolateral margins of postmentum of labium acute. (Western U. S.). . . . . **Dysmicohermes crepusculus** Chandler

4b   Spiracles of eighth abdominal segment larger and posteromesad of lateral appendages; anterolateral margins of postmentum emarginate.   5

5a **(4)** Lateral abdominal appendages short, to ½ width of abdominal segments in length (longer in young larvae); spiracles of eighth abdominal segment large, on short dorsal respiratory tubules, about as long as wide. (Western U. S.) . . . . **Dysmicohermes** Munroe

5b   Lateral abdominal appendages longer, may be as long as or longer than width of abdominal segments; spiracles of eighth abdominal segment smaller   6

6a **(5)** Each spiracle of eighth abdominal segment at end of short tapered dorsal respiratory tubule, about 1½ times as long as wide. (Eastern and central U. S.). . . . . **Nigronia** Banks

6b   Each spiracle of eighth abdominal segment on posterior edge of eighth segment dorsally, not on respiratory tubule. (Eastern and western U. S., Canada) . . . . **Neohermes** Banks
       (Western U. S., Canada, S. A.) . . . . **Protochauliodes** Weele

Suborder Planipennia

The only aquatic family of the more highly specialized Planipennia is the Sisyridae, or spongilla-flies, the larvae of which are parasitic on fresh-water sponges. Two Nearctic genera occur, each having at least three species: *Sisyra apicalis* Banks, *S. fuscata* (Fabricius), and *S. vicaria* (Walker); and *Climacia areolaris* (Hagen), *C. californica* Chandler, and *C. chapini* Parfin and Gurney. There are comparatively few records from the western half of the United States.

The larvae are aquatic, and may be as small as 0.5 mm when newly hatched, and as large as 8 mm when full grown. The body color varies from pale greenish to light brown, and as it is usually similar to that of the sponge, close examination is necessary to detect these insects. The straight threadlike mouthparts and antennae (Fig. 37.2a) readily separate them from other insects. The mouthparts include two slender piercing and sucking tubes, each formed by the close apposition of the ventrally grooved mandible over the dorsally grooved maxilla; those of the first instar are proportionally much shorter in comparison with the body length. Antennae of full-grown larvae consist of fourteen to sixteen segments. There are no labial palpi, and the eyes are reduced to about six simple dark spots. The dorsal surface of the body has many setae, which are borne in clusters on tubercles or chalazae, and become progressively longer posteriorly. Legs are slender and always bear a *single* claw (Fig. 37.2d,cl). On second- and third-instar larvae (approximately 1 to 8 mm), each of the first seven abdominal segments bears a pair of two- or three-segmented, transparent, ventral tracheal gills (Fig. 37.2e), which are folded and taper posteriorly, but become blunter and shrink as the larva approaches pupation; in living individuals, these usually vibrate rapidly. The first-instar larvae (approximately 0.5 to 1.0 mm) (Fig. 37.3b,c), which do not have gills, resemble *Cyclops* in their manner of move-

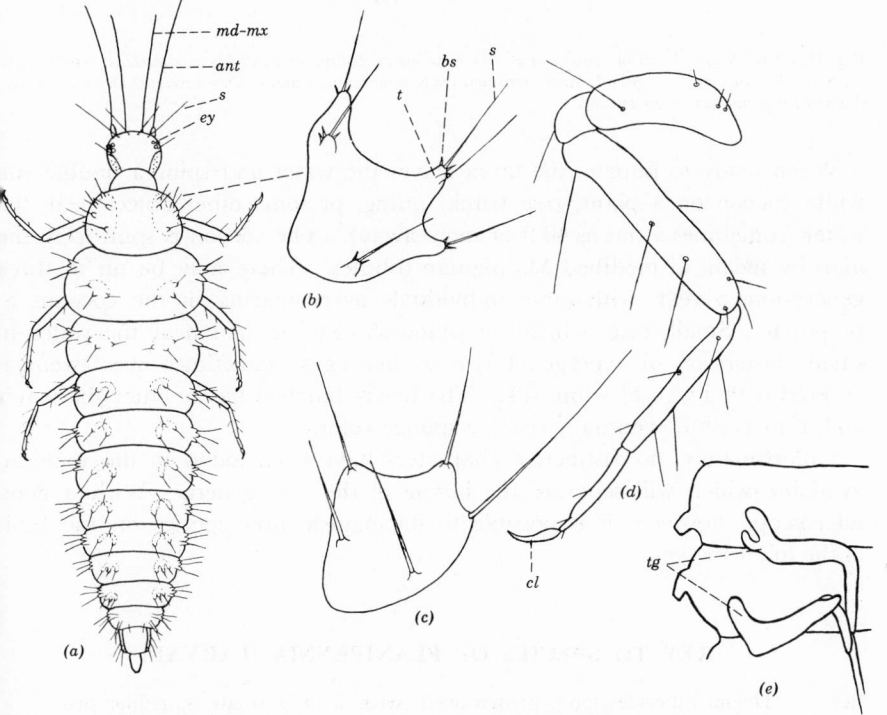

Fig. 37.2. (a) Dorsal view of larva of *Climacia areolaris*. (b) Dorsolateral plate of pronotum (greatly enlarged). (c) Same plate of *Sisyra vicaria*. (d) Front leg of *Climacia*. (e) Tracheal gills on right side of first and second abdominal segments of *Sisyra vicaria*, ventral view.

ant, antenna; *bs*, basal spine; *cl*, claw; *ey*, eye; *md-mx*, combined mandible and maxilla; *s*, seta; *t*, tubercle; *tg*, tracheal gill.

ment as they wander about in search of a sponge host, and probably have been mistaken for them in plankton studies.

The larvae crawl over the sponges, sometimes through the osteoles, into the cavities within, and pierce the sponges with their mouthparts, sucking the juices as though through two drinking straws. In North America, sisyrid larvae have been reported from *Spongilla fragilis* Leidy, and they are undoubtedly found on other fresh-water sponges.

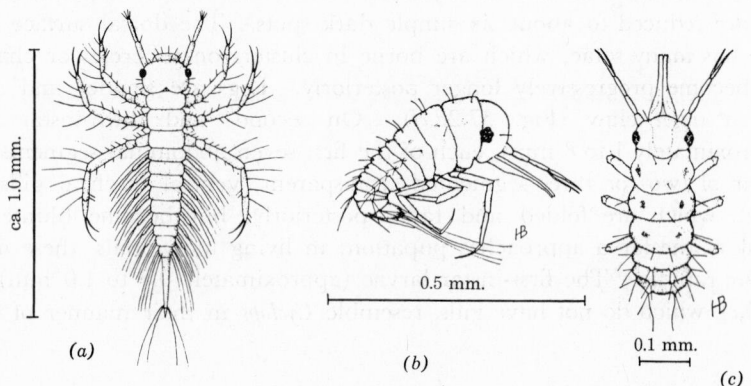

Fig. 37.3. (*a*) Young larva of *Sialis lutaria*. (*b*) First instar drifting larva of *Climacia areolaris*, lateral view (*c*) Same, dorsal view. (*a* after Lestage, redrawn by Sommerman; *b* and *c* after Brown. Drawings reproduced with permission of the authors.)

When ready to pupate, the larva leaves the water and spins a double silk white cocoon on a plant, tree trunk, piling, or some other object near the water (sometimes as far as 40 ft or more away). The cocoon is spun from the anus by means of modified Malpighian tubules. There may be up to three generations a year, with some individuals overwintering in the cocoons as prepupae. Small, oval, whitish to yellowish eggs are laid near the water in small clusters of an average of two to five eggs (sometimes over twenty), covered with a web of white silk. The newly hatched larvae enter the water and, if successful, eventually reach a sponge colony.

Unfortunately, no distinctive characters have been found in the material available which will separate the larvae of the two genera. With a good microscope, however, it is possible to distinguish three species on the basis of the following key.

KEY TO SPECIES OF PLANIPENNIA (LARVAE)

1a   Dorsal tubercles long, pronounced, with 2 or 3 small, spinelike projections at bases of setae (Fig. 37.2*b,bs*). Detection of basal spines requires a magnification of about 70. (Eastern and central U. S., Canada) . . . . *Climacia areolaris* (Hagen)

1b   Dorsal tubercles shorter, with no small, spinelike projections at bases of setae (Fig. 37.2*c*) 2

2a   Ventral median setae of eighth abdominal segment much closer to-
           gether than those of ninth segment. (U. S., Canada). . . . .
                                                      Sisyra vicaria (Walker)
2b   Ventral median setae of eighth segment only slightly closer together
           than those of ninth. (Western U. S.) . **Climacia californica** Chandler

KEY TO PUPAE AND COCOONS OF PLANIPENNIA

a   Apical segment of maxillary palpus broadly triangular-shaped (Fig.
           37.4*b*); cocoon usually close woven, appearing almost single layered,
           but sometimes with an outer layer of irregularly and closely spaced
           open hexagonal mesh, separate from the inner layer. (Fig. 37.4*a*).
           (Cosmopolitan) . . . . **Sisyra** Burmeister
b   Apical segment of maxillary palpus more cylindrical in shape (Fig.
           37.4*d*); cocoon sometimes with outer layer of a striking hexagonal
           mesh (Fig. 37.4*c*), but sometimes without this tentlike layer. (Western
           Hemisphere). . . . . **Climacia** McLachlan

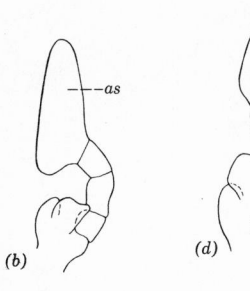

◄ **Fig. 37.4.** (*a*) Pupa and cocoon of *Sisyra*
vicaria. (*b*) Maxilla of pupa of same.
(*c*) Pupa and cocoon of *Climacia areolaris*.
(*d*) Maxilla of pupa of same.
  as, apical segment; *ic*, inner cocoon;
ls, last larval skin; *oc*, outer cocoon.

References

Banks, Nathan. 1908. On the classification of the Corydalinae, with description of a new
species. *Proc. Entomol. Soc. Wash.*, 10:27–30.   **Brown, Harley P. 1952.** The life history of
Climacia areolaris (Hagen), a neuropterous 'parasite' of fresh-water sponges. *Am. Midland Naturalist*, 47:130–160.   **Caudell, A. N. 1933.** *Neohermes infuscatus*, a new Sialid from California. *Pan-Pacific Entomologist*, 9:125–126.   **Chandler, H. P. 1954.** Four new species of
dobsonflies from California. *Pan-Pacific Entomologist*, 20:105–111.   **1956.** Megaloptera,
Chapter 8. In: R. L. Usinger (ed.). *Aquatic Insects of California, with Keys to North American
Genera and California Species*, pp. 229–236. University of California Press, Berkeley and Los

Angeles. **Davis, K. C. 1903.** Sialididae of North and South America. *N. Y. State Museum Bull.*, No. 68:442–487. **Kimmins, D. E. 1954.** A new genus and some new species of the Chauliodini (Megaloptera). *Bull. British Museum*, 3:417–444. **Maddux, D. E. 1954.** A new species of dobsonfly from California (Megaloptera: Corydalidae). *Pan-Pacific Entomologist*, 30:70–71. **Munroe, E. G. 1953.** *Chauliodes disjunctus* Walker: a correction, with the descriptions of a new species and a new genus (Megaloptera: Corydalidae). *Can. Entomologist*, 85:190–192. **Needham, J. G. 1901.** Neuroptera. In: Needham and Betten. Aquatic insects in the Adirondacks. *N. Y. State Museum Bull.*, 47:540–560. **Parfin, S. I. and A. B. Gurney. 1956.** The spongilla-flies, with special reference to those of the Western Hemisphere (Sisyridae, Neuroptera). *Proc. U. S. Natl. Museum*, 105:421–529. **Ross, H. H. 1937.** Nearctic alder flies of the genus *Sialis* (Megaloptera, Sialidae). In: Studies of Nearctic aquatic insects. *Bull. Illinois Nat. Hist. Survey*, 21:57–78. **Weele, H. W. van der. 1910.** Megaloptera, monographic revision. *Coll. Zool. Selys Longch.*, 5:1–93.

38

Coleoptera

HUGH B. LEECH
MILTON W. SANDERSON

One or more stages of species in at least twenty-one families of North American beetles are aquatic or semiaquatic. The water-inhabiting forms of the Carabidae, Staphylinidae, Melyridae, and Eurystethidae occur principally in the intertidal zone of the Pacific Coast sea beaches, and are not treated here. If beetles actually found in fresh water cannot be run through the following keys satisfactorily, it is probable that they are not true aquatics. A great many flying insects fall into the water, and others which live on the shore may be washed or knocked in by accident.

In general, the phylogenetically more primitive water-beetle groups are structurally adapted for an aquatic life, both as larvae and adults. Species of the more advanced groups enter the water only occasionally, or may have but one aquatic stage. Families in which the adults are specialized for life in the water are also those in which almost all the species, as larvae and adults, are aquatic. They are not all closely related, however, and result from several independent invasions of the habitat.

Two of the three recognized suborders of Coleoptera include water beetles. The Families Amphizoidae, Haliplidae, Dytiscidae, Noteridae, and Gyrinidae belong to the Adephaga. All but the crawling amphizoids swim fairly well

981

to very strongly; except for the largely vegetarian haliplids, adults and larvae of virtually all species are predacious. Nearly all adults and larvae have to come to the surface at intervals to renew their air supply, except for the Gyrinidae (larvae with gills, adults *on* the water), larval noterids (known forms obtain air by piercing cells in the underwater roots of plants), larval haliplids, and the gilled *Coptotomus* spp. (dytiscid) larvae.

The remaining families belong to the Suborder Polyphaga. Though the larvae of many species of Hydrophilidae are predators, the family as a whole is the most nearly polyphagous of the water beetles. The hydraenid adults are largely vegetarian, but their larvae are predacious, and the hydroscaphids feed on algae in both stages. The known larvae of the Helodidae are aquatic; their mouthparts are suited for scraping detritus from the surface of underwater plants and debris. The larvae of the Psephenidae, Ptilodactylidae, and Limnichidae are perhaps also detritus feeders. Both adults and larvae of the Dryopidae and Elmidae are detritus feeders; some feed in calcareous incrustations, others are phytophagous, specializing on plant roots. The few aquatic chrysomelid larvae are also root eaters, but at the same time are remarkably adapted for obtaining air from cells in the roots on which they feed. The larvae of most of the aquatic Curculionidae live within and eat the underwater parts of plants, but can neither swim nor live freely in the water.

The most truly aquatic families are the Dryopidae and Elmidae. Except for the semiaquatic adults of the elmid genera *Lara* and *Phanocerus*, both larvae and adults cling to underwater objects, and some species do not come to the surface at all, although *Stenelmis* spp. are commonly attracted to light, possibly following pupation, and before entering water. Except for adult Curculionidae, most of the aquatic Polyphaga are able to stay submerged for long periods, even for the entire aquatic stage.

Following is an annotated list of the Nearctic families having one or more aquatic or semiaquatic species. A species of one of them occurs only in the intertidal zone of the Pacific Coast seashore (Hydraenidae: *Ochthebius vandykei* Knisch). All the rest are found in fresh water, though of course this term includes saline waters on land areas and mineralized springs.

Amphizoidae. (Fig. 38.1). Crawling water beetles, 11–15 mm long; larvae and adults in streams and rivers, rarely in lakes. Both stages have to come to the surface at intervals to renew their air supply. They feed exclusively on stone-fly nymphs (Plecoptera). Eggs are laid in cracks in floating or water-logged wood. Pupal cells are made in debris-filled crevices and in damp soil near the water. Western U. S. and Canada.

Haliplidae. (Fig. 38.2). Small beetles, 4.5 mm or less in length, with greatly expanded hind coxal plates. Larvae eat algae, adults both algae and animal material; they are commonest in the shallow water of lakes, ponds, and streams. The adults, but not the larvae, must come to the surface for fresh air. Eggs are laid on and in aquatic vegetation. Mature larvae leave the water to form pupal cells in damp soil, under stones, etc. Widespread.

Dytiscidae. (Fig. 38.3). Adults are known as the diving water beetles; larvae of the larger forms are called water tigers. Both stages are predacious. The adults of some species are the most powerful swimmers of the water beetles; the hind legs are moved in unison in a "rowing" fashion. All adults must break the surface film to

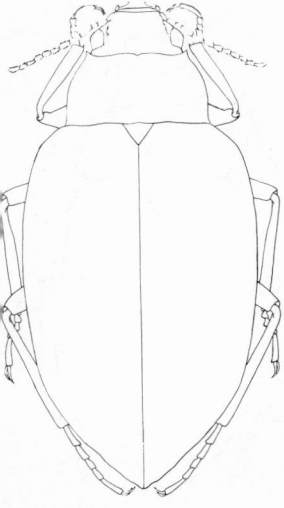

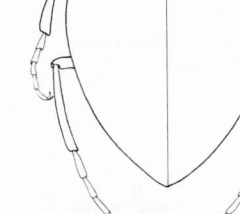

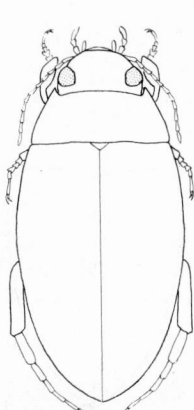

Fig. 38.1. Amphizoidae. *Amphizoa lecontei* Matthews.

Fig. 38.2. Haliplidae. *Peltodytes* sp.

Fig. 38.3. Dytiscidae. *Copelatus* sp.

obtain air (either at the surface of the water or in trapped underwater bubbles), and they do this tail first, with the tip of the abdomen, in contrast to the head-first method of the Hydrophilidae and allies. The larvae, except for those of *Coptotomus* spp. and some of the Hydroporinae, must also come up for air. The eggs are laid on or in aquatic plants. Pupal cells are made in damp soil near the water. Widespread.

Noteridae. (Fig. 38.4). Small beetles resembling dytiscids; all but the tiny *Notomicrus* spp. have a curved hook or spine at the apex of the foretibiae; the scutellum is always covered. The adults come to the surface for air, and are predacious; the larvae of at least some species live under several feet of water, obtain air by piercing plant roots, and are thought to eat both plant and animal matter. They pupate in underwater cocoons which are attached to plant roots and filled with air from their cells. Presumably the eggs are laid on the roots of aquatic plants, or in the mud close by.

Gyrinidae. (Fig. 38.5). These are the whirligig beetles. With the aid of their peculiar fanlike middle and hind legs the predacious adults skim over the surface film, but can dive and swim well too. Each eye is completely divided, the lower part being against the surface film and of use for underwater vision, while the upper section is for use in the air. Adults tend to congregate in large swarms or "schools." The larvae are predacious, stay beneath the surface, and breathe by means of abdominal gills. They build cocoons of mud and debris on emergent vegetation or on objects close to the water's edge. Eggs are attached to the submerged parts of plants, in rows or clusters.

Hydrophilidae. (Fig. 38.6). The so-called water-scavenger beetles, comparable to the Dytiscidae in size and in number of species. The adults have club-shaped antennae, move their legs alternately when swimming, and break the surface film with an antenna when renewing their air supply; they are largely herbivorous. The larvae are carnivorous; they come to the surface for air, except those of *Berosus* spp., which have abdominal gills. Pupal cells are made in damp soil near the water. The eggs may be embedded in a loose web (*Cymbiodyta*, *Paracymus*) in wet places, completely enclosed in a silken case which has a vertical mast or flexible ribbon

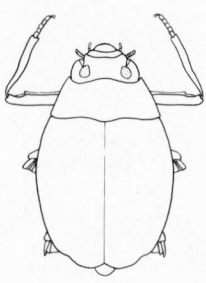

Fig. 38.5. Gyrinidae. *Dineutus* sp.

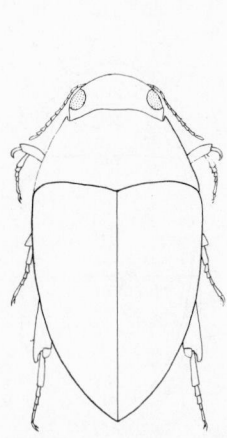

Fig. 38.4. Noteridae. *Hydrocanthus iricolor* Say.

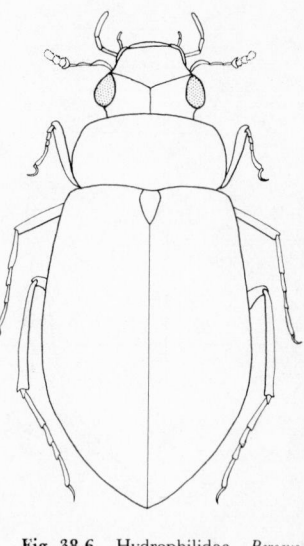

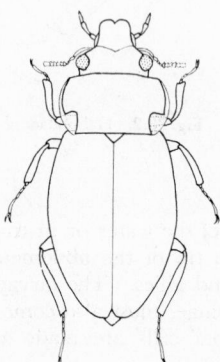

Fig. 38.6. Hydrophilidae. *Berosus dolerosus* Leech.

Fig. 38.7. Hydraenidae. *Ochthebius* sp.

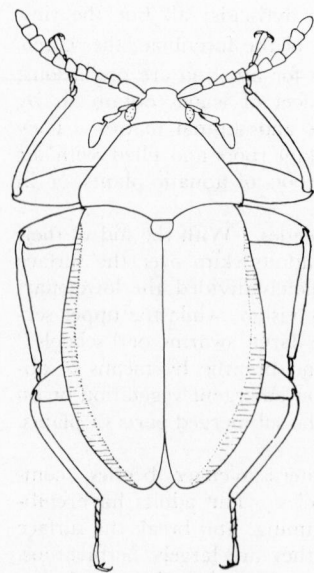

Fig. 38.8. Psephenidae. *Psephenus herricki* De Kay.

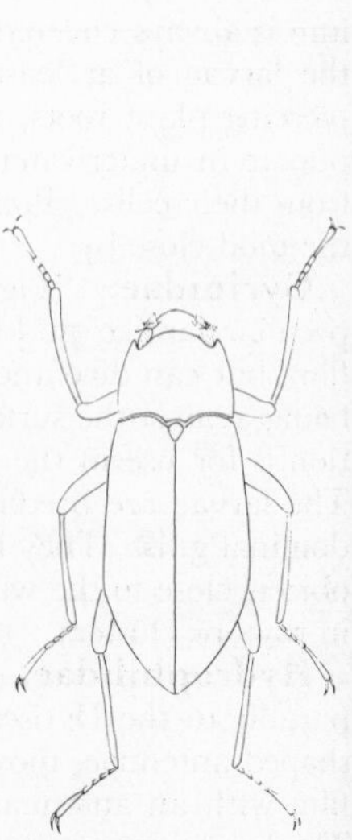

Fig. 38.9. Dryopidae. *Helichus lithophilus* Germar.

attached to it (*Hydrophilus, Laccobius, Helophorus*, etc.), or carried by the female beneath her abdomen in a nearly transparent bag-shaped case (*Helochares, Epimetopus*).

Hydraenidae. (Fig. 38.7). Tiny (1–2.5 mm long) crawling water beetles, resembling small hydrophilids. The adults are largely vegetarians, and aquatic; the larvae are predacious and almost terrestrial, occurring in the damp sand and mud at the water's edge. Eggs are laid singly, on stones or algae in shallow water or out of the water but in damp places; they are usually covered with silk. Adults come to the surface to renew their air supply and break the film with an antenna.

Hydroscaphidae. Minute (1 mm) beetles with truncated elytra exposing the conical abdomen. Both larvae and adults are aquatic and eat filamentous algae. The egg is very large in proportion to the female's abdomen, and only 1 is developed at a time; it is laid on the algae.

Psephenidae. (Fig. 38.8). Five Nearctic genera containing about 10 species belong in this family. Adults are small, depressed, black or brown, sometimes variegated, terrestrial beetles which prefer wave-splashed stones in streams but may occur around lakes. Often they congregate in numbers at the water's edge, and so far as known, enter the water only to oviposit. Their aquatic larvae, called "water pennies," can cling tightly to stones in currents. One species of *Psephenus* occurs widely over the eastern U. S., all others being found in Calif., Wash., and Ida. *Eubrianax* occurs in Calif., Ore., and Nev., other species of this genus are found in the Old World. *Ectopria* and *Dicranopselaphus* are found in eastern N. A. *Acneus* is recorded from Calif. and Ore. *Ectopria* is the only genus in this family in N. A. that has been taken at lights. The Old World genus *Psephenoides* has one of the few truly aquatic pupae known in the Coleoptera.

Ptilodactylidae. This family contains about half a dozen North American genera of which 3 are known to have aquatic larvae. The terrestrial adults of these genera are brown to black in color and usually are found near water. Larvae of the aquatic species are elateroid in shape and have gills on the abdomen on segments 1 to 7 or in the anal region. The genus *Anchytarsus*, with one species, occurs in the eastern U. S. from N. Y. to Ga., and *Stenocolus* and *Anchycteis*, each with one species, are known from Calif. The larvae are found in streams and springs.

Chelonariidae. This family is comprised of the single genus *Chelonarium* with species in both the Old and New World tropics. One species is known to occur in Fla., Ala., and Tenn. The oval compact adults, about 5 mm in length, are terrestrial; they occur on vegetation and have been taken at lights. The larvae occur in damp moss and resemble larvae of the family Elmidae but perhaps are not as fully aquatic.

Limnichidae. Several genera of small convex hairy or scaly byrrhidlike beetles belong to this family. The larva of the genus *Lutrochus* only is known to be aquatic, occurring with elmid larvae on submerged stones and debris in rapid streams. The larva has similar retractile anal gills. Unlike most Elmidae, *Lutrochus* adults are riparian, occupying shaded surfaces of stones over water and taking flight when disturbed. Rarely are they attracted to lights. One or two short chunky species are found in the Mississippi Valley area and in N. M.

Dryopidae. (Fig. 38.9). This family as now restricted in N. A. comprises 3 genera of elongate, dull-colored brown or black and generally hairy beetles, usually less than 10 mm in length. The tarsal claws are large as in the Elmidae, and the antenna usually is short; the first 2 segments are strongly enlarged, the remaining segments often very short and pectinate. In North America the adults and larvae of *Helichus* only are known to be truly aquatic, but it is not certain whether at least the larvae of *Pelonomus* and *Dryops* are aquatic. Adults and larvae of 1 of the 2 southeastern species of *Pelonomus* occur in damp places in swampy areas although Darlington (1936) recorded 2 West Indian species from aquatic vegetation and waterlogged trash in ponds or slow streams. The larvae and habits of the single North American species of *Dryops*, recorded from Ariz., are unknown. *Helichus* is transcontinental in distribution, and probably one or more species occur in each of the

United States; adults have been collected in streams of various sizes, in ponds, and ir an artificial tank located many miles from the nearest stream (Leech and Chandler, 1956). A few species of *Helichus*, such as the eastern *lithophilus* Germar, are strongly attracted to lights, a characteristic also of *Dryops* and *Pelonomus*. Present knowledge of the larvae of *Helichus* in N. A. is restricted to two specimens (Leech and Chandler, 1956) collected in two small turbulent streams in northern Calif. These larvae are elateroid, and they have no anal gills as do the Elmidae. Food habits of North American Dryopidae are not definitely known but probably they are vegetable feeders (Leech and Chandler, 1956; Hinton, 1955). *Helichus* and *Dryops* occur also in the Old World.

Elmidae. (Fig. 38.10). This is the largest family of Dryopoidea in N. A., containing approximately 75 species distributed among 24 genera. The family is included by some authors with the Dryopidae (Bertrand, 1955) and together with other families of Dryopoidea has been known under the old name Parnidae. Adults of all our elmid genera, except *Lara* and *Phanocerus*, are aquatic, and they rarely exceed 3 mm in length. The color usually is black or brown, and occasionally some species are marked with red or yellow. The family is remarkable for the long adult tarsal claws enabling the beetles to cling to rocks, water-soaked wood, and other objects in swift streams. Adults are equally remarkable in their adaptation to aquatic existence in that they obtain their oxygen principally by diffusion through a hydrofuge tomentum or plastron on the underside of the body. Consequently it appears to be unnecessary for most elmid adults to surface for respiratory purposes as do many other aquatic beetles. However, most species of *Stenelmis* and one of *Microcylloepus* in N. A., as well as several Neotropical genera, fly to lights, but it is possible that these flights occur following pupation outside water (Sanderson, 1953).

The larval stages of all except one North American genus are known, and they occur in the same habitats as adults. Identification of our larvae is based largely on association and the process of elimination (Sanderson, 1953; Leech and Chandler, 1956; Hinton, 1940). Most larvae are slender, shaped somewhat like some larval Dermestidae, and have retractile filamentous gills in a ventral operculate chamber in the last abdominal segment.

The Elmidae are distributed throughout N. A. north to Great Slave Lake at about 60 degrees north latitude. No species has been recorded from Alaska. The family has a preference for rapidly flowing clear rocky streams, but a few species may occur in ponds, lakes, or slowly moving streams. Several are confined to warm springs. Genera and species of Elmidae are particularly abundant in the Appalachian Mountains in eastern N. A., in the Ozark Plateau and adjacent areas in Mo., Ark., and Okla., in the southwestern U. S., and in the mountainous areas in western N. A. especially Calif.

With the exception of one species and a subspecies of *Stenelmis* occurring in warm springs in Nev., all other species of this genus are found in eastern N. A., generally east of the 100th meridian; many of these species are stream inhabitants occurring in the Ozark region; 2 species appear to be confined to lakes in the northeastern states. *Microcylloepus* occurs throughout the U. S.; all our species inhabit streams, but 2 are confined to warm springs in Nev. *Zaitzevia* is found through the mountainous regions in the West, generally in cold streams, but also in warm springs. *Dubiraphia* is found throughout the U. S., chiefly on aquatic vegetation in lakes and ponds and at the margins of slow-moving streams. *Optioservus* probably is the most widely distributed of all our Elmidae, occurring throughout the U. S. and in parts of Canada; it appears to be limited to streams. *Macronychus* is distributed from the Mississippi Valley region to the east coast, and usually is found in streams but has been taken in lakes.

The remaining genera of the family occur in streams and are more restricted in distribution than the ones listed above. Several genera are northern extensions of a larger Mexican fauna and are confined to the southwestern states. These are: *Cyl-*

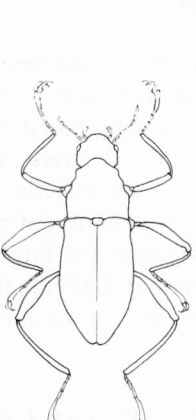

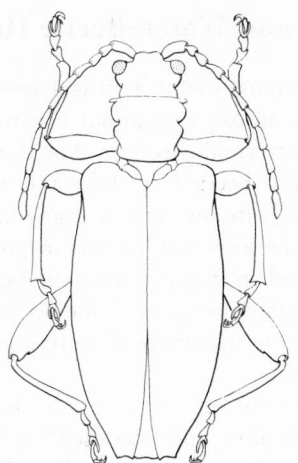

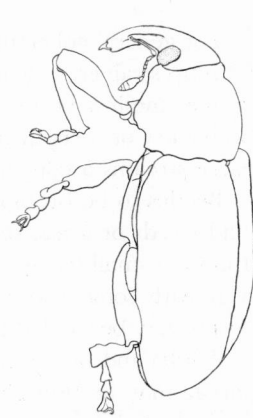

Fig. 38.10. Elmidae.
Ancyronyx variegatus
Germar.

Fig. 38.11. Chrysomelidae. *Don-*
acia hirticollis Kirby.

Fig. 38.12. Curculionidae.
Endalus sp.

loepus, Elsianus, Heterelmis, Hexacylloepus, Neoelmis, and *Phanocerus.* Five genera are known to occur only in the eastern and southeastern states: *Ancyronyx, Macronychus, Promoresia, Gonielmis,* and *Limnius.* Genera limited to the western states, particularly mountainous regions, are: *Lara, Heterlimnius, Cleptelmis, Zaitzevia, Narpus, Ordobrevia, Ampumixis, Atractelmis,* and *Rhizelmis.* Leech and Chandler (1956) record the food of adult and larval Elmidae as algae, moss, and other vegetable matter. Adults of *Stenelmis* have been observed feeding on vegetable growth and deposits on stones and on the elytra of other *Stenelmis* beetles (Sanderson, 1938). Five North American genera of Elmidae occur in the Old World: *Stenelmis, Limnius, Macronychus, Zaitzevia,* and *Ancyronyx.*

Helodidae. Small, rather soft beetles occurring near water. The adults are terrestrial (those of *Elodes* spp. sometimes enter the water); the larvae are aquatic, in shallows or on floating vegetation, etc. Species of several genera breed in water trapped in cavities in stumps and trees, and in bromeliads. The larvae are unique in the Coleoptera for their long multiarticulate antennae; they also have terminal retractile abdominal gills, and their mouthparts are specialized for scraping detritus from underwater surfaces.

Chrysomelidae. (Fig. 38.11). A large family of predominantly terrestrial beetles. Adults of the subfamily Donaciinae are semiaquatic, and their larvae are truly aquatic. The small, fat, white or greenish larvae feed on the submerged underground stems and roots of aquatic plants (*Nymphaea, Sagittaria, Potamogeton,* etc.). They obtain air by piercing the plant cells with the spurlike spiracles of the eighth abdominal segment; cocoons are spun near the feeding sites, and filled with air in the same manner.

Curculionidae. (Fig. 38.12). Weevils, or snout beetles; adults are recognized by the beaklike prolongation of the head, with mouthparts at the tip; the larvae are legless. All stages of most species are terrestrial, but adults of some (chiefly of the Tribe Hydronomini) crawl down plants growing in water to feed and oviposit. Except for *Lissorhoptrus* their larvae are not truly aquatic either; most burrow in the underwater sections of plants, though some live between the leaf sheath and stem, and have piercing dorsal spiracles which enable them to get air from the plant cells.

Collecting Methods and Water-Beetle Habitats

For general collecting the most useful item is a strongly made kitchen sieve or soup strainer, $6\frac{1}{2}$ or 7 in. across, with about 17 wires to the inch; one with a finer mesh is better for very small beetles. For deep water a long-handled water net or a small dredge is needed, and for the swifter parts of rivers or large streams a copper mesh window screen, frame and all, is very effective.

Beetles to be brought home alive can be put in jars half filled with water and weeds or grass, or placed in damp moss; it is best to separate the large from the small forms. Any predacious larva should be put into an individual vial with some water weed or dampened shore-line debris; nonpredacious forms may be carried together.

Adults and larvae may be put into 80 per cent alcohol in the field, or the larvae may be brought back alive and "coddled" in boiling water, then put into alcohol or into one of the special preserving fluids. If field collected material is to be stored, the alcohol should be changed at least once, a day or so after collecting.

The greatest number of water beetles are to be found in weedy pools and ponds. Try working a sieve back and forth a number of times in shallow weedy places, stirring the water up thoroughly. Dump the debris from the net onto a sheet of white canvas or rubberized cloth, and pick out the wanted specimens as they move; they will do so more quickly if the canvas is in the hot sun and on a slight slope. If you have no canvas, use a flat rock or a patch of bare ground.

Use the same system at the margins of lakes and quiet pools in streams and rivers. If weeds are absent, try pushing some of the shore line out into the water before using the sieve. As the water is roiled the dislodged insects swim about and can be caught in the net. In running water place the sieve or screen against the bottom, nearly upright, and turn stones, etc. upstream from it, letting the current wash the insects onto the netting.

More specialized habitats include swamps; springs, both hot and cold; seepage areas and wet mosses; algal mats; the roots and underground stems of aquatic plants; cracks and crevices in waterlogged wood; water trapped in stumps, tree holes, and certain plants; and in cracks in the rocks or among barnacles in the intertidal zone of the seashore. If waters dry up during the summer, some beetles fly to new places, but others bury themselves in the mud or gravel and aestivate, for not all species have functional wing muscles. Where winters are severe adults hibernate; in warmer places many are active under the ice; in mild climates with winter rains, both larvae and adults may be present and active.

Most water beetles hide during the day, but may be seen feeding and swimming or crawling about after dark, if one goes collecting with a flashlight (headlamp). This applies especially to forms such as *Chaetarthria* (Hydrophilidae) which live in the wet sand at the water's edge. Many species fly well and are attracted to lights at night; others fly during the day and "rain" down on shiny car tops, greenhouses, etc., mistaking the metal or glass for water surfaces.

KEY TO GENERA (ADULTS)
(Larvae on p. 1009)

1a   First visible abdominal sternite completely divided by hind coxal cavities (Fig. 38.13); in Gyrinidae, first apparent sternite is actually the second, but note 2 pairs of eyes and short, irregular antennae (Fig. 38.14); in Haliplidae, first 2 or 3 sternites hidden by expanded hind coxal plates (Fig. 38.15) **2**

1b   First visible abdominal sternite extending for its entire breadth behind coxal cavities, not divided by them (Fig. 38.33); if both a dorsal and a ventral pair of eyes present, see Gyrinidae, above . . .   **53**

2a   (1)   Eyes divided by sides of head, appearing as 4; antennae short, stout (Fig. 38.14); middle and hind legs short, flattened, tarsi folding fanwise . . . . Family **Gyrinidae**   **3**

2b   Eyes 2; antennae elongate, slender; hind legs suited for crawling or swimming, tarsi never folded fanwise **5**

3a   (2)   Dorsum glabrous; last abdominal segment rounded, its sternite without a median longitudinal line of hairs; scutellum visible or not. **4**

3b   Sides of pronotum and elytra pubescent; last abdominal segment elongate, conical, its sternite with a median line of golden hairs; scutellum not visible. (3 species) . . . . *Gyretes* Brullé

4a   (3)   Scutellum visible; elytral striae punctate, suture margined; smaller, more convex species, 3–8 mm long. (35 species) . . . .
Gyrinus (Geoffroy *in*) Müller

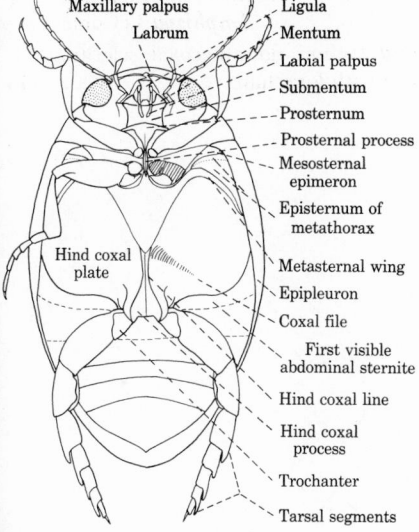

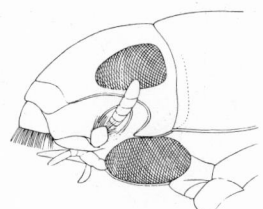

Fig. 38.13. Dytiscidae. Underside of *Laccophilus terminalis* Sharp, to show parts.

Fig. 38.14. Gyrinidae. *Dineutus* sp., lateral view of head of adult, to show divided eyes.

4b Scutellum not visible; elytral striae not punctate, suture not margined; larger, flatter species, 9–16 mm long. (Figs. 38.5, 38.14). (15 species) . . . . ***Dineutus*** MacLeay

5a (2) Hind coxae expanded into large plates (Fig. 38.15); small beetles, 5.5 mm or less in length (Fig. 38.2) . . . . . . Family **Haliplidae** 6

5b Hind coxae not expanded into plates, not covering much of hind femora nor more than first abdominal sternite 9

6a (5) Last segment of maxillary palpi as wide and as long as or longer than, next to last; hind coxal plates large, only last abdominal sternite completely exposed. (Figs. 38.15, 38.16). (15 species) . . . ***Peltodytes*** Regimbart

6b Last palpal segment narrow, shorter than next to last (Fig. 38.17); last 3 abdominal sternites showing beyond hind coxal plates 7

7a (6) Pronotum with sides of basal two-thirds nearly parallel; epipleura broad, extending almost to tips of elytra, which are never truncate. (4 species) . . . . ***Brychius*** Thomson

7b Pronotum with sides widest at base, convergent anteriorly; epipleura evenly narrowed, usually ending near base of last abdominal sternite, never reaching elytral apices 8

8a (7) Median portion of prosternum and base of prosternal process forming a plateaulike elevation, at least in part angularly separated from sides of prosternum. (Fig. 38.17). (42 species) . . . . ***Haliplus*** Latreille

8b Prosternum evenly rounded from side to side. (1 species) . . . . . ***Apteraliplus*** Chandler

9a (5) Metasternum with a transverse, triangular antecoxal sclerite separated by a well-marked suture (as in Fig. 38.15); hind tarsi not flattened or fringed with hairs, but simple and carabidlike (Fig. 38.1) . . . . · . . . . Family **Amphizoidae** (4 species) . . . . ***Amphizoa*** LeConte

9b Metasternum without a transverse suture, no antecoxal sclerite; hind tarsi flattened, usually fringed with long hairs 10

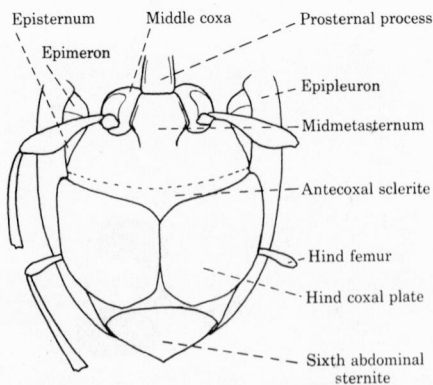

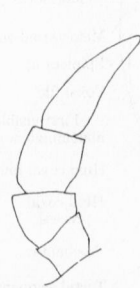

Fig. 38.15. Haliplidae. Underside of *Peltodytes* sp., to show parts.

Fig. 38.16. Haliplidae *Peltodytes dispersus* Roberts, ventral view of right maxillary palpus of adult.

Fig. 38.17. Haliplidae. *Haliplus gracilis* Roberts, ventral view of right maxillary palpus of adult.

10a (9) Middle of prosternum and its postcoxal process (Fig. 38.18) in same plane; front and middle tarsi distinctly 5-segmented, segment 4 as long as 3. . . . . **11**

10b Middle of prosternum not in same plane as its process (Fig. 38.19); front and middle tarsi 4- or 5-segmented, with fourth very small and almost concealed between lobes of third . . Family **Dytiscidae** Subfamily **Hydroporinae** **12**

11a (10) Scutellum fully visible or, if concealed, hind tarsi each have a single straight claw and segments lobed behind on outer side (Fig. 38.13) . . . . (in part) Family **Dytiscidae** **27**

11b Scutellum covered by bases of elytra and hind margin of pronotum; hind tarsi with 2 slender, curved claws of equal length, sides of tarsal segments nearly parallel; front tibiae (except in *Notomicrus* spp., beetles 1.5 mm or less in length) usually with a curved spur or hooked apex. (Figs. 38.4; 38.32) . . . . Family **Noteridae** **49**

12a (10) Scutellum fully visible; apices of elytra and last abdominal sternite produced, acuminate. (4 species). . . . . *Celina* Aubé

12b Scutellum covered by pronotum; apices of elytra rounded, sub-truncate, or acute **13**

13a (12) Prosternal process short, broad, not reaching metasternum, its tip ending at front of the contiguous middle coxae. (1 species) . . . . *Derovatellus* Sharp

13b Prosternal process more elongate, separating middle coxae and contacting metasternum. **14**

14a (13) Broad apex of hind coxal processes conjointly divided into 3 parts, i.e., 2 widely separated narrow lateral lobes and a broad depressed middle region (Fig. 38.20); small, broadly ovate beetles about 2.5 mm long. (6 species) . . . . *Hydrovatus* Motschulsky

14b Hind coxal processes not divided into 3 parts as above, but either without lateral lobes, or with these lobes covering bases of trochanters. **15**

15a (14) Hind coxal processes without lateral lobes, bases of hind trochanters entirely free **16**

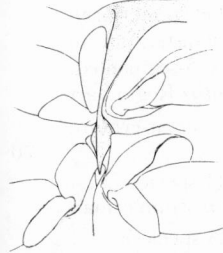

Fig. 38.18. Dytiscidae. Pro- and mesosternal area of *Laccophilus* sp. Prosternum and its process (stippled) in the same plane.

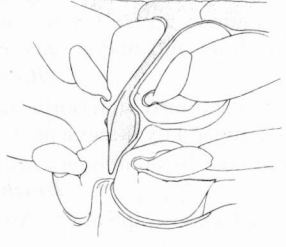

Fig. 38.19. Dytiscidae. Pro- and mesosternal area of *Hygrotus* sp. Prosternal process bent, not in the same plane as prosternum.

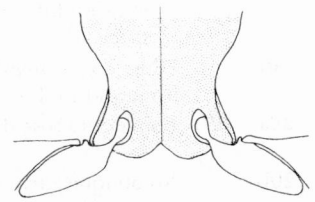

Fig. 38.20. Dytiscidae. *Hydrovatus pustulatus* Melsheimer, hind coxal process of adult.

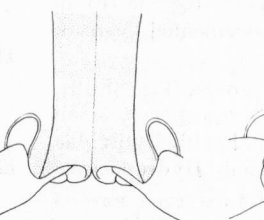

Fig. 38.21. Dytiscidae. *Hydroporus pilatei* Fall, hind coxal process of adult.

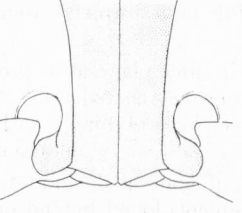

Fig. 38.22. Dytiscidae. *Hydroporus superioris* Balfour-Browne, hind coxal process of adult.

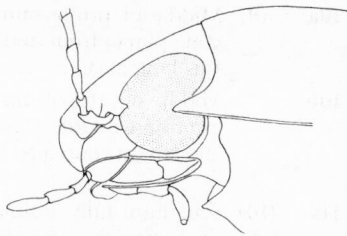

Fig. 38.23. Dytiscidae. Head of *Agabus disintegratus* Crotch, showing emargination of front of eye above antennal base.

15b	Sides of hind coxal processes divergent, more or less produced into lobes which cover bases of hind trochanters (as in Fig. 38.24) . . .	**21**
16a	(15) Hind tibiae straight, of almost uniform width from near base to apex; hind tarsal claws unequal; prosternal process short and broad, or rhomboid; epipleura with a diagonal carina crossing near base.	**17**
16b	Hind tibiae slightly arcuate, narrow at base, gradually widening to apex; hind tarsal claws equal; prosternal process oblong; epipleura without a diagonal carina near base (except in *Brachyvatus*).	**18**
17a	(16) Middle coxae separated by about width of a middle coxa; prosternal process short and broad, apex obtuse. (1 species). . . . . *Pachydrus* Sharp
17b	Middle coxae separated by only $\frac{1}{2}$ the width of a middle coxa; prosternal process rhomboid, apex acute. (7 species) . . . . *Desmopachria* Babington
18a	(16) Each elytron with a sharp narrow carina, starting at base opposite pronotal plica and fading out at declivity; pronotum transversely impressed at base. (1 species) . . . . . . *Anodocheilus* Babington
18b	Elytra without sharp narrow carinae, though often with a short basal groove opposite each pronotal plica; pronotum not transversely impressed at base	**19**
19a	(18) Hind coxal lines strongly sulcate-impressed, parallel posteriorly, converging as they continue forward across mid-metasternum to meet at middle coxae; front and middle tarsi clearly 5-segmented. (3 species) . . . . *Bidessonotus* Régimbart
19b	Hind coxal lines not continued anteriorly across metasternum; front and middle tarsi apparently 4-segmented.	**20**
20a	(19) Epipleura crossed near base by an oblique carina. (1 species) . . . *Brachyvatus* Zimmerman
20b	No oblique carina near base of epipleura. (About 25 species) . . . *Bidessus* Sharp
21a	(15) Bases of hind femora touching hind coxal lobes. (5 species) . . . . *Laccornis* Des Gozis
21b	Hind femora separated from hind coxal lobes by basal part of trochanters.	**22**

22a (21) A diagonal carina crossing epipleura near base; front and middle tarsi 4-segmented. (49 species) . . . . ***Hygrotus*** Stephens

22b No carina crossing epipleura; front and middle tarsi actually 5-segmented, though fourth is usually very small and hidden between lobes of third **23**

23a (22) Hind margin of hind coxal processes (best viewed with head of insect towards observer) together virtually straight across, or sinuate and angularly prominent at middle (Fig. 38.21), or obtusely angulate (Fig. 38.22), but never triangularly incised at middle, therefore median line as long as or longer than lateral coxal lines **24**

23b Hind margins of hind coxal processes slightly to deeply and more or less triangularly incised at middle, median line thus shorter than lateral coxal lines. **26**

24a (23) Hind margins of hind coxal processes together either truncate or angularly prominent (Fig. 38.22). (About 100 species). . . . . . . (in part) ***Hydroporus*** Clairville

24b Hind margins of hind coxal processes conjointly sinuate and somewhat angularly prominent at middle (Fig. 38.21) **25**

25a (24) Hind angles of pronotum rectangular or obtuse . . . . (in part) ***Hydroporus*** Clairville

25b Hind angles of pronotum acute . . . . (in part) ***Deronectes*** Sharp

26a (23) Pronotum with a longitudinal impressed line or crease on each side, and usually with a shallow transverse impression near base; hind femora with a median line of setiferous punctures, otherwise sparsely punctate or nearly smooth; body beneath densely finely punctate or shagreened, with scattered or numerous coarser punctures. (16 species) . . . . ***Oreodytes*** Seidlitz

26b Pronotum without sublateral impressed lines, usually without basal impression; hind femora usually densely punctate over entire surface; body beneath densely finely punctate to subgranulate, usually without scattered large punctures. (17 species) . . . . (in part) ***Deronectes*** Sharp

27a (11) Scutellum covered by hind margin of pronotum, or rarely a small tip visible; hind tarsi each with a single straight claw. (Fig. 38.13). **28**

27b Scutellum entirely visible **29**

28a (27) Spines of hind tibiae notched or bifid at tip; apical third of prosternal process lanceolate, only moderately broad (Fig. 38.13); larger species, 2.5–6.5 mm long. (18 species) . ***Laccophilus*** Leach

28b Spines of hind tibiae simple, acute at tip; apical third of prosternal process somewhat diamond-shaped, dilated behind front coxae and with tip acute. (1 species) . . . . ***Laccodytes*** Régimbart

29a (27) Eyes emarginate above bases of antennae (Fig. 38.23); first 3 segments of front tarsi of male widened and with adhesion discs or not, but never together forming a nearly rounded plate **30**

29b Front margin of eyes not indented above bases of antennae; first 3 segments of front tarsi of male greatly broadened, forming a nearly round or an oval plate with adhesion discs. **42**

30a (29) Hind femora with a linear group of ciliae near posterior apical angle (Fig. 38.24) **31**

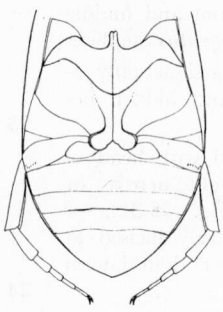

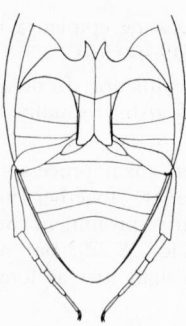

Fig. 38.24.   Dytiscidae.
Underside of *Agabus* sp.

Fig. 38.25. Dytiscidae.
Underside of *Agabinus
glabrellus* Motschulsky.

Fig. 38.26. Dytiscidae.
Tip of left hind tarsus of
Ilybius sp., to show short
outer claw.

30b		Hind femora without such a group of ciliae	**35**
31a	(30)	Hind coxal processes in form of rounded lobes (Fig. 38.24)	**32**
31b		Sides of hind coxal processes parallel, lateral margins straight to apices (Fig. 38.25). (2 species) . . . . *Agabinus* Crotch	
32a	(31)	Hind tarsal claws of equal length; if slightly unequal then both are very short, only ⅓ the length of fifth tarsal segment	**33**
32b		Hind tarsal claws obviously unequal, outer one of each pair ⅔ or less length of inner claw (Fig. 38.26)	**34**
33a	(32)	Labial palpi very short (Fig. 38.27), terminal segment subquadrate. (1 species) . . . . *Hydrotrupes* Sharp	
33b		Labial palpi approximately as long as maxillary palpi (Fig. 38.23), terminal segment linear, not subquadrate. (76 species) . . . . . . *Agabus* Leach	
34a	(32)	Labial palpi with penultimate segments enlarged, triangular in cross section, the faces concave and unequal (Fig. 38.28); genital valves of female dorsoventrally flattened, unarmed. (1 species) . . *Carrhydrus* Fall	
34b		Labial palpi with penultimate segments linear, not enlarged and triangular; genital valves of female laterally compressed, sawlike. (15 species) . . . . *Ilybius* Erichson	
35a	(30)	Prosternum with a median longitudinal furrow, from near front margin to apex of prosternal process. (3 species) . . *Matus* Aubé	
35b		Prosternum not longitudinally furrowed	**36**
36a	(35)	Hind coxal lines divergent anteriorly (Fig. 38.29), coming so close together posteriorly as almost to touch median line, thence turning outward almost at right angles onto hind coxal processes; hind tarsal claws unequal; pronotum clearly but narrowly margined laterally. (6 species) . . . . *Copelatus* Erichson	
36b		Hind coxal lines never almost touching median line and thence turning outward almost at right angles onto coxal processes; hind tarsal claws equal or not; pronotum margined or not	**37**
37a	(36)	Hind claws of same length or virtually so; smaller species, 6–9 mm long.	**38**

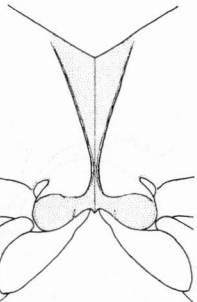

Fig. 38.27. Dytiscidae. Labial palpus of *Hydrotrupes palpalis* Sharp.

Fig. 38.28. Dytiscidae. Labial palpus of *Carrhydrus crassipes* Fall.

Fig. 38.29. Dytiscidae. *Copelatus glyphicus* Say, adult, to show hind coxal lines almost touching median line.

37b		Hind claws obviously unequal, outer ones only ⅓ to ⅔ the length of inner ones; larger species, 9–20 mm long	**39**
38a	**(37)**	Terminal segment of palpi (especially of labial palpi) notched or emarginate at apex; pronotum clearly though narrowly margined laterally. (3 species) . . . . **Coptotomus** Say	
38b		Terminal segment of palpi not emarginate at apex; pronotum with an exceedingly fine line along lateral edge, but not margined. (1 species) . . . . **Agabetes** Crotch	
39a	**(37)**	Anterior point of metasternum (between middle coxal) clearly triangularly split to receive tip of prosternal process, the triangular channel usually deep, with its apex about on a line with hind margins of middle coxae; pronotum usually margined laterally . . .	**40**
39b		Anterior tip of metasternum depressed, with a shallow pit or broad notch to receive tip of prosternal process, never with a sharply outlined triangular excavation; pronotum not margined	**41**
40a	**(39)**	Prosternal process flat. (1 species) . . . . **Hoperius** Fall	
40b		Prosternal process convex or cariniform. (15 species) . . . . **Rhantus** Dejean	
41a	**(39)**	Elytral sculpture consisting of numerous parallel transverse grooves. (10 species) . . . . **Colymbetes** Clairville	
41b		Elytra coarsely reticulate, without transverse grooves. (2 species) . . . . **Neoscutopterus** J. Balfour-Browne	
42a	**(29)**	Inferior spur at apex of hind tibiae dilated, *much* broader than the other large spur; first 3 segments of front tarsi of male forming an oval plate; large beetles, 20–32 mm long	**48**
42b		Inferior spur not or but little broader than its fellow; first 3 segments of front tarsi of male forming a nearly round plate; medium-size to large beetles	**43**
43a	**(42)**	Posterior margins of first 4 hind tarsal segments beset with a dense fringe of flat golden cilia; smaller beetles, about 8–15 mm long	**44**
43b		Posterior margins of first 4 hind tarsal segments bare; large beetles, about 20–38 mm long. (11 species) . . **Dytiscus** Linnaeus	

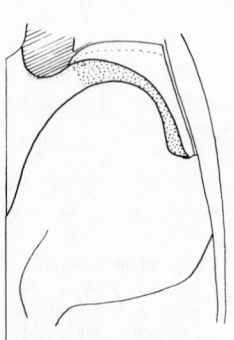

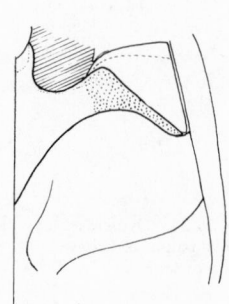

Fig. 38.30. Dytiscidae. Underside of *Thermonectus* sp., metasternal side wing stippled.

Fig. 38.31. Dytiscidae. Underside of *Hydaticus* sp., metasternal side wing stippled.

Fig. 38.32. Noteridae. Front leg of *Colpius inflatus* LeConte, to show curved tibial spur.

44a	**(43)**	Apex of prosternal process sharply pointed, pronotum margined laterally. (1 species). . . . . ***Eretes*** Laporte
44b		Apex of prosternal process rounded; pronotum not margined laterally **45**
45a	**(44)**	Outer margin of metasternal side wings arcuate (Fig. 38.30); outer (shorter) spur at apex of hind tibiae blunt, more or less emarginate. **46**
45b		Outer margin of metasternal wings straight (Fig. 38.31); outer spur at apex of hind tibiae acute. (4 species) . . ***Hydaticus*** Leach
46a	**(45)**	Elytra densely punctate, and in addition usually fluted and hairy in female. (3 species). . . . . ***Acilius*** Leach
46b		Elytral punctation extremely fine or absent; some females with a superimposed sexual sculpture of small elongate grooves, or granulate. **47**
47a	**(46)**	Elytra basically yellowish, uniformly speckled or vermiculate with black; hind margin of middle femora with a series of stiff setae which are only about $\frac{1}{2}$ as long as femora are wide. (5 species) . . . ***Graphoderus*** Sturm
47b		Elytra black with yellow maculae or transverse bands, or yellow with black spots, or irrorate; hind margin of middle femora with series of stiff setae which, if unbroken, are as long as or longer than femora are wide. (6 species). . . . . . ***Thermonectus*** Dejean
48a	**(42)**	Apex of hind tarsi of males with 2 claws, of females with a long outer and a rudimentary inner claw. (2 species) . . . . ***Megadytes*** Sharp
48b		Apex of hind tarsi of males always, of females usually, with only one claw. (5 species) . . . . ***Cybister*** Curtis
49a	**(11)**	Apex of front tibiae bearing more or less conspicuous curved spurs or hooks. (Fig. 38.32) **50**

Key to the genera of Noteridae supplied by Dr. F. N. Young.

49b　　　No spurs or hooks on front tibiae; small beetles, rarely exceeding 1.5 mm in length. (2 species). . . . . ***Notomicrus*** Sharp

50a　(49)　Front tibial spurs strong and conspicuous (Fig. 38.32); hind femora with angular cilia; prosternal process truncate behind (or if rounded in male, form very broad, almost hemispherical) **51**

50b　　　Front tibial spurs weak and inconspicuous; hind femora usually without angular cilia; prosternal process rounded behind in both sexes. (2 species) . . . . ***Pronoternus*** Sharp

51a　(50)　Laminate inner plates of hind coxae truncate at apex with an arcuate emargination on each side of the depressed middle; hind coxal cavities separated. (1 species) . . . . ***Colpius*** LeConte

51b　　　Laminate inner plates of hind coxae with a broad and deep excision at apex, leaving on each side a diverging triangular process; hind coxal cavities contiguous. **52**

52a　(51)　Apex of prosternal process at least twice as wide as its breadth between front coxae, not broader than long; pronotum with lateral marginal lines originating at hind angles but disappearing at about middle; beetles usually less than 3 mm in length. (8 species) . . . . ***Suphisellus*** Crotch

52b　　　Apex of prosternal process very broad, at least 2½ to 3 times its breadth between coxae, broader than long; pronotum with lateral marginal lines the entire length of margins and joining front margin anteriorly; beetles usually between 4 and 5.5 mm in length. (5 species) . . . . ***Hydrocanthus*** Say

53a　(1)　Tiny beetles, from 0.5–1 mm long; hind coxae laminate, widely separated (Fig. 38.33); Staphylinidlike beetles with truncate elytra and exposed conical abdomen . . . . . . Family **Hydroscaphidae** (1 species) . . . . ***Hydroscapha*** LeConte

53b　　　Beetles of various sizes, from nearly 1 mm to 40 mm long; in very small dorsally glabrous forms, never with hind coxae laminate . . **54**

54a　(53)　Antennae short, true segment 6 modified to form a cupule (Fig. 38.34), segments 7 to 11 (often reduced in number to 3) forming a differentiated pubescent club, segments 1 to 5 (sometimes reduced in number to 3 or 4) simple and glabrous; maxillary palpi nearly always longer than antennae; head usually with a Y-shaped (Fig. 38.6) impressed line on vertex. **55**

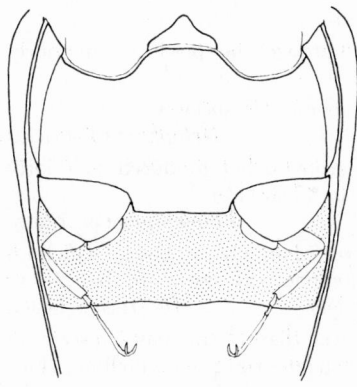

Fig. 38.33. Hydroscaphidae. Underside of *Hydroscapha natans* LeConte, first abdominal sternite stippled.

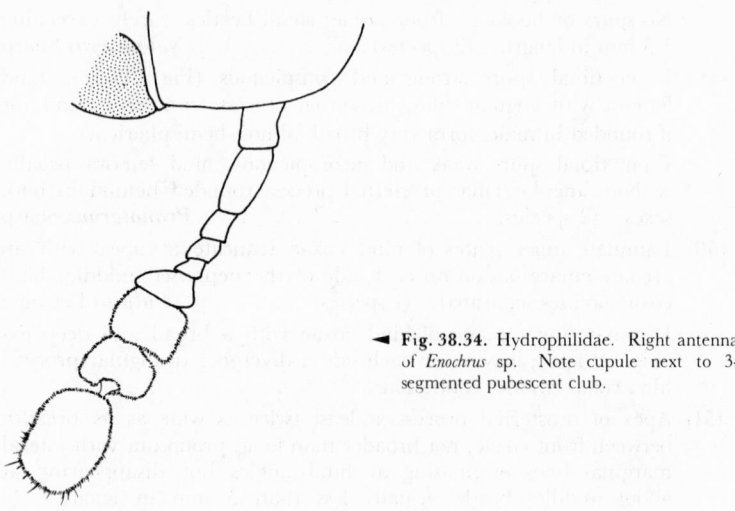

◄ **Fig. 38.34.** Hydrophilidae. Right antenna of *Enochrus* sp. Note cupule next to 3-segmented pubescent club.

54b   Antennae not so constructed, longer than maxillary palpi; head without Y-shaped impressed line on vertex **81**

55a (54) Antennal club of 5 pubescent segments; tiny beetles, not over 2.5 mm in length . . . . Family **Hydraenidae**   **56**

55b   Antennae with only 3 pubescent segments beyond cupule (Fig. 38.34); size varied, but species approaching 1 mm in length are convex and rounded . . . . Family **Hydrophilidae**   **58**

56a (55) Pronotum smooth, not coarsely punctate or sculptured, sides evenly arcuate. (6 species) . . . . *Limnebius* Leach

56b   Pronotum with surface uneven, coarsely punctate and/or with a transparent lateral border (Fig. 38.7), sides sinuate or irregular . .   **57**

57a (56) Maxillary palpi very long, much longer than antennae, pronotum coarsely, closely punctate, sides without a transparent border. (4 species) . . . . *Hydraena* Kugelann

57b   Maxillary palpi shorter than antennae (Fig. 38.7); pronotum variously sculptured, often with deep fossae and grooves, almost always with a transparent border in at least basal half. (22 species) . . . . *Ochthebius* Leach

58a (55) Pronotum with 5 longitudinal grooves, or produced anteriorly at middle so as to hide much of head **59**

58b   Pronotum not with 5 longitudinal grooves, not produced anteriorly to hide much of head **60**

59a (58) Pronotum with 5 longitudinal grooves. (18 species) . . . .
Helophorus Fabricius

59b   Pronotum without longitudinal grooves, but produced anteriorly at middle, covering much of head. (2 species) . . . .
Epimetopus Lacordaire

60a (58) Pronotum conspicuously narrower than elytral bases; scutellum very small; eyes protuberant; antennae with not more than 3 segments *before* the cupule. (16 species) . . . . . . *Hydrochus* Leach

60b   Pronotum not appreciably narrower than elytral bases (except in some *Berosus* ssp., but there note elongate triangular scutellum, (Fig.

38.6); antennae usually with 5 well-developed segments before the
cupule **61**

61a   **(60)** Hind tarsi with basal segment (which may be oblique and very
small, see Figs. 38.6, 38.35) shorter than second; antennae about as
long as or shorter than maxillary palpi; second segment of
maxillary palpi not, or very little, thicker than third or fourth;
aquatic . . . . Family **Hydrophilidae**   **62**

61b   Basal segment of hind tarsi longer than second; antennae usually
longer than maxillary palpi, which are never very long; second
segment of maxillary palpi (first is very small) *much* thicker than
third or fourth. Species terrestrial, in damp places. Not treated
further here . . . . Family **Hydrophilidae**
Subfamily **Sphaeridiinae**

62a   **(61)** Meso- and metasternum with a continuous median longitudinal
keel, which is prolonged posteriorly into a spine between hind
coxae **63**

62b   Meso- and metasternum without a continuous keel in common . . .   **67**

63a   **(62)** Prosternum sulcate to receive anterior part of mesosternal keel . . .   **64**

63b   Prosternum carinate, not sulcate **66**

64a   **(63)** Larger species, 30–45 mm long **65**

64b   Smaller species, 6–15 mm long. (14 species) . *Tropisternus* Solier

65a   **(64)** Prosternum bifurcate at middle so that anterior tip of mesosternal
keel could contact head. (1 species). . . . . . . *Dibolocelus* Bedel

65b   Prosternum sulcate at middle posteriorly, to receive mesoternal
keel, but closed anteriorly. (3 species) . . . *Hydrophilus* Geoffroy

66a   **(63)** Antennal club compact, almost symmetrical; front margin of
clypeus simply truncate or arcuate. (5 species) . . . .
Hydrochara Berthold

66b   Antennal club perfoliate, of very asymmetrical segments; front
margin of clypeus arcuate, and emarginate at middle to expose a
preclypeus. (1 species). . . . . . . *Neohydrophilus* d'Orchymont

67a   **(62)** First 2 abdominal sternites with an excavation in common, which
is large and spectacle-shaped and usually filled with a hyaline mass
supported by golden hairs; small beetles (1–2.5 mm long) with
ability to roll up partially. (5 species) . . *Chaetarthria* Stephens

67b   First 2 abdominal sternites without a broad excavation in common; no fringe of long stiff golden hairs from anterior margin of
first sternite. **68**

68a   **(67)** Head markedly deflexed, often with a deep transverse groove
delimiting a postoccipital region; antennae usually with only 3
segments before cupule, hence 7-segmented; scutellum a long
triangle (Fig. 38.6); middle and hind tibiae fringed with long
swimming hairs **69**

68b   Head not strongly deflexed, without a transverse occipital groove;
antennae normally 9-segmented; scutellum not or not much longer
than its basal width; middle and hind tibiae without natatory
fringes **70**

69a   **(68)** Eyes protuberant (Fig. 38.6); front tibiae slender, linear; labrum
prominent. (23 species) . . . . *Berosus* Leach

69b   Eyes not prominent; front tibiae triangular; labrum very short,
not prominent. (1 species) . . . . *Derallus* Sharp

70a (68) Maxillary palpi robust and short, shorter or not much longer than antennae, ultimate segment as long as or longer than penultimate   71

70b   Maxillary palpi more slender, longer than antennae, with ultimate segment usually shorter than penultimate. (Figs. 38.36, 38.37) . .   77

71a (70) Elytra with sutural striae in at least apical half; hind tibiae not arcuate   72

71b   Elytra with rows of punctures but no sutural striae; hind tibiae arcuate. (7 species) . . . . *Laccobius* Erichson

72a (71) Larger species, at least 4.5 mm long; elytra striate or with pronounced rows of punctures   73

72b   Smaller species, not over 3 mm long; elytra impunctate or confusedly punctate, never striate, at most with punctures subserially arranged   75

73a (72) Segments 2 to 5 of middle and hind tarsi with a fringe of long, fine swimming hairs, which arise from a series of punctures or a narrow groove (Fig. 38.35) along upper inner edge of tarsi; lateral margins of elytra even. (3 species) . . . . *Hydrobius* Leach

73b   Middle and hind tarsi completely without groove and fringe of fine swimming hairs along upper inner edge; lateral margins of elytra weakly serrate at least basally   74

74a (73) Form strongly convex, almost hemispherical in profile; clypeus more deeply emarginate, median part nearly truncate. (1 species) . . . . *Sperchopsis* LeConte

74b   Form oval; clypeus evenly, shallowly, arcuately emarginate. (2 species) . . . . *Ametor* Semenov

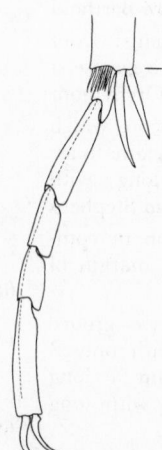

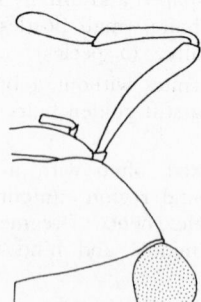

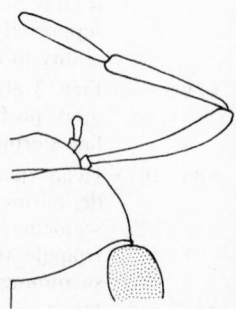

Fig. 38.35. Hydrophilidae. *Hydrobius fuscipes* Linnaeus, end of left hind leg, showing line on inner face of tarsal segments 2 to 5, along which a fringe of fine golden swimming hairs arises.

Fig. 38.36. Hydrophilidae. Head of *Enochrus* sp., right maxillary palp, showing forward curvature of long pseudobasal segment.

Fig. 38.37. Hydrophilidae. Head of *Helocombus bifidus* LeConte, right maxillary palp, long pseudobasal segment with convexity to the rear.

75a (72) Prosternum longitudinally carinate at middle; mesosternum usually with a longitudinal median carina behind the anterior transverse ∧ -shaped protuberance. (8 species) . . . . ***Paracymus*** Thomson

75b Prosternum not carinate; mesosternum simple, or with only a transverse ∧ -shaped or arcuate carina or protuberance 76

76a (75) Mesosternum simple, either noncarinate or with a small transverse protuberance before the middle coxae. (11 species) . . . .
Crenitis Bedel

76b Mesosternum with a prominent angularly elevated or dentiform protuberance before middle coxae. (3 species) . . . .
Anacaena Thomson

77a (70) All tarsi 5-segmented, though basal segment may be very small (even smaller than in Fig. 38.35), oblique, and best seen from an inner or ventral view 78

77b Middle and hind tarsi definitely 4-segmented 79

78a (77) Long curved pseudobasal segment of maxillary palpi (Fig. 38.36) with convexity to front; mesosternum with projecting longitudinal lamina. (20 species) . . . . ***Enochrus*** Thomson

78b Curved pseudobasal segment with convexity to rear (Fig. 38.37); mesosternum feebly protuberant at most. (3 species) . . . .
Helochares Mulsant

79a (77) Anterior coxal cavities closed behind; labrum concealed beneath projecting nonemarginate clypeus, which extends around to about the middle of each eye and outward for a distance equal to about the width of an eye. (1 species) . . . . ***Helobata*** Bergroth

79b Anterior coxal cavities open behind; labrum fully exposed; clypeus truncate or emarginate, not extending laterally in front of eyes . . . 80

80a (79) Maxillary palpi long and slender (Fig. 38.37), pseudobasal segment at least ⅔ as long as a front tibia. (1 species). . . . .
Helocombus Horn

80b Maxillary palpi shorter, stouter; pseudobasal and following segment together subequal in length to a front tibia. (12 species) . . .
Cymbiodyta Bedel

81a (54) Head beyond eyes produced into a distinct beak (Fig. 38.12), on the end of which are the mandibles and trophi. . . . .
Family **Curculionidae** 115

81b Head not produced into a distinct beak 82

82a (81) All tarsi actually 5-segmented, but segment 4 very small and nearly concealed within lobes of third (Fig. 38.38); or solidly joined to segment 5 so that tarsus actually is 4-segmented; first 3 segments dilated, with adhesive (hairy) pads beneath; beetles of 5–10 mm in length . . . . Family **Chrysomelidae** 114

82b Tarsi with 5 or fewer segments, but not fitting above description . . 83

83a (82) Front coxae more or less conically projecting; hind margin of prothorax never crenulate. (Adults of some species of *Elodes* dive, and swim below the surface.) . . . . Family **Helodidae**

83b Front coxae variously formed; if they are projecting, then hind margin or prothorax is crenulate 84

84a (83) Six or 7 abdominal sternites. (Adults largely terrestrial, but enter water at least to lay eggs) . . . . . . Family **Psephenidae** 85

84b Five abdominal sternites. 88

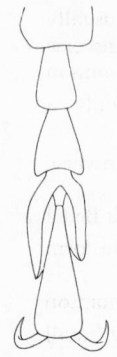

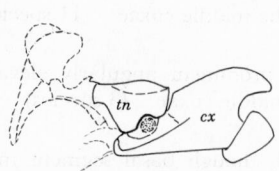

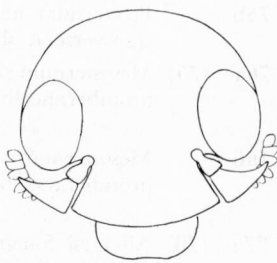

Fig. 38.38. Chryso-
melidae. Hind tarsus
of *Donacia* sp., showing
small fourth segment.

Fig. 38.39. Dryopidae. *Helichus lithophilus* Germar, right
coxa (*cx*) and trochantin (*tn*) of
adult.

Fig. 38.40. Dryopidae. *Helichus
lithophilus* Germar, head of adult.

85a	**(84)**	Hind margin of prothorax simple, smooth	**86**
85b		Hind margin of prothorax finely beaded or crenulate	**87**
86a	**(85)**	Head hidden beneath the expanded pronotum. (1 species).	
		Eubrianax Kiesenwetter	
86b		Head (Fig. 38.8) visible from above. (4 species)	
		Psephenus Haldeman	
87a	**(85)**	Prosternum narrow, depressed between front coxae; antennae of male flabellate. (2 species) . . . . *Acneus* Horn	
87b		Prosternum of moderate width, not depressed between coxae; antennae simple. (1 species) . . . . *Ectopria* LeConte	
88a	**(84)**	Anterior coxae (Fig. 38.39) transverse, with exposed trochantin; antennae generally short and serrate (Figs. 38.40, 38.41)	
		Family **Dryopidae**	**89**
88b		Anterior coxae globular, nearly always without trochantin; antennae usually slender . . . . Family **Elmidae**	**91**
89a	**(88)**	Dorsum with long erect silky hairs; eyes at least in part hairy; antennae closely approximate, separated by a distance equal to or less than length of first antennal segment	**90**
89b		Dorsum with hairs recumbent; eyes bare; antennae (Fig. 38.40) widely separated, the distance between them equal to about 3 times length of first segment. (11 species) . . . *Helichus* Erichson	
90a	**(89)**	Pronotum with a narrow groove on each side extending in a curved diagonal line from the base to the anterior margin near the apical angle. (1 species). . . . . *Dryops* Olivier	
90b		Pronotum with a shallow poorly defined depression at base between middle and lateral angle. (2 species) . . . . *Pelonomus* Erichson	
91a	**(88)**	Antennae (Fig. 38.41) 11-segmented, very short, strongly clubbed segments 6 to 10 are 2 to 3 times wider than long; body densely hairy above; length 2.5–3.5 mm. (1 species). . *Phanocerus* Sharp	
91b		Antennae usually slender, usually 10- or 11-segmented but with some segments wider than long if only 7- to 8-segmented	**92**
92a	**(91)**	Body densely hairy above; length about 5–8 mm. (3 species) . . .	
		Lara LeConte	

Fig. 38.41. Dryopidae. *Phanocerus clavicornis* Sharp, antenna of adult.

Fig. 38.42. Elmidae. *Ancyronyx variegatus* Germar, tarsal claw of adult.

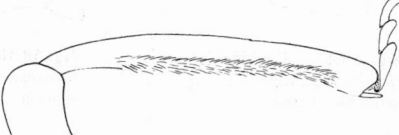

Fig. 38.43. Elmidae. *Elsianus texanus* Schaeffer, anterior tibia of adult.

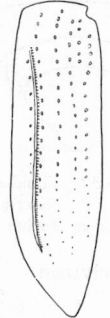

Fig. 38.44. Elmidae. *Ordobrevia nubifera* Fall, left elytron of adult.

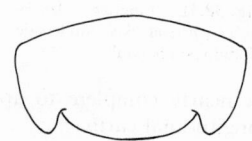

Fig. 38.45. Elmidae. *Elsianus texanus* Schaeffer, last abdominal sternite of adult.

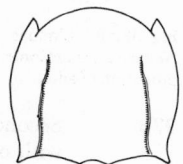

Fig. 38.46. Elmidae. *Elsianus texanus* Schaeffer, pronotum of adult.

92b		Body at most thinly hairy above; length not over 4 mm	93
93a	(92)	Tarsal claw (Fig. 38.42) with a small basal tooth; basal elytral spot angulate near suture (1 species) . . . . *Ancyronyx* Erichson	
93b		Tarsal claw without basal tooth.	94
94a	(93)	Anterior tibiae without fringe of tomentum	95
94b		Anterior tibiae (Fig. 38.43) with fringe of tomentum	96
95a	(94)	Second elytral stria (Fig. 38.44) incomplete, terminating at about basal fifth. (1 species) . . . . *Ordobrevia* Sanderson	
95b		Second elytral stria complete, extending nearly to elytral apex. (27 species) . . . . *Stenelmis* Dufour	
96a	(94)	Last abdominal sternite (Fig. 38.45) produced on each lateral margin as a distinct lobe	97
96b		Last sternite slightly emarginate, nearly evenly convex, or straight on lateral margin. (Figs. 38.51, 38.52)	102
97a	(96)	Second elytral stria incomplete, terminating at about basal fifth; elytron without longitudinal carinae; pronotum (Fig. 38.46) with a complete or nearly complete but sometimes faint carina between meson and lateral margin. (2 species) . . . . *Elsianus* Sharp	

Fig. 38.47. Elmidae. *Heterelmis* sp., right mandible of adult.

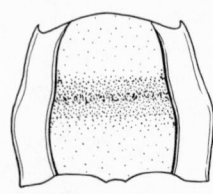

Fig. 38.48. Elmidae. *Heterelmis* sp., pronotum of adult.

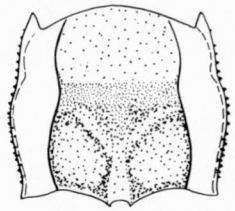

Fig. 38.49. Elmidae. *Microcylloepus* sp., pronotum of adult.

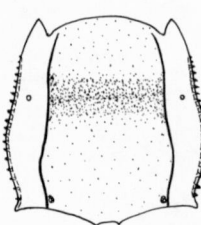

Fig. 38.50. Elmidae. *Neoelmis caesa* LeConte, pronotum of adult.

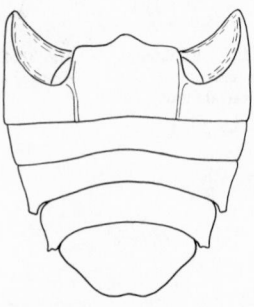

Fig. 38.51. Elmidae. *Macronychus glabratus* Say, underside of abdomen of adult.

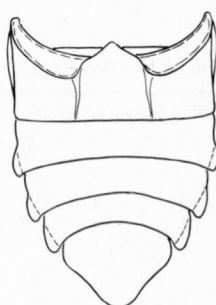

Fig. 38.52. Elmidae. *Zaitzevia* sp., underside of abdomen of adult.

97b Second elytral stria nearly complete to apex of elytron; elytron with one or more longitudinal carinae **98**

98a (97) Each mandible (Fig. 38.47) with a narrow lobe on lateral margin **99**

98b Mandible without narrow lateral lobe **100**

99a (98) Pronotum (Fig. 38.48) transversely depressed at middle. (3 species) . . . . *Heterelmis* Sharp

99b Pronotum (Fig. 38.49) transversely depressed at anterior third or two-fifths. (6 species) . . . . *Microcylloepus* Hinton

100a (98) Pronotum (Fig. 38.50) transversely depressed at or slightly before middle, with a deep fovea at each end of depression outside lateral carina, and another fovea on inside of carina near base; a serrate carina parallel to lateral margin. (1 species). *Neoelmis* Musgrave

100b Pronotum not conspicuously transversely depressed, without a serrate carina near lateral margin and without foveae. **101**

101a (100) Prothoracic hypomera with a complete transverse band of flat tomentum; last 2 rows of elytral punctures between lateral carina and lateral elytral margin confused and indistinct so that the elytron appears to have but 7 complete rows. (1 species). . . . . . *Hexacylloepus* Hinton

101b Prothoracic hypomera without a complete band of flat tomentum; 9 distinct rows of elytral punctures. (1 species) . . . . *Cylloepus* Erichson

102a (96) Antennae 7- or 8-segmented. **103**

102b Antennae 10- or 11-segmented **104**

103a (102) Antennae 7-segmented; posterior coxae widely separated by a

Fig. 38.53. Elmidae. *Limnius latiusculus* LeConte, dorsal view of adult.

Fig. 38.54. Elmidae. *Narpus* sp., right maxillae of adult, palpal segments numbered.

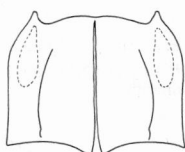

Fig. 38.55. Elmidae. *Ampumixis dispar* Fall, pronotum of adult.

quadrate lobe (Fig. 38.51) of the first abdominal sternite; length 2.5–3.5 mm. (1 species) . . . . **Macronychus** Müller

103b   Antennae 8-segmented; posterior coxae more narrowly separated by the more pointed lobe (Fig. 38.52) of the first abdominal sternite; length about 2 mm. (3 species) . . **Zaitzevia** Champion

104a (102)   Antennae 10-segmented; pronotum with a short basal carina between meson and lateral margin. (2 species) . . . .
     Heterlimnius Hinton

104b   Antennae 11-segmented **105**

105a (104)   Maxillary palpi 3-segmented (Fig. 38.54) **106**

105b   Maxillary palpi 4-segmented **111**

106a (105)   Body (Fig. 38.53) very minute, about 1 mm long; each side of pronotum with a distinct carina extending full length of pronotum. (1 species) . . . . **Limnius** Erichson

106b   Body larger, measuring 1.5 mm or more in length. **107**

107a (106)   Pronotum evenly punctured without carinae or swellings; length 3 to 4 mm. (2 species). . . . . **Narpus** Casey

107b   Pronotum with sublateral carina or swellings **108**

108a (107)   Lateral pronotal carina not reaching anterior margin of pronotum; base of pronotum between carinae broadly depressed. (1 species) . .
     Atractelmis Chandler

108b   Lateral pronotal carinae reaching or nearly reaching anterior margin of pronotum; base of pronotum between carinae flat or convex **109**

109a (108)   Pronotum (Fig. 38.55) with a very narrow median longitudinal carina or narrow depression **110**

109b   Pronotum (Fig. 38.56) without median carina or narrow depression. (2 species) . . . . **Cleptelmis** Sanderson

110a (109)   Prosternal ridge (Fig. 38.57) perpendicular to anterior prosternal margin. (1 species). . . . . **Ampumixis** Sanderson

110b   Prosternal ridge oblique. (1 species). . . . . . **Rhizelmis** Chandler

111a (105)   Pronotum without distinct basal carinae **112**

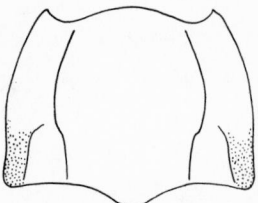

Fig. 38.56. Elmidae. *Cleptelmis ornata* Schaeffer, pronotum of adult.

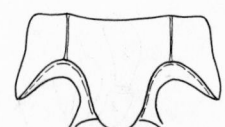

Fig. 38.57. Elmidae. *Ampumixis dispar* Fall, prosternum of adult.

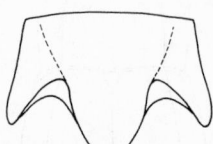

Fig. 38.58. Elmidae. *Dubiraphia* sp., prosternum of adult.

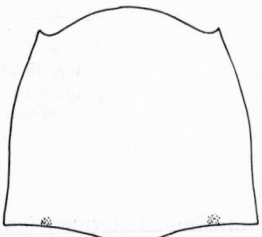

Fig. 38.59. Elmidae. *Gonielmis dietrichi* Musgrave, pronotum of adult.

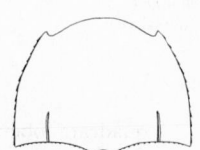

Fig. 38.60. Elmidae. *Optioservus trivittatus* Brown, pronotum of adult.

111b		Pronotum with a short sublateral carina in basal fourth to basal half **113**
112a	**(111)**	Pronotal disc entirely smooth; prosternum (Fig. 38.58) with a faint oblique carina on each side not reaching anterior margin; elytral spots, if present, longitudinal or circular. (2 species) . . . . ***Dubiraphia*** Sanderson
112b		Pronotum (Fig. 38.59) with a faint rugose area at base between meson and posterior angle; prosternum without carinae; elytral spots oblique. (1 species) . . . . ***Gonielmis*** Sanderson
113a	**(111)**	Posterior prosternal lobe broadly rounded at apex; lateral margins of pronotum (Fig. 38.60) sometimes slightly serrate, posterior margin with many small closely placed teeth. (11 species) . . . . . . ***Optioservus*** Sanderson
113b		Posterior prosternal lobe more triangular in outline; lateral and posterior margins of prothorax smooth. (2 species) . . . . ***Promoresia*** Sanderson
114a	**(82)**	Tarsi slender, nearly glabrous below; third segment entire, much shorter than second, last nearly as long as the rest together. (3 species) . . . . ***Neohaemonia*** Szekessy
114b		Tarsi dilated, spongy beneath; third segment deeply bilobed (Fig. 38.38), never much shorter than second, last short, rarely as long as the two preceding together. (44 species) . . . ***Donacia*** Fabricus
115a	**(81)**	Prosternum short, not continuing between and separating front coxae, the coxae contiguous **116**
115b		Prosternum continuing between front coxae, narrowly but completely separating them **129**

116a (115) Tibiae ending in a large inwardly curving claw, which is fully as
long as first tarsal segment. (4 species). . . . *Steremnius* Schönherr

116b Tibiae unarmed, or at most with a short tooth or spur 117

117a (116) Grooves for antennae starting near tip of beak, usually near bases
of mandibles 118

117b Grooves for antennae beginning further from tip of beak, at apical
quarter or third (Fig. 38.12), or even near mid-point 120

118a (117) Front margin of prothorax lobate adjacent to eyes (as in Fig.
38.12) and partially covering them 119

118b Front margin of prothorax straight, not at all lobed at sides . . . .
Phytonomus Schönherr

119a (118) Second segment of antennal funicle much longer than first; larger
species, usually over 5 mm long . . . . *Listronotus* Jekel

119b Second segment of antennal funicle as long as or but little longer
than first; smaller species, seldom over 4.5 mm long . . . .
Hyperodes Jekel

120a (117) Beak very short and broad, not longer than head; tarsi narrow,
third segment deeply emarginate . . . . . *Stenopelmus* Schönherr

120b Beak cylindrical, much longer than head 121

121a (120) Third segment of hind tarsi emarginate or bilobed (Fig. 38.12). . . 122

121b Third segment of hind tarsi simple; legs long and slender 127

122a (121) Beak curved; antennal funicle of 6 segments, the second short, as
is last tarsal segment 123

122b Beak straight; second segment of funicle long, as is last tarsal
segment 126

123a (122) Each tarsus with a single claw . . . . *Brachybambus* Germar

123b Tarsi with 2 claws 124

124a (123) Last segment of tarsi broad, claws well separated (Fig. 38.12) . . . 125

124b Last tarsal segment narrow, projecting beyond lobes of third, claws
slender . . . . *Onychylis* LeConte

125a (124) Elytra but slightly if at all wider than prothorax; length usually
2 mm or more . . . . *Endalus* Laporte

125b Elytra much wider than prothorax; body length less than 1.5 mm . .
Tanysphyrus Germar

126a (122) Front and middle tibiae serrate on inner side; third tarsal segment
narrow, slightly emarginate . . . . *Lixellus* LeConte

126b Tibiae not serrate on inner side; third tarsal segment broad,
deeply bilobed . . . . *Anchodemus* LeConte

127a (121) Antennal club partly smooth and shining; prosternum not excavated . . . . *Lissorhoptrus* LeConte

127b Antennal club entirely pubescent and sensitive; prosternum
broadly and deeply excavated in front of coxae 128

128a (127) Pronotum feebly constricted in front . . . . *Bagous* Germar

128b Pronotum very strongly constricted and tubulate in front . . . . . .
Pnigodes LeConte

129a (115) Antennae inserted at mid-point or apical third of beak; tiny squat
beetles covered with small, flat scales . . . . *Phytobius* Schönherr

129b Antennae inserted at base of beak, near eyes. (Billbugs, not
further treated here) . . . . Subfamily **Rhynchophorinae**

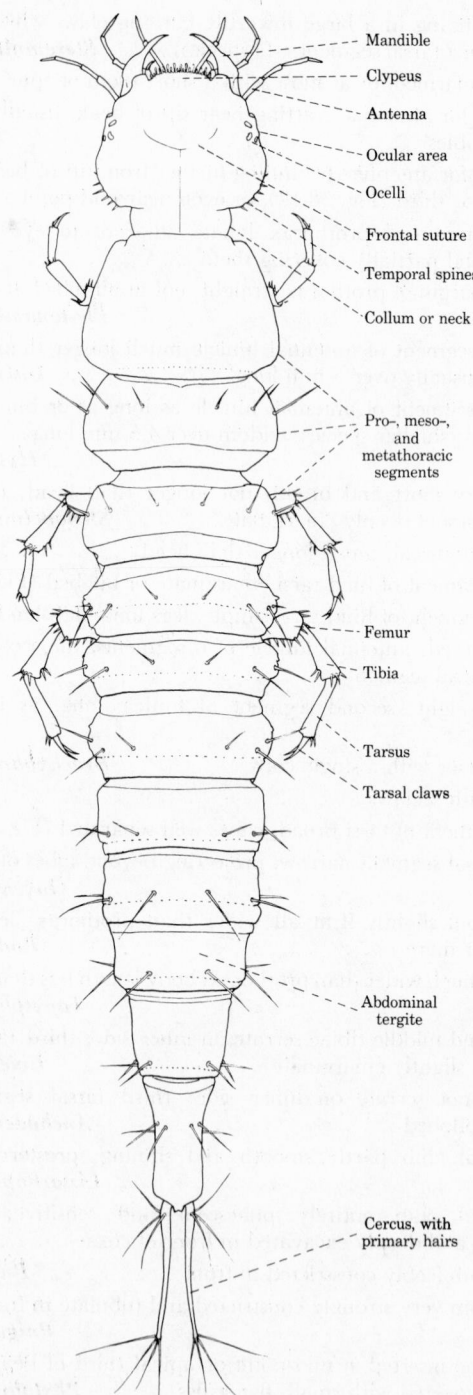

Fig. 38.61. Dytiscidae. Larva of *Agabinus glabrellus* Motschulsky, dorsal view, to show parts.

KEY TO GENERA (LARVAE)
(Adults on p. 989)

1a   With legs. **2**

1b   Without legs **91**

2a (1) Legs apparently 5-segmented (Fig. 38.62), tarsi with 2 movable
claws (except Haliplidae which are 1-clawed) **3**

2b   Legs apparently 4-segmented, each tarsus united with its single
claw **32**

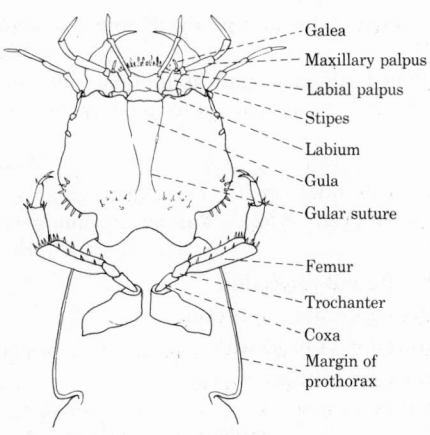

Fig. 38.62. Dytiscidae. Underside of head and prothorax of larva of *Agabinus glabrellus* Motschulsky, to show parts.

3a (2) Tenth abdominal segment armed with 4 hooks; lateral gills present
on all abdominal segments. . . . . Family **Gyrinidae**   **4**

3b   Tenth segment not armed with 4 hooks; abdominal gills present
or not. **6**

4a (3) Head subcircular with collum narrow and distinct; mandible
falcate, without retinaculum . . . . ***Dineutus*** MacLeay

4b   Head elongate, with collum about as wide as rest of head and not
distinct; mandible with retinaculum **5**

5a (4) Nasale without teeth . . . . ***Gyretes*** Brullé

5b   Nasale with 2 to 4 teeth in a transverse row . . . .
                          Gyrinus (Geoffroy *in*) Müller

6a (3) Abdomen with 9 or 10 segments . . . . Family **Haliplidae**   **7**

6b   Abdomen with 8 visible segments (Fig. 38.61) **10**

7a (6) Abdomen with 9 segments; each body segment with 2 or more
erect, segmented, hollow, spine-tipped filaments, each ½ as long
as body . . . . ***Peltodytes*** Régimbart

7b   Tenth abdominal segment produced posteriorly in a forked or un-
forked horn; body spines, except in first instar, never stalked or
much longer than length of 1 body segment **8**

8a (7) Third antennal segment shorter than second. . ***Brychius*** Thomson

8b   Third antennal segment 2 to 3 times as long as second **9**

9a   (8) Third segment of front legs produced and edged by 2 blunt teeth
                                                    Apteraliplus Chandler
9b   Front legs weakly to moderately chelate, fourth segment more or
             less produced, usually bearing 2 or 3 spines . . . *Haliplus* Latreille
10a   (6) Legs fossorial; larval form elateroid (wireworm-like). Larvae of
             most genera still unknown . . . . Family **Noteridae**   11
10b   Legs ambulatory or natatory; larval form not elateroid (Fig.
             38.61).   12
11a   (10) Mandibles slender, narrowly sulcate at tip, simple; third antennal
             segment more than twice as long as fourth . . . *Hydrocanthus* Say
11b   Mandibles stout, bifid at tip; third antennal segment no longer
             than fourth. . . . . *Suphisellus* Crotch
12a   (10) Larva flattened, thoracic and abdominal sides greatly expanded
             into thin lateral plates; gular suture median and simple. . . . . . .
                                                    Family **Amphizoidae**
                                                    Amphizoa LeConte
12b   Larvae not with sides greatly expanded; gular suture double, at
             least anteriorly (Fig. 38.62). Larvae of some genera are still un-
             known . . . . Family **Dytiscidae**   13
13a   (12) Head with a frontal projection   14
13b   Head without a frontal projection   18
14a   (13) Frontal projection of head with a notch on each side   15
14b   Frontal projection without notches   16
15a   (14) Cerci with only primary hairs, 6 or 7 in number (as in Fig. 38.61).
                                    Hydroporus Clairville, *Hygrotus* Stephens
15b   Cerci with additional secondary hairs. . . . .
                                    (in part) *Oreodytes* Seidlitz, *Deronectes* Sharp
16a   (14) Larvae not greatly widened; last abdominal segment long and
             tapering; cerci with only primary hairs. . . . . *Bidessus* Sharp
16b   Larvae greatly widened in middle; last abdominal segment long or
             short; cerci long with secondary hairs, or short with primary hairs
             only.   17
17a   (16) Last abdominal segment long and tapering; cerci short, arising
             beneath segment, projecting beyond it, and having primary hairs
             only . . . . *Hydrovatus* Motschulsky
17b   Last abdominal segment short; cerci long, with secondary hairs . .
                                                    (in part) *Oreodytes* Seidlitz
18a   (13) Maxillary stipes broad, suboval (Fig. 38.62), usually with 1 or 2
             strong inner marginal hooks.   19
18b   Maxillary stipes slender, long, without inner marginal hooks   30
19a   (18) Abdominal segments 7 and 8 without lateral fringe of long swim-
             ming hairs; labium simple anteriorly (Fig. 38.62).   20
19b   Abdominal segments 7 or 8 or both with even fringe of long
             swimming hairs; labium with a single or bilobed median anterior
             projection, the ligula.   26
20a   (19) Fourth (last) segment of antennae less than ⅔ the length of
             third   21
20b   Fourth segment more than ⅔ the length of third . . . .
                                    Rhantus Dejean, *Colymbetes* Clairville

21a	(20)	Mandibles toothed along inner edge; fourth antennal segment double (biramous). . . . . ***Copelatus*** Erichson	
21b		Mandibles not toothed; fourth antennal segment not double	22
22a	(21)	Cerci with numerous secondary hairs; fourth antennal segment less than $\frac{1}{4}$ as long as third . . . . ***Laccophilus*** Leach	
22b		Cerci usually with 7 primary hairs in 2 whorls, 3 near middle and 4 apically; fourth antennal segment about $\frac{1}{2}$ as long as third	23
23a	(22)	Thorax as wide as long, about equal in length to abdomen; head without neck area set off by shallow groove . . ***Hydrotrupes*** Sharp	
23b		Thorax longer than wide, never as long as abdomen; head with neck area set off by occipital suture or shallow groove.	24
24a	(23)	Tibiae and tarsi with conspicuous spines confined to apical half, mostly terminal (Fig. 38.61). . . . . ***Agabinus*** Crotch	
24b		Tibiae and tarsi with spines not confined to apical half	25
25a	(24)	Lateral margin of head more or less compressed or keeled, with temporal spines on a line which would intersect the ocelli or pass just below them. . . . . ***Ilybius*** Erichson	
25b		Lateral margin of head not keeled, temporal spines on a line which would run well below the ocelli . . . . ***Agabus*** Leach	
26a	(19)	With a pair of long lateral gills on each of the 6 anterior abdominal segments. . . . . ***Coptotomus*** Say	
26b		No gills on abdominal segments.	27
27a	(26)	Ligula (median anterior projection of labium) very short, armed with 4 spines . . . . ***Eretes*** Laporte	
27b		Ligula long, simple or bifid but without 4 spines.	28
28a	(27)	Ligula apically bifid. . . . . ***Acilius*** Leach	
28b		Ligula simple.	29
29a	(28)	Ligula not as long as first segment of labial palpus . . . . ***Thermonectus*** Dejean	
29b		Ligula nearly equal to or exceeding length of first segment of labial palpus . . . . ***Graphoderus*** Sturm	
30a	(18)	Head anteriorly dentate; abdominal cerci absent . ***Cybister*** Curtis	
30b		Head not dentate anteriorly; cerci present	31
31a	(30)	Cerci with lateral fringes; labium without projecting lobes . . . . . ***Dytiscus*** Linnaeus	
31b		Cerci without lateral fringes; labium with 2 projecting lobes . . . . ***Hydaticus*** Leach	
32a	(2)	Urogomphi (cerci) segmented usually individually movable (often retracted into a terminal breathing pocket in eighth abdominal segment in Hydrophilidae)	33
32b		Urogomphi solidly united at base, or absent	51
33a	(32)	Maxillary palpiger free and segmentlike, usually carrying a finger-shaped galea; spiracles biforous. Larvae of some genera are unknown . . . . Family **Hydrophilidae**	34
33b		Maxillary palpiger closely united with stipes, without finger-shaped galea; spiracles annular or absent.	47
34a	(33)	Nine complete abdominal segments, tenth reduced but distinct . . . ***Helophorus*** Fabricius	

34b   Eight complete abdominal segments, 9 and 10 reduced and form-
             ing a stigmatic atrium (except in *Berosus* in which atrium is not
             developed)   35

35a   (34)   Antennae with their points of insertion nearer the externofrontal
             angles than are those of the mandibles; labium and maxillae inserted
             in a furrow on under side of head . . . . **Hydrochus** Leach

35b   Antennae with their points of insertion farther from externofrontal
             angles than are those of mandibles; labium and maxillae inserted at
             anterior margin of underside of head   36

36a   (35)   First 7 abdominal segments with long lateral tracheal gills; seg-
             ments 9 and 10 reduced but no stigmatic atrium present . . . . . .
                                                              Berosus Leach

36b   Tracheal gills not nearly·so prominent, or else absent; abdominal
             segments 9 and 10 reduced, forming a stigmatic atrium   37

37a   (36)   Ocular area round, ocelli so minute and closely aggregated as
             often to appear as one on each side of head; legs absent, or much
             reduced and without claw-bearing segment. Terrestrial. (Not
             treated here) . . . . Family **Hydrophilidae**

37b   Ocular area oval, larger, ocelli more distant; legs well developed,
             with distinct claw-bearing segment   38

38a   (37)   First segment of antennae not distinctly longer than following 2
             taken together, fingerlike antennal appendage present; femora
             without fringes of swimming hairs   39

38b   First segment of antennae distinctly longer than next 2 taken to-
             gether, fingerlike antennal appendage absent; femora with fringes
             of long swimming hairs   45

39a   (38)   Frontal sutures parallel, not uniting to form an epicranial suture;
             left expansion of epistoma much more prominent than the right
             and with a row of stout setae . . . . **Laccobius** Erichson

39b   Frontal sutures not parallel, and they may or may not unite to
             form an epicranial suture; lateral expansions of epistoma similar
             and usually in line with anterior margin of labro-clypeus   40

40a   (39)   Antennae shorter and antennal appendages more prominent;
             epicranial suture absent; legs reduced; abdomen more truncate;
             cercus with long terminal seta.   41

40b   Antennae longer and antennal appendages less prominent;
             epicranial suture present but usually short; legs fairly long, not
             reduced; abdomen narrowed caudally; cercus with a shorter
             terminal seta   42

41a   (40)   Frons truncate behind; labrum tridentate, lateral teeth bifid . . . .
                                                          Paracymus Thomson

41b   Frons rounded behind; labrum quadridentate . . . .
                                                           Anacaena Thomson

42a   (40)   Mandibles asymmetrical, the right with 2 teeth, the left with only
             1; abdomen with prolegs on segments 3 to 7. . **Enochrus** Thomson

42b   Mandibles symmetrical, each with 2 or 3 inner teeth; abdomen
             without prolegs.   43

43a   (42)   Labro-clypeus with 5 distinct teeth, outer left tooth a little distant
             from rest; each mandible with 3 inner teeth . . . **Hydrobius** Leach

43b   Labro-clypeus with at least 6 teeth; each mandible with 2 inner
             teeth   44

44a **(43)** Labro-clypeus with 6 distinct teeth in 2 groups, 2 on left and 4 on right . . . . *Helochares* Mulsant

44b Labro-clypeus with more than 6 teeth, those at right not clearly defined and with several smaller teeth . . . . . *Cymbiodyta* Bedel

45a **(38)** Head subspherical; labro-clypeus without teeth; each mandible with a single inner tooth . . . . *Hydrophilus* Geoffroy

45b Head subquadrangular, narrowed behind; labro-clypeus with inconspicuous teeth; each mandible with more than 1, usually with 2 inner teeth **46**

46a **(45)** Mentum with sides nearly straight; its fronto-external angles very prominent; pleural gills rudimentary but indicated by tubercular projections, each with several terminal setae . . *Tropisternus* Solier

46b Mentum with sides convergent basally; its frontoexternal angles less prominent; pleural gills fairly well developed and pubescent . . . . . *Hydrochara* Berthold

47a **(33)** Mandible lacking asperate or tubercular molar portion, usually without molar parts . . . . Family **Staphylinidae**
    A few small species of aleocharine Staphylinidae occur in cracks in rocks in the intertidal zone of the Pacific Coast seashore.

47b Mandible with molar portion usually large, asperate or tuber-cular **48**

48a **(47)** Tenth abdominal segment with a pair of recurved hooks; spiracles present, annuliform; no balloonlike appendices on prothorax or abdomen . . . . Family **Hydraenidae** **49**

48b Tenth abdominal segment without terminal hooks; spiracles absent, but a small balloonlike appendix at each side of prothorax and first and eighth abdominal segments . . . .
    Family **Hydroscaphidae**
    Hydroscapha LeConte

49a **(48)** Setae on clypeus not placed at anterior margin, two median ones distant from each other; cerci nearly contiguous proximally and divergent . . . . *Ochthebius* Leach

49b Setae on clypeus placed at anterior margin, equidistant **50**

50a **(49)** A pectinate seta present at each side of frontal margin of labrum . . .
    Hydraena Kugelann

50b All setae at frontal margin of labrum without ramifications, pointed and uniformly shaped . . . . *Limnebius* Leach

51a **(32)** Body rounded or oval and depressed (Fig. 38.63); lateral margins of body segments greatly expanded, the head completely concealed from a dorsal view by the rounded anterior pronotal margin (water pennies and allies) often oval. . . . . . Family **Psephenidae** **52**

51b Body usually slender and round or triangular in cross section, the head exposed from a dorsal view. (Figs. 38.79, 38.88) **55**

52a **(51)** Lateroposterior margins of the tergites overlapping the succeeding tergites so that the margins appear to be joined (Fig. 38.63); abdomen with 4 or 5 pairs of branched gills (Fig. 38.65) the first pair arising from the posterior margin of the second sternite **53**

52b Lateral margins of tergites well separated (Fig. 38.64); ventral gills absent, retractile anal gills present **54**

53a **(52)** Two pairs of gills each on abdominal sternites 2 to 6, about 12 filaments in each gill (Fig. 38.65); median pronotal suture (Fig.

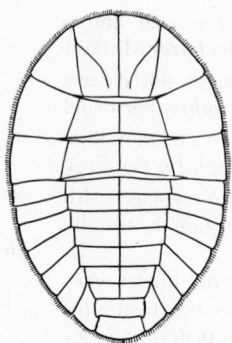

Fig. 38.63. Psephenidae. *Psephenus herricki* De Kay, dorsal view of larva.

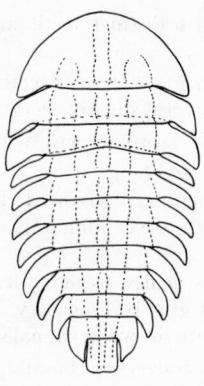

Fig. 38.64. Psephenidae. *Ectopria* sp., dorsal view of larva.

Fig. 38.65. Psephenidae. *Psephenus herricki* De Kay, abdominal gill of larva.

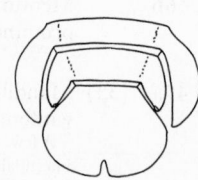

Fig. 38.66. Psephenidae. *Acneus* sp., seventh, eighth, and ninth abdominal tergites of larva.

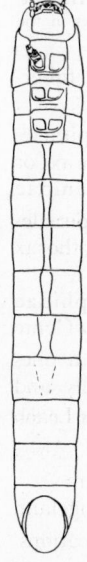

Fig. 38.67. Dryopidae. *Pelonomus* sp., ventral surface of larva.

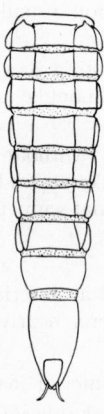

Fig. 38.68. Elmidae. *Optioservus* sp., ventral view of abdomen of larva, showing 7 lateral pleura.

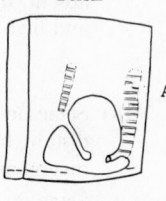

Fig. 38.69. Dryopidae. *Helichus* sp., third abdominal segment of larva, lateral view, anterior end to right.

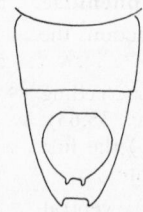

Fig. 38.70. Dryopidae. *Helichus* sp., last 2 abdominal sternites of larva.

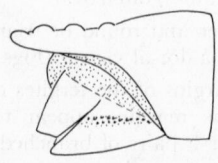

Fig. 38.71. Chelonariidae. *Chelonarium* sp., lateral view of last abdominal segment of larva, showing operculum at bottom, and movable sclerite above it.

Fig. 38.72. Elmidae. *Ancyronyx variegatus* Germar, operculum of last abdominal segment of larva, showing internal hooks.

           38.63) without expansion at middle; eighth abdominal tergite without lateral expansion; last tergite wider than long . . . .
                                                    Psephenus Haldeman

53b   Two pairs of gills each on abdominal sternites 2 to 5, about 30 filaments in each gill; pronotum with an oval or angulate expansion at middle on median suture; eighth abdominal tergite with lateral expansion reaching lateral margin of body; last tergite longer than wide. . . . . ***Eubrianax*** Kiesenwetter

54a **(52)** Lateral expansion of eighth abdominal segment (Fig. 38.64) expanded toward apex and reaching lateral body marginal outline; last tergite longer than wide, truncate at apex . ***Ectopria*** LeConte

54b   Lateral expansion of eighth abdominal segment (Fig. 38.66) short and narrowed toward apex, not reaching lateral body marginal outline; last tergite wider than long, rounded on sides and narrowly emarginate at apex. . . . . ***Acneus*** Horn

55a **(51)** Ninth abdominal segment with a ventral movable operculum closing a caudal chamber (Figs. 38.67, 38.68, 38.71) **56**

55b   Ninth abdominal segment without operculum **82**

56a **(55)** First 5 or 8 abdominal sternites (Fig. 38.67) each with a median fold . . . . Family **Dryopidae**   **57**

56b   Abdominal sternites flattened or evenly convex (Fig. 38.68); operculum present; cloacal chamber with 3 tufts of retractile gills. **59**

57a **(56)** Mesothorax and each of first 7 abdominal segments (Fig. 38.69) with a swollen area adjacent to spiracle; tergites, especially on sides, transversely grooved; eighth tergite with 2 tuberculate swellings on posterior margin; ninth tergite (Fig. 38.70) biangulate at apex, operculum with 2 rounded tubercles at apex . . .
                                                 Helichus Erichson

57b   Spiracles without adjacent swollen areas; tergites not transversely grooved; eighth and ninth tergites without tubercles or swellings; operculum evenly rounded at apex. (Fig. 38.67) **58**

58a **(57)** All dorsal tergites except pronotum with many longitudinal carinae arising near each anterior margin. . . . . . . ***Pelonomus*** Erichson

58b   All dorsal tergites smooth at anterior margins . . . ***Dryops*** Olivier

59a **(56)** Operculum without internally attached hooks but with a flat movable sclerite (Fig. 38.71) attached to each lateral margin; thoracic segments and first 8 abdominal segments each with a dorsolateral flattened projection bearing about 15 long minutely hairy filaments. . . . . Family **Chelonariidae**
                                              Chelonarium Fabricius

59b   Operculum with a pair of internally attached hooks (Fig. 38.72) . . .   **60**

60a **(59)** Apex of last abdominal segment (Fig. 38.73) evenly rounded; each thoracic segment with an undivided slender pleuron margined internally with erect hairs; 5 ocelli; antenna deeply retractile. . . .
                                         Family **Limnichidae**
                                         Lutrochus Erichson

60b   Apex of last abdominal segment shallowly to deeply emarginate (Fig. 38.84); each mesothoracic and metathoracic segment with pleuron of 1 short part (Fig. 38.74) or subdivided into 2 or 3 parts (Figs. 38.75, 38.76); antenna generally slender . . . .
                                        Family **Elmidae**   **61**

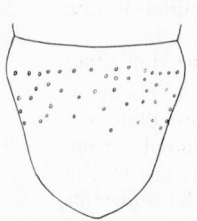

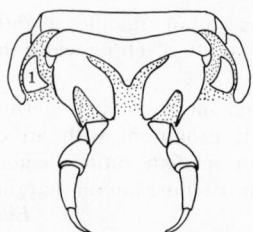

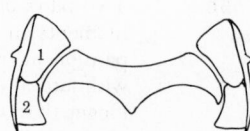

Fig. 38.73. Limnichidae. *Lutrochus* sp., dorsal view, last abdominal segment of larva.

Fig. 38.74. Elmidae. *Optioservus* sp., mesosternum of larva, showing 1 pleural part.

Fig. 38.75. Elmidae. *Macronychus glabratus* Say, mesosternum, pleural parts numbered.

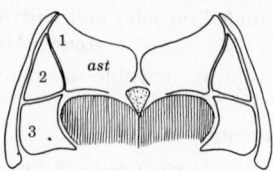

Fig. 38.76. Elmidae. *Heterlimnius* sp., prosternum of larva; pleural parts numbered. *ast*, anterior sternum.

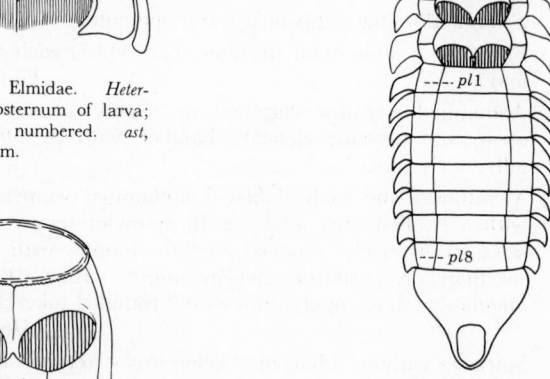

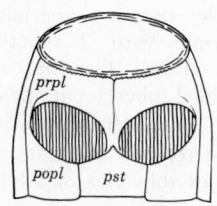

Fig. 38.77. Elmidae. *Stenelmis* sp., prosternum of larva. *pst*, posterior sternum; *prlp*, prepleurite; *popl*, postpleurite.

Fig. 38.78. Elmidae. *Phanocerus clavicornis* Sharp, ventral view of larva. *pst*, posterior sternum; *pl* 1, *pl* 8, abdominal pleura 1 and 8.

61a	**(60)**	Prothorax with a posterior sternum (Fig. 38.77) behind the middle coxae, usually separated from lateral pleural sclerite by a distinct suture. **62**
61b		Posterior sternum absent (Fig. 38.76) **71**
62a	**(61)**	Eight complete abdominal pleura on each side of venter (Fig. 38.78). **63**
62b		Seven abdominal pleura on each side. **64**
63a	**(62)**	Body depressed, the lateral margins of the thoracic segments and the first 8 abdominal segments produced into curved flattened lobes; 2 mesothoracic and 2 metathoracic pleurites (Fig. 38.78); propleuron of 2 parts . . . . *Phanocerus* Sharp
63b		Body cylindrical (Fig. 38.79), the lateral margins of thoracic and abdominal segments not produced; propleuron, and metapleuron each of 1 part . . . . *Cylloepus* Erichson

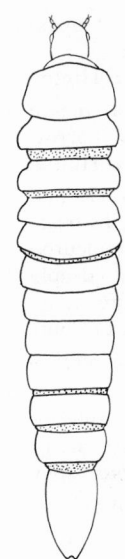

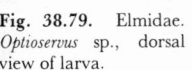

Fig. 38.80. Elmidae. *Stenelmis* sp., head of larva. Arrow points to frontal tooth.

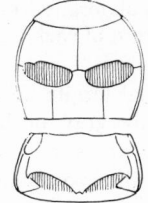

Fig. 38.81. Elmidae. *Elsianus graniger* Sharp, prosternum and mesosternum of larva. (After Hinton.)

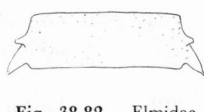

Fig. 38.82. Elmidae. *Ancyronyx variegatus* Germar, third abdominal tergite of larva.

Fig. 38.79. Elmidae. *Optioservus* sp., dorsal view of larva.

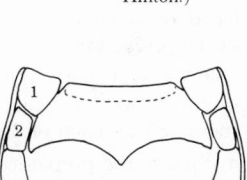

Fig. 38.83. Elmidae. *Heterelmis* sp., mesosternum of larva, pleural parts numbered.

Fig. 38.84. Elmidae. *Heterelmis* sp., dorsal view of last abdominal segment of larva.

64a	**(62)**	Front of head at each anterolateral angle produced. (Fig. 38.80). **65**
64b		Front of head not produced **68**
65a	**(64)**	Abdominal segments 2 to 7 each with tergal (dorsolateral) sutures; mesopleuron and metapleuron (Fig. 38.81) each of 2 parts **66**
65b		Abdominal segments without tergal sutures **67**
66a	**(65)**	Propleuron of 1 continuous part (Fig. 38.77) . . ***Stenelmis*** Dufour
66b		Propleuron of 2 parts (Fig. 38.81). . . . . ***Elsianus*** Sharp
67a	**(65)**	Propleuron of 2 parts (Fig. 38.81) . . . . ***Neoelmis*** Musgrave
67b		Propleuron of 1 part (Fig. 38.77), the prepleurite and postpleurite untied to form a continuous sclerite adjoining the lateral suture. . . . . ***Ordobrevia*** Sanderson
68a	**(64)**	Posterior angles of first abdominal segments each (Fig. 38.82) produced on lateral margin; propleuron of 1 part; mesopleuron and metapleuron of 2 parts . . . . ***Ancyronyx*** Erichson
68b		Posterior angles of abdominal segments not produced (Fig. 38.68); propleuron of 2 parts; mesopleuron and metapleuron each divided into 3 parts (Fig. 38.83) **69**
69a	**(68)**	First 8 abdominal segments each with 4 diagonal dorsal rows of blunt spines; mesonotal and metanotal rows similar but each lateral row divided at apical half; last abdominal segment (Fig. 38.84) with a median dorsal group and a lateral row of spines . . . ***Heterelmis*** Sharp
69b		First 8 abdominal segments each with 4 rows of flat tubercles; median ridge of last segment (Fig. 38.85) with an irregular row of flat tubercles. **70**

70a (69) Last abdominal tergite (Fig. 38.85) slightly emarginate at apex; posterior margin of each tergite with a pale spot at middle . . . . .
           Microcylloepus Hinton

70b   Last abdominal tergite nearly straight at apex; tergites at meson each uniformly gray or brownish except for the very narrow median line . . . . **Hexacylloepus** Hinton

71a (61) Body irregularly quadrangular in cross section, the thoracic and first 8 abdominal tergites with lateral and posterior margins bearing long, soft fingerlike projections; mesopleuron and metapleuron each of 1 part (Fig. 38.74); last abdominal segment deeply emarginate at apex; length of full-grown larva about 15 mm . . . .
           Lara LeConte

71b   Body more cylindrical or triangular in cross section, the tergites without long, fingerlike projections   72

72a (71) Postpleurite (Fig. 38.86) of 1 part, the prepleurite and anterior sternum absent; pronotum with 2 large, transverse, smooth areas; mesopleuron and metapleuron each of 2 parts . . . **Narpus** Casey

72b   Postpleurite of 1 or 2 parts, the prepleurite present (Fig. 38.87) . .   73

73a (72) Postpleurite of 2 parts (Fig. 38.87)   74

73b   Postpleurite of 1 part   80

74a (73) Mesopleuron of 2 parts (Fig. 38.75)   75

74b   Mesopleuron of 1 part (Fig. 38.74)   79

75a (74) Six abdominal pleura; last abdominal segment with 2 long, acute, narrowly separated apical processes . . . . . **Macronychus** Müller

75b   Seven abdominal pleura   76

76a (75) Last abdominal segment with lateral angles of apical emargination produced, the lateral margin before angle distinctly serrate; body more evenly cylindrical in cross section . . . **Zaitzevia** Champion

76b   Apex of last abdominal segment very shallowly emarginate, angles not so acute as in Fig. 38.79; body (Fig. 38.88) more nearly triangular in cross section   77

77a (76) Anterior pleurite of mesopleuron and metapleuron each long and narrow averaging about ½ the width of posterior pleurite; abdominal tergites tumid on meson somewhat as in Fig. 38.90, each tumidity covered with scalelike hairs; 2 longitudinal dark marks on each thoracic tergite . . . . **Gonielmis** Sanderson

77b   Anterior pleurite of mesopleuron and metapleuron short and broad

Fig. 38.85. Elmidae. *Microcylloepus* sp., dorsal view of last abdominal segment of larva.

Fig. 38.86. Elmidae. Prosternum of larva of *Narpus* sp., legs removed.

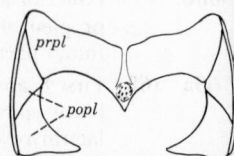

Fig. 38.87. Elmidae. *Macronychus glabratus* Say larva, prosternum. *prpl*, prepleurite; *popl*, 2 parts of post-pleurite.

(Fig. 38.75); abdominal tergites not tumid on meson; thoracic
tergites without dark markings **78**

78a **(77)** Length of full-grown larva about 4 mm. Western states . . . . . .
Heterlimnius Hinton

78b Length of full-grown larva about 2 mm. Appalachian area . . . .
Limnius Erichson

79a **(74)** Abdominal tergites on meson strongly humped at posterior
margins, more strongly so than in Fig. 38.89; each abdominal
tergite except last with a large swelling between median hump
and lateral margin; last tergite slightly tumid at middle. . . . . . .
Promoresia Sanderson

79b Abdominal tergites not conspicuously humped at middle and without large swellings (Figs. 38.79, 38.88) . . *Optioservus* Sanderson

80a **(73)** Eight abdominal segments with pleura; body (Fig. 38.89) very
slender, the last segment about 4 times longer than wide; operculum
with apex acute . . . . *Dubiraphia* Sanderson

80b Six or 7 abdominal segments with pleura (Fig. 38.68). **81**

81a **(80)** First 8 abdominal segments (Fig. 38.90) each tuberculate at middle
on posterior half; each tubercle covered with a dense group of
scales; lateral surface of tergite swollen; last abdominal segment
strongly tuberculate on meson and with a dense group of scales . .
Ampumixis Sanderson

81b Abdominal segments not tuberculate at middle or on sides; body
nearly evenly convex above . . . . *Cleptelmis* Sanderson

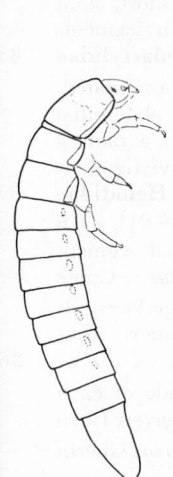

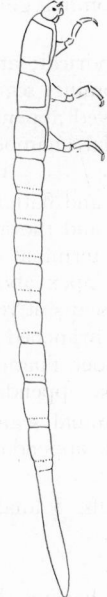

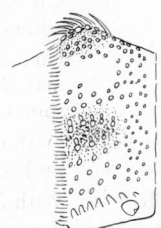

Fig. 38.88. Elmidae.
Heterlimnius sp., lateral
view of larva.

Fig. 38.89. Elmidae.
Larva of *Dubiraphia* sp.,
lateral view.

Fig. 38.90. Elmidae.
Ampumixis dispar Fall,
lateral view of fourth
abdominal segment of
larva.

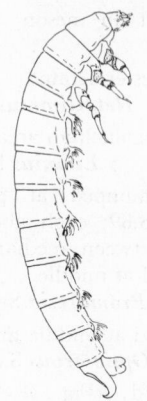

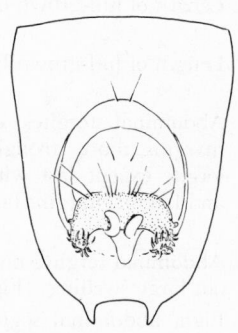

Fig. 38.91. Ptilodac-
tylidae. Larva of *Steno-*
colus sp.?, lateral view.

Fig.38.92. Ptilodactylidae.
Ventral view of last ab-
dominal segment of larva.

82a	**(54)**	Four or more of first 7 abdominal segments bearing 2 ventral tufts of filamentous gills . . . . . . (in part) Family **Ptilodactylidae**	**85**
82b		Abdomen without such gills on 4 or more segments	**83**
83a	**(82)**	With gills restricted to anal region, more or less in a caudal chamber (may be retracted out of sight into caudal chamber; open and examine apex of abdomen).	**84**
83b		Without conspicuous abdominal gills (see note above). . . . . Family **Chrysomelidae**	**90**
84a	**(83)**	Body form elongate, cylindrical; antennae 3-segmented, rarely as long as head; ninth abdominal segment with 5 to 21 mammillate gills and 2 prehensile curved appendages covered with short, stout spines in a shallow caudal chamber; first 8 abdominal segments with lateral spiracles . . . . . (in part) Family **Ptilodactylidae**	**85**
84b		Body form more elliptical and flattened; antennae multiarticulated, usually longer than head and thorax combined; eighth abdominal segment with gills in the terminal caudal chamber and a pair of enlarged spiracles on the apex above, other spiracles vestigial or absent; ninth abdominal segment vestigial . Family **Helodidae**	**87**
85a	**(82)**	Abdomen with 7 pairs of branched ventral gills (Fig. 38.91), each gill having about 10 slender filaments; ninth abdominal segment without prehensile hooklike appendages . . . . *Stenocolus* LeConte	
85b		Anal area of ninth abdominal segment with short fingerlike gills and 2 prehensile curved appendages covered with short stout spines (Fig. 38.92).	**86**
86a	**(85)**	With only 5 fingerlike gills, 3 anal and 1 on outer side of each prehensile appendage. . . . . *Anchycteis* Horn	
86b		With 21 fingerlike gills . . . . *Anchytarsus* Guérin	
87a	**(84)**	Anterior margin of hypopharynx (Fig. 38.94) with a central cone bearing 2 pairs of flat, usually serrate, spines; head with 1 or 2 ocelli on each side	**88**

Written with the aid of unpublished data supplied by W. E. Snow and F. van
Emden. Since the larvae of so few species of Helodidae have been described, the
characters used here may not all be satisfactory.

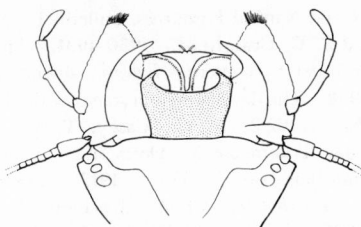

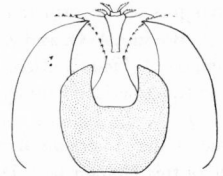

Fig. 38.93. Helodidae. Dorsal view of head of larva of *Elodes* sp. Labrum (stippled) with anterior angles bent under; anterior margin of hypopharynx (center) with a small central cone and a pair of simple, curved spines.

Fig. 38.94. Helodidae. Part of head of larva of *Cyphon* sp., showing labrum (stippled), mandibles, maxillae, and 2 pairs of serrate spines on hypopharynx.

87b	Cone of hypopharynx with 1 pair of spines; head with 3 ocelli on each side (Fig. 38.93). . . . . **Elodes** Latreille	
88a	(87) Abdominal segments 3 to 6 with a regular series of short, flattened, differentiated setae along lateral margins.	89
88b	Sides of abdominal segments with only scattered, thin setae, as on dorsum . . . . **Cyphon** Paykull	
89a	(88) Anterior angles of labrum bent under, inner margin projecting from under transverse front margin somewhat as in Fig. 38.93 . . **Prionocyphon** Redtenbacher	
89b	Anterior angles of labrum on same plane as rest (Fig. 38.94), labrum thus simply emarginate . . . . **Scirtes** Illiger	
90a	(83) Color green; tergum of seventh abdominal segment broadly pointed posteriorly and composed of dorsal and ventral portions . . . **Neohaemonia** Szekessy	
90b	Color white or cream; tergum of seventh abdominal segment smoothly rounded laterally and not composed of dorsal and ventral portions . . . . **Donacia** Fabricius	
91a	(1) Mouthparts terminal, body chiefly membranous with pro-, meo-, and metatergum and propleurae and sternum chitinized; eighth abdominal segment more or less truncate, terminated by a flat plate, below which is ninth segment bearing 2 large spiracles. . . . Certain terrestrial species of the Family **Hydrophilidae** **Cercyon** Leach	
91b	Mouthparts ventral, body membranous except for protergum; no dorsal plate on eighth abdominal segment. (Larvae mostly unknown, not further treated here) . . . . . . Family **Curculionidae**	

References

Bertrand, H. **1928.** Les larves et nymphes des Dytiscides, Hygrobiides, Haliplides. *Encyclopédie entomologique*, A (10). Lechevalier, Paris. **1955.** Notes sur les premiers etats des Dryopides d'Amerique. *Ann. soc. entomol. France*, 124:97–139. **Blackwelder, R. E.** **1931.** The Sphaeridiinae of the Pacific Coast. *Pan-Pacific Entomol.*, 8:19–32. **Blatchley, W. S.** **1910.** *An Illustrated Descriptive Catalogue of the Coleoptera or Beetles (Exclusive of the Rhyncophora)*

Known to Occur in Indiana. Indiana Dept. of Geology and Natural Resources, Bulletin 1. Nature Publishing Co., Indianapolis. **Böving, A. G. and F. C. Craighead. 1930–1931.** An illustrated synopsis of the principal larval forms of the order Coleoptera. *Entomol. Americana,* 11: 1–351. **Böving, A. G. and Kai L. Henriksen. 1938.** The developmental stages of the Danish Hydrophilidae. *Vidensk. Medd. Dansk Naturhist. Foren.,* 102:27–162. **Casey, T. L. 1900.** Review of the American Corylophidae, Cryptophagidae, Tritomidae and Dermestidae, with other studies. *J. N. Y. Entomol. Soc.,* 8:51–172. [Limnebius, pp. 51–53.] **1912.** Descriptive catalogue of the American Byrrhidae. *Memoirs on the Coleoptera,* 3:1–69. [Includes the Limnichidae of this chapter.] **Darlington, P. J. 1936.** A list of the West Indian Dryopidae, with a new genus and eight new species, including one from Colombia. *Psyche,* 43:65–83. **Edwards, J. G. 1951.** Amphizoidae (Coleoptera) of the world. *Wasmann J. Biol.,* 8:303–332. **Fall, H. C. 1919.** The North American Species of Coelambus [=Hygrotus of this chapter]. J. D. Sherman, Jr., Mount Vernon, N. Y. **1922a.** The North American species of Gyrinus. *Trans. Am. Entomol. Soc.,* 47:269–306. **1922b.** A Revision of the North American Species of Agabus Together with a Description of a New Genus and Species of the Tribe Agabini. J. D. Sherman, Jr., Mount Vernon, N. Y. **1923.** A revision of the North American species of Hydroporus and Agaporus. Privately printed. **Hatch, M. H. 1929.** Studies on Dytiscidae. *Bull. Brooklyn Entomol. Soc.,* 23:217–229. **1930.** Records and new species of Coleoptera from Oklahoma and western Arkansas, with subsidiary studies. *Univ. of Oklahoma Publ. Biol. Survey,* 2:15–26. [Key to species of Dineutus.] **1953.** The beetles of the Pacific Northwest. Part 1: Introduction and Adephaga. *Univ. Wash. Publs. Biol.,* 16:1–340. **Hinton, H. E. 1935.** Notes on the Dryopidae. *Stylops,* 4:169–179. **1939.** An inquiry into the natural classification of the Dryopoidea, based partly on a study of their internal anatomy. *Trans. Roy. Entomol. Soc. London,* 89:133–184. **1940.** A monographic revision of the Mexican water beetles of the family Elmidae. *Novitates Zoologicae,* 42:19–396. **1955.** On the respiratory adaptations, biology, and taxonomy of the Psephenidae, with notes on some related families. *Proc. Zool. Soc. London,* 125:543–568. **Hoffman, C. E. 1939.** Morphology of the immature stages of some northern Michigan Donaciini. *Pap. Mich. Acad. Sci.,* 25:243–290. **Horn, G. H. 1873.** Revision of the genera and species of the tribe Hydrobiini. *Proc. Am. Phil. Soc.,* 13:118–137. **1880.** Synopsis of the Dascyllidae of the United States. *Trans. Am. Entomol. Soc.,* 8:76–114. **1890.** Notes on the species of Ochthebius of Boreal America. *Trans. Am. Entomol. Soc.,* 17:17–26. **La Rivers, I. 1950.** The staphylinoid and dascilloid aquatic Coleoptera of the Nevada area. *Great Basin Naturalist,* 10:66–70. (Also an erratum page, *loc. cit.,* 11:52.) **Leech, H. B. 1940.** Description of a new species of Laccornis, with a key to the Nearctic species. *Canadian Entomol.,* 72:122–128. **1941a.** Note on the species of Agabinus. *Can. Entomol.,* 73:53. **1941b.** The species of Matus, a genus of carnivorous water beetles. *Can. Entomol.,* 73:77–83. **1948.** Coleoptera: Haliplidae, Dytiscidae, Gyrinidae, Hydrophilidae, Limnebiidae. No. 11 in Contributions toward a knowledge of the insect fauna of Lower California. *Proc. Calif. Acad. Sci.,* Ser. 4, 24:375–484. **Leech, H. B., and H. P. Chandler. 1956.** Chapter 13, Aquatic Coleoptera. In: R. L. Usinger (ed.). *Aquatic Insects of California, with Keys to North American Genera and California Species,* pp. 293–371. University of California Press, Berkeley and Los Angeles. **Marx, E. J. F. 1957.** A review of the subgenus Donacia in the Western Hemisphere (Coleoptera, Donaciidae). *Bull. Am. Museum Nat. Hist.,* 112:191–278. **McGillivray, A. D. 1903.** Aquatic Chrysomelidae and a table of families of coleopterous larvae. *N. Y. State Museum Bull.,* No. 68:288–331. **Musgrave, P. N. 1935.** A synopsis of the genus Helichus Erichson in the United States and Canada, with description of a new species. *Proc. Entomol. Soc. Wash.,* 37:137–145. **d'Orchymont, A. 1921.** Le genre Tropisternus. I. *Ann. soc. entomol. Belg.,* 61:349–374. **1922.** Le genre Tropisternus. II. *Ann. soc. entomol. belg.,* 62:11–47. **1942.** Revision des Laccobius americains. *Bull. musée roy. hist. nat. Belg.,* 18: 1–18. **1946** Notes on some American Berosus (s. str.) *Bull. musée roy. hist. nat. Belg.,* 22:1–20. **Richmond, E. A. 1920.** Studies on the biology of the aquatic Hydrophilidae. *Bull. Am. Museum Nat. Hist.,* 42:1–94. **Roberts, C. H. 1895.** The species of Dineutes of America north of Mexico. *Trans. Am. Entomol. Soc.,* 22:279–288. **1913.** Critical notes on the species of Haliplidae of America north of Mexico with descriptions of new species. *J. N. Y. Entomol. Soc.,* 21:91–123. **Sanderson, M. W. 1938.** A monographic revision of the North American species of Stenelmis. *Univ. Kansas Sci. Bull.*, 25:635–717. **1953.** A revision of the Nearctic genera of Elmidae. Part I. *J. Kansas Entomol. Soc.*, 26:148–163. **1954.** A revision of the Nearctic genera of Elmidae. Part II. *J. Kansas Entomol. Soc.*, 27:1–13. **Schaeffer, C. 1908.** On North American and some Cuban Copelatus. *J. N. Y. Entomol. Soc.*, 16:16–18. **1925.** Revision of the New World species of the tribe Donaciini of the coleopterous family Chrysomelidae. *Brooklyn Museum Sci. Bull.*, 3:45–165. **Tanner, V. M. 1943.** Studies of the subtribe Hydronomi with descriptions of new species. *Great Basin Naturalist*, 4:1–38. **Wallis, J. B. 1933.** Revision of the North American species, (north of Mexico), of the genus *Haliplus*, Latreille. *Trans. Roy. Can. Inst.*, 19:1–76. **1939a.** The genus Graphoderus Aubé in North America (north of Mexico). *Can. Entomol.*, 71:128–130. **1939b.** The genus *Ilybius* Er. in North America. *Can. Entomol.*, 71:192–199. **Wilson, C. B. 1923.** Water beetles in relation to pondfish culture, with life histories of those found in fishponds at Fairport, Iowa. *Bull. Bur. Fish.*, 39:231–345. (Document No. 953.) **Winters, F. E. 1926.** Notes on the Hydrobiini (Coleoptera, Hydrophilidae) of Boreal America. *Pan-Pacific Entomologist*, 3:49–58. **1927.** Key to the subtribe Helocharae Orchym. (Coleoptera-Hydrophilidae) of Boreal America. *Pan-Pacific Entomologist*, 4:19–29. **Young, F. N. 1951.** A new water beetle from Florida, with a key to the species of Desmopachria of the United States and Canada. *Bull. Brooklyn Entomol. Soc.*, 46:107–112. **1954.** *The Waterbeetles of Florida.* University of Florida Press, Gainesville.

39

Trichoptera

HERBERT H. ROSS

The caddisflies have complete metamorphosis and are close relatives of the Lepidoptera. In a typical caddisfly life history the gravid female crawls into the water and lays the eggs on or under sticks or stones; the eggs hatch into larvae, which make cases or fixed netlike retreats, or simply lay down a ground line of silk (Fig. 39.1). When full grown the larva spins a cocoon (inside the case if it makes a portable one) and pupates in it; when mature, the pupa cuts its way out of the case or cocoon, swims to the surface, crawls up on and attaches firmly to a support, and the adult emerges there. The adults vary in length from 1.5 to 35 mm and resemble somber moths. The pupae are unusual in possessing segmentally arranged, dorsal pairs of sclerous plates bearing hooks or teeth (Fig. 39.2) and in having hard sharp mandibles for cutting an opening in the cocoon or case. The larvae are essentially wormlike.

Keys for identification of adults are contained in Betten (1934), Denning (1956), and Ross (1944, 1956); keys for identification of pupae are contained in Ross (1944). The larva of *Yphria* Milne (1 species), our only representative of the family Kitagamiidae or Limnocentropidae, is unknown.

Larvae of this order possess the following distinctive combination of characters: head with sclerotized capsule, antenna with one principal segment, eye with a single lens; thorax with three pairs of segmented legs, distinct pleurites, sclerotized pronotum and three pairs of primary setae or sclerites on the meso- and metanotum; abdomen with nine annular segments, sometimes bearing gills, a tenth segment which is divided or bulbous, and a pair of anal legs each bearing a terminal claw.

The mesontum, metanotum, and most of the abdominal tergites have a primitive basic pattern of three pairs of simple primary setae. In many genera this condition is modified, and the sites of these primary setae bear clusters of setae or sclerotized areas bearing setae. In these keys the primary setae or setal areas have been identified by the designations *sa*1, *sa*2, and *sa*3.

The approximate number of North American species has been indicated for each genus.

The great bulk of the illustrations are by courtesy of the Illinois Natural History Survey.

KEY TO GENERA

1a	Three thoracic segments each covered with a single dorsal plate (Fig. 39.1)	**2**
1b	Metanotum mostly membranous, bearing only scattered hairs or small plates (Figs. 39.5*a*, 39.16*m*) or divided into at least 2 sclerites (Fig. 39.16*k, l*)	**3**
2a	(1) Abdomen with rows of branched gills, and with a large fan of long hairs at base of anal claw (*f*, Fig. 39.1*c*); living in a retreat or nest . . . . (p. 1033) Family **Hydropsychidae**	
2b	Abdomen without gills, and with only 2 or 3 long hairs at base of anal claw (Fig. 39.1*a, b*); minute forms making barrel- or purselike cases (Fig. 39.11) . . . . (p. 1036) Family **Hydroptilidae**	
3a	(1) Ninth segment of abdomen with dorsum entirely membranous	**4**
3b	Ninth segment of abdomen bearing a sclerotized dorsal plate (Fig. 39.2*b*).	**5**
4a	(3) Labrum membranous and T-shaped (Fig. 39.2*h*) . . . . (p. 1030) Family **Philopotamidae**	
4b	Labrum sclerotized and widest near base (Fig. 39.2*i*) . . . . (p. 1031) Family **Psychomyiidae**	
5a	(3) Anal leg with sclerite *s* more or less rectangular and longer than deep (Fig. 39.2*c, d, f*); anal legs either projecting free from last segment for at least ½ their length (Fig. 39.2*g*) or angling down at nearly a right angle from linear axis of abdomen (Fig. 39.2*d*); mesonotum and metanotum membranous with *sa*3 consisting of a single seta (Fig. 39.2*a*).	**6**
5b	Anal leg with sclerite *s* triangular and short (Fig. 39.2*e*) with the claw projecting chiefly laterad; only part of the claws project beyond the last segment, the anal legs being embedded in its side, and projecting downward little if at all; mesonotum and metanotum with *sa*3 consisting of a cluster of setae (Fig. 39.13*e*), or on a plate (Fig. 39.16*k*), or entire dorsum sclerotized	**7**
6a	(5) Anal claw large, nearly as long as sclerite *s* (Fig. 39.2*c, f*); trochantin of foreleg projecting forward and conspicuous (Fig. 39.3*h*). Free-living larva without case . . . . . (p. 1035) Family **Rhyacophilidae**	
6b	Anal claw small, retractile (Fig. 39.2*d*); fore trochantin difficult to	

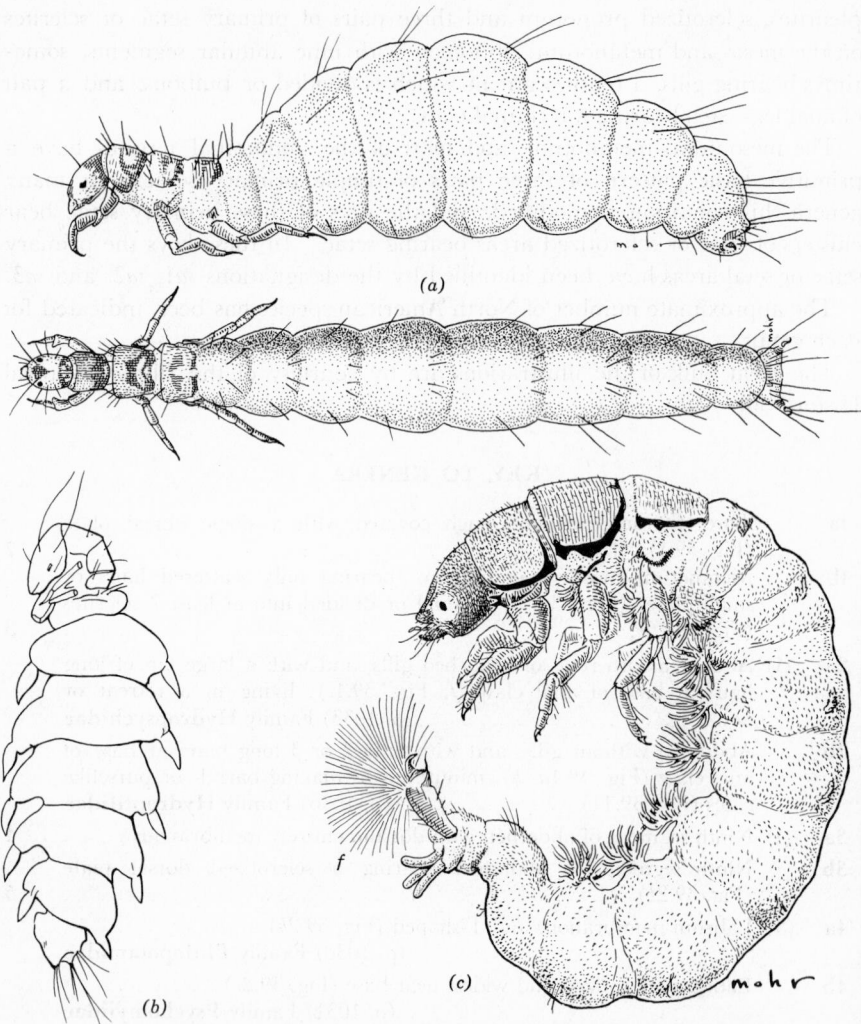

Fig. 39.1. Larvae of Hydroptilidae and Hydropsychidae. (*a*) *Hydroptila*, lateral aspect above, dorsal aspect below. (*b*) *Ithytrichia*, lateral aspect. (*c*) *Hydropsyche*, lateral aspect. *f*, fan of hairs at base of anal claw.

		distinguish, minute (Fig. 39.3*g*). Living in saddle-shaped case (Fig. 39.4*d*) . . . . (p. 1035) Family **Glossosomatidae**
7a	(5)	Claws of hind legs very small, those of middle and front legs large (Fig. 39.2*k, l*) . . . . (p. 1049) Family **Molannidae**
7b		Claws of hind legs as long as those of middle legs 8
8a	(7)	Antennae long, at least 8 times as long as wide, and arising at base of mandibles (Fig. 39.3*a*) . . . . . . (p. 1047) Family **Leptoceridae**
8b		Antennae much shorter (Fig. 39.3*c*) not more than 3 or 4 times as long as wide, often very inconspicuous, and arising at various points. 9
9a	(8)	Mesonotum submembranous except for a pair of parenthesislike, sclerotized bars (Fig. 39.17*a*) . . . . (p. 1047) Family **Leptoceridae**
9b		Mesonotum without such bars 10

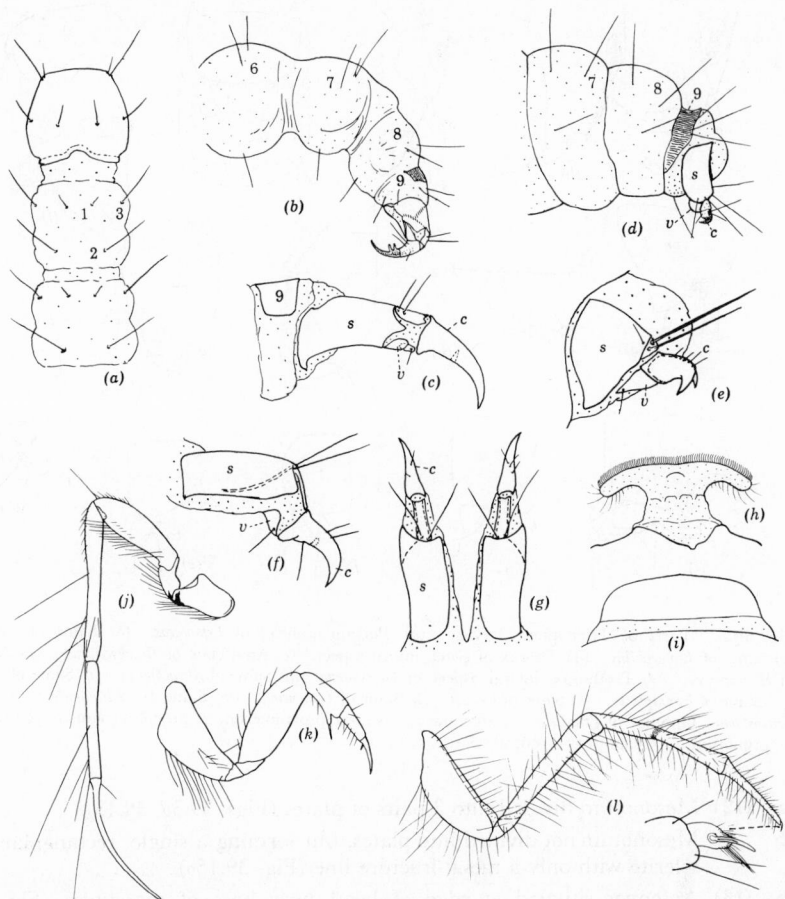

Fig. 39.2. Parts of Trichoptera larvae. (*a*) Thorax of *Rhyacophila*, dorsal aspect, showing numbered primary setae 1, 2, and 3. (*b*) Apex of abdomen of *Rhyacophila*, lateral aspect. (*c*) Anal leg of *Rhyacophila*, lateral aspect. (*d*) Apex of abdomen of *Agapetus*, lateral aspect. (*e*) Anal leg of *Limnephilus*, lateral aspect. (*f*) Anal leg of *Atopsyche*, lateral aspect. (*g*) Same, dorsal aspect. (*h*) Labrum of *Chimarra*. (*i*) Labrum of *Polycentropus*. (*j*) Leg of *Beraea*. (*k*) Front legs of *Molanna*. (*l*) Hind leg of *Molanna*. *c*, anal claw; *s*, *v*, sclerites of anal leg. (*j* after Ulmer.)

10a	**(9)**	Meso- and metanotum entirely membranous or with only minute sclerites (Fig. 39.13) . . . . (p. 1040) Family **Phryganeidae**
10b		Mesonotum and sometimes metanotum with some conspicuous sclerotized plates **11**
11a	**(10)**	Pronotum with sharp furrow extending across middle, much as in Fig. 39.14*a*, and hind tarsal claws long and with basal tooth reduced to a fine hair (Fig. 39.2*j*) . . . . (p. 1049) Family **Beraeidae**
11b		Either pronotum without such a transverse furrow or hind tarsal claws with a basal tooth (Fig. 39.14*c, d*) **12**
12a	**(11)**	Pronotum in side view with a suturelike furrow starting in line with the pleural suture of the propleurae and running in front of a ridge along the posterior margin of the pronotum (Fig. 39.3*j*) **13**
12b		Pronotum in side view with no such suture, or with a suture running directly into posterior ridge (Fig. 39.3*k*) **17**

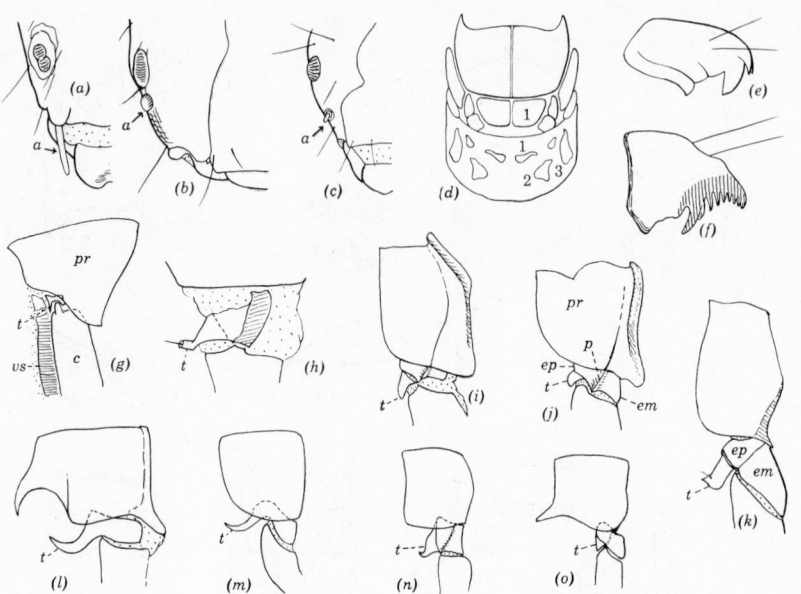

Fig. 39.3. Parts of Trichoptera larvae. (*a*) Portion of head of *Leptocerus*. (*b*) Same of *Lepidostoma* (*c*) Same of *Limnephilus*. (*d*) Thorax of *Goera*, dorsal aspect. (*e*) Anal claw of *Brachycentrus*. (*f*) Anal claw of *Helicopsyche*. (*g*) Prothorax, lateral aspect of *Glossosoma*. (*h*) Same of *Rhyacophila*. (*i*) Same of *Phryganea* (*j*) Same of *Limnephilus*. (*k*) Same of *Oecetis*. (*l*) Same of *Ganonema*. (*m*) Same of *Helicopsyche*. (*n*) Same o *Sericostoma*. (*o*) Same of *Psilotreta*. *a*, antenna; *c*, coxa; *em*, proepimeron; *ep*, proepisternum; *p*, pleural suture *pr*, pronotum; *t*, trochantin; *vs*, ventral sclerite.

13a	(12)	Mesonotum divided into 2 pairs of plates (Figs. 39.3*d*, 39.15*g*)	**14**
13b		Mesonotum not divided into plates, but forming a single, rectangular sclerite with only a mesal fracture line (Fig. 39.15*o*).	**16**
14a	(13)	Antennae situated at edge of head near base of mandibles (Fig. 39.16*j*), or difficult to distinguish (Fig. 39.14*a*, *b*)	**15**
14b		Antennae situated between edge of head and eye, or near eye (Fig. 39.3*b*, *c*)	**16**
15a	(14)	Pronotum with a sharp furrow running across it near middle; plates of mesonotum rectangular (Fig. 39.14*a*, *b*); metanotum with *sa*1 absent; no horn between base of front legs . . . . (p. 1041) Family **Brachycentridae**	
15b		Pronotum with no transverse furrow but always with projecting anterolateral points; plates of mesontum trianguloid (Fig. 39.3*d*); metanotum with *sa*1 represented by a small plate with several setae; between base of front legs and close to head arises a curved, small membranous horn . . . . (p. 1041) Family **Goeridae**	
16a	(13, 14)	Antennae situated very close to eye (Fig. 39.3*b*); first abdominal tergite without a hump; metanotum with *sa*1 and *sa*2 single hairs (Fig. 39.5*a*), sometimes each with a minute extra seta. . . . . (p. 1045) Family **Lepidostomatidae**	
16b		Antennae situated either midway between eye and margin of head or closer to margin of head than to eye (Fig. 39.3*c*); first abdominal tergite with a hump; metanotum usually with *sa*1 and *sa*2 represented by a cluster of setae, or by small plates each bearing one to several setae (Fig. 39.15*j*, *n*) . . . . (p. 1042) Family **Limnephilidae**	

17a **(12)** Metanotum with *sa*1 forming a wide plate, sclerotized but light colored and sometimes difficult to see at first glance, bearing a row of hairs (Fig. 39.16*k*, *l*); fore trochantin with a basal suture and with a short anterior projection, (Fig. 39.3*n*) or none (Fig. 39.3*o*); antennae on or under a ridge at anterior margin of head (Fig. 39.16*j*) **18**

17b Metanotum with *sa*1 a single hair; fore trochantin fused completely with episternum, forming a long, sharp, curved projection (Fig. 39.3*l*, *m*); antennae nearly midway between head margin and eye, and not on a ridge (Fig. 39.16*i*) **19**

18a **(17)** Gills elongate and single; metanotum with *sa*2 a single seta on a small plate (Fig. 39.16*e*); fore trochantin produced as a short, curved point (Fig. 39.3*n*) . . . . (p. 1046) Family **Sericostomatidae**

18b Gills composed of tufts of fine threads; metanotum with *sa*2 a row of hairs on a thin, faint, linear plate (Fig. 39.16*k*, *m*); fore trochantin not produced beyond edge of coxa (Fig. 39.3*o*) . . . .
(p. 1046) Family **Odontoceridae**

19a **(17)** Anal hooks with a long comb of teeth (Fig. 39.3*f*); larva living in a case shaped exactly like a snail shell (Fig. 39.4*c*). . . . .
(p. 1046) Family **Helicopsychidae**

19b Anal hooks with 2 accessory teeth, but these not forming a comb (Fig. 39.3*e*); case not at all snail-like . . . .
(p. 1046) Family **Calamoceratidae**

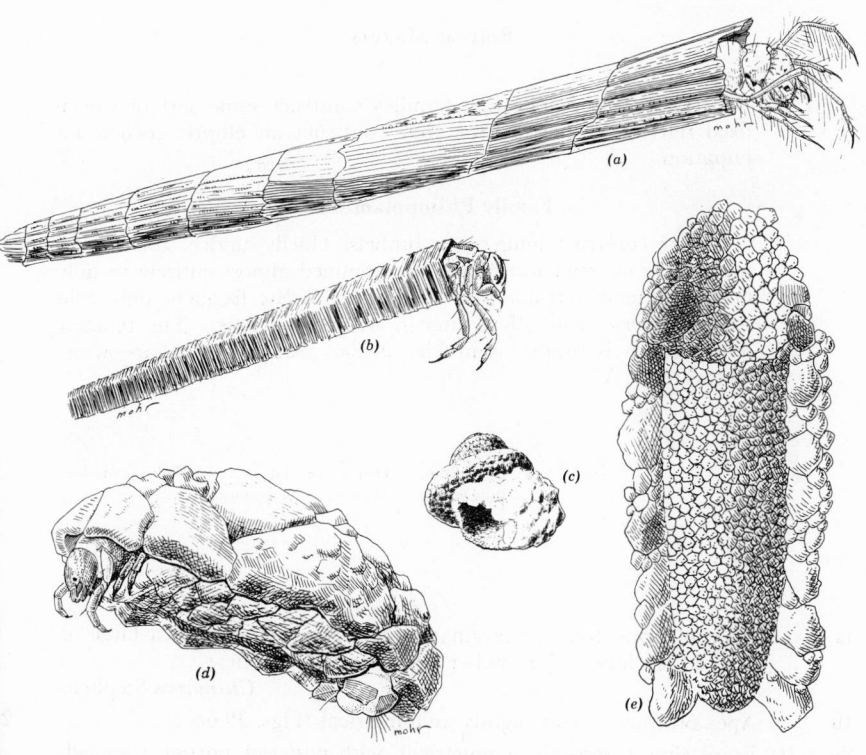

Fig. 39.4. Cases of Trichoptera. (*a*) *Triaenodes*. (*b*) *Brachycentrus*. (*c*) *Helicopsyche*. (*d*) *Glossosoma*. (*e*) *Molanna*. (Courtesy Illinois Natural History Survey.)

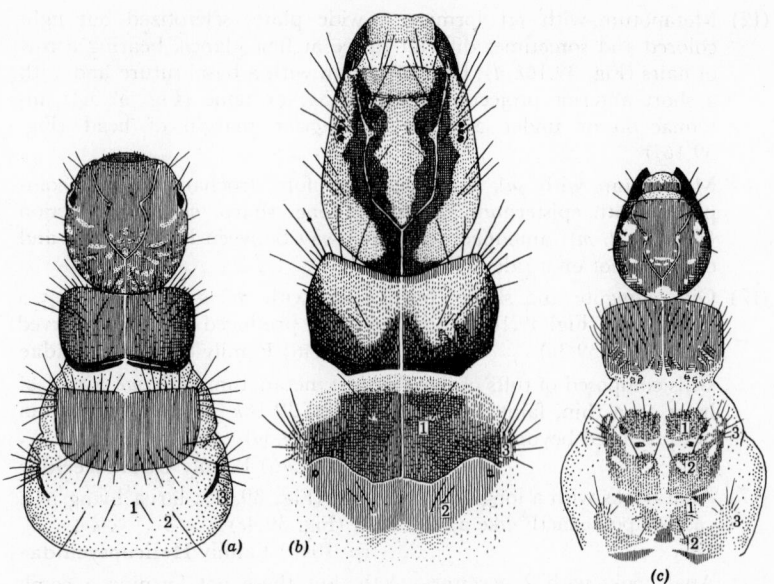

Fig. 39.5. Head and thorax, dorsal aspect, of larvae of Trichoptera. (*a*) *Lepidostoma*. (*b*) *Molanna*. (*c*) *Helicopsyche*.

Retreat Makers

Larvae of the 3 following families construct some sort of woven fixed retreat. When mature they construct an elliptic cocoon for pupation.

Family **Philopotamidae**

Larvae construct long silken tunnels, chiefly under stones. All species live in rapid water, and are confined almost entirely to hilly or mountainous terrain. *Wormaldia* and *Sortosa* frequent only cold streams; *Chimarra* usually occurs in warmer streams. The thoracic setal pattern is simple as in Fig. 39.2*a*. All 3 genera are widespread in N. A.

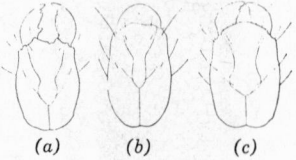

◄ **Fig. 39.6.** Heads of larvae of Philopotamidae. (*a*) *Chimarra*. (*b*) *Sortosa*. (*c*) *Wormaldia*.

1a	Apex of frons deeply emarginate (Fig. 39.6*a*), often with a large or pointed left lobe and a smaller right one. (12 species) . . . .

Chimarra Stephens

1b	Apex of frons at most slightly asymmetrical (Figs. 39.6*b*, *c*) 2
2a	(1) Frons almost perfectly symmetrical, with posterior portion widened, separated by a constriction from anterior portion (Fig. 39.6*c*). (11 species) . . **Wormaldia** McLachlan (=**Dolophilus** McLachlan)

2b   Frons slightly asymmetrical, without constriction, posterior portion uniform in width (Fig. 39.6b). (5 species) . . . .
            Sortosa Navas (= **Dolophilodes** Ulmer, **Trentonius** Betten and Mosely)

Family Psychomyiidae

Larvae build retreats such as silken tunnels on aquatic plants or burrow into sandy stream beds. Most of the species live in rapid water but some live in lakes also. The retreats built on stones, sticks, or plants, collapse into silky webs when taken out of water. Larvae are unknown for **Nyctiophylax** Brauer and **Cernotina** Ross. In the Nearctic genera the thoracic setal pattern is simple, as in Fig. 39.2a.

1a   Fore trochantin fused completely with episternum, represented by a long, sharp point (Fig. 39.7b), but with no suture at base . . . . . .
                       Subfamily **Polycentropinae**   **2**

1b   Fore trochantin marked off from episternum by an internal sclerotized ridge visible externally as a black line, trochantin squarish and twisted slightly mesad (Fig. 39.7a). . Subfamily **Psychomyiinae**   **7**

2a   (1) Mandibles short and triangular, each with a large, thick brush on the mesal side (Fig. 39.8c). Larvae burrow in sand bottom and make branched retreats. Cold water, eastern forms. (4 species). . .
                           Phylocentropus Banks

2b   Mandibles longer (Fig. 39.8g–i), with only a thin brush on left mandible, none on right. **3**

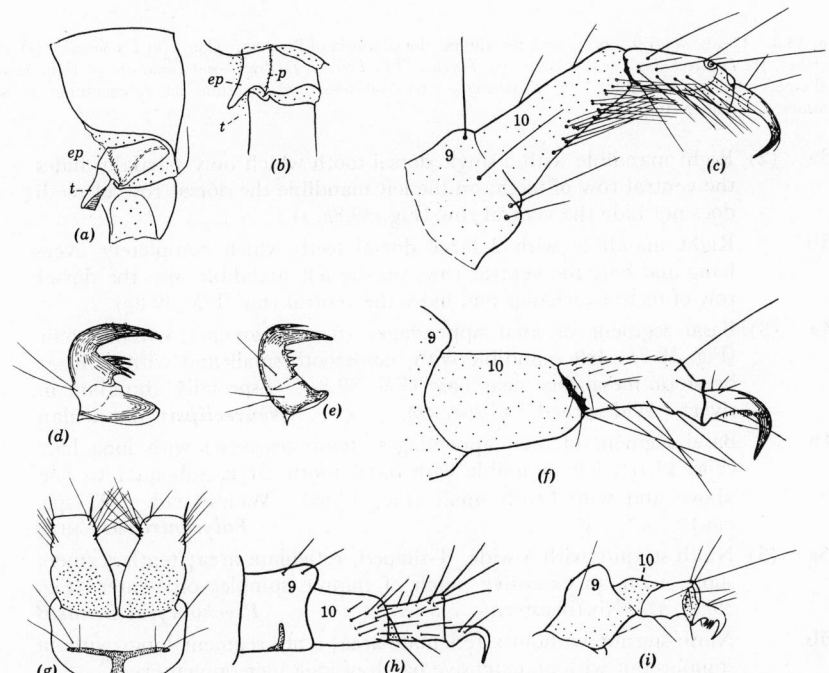

Fig. 39.7. Parts of Psychomyiidae larvae. (a) Prothorax of *Psychomyia*, lateral aspect. (b) Same of *Phylocentropus*. (c) Apex of abdomen of *Polycentropus*. (d) Anal claw of *Psychomyia*. (e) Same of *Psychomyiid Genus A*. (f) Apex of abdomen of *Neureclipsis*. (g, h) Apex of abdomen of *Psychomyiid Genus B*, ventral and lateral aspects, respectively. (i) Apex of abdomen of *Psychomyia*. *ep*, episternum; *p*, pleural suture; *t*, trochantin.

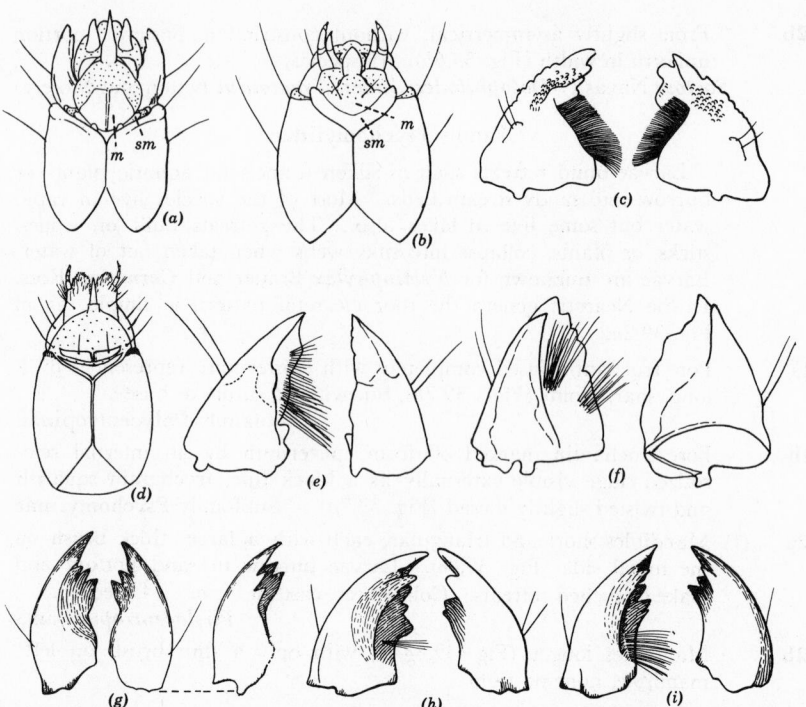

Fig. 39.8. Heads, ventral aspect, and mandibles, dorsal aspect of Psychomyiidae. (*a*) *Psychomyia*. (*b*) *Polycentropus*. (*c*) *Phylocentropus*. (*d*) *Tinodes*. (*e*) *Tinodes*. (*f*) *Lype*. (*g*) *Psychomyiid Genus A* (to right is ventral aspect of right mandible). (*h*) *Polycentropus*. (*i*) *Neureclipsis*. *m*, mentum; *sm*, submentum. (*c* after Vorhies; *d, e, f*, after Hickin.)

3a	**(2)**	Right mandible with a single dorsal tooth which only partially hides the ventral row of teeth; on the left mandible the dorsal row of teeth does not hide the ventral row (Fig. 39.8*h, i*) **4**
3b		Right mandible with 2 large dorsal teeth which completely overhang and hide the ventral row; on the left mandible also the dorsal row of teeth overhangs and hides the ventral row (Fig. 39.8*g*) **5**
4a	**(3)**	Basal segment of anal appendages (tenth segment) without hair (Fig. 39.7*f*); left mandible with basal tooth small and with a linear brush on mesal face near base (Fig. 39.8*i*). Especially abundant in rivers. Widespread. (4 species) . . . . . ***Neureclipsis*** McLachlan
4b		Basal segment of anal appendages (tenth segment) with long hair (Fig. 39.7*c*); left mandible with basal tooth large, subequal to one above and with brush small (Fig. 39.8*h*). Widespread. (27 species) . . . . ***Polycentropus*** Curtis
5a	**(3)**	Ninth sternite with a wide, T-shaped, reticulate area; tenth segment short, with an extensive patch of minute spinules on venter (Fig. 39.7*g, h*). Northeastern . . . . ***Psychomyiid Genus B***
5b		Ninth sternite without a reticulate area; tenth segment long, without spinules but with an extensive patch of long hair on venter **6**
6a	**(5)**	Anal claw with apex sharply and acutely angled, the basal part of the claw with a series of several large teeth (Fig. 39.7*e*). Central. In streams. . . . . ***Psychomyiid Genus A***

6b Anal claw with apex more rounded and less acute, and basal part of claw with only fine, hairlike teeth (as in Fig. 39.7*f*). In lakes and streams. Mexico to Northeast (1 species) . . . . . . ***Cyrnellus*** Banks

7a (1) Mentum formed of a pair of high, trianguloid sclerites (Fig. 39.8*a*); anal claw with a cluster of fine, long, preapical teeth (Fig. 39.7*d*). Cold streams, widespread. (3 species) . . . . . . ***Psychomyia*** Pictet

7b Mentum formed of a pair of wide, short sclerites (Fig. 39.8*d*); anal claw with no inner teeth, much as in Fig. 39.7*f* **8**

8a (7) Mandibles short and triangular (Fig. 39.8*f*), each with a dorsal dark stout tooth before the apex. Larva makes irregular long, narrow web tunnel on submerged sticks. Eastern and northcentral. (1 species) . . . . ***Lype*** McLachlan

8b Mandible narrow and pointed at apex, often quite long, with no dorsal preapical tooth (Fig. 39.8*e*). Larva makes similar long, narrow tunnels, usually on rocks. Western. (5 species) . . . .
Tinodes Stephens

Family **Hydropsychidae**

A very abundant family in rivers and streams of moderate current. Larvae live in retreats connecting with a net spun in the current. In N. A. the eastern genera ***Oropsyche*** Ross (1 species) and ***Aphropsyche*** Ross (1 species) and the western genera ***Homoplectra*** Ross (4 species) and ***Smicridea*** subgenus ***Rhyacophylax*** Müller (1 species), have not been reared.

It is especially important to have well-cleaned larvae of this family before attempting identification. An excellent cleaning tool is a camel's hair brush with the bristles cut short.

1a Head with a broad, flat, dorsal area set off by an extensive arcuate carina (Fig. 39.9*a*). Frequents large streams or rivers. Eastern. (3 species). . . . . ***Macronemum*** Burmeister

1b Head without such a carina (Fig. 39.9*f, g*) **2**

2a (1) Left mandible with a high, thumblike dorsolateral projection on basal portion (Fig. 39.9*s, t*). Occurs in springs in central and eastern U. S. . . . . ***Hydropsychid Genus A***

2b Left mandible without a high projection on the basal portion, sometimes with a low carina which is chiefly lateral (Fig. 39.9*v*). **3**

3a (2) Fore trochantin forked (Fig. 39.9*b*) **4**

3b Fore trochantin simple (Fig. 39.9*c*) **5**

4a (3) Prosternal plate with a pair of detached, moderate-sized, posterior sclerites (Fig. 39.9*d*); basal tooth on mandibles single (Fig. 39.9*r*). Predominant in colder, larger streams and rivers. Widespread. (55 species) . . . . ***Hydropsyche*** Pictet

4b Prosternal plate with only 1 pair of minute sclerotized dots posterior to it (Fig. 39.9*e*); basal tooth on mandibles double (Fig. 39.9*q*). Predominant, on the average, in smaller and warmer streams than the preceding. Widespread. (20 species) . . . .
Cheumatopsyche Wallengren

5a (3) Gula rectangular and long, separating genae completely (Fig. 39.9*j*); each branched gill with all its branches arising from the top of the

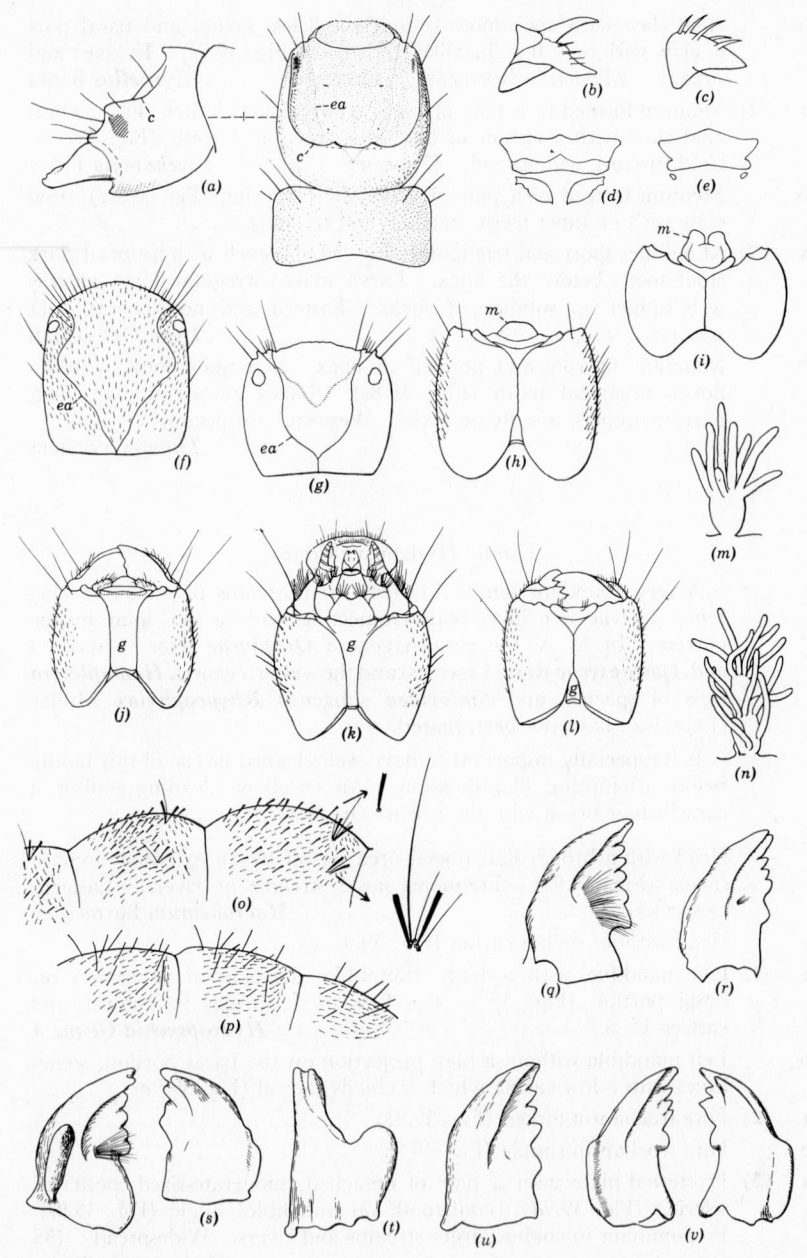

Fig. 39.9. Parts of Hydropsychidae larvae. (*a*) Head of *Macronemum*. (*b*) Fore trochantin of *Hydropsyche*. (*c*) Same of *Smicridea*. (*d*) Prosternum of *Hydropsyche*. (*e*) Same of *Cheumatopsyche*. (*f*) Head of *Diplectrona*, dorsal aspect. (*g*) Same of *Smicridea*. (*h*) Head of *Diplectrona*, ventral aspect. (*i*) Same of *Potamyia*. (*j*) Same of *Parapsyche*. (*k*) Same of *Arctopsyche*. (*l*) Same of *Diplectrona* with mentum infolded. (*m*) Abdominal gill of *Parapsyche*. (*n*) Same of *Diplectrona*. (*o*) Portion of abdomen of *Parapsyche*. (*p*) Same of *Arctopsyche*. (*q*) Left mandible of *Cheumatopsyche*. (*r*) Same of *Hydropsyche*. (*s*) Mandibles of *Hydropsychid Genus A*. (*t*) Lateral aspect of left mandible of same. (*u*) Left mandible of *Diplectrona*. (*v*) Mandibles of *Potamyia*. *ea*, epicranial arms; *c*, carina; *g*, gula; *m*, mentum.

central stalk of the gill (Fig. 39.9*m*). Confined to very cold or montane streams. Eastern, northern and western 6

5b   Gula triangular and short, genae fused for most of their length (Fig. 39.9*l*); each branched gill with branches arising both from sides and top of stalk (Fig. 39.9*n*). In a wide variety of streams . . . 7

6a   (5)   Gula rectangular and of even width (Fig. 39.9*j*); abdomen with stout, short, black, scalelike hairs arranged in tufts along dorsum near sides, frequently with broad scales scattered between them (Fig. 39.9*o*). (5 species) . . . . ***Parapsyche*** Betten

6b   Gula narrowed posteriorly (Fig. 39.9*k*); abdomen without distinct setal tufts, with coarse hairs of varying lengths, some of them scale-like but narrow and long (Fig. 39.9*p*). (5 species) . . . .
                                      Arctopsyche McLachlan

7a   (5)   Mandibles with winglike dorsolateral flanges along basal half (Fig. 39.9*v*); mentum cleft (Fig. 39.9*i*). Frequenting chiefly warmer streams and rivers. Eastern and southern. (1 species) . . . .
                                        Potamyia Banks

7b   Mandibles without distinct dorsolateral flanges (Fig. 39.9*u*); sub-mentum subconical, not cleft (Fig. 39.9*h*) 8

8a   (7)   Frons expanded laterad, its lateral extensions sharp (Fig. 39.9*f*). Occurring in cold streams and springs. Widespread. (5 species) . .
                                     Diplectrona Westwood

8b   Frons not expanded laterad, its lateral extensions scarcely produced (Fig. 39.9*g*). Occurring chiefly in warm streams and springs. Southwestern. (1 species) . . . . ***Smicridea*** McLachlan

Free-Living Larvae

Family **Rhyacophilidae**

Free-living larvae found in cold running water, especially abundant in mountain streams. The mature larvae construct stone cases for pupation. Often found as glacial relicts in cold springs in eastern U. S. Two genera in N. A.

1a   Front leg simple, primarily same as others (Fig. 39.10*b*); prosternum membranous. Over all N. A. except plains region. (70 species). . .
                                     Rhyacophila Pictet

1b   Front leg chelate (Fig. 39.10*a*), other legs simple; prosternum with a large sclerotized plate. Southwestern U. S. (3 species) . . . . . .
                                     Atopsyche Banks

Saddle-Case Makers

Family **Glossosomatidae**

Larvae of this family construct a stone case shaped like a tortoise shell, and having a stone bridge across the center of the ventral opening (Fig. 39.4*d*). For pupation this bridge is removed and the case is cemented to a rock. The saddle-case makers all have a sclerotized strap narrowed in the middle across the posternal region. All species are denizens of cold running water, chiefly confined to

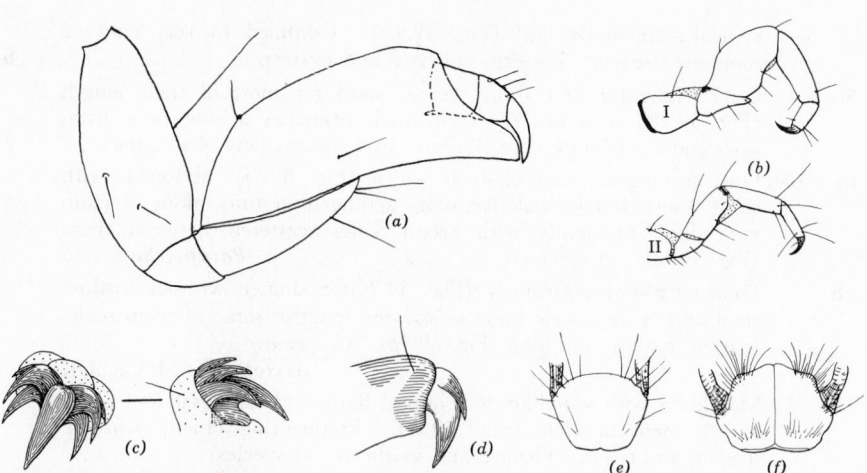

Fig. 39.10. Parts of Rhyacophilidae and Glossosomatidae larvae. (*a*) Front leg of *Atopsyche*. (*b*) Front (upper) and hind (lower) legs of *Rhyacophila*. (*c*) Anal hooks of *Protoptila*. (*d*) Anal hooks of *Agapetus*. (*e*) Pronotum of *Glossosoma*. (*f*) Same of *Agapetus*.

springs or semipermanent streams in the southern part of the families' range.   Four genera occur in the Nearctic region.

1a   Anal claw divided into many teeth (Fig. 39.10*c*); meso- and metanotum with only a single dorsal pair of hairs in addition to *sa*3.   Widespread, frequently in rivers.   (15 species). . . . .
                                          Protoptila Banks

1b   Anal claw with 1 large tooth, and 1 or 2 small ones (Fig. 39.10*d*); mesonotum, usually also metanotum, with both *sa*1 and *sa*2 present .   **2**

2a (1) Pronotum notched only at extreme anterolateral angle, at which point the legs are attached (Fig. 39.10*e*); only *sa*2 and *sa*3 on abdominal tergites.   Widespread.   (20 species) . . . ***Glossosoma*** Curtis

2b   Pronotum narrow from anterior margin to middle, the legs attached at this central point (Fig. 39.10*f*); several abdominal tergites with *sa*1, 2, and 3 all present **3**

3a (2) *Sa*1 usually present on abdominal tergites 3 and 6.   Rocky Mountain region only, in small streams.   (4 species) . . . . . ***Anagapetus*** Ross

3b   *Sa*1 usually absent on abdominal tergites 3 and 6.   Widespread including Rocky Mountains, usually in small streams.   (15 species) . . . . ***Agapetus*** Curtis

Family **Hydroptilidae**

These are the "micro" caddisflies, and seldom reach a length of 5 mm.   The larvae are unique among Trichoptera in possessing a type of hypermetamorphosis.   The first 4 instars build no case, and have long, curved anal claws; the fifth and last instar builds a barrel or purselike case (Fig. 39.11), and has short claws.   Pupation takes place within the case.   Only the case-making stage is keyed here.   See Neilsen (1948) for an excellent, comprehensive account of this phenomenon.   Larvae are unknown for **Metrichia** Ross (1 species) and **Dibusa** Ross (1 species), both of which are widespread but local in the southern part of the U. S.

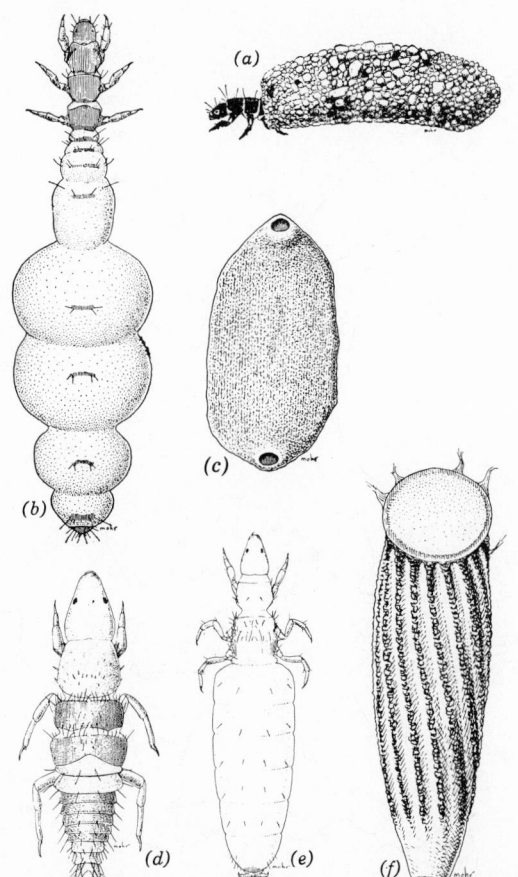

Fig. 39.11. Larvae and cases of Hydroptilidae. (*a*) *Ochrotrichia.* (*b, c*) *Leucotrichia.* (*d*) Pre-casemaking form of *Mayatrichia.* (*e*) Casemaking form of same. (*f*) Case of same, sealed for pupation.

1a		Abdomen enlarged, at least some part of it much thicker than thorax (Figs. 39.1*a*, 39.11*b*, *e*); living in case (last instar)	**2**
1b		Abdomen slender, not appreciably thicker than thorax (Fig. 39.11*d*); not with case (early instars) . . . . **Not keyed**	
2a	(1)	Each segment of abdomen with a dark, sclerotized dorsal area (Figs. 39.11*b*, 39.12*a*)	**3**
2b		Abdomen with at least segments 2 to 7 without dark, sclerotized dorsal areas, at most with a small delicate ring (Fig. 39.12*b*)	**4**
3a	(2)	Abdomen with dorsal sclerites solid; segments 1 and 2 small, 3 to 6 greatly expanded (Fig. 39.11*b*). Case translucent, ovoid and irregularly water-penny-shaped (Fig. 39.11*c*). In cold streams. Widespread. (4 species) . . . . *Leucotrichia* Mosely	
3b		Abdomen with dorsal sclerites membranous across middle (Fig. 39.12*a*); segments 1 to 6 evenly expanded, as in Fig. 39.1*a*. In a wide variety of small streams, including semipermanent types. Widespread. (20 species) . . . . *Ochrotrichia* Mosely	

Fig. 39.12. Parts of Hydroptilidae larvae. (*a, b*) Abdominal rings of two species of *Ochrotrichia*. (*c*) Pronotum of *Neotrichia*. (*d*) Same of *Mayatrichia*. (*e*) Apical tergites of *Mayatrichia*. (*f*) Legs of *Oxyethira*. (*g*) Legs of *Neotrichia*. (*h*) Legs of *Orthotrichia*. (*i*) Legs of *Hydroptila*. (*j*) Legs of *Ochrotrichia*. (*k*) Tarsi and claws of *Stactobiella*. (*l*) Same of *Ochrotrichia*. (*m*) Metanotum of *Ochrotrichia*. (*n*) Same of *Hydroptila*. (*o*) Legs of *Agraylea*. (*p*) Legs of *Mayatrichia*.

4a (2) Abdominal segments with dorsal and ventral projections (Fig. 39.1*b*). In streams and possibly lakes also. Widespread. (2 species) . . . . .
Ithytrichia Eaton

4b Abdominal segments without dorsal and ventral projection (Fig. 39.1*a*). **5**

5a (4) Middle and hind legs almost 3 times as long as front legs (Fig. 39.12*f*). Very wide ecological tolerance. Streams and lakes. Widespread. (20 species) . . . . *Oxyethira* Eaton

5b Middle and hind legs not more than 1½ times as long as front legs (Fig. 39.12*g*). **6**

6a (5) Tarsal claws about same length as tarsus (Fig. 39.12*i, j, k, o*); case purselike (Fig. 39.11*a*). **7**

6b Tarsal claws much shorter than tarsus (Fig. 39.12*g, h, p*); case not purselike, more barrel-shaped, such as in Fig. 39.11*f* **10**

7a (6) Tarsal claw with long, stout inner tooth (Fig. 39.12*k*); case purselike, robust. In cold streams. Widespread. (3 species) . . . .
Stactobiella Martynov (= *Tascobia* Ross)

7b Tarsal claw without prominent inner tooth (Fig. 39.12*l*); case either purselike or cylindrical **8**

8a (7) Tibia twice as long as deep (Fig. 39.12*o*). (3 species). . . . .
Agraylea Curtis

8b Tibia about as long as deep (Fig. 39.12*i, j*) **9**

9a (8) Metanotum with a distinct, widened ventrolateral area (Fig. 39.12*m*). In a wide variety of small streams, including semipermanent types. Widespread. (20 species). . . . . *Ochrotrichia* Mosely

9b Metanotum without a widened lateral area (Fig. 39.12*n*). Common in streams, less so in lakes. Widespread. (45 species) . . . .
Hydroptila Dalman

10a (6) Anal legs apparently combined with body mass and only the claws projecting as in Fig. 39.1*a*; eighth abdominal tergite with only 1 or 2 pairs of weak setae. (3 species) . . . . *Orthotrichia* Eaton

10b Anal legs distinctly projecting from body mass (Fig. 39.11*e*) eighth abdominal tergite with a brush of setae (Fig. 39.12*e*) **11**

11a (10) Thoracic tergites clothed with long, slender, erect, inconspicuous setae (Fig. 39.12*c*); case of sand grains, evenly tapered and without posterior slit. In warmer streams and rivers. Widespread, rare in Canada. (15 species) . . . . *Neotrichia* Morton

11b Thoracic tergites clothed with shorter, stout, black setae which are conspicuous and appressed to the surface of the body (Fig. 39.12*d*); case semitranslucent, evenly tapered and with dorsal side either ringed or fluted with raised ridges (Fig. 39.11*f*). In rapid and clear streams. Widespread. (3 species) . . . . *Mayatrichia* Mosely

Tube-Case Makers

Larvae make cases which have posterior end closed, the closure perforated by a network of cross strands or a single, small slit or circle. Certain genera make cases of particular shape and texture, others may use a variety of materials depending on availability. Pupation occurs within the case.

Family **Phryganeidae**

Cases made as shown in Fig. 39.4a, composed of pieces of grass stems usually arranged in spiral pattern. Most of the genera are cold-water marsh inhabitants, but may occur also in stream backwaters. The family is most abundant in the cool temperate belt and northward. Few characters have been found to date for larval identification, and the following key must be considered provisional. Larvae have not been reared of **Fabria** Milne (2 species) and **Oligotricha** Rambur (1 species).

1a	Frons with a median black line (Fig. 39.13a)	**2**
1b	Frons without median black line (Fig. 39.13c)	**3**

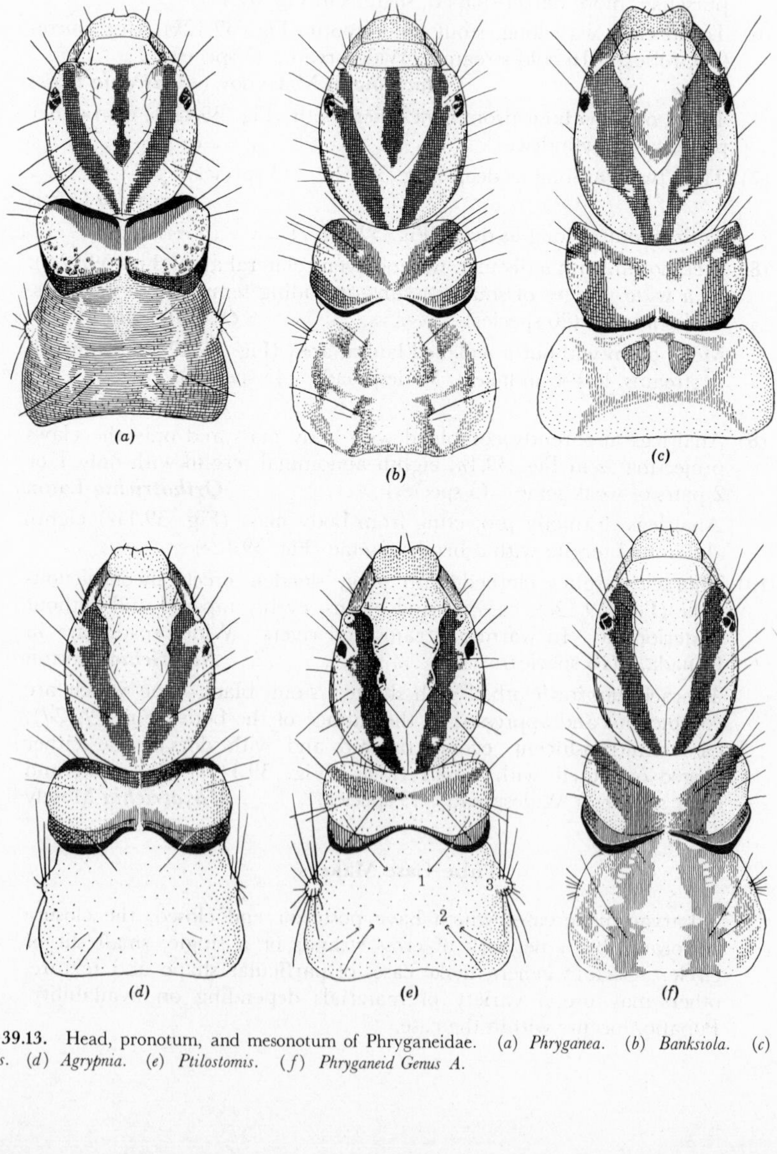

Fig. 39.13. Head, pronotum, and mesonotum of Phryganeidae. (a) *Phryganea.* (b) *Banksiola.* (c) *Oligostomis.* (d) *Agrypnia.* (e) *Ptilostomis.* (f) *Phryganeid Genus A.*

2a (1) Pronotum with anterior margin black (Fig. 39.13*a*), and without a diagonal black line. Marsh species. Widespread. (3 species). . .

Phryganea Linnaeus

2b Pronotum with a diagonal black line (Fig. 39.13*b*), anterior margin mostly yellow. Some species of the widespread genera . . . .
(5 species) ***Banksiola*** Martynov, (8 species) ***Agrypnia*** Curtis

3a (1) Mesonotum with a pair of small sclerites near anterior margin (Fig. 39.13*c*) **4**

3b Mesonotum without a pair of sclerites (Fig. 39.13*d*), sometimes with a very small sclerotized area around the base of 1 seta **5**

4a (3) Mature larvae attaining length of 30 mm. Northeastern. (1 species) . . . . ***Eubasilissa*** Martynov

4b Mature larvae attaining length of only 20 mm. Northeastern. (1 species) . . . . ***Oligostomis*** Kolenati

5a (3) Pronotum with anterior margin black (Fig. 39.13*d*) and without a diagonal black line. (10 species.) Some species of . . . .
Agrypnia Curtis

5b Pronotum with a diagonal black line (Fig. 39.13*e, f*) anterior margin mostly yellow **6**

6a (5) Diagonal marks on pronotum meeting at posterior margin to form a V-mark (Fig. 39.12*f*) . . . . ***Phryganeid Genus A***

6b Diagonal marks on pronotum not reaching posterior margin but joining each other on meson to form an arcuate mark (Fig. 39.13*e*). Larvae frequent both streams and lakes. Widespread. (4 species). . .
Ptilostomis Kolenati

Family **Goeridae**

Larvae make very solid stone cases, exactly like those of *Neophylax*, in Limnephilidae. Goeridae inhabit cold, fast streams, and are widespread in distribution but relatively rare. Three genera are known in N. A., ***Goera*** Curtis (4 species), ***Goerita*** Ross (2 species), and ***Pseudogoera*** Carpenter (1 species). Larvae of only the first are known (Fig. 39.3*d*).

Family **Brachycentridae**

Larvae make either cylindrical cases, chiefly of spun silk, or square cases (Fig. 39.4*b*) using small bits of woody material. In *Brachycentrus* the same case may be partly one type, partly another. Favorite haunts are cold, fast streams and rivers. Of the 3 North American genera, the larva of the western ***Oligoplectrum*** McLachlan (3 species) is unknown. The other 2 genera are widespread in the montane areas and northward.

1a Middle and hind tibiae with an inner, apical, seta-bearing spur (Fig. 39.14*d*); hind coxae with a ventral, semicircular lobe bearing a row of long setae; mesonotum with sclerites long and narrow, plates of metanotum heavily sclerotized (Fig. 39.14*b*). (12 species) . . . . . .
Brachycentrus Curtis

1b Middle and hind tibiae without an apical spur (Fig. 39.14*c*); hind coxae without a ventral lobe; mesonotum with sclerites short and very wide, plates of metanotum only lightly sclerotized but recognized chiefly by their cluster of setae (Fig. 39.14*a*). (15 species) . . .
Micrasema McLachlan

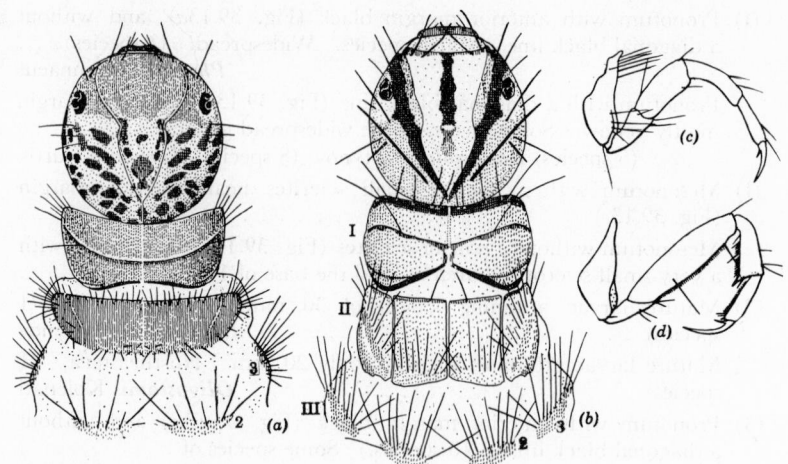

Fig. 39.14. Parts of Brachycentridae larvae. (*a*) Head and thorax of *Micrasema*. (*b*) Same of *Brachycentrus*. (*c*) Hind leg of *Micrasema*. (*d*) Same of *Brachycentrus*.

Family **Limnephilidae**

Larvae making a great variety of cases and occupying many diverse habitats. The following key contains only about half the North American genera, the larvae of the remainder being un-associated with their respective adults. In addition, the generic diagnoses for some large genera are based on only the few species that have been reared. Some of these diagnoses may need to be emended when larvae are known for more species. Much more rearing needs to be done to improve our knowledge of the immature stages of this family.

1a   Abdomen without gills; mesonotal plates incised to form a deep angle on meson; metanotum with *sa*3 forming a large and quad-rangular sclerite. A small larva making a thin, long, smooth, sand case. In rapid, cold streams. Western. (2 species) . . . .

                                                   Neothremma Banks

1b   Abdomen with gills; mesonotal plates little or not incised; metanotum with *sa*3 bean-shaped, narrow. **2**

2a (1) Each gill single (Fig. 39.15*c*) . . . ., ... **3**

2b   Many gills with 3 or more branches (Fig. 39.15*d-f*) **15**

3a (2) Head with short, stout spines each on a raised base (Fig. 39.15*h*); mesonotum with 2 pairs of plates (Fig. 39.15*g*); metanotum and first 2 abdominal tergites covered with short, fine hair. In streams. Western. (1 species) . . . . **Pedomoecus** Ross

3b   Head with principal setae long and slender; mesonotum with only 1 pair of plates (Fig. 39.15*j*); at least second abdominal segment not hairy. **4**

4a (3) Anterior margin of mesonotum with a mesal, rectangular emargina-tion (Fig. 39.15*j*); at this point it is connected to pronotum by a short, sclerotized strap. Always in cold, rapid streams, including semipermanent types. Widespread. (15 species) . . . .

                                                 Neophylax McLachlan

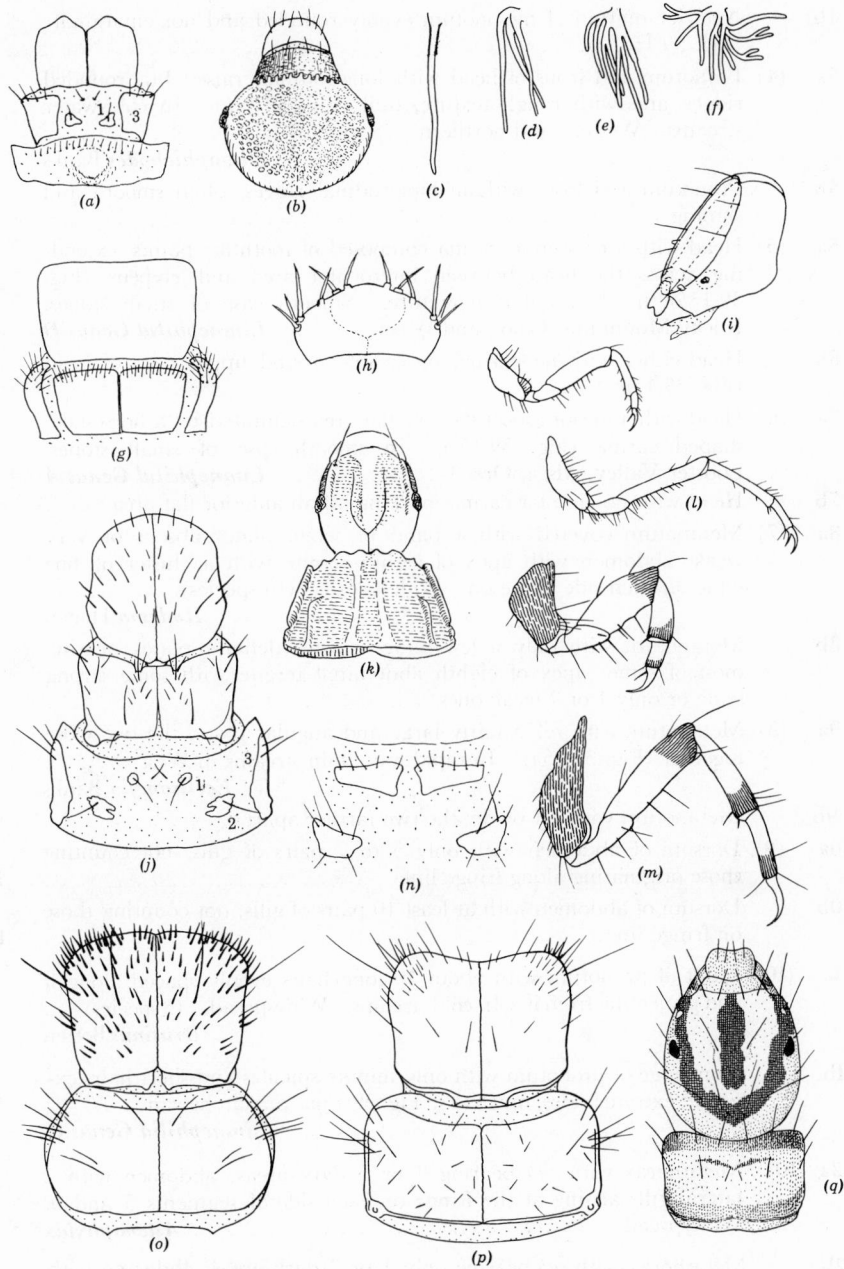

Fig. 39.15.   Parts of Limnephilidae larvae.   (*a*) Meso- and metanotum and abdomen of *Neothremma*. (*b*) Head of *Genus D*. (*c*) Abdominal gill of *Pycnopsche*. (*d*) Same of *Limnephilus*. (*e*) Same of *Hesperophylax*. (*f*) Same of *Caborius*. (*g*) Pro- and mesonotum of *Pedomoecus*. (*h*) Head of *Pedomoecus*. (*i*) Head of *Genus A*. (*j*) Thorax of *Neophylax*. (*k*) Head and pronotum of *Oligophlebodes*. (*l*) Front and hind legs of *Dicosmoecus*. (*m*) Same of *Glyphopsyche*. (*n*) Metanotum of *Ecclisomyia;* (*o*) Pro- and mesonotum of *Glyphopsyche*. (*p*) Same of *Frenesia*. (*q*) Head and pronotum of *Glyphotaelius*.

4b   Anterior margin of mesonotum evenly rounded and not emarginate
(Fig. 39.15*o*, *p*). **5**

5a   **(4)** Pronotum and frons of head with longitudinal, raised but rounded
ridges, and with rough texture, dull (Fig. 39.15*k*). In cold-water
streams. Western and northern. (5 species) . . . .
Oligophlebodes Banks

5b   Pronotum and frons without longitudinal ridges, often smooth and
shining **6**

6a   **(5)** Head with a crescentic carina composed of toothlike points, extending across the head between interocular area and clypeus (Fig.
39.15*b*); head granular in texture. Smooth case of small stones.
Rocky Mountains, Colo., and Wash. . . . . . *Limnephilid Genus D*

6b   Head either with no carina, or carina around upper part of head
(Fig. 39.15*i*) **7**

7a   **(6)** Head with anterior aspect flat, the flat area delimited by a horseshoe-
shaped carina (Fig. 39.15*i*). In smooth case of small stones.
Lobster Valley, Alsea, Ore. . . . . *Limnephilid Genus A*

7b   Head with no circular carina marking off an anterior flat area **8**

8a   **(7)** Metanotum covered with a band of setae, plates absent or very
weak; abdomen with apex of eighth tergite with a cluster of fine
setae on each side of meson. Widespread. (10 species). . . . . . .
Radema Hagen

8b   Metanotum with only a few setae or with definite plates bearing
most of setae; apex of eighth abdominal tergite with some strong
setae or only 1 or 2 weak ones **9**

9a   **(8)** Metanotum with *sa*1 a fairly large and angular plate, the two close
together (Fig. 39.15*n*). In rapid mountain streams in West . . . . .
Ecclisomyia Banks

9b   Metanotum with *sa*1 ovate, the two farther apart **10**

10a **(9)** Dorsum of abdomen with only 5 to 7 pairs of gills, not counting
those originating along fringe line **11**

10b   Dorsum of abdomen with at least 10 pairs of gills, not counting those
on fringe line. **12**

11a **(10)** Front of pronotum with about 10 long hairs evenly spaced along it
and projecting from it. In cold streams. Widespread. (5 species) . .
Drusinus Betten

11b   Front edge of pronotum with only minute spicules and short hairs except at extreme sides, much as in Fig. 39.15*p*. Saskatchewan. . . . . .
Limnephilid Genus C

12a **(10)** Metathorax with *sa*3 bearing 4 or 5 dark areas; abdomen with 2
lateral gills arising at the fringe on each side of segments 5 and 6.
Widespread . . . . *Astenophylax*

12b   Metathorax with *sa*3 bearing only 1 or 2 dark areas; abdomen with
no fringe gills on segment 6 and at most 1 on segment 5 . . . .   **13**

13a **(12)** First segment of abdomen with only a few pairs of setae, all close to
sides of hump, including both dorsum and venter. Western. (2
species) . . . . *Clostoeca* Banks

13b   First segment of abdomen with a band of scattered setae across most
of tergite and sternite **14**

14a (13) Abdominal gills long, stout and overlapping considerably on meson, especially on segments 2 and 3. Eastern. (18 species) . . . .
Pycnopsyche Banks

14b Abdominal gills shorter, scarcely or not overlapping. Western. (4 species) . . . . *Ecclisomyia* Banks

15a (2) Front femora slender, the apical margin short (Fig. 39.15*l*); pronotum slightly incised along anterior margin; always with sclerotized portions of head and body black or nearly so. Rapid streams. Western and northern. (15 species) . . . . *Dicosmoecus* McLachlan

15b Front femora somewhat chelicerate, widened, with the apical margin very oblique and nearly as long as the lower margin (Fig. 39.15*m*) . . . **16**

16a (15) Pronotum with dense, short, black spines, especially along anterior margin (Fig. 39.15*o*) legs banded with red and black (Fig. 39.15*m*). Widespread. (2 species) . . . . *Glyphopsyche* Banks

16b Pronotum without dense black spines, clothed primarily with long setae (Fig. 39.15*p*); legs not banded **17**

17a (16) Anal legs with a group of about 10 setae on the bulbous, ventral, membranous portion between the two claws. Eastern. (2 species) . *Frenesia* Betten and Mosely

17b Anal legs with no setae on bulbous, membranous, ventral portion . . **18**

18a (17) Dorsal gills of first few abdominal segments with 6 to 12 branches, forming a fan-shaped spread (Fig. 39.15*e, f*) **19**

18b Dorsal gills of first few abdominal segments with 2 to 4 branches at the most, not spreading fanlike (Fig. 39.15*d*). Sometimes a few ventral gills will have 6 or 7 branches **20**

19a (18) Dorsal gills at base of abdomen with about 12 branches each (Fig. 39.15*f*). Ponds or streams. Eastern. (3 species) . . . .
Caborius Navas

19b Dorsal gills at base of abdomen with about 6 branches each (Fig. 39.15*e*). Streams. Widespread. (6 species). . . . .
Hesperophylax Banks

20a (18) Head with a narrow, dark line along meson of frons and with a dark, U-shaped line running through eyes and above frons (Fig. 39.15*q*). Northern. (1 species) . . . . . . *Glyphotaelius* Stephens

20b Head either with wider dark areas, or without indication of dark lines. **21**

21a (20) Prosternal horn short, not projecting beyond apexes of front coxae . . . **22**

21b Prosternal horn projecting distinctly beyond apexes of front coxae. Northern. (5 species) . . . . *Platycentropus* Ulmer

22a (21) Segments 2 and 3 of abdomen with some dorsal gills 4-branched and some ventral gills 6- or 7-branched; sternite of segment 1 of abdomen with a band of fairly dense, thin, but sharp, stiff black setae. Western. (2 species) . . . . *Psychoronia* Banks

22b Abdomen with dorsal gills at most 3-branched, ventral gills not more than 4-branched; sternite of segment 1 with various types of vestiture. Ponds, lakes, or streams. Common and widespread. (100 species) . . *Limnephilus* Leach

Family Lepidostomatidae

Larvae construct either log cabin cases of little sticks, irregular cases of varied materials, or slender, smooth, sand cases. They inhabit cold water, are often abundant in springs or small streams, and also occur in rivers and small lakes. Of the 2 Nearctic genera, the larva of **Theliopsyche** Banks (4 species) is unknown. The larva of **Lepidostoma** Rambur (30 species) is illustrated in Fig. 39.5*a;* the metanotum has *sa*1 and *sa*2 single.

Family Calamoceratidae

Larvae make cases from sticks or leaf fragments. Leaf cases are usually composed of three large pieces plus other small bits, and the whole has a triangular cross section. All genera inhabit springs or rapid streams. The larvae have the antenna situated midway between the margin of the head and the eye, as in Fig. 39.16*i*. The 3 genera listed below all have *sa*3 of the mesonotum straplike and separate from the central sclerite composed of *sa*1 and *sa*2. The larva of *Notiomyia* Banks is identified only on circumstantial evidence.

1a   Anterolateral corners of pronotum produced into sharp, down-curved hooks (Fig. 39.16*b*). Larva makes a leaf case   2
1b   Pronotum almost rectangular, its anterior corners not produced (Fig. 39.16*a*). Larva hollows out a small stick and uses it for a case. Eastern and western. (2 species). . . . **Heteroplectron** Banks
2a   Tarsal claw of fore and middle legs similar, short and stout (Fig. 39.16*c*). Southeastern. (1 species) . . . **Anisocentropus** McLachlan
2b   Tarsal claw of middle leg elongate, twice length of that of foreleg (Fig. 39.16*d*). Southwestern. (2 species) . . . . . **Notiomyia** Banks

Family Helicopsychidae

Larvae make a spiral stone case resembling a snail shell (Fig. 39.4*c*), and live in springs and fast streams over most of the continent. Larvae have antenna midway between margin of head and eye (Fig. 39.16*i*). Only the genus **Helicopsyche** Hagen (4 species) has been collected north of central Mexico (Figs. 39.5*c* and 39.16*h, i*).

Family Odontoceridae

Larvae make elongate, extremely strong, smooth cases of minute stones, and live in cold, rapid streams. Of the 5 known Nearctic genera, only 2 are definitely reared; larvae of the western genera **Namamyia** Banks (1 species), **Nerophilus** Banks (1 species) and **Parthina** Denning (1 species) have not been identified.

1a   Mesonotum consisting of a single, undivided plate resembling the pronotum in general shape; metanotum with each *sa*1 a small sclerite (Fig. 39.16*m*). Lake County, Calif. . . . .
                                              Odontocerid Genus A
1b   Mesonotum divided by linear sutures into several sclerites; metanotum with the two *sa*1 forming large, contiguous sclerites (Fig. 39.16*l*) . . .   2
2a   Anterolateral corner of pronotum evenly rounded (Fig. 39.16*l*). Widespread. Southern. (2 species) . . . . **Marilia** Müller
2b   Anterolateral corner of pronotum produced into a sharp point (Fig. 39.16*k*). Eastern. (6 species) . . . . **Psilotreta** Banks

Family Sericostomatidae

Larvae construct strong, robust cases of minute stones and live in cold, rapid streams. Only 1 genus is recognized in N. A., **Seri-**

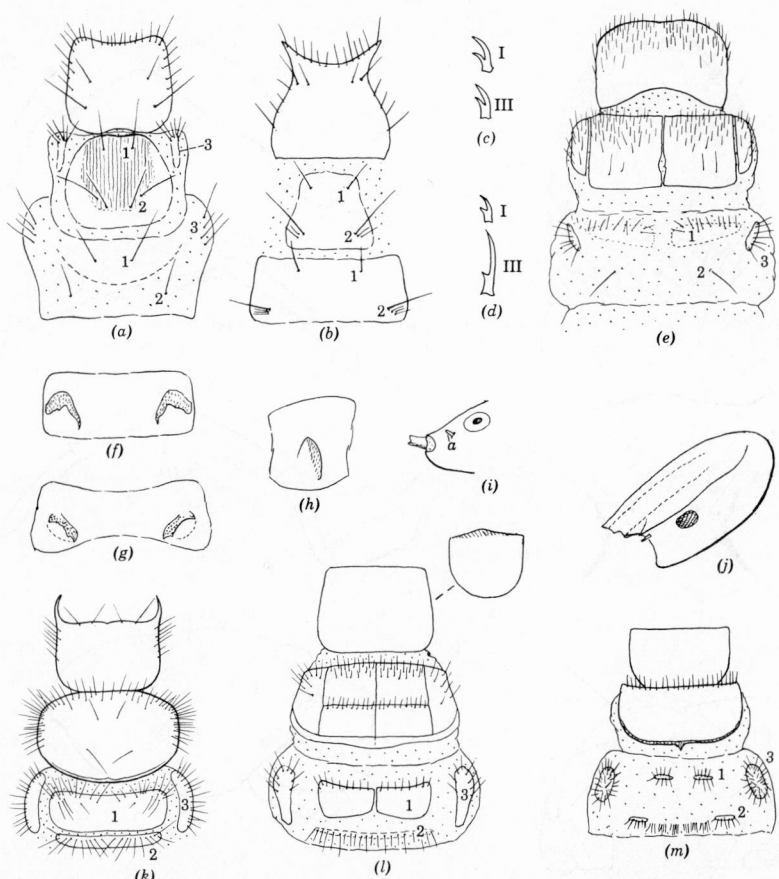

Fig. 39.16. Parts of Trichoptera larvae. (*a*) Thorax of *Heteroplectron*. (*b*) Same of *Anisocentropus*. (*c*) Front and hind tarsal claws of *Anisocentropus*. (*d*) Same of *Notiomyia*. (*e*) Thorax of *Sericostoma*. (*f*) First abdominal sternite of *Heteroplectron*. (*g*) Same of *Anisocentropus*. (*h*) Same (lateral aspect) of *Helicopsyche*. (*i*) Head of *Helicopsyche*. (*j*) Head of *Odontocerid Genus A*. (*k*) Thorax of *Psilotreta*. (*l*) Same of *Marilia*, with inset showing lateral aspects of pronotum. (*m*) Same of *Odontocerid Genus A*. *a*, antenna.

costoma Berthold (10 species), occurring in the eastern and western montane regions and across the north (Figs. 39.3*n*, 39.16*e*).

Family **Leptoceridae**

Larvae construct a wide variety of cases and are found in an equally wide range of habitats. Each genus occurs in both lakes and streams.

1a   Middle legs with claw stout and hook-shaped, tarsus bent (Fig. 39.17*d*); case slender and transparent. Northern and eastern. (1 species) . . . . **Leptocerus** Leach

1b   Middle legs with claw slender, slightly curved, tarsus straight (Fig. 41.17*e*); case seldom transparent **2**

2a   (1) Maxillary palpi nearly as long as stipes (Fig. 39.17*h*); mandibles

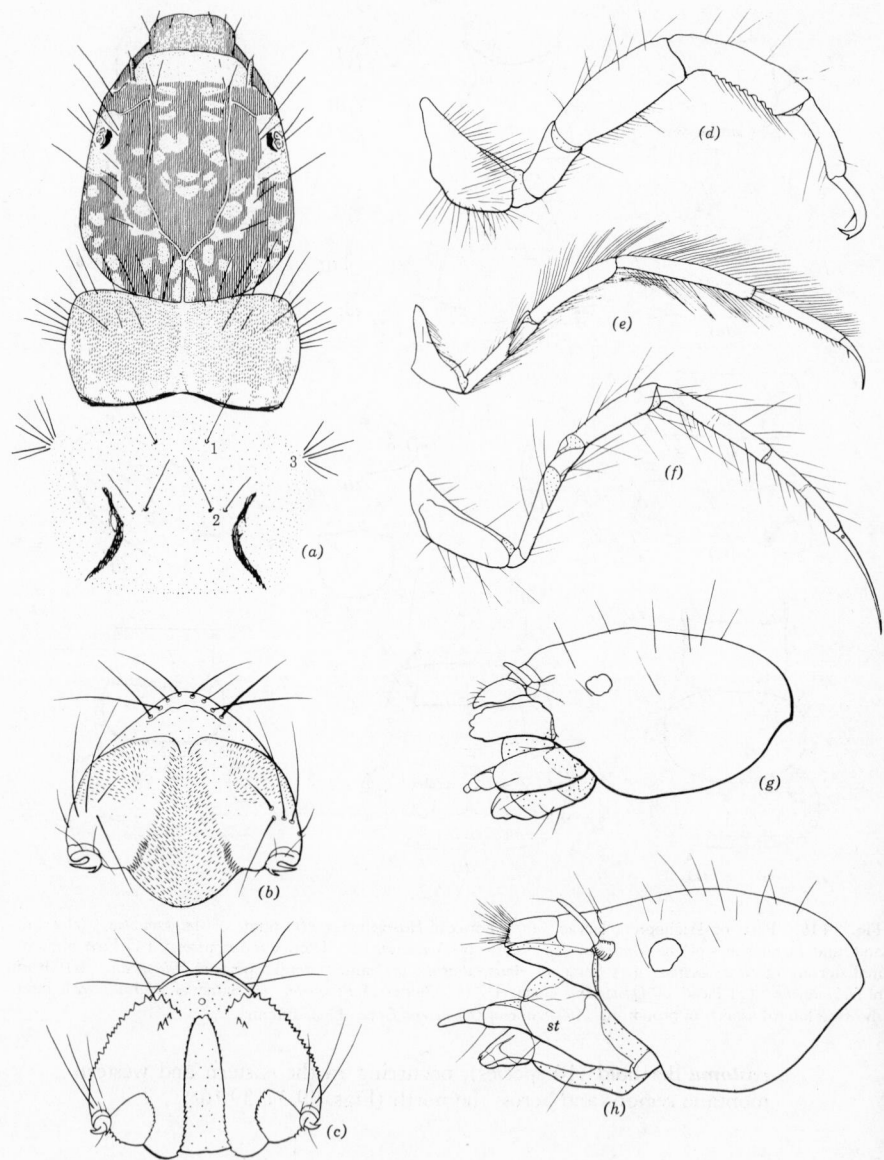

Fig. 39.17. Parts of Leptoceridae larvae. (*a*) Head, pronotum, and mesonotum of *Athripsodes*. (*b*) Last segment, posterior aspect, of *Mystacides*. (*c*) Same of ?*Setodes*. (*d*) Middle leg of *Leptocerus*. (*e*) Hind leg of *Leptocella*. (*f*) Same of *Mystacides*. (*g*) Head of *Leptocella*. (*h*) Same of *Oecetis*. *st*, stipes.

long, sharp at apex, the teeth considerably below apex. Cases of several types. Common and widespread. (15 species) . . . .
Oecetis McLachlan

2b   Maxillary palpi short, about ½ the length of stipes (Fig. 39.17*g*); mandibles shorter, blunt at apex, the teeth near or at apex **3**

3a (2) Head with suturelike line or pale area paralleling the epicranial arms (Fig. 39.17*a*). (30 species.) Some species of . . . . *Athripsodes* Billberg

3b Head without a suturelike line in addition to the epicranial arms. . . 4

4a (3) Mesonotum membranous with a pair of sclerotized, narrow, curved or angled bars (Fig. 39.17*a*). Cases ovate, convex, or like a water penny. Widespread. (30 species) . . . . . . *Athripsodes* Billberg

4b Mesonotum without such a pair of sclerotized bars 5

5a (4) Anal segment developed into a pair of sclerotized, concave plates, with spinose dorsolateral and mesal carinae, and an overhanging ventral flap (Fig. 39.17*c*). Case slender. Eastern. (5 species). . . . *?Setodes* Rambur

5b Anal segments convex and without carinae between anal hooks (Fig. 39.17*b*) 6

6a (5) Hind tibiae entirely sclerotized, without a fracture in middle (Fig. 39.17*e*); abdomen without gills. Case long, slender, of various materials. Widespread. (15 species) . . . . . . *Leptocella* Banks

6b Hind tibiae with a fracture near middle which appears to divide tibiae into 2 segments (Fig. 39.17*f*); abdomen with at least a few gills 7

7a (6) Hind tibiae with a regular fringe of long hair (Fig. 39.17*e*). Case elongate, made of spirally arranged bits of grass (Fig. 39.4*a*). Widespread. (20 species) . . . . *Triaenodes* McLachlan

7b Hind tibiae with only irregularly placed hairs (Fig. 39.17*f*). Case elongate, of various materials. (3 species) . . . *Mystacides* Berthold

Family **Molannidae**

Larvae make a sand case shaped like a water penny (Fig. 39.4*e*); this is very like the cases of a few species of *Athripsodes* (Leptoceridae). Molannidae live in lakes or streams, and are common in northern lakes. The genus *Molanna* Curtis (5 species) (Fig. 39.5*b*) is widespread. The larva of the Alaskan and Palearctic genus *Molannodes* McLachlan (1 species) is unknown.

Family **Beraeidae**

Larvae make a curved, tapering case of sand grains. Only one genus, *Beraea* Stephens (3 species), occurs in N. A. (Wiggins 1954).

References

Betten, Cornelius. 1934. The caddisflies or Trichoptera of New York State. *N. Y. State Museum Bull.*, No. 292:1–576. Betten, Cornelius and Martin E. Mosely. 1940. *The Francis Walker Types of Trichoptera in the British Museum.* London. Denning, D. G. 1956. Trichoptera. *In:* R. L. Usinger (ed.) *Aquatic Insects of California, with Keys to North American Genera and California Species*, pp. 237–270. University of California Press, Berkeley and Los Angeles. Ross, Herbert H. 1944. The Caddisflies or Trichoptera of Illinois. *Bull. Illinois Nat. Hist. Survey*, 23:1–326. 1956. *The Evolution and Classification of the Mountain Caddisflies.* University of Illinois Press, Urbana. Wiggins, Glenn B. 1954. The caddisfly genus *Beraea* in North America. *Contrib. Royal Ontario Mus. Zool. Pal.*, 39:1–18. 1956. The Kitagamiidae, a family of caddisflies new to North America. *Contrib. Royal Ontario Mus. Zool. Pal.*, 44:1–10.

40

Lepidoptera

PAUL S. WELCH

The Lepidoptera are primarily terrestrial, but scattered irregularly throughout the suborder Heterocera (moths) single species or small groups of species occupy water in one or more stages of the life cycle. So dispersed is this habit that in only a few instances have all of the species of any genus invaded the water. No definite dividing line between terrestrial and water-inhabiting Lepidoptera exists, since relation to water varies from the merely incidental to complete aquatic existence. Those species that make a major use of water as a normal environment and possess special adaptations for such a life may properly be called aquatic; others manifesting a less intimate but some definite relation to water are referred to as semiaquatic. Occupancy of water is almost wholly confined to the immature stages. In Europe, *Acentropus niveus*, now known to occur also in North America, produces different forms of the adult, one of which spends its entire life under water. Chief among the special adaptations for life in water are those concerned with respiration and locomotion. Representatives of certain genera are widely distributed over much of North America wherever environmental conditions are favorable; all are related to aquatic plants in one or more essential ways; and locally, on occasion, they may occur in such abundance that the food plants are devastated. Most species are confined largely to quiet waters.

The keys that follow include genera containing certain species which occur n the United States and Canada and which are sufficiently known to justify consideration here. Detailed information on the life cycles of some of these insects is scanty. Therefore, any keys to the larvae and pupae attempted at this time must be regarded as tentative.

The first key deals only with the full-grown larvae (last larval instar). Younger larvae which in some species differ, sometimes markedly, from their last larval instars, are mostly too little known to make their inclusion possible.

As known at present, the geographical distribution within the United States and Canada is, in general, as follows:

Bellura, Arzama, Nymphula, and *Cataclysta.* Widely, but often locally, over eastern half of U. S. and northeastern Canada. *Nymphula* and *Cataclysta* reported also from Calif.

Chilo. Eastern U. S.

Pyrausta. Eastern, central, and southern U. S.

Occidentalia. N. Y., Ill., Mich., and southward.

Schoenobius. Northern U. S. and eastern Canada.

Nepticula. Mich., N. Y.

Acentropus. Ontario, Quebec, N. Y., and Mass.

The classification and nomenclature used here follow McDunnough's Check List (1938, 1939). Recently, Lange (1956) published a revision of the Nymphulinae of North America in which certain generic changes are proposed.

Explanation of terms as used in this chapter:

ADFRONTAL SUTURES. The two outermost branches of the epicranial suture.

ANTENNAE. Elongated, slender appendages on ventral side of a pupa, each lying along ventral edge of wing (Fig. 40.3).

CREMASTER. Attachment mechanism on posterior end of pupa; variously constructed; often provided with hooks. Absent in some pupae.

HEAD EMARGINATION. V-shaped depression on dorsal region of head; at upper end of stem of epicranial suture; absent in some larvae.

EPICRANIAL SUTURE. Suture on front of head, shaped like an inverted Y; divides face into right and left halves.

FRONS. Region of head between the two arms of the epicranial suture.

MESONOTUM. Dorsal wall of second thoracic segment (mesothorax).

MESOTHORACIC SHIELD. Hardened (sclerotized) area on dorsal surface of mesothorax.

METALEG. Leg on third thoracic segment.

PROTHORACIC SHIELD. Hardened (sclerotized) area on dorsal surface of prothorax.

WING. Large, flat, somewhat triangular structure on each lateroventral side of pupa (Fig. 40.3).

KEY TO GENERA (LARVAE)

1a		Thoracic legs present; prolegs well developed or somewhat reduced	**2**
1b		Legless; on *Scirpus*, sometimes on *Eleocharis;* in serpentine mine; usually less than ½ of mine below water surface; length, about 7 mm.	***Nepticula*** von Heyden
2a	(1)	Filamentous lateral gills present	**3**
2b		Gills absent	**4**

3a (2) Gills branched, ranging up to about 400 gill filaments; body slightly flattened; mandibles small; larval case oval to oblong, cut from leaf of food plant, filled with water; spiracles on segments 1, 4, and 8 to 11 reduced; on *Nuphar, Potamogeton, Vallisneria,* or certain other aquatic plants; in quiet waters; length, about 15–20 mm . . . .
Nymphula (= Paraponyx) Schrank

3b Gills unbranched; about 120 gill filaments; mandibles large, protruding; body strongly depressed; spiracles uniform in size; under silken web on stones in rapid streams; length, 10–11 mm. (Fig. 40.1). . . . . *Cataclysta (=Elophila)* Hübner

4a (2) Dorsal half of last abdominal segment greatly depressed; next to last abdominal segment with 2 large spiracles on dorsal posterior margin 5

4b Last abdominal segment not depressed dorsally; spiracles absent on dorsal posterior margin of next to last abdominal segment 6

5a (4) Larva in long, water-filled burrow in petiole of *Nuphar,* sometimes in *Pontederia* or *Typha;* head light brown; prothoracic shield dark brown; body color deep gray to dark brown; length, 45–55 mm . . . *Bellura* Walker

5b Larva in long, partly water-filled burrow in stem of *Typha,* sometimes in *Pontederia;* head and prothoracic shield very dark brown; body color shiny muddy black; length, 50–60 mm. (Fig. 40.2) . . . *Arzama* Walker

6a (4) Larva in case made from part of food plant; burrowing habit absent 7

6b Larval case usually absent; larva burrowing in food plant 9

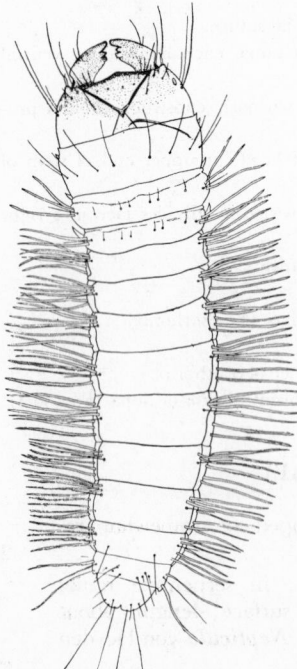

◄ **Fig. 40.1.** Dorsal view of full-grown larva of *Cataclysta (Elophila) fulicalis* Clemens. (After Lloyd.)

Fig. 40.2. Lateral view of full-grown larva of *Arzama obliqua* Walker. (After Claassen.)

7a (6) Spiracles not reduced, distinct. **8**

7b Spiracles reduced and inconspicuous; body slender, cylindrical; larva surrounded by dense silken web; on *Chara, Ceratophyllum, Potamogeton*, and certain other aquatic plants; length, about 12–14 mm. . . . . *Acentropus* Curtis

8a (7) Body stout, somewhat flattened; not moniliform; larval case subcircular to oblong, not filled with water; head darker than body; spiracles on abdominal segment 2 smaller than those on abdominal segment 3; on *Potamogeton, Lemna, Nuphar*, sometimes on certain other aquatic plants; length, about 13–18 mm . . . .
Nymphula (= *Hydrocampa*) Schrank

8b Body cylindrical, moniliform; larval case ovate; on *Lemna;* head yellowish, paler than grayish body; spiracles on abdominal segments 3 to 5 enlarged; length, about 12 mm . . . .
Cataclysta (= *Elophila*) Hübner

9a (6) Larva in short burrow in top of petiole of *Nelumbo*, sometimes in certain other aquatic plants; closing cap on top of burrow; adfrontal sutures extending to dorsal emargination of head; frons large, reaching beyond middle of head; length, about 25–35 mm . . . .
Pyrausta Schrank

9b Larva in long burrow in culm of food plant; no closing cap on top of burrow; adfrontal sutures extending to, or little beyond, ½ the distance to dorsal emargination of head; frons not reaching middle of head **10**

10a (9) Prothoracic shield fused anteriorly; mesothoracic shield absent; spiracles small; usually on *Eleocharis;* length, 18–24 mm. . . . .
Schoenobius Duponchel

10b Prothoracic shield divided; mesothoracic shield present; spiracles distinct **11**

11a (10) Larva usually in *Scirpus;* cuticula rough, covered with minute spines; mesothoracic shield less than ½ the width of mesonotum, setae absent; prothoracic spiracles larger than those on abdominal segments 1 to 7; length, about 23 mm . . . . *Chilo* Zincken

11b Larva in *Scirpus;* cuticula without minute spines; mesothoracic shield ½ as wide as mesonotum and bearing setae; prothoracic spiracles but slightly larger than those on abdominal segments 1 to 7; length about 23 mm . . . . *Occidentalia* Dyar and Heinrich

KEY TO GENERA (PUPAE)

1a Pupal case usually present and on outside of support **2**

1b Pupal case absent; pupa in burrow in plant **5**

2a (1) Pupal case of leaves or parts of leaves of plants; case lined with firm silken web; posterior end of abdomen without large hooks; spiracles on abdominal segments 2, 3, and 4 large, either equal or those on segment 2 smaller, on short fleshy tubercles, not tubular **3**

2b Pupal case of thick double layer of silk; 2 large oppositely directed hooks on posterior end of abdomen; spiracles on abdominal segments 3 and 4 large, tubular; spiracles greatly reduced on abdominal segments 5, 6, and 7; on rocks in rapid streams; length, 6–7 mm . . . .
Cataclysta (= *Elophila*) Hübner

3a   **(2)** Antennae short, extending only to or near level of abdominal segment 2; length, 6–7 mm . . . . (female) *Acentropus* Curtis

3b   Antennae long, extending to or near abdominal segment 5 . . . . . .   **4**

4a   **(3)** Wings medium size, extending over about ½ of ventral side; pupal case usually made from leaves of *Nuphar*, *Potamogeton*, or certain other aquatic plants; length, 6–15 mm. (Fig. 40.3) . . . .
Nymphula (= *Paraponyx* and *Hydrocampa*) Schrank
An exception occurs in *Nymphula serralineatis* Barnes and Benjamin in which pupation takes place in pupal chambers in petioles of white water lilies.

4b   Wings large, extending over ⅔ to ¾ of ventral side; length, about 6 mm . . . . (male) *Acentropus* Curtis

5a   **(1)** Posterior portions of wings contiguous at mid-ventral line (Fig. 40.4); legs short **6**

5b   Posterior portions of wings not contiguous at mid-ventral line (Fig. 40.3); legs long, extending between tips of wings. **7**

6a   **(5)** Intersegmental grooves of abdomen deeply impressed; each mid-abdominal segment crossed at middle by transverse, finely toothed ridge; tip of abdomen with 2 pairs of stiff, curved spines; in *Nuphar*, sometimes in *Pontederia* or *Typha*; length, about 25–35 mm. (Fig. 40.4) . . . . *Bellura* Walker

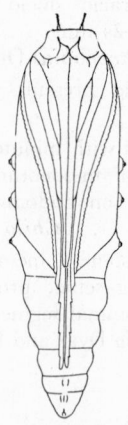

Fig. 40.3. Ventral view of pupa of *Nymphula maculalis* Clemens. Posterior tips of wings not contiguous; separated by appendages. Metalegs long, extending to posterior region of body. Antennae lie along ventral edges of wings, terminating at wing tips. (After Welch.)

Fig. 40.4. Ventral view of pupa of *Bellura gortynoides* Walker. Posterior tips of wings contiguous; appendages confined to anterior region of body. (After Robertson-Miller.)

6b   Intersegmental grooves of abdomen shallow; each mid-abdominal segment crossed near posterior margin by transverse row of tubercles; tip of abdomen with 4 short, equal setae; in *Typha*, sometimes in *Pontederia;* length, 28–35 mm . . . . *Arzama* Walker

7a (5) Cremaster stout, broader than long; antennae long, extending to posterior tips of wings; in *Nelumbo, Polygonum*, or certain other plants; length, about 12–16 mm . . . . ***Pyrausta*** Schrank

7b Cremaster absent; antennae not reaching posterior tips of wings . . . 8

8a (7) Metalegs long, extending almost to or slightly beyond caudal end of abdomen; body slender, elongate; in *Eleocharis;* length, about 13 mm . . . . ***Schoenobius*** Duponchel

8b Metalegs extending little beyond tips of wings 9

9a (8) Dorsal furrow between abdominal segments 9 and 10 absent; tip of abdomen rounded, smooth; in *Scirpus;* length, 11–18 mm . . . .
Occidentalia Dyar and Heinrich

9b Dorsal furrow between abdominal segments 9 and 10 present; tip of abdomen blunt, faintly toothed; in *Scirpus;* length, 12–20 mm . . . .
Chilo Zincken

References

Ainslie, G. G. and W. B. Cartwright. 1922. Biology of the Lotus Borer (Pyrausta penitalis Grote). *U. S. Dept. Agr. Bull.* No. 1076. **Berg, C. O. 1950.** Biology of certain aquatic caterpillars (Pyralididae:Nymphula spp.) which feed on *Potamogeton. Trans. Am. Microscop. Soc.*, 69:254–266. **Berg, Kaj. 1941.** Contributions to the biology of the aquatic moth *Acentropus niveus* (Oliv.). *Vidensk. Medd. Dansk Naturhist. Foren.*, 105:59–139. **Claasen, P. W. 1921.** Typha Insects: Their biological relationships. *Cornell Univ. Agr. Exp. Sta., Mem.*, 47:463–531. **Flint, W. P. and J. H. Malloch. 1920.** The European corn-borer and some similar native insects. *Bull. Illinois Nat. Hist. Survey*, 13:287–305. **Forbes, W. T. M. 1910.** The aquatic caterpillars of Lake Quinsigmond. *Psyche*, 17:219–228. **1923.** The Lepidoptera of New York and neighboring states. *Cornell Univ. Agr. Exp. Sta., Mem.*, 68:1–729. **Frohne, W. C. 1939a.** Biology of *Chilo forbesellus* Fernald, an hygrophilous crambine moth. *Trans. Am. Microscop. Soc.*, 58:304–326. **1939b.** Observations on the biology of three semiaquatic lacustrine moths. *Trans. Am. Microscop. Soc.*, 58:327–348. **Hart, C. A. 1895.** On the entomology of the Illinois River and adjacent waters. *Bull. Ill. State Lab. Nat. Hist.*, 4:149–273. **Judd, W. W. 1950.** *Acentropus niveus* (Oliv.) (Lepidoptera:Pyralidae) on the north shore of Lake Erie with a consideration of its distribution in North America. *Can. Entomologist*, 82:250–252. **1953.** A study of the population of insects emerging as adults from the Dundas Marsh, Hamilton, Ontario, during 1948. *Am. Midland Naturalist*, 49:801–824. **Lange, W. H. 1956a.** A generic revision of the aquatic moths of North America (Lepidoptera: Pyralidae, Nymphulinae). *Wasmann J. Biol.*, 14:59–144. **1956b.** Aquatic Lepidoptera. In: R. L. Usinger (ed.). *Aquatic Insects of California, with Keys to North American Genera and California Species*, pp. 271–288. University of California Press, Berkeley and Los Angeles. **Lloyd, J. T. 1914.** Lepidopterous larvae from rapid streams. *J. N. Y. Entomol. Soc.*, 22:147–152. **McDunnough, J. 1938.** Check list of the Lepidoptera of Canada and the United States of America. Part I. Macrolepidoptera. *Mem. So. Calif. Acad. Sci.*, I:1–272. **1939.** Check list of Lepidoptera of Canada and the United States of America. Part II. Microlepidoptera. *Mem. So. Calif. Acad. Sci.*, 2:1–171. **McGaha, Y. J. 1952.** The limnological relations of insects to certain aquatic flowering plants. *Trans. Am. Microscop. Soc.*, 71:355–381. **1954.** Contribution to the biology of some Lepidoptera which feed on certain aquatic flowering plants. *Trans. Am. Microscop. Soc.*, 73:167–177. **Poos, F. W. 1927.** Biology of the European corn borer (*Pyrausta nubilalis* Hübn.) and two closely related species in northern Ohio. *Ohio J. Sci.*, 27:47–94. **Robertson-Miller, Ellen. 1923.** Observation on the Bellura. *Ann. Entomol. Soc. Am.*, 16:374–383. **Welch, P. S. 1914.** Habits of the larva of *Bellura melanopyga* Grote (Lepidoptera). *Biol. Bull.*, 27:97–114. **1916.** Contribution to the biology of certain aquatic Lepidoptera. *Ann. Entomol. Soc. Am.*, 9:159–190. **1919.** The aquatic adaptations of *Pyrausta penitalis* Grt. (Lepidoptera). *Ann. Entomol. Soc. Am.*, 12:213–226. **1922.** The respiratory mechanism in certain aquatic Lepidoptera. *Trans. Am. Microscop. Soc.*, 41:29–50. **1924.** Observations on the early larval activities of *Nymphula maculalis* Clemens (Lepidoptera). *Ann. Entomol. Soc. Am.*, 17:395–402. **Welch, P. S. and G. L. Sehon. 1928.** The periodic vibratory movements of the larva of *Nymphula maculalis* Clemens (Lepidoptera) and their respiratory significance. *Ann. Entomol. Soc. Am.*, 21:243–258.

41

Diptera

MAURICE T. JAMES

This order, one of the four largest of the Insecta, contains many members that in the larval stage have become adapted to fresh water and a great many more that occupy habitats that may be considered semiaquatic. A few are marine; others occupy a wide variety of habitats, many of them being terrestrial and many others being scavengers or parasites on insects and other animals. A problem arises, consequently, as to where to draw the line between those which are to be included in this work and those which are to be excluded. In some cases the decision is necessarily more or less arbitrary, but in general, all truly aquatic forms, semiaquatic forms which are likely to be encountered under aquatic conditions, parasites of aquatic insects and snails, and scavengers that might be encountered in the water, are included. Families like the Tipulidae, in which, even within a single genus, habitats may range from terrestrial to semiaquatic or even aquatic, pose an especially difficult problem. Our incomplete knowledge of many groups, especially the Muscoidea, further complicates the problems of what forms to include and presenting satisfactory distinguishing characters.

Adults are never truly aquatic and, consequently, are omitted from consideration here; for determination of North American forms the student is referred to Curran (1934), and to such special monographic works as Matheson

(1944), Stone (1938), Brennan (1935), Alexander (1919, 1942), Townes (1945) and Malloch (1915). Especially valuable larval treatments are found in Matheson (1944), Carpenter and LaCasse (1955), Alexander (1920), Malloch (1917), Johannsen (1934–1937) and Hennig (1948–1950).

It is difficult to say just what is a "typical" dipterous larva, since so much variation occurs in the order. All are legless, however, though pseudopods may be developed on the thorax or abdomen, or, when not developed, creeping welts or at least spinulose areas or rings may occur. The term "pseudopod" seems preferable to "proleg," since the dipterous pseudopod is structurally quite different and probably not homologous with the lepidopterous or hymenopterous prolegs.

The head capsule is more or less well developed in the lower Diptera; it may be quite prominent, strongly sclerotized, and nonretractile, or, on the other hand, it may be feebly developed and almost wholly retractile into the thorax. In the higher Diptera, the head capsule is not developed, the larvae, from external appearances, being headless. In some cases, for example some Tipulidae, larvae with head capsules may at first examination appear headless, though the parts of the head capsule, including the opposed (rather than parallel) mouthparts, may be determined by forcing the head capsule out or by dissection of the anterior part of the head. In the higher Diptera, the opposed mandibles are replaced, functionally though not morphologically, by the parallel mouth hooks or, occasionally, by a single unpaired hook.

Respiratory structures vary. Some aquatic larvae are provided with blood gills, usually on the ventral side of the body. A few larvae, the "bloodworms" (some Tendipedidae) contain haemoglobin in their blood. Larvae breathing through spiracles may have these structures at the posterior end of the body (metapneustic), at the anterior and posterior ends (amphipneustic), or laterally on most segments (holopneustic), and some pupae may breathe through anterior spiracles only (propneustic). Many aquatic forms are metapneustic as larvae, the posterior spiracles often being situated at the ends of tubes which, in some forms, are greatly elongated. The posterior spiracular disc, when sessile, with its surrounding protuberances, often furnishes important taxonomic characters.

An unfortunate duality of generic and family nomenclature in the Diptera has arisen because of the publication of two sets of names by Meigen in 1800 and 1803. The controversy cannot be discussed here, but in accordance with the ruling of the International Committee on Zoological Nomenclature the 1800 names are used here, the 1803 equivalents being given in synonymy. There is at present a move under way to establish the 1803 names as *nomina conservanda*, in which case they would have to be used in place of the 1800 names.

All the new figures in this chapter were drawn by Joan Laval and the author; these illustrations are identified by the initials J. L. and M. T. S., respectively.

KEY TO GENERA (LARVAE)

Pupal characters included in this key are intended merely to aid the user in his determination of material in which larvae and pupae or pupal cases are associated, or in making the association of such material; the key will not work satisfactorily to determine pupae.

1a   Mouthparts vestigial or very indistinct; head capsule poorly formed, especially posteriorly; body behind head apparently 13-segmented, the second apparent segment, in the last instar, bearing a ventral sclerotized spatula-shaped plate, the "breastbone." In galls of aquatic plants . . . . (in part) Suborder **Nematocera** Family **Itonididae** (=**Cecidomyiidae**)

1b   Mouthparts in aquatic forms well developed; body behind head composed of fewer than 13 segments (except that tegumentary folds or intercalary segments may make the number appear greater); no "breastbone" present **2**

2a   (1)   Mandibles and other sclerotized mouthparts opposed, moving in a horizontal plane. Pupa free (except in the Scatopsidae); prothoracic spiracles usually located at the ends of a pair of processes; antennal sacs elongated, extending over the compound eyes, thence to or beyond the bases of the wing sheaths . . . . (major part) Suborder **Nematocera**   **3**

2b   Mandibles replaced by mouth hooks which either lie parallel to each other, are never opposed, and move in a vertical plane, or which are fused into a single structure. Pupa either remaining in the last larval skin, or free; if free, the prothoracic respiratory organs are rudimentary or lacking and the antennal sacs are short, directed caudad and laterad, and not over the eyes **72**

3a   (2)   Body flattened, the abdominal segments with ventral suckers serially arranged either in a median row of 6 or in 2 rows at the apices of lateral prominences. Pupa free, strongly flattened ventrally, the legs lying alongside one another and far exceeding the wing tips . . .   **4**

3b   Body either not flattened and with serially arranged ventral suckers, or if body flattened (*Maruina*), the suckers are 8 in number. Pupa either not flattened or with the legs scarcely exceeding the wing tips.   **8**

4a   (3)   Thorax distinct from head and first abdominal segment; ventral suckers in 2 series, at apices of pseudopods. Pupal respiratory organ terminating in 3 or 4 filamentous processes . . . . Family **Deuterophlebiidae** ***Deuterophlebia*** Edwards

Two species: *D. coloradensis* Pennak and *D. shasta* Wirth.

4b   Thorax fused with head and first abdominal segment, the abdominal segments beyond the seventh likewise a composite; the body, therefore, apparently 7-segmented; first 6 abdominal segments each with a prominent unpaired ventral sucker. Pupa free; each respiratory organ consisting of 4 simple lamellae. . . . . Family **Blepharoceridae**   **5**

5a   (4)   Intercalary segments absent. . (tropical) Subfamily **Paltostomatinae** ***Paltostoma*** Schiner

5b   Intercalary segments present between the abdominal segments . . . . Subfamily **Blepharocerinae**   **6**

6a   (5)   Lateral body processes absent (4 species) . . ***Blepharocera*** Macquart

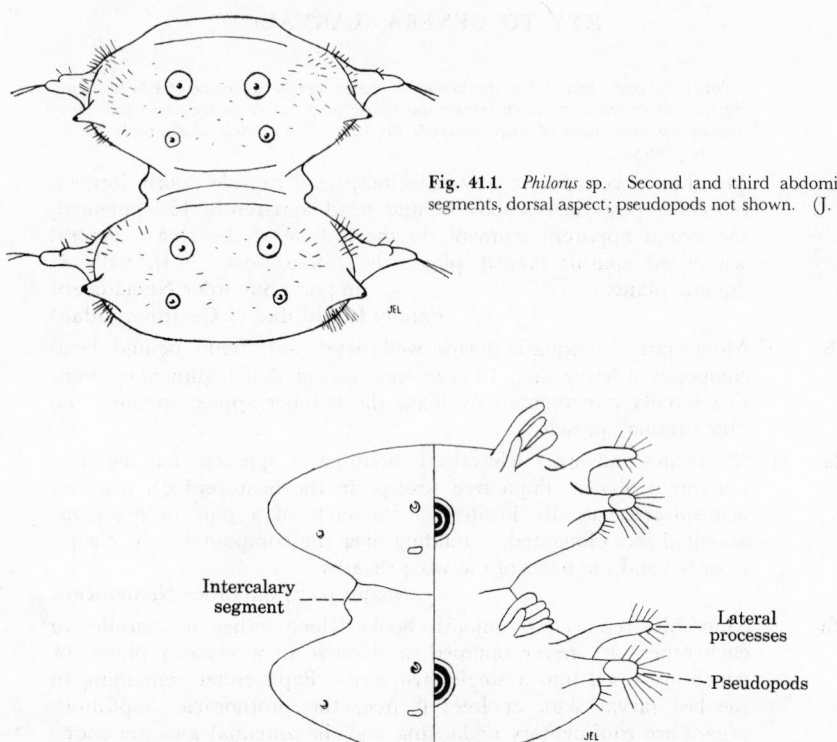

Fig. 41.1. *Philorus* sp. Second and third abdominal segments, dorsal aspect; pseudopods not shown. (J. L.)

Fig. 41.2. *Bibiocephala* sp. Second and third abdominal segments: right, ventral aspect, showing ventral suckers; left, outline of dorsal aspect, showing rudimentary spines. (J. L.)

6b   Lateral body processes present **7**

7a **(6)** First 7 abdominal segments each with 6 stout thorns or spines dorsally, 2 of these located on each side of the median line and 1 above each pseudopod and lateral process (5 species.) (Fig. 41.1.) . . . . . .
                                                          ***Philorus* Kellogg**

7b   Abdomen with at most traces of such spines (Several species.) (Fig. 41.2). . . . . ***Bibiocephala* Osten Sacken, *Agathon* Roeder**

8a **(3)** Head capsule incomplete, nonsclerotized posteriorly and often also ventrally, retractile into the prothorax. Pupa not flattened; legs extending far beyond wing tips; respiratory tubes never exceding the body in length . . . . Family **Tipulidae**   **9**
           Most tipulid larvae are terrestrial or semiaquatic; only those likely to be met in aquatic situations and the more common semiaquatic groups are keyed here. It has been impossible to key the aquatic genera *Cryptolabis* and *Lipsothrix*, neither of which is common.

8b   Head capsule complete or at least well developed posteriorly, not divided into lobes or rods, not retractile **19**

9a **(8)** Body provided with numerous long filiform (*Phalacrocera* Schiner) or leaflike (*Triogma*) processes. Pupa with similar processes, at least on the abdomen. (Several species) . . . .
                                                Subfamily **Cylindrotominae**
                                     ***Phalacrocera* Schiner, *Triogma* Schiner**

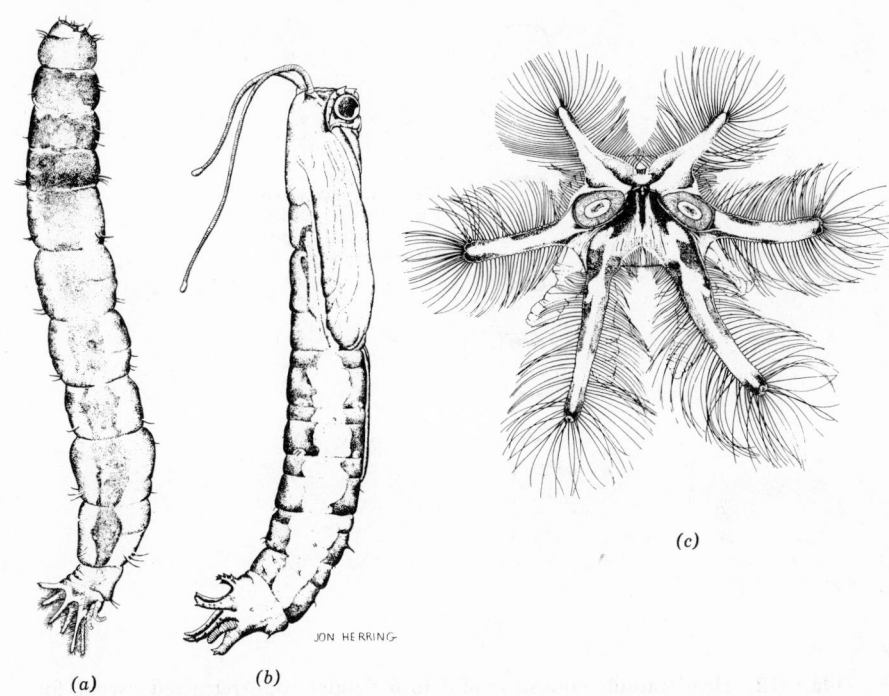

JON HERRING

(a) *(b)* *(c)*

Fig. 41.3. *Megistocera longipennis* (Macquart). (*a*) Partly extended fully grown larva, side view. (*b*) Female pupa, side view. (*c*) Caudal disc. (After Rogers.)

9b	Body without filiform or leaflike processes	**10**
10a	(**9**) Spiracular disc surrounded by 6 (rarely 8) lobes; head capsule broad and massive. Pupa with maxillary palpus usually recurved backwards or inwards, or if straight (*Longurio* Loew) the pupa is 30 mm. or more in length . . . . Subfamily **Tipulinae**	**11**
10b	Spiracular disc surrounded by at most 5 lobes; head capsule slender. Pupa with maxillary palpus straight . . . . . Subfamily **Limoniinae**	**12**

11a (**10**) Anal gills pinnately branched. (3 species). . . . .
                                     Longurio Loew (=***Aeshnasoma*** Johnson)

11b Anal gills not pinnately branched. (Fig. 41.3) . . . .
                             Megistocera Wiedemann, ***Tipula*** Linnaeus

    One species of the essentially tropical *Megistocera*, **M. longipennis** (Macquart), occurs in the neuston fauna of Fla.; about 30 species of the large genus *Tipula* are aquatic or semiaquatic.

12a (**10**) Spiracular disc with the 2 ventral lobes elongated, the others reduced or vestigial **13**

12b Spiracular disc with either more than the 2 ventral lobes developed, or with all the lobes reduced **14**

13a (**12**) Spiracles lacking or vestigal. Larva and pupa enclosed in a silken case; pupal respiratory tube 6- to 8-branched. (About 7 species.) (Fig. 41.4) . . . . ***Antocha*** Osten Sacken

13b Spiracles well developed. Larva and pupa naked; pupal respiratory tube simple. (About 30 species each) . . . .
                             Pedicia Latreille, ***Dicranota*** Zetterstedt

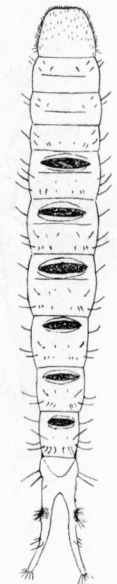

Fig. 41.4. *Antocha saxicola* Osten Sacken. Larva, dorsal aspect. (After Alexander.)

14a (12) Head capsule consisting of 4 to 6 slender rods, retracted except for the tips of the greatly produced maxillary blades, which usually protrude at least a short distance **15**

14b Head capsule usually massive, or, if reduced to slender rods and wholly retractile, the maxillae do not bear elongated, protruding blades. **16**

15a (14) Mentum in the form of a sclerotized transverse bar. (More than 40 species.) (Fig. 41.5). . . . . *Limnophila* Macquart

15b Mentum not sclerotized. (About 15 species). . . . .
Hexatoma Latreille (including *Eriocera* Macquart)

16a (14) Abdominal segments without distinct creeping welts. All semi-aquatic Eriopterini. . . . .
Erioptera Meigen, *Molophilus* Curtis, *Trimicra* Osten Sacken, *Gonomyia* Meigen, etc.

16b Abdominal segments with basal creeping welts. **17**

17a (16) Body depressed. (Several species) . . . .
Dactylolabis Osten Sacken, *Elliptera* Schiner

17b Body cylindrical **18**

18a (17) Abdomen with ventral segmental welts only. (2 species) . . . .
Helius St. Fargeau and Serville

18b Abdomen with both dorsal and ventral segmental welts. (About 100 species, many aquatic (mostly fauna hydropetrica.) . . . .
Limonia Meigen (including *Geranomyia* Haliday)

19a (8) Pseudopods lacking **20**

19b Pseudopods present, either at one or both ends of the body, or on the intermediate segments; rarely (*Dasyhelia*) partially or wholly withdrawn and evident only as terminal hooks on the last segment.   **40**

20a (19) Thoracic segments fused into an enlarged complex which is dis-

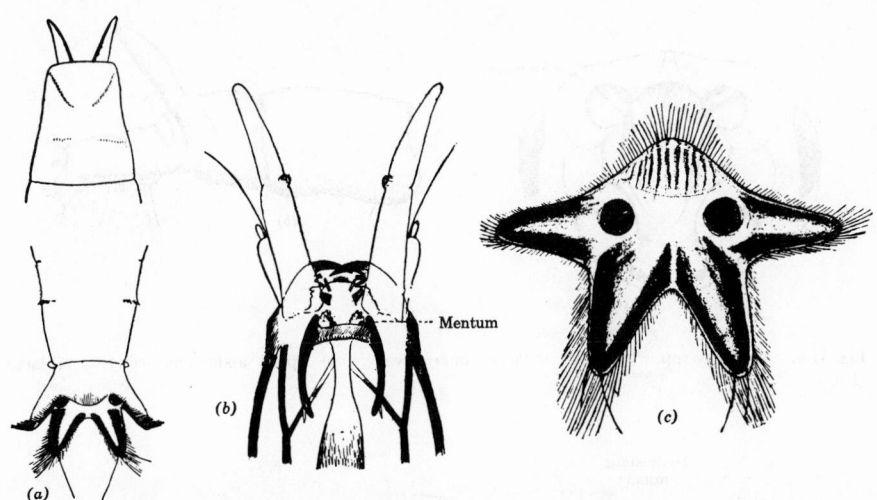

Fig. 41.5. *Limnophila fuscovaria* Osten Sacken. (*a*) Larva, anterior and posterior ends, dorsal. (*b*) Larval head capsule, ventral. (*c*) Spiracular disc of larva. (After Alexander.)

tinctly broader than the abdomen; metapneustic, the spiracles either sessile or at the end of a long or short respiratory siphon. Pupa active; when at rest the abdomen curved under the bulky cephalothorax, terminating in 2 or 4 paddlelike plates . . . .
Family **Culicidae**   21

20b   Thoracic segments usually distinct from one another, not noticeably broader than the abdomen   32

21a (**20**) Antenna developed into a prehensile organ with long, strong apical spines. Pupa either with the swimming paddles fused basally, or the respiratory horn closed or almost so at the tip . . . .
Subfamily **Chaoborinae**   22

21b   Antenna not prehensile and lacking strong apical spines. Pupa with swimming paddles free and with the respiratory horn open at the tip . . . . Subfamily **Culicinae**   25

22a (**21**) Eighth abdominal segment with a respiratory siphon which is much longer than broad   23

22b   Eighth abdominal segment without an elongated respiratory siphon, the spiracles either sessile or located at the apex of a short tube that is no longer than it is wide. Pupa with paddles movable   24

23a (**22**) Antennae inserted close together; anal brush not developed. Pupa with swimming paddles not movable. (2 species) . . . .
Corethrella Coquillett

23b   Antennae inserted far apart; anal brush well developed. Pupa with swimming paddles movable. (3 species.) (Fig. 41.6) . . . . . .
Mochlonyx Loew

24a (**22**) Hydrostatic organs present in thorax and seventh abdominal segment. Pupal prothoracic horn with an apical, almost closed spiracle. (About 8 species.) (Fig. 41.7) . . . . ***Chaoborus*** Lichtenstein

24b   Hydrostatic organs absent. Pupal prothoracic horn with an open spiracle near its middle. . . . . ***Eucorethra*** Underwood
One species, ***E. underwoodi*** Underwood.

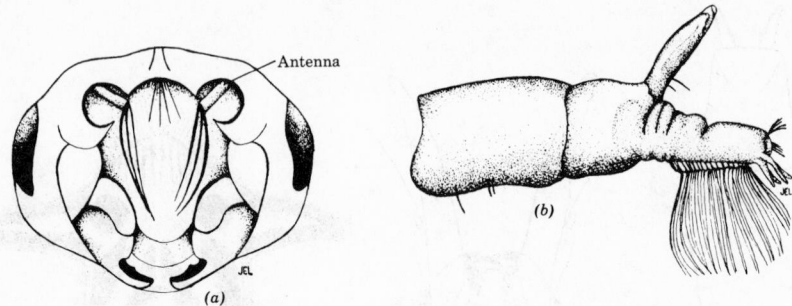

Fig. 41.6. *Mochlonyx* spp. (*a*) Head of larva, anterior view. (*b*) Apical abdominal segments of larva. (J. L.)

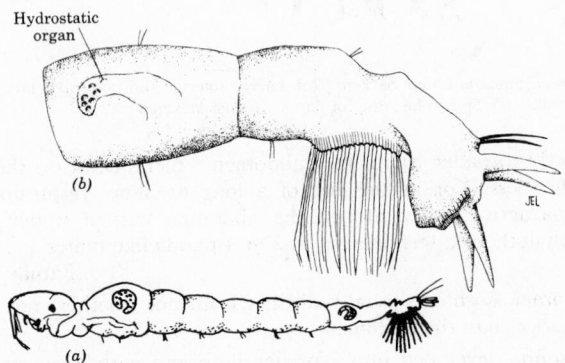

Fig. 41.7. *Chaoborus* sp. (*a*) Lateral view of entire larva. (*b*) Apical abdominal segments of larva, (*a* after unpublished drawing by R. A. Main., *b* J. L.)

25a (21) Eighth abdominal segment without a respiratory siphon. Pupa with lateral apical hairs of the abdominal segments placed almost exactly at the corners. (About 12 species.) (Fig. 41.8) . . . .
Anopheles Meigen

25b Eighth abdominal segment with a respiratory siphon which is at least as long as broad, usually much longer. Pupa with lateral apical hairs of the abdominal segments placed well before the corners 26

26a (25) Mouth brushes prehensile, each composed of 10 stout rods. Pupa with outer parts of paddle produced beyond tip of midrib. (2 species) . . . . *Toxohynchites* Theobald

26b Mouth brushes rarely prehensile, composed of 30 or more hairs. Pupa with outer parts of paddle not produced beyond tip of midrib. 27

27a (26) Respiratory siphon without a pecten. (Fig. 41.9) . . . .
Mansonia Blanchard, *Wyeomyia* Theobald, *Orthopodomyia* Theobald

Three Neotropical genera, poorly represented in the Nearctic region. Of these, *Mansonia* (2 species) has the respiratory siphon modified for piercing stems and roots of aquatic plants, *Wyeomyia* (3 species) has only a pair of ventral hairs instead of the anal brush, and *Orthopodomyia* (2 species) shows neither of these peculiarities.

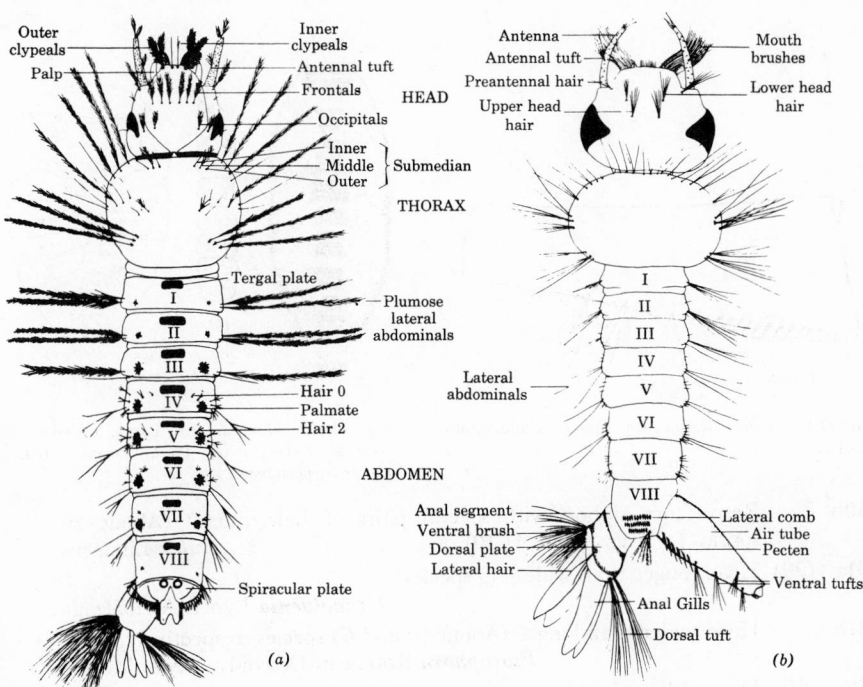

Fig. 41.8. Mosquito larvae. (a) Anopheles. (b) Culex. (U. S. Public Health Service.)

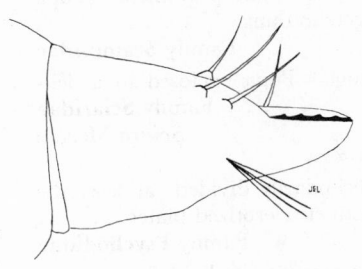

Fig. 41.9. *Mansonia*. Respiratory siphon, semidiagrammatic. (J. L.)

27b	Respiratory siphon with a pecten	28
28a	(27)	Head with a prominent triangular pouch on each side. (2 species). **Deinocerites** Theobald
28b		Head without prominent lateral pouches	29
29a	(28)	Respiratory siphon either with several pairs of ventral hair tufts or with a single pair located near its base	30
29b		Respiratory siphon with 1 pair of ventral hair tufts located near its middle, or the tufts vestigial or absent	31
30a	(29)	Respiratory siphon greatly elongated, with a single pair of basal tufts, which may, however, be followed by a median row of unpaired smaller tufts which extend to the apex of the siphon. (7 species). . . . . **Culiseta** Felt (= **Theobaldia** Neveu-Lemaire)

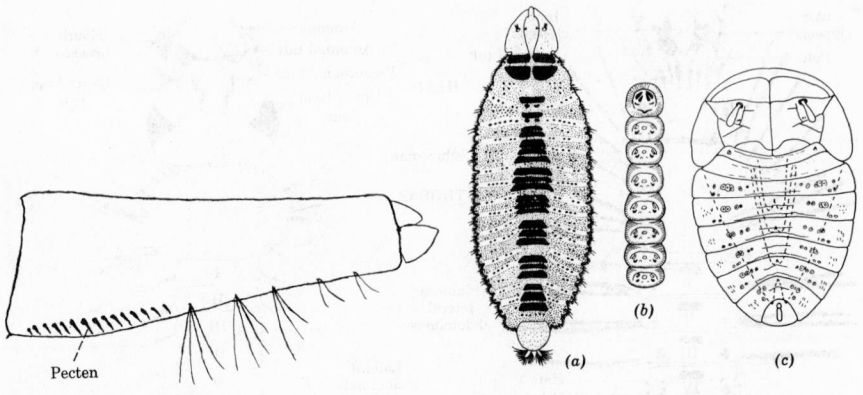

Fig. 41.10. *Culex.* Respiratory siphon, semidiagrammatic.

Fig. 41.11. *Maruina.* (*a*) Larva, dorsal. (*b*) Ventral suckers. (*c*) Pupa, dorsal. (After Quate by permission.)

30b Respiratory siphon with several pairs of hair tufts. (About 20 species.) (Figs. 41.8*b*, 41.10). . . . . ***Culex*** Linnaeus

31a (29) Head longer than wide. (3 species) . . . .
Uranotaenia Lynch-Arribálzaga

31b Head wider than long. (About 10 and 60 species, respectively.) . . .
Psorophora Robineau-Descoidy, ***Aedes*** Meigen

32a (20) Intermediate body segments provided with spiracles; larvae without truly aquatic adaptations, and rarely occurring under aquatic conditions **33**

32b Intermediate body segments without spiracles **34**

33a (32) Body covered with short, coarse hairs; antennae prominent. Pupa in last larval skin. Normally scavengers in dung . . . .
Family **Scatopsidae**

33b Body glabrous; antennae inconspicuous. Pupa enclosed in a delicate cocoon . . . . Family **Sciaridae**
Sciara Meigen
S. macfarlanei Jones breeds in pitchers of *Sarracenia.*

34a (32) Thoracic and abdominal segments secondarily divided, at least the terminal abdominal segments with distinct sclerotized plates . . . . .
Family **Psychodidae** **35**

34b Thoracic and abdominal segments not secondarily divided . . . . . .
(in part) Family **Heleidae** (=**Ceratopogonidae**) **37**

35a (34) Body flattened, depressed, with a series of 8 ventral sucker discs. Pupa oval, flattened. (Fig. 41.11) . . . . ***Maruina*** Müller
One species, *M. lanceolata* (Kincaid).

35b Body more or less cylindrical, without sucker discs. Pupa not flattened **36**

36a (35) Anal region with an unpaired preanal plate and with paired adanal and prosternal plates; dorsal plates 26 in number. (Many species.) (Fig. 41.12). . . . . ***Pericoma*** Walker, ***Telmatoscopus*** Eaton

36b No preanal or prosternal plates, the adanal region with a transverse unpaired plate only; dorsal plates usually less than 26, sometimes as few as 6. (Many species.) (Fig. 41.13) . . . .
Psychoda Latreille

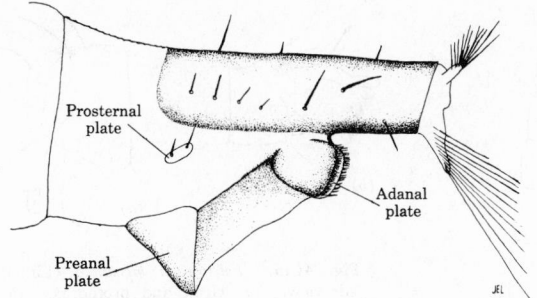

Fig. 41.12. *Pericoma.* Apex of abdomen, semidiagrammatic. (J. L.)

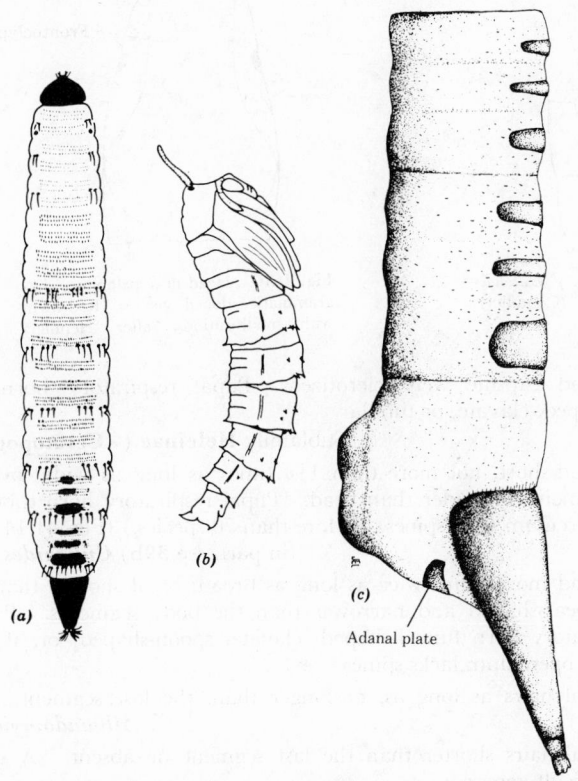

Fig. 41.13. *Psychoda* sp. (*a*) Larva, dorsal. (*b*) Pupa, lateral. (*c*) Terminal abdominal segments, (*a*, *b* after Quate.)

37a (34) Head capsule not sclerotized, but provided with an internal system of sclerotized rods and levers. Pupal respiratory horn oval or barrel-shaped with about 10 spiracles . . Subfamily **Leptoconopinae** (About 4 species). . . . . **Leptoconops** Skuse

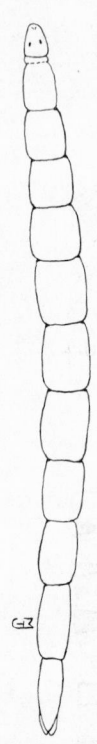

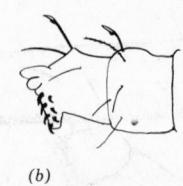

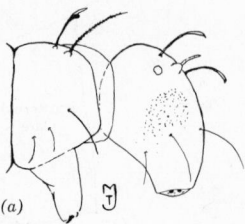

Fig. 41.15. *Forcipomyia bipunctata* (Linnaeus). Larva, side view. (*a*) Head and prothorax. (*b*) Last 2 segments of the abdomen. (M. T. J.)

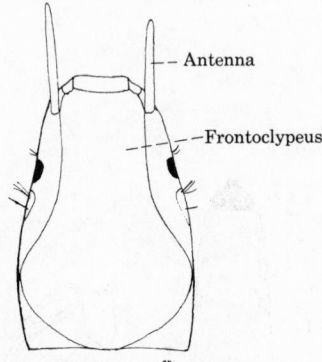

Fig. 41.14. *Culicoides variipennis.* (Coquillet). (M. T. J.)

Fig. 41.16. Head of a pelopiine, diagrammatic, dorsal view. (Redrawn, with modifications, after Hennig.)

37b		Head capsule well sclerotized. Pupal respiratory horn funnel-shaped, clavate, or tubular . . . .
		Subfamily **Heleinae** (=**Ceratopogoninae**) 38
38a	(37)	Head short, not more than 1½ times as long as wide, oval; body segments no wider than head. Pupal respiratory horn tubular, the operculum with spines. (More than 70 species.) (Fig. 41.14). . . . .
		(in part, see **39b**) *Culicoides* Latreille
38b		Head more than twice as long as broad, or, if shorter than this, it is pear-shaped and narrower than the body segments. Pupal respiratory horn funnel-shaped, clavate, spoon-shaped, or, if tubular, the operculum lacks spines 39
39a	(38)	Anal hairs as long as, or longer than, the last segment. (7 species). . . . . *Alluaudomyia* Kieffer
39b		Anal hairs shorter than the last segment or absent. A group of difficult genera. . . . .
		Palpomyia Meigen, *Bezzia* Kieffer, *Probezzia* Kieffer, *Johannsenomyia* Malloch, *Sphaeromias* Curtis, (in part, see **38a**) *Culicoides* Latreille
40a	(19)	Pseudopods present on the anal and usually on the prothoracic segment, otherwise lacking 41
40b		Pseudopods present on the intermediate body segments, or confined to the prothorax 66

41a (**40**) Pseudopods unpaired **42**

41b Pseudopods paired **45**

42a (**41**) Amphipneustic, a pair of respiratory tubes present on the prothorax and a single respiratory opening posteriorly . . . .
Family **Thaumaleidae** (=**Orphnephilidae**)
Thaumalea Ruthé

42b Neither prothoracic nor caudal spiracles developed . . . .
(in part) Family **Heleidae** (=**Ceratopogonidae**) **43**

43a (**42**) Last abdominal segment with retractile pseudopod bearing 10 to 12 hooks; prothoracic pseudopod lacking. Pupa free from larval exuviae; respiratory horn elongated . . . . Subfamily **Dasyheliinae** (About 25 species). . . . . *Dasyhelia* Kieffer

43b Prothoracic pseudopod present. Pupa with larval exuviae attached to the last 3 segments; respiratory horn short, knoblike . . . .
Subfamily **Forcipomyiinae** **44**

44a (**43**) Body flattened, oval in cross section; lateral body processes at least as long as the segments. Pupa: abdomen with branched or setaceous processes on the first 5 segments. (About 15 species) . . . . . .
Atrichopogon Kieffer

44b Body circular in cross section, or, if flattened, segments with processes less than ½ their length. Pupa: abdomen with spines or stumplike processes on all but last segment. (About 40 species.) (Fig. 41.15) . . . . *Forcipomyia* Meigen

45a (**41**) Caudal end of body with 6 long filaments and 2 pseudopods . . . . .
Family **Tanyderidae**
Protoplasa Osten Sacken

A rare species, *P. fitchii* Osten Sacken.

45b Caudal end of body lacking such filaments. . . . .
Family **Tendipedidae** (=**Chironomidae**) **46**

46a (**45**) Antenna retractile, usually elongated; frontoclypeus almost as broad as head, rounded behind; at least the anal pseudopods elongated, stiltlike. (Fig. 41.16) . . . Subfamily **Pelopiinae** (=**Tanypodinae**) **47**

46b Antenna not retractile; frontoclypeus narrowed behind **51**

47a (**46**) Three pairs of anal gills . . . . *Pelopia* Meigen (=*Tanypus* Meigen)

47b Two pairs of anal gills. **48**

48a (**47**) Both pairs of anal gills close to the anal opening; body without lateral hair fringes, with only scattered bristles. (Fig. 41.17) . . . . .
Pentaneura Philippi

48b Ventral pair of anal gills remote from anal opening, located on the anal pseudopods; body with a fringe of hair on each side **49**

49a (**48**) Labium with a paralabial comb; antenna ¼ to ⅓ as long as the head . . . .
Procladius Skuse, *Psilotanypus* Kieffer, *Anatopynia* Johannsen

49b Labium without a paralabial comb, with only separate teeth in a row on each side; antenna ½ to ¾ as long as the head **50**

50a (**49**) Antenna about ¾ head length; mandible hooklike . . . .
Clinotanypus Kieffer

50b Antenna about ½ head length; mandible gently curved. . . . .
Coelotanypus Kieffer

51a (**46**) Tormae ("premandibles") rudimentary . . Subfamily **Podonominae**
Podonomus Philippi

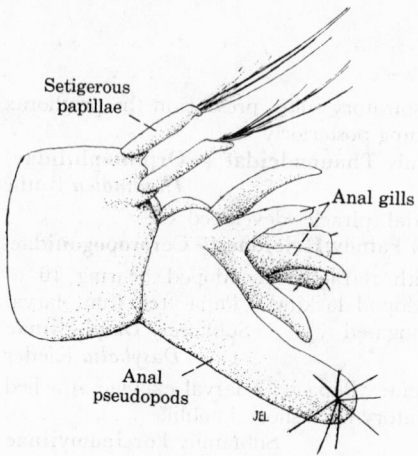

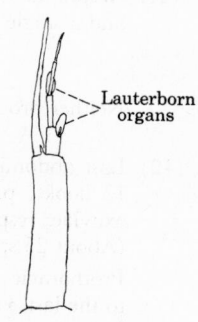

Fig. 41.17. *Pentaneura*, anal segment, dorsolateral view. (J. L.)

Fig. 41.18. *Microtendipes pedellus aberrans* (Johannsen). Antenna. (After Johannsen.)

51b   Tormae well developed **52**

52a **(51)** Paralabial plates present and radially striated; third antennal segment not annulated . . . .
                        Subfamily **Tendipedinae** (=**Chironominae**)   **53**

52b   Paralabial plates usually absent; if present they may be bearded, but never radially striated . . . .
                        Subfamily **Hydrobaeninae** (=**Orthocladiinae**)   **57**

53a **(52)** Antenna elongated, mounted on a prominent tubercle; abdominal segments 2 to 6 each with a bifid plumose bristle at each latero-posterior angle . . . . Tribe **Calopsectrini** (=**Tanytarsini**)
                        Calopsectra sens. lat. (=***Tanytarsus*** Auctt., non Wulp)
        Dryadontanytarsus Andersen (1943) known as microfossils from bogs will key here.
        It may be distinguished from *Calopsectra* by the peculiar form of the labium which is
        rolled up dorsally in the form of a tube which encloses the hypopharynx.

53b   Antenna shorter, not mounted on a tubercle; no bifid plumose bristle laterally on abdomen. . . . .
                        (in part) Tribe **Tendipedini** (=**Chironomini**)   **54**

54a **(53)** Antenna 6-segmented, a Lauterborn organ present apically or pre-apically on the second and third segments. (Fig. 41.18) **55**

54b   Antenna 5-segmented, without distinct Lauterborn organs **56**

55a **(54)** The 2 middle teeth of the labial plate pale, the others dark.   (3 species.) (Fig. 41.18) . . . . ***Microtendipes*** Kieffer

55b   Four middle teeth of the labial plate pale, the others dark.   (6 species) . . . . ***Paratendipes*** Kieffer

56a **(54)** Eighth abdominal segment with fingerlike ventral gills.   (Fig. 41.19). . . . . . . (in part) ***Tendipes*** Meigen (***Chironomus*** Meigen)

56b   Eighth abdominal segment without ventral gills.   A large complex, including *Tendipes* in part, difficult to separate further. . . . .
                        (most) Tribe **Tendipedini**

57a **(52)** Antenna at least ½ as long, often as long, as the head. . . . .
                        Corynoneura Winnertz

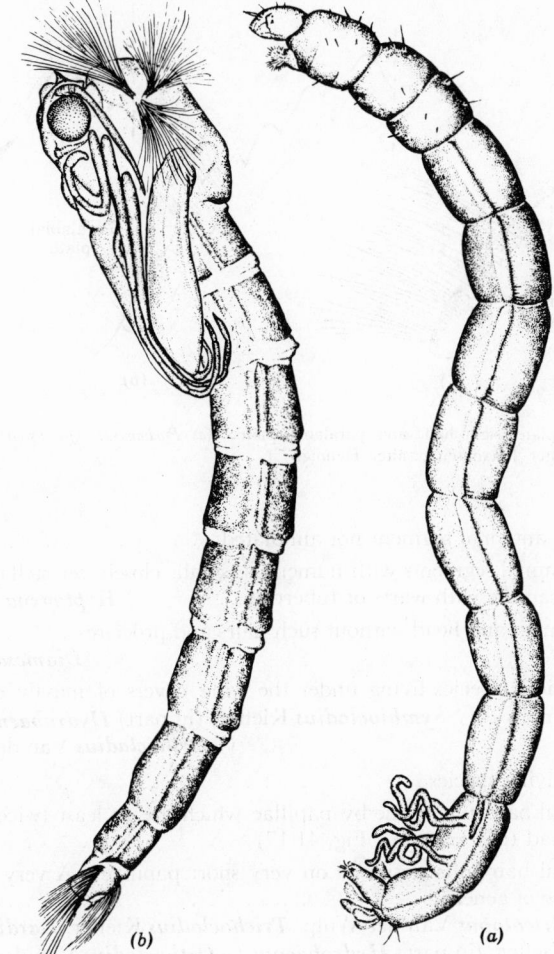

(b)   *(a)*

Fig. 41.19. *Tendipes tentans.* (*a*) Larva. (*b*) Pupa. (From
Johannsen by permission.)

57b	Antenna less, usually much less, than ½ as long as the head	**58**
58a (**57**)	Paralabial plates present, though often small, in the Nearctic species fringed with hairs	**59**
58b	Paralabial plates wholly absent	**61**
59a (**58**)	Paralabial plates of moderate size, bearded with black bristles. (Fig. 41.20*a*). . . . . *Prodiamesa* Kieffer	
59b	Paralabial plates small, inconspicuous	**60**
60a (**59**)	Paralabial plates bare (not Nearctic). (Fig. 41.20*c*). . . . . *Trissocladius* Kieffer	
60b	Paralabial plates bearded. (Fig. 41.20*b*). . . . . *Diplocladius* Kieffer, *Psectrocladius* Kieffer, (in part) *Hydrobaenus* Fries (=*Orthocladius*) Van der Wulp	
61a (**58**)	Third antennal segment annulated	**62**

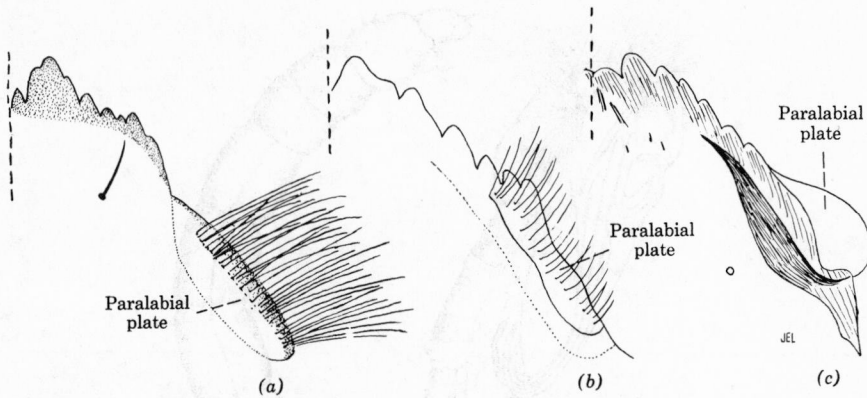

Fig. 41.20. Labial plate, right half, and paralabial plate. (*a*) *Prodiamesa*. (*b*) *Psectrocladius*. (*c*) *Trissocladius*. (*a* and *b* after Johannsen; *c* after Hennig.)

61b   Third antennal segment not annulated **63**

62a **(61)** Abdominal segments with numerous, small, closely set stellate hairs; head capsule with warts or tubercles . . . . **Heptagyna** Philippi

62b   Abdomen and head without such hairs and processes . . . .
                                                                    Diamesa Meigen

63a **(61)** Symbiotic species living under the wing covers of mayfly or stone-fly nymphs . . . **Symbiocladius** Kieffer, (in part) **Hydrobaenus** Fries
                                          (=**Orthocladius** Van der Wulp)

63b   Free-living species **64**

64a **(63)** Caudal hair tufts borne by papillae which are at least twice as long as broad (cf. *Pentaneura*, Fig. 41.17) **65**

64b   Caudal hair tufts sessile or on very short papillae. A very difficult residue of genera . . . .
                Cricotopus Van der Wulp, **Trichocladius** Kieffer, **Cardiocladius**
                Kieffer, (in part) **Hydrobaenus** (=**Orthocladius** Van der Wulp),
                                          and probably others.

65a **(64)** Basal antennal segment slightly but distinctly bent . . **Brillia** Kieffer

65b   Basal antennal segment straight . . . . **Metriocnemus** Van der Wulp

66a **(40)** Pseudopods present only on the prothorax, the apex of the abdomen with an adhesive disc; species living in rapidly flowing water. Pupa within a cornucopia- or slipper-shaped cocoon from which the fila-mentous respiratory organs project. (Fig. 41.21) . . . .
                          Family **Simuliidae** (=**Melusinidae**)
          Though good larval characters exist, it is not yet possible to formulate a good
          key for the separation of genera.

66b   Pseudopods on the intermediate body segments; apex of abdomen without an adhesive disc. Pupa not enclosed in a cocoon **67**

67a **(66)** Abdominal segments 1 to 3 each with a pair of pseudopods; abdo-men terminating in a long, segmented respiratory tube. Pupa with 2 respiratory processes of markedly unequal length, one short, the other longer than the body . . . .
                      Family **Liriopeidae** (=**Ptychopteridae**)   **68**

67b   Abdominal segments 1 and 2 each, or 1 only, with a pair of pseu-

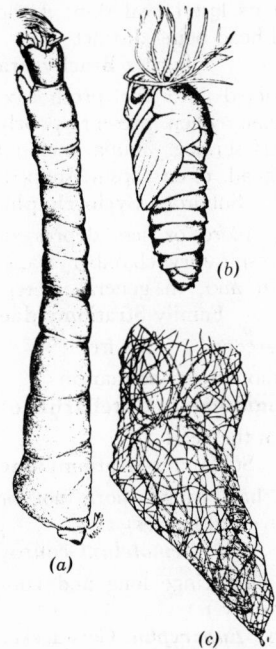

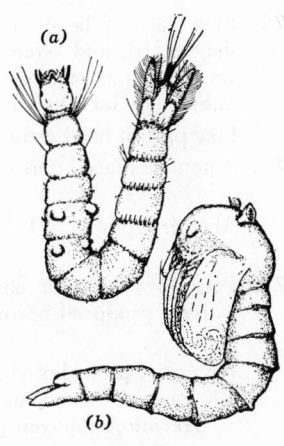

Fig. 41.21. *Simulium pictipes* Hagen. (*a*) Larva. (*b*) Pupa. (*c*) Pupal case.

Fig. 41.22. (*a*) Larva, and (*b*) pupa of a dixid. (After Johannsen.)

dopods; abdomen without a long respiratory tube. Pupa with short respiratory processes of equal length . . . . . . Family **Dixidae**   70

68a (67) Mandible with 3 large outer teeth; pseudopods small . . . .
Subfamily **Liriopeinae** (=**Ptychopterinae**)
Liriope Meigen (=*Ptychoptera* Meigen)

68b   Mandible with a single outer tooth; pseudopods prominent . . . . . .
Subfamily **Bittacomorphinae**   69

69a (68) Body black; integument covered with long projections which are incased in a black, horny substance . . *Bittacomorphella* Alexander

69b   Body mostly red; integument covered with transverse rows of short, stellate tubercles . . . . *Bittacomorpha* Westwood

70a (67) Abdomen dorsally with coronae or rosettes of hairs on abdominal segments 2 to 7 or 3 to 7. (About 22 species.) (Fig. 41.22) . . . . .
Dixa Meigen

70b   Abdomen dorsally bare or almost so   71

71a (70) Ventral pseudopods on first 2 abdominal segments. (About 18 species) . . . . *Paradixa* Tonnoir

71b   Ventral pseudopods on first abdominal segment only . . . .
Meringodixa Nowell
One species, *M. chalonensis* Nowell

72a (2) Head capsule well developed, usually sclerotized dorsally though often retractile; maxillae usually well developed, their palpi usually distinct; antennae well developed, situated on the sclerotized dorsal

plate. Pupa either free or remaining in its last larval skin; if the latter, it is unchanged in shape, the larval head being distinct . . . .
Suborder **Brachycera**   73

72b   Head nonsclerotized, permanently retracted into the prothorax; maxillae and their palpi absent; antennae, when present, poorly developed and situated on a membranous surface. Pupa enclosed in its last larval skin, without a distinct head, often capsule-shaped.
Suborder **Cyclorrhapha**   82

73a (72) Free part of head not retractile; body more or less depressed, shagreened, and often striated; cleft of respiratory chamber transverse. Pupa enclosed in last larval skin and, in general, determinable by larval characters . . . . Family **Stratiomyidae**   74

73b   Free part of head retractile; body not shagreened. Pupa free   77

74a (73) Antennae placed dorsally on the head, remote from the margin . . .
Subfamily **Potamidinae** (=**Clitellariinae**)   75

74b   Antennae placed at lateroanterior angles of the head . . . .
Subfamily **Stratiomyidae**   76

75a (74) Posterior spiracular chamber dorsal, its hair fringe short, not or scarcely produced beyond its margin. (Over 20 species) . . . . . .
Nemotelus Geoffroy

75b   Posterior spiracular chamber apical, its hair fringe long and conspicuous. (About 5 and 25 species respectively.) . . . .
Hermione Meigen (=***Oxycera*** Meigen), ***Euparyphus*** Gerstaecker

76a (74) Last abdominal segment not more than twice as long as wide, only moderately tapering. (Fig. 41.23*a, b*) . . . .
Eulalia Meigen (=***Odontomyia*** Meigen) sens. lat.

76b   Last abdominal segment usually 3 or more times as long as wide, strongly tapering. (Fig. 41.23*c*). . . . . ***Stratiomys*** Geoffroy

77a (73) Body cylindrical, each abdominal segment with a girdle of pseudopods which may bear hooks or which may be reduced to fleshy swellings; posterior swellings close together, situated in a vertical cleft. . . . . Family **Tabanidae**   78

77b   Body variable, but never with more than 1 pair of pseudopods on each abdominal segment; posterior spiracles not in a vertical cleft . .   80

78a (77) Second and third antennal segments subequal, or the distal one the longer; mature larvae usually shorter than 20 mm. . . . .
Subfamily **Pangoniinae**
(About 70 species). . . . . ***Chrysops*** Meigen

78b   Terminal antennal segment much the shortest and smallest; mature larvae usually longer than 20 mm. . . . . . . Subfamily **Tabaninae**   79

79a (78) Anal segment rounded, the siphon short, hardly exsertile. (3 species) . . . . ***Chrysozona*** Meigen (=***Haematopota*** Meigen)

79b   Anal segment usually tapering, the siphon more or less elongated when exserted. (About 150 species.) (Fig. 41.24) . . . .
Tabanus Linnaeus sens. lat.

80a (77) Apex of abdomen with a pair of caudal processes which are definitely longer than the pseudopods; each abdominal segment with a pair of ventral pseudopods. . . . Family **Rhagionidae** (=**Leptidae**) (Fig. 41.25) . . . . ***Atherix*** Meigen
One species, *A. variegata* Walker

80b   Apex of abdomen without caudal processes, or with such processes shorter than the pseudopods, or the latter wanting   81

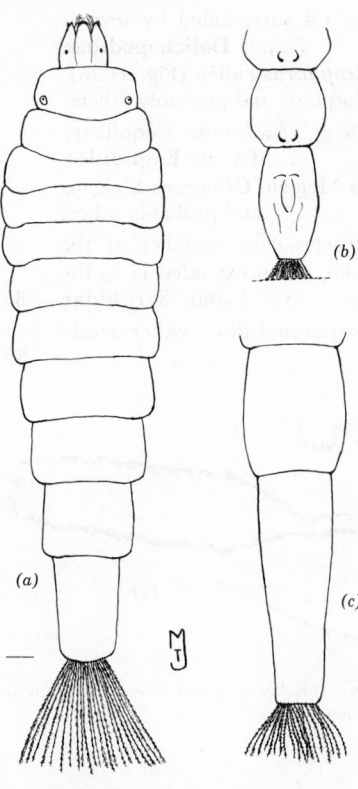

Fig. 41.23. (*a*) *Eulalia* sp., probably *communis* James; Puparium in outline, dorsal view. (*b*) Same; last 2½ segments of larva, ventral view (anal fringe abbreviated). (*c*) *Stratiomys* sp.; last two segments of larva, in outline, dorsal view. (M. T. J.)

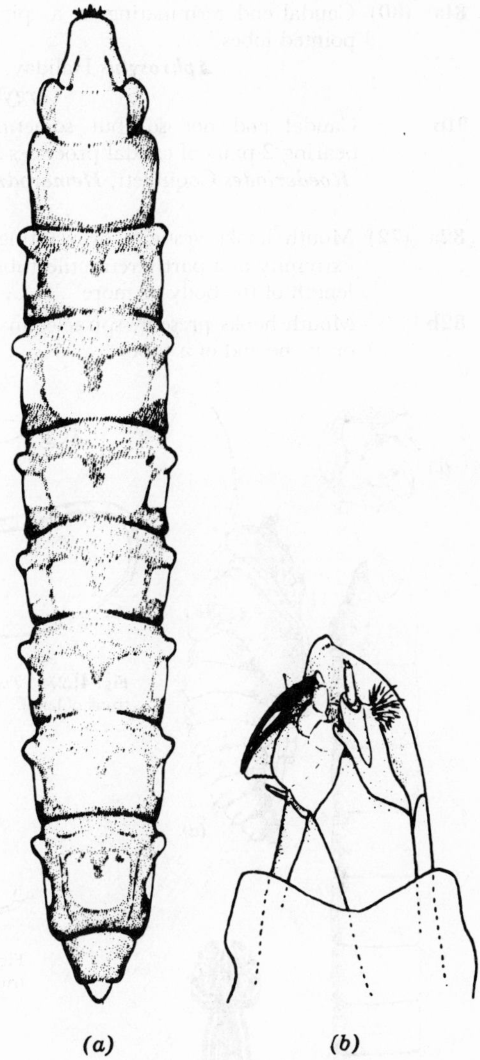

(*a*)   (*b*)

Fig. 41.24. (*a*) *Tabanus atratus* Fabricius, larva, dorsal view. (*b*) *Tabanus* sp. Anterior part of head of larva. (*a* after Johannsen; *b* after Webb and Wells.)

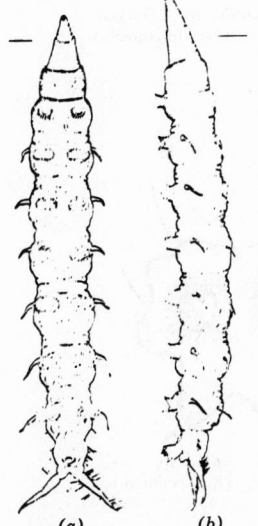

Fig. 41.25. *Atherix variegata* Walker. Larva. (*a*) Ventral. (*b*) Lateral. (After Johannsen.)

81a (80) Caudal end terminating in a spiracular pit surrounded by several pointed lobes . . . . Family **Dolichopodidae**

Aphrosylus Haliday, ***Hydrophorus*** Fallén (Fig. 41.26), *Argyra* Maquart, and probably others.

81b   Caudal end not so, but sometimes (e.g., *Roederiodes* Coquillett) bearing 2 pairs of caudal processes . . . . Family **Empididae**

Roederiodes Coquillett, ***Hemerodromia*** Meigen, ***Clinocera*** Meigen, and probably others

82a (72) Mouth hooks vestigial or wanting; spiracles close together at the extremity of a partly retractile tube which, when extended is ½ the length of the body or more . . . . Family **Syrphidae**   83

82b   Mouth hooks present; spiracles in well-separated discs, either sessile or at the end of a tube.   85

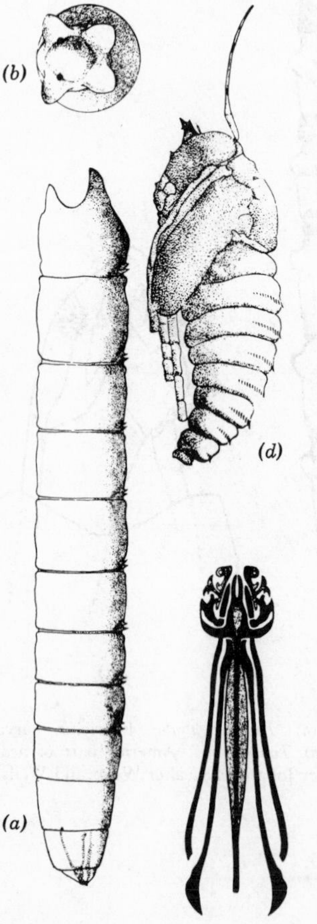

(b)

(b)

(a)

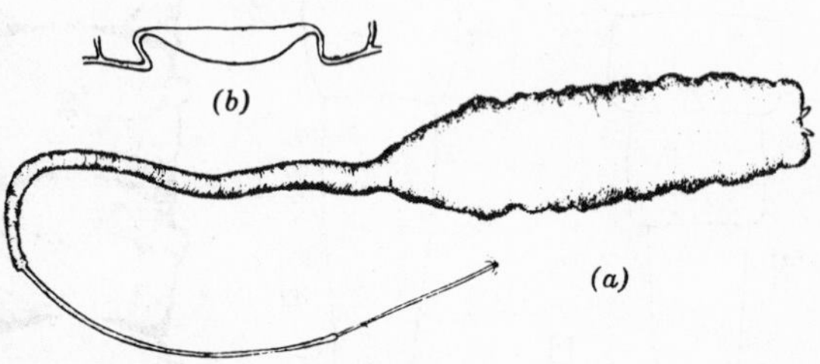

Fig. 41.27. *Tubifera* sp. (*a*) Larva, dorsal view. (*b*) Tracheal trunk of larva. (After Johannsen.)

(d)

Fig. 41.28. *Elophilus* sp. Tracheal trunk of larva. (After Johannsen.)

(a)

(c)

Fig. 41.26. *Hydrophorus agalma.* (*a*) Larva, lateral view. (*b*) Posterior end and posterior spiracles of larva. (*c*) Cephalopharyngeal skeleton, dorsal view. (*d*) Pupa, lateral view. (After Greene.)

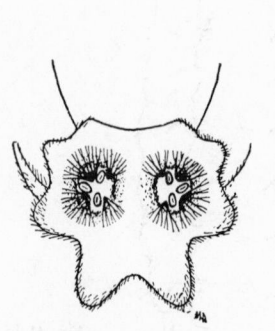

Fig. 41.29. *Sepedon* sp., spiracular disc. (By Needham.)

83a (82) Caudal respiratory tube, when extended, about ½ the length of the body. (Several species). . . . . ***Chrysogaster*** Meigen

83b Caudal respiratory tube, when extended, much longer than the body **84**

84a (83) The 2 longitudinal tracheal trunks straight. (About 30 species.) (Fig. 41.27) . . . . ***Tubifera*** Meigen (= ***Eristalis*** Latreille)

84b The 2 longitudinal tracheal trunks undulating. (About 25 species.) (Fig. 41.28) . . . .
Elophilus (= ***Helophilus*** Meigen), ***Lejops*** Rondani, ***Parhelophilus*** Girschner

85a (82) Parasitic on aquatic larvae . . . .
Family **Larvaevoridae** (= **Tachinidae**)
Euadmontia pergandei (Coquillett) is known to parasitize the larva of the tipulid *Tipula abdominalis* (Say), and *Ginglymyia acrirostris* Townsend the larva of the pyralid moth *Cataclysta fulicalis* Clem.

85b Not parasitic on insects **86**

86a (85) Spiracular disc surrounded by several lobes . . . .
Family **Sciomyzidae** (= **Tetanoceridae, Tetanoceratidae**) **87**

86b Spiracular disc without such lobes **88**

87a (86) Caudal spiracular plate with palmate hairs . . . .
(most) Family **Sciomyzidae**
Sepedon Latreille (Fig. 41.29), ***Hedroneura*** Hendel, ***Tetanocera*** Latreille, ***Dictya*** Meigen, and probably others

87b Caudal spiracular plate without palmate hairs. . . . .
Poecilographa Melander

88a (86) Mouth hooks serrate, palmate, or digitate; if simple, (some *Hydrellia* Robineau-Desvoidy, posterior respiratory openings consisting of 2 small slits at the apices of 2 sharp, hollow spines; some forms with respiratory openings at the apices of 2 long, retractile tubes. . . . . .
Family **Ephydridae** **89**

88b Mouth hooks simple; posterior respiratory organs never as in *Hydrellia* Robineau-Desvoidy nor located on long, retractile tubes, usually at the apices of short tubercles **93**

89a (88) Posterior respiratory openings small, slitlike, situated at the apices of 2 hollow spines which, in life, are inserted into plant tissues **90**

89b Posterior respiratory openings at the apices of tubercles or of long retractile tubes **91**

90a (89) Creeping welts present; leaf miners . . . .
Lemnaphila Cresson, ***Hydrellia*** Robineau-Desvoidy
Lemnaphila (one species, *L. scotlandae* Cresson) breeds in the thallus of *Lemna;* some *Hydrellia* (about 35 species) breed in Potamogeton, water cress, and other aquatic plants.

90b Creeping welts absent; mud-dwellers attached to aquatic roots. (About 25 species.) (Fig. 41.30) . . . . ***Notiphila*** Fallén

91a (89) Eight pairs of pseudopods present, each with strong claws. (About 20 species.) . . . . . . ***Ephydra*** Fallén (Fig. 41.31), ***Setacera*** Cresson

91b Pseudopods absent **92**

92a (91) Caudal respiratory organ, with sheath, about ½ the length of the body. (Several species) . . . . ***Ochthera*** Latreille

92b Caudal respiratory organ short . . . .
Brachydeutera Loew, ***Paralimna*** Loew

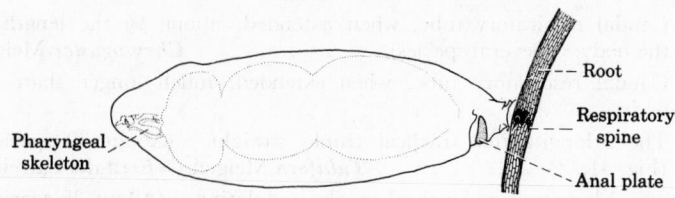

Fig. 41.30. *Notiphila loewi* Cresson. Lateral view of puparium, attached to root. (After Berg.)

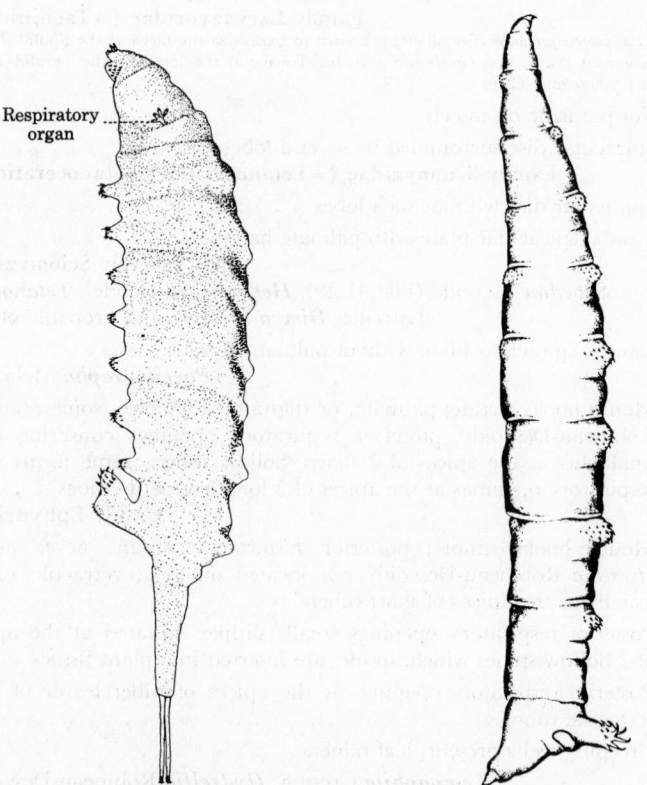

Fig. 41.31. *Ephydra sobopaca.* Loew. Larva. (After Johannsen by permission.)

Fig. 41.32. *Limnophora aequifrons.* Stein. Larva. (After Johannsen by permission.)

93a (88) Posterior spiracle with 2 slits of the usual muscoid type and with a third situated at the apex of a sharp spine similar to the posterior respiratory mechanism of *Hydrellia;* miners in stems of pond lilies. . .
Family **Scopeumatidae** (=**Scatophagidae, Cordyluridae**)
Hydromyza Fallén

One species, *H. confluens* Loew.

93b   Posterior spiracle without a spinelike slit-bearing process . . . .
Family **Musidae** including **Anthomyiidae**
Lispe Latreille (=*Lispa*), *Mydaeina* Malloch, *Limnophora*
Robineau-Desvoidy (Fig. 41.32), *Hydrophoria* Robineau-Desvoidy

References

Alexander, C. P. 1919. The Crane-flies of New York. Part I. Distribution and taxonomy of the adult flies. *Cornell Univ. Agr. Exp. Sta. Mem.*, 25:767–993. **1920.** The Crane-flies of New York. Part II. Biology and phylogeny. *Cornell Univ. Agr. Exp. Sta. Mem.*, 38:699–1133. **1942.** The Diptera or true flies of Connecticut. Fasc. 1. *Conn. State Geol. and Nat. Hist. Survey Bull.*, 64:183–486. **Andersen, F. S. 1943.** *Dryadotanytarsus edentulus* n.g. et sp. from the late glacial period in Denmark. *Entomol. Medd.*, 23:174–178. **Brennan, J. M. 1935.** The Pangoniinae of Nearctic America. *Univ. Kansas Sci. Bull.*, 22:249–401. **Carpenter, S. J. and W. J. LaCasse. 1955.** *Mosquitoes of North America (North of Mexico).* University of California Press, Berkeley. **Cook, E. F. 1956.** The nearctic Chaoborinae (Diptera: Culicidae). *Univ. Minn. Agr. Exp. Sta. Tech. Bull. 218.* **Curran, C. H. 1934.** *The Families and Genera of North American Diptera.* Ballou Press, New York. **Hennig, Willi. 1948–1952.** *Die Larvenformen der Dipteren,* 3 vols. Berlin. **Johannsen, O. A. 1934.** Aquatic Diptera. Part I. Nemocera, exclusive of Chironomidae and Ceratopogonidae. *Cornell Univ. Agr. Exp. Sta. Mem.*, 164:1–71. **1935.** Aquatic Diptera. Part II. Orthorrhapha-Brachycera and Cyclorrhapha. *Cornell Univ. Agr. Exp. Sta. Mem.*, 177:1–62. **1937a.** Aquatic Diptera. Part III. Chironomidae: Subfamilies Tanypodinae, Diamesinae, and Orthocladiinae. *Cornell Univ. Agr. Exp. Sta. Mem.*, 205:1–84. **1937b.** Aquatic Diptera. Part IV. Chironomidae: Subfamily Chironominae, and Part V. Ceratopogonidae (by Lilian C. Thomsen). *Cornell Univ. Agr. Exp. Sta. Mem.*, 210:1–80. **Malloch, John R. 1915.** The Chironomidae, or midges, of Illinois, with particular reference to the species occurring in the Illinois River. *Bull. Illinois State Lab. Nat. Hist.*, 10:275–543. **1917.** A preliminary classification of Diptera, exclusive of Pupipara, based upon larval and pupal characters, with keys to imagines in certain families. Part I. *Bull. Illinois State Lab. Nat. Hist.*, 12:161–409. **Matheson, Robert. 1944.** *Handbook of the Mosquitoes of North America,* 2nd ed. Comstock, Ithaca, New York. **Quate, Larry W. 1955.** A revision of the Psychodidae (Diptera) in America North of Mexico. *Univ. Calif., Berkeley Publs. Entomol.*, 10:103–273. **Stone, Alan. 1938.** The horseflies of the subfamily Tabaninae of the Nearctic region. *U. S. Dept. Agr. Misc. Pub.*, 305:1–171. **Townes, Henry K., Jr. 1945.** The Nearctic species of Tendipedini. *Am. Midland Naturalist,* 34:1–206. **Wirth, W. E. 1952.** The Heleidae of California. *Univ. Calif. Berkeley Publs. Entomol.*, 9:95–266. **Wirth, W. E. and A. Stone. 1956.** Aquatic Diptera. In R. L. Usinger (ed.) Aquatic insects of California with keys to North American Genera and California species. *University of California Press,* Berkeley and Los Angeles.

42

Acari

IRWIN M. NEWELL

PARASITENGONA

The mites restricted to habitats in or immediately around fresh water include representatives of the three major groups of Acari, namely the Parasitiformes, Sarcoptiformes, and Trombidiformes. By far the majority, however, belong in the Trombidiformes, and to the Parasitengona[1] in particular. It is these that are commonly spoken of as "water mites" or "aquatic mites," and it is this group alone which is treated in the present section of this chapter. Water mites occupy a vast range of habitats, from stagnant ponds to glacial torrents, from hot springs to tundra pools frozen solid for most of the

---

[1]The group *Parasitengona* Oudemans 1909 includes those *Trombidiformes* in which the adults and larvae differ markedly in form and habit. The larvae of the water mites are parasitic upon aquatic insects, and the adults are almost exclusively predaceous. The *Parasitengona* differ from the group *Cursoria* Grube 1851 (=*Eleutherengona* Oudemans 1909) in which the larvae closely resemble the adults in general form as well as habit (that is, if the larvae are parasitic, the adults are also parasitic, etc.). The closest terrestrial relatives of the water mites are the polytrichous trombidiform mites such as the *Johnstonianidae* and *Trombidiidae*. In marine habitats, the predominant aquatic mites are the *Halacaridae*. Although the majority of this family are marine, quite a number of species and genera are found in fresh water exclusively. The *Halacaridae* belong to the *Cursoria* and are only remotely related to the water mites proper. The names *Hydrachnellae* and *Hydracarina*, when applied to the fresh-water mites and *Halacaridae* collectively, are meaningless from a systematic standpoint and should be abandoned.

year, from the open waters of the largest lakes to water-soaked moss and algae on the vertical faces of cliffs. A few species are marine. Nevertheless, the ecological distribution of individual species, or even genera, is usually limited to one or a very few types of habitat, so that the mite fauna of a swift mountain stream is markedly different from that of a lowland lake or an estuary. There is also a rich subterranean fauna.

For nearly 35 years, the only available summary of the genera of water mites of North America has been that of Wolcott (1918) who included 33 genera in his key.[2] Since that time, numerous genera have been discovered, and many of the older ones subdivided, until now the number of genera and subgenera known from North America has reached 100. The number of named species, subspecies and varieties is roughly 325, or approximately double what it was in 1918. With such a growth in the number of known groups in our fauna, it is difficult to treat the water mites in a short chapter. The accompanying keys permit identification to subgenus only, and if specific determinations are desired, these must be obtained by consulting the literature, or from specialists in the group.

Subsequent reviewers of the group should not regard the inclusion here of any particular genus, subgenus, or even family as an indication of the author's unqualified acceptance of that group. For the benefit of other students, the writer has attempted to present all current names pertinent to our fauna, and in doing so several of questionable status have been included. Some of these have been pointed out, but not all, since it is not feasible to review the systematics and nomenclature of the water mites in the space available here.

Preservation and Identification

Anyone unfamiliar with the water mites will find it impossible to key out specimens without making dissections. It is especially important to have at least the palpi dissected and visible in both lateral and medial view. Specimens should first be treated to hydrolyze the soft parts. For cursory studies, they may be treated with 2 percent KOH in water for 24 hours, but for research purposes it is certainly preferable to use enzymatic hydrolysis (see Newell, 1947, Reference 46 p. 7). In any case, formalin specimens cannot be hydrolyzed with either alkali or enzymes, hence only alcohol (60–75 percent) should be used as a preservative. After hydrolysis, the specimens are rinsed at least three times, placed in a very small dish or vial (about 1 to 2 cc volume) and dehydrated with 95 percent alcohol, the alcohol concentration being increased in about three steps over a period of 10 or 15 minutes. If excessive oil is present, the specimens should be left in 95 percent alcohol

---

[2]A recent synopsis of the families of water mites is found in *An Introduction to Acarology* by Edward W. Baker and G. W. Wharton, 1952, Macmillan, New York. The reader's attention is also called to the recent works of Habeeb, Mitchell, and Cook, most of which appeared after the original manuscript for this section was completed. So far as possible, the new generic records of these workers have been included in this chapter, but a few omissions were unavoidable. Another recent major work is the comprehensive catalogue, nomenclator, and bibliography published by Viets (references P and Q).

until the oil is dissolved, after which the alcohol is pipetted off and replaced. Sufficient glycerine is then added to make a layer 1 to 2 mm deep in the bottom of the dish or vial (depending on the size of the specimens). This is mixed with the alcohol, and the container is set in a warm place to permit evaporation of the alcohol. Temporary mounts of dissected mites can be made in glycerine, or permanent glycerine preparations can be made by the methods described elsewhere by the author (1947, pp. 10–14). Undissected mites can be examined in depression slides and stored in small vials, or they can be mounted on slides. Only glycerine should be used as a mountant, for the mites are too clear in the standard resinous media.

The structures used to identify water mites can be learned easily by keying out specimens, so only the major landmarks will be pointed out here (Fig. 42.1). The body shows no sign of subdivision, except in the males of some species of *Arrenurus* which have a posterior caudiform projection (Fig. 42.84). The dorsum may be armored or not, but there are always several pairs of *dorsoglandularia* (dgl.), each consisting of a gland pore and an associated seta (Fig. 42.50). Even in unarmored species the glandularia are often borne on minute sclerites. The venter bears four pairs of *epimera* (ep. I–IV) which are the coxal portions of the legs. These may be discrete (Fig. 42.32) or variously modified by fusions (Fig. 42.36) or expansions. In some genera the body is encased in a nearly continuous shell-like armor, with the limits of the epimera indistinct or partly obliterated.

There are paired glandularia on the ventral side also, some of which, the *epimeroglandularia* (epg.), are highly useful diagnostic structures. Typically there are three pairs of these, with epg. 1 lying between ep. II and III (Figs. 42.2, 42.55, 42.58), in ep. II (Figs. 42.49, 42.85), or in the anterior margin of ep. III (Fig. 42.36). Epg. 2 is the most variable in position, lying just lateral to the genital opening (Fig. 42.17), just posterior to ep. IV (Figs. 42.66, 42.74), within ep. IV (Fig. 42.49), or sometimes apparently absent. Epg. 3 is usually found just behind ep. IV, and lateral to the position of epg. 2 (Fig. 42.74). The reader should soon learn to identify the epimeroglandu-

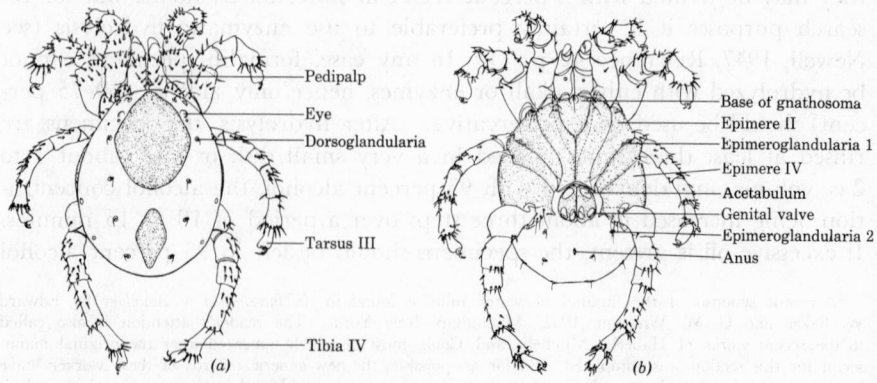

Fig. 42.1. Morphology of a water mite, *Tyrrelia*. (*a*) Dorsal. (*b*) Ventral.

laria, for the position of these small but important morphological landmarks is considered at several points in the following keys.

The *genital opening* lies between or posterior to the epimera (Figs. 42.1, 42.2, 42.49). Variation in the number, form, and arrangement of the genital *acetabula* (Figs. 42.26, 42.48), the *valves* or *plates* guarding the opening (see footnote, p. 1085), and the position of the opening, are all important diagnostic features.

At the anterior end of the body, the *gnathosoma* (Fig. 42.57) lies in a *camerostome* of variable depth (Fig. 42.45). The gnathosoma has also been called *capitulum*. In a few genera, the gnathosoma is borne at the end of a long, protrusible tube (Fig. 42.89). The base of the gnathosoma bears a pair of *palpi* and a pair of *chelicerae*, and in some forms is produced anteriorly as a pointed or rounded *rostrum*. In all known North American genera, the palpi have five segments (trochanter, femur, patella, tibia, tarsus, numbered P–1 to P–5), with the first one short and ringlike. Variations in the form of P–2, P–4, and P–5 provide some of the major key characters in the fresh-water mites and it is this which makes it imperative to have the palpi dissected and clearly visible in lateral and medial view. The legs (I–IV) contain six segments each, namely the trochanter, basifemur, telofemur (or femur), patella, tibia, and tarsus. For convenience these are often designated by number, I–5 being the tibia of leg I, IV–6 being the tarsus of leg IV, etc. Many water mites, especially those which swim, bear swimming setae on some segments of the legs (swimming "hairs" of authors; see footnote, p. 1084). These are very long, slender, flexible setae, usually borne in clumps or rows. Occasionally they are very few, numbering only one or two on a segment. Swimming setae are generally confined to segments 4 and 5 of legs II, III, and IV, although they are also common on IV–3. The long, stiff, swordlike setae found in *Unionicola* and some other genera are not to be confused with swimming setae. Species that crawl on the bottom or on submerged plants do not have swimming setae. Usually it is possible to distinguish the swimmers from the crawlers, although a few of the former seem to manage very well without specialized setae.

The larvae of water mites are six-legged, and often brightly colored (red, orange, yellow). They are parasitic on insects, and are familiar to all who have collected aquatic insects. Host preference is moderately developed, the larvae of most genera of water mites attacking members of only certain genera or families of insects. The nymphs have eight legs and are predaceous, with rare exceptions, upon small aquatic organisms. Nymphs can be distinguished from adults with little difficulty on the basis of the stage of development of the external genitalia and a reduced number of acetabula.

Acknowledgments

The writer wishes to acknowledge the substantial assistance of several colleagues who provided material for this work. Melville H. Hatch and W. T. Edmondson of the Department of Zoology of the University of Washington made available the collection of the late C. H. Lavers, Jr., R. L. Usinger, D. E. Hardy, and Borys Malkin also provided valuable collections which have been utilized in preparing the accompanying figures. J. L. Stannard and the Illinois Natural History Survey loaned representatives of several genera that could not be obtained from other sources. Rodger Mitchell and David Cook provided valuable information based on their experience in the central states.

Arrangement of the Key

The arrangement of the key will be unfamiliar to most students, but it was decided upon only after long study. It was designed to reduce ambiguity by eliminating as much as possible the subjective factor of interpretation of the meaning of words. There is no key to families, for the variation within families is so great that such a key would be too lengthy and would include so many involved qualifying statements that the beginner would soon be discouraged. The genera are first divided into nine key groups which have no systematic significance whatever. These groups are introduced to simplify backtracking, a process which, while not inevitable, is never entirely avoided by users of keys. Some genera appear in more than one group, owing to variation within the genus, or, in a few cases, the possibility of an alternative interpretation of one of the key characters. Keys to subgenera, or brief diagnoses of these are given in the systematics section, in which the genera are arranged according to families. Specific names have not been applied to any of the figures, chiefly because positively determined material was not available in most cases. Nearly all figures were drawn from actual specimens studied by the author. In the scales accompanying the figures, each subdivision represents 10μ.

KEY TO GROUPS (ADULTS)

1a   Tarsus of palp (P–5) inserted at distal end of tibia (P–4) (Figs. 42.10, 42.37, 42.62). Distal end of tibia not produced to form a dorsal spine, although a seta or a medial process may be present here. Distal end of the tibia not especially higher than base of tarsus (i.e., not more than twice as high). Tarsus of palp not opposable to distal end of tibia. (See also 1c)   2

> The absence of a distal spine on the tibia is one of the most constant and reliable key characters, but care must be taken to avoid confusing the large dorsomedial seta, found in some species, with a spine. A spine is a projection of a segment, or of the body wall, and in most cases is probably the product of a number of cells. There is no alveolus, and the function, insofar as this can be ascertained from its structure, must be mechanical rather than sensory. A seta is generally a sensory organ, formed by a single cell, and inserted in an alveolus (Fig. 42.50). The alveolus may be ringlike or pitlike, the latter type being common in the water mites. Setae are occasionally superimposed upon spines (Fig. 42.33). Authors are not consistent in the use of these terms, heavy setae often being called "spines," long slender setae "hairs," etc. The writer prefers to apply the term "seta" to all types of setae, using appropriate modifying terms (spiniform seta, pectinate seta, swimming seta, etc.). The term "spine" is in no case applied to a true seta in this key. The user of the key can learn quickly to differentiate setae and spines, and should experience little difficulty from this source.

1b Tarsus of palp (P–5) attached at distal end of tibia as above, but end of tibia (P–4) 2 to 3 times as high as base of tarsus (Fig. 42.83). Tarsus opposable to distal end of tibia . . . (p. 1091) **Key Group 7**

1c Tibia of palp (P–4) produced in form of a distidorsal spine (not a seta) which projects beyond the insertion of the tarsus, so that the tarsus is inserted on the distiventral surface of the tibia (Figs. 42.19, 42.27). 5

2a (1) With 3 pairs of genital acetabula, variously arranged (Figs. 42.26, 42.29, 42.44), sometimes hidden under the genital valves 3

2b With 4, 5, or 6 pairs of acetabula, variously arranged. . . . . (p. 1088) **Key Group 3**

2c With more than 6 pairs of genital acetabula, the acetabula sometimes very small (Figs. 42.6, 42.18, 42.77). In the large, soft-bodied *Eylais* (epimera III and IV as in Fig. 42.12) and *Limnochares* (epimera as in Fig. 42.13), the acetabula are only 5 to 10μ in diameter, fungiform, and scattered widely over portions of the venter. In both genera the genital area is weakly developed 4

2d Genital acetabula absent. A single genus, *Hydrovolzia*, with heavily armored body (Fig. 42.7), and living in cold, running streams . . . . (p. 1091) **Key Group 6**

3a (2) P–2 with a ventral seta or setae which may be borne on a spine or tubercle or at the base of a spine (Fig. 42.33), or there may be no spine whatever (Figs. 42.31, 42.37, 42.46). . (p. 1085) **Key Group 1**

3b P–2 without a ventral seta or setae although a spine (Fig. 42.80) or tubercle may be present . . . . (p. 1086) **Key Group 2**

4a (2) Dorsum hard, armored, that is, with a central unpaired plate which usually occupies virtually the entire dorsum (Fig. 42.79) or which may be smaller, surrounded by a number of small paired plates (Fig. 42.43). . . . . (p. 1088) **Key Group 4**

4b Dorsum mostly membranous, not armored, with no large, unpaired central plate. At most with an anterior median ocular bridge (Figs. 42.14, 42.15); or with a series of paired plates bearing the individual dorsoglandularia . . . . (p. 1089) **Key Group 5**

5a (1) With 3 pairs of genital acetabula . . . . . (p. 1091) **Key Group 8**

5b With more than 3 pairs of genital acetabula . . . . (p. 1092) **Key Group 9**

KEY GROUP 1

1a Genital acetabula discoidal or pitlike, inseparably imbedded in the genital *valves*, *plates*, or *plate*. (Figs. 42.44, 42.48) 7

1b Genital acetabula not inseparably imbedded in the genital plate, plates, or valves, but lying free, often largely concealed under paired valves (Figs. 42.29, 42.30) 2

The term *valve* is applied to paired structures which lie directly alongside the genital opening, which are hinged to the body wall along their lateral margins, and which are capable of opening outward like a double door (Fig. 42.35). The term *plate* is applied to paired (Fig. 42.77) or fused (Fig. 42.49) structures which are quite immovably imbedded in the body wall. When fused, the plate surrounds the genital opening (Fig. 42.44); when paired they may lie at some distance from it (Fig. 42.58). The distinction between valves and plates is usually, but not always sharp; furthermore, paired plates may be found in one sex (almost invariably the female) and a fused plate in the other (the male) of a single species. The form of the acetabula, however, does not vary within a genus, or in the two sexes of any given species.

2a (1) Ep. II separated from III by a distinct space, suture, or carina which is complete (Figs. 42.36, 42.32). Suture separating I and II also complete, extending to the inner margin of the epimeral area (Fig. 42.32), but not joining with the suture of the opposite side to make a cordiform configuration. 3

2b Ep. II and III partially fused, the suture or carina between them incomplete (Fig. 42.30). Sutures separating I and II coalescing posteriorly to make a cordiform configuration behind which ep. II are fused medially 6

3a (2) Ep. III of right and left sides separated from each other by a distinct interval of membranous cuticle (Fig. 42.29). P–2 with a spine or a spine and setae ventrally (Fig. 42.33). Ep. II and III separated by membranous cuticle. Body wall outside of epimera only partially sclerotized at most 4

3b Ep. III of right and left sides contiguous, although not fused (Fig. 42.30). Ep. II and III not separated by membranous cuticle. P–2 with a seta ventrally, but no spine. Claws of tarsus IV reduced to peglike rudiments (Fig. 42.34). Body wall almost entirely sclerotized; but dorsal plate surrounded by striated membranous cuticle containing 6 or 7 pairs of small plates . . . .
(p. 1098) *Mamersella*

4a (3) IV–6 with claws 5

4b IV–6 without claws; with only a few setae at the tip . . . .
(p. 1097) *Teutonia*

5a (4) Basal third of P–4 with 2 fine setae borne at the end of a truncate, conical tubercle. . . . . (p. 1097) *Sperchonopsis*

5b Without such a tubercle in this position. Ventral margin of P–4 with 2 small setae, usually widely spaced and spiniform, but occasionally close together . . . . (p. 1097) *Sperchon*

6a (2) Dorsum almost entirely covered by a large central plate and several smaller marginal plates (Fig. 42.43). . . (p. 1100) *Testudacarus*

6b Dorsum devoid of plates or with only a few minute sclerites at the muscle insertions. Cuticle often leathery . . . . . (p. 1098) *Lebertia*

7a (1) IV–6 with claws . . . . (p. 1101) *Tyrrellia*

7b IV–6 without claws (Fig. 42.47) . . . . (p. 1100) *Limnesia*

KEY GROUP 2

1a Leg insertions displaced far anteriorly, confined to anterior quarter of body. Body elongate when seen in dorsal view, usually 1½ or more than 1½ times as long as broad (Figs. 42.39, 42.40). Tarsus IV without claws (Fig. 42.42). 2

1b Leg insertions normally disposed, the insertion of IV being near or behind the middle of the body (Figs. 42.85, 42.49). Body less than 1½ times as long as broad. Tarsus IV with claws except in the genus *Teutonia* 5

2a (1) Sclerotization of epimera extending well up onto dorsal surface of body, leaving only a narrow median strip of cuticle (Fig. 42.39). 3

2b Sclerotization not so extensive, dorsum almost or entirely covered by membranous cuticle (Fig. 42.40) 4

3a (2) Dorsomedian strip of membranous cuticle with a row of narrow sclerites (Fig. 42.39). . . . . (p. 1099) *Frontipoda*

3b With no such row of sclerites . . . . . (males, p. 1099) *Gnaphiscus*

4a (2) Epimera extending to dorsal surface, where they are visible at the sides of the body in dorsal view, but the greater portion of the dorsum is covered by membranous cuticle (Fig. 42.40). Body usually higher than broad . . . . . . (females, p. 1099) *Gnaphiscus*

4b Epimera not visible in dorsal view. Body somewhat broader, or at least as broad as high, cigar-shaped . . . . (p. 1099) *Oxus*

5a (1) Genital acetabula enclosed within the genital opening itself (Figs. 42.85, 42.29), not lying on plates or in the body wall surrounding the genital opening 6

5b Genital acetabula lying on plates outside the genital opening (Figs. 42.49, 42.66), or in body wall. 7

6a (5) Tarsus IV with claws. Body entirely encased in a hard, porous armor, except for a suture or narrow band of soft cuticle around the dorsal plate (Fig. 42.86). Epg. 2 just behind IV (Fig. 42.85). (p. 1106) *Mideopsis*

6b Tarsus IV without claws. Body not armored, except for the epimera. Epg. 2 enclosed within ep. IV, or in a deep recess in IV (Fig. 42.29) . . . . (p. 1097) *Teutonia*

7a (5) Sutures between epimera II and III considerably reduced, especially those between II and III, III and IV (Fig. 42.78) 15

7b All epimera quite distinct although some of the sutures may not be complete (Fig. 42.49). Separation of II and III usually complete, often with a distinct interval of membranous cuticle. 8

8a (7) Epg. 2 in the anterior half of IV, often in close contact with the suture between III and IV (Fig. 42.49) 9

8b Epg. 2 behind IV, or in the very posterior margin of IV (Fig. 42.67). 11

9a (8) I–5 distiventrally with 2 greatly enlarged setae; I–6 usually curved, rarely straight (Fig. 42.52). Gnathosoma not fused with ep. I. P–4 with many (more than 20) dorsal setae . . (p. 1101) *Atractides*

9b I–5 without such setae, I–6 straight. Gnathosoma usually, but not always, fused with ep. I (Fig. 42.49). P–4 with fewer than 10 dorsal setae. 10

10a (9) Epg. 2 in extreme anterior margin of ep. IV, contiguous with the suture between III and IV, or separated from the suture by a distance no greater than the diameter of epg. 2 (Fig. 42.49). P–IV with no prominent ventral spine or tubercle. . (p. 1101) *Hygrobates*

10b Epg. 2 much nearer the middle of ep. IV, separated from the suture by a distance greater than twice the diameter of epg. 2 (Fig. 42.51). P–4 often with a prominent ventral spine or tubercle . . . . (p. 1101) *Corticacarus*

11a (8) Carina or suture separating ep. III and IV intersecting medial margin of epimeral area (Fig. 42.67). 12

11b This carina or suture intersecting posteromedial angle, or even the posterior margin of the epimeral area (Fig. 42.66). . . . . (p. 1103) *Wettina*

12a (11) P–4 with only 1, 2, or 3 setae ventrally (Fig. 42.65). Male without a petiolus at the posterior end of the body 13

12b P–4 very long and slender, with 6 to 12 ventral setae (Fig. 42.68). Male with a small petiolus at the posterior end of the body . . . . . . (p. 1104) *Hydrochoreutes*

13a (12) Epg. 1 in the space or suture between ep. II and III (Fig. 42.66).
Males with IV–6 normal, straight **14**

13b   Epg. 1 in ep. II.   Males with IV–6 grotesquely curved, with
numerous spiniform setae on the concave side . . . .
(p. 1104) *Pionacercus*

14a (13) Males with IV–4 flattened, broad, and with many swimming setae
Fig. 42.64) . . . . (p. 1104) *Tiphys*
<small>Thus far, no characters have been advanced to separate the females of *Acercus*
and *Pionopsis*.</small>

14b   Males with IV–4 not flat and broad, but shortened, cylindrical,
and showing only moderate sexual differences in chaetotaxy . . . . .
(p. 1104) *Pionopsis*

15a (7) P–2 with a ventral spine or tubercle **16**

15b   P–2 without a ventral spine or tubercle . . . . (p. 1105) *Ljania*

16a (15) Some segments of legs with elongate, slender, typical swimming
setae. . . . . (p. 1105) *Brachypoda*

16b   Legs without typical swimming setae . . . . (p. 1105) *Neoaxonopsis*

KEY GROUP 3

1a   With 4 pairs of pitlike genital acetabula **2**

1b   With 5 or 6 pairs of pitlike genital acetabula (Fig. 42.54) . . . .
(p. 1102) *Unionicola*

1c   With 6 pairs of acetabula, not pitlike, but lying free under the
paired genital valves (Fig. 42.41) . . . . (p. 1100) *Torrenticola*

2a   (1) Tarsus IV with claws **3**

2b   Tarsus IV without claws (Fig. 42.47) . . . . . . (p. 1100) *Limnesia*

3a   (2) Epimera distinct, well separated by sutures or spaces; majority of
body covered by membranous cuticle. Genital area not confined to
posterior quarter of body. (Fig. 42.49) . . . . (p. 1101) *Hygrobates*

3b   Epimeral boundaries extensively obliterated by fusions so that venter
is virtually a single plate; body well armored. Genital area confined to posterior quarter of body . . . . (p. 1105) *Axonopsis*

KEY GROUP 4

1a   Genital acetabula arranged lineally along the posterior and posterolateral margin of the composite epimeral area (Fig. 42.76) **2**

1b   Genital acetabula arranged otherwise **3**

2a   (1) P–2 with or without a spiniform process ventrally, but if one is
present, it is distal in position (Fig. 42.80.) Body sometimes cleft
posteriorly (Fig. 42.76). . . . . (p. 1105) *Aturus*

2b   P–2 with 1 or 2 processes ventrally, at least one of which is at or
behind the middle of the ventral margin . . . (p. 1105) *Kongsbergia*

3a   (1) Leg IV with claws **4**

3b   Leg IV ending in a clawless point (Fig. 42.47) . (p. 1100) *Limnesia*

4a   (3) Acetabula borne on paired, movable valves, or enclosed within the
genital opening (Fig. 42.22), or arranged in a ring about the genital
opening (Fig. 42.82) **8**

4b   Acetabula borne on winglike areas of the body wall extending laterally from the genital opening (Fig. 42.63), or on plates (Fig. 42.77) oriented laterally from the genital opening 5

5a (4) Claws II and III with a basal lamella (Fig. 42.61), distal portion of claw variously modified 6

5b Claws II and III simple, scythe-shaped, either smooth or with a fine accessory tooth on the outer margin, but with no basal lamella. . . . (p. 1103) *Koenikea*

6a (5) Legs with swimming setae on some segments (Fig. 42.69) 7

6b Legs without swimming setae . . . . (p. 1103) *Feltria*

7a (6) Ep. II and III separated by a distinct, although sometimes narrow space which includes epg. 1. Sutures separating ep. I and II not uniting medially to make a cordiform configuration (Fig. 42.74) . . . (p. 1104) *Forelia*

7b Ep. II and III separated only by a carina, which is not complete. Epg. 1 in ep. II, or in the carina between II and III. Sutures separating ep. I and II uniting medially to make a cordiform configuration (Fig. 42.77). . . . . (p. 1105) *Albia*

8a (4) Legs with swimming setae. P–4 without a distal spiniform seta. Genital acetabula of female about 15 to 25 in number, solidly imbedded in the body wall around the genital opening (Fig. 42.82). Those of male on movable valves which bear winglike projections (Fig. 42.81) . . . . (p. 1106) *Midea*

8b Legs without swimming setae. Genital acetabula more than 30, borne on (or under) paired, movable valves 9

9a (8) P–4 with a distal spiniform seta. Gnathosoma not borne on a protrusible tube. Genital acetabula as in Fig. 42.22. Living in hot springs. . . . . (p. 1097) *Thermacarus*

9b P–4 with only a small, slender seta distally. Gnathosoma borne on a protrusible tube. Not living in hot springs. . . . . (p. 1096) *Clathrosperchon*

KEY GROUP 5

1a Tarsus IV with claws (these are sometimes retracted into the claw fossa) 2

1b Tarsus IV ending in a point, with neither claws nor claw fossa (Fig. 42.47) . . . . (p. 1100) *Limnesia*

2a (1) Eyes widely spaced, not borne on an anteromedian ocular plate or bridge 3

2b Eyes borne on a median longitudinal ocular plate of the type shown in Fig. 42.15. Ep. III and IV as shown in Fig. 42.13. . . . . (p. 1094) *Limnochares*

2c Eyes borne on a median transverse ocular bridge of the type shown in Fig. 42.14. Ep. III and IV contiguous only medially, diverging laterally (Fig. 42.12) . . . . (p. 1094) *Eylais*

3a (2) Epg. 2 in the anterior margin of ep. IV (Fig. 42.49) . . . . (p. 1101) *Hygrobates*

3b Epg. 2 not in this position, but behind, or in the very posterior margin of ep. IV (Figs. 42.55, 42.72). 4

4a (3) Ep. IV with medial portion roughly blocklike, or rectangular in form, with a distinct medial margin (Figs. 42.55, 42.72) 6

4b   Ep. IV with medial portion angular (Fig. 42.74) **5**

5a (4) With 12 to 20 moderate-sized acetabula on each side of the genital opening. With a pair of epimeroglandularia between the genital area and coxae IV (Fig. 42.74). Coxae I and II with small epidesmids. Swimming hairs present . . . . (p. 1104) *Forelia*

5b   With 50 to 160 small acetabula on each side of the genital opening, borne on a pair of distinctly bipartite genital plates. With no epimeroglandularia between the genital area and coxae IV. Coxae I and II without epidesmids. Swimming hairs absent . . . .

                                               (p. 1095) *Piersigia*

        In order to identify the remaining genera, it is necessary to determine the sex of the mite. Males can be identified by the penis (indicated by the dashed lines in Fig. 42.72), a complex, usually refractile structure lying under (ventral view) and usually extending somewhat anterior to the genital area. The presence of eggs is positive proof that the specimen is a female, but not all females contain mature eggs. Also, the genital plates of the right and left sides are usually separate in the females (Fig. 42.73), but fused in the males (Fig. 42.72). Specimens showing sexual modifications in legs III and IV are males.

6a (4) Females. **7**

6b   Males. **11**

7a (6) Genital acetabula borne on 2 or 4 plates which are rather distantly removed from the slitlike genital opening, leaving a distinct interval of membranous cuticle (Figs. 42.55, 42.73); in certain species of *Piona*, some of the acetabula may lie in the membranous body wall itself (Fig. 42.73). **8**

7b   Acetabula borne on 2 or 4 plates which are closely adjacent to the slitlike genital opening (Figs. 42.53, 42.54). All acetabula borne on plates. Some species parasitic in mussels . . . (p. 1102) *Unionicola*

8a (7) Epidesmids usually very long, reaching to or beyond the division between ep. III and IV (Fig. 42.55). Not parasitic in mussels . . . . .

                                           (p. 1102) *Neumania*

8b   Epidesmids much shorter, never reaching as far as ep. IV, and seldom to the middle of ep. III (Fig. 42.58). Some species living in mussels. **9**

9a (8) Posterior margin of ep. IV straight or convex. Acetabula borne on a pair of long, laterally oriented plates (Fig. 42.59). Living in mussels . . . . (p. 1102) *Najadicola*

9b   Posteromedial margin of ep. IV concave (Fig. 42.75). Genital plates crescentric to hemi-ellipsoid, oval, or largely absent (Figs. 42.58, 42.73). Not living in mussels **10**

10a (9) P–4 distally with a heavy spiniform seta on the medial aspect (Fig. 42.62). P–3 with a long pectinate seta about the middle of the lateral side, the seta twice as long as P–3 . . . (p. 1103) *Huitfeldtia*

10b   P–4 without such a seta. P–3 with only a short seta in this position, its length less than or scarcely equal to the dorsal length of P–3 . . . . (p. 1104) *Piona*

11a (6) Legs III or IV, or both, exhibiting sexual dimorphism, and differing markedly from the other legs. III–6 (Fig. 42.71) modified for sperm transfer, usually expanded distally, and with the claw considerably modified. IV–4 with a prominent concavity, with stiff, short setae (Fig. 42.70). Tarsi III are often found inserted into the genital opening (Fig. 42.72) . . . . (p. 1104) *Piona*

11b   Legs III and IV not especially different from others in the structure of III–6 and IV–6. . . . . **12**

12a	**(11)**	Epidesmids usually very long, reaching to or beyond the division between ep. III and IV (Fig. 42.55) . . . . . . (p. 1102) **Neumania**
12b		Epidesmids much shorter, never reaching as far as ep. IV, and seldom to the middle of ep. III (Fig. 42.58) **13**
13a	**(12)**	Genital plate extending winglike from genital opening (Fig. 42.60). Parasitic in mussels . . . . (p. 1102) **Najadicola**
12b		Genital plate circular or oval in form, divided or not **14**
14a	**(13)**	Ep. IV rather blocklike or rectangular in form, posterior margin straight, or with a small projection at most; not concave posteromedially. Genital plates nearer to end of body than to epimera. Often living in mussels . . . . (p. 1102) **Unionicola**
14b		Ep. IV with posteromedial margin concave (Fig. 42.58) female. Genital plates nearer to epimera than to end of body . . . . (p. 1103) **Huitfeldtia**

KEY GROUP 6

A single genus. Legs III and IV inserted at sides of body and visible in dorsal view. Dorsum and venter with heavy plates (Fig. 42.7). Legs with spiniform setae, but no swimming setae . . . . . . (p. 1093) **Hydrovolzia**

KEY GROUP 7

1a		With 3 or 4 pairs of genital acetabula **2**
1b		With more than 6 pairs of acetabula . . . . . . (p. 1107) **Arrenurus**
2a	**(1)**	Gnathosoma $2\frac{1}{2}$ to 3 times as long as broad (Fig. 42.89); borne at the end of long, protrusible, membranous tube. . . (p. 1106) **Geayia**

<small>Evidently the tube is not normally protruded, but becomes so after preservation in alcohol and probably other media as well. The correlation between the presence of a tube and an elongate capitulum is unmistakable, however, and can be relied on in cases in which the tube is not extended.</small>

2b		Gnathosoma less than twice as long as broad, not borne at the end of a protrusible tube . . . . (p. 1106) **Krendowskia**

KEY GROUP 8

1a		Swimming setae present on at least some segments of some legs. . . . **2**
1b		Swimming setae absent; legs with spiniform setae **4**
2a	**(1)**	Dorsum with a rather large plate, variable in form, lying between and behind the eyes (Fig. 42.21). First and third pairs of acetabula attached to the genital valves, and knoblike in form . . . . (p. 1097) **Hydryphantes**
2b		Without a dorsal plate. Genital acetabula all lying free in the area between the valves, or on plates **3**
3a	**(2)**	Genital valves present; the acetabula not pitlike, but lying in the space between the valves . . . . . . (p. 1097) **Pseudohydryphantes**
3b		Genital plates present, the acetabula pitlike, imbedded in the plates . . . . (p. 1104) **Tiphys**

<small>The structure of P-4 of *Tiphys* is actually unlike that of the other genera in Key Group 8, and most species will run to Group 2. However, exceptional species, e.g., *T. torris* (Müller) 1776, could understandably run to group 8, so the genus is included here also.</small>

4a   **(1)**   Each genital valve with an anterior setigerous process which extends beyond the level of the anterior acetabulum and bears 1 to several setae (Fig. 42.26) **5**

4b   Anterior setigerous process absent, or if present, it at least does not extend beyond the acetabulum; no setae anterior to the acetabulum (Fig. 42.16) **6**

5a   **(4)**   Anterior setigerous process of genital valve bearing 4 to 8 stout setae which usually curve backward over the anterior acetabulum. A posterior setigerous process also present, lying between the second and third acetabula (Fig. 42.26). . . . . (p. 1096) **Panisopsis**

> *Marshallothyas* Cook 1953 also keys out here. In *Panisopsis*, the anterior dorsal shield forms a complete ring enclosing the median eye; in *Marshallothyas* this ring is open anteriorly, and usually posteriorly as well. Whether or not the differences in degree of development of dorsal plates, which have been utilized to differentiate certain of the genera in the Thyasidae, are of real generic import is questionable. The entire family requires a critical reevaluation.

5b   Anterior setigerous process with a single seta (in known cases); posterior setigerous process absent . . . . **Thyasella***

6a   **(4)**   Dorsum covered with a single reticulate shield in which the dorso-glandularia are imbedded (Fig. 42.20) . . . . . . (p. 1096) **Thyopsis**

6b   Dorsum largely membranous, sometimes scalelike in appearance, or only partially covered with plates. When plates are present they are often irregular in form and extent **7**

7a   **(6)**   Median eye surrounded by a very small ring or elongate sclerite, the width of which is only $\frac{1}{5}$ to $\frac{1}{15}$ the distance between the lateral eyes (Fig. 42.23) **8**

7b   Median eye enclosed in a plate of appreciable size, usually occupying $\frac{1}{3}$ to $\frac{1}{2}$ of the distance between the lateral eyes. **10**

8a   **(7)**   Second and third acetabula close together, at or near the posterior end of the genital sclerites. **9**

8b   Second pair of acetabula reduced in size, lying about halfway between the first and third pairs . . . . (p. 1097) **Euthyas**

9a   **(8)**   With 4 dorsal longitudinal rows of distinct but small sclerites. . . . . (p. 1097) **Thyas**

9b   Dorsum without such sclerites . . . . (p. 1097) **Zschokkea**

10a   **(7)**   Second and third acetabula on either side separated by a pronounced setigerous process (Fig. 42.16) . . . . . . (p. 1096) **Panisus**

10b   No such setigerous process here . . . . **Parathyas***

> *The genera *Parathyas* and *Thyasella* are not yet known from N. A., but their distribution indicates that they probably will be recorded from this continent eventually, along with other genera of the Hydryphantidae.

KEY GROUP 9

1a   Many (usually most) of genital acetabula stalked (Fig. 42.18). Legs without swimming setae. **2**

1b   Acetabula pitlike or knoblike, but not stalked. Legs usually (but not always) with swimming setae **3**

2a   **(1)**   Claws pectinate . . . . (p. 1095) **Protzia**

2b   Claws smooth. . . . . (p. 1096) **Partnunia**

3a   **(1)**   Genital opening covered by a single operculum bearing numerous (often 200 to 300) minute, pitlike acetabula; the operculum often largely enclosed by ep. III and IV (Fig. 42.2)   (p. 1093) **Hydrachna**

3b Genital opening not covered by such an operculum; but guarded by paired valves. With fewer than 100 acetabula on each side of opening. **4**

4a (3) Genital valves each bearing fewer than 20 acetabula. Distidorsal process of P-4 short, thick, less than ½ as long as the basal portion of the segment (Fig. 42.19) **5**

4b Genital valves each bearing 30 to 60 pitlike acetabula. Distidorsal process of P-4 very long, slender, more than ½ as long as the basal portion of the segment (Fig. 42.27) . . . . . . (p. 1097) ***Hydrodroma***

5a (4) With a dorsal plate between the eyes (Fig. 42.21). Legs with swimming setae . . . . (p. 1097) ***Hydryphantes***

5b With no such plate here. Legs without swimming setae . . . .
                   (p. 1096) ***Partnuniella***

Systematics

References treating endemic species are indicated by number; those pertaining to species in other parts of the world are indicated by letter. The following abbreviations are used in giving the world distribution of each genus: N. A., North America; S. A., South America; Eu., Europe; Af., Africa; As., Asia; Aus., Australia; W. I., West Indies; E. I., East Indies.

Superfamily **Hydrovolziae** Viets 1931

Family **Hydrovolziidae** Thor 1905

Genus ***Hydrovolzia*** (s. str.) Thor 1905 (Figs. 42.7, 42.10). Living chiefly on plants growing in cold mountain streams. Two species have been recorded from N. A. Distr.: N. A., Eu., Af., As., E. I. Refs.: 61, 64, 65 A, M.

Superfamily **Hydrachnae** Viets 1931

Family **Hydrachnidae** Leach 1815

Genus ***Hydrachna*** O. F. Müller 1776 (= ***Hydrarachna*** Hermann 1804) (Figs. 42.2, 42.6, 42.8, 42.9). Often very large mites, up to 8000 μ long. Actively swimming mites living principally in standing water. About 20 species, subspecies, and varieties recorded from N. A. in 5 subgenera (below). Distr.: N. A., S. A., Eu., Af., As., E. I., Aus. Refs.: 3, 7, 26, 29, 31, 33, C, M.

Subgenus ***Anohydrachna*** Thor 1916. Dorsum entirely soft, membranous, with no large plates or long, narrow sclerites in anterior portion of body. Cuticle minutely denticulate in known forms. The subgenus has not been recorded previously from N. A., but the writer has seen one species from the state of Conn. It is also known from the W. I. (Aruba), Eur., and As.

Subgenus ***Rhabdohydrachna*** Viets 1931. Dorsum with a pair of long, slender sclerites, each bearing a seta at both ends (Fig. 42.9), or with each sclerite divided into two parts, the anterior and posterior portions of approximately equal width. See also *Scutohydrachna*.

Subgenus ***Diplohydrachna*** Thor 1916. Anterodorsal plates much thicker in anterior half (Fig. 42.8), and sometimes narrowly fused in the vicinity of the unpigmented median eye (Fig. 42.4).

Subgenus ***Hydrachna*** (s. str.) O. F. Müller 1776. Anterodorsal plate moderately large, but not covering the entire dorsum. Posterior margin of plate entire (Fig.

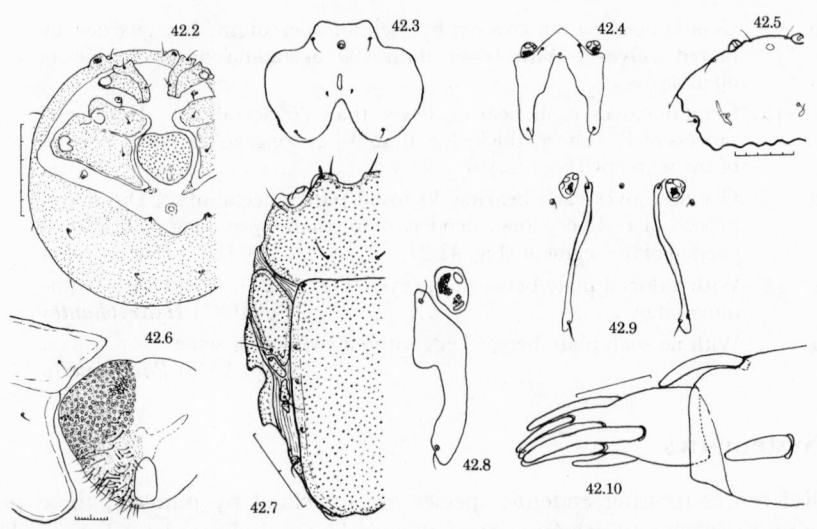

HYDROVOLZIIDAE, HYDRACHNIDAE
Fig. 42.2. *Hydrachna (Scutohydrachna)*, male, venter. **Fig. 42.3.** *Hydrachna* (s.s.), dorsal plate. **Fig. 42.4.** *Hydrachna (Diplohydrachna)*, dorsal plate. **Fig. 42.5.** *Hydrachna* (s.s.), female, dorsal plate. **Fig. 42.6.** *Hydrachna* (s.s.), male, genital operculum, ep. III and IV. **Fig. 42.7.** *Hydrovolzia*, male, dorsum. **Fig. 42.8.** *Hydrachna (Diplohydrachna)*, dorsal plate. **Fig. 42.9.** *Hydrachna (Rhabdohydrachna)*, dorsal plate. **Fig. 42.10.** *Hydrovolzia*, female, tibia and tarsus of palp. (Figs. 42.3, 42.4, 42.8, and 42.9 after Viets.)

42.5), or notched (Fig. 42.3), in which case the two sides of the plate are broadly joined.

Subgenus *Scutohydrachna* Viets 1933 (= *Tetrahydrachna* Lundblad 1934). Male with entire dorsum covered, the sclerotization extending also onto the ventral surface (Fig. 42.2). Female with 4 plates as in some species of the subgenus *Rhabdohydrachna*, but the plates are trapezoidal rather than rodlike or circular and the anterior ones are wider than the posterior ones.

Superfamily **Limnocharae** Viets 1926

Family **Limnocharidae** Kramer 1877

Genus *Limnochares* Latreille 1796 (Figs. 42.13, 42.15, 42.17). Soft-bodied red mites, often of considerable size (up to 5 mm), living in ponds, lakes, or slowly flowing water. Three species known from N. A. in 2 subgenera (below). Distr.: N. A., S. A., Eu., Af., As., Aust. Refs.: 5, 9, 33, 58, M.

Subgenus *Limnochares* (s. str.) Latreille 1796. Creeping forms, legs provided with heavy pectinate setae.

Subgenus *Cyclothrix* Wolcott 1905. Swimming forms, legs with numerous long swimming setae.

Family **Eylaidae** Leach 1815

Genus *Eylais* Latreille 1796 (Figs. 42.12, 42.14). Soft-bodied red mites, often of considerable size (up to 7 mm), living in standing or slowly flowing water. Excellent swimmers, the fourth pair of legs trailing. About 6 species known from N. A. in 2 subgenera (below). Distr.: N. A., S. A., Eu., Af., As., Aus., E. I. Refs.: 3, 4, 9, 29, 33, B, M.

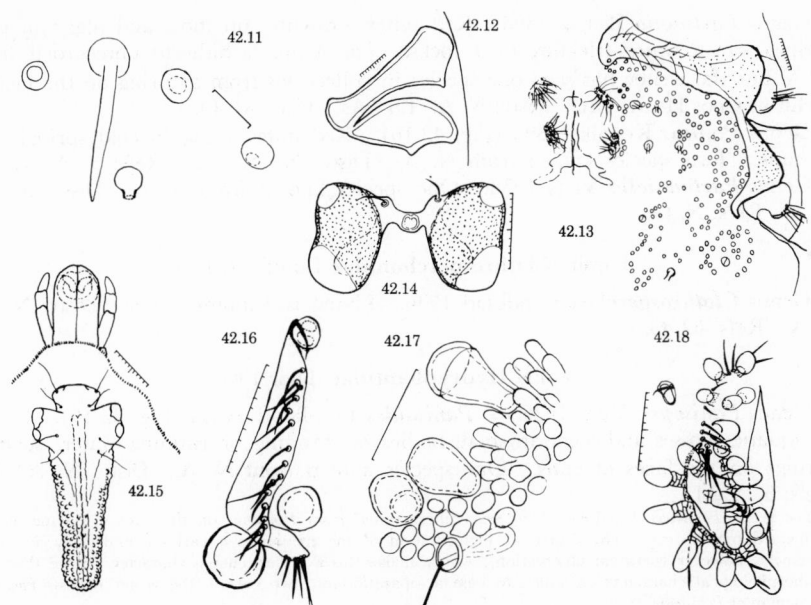

LIMNOCHARIDAE, EYLAIDAE, PROTZIIDAE
Fig. 42.11. *Eylais* (s.s.), female, seta and acetabula. **Fig. 42.12.** *Eylais* (s.s.), ep. III and IV. **Fig. 42.13.** *Limnochares*, male genital area, ep. III and IV. **Fig. 42.14.** *Eylais* (s.s.), ocular bridge. **Fig. 42.15.** *Limnochares*, ocular plate. **Fig. 42.16.** *Panisus*, genital plate and acetabula of right side. **Fig. 42.17.** *Limnochares*, male, genital acetabula and epg. **Fig. 42.18.** *Protzia*, male genital area, right side.

Subgenus **Eylais** (s. str.) Latreille 1796. Genital opening of male bordered by a pair of crescentic plates.

Subgenus **Syneylais** Lundblad 1936. Genital opening of male in the posterior portion of a single pyriform or oblong plate.

Family **Piersigiidae** Wolcott 1905

Genus **Piersigia** Protz 1896. Red mites of moderate size (1300–2000 μ), crawling on the bottom; inhabiting swamps, etc. A single species recorded from N. A. Distr: N. A., Eu. Refs: 62, 66, M.

Superfamily **Hydryphantae** Viets 1931

Family **Protziidae** Viets 1926

Both Lundblad and Viets placed the Protziidae in the Limnocharae, a position which appears to be contrary to most of the morphological characteristics of the family. The writer believes that the family is very closely related to the Hydryphantidae, has only a remote relationship to the Limnocharidae, and therefore the Protziidae are placed in the Hydryphantae.

Genus **Protzia** Piersig 1896 (= **Sporadoporus** Wolcott 1905 = **Calonyx** Walter 1907) (Fig. 42.18). Soft-bodied red mites living mostly in cold mountain streams. Two North American species. Distr: N. A., Eu., Af., As., E. I. Refs: 33, 44, 59, M.

The characters which have been employed to set off *Calonyx* from *Protzia* appear to be of little consequence, and the writer therefore considers the two groups as comprising a single genus which must bear the earlier name proposed by Piersig.

Genus **Partnunia** Piersig 1896. Red mites, crawling on moss and algae in cold springs, or in water trickling over rocks. The genus is hitherto unrecorded from N. A. but the writer has seen one species in collections from a spring on the face of a cliff at Coos Head, Ore. Distr: N. A., Eu., As. Refs: M, O.

Genus **Panisus** Koenike 1896 (Fig. 42.16). Red mites living in cold springs and streams. Two species known from N. A. Distr: N. A., Eu. Refs: 3, A, M, O.

Genus **Partnuniella** Viets 1938. One species known from N. A. Distr: N. A., S. A. Refs: N, O.

Family **Clathrosperchonidae** Lundblad 1936

Genus **Clathrosperchon** Lundblad 1936. Living in running water. Distr: N. A., S. A. Refs: 62, C, O.

Family **Hydryphantidae** Thor 1900

Genus **Panisopsis** Viets 1926 (= **Panisoides** Lundblad 1926) (Fig. 42.19). Living in aquatic mosses and algae in small bodies of standing or running water, such as springs on the faces of cliffs. One species known from N. A. Distr: N. A., Eu. Refs: 3, A, M.

The genus *Panisoides* Lundblad 1926 was differentiated from *Panisopsis* on the basis of having an unpigmented median eye. The degree of pigmentation of the median eye varies considerably in living specimens, however (personal observation), which makes this a rather tenuous character. Since there are no morphological characters on which to base a separation into two genera, the writer regards *Panisoides* a synonym of *Panisopsis*.

Genus **Thyopsis** Piersig 1899 (Fig. 42.20). Our own species of this genus is not sufficiently known to say anything of its habits. However, the European *T. cancellata* (Protz) 1896 is found in a wide variety of habitats including brackish water, cold springs, small ponds, lakes, etc. The genus is hitherto unrecorded from N. A., but the writer has seen it in collections from Calif. and Ore. Distr: N. A., Eu., Af. (Madeira). Refs: A, D, M.

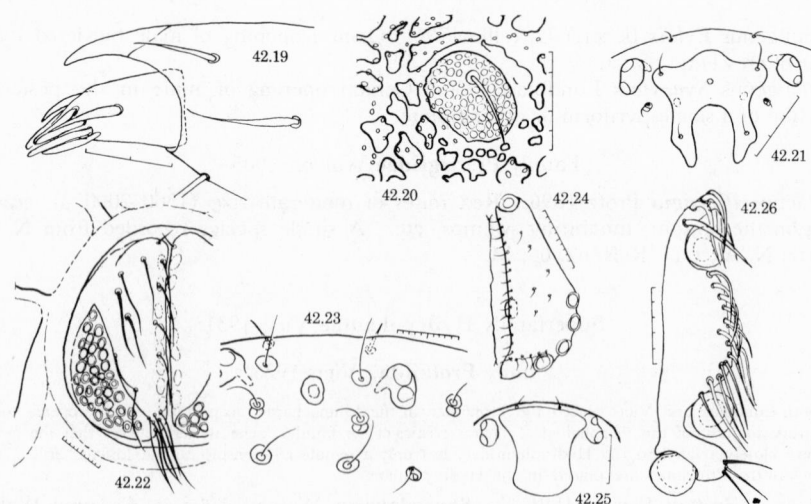

HYDRYPHANTIDAE, THERMACARIDAE
Fig. 42.19. *Panisopsis,* P–4 and P–5. **Fig. 42.20.** *Thyopsis,* portion of reticulate dorsal shield. **Fig. 42.21.** *Hydryphantes* (*Polyhydryphantes*), dorsal shield. **Fig. 42.22.** *Thermacarus,* female genital area. **Fig. 42.23.** *Thyas,* lateral and median eyes. **Fig. 42.24.** *Hydryphantes* (*Polyhydryphantes*), female genital area. **Fig. 42.25.** *Hydryphantes* (s.s.), dorsal plate. **Fig. 42.26.** *Panisopsis,* male genital area.

Genus *Thyas* Koch 1835 (Fig. 42.23). Red mites of moderate size (up to 2000μ); creeping forms living in lakes, ponds, springs, or streams. Two species known from N. A. *T. cataphracta* Koenike 1895 is a synonym of *Panisus cataphracta* (Koenike) 1895, and *T. pedunculata* Koenike 1895 is a synonym of *Panisopsis pedunculata* (Koenike) 1895. Distr.: N. A., Eu., Af. (Madeira), As. Refs.: 3, A, D, M, O.

Genus *Zschokkea* Koenike 1892. Living in either standing or running water. First record from N. A. by David Cook (unpublished). Distr.: N. A., Eu., As. Ref.: A.

Genus *Euthyas* Piersig 1898. Living in either standing or running water. First records from N. A. by Habeeb. Distr.: N. A., Eu. Refs.: A, P, Q.

Genus *Hydryphantes* C. L. Koch 1841 (Figs. 42.21, 42.24, 42.25). Red, soft-bodied mites of wide ecological distribution, occasionally in brackish tide pools. About 5 recorded species from N. A., in 2 subgenera (below). Distr.: N. A., S. A., Eu., Af., As., Aus. Refs.: 24, 29, 31, C, M.

Subgenus *Hydryphantes* (s.s.) Koch 1841. Three pairs of genital acetabula.
Subgenus *Polyhydryphantes* Viets 1926. More than 4 pairs of acetabula.

Species with 4 pairs of acetabula go in the subgenus *Octohydryphantes* Lundblad 1927, not yet recorded from N. A.

Family **Thermacaridae** Sokolow 1927

Genus *Thermacarus* Sokolow 1927 (Fig. 42.22). Living in hot springs. A single species recorded from N. A., *T. nevadensis* Marshall 1928. Distr.: N. A., As. Refs.: 28, I.

Family **Hydrodromidae** Viets 1936

Genus *Hydrodroma* Koch 1837 (= *Diplodontus* auct., *nec* Dugès 1833) (Fig. 42.27). Red-bodied swimming mites, widely distributed in standing waters. Two species recorded from N. A. Distr.: N. A., S. A., Eu., Af., As., Aus., W. I., E. I. Refs.: 29, 33, C.

Family **Pseudohydryphantidae** Viets 1926

Genus *Pseudohydryphantes* Viets 1907. Excellent swimmers living in ponds, lakes, swamps or very slowly flowing water. Two species recorded from N. A. Distr.: N. A., Eu. Refs.: 22, 29, M.

Family **Teutoniidae** Lundblad 1927

Genus *Teutonia* Koenike 1899 (Fig. 42.29). Very rapid swimmers, living in a wide variety of habitats in standing water. A single species recorded from N. A. Distr.: N. A., Eu., Af., As. Refs.: 22, M.

Family **Sperchonidae** Thor 1900

Genus *Sperchonopsis* Piersig 1896 (= *Pseudosperchon* Piersig 1901). Found in running water. Two species recorded from N. A. Distr.: N. A., Eu., Af., As. Refs.: 44, M.

Genus *Sperchon* Kramer 1877 (Figs. 42.28, 42.32, 42.33). A group with rather wide ecological range, but most commonly found in cold running water. Five species recorded from N. A. Distr.: N. A., S. A., Eu., Af., As., Aus., E. I. Refs.: 3, 29, 35, M.

A number of subgenera have been established in the genus, based solely on cuticular patterns. It is very probable that these are not all natural groups, and unless other morphological characters can be found to augment the cuticular characters, it will be advisable to abandon the present subgenera. They are included here, with great reservation, for the information of the student.

Subgenus *Sperchon* (s. str.) Kramer 1877. Cuticle of dorsum papillose (Fig. 42.28) or marked with coarse, wavy lines.

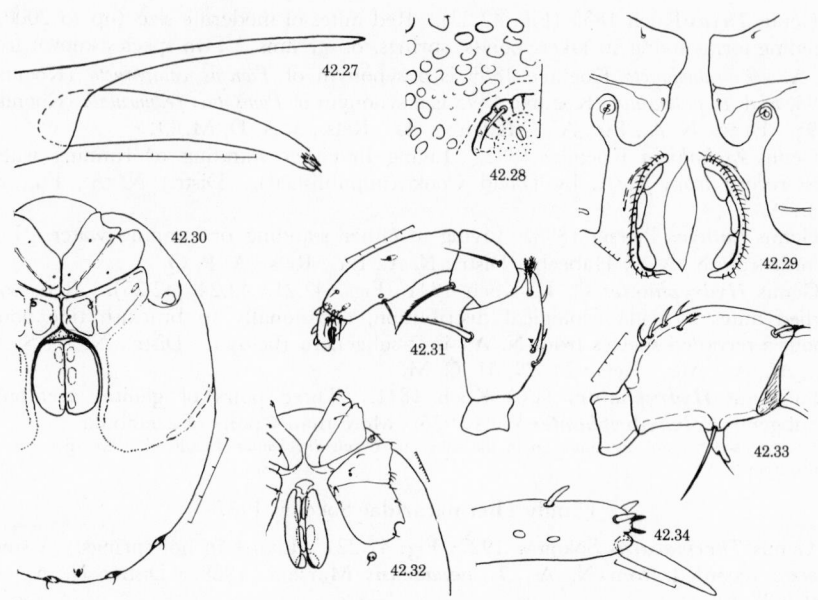

HYDRODROMIDAE, TEUTONIIDAE, SPERCHONIDAE, ANISITSIELLIDAE
Fig. 42.27. *Hydrodroma*, male, P–4 and P–5. **Fig. 42.28.** *Sperchon* (s.s.), dorsal cuticle and dgl. **Fig. 42.29.** *Teutonia*, female genital area. **Fig. 42.30.** *Mamersella*, female, venter. **Fig. 42.31.** *Mamersella*, female, palp. **Fig. 42.32.** *Sperchon* (*Scutosperchon*), female, venter. **Fig. 42.33.** *Sperchon* (s.s.), P–1 and P–2. **Fig. 42.34.** *Mamersella*, IV–6.

Subgenus **Hispidosperchon** Thor 1901. Cuticle of dorsum reticulately paneled, the walls of the panels made up of short, hairlike processes of uniform size.

Subgenus **Mixosperchon** Viets 1926. Cuticle of dorsum reticulately paneled, but with small tubercles or coniform processes interspersed among the panels. No known North American species are referable with certainty to this group.

Subgenus **Scutosperchon** Viets 1926. Cuticle of dorsum armored (**Sperchon parmatus** Koenike 1895, with 2 unpaired dorsal plates, probably goes here).

Family **Anistisiellidae** Viets 1929

Genus **Mamersella** Viets 1929 (Figs. 42.30, 42.34). Living in springs and streams. A single undescribed species from Ore. The genus has not been recorded previously from N. A. Distr.: N. A., E. I. Ref.: J.

Superfamily **Lebertiae** Viets 1935

Family **Lebertiidae** Thor 1900

Genus **Lebertia** Neuman 1880 (Figs. 42.35–42.37). Living in cold springs, creeks and lakes, the various subgenera showing moderate ecological specificity. About 12 species recorded from N. A. in 3 subgenera (below). Distr.: N. A., Eu., Af., As. Refs.: 3, 4, 15, 16, 28, 29, 35, 44, M.

KEY TO SUBGENERA OF *LEBERTIA*

(Both palpi should be examined for variability)

1a   P–3 with 5 setae medially (Fig. 42.37) **2**

1b   P–3 with 6 (sometimes 7) setae medially . . . .
Subgenus *Mixolebertia* Thor 1906

The writer can find no reliable character on which to separate *Hexalebertia* Thor 1907 from *Mixolebertia* Thor 1906, and regards them as identical. *Mixolebertia* has priority.

2a   (1)   All of the small dorsal setae of P–4 are confined to the distal third of the segment (Fig. 42.37). Legs II–IV usually with swimming setae, often few on II . . . . . . Subgenus *Pilolebertia* Thor 1900

2b   One or 2 of the small dorsal setae of P–4 are in the middle, or even basal third of the segment   3

3a   (2)   IV–1 with 3 to 4 setae dorsally, swimming setae few or absent, especially on II. Dorsal cuticle smooth . . . .
Subgenus *Lebertia* (s. str.) Neuman 1880

3b   IV–1 usually with 6 setae dorsally. Swimming setae absent. Dorsal cuticle papillose or striate. (Not yet recorded from N. A.) . . . .
Subgenus *Pseudolebertia* Thor 1897

Genus *Frontipoda* Koenike 1891 (Figs. 42.39, 42.42). Principally in standing and slowly running water. One species recorded from N. A. Distr.: N. A., S. A., Eu., Af., As. Refs.: 22, 34, C, M.

Genus *Gnaphiscus* Koenike 1898 (Fig. 42.40). In lakes and ponds. One species recorded from N. A. Distr.: N. A., S. A., Eu., As. Refs.: 22, C, M.

Genus *Oxus* Kramer 1877. Found in lakes and ponds. Three species described from N. A. Distr.: N. A., Eu., Af., As., E. I. Refs.: 29, 34, C, M.

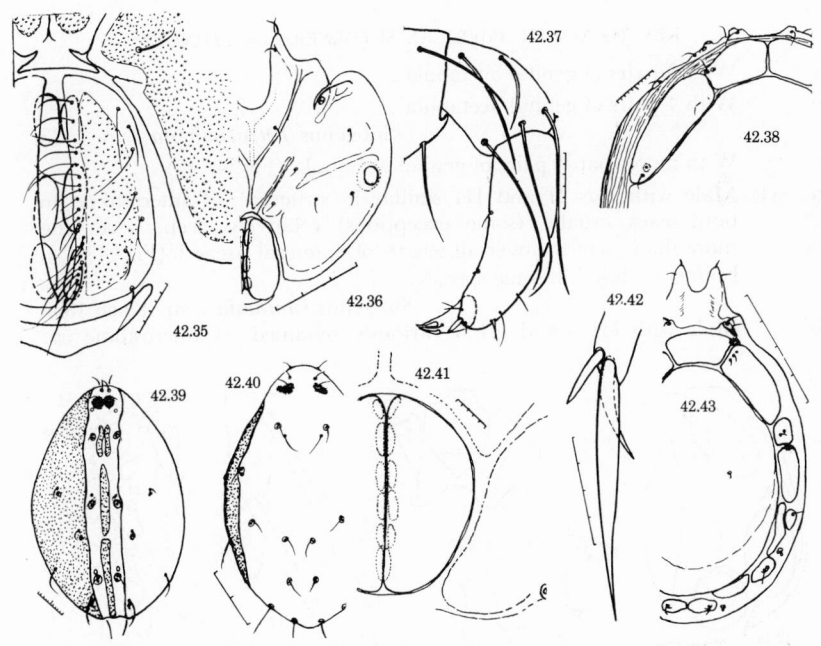

LEBERTIIDAE, TORRENTICOLIDAE
Fig. 42.35. *Lebertia* (*Pilolebertia*), male, genital area. **Fig. 42.36.** *Lebertia* (*Pilolebertia*), female, ep. **Fig. 42.37.** *Lebertia* (*Pilolebertia*), P–4 and P–5. **Fig. 42.38.** *Torrenticola* (s.s.), dorsum. **Fig. 42.39.** *Frontipoda*, female, dorsum. **Fig. 42.40.** *Gnaphiscus*, female, dorsum. **Fig. 42.41.** *Torrenticola*, female, genital area. **Fig. 42.42.** *Frontipoda*, IV–5 and IV–6. **Fig. 42.43.** *Testudacarus*, female, dorsum.

Family **Torrenticolidae** Piersig 1902

Genus **Torrenticola** Piersig 1896 (Figs. 42.38, 42.41). Living in mountain streams for the most part. About 15 species recorded from N. A., in 2 subgenera (below). The majority of species of *Torrenticola* heretofore described or listed have been erroneously assigned to the genus *Atractides*, while the majority of true species of *Atractides* have been assigned to the genus *Megapus*. The latter is a synonym of *Atractides* (see footnote to list of synonyms on page 1107). Distr: N. A., S. A., Eu., Af., As., E. I. Refs: 28, 31, 34, 35, 43, J, K, M, O.

Subgenus **Torrenticola** (s.str.) Piersig 1896. With the 4 anterior plates discrete, separated not only from each other but from the main dorsal plate as well (Fig. 42.38).

Subgenus **Rusetria** Thor 1897. With the anterolateral plates fused to the main dorsal plate. This subgenus is of questionable validity, and certainly the character is difficult to interpret in many specimens.

Genus **Testudacarus** Walter 1928 (Fig. 42.43). Living chiefly in cold streams. Widely distributed in western N. A., 2 species described. Distr: N. A., As. Ref: 43.

Superfamily **Pionae** Viets 1930

Family **Limnesiidae** Thor 1900

Genus **Limnesia** C. L. Koch 1836 (Figs. 42.44–48). Usually excellent swimmers, inhabiting a wide range of habitats in standing and slowly flowing water. About 20 named species from N. A. in 5 subgenera (see below). Distr: N. A., S. A., Eu., Af., As., Aus., E. I., W. I. Refs: 8, 9, 22, 28, 34, 47, 50, 57, 65, C, M, P, Q.

KEY TO NORTH AMERICAN SUBGENERA OF *LIMNESIA*

1a   With 3 pairs of genital acetabula **2**

1b   With 4 pairs of genital acetabula . . . .
                                    Subgenus **Tetralimnesia** Thor 1923

1c   With more than 4 pairs of genital acetabula (Fig. 42.48) **3**

2a   (1) Male with legs II and III similar in structure. Camerostome in both sexes usually (some exceptions) relatively deep, occupying more than ½ of the over-all length of epimeral areas I (Fig. 42.45). Epimera I fused in some species . . . .
                                    Subgenus **Limnesia** s. str. Koch 1836

2b   Male with III–5 and III–6 variously modified. Camerostome not

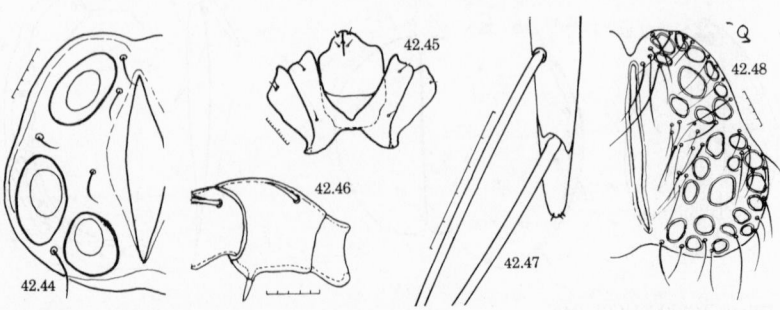

LIMNESIIDAE
Fig. 42.44. *Limnesia* (s.s.), male, genital area. **Fig. 42.45.** *Limnesia* (s.s.), female, ep. I and II. **Fig. 42.46.** *Limnesia* (s.s.), male, P–1 and P–2. **Fig. 42.47.** *Limnesia* (s.s.), tarsus IV. **Fig. 42.48.** *Limnesia* (*Limnesiopsis*), male, genital area.

deep, occupying no more than ½ of over-all length of epimeral areas I (Females which go here should not be assigned to this subgenus unless the male is also known.) Epimera I not fused in any known species. Thus far known only from Haiti and S. A. . . . . .

Subgenus *Centrolimnesia* Lundblad 1935

3a　(1) Claws pectinate, or with a large accessory tooth . . . .

Subgenus *Limnesiopsis* Piersig 1896

3b　　　Claws not pectinate. . . . . Subgenus *Limnesiella* Daday 1905

Family **Tyrrelliidae** Viets 1935

Genus *Tyrrellia* Koenike 1895 (Fig. 42.1). Living in ponds, or in water running down cliffs, and in warm springs. Two forms known from N. A. Distr: N. A., S. A. Refs: 3, 34, 42, C.

Family **Hygrobatidae** Koch 1842

Genus *Hygrobates* Koch 1837 (Fig. 42.49). Chiefly inhabitants of rivers and other running waters. About 7 species from N. A. in 4 subgenera (below). Distr: N. A., S. A., Eu., Af., As., Aus., E. I. Refs: 3, 22, 29, 34, 44, E, M.

Subgenus *Hygrobates* (s. str.) Koch 1837. Three pairs of genital acetabula.
Subgenus *Tetrabates* Thor 1923. Four pairs of genital acetabula.
Subgenus *Dekabates* Thor 1923. Five pairs of genital acetabula.
Subgenus *Rivobates* Thor 1897. More than 6 pairs of genital acetabula.

Genus *Corticacarus* Lundblad 1936 (Fig. 42.51). Inhabitants of running water. The genus has never been recorded from N. A., but the writer has seen 1 species of uncertain subgeneric position from Utah. A number of related genera and subgenera have been described from S. A. The status of some of these will undoubtedly require revision when the group is better known. Distr: N. A., S. A. Ref: E.

Genus *Atractides* Koch 1837 (= *Megapus* Neuman 1880) (Figs. 42.50, 42.52). Principally inhabitants of running water, especially of mountain streams. Four or 5 species known from N. A., all in the subgenus *Atractides* s. str. The majority of species heretofore referred to *Atractides* actually belong in the genus *Torrenticola*, while the majority of true species of *Atractides* have been assigned erroneously to the genus *Megapus*. See footnote to list of generic synonyms on page 1107. Distr: N. A., S. A., Eu., As., Af. Refs: 17, 26, 34, 44, E, M, O.

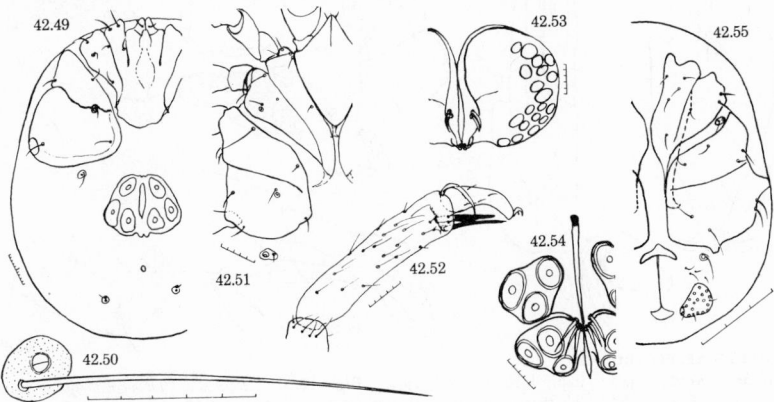

HYGROBATIDAE, UNIONICOLIDAE
Fig. 42.49. *Hygrobates* (s.s.), female, venter. **Fig. 42.50.** *Atractides* (s.s.), dgl. **Fig. 42.51.** *Corticacarus*, male, venter. **Fig. 42.52.** *Atractides* (s.s.), I–5 and I–6. **Fig. 42.53.** *Unionicola* (s.s.), female genital area. **Fig. 42.54.** *Unionicola* (*Hexatax*), female genital area. **Fig. 42.55.** *Neumania* (s.s.), female venter.

Family **Unionicolidae** Oudemans 1909

Genus *Unionicola* Haldeman 1842 (= *Atax* Fabricius 1805) (Figs. 42.53, 42.54). Occupying a wide variety of habitats in standing and running water. Many, but not all of the species, are parasitic in the branchial chamber of Unionidae. About 20 named species from N. A., in 5 subgenera (below). Distr: N. A., S. A., Eu., Af., As., Aus., E. I. Refs: 2, 3, 10, 36, 37, 50, 54, E, M.

KEY TO NORTH AMERICAN SUBGENERA OF *UNIONICOLA*

1a　　　With 5 pairs of genital acetabula (both sexes); female with 4 genital plates . . . . Subgenus *Pentatax* Thor 1923

1b　　　With 6 pairs of acetabula; female with 4 genital plates (Fig. 42.54).
　　　　　　　　　　　　　　　　Subgenus *Hexatax* Thor 1926

1c　　　With more than 6 pairs of acetabula . . . .　**2**

2a　(1) Female with only 2 genital plates (Fig. 42.53). Leg IV of male showing no sexual dimorphism . . . .
　　　　　　　　　Subgenus *Unionicola* (s. str.) Haldeman 1842

2b　　　Female with 4 genital plates (Fig. 42.54). Leg IV of male showing sexual dimorphism or not . . . .　**3**

3a　(2) Leg IV of male not showing sexual dimorphism . . . .
　　　　　　　　　　　　　　Subgenus *Polyatax* Viets 1933

3b　　　Leg IV of male showing sexual dimorphism . . . .
　　　　　　　　　　　　Subgenus *Neoatax* Lundblad 1941

Genus *Neumania* Lebert 1879 (Fig. 42.55). Largely restricted to standing or slowly flowing water. About 15 species in N. A. in 2 subgenera (below). Distr.: N. A., S. A., Eu., Af., As., E. I. Refs.: 20, 22, 28, 36, 39, E, K, M.

Subgenus *Neumania* (s. str.) Lebert 1879. Genital acetabula of female borne on 2 plates (Fig. 42.55).

Subgenus *Tetraneumania* Lundblad 1930. Genital acetabula of female borne on 4 plates.

Genus *Najadicola* Piersig 1897 (Figs. 42.59, 42.60). Parasitic in species of *Unio*

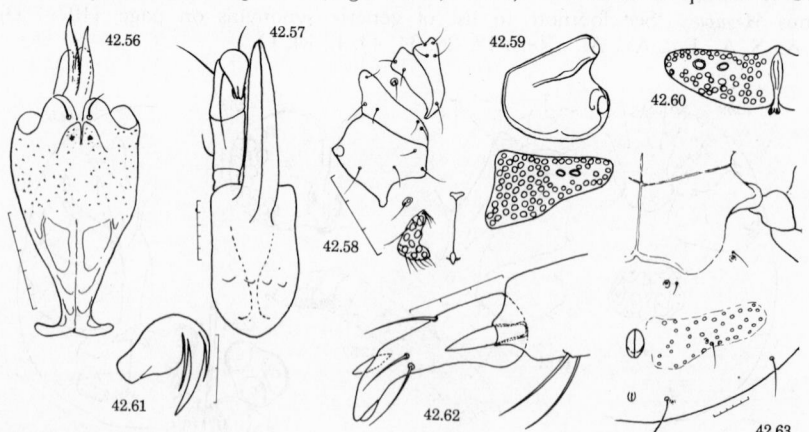

UNIONICOLIDAE, FELTRIIDAE

Fig. 42.56. *Koenikea* (s.s.), capitulum, ventral view. **Fig. 42.57.** *Koenikea* (*Tanaognathella*), capitulum, ventral view. **Fig. 42.58.** *Huitfeldtia,* female genital area, ep. **Fig. 42.59.** *Najadicola,* female genital area and posterior ep. **Fig. 42.60.** *Najadicola,* male genital area. **Fig. 42.61.** *Feltria* (*Feltriella*), male, claw III. **Fig. 42.62.** *Huitfeldtia,* P-4 and P-5. **Fig. 42.63.** *Koenikea* (s.s.), male, venter. (Figs. 42.58 and 42.60 after Wolcott.)

and *Anodonta.* A single known species, *N. ingens* (Koenike) 1895. Distr.: N. A. Refs: 3, 54.

Genus *Huitfeldtia* Thor 1898 (Figs. 42.58, 42.62). In lakes, commonly at depths below 10 meters. A single North American species, *H. rectipes* Thor 1898. Refs: 38, M.

Genus *Koenikea* Wolcott 1900 (Figs. 42.56, 42.57, 42.63). Found principally in rivers and creeks, rarely in springs. About 8 species described from N. A. in 4 subgenera (below). Distr.: N. A., S. A., Af., As., E. I., W. I. Refs: 8, 39, 52, 55, F.

KEY TO NORTH AMERICAN SUBGENERA OF *KOENIKEA*

1a		Base of gnathosoma when seen in ventral view anchor-shaped at posterior end (Fig. 42.56) . . . . Genus *Koenikea* (s. str.)
1b		Base of gnathosoma rounded posteriorly, when seen in ventral view (Fig. 42.57) 2
2a	(1)	Rostrum very long (Fig. 42.57) . . . . Subgenus *Tanaognathella* Lundblad 1941
2b		Rostrum much shorter. 3
3a	(2)	Males with legs showing sexual modifications . . . . Subgenus *Tanaognathus* Wolcott 1900
3b		Males without sexual modifications of legs . . . . Subgenus *Pseudokoenikea* Lundblad 1941

Family **Feltriidae** Thor 1929

Genus *Feltria* Koenike 1892 (Fig. 42.61). Living in moss and algae growing in running water. Two North American species in 2 subgenera (below). Distr.: N. A., Eu., As. Refs.: 1, M.

Subgenus *Feltria* ((s. str.) Koenike 1892. IV–6 of male with a prominent ventral process which bears several setae.

Subgenus *Feltriella* Viets 1930. IV–6 normal, like III–6 in form. (New record, based on original material from Coos Head, Ore.)

Family **Pionidae** Thor 1900

Genus *Wettina* Piersig 1892 (Fig. 42.66). Living in running water, especially in small creeks. A single species, *W. primaria* Marshall 1929, described from N. A. Distr.: N. A., Eu. Refs.: 29, M.

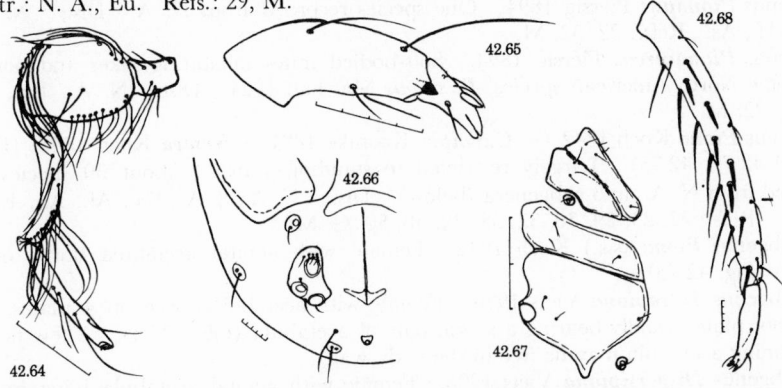

PIONIDAE

Fig. 42.64. *Tiphys* (s.s.), male, IV–3 to IV–6. **Fig. 42.65.** *Tiphys*, female, P–4 and P–5. **Fig. 42.66.** *Wettina*, female genital area. **Fig. 42.67.** *Tiphys* (s.s.), ep. **Fig. 42.68.** *Hydrochoreutes*, female, P–3 and P–4.

Genus **Hydrochoreutes** Koch 1842 (Fig. 42.68). Living in standing or slowly flowing waters. A single widely distributed species, **H. ungulatus** (Koch) 1826, recorded from N. A. Distr.: N. A., Eu., Af., As. Refs.: 40, M.

Genus **Tiphys** Koch 1837 (=**Laminipes** Piersig 1901; **Acercus** Koch 1842) (Figs. 42.64, 42.65, 42.67). Swimming mites, inhabitants of running water and tundra pools. Three forms recorded from N. A. Distr.: N. A., Eu., Af., As. Refs.: 22, 40, M, O.

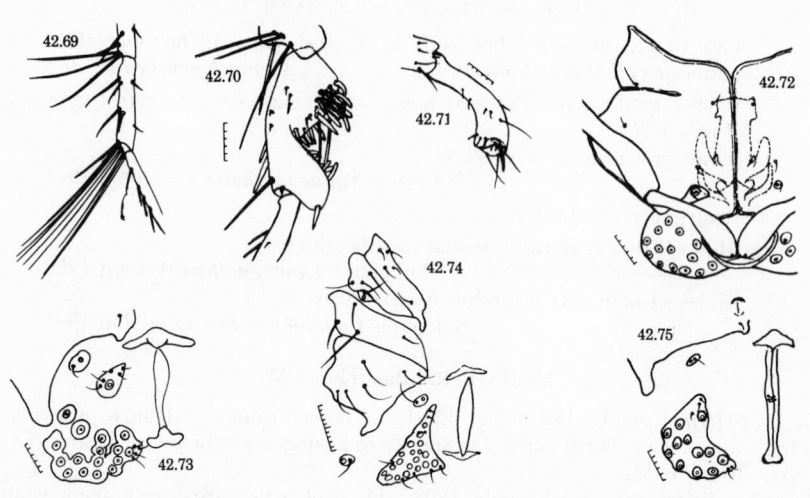

PIONIDAE
Fig. 42.69. *Forelia*, female, IV-5. **Fig. 42.70.** *Piona* (s.s.), male, IV-4. **Fig. 42.71.** *Piona* (s.s.), male, III-6. **Fig. 42.72.** *Piona* (s.s.), male genital area and ep. III and IV. **Fig. 42.73.** *Piona* (*Tetrapiona*), female genital area. **Fig. 42.74.** *Forelia*, female genital area and ep. **Fig. 42.75.** *Piona* (s.s.), female genital area.

Genus **Pionopsis** Piersig 1894. One species recorded from N. A. Distr.: N. A., Eu., Af., As. Refs.: 22, 32, M.

Genus **Pionacercus** Piersig 1894. Soft-bodied mites inhabiting lakes and ponds. A single North American species, **P. novus** Marshall 1924. Distr.: N. A., Eu., As. Refs.: 22, M.

Genus **Piona** Koch 1842 (= **Curvipes** Koenike 1891 = **Nesaea** Koch 1842) (Figs. 42.70–42.73, 42.75). Largely restricted to standing water. About 30 species recorded from N. A., in 3 subgenera (below). Distr.: N. A., S. A., Eu., Af., As., E. I., W. I. Refs.: 22, 23, 29, 30, 32, 38, 39, 40, 50, G, M.

Subgenus **Piona** (s.s.) Koch 1842. Female with genital acetabula borne on 2 plates (Fig. 42.75).

Subgenus **Tetrapiona** Viets 1926. Female with acetabula borne on 4 plates, the anterior plates usually bearing a single pair of acetabula (Fig. 42.73). A few of the remaining acetabula may lie free in the body wall.

Subgenus **Dispersipiona** Viets 1926. Female with genital acetabula lying in the body wall; plates absent.

Genus **Forelia** Haller 1882 (Figs. 42.69, 42.74). Principally inhabitants of standing or slowly flowing water. Two species recorded from N. A. Distr.: N. A., Eu., Af., As. Refs.: 29, 38, M.

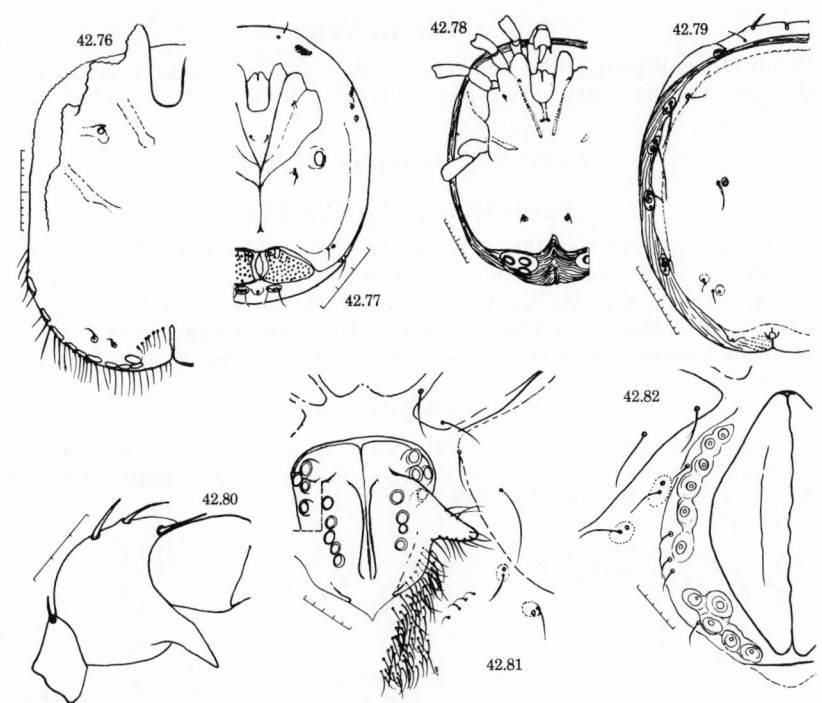

AXONOPSIDAE, MIDEIDAE
Fig. 42.76. *Aturus*, male venter. **Fig. 42.77.** *Albia*, female venter. **Fig. 42.78.** *Brachypoda*, female venter.
Fig. 42.79. *Aturus*, female dorsum. **Fig. 42.80.** *Aturus*, female, P-2. **Fig. 42.81.** *Midea*, male genital
area (portion of right valve deleted to show acetabula). **Fig. 42.82.** *Midea*, female genital area.

Superfamily **Axonopsae** Viets 1931

Family **Axonopsidae** Viets 1929

Genus *Axonopsis* Piersig 1893. Principally found in lakes and ponds. A single
species recorded from N. A. Distr: N. A., Eu., Af., As., E. I. Refs: 59, K, M.

Genus *Albia* Thon 1899 (Fig. 42.77). Living in lakes, ponds, and slowly flowing
streams. Two species recorded from N. A. Distr: N. A., S. A., Af., As., E. I.
Refs: 26, 41, 58, K, M.

Genus *Aturus* Kramer 1875 (= *Subaturus* Viets 1916) (Figs. 42.76, 42.79, 42.80).
Principally inhabitants of mountain streams. Two species recorded from N. A.
Distr: N. A., Eu., Af., As., E. I. Refs: 3, K, M.

A. (s.s.) *obtusisetus* Viets 1935, from Java and Sumatra (K), is represented by 2 forms, one of which
lacks the ventral spine of P-2. Since this is the one character on which the subgenus *Subaturus* is based,
then this species would occupy two subgenera. Therefore, *Subaturus* is invalid.

Genus *Kongsbergia* Thon 1899. Living chiefly in cold mountain streams. Four
species known from N. A. Distr: N. A., Eu., Af., As., E. I. Refs: K, M.

Genus *Brachypoda* Lebert 1879 (Fig. 42.78). In standing water or small creeks.
One species known from N. A. Distr: N. A., Eu., As. Ref: N.

Genus *Ljania* Thor 1898. In creeks. Distr: N. A., Eu. Refs: 62, 65, N.

Genus *Neoaxonopsis* Lundblad 1938. Two species listed from N. A. Distr:
N. A., S. A. Refs: 62, 65, G.

Family **Mideidae** Viets 1929

Genus *Midea* Bruzelius 1854 (Figs. 42.81, 42.82). Widely distributed in standing and slowly running water. Two species recorded from N. A. Distr: N. A., Eu. Refs: 29, 41.

Superfamily **Mideopsae** Viets 1931

Family **Mideopsidae** Thor 1928

Genus *Mideopsis* Neuman 1880 (Figs. 42.85, 42.86). Chiefly inhabitants of standing or slowly flowing water. About 7 species in 2 subgenera (below) in N. A. Distr.: N. A., S. A., Eu., Af., As., W. I. Refs.: 9, 23, 41, 44, 55, G, M.
Subgenus *Mideopsis* (s. str.) Neuman 1880. Legs with swimming setae.
Subgenus *Xystonotus* Wolcott 1900. Legs without swimming setae.

Family **Krendowskijidae** Lundblad 1930

Genus *Krendowskia* Piersig 1895. Inhabitants of standing water and slowly flowing streams. One species recorded from N. A., *K.* (s. str.) *similis* Viets 1931. Distr.: N. A., S. A., Eu., Af., As. Refs.: 41, 53, G.

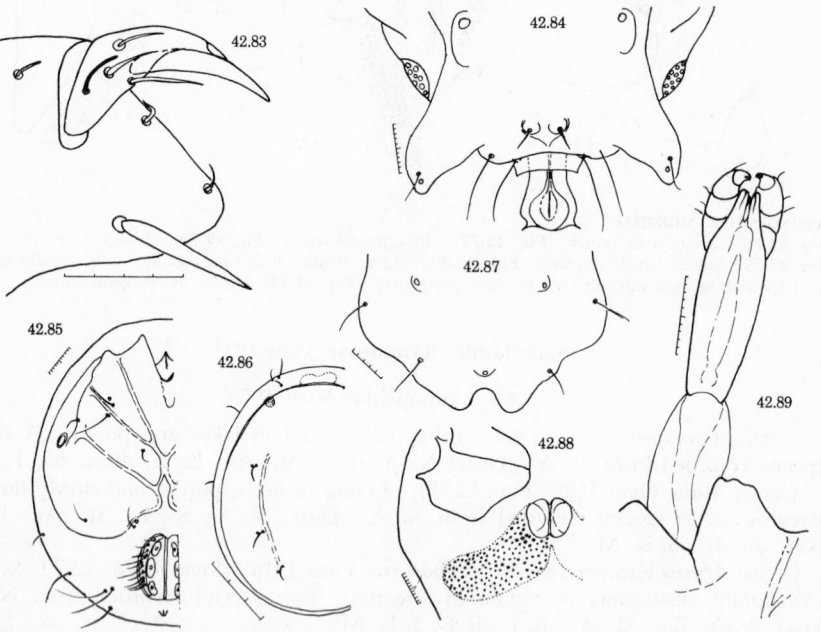

MIDEOPSIDAE, KRENDOWSKIJIDAE, ARRENURIDAE
Fig. 42.83. *Arrenurus*, female, P-4 and P-5. **Fig. 42.84.** *Arrenurus* (s.s.), male cauda, dorsal view.
Fig. 42.85. *Mideopsis* (*Xystonotus*), venter. **Fig. 42.86.** *Mideopsis* (*Xystonotus*), dorsum. **Fig. 42.87.** *Arrenurus* (*Truncaturus*), male cauda, ventral view. **Fig. 42.88.** *Arrenurus*, female, genital area. **Fig. 42.89.** *Geayia* (s.s.), female, capitulum.

Genus *Geayia* Thor 1897 (Fig. 42.89). Living in running water. One described species from N. A., *G.* (s.s.) *ovata* (Wolcott) 1900. It is probable that others exist, and that there are at least 2 subgenera (below). Distr.: N. A., S. A. Refs.: 10, 41, 55, G.
Subgenus *Geayia* (s.s.) Thor 1897. With 4 pairs of genital acetabula.
Subgenus *Geayidea* Lundblad 1941. With 3 pairs of genital acetabula.

The male of *Geayia ovata* (Wolcott) was described as having 3 pairs of acetabula although the female had 4. From this and other considerations, it is evident that Wolcott had 2 species, the other one probably a *Geayidea*.

Superfamily Arrenurae Oudemans 1902

Family Arrenuridae Thor 1900

Genus *Arrenurus* Dugès 1834 (= *Stegnaspis* Wolcott 1901 = *Arrhenurus* auct.) (Figs. 42.83, 42.84, 42.87, 42.88). Chiefly inhabitants of lakes, ponds, and slowly flowing streams. About 85 forms described from N. A. in 2 subgenera. Distr.: N. A., S. A., Eu., Af., As., Aus., E. I., W. I. Refs.: 3, 6, 11, 12, 13, 14, 16, 18, 19, 21, 29, 39, 41, 45, H, L, M.

The author is in agreement with Lundblad (H, p. 34) in the view that no more than 2 subgenera, at the most, are valid in *Arrenurus*. One of these is *Arrenurus* (s. str.); the other must be either *Truncaturus* [Type *A. knauthei* Koenike 1895] or *Megaluracarus* [Type *A. globator* (Mull.) 1776]. Since *Truncaturus* Thor 1900 has obvious priority over *Megaluracarus* Viets 1911, the author uses the former name, and regards *Megaluracarus* as a synonym.

While it may be possible to differentiate more than two subgeneric groups on the basis of body form in limited faunules, the artificiality of this system becomes increasingly apparent as more and more species become known. Inevitably cases are encountered in which naturally related species are sundered and placed in separate subgenera along with species with which they have little in common.

Subgenus *Arrenurus* (s. str.) Dugès 1834. Cauda of males with a pronounced petiolus, usually projecting well beyond posterior margin. Pygal lobes usually, but not always, present. Cauda (exclusive of petiolus) broader than long (Fig. 42.84).

Subgenus *Truncaturus* Thor 1900. Petiolus small or absent, never projecting well beyond margin of cauda. Pygal lobes absent. Cauda usually, but not always, longer than broad (Fig. 42.87).

Generic Synonyms Appearing in Works Based on American Species

Acercus Koch 1842. Syn. of *Tiphys* Koch 1837.[3]
Arrhenurus auct. Syn. of *Arrenurus* (s. str.) Dugès 1834.
Atax Fabricius 1805. Syn. of *Unionicola* Haldeman 1842.
Atractides auct., nec Koch 1837. Most species heretofore referred to this genus actually belong in *Torrenticola* Piersig 1897. Both genera are valid.[3]
Calonyx Walter 1907. Syn. of *Protzia* Piersig 1896.
Curvipes Koenike 1891. Syn. of *Piona* Koch 1842.
Diplodontus auct., nec Dugès 1833. Syn. of *Hydrodroma* Koch 1837.
Hydrarachna auct. Syn. of *Hydrachna* O. F. Müller 1776.
Laminipes Piersig 1901. Syn. of *Tiphys* Koch 1837.[3]
Megaluracarus Viets 1911. Syn. of *Truncaturus* Thor 1900.
Megalurus Thon 1900. Syn. of *Truncaturus* Thor 1900.
Megapus Neuman 1880. Syn. of *Atractides* Koch 1837.[3]
Micruracarus Viets 1911. Syn. of *Truncaturus* Thor 1900.
Micrurus Thon 1900. Syn. of *Truncaturus* Thor 1900.
Nesaea Koch 1842. Syn. of *Piona* Koch 1842.
Panisoides Lundblad 1926. Syn. of *Panisopsis* Viets 1926.
Pseudosperchon Piersig 1901. Syn. of *Sperchonopsis* Piersig 1896.
Sporadoporus Wolcott 1905. Syn. of *Protzia* Piersig 1896.
Steganaspis Wolcott 1901. Syn. of *Arrenurus* Dugès 1833.
Subaturus Viets 1916. Syn. of *Aturus* Kramer 1875.
Tetrahydrachna Lundblad 1934. Syn. of *Scutohydrachna* Viets 1933.

---

[3] Prior to 1950, the generic names *Atractides*, *Megapus*, *Torrenticola*, *Acercus*, and *Tiphys* were almost universally used in an incorrect manner. Their status was clarified by Viets in 1949 (Reference O, page 1116).

HALACARIDAE

Generally speaking the term water mite is applied only to aquatic members of the Parasitengona, a practice which has much to recommend it and will be followed here. However, other mites are frequently encountered in fresh-water studies, most notably certain of the Oribatei and the Halacaridae. The latter family is a predominantly marine family with most of the genera and species restricted to marine habitats of one kind or another. But several genera have evolved in fresh water and their representatives are often found crawling about on the surface of aquatic plants or on the bottom of lakes, streams, etc., or even in interstitial water of coarse sand. Unlike some of the Parasitengona, they do not swim. Rarely found in large numbers, the fresh-water Halacaridae are nevertheless a highly characteristic element of the fresh-water fauna, and in certain studies may be encountered more frequently than the aquatic Parasitengona themselves.

The invasion of fresh water by the Halacaridae has obviously occurred at more than one time in their history; that is, the fresh-water Halacaridae are polyphyletic. For example, the fresh-water genera *Porolohmannella* and *Porohalacarus* are much more closely related to the marine genera *Lohmannella* and *Halacarus*, respectively, than they are to each other. Viets (1936, pp. 520–521, and other sources) established a separate family, the Porohalacaridae Viets 1933, for the Halacaridae found in fresh water, but as the writer pointed out (Newell 1947, pp. 22–25, 38) this was a highly artificial arrangement cutting sharply across natural lines. Some revisional work remains to be done in order to express properly the true relationships of the fresh-water Halacaridae to the marine genera, but it is already obvious that the separation of the fresh-water genera into a distinct family is untenable on morphological grounds. The only thing that all these genera have in common, outside of habitat, is the possession of external rather than internal genital acetabula, structures which ecologically are extremely labile, and which have little significance above the generic level.

Three of the ten generic or subgeneric groups of fresh-water Halacaridae, namely *Caspihalacarus* Viets 1928 (Caspian Sea), *Troglohalacarus* Viets 1937 (Spain), and *Stygohalacarus* Viets 1934 (Yugoslavia) appear to be rather improbable inhabitants of North America. *Parasoldanellonyx* Viets 1929 (Europe, Great Britain), has not yet been recorded from North America although its distribution would strongly suggest that it will eventually be found here. *Hamohalacarus* Walter 1931 is known only from caves in Indiana. The remaining five genera are rather widely distributed on the North American continent or in Hawaii. Some of our North American species appear to be identical with European forms but others will probably prove to be distinct.

Most of the fresh-water Halacaridae are small, having a body length less than 500 μ. Some of them, such as *Porolohmannella*, appear to be rare and show up in collections only in small numbers. *Porohalacarus* and *Soldanellonyx* have been taken in considerable numbers but this appears to be exceptional. Little is known of the habits of these mites but they are presumed to be predaceous.

The illustrations for this section were drawn by Mari Riess of the University of California, Riverside. The scales provided are marked off in 10-μ units.

KEY TO GENERA (ADULTS)

1a   Palpi attached to gnathosoma dorsally; the openings in which they are inserted separated by an interval no greater than the width of the base of the first segment (trochanter) of the palp, so that the trochanters are not visible in ventral view, except by transparency (Figs. 42.92, 42.95) **2**

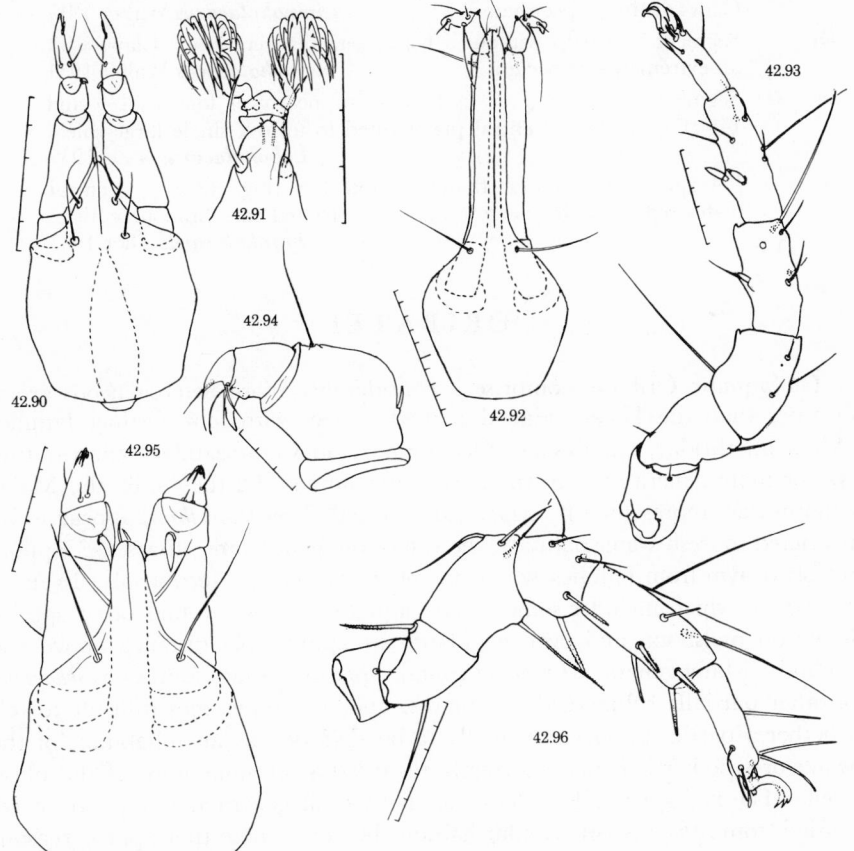

HALACARIDAE
Fig. 42.90. *Porohalacarus*, gnathosoma, ventral view. **Fig. 42.91.** *Soldanellonyx*, tarsal claws I. **Fig. 42.92.** *Porolohmannella*, gnathosoma, ventral view. **Fig. 42.93.** *Lobohalacarus*, leg I, anterior aspect. **Fig. 42.94.** *Limnohalacarus*, palp. **Fig. 42.95.** *Soldanellonyx*, gnathosoma, ventral view. **Fig. 42.96.** *Porohalacarus*, leg I, anterior aspect.

1b   Palpi attached laterally, more widely separated at their insertions, so that the trochanters of the palpi are clearly visible in ventral view (Fig. 42.90). . . . . Subfamily **Porohalacarinae**   5

2a   (1) Rostrum very long, slender, parallel-sided; distance from posterior margin of gnathosoma to posterorostral setae less than distance from posterorostral setae to tip of rostrum (Fig. 42.92) . . . .
Subfamily **Lohmannellinae**
Porolohmannella Viets 1933

2b   Rostrum much shorter and more triangular in outline; distance from posterior margin of gnathosoma to posterorostral setae greater than distance from that level to tip of rostrum (Fig. 42.95) . . . . . .
Subfamily **Limnohalacarinae**   3

3a   (2) Claws on I-6 with a distal mushroomlike expansion bearing a number of teeth which curve back sharply in direction of base of claw (Fig. 42.91) . . . . *Soldanellonyx* Walter 1917

3b   Claws pectinate, but the pecten is a straight comb.   4

4a   (3) Segment 3 of palp with a heavy seta distiventrally (Fig. 42.94). Claws distinctly pectinate . . . . . . *Limnohalacarus* Walter 1917

4b   Segment 3 of palp without a heavy seta distiventrally. Claws with an extremely fine pecten . . . . *Hamohalacarus* Walter 1931

5a   (1) Segment 4 of leg I (I-4) as long as or nearly as long as I-3 and I-5 (Fig. 42.93). Ventral plates fused to form a single large shield
Lobohalacarus Viets 1939

5b   I-4 appreciably shorter than I-3 and I-5 (Fig. 42.96). Ventral plates separated by narrow bands of striated membranous cuticle.
Porohalacarus Thor 1923

ORIBATEI

The aquatic Oribatei comprise a considerably more heterogeneous group of mites than the Halacaridae described above, with four distinct families (Malaconothridae, Camisiidae, Oribatidae, and Ceratozetidae) contributing two or more genera each to the fresh-water fauna. Of these only the Malaconothridae, represented by *Malaconrothrus* and *Trimalaconothrus*, appear to be restricted to fresh-water habitats; the rest of the genera are, in a sense, opportunists drawn from families whose members are chiefly terrestrial. Even in *Hydrozetes*, which includes some strictly aquatic species, we find other species living on moist soil and even on plants an appreciable distance above the ground. Many genera of Oribatei contain species that are found on *Sphagnum* or other partially submerged bog plants, and it is sometimes difficult to tell whether a particular species is really to be regarded as an inhabitant of the water or whether it is more correctly regarded as an inhabitant of the plant itself. The key given below does not include all genera that have been recorded from *Sphagnum* and similar habitats but only those that appear reasonably certain to contain species characteristic of the aquatic fauna. With a single exception (*Limnozetes* Hull 1916) all are known on the basis of the writer's experience to exist in North America and to have at least one or more species characteristically found in aquatic or subaquatic situations.

Limnozetes is included in the key because its species appear to be strictly aquatic and its distribution in Europe and Britain indicates that it might eventually turn up in North America as well. All of our North American genera that the writer has seen in fresh water are also known from Europe and other parts of the world.

Although the food habits of only a few genera are known there is little doubt that all are herbivorous, feeding on fungi or green plant material. The writer has seen *Lemna* severely damaged by the feeding of one of our eastern species of *Hydrozetes*. Most of the damage was confined to the upper surface of the leaves.

Two interesting evolutionary trends appear in the aquatic Oribatei, namely the loss of the tracheae and the development of viviparity. In forms such as *Trimalaconothrus* both of these conditions are to be found. In *Hydrozetes* the tracheae are still present but oviparity has been replaced by viviparity (or at least ovoviviparity) in some species. *Hydrozetes* is also exceptional among genera of Oribatei in that external sexual differences are known in the two sexes. Moreover, there is a great variation in sex ratio, with populations of some species having less than 1 per cent males, although in other species the sex ratio is normal.

KEY TO GENERA (ADULTS)

1a	Genital and anal openings contiguous or nearly so, separated by an interval less than ¼ the length of the genital opening (Figs. 42.97–42.99).	2
1b	These openings separated by a distinct interval of the ventral plate, this interval at least ⅓ as long as the genital opening (Figs. 42.102, 42.104)	8
2a	(1) Ventral plate of opisthosoma broad, extending nearly to lateral margin; suture between dorsal and ventral plates marginal in position so that only the rim of the dorsal plate is seen at the sides of the body in ventral view (Fig. 42.99) . . . **Hermannia** Nicolet 1865	

The tagmation of most Acari is unique in that the principal division of the body lies between legs II and III. The portion of the body anterior to this division includes the gnathosoma ("capitulum"), and the propodosoma, including the first two pairs of legs. The portion of the body behind this division is the hysterosoma, and the portion behind legs IV is the opisthosoma. As seen in dorsal view, the propodosoma of the Oribatei is a somewhat triangular tagma, inserted on the anterior end of the globular or elliptical hysterosoma (Fig. 42.101).

2b	Ventral plate of opisthosoma compressed laterally by encroachment of dorsal plate over ventral surface of body. Suture between dorsal and ventral plates not marginal in position, but closer to the median line so that much of the dorsal plate can be seen in ventral view (Figs. 42.97, 42.98)	3
3a	(2) Alveoli of propodosomal sensilla very simple, superficial, like those of ordinary setae; sensilla thickest at base, setiform or uniformly tapering, never clavate (Fig. 42.100)	4

One of the most conspicuous features of most Oribatei is the pair of specialized setae, the propodosomal sensilla, found near the posterolateral angles of the propodosoma. These assume a variety of forms, from tapering to clavate, and are quite frequently referred to as the pseudostigmatic organs. They usually arise from a pair of deep pits. In Fig. 42.101, the propodosomal sensilla are directed laterally from their origins near the posterolateral corner of the propodosoma.

3b   Alveoli of sensilla of the typical complex form found in most Orib-
              atei, extending deep into surface of propodosoma in form of a tor-
              tuous passage; sensilla generally not uniformly tapering, but clavate,
              or at least of fairly uniform diameter throughout **5**

4a   (3) With 1 claw on each tarsus. Genital sclerites each with about 20
              or more minute setae in a single straight row down the medial
              margin (Fig. 42.97) . . . . ***Malaconothrus*** Berlese 1905

4b   With 3 claws on each tarsus. Genital sclerites each with only 6 to
              8 setae of moderate size in a row near the medial margin . . . .
                                           Trimalaconothrus Berlese 1916

5a   (3) Each genital sclerite with about 9 setae. **6**

5b   Genital sclerites each with about 12 or more setae, all of which are
              along the medial margin of the plate **7**

6a   (5) Width of right and left genital sclerites together as great as length
              of sclerites. One of setae lying at or near the posterolateral angle
              of the plate (Fig. 42.98) . . . . ***Nothrus*** Koch 1840

6b   Genital sclerites more rectangular, their length about 1½ times as
              great as their combined width (as in *Trimalaconothrus*, Fig. 42.97).
              All of 9 or so setae along medial margin of genital plate . . . . . .
                                           Trhypochthonius Berlese 1905

7a   (5) Propodosoma with anterior end (rostrum) forming a small rectang-
              ular lobe bearing the rostral setae (Fig. 42.103) . . . .
                                           Platynothrus Berlese 1913

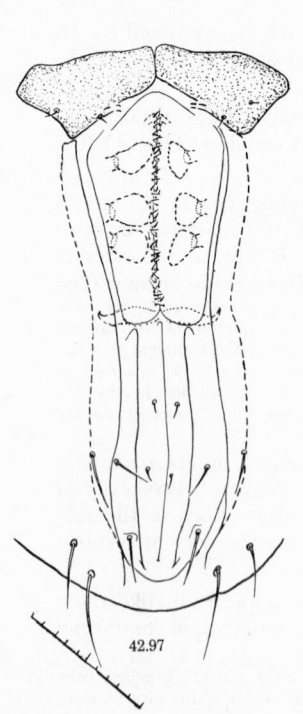

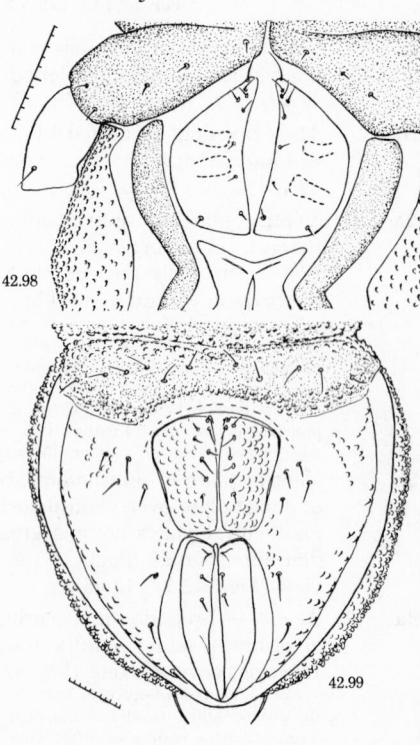

ORIBATEI

Fig. 42.97. *Malaconothrus*, genitoanal area and coxae IV. **Fig. 42.98.** *Nothrus*, genital opening and surrounding structures. **Fig. 42.99.** *Hermannia*, opisthosoma, ventral.

7b   Without such a lobe here, anterior end of propodosoma more uniformly convex . . . . **Heminothrus** Berlese 1913

8a   **(1)** Hysterosoma rather soft, dorsum coarsely wrinkled, even in adult. Genital and anal openings separated by an interval about 0.5 to 0.6 the length of the genital opening. Common in weakly saline waters of estuaries, or in spray pools above the tide zone; also intertidal . . . . **Ameronothrus** Berlese 1896

8b   Hysterosoma of adults brittle, cuticle smooth, usually shining. Genital and anal openings separated by a distance equal to or greater than the length of the genital opening. Usually in fresh water, occasionally in brackish situations. **9**

9a   **(8)** Anterolateral margins of hysterosoma (abdomen) with small to moderate winglike expansions (pteromorphae) (Fig. 42.102) **10**

9b   Anterolateral margins of hysterosoma without pteromorphae (Fig. 42.104) **12**

10a **(9)** Tarsus I with 1 claw, II–IV with 3 claws . . . . *Heterozetes* Willman 1917

10b   All tarsi with equal numbers of claws (1 or 3) **11**

11a **(10)** Pteromorphae extending back to or beyond the middle of the hysterosoma, where their outlines blend imperceptibly with the hysterosomal margin (Fig. 42.102) . . . . **Ceratozetes** Berlese 1908

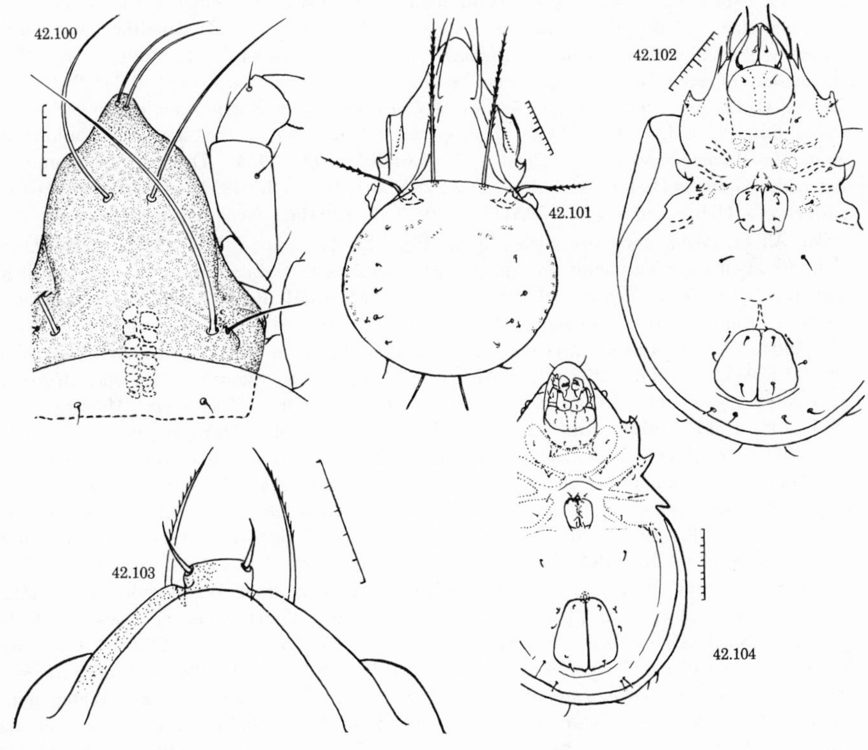

ORIBATEI
Fig. 42.100. *Malaconothrus*, propodosoma, dorsal view. **Fig. 42.101.** *Ceratoppia*, male, dorsum. **Fig. 42.102.** *Ceratozetes*, female, venter. **Fig. 42.103.** *Platynothrus*, rostrum, ventral view. **Fig. 42.104.** *Hydrozetes*, female, venter.

11b   Pteromorphae shorter, and distinctly protruding beyond the hysterosomal margin . . . . *Limnozetes* Hull 1916

12a   (9) Sensilla and interlamellar setae of propodosoma very long, straight, slender, extending well beyond margins of body. Lamellae ending in a long spine bearing the straight, stiff lamellar setae (Fig. 42.101). Carina between coxae II and III well developed; others weakly developed or absent, and not reaching the mid-line . . . . . .
               Ceratoppia Berlese 1908

12b   Sensilla short, clavate, interlamellar setae very short, simple; neither reaching to or beyond margins of body. Lamellae not ending in spines, the lamellar setae borne on the surface of the propodosoma. All coxal areas well defined by carinae in mature adults (Fig. 42.104), carina between coxae II and III strongly arcuate, not straight . . . . *Hydrozetes* Berlese 1902

References

PARASITENGONA (NORTH AMERICAN)

1. Banks, Nathan. 1907. A catalogue of the Acarina, or mites, of the United States. *Proc. U. S. Natl. Mus.*, 32:595–625.   **2. Haldeman, S. S. 1842.** On some American species of Hydrachnidae. *Zoological Contributions*, No. 1. Philadelphia.   **3. Koenike, F. 1895.** Nordamerikanische Hydrachniden. *Abhandl. Naturw. Ver. Bremen*, 13:167–226.   **4. 1912.** A revision of my "Nordamerikanische Hydrachniden" *Trans. Can. Inst.*, 9:281–293.   **5. Lavers, C. H., Jr. 1941.** A new species of *Limnochares* from North America. *Univ. Wash. Publs. Biol.*, 12:1–6.   **6. 1945.** The species of *Arrenurus* of the state of Washington. *Trans. Am. Microscop. Soc.*, 64:228–264.   **7. Lundblad, O. 1934.** Die nordamerikanischen Arten der Gattung Hydrachna. *Arkiv Zool.*, 28A(3):1–44.   **8. 1935.** Über einige Hydracarinen aus Haiti. *Arkiv Zool.*, 28A(13):1–30.   **9. 1941a.** Neue Wassermilben aus Amerika, Afrika, Asien, und Australien. *Zool. Anz.* 133:155–160.   **10. 1941b.** Eine Übersicht des Hydrachnellensystems und der bis jetzt bekannten Verbreitung der Gattungen dieser Gruppe. *Zool. Bidrag Uppsala*, 20:359–379.   **11. Marshall, Ruth. 1903.** Ten species of Arrenuri belonging to the subgenus *Megalurus* Thon. *Trans. Wisconsin Acad. Sci.*, 14:145–172.   **12. 1904.** A new *Arrenurus* and notes on collections made in 1903. *Trans. Wisconsin Acad. Sci.*, 14:520–526.   **13. 1908.** The Arrhenuri of the United States. *Trans. Am. Microscop. Soc.*, 28:85–140.   **14. 1910.** New studies of the Arrhenuri. *Trans. Am. Microscop. Soc.*, 29:97–108.   **15. 1912.** Some American *Lebertia*. *Trans. Am. Microscop. Soc.*, 31:225–230.   **16. 1914.** Some new American water mites. *Trans. Wisconsin Acad. Sci.*, 17:1300–1304.   **17. 1915.** American species of the genus *Atractides*. *Trans. Am. Microscop. Soc.*, 34:185–188.   **18. 1919.** New species of water mites of the genus *Arrhenurus*. *Trans. Am. Microscop. Soc.*, 38:275–281.   **19. 1921.** New species and collections of *Arrhenuri*. *Trans. Am. Microscop. Soc.*, 40:168–176.   **20. 1922.** New American water mites of the genus *Neumania*. *Trans. Wisconsin Acad. Sci.*, 20:205–213.   **21. 1924a.** Arrhenuri from Washington and Alaska. *Trans. Wisconsin Acad. Sci.*, 21:214–218.   **22. 1924b.** Water mites of Alaska and the Canadian Northwest. *Trans. Am. Microscop. Soc.*, 43:236–255.   **23. 1926a.** Collecting water mites in Cuba. *Trans. Illinois Acad. Sci.*, 19:197–199.   **24. 1926b.** Water mites of the Okoboji region. *Univ. Iowa Studies Nat. Hist.*, 11:28–35.   **25. 1927a.** Water mites from Cuba. *Trans. Am. Microscop. Soc.*, 46:60–65.   **26. 1927b.** Hydracarina of the Douglas Lake region. *Trans. Am. Microscop. Soc.*, 46:268–285.   **27. 1928.** A new species of water mite from thermal springs. *Psyche*, 35:92–97.   **28. 1929a.** The water mites of Lake Wawasee. *Proc. Indiana Acad. Sci.*, 38:315–320.   **29. 1929b.** Canadian Hydracarina. *Univ. Toronto Studies Biol. Ser.*, 33:57–93.   **30. 1929c.** The morphology and develop-

mental stages of a new species of *Piona. Trans. Wisconsin Acad. Sci.*, 24:401–404. **31.**
1930a. The water mites of the Jordan Lake region. *Trans. Wisconsin Acad. Sci.*, 25:245–253.
32. 1930b. Hydracarina from Glacier National Park. *Trans. Am. Microscop. Soc.*, 49:342–
345. **33. 1931.** Preliminary list of the Hydracarina of Wisconsin. I. The red mites.
Trans. Wisconsin Acad. Sci., 26:311–319. **34. 1932.** Preliminary list of the Hydracarina
of Wisconsin. Part II. *Trans. Wisconsin Acad. Sci.*, 27:339–358. **35. 1933a.** Water
mites from Wyoming as fish food. *Trans. Am. Microscop. Soc.*, 52:34–41. **36. 1933b.**
Preliminary list of the Hydracarina of Wisconsin. Part III. *Trans. Wisconsin Acad. Sci.*,
28:37–61. **37. 1935a.** A new parasitic *Unionicola. Univ. Toronto Studies Biol. Ser.*, 39:97–
102. **38. 1935b.** Preliminary list of the Hydracarina of Wisconsin. Part IV. *Trans.
Wisconsin Acad. Sci.*, 29:273–297. **39. 1936.** Hydracarina of Yucatan. *Carnegie Inst.
Wash. Publ.*, No. 457:133–137. **40. 1937.** Preliminary list of the Hydracarina of Wisconsin. Part V. *Trans. Wisconsin Acad. Sci.*, 30:225–252. **41. 1940a.** Preliminary list of
the Hydracarina of Wisconsin. Part VI. *Trans. Wisconsin Acad. Sci.*, 32:135–165. **42.
1940b.** The water mite genus *Tyrrellia. Trans. Wisconsin Acad. Sci.*, 32:383–389. **43.
1943a.** Hydracarina from California. Part I. *Trans. Am. Microscop. Soc.*, 62:306–324.
44. 1943b. Hydracarina from California. Part II. *Trans. Am. Microscop. Soc.*, 62:404–415.
45. 1944. New species and notes on the Arrenuri. *Am. Midland Naturalist*, 31:631–637.
46. Newell, I. M. 1947. A systematic and ecological study of the Halacaridae of eastern
North America. *Bull. Bingham Oceanog. Collection*, 10:1–232. **47. Piersig, R. 1894.**
Sachsens Wassermilben. *Zool. Anz.*, 17:213–216. **48. 1904.** (Referat). *Zool. Cent.*,
11:210–211. **49. 1905.** (Referat). *Zool. Cent.*, 12:185. **50. Stoll, Otto. 1893.**
Arachnida Acaridea. Biologia Centrali-Americana. Zoologia. Godman and Salvin, London. **51.
Viets, Karl. 1907.** Neue Hydrachniden. *Abhandl. Naturw. Ver. Bremen*, 19:142–146. **52.
1930.** Über nordamerikanische *Koenikea*-Arten (Hydracarina) *Zool. Anz.*, 92:266–272. **53.
1931.** Über einige Gattungen und Arten der Axonopsae, Mideopsae und Arrhenurae (Hydracarina). *Zool. Anz.*, 93:33–48. **54. Wolcott, R. H. 1899.** On the North American
species of the genus *Atax* (Fabr.) Bruz. *Trans. Am. Microscop. Soc.*, 20:193–259. **55. 1900.**
New genera and species of North American Hydrachnidae. *Trans. Am. Microscop. Soc.*,
21:177–200. **56. 1902.** On the North American species of the genus *Curvipes. Trans.
Am. Microscop. Soc.*, 23:201–256. **57. 1903.** On the North American species of the genus
Limnesia. Trans. Am. Microscop. Soc., 24:85–107. **58. 1905.** A review of the genera of
the water mites. *Trans. Am. Microscop. Soc.*, 26:161–243. **59. 1918.** Chapter 26, The
Water-Mites (Hydracarina). In: Ward and Whipple (eds.). *Fresh-Water Biology*, pp. 851–
875. Wiley, New York.

Additional References to North American Species and Genera

The following references were added subsequent to the completion of the original manuscript.
60. Cook, David R. 1953. *Marshallothyas*, a new genus belonging to the subfamily Thy-
asinae. *Proc. Entomol. Soc. Wash.*, 55:305–308. **61. Habeeb, Herbert. 1950.** Three
interesting water mites. *Naturaliste can.*, 77(3–4):112–117. **62. 1953.** North American
Hydrachnellae, Acari I–V. *Leaflets of Acadian Biol.*, No. 1:1–16. **63. Mitchell, Rodger D.
1953.** A new species of *Lundbladia* and remarks on the family Hydryphantidae. *Am. Midland
Naturalist*, 49:159–170. **64. 1954.** A description of the water-mite, *Hydrovolzia gerhardi*
new species, with observations on the life history and ecology. *Nat. Hist. Miscellanea*,
135:1–9. **65. 1954.** Check list of North American water mites. *Fieldiana, Zool.*, 35:29–
70. **66. 1955.** Two water mites from Illinois. *Trans. Am. Microscop. Soc.* 74:333–342.

PARASITENGONA (OTHER PARTS OF THE WORLD)

A. Lundblad, O. 1927. Die Hydracarinen Schwedens I. Beitrag zur Systematik, Embry-
ologie, Ökologie und Verbreitungsgeschichte der Schwedischen Arten. *Zool. Bidrag Uppsala*,
11:185–540. **B. 1936.** Schwedisch-chinesische wissenschaftliche Expedition nach den

nordwestlichen Provinzen Chinas. Wassermilben. *Arkiv. Zool.*, 29A:1–40. **C. 1941a.** Die Hydracarinenfauna Südbrasiliens und Paraguays. Erster Teil. *Kgl. Svenska Vetenskapsakad. Handl., Ser. 3*, 19:1–183. **D. 1941b.** Neue Wassermilben aus Madeira. *Entomol. Tidskr.*, 62:93–96. **E. 1942.** Die Hydracarinenfauna Südbrasiliens und Paraguays. Zweiter Teil. *Kgl. Svenska Vetenskapsakad. Handl., Ser. 3*, 20:1–175. **F. 1943a.** Die Hydracarinenfauna Südbrasiliens and Paraguays. Dritter Teil. *Kgl. Svenska Vetenskapsakad. Handl., Ser. 3*, 20:1–148. **G. 1943b.** Die Hydracarinenfauna Südbrasiliens und Paraguays. Vierter Teil. *Kgl. Svenska Vetenskapsakad. Handl., Ser. 3*, 20:1–171. **H. 1944.** Die Hydracarinenfauna Südbrasiliens und Paraguays. Fünfter Teil. *Kgl. Svenska Vetenskapsakad. Handl., Ser. 3*, 20:1–182. **I. 1927.** *Thermacarus thermobius* n. gen. n. sp., eine Hydracarine aus heisser Quelle. *Zool. Anz.*, 73:11–20. **J. Viets, K. 1935.** Die Wassermilben von Sumatra, Java und Bali nach den Ergebnissen der Deutschen Limnologischen Sunda-Expedition. *Arch. Hydrobiol. Suppl.*, 13:484–594. **K. 1935.** Die Wassermilben von Sumatra, Java und Bali nach den Ergebnissen der Deutschen Limnologischen Sunda-Expedition. *Arch. Hydrobiol. Suppl.*, 13:595–738. **L. 1935.** Die Wassermilben von Sumatra, Java und Bali nach den Ergebnissen der Deutschen Limnologischen Sunda-Expedition. *Arch. Hydrobiol. Suppl.*, 14:1–113. **M. 1936.** Spinnentiere oder Arachnoidea, VII. Wassermilben oder Hydracarina (Hydrachnellae und Halacaridae). *Tierwelt Deutschlands*, 31:1–288. **N. 1938.** Über die verschiedenen Biotope der Wassermilben, besonders über solche mit anormalen Lebensbedingungen und über einige neue Wassermilben aus Thermalgewässern. *Verhandl. Intern. Ver. Limnol. Paris*, 1937 (1938). 8:209–224. **O. 1949.** Nomenklatorische und taxonomische Bemerkungen zur Kenntnis der Wassermilben (Hydrachnellae, Acari). 1–10. *Abhandl. naturw. Ver. Bremen*, 32:292–331. **P. 1955.** *Die Milben des Süsswassers und des Meeres*, erster Teil. G. Fischer, Jena. **Q. 1956.** *Die Milben des Süsswassers und des Meeres*, zweiter und dritter Teil. G. Fischer, Jena.

HALACARIDAE

1. Newell, I. M. 1947. A systematic and ecological study of the Halacaridae of eastern North America. *Bull. Bingham Oceanogr. Collection*, 10:1–232. **2. Viets, Karl. 1934.** Siebente Mitteilung über Wassermilben aus unterirdischen Gewässern. *Zool. Anz.*, 106:118–124. **3. 1936.** Spinnentiere oder Arachnoidea. VII. Wassermilben oder Hydracarina (Hydrachnellae und Halacaridae). In: Friedrich Dahl. *Tierwelt Deutschlands*, 31:1–288; 32:289–574. **4. Walter, C. 1931.** Biospeologica. LVI. Campagne speologique de C. Bolivar et R. Jeannel dans l'Amerique du Nord (1928). 6. Arachnides halacariens. *Arch. zool. exp. et gén.*, 71: 375–381.

AQUATIC ORIBATEI

1. Grandjean, F. 1948. Sur les *Hydrozetes* (Acariens) de l'Europe occidentale. *Bull. muséum natl. hist. nat. Paris*, 2nd Ser., 20:328–335. **2. Newell, I. M. 1945.** *Hydrozetes* Berlese (Acari, Oribatoidea): the occurrence of the genus in North America and the phenomenon of levitation. *Trans. Conn. Acad. Arts and Sci.*, 36:253–275. **3. Sellnick, Max, and Karl-Herman Forsslund. 1955.** Die Camisiidae Schwedens (Acar. Oribat.) *Arkiv Zool.*, Ser. 2, 8:473–530. **4. Willmann, C. 1931.** Moosmilben oder Oribatiden (Oribatei). In: Friedrich Dahl. *Tierwelt Deutschlands*, 22:79–200.

43

Mollusca

WILLIAM J. CLENCH

The fresh-water Mollusca include univalves (snails) and bivalves (clams or mussels). In the univalve mollusks or Gastropoda, the shell may be coiled obliquely or horizontally, or it may be conical and somewhat tent-shaped. In the Class Gastropoda the fresh-water forms possess a distinct head with a pair of contractile tentacles at the base of which are the eyes. The mouth is located on the lower portion of the head between the tentacles. The upper portion of the mouth is usually provided with a chitinous jaw which may consist of one to three pieces. The lower portion of the mouth is provided with a radula, an organ peculiar to the mollusks. This is a chitinous ribbon provided with transverse rows of teeth. It can be extended and then pulled back and forth rapidly, rasping off food which is then carried back into the mouth. The jaw, besides cutting the food, also aids in holding the food in a firm position so the radula can work on it.

The fresh-water Gastropoda are divided into two main groups or subclasses, the Prosobranchia, which possess a gill for respiration under water, and the Pulmonata, which have a lung for obtaining air directly. Many Pulmonata, however, are able to remain submerged in water for indefinite periods of time, perhaps for their entire existence, for all mollusks are capable of carrying on some gas exchange through most parts of the body. These animals progress by crawling on the ventral surface of the body, which is modified to form a flat, muscular organ called the foot.

The bodies of bivalve mollusks, or Pelecypoda (clams or mussels), are protected by two symmetrical and opposing valves which are united above by an elastic tissue called the ligament. They have no head, tentacles, eyes, jaws, or radula. The mouth is an orifice at the anterior end of the body, and on each side there is a flap or labial palp which assists in guiding the food to the mouth. The foot is an axe-shaped organ of muscular tissue which can be extended from the anterior portion of the animal and by lodging in the mud or sand, pulls the animal forward. The Pelecypoda breathe by means of two gills suspended on each side of the body. These gills are divided into a series of water tubes by septa or lamellae, through which the water circulates by means of cilia. The body is enclosed by a tissue, the mantle, which secretes the shell. Posteriorly the mantle has two openings, the siphons, the lower one taking in water which aerates the gills and brings in food, and the dorsal one through which the water flows out carrying away waste products.

The North American Fauna

The distribution of the various families, genera, and species represented in our fresh-water fauna varies greatly in the different sections of the continent. Most of the families have representatives in nearly all portions of the country where suitable conditions of environment are to be found. There are, of course, some notable exceptions. The Viviparidae, which form a very conspicuous element in the fauna of the eastern states, are not to be found west of the Mississippi valley. However, two species of oriental Viviparus have been introduced into North America, both of which have been recorded from California. The Pilidae, a family limited mainly to the tropics, are to be found only in Florida and southern Georgia. The Lancidae occur only in the northwestern states and the Lepyriidae with but a single genus and species is to be found, so far as is known, only in the Cahawba River of south central Alabama. Many genera have a general distribution in all parts of the continent, that is, some representatives may be found wherever suitable conditions are available. However, very few species have a general distribution. Many have an exceedingly limited range—a river system, a single river, and in a few cases only a very small portion of a river. Many genera are likewise restricted to certain portions of the continent, many to a single river system, and others to a single river. The Coosa River in Alabama in this respect has a most remarkable fauna; no less than six genera and a multitude of species are known to occur only in this river and its tributaries. This, of course, is an extreme example, but many others, such as the Tennessee, Green, Cumberland, and Altamaha rivers in the eastern states and the Columbia in the Northwest have many species and species groups that are to be found only in these rivers and their tributaries.

On the other hand, our North American lakes have but few endemic species. These lakes are either geologically young, or because of subsequent modification such as glaciation, the fauna is relatively recent and consequently has not developed any significant modifications. Nothing in the lakes of North America is comparable to the remarkable mollusk fauna of Lake Tanganyika in central Africa or to that of Lake Titicaca in Peru or Lake Baikal in Siberia, to mention only a few of several outstanding examples (Brooks, 1950).

No comparable area in the world exceeds North America in the richness of its fresh-water mollusks. This comparison is of even greater interest when it is considered that probably 75 per cent of this fauna occurs in the Mississippi drainage system and the many rivers that drain independently into the Atlantic Ocean and the Gulf of Mexico. Certainly there are many other areas that are rich in species, such as Central and South America, equatorial Africa, and southern Asia, but none has a mollusk fauna equal numerically to that of North America.

Two families, the Pleuroceridae and the Unionidae are preeminent in our fauna because of the large number of genera, species, and subspecies of which they are composed. Close to half of all the known fresh-water mollusks of the North American fauna are included in these two families. Both reach their greatest development in the vast Mississippi River system and a few of the larger rivers that drain directly into the Gulf of Mexico. Curiously enough, the Mississippi River proper below the mouth of the Missouri River is comparatively poor in mollusks. This is due, in a large measure, to the quantities of sand and silt brought in by the Missouri River. The richness of the river system, however, remains in the many tributaries, several of which, like the Ohio, are themselves great rivers.

The number of species of mollusks that compose our fresh-water fauna is not known. It is, perhaps, close to 3000 which is nearly double the number of land mollusks for the same area.

A large amount of literature has been written regarding fresh-water mollusks, but so far nothing comprehensive exists that is at all recent, embracing the region as a whole. As a consequence, it is impossible to include in this report more than a selected list of some of the more important papers. Students interested in the fauna of localized areas should consult the Zoological Record and the bibliographies of the catalogs and monographs cited at the end of this chapter.

Collection and Preparation of Specimens

Nearly every permanent body of water has its mollusks, which vary according to its character. Some species are found only in rapidly flowing water and others only in ponds and still water. Ditches and other stagnant waters are usually good collecting grounds for many small species. Low places in the woods, which dry up in the summer time, have a number of species which bury themselves in the mud when the water disappears. Sand banks along rivers and lakes are favorite resorts of many of the smaller species. The under side of the lily pads should be scrutinized, and species of the Ancylidae should be looked for on stones and dead clam shells. With a little training and practice in the field, one soon learns the kind of localities that are frequented by mollusks.

The field equipment needed is rather limited. A few cloth sacks, 9 by 18 in. may be used for Unionidae unless collecting is attempted on many of the southern rivers where this group is exceedingly abundant. Much larger sacks should then be used, especially if the catch has to be carried any distance. A few vials, some empty and some with 75 per cent alcohol, will provide containers for the small species. One or two wide-mouth bottles will serve for the more or less fragile gastropods such as *Lymnaea* and *Helisoma*. A dip net can be of considerable aid in deep-water collecting. Perhaps the handiest tool of this sort is the simple soup strainer. When the collecting grounds are reached, a pole of 6 to 8 ft. can be cut and the strainer can be wired or tied on with string; the pole can then be discarded when the day's collecting is over. A small dredge can be used if waters of any depth are to be collected.

After returning to the laboratory, Unionidae can be boiled, the soft parts extracted and studied immediately, or both soft parts and shell can be placed in alcohol for future research. Large gastropods can be treated the same way, and after boiling, the soft parts are easily removed by means of a bent pin or wire. Minute species can remain in alcohol or can be dried out after 24 to 36 hours in 70 per cent alcohol.

Parasitic Stages

With very few known exceptions, species of the Unionidae are parasitic on fish in their larval stages and, in a single case, on a mud puppy (*Necturus maculosus* Rafinesque). There is a considerable degree of host selectivity; certain species of these fresh-water clams are capable of parasitizing several species of fish, and others are limited to a few and many to a single species of fish. The unionid larvae, known as glochidia, are extruded through the siphon when the glochidial sac is ruptured, and these minute clams then attach themselves to the fins and gills of their host. They exist as parasites from a few to many weeks depending on the species, subsisting on the body fluids of their hosts. When this stage in their development is over the encystment sloughs off, the young clams drop to the bottom and then begin their own independent existence. The interested student can obtain much detailed information on this subject from an excellent report by G. Lefevre and W. C. Curtis (1912).

Many snails are of medical importance, because they are intermediate hosts of parasitic worms. Many vertebrates, including man, are the primary hosts. In North America we are quite fortunate, as man is seldom affected by the fluke diseases, although the liver fluke of sheep and cattle cause considerable damage to livestock in some areas. Certain of the blood flukes of birds can affect man, but they cause only a superficial skin irritation known as swimmers' itch. Elsewhere, particularly in Africa, southeastern Asia, and portions of the West Indies and South America, this disease, known as schistosomiasis, is exceedingly serious. The intermediate host for this type of blood fluke is a snail, and the primary host is a mammal. Man, domesticated animals, and many wild mammals are involved in the life history of these parasites. Generally, when the life history of the blood fluke is known and the secondary host has been determined to be a particular species of snail, control measures for destroying the snail can be undertaken.

Measurements and Descriptive Terms

The *length* or *height* of a univalve shell is the distance from the apex to the basal edge of the lip, measured along a line drawn through the axis. The *diameter* is the greatest width, including the lip, measured on a line drawn at right angles to the axis.

Univalve mollusks are *dextral* or *sinistral* according to whether the aperture is on the right or left of the axis when the shell is held with the apex uppermost and with the aperture facing the observer.

The term parietal refers to the inner wall of the aperture in univalves. Body whorl refers to the last or ultimate whorl produced by the snail. The operculum is the "door" to the aperture. This is produced by a gland located

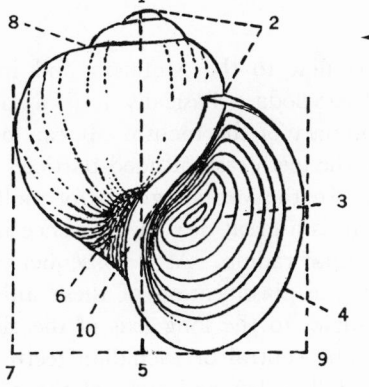

Fig. 43.1. The shell of a dextral univalve. 1, apex; 2, spire; 3, operculum; 4, lip; 6, umbilicus; 8, suture; 10, columella; 1–5, height; 7–9, greatest diameter. (By Walker.)

upon the back or top of the foot, and when the animal retracts within the shell the foot infolds about midway and at right angles to the long axis of the foot. This procedure brings the operculum in complete alignment with the aperture, and as the animal withdraws well within the aperture, the operculum fits tightly to its walls. The operculum is formed either entirely of chitin or of calcium carbonate with a chitinized base. All of our fresh-water gastropods, except members of the order Pulmonata, possess an operculum.

In bivalve mollusks, the *length* is the distance from the anterior to the posterior end of the valve; the *height* is the distance from the umbo to the

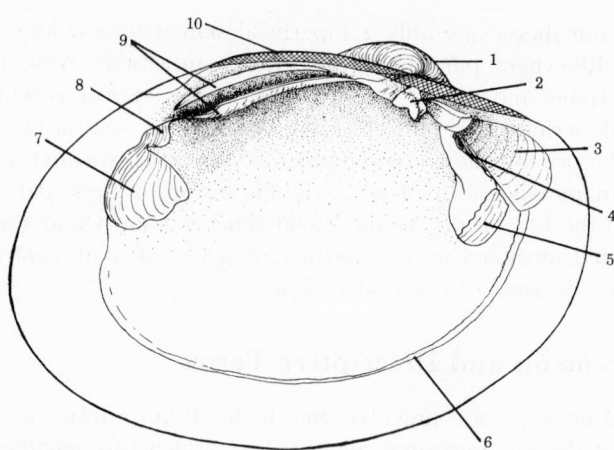

Fig. 43.2. The shell of a bivalve. 1, beak; 2, pseudocardinal teeth; 3, scar of anterior adductor; 4, scar of anterior retractor; 5, scar of foot protractor; 6, pallial line; 7, scar of posterior adductor; 8, scar of posterior retractor; 9, lateral teeth; 10, ligament. (After Turner.)

ventral margin of the shell; the *width* is the distance through the middle of the shell including the thickness of both valves. The umbos or beaks are generally the highest portion of a bivalve and represent the earliest stage in the development of the shell.

The Radula

The radula is an organ which is peculiar to the Mollusca and in this phylum is absent only in the Class Pelecypoda. Basically it is a ribbon studded with teeth which lies in the forepart of the mouth. It can be extended, and when pulled back and forth, the teeth rasp off food particles from the substance upon which it is feeding. In the various groups of mollusks, particularly in the Class Gastropoda, there is a great deal of difference in the shape and number of teeth on the individual radula. As a consequence the radula is a very important character in the classification of these animals. The teeth are in rows, either at right angles to the long axis of the ribbon or at angles similar to an inverted V. The central or rachidian teeth form the long axis of the radula, the laterals follow left and right of the central teeth, and the marginal teeth, when present, are left and right of the lateral teeth.

In the figures of the radulae which follow (Fig. 43.3), only a portion of a row may be given (*Planorbis*) with teeth selected from the row to show variation in the shape and position of the denticles. In other cases, such as in the Valvatidae, the entire row is given because in this group the number of teeth per row is limited to the central, two laterals, and four marginals (two on each side).

The radula can be extracted by boiling the head of the mollusk or, if small, the entire animal in 40 per cent solution of potassium or sodium hydroxide.

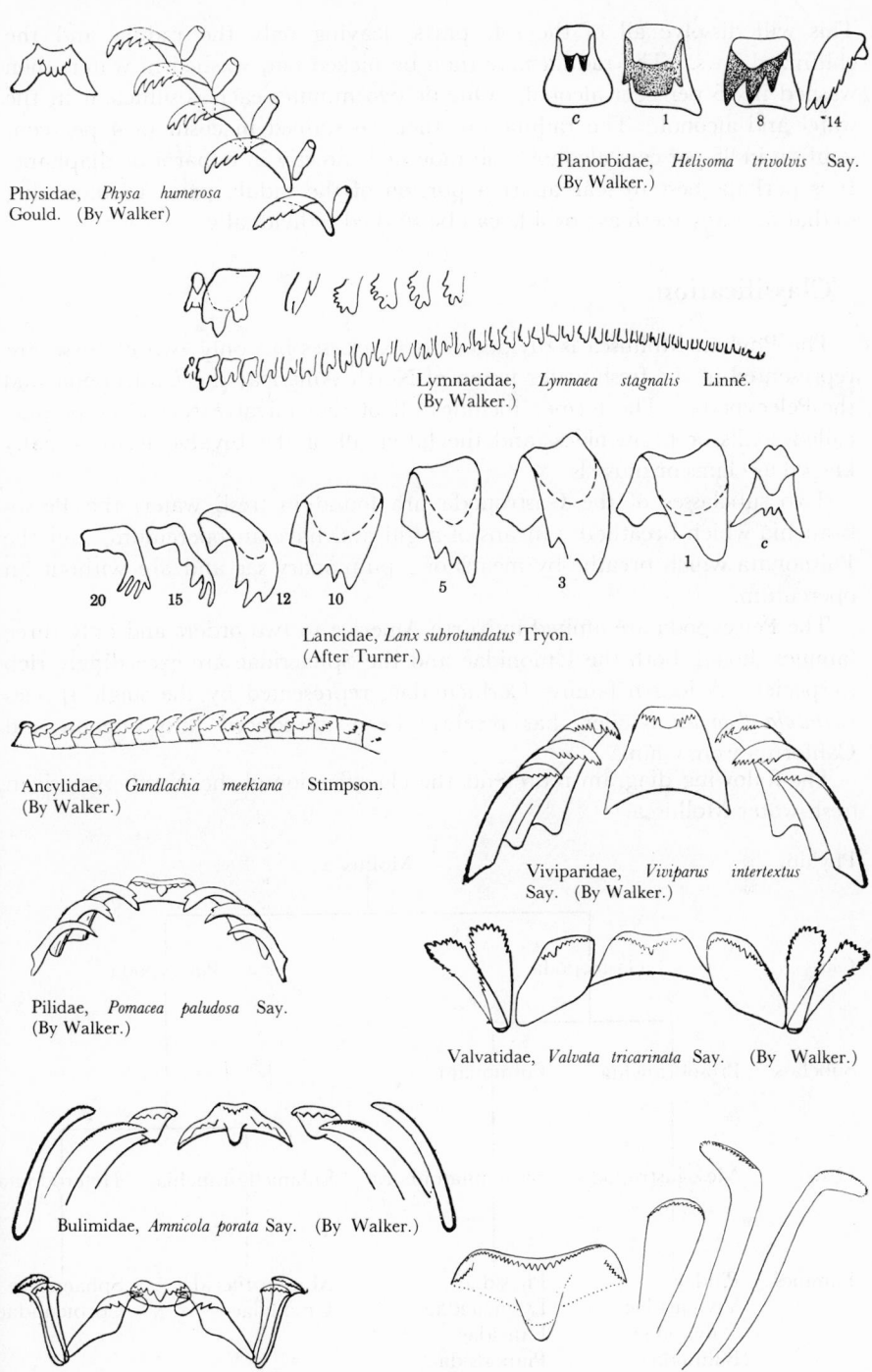

Physidae, *Physa humerosa* Gould. (By Walker)

Planorbidae, *Helisoma trivolvis* Say. (By Walker.)

Lymnaeidae, *Lymnaea stagnalis* Linné. (By Walker.)

Lancidae, *Lanx subrotundatus* Tryon. (After Turner.)

Ancylidae, *Gundlachia meekiana* Stimpson. (By Walker.)

Viviparidae, *Viviparus intertextus* Say. (By Walker.)

Pilidae, *Pomacea paludosa* Say. (By Walker.)

Valvatidae, *Valvata tricarinata* Say. (By Walker.)

Bulimidae, *Amnicola porata* Say. (By Walker.)

Pleuroceridae, *Leptoxis dissimilis* Say. (By Walker.)

Lepyriidae, *Lepyrium showalteri* Lea. (After Turner.)

Fig. 43.3 Radulae.

This will dissolve all of the soft parts, leaving only the radula and the chitinized jaws. The radula may then be picked out, washed in water, then washed in 95 per cent alcohol. One or two minutes each is sufficient in the water and alcohol. The radula can then be stained in eosin (a 4 per cent solution in 95 per cent alcohol) and mounted directly in euparol or diaphane. It is perhaps best to tear apart a portion of the radula prior to mounting so that as many teeth as possible can be studied individually.

Classification

The Phylum Mollusca is divided into six classes but only two of these are represented in the fresh-water fauna of North America; the Gastropoda and the Pelecypoda. The former includes all of the univalve species commonly called snails or periwinkles, and the latter all of the bivalve forms usually known as clams or mussels.

Two subclasses of the Gastropoda are found in fresh water, the Prosobranchia which breathe by means of a gill and have an operculum, and the Pulmonata which breathe by means of a pulmonary sac and are without an operculum.

The Pelecypoda are limited in North America to two orders and only three families though both the Unionidae and the Sphaeridae are exceedingly rich in species. A fourth family, Corbiculidae, represented by the single species *Corbicula fluminea* Müller, has recently been introduced into Oregon and California from China.

The following diagram represents the classification of the North American fresh-water Mollusca.

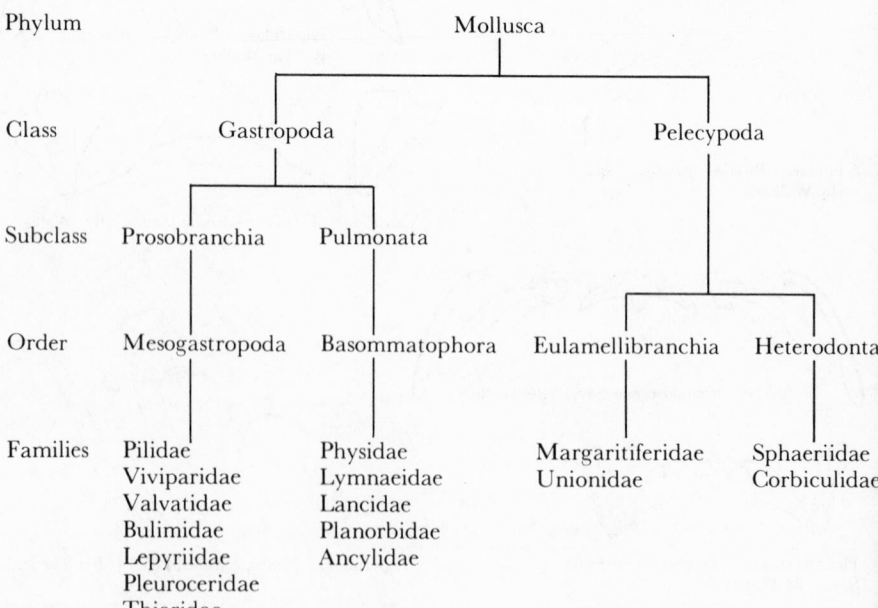

Phylum				Mollusca		
Class		Gastropoda			Pelecypoda	
Subclass	Prosobranchia		Pulmonata			
Order	Mesogastropoda		Basommatophora	Eulamellibranchia		Heterodonta
Families	Pilidae		Physidae	Margaritiferidae		Sphaeriidae
	Viviparidae		Lymnaeidae	Unionidae		Corbiculidae
	Valvatidae		Lancidae			
	Bulimidae		Planorbidae			
	Lepyriidae		Ancylidae			
	Pleuroceridae					
	Thiaridae					

Acknowledgements

I am deeply indebted to my colleagues, Dr. Merrill E. Champion and Dr. Ruth Turner, for much critical advice and substantial aid in the preparation of this chapter.

KEY TO FAMILIES

A key is given to all the families known to occur in North America. Once the family has been identified, identification of the genus is to be made by comparing the specimen with the illustrations given of each genus in the family. In this way all but one or two genera will be eliminated as possibilities, and identification can be made by recourse to the printed description.

In the descriptions of the genera, the terms *small*, *medium*, and *large* are used. These are purely comparative terms but we have set approximate limitations on these for convenience. In the Gastropoda small means ½ in. or under; medium, ½ to 1½ in.; large, over 1½ in. In the Pelecypoda, small is 2 in. or under; medium, 2 to 4 in.; large, over 4 in.

1a		Animal having a shell consisting of one piece (Univalve)	
		Class **Gastropoda**	2
1b		Animal having a shell consisting of two pieces (bivalve).	
		Class **Pelecypoda**	13
2a	(1)	Animal without an operculum	3
2b		Animal with an operculum	7
3a	(2)	Shell spirally coiled	4
3b		Shell a flattened cone	6
4a	(3)	Shell extended, coiled in two planes	5
4b		Shell flattened, coiled in a single plane or nearly so	
		(p. 1128) Family **Planorbidae**	
5a	(4)	Shell coiled dextrally . . . . (p. 1126) Family **Lymnaeidae**	
5b		Shell coiled sinistrally. . . . . (p. 1126) Family **Physidae**	
6a	(3)	Length usually under ¼ inch, corneous, thin	
		(p. 1130) Family **Ancylidae**	
6b		Length usually over ¼ inch, limy, fairly thick	
		(p. 1128) Family **Lancidae**	
7a	(2)	Operculum with concentric growth lines	8
7b		Operculum with spiral growth lines	9
8a	(7)	Animal with both gill and lung . . . . . . (p. 1131) Family **Pilidae**	
8b		Animal with gill only . . . . (p. 1132) Family **Viviparidae**	
9a	(7)	Operculum circular, multispiral . . . . (p. 1133) Family **Valvatidae**	
9b		Operculum ovate, paucispiral.	10
10a	(9)	Shell depressed and having a very broad parietal area	
		(p. 1136) Family **Lepyriidae**	
10b		Shell rounded or extended with a narrow parietal area	11
11a	(10)	Animal with external verge (male organ), central tooth of radula with basal denticles . . . . (p. 1133) Family **Bulimidae**	
11b		Animal without external verge, central tooth of radula without basal denticles	12

12a (11) Mantle border not fringed; oviparous . . . .
                                    (p. 1136) Family **Pleuroceridae**

12b   Mantle border fringed; viviparous . . . (p. 1138) Family **Thiaridae**

13a (1) Shell non-nacreous; hinge with cardinal and lateral teeth **14**

13b   Shell nacreous; hinge with lateral and pseudocardinal teeth or
                without teeth **15**

14a (13) Hinge with cardinal and smooth anterior and posterior lateral
                teeth . . . . (p. 1158) Family **Sphaeridae**

14b   Hinge with cardinal and serrated anterior and posterior lateral
                teeth . . . . (p. 1159) Family **Corbiculidae**

15a (13) Gills with distinct interlamellar septa, parallel with the gill fila-
                ments . . . . (p. 1138) Family **Unionidae**

15b   Gills without distinct interlamellar septa, or when present, oblique
                to the gill filaments . . . . (p. 1138) Family **Margaritiferidae**

SUBCLASS PULMONATA

Mollusks with a pulmonary sac and without an operculum.

Family **Physidae**

Shells small to medium in size, sinistral, attenuate, usually shining, brown to light
amber in color, and imperforate. Aperture elongate and possessing a simple lip
which may be slightly thickened a little below the edge. Radula having the central
tooth with a few denticles and the lateral teeth denticulate and having an apophysis.

Genus **Physa** Draparnaud. Shells somewhat inflated and sinistral. Inner edge
of mantle usually digitate or lobed and extending over the parietal area.
Widely distributed throughout N. A.

◄ **Fig. 43.4.** *Physa gyrina* Say. × 1½. (By Walker.)

Genus **Aplexa** Fleming. Shell attenuate, sinistral, thin, and not inflated to any
extent. Mantle digitations weak or absent.
Widely distributed throughout northern U. S. and Canada.

◄ **Fig. 43.5.** *Aplexa hypnorum* Linné. × 1½. (By Walker.)

Family **Lymnaeidae**

Shells small to medium in size, globose to attenuate, smooth to finely sculptured,
usually imperforate and grayish-white to dark brown in color. Aperture rounded to
lengthened with the lip simple or with a thickening just below the inner edge. Radula
with a unicuspid central tooth, normally bicuspid lateral teeth and serrated marginal
teeth.

The genus **Lymnaea** Lamarck is the only genus of this family found in N. A. Eight
subgenera are considered here. They are not clear-cut divisions but they are of value
in differentiating groups of closely related species.

Subgenus **Lymnaea** Lamarck. Shells medium in size, usually attenuate, rather thin, imperforate, columella twisted, and generally with the body whorl inflated. Widely distributed in N. A., Europe, and northern Asia.

◀ **Fig. 43.6.** *Lymnaea (Lymnaea) stagnalis* Linné. × $\frac{5}{7}$. (By Walker.)

Subgenus **Bulimnaea** Haldeman. Shell medium in size, rather solid, spire somewhat convex, body whorl inflated, olivaceous brown in color, and imperforate. The surface of the shell may be smooth, faintly and axially costate, and occasionally malleated.

Limited to northern U. S. and southern Canada, west to Iowa and Manitoba.

◀ **Fig. 43.7.** *Lymnaea (Bulimnaea) megasoma* Say. × $\frac{3}{4}$. (By Walker.)

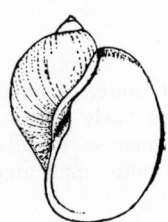

Subgenus **Radix** Montfort. Shell medium in size, thin, generally short-spired, and with an exceedingly large and inflated body whorl with an expanded lip. Usually imperforate, occasionally subperforate.

European in distribution. Introduced into eastern N. A. and spreading.

◀ **Fig. 43.8.** *Lymnaea (Radix) auricularia* Linné. (By Walker.)

Subgenus **Pseudosuccinea** Baker. Shells small to medium in size, thin, spire rather short, imperforate, generally light amber in color, with a large body whorl, and with the lip generally not expanded.

Generally distributed throughout eastern U. S. and Canada.

◀ **Fig. 43.9.** *Lymnaea (Pseudosuccinea) columella* Say. (By Walker.)

Subgenus **Acella** Haldeman. Shell small to medium in size, very attenuate, with a long narrow body whorl, and smooth columella. Color generally grayish-white to light amber.

Occurs in the upper Mississippi and St. Lawrence drainage areas.

◀ **Fig. 43.10.** *Lymnaea (Acella) haldemani* Binney. (By Walker.)

Subgenus **Stagnicola** Jeffreys. Shell small to medium in size, attenuate, elongate to ovate, outer lip usually somewhat thickened within, columella twisted. Color light amber to dark brown. The surface of the shell may be smooth, malleated, or occasionally may have spirally impressed lines.

Widely distributed in the northern states and Canada.

◀ **Fig. 43.11.** *Lymnaea (Stagnicola) palustris* Müller. (By Walker.)

Subgenus *Polyrhytis* Meek. Shell small, subglobose with a rounded body whorl, grayish-white in color, and generally axially ribbed or costate.

Only a single recent species in this subgenus is known, limited to the Bonneville Lake Basin of Utah.

◄ **Fig. 43.12.** *Lymnaea (Polyrhytis) utahensis* Call. (By Walker.)

Subgenus **Galba** Schrank. Shell small, with a somewhat elevated spire, generally smooth, with a straight columella, and with the inner lip reflected over the umbilical area.

Widely distributed throughout most of N. A.

◄ **Fig. 43.13.** *Lymnaea (Galba) abrussa* Say. × 1½. (By Walker.)

Family **Lancidae**

Shell small, limpet-shaped, rather solid in structure, with a smooth apex which is usually located near the center of the shell. Lung consisting of an open furrow between the mantle and the foot. Radula and jaws similar to those of the Lymnaeidae.

Lanx Clessin is the only genus in the family. Occurs from northern Calif. north to Wash. and Ida.

◄ **Fig. 43.14.** *Lanx newberryi* Lea. (By Walker.)

Family **Planorbidae**

Shells small to medium in size, usually discoidal, a few with moderate spire. Animal sinistral in a dextral shell (ultradextral). Surface of shell smooth to finely costate, whorls rounded to strongly keeled. Tentacles cylindrical. Jaw in 3 segments. Radula with numerous teeth arranged in nearly horizontal rows. Central tooth small and bicuspid, marginals tricuspid and lateral teeth multicuspid.

Genus **Gyraulus** Charpentier. Shell small and discoidal, with the whorls rounded to carinate. Aperture oblique and somewhat deflected. In many species the surface of the shell is covered with hairlike processes of periostracum and in addition is spirally striate.

Occurs throughout N. A. The subgenus *Gyraulus* occurs in the eastern and northern states and eastern Canada.

◄ **Fig. 43.15.** *Gyraulus hirsutus* Gould. × 3. (By Walker.)

Subgenus **Torquis** Dall. Shell small, with the whorls indistinctly spirally striate. Not hirsute. Base more or less concave. Lip often slightly thickened within.

Widely distributed in N. A.

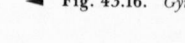
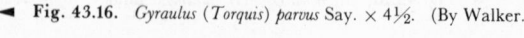

◄ **Fig. 43.16.** *Gyraulus (Torquis) parvus* Say. × 4½. (By Walker.)

Genus **Armiger** Hartman. Shell small and discoidal with the whorls rather strongly axially costate, the costae projecting at the periphery.

Only a single species in N. A., found from Me. west to Ill. and north into Canada.

◄ **Fig. 43.17.** *Armiger crista* Linné. × 7. (By Walker.)

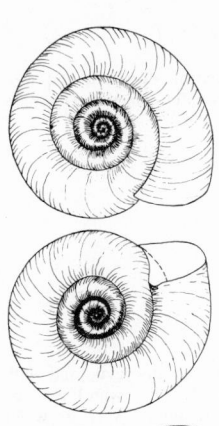

Genus **Tropicorbis** Pilsbry and Brown. Shell small, discoidal and smooth, usually having the body whorl rapidly increasing in size. Aperture oblique and with or without lamellae.

Occurs in La. and Tex., and south into Central and South America.

◄ **Fig. 43.18.** *Tropicorbis havanensis* Pfeiffer. (After Turner.)

Genus **Drepanotrema** Fischer and Crosse. Shell small, ultradextral, discoidal, the last whorl enlarged and expanded. Whorls rounded or carinate. There are no lamellae within the aperture.

Occurs from southern Tex. and south through Central and South America and the West Indies.

◄ **Fig. 43.19.** *Drepanotrema cultratum* d'Orbigny. × 2. (By Walker.)

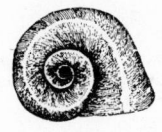

Genus **Helisoma** Swainson. Shell small to medium, discoidal, sinistral, with comparatively few rounded whorls, which may be carinated in certain species. Spire and umbilicus funicular. Aperture expanded, the outer lip somewhat thickened.

Widely distributed throughout all of N. A., the West Indies, Central and South America.

◄ **Fig. 43.20.** *Helisoma anceps* Menke. × 1½. (By Walker.)

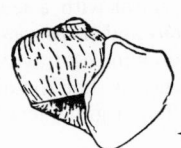

Genus **Carinifex** Binney. Shell small to medium in size, ultradextral, and the body whorl angled. The spire is depressed to elevated, the whorls terraced and angular. Aperture triangular with the outer lip thin and the inner lip with a thin callus, umbilicus funicular.

Occurs from Calif. and Ore. east to Wyo.

◄ **Fig. 43.21.** *Carinifex newberryi* Lea. (By Walker.)

Genus **Parapholyx** Hanna. Shell small, imperforate, ultradextral, globose, the spire very short and raised but little above the body whorl. Aperture wide and greatly expanded. Inner lip thickened, outer lip thin and acute.

Known to occur only in Wash., Ore., Calif., and Nev.

◄ **Fig. 43.22.** *Parapholyx effusa* Lea. × 2⅔. (By Walker.)

Genus *Planorbula* Haldeman. Shell small, ultradextral, discoidal, and with a few closely coiled whorls. The body whorl is usually somewhat carinate. Aperture with 6 lamellae which are situated well within.

Occurs throughout most of the eastern half of N. A.

◄ **Fig. 43.23.** *Planorbula armigera* Say. × 2. (By Walker.)

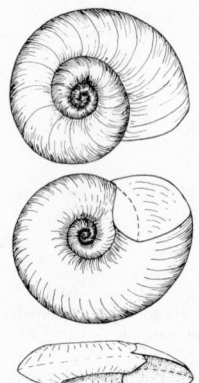

Genus *Promenetus* Baker. Shell small, ultradextral, discoidal, and with few whorls which rapidly increase in diameter. The last whorl may be rounded or carinate. All whorls exposed from both sides. Aperture wider than high with the outer lip thin.

Occurs throughout N. A.

◄ **Fig. 43.24.** *Promenetus exacuous* Say. × 15. (After Turner.)

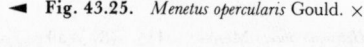
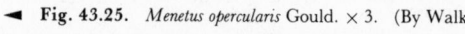

Genus *Menetus* H. and A. Adams. Shell small, ultradextral, discoidal, and having a few rapidly enlarging whorls. Shoulder of the body whorl more or less carinated. Aperture wide and somewhat expanded with the outer lip thin. Limited in its distribution to northern Calif. and Vancouver.

◄ **Fig. 43.25.** *Menetus opercularis* Gould. × 3. (By Walker.)

Family **Ancylidae**

Shell small, usually depressed cone-shaped, occasionally dextrally spiral with a few species having a small ledge or septum within. Two genera, *Neoplanorbis* and *Amphigyra*, are coiled. The foot is large and oval; tentacles are short, blunt, and cylindrical, with eyes located at their inner bases. Radula with teeth arranged in nearly horizontal rows. Central tooth small and usually unicuspid, laterals bicuspid, marginal teeth comblike.

This family is found throughout N. A.

Genus *Ferrissia* Walker. Shell small, conical, thin, with the apex posterior and slightly inclined to one side. The shell may be radially striate or smooth.

There are 2 subgenera, *Ferrissia* having the shell elevated with the apex radially striate and **Laevapex** having the shell depressed with a smooth apex.

Widely distributed throughout N. A.

◄ **Fig. 43.26.** *Ferrissia rivularis* Say. × 3. (By Walker.)

Genus **Gundlachia** Pfeiffer. Shell small, thin, obliquely conical, with the apex posterior and inclined to the right, smooth or radially striate, and having the apical portion of the interior more or less closed by a flat, horizontal septum.

There are 2 subgenera, **Gundlachia** with the apex smooth or only concentrically wrinkled and **Kincaidella** with the apex radially striate.

The genus is widely though locally distributed in N. A.

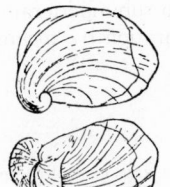

◄ **Fig. 43.27.** *Gundlachia meekiana* Stimpson. × 6. (By Walker.)

Genus **Rhodacmaea** Walker. Shell small, conic, elevated or depressed, smooth or radially striate, and having the apex tinged with pink.

There are 2 subgenera, **Rhodacmea** having the shell depressed and smooth, and **Rhodocephala** having the shell elevated and radially striate.

Limited to rivers, Ill. south to Ala.

◄ **Fig. 43.28.** *Rhodacmea filosus* Conrad. × 16. (After Turner.)

Genus **Amphigyra** Pilsbry. Shell small, spiral, dextral, and with a broad, thin columella plate projected across the end of the aperture next to the spire.

A single species in this genus, limited to the Coosa River, Ala.

◄ **Fig. 43.29.** *Amphigyra alabamensis* Pilsbry. × 10. (By Walker.)

Genus **Neoplanorbis** Pilsbry. Shell minute, dextral, spiral, subdiscoidal, and having the columellar margin broadly flattened.

Only 4 species in this genus, all limited to the Coosa River, Ala.

◄ **Fig. 43.30.** *Neoplanorbis tantillus* Pilsbry. × 10. (By Walker.)

SUBCLASS PROSOBRANCHIA

Mollusks with a gill and an operculum.

Family **Pilidae**

Shell large, globose-turbinate, umbilicate, and greenish in color. Operculum present. Respiratory chamber divided into two parts, one being the lung and the other containing the gill.

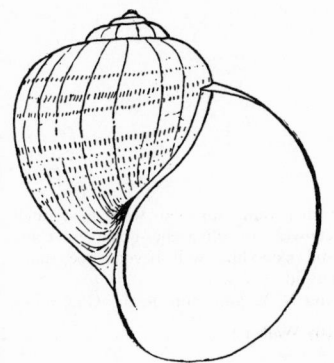

Genus **Pomacea** Perry. The characters of the family apply equally to the genus. These are the largest of our fresh-water snails. They occur in Fla. and southern Ga. along the margins of rivers, in swamps and drainage ditches. The animals crawl about mainly at night. They lay clusters of white eggs on the grass stems just above the water line to protect the eggs from predators. The young drop into the water upon hatching.

◄ **Fig. 43.31a.** *Pomacea paludosa* Say. (By Walker.)

Family **Pilidae**

Genus *Marisa* Gray. Shell medium in size, discoidal and with the aperture somewhat flaring. Colored yellowish-brown to greenish-brown and banded. Aperture fitted with a corneus operculum.

This northern South American genus has recently been introduced in the canals at Coral Gables, Florida, and apparently is spreading rapidly. It has been introduced also in Cuba and Puerto Rico. In the aquarium trade it is known as the "Columbian Snail" and was possibly introduced by someone discarding the contents of an aquarium. (See *Nautilus*, 72:53–55.)

◄ **Fig. 43.31b.** *Marisa cornuarietis* Linné. (After Turner.)

Family **Viviparidae**

Shell medium to large, globose to globose-turbinate, imperforate to subimperforate, usually green to nearly black in color. Operculum present. Respiration by means of gills.

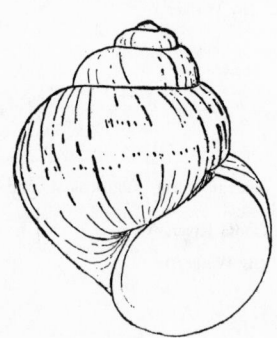

Genus *Viviparus* Montfort. Shell medium to large, generally thin with strongly convex whorls, imperforate to subimperforate, and usually light to dark green in color and somewhat banded. Outer or palatal margin or lip simple. Foot of moderate size and not extended beyond the snout. Teeth on the radula ribbon multicuspid.

Several species occur between the Atlantic states and the Mississippi Valley. They are rare north of Ill., Ohio, and N. Y. Two Oriental species, introduced into N. A., have become widely distributed. Both are much larger than our native species, reaching over 2 in. in length.

◄ **Fig. 43.32.** *Viviparus intertextus* Say. (By Walker.)

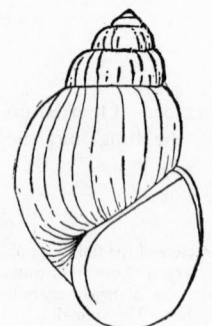

Genus *Campeloma* Rafinesque. Shell rather thick and solid with moderately convex whorls, imperforate, and light green to olivaceous in color. Outer lip thin. Foot large and extending beyond the snout. Teeth on the radula not multicuspid but simple or only minutely crenulate.

Known only in N. A., from the Atlantic states west to the Mississippi Valley and from the Great Lakes—St. Lawrence area to the Gulf states.

◄ **Fig. 43.33.** *Campeloma subsolidum* Anthony. (By Walker.)

Genus *Lioplax* Trochel. Shell rather thin, turreted with moderately convex whorls, imperforate, and yellowish to olivaceous-green in color. Outer lip thin. Foot very large and extending well beyond the snout. Teeth on the radula ribbon not multicuspid.

Known only in N. A., from Wis. and N. Y. and south to the Gulf states.

◄ **Fig. 43.34.** *Lioplax subcarinata* Say. (By Walker.)

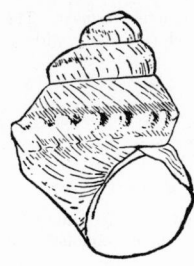

Genus *Tulotoma* Haldeman. Shell medium to large, rather thick and strong, and with moderately convex whorls. Young specimens are thin, usually carinate, and smooth. Mature specimens are very strong and heavy, nodulose, and without the carina.

Known only from the Alabama-Coosa Rivers in Ala.

◄ **Fig. 43.35.** *Tulotoma magnifica* Conrad. (By Walker.)

Family **Valvatidae**

Shell small, turbinate to subdiscoidal and widely umbilicate. Aperture holostomatous and simple. Operculum circular and multispiral. Respiration by means of gills. Radula without basal denticles on the central tooth.

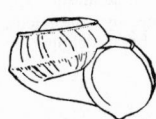

Genus *Valvata* Müller. Characters for the family apply as well to the genus. Widely distributed in N. A., especially in the northern states and Canada, in lakes and ponds and along river margins where there is ample vegetation.

◄ **Fig. 43.36.** *Valvata tricarinata* Say. × 4. (By Walker.)

Family **Bulimidae**

Shell small, spiral, conical, imperforate or umbilicate, and usually unicolored. Aperture entire with a simple and thin lip. Operculum generally paucispiral. Animal with a long snout and having the tentacles long and cylindrical with eyes at their bases. The foot is oblong, truncate before and behind. Gills internal. Verge exserted, placed on the back and some distance behind the right tentacle. There are two jaws. Radula with the central, lateral, and marginal teeth multicuspid. There are one or more basal denticles on the central tooth.

Widely distributed over much of N. A.

Genus *Bulimus* Scopoli. Shell small, somewhat extended, and subperforate. Aperture oval with a thin continuous and simple lip. Operculum calcareous and concentric. Radula with a broad denticulate central tooth and with a tonguelike process below.

A single species, occurring from the Hudson River in N. Y., west to Ill. and north into Ontario.

◄ **Fig. 43.37.** *Bulimus tentaculata* Linné. × 2. (By Walker.)

Genus *Amnicola* Gould and Haldeman. Shell small, ovate-conic to elongate; spire subacute. Whorls 4 to 6 and convex. Aperture rounded-ovate and with a continuous lip. Umbilicus narrow to wide. Operculum paucispiral, thin, and corneous. Central tooth of the radula multicuspid and with a tonguelike process below. Verge bifid and with a globular base.

Widely distributed throughout most of N. A.

◄ **Fig. 43.38.** *Amnicola limosa* Say. × 4. (By Walker.)

Genus *Fontigens*. Pilsbry. Shell similar to that of *Amnicola* but more attenuate. Central tooth of the radula high and wide, a single basal denticle and a distinct tongue-shaped projection on the lower margin. Verge trifid.

Widely distributed from the Mississippi Valley to the Atlantic states.

◄ **Fig. 43.39.** *Fontigens nickliniana* Lea. × 6. (By Walker.)

Genus **Paludestrina** d'Orbigny. Shell very similar to that of *Fontigens*. The central tooth of the radula has but one basal denticle on each side and is without the tongue-shaped process. The verge is bifid.

Widely distributed throughout most of N. A.

◄ **Fig. 43.40.** *Paludestrina minuta* Totten. (By Walker.)

Genus **Tryonia** Stimpson. Shell small, umbilicate, elongate, turreted, and with the surface longitudinally ribbed or costate. Aperture small, oblique, and subovate. Lip continuous, thin, and simple.

Found in Nev. and probably south through Central America.

◄ **Fig. 43.41.** *Tryonia clathrata* Stimpson. × 2⅔. (By Walker.)

Genus **Pyrgulopsis** Call and Pilsbry. Shell small, elongate, imperforate, and having a single strong carina at the periphery. Aperture ovate, lip continuous, thin, and simple. Central tooth of the radula with one basal denticle on each side.

Scattered distribution from the Mississippi Valley west to Nev.

◄ **Fig. 43.42.** *Pyrgulopsis nevadensis* Stearns. × 3. (By Walker.)

Genus **Lyrodes** Doering. Shell small, ovate-conic, and imperforate. Whorls angulated and usually coronated with spines. Aperture ovate with a simple lip. Verge with appendages. Central tooth of radula trapezoidal with very small basal denticles.

In N. A., limited to Fla. and Tex.

◄ **Fig. 43.43.** *Lyrodes coronatus* Pfeiffer. × 3¾. (By Walker.)

Genus **Littoridina** Souleyet. Shell small, narrowly perforate, somewhat extended, solid, and opaque. Body whorl subangulate at the periphery. Aperture pyriform, angulate above, lip simple but not continuous. Radula similar to that in *Amnicola*.

Fla. and Tex., and south through Central and South America.

◄ **Fig. 43.44.** *Littoridina monroensis* Frauenfeld. × 7. (By Walker.)

Genus **Cochliopa** Stimpson. Shell small, depressed-conic, base concave, and possessing a large and deep umbilicus. Aperture subcircular, oblique, and with a simple lip. Central tooth of the radula with 2 or 3 basal denticles. Verge elongate, compressed, and bifid.

Tex. and south through Central America.

◄ **Fig. 43.45.** *Cochliopa riograndensis* Pilsbry and Ferriss.× 6. (By Walker.)

Genus **Clappia** Walker. Shells small, globose-turbinate, narrowly and deeply umbilicate. Aperture subcircular and having a simple lip. Operculum paucispiral. Central tooth of the radula broad and possessing several basal-denticles. Lateral tooth with a tonguelike process in the blade.

A single species, restricted to the Coosa River in Ala.

◄ **Fig. 43.46.** *Clappia clappi* Walker. × 6¼. (By Walker.)

Genus **Fluminicola** Stimpson. Shell small, solid, ovate, and imperforate. Aperture subcircular with a simple lip. Central tooth of the radula with several denticles on each side of the base. Verge winged.

Occurs in the northwestern states and British Columbia.

◄ **Fig. 43.47.** *Fluminicola nuttalliana* Lea. × 2. (By Walker.)

Genus **Somatogyrus** Gill. Shell small, usually rather solid, smooth, imperforate, or very narrowly perforate. Aperture oblique with the lip thin and projecting above. Columella area thickened with a callus. Central tooth of the radula with 3 to 4 basal denticles. Verge broad, compressed, and bifid.

Occur mainly south of the Ohio River and east of the Mississippi River.

◄ **Fig. 43.48.** *Somatogyrus subglobosus* Say. × 2. (By Walker.)

Genus **Gillia** Stimpson. Shell small, smooth, imperforate, obtuse. Aperture large, oblique, the lip thin and continuous on the same plane. Central tooth of the radula with 2 basal denticles on each side. Verge small, simple, and lunate.

Restricted to the Atlantic Coast states from N. J. south to Va.

◄ **Fig. 43.49.** *Gillia altilis* Lea. × 2. (By Walker.)

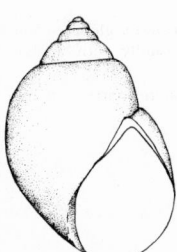

Genus **Notogillia** Pilsbry. Shell smooth, imperforate, subglobose, and solid. Aperture subcircular, slightly oblique, lip simple but thickened and continuous on the same plane. Central tooth of the radula with 2 basal teeth set high on the plate, the upper row consisting of 8 small teeth and 1 large, central tooth.

Restricted to rivers of northern Fla. from the upper St. Johns west to the Apalachicola River system.

◄ **Fig. 43.50.** *Notogillia wetherbyi* Dall. × 4. (After Turner.)

Genus **Lyrogyrus** Gill. Shell very small, smooth, perforate, elongate-ovate, and rather thin. Aperture nearly circular and with a thin and continuous lip. Operculum corneous, circular, and multispiral. Central tooth of the radula with 2 denticles on each side of the base.

A genus of very small species restricted to the Atlantic Coast states.

◄ **Fig. 43.51.** *Lyrogyrus pupoideus* Gould. × 6. (By Walker.)

Genus **Horatia** Bourguignat. Shell exceedingly small, depressed, umbilicate, and with a deep suture. Aperture moderately oblique, subcircular, with a thin simple and continuous lip.

In N. A., confined to Tex. It also occurs in Europe.

◄ **Fig. 43.52.** *Horatia micra* Pilsbry and Ferriss. × 20. (After Turner.)

Genus **Pomatiopsis** Tryon. Shell small, turreted or extended, and umbilicate. Aperture subcircular, expanded, and with a thin, simple lip. Central tooth of the radula with one basal denticle on each side near the lower margin. Verge large, simple, and convoluted. Foot divided by a transverse sulcus at about its anterior third.

Species in this genus are semiamphibious, living on damp or wet soil and rocks. Rather widely distributed in central and eastern N. A.

◄ **Fig. 43.53.** *Pomatiopsis lapidaria* Say. × 4. (By Walker.)

Family **Lepyriidae**

Shell small and depressed with a broad parietal area, imperforate. Aperture auriculate with the outer lip flaring. Operculum thin and paucispiral. Radula with the central tooth broad and with numerous denticles along its upper margin.

Genus **Lepyrium** Dall. The characters of the family apply as well to the genus. The family contains but a single genus and species, limited to the Cahawba River and possibly to the Coosa River in Ala.

◄ **Fig. 43.54.** *Lepyrium showalteri* Lea. × 3½. (By Walker.)

Family **Pleuroceridae**

Shells small to large, attenuate to globose, smooth to nodulose, generally dark green to brown in color and occasionally with darker bands of color. Radula with a broad and denticulate central tooth. Aperture rounded to lengthened and occasionally canaliculate below. Operculum paucispiral and thin. So far as we know all species in this family are oviparous.

The family occurs in N. A. mainly in the Mississippi Valley, east to the Atlantic states and north to the Great Lakes. It again occurs in Calif. and north to Wash.

Genus **Lithasia** Haldeman. Shells of medium size, imperforate, globose-conic, smooth to tuberculate, and solid. Aperture rhomboidal usually with a short canal at the base.

Occurs mainly in Ky., Tenn., and Ala. A few species extend northward into Ill., Ind., and Ohio.

◄ **Fig. 43.55.** *Lithasia geniculata* Conrad. (By Walker.)

Genus **Io** Lea. Shell medium to large, nearly smooth to nodulose, and solid. Aperture rhomboidal, extended below to form a siphonal canal.

Confined to the Tennessee River system in eastern Tenn. and Va.

◄ **Fig. 43.56.** *Io fluvialis* Say. (By Walker.)

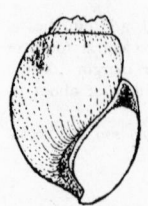

Genus **Eurycaelon** Lea. Shell small to medium is size, solid, ovate, with a very short spire, and large body whorl. Aperture subovate with a short canal at the base.

Occurs only in the Tennessee River drainage system in eastern Tenn.

◄ **Fig. 43.57.** *Eurycaelon anthonyi* Budd. (By Walker.)

Genus *Pleurocera* Rafinesque. Shell small to medium in size, attenuate, imperforate, smooth, and nodulose or carinate. Aperture subrhomboidal, prolonged into a short canal below. Columella twisted but not thickened.

Widely distributed in the eastern half of the U. S. This genus is particularly rich in species in Ala., Tenn., and Ky.

◄ **Fig. 43.58.** *Pleurocera acuta* Rafinesque. (After Turner.)

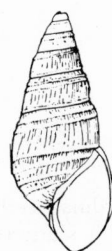

Genus *Goniobasis* Lea. Shell small to medium in size, attenuate, imperforate, smooth, carinate, and occasionally tuberculate. Aperture subrhomboidal, subangular at the base but not canaliculate. Columella smooth, not twisted.

Widely distributed from the Mississippi Valley and east to the Atlantic states and from Fla. north to the Great Lakes. A small group of species occurs in northern Calif. north to Wash.

◄ **Fig. 43.59.** *Goniobasis virginica* Gmelin. (By Walker.)

Genus *Apella* Anthony. Shell small to medium in size, somewhat extended, imperforate, and smooth to carinate. Aperture subrhomboidal, rounded or slightly angulate below, with a slit or fissure above.

Known only from the Coosa River, Ala.

◄ **Fig. 43.60.** *Apella demissum* Lea. (By Walker.)

Genus *Leptoxis* Rafinesque. Shell small in size, globose, solid, imperforate, and smooth. Aperture oval to subcircular, entire below, and having the columella thickened with a callus.

Distributed from the Ohio River south into Ala. and Ga.

◄ **Fig. 43.61.** *Leptoxis praerosa* Say. (By Walker.)

Genus *Mudalia* Haldeman. Shell small in size, conical, rather thin, imperforate, and smooth to carinate. Aperture subrhomboidal with the outer lip thin.

Occurs in rivers from Pa. west to Ohio and south to northern Ala.

◄ **Fig. 43.62a.** *Mudalia carinata* Bruguiere. (After Turner.)

Family **Thiaridae**

Genus *Tarebia* Adams.   Shell medium in size, attenuate, imperforate and colored a yellowish-brown.   Aperture subovate and flaring below.   Columella slightly thickened.   Radula similar to that of the Pleuroceridae (Fig. 43.3).

Known only at present from Lithia Spring, Hillsborough Co., Florida.   This species was introduced about 1940 from Hawaii or some western Pacific island where this species normally occurs.

◄   **Fig. 43.62b.**   *Tarebia granifera* Lamarck.   (After Turner.)

CLASS PELECYPODA

Family **Margaritiferidae**

Shell elongate, laterally compressed; hinge usually with only pseudocardinal teeth, laterals when present very obscure.   Gills without water tubes and with scattered interlamellar connections which in certain places form irregular rows.

Genus *Margaritifera* Schumacher.   The characters of the family apply equally to the genus.

The typical species, **M. margaritifera** Linné, is circumboreal but in N. A. is found only in the northeastern section and Pacific Northwest.   Another species is found in the Tennessee and Ohio drainage systems and 2 more have been described from the Gulf drainage.

Fig. 43.63.   *Margaritifera margaritifera* Linné.   × ³/₅.   (By Walker.)

Family **Unionidae**

Shell subcircular, oval, subtriangular, or elongate; hinge edentulous or with pseudocardinals only or with both pseudocardinals and laterals.   Gills with water tubes and distinct, continuous, interlamellar septa running parallel to the filaments.

Subfamily **Unioninae.**   Marsupia formed by all 4 gills or by the outer gills only. The edges of the marsupia always sharp and not distending.   Shell usually heavy and solid, rounded to elongate, and generally with a dull-colored periostracum. Beak sculpture rather indistinct and consisting of concentric or double-looped bars. Hinge always complete and having well-developed teeth.   There is little or no difference of sex indicated by the shell.

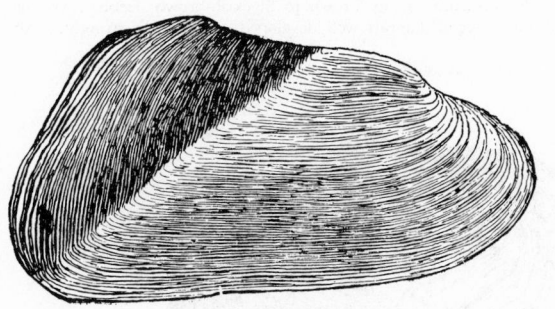

Genus **Gonidea** Conrad. Shell smooth, elongate, subtriangular, with usually a high and sharp posterior ridge. Hinge with rudimentary pseudocardinal and lateral teeth in each valve. This genus, represented by a single species, is remarkable for the sharp posterior ridge and more or less flattened posterior region.

A West Coast species ranging from central Calif. north to British Columbia and east to Ida.

◄ **Fig. 43.64.** *Gonidea angulata* Lea. × ¾. (By Walker.)

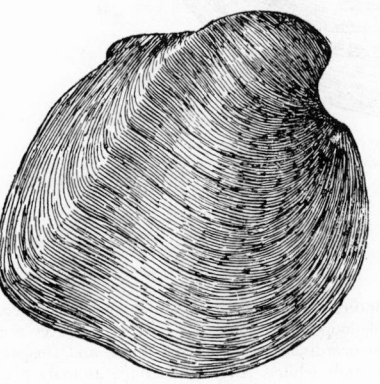

Genus **Fusconaia** Simpson. Shell rounded, rhomboid, triangular or short elliptical, and with a moderate posterior ridge. Beaks high and full, curved inward and forward, sculptured with a few coarse, parallel ridges. Periostracum dark, surface not sculptured, and having a hinge plate of moderate width. Pseudocardinal teeth strong. Nacre white, salmon, or purple. All 4 gills serving as marsupia.

The majority of the species in this genus are found in the southern states. However, a few species range well north into Mich. and the upper Mississippi River.

◄ **Fig. 43.65.** *Fusconaia undata* Barnes. (By Walker.)

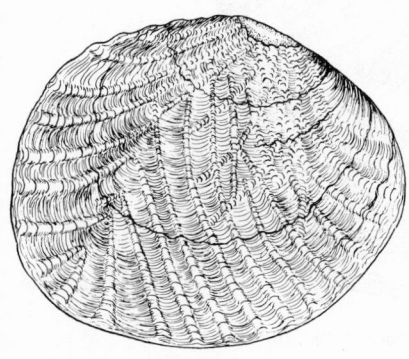

Genus **Quincuncina** Ortmann. Shell subelliptical to rounded, compressed, and somewhat solid. Beaks sculptured with subconcentric ridges. There is a rather complex zigzag sculpture over most of the disc of the shell and the posterior ridge is moderately to well developed. Periostracum a blackish-brown. All 4 gills serving as marsupia.

Known only from the Choctowhatchee River, Ala. east to the Suwanee River, Fla.

◄ **Fig. 43.66.** *Quincuncina infucata* Conrad. (After Turner.)

Genus *Quadrula* Rafinesque. Shell subcircular to oblong, solid, usually inflated, and having a well-developed posterior ridge. Surface of the shell usually well sculptured with flattened knobs or ridges, generally heavier towards the posterior portion of the shell. Beaks prominent, sculptured with a few coarse and irregular subparallel ridges. Periostracum usually brown to blackish-brown, feebly rayed or without rays. Pseudocardinal teeth solid and ragged, laterals well developed and nearly straight. All 4 gills serving as marsupia.

Widely distributed from Mich. west to Minn. and south to Ark. and Ala.

Fig. 43.67.   *Quadrula quadrula* Rafinesque.   (By Walker.)

Genus *Crenodonta* Schlüter. Shell large, solid, and sculptured with strong diagonal plicae. Beaks prominent, nearly smooth, or sculptured with coarse, double-looped corrugations which extend over the upper portion of the disc. Periostracum dark brown. Pseodocardinal teeth relatively large and ragged, laterals long and developed upon a broad plate. Nacre generally white. All 4 gills serving as marsupia.

Widely distributed from the St. Lawrence system west to Lake Winnipeg and south to Fla. and La.

Fig. 43.68.   *Crenodonta costata* Rafinesque.   × ½.   (By Walker.)

Genus *Tritigonia* Agassiz. Shell large, solid, rhomboid in shape, and possessing a well-defined posterior ridge. Sexes dissimilar in shape, the shell in the male truncated posteriorly, rounded and subcompressed in the female. Surface of the shell pustulose except on the extended portion of the female. Both pseudocardinal teeth and laterals well developed. All 4 gills serving as marsupia.

Widely distributed throughout most of the Mississippi system and the Gulf drainage from Ala. to Tex.

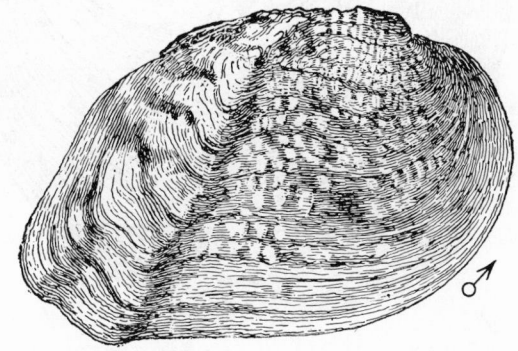

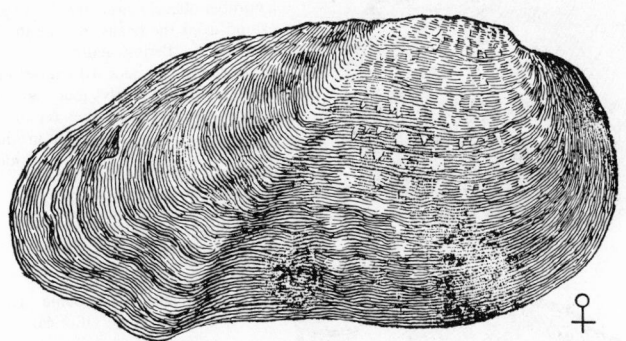

Fig. 43.69. *Tritigonia verrucosa* Rafinesque. × ½. (By Walker.)

Genus *Cyclonaias* Pilsbry.   Shell rounded, solid, and having a moderately well-defined posterior ridge. Surface of the shell pustulose.   Beaks prominent and sculptured with numerous fine and irregular corrugations.   Periostracum a dark brown.   Pseudocardinal teeth massive and ragged, laterals poorly defined and built upon a broad and irregular hinge plate.   Nacre pinkish.   Only the outer gills serving as marsupia.

Widely distributed from southern Mich. west to Ia. and south to Ala. and Tex.

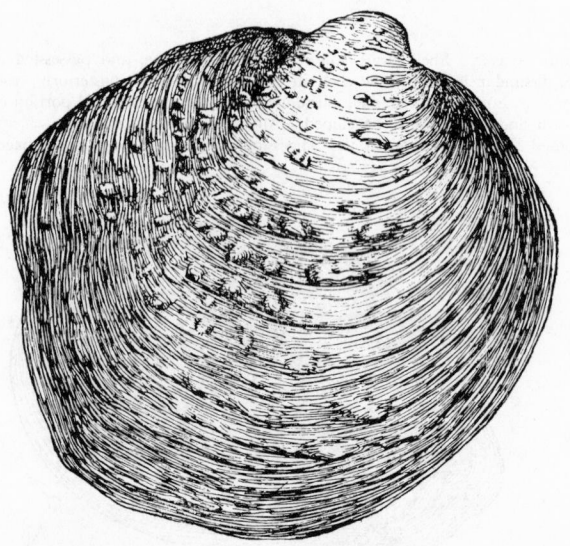

Fig. 43.70.   *Cyclonaias tuberculata* Rafinesque.   (By Walker.)

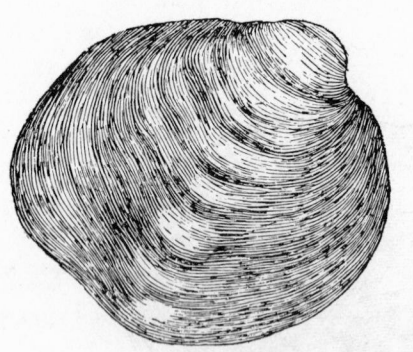

Genus **Plethobasus** Simpson.   Shell large, irregularly oval, inflated, and with a poorly defined posterior ridge. Surface of the shell sculptured with a number of fairly large and flattened tubercles which extend from the beaks to near the lower margin of the valves. Periostracum a brownish-yellow to a dark brown. Pseudocardinal teeth rather small and ragged, laterals fairly long and slightly curved. Only the outer gills serving as marsupia.

There are only 2 species in this genus, which ranges from Ohio to Minn. and south to the Tenn. River system.

◄ **Fig. 43.71.**   *Plethobasus cyphus* Rafinesque. × ⅔. (By Walker.)

Genus **Pleurobema** Rafinesque.   Shell triangular to elliptical, solid, moderately inflated, and with a fairly well-defined posterior ridge. Surface of the shell usually smooth.   Beaks anterior.   Periostracum brown to yellowish and frequently rayed. Pseudocardinal teeth small and somewhat ragged, laterals long and slightly curved. The outer gills only serving as marsupia.

Contains a large number of species, and ranges widely from Ohio and Ill. south to Ga. and Miss.

◄ **Fig. 43.72.**   *Pleurobema mytiloides* Rafinesque. (By Walker.)

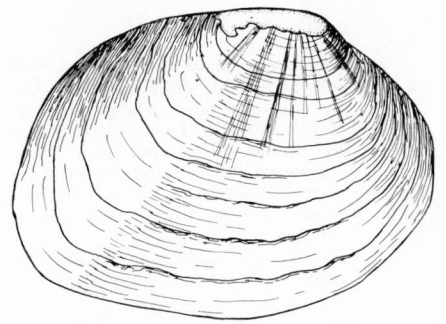

Genus **Lexingtonia** Ortmann.   Shell subquadrate with slightly elevated beaks and well-
developed hinge teeth.   Beaks slightly. anterior.
Outer surface of the shell without sculpture.
Periostracum light brown to dark brown with
rather indistinct rays.   Beak sculpture consisting
of 6 to 8 rather crowded subconcentric ridges.
Nacre white or pink.   Outer gills only serving
as marsupia.

A genus consisting of a few species, limited in
distribution mainly to N. C. and Va.

◄ **Fig. 43.73.**   *Lexingtonia   subplana*   Conrad.
(After Turner.)

Genus **Elliptio** Rafinesque.   Shell
elongate, rhomboidal to oval in shape,
and usually having a well-defined posterior ridge.   Surface of shell smooth or
feebly corrugated.   Beak sculpture consisting of a few rather strong ridges
which are nearly parallel to the growth
lines.   Pseudocardinal teeth relatively
small and ragged, laterals long and
slightly curved.   Periostracum dark
brown and occasionally rayed.   Nacre
white to pink.   Outer gills only serving
as marsupia.

This genus, very numerous in species,
is widely distributed throughout most of
eastern N. A.   The various species are
exceedingly difficult to separate from
one another.

◄ **Fig. 43.74.**   *Elliptio crassidens* Lamarck.
× ½.   (By Walker.)

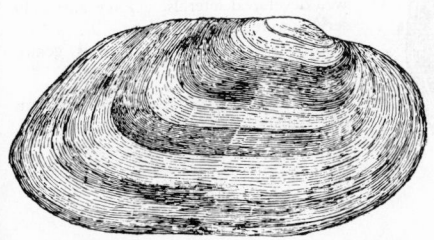

Genus **Uniomerus** Conrad.   Shell trapezoidal
in outline, with a rounded posterior ridge, and
somewhat pointed posteriorly.   Beaks not prominent and sculptured with curved concentric
ridges.   Surface of the shell sculptured with fine
concentric lines.   Pseudocardinal teeth usually
compressed, laterals delicate and slightly curved.
Outer gills only serving as marsupia.

Only a few species in this genus, which ranges
from Ohio south to Fla. and Tex.

◄ **Fig. 43.75.**   *Uniomerus tetralasmus* Say.   × ½.
(By Walker.)

Subgenus **Canthyria** Swainson. Shell inflated, suboval, with a high and rather sharp posterior ridge. The shell is sculptured with a series of spines which more or less parallel the posterior ridge. Beaks compressed. Hinge sharply curved at the center. Pseudocardinal teeth rather compressed, laterals short. Beak cavities rather deep.

Only a single species in this remarkable subgenus which is limited to the Altamaha River in Ga.

◄ **Fig. 43.76.** *Elliptio (Canthyria) spinosa* Lea. × ⅔. (By Walker.)

Genus **Hemistena** Rafinesque. Shell elongate, subsolid, inequilateral, rounded anteriorly, and pointed posteriorly. Posterior ridge low with one or more secondary ridges above it. Beaks low and sculptured with a few coarse, irregular, longitudinal folds. Periostracum shining and often rayed. Pseudocardinal teeth limited to one in each valve, laterals usually vestigial. Nacre purplish, shading toward blue at the margin of the shell. The middle portion of the outer gills serving as marsupia.

Only a single species in this genus, in the Ohio River system.

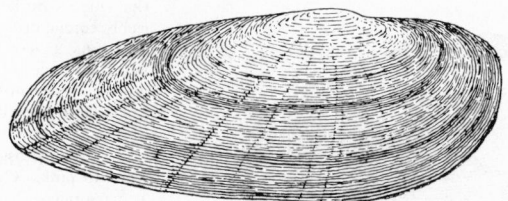

Fig. 43.77. *Hemistena lata* Rafinesque. (By Walker.)

Subfamily **Anodontinae.** Marsupia formed by the entire outer gills, distending transversely when charged. Water tubes in the gravid female divided longitudinally into 3 tubes, of which the center one is used as an ovisac. Hinge rarely complete, the lateral or both the lateral and the pseudocardinal teeth often missing. Little or no difference between the shells of the two sexes.

Genus **Arkansia** Ortmann and Walker. Shell large, subcircular, solid, and inflated. Beak sculpture weak, consisting of 2 or 3 double-looped bars. Disc sculptured with irregular and oblique folds which are occasionally indistinct. Hinge well developed with strong pseudocardinal teeth and well-developed laterals. Outer gills only serving as marsupia.

Only a single species in this genus, limited to the Ouachita River, Ark.

◄ **Fig. 43.78.** *Arkansia wheeleri* Ortmann and Walker. × ¾. (By Walker.)

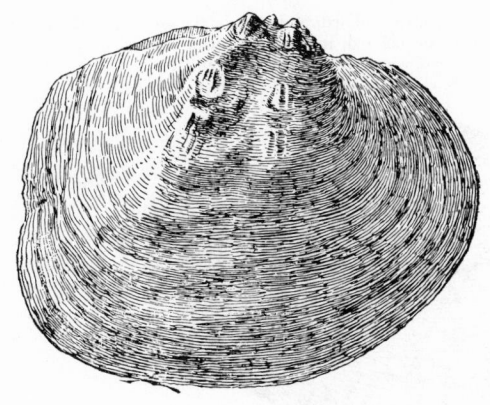

Genus **Arcidens** Simpson. Shell large, subsolid, inflated, subrhomboidal, with full and high beaks. Beak sculpture very strong, consisting of irregular corrugations. Disc of shell sculptured with oblique folds and wrinkles. Periostracum dark brown to blackish. Outer gills only serving as marsupia.

Only a single species in this genus, ranging widely throughout the Mississippi and Ohio river systems as far west as eastern Tex.

◄ **Fig. 43.79.** *Arcidens confragosa* Say. × ½. (By Walker.)

Genus **Alasmidonta** Say. Shell medium to large in size, generally rhomboidal in shape, inflated, and with a well-developed posterior ridge. Beaks full and high, with coarse, concentric or slightly double-looped bars. Periostracum rayed and shining. Hinge with 2 pseudocardinal teeth in the left valve and 1 in the right, laterals imperfect or wanting (present in *Prolasmidonta*). Beak cavities deep. Nacre bluish. Outer gills only serving as marsupia.

This genus has several well-defined subgenera. The many species range from the Mississippi River east to the Atlantic states.

Subgenus **Alasmidonta** Say. Shell medium to large, ovate-rhomboid, rather thin, and inflated. Beak sculpture of strong concentric bars. Periostracum dark green to dark brown and generally rayed. Pseudocardinal teeth solid, short, and with ridges; laterals short, imperfect, or wanting. Beak cavities deep and compressed.

A single species in this subgenus, occurring from Nova Scotia south to N. C.

◄ **Fig. 43.80.** *Alasmidonta* (*Alasmidonta*) *undulata* Say. × ⅔. (By Walker.)

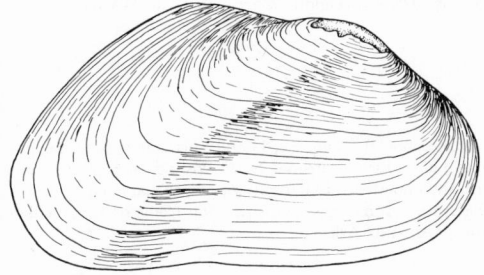

Subgenus **Prolasmidonta** Ortmann. Shell small to medium in size and rhomboidal in shape. Posterior ridge well defined. Beak sculpture moderately heavy. Pseudocardinal teeth small, laterals present but reversed, 2 in the right valve and 1 in the left.

Only a single species in this subgenus, limited in distribution to Va. and north to New Brunswick, Canada.

◄ **Fig. 43.81.** *Alasmidonta* (*Prolasmidonta*) *heterodon* Lea. (After Turner.)

Subgenus **Decurambis** Rafinesque.   Shell large, elongate, inflated, and rhomboidal in shape.   Posterior slope slightly corrugated.   Periostracum generally highly colored, mainly green with rays which are often broken into a dappled or splashed pattern.   Hinge weak with teeth imperfect, lateral teeth absent.

Widely distributed from New England south to S. C. and west to the Mississippi Valley.

Fig. 43.82.   *Alasmidonta* (*Decurambis*) *marginata* Say.   (By Walker.)

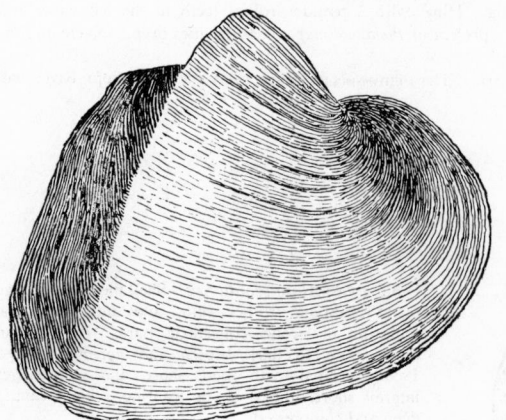

Subgenus **Bullella** Simpson.   Shell large, somewhat triangular, thin, inflated, and having a sharp posterior ridge.   Beaks full and possessing an exceedingly strong concentric sculpture.   Pseudocardinal teeth reflexed and compressed.

Only 2 species in this subgenus, limited to S. C. and Ga.

◄ Fig. 43.83.   *Alasmidonta* (*Bullella*) *arcula* Lea. × ⅚.   (By Walker.)

Subgenus **Pegias** Simpson.   Shell small, thickened in front, and with a sharp posterior ridge, which has a wide radial impression ending in a basal sinus.   Beak sculpture consisting of subconic corrugations.   Periostracum dark and usually showing a few radial rays at the base.   Pseudocardinal teeth rather solid, laterals wanting.

Only a single species in this subgenus, limited to the Cumberland and Tennessee river systems.

◄ Fig. 43.84.   *Alasmidonta* (*Pegias*) *fabula* Lea.   (By Walker.)

Genus **Lasmigona** Rafinesque.   Shell large, subrhomboidal, compressed, and corrugated over the posterior slope.   Beaks low and sculptured with strong bars.   Pseudocardinal teeth existing as 1 in the right valve and 2 in the left valve.   Lateral teeth generally imperfect.   Outer gills serving as marsupia.

This genus contains 4 rather well-defined subgenera.   It has a wide distribution from the Great Lakes, St. Lawrence, and upper Mississippi rivers south to Ala. and Ark.

Subgenus **Lasmigona** Rafinesque. Shell large, compressed, and generally corrugated on the posterior slope. Beaks low with shallow cavities within. Beak sculpture consisting of coarse ridges or bars which form slight loops. Periostracum brownish-green with radiating greenish rays.

Only a single species in this subgenus, with a wide distribution in the St. Lawrence and Mississippi river systems.

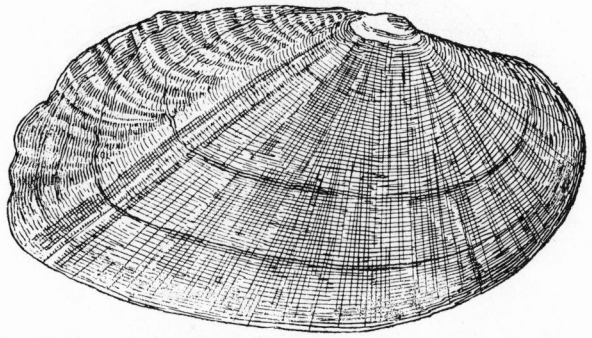

Fig. 43.85. *Lasmigona* (*Lasmigona*) *costata* Rafinesque. × ⅔. (By Walker.)

Subgenus **Pterosyna** Rafinesque. Shell very large, ovate-rhomboidal, somewhat inflated and with a rather well-defined posterior ridge. Beaks much compressed, sculptured with double-looped bars and ridges. Periostracum blackish-brown. Nacre white. Hinge plate very heavy, lateral teeth imperfectly developed.

Only a single species in this subgenus, widely distributed from the Great Lakes system and upper Mississippi River south to Ala. and Ark.

Fig. 43.86. *Lasmigona* (*Pterosyna*) *complanata* Rafinesque. × ⅓. (By Walker.)

Subgenus *Platynaias* Walker.   Shell large, compressed, rather thin, smooth, and generally rayed. Beaks compressed and sculptured with double-looped bars and ridges.   Hinge teeth delicate.

The few species in this subgenus are widely distributed from Vt. west to Neb. and south to northern Ala.

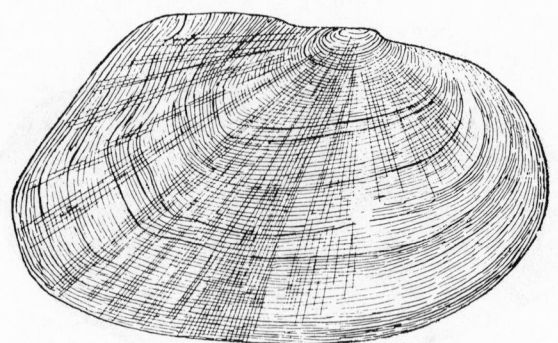

Fig. 43.87.   *Lasmigona (Platynaias) compressa* Lea. × ⅔.   (By Walker.)

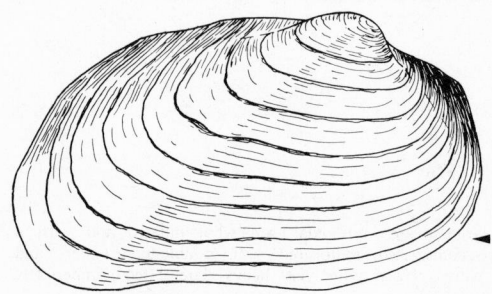

Subgenus *Sulcularia* Rafinesque.   Shell small, elliptical, and smooth.   Beaks sculptured with 4 to 6 rather fine bars, the first 1 or 2 subconcentric, the remainder double-looped.   Pseudocardinal teeth delicate, laterals obsolete or wanting entirely.

Only a single species in this subgenus, occurring mainly in the smaller rivers and streams of Tenn. and Ga.

◄ **Fig. 43.88.**   *Lasmigona (Sulcularia) holstonia* Lea.   (After Turner.)

Genus *Simpsoniconcha* Frierson.   Shell small, thin, elongate, and rounded in front and behind.   Beaks sculptured with fine, parallel ridges which are looped in the middle. Periostracum yellowish or greenish-brown and without rays. Pseudocardinal teeth irregular and compressed, laterals obsolete or lacking.   Outer gills only serving as marsupia.

There is but a single species in this genus, which ranges from Ohio and Mich. west to Ia. and south to Tenn. and Ark.

◄ **Fig. 43.89.**   *Simpsoniconcha ambigua* Say.   (By Walker.)

Genus *Anodontoides* Simpson.   Shell elliptical, smooth, inflated, thin, and having a faint posterior ridge.   Beaks full and sculptured with a few subparallel concentric ridges.   Periostracum smooth, shining and frequently rayed.   Hinge line slightly incurved in front of the beaks.   Teeth lacking or only indicated by rudiments.   Nacre bluish-white.   Outer gills only serving as marsupia.

Widely distributed from the St. Lawrence system south through much of the Mississippi system.

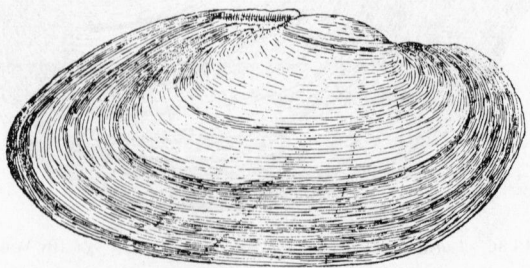

Fig. 43.90.   *Anadontoides ferussaciana* Lea. × ¾.   (By Walker.)

Genus **Anodonta** Lamarck.   Shell thin, smooth, elliptical, inflated, and occasionally slightly winged posterior to the beaks.   Posterior ridge usually well defined.   Beaks sculptured with rather numerous and nearly parallel ridges, usually double-looped.   Periostracum thin, smooth, and occasionally shining. Hinge without teeth, the hinge plate reduced to a narrow and regularly curved ridge.   Outer gills only serving as a marsupia.

Widely distributed throughout most of N. A.

Fig. 43.91.   *Anodonta grandis* Say. × ½.   (By Walker.)

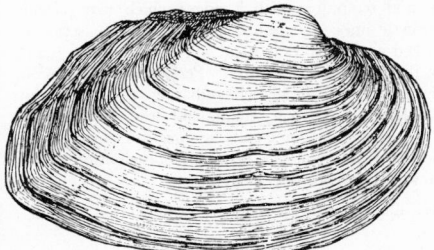

Genus **Strophitus** Rafinesque.   Shell elliptical, inflated, rather thin, generally pointed posteriorly, and having a low posterior ridge.   Beaks full and sculptured with a few strong concentric ridges which curve upwards posteriorly.   Periostracum thin, smooth, shining and occasionally rayed. Teeth rudimentary.   Outer gills only serving as marsupia.

Only a few species in this genus, ranging widely from New England west to Minn. and south to N. C. and Tenn.

Fig. 43.92.   *Strophitus undulatus* Say. × ¾.   (By Walker.)

Subfamily **Lampsilinae.**   Marsupia formed from the outer gills and usually only the posterior portion is utilized.   Edge of marsupia when charged distending beyond the original edge of the gills.   Water tubes simple in the gravid female.   Hinge complete with both lateral and pseudocardinal teeth.   Male and female shells showing moderate differences.

Genus **Ptychobranchus** Simpson.   Shell triangular, solid, and with a well-developed rounded posterior ridge.   Periostracum brownish-yellow with greenish, wavy, hairlike rays, or broken radiating bars.   Hinge plate wide and flat.   Pseudocardinal teeth small and triangular in shape, laterals club-shaped and distant.

The few species in this genus range from Ohio and Mich. south to Ala. and La.

Fig. 43.93.   *Ptychobranchus fasciolare* Rafinesque. × ¾.   (By Walker.)

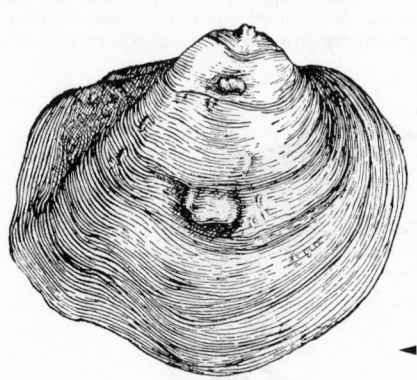

Genus *Obliquaria* Rafinesque. Shell inflated, solid, oval, and somewhat pointed at the ventral end of the well-defined posterior ridge. The shell is sculptured with 2 or more large, compressed knobs which extend from the beaks to the ventral margin. In profile the position of the knobs on one valve alternate with the interspaces between the knobs on the other valve. The posterior slope is sculptured with well-defined corrugations. Beaks prominent, with 4 or 5 heavy parallel ridges. Periostracum smooth, usually shining and generally rayed. Pseudocardinal teeth distinct and ragged, laterals short and nearly straight. Outer gills serving as marsupia, the ovisacs positioned just behind the center of the gills and projecting below the rest of the branchiae.

Only a single species in this genus, ranging from Mich. south to Ala. and Tex.

◀ **Fig. 43.94.** *Obliquaria reflexa* Rafinesque. (By Walker.)

Genus *Conchodromus* Haas. Shell solid, rounded to ovate, and with a poorly defined posterior ridge. Beaks forward, rather high, and sculptured with 5 ridges which run parallel with the growth lines. Sculpture consisting of a series of strongly marked concentric ridges and a few low humps which extend from the beaks to the margin. Periostracum yellowish-brown with a series of undulated, radiating rays of green. Pseudocardinal teeth triangular, small, and low, laterals low, short, and club-shaped. Outer gills serving as marsupia.

The 2 species in this genus are limited in distribution to the Cumberland and Tennessee river systems.

◀ **Fig. 43.95.** *Conchodromus dromas* Lea. × ½. (By Walker.)

Genus *Cyprogenia* Agassiz. Shell solid, inflated, subcircular, and having a well-developed posterior ridge. Beaks curved inward and forward, sculpture with faint double-looped ridges. Surface of the shell nodulose and with a few irregular, concentric ridges. Periostracum yellowish-brown and rayed with groups of greenish, interrupted lines. Pseudocardinal teeth heavy and triangular, laterals short and obliquely striated. Nacre white and silvery. Outer gills serving as marsupia. These consist of several long, purple ovisacs which are pendent from near the central base of the gills.

Only a single species in this genus and it is widely distributed in the Ohio, Cumberland, and Tennessee river systems.

◀ **Fig. 43.96.** *Cyprogenia irrorata* Lea. (By Walker.)

Genus *Plagiola* Rafinesque. Shell solid, subelliptical, the surface concentrically ridged and having a well-defined posterior ridge. Periostracum yellowish-green and rayed with rather fine, greenish, interrupted lines. Beaks well forward and sculptured with a few irregular, double-looped ridges. Pseudocardinal teeth large and triangular in shape, laterals slightly curved.

Only a single species in this genus and it is widely distributed from Ohio west to Ia. and south to Ark. and Ala.

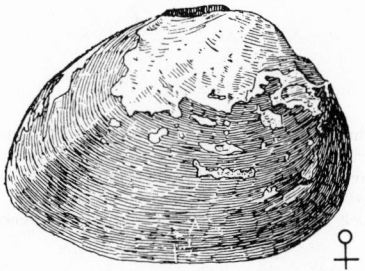

Fig. 43.97. *Plagiola securis* Lea. × ½. (By Walker.)

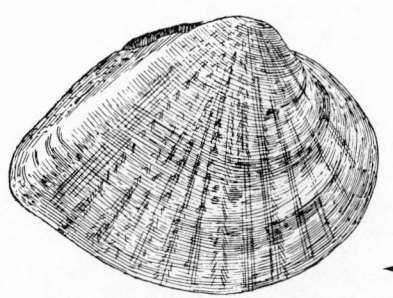

Genus *Truncilla* Rafinesque. Shell inflated, rather thin, rhomboidal, and possessing a well-defined posterior ridge. Sculpture consisting of a few well-marked concentric ridges. Beaks nearly central and sculptured with delicate and double-looped ridges. Periostracum yellowish-brown and strongly rayed with broken lines of green. Pseudocardinal teeth compressed and high, laterals narrow and slightly curved. Outer gills serving as marsupia. In most of the species the male and female shells differ in size and proportions.

The few species in this genus range widely in the Mississippi River system and south into Tex.

◄ **Fig. 43.98.** *Truncilla truncata* Rafinesque. (By Walker.)

Genus *Medionidus* Simpson. Shell small, somewhat elongate, arcuate when adult, and having the posterior ridge somewhat obscure and usually finely plicate. Periostracum smooth, usually shining, and possessing broken green rays. Beaks small and sculptured with subparallel, often broken ridges. Pseudocardinal teeth small, stumpy, and somewhat roughened; laterals short, slightly curved, and club-shaped. Marsupia occupying the central portion of the outer gills.

Occurs in the southeastern states from Tenn. to Ala. and northern Fla.

Fig. 43.99. *Medionidus conradicus* Lea. (By Walker.)

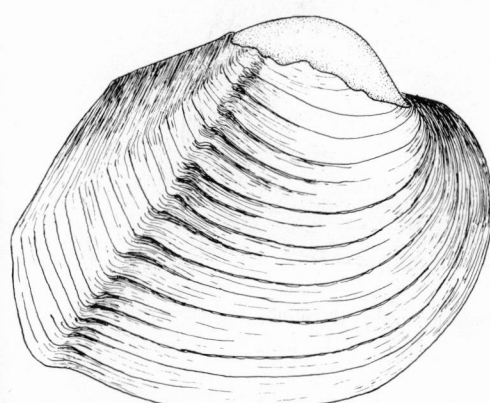

Genus *Glebula* Conrad. Shell solid, inflated, elliptical, bluntly pointed posteriorly and with a low but well-defined posterior ridge. Beaks compressed and smooth. Periostracum brownish and dull. Pseudocardinal teeth small with the top surface ridged; laterals rather short and low. Marsupia developed in the hinder portion of the outer gills.

Only a single species in this genus and it occurs from northern Fla. west to Tex.

◄ **Fig. 43.100.** *Glebula rotundata* Lamarck. (After Turner.)

Genus *Proptera* Rafinesque. Shell large, ovate, compressed, usually winged, and having a fairly well-defined posterior ridge. Periostracum dark brown and usually smooth. Beaks well forward, compressed, and weakly sculptured. Pseudocardinal teeth imperfect or nearly wanting, laterals remote. Marsupia occupying the posterior portion of the outer gills. Nacre generally pink to a rather dark purple.

Widely distributed throughout the St. Lawrence and Mississippi river systems and south to Ala. and Tex.

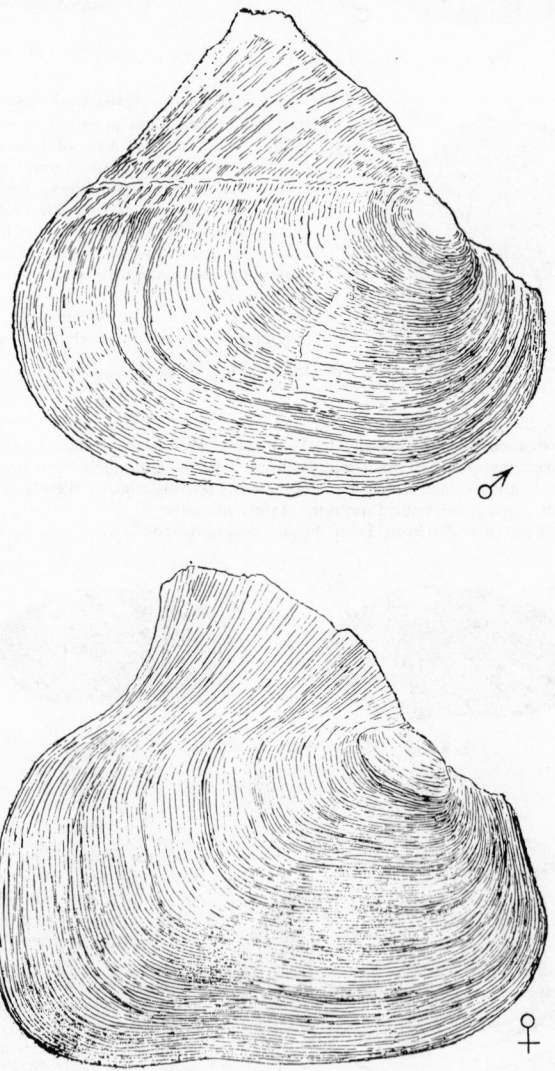

Fig. 43.101. *Proptera alata* Say. × ½. (By Walker.)

Genus **Leptodea** Rafinesque. Shell large, ovate to elliptical, thin, compressed, winged and with a poorly defined posterior ridge. Periostracum yellowish-brown, smooth, and with but few faint, narrow rays. Pseudocardinal teeth feebly and often imperfectly developed, laterals fairly long and narrow. Marsupia limited to the posterior portion of the outer gills. Nacre highly opalescent and usually tinged with purple.

Only two species in this genus, ranging widely from the Great Lakes south to Ala. and the lower Mississippi Valley.

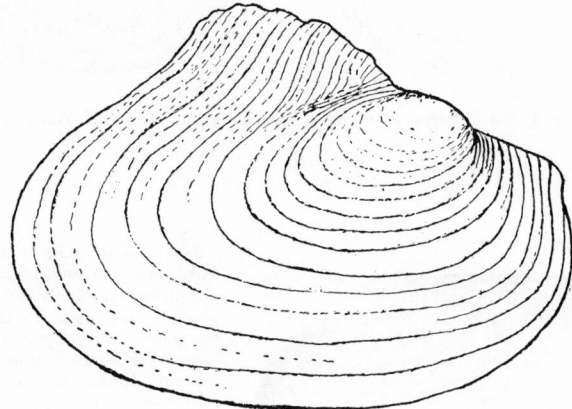

Fig. 43.102. *Leptodea fragilis* Rafinesque. × ⁴/₅. (By Walker.)

Genus **Obovaria** Rafinesque. Shell elliptical to ovate, usually very solid and inflated. Beaks high, sculptured with very faint, irregular, slightly nodulose ridges. Periostracum dull brownish and rarely rayed. Posterior slope faintly indicated or wanting. Pseudocardinal teeth solid and somewhat stumpy, laterals rather short. Marsupia developed on the posterior portion of the outer gills.

There are 2 subgenera in this genus which is widely distributed from the Great Lakes south to La. and Ala.

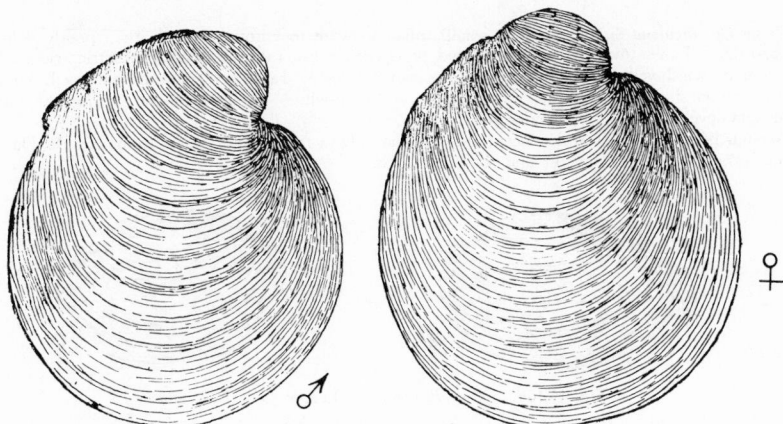

Fig. 43.103. *Obovaria (Obovaria) retusa* Lamarck. × ⅔. (By Walker.)

Subgenus **Pseudoon** Simpson. Shell ovate to elliptical, solid, and inflated. Beaks rather high and anterior, beak cavities rather shallow.

Two species in this subgenus, ranging from the upper Mississippi and the lower Great Lakes south to La. and Ala.

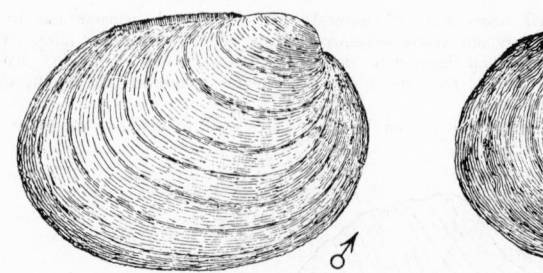
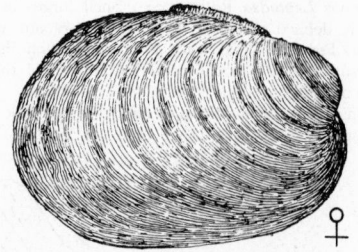

Fig. 43.104. *Obovaria (Pseudoon) olivaria* Rafinesque. × ⅔. (By Walker.)

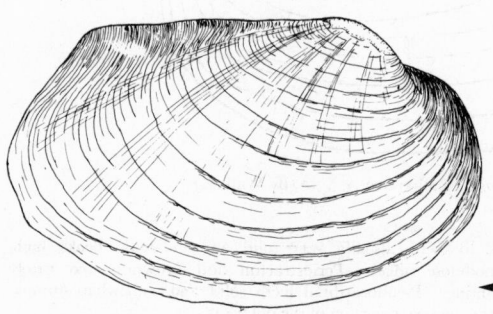

Genus *Actinonaias* Fischer and Crosse. Shell ovate to subelliptical, rather solid, compressed to slightly inflated, and with a poorly defined posterior ridge. Beaks toward the anterior end sculptured with a few faint bars. Periostracum yellowish to greenish and with distinct rays. Pseudocardinal teeth rather small and rough, laterals somewhat remote and rarely straight. Nacre white. Marsupia developed only on the outer gills.

Several species in this genus, which is widely distributed in the upper Mississippi and Great Lakes area south to Ala.

◄ **Fig. 43.105.** *Actinonaias carinata* Barnes. (After Turner.)

Genus *Corunculina* Simpson. Shell small, inflated, ovate to elliptical, and with a poorly defined posterior ridge. Beaks toward the anterior end sculptured with rather strong concentric ridges. Periostracum brownish-yellow to greenish and occasionally feebly rayed. Pseudocardinal teeth compressed, laterals rather short and straight. Nacre white or bluish-white to pinkish and frequently opalescent. Marsupia developed only on the outer gills.

A genus of small species which occur most abundantly in the southern states from Tex. to Fla. A few species range as far north as Ill. and southern Mich.

Fig. 43.106. *Corunculina texasensis* Lea. × ¾. (By Walker.).

Genus *Ligumia* Swainson. Shell oval to oblong, small to large and having a poorly defined posterior ridge. Beaks low and sculptured with delicate double-looped ridges. Pseudocardinal teeth small but well developed, laterals rather long and slightly curved. Marsupia occupying only the posterior part of the outer gills.

Genus **Ligumia** Swainson. Shell elongate, generally pointed posteriorly, thin to rather thick and solid, and having a poorly defined posterior ridge. Beaks low and sculptured with double-looped ridges. Periostracum usually a dark blackish-brown or greenish-brown. Nacre white, occasionally tinged with pink.

Widely distributed from the Red River of the North, the Great Lakes and St. Lawrence system, south to Ala.

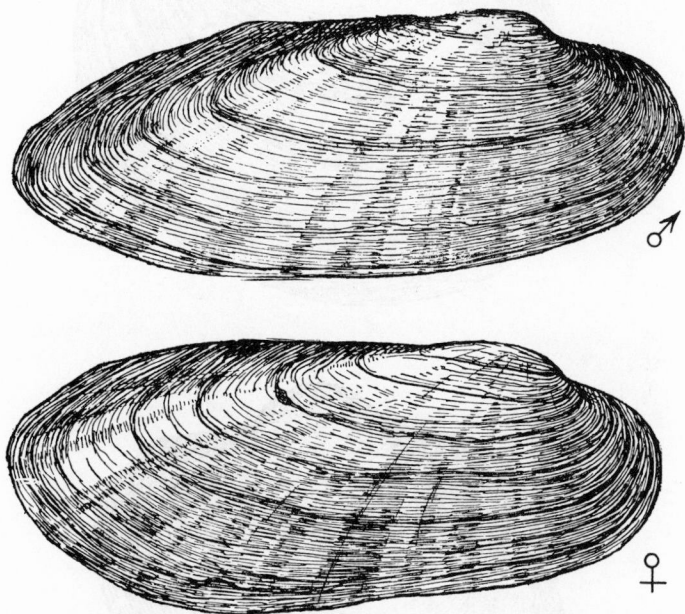

Fig. 43.107. *Ligumia latissima* Rafinesque. × ⅔. (By Walker.)

Genus **Villosa** Frierson. Shell small to medium in size, suboval to subelliptical, and with a poorly defined posterior ridge. Beaks low and sculptured with rather distinct double-looped ridges. Periostracum dark brown or dark green and often rayed.

Widely distributed from the Great Lakes area south to Ga. and Ala.

Fig. 43.108. *Villosa fabalis* Lea. (After Turner.)

Genus **Lampsilis** Rafinesque. Shell usually large, oval to elliptical, generally inflated, and with or without a well-defined posterior ridge. Beaks usually high and sculptured with rather coarse parallel ridges. Pseudocardinal teeth small but well developed, laterals short and curved. Periostracum yellowish to dark brown, occasionally greenish and often rayed. Nacre generally white.

Widely distributed in all of eastern N. A. from New England south to northern Fla. and west to the Mississippi Valley.

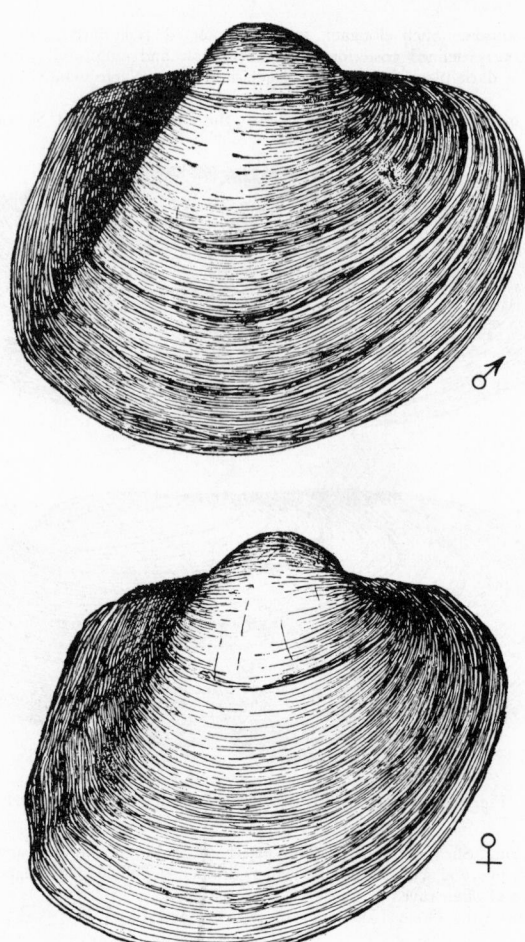

Fig. 43.109. *Lampsilis ovata* Say. × ⁵⁄₈. (By Walker.)

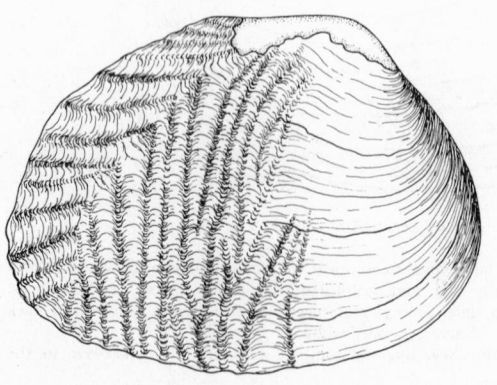

Genus **Lemiox** Rafinesque. Shell small, compressed, ovate, and having a low, rounded posterior ridge. Beaks high, anterior, and sculptured with double-looped ridges. Surface of the shell sculptured with strong subradial corrugations on the posterior half, divaricate on the posterior ridge. Periostracum dull greenish-brown and feebly rayed. Pseudocardinal teeth low and ragged, laterals heavy and curved. Nacre white.

A single species only in this genus, limited to the Tennessee River system.

◄ **Fig. 43.110.** *Lemiox rimosus* Rafinesque. (After Turner.)

Genus *Epioblasma* Rafinesque. Shell medium in size, rounded, oval or subtriangular, thin to solid, and the shells on the two sexes quite different. The female possesses a decided inflation at the basal area of the posterior ridge. Beaks full and sculptured weakly with double-looped ridges. Periostracum dark greenish or brownish and frequently rayed. Marsupia occupying the posterior portion of the outer gills.

The various species in this genus have a wide range, from N. Y. west to Neb. and south to Ga. and Ala. Four well-defined subgenera in this genus.

Subgenus *Truncillopsis* Rafinesque. Male and female shells different, the female having a decided inflation at the posterior basal area. In both sexes the shell is subtriangular and the periostracum is marked with broken rays.

Distributed from N. Y. west to Neb. and south to Kan. and northern Ala.

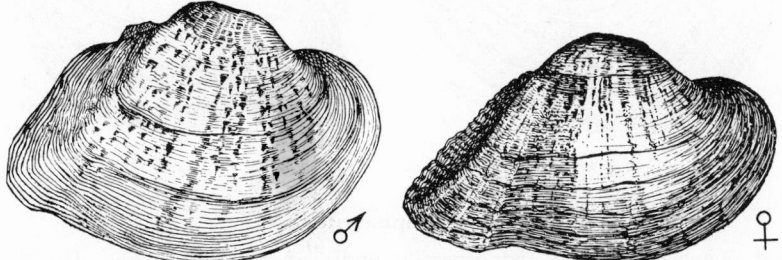

Fig. 43.111. *Epioblasma (Truncillopsis) triquetrum* Rafinesque. (By Walker.)

Subgenus *Pilea* Simpson. Male and female shells different, male shell with a wide, shallow, radiating depression in front of the posterior ridge. Female shell with a rounded swelling at the posterior basal area. The several species in this subgenus range from southern Mich. and south to Ark. and Tenn.

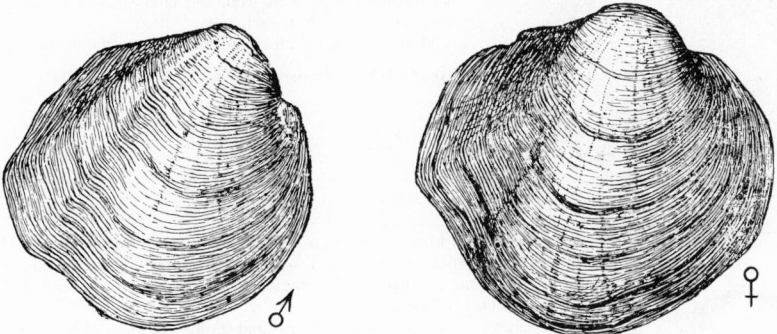

Fig. 43.112. *Epioblasma (Pilea) personatum* Lea. (By Walker.)

Subgenus *Epioblasma* Rafinesque. Male and female shells different, male shell with posterior and central radiating ridges with a furrow between. Female shell with a greatly produced inflation a little behind the center of the base.

The 3 species in this subgenus are found in the Ohio, Cumberland, and Tenn. rivers.

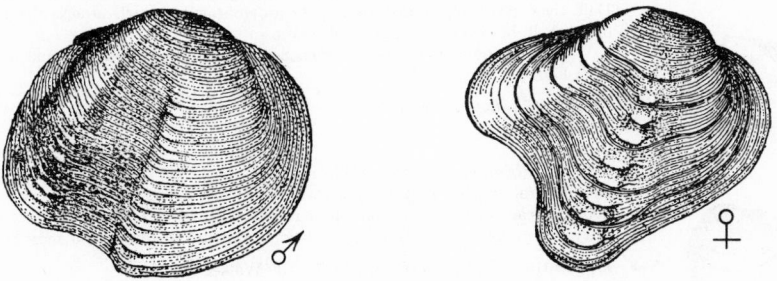

Fig. 43.113. *Epioblasma (Epioblasma) bilobum* Rafinesque. × ½. (By Walker.)

Subgenus *Scalenilla* Ortmann and Walker. Male and female shells different, male shell with a wide, radiating shallow depression in front of the posterior ridge. Female shell with a small, rounded and radial basal swelling.

The 3 species in this subgenus range from southern Mich. south to Tenn. and Ga.

Fig. 43.114. *Epioblasma* (*Scalenilla*) *sulcatum* Lea. (By Walker.)

Family **Sphaeriidae**

Shell nonnacreous, thin and generally under an inch in length. Hinge with cardinal teeth and having both anterior and posterior lateral teeth but no hinge plate. Pallial line simple.

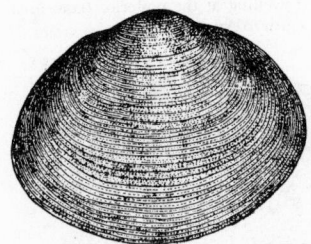

Genus *Sphaerium* Scopoli. Shell oval, equilateral, and having the beaks subcentral. Nepionic valves (young stages) not distinctly separated from the later growth stages. There are 2 cardinal teeth in each valve.

Many species, which range widely throughout most of N. A.

◀ **Fig. 43.115.** *Sphaerium simile* Say. × 2. (By Walker.)

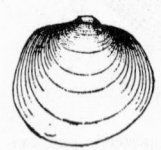

Genus *Musculium* Link. Shell thin, delicate, subcircular to oblong, and having prominent beaks which are more or less centrally located. Usually the nepionic valves or early stages have been retained. Cardinal teeth minute and often obsolete, lateral teeth present.

Many species in this genus, widely distributed throughout most of N. A.

◀ **Fig. 43.116.** *Musculium partumeium* Say. × 2. (By Walker.)

Eupera Bourguignat. Shell subrhomboidal, thin, moderately inflated, and having the beaks subcentral but located in the anterior half of the valves. Cardinal teeth feeble, anterior and posterior laterals well developed. A mottled coloration of brownish-red exists in all the species so far seen. According to Dall, these spots are caused by a parasitic infusorian which attacks the interior of the shell; however, this needs confirmation.

Widely distributed in the West Indies, South and Central America, and in N. A. from Fla. west to Tex.

◀ **Fig. 43.117.** *Eupera singleyi* Pilsbry. × 3. (By Walker.)

Genus *Pisidium* Pfeiffer. Shell rounded, oval, or obliquely wedge-shaped, inequilateral and having the umbos well posterior to the center. Cardinal teeth double in each valve, anterior and posterior laterals well developed.

The many species in this genus are widely distributed throughout most of N. A.

◀ **Fig. 43.118.** *Pisidium dubium* Say. × 2. (By Walker.)

Family **Corbiculidae**

Shell nonnacreous, rather thick, ranging in size from less than 1 in. to 2½ in. in length. The shape is triangular to subcircular, and the outer surface usually sculptured with concentric ridges and covered with a greenish and shining periostracum.

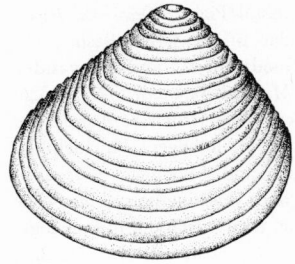

Genus *Corbicula*. Shell triangular in shape, equilateral, and having the beaks subcentral. Outer surface with rather heavy concentric ridges. Two cardinal teeth in each valve. Lateral teeth serrated.

Found in northern Calif. and Ore. Introduced from China.

◀ **Fig. 43.119.** *Corbicula fluminea* Müller. × 1½. (After Turner.)

References

The following bibliography of North American fresh-water mollusks covers the leading studies in this group. Most of the references contain extensive bibliographies of their own.

Reference should also be made to *The Nautilus*, Philadelphia. This journal numbers over 70 volumes and has a great deal of information on fresh-water mollusks. *Occasional Papers on Mollusks, Museum of Comparative Zoology, Harvard University*, with one volume completed, plans to devote much of its effort to papers on North American fresh-water mollusks.

Baker, F. C. 1898–1902. The Mollusca of the Chicago area. *Chicago Acad. of Sci. Bull.* Part 1, Pelecypoda, No. 3:1–130. Part 2, Gastropoda, No. 3:137–410. **1906.** A catalogue of the Mollusca of Illinois. *Bull. Illinois State Lab. Nat. Hist.*, 7:53–136. **1910.** The ecology of the Skokie Marsh area, with special reference to the Mollusca. *Bull. Illinois State Lab. of Nat. Hist.*, 8:441–499. **1911.** The Lymnaeidae of North and Middle America recent and fossil. *Chicago Acad. Sci., Spec. Publ.*, No. 3:16–539. **1928.** The fresh-water Mollusca of Wisconsin. *Wisc. Acad. Sci. Bull.* Part 1, Gastropoda, 70:1–505. Part 2, Pelecypoda, 70:1–495. **1945.** *The Molluscan Family Planorbidae.* University of Illinois Press, Urbana. **Binney, W. G. 1865.** Land and freshwater shells of North America. *Smithsonian Inst. Publs. Misc. Collections.* Part II, No. 143:1–161. Part III, No. 3:1–120. **Brooks, J. L. 1950.** Speciation in ancient lakes. *Quart. Rev. Biol.*, 25:30–60, 131–176. **Call, R. E. 1898.** A descriptive illustrated catalogue of the Mollusca of Indiana. *Twenty-fourth Ann. Rept. Dept. Geol. Natural Resources Indiana*, 335–535. **Chamberlain, R. V. and D. T. Jones. 1929.** A descriptive catalogue of the Mollusca of Utah. *Bull. Univ. Utah*, 19:1–203. **Clench, W. J. and R. D. Turner. 1955.** The North American genus *Lioplax* in the family Viviparidae. *Occasional Papers Mollusks, Harvard Univ.*, 2:1–20. **1956.** Freshwater mollusks of Alabama, Georgia and Florida from the Escambia to the Suwannee River. *Bull. Florida State Museum*, 1:97–220. **Dall, W. H. 1905.** *Land and Fresh Water Mollusks. Harriman Alaska Expedition*, Vol. 13. Doubleday, Page, New York. **Frierson, L. S. 1927.** *A Classified and Annotated Check List of the North American Naiades.* Baylor University Press, Waco, Texas. **Goodrich, C. 1922.** The Anculosae of the Alabama River drainage. *Museum Zool. Univ. of Mich. Misc. Publ.*, No. 7:1–57. **1924.** The genus *Gyrotoma. Museum Zool. Univ. Mich. Misc. Publ.*, No. 12:1–30. **1930.** Goniobasis of the vicinity of muscle shoals. *Occasional Papers Museum Zool. Univ. of Mich.*, No. 209:1–25. **1931.** The Pleurocerid genus *Eurycaelon. Occasional Papers Museum Zool. Univ. of Mich.*, No. 223:1–10. **1932.** The Mollusca of Michigan. *Univ. of Mich. Handbook Series*, No. 5:1–120. **1936.** Goniobasis of the

Coosa River, Alabama. *Museum Zool. Univ. of Mich. Misc. Publ.*, No. 31:1–60. **1939.** Pleuroceridae of the Mississippi River Basin exclusive of the Ohio River system. *Occasional Papers Museum of Zool. Univ. of Mich.*, No. 406:1–4. **1940.** The Pleuroceridae of the Ohio River drainage system. *Occasional Papers Museum of Zool. Univ. Mich.*, No. 417:1–21. **1941.** Pleuroceridae of the small streams of the Alabama River system. *Occasional Papers Museum Zool. Univ. Mich.*, No. 427:1–10. **1942.** Pleuroceridae of the Atlantic Coastal Plain. *Occasional Papers Museum Zool. Univ. of Mich.*, No. 456:1–6. **1944.** Pleuroceridae of the Great Basin. *Occasional Papers Museum Zool. Univ. of Mich.*, No. 485:1–11.   **Goodrich, C. and H. vander Schalie. 1944.** A revision of the Mollusca of Indiana. *Am. Midland Naturalist,* 32:257–326. **Hartman, W. D. and E. Michener. 1874.** Conchologia Cestrica. *The Molluscous Animals and their Shells of Chester County, Pennsylvania.* Claxton, Remson, and Haffelfinger, Philadelphia. **Henderson, J. 1924.** Mollusca of Colorado, Utah, Montana, Idaho and Wyoming. *Univ. Colo. Studies,* 13:65–223. **1929.** Non-marine Mollusca of Oregon and Washington. *Univ. Colo. Studies,* 17:47–190.   **Hubendick, B. 1951.** Recent Lymnaeidae, their variation, morphology, taxonomy, nomenclature and distribution. *Kgl. Svenska Vetenskapsakad. Handl.,* 3:1–222. **Johnson, C. W. 1915.** Fauna of New England: List of Mollusca. *Occasional Papers Boston Soc. Nat. Hist.,* 7:1–231.   **La Rocque, A. 1953.** Catalogue of the recent Mollusca of Canada. *Bull. Natl. Museum Can.,* No. 129:9–406.   **Lefevre, G. and W. C. Curtis. 1912.** Studies on the reproduction and artificial propagation of fresh-water mussels. *Bull. U. S. Bur. Fisheries,* 30:103–201.   **Mazyck, W. G. 1913.** Catalog of Mollusca of South Carolina. *Contr. Charleston Museum,* No. 2:10–39.   **Ortmann, A. E. 1919.** A monograph of the Naiades of Pennsylvania. *Mem. Carnegie Museum,* 8:1–384.   **Robertson, I. C. S. and C. L. Blakeslee. 1948.** The Mollusca of the Niagara Frontier region. *Buffalo Soc. Nat. Sci.,* 19:1–191.   **Simpson, C. T. 1900.** Synopsis of the Naiades, or pearly fresh-water mussels. *Proc. U. S. Natl. Museum,* 22:501–1044. (This report contains a very fine bibliography.)   **1914.** *A Descriptive Catalogue of the Naiades, or Pearly Fresh-water Mussels.* Detroit. (Can be obtained from the Museum of Zoology, University of Michigan, Ann Arbor, Michigan.)   **Stimpson, W. 1865.** Researches upon the Hydrobiinae and allied forms. *Smithsonian Inst. Publs. Misc. Collections,* No. 201:1–59.   **Thiele, J. 1929–1935.** *Handbuch der Systematischen Weichtierkunde,* 2 vols. G. Fischer, Jena.   **Tryon, George W. 1873.** Land and freshwater shells of North America. Part 4, Strepomatidae. *Smithsonian Inst. Publs. Misc. Collections,* No. 253:1–435.   **Walker, B. 1918.** A Synopsis of the classification of the freshwater Mollusca of North America north of Mexico. *Misc. Publ. Mus. Zool. Univ. Mich.,* No. 6:1–213.

44

Bryophyta

HENRY S. CONARD

The group Bryophyta (Atracheata) includes mostly small plants with numerous round chloroplasts in each cell, entirely lacking the thick-walled (spiral, annular, reticulate, or pitted) water-conducting cells of vascular plants (Tracheata). The life cycle is in two alternating phases:

1. *Gametophyte.* The green plant, scalelike, ribbonlike, or leafy, bearing eggs in flask-shaped archegonia and spiral, ciliated sperms in spherical or cylindrical antheridia.

2. *Sporophyte.* The fertilized egg remains in the archegonium, giving rise ultimately to a capsule containing spores. Dissemination is by fragments of the gametophyte or by spores. Aquatic forms rarely produce spores.

Among Bryophytes we find every possible degree of hygrophily. The choice of forms to be listed here is, therefore, purely arbitrary.

Class 1. Hepaticae. Liverworts (p. 1162). Plant body flat, without distinction of stem and leaf, 1 to 16 cells thick; *or* stems with 2 dorsolateral rows of leaves which are 1 cell thick, the cells all essentially alike and isodiametric (no midrib); sometimes a row of small ventral leaves (underleaves). Sporophytes evanescent.

Class 2. Musci. Mosses (p. 1163). Stems, with leaves many-ranked (2 in Fissidens, etc.), the cells differing in different parts of the leaf, often very

Fig. 44.1. *Riccia flui-tans.* Thallus.

Fig. 44.2. *Ricciocarpus natans.* Thallus.

Fig. 44.3. *Riella.* Thallus.

Fig. 44.4. *Riccardia multifida.*

long and narrow. Midrib often present. Sporophytes persisting for weeks or months.

Illustrations are from H. S. Conard, *How to Know the Mosses and Liverworts*, Wm. C. Brown Co., Dubuque, Iowa, by permission of the editor, H. E. Jaques.

KEY TO HEPATICAE

1a		Plant body scalelike, ribbonlike, or in rosettes (a thallus)	2
1b		Plant with stem and leaves . . . . Order **Jungermanniales**	6
2a	(1)	Thallus opaque because of air chambers	3
2b		Thallus translucent, without air chambers	4
3a	(2)	Narrow (1 mm), forked, floating singly or in tangled mats. (Fig. 44.1) . . . . ***Riccia*** L.	
3b		Triangular or semicircular floating scales to 1 cm across, with a few shallow, radiating furrows, and large air chambers. Underside beset with blackish hairs and scales. (Fig. 44.2). . . . . ***Ricciocarpus*** Corda	
4a	(2)	A cordlike stem with 1-layered wing along one side. Very rare, southwestern. (Fig. 44.3) . . . . ***Riella*** Mont.	
4b		Without stem. Several cells thick	5
5a	(4)	Variously branched, 0.5 to 2 mm wide. (Fig. 44.4). . . . . ***Riccardia*** S. F. Gray	

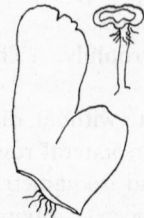
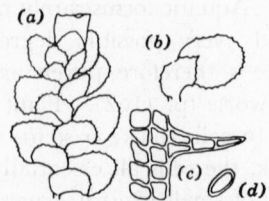
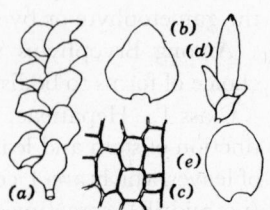

Fig. 44.5. *Dumortiera hirsuta.* Thallus and spore-bearing shoot.

Fig. 44.6. *Scapania nemorosa.* (*a*) Shoot with perianth. (*b*) Leaf. (*c*) Margin of leaf. (*d*) Gemma.

Fig. 44.7. *Jungermannia cordifolia.* (*a*) Plant. (*b*) Leaf. (*c*) Cells of leaf. (*d*) Perianth. (*e*) Leaf of *J. jumila.*

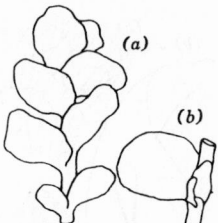

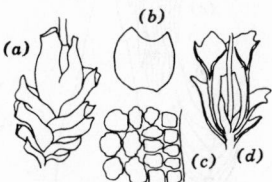

Fig. 44.8. *Porella pinnata.*
(*a*) Plant. (*b*) Leaf, under-
leaf, and underlobe.

Fig. 44.9. *Chiloscyphus rivularis.*

Fig. 44.10. *Marsupella emarginata.*
(*a*) Perianth with bracts. (*b*) Leaf.
(*c*) Cells of leaf. (*d*) Section of
perianth and bracts.

5b   Branching, if any, not noticeable; about 1 cm wide.   Southeastern.
(Fig. 44.5). **Dumortiera** Reinw.

6a   (1) Leaves in 2 parts folded together; the upper smaller than the lower,
appearing as 4 rows of leaves.   Margins toothed or entire.   (Fig.
44.6) . . . . **Scapania** Dum.

6b   Leaves appearing in 2 rows only **7**

7a (6) Leaves entire, not lobed **8**

7b   Leaves 2-lobed **10**

8a (7) Small plants without underleaves.   (Fig. 44.7) . . **Jungermannia** L.

8b   Large.   Underleaves present, at least near ends of shoots **9**

9a (8) Underleaves tongue-shaped, conspicuous.   A narrow lobe (underlobe) of the leaf is close to underside of stem.   (Fig. 44.8). . . . . . .
Porella L.

9b   Underleaves fugacious, best seen on youngest parts.   (Fig. 44.9) . . .
Chiloscyphus Loeske

10a (7) Shoots 1 to 2 mm wide.   Stems erect, crowded.   Leaves concave.
Emergent.   (Fig. 44.10) . . . . **Marsupella** Dum.

10b   Shoots 0.5–0.7 mm wide, submersed.   Leaves flat.   (Fig. 44.11) . . .
Cladopodiella Joerg.

KEY TO MUSCI

1a   Large soft plants, whitish when dry.   Stems to 2 dm long, with
alternate clusters of branches, terminating in a dense head (to 3
cm across).   Leaf cells in 1 layer, of 2 kinds: large empty cells
surrounded by narrow green cells.   (Fig. 44.12) . . . . **Sphagnum** L.

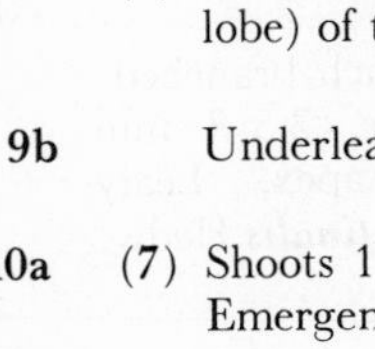

Fig. 44.11. *Cladopodiella fluitans.*

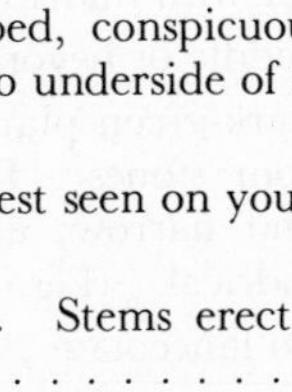

Fig. 44.12. *Sphagnum.*
End of stem with cap-
sules.

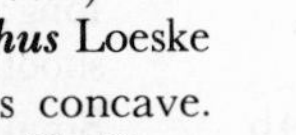

Fig. 44.13. *Fissidens.*
Leaf.

Fig. 44.14. *Fontinalis antipyretica.* (*a*) Shoot. (*b*) Leaf. (*c*) Cells of leaf. (*d*) Section of leaf.

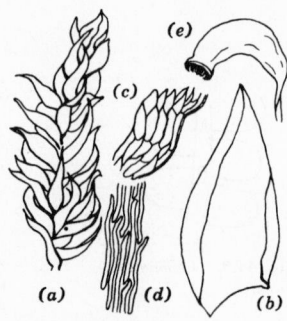

Fig. 44.15. *Scorpidium scorpioides.* (*a*) Shoot. (*b*) Leaf. (*c*) Angle cells. (*d*) Median cells. (*e*) Capsule.

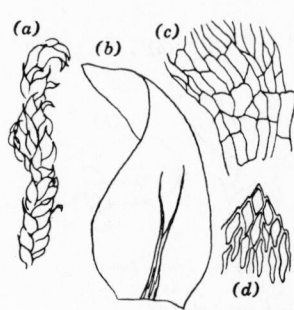

Fig. 44.16. *Hygrohypnum ochraceum.* (*a*) Shoot. (*b*) Leaf. (*c*) Angle cells. (*d*) Apex.

1b	Branches, if any, single and similar to stems. All cells above base of leaf contain chloroplasts	**2**
2a	(1) Leaves in 2 *opposite* lateral rows, each leaf split at base and sitting astride of stem and next leaf above (equitant). Stems 2–20 cm long. (Fig. 44.13) . . . . **Fissidens** Hedw.	
2b	Leaves flat, rolled, or keeled, but not equitant	**3**
3a	(2) Leaves without midrib, or with midrib short and double	**4**
3b	Leaves with midrib to middle or beyond	**8**
4a	(3) Large (to 3 dm long), dark-green plants, crowded, much branched, floating or dangling from stones. Leaves lanceolate, 2.5–8 mm long; cells very long and narrow; entire except at apex. Leafy shoots triangular or cylindrical. (Fig. 44.14) . . . **Fontinalis** Hedw.	
4b	Smaller. Leaves ovate to lanceolate	**5**
5a	(4) Leaves all bent to one side (falcate).	**6**
5b	Leaves straight or bent backward.	**7**
6a	(5) Leaves to 4 mm long, entire, wrinkled when dry. Stems almost unbranched, to 3 dm long. (Fig. 44.15) . . . . . . **Scorpidium** BSG	
6b	Leaves smaller, entire or toothed. (Fig. 44.16) . . . . **Hygrohypnum** Lindb.	
7a	(5) Shoots ending in a slender acute bud. (Fig. 44.17) . . . . **Calliergonella** Loeske	
7b	End bud loose . . . . **Hygrohypnum** Lindb.	
8a	(3) Leaves strongly bent to one side (falcate).	**9**
8b	Leaves straight or bent backward.	**13**
9a	(8) Leaves sharply folded along the middle (keeled). (Fig. 44.18) . . . . **Dichelyma** Myr.	
9b	Leaves flat or concave, not keeled	**10**
10a	(9) Stems erect. Leaves papillose. (Fig. 44.19) . . . . **Philonotis** Brid.	
10b	Not papillose.	**11**
11a	(10) Stems beset with green filaments (paraphyllia). (Fig. 44.20) . . . . . **Cratoneuron** (Sull.) Roth	
11b	Without such filaments	**12**

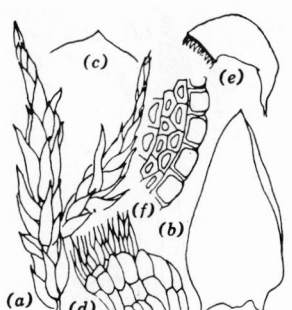

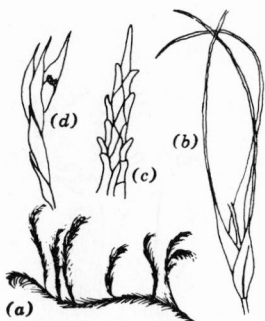

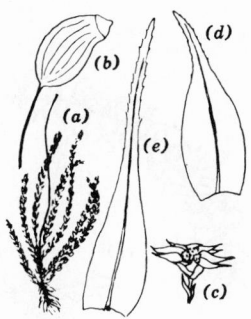

Fig. 44.17. *Calliergonella cuspidata.* (a) Shoot. (b) Leaf. (c) Apex of leaf. (d) Angle cells. (e) Capsule. (f) Section of stem.

Fig. 44.18. *Dichelyma capillaceum.* (a) Shoot. (b) Leaves. (c) Apex of leaf. (d) Capsule and perichaetium.

Fig. 44.19. *Philonotis fontana.* (a) Plant. (b) Capsule. (c) Antheridial head. (d) Leaf. (e) Leaf of *P. longiseta.*

12a	**(11)**	Leaves 4 to 6 times longer than wide, slenderly taper pointed. (Fig. 44.21) . . . . **Drepanocladus** (C. Muell.) Roth
12b		Leaves less than 2 times longer than wide, broadly pointed . . . . . . **Hygrohypnum** Lindb.
13a	**(8)**	Stems and branches beset with green filaments (paraphyllia) between the leaves **14**
13b		Without paraphyllia **16**
14a	**(13)**	A large cluster of swollen clear cells at basal angles of leaf; midrib very strong . . . . **Cratoneuron** (Sull.) Roth
14b		Without inflated alar cells **15**
15a	**(14)**	Leaves auricled at base, broadly acute or obtuse at apex. (Fig. 44.22) . . . . **Climacium** W. and M.
15b		Leaves contracted at base to a kind of petiole, slenderly acute at apex. Cells usually slightly papillose. (Fig. 44.23) . . . . **Helodium** (Sull.) Warnst.
16a	**(13)**	Leaf cells papillose **17**
16b		Leaf cells smooth **19**

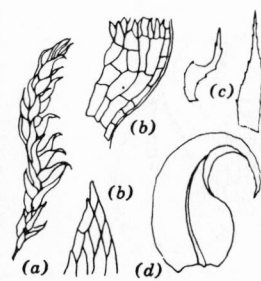

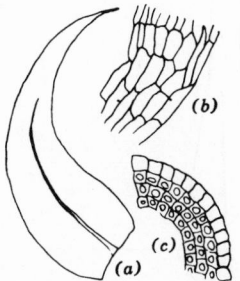

Fig.44.20. *Cratoneuron filicinum.* (a) Shoot. (b) Angle cells and apex. (c) Paraphyllia. (d) Leaf of *C. commutatum.*

Fig. 44.21. *Drepanocladus intermedius.* (a) Leaf. (b) Angle cells. (c) Section of stem.

Fig. 44.22. *Climacium americanum.* Leaf.

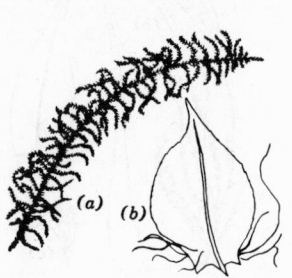

Fig. 44.23. *Helodium blandowii.*
(a) Plant. (b) Leaf.

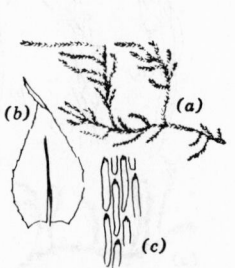

Fig. 44.24. *Bryhnia novae-angliae.* (a) Plant. (b) Leaf.
(c) Cells with papillae.

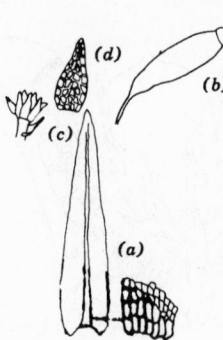

Fig. 44.25. *Aulocomnium palustre.* (a) Leaf with swollen base cells. (b) Capsule. (c) Cluster of gemmae. (d) Single gemma.

17a (16) Stems crowded, matted together, erect, and parallel. Papillae over the cell cavity 18

17b Stems spreading, much branched. Papillae over end walls of cells on back of leaf only. Northeastern. (Fig. 44.24) . . . .
Bryhnia (Lesq.) Grout

18a (17) One large papilla at middle of cell on each side of leaf. Pale green plants, often with naked stalks bearing gemmae. (Fig. 44.25) . . . .
Aulacomnium Schwaegr.

18b One papilla near end of cell, distal or proximal or both, on back of leaf or on both sides. Dark green . . . . ***Philonotis*** Brid.

19a (16) Stems with many short branches, beset with matted, reddish-brown tomentum. Leaves pleated lengthways. Northern. (Fig. 44.26) . .
Camptothecium Schimp.

19b Leaves not plicate; stems not tomentose 20

20a (19) Leaves small, bordered with elongated cells. Southeastern. (Fig. 44.27) . . . . ***Sciaromium*** Broth.

20b Border cells superficially like median cells 21

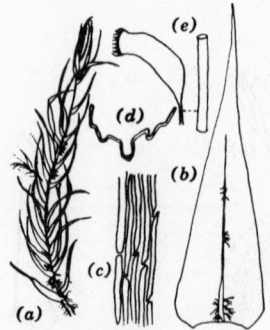

Fig. 44.26. *Camptothecium nitens.*
(a) Shoot. (b) Leaf. (c) Median cells. (d) Section of leaf.
(e) Capsule and seta.

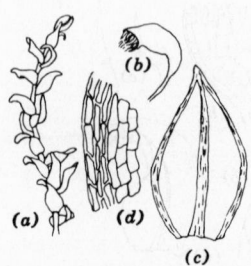

Fig. 44.27. *Sciaromium lescurii.* (a) Shoot. (b) Capsule. (c) Leaf. (d) Margin of leaf.

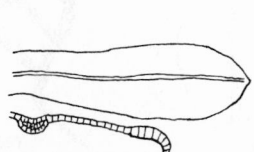

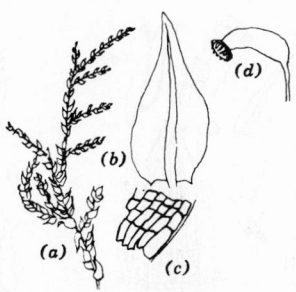

Fig. 44.28. *Merceya latifolia.*
Leaf and section.

Fig. 44.29. *Hygroamblystegium irriguum.* (a) Plant. (b) Leaf.
(c) Angle cells. (d) Capsule.

21a (20) Margins of leaves thickened by deeper cells. Rocky Mountains and
           westward. (Fig. 44.28) . . . . **Merceya** Kindb.

21b   Margins not thicker than lamina **22**

22a (21) Median cells of leaf less than 7 times longer than wide. (Fig.
           44.29). **Hygroamblystegium** Loeske

22b   Median cells 6 to 15 times longer than wide **23**

23a (22) Leaf margins entire, or finely toothed at apex **24**

23b   Leaf margins toothed nearly or quite to base. **28**

24a (23) Leaf margins absolutely entire throughout **25**

24b   Margins with fine teeth at apex. **27**

25a (24) Leaves long-lanceolate; apex slenderly acuminate. (Fig. 44.30) . . .
                                                         Leptodictyum Warnst.

25b   Leaves rounded at apex, or with short, broad apex **26**

26a (25) Cells at basal angles of leaf large and clear. (Fig. 44.31). . . . . . .
                                                            Calliergon Kindb.

26b   Cells at basal angles only slightly larger than above. (Fig. 44.32) . .
                                                          Hygrohypnum Loeske

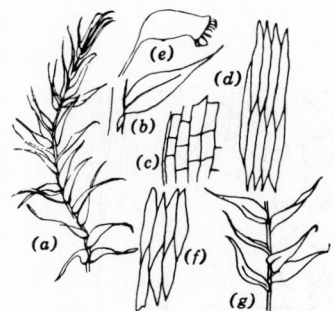

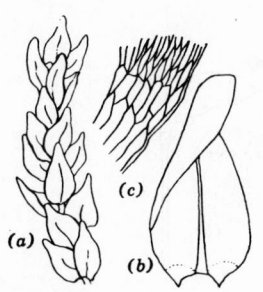

Fig. 44.30. *Leptodictyum riparium.*
(a) Shoot. (b) Leaf. (c) Angle cells.
(d) Median cells. (e) Capsule.
(f) Median cells of *L. lascirete.*
(g) Shoot of *L. riparium* forma *fluitans.*

Fig. 44.31. *Calliergon cordi-
folium.* (a) Shoot. (b) Leaf.
(c) Angle cells.

Fig. 44.32. *Hygro-
hypnum palustre.* Leaf
and apex.

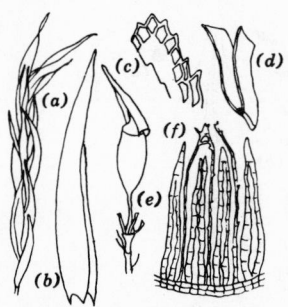

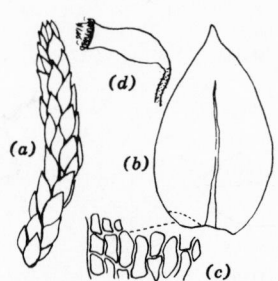

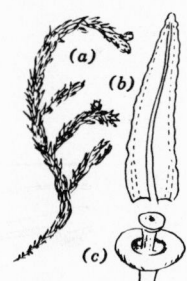

Fig. 44.33. *Brachelyma subulatum.* (a) Shoot. (b) Leaf. (c) Apex of leaf. (d) Section of leaf. (e) Seta, capsule, and calyptra. (f) Peristome.

Fig. 44.34. *Scleropodium obtusifolium.* (a) Shoot. (b) Leaf. (c) Alar region. (d) Capsule.

Fig. 44.35. *Scouleria aquatica.* (a) Plant. (b) Leaf. (c) Capsule.

27a (24) Leaves long and narrowly lanceolate (to 4 mm). (Fig. 44.33) . . . . **Brachelyma** Sch.

27b   Leaves ovate, rounded or acute at apex. Cells at basal angles large and clear. Calif. to British Columbia. (Fig. 44.34) . . . . **Scleropodium** Kindb.

28a (23) Thin, dense blackish pads on rocks. Teeth of leaf distant, blunt. Capsules at ends of short branches among the leaves. (Fig. 44.35) . . . . **Scouleria** Hook.

28b   Without the above combination of characters 29

29a (28) Cells at basal angles of leaf greatly enlarged, clear or colored. 30

29b   Cells at basal angles not larger or clearer. 31

30a (29) Enlarged cells extending down along stem (decurrent). (Fig. 44.36). **Brachythecium** BSG

30b   Angle cells not decurrent, often colored. . . . **Hygrohypnum** Lindb.

31a (29) Leaves lanceolate, slightly toothed in lower half . . . . **Hygroamblystegium** Loeske

31b   Leaves broadly ovate to orbicular, finely toothed all around. (Fig. 44.37) . . . . **Eurhynchium** BSG

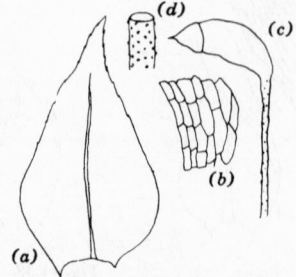

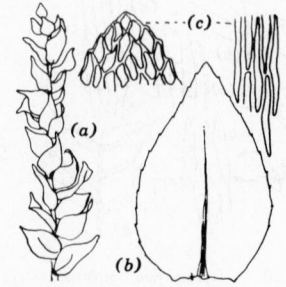

Fig. 44.36. *Brachythecium rivulare.* (a) Leaf. (b) Alar region. (c) Capsule. (d) Seta.

Fig. 44.37. *Eurhychium riparioides.* (a) Short. (b) Leaf. (c) Cells of apex and middle of leaf.

Habitat Lists

Plants of Swift Running Waters

Chiloscyphus rivularis, *Jungermannia* spp., *Scapania* spp., *Dichelyma capillaceum*, *Brachythecium rivulare*, *Eurhynchium riparioides*, *Fissidens grandifrons*, *debilis*, *Fontinalis* spp., *Hygroamblystegium* spp., *Hygrohypnum* spp., *Leptodictyum* forms, *Sciaromium lescurii*, *Scleropodium obtusifolium*, *Scouleria* spp., *Philonotis* (rarely).

Plants of Slow or Stagnant Waters

Submerged:

Cladopodiella fluitans, *Porella pinnata*, *Riccia fluitans*, *Ricciocarpus natans* (floating), *Riella* spp., *Brachelyma* spp., *Brachythecium rivulare*, *Calliergon* spp., *Dichelyma* spp., *Drepanocladus* spp., *Fontinalis* spp., *Leptodictyum riparium* forms, *Merceya latifolia*, *Scleropodium obtusifolium*.

Emergent:

Dumortiera hirsuta (springs), *Jungermannia* spp., *Marsupella* spp., *Riccardia* spp., *Scapania* spp., *Aulacomnium palustre*, *Brachythecium rivulare*, *Bryhnia novae-angliae*, *Calliergonella cuspidata*, *Camptothecium nitens*, *Climacium kindbergii*, *Cratoneuron* spp., *Drepanocladus* spp., *Helodium* spp., *Hygroamblystegium irriguum*, *Hygrohypnum* spp., *Leptodictyum* forms, *Merceya latifolia*, *Philonotis fontana*, *americana*, *Scleropodium obtusifolium*, *Scorpidium scorpioides*, *Sphagnum* spp.

References

Andrews, A. LeRoy. **1913.** Sphagnaceae (in *N. Amer. Flora*). 15(1):1–75. N. Y. Bot. Garden. **Conard, H. S.** **1956.** *How to Know the Mosses and Liverworts.* Wm. C. Brown, Dubuque, Iowa. **Frye, T. C. and Lois Clark.** **1937.** *Hepaticae of North America.* University of Washington Press, Seattle. **Grout, A. J.** **1903–1910.** *Mosses with Hand Lens and Microscope.* Privately printed. New York. **1905.** *Mosses with a Hand Lens,* 4th ed. O. T. Louis, New York. **1928–1940.** *Moss Flora of North America North of Mexico.* 3 vols. Privately printed. New York. **Jennings, O. E.** **1951.** *A Manual of the Mosses of Western Pennsylvania,* 2nd ed. University of Notre Dame Press, Notre Dame, Indiana. **Schuster, R. M.** **1949.** The ecology and distribution of Hepaticae in Central and Western New York. *Am. Midland Naturalist,* 42:513–712. **1956.** Boreal Hepaticae. *Am. Midland Naturalist,* 49:257–684. **Welch, W. H.** **1957.** *Manual of the Mosses of Indiana.* Indiana Dept. of Conservation. The Bookwalter Co., Indianapolis.

45

Vascular Plants

W. C. MUENSCHER

The higher aquatic plants included in this chapter normally grow, or at least start their life cycle, in the water. Because they require light they are mostly limited to shallow water where they grow toward the surface and often produce floating leaves. Many grow completely submersed throughout their life and rarely reach the surface when they flower. They nearly all grow anchored in the muddy or silty bottom and through their roots and root hairs absorb mineral nutrients from the soil to be utilized in their metabolism and growth. Some absorb nutrients through their leaves. When their tops die part of the nutrients in organic combination are released to the water. A body of water that produces many aquatic plants is usually considered rich for many forms of life because it furnishes shelter and food.

The present chapter deals with some of the more common true aquatics and their recognition. The distribution of aquatic plants is frequently considered rather cosmopolitan. Compared to a mesophytic habitat, the hydrophytic habitat is often less subject to fluctuations in temperature and water supply; but the dissolved salts and nutrients and the color and transparency of the water, as well as the physical and chemical properties of the water and also the bottom, are subject to much variation even in waters with but slight differences in altitude or latitude. Some species of aquatic plants tolerate a wide range of variation in habitat. This is illustrated by the following cosmopolitan species: *Potamogeton richardsonii*, *P. gramineus*, *P. natans*, *P. epihydrus*, *P. pectinatus*, *P. pusillus*, *P. zosteriformis*, *Najas flexilis*, *Alisma plantago-aquatica*, *Sagittaria latifolia*, *Lemna minor*, *L. trisulca*, *Spirodela polyrhiza*, *Ceratophyllum demersum*, *Myriophyllum exalbescens*, *Utricularia vulgaris*, and *Bidens beckii*, all of which occur in all of the following widely separated Lakes: Champlain (Vt.-N. Y.), Cayuga (N. Y.), Erie (Ohio), Flathead (Mont.), Pend d'Oreille (Ida.), and Ozette (Wash.).

Certain species of aquatic plants, however, are rather exacting in their requirements, and may be somewhat restricted in their range. Their limited distribution may be due to such factors as the temperature or depth of the water, the physical properties of the bottom, the reaction of the water or bottom, the quantity or quality of the salts dissolved in the water, the competition of other plants. Depending upon their tolerance and aggressiveness and also upon their mobility, aquatic plants, like terrestrial plants, contain many restricted species as well as many cosmopolitan ones.

In their natural environment, angiospermous aquatic plants reproduce and spread by seeds and in many species also by vegetative propagation. Many of them blossom and fruit in abundance in shallow water but seldom produce mature seeds in deeper water or where they are continuously submersed, so that it is frequently desirable and often necessary to identify plants in the

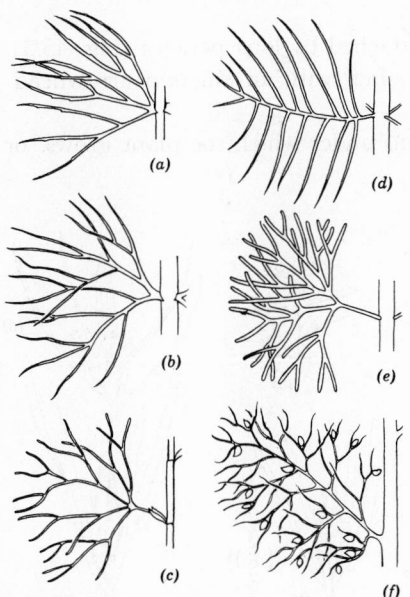

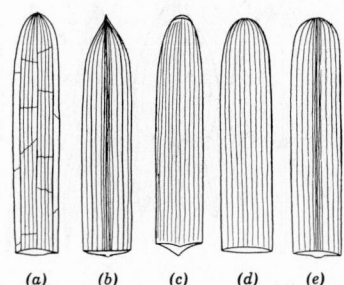

Fig. 45.1. Leaves with blades dissected into linear segments. (*a*) *Ceratophyllum echinatum*. (*b*) *Bidens beckii*. (*c*) *Ranunculus aquatilis*. (*d*) *Myriophyllum exalbescens*. (*e*) *Cabomba caroliniana*. (*f*) *Utricularia vulgaris*.

Fig. 45.2. Leaves linear or tapelike. (*a*) *Vallisneria americana*. (*b*) *Potamogeton zosteriformis*. (*c*) *Sparganium fluctuans*. (*d*) *Potamogeton epihydrus*. (*e*) *Heteranthera dubia*.

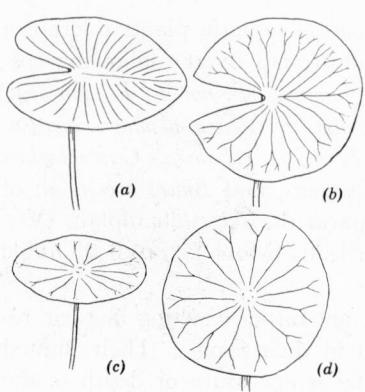

Fig. 45.3. Leaves with broad, floating blades. (a) *Nuphar variegatum.* (b) *Nymphaea odorata.* (c) *Braserma schreberi.* (d) *Nelumbo lutea.*

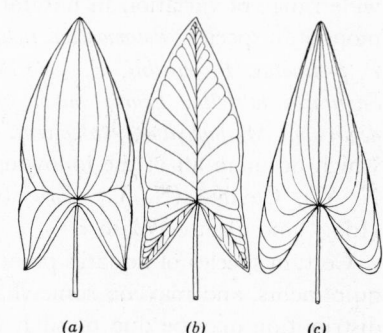

Fig. 45.4. Leaves with arrowhead shape of blade. (a) *Sagittaria latifolia.* (b) *Peltandra virginica.* (c) *Pontederia cordata.*

vegetative condition.   This can sometimes be done by the leaves which are usually helpful even though they are often quite variable.

The leaves of most aquatic plants belong to a few general types, the most common of which are:

1. Leaves submersed, with blades dissected into linear segments (Fig. 45.1).

2. Leaves submersed, linear, or tapelike, often the upper part partly floating (Fig. 45.2).

3. Leaves with broad floating blades attached by long petioles (Fig. 45.3).

4. Leaves with streamlined shape and often with an emersed arrowhead-shaped blade (Fig. 45.4).

The last vary greatly with the conditions under which the plant grows, or even on a single plant (Fig. 45.5).

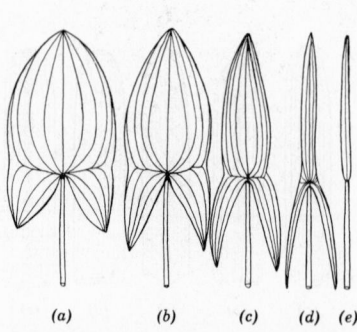

Fig. 45.5. Leaf variation in shape of blade on one plant of *Sagittaria latifolia.*

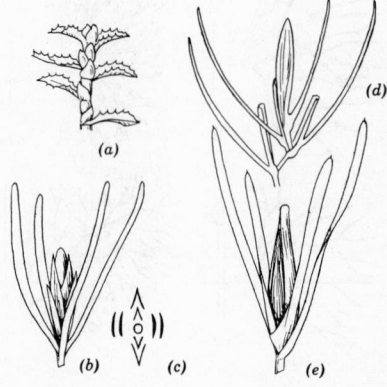

Fig. 45.6. Winter buds or turions of *Potamogeton.* (a) *P. crispus.* (b, c) *P. friesii.* (d) *P. pusillus.* (e) *P. zosteriformis.*

In numerous species vegetative propagation is accomplished by special organs; in others any part of the stem may break off and take root. The most common propagating parts consist of rhizomes, runners, tubers, corms, and shortened axillary or terminal leafy axes, the turions or winter buds (Fig. 45.6).

The tubers, common to some species, usually are rich in stored materials which are often the source of food for animal life, especially water fowl (Fig. 45.7). These tubers are sometimes very characteristic and supply diagnostic features for identification of the plants that produced them. Most aquatic

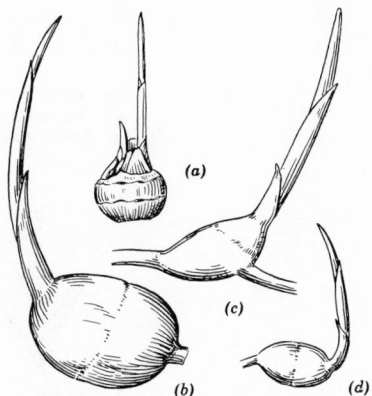

Fig. 45.7. Tubers. (*a*) *Eleocharis tuberosa.*
(*b*) *Sagittaria latifolia.* (*c*) *Vallisneria americana.*
(*d*) *Potamogeton pectinatus.*

plants produce seeds in abundance. When these germinate, the seedlings that they produce at first often look rather similar and are not recognizable, except by a specialist, until they are developed into mature plants. Some seedlings of a number of common species are illustrated in Figs. 45.8–45.14.

The seeds of some aquatic plants are rich in stored food such as starch and are frequently eagerly sought by animals. One of the most common foods of water fowl is the seeds of *Potamogeton* (Fig. 45.15).

The more common of the true aquatic plants of the United States are included in the following artificial key to species based largely upon vegetative characteristics. Flowers and fruits are used only where they seem indispensable for identification. About a quarter of the species known from the United States are included, but since the key includes the commonest species, the majority of specimens collected will be covered. Details of the classification of these plants can be found in several of the works cited in the References.

I am indebted to Miss Elfriede Abbe who adapted most of the illustrations from *Aquatic Plants of the United States* by W. C. Muenscher. They are here used with permission of Comstock Publishing Co., a division of the Cornell University Press.

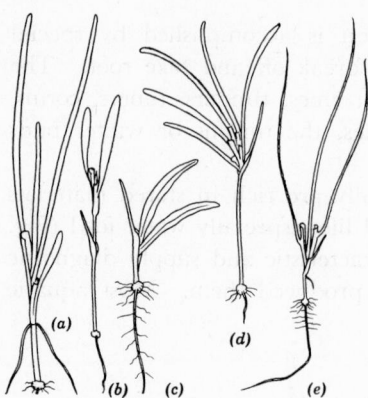

Fig. 45.8. Seedlings of *Potamogeton*.
(a) *natans*. (b) *obtusifolius*. (c) *amplifolius*.
(d) *epihydrus*. (e) *capillaceus*.

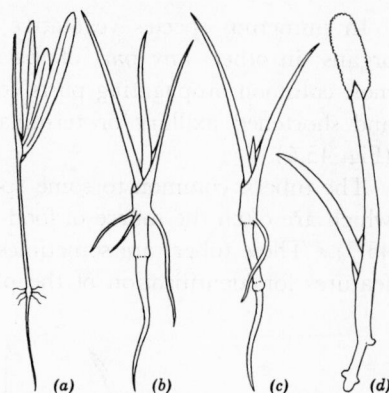

Fig. 45.9. Seedlings. (a) *Najas flexilis*. (b)
Alisma plantago-aquatica. (c) *Sagittaria latifolia*.
(d) *Butomus umbellatus*.

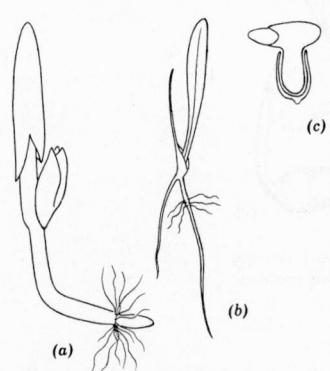

Fig. 45.10. Seedlings. (a) *Elodea
occidentalis*. (b) *Vallisneria americana*.
(c) *Lemna minor*.

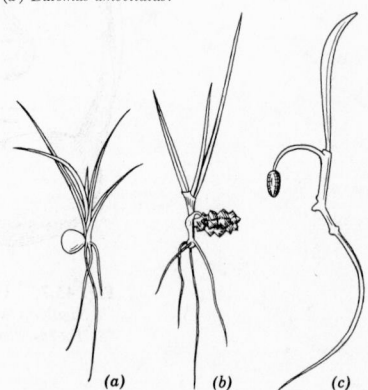

Fig. 45.11. Seedlings. (a) *Orontium aquaticum*. (b) *Pontederia cordata*. (c) *Heteranthera dubia*.

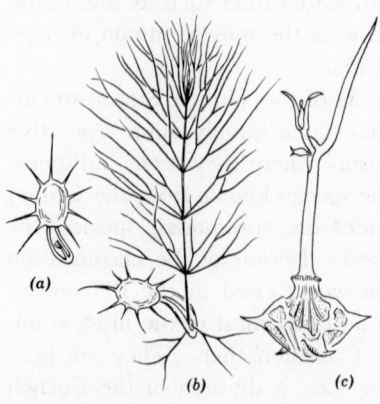

Fig. 45.12. Seedlings. (a, b) *Ceratophyllum
echinatum*. (c) *Trapa natans*.

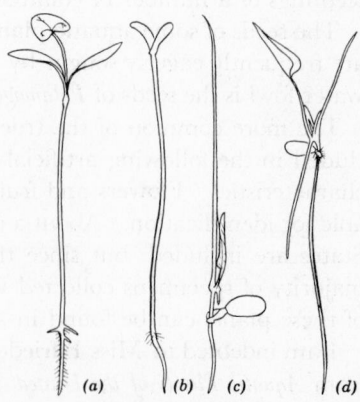

Fig. 45.13. Seedlings. (a) *Nymphoides
cordatum*. (b) *Lobelia dortmanna*. (c)
Nuphar variagatum. (d) *Nymphaea tuberosa*.

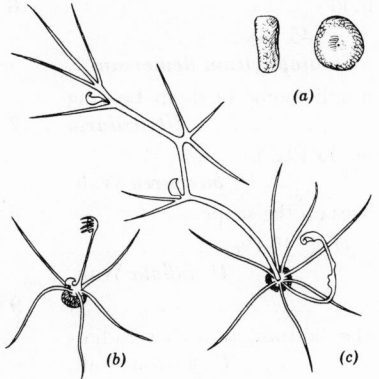

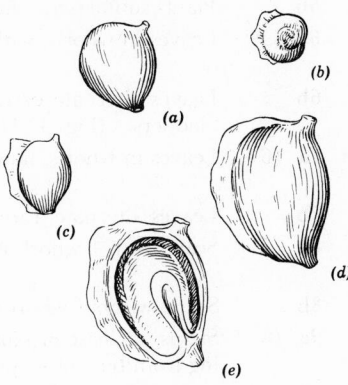

Fig. 45.14. *Utricularia geminiscapa.* (*a*) Seeds. (*b*, *c*) Seedlings.

Fig. 45.15. Seeds of *Potamogeton.* (*a*) *pectinatus.* (*b*) *spirillus.* (*c*) *epihydrus.* (*d*, *e*) *praelongus.*

KEY TO SPECIES

1a		Plants without roots, floating or submerged	**2**
1b		Plants with roots (but see **5a**).	**11**
2a	(1)	Stems not developed, plant reduced to a small, undifferentiated, flat globular or tubular floating frond	**3**
2b		Stems slender, leafy	**5**
3a	(2)	Frond thin, sickle-shaped or much elongated. (Fig. 45.16*f*) . . . . ***Wolffiella floridana*** (Smith) Thompson	
3b		Fronds thick, globular or ellipsoidal	**4**
4a	(3)	Globular. (Fig. 45.16*d*) . . . . ***Wolffia columbiana*** Karst.	
4b		Ellipsoidal. (Fig. 45.16*e*) . . . . ***W. punctata*** Griseb.	
5a	(2)	Plants floating; leaves 2, at a node, rotund, and a third modified into filiform submerged rootlike segments. (Fig. 45.17*a*) . . . . . . ***Salvinia rotundifolia*** Willd.	

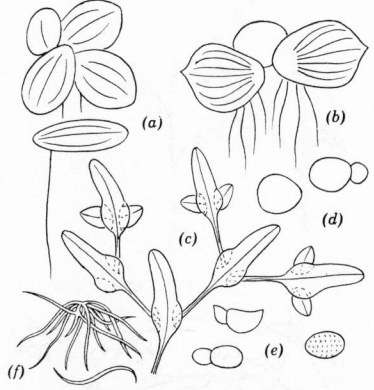

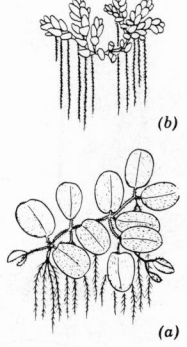

Fig. 45.16. Lemnaceae. (*a*) *Lemna minor.* (*b*) *Spirodela polyrhiza.* (*c*) *Lemna trisulca.* (*d*) *Wolffia columbiana.* (*e*) *Wolffia punctata.* (*f*) *Wolffiella floridana.*

Fig. 45.17. (*a*) *Salvinia rotundifolia.* (*b*) *Azolla caroliniana.*

5b		Plants submersed, often anchored in mud	**6**
6a	**(5)**	Leaves in whorls, without bladders. (Fig. 45.18)	
		Ceratophyllum demersum L.	
6b		Leaves alternate or rarely whorled; usually some of them bearing bladders. (Fig. 45.1*f*). . . . . *Utricularia*	**7**
7a	**(6)**	Leaves in whorls, flowers purple. (Fig. 45.19*a*, *b*).	
		U. purpurea Walt.	
7b		Leaves alternate, rarely 1 whorl of leaves on the scape	**8**
8a	**(7)**	Scape with 1 whorl of inflated leaves. (Fig. 45.19*e*).	
		U. inflata Walt.	
8b		Scape without whorl of inflated leaves	**9**
9a	**(8)**	Stems, at least in part, creeping on the bottom, branches radiating from base of scape. (Fig. 45.19*f*) . . . . *U. fibrosa* Walt.	
9b		Stems floating, at least some of them, draped in the water	**10**
10a	**(9)**	Leaf segments minutely serrate along the margin; scape 6 to 12 flowered. (Figs. 45.1*b*, 45.19*g*) . . . . *U. vulgaris* L.	
10b		Leaf segments not serrate on the margin; scape 1 to 5 flowered. (Fig. 45.19*c*, *d*) . . . . *U. geminiscapa* Benj.	

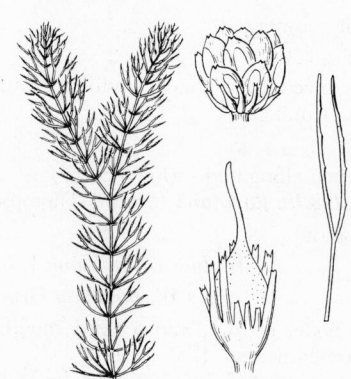

Fig. 45.18.   *Ceratophyllum demersum.*

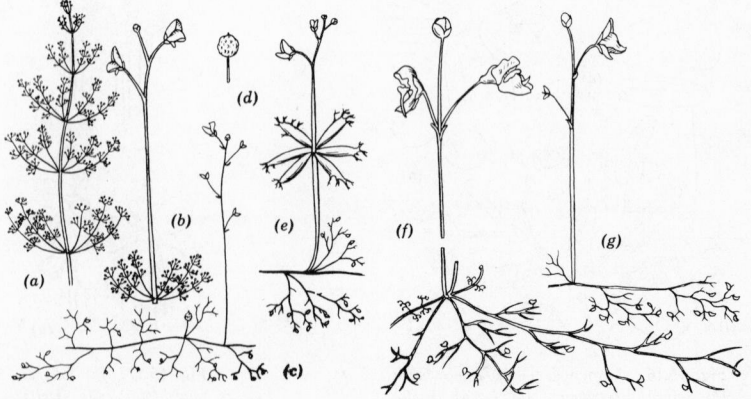

Fig. 45.19.   *Utricularia.*   (*a*, *b*) *purpurea.*   (*c*, *d*) *geminiscapa.*   (*e*) *inflata.*   (*f*) *fibrosa.*   (*g*) *vulgaris.*

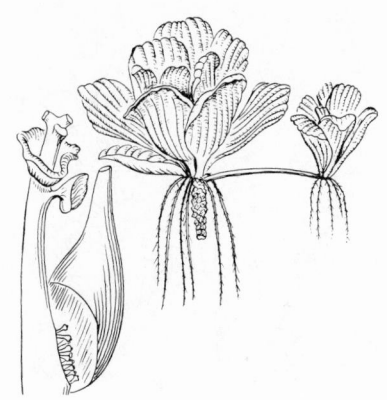

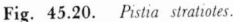

Fig. 45.20. *Pistia stratiotes.*

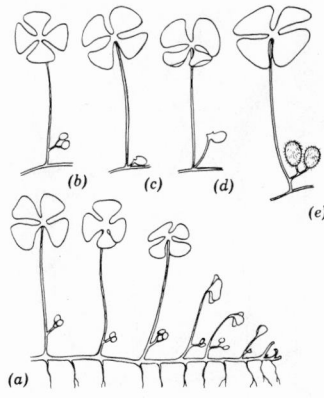

Fig. 45.21. *Marsilea.* (*a, b*) *quadrifolia.*
(*c*) *vestita.* (*d*) *uncinata.* (*e*) *macropoda.*

11a **(1)** Plants free-floating. **12**

11b   Plants rooted on the bottom, submersed or rarely emersed **16**

12a **(11)** Plants reduced to 1 or a few small flat floating fronds **13**

12b   Plants with several to many leaves inserted on an axis. **15**

13a **(12)** Fronds with 2 or more roots. (Fig. 45.16*b*) . . . .
                       Spirodela polyrhiza (L.) Schleid.

13b   Fronds with 1 root. **14**

14a **(13)** Fronds 6–12 mm long, oblong, often stalked, several connected.
             (Fig. 45.16*c*) . . . . ***Lemna trisulca*** L.

14b   Fronds 2–5 mm long, round or ovate. (Fig. 45.16*a*) . **L. minor** L.

15a **(12)** Axis vertical, leaves 4–10 cm long, erect, in a dense rosette. (Fig.
             45.20). . . . . ***Pistia stratiotes*** L.

15b   Axis horizontal, leaves overlapping, not over 5 mm long, not in a
             rosette. (Fig. 45.17*b*) . . . . ***Azolla caroliniana*** Willd.

16a **(11)** Leaves compound, with 4 broad leaflets. . . . . ***Marsilea***   **17**

16b   Leaves not as above, simple or compound **20**

17a **(16)** Peduncles with 2–6 sporocarps, adnate to the base of the petiole . .   **18**

17b   Peduncles solitary, not adnate to the petiole **19**

18a **(17)** Leaflets glabrous. (Fig. 45.21*a, b*) . . . . ***Marsilea quadrifolia*** L.

18b   Leaflets silky with white hairs. (Fig. 45.21*e*) . . . .
                       M. macropoda Engelm.

19a **(17)** Peduncle about as long as the sporocarp. (Fig. 45.21*c*). . . . . . .
                       M. vestita Hook and Grev.

19b   Peduncle much longer than the sporocarp. (Fig. 45.21*d*). . . . . .
                       M. uncinata A. Br.

20a **(16)** Leaves opposite or whorled **21**

20b   Leaves alternate **42**

21a **(20)** Leaves simple **22**

21b   Leaves compound **35**

22a **(21)** Leaves whorled **23**

22b   Leaves opposite **26**

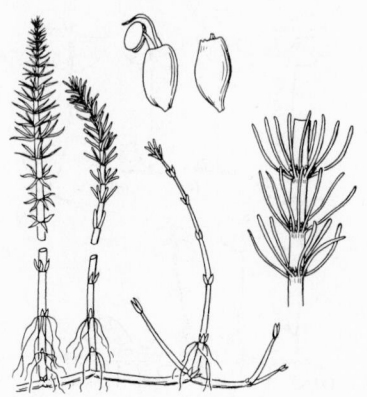

Fig. 45.22. *Hippuris vulgaris.*

Fig. 45.23. *Anacharis.* (*a*) *canadensis.* (*b*) *densa.*

23a (22) Stems unbranched, from hollow underground rhizomes. (Fig.
45.22) . . . . **Hippuris vulgaris** L.

23b   Stems branched above ground, solid . . . . . **Anacharis** (=**Elodea**)   **24**

24a (23) Leaves 6–9 mm wide. (Fig. 45.23*b*) . . . .
A. densa (Planch.) Caspary

24b   Leaves to 5 mm wide **25**

25a (24) Leaves 0–1.5 mm wide. (Fig. 45.10*a*) . . . **A. occidentalis** Pursh.

25b   Leaves 1–5 mm wide. (Fig. 45.23*a*). . . . . **A. canadensis** Michx.

26a (22) Leaves glandular-dotted, usually less than 1 cm long, blunt and
rounded, fruits with several seeds. (Fig. 45.24*a, b*) . . . .
Elatine americana Pursh.

26b   Leaves not glandular-dotted. **27**

27a (26) Leaves large, more than 2 cm long, broad and net-veined **28**

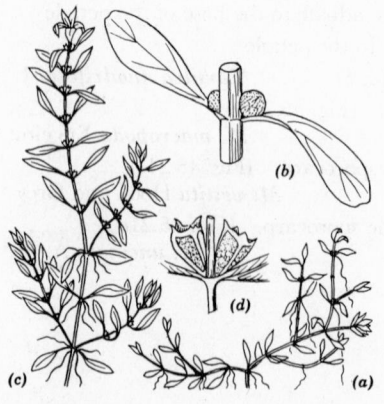

Fig. 45.24. (*a, b*) *Elatina americana.* (*c, d*)
Ludvigia palustris.

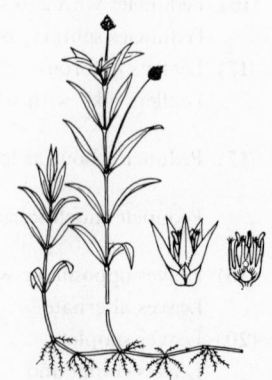

Fig.   45.25.   *Alternanthera
philoxeroides.*

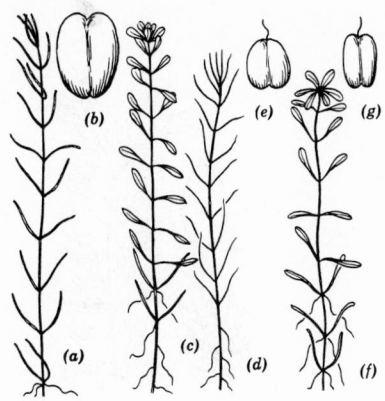

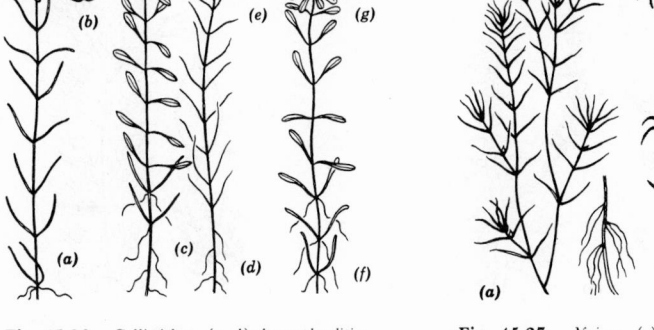

Fig. 45.26. *Callitriche.* (*a, b*) *hermaphroditica.* (*c–e*) *heterophylla.* (*f, g*) *palustris.*

Fig. 45.27. *Najas.* (*a*) *flexilis.* (*b*) *minor.*

27b		Leaves small, linear to spatulate	**29**
28a	**(27)**	Flowers axillary. (Fig. 45.24*c, d*). . . ***Ludvigia palustris*** (L.) Éll.	
28b		Flowers in capitate clusters. (Fig. 45.25)	
		Alternanthera philoxeroides (Mart.) Griseb.	
29a	**(27)**	Fruits axillary, 4-celled, forming 4 nutlets, leaves often forming floating rosettes	**30**
29b		Fruits axillary, solitary, 1-seeded, leaves linear, often with a dilated base	**32**
30a	**(29)**	All leaves linear. (Fig. 45.26*a, b*)	
		Callitriche hermaphroditica L.	
30b		The upper leaves often spatulate, in floating rosettes	**31**
31a	**(30)**	Fruits higher than wide. (Fig. 45.26*f, g*) . . . . . ***C. palustris*** L.	
31b		Fruits as high as wide. (Fig. 45.26*c–e*) . . ***C. heterophylla*** Pursh.	
32a	**(29)**	Leaf bases broadly and truncately lobed	**33**
32b		Leaf base neither broadly nor truncately lobed, but little enlarged.	**34**
33a	**(32)**	Leaves stiff, recurved, spiny. (Fig. 45.27*b*) . . . . ***Najas minor*** L.	
33b		Leaves flaccid, not recurved . . . ***N. gracillima*** (A. Br.) Morong	
34a	**(32)**	Seed coat smooth and glossy. (Fig. 45.27*a*)	
		N. flexilis (Willd.) R. and S.	
34b		Seed coat coarsely reticulate, not glossy	
		Najas guadalupensis (Spreng.) Morong	
35a	**(21)**	Leaves pinnately divided	**36**
35b		Leaves palmately divided	**40**
36a	**(35)**	Flowers in axils of ordinary leaves	
		Myriophyllum brasiliense Comb.	
36b		Flowers in axils of bracts on terminal spikes	**37**
37a	**(36)**	Bracts alternate . . . . ***M. alterniflorum*** D.C.	
37b		Bracts in whorls	**38**
38a	**(37)**	Bracts pinnately dissected or lobed . . . . . . ***M. verticillatum*** L.	
38b		Bracts entire or toothed	**39**

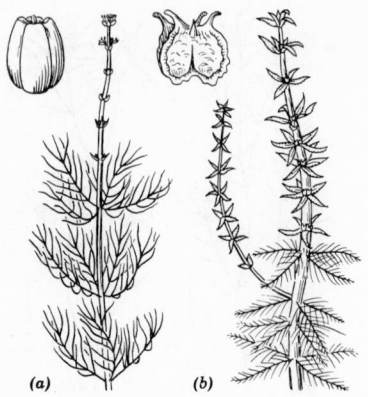

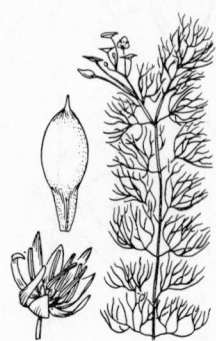

Fig. 45.28. *Myriophyllum.* (a) *exalbescens.*
(b) *heterophyllum.*

Fig.   45.29.   *Cabomba
caroliniana.*

39a (38) Bracts oval, not exceeding the fruit. (Fig. 45.28a) . . . .
                                    M. exalbescens Fernald
39b   Bracts oblanceolate, toothed, much exceeding the fruit. (Fig.
            45.28b) . . . . ***M. heterophyllum*** Michx.
40a (35) Leaves petioled. (Figs. 45.1e, 45.29) . . . .
                                    Cabomba caroliniana Gray
40b   Leaves sessile **41**
41a (40) Leaves crisp, margin of lobes serrate (roots lacking). (Fig. 45.18;
            see also **6a**) . . . . **Ceratophyllum demersum** L.
41b   Leaves flaccid, margins of lobes entire. (Figs. 45.1b, 45.30) . . . .
                                    Bidens beckii Torr.
42a (20) Leaves compound **43**
42b   Leaves simple **47**
43a (42) Leaves palmately compound, stipules fused to the base of petiole . .   **44**

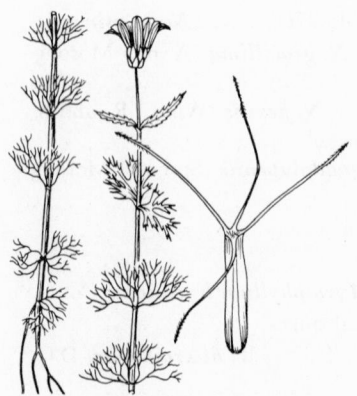

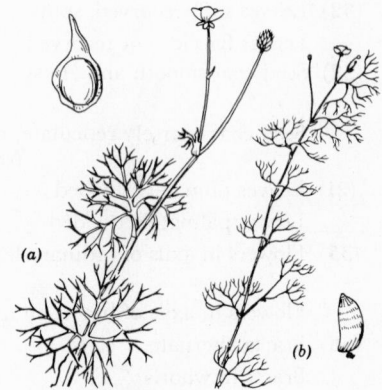

Fig. 45.30. *Bidens beckii.*

Fig. 45.31. *Ranunculus.* (a) *flabellaris.*   (b)
aquatilis.

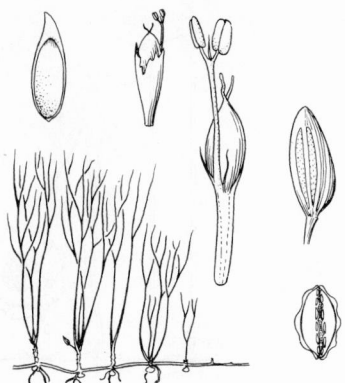
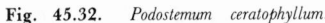

Fig. 45.32.   *Podostemum   ceratophyllum.*

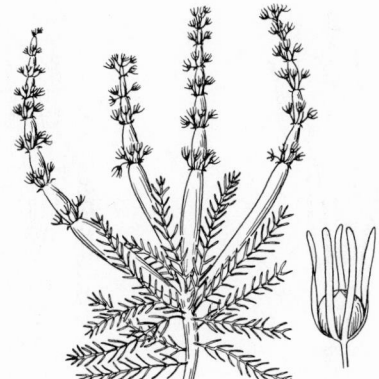

Fig. 45.33.   *Hottonia inflata.*

43b   Leaves pinnately compound or divided, without stipules **46**

44a   (**43**)   Flowers yellow; some leaves floating.   (Fig. 45.31*a*) . . . .
                                                                    Ranunculus flabellaris Raf.

44b   Flowers white; leaves all submersed. **45**

45a   (**44**)   Petiole sheath extending to base of blade; leaves rigid . . . .
                                                                    R. circinatus Sibth.

45b   Petiole sheath much shorter than petiole; leaves limp and col-
                  lapsing when lifted out of water.   (Figs. 45.1*c*, 45.31*b*) . . . .
                                                                    R. aquatilis L.

46a   (**43**)   Leaf segments cartilaginous; flowers axillary.   (Fig. 45.32) . . . . .
                                                                    Podostemon ceratophyllum Michx.

46b   Leaf segments not cartilaginous; flowers in whorls, on erect inflated
                  stems.   (Fig. 45.33) . . . . *Hottonia inflata* Ell.

47a   (**42**)   Leaves clustered in dense, mostly basal, rosettes **48**

47b   Leaves cauline, on branched stems **70**

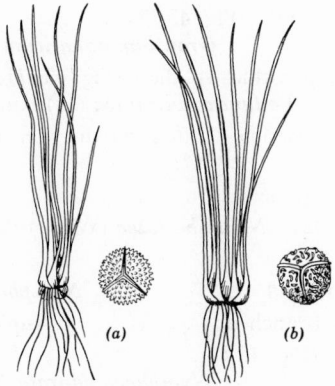

Fig. 45.34.   *Isoetes.*   (*a*) *muricata.*   (*b*)
engelmanni.

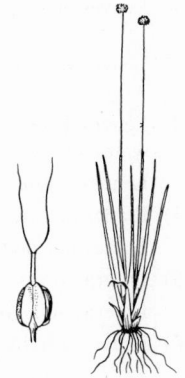

Fig. 45.35.   *Eriocaulon
septangulare.*

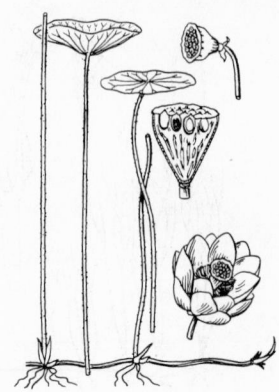

Fig. 45.36.   *Pontederia*   Fig. 45.37.   *Orontium*   Fig.   45.38.   *Nelumbo   lutea.*
cordata.   *aquaticum.*

48a	(47)	Leaf rosettes on a short vertical axis, leaves in more than 2 ranks. .	**49**
48b		Leaf rosettes from the nodes of creeping rhizomes or rootstocks. . .	**55**
49a	(48)	Plants producing spores on quill-like leaves in basal sporangia . . .	**50**
49b		Plants producing seeds from flowers on a scapose stem	**51**
50a	(49)	Megaspores spiny on surface. (Fig. 45.34*a*)	

                                                          Isoetes muricata Dur.

50b		Megaspores reticulated on surface. (Fig. 45.34*b*)	

                                                          I. engelmanni A. Br.

51a	(49)	Flowers in a solitary terminal head or raceme	**52**
51b		Flowers in a spadix or in a spike . . . . , ...	**53**
52a	(51)	Leaves erect; flowers in a head. (Fig. 45.35)	

                                                          Eriocaulon septangulare With.

52b		Leaves recurved; flowers in a raceme. (Fig. 45.13*b*)	

                                                          Lobelia dortmanna L.

53a	(51)	Flowers in a spadix	**54**
53b		Flowers in a spike. (Figs. 45.11*b*, 45.36).	

                                                          Pontederia cordata L.

54a	(53)	Leaves oblong, spadix without spathe. (Fig. 45.37).	

                                                          Orontium aquaticum L.

54b		Leaves sagittate, spadix covered by a thick spathe. (Fig. 45.4*b*) . .	

                                                          Peltandra virginica (L.) Kunth

55a	(48)	Leaves broad, terminating long petioles, often floating on the water surface.	**56**
55b		Leaves sessile or on short inflated petioles	**61**
56a	(55)	Blades peltate. (Figs. 45.3*d*, 45.38) . **Nelumbo lutea** (Willd.) Pers.	
56b		Blades not peltate	**57**
57a	(56)	Lateral veins dichotomously branched . . . . **Nymphaea**	**58**
57b		Lateral veins not dichotomously branched . . . . **Nuphar**	**59**
58a	(57)	Leaves usually purple beneath. (Fig. 45.3*b*).	

                                                          Nymphaea odorata Ait.

58b		Leaves usually not purple beneath. (Fig. 45.39)	

                                                          N. tuberosa Paine

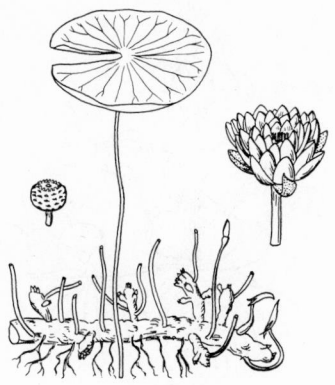

Fig. 45.39. *Nymphaea tuberosa.*

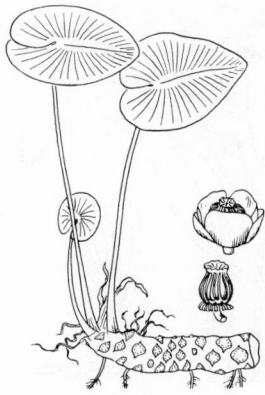

Fig. 45.40. *Nuphar variegatum.*

59a (*57*) Blade of leaf less than ½ as wide as long . . . .
Nuphar sagittifolium Pursh.

59b Blade of leaf more than ½ as wide as long **60**

60a (*59*) Sepals 9, rarely 7 . . . . *N. polysepalum* Engelm.

60b Sepals mostly 6. (Figs. 45.3*a*, 45.40) . . . *N. variegatum* Engelm.

60c Sepals 3, persistent. (Fig. 45.41) . . . .
Hydrocleis nymphoides Buchenau

61a (*55*) Leaves in 2 ranks, linear **62**

61b Leaves in more than 2 ranks **67**

62a (*61*) Flowers imperfect, in heads on a stem with few leaves . . . .
Sparangium **63**

62b Flowers dioecious, in leaf axils. (Fig. 45.2*a*, 45.42) . . . .
Vallisneria americana Michx.

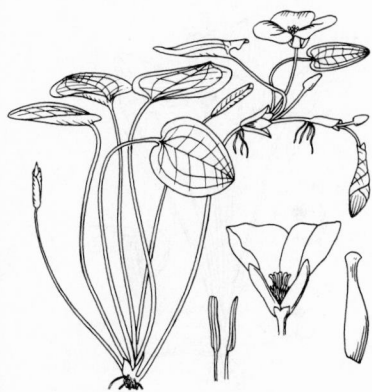

Fig. 45.41. *Hydrocleis nymphoides.*

Fig. 45.42. *Vallisneria americana.* (*a*) Staminate. (*b*) Pistillate.

Fig. 45.43. *Sparganium fluctuans.*

Fig. 45.44. *Eichornia crassipes.*

63a	**(62)**	Stigmas mostly 2; leaves erect, emersed. . . . .
		Sparganium eurycarpum Engelin.
63b		Stigmas solitary; leaves erect or floating **64**
64a	**(63)**	Beaks strongly curved; achenes reddish-brown. (Fig. 45.43) . . . .
		S. fluctuans (Morong) Robinson
64b		Beaks straight or slightly curved; achenes greenish. **65**
65a	**(64)**	Pistillate heads or branches all axillary; leaves without scarious
		margins . . . . ***S. americanum*** Nutt.
65b		Pistillate heads, at least some, supra-axillary. **66**
66a	**(65)**	Leaves 3–4 mm wide, not scarious-margined. . . . .
		S. angustifolium Michx.
66b		Leaves 3–9 mm wide, with scarious margin near base. . . . .
		S. chlorocarpum Rydb.
67a	**(61)**	Leaf rosettes mostly floating, leaves broad **68**
67b		Leaf rosettes basal, rooted in the mud **69**
68a	**(67)**	Leaves sessile; flowers axillary. (Fig. 45.20; see also **15a**) . . . . . .
		Pistia stratiotes L.

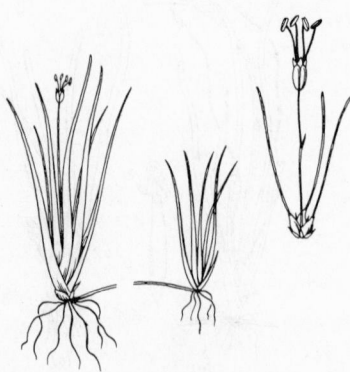

Fig. 45.45. *Littorella americana.*

Fig. 45.46. *Sagittaria lati-folia.*

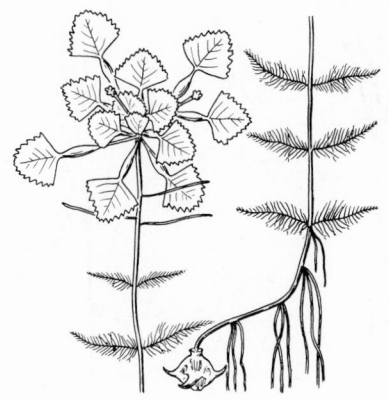

Fig. 45.47.   *Trapa natans.*   **Fig. 45.48.**   *Polygonum amphibium.*

68b		Leaves on inflated petioles; flowers in terminal clusters. (Fig. 45.44) . . . . ***Eichhornia crassipes*** (Mart.) Solms.
69a	**(67)**	Leaves linear. (Fig. 45.45) . . . . . ***Littorella americana*** Fernald
69b		Leaves broader, often sagittate. (Figs. 45.4*a*, 45.5, 45.7*b*, 45.9*c*, 45.46) . . . . ***Sagittaria latifolia*** (Willd.)
70a	**(47)**	Dicotyledonous plants with large broad leaves **71**
70b		Monocotyledonous plants, mostly with narrow leaves (except when floating) **76**
71a	**(70)**	Petioles inflated, leaves mostly floating, in terminal rosettes. (Figs. 45.12*c*, 45.47) . . . . ***Trapa natans*** L.
71b		Petioles not inflated **72**
72a	**(71)**	Petioles with fused stipules (acreae) surrounding the jointed stem. (Fig. 45.48) . . . . ***Polygonum amphibium*** L.
72b		Petioles without stipules surrounding the stems. **73**
73a	**(72)**	Leaves peltate. (Figs. 45.3*c*, 45.49) . . . ***Brasenia schreberi*** Gmel.

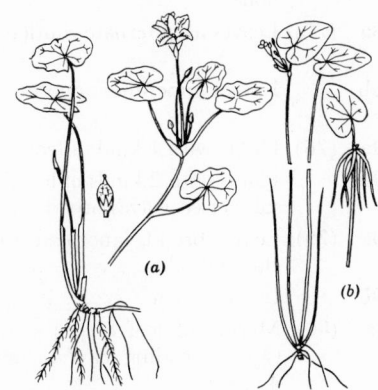

Fig. 45.49.   *Brasenia schreberi.*   **Fig. 45.50.**   *Nymphoides.* (*a*) *peltatum.* (*b*) *cordatum.*

Fig. 45.51. *Heteranthera dubia.*

Fig. 45.52. *Ruppia maritima.*

Fig. 45.53. *Zannichellia palustris.*

73b		Leaves not peltate, mostly cordate, many of them basal . . . .
		Nymphoides **74**
74a	(73)	Petioles slender, mostly provided with clusters of roots, often also with white flowers **75**
74b		Petioles without roots, flowers yellow. (Fig. 49.50*a*) . . . .
		Nymphoides peltatum (Gmel.) B. and R.
75a	(74)	Leaf blade ovate, mostly less than 6 cm long. (Figs. 45.13*a*, 45.50*b*) . . . . *N. cordatum* (Gmel.) B. and R.
75b		Leaf blade orbicular to reniform, mostly 8–15 cm long . . . . . .
		N. aquaticum (Walt.) Fernald
76a	(70)	Flowers axillary; leaves linear. **77**
76b		Flowers in spikes, each with 4 separate sepals, carpels and stamens . . . . *Potamogeton* **79**
77a	(76)	Pistil solitary, several-seeded; corolla yellow. (Fig. 45.51) . . . . .
		Heteranthera dubia (Jacq.) MacM.
77b		Pistils several in each flower, forming curved nutlets; corolla none **78**
78a	(77)	Leaves all alternate; nutlets on long stalks. (Fig. 45.52) . . . . . .
		Ruppia maritima L.
78b		Leaves sometimes opposite; nutlets sessile. (Fig. 45.53) . . . .
		Zannichellia palustris L.
79a	(76)	Plants with 1 kind of leaves, all submersed **80**
79b		Plants with 2 kinds of leaves; floating leaves broad and coriaceous; submersed leaves broad and membranous or linear **99**
80a	(79)	Leaves broad, lanceolate to elliptical or ovate, never linear, often clasping. **81**
80b		Leaves linear. **86**
81a	(80)	Margin of leaf blades serrulate; winter buds hard, with serrate, rigid, spreading leaves; fruit with long, slender beak. (Figs. 45.6*a*, 45.54*a*) . . . . *Potamogeton crispus* L.
81b		Margin of leaf blades entire, rarely serrulate at tip **82**
82a	(81)	Base of blade tapering, not clasping **83**

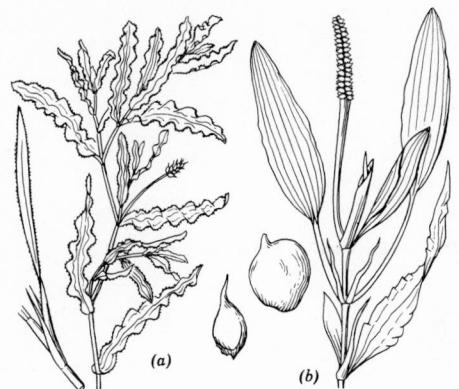
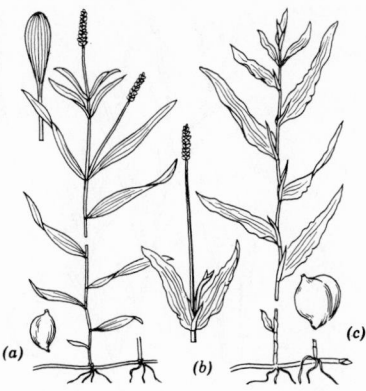

Fig. 45.54. *Potamogeton.* (*a*) *crispus.* (*b*) *illinoensis.*

Fig. 45.55. *Potamogeton.* (*a, b*) *alpinus.* (*c*) *praelongus.*

82b		Base of blade clasping **84**
83a	(82)	Upper leaves petioled; blades serrulate near apex; plant green. (Fig. 45.54*b*) . . . . ***P. illinoensis*** Morong
83b		Upper leaves sessile or nearly so; blades entire; plant reddish. (Fig. 45.55*a, b*). . . . . ***P. alpinus*** Balbis
84a	(82)	Blade 10–30 cm long, with cucullate apex; stipules 2–8 cm long, persistent; stem whitish; fruit 4–5 mm long, sharply 3-keeled; embryo with straight apex. (Figs. 45.15*d, e*, 45.55*c*) . . . . ***P. praelongus*** Wulfen
84b		Blade 1–12 cm long, apex not cucullate; stem green; fruit 2–4 mm long, obscurely 3-keeled; embryo with apex curved inward **85**
85a	(84)	Leaves short, with rounded apex and plain margin, drying dark green or olive; stipules small or wanting; peduncle slender. (Fig. 45.56*a*) . . . . ***P. perfoliatus*** L.
85b		Leaves narrowly ovate, with tapering apex and crinkly margin, drying light green; stipules conspicuous, persisting as shreds; peduncle spongy. (Fig. 45.56*b*) . . . . . ***P. richardsonii*** A. Benn.

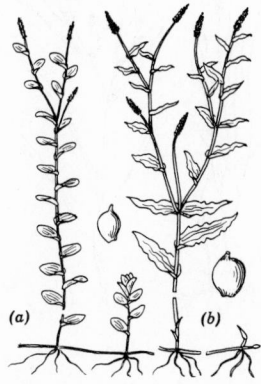

Fig. 45.56. *Potamogeton.* (*a*) *perfoliatus.* (*b*) *richardsonii.*

Fig. 45.57. *Potamogeton latifolius.*

86a	(80)	Stipules fused with the lower part of the leaf to form a sheath at least 1 cm long **87**
86b		Stipules free from the leaf or, rarely, fused to the base for 1 or 2 mm **91**
87a	(86)	Leaves 4–8 mm wide, auricled at base, oriented on the axis into a rigid, flattened spray **88**
87b		Leaves filiform, rarely up to 3 mm wide, not auricled, entire, oriented into a lax, diffuse, branched spray **89**
88a	(87)	Leaves serrulate, pointed at apex. (Fig. 45.58*c*) . . . . *P. robbinsii* Oakes
88b		Leaves entire, rounded at apex. (Fig. 45.57) . . . . *P. latifolius* (Robbins) Morong
89a	(87)	Stigmas raised on a minute style, capitate; leaves gradually acuminate; rhizomes tuber-bearing. (Figs. 45.7*d*, 45.15*a*, 45.58*a*) . . *P. pectinatus* L.
89b		Stigmas inconspicuous, broad and sessile; leaves retuse, blunt, or shortly apiculate **90**

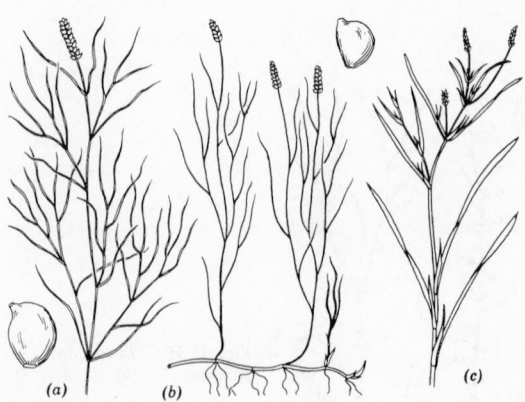

Fig. 45.58. *Potamogeton.* (*a*) *pectinatus.* (*b*) *filiformis.* (*c*) *robbinsii.*

90a (89) Plants short, slender; leaves all filiform; sheaths close around stem; spike with 2 to 5 whorls of flowers. (Fig. 45.58*b*) . . . .
P. filiformis Pers.

90b Plants coarse, 2–5 mm long; leaves on main stem short, flat, their sheaths enlarged to 2 to 5 times the diameter of the stem; spikes with 5 to 12 whorls of flowers . . . . *P. vaginatus* Turcz.

91a (86) Plants with slender, creeping rhizomes; leaves without basal glands   92

91b Plants with short rhizomes or none at all (often rooting at the lower nodes of the stem).   93

92a (91) Peduncles terminal, mostly 5–25 cm long; leaves narrower than the stems, flaccid, filiform, with long, tapering apex. (Fig. 45.59*a*) . .
P. confervoides Reichenb.

92b Peduncles axillary, less than 3 cm long; leaves broader than the stems, acute or cuspidate at apex. (Fig. 45.59*c*, See **95a**). . . . . .
P. foliosus Raf.

93a (91) Leaves 9- to 35-nerved, subrigid; prominent winter buds with imbricated stipules and ascending blades.   94

93b Leaves 1- to 7- nerved.   95

94a (93) Stems much flattened and winged, about as wide as the leaves; leaves 2 to 5 mm wide, without basal glands. (Figs. 45.2*b*, 45.6*e*, 45.60*b*) . . . . *P. zosteriformis* Fernald

94b Stems somewhat flattened, not winged; leaves mostly less than 2 mm wide, bristle-tipped, with a pair of basal glands. . . . .
P. longiligulatus Fernald

95a (93) Leaves without basal glands. (Fig. 45.59*c*, see **92b**) . . . .
P. foliosus Raf.

95b Leaves, at least some of them, with a pair of basal glands.   96

96a (95) Leaves with 5 to 7 nerves, thin; winter buds composed largely of overlapping, whitish, fibrous stipules and blades. (Figs. 45.6*b*, *c*, 45.59*b*). . . . . *P. friesii* Rupr.

96b Leaves with 3 (rarely 1 or 5) nerves, obtuse or acute   97

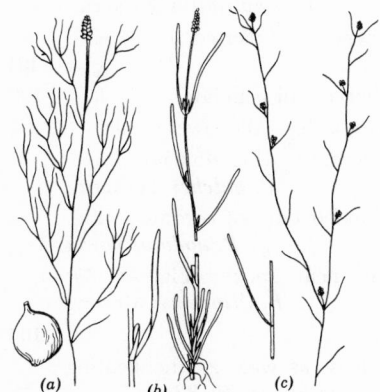

Fig. 45.59. *Potamogeton.* (*a*) *confervoides.*
(*b*) *friesii.* (*c*) *foliosus.*

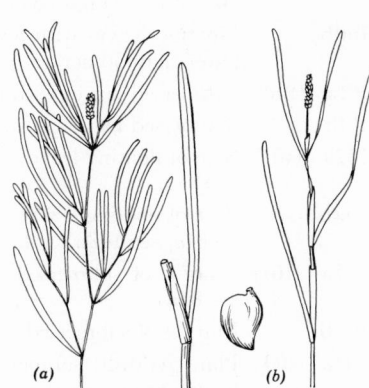

Fig. 45.60. *Potamogeton.* (*a*) *obtusifolius.*
(*b*) *zosteriformis.*

97a (96) Body of winter bud 2–4 cm long, covered with scarious stipules. (Fig. 45.8*b*, 45.60*a*) . . . . ***P. obtusifolius*** Mert. and Koch

97b Body of winter bud less than 2 cm long, solid; leaves green, rarely reddish (except *P. strictifolius*, winter bud 1.0–2.5 cm long, see below) 98

98a (97) Stipules not fibrous, in early stages with edges at least in part connate; peduncle filiform, mostly 3–8 cm long. (Figs. 45.6*d*, 45.61*c*) . . . . ***P. pusillus*** L.

98b Stipules not fibrous or connate, but flat or convolute; peduncle mostly 0.5–3 cm long. (Fig. 45.61*a, b*) . . . ***P. berchtoldii*** Fieber

98c Stipules strongly fibrous, connate when young, soon splitting; peduncle 1–9 cm long. (Fig. 45.61*d*) . . . ***P. strictifolius*** A. Benn.

Fig. 45.61. *Potamogeton.* (*a, b*) *berchtoldii.* (*c*) *pusillus.* (*d*) *strictifolius.*

99a (79) Submersed leaves broad, never linear. 100

99b Submersed leaves (or phyllodia) linear 105

100a (99) Floating leaves with 30 to 55 nerves; submersed leaves with 30 to 40 nerves. (Figs. 45.8*c*, 45.62) . . . . . . ***P. amplifolius*** Tuckerm.

100b Floating leaves with fewer than 30 nerves; submersed leaves with fewer than 30 nerves. 101

101a (100) Submersed leaves with more than 7 nerves, all petiolate. 102

101b Submersed leaves mostly with 7 nerves, at least the lower sessile . . 103

102a (101) Base of floating leaves cordate or subcordate. (Fig. 45.63*b*) . . . .
P. pulcher Tuckerm.

102b Base of floating leaves tapering or rounded but not cordate. (Fig. 45.64*a;* see **108a**) . . . . ***P. nodosus*** Poiret

103a (101) Margin of submersed leaves serrulate near apex. (Fig. 45.54*b;* see **83a**) . . . . ***P. illinoensis*** Morong

103b Margin of submersed leaves entire 104

104a (103) Plant reddish; submersed leaves at least as wide as the floating leaves, mostly on the main stem. (Fig. 45.55*a, b;* see **83b**) . . . . .
P. alpinus Balbis

Fig. 45.62.   *Potamogeton amplifolius.*

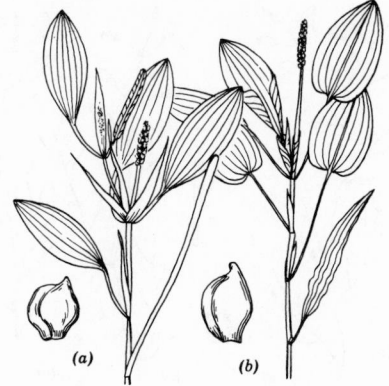

Fig. 45.63.   *Potamogeton.*   (*a*) *epihydrus.*   (*b*) *pulcher.*

104b   Plant green; submersed leaves narrower than the floating leaves, often numerous on short, axillary branches. (Fig. 45.64*b*) . . . . .
                                                          P. gramineus L.

105a **(99)** Stipules all free from the leaf bases; spikes of 1 kind only, fruits not at all or but slightly compressed **106**

105b   Stipules of all, or at least of some of the lower leaves fused with the leaf base; winter buds rare; spikes of 2 kinds, those in the axils of the lower submersed leaves globose, submersed on short peduncles; those in the axils of the upper or floating leaves cylindrical, often emersed on longer peduncles; fruit laterally compressed, 3-keeled, with spirally coiled embryo **111**

106a **(105)** Floating leaves more than 1 cm wide and more than 2 cm long; winter buds usually wanting **107**

106b   Floating leaves less than 1 cm wide and less than 2 cm long, 5- to 9-nerved **110**

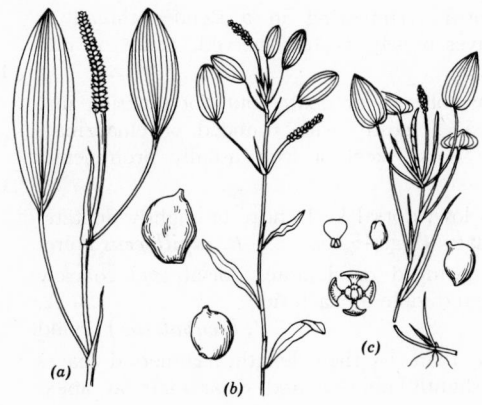

Fig. 45.64.   *Potamogeton.*   (*a*) *nodosus.* (*b*) *gramineus.*   (*c*) *natans.*

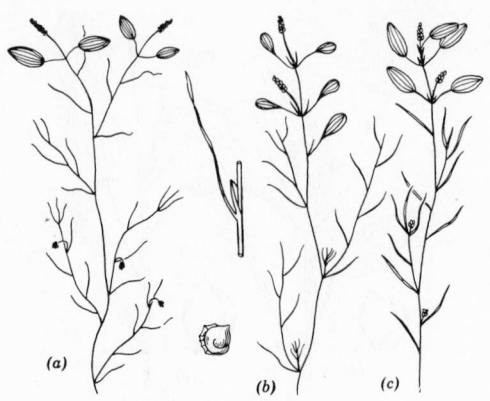

Fig. 45.65. *Potamogeton.* (*a*) *capillaceus.* (*b*) *vaseyi.* (*c*) *spirillus.*

107a (106) Submersed leaves tapelike, 2–10 mm wide, with a prominent cellular median band; fruit 3-keeled. (Figs. 45.63a, 45.2d, 45.8d, 45.15c). . . . . **P. epihydrus** Raf.

107b Submersed leaves terete, often reduced to petiole, mostly less than 1.5 mm thick, without a median band 108

108a (107) Blade of floating leaves elliptical, with tapering base; fruit 3-keeled, without lateral dimple. (Fig. 45.64a, see **102b**) . . . .
                                                                    P. nodosus Poiret

108b Blade of floating leaves ovate to subcordate; fruit scarcely keeled, with a dimple on each side 109

109a (108) Fruits with concave sides; spikes 3–6 cm long; floating leaves mostly 3–10 cm long. (Figs. 45.8a, 45.64c) . . . . . . **P. natans** L.

109b Fruits with plane sides; spikes 1–3 cm long; floating leaves 2–5 cm long. **P. oakesianus** Robbins

110a (106) Submersed leaves filiform, tapering; floating leaves 3–8 mm wide, in marginless petioles; winter buds nearly sessile on short, axillary branches. (Fig. 45.65b) . . . . **P. vaseyi** Robbins

110b Submersed leaves linear, acute; floating leaves 2–4 mm wide, tapering to margined petioles; winter buds terminating upper branches. . . . . **P. lateralis** Morong

111a (105) Submersed leaves filiform, terminating in a slender thread or bristle tip; floating leaves mostly 3- to 7-nerved, acute or mucronate 112

111b Submersed leaves linear, obtuse, or acute, but not tapering into bristle tips; floating leaves mostly with rounded or emarginate apex, 5- to 15-nerved; dorsal keel of fruit usually prominently toothed 113

112a (111) Fruit with lateral keels low; dorsal keel entire or slightly dentate; fruit with sides nearly flat. (Fig. 45.65a). . . **P. capillaceus** Poiret

112b Fruit with lateral keels winged or dentate; dorsal keel coarsely dentate; fruit with a deep dimple on each side . . . .
                                                                    P. bicupulatus Fernald

113a (111) Stipules fused for more than ½ their length; submersed leaves blunt; floating leaves slightly oblique and emarginate at apex;

fruit with obsolete beak and sides rounded instead of keeled. (Figs. 45.15*b*, 45.65*c*) . . . . ***P. spirillus*** Tuckerm.

113b   Stipules fused about ½ their length; submersed leaves pointed; floating leaves rounded, not emarginate at apex; fruit with minute beak and low lateral keels . . . . ***P. diversifolius*** Raf.

References

Benson, Lyman. 1957. *Plant Classification.* Heath, Boston.   **Britton, N. L. and A. Brown.** 1952. *An Illustrated Flora of the Northern United States, Canada and British Possessions,* 3 vols. Ed. H. A. Gleason. N. Y. Botan. Gardens, Lancaster Press, Lancaster, Pa.   **Clausen, R. T. 1936.** Studies in the genus Najas in the Northern United States. *Rhodora,* 38:333–345.   **Fassett, Norman C.** 1957. *A Manual of Aquatic Plants.* University of Wisconsin Press, Madison. **Fernald, M. L.** 1932. The linear-leaved North American species of Potamogeton, Section Axillares. *Mem. Am. Acad. Arts and Sci.,* 17:1–183.   **1950.** *Gray's Manual of Botany,* 8th ed. American Book Co. New York.   **Hitchcock, A. S.** 1935. Manual of grasses of the United States. *U. S. Dept. Agr. Misc. Publ.,* 200.   **Lawrence, George H. M. 1951.** *Taxonomy of Vascular Plants.* Macmillan, New York.   **Martin, A. C. and F. M. Uhler.** 1939. Food of game ducks in the United States and Canada. *U. S. Dept. Agr. Tech. Bull.,* 634:1–156.   **Mason, Herbert L.** 1957. *A Flora of the Marshes of California.* University of California Press, Berkeley and Los Angeles.   **Muenscher, W. C. 1944.** *Aquatic Plants of the United States.* Comstock, Ithaca, New York.   **Ogden, E. C.** 1943. The broad-leaved species of Potamogeton of North America north of Mexico. *Rhodora,* 45:57–105, 109–163, 171–214.   **Pfeiffer, N. E. 1922.** Monograph of the Isoetaceae. *Ann. Missouri Botan. Garden,* 9:99–232.   **Rossbach, G. B.** 1939. Aquatic Utricularias. *Rhodora,* 41:113–128.

46

Methods and Equipment

W. T. EDMONDSON

The purpose of this final chapter is to suggest in very brief form materials and techniques useful in collecting and handling organisms, to indicate sources of more detailed information, and to give addresses of suppliers of certain materials. Other suppliers of limnological equipment are listed in a special publication by the American Society of Limnology and Oceanography, cited in the references. The suggestions made in this chapter are applicable to a number of groups. Special techniques of more or less limited application have already been given in connection with the individual chapters.

Collection and Concentration of Material

The most direct way to collect fresh-water organisms is for the collector to enter the water and pick up specimens. The range in which this technique is usable is currently being extended by increased use of self-contained diving apparatus. However, because of small size, cryptic habits, or other features, most organisms are not amenable to this simple technique and a variety of aids have been developed to catch and concentrate them. Specific methods are generally limited to particular types of habitats, and collecting techniques, therefore, are conveniently described for general habitat types—the pelagial, littoral, and profundal of standing water, and running water.

Since most pelagic or planktonic organisms living in the open water are small and dispersed, the major problem is concentration. For this purpose a plankton net made of fine silk or nylon is most useful. After the net is drawn through the water, the water is allowed to drain until the catch has been concentrated in a small volume at the bottom of the net. Such nets commonly terminate in a small vial, but a more useful arrangement is a short piece of wide-bore rubber tubing inside the tip, closed with a spring clamp; this permits the material to be easily drained from the net into a bottle. A No. 25 net has holes of approximately 0.05 mm in diameter; such a net will catch most planktonic metazoa and many kinds of algae and protozoa. A plankton haul, live or preserved, may be poured into a Syracuse dish or petri plate for examination with a binocular dissecting microscope. Individual organisms may then be removed with a pipette of appropriate bore, and a steady hand, for examination with high magnification on a slide.

Many planktonic organisms such as algae and protozoa are so small they pass through the mesh of even the finest net. Such organisms can be concentrated by filtration of water samples through a membrane filter (e.g., Millepore) from which they may be washed or examined in place (see p. 26, also Richards and Krabek and Millepore Company brochure, 1954). Concentration can be achieved by centrifuging with a clinical centrifuge or a Foerst continuous centrifuge; the utility of the latter instrument is limited by the fact that many organisms are damaged by its great force. Organisms can also be concentrated by adding preservative to water samples and permitting the material to settle to the bottom of a tall vessel. Minute algae are more abundant than is commonly realized. They are well preserved with acid Lugol's solution (p. 1200). Most of the liquid is removed by siphoning, leaving the organisms in a small volume of water.

Attraction by light at night has been very little used by limnologists, although this method is often used by marine biologists. The Daphnia trap of Baylor and Smith (1953) represents a potentially useful apparatus of this sort. Aerial light traps may be used in collecting adult insects for help in identification of immature stages.

Ponds and the littoral region of lakes are often characterized by the presence of masses of vascular plants. Such plants are easy to collect by hand or by a variety of grapples and rakelike devices. They are inhabited by numbers of macroscopic animals clinging to or moving around on the surface, notably amphipods, insects, mites, molluscs, flatworms, etc. Such organisms can be collected by shaking masses of vegetation in buckets or in nets, and the material can be sorted out in white enameled pans. When a plankton net is used in such locations, it may become clogged with debris. The Birge net or cone dredge has a protective cone of screening which diverts large pieces from the mouth of the net. This instrument is especially useful for littoral ostracods, copepods, and cladocerans. Sessile and other firmly attached animals (sponges, bryozoa, hydra, some molluscs) must be sought by examining the surfaces of the vegetation.

Many of the animals and algae associated with aquatic vegetation are too small to concentrate and sort by hand. Some of these may be most conveniently concentrated by a method especially useful with the rotifers in which the animals are permitted to accumulate at the top of the illuminated side of a jar containing plants in water. Individual organisms may then be picked out of a concentrate with a fine pipette and placed on a slide or in a compressor (p. 1198) for study with high magnification. It pays to sample the jar from time to time over a period of an hour or two, since some organisms take more time to congregate than others. This method will be found extremely useful for rotifers, small crustaceans, some flatworms, motile algae and many others. In the same jars, some organisms will fall to the bottom where they can be collected in the sediment that accumulated there. Included will be the less motile rotifers, tardigrades, some protozoa, gastrotrichs, insects, and crustaceans. Minute sessile organisms can be located by examining parts of leaves in water in a Syracuse dish or a petri dish with a binocular dissecting microscope. Broad leaves can be cut into thin strips and examined edgewise. Finely divided leaves may be examined entire or in small sections. When particularly interesting organisms are located, small bits of leaf may be snipped off with fine scissors and mounted on a compressor (p. 1198) or on a slide under a coverglass. Iridectomy scissors are especially useful for this purpose. Many sessile rotifers, protozoa, and algae will be found in this way which might easily escape detection otherwise. An entire *Utricularia* or *Myriophyllum* leaf may be mounted under a large coverglass on a slide and examined with a high power microscope. The sessile microfauna and flora on *Utricularia vulgaris americana* are especially rich.

Small sessile organisms may also be collected by submerging glass microsope slides in holders in the pond or lake for several days or weeks. Many sessile rotifers, protozoa, and algae will be found in this way.

The damp sand above water level in a sandy beach contains a specialized biota (called the psammon) which may be easily collected by scooping up sand in glass vials. In the laboratory, filtered lake water is added, the whole mass shaken, and the sand briefly allowed to settle out. The supernatant water is decanted and examined. Such samples are often rich in algae, protozoa, nematodes, rotifers, tardigrades, and harpacticoid copepods. Larger sand-dwelling organisms, such as insects, may be separated by screening sand samples.

In littoral regions which do not have a massive growth of vegetation rocks may bear a biota of attached as well as motile organisms. Small rocks may be examined in a deep bowl of water under a dissecting microscope and organisms scraped off. If rocks are brushed in buckets of water many of the sessile organisms will fall into the bucket and can be further concentrated by pouring the water through a plankton net.

The profundal region of lakes is generally characterized by soft sediments which may easily be sampled with an Ekman dredge. The material may then be treated in a variety of ways. Organisms more than a few millimeters long may be removed by straining through fairly coarse silk, nylon, or brass meshes. The microfauna ordinarily must be sought in unscreened samples by suspending material in water. Rawson's ooze sucker may be used for collecting microfauna (1930). Gravel and stone bottoms, whether shallow or deep, are not well sampled by the Ekman dredge; the heavier Peterson dredge will sometimes be effective. In relatively shallow water successful samples of hard material may be obtained with a sampler which is pushed into the bottom with a pole, the jaws being closed by ropes (Deevey and Bishop, 1948).

By and large, collecting in small streams has been mostly confined to bottom dwelling organisms. Again, stones can be lifted from the water and brushed in buckets of water. In some cases it is convenient to place a net of fine-meshed screen in the water and disturb the bottom material upstream from the net. Dislodged organisms will then be swept into the net. Organisms of the open water can be collected with a plankton net. Some regions of very slowly flowing waters will have many of the characteristics of the littoral of lakes, and will be sampled in the same way.

Detailed descriptions and illustrations of limnological collecting instruments have been presented by Welch (1948) and Pennak (1953). Quantitative methods are given in some detail by the former author.

Many animals are killed by a sudden rise in temperature, especially when crowded, and provision should be made for keeping collections cool until they can be examined or fixed. Thermos bottles and jugs are useful for this purpose, but usually simpler means are adequate

Examination of Live Material

Anyone who wants to attempt to identify preserved material should be thoroughly familiar with the appearance and activities of live organisms. The examination of active animals, especially microscopic ones, may be difficult, since it is necessary to restrain motion without damaging or distorting the animal. For close examination of rotifers and other small animals, either alive or dead, nothing can fully substitute for a compressor of the type which was formerly made by Watson & Sons, London (Fig. 46.1). The information given in the legend of the figure should be enough to enable a competent mechanic to construct one. With this instrument, individual rotifers can be compressed just enough to hold them still without serious distortion. The author would not have been able to make Fig. 18.11 without this type of compressor. Preserved individuals can be oriented with a precision and ease not possible with any other method. Examination of many other kinds of organisms can be facilitated by the compressor—gastrotrichs, tardigrades, some protozoa, small arthropods, nematodes, etc. As far as the author knows, such an instrument is not now being manufactured by anyone, although the Rotocompressor has a similar function, but is without the valuable swinging arm feature.

In the absence of a compressor, resort must be made to less satisfactory but quite effective methods. Many littoral crustaceans and rotifers can be kept within sight of the high-power microscope by gentle compression under a coverglass, the cover being supported by bits of paper, chips of coverglass, cotton fibers, or other material. Much can be seen in wet mounts of finely divided leaves of aquatic plants, since many rotifers and crustaceans become trapped between leaves, or fasten themselves to surfaces. Another method is to use a water suspension of methyl cellulose. This very viscous material will entangle motile algae, protozoa, rotifers, nematodes, and other animals without greatly modifying their appearance or general behavior. An effective way to use the material is to make a ring about a half inch in diameter on a

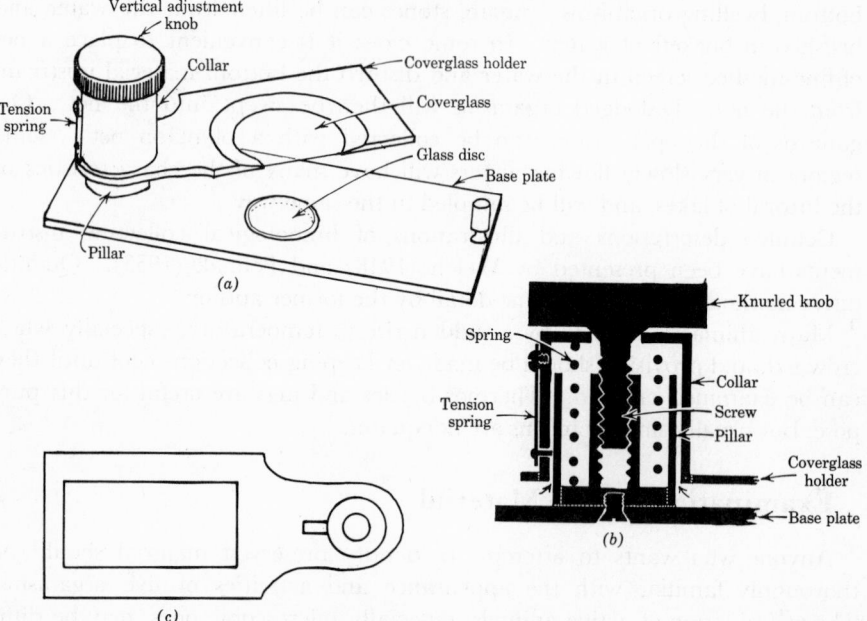

Fig. 46.1. Compressor of the type formerly made by Watson & Sons. (a) General view. In use the adjustment knob would be to the right for a right-handed person. (b) Vertical section of adjustment mechanism. (c) Suggested improved form of coverglass holder.

The essential pieces are a glass disc on which an organism is placed, and a coverglass which compresses the organism. The metal parts are arranged to hold the glass disc and to permit a precise adjustment of the coverglass. The disc is placed in a hole through a base plate (3 by 1 in.), its surface elevated somewhat above the base plate. The coverglass holder is attached to a collar which fits closely on a hollow pillar. The coverglass holder is lowered by turning a vertical adjustment knob with a screw which works in an element in the center of the pillar (b). When the knob is turned in a reverse direction, the coverglass holder is raised by a spring within the pillar. The collar turns on the pillar, being kept from turning too easily by a tension spring. In b, the space between collar and pillar is exaggerated (arrows). The space is filled by a film of oil. Soldered joints are cross-hatched. Myers (1936) described a more elaborate compressor with double collar which permits somewhat smoother action.

To use the compressor, an organism is placed on the disc in a small drop of water, and the coverglass is lowered by turning the knob until it touches the water. The coverglass is lowered further while the organisms are being observed with a microscope, until just the right degree of compression is achieved. If the coverglass holder is moved in the horizontal plane, the organism will be rolled over. After sufficient practice, one can learn to orient organisms with great precision, even flattened ones to some extent. Obviously, for best results, the coverglass must be parallel to the disc, and operation of the knob should not turn the coverglass holder in the horizontal direction.

One difficulty in using this compressor with a microscope that has a rotating nosepiece for objectives, is that the body of the microscope must be turned up to permit changing objectives; otherwise an objective will strike the adjustment knob. A suggested improvement (c) to eliminate this difficulty would involve making the coverglass holder longer, and adding a brace to keep the coverglass holder perpendicular to the collar.

slide, add a drop of water containing a concentration of organisms, and top with a coverglass. In time, many animals will be found trapped near the ring.

To pick up individual microscopic organisms requires a fine pipette. While a medicine dropper, drawn out to a fine point in a flame can be used, much more precise control is obtainable with one made as follows: A small (2 to 3 mm diameter) glass tube is drawn out to a fine point, the exact size being determined by the size range of the organisms to be handled. The other end of the tube is fitted with a 1-inch length of tightly fitting rubber tubing which is closed at its other end by a short length of glass rod. A battery of such pipettes of different sizes will permit easy manipulation of material. Another useful instrument is a fine, flattened wire loop about 0.5 mm in diameter, sold as Irwin loops. Organisms of moderate size are brought up to the surface and lifted out on the loop.

Relaxation

Many aquatic animals contract or become distorted when placed directly in preservatives. It is necessary to treat these animals with a relaxing agent that will inhibit sensitivity or muscular contraction. Cooling with ice water is effective with some animals, such as planarians.

To prevent contraction there are a number of substances that may be used with varying degrees of success. Use of these materials seems to be as much an art as a science, and one must be prepared to spend time in practicing varied methods. Time of action, condition of organisms, temperature, and other conditions will affect the reaction of a population to relaxing agents. In general, one tries to work with a large number of individuals, so that even a small fractional success will result in a satisfactory number of usable specimens. Usually the material is added dropwise to a small volume of water containing the animals. The animals are examined until they are observed to have stopped moving; before death, fixative is added. Relaxation may take minutes or hours, depending on the material. After fixation, the material should be stored in formalin or other preservative. An alternative method is to place samples of the population in a number of dishes and add various concentrations of narcotic. When animals have begun to die in the highest concentration, fixative is added to all the dishes. The following materials have been used much, and further information about them and others will be found in publications by Lee (1950), Pennak (1953), and Welch (1948). Some of them are narcotic or poisonous substances, and should be handled cautiously.

Cocaine. This material is frequently used in a concentration of 0.25 to 1 per cent in water or 10 per cent ethyl alcohol. The modification for gastrotrichs, using powdered cocaine (p. 408, footnote) may be adaptable to other groups.

Neosynephrin. This material, obtainable as the hydrochloride in 1 per cent solution at drugstores is best used as 0.1 to 0.5 per cent. Many rotifers, completely intractable to cocaine are easily fixed extended by proper use of neosynephrin. It seems to work best in acid waters, rather poorly in alkaline waters.

Acetone. Pure acetone added dropwise to small volumes of water has been found to be useable with some kinds of animals, such as rotifers and bryozoans.

Chloral hydrate. Used as a 10 per cent solution.

Chloretone. Used as 0.1 to saturation, about 0.8 per cent.

Clove oil. A few drops are scattered on the surface of water in a finger bowl. For smaller vessels, use less. After fixation, the animals should be rinsed to remove the oil.

Magnesium chloride. Used as 2.5 per cent solution of the hexahydrate in tap water.

Magnesium sulfate. Used as 20 per cent of the hydrated form.

Menthol. A saturated solution is added dropwise to animals in a small amount of water. A mixture of equal parts of menthol and chloral hydrate may also be used.

Nickel sulfate. A 1 per cent solution paralyzes the cilia of protozoa.

Potassium or Sodium iodide. A 1 per cent solution prevents contraction of the myonemes of *Stentor* and presumably other ciliates.

Urethane. This material may be applied in solutions or by scattering crystals on the water.

For further description of methods of relaxing, see Chapter 13, 18, 19, and 23.

Fixation and Preservation

Although one must study live organisms, most identification is done from preserved material since it is generally impracticable to keep an entire collection alive until it is thoroughly examined, and since many organisms must be dead before the necessary dissections can be made. The foundation of a museum collection is properly preserved specimens. Many animals can be satisfactorily preserved for taxonomic work by placing them in 4 per cent neutralized formaldehyde (10 per cent commerical formalin) or 70 per cent alcohol. Special preservatives are used for material which is to be sectioned (Lee, 1950). Although alcohol is better for most arthropods, most plankton collections are well preserved simply by adding formaldehyde. Alcohol collections are too easily stirred by convection currents when examined in shallow dishes under a microscope. A 0.25 per cent solution of osmic acid is often used after relaxation when especially fast fixation is needed, as with rotifers.

The hot water method (p. 433) works very well with many rotifers, although a certain amount of practice is required before full success will be achieved. Animals fixed with hot water should be preserved in formalin, or in some other fixative appropriate to the purpose for which the material is desired. This method may be useful with other organisms as well, such as contractile protozoa, gastrotrichs, and the like.

Acid Lugol's solution preserves algae and other microorganisms very well (10 gm iodine, 20 gm potassium iodide, 20 gm glacial acetic acid, 200 gm distilled water. This formula was kindly supplied by Wilhelm Rodhe). About 1 ml of preservative is added to each 100 ml of sample.

If material is to be stored in vials, a little glycerine should be added to reduce or prevent damage if the preservative evaporates.

Mounting

Often permanent whole mounts are made in balsam, damar, euparol, Hoyer's medium or other mounting media, and the organisms are generally cleared and stained before mounting. These techniques require explanation in a detail beyond the scope of this book. Fortunately they are well described by Lee (1950) and Pantin (1948).

Many small organisms such as rotifers, gastrotrichs, tardigrades, small cladocera, and copepods are easily mounted permanently in glycerine as follows: First put the specimens into 10 per cent glycerine in a watch glass or similar vessel and allow the water to evaporate, which will leave the objects in pure glycerine. For animals with especially impenetrable cuticle, such as nematodes, use 5 per cent glycerine. To dehydrate the glycerine completely, the vessel should be warmed or placed in a desiccator. Then place a small drop of pure glycerine in the center of a clean slide. Transfer to it whatever objects are to be mounted. The organisms can be lifted out of glycerine on the end of a fine (No. 00) insect pin or an Irwin loop. Glycerine material should not be handled with a pipette because it will stick inside. With the pin, push the organisms down against the glass, or they will be displaced when the coverglass is put on. Place three small pieces of broken coverglass around the drop to support the cover. (Small pieces of gummed paper may also be used for thinner mounts. They can be attached to the slide ahead of time.) Lower the coverglass onto the drop gently, using a pin. It is best to use a small round cover. The mount is completed by allowing cement to run in under the cover from a glass rod, a small amount at a time. The slides should be stored in a horizontal position.

The use of glycerine as a medium in which to make dissections of small arthropods is widespread. Another useful medium is pure lactic acid, in which an appropriate amount of methyl blue is dissolved (approximately 2 to 4 mgm in 10 ml). Lactic acid is a syrupy liquid which has just the right viscosity to facilitate dissection of small parts. Organisms may be transferred directly from aqueous media. The muscles are cleared while the cuticle becomes stained, rendering the details of segmentation and setation plainly visible. Lactic acid will eventually evaporate, and to make a permanent mount, the preparation must be very securely sealed, or the parts must be transferred to some other medium, such as Hoyer's. For sealing, Lactoseal or certain fingernail polishes are suitable. For information on mounting media, in addition to the general works cited at the end of this chapter, see Baker and Wharton (1952), cited on page 1081.

Murrayite cement has been much used for sealing mounts. It eventually sets to a very hard, brittle, but strong consistency. Another useful cement is Zut slide ringing compound. This material is said never to get brittle. A method using two coverglasses and glycerine jelly is described in Chapter 27, and further directions for the use of glycerine are given in Chapter 15. Organisms which must be viewed from both dorsal and ventral sides may be mounted on a 25 mm square coverglass instead of on a slide, and held in a thin sheet-metal or heavy cardboard frame.

Sectioning

A discussion of the techniques of sectioning is out of place in this brief outline; see the books by Lee (1950), Pantin (1948), and Gray (1954).

Measurement

In some groups, size must be measured. For microscopic organisms an eyepiece micrometer is usually used. In the absence of this instrument, size can be estimated closely enough for many purposes by measuring the diameter of the microscope field with a stage micrometer and using it as a standard of reference to which organisms are compared.

References

American Society of Limnology and Oceanography. 1949. *Sources of Limnological and Oceanographic Apparatus and Supplies,* Special Publication No. 1, Revised. (Obtainable from the secretary of the Society.) **Baylor, E. R. and F. E. Smith. 1953.** A physiological light trap. *Ecol.,* 34:223–224. **Creitz, G. L. and F. A. Richards. 1955.** The estimation and characterization of plankton populations by pigment analysis. III. A note on the use of "Millipore" membrane filters in the estimation of plankton pigments. *J. Marine Research Sears Foundation,* 14:211–216. **Deevey, Edward S. Jr. and James S. Bishop. 1948.** Limnology. In: A fishery survey of important Connecticut Lakes. *Conn. State Geol. and Nat. Hist. Survey. Bull.,* 63:69–121. **Galigher, A. E. 1934.** *The essentials of Practical Microtechnique in Animal Biology.* Published by the author, Berkeley. **Gray, P. 1954.** *The Microtomist's Formulary and Guide.* Blakiston, New York. **Guyer, M. F. 1953.** *Animal Micrology,* 5th ed. University of Chicago Press, Chicago. **Lee, A. B. 1950.** *The Microtomist's Vade Mecum,* 11th ed. J. B. Gatenby and H. W. Beams (eds.). London, J. and A. Churchill. **Myers, F. J. 1936.** Mounting rotifers in pure glycerine. *J. Quekett Microscop. Club,* Ser. 3, 1:1–9. (Ed. Note: dioxane is toxic and should be used only with great caution.) **Pantin, C. F. A. 1948.** *Notes on Microscopical Technique for Zoologists.* Cambridge University Press, Cambridge. **Pennak, R. W. 1953.** *Freshwater Invertebrates of the United States.* Ronald, New York. **Rawson, D. S. 1930.** The bottom fauna of Lake Simcoe and its role in the ecology of the lake. *Publ. Art. Fisheries Research Lab.,* 40:1–183. **Richards, O. W. and W. B. Krabek. 1954.** Visibilizing microorganisms on membrane filter surface. *J. Bact.,* 67:613. **Tartar, V. 1950.** Methods for the study and cultivation of Protozoa. In: *Studies Honoring Trevor Kincaid,* pp. 104–107. University of Washington Press, Seattle. **1957.** Reactions of Stentor coeruleus to certain substances added to the medium. *Exp. Cell. Res.* 13:317–332. **Wagstaffe, R. and J. H. Fidler. 1955.** *The Preservation of Natural History Specimens.* Vol. 1, *Invertebrates.* H. F. & G. Witherby, London. **Welch, P. S. 1948.** *Limnological Methods.* Blakiston, Philadelphia.

SOURCES OF MATERIAL MENTIONED

Hoyer's medium. Ward's Natural Science Establishment, P. O. Box 24, Beechwood Station, Rochester 9, New York. (Various modifications have been published.)
Irwin loops. W. M. Welch Scientific Co., Chicago. Catalog No. 8190.
Lactoseal. Edward Gurr, Ltd., 42 Upper Richmond Road West, London, SW14, England.
Millipore filters. Millipore Filter Corporation, Watertown 72, Massachusetts. (Also a brochure with extensive bibliography.)
Murrayite Cement. Arthur H. Thomas Co., Philadelphia.
Rotocompressor. Biological Institute, 2018 North Broad Street, Philadelphia 21, Pennsylvania.
Zut. Bennett Glass and Paint Co., 2131 South Second West, Salt Lake City, Utah.

Index

The main purpose of this Index is to aid in locating the names of organisms in the keys. All taxa identified in the keys are indexed. In keys to genera, species cited merely as examples are not indexed, but if all the species in a certain genus are cataloged, they are indexed. Species are listed under the genus to which they belong, and are not listed independently.

The first number after the name of an organism gives the page on which it keys out, and if it keys out on more than one page, all are listed. If the illustration in the key is on a different page, that is given next; other references follow in numerical order. Page numbers of references of secondary importance are italicized.

References to organisms in the introductory text sections are not indexed unless they present material not included in the keys that will help in identification. All generic names in illustration legends are indexed.

In general, the only morphological terms indexed are names of specialized structures that occur in relatively few of the groups of organisms included in the book, or that have special significance in the identification of members of the groups. These terms are indexed only when the references aid in the use of the key.

Abdominal processes, Cladocera, 593
Abdominal setae, Cladocera, 594
Abedus, 966, *960*
Abreptor, Cladocera (postabdomen), 593
Abrochtha, 489, 486, *437*
Acanthocephala, *368*
Acanthocyclops, subgenus of *Cyclops*, 802
Acanthocystis, 262
Acanthodiaptomus, 756: *denticornis*, 783, 781, *739*
Acantholeberis, 628: *curvirostris*, 628
Acanthometropus, 915
Acanthomysis awatchensis, 879
Acanthosphaera, 133, 132
Acari, 1080

Accessory flagellum of first antenna, Malacostraca, 870
Acella, 1127
Acentropus, 1053, 1054, *1051: niveus*, 1050
Acenus, 1002
Acercus, 1104, 1107
Acetabula, Acari, 1083
Achlya, 79, 77
Achlyogeton, 53
Achlyogetonaceae, 51
Achnanthaceae, 174
Achnanthes, 179, *174*
Achnanthoideae, 174
Achromadora, 393

Achromatium, 36, 37, 45: oxaliferum, 37, 43; volutans, 37, 43

Achtheres, 867, 864: ambloplitis, 864; beani, 864; bicauliculata, 864; californiensis, 864; carpenteri, 864; coregoni, 864; corpulentus, 864; edwardsii, 864; extumescens, 864; falculata, 864; gibber, 864; inermis, 864; lacae, 864; micropteri, 864; oquassa, 864; pimelodi, 864; salmonea, 864; salvelini, 864; siscowet, 864; thymalli, 864; wisconsinensis, 864

Acilius, 996, 1011

Acineria, 273, 272

Acineta, 296, 295

Acinetidae, 294, 295

Acneus, 1015, 1014, 985

Aconchulina, 251

Acroneuria, 953, 948, 946

Acroperus, 638, 637: harpae, 638

Acropisthium, 271

Actidesmium, 135, 134

Actinastrum, 139, 138

Actinella, 179, 174

Actinobdella, 551: annectens, 551; inequiannulata, 551; triannulata, 551

Actinobolina, 270

Actinobolinidae, 268, 270

Actinolaiminae, 398

Actinolaimus, 398

Actinomonas, 193, 197

Actinomycetes, 12, 14, 19, 44

Actinonaias, 1154

Actinophrydia, 260

Actinophrys, 260

Actinopoda, 232, 233

Actinosphaerium, 260

Actinospora, 91

Acumen, 871

Acyclus, 480, 481, 484, 438

Adenodactyls, Turbellaria, 327

Adenophorea, 380

Adephaga, 981

Adfrontal sutures, Lepidoptera, 1051

Adhesive organ, Turbellaria, 326

Adineta, 489, 439, 437

Adinetidae, 485, 437

Adoral zone, Ciliophora, 265

Adorus, 397

Aedes, 1066

Aeolosoma, 528: headleyi, 528; hemprichi, 528, 529; leidyi, 528; niveum, 528; quaternarium, 528; tenebrarum, 528; variegatum, 528

Aeolosomatidae, 528

Aeshna, 931, 934, 929, 930

Aeshnasoma, 1061

Aeshnidae, 919, 922, 931

Agabetes, 995

Agabinus, 994, 1011, 1008, 1009

Agabus, 994, 1011

Agapetus, 1036, 1027

Agathon, 1060

Aglaodiaptomus, subgenus of Diaptomus, 739, 756, 772

Agmenellum, 98: quadruplicatum, 98; thermale, 98; wichurae, 98

Agraylea, 1039, 1038

Agrion, 926, 923, 925

Agrionidae, 922, 924

Agrypnia, 1041, 1040

Alaimidae, 396

Alaimus, 396

Alaocharis, subgenus of Palaemonetes, 880

Alasmidonta, 1145

Alatospora, 90

Albertia, 460, 437

Albia, 1089, 1105

Alderflies, 904

Algae, 8, 14, 115, 190, 165: classification, 117; color, 116

Allocapnia, 950, 951, 953

Allocrangonyx, 877: pellucidus, 877

Alloeocoela, 325, 326, 359

Allogromiidae, 259

Alloionema, 386

Alloinematinae, 386

Allomyces, 72,; life cycle, 49

Allonais, 531: paraguayensis, 531

Allonarcys, 947, 948

Alloperla, 956, 952, 953, 955

Alluaudomyia, 1068

Alona, 639: affinis, 642; costata, 642, 641; guttata, 640, 641, 642; intermedia, 643, 642; karau, 641, 653; monacantha, 641; quadrangularis, 642; rectangula, 643, 642; rectangula var. pulchra, 642

Alonella, 643, 645, 649: acutirostris, 654; dadayi, 654; dentifera, 644, 653; diaphana, 640, 653; excisa, 635, 643; exigua, 635, 643; globulosa, 652; nana, 654, 648, rostrata, 654

Alonopsis, 637: aureola, 637; elongata, 637

Aloricata, 291

Alternanthera philoxeroides, 1179, 1178

Alula, Rotifera, 428, 429

Alveola, Bacillariophyceae, 175

Amastigogenina, 234

Amastigomonas, 196

Ambrysus, 967, 961

Ameletus, 915, 910

Ameridae, 816

Ameronothrus, 1113

Ametor, 1000

Ametropodinae, 913
Ametropus, 913, 910
Amnicola, 1133
Amnicolidae, 1123
Amoeba, 235, 232
Amoebaea, 234
Amoebobacter, 32, 44
Amoebochytrium, 62, 63
Amoeboid protozoa, 14
Amphiagrion, 924, *926*, 923
Amphibolella, 343: *virginiana*, 343
Amphicampa, 177, *174*
Amphichaeta, 530: *americana*, 530
Amphichrysis, 155, 154
Amphidelus, 397
Amphidial glands, Nemata, 369
Amphidinium, 163, 161
Amphids, Nemata, 369
Amphigyra, 1131
Amphileptidae, 272
Amphileptus, 273, 272
Amphimonas, 218
Amphinemura, 948, 949
Amphipleura, 184, *174*
Amphipoda, 872, 873
Amphipods, 871
Amphiprora, 182, *174*
Amphiproroideae, 174
Amphisiella, 288, 289
Amphithrix, 103: *janthina*, 103
Amphitrema, 252
Amphitremidae, 252
Amphizoa, 990, 1010
Amphizoidae, 982, 983, 990, 1010
Amphizonella, 237
Amphora, 186, *174*
Ampulla, Nemata, 370
Ampullariidae, 1123
Ampumixis, 1005, 1006, 1019, *987*
Anabaena, 102, 103: *bornetiana*, 103; *catenula*, 103; *circinalis*, 103; *flos-aquae*, 103; *inaequalis*, 103; *oscillarioides*, 103, 102; *sphaerica*, 103; *unispora*, 103; *variabilis*, 103
Anacaena, 1001, 1012
Anacharis, 1178: *canadensis*, 1178; *densa*, 1178; *occidentalis*, 1178, 1174
Anacroneuria, 952
Anacystis, 97: *cyanea*, 97; *dimidiata*, 97; *incerta*, 97; *marina*, 97; *montana*, 97; *thermalis*, 97
Anadontoides, 1148
Anagapetus, 1036
Anal operculum, Harpacticoida, 819
Anal spines, Cladocera, 594
Anaplectus, 392, 391
Anapus, 442

Anarma, 295
Anarthra, 440, 439
Anatonchus, 395
Anatopynia, 1069
Anax, 931, *934*
Anchistrops, 648: *minor*, 648
Anchodemus, 1007
Anchycteis, 1020, *985*
Anchytarsus, 1020, *985*
Ancylidae, 1125, 1130, *1123*
Ancylistaceae, 50
Ancylistes, 50, 51, *49*
Ancyromonas, 198
Ancyronyx, 1003, 1017, 1014, *987*
Ancystropodium, 289
Anepeorus, 914
Anguillospora, 93, 92
Animal, definition, 7
Anisocentropus, 1046, 1047
Anisogammarus, 877: *oregonensis*, 877; *ramellus*, 877
Anisonema, 127
Anisoptera, 920, 917, 918–919
Anistisiellidae, 1098
Ankistrodesmus, 137, 136
Annulus ventralis, 871
Anodocheilus, 992
Anodonitinae, 1144
Anodonta, 1149
Anohydrachna, 1093
Anomalagrion, 926
Anomoeoneis, 184, *174*
Anomopoda, 599, *588*
Anomopus, 484, *437*
Anonchus, 389
Anopheles, 1064, 1065
Anostraca, *558*, *559*, 587
Antarcella, 238
Antenna: Anostraca, 559; Cladocera, 590; Conchostraca, 577, 578; Copepoda, 735; Crustacea, 558; Insecta, 903; Lepidoptera, 1051; Malacostraca, 870; Ostracoda, 659; Rotifera, 422
Antennal appendage, Anostraca, 559
Antennule (see Antenna)
Anthomyiidae, 1079
Anthonema, 390
Anthophysa, 227, 40
Antocha, 1061, 1062
Antrum, Turbellaria, 327
Anuraeopsis, 441, *437: fissa*, 441
Apella, 1137
Aphanocapsa, 97
Aphanochaete, 147, 146
Aphanodictyon, 79, 78

Aphanolaiminae, 386
Aphanolaimus, 386, 387
Aphanomyces, 79, 78
Aphanomycopsis, 77
Aphanothece, 96
Aphanizomenon, 102: *flos-aquae*, 102; *holsaticum*, 102; *ovalisporum*, 102
Aphelenchoidea, 380
Aphelenchoides, 381
Aphelenchus, 381
Aphrosylus, 1076
Aphrothoraca, 260
Apical axis, Bacillariae, 175
Apiocystis, 131, 130
Aplexa, 1126
Apobaetis, 916
Apocrangonyx, 878: *lucifugus*, 878; *subtilis*, 878
Apodachlya, 82, 80
Apodachlyella, 82, 81
Apophyses, cuticular, Tardigrada, 509
Apostemidium, 87, 86: asci, 49
Apsilus, 447, *438*
Apteraliplus, 990, 1010
Apus, 574 (see also *Triops*): *aequalis*, 574; *biggsi*, 574; *glacialis*, 574; *lucasanus*, 574; *newberryi*, 574; *oryzaphagus*, 574
Arachnochloris, 153, 150
Araiospora, 82
Araphidineae, 174
Arcade, Nemata, 370
Arcella, 238: *artocrea*, 239; *dentata*, 239; *discoides*, 239; *megastoma*, 238; *mitrata*, 239; *polypora*, 238; *vulgaris*, 239
Arcellidae, 236
Archilestes, 924, 923
Arcidens, 1145
Arcticocamptus, subgenus of *Bryocamptus*, 851
Arctocorisa, 970, *961*
Arctodiaptomus, subgenus of *Diaptomus*, 739, 772, 773, 776
Arctopsyche, 1035, 1034
Arcynopteryx, 954, 947, 948, 954, *946*
Areola: crayfish, 870; Hemiptera, 964
Argia, 924, 921, 925, *926*
Argulus, 862, 863: *japonicus*, 863; *trilineatus*, *863*
Argyra, 1076
Arigomphus, 928, *929*, 933
Aristerostoma, 280, 279
Arkansia, 1144
Armiger, 1128
Arrenurae, 1107
Arrenuridae, 1107
Arrenurus, 1091, 1107, 1106
Artemia, 566, *561*: *salina*, 560, 566, 567
Artemiidae, 566, *561*

Artemiopsis, *561*: *bungei*, 567; *stephanssoni*, 566, 567; *stefanssoni* var. *groenlandicus*, 567
Arthrodesmus, 143, 142
Arthroplea, 914, 910
Arthropoda, 558
Arthrospira, 114: *gomontiana*, 114; *jenneri*, 114; *khannae*, 114; *platensis*, 114
Arthrotardigrada, 513
Articulospora, 91, 90
Arzama, 1052, 1054
Aschelminthes, 368, 406, 420
Asci, Ascomycetes, 49
Ascocarps, Ascomycetes, 49
Ascoglena, 125, 124
Ascomorpha, 442, 445, 446, *437*: *ecaudis*, 442, 446; *minima*, 442
Ascomorphella, 445, 446, 437: *volvocicola*, 445
Ascomycetes, *11*, *14*, 49, 87
Ascophorinae, 348
Asellidae, 873
Asellus, 875: *acuticarpus*, 875; *adentus*, 875; *alabamensis*, 875; *antricolus*, 875; *attenuatus*, 875; *brevicaudus*, 875; *californicus*, 875; *communis*, 875; *dentadactylus*, 875; *dimorphus*, 875; *hobbsi*, 875; *intermedius*, 875; *macropropodus*, 875; *militaris*, 875; *montanus*, 875; *nickajackensis*, 875; *oculatus*, 875; *packardi*, 875; *pricei*, 876; *smithi*, 875; *spatulatus*, 876; *stiladactylus*, 876; *stygius*, 875; *tomalensis*, 875; *tridentatus*, 875
Askenasia, 271
Askenasyella, 131, 130
Aspelta, 461, *437*
Aspidiophorus, 413
Aspidisca, 290
Aspidiscidae, 286, 290
Asplanchna, 446, *437*: *brightwelli*, 446, 447; *girodi*, 446; *herricki*, 446; *intermedia*, 446; *priodonta*, 446; *sieboldi*, 446; *silvestri*, 446
Asplanchnidae, 437
Asplanchnopus, 463, 462, *437*
Assulina, 256: *muscorum*, 256; *seminulum*, 256
Astacidae, 879
Astacinae, 883
Astasia, 125, 124, *10*
Astenophylax, 1044
Asterionella, 178, *174*, 54
Asterococcus, 131, 130
Asterocytis, 167, 168
Asteromeyenia plumosa, 311, 312: *radiospiculata*, 311, 312
Asterophlyctis, 62, 61
Astramoeba, 235
Astrodisculus, 261
Astrophrya, 297, 296

Astrosiga, 210
Astylozoon, 291
Astylozoonidae, 291
Atax, 1102, 1107
Athalamia, 259
Atherix, 1074, 1075
Athripsodes, 1049, 1048
Atoperla, 951, 952
Atopodinium, 286
Atopsyche, 1035, 1036, 1027
Atracheata, 1161
Atracterlmis, 1005, *987*
Atractides, 1087, 1101, 1107
Atrichopogon, 1069
Atrochus, 482, 484, *438*, *446*
Attaneuria, 953
Attheya, 186, *174*
Attheyella, 816, 840, 844: *alaskaensis*, 845; *americana*, 849; *bicolor*, 838; *carolinensis*, 847; *dentata*, 848; *dogieli*, 848; *hyperboreus*, 845; *idahoensis*, 844; *illinoisensis*, 846, 845; *nordenskioldii*, 845; *northumbrica*, 848; *obatogamensis*, 849; *pilosa*, 847; *subarctica*, 852; *wierzejskii*, 849
Attheyella, subgenus of *Atthyella*, 844
Aturus, 1088, 1105
Atyidae, 879
Atylenchus, 384
Audouinella, 169, 168
Aulacomnium, 1166
Aulolaimoides, 398
Aulolaimus, 392
Aulomonas, 199
Aulophorus, 530: *furcatus*, 530, 531; *vagus*, 531
Aulosira, 105: *implexa*, 105, 104; *laxa*, 105
Auricles: Rotifera, 426; Turbellaria, 326
Awerintzewia, 248
Axial area, Bacillariophydae, 175
Axonolaimidae, 392
Axonopsae, 1105
Axonopsidae, 1105
Axonopsis, 1088, 1105
Azolla caroliniana, 1177, 1175
Azotobacter, 29, 44

Bacillaria, 181, 180, *174*
Bacillariales, 174
Bacillariophyceae, *117*, 122, 171, 174
Bacilli, 17
Bacillus, 44
Back swimmer, 966
Bacteria, 8, 16: ecology, 27 ff.; iron, 38 ff.; relationships, 12, 22; stalked, 30; sulfur, 14, 31–37; unstalked, 31
Bacteriochlorophyll, 14, 33
Bacterium, 17

Bactrurus, 878: *brachycaudus*, 878; *hubrichti*, 878; *mucronatus*, 878
Baetidae, 913, *908*
Baetis, 916
Baetisca, 911, 909
Baetiscidae, 911
Baetodes, 916
Bagous, 1007
Balanonema, 281
Balantidioides, 283
Balatro, 460, *437*
Balladyna, 288
Bangia, 167, 168
Bangiaceae, 167
Bangiales, 167
Bangioideae, 167
Banksiola, 1041, 1040
Basal spines, Cladocera, 594
Basiaeschna, 931, 930, 932, 933, *934*
Basicladia, 149, 148
Basidiomycetes, 11, 14
Basipod, copepod, 736
Basipodite, copepod, 736
Bastiania, 386, 388
Bastianiidae, 386
Bathyodontus, 394
Batrachobdella, 548: *paludosa*, 548; *phalera*, 548; *picta*, 548, 549
Batrachospermum, 167, 168
Bdellodrilus, 527: *illuminatus*, 527
Bdelloida, 437, 483, 484
Bdelloidea, 438, *437*
Beatogordius, 404
Beauchampia, 476, 477, *438*
Beetles, 903, 981
Beggiatoa, *20*, 35, 36, 44: relation to Oscillatoria, 12
Behningiidae, 911
Bellura, 1052, 1054, *1051*
Belondiridae, 398
Beloneuria, 953
Belonia, 939
Belonolaimus, 381
Belostoma, 966, 960
Belostomatidae, 965, *960*
Benacus, 965, *960*
Beraea, 1049, 1027
Beraeidae, 1027, 1049
Bernardinium, 163, 161
Berosus, 999, 1012, *983*
Bezzia, 1068
Bibiocephala, 1060
Bicoeca, 201
Biddulphia, 187, 188, *174*
Biddulphiaceae, 174

Biddulphineae, 174
Biddulphioideae, 174
Bidens beckii, 1180, 1171
Bidessonotus, 992
Bidessus, 992, 1010
Billbugs, 1007
Binuclearia, 145, 144
Bipalpus, 453
Biraphidineae, 174
Birgea, 465, *437*
Birgeidae, 437
Bitrichia, 159, 158
Bittacomorphella, 1073
Bittacomorphinae, 1073
Bivalves, 1117
Bizone, 280, 279
Blastocaulis, 18, *30*
Blastocladia, 72, 71
Blastocladiaceae, 71
Blastocladiales, 70, *11:* habitat, 48
Blastocladiella, 71
Blastocladiopsis, 72, 71
Blastodiniaceae, 165
Blepharisma, 284
Blepharocera, 1059
Blepharoceridae, 1059
Blepharocerinae, 1059
Blepharoplast, 191
Blood worms, 1058
Blue-green algae, 118, *8, 12*, 95, *117*
Blyttiomyces, 55, 54
Bodo, 217
Bodopsis, 195
Bohlinia, 137, 138
Bopyridae, 873
Bosmina, 624: *coregoni*, 625; *longirostris*, 624, 625
Bosminidae, 604, *588*, 589
Bosminopsis, 624: *deitersi*, 624
Bothrioplana, 364: *semperi*, 364
Bothrioplanidae, 364
Bothromesostoma, 353: *personatum*, 353
Botrydiaceae, 155
Botrydiopsis, 151, 150
Botrydium, 155, 152
Botryococcus, 155, 152
Boyeria, 934, 930, 933
Brachelyma, 1168, *1169*
Brachionidae, 437
Brachioninae, 437
Brachionus, 451, *437*, 439, 454, 469: *dimidiatus* var. *inermis*, 454; *tridens*, 454
Brachybambus, 1007
Brachycentridae, 1028, 1041
Brachycentrus, 1041, 1028, 1029, 1042
Brachycera, 1074

Brachycercus, 913
Brachydeutera, 1077
Brachymesia, 937, *939*
Brachymetra, 963, 959
Brachypoda, 1088, 1105
Brachyptera, 948, *946*
Brachythecium, 1168, *1169*
Brachyvatus, 992
Bradyscela, 489, 437
Branchinecta, *564, 561:* coloradensis, 566, 565, 567; *cornigera*, 566; *gigas*, 565, *560; lindahli*, 565, 566; *mackini*, 565; *occidentalis*, 566; *packardi*, 565; *paludosa*, 564; *shantzi*, 565, *561*
Branchinectidae, 564, *561*
Branchinella, 566, *561*: *alachua*, 566; *gissleri*, 568; *lithaca*, 566, 567
Branchiobdella, 524: *americana*, 524; *tetradonta*, 524
Branchiobdellidae, 524, *543*
Branchioecetes, 273
Branchiopoda, 558, 587, 589, *572*
Branchipus gelidus, 567
Branchiura (Crustacea), 862, *559*
Branchiura (Oligochaeta), 534, 535: *sowerbyi*, 534
Braseria schreberi, 1185, 1172
Brebissonia, 185, 174
Brechmorhoga, 936
Bresslaua, 277
Brevilegnia, 81, 80
Brillia, 1072
Bristles tactile, Gastrotricha, 406
Brood chamber, Cladocera, 594
Brood pouch, mysid, 869
Brown algae, *117*
Bryceela, 465, *437*
Brychius, 990, 1009
Bryhnia, 1166
Bryocamptus, 830, 833, 838, 840, 851, *816:* alleganiensis, 850; *arcticus*, 852, 853; *australis*, 855; *cuspidatus*, 851; *douwei*, 854; *frigidus*, 850; *hiatus*, 855; *hiemalis*, 853; *hiemalis brevifurca*, 854; *hutchinsoni*, 857; *minnesotensis*, 857; *minusculus*, 856; *minutiformis*, 856; *minutus*, 856, 857; *morrisoni*, 853; *newyorkensis*, 855; *nivalis*, 854; *pygmaeus*, 851; *simplicidentata*, 857; *subarcticus*, 851; *tikchikensis*, 852; *umiatensis*, 855, 856; *vejdovskyi*, 856; *washingtonensis*, 857; *zschokkei*, 850, 851
Bryocamptus, subgenus of *Bryocamptus*, 840
Bryochoerus, subgenus of *Echiniscus*, 514
Bryodelphax, subgenus of *Echiniscus*, 514
Bryometopus, 283
Bryophrya, 277
Bryophyllum, 273, 272
Bryophyta, 1161
Bryophytes, *8*, 1161

Bryozoa, 495
Buccal capsule, Nemata, 370
Buccal cavity, Ciliophora, 265
Buds, Bryozoa, 497
Buenoa, 967, *960*
Bugs, 958, *903*
Bulbochaete, 149, 148
Bulimidae, 1125, 1133
Bulimnaea, 1127
Bulimus, 1133
Bullella, 1146
Bullinularia, 241
Bumillaria, 154, 152
Bumilleriopsis, 153, 150
Bunonema, 385
Bunonematinae, 385
Bunops, 630: *serricaudata*, 630
Bursaria, 285, 284
Bursaridium, 285, 284
Bursariidae, 284, 283

Cabomba caroliniana, 1180, 1171
Caddisflies, *903*, 1024
Caecum, Cladocera, 593
Caenestheria, *579*, *580*
Caenestheriella, 583, *579: belfragei* 584; *gynecia*, 584; *setosa*, 584
Caenestheriidae, 579
Caenidae, 912
Caenis, 913, 910
Caenomorpha, 283
Calacanthia, 964
Calamoceratidae, 1029, 1046
Calanoida, 738, 735: characteristics, 737
Calineuria, 953, 952
Callidina, 485
Callibaetis, 916, 910
Callicorixa, 970, *961*
Calliergon, 1167, *1169*
Calliergonella, 1164, 1165, *1169*
Calliperla, 957
Callitriche, 1179: *hermaphroditica*, 1179; *heterophylla*, 1179; *palustris*, 1179
Calohypsibius, subgenus of *Hypsibius*, 517
Caloneis, 184, *174*
Calonyx, 1095, 1107
Calopsectra, 1070
Calopsectrini, 1070
Calopterygidae, 920, 923
Calopteryx, 923, 921
Calothrix, 104: *adscendens*, 104; *juliana*, 104; *parietina*, 104
Calyptomera, *588*
Calyptotricha, 282
Calyptralegnia, 80, 78

Camacolaimidae, 386
Cambarellus, 883: *diminutus*, 884; *ninae*, 883, 884; *puer*, 883, 884; *schmitti*, 884; *shufeldtii*, 883, 884
Cambarinae, 883
Cambarincola, 526, 525: *branchiophila*, 526; *chirocephala*, 526; *elevata*, 526; *floridana*, 526; *gracilis*, 526; *inversa*, 526; *macbaini*, 526; *macrocephala*, 526; *macrodonta*, 526; *philadelphica*, 526; *vitrea*, 526
Cambarus, 883: *asperimanus*, 897; *bartonii*, 898, 896; *brachydactylus*, 897; *byersi*, 896; *cahni*, 895; *carolinus*, 897; *cornutus*, 895; *cristatus*, 897; *cryptodytes*, 895; *diogenes*, 897; *dissitus*, 896; *distans*, 898; *extraneus*, 896; *floridanus*, 897; *fodiens*, 896; *friaufi*, 897; *hamulatus*, 895; *hedgpethi*, 896; *hubbsi*, 895; *hubrichti*, 895; *latimanus*, 897; *longulus*, 898; *mononglaensis*, 897; *montanus*, 898; *obeyensis*, 897, 896, *883*; *ortmanni*, 896; *parvoculus*, 897; *reduncus*, 897; *robustus*, 898; *rusticiformis*, 895; *sciotensis*, 898; *setosus*, 895; *spicatus*, 896; *striatus*, 897; *tenebrosus*, 898; *uhleri*, 896
Camerostome, Acari, 1083
Camisiidae, 1110
Campanella, 292
Campascus, 252
Campeloma, 1132
Campsopogan, 167, 168
Campsurinae, 911
Campsurus, 913, 910
Camptocercus, 636: *macrurus*, 636; *oklahomensis*, 636; *rectirostris*, 636, 637
Camptothecium, 1166, 1169
Campydora, 398
Campydorinae, 398
Campylodiscus, 181, 180, *174*
Canal raphe, Bacillariophyceae, 175
Candelabrum, 93
Candocypria, 665: *osborni*, 665
Candocyprinae, 664
Candona, 677: *acuta*, 684; *albicans*, 682, 683; *annae*, 681, 682; *balatonica*, 682; *biangulata*, 682, 683; *candida*, 688; *caudata*, 686; *crogmaniana*, 679, 678; *decora*, 687, 688; *delawarensis*, 682; *distincta*, 687; *elliptica*, 680, 681; *eriensis*, 684, 685; *exilis*, 680; *fluviatilis*, 679; *fossulenis*, 687; *foveolata*, 681; *hyalina*, 677; *indigena*, 684; *inopinata*, 686; *intermedia*, 679, 678; *jeanneli*, 680, 681; *marengoensis*, 680; *ohioensis*, 678; *parvula*, 679; *peirci*, 680; *punctata*, 682; *recticauda*, 688; *reflexa*, 680; *scopulosa*, 683; *sharpei*, 685; *sigmoides*, 687; *simpsoni*, 680, 681; *stagnalis*, 687, 686; *subgibba*, 683; *suburbana*, 685; *truncata*, 688; *uliginosa*, 678

Candonini, 665

Candonocypris, 701: pugionis, 701; serrato-marginata, 702

Candonopsis, 676: kingsleii, 676, 677

Cannacria, 937, 935, 939

Canthocamptidae, 816

Canthocamptus, 840, 816: assimilis, 843; douweanus, 854; oregonensis, 841; robertcokeri, 842; sinuus, 842; staphylinoides, 841, 842; staphylinus, 840; vagus, 844

Canthyria, 1144

Capitulum, Acari, 1083

Capnia, 950, 953, 946

Capniinae, 949

Capsellina, 258

Capsosira, brebissonii, 99

Capsules, bacterial, 18

Carapace, Cladocera, 589, 591: Malacostraca, 870; measurement of length, crayfish, 871

Carborius, 1045, 1043

Carchesium, 293

Cardiocladius, 1072

Caridea, 879

Caridean shrimps, 871

Carinifex, 1129

Carpus, Malacostraca, 870

Carrhydrus, 994

Carteria, 129, 126

Carterius, 303: latitentus, 310; tentaspermus, 311; tubispermus, 310

Caryophanon, 19

Caspihalacarus, 1108

Castrada, 346: hofmanni, 346; lutheri, 347; vir-giniana, 346, 347

Castrella, 342: graffi, 342, 343; marginata, 343; pinguis, 342

Cataclysta, 1052, 1053, 1051

Cataractocoris, 967

Catenaria, 71, 70

Catenariaceae, 71

Catenochytridium, 66, 67

Catenomyces, 64

Catenula, 335: confusa, 336; lemnae, 335, 336; leptocephala, 336, 337; sekerai, 337; virginia, 336

Catenulida, 325, 334

Catenulidae, 335

Caudal furca, Ostracoda, 658

Caudal glands, Nemata, 370

Caudal ramus, Harpacticoida, 819

Caulicola, 293, 294

Caulobacter, 18, 19, 30, 31

Cecidomyiidae, 1059

Cecum (see Caecum)

Celina, 991

Celithemis, 937, 939, 932, 935

Cenocorixa, 970, 961

Central area, Bacillariophyceae, 175

Centrales, 174

Central nodule, Bacillariophyceae, 175

Central pores of raphe, Bacillariophyceae, 175

Centritractaceae, 153

Centritractus, 153, 150

Centrocorisa, 968, 961

Centrohelidia, 260

Centrolimnesia, 1101

Centronella, 178, 174

Centropagidae, 752, 756, 738

Centroptilum, 916, 910

Centropyxidae, 240

Centropyxis, 241: aculeata, 241; aërophila, 242; arcelloides, 243; constricta, 242; ecornis, 242; hemisphaerica, 242; stellata, 243

Cephalic appendages; Tardigrada, 509

Cephalic segment: Copepoda, 736; Harpac-ticoida, 819

Cephalobidae, 384, 385

Cephalobinae, 386

Cephalobus, 386

Cephalodella, 454, 455, 473, 437

Cephalomonas, 129, 128

Cephalosiphon, 476

Cephalothamnion, 227

Cephalothorax, Copepoda, 735

Cerasterias, 137, 136

Ceratiaceae, 163

Ceratium, 163, 162

Ceratoneis, 177, 174

Ceratophyllum demersum, 1176, 1180

Ceratopogonidae, 1066, 1069

Ceratopogoninae, 1068

Ceratoppia, 1114, 1113

Ceratotrocha, 485, 437

Ceratozetes, 1113

Ceratozetidae, 1110

Cerci: Odonata, 918; Plecoptera, 943

Cercobodo, 215

Cercomastix, 214

Cercyon, 1020

Ceriodaphnia, 617: acanthina, 618; lacustris, 619; laticaudata, 620; megalops, 619; pulchella, 619; quadrangula, 620; reticulata, 618; rigaudi, 618; rotunda, 620

Ceriospora, 89, 88: asci, 50

Cernotina, 1031

Ceropods, Anostraca, 560

Cervical groove, Malacostraca, 870

Cervical notch, Cladocera, 589

Cervical sinus, Cladocera, 589

Cestoda, 369

Chaemisiphon, 98

Chaenea, 269, 268
Chaetarthria, 999
Chaetoceraceae, 174
Chaetoceroideae, 174
Chaetoceros, 187, *174*
Chaetogaster, 529: *cristallinus*, 529; *diaphanus*, 529; *diastrophus*, 529; *langi*, 529; *limnaei*, 529
Chaetonema, 147, 146
Chaetonodidae, 410
Chaetonotus, 413: *acanthophorus*, 417; *anomalus*, 417; *brevispinosus*, 414, 415; *formosus*, 417; *gastrocyaneus*, 417, 418; *longispinosus*, 416; *octonarius*, 416; *robustus*, 414, 415; *similis*, 414, 415; *spinulosus*, 416; *tachyneusticus*, 418; *trichodrymodes*, 416; *trichostichodes*, 416; *vulgaris*, 418
Chaetopeltis, 133, 132
Chaetophora, 147, 146
Chaetophoraceae, 147
Chaetosphaeridium, 147, 146
Chaetospira, 287, 288
Chalarothoraca, 261
Chamaesiphonaceae, 98, *96*
Chantransiaceae, 169
Chaoborinae, 1063
Chaoborus, 1063, 1064
Chaos, 235
Chaosidae, 234
Chara, 149, 150
Characeae, 149
Characiaceae, 135, 134
Characiopsidaceae, 153
Characiopsis, 153, 152
Characium, 135, 134
Charales, 149
Charophyceae, 149, *117*
Chauliodes, 976, 975, *974*
Cheilorhabdions, Nemata, 370
Cheilostom, Nemata, 370
Chela, Malacostraca, 870
Chelicerae, Acari, 1083
Chelipeds, Malacostraca, 871
Chelonariidae, 1015, *985*
Chelonarium, 1015, 1014, *985*
Chernokrilus, 957
Cheumatopsyche, 1033, 1034
Chilo, 1053, 1055, *1051*
Chilodonella, 275, 274
Chilodontopsis, 274
Chilomonas, 167, 166
Chilophrya, 268
Chiloscyphus, 1163, *1169*
Chiloxanthus, 965
Chimarra, 1030, 1027
Chirocephalidae, 566, *561*

Chirocephalopsis, 566, 561: *bundyi*, 566, 569
Chirocephalus lithacus, 566
Chirodrilus, 534
Chironomidae, 1069
Chironominae, 1070
Chironomus, 1070
Chlamydodontidae, 274
Chlamydomonadaceae, 127
Chlamydomonas, 127, 126, 10
Chlamydophora, 260
Chlamydophrys, 257
Chlamydotheca, 695: *arcuata*, 706; *azteca*, 705; *flexilis*, 703; *mexicana*, 706; *speciosa speciosa*, 704; *texasiensis*, 704, 705; *unispinosa*, 704
Chlorallanthus, 153, 150
Chlorangiaceae, 131, 130
Chlorangium, 131, 130
Chlorella, 135, 136
Chlorobium, 43, 34, 35; chlorophyll, 14
Chlorobotrys, 151, 150
Chlorochromatium, 44: *aggregatum*, 35
Chlorochytrium, 133, 132
Chlorocloster, 153, 150
Chlorococcacceae, 133
Chlorococcales, 133
Chlorococcum, 133, 132
Chlorogibba, 151, 150
Chlorogonium, 127, 126
Chlorohydra viridissima, 317
Chloromonadales, 165
Chloromonadophyceae, 118
Chloronium mirabile, 35
Chloropediaceae, 153
Chloroperla, 956, *946*
Chloroperlidae, 954
Chloroperlinae, 956
Chlorophyceae, 127, *117*: colorless, 118
Chlorophyll *a*, 34
Chlorophyta, 127, *117*, *10*
Chlorosaccaceae, 151
Chlorosarcina, 133, 132
Chlorotheciaceae, 153
Chlorothecium, 153, 152
Chlorotylium, 147, 144
Choanoeca, 206
Choanophrya, 297, 296
Chodatella, 137, 136
Chonotrichida, 267, 294
Chordodes, 404
Chordodidae, 403, 404
Chordodiolinae, 403
Chordodiolus, 403
Chords, Nemata, 369
Choroterpes, 914, 910
Chromadora, 393, 392

Chromadorida, 386
Chromadoridae, 393
Chromadorita, 393
Chromagrion, 924, *926*
Chromatium, 44, 32, 33
Chromogaster, 442, 443, *437*
Chromulina, 155, 154
Chromulinaceae, 155
Chrondrymyces, 20
Chronogaster, 390, 391
Chroococcaceae, 96
Chroococcus, 97
Chroomonas, 165, 164
Chrysamoeba, 159, 158, *11*
Chrysapsis, 155, 154
Chrysidiastrum, 159, 158
Chrysoamphitrema, 159, 158
Chrysocapsa, 159, 158
Chrysocapsaceae, 159
Chrysocapsales, 159
Chrysochromulina, 155, 156
Chrysococcus, 155, 154
Chrysogaster, 1077
Chrysomelidae, 1001, 1020, 987
Chrysomonadales, 155
Chrysomonadida, 155, *190*
Chrysomonadines, *191*
Chrysomonads, *9*, *11*
Chrysophyceae, 155, 117: colorless, 118
Chrysophyta, 151, *174*, *117*
Chrysophyxis, 159, 158
Chrysopinae, 1074
Chrysops, 1014
Chrysosphaeraceae, 159
Chrysosphaerales, 159
Chrysosphaerella, 155, 154
Chrysostephanosphaera, 160
Chrysotrichales, 160
Chrysozona, 1074
Chydoridae, 604, *588*, 589
Chydorinae, 634
Chydoroidea, 589, 599
Chydorus, 461, 648: *barroisi*, 652, 648, *649; bicornutus*, 650, *649; faviformis*, 649; *faviformis* group, 649; *gibbus*, 650, *649; globosus*, 649; *hybridus*, 652, 648, *649; latus*, 650, *649; ovalis*, 651, *649; piger*, 651, *649; poppei*, 652, *649; sphaericus*, 651, *649; sphaericus* group, 649
Chytridiaceae, 65
Chytridiales, 50, *11*
Chytridium, 65
Chytriomyces, 66
Ciliata, 266
Ciliates, *14*
Ciliophora, 265

Cinetochilum, 281
Cinygma, 915, 910
Cinygmula, 915, 909, 910
Cirolanidae, 873
Cirolanides texensis, 873
Cirrodrilus, 525: *thysanosomus*, 525
Claassenia, 953, 952, *946*
Cladocera, 559, 587, 589
Cladocera, classification, 589
Cladochytrium, 64, 63
Cladocopa, 663
Cladomonas, 224
Cladonema, 225
Cladophora, 149, 148, *53*
Cladophoraceae, 149
Cladophorales, 149
Cladopodiella, 1163, *1169*
Cladospongia, 212
Cladothrix, 41
Clams, 1117
Clam shrimps, 559
Clappia, 1134
Classification: Actinopoda, 232; Algae, 117–118; Bacillariales, 174; Bacillariophyceae, 174; Bacteria, *16–22;* Bryophyta, *1161;* Bryozoa, 499; Calanoida, 738–740; Cladocera, 587–589; Copepoda, 737 (see also Calanoida, Cyclopoida, Harpacticoida); Crustacea, 558–559, *587–588*, 904; Cyclopoida, *795;* Harpacticoida, 818; Megaloptera, *973;* Mollusca, *1117*, 1124; Neuroptera, *973;* Polychaeta, *538–539;* Protista, *2–14;* Protozoa (see Actinopoda, Algae, Ciliophora, Rhizopoda, Zooflagellates); Raphidioidea, *973;* Rhizopoda, *232;* Rotifera, 437–438; Zooflagellates, *190–191*
Clastidiaceae, 98, *96*
Clastidium, 98: *setigerum*, 98
Clathrochloris, 35, 34, 43
Clathrosperchon, 1089, 1096
Clathrosperchonidae, 1096
Clathrosphaerina, 93
Clathrostoma, 276
Clathrostomidae, 275, 276
Clautriavia, 197
Claval suture, Hemiptera, 969
Clavariopsis, 90
Clavus, Hemiptera, 969
Claws, Tardigrada, 509
Claw shrimps, 559
Cleptelmis, 1005, 1006, 1019, *987*
Cletocamptus, 837, 838, *816;* bicolor 838; *brevicaudata*, 838; *deitersi*, 838
Cletodidae, 816
Climacia, 977, 978, 979, *976*

Climacium, 1165, *1169*
Climacostomum, 284
Clinocera, 1076
Clinotanypus, 1069
Clioperla, *946*
Clitellariinae, 1074
Cloeon, 916
Clonothrix, 41
Closteridium, 137, 136
Closteriopsis, 137, 136
Closterium, 141, 140, *51*
Clostoeca, 1044
Clostridium, 44
Chidaria, 313
Cocaine, use of powdered, 408
Cocci, 17
Coccochloris, 96
Coccoid Myxophyceae, 96
Coccolithophoridaceae, 155
Coccomonas, 129, 128
Coccomyxa, 133, 132
Coccomyxaceae, 133
Cocconeioideae, 174
Cocconeis, 179, *174*
Cochliopa, 1134
Cochliopodiidae, 236
Cochliopodium, 236
Codomonas, 200
Codoneca, 200
Codonella, 285
Codonobotrys, 223
Codonocladium, 209
Codonodendron, 203
Codonosiga, 209
Codonosigopsis, 203
Coelastraceae, 135, 134
Coelastrum, 135, 134
Coelenterata, 313
Coeleochaetaceae, 147
Coelomomyces, 70
Coelomomycetaceae, 70
Coelospherium, 97
Coelotanypus, 1069
Coenagrion, 926, 925
Coenagrionidae, 922, 924
Cohnilembidae, 278, 281
Cohnilembus, 281
Colaciaceae, 123
Colaciales, 123
Colacium, 123
Coleochaete, 147, 146
Coleoptera, 905, 906, 981: key to adults, 989; key to larvae, 1009
Colepidae, 268, 270
Coleps, 270

Collembola, 902, 904
Collodictyon, 228
Collotheca, 480, 483, *438*
Collothecaceae, 473, *438*
Collothecidae, 438
Colpidium, 280, 279
Colpius, 997
Colpoda, 277
Colpodidae, 275, 277
Colponema, 214
Colurella, 448, 449, *437*
Colurinae, 437
Colurus, 449
Colymbetes, 994, 1010
Compressor, 1196, 1197, 1198
Conchodromus, 1150
Conchophthiridae, 275, 277
Conchophthirus, 277
Conchostraca, 577, *559*: relation to Cladocera, 588
Condylostoma, 284
Condylostomidae, 283, 284
Conidia: *Ancylistis*, 49; Fungi Imperfecti, 50
Conidiophore, Fungi Imperfecti, 50
Conochaete, 147, 146
Conochilidae, 438
Conochiloides, 475, *438*: *coenobasis*, 475; *dossuarius*, 475; *exiguus*, 475; *natans*, 475
Conochilus, 475, 474, *438*: *hippocrepis*, 475, 474; *norvegicus*, 475, 474; *unicornis*, 475, 474
Copelatus, 994, 1011
Copepoda, 559: free-living, 735; parasitic, 862
Copepodid stages, copepod, 738, 745–747
Copepods (see Copepoda)
Copeus, 473
Coptotomus, 995, 1011, *983*
Copulatory bursa, Turbellaria, 327
Copulatory sac, Turbellaria, 327
Coralliochytrium, 57, 56
Corbicula, 1159
Corbiculidae, 1125, 1159
Cordulegaster, 934, 930, *919*
Cordulegastridae, 922, 934, *919*
Cordulia, 939, *934*
Corduliinae, 934, *919*
Cordylophora lacustris, 313, 316
Cordylosoma perlucidum, 469, 489, *438*
Cordyluridae, 1078
Corethra (see *Chaoborus*)
Corethrella, 1063
Corisella, 968
Corium, Hemiptera, 969
Corixidae, 967, *961*
Corona, Rotifera, *420*, 425
Coronastrum, 139, 138

Corpus, Nemata, 371
Corticacarus, 1087, 1101
Corunculina, 1154
Corvomyenia, 303: *discoides*, 303; *everetti*, 303
Corvospongilla, 302: *novae-terrae*, 302
Corydalidae, 976, *904*, *905*
Corydalus, 976, 975, *974*
Corynoneura, 1070
Coryphaeschna, 931, *934*, 930
Corythion, 256
Coscinodiscaceae, 174
Coscinodiscoideae, 174
Coscinodiscus, 188, *174*
Cosmarium, 143, 142
Cosmocladium, 143, 142
Costa, Bacillariophyceae, 175
Cothurnia, 293, 294
Coxal plate, Malacostraca, 870
Cranotheridium, 272
Crangonyx, 877: *anomalus*, 877; *antennatus*, 877; *dearolfi*, 877; *forbesi*, 877; *gracilis gracilis*, 877; *gracilis packardii*, 878; *hobbsi*, 877; *obliquus*, 877; *occidentalis*, 877; *serratus*, 877; *shoemakeri*, 877
Craspedacusta sowerbyi, 314, 316
Cratoneuron, 1164, 1165, *1169*
Crawfishes, 871
Crayfishes, 871
Cremaster, Lepidoptera, 1051
Crenitis, 1001
Crenodonta, 1140
Crenothrix, 44, 45, *21*: *polyspora*, 30, 39
Criconema, 381, 383
Criconematidae, 381
Criconematinae, 381
Criconemoides, 381
Cricotopus, 1072
Cristatella, *499*: *mucedo*, 503
Cristatellidae, *499*
Cristigera, 282
Crucigenia, 139, 138
Crustacea, 558: classification, 558
Cryphocricos, 966
Cryptochrysidaceae, 165
Cryptochrysis, 166
Cryptococcaceae, 165
Cryptococcales, 165
Cryptodifflugia, 251
Cryptoglena, 125, 123
Cryptolabis, 1060
Cryptomonadaceae, 166
Cryptomonadales, 165
Cryptomonadida, *190*
Cryptomonads, *9*
Cryptomonas, 166

Cryptonchinae, 396
Cryptonchus, 396
Cryptophyceae, 165, *118:* colorless, 118
Ctedoctema, 282
Ctenodaphnia, subgenus of *Daphnia*, 605
Ctenopoda, *588*, 589, 599
Ctenostomina, 282, 285
Cubitus, Hemiptera, 969
Cucurbitella, 244
Culex, 1066, 1065
Culicidae, 1063
Culicinae, 1063
Culicoides, 1068
Culiseta, 1065
Cultus, 956, 955
Cumulata, 362
Cupelopagis, 447, 448, *438*
Cura foremanii, 331
Curculionidae, 1001, 1020, 987
Curicta, 965, *960*
Cursoria, 1080
Curvipes, 1104, 1107
Cyanomastix, 165, 164
Cyanomonas, 165, 164
Cyanophyta, 95, *117*
Cyatholaimidae, 393
Cyatholaiminae, 393
Cyathomonas, 166
Cybister, 996, 1011
Cyclestheria, *579*
Cyclestheriidae, *579*
Cyclidium, 282
Cyclochaeta, 290, 291
Cyclocypria, 666: *kinkaidia*, 666
Cyclocyprini, 665
Cyclocypris, 666: *ampla*, 667; *cruciata*, 666; *forbesi*, 668, 669; *globosa*, 666, 667; *laevis*, 668, 669; *nahcotta*, 667; *ovum*, 668, 669; *serena*, 668; *sharpei*, 668, 669; *washingtonensis*, 668
Cyclogramma, 274
Cyclonaias, 1142
Cyclonexis, 156, 157
Cyclopidae, 795
Cyclopoida, *735*, 737, 795
Cyclops, 797, 800, 801, 802, 977: *agilis*, 799; *agilis montanus*, 799; *bicolor*, 810, 811, *795;* *bicuspidatus*, 807, *808;* *bicuspidatus lubbocki*, 808; *bicuspidatus thomasi*, 809, 807; *bisetosus*, 806, 807; *bissextilis*, 808; *bistriatus*, 815; *capillatus*, 805; *carolinianus*, 805; *crassicaudis brachycercus*, 808; *dentatimanus*, 810, 811; *dimorphus*, 810; *donnaldsoni*, 804; *exilis*, 805; *gigas*, 803; *haueri*, 807; *ingens*, 804; *insignis*, 800; *jeanneli*, 806, 807; *jeanneli putei*, 809; *languidoides*, 809; *latipes*, 803; *magnus*, 802;

nanus, 809; *navus*, 806, 807; *nearcticus*, 807; *panamensis*, 810; *pilosus*, 806; *scutifer*, 800, 801; *serrulatus*, 799; *serrulatus elegans*, 799; *strenuus*, 801; *varicans*, 795; *varicans rubellus*, 810; *venustoides*, 806; *venustoides bispinosus*, 806; *venustus*, 805; *vernalis*, 804; *vicinus*, 800; *viridis*, 803
Cyclops, subgenus of *Cyclops*, 800
Cyclorrhapha, 1074
Cyclotella, 188, 172, *174*
Cyclothrix, 1094
Cylindrocapsa, 145, 144
Cylindrocapsaceae, 145
Cylindrochytridium, 66, 67
Cylindrocorporidae, 384
Cylindrocorpus, 384
Cylindrocystis, 141, 140
Cylindrolaiminae, 392
Cylindrolaimus, 392
Cylindropsyllidae, 816
Cylindrospermum, 103: *catenatum*, 103; *licheniforme*, 103, 102; *majus*, 103; *muscicola*, 103; *stagnale*, 103
Cylindrotheca, 181, 180, *174*
Cylindrotominae, 1060
Cylloepus, 1004, 1016, *986*
Cymatia, 967, *961*
Cymatopleura, 181, *174*
Cymbella, 186, *174*
Cymbellaceae, 174
Cymbiodyta, 1001, 1013, *985*
Cyphoderia, 252: *ampulla*, 253; *ampulla* var. *papillata*, 253; *trochus*, 252
Cyphoderiidae, 252
Cyphon, 1020, 1021
Cypretta, 695: *bilicis*, 717; *brevisaepta*, 717; *intonsa*, 715, 716; *nigra*, 716, 717; *turgida*, 715, 716
Cypria, 669: *exculpta*, 669; *inequivalva*, 671, 670; *lacustris*, 673, 672; *maculata*, 671; *mediana*, 671, 670; *mons*, 673, 672; *obesa*, 670; *ophthalmica*, 673, 672; *palustera*, 673, 672; *pellucida*, 671; *pseudocrenulata*, 672; *turneri*, 669, 670
Cypricercus, 706: *affinis*, 712, 713; *columbiensis*, 712, 713; *dentifera*, 712; *elongata*, 712; *fuscatus*, 709, 710; *hirsutus*, 710; *horridus*, 709; *mollis*, 711; *obliqvus*, 712, 713; *passaica*, 711; *reticulatus*, 708; *serratus*, 712, 713; *splendida*, 709, 710; *tincta*, 709; *tuberculatus*, 708
Cypriconcha, 701: *alba*, 703; *barbata*, 702; *gigantea*, 702, 703
Cypridae, 664
Cypridicola, 446, 438: parasitica, 438, 489
Cypridopsis, 717: *aculeata*, 717, 718; *helvetica*, 721, 720; *inaudita*, 721; *mexicana*, 717; *niagrensis*, 721, 720; *okeechobei*, 719; *rhomboidea*,

719; *vidua*, 720; *viduella*, 720; *yucatanensis*, 717
Cyprinae, 664
Cyprinotus, 690: *americanus*, 692; *aureus*, 690; *crenatus*, 692; *dentatus*, 691; *fluviatilis*, 693, 692; *fretensis*, 694; *glaucus*, 693; *incongruens*, 690, 691; *inconstans*, 690, 691; *pellucidus*, 693; *putei*, 694; *salinus*, 694; *scytoda*, 692; *symmetricus*, 695, 694; *unispinifera*, 691
Cypris, 696: *pubera*, 701, 700; *subglobosa*, 701, 700
Cyprogenia, 1150
Cyprois, 689: *marginata*, 689, 688; *occidentalis*, 689
Cyrnellus, 1033
Cyrtolophosis, 280, 279
Cyrtonia, 469, *437*
Cystobranchus verrilli, 552
Cystodinium, 165, 164
Cysts, Myxobacteria, 20
Cytheridae, 663
Cytophaga, 20, 21
Cytostome, Ciliophora, 266
Cyzicidae, 583, *579*
Cyzicus, 583, *579*: *californicus*, 584, 585; *elongatus*, 584; *mexicanus*, 584, 578, *579*; *morsei*, 584

Dactylococcus, 137, 138
Dactylolabis, 1062
Dactylopusia brevicornis, 827
Dactylothece, 133, 132
Dadaya, 635: *macrops*, 635
Dallingeria, 218
Dalyellia, 343: *viridis*, 343
Dalyelliidae, 342
Dalyellioida, 341
Damselflies, *903*, 919
Dangeardia, 55, 54
Daphnia, 605: *ambigua*, 607, 608; *catawba*, 615; *dentifera*, 610; *dubia*, 609; *galeata mendotae*, 610; *laevis*, 608, 609; *longiremis*, 607; *magna*, 605, 606; *middendorffiana*, 613; *parvula*, 611, 612; *pulex*, 613, 614, var. *tenebrosa*, 613; *retrocurva*, 611, 612; *rosea*, 610; *schødleri*, 614; *similis*, 606; *thorata*, 611, 612
Daphnia, subgenus of *Daphnia*, 605
Daphnia trap, 1195
Daphnidae, 604, *588*, 589
Dapidia, 458, *437*
Darwinula, 663: *stevensoni*, 663, 664
Darwinulidae, 663
Dasycorixa, 967, *961*
Dasydytes, 409: *oöeides*, 409, 410; *saltitans*, 409, 410
Dasydytidae, 409
Dasyhelia, 1069
Dasyheliinae, 1069

Daubaylia, 384
Daubayliinae, 384
Debarya, 145, 142
Decapoda, 872, 878
Decapods, 871
Decurambis, 1146
Deinocerites, 1065
Dekabates, 1101
Demoiselle, 919
Dendrocoelidae, 329, 326
Dendrocoelopsis: alaskensis, 334; *piriformis*, 333; *vaginatus*, 334
Dendrocometes, 297, 296
Dendrocometidae, 294, 297
Dendromonas, 226
Dendrosoma, 297, 296
Dendrosomidae, 294, 297
Dendrospora, 93, 92
Denticles, Nemata, 370
Denticula, 182, *174*
Derallus, 999
Derepyxis, 156
Dermatophyton, 147, 146
Dermocarpa, 98
Dero, 530: *digitata*, 530, 531; *limosa*, 530; *obtusa*, 530
Deronectes, 993, 1010
Derovatellus, 991
Desmarella, 210
Desmatractum, 133, 132
Desmidiaceae, 141, 140
Desmidium, 143, 142
Desmokontae, 160, *117*
Desmomonadales, 160
Desmonema, 105: *wrangelii*, 105, 104
Desmonemes (see Nematocysts)
Desmopachria, 992
Desmothoraca, 263
Despaxia, 950
Desulfovibrio, 42, 44
Deuterophlebia, 1059
Deuterophlebiidae, 1059
Devil's darningneedle, 919
Dexiotrichides, 281
Dextral, Gastropoda, 1121
Diachros, 151, 150
Diacyclops, subgenus of *Cyclops*, 802
Diamesa, 1072
Diaphanosoma, 601: *brachyurum*, 601; *leuchtenbergianum*, 601
Diaphoropodon, 258
Diaptomidae, 756, 738
Diaptomids, 738
Diaptomus, 756, 762, 764, 774, 776, 783, 791, *738, 739*, 742, 743, 744: *alaskaensis*, 773, *739;*

albuquerquensis, 764, *739; amatitlanensis*, 763, *739; arapahoensis*, 773, 764, *739; arcticus*, 769, *739; ashlandi*, 786, 787, 782, *739; asymmetricus*, 777, *739; augustaensis*, 764, 765, *739; bacillifer*, 774, *739; bakeri*, 775; *birgei*, 790, 783, *739; bogalusensis*, 794, 782, *739; breweri*, 769, *739; caducus*, 770, 748, *739; castor*, 774, 775, *739; cokeri*, 780, *739; clavipes*, 761, *739; clavipoides*, 762, *739; colombiensis*, 775, 776, *739; coloradensis*, 784, 781, *739; conipedatus*, 760, *739; connexus*, 788, 782, *739; cuauhtemoci*, 788; *dampfi*, 778, *739; dentipes*, 766; *dilobatus*, 758, *739; dorsalis*, 778, *739; eiseni*, 769, *739; eiseni occidentalis*, 770; *floridanus*, 779, *739; forbesi*, 757, *739; franciscanus*, 775, *739; garciai*, 789, *glacialis*, 774, *739; gracilis*, 777, *739; hesperus*, 790, 783, *739; hirsutus*, 770, 771, *739; insularis*, 787, 782, *739; judayi*, 787, 782, *739; kenai*, 765, *739; kiseri*, 768, *739; kurilensis*, 789, 778, *739; lehmeri*, 764; *leptopus*, 760, *739; lighti*, 784; *lintoni*, 757, *739; louisianensis*, 792, 783, *739; manitobensis*, 761; *marshi*, 776; *marshianus*, 758, *739; mexicanus*, 785, *739; minutus*, 779, *739; mississippiensis*, 782, 793, *739; montezumae*, 764, 765, *739; moorei*, 789, 782, *739; natriophilus*, 786; *nebraskensis*, 761; *nevadensis*, 766, *739; novemdecimus*, 771, 772, *739; novamexicanus*, 789, 781, 782, *739; nudus*, 789, 782, *739; oregonensis*, 792, 782, *739; pallidus*, 792, 782, *739; patzcuarensis*, 764; *piscinae*, 761; *pribilofensis*, 785, 781, *739; proximus*, 778; *pseudosanguineus*, 758, *739; pugetensis*, 790; *purpureus*, 762, *739; pygmaeus*, 793, 782, *739; reighardi*, 794, 782, *739*, *793; saltillinus*, 779, *739; sanguineus*, 791, 783, *739; sarsi*, 776, 777, *739; saskatchewanensis*, 759, *739; schefferi*, 767, *739; shoshone*, 770, *739; shoshone beringianus*, 768; *sicilis*, 786, 782, *739; siciloides*, 788, 782, *739; signicauda*, 789, 782, *739; sinuatus*, 794, 782, *739; spatulocrenatus*, 758, *739; spinicornis*, 786, 781, *739; stagnalis*, 772, *739; tenuicaudatus*, 786; *texensis*, 763, *739; theeli*, 773, *739; trybomi*, 780, *739; tyrrelli*, 783, 784, 781, *739; victoriaensis*, 768, *739; virginiensis*, 792, 783, *739; wardi*, 766, *739; washingtonensis*, 789; *wilsonae*, 767, *739*

Diaptomus, subgenus of *Diaptomus*, 739, 774
Diaptomus, table of subgenera and species, 739
Diaschiza, 454
Diatoma, 177, 176, *174*
Diatomella, 183, *174*
Diatoms, *14, 117*, 171 (see also Bacillariophyceae)
Dibolocelus, 999
Dibusa, 1036

Diceras, 272, 271
Dichaetura, 410
Dichaeturidae, 410
Dichelyma, 1164, 1165, *1169*
Dichilum, 280, 279
Dichothrix, 104: *baueriana*, 105; *gypsophila*, 104; *hosfordii*, 105; *inyoensis*, 105; *orsiniana*, 105
Dichotomococcus, 137, 138
Dichotomosiphon, 135, 134
Dichotomosiphonaceae, 135
Dicosmoecus, 1045, 1043
Dicranochaete, 147, 146
Dicranophoridae, 454, 458, *437*
Dicranophorus, 460, 461, *437*
Dicranopselaphus, *985*
Dicranota, 1061
Dicraspedella, 205
Dictya, 1077
Dictyopterygella, *946*
Dictyosphaerium, 135, 134
Dictyuchus, 80, 79
Didiniidae, 268, 271
Didinium, 271
Didymodactylus, 485, 486, *437*
Didymops, 934, 935
Difflugia, 244: *acuminata*, 246; *bacillifera*, 245; *corona*, 245; *lebes*, 244; *lobostoma*, 245; *oblonga*, 245; *rubescens*, 245; *tuberculata*, 244; *urceolata*, 244
Difflugiella, 251
Difflugiidae, 240
Digestive system, Nemata, 370
Diglena, 460
Digononta, 438, 439
Dileptus, 273
Dimorpha, 192
Dimorphism, zoospores of Phycomycetes, 49
Dimorphococcus, 135, 134
Dina, 556: *anoculata*, 556; *dubia*, 556; *lateralis*, 556; *parva*, 556
Dinamoeba, 235: *horrida*, 235; *mirabilis*, 235
Dinema, 127
Dineutus, 990, 1009
Dinobryon, 157
Dinocapsales, 163
Dinocharis, 449
Dinoflagellates, *9, 10, 117*
Dinoflagellida, *190*
Dinomonas, 216
Dinophyceae, 160, *10, 117*
Dinopodiella, 165, 164
Dinotrichales, *10*
Dioctophymatida, 397
Diosaccidae, 816
Dioxys, 153, 152

Diphanoeca, 208
Diphascon, subgenus of *Hypsibius*, 519
Diphtherophora, 398
Diphtherophoridae, 398
Diplanetism, zoospores of Phycomycetes, 49
Diplectrona, 1035, 1034
Dipleuchlanis, 457, 458, *437*
Diplochlamys, 237
Diplocladius, 1071
Diplodontus, 1097, 1107
Diploeca, 205
Diplogaster, 384, 385
Diplogasteridae, 384
Diplohydrachna, 1093, 1094
Diplois, 455, *437*
Diplomita, 222
Diploneis, 183, *174*
Diploperla, 955, *946*
Diplophlyctis, 59, 58
Diplophrys, 260
Diploscapter, 385
Diploscapterinae, 385
Diplosiga, 204
Diplosigopsis, 206
Diplostraca (= Cladocera + Conchostraca), 588, 589
Diptera, 904, 906, 1057
Discineae, 174
Discomorpha, 286
Discomorphidae, 286
Discophrya, 296
Discophryidae, 294, 296
Disematostoma, 279
Dispersipiona, 1104
Dispora, 133, 132
Dissotrocha, 488, 486, *437*
Distichodont teeth, Hirudinea, 544
Distigma, 125, 124
Distribution: Acari, 1093–1114 *passim;* Algae, 115–116 (see also Bacillariae); Anostraca, 560; Bacillariophyceae, 173; Bacteria, 27–43; Bryozoa, 498; Calanoida, 740–741; Cladocera, 595; Coleoptera, 982–987; Conchostraca, 579; Copepoda (see Calanoida, Harpacticoida); Fungi (substrates), 47–48; Gastrotricha, 407–408; Harpacticoida, 817–818; Hemiptera, 959–962; Insecta, general, 903 (see also the various orders); Lepidoptera, 1051; Malacostraca, 871–872; Mollusca, 1118; Nemata, 374; Odonata, 923, 924, 926, 929, 930, 934, 939; Oligochaeta, 525; Plecoptera, 946–947; Polychaeta, 539; Rotifera, 433; Tardigrada, 509
Ditrema, 252
Ditylenchus, 384

Diura, 955, *946*
Diurella, 447, *437*
Dixa, 1073
Dixidae, 1073
Dobsonflies, *904*, 974
Docidium, 141, 140
Dolania, 911, 909
Dolichodorus, 381, 382
Dolichopodidae, 1076
Dolichorinae, 381
Dolkrila, 955
Dolophilodes, 1031
Dolophilus, 1030
Domatomonas, 199
Domorganus, 389
Donacia, 1006, 1020, 1002
Dorocordulia, 936, 937, *934*
Dorria, 471, 470, 460, *437*
Dorsal ridge, Harpacticoida, 819
Dorsal seta, Harpacticoida, 819
Dorsoglandularia, Acari, 1082
Dorylaimida, 397
Dorylaimidae, 398
Dorylaiminae, 399
Dorylaimoidea, 398
Dorylaimoides, 398
Dorylaimus, 399
Doryllium, 398
Dorystoma, 471, 470, *437*
Dosalia, 311: *palmeri*, 311; *plumosa*, 311
Dragonflies, *903*, 919
Draparnaldia, 147, 146
Draparnaldiopsis, 147, 144
Drepanocladus, 1165, *1169*
Drepanomonas, 278, 277
Drepanothrix, 627: *dentata*, 627
Drepanotrema, 1129
Drilophaga, 466, *437*
Dromogomphus, 928, *929*, 930, 932
Drusinus, 1044
Dryadontanytarsus, 1070
Dryopidae, 1002, 1015, 984, *985*
Dryops, 1002, 1015, *985*, *986*
Dubiraphia, 1006, 1018, 1019, *986*
Dugesia, 329: *agilis*, 330; *microbursalis*, 330; *dorotocephala*, 330; *tigrina*, 330
Dumortiera, 1163, 1162, *1169*
Dunhevedia, 644: *crassa*, 644; *serrata*, 644
Dysmicohermes, 976
Dysmorphococcus, 129, 128
Dysteria, 275, 274
Dysteriidae, 275, 274
Dythemis, 937, *939*
Dytiscidae, 991, 1010, 982, 983
Dytiscus, 995, 1011

Earthworms, 532
Ecclisomyia, 1045, 1043, 1044
Eccoptura, 953
Echiniscoidea, 513
Echiniscus, 513: *calvus*, 514; *gladiator*, 514; *granulatus*, 514; *mauccii*, 514; *merokensis*, 514; *merokensis suecica*, 514; *oinonnae*, 515; *phocae*, 514; *quadrispinossus cribrosa*, 515; *quadrispinosus fissispinosa*, 514; *reticulatus*, 514; *spiniger*, 514; *spitsbergensis*, 515; *tympanista*, 515; *viridis*, 514; *wendti*, 514
Echiniscus, subgenus of *Echiniscus*, 514
Echinodera, *368*
Echinospaerella, 137, 136
Eclipidrilus, 533: *frigidus*, 533
Ectocyclops, 795: *phaleratus*, 796
Ectopria, 1002, 1015, 1014, *985*
Ectoprocta, 500, 495, *499*
Ectrogella, 77, 76
Ectrogellaceae, 77
Edmundsius, 915
Eichhornia crassipes, 1185, 1184
Eiseniella, 534: *tetraedra*, 534
Ejaculatory apparatus, Ostracoda, 661
Ejaculatory duct, Turbellaria, 327
Ekman dredge, 1196, 1197
Elaeorhanis, 261
Elakatothrix, 133, 132
Elaphoidella, 840, *816*: *bidens coronata*, 849; *subgracilis*, 850
Elatine americana, 1178
Electric light bug, 960
Eleutherengona, 1080
Elliptera, 1062
Elliptio, 1143
Elmidae, 1002, 1015, *986*, 987
Elodea, 1178
Elodes, 1020, 1021, *987*
Elophila, 1052, 1053, *1051*
Elophilus, 1077, 1076
Elosa, 445, 444, *437*
Elsianus, 1003, 1017, 987
Embata, 488, 486, *437*
Embolium, Hemiptera, 967, 969
Empididae, 1076
Enallagma, 926, 921, 925
Encentrum, 460, 459, *437*
Enchelydium, 272, 271
Enchelyodon, 270
Enchelys, 270
Enchodelus, 399
Enchytraeidae, 533
Enchytraeus, 534
Endalus, 1007
Endochytrium, 68

Endocoenobium, 62
Endodesmidium, 52, 53
Endopod, copepod, 736
Endopodite, copepod, 736
Endosphaera, 295
Endosphaeraceae, 133, 132
Endospore: bacteria 18; Chamaesiphonaceae, 95
Enochrus, 1001, 1012, 1000
Enoplida, 393
Enoplochilus, 399
Enteromorpha, 149, 148
Enteroplea, 466, 467, *437, 473*
Entocladia, 147, 144
Entocythere, 724: *cambaria*, 728; *claytonhoffi*, 727; *columbia*, 728; *copiosa*, 731, 730; *dobbinae*, 729; *donaldsonensis*, 733, 732; *dorsorotunda*, 733, 732; *elliptica*, 728; *equicurva*, 729; *heterodonta*, 731, 730; *hobbsi*, 732; *humesi*, 733, 732; *illinoisensis*, 727; *insignipes*, 731, 730; *mexicana*, 729; *riojai*, 730; *serrata*, 726; *sinuosa*, 731, 730; *talulus*, 731, 730
Entomophthorales, 50, 49
Entophlyctis, 57
Entophysalis, 98: *lemaniae*, 98; *rivularis*, 98
Entoprocta, 499, 495, *499*
Entosiphon, 127
Eobates, 962
Eocyzicus, 583, *579:* *concavus*, 585; *digueti*, 585
Eosphora, 472, 421, *437*
Eothinia, 471, *437*
Epactophanes, 832, *816; richardi*, 832
Epalcidae, 286
Epalxis, 286
Epeorus, 914
Ephemera, 913, 910
Ephemerella, 912, 909
Ephemerellidae, 912
Ephemeridae, 911
Ephemeroptera, 906, 907, 908
Ephippium, Cladocera, 595
Ephoron, 911, 910
Ephydra, 1077
Ephydridae, 1077
Epiaeschna, 931, *934*
Epichrysis, 159, 158
Epicordulia, 936, *934*, 927, 930
Epicranial suture, Lepidoptera, 1051
Epimera, Acari, 1082
Epimeroglandularia, Acari, 1082
Epimetopus, 998, 985
Epioblasma, 1157
Epiphanes, 469, 437, *454*
Epipharynx, Rotifera, 428
Epiproct, Odonata, 918

Epipyxis, 157
Epischura, 750, *738: fluviatilis*, 750; *lacustris*, 751; *massachusettsensis*, 751, 752; *nevadensis*, 751; *nordenskiöldi*, 752
Epistylidae, 291, 292
Epistylis, 292
Epithemia, 182, *174*
Epithemiaceae, 174
Epithemiodeae, 174
Eremosphaera, 137, 136
Eretes, 996, 1011
Ergasilidae, 865
Ergasilus, 864, 865: *caeruleus*, 865, 864; *centrarchidarum*, 865; *chautauguaensis*, 864, 865; *confusus*, 864; *cotti*, 864; *elegans*, 864; *elongatus*, 864; *lanceolatus*, 864; *luciopercarum*, 864; *megaceros*, 864; *nigritus*, 864; *osburni*, 864; *versicolor*, 865, 864
Erignatha, 460, 461, *437*
Eriocaulon septangulare, 1182, 1181
Eriocera, 1062
Erioptera, 1062
Eristalis, 1077
Erpetogomphus, 928, *929*
Erpobdella, 556: *annulata*, 556; *punctata*, 556
Erpobdellidae, 552, *543, 544*
Errerella, 135, 132
Erythemis, 938, *939*, 935
Erythrodiplax, 938, *934, 939*
Erythrotrichiaceae, 167
Eschaneustyla, 287
Escherichia coli, 40
Espejoia, 280, 279
Estheria, 583, 579
Estheriidae, *579*
Estheriids, *579*
Ethmolaiminae, 393
Ethmolaimus, 393
Euadmontia, 1077
Euastropsis, 135, 134
Euastrum, 141, 142
Eubacteria, 16
Eubasilissa, 1041
Eubranchipus, 566, *561: floridanus*, 568, 569, *560; holmani*, 568, 569; *neglectus*, 568, 569; *oregonus*, 568, 569; *ornatus*, 568, 569; *serratus*, 568, 569; *vernalis*, 568, 569
Eubrianax, 1002, 1015, 985
Eucapnopsis, 950
Eucapsis, 97
Eucephalobus, 386
Euchlanis, 453, 456, 457, 458, *437*
Euchordodes, 404
Euciliata, 266
Eucladocera, 598, 589

Eucocconeis, 179, *174*
Eucorethra, 1063
Eucyclops, 797: *macrurus*, 798; *prionophorus*, 798; *speratus*, 799, 798
Eucypris, 696: *affinis hirsuta*, 699; *arcadiae*, 698; *cisternina*, 698, 699; *crassa*, 696; *fuscatus*, 697, 696; *hystrix*, 697; *rava*, 698; *reticulata*, 698, 699; *virens*, 697
Eudactylota, 449
Eudiaptomus, 777, *739*
Eudorina, 129, 128
Euglena, 125, 123, 124, *10, 52, 56*
Euglenaceae, 125
Euglenales, 125
Euglenamorpha, 125, 124
Euglenida, *190*
Euglenids, *9, 10*
Euglenocapsales, *10*
Euglenoids, *117*
Euglenophyta, 123, 117
Euglenopsis, 125, 124
Euglypha, 254: *brachiata*, 254; *ciliata*, 255; *compressa*, 255; *cristata*, 254; *laevis*, 255; *mucronata*, 254; *tuberculata*, 254
Euglyphidae, 253
Eukalyptorhynchia, 341
Eulalia, 1074, 1075
Eulimnadia, 581, *579:agassizii*, 581; *alineata*, 581; *antillarum*, 581; *antlei*, 582; *diversa*, 581; *francesae*, 581, 582; *inflecta*, 582; *oryzae*, 583; *stoningtonensis*, 583; *texana*, 582; *thompsoni*, 583; *ventricosa*, 583
Eulobosa, 236
Eunotia, 179, *174*
Eunotiaceae, 174
Eunotioideae, 174
Euparyphus, 1074
Eupera, 1158
Euphyllopoda, 558
Euplotes, 290
Euplotidae, 286, 290
Eurhynchium, 1168, *1169*
Euryalona, 637: *occidentalis*, 637, 638
Eurycaelon, 1136
Eurycercinae, 634
Eurycercus, 634: *glacialis*, 635; *lamellatus*, 634
Eurytemora, 752, *738*: *affinis*, 753; *canadensis*, 752, 753; *composita*, 754; *hirundoides*, 753; *tolli*, 753; *yukonensis*, 753, 754
Eutardigrada, 513
Euthyas, 1092, 1097
Eutreptia, 125, 124
Eutylenchus, 384
Evolution of flagellates, 10
Excentrosphaera, 135, 136

Excretory system, Nema, 373
Exopod: Copepod, 736; Malacostraca, 870
Exopodite (see exopod)
Exosphaeroma, 874: *insulare*, 874; *thermophilum*, 874
Exuviaella, 160, 161
Eye, Cladocera, 590
Eylaidae, 1094
Eylais, 1089, 1094, 1095

Faago, 377
Fabria, 1040
Fadeewella, 440, 441, *438: minuta*, 441
Fairy shrimps, 559
Faxonella, subgenus of *Orconectes*, 890
Feltria, 1089, 1103
Feltriella, 1103
Feltriidae, 1103
Ferrissia, 1130
Filamentous blue-green algae, 14
Filamentous true bacteria, 14
Filinia, 440, 446, *438: aseta*, 440; *brachiata*, 440; *camascela*, 440; *cornuta*, 440; *limnetica*, 441; *longiseta*, 440, 441; *major*, 440, 441; *passa*, 441; *terminalis*, 441
Filosa, 233
Filter, Millepore, 1195, 1202
First antenna (see Antenna)
First maxilla, Ostracoda, 659
Fischerella, 100: *ambigua*, 100; *thermalis*, 100
Fishflies, 974
Fish lice, *559*
Fissidens, 1164, 1163, *1169*
Flagellata, relationships, 9
Flagellate protozoa, *14*, 190
Flagellospora, 89, 88
Flagellum: bacterial, 9; structure, 17
Flatworms, 326
Flies (see Diptera)
Floatoblasts, Bryozoa, 497
Florideae, 167
Floscularia, 478, 480, *438*
Flosculariaceae, 473, *438*
Flosculariidae, 438
Fluminicola, 1135
Fontigens, 1133
Fontinalis, 1164, *1169*
Foot, Rotifera, 425
Forcipomyia, 1069, 1068
Forcipomyiinae, 1069
Forelia, 1089, 1090, 1104
Fornix, Cladocera, 592
FPA, 117
Fragilaria, 177, *174*
Fragilariaceae, 174

Fragilarioideae, 174
Franceia, 137, 136
Fredericella, 501, *499: australiensis* subsp. *browni*,
  501; *sultana*, 501
Fredericellidae, 499
Fremyella, 107: *diplosiphon*, 107; *tenera*, 107
Frenesia, 1045, 1043
Fridaea, 145, 144
Frisonia, 954
Frons: Hemiptera, 969; Lepidoptera, 1051
Frontal appendage, Anostraca, 559
Frontipoda, 1086, 1099
Frontonia, 280, 279
Frontoniella, 279, 278
Frontoniidae, 278
Fruiting bodies, Myxobacteria, 20
Frustule, Bacillariophyceae, 175
Frustulia, 184, *174*
Fungi, 47, *8*, *11:* Imperfecti, 89, 50; origins, 11
Furca, caudal; Gastrotricha, 406
Furcilla, 219
Furcularia, 454, 465
Fusconaia, 1139

Galba, 1128
Gallionella, 44, 45, *19*, *31*, 38, 39
Gametes, 11
Gammaridae, 876
Gammarus, 877: *acherondytes*, 877; *fasciatus*, 877;
  lacustris, 877; *limnaeus*, 877; *minus*, 877; *pro-
  pinquus*, 877; *pseudolimnaeus*, 877; *purpurascens*,
  877; *troglophilus*, 877
Ganonema, 1028
Gastronauta, 275, 274
Gastropidae, 437
Gastropoda, 1125
Gastropus, 453, 452, *437*, *439: hyptopus*, 453;
  minor, 453; *stylifer*, 453
Gastrostyla, 289
Gastrotricha, 406, *368*
Geayia, 1091, 1106, 1107
Geayidea, 1106
Gelastocoridae, 965, *960*
Gelastocoris, 965, *960*
Gelatinous matrix, list of algal genera, 118
Geleiella, 291
Geminella, 145, 144
Gemmule, Porifera, 298
Gena, Hemiptera, 969
Genital acetabula, Acari, 1082
Genital opening, Acari, 1082
Genital segment, Copepoda, 736
Geocentrophora, 360: *applanata*, 361; *baltica*, 362;
  sphyrocephala, 361; *tropica*, 362
Geolegnia, 81, 80

Geranomyia, 1062
Gerridae, 962, *959*
Gerrinae, 962
Gerris, 962, *959*
Giant water bug, 900, 965
Gillia, 1135
Gills: Odonata, 917; Plecoptera, 943–945
Ginglymyia, 1077
Girdle, Bacillariaceae, 175
Glaenocorisa, 967, *961*
Glaucoma, 280, 279
Glebula, 1151
Glenodiniaceae, 163
Glenodinium, 163, 162
Gloeoactinium, 137, 138
Gloeobotrydiaceae, 151
Gloeobotrys, 151, 150
Gloeocapsa, 97
Gloeochloris, 151, 150
Gloeocystis, 131, 130
Gloeodiniaceae, 163
Gloeodinium, 163, 162
Gloeomonas, 127, 126
Gloeotaenium, 139, 138
Gloeothece, 96
Gloeotrichia, 105: *echinulata*, 105, 104; *natans*,
  105; *pisum*, 105
Glossa, Plecoptera, 942
Glossatella, 291, 292
Glossiphonia, 548: *complanata*, 548; *fusca*, 549;
  heteroclita, 548; *hepheloidea*, 548
Glossiphoniidae, 547, *544*, *545*
Glossoscolecidae, 534
Glossosoma, 1036, 1028, 1029
Glossosomatidae, 1026, 1035, 1036
Glyphopsyche, 1045, 1043
Glyphotaelius, 1045, 1043
Gnaphiscus, 1087, 1099
Gnathobdellida, 552
Gnathopods, Malacostraca, 870
Gnathosoma, Acari, 1083
Goera, 1041
Goeridae, 1028, 1041
Goerita, 1041
Golden algae, *117*
Golenkinia, 133, 132
Gomontia, 145, 144
Gomphaeschna, 931, *934*
Gomphidae, 922, *919*
Gomphoides, 928, *929*
Gomphoneis, 185, *174*
Gomphonema, 186, *174*
Gomphonemaceae, 174
Gomphonemoideae, 174
Gomphosphaeria, 97: *aponina*, 98; *lacustris*, 98

Gomphus, 928, *929*, 933
Gonapodya, 70
Gonatozygon, 141, 140
Gongrosira, 147, 144
Gonidea, 1139
Gonielmis, 1006, 1018, *987*
Goniobasis, 1137
Goniochloris, 153, 150
Goniotrichaceae, 167
Gonium, 129, 128
Gonomyia, 1062
Gonopore, Turbellaria, 327
Gonostomum, 289
Gonyaulaceae, 163
Gonyaulax, 163, 162
Gonyostomum, 165, 164
Gordiida, 402
Gordiidae, 403
Gordionus, 404
Gordius, 403
Granulo-reticulosa, 233, 234
Graphoderus, 996, 1011
Graptocorixa, 968, *961*
Graptoleberis, 639: *testudinaria*, 639
Grass-green algae, 117
Green algae, *10*
Green bacteria, *12*, 19
Green sulfur bacteria, *31*, 33
Grimaldina, 628: *brazzai*, 628
Gromia, 257
Gromiidae, 252
Gubernaculum, Nemata, 374
Gullet: Ciliophora, 266; Tardigrada, 509
Gundlachia, 1131
Gymnodiniaceae, 161
Gymnodiniales, 161
Gymnodinium, 161
Gymnolaemata, 500, *499*
Gymnomera, *588*
Gymnostomina, 267
Gymnozyga, 143, 142
Gynacantha, 931, *934*
Gyratricidae, 357
Gyratrix, 357: *hermaphroditus*, 357
Gyraulus, 1128
Gyretes, 989, 1009
Gyrinidae, 989, 1009, *905*, 983, 984
Gyrinus, 989, 1009
Gyrodinium, 161
Gyromonas, 229
Gyrosigma, 183, *174*

Habrophlebia, 914, 910
Habrophlebiodes, 914, 910
Habrotrocha, 485, *484*, 486, *437*

Habrotrochidae, 484, *437*
Haematococcaceae, 131
Haematococcus, 131, 130
Haematopota, 1074
Haemopis, 554: *grandis*, 554; *kingi*, 554; *lateralis*, 554; *marmorata*, 554, 555; *plumbea*, 554
Hagenius, 928, *929*, 927
Hairs, Acari, 1083, 1084
Halacaridae, 1080, 1108
Halacarus, *1108*
Halicyclops, 797: *aequoreus*, 797; *magniceps*, 797
Haliplectinae, 389
Haliplectus, 389
Haliplidae, 990, 1009, 982, 983
Haliplus, 990, 1010
Hallezia, 295, 296
Halobates, 962
Halobatinae, 962, *959*
Halolindia, 467
Halteria, 285
Halteriidae, 285
Hamohalacarus, 1110, *1108*
Hantzchia, 181, 180, *174*
Hapalosiphon, 100: *fontinalis*, 100; *laminosus*, 100
Haplomacrobiotus, 515
Haplopoda, 598, *588*, 589
Haplotaxidae, 531
Haplotaxis, 531: *forbesi*, 532; *gordioides*, 532
Harpacticidae, 816
Harpacticoida, 815, *735*: characteristics, 737; tabulation of families and genera, 816
Harpacticoid copepods, 815
Harpacticus, 824, 816: *chelifer*, 824; *gracilis*, 824
Harringia, 463, *437*
Hartmannellidae, 234
Hastaperla, 956, 953, *946*
Hastatella, 291
Haustoriidae, 876
Head emargination, Lepidoptera, 1051
Hebridae, 963, *959*
Hebrus, 964
Hedroneura, 1077
Heleidae, 1066, 1069
Heleinae, 1068
Heleocoris, 967
Heliapsis, 159, 158
Helichus, 1002, 1015, 1014, *985*, *986*
Helicodendron, 93, 92
Helicoon, 93, 92
Helicopsyche, 1046, 1028, 1029, 1030, 1041
Helicopsychidae, 1029, 1046
Helicotylenchus, 383
Heliozoa, 233
Heliscus, 89, 88
Helisoma, 1129

Helius, 1062
Helkesimastix, 214
Hellgrammites, 974
Helobata, 1001
Helobdella, 548: *elongata*, 548; *fusca*, 548; *lineata*, 549; *papillata*, 549; *punctata-lineata*, 548; *stagnalis*, 548
Helochares, 1001, 1013, 985
Helocombus, 1001, 1000
Helocordulia, 936, *934*
Helodidae, 1001, 1020, 987
Helodium, 1165, 1166, *1169*
Helopera, 248: *petricola*, 249; *rosea*, 249; *spagni*, 248
Helophilus, 1077
Helphorus, 998, 1011, 985
Helopicus, 954
Hemerodromia, 1076
Hemicycliophora, 381
Hemicycliostyla, 287
Hemidinium, 163, 161
Heminothrus, 1113
Hemiptera, 906, 907, *904*, 958
Hemistena, 1144
Henlea, 534
Henoceros, 489, 486, *437*
Hepaticae, 1161, 1162
Hepatic caeca, Cladocera, 593
Hepatic spine, Malacostraca, 870
Heptagenia, 915, 909
Heptageniidae, 913
Heptageniinae, 913
Heptagyna, 1071
Heribaudiella, 160
Hermannia, 1111
Hermione, 1074
Herpetocypris, 701: *chevreuxi*, 714; *meridana*, 714; *reptans*, 714, 715; *testudinaria*, 714, 715
Hertwigia, 445
Hesperagrion, 925, *924*, 926
Hesperocorixa, 970, 969, *961*
Hesperodiaptomus, subgenus of *Diaptomus*, 739, 762, 764, 766, 770, 775
Hesperoperla, 953, 952
Hesperophylax, 1045, 1043
Hetaerina, 923
Heteranthera dubia, 1186
Heterelmis, 1004, 1017, *987*
Heterlimnius, 1005, 1019, 1016, *987*
Heterocapsales, 151
Heterococcales, 151
Heterocope, 738: *septentrionalis*, 750
Heterocysts, Myxophyceae, 95
Heterodera, 383
Heteroderinae, 381

Heterolaophonte, *816*: *strömi*, 826, *816*
Heterolepidoderma, 413
Heteromastix, 127, 126
Heteromeyenia, 304, 303: *argyrosperma*, 308, 304; *baileyi*, 309; *biceps*, 309; *conigera*, 308; *pictouensis*, 307; *repens*, 309; *ryderi*, 308
Heteronema, 127
Heterophrys, 261
Heteroplectron, 1046, 1047
Heterosiphonales, 155
Heterotardigrada, 513
Heterotrichales, 154
Heterotrichina, 282
Heterozetes, 1113
Hexacylloepus, 1004, 1018, *987*
Hexagenia, 913, 910
Hexalebertia, 1099
Hexamitus, 230
Hexarthra, 441, *438*
Hexarthridae, 438
Hexatax, 1102
Hexatoma, 1062
Hibernacula, Bryozoa, 497
Hippurus vulgaris, 1178
Hirudidae, 552, *543*, *544*, *545*
Hirudinea, 542
Hirudo medicinalis, 552
Hispidosperchon, 1098
Histiobalantium, 282
Histiona, 202
Histrio, 290
Holocarpic thalli, Fungi, Phycomycetes, 48
Holocoela, 362
Holognatha, 947, *942*
Holopedidae, 599, *588*, 589
Holopedium, 599: *amazonicum*, 603, 604; *gibberum*, 603
Holophrya, 270
Holophryidae, 268
Holosticha, 288, 289
Holotrichida, 267
Homalogastra, 281
Homalozoon, 271
Homoeoneuria, 913
Hoperius, 995
Hoplolaimidae, 381
Hoplolaiminae, 383
Hoplolaimus, 383
Horaella, 446, *438*
Horatia, 1135
Hormidium, 145, 144
Hormogonia, Myxophyceae, 95
Hormotila, 131, 130
Hormotilopsis, 131, 130
Horsehair worms, 402

Horse stinger, 920
Hottonia inflata, 1181
Huitfeldtia, 1090, 1091, 1103, 1102
Huntemannia, 832, *816: lacustris*, 832
Hyalella: azteca, 876; *knickerbockeri*, 876
Hyalinella, *499: punctata*, 503, 504
Hyalobryon, 157
Hyalocephalus trilobus, 480, *438*
Hyalodiscidae, 234
Hyalosphenia, 246: *cuneata*, 247; *elegans*, 247; *papilio*, 247
Hyalotheca, 143, 142
Hydaticus, 996, 1011
Hydatina, 469
Hydra, 317: *americana*, 317; *canadensis*, 322; *carnea;* 320; *cauliculata*, 321; *hymanae*, 318; *littoralis*, 321; *oligactis*, 318, 319, *314; oregona*, 322; *pseudoligactis*, 318, 319; *utahensis*, 320
Hydracarina, 1080
Hydrachna, 1092, 1093, 1094
Hydrachnae, 1093
Hydrachnidae, 1093
Hydraena, 998, 1013
Hydraenidae, 998, 1013, 984, 985
Hydrarachna, 1093, 1107
Hydras, 313, 316
Hydrellia, 1077
Hydrobaeninae, 1070
Hydrobaenus, 1071, 1072
Hydrobius, 1000, 1012
Hydrocampa, 1053, 1054
Hydrocanthus, 997, 1010
Hydrochara, 999, 1013
Hydrochoreutes, 1087, 1104, 1103
Hydrochus, 998, 1012
Hydrocleis nymphoides, 1183
Hydrocoleum, 107: *groesbeckianum*, 108; *homoeotrichum* 107, 106
Hydrocoryne, 100: *spongiosa*, 100
Hydrodictyaceae, 135, 134
Hydrodictyon, 135, 134
Hydrodroma, 1093, 1097, 1098
Hydrodromidae, 1097
Hydroida, 313
Hydrolimax, 363: *grisea*, 363
Hydrometra, 962, *959*
Hydrometridae, 962, *959*
Hydromyza, 1078
Hydroperla, 954, 949, *946*
Hydrophilidae, 998, 999, 1011, 1012, 1021, 983, 984
Hydrophilus, 999, 1013, 985
Hydrophoria, 1079
Hydrophorus, 1076
Hydroporinae, 991

Hydroporus, 993, 1010
Hydropsyche, 1033, 1026, 1034
Hydropsychidae, 1025, 1033, 1026, 1034
Hydropsychid Genus A, 1033, 1034
Hydroptila, 1039, 1026, 1038
Hydroptilidae, 1025, 1036, 1026, 1037, 1038
Hydroscapha, 997, 1013
Hydroscaphidae, 997, 1013, 985
Hydrosera, 188, *174*
Hydrotrupes, 994, 1011
Hydrovatus, 991, 1010
Hydrovolzia, 1091, 1093, 1094
Hydrovolziae, 1093
Hydrovolziidae, 1093
Hydrozetes, 1114, *1110, 1111*
Hydrozoa, 313
Hydruraceae, 159
Hydrurus, 159, 158
Hydryphantae, 1095
Hydryphantes, 1091, 1093, 1096, 1097
Hydryphantidae, 1096
Hyella, 98
Hygroamblystegium, 1167, 1168, *1169*
Hygrobates, 1087, 1088, 1089, 1101
Hygrobatidae, 1101
Hygrohypnum, 1164, 1165, 1167, 1168, *1169*
Hygrotus, 993, 991, 1010
Hymanella retenuova, 331
Hymenomonas, 155, 156
Hymenoptera, 906, 907, *904*
Hymenostomina, 267, 278
Hypechiniscus, subgenus of *Echiniscus*, 514
Hyperodes, 1007
Hyphae, Phycomycetes, 47
Hypal thalli, Phycomycetes, 48
Hyphochytriaceae, 72
Hyphochytriales, 50
Hyphochytrium, 72, 73
Hyphomicrobium, 21, *30*
Hypnodinium, 165, 164
Hypocular suture, Hemiptera, 969
Hyponeura, 924, *926*
Hypopharyngeal muscle, Rotifera, 428
Hypostomata, 267, 274
Hypostome, 314
Hypotrichidium, 287, 288
Hypotrichina, 282, 286
Hypsibius, 517: *alpinus*, 520; *angustatus*, 520; *annulatus*, 518; *arcticus*, 519; *asper*, 518; *augusti*, 518; *baldii*, 518; *chilenensis*, 520; *conjungens*, 520; *convergens*, 519; *dujardini*, 519; *evelinae*, 519; *granulifer*, 518; *myrops*, 519; *nodosus*, 518; *oberhaeuseri*, 519; *oculata* forma *vancouverensis*, 520; *ornatus*, 518; *ornatus caelata*, 518; *pallidus*, 519; *papillifer*, 518; *papillifer bulbosa*, 518;

prorsirostris, 520; *prosostomus*, 519; *recamieri*, 520; *sattleri*, 518; *schaudinni*, 519; *scoticus*, 520; *spitzbergensis*, 520; *tetradactyloides*, 519; *trachydorsatus*, 519; *tuberculatus*, 518; *verrucosus*, 518; *zetlandicus*, 519
Hypsibius, subgenus of *Hypsibius*, 519

Ichthydium, 41: *auritum*, 412; *brachykolon*, 411; *cephalobares*, 411; *leptum*, 413; *macropharyngistum*, 412; *minimum*, 412, 413; *monolobum*, 411; *podura*, 412; *sulcatum*, 412
Idiataphe, 937, *939*
Ileonema, 269, 268
Illinobdella, 552: *alba*, 552; *moorei*, 552, 553
Ilybius, 994, 1011
Ilyocryptus, 630, *635: acutifrons*, 631; *sordidus*, 631; *spinifer*, 632
Ilyocyprinae, 664
Ilyocypris, 664: *biplicata*, 665; *bradyii*, 665, 664; *gibba*, 664
Ilyodrilus, 536: *fragilis*, 536; *perrierii*, 536; *sodalis*, 536
Ilyodromus, 696: *pectinatus*, 696
Imperfect fungi, 22
Inner caudal seta, Harpacticoida, 819
Inoperculatae, 51
Inoperculate sporangia, Chytridales, 48
Insecta, 902
Intercalary bands, Bacillariophyceae, 175
Intranstylum, 292, 293
Io, 1136
Ioscytus, 965
Iotonchus, 395
Iron, 914
Ironidae, 395
Ironinae, 396
Ironodes, 914
Ironopsis, 914
Ironus, 396
Irwin loops, 1199, 1201
Ischium, Malacostraca, 870
Ischnura, 926, 918, 923, 925
Isoachlya, 79, 77
Isocapnia, 949, 950, 951
Isoetes: muricata, 1182, 1181; *engelmanni*, 1182, 1181
Isogenoides, 954
Isogenus, 954, 955, 956, *946*, 949
Isohypsibius, subgenus of *Hypsibius*, 518
Isonychia, 912, 910
Isonychiinae, 912
Isoperla, 956, 953, 955, 946
Isopoda, 872
Isopods, 870
Isorhizas (see Nematocysts)

Isthmus, Nemata, 371
Itaquascon, 517
Ithytrichia, 1039, 1026
Itonididae, 1059
Itura, 461, 471, *437*, 460, 470

Jaws: Nemata, 370; Rotifera, 422
Jenningsia, 125, 124
Johannesbaptistia 97: *pellucida*, 97
Johannsenomyia, 1068
Jungermannia, 1163, 1162, *1169*
Jungermanniales, 1162

Kahlia, 287
Kalyptorhynchia, 341
Karlingia, 59
Kathroperla, 956, 952, 953, 955
Keel, Bacillariophyceae, 175
Keel punctae, Bacillariophyceae, 175
Kellicottia, 442, 443, *437: bostoniensis*, 442, 443; *longispina*, 442, 443
Kenkiidae, 329, 326
Kentrosphaera, 133, 132
Kephyrion, 155, 154
Keratella, 442, 443, 444, *437*
Keriochlamys, 137, 136
Kerona, 288
Keronopsis, 288, 289
Kincaidella, 1131
Kincaidiana, 533: *hexatheca*, 533
Kinetonucleus, 191
Kirchneriella, 139, 138
Kitagamiidae, 1024
Klattia, 359: *virginiensis*, 359
Koenikea, 1089, 1103, 1102
Kogotus, 555
Koinocystidae, 357
Koinocystis, 357
Kongsbergia, 1088, 1105
Krendowskia, 1091, 1106
Krendowskijidae, 1106
Krizousacorixa, 968, *962*
Krumbachia, 345: *minuta*, 345; *virginiana*, 346
Kurzia, 638, *637: latissima*, 638
Kybotion, 159, 158
Kyliniella, 167, 168

Labium, Ostracoda, 659
Labrum, Ostracoda, 659
Laccobius, 1000, 1012, 985
Laccodytes, 993
Laccophilus, 993, 1011
Laccornis, 992
Lacewing flies, 904
Lachlania, 913

Lacinia, Plecoptera, 942
Lacinularia, 476, 479, *438*, *478*, 470: *flosculosa*, 479, 480; *ismailoviensis*, 476, 480
Lacrymaria, 269, 268
Ladona, 937, *939*
Laeonereis culveri, 540, 539
Laevapex, 1130
Lagenidiales, 83, *11*
Lagenidium, 85, 84
Lagenoeca, 207
Lagenophryidae, 293, 294
Lagenophrys, 294
Lagynion, 159, 158
Laminipes, 1104, 1107
Lampracanthia, 965
Lamprocystis, 44, 32
Lampsilinae, 1149
Lampsilis, 1155
Lancidae, 1125, 1128, *1118*
Langenidiaceae, 83
Lanthus, 928, 929
Lanx, 1128
Laophonte: mohammed, 825; *proxima*, 827: *strömi*, 826
Laophontidae, 816
Lara, 1002, 1018, 986, 987
Larva, Insecta, 903
Larvaevoridae, 1077
Lasmigona, 1146, 1147
Lateral setae, Harpacticoida, 819
Lateral spines, Cladocera, 594
Lathonura, 630: *rectirostris*, 630
Latona, 600: *parviremis*, 600, 601; *setifera*, 600
Latonopsis, 602: *fasciculata*, 603; *occidentalis*, 602
Latrostium, 74, 73
Lebertia, 1086, 1098, 1099
Lebertiae, 1098
Lebertiidae, 1098
Lecane, 456, 457, *437*
Lecanidae, 437
Lecithoepitheliata, 360
Lecythium, 257: *hyalinum*, 257; *mutabile*, 257
Leeches, 542
Legendrea, 272, 271
Legs: Cladocera, 592; Ostracoda, 659; Copepoda, 736
Lejops, 1077
Lemanea, 169, 168
Lemaneaceae, 169
Lembadion, 278
Lemiox, 1156
Lemna: minor, 1177, 1175; *trisulca*, 1177, 1175
Lemnaphila, 1077
Lemonniera, 89, 88
Lepadella, 451, 452, *437*, *78*
Lepidocaris rhyniensis, 587

Lepidodermella, 413: *squamatum*, 413; *trilobum*, 414, 415
Lepidoptera, 905, 906, 907, 1050
Lepidostoma, 1046, 1028, 1030
Lepidostomatidae, 1028, 1045
Lepidurus, 575, *573*: *arcticus*, 575; *bilobatus*, 575; *couesii*, 575, *572*; *lynchi*, 575; *lynchi* var. *echinatus*, 576; *macrurus*, 575; *packardi*, 575; *patagonius*, 575
Lepocynclis, 125, 123
Leptestheria, 583, 580: *compleximanus*, 583
Leptestheriidae, 583, *579*, *580*
Lepthemis, 938, *939*
Leptidae, 1074
Leptocella, 1049, 1048
Leptoceridae, 1026, 1047, 1048
Leptocerus, 1047, 1028, 1048
Leptoconopinae, 1067
Leptoconops, 1067
Leptodea, 1153
Leptodiaptomus, subgenus of *Diaptomus*, 739, 780, 781, 783
Leptodictyum, 1167, *1169*
Leptodora, 598, *588*, 589: *kindtii*, 598, 589
Leptodoridae, 598, 589
Leptohyphes, 912
Leptolaimidae, 389
Leptolegnia, 79, 76
Leptomitales, 75, 11, 48
Leptomitus, 81
Leptomycetaceae, 81
Leptonchidae, 398
Leptonchinae, 398
Leptonchus, 398
Leptophlebiidae, 911, 912, 913, 914, 910
Leptorhynchus dentifer, 654: *excisa*, 655; *exigua*, 655
Leptosira, 147, 144
Leptospira, 44, 21
Leptothrix, 44, 45, *19*, 31, 39, 41: *crassa*, 38, 39; *ochracea*, 38, 39; *trichogenes*, 39
Leptoxis, 1137
Lepyriidae, 1125, 1136, *1118*
Lepyrium, 1136
Lesquereusia, 246: *epistomium*, 246; *modesta*, 246; *spiralis*, 246
Lernaea, 866: *anomala*, 864; *carassii*, 864; *catostomi*, 864; *cruciata*, 864; *cyprinacea*, 864; *dolabrodes*, 864; *insolens*, 864; *pectoralis*, 864; *pomotidis*, 864; *ranae*, 864; *tenuis*, 864; *tortua*, 864; *tortua coquae*, 864; *variabilis*, 864
Lernaeidae, 866
Lernaeopodidae, 866
Lestes, 924, 921, 923, 925
Lestidae, 921, 924
Lethocerus, 966, *960*

Leucophrydium, 279, 278
Leucophrys, 278
Leucophytes, 10, *14*
Leucorrhinia, 938, *939*, 932
Leucosin, 191
Leucotrichia, 1037
Leuctra, 950, 953
Leuctrinae, 949
Leuvenia, 151, 150
Lexingtonia, 1143
Leydigia, 639: *acanthocercoides*, 640; *quadrangularis*, 640
Libellula, 938, 918, *939*, 930, 932, 935
Libellulidae, 922, 934, *919*
Libellulinae, 934
Light: absorption by pigments, 34; attraction by, 1195, 1196 (see also *Daphnia* trap)
Ligumia, 1154, 1155
Limnadia, 581: *americana*, 581; *lenticularis*, 581
Limnadiidae, 581, *579*
Limnebius, 998, 1013
Limnephilidae, 1028, 1042, 1043
Limnephilid Genus A, 1044, 1043
Limnephilid Genus C, 1044
Lemnephilid Genus D, 1044, 1043
Limnephilus, 1045, 1027, 1028, 1043
Limnesia, 1086, 1088, 1089, 1100
Limnesiella, 1101
Limnesiidae, 1100
Limnesiopsis, 1101
Limnetis, 580, *579*: *brachyurus*, 580; *brevifrons*, 580; *gouldi*, 580; *gracilicornis*, 580; *mucronatus*, 581
Limnias, 476, 477, *438*
Limnichidae, 1015, *985*
Limniids, *579*
Limnius, 1005, 1018, *987*
Limnobatodes, 962, *959*
Limnocalanus, 752, *738*: *johanseni*, 755; *macrurus*, 755
Limnocamptus, subgenus of *Bryocamptus*, 830
Limnocentropidae, 1024
Limnocharae, 1094
Limnochares, 1089, 1094, 1095
Limnocharidae, 1094
Limnocoris, 967
Limnocythere, 724: *glypta*, 726, 727; *illinoisensis*, 725, 724; *inopinata*, 726; *ornata*, 725; *reciculata*, 724; *sancti-patrici*, 725; *verrucosa*, 725, 726
Limnodrilus, 535: *claparedianus*, 536; *gracilis*, 535; *hoffmeisteri*, 536; *udekemianus*, 536
Limnogonus, 962
Limnohalacarinae, 1110
Limnohalacarus, 1110, 1109
Limnometra, 959

Limnophila, 1062, 1063
Limnophora, 1079
Limnozetes, 1114, *1110*
Limonia, 1062
Limoniinae, 1061
Lindia, 467, *437*
Lindiidae, 437
Lionotus, 273
Lioplax, 1132
Lipostraca, 587, 588
Lipsothrix, 1060
Lirceus, 875: *alabamae*, 875; *bicuspidatus*, 875; *bidentatus*, 875; *brachyurus*, 875; *fontinalis*, 875; *garmani*, 875; *hargeri*, 875; *hoppinae*, 875; *louisinae*, 875; *megapodus*, 875; *richardsonae*, 875; *trilobus*, 875
Liriope, 1073
Liriopeidae, 1072
Liriopeinae, 1073
Lispa, 1079
Lispe, 1079
Lissorhoptrus, 1007, *987*
Listronotus, 1007
Lithasia, 1136
Littorella americana, 1185, 1184
Littoridina, 1134
Liverworts, 1161
Lixellus, 1007
Ljania, 1088, 1105
Lobelia dortmanna, 1182, 1174
Lobocystis, 139, 138
Lobohalacarus, 1110, 1109
Lobomonas, 127, 126
Loborhiza, 55, 54
Lobosa, 233
Locule, Bacillariophyceae, 175
Lohmannella, 1108
Lohmannellinae, 1110
Longurio, 1061
Lophocharis, 454, *437*
Lophopodella, 499: *carteri*, 502
Lophopodidae, 499
Lophopus, 499: *crystallinus*, 502
Loramyces, 89, 87: *asci*, 50
Lorica: Rotifera, 424; Zooflagellate, *191*
Loricata, 291, 293
Lower protists, 12
Loxocephalus, 281
Loxodes, 273
Loxodidae, 272, 273
Loxophyllum, 273, 272
Ludvigia palustris, 1179, 1178
Lumbricidae, 534
Lumbricillus, 534
Lumbriculidae, 532

Lumbriculus, 533: *inconstans*, 533; *variegatus*, 533
Lunulospora, 93, 92
Lutherella, 153, 152
Lutrochus, 1015, 1016, *985*
Lycastoides alticola, 540, 539
Lycastopsis, 540, 539
Lymnaea, 1126, 1127
Lymnaeidae, 1125, 1126
Lynceidae, 580
Lynceus, 580, *579: brachyurus*, 580, 579; *brevifrons*, 580; *gouldi*, 580; *gracilicornis*, 580; *mucronatus*, 581
Lyngbya, 110: *aestuarii*, 110, 111; *birgei*, 110; *contorta*, 110; *diguetii*, 110; *epiphytica*, 110; *giuseppei*, 110; *patrickiana*, 110; *putealis*, 110; *taylorii*, 110; *thermalis*, 111; *versicolor*, 110
Lype, 1033, 1032
Lyrodes, 1134
Lyrogyrus, 1135

Macilla, Plecoptera, 942
Macrobdella, 553: *decora*, 553; *ditetra*, 553; *sestertia*, 553
Macrobiotidae, 515
Macrobiotus, 515: *ambiguus*, 517; *ampullaceus*, 517; *dispar*, 515, 517; *dubius*, 516, 517; *echinogenitus*, 516, 517; *furcatus*, 516; *furciger*, 516, 517; *grandis*, 517; *harmsworthi*, 516; *harmsworthi* var. *coronata*, 516; *hastatus*, 516; *hufelandii*, 516, 517; *intermedius*, 516; *islandicus*, 516, 517; *macronyx*, 516, 517; *montanus*, 516; *occidentalis*, 516, 517; *pullari*, 516, 517; *richtersii*, 516; *tonollii*, 516
Macrobrachium, 880: *acanthurus*, 882; *carcinus*, 881; *jamaicensis*, 881; *ohione*, 882; *olfersii*, 882
Macrochaetus, 449, 450, *437*
Macrochytrium, 66
Macrocotyla glandulosa, 334
Macrocyclops, 797: *ater*, 796
Macrodiplax, 938, *934, 939*
Macrolaimus, 386
Macromastix, 218
Macromia, 934, 932, 935
Macromiidae, 922, 934, *919*
Macromonas, 44, 45, 37
Macronemum, 1033, 1034
Macronychus, 1005, 986, 987, 1016, 1018
Macrostomida, 325, 338, 339
Macrothemis, 936, *939*
Macrothricidae, 604, *588*, 589
Macrothrix, 629: *borysthenica*, 633; *hirsuticornis*, 634; *laticornis*, 633; *montana*, 633; *rosea*, 632
Macrotrachela, 488, 486, *437*
Macrovelia, 963, *962*
Malaconothridae, 1110

Malaconothrus, 1112, 1113, *1110*
Malacophrys, 280, 279
Malacostraca, 559, 869
Malenka, 948
Males, Cladocera, 595
Malirekus, 955, 949, 954
Malleochloris, 131, 130
Malleodendraceae, 151
Malleodendron, 151, 150
Mallomonadaceae, 155
Mallomonas, 155, 154
Mamersella, 1086, 1098
Manayunkia, 540, 544, 539: *speciosa*, 540, 539
Mancasellus, 875
Mandibles: Copepoda, 736; Ostracoda, 659
Mandibulata, 558
Manfredium, 449; *437*
Mansonia, 1064, 1065
Manubrium, Rotifera, 428, 429
Maraenobiotus, 832, *816: brucei*, 833; *insignipes*, 833
Margaritifera, 1138
Margaritiferidae, 1125, 1138
Margaritispora, 89, 88
Marilia, 1046, 1047
Marionina, 534
Marisa, 1132
Marsh treader, 962
Marshallothyas, 1092
Marshia, 837: *albuquerquensis*, 838; *dominicanus*, 838
Marsilea, 1177: *macropoda*, 1177; *quadrifolia*, 1177; *uncinata*, 1177; *vestita*, 1177
Marssaniella, 97
Marsupella, 1163, *1169*
Marsupium, Malacostraca, 869
Martarega, 967
Maruina, 1066, *1059*
Maryna, 276
Marynidae, 275, 276
Massartia, 163, 161
Mastax, Rotifera, *420*, 428
Mastigamoeba, 194, 236
Mastigamoebidae, 236
Mastigella, 194
Mastigodiaptomus, subgenus of *Diaptomus*, 739, 762
Mastigogenina, 234
Mastigophora, 9, 10, 11, *14*, 115, 123
Mastogloia, 183, *174*
Matus, 994
Maxillae: Copepoda, 736; Ostracoda, 659
Maxilliped, Malacostraca, 870
Maxillules, Copepoda, 736
Mayatrichia, 1039, 1037, 1038

Mayflies, *903*
Mayorellidae, 234
Medionidus, 1151
Megachytriaceae, 65
Megachytrium, 68
Megacyclops, subgenus of *Cyclops*, 802
Megadytes, 996
Megaleuctra, 951
Megaloptera, 905, 906, 907, 904, 973
Megalotrocha, 476, 479
Megaluracurus, 1107
Megalurus, 1107
Megapus, 1101, 1107
Megarcys, 954, 948, 954
Megarhinus, 1064
Megistocera, 1061
Melicerta, 478, 480
Meloidodera, 383
Meloidogyne, 383
Melosira, 187, *174*
Melosiroideae, 174
Melusinidae, 1072
Membrane: Hemiptera, 969; undulating, Ciliophora, 265
Membranelles, Ciliophora, 265
Menetus, 1130
Menoidium, 125, 124
Mentum, Odonata, 917
Merceya, 1167, *1169*
Mercierella enigmatica, 540, 539
Meridion, 177, *174*
Meridionoideae, 174
Meringodixa, 1073
Merismopoedia, 98
Mermithoidea, 398
Meromyarian, Nemata, 370
Merotrichia, 165, 164
Merragata, 964
Mesenchytraeus, 534
Mesochra, 838, 816: *alaskana*, 840; *lilljeborgi*, 839; *rapiens*, 839
Mesocyclops, 797: *albidus*, 815; *distinctus*, 814; *dybowskii*, 813; *edax*, 811; *fuscus*, 814; *hyalinus*, 812; *inversus*, 814; *leuckarti*, 812; *oithonoides*, 812; *tenuis*, 813
Mesodinium, 271
Mesonotum, Lepidoptera, 1051
Mesoporodrilus, 533: *asymmetricus*, 533; *lacustris*, 533
Mesorhabdions, Nemata, 370
Mesostigma, 127, 126
Mesostom; Nemata, 370
Mesostoma, 353: *andrewsi*, 356; *arcticum*, 354; *californicum*, 354; *columbianum*, 356; *curvipenis*, 356; *ehrenbergii*, 354, 355; *macropenis*, 354;

macroprostatum, 355; *vernale*, 356; *virginianum*, 355
Mesostominae, 353
Mesotaeniaceae, 139
Mesotaenium, 141, 140
Mesotardigrada, 513
Mesothoracic shield, Lepidoptera, 1051
Mesovelia, 964, 959
Mesoveliidae, 963, *959*
Mesoveloidea, 964
Metacineta, 295
Metacorpus, Nemata, 371
Metacypris, 723: *americana*, 724; *maracaoensis*, 723
Metacystidae, 268, 271
Metacystis, 271, 270
Metadiaschiza, 454
Metaleg, Lepidoptera, 1051
Metamorphosis, Insecta, 903
Metarhabdions, Nemata, 370
Metasome, Copepoda, 735
Methane fermentation, 42
Methanobacterium, 42
Methanococcus, 42
Methanosarcina, 42
Methods, 1194:
   Collection, 1194–1197: algae, 116; bacteria, 26; Coleoptera, 988; fungi, 48; Gastrotricha, 408; Mollusca, 1119; Nemata, 375; Ostracoda, 661; planarians, 328; Plecoptera, 941
   Concentration, 1194: bacteria 26
   Culturing: bacteria, 22; fungi, 48; Hydra, 315
   Dissection, 1201: Calanoida, 748; Cyclopoida, 795, Harpacticoida, 818; Ostracoda, 662
   Fixation (see also preservation): Cyclopoida, 795; Gastrotricha, 408n; Hirudinea, 547; planarians, 328; Rotifera, 433
   Measurement, 1202
   Mounting, 1200: algae, 96, 117; Bacillariae, 171; bacteria, 23; Calanoida, 748; Cladocera, 597; Cyclopoida, 795; Gordiida, 403; Nemata, 377; Ostracoda, 662
   Observation, 1197: bacteria, 23
   Preservation, 1200: Acari, 1081, 1108, 1110; algae, 96, 117; Bacillariae, 171; Bryozoa, 498; Cladocera, 597; Ephemeroptera, 911; Hirudinea, 546; Mollusca, 1120; Ostracoda, 662; Rotifera, 433
   Relaxation, 1199: Bryozoa, 498; Gastrotricha, 408n; Hirudinea, 546; Rotifera, 433
   Sectioning, 1202
   Staining, iron bacteria, 24
Metidae, 816
Metis, 831, *816: jousseaumei*, 831; *sarsi*, 831

Metopidae, 283
Metopidia, 451
Metopus, 283
Metostom, Nemata, 370
Metretopodinae, 913
Metretopus, 915
Metrichia, 1036
Metriocnemus, 1072
Metrobates, 963, *959*
Meyenia, 304, *303: crateriformis*, 306; *fluviatilis*, 305, *304; millsii*, 304; *mülleri*, 305; *robusta*, 306; *subdivisa*, 304; *subtilis*, 306
Meyeninae, 300
Miathyria, 937, *939*
Michrochloris, 34
Micracanthia, 965
Micractiniaceae, 133, 132
Micractinium, 135, 132
Micrasema, 1041, 1042
Micrasterias, 141, 140
Micrathyria, 938, *939*
Microarthridion, 828, *816: littorale*, 828
Microchlamys, 237
Microchloris, 35, 44
Micrococcus, 44, 17
Microcodides (see *Mikrocodides*)
Microcodon, 464, *438*
Microcodonidae, 438
Microcoleus, 107: *acutissimus*, 107; *lacustris*, 107; *paludosus*, 107; *rupicola*, 107; *vaginatus*, 107, 106
Microcometes, 259
Microcometesidae, 259
Microcorycia, 236
Microcoryciidae, 236
Microcrocis geminata, 98
Microcyclops, subgenus of *Cyclops*, 801
Microcylloepus, 1004, 1018, 986
Microcystis, 97
Microcysts, Myxobacteria, 20
Microdalyellia, 343: *gilesi*, 343
Microdiaptomus, 780, *739*
Microdina, 489, *437*
Microgromiidae, 259
Microhydra ryderi, 314, 316
Microkalyptorhynchus, 357: *virginianus*, 357
Microlaimidae, 392
Microlaimoides, 393
Microlaimus, 393, 392
Micromonospora, 44
Micromyces, 52
Micromycopsis, 52
Microregma, 269
Microspora, 145, 144
Microsporaceae, 145

Microstomidae, 339
Microstomum, 339: *bispiralis*, 340; *caudatum*, 340; *lineare*, 339
Microtendipes, 1070
Microthamnion, 147, 144
Microthorax, 278, 277
Microvelia, 963, *959, 962*
Micruracarus, 1107
Micrurus, 1107
Middle caudal seta, Harpacticoida, 819
Midea, 1089, 1106, 1105
Mideidae, 1106
Mideopsae, 1106
Mideopsidae, 1106
Mideopsis, 1087, 1106
Mikrocodides, 464, 454, *437*
Milnesiidae, 515
Milnesium tardigradum, 515
Mindeniella, 82, 81
Mischococcaceae, 153
Mischococcus, 153, 152
Mites, 1080
Mitochytridium, 57
Mitrula, 87, 86: *asci*, 49
Mixodiaptomus, 773, *739*
Mixolebertia, 1099
Mixosperchon, 1098
Mniobia, 485, 486, *437*
Mobilia, 290
Mochlonyx, 1063, 1064
Moina, 621: *affinis*, 624, 623; *brachiata*, 623; *hutchinsoni*, 622; *irrasa*, 622; *macrocopa*, 622, 623; *micrura*, 621; *rectirostris*, 623
Moinodaphnia, 621: *macleayii*, 621
Molanna, 1049, 1027, 1029, 1030
Molannidae, 1026, 1049
Molannodes, 1049
Mollusca, 1117
Mollusks, 1117
Molophilus, 1062
Monallantus, 151, 150
Monads, 118
Monas, 220
Monera, 9
Monhystera, 393
Monhysteridae, 393
Monhysteroidea, 392
Monhystrella, 393
Monoblepharella, 72, *70*
Monoblepharidales, 70, 11: habitat, 48
Monoblepharis, 72, 73
Monocentric thalli, Phycomycetes, 48
Monochilum, 280, 279
Monochromadora, 393
Monochus, 395

Monocilia, 154, 152
Monociliaceae, 154
Monogononta, 438, 437
Monohysterida, 386
Monomastix, 166, 164
Monommata, 465, 464, *437*
Mononchidae, 394
Mononchulus, 395
Mononyx, 965
Monopytephorus, 535
Monoraphidineae, 174
Monosiga, 204
Monospilus, 635: *dispar*, 635
Monostichodont, teeth, Hirudinea, 544
Monostroma, 149, 148
Monostyla, 456, 457, *437*
Mooreobdella, 555: *bucera*, 554; *fervida*, 554, 556; *microstoma*, 544
Mopsechiniscus, 515
Moraria, 833, *816: affinis*, 836; *americana*, 836; *cristata*, 835; *duthiei*, 835; *laurentica*, 836; *mrazeki*, 837; *virginiana*, 835
Morphocorixa, 970, *961*
Moselia, 950
Mosquito-hawk, 920
Mosses, 1161
Mougeotia, 145, 142
Mougeotiopsis, 145, 142
Mrazekiella, subgenus of *Atthyella*, 844
Mudalia, 1137
Muticellular algae, 14
Multicilia, 192
Multifasciculatum, 296, 295
Musci, 1161, 1163
Musculium, 1158
Musidae, 1079
Mussels, 1117
Mycelium, Phycomycetes, 47
Mycetozoa, 48
Mycterothrix, 276
Mydaeina, 1079
Myelostoma, 286
Myelostomidae, 286
Myersinella, 461, *437*
Mylonchulus, 395
Myodocopa, 663
Myriophyllum: alterniflorum, 1179; *brasiliense*, 1179; *exalbescens*, 1180; *heterophyllum*, 1180; *verticillatum*, 1179
Myrmecia, 133, 132
Mysidacea, 872, 878
Mysids, 869
Mysis relicta, 879
Mystacides, 1049, 1048
Mytilina, 456, *437*, 449

Myxobacteria, 14, 20, 44
Myxophyceae, 118, 95, *117*
Myzocytium, 85, 84

Naegeliella, 160
Naegeliellaceae, 160
Naididae, 527
Nais, 531: *barbata*, 531; *communis*, 531; *elinguis*, 531; *pseudobtusa*, 531; *simplex*, 531; *variabilis*, 531
Najadicola, 1090, 1091, 1102
Najas: flexilis, 1179; *gracillima*, 1179; *minor*, 1179; *guadalupensis*, 1179
Namamyia, 1046
Namanereis hawaiiensis, 540, 541, 539
Nannochloris, 133, 132
Nannopus, 837, *816: littoralis*, 837; *palustris*, 837
Nannothemis, 939
Narpus, 1005, 1018, 987
Nasiaeschna, 931, *934*, 932, 933
Nassula, 274
Nassulidae, 274
Natatory setae, Ostracoda, 659
Naucoridae, 966, *961*
Nauplius larva, *Leptodora*, 589, 594
Nauplius stages, copepod, 738, 745–747
Navicula, 185, *174*
Naviculoideae, 174
Neanthes: lighti, 540, 539; *limnicola*, 540, 541, 539; *saltoni*, 540, 539; *succinea*, 540, 541, 539
Nebela, 248, 247: *caudata*, 249; *collaris*, 250; *dentistoma*, 249; *flabellulum*, 250; *lageniformis*, 250; *militaris*, 250
Nebelidae, 240
Nehalennia, 925, *926*
Neidium, 185, *174*
Nelumbo lutea, 1182, 1172
Nema, 368
Nemalionales, 167
Nemata, 368
Nemates, 368
Nemathelminthes, 368
Nematocera, 1059
Nematocysts, 315
Nematoda, 368
Nematodes, 368
Nematoidea, 368
Nematomorpha, 402, *368*
Nemertea, 366
Nemocapnia, 950, *946*
Nemotelus, 1074
Nemoura, 948, 949, 952, 953
Nemouridae, 948
Nemourinae, 948
Neoatax, 1102, 1101

Neoaxonopsis, 1088, 1105
Neochordodes, 404
Neocloeon, 916
Neocorixa, 968, *961*
Neoelmis, 1004, 1017, 987
Neoephemera, 912, 909
Neoephemeridae, 912
Neogossea, 409: *fasciculata*, 409; *sexiseta*, 409
Neogosseidae, 409
Neohaemonia, 1006, 1020
Neohermes, 976, *974*
Neohydrophilus, 999
Neomysis mercedis, 879
Neoneura, 925, *924, 926*
Neoperla, 952, **948**
Neophasganophora, *946*
Neophylax, 1042, 1043
Neoplanorbis, 1131
Neorhabdocoela, 325, 341
Neoscutopterus, 995
Neothremma, 1042, 1043
Neotrichia, 1039, 1038
Nepa, 965, *960*
Nephelopsis obscura, 555, 556
Nephrochytrium, 66, 67
Nephrocytium, 139, 138
Nephropsidae, 879
Nephroselmidaceae, 167
Nephroselmis, 167, 166
Nepidae, 965, *960*
Nepticula, 1051
Nerophilus, 1046
Nerthra, 965, *960*
Nesaea, 1104, 1107
Netrium, 141, 140
Neumania, 1090, 1091, 1102
Neureclipsis, 1032, 1031, 1032
Neurocordulia, 936, *934*, 932, 935: *molesta*, 922
Neuroptera, 904, 905, 906, 907, 973
Nevskia, *31*
Nigronia, 976, *974*
Nitella, 149, 150
Nitocra, 829, *816*: *hibernica*, 830; *lacustris*, 829; *spinipes*, 829, 830; *typica*, 830
Nitocrella, 829, *816*: *incerta*, 829; *subterranea*, 829
Nitrosomonas, 27
Nitzchia, 181, 180, *174*
Nitzchiaceae, 174
Nitzchioideae, 174
Nocardia, 44
Nodal furrow, Hemiptera, 969
Nodularia, 102: *harveyana*, 103; *sphaerocarpa*, 102; *spumigena*, 102
Nordodiaptomus, 774, *739*
Nostoc, 100: *amplissimum*, 101; *caeruleum*, 101;
carneum, 101; *commune*, 102; *cuticulare*, 100; *ellipsosporum*, 102; *entophytum*, 101; *hederulae*, 101; *humifusum*, 102; *linckia*, 101; *macrosporum*, 102; *maculiforme*, 101; *microscopicum*, 102; *muscorum*, 102; *paludosum*, 101; *parmelioides*, 101; *piscinale*, 101; *pruniforme*, 101; *rivulare*, 101; *sphaericum*, 101; *spongiiforme*, 101; *verrucosum*, 101
Nostocaceae, 100
Nostochopsis, 99: *lobatus*, 99
Noteridae, 991, 1010, 983, 984
Noteus, 451
Notholca, 445, 444, *437*, 442
Nothotylenchinae, 384
Nothotylenchus, 384
Nothrus, 1112
Notiomyia, 1046, 1047
Notiphila, 1077
Notodromus monacha, 689
Notogillia, 1135
Notomicrus, 997, 983, *991*
Notommata, 473, *437*
Notommatidae, 437
Notonecta, 967, *960*
Notonectidae, 967, *960*
Notops, 469
Notosolenus, 125, 124
Notostraca, 559, 572, *587*
Nowakowskia, 61, 59
Nowakowskiella, 68, 69
Nuchal organ, Notostraca, 573
Nuclearia, 263
Nudechiniscidae, 513
Nuphar, 1182: *polysepalum*, 1183; *sagittifolium*, 1183; *variegatum*, 1183, 1172
Nyctiophylax, 1031
Nygolaiminae, 398
Nygolaimus, 398
Nymphaea, 1182: *odorata*, 1182, 1172; *tuberosa*, 1182
Nymphoides, 1186: *aquaticum*, 1186; *cordatum*, 1186, 1185, *1174*; *peltatum*, 1186
Nymphs: Insecta, 903; Odonata, 917
Nymphula, 1052, 1053, 1054, *1051*

Obelidium, 61, 60
Obliquaria, 1150
Obovaria, 1153
Occidentalia, 1053, 1055, *1051*
Ocellus, Cladocera, 590
Ochromonadaceae, 156
Ochromonas, 156
Ochrotrichia, 1037, 1039, 1038
Ochteridae, 965, *960*
Ochterus, 960

Ochthebius, 998, 1013
Ochthera, 1077
Octogomphus, 928, 929, 932
Octomyxa, 74
Octotrocha, 478, *479*, *438: speciosa*, 478, 479, 476
Oculobdella lucida, 551
Odonata, 906, 907, *904*, 917
Odontoceridae, 1029, 1046
Odontocerid Genus A, 1046, 1047
Odontomyia, 1074
Oecetis, 1047, 1026, 1048
Oecistes, 476
Oedocladium, 149, 148
Oedogoniaceae, 149
Oedogonium, 149, 148
Oicomonas, 201
Oionchus, 398
Oligobdella biannulata, 550, 551
Oligochaeta, 522
Oligoneuriidae, 912
Oligophlebodes, 1044, 1043
Oligoplectrum, 1041
Oligostomis, 1041, 1040
Oligotricha, 1040
Oligotrichina, 282, 285
Olisthanella, 349
Olisthanellinae, 349
Olpidiaceae, 51
Olpidiopsidaceae, 83
Olpidiopsis, 85, 83
Olpidium, 51
Onychocamptus, 825, *816: calamorum*, 825; *mohammed*, 825; *talipes*, 825
Onychodiaptomus, subgenus of *Diaptomus*, 739, 790, 791
Onychodromopsis, 289
Onychodromus, 289
Onychonema, 143, 142
Onychopoda, *588*, 589, 598
Onychylis, 1007
Oöcardium, 143, 142
Oöcystaceae, 135
Oöcystis, 137, 139, 136
Oödinium, 165, 162
Oöphila, 133, 132
Opephora, 178, *174*
Opercularia, 292
Operculatae, 51
Operculate sporangia, Chytridiales, 48
Ophidonais, 530: *serpentina*, 530, 531
Ophiobolus, 89, 88: asci, 50
Ophiocytium, 153, 152
Ophiogomphus, 928, *929*, 927
Ophrydiidae, 291, 293
Ophrydium, 293

Ophryoglena, 281, 282
Ophryoglenidae, 278, 281
Ophryoxus, 626: *gracilis*, 626
Opisthocysta flagellum, 528, 529
Opisthocystidae, 528
Opisthonecta, 290
Opisthotricha, 290
Opistominae, 353
Opistomum, 353: *pallidum*, 353
Oplonaeschna, 931, *934*
Opposum shrimps, 869
Optioservus, 1019, 986, 1006, 1014, 1016, 1017, 1018
Oral groove, Ciliophora, 265
Orconectes, 883: *alabamensis*, 894; *beyeri*, 890; *clypeatus*, 890, 891; *compressus*, 895; *difficilis*, 892; *eupunctus*, 894; *harrisoni*, 894; *hathawayi*, 892; *hobbsi*, 895; *hylas*, 892; *immunis*, 894; *indianensis*, 893; *inermis*, 890; *juvenilis*, 892; 891; *kentuckiensis*, 894; *lancifer*, 890; *leptogonopodus*, 892; *limosus*, 893, 891; *longidigitus*, 894; *luteus*, 893; *marchandi*, 895; *medius*, 893; *meeki*, 895; *menae*, 893; *mississippiensis*, 892; *nais*, 894; *nana*, 892; *neglectus*, 893; *obscurus*, 893; *ozarkae*, 892; *palmeri*, 892; 891; *pellucidus*, 890; *peruncus*, 892; *propinquus*, 894, 891; *punctimanus*, 895; *rafinesquei*, 893; *rhoadesi*, 895; *rusticus*, 893; *shoupi*, 893; *sloani*, 894, 891; *tricuspis*, 894; *validus*, 895; *virginiensis*, 893; *virilis*, 894, 891; *wrighti*, 893
Orconectes, subgenus of *Orconectes*, 890
Ordobrevia, 1003, 1017, 987
Oreella, 513
Oreodytes, 993, 1010
Oribatei, 1108, 1110
Oribatidae, 1110
Orontium aquaticum, 1182
Oroperla, 954, 948
Oropsyche, 1033
Orphnephilidae, 1069
Orthemis, 938, *939*
Orthocladiidae, 1070
Orthocladius, 1071, 1072
Orthocyclops, 796: *modestus*, 796
Orthodon, 274
Orthopodomyia, 1064
Orthotrichia, 1039
Oscillatoria, 112, 12: *acuminata*, 113; *agardhii*, 112; *amoena*, 113; *anguina*, 113; *animalis*, 113; *articulata*, 113; *boryana*, 113; *brevis*, 113; *chalybea*, 113; *chlorina*, 113; *cortiana*, 113; *curviceps*, 113; *formosa*, 113; *geminata*, 113; *granulata*, 114; *irrigua*, 113; *limosa*, 113; *okeni*, 113; *ornata*, 113; *princeps*, 112; *proboscidea*, 112; *prolifica*, 112; *rileyi*, 112; *rubescens*, 112;

sancta, 113; *simplicissima*, 113; *splendida*, 113; *tenuis*, 113; *terebriformis*, 114
Oscillatoriaceae, 105
Osobenus, 956
Osphranticum, 756, *738*: *labronectum*, 756
Ostracoda, 559, 657
Otomesostoma, 363: *auditivum*, 363
Otomesostomidae, 363
Otostephanos, 485, *437*
Ourococcus, 133, 132
Outer caudal seta, Harpacticoida, 819
Ovovitelline ducts, Turbellaria, 327
Oxus, 1087, 1099
Oxycera, 1074
Oxyethira, 1039, 1038
Oxytricha, 290
Oxytrichidae, 286
Oxyurella, 639, 641: *longicauda*, 640; *tenuicaudis*, 640, 642

Pachycladon, 137, 136
Pachydiplax, 938, *939*, 929, 932, 935
Pachydrus, 992
Pachysoeca, 208
Pachytrocha, 293, 294
Pacifastacus, 883: *gambelii*, 883, 886; *klamathensis*, 883; *leniusculus*, 883; *nigrescens*, 883; *trowbridgii*, 883
Pala, Hemiptera, 969
Palaemonetes, 880; *antrorum*, 880; *cummingi*, 880; *exilipes*, 881; *kadiakensis*, 881; *paludosus*, 880
Palaemonias ganteri, 879
Palaemonidae, 879
Palmella, 131, 130
Palmellaceae, 131, 136
Palmellococcus, 135, 136
Palmocorixa, 968, *961*
Palmodictyon, 131, 130
Palpi, Acari, 1083
Palpomyia, 1068
Paltostoma, 1059
Paltostomatinae, 1059
Paltothemis, 937, 932
Paludestrina, 1134
Paludicella, 499: *articulata*, 500
Paludicellidae, *499*
Panagrolaiminae, 386
Panagrolaimus, 386, 387
Pandorina, 129, 128
Panisoides, 1096, 1107
Panisopsis, 1092, 1096, 1097
Panisus, 1092, 1096, 1095, 1097
Pantala, 937, *939*, 930, 935
Papillae, Nemata, 370

Parabasal body, 191
Parabodo, 216
Paracamptus, 827, *816: reductus*, 833; *reggiae*, 834
Paracandona, 676: *euplectella*, 676, 677
Paracapnia, 950, 951, *946*
Parachordodes, 404
Parachordodinae, 404
Paracineta, 294, 295
Paracloeodes, 916
Paracolurella, 449, *437*
Paracyatholaimus, 393
Paracyclops, 797: *affinis*, 797; *fimbriatus*, 797; *fimbriatus poppei*, 798, 797
Paracymus, 1001, 1012, 983
Paradactylopodia, 827, *816: brevicornis*, 827
Paradicranophorus, 460, 461, *437*
Paradileptus, 273
Paradixa, 1073
Paraglossa, Plecoptera, 942
Paragnetina, 953, *946*
Paragordiinae, 403
Paragordionus, 404
Paragordius, 403
Paraholosticha, 282
Paraleptophlebia, 911, 914, 910
Paraleuctra, 950
Paralimna, 1077
Paralonella, 649
Paramastix, 229
Paramecidae, 275, 276
Paramecium, 276
Parameletus, 915, 910
Parameyenia, 302: *discoides*, 302
Paramylum, 191
Paranais, 530: *litoralis*, 530
Paraperla, 956, 952, 955
Paraperlinae, 956
Paraphanolaimus, 386
Paraphelenchus, 381
Parapholyx, 1129
Paraplectonema, 389
Paraponyx, 1052, 1054
Paraprocts, Odonata, 918
Parapsyche, 1035, 1034
Paraquadrula, 248
Parasitengona, 1080, *1108*
Parasitiformes, 1080
Parasoldanellonyx, *1108*
Parastenocaridae, 816
Parastenocaris, 823, *816: brevipes*, 823; *delamarei*, 823; *lacustris*, 823; *starretti*, 823; *wilsoni*, 823
Parasynchaeta, 438
Paratendipes, 1070
Parathyas, 1092
Paratylenchinae, 381

Paratylenchus, 381
Paravorticella, 292
Parechiniscus, 513
Parencentrum, 460, *437*
Pareuglypha, 253
Parhelophilus, 1077
Parmulina, 237
Parophryoxus, 626: *tubulatus*, 626
Parthina, 1046
Partnunia, 1092, 1096
Partnuniella, 1093, 1096
Paruroleptus, 288, 289
Pascheriella, 129, 130
Paulinella, 253
Paulinellidae, 253
Pecten, Cladocera, 594
Pectinatella, 499: *magnifica*, 502, 503
Pectinelles, Ciliophora, 265
Pectodictyon, 139, 138
Pedalia, 441
Pedalion, 441
Pediastrum, 135, 134
Pedicia, 1061
Pedinomonas, 124
Pedinopera, 129, 128
Pedipartia, 459, *437*
Pedomoecus, 1042, 1043
Pegias, 1146
Pelatractus, 271, 270
Pelecypoda, 1125, 1138, *1118*
Pelocoris, 967, *961*
Pelodictyon, 44
Pelodinium, 286
Pelogloea, 35
Pelomyxa, 235
Pelonomus, 1015, 1014, *985*, 986, 1002
Pelopia, 1069
Pelopiinae, 1069
Peloscolex, 534: *multisetosus*, 535; *variegatus*, 534
Peltandra virginica, 1182, 1172
Peltodytes, 990, 1009
Peltoperla, 948, 952
Peltoperlidae, 948
Penardia, 251
Penardiella, 272, 271
Penardochlamys, 237
Penis bulb, Turbellaria, 327
Penis papilla, Turbellaria, 327
Penium, 141, 140
Pennales, 174
Pentacora, 965
Pentagenia, 913, 910
Pentaneura, 1069
Pentatax, 1102
Peracantha, 645

Peranema, 125, 127, 124
Pereiopods, Malacostraca, 870
Periacineta, 296, 295
Pericoma, 1066, 1067
Peridiniaceae, 163
Peridiniales, 163
Peridinium, 163, 162
Perispira, 272, 271
Peristome, Ciliophora, 265
Peristomium, 522
Perithecium, Ascomycetes, 50
Perithemis, 937, *939*, 935
Peritricha, 290
Peritrichida, 267
Perlesta, 952, 947
Perlidae, 951
Perlinella, 951
Perlinodes, 954, 949
Perlodes, *946*
Perlodidae, 953
Perlomyia, 951
Peronia, 179, 178, *174*
Peroniella, 153, 152
Peronioideae, 174
Peronosporales, 83, *11*
Petalomonas, 125, 124
Petaluridae, 922, 929, *919*
Petersenia, 85, 84
Phacodinium, 284
Phacotaceae, 129
Phacotus, 129, 128
Phacus, 125, 123
Phaenocora, 350: *agassizi*, 351; *falciodenticulata*, 350; *highlandense*, 351; *kepneri*, 352; *lutheri*, 352; *virginiana*, 350
Phaenocorinae, 349
Phaeophyta, 160, *117*
Phaeoplaca, 159, 158
Phaeosphaera, 159, 158
Phaeothamniaceae, 160
Phaeothamnion, 160
Phagocata, 331: *gracilis gracilis*, 332; *gracilis monopharyngea*, 332; *gracilis woodworthi*, 332; *morgani*, 332, *333*; *morgani polycelis*, 333; *nivea*, 332; *subterranea*, 332, *329*; *velata*, 332, *333*; *vernalis*, 332, 333
Phalacrocera, 1060
Phalansterium, 212
Phanerobia, 213
Phanocerus, 1002, 1003, 1016, 986, 987
Pharyngobdellida, 552, *545*
Pharynx: Nemata, 370; Turbellaria, 324
Phascolodon, 275, 274
Phasganophora, 953, *946*
Phasmids, Nemata, 369

Philasteridae, 278, 281
Philasterides, 281
Philobdella, 554: floridana, 554; gracilis, 554
Philodina, 488, 486, 437
Philodinavidae, 485, 437
Philodinavus, 489, 437
Philodinidae, 484, 437
Philonotis, 1164, 1166, 1165, 1169
Philopotamidae, 1025, 1030
Philorus, 1060
Phlycitidum, 55, 54
Phlyctidiaceae, 51
Phlyctochytrium, 57, 55
Phlyctorhiza, 57, 59
Phormidium, 111: ambiguum, 112; anabaenoides, 111; autumnale, 111; corium, 112; favosum, 111; groesbeckianum, 112; incrustatum, 111; inundatum, 111; laminosum, 111; luridum, 112; minnesotense, 112; molle, 112; mucicola, 112; papyraceum, 112; retzii, 112; richardsii, 111; setchellianum, 111; subfuscum, 111; tenue, 111; treleasei, 112; uncinatum, 111; valderianum, 112
Phryganea, 1041, 1028, 1040
Phryganeidae, 1027, 1040
Phryganeid Genus A, 1041, 1040
Phryganella, 250: hemisphaerica, 251; nidulus, 251
Phycomycetes, 11, 14, 47, 50
Phylactolaemata, 500, 499
Phyllobium, 133, 132
Phyllognathopidae, 816
Phyllognathopus, 830, 816: viguieri, 830
Phyllomitus, 217
Phyllopoda, 587
Phyllopods, 558, 577
Phylocentropus, 1031, 1032
Phymatodocis, 143, 142
Physa, 1126
Physalophrya, 276
Physidae, 1125, 1126, 1123
Physocladia, 64
Physocypria, 669: denticulata, 674; dentifera, 673; exquisita, 675; fadeewi, 675, 674; gibbera, 676; globula, 675, 676; inflata, 675, 674; posterotuberculata, 674; pustulosa, 675; xanabanica, 676, 677
Physomonas, 221
Physorhizophidium, 57, 56
Phytobius, 1007
Phytodiniaceae, 165
Phytodinium, 165, 164
Phytomastigina, 9
Phytomastigophorea, 191
Phytomonadida, 190
Phytonomus, 1007
Phytophthora, 87, 86

Pictetia, 955, 949
Piersigia, 1090, 1095
Piersigiidae, 1095
Pilea, 1157
Pilidae, 1125, 1131, 1118
Pilolebertia, 1099
Pinnularia, 184, 172, 174
Pionacercus, 1088, 1104
Pionae, 1100
Pionidae, 1103
Pionopsis, 1088, 1104
Piricularia, 93, 92
Piscicolaria reducta, 552
Piscicolidae, 548, 543, 545
Pisicola, 552: geometra, 552; milneri, 552; punctata, 552; salmonsitica, 552
Pisidium, 1158
Pistia stratiotes, 1177, 1184
Pithophora, 149, 148
Pithothorax, 269
Placobdella, 549: hollensis, 550, 547, 548; montifera, 549; multilineata, 550; ornata, 550; parasitica, 550, 549; pediculata, 550; phalera, 548; rugosa, 550
Placocista, 256
Placoids, Tardigrada, 509
Placus, 269
Plagiocampa, 269
Plagiola, 1150, 1151
Plagiophrys, 258
Plagiopyla, 277
Plagiopylidae, 275, 277
Plagiopyxis, 241: callida, 241; labiata, 241
Plagiostomidae, 362
Planaria dactyligera, 331: maculata, 330; simplicissima, 331
Planariidae, 329
Planipennia, 904, 973, 976: cocoons, 979; larvae, 978; pupae, 979
Planktosphaeria, 137, 138
Planorbidae, 1125, 1128, 1123
Planorbula, 1130
Plant, definition, 7
Plants, vascular, 1170
Plasmodiophorales, 74: habitat, 48
Plathemis, 938, 939, 930, 932
Platycentropus, 1045
Platychloris, 127, 126
Platycola, 293, 294
Platycopa, 663
Platydorina, 129, 128
Platygerris, 963
Platyhelminthes, 323, 369
Platyias, 451, 452, 437: patulus, 452; polyacanthus, 452; quadricornis, 451, 452

Platymonas, 129, 128
Platynaias, 1148
Platynematum, 281
Platynothrus, 1112, 1113
Platyophrya, 269
Platytheca, 200
Plea, 966, *960*
Plecoptera, 906, 907, 941
Plectidae, 389
Plectids, Nemata, 389
Plectinae, 389
Plectoidea, 386
Plectonema, 109: *cloverianum*, 110; *nostocorum*, 110; *purpureum*, 110; *terebrans*, 110; *tomasinianum*, 110; *wollei*, 109, 108
Plectospira, 79, 78
Plectus, 392
Pleidae, 966, *960*
Pleodorina, 129, 128
Pleopods, Malacostraca, 870
Plethobasus, 1142
Pleuretra, 489, 486, *437*
Pleurobema, 1142
Pleurocapsa, 98
Pleurocera, 1137
Pleuroceridae, 1125, 1135, *1119*, 1123
Pleurochloridaceae, 151
Pleurodiscus, 143, 142
Pleurogaster, 153, 150
Pleuromonas, 213
Pleuronema, 282
Pleuronematidae, 278, 282
Pleurostomata, 267, 272
Pleurotaenium, 141, 140
Pleurotricha, 289
Pleurotrocha, 473, *437*, 464
Pleuroxalonella, 649
Pleuroxus, 644: *aduncus*, 648; *denticulatus*, 646, *647*; *hamulatus*, 647, 635; *hastatus*, 646; *procurvus*, 645, 644; *striatus*, 646, 644, *635*, *647*; *trigonellus*, 647; *truncatus*, 645, 644; *uncinatus*, 645
Ploesoma, 453, *438*, *439*: *hudsoni*, 453; *lenticulare*, 453; *triacanthum*, 453; *truncatum*, 453
Ploima, 437
Plumatella, 504, *499*: *casmiana*, 504, 505; *emarginata*, 505, 506; *fruticosa*, 505; *fungosa*, 506; *repens*, 506
Plumatellidae, 499
Pnigodes, 1007
Podochytrium, 55, 57, 54
Podocopa, 663
Podonominae, 1069
Podonomus, 1069
Podophrya, 294, 295

Podophryidae, 294
Podostemon ceratophyllum, 1181
Poecilographa, 1077
Pole of valve, Bacillariophyceae, 175
Polyatax, 1102, 1101
Polyartemiella, *562*, 561: *hazeni*, 562, 563; *judayi*, 562, 563
Polyartemiidae, 562, *561*
Polyarthra, 439, 446, *438*: *bicera*, 440; *dissimulans*, 440; *dolichoptera*, 440; *euryptera*, 440; *longiremis*, 440; *major*, 440; *minor*, 440; *platyptera*, 439; *proloba*, 440; *remata*, 440; *trigla*, 439; *vulgaris*, 440
Polyblepharidaceae, 127
Polyblepharides, 127, 126
Polycelis, 333: *borealis*, 333; *coronata*, 333
Polycentric thalli, Phycomycetes, 48
Polycentropinae, 1031
Polycentropus, 1032, 1031, 1027, 1032
Polychaeta, 538
Polychaetus, 449
Polychytrium, 64
Polycystidae, 358
Polycystis, 358: *goettei*, 358
Polyedriopsis, 137, 136
Polygonum amphibium, 1185
Polyhydryphantes, 1097, 1096
Polymerurus, 410: *callosus*, 411; *rhomboides*, 410
Polymitarcidae, 911
Polymitarcinae, 911
Polymyarian, Nemata, 370
Polyoeca, 210
Polyphaga, 982
Polyphagus, 62
Polyphemidae, 598, *588*, *589*
Polyphemoidea, 598, *589*
Polyphemus, 599: *pediculus*, 599
Polyrhytis, 1128
Polytoma, 127, 126, *10*, *190*
Polytomella, 127, 126
Pomacea, 1131
Pomatiopsis, 1135
Pompholyx, 445, 444, *438*: *complanata*, 445; *sulcata*, 445, 444; *trilobata*, 445
Pompholyxophrys, 262
Pontederia cordata, 1182, 1174
Pontigulasia, 243
Pontoporeia, 876: *affinis*, 876; *filicornis*, 876; *hoyi*, 876
Porella, 1163, *1169*
Porifera, 298
Porohalacaridae, *1108*
Porohalacarinae, 1110
Porohalacarus, 1109, 1110, *1108*
Porolohmannella, 1110, 1109, *1108*

Porphrydium, 167, 168

Porphyrosiphon, 107, 110: *fuscus*, 107; *notarisii*, 107, 106

Postabdomen, Cladocera, 592, 593

Postclausa, 453

Postmentum, Odonata, 917

Potamanthidae, 911

Potamanthus, 911

Potamidinae, 1074

Potamobates, 962, *959*

Potamocoris, 966

Potamocypris, 718: *comosa*, 722; *elegantula*, 723; *illinoisensis*, 722; *islagrandensis*, 721, 722; *pallida*, 721; *smaragdina*, 722; *variegata*, 723

Potamogeton, 1186: *alpinus*, 1187, 1190; *amplifolius*, 1190, 1174; *berchtoldii*, 1190; *bicupulatus*, 1192; *capillaceus*, 1192; *confervoides*, 1189; *crispus*, 1186, 1187; *diversifolius*, 1193; *epihydrus*, 1192, 1191, 1171, 1174; *filiformis*, 1189; *foliosus*, 1189; *friesii*, 1189, 1172; *gramineus*, 1191; *illinoensis*, 1187, 1190; *lateralis*, 1192; *latifolius*, 1188; *longiligulatus*, 1189; *natans*, 1192, 1174, 1191; *nodosus*, 1190, 1191, 1192; *oakesianus*, 1192; *obtusifolius*, 1190, 1174, 1189; *pectinatus*, 1188, 1173, 1175; *perfoliatus*, 1187; *praelongus*, 1187, 1175; *pulcher*, 1190, 1191; *pusillus*, 1190, 1172; *richardsonii*, 1187; *robbinsii*, 1188; *spirillus*, 1192, 1174; *strictifolius*, 1190; *vaginatus*, 1189; *vaseyi*, 1192; *zosteriformis*, 1189, 1171, 1172

Potamyia, 1035, 1034

Poteriodendron, 202

Pottsiella, 499: *erecta*, 500

Prasiola, 149, 148

Pratylenchinae, 383

Pratylenchus, 383

Prawns, 871

Prehensile palp, Ostracoda, 661

Prementum, Odonata, 917

Premnodrilus, 533: *palustris*, 533

Prionchulus, 395

Prionocyphon, 1020

Prionocypris, 699: *canadensis*, 700; *longiforma*, 699

Prionodiaptomus, 776, *739*

Prismatolaimus, 397

Pristicephalus, 566, *561*: *occidentalis*, 566, 568

Pristina, 530, *528*: *aequiseta*, 530; *bilongata*, 530; *breviseta*, 530; *longiseta leidyi*, 530; *osborni*, 530; *plumiseta*, 530; *schmiederi*, 530

Proales, 469, 464, *437*

Proalidae, 437

Proalides, 445, *437*

Proalinopsis, 469, *437*

Probezzia, 1068

Probolae, Nemata, 370

Probopyrus bithynis, 873

Procambarus, 883: *acherontis*, 884; *acutissimus*, 890; *advena*, 887, 886; *alleni*, 889; *angustatus*, 889; *apalachicolae*, 888; *barbatus*, 887, 886; *bivittatus*, 889; *blandingii*, 890, 886; *clarkii*, 885, 886; *dupratzi*, 885; *echinatus*, 885; *econfinae*, 888; *enoplosternum*, 889, 886; *escambiensis*, 887; *evermanni*, 890; *fallax*, 889; *geodytes*, 885; *gracilis*, 887, 886; *hagenianus*, 887; *hayi*, 890; *hinei*, 888; *howellae*, 885; *hubbelli*, 887; *hybus*, 888; *jaculus*, 887; *kilbyi*, 888; *latipleurum*, 888; *lecontei*, 890; *leonensis*, 889; *lepidodactylus*, 889; *litosternum*, 889; *lucifugus*, 884; *lunzi*, 889; *mancus*, 887; *natchitochae*, 885; *okaloosae*, 885; *paeninsulanus*, 885; *pallidus*, 884; *pearsei*, 888; *penni*, 885; *pictus*, 889; *planirostris*, 888; *pubescens*, 889; *pubischelae*, 887; *pycnogonopodus*, 889; *pygmaeus*, 887; *raneyi*, 885; *rathbunae*, 888; *rogersi*, 885; *seminolae*, 889; *shermani*, 887; *simulans*, 888; *spiculifer*, 884, 886; *suttkusi*, 885; *tenuis*, 888; *troglodytes*, 885; *truculentus*, 887; *tulanei*, 887; *verrucosus*, 890; *versutus*, 884; *viaeviridis*, 888; *vioscai*, 885; *youngi*, 888

Prochromadorella, 393

Procladius, 1069

Procorpus, Nemata, 371

Procotyla, 333: *fluviatilis*, 334; *typhlops*, 334

Prodesmodora, 393

Prodiamesa, 1071

Prolasmidonta, 1145

Proleg, 1058

Promenetus, 1130

Promoresia, 1006, 1018, *987*

Prongs, caudal, Gastrotricha, 406

Pronotal disc, Hemiptera, 969

Pronoterus, 997

Propleurae, Hemiptera, 965

Proptera, 1152

Prorhabdions, Nemata, 370

Prorhynchella, 345: *minuta*, 345

Prorhynchidae, 360

Prorhynchus, 360: *stagnalis*, 360

Prorocentraceae, 160

Prorodon, 270

Prosobranchia, 1117, 1131

Prostom, Nemata, 370

Prostoma rubrum, 366

Prostomata, 267

Proteomyxida, 262

Proterhabdions, Nemata, 370

Proterospongia, 211

Prothoracic shield, Lepidoptera, 1051

Protista, 7, 8, 115: interrelationships, 14

Protoascus, 348: *wisconsinensis*, 348, 349

Protochauliodes, 976

Protochrysis, 167, 166

Protociliata, 266

Protococcaceae, 149
Protococcus, 149, 148
Protocrucia, 283, 284
Protoderma, 147, 146
Protoplanellinae, 343
Protoplasa, 1069
Protoptila, 1036
Protosiphon, 135, 134
Protosiphonaceae, 135, 134
Protospongia, 211
Protostom, Nemata, 370
Protozoa, 8, 9 (see also Actinopoda, Algae, Ciliophora, Rhizopoda, Zooflagellates)
Protzia, 1092, 1095
Protziidae, 1095
Provortex, 342: *virginiensis*, 342
Provorticidae, 342
Prymnesiaceae, 155
Psectrocladius, 1071
Psephenidae, 1001, 1013, 984, 985
Psephenoides, 985
Psephenus, 1002, 1015, 1014, 985
Pseudechiniscus, 515: *cornutus*, 515; *suillus*, 515; *tridentifer*, 515
Pseudiron, 913, 910
Pseudironinae, 913
Pseudoblepharisma, 284
Pseudocalanidae, 749, *738*
Pseudochordodes, 404
Pseudocloeon, 916
Pseudocorixa, 970, *961*
Pseudodifflugia, 258
Pseudoecistes rotifer, 476, *438*
Pseudoglaucoma, 280, 279
Pseudogoera, 1041
Pseudoharringia, 473, 466, *437*
Pseudohydryphantes, 1091, 1097
Pseudohydryphantidae, 1097
Pseudokoenikea, 1103
Pseudolebertia, 1099
Pseudoleon, 939
Pseudolpidium, 85
Pseudomicrothorax, 277
Pseudomonas, 27, *29*, *30*
Pseudonychocamptus, 816: *proximus*, 827
Pseudoon, 1153, 1154
Pseudophaenocora, 349: *sulfophila*, 349
Pseudoploesoma, 453, *437*
Pseudopod: Diptera, 1058; Rhizopoda, 232
Pseudoprorodon, 270
Pseudoraphe, Bacillariophyceae, 175
Pseudosida, 602: *bidentata*, 602
Pseudosperchon, 1097, 1107
Pseudosphaerita, 83
Pseudosuccinea, 1127
Pseudoulvella, 147, 146

Pseudovacuoles, Myxophyceae, 95
Psilenchus, 384
Psilotanypus, 1069
Psilotreta, 1046, 1047
Psilotricha, 287
Psorophora, 1066
Psychoda, 1066, 1067
Psychodidae, 1066
Psychomyia, 1033, 1031, 1032
Psychomyiidae, 1025, 1031, 1032
Psychomyiid Genus A, 1032, 1031, 1032
Psychomyiid Genus B, 1032, 1031
Psychomyiinae, 1031
Psychoronia, 1045
Psythiopsis, 77, 76
Pteridomonas, 193
Pterodina, 473
Pterodrilus, 525: *alcicornus*, 525; *distichus*, 525; *durbini*, 525; *mexicanus*, 525
Pteromonas, 129, 128
Pteronarcella, 947
Pteronarcidae, 947
Pteronarcys, 947, 948
Pterosyna, 1147
Ptilodactylidae, 985, 1020
Ptilostomis, 1041, 1040
Ptychobranchus, 1149
Ptychoptera, 1073
Ptychopteridae, 1072
Ptychopterinae, 1073
Ptygura, 476, 477, *479*, *438*
Pulmonata, 1117, 1126
Punctae, Bacillariophyceae, 175
Punctodora, 393
Pupa, Insecta, 903
Purple bacteria, *14*, 19
Purple sulfur bacteria, *31*
Pycnopsyche, 1045, 1043
Pygidium, Cladocera, 592
Pyramidomonas, 127, 126
Pyrausta, 1053, 1055, *1051*
Pyrgulopsis, 1134
Pyrobotrys, 129, 130
Pyrrophyta, 160, *117*
Pythiella, 77, 76
Pythiogeton, 87, 85
Pythium, 87, 86
Pyxicola, 293, 294
Pyxidicula, 238: *cymbalum*, 240; *operculata*, 240; *scutella*, 240
Pyxidium, 292

Quadricoccus, 139, 138
Quadrigula, 137, 138
Quadrula, 1140

Quadrulella, 247
Quincuncina, 1139

Raciborskia, 165, 164
Radaisia, 98
Radema, 1044
Radiococcus, 137, 138
Radiofilum, 145, 144
Radopholus, 383
Radula, Gastropoda, 1117, 1122
Ramphocorixa, 970, *961*
Ramus, Rotifera, 428, 429
Ranatra, 965, *960*
Ranunculus, 1181: *aquatilis*, 1181, 1171, 1180; *circinatus*, 1181; *flabellaris*, 1181, 1180
Raphe, Bacillariophyceae, 175
Raphidiodea, 973
Raphidioidineae, 174
Raphidionema, 145, 144
Raphidiophrys, 262
Raphidiopsis, 103: *curvata*, 103, 104
Rattulus, 447
Reagents (see Methods)
Reckertia, 196
Rectocephala exotica, 333
Rectum, Cladocera, 593
Red algae, 14, *117*
Red sulfur bacteria, *31*
Reichenowella, 283
Reichenowellidae, 283
Remenus, 555
Rennette, Nemata, 373
Resticula, 472, 471, *437*
Reticulo-lobosa, 236
Retreat matters, Trichoptera, 1030
Retrocerebral organ, Rotifera, 421
Retrocerebral sac, Rotifera, 421
Rhabdions, Nemata, 370
Rhabdites, Turbellaria, 323
Rhabditida, 380
Rhabditidae, 385
Rhabditinae, 386
Rhabditis, 386
Rhabdochromatium, 32, 44: relation to *chromatium*, 33
Rhabdocoela, 325
Rhabdoderma, 96: *aeruginosa*, 96; *elabens*, 97; *peniocystis*, 97; *stagina*, 96, 97
Rhabdohydrachna, 1093
Rhabdoids, Turbellaria, 323
Rhabdolaiminae, 389
Rhabdolaimus, 389, 388
Rhabdomonas, 125, 124
Rhabdostyla, 292
Rhagionidae, 1074

Rhagovelia, 963, *959*
Rhammites, Turbellaria, 323
Rhantus, 995, 1010
Rheumatobates, 962, 959
Rhinoglena, 468, 467, 437
Rhinops, 468
Rhipidiaceae, 81
Rhipidium, 82, 83
Rhipidodendron, 224
Rhithrogena, 914
Rhizelmis, 1005, 987
Rhizidiaceae, 51
Rhizidiomyces, 74, 73
Rhizidiopsis, 55
Rhizidium, 59, 58
Rhizochloridales, 151
Rhizochrysidaceae, 159
Rhizochrysidales, 159
Rhizochrysis, 159, 158
Rhizoclonium, 149, 148
Rhizoclosmatium, 62, 61
Rhizodrilus, 535: *lacteus*, 535
Rhizoidal thalli, Phycomycetes, 48
Rhizoids, Phycomycetes, 47
Rhizophlyctis, 59, 58
Rhizophydium, 55, 54
Rhizopoda, 232, 233, *11*
Rhizosiphon, 57, 56
Rhizosolenia, 185, *174*
Rhizosoleniaceae, 174
Rhizosolenioideae, 174
Rhodacmaea, 1131
Rhodocephala, 1131
Rhodomicrobium, 44, *19*, 21
Rhodomonas, 166
Rhodophyceae, 167
Rhodophyta, 167, *117*
Rhodopseudomonas, 44
Rhodospirillum, 44
Rhoicosphenia, 179, *174*
Rhopalodia, 181, *174*
Rhopalodoideae, 174
Rhopalophlyctis, 65
Rhopalophrya, 269
Rhyacophila, 1035, 903, 1027, 1028, 1036
Rhyacophilidae, 1025, 1035, 1036
Rhyacophylax, 1033
Rhynchelmis, 533: *elrodi*, 533; *glandula*, 533
Rhyncheta, 297, 296
Rhynchobdellida, 547, *543*, *544*, *545*
Rhynchomesostoma, 346: *rostrata*, 346
Rhynchomesostominae, 343
Rhynchomonas, 197
Rhynchophora, 297, 296
Rhynchophorinae, 1007

Rhynchoscolex simplex, 337, 338
Rhynchotalona, 643: *falcata*, 643
Riccardia, 1162, *1169*
Riccia, 1162, *1169*
Ricciocarpus, 1162, *1169*
Rickera, 955
Riella, 1162, *1169*
Rivobates, 1101
Rivularia, 105: *bornetiana*, 105; *dura*, 105; *haematites*, 105; *minutula*, 105, 104
Rivulariaceae, 103
Roederiodes, 1076
Rogerus, 393
Rostrum: Acari, 1083; Cladocera, 592; Hemiptera, 969; Malacostraca, 870; Rotifera, 427
Rotaria, 488, 486, *437*
Rotatoria, 420
Rotifer, 84, 420
Rotifer, 488, *437*
Rotifera, 420, *368*
Rotylenchus, 383
Rousseletia, 471, *437*
Roya, 141, 140
Rozella, 52
Rozellopsis, 83
Rugae, circumoral, Nemata, 370
Rugipes, 234
Ruppia maritima, 1186
Rusetria, 1100

Sabellidae, 539
Saccomyces, 57, 56
Sagittaria latifolia, 1185, 1172, 1173, 1174, 1184
Salda, 964
Saldidae, 964, *959*
Saldoida, 964, *960*
Saldula, 965, *959*
Salmincola, 866, 867
Salpina, 456
Salpingoeca, 207
Salpingorhiza, 197
Salvinia rotundifolia, 1175
Saprodinium, 286
Saprolegnia, 79, 76
Saprolegniaceae, 77
Saprolegniales, 75, *11*
Sapromyces, 82
Saprophilus, 281
Sarcina, 44, 17
Sarcodina, *11*
Sarcoptiformes, 1080
Scalenilla, 1158
Scales, Gastrotricha, 406
Scapania, 1163, 1162, *1169*
Scapholeberis, 616: *kingi*, 617

Scaridium, 451, 450, *437*
Scatophagidae, 1078
Scatopsidae, 1066
Scenedesmaceae, 139
Scenedesmus, 139, 138
Scepanotrocha, 485, *437*
Scherffelia, 129, 128
Scherffeliomyces, 57, 56
Schizocerca, 451, *437*
Schizochlamys, 131, 130
Schizodictyon, 131, 130
Schizogoniaceae, 149
Schizogoniales, 149
Schizogonium, 149, 148
Schizomeridaceae, 149
Schizomeris, 149, 148
Schizopera, 827, 830, *816*: *haitiana*, 827; *triacantha*, 827
Schizothrix, 107: *acuminata*, 109; *acutissima*, 109; *aikenensis*, 108; *calcicola*, 108; *californica*, 109; *coriacea*, 108; *dailyi*, 108; *fasciculata*, 108; *fragilis*, 109; *friesii*, 109, 108; *fuscescens*, 109; *giuseppei*, 109; *heufleri*, 109; *lacustris*, 108; *lamyi*, 109; *lardacea*, 108; *lateritia*, 108; *longiarticulata*, 109; *macbridei*, 109; *muelleri*, 109; *penicillata*, 108; *pulvinata*, 108; *purcellii*, 109; *purpurascens*, 109; *richardsii*, 109; *rivularis*, 108; *roseola*, 109; *stricklandii*, 108; *taylorii*, 109; *thelephoroides*, 109; *tinctoria*, 108; *vaginata*, 108; *wollei*, 109
Schoenobius, 1053, 1055, *1051*
Schroederia, 135, 134
Sciara, 1066
Sciaridae, 1066
Sciaromium, 1166, *1169*
Sciomyzidae, 1077
Scirtes, 1020
Scleropodium, 1168, 1169
Scolecida, *369*
Scopeumatidae, 1078
Scorpidium, 1164, *1169*
Scotiella, 137, 136
Scouleria, 1168, *1169*
Scuds, 871
Scutechiniscidae, 513
Scutohydrachnà, 1094, 1107
Scutosperchon, 1098
Scyphidia, 292
Scyphidiidae, 291
Scytonema, 106: *alatum*, 107; *coactile*, 106; *crispum*, 106; *crustaceum*, 106; *densum*, 107; *guyanense*, 106; *hofmannii*, 106; *mirabile*, 106; *myochrous*, 106; *ocellatum*, 106; *tolypotrichoides*, 106
Scytonemataceae, 105
Secernentea, 380
Second antenna (see Antenna)

Seed shrimps, 559
Seinura, 381
Seison, 436, 439
Seisonida, 436
Seisonidae, 436, 439
Selenastrum, 137, 138
Semilorica, Rotifera, 424
Seminal receptacle, Turbellaria, 327
Senecella, *738: calanoides*, 749
Sensory papillae, Tardigrada, 509
Sepedon, 1077
Septochytrium, 70, 69
Septolpidium, 53
Septosperma, 53
Septum, Bacillariophyceae, 175
Seriata, 362
Sericostoma, 1046, 1028, 1047
Sericostomatidae, 1029, 1046
Serpulidae, 539
Sessilia, 290, 291
Sessoblasts, Bryozoa, 497
Seta: Acari, 1083, 1084; Bacillariae, 175;
    Nemata, 370; Oligochaeta, 522, 523; tactile,
    Cliophora, 265
Setacera, 1077
Setation, diaptomid first antenna, 748
Setodes, 1049, 1048
Setvena, 954
Sewage, 40
Sexangularia, 243
Sheaths: bacterial, 18; myxophyceae, 95
Shell, Ostracoda, 658
Shore bug, 964
Shrimps, 871
Sialidae, 905, 904, 975
Sialis, 975, 978
Sialodea, 973, 974–975
Sialoidea, 904
Sida, 599: *crystallina*, 599, 600
Siderocapsa, 43, *31*, 40: *monas*, 45
Siderocelis, 137, 136
Sideromonas, 44, *31*, 40
Sididae, 599, *588*, 589
Sidoidea, 599, *589*
Sierraperla, 948
Sigara, 970, *961*
Simocephalus, 616: *aurita*, 617; *exspinosus*, 616;
    serrulatus, 617, 616; *vetulus*, 617, 616
Simplex stages, Tardigrada, 513
Simpsoniconcha, 1148
Simuliidae, 1072
Sinantherina, 476, 479, 481, *438: ariprepes*, 479,
    481; *procera*, 479; *semibullata*, 476; *socialis*,
    479, 481; *spinosa*, 476
Sinistral, Gastropoda, 1121

Sinodiaptomus, 776, *739*
Siphlonisca, 915
Siphlonuridae, 912, 913
Siphlonurus, 915, 910
Siphloplecton, 915, 910
Siphonaria, 61, 60
Sirodotia, 169, 168
Sirogonium, 143, 142
Sisyra, 979, *976*, 977
Sisyridae, 905
Skadovskiella, 156
Skistodiaptomus, subgenus of *Diaptomus*, 739, 791
Skwala, 951
Slavina, 531: *appendiculata*, 531
Smicridea, 1035, 1034, *1033*
Snails, 1117
Snake doctor, 920
Snake feeder, 920
Soldanellonyx, 1110, *1108*, 1109
Soleniineae, 174
Solenophrya, 296, 295
Soliperla, 948
Solutoparies, 61, 60
Somatochlora, 936, 939, 934, *932*
Somatogyrus, 1135
Sommerstorffia, 79, 78
Sorastrum, 135, 134
Sorocelis americana, 333
Sorodiscus, 75, 74
Sortosa, 1031, 1030
Sow-bugs, 870
Sparangium, 1183: *americanum*, 1184; *angusti-
    folium*, 1184; *chlorocarpum*, 1184; *eurycarpum*,
    1184; *fluctuans*, 1184
Sparganophilus, 534
Spasmostoma, 268
Spathidiidae, 268, 271, 267
Spathidioides, 272
Spathidium, 272, 271
Spear, Nemata, 370
Sperchon, 1086, 1097, 1098
Sperchonidae, 1097
Sperchopsis, 1000, 1086, 1097
Spermatozoopsis, 127, 126
Spermiducal vesicles, Turbellaria, 326
Sphaerellopsis, 127, 126
Sphaeridiinae, 999
Sphaeriidae, 1126, 1158
Sphaerita, 52
Sphaerium, 1158
Sphaerocystis, 131, 130
Sphaerodinium, 163, 162
Sphaeroeca, 211
Sphaeroma, 874
Sphaeromias, 1068

Sphaeromidae, 873
Sphaerophrya, 294, 295
Sphaeroplea, 147, 148
Sphaeropleaceae, 147
Sphaerotilus, 44, 45, *13*, 40, 41: *natans*, 41
Sphaerozosma, 143, 142
Sphagnum, 1163, *1169*
Sphenoderia, 255: *lenta*, 255; *macrolepis*, 255
Sphenomonas, 125, 124
Sphyrias, 465, 464, *437*
Spicule: Nemata, 373; sponge, 298–299
Spiculum, Nemata, 373
Spine: Bacillariophyceae, 175; Cladocera, 592; Gastrotricha, 406
Spinneret, Nemata, 370
Spinoclosterium, 141, 140
Spirilla, 17
Spirillum, 44, 17, 42
Spirochaeta, 44: *plicatilis*, 21, 42
Spirochaetes, 14; 21, 42
Spirochona, 294
Spirochonidae, 294
Spirodela polyrhiza, 1177, 1175
Spirogyra, 143, 142
Spiromonas, 215
Spirostomidae, 283
Spirostomum, 284
Spirotaenia, 139, 140
Spirotrichida, 266, 282
Spirozona, 276
Spirozonidae, 275, 276
Spirulina, 114: *caldaria*, 114; *labyrinthiformis*, 114; *major*, 114; *princeps*, 114; *subsalsa*, 114; *subtilissima*, 114
Spondylomorum, 129, 130
Spondylosium, 143, 142
Spongeflies, *903*
Sponges, *8*, *14*, 298
Spongilla, 300: *aspinosa*, 300; *discoides*, 302; *fragilis*, 301; *heterosclerifera*, 301; *igloviformis*, 302; *johanseni*, 302; *lacustris*, 300; *lacustris* var *montana*, 300; *mackayi*, 301, 302
Spongilla-flies, 976
Spongillinae, 300
Spongomonas, 225
Sporadoporus, 1095, 1107
Sporangium, Phycomycetes, 47
Spores: Myxophyceae, 95; Phycomycetes, 48–49
Sporophlyctidium, 59, 58
Sporophlyctis, 62, 63
Sporozoa, 14
Springtails, 903
Squalorophrya, 294
Squatinella, 453, 454, *437*

Stactobiella, 1039, 1038
Stagnicola, 1127
Stalks, bacterial, 18
Staphylinidae, 1013
Statoblasts, Bryozoa, 497
Staurastrum, 141, 140, 142
Stauroneis, 185, *174*
Staurophrya, 297, 296
Stauros, Bacillariophyceae, 175
Stegnaspis, 1107
Steinia, 290
Stelexomonas, 211
Stenelmis, 1003, 1017, 1016, *986*, *987*
Stenocaris, 825, *816: minor*, 826
Stenocodon, 222
Stenocolus, 1020, 985
Stenocypria, 695: *longicomosa*, 695
Stenocypris, 696: *fontinalis*, 714, 715; *malcolmsoni*, 714, 715
Stenopelmus, 1007
Stenonema, 915, 910
Stenostomidae, 335
Stenostomum, 338: *tenuicaudatum*, 338; *virginianum*, 338
Stenotels (see Nematocysts)
Stentor, 284
Stentoridae, 283, 284
Stephanoceros, 480, *438: fimbriatus*, 482
Stephanocodon, 202
Stephanodiscus, 188, *174*, *56*
Stephanodrilus, 527
Stephanoeca, 208
Stephanops, 453
Stephanosphaera, 131, 130
Steremnius, 1007
Sterromonas, 220
Stichococcus, 145, 144
Stichosiphon, 99: *sansibaricus*, 99
Stichotricha, 287, 288
Stigeoclonium, 147, 146
Stigonema, 99: *hormoides*, 99; *informe*, 99; *mamillosum*, 99; *minutum*, 99; *ocellatum*, 99; *panniforme*, 99; *turfaceum*, 99
Stigonemataceae, 99
Stipitococcaceae, 151
Stipitococcus, 151, 150
Stokesia, 279
Stokesiella, 222
Stolella, 499: *indica;* 504
Stoma, Nemata, 370
Stomatochone, 221
Stoneflies, *903*, 941
Stoneworts, 117
Strandesia, 707: *bicuspis bicuspis*, 707; *intrepida*, 708; *obtusata*, 707

Stratiomyidae, 1074
Stratiomys, 1074
Streblocerus, 627: *pygmaeus*, 627; *serricaudatus*, 627
Streptocephalidae, 562, *561*
Streptocephalus, 562, *561: antillensis*, 562, 563; *dorothae*, 563, 564; *seali*, 562, 563; *similis*, 562, 563; *texanus*, 563, 564
Streptococcus, 44, 17
Streptognatha, 460, 461, *437*
Streptomonas, 218
Streptomyces, 44
Stria, Bacillanophyceae, 175
Strigil, Hemiptera, 969
Strobilidiidae, 285
Strobilidium, 285
Strombidinopsis, 285
Strombidium, 285
Strongylidium, 287, 288
Strongylostoma, 348: *gonocephalum*, 348
Strophitus, 1149
Strophopterys, *946*
Stygobromus, 878: *exilis*, 878; *heteropodus*, 878; *hubbsi*, 878; *iowae*, 878; *mackini*, 878; *onondagaensis*, 878; *putealis*, 878; *smith*, 878; *spinosus*, 878; *vitreus*, 878
Stygohalacarus, *1108*
Stygonectes, 878: *balconis*, 878; *flagellatus*, 878
Stylaria, 531: *fossularis*, 531; *lacustris*, 531
Stylet: Nemata, 370; Tardigrada, 509
Stylobryon, 223
Stylochaeta, 409: *scirteticus*, 409
Stylocometes, 297, 296
Stylodinium, 165, 164
Stylonychia, 290
Stylosphaeridium, 131
Stylurus, 928, *929*
Subanal lobes, Plecoptera, 943
Subaturus, 1105, 1107
Subcerebral glands, Rotifera, 421
Submentum, Odonata, 917
Suctoria, 266, 294
Sulcularia, 1148
Sulfate reduction, 42
Suomina turgida, 335
Suphisellus, 997, 1010
Surirella, 181, *174*
Surirellaceae, 174
Surirelloideae, 174
Sutroa, 533: *alpestris*, 533; *rostrata*, 533
Symbiocladius, 1072
Sympetrum, 937, 935, *939*
Symploca, 110: *dubia*, 110; *kieneri*, 110; *muralis*, 110; *muscorum*, 110, 111
Syncaris, 879: *pacifica*, 880; *pasadenae*, 880
Synchaeta, 463, 462, *438*

Synchaetidae, 438
Synchytriaceae, 51
Syncilium, Ciliophora, 265
Syncrypta, 156
Syncryptaceae, 156
Synechococcus, 96
Synedra, 177, *174*
Syneylais, 1095
Synkentronia, 484, *437*
Synpleonia, 878: *alabamensis*, 878; *americana*, 878; *clantoni*, 878; *emarginata*, 878; *hayi*, 878; *pizzinii*, 878; *tenuis*, 878
Synura, 156
Synuraceae, 156
Synurella, 878: bifurca, 878; chamberlaini, 878; dentata, 878; *johanseni*, 878
Syrphidae, 1076
Systellognatha, 947, *942*
Systylis, 292

Tabanidae, 1074
Tabaninae, 1074
Tabanus, 1074, 1075
Tabellaria, 176, *174, 66*
Tabellarioideae, 174
Tachidiidae, 816
Tachidius, 828, *816: brevicornis*, 828; *discipes*, 828; *littoralis*, 828; *spitzbergensis*, 828
Tachinidae, 1077
Tachopteryx, 930, 929, 933
Tachygerris, 962, 959
Tachysoma, 290
Tadpole shrimps, 559
Taeniopteryginae, 948
Taeniopteryx, 949, 947, 948, *946*
Talitridae, 876
Tanaognathella, 1103
Tanaognathus, 1103
Tanyderidae, 1069
Tanypodinae, 1069
Tanypteryx, 930, 929
Tanypus, 1069
Tanysphyrus, 1007
Tanytarsini, 1070
Tanytarsus, 1070
Taphrocampta, 465, *437*
Taphromysis louisianae, 879
Tardigrada, 508
Tarebia, 1138
Tarnetrum, 939
Tascobia, 1039
Tauriphila, 936, *939*
Teeth, Nemata, 370
Teleallagma, 926
Telebasis, 926

Telerhabdions, Nemata, 370
Telmatodrilus, 535: *mcgregori*, 536; *vejdovskyi*, 535
Telmatometra, 963, *959*
Telmatoscopus, 1066
Teloleuca, 965
Telostom, Nemata, 370
Telson, Malacostraca, 870
Temoridae, 750, 752, 738
Tenagogonus, 962, 959
Tendipedidae, 1069
Tendipedinae, 1070
Tendipedini, 1070
Tendipes, 1070
Tenegobia, 967, *962*
Tentaculiferida, 294
Teratocephalus, 389
Terminal claws, Cladocera, 594
Terminal or polar nodule, Bacillariophyceae, 176
Terpsinioideae, 174
Terpsinöe, 187, *174*
Testaceafilosa, 252, 233
Testacealobosa, 234, 233
Testaceous Rhizopoda, 233
Testudacarus, 1086, 1100, 1099
Testudinella, 473, *438, 439*
Testudinellidae, 438
Tetanocera, 1077
Tetanoceratidae, 1077
Tetanoceridae, 1077
Tetmemorus, 141, 140
Tetrabates, 1101
Tetrachaetum, 91
Tetracladium, 91, 90
Tetracyclus, 176, *174*
Tetradesmus, 139, 138
Tetradinium, 165, 164
Tetraedriella, 153, 150
Tetraëdron, 137, 136
Tetragoneuria, 936, *934*
Tetragonidium, 165, 164
Tetrahydrachna, 1094, 1107
Tetrahymena, 280, 279
Tetralimnesia, 1100
Tetrallantos, 139, 138
Tetramastix (Rotifera), 441, *438: opoliensis*, 441
Tetramastix (Zooflagellates), 228
Tetramitus, 228
Tetraneumania, 1102
Tetrapiona, 1104
Tetrasiphon, 464, 463, *437*
Tetrasiphonidae, 437
Tetraspora, 131, 130
Tetrasporaceae, 131
Tetrasporales, 131, 130

Tetrastrum, 139, 138
Tetylenchus, 384
Teuthophrys, 273, 267
Teutonia, 1086, 1087, 1097, 1098
Teutoniidae, 1097
Thalamia, 259, 233
Thalestridae, 816
Thamniochaete, 147, 146
Thamnocephalidae, 566, *561*
Thamnocephalus, 566, *561: platyurus*, 566, 567
Thaumalea, 1069
Thaumaleidae, 1069
Thaumatomastix, 196
Thaumatomonas, 195
Thecacineta, 296, 295
Thecamoeba, 234
Thecamoebidae, 234
Theliopsyche, 1046
Theobaldia, 1065
Thermacaridae, 1097
Thermacarus, 1089, 1097, 1096
Thermonectus, 996, 1011
Thermozoidium esakii, 513
Theromyzon, 548, 549: *meyeri*, 548; *rude*, 548; *tessulatum*, 548
Thiaridae, 1125, 1138
Thiocapsa, 43, 32
Thiocystis, 43, 32
Thiodictyon, 44, 32, 33
Thiogloea, 43, 37
Thiopedia, 43, 31, 32
Thiophysa, 43, 37
Thioploca, 44, 36
Thiopolycoccus, 43, 44, 32
Thiosarcina, 43, 32
Thiospirillopsis, 44, 36
Thiospirillum, 44, 32, 33
Thiothece, 43, 32
Thiothrix, 44, 33, 35, 36
Thiovulum, 43, 36, 37
Thoracic appendages, Ostracoda, 659
Thoracomonas, 129, 128
Thorea, 169, 168
Thoreaceae, 169
Thraulodes, 914, 910
Thraustotheca, 80, 79
Thuricola, 293, 294
Thyas, 1092, 1097
Thyasella, 1092
Thylacidium, 285, 284
Thylacomonas, 198
Thyopsis, 1092, 1096
Thysanura, 902
Tigriopus, 825, *816: californicus*, 825; *triangulus*, 825

Tillina, 277
Tinodes, 1033, 1032
Tintinnidae, 285
Tintinnidium, 285
Tintinnopsis, 285
Tiphys, 1088, 1091, 1103, 1104, 1107
Tipula, 1061, *1077*
Tipulidae, 1060
Tipulinae, 1061
Toad bug, 960, 965
Toe: Rotifera, 421, 425; Tardigrada, 509
Toe biter, 960, 965
Togoperla, *946*
Tokophrya, 296, 295
Tolypella, 149, 150
Tolypothrix, 107: *distorta*, 107; *lanata*, 107, 106; *tenuis*, 107
Tomaculum, 139, 138
Torquis, 1128
Torrenticola, 1088, 1099, 1100, 1101, 1107
Torrenticolidae, 1100
Tortopus, 913, 910
Toxicysts, Ciliophora, 266
Tracheata, 1161
Tracheleuglypha, 254
Tracheliidae, 272, 273
Trachelius, 273
Trachelomonas, 125, 123, 40
Trachelophyllum, 299, 268
Trachychloron, 153, 150
Tramea, 938, 930, 935, *937*, *939*
Transverse axis, Bacillariophyceae, 176
Trapa natans, 1185, 1174
Traverella, 913, 910
Trentonius, 1031
Trepobates, 963, *959*
Trepomonas, 220
Treubaria, 137, 136
Trhypochthonius, 1112
Triacanthagyna, 931, *934*
Triaenodes, 1049, 1029
Triannulata, 527: *magna*, 527; *montana*, 527
Triarthra, 441
Tribonema, 154, 152
Tribonemataceae, 154
Trichites, Ciliophora, 266
Trichocerca, 447, 448, *437*
Trichocercidae, 437
Trichocladius, 1072
Trichocorixa, 968, *961*
Trichocorixella, 968, *962*
Trichodina, 270, 290
Trichodrilus, 534: *allobrogum*, 534
Trichome, Myxophyceae, 95
Trichopelma, 278, 277

Trichopelmidae, 275, 277
Trichophrya, 297, 296
Trichoptera, 905, 906, 907, 1024
Trichospira, 276
Trichospiridae, 275, 276
Trichostomina, 267, 275
Trichotaxis, 288, 289
Trichotria, 449, 450, 454, *437*
Trichuroidea, 397
Tricladida, 326, 325
Tricladium, 93, 92
Triclads, Turbellaria, 326
Tricorythidae, 912
Tricorythodes, 912, 910
Trigonomonas, 229
Trigonopyxis, 240
Trilobus, 397
Trimalaconothrus, 1112, *1110*, *1111*
Trimastigamoeba, 217
Trimastix, 228
Trimicra, 1062
Trimyema, 276
Trimyemidae, 275, 276
Trinema, 256: *complanatum*, 256; *enchelys*, 257; *lineare*, 257
Triogma, 1060
Triops, 574, *573: longicaudatus*, *573*
Triphylus, 466
Tripleuchlanis, 456, 457, 458, *437*
Triploceras, 141, 140
Tripyla, 397
Tripylidae, 396
Tripyloidea, 393
Triscelophorus, 91
Trischistoma, 397
Trissocladius, 1071
Tritigonia, 1141
Trochilia, 275, 274
Trochilioides, 275, 274
Trochiscia, 137, 136
Trochosphaera, 446, *438: aequatorialis*, 446; *solstitialis*, 446
Trochospongilla 304: *erenaceus*, 304; *horrida*, 307; *leidyi*, 306
Troglocambarus, 883: *maclanei*, 883
Troglohalacarus, 1108
Trombidiformes, 1080
Trophi, Rotifera, 428–432, *420*
Tropicorbis, 1129
Tropidoatractus, 283
Tropisternus, 999, 1013
Tropocyclops, 797: *prasinus*, 799; *prasinus mexicanus*, 800, 799
Truittella, 68, 67
Truncaturus, 1107

Truncilla, 1151
Truncillopsis, 1157
Tryonia, 1134
Tubella, 304: *paulula*, 304; *pennsylvanica*, 304
Tubifera, 1077, 1076
Tubifex, 536: *tubifex*, 536
Tubificidae, 534
Tufts, tactile ciliary, Gastrotricha, 407
Tulotoma, 1133
Tuomeya, 169, 168
Turania, 279, 278
Turbellaria, 323
Tylenchida, 380
Tylenchidae, 384
Tylenchinae, 384
Tylenchoidea, 380
Tylencholaimellus, 398
Tylencholaiminae, 399
Tylenchlaimus, 399
Tylenchorhynchus, 384
Tylenchulidae, 384
Tylenchulinae, 384
Tylenchulus, 384
Tylenchus, 384
Tyleptus, 398
Tylocephalus, 390
Tylotrocha, 463, *437*
Tylotrochidae, 437
Typhloplana, 348: *viridata*, 348
Typhloplanidae, 341
Typhloplaninae, 346
Typhloplanoida, 341
Tyrrellia, 1086, 1101, 1082
Tyrrelliidae, 1101

Ulothrix, 145, 144
Ulotrichales, 145
Ulotrichaceae, 145
Ulvaceae, 149
Ulvales, 149
Uncus, Rotifera, 428, 429
Unicellular blue-green algae, 14
Unicellular true bacteria, *14*
Uniomerus, 1143
Unionicola, 1088, 1090, 1091, 1101, 1102
Unionicolidae, 1102
Unionidae, 1125, 1138, 1118
Unioninae, 1138
Univalves, 1117
Uranotaenia, 1066
Urceolaria, 290, 291
Urceolariidae, 290
Urceolus, 125, 124
Urnatella, *499*: *gracilis*, 499
Urnatellidae, 499

Urocentrum, 280, 281
Urochaenia, 269, 268
Urococcus, 163, 162
Uroglena, 156, 157
Uroglenopsis, 156, 157
Uroleptus, 287
Uronema, 281, 145
Uronemopsis, 281
Urophagus, 230
Uropod, Malacostraca, 870
Urosoma, 289
Urosome: Copepoda, 735, 736; Harpacticoida, 818; Malacostraca, 870
Urostyla, 288, 289
Urotricha, 268
Urozona, 280, 281
Usingerina, 967
Utaperla, 957
Utricularia, 1176: *fibrosa*, 1176; *geminiscapa*, 1176; *inflata*, 1176; *purpurea*, 1176; *vulgaris*, 1176, 1171

Vacuoles, Myxophyceae, 95
Vaginicola, 293, 294
Vaginicolidae, 293
Vahlkampfiidae, 236
Vallisneria americana, 1183
Valvata, 1133
Valvatidae, 1125, 1133
Valve: Bacillariophyceae, 176; esophago-intestinal, Nemata, 371
Vampyrella, 263
Vanoyella globosa, 438, 489
Varicosporium, 89, 88
Vas deferens, Nemata, 373
Vasicola, 271, 270
Vaucheria, 155, 152
Vaucheriaceae, 155
Vejdovskyella, 531: *comata*, 531
Velia, 963, *959*
Veliidae, 962, *959*
Veloidea, 963
Vertebrates, 5
Vertex, Cladocera, 591
Vibrio, 44, 17, *30*
Vibrios, 17
Vibrissea, 87, 86: asci, 49
Victorellidae, 499
Viehoperla, 948
Villosa, 1155
Viruses, 22
Visoka, 948, 949
Vitreoscilla, 44, *20*, *31*
Vitreoscillaceae, 13, 14
Viviparidae, 1125, 1132, 1178, 1123

Viviparus, 1132, *1118*
Volutin, 191
Volvocaceae, 129
Volvocales, 127
Volvocine flagellates, 9, 10 (see also Volvocales)
Volvox, 129, 128
Volvulina, 129, 128
Voronkowia, 476, 475, *438*
Vorticella, 292, 293
Vorticellidae, 291, 292
Vulva, Nemata, 373

Wailesella, 250
Wasps, *904*
Water bear, 509
Water boatman, 967
Water bug, creeping, 966
Water fleas, 559
Water mites, 1080, 1108
Water penny, 1013, 1049
Water scorpion, 965
Water-striders, *904*: broad-shouldered, 962
Westella, 139, 138
Wettina, 1087, 1103
Wheel animalcules, 420
Wierzejkiella, 459, *437*
Wigrella, 459, *437*
Williamsonia, 934
Wilsonema, 390
Wilsonematinae, 389
Wing, Lepidoptera, 1051
Wing pads, Plecoptera, 942
Wislouchiella, 129, 128
Wlassicsia, 629: *kinistinensis*, 629
Wolffia columbiana, 1175: *punctata*, 1175
Wolffiella floridana, 1175
Wolga, 449, *437*, 454
Wollea saccata, 100
Woloszynskia, 163, 161
Wormaldia, 1030
Woronina, 75

Wulfertia, 466, 469, *437*
Wyeomyia, 1064

Xanthidium, 143, 142
Xanthophyceae, 151, *117:* doubtful, 155
Xiphinema, 399
Xironodrilus, 527: *appalachius*, 528; *dentatus*, 528; *formosus*, 528; *pulcherrimus*, 527
Xironogiton, 526: *instabilius instabilius*, 527; *instabilius oregonensis*, 527; *occidentalis*, 526
Xystonotus, 1106

Yellow-brown algae, *117*
Yellow-green algae, *117*
Yolk glands, Turbellaria, 327
Yoraperla, 948
Yphria, 1024
Yugus, 955

Zaitzevia, 1005, 1004, 1018, 987
Zannichellia palustris, 1186
Zapada, 948, 949
Zealeuctra, 950
Zelinkiella, 484, *437*
Zoniagrion, 926
Zonomyxa, 237
Zooflagellates, 190
Zooflagèlles, *191*
Zoogloea ramigera, 45, 41
Zoomastigophorea, 191
Zoophagus, 85, 84
Zoospores, 11: types in Phycomycetes, 48
Zoothamnium, 292
Zschokkea, 1092, 1097
Zut, 378
Zygnema, 145, 142
Zygnemataceae, 143, 142
Zygnematales, 139
Zygnemopsis, 145, 142
Zygogonium, 145, 142
Zygoptera, 920, 922, 917, 918, 919
Zygorhizidium, 65